Langenscheidt
Fachwörterbücher

Langenscheidt

Dictionary of **Chemistry** Concise Edition English

English – German
German – English

Second enlarged edition

Edited by
Technische Universität Dresden

Langenscheidt

Berlin · München · Wien · Zürich · New York

Vorwort zur 2. Auflage

In der zweiten Auflage der Kompaktausgabe des Fachwörterbuches Chemie wurde den rasanten Entwicklungen auf den Gebieten der Nanotechnologie, der Biotechnologie und funktionalen Polymere sowie der stark interdisziplinär ausgerichteten Forschungsrichtung der Lebenswissenschaften (life sciences) Rechnung getragen. Dementsprechend wurde die erste Auflage um etwa 1800 Einträge, vorwiegend aus den genannten Themenbereichen, erweitert. Damit steht mit dieser Ausgabe ein kompaktes Werk zur Verfügung, mit dem sich nun auch die Fachwörter der modernen Chemie, die den Nutzern heutzutage in vielen Berufszweigen sowie zahlreichen Bereichen des täglichen Lebens begegnen, übersetzen lassen.

Ich möchte mich bei Frau A. Jiranek, Herrn J. Knepper, Prof. K.-F. Arndt und Prof. A. Satzger für die zahlreichen Hinweise und für die Unterstützung bedanken. Den Lektorinnen des Langenscheidt Fachverlages danke ich sehr für die Hilfe bei der Anfertigung des Manuskripts.

Franziska Eckert

Vorwort zur 1. Auflage

Die vorliegende Kompaktausgabe „Chemie“ basiert auf den im gleichen Verlag erschienenen Fachwörterbüchern „Chemie und chemische Technik“ Englisch-Deutsch (6. Auflage) und Deutsch-Englisch (5. Auflage). Die Kompaktausgabe richtet sich vor allem an Studenten und Schüler sowie an Fachleute, die mit der Chemie als Randgebiet in Berührung kommen, aber auch an Nutzer aus verschiedensten Berufszweigen, die gelegentlich Übersetzungen für chemische Fachwörter benötigen.

Bei der Bearbeitung der Kompaktausgabe wurden neben dem Grundwortschatz der Chemie und chemischen Technik besonders Fachwörter aus den Gebieten physikalische Chemie, organische Chemie, Biochemie und Biotechnologie berücksichtigt. Der Herausgeber und die Bearbeiter waren bemüht, mit der vorliegenden Bearbeitung den Erwartungen der Nutzer hinsichtlich Genauigkeit und Zuverlässigkeit der Wörterbucheinträge zu entsprechen.

Den Lektorinnen des Langenscheidt Fachverlags danken wir für die gute Zusammenarbeit, die Hilfe und die Geduld bei der Fertigstellung des Manuskripts. Helfende Kritik und Vorschläge zur Verbesserung des Wörterbuchs nehmen die Autoren gern entgegen und bitten, diese an den Langenscheidt Fachverlag, Postfach 40 11 20, D-80711 München, zu richten.

J. Knepper

Preface to the 2nd edition

This second, compact edition of the Chemistry Dictionary takes account of the burgeoning developments in the fields of nanotechnology, biotechnology and functional polymers as well as of the highly interdisciplinary nature of research in the life sciences. As a result the first edition has been expanded by about 1,800 entries, mainly from the above-mentioned fields. This edition thus constitutes a compact work which enables users from a wide range of professions to translate the technical terms of modern chemistry which they may encounter in the course of their work and in numerous situations of daily life.

I should like to thank Ms. A. Jiranek, Mr. J. Knepper, Prof. K.-F. Arndt and Prof. A. Satzger for their many helpful contributions and general support. My thanks also to the editorial staff of the publisher, Langenscheidt, for their help in the preparation of the manuscript.

Franziska Eckert

Preface to the 1st Edition

The Compact dictionary of *Chemistry* is based on the English-German (6th edition) and German-English (5th edition) *Chemistry and Chemical Engineering* dictionaries from the same publisher. This Compact version is intended mainly for students and pupils, and specialists for whom chemistry is a marginal subject, but also for users from a wide range of occupations who occasionally need translations for chemical terms.

Besides the basic vocabulary of chemistry and chemical engineering, special attention was devoted to terms from the fields of physical and organic chemistry, biochemistry, and biotechnology in preparing the Compact edition. The editor and the editorial team were concerned to meet the users' expectations with regard to the precision and reliability of the dictionary entries in this revision.

We thank the publisher's readers of Langenscheidt Fachverlag for their co-operation, assistance, and patience during the preparation of the manuscript. The authors will be grateful for constructive criticism and suggestions for improving the dictionary. Please send these to:
Langenscheidt Fachverlag, Postfach 40 11 20, D-80711 München, Germany.

J. Knepper

Benutzungshinweise • Directions for Use

1. Beispiele für die alphabetische Ordnung • Examples of Alphabetization

bulk (...) • evaporate to low [small] bulk
bulk blend
bulk material
bulk solution
bulk volume
bulking agent
bunker fuel (oil)
bunker oil
Bunte (gas)
Bunte salt
burn *v*
burn *v* off
burn
burn-off
burn-up
burnable
burner
butadiene
butane
1,4-butanedicarboxylic acid
butanedioic acid
1,4-butanediol
2,3-butanedione
butanoic acid
1-butanol
2-butanol
cis-butenedioic acid
trans-butenedioic acid
butyl group
sec-butyl alcohol
tert-butyl alcohol
butylation

diatherm(an)
Diathermansie
Dibasizität
Dibenzo-1,4-dioxin
Dibenzo-1,4-dioxin/polychloriertes
Dibenzo-*p*-dioxin
Dibromid
2,6-Di-*tert*-butyl-*p*-cresol
Dichte
Dichte/optische
Dichte/relative
Dichtebestimmung
Dichtebestimmung von Flüssigkeiten
Dichtegradientenzentrifugation
dichten
dickflüssig (...) • dickflüssig werden
Dicyan
1,1-Dicyanethen
Dicyanoaurat(I)
Diffusat
difunktionell
m-Digallussäure
dimerisieren
Dimerisierung
Dimerisierung/oxidative
Dimerisierung/reduktive [reduzierende]
Di-π-methan-Umlagerung
Dimethylbenzen
2,5-Dimethylbenzoesäure
α,α-Dimethylbenzylalkohol
5,5-Dimethylcyclohexan-1,3-dion
Dimethyldichlorsilan

Zahlen, kursiv gesetzte Stellungs- und Konfigurationssymbole wie *asym-*, *sym-*, *prim-*, *sec-*, *tert-*, *cis-*, *trans-*, *m-*, *o-*, *p-*, *N-*, *d-*, *l-* sowie griechische Buchstaben bleiben bei der alphabetischen Einordnung unberücksichtigt. Wortzusammensetzungen mit derartigen Elementen als Vorsätze sind voll ausgeschrieben.

Numbers and italicized symbols denoting position or configuration like *asym-*, *sym-*, *prim-*, *sec-*, *tert-*, *cis-*, *trans-*, *m-*, *o-*, *p-*, *N-*, *d-*, *l-* and Greek letters are disregarded in alphabetization. Terms with such elements are given in full spelling.

2. Bedeutung der Zeichen • Meaning of Signs

/	Abbau / biologischer = biologischer Abbau vacuo / in = in vacuo
( )	Aminosäure(reihen)folge = Aminosäurefolge *oder* Aminosäurereihenfolge Hooker (diaphragm) cell = Hooker cell *or* Hooker diaphragm cell Diese Bedeutung der runden Klammern gilt nicht für Wertigkeits- und Stellungsangaben in Formeln und chemischen Benennungen. This meaning of the round brackets does not apply to valency and position indications in formulae and chemical terms.
[ ]	Enzym / eiweißabbauendes [eiweißspaltendes] = eiweißabbauendes Enzym *oder* eiweißspaltendes Enzym ketonic cleavage [fission] = ketonic cleavage *or* ketonic fission
( )	Kursive Klammern enthalten Erklärungen – Italicized brackets contain explanations
*	kennzeichnet Benennungen nach IUPAC – marks IUPAC names
•	kennzeichnet Wendungen – marks phrases
=	Abkürzung für – Abbreviation for

3. Abkürzungen • Abbreviations

allg.	allgemein / generally
Am	amerikanisches Englisch / American English
bzw.	beziehungsweise / respectively
f	Femininum / feminine noun
i. e. S.	im engeren Sinne / in a narrower sense
i. w. S.	im weiteren Sinne / in a broader sense
m	Maskulinum / masculine noun
n	Neutrum / neuter noun
pl	Plural / plural
s.	siehe / see
U.K.	in Großbritannien / in the United Kingdom
U.S.	in den Vereinigten Staaten von Amerika / in the United States
v	Verb / verb
z.B.	zum Beispiel / for example

Am	American English / amerikanisches Englisch
e.g.	(exempli gratia) for example / zum Beispiel
esp	especially / besonders
f	feminine noun / Femininum
m	masculine noun / Maskulinum
n	neuter noun / Neutrum
pl	plural / Plural
s.	see / siehe
specif	specifically / im engeren Sinne
U.K.	in the United Kingdom / in Großbritannien
U.S.	in the United States / in den Vereinigten Staaten von Amerika
v	verb / Verb

Fachgebietskürzel • Subject-area labels

Anal	analytische Chemie / analytical chemistry
Anstr	Chemie und Technologie der Anstrichstoffe / chemistry and technology of organic coating materials
Aufb	Aufbereitungstechnik / mineral technology
Bau	Chemie und Technologie der Baustoffe / chemistry and technology of building materials
Bgb	Bergbau / mining
Bio	Biologie / biology
Bioch	Biochemie / biochemistry
Biot	Biotechnologie / biotechnology
Boden	Bodenchemie und Bodenkunde / soil chemistry and soil science
Bot	Botanik / botany
Carb	Carbochemie / coal chemistry
Ch	Chemie / chemistry
Chromat	Chromatographie / chromatography
Dest	Destillationstechnik / distillation apparatus and practice
Ech	Elektrochemie / electrochemistry
Erdöl	Chemie und Technologie des Erdöls / chemistry and technology of petroleum
Farb	Chemie und Technologie der organischen Farbstoffe / chemistry and technology of organic dyes
Filtr	Filtriervorrichtungen und Filtriertechnik / filtration equipment and practice
Foto	fotografische Chemie / photographic chemistry
Gär	Gärungschemie und Gärungstechnologie / chemistry and technology of fermentation
Geoch	Geochemie / geochemistry
Geol	Geologie / geology
Gerb	Gerbereichemie und Gerbereitechnologie / chemistry and technology of leather manufacture
Glas	Glasherstellung / glass technology
Gum	Technologie der Gummiherstellung / rubber technology
Hyd	Hydrochemie, Wasseraufbereitung und Abwasserbehandlung / hydrochemistry, water and waste-water treatment
Ker	Keramik / ceramics
Kern	Kernchemie, Kernphysik und Kerntechnik / nuclear science and technology
Koll	Kolloidchemie / colloid chemistry
Kosmet	Chemie und Technologie der Kosmetika und Riechstoffe / chemistry and technology of cosmetics and perfumes
Krist	Kristallographie und Kristallchemie / crystallography and crystal chemistry
Kst	Chemie und Technologie der Kunststoffe / chemistry and technology of plastics

Lab	Labortechnik und Laborgeräte / laboratory technique and apparatus
Landw	land- und forstwirtschaftliche Chemie / agricultural and silvicultural chemistry
Lebm	Lebensmittelchemie und -technologie / food chemistry and technology
Med	Medizin / medicine
Met	Metallurgie und Metallkunde / metallurgy and science of metals
Min	Mineralogie und Petrographie / mineralogy and petrography
Nom	chemische Nomenklatur / chemical nomenclature
org Ch	organische Chemie / organic chemistry
Pap	Zellstoff- und Papierchemie und -technologie / pulp and paper chemistry and technology
Pharm	pharmazeutische Chemie / pharmaceutical chemistry
phys Ch	physikalische Chemie / physical chemistry
Spektr	Spektroskopie / spectroscopy
Tech	Technik und Technologie / technology
Text	Textilchemie / textile chemistry
Tox	Toxikologie / toxicology
Zucker	Technologie der Zuckerherstellung / sugar technology

agric	agricultural and silvicultural chemistry / land- und forstwirtschaftliche Chemie
anal	analytical chemistry / analytische Chemie
bioch	biochemistry / Biochemie
biol	biology / Biologie
biot	biotechnology / Biotechnologie
bot	botany / Botanik
build	chemistry and technology of building materials / Chemie und Technologie der Baustoffe
ceram	ceramics / Keramik
ch	chemistry / Chemie
chromat	chromatography / Chromatographie
coal	coal chemistry / Carbochemie
coat	chemistry and technology of organic coating materials / Chemie und Technologie der Anstrichstoffe
coll	colloid chemistry / Kolloidchemie
cosmet	chemistry and technology of cosmetics and perfumes / Chemie und Technologie der Kosmetika und Riechstoffe
cryst	crystallography and crystal chemistry / Kristallographie und Kristallchemie
distil	distillation apparatus and practice / Destillationstechnik
dye	chemistry and technology of organic dyes / Chemie und Technologie der organischen Farbstoffe
el ch	electrochemistry / Elektrochemie
ferm	chemistry and technology of fermentation / Gärungschemie und Gärungstechnologie

Englisch-Deutsch

A

a 1. *s.* activity; 2. *s.* antarafacial; 3. = ampere
a. *s.* acid
A 1. *s.* absolute temperature; 2. *s.* angstrom (unit); 3. *s.* atomic weight; 4. *s.* anticlinal conformation; 5. = ampere
A-stage resin Harz *n* im A-Zustand, A-Harz *n*, *(bei Phenolharzen auch)* Resol *n*
Å *s.* angstrom
aa *s.* amino acid
AA 1. *s.* allyl alcohol; 2. = adenylic acid
AAS *s.* atomic absorption spectrometry
ABA *s.* abscisic acid
Abbe number abbesche Zahl *f*, ν-Wert *m*, reziproke relative Dispersion *f*
Abbe refractometer (Zeiß-)Abbe-Refraktometer *n*
aberration Aberration *f*, Abweichung *f*, Abbildungsfehler *m*
abherent Trennmittel *n*
abhesion Abhäsion *f*
abhesive abhäsiv, antiadhäsiv
abhesiveness abhäsive Eigenschaft *f*, Trennvermögen *n*
abietate Abietat *n (Salz oder Ester der Abietinsäure)*
abietic acid Abietinsäure *f (Harzsäure, Diterpenderivat)*
ability of being wetted Benetzbarkeit *f*
ability of wetting Benetzungsfähigkeit *f*, Netzfähigkeit *f*, Benetzungsvermögen *n*, Netzvermögen *n*
ability to swell Quellfähigkeit *f*, Quellvermögen *n*, *(passiv auch)* Quellbarkeit *f*
ablate *v* abtragen; abgetragen werden *(durch Schmelzen, Verdampfen oder Zersetzen bei hohen Temperaturen)*
ablation Ablation *f (Materialabtrag durch Schmelzen, Verdampfen oder Zersetzen bei hohen Temperaturen)*
ablative ablativ
ablative coating Ablationsschicht *f*
abnormal transference number *(phys Ch)* anomale Überführungszahl *f*
abrade *v* abreiben; abschleifen; verschleißen
abrasion 1. Abreiben *n*; 2. *(Text)* Scheuern *n*; Abschleifen *n*; Abrieb *m*, Abriebsabnutzung *f*, (reibender) Verschleiß *m*, Reibungsverschleiß *m*; 3. *(Text)* Scheuerverschleiß *m*
abrasion resistance 1. Abriebwiderstand *m*, Abriebbeständigkeit *f*, Abriebfestigkeit *f*, Verschleißfestigkeit *f*; 2. *(Text)* Scheuerbeständigkeit *f*, Scheuerfestigkeit *f*
abrasion-resistant 1. abriebbeständig, abriebfest, verschleißfest; 2. *(Text)* scheuerbeständig, scheuerfest
abrasion test 1. Abriebprüfung *f*, Abriebversuch *m*, Abnutzungsprüfung *f*; 2. *(Text)* Scheuerprüfung *f*
abrasive abreibend, reibend, schleißend; abschleifend, Schleif...
abrasive Abrasivmittel *n*, Abrasivwerkstoff *m*, abrasiver Werkstoff *m (z. B. Schleifmittel)*
abrasive wear Abnutzung *f* durch Abrieb, Abrieb *m*, Abriebsabnutzung *f*, (reibender) Verschleiß *m*, Reibungsverschleiß *m*
ABS *s.* alkylbenzene sulphonate
ABS polymer ABS-Polymer(e) *n*, Acrylnitril-Butadien-Styrol-Copolymerisat *n*
abscisic acid Abscisinsäure *f*, ABS *(ein Phytohormon)*
absence of oxygen Sauerstoffabwesenheit *f*
absence of reaction Reaktionslosigkeit *f*
absolute 1. absolut, rein, unvermischt, *(oft auch)* wasserfrei, abs.; 2. *(Kosmet)* absolut, abs. *(in flüssiger Form)*
absolute alcohol absoluter [wasserfreier] Alkohol *m* [Ethylalkohol *m*], Absolutalkohol *m*
absolute atomic mass absolute Atommasse *f*
absolute detection limit *(Anal)* absolute Nachweisgrenze *f*, absolute Erfassungsgrenze *f*
absolute humidity absolute Feuchte *f* [Feuchtigkeit *f*]
absolute temperature absolute [thermodynamische] Temperatur *f*, Kelvin-Temperatur *f*
absolute viscosity absolute [dynamische] Viskosität *f*
absolute zero absoluter Nullpunkt *m*
absorb *v* absorbieren, aufsaugen, einsaugen, aufnehmen; absorbieren,

(Strahlen) (ver)schlucken; aufziehen *(Farbstoffe)*
absorbability Absorbierbarkeit *f*, Aufsaugbarkeit *f*
absorbable absorbierbar, aufsaugbar
absorbance *(Spektr)* Extinktion *f*, Absorbanz *f*
absorbance coefficient Extinktionskoeffizient *m*, Extinktionskonstante *f*
absorbancy *s.* absorbance
absorbate 1. Absorbat *n*, zu absorbierender Stoff *m*; 2. Absorpt *n*, absorbierter Stoff *m*
absorbefacient absorptionsfördernd, absorptionsbeschleunigend
absorbefacient absorptionsförderndes [absorptionsbeschleunigendes] Mittel *n*
absorbency 1. Absorptionsvermögen *n*, Absorptionsfähigkeit *f*, Aufnahmevermögen *n*, Aufnahmefähigkeit *f*; 2. *(Pap)* Saugvermögen *n*, Saugfähigkeit *f*; 3. *s.* absorbance
absorbency index *s.* absorptivity
absorbent absorptionsfähig, absorbierend, absorptiv, Absorptions..., *(i. w. S.)* aufnahmefähig, (auf)saugend, Aufnahme..., Saug...
absorbent absorbierender Stoff *m*, Absorbens *n*, Absorptionsmittel *n*, *(i. w. S.)* aufnehmender [aufsaugender] Stoff *m*
absorbent paper Saugpapier *n*, Fließpapier *n*
absorber 1. Absorber *m*, Absorptionsapparat *m*; 2. *s.* absorbent
absorbing agent *s.* absorbent
absorbing capacity 1. *s.* absorbency 1.; 2. *(Farb, Text)* Aufziehvermögen *n*, Ziehvermögen *n*
absorbing region *(Spektr)* Absorptionsbereich *m*, Absorptionsgebiet *n*
absorptance Absorptionsgrad *m*, Absorptionsvermögen *n*
absorptiometer Absorptiometer *n*
absorptiometric absorptiometrisch
absorptiometry Absorptiometrie *f*, Absorptionsanalyse *f*
absorption Absorption *f*, Absorbieren *n*, Aufnahme *f*, Aufnehmen *n*, *(von Flüssigkeiten auch)* Aufsaugen *n*, Einsaugen *n*; Absorption *f (von Strahlung)*; Aufziehen *n (von Farbstoffen)*
absorption band Absorptionsbande *f*
absorption bottle Waschflasche *f*, Absorptionsflasche *f*
absorption cell *(Anal)* Absorptionsküvette *f*
absorption column Absorptionskolonne *f*, Absorptionssäule *f*
absorption cross section Absorptionsquerschnitt *m*
absorption line Absorptionslinie *f*
absorption-line spectrum Linienabsorptionsspektrum *n*
absorption liquid Absorptionsflüssigkeit *f*
absorption pathlength *(Spektr)* durchstrahlte Schichtdicke *f*
absorption pipette Absorptionspipette *f*
absorption rate Absorptionsgeschwindigkeit *f*
absorption refrigeration system Absorptionskältemaschine *f*, Absorptionskälteanlage *f*
absorption spectrometer Absorptionsspektrometer *n*
absorption spectroscopy Absorptionsspektroskopie *f*
absorption spectrum Absorptionsspektrum *n*
absorption train Absorptionsbatterie *f (für Gase)*
absorption tube Absorptionsrohr *n*, Absorptionsröhrchen *n*
absorptive absorptionsfähig, absorbierend, absorptiv, Absorptions..., *(i. w. S.)* aufnahmefähig, (auf)saugend, Aufnahme..., Saug...
absorptivity *(Spektr)* Extinktionskoeffizient *m*, Absorptionskoeffizient *m*; Absorptionszahl *f (Wärmestrahlung)*
abstract *v* 1. abtrennen, absondern, abspalten; abscheiden, separieren; extrahieren, (her)ausziehen; entziehen; 2. zusammenfassen, referieren *(Fachveröffentlichungen)*
abstract Zusammenfassung *f*, Referat *n (einer Fachveröffentlichung)*
abstract journal Referatezeitschrift *f*, referierende Zeitschrift *f*
abstraction Abtrennung *f*, Absonderung *f*, Abspaltung *f*; Abscheidung *f*; Extraktion *f*; Entzug *m*, Entziehung *f*
abstraction reaction Abspaltungsreaktion *f*

abundance Häufigkeit *f (eines Elements oder Isotops)*
abundance ratio Mischungsverhältnis *n (von Isotopen)*; Isotopenverhältnis *n*
ac. *s.* acid
AC. *(alternating current)* Wechselstrom *m*
a.c. *(alternating current)* Wechselstrom *m*
a.c. polarography Wechselstrompolarographie *f*
accelerate *v* beschleunigen
accelerated ageing beschleunigte [künstliche] Alterung *f*, Kurzalterung *f*
accelerated exposure Schnellprüfung *f*, beschleunigte Prüfung *f (von Werkstoffen gegen Umwelteinflüsse)*
accelerated test Kurzversuch *m*, Kurzprüfung *f*
accelerating potential Beschleunigungsspannung *f*
acceleration Beschleunigung *f*
accelerator 1. Beschleuniger *m*; 2. *(Kst)* Akzelerator *m*; 3. *(Bau)* Abbindebeschleuniger *m*
accept *v* aufnehmen *(z. B. Elektronen)*; annehmen *(Farbstoffe)*
acceptable daily intake *(Tox)* duldbare [zulässige] Tagesdosis *f (in mg/kg Körpermasse)*
acceptable for discharge to a receiving stream *(Hyd)* vorflutergerecht, vorfluterwürdig
acceptor Akzeptor *m*
acceptor atom Akzeptoratom *n*
acceptor compound Akzeptorverbindung *f*
acceptor site *(Bioch)* Akzeptorort *m*, A-Ort *m*, Akzeptorstelle *f*, Decodierungsort *m*, Aminoacylort *m*, Erkennungsregion *f*
access of air Luftzutritt *m*
accessibility Akzessibilität *f*, Zugänglichkeit *f*
accessory agent Zusatzstoff *m*
accidental degeneracy zufällige Entartung *f (Quantenmechanik)*
accidental error zufälliger [subjektiver] Fehler *m*, Zufallsfehler *m*
accommodate *v* angleichen, anpassen; sich angleichen [anpassen]
accommodation Angleichung *f*, Anpassung *f*, Akkommodation *f*, Akkommodierung *f*
accretion Vergrößerung *f (von Aerosolen)*; Akkretion *f*
accumulate *v* (an)sammeln, anhäufen, (auf)speichern, anreichern; sich ansammeln [sammeln], sich anhäufen [speichern, anreichern]
accumulation Ansammlung *f*, Anhäufung *f*, Aufspeicherung *f*, Speicherung *f*, Anreicherung *f*
accumulator Speicher *m*, Akkumulator *m*, Energiespeicher *m*; (elektrischer) Akkumulator *m*, Akku *m*; Hydraulikspeicher *m*, hydraulischer Druckspeicher *m*
accuracy *(Anal)* Richtigkeit *f*
acenaphthene *(org Ch)* Acenaphthen *n*
acene *(org Ch)* Acen *n*
acet. *s.* acetone
acet. a. *s.* acetic acid
acetal $R–CH(OR')_2$ Acetal *n*, *(i. e. S.)* $CH_3CH(OC_2H_5)_2$ Acetal *n*, 1,1-Diethoxyethan *n*
acetal resin *(Kst)* Acetalharz *n*
acetalate *v (org Ch)* acetalisieren
acetalation *(org Ch)* Acetalisierung *f*
acetaldehyde Acetaldehyd *m*, *Ethanal *n*
acetaldol Acetaldol *n*, *3-Hydroxybutanal *n*
acetalize *v (org Ch)* acetalisieren
acetamide Acetamid *n*
acetamidine Acetamidin *n*, Methylcarboxamidin *n*
acetamido group $–NHCOCH_3$ Acetamidogruppe *f*, Acetylaminogruppe *f*
acetate 1. Acetat *n*; 2. Acetatfaserstoff *m*, AZ
acetate fibre Acetatfaser *f*; Acetatfaserstoff *m*, AZ
acetate rayon Acetatseide *f*, Acetatrayon *m(n)*
acetic acid Essigsäure *f*, *Ethansäure *f*
acetic-acid fermentation Essig(säure)gärung *f*
acetic anhydride Essigsäureanhydrid *n*, Ethansäureanhydrid *n*, Acetanhydrid *n*
acetification Essigsäurebildung *f*, Essigsäuregärung *f*; Essigherstellung *f*
acetoacetate *(Bioch)* Acetacetat *n*
acetoacetic acid Acetessigsäure *f*, *(Ketoform)* 3-Oxobutansäure *f*, *(Enolform)* *3-Hydroxybut-2-ensäure *f*

acetoacetic ester Acetessigsäureethylester *m*, Acetessigester *m*
acetoacetylation *(org Ch)* Acetoacetylierung *f*
acetoethylation *(org Ch)* Acetoethylierung *f*
acetogenin Polyketid *n*, Acetogenin *n*
acetoin Acetoin *n*, *3-Hydroxybutan-2-on *n*
acetolysis Acetolyse *f*
acetonation *(org Ch)* Acetonierung *f*
acetone Aceton *n*, Dimethylketon *n*, *Propan-2-on *n*
acetone-butanol fermentation *(Biot)* Aceton-Butanol-Gärung *f*
acetone resin *(Kst)* Acetonharz *n*
acetonedicarboxylic acid Acetondicarbonsäure *f*, 3-Ketoglutarsäure *f*, *3-Oxopentandisäure *f*
acetonitrile Acetonitril *n*
acetophenone Acetophenon *n*, *Methylphenylketon *n*, Acetylbenzen *n*
acetous fermentation Essig(säure)gärung *f*
acetoxy group $-CH_3CO_2$ Acetoxygruppe *f*
acetoxylation *(org Ch)* Acetoxylierung *f*
acetoxymethylation *(org Ch)* Acetoxymethylierung *f*
acetyl CoA [coenzyme A] *(Bioch)* Acetyl-Coenzym *n* A, aktivierte Essigsäure *f*, aktiviertes Acetat *n*
acetyl group $-COCH_3$ Acetylgruppe *f*, Acetylrest *m*
acetyl number Acetylzahl *f (von Fetten und Ölen)*
acetylable acetylierbar
acetylatable acetylierbar
acetylate *v* acetylieren
acetylated cellulose *(Text)* Acetylcellulose *f*, Celluloseacetat *n*
acetylating agent Acetylierungsmittel *n*
acetylation Acetylieren *n*, Acetylierung *f*
acetylcholine *(Bioch)* Acetylcholin *n*
acetylene Acetylen *n*, *Ethin *n*, *(i. w. S.)* Acetylen *n*, Acetylenkohlenwasserstoff *m (acyclisch oder cyclisch, mit einer oder mehreren C≡C-Dreifachbindungen)*
acetylene gas generator Acetylenentwickler *m*
acetylene hydrocarbon *s.* acetylenic hydrocarbon
acetylene lime Carbidkalk *m*
acetylenedicarboxylic acid Acetylendicarbonsäure *f*, *Ethindicarbonsäure *f*, *Butindisäure *f*
acetylenic acetylenisch; Acetylen..., *Ethin...
acetylenic acid Acetylencarbonsäure *f*, *Alkinsäure *f (allg. für eine Carbonsäure mit Dreifachbindung)*
acetylenic alcohol Alkinol *n*
acetylenic hydrocarbon Acetylen *n*, Acetylenkohlenwasserstoff *m (acyclisch oder cyclisch, mit einer oder mehreren C≡C-Dreifachbindungen)*
acetylenic linkage C≡C–Bindung *f*, C–C-Dreifachbindung *f*
acetylide Acetylid *n*
acetylsalicylic acid Acetylsalicylsäure *f*, 2-Acet-oxybenzoesäure *f*
acetylurea Acetylharnstoff *m*, Acetylureid *n*, Essigsäureureid *n*
AChE = acetylcholinesterase
Acheson process Acheson-Verfahren *n (zur Herstellung von Kunstgraphit)*
achiral achiral *(Stereochemie)*
achirality Achiralität *f (Stereochemie)*
achromatic lens achromatische Linse *f*
aci form *aci*-Form *f*, Säureform *f (von Nitroverbindungen)*
acicular nadelförmig, nadelartig, nadelig
acid sauer
acid Säure *f*, Sre. • **be acid to** sauer reagieren gegen
acid amide Säureamid *n*
acid anhydride Säureanhydrid *n*
acid-base catalysis Säure-Base-Katalyse *f*
acid-base dissociation curve Säure-Base-Dissoziationskurve *f*
acid-base equilibrium Säure-Base-Gleichgewicht *n*
acid-base indicator Säure-Base-Indikator *m*
acid-base pair Säure-Base-Paar *n*
acid-base reaction Säure-Base-Reaktion *f*, Neutralisationsreaktion *f*
acid-base titration Säure-Base-Titration *f*, Neutralisationstitration *f*, Protolysetitration *f*
acid Bessemer process Bessemer-Verfahren *n*, Bessemer-Konverterverfahren *n*, saures Windfrischverfahren *n*

acid carbonate Me^IHCO_3 Hydrogencarbonat *n*
acid carboy Säure(glas)ballon *m*
acid catalysis Säurekatalyse *f*, Reaktionsbeschleunigung *f* durch Säuren
acid chloride Säurechlorid *n*
acid color *(Am) s.* acid dye
acid conversion *(Lebm)* Verzuckerung *f* [Konversion *f*] mit Säure
acid corrosion Korrosion *f* in saurem Medium
acid denaturation Säuredenaturierung *f* *(von Proteinen)*
acid deposition saure Ablagerung *f* [Deposition *f*]
acid dissociation constant Säuredissoziationskonstante *f*
acid dye saurer Farbstoff *m*, Säurefarbstoff *m*, anionischer Farbstoff *m*
acid equivalent Säureäquivalent *n*
acid exchange Säureaustausch *m*
acid-fast *(Farb)* säureecht
acid fastness *(Farb)* Säureechtheit *f*
acid fermentation saure Gärung *f*
acid fog saurer Nebel *m*
acid fume Säuredampf *m*
acid gas saures Gas *n*, Sauergas *n* *(Kohlendioxid und Schwefelwasserstoff)*
acid halide Säurehalogenid *n*
acid hardening *(Kst)* Säurehärtung *f*
acid hydrogen Säurewasserstoff *m*
acid hydrolysis saure Hydrolyse *f*, Säurehydrolyse *f*
acid indicator saurer Indikator *m*, Indikatorsäure *f*
acid-labile säurelabil, säureempfindlich, säurezersetzlich, säureunbeständig
acid mist Säurenebel *m*
acid number Säurezahl *f*, SZ, Neutralisationszahl *f*
acid open-hearth process saures Herdfrischverfahren *n* [Siemens-Martin-Verfahren *n*, SM-Verfahren *n*]
acid phosphatase *(Bioch)* saure Phosphatase *f*
acid phosphate $M^IH_2PO_4$ Dihydrogenphosphat *n*; $M^I_2HPO_4$ Hydrogenphosphat *n*
acid phosphite Hydrogenphosphit *n*
acid potassium sulphate Kaliumhydrogensulfat *n*
acid precipitation Säurefällung *f*
acid radical Säurerest *m*
acid rain saurer Regen *m*
acid residue Säurerest *m*
acid resistance Säurebeständigkeit *f*, Säurefestigkeit *f*, Säureresistenz *f*
acid-resistant säurebeständig, säurefest
acid-resisting säurebeständig, säurefest
acid salt saures Salz *n*, Hydrogensalz *n*
acid site saures Zentrum *n*, Säurezentrum *n (an Molekülen, Katalysatoren)*
acid slag saure Schlacke *f*, *(i. e. S.)* Bessemer-Schlacke *f*
acid smut *Luftverunreinigung in Form säurehaltiger, durch Luftfeuchte zusammengeballter schwebender Kohleteilchen*
acid solution saure Lösung *f*; Säurelösung *f*
acid soot *s.* acid smut
acid steel saurer Stahl *m*, *(i. e. S.)* Bessemer-Stahl *m*, Bessemer-Konverterstahl *m*
acid strength Säurestärke *f*
acid sulphide Hydrogensulfid *n*
acid sulphite Hydrogensulfit *n*
acid tower Säureturm *m*
acid treatment Säurebehandlung *f*, *(bei Ölen auch)* Säureraffination *f*
acid value Säurezahl *f*, SZ, Neutralisationszahl *f*
acid washing Säurewaschen *n*, Säurewäsche *f*
acidic 1. azid *(mit Säurecharakter)*; 2. sauer, säurehaltig; Säure... • **make acidic** (an)säuern
acidic catalyst Säurekatalysator *m*
acidic character Säurecharakter *m*
acidic cleavage Säurespaltung *f*, Spaltung *f* unter Entstehung von Säure
acidic function Säurefunktion *f*
acidic soil saurer Boden *m*
acidiferous säurehaltig
acidification 1. Ansäuern *n (von Lösungen)*; 2. *(Lebm)* Säuern *n*, Ansäuern *n*; 3. *(Boden)* Versauerung *f (durch Düngemittel)*
acidify *v* 1. ansäuern, sauer einstellen *(Lösungen)*; 2. *(Lebm)* (an)säuern; 3. *(Boden)* versauern *(durch Düngemittel)*; 4. azidifizieren *(ein H-Atom)*
acidimeter Azidimeter *n*

acidimetric azidimetrisch
acidimetry Azidimetrie *f*
acidity Azidität *f*, *(quantitativ auch)* Säuregrad *m*
acidity constant Aziditätskonstante *f*, Säurekonstante *f*
acidity function Aziditätsfunktion *f*, Säurefunktion *f*
acido complex Azidokomplex *m*
acidogen *(Lebm)* Säurebildner *m*
acidolysis Azidolyse *f*, Säurespaltung *f*
acidolytic azidolytisch
acidometer *s.* acidimeter
acidulant 1. Ansäuerungsmittel *n*; 2. *(Lebm)* Säuerungsmittel *n*
acidulate *v* 1. (schwach) ansäuern; 2. *(Lebm)* säuern
acidulation 1. (schwaches) Ansäuern *n*; 2. *(Lebm)* Säuern *n*, Säuerung *f*
aconite alkaloid Aconit(um)alkaloid *n*
aconitic acid Aconitsäure *f*, *Propen-1,2,3-tricarbonsäure *f*
aconitum alkaloid Aconit(um)alkaloid *n*
acorn condenser Einhängekühler *m*, Kühlfinger *m*
acoxylation *(org Ch)* Acoxylierung *f*, Acyloxylierung *f*
ACP *s.* 1. acyl-carrier protein; 2. acetophenone; 3. a.c. polarography
acquire *v* annehmen, erreichen; aufnehmen
acquisition Annahme *f*, Erreichung *f*; Aufnahme *f*
acrid *(Lebm)* scharf, beißend, herb
acridine dye Acridinfarbstoff *m*
acridity *(Lebm)* Schärfe *f*, Herbheit *f*, Herbe *f*
acrolein Acrolein *n*, *Propenal *n*, Acrylaldehyd *m*
acrylaldehyde *s.* acrolein
acrylamide Acrylamid *n*, Propensäureamid *n*
acrylate Acrylat *n*
acrylate elastomer Acrylelastomer(e) *n*, Polyacrylatelastomer(e) *n*
acrylate resin *s.* acrylic resin
acrylic Acryl...
acrylic acid Acrylsäure *f*, *Propensäure *f*
acrylic aldehyde *s.* acrolein
acrylic amide *s.* acrylamide
acrylic ester Acrylsäureester *m*
acrylic fibre Acrylfaser *f*, Polyacrylnitrilfaser *f*; Acrylfaserstoff *m*, Polyacrylnitrilfaserstoff *m*, PVY
acrylic glass Acrylglas *n*, Kunstglas *n*, Plexiglas® *n*
acrylic lacquer Acryllack *m*
acrylic resin Acrylharz *n*, Polyacrylat *n*
acrylonitrile Acrylnitril *n*, *Propennitril *n*, Vinylcyanid *n*
acrylonitrile-butadiene rubber Acrylnitril-Butadienkautschuk *m*, Nitrilkautschuk *m* *(Butadien-Acrylnitril-Copolymerisat)*
acrylonitrile-butadiene-styrene plastic Acrylnitril-Butadien-Styrol-Kunststoff *m*, ABS-Kunststoff *m*
ACS = American Chemical Society
act *v* (ein)wirken
actinic aktinisch, photochemisch wirksam
actinide *s.* actinoid
actinium Ac Actinium *n*
actinium decay series Actinium(zerfalls)reihe *f*, Uran-Actinium-Zerfallsreihe *f*, Uran-Actinium-Reihe *f*
actinoid Actinoid *n*, Actinoidenelement *n* *(Element der Actiniumreihe)*
actinoid contraction Actinoidenkontraktion *f*
actinoid group [series] Actinoidenreihe *f*, Actiniumreihe *f*
action Einwirken *n*, Einwirkung *f*, Wirkung *f*
action constant Planck-Konstante *f*, plancksches Wirkungsquantum *n*
activable aktivierbar
activate *v* aktivieren, anregen, erregen
activated adsorption aktivierte Adsorption *f* *(häufigster Fall der chemischen Adsorption)*
activated alumina (künstlich) aktiviertes Aluminiumoxid *n*
activated atom angeregtes Atom *n*
activated carbon Aktivkohle *f*, A-Kohle *f*
activated carbon (bed) filter *(Hyd)* Aktivkohlefilter *n*, A-Kohle-Filter *n*
activated charcoal *s.* activated carbon
activated clay aktivierter Ton *m*
activated complex aktivierter Komplex *m*, Übergangszustand *m*
activated earth aktivierte Erde *f* [Bleicherde *f*], Aktiverde *f*, Edelerde *f*
activated molecule aktiviertes [aktives, energiereiches] Molekül *n*

activated sludge *(Hyd)* belebter [aktivierter, biologischer] Schlamm *m*, Belebtschlamm *m*, Bioschlamm *m*
activated-sludge waste treatment Abwasserbehandlung *f* nach dem Belebtschlammverfahren
activating collision reaktiver [wirksamer] Stoß *m (Kinetik)*
activating group aktivierende Gruppe *f*, Substituent *m* mit aktivierender Wirkung
activation 1. Aktivieren *n*, Aktivierung *f*; 2. *(Tox)* Aktivation *f*, Aktivatoreffekt *m (eine Form der Verbundwirkung)*; 3. *(Aufb)* Beleben *n*, Aktivieren *n*, Aktivierung *f*
activation analysis Aktivierungsanalyse *f*
activation cross section Wirkungsquerschnitt *m*, Aktivierungsquerschnitt *m (Aktivierungsanalyse)*
activation energy Aktivierungsenergie *f*
activation Gibbs function freie Aktivierungsenthalpie *f*
activator Aktivator *m*, Aktivierungsmittel *n*; Aktivator *m*, Promotor *m*, (synergetischer) Verstärker *m*
active 1. aktiv, wirksam; 2. unedel *(Metall)*; 3. radioaktiv
active acetate aktiviertes Acetat *n*, aktivierte [aktive] Essigsäure *f*, Acetyl-Coenzym *n*, Acetyl-CoA
active acidity aktuelle Azidität *f*
active alkalinity aktuelle Basizität *f*
active centre *s.* active site
active chlorine aktives wirksames Chlor *n*, Aktivchlor *n*; Aktivchlorgehalt *m*
active group 1. *(Bioch)* aktive Gruppe *f*, Wirk(ungs)gruppe *f*; 2. *(Hyd)* aktive [austauschaktive] Gruppe *f*, Ankergruppe *f (Ionenaustausch)*
active hydrogen aktiver Wasserstoff *m*
active molecule aktives [aktiviertes, energiereiches] Molekül *n*
active oxygen aktiver Sauerstoff *m*, Aktivsauerstoff *m (z. B. beim Bleichen)*
active site Aktivstelle *f*, Aktivzentrum *n*, aktive Stelle *f*, aktives Zentrum *n*
active solvent aktiver [echter] Löser *m*, aktives Lösungsmittel *n*
active substance aktiver [wirksamer] Stoff *m*, wirksames Agens *n*, aktive [wirksame] Substanz *f*, Wirkstoff *m*
activity 1. Aktivität *f*, Wirksamkeit *f (eines Katalysators)*; 2. Aktivität *f (korrigierter Molenbruch eines Stoffes in einer Mischphase)*; 3. Aktivität *f*, Umwandlungsrate *f (eines radioaktiven Stoffes)*
activity coefficient Aktivitätskoeffizient *m*
actor Aktor *m (bei induzierten Reaktionen)*
actual tatsächlich, wirklich, real, praktisch *(z. B. Versuch)*; praktisch erzielbar *(z. B. Temperatur)*
actual gas reales Gas *n*
actual plate *(Dest)* praktischer Boden *m*
actuator 1. Betätigungsorgan *n*; Stellantrieb *m*, Stellmotor *m (Regelungstechnik)*; 2. *(Anal)* Aktuator *m*
acute toxicity akute Toxizität *f* [Giftigkeit *f*]
acyclic acyclisch, nicht cyclisch
acyclic diene metathesis acyclische Dienmetathese *f*
acyl-carrier protein acylübertragendes Enzym *n* [Protein *n*], Acyl-Carrier-Protein *n*, ACP
acyl-carrying protein acylübertragendes Enzym *n* [Protein *n*], Acyl-Carrier-Protein *n*, ACP
acyl cation Acylkation *n*, Acylium-Ion *n*
acyl group Acylgruppe *f*, Acylrest *m*
acyl halide Acylhalogenid *n (Säurehalogenid von Carbonsäuren)*
acyl radical 1. (freies) Acylradikal *n*; 2. *s.* acyl group
***N*-acyl amino carboxylic acid** *N*-Acylaminocarbonsäure *f*
acylable acylierbar
acylaminoalkylation *(org Ch)* Acylaminoalkylierung *f*; Amidoalkylierung *f*
acylate *v* acylieren
acylating agent Acylierungsmittel *n*
acylation Acylierung *f*
acylglycerol Glycerid *n*, Acylglycerol *n (Ester des Glycerols mit Fettsäuren)*
acylium ion Acylkation *n*, Acylium-Ion *n*
acyloin Acyloin *n (Ketonalkohol der allgemeinen Formel RCOCHR'OH)*
acyloin condensation Acyloinkondensation *f*, Acyloinsynthese *f*
acyloin rearrangement Acyloin-Umlagerung *f*
acyloxylation *(org Ch)* Acoxylierung *f*, Acyloxylierung *f*

A.D. *s.* air-dried
adamantane *(org Ch)* Adamantan *n*
adamantine diamanten, Diamant...; diamantartig; diamanthart
adapter Übergangsstück *n*, Anpassstück *n*, Passstück *n*; Reduzierstück *n*; Verlängerungsstück *n*, Zwischenstück *n*; Destilliervorstoß *m*, Vorstoß *m*
adaptive metallurgy verarbeitende Metallurgie *f*
adaptor 1. *(Bioch)* Adaptor *m*; 2. *s.* adapter
adatom adsorbiertes Atom *n*, ad-Atom *n*
add *v* (hin)zufügen, (hin)zugeben, zusetzen, beimengen, beimischen, versetzen mit; anlagern, addieren; sich anlagern [addieren]
add *v* **to form a lower layer** unterschichten
addend Addend *m*, Ligand *m*
addictive *(Pharm)* suchterregend
addition 1. Zufügen *n*, Zufügung *f*, Zugeben *n*, Zugabe *f*, Zusatz *m*, Beimengung *f*, Beimischung *f*; Addition *f*, Anlagerung *f*, Anlagern *n*; Zufuhr *f (z. B. von Luft)*; 2. *s.* addition reaction • **undergo addition** sich anlagern
addition agent Zusatzmittel *n*, Zusatzstoff *m*, Zusatz *m*, *(manchmal auch)* Hilfsmittel *n*, Hilfsstoff *m*
addition compound Additionsverbindung *f*, Anlagerungsverbindung *f*; Molekülverbindung *f*
addition polymer Additionspolymer(e) *n*, Polyaddukt *n*
addition polymerization Additionspolymerisation *f*, Polyaddition *f*
addition reaction Additionsreaktion *f*, Anlagerungsreaktion *f*
addition solid solution Additionsmischkristall *m*
additive additiv
additive Zusatz(stoff) *m*, Zusatzmittel *n*, *(manchmal auch)* Hilfsmittel *n*, Hilfsstoff *m*; Additiv(e) *n (Zusatzstoff in Mineralölprodukten, der in Mengen von über 1 bis 10 % benutzt wird)*
additive property additive Eigenschaft *f*
additivity principle Additivitätsprinzip *n (zur Berechnung von Moleküleigenschaften aus Beiträgen einzelner Bindungen)*
adduct Addukt *n*
adduct rubber Adduktkautschuk *m*
π-adduct π-Komplex *m*
σ-adduct σ-Komplex *m*
adenine Adenin *n*, 6-Aminopurin *n*
adenohypophyseal hormone Hormon *n* der Adenohypophyse, Hypophysenvorderlappenhormon *n*
adenosine diphosphate Adenosindiphosphat *n*, ADP
adenosine monophosphate Adenosinmonophosphat *n*, AMP, Adenylsäure *f*
adenosine triphosphate Adenosintriphosphat *n*, ATP
adenylating agent Adenylierungsmittel *n*
adenylation Adenylierung *f*
adenylic acid *s.* adenosine monophosphate
adenylosuccinate *(Bioch)* Adenylosuccinat *n*, sAMP
adenylosuccinic acid *(Bioch)* Adenylosuccinat *n*, sAMP
adenylylate *v (Bioch)* adenylylieren
adenylylation *(Bioch)* Adenylylierung *f*
adermin(e) Adermin *n*, Vitamin *n* B_6, Pyridoxin *n*
ADH *s.* antidiuretic hormone
adhere *v* (an)haften, adhärieren; (an)kleben
adherence 1. Adhäsion *f*, Haften *n*, Anhaften *n*; Kleben *n*, Ankleben *n*; 2. *s.* adhesiveness
adherend Adhärend *m*, Packstoff *m*, Schichtträger *m (zu verklebender Stoff)*
adherent (an)haftend, adhärierend, adhäsionsfähig, haftfähig, adhäsiv; klebend, anklebend, klebfähig
adhesion Adhäsion *f*, Haften *n*, Anhaften *n*; Kleben *n*, Ankleben *n*
adhesive adhäsiv, adhäsionsfähig, haftfähig, (an)haftend, adhärierend, Adhäsions..., Haft...; klebfähig, klebend, anklebend, Kleb(e)...
adhesive Klebstoff *m*, Kleber *m*
adhesive agent Haftmittel *n*, Haftstoff *m*, Haftvermittler *m*, Haftlösung *f*; Klebstoff *m*; Kleber *m*; Bindemittel *n*
adhesive bonding *(Kst)* Aufkleben *n*
adhesive constituent *(Kst)* Klebstoffbestandteil *m*
adhesive dissolver *(Kst)* Klebstofflöser *m*
adhesive force Adhäsionskraft *f*

adhesive liquid *(Kst)* Klebflüssigkeit *f*
adhesive strength Adhäsionsfestigkeit *f*, Haftfestigkeit *f*
adhesive technology *(Kst)* Klebstofftechnologie *f*
adhesiveness Adhäsionsfähigkeit *f*, Adhäsionsvermögen *n*, Adhäsionskraft *f*, Haftfähigkeit *f*, Haftvermögen *n*; Klebvermögen *n*, Klebfähigkeit *f*, Klebkraft *f*
ADI *s.* acceptable daily intake
adiabat Adiabate *f*
adiabatic *(phys Ch)* adiabat(isch)
adiabatic *(phys Ch)* Adiabate *f*
adiabatic change adiabatische Zustandsänderung *f*
adiabatic curve *s.* adiabatic line
adiabatic expansion adiabatische Ausdehnung *f* [Expansion *f*]
adiabatic line Adiabate *f*
adiabatic operation adiabatische [adiabate] Fahrweise *f*
adiabatic reactor adiabat(isch)er Reaktor *m*
adiabaticity *(phys Ch)* Adiabasie *f*
adion Adion *n (adsorbiertes Ion)*
adipamide Adipinsäurediamid *n*, Adipamid *n*
adipate Adipat *n*
adipic acid Adipinsäure *f*, *Butan-1,4-dicarbonsäure *f*, *Hexandisäure *f*
adiponitrile Adipinsäuredinitril *n*, Adiponitril *n*
adjacent angrenzend, benachbart; nachbarständig *(Atome im Molekül)*
adjacent atom Nachbaratom *n*
adjacent position Nachbarstellung *f*
adjust *v* einstellen *(z. B. pH-Wert)*; justieren *(z. B. Waagen)*
adjusting screw Justierschraube *f*, Stellschraube *f (z. B. einer Waage)*
adjustment Einstellung *f (z. B. des pH-Werts)*; Justierung *f (z. B. von Waagen)*
adjuvant 1. Hilfsmittel *n*, Hilfsstoff *m*, Zusatzmittel *n*, Zusatzstoff *m*, Zusatz *m*; 2. *(Pharm)* Adjuvans *n (unterstützendes Mittel)*
ADMET *s.* acyclic diene metathesis
administer *v (Pharm)* applizieren, verabreichen
administration *(Pharm)* Applikation *f*, Verabreichung *f*
admittance *(Anal)* Admittanz *f*, Scheinleitwert *m*, Wechselstromleitwert *m*
admix *v* beimischen, beimengen, zumischen, zusetzen
admixture 1. Beimischen *n*, Beimengen *n*, Zumischen *n*, Zusetzen *n*, Zusatz *m*; 2. Beimischung *f*, Beimengung *f*, Begleitstoff *m*, Fremdstoff *m*, Fremdbestandteil *m*, Zusatz *m*
ADP *s.* adenosine diphosphate
adrenaline Adrenalin *n*, 1-(3′,4′-Dihydroxyphenyl)-2-methylaminoethanol *n*
adrenocortical hormone Nebennierenrindenhormon *n*, NNR-Hormon *n*
ADSA *s.* asymmetric drop shape analysis
adsorb *v* adsorbieren; sich adsorptiv anlagern
adsorbability Adsorbierbarkeit *f*
adsorbable adsorbierbar
adsorbate 1. Adsorpt *n*, adsorbierter [aufgenommener] Stoff *m*; 2. *s.* adsorption complex
adsorbed layer Adsorptionsschicht *f*
adsorbed material [substance] *s.* adsorbate 1.
adsorbent adsorbierend, adsorptionsfähig, adsorptiv
adsorbent Adsorbens *n*, Adsorptionsmittel *n*, Adsorber *m*, adsorbierender [aufnehmender] Stoff *m*
adsorbent-adsorbate complex [system] *s.* adsorption complex
adsorber Adsorber *m*, Adsorptionsapparat *m*
adsorption Adsorption *f*
adsorption analysis Adsorptionsanalyse *f*
adsorption chromatography Adsorptionschromatographie *f* • **by adsorption chromatography** adsorptionschromatographisch
adsorption complex Adsorbat *n (System aus Adsorptiv plus Adsorbens)*
adsorption equilibrium Adsorptionsgleichgewicht *n*
adsorption filtration Adsorptionsfiltration *f*
adsorption indicator Adsorptionsindikator *m*
adsorption isotherm Adsorptionsisotherme *f*
adsorption site Adsorptionszentrum *n*

adsorptive adsorptiv, adsorptionsfähig, adsorbierend, Adsorptions...
adsorptive Adsorptiv *n*, zu adsorbierender Stoff *m*
advanced waste treatment (stage) *(Hyd)* weitgehende Abwasserbehandlung *f*, dritte Reinigungsstufe *f*, Nachreinigung *f*
advancement of reaction Reaktionslaufzahl *f*
aeolotropic anisotrop
aeolotropism Anisotropie *f*
aerate *v* mit Luft versetzen [vermischen], belüften, begasen
aerated concrete Porenbeton *m*, Gasbeton *m*
aerated fermentation *(Biot)* belüftete Fermentation *f*
aeration Lüftung *f*, Luftbegasung *f*, Begasung *f*
aeration basin *s.* aeration tank
aeration cell *s.* differential aeration cell
aeration tank *(Hyd)* Belüftungsbecken *n*, Lüftungsbecken *n*, Belebungsbecken *n*, Belebtschlammbecken *n*
aerator *(Hyd)* Belüftungseinrichtung *f*, Belüfter *m*, Begaser *m*
aerobe *(Bio)* Aerobier *m*, Aerobiont *m*
aerobic aerob • **under aerobic conditions** unter aeroben Bedingungen [Verhältnissen], im aeroben Medium, unter Luftzutritt, aerob
aerobic digestion *(Hyd)* aerober biologischer Abbau *m*
aerobic fermentation aerobe [oxidative] Gärung *f*; aerobe Fermentation *f*
aerobic oxidation aerobe Oxidation *f*
aerobic reactor *(Biot)* Reaktor *m* für aerobe Verfahren
aerobiotic *(Biot)* aerob, oxybiontisch *(in Gegenwart von Luftsauerstoff lebend)*
aerogel Aerogel *n (Gel mit gasförmigem Dispersionsmittel)*
aerosol Aerosol *n*
aerosol mist Aerosolnebel *m*; Wirkstoffnebel *m*
aerosol propellant Aerosoltreibmittel *n*, Treibgas *n* für Aerosole
aerosolization Aero(so)lisierung *f*, Vernebelung *f*
aerosolize *v* aero(so)lieren, vernebeln
aerugo Grünspan *m (Gemisch aus basischen Kupfer(II)-acetaten)*
AES 1. *s.* Auger (electron) spectroscopy; 2. *s.* atomic-emission spectroscopy; 3. = American Electrochemical Society
AFC *s.* alkaline fuel cell
affect *v* angreifen, beeinflussen, wirken auf, einwirken auf
affinity 1. Affinität *f*; 2. *s.* chemical affinity
affinity chromatography Affinitätschromatographie *f*, Bioaffinitätschromatographie *f* • **by affinity chromatography** affinitätschromatographisch
affinity for oxygen Sauerstoffaffinität *f*, Affinität *f* zu(m) Sauerstoff
affix *v* fixieren, festheften
afflux Zufluss *m*
AFM *s.* 1. atomic force microscope; 2. atomic force microscopy
AFS *s.* atomic fluorescence spectroscopy
afterbake *v (Kst)* nachhärten
afterbake *(Kst)* Nachhärten *n*, Nachhärtung *f*
afterburner Nachverbrennungsanlage *f (Müllverbrennung)*
afterburning Nachverbrennung *f (von Schadstoffen in Abgasen)*
afterchlorinate *v* nachchlorieren
afterchrome dye Nachchromier(ungs)farbstoff *m*, Chromentwicklungsfarbstoff *m*
aftercrystallization Nachkristallisation *f*
afterglow Nachleuchten *n*; Nachglühen *n*, Nachglimmen *n*
afterprecipitation Nachfällung *f*
aftertreat *v* nachbehandeln
aftertreating Nachbehandeln *n*, Nachbehandlung *f*
aftertreatment Nachbehandeln *n*, Nachbehandlung *f*
agar(-agar) Agar(-Agar) *m(n)*, Gelose *f*
agar gel electrophoresis *(Anal)* Agar-Gel-Elektrophorese *f*
agate *(Min)* Achat *m*
age *v* 1. altern; 2. *(Lebm)* ablagern; reifen *(Viskose)*; 3. *(Pap)* vergilben, altern
age-harden *v* aus(scheidungs)härten
age determination Altersbestimmung *f*
age hardening Aus(scheidungs)härten *n*
ageing 1. Altern *n*, Alterung *f*; 2. *(Lebm)* Ablagern *n*; Reifen *n (von Viskose)*; 3. *(Pap)* Vergilben *n*, Altern *n*
ageing resistance Alterungsbeständigkeit *f*, Alterungswiderstand *m*

agency treibende Kraft *f*, Triebkraft *f*, Faktor *m*
agent Agens *n*, (wirksames) Mittel *n*
agent of damage Schadstoff *m*
agglomerate *v* agglomerieren, zusammenbacken, zusammenballen, stückig machen, *(mittels Flüssigkeit:)* granulieren, krümeln, *(mittels Hitze:)* sintern, fritten; agglomerieren, sich zusammenballen, zusammenbacken, klumpen, *(bei Hitzeeinwirkung:)* sintern, (zusammen)fritten
agglomerate *v* **into floc** Flocken bilden
agglomerate Agglomerat *n*, Anhäufung *f*; Granulatkorn *n*, Granalie *f*; Sinterstoff *m*, Sintererzeugnis *n*
agglomeration Agglomerieren *n*, Agglomeration *f*, Agglomerierung *f*, Zusammenbacken *n*, Zusammenballen *n*, Stückigmachen *n*, *(mittels Flüssigkeit:)* Granulieren *n*, Granulation *f*, Granulatformen *n*, Krümeln *n*, *(mittels Hitze:)* Sintern *n*, Fritten *n*, Zusammenfritten *n*
agglutinant agglutinierend, klebend
agglutinant Bindemittel *n*, Klebemittel *n*
agglutinate *v* agglutinieren, verkleben, sich zusammenballen, klumpen; agglutinieren, (ver)kleben, (ver)kitten, binden
agglutination Agglutination *f*, Verklebung *f*, Zusammenballung, Verklumpung *f*
agglutinogen Agglutinogen *n*, agglutinables Antigen *n*
aggregate *v* 1. (sich) aggregieren, ein Aggregat bilden, sich zu einem Aggregat vereinigen, zu einem Aggregat zusammentreten; 2. aggregieren, anhäufen
aggregate 1. Aggregat *n (Verband von Atomen, Molekülen oder größeren Teilchen)*; 2. *(org Ch)* Ringkomplex *m*, polycyclische Verbindung *f*; 3. *(Bau)* Zuschlag(stoff) *m*
aggregate of molecules Molekülaggregat *n*, Molekularaggregat *n*, molekulares Aggregat *n*
aggregate of particles Teilchenaggregat *n*
aggregation Aggregation *f*, Aggregierung *f*
aggregation of molecules Molekülaggregation *f*, Molekularaggregation *f*
aggressive aggressiv *(Chemikalien)*
aggressiveness Aggressivität *f*, aggressiver Charakter *m (von Chemikalien)*
aggressivity *s.* aggressiveness
aging *s.* ageing
agitate *v* bewegen; rühren, umrühren, durchrühren; schütteln; auflockern
agitated batch reactor diskontinuierlicher Rührkessel(reaktor) *m*
agitated bioreactor *(Biot)* Rühr(kessel)-fermenter *m*, gerührter Bioreaktor *m*
agitated extraction tower Rührkolonne *f*
agitated fermenter *s.* agitated bioreactor
agitated tank Rührkessel *m*
agitation Bewegung *f*; Rühren *n*, Umrühren *n*, Durchrühren *n*, Aufrühren *n*; Schütteln *n*
agitation leaching *(Aufb)* Rührlaugung *f*
agitator 1. Rührer *m*; Rührwerk *n (Rührer mit Motor)*; Rührmaschine *f*, Rührapparat *m*; 2. *s.* agitator blade
agitator blade Rühr(er)schaufel *f*, Rühr(er)flügel *m*, Rühr(er)blatt *n*, Mischflügel *m*
aglucone Aglucon *n (zuckerfreier Bestandteil eines Glucosids)*
aglycon(e) Aglycon *n*, zuckerfreier [nicht zuckerartiger, zuckerfremder] Bestandteil *m*, Nichtzuckeranteil *m*
agon *s.* coenzyme
agonist Agonist *m (sich mit einem Rezeptor verbindendes Arzneimittel, das Zelleigenschaften verändert)*
agricultural chemical Agrochemikalie *f*
agricultural chemist Agrochemiker *m*
agricultural chemistry Agrarchemie *f*, Agrochemie *f*, landwirtschaftliche Chemie *f*, Landwirtschaftschemie *f*
agrochemistry *s.* agricultural chemistry
AHF *s.* antihaemophilic factor
AIBN *s.* azobisisobutyronitrile
AIC 1. = American Institute of Chemists; 2. = Associate of the Institute of Chemistry
AICAR = 5-amino-4-imidazole carboxamide ribotide
AICE 1. = American Institute of Chemical Engineers; 2. = American Institute of Consulting Engineers
aid Hilfsstoff *m*, Hilfsmittel *n*, Hilfe *f*

AIR = 5-aminoimidazole ribotide
air *v* lüften, der Luft aussetzen
air-dry *v* an der Luft trocknen
air Luft *f* • **in the absence of air** unter Luftabschluss • **sealed from the air** unter Luftabschluss
air agitation pneumatisches Rühren *n*
air bath Luftbad *n*
air-blown asphalt Blasbitumen *n*, geblasenes Bitumen *n*
air-borne contaminant *s.* air pollutant
air bubble Luftblase *f*, Luftbläschen *n*
air-carbon reaction Reaktion *f* des Kohlenstoffs mit Luft, Generatorgasreaktion *f*, Boudouard-Reaktion *f*
air classifier Sichter *m*, Windsichter *m*
air classifying Sichten *n*, Windsichten *n*, Aeroklassieren *n*
air condenser *(Lab)* Luftkühler *m*
air contaminant *s.* air pollutant
air conveying pneumatisches Fördern *n*
air conveyor pneumatischer Förderer *m*, Luftförderer *m*
air cooling Luftkühlung *f*
air-damped balance Waage *f* mit Luftdämpfung
air dew point Lufttaupunkt *m*
air-dried luftgetrocknet, lufttrocken, lutro
air-dried moisture Feuchte *f* der lufttrockenen Probe, hygroskopische Feuchte *f*
air-dry lufttrocken, luftgetrocknet, lutro
air dryer Lufttrockner *m*, atmosphärischer Trockner *m*
air drying Lufttrocknen *n*, Lufttrocknung *f*, Freilufttrocknen *n*, natürliches Trocknen *n*, atmosphärisches Trocknen *n*, Trocknen *n* an der Luft
air emission standard Emissionsgrenzwert *m* für Luftverunreinigungen
air entrainment 1. Mitreißen *n* von Luft *(in strömenden Flüssigkeiten)*; 2. *(Hyd)* Einleiten *n* [Einführung *f*] von Luft *(in Wasser)*; Lufteintrag *m*, Lufteintragung *f*, Luftzufuhr *f*, Luftzuführung *f*
air filter Luftfilter *n*
air floating Sichten *n*, Windsichten *n*, Aeroklassieren *n*
air flow Luftstrom *m*
air furnace Flammofen *m*
air-hardening lime Luftkalk *m*
air-hardening steel lufthärtender Stahl *m*, Lufthärtestahl *m*, Lufthärter *m*
air heater Lufterhitzer *m*; Luftvorwärmer *m*; Winderhitzer *m*
air humidity Luftfeuchte *f*
air input *s.* air entrainment 2.
air lift Drucklufttheber *m*, Lufttheber *m*, Airlift *m*, Mammutpumpe *f*; Airlift *m* *(Kontakttheber beim Airliftkracken)*
air liquefaction Luftverflüssigung *f*
air-liquid interface Grenzfläche *f* Luft gegen Flüssigkeit
air monitoring Luft(qualitäts)überwachung *f*
air-operated pneumatisch betätigt, druckluftbetätigt, luftbewegt; luftgesteuert
air oven *(Lab)* Trockenschrank *m*
air oxidation Oxidation *f* durch Luft(-sauerstoff), Luftoxidation *f*
air pocket Lufteinschluss *m*
air pollutant Schmutzstoff *m* in der Luft, luftverunreinigender Stoff *m*, Luftverunreinigung *f*, *(i. e. S.)* Luftschadstoff *m*
air pollution Luftverunreinigung *f*, Luftverschmutzung *f*
air-pollution index Luftgüteindex *m*, Luftqualitätsindex *m*, Luftqualitätskennziffer *f*
air preheater Luftvorwärmer *m*, Luvo *m*
air pressure Luftdruck *m*, Atmosphärendruck *m*
air-quality index *s.* air-pollution index
air rate Luftdurchsatz *m*, Luftmassestrom *m*; Luftvolumenstrom *m*
air recirculation Luft(rück)führung *f* im Kreislauf, Luftführung *f* im Umluftbetrieb; Luftzirkulation *f*
air refining *(Met)* Windfrischen *n*
air relief Entlüftungseinrichtung *f*, Entlüftungsöffnung *f*, Lüftungsöffnung *f*
air sample Luftprobe *f*
air-sensitive luftempfindlich
air separation 1. Sichten *n*, Windsichten *n*, Aeroklassieren *n*; 2. Luftzerlegung *f*, Lufttrennung *f*
air separator Sichter *m*, Windsichter *m*
air sizing Sichten *n*, Windsichten *n*, Aeroklassieren *n*
air sparger *(Biot)* Luftverteiler *m* *(Fermenterbelüftung)*
air sweeping Sichten *n*, Windsichten *n*, Aeroklassieren *n*
air vent Entlüftung(seinrichtung) *f*, Entlüftungsöffnung *f*, Lüftungsöffnung *f*

air void Lufteinschluss *m*, Luftpore *f*, Luftblase *f*
air-water interface Grenzfläche *f* Luft gegen Wasser
aircraft turbine fuel Turbinentreibstoff *m* [Düsentreibstoff *m*] für Flugturbinen
airfree luftfrei
airlift *s.* air lift
Aitken particle Aitken-Partikel *n*, Aitken-Teilchen *n* *(Aerosolteilchen unter 0,1 µm Durchmesser)*
alanate Alanat *n* *(Metall-Mischhydrid der allgemeinen Formel* $M(AlH_4)_n$*)*
alane $(AlH_3)_x$ Alan *n*, Aluminiumwasserstoff *m*
alanine Alanin *n*, Ala, 2-Aminopropionsäure *f*
ALAT = L-alanine-2-oxoglutarate aminotransferase
albedo Oberflächenhelligkeit *f*
albumen 1. Eiweiß *n*, Eiklar *n*, Weißei *n*; 2. *s.* albumin
albumin *(Bioch)* Albumin *n*
albuminoid Skleroprotein *n*, Gerüsteiweißstoff *m*, Albuminoid *n*; *(früher i. w. S.)* Eiweißstoff *m*
albuminous Eiweiß...
albuminous matter Eiweißstoff *m*, Eiweiß *n*
alc. *s.* alcohol
alchemist Alchimist *m*, Alchemist *m*
alchemistic(al) alchimistisch, alchemistisch
alchemy Alchimie *f*, Alchemie *f*
alcohol Alkohol *m*, *(i. e. S.)* Ethylalkohol *m*, *Ethanol *n*
alcohol fuel Alkoholkraftstoff *m*
alcohol-in-glass thermometer Alkoholthermometer *n*
alcohol-soluble alkohollöslich
alcoholate Alkoholat *n*
alcoholic alkoholisch, spirituos, spirituös, geistig, Alkohol..., Spiritus..., Sprit...
alcoholic beverage alkoholisches [geistiges] Getränk *n*
alcoholic extract alkoholischer Auszug *m*, Alkoholauszug *m*
alcoholic fermentation alkoholische Gärung *f*, Alkoholgärung *f*
alcoholic solution alkoholische Lösung *f*
alcoholization 1. Versetzen *n* mit Alkohol; 2. *(Lebm)* Alkoholisierung *f* *(z. B. von Wein)*; Sättigung *f* mit Alkohol; 3. Überführung *f* in Alkohol
alcoholize *v* 1. mit Alkohol versetzen; 2. *(Lebm)* alkoholisieren *(z. B. Wein)*; mit Alkohol sättigen; 3. in Alkohol überführen
alcoholysis Alkoholyse *f*, *(meist)* Umesterung *f*
alcoholyze *v* einer Alkoholyse unterwerfen, *(meist)* umestern
ALD *s.* atom layer deposition
aldaric acid Aldarsäure *f*, Zuckerdicarbonsäure *f*
aldazine *(org Ch)* Aldazin *n*
aldehyde Aldehyd *m*
aldehyde condensation Aldehydkondensation *f*
aldehyde group –CHO Aldehydgruppe *f*, Formylgruppe *f*
aldehyde-ketone-rearrangement Aldehyd-Keton-Umlagerung *f*
aldehyde sugar Aldehydzucker *m*
aldehydic aldehydig; Aldehyd...
aldehydic acid Aldehyd(carbon)säure *f*
aldehydic group *s.* aldehyde group
aldimine Aldimin *n*
aldohexose Aldohexose *f* *(Hexose mit einer Aldehydgruppe)*
aldoketene RCH=C=O Aldoketen *n*
aldol Aldol *n* *(allg. für einen 3-Hydroxyaldehyd)*; *(i. e. S.)* $CH_3CH(OH){=}CH_2CHO$ Aldol *n*, Acetaldol *n*, *3-Hydroxybutanal *n*, 3-Hydroxybutyraldehyd *m*
aldol condensation 1. Aldolkondensation *f* *(unter Bildung einer α,β-ungesättigten Carbonylverbindung)*; 2. *s.* aldol reaction
aldol reaction Aldolreaktion *f*, Aldoladdition *f*, Aldolisierung *f* *(Dimerisierung von Aldehyden)*
aldolization *s.* aldol condensation
aldonic acid Aldonsäure *f*, Onsäure *f* *(allg. für ein Aldose-Zuckerderivat der Formel* $HOCH_2(CHOH)_nCOOH$*)*
aldose Aldose *f* *(allg. für ein Monosaccharid mit der Gruppe –CHO)*
aldoxime R–CH=NOH Aldoxim *n*
algae *(Bio)* Algen *fpl*
algaecide *s.* algicide
algal biomass *(Biot)* Algenbiomasse *f*, Algenrohmasse *f*
algal culture *(Biot)* Algenkultur *f*

algicide Algizid *n*, Algenbekämpfungsmittel *n*
alginate Alginat *n (1. Salz oder Ester der Alginsäure*; *2. aus Meeresalgen gewonnenes Handelsprodukt)*
alginate fibre *(Text)* Alginatfaser *f*; Alginatfaserstoff *m*, AL
alginate yarn Alginatseide *f*, ALS
alginic acid Alginsäure *f*, Algensäure *f*
alicyclic compounds alicyclische [cycloaliphatische] Verbindungen *fpl*, Alicyclen *pl*
alicyclics *s.* alicyclic compounds
align *v* ausrichten; sich ausrichten, sich (regelmäßig) anordnen
alignment Ausrichtung *f*, Orientierung *f*, (regelmäßig ausgerichtete) Anordnung *f*, Ordnung *f (von Kettenmolekülen)*
aliphatic aldehyde aliphatischer Aldehyd *m*, Alkanal *n*
aliphatic chemistry aliphatische Chemie *f*, Aliphatenchemie *f*
aliphatic compounds *s.* aliphatics
aliphatic diol Glykol *n*, zweiwertiger Alkohol *m*
aliphatic fluorocarbon Fluoralkan *n*
aliphatic hydrocarbon aliphatischer [acyclischer, kettenförmiger, offenkettiger] Kohlenwasserstoff *m*
aliphatics Aliphaten *pl*, aliphatische [acyclische] Verbindungen *fpl*, Verbindungen *fpl* der Fettreihe
aliquot *v* aliquotieren, in gleiche Anteile aufteilen
aliquot (part) aliquoter Teil *m*
aliquot (portion) aliquoter Teil *m*
alitizing Alitieren *n (Metall-Veredelungsverfahren)*
alizarin dye Alizarinfarbstoff *m*
alk. *s.* alkali
alkadiene Alkadien *n*, Dien *n*
alkadiyne *(org Ch)* Alkadiin *n*
alkalescent schwach alkalisch
alkali 1. Alkali *n*; 2. *im Handel ungenau für Alkalihydroxide und -carbonate*
alkali alkoxide Alkalialkoholat *n*
alkali borate glass Alkaliboratglas *n*
alkali cellulose Alkalicellulose *f*, *(meist)* Natroncellulose *f*
alkali-chlorine cell Chloralkali-Elektrolysezelle *f*
alkali cleavage Alkalispaltung *f*
alkali error Alkalifehler *m (Elektroanalyse)*
alkali-fast *(Farb)* alkaliecht
alkali feldspar Alkalifeldspat *m*
alkali fusion Alkalischmelze *f*, Schmelzen *n* mit Alkali
alkali glass Alkaliglas *n*, alkalihaltiges Glas *n*
alkali halide Alkali(metall)halogenid *n*
alkali hydroxide Alkali(metall)hydroxid *n*
alkali-insoluble alkaliunlöslich
alkali ion Alkaliion *n*
alkali-labile alkalilabil, alkaliunbeständig
alkali metal Alkalimetall *n*
alkali metal ion Alkalimetallion *n*
alkali metal salt Alkali(metall)salz *n*
alkali oxide Alkalimetalloxid *n*
alkali phosphate glass Alkaliphosphatglas *n*
alkali-proof alkalibeständig, alkalifest
alkali pulp alkalisch gekochter [erkochter, aufgeschlossener] Zellstoff *m*
alkali resistance Alkalibeständigkeit *f*, Alkalistabilität *f*, Laugenbeständigkeit *f*
alkali-resistant alkalibeständig, alkalifest, laugenbeständig
alkali sensitivity Alkaliempfindlichkeit *f*
alkali silicate glass Alkalisilikatglas *n*
alkali soluble alkalilöslich
alkali-unstable alkaliunbeständig, nicht alkalibeständig, alkalifest
alkalies Alkalien *npl*
alkalify *v* alkalisieren, alkalisch [basisch] stellen; alkalisch werden
alkalimeter 1. Alkalimeter *n (zur Bestimmung des Laugengehalts einer Flüssigkeit)*; 2. Kaliapparat *m (zur Kohlensäurebestimmung)*
alkalimetric alkalimetrisch
alkalimetry Alkalimetrie *f*
alkaline alkalisch, basisch • **highly alkaline** stark alkalisch • **mildly alkaline** schwach alkalisch • **be alkaline to** alkalisch reagieren gegen • **make alkaline** 1. alkalisch machen, alkalisieren; 2. *(Farb)* alkalisch stellen
alkaline cell *(Ech)* Alkalibatterie *f*
alkaline earth Erdalkali *n*
alkaline-earth metal Erdalkalimetall *n*, Metall *n* der Erdalkaligruppe
alkaline earths Erdalkalien *npl*
alkaline error Alkalifehler *m (Elektroanalyse)*

alkaline fuel cell alkalische Brennstoffzelle *f*
alkaline hydrolysis alkalische Hydrolyse *f*
alkaline pulping *(Pap)* alkalischer Aufschluss *m*
alkaline reaction alkalische [basische] Reaktion *f*
alkaline saponification alkalische Verseifung *f*
alkaline solution alkalische Lösung *f*
alkalinity 1. Alkalität *f*, Basizität *f*; 2. *(Boden)* Basengehalt *m*
alkalinization Alkalisierung *f*, Alkalischmachen *n*
alkalinize *v* 1. alkalisieren, alkalisch machen; 2. *(Farb)* alkalisch stellen
alkalizable alkalisierbar
alkaloid Alkaloid *n*
alkaloidal alkaloidisch, Alkaloid...
alkamine Alkamin *n*, Alkanolamin *n*, Aminoalkohol *m*
alkanal *Alkanal *n*, aliphatischer Aldehyd *m*
alkanation Alkylation *f*, Alkylieren *n*, Alkylierung *f*
alkane *Alkan *n*, Paraffinkohlenwasserstoff *m*, gesättigter Kohlenwasserstoff *m*
alkane family Klasse *f* der Alkane, Alkanreihe *f*, Paraffinreihe *f*
alkane fermentation *(Biot)* Alkanfermentation *f*
alkanediyl group Alkandiylgruppe *f*, Alkandiylrest *m*
alkanephosphonous acid *(Am)* $RP(OH)_2$ Alkanphosphonigsäure *f*
alkanoic acid Alkansäure *f (allg. für eine gesättigte aliphatische Carbonsäure)*
alkanolamine *s.* alkamine
alkanone *Alkanon *n*, aliphatisches Keton *n*
alkatriene *(org Ch)* Alkatrien *n*
alkatriyne *(org Ch)* Alkatriin *n*
alkene *Alken *n*, Olefin *n*, Ethylenkohlenwasserstoff *m (acyclischer Kohlenwasserstoff mit einer C=C-Bindung)*
alkene family Klasse *f* der Alkene, Alkenreihe *f*
alkenoic acid Alkensäure *f (allg. für eine einfach ungesättigte Fettsäure)*
alkenylation *(org Ch)* Alkenylierung *f*
alkenylbenzene Alkenylbenzen *n*
alkine *s.* alkyne
alkoxide Alkoholat *n*
alkoxy group $C_nH_{2n+1}O$– Alkoxygruppe *f*, Alkoxylgruppe *f*
alkoxyalkyliminophosphine (RO)P=NR Phosphorigsäurealkylesteralkylimid *n*
alkoxyboration *(org Ch)* Alkoxyborierung *f*
alkoxycarbonylation *(org Ch)* Alkoxycarbonylierung *f*, Carbalkoxylierung *f*
alkoxydealkylation *(org Ch)* Alkoxydealkylierung *f*
alkoxyhalogenation *(org Ch)* Alkoxyhalogenierung *f*
alkoxyl (radical) *s.* alkoxy group
alkoxylalkylation *(org Ch)* Alkoxylalkylierung *f*
alkoxylation *(org Ch)* Alkoxylierung *f*
alkyd *s.* alkyd resin
alkyd paint Alkydharzanstrichstoff *m*, Alkydharzfarbe *f*
alkyd resin Alkyd(harz) *n*
alkyl Alkyl *n*
alkyl bridge Alkylbrücke *f*
alkyl compound Alkylverbindung *f*
alkyl cyanide Alkylcyanid *n*, Nitril *n* der aliphatischen Reihe
alkyl derivative Alkylderivat *n*
alkyl group Alkylgruppe *f*, Alkylrest *m*
alkyl halide Alkylhalogenid *n*, *Halogenalkan *n*
alkyl peroxide Alkylperoxid *n*
alkyl phenol resin Alkylphenolharz *n*
alkyl radical 1. (freies) Alkylradikal *n*; 2. *s.* alkyl group
alkyl shift 1,2-Alkylcarbonium-Umlagerung *f*
alkyl silicone *Alkylsiloxan *n*, Alkylsilikon *n*
alkyl sulphonate Alkylsulfonat *n*
alkylamine Alkylamin *n*
alkylate *v* alkylieren
alkylate Alkylat *n*
alkylating agent Alkylierungsmittel *n*, alkylierendes Mittel *n*, alkylierende Verbindung *f*, Alkylans *n*
alkylation Alkylieren *n*, Alkylierung *f*
alkylbenzene Alkylbenzen *n*
alkylbenzene sulphonate Alkylbenzensulfonat *n (Detergens)*
alkylene Alkandiylgruppe *f*, Alkandiylrest *m*

alkylidene group Alkylidengruppe *f*, Alkylidenrest *m*
alkylideneazinic acid Nitronsäure *f*, *aci*-Nitroverbindung *f*
alkylsulphonic acid Alkylsulfonsäure *f*
alkylsulphuric acid Alkylschwefelsäure *f*
alkyne *Alkin *n*, Acetylen *n*, Acetylenkohlenwasserstoff *m (acyclischer Kohlenwasserstoff mit einer C≡C-Bindung)*
alkyne family Klasse *f* der Alkine, Alkinreihe *f*
alkyne metathesis *(org Ch)* Alkinmethatese *f*
alkynoic acid Alkinsäure *f*
all-cis isomer all-cis-Isomer(e) *n*
all-trans isomer all-trans-Isomer(e) *n*
allantoic acid Allantoinsäure *f*, Glyoxylsäurediureid *n*, Diureidoacetat *n*
allantoin Allantoin *n*, 5-Ureido-hydantoin *n*
allelochemical Allelochemikalie *f (von Organismen erzeugter, auf andere Organismen einwirkender Stoff)*
allelotropic allelotrop
allelotropism Allelotropie *f (Koexistenz tautomerer Formen)*
Allen cone classifier Allen-Konus *m*, Allen-Kegel *m (Klassierapparat)*
allene Allen *n*, *(i. e. S.)* $CH_2=C=CH_2$ Allen *n*, *Propadien *n*
allene isomerism Allen-Isomerie *f*
allenic *(org Ch)* allenisch
allergen *(Med)* Allergen *n*
allied verwandt *(chemische Stoffe)*
Allihn condenser Allihn-Kühler *m*, allihnscher Kühler *m*, Kugelkühler *m* nach Allihn
allocatalysis Allokatalyse *f (Gegenteil: Autokatalyse)*
allomerization *(Bioch)* Allomerisierung *f*
allose Allose *f (eine Aldohexose)*
allosteric *(Bioch)* allosterisch
allosteric enzyme allosterisches (kooperatives oligomeres) Enzym *n*
allosterism *(Bioch)* Allosterie *f*
allotriomorphic *(Krist)* allotriomorph, xenomorph, fremdgestaltig
allotriomorphic transition *s.* allotropic transition
allotrope allotrope Modifikation *f*, Allotrop *n (eines Elements)*
allotropic allotrop
allotropic form [modification] *s.* allotrope
allotropic transition allotrope Umwandlung *f (eines reinen Elements)*
allotropism Allotropie *f (Auftreten eines chemischen Elements in mehreren Modifikationen)*
allotropy Allotropie *f (Auftreten eines chemischen Elements in mehreren Modifikationen)*
allow *v* **to cool** abkühlen lassen
allow *v* **to settle** *(Lab)* absitzen lassen
allow *v* **to stand** stehen lassen
allowed band erlaubtes Band *n*, erlaubter [zugelassener] Energiebereich *m*
allowed transition *(phys Ch)* erlaubter Übergang *m*
alloy *v* legieren, zusammenschmelzen; sich legieren lassen
alloy Legierung *f*
alloy skeleton catalyst Legierungsskelettkatalysator *m*, Raney-Katalysator *m*
alloy steel legierter Stahl *m*
alloying Legieren *n*
alloying element Legierungsbestandteil *m*, Legierungskomponente *f*, Legierungselement *n*, Legierungszusatz *m*
allyl alcohol Allylalkohol *m*, *Prop-2-en-1-ol *n*
allyl anion Allylanion *n*
allyl cation Allylkation *n*
allyl chloride Allylchlorid *n*, *3-Chlorprop-1-en *n*
allyl plastic Allylharz-Kunststoff *m*, Allylester-Kunststoff *m*
allyl resin Allylharz *n*
allylation Allylierung *f*
allylic allylisch, allylständig, Allyl...
allylic chlorination Allylchlorierung *f*
allylic condensation Allylkondensation *f*
allylic group Allylgruppe *f*, Allylrest *m*
allylic hydrogen allylständiges Wasserstoffatom *n*
allylic rearrangement Allylumlagerung *f*
allylic substitution Allylsubstitution *f*
alpha addition (reaction) α-Addition *f*
alpha cellulose Alphacellulose *f*, α-Cellulose *f*
alpha cleavage α-Spaltung *f*
alpha decay [disintegration] Alpha-Zerfall *m*, α-Zerfall *m*
alpha elimination α-Eliminierung *f*

alpha emitter Alphastrahler *m*, α-Strahler *m*
alpha-halogenated α-halogeniert, in α-Stellung halogeniert
alpha helix *(Bioch, Kst)* α-Helix *f*
alpha hydroxy acid 2-Hydroxysäure *f*, α-Hydroxysäure *f*
alpha particle Alphateilchen *n*, α-Teilchen *n*
alpha radiation Alphastrahlung *f*, α-Strahlung *f*
alternating copolymer alternierendes Copolymer(e) *n*
alternating copolymerization alternierende Copolymerisation *f*
alternating current Wechselstrom *m*
alternating voltage sinusförmige Wechselspannung *f*
alum 1. $M^IM^{III}(SO_4)_2 \cdot 12\ H_2O$ Alaun *m*; 2. technisches [handelsübliches] Aluminiumsulfat *n*
alumetize *v* alumetieren, spritzalitieren, spritzaluminieren
alumina Aluminiumoxid *n*, Tonerde *f*
alumina boat Tonerdeschiffchen *n*, Substanzschiffchen *n* aus Tonerde
alumina gel Aluminogel *n*, Tonerdegel *n* *(Aluminiumoxidhydrat)*
aluminate Aluminat *n*
aluminic Aluminium..., Alu...
aluminide Aluminid *n*
aluminiferous 1. aluminiumhaltig; 2. tonerdehaltig, aluminiumoxidhaltig; 3. alaunhaltig
aluminium Al Aluminium *n*
aluminium acetate Aluminiumacetat *n*
aluminium alkyl Trialkylaluminium *n*
aluminium alloy Aluminiumlegierung *f*
aluminium bronze Aluminiumbronze *f*
aluminium chloride Aluminiumchlorid *n*
aluminium fluoride Aluminiumfluorid *n*
aluminium fluorosilicate Aluminiumhexafluorosilicat *n*
aluminium nitrate Aluminiumnitrat *n*
aluminium oxide Aluminiumoxid *n*
aluminium silicofluoride *s.* aluminium fluorosilicate
aluminium soap Aluminiumseife *f*
aluminium sulphate Aluminiumsulfat *n*
aluminium trialkyl Trialkylaluminium *n*
aluminize *v* alitieren; (ver)aluminieren *(ohne Ausbildung von Legierungsschichten)*
aluminosilicate Alumosilicat *n*
aluminothermic aluminothermisch
aluminothermics Aluminothermie *f*, Thermitverfahren *n*, Goldschmidt-Verfahren *n*
aluminothermy Aluminothermie *f*, Thermitverfahren *n*, Goldschmidt-Verfahren *n*
aluminous 1. aluminiumhaltig, Aluminium...; 2. alaunhaltig, Alaun...
aluminum *s.* aluminium
am. = amyl
Am = amyl
Amadori reaction [rearrangement] Amadori-Umlagerung *f (von Glykosylaminen in 1-Aminoketosen)*
amalgam *v s.* amalgamate
amalgam Amalgam *n (Quecksilberlegierung)*
amalgamate *v* amalgamieren; sich amalgamieren
amalgamation Amalgamation *f*, Amalgamieren *n*, Amalgamierung *f*
amalgamator Amalgamator *m*, Amalgamier(ungs)apparat *m*, Amalgamier(ungs)vorrichtung *f*
ambident ambident, ambifunktionell, ambivalent *(mit zwei reaktiven Zentren)*
ambient umgebend, Umgebungs...
ambient temperature Umgebungstemperatur *f*
amebicidal *s.* amoebicidal
americium Am Americium *n*
Ames test Ames-Test *m*, Rückmutationstest *m (zur Prüfung der Mutagenität von Chemikalien)*
amic acid Carbamoylcarbonsäure *f*
amicron Amikron *n (mikroskopisch unsichtbares Schwebeteilchen)*
amidate *v* amidieren
amidation Amidierung *f*
amide Amid *n*, Säureamid *n*; Amid *n*, Metallamid *n*
amide hydrazone Amidhydrazon *n*
amide nitrogen Amidstickstoff *m*
amide oxime Amidoxim *n*
amidic amidisch, Amid...
amidine Amidin *n*, *(i. e. S.)* $RC(=NH)NH_2$ Amidin *n*, Carboxamidin *n*
amidine group $-C(=NH)NH_2$ Amidinogruppe *f*, Amidinocarbonylgruppe *f*, Guanylgruppe *f*
amido group $-CONH_2$ Amidogruppe *f*

amidoalkylation *(org Ch)* Amidoalkylierung *f*, Acylaminoalkylierung *f*
amidrazone Amidrazon *n*
aminal Aminal *n*
aminate *v* aminieren
amination Aminierung *f*
amine Amin *n*
amine oxide Aminoxid *n*, N-Oxid *n*
amine ylide Ammoniumylid *n*
aminediyl group –NH– Iminogruppe *f*
aminize *v s.* aminate
amino acid Amino(carbon)säure *f*
amino-acid analyser Aminosäureanalysator *m*
amino-acid residue Aminosäurerest *m*
amino-acid sequence Aminosäuresequenz *f*, Aminosäure(reihen)folge *f*
amino-acid unit Aminosäureeinheit *f*
amino alcohol Aminoalkohol *m*, Alkanolamin *n*, Alkamin *n*
amino compound Aminoverbindung *f*
amino dicarboxylic acid Aminodicarbonsäure *f*
amino group $-NH_2$ Aminogruppe *f*
amino moiety Amino(an)teil *m*, Aminokomponente *f*
amino polycarboxylic acid Aminopolycarbonsäure *f*
amino resin Aminoplastharz *n*
amino succinic acid Aminobernsteinsäure *f*, Asparaginsäure *f*, *Aminobutandisäure *f*
amino sugar Aminozucker *m*
α-**amino alcohol** Halbaminal *n*, α-Aminoalkohol *m*
α-**amino carboxylic acid** α-Aminocarbonsäure *f*
aminoacetal Aminoacetal *n*, Aminoacetaldehyddiethylacetal *n*
aminoacetanilide Aminoacetanilid *n*
aminoacetic acid Aminoessigsäure *f*, Glycin *n*, Glykokoll *n*, Gly
aminoacylation Aminoacylierung *f*
α-**aminoadipic acid pathway** *(Bioch)* α-Aminoadipinsäureweg *m*
aminoanthraquinone Aminoantrachinon *n*
aminoazo compound Aminoazoverbindung *f*
aminobenzene Aminobenzen *n*, *Phenylamin *n*, Anilin *n*
aminobenzoic acid Aminobenzoesäure *f*
aminobenzylation Aminobenzylierung *f*
aminobutyric acid Aminobuttersäure *f*
2-amino-2-deoxy-D-α-glucopyranose D-Glucos-amin *n*, 2-Amino-2-deoxy-α-D-glucopyranose *f*
aminoethane *Etylamin *n*
α-**aminoethyl alcohol** 1-Aminoethylalkohol *m*, *1-Aminoethanol *n*
β-**aminoethyl alcohol** 2-Aminoethylalkohol *m*, *2-Aminoethanol *n*, Colamin *n*
aminolysis Aminolyse *f*
aminomethane Methylamin *n*, Aminomethan *n*
aminomethylate *v* aminomethylieren
aminomethylation Aminomethylierung *f*
***p*-aminophenol** *p*-Aminophenol *n*, 4-Amino-1-hydroxybenzen *n*
aminoplastic Aminoplast *m*
aminosulphate Amidosulfat *n*, Sulfamidat *n*, Sulfamat *n*
aminothiourea Thiosemicarbazid *n*, Thiocarbazidsäureamid *n*, Hydrazinthiocarbonsäureamid *n*
aminotoluene Aminotoluen *n*, Toluidin *n*
ammeter Am(pere)meter *n*, Strommesser *m*
ammine Amminverbindung *f*, Ammin(-salz) *n*
ammonia Ammoniak *n*
ammonia alum Ammoniakalaun *m*, Ammoniumalaun *m (Ammoniumsulfat-12-Wasser)*
ammonia gas (gasförmiges) Ammoniak *n*, Ammoniakgas *n*
ammonia nitrogen Ammoniumstickstoff *m*, Ammoniakstickstoff *m*
ammonia oxidation Ammoniakoxidation *f*, Ammoniakverbrennung *f*
ammonia oxidizers *s.* nitrite-forming bacteria
ammonia scrubber Ammoniakwäscher *m*, *(hohe Bauform auch)* Ammoniakwaschturm *m*
ammonia soda process Ammoniak-Soda-Verfahren *n*, Solvay-Verfahren *n*
ammonia solution [spirit] (wässrige) Ammoniaklösung *f*, wässriges Ammoniak *n*, Ammoniakwasser *n*, Salmiakgeist *m*
ammonia synthesis Ammoniaksynthese *f*
ammoniacal ammoniakalisch, ammoniakhaltig

ammoniate *v* ammonisieren *(Düngemittel)*
ammoniate Ammoniakat *n*
ammoniation Ammonisierung *f (von Düngemitteln)*
ammonification 1. *(Boden)* Ammonifizierung *f*, Ammonifikation *f (Ammoniakbildung durch bakterielle Zersetzung organischer Stoffe)*; 2. Ammonisierung *f (von Düngemitteln)*
ammonifiers Ammoniakbildner *mpl*
ammonifying bacteria Ammoniakbildner *mpl*
ammonium acetate Ammoniumacetat *n*
ammonium acid carbonate *s.* ammonium hydrogencarbonate
ammonium alum Ammoniakalaun *m*, Ammoniumalaun *m (Ammoniumaluminiumsulfat-12-Wasser)*
ammonium benzoate Ammoniumbenzoat *n*
ammonium bicarbonate *s.* ammonium hydrogencarbonate
ammonium bichromate *s.* ammonium dichromate
ammonium carbamate Ammoniumcarbamat *n*, Ammoniumamidocarbonat *n*
ammonium carbonate 1. Ammoniumcarbonat; 2. *(Lebm)* (gewöhnliches) Ammoniumcarbonat *n*, Hirschhornsalz *n (Gemisch aus Ammoniumhydrogencarbonat und Ammoniumcarbamat)*
ammonium chloride Ammoniumchlorid *n*, Salmiak *m*
ammonium dichromate Ammoniumdichromat(VI) *n*
ammonium ferrocyanide Ammoniumhexacyanoferrat(II) *n*
ammonium fluorosilicate Ammoniumhexafluorosilicat *n*
ammonium formate Ammoniumformiat *n*
ammonium group Ammoniumgruppe *f*, Ammoniumrest *m*
ammonium hydrogencarbonate Ammoniumhydrogencarbonat *n*
ammonium hydrogensulphide Ammoniumhydrogensulfid *n*
ammonium hydroxide Ammoniakhydrat *n*, *(hypothetisch:)* Ammoniumhydroxid *n*
ammonium molybdate 1. $(NH_4)_2MoO_4$ (normales) Ammoniummolybdat(VI); 2. *s.* ammonium paramolybdate
ammonium nitrate Ammoniumnitrat *n*
ammonium nitrate explosive Ammonsalpetersprengstoff *m*
ammonium paramolybdate Ammoniumparamolybdat *n*, Ammoniummolybdat *n* des Handels, Hexammoniumheptamolybdat-4-Wasser *n*
ammonium peroxodisulphate Ammoniumper(oxodi)sulfat *n*
ammonium per(oxydi)sulphate Ammoniumper(oxodi)sulfat *n*
ammonium picrate Ammoniumpikrat *n*
ammonium radical Ammoniumgruppe *f*, Ammoniumrest *m*
ammonium salt Ammoniumsalz *n*
ammonium sulphamate Ammoniumsulfamat *n*
ammonium sulphate Ammoniumsulfat *n*, Ammonsulfat *n*
ammonium sulphide Ammoniumsulfid *n*
ammonium thiocyanate Ammoniumthiocyanat *n*
ammonium thiosulphate Ammoniumthiosulfat *n*
ammonium tungstate Ammoniumwolframat *n*
ammonium ylide Ammoniumylid *n*
ammonolysis Ammonolyse *f*
ammonolytic ammonolytisch
ammonotelic *(Bioch)* ammonotel *(Stickstoff als Ammoniak ausscheidend)*
ammonotelism *(Bioch)* Ammonotelie *f (Ausscheidung von Stickstoff als Ammoniak)*
ammoxidation Ammonoxidation *f*
amoebicidal *(Pharm)* amöbizid
amoebicide *(Pharm)* Amöbizid *n*
amorphicity *s.* amorphousness
amorphous amorph, nicht kristallin
amorphousness Amorphie *f*
amount adsorbed adsorbierte Menge *f*, *(i. e. S.)* Sättigungsmenge *f*
amount of charge Ladungsmenge *f*
amount of heat Wärmemenge *f*
amount of substance Stoffmenge *f*, Substanzmenge *f*
AMP *s.* adenosine monophosphate
amperometric amperometrisch

amperometric titration amperometrische Titration *f*
amperometry Amperometrie *f*
amphetamine Amphetamin *n*, 1-Methyl-2-phenylethylamin *n*
amphi-position amphi-Stellung *f*
amphibolic pathway *(Bioch)* amphibolische Sequenz *f*, Zentralbahn *f*, Hauptstoffwechselweg *m*
amphibolism *(Bioch)* Amphibolismus *m*
amphion Ampho-Ion *n*, Zwitterion *n*, inneres [intramolekulares] Salz *n*
amphipathic *s.* amphiphilic
amphiphilic *(Bioch)* amphiphil, amphipathisch, lyobipolar *(mit polaren und apolaren Gruppen)*
amphiprotic amphoter
ampholyte Ampholyt *m*, amphoterer Stoff *m* [Elektrolyt *m*]
ampholyte ion *s.* amphion
amphoteric amphoter
amphoteric ion *s.* amphion
amphotericity Amphoterie *f*, amphoterer Charakter *m*
amphoterism 1. Amphoterie *f (als Erscheinung)*; 2. *s.* amphotericity
amplitude Amplitude *f*, Schwing(ungs)weite *f*
ampoule Ampulle *f*
ampul(e) *s.* ampoule
AMS 1. *s.* ammonium sulphamate; 2. = α-methylstyrene
amu *s.* atomic mass unit
a. m.u. *s.* atomic mass unit
amyl acetate Essigsäureamylester *m*, Amylacetat *n*
amyl group Amylgruppe *f*, Amylrest *m*, *Pentylgruppe *f*, Pentylrest *m*
amyl nitrite Salpetrigsäureamylester *m*, *(i. e. S. meist)* Isoamylnitrit *n*, *Isopentylnitrit *n*
***sec act*-amyl alcohol** optisch aktiver *sec*-Amylalkohol *m*, *Pentan-2-ol *n*
***n*-amyl chloride** *n*-Amylchlorid *n*, *1-Chlorpentan *n*
amylaceous Stärke...; stärkeartig
amyliferous stärkehaltig
amyloid(al) amyloid, stärkeähnlich
amylolytic enzyme stärkespaltendes Enzym *n*, Amylase *f*
amylopectin Amylopektin *n (Hauptbestandteil der Stärke)*
AN *s.* acetonitrile
A.N. *s.* atomic number
an. *s.* anhydrous
anabolic *(Bioch)* anabol(isch), synthetisch
anabolic route Aufbauweg *m*, anabolische Bahn *f*, Bildungsweg *m*
anabolism *(Bioch)* Anabolismus *m*, Aufbau *m*
anaerobe Anaerobier *m*, Anaerobiont *m*
anaerobic anaerob • **under anaerobic conditions** unter anaeroben Bedingungen, im anaeroben Medium, unter Luftabschluss, anaerob
anaerobic decomposition *s.* anaerobic digestion
anaerobic digester *(Hyd)* Faulbehälter *m*, Faulraum *m*
anaerobic digestion anaerober biologischer Abbau *m*, Faulung *f*
anaerobic fermentation anaerobe [eigentliche] Gärung *f*; anaerobe Fermentation *f*
anaerobic reactor *(Biot)* Reaktor *m* für anaerobe Verfahren
anaesthetic Anästhetikum *n*, Anästhesie(rungs)mittel *n*
anaesthetic action anästhesierende [betäubende] Wirkung *f*
analar *s.* analytical-reagent-quality
analar-grade *s.* analytical-reagent-quality
analeptic *(Pharm)* Analeptikum *n*, (das Zentralnervensystem) anregendes Mittel *n*
analgesic analgetisch, schmerzstillend, schmerzlindernd
analgesic *(Pharm)* Analgetikum *n*, schmerzstillendes [schmerzlinderndes] Mittel *n*, Schmerzlinderungsmittel *n*
analgetic *s.* analgesic
analog(ue) Analogon *n*, Analog(e) *n*, analoge Verbindung *f*
analysable analysierbar
analysand zu analysierender Stoff *m*, Analysensubstanz *f*
analyse *v* 1. analysieren; 2. enthalten; 3. sich erweisen
analyser Analysiergerät *n*, Analysator *m*
analysis 1. Analyse *f*; 2. Zusammensetzung *f*; 3. Gehalt *m* • **perform [run] an analysis** eine Analyse durchführen

analysis by weight Gewichtsanalyse *f*, gravimetrische Analyse *f*
analysis report Analysenprotokoll *n*
analysis scheme Analysengang *m*, Analysenweg *m*
analyst Analytiker *m*, analytisch arbeitender Chemiker *m*
analyte Analyt *m*
analytic *s.* analytical
analytical analytisch, Analysen...
analytical accuracy *(Anal)* *Richtigkeit *f*
analytical balance Analysenwaage *f*, analytische Waage *f*
analytical chemist *s.* analyst
analytical chemistry analytische Chemie *f*, Analytik *f*
analytical column *(Chromat)* Trennsäule *f*, Hauptsäule *f*
analytical enzyme *(Biot)* analytisches Enzym *n*, Enzym *n* für die Analytik
analytical error Analysenfehler *m*
analytical factor Faktor *m (Masseanalyse)*
analytical instrument Analysengerät *n*
analytical method Analysenmethode *f*, Analysenverfahren *n*
analytical procedure Analysenverfahren *n*; Analysendurchführung *f*
analytical reagent Nachweisreagens *n*, (chemisches) Nachweismittel *n*
analytical-reagent-quality analysenrein, p. a.
analytical result Analysenbefund *m*, Analysenergebnis *n*
analytical sample Analysenprobe *f*
analytical specimen *s.* analytical sample
analytical technique Analysenmethodik *f*, Analysenverfahren *n*
analytical value Analysenwert *m*
analytically pure analysenrein
analyze *v s.* analyse
anaphoresis *(phys Ch)* Anaphorese *f*
anatoxin Anatoxin *n (entgiftetes Toxin)*
anchimeric assistance anchimere Unterstützung *f* [Beschleunigung *f*] *(der Reaktionsgeschwindigkeit durch Nachbargruppen)*
anchor *v* verankern, befestigen
ancillary equipment Hilfseinrichtung *f*, Hilfsausrüstung *f*
androgen *(Med)* Androgen *n*
anellate *v (org Ch)* anellieren, kondensieren
anellated system anelliertes [kondensiertes] Ringsystem *n*
anellation *(org Ch)* Anellierung *f*, Kondensation *f*, Ringkondensation *f*
anemometer Anemometer *n*, Windmesser *m*; Gasdurchflussmesser *m*
anerobic *s.* anaerobic
anesthetic *s.* anaesthetic
aneurin(e) Aneurin *n*, Thiamin *n*, Vitamin *n* B_1, antineuritisches Vitamin *n*
angle connector Winkelstück *n*
angle of contact *(phys Ch)* Kontaktwinkel *m*, Randwinkel *m*
angle of diffraction Beugungswinkel *m*
angle of incidence Einfall(s)winkel *m*
angle of polarization Polarisationswinkel *m*
angle of reflection Reflexionswinkel *m*
angle of refraction Brechungswinkel *m*
angle of rotation Drehwert *m*, (optischer) Dreh(ungs)winkel *m*; Torsionswinkel *m*, Konformationswinkel *m*, Rotationswinkel *m*, Diederwinkel *m*
angle of scattering Streu(ungs)winkel *m*
angle of wetting Randwinkel *m*, Benetzungswinkel *m*
angle strain Winkelspannung *f*, Ringspannung *f*, Baeyer-Spannung *f*
angle thermometer Winkelthermometer *n*
angle valve Eckventil *n*
angstrom (unit) Ångström *n*, Å, Ångström-Einheit *f (inkohärente, SI-fremde Einheit der Länge; 1 Å = 10^{-1} nm)*
Ångström (unit) Ångström *n*, Å, Ångström-Einheit *f (inkohärente, SI-fremde Einheit der Länge; 1 Å = 10^{-1} nm)*
angular frequency Winkelfrequenz *f*, Kreisfrequenz *f*
angular momentum Drehimpuls *m*, Impulsmoment *n*
angular momentum rule *(Spektr)* 2. [zweite] hundsche Regel *f*
angular orientation V-förmige Anordnung *f (von Liganden)*
angular rotor Winkelrotor *m (einer Zentrifuge)*
angular velocity Winkelgeschwindigkeit *f*
anh. *s.* anhydrous

anharmonic vibration anharmonische Schwingung *f*
anharmonicity Anharmonizität *f*
anhedral *(Krist)* xenomorph, allotriomorph, fremdgestaltig
anhedron xenomorpher [allotriomorpher] Kristall *m*
anhidrotic *s.* antiperspirant
anhyd. *s.* anhydrous
anhydride Anhydrid *n*
anhydridization Anhydri(di)sierung *f*
anhydridize *v* anhydri(di)sieren
anhydrite *(Min)* Anhydrit *m (wasserfreies Calciumsulfat)*
anhydrize *v s.* anhydridize
anhydro base Anhydrobase *f*
anhydro sugar Anhydrozucker *m*
anhydrotic *s.* antiperspirant
anhydrous wasserfrei, *(i. e. S.)* kristallwasserfrei
anhydrous alcohol absoluter [wasserfreier] Alkohol *m*, Absolutalkohol *m*, wasserfreier Ethylalkohol *m*, wasserfreies Ethanol *n*
anhydrous sodium carbonate wasserfreies Natriumcarbonat *n*, kalzinierte [kristallwasserfreie] Soda *f*
anil Anil *n (eine sich vom Anilin oder einem anderen primären aromatischen Amin ableitende schiffsche Base)*
anilide 1. Anilid *n*, *N*-Phenylamid *n*; 2. Anilid *n (Salz des Anilins)*
aniline Anilin *n*, Aminobenzen *n*, *Phenylamin *n*
aniline dye Anilinfarbstoff *m*, Anilinfarbe *f*
aniline-formaldehyde resin Anilin-(formaldehyd)harz *n*
aniline hydrochloride *Aniliniumchlorid *n*, Anilinhydrochlorid *n*
aniline poisoning Anilinvergiftung *f*
aniline resin 1. Anilinharz *n*; 2. *s.* aniline-formaldehyde resin
aniline yellow Anilingelb *n*, *p*-Aminoazobenzen *n*
***p*-anilinesulphonic acid** Anilin-*p*-sulfonsäure *f*, *p*-Aminobenzensulfonsäure *f*, Sulfanilsäure *f*
anilino group C_6H_5NH- Anilinogruppe *f*
animal tierisch, animalisch, Tier…
animal assay [bioassay] *s.* animal test
animal char(coal) Tierkohle *f*, *(i. e. S.)* Knochenkohle *f*
animal fat tierisches Fett *n*, Tierfett *n*
animal feed additive [supplement] Futtermittelzusatz(stoff) *m*, Futterzusatz *m*, Futtersupplement *n*
animal fibre tierische Faser *f*, Tierfaser *f*
animal glue tierischer Leim *m*, Tierleim *m*
animal health product veterinärmedizinisches [tiermedizinisches] Präparat *n*, Tierarzneimittel *n*
animal protein tierisches Protein *n*, Tiereiweiß *n*
animal starch tierische Stärke *f*, Tierstärke *f*, Leberstärke *f*, Glykogen *n*
animal test Tierversuch *m*, Tierexperiment *n*
anion Anion *n*, negatives [negativ geladenes] Ion *n*
anion acid Anionsäure *f*
anion-active anion(en)aktiv, anionisch, Anion(en)…
anion base Anionbase *f*
anion exchange Anionenaustausch *m*
anion exchanger Anionenaustauscher *m*, Anionit *m*
anion radical Radikalanion *n*
anion scheme *(Anal)* Anionentrennungsgang *m*
anion vacancy Anionenleerstelle *f*, Anionenfehlstelle *f*, Anionenlücke *f*
anionic anionisch, anion(en)aktiv, Anion(en)…
anionic ligand anionischer Ligand *m*, Acidoligand *m*
anionic polymerization anionische Polymerisation *f*
anionic surfactant anion(en)aktiver Stoff *m*, anionisches Tensid *n*, Aniontensid *n*
anionoid anionoid, nukleophil
anionotropic anionotrop
anionotropy Anionotropie *f*
anisodesmic *(Krist)* anisodesmisch
anisole Anisol *n*, *Methylphenylether *m*, *Methoxybenzen *n*
anisotropic anisotrop
anisotropy Anisotropie *f*
ankerite *(Min)* Braunspat *m*
anneal *v* 1. *(Met)* glühen, *(i. e. S.)* stabilglühen, spannungsfrei glühen; 2. *(Kst)* tempern, spannungsfrei machen; 3. *(Glas)* tempern, entspannen *(durch gelenktes Kühlen)*
anneal *s.* annealing
annealing 1. *(Met)* Glühen *n*, *(i. e. S.)* Stabilglühen *n*, Spannungsarmglühen

n, Spannungsfreiglühen *n*; 2. *(Kst)* Tempern *n*, Spannungsfreimachen *n*; 3. *(Glas)* Tempern *n*, Entspannen *n* *(durch gelenktes Kühlen)*; 4. *(Bioch)* Aufschmelzen *n*, Schmelzen *n* *(der Desoxyribonucleinsäure)*

annealing colour Anlassfarbe *f*

annealing point *(Glas)* oberer Kühlpunkt *m*, obere Kühltemperatur *f*, 15-Minuten-Entspannungstemperatur *f*

annealing temperature 1. *(Met)* Glühtemperatur *f*; 2. *s.* annealing point

annihilate *v* *(phys Ch)* vernichten

annihilation *(phys Ch)* Annihilation *f*, Paarvernichtung *f*, Zerstrahlung *f*

annihilation radiation Vernichtungsstrahlung *f*

annular ringförmig, Ring...

annular die *(Kst)* Ringdüse *f*

annular electrode ringförmige Elektrode *f*, Ringelektrode *f*

annular structure ringförmige Struktur *f*, Ringstruktur *f*

annulene *(org Ch)* Annulen *n*

[8]annulene [8]Annulen *n*, Cycloocta-1,3,5,7-tetraen *n*

annulus Ring *m*; Ringrohr *n*, Mantelrohr *n*

anode Anode *f*

anode compartment Anodenraum *m*

anode potential Anodenpotenzial *n*

anode reaction Anodenreaktion *f*, anodische Reaktion *f*

anode slime [sludge] Anodenschlamm *m*

anode voltage Anodenspannung *f*

anodic anodisch, Anoden...

anodic coating Anodisier(ungs)schicht *f*, anodisch hergestellte Schicht *f*

anodic current Anodenstrom *m*

anodic oxidation 1. anodische Oxidation *f* *(Vorgang)*; 2. *s.* anodizing

anodic stripping (analysis) *(Anal)* Inversvoltammetrie *f* [inverse Voltammetrie *f*] an der Anode, anodische Inversvoltammetrie *f* • **by anodic stripping analysis** inversvoltammetrisch

anodization *s.* anodizing

anodize *v* *(Met)* anodisieren, anodisch behandeln, anodisch [elektrochemisch, elektrolytisch] oxidieren

anodizing *(Met)* Anodisieren *n*, anodische Behandlung *f*, anodische [elektrochemische, elektrolytische] Oxidation *f*

anolyte Anolyt *m* *(Elektrolyt im Anodenraum)*

anomalous 1. anomal, abnorm, regelwidrig, unregelmäßig; 2. *(Tech)* normwidrig

anomaly Anomalie *f*, Regelwidrigkeit *f*, Unregelmäßigkeit *f*

anomer Anomer(e) *n* *(Stereochemie)*

anomeric anomer *(Stereochemie)*

anomerization Anomerisierung *f* *(Stereochemie)*

anoxic reactor *(Biot)* *Reaktor für aerobe Mikroorganismen ohne Belüftung*

ansa-compound Ansa-Verbindung *f*

Anschütz head *(Dest)* Anschütz-Aufsatz *m*

ANSI = American National Standards Institute

antacid säurewidrig

antacid *(Pharm)* Antazidum *n* *(Mittel gegen Superazidität)*

antagonism Antagonismus *m*

antagonist Antagonist *m*, antagonistisch wirkender Stoff *m*

antagonistic action antagonistische Wirkung *f*

antagonize *v* entgegenwirken

antarafacial antarafacial, a *(Reaktion)*

anterior-pituitary hormone Hypophysenvorderlappenhormon *n*, Hormon *n* der Adenohypophyse

anthanthrone dye Anthanthronfarbstoff *m*

anthelmint(h)ic anthelminthisch, wurmvertreibend, wurmwidrig, gegen Würmer wirksam

anthelmint(h)ic *(Pharm)* Anthelminthikum *n*, Wurmmittel *n*, wurmwidriges Mittel *n*

anthocyan *s.* anthocyanin

anthocyanin Anthocyan(in) *n* *(Vertreter einer Gruppe von Pflanzenfarbstoffen)*

anthracene *(org Ch)* Anthracen *n*

anthracite *(Carb)* Anthrazit *m*

anthracitic coal anthrazitische Kohle *f*

anthranilate Anthranilat *n*, Anthranilsäureester *m*

anthranilic acid Anthranilsäure *f*, *o*-Aminobenzoesäure *f*

anthraquinone Anthrachinon *n*, 9,10-Dioxodihydroanthracen *n*

anthraquinone dye Anthrachinonfarbstoff *m*, anthrachinoider Farbstoff *m*
anthraquinonoid anthrachinoid, Anthrachinon...
anthrimide *(Farb)* Anthrimid *n*
anthrone Anthron *n*, 9,10-Dihydro-9-oxoanthracen *n*
anthropogenic input anthropogener Eintrag *m*
anti *anti-*..., in *anti*-Stellung befindlich
• **be anti** in *anti*-Stellung stehen
anti-bumping stone Siedestein *m*
anti conformation 1. *anti*-Konformation *f (bei Makromolekülen, Torsionswinkel ± 60°)*; 2. *s.* anticlinal conformation
anti form Antiform *f*, *anti*-Form *f*, *trans*-Form *f*
anti isomer *anti*-Isomer(e) *n*
anti-Markovnikov addition Anti-Markovnikov-Addition *f (durch Peroxide ausgelöste Addition)*
anti-periplanar ± anti-periplanar, gestaffelt, auf Lücke *(Stereochemie)*
anti position *anti*-Stellung *f*
anti-static bars Entionisierstäbe *mpl*
anti-Stokes line anti-Stokes-Linie *f*, anti-stokessche Linie *f (im Raman-Spektrum)*
anti-syn isomerism *anti-syn*-Isomerie *f*, *syn-anti*-Isomerie *f*
antiacid säurewidrig
antiaromatic compound *(org Ch)* antiaromatische Verbindung *f*
antiarthritic *(Pharm)* Antiarthritikum *n*
antibacterial antibakteriell (wirksam)
antibacterial (agent) *s.* bactericide
antibiosis Antibiose *f*
antibiotic *(Pharm)* antibiotisch (wirksam)
antibiotic (agent) *(Pharm)* Antibiotikum *n*, antibiotischer Stoff *m* [Wirkstoff *m*], antibiotisches Heilmittel *n*
antibiotic residue Antibiotikarückstand *m*
antibiotic resistance Antibiotikaresistenz *f*
antibiotic-resistant antibiotikaresistent
antibody Antikörper *m*, Immunkörper *m*
antibody globulin Immunglobulin *n*, Ig
antibody protein Abwehrprotein *n*
antibond Antibindung *f*
antibonding antibindend, lockernd
antibonding Antibindung *f*, Bindungslockerung *f*, Lockerung *f*
antibonding electron antibindendes [lockerndes] Elektron *n*
antibonding orbital antibindendes [lockerndes] Orbital *n*
anticatalysis Antikatalyse *f*, negative Katalyse *f*, *(i. w. S.)* Inhibition *f*, Hemmung *f*, Verzögerung *f*
anticatalyst Antikatalysator *m*, negativer Katalysator *m*, *(i. w. S.)* Inhibitor *m*, Hemmstoff *m*, Verzögerer *m*, Passivator *m*; Katalysatorgift *n*, Kontaktgift *n*
anticatalytic antikatalytisch
anticlinal conformation *anti*-Konformation *f*, anticlinale Konformation *f (Torsionswinkel ±120°)*
anticlined chain *(Kst)* anticlinale Kette *f*
anticlockwise entgegen dem Uhrzeigersinn, im Gegen(uhr)zeigersinn
anticoagulant gerinnungshemmend, gerinnungsverzögernd
anticoagulant *(Pharm)* Antikoagulans *n*, gerinnungshemmendes [gerinnungsverzögerndes] Mittel *n*
anticodon *(Bioch)* Anticodon *n (Sequenz dreier Nucleotide)*
anticompetitive inhibition *(Bioch)* unkompetitive Hemmung *f*
anticonvulsant krampflindernd
anticonvulsant (drug) *(Pharm)* Antikonvulsivum *n*, Antikonvulsans *n*, Antiepileptikum *n*
anticonvulsive *s.* anticonvulsant
anticorrosive korrosionshemmend, korrosionshindernd, korrosionsmindernd, antikorrosiv
anticorrosive agent Korrosionsschutzmittel *n*, Korrosionsinhibitor *m*, Korrosionshemmer *m*, Korrosionshemmstoff *m*, Korrosionsverzögerer *m*
antidepressant *(Pharm)* Antidepressivum *n*, Mittel *n* gegen depressive Verstimmung
antidiabetic *(Pharm)* Antidiabetikum *n*, Antihyperglykämikum *n*
antidiabetic antidiabetisch
antidiarrhoeal antidiarrhöisch
antidiarrhoeic *(Pharm)* Antidiarrhoikum *n*
antidiuretic antidiuretisch, die Harnsekretion hemmend
antidiuretic (agent) *(Pharm)* Antidiuretikum *n*, die Harnsekretion hemmendes Mittel *n*

antidiuretic hormone Vasopressin *n*, antidiuretisches Hormon *n*, ADH
antidote Gegengift *n*, Antidot *n*
antielectron Positron *n*, positives Elektron *n*
antiemetic *(Pharm)* Antiemetikum *n*
antienzyme Antienzym *n*
antiepileptic *s.* anticonvulsant
antifermentative gärungshemmend, gärungsverhindernd
antiferroelectric transition antiferroelektrischer Übergang *m* [Phasenübergang *m*]
antiferromagnetic transition antiferromagnetischer Übergang *m* [Phasenübergang *m*]
antiferromagnetism Antiferromagnetismus *m*
antifertility agent *(Pharm)* Ovulationshemmer *m*
antifoam (agent) *s.* antifoaming agent
antifoam compound *s.* antifoaming agent
antifoaming agent Antischaummittel *n*, Schaumverhinderungsmittel *n*, Schaumzerstörungsmittel *n*, Schaumzerstörer *m*, Entschäumungsmittel *n*, Entschäumer *m*, Schaumdämpfungsmittel *n*, Schaumdämpfer *m*
antifouling (marine) paint Antifoulingfarbe *f*, bewuchsverhindernde Farbe *f*, anwuchsverhindernde Farbe *f*
antifreeze agent Frostschutzmittel *n*, Gefrierschutzmittel *n*
antifroth(ing) agent *s.* antifoaming agent
antifungal 1. antifungal, pilzwidrig, pilzwirksam; 2. *(Pharm)* antimykotisch
antifungal agent [drug] *(Pharm)* Antimykotikum *n*
antigen Antigen *n*
antigenic Antigen...
antigenicity *(Med, Tox)* Antigenität *f*, Antigenwirkung *f*
antihaemophilic factor *(Biot)* antihämophiles Globulin *n* (A), AHG, Faktor *m* VIII
antihaemorrhagic vitamin antihämorrhagisches Vitamin *n*, Koagulationsvitamin *n*, Vitamin *n* K_1, Phyllochinon *n*
antihidrotic *s.* antiperspirant
antihistamine *(Pharm)* Antihistaminikum *n*, Antihistaminpräparat *n*
antihistaminic die Histaminwirkung abschwächend [aufhebend]
antihistaminic (agent) *s.* antihistamine
antihormone Antihormon *n*
antihypertensive blutdrucksenkend, antihypertonisch
antihypertensive drug *(Pharm)* Hypotensivum *n*, Antihypertonikum *n*, blutdrucksenkendes Mittel *n*, Mittel *n* gegen Bluthochdruck
antiinfective vitamin Epithelschutzvitamin *n*, Vitamin *n* A, Retinol *n*
antiinflammatory *s.* antiphlogistic
antiknock (additive) Antiklopfmittel *n*, Klopfbremse *f*
antiknock agent Antiklopfmittel *n*, Klopfbremse *f*
antiknock rating 1. Klopfwertbestimmung *f*, Klopf(wert)prüfung *f*, Ermittlung *f* der Klopffestigkeit; 2. *s.* antiknock value
antiknock value Klopfwert *m*, Klopffestigkeit *f (quantitativ)*
antileucocytic *(Pharm)* antileukozytär
antimalarial gegen Malaria wirksam, Antimalaria...
antimalarial (drug) *(Pharm)* Malaria(bekämpfungs)mittel *n*, Malariaheilmittel *n*, Malariapräparat *n*, Mittel *n* gegen Malariainfektion
antimatter Antimaterie *f*
antimer *s.* optical isomer
antimetabolite Antimetabolit *m (Enzyminhibitor, meist bei kompetitiver Hemmung wirkend)*
antimicrobial agent antimikrobieller Wirkstoff *m*, Antimikrobiotikum *n*
antimicrobial spectrum *(Biot)* antimikrobielles Wirkungsspektrum *n (eines Antibiotikums)*
antimildew agent *(Text)* Antiseptikum *n* gegen Schimmel, verrottungshemmendes Ausrüstungsmittel *n*, Verrottungsschutzmittel *n*
antimonate Antimonat *n*
antimonial lead Hartblei *n (Bleilegierung mit bis zu 15 % Antimon)*
antimonial speiss *(Met)* Antimonspeise *f*
antimonic Antimon..., *(i. e. S.)* Antimon(V)-...
antimonic acid Antimon(V)-säure *f*
antimonic chloride Antimon(V)-chlorid *n*, Antimonpentachlorid *n*

antimonic oxide Antimon(V)-oxid *n*, Diantimonpentoxid *n*, Antimonpentoxid *n*
antimonide Antimonid *n*
antimonious *s.* antimonous
antimonite 1. Antimonat(III) *n*; 2. *(Min)* Antimonit *m*, Stibnit *m*, Antimonglanz *m*, Grauspießglanz *m* *(Antimon(III)-sulfid)*
antimonous Antimon..., *(i. e. S.)* Antimon(III)-...
antimonous chloride Antimon(III)-chlorid *n*, Antimontrichlorid *n*
antimonous oxide Antimon(III)-oxid *n*, Diantimontrioxid *n*, Antimontrioxid *n*
antimony Sb Antimon *n*
antimony hydride Antimonhydrid *n*, *(i. e. S.)* SbH_3 Antimonwasserstoff *m*, Stibin *n*, Antimon(III)-hydrid *n*
antimony pentachloride Antimonpentachlorid *n*, Antimon(V)-chlorid *n*
antimony pentasulphide Antimonpentasulfid *n*, Antimon(V)-sulfid *n*
antimony pentoxide Antimonpentoxid *n*, Diantimonpentoxid *n*, Antimon(V)-oxid *n*
antimony tetroxide Antimontetroxid *n*, Antimon(III,V)-oxid *n*
antimony trichloride Antimontrichlorid *n*, Antimon(III)-chlorid *n*
antimony trioxide Antimontrioxid *n*, Diantimontrioxid *n*, Antimon(III)-oxid *n*
antimony trisulphide Antimontrisulfid *n*, Antimon(III)-sulfid *n*
antimony white Antimonweiß *n*, Antimon(III)-oxid *n*
antimony yellow Antimongelb *n*, Neapelgelb *n*, Bleiantimonat(V) *n*
antimonyl SbO– Antimonoxid..., *(veraltet:)* Antimonyl...
antimonyl chloride Antimon(III)-oxidchlorid *n*
antimutagen *(Bioch)* Antimutagen *n* *(die Mutationsrate senkender Stoff)*
antimycotic *(Pharm)* Antimykotikum *n*, pilztötendes Mittel *n*
antineuralgic antineuralgisch, gegen Kopf- und Nervenschmerzen wirksam
antineuralgic *(Pharm)* Antineuralgikum *n*
antineutrino Antineutrino *n*
antineutron Antineutron *n*
antinucleon *(Kern)* Antinukleon *n* *(Antineutron oder Antiproton)*
antioxidant oxidationshemmend
antioxidant (agent) Antioxidationsmittel *n*, Antioxidans *n*, Antioxygen *n*, Oxidationsverhinderer *m*
antioxidizing agent *s.* antioxidant
antioxygen *s.* antioxidant
antiozidant Ozonschutzmittel *n*, Antiozonans *n*, Antiozonisator *m*
antiozonant Ozonschutzmittel *n*, Antiozonans *n*, Antiozonisator *m*
antiparallel spins antiparallele [entgegengesetzte] Spins *mpl*
antiparticle Antiteilchen *n*, Antipartikel *n*
antiperiplanar conformation *trans*-Konformation *f*, antiperiplanare Konformation *f* *(Torsionswinkel ±180°)*
antiperspirant transpirationshemmend, schweißhemmend, schweißlindernd
antiperspirant Antiperspirans *n*, Antihidrotikum *n*, Antitranspirationsmittel *n*, Schweiß(hemmungs)mittel *n*, schweißhemmendes [schweißlinderndes] Mittel *n*
antiphlogistic antiphlogistisch, antiinflammatorisch, entzündungshemmend
antiphlogistic (agent) *(Pharm)* Antiphlogistikum *n*, Antiinflammatorikum *n*, entzündungshemmendes Mittel *n*
antipode (optischer) Antipode *m*, enantiomorphe Form *f*, Enantiomer(e) *n*, Spiegelbildisomer(e) *n*
antiproton Antiproton *n*
antipsychotic *(Pharm)* Neuroleptikum *n*, Antipsychotikum *n*, Psycholeptikum *n*, Antihalluzinogen *n*
antipyretic antipyretisch, fiebersenkend
antipyretic *(Pharm)* Antipyretikum *n*, fiebersenkendes Mittel *n*, Fiebermittel *n*
antirachitic antirachitisch
antirachitic (agent) *(Pharm)* Antirachitikum *n*
antirachitic vitamin antirachitisches Vitamin *n*, Vitamin *n* D
antirheumatic antirheumatisch
antirheumatic *(Pharm)* Antirheumatikum *n*
antirust rostschützend, Rostschutz...
antirust agent Rostschutzmittel *n*, Rostverhütungsmittel *n*, Rostverhinderer *m*
antirust paint Rostschutzfarbe *f*, Rostschutzanstrichstoff *m*

antiscorbutic vitamin antiskorbutisches Vitamin *n*, Vitamin *n* C
antisense RNA *(Bioch)* Gegensinn-RNA *f*
antiseptic antiseptisch
antiseptic (agent) *(Pharm)* Antiseptikum *n*, antiseptisches Mittel *n*
antisettling agent Absetzverhütungsmittel *n*, Absetzverhinderungsmittel *n*, Antiabsetzmittel *n*, Schwebemittel *n*, absetzverhinderndes Mittel *n*
antispasmodic *(Pharm)* Antispasmodikum *n*, Spasmolytikum *n*, krampflösendes Mittel *n*
antistatic antistatisch
antistatic (additive) Antistatikum *n*, Antistatikmittel *n*, antistatisches Mittel *n*
antistatic agent Antistatikum *n*, Antistatikmittel *n*, antistatisches Mittel *n*
antisterility vitamin Antisterilitätsvitamin *n*, Fertilitätsvitamin *n*, Vitamin *n* E
antisymmetric antisymmetrisch *(Orbital, Schwingung)*
antithixotrophy *(Koll)* rheopexes Verhalten *n*
antitoxic antitoxisch
antitoxin Antitoxin *n*
antitubercular drug *(Pharm)* Tuberkulosemittel *n*, Tuberkulostatikum *n*
antitumour antitumoral, tumorhemmend, Antitumor...
antitumour agent *(Pharm)* Antitumormittel *n*, Antikrebsmittel *n*, Zytostatikum *n*
antitussive *(Pharm)* Antitussivum *n*, Hustenmittel *n*
antiviral antiviral, virentötend, viruzid
antivirus antiviral, virentötend, viruzid
antivitamin Antivitamin *n*, Vitaminantagonist *m*
antixerophthalmic vitamin antixerophthalmisches [antiinfektiöses] Vitamin *n*, Vitamin *n* A, Retinol *n*, Axerophthol *n*
Antu ANT *n*, Antu *n*, α-Naphthylthioharnstoff *m* *(Rodentizid)*
ANTU ANT *n*, Antu *n*, α-Naphthylthioharnstoff *m* *(Rodentizid)*
anxiolytic (sedative) *(Pharm)* Tranquilizer *m*, Ataraktikum *n*, Anxiolytikum *n*, Psychosedativum *n*
AO *s.* atomic orbital
AP = American Patent
A.P. *s.* annealing point
apatite *(Min)* Apatit *m*
APCI *s.* atmospheric pressure chemical ionization
aperiodic aperiodisch
aperture 1. Öffnung *f*, *(bei Sieben auch)* lichtes Lochmaß *n*; 2. Apertur *f* *(Maß für die Güte eines optischen Systems)*
aperture angle Öffnungswinkel *m* *(eines Strahlenkegels)*
apex Scheitel(punkt) *m*; Apex *m* *(eines Hydrozyklons)*
aphrodisiac den Geschlechtstrieb steigernd [anregend]
aphrodisiac (agent) Aphrodisiakum *n*
API 1. = American Petroleum Institute; 2. *s.* atmospheric pressure ionization
APIC = American Petroleum Industries Committee
aplanatic lens aplanatische Linse *f*
apo rearrangement Apo-Umlagerung *f*
apodization *(Spektr)* Apodisation *f* *(ein Multiplikationsverfahren)*
apoenzyme Apoenzym *n*, Enzymprotein *n*
apolar apolar, nicht polar, unpolar
apoprotein *(Bioch)* Apoprotein *n*
aporphine alkaloid Aporphinalkaloid *n*
apparatus Apparat *m*, Gerät *n*, Vorrichtung *f*, Einrichtung *f*; Apparatur *f*
apparatus constant Apparat(e)konstante *f*, Gerätekonstante *f*
apparent density scheinbare Dichte *f*, Schüttdichte *f*, Korndichte *f*
apparent solubility scheinbare Löslichkeit *f*, Scheinlöslichkeit *f*
appearance potential Auftrittsenergie *f*, Auftrittspotenzial *n*, appearance-Potenzial *n* *(Mindestenergie für Auftreten einer Massenlinie im Spektrum)*
APPI *s.* atmospheric pressure photoionization
apple acid Apfelsäure *f*, *Hydroxybutandisäure *f*
applicability Anwendbarkeit *f*, Verwendbarkeit *f*, Einsetzbarkeit *f*, Anwendungsmöglichkeit *f*, Verwendungsmöglichkeit *f*
application 1. Anwendung *f*, Einsatz *m*, Auftragen *n*, Aufbringen *n* *(auf Oberflächen)*; 2. *(Landw)* Ausbringen *n*; 3. *(Med)* Applikation *f*, Verabreichung *f*
application-specified anwendungsspezifisch

applied chemistry angewandte Chemie *f*
apply *v* 1. anwenden, verwenden, einsetzen; auftragen, aufbringen *(auf Oberflächen)*; 2. *(Landw)* ausbringen; 3. *(Med)* verabreichen
apportion *v* zudosieren
appreciably coloured deutlich gefärbt
approach Methode *f*, Verfahren *n*
approval (staatliche) Anerkennung *f (z. B. von Pflanzenschutzmitteln)*
approval scheme Anerkennungsverfahren *n (z. B. für Pflanzenschutzmittel)*
approved name offizieller Name *m (einer organischen Verbindung)*
approximate formula angenäherte Formel *f*, Näherungsformel *f*
approximate procedure Näherungsverfahren *n*
approximate value Näherungswert *m*
approximation 1. Näherung *f (Mathematik)*; 2. Näherungsverfahren *n*
approximation of linear combinations of atomic orbitals LCAO-Näherung *f*
appurtenance Zubehör *n*, Geräte *npl*
apron Band *n*, Förderband *n*
apron conveyor Platten(band)förderer *m*, Gliederbandförderer *m*
aprotic aprot(on)isch
APS = American Physical Society
aq. *s.* aqueous
aq. reg. *s.* aqua regia
aqua regia Aqua regia, Königswasser *n*, Kö., Kw. *(Salzsäure-Salpetersäure-Gemisch)*
aquametry Aquametrie *f*
aquasol Hydrosol *n*
aquate *v* aquotisieren *(Komplexverbindungen)*
aquate Aquat *n (Salz mit Hydratationswasser in nicht stöchiometrischem Verhältnis)*
aquated ion Aquoion *n*
aquatic chemistry Wasserchemie *f*, Chemie *f* des Wassers, Hydrochemie *f*
aquat(izat)ion Aquation *f*, Aquotisierung *f (Komplexchemie)*
aqueous wässrig, wässerig, wssr., wss., Wasser...; wasserartig
aqueous ammonia (wässrige) Ammoniaklösung *f*, Ammoniak *n* in wässriger Lösung, wässriges Ammoniak *n*, Ammoniakwasser *n*, Salmiakgeist *m*
aqueous emulsion wässrige Emulsion *f*
aqueous phase wässrige Phase *f*, Wasserphase *f*
aqueous solution wässrige Lösung *f*
aquifer protection zone Trinkwasserschutzzone *f*
aquo complex [compound] Aquakomplex *m*, Aquokomplex *m*
aquo ion Aquoion *n*
aquotization *s.* aquat(izat)ion
aquotize *v* der Aquation unterliegen
A.R. *s.* analytical reagent
Ar *s.* aryl group
arach(id)ic acid Arachinsäure *f*, *Eicosansäure *f*
arachidonic acid Arachidonsäure *f (eine Eicosatetraensäure)*
araliphatic araliphatisch
aralkane arylsubstituiertes Alkan *n*
aralkyl Aralkyl *n*, arylsubstituierter Alkylrest *m*
aralkyl sulphonate Alkylarensulfonat *n (Waschrohstoff)*
aralkylation Arylalkylierung *f*
arc black Lichtbogenruß *m*
arc discharge Lichtbogenentladung *f*, Bogenentladung *f*
arc excitation *(Anal)* Bogen-Anregung *f*
arc (heated) furnace Lichtbogenofen *m*
arc lamp Lichtbogenlampe *f*, Bogenlampe *f*
arc melting Lichtbogenschmelzen *n*
arc spectrum *(Anal)* Bogenspektrum *n*
arc welding Lichtbogenschweißen *n*
area of filtration Filterfläche *f*
arene epoxide [oxide] Arenepoxid *n*, Arenoxid *n*
arenediyl group Arylengruppe *f*, Arylenrest *m*, Arylen *n (allg. für einen zweiwertigen aromatischen Rest)*
arenium-σ-adduct Wheland-Intermediat *m*
arenium ion Arenenium-Ion *n*
arenonium ion Arenenium-Ion *n*
areometer Aräometer *n*, Senkwaage *f*, Senkspindel *f*, Spindel *f*
areometric aräometrisch
areometry Aräometrie *f*
argentian silberhaltig
argentic Silber..., *(i. e. S.)* Silber(II)-...
argentiferous silberhaltig, silberführend
argentocyanide Cyanoargentat(I) *n*
argentometric argentometrisch

argentometry Argentometrie *f*
argentous Silber..., *(i. e. S.)* Silber(I)-...
argil 1. Ton *m*, *(i. e. S.)* Töpferton *m*; 2. Tonerde *f*, Aluminiumoxid *n*
argilla Ton *m*; Kaolin *m*, Porzellanerde *f*
argillaceous tonig, tonhaltig, Ton...; lehmig, lehmhaltig, Lehm...
argilliferous tonhaltig; lehmhaltig
arginine Arginin *n*, Arg, α-Amino-δ-guanidovaleriansäure *f*
argon Ar Argon *n*
argon (ion) laser *(Spektr)* Argonionen-Laser *m*, Argon-Laser *m*, Ar^+-Laser *m*
arithmetic mean *(Anal)* arithmetisches Mittel *n*
arm Schaufel *f*, Arm *m*, Flügel *m*, Paddel *n*
aroma Aroma *n*, Geruch *m*, Wohlgeruch *m*, Duft *m*; Würze *f*, Blume *f (des Weins)*
aroma body *s.* aroma substance
aroma development Aromabildung *f*
aroma substance Aromastoff *m*
aromatic 1. *(org Ch)* aromatisch; 2. aromatisch, wohlriechend, duftend, würzig
aromatic character *(org Ch)* aromatischer Charakter *m*, Aromatizität *f*
aromatic compounds aromatische Verbindungen *fpl*, Aromaten *mpl (benzoid und nicht benzoid)*
aromatic ether aromatischer Ether *m*, Arylether *m*
aromatic hydrocarbons aromatische Kohlenwasserstoffe *mpl*, Benzenkohlenwasserstoffe *mpl*, Aromaten *pl*
aromatic nucleus aromatischer Kern *m*
aromatic rearrangement Umlagerung *f* am Aromaten
aromatic ring Benzenring *m*, Benzenkern *m*
aromatic substance Aromastoff *m*
aromaticity 1. *(org Ch)* Aromatizität *f*, aromatischer Charakter *m*; 2. *(Carb)* Aromatizität *f (Maß für den Gehalt an aromatischem Kohlenstoff)*; 3. Aromatizität *f*, Wohlgeruch *m*, Würzigkeit *f*
aromatics *s.* aromatic hydrocarbons
aromatization 1. *(org Ch)* Aromatisieren *n*, Aromatisierung *f*, Aromatensynthese *f*; 2. *(Lebm)* Aromatisieren *n*, Aromatisierung *f*
aromatize *v* 1. *(org Ch)* aromatisieren *(in aromatische Kohlenwasserstoffe überführen)*; 2. *(Lebm)* aromatisieren
aroyl group *(org Ch)* Aroyl-Gruppe *f*, Aroyl-Rest *m*
aroylbenzoic acid Aroylbenzoesäure *f*
arrange *v* anordnen
arrangement Ordnung *f*, Anordnung *f*, Einrichtung *f*, Einteilung *f*, Aufstellung *f*, Zusammenstellung *f*; System *n*; (räumliche) Anordnung *f*, Konfiguration *f*
array Anordnung *f (im Molekül)*
arrest *v* arretieren, feststellen *(Waage)*; zum Stehen bringen, abbrechen *(Reaktion)*
arrestment Arretierung *f*
Arrhenius activation energy Arrhenius-Aktivierungsenergie *f*, arrheniussche Aktivierungsenergie *f*
Arrhenius equation Arrhenius-Gleichung *f*, arrheniussche Gleichung *f*
arsane Arsan *n*, Arsenwasserstoff *m*, Arsenhydrid *n (allgemeine Formel* As_nH_{n+2}*)*
arsenate Arsenat(V) *n*
arsenian arsenhaltig
arsenic 1. Arsen..., *(i. e. S.)* Arsen(V)-...; 2. Arsenik...
arsenic 1. As Arsen *n*; 2. *s.* arsenic trioxide
arsenic acid Arsen(V)-säure *f*, Arsensäure *f*
arsenic pentasulphide Arsenpentasulfid *n*, Diarsenpentasulfid *n*, Arsen(V)-sulfid *n*
arsenic pentoxide Arsenpentoxid *n*, Diarsenpentoxid *n*, Arsen(V)-oxid *n*
arsenic poisoning Arsenvergiftung *f*; Arsenikvergiftung *f*
arsenic trichloride Arsentrichlorid *n*, Arsen(III)-chlorid *n*
arsenic trioxide Arsentrioxid *n*, Diarsentrioxid *n*, Arsen(III)-oxid *n*, Arsenik *n*
arsenic trisulphide Arsentrisulfid *n*, Diarsentrisulfid *n*, Arsen(III)-sulfid *n*
α-arsenic gelbes Arsen *n*
β-arsenic schwarzes Arsen *n*
γ-arsenic graues [metallisches] Arsen *n*
arsenical arsenhaltig, Arsen...
arsenical Arsenikalie *f*, arsenhaltiger Stoff *m*, arsenhaltige Substanz *f*; Arsenverbindung *f*
arsenical speiss *(Met)* Arsenspeise *f*

arsenicalism chronische [gewerbliche] Arsenvergiftung *f*
arsenicate *v* mit Arsen behandeln
arsenide Arsenid *n*
arseniferous arsenhaltig
arsenious Arsen..., *(i. e. S.)* Arsen(III)-...
arsenious acid 1. Arsen(III)-säure *f*, arsenige Säure *f*; 2. *s.* arsenic trioxide
arsenious chloride Arsen(III)-chlorid *n*, Arsentrichlorid *n*
arsenious oxide Arsen(III)-oxid *n*, Diarsentrioxid *n*, Arsentrioxid *n*, Arsenik *n*
arsenious sulphide Arsen(III)-sulfid *n*, Diarsentrisulfid *n*, Arsentrisulfid *n*
arsenite Arsenat(III) *n*
arsenous *s.* arsenious
arsine Arsan *n*, Monoarsan *n*, Arsenwasserstoff *m*, *(unsubstituiertes AsH_3 auch:)* Arsin *n*
arsinic acid $R_2As(O)OH$ Arsinsäure *f*
arsonation Arsonylierung *f*
arsonic acid $R{-}As(O)(OH)_2$ Arsonsäure *f*
arsonium compound Arsoniumverbindung *f*
arsonium group $AsH^{+}{}_4$ Arsoniumgruppe *f*
artifact Artefakt *n*
artificial künstlich, Kunst...
artificial ageing künstliche [beschleunigte] Alterung *f*, Kurzalterung *f*
artificial carbon article [body] Kunstkohlenstoff-Formkörper *m*
artificial cell künstliche Zelle *f*
artificial fibre *s.* man-made fibre
artificial graphite künstlicher Graphit *m*
artificial isotope künstlich erzeugtes Isotop *n*
artificial leather Kunstleder *n*
artificial radioactivity künstliche [induzierte] Radioaktivität *f*
artificial resin Kunstharz *n*
artificial rubber künstlicher [synthetischer] Kautschuk *m*, Kunstkautschuk *m*, Synthesekautschuk *m*, SK
artificial silk Regeneratseide *f*, Kunstseide *f*, Chemieseide *f*
artificial skin künstliche Haut *f*
artificial transmutation (künstliche) Kernumwandlung *f*
aryl cation Arylkation *n*
aryl group Arylgruppe *f*, Arylrest *m*
aryl halide Arylhalogenid *n*
aryl siloxane Arylsiloxan *n*
arylalkylation Arylalkylierung *f*
arylate *v* arylieren
arylation Arylierung *f*
arylazo group Ar–N=N– Arylazogruppe *f*, Arylazorest *m*
arylene (group) Arylengruppe *f*, Arylenrest *m*, Arylen *n* *(allgemein für einen zweiwertigen aromatischen Rest)*
arylmethyl group $ArCH_2{-}$ Arylmethylgruppe *f*, Arylmethylrest *m*
arylsulphonic acid Arylsulfonsäure *f*
aryne Arin *n*, Aryn *n*, 1,2-Dehydroaromat *m*, *(i. e. S.)* Dehydrobenzen *n*, Benzin *n*, Benzyn *n*, Arin *n*
as-received im Anlieferungszustand
ASA = American Standards Association
ASAT = aspartate aminotransferase
asbestos Asbest *m*
asbestos cement Asbestzement *m*
asbestos diaphragm Asbestdiaphragma *n*, Asbestmembran(e) *f*
asbestos dust Asbeststaub *m*
asbestos fibre Asbestfaser *f*
asbestos filter Asbestfilter *n*
asbestos gauze Asbestdrahtnetz *n*
asbestosis *(Med)* Asbestose *f*
ascending current aufwärts gerichtete Strömung *f*, aufsteigender Strom *m*, Auf(wärts)strom *m*
ascending development *(Chromat)* aufsteigende Entwicklung *f*
ascorbic acid Ascorbinsäure *f*
ascorbigen Ascorbigen *n* *(Ascorbinsäure-Eiweiß-Verbindung)*
asepsis Asepsis *f*, Keimfreiheit *f*, Sterilität *f*
aseptic aseptisch, keimfrei, steril
ash *v* veraschen
ash 1. Asche *f*; 2. kristallwasserfreie [wasserfreie, kalzinierte] Soda *f*, wasserfreies Natriumcarbonat *n*
ash content Aschegehalt *m*
ash-free aschefrei, af
ash-fusion temperature Ascheschmelzpunkt *m*
ashing Veraschen *n*, Veraschung *f*
ashless aschefrei, af; aschearm, nahezu aschefrei
ASLE = American Society of Lubrication Engineers
Asp *s.* aspartic acid
asparagic acid *s.* aspartic acid
asparagine Asparagin *n*, $Asp{-}NH_2$

asparaginic acid *s.* aspartic acid
aspartic acid Asparaginsäure *f*, Asp, *Aminobutandisäure *f*
asphalt 1. Asphalt *m*, Erdpech *n (Naturasphalt)*; 2. Erdölasphalt *m*, Petrol(eum)asphalt *m (künstlicher Asphalt)*; 3. Asphalt *m (Gemisch aus Bitumen und Mineralstoffen)*
asphalt-base petroleum asphaltbasisches [asphaltisches, asphaltöses] Erdöl *n*, Asphalt(basis)öl *n*, Erdöl *n* auf Asphaltbasis
asphaltene Asphalten *n*
asphaltic asphaltisch, asphalthaltig, Asphalt...
asphaltic bitumen Asphaltbitumen *n*, *(i. e. S.)* Bitumen *n*
asphaltic residue Erdölasphalt *m*, Petroleumasphalt *m (künstlicher Asphalt)*
asphaltite Asphaltit *m (Naturasphalt)*
aspirate *v* ansaugen
aspiration Ansaugen *n*
aspirator 1. Sauggebläse *n*, Saugventilator *m*, Absauger *m*, Aspirator *m*; 2. *(Lab)* Stutzenflasche *f*, mariottesche Flasche *f*, Abklärflasche *f*; 3. Staubschutzmaske *f*
aspirator bottle *(Lab)* Saugflasche *f*
assay *v* 1. prüfen, untersuchen; auf Gehalt prüfen *(Erze, Legierungen)*; 2. enthalten *(aufgrund des Analysenbefunds)*
assay 1. Versuch *m*; 2. Gehaltsbestimmung *f*; 3. Untersuchungsmaterial *n*, Probe *f*
assay at source vom Hersteller vorgenommene Bestimmung *f*
assay balance chemische Waage *f*, *(i. e. S.)* analytische Waage *f*, Analysenwaage *f*
assemble *v* zusammenbauen, vereinigen
assembly Zusammenbau *m*, Vereinigung *f*; Aggregat *n (Apparate-, Geräte- oder Maschinensatz)*
assignment of bands *(Spektr)* Bandenzuordnung *f*
assimilability *(Bioch)* Assimilierbarkeit *f*
assimilable *(Bioch)* assimilierbar
assimilate *v (Bioch)* assimilieren
assimilation 1. *(Bioch)* Assimilation *f*; 2. *(Geol)* Assimilation *f (Einschmelzung in Magma)*
assistent chemist Chemieassistent *m*
associate *v* 1. assoziieren; sich assoziieren; 2. *(Min)* vergesellschaften, assoziieren; sich vergesellschaften [assoziieren]
association 1. *(Ch)* Assoziation *f*; 2. *(Min)* Vergesellschaftung *f*, Assoziation *f*
association colloid Assoziationskolloid *n*, Mizellkolloid *n*
association reaction Assoziationsreaktion *f*
AST *s.* ASAT
astatine At Astat *n*
asterisk conditions Standardbedingungen *fpl*, Normalbedingungen *fpl*, Normbedingungen *fpl*
ASTM = American Society for Testing and Materials
Aston mass spectrograph astonscher Massenspektrograph *m*, aston-Massenspektrograph *m*
astringency Adstringenz *f*
astringent adstringierend, adstringent, zusammenziehend
astringent *(Pharm)* Adstringens *n*, adstringierendes Mittel *n*, adstringierende Substanz *f*
ASV *(anodic stripping voltammetry) s.* anodic stripping
asymmetric asymmetrisch, unsymmetrisch
asymmetric carbon atom asymmetrisches Kohlenstoffatom *n*
asymmetric centre Asymmetriezentrum *n*
asymmetric drop shape analysis asymmetrische Tropfenformanalyse *f*
asymmetric synthesis asymmetrische Synthese *f*
asymmetrical *s.* asymmetric
asymmetry Asymmetrie *f*, Unsymmetrie *f*
asymmetry parameter *(Glas)* Asymmetrieparameter *m*
asymmetry potential Asymmetriepotenzial *n*
at *s.* atmosphere 2.
atactic *(Kst)* ataktisch
atactic polymer ataktisches Polymer(e) *n*

ATE *(aluminium triethyl)* *s.* triethyl aluminium
atherm(an)ous adiatherman, atherman, keine Wärmestrahlen durchlassend, wärmeundurchlässig
atm *s.* atmosphere 2.
ATM *(aluminium trimethyl)* *s.* trimethylaluminium
atmophile element atmophiles Element *n*, Element *n* der Lufthülle
atmosphere 1. Atmosphäre *f*, Lufthülle *f*; 2. (physikalische) Atmosphäre *f*, atm *(1 atm = 101 325 N/m²)*; (technische, metrische, neue) Atmosphäre *f*, at *(1 at = 10⁴ kp/m²)*; 3. Atmosphäre *f*, Medium *n*, Mittel *n*
atmospheric atmosphärisch
atmospheric chemistry Chemie *f* der Atmosphäre, Atmosphärenchemie *f*
atmospheric contaminant Schmutzstoff *m* in der Luft, luftverunreinigender Stoff *m*, Luftverunreinigung *f*, *(i. e. S.)* Luftschadstoff *m*
atmospheric corrosion atmosphärische Korrosion *f*
atmospheric moisture Luftfeuchte *f*
atmospheric nitrogen Luftstickstoff *m*, Stickstoff *m* der Luft, atmosphärischer Stickstoff *m*
atmospheric oxidation Luftoxidation *f*, Oxidation *f* durch Luft(sauerstoff)
atmospheric oxygen Luftsauerstoff *m*, Sauerstoff *m* der Luft, atmosphärischer Sauerstoff *m*
atmospheric pollution Luftverunreinigung *f*, Luftverschmutzung *f*
atmospheric pressure atmosphärischer Druck *m*, Atmosphärendruck *m*, Luftdruck *m*
atmospheric pressure chemical ionization *(Spektr)* chemische Ionisation *f* bei Atmosphärendruck
atmospheric pressure ionization *(Spektr)* Ionisation *f* bei Atmosphärendruck
atmospheric pressure photoionization *(Spektr)* Photoionisation *f* bei Atmosphärendruck
atmospheric temperature and pressure technischer Normzustand *m* *(20 °C und 98,0665 kPa)*
atom Atom *n*
atom % Atomprozent *n*, Atom-%
atom arrangement Atomanordnung *f*
atom disintegration Atomzerfall *m*
atom lattice *(Krist)* Atomgitter *n*
atom layer deposition Atomlagenabscheidung *f*
atom model Atommodell *n*
atom polarization Atompolarisation *f*
atom tagging method Tracermethode *f*, Leitisotopenmethode *f*, Isotopenmethode *f*
atom transfer radical polymerization Atomübertragungsreaktion *f*, radikalische Polymerisation *f* durch Atomübertragung
atomic 1. atomar, einatomig, aus einem Atom bestehend; atomar *(in der Größenordnung eines Atoms)*; 2. *s.* atomistic
atomic absorption spectrometry *(Anal)* Atomabsorptionsspektralphotometrie *f*, Atomabsorptionsspektrometrie *f*, AAS
atomic absorption spectroscopy Atomabsorptionsspektroskopie *f*, AAS
atomic aggregate Atomaggregat *n*
atomic arrangement Atomanordnung *f*
atomic bond Atombindung *f*, homöopolare [kovalente, unpolare] Bindung *f*, Kovalenz *f*, Elektronenpaarbindung *f* *(Zustand)*
atomic combining ratio *s.* atomic ratio
atomic compound Atomverbindung *f*, homöopolare [unpolare] Verbindung *f*
atomic configuration Atomkonfiguration *f*, Konfiguration *f*
atomic core Atomrumpf *m*
atomic disintegration Atomzerfall *m*
atomic distance Atomabstand *m*, Kernabstand *m*
atomic-emission spectroscopy Atomemissionsspektroskopie *f*
atomic energy 1. Atom(kern)energie *f*, Kernenergie *f*; 2. Bindungsenergie *f*
atomic energy level Energieniveau *n* des Atoms, Atomniveau *n*
atomic excitation Anregung *f* von Atomen
atomic fission *s.* nuclear fission
atomic fluorescence spectroscopy Atomfluoreszenzspektroskopie *f*, AFS
atomic force microscope Rasterkraftmikroskop *n*

atomic force microscopy Rasterkraftmikroskopie *f*
atomic fusion *s.* nuclear fusion
atomic ground state Grundzustand *m*, Normalzustand *m*, Grundniveau *n*, Grundterm *m*
atomic hydrogen atomarer [einatomiger] Wasserstoff *m*, Monowasserstoff *m*
atomic kernel *s.* atomic core
atomic lattice *(Krist)* Atomgitter *n*
atomic magnetic moment atomares magnetisches Moment *n*
atomic mass Atommasse *f*, atomare Masse *f*
atomic mass number Atommasse(n)zahl *f*, atomare Masse(n)zahl *f*
atomic mass unit Atommasseeinheit *f*, atomare Masseeinheit *f*
atomic nucleus Atomkern *m*
atomic number Kernladungszahl *f*, Z, Ordnungszahl *f*, Atomnummer *f*
atomic orbital Atomorbital *n*, atomares Orbital *n*, AO
atomic oxygen atomarer Sauerstoff *m*, Monosauerstoff *m*
atomic percent(age) Atomprozent *n*, Atom-%
atomic polarization Atompolarisation *f*
atomic proportion *s.* atomic ratio
atomic ratio *(Anal)* Atomverhältnis *n*, atomares Verhältnis *n* [Verbindungsverhältnis *n*]
atomic refraction Atomrefraktion *f*
atomic shift Atomverschiebung *f*
atomic spectroscopy Atomspektroskopie *f* • **by atomic spectroscopy** atomspektroskopisch
atomic spectrum Atomspektrum *n*
atomic spin Atomspin *m*
atomic state atomarer Zustand *m*, Atomzustand *m*, Energiezustand *m* des Atoms
atomic structure Atombau *m*, Atomstruktur *f*
atomic susceptibility Atomsuszeptibilität *f*
atomic transmutation Atomumwandlung *f*
atomic unexcited state *s.* atomic ground state
atomic vibration Atomschwingung *f*
atomic volume Atomvolumen *n*
atomic weight relative Atommasse *f*
atomicity 1. Wertigkeit *f*; 2. Atomigkeit *f*, Atomizität *f (Anzahl der Atome eines Moleküls)*
atomistic atomistisch *(die Atomlehre betreffend)*
atomization Versprühen *n*, Zersprühen *n*; Verdüsen *n*; Verstäuben *n*, Zerstäuben *n (von Feststoffen)*
atomize *v* versprühen, zersprühen; verdüsen; verstäuben, zerstäuben *(Feststoffe)*
atomizer 1. Sprüher *m*, Versprüher *m*, Atomisator *m*; Zerstäuber *m (für Feststoffe)*; 2. *(Anal)* Atomisator *m*
atomizing nozzle Sprühdüse *f*; Zerstäuberdüse *f (für Feststoffe)*
ATP *s.* 1. adenosine triphosphate; 2. atmospheric temperature and pressure
ATPD = atmospheric temperature and pressure, dry
ATPS = atmospheric temperature and pressure, saturated
ATR *s.* attenuated total reflectance
atropina Atropin *n*, DL-Hyoscyamin *n*
atropine Atropin *n*, DL-Hyoscyamin *n*
atropisomer Atropisomer(e) *n*
atropisomerism Atropisomerie *f*, Behinderungsisomerie *f*
ATRP *s.* atom transfer radical polymerization
attach *v* 1. befestigen, anbringen *(mechanisch)*; 2. anlagern; sich anlagern *(chemisch)*; 3. aufziehen *(Farbstoffe)*
attached *v* **to/be** gebunden sein an, hängen [sitzen] an
attached *v* **to/become** sich anlagern an
attachment 1. Befestigung *f*, Anbringung *f (mechanisch)*; 2. Anlagerung *f (chemisch)*; 3. Aufziehen *n (von Farbstoffen)*; 4. Zusatzgerät *n*
attachment to a carrier *(Biot)* Trägerbindung *f (eines Enzyms)*
attack *v* angreifen
attack Angriff *m*
attacking agent angreifendes Agens *n*
attenuate *v* 1. verdünnen *(Lösungen)*; 2. dämpfen, (ab)schwächen *(z. B. Wirkungen)*
attenuated total reflectance abgeschwächte Totalreflexion *f*, ATR
attenuation 1. Verdünnen *n*, Verdünnung *f (von Lösungen)*; 2. Dämpfung *f*,

Schwächung *f*, Abschwächung *f (z. B. von Wirkungen)*
attenuator 1. *(Spektr)* Abgleichsblende *f*; 2. *(Bioch)* Attenuator *m (als Kontrollelement fungierender DNA-Abschnitt)*
attract *v* anziehen; aufnehmen, ziehen *(Elektronen)*
attractant Lockmittel *n*, Lockstoff *m*, Anlockstoff *m*, Attraktans *n*
attraction Anziehung *f*, Attraktion *f*
attractive force [power] Anziehungskraft *f*, Attraktionskraft *f*
attrition Abreiben *n*, Abrieb *m*, Abtragen *n*, Abtragung *f*; Abnutzung *f* durch Reibung, Abriebsabnutzung *f*, Schleifabnutzung *f*, Reibungsverschleiß *m*
Å.U. *s.* angstrom (unit)
A.U. *s.* angstrom (unit)
aufbau principle Aufbauprinzip *n (der Elektronenkonfiguration)*
aufbau reaction Aufbaureaktion *f (Polymerisation)*
auger 1. *(Ker)* Schneckenpresse *f (eine Art Strangpresse)*; 2. *s.* auger sampler
auger sampler Probenbohrer *m*
Auger effect Auger-Effekt *m*, innere Absorption *f*
Auger electron Auger-Elektron *n*
Auger (electron) spectroscopy Auger-Elektronenspektroskopie *f*, AES
augment *v* verstärken *(z. B. eine Farbe)*
aurate Aurat *n*
auric Gold..., *(i. e. S.)* Gold(III)-...
auric chloride Gold(III)-chlorid *n*, Goldtrichlorid *n*
auric cyanide Gold(III)-cyanid *n*, Goldtricyanid *n*
auric hydroxide Gold(III)-hydroxid *n*
auric oxide Gold(III)-oxid *n*, Goldtrioxid *n*
aurichloride Chloroaurat(III) *n*
auricyanide Cyanoaurat(III) *n*, Tetracyanoaurat(III) *n*
auride Aurid *n*
auriferous *(Geoch)* goldhaltig, goldführend
aurin(e) Aurin *n*, *p*-Rosolsäure *f*
aurochloride Chloroaurat(I) *n*
aurocyanide Cyanoaurat(I) *n*, Dicyanoaurat(I) *n*
auroso-auric chloride Gold(I,III)-chlorid *n*
aurous Gold..., *(i. e. S.)* Gold(I)-...
aurous chloride Gold(I)-chlorid *n*, Goldmonochlorid *n*
aurous cyanide Gold(I)-cyanid *n*
aurous hydroxide Gold(I)-hydroxid *n*
aurous oxide Gold(I)-oxid *n*, Goldmonoxid *n*
austenite *(Met)* Austenit *m*, γ-Eisen *n (kubisch-flächenzentrierte Fe-C-Mischkristalle)*
austenitic austenitisch
austenitization *(Met)* Austenitisierung *f*
austenitize *v (Met)* austenitisieren
austenitizing temperature *(Met)* Austenitisierungstemperatur *f*
auto-oxidaton-reduction Disproportionierung *f*
autoacceleration Selbstbeschleunigung *f*
autocatalysis Autokatalyse *f*
autocatalytic autokatalytisch
autocatalytic agent Autokatalysator *m*
autoclave *v* 1. im Autoklaven behandeln [kochen], autoklavieren; 2. *(Med)* sterilisieren
autoclave 1. Autoklav *m*, Druckgefäß *n*, Druckkessel *m*; 2. *(Gum)* Vulkanisierkessel *m*, Vulkanisationskessel *m*; 3. *s.* autoclave sterilizer
autoclave sterilizer *(Med)* Autoklav *m*, Drucksterilisator *m*, Drucksterilisierapparat *m*
autoclaving Autoklavenbehandlung *f*, Autoklavieren *n*
autocondensation Selbstkondensation *f*
autocorrelation function Autokorrelationsfunktion *f*
autogenous grinding autogenes Mahlen *n*, autogene Mahlung *f*
autogenous ignition temperature Selbstentzündungstemperatur *f*, Selbstzünd(ungs)temperatur *f*, SZT, Selbstzündpunkt *m*, Zündpunkt *m*
autogenous mill Autogenmühle *f*, Mühle *f* für autogene Mahlung, autogen arbeitende Mühle *f*
autogenous smelting *(Met)* autogenes Schmelzen *n*, Autogenschmelzen *n*
autogenous welding autogenes Schweißen *n*, Autogenschweißen *n*, Autogenschweißung *f*, Gas(schmelz)-schweißen *n*

autoignition Selbst(ent)zündung *f*, Spontanzündung *f*
autoionization 1. Autoionisation *f*, Selbstionisation *f*, Eigenionisation *f*, Präionisation *f*; 2. *s.* autoprotolysis
autolysate Autolysat *n*
autolysis *(Bioch)* Autolyse *f (Selbstauflösung)*
autolytic *(Bioch)* autolytisch
autolyze *v (Bioch)* Autolyse erleiden, der Autolyse unterliegen, sich selbst auflösen; Autolyse auslösen *(Enzyme)*
automatic control automatische [selbsttätige] Steuerung *f*
automatic titrator Titrierautomat *m*
automatic zero burette Bürette *f* mit automatischer Nullpunkteinstellung
automorphic *(Min)* automorph, idiomorph *(in der typischen Kristallform vorliegend)*
automorphic-granular *(Geol)* panidiomorph, panidiomorphkörnig
automotive fuel Kraftstoff *m*
autooxidation *s.* autoxidation
autoprotolysis Autoprotolyse *f*, Eigendissoziation *f*
autoprotolysis constant Autoprotolysekonstante *f*
autoprotolysis equilibrium Autoprotolysegleichgewicht *n*
autoracemization Autoracemisierung *f*, Autoracemisation *f*
autoradiograph Autoradiogramm *n*, autoradiographische Aufnahme *f*
autoradiography Autoradiographie *f*, Radioautographie *f*
autosampler *(Chromat)* automatisches Probendosiersystem *n*, automatischer Probengeber *m*
autotitrator *s.* automatic titrator
autotrophic *(Bioch)* autotroph
autotrophs autotrophe Mikroorganismen *mpl*
autotrophy *(Bioch)* Autotrophie *f*, autotrophe Ernährung *f*
autoxidation Autoxidation *f*, Selbstoxidation *f*
autoxidator Autoxidator *m*
autozero automatische Nullpunkteinstellung *f* [Nullpunktkorrektur *f*]
auxiliary zusätzlich, Zusatz..., Hilfs..., Neben...
auxiliary electrode Hilfselektrode *f*
auxiliary material Hilfsstoff *m*
auxochrome Auxochrom *n*, auxochrome [farbverstärkende] Gruppe *f*
auxochromic auxochrom, farbverstärkend, farbvermehrend
auxotrophic *(Bioch)* auxotroph
av. = average
A.V. *s.* acid value
availability 1. Verfügbarkeit *f*; 2. *(Landw)* Pflanzenverfügbarkeit *f*, Aufnehmbarkeit *f (von Nährstoffen)*
available 1. verfügbar; 2. *(Landw)* pflanzenverfügbar, zugänglich, aufnehmbar *(Nährstoffe)*
available chlorine 1. *(Pap, Text)* Aktivchlor *n*, wirksames [aktives] Bleichchlor *n* [Chlor *n*]; 2. *(Hyd)* wirksames [aktives] Chlor *n*
available oxygen aktiver Sauerstoff *m*, Aktivsauerstoff *m (z. B. beim Bleichen)*
available sulphur dioxide *(Pap)* freies [loses] SO_2 *n*
avalanche of electrons Elektronenlawine *f*
avalent nullwertig
average analysis mittlere Zusammensetzung *f*
average life (period) *(phys Ch)* (mittlere) Lebensdauer *f*
average linear expansion coefficient mittlerer linearer Ausdehnungskoeffizient *m*
average molecular weight mittlere Molmasse *f*, mittlere relative Molekülmasse *f*
average residence [retention] time mittlere Verweilzeit *f*
average sample Durchschnittsprobe *f*
average yield Durchschnittsausbeute *f*
aviation fuel Flugkraftstoff *m*
avicide Avizid *n*, Vogelbekämpfungsmittel *n*
AVLIS = atomic vapour laser isotope separation
Avogadro constant Avogadro-Konstante *f*, N_A *(Anzahl der Teilchen in 1 mol,* $N_A = 6{,}022045 \cdot 10^{23}\ mol^{-1}$; *früher Loschmidt-Konstante* N_L*)*
Avogadro hypothesis [law] avogadrosche Hypothese *f* [Regel *f*], avogadrosches Gesetz *n*, Satz *m* des Avogadro
Avogadro number 1. Avogadro-Zahl *f*,

N_0 *(dimensionsloser Zahlenwert der Avogadro-Konstanten)*; 2. *s.* Avogadro constant
A.W. *s.* atomic weight
awu *(atomic weight unit) s.* atomic mass unit
axial axial
axial angle *(Krist)* Achsenwinkel *m*
axial bond axiale Bindung *f*, α-Bindung *f*
axial conformation axiale Konformation *f*
axial diffusion axiale Diffusion *f*, Longitudinaldiffusion *f*, Längsdiffusion *f*
axial dispersion 1. axiale Dispersion *f* *(im Rohrreaktor)*; 2. *(Chromat)* Bandenverbreiterung *f*
axial flow axiale Strömung *f*
axial mixing Axialvermischung *f*, Längsvermischung *f*, axiale Rückvermischung *f*
axial ratio (kristallographisches) Achsenverhältnis *n*
axial symmetry Achsensymmetrie *f*
axis of rotation Rotationsachse *f*, Drehachse *f*
axis of symmetry *(Krist)* Symmetrieachse *f*, Drehachse *f*
axis of threefold symmetry *(Krist)* dreizählige Achse *f* [Drehachse *f*]
axisymmetric(al) achsensymmetrisch, axialsymmetrisch
azane Azan *n (Stickstoffhydrid der allgemeinen Formel N_nH_{n+2})*
azanide Amid *n*, Metallamid *n*
azelaic acid Azelainsäure *f*, *Heptan-1,7-dicarbonsäure *f*, *Nonandisäure *f*
azelaic aldehyde Azelainaldehyd *m*
azelate Azelat *n (Salz oder Ester der Azelainsäure)*
azeotrope Azeotrop *n*, azeotropes Gemisch *n*, azeotrope Mischung *f*, Azeotropmischung *f*
azeotrope agent [former] *(Dest)* Azeotropbildner *m*, Schleppmittel *n*, Mitnehmer *m*
azeotropic azeotrop
azeotropic copolymerization azeotrope Copolymerisation *f*
azeotropic distillation azeotrope Destillation *f*, Azeotropdestillation *f*
azeotropic mixture *s.* azeotrope
azeotropic point azeotroper Punkt *m*, Azeotroppunkt *m*
azeotropy Azeotropie *f*
azide Azid *n (Salz M^IN_3 oder organische Verbindung RN_3 der Stickstoffwasserstoffsäure)*; RCO–N=N=N Säureazid *n*
azido group $-N_3$ Azidogruppe *f*
aziminobenzene Aziminobenzen *n*, Benzenazimid *n*, 1,2,3-Benzotriazol *n*
azimuthal quantum number Bahndrehimpulsquantenzahl *f*, Drehimpulsquantenzahl *f*, Azimut(al)quantenzahl *f*, azimutale Quantenzahl *f*, Nebenquantenzahl *f*, Orbitaldrehimpulsquantenzahl *f*
azine Azin *n (Kondensationsprodukt aus Carbonylverbindungen mit Hydrazin)*
azine dye Azinfarbstoff *m*, Phenazinfarbstoff *m*
aziridine *(org Ch)* Aziridin *n*
azlactone Azlacton *n*, Oxazolon *n*, Oxazolinon *n*
azo color *(Am) s.* azo dye
azo compound R–N=N–R′ Azoverbindung *f*, Azokörper *m*
azo coupling Azokupplung *f*, Diazokupplung *f*
azo dye Azofarbstoff *m*
azo group –N=N– Azogruppe *f*
azo link(age) Azobindung *f*, Azobrücke *f*
azobenzene Azobenzen *n*, Diphenyldiimid *n*
azobisisobutyronitrile Azobisisobutyronitril *n*, 2,2′-Azo-bis-isobutyronitril *n*, AIBN
azodicarbonamide Azodicarbonamid *n*
azofication *(Landw)* nicht symbio(n)tische Stickstoffbindung *f* [Fixierung *f* von Luftstickstoff], Stickstoffbindung *f* durch freilebende Mikroorganismen
azoic dye (auf der Faser erzeugter) unlöslicher Azofarbstoff *m*, Eisfarbe *f*
azoimide Azoimid *n*, Stickstoffwasserstoffsäure *f*
azole Azol *n*, *(i. e. S.)* Pyrrol *n*, Imidol *n*
azomethine (compound) $R_2C{=}NR'$ Azomethin *n*; RHC=NR′ Azomethin *n*, schiffsche Base *f*
azomethine dye Azomethinfarbstoff *m*
azomethine oxide Nitron *n (N-Oxid einer schiffschen Base)*
azophenylene Azophenylen *n*, Phenazin *n*, Dibenzopyrazin *n*
azotometer Azotometer *n*, Nitrometer *n*
azoxy compound Azoxyverbindung *f*

azoxy group –N(O)N– Azoxygruppe *f*
azulene Azulen *n (bicyclischer Kohlenwasserstoff mit 10 p-Elektronen)*

B

B group Nebengruppe *f (des Periodensystems)*
B-stage resin Harz *n* im B-Zustand, B-Harz *n*, *(bei Phenolharzen auch)* Resitol *n*
B strain effect B-strain-Effekt *m*, sterische Beschleunigung *f*
BAC = British Association of Chemists
back Rückstand *m*
back-diffusion Rückdiffusion *f*
back-donation bond Rückgabebindung *f*
back e.m.f. gegenelektromotorische Kraft *f*, Gegen-EMK *f*, Gegenurspannung *f*
back end *(Dest)* Nachlauf *m*
back flow Rückfluss *m*, Rücklauf *m*, Rückstrom *m*
back-flow valve Rückschlagventil *n*, Rückstromventil *n*, Rückstromsperre *f*
back-mixing Rück(ver)mischung *f*
back pressure Gegendruck *m*, Rückdruck *m*
back reaction Rückreaktion *f*
back reflection method Rückstrahltechnik *f (zur Kristallstrukturuntersuchung)*
back-side attack Angriff *m* von der Rückseite her *(z. B. bei Umlagerungen)*
back titration Rücktitration *f*, Rücktitrieren *n*, Zurücktitrieren *n*
backbone (chain) Hauptkette *f*, Rückgratkette *f*, Verbindungsstamm *m (eines verzweigten Moleküls)*
backflush *v s.* backwash
background 1. *(Anal)* Untergrund *m*, Hintergrund *m*, Nulleffekt *m*, Nullrate *f*, Leerwert *m*; 2. *(Kern)* Fond *m*
background interference Störung *f* durch Untergrundeffekte, Untergrundstörung *f*
background noise Untergrundrauschen *n*
background spectrum Untergrundspektrum *n*
backscatter method [radiography] *(Anal)* Rückstreu(mess)verfahren *n*
backscattering *(Kern)* Rückstreuung *f*
backtitrate *v* zurücktitrieren
backward reaction Rückreaktion *f*
backwash *v* rückspülen, durch Rückspülung waschen
backwash(ing) Rückspülen *n*, Rückspülung *f*, Rückwäsche *f*
bacteria-retaining filter Bakterienfilter *n*, bakteriendichtes Filter *n*
bacteria-tight bakteriendicht
bacterial bakteriell, Bakterien…
bacterial attack bakterieller Angriff *m*, Bakterieneinwirkung *f*
bacterial count *(Lebm, Hyd, Med)* Keimzahl *f (pro Probenmenge)*
bacterial decomposition [degradation] bakterieller Abbau *m*
bacterial fermentation *(Biot)* bakterielle Fermentation *f*; bakterielle Gärung *f*
bacterial number *s.* bacterial count
bacterial protein bakterielles Protein *n*, Bakterieneiweiß *n*
bactericidal bakterizid, bakterientötend, keimtötend
bactericide Bakterizid *n*, antibakterieller Wirkstoff *m*, antibakterielles Arzneimittel *n*, bakterizide [bakterientötende] Substanz *f*
bacteriochlorophyll *(Bioch)* Bakteriochlorophyll *n*
bacteriocidal *s.* bactericidal
bacteriological count *s.* bacterial count
bacteriological filter Bakterienfilter *n*, bakteriendichtes Filter *n*
bacteriostat Bakteriostatikum *n*, bakteriostatisches Mittel *n*
bacteriostatic bakteriostatisch, das Bakterienwachstum hemmend, bakterienwachstumshemmend
bacteriostatic *s.* bacteriostat
Baeyer (angle) strain theory baeyersche Spannungstheorie *f*
Baeyer-Villiger oxidation [reaction] Baeyer-Villiger-Oxidation *f*, Baeyer-Villiger-Reaktion *f (von Ketonen zu Estern)*
Baeyer's angle strain Winkelspannung *f*, Ringspannung *f*, Baeyer-Spannung *f*
baffle *v* mit Prallblechen versehen [ausstatten], mit Schikanen [Leitblechen] versehen

baffle Prallblech *n*, Prallfläche *f*, Prallwand *f*, Prallplatte *f*, Schikane *f*, Leitblech *n*, Lenkblech *n*, Umlenkblech *n*, Umlenkplatte *f*
baffle plate *s.* baffle
baffle-plate column Kaskadenbodenkolonne *f (Solventextraktion)*
bag *v* absacken
bag Beutel *m*; Sack *m*; Schlauch *m (z. B. im Schlauchfilter)*
bag collector Schlauch(kammer)filter *n*, Schlauchabscheider *m (zur Gasreinigung)*
bag filter Schlauchabscheider *m*, Schlauchfilter *n*, Beutelfilter *n*, Seihfilter *n*, Sackfilter *n*
bagasse Bagasse *f*, Begasse *f*, Zuckerrohrbagasse *f (Zuckerrohrabfälle)*
bainite (structure) *(Met)* Bainit *m*, Zwischenstufengefüge *n*, bainitisches Gefüge *n*, Gefüge *n* der Zwischenstufe
bainite transition Bainitumwandlung *f*
bainitic structure *s.* bainite (structure)
bait Köder *m*, Ködermittel *n*
bake *v* 1. wärmebehandeln, thermisch [mit Hitze] behandeln; mittels Wärme trocknen; 2. (ver)backen *(Amine zwecks Sulfonierung)*; 3. *(Ker)* brennen; glühen, schrühen; einbrennen *(Aufglasurfarben)*; 4. *(Anstr)* einbrennen, im Ofen trocknen [härten]; 5. *(Kst)* härten; 6. *(Lebm)* backen
baked clay gebrannter Ton *m*, Töpferware *f*
Baker-Venkataraman rearrangement Baker-Venkataraman-Umlagerung *f (von 2-Acyloxy-aceto-phenonen zu 1,3-Diketonen)*
baking 1. Wärmebehandlung *f*, thermische Behandlung *f*, Hitzebehandlung *f*, Trockenhitzebehandlung *f*; Brennen *n (von grünen Formlingen zur Herstellung von Kohlenstoffkörpern)*; Trocknen *n*, Trocknung *f (mittels Wärme)*; 2. Backen *n*, Verbacken *n (von Aminen zwecks Sulfonierung)*; 3. *(Ker)* Brennen *n*; Glühen *n*, Schrühen *n*; Einbrennen *n*, Einbrand *m (von Aufglasurfarben)*; 4. *(Anstr)* Einbrennen *n*, Ofentrocknung *f*, Ofenhärtung *f*; 5. *(Kst)* Härten *n*, Härtung *f*; 6. *(Lebm)* Backen *n*
baking agent Backtriebmittel *n*
baking finish 1. Einbrennlack *m*, ofentrocknender Lack *m*; Einbrenndecklack *m*; 2. Einbrennlackierung *f*
baking oven 1. *(Anstr)* Einbrennofen *m*, Trockenofen *m*; 2. *(Lebm)* Backofen *m*
baking powder *(Lebm)* Backpulver *n*
baking soda Natriumhydrogencarbonat *n*
baking varnish Einbrennlack *m*, ofentrocknender Lack *m*
balance *v* 1. austarieren; ausgleichen; 2. bilanzieren *(Reaktionsgleichungen)*
balance *v* **out** *s.* balance 1.
balance 1. Waage *f*; 2. Gleichgewicht *n*; 3. Bilanz *f*; 4. Rest *m (Differenz zu 100 % in Analysentabellen)*
balance beam Waag(e)balken *m*
balance dish Wägeschiffchen *n*
balance method Abgleichmethode *f*, Kompensationsmethode *f*, Nullmethode *f*
balance rider Reiter *m (der Analysenwaage)*
balance weight Wägestück *n*, Massestück *n*, Vergleichsmassestück *n*
balanced equation stöchiometrische Gleichung *f*, Bilanzgleichung *f*
balanced reaction reversible [umkehrbare] Reaktion *f*; Gleichgewichtsreaktion *f*
balanced salt solution *(Biot)* Mineralsalzmedium *n (eine synthetische Nährlösung, z. B. nach Tyrode)*
ball *v* 1. zusammenballen; sich zusammenballen; 2. *(Met)* Luppen bilden, zu Luppen zusammenbacken
ball *v* **up** *s.* ball
ball-mill *v* kugelmahlen, in der Kugelmühle vermahlen, mit der Kugelmühle mahlen
ball 1. Schliffkugel *f*, Schliffball *m (einer Schliffverbindung)*; Stahlkugel *f (Mahlkörper)*; 2. *(Met)* Luppe *f*
ball-and-socket joint Kugelschliffverbindung *f*
ball-and-stick model Kugel-Stäbchen-Modell *n (von Molekülen)*
ball condenser Kugelkühler *m*
ball impact test Kugelfallversuch *m*
ball mill Kugelmühle *f*
ball milling Zerkleinern *n* in der Kugelmühle
ball-puncture resistance Kugeldruckhärte *f*

ball valve Kugelventil *n*
balling 1. Zusammenballen *n*, Kugelbildung *f*; Pellet(is)ieren *n*; 2. *(Met)* Luppenbildung *f*
balling drum Pellet(is)iertrommel *f*
ballistic galvanometer ballistisches Galvanometer *n*, Stromstoßgalvanometer *n*, Stoßgalvanometer *n*
balloon flask Kurzhalsrundkolben *m*
balm 1. Balsam *m*; 2. *s.* bee glue
Balmer series Balmer-Serie *f*
balsam Balsam *m*
balsam of fir Kanadabalsam *m* *(von nordamerikanischen Abies-Arten)*
Bamford-Stevens reaction Bamford-Stevens-Reaktion *f* *(zur Darstellung von Diazoalkanen)*
banana bond Bananenbindung *f*, gebogene Bindung *f*
Banbury (mixer) Banbury-Mischer *m*, Banbury-Innenmischer *m*, Banbury-Kneter *m*
band 1. Band *n*, Streifen *m*; 2. *(Anal)* Bande *f*; 3. *(Chromat)* Substanzbereich *m*; 4. Band *n*, Energieband *n*, Energiebereich *m*, Zone *f*
band broadening *(Anal)* Bandenverbreiterung *f*
band conveyor Gurt(band)förderer *m*, Bandförderer *m*
band edge *s.* band head
band filter Bandfilter *n*, Umlauffilter *n*
band gap Bandlücke *f*
band head *(Spektr)* Bandenkopf *m*, Bandenkante *f*
band heater Bandheizkörper *m*, Heizband *n*
band-pass filter *(Spektr)* Bandpassfilter *n*, Bandpass-Filter *n*
band spectrum Bandenspektrum *n*, *(i. e. S.)* Molekülspektrum *n*
band splitting *(Spektr)* Bandenaufspaltung *f*
band structure Bandstruktur *f*
band theory Bändermodell *n*
band theory of solids Bandtheorie *f* [Bändertheorie *f*] der Festkörper
banded 1. gestreift, streifig, streifenartig (ausgebildet), Streifen...; 2. *(Geol)* gebändert
bandwidth *(Spektr)* Bandenbreite *f*
bank *(Tech)* Reihe *f*, Gruppe *f*, Serie *f*, Satz *m*, Anordnung *f*
bar 1. Bar *n*, bar *(inkohärente, SI-fremde Einheit des Drucks; 1 bar = 10^5 Pa)*; 2. Stange *f*, Stab *m*; Barren *m*; Strang *m*; Riegel *m* *(z. B. aus Seife)*
bar grizzly Stangen(sieb)rost *m*, Stab(sieb)rost *m*, Siebrost *m*
bar mill Desintegrator *m*, Schleudermühle *f*, Schlagkorbmühle *f*
bar screen Stangenrostsieb *n*, Stabrostsieb *n*, Rostsieb *n*
Barbier-Wieland degradation [reaction] Barbier-Wieland-Abbau *m*, barbier-wielandscher Carbonsäureabbau *m*, Barbier-Wieland-Reaktion *f*
barbital Barbital *n*, 5,5-Diethylbarbitursäure *f*
barbiturate *(org Ch)* Barbiturat *n*
barbituric acid Barbitursäure *f*, *N,N'*-Malonylharnstoff *m*, 2,4,6-Trioxo-hexahydropyrimidin *n*
barium Ba Barium *n*
barium crown (glass) Baritkron(glas) *n*, Bariumkronglas *n*
barium dichromate Bariumdichromat(VI) *n*
barium flint (glass) Baritflint(glas) *n*, Bariumflintglas *n*
barium hydride Bariumhydrid *n*
barium hydrogensulphide Bariumhydrogensulfid *n*
barium hydroxide Bariumhydroxid *n*
barium lake Bariumlack *m*
barium oxide Bariumoxid *n*
barium peroxide Bariumperoxid *n*
barium peroxydisulphate Bariumperoxodisulfat *n*
barium sulphate Bariumsulfat *n*
barium sulphide Bariummonosulfid *n*, Bariumsulfid *n*
bark *v* *(Pap)* entrinden, schälen
bark 1. Rinde *f*; 2. *(Gerb)* Lohe *f*
bark leather pflanzlich gegerbtes Leder *n*
bark tannage [tanning] pflanzliche Gerbung *f*, *(i. e. S.)* Lohgerbung *f*
barley malt Gerstenmalz *n*
barley starch Gerstenstärke *f*
barm Kulturhefe *f*, Zuchthefe *f*
barn Barn *n*, b *(inkohärente Einheit für den Wirkungsquerschnitt des Atomkerns; 1 b = 10^{24} cm^2)*
barometric formula barometrische Höhenformel *f*

barometric pressure Barometerdruck *m*, barometrischer Druck *m*
barrel 1. Fass *n*, Tonne *f*; Barrel *n (1 barrel U.K. = 163,66 l, U.S. = 119,23 l, bei Erdöl 158,97 l)*; 2. Brenn(er)rohr *n (z. B. am Bunsenbrenner)*; 3. Hahnhülse *f (des Glashahns)*
barrel amalgamation *(Aufb)* Fassamalgamation *f*, Fässeramalgamation *f*, europäische Amalgamation *f*
barrel blender Trommelmischer *m*
barrel furnace Diffusionsofen *m*
barrel mixer Trommelmischer *m*
barren 1. *(Boden)* unfruchtbar, steril; 2. *(Geol)* taub, unhaltig
barren liquor [solution] *(Aufb)* Armlauge *f (Cyanidlaugerei)*
barrier 1. Schwelle *f*, Schranke *f*, Barriere *f*; Potenzialbarriere *f*, Potenzialschwelle *f*, Potenzialberg *m*, Potenzialwall *m*; 2. *(Hyd)* Trennwand *f*, Membran *f (Umkehrosmose)*
barrier cream Gewerbeschutzsalbe *f*, Arbeitsschutzsalbe *f*
barrier-layer cell Sperrschicht(photo)zelle *f*, Sperrschicht(photo)element *n*, Photoelement *n*
barrier polymer Barrierekunststoff *m (Verpackungstechnologie)*
baryon *(Kern)* Baryon *n*
barysphere *(Geoch)* Barysphäre *f*, Erdkern *m*
baryta Bariumoxid *n*
baryta water *(Anal)* Barytwasser *n (wässrige Lösung von Bariumhydroxid)*
baryta yellow Barytgelb *n*, gelbes Ultramarin *n (Bariumchromat)*
baryte *(Min)* Baryt *m*, Schwerspat *m (Bariumsulfat)*
barytic 1. Schwerspat..., Baryt...; 2. bariumhaltig; Barium...
basal medium *(Biot)* Grundsubstrat *n*, Ausgangssubstrat *n*, Hauptkulturmedium *n*
basal metabolism *(Bioch)* Grundumsatz *m*, Ruhestoffwechsel *m*
basalt Basalt *m*
basaltic basaltisch, Basalt...; basalthaltig
base geringwertig; unedel *(Metall)*
base 1. *(Ch)* Base *f*; 2. Basis *f*, Grundlage *f*; 3. Grundbestandteil *m*, Hauptbestandteil *m*; 4. *(Erdöl)* Basis *f*, Charakter *m (eines Rohöls)*
base-binding capacity [power] Basenbindungsvermögen *n*
base catalysis Basekatalyse *f*, Reaktionsbeschleunigung *f* durch Basen
base coat Grundschicht *f*, Grundierung *f (Oberflächentechnologie)*
base electrolyte Grundelektrolyt *m*, Leitelektrolyt *m*, Zusatzelektrolyt *m*, indifferenter Elektrolyt *m*
base equivalence Basenäquivalenz *f (in DNA)*
base exchange 1. Basenaustausch *m*; 2. *(Boden)* Basenumtausch *m*, *(i. e. S.)* Basenaustausch *m (Gegensatz: Baseneintausch)*
base excision repair *(Bioch)* Basen-Excisionsreparatur *f*
base hydrolysis alkalische Hydrolyse *f*
base-insoluble alkaliunlöslich
base line *(Anal)* *Basislinie *f*, Grundlinie *f*
base-line concentration *(Anal)* Hintergrundkonzentration *f*, Backgroundkonzentration *f*
base-line noise *(Anal)* Untergrundrauschen *n*
base material Ausgangsmaterial *n*
base metal 1. unedles Metall *n*, Unedelmetall *n*, Nichtedelmetall *n*; 2. Grundmetall *n (einer Legierung)*
base molecule Grundmolekül *n (für Polymerisationen)*
base pair *(Bioch)* Basenpaar *n*
base pairing *(Bioch)* Basenpaarung *f*
base sequence *(Bioch)* Basensequenz *f*
base-soluble alkalilöslich
base stacking *(Bioch)* Basenstapelung *f*
base stock *(Gum)* Grundmischung *f*
base strength Basenstärke *f*
base unit 1. Basiseinheit *f*, Grundeinheit *f (eines Einheitensystems)*; 2. *s.* basic unit
baseline *s.* base line
basic 1. basisch, alkalisch; 2. *(Met)* basisch zugestellt [ausgekleidet]; 3. grundlegend, Grund...
basic anhydride Basenanhydrid *n*
basic Bessemer process basisches Windfrischverfahren *n* [Bessemer-Verfahren *n*], Thomas-Verfahren *n*, Thomas-Konverterverfahren *n*

basic Bessemer steel Thomas-Stahl *m*, Thomas-Konverterstahl *m*
basic building block Grundbaustein *m*, grundlegender [fundamentaler] Baustein *m*, Fundamentalbaustein *m*
basic capacity Basizität *f (von Säuren)*
basic catalysis Basekatalyse *f*, Reaktionsbeschleunigung *f* durch Basen
basic chemical (material) Grundchemikalie *f*, Chemiegrundstoff *m*, chemischer Grundstoff *m*
basic chemical reaction chemische Grundreaktion *f*
basic dye basischer Farbstoff *m*
basic indicator basischer Indikator *m*, Indikatorbase *f*
basic lining basische Zustellung *f* [Auskleidung *f*], basisches Futter *n (eines metallurgischen Ofens)*
basic molecule Grundmolekül *n*
basic open-hearth process basisches Herdfrischverfahren *n* [Siemens-Martin-Verfahren *n*, SM-Verfahren *n*]
basic open-hearth steel basischer Siemens-Martin-Stahl *m*
basic operation Grundoperation *f*, grundlegende Operation *f*
basic oxide basisches Oxid *n*
basic oxygen converter process Sauerstoff(auf)blasverfahren *n*, Sauerstoff-Frischverfahren *n*, (Sauerstoff-) Blasstahlverfahren *n*, Oberwindfrischverfahren *n*
basic oxygen steel Sauerstoffblasstahl *m*, Blasstahl *m*, sauerstoffgefrischter Stahl *m* [Konverterstahl *m*]
basic process 1. Grundvorgang *m*; Grundverfahren *n*; 2. basisches Verfahren *n*; 3. *s.* basic Bessemer process
basic reaction 1. Grundreaktion *f*, grundlegende [fundamentale] Reaktion *f*; 2. basische Reaktion *f*
basic refractory (material) basisches Futter *n* [Ofenfutter *n*], *(i. w. S.)* basisches feuerfestes Material *n*
basic repeating unit *s.* basic molecule
basic research Grundlagenforschung *f*
basic salt Hydroxidsalz *n*; Oxidsalz *n*
basic slag basische Schlacke *f*, *(i. e. S.)* Thomas-Schlacke *f*
basic steel basischer Stahl *m*, *(i. e. S.)* Thomas-Stahl *m*
basic strength Basenstärke *f*
basic structural unit strukturelle Grundeinheit *f*
basic structure Grundstruktur *f*
basic substance 1. Grundstoff *m*, Grundsubstanz *f*; 2. basischer Stoff *m*
basic unit Grundeinheit *f (eines Moleküls)*
basicity 1. Basizität *f*, basischer Charakter *m*; 2. Basizität *f*, Alkalität *f*, Basengehalt *m (einer Lösung)*; 3. Basizität *f (einer Säure)*
basification Basischstellen *n*
basify *v* alkalisieren, alkalisch [basisch] stellen; alkalisch werden
basin Becken *n*, Schale *f*, Schüssel *f*, Behälter *m*, Bassin *n*
basis Basis *f*, Grundlage *f*; Grundbestandteil *m*, Hauptbestandteil *m*
basis solution Grundlösung *f*
basket Trommel *f (einer Zentrifuge)*
basket centrifugal [centrifuge] Trommelzentrifuge *f*
basoid Basoid *n (mit OH^--Ionen gesättigte kolloidale Substanz)*
basophile *s.* 1. basophilous; 2. basophilic
basophilia Basophilie *f (leichte Färbbarkeit durch basische Farbstoffe)*
basophilic basophil *(durch basische Farbstoffe leicht färbbar)*
basophilous 1. *(Bot)* basiphil, basophil *(auf Böden alkalischer Reaktion gedeihend)*; 2. *s.* basophilic
batch *v* 1. einsetzen, zuteilen *(eine bestimmte Menge)*; dosieren, abmessen; 2. durchsetzen *(eine bestimmte Menge)*
batch diskontinuierlich [chargenweise] arbeitend, periodisch [unstetig, absatzweise, satzweise] arbeitend, diskontinuierlich betrieben *(Anlage)*; diskontinuierlich, chargenweise, periodisch, unstetig, absatzweise, satzweise *(Betrieb)*
batch 1. Charge *f*, Partie *f*, Posten *m*, Füllung *f*, Ladung *f*, Einsatz *m*, Eintrag *m*, Beschickung *f*, Satz *m*, Ansatz *m*; 2. *(Glas)* Gemenge *n*, Glassatz *m*; 3. *(Ker)* Versatz *m*, Masseversatz *m*; 4. *(Gum)* Vormischung *f*, Batch *m*
batch analysis diskontinuierliche Analyse *f*

batch distillation diskontinuierliche Destillation *f*, Chargendestillation *f*
batch dryer Chargentrockner *m*
batch fermentation *(Biot)* Batch-Fermentation *f*, diskontinuierliche [statische] Fermentation *f*, Satzfermentation *f*
batch fermenter *(Biot)* Batch-Fermenter *m*, Fermenter *m* für diskontinuierliches Verfahren
batch mixer 1. Chargenmischer *m*, satzweise arbeitender Mischer *m*; 2. *(Glas)* Gemengemischer *m*
batch nitration Chargennitrierung *f*, Nitrierung *f* in Ansätzen, diskontinuierliche Nitrierung *f*
batch-operated tank reactor diskontinuierlicher Rührkessel(reaktor) *m*
batch operation diskontinuierlicher [periodischer] Betrieb *m*, diskontinuierliche Arbeitsweise *f* [Fahrweise *f*], Chargenbetrieb *m*, Satzbetrieb *m*
batch process diskontinuierliches [periodisches] Verfahren *n*, Chargenverfahren *n*
batch reactor diskontinuierlicher Reaktor *m*
batchwise diskontinuierlich, periodisch, unstetig, absatzweise, satzweise, chargenweise
bath 1. Bad *n*; 2. *(Text)* Flotte *f*
bath carburizing Badaufkohlen *n*, Salzbadaufkohlen *n*, flüssiges Aufkohlen *n*, Badzementieren *n*, Salzbadzementieren *n*, Badeinsetzen *n*, Aufkohlen *n* [Zementieren *n*, Einsetzen *n*] im Salzbad, Aufkohlen *n* [Zementieren *n*, Einsetzen *n*] in flüssigen Mitteln
bath ratio *(Text, Farb)* Flottenverhältnis *n*
bath-carburize *v* badaufkohlen, badzementieren, badeinsetzen, im Salzbad aufkohlen [zementieren, einsetzen], in flüssigen Mitteln aufkohlen [zementieren, einsetzen]
bathochrome *s.* bathochromic group
bathochromic group bathochrome [farbvertiefende] Gruppe *f*
bathochromic shift bathochrome Verschiebung *f* [Farbtonverschiebung *f*], Rotverschiebung *f*
bathychrome *s.* bathochromic group
bats-wing burner *(Lab)* Schnittbrenner *m*, Fischschwanzbrenner *m*, Schlitzbrenner *m*, Breitbrenner *m*
battery 1. Batterie *f*, Sammler *m*, Akku(mulator) *m*; 2. Gruppe *f*, Serie *f*, Satz *m*, Batterie *f*
battery acid Akkumulatorensäure *f*, Akku-Säure *f*
batting Watte *f* in Lagen
Baumé scale Baumé-Skale *f*
bauxite Bauxit *m*
bauxitic bauxitisch
bay salt Seesalz *n*, Meersalz *n (durch solare Eindunstung gewonnen)*
Bayer process Bayer-Verfahren *n (Aufschluss von Bauxit mit Natronlauge)*
bayonet coupling [joint] Bajonettkupplung *f*
bayonet-tube heat exchanger Nadelwärmeaustauscher *m*, Nadelwärmeübertrager *m*
BCCP = biotin carboxyl-carrier protein
BD *s.* 1,4-butanediol
B.D. *s.* 1. bone-dry; 2. bulk density
bead 1. Perle *f*, Kügelchen *n*; Korn *n (bei Ionenaustauschverfahren)*; 2. Wulst *m(f)*, Rand *m*
bead catalyst perlförmiger Katalysator *m* [Kontakt *m*], Perlkontakt *m*
bead polymerizate Perlpolymerisat *n*
bead polymerization Perlpolymerisation *f*, Suspensionspolymerisation *f*, Kornpolymerisation *f*
bead test Perlenreaktion *f*; Perlenprobe *f (mit Phosphorsalz oder Borax)*
beaded perlenförmig
beak Tubus *m (einer Retorte)*
beaker Becherglas *n*; Becher *m*
beaker cover *(Lab)* Uhrglasschale *f*
beaker flask weithalsiger Erlenmeyer-Kolben *m*
beam 1. Strahl *m*, Lichtstrahl *m*; Strahlenbündel *n*; 2. Waag(e)balken *m*
beam balance Balkenwaage *f*
beam of electrons Elektronenstrahl *m*
beam splitter *(Spektr)* Strahlungsteiler *m*
bearding Bartbildung *f*, *Fronting *n (in Chromatogrammen)*
bearing Lager *n*, Lagerung *f*, Lagerstelle *f*; Halterung *f*
bearing metal Lagermetall *n*
beater 1. Schläger *m*, Stampfer *m*, Hammer *m (der Hammermühle)*; Flügel *m (des Rührwerks)*; 2. *(Text)* Walk-

hammer *m*, Walkmaschine *f*; 3. *(Pap)* Mahlmaschine *f*, Stoffmühle *f*, *(i. e. S.)* Holländer *m*, Mahlholländer *m*, Messerholländer *m*, Ganzzeugholländer *m*
beater mill Schlägermühle *f*, Schlagmühle *f*
beating *(Pap)* Mahlen *n*, Mahlung *f*, Stoffmahlung *f*
Beattie-Bridgeman equation (of state) *(phys Ch)* Beattie-Bridgeman-Zustandsgleichung *f*, Beattie-Bridgeman-Gleichung *f*
Béchamp process Béchamp-Verfahren *n (zur Herstellung primärer aromatischer Amine)*
Béchamp reaction [reduction] Béchamp-Reaktion *f*, Béchamp-Reduktion *f (von Nitroarylverbindungen)*
beck Kufe *f*, Färbekufe *f*, Bottich *m*, Wanne *f*, Trog *m*
Beckmann (oxime-amide) rearrangement Beckmann-Umlagerung *f (eines Ketoxims in ein Amid)*
Beckmann thermometer Beckmann-Thermometer *n*, beckmannsches Thermometer *n*
bed 1. Bett *n*; Grundplatte *f*; 2. Bett *n*, Schicht *f*, Schüttgutbett *n*, Schütt(gut)schicht *f*, Schüttung *f (z. B. eines Filters)*
bed filter Schüttgutfilter *n*, Haufwerkfilter *n*
bed filtration *(Hyd)* Filtration *f* über körniges [gekörntes] Filtermaterial
bed material Schüttgut *n*
bed of crystals Kristallbett *n*, Kristallpolster *n*, Kristallschüttung *f*, Kristall(schütt)schicht *f*
bed of granular solids Körnerhaufwerk *n*, körnige Schüttschicht *f*, Körnerschüttung *f*
bee glue Bienenharz *n*, Kittharz *n*, Propolis *f*
bee venom Bienengift *n*
beer Bier *n*, *(i. e. S.)* untergäriges Bier *n*
beer wort Bierwürze *f*, Würze *f*
beer yeast Bierhefe *f*
Beer-Lambert(-Bouguer) law lambert-beersches Gesetz *n* [Absorptionsgesetz *n*], bouguer-lambert-beersches Gesetz *n (der Lichtabsorption)*
beeswax Bienenwachs *n*
beet cossettes Zuckerrübenschnitzel *npl*, Rübenschnitzel *npl*
beet diffuser Diffuseur *m*
beet juice Rübensaft *m*
beet molasses Zuckerrübenmelasse *f*, Rüben(zucker)melasse *f*
beet pulp ausgelaugte Zuckerrübenschnitzel *npl* [Rübenschnitzel *npl*]
beet sugar Rübenzucker *m*, Saccharose *f*
begass(e) *s.* bagasse
behave *v* sich verhalten
behave *v* **perfectly** sich ideal verhalten
behaviour Verhalten *n*, Reaktion *f (eines chemischen Stoffs)*
behenic acid Behensäure *f*, *Docosansäure *f*
Beilstein's test Beilsteinprobe *f*
bell Glocke *f*, Verschlussglocke *f*, Gichtglocke *f (eines Hochofens)*
bell bronze Glockenbronze *f*
bell cap tray Glockenboden *m*
bell-jar Glasglocke *f*, Rezipientenglocke *f*, Glocke *f*
bell metal Glockenmetall *n*
belladonna alkaloid Belladonnaalkaloid *n*
bellows-type expansion joint Wellrohrkompensator *m*, Wellrohr(dehnungs)ausgleicher *m*
Belousov-Zhabotinskii reaction Belousov-Žabotinskij-Reaktion *f (eine oszillierende Reaktion)*
below-cloud scavenging Washout *n*, Auswaschen *n (von Luftschadstoffen durch atmosphärische Niederschläge)*
belt Band *n*, Gurt *m*; Riemen *m*, Treibriemen *m*
belt conveyor Gurt(band)förderer *m*, Bandförderer *m*
belt dryer Bandtrockner *m*, Förderbandtrockner *m*, Bahnentrockner *m*
belt filter Bandfilter *n*, Umlauffilter *n*
belt (filter) press Bandfilterpresse *f*
belt-type vacuum filter Vakuumbandfilter *n*, Bandsaugfilter *n*, Belt-Filter *n*, Trommelfilter *n* mit ablaufendem Filtertuch
belt weigher Bandwaage *f*
bench Batterie *f (z. B. Verkokungsbatterie)*; Bühne *f*, Bedienungsbühne *f*; Rampe *f*; Prüfstand *m*

bench blowpipe Gebläselampe *f*, Gebläsebrenner *m*
bench-scale reactor Laborreaktor *m*
bench-scale study Laboruntersuchung *f*
bench sink Ausgussbecken *n*, Spülbecken *n*, Abflussbecken *n*
bending endurance Dauerbiegefestigkeit *n*
bending strength Biegefestigkeit *f*
bending test *(Kst)* Biegeversuch *m*, Biegeprüfung *f*
bending vibration Deformationsschwingung *f*, Knickschwingung *f*, *(i. e. S.)* Scherenschwingung *f*, δ-Schwingung *f*
Benedict's reagent [solution] benedictsches Reagens *n*, benedictsche Lösung *f (zum Nachweis reduzierender Zucker)*
beneficiate *v* aufbereiten, anreichern *(Erze)*
beneficiation Aufbereitung *f*, Anreicherung *f (von Erzen)*
Bengough-Stuart process Bengough-Stuart-Verfahren *n*, Chromsäureverfahren *n (zur anodischen Oxidation)*
benign unschädlich, harmlos
bent bond Bananenbindung *f*, gebogene Bindung *f*
bent orientation V-förmige Anordnung *f (von Liganden)*
bent-tube (connection) *(Lab)* Winkelstück *n*, Winkelrohr *n*, Krümmer *m*
bentonite (clay) Bentonit *m*, Bentonitton *m*
benzal acetone Benzalaceton *n*
benzal chloride *s.* benzylidene chloride
benzal green Malachitgrün *n*, Neugrün *n*, Benzalgrün *n*, Benzoylgrün *n*, Bittermandelölgrün *n (ein Triphenylmethanfarbstoff)*
benzaldehyde Benzaldehyd *m*
benzamide Benzamid *n*
benzanthrone dye Benzanthronfarbstoff *m*
benzene Benzen *n*, Benzol *n*
benzene azimide Benzenazimid *n*, Aziminobenzen *n*, 1,2,3-Benzotriazol *n*
benzene carboxylic acid Benzencarbonsäure *f*, Benzoesäure *f*
benzene derivative Benzenderivat *n*, Benzenabkömmling *m*
benzene dicarboxylic acid Benzendicarbonsäure *f*
benzene hexachloride Hexachlorcyclohexan *n*, HCH, Benzenhexachlorid *n*
benzene hydrocarbon Benzenkohlenwasserstoff *m*, aromatischer Kohlenwasserstoff *m*
benzene nucleus Benzenkern *m*
benzene ring Benzenring *m*
benzene series Benzenreihe *f*
benzene sulphonamide Benzensulfonamid *n*
benzene-toluene-xylenes BTX-Aromaten *mpl (Benzen, Toluen und Xylen)*
1,4-benzene diamine *p*-Phenylendiamin *n*
1,2,4,5-benzene tetracarboxylic acid 1,2,4,5-Benzentetracarbonsäure *f*, Pyromellit(h)säure *f*
1,2,3-benzene tricarboxylic acid 1,2,3-Benzentricarbonsäure *f*, Hemimellit(h)säure *f*
α-benzenesulphonate *(Hyd)* lineares Alkylbenzolsulfonat *n*, LAS
benzenesulphonic acid Benzensulfonsäure *f*, *(i. e. S.)* Benzenmonosulfonsäure *f*
benzenium ion $C_6H_7^+$ Benzeniumion *n*, Benzenenium-Ion *n*
benzenoid benzoid
benzidine Benzidin *n*, 4,4′-Diaminobiphenyl *n*
benzidine rearrangement Benzidin-Umlagerung *f*
benzil Benzil *n*, *Diphenyldiketon *n*
benzilic acid Benzilsäure *f*, Diphenylglykolsäure *f*, α-Hydroxydiphenylessigsäure *f*
benzilic acid rearrangement Benzilsäureumlagerung *f*
benzin *s.* benzine
benzine Benzin *n*; Leichtbenzin *n*, leichtes Benzin *n (Siedebereich 20 bis 135 °C)*; Petrol(eum)äther *m (Siedebereich 40 bis 70 °C)*
benzoate Benzoat *n*
benzofuran Benzofuran *n*, Cumaron *n*
benzoic acid Benzoesäure *f*, Benzencarbonsäure *f*
benzoic anhydride Benzoesäureanhydrid *n*
benzoin 1. Benzoin *n*, α-Hydroxy-α-

phenylacetophenon *n*, Bittermandelölcampher *m*; 2. *s.* benzoin gum
benzoin condensation Benzoinkondensation *f*
benzoin gum [resin] Benzoeharz *n*, Benzoe *f (von Styrax-Arten)*
benzol 1. Benzol *n (als unreines Handelsprodukt)*; 2. *s.* solvent naphtha; 3. *s.* benzene
benzole *s.* benzol
benzolize *v* mit Benzen beladen [anreichern, sättigen]
benzonitrile Benzonitril *n*
benzophenol Phenol *n (i. e. S.)*
benzophenone Benzophenon *n*, *Benzoylbenzen *n*, *Diphenylketon *n*
benzo[*b*]pyridine Benzo[b]pyridin *n*, 2,3-Benzopyridin *n*, 1-Benzazin *n*, Chinolin *n*
benzo[*c*]pyridine Benzo[c]pyridin *n*, 3,4-Benzopyridin *n*, 2-Benzazin *n*, Isochinolin *n*
***p*-benzoquinone** *p*-Benzochinon *n*, *Cyclohexa-2,5-dien-1,4-dion *n*
1,2,3-benzotriazole 1,2,3-Benzotriazol *n*, Azimidobenzen *n*, Benzenazimid *n*
benzoyl group C_6H_5CO– Benzoylgruppe *f*, Benzoylrest *m*
benzoyl peroxide Benzoylperoxid *n*, Dibenzoylperoxid *n*
benzoylate *v* benzoylieren
benzoylation Benzoylierung *f*
benzoyloxylation Benzoyloxylierung *f*
3,4-benzpyrene Benzo(a)pyren *n*, 1,2-Benzpyren *n*, Benzo(def)chrysen *n*
benzyl alcohol Benzylalkohol *m*
benzyl cellulose Benzylcellulose *f*
benzyl chloride Benzylchlorid *n*, α-Chlortoluen *n*
benzyl cyanide Benzylcyanid *n*, Phenylacetonitril *n*, Phenylessigsäurenitril *n*
benzyl ether Dibenzylether *m*
benzyl group $C_6H_5CH_2$– Benzylgruppe *f*, Benzylrest *m (i. w. S. auch für substituierte Derivate* $ArCR_2$*–)*
benzyl radical 1. (freies) Benzylradikal *n*; 2. *s.* benzyl group
benzylation Benzylierung *f*
benzylic group *s.* benzyl group
benzylidene chloride Benzylidendichlorid *n*, α,α-Dichlortoluen *n*
benzylpenicillin Benzylpenicillin *n*, Penicillin *n* G, Penicillin *n* II
benzyne Benzin *n*, Benzyn *n*, 1,2-Dehydrobenzen *n*
Bergius process *(Carb)* Bergius-Hydrierverfahren *n*, Bergius-Verfahren *n*, Bergius-Hochdruckverfahren *n*, (katalytisches) Hochdruckhydrier(ungs)verfahren *n* nach Bergius, Bergius-Pier-Verfahren *n*, Berginverfahren *n*, Leuna-Verfahren *n*
berkelium Bk Berkelium *n*
Berlin blue Berliner Blau *n*, Pariser Blau *n*, Preußisch Blau *n*
Berthelot calorimeter Berthelot-Kalorimeter *n*
berthollide compounds *s.* berthollid(e)s
berthollid(e)s Berthollide *npl*, berthollide [nicht daltonide, nicht stöchiometrische] Verbindungen *fpl*
beryl *(Min)* Beryll *m (Berylliumaluminiumsilicat)*
beryllate Beryllat *n*
beryllia *s.* beryllium oxide
beryllide Beryllid *n*
berylliosis *s.* beryllium disease
beryllium Be Beryllium *n*
beryllium disease Berylliumkrankheit *f*, Beryll(i)ose *f (durch Berylliumstäube verursachte Staublungenkrankheit)*
beryllium halide Berylliumhalogenid *n*
beryllium hydride Berylliumhydrid *n*
beryllium hydroxide Berylliumhydroxid *n*
beryllium oxide Berylliumoxid *n*
beryllosis *s.* beryllium disease
Bessemer converter Windfrischkonverter *m*; Bessemer-Konverter *m*, Bessemer-Birne *f*
Bessemer process 1. Windfrischverfahren *n*, Konverter(frisch)verfahren *n*, Bessemer-Verfahren *n*; 2. *s.* acid Bessemer process
Bessemer steel Windfrischstahl *m*, windgefrischter Stahl *m*, Konverterstahl *m*; Bessemer-Stahl *m*, Bessemer-Konverterstahl *m*
bessemerize *v* windfrischen, bessemern, im Konverter verblasen; (Kupferstein) im Konverter verblasen, kupferbessemern, bessemern
bessemerizing Windfrischen *n*, Bessemern *n*, Verblasen *n* im Konverter; Verblasen *n* (von Kupferstein) im Kon-

verter, Kupferbessemern *n*, Bessemern *n*
BET isotherm BET-Adsorptionsisotherme *f*, Brunauer-Emmett-Teller-Adsorptionsisotherme *f*, BET-Isotherme *f*
beta backscatter *(Anal)* β-Rückstreuung *f*, Betarückstreuung *f*
beta cellulose β-Cellulose *f*, Betacellulose *f*
beta decay [disintegration] β-Zerfall *m*, Betazerfall *m*, Beta-Umwandlung *f*
beta-emitter *(Kern)* β-Strahler *m*, Betastrahler *m*
beta-hydroxy (carboxylic) acid 3-Hydroxysäure *f*, β-Hydroxysäure *f*, 3-Hydroxycarbonsäure *f*
beta particle β-Teilchen *n*, Betateilchen *n*, Elektron *n*
beta-ray β-Strahl *m*, Betastrahl *m*
beta-ray spectrum β-Spektrum *n*, Betaspektrum *n*
betaine Betain *n (allg. für eine Klasse zwitterionischer Verbindungen)*; *(i. e. S.)* Betain *n*, Carboxy-methyltrimethylammonium *n*, Trimethylammonioacetat *n*, Trimethylglycin *n*, Oxyneurin *n*, Lycin *n*
Bettendorf's test (for arsenic) bettendorfsche Arsenprobe *f* [Probe *f*]
betweenanene *(org Ch)* Betweenanen *n*
beverage alcohol Trinkbranntwein *m*
B.F.W. *s.* boiler feed water
BFW *s.* boiler feed water
BHA *s.* butylated hydroxyanisole
BHC *s.* benzene hexachloride
BHT *(butylated hydroxytoluene) s.* 2,6-di-*tert*-butyl-*p*-cresol
biacid zweisäurig *(Basen)*
biaryl Biaryl *n*
bias 1. *(Anal)* Vorspannung *f*; 2. *s.* bias error
bias error *(Anal)* systematischer Fehler *m*
biaxial *(Krist)* biaxial, (optisch) zweiachsig
biaxially oriented film biaxial gereckte Folie *f*
bicarbonate Hydrogencarbonat *n*
bicarbonate hardness *s.* carbonate hardness
bicarbonate of soda Natriumhydrogencarbonat *n*
bichloride Dichlorid *n*
bichromate Dichromat *n*
bicomponent fibre Bi(komponenten)faser *f*
biconcave lens bikonkave Linse *f*
biconvex lens bikonvexe Linse *f*
bicycle *s.* bicyclic compound
bicyclic compound *(org Ch)* bicyclische Verbindung *f*, Bicyclus *m*
bidentate zweizähnig, zweizählig *(Komplexchemie)*
bifluoride Hydrogenfluorid *n*
bifocal lens bifokale Linse *f*
bifunctional bifunktionell, difunktionell
bifunctional catalysis bifunktionelle Katalyse *f*
bifunctional unit difunktionelle Einheit *f*, D-Einheit *f*, Grundbaustein *m* D *(zum Aufbau von Polymeren)*
biguanide *(org Ch)* Biguanid *n*
bihelical *s.* double-helical
bilayer *(Bioch)* Doppelschicht *f*
bilayer membrane Doppelmembran *f*
bile acid Gallensäure *f*
biliary calculus Gallenstein *m*
bilirubin Bilirubin *n*
biliverdin Biliverdin *n*, Dehydrobilirubin *n*
Billiter cell Billiter-Zelle *f*, Horizontalzelle *f (Elektrolyse)*
bimetal Bimetall *n*
bimetallic bimetallisch, Bimetall...
bimetallic thermometer Bimetallthermometer *n*
bimodal distribution *(Anal)* bimodale [zweigipflige] Verteilung *f*
bimolecular collision bimolekularer Stoß *m*, Zweierstoß *m*
bimolecular rate constant Geschwindigkeitskonstante *f* der bimolekularen Reaktion
bimolecular reaction bimolekulare Reaktion *f*
bin Silo *n(m)*, Bunker *m*, Behälter *m*, Zellenspeicher *m*
binary alloy binäre Legierung *f*, Zweistofflegierung *f*
binary collision *(phys Ch)* Zweierstoß *m*
binary compound binäre Verbindung *f*
binary mixture binäres Gemisch *n*, Zweistoffgemisch *n*
bind *v* binden; verbinden; sich binden *(an etwas)*
binder 1. Bindemittel *n*, Binder *m*; 2.

(Anstr) Bindekörper *m*; 3. *(Kst)* Harzträger *m*; 4. *(Geol)* Zement *m*

binderless bindemittellos, bindemittelfrei, ohne Bindemittel(zusatz)

binding Binden *n*; Verbinden *n*; (chemische) Bindung *f (Vorgang)*

binding agent 1. Bindemittel *n*; 2. *(Bau)* Binder *m*

binding site *(Bioch)* Bindungsort *m*, Bindungsstelle *f*, Bindungsregion *f*

binegative zweifach negativ

Bingham body *s.* Bingham plastic

Bingham flow binghamsches Fließen *n*

Bingham plastic Bingham-Körper *m*, Bingham-plastischer Körper *m*

binodal curve Binodal-Kurve *f*, binodale Kurve *f*

binuclear zweikernig

binucleate zweikernig

bioassay (test) *s.* biological assay

bioavailability *(Pharm)* Bioverfügbarkeit *f*

bioavailable metal level biologisch verfügbarer Metallgehalt *m*

biocatalysis Biokatalyse *f*

biocatalyst Biokatalysator *m*, Ergin *n*

biocatalytic biokatalytisch

biochemical biochemisch

biochemical *(Biot)* Biochemikalie *f*, biochemisches [biogenes] Reagens *n*, Bioreagens *n*

biochemical conversion 1. biochemische Umwandlung *f*; 2. *s.* bioconversion

biochemical decomposition [degradation] biochemischer Abbau *m*

biochemical engineering biochemische Verfahrenstechnik *f*

biochemical map topographische Karte *f (z. B. der Ribonucleinsäure)*

biochemical oxygen demand *(Hyd)* biochemischer Sauerstoffbedarf *m*, BSB *m*

biochemist Biochemiker *m*

biochemistry Biochemie *f*, biologische Chemie *f*

biochip *(Biot)* Biochip *m*

biochrome Naturfarbstoff *m*, natürlicher Farbstoff *m*

biocidal biozid, Organismen (ab)tötend

biocide Biozid *n*, Lebensgift *n (Organismen tötende Chemikalie)*; *(i. e. S.)* Pestizid *n*, Schädlingsbekämpfungsmittel *n*

biocompatibility biologische Verträglichkeit *f*, Biokompatibilität *f*

biocompatible biologisch verträglich

bioconversion *(Biot)* Biotransformation *f*, Biokonversion *f*, mikrobielle Transformation *f* [Umsetzung *f*, Stoffumwandlung *f*]

biocorrosion Biokorrosion *f*, mikrobiell induzierte Korrosion *f*

biodegradability biologische Abbaubarkeit *f*

biodegradable biologisch abbaubar, bioabbaubar

biodegradation biologischer Abbau *m*

biodegrade *v* biologisch abbauen

biodeterioration of materials Materialzerstörung *f* durch Mikroorganismen, biologische Zerstörung *f* von Werkstoffen

biodiesel Biodiesel *m (Biokraftstoff)*

bioelectricity Bioelektrizität *f*

bioenergy Bioenergie *f*

bioengineering Bioverfahrenstechnik *f*, Bioingenieurtechnik *f*, Bioingenieurwesen *n*

bioethanol Bioethanol *m (Biokraftstoff)*

biofilm Biofilm *m*

biofilter 1. Biofilter *n*, biologisch arbeitendes Filter *n*; 2. *s.* trickling filter

biofouling Bewuchs *m*

biofuel Biokraftstoff *m*

biogas Biogas *n*

biogas digester Biogasreaktor *m*

biogenesis Biogenese *f*, Biosynthese *f*, biologische Synthese *f*

biogenetic(al) *s.* biogenic

biogenic biogen

biogenic amines biogene Amine *npl*, Protoalkaloide *npl*

biogenous biogen

biogeochemical biogeochemisch

bioindication Bioindikation *f*

bioindicator Bioindikator *m*, Indikatororganismus *m*

bioinorganic bioanorganisch

bioinsecticide *(Biot)* Bioinsektizid *n*, mikrobielles Insektizid *n*

biokinetic Biokinetik *f*

biological activity biologische Aktivität *f* [Wirksamkeit *f*], Bioaktivität *f*

biological assay 1. Biotest *m*, biologi-

scher Test *m*; 2. *(Pharm)* biologische Wertbestimmung *f*
biological chemistry biologische Chemie *f*, Biochemie *f*
biological cycle biologischer Kreislauf *m*
biological digestion biologischer [biochemischer, mikrobieller] Abbau *m (organischer Abwasserinhaltsstoffe)*
biological engineering biologische Verfahrenstechnik *f*
biological equilibrium biologisches Gleichgewicht *n*
biological exposure index biologischer Arbeitsplatztoleranzwert *m*, BAT
biological film reactor *(Biot)* Filmfermenter *m*
biological filter 1. Biofilter *n*, biologisch arbeitendes Filter *n*; 2. *s.* trickling filter
biological half-life *(Tox)* Rückstands-Halbwertszeit *f*, RL_{50}
biological membrane Biomembran *f*, biologische Membran *f*
biological oxygen demand *s.* biochemical oxygen demand
biological pollutant biologischer Schadstoff *m*
biological pollution biologische Verunreinigung *f*
biological populations biologische Kulturen *fpl*
biological production process biotechnologisches Verfahren *n* der Stoffproduktion, biotechnologisches [biochemisches, biosynthetisches] Produktionsverfahren *n*
biological sewage treatment biologische [biochemische] Abwasserreinigung *f*, biologische [biochemische] Abwasserbehandlung *f*
biological signal Biosignal *n*
biological sludge *s.* 1. activated sludge; 2. sewage sludge
biological (waste) treatment *s.* biological sewage treatment
biologically inert refractory matter *s.* refractory organics
bioluminescence Biolumineszenz *f*
bioluminescent biolumineszent
biomacromolecule Biomakromolekül *n*, Biopolymer(e) *n*
biomass 1. Biomasse *f*; 2. *s.* microbial mass
biomass fuel Biokraftstoff *m*
biomimetic *(org Ch)* biomimetisch
biomimetics *(org Ch)* Biomimetik *f*
biomineralization *(Bioch)* Biomineralisation *f*
biomolecule *(Bioch)* Biomolekül *n*
bioorganic bioorganisch
bio-oxidation *(Hyd)* biologische Oxidation *f*, Biooxidation *f*
biophile element biophiles Element *n*
biopolymer Biopolymer(e) *n*, Biomakromolekül *n*
bioprocess 1. biotechnologisches Verfahren *n* der Stoffproduktion und Stoffwandlung, *(i. e. S.)* biotechnologisches Fermentationsverfahren *n*; 2. Bioprozess *m*, biologischer Prozess *m*, biologische Umsetzung *f (innerhalb eines Organismus)*
bioprocess engineering biotechnologische Verfahrenstechnik *f*
bioprocess industry biotechnologische Industrie *f*, *(i. e. S.)* Fermentationsindustrie *f*
bioprocess production biotechnologische Stoffproduktion *f*
bioprocess technology Bioprozesstechnik *f*
bioprocessing Durchführung *f* biotechnologischer Produktionsverfahren, *(i. e. S.)* Durchführung *f* biotechnologischer Fermentationsverfahren
bioprotein *s.* microbial protein
bioreactor *(Biot)* Bioreaktor *m*, Biotank *m*, biologischer [biochemischer] Reaktor *m* [Reaktionstank *m*]
biorefractory compound *(Tox)* persistente Verbindung *f*
biose 1. Biose *f (Monosaccharid mit zwei Kohlenstoffatomen)*; 2. *s.* disaccharide
biosensor *(Biot)* Biosensor *m*
biosorption Biosorption *f*
biosurfactant *(Biot) mikrobiell hergestellter grenzflächenaktiver Stoff*
biosynthesis Biosynthese *f*, biologische Synthese *f*, Biogenese *f*
biosynthetic biosynthetisch
biosynthetic pathway [route] Biosyntheseweg *m*
biotechnological biotechnologisch, biotechnisch
biotechnology Biotechnologie *f*, Biotechnik *f*
biotic biotisch

biotic balance *s.* biological equilibrium
biotic equilibrium *s.* biological equilibrium
biotransformation *s.* bioconversion
biphenyl Biphenyl *n*
biphotonic excitation Zweiphotonenanregung *f*
bipolar cell Bipolarzelle *f*, bipolare Zelle *f*
biradical Diradikal *n*, Biradikal *n*
birch oil Birkenöl *n*; Birkenteeröl *n*
Birch reduction Birch-Hückel-Reaktion *f* *(Hydrierung aromatischer Verbindungen mit Natrium und Ammoniak)*
bird repellent Vogelabschreckmittel *n*, Vogelfraß-Abwehrmittel *n*, Saatschutzmittel *n (gegen Vögel)*
birefringence Doppelbrechung *f*
birefringent doppelbrechend
Birkeland-Eyde process Birkeland-Eyde-Verfahren *n (Herstellung von Stickoxiden durch Luftoxidation)*
bis-axial bis-axial
bis-substituted *s.* disubstituted
bisazo dye Disazofarbstoff *m*
Bischler-Napieralski reaction Bischler-Napieralski-Reaktion *f (Isochinolinsynthese)*
BISCT = Biotechnology Institute and Studies Centre Trust *(U.K.)*
bishydrazone Dihydrazon *n*
bismuth Bi *Bismut *n*, Wismut *n*
bismuth glance *(Min)* Wismutglanz *m*, Bismuthinit *m*, Bismut(h)in *m (Bismut(III)-sulfid)*
bismuth ochre *(Min)* Wismutocker *m* *(Bismut(III)-oxid)*
bismuth oxychloride Bismutoxidchlorid *n*
bismuth pentoxide Bismutpentoxid *n*, Bismut(V)-oxid *n*
bismuth trihydroxide Bismuttrihydroxid *n*, Bismut(III)-hydroxid *n*
bismuth trioxide Bismuttrioxid *n*, Bismut(III)-oxid *n*
bismuth trisulphide Bismuttrisulfid *n*, Bismut(III)-sulfid *n*
bismuthal Bismut...
bismuthane Bismutan *n*, Bismutwasserstoff *m*, Bismuthydrid *n (allgemeine Formel* Bi_nH_{n+2}*)*
bismuthate M^IBiO_3 Bismutat(V) *n*
bismuthic Bismut..., *(i. e. S.)* Bismut(V)-...
bismuthic acid Bismutsäure *f*
bismuthiferous bismuthaltig
bismuthine 1. BiH_3 Bismutan *n*, Bismutwasserstoff *m*; 2. *(Min)* Bismuthinit *m*, Wismutglanz *m (Bismuttrisulfid)*
bismuthous Bismut..., *(i. e. S.)* Bismut(III)-...
bismuthyl Bismutoxid...
bismutite *(Min)* Bismutit *m*, Wismutspat *m (Bismutoxidcarbonat)*
bisphenol Bisphenol *n*, Methylendiphenol *n*
bistability Bistabilität *f (eines Reaktionssystems)*
bistable bistabil *(Reaktionssystem)*
bisubstrate reaction *(Bioch)* Zweisubstratreaktion *f*
bisulphate Hydrogensulfat *n*
bisulphide 1. M^IHS Hydrogensulfid *n*; 2. $M^I_4S_2$ Disulfid *n*
bisulphite Hydrogensulfit *n*
bisulphite addition compound Bisulfit-Additionsverbindung *f*, Bisulfitaddukt *n*
bisulphite (cooking) liquor *(Pap)* Sulfit(koch)säure *f*, Kochsäure *f*
bitartrate Hydrogentartrat *n*
bitter almond oil Bittermandelöl *n (von Prunus amygdalus Batsch var. amara)*
bitter principle *(Pharm)* Bitterstoff *m*
bitter salt 1. Bittersalz *n*; 2. *(Min)* Epsomit *m (Magnesiumsulfat-7-Wasser)*
bitter substance Bitterstoff *m*
bittern 1. Bitterstoff *m*; 2. restliche Mutterlauge *f (bei Speisesalzgewinnung aus Meerwasser)*
bitumen Bitumen *n*
bituminous bituminös, bitumenhaltig
bituminous coal bituminöse Kohle *f*
bituminous paint Bitumenanstrichstoff *m*
bituminous sand bituminöser [bitumenhaltiger] Sand *m*, Bitumensand *m*
bituminous shale bituminöser Schiefer *m*, Bitumenschiefer *m*
biurate Hydrogenurat *n*
biurea Biharnstoff *m*, Bis-harnstoff *m*, Hydrazin-*N,N'*-bis-carbonsäureamid *n*
biuret Biuret *n*, Allophansäureamid *n*
biuret reaction Biuretreaktion *f*
bivalency Zweiwertigkeit *f*, Bivalenz *f*
bivalent zweiwertig, bivalent
bivariant bivariant, divariant, zweifachfrei

bixin *(Lebm)* Bixin *n (Farbstoff von Bixa orellana L.)*
black 1. Schwarz *n (Farbempfindung)*; 2. schwarzfärbender [schwarzer] Farbstoff *m*, Schwarz *n*; Schwärze *f*; 3. Ruß *m*
black body schwarzer Körper *m*, Schwarzkörper *m*
black-body radiation schwarze Strahlung *f*, Hohlraumstrahlung *f*
black-body radiator schwarzer Strahler *m*, Hohlraumstrahler *m*
black phosphorus schwarzer Phosphor *m*
black powder Schwarzpulver *n*, Sprengpulver *n*
blacken *v* schwärzen; schwarz färben; sich schwärzen
blackening Schwärzen *n*, Schwarzfärben *n*; Schwärzung *f*, Schwarzfärbung *f (Vorgang)*
blackish schwärzlich
blade 1. Schaufel *f*, Flügel *m (z. B. einer Mischmaschine)*; 2. Messer *n*; 3. *(Krist)* flacher Stängel *m*
blade mixer Schaufelmischer *m*, Flügelmischer *m*, Schaufelrührer *m*, Schaufelrührwerk *n*, Paddelrührer *m*
blade wheel reactor *(Biot)* Schaufelradreaktor *m*, Schaufelradfermenter *m*
bladed *(Krist)* plattstängelig
blanc fixe Blanc fixe *n*, Blankfix *n*, Permanentweiß *n*, Barytweiß *n (gefälltes Bariumsulfat)*
Blanc chloromethylation reaction blancsche Reaktion *f (Chlormethylierung)*
blanch *v* 1. *(Lebm)* blanchieren; 2. *s.* bleach
blancher *(Lebm)* Blanchierapparat *m*, Blanchiermaschine *f*
blank 1. Rohling *m*, Rohteil *n*, Rohstück *n*, Formling *m*, Formteil *n*; 2. *(Tech)* Blindflansch *m*; 3. *s.* blank determination • **perform [run] a blank (test)** einen Blindversuch durchführen [anstellen]
blank determination Blindbestimmung *f*
blank reading *(Anal)* Blindwert *m*, Leerwert *m*
blank solution *(Anal)* Blindlösung *f*
blank value *s.* blank reading
blanket *v* abdecken, abschirmen; ersticken *(Feuer)*
blanket-type clarifier *(Hyd)* Schwebefilter *n*, Schwebefilteranlage *f*
blast 1. Wind *m*, Gebläsewind *m*, Gebläseluft *f*; 2. Explosion *f*, Sprengung *f*; Druckwelle *f*
blast burner Gebläsebrenner *m*, Gebläselampe *f*
blast furnace Gebläse(schacht)ofen *m*, Blashochofen *m*, *(i. e. S.)* Hochofen *m (zur Gewinnung von Eisen)*
blast-furnace coke Hochofenkoks *m*, Hüttenkoks *m*
blast-furnace gas Gichtgas *n*, Hochofen(gicht)gas *n*
blast main Windleitung *f*
blast of air Luftstrom *m*
blasting Sprengen *n*, Sprengung *f*
blasting agent Sprengmittel *n*
blasting gelatin Sprenggelatine *f*
blasting powder Sprengpulver *n*, Schwarzpulver *n*
blazed grating *(Spektr)* Echelle-Gitter *n*, Blaze-Gitter *n*, geblaztes Gitter *n (mit sägezahnartigem Muster)*
bleach *v* bleichen, entfärben, aufhellen
bleach *v* **out** ausbleichen, sich entfärben
bleach 1. Bleichen *n*, Bleichung *f*, Bleiche *f*; 2. Bleichmittel; 3. *s.* bleach liquor
bleach-fastness Bleichechtheit *f*
bleach liquor Bleichlauge *f*, Bleichflüssigkeit *f*, Bleichlösung *f*, Bleichflotte *f*, *(i. e. S.)* Calciumhypochloritbleichlauge *f*
bleachability Bleichfähigkeit *f*, Bleichbarkeit *f*
bleachable bleichfähig, bleichbar
bleacher Bleicher *m*, Bleichapparat *m*; Bleichholländer *m*; Bleichturm *m*
bleaching Bleichen *n*, Bleichung *f*, Bleiche *f*, Entfärben *n*, Entfärbung *f*
bleaching agent Bleichmittel *n*
bleaching assistant Bleichhilfsmittel *n*
bleaching bath Bleichbad *n*, Bleichflotte *f*
bleaching clay Bleichton *m*, *(bei der Mineralölraffination auch)* Bleicherde *f*, Entfärbungserde *f*
bleaching powder Bleichpulver *n*, *(i. e. S.)* Bleichkalk *m*, Chlorkalk *m*
bleaching vat Bleichbottich *m*, Bleichkessel *m*, Bleichkufe *f*

bleed *v* **(off)** 1. (aus)bluten, durchbluten, (farbig) durchschlagen, abschmutzen *(z. B. Farb- oder Gerbstoffe)*; 2. *(Tech)* Luft ablassen, entlüften
bleed *v* **through** *s.* bleed
bleeding 1. Bluten *n*, Ausbluten *n*, Durchbluten *n*, Durchschlagen *n*, Abschmutzen *n (z. B. von Farb- oder Gerbstoffen)*; 2. Schwitzen *n (z. B. des Zements)*; 3. *(Filtr)* Trüblauf *m*, Durchlaufen *n* der Feststoffpartikel; 4. *(Chromat)* Abdampfen *n*, Ausbluten *n (der Trennflüssigkeit)*
blend *v* 1. (ver)mischen, zusammenmischen, *(i. e. S.)* innig (ver)mischen, *(Schüttgut auch)* (ver)mengen; sich (ver)mischen; 2. *(Lebm, Gum)* verschneiden; 3. *(Lebm)* kupieren; 4. *(Erdöl)* blenden *(Basisprodukte)*
blend 1. Mischung *f*, *(i. e. S.)* inniges Gemisch *n*; 2. *(Lebm, Gum)* Verschnitt *m*
blende *(Min)* Blende *f*, *(i. e. S.)* Blende *f*, Zinkblende *f*, Sphalerit *m (Zinksulfid)*
blender Mischer *m*, Mischmaschine *f*
blending 1. Mischen *n*, Vermischen *n*, Mischung *f*, Mengen *n*, Vermengen *n*; 2. *(Gum)* Verschneiden *n*; 3. *(Lebm)* Verschneiden *n*, Verschnitt *m*, Kupieren *n*, Coupage *f*; 4. *(Erdöl)* Blenden *n*, Blending *n (von Basisprodukten)*
blending agent Verschnittmittel *n*, Zusatzmittel *n*, Zusatzstoff *m*
blending octane number Mischoctanzahl *f*, Misch-OZ *f*
blind *v* 1. zusetzen, versetzen, verstopfen *(Siebe, Filtertücher)*; 2. *(Text)* mattieren; matt [glanzlos] werden
blinding 1. Zusetzen *n*, Versetzen *n*, Versatz *m*, Verstopfen *n (von Sieben, Filtertüchern)*; 2. *(Text)* Mattieren *n*; Mattwerden *n*, Glanzloswerden *n (unerwünscht)*
blister *v* Blasen bilden, sich mit Blasen bedecken; Blasen ziehen
blister 1. Blase *f*, Gasblase *f*, Gaseinschluss *m (Materialfehler)*; 2. *(Glas)* Blatter *f*; 3. *(Met)* Gussblase *f*
blister agent blasenziehendes Mittel *n*; blasenziehender Kampfstoff *m*, *(i. w. S.)* hautschädigender Kampfstoff *m*
blistery blasig
Bloch wall *(Krist)* Bloch-Wand *f*, blochsche Wand *f*, Domänenwand *f*, Antiphasengrenze *f*
block *v* blockieren, inaktivieren, unwirksam machen *(eine reaktionsfähige Gruppe oder Position)*; blockieren *(eine Reaktion)*
block 1. *(Tech)* Block *m*; 2. *(Bioch)* Block *m (Vorgang, Zustand oder Ort)*
block copolymer Block(co)polymer(e) *n*, Block(co)polymerisat *n*, Segmentcopolymerisat *n*
block diagram Block(schalt)bild *n*
block valve Absperrventil *n*
blockage *s.* blocking
blocking Blockierung *f*, Inaktivierung *f (einer reaktionsfähigen Gruppe oder Position)*; Blockierung *f (einer Reaktion)*
blocking agent *(Bioch)* Blocker *m*
blocky blockig, großstückig
blood coagulation Blutgerinnung *f*
blood glucose Blutzucker *m*
blood pigment Blutfarbstoff *m*, Hämoglobin *n*, Hb
blood plasma extender Plasmaexpander *m*, Blutplasmaersatz *m*
blood poison Blutgift *n*
blood-pressure depressant blutdrucksenkendes Mittel *n*, Mittel *n* gegen Bluthochdruck, Antihypertonikum *n*, Hypotensivum *n*
blood serum Blutserum *n*, Serum *n*
blood sugar Blutzucker *m*
bloom *v* **(out)** ausblühen
bloom Ausblühung *f*, Beschlag *m*
blot *v* abtupfen *(mit Fließpapier)*
blotting paper Löschpapier *n*
blow *v* 1. blasen; drücken *(z. B. eine Flüssigkeit in einen Behälter)*; 2. *(Carb)* (heiß)blasen, warmblasen *(bei der Gaserzeugung)*; 3. *(Met)* frischen
blow *v* **in** einblasen; eindrücken; anblasen *(Hochofen)*
blow *v* **out** ausblasen; ausdrücken, herausdrücken, hinausdrücken; eruptieren *(z. B. Erdgas)*; abblasen *(Hochofen)*
blow *v* **up** explodieren, zerplatzen; explodieren lassen
blow 1. *(Carb)* Blasen *n*, Heißblasen *n*, Warmblasen *n (Gaserzeugung)*; 2. Schlag *m*, Stoß *m*
blow air Blasluft *f*
blow gas Blasegas *n*

blow-head 1. Blaskopf *m*; 2. *(Kst)* Folienblaskopf *m*
blow-hole *(Met)* Blase *f*, Gasblase *f*, Gaseinschluss *m*, Gussblase *f*
blow mould *(Kst)* Blaswerkzeug *n*
blow moulding 1. *(Kst)* Blasformen *n*, Hohlkörperblasen *n*; 2. *(Kst)* Blasformteil *n*
blow off pipe Abblaserohr *n*
blow off points Abblasstellen *fpl*
blowback Rückstoß *m*, Rückschlag *m*
blowcase Druckfass *n*, Druckbirne *f*, Montejus *n*; Pulsometer *n*
blower Lüfter *m*, Ventilator *m*; Gebläse *n*
blowing 1. Blasen *n*; 2. *(Carb)* Blasen *n*, Heißblasen *n*, Warmblasen *n (Gaserzeugung)*
blowing agent *(Gum, Kst)* Treibmittel *n*, Blähmittel *n*
blowing torch Lötrohr *n*
blowlamp Gebläsebrenner *m*, Gebläselampe *f*; Lötlampe *f*
blown bitumen Blasbitumen *n*, geblasenes Bitumen *n*
blown tubing *(Kst)* Blasfolie *f*, Schlauchfolie *f*
blowpipe 1. Lötrohr *n*; 2. *(Glas)* Glasmacherpfeife *f*, Pfeife *f*; 3. *(Tech)* Entlüftungsrohr *n*; Ausblasrohr *n*, Blasrohr *n*, Abblasleitung *f*; 4. Schweißbrenner *m*
blowpipe analysis Lötrohranalyse *f*, Lötrohrprobierkunde *f*
blowtorch Gebläselampe *f*, Gebläsebrenner *m*; Lötlampe *f*
blue *v* bläuen, anbläuen, blau färben; blauen, blau werden
blue biotechnology blaue Biotechnologie *f (Produkte aus dem Meer)*
blue iron ore Blaueisenerz *n*
blue-sensitive blausensibilisiert, blauempfindlich
blue-sensitized blausensibilisiert, blauempfindlich
blue shade Blauton *m*, blauer Farbton *m*, blaue Farbnuance *f*
blue-shift Blauverschiebung *f*, hypsochrome Verschiebung *f*
blue staining Blaufärbung *f*
blueing 1. Bläuen *n*, Bläuung *f*, Anbläuen *n*; Blauanlaufen *n (z. B. eines Werkzeugs)*; 2. *(Text)* Bläuungsmittel *n*, blauer Farbstoff *m (zum Verbessern des Weißgrads)*
blueness Bläue *f*, blaue Färbung *f*
bluish bläulich; blaustichig
blunt *v* abstumpfen *(1. Säuren oder Basen; 2. Glasschnittflächen)*
BM = barrels per month
BMR *(basal metabolic rate)* s. basal metabolism
board 1. Platte *f*; 2. Pappe *f*; Karton *m*
boat 1. *(Lab)* Schiffchen *n*, Substanzschiffchen *n*, Boot *n*; 2. *s.* boat form
boat conformation Bootkonformation *f*, Wannenkonformation *f (Stereochemie)*
boat form Bootform *f*, Wannenform *f (Stereochemie)*
bobby Gummiwischer *m (für Rührstäbe)*
BOD *(biochemical oxygen demand) (Hyd)* BSB *m*, biochemischer Sauerstoffbedarf *m*
B.O.D. *(biochemical oxygen demand) (Hyd)* BSB *m*, biochemischer Sauerstoffbedarf *m*
BOD loading BSB-Belastung *f*, BSB-Last *f*
Bodenstein (steady-state) approximation *(phys Ch)* (bodensteinsches) Stationaritätsprinzip *n*, Bodenstein-Prinzip *n*
body *v* 1. *(Anstr)* Körper geben, füllend wirken; 2. *(Anstr)* verdicken, eindicken
body 1. Stoff *m*, Substanz *f*, Körper *m*; 2. *(Anstr)* Körper *m*, Konsistenz *f*, Festigkeit *f (der Farbmasse)*; 3. *(Ker)* Masse *f*; Scherben *m*; 4. *(Tech)* Konvertermittelstück *n*; Brüdenraum *m (eines Verdampfers)*; Verdampfereinheit *f*, Verdampferkörper *m*; Körper *m*, Gehäuse *n (z. B. eines Ventils)*
body-centred lattice *(Krist)* raumzentriertes [innenzentriertes] Gitter *n*
body fluid Körperflüssigkeit *f*
bog iron (ore) *(Min)* Raseneisenerz *n*, Raseneisenstein *m*
Bohn-Schmidt reaction Bohn-Schmidt-Reaktion *f (zur Darstellung von Polyhydroxyanthrachinonen)*
Bohr atom (model) bohrsches [bohr-rutherfordsches] Atommodell *n*
Bohr magneton bohrsches Magneton *n*
Bohr Sommerfeld atom (model) bohr-sommerfeldsches Atommodell *n*
bohrium Bh Bohrium *n (Element 107)*

boil *v* kochen (lassen); kochen, sieden
boil *v* **away** abdampfen, austreiben, verkochen, wegkochen
boil *v* **down** eindampfen, einengen; eindicken; eingeengt [konzentrierter] werden; eindicken, dick [zähflüssig] werden
boil *v* **out** auskochen
boil *v* **over** überkochen, überlaufen, überschäumen
boil *v* **up** aufkochen (lassen)
boil 1. *s.* boiling point; 2. *(Pap)* Kochung *f*; 3. *(Met)* Kochen *n (der gefrischten Schmelze)* • **be on the boil** kochen, sieden • **bring to the boil** zum Kochen [Sieden] bringen • **come to the boil** zu kochen [sieden] beginnen • **keep at the boil** am Kochen [Sieden] halten
boildown Eindampfen *n*, Einengen *n*; Eindicken *n*
boiled oil Ölfirnis *m*, Firnis *m*, *(i. e. S.)* Leinölfirnis *m*
boiler Destillationsblase *f*, Destillierblase *f*, Blase *f*, Destillationsgefäß *n*; Dampferzeuger *m*, Dampfkessel *m*, Kessel *m*
boiler feed water Kesselspeisewasser *n*, Speisewasser *n* für Dampferzeuger, Dampferzeugerspeisewasser *n*
boiler make-up Zusatz(kesselspeise)-wasser *n*, Zusatzspeisewasser *n*
boiler scale Kesselstein *m*
boiler water Kesselwasser *n*, Inhaltswasser *n* (für Dampferzeuger), Dampferzeugerinhaltswasser *n*, DE-Inhaltswasser *n*
boiler water treatment Kesselwasseraufbereitung *f*
boilfast kochecht, kochfest, kochbeständig
boiling bed brodelnde [kochende, blasenbildende] Wirbelschicht *f*
boiling curve *(Dest)* Siedelinie *f*, Siedekurve *f*
boiling flask Kochkolben *m*, Kochflasche *f*, Langhalskolben *m*, *(i. e. S.)* Langhals-Stehkolben *m*
boiling heat Siedehitze *f*
boiling point Kochpunkt *m*, Kp., K, Siedepunkt *m*, Siedetemperatur *f*
boiling-point constant ebullioskopische Konstante *f*, molale Siedepunktserhöhung *f*
boiling-point diagram Siedediagramm *n*, Siedeschaubild *n*
boiling-point elevation Siedepunktserhöhung *f*
boiling reactor Siede(wasser)reaktor *m*, Verdampferreaktor *m*, Kochendwasserreaktor *m*
boiling stone Siedesteinchen *n*, Siedestein *m*
boiling temperature Siedetemperatur *f*, Siedepunkt *m*, Kochpunkt *m*, Kp., K
boiling tube Siederohr *n*
boiling-up Aufkochen *n*
boiling water reactor *s.* boiling reactor
boilout Auskochen *n*
boilover Überkochen *n*, Überlaufen *n*, Überschäumen *n*
bolometer Bolometer *n*
bolometric bolometrisch
bolt *v* sieben, (durch)beuteln
bolt-head flask *(Lab)* Kurzhalsrundkolben *m*
Boltzmann constant boltzmannsche Konstante *f*, Boltzmann-Konstante *f*, k
Boltzmann distribution Boltzmann-Verteilung *f*
Boltzmann statistics boltzmannsche [klassische] Statistik *f*, Boltzmann-Statistik *f*
bolus alba *(Min)* Kaolin *m*, Porzellanerde *f*, weißer [reiner] Ton *m*, China Clay *m(n)*, Bolus alba
bomb calorimeter Bombenkalorimeter *n*, Kalorimeterbombe *f*, kalorimetrische Bombe *f*, Verbrennungsbombe *f*
bomb tube *(Lab)* Bombenrohr *n* [Einschmelzrohr *n*, Schießrohr *n*] nach Carius, Carius-Rohr *n*
bombard *v (Kern)* beschießen, bestrahlen, bombardieren
bombardment *(Kern)* Beschuss *m*, Beschießen *n*, Bombardement *n*, Bombardierung *f*
BON 1. *s.* blending octane number; 2. = beta-oxy-naphthoic acid
bond *v* 1. *(Ch)* verbinden; sich verbinden, eine Bindung eingehen; 2. binden; (ver)kleben
bond *v* **together** verbinden, verknüpfen
bond 1. (chemische) Bindung *f (Zustand)*; 2. Verbindung *f*; Klebung *f*, Verklebung *f*

bond angle Bindungswinkel *m*, Valenzwinkel *m*
bond cleavage Bindungsspaltung *f*
bond dipole moment Bindungsdipolmoment *n*
bond dissociation Bindungsspaltung *f*, *(i. e. S.)* homolytische Bindungsspaltung *f*, Homolyse *f*
bond-dissociation energy Bindungsspaltungsenergie *f*, Bindungsdissoziationsenergie *f*, Dissoziationsenergie *f*
bond distance *s.* bond length
bond energy (mittlere) Bindungsenergie *f*
bond fission *s.* bond cleavage
bond length Bindungslänge *f*, Atom(kern)abstand *m*, Kernabstand *m*
bond number Bindungszahl *f*, Bindigkeit *f*, Bindungswertigkeit *f*; Bindungsgrad *m*, Bindungsordnung *f*
bond orbital Bindungsorbital *n*; Valenzorbital *n*
bond order Bindungsgrad *m*, Bindungsordnung *f*
bond rearrangement Bindungsumgruppierung *f*
bond rupture Sprengung *f* der Bindung
bond scission *s.* bond cleavage
bond specificity *(Bioch)* Bindungsspezifität *f (eines Enzyms)*
bond strength 1. Bindefestigkeit *f*, Haftfestigkeit *f*; 2. Bindungsstärke *f*, Bindungsfestigkeit *f*, Bindefestigkeit *f* *(Bindungstheorie)*
bond vibration Bindungsschwingung *f*
δ **bond** δ-Bindung *f*
π **bond** π-Bindung *f*, Pi-Bindung *f*, π-Elektronenpaarbindung *f*
σ **bond** σ-Bindung *f*, Sigma-Bindung *f*, σ-Elektronenpaarbindung *f*
bonding 1. (chemische) Bindung *f (Vorgang)*; 2. Binden *n*, Verbinden *n*; Kleben *n*, Verkleben *n*; 3. Bindemittel *n*, bindender Zusatz *m*
bonding agent 1. Bindemittel *n*; 2. *(Gum)* Haftmittel *n*, Bindemittel *n*
bonding electron Bindungselektron *n*
bonding electron pair bindendes Elektronenpaar *n*, Bindungselektronenpaar *n*, Bindungsdublett *n*
bonding force bindende Kraft *f*, Bindungskraft *f*, Bindekraft *f*
bonding molecular orbital bindendes Molekülorbital *n*
bonding pair of electrons *s.* bonding electron pair
bonding strength *s.* bond strength
bone char(coal) Knochenkohle *f*, Spodium *n*
bone-dry knochentrocken, absolut trocken, atro
bone-dry weight Trocken(substanz)masse *f*, Trockenstoffmasse *f*
boost *v (Gum)* aktivieren *(einen Beschleuniger)*
booster 1. Verstärker *m*, Aktivator *m*; 2. *(Gum)* aktivierender Zusatzbeschleuniger *m* [Sekundärbeschleuniger *m*, Zweitbeschleuniger *m*]
borane Boran *n*, *(i. w. S.)* Borhydrid *n*
borate Borat *n*
borax 1. Borax *m*, Natriumtetraborat-Dekahydrat *n*; 2. *(Min)* Borax *m*, Tinkal *m (Natriumtetraborat-10-Wasser)*
borax bead *(Anal)* Boraxperle *f*
borazene Borazin *n*, *(i. e. S.)* $B_3N_3H_6$ Borazin *n*, *Cyclotriborazan *n*
borazine Borazin *n*, *(i. e. S.)* $B_3N_3H_6$ Borazin *n*, *Cyclotriborazan *n*
boric Bor...
boric acid Orthoborsäure *f*, Trioxoborsäure *f*, Borsäure *f*
boric-acid ester Borsäureester *m*
boride Borid *n*
borine Boran *n (der allgemeinen Formel B_nH_{n+2})*; *(i. e. S.)* BH_3 Boran *n*, Monoboran *n*
Born-Haber (thermochemical) cycle born-haberscher Kreisprozess *m*, Born-Haber-Kreisprozess *m (zur Berechnung von Gitterenergien)*
Born-Oppenheimer approximation Born-Oppenheimer-Näherung *f (zur Berechnung von Elektronenübergängen)*
bornane *(org Ch)* Bornan *n*
borneol Borneol *n*, Bornylalkohol *m*
borofluoric acid Borfluorwasserstoffsäure *f*, Fluoroborsäure *f*
borofluoride Fluoroborat *n*
borohydride Tetrahydridoborat *n*, Boronat *n*
boron B Bor *n*
boron carbide Borcarbid *n*
boron hydride *s.* borane

boron mononitride Bormononitrid *n*, Bornitrid *n*
boron steel Borstahl *m*
boron trichloride Bortrichlorid *n*, Borchlorid *n*
boron trioxide Bortrioxid *n*, Bor(III)-oxid *n*, Boroxid *n*
boronate *v* mit Borverbindungen versetzen *(Düngemittel)*
borosilicate glass Borosilikatglas *n*
Bose-Einstein statistics Bose-Einstein-Statistik *f*, Bose-Statistik *f*
boson Boson *n (Elementarteilchen)*
bosshead Doppelmuffe *f (am Stativ)*
botanical pflanzliches Insektizid *n*, insektizider Naturstoff *m*
botryoidal *(Krist)* traubenförmig, traubig
bottle *v* auf [in] Flaschen (ab)füllen, auf Flaschen (ab)ziehen
bottle 1. Flasche *f*; 2. *(Lebm)* Einkochglas *n*, Einweckglas *n*
bottle centrifuge Flaschenzentrifuge *f*
bottle fermentation Flaschengärung *f*
bottle glass Flaschenglas *n*
bottled gas Flüssiggas *n* in Stahlflaschen *(besonders Propan und Butan)*
bottling Abfüllen *n* auf Flaschen, Flaschenfüllung *f*, Flaschenabzug *m*
bottling machine Flaschen(ab)füllmaschine *f*, Flaschenfüller *m*
bottom 1. Boden *m*, unterster Teil *m*; 2. *(Met)* Sohle *f*; Boden *m (beim Kopf- und Bodenschmelzen)*
bottom-blown converter bodenblasender [normal blasender] Konverter *m*, Unterwindfrischkonverter *m*
bottom discharge Untenaustrag *m*, Untenentleerung *f*, Bodenentleerung *f*, Bodenablass *m*
bottom-driven centrifuge Stehzentrifuge *f*, stehende Zentrifuge *f*
bottom fermentation Untergärung *f*
bottom-fermented untergärig
bottom fraction Bodenfraktion *f*
bottom layer untere Schicht *f*, Bodenschicht *f*
bottom product *s.* bottoms 3.
bottom sediment(s) *s.* bottom settlings
bottom settlings Bodensatz *m*, Satz *m*, Sediment *n*, Ausscheidung *f*, Ausscheidungsprodukt *n*, Ablagerung *f*, Niederschlag *m*
bottom valve Bodenventil *n*
bottoms 1. *s.* bottom settlings; 2. *(Lebm)* Fassgeläger *n*; 3. *(Dest)* Bodenprodukt *n*, Sumpfprodukt *n*, Bodenkörper *m*
Bouguer-Lambert law (of absorption) bouguer-lambertsches Gesetz *n* [Absorptionsgesetz *n*], lambertsches Gesetz *n* [Absorptionsgesetz *n*]
bound carbon dioxide *(Hyd)* gebundene Kohlensäure *f*, *(i. e. S.)* festgebundene [ganzgebundene] Kohlensäure *f*
bound enzyme immobilisiertes [trägerfixiertes] Enzym *n*
bound moisture gebundene Feuchte *f* [Feuchtigkeit *f*], gebundenes Wasser *n*
boundary Grenze *f*
boundary condition Randbedingung *f*
boundary diffusion Grenzflächendiffusion *f*
boundary layer Grenzschicht *f*
boundary reaction Phasengrenzreaktion *f*
boundary surface Grenzfläche *f*; Bindungssphäre *f (Bindungstheorie)*
Bouveault-Blanc reduction Bouveault-Blanc-Reduktion *f*, Bouveault-Blanc-Methode *f*, bouveault-blancsche Reduktion *f (von Carbonsäureestern zu primären Alkoholen)*
b.o.v. *s.* brown oil of vitriol
B.O.V. *s.* brown oil of vitriol
bovine serum albumin Rinderserumalbumin *n*
bowk *v* beuchen, abkochen, brühen
bowl 1. Schale *f*, Schüssel *f*; Mahlschüssel *f*, Schüssel *f (einer Schüsselmühle)*; Speiserbecken *n*, Speiserkopf *m*; 2. Walze *f*, *(oft)* Kalanderwalze *f*; Trommel *f (einer Zentrifuge)*
bowl centrifuge Trommelzentrifuge *f*
box 1. Kasten *m*, Kammer *f*; Buchse *f*; Büchse *f*, Packung *f*; 2. *(Met)* Kasten *m*, Kiste *f (zur Wärmebehandlung)*
box annealing Kastenglühen *n*, Kistenglühen *n*
box carburizing Kastenaufkohlen *n*, Kastenzementieren *n*, Kasteneinsetzen *n*, Einsetzen *n* in Zementationskästen
box kiln Kammerofen *m*
box press Etagenpresse *f*
box-anneal *v* kastenglühen, kistenglühen
box-carburize *v* kastenaufkohlen, kas-

tenzementieren, kasteneinsetzen, in Zementationskästen einsetzen
Boyle law boyle-mariottesches Gesetz *n*
bp *s.* boiling point
b.p. *s.* boiling point
BP 1. = British Patent; 2. = British Pharmacopoeia
bpd = barrels per day
BPE *s.* boiling-point elevation
BPh = British Pharmacopoeia
BPL *s.* β-propiolactone
bpm = barrels per month
b.pt. *s.* boiling point
BR *s.* 1. butyl rubber; 2. butadiene rubber
Brackett series *(Spektr)* Brackett-Serie *f*
brackish brackig, etwas [leicht] salzig, Brack...
brackish water Brackwasser *n*
Bradford assay Bradford-Bestimmung *f (Proteinchemie)*
bradykinin Bradykinin *n (ein Nonapeptid)*
Bragg angle *(Krist)* braggscher Winkel *m*, Glanzwinkel *m*
Bragg method (of crystal analysis) Bragg-Methode *f*, Bragg-Verfahren *n*, Drehkristallmethode *f* [Drehkristallverfahren *n*] von Bragg
branch *v* sich verzweigen *(Moleküle und Reaktionsketten)*
branch *v* **off** abzweigen *(Rohrleitungen)*
branch Verzweigung *f*, Seitenkette *f*, Seitenzweig *m (eines Moleküls)*; Zweig *m (einer Brückenringverbindung)*; Abzweig *m*, Abzweigung *f (einer Rohrleitung)*
branch point 1. Verzweigungsstelle *f (eines Moleküls)*; 2. *(Bioch)* Verzweigungsstelle *f*, Verzweigungspunkt *m*, Knotenpunkt *m (des Stoffwechsels)*
branched verzweigt
branched biosynthetic pathway *(Biot)* verzweigter Biosyntheseweg *m*
branched-chain hydrocarbon verzweigtkettiger Kohlenwasserstoff *m*, Kohlenwasserstoff *m* mit verzweigter Kette
branched-chain mechanism Kettenverzweigungsmechanismus *m*, verzweigter Kettenmechanismus *m*
branched polyethylene verzweigtes Polyethylen *n*, Hochdruckpolyethylen *n*, Polyethylen *n* niedriger Dichte
branched polymer verzweigtes Polymer *n*
branching Verzweigung *f (von Molekülen)*
branching enzyme Verzweigungsenzym *n*, Branching-Enzym *n*
branching of chains Kettenverzweigung *f*
brand name Markenname *m*, Markenbezeichnung *f*
brass Messing *n*, Ms
Braun degradation [reaction] (von-) braunscher Abbau *m*, von-Braun-Reaktion *f (tertiärer Amine mit Bromcyanid)*
Bravais lattice *(Krist)* Bravais-Gitter *n*, bravaissches Translationsgitter *n*
braze *v* hartlöten
brazing Hartlöten *n*, Hartlötung *f*
brazing alloy Hartlot *n*
brazing metal Hartlot *n*
brazing solder Hartlot *n*
break *v* (zer)brechen; (auf)spalten, sprengen, lösen *(chemische Bindungen)*; abbrechen, beenden *(Reaktionsketten)*; brechen, spalten, entmischen *(Emulsionen)*; brechen, sich entmischen *(Emulsionen)*; brechen *(Speiseöl durch Erhitzen entschleimen)*
break *v* **apart** auseinanderbrechen
break *v* **down** 1. zerlegen *(mechanisch)*; 2. *(org Ch)* abbauen, zerlegen; zerfallen; 3. *(Ech)* zusammenbrechen, zerstört werden *(Passivität, Passivschicht)*; zusammenbrechen *(Potenzial)*
break *v* **irregularly** *(Min)* einen unebenen Bruch haben
break *v* **through** durchbrechen *(Adsorption, Ionenaustausch, Filtration)*
break *v* **up** aufbrechen, aufreißen
break 1. Umschlag *m (z. B. bei der Titration)*; 2. Knickpunkt *m (z. B. einer Kurve)*
breakage 1. Brechen *n*, Bruch *m*; 2. *(Ker, Glas)* Bruch *m (zerbrochene Ware)*
breakdown 1. Zerlegung *f (mechanisch)*; 2. *(org Ch)* Abbau *m*, Zerlegung *f*; Zerfall *m*; 3. *(Ech)* Zusammenbruch *m*, Zerstörung *f (der Passivität oder Passivschicht)*; Zusammenbruch *m (des Potenzials)*
breakdown product *(org Ch)* Abbau-

produkt *n*, Zerlegungsprodukt *n*; Zerfallsprodukt *n*
breakdown reaction Abbaureaktion *f*, abbauende Reaktion *f*
breakdown strength Durchschlagfestigkeit *f*, dielektrische [elektrische] Festigkeit *f*
breakdown voltage Durchschlag(s)-spannung *f*
breaker 1. Zerteiler *m*; 2. *s.* breaking mill
breaking-apart Auseinanderbrechen *n*
breaking length Reißlänge *f*
breaking load Bruchkraft *f (Biegeversuch)*
breaking mill Brecherwalzwerk *n*
breaking of emulsions Brechen *n* [Spalten *n*, Trennen *n*, Zerstören *n*] von Emulsionen, Dismulgieren *n*, Entemulsionieren *n*; Brechen *n* [Entmischung *f*, Spaltung *f*, Zerfallen *n*] einer Emulsion, Emulsionsentmischung *f*, Emulsionsspaltung *f*
breaking of foams Schaumzerstörung *f*, Entschäumen *n*
breaking-off Abbrechen *n*, Abbruch *m*
breaking point *(Erdöl)* Brechpunkt *m*; Bruchpunkt *m*, Zerreißpunkt *m*
breaking strength Bruchfestigkeit *f*; Reißfestigkeit *f*
breaking tenacity *(Text)* Reißfestigkeit *f*
breaking-up Aufbrechen *n*, Aufreißen *n*
breakpoint 1. Durchbruchpunkt *m (Adsorption, Ionenaustausch, Filtration)*; 2. *(Hyd)* Brechpunkt *m*, Knickpunkt *m (Chlorung)*
breakpoint chlorination *(Hyd)* Brechpunktchlorung *f*, Knickpunktchlorung *f*
breakthrough Durchbruch *m (Adsorption, Ionenaustausch, Filtration)*
breakthrough point Durchbruchpunkt *m*
breakup Aufbrechen *n*, Aufreißen *n*
breathability Atmungsfähigkeit *f*, Atmung *f (eines Überzugs)*
breathe *v* 1. atmen; 2. arbeiten *(sich ausdehnen oder zusammenziehen)*; 3. *(Kst)* entlüften, entgasen *(Werkzeug, Form)*
breathing apparatus Atemschutzgerät *n*
breathing vibration *(Spektr)* Pulsationsschwingung *f*, Atemschwingung *f*, Breathing-Schwingung *f*
breeching Rauchgaskanal *m*
breeding reactor *(Kern)* Brutreaktor *m*, Brüter *m*
breeze 1. Grus *m (von Kohle oder Koks)*; Kohlenstaub *m*; 2. Asche *f*, Lösche *f*
Bremen blue Bremer Blau *n*, Braunschweiger Blau *n*, Kalkblau *n*, Neuwieder Blau *n (Kupfer(II)-hydroxid)*
bremsstrahlung Bremsstrahlung *f*, kontinuierliche Röntgenstrahlung *f*
brew *v* brauen
brew Sud *m*, Gebräu *n*; Brühe *f*
brew kettle Braukessel *m*, Braupfanne *f*, Würze(koch)kessel *m*, Würzepfanne *f*
brewers' malt Brau(erei)malz *n*
brewers' wort Bierwürze *f*
brewery Brauerei *f*, Bierbrauerei *f*
brewhouse Brauhaus *n*, Brauerei *f*, *(i. e. S.)* Sudhaus *n*
brewing 1. Brauen *n*, Bierbrauen *n*; 2. Sud *m*
brewing barley Braugerste *f*; BG
brewing chemistry Brauereichemie *f*
brewing water Brau(erei)wasser *n*
brewing yeast Bierhefe *f*
Brewster angle Brewster-Winkel *m*, Polarisationswinkel *m*
brick Ziegel *m*, Ziegelstein *m*; Stein *m (aus geformten und gebrannten keramischen Rohstoffen)*
brick clay Ziegelton *m*
brick dust Ziegelmehl *n*
brick kiln Ziegelofen *m*
brick-line *v* ausmauern, aussteinen; mit Ziegeln ausmauern
brickware Ziegeleierzeugnis *n*
brickwork Mauerwerk *n*
bridge *v* überbrücken, untereinander verbinden
bridge *(Tech, Ch)* Brücke *f*
bridge atom Brückenatom *n*
bridge bicycle *(org Ch)* Brücken-Bicyclus *m*
bridge bond Brückenbindung *f*
bridged hydrocarbon Brücken-Kohlenwasserstoff *m*
bridged ring compound Brückenringverbindung *f*
bridgehead Brückenkopf *m (einer Brückenringverbindung)*
bridgehead atom Brückenkopfatom *n*
bridging brückenbildend

bridging Brückenbildung *f*; Überbrückung *f*
bridging oxygen Brückensauerstoff *m*
bright glänzend, Glanz...; leuchtend, lebhaft; hell; klar; blank
bright-anneal *v* blankglühen, in Schutzgas glühen
bright anneal(ing) Blankglühen *n*, Schutzgasglühen *n*, Glühen *n* [Glühung *f*] in Schutzgas
bright coal Glanzkohle *f*, glänzende Kohle *f*
bright nickel Glanznickel *m*
bright plating Glanzabscheidung *f* *(Galvanotechnik)*
bright plating solution [bath] Glanzbad *n*
bright silver Glanzsilber *n*
bright stock Brightstock *m*, Brightstock-Öl *n* *(hochviskoser Schmieröldestillationsrückstand)*
brightener (optischer) Aufheller *m*, (optisches) Aufhellungsmittel *n*, optisches Bleichmittel *n*, Weißtöner *m*, *(in Waschmitteln auch)* Weißmacher *m*; Glanzbildner *m* *(Galvanotechnik)*
brightening 1. Aufhellen *n*; 2. *(Pap)* Bleichen *n*, Entfärben *n*, Schaffung *f* [Herstellung *f*] eines hohen Weißgehalts
brightening agent *s.* brightener
brightness Glanz *m*; Leuchtkraft *f*, Lebhaftigkeit *f*; Helligkeit *f*; Klarheit *f*; Leuchtdichte *f*
brilliance Brillanz *f*, Farbbrillanz *f*, Leuchtkraft *f*, Farb(en)schönheit *f*
brilliancy Brillanz *f*, Farbbrillanz *f*, Leuchtkraft *f*, Farb(en)schönheit *f*
brilliant brillant, leuchtend, farbstark, farbkräftig
brilliant dye Brillantfarbstoff *m*
brilliant lustre Diamantglanz *m*
brilliant nickel *s.* bright nickel
Brillouin polyhedron *(Krist)* Brillouin-Polyeder *n*
brim Rand *m*
brimstone gediegener [natürlicher] Schwefel *m*; Rohschwefel *m*, Schwefel *m* *(in Stücken oder Stangen als Handelsprodukt)*
brine *v* mit Salzlake [Salzlösung, Lake] behandeln; nasspökeln; nasssalzen
brine 1. Sole *f*, Lake *f*, Salzsole *f*, Salzlake *f*, Salzlösung *f*, Salzlauge *f*; 2. Kälteträger *m*, *(i. e. S.)* Kältesole *f*, Kühlsole *f*, Sole *f*; 3. *s.* brine regenerant
brine curing 1. *(Gerb)* Lakenkonservierung *f*, Salzlakenkonservierung *f*, Konservierung *f* durch Salzlakenbehandlung; 2. *(Lebm)* Nasspökeln *n*, Nasspökelung *f*
brine regenerant *(Hyd)* Natriumchloridlösung *f* [Kochsalzlösung *f*, Kochsalzsole *f*] zum Regenerieren *(Ionenaustausch)*
brine salting *(Lebm)* Nasspökeln *n*, Nasspökelung *f*
Brinell hardness Brinellhärte *f*, HB
bring *v* **about** bewerkstelligen, bewirken, herbeiführen
bring *v* **into solution** in Lösung bringen, lösen
bring *v* **to** einstellen auf *(z. B. die gewünschte Stärke)*
bring *v* **to the boil** zum Sieden [Kochen] bringen
brining 1. Behandlung *f* mit Salzlake [Salzlösung, Lake]; 2. *(Lebm)* Nasspökeln *n*, Nasspökelung *f*; Nasssalzung *f*; 3. *(Hyd)* Regenerierung *f* mit Natriumchloridlösung [Kochsalzlösung, Kochsalzsole] *(Ionenaustausch)*
briny salzig, salzhaltig, solehaltig, Salz...
briquet *s.* briquette
briquettability Brikettierbarkeit *f*, Verpressbarkeit *f*, Brikettierfähigkeit *f*
briquettable brikettierbar, verpressbar, brikettierfähig
briquette *v* brikettieren, verpressen, zu Briketts pressen
briquette Pressling *m*, Pressstück *n*, Presskörper *m*; Brikett *n*
briquetting Brikettieren *n*, Brikettierung *f*, Verpressen *n*, Verpressung *f*
briquetting press Brikett(ier)presse *f*, Brikettiermaschine *f*; Pastillenpresse *f* *(einer Kalorimetereinrichtung)*
brisance Brisanz *f*, Sprengkraft *f*
brisk 1. lebhaft; 2. schäumend, perlend, prickelnd; 3. *(Lebm)* pikant
bristle Borste *f*
Brit. Pat. = British Patent
Brit. Pharm. = British Pharmacopoeia
British thermal unit *veraltete Einheit der Wärmemenge mit geringfügig diver-*

gierenden Bezugsbasen; *1 Btu ≈ 1055 J*
brittle spröde, zerbrechlich; brüchig
brittle failure [fracture] Sprödbruch *m*
brittle temperature Sprödbruchtemperatur *f*, Versprödungstemperatur *f*, Kältebruchtemperatur *f*
brittleness Spröde *f*, Sprödigkeit *f*; Brüchigkeit *f*
broad-band decoupling *(Spektr)* Breitbandentkopplung *f*, Rauschentkopplung *f*
broad-band spectrum *(Anal)* Breitbandspektrum *n*
broad-spectrum Breitspektrum..., Breitband..., mit breitem Wirkungsspektrum, mit großer Wirkungsbreite [Anwendungsbreite]
broad-spectrum antibiotic Breitspektrumantibiotikum *n*, Breitbandantibiotikum *n*
bromalide *(org Ch)* Bromalid *n*
bromate *v* bromieren, mit Brom behandeln
bromate M^IBrO_3 Bromat *n*
bromatometric bromatometrisch
bromatometry Bromatometrie *f*
bromic Brom..., *(i. e. S.)* Brom(V)-...
bromic acid Bromsäure *f*
bromide M^IBr Bromid *n*
brominate *v* bromieren, mit Brom behandeln
brominating agent Bromierungsmittel *n*
bromination Bromieren *n*, Bromierung *f*
bromine Br Brom *n*
bromine number *(Anal)* Bromzahl *f*
bromine pentafluoride Brom(V)-fluorid *n*, Brompentafluorid *n*
bromine trifluoride Brom(III)-fluorid *n*, Bromtrifluorid *n*
bromine water Bromwasser *n*
brominolysis Bromolyse *f*
bromism Bromismus *m (Symptomenkomplex der Bromvergiftung)*
bromoacetic acid Bromessigsäure *f*, Bromethansäure *f*
bromoacetone Bromaceton *n*, *1-Brompropan-2-on *n*
bromoacetoxylation Bromacetoxylierung *f*
bromoacetylation Bromacetylierung *f*
bromoacylation Bromacylierung *f*
bromoalkane Bromalkan *n*, Alkylbromid *n*
bromoalkylation Bromalkylierung *f*
bromobenzene Brombenzen *n*, *(i. e. S.)* C_6H_5Br Brombenzen *n*, Monobrombenzen *n*
bromobenzoic acid Brombenzoesäure *f*
bromocarbonylation Carbobromierung *f*
bromocresol green Bromcresolgrün *n*, 3′,3″,5′,5″-Tetrabrom-*m*-cresolsulfonphthalein *n*
bromocresol purple Bromcresolpurpur *m*, 5′,5″-Dibrom-*o*-cresolsulfonphthalein *n*
bromodecarboxylation bromierende Decarboxylierung *f*
bromodesulphonation Bromdesulfonierung *f*
bromoethane Bromethan *n*, Ethylbromid *n*
bromoethene *Bromethen *n*, Vinylbromid *n*
bromoform Bromoform *n*, *Tribrommethan *n*
bromoformylation Carbobromierung *f*
bromolactonization Bromlactonisierung *f*
bromomethane Brommethan *n*, Monobrommethan *n*, Methylbromid *n*
bromomethylation Brommethylierung *f*
bromometric bromometrisch
bromometry Bromometrie *f*
bromonium compound Bromoniumverbindung *f*
bromophenol blue Bromphenolblau *n*, 3′,3″,5′,5″-Tetrabrom-phenolsulfonphthalein *n*
3-bromopropene 3-Brompropen *n*, Allylbromid *n*
***N*-bromosuccinimide** *N*-Bromsuccinimid *n*, NBS, *N*-Brombernsteinsäureimid *n*
bromothymol blue Bromthymolblau *n*, 3′,3″-Dibromthymolsulfonphthalein *n*
bromous Brom..., *(i. e. S.)* Brom(III)-...
bromous acid bromige Säure *f*
Brönsted catalysis law Brönsted-Katalysegesetz *n*, brönstedsches Katalysegesetz *n*
Brönsted-Lowry acid Brönsted-Säure *f*, Protonsäure *f*, Protonendonator *m*, Dysprotid *n*

Brönsted-Lowry base Brönsted-Base *f*, Protonenakzeptor *m*, Emprotid *n*
Brönsted-Lowry concept of acids and bases brönstedscher Säure- und Basebegriff *m*
bronze *v* bronzieren
bronze Bronze *f*
broth 1. *(Lebm)* Fleischbrühe *f*, Brühe *f*, Bouillon *f*; 2. Nährbouillon *f*, Bouillon *f*, Nährflüssigkeit *f*, Nährlösung *f*, Kulturflüssigkeit *f*, Kulturlösung *f (für Mikroorganismen)*; 3. *(Biot)* Fermentationsbrühe *f*, Fermenterbrühe *f*, Fermentationslösung *f*, Fermentationsflüssigkeit *f*, Fermenterflüssigkeit *f*
broth culture Bouillonkultur *f*
brown *v* 1. *(Lebm)* bräunen; braun werden; 2. *(Met)* brünieren
brown coal Braunkohle *f (i. e. S. minderwertige, lockere Braunkohle)*
brown-coal coke Braunkohlenkoks *m*
brown-coal tar Braunkohlenteer *m*
brown humic acid *(Boden)* Braunhuminsäure *f*
brown iron ore [stone] *(Min)* Limonit *m*, Brauneisenerz *n*, Brauneisenstein *m*, brauner Glaskopf *m (wasserhaltiges Eisenoxidhydroxid)*
brown oil of vitriol Glover-Säure *f*, Glover-Turmsäure *f (konzentrierte Schwefelsäure mit 77 bis 80 % H_2SO_4)*
brown-ring test Ringprobe *f (zum NO_3^--Nachweis)*
brown soil *(Boden)* Braunerde *f*
brown sugar brauner Rohzucker *m* [Zucker *m*]; Farin *m*, Farinzucker *m*
brown ware *(Ker)* braunes Steinzeug *n*
Brownian motion [movement] brownsche Bewegung *f* [Molekularbewegung *f*]
browning 1. *(Lebm)* Bräunen *n*, Bräunung *f*; Braunwerden *n*; 2. *(Met)* Brünieren *n*
brucine paper Brucinpapier *n (zum HNO_2-Nachweis)*
Brunauer-Emmett-Teller relationship Gleichung *f* von Brunauer, Emmett und Teller, BET-Gleichung *f (der Adsorption)*
brush *v* bürsten; streichen, anstreichen
brush *v* **on** aufstreichen, mit dem Pinsel auftragen [aufbringen]
brush *v* **with the free flame** mit fächelnder Flamme erwärmen [erhitzen]
brush Bürste *f*; Pinsel *m*
brush aeration *(Hyd)* Bürstenbelüftung *f*
brush application Streichauftrag *m (Bürsten- oder Pinselauftrag)*
brushing Bürsten *n*; Streichen *n*, Anstreichen *n*
brushing paint Streichfarbe *f*
brushkiller Gehölzvernichtungsmittel *n*, Arborizid *n*
BS 1. = British Standard; 2. = Biochemical Society
B.S. *s.* bottom settlings
BSA *s.* bovine serum albumin
BSI = British Standards Institution
BSS *s.* balanced salt solution
BTHM = bromine-containing trihalomethanes
Btu *s.* British thermal unit
BTX *s.* benzene-toluene-xylenes
Bu *s.* butyl 1.
bubble *v* 1. in Blasen aufsteigen lassen, hindurchperlen lassen *(Gas in Flüssigkeiten)*; 2. Blasen bilden, sprudeln, perlen; 3. *s.* bubble up
bubble *v* **up** in Blasen aufsteigen
bubble Blase *f*, Bläschen *n*
bubble aeration *(Hyd)* Druck(luft)belüftung *f*
bubble-cap (distillation) column *(Dest)* Glocken(boden)kolonne *f*
bubble-cap plate Glockenboden *m*
bubble chamber *(Kern)* Blasenkammer *f*
bubble column (fermenter) *(Biot)* Blasensäule *f*, Blasensäulenfermenter *m*, Blasensäulenreaktor *m*
bubble counter *(Lab)* Blasenzähler *m*
bubble plate Glockenboden *m*
bubble tower fermenter *s.* bubble column (fermenter)
bubble tray Glockenboden *m*
bubble-tray column Glockenbodenkolonne *f*
bubbler 1. Barboteur *m*, Druckmischer *m*, pneumatisches Rührwerk *n*; 2. Waschflasche *f*, Gasspüler *m*
bubbling Blasenbildung *f*, Bläschenbildung *f*
bubbling bed brodelnde [kochende, blasenbildende] Wirbelschicht *f*
Bucherer reaction Bucherer-(Lepetit-)Reaktion *f (Naphthol-Naphthylamin-Umwandlung)*
Büchner filter *(Lab)* Büchner-Trichter *m*,

Büchner-Nutsche *f*, Filternutsche *f* nach Büchner
Büchner flask *(Lab)* Saugflasche *f*
Büchner funnel *(Lab)* Büchner-Trichter *m*, Büchner-Nutsche *f*, Filternutsche *f* nach Büchner
bucket Becher *m*, Eimer *m*, Kübel *m*, Behälter *m*; Löffel *m*, Kelle *f*
bucket elevator Becherwerk *n*, Becheraufzug *m*
buckytubes Buckminster-Fullerene *npl*
buffer *v* (ab)puffern
buffer Puffer *m*, Pufferlösung *f*
buffer acid Puffersäure *f*
buffer base Pufferbase *f*
buffer mixture Puffergemisch *n*, Puffermischung *f*
buffer reagent Puffersubstanz *f*
buffer region Pufferbereich *m*, Pufferzone *f*
buffer solution Pufferlösung *f*, Puffer *m*
buffered solution gepufferte Lösung *f*
buffering Abpuffern *n*, Puffern *n*, Pufferung *f*
buffering agent Puffersubstanz *f*
buffering capacity Puffer(ungs)kapazität *f*, Puffer(ungs)vermögen *n*
build *v* **up** 1. aufbauen *(z. B. eine chemische Verbindung)*; 2. sich bilden, zustande kommen
build Bau *m*, Aufbau *m*, Bauart *f*
build-up 1. Aufbau *m*, Struktur *f (z. B. einer chemischen Verbindung)*; 2. Bildung *f*, Zustandekommen *n*; 3. *s.* building-up
build-up of pressure Druckaufbau *m*
build-up of resistance Resistenzentwicklung *f*
build-up reaction Aufbaureaktion *f*, aufbauende Reaktion *f*
builder Builder *m*, Gerüststoff *m*, Aufbaustoff *m (in Waschmitteln)*
building block Baustein *m (z. B. einer chemischen Verbindung)*
building cement Bauzement *m*
building ceramics Baukeramik *f*
building constituent Baubestandteil *m*, Aufbaubestandteil *m*
building lime Baukalk *m (Hauptanteil meist Calciumhydroxid, seltener Calciumoxid)*
building unit Baueinheit *f*
building-up Aufbau *m*, Aufbauen *n (z. B. einer chemischen Verbindung)*
building-up principle *(phys Ch)* Aufbauprinzip *n*
buildup *s.* build-up
built-in eingebaut, Einbau...
bulb 1. Kolben *m*, Ballon *m*, (kugelförmiges) Gefäß *n*, Kugel *f*; 2. Thermometergefäß *n*; 3. Zwiebel *f*, Spinnzwiebel *f*, Ziehzwiebel *f (für Glasseide)*
bulb condenser Kugelkühler *m*, Allihn-Kühler *m*, allihnscher Kühler *m*
bulb pipette Vollpipette *f*
bulk lose
bulk Masse *f*, Menge *f*; Volumen *n*; Hauptmasse *f*, Hauptmenge *f*, Großteil *m* • **evaporate to low [small] bulk** einengen
bulk blend Gemenge *n*
bulk density Rohdichte *f (von porigen Stoffen)*; Schüttdichte *f (von Schüttgütern)*
bulk liquid Hauptteil *m* [Inneres *n*] der flüssigen Phase, Flüssigkeitsraum *m (an der Phasengrenzschicht)*
bulk material Schüttgut *n*
bulk modulus Kompressionsmodul *m*
bulk polymer *(Tech)* Massepolymerisat *n*
bulk polymerization Substanzpolymerisation *f*, Massepolymerisation *f*, Polymerisation *f* in Masse [Substanz]
bulk solution Gesamtlösung *f*
bulk volume 1. Schüttvolum(en) *n*; 2. *(Ker)* Rohvolumen *n*, Volumen *n*, Rauminhalt *m*
bulking agent Füllstoff *m*, Füllmittel *n*, Füllmaterial *n*
bulky groß, dick, umfangreich, voluminös; sperrig
bullet Druck(gas)flasche *f*, Bombe *f*
bump *v* stoßen *(beim Siedeverzug)*
bumping stick *(Lab)* Siedestab *m*
buna rubber Bunakautschuk *m*, Buna *m(n)*
bunker coal Bunkerkohle *f*
bunker fuel (oil) Bunkeröl *n*
bunker oil Bunkeröl *n*
Bunsen burner Bunsenbrenner *m*
Bunsen coefficient bunsenscher Absorptionskoeffizient *m*
Bunsen funnel Bunsentrichter *m*, Trichter *m* mit langem Stiel

Bunte (gas) burette buntesche Bürette *f*, Bunte-Bürette *f*
Bunte salt Bunte-Salz *n*, buntesches Salz *n* *(Salze, meist Natriumsalze, der S-Alkylthioschwefelsäure)*
buoy *v* **up** schwimmend erhalten, emporheben, an die Oberfläche treiben
buoyancy (statischer) Auftrieb *m* *(in einer Flüssigkeit oder einem Gas)*
buoyant density Schwebedichte *f*, Schwimmdichte *f* *(bei der Dichtegradientenzentrifugation)*
buoyant force Auftriebskraft *f*, Auftrieb *m*
burden *v* *(Met)* beschicken, begichten, möllern
burden *(Met)* Möller *m*, Hochofenmöller *m*, Beschickung(smenge) *f*, Gicht *f*
buret *s.* burette
burette Bürette *f*
burette clamp Bürettenklemme *f*, Bürettenhalter *m*
burette stopcock Büretten(sperr)hahn *m*
burial 1. Erdverlegung *f*, Einerdung *f* *(von Leitungen)*; 2. Vergraben *n* *(Materialprüfung)*; 3. Vergraben *n* *(von Abprodukten)*; Versenken *n* *(ins Meer)*
buried 1. unterirdisch, erdverlegt, Erd...; 2. abgeschirmt *(Atomgruppe)*
burn *v* 1. brennen; verbrennen; 2. *(Tech)* brennen *(Kalkstein)*; kalzinieren *(Soda)*; (ab)rösten *(Erze)*; verkohlen *(Holz)*; 3. *(Ker)* brennen; 4. *(Landw)* verätzen, verbrennen *(Pflanzenteile durch Chemikalien)*; 5. *(Med)* verätzen *(z. B. Haut durch Chemikalien)*
burn *v* **in** einbrennen *(z. B. Farben)*
burn *v* **off** 1. wegbrennen, abbrennen, ausbrennen; 2. *(Filtr)* veraschen
burn *v* **out** ausbrennen
burn *v* **to ashes** veraschen
burn 1. *(Landw)* Verätzung *f*, Verbrennung *f* *(an Pflanzen durch Chemikalien)*; 2. *(Med)* Verbrennung *f*; Verätzung *f* *(durch Chemikalien)*; 3. *(Ker)* Brennen *n*, Brand *m*
burn-off Abbrennen *n* *(alter Anstriche)*
burn-out Ausbrand *m*
burn-up *(Kern)* Abbrand *m*
burnable (ver)brennbar
burned lime Branntkalk *m*, gebrannter Kalk *m*, Ätzkalk *m* *(Calciumoxid)*
burner Brenner *m*; Ofen *m*, Verbrennungsofen *m*
burner flame spreader *(Lab)* Breitbrenneraufsatz *m*, Schnittbrenneraufsatz *m*, Schlitzaufsatz *m*
burner gas Verbrennungs(ab)gas *n*, Ofengas *n*; Röstgas *n*
burner wingtop *s.* burner flame spreader
burning-in Einbrennen *n* *(z. B. von Farben)*
burning of pyrites Kiesabröstung *f*
burning of wet fuels Nassverbrennung *f*, Nass-Luft-Oxidation *f* *(von Abprodukten)*
burning-off 1. Wegbrennen *n*, Abbrennen *n*, Ausbrennen *n*; 2. *(Filtr)* Veraschen *n*
burning oil Brennöl *n*
burning-out Ausbrennen *n*
burning point Brennpunkt *m* *(flüssiger Brennstoffe)*
burning rate Verbrennungsgeschwindigkeit *f*, Brenngeschwindigkeit *f*, Abbrandverhältnis *n*
burnt borax gebrannter [kalzinierter] Borax *m*, wasserfreies Natriumtetraborat *n*
burnt lime gebrannter Kalk *m*, Branntkalk *m*, Ätzkalk *m* *(Calciumoxid)*
burnt-out material abgebranntes Material *n*
burr *(Tech)* Grat *m*, Bart *m*
burst *v* **into flame** entflammen; aufflammen *(glimmender Span)*
bursting disk Berstscheibe *f*, Reißscheibe *f*, Platzscheibe *f*
bursting strength *(Kst, Pap)* Berst(druck)festigkeit *f*, Berstwiderstand *m*
butadiene Butadien *n*, *(i. e. S.)* Buta-1,3-dien *n*, Butadien *n*
butadiene-acrylonitrile copolymer Butadien-Acrylnitril-Copolymerisat *n*
butadiene rubber Butadienkautschuk *m*
butadiene-styrene copolymer Butadien-Styren-Copolymerisat *n*
butadiene-styrene rubber Butadien-Styren-Kautschuk *m*, Styren-Butadien-Kautschuk *m*, SBK, Styrenkautschuk *m* *(Butadien-Styren-Copolymerisat)*
butadiyne Butadiin *n*
***n*-butanal** *Butanal *n*, Butyraldehyd *m*
butane Butan *n* *(allg. für die beiden Al-*

kane C_4H_{10}); (i. e. S.) $CH_3CH_2CH_2CH_3$ *Butan *n*
1,4-butanedicarboxylic acid Butan-1,4-dicarbonsäure *f*, *Hexandisäure *f*, Adipinsäure *f*
butanedioic acid *Butandisäure *f*, Ethan-1,2-dicarbonsäure *f*, Bernsteinsäure *f*
1,4-butanediol Butan-1,4-diol *n*
2,3-butanedione *Butan-2,3-dion *n*, Biacetyl *n*, Dimethylglyoxal *n*
butanoic acid *Butansäure *f*, Buttersäure *f*
1-butanol *Butan-1-ol *n*, Butylalkohol *m*
2-butanol *Butan-2-ol *n*, *sec*-Butylalkohol *m*
2-butanone *Butan-2-on *n*, Methylethylketon *n*
2-butenal *But-2-enal *n*, Crotonaldehyd *m*
1-butene But-1-en *n*
2-butene But-2-en *n*
***cis*-butenedioic acid** **cis*-Butendisäure *f*, *cis*-Ethylen-1-2-dicarbonsäure *f*, Maleinsäure *f*
***trans*-butenedioic acid** **trans*-Butendisäure *f*, *trans*-Ethylendicarbonsäure *f*, Fumarsäure *f*
***cis*-2-butenoic acid** **cis*-But-2-ensäure *f*, Isocrotonsäure *f*
***trans*-2-butenoic acid** **trans*-But-2-ensäure *f*, Crotonsäure *f*
butter colour Butterfarbe *f (Farbstoff)*
butter fat Butterfett *n*
butter of antimony Antimonbutter *f*, Antimon(III)-chlorid *n*
butter yellow Buttergelb *n*, Dimethylgelb *n*, *p*-Dimethylaminoazobenzen *n*
butterfly valve Drosselklappe *f*, Drossel(klappen)ventil *n*
buttery butterartig, butterähnlich, butt(e)rig, Butter...
butyl 1. Butyl *n*; 2. *s.* butyl rubber
butyl group Butylgruppe *f*, Butylrest *m*
butyl halide Buthylhalogenid *n*
butyl rubber Butylkautschuk *m*
***n*-butyl acetate** Essigsäurebutylester *m*, Butylacetat *n*
***n*-butyl alcohol** Butylalkohol *m*, *Butan-1-ol *n*
***prim*-butyl alcohol** *s.* *n*-butyl alcohol
***sec*-butyl alcohol** *sec*-Butylalkohol *m*, *Butan-2-ol *n*
***tert*-butyl alcohol** *tert*-Butylalkohol *m*, *2-Methyl-propan-2-ol *n*
butylated hydroxyanisole *(Lebm)* Butylhydroxyanisol *n*, BHA *(Antioxidans)*
butylation Butylierung *f*
butylene polymer Butenpolymer(e) *n*
butylic fermentation *(Biot)* Butanol-Aceton-Gärung *f*
1-butyne But-1-in *n*
2-butyne But-2-in *n*
butynedioic acid *Butindisäure *f*, Ethindicarbonsäure *f*
***n*-butyraldehyde** Butyraldehyd *m*, *Butanal *n*
butyrate Butyrat *n*
butyric(-acid) fermentation Buttersäuregärung *f*
***n*-butyric acid** Buttersäure *f*, *Butansäure *f*
by-product Nebenprodukt *n*, Beiprodukt *n*; Nebenerzeugnis *n*
by-product plant Nebenproduktenanlage *f*, Wertstoffanlage *f*
by-product recovery Nebenproduktengewinnung *f*, Wertstoff(rück)gewinnung *f*
by-product utilization Abproduktverwertung *f*, Abproduktnutzung *f*
bypass *v* umgehen; vorbeiführen, herumführen
bypass Kurzschlussströmung *f*, Kurzschluss *m (im Reaktor)*
bypassing Ausbildung *f* [Auftreten *n*] von Kurzschlussströmungen, Auftreten *n* eines Kurzschlusses *(im Reaktor)*
byproduct *s.* by-product
BZ reaction *s.* Belousov-Zhabotinskii reaction

C

c *s.* 1. cold; 2. concentrate; 3. concentrated; 4. concentration
C *s.* 1. cis conformation; 2. cold; 3. concentrate; 4. concentrated; 5. concentration
C_4 (dicarboxylic acid) cycle *(Bioch)* Hatch-Slack-Kortschak-Zyklus *m*, C_4-Säurenzyklus *m*, C_4-Dicarbonsäure-Weg *m*, C_4-Zyklus *m*
C^{14} method *s.* radiocarbon dating

C_4 pathway *s.* C_4 (dicarboxylic acid) cycle
C_3 pentose phosphate cycle *(Bioch)* Calvin-(Bassham-)Zyklus *m*, C_3-Zyklus *m*, reduktiver Pentosephosphatzyklus *m*, Ribulose-Diphosphat-Zyklus *m*
C_3 plant C_3-Pflanze *f*, Calvinpflanze *f (mit Dreikohlenstoffverbindungen als ersten Assimilationsprodukten)*
C_4 plant C_4-Pflanze *f (mit Vierkohlenstoffverbindungen als ersten Assimilationsprodukten)*
C-stage resin Harz *n* im C-Zustand, C-Harz *n*, *(bei Phenolharzen auch)* Resit *n*
C-terminal group [residue] C-terminaler Rest *m*, C-terminale Endgruppe *f*, Carboxylende *n (in Proteinen)*
CA *s.* 1. cellulose acetate; 2. collision activation
C.A. = Chemical Abstracts
CAB *s.* cellulose acetate butyrate
cabinet dryer Kammertrockner *m (mit einer Kammer)*
cacao butter Kakaobutter *f*, Kakaoöl *n*, Kakaofett *n*
cacodyl Kakodyl *n*, *Tetramethyldiarsan *n*
cacodylic acid Kakodylsäure *f*, *Dimethylarsinsäure *f*
cadmic Cadmium...
cadmium Cd Cadmium *n*
cadmium-bearing cadmiumhaltig
cadmium-helium laser Helium-Cadmium-Laser *m*, He-Cd^+-Laser *m*
cadmium hydroxide Cadmiumhydroxid *n*
cadmium-nickel storage cell Nickel-Cadmium-Akkumulator *m*
cadmium oxide Cadmium(II)-oxid *n*
cadmium poisoning Cadmiumvergiftung *f*
cadmium red Cadmiumrot *n*
cadmium sulphate Cadmiumsulfat *n*
cadmium sulphide Cadmiumsulfid *n*
cadmium yellow Cadmiumgelb *n*, Cadmiumsulfid *n*
caesium Cs Caesium *n*
caesium chloride Caesiumchlorid *n*
caesium hydride Caesiumhydrid *n*
caesium hydroxide Caesiumhydroxid *n*
caesium iodide Caesiumiodid *n*
caesium oxide Caesiumoxid *n*
caesium sulphate Caesiumsulfat *n*
caesium sulphide Caesiumsulfid *n*
caffeine Coffein *n*, Thein *n*, 1,3,7-Trimethylxanthin *n*
caffeinic coffeinhaltig
cage 1. Käfig *m (aus Molekülen in flüssigen und festen Phasen)*; 2. Seiher *m (an der Seiherpresse)*
cage compound Käfig(einschluss)verbindung *f*, Clathratverbindung *f*, Clathrat *n*
cage disintegrator [mill] Schlagkorbmühle *f*, *(i. e. S.)* Desintegrator *m*, *(i. e. S.)* Desintegratormühle *f*, Schleudermühle *f*
caged ring system endocyclisches System *n*
Cahn-Ingold-Prelog system Cahn-Ingold-Prelog-System *n*, CIP-System *n*, R,S-System *n (zur Bezeichnung der absoluten Konfiguration)*
cake *v* zusammenbacken; backen *(Kohle)*
cake 1. Kuchen *m*, Filterkuchen *m*; Zentrifugenkuchen *m*, Presskuchen *m*; 2. *(Lebm)* Ölkuchen *m*; 3. *(Kosmet)* Seifenstück *n*
cake filter Rückstandsfilter *n*, Schlammfilter *n*
cake filtration Schlammfiltration *f*, Scheidefiltration *f*, Kuchenfiltration *f*
caking Zusammenbacken *n*; Backen *n (z. B. der Kohle)*
calabash-curare alkaloid Calebassen-curare-Alkaloid *n*, Calebassenalkaloid *n*, C-Alkaloid *n*
calandria Heizkammer *f*, Heizraum *m (des Vertikalrohrverdampfers)*
calc-spar *s.* calcite
calcareous kalkartig, kalkig; kalkhaltig, kalkreich; Kalk...
calcareous rock Kalkgestein *n*
calcareous tufa Kalktuff *m*
calcination Kalzinieren *n*, Kalzinierung *f*, Kalzination *f*, Brennen *n*; Rösten *n*, Röstung *f (von Erzen)*; Glühaufschluss *m (Düngemittelindustrie)*
calcine *v* kalzinieren, brennen, glühen; rösten *(Erze)*
calcine 1. Abbrand *m*, abgeröstetes Gut *n* [Material *n*], Röstgut *n (von Erzen)*; 2. *(Ker)* Kalzinationsprodukt *n*
calcined limestone Branntkalk *m (Calciumoxid)*

calcined soda kalzinierte [kristallwasserfreie] Soda *f*, wasserfreies Natriumcarbonat *n*
calciner Kalzinierofen *m*; Röstofen *m (für Erze)*
calcining furnace [kiln] *s.* calciner
calcite *(Min)* Calcit *m*, Kalkspat *m (Calciumcarbonat)*
calcium Ca Calcium *n*
calcium carbide Calciumcarbid *n*, Calciumacetylid *n*, Carbid *n*
calcium carbonate Calciumcarbonat *n*
calcium carbonate hardness *(Hyd)* Calciumcarbonathärte *f*, Calcium-Karbonathärte *f*, Ca-KH
calcium carbonate scale *(Hyd)* Calciumcarbonatstein *m*, *(i. w. S.)* Kesselstein *m*
calcium chloride Calciumchlorid *n*
calcium cyanamide Calciumcyanamid *n*
calcium fluoride Calciumfluorid *n*
calcium hardness *(Hyd)* Kalkhärte *f*, durch Calcium verursachte Härte *f*, Calciumhärte *f*
calcium hydride Calciumhydrid *n*
calcium hydrogencarbonate Calciumhydrogencarbonat *n*
calcium hydroxide Calciumhydroxid *n*
calcium hypochlorite Calciumhypochlorit *n*
calcium level Calciumspiegel *m (des Bluts)*
calcium non-carbonate hardness *(Hyd)* Calcium-Nichtcarbonathärte *f*, Calcium-Nichtkarbonathärte *f*, Ca-NKH, Kalk-Nichtkarbonathärte *f*
calcium oxide Calciumoxid *n*
calcium peroxide Calciumperoxid *n*
calcium soap Calciumseife *f*
calcium stearate Calciumstearat *n*
calcium sulphate Calciumsulfat *n*
calcium sulphate scale *(Hyd)* Gipsstein *m*
calcium sulphide Calciumsulfid *n*
calender *v* 1. kalandern, kalandrieren; 2. *(Pap)* glätten, satinieren; 3. *(Kst)* auswalzen
calender 1. Kalander *m*, Glätt(walzen)werk *n*, Walzenglättwerk *n*, Glättmaschine *f*, Maschinenglättwerk *n*, Maschinenkalander *m*; 2. *(Pap)* Papierkalander *m*, Trockenglättwerk *n*, Satinierkalander *m*, Satinier(walz)werk *n*
calendering 1. Kalandrieren *n*, Kalandern *n*; 2. *(Pap)* Glätten *n*, Glättung *f*, Satinieren *n*, Satinierung *f*, Satinage *f*; 3. *(Kst)* Auswalzen *n*
calibrate *v* kalibrieren, graduieren, mit genauer Einteilung versehen *(Messgeräte)*; eichen *(Glaselektroden, Kalorimeter)*
calibrated to contain auf Einguss geteilt [graduiert, justiert]
calibrated to deliver auf Auslauf geteilt [graduiert, justiert]
calibrating solution Eichlösung *f*
calibration Kalibrieren *n*, Kalibrierung *f*, Graduieren *n*, Graduierung *f (von Messgeräten)*; Eichen *n*, Eichung *f (von Glaselektroden, Kalorimetern)*
calibration curve *(Anal)* Kalibrierkurve *f*; Eichkurve *f*
calibration function *(Anal)* Kalibrierfunktion *f*; Eichfunktion *f*, Analyseneichfunktion *f*
calibration line *(Anal)* Kalibriergerade
calibration mark Eichmarke *f*, Eichstrich *m*
calibration sample *(Anal)* Kalibrierprobe *f*
calibration standard *(Anal)* Eichstandard *m*
californium Cf Californium *n*
calixarene *(org Ch)* Calixaren *n*
calomel Kalomel *n*, Quecksilber(I)-chlorid *n*
calomel electrode [half-cell] Kalomelelektrode *f*
caloric kalorisch, Kalorien...
calorie Kalorie *f (SI-fremde Einheit der Wärmemenge; 1 cal = 4,1868 J)*
calorific wärmeerzeugend, Wärme...
calorific value 1. Heizwert *m*; 2. *(Lebm)* kalorischer Wert *m* • **of high calorific value** heizwertreich, heizkräftig, mit hohem Heizwert • **of low calorific value** heizwertarm, mit niedrigem Heizwert
calorigenic wärmeerzeugend
calorimeter Kalorimeter *n*
calorimetric kalorimetrisch
calorimetric bomb Kalorimeterbombe *f*, kalorimetrische Bombe *f*, Bombenkalorimeter *n*, Verbrennungsbombe *f*
calorimetrical kalorimetrisch
calorimetry Kalorimetrie *f*

calorize *v* 1. *(Met)* kalorisieren; 2. durch Hitze eindicken *(z. B. Leinöl)*
calorizing *(Met)* Kalorisieren *n*, Kalorisierung *f (Veredlungsverfahren)*
Calvin(-Bassham) cycle *(Bioch)* Calvin-(Bassham-)Zyklus *m*, C_3-Zyklus *m*, reduktiver Pentosephosphatzyklus *m*, Ribulose-Diphosphat-Zyklus *m*
camouflage of a trace element Tarnung *f* eines Spurenelements
cAMP = cyclic adenosine 3′,5′-monophosphate
CAMP = continuous air monitoring program *(Luftüberwachungsprogramm in den USA)*
camphene Camphen *n*, 2,2-Dimethyl-3-methylen-norbornan *n (bicyclischer Terpenkohlenwasserstoff)*
camphor Campher *m*, Kampfer *m*, Bornan-2-on *n*
camphor oil Campheröl *n (von Cinnamomum camphora L. Sieb.)*
camphoraceous campherartig
camphoric acid Camphersäure *f*, *cis*-1,2,2-Trimethylcyclopentan-1,3-dicarbonsäure *f*
CAMR = Centre for Applied Microbiology and Research *(U.K.)*
can *v (Lebm)* eindosen, in Dosen [Büchsen] konservieren, *(Am, i. w. S.)* konservieren *(auch in Gläsern)*
can 1. Dose *f*, Büchse *f*; Kanne *f*; Kanister *m*; Konservendose *f*, Konservenbüchse *f*; 2. Walze *f*, Zylinder *m (des Walzenbandtrockners)*; 3. *(Text)* Kanne *f*, Spinnkanne *f*, Topf *m*
canal ray *(Kern)* Kanalstrahl *m*, positiver Strahl *m*
cancer-causing *s.* cancerogenic
cancer-producing *s.* cancerogenic
cancerigenic *s.* cancerogenic
cancerogenic kanzerogen, karzinogen, krebserzeugend, krebserregend, krebsauslösend
cane 1. Rohr *n*, Zuckerrohr *n*; 2. Glasstab *m*
cane molasses Rohr(zucker)melasse *f*, Zuckerrohrmelasse *f*
cane sugar Rohrzucker *m*, Saccharose *f*
cannel coal Cannelkohle *f*, Kännelkohle *f*, Candelkohle *f*, Kerzenkohle *f*, Blätterkohle *f*, Gasschiefer *m*, Kandelit *m*
Cannizzaro reaction Cannizzaro-Reaktion *f*, cannizzarosche Reaktion *f (Aldehyd-Dismutation)*
cantilever Cantilever *m*
caoutchouc Kautschuk *m(n)*
cap *v* überdecken, abdecken
cap 1. Deckel *m*, Haube *f*, Klappe *f*; 2. *(Dest)* Glocke *f*; 3. *(Geol)* Deckschicht *f*, Hut *m*
CAP 1. = chloramphenicol; 2. = 6-chloro-6-dehydro-17α-acetoxyprogesterone; 3. = carbamyl phosphate; 4. = cellulose acetopropionate; 5. = cystine aminopeptidase; 6. = catabolite (gene) activator protein
capable of being vaporized verdampfbar
capable of oxidation oxidierbar, oxidabel
capable to be regenerated *(Hyd)* regenerierbar
capacitance Kapazität *f*, kapazitiver Widerstand *m*
capacitively coupled microwave plasma kapazitiv gekoppeltes Mikrowellenplasma *n*
capacitor Kondensator *m*
capacity 1. Fassungsvermögen *n*, Kapazität *f*; Aufnahmevermögen *n*, Aufnahmefähigkeit *f*; 2. Inhalt *m*, Volumen *n*; 3. Leistungsfähigkeit *f*, Leistungsvermögen *n*, Produktionsvermögen *n*; Durchsatzleistung *f*
capacity current Kapazitätsstrom *m*, kapazitiver Strom *m*, Ladestrom *m (Polarographie)*
capillarity Kapillarität *f*, Kapillarattraktion *f*, Kapillareffekt *m*, Kapillarwirkung *f*, Haarröhrchenwirkung *f*
capillary kapillar, Kapillar..., haarfein, sehr eng
capillary Kapillare *f*, Kapillarrohr *n*, Kapillarröhrchen *n*, Haarröhrchen *n*
capillary action *s.* capillary action
capillary-active kapillaraktiv
capillary air bleed Siedekapillare *f*
capillary column *(Anal)* Kapillarsäule *f*
capillary condensation Kapillarkondensation *f*, kapillare Kondensation *f*
capillary depression Kapillardepression *f*
capillary electrophoresis *(Anal)* Kapillarelektrophorese *f*
capillary force Kapillarkraft *f*

capillary gas chromatography Kapillar-Gaschromatographie *f*, Kapillar-GC *f*
capillary-inactive kapillarinaktiv
capillary pressure Kapillardruck *m*
capillary rise Kapillaraszension *f*
capillary salt bridge *(phys Ch)* lugginsche Kapillare *f*, (Haber-)Luggin-Kapillare *f*
capillary tube Kapillarrohr *n*, Kapillarröhrchen *n*, Haarröhrchen *n*, Kapillare *f*
capillary viscometer Kapillarviskosimeter *n*
caprate *Decanoat *n*, *(veraltet:)* Caprinat *n (Salz oder Ester der Decansäure)*
capric acid *Decansäure *f*, *(veraltet:)* Caprinsäure *f*
caproate *Hexanoat *n*, *(veraltet:)* Capronat *n (Salz oder Ester der Hexansäure)*
caproic acid *Hexansäure *f*, *(veraltet:)* Capronsäure *f*
caprolactam Caprolactam *n*, *Hexan-6-lactam *n*
caprylate *Octanoat *n*, *(veraltet:)* Caprylat *n (Salz oder Ester der Octansäure)*
caprylic acid *Octansäure *f*, *Heptan-1-carbonsäure *f*, *(veraltet:)* Caprylsäure *f*
capsid protein Virushüllprotein *n*, Hüllprotein *n*, Kapsid *n*
capsomere *(Bioch)* Capsomer *n (Untereinheit der Proteinhülle von Viren)*
capsule 1. Kapsel *f*; 2. *(Lab)* Glühschale *f*; Glühschiffchen *n*
capsule membrane Kapselmembran *f*
capsule surface Kapseloberfläche *f*
captive drop liegender Tropfen *m*
capture *v* einfangen
capture 1. *(Kern)* Einfang *m*, Einfangen *n (z. B. von Elektronen)*; 2. *(Krist)* Abfangen *n (z. B. eines Spurenelements)*
capture cross section *(Kern)* Einfangquerschnitt *m*, Absorptionsquerschnitt *m*
caramel Karamell *m*, gebrannter Zucker *m*, *(als Lösung)* Zuckercouleur *f*, Kulör *f*
caramelization Karamellisierung *f*
caramelize *v* karamellisieren
carbaborane *(org Ch)* Carboran *n*
carbalkoxylation Alkoxycarbonylierung *f*, Carbalkoxylierung *f*
carbalumination Carbaluminierung *f*
carbamate Carbamat *n (Salz oder Ester der Carbamidsäure)*; *(bei Estern, besonders bei Ethylestern auch)* Urethan *n*
carbamic acid Carbamidsäure *f*, Amidokohlensäure *f*, Kohlensäuremonamid *n*
carbamide Carbamid *n*, Kohlensäurediamid *n*, Harnstoff *m*
carbamoyl carboxylic acid Carbamoylcarbonsäure *f*
carbamoylation Carbam(o)ylierung *f*
carbamoylethylation Carbamoylethylierung *f*
carbamylation Carbam(o)ylierung *f*
carbamylethylation Carbamoylethylierung *f*
carbanion Carbanion *n (durch Abspaltung eines Protons aus einem C-Atom entstandenes Anion)*
carbazide RNH–NH–CO–NH–NHR′ *Carbono-hydrazid *n*, Carbazid *n*, *(i. e. S.)* H_2N–NH–CO–NH–NH_2 *Carbonohydrazid *n*
carbazole Carbazol *n*, Dibenzopyrrol *n*, Diphenylimid *n*
carbazone R–N=N–CO–NH–NHR′ Carbazon *n*
carbene Carben *n*, Methylen *n*, *(i. e. S.)* :CH_2 Carben *n*, Methylen *n*
carbenium ion Carbeniumion *n*, Carbeniumkation *n (mit dreibindigem sp^2-hybridisiertem Kohlenstoff)*
carbenoid Carbenoid *n*
carbethoxylation Carbethoxylierung *f*
carbethoxymethylation Carbethoxymethylierung *f*
carbide Carbid *n*, *(i. e. S.)* Calciumcarbid *n*, Calciumacetylid *n*, Carbid *n*
carbidic carbidisch, Carbid...
carbinolamine Halbaminal *n*, α-Aminoalkohol *m*
carbobenzoxylation Carbobenzoxylierung *f*
carbocation Carbokation *n (positiv geladenes Kohlenstoffion)*
carbocupration Carbocuprierung *f*
carbocyclic compound Carbocyclus *m*, Isocyclus *m*, carbocyclische [isocyclische, homocyclische] Verbindung *f*
carbodiimide Carbodiimid *n*
carboethoxylation Carbethoxylierung *f*
carbohalogenation Carbohalogenierung *f*

carbohydrate Kohlenhydrat *n*
carbolic acid C_6H_5OH Karbolsäure *f*, Phenol *n*, Hydroxybenzen *n*, *(i. w. S. allg. für ein)* Phenol *n*
carbolithiation Carbolithiierung *f*
carbometallation Carbmetallierung *f*
carbomethoxylation Carbomethoxylierung *f*
carbon 1. C Kohlenstoff *m*, Carbon *n*; 2. Kohle(elektrode) *f*; 3. *s.* activated carbon
carbon arc Kohlebogen *m*
carbon balance Kohlenstoffbilanz *f*
carbon black Ruß *m*, Kohleschwarz *n*, Rußschwarz *n*, Carbon-Black *n*
carbon brush 1. Kollektorbürste *f*; 2. Schleifringbürste *f*
carbon canister Aktivkohle-Filter *m*
carbon-carbon bond Kohlenstoff-Kohlenstoff-Bindung *f*, C–C-Bindung *f*
carbon-carbon composite *s.* carbon fibre-reinforced carbon
carbon-carbon double bond Kohlenstoff-Kohlenstoff-Doppelbindung *f*, C–C-Doppelbindung *f*, C=C-Bindung *f*
carbon-carbon scission Spaltung *f* einer C–C-Bindung
carbon-carbon single bond Kohlenstoff-Kohlenstoff-Einfachbindung *f*, C–C-Einfachbindung *f*, (einfache) C–C-Bindung *f*
carbon-carbon triple bond Kohlenstoff-Dreifachbindung *f*, C–C-Dreifachbindung *f*, C≡C-Bindung *f*
carbon-carburizing steel Kohlenstoff-einsatzstahl *m*
carbon-casehardening steel Kohlenstoffeinsatzstahl *m*
carbon-centred radical Kohlenstoffradikal *n*
carbon chain Kohlenstoffkette *f*
carbon combustion *(Tech)* Kohlenstoffausbrand *m*
carbon compound Kohlenstoffverbindung *f*
carbon dating Radiokohlenstoffdatierung *f*, Altersbestimmung *f* mit Hilfe der Radiocarbonmethode [Kohlenstoff-14-Methode, ^{14}C-Methode]
carbon deposit Rußniederschlag *m*
carbon dioxide Kohlendioxid *n*
carbon dioxide assimilation Kohlendioxidassimilation *f*
carbon dioxide ice Kohlendioxidschnee *m*, Kohlensäureschnee *m*, Trockeneis *n*
carbon dioxide removal Kohlendioxidentfernung *f*, CO_2-Entfernung *f*
carbon dioxide snow *s.* carbon dioxide ice
carbon disulphide Kohlendisulfid *n*, Kohlenstoffdisulfid *n*, Schwefelkohlenstoff *m*
carbon electrode Kohleelektrode *f*
carbon fibre Kohlenstofffaser *f*
carbon fibre-reinforced carbon Kohlenstofffaser-Kohlenstoff-Verbundwerkstoff *m*, C/C-Verbundwerkstoff *m*, CFC-Verbundwerkstoff *m*, kohlenstofffaserverstärkter Kohlenstoff *m*, KK
carbon fibre type HM [high modulus] Hochmodul-Kohlenstofffaser *f (Elastizitätsmodul > 300 GPa)*
carbon fibre type HT Hochfest-Kohlenstofffaser *f (Elastizitätsmodul 150 bis 300 GPa)*
carbon filament Kohlenstofffaden *m*
carbon filter Kohlefilter *n*, Aktivkohlefilter *n*, A-Kohle-Filter *n*
carbon filtration Aktivkohlefiltration *f*, Filtration *f* über A-Kohle
carbon-functional carbofunktionell, organofunktionell
carbon-hydrogen bond Kohlenstoff-Wasserstoff-Bindung *f*, C–H-Bindung *f*
carbon material Kohlenstoffmaterial *n*, Werkstoff *m* aus Kohlenstoff
carbon monoxide Kohlenmonoxid *n*, Kohlenoxid *n*
carbon-monoxide-shift reaction *(Carb)* Konvertierungsreaktion *f* (des Kohlenmonoxids), homogene Wassergasreaktion *f*
carbon monoxy haemoglobin Kohlenmonoxidhämoglobin *n*, CO-Hämoglobin *n*, Hb–Co
carbon nanofibre Kohlenstoff-Nanofaser *f*
carbon nanotube Kohlenstoff-Nanoröhrchen *n*
carbon restoration 1. Kohlenstoffaufnahme *f*; 2. *(Met)* Wiederaufkohlen *n*, Rückkohlen *n*, Wiederaufkohlung *f*
carbon ring Kohlenstoffring *m*
carbon-12 scale Kohlenstoff-12-Skala *f*, ^{12}C-Skala *f (der relativen Atommassen)*

carbon skeleton Kohlenstoffgerüst *n*
carbon source *(Biot)* C-Quelle *f*, Kohlenstoffquelle *f*
carbon steel Kohlenstoffstahl *m*, unlegierter Stahl *m (mit bis zu 2 % Kohlenstoff)*
carbon tetrachloride Kohlenstofftetrachlorid *n*, Tetrachlorkohlenstoff *m*, *Tetrachlormethan *n*, Tetra *n*
carbon whisker Kohlenstoff-Whisker *m*
carbon yarn Kohlenstoffgarn *n*
3-carbon-tautomerism Dreikohlenstofftautomerie *f*
carbonaceous kohlenstoffhaltig; kohlenstoffreich
carbonate *v* 1. carbonisieren, mit CO_2 sättigen, *(Getränke auch)* mit CO_2 imprägnieren; 2. saturieren *(Zuckerrübensaft)*; 3. in ein Carbonat überführen [umwandeln]; 4. *s.* carboxylate
carbonate *Carbonat *n*, Karbonat *n*
carbonate hardness *(Hyd)* Carbonathärte *f*, Karbonathärte *f*, KH, temporäre [vorübergehende] Härte *f*
carbonate scale *(Hyd)* Carbonatstein *m*
carbonation 1. Carbonisieren *n*, Carbonisation *f*, Sättigung *f* mit CO_2, *(bei Getränken auch)* Imprägnieren *n* mit CO_2; 2. Saturation *f (Entkalkung des Zuckerrübensafts mittels CO_2)*; 3. Überführung *f* [Umwandlung *f*] in ein Carbonat; 4. *s.* carboxylation
carbonatization 1. Umwandlung *f* in ein Carbonat; 2. *(Geoch)* Carbonatisierung *f*, Carbonatisation *f*
carbonic 1. Kohlenstoff...; 2. Kohlensäure...; Kohlendioxid...
carbonic acid Kohlensäure *f*
carbonitride *v* carbonitrieren, cyanieren, *(i. e. S.)* gascyanieren, trockencyanieren
carbonitrile Carbonitril *n*
carbonium ion 1. Carboniumion *n (mit pentakoordiniertem Kohlenstoffatom)*; 2. *s.* carbenium ion
carbonization 1. Umwandlung *f* in Kohlenstoff; 2. *(i. w. S.)* Verkokung *f*, Entgasung *f*; *(i. e. S.)* Verkokung *f*, Hochtemperaturverkokung *f*, HT-Verkokung *f*, Normalverkokung *f*, Vollverkokung *f*, Hochtemperaturentgasung *f*, HT-Entgasung *f*; Schwelung *f*, Verschwelung *f*, Tieftemperaturverkokung *f*, Tieftemperaturentgasung *f*; 3. Verkohlung *f*, Entgasung *f (z. B. von Holz)*; 4. Karbonisieren *n*, Karbonisation *f*, Auskohlen *n*, Entkletten *n (von Wolle)*; 5. *s.* carburization
carbonization coal Kokskohle *f*
carbonization gas Verkokungsgas *n*, Entgasungsgas *n*; Schwelgas *n*
carbonize *v* 1. in Kohlenstoff umwandeln; 2. verkoken, entgasen; (ver-) schwelen; 3. verkohlen *(z. B. Holz)*; 4. karbonisieren, auskohlen, entkletten *(Wolle)*; 5. *s.* carburize
carbonizer Schweler *m*, Schwelofen *m*
carbonizing generator Schwelgenerator *m*, Doppelgasgenerator *m*, Doppelgaserzeuger *m*
carbonyl 1. Metallcarbonyl *n*, Carbonyl *n*; 2. *s.* carbonyl group
carbonyl chloride Carbonylchlorid *n*, Kohlenoxidchlorid *n*, Phosgen *n*
carbonyl double bond C=O-Doppelbindung *f*
carbonyl group Carbonylgruppe *f*, C=O-Gruppe *f*, CO-Gruppe *f*
carbonyl oxide Carbonyloxid *n*
carbonyl sulphide Carbonylsulfid *n*, Kohlenoxidsulfid *n*
carbonylate *v* carbonylieren
carbonylation Carbonylierung *f*
carborane *(org Ch)* Carboran *n*
carbothermal *(Met)* karbothermisch
carbothermic *(Met)* karbothermisch
carboxamidation Carbamoylierung *f*, Carbamylierung *f*
carboxamide Carboxamid *n*, Carbonsäureamid *n*
carboxamidine $RC(=NH)NH_2$ Carboxamidin *n*, Amidin *n*
carboxazylic acid *s.* carboxamide
carboxy group *s.* carboxyl group
carboxyalkylation Carboxyalkylierung *f*
carboxyethylation Carboxyethylierung *f*
carboxyhaemoglobin Kohlenmonoxidhämoglobin *n*, CO-Hämoglobin *n*, Hb–CO
carboxyl group Carboxylgruppe *f*, Carboxyl *n*, COOH-Gruppe *f*
carboxylate *v* carboxylieren *(in organische Verbindungen die Carboxylgruppe einführen)*
carboxylate Carboxylat *n (Salz oder Ester einer Carbonsäure)*

carboxylation Carboxylierung *f*
carboxylic acid Carbonsäure *f*
carboxylic acid amide Carbonsäureamid *n*, Carboxamid *n*
carboxymethylation Carboxymethylierung *f*
carboxymethylcellulose Carboxymethylcellulose *f*, CM-Cellulose *f*, Celluloseglykolsäure *f*
carboy Ballon *m (umkleidet)*; Korbflasche *f (besonders für Säuren)*
carburant Vergaserkraftstoff *m*, VK
carburet *v* karburieren *(Gase)*
carburetted water gas karburiertes Wassergas *n*
carburetting fuel Vergaserkraftstoff *m*, VK
carburettor 1. Karburator *m (zum Karburieren von Gasen)*; 2. Vergaser *m (eines Motors)*
carburization *(Met)* Aufkohlen *n*, Kohlen *n*, Zementieren *n*, Einsetzen *n*
carburize *v (Met)* (auf)kohlen, zementieren, einsetzen
carburized steel aufgekohlter [zementierter] Stahl *m*
carburizer *(Met)* Aufkohl(ungs)mittel *n*, Kohlungsmittel *n*, Zementationsmittel *n*, Einsatzmittel *n*, (auf)kohlendes Mittel *n*
carburizing *(Met)* Aufkohlen *n*, Kohlen *n*, Zementieren *n*, Einsetzen *n*
carburizing bath Aufkohlungsbad *n*, Kohlungsbad *n*, aufkohlendes Salzbad *n*, Zementierbad *n*, Zementationsbad *n*, Einsatzbad *n*
carburizing depth Aufkohlungstiefe *f*, Kohlungstiefe *f*, Zementationstiefe *f*, Einsatztiefe *f*, Tiefe *f* der Einsatzschicht
carburizing material [medium] Aufkohl(ungs)mittel *n*, Kohlungsmittel *n*, Zementationsmittel *n*, Einsatzmittel *n*, (auf)kohlendes Mittel *n*
carburizing oven Aufkohlungsofen *m*, Kohlungsofen *m*, Zementier(ungs)ofen *m*
carburizing steel Stahl *m* für Einsatzhärtung, Einsatzstahl *m*
carbylic acids *allg. für Carbonsäuren, Carbonsäureamide und Amidine*
carcinogen Karzinogen *n*, Kanzerogen *n*, karzinogener [krebserzeugender] Stoff *m*
carcinogenic karzinogen, kanzerogen, krebserzeugend, krebserregend, krebsauslösend
carcinogenicity karzinogene [kanzerogene, krebserzeugende] Wirkung *f*
cardiac-active *(Pharm)* herzaktiv, herzwirksam
cardiac glycoside Herzglykosid *n*, herzaktives Glykosid *n*
cardioactive *s.* cardiac-active
cardiotonic cardiotonisch
cardiotonic *(Pharm)* Kardiotonikum *n (die Herzmuskulatur kräftigendes Mittel)*
carminative *(Pharm)* Carminativum *n*, blähungstreibendes Mittel *n*
Carnot cycle carnotscher Kreisprozess *m*, Carnot-Prozess *m (Thermodynamik)*
Carnot efficiency carnotscher Wirkungsgrad *m*, Wirkungsgrad *m* des (reversiblen) carnotschen Kreisprozesses
Caro's acid carosche Säure *f*, Peroxomonoschwefelsäure *f*
carotene *(Bioch)* Caroten *n*, Carotin *n*
carotenoid *(Bioch)* *Carotenoid *n*, Carotenoid-Kohlenwasserstoff *m*
carrier 1. Träger *m*, Überträger *m*, Trägerstoff *m*, Trägersubstanz *f*, Trägermittel *n*, Carrier *m*; 2. *(Kern)* Trägerelement *n*; Trägergas *n*; 3. *(Bioch)* Transportsystem *n*
carrier-bound enzyme *(Biot)* trägergebundenes [trägerfixiertes, matrixgebundenes] Enzym *n*
carrier distillation Trägerstoffdestillation *f*
carrier fluid Trägerflüssigkeit *f*; Trägergas *n*
carrier sublimation Trägergas-Sublimation *f*
carry *v* **down** mitreißen *(im Niederschlag)*
carry *v* **off** mitreißen *(z. B. in einem Gasstrom)*
carry *v* **out** ausführen, durchführen, anstellen *(ein Experiment)*
carry *v* **over** 1. verschleppen *(eine Substanz in die andere)*; mitreißen *(z. B. in einem Gasstrom)*; 2. überleiten, übertreiben *(Dämpfe)*
carry *v* **through** 1. durchführen; 2. verschleppen *(eine Substanz in die an-*

dere); verschleppt werden *(z. B. im Analysen- oder Synthesegang)*
carry-over 1. Verschleppen *n (einer Substanz in die andere)*; Mitreißen *n (z. B. in einem Gasstrom)*; 2. Überleiten *n*, Übertreiben *n (von Dämpfen)*
carryover *s.* carry-over
CARS *s.* coherent anti-Stokes Raman spectroscopy
cascade 1. Kaskade *f (Art von Mehrstufensystemen)*; 2. *(Hyd)* Kaskadenreaktor *m*
cascade dryer Kaskadentrockner *m*
cascade evaporator Kaskadenverdampfer *m*, Mehrstufenverdampfer *m*, Mehrkörperverdampfer *m*
cascade tray Kaskadenboden *m*
case *v* mit einem Gehäuse umgeben
case-carburize *v (Met)* rand(auf)kohlen, in der Randschicht aufkohlen [zementieren]
case-harden *v (Met)* einsatzhärten, im Einsatz(verfahren) härten
case 1. Kasten *m*, Kiste *f*; Gehäuse *n*; 2. *(Met)* Randschicht *f*, Oberflächenschicht *f*, aufgekohlte [gekohlte, zementierte, eingesetzte] Randschicht *f*, Aufkohlungsschicht *f*, Zementationsschicht *f*, Einsatzschicht *f*; gehärtete Randschicht *f* [Einsatzschicht *f*], Härteschicht *f*
case carburization *(Met)* Randaufkohlung *f*, Aufkohlen *n* in der Randschicht
case carburizing *(Met)* Randaufkohlung *f*, Aufkohlen *n* in der Randschicht
case depth *(Met)* Einsatztiefe *f*, Tiefe *f* der Einsatzschicht, Zementationstiefe *f*, Aufkohlungstiefe *f*, Kohlungstiefe *f*; Einsatzhärtungstiefe *f*, Einsatzhärtetiefe *f*
case-hardened glass vorgespanntes Glas *n*, *(i. e. S.)* Einscheibensicherheitsglas *n*, Einschichtensicherheitsglas *n*
case-hardened steel einsatzgehärteter Stahl *m*, Einsatzstahl *m*
case-hardening *(Met)* Einsatzhärten *n*, Zementationshärten *n*, *(i. w. S.)* Oberflächenhärten *n*
case-hardening carburizer *(Met)* Aufkohl(ungs)mittel *n*, Kohlungsmittel *n*, Zementationsmittel *n*, Einsatzmittel *n*, (auf)kohlendes Mittel *n*
case-hardening steel Stahl *m* für Einsatzhärtung, Einsatzstahl *m*
case hardness *(Met)* Einsatzhärte *f*, Randhärte *f*, Härte *f* der (gehärteten) Randschicht
casein Casein *n*, Kasein *n*
casein fibre *(Text)* Caseinfaser *f*; Caseinfaserstoff *m*, KA
casein plastic Kunsthorn *n*
caseinate Caseinat *n*
caseous käseartig, käsig
casing Gehäuse *n*; Behälter *m*, Behältnis *n*
casing-head gas Naturgas *n*, Bohrlochkopfgas *n*, Casinghead-Gas *n*
casing-head gasoline [spirit] Rohrkopfbenzin *n*, Natur(gas)benzin *n*, Erdgasbenzin *n*, Casinghead-Benzin *n*
casing rubber Karkassengummi *m*
cask Fass *n*, Tonne *f*; Bottich *m*
cast *v* gießen
cast alloy Gusslegierung *f*
cast bronze Gussbronze *f*
cast film *(Kst)* gegossene Folie *f*, Gießfolie *f*, Gießfilm *m*
cast iron Gusseisen *n*
cast moulding 1. Gießen *n*; 2. gegossenes Formteil *n*
cast resin Gießharz *n*, Vergussharz *n*, Schmelzharz *n*
cast sheet *(Kst)* gegossene Folie *f*, Gießfolie *f (als Stück und Dicke > 0,25 mm)*
cast steel Gussstahl *m*, Stahl(form)guss *m*
castability Gießbarkeit *f*
casting 1. *(Tech)* Gießen *n*, Guss *m*, Vergießen *n*, Gussverfahren *n*; 2. Gussstück *n*, Gussteil *n*, Formgussstück *n*, Gießling *m*
casting alloy Gusslegierung *f*
casting mould Gießform *f*
casting resin Gießharz *n*, Vergussharz *n*, Schmelzharz *n*
Castner cell Castner-Zelle *f (Elektrolyse)*
castor oil Ricinusöl *n*, Kastoröl *n*
cat *s.* 1. catalyst; 2. catalytic
cat-crack *v* katalytisch kracken [spalten]
cat-cracked gasoline katalytisches Krackbenzin *n*, Cat-Benzin *n*
cat cracking *s.* catalytic cracking
catabolic *(Bioch)* katabol(isch), dissimilatorisch, abbauend

catabolic route Abbauweg *m*, katabolische Bahn *f*, kataboler Stoffwechselweg *m*
catabolism *(Bioch)* Katabolismus *m*, Abbau *m*, Zerlegung *f*, Dissimilation *f*
catabolite *(Bioch)* Katabolit *m*
catabolize *v (Bioch)* katabolisieren, abbauen, zerlegen; katabolisiert [abgebaut, zerlegt] werden
catalase activity *(Bioch)* Katalaseaktivität *f*
catalyse *v* katalysieren
catalysed by acid(s) säurekatalysiert, sauer [durch Säure] katalysiert
catalysed by base(s) basekatalysiert, basisch [durch Basen] katalysiert
catalysis Katalyse *f*
catalysis law Brönsted-Katalysegesetz *n*, brönstedsches Katalysegesetz *n*
catalyst Katalysator *m*, *(bei heterogener Katalyse auch)* Kontakt(stoff) *m*
catalyst activity Katalysatoraktivität *f*, Katalysatorwirksamkeit *f*
catalyst bed Katalysatorbett *n*, Kontaktbett *n*, Katalysatorschicht *f*, Kontaktschicht *f*
catalyst deactivation Katalysatordesaktivierung *f*
catalyst pellet Katalysatorpellet *n*, Katalysatorpille *f*, Kontaktpille *f*
catalyst poison Katalysatorgift *n*, Kontaktgift *n*
catalyst-reactant complex Katalysatorkomplex *m*
catalyst recycling Katalysatorrückgewinnung *f*
catalyst regeneration Katalysatorregenerierung *f*
catalyst selectivity Katalysatorselektivität *f*
catalyst support Katalysatorträger *m*, Kontaktträger *m*
catalyst surface Katalysatoroberfläche *f*
catalytic katalytisch
catalytic activity katalytische Aktivität *f* [Wirksamkeit *f*]
catalytic coefficient Katalysekonstante *f*, katalytische Konstante *f*
catalytic cracking katalytisches [ionisches] Kracken *n* [Spalten *n*], katalytische [ionische] Krackung *f* [Spaltung *f*], Katkracken *n*, Catcracking *n*, Kracken *n* am Katalysator, Kracken *n* über Ionen
catalytic fixed-bed reactor katalytischer Festbettreaktor *m*, festbettkatalytischer Reaktor *m*
catalytic hydrogenation katalytische Hydrierung *f*
catalytic reaction katalytische [katalysierte] Reaktion *f*
catalytic reactor katalytischer Reaktor *m*, Reaktor *m* für katalytische Reaktionen, Katalyseofen *m*, Kontaktofen *m*
catalytic reforming katalytisches Reform(ier)en *n*, katalytische Reformierung *f*, katalytisches Reforming *n*
catalyze *v s.* catalyse
cataphoresis Kataphorese *f*
cataphoretic kataphoretisch
catch *v* abfangen, auffangen
catcher Auffänger *m*, Fänger *m*, Fang *m*, Abscheider *m*
catecholamine Catecholamin *n*
catena compound Catena-Verbindung *f*
catenane Catenan *n*
catenated mit Catena-Struktur
catenation Verknüpfung *f (gleichartiger Atome zu Ketten oder Ringsystemen)*
catenation compound Catena-Verbindung *f*
catforming Catformen *n*, Catforming *n*, katalytisches Reform(ier)en *n*, katalytische Reformierung *f*
cathartic *(Pharm)* Kathartikum *n*, Abführmittel *n*
cathode Katode *f*
cathode compartment Katodenraum *m*
cathode potential Katodenpotenzial *n*
cathode ray Katodenstrahl *m*
cathode reaction Katodenreaktion *f*, katodische Reaktion *f*
cathode sputtering Katodenzerstäuben *n*, Vakuumzerstäuben *n*, Ionen-Plasma-Zerstäuben *n*, Sputtern *n*; katodisches Bestäuben *n*, Vakuumbestäuben *n*
cathodic katodisch, Katoden...
cathodic current Katodenstrom *m*
cathodic deposition katodisches Abscheiden *n (von Metallen)*
cathodic dip painting katodische Tauchlackierung *f (Automobillacke)*
cathodic protection katodischer Schutz *m* [Korrosionsschutz *m*], Katodenschutz *m*
cathodic stripping voltammetry *(Anal)*

Inversvoltammetrie *f* [inverse Voltammetrie *f*] an der Katode, katodische Inversvoltammetrie *f*
catholyte Katolyt *m (Elektrolyt des Katodenraums)*
cation Kation *n*, positives [positiv geladenes] Ion *n*
cation acid *(phys Ch)* Kationsäure *f*
cation-active kation(en)aktiv, kationkapillaraktiv, kationisch, Kation(en)...
cation exchange 1. Kationenaustausch *m*; 2. *(Boden)* Kationenumtausch *m*
cation exchanger Kationenaustauscher *m*, Kationit *m*
cation radical Radikalkation *n*
cation scheme *(Anal)* Kationentrennungsgang *m*
cation vacancy Kationenleerstelle *f*, Kationenfehlstelle *f*
cationic kationisch, Kation(en)..., kation(en)aktiv
cationic kationisches Tensid *n*
cationic polymerization kationische Polymerisation *f*
cationic surfactant kationisches Tensid *n*
cationoid kationoid
cationoid dye kationischer Farbstoff *m*
cationotropic kationotrop
cationotropy Kationotropie *f*
caulk *v* (ab)dichten
caustic ätzend, brennend, kaustisch
caustic 1. Alkali *n*, *(meist)* Natriumhydroxid *n*, Ätznatron *n*; 2. *s.* caustic agent
caustic agent 1. Ätzmittel *n*; 2. *(Med)* Kaustikum *n*
caustic alkalies Ätzalkalien *npl*
caustic cracking [embrittlement] *(Met)* Laugensprödigkeit *f*, Laugenbrüchigkeit *f (interkristalline Spannungsrisskorrosion)*
caustic fusion Alkalischmelze *f*, Schmelzen *n* mit Alkali
caustic lime Ätzkalk *m*, gebrannter Kalk *m*, Branntkalk *m*, Calciumoxid *n*
caustic potash Ätzkali *n*, Kaliumhydroxid *n*
caustic soda Ätznatron *n*, Natriumhydroxid *n*, *(als Produkt des Kalk-Soda-Verfahrens auch)* kaustische Soda *f*, Sodastein *m*
caustic-soda fusion Ätznatronschmelze *f*
caustic solution ätzende Lösung *f*, Ätzlösung *f*
causticity Ätzwirkung *f*, ätzende Wirkung *f*
causticization 1. Kaustifizierung *f*; 2. *(Pap)* Aussüßen *n (Überführen von Soda oder Pottasche in Natron- bzw. Kalilauge)*; 3. *(Text)* Merzerisieren *n*
causticize *v* 1. kaustifizieren; 2. *(Pap)* aussüßen *(Soda oder Pottasche in Natron- bzw. Kalilauge überführen)*; 3. *(Text)* merzerisieren
causticproof laugenbeständig, laugenfest
cauterant *(Med)* ätzend *(zum Kauterisieren verwendbar)*
cauterant *(Med)* Ätzmittel *n (zum Kauterisieren)*
cauterization *(Med)* Kauterisation *f*, Verätzung *f (mittels Chemikalien oder Hitze)*
cauterize *v (Med)* kauterisieren, verätzen *(mittels Chemikalien oder Hitze)*
cavitand *(org Ch)* Cavitand *m*
cavitation Kavitation *f*, Hohlraumbildung *f*, *(in Flüssigkeiten auch)* Hohlsog *m*
cavity 1. Hohlraum *m*, Aushöhlung *f*; 2. *(Kst)* Werkzeughohlraum *m*, Werkzeughöhlung *f*, Formhöhlung *f*, Formnest *n*
cb. *s.* cubic
cc *s.* column chromatography
C–C bond C–C-Bindung *f*, Kohlenstoff-Kohlenstoff-Bindung *f*
CCC *s.* countercurrent chromatography
CCH *s.* calcium hypochlorite
ccp *s.* 1. cubic close-packed; 2. cubic close packing
CCP = carboxyl carrier protein
CCR = catalyst circulation rate
CD *s.* circular dichroism
C.D. *s.* current density
cDNA *s.* 1. complementary DNA; 2. copy DNA
CE *s.* 1. charge exchange; 2. chemical engineer
C.E.D. *s.* cohesive energy density
cell 1. Zelle *f*, Kammer *f*, Raum *m*; 2. Küvette *f*; 3. *(Ech)* Elektrolyse(n)zelle *f*, Elektrolysierzelle *f*, elektrolytische Zelle *f*; 4. *(phys Ch)* (galvanische, elektrochemische) Zelle *f*, (galvanisches,

elektrochemisches) Element *n*; 5. *(Bio)* Zelle *f*
cell attachment *(Bioch)* Zellhaftung *f*
cell-bound *(Biot)* zellgebunden
cell breakage by sonication *(Bioch)* Ultraschallaufschluss *m*
cell changer Küvettenwechsler *m*
cell culture Zellkultur *f*
cell death *(Bio)* Zelltod *m*
cell detachment *(Bioch)* Zellabtrennung *f*
cell disruption *(Bioch)* Zellaufschluss *m*
cell division *(Bio)* Zellteilung *f*
cell doubling time *(Biot)* Verdopplungszeit *f (Mikroorganismenwachstum)*
cell dry weight *(Biot)* Zelltrockenmasse *f*, Zelltrockensubstanz *f*, Zell-TS *f*
cell growth *(Bio)* Zellwachstum *m*
cell immobilization *(Biot)* Zellimmobilisierung *f*
cell lysis *(Bio, Biot)* Lyse *f*, Zellauflösung *f*, Zytolyse *f*
cell metabolism Zellstoffwechsel *m*
cell motility Zellbeweglichkeit *f*
cell proliferation *(Bio)* Zellvermehrung *f*
cell protein Zellprotein *n*
cell reaction *(Ech)* Zellreaktion *f*
cell separation *(Biot)* Abtrennung *f* der Zellen *(aus der Fermentationslösung)*
cell slurry *(Biot)* Zellsuspension *f*
cell voltage *(Ech)* Zell(en)spannung *f*
cell yield *(Biot)* Zellausbeute *f*, Masseausbeute *f*
cellobiose Cellobiose *f (ein Disaccharid)*
cellophane Zellglas *n*, Zellhaut *f*
cellular zellig, zellenartig, zellenähnlich, zellulär, Zell(en)..., *(manchmal auch)* blasig, porig, Schaum...
cellular enzyme Zellenzym *n*
cellular filter Zellenfilter *n*
cellular plastic Schaum(kunst)stoff *m*
cellular PVC PVC-Schaumstoff *m*, Polyvinylchloridschaumstoff *m*
celluloid Zelluloid *n*, Zellhorn *n*
cellulolytic microorganisms *(Biot)* celluloseabbauende [cellulolytische] Mikroorganismen *mpl*
cellulose 1. Cellulose *f*; 2. *(Tech)* Zellulose *f*; Zellstoff *m (Hauptbestandteil Cellulose)*
cellulose acetate Celluloseacetat *n*, CA, Acetylcellulose *f*
cellulose acetate butyrate Celluloseacetobutyrat *n*, CAB, Acetylbutyrylcellulose *f*
cellulose acetate fibre *(Text)* Celluloseacetatfaser *f*; Celluloseacetatfaserstoff *m*
cellulose acetate propionate Celluloseacetatpropionat *n*, CAP
cellulose acetate rayon *(Text)* Celluloseacetatseide *f*, Acetatseide *f*
cellulose ester Celluloseester *m*
cellulose ether Celluloseether *m*
cellulose fibre *(Text)* Cellulosefaser *f*; Cellulosefaserstoff *m*
cellulose nitrate Cellulosenitrat *n*, Nitratcellulose *f*, Cellulosesalpetersäureester *m*
cellulose nitrate lacquer Cellulosenitratlack *m*, Nitrolack *m*
cellulose propionate Cellulosepropionat *n*, CP
cellulose triacetate fibre *(Text)* Triacetatfaser *f*; Triacetatfaserstoff *m*, TA
cellulose xanthate Cellulosexanthogenat *n*
cellulosic cellulosisch, Cellulose...
cellulosic Celluloseerzeugnis *n*; Cellulosederivat *n*
cellulosic fibre *(Text)* Cellulosechemiefaser *f*; Cellulosechemiefaserstoff *m*, CZ
Celsius (temperature) scale Celsius-Skala *f*, Celsius-Temperaturskala *f*
cement *v* 1. (ver)kitten, (ver)kleben; einstreichen *(mit Gummilösung)*; 2. *(Bau)* zementieren; 3. *(Met)* oberflächenhärten *(durch Diffusion)*; *(mit Kohlepulver auch)* zementieren; 4. (aus)zementieren, ausfällen *(ein Metall durch ein unedleres Metall aus seiner Lösung)*
cement *v* **together** zusammenkitten
cement 1. Kleb(e)stoff *m*, Kleber *m*, *(i. e. S.)* Kitt *m*; 2. *(Gum)* Klebelösung *f*, Lösung *f*, Konfektionierlösung *f*, Klebezement *m*, Zement *m*; 3. *(Bau)* Zement *m*; 4. *(Med)* Zahnzement *m*, Zement *m*; 5. *(Geol)* Bindemittel *n*, Zement *m*
cement-asbestos Asbestzement *m*
cement copper Zementkupfer *n*, Fällkupfer *n*
cementation 1. Kitten *n*, Verkitten *n*, Kleben *n*, Verkleben *n*; 2. *(Bau)* Ze-

mentieren *n*; 3. *(Met)* Oberflächenhärten *n* durch Diffusion, *(mit Kohlepulver auch)* Aufkohlen *n*, Einsetzen *n*, Einsatzhärten *n*, Zementieren *n*; 4. *(Met, Geol)* Zementation *f*, Auszementierung *f (Ausfällen oder Ausfallen eines Metalls aus seiner Lösung durch Einbringen oder Anwesenheit eines unedleren)*; 5. *(Geol)* Verkittung *f*

cemented carbide 1. Sinterhartmetall *n*, gesintertes Hartmetall *n*, Hartmetall *n*, HM; 2. *s.* cementite

cemented hard carbide [metal] *s.* cemented carbide

cemented tungsten carbide gesintertes Wolframcarbid *n*, Wolframsintercarbid *n*, Wolframcarbidhartmetall *n*

cementite Eisencarbid *n*, Zementit *m*

cementitious 1. zementartig; 2. verkittend

center *v s.* centre

centigrade heat unit *Einheit der Wärmemenge in U.K. und U.S.; 1 chu [ctu] = 1,9 kJ*

centigrade scale *s.* Celsius (temperature) scale

centigrade thermal unit *s.* centigrade heat unit

centigram(me) analysis Halbmikroanalyse *f*

centigram(me) range *(Anal)* Zentigrammbereich *m (0,01 bis 0,1 g)*

centimolar solution 0,01m-Lösung *f*, m/100-Lösung *f*

central area of metabolism *(Bioch)* Stoffwechselpool *m*, Sammelbecken *n* des Stoffwechsels, labile Mischphase *f*

central atom Zentralatom *n*

central nervous system stimulant *(Pharm)* Analeptikum *n*, (das Zentralnervensystem) anregendes Mittel *n*

central pathway *(Bioch)* Zentralbahn *f*, amphibolische Sequenz *f*

centrally located mittelständig

centre *v* zentrieren, (ein)mitten

centre 1. Zentrum *n*, Mitte *f*, Mittelpunkt *m*; 2. *(Krist)* Zentrum *n*, Keim *m*

centre band *(Spektr)* Hauptbande *f*

centre-feed clarifier *(Hyd)* Rund(absetz)becken *n (mit zentraler Verteilung des Rohwassers)*

centre of asymmetry Asymmetriezentrum *n*

centre of charge Ladungsschwerpunkt *m*

centre of crystallization Kristallisationszentrum *n*

centre of symmetry 1. Symmetriezentrum *n*; 2. *(Krist)* Inversionszentrum *n*

centrifugal zentrifugal, Fliehkraft...

centrifugal 1. Zentrifuge *f*, Trennschleuder *f*, Schleuder *f*; 2. *s.* centrifugal filter; 3. *s.* centrifuge basket

centrifugal atomizer Zentrifugalzerstäuber *m*, Fliehkraftzerstäuber *m*; Zentrifugalversprüher *m*, Fliehkraftversprüher *m*

centrifugal casting 1. Schleudergussteil *n*; 2. Schleuderguss *m*, Zentrifugalguss *m*, Zentrifugalgießen *n*

centrifugal clarifier Klärzentrifuge *f*

centrifugal classifier Zentrifugal(kraft)sichter *m*, Fliehkraftsichter *m*, Zentrifugal(kraft)klassierer *m*, Fliehkraftklassierer *m*

centrifugal collector Zentrifugal(ab)scheider *m*, Fliehkraft(ab)scheider *m*, *(meist)* Zyklon(ab)scheider *m*, Wirbel(ab)scheider *m*, Abscheidezyklon *m*, Zyklon *m*; Spiral(ab)scheider *m (mit umlaufender Abscheidefläche)*

centrifugal compressor Kreiselverdichter *m*

centrifugal decanter Sedimentierzentrifuge *f*, Absetzzentrifuge *f*, Vollmantelzentrifuge *f*, Dekantierzentrifuge *f*

centrifugal dryer Trockenzentrifuge *f*

centrifugal extractor Zentrifugalextraktor *m*

centrifugal filter Filterzentrifuge *f*, Filtrierzentrifuge *f*, Siebzentrifuge *f*, Zentrifugal(kraft)filter *n*

centrifugal force Zentrifugalkraft *f*, Fliehkraft *f*

centrifugal mill Fliehkraftmühle *f*, Schleudermühle *f*

centrifugal moulding *(Kst)* Schleuderguss *m*, Rotationsguss *m*

centrifugal pump Kreiselpumpe *f*, Zentrifugalpumpe *f*

centrifugal separation Fliehkraft(ab)scheidung *f*

centrifugal separator *s.* centrifugal collector

centrifugal still Rotationskolonne *f*

centrifugal washer Kreiselwäscher *m*

centrifugate *v* zentrifugieren, schleudern
centrifugate Zentrifugat *n*, Zentrifugenablauf *m*, Zentrifugenüberlauf *m*, Fugat *n*
centrifugation Zentrifugieren *n*, Zentrifugierung *f*, Schleudern *n*, Abschleudern *n*
centrifuge *v* zentrifugieren, schleudern
centrifuge *v* **off** abzentrifugieren, abschleudern
centrifuge Zentrifuge *f*, Trennschleuder *f*, Schleuder *f*
centrifuge basket Siebtrommel *f* einer Zentrifuge, Zentrifugentrommel *f*
centrifuge cup Schleuderbecher *m*
centrifuge head Zentrifugenrotor *m (bei Laborzentrifugen)*
centring Zentrierung *f*, Zentrieren *n*, Einmitten *n*, Mitten *n*
centripetal force Zentripetalkraft *f*
centrosymmetric zentrosymmetrisch, zentralsymmetrisch
cephalin Kephalin *n*, Cephalin *n (ein Glycerophospholipid)*
cephalosporin Cephalosporin *n (Antibiotikum)*
cerametallic metallkeramisch
ceramic keramisch, Keramik...
ceramic fibre keramische Faser *f*, Keramikfaser *f*; keramischer Faserstoff *m*
ceramic filter Keramikfilter *n*
ceramic material keramischer Werkstoff *m*
ceramics 1. Keramik-, Glas- und Zementindustrie *f*; 2. Keramik-, Glas- und Zementherstellung *f*; 3. Keramik-, Glas- und Zementerzeugnisse *npl*
ceramide *(Bioch)* Ceramid *n*, *N*-Acylsphingosin *n*
cereal starch Getreidestärke *f*
cerebroside *(Bioch)* Cerebrosid *n*
cerebrospinal fluid cerebrospinale Flüssigkeit *f*, Hirn-Rückenmarks-Flüssigkeit *f*, Liquor *m* cerebrospinalis
ceresin (wax) Ceresin *n*, Hartparaffin *n*, gereinigtes Erdwachs *n*
ceria *s.* ceric oxide
ceric Cer..., *(i. e. S.)* Cer(IV)-...
ceric fluoride Cer(IV)-fluorid *n*, Certetrafluorid *n*
ceric oxide Cer(IV)-oxid *n*, Cerdioxid *n*
cerimetry *(Anal)* Cerimetrie *f*
cerite earth Ceriterde *f (allg. für ein Oxid der leichteren Lanthanidenelemente)*
cerium Ce Cer *n*
cerium carbide Cercarbid *n*
cerium dioxide Cerdioxid *n*, Cer(IV)-oxid *n*
cermet Cermet *n*, Kerametall *n*, mischkeramischer [metallkeramischer, kerametallischer] Werkstoff *m*, Mischkeramik *f*
cerotic acid Cerotinsäure *f*, *Hexacosansäure *f*
cerous Cer..., *(i. e. S.)* Cer(III)-...
cerous fluoride Cer(III)-fluorid *n*, Certrifluorid *n*
cerous oxide Cer(III)-oxid *n*
cesium *s.* caesium
cetane Cetan *n*, *Hexadecan *n*
cetane number [rating] Cetanzahl *f*, Cetanziffer *f*, CaZ
CF *s.* carbon filtration
CFC *s.* 1. carbon fibre-reinforced carbon; 2. chlorofluorocarbon
CFR = Cooperative Fuel Research
CFR motor method CFR-Motormethode *f (zur Bestimmung der Octanzahl)*
cGMP = cyclic guanosine 5′-monophosphate
C/H ratio C/H Verhältnis *n*, Kohlenstoff-Wasserstoff-Verhältnis *n*
chain 1. Kette *f*; Atomkette *f*; Reaktionskette *f*; 2. *(Kern)* Zerfallsreihe *f*
chain branching Kettenverzweigung *f*
chain breakage [breaking] Kettenabbruch *m*
chain carrier Kettenträger *m*
chain compound Kettenverbindung *f*
chain conformation Kettenkonformation *f*
chain conveyor Kettenförderer *m*
chain-end distance Kettenendenabstand *m*, Fadenendenabstand *m*, Endpunktabstand *m*
chain formation *(Kst)* Kettenbildung *f*
chain growth Kettenwachstum *n*, Wachstumsperiode *f (Polymerisation)*
chain-growth polymerization Ketten(wachstums)polymerisation *f*
chain initiation Kettenstart *m*, Startreaktion *f*
chain isomer Skelettisomer(e) *n*, Kettenisomer(e) *n*

chain isomerism Skelettisomerie *f*, Kettenisomerie *f*
chain molecule Kettenmolekül *n*
chain of reactions Reaktionskette *f*
chain polymerization Kettenpolymerisation *f*
chain-propagating reaction Kettenfortpflanzungsreaktion *f*
chain propagation Kettenwachstum *n*, Wachstumsperiode *f (Polymerisation)*
chain reaction Kettenreaktion *f*
chain-reaction mechanism Ketten(reaktions)mechanismus *m (bei der Polymerisation)*
chain scission Kettenspaltung *f*
chain segment Kettensegment *n*, Kettenbruchstück *n*
chain silicate Kettensilicat *n*, Inosilicat *n*
chain stopper *s.* chain terminator
chain termination Kettenabbruch *m*
chain terminator Kettenabbrecher *m*, Kettenabbruchmittel *n*, Stopper *m*
chain transfer Kettenübertragung *f*
chair *s.* chair form
chair conformation Sesselkonformation *f (Stereochemie)*
chair form Sesselform *f*, Sessel *m*, starre Form *f (Stereochemie)*
chalcogen Chalkogen *n (Sauerstoff, Schwefel, Selen, Tellur)*
chalcogenide Chalkogenid *n (binäre Verbindung eines Chalkogens)*
chalcone Chalkon *n*, 1,3-Diarylpropenon *n*, *(i. e. S.)* Benzylidenacetophenon *n*, Benzalacetophenon *n*, 1,3-Diphenylpropenon *n*, Chalkon *n*
chalcophile element chalkophiles Element *n*
chalk lime Kreidekalk *m*
chamber Kammer *f*, Zelle *f*, Raum *m*
chamber acid *s.* chamber sulphuric acid
chamber dryer Kammertrockner *m*
chamber filter press Kammer(filter)presse *f*
chamber furnace *(Met)* Kammerofen *m*
chamber process Bleikammerverfahren *n*, Kammerverfahren *n (zur Schwefelsäureherstellung)*
chamber sulphuric acid Kammer(schwefel)säure *f*
chamotte Schamotte *f*
chance contaminant zufällige Verunreinigung *f*
change *v* 1. (ver)ändern, wechseln *(Farbe, Gestalt, Ladung)*; verwandeln, umwandeln, umformen *(Struktur)*; 2. sich (ver)ändern, wechseln *(Farbe)*; sich verwandeln [umwandeln, umformen, umlagern] *(Struktur)*; übergehen; umschlagen *(Farbe, Reaktion)*
change Änderung *f*, Veränderung *f*, Wechsel *m*, Übergang *m*, Umschlag *m*; Umwandlung *f*, Umformung *f*, Umlagerung *f* • **undergo change** *s.* change 2.
change in ... *siehe auch: change of ...*
change in colour Farbänderung *f*, Farbumschlag *m*, Farbwechsel *m*
change in structure Strukturänderung *f*
change in volume Volumenänderung *f*
change of ... *siehe auch: change in ...*
change of concentration Konzentrationsänderung *f*
change of material stoffliche Veränderung *f* [Änderung *f*], Stoff(ver)änderung *f*, Material(ver)änderung *f*
change of state Zustandsänderung *f*, *(i. e. S.)* Aggregatzustandsänderung *f*
changeover Umschlag *m*, Wechsel *m (z. B. bei Titration)*
channel Kanal *m*, Rinne *f*
channel inclusion compound Kanaleinschlussverbindung *f*
channel(l)ing 1. Kanalbildung *f (z. B. in Staubfließsystemen und Filterkuchen)*; Bachbildung *f (z. B. in Füllkörperschüttungen)*; 2. Kanalisierung *f (bei der Ionenbeugung)*
chaotropic chaotrop *(H_2O-Struktur zerstörend)*
char *v* (ver)kohlen, ankohlen, verschwelen
char 1. künstliche Kohle *f (z. B. Holz- oder Knochenkohle)*; *(i. e. S.)* Entfärbungskohle *f*; 2. Halbkoks *m*, Schwelkoks *m*, Tieftemperaturkoks *m*
char filter Kohlefilter *n*
characteristic Kennzeichen *n*; Kennziffer *f*
characteristic curve Kennlinie *f*, Charakteristik *f*
characteristic number *s.* eigenvalue
characteristic temperature charakteristische Temperatur *f*, Debye-Temperatur *f*

characteristics Verhalten *n*, Charakteristik *f*
charcoal künstliche Kohle *f (z. B. Holz- oder Knochenkohle)*; *(i. e. S.)* Holzkohle *f*
charge *v* 1. beschicken, besetzen, speisen, füllen, chargieren *(einen chemischen Reaktor)*; begichten *(einen Hochofen)*; eintragen, einsetzen, einlegen, einspeisen, einfüllen, einschütten, aufgeben *(das Beschickungsgut)*; chargenweise [portionsweise] zugeben, zuteilen, (zu)dosieren; 2. *(phys Ch)* (auf)laden
charge *v* **by pumping** vollpumpen *(Gefäße)*
charge 1. Charge *f*, Beschickungsgut *n*, Beschickungsmaterial *n*, Füllung *f*, Schüttung *f*, Einsatz *m*, Einsatzgut *n*, Einsatzmaterial *n*, Eintrag *m*, Ansatz *m*, Aufgabegut *n*; Gicht *f (des Hochofens)*; 2. *(phys Ch)* Ladung *f*
charge carrier *(phys Ch)* Ladungsträger *m*
charge carrier density Ladungsträgerdichte *f*
charge centre *(phys Ch)* Ladungsschwerpunkt *m*
charge cloud *(phys Ch)* Ladungswolke *f*
charge compensation Ladungsausgleich *m*, Ladungskompensation *f*
charge density *(phys Ch)* Ladungsdichte *f*
charge distribution *(phys Ch)* Ladungs(dichte)verteilung *f*
charge exchange Ladungsaustausch *m*
charge injection Ladungsinjektion *f*
charge neutrality Ladungsneutralität *f*
charge number *(phys Ch)* Ladungszahl *f*
charge sign Ladungsvorzeichen *n*
charge stock *s.* charge 1.
charge transfer *(phys Ch)* Ladungsübertragung *f*
charge-transfer coefficient Ladungsdurchtrittsfaktor *m*
charge-transfer complex Ladungsübertragungskomplex *m*, Elektronen-Donor-Akzeptor-Komplex *m*, EDA-Komplex *m*, Charge-transfer-Komplex *m*, CT-Komplex *m*
charge transport Ladungstransport *m*
charge transport rate Ladungstransportrate *f*
charging 1. Beschicken *n*, Beschickung *f*, Besetzen *n*, Speisen *n*, Füllen *n*, Chargieren *n (eines chemischen Reaktors)*; Begichten *n*, Begichtung *f (eines Hochofens)*; Eintragen *n*, Eintrag *m*, Einsetzen *n*, Einsatz *m*, Einlegen *n*, Einlage *f*, Einspeisen *n*, Einfüllen *n*, Einschütten *n*, Aufgeben *n*, Aufgabe *f (des Beschickungsguts)*; chargenweises [portionsweises] Zugeben *n*, Zuteilen *n*, Dosieren *n*, Zudosieren *n*; 2. *(phys Ch)* Aufladen *n*, Laden *n*
charging hole Füllöffnung *f*, Füllloch *n*, Beladeöffnung *f*, Eintragöffnung *f*, Beschickungsöffnung *f*, Einwurföffnung *f*, Aufgabeöffnung *f*
charging pipe Zuflussrohr *n*, Zulaufrohr *n*, Füllrohr *n*; Zuflussstutzen *m*, Zulaufstutzen *m*, Füllstutzen *m*
charging point Guteintrag *m*, Gutaufgabe *f*
chart Diagramm *n*, Schaubild *n*; Tabelle *f*; Tafel *f*
Chatelier-Braun principle *s.* Le Châtelier(-Braun) principle
Ch D = Doctor of Chemistry
Ch E *s.* chemical engineer
cheapener *(Gum)* Streckmittel *n*; inaktiver Füllstoff *m*
check *v* prüfen, kontrollieren
check *v* **for purity** auf Reinheit prüfen
check 1. Riss *m*, Spalt *m*, Spalte *f*; 2. *s.* check valve
check analysis Kontrollanalyse *f*
check valve Rückschlagventil *n*
chelate *v* Chelate bilden, Chelatbindung [Wasserstoffbrückenbindung] eingehen, chelatisieren; in einem Chelat binden
chelate Chelat *n*, Chelatverbindung *f*, Scherenverbindung *f*, Chelatkomplex *m*
chelate complex [compound] *s.* chelate
chelate effect Chelateffekt *m (Stabilitätsgewinn)*
chelate formation constant Chelatbildungskonstante *f*
chelate ring Chelatring *m*
chelate stability constant Chelat-Stabilitätskonstante *f*
chelated chelatgebunden
chelating agent Chelatbildner *m*, Che-

lator *m*, chelatbildende Substanz *f*, chelatbildendes Reagens *n*
chelating ion-exchange resin Chelataustauscher *m*, chelatbildendes Austauscherharz *n*, Chelatharz *n*
chelating ligand Chelatligand *m*
chelation Chelatbildung *f*, Scherenbildung *f*
chelatometric chelatometrisch
chelatometric titration *s.* chelatometry
chelatometry Chelatometrie *f*, chelatometrische Titration *f*
cheletropic reaction cheletrope Reaktion *f*
chelotropic reaction cheletrope Reaktion *f*
chem *s.* 1. chemical; 2. chemist; 3. chemistry
ChemFET *s.* chemically sensitive field effect transistor
CHEMFET *s.* chemically sensitive field effect transistor
chemical chemisch, Chemie... • **by a chemical route** auf chemischem Wege • **by chemical means** auf chemischem Wege
chemical Chemikalie *f*
chemical activity chemische Aktivität *f* [Wirksamkeit *f*]
chemical adsorption chemische Adsorption *f*, *(i. w. S.)* Chemisorption *f*
chemical affinity (chemische) Affinität *f* *(eines Stoffs)*; (chemische) Triebkraft *f*, (chemische) Affinität *f*, Reaktionsaffinität *f*
chemical agent chemisches [chemisch wirksames] Mittel *n* [Agens *n*], chemischer Wirkstoff *m*
chemical analysis chemische Analyse *f*
chemical balance chemische Waage *f*, *(i. e. S.)* Analysenwaage *f*, analytische Waage *f*
chemical bond chemische Bindung *f* *(Zustand)*
chemical bonding chemische Bindung *f* *(Vorgang)*
chemical cellulose Alphacellulose *f*, α-Cellulose *f*
chemical ceramics chemische Keramik *f*, Chemiekeramik *f*
chemical change chemische Veränderung *f*, *(i. e. S.)* chemische Umwandlung *f*
chemical combination chemische Vereinigung *f* [Verbindung *f*]
chemical composition chemische Zusammensetzung *f*
chemical compound chemische Verbindung *f*
chemical conversion chemische Umwandlung *f* [Umsetzung *f*]
chemical dating physikalische [radiometrische, absolute, physikalisch-chemische] Altersbestimmung *f*, Altersbestimmung *f* durch Radionuklide
chemical degradation chemischer Abbau *m*
chemical displacement *s.* chemical shift
chemical energy chemische Energie *f*
chemical engineer Chemieingenieur *m*, *(i. e. S.)* Verfahrensingenieur *m*, Verfahrenstechniker *m*, Verfahrenschemiker *m*
chemical engineering Chemieingenieurwesen *n*, chemisches Ingenieurwesen *n*, *(i. e. S.)* chemische Verfahrenstechnik *f*
chemical engineering science Chemieingenieurwissenschaft *f*
chemical engineering technology Chemieingenieurtechnik *f*, chemische Ingenieurtechnik *f*
chemical-engineering unit operation Grundoperation *f* der chemischen Verfahrenstechnik
chemical entity chemisches Objekt *n* [Gebilde *n*]
chemical equation chemische Gleichung *f* [Umsatzgleichung *f*], (chemische) Reaktionsgleichung *f*
chemical exposure chemische Beanspruchung *f*
chemical feed zugesetzte [dosierte] Chemikalien *fpl*, Chemikalienzusatz *m*
chemical feeder Aufgabeapparat *m* für Chemikalien, *(mit Mengenmessung:)* Chemikalienzuteilvorrichtung *f*, Dosierapparat *m* [Dosiervorrichtung *f*] für Chemikalien
chemical feeding Chemikalienzugabe *f*, Chemikalienzusatz *m*, Chemikaliendosierung *f*
chemical individual chemisches Individuum *n*

chemical industry chemische Industrie *f*, Chemieindustrie *f*
chemical inertness chemische Trägheit *f* [Reaktionsträgheit *f*], chemische Inertie *f* [Indifferenz *f*]
chemical ionization chemische Ionisation *f*
chemical ionization mass spectrometer CI-Massenspektrometer *n*
chemical kinetics chemische Kinetik *f*, Reaktionskinetik *f*
chemical laser chemischer Laser *m*
chemical loss Chemikalienverlust *m*
chemical nomenclature chemische Nomenklatur *f*, Nomenklatur *f* der chemischen Stoffe
chemical notation chemische Notation *f*, Notation *f* chemischer Strukturen
chemical oxygen demand *(Hyd)* chemischer Sauerstoffbedarf *m*, CSB, chemischer Sauerstoffverbrauch *m*, CSV
chemical passivity chemische Passivität *f*
chemical physics chemische Physik *f*
chemical plant chemische Fabrik *f*, chemisches Werk *n*, Chemiewerk *n*, chemischer Betrieb *m*, Chemiebetrieb *m*; chemische Anlage *f*, Chemieanlage *f*
chemical-plant construction Chemieanlagenbau *m*
chemical plasticization *(Gum)* chemische Plastizierung *f* [Weichmachung *f*]
chemical plasticizer *(Gum)* chemisches Plastizierungsmittel *n* [Plastiziermittel *n*, Abbaumittel *n*], Plastifikator *m*, Peptisier(ungs)mittel *n*
chemical porcelain chemisches Porzellan *n*
chemical potential chemisches Potenzial *n*
chemical preservative *(Lebm)* chemisches Konservierungsmittel *n*, chemischer Konservierungsstoff *m*
chemical probe chemische Sonde *f* *(Molekül, das in bestimmter Weise mit Teilen eines anderen Moleküls reagiert)*
chemical processing industry chemische Industrie *f*, Chemieindustrie *f*
chemical pulp (chemisch vollkommen aufgeschlossener) Zellstoff *m*, Vollzellstoff *m*, klassischer Zellstoff *m*
chemical pulping Zellstoffaufschluss *m*, chemischer Aufschluss *m*
chemical pump Chemiepumpe *f*
chemical raw material Chemierohstoff *m*, chemischer Rohstoff *m*
chemical reaction chemische Reaktion *f*
chemical reaction engineering chemische Reaktionstechnik *f*
chemical-reaction equation *s.* chemical equation
chemical reactor chemischer Reaktor *m* [Reaktionsapparat *m*], Stoffumsetzer *m*
chemical recovery Chemikalienrückführung *f*, Chemikalienrückgewinnung *f*
chemical reprocessing *(Kern)* chemische Wiederaufbereitung *f*
chemical resistance chemische Widerstandsfähigkeit *f* [Beständigkeit *f*, Resistenz *f*], Beständigkeit *f* gegen chemische Einwirkungen; Chemikalienbeständigkeit *f*, Widerstandsfähigkeit *f* [Resistenz *f*] gegen Chemikalien
chemical-resistant chemikalienbeständig, chemikalienfest, chemikalienresistent
chemical rubber synthetischer [künstlicher] Kautschuk *m*, Synthesekautschuk *m*, SK, Kunstkautschuk *m*
chemical shift chemische Verschiebung *f* *(NMR-Spektroskopie)*
chemical sign chemisches Zeichen *n*, Elementsymbol *n*, Symbol *n*
chemical species chemische Spezies *f*
chemical stability *s.* chemical resistance
chemical stoneware chemisches Steinzeug *n*, Steinzeug *n* für chemische Industrie
chemical symbol *s.* chemical sign
chemical technician Chemotechniker *m*
chemical technique chemische Arbeitstechnik *f*, chemisches Arbeitsverfahren *n*
chemical-technological chemisch-technisch, chemietechnisch; chemisch-technologisch
chemical technologist Chemietechnologe *m*
chemical technology chemische Technik *f*, Chemietechnik *f*; chemische Technologie *f*

chemical thermodynamics chemische Thermodynamik *f*
chemical tracer chemischer Tracer *m*
chemical transformation chemische Umwandlung *f*
chemical transport chemischer Transport *m*
chemical uniformity chemische Einheitlichkeit *f*
chemical union chemische Vereinigung *f*
chemical vapour deposition chemische Abscheidung *f* aus der Gasphase, chemisches Abscheiden *n* aus der Gasphase, CVD-Verfahren *n*, CVD-Beschichtungstechnik *f (über zersetzliche gasförmige Zwischenstufen, meist Halogenide)*
chemical waste(s) chemische Abprodukte *npl*, Chemiemüll *m*
chemical works chemisches Werk *n*, Chemiewerk *n*, chemischer Betrieb *m*, Chemiebetrieb *m*, chemische Fabrik *f*
chemically bound chemisch gebunden
chemically homogeneous chemisch homogen
chemically induced dynamic electron polarization chemisch induzierte dynamische Elektronenpolarisation *f*, CIDEP
chemically induced dynamic nuclear polarization chemisch induzierte dynamische Kernpolarisation *f*, CIDKP
chemically inert reaktionsträge, chemisch träge [inert, indifferent]
chemically pure chemisch rein
chemically sensitive field effect transistor chemisch sensitiver Feldeffekttransistor *m*
chemically uniform chemisch einheitlich
chemiluminescence Chemilumineszenz *f*
chemiluminescent chemilumineszierend
chemism Chemismus *m*
chemisorb *v* chemisch adsorbieren, chemisorbieren, chemosorbieren
chemisorption Chemisorption *f*, Chemosorption *f*, chemische Adsorption *f*, *(i. e. S.)* irreversible chemische Adsorption *f*
chemist Chemiker *m*
chemist-analyst Analytiker *m*
chemistry Chemie *f*
chemistry laboratory chemisches Laboratorium *n*, Chemielabor *n*
chemistry of atmosphere Chemie *f* der Atmosphäre, Atmosphärenchemie *f*
chemistry of coordination compounds Chemie *f* der Koordinationsverbindungen, Koordinationschemie *f*
chemistry of exotic atoms Chemie *f* exotischer [neuer] Atome
chemistry of fats Fettchemie *f*
chemistry of fibres Faserstoffchemie *f*
chemistry of natural products Naturstoffchemie *f*
chemistry of photography Chemie *f* der Fotografie [fotografischen Prozesse], fotografische Chemie *f*
chemistry of surfaces Oberflächenchemie *f*
chemistry of tanning Gerbereichemie *f*
chemistry of the carbon compounds Chemie *f* der Kohlenstoffverbindungen, Kohlenstoffchemie *f*
chemistry of water Wasserchemie *f*, Chemie *f* des Wassers, Hydrochemie *f*
chemoautotrophic chemo(auto)troph
chemoautotrophs chemo(auto)trophe Organismen *mpl*
chemoautotrophy *(Bio)* Chemo(auto)-trophie *f (Gewinnung von Assimilationsenergie aus Oxidationsvorgängen)*
chemolithotrophic chemolithotroph
chemolithotrophs chemolithotrophe Organismen *mpl*
chemoorganotrophs chemoorganotrophe Organismen *mpl*
chemophobia Chemophobie *f*, Chemiefeindlichkeit *f*
chemoselective chemoselektiv
chemoselectivity Chemoselektivität *f*
chemosorb *v s.* chemisorb
chemostat *(Biot)* Chemostat *m*
chemosterilant Chemosterilans *n*, Chemosterilisierungsmittel *n*
chemosynthesis Chemosynthese *f* *(Mikrobiologie)*
chemosynthetic chemosynthetisch *(Mikrobiologie)*
Chemotaxis *f (Bio)* chemotaxis
chemotherapeutant Chemotherapeutikum *n*, chemotherapeutisches Mittel *n*
chemotherapeutic chemotherapeutisch

chemotherapeutic agent *s.* chemotherapeutant
chemotherapy Chemotherapie *f*
chemurgy Chemurgie *f (Gewinnung chemischer Produkte aus pflanzlichen oder tierischen Rohstoffen)*
chequerwork Steingitterwerk *n*, Gitter(mauer)werk *n (z. B. im Winderhitzer)*
cherry-red heat Kirschrotglut *f*
Chichibabin hydrocarbon Čičibabin-Kohlenwasserstoff *m*, Tschitschibabin-Kohlenwasserstoff *m*
Chichibabin reaction [synthesis] Čičibabin-Reaktion *f*, Čičibabin-Synthese *f*
chief component [constituent] Hauptbestandteil *m*, Hauptkomponente *f*, Grundkomponente *f*
chief product Hauptprodukt *n*
Chile saltpetre Chilesalpeter *m*, Natronsalpeter *m (Natriumnitrat)*
chill *v* (ab)kühlen, erkalten lassen; abschrecken; abkühlen, erkalten; abgeschreckt werden
chill Abschrecken *n*, Abschreckung *f*
chill-cast iron Schalen(hart)guss *m*, Hartguss *m (Produkt)*
chill roll Kühlwalze *f*, Kühltrommel *f*
chiller Chiller *m*, Kühlapparat *m (für Temperaturen, die mit Wasser nicht mehr erreichbar sind)*
chilling Kühlen *n*, Kühlung *f*, Abkühlen *n*, Abkühlung *f*; Abschrecken *n*, Abschreckung *f*; Erkalten *n*, Kaltwerden *n*
chimney 1. Schornstein *m*, Esse *f*, Schlot *m*, Kamin *m*; 2. *(Dest)* Kamin(stummel) *m*, Dampfkamin *m*, Dampfhals *m*, Dämpfestutzen *m (einer Glocke)*
china (nicht technisches) Porzellan *n*
china clay 1. *(Min)* Kaolin *m*, Porzellanerde *f*, weißer [reiner] Ton *m*, China Clay *m(n)*; *(i. e. S.)* Kaolin *m* aus Cornwall; 2. *(Tech)* Kaolin *m*, Schlämmkaolin *m*, geschlämmte Porzellanerde *f*
chinoline Chinolin *n*, 2,3-Benzopyridin *n*
chinone *s.* quinone
chipped ice Splittereis *n*
chiral chiral *(Stereochemie)*
chiral centre Chiralitätszentrum *n (Stereochemie)*
chirality Chiralität *f (Stereochemie)*
chitin Chitin *n (ein Polysaccharid)*
chitinous chitinös, chitinig, Chitin...
chl. *s.* chloroform
chlor-alkali cell Chloralkali-Elektrolysezelle *f*
chlor-alkali electrolysis Chloralkalielektrolyse *f*, Alkalichloridelektrolyse *f*
chloracne *(Med)* Chlorakne *f*
chloral Chloral *n*, Trichloracetaldehyd *m*, *Trichlorethanal *n*
chloral hydrate Chloralhydrat *n*, *1,1,1-Trichlor-2,2-dihydroxyethan *n*
chloramine 1. Chloramin *n*, *N*-Chloramin *n*, *(i. e. S.)* NH_2Cl Chloramin *n*; 2. *s.* chloramine-T
chloramine-T Monochloramin *n* T, Toluensulfonsäurechloramidnatrium *n*
chloranil Chloranil *n*, Tetrachlor-*p*-benzochinon *n*
chloraniline Chloranilin *n*
chlorate M^IClO_3 Chlorat *n*
chloric acid Chlorsäure *f*
chloride M^ICl Chlorid *n*
chloride brine Chloridsole *f*
chloride hardness *(Hyd)* Chloridhärte *f*
chloride melt Chloridschmelze *f*
chloride of lime Chlorkalk *m*, Bleichkalk *m*, Bleichpulver *n*
chloridize *v* mit Chlor [Chlorid] behandeln, chlorieren
chlorinate *v* 1. *(org Ch, Pap, Text)* chlorieren; 2. *(Hyd)* chloren
chlorinate Chlorierungsprodukt *n*
chlorinated hydrocarbon Chlorkohlenwasserstoff *m*, chlorierter Kohlenwasserstoff *m*
chlorinated product Chlorierungsprodukt *n*
chlorinated rubber Chlorkautschuk *m*
chlorinating agent 1. Chlorierungsmittel *n*; 2. *(Hyd)* Chlorungsmittel *n*
chlorinating vessel Chlorierungsgefäß *n*, Chlorierungsbehälter *m*, Chlorierungskessel *m*, Chlorierungsapparat *m*, Chlorierer *m*, Chlorator *m*
chlorination 1. *(org Ch, Pap, Text)* Chlorierung *f*; 2. *(Hyd)* Chlorung *f*, *(i. e. S.)* Chlorgasverfahren *n*
chlorinator 1. *s.* chlorinating vessel; 2. Chlor(ierungs)turm *m*; 3. *(Hyd)* Chlor(gas)anlage *f*
chlorine Cl Chlor *n*
chlorine bleaching Chlorbleiche *f*
chlorine cyanide Chlorcyan *n*, Cyanchlorid *n*

chlorine dioxide Chlordioxid *n*, Chlor(IV)-oxid *n*
chlorine dioxide bleaching *(Pap)* Chlordioxidbleiche *f*
chlorine gas Chlorgas *n*
chlorine heptoxide Chlorheptoxid *n*, Dichlorheptoxid *n*, Chlor(VII)-oxid *n*
chlorine monoxide Chlor(mon)oxid *n*, Dichloroxid *n*, Chlor(I)-oxid *n*
chlorine residual *(Hyd)* (freies wirksames) Restchlor *n*, Restgehalt *m* an freiem (wirksamem) Chlor, Restchlorgehalt *m*, Chlorüberschuss *m*
chlorine tetroxide Chlortetroxid *n*
chlorine water Chlorwasser *n*
chlorinity Chlorinität *f*, Cl%
chlorinize *v s.* chlorinate
chlorinolysis Chlorolyse *f*
chlorinous taste *(Hyd)* Chlorgeschmack *m*
chlorite $H^{I}ClO_2$ Chlorit *n*
chlorite bleaching *(Text)* Chloritbleiche *f*, Natriumchloritbleiche *f*
chloroacetal Chloracetal *n*, Chloracetaldehyddiethyl-acetal *n*
chloroacetic acid Chloressigsäure *f*, Chlorethansäure *f*
chloroacetone Chloraceton *n*, *Chlorpropanon *n*
chloroacetophenone Chloracetophenon *n*, *(i. e. S.)* $C_6H_5COCH_2Cl$ α-Chloracetophenon *n*, ω-Chloracetophenon *n*, Phenacylchlorid *n*
chloroacetylation Chloracetylierung *f*
chloroamination Chloraminierung *f*
***N*-chloroamine** Chloramin *n*, *N*-Chloramin *n*
chloroarylation Chlorarylierung *f*
chlorobenzene Chlorbenzen *n*, *(i. e. S.)* C_6H_5Cl Chlorbenzen *n*, Monochlorbenzen *n*
chlorobutyl rubber Chlorbutylkautschuk *m*
chlorocarbonate Chlorameisensäureester *m*, Chlorkohlensäureester *m*
chlorocarbonic acid Chlorameisensäure *f*, Chlorkohlensäure *f*
chlorocomplex Chlorokomplex *m*
chlorodecarboxylation chlorierende Decarboxylierung *f*
chlorodehydroxylation Chlordehydroxylierung *f*
chloroethane *Chlorethan *n*, Monochlorethan *n*, Ethylchlorid *n*
chloroethanoic acid Chlorethansäure *f*, Chloressigsäure *f*
2-chloroethanol *2-Chlorethanol *n*, 2-Chlorethylalkohol *m*
chloroethyl group $ClCH_2CH_2$– Chlorethylgruppe *f*, Chlorethylrest *m*
chloroethylation Chlorethylierung *f*
chloroethylene *Chlorethen *n*, Vinylchlorid *n*
chlorofluorination Chlorfluorierung *f*
chlorofluorocarbon Fluorchlorkohlenwasserstoff *m*, FCKW
chloroform Chloroform *n*, Trichlormethan *n*
chloroformate Chlorameisensäureester *m*, Chlorkohlensäureester *m*
chlorohydrin 1. Chlorhydrin *n (allg. für eine Verbindung* >C(OH)C(Cl)<*)*; 2. *s.* α-chlorohydrin
α-chlorohydrin α-Monochlorhydrin *n*, Glycerol-1-chlorhydrin *n*, *3-Chlorpropan-1,2-diol *n*
chloroiodination Chloriodierung *f*, Iodchlorierung *f*
chlorometallic acid Chlorosäure *f*
chloromethane Chlormethan *n*, Methylchlorid *n*
chloromethoxylation Chlormethoxylierung *f*
chloromethylation Chlormethylierung *f*
chloronium compound Chloroniumverbindung *f*
chlorophenol Chlorphenol *n*
chlorophenylation Chlorphenylierung *f*
chlorophosphon(yl)ation Chlorphosphon(yl)ierung *f*
chlorophyl(l) Chlorophyll *n*, Blattgrün *n*
chlorophyllide *(Bioch)* Chlorophyllid *n*
chloroplatinate Chloroplatinat *n*
chloroplatinic acid Chloroplatin(IV)-säure *f*, Hexa-chloroplatin(IV)-säure *f*
chloroprene Chloropren *n*, *2-Chlorbuta-1,3-dien *n*
chloroprene rubber Chloroprenkautschuk *m*
chloropropane Chlorpropan *n*
chloropropene Chlorpropen *n*
chlorosil(ic)ane Chlorsilan *n*, Monochlorsilan *n*
chlorosulphonate 1. Chlorosulfat *n*; 2. *(org Ch)* Sulfochlorid *n*

chlorosulphonated polyethylene chlorosulfoniertes Polyethylen *n*
chlorosulphonation Chlorsulfonierung *f*
chlorosulphonic acid Chloroschwefelsäure *f*
α-chlorotoluene α-Chlortoluen *n*, Benzylchlorid *n*
chlorotrifluoroethylene *Chlortrifluorethen *n*, Chlortrifluorethylen *n*
chlorous acid chlorige Säure *f*
choke *v* 1. verstopfen, versperren, verschmutzen, verschmieren; drosseln; hemmen, ersticken *(Feuer)*; sich verstopfen; 2. *(Dest)* stauen; erstickend wirken *(Gase)*
choke *v* **up** verstopfen, versperren, verschmutzen, verschmieren
cholagogic *(Pharm)* gallentreibend
cholagogue *(Pharm)* galle(n)treibendes Mittel *n*
cholesteric cholester(in)isch
cholesteric liquid crystal cholester(in)ischer Flüssigkristall *m*
cholesteric phase cholester(in)ische Phase *f*
cholesterin *s.* cholesterol
cholesterol *(org Ch)* Cholesterol *n*
cholic acid Cholsäure *f*, 3α,7α,12α-Trihydroxy-5β-cholansäure *f*
chop *v* (zer)hacken, zerkleinern, schneiden, schnitzeln; häckseln *(Stroh)*; stoßen *(Eis)*
chopper 1. Hackmaschine *f*, Hacke *f*, Hacker *m*; 2. *(Pap)* Häckselmaschine *f* *(Strohaufschluss)*; 3. *(Spektr)* Chopper *m*, Zerhacker *m*
CHP *s.* cumene hydroperoxide
chromable chromierbar
chromate *v* chromatieren
chromate $M^I_3CrO_4$ Chromat *n*
chromatic aberration chromatische Aberration *f*, Farbenabweichung *f*, Farb(en)fehler *m*
chromatize *v s.* chromate
chromatogram Chromatogramm *n*
chromatograph *v* chromatographieren
chromatographic chromatographisch
chromatographic column Chromatographiesäule *f*, chromatographische Säule *f*
chromatographic packing Füllkörper *mpl*, Füllmaterial *n*, Füllstoff *m*, Packungsmaterial *n*, Packung *f*
chromatographic paper chromatographisches Papier *n*, Chromatographiepapier *n*
chromatographically pure chromatographisch rein
chromatography Chromatographie *f*
• **by affinity chromatography** affinitätschromatographisch • **by column chromatography** säulenchromatographisch • **by gas chromatography** gaschromatographisch • **by gel chromatography** gelchromatographisch • **by gel permeation chromatography** gelpermeationschromatographisch • **by liquid chromatography** flüssig(keits)chromatographisch • **by partition chromatography** verteilungschromatographisch
chromatography chamber (chromatographische) Entwicklungskammer *f*, Chromatographie(r)kammer *f*, Trennkammer *f*
chromatoplate Chromatographieplatte *f*, Fertigplatte *f*
chrome *v* 1. *(Farb)* chromieren; 2. mit Chromsalzen gerben, chromgerben; 3. verchromen *(Metalle)*
chrome *s.* 1. chrome yellow; 2. chrome leather
chrome alum Chromalaun *m*, *(i. e. S.)* Kaliumchromalaun *m*, Kaliumchrom(III)-sulfat-12-Wasser *n*
chrome-developed dye Chromentwicklungsfarbstoff *m*, Nachchromier(ungs)farbstoff *m*
chrome dye Chromfarbstoff *m*, Chromier(ungs)farbstoff *m*
chrome green Chromgrün *n*, Deckgrün *n*, Druckgrün *n*, Ölgrün *n* *(Gemisch aus Chromgelb und Berliner Blau)*; $Cr_2O_3 \cdot \times H_2O$ Chromoxidhydratgrün *n*, Brillantgrün *n*, Guignetgrün *n*; C_2O_3 Chromoxidgrün *n*, Laubgrün *n*, grüner Zinnober *m*
chrome leather Chromleder *n*
chrome liquor Chrom(gerb)brühe *f*
chrome molybdenum steel Chrom-Molybdän-Stahl *m*
chrome nickel steel Chrom-Nickel-Stahl *m*
chrome oxide green Chromoxidgrün *n*, grüner Zinnober *m*, Laubgrün *n* *(Chrom(III)-oxid)*

chrome red Chromrot *n*, Chromzinnober *m*, Persischrot *n (basisches Bleichromat schwankender Zusammensetzung)*
chrome steel Chromstahl *m*
chrome tanning Chromgerbung *f*
chrome yellow Chromgelb *n (Gemisch und Mischkristalle aus Bleichromat und Bleisulfat)*
chromic Chrom..., *(i. e. S.)* Chrom(III)-...
chromic acid Chromsäure *f*
chromic acid cleaning mixture *(Lab)* Chromschwefelsäure *f*, Dichromatschwefelsäure *f*
chromic chloride Chrom(III)-chlorid *n*, Chromtrichlorid *n*
chromic hydroxide Chrom(III)-hydroxid *n*
chromic oxide Chrom(III)-oxid *n*, Dichromoxid *n*
chromic sulphate Chrom(III)-sulfat *n*
chroming 1. *(Farb)* Chromierung *f*; 2. Chromgerbung *f*; 3. Verchromen *n (von Metallen)*
chromite 1. Chromat(III) *n*; 2. *(Min)* Chromit *m*, Chromeisenerz *n (Eisen(II)-chrom(III)-oxid)*
chromium Cr Chrom *n*
chromium carbonyl Chromcarbonyl *n*, Chromhexacarbonyl *n*
chromium dioxide Chromdioxid *n*, Chrom(IV)-oxid *n*
chromium mordant Chrombeize *f*
chromium oxychloride Chrom(IV)-oxidchlorid *n*, Chromylchlorid *n*
chromium plating 1. Verchromen *n*; 2. Chrom(schutz)schicht *f*
chromium steel Chromstahl *m*
chromium trioxide Chromtrioxid *n*, Chrom(VI)-oxid *n*
chromize *v* inchromieren, chromdiffundieren, diffusionsverchromen, diffusionschromieren, einsatzverchromen *(mittels Chrompulvers oder Chromschmelzen)*
chromized steel Inchromierstahl *m*, inchromierter Stahl *m*
chromogen Chromogen *n (Verbindung mit chromophorer Gruppe)*
chromogenic reagent Farbreagens *n*
chromoisomer Chromoisomer(e) *n*
chromoisomerism Chromoisomerie *f*
chromophore Chromophor *m*, *(i. e. S.)* Farbträger *m*, chromophore [farbtragende, farbgebende] Gruppe *f*, *(i. w. S.)* farbbestimmendes Bindungssystem *n*
chromophoric chromophor, farbtragend, farbgebend
chromophoric group *s.* chromophore
chromous Chrom..., *(i. e. S.)* Chrom(II)-...
chromous chloride Chrom(II)-chlorid *n*, Chromdichlorid *n*
chromous hydroxide Chrom(II)-hydroxid *n*
chromous oxide Chrom(II)-oxid *n*, Chrommonoxid *n*
chromous sulphate Chrom(II)-sulfat *n*
chromyl chloride Chromylchlorid *n*, Chrom(VI)-oxidchlorid *n*
chronic toxicity chronische Toxizität *f* [Giftigkeit *f*]
chronopotentiometry *(Anal)* Chronopotenziometrie *f*
chrysene Chrysen *n*, Benzophenanthren *n*
Chu *s.* centigrade heat unit
CHU *s.* centigrade heat unit
churn *v (Text)* sulfidieren, xanthogenieren
churn *(Text)* Baratte *f*, Sulfidiertrommel *f*, Xanthatkneter *m*
chute Rutsche *f*, Schurre *f*, Rinne *f*, Schütte *f*
CI 1. = Colour Index; 2. = Chlorine Institute
CIDEP *s.* chemically induced dynamic electron polarization
CIDNP *s.* chemically induced dynamic nuclear polarization
cinchona alkaloid Cinchonaalkaloid *n*, China(rinden)alkaloid *n*
cinder Schlacke *f*; Zunder *m*
cine-substitution cine-Substitution *f*
cinebarite Zinnober *m*
cinnamaldehyde Zimtaldehyd *m*, *3-Phenylpropenal *n*
cinnamate Cinnamat *n (Salz oder Ester der Zimtsäure)*
cinnamic acid (gewöhnliche) Zimtsäure *f*, *trans*-Zimtsäure *f*, *trans*-3-Phenylpropensäure *f*
cinnamic alcohol Zimtalkohol *m*, *3-Phenylprop-2-en-1-ol *n*
CIP system *s.* Cahn-Ingold-Prelog system
circuit water Kreislaufwasser *n*

circular chromatography Zirkularchromatographie *f*, Ringchromatographie *f*, *(auf Papier auch)* Rundfilterchromatographie *f*
circular clarifier *(Hyd)* Rund(absetz)becken *n*, Rundklärbecken *n*
circular dichroism *(Anal)* Zirkulardichroismus *m*
circular die Ringdüse *f* *(Strangpresswerkzeug)*
circular piston pump Kreiskolbenpumpe *f*
circular polarimeter Kreispolarimeter *n*
circular polarization zirkulare Polarisation *f*, Zirkularpolarisation *f*
circularly polarized zirkularpolarisiert, zirkular polarisiert
circulate *v* im Kreis(lauf) führen, zirkulieren [umlaufen] lassen, umwälzen
circulating pump Umwälzpumpe *f*, Umlaufpumpe *f*, Kreislaufpumpe *f*, Zirkulationspumpe *f*
circulating water Kreislaufwasser *n*
circulation Kreislauf *m*, Zirkulation *f*, Umlauf *m*; Umwälzung *f*
circulation evaporator Umlaufverdampfer *m*
circulation system Kreislaufsystem *n*, Zirkulationssystem *n*, Umlaufsystem *n*
circumambient medium umgebendes Medium *n*
circumneutral pH range Neutralbereich *m*
cis *cis-...*, *cis*-ständig, in *cis*-Stellung befindlich • **be cis** in *cis*-Stellung stehen, *cis*-ständig (angeordnet) sein
cis conformation *cis*-Konformation *f*, synperiplanare Konformation *f* *(Torsionswinkel 0°)*
cis-trans conversion *cis-trans*-Umwandlung *f*
cis-trans isomerism *cis-trans*-Isomerie *f*, geometrische Isomerie *f*
cisoid cisoid
cistactic *cis*-taktisch, ct
cistacticity *cis*-Taktizität *f*
citral a Citral *n* A, *trans*-Citral *n*, Geranial *n*
citral b Citral *n* B, *cis*-Citral *n*, Neral *n*
citrate *v* Citrat zusetzen, mit Citrat versetzen
citrate Citrat *n* *(Salz oder Ester der Citronensäure)*
citrate-cleavage enzyme citratspaltendes Enzym *n*, ATP-Citratlyase *f*
citrate-condensing enzyme citratkondensierendes Enzym *n*, Citratsynthase *f*, Citrogenase *f*
citric acid Citronensäure *f*, *2-Hydroxypropan-1,2,3-tricarbonsäure *f*
citric-acid cycle Citronensäurezyklus *m*, Citratzyklus *m*, Tricarbonsäurezyklus *m*, (Szent-Györgyi-)Krebs-Zyklus *m*
citric acid fermentation Citronensäuregärung *f*, Citronensäurefermentation *f*
city gas Stadtgas *n*
clad *v* plattieren *(Metall)*; *(i. w. S.)* verkleiden, umhüllen, überziehen
cladding 1. Plattieren *n* *(von Metall)*; *(i. w. S.)* Verkleiden *n*, Umhüllen *n*, Überziehen *n*; 2. Plattierung *f*, Plattierüberzug *m*, *(i. w. S.)* Verkleidung *f*, Umhüllung *f*, Überzug *m*, Auflage *f*
Claisen condensation Claisen-Kondensation *f*, claisensche Kondensation *f* [Esterkondensation *f*]
Claisen distillation flask Claisen-Kolben *m*, Claisen-Destillierkolben *m*
Claisen distilling flask Claisen-Kolben *m*, Claisen-Destillierkolben *m*
Claisen flask Claisen-Kolben *m*, Claisen-Destillierkolben *m*
Claisen rearrangement (Oxo-)Claisen-Umlagerung *f* *(O-Allyl-C-Allyl-Orthoumlagerung)*
clamp *v* festklemmen, einspannen
clamp Haltevorrichtung *f*, Halter *m*, Klemmvorrichtung *f*, Klemme *f*, Zwinge *f*, Klammer *f*, Schelle *f*, Feststellvorrichtung *f*; Quetschhahn *m*
Clapeyron-Clausius equation clausius-clapeyronsche Gleichung *f*
clarificant Klär(hilfs)mittel *n*
clarification 1. Klären *n*, Klärung *f*, Abklärung *f*, Klarifikation *f*; Abschlämmen *n*; Läutern *n*, Läuterung *f*; Klärfiltration *f*, Blankfiltration *f*, Polierfiltration *f*, Feinfiltration *f*; Reinigen *n*, Reinigung *f*; Raffinieren *n*, Raffination *f*; 2. *(Hyd)* (mechanische) Klärung *f*
clarification-flocculation *(Hyd)* Flockung *f* und Sedimentation *f* *(in einem Arbeitsgang)*
clarification plant *(Hyd)* Kläranlage *f*, KA, Absetz(klär)anlage *f*, Klärwerk *n*

clarification sludge *(Hyd)* Klärschlamm *m*, Absetzschlamm *m*, Sedimentationsschlamm *m*
clarifier 1. Klär(hilfs)mittel *n*; 2. Klärvorrichtung *f*, Klarifikator *m*, Klärseparator *m*, Klärapparat *m*, Klärgefäß *n*; 3. *(Hyd)* Klärbecken *n*, Absetzbecken *n*
clarifier-thickener *(Hyd)* Kläreindicker *m*
clarify *v* (ab)klären; abschlämmen; reinigen; raffinieren; sich (ab)klären, klar werden; sich reinigen
clarifying agent Klär(hilfs)mittel *n*
clarifying centrifuge Klärzentrifuge *f*
clarifying filter Klärfilter *n*, Blankfilter *n*, Polierfilter *n*, Feinfilter *n*
clarity Klarheit *f*, Durchsichtigkeit *f*, Lichtdurchlässigkeit *f*; Reinheit *f*
Clark degree *s.* English degree of hardness
class *v s.* classify
class Klasse *f (z. B. chemischer Verbindungen)*
class [a] metal ion A-Metallion *n*, A-Kation *n*
class [b] metal ion B-Metallion *n*, B-Kation *n*
class of compounds Verbindungsklasse *f*; Stoffgruppe *f*
class of crystal symmetry Kristallklasse *f*, Symmetrieklasse *f*, Punkt(symmetrie)gruppe *f*
class of substances Stoffklasse *f*, Substanzklasse *f*; Stoffgruppe *f*
classical distribution *(phys Ch)* klassische Verteilung *f*, *(i. e. S.)* Boltzmann-Verteilung *f*
classification 1. Klassifizierung *f*, Klassifikation *f*, Einteilung *f*, Unterteilung *f*, Eingruppierung *f*, Einstufung *f*, Einordnung *f*; 2. *(Tech)* Klassieren *n*, Klassierung *f*
classifier Klassierer *m*, Klassierapparat *m*, Klassiergerät *n*
classify *v* 1. klassifizieren, einteilen, unterteilen, eingruppieren, einstufen, einordnen; 2. *(Tech)* klassieren
classifying chamber Sichtkammer *f*
clathrate (compound) Einschlussverbindung *f*, *(i. e. S.)* Clathrat *n*, Clathratverbindung *f*, Käfigeinschlussverbindung *f*
Claude process Claude-Verfahren *n* *(Gewinnung von NH_3 aus der Luft)*
Claus plant Claus-Anlage *f (zur Schwefelgewinnung)*
clausius Clausius *n (SI-fremde Einheit der Entropie; 1 Cl = 4,1868 J/K)*
Clausius-Clapeyron equation clausius-clapeyronsche Gleichung *f*
clay *v* 1. mit Ton [Lehm] behandeln; 2. durch Ton filtrieren
clay 1. Ton *m*; Lehm *m*; Bleicherde *f (zum Raffinieren)*; 2. *(Pap)* Kaolin *m*, Schlämmkaolin *m*, geschlämmte Porzellanerde *f*
clay catalyst Bleicherdekontakt *m*
clay combustion boat Glühschiffchen *n* [Verbrennungsschiffchen *n*, Schiffchen *n*] aus Ton
clay contact *s.* clay catalyst
clay crucible Tontiegel *m*
clay mineral Tonmineral *n*
clay treating [treatment] *(Erdöl)* Bleicherdebehandlung *f*, Bleicherderaffination *f*, Behandlung *f* mit Bleicherde
clayey tonhaltig, tonig; tonartig; lehmhaltig, lehmig; lehmartig
clayish tonhaltig, tonig; tonartig; lehmhaltig, lehmig; lehmartig
clayware Tonware *f*
clean *v* **(up)** 1. reinigen, säubern *(z. B. Oberflächen)*; 2. *(Hyd)* reinigen
clean bench *(Bio)* Sterilbank *f*, Sterilbox *f*
clean box *(Anal)* reiner Arbeitsplatz *m (von gefilterter Luft durchströmt)*
clean fuel umweltfreundlicher Brennstoff *m*
clean gas Reingas *n*
clean room *(Anal)* Reinraum *m*, reiner Raum *m (von gefilterter Luft durchströmt)*
clean room technique *(Lab)* Reinraumtechnik *f*
clean-up Reinigen *n*, Reinigung *f*, Säubern *n*, Säuberung *f*
clean water Reinwasser *n*
cleaned gas Reingas *n*
cleaner 1. Reiniger *m*, Reinigungseinrichtung *f*, Reinigungsanlage *f*; 2. Reinigungsmittel *n*, Reiniger *m*
cleaning agent Reinigungsmittel *n*, Reiniger *m*
cleaning mixture *(Lab)* Chromschwefelsäure *f*, Dichromatschwefelsäure *f*
cleanliness Reinheit *f*, Sauberkeit *f*; Reinheitsgrad *m*

cleanse *v* reinigen, säubern *(Oberflächen)*; reinigen *(Gase)*
cleanser Reiniger *m*, Reinigungsmittel *n*
cleansing agent Reinigungsmittel *n*, Reiniger *m*
cleanup *s.* clean-up
clear *v* 1. klären; reinigen; 2. aufhellen, bleichen
clear filtrate Klarfiltrat *n*
clear lacquer Klarlack *m*, Lack *m (physikalisch trocknend)*
clear opening lichte Weite *f*, Maschenweite *f (eines Siebs)*
clear varnish Klarlack *m*, Lack *m*, *(i. e. S.)* oxidativ trocknender Klarlack *m* [Lack *m*]
clear water Klarwasser *n*
clearance 1. *(Tech)* freier Raum *m*, Spielraum *m*, Spiel *n*, Luft *f*; relativer Schadraum *m (an der Hubkolbenmaschine)*; 2. *(Med)* Clearance *f (Maß für die Ausscheidung rückresorbierbarer Blutinhaltsstoffe)*
clearcoat *(Anstr)* Überlack *m*, Glanzschutzlack *m*, Klarlack *m*
cleavability *(Min)* Spaltbarkeit *f*
cleavable *(Min)* spaltbar
cleavage 1. Spalten *n*, Spaltung *f*, Aufspalten *n*, Aufspaltung *f*; Abspalten *n*, Abspaltung *f*; Aufblättern *(von Schichtstoffen)*; 2. *(Min)* Bruch *m*; Spaltbarkeit *f*; Richtung *f* der Spaltebene; 3. *(Geol)* Schieferung *f*
cleavage by ozone Ozon(id)spaltung *f*, Ozonolyse *f*
cleavage face [plane] Spaltfläche *f*, Spaltebene *f*
cleavage product Spaltprodukt *n*
cleavage reaction Spaltungsreaktion *f*
cleave *v* (auf)spalten; abspalten; sich (auf)spalten; sich abspalten; aufblättern *(Schichtstoffe)*
cleave *v* **off** abspalten *(z. B. Atome von Molekülen)*; sich abspalten
Clemmensen method [reduction] Clemmensen-Reduktion *f*, Clemmensen-Methode *f (Carbonyl-Methylen-Reduktion)*
click chemistry *(org Ch)* Click-Chemie *f*
click reaction *(org Ch)* Click-Reaktion *f*
climatic chamber Klimakammer *f*
climbing-film evaporator Kletterfilmverdampfer *m*
clinching *(Tech)* Druckfügen *n*, Durchsetzfügen *n*
clinical chemistry klinische Chemie *f*
clinker 1. *(Bau)* Klinker *m*; 2. Schlacke *f*; 3. Hammerschlag *m (Eisen(II,III)-oxid)*
clinkering Schlackenbildung *f*, Verschlackung *f*
clip Klemmvorrichtung *f*, Klemme *f*, Klammer *f*, Haltevorrichtung *f*, Halter *m*, Zwinge *f*, Feststellvorrichtung *f*
clock glass Uhrglas *n*, Uhrglasschale *f*
clock reaction Zeitreaktion *f*
clog *v* **(up)** 1. verstopfen, zusetzen, versetzen, verlegen; sich verstopfen [zusetzen]; 2. (ver)klumpen, klumpig werden, sich zusammenballen, Klumpen bilden
cloning *(Biot)* Klonierung *f*
close boiling *(Dest)* engsiedend
close-grain(ed) feinkörnig, Feinkorn..., dicht *(Gefüge)*
close-packed *(Krist)* dicht gepackt
close packing *(Krist)* dichteste Kugelpackung *f* [Packung *f*], d. P.
closed-cell foam geschlossenzelliger Schaum(kunst)stoff *m*
closed-chain geschlossenkettig, mit geschlossener [ringförmiger, cyclischer] Kette
closed circuit geschlossener Kreislauf *m*
closed-cycle system plant Anlage *f* mit geschlossenem Kreislauf
closed-loop control system Regelkreis *m*, Regelsystem *n*
closed mixer geschlossener Mischer *m* [Kneter *m*], Innenmischer *m*
closed shell abgeschlossene Schale *f*, vollbesetztes Niveau *n*
closely graded engklassiert
closing cap Abschlusskappe *f*
closing of a ring Ringschluss *m*, Zyklisierung *f*
closing valve Abschlussventil *n*
clot *v* (ver)klumpen, klumpig werden, Klumpen bilden, sich zusammenballen, ausflocken, gerinnen, koagulieren
clot Klumpen *m*, Klümpchen *n*; Koagulat *n*, Koagulum *n*, Gerinnsel *n*
cloth Gewebe *n*, Stoff *m*, Tuch *n*, Zeug *n*
cloth filter Tuchfilter *n*, Stofffilter *n*, Gewebefilter *n*
cloth-tube collector Schlauch(kam-

mer)filter *n*, Schlauchabscheider *m (zur Gasreinigung)*
clotting Klumpen *n*, Verklumpen *n*, Gerinnen *n*, Koagulieren *n*, Koagulation *f*
cloud *v* matt [glanzlos] werden
cloud 1. Wolke *f*; Nebel *m*; 2. *(Ch)* wolkiger Niederschlag *m*
cloud chamber (wilsonsche) Nebelkammer *f*, Wilson-Kammer *f*
cloud point Trübungspunkt *m*
cloud track *(phys Ch)* Nebelspur *f*
cloudiness Trübung *f*, Trübheit *f*, Trübsein *n*, Trübe *f*
cloudy wolkig *(Niederschlag)*
cloverleaf arrangement [conformation] *(Bioch)* Kleeblattstruktur *f (der tRNA)*
Clusius column Clusius-Trennrohr *n (zur Trennung gasförmiger Isotope)*
cluster 1. *(Krist)* Cluster *m*, Subkeim *m*, Unterkeim *m*, Agglomerat *n (Atom- oder Molekülverband)*; 2. *(Hyd)* Cluster *m (Aggregation aus Wassermolekülen)*
clustering 1. *(Krist, Hyd)* Clusterbildung *f*; 2. *(Krist)* Nahentmischung *f*
CM cellulose CM-Cellulose *f*, Carboxymethylcellulose *f*
CMC CM-Cellulose *f*, Carboxymethylcellulose *f*
c.m.c. *s.* critical micelle concentration
CMP *s.* capacitively coupled microwave plasma
CNF *s.* carbon nanofibre
CNS-depressant *(Pharm)* Neuroleptikum *n (das Zentralnervensystem dämpfendes Mittel)*
CNT *s.* carbon nanotube
CO_2 fixation CO_2-Fixierung *f*
CO shift *(Carb)* CO-Konvertierung *f*, Kohlenmonoxid-Konversion *f*
coacervate Koazervat *n*
coacervation Koazervation *f*, Koazervierung *f*
coagel Koagel *n*
coagulability Koagulierbarkeit *f*, Gerinnbarkeit *f*
coagulable koagulierbar, gerinnbar
coagulant 1. Koagulans *n*, Koagulator *m*, Koagulationsmittel *n*, Flockungsmittel *n*; 2. *(Hyd)* Metallsalzflockungsmittel *n*, „klassisches" Flockungsmittel *n*, Koagulant *m*
coagulate *v* 1. koagulieren, gerinnen lassen, zur Ausflockung bringen; 2. *(Hyd)* (aus)flocken *(Wasserinhaltsstoffe)*; 3. *(Hyd)* flocken *(Roh- und Abwasser)*; 4. *(Text)* fällen; koagulieren, (aus)flocken, gerinnen
coagulate Koagulat *n*
coagulating agent *s.* coagulant
coagulation 1. Koagulation *f*, Ausflockung *f*, Gerinnung *f*; 2. *(Hyd)* Flockung *f*, Ausflockung *f (von Wasserinhaltsstoffen mit Flockungsmitteln)*; *(i. e. S.)* Entstabilisierung *f*; 3. *(Text)* Fällung *f*, Fällen *n*; 4. *s.* coagulation treatment; 5. Koagulat *n*
coagulation aid *(Hyd)* Flockungshilfsmittel *n*, Flockungsbeschleuniger *m*
coagulation-filtration *(Hyd)* Koagulationsfiltration *f*, Flockungsfiltration *f*
coagulation point Gerinnungspunkt *m*
coagulation treatment *(Hyd)* Flockung(sbehandlung) *f (von Roh- und Abwasser)*
coagulation value Flockungswert *m*
coagulator *s.* coagulant
coagulatory Koagulator..., Koagulations..., Ausflockungs...
coagulum Koagulat *n*
coal Kohle *f*, *(i. e. S.)* Steinkohle *f*
coal carbonization Kohlenverkokung *f*, Kohlenentgasung *f*, *(i. e. S.)* Steinkohlenverkokung *f*, Steinkohlenentgasung *f*; Kohlen(ver)schwelung *f*, *(i. e. S.)* Steinkohlen(ver)schwelung *f*
coal char Halbkoks *m*
coal chemical Kohle(n)wertstoff *m*
coal chemistry Kohlechemie *f*, Carbochemie *f*
coal coke Steinkohlenkoks *m*
coal conversion Kohle(um)wandlung *f*, Kohleveredlung *f*
coal-derived kohlestämmig
coal-derived chemical product *s.* coal chemical
coal desulphurization Kohleentschwefelung *f*
coal dressing Kohle(n)aufbereitung *f*, *(i. e. S.)* Steinkohle(n)aufbereitung *f*
coal dust Kohlenstaub *m*
coal extraction Kohleextraktion *f*
coal fines Feinkohle *f*
coal gas Kohlengas *n*, *(i. e. S.)* Steinkohlengas *n*
coal gasification Kohle(n)vergasung *f*

coal hydrogenation Kohle(n)hydrierung *f*, *(i. e. S.)* Steinkohle(n)hydrierung *f*
coal-like kohle(n)ähnlich, kohleartig
coal liquefaction Kohle(n)verflüssigung *f*
coal-ore briquette Kohle-Erz-Brikett *n*
coal preparation Kohle(n)aufbereitung *f*, *(i. e. S.)* Steinkohlenaufbereitung *f*
coal product Kohlefolgeprodukt *n*
coal pyrolysis Kohlepyrolyse *f*
coal refinery carbochemische [kohlechemische] Raffinerie *f*, Kohleraffinerie *f*
coal slurry Kohle(n)brei *m*, Kohle(n)maische *f*
coal tar Kohlenteer *m*, *(i. e. S.)* Steinkohlenteer *m*
coal-tar dye Teerfarbstoff *m*, Teerfarbe *f*
coal-tar pitch Kohlenteerpech *n*, *(i. e. S.)* Steinkohlen(teer)pech *n*
coal water gas Kohlenwassergas *n*
coalesce *v* sich vereinigen, miteinander verschmelzen, zusammenfließen, koal(is)ieren, koaleszieren
coalescence Koaleszenz *f*, Vereinigung *f*, Verschmelzung *f*, Zusammenfließen *n (von Flüssigkeitströpfchen)*
coalescer Flüssigkeitsabscheider *m*, Absetzer *m (für Flüssigkeitströpfchen)*
coarse grob, Grob...; grobteilig
coarse adjustment Grobeinstellung *f*
coarse-crystalline grobkristallin
coarse filter *(Hyd)* Grobfilter *n*
coarse-grained 1. grobkörnig; grobkristallin; grobfas(e)rig; 2. *(Gerb)* grobnarbig
coarse grinding Grobmahlen *n*, Grob(aus)mahlung *f*, grobe Mahlung *f* [Ausmahlung *f*]
coarse-mesh(ed) grobmaschig, weitmaschig
coarse-to-fine filtration Mehrschichtfiltration *f (über gekörntes Filtermaterial mit abnehmendem Korndurchmesser)*
coarsely disperse(d) grobdispers
coarseness Grobkörnigkeit *f*
coat *v* beschichten
coat *v* **onto a support** *(Lab)* auftragen
coat Schicht *f*, Schutzschicht *f*; Anstrich *m*
coat protein Hüllprotein *n*, Virushüllprotein *n*, Kapsid *n*
coated beschichtet • **become coat** sich überziehen
coated fabric gestrichenes [beschichtetes, kaschiertes] Gewebe *n*
coater *s.* coating machine
coating 1. Beschichten *n*; Anstreichen *n*; 2. Schicht *f*, Schutzschicht *f*; Anstrich *m*; Deckschicht *f (spontan entstanden)*; 3. Beschichtungsstoff *m*, Beschichtungsmaterial *n*, Schutzschichtstoff *m*; Anstrichstoff *m*
coating calender Kalander *m* zum Belegen von Geweben
coating machine Auftragmaschine *f*
coating metal Schutzschichtmetall *n*, Beschichtungsmetall *n*
coating of rubber Gummibelag *m*, Gummierung *f*
coating resin Lackharz *n*
coating system 1. Beschichtungssystem *n*, Schutzschichtsystem *n*, Schichtsystem *n*; 2. *(Anstr)* Anstrichsystem *n*, Anstrichaufbau *m*
coating technique *(Anstr)* Beschichtungstechnik *f*
coating technology *(Anstr)* Beschichtungstechnologie *f*
coating vehicle Anstrichbindemittel *n*
coaxial cylinder viscometer Koaxialzylinderviskosimeter *n*
cobalamine Corrinoid *n*, Corrin *n*, Cobalamin *n*, Verbindung *f* der Vitamin-B_{12}-Gruppe, *(i. e. S.)* Cyanocobalamin *n*, Cobalamin *n*, Vitamin *n* B_{12}
cobalt Co *Cobalt *n*
cobalt blue Kobaltblau *n*, Kobaltultramarin *n*, Thénards Blau *n (Cobaltaluminat)*
cobalt catalyst Cobaltkatalysator *m*, Cobaltkontakt *m*, Co-Kontakt *m*
cobalt drier *(Anstr)* Kobalttrockner *m*, Kobalttrockenstoff *m*, Kobaltsikkativ *n*
cobalt green Kobaltgrün *n*, Rinmans Grün *n*
cobalt monoxide Cobalt(mon)oxid *n*, Cobalt(II)-oxid *n*
cobalt yellow Kobaltgelb *n*, Aureolin *n*, Indischgelb *n (Kaliumhexanitrocobaltat)*
cobaltammine Cobaltammin *n*, *(i. e. S.)* Cobalt(III)-ammin *n*, Cobaltiak *n*
cobaltate Cobaltat *n*
cobaltic Cobalt..., *(i. e. S.)* Cobalt(III)-...
cobaltic ammine Cobalt(III)-ammin *n*

cobaltic oxide Cobalt(III)-oxid *n*, Dicobalttrioxid *n*, Cobalttrioxid *n*
cobaltic sulphate Cobalt(III)-sulfat *n*
cobalticyanide Cyanocobaltat(III) *n*, Hexacyanocobaltat(III) *n*
cobaltocyanide Cyanocobaltat(II) *n*, Hexacyanocobaltat(II) *n*
cobaltous Cobalt..., *(i. e. S.)* Cobalt(II)-...
cobaltous acetate Cobalt(II)-acetat *n*
cobaltous ammine Cobalt(II)-ammin *n*
cobaltous chelate Cobalt(II)-chelat *n*
cobaltous chloride Cobalt(II)-chlorid *n*, Cobaltdichlorid *n*
cobaltous fluorosilicate Cobalt(II)-hexafluorosilicat *n*
cobaltous nitrate Cobalt(II)-nitrat *n*
cobaltous oxide Cobalt(II)-oxid *n*, Cobaltmonoxid *n*
cobaltous sulphate Cobalt(II)-sulfat *n*
coca alkaloid Cocaalkaloid *n*
cocaine Cocain *n*, Kokain *n*, Benzoylmethylecgonin *n*, Benzoylecgoninmethylester *m*
cocatalysis Cokatalyse *f*
cocatalyst Cokatalysator *m*
cochromatograph *v* mitlaufen lassen; mitlaufen
cock Hahn *m*, Hahnventil *n*
cock plug Hahnkegel *m*, Hahnküken *n*
cocondensation *(Kst)* Mischkondensation *f*, Cokondensation *f*
co-current im Gleichstrom [Parallelstrom], nach dem Gleichstromprinzip [Parallelstromprinzip], im Gleichstrom [Parallelstrom] geführt
co flow Gleichstrom *m*, Parallelstrom *m*
co operation Gleichstromfahrweise *f*
COD *(chemical oxygen demand)* chemischer Sauerstoffbedarf *m*, CSB
C.O.D. *s.* chemical oxygen demand
code *v* codieren, determinieren, bestimmen *(Aminosäurefolge)*
code (of life) *(Bioch)* (genetischer) Code *m*
codeine Codein *n*, Morphinmethylether *m*
co-deposit *v* gemeinsam abscheiden, mitabscheiden; sich gemeinsam abscheiden
co-deposition Mitabscheidung *f*
coding triplet *s.* codon
codistillant *(Dest)* Zusatzstoff *m*, Zusatzkomponente *f*
codistillation Destillation *f* mit Zusatzstoff(en), Codestillation *f*, *(i. e. S.)* Trägerdampfdestillation *f*
codon *(Bioch)* Codon *n*, Codeeinheit *f*, Codetriplett *n*, Basentriplett *n*
coefficient of absorption Absorptionskoeffizient *m*
coefficient of cubic(al) expansion kubischer [räumlicher] Ausdehnungskoeffizient *m*, Volumenausdehnungskoeffizient *m*
coefficient of diffusion Diffusionskoeffizient *m*, Diffusionskonstante *f*
coefficient of elongation linearer thermischer Ausdehnungskoeffizient *m*
coefficient of hardness Härtezahl *f*
coefficient of heat transfer Wärmeübergangszahl *f*, Wärmetransmissionskoeffizient *m*
coefficient of linear expansion linearer Ausdehnungskoeffizient *m*
coefficient of mass transfer Stoffübergangszahl *f*
coefficient of solubility Löslichkeitskoeffizient *m*, Löslichkeitskonstante *f*
coefficient of thermal conduction [conductivity] Wärmeleitzahl *f*, (spezifisches) Wärmeleit(ungs)vermögen *n*, Wärmeleitfähigkeit *f*
coefficient of viscosity Viskositätskoeffizient *m*, Viskositätskonstante *f*, Konstante *f* der inneren Reibung
coenzymatic coenzymatisch
coenzyme Coenzym *n*, Wirk(ungs)gruppe *f*, prosthetische [aktive] Gruppe *f*
coesite Coesit *n (Hochdruckmodifikation von* SiO_2*)*
coexist *v* koexistieren, nebeneinander bestehen *(z. B. zwei Phasen)*
coexistence Koexistenz *f (z. B. zweier Phasen)*
coexistence curve Binodal-Kurve *f*, binodale Kurve *f*
coextract *v* gemeinsam extrahieren
cofactor *(Biot)* Cofaktor *m (Enzymtechnik)*
coffeine *s.* caffeine
Coffey still Rektifizierapparat *m*, Rektifikationsapparat *m*
coflocculation gemeinsames Fällen *n*
cohere *v* zusammenhalten, kohärieren
coherence Zusammenhalt *m*, Kohärenz *f*

coherency Zusammenhalt *m*, Kohärenz *f*
coherent kohärent
coherent anti-Stokes Raman spectroscopy kohärente anti-Stokes-Raman-Spektroskopie *f*, CARS
coherent radiation kohärente Strahlung *f*
cohesion Kohäsion *f*
cohesion pressure *(phys Ch)* Kohäsionsdruck *m*, Binnendruck *m*
cohesive kohäsiv, zusammenhaltend, Kohäsions...
cohesive energy density Kohäsionsenergiedichte *f*, kohäsive Energiedichte *f*, CED
cohesive force Kohäsionskraft *f*
cohobate *v (Dest)* kohobieren
cohobation *(Dest)* Kohobation *f*
coil 1. Rohrschlange *f*, Schlangenrohr *n*, Schlange *f*, Rohrspirale *f*, Spirale *f*; 2. Knäuel *n (z. B. von Fadenmolekülen)*
coiled-coil Superhelix *f (Feinstruktur von Polypeptidketten)*
coiled molecule geknäueltes Molekül *n*, Knäuelmolekül *n*, Molekülknäuel *n*
coiled tube *s.* coil 1.
coiled-tube condenser *(Dest)* Schlangenkühler *m*
coiled tubular (heat) exchanger Schlangenrohr-Wärmeübertrager *m*
coiling Knäuelung *f (z. B. einer Molekülkette)*
coincide *v* zusammenfallen *(zeitlich oder räumlich)*; koinzidieren
coincidence Koinzidenz *f*, zeitliches [räumliches] Zusammenfallen *n*
coincident koinzident, zeitlich [räumlich] zusammenfallend; einander deckend
co-ion *(phys Ch)* Coion *n*
coke *v* verkoken, entgasen; (ver)koken, zu Koks werden, entgasen
coke Koks *m*, *(i. e. S.)* Hochtemperaturkoks *m*
coke-contaminated koksbeladen, mit Koks beladen *(Katalysator)*
coke gasification Koksvergasung *f*
coke number Verkokungszahl *f*, Verkokungswert *m*
coke oven Kokereiofen *m*, Verkokungsofen *m*, Kokungsofen *m*, Koksofen *m*
coke-oven battery Koks(ofen)batterie *f*, Verkokungsbatterie *f*
coke-oven benzol Kokereibenzol *n*
coke-oven gas Koks(ofen)gas *n*, Kokereigas *n*
coke-oven tar Koksofenteer *m*, Kokereiteer *m*
coke plant waste (water) Kokereiabwasser *n*, Abwasser *n* der Kokereien
coke producer Koksgenerator *m*
coke residue Verkokungsrückstand *m*, Koksrückstand *m*, Kohlenstoffrückstand *m (beim Verkokungstest nach Conradson)*
coking 1. Verkokung *f*, Verkoken *n*, Entgasung *f*, Entgasen *n*; 2. *(Erdöl)* Verkokung *f*, Kracken *n* auf Koks *(beim thermischen Kracken)*
coking chamber 1. Verkokungskammer *f*, Koks(ofen)kammer *f*; 2. *(Erdöl)* Verkokungskammer *f*, Kokskammer *f (beim Kracken auf Koks)*
coking coal Kokskohle *f*, Verkokungskohle *f*, *(i. w. S.)* kokende [verkokbare] Kohle *f*
coking still Kokungsdestillationsanlage *f*, Verkokungsblase *f*
col. *s.* colourless
colation Durchseihen *n*, Abseihen *n*, Kolieren *n*
colature Seihflüssigkeit *f*, Kolatur *f*
cold-draw *v* 1. *(Met)* kaltziehen; 2. *(Kst)* kalt(ver)strecken, kaltrecken; 3. *(Lebm)* kaltpressen, kaltschlagen *(z. B. Öle)*
cold-press *v* kaltpressen, kaltverpressen; kaltpressen, kaltschlagen *(z. B. Öle)*
cold-vulcanize *v* kaltvulkanisieren
cold-work *v* kaltverarbeiten
cold 1. kalt; 2. nicht (radio)aktiv, aktivitätsfrei
cold bath Kältebad *n (eines Gefriertrocknungsapparats)*
cold brine Kühlsole *f*
cold-cathode ionization gauge Kaltkatoden-Ionisationsvakuummeter *n*, Penning-Vakuummeter *n*, Philips-Vakuummeter *n*
cold-curing *(Anstr, Kst)* kalthärtend, kaltabbindend; kaltvulkanisierend
cold curing *(Anstr, Kst)* Kalthärtung *f*; Kaltvulkanisation *f*
cold drawing 1. *(Met)* Kaltziehen *n*; 2. *(Kst)* Kalt(ver)strecken *n*, Kaltrecken; 3. *(Lebm)* Kaltpressen *n*, Kaltschlagen *n (von Ölen)*

cold-finger (condenser) *(Dest)* Kühlfinger *m*, kalter Finger *m*, Einhängekühler *m*
cold-finger trap *(Dest)* Kühlfinger *m*, kalter Finger *m*, Einhängekühler *m*
cold flow *(Kst)* kalter Fluss *m*, kaltes Fließen *n*
cold forming Kalt(um)formen *n*, Kaltformgebung *f*
cold galvanizing 1. elektrochemisches [galvanisches] Verzinken *n*; 2. Kaltverzinken *n (mit zinkstaubpigmentierten Anstrichstoffen)*
cold hardening Kalthärten *n*
cold-insoluble in der Kälte unlöslich
cold light kaltes Licht *n*
cold moulding 1. *(Kst)* Kaltpressen *n*; 2. *(Kst)* Kaltformteil *n*
cold-moulding compound [material] Kaltpressmasse *f*
cold neutron kaltes [unterthermisches] Neutron *n*
cold polymer Tieftemperaturpolymerisat *n*, Tieftemperaturpolymer(e) *n*
cold polymerization Tieftemperaturpolymerisation *f*, Kaltpolymerisation *f*
cold pressing Kaltpressen *n*
cold resistance Kältebeständigkeit *f*, Tieftemperaturbeständigkeit *f*, Kältefestigkeit *f*
cold-resistant kältebeständig, tieftemperaturbeständig, kältefest
cold rolling Kaltwalzen *n*
cold rubber Kaltkautschuk *m*, Tieftemperaturkautschuk *m*, kalt polymerisierter Kautschuk *m*, Coldrubber *m*
cold-setting *(Anstr, Kst)* kalthärtend, kaltabbindend
cold test Kälte(beständigkeits)prüfung *f*
cold trap Kühlfalle *f*
cold vulcanizate Kaltvulkanisat *n*
cold vulcanization Kaltvulkanisation *f*
cold working 1. Kaltverarbeitung *f*; 2. *s.* cold forming
coliform bacteria Colibakterien *npl*
collagen Kollagen *n (Eiweißstoff)*
collect *v* sammeln, auffangen *(Destillat, Gase)*; sich (an)sammeln, sich anhäufen
collect *v* **samples** Proben entnehmen
collecting electrode 1. Niederschlagselektrode *f*, Sammelelektrode *f*; 2. *s.* collector 2.
collecting main Sammelleitung *f*
collection Sammeln *n*, Sammlung *f*; Ansammlung *f (z. B. von Erdöl)*
collection chamber *(Biot)* Sammelraum *m*, Sammelbehälter *m*, Maischesammelraum *m*, Maischesammler *m*
collection efficiency Abscheidegrad *m*, Ausscheidungsgrad *m*, Entstaubungsgrad *m*
collection of samples Proben(ent)nahme *f*, Probenahme *f*
collection vessel Sammelgefäß *n*, Sammelbehälter *m*, Auffanggefäß *n*
collective floatation kollektive Flotation *f*, Kollektivflotation *f*
collector 1. Sammler *m*, Kollektor *m (bei der Flotation)*; Abscheider *m*, Entstauber *m*; Abgasreiniger *m*; Abluftreiniger *m*; 2. Kollektor *m*, Kollektorelektrode *f*, Sammelelektrode *f*, Auffangelektrode *f*, Fangelektrode *f (bei Röhren, Transistoren)*; 3. Probenehmer *m*, Probenahmegerät *n*, Probenahmevorrichtung *f*
collide *v* zusammenstoßen, zusammentreffen, kollidieren
colligative property kolligative Eigenschaft *f (nur von der Teilchenanzahl abhängig)*
collimating lens *(Spektr)* Kollimatorlinse *f*
collision Zusammenstoß *m*, Stoß *m*, Zusammentreffen *n*, Zusammenprall *m*, Kollision *f*
collision activation *(Spektr)* Stoßaktivierung *f*
collision broadening Stoßverbreiterung *f*, Druckverbreiterung *f*, Lorentz-Verbreiterung *f (von Spektrallinien)*
collision complex Stoßkomplex *m*
collision cross section Stoßquerschnitt *m*, Streuquerschnitt *m*
collision excitation Stoßanregung *f*
collision frequency 1. Stoßfrequenz *f (eines einzelnen Moleküls)*; 2. Stoßhäufigkeit *f*, Stoßzahl *f* (je Zeiteinheit), Stoßgeschwindigkeit *f*, Stoßfrequenz *f (vieler Moleküle)*
collision-induced reaction stoßinduzierte Reaktion *f*
collision ionization Stoßionisation *f*, Stoßionisierung *f*

collision number *s.* collision frequency 2.
collision of the first kind Stoß *m* erster Art, anregender [aktivierender] Stoß *m*
collision of the second kind Stoß *m* zweiter Art, desaktivierender Stoß *m*
collision rate *s.* collision frequency
collodion cotton Kollodiumwolle *f*, Kolloxylin *n (niedrignitrierte Cellulose)*
collodion membrane Kollodiummembran *f*
colloid *s.* colloidal
colloid Kolloid *n*
colloid-chemical kolloidchemisch
colloid chemist Kolloidchemiker *m*
colloid chemistry Kolloidchemie *f*
colloid-disperse kolloiddispers
colloid mill Kolloidmühle *f*
colloidal kolloid(al), fein verteilt, kolloiddispers, Kolloid...
colloidal dispersion kolloide Dispersion *f*
colloidal gold kolloidales Gold *n*
colloidal particle Kolloidteilchen *n*, Kolloidpartikel *n*, kolloides Teilchen *n*
colloidal silver kolloidales Silber *n*
colloidal solution kolloide Lösung *f*, Kolloidlösung *f*
colloidal stability Kolloidstabilität *f*
colloidal sulphur kolloider Schwefel *m*, Kolloidschwefel *m*
colloidally dispersed kolloiddispers
colophony Kolophonium *n*, Geigenharz *n (von Pinus-Arten)*
color *v s.* colour
color *(Am)* 1. Farbe *f (Qualität der Lichtempfindung)*; 2. Färbung *f (Zustand)*; 3. Farbstoff *m*; 4. Pigment *n*
colorant *s.* colouring agent [matter]
coloration 1. Färbung *f (Erscheinungsbild)*; 2. Färben *n*
colorimeter Kolorimeter *n*, Tintometer *n*
colorimetric kolorimetrisch
colorimetric analysis kolorimetrische Analyse *f* [Bestimmung *f*]
colorimetry Kolorimetrie *f*
coloring *s.* colouring
coloristic farblich
colorless *s.* colourless
colour *v* färben; sich (ver)färben, Farbe annehmen • **appreciably coloured** deutlich gefärbt
colour 1. Farbe *f (Qualität der Lichtempfindung)*; 2. Färbung *f (Zustand)*; 3. Künstlerfarbe *f (Pigment)*; 4. *s.* colouring agent [matter]
colour base Farb(stoff)base *f*, Färbebase *f*
colour-bearing group chromophore [farbtragende, farbgebende] Gruppe *f*, Chromophor *m*, Farbträger *m*
colour change Farbänderung *f*, Farbwechsel *m*, *(bei plötzlichem Eintreten:)* Farbumschlag *m*
colour change interval Umschlagsintervall *n*, Umschlagsgebiet *n*, Umschlagsbereich *m (eines pH-Indikators)*
colour fastness Farbechtheit *f*
colour filter Farbfilter *n*
colour indicator Farbindikator *m*
colour-indicator paper Indikatorpapier *n*
colour lake *(Anal, Text)* Farblack *m*
colour reaction Farbreaktion *f*
colour reagent Farbstoffreagens *n*
colour salt Farbsalz *n*
colour tint Farbtönung *f*
Colour Index Colour Index *m*, CI *(von der Society of Dyers and Colourists und der American Association of Textile Chemists and Colorists herausgegebenes Nachschlagewerk für Handelsfarbstoffe)*
colourant *s.* colouring agent [matter]
colouration *s.* coloration
colouration glass farbiges Glas *n*, Farbglas *n*, Buntglas *n*
colouration lake *(Anal, Text)* Farblack *m*
colouring 1. Färben *n*; 2. *s.* colouring agent
colouring agent [matter] färbender [farbgebender] Stoff *m*, Färbemittel *n*, Farbmittel *n*
colouring strength Farbintensität *f*
colourless farblos, farbfrei, nicht farbig
colourless oil of vitriol hochkonzentrierte Schwefelsäure *f (enthält 93 bis 98 % H_2SO_4)*
colourlessness Farblosigkeit *f*
column 1. *(Dest, Tech)* Kolonne *f*, Säule *f*; 2. *(Anal)* Säule *f*; 3. *(Krist)* Stängel *m*
column chromatography Säulenchromatographie *f*, SC • **by column chromatography** säulenchromatographisch

column distillation Kolonnendestillation *f*
column hold-up *(Dest)* Betriebsinhalt *m* der Kolonne
column of fluid [liquid] Flüssigkeitssäule *f*
column packing 1. Säulenfüllung *f*, Säulenpackung *f*; 2. *(Dest)* Kolonnenfüllung *f*
column reactor *(Biot)* Säulenreaktor *m*, Biosäule *f*
column temperature *(Anal)* Säulentemperatur *f*, Temperatur *f* der Säule *(Parameter bei chromatographischen Methoden)*
columnar *(Krist)* säulenförmig, stängelförmig, stängelkristallinisch, palisadenartig, kolumnar
columnar crystal Stängelkristall *m*
comb attenuator *(Spektr)* Kammblende *f*
combination Verbindung *f*, Vereinigung *f*, Kombination *f*
combination band Kombinationsschwingungsbande *f*
combination reaction Kombinationsreaktion *f*, Vereinigungsreaktion *f*
combination vibration Kombinationsschwingung *f*
combine *v* 1. zusammensetzen, kombinieren; 2. *(Ch)* verbinden, vereinigen; sich verbinden [vereinigen], eine Verbindung eingehen
combine *v* **with ... to form ...** sich unter Bildung von ... mit ... verbinden
combined (chemisch) gebunden • **in combined form** gebunden, in gebundener Form • **in the combined state** gebunden, in gebundener Form
combined chlorine *(Hyd)* gebundenes Chlor *n*
combined water chemisch gebundenes Wasser *n*
combining ability [capacity] *s.* combining power
combining force Bindungskraft *f*, Bindekraft *f*, bindende Kraft *f*
combining power Bindungsvermögen *n*, Bindungsfähigkeit *f*, Bindungskraft *f*
combining weight (relative) Äquivalentmasse *f*
combustibility Verbrennbarkeit *f*, Brennbarkeit *f*, Verbrennlichkeit *f*
combustible (ver)brennbar, verbrennlich
combustible(s) Brennbares *n*, Verbrennliches *n*
combustion Verbrennen *n*, Verbrennung *f* • **sustain combustion** die Verbrennung unterhalten
combustion analysis *(org Ch)* quantitative Elementaranalyse *f*, Verbrennungsanalyse *f*
combustion boat *(Lab)* Glühschiffchen *n*, Schiffchen *n*
combustion chamber Verbrennungskammer *f*, Verbrennungsraum *m*, Brennkammer *f*, Brennraum *m*, Feuerraum *m*
combustion fuel Brennstoff *m*
combustion furnace Verbrennungsofen *m*
combustion gas Verbrennungsgas *n*
combustion of sludge *(Hyd)* Schlammverbrennung *f*
combustion rate Verbrennungsgeschwindigkeit *f*
combustion reaction Verbrennungsreaktion *f*
combustion residue Verbrennungsrückstand *m*
combustion train Verbrennungsapparat *m*, Elementaranalysenapparat *m*
combustion tube Verbrennungsrohr *n*, Verbrennungsröhre *f*
come *v* **down** ausfallen *(als Niederschlag)*
come *v* **over** übergehen, überdestillieren
come *v* **to the boil** zu kochen [sieden] beginnen
coming-up time Anwärmzeit *f*, Anheizzeit *f*, Aufheizzeit *f*
commercial 1. handelsüblich, Handels...; 2. technisch, industriell • **bring to the commercial stage** zur Betriebsreife entwickeln, fabrikreif gestalten
commercial analysis technische Analyse *f*
commercial catalyst technischer Katalysator *m*
commercial chemist Handelschemiker *m*
commercial fermenter *(Biot)* Industriefermenter *m*
commercial fertilizer Handelsdünger *m*

commercial grade handelsübliche Qualität *f*, Handelsqualität *f*
commercial name Handelsname *m*, kommerzieller Name *m*
commercial plant Produktionsanlage *f*
commercial process technisches Verfahren *n*
commercial reactor Produktionsreaktor *m*, Betriebsreaktor *m*, technischer Reaktor *m*
commercial-scale reactor *s.* commercial reactor
commercial spirit technischer Alkohol *m*, Industriealkohol *m*
commercial synthesis technische Synthese *f*, technisch brauchbare [verwertbare, erfolgreiche] Synthese *f*
commercial waste 1. gewerbliche Abprodukte *npl*; 2. *(Hyd)* gewerbliches Abwasser *n*, Abwasser *n* gewerblicher Betriebe, Abwasser *n* von Gewerbebetrieben
commercially available im Handel erhältlich, handelsüblich
commercially pure handelsüblich
comminute *v* (fein) zerkleinern, pulverisieren
comminution Feinzerkleinern *n*, Zerkleinern *n*
commodity chemicals *in großem Umfang hergestellte Chemikalien*; *etwa: Grundchemikalien*
commodity plastic 1. Massenkunststoff *m*; 2. Standardkunststoff *m*
common electrical connection galvanische Kopplung *f*
common ion effect Wirkung *f* gleichioniger Zusätze, Löslichkeitsverminderung *f*, Löslichkeitsverringerung *f*, Aussalzeffekt *m*
common name 1. Trivialname *m*; 2. Gruppenbezeichnung *f (für Schädlingsbekämpfungsmittel)*
common salt Kochsalz *n*
comonomer Comonomer(e) *n*, Gegenmonomer(e) *n*
compact *v* kompaktieren, (press)verdichten, zusammenpressen; sich verfestigen
compact 1. kompakt, pressdicht; dicht; 2. *(Geol)* massiv
compact Pressling *m*, Presskörper *m*, Pressstück *n*, Pulverpressling *m*, Pulverpresskörper *m*, Metallpulverpressling *m (in der Pulvermetallurgie)*
compacting Verdichten *n*, *(i. e. S.)* Kompaktieren *n*, Pressverdichten *n*; Komprimieren *n*, Pressen *n*, Zusammenpressen *n*; Verfestigen *n*
compaction Verdichten *n*, *(i. e. S.)* Kompaktieren *n*, Pressverdichten *n*; Komprimieren *n*, Pressen *n*, Zusammenpressen *n*; Verfestigen *n*
companion Begleiter *m*, Begleitstoff *m*, Begleitsubstanz *f*
companion alkaloid [base] Nebenalkaloid *n*, Nebenbase *f*, Begleitalkaloid *n*
companion material [substance] *s.* companion
comparator Komparator *m (Kolorimetrie)*
comparison compound Vergleichsverbindung *f*
comparison electrode Vergleichselektrode *f*, Bezugselektrode *f*
comparison line *(Spektr)* Bezugslinie *f*
comparison solution Vergleichslösung *f*
comparison spectrum Vergleichsspektrum *n*
compartment *v* unterteilen
compartment 1. Kammer *f*; Zelle *f*; 2. *(Bioch)* Kompartiment *n*, Zellkompartiment *n*, Stoffwechselraum *m (einer Zelle)*
compartment dryer Kammertrockner *m (mit mehreren Kammern)*
compartment(aliz)ation *(Bioch)* Kompartimentierung *f*
compatibility Verträglichkeit *f*, Kompatibilität *f*
compatibilizer Verträglichkeitsmacher *m*, Verträglichkeitsvermittler *m*
compatible verträglich, kompatibel
compensation Ausgleich *m*, Kompensation *f*
compensation of errors Fehlerkompensation *f*, Fehlerausgleich *m*
compensation pH-meter Kompensations-pH-Meter *n*
compensator Kompensator *m*
competing reaction *s.* competitive reaction
competition *s.* competitive reaction
competitive inhibition kompetitive [konkurrierende] Hemmung *f*, Ver-

drängungshemmung *f*, Konkurrenzhemmung *f (einer Enzymreaktion)*
competitive inhibitor *(Bioch)* kompetitiver Inhibitor *m* [Hemmstoff *m*]
competitive reaction Konkurrenzreaktion *f*, Parallelreaktion *f*
complementariness Komplementarität *f*
complementarity Komplementarität *f*
complementary colour Komplementärfarbe *f*
complementary DNA *(Bioch)* komplementäre DNA *f*, cDNA
complete analysis Vollanalyse *f*
complete biological treatment *(Hyd)* vollbiologische Reinigung *f*
complete combustion vollständige Verbrennung *f*
complete deionization *(Hyd)* Vollentsalzung *f*
complete mixing vollständige Vermischung *f* [Durchmischung *f*]
complete precipitation vollständige Fällung *f*
complete reaction Gesamtreaktion *f*
complete softening *(Hyd)* Vollenthärtung *f*
complete treatment *(Hyd)* Vollreinigung *f*
completely mixed fermenter *(Biot)* vollständig gemischter Fermenter *m*
completely mixed reactor *s.* 1. completely mixed tank; 2. completely mixed fermenter
completely mixed tank *(Hyd)* Rühr(kessel)reaktor *m*, Mischreaktor *m*, ideal durchmischter Rührkessel *m*, vollständiges Mischbecken *n*
completing step letzte Stufe *f (einer Reaktion)*
complex *v* einen Komplex bilden, komplex [im Komplex] binden; in einen Komplex überführen, komplexieren
complex 1. Komplex *m*, komplexe Gruppe *f*; 2. Assoziat *n (aus mehreren Molekülen)*
complex compound Komplexverbindung *f*, Koordinationsverbindung *f*
complex-formation constant Komplexbildungskonstante *f*, Komplexstabilitätskonstante *f*
complex-forming komplexbildend
complex ion komplexes Ion *n*, Komplexion *n*
complex lipid komplexes Lipid *n*, Lipoid *n*
complex salt Komplexsalz *n*, komplexes Salz *n*
π-complex π-Komplex *m*
σ-complex σ-Komplex *m*
1:1 complex 1:1-Komplex *m*
2:1 complex 2:1-Komplex *m*
complexation Komplexbildung *f*; Komplexierung *f*
complexation analysis [titration] Komplexbildungstitration *f*, komplexometrische Titration *f*
complexing agent Komplexbildner *m*, komplexbildender Stoff *m*
complexing catalyst Komplexkatalysator *m*, komplexkoordinativer Katalysator *m*
complexometric komplexometrisch
complexometric titration *s.* complexation analysis
complexometry Komplexometrie *f*
component am Aufbau beteiligt
component Komponente *f*, Bestandteil *m*, Anteil *m*; Komponente *f (z. B. die Molekülart als Teil eines Systems)*
component glyceride gemischtes [gemischtsäuriges] Glycerid *n*
composite Verbund(werk)stoff *m*
composite reaction komplexe [zusammengesetzte] Reaktion *f*
composited sample *(Anal)* Sammelprobe *f*
composition 1. Zusammensetzung *f*; 2. *(Gum)* Mischung *f*
compositional analysis qualitative Analyse *f*
compound *v* 1. *(Gum)* mischen, eine Mischung herstellen; ein Mischungsrezept aufbauen [aufstellen]; 2. compoundieren, fetten *(Öl)*
compound 1. *(Ch)* Verbindung *f*; 2. Kompound *n*, Kompoundmasse *f (bituminöser Isolierstoff)*; 3. *(Gum)* Mischung *f*, Mischungsrezept *n*
compound fertilizer Mehrnährstoffdünger *m*, Kombinationsdünger *m*
compound lipid komplexes Lipid *n*, Lipoid *n*
compound substance zusammenge-

setzter Stoff *m*, zusammengesetzte Substanz *f*
compounding 1. *(Gum)* Mischen *n*, Mischungsherstellung *f*; Rezeptaufstellung *f*, Aufstellung *f* eines Mischungsrezepts; 2. Compoundieren *n (von Ölen)*
compress *v* komprimieren, zusammenpressen, zusammendrücken, verdichten
compressed air Druckluft *f*, Pressluft *f*
compressed air aeration Druckbelüftung *f (Abwasser)*
compressed-air supply Druckluftversorgung *f*
compressibility Kompressibilität *f*, Zusammendrückbarkeit *f*, Komprimierbarkeit *f*, Verdichtbarkeit *f*, Pressbarkeit *f*
compressibility coefficient (kubischer) Kompressibilitätskoeffizient *m*
compressibility factor Kompressibilitätsfaktor *m*
compressible kompressibel, zusammendrückbar, komprimierbar, verdichtbar, pressbar
compression Kompression *f*, Komprimieren *n*, Zusammenpressen *n*, Zusammendrücken *n*, Pressen *n*, Verdichten *n*, Verdichtung *f*
compression moulding *(Kst)* Formpressen *n*, Pressen *n*
compression-moulding material Formpressstoff *m*, Pressmasse *f*
compression-moulding resin Pressharz *n*
compression refrigerating machine Kompressionskältemaschine *f*
compressive modulus Kompressionsmodul *m*
compressive strength Druckfestigkeit *f*
compressor Verdichter *m*, Kompressor *m*
comproportionation Komproportionierung *f*, Symproportionierung *f*
Compton effect Compton-Effekt *m*
Compton (recoil) electron Compton-Elektron *n*, Rückstoßelektron *n*
Compton scattering Compton-Streuung *f*
conc. *s.* 1. concentration; 2. concentrated
concatenation Verkettung *f*, Verknüpfung *f (von Reaktionen)*
concd. *s.* concentrated
concentrate *v* 1. einengen, eindicken *(Lösungen)*; anreichern, konzentrieren *(Erze)*; 2. bündeln *(Strahlen)*
concentrate *v* **by evaporation** eindampfen
concentrate *v* **to small bulk** einengen
concentrate 1. Konzentrat *n*; 2. *(Aufb)* Aufbereitungskonzentrat *n*
concentrated konzentriert, k, konz.
concentrated liquor Dicklauge *f*
concentrated sludge *(Hyd)* Dickschlamm *m*, eingedickter Schlamm *m*
concentration 1. Einengen *n*, Eindicken *n (einer Lösung)*; Anreichern *n*, Anreicherung *f*, Konzentrierung *f (von Erz)*; Bündelung *f (von Strahlen)*; 2. Konzentration *f*, *(manchmal auch)* Gehalt *m*
concentration cell *(Ech)* Konzentrationskette *f*, Konzentrationszelle *f*, Konzentrationselement *n*
concentration-cell corrosion konzentrationsbedingte Korrosion *f*, (örtliche) Korrosion *f* durch Konzentrationsketten
concentration dependence Konzentrationsabhängigkeit *f*
concentration-dependent konzentrationsabhängig
concentration gradient Konzentrationsgefälle *n*, Konzentrationsgradient *m*
concentration limits 1. Konzentrationsgrenzwerte *mpl*, Konzentrationsgrenze *f*; 2. *s.* discharge limits
concentration of active constituents Wirkstoffkonzentration *f*
concentration overpotential [overvoltage] Konzentrationsüberspannung *f*
concentration polarization Konzentrationspolarisation *f*
concentrator 1. Eindicker *m*, Eindickapparat *m*, Eindickzylinder *m*, Eindickmaschine *f*, Entwässerungsmaschine *f*, Entwässerungsanlage *f*; 2. *(Aufb)* Anreicher(ungs)gerät *n*, Anreicherungsapparat *m*, Anreicherungsmaschine *f*, Anreicherungsanlage *f*
concentric-tube heat exchanger Dop-

pelrohrwärmeübertrager *m*, Doppelrohrwärmeaustauscher *m*
concerted inhibition konzertierte [multivalente] Hemmung *f* [Endprodukthemmung *f*]
concerted transition konzertierter Übergang *m*, Alles-oder-Nichts-Übergang *m (Enzymkinetik)*
conchoidal *(Min)* musch(e)lig *(z. B. Bruch)*
concn. *s.* concentration
concrete Beton *m*
concrete aggregate Betonzuschlag(-stoff) *m*
concrete disintegration Betonzerfall *m*
concurrent flow Parallelstrom *m*, Gleichstrom *m*
concurrent reaction Konkurrenzreaktion *f*, Parallelreaktion *f*; Nebenreaktion *f*, Begleitreaktion *f*
condensability Kondensierbarkeit *f*, Niederschlagbarkeit *f (von Gasen, Dämpfen)*
condensable kondensierbar, niederschlagbar *(Gase, Dämpfe)*
condensate 1. Kondensat *n*, Niederschlag *m*; Kondenswasser *n*, Niederschlagswasser *n*, Schwitzwasser *n*; 2. *s.* condensation product
condensation 1. Kondensation *f*, Kondensierung *f*, Niederschlagung *f (von Gasen, Dämpfen)*; 2. *(org Ch)* Kondensation *f*; 3. *(Lebm)* Eindampfung *f*, Eindickung *f*, Kondensation *f*
condensation curve *(Dest)* Kondensationskurve *f*, Taukurve *f*, Kondensationslinie *f*, Taulinie *f*
condensation nucleus Kondensationskeim *m*, Kondensationskern *m*
condensation polymer Kondensationspolymer(e) *n*, Polykondensat *n*
condensation polymerization Kondensationspolymerisation *f*, Polykondensation *f*
condensation product *(org Ch)* Kondensationsprodukt *n*, Kondensat *n*
condensation reaction Kondensationsreaktion *f*
condensation resin Kondensationsharz *n*
condense *v* 1. kondensieren, niederschlagen *(Gase, Dämpfe)*; (sich) kondensieren, sich niederschlagen [verdichten]; 2. *(org Ch)* kondensieren; 3. *(Lebm)* eindampfen, eindicken, kondensieren
condense *v* **together** (miteinander) kondensieren *(Ringe)*
condensed-ring system kondensiertes Ringsystem *n*
condenser 1. *(Tech)* Kondensatorkühler *m*, Kühler *m*; 2. *(Lab)* Kühler *m*; Verflüssiger *m*, Kondensator *m (einer Kältemaschine)*; Sublimiervorlage; 3. Kondensator *m (Elektrotechnik)*; 4. Kondensor *m (Optik)*
condition *v* 1. konditionieren *(Stoffe vor der Weiterverarbeitung)*; 2. klimatisieren *(Räume)*
condition Bedingung *f*; Zustand *m*
conditional acceptable daily intake *(Tox)* bedingt duldbare Tagesdosis *f*
conditioner 1. Zusatzstoff *m (zur besseren Handhabung einer Chemikalie)*; *(i. e. S.)* Umhüllungsmittel *n*, Füllstoff *m*, Trägerstoff *m*, Lockerungsmittel *n*; 2. Konditionierapparat *m*; Klimagerät *n*
conduct *v* 1. leiten *(Elektrizität, Wärme)*; 2. durchführen *(Versuche)*
conductance (elektrische) Leitfähigkeit *f*, *(quantitativ)* elektrischer Leitwert *m*, Wirkleitwert *m*, Konduktanz *f*; Wärmeleitfähigkeit *f*
conduction Leitung *f (von Elektrizität oder Wärme)*
conduction band Leitungsband *n*, Leitfähigkeitsband *n*, L-Band *n*
conduction drying Kontakttrocknen *n*, Kontakttrocknung *f*, Berührungstrocknen *n*, Berührungstrocknung *f*
conduction electron Leitungselektron *n*
conductive leitfähig, leitend
conductive polymer leitfähiges Polymer *n*
conductivity 1. (spezifische elektrische) Leitfähigkeit *f (Materialkenngröße)*; 2. *s.* conductivity of heat
conductivity bridge Leitfähigkeitsmessbrücke *f*
conductivity cell Leitfähigkeits(mess)-zelle *f*
conductivity of heat (spezifische) Wärmeleitfähigkeit *f*, (spezifisches) Wärmeleitvermögen *n*, *(quantitativ auch)* Wärmeleitzahl *f*

conductometer Leitfähigkeitsmesser *m*, Leitfähigkeitsmessgerät *n*
conductometric titration konduktometrische Titration *f*, Leitfähigkeitstitration *f*
conductometry Konduktometrie *f*
conductor Leiter *m*, *(i. e. S.)* (elektrischer) Leiter *m*, Elektrizitätsleiter *m*, Stromleiter *m*
conductor of electricity (elektrischer) Leiter *m*, Elektrizitätsleiter *m*, Stromleiter *m*
conductor of heat Wärmeleiter *m*
cone Kegel *m*, Konus *m*; Kern *m*, Schliffkern *m (einer Kegelschliffverbindung)*
cone-and-plate viscometer Konus-Platte-Viskosimeter *n*
cone classifier Klassierkegel *m*, (konische) Klärspitze *f*, Spitztrichter *m*
cone-shaped kegelförmig, keg(e)lig, konusartig, konisch
confidence interval Vertrauensbereich *m*, Vertrauensintervall *n*
confidence limit Vertrauensgrenze *f*
configuration Konfiguration *f*, Atomkonfiguration *f*, Atomanordnung *f*; Elektronenkonfiguration *f*, Elektronenanordnung *f*
configuration retention Konfigurationserhalt *m*, Retention *f* (der Konfiguration)
configurational formula Konfigurationsformel *f*, Raumformel *f*, Stereoformel *f*
configurational isomer Konfigurationsisomer(e) *n*
configurational isomerism Konfigurationsisomerie *f*
confirmatory test Nachweis *m*
conformation 1. Konformation *f*, Konstellation *f (Stereochemie)*; 2. *(Bioch)* Konformation *f (Oberbegriff für Tertiär- und Quartärstruktur der Proteine)*
conformational konformer, Konformations...
conformational equilibrium Konformationsgleichgewicht *n*
conformational formula Konformationsformel *f*, Konstellationsformel *f*
conformational isomerism Konformationsisomerie *f*
conformer Konformationsisomer(e) *n*, Konformer(e) *n*
congeal *v* erstarren, fest [hart, starr] werden; gerinnen *(eiweißhaltige Flüssigkeiten)*; gefrieren *(wässrige Lösungen)*; erstarren lassen, fest [hart, starr] werden lassen; gerinnen lassen; gefrieren lassen
congelation Erstarren *n*, Festwerden *n*; Gerinnen *n (eiweißhaltiger Flüssigkeiten)*; Gefrieren *n (wässriger Lösungen)*; Erstarrenlassen *n*; Gerinnenlassen *n*; Gefrierenlassen *n*
conglomerate Konglomerat *n*
Congo red Kongorot *n*
Congo rubin(e) Kongorubin *n*
Congo rubine number Rubinzahl *f (zur Messung der Schutzkolloidwirkung)*
congregate *v* sich zusammenlagern
congruent transition kongruenter Übergang *m* [Phasenübergang *m*]
conical flask Erlenmeyer-Kolben *m*
conical ground (glass) joint Kegelschliffverbindung *f*
conical mixer Konusmischer *m*, Kegelmischer *m*
conical-spherical adapter Übergangsstück *n* mit Normschliffhülse [NS-Hülse] und Kugelschliffball [KS-Ball]; Übergangsstück *n* mit Kugelschliffschale [KS-Schale] und Normschliffkern [NS-Kern]
coniferous wood *(Pap)* Nadelholz *n*
coni(i)ne Coniin *n*, 2-Propylpiperidin *n*
coning and quartering *(Anal)* Aufkegeln *n* und Vierteln *n*, Vierteln *n*, Viertelung *f*, Quartieren *n*
coning plume kegelförmige [konische] Rauchfahne *f*
conjugate 1. konjugierte Verbindung *f*; 2. *(Bioch)* Paarling *m*
conjugate acid-base pair konjugiertes [korrespondierendes] Säure-Base-Paar *n*
conjugate group prosthetische Gruppe *f (Eiweißchemie)*
conjugated 1. konjugiert *(Doppelbindung)*; mit konjugierten Doppelbindungen *(Molekül)*; 2. *(Bioch)* konjugiert, gepaart *(mit einer anderen chemischen Verbindung zur besseren Ausscheidbarkeit)*; zusammengesetzt, konjugiert *(Eiweißstoff)*

conjugated double bond konjugierte Doppelbindung *f*
conjugated protein zusammengesetztes [konjugiertes] Protein *n*, Proteid *n*
conjugation 1. Konjugation *f (zweier Doppelbindungen)*; 2. *(Bioch)* Konjugation *f*, Paarung *f (zweier chemischer Verbindungen zur besseren Ausscheidbarkeit)*
connect *v* **to** anschließen an
connecting peptide *(Bioch)* C-Peptid *n*, Verbindungspeptid *n*
connection 1. Verbindung *f*, Anschluss *m*; 2. *s.* connector
connector (adapter) Zwischenstück *n*, Verbindungsstück *n*
conodal curve Binodal-Kurve *f*, binodale Kurve *f*
conode Gleichgewichtslinie *f*, Ko(n)node *f (im Gleichgewichtsdiagramm)*
Conrad-Limpach synthesis conrad-limpachsche Synthese *f (von 4-Hydroxychinolin)*
conrotatory konrotatorisch
consecutive reaction Folgereaktion *f*
conservation law Erhaltungssatz *m*
conservation of energy Energieerhaltung *f*, Erhaltung *f* der Energie
conservation of mass [matter] Massenerhaltung *f*, Erhaltung *f* der Masse
conserve *v* erhalten, aufrechterhalten
consistence *s.* consistency
consistency Konsistenz *f*
consistent konsistent
consolidate *v* verfestigen, verdichten *(Schüttgüter)*
consolidation Verfestigung *f*, Verdichtung *f (von Schüttgütern)*
consolute point kritischer Lösungspunkt *m* [Mischungspunkt *m*] *(bezüglich der Zusammensetzung)*
consolute temperature kritische Lösungstemperatur *f* [Mischungstemperatur *f*], kritischer Lösungspunkt *m* [Mischungspunkt *m*], *(i. e. S.)* obere kritische Lösungstemperatur *f*
constant konstant • **hold constant** konstant halten • **keep [maintain] constant** konstant halten • **with constant shaking** unter ständigem Schütteln
constant Konstante *f*, konstante Größe *f*
constant-boiling konstant siedend, konstantsiedend
constant-boiling(-point) mixture azeotropes Gemisch *n*
constant error systematischer Fehler *m*
constant-temperature bath Temperierbad *n*, Thermostat *m*
constant-temperature chamber Thermostat *m*
constant-temperature transformation Umwandlung *f* bei gleich bleibender Temperatur, isotherme Umwandlung *f*
constant-volume heat capacity Wärmekapazität *f* bei konstantem Volumen
constant weight Massekonstanz *f*; Gewichtskonstanz *f*
σ-constant Substituentenkonstante *f*
constituent konstituierend, am Aufbau beteiligt
constituent Bestandteil *m*, Teil *m*, Anteil *m*; Inhaltsstoff *m*
constitution Konstitution *f*; Anordnung *f*, Bau *m*, Beschaffenheit *f*
constitution diagram Zustandsdiagramm *n*, Phasendiagramm *n*, Zustandsschaubild *n*
constitutional formula Konstitutionsformel *f*, Strukturformel *f*, Bauformel *f*
constitutional repeating unit Strukturelement *n*, sich wiederholende Struktureinheit *f*, WSE *(von Polymeren)*
constitutional water konstitutiv gebundenes Wasser *n*, Konstitutionswasser *n*
constitutive property konstitutive Eigenschaft *f*
constriction 1. Verengung *f*, Einschnürung *f*; 2. Einschnürung(sstelle) *f*, Drosselstelle *f*
consume *v* verbrauchen; aufzehren, aufbrauchen
consumption Verbrauch *m*
contact *v* berühren, mit etwas in Kontakt kommen; in Kontakt bringen, andrücken, anlegen; sich berühren, miteinander in Kontakt kommen
contact Kontakt *m* • **out of contact with air** unter Luftabschluss
contact adhesive Kontaktklebstoff *m*, Kontaktkleber *m*, Haftkleber *m*
contact angle Kontaktwinkel *m*, Randwinkel *m*, Benetzungswinkel *m*

contact area Berührungsfläche *f*, Kontaktfläche *f*
contact catalysis Kontaktkatalyse *f*, Oberflächenkatalyse *f*, heterogene Katalyse *f*
contact catalyst Kontaktkatalysator *m*, heterogener Katalysator *m*, Kontakt *m*
contact corrosion Kontaktkorrosion *f*, Berührungskorrosion *f (durch Kontakt zweier Metalle)*; galvanische Korrosion *f*
contact drying Kontakttrocknen *n*, Kontakttrocknung *f*, Berührungstrocknen *n*, Berührungstrocknung *f*
contact ion pair Kontaktionenpaar *n*, inneres Ionenpaar *n*
contact poison Kontaktgift *n*, Berührungsgift *n*
contact process Kontaktverfahren *n*, *(i. e. S.)* Schwefelsäurekontaktverfahren *n*
contact sulphuric acid Kontaktschwefelsäure *f*, Kontaktsäure *f*
contact thermometer Kontaktthermometer *n*
contact time Kontaktzeit *f*, Einwirkungszeit *f*, Reaktionszeit *f*
contactor Kontaktor *m*, Extraktor *m*, Extrakteur *m*, Extraktionskolonne *f*, Extraktionsapparat *m*, Extraktionsmaschine *f (zur Solventextraktion)*; Kontaktor *m*, Kontaktautoklav *m*
container Gefäß *n*, Behälter *m*; (großer) Transportbehälter *m*, Container *m*
contaminant 1. Schmutzstoff *m*, Verschmutzung *f*, Verunreinigung *f*, Kontaminant *m*; 2. *(Hyd)* Wasserinhaltsstoff *m*, *(i. e. S.)* Schadstoff *m*, schädlicher Wasserinhaltsstoff *m*, Wasserschadstoff *m*
contaminant load(ing) 1. Schmutz(stoff)last *f*, Schmutz(stoff)belastung *f*, *(i. e. S.)* Schadstoffbelastung *f*; 2. *(Hyd)* Abwasserlast *f*, Abwasserbelastung *f*, Belastung *f* mit Abwasserinhaltsstoffen
contaminant microorganisms *(Biot)* Infektionsmikroorganismen *mpl*, Kontaminanten *mpl*
contaminate *v* verunreinigen, verschmutzen, kontaminieren
contamination 1. Verunreinigung *f*, Verschmutzung *f*, Kontamination *f*; 2. *(Biot)* Infektion *f (durch fremde Mikroorganismen)* • **have a high risk of contamination** 1. sehr kontaminationsgefährdet sein; 2. *(Biot)* sehr infektionsgefährdet sein
content Gehalt *m*
content of active component Wirkstoffgehalt *m*
continuity equation Massenbilanzgleichung *f*, Kontinuitätsgleichung *f*
continuous kontinuierlich [stetig] arbeitend [betrieben] *(Gerät)*; kontinuierlich, stetig *(Betrieb)*; anhaltend, ständig *(z. B. Schütteln)* • **by a continuous process** in laufender Betriebsanordnung, im Fließbetrieb [Dauerbetrieb]
continuous casting *(Tech)* Stranggießen *n*, Strangguss *m*
continuous culture *(Hyd, Biot)* kontinuierliche Kultur *f*, Dauerkultur *f (von Mikroorganismen)*
continuous development *(Chromat)* Durchlaufentwicklung *f*, Durchlauftechnik *f*
continuous dialysis Fließdialyse *f*
continuous distillation kontinuierliche Destillation *f*
continuous electrode Dauerelektrode *f*
continuous equilibrium vaporization (kontinuierliche) Entspannungsdestillation *f*, geschlossene [integrale] Destillation *f*, Gleichgewichtsdestillation *f*
continuous fermenter *(Biot)* kontinuierlicher Reaktor *m*, Durchflussfermenter *m*, Fermenter *m* für kontinuierliches Verfahren
continuous fibre [filament] Elementarfaden *m*, Endlosfaser *f*, Filament *n*
continuous-flow chromatography Durchlaufchromatographie *f*
continuous-flow reactor Durchflussreaktor *m*, kontinuierlicher Reaktor *m*
continuous-flow stirred-tank reactor kontinuierlicher Rührkesselreaktor *m*, Durchflussmischreaktor *m*
continuous furnace Durchlaufofen *m*, Ofen *m* für durchlaufenden [kontinuierlichen] Betrieb, kontinuierlicher [kontinuierlich arbeitender] Ofen *m*
continuous heat resistance Dauerwärmebeständigkeit *f*
continuous operation 1. kontinuierlicher [stetiger, ununterbrochener] Betrieb *m*, kontinuierliche Arbeitsweise *f*

[Fahrweise f], Fließbetrieb *m*, Dauerbetrieb *m*; 2. *(Farb)* Kontinuebetrieb *m*
continuous phase zusammenhängende [geschlossene] Phase *f*, Dispersionsphase *f*
continuous process kontinuierliches [ununterbrochenes] Verfahren *n*, Durchlaufverfahren *n*, Kontinueverfahren *n*
continuous reactor 1. kontinuierlicher Reaktor *m*, Durchflussreaktor *m*; 2. s. continuous fermenter
continuous rectification stetige [kontinuierliche] Gegenstromdestillation *f* [Rektifikation f]
continuous spectrum kontinuierliches Spektrum *n*
continuous stirred tank Durchflussrührkessel *m*
continuous stirred-tank fermenter *(Biot)* kontinuierlicher Rührkesselreaktor *m*, Durchflussrühr(kessel)fermenter *m*
continuous transition Umwandlung *f* [Phasenumwandlung f] 2. Ordnung, Phasenübergang *m* 2. Ordnung, kontinuierliche Umwandlung *f*, kontinuierlicher Übergang *m*, *(bei Polymeren auch)* Glasumwandlung *f*, Glasübergang *m*, Gamma-Umwandlung *f*, α-Anomalie *f*
continuous-tubular-flow reactor Strömungsrohrreaktor *m*, Rohrreaktor *m*, Strömungsrohr *n*
continuous-wave laser kontinuierlicher Laser *m*
continuous-wave technique *(Spektr)* Continuous-wave-Technik *f*, CW-Technik *f*, Dauerstrichbetrieb *m*, variable Frequenzmethode *f*
continuous working *s.* continuous operation
continuum Kontinuum *n*
continuum source *(Anal)* Kontinuumstrahler *m*, Kontinuumlampe *f*, Kontinuumlichtquelle *f*
contour length Konturlänge *f*, Gesamtlänge *f (einer Molekülkette)*
contraceptive antikonzeptionell, empfängnisverhütend, schwangerschaftsverhütend
contraceptive antikonzeptionelles [empfängnisverhütendes, schwangerschaftsverhütendes] Mittel *n*, Antikonzipiens *n*, Antikonzeptivum *n*
contract *v* zusammenziehen, kontrahieren; sich zusammenziehen, (ein-)schrumpfen, schwinden, sich verkleinern, kleiner werden
contraction Kontraktion *f*, Zusammenziehung *f*, Schrumpfung *f*, Einschrumpfung *f*, Schwindung *f*, Schwund *m*
contrast enhancement *(Spektr)* Kontrastverstärkung *f*, Kontrastverbesserung *f*
contribute *v* beisteuern, liefern; beitragen, beteiligt sein, teilhaben
contributing structure mesomere Grenzstruktur *f*
control *v* 1. lenken, steuern, leiten, kontrollieren, zügeln *(eine chemische Reaktion)*; steuern; regeln; 2. bekämpfen *(z. B. Schädlinge, Korrosion)*
control 1. Lenkung *f*, Steuerung *f*, Leitung *f*, Zügelung *f (einer chemischen Reaktion)*; Steuerung *f*; Regelung *f*; Kontrolle *f*; 2. Bekämpfung *f (z. B. von Schädlingen, Korrosion)*; 3. Vergleichsprobe *f* • **get out of control** außer Kontrolle geraten
control agent Schädlingsbekämpfungsmittel *n*; Pflanzenschutzmittel *n*
control volume Bilanzraum *m (Energiebilanz)*
controllability Steuerbarkeit *f (eines chemischen Reaktors)*
controlled atmosphere 1. kontrollierte Atmosphäre *f (z. B. eines Ofens)*; (kontrollierte) Schutzgasatmosphäre *f*, Schutzatmosphäre *f*; 2. Prüfgas *n*
controlled biosynthesis *(Biot)* gesteuerte [gelenkte] Biosynthese *f*
controlled-potential electrolysis *(Anal)* Elektrolyse *f* bei kontrolliertem Potenzial
controller Steuergerät *n*, Steuereinrichtung *f*, Steuerung *f*; Regeleinrichtung *f*, Regelanlage *f*, Regler *m*; Stellglied *n*
convection Konvektion *f*
convection dryer Konvektionstrockner *m*
convection heating Konvektionsheizung *f*
convective konvektiv

convective mixing Konvektions(ver)mischen *n*
conversion 1. Umwandlung *f*, Überführung *f*, Konvertierung *f*, Konversion *f*, Umsetzung *f*; 2. Umrechnung *f*; 3. Übergang *m (in eine andere chemische Verbindung)*; 4. Umsatz *m (quantitativ)*
conversion curve Umsatzkurve *f*
conversion efficiency *(Biot)* Stoffwandlungsleistung *f*
conversion factor Umrechnungsfaktor *m*
conversion of materials Stoff(um)wandlung *f*, Stoffumsetzung *f*, stoffliche Umwandlung *f* [Umsetzung *f*]
conversion product 1. Umwandlungsprodukt *n*, Umsetzungsprodukt *n*; 2. *(Biot)* Fermentationsprodukt *n*
conversion reaction Umwandlungsreaktion *f*, Umsetzungsreaktion *f*, Konvertierungsreaktion *f*
conversion table Umrechnungstabelle *f*
conversion yield *(Biot)* Produkt(ions)-ausbeute *f*
convert *v* 1. umwandeln, überführen, umsetzen; sich umwandeln [umsetzen], übergehen *(in eine andere chemische Verbindung)*; 2. *(Met)* windfrischen, im Konverter verblasen, bessemern; 3. umrechnen • **be converted into** übergehen in
convert back rückumwandeln, zurückverwandeln, rück(ver)wandeln; sich rückumwandeln [zurückverwandeln], sich rück(ver)wandeln
converter 1. Reaktor *m*, Reaktionsofen *m*, Kontaktofen *m*, *(bei der Kohlenmonoxidhydrierung auch)* Hydrierungsofen *m*; 2. *(Met)* Konverter *m*
converter process Windfrischverfahren *n*, Konverter(frisch)verfahren *n*
converter steel Konverterstahl *m*, Windfrischstahl *m*, windgefrischter Stahl *m*
convertible 1. umwandelbar; 2. *(Anstr)* chemisch [durch chemische Reaktion] trocknend
conveying Fördern *n*
conveyor Fördergerät *n*, Förderanlage *f*, Fördereinrichtung *f*, Förderer *m*, *(i. e. S.)* Stetigförderer *m*
conveyor belt Förderband *n*, Fördergurt *m*, Transportband *n*
conveyor screw [worm] Förderschnecke *f*, Transportschnecke *f*, Schneckenförderer *m*
convulsant Konvulsionen [Schüttelkrämpfe] auslösend
convulsant Konvulsionen [Schüttelkrämpfe] auslösendes Mittel *n*
cook *v* 1. kochen; 2. *(Pap)* kochen, aufschließen *(in Gegenwart von Chemikalien)*; erkochen *(Zellstoff)*
cook *v* **thoroughly** durchkochen
cook 1. *(Pap)* Kochung *f*; 2. *(Pap)* Kocherinhalt *m*, Kochereintrag *m*, Kocherfüllung *f*, Kochgut *n*, Endkochgut *n*, Kochstoff *m*
cooker Kocher *m*
cooking acid *(Pap)* Kochsäure *f*, Sulfit(koch)säure *f*
cooking kettle Kochkessel *m*
cooking salt Kochsalz *n (grob)*
cooking temperature Kochtemperatur *f*
cool *v* (ab)kühlen; sich abkühlen, kalt werden, erkalten
cool *v* **down** herunterkühlen, abkühlen; sich abkühlen, kalt werden, erkalten
cool *v* **rapidly** abschrecken
coolant Kühlmittel *n*, Kältemittel *n*, Kühlmedium *n*
cooler *(Tech)* Kühler *m*, Kühlvorrichtung *f*
cooling Kühlen *n*, Abkühlen *n*; Abkühlung *f*, Kaltwerden *n*, Erkalten *n*
cooling agent *s.* coolant
cooling bath Kühlbad *n*
cooling brine Kälteträger *m*, *(i. e. S.)* Kältesole *f*, Kühlsole *f*, Sole *f*
cooling coil Kühlschlange *f*
cooling curve Abkühlungskurve *f*
cooling cycle Abkühlzyklus *m*
cooling jacket Kühlmantel *m*
cooling liquid Kühlflüssigkeit *f*
cooling loop Kühlwasserkreislauf *m*
cooling lubricant Kühlschmierstoff *m*
cooling rate Kühlgeschwindigkeit *f*
cooling tower Kühlturm *m*, Turmkühler *m*
cooling-water circuit [loop] Kühlwasserkreislauf *m*
cooperative *(Bioch)* kooperativ
cooperative behaviour Kooperativitätsverhalten *n*
cooperative diffusion coefficient kooperativer Diffusionskoeffizient *m*
cooperativity *(Bioch)* Kooperativität *f*,

kooperativer Substratbindungsmechanismus *m*
coordinate *v* koordinativ anlagern [binden] *(z. B. Moleküle)*; sich (koordinativ) anlagern, koordinativ gebunden werden
coordinate 1. koordinativ; koordinativ gebunden, komplexgebunden; 2. koordiniert *(Reaktion, Komplex)*
coordinate Koordinate *f*
coordinate bond koordinative [dative] Bindung *f*, Koordinationsbindung *f*, Donator-Akzeptor-Bindung *f*
coordinate repression koordinierte Repression *f*, Endprodukthemmung *f* *(von Enzymgruppen)*
coordination 1. Koordination *f*, koordinative Anlagerung *f* [Bindung *f*]; 2. *s.* coordinate bond; 3. *s.* coordination number
coordination catalysis Koordinationskatalyse *f*, Komplexkatalyse *f*
coordination catalyst komplexkoordinativer Katalysator *m*, Komplexkatalysator *m*
coordination chemistry Koordinationschemie *f*, Komplexchemie *f*, Koordinationslehre *f*
coordination complex Koordinationskomplex *m*, koordinierter Komplex *m*
coordination compound Koordinationsverbindung *f*, Komplexverbindung *f*
coordination lattice *(Krist)* Koordinationsgitter *n*
coordination number Koordinationszahl *f*, KZ, Zähligkeit *f*, koordinative Wertigkeit *f*
coordination polymerization Koordinationspolymerisation *f*, koordinative Polymerisation *f*
coordination sphere Koordinationssphäre *f*
copal (resin) Kopal *m* *(allg. für ein hochschmelzendes pflanzliches Harz)*
Cope elimination Cope-Eliminierung *f* *(Pyrolyse von Aminoxiden)*
Cope rearrangement *(org Ch)* Cope-Umlagerung *f*
copigment *(Bioch)* Copigment *n*
coplanar koplanar *(Bindungssystem)*
coplanarity Koplanarität *f* *(eines Bindungssystems)*
copolyaddition *(Kst)* Copolyaddition *f*
copolycondensation *(Kst)* Copolykondensation *f*
copolymer Copolymer(e) *n*, Copolymerisat *n*
copolymerization Copolymerisation *f*
copolymerize *v* copolymerisieren
copper *v* verkupfern
copper Cu Kupfer *n*
copper alloy Kupferlegierung *f*
copper amalgam electrode Kupferamalgamelektrode *f*
copper anode slime Kupfer-Anodenschlamm *m*
copper-bearing kupferhaltig
copper blossom Kupferblüte *f*
copper carbide Kupfer(I)-acetylid *n*
copper converter Kupfer(stein)konverter *m*
copper converting Kupfersteinverblasen *n*, Verblasen *n* von Kupferstein im Konverter, Kupferbessemern *n*
copper coulometer Kupfercoulometer *n*
copper dichloride Kupferdichlorid *n*, Kupfer(II)-chlorid *n*
copper dichromate Kupfer(II)-dichromat(VI) *n*
copper electrode Kupferelektrode *f*
copper emerald Kupfersmaragd *m*
copper glance Kupferglanz *m* *(Mineralgruppe)*
copper green Kupfergrün *n*
copper halide Kupferhalogenid *n*
copper hydroxide Kupferhydroxid *n*
copper leaching *(Biot)* (mikrobielle) Laugung *f* von Kupfererzen, Kupfer-Leaching *n*
copper-lean kupferarm
copper-liquor scrubbing Kupfersalzlösungswäsche *f* *(zur Entfernung von Kohlenmonoxid aus Gasen)*
copper matte *(Met)* armer Kupferstein *m*
copper mica Kupferglimmer *m*
copper mineral Kupfermineral *n*
copper monoxide Kupfermonoxid *n*, Kupfer(II)-oxid *n*
copper ore Kupfererz *n*
copper phthalocyanine Kupferphthalocyanin *n*
copper-plated verkupfert
copper plating 1. Verkupfern *n*; 2. Kupfer(schutz)schicht *f*
copper-plating bath Verkupferungselektrolyt *m*, Kupferelektrolyt *m*

copper pyrites *(Min)* Kupferkies *m*, Chalkopyrit *m (Kupfer(I)-Eisen(III)-sulfid)*
copper rayon Kupfer(kunst)seide *f*, Kupfer-Reyon *n*, Glanzstoff *m*
copper red Kupferrot *n*
copper sheet Kupferblech *n*
copper tapping Kupferstich *m*
copper-tin alloy Kupfer-Zinn-Legierung *f*
copper-vapour laser Kupferdampflaser *m*
copper vitriol 1. Kupfersulfat *n*, *(i. e. S.)* Kupfervitriol *n*, Kupfer(II)-sulfat-5-Wasser *n*; 2. *(Min)* Kupfervitriol *m*, Chalkanthit *m*
coprecipitate *v* mitfällen
coprecipitation (induzierte) Mitfällung *f*, Mitfallen *n*
coproduct Nebenprodukt *n*, Beiprodukt *n*
copy DNA *(Biot)* DNA-Kopie *f*
cordierite ceramics Cordieritkeramik *f*
core 1. Kern *m*, *(beim Einsatzhärten auch)* Kernzone *f*; 2. Seele *f*, Kern *m (eines Kabels oder Seils)*; 3. Dorn *m (bei Strangpressdüsen)*; 4. *(phys Ch)* Atomrumpf *m*, Rumpf *m*; 5. *(Kern)* Spaltzone *f*
core activity Kernaktivität *f*
core enzyme *(Bioch)* Kernenzym *n*
core particles *(Bioch)* Core-Partikel *npl*, Ribonucleoproteinkerne *mpl (der Ribosomen)*
core protein *s.* core particles
core-shell particle Hülle-Kern-Teilchen *n*, Kern-Schale-Teilchen *n*
coreactant Reaktionspartner *m*
coreless induction furnace kernloser Induktionsofen *m*, Mittelfrequenzinduktionsofen *m*, MF-Induktionsofen *m*
corepressor Corepressor *m (Enzymkinetik)*
cork *v* zukorken, verkorken, *(i. w. S.)* zustöpseln, verstöpseln, verschließen *(Flaschen)*
cork 1. Kork *m*, Korkrinde *f (von Quercus suber L.)*; Kork *m*, Korkgewebe *n*, Phellem *n*; 2. Kork(en) *m*, Korkstopfen *m*, *(i. w. S.)* Stopfen *m*, Pfropf(en) *m*, Stöpsel *m*
cork gasket Korkdichtung *f*
cork powder Korkmehl *n*
corky korken, Kork...
corn Samenkorn *n*, *(i. e. S.)* Getreidekorn *n*; Körnerfrüchte *fpl*, Getreide *n*, *(in England:)* Weizen *m*, *(in Schottland und Irland:)* Hafer *m*, *(Am)* Mais *m*
corn fermentation *(Biot)* Maisvergärung *f*
corn oil Mais(keim)öl *n*
corn starch Maisstärke *f*
Cornu prism *(Spektr)* (Quarz-)Cornu-Prisma *n*
corona (discharge) Koronaentladung *f*
coronand Coronand *m (monocyclische Kronenverbindung)*
coronate Coronat *n (Chelatkomplex eines Coronanden)*
corpuscular radiation Korpuskularstrahlung *f*, Teilchenstrahlung *f*
correction factor Korrekturfaktor *m*, Korrekturkoeffizient *m*
corrective 1. Hilfsstoff *m*, Beistoff *m (bei der Konfektionierung von Wirkstoffgemischen)*; 2. *(Pharm)* Korrigens *n*, geschmacksverbessernder Zusatz *m*
correlation Korrelation *f*, Wechselbeziehung *f*
$\rho\sigma$**-correlation** Hammett-Beziehung *f*, Hammett-Korrelation *f*
corresponding states *(phys Ch)* übereinstimmende Zustände *mpl*
corrigent *(Pharm)* Korrigens *n*, geschmacksverbessernder Zusatz *m*
corrin Corrin *n*, Corrinoid *n*, Cobalamin *n*, Verbindung *f* der Vitamin-B12-Gruppe
corrinoid *s.* corrin
corrode *v* korrodieren, zerfressen, *(im Anfangsstadium auch)* anfressen; der Korrosion unterliegen, korrodiert werden, korrodieren, *(bei Eisen und Stahl auch)* rosten
corrodibility Korrosionsanfälligkeit *f*, Korrosionsempfindlichkeit *f*, Angreifbarkeit *f*
corrodible korrosionsanfällig, korrosionsempfindlich, angreifbar
corrosion Korrosion *f*, *(im Anfangsstadium auch)* Anfressung *f*, *(bei Eisen und Stahl auch)* Rosten *n*
corrosion cell Korrosionselement *n*
corrosion control Korrosionsschutz *m*
corrosion current Korrosionsstrom *m*
corrosion fatigue Ermüdungskorrosion *f*, Korrosionsermüdung *f*, Dauer-

schwingkorrosion *f*, Schwingungs(riss)korrosion *f*
corrosion inhibitor Korrosionsinhibitor *m*, Korrosionshemmstoff *m*, Korrosionshemmer *m*, Korrosionsverzögerer *m*
corrosion protection oil Korrosionsschutzöl *n*
corrosion rate Korrosionsgeschwindigkeit *f*, Korrosionsrate *f*
corrosion resistance Korrosionsbeständigkeit *f*, Korrosionsfestigkeit *f*, Korrosionssicherheit *f*; Korrosionswiderstand *m*, Korrosionsbeständigkeit *f* *(quantitativ)*
corrosion-resistant korrosionsbeständig, korrosionsfest, korrosionssicher
corrosion-resisting korrosionsbeständig, korrosionsfest, korrosionssicher
corrosion test Korrosionsversuch *m*, Korrosionsprüfung *f*, Korrosionstest *m*
corrosive korrosiv, korrodierend (wirkend), korrosionswirksam, zerfressend, angreifend, Korrosions...
corrosive agent Korrosionsmittel *n*, Korrosionsmedium *n*, korrodierendes Mittel *n*
corrosive environment Korrosionsmilieu *n*, Korrosionsumgebung *f*, *(i. e. S.)* Korrosionsklima *n*
corrosiveness Korrosivität *f*, korrodierende Wirkung *f*, Aggressivität *f*
corrosivity Korrosivität *f*, korrodierende Wirkung *f*, Aggressivität *f*
corrugated asbestos Wellasbest *m*
corrugated bellows Faltenbalg *m*
corrugated expansion joint Wellrohrkompensator *m*, Wellrohr(dehnungs)ausgleicher *m*
corrugated roll geriffelte Walze *f*, Riffelwalze *f*
corrugated sheet *(Kst)* Wellplatte *f*
corticoid (hormone) Nebennierenrindenhormon *n*, Corticoid *n*, Corticosteroid *n*
corticosterone Corticosteron *n* *(Hormon der Nebennierenrinde)*
corticotrop(h)in *Corticotropin *n*, adrenocorticotropes Hormon *n*, ACTH *n*
cortisol Cortisol *n*, Hydrocortison *n*, 17α-Hydroxycorticosteron *n*
cortison Cortison *n*, 11-Dehydro-17α-hydroxycorticosteron *n*
corundum *(Min)* Korund *m* *(α-Aluminiumoxid)*
coslettize *v* nach dem Coslett-Verfahren phosphatieren *(Metalle)*
cosmetic kosmetisches Präparat *n*, Kosmetikum *n*, Körperpflegemittel *n*, *(i. e. S.)* Schönheitsmittel *n*
cosmetic chemist Kosmetikchemiker *m*
cosmetic chemistry kosmetische Chemie *f*
cosmochemistry Kosmochemie *f*
cosolvent Hilfslöser *m*, Hilfslösungsmittel *n*, latenter [mittelbarer] Löser *m*, latentes Lösungsmittel *n*
cost of maintenance [upkeep] Unterhaltungskosten *pl*
cosubstrate *(Bioch)* Cosubstrat *n*, Transportmetabolit *m*
cosynthesis *(Bio)* Cosynthese *f*
cotton Baumwolle *f*
cotton dye Baumwollfarbstoff *m*
cotton fibre Baumwollfaser *f*
cotton wax Baumwollwachs *n*
cotton wool Watte *f*
cottonseed oil Baumwollsamenöl *n*, Baumwollkernöl *n*, Cottonöl *n*
Cottrell precipitator Cottrell-Elektroabscheider *m*, Cottrell-Abscheider *m*, Cottrell-Staubfilter *n*, Cottrell-Filter *n*
Couette flow Couette-Strömung *f* *(Rheologie)*
coulomb Coulomb *n*, C, Amperesekunde *f*, As *(SI-Einheit der Elektrizitätsmenge)*
Coulomb attraction Coulomb-Anziehung *f*; coulombsche [elektrostatische, elektrische] Anziehung *f*
Coulomb force coulombsche [elektrostatische] Kraft *f*
Coulomb law Coulomb-Gesetz *n*, coulombsches Gesetz *n*
Coulomb repulsion Coulomb-Abstoßung *f*; coulombsche Abstoßung *f*
coulombic force *s.* Coulomb force
coulombic interaction Coulomb-Wechselwirkung *f*
coulombmeter *s.* coulometer
coulometer Coulometer *n*
coulometric coulometrisch
coulometric titration coulometrische Titration *f*, galvanostatische Coulometrie *f*, Coulometrie *f* bei konstanter Stromstärke

coulometry Coulometrie *f*
coulometry at controlled potential Coulometrie *f* bei konstantem Potenzial, potenziostatische [potenzialkontrollierte] Coulometrie *f*
Coulter-method *(Anal)* Coulter-Verfahren *n (zur Partikelgrößenbestimmung)*
coumarin Cumarin *n*, 2*H*-Chromen-2-on *n*, α-Chromon *n*, 1,2-Benzopyron *n*, Tonkabohnencampher *m*, 2-Hydroxyzimtsäurelacton *n*
coumarone Benzo[b]furan *n*, *(veraltet:)* Cumaron *n*
coumarone(-indene) resin Cumaron-Inden-Harz *n*, Cumaronharz *n*
counter electrode Gegenelektrode *f*
counter emf gegenelektromotorische Kraft *f*, Gegen-EMK *f*, Gegenurspannung *f*
counter-ion Gegenion *n*
counter-rotating gegenläufig
counter tube *(Kern)* Zählrohr *n*, Zähler *m*
counteract *v* entgegenwirken
counterbalance *v* ausgleichen, kompensieren; ausbalancieren, austarieren
counterclockwise entgegen dem Uhrzeigersinn, gegen den Uhrzeigersinn, im Gegen(uhr)zeigersinn
countercurrent Gegenstrom...
countercurrent Gegenstrom *m*, Gegenfluss *m* • **in countercurrent** *s.* countercurrently
countercurrent chromatography Gegenstromchromatographie *f*
countercurrent distribution Gegenstromverteilung *f*
countercurrent extraction Gegenstromextraktion *f*, Jantzen-Verteilung *f*, fraktionierte Verteilung *f*
countercurrently im Gegenstrom, nach dem Gegenstromprinzip
counterelectrode Gegenelektrode *f*
counterflow Gegenfluss *m*, Gegenstrom *m*
counterflow heat exchanger Gegenstromwärmeaustauscher *m*, Gegenströmer *m*
counterion Gegenion *n*
counterpoise Gegenmasse *f*, Gegengewicht *n*
counterweight Gegenmasse *f*, Gegengewicht *n*
counting tube *(Kern)* Zählrohr *n*, Zähler *m*
couple *v* 1. verknüpfen; 2. *(Farb)* (an-) kuppeln, kombinieren; kuppeln, sich kombinieren
couple *v* **in acid solution** sauer kuppeln, in saurer Lösung kuppeln, in saurem Medium kuppeln
couple *v* **in alkaline solution** alkalisch kuppeln, in alkalischer Lösung kuppeln, in alkalischem Medium kuppeln
coupled reaction gekoppelte Reaktion *f*
coupling 1. Verknüpfung *f*; 2. *(Farb)* Kuppeln *n*, Kupp(e)lung *f*, Kopp(e)lung *f*, Kombinieren *n*; 3. *(Tech)* Kupplung *f (starr oder elastisch)*
coupling component *(Farb)* Kupplungskomponente *f*, passive Komponente *f*, Entwickler *m*
coupling constant *(Spektr)* Kopplungskonstante *f*
coupling factor *(Bioch)* Kopplungsfaktor *m*
coupling termination Rekombinationsabbruch *m*, Kombinationsabbruch *m*, Kettenabbruch *m* durch Rekombination [Kombination]
course Verlauf *m*, Lauf *m*, Ablauf *m*, Gang *m*
course of analysis Analysengang *m*
course of reaction Reaktionsverlauf *m*, Reaktionsablauf *m*
covalence 1. Bindigkeit *f*, Bindungswertigkeit *f*, Bindungszahl *f*, Kovalenz *f*; 2. *s.* covalent bond
covalence number Koordinationszahl *f*, Zähligkeit *f*, koordinative Wertigkeit *f*
covalency *s.* covalence
covalent kovalent, homöopolar, unpolar
covalent bond kovalente [unpolare, homöopolare] Bindung *f*, Elektronenpaarbindung *f*, Atombindung *f*, Kovalenz *f (Zustand)*
covalent bonding kovalente [unpolare, homöopolare] Bindung *f*, Elektronenpaarbindung *f*, Atombindung *f (Vorgang)*
covalent catalysis *(Bioch)* kovalente Katalyse *f*
covalent crystal kovalenter [homöopolarer] Kristall *m*, Atomkristall *m*
covalent radius (of atoms) kovalenter Atomradius *m*

covalently bonded kovalent [homöopolar, unpolar] gebunden

cover *v* 1. (be)decken, abdecken; belegen, überziehen, umhüllen, ummanteln; 2. bedecken, einnehmen, umfassen, sich erstrecken auf

cover Decke *f*, Belag *m*, Schicht *f*, Überzug *m*, Bezug *m*, Hülle *f*, Umhüllung *f*, Mantel *m*, Ummantelung *f*; Deckel *m*, Deckplatte *f*, Abdeckplatte *f*, Verschluss *m*

cover coat(ing) 1. Deckschicht *f*; 2. *(Anstr)* Deckanstrich *m*, Schlussanstrich *m*

cover glass Deckglas *n*, Deckgläschen *n* *(Mikroskopie)*

coverage 1. *(Anstr)* Deckvermögen *n*, Deckfähigkeit *f*, Deckkraft *f*; 2. *(Text)* Decken *n*, Egalfärben *n*; 3. behandelte [behandelbare] Fläche *f (mit einer gegebenen Chemikalienmenge)*; 4. deckender Belag *m (von Insektiziden)*

covolume *(phys Ch)* Covolumen *n*

CP *s.* 1. cross polarization; 2. core particles; 3. chemically pure

cp [c.p., C.P.] *s.* chemically pure

CPA = coulostatic-pulse amperometry

cr. *s.* 1. crystal; 2. crystalline

CR *s.* chloroprene rubber

crab's claw complex Chelatkomplex *m*, Chelat *n*, Chelatverbindung *f*, Scherenverbindung *f*

crack *v* 1. sprengen, (auf)spalten, lösen *(eine chemische Bindung)*; spalten *(in einfachere Verbindungen)*; 2. *(Erdöl)* kracken; sich (auf)spalten *(Verbindungen)*; brechen *(Emulsionen)*; 3. (zer-)platzen, (zer)springen, bersten, aufspringen, (auf)reißen, rissig werden, zerreißen

crack *v* **up** aufspalten

crack Sprung *m*, Riss *m*, Spalt *m*, Bruch *m*

crack growth Risswachstum *n*

crack initiation Rissauslösung *f*, Risskeimbildung *f*

cracked distillate Krackdestillat *n*

cracked fraction Krackfraktion *f*

cracked gasoline [naphtha] Krackbenzin *n*, Spaltbenzin *n*

cracker Krackapparatur *f*, Krackanlage *f*

cracker gas Krackgas *n*

cracking 1. Sprengen *n (einer Bindung)*; Spalten *n*, Spaltung *f (in einfachere Verbindungen)*; 2. *(Erdöl)* Kracken *n*, Krackung *f*; Brechen *n (von Emulsionen)*; 3. Platzen *n*, Zerplatzen *n*, Springen *n*, Zerspringen *n*, Bersten *n*, Aufspringen *n*, Reißen *n*, Aufreißen *n*, Rissigwerden *n*, Rissbildung *f*, Rissebildung *f*, Zerreißen *n* • **give a mild cracking treatment** mild [gelinde] kracken

cracking catalysts *(Erdöl)* Krackkatalysatoren *mpl*

cracking conditions *(Erdöl)* Krackbedingungen *fpl*

cracking distillate *(Erdöl)* Krackdestillat *n*

cracking feed(stock) *(Erdöl)* Krackgut *n*, Krackansatz *m*, Krackrohstoff *m*, Krackeinsatz *m*

cracking fraction *(Erdöl)* Krackfraktion *f*

cracking fuel *(Erdöl)* Krackbenzin *n*

cracking furnace *(Erdöl)* Krackofen *m*

cracking gas *(Erdöl)* Krackgas *n*

cracking glaze *(Erdöl)* Krackglasur *f*

cracking plant Krackanlage *f*, Spaltanlage *f*

cracking process *(Erdöl)* Krackverfahren *n*, Spaltverfahren *n*; Krackvorgang *m*, Spaltvorgang *m*

cracking product *(Erdöl)* Krackgut *n*

cracking raw material *(Erdöl)* Krackrohstoff *m*

cracking reaction *(Erdöl)* Krackreaktion *f*

cracking reactor *(Erdöl)* Krackreaktor *m*

cracking sharpness *(Erdöl)* Krackschärfe *f*

cracking treatment *(Erdöl)* Krackbehandlung *f*

cracking tube heater *(Erdöl)* Krackröhrenerhitzer *m*

cracking zone *(Erdöl)* Krackzone *f*

crankshaft Kurbelwelle *f*

craze *v* reißen *(Oberfläche)*

craze Haarriss *m (Oberflächenfehler)*

crazing 1. Haarrissbildung *f*, Haarrissigwerden *n (von Oberflächen)*; 2. Haarrisse *mpl (Oberflächenfehler)*

creaming *(Lebm, Gum, Kst)* Aufrahmen *n*, Aufrahmung *f*, Emulsionsverdichtung *f*

creamy cremig, cremeförmig, cremeartig

creep *v* kriechen
creep *v* **up** emporkriechen, hochkriechen *(an der Gefäßwandung)*
creep Kriechen *n*, Kriechvorgang *m*
creep resistance [strength] Kriechfestigkeit *f*, Kriechwiderstand *m*, Zeitstandfestigkeit *f*
creep rheometer Kriechrheometer *n*
creep test Kriechversuch *m*, Zeitstandversuch *m*
cresol Cresol *n*, Methylphenol *n*
cresol red Cresolrot *n*, *o*-Cresolsulfophthalein *n*, 3′,3″-Dimethylphenolsulfophthalein *n*
cresol resin Cresolharz *n*
cresylic acid rohe Carbolsäure *f (Gemisch aus Phenolen, Cresolen und Xylenolen)*; *(i. e. S.)* Teercresol *n*, Cresylsäure *f* (*Gemisch aus o-, m- und p-Cresol*)
cresylic resin Cresolharz *n*
crevice Spalt *m*, Riss *m*
crevice corrosion Spaltkorrosion *f*, *(i. w. S.)* Berührungskorrosion *f*
Criegee reaction criegeesche Glykolspaltung *f*, Criegee-Oxidation *f*
crimped fibre Kräuselfaser *f*
critical constants [data] kritische Konstanten *fpl* [Daten *pl*, Größen *fpl*]
critical humidity *s.* critical moisture content
critical micelle concentration *(Koll)* kritische Mizell(bildungs)konzentration *f*, KMK
critical moisture content Knickpunktfeuchte *f*, Knickpunkt-Feuchtebeladung *f*, kritische Feuchte(beladung) *f*
critical molar volume kritisches Molvolumen *n* [Volumen *n*]
critical point kritischer Punkt *m (mehrphasiger Systeme)*; *(bei Zweikomponentensystemen auch)* kritischer Mischungspunkt *m*
critical pressure kritischer Druck *m*
critical temperature kritische Temperatur *f*
cRNA = complementary ribonucleic acid
crop *v* **(out)** *(Geol)* ausstreichen, ausgehen, zutage treten
crop (of crystals) Kristallisationsprodukt *n*, Kristallisat *n*
crop protection chemical Pflanzenschutzmittel *n*
cross-link *v* (quer)vernetzen
cross Kreuzstück *n*
cross-belt controller Kreuzbalgregler *m*
cross-belt mixer Kreuzbalgrührer *m*
cross-belt separator Kreuzbalgscheider *m*
cross coupling *(org Ch)* Kreuzkopplung *f*
cross flow Querströmung *f*, Querstrom *m*, Kreuzströmung *f*, Kreuzstrom *m*
cross-flow filtration *(Biot)* Cross-Flow-Filtration *f*
cross-flow tray Kreuzstromboden *m*
cross-laminate *(Kst)* Kreuzschichtstoff *m*
cross-link Vernetzungsstelle *f*, Brücke *f*, Querverbindung *f*, Quervernetzung *f*
cross-linkable vernetzbar
cross-linkage 1. Vernetzung *f*, Quervernetzung *f (Vorgang)*; 2. *s.* cross-link
cross-linked polymer vernetztes Polymer(e) *n*
cross-linking agent Vernetzungsmittel *n*, Vernetzer *m*
cross metathesis Kreuzmetathese *f*
cross polarization *(Spektr)* Kreuzpolarisation *f*, Kreuzpolarisierung *f*
cross-reaction Kreuzreaktion *f*
cross-sectional area Querschnittsfläche *f*, Querschnitt *m*
crossed aldol condensation gekreuzte Aldolkondensation *f*
crossed double bond gekreuzte Doppelbindung *f*
crossed molecular beams gekreuzte Molekularstrahlen *mpl (zur Untersuchung von Elementarreaktionen)*
crosslink *v s.* cross-link
crossroads element Halbmetall *n*
crotonaldehyde Crotonaldehyd *m*, *But-2-enal *n*
crotonic acid Crotonsäure *f*, **trans*-But-2-ensäure *f*
crotonization Crotonisierung *f*
crowded spectrum linienreiches Spektrum *n*
crown compound Kronenverbindung *f*
crown ether Kronenether *m (cyclischer Polyether)*
crown (optical) glass Kronglas *n*
18-crown 6 (18)Krone-6 *f*, 1,4,7,10,13,16-Hexaoxacyclooctadecan *n*
CRP = cyclic AMP receptor protein

CRU *s.* constitutional repeating unit
crucial intermediate *(Bioch)* Schlüsselintermediäres *n*, Schlüsselverbindung *f*
crucible Tiegel *m*, Schmelztiegel *m*
crucible adapter *(Lab)* Tulpe *f*, Einsatztulpe *f*, Gooch-Tiegel-Halter *m*, Vorstoß *m* für Frittentiegel; Gummimanschette *f* für Frittentiegel
crucible coke Tiegelkoks *m*
crucible furnace Tiegelofen *m*
crucible steel Tiegelstahl *m*
crucible tongs *(Lab)* Tiegelzange *f*, Schmelztiegelzange *f*
crude roh, Roh...
crude 1. Rohgut *n*, Rohprodukt *n*; 2. *s.* crude oil
crude acid Rohsäure *f*
crude alcohol Rohspiritus *m*, Rohsprit *m*
crude distillate Rohdestillat *n*
crude extract Rohextrakt *m*
crude fibre *(Bioch, Landw)* Rohfaser *f*
crude gas Rohgas *n*
crude lead Rohblei *n*, Werkblei *n*
crude oil Rohöl *n*, rohes Öl *n*; Roh(erd)öl *n*, rohes Erdöl *n*, Crude(oil) *n*, Naphta *n*
crude ore Roherz *n*, Fördererz *n*
crude petroleum rohes Erdöl *n*, Roh(erd)öl *n*
crude product Rohprodukt *n*
crude refuse Rohmüll *m*
crude rubber Rohkautschuk *m*
crude salt Rohsalz *n*
crude sulphur Rohschwefel *m*
crude wax Rohwachs *n*, rohes Wachs *n*
crumble *v* zerkrümeln, zerbröckeln; zerbröckeln, zerfallen
crumble *v* **away** zerbröckeln, zerfallen
crumble *v* **to pieces** in Stücke zerfallen
crumbly krüm(e)lig, bröck(e)lig, leicht bröckelnd
crumbs Krümel *mpl(npl)*
crush *v* brechen, grobzerkleinern; zerquetschen; zerstoßen; stoßen *(Eis)*
crusher Brecher *m*
crushing and grinding Zerkleinern *n*, Zerkleinerung *f*
crushing plate Brechplatte *f*
crushing roll Brechwalze *f*
crust 1. Kruste *f*, Ansatz *m (z. B. auf Heizflächen von Trocknern)*; 2. Kruste *f*, Rinde *f (der Erde)*
cry of tin Zinngeschrei *n*, Zinnschrei *m*
cryo storage Kryospeicherung *f*
cryogenic Tieftemperatur...
cryogenic engineering Tieftemperaturtechnik *f*, Kryotechnik *f*
cryogenic pump Kryopumpe *f*
cryogenic trap Kühlfalle *f*
cryogenics 1. Kryogenik *f (Wissenschaft von den Erscheinungen bei sehr tiefen Temperaturen)*; 2. *s.* cryogenic engineering
cryohydrate Kryohydrat *n*
cryohydric kryohydratisch
cryoscope Kryoskop *n*, Gefrierpunktmesser *m*
cryoscopic constant kryoskopische Konstante *f*, molare Gefrierpunktserniedrigung *f*
cryoscopy Kryoskopie *f*
cryptand Kryptand *m (sphärische Kronenverbindung)*
cryptate Kryptat *n (Chelatkomplex eines Kryptanden)*
crypto-ion Krypto-Ion *n*
crypto-ionic kryptoionisch
crypto-ionic reaction Kryptoionenreaktion *f*
cryptobase Kryptobase *f*
cryptocrystalline *(Min)* kryptokristallin
cryst. *s.* 1. crystal; 2. crystalline
crystal Kristall *m*
crystal analysis Kristallstrukturanalyse *f*, Feinstrukturanalyse *f*
crystal axis Kristallachse *f*, kristallographische Achse *f*
crystal chemistry Kristallchemie *f*
crystal class Kristallklasse *f*, Symmetrieklasse *f*, Punkt(symmetrie)gruppe *f*
crystal defect Kristall(bau)fehler *m*, Gitter(bau)fehler *m*, Kristallstörung *f*, Gitterstörung *f*, Gitterdefekt *m*, Störstelle *f*
crystal diffraction Beugung *f* am Kristall(gitter)
crystal dislocation Kristallversetzung *f*, Versetzung *f*, Dislokation *f*
crystal face Kristallfläche *f*, Kristallebene *f*
crystal-field theory Kristallfeldtheorie *f*
crystal gliding Gleiten *n*, Gleitung *f*, Abgleiten *n*, Translation *f (entlang kristallographischer Ebenen)*
crystal grating Kristallgitter *n*
crystal growth Kristallwachstum *n*; Kristallzüchtung *f*

crystal habit Kristallhabitus *m*
crystal imperfection *s.* crystal defect
crystal lattice Kristallgitter *n*
crystal nucleus Kristall(isations)keim *m*, Eigenkeim *m*; Kristall(isations)kern *m*, Fremdkeim *m*
crystal plane Kristallebene *f*, Kristallfläche *f*
crystal pulling Ziehen *n* aus der Schmelze *(von Einkristallen)*
crystal shape Kristallform *f*
crystal structure Kristallstruktur *f*, Kristallbau *m*
crystal-structure analysis *s.* crystal analysis
crystal symmetry Kristallsymmetrie *f*
crystal system Kristallsystem *n*, kristallographisches System *n*
crystal twin Kristallzwilling *m*, Zwilling(skristall) *m*
crystal violet Kristallviolett *n*, Hexamethylpararosanilinchlorhydrat *n*
crystal whisker Whisker *m*, Haarkristall *m*, Nadelkristall *m*, Faserkristall *m*, Einkristallfaser *f*
crystalline kristallin
crystalline aggregate Kristallaggregat *n*, Kristallhaufwerk *n*
crystalline area kristalliner Bezirk *m* [Bereich *m*]
crystalline crop Kristallisationsprodukt *n*, Kristallisat *n*
crystalline liquid kristalline [anisotrope, mesomorphe] Flüssigkeit *f*, flüssiger Kristall *m*, Flüssigkristall *m*, Mesophase *f*
crystalline phase kristalline Phase *f*, Kristallphase *f*
crystalline region *s.* crystalline area
crystalline solid kristalliner Feststoff *m*
crystallinity Kristallinität *f*
crystallite 1. *(Krist)* Kristallit *m*; Mosaikblöckchen *n*, Mosaikblock *m*; 2. *(Bioch)* Mizelle *f*, Mizell *n*
crystallizability Kristallisierbarkeit *f*, Kristallisationsvermögen *n*, Kristallisationsfähigkeit *f*
crystallizable kristallisierbar, kristallisationsfähig
crystallizate Kristallisat *n*, Kristallisationsprodukt *n*
crystallization Kristallisation *f* • **induce crystallization** die Kristallisation anregen [auslösen]
crystallization centre Kristallisationszentrum *n*
crystallization column Kristallisationskolonne *f*
crystallization conveyor Kristallisierband *n*
crystallization kinetics Kristallisationskinetik *f*
crystallization rate Kristallisationsgeschwindigkeit *f*
crystallization roll Kristallisierwalze *f*
crystallization screw conveyor Kristallisierschnecke *f*
crystallize *v* **(out)** (aus)kristallisieren, Kristalle bilden; zur Kristallisation bringen
crystallizer Kristallisator *m*, Kristallisierapparat *m*, Kristaller *m*, Kristallierer *m*
crystallizing dish Kristallisierschale *f*
crystallizing evaporator Verdampfungskristallisator *m*
crystallizing starch Kristallstärke *f*
crystalloblastic texture *(Geol)* kristalloblastisches Gefüge *n*
crystallochemical kristallchemisch
crystallogram Einkristallaufnahme *f*, Röntgendiagramm *n* eines Einkristalls
crystallographic axial ratio (kristallographisches) Achsenverhältnis *n*
crystallographic axis kristallographische Achse *f*, Kristallachse *f*
crystallographic plane Kristallebene *f*, Kristallfläche *f*
crystallographic system Kristallsystem *n*, kristallographisches System *n*
crystallography Kristallographie *f*
crystalloid Kristalloid *n*
crystallyzand Kristallisiergut *n*
crystn. *s.* crystallization
CS 1. CS-Kampfstoff *m*, *O*-Chlorbenzylidenmalonitril *n (Tränenreizstoff)*; 2. = Chemical Society
CSV *s.* cathodic stripping voltammetry
CT *s.* charge transfer
ct product *(Tox)* habersches Produkt *n* [Tödlichkeitsprodukt *n*, Wirkungsprodukt *n*], Tödlichkeitsindex *m*
CTC = chlorotetracycline
CTFE *s.* chlorotrifluoroethylene
CTP = cytidine 5′-triphosphate

CTU *(centigrade thermal unit) s.* centigrade heat unit
cub. *s.* cubic
cube 1. Würfel *m*, Kubus *m*; 2. *(Krist)* Elementarwürfel *m*
cubic kubisch, würfelförmig, Würfel...
cubic close-packed *(Krist)* kubisch dicht gepackt
cubic close packing *(Krist)* kubisch dichteste Kugelpackung *f*
cubic expansion kubische [räumliche] Ausdehnung *f*, Volumenausdehnung *f*
cubic lattice *(Krist)* kubisches Gitter *n*, Würfelgitter *n*
cubic system *(Krist)* kubisches System *n*
cubical *s.* cubic
culinary acid *(Lebm)* Genusssäure *f*
cull *v* aussuchen, wählen, aussortieren
cull *(Kst, Glas)* Abfall *m*
cultivation *(Biot)* Kultivierung *f*, Kultivation *f*, Züchtung *f*, Zucht *f*, Anzucht *f*, Inkulturnahme *f (von Mikroorganismen)*
culture 1. *(Bioch, Biot)* Kultur *f*; 2. *s.* cultivation
culture bottle Kulturkolben *m*
culture broth *s.* culture solution
culture dish Petrischale *f*
culture fermenter *(Biot)* Impffermenter *m*, Impftank *m*, Impfkessel *m*
culture flask Kulturkolben *m*
culture medium Kulturmedium *n*, Nährmedium *n*, Nährsubstrat *n*
culture of microorganisms Mikroorganismenkultur *f*
culture solution Kulturlösung *f*, Kulturbrühe *f*, Kulturflüssigkeit *f*
culture yeast Kulturhefe *f*, Zuchthefe *f*
cumarin *s.* coumarin
cumene Cumen *n*, Cumol *n*, *Isopropylbenzen *n*
cumene hydroperoxide Cumenhydroperoxid *n*
cumol *s.* cumene
cumulated double bond kumulierte [angehäufte, gehäufte] Doppelbindung *f*
cumulative kumulativ, kumulierend
cumulative action *(Tox)* kumulative Wirkung *f*
cumulative inhibition *(Bioch)* kumulative Hemmung *f*
cumulative ionization lawinenartige Ionisation *f* [Ionisierung *f*]
cumulene Kumulen *n (allg. für eine Verbindung mit drei oder mehr kumulierten Doppelbindungen)*
cupel *(Met)* Kapelle *f*
cupellation 1. Kupellation *f*, Kupellieren *n*, Treiben *n*, Abtreiben *n*, Treibarbeit *f* *(Edelmetallgewinnung)*; 2. *(Met)* Kapellenprobe *f*
cupferron Cupferron *n*, *N*-Nitroso-phenylhydroxylamin-Ammonium *n*
cupola (furnace) Kupolofen *m*, Kuppelofen *m*
cuprammonia *s.* cuprammonium hydroxide solution
cuprammonia process Kupferoxidammoniak-Spinnverfahren *n*, Kuoxam-Spinnverfahren *n*
cuprammonium cellulose Kupferoxidammoniakcellulose *f*, Kuoxamcellulose *f*, Blaumasse *f*
cuprammonium hydroxide solution Kupfertetramminhydroxidlösung *f*, Kuoxamlösung *f*, ammoniakalische Kupferhydroxidlösung *f*, Schweizers Reagens *n*
cuprammonium rayon Kuoxamfaserstoff *m*, KU
cuprammonium sulphate Tetrammin-kupfer(II)-sulfat *n*
cuprate Cuprat *n*
cupreous kupferhaltig; kupferähnlich
cupric Kupfer..., *(i. e. S.)* Kupfer(II)-...
cupric acetate Kupfer(II)-acetat *n*
cupric ammonium sulphate Ammoniumkupfer(II)-sulfat *n*
cupric carbonate 1. Kupfer(II)-dihydroxidcarbonat *n*; 2. Kupfer(II)-dihydroxiddicarbonat *n*
cupric chelate Kupfer(II)-chelat *n*
cupric chloride Kupfer(II)-chlorid *n*
cupric cyanide Kupfer(II)-cyanid *n*, Kupferdicyanid *n*
cupric ferricyanide Kupfer(II)-hexacyanoferrat(III) *n*
cupric ferrocyanide Kupfer(II)-hexacyanoferrat(II) *n*
cupric hydroxide Kupfer(II)-hydroxid *n*, Kupferdihydroxid *n*
cupric nitrate Kupfer(II)-nitrat *n*
cupric oxide Kupfer(II)-oxid *n*
cupric oxychloride Kupfer(II)-oxidchlorid *n*
cupric sulphate Kupfer(II)-sulfat *n*

cupric sulphide Kupfer(II)-sulfid *n*
cupriethylenediamine *(Text)* Kupferethylendiamin *n*, Cuen *n*
cupriferous kupferhaltig, Cu-haltig
cuprite *(Min)* Kuprit *m*, Rotkupfererz *n* *(Kupfer(I)-oxid)*
cupro Kupferseide *f*, KUS
cupro filament Kupferseidenelementarfaden *m*
cupro-nickel Kupfer-Nickel-Legierung *f*, Kupfernickel *n*
cupro staple Kupferfaser *f*, KUF
cuprous Kupfer..., *(i. e. S.)* Kupfer(I)-...
cuprous acetylide Kupfer(I)-acetylid *n*
cuprous chloride Kupfer(I)-chlorid *n*
cuprous cyanide Kupfer(I)-cyanid *n*
cuprous ferricyanide Kupfer(I)-hexacyanoferrat(III) *n*
cuprous ferrocyanide Kupfer(I)-hexacyanoferrat(II) *n*
cuprous oxide Kupfer(I)-oxid *n*
cuprous sulphate Kupfer(I)-sulfat *n*
curable 1. *(Anstr, Kst)* härtbar; 2. *(Gum)* vulkanisierbar, vernetzbar
curare alkaloid Curarealkaloid *n*
curative action heilende Wirkung *f*, Heilwirkung *f*
curative drug Heilmittel *n*
curd *v* gerinnen, dick werden, koagulieren, ausflocken, (ver)klumpen; gerinnen lassen, koagulieren, zur Ausflockung bringen
curd Gerinnsel *n*, Klumpen *m*
curdle *v s.* curd
curdly 1. leicht gerinnbar; 2. geronnen, käsig
curds käsiger Niederschlag *m*
curdy geronnen, dick; klumpig; flockig; quarkig, käsig
curdy precipitate käsiger Niederschlag *m*
cure *v* 1. *(Gum)* vulkanisieren, heizen, vernetzen; 2. *(Anstr, Kst)* (aus)härten; 3. *(Lebm)* konservieren, haltbar machen *(durch Trocknen, Salzen, Räuchern oder Säuern)*; *(i. e. S.)* (ein)pökeln
cure *(Gum)* Vulkanisation *f*, Vernetzung *f*, Heizung *f*
Curie point [temperature] Curie-Punkt *m*, Curie-Temperatur *f*
curing agent 1. *(Gum)* Vulkanisationsmittel *n*, Vulkanisationsagens *n*, Vulkanisiermittel *n*, Vernetzungsmittel *n*, Vernetzer *m*; 2. *(Anstr, Kst)* Härtungsmittel *n*, Härtemittel *n*, Härter *m*; 3. *(Lebm)* Konservierungsmittel *n*, *(i. e. S.)* Pökelsalz *n*
curing catalyst *(Kst)* Härtekatalysator *m*, Härtungskatalysator *m*
curing curve *(Gum)* Vulkanisationskurve *f*
curing mould *(Gum)* Vulkanisierform *f*
curing press *(Gum)* Vulkanisierpresse *f*
curing temperature 1. *(Gum)* Vulkanisationstemperatur *f*, Heiztemperatur *f*; 2. *(Anstr, Kst)* Härtetemperatur *f*
curing time 1. *(Gum)* Vulkanisationszeit *f*, Heizzeit *f*, Gesamtheizzeit *f*; 2. *(Anstr, Kst)* Härtezeit *f*, Härtungszeit *f*, Aushärtungszeit *f*
curium Cm Curium *n*
current 1. Strom *m*, Strömung *f*, Fluss *m*; 2. (elektrischer) Strom *m*; 3. *s.* current strength
current-carrying stromführend, stromdurchflossen, unter Strom
current density Stromdichte *f*
current efficiency Stromausbeute *f*
current flow Stromfluss *m*
current-potential curve Strom-Potenzial-Kurve *f*, SPK, Strom-Potenzial-Kennlinie *f*, Strom-Spannungs-Kurve *f*, I/U-Kurve *f*, I-U-Kennlinie *f*
current source Stromquelle *f*
current strength Stromstärke *f*
Curtius degradation Curtius-Abbau *m*, curtiusscher Abbau *m* *(von Carbonsäureaziden zu primären Aminen)*
curvature Krümmung *f*, Biegung *f*; Kurvenform *f*
curve fitting *(Anal)* Kurvenanpassung *f*
cut *v* schneiden *(z. B. Glas, Papier, Hadern)*; schleifen *(Glas)*; häckseln *(Stroh)*
cut *v* **back** verschneiden, fluxen *(hochsiedende Erdölfraktionen)*
cut *v* **out of service** außer Betrieb setzen
cut Schnitt *m*, Trennschnitt *m* • **make a cut** trennen, scheiden *(z. B. beim Klassieren)*
cut-off filter *(Spektr)* Kantenfilter *n*
cut point [size] Trennkorngröße *f*, Kornscheide *f*
cutback (bitumen) Verschnittbitumen *n*, Cutbackbitumen *n*
cutback product Verschnittprodukt *n*

cuticular residue kutikulärer Rückstand *m (von Pflanzenschutzmitteln)*
cutter Schneider *m*; Häckselmaschine *f (Strohaufschluss)*; Schneid(e)apparat *m*, Schneid(e)maschine *f*
cutting ceramics Schneidkeramik *f*
cutting fluid Schneidflüssigkeit *f*
cutting oil Schneidöl *n*
cuvette Küvette *f*
cuvette test *(Spektr)* Küvettentest *m*
c.v. *s.* calorific value
CVD *s.* chemical vapour deposition
CW laser kontinuierlicher Laser *m*
CW technique *s.* continuous-wave technique
C.W.G. *s.* carburetted water gas
cyanamide Cyanamid *n*
cyanate M^IOCN Cyanat *n*; ROCN Cyanat *n*, Cyansäureester *m*
cyanation Cyanierung *f*
cyanhydrin Cyanhydrin *n*, α-Hydroxynitril *n*
cyanic acid Cyansäure *f*
cyanidation *(Met)* Cyanidlaugung *f*
cyanide *v (Met)* cyanieren, im Cyan(salz)bad härten
cyanide M^ICN Cyanid *n*; RCN Cyanid *n (Derivat des Cyanwasserstoffs)*
cyanide (case-)hardening *(Met)* Carbonitrierhärten *n*
cyaniding *s.* 1. cyanide (case-)hardening; 2. cyanidation
cyanine 1. Cyanin *n*, Cyaninfarbstoff *m*; 2. Cyaninblau *n*, Chinolinblau *n*, 1,1′-Diethyl-4,4′-cyaniniodid *n*
cyano group CN– Cyangruppe *f*, Cyanrest *m*
cyanoalkylation Cyanalkylierung *f*
cyanoaurate(I) $M^I[Au(CN)_2]$ Cyanoaurat(I) *n*, Dicyanoaurat(I) *n*
cyanoaurate(III) $M^I[Au(CN)_4]$ Cyanoaurat(III) *n*, Tetracyanoaurat(III) *n*
cyanoethylation Cyanethylierung *f*
cyanoferrate(II) $M^I_4[Fe(CN)_6]$ Cyanoferrat(II) *n*, Hexacyanoferrat(II) *n*
cyanoferrate(III) $M^I_3[Fe(CN)_6]$ Cyanoferrat(III) *n*, Hexacyanoferrat(III) *n*
cyanogen 1. CN– Cyangruppe *f*, Cyanrest *m*; 2. $(CN)_2$ Dicyan *n*, Cyan *n*
cyanogen chloride Cyanchlorid *n*, Chlorcyan *n*
cyanogenation Cyanierung *f*
cyanogenetic glycoside cyanogenes Glykosid *n*, Blausäureglykosid *n*
cyanohydrin Cyanhydrin *n*, α-Hydroxynitril *n*
cyanomethylation Cyanmethylierung *f*
cyanonickelate Cyanoniccolat *n*
cyanoplatinate(II) $M^I_2[Pt(CN)_4]$ Cyanoplatinat(II) *n*, Tetracyanoplatinat(II) *n*
cyanoplatinate(IV) $M^I_2[Pt(CN)_6]$ Cyanoplatinat(IV) *n*, Hexacyanoplatinat(IV) *n*
cyanuric acid Cyanursäure *f*, 2,4,6-Trioxohexahydro-1,3,5-triazin *n*
cyanuric chloride Cyanurchlorid *n*, Cyanursäurechlorid *n*, Trichlor-*sym*-triazin *n*
cyanylation Cyanierung *f*
cybotactic *(phys Ch)* zybotaktisch
cybotaxis zybotaktische Struktur *f*, zybotaktischer Zustand *m*
cyclane *s.* cycloalkane
cycle Kreis *m*; Kreislauf *m*, Zyklus *m*, Kreisprozess *m*
cycle time Dauer *f* eines Arbeitszyklus [Arbeitstakts] *(einer Anlage)*
cyclic cyclisch, ringförmig, Ring...
cyclic acid anhydride cyclisches Säureanhydrid *n* [Anhydrid *n*]
cyclic aliphatic hydrocarbon cycloaliphatischer [alicyclischer] Kohlenwasserstoff *m*
cyclic alkane *Cycloalkan *n*, Cycloparaffin *n*, Naphthen *n*
cyclic alkene *Cycloalken *n*, Cycloolefin *n*
cyclic anhydride *s.* cyclic acid anhydride
cyclic compound cyclische [ringförmige] Verbindung *f*, Ringverbindung *f*
cyclic hydrocarbon cyclischer [ringförmiger] Kohlenwasserstoff *m*, Cyclokohlenwasserstoff *m*, Ringkohlenwasserstoff *m*
cyclic olefin Cycloalken *n*, Cycloolefin *n*
cyclic peptide cyclisches Peptid *n*, Cyclopeptid *n*, Peptid *n* mit ringartiger Struktur
cyclic process Kreisprozess *m*, Kreislauf *m*
cyclic voltammetry *(Anal)* zyklische Voltammetrie *f*
cycling Kreislaufführung *f*
cyclization Cyclisierung *f*, Ringschließung *f*, Ringschluss *m*, Ringbildung *f*

cyclization reaction Cyclisierungsreaktion *f*, Ringschlussreaktion *f*
cyclize *v* cyclisieren, einem Ringschluss unterwerfen, „ringschließen"; sich zum Ring schließen
cyclized rubber cyclisierter Kautschuk *m*, Cyclokautschuk *m*
cycloaddition *(org Ch)* Cycloaddition *f*
cycloaliphatic cycloaliphatisch, alicyclisch
cycloalkadiene Cycloalkadien *n*
cycloalkane *Cycloalkan *n*, Cycloparaffin *n*, Naphthen *n*
cycloalkatriene Cycloalkatrien *n*
cycloalkene *Cycloalken *n*, Cycloolefin *n*
cycloalkenylation Cycloalkenylierung *f*
cycloalkyl group Cycloalkylgruppe *f*, Cycloalkylrest *m*
cycloalkylation Cycloalkylierung *f*
cycloalkyne Cycloalkin *n*
cyclocondensation Cyclokondensation *f*
cyclodehydration Cyclodehydratisierung *f*
cyclodehydrogenate *v* cyclodehydrieren
cyclodehydrogenation Cyclodehydrierung *f*
cyclodextrin Cyclodextrin *n*, Schardinger-Dextrin *n*
cyclodimerization Cyclodimerisierung *f*
cycloelimination Cycloeliminierung *f*
cycloheptatrienylium ion Cycloheptatrienylium-Ion *n*, Tropylium-Ion *n*
cyclohexane Cyclohexan *n*
cyclohexanol *Cyclohexanol *n*, Hexahydrophenol *n*
cyclohexene *Cyclohexen *n*, 1,2,3,4-Tetrahydrobenzen *n*
cyclohexylation Cyclohexylierung *f*
cyclone (collector) Zyklon(ab)scheider *m*, Wirbel(ab)scheider *m*, Abscheidezyklon *m*, Zyklon *m*
cyclone separation Zyklonieren *n*, Fliehkraftabscheiden *n*
cyclone separator *s.* cyclone (collector)
cycloocta-1,3,5,7-tetraene [8]Annulen *n*, Cycloocta-1,3,5,7-tetraen *n*
cyclooligomerization Cyclooligomerisierung *f*
cycloparaffin Cycloparaffin *n*, *Cycloalkan *n*, Naphthen *n*
cyclopentadiene Cyclopentadien *n*
cyclopentadienyl anion Cyclopentadienyl-Anion *n*
cyclopentane Cyclopentan *n*
cyclopentylation Cyclopentylierung *f*
cyclophane Cyclophan *n*, Cyclo-phenylalkan *n (überbrücktes aromatisches System)*
cyclopolymerization cyclisierende Polymerisation *f*
cyclopropane Cyclopropan *n*
cyclopropenyl cation Cyclopropenylkation *n*
cycloreversion Cycloeliminierung *f*
cyclorubber Cyclokautschuk *m*, cyclisierter Kautschuk *m*
cyclosilicate *(Min)* Cyclosilicat *n*
cyclotrimerization Cyclotrimerisierung *f*
cyclotron radiation Cyclotronstrahlung *f*
cyclovoltammetric *(Ech)* cyclovoltammetrisch
cyclovoltammetry *(Ech)* Cyclovoltammetrie *f*
cylinder Zylinder *m*, Walze *f*, Trommel *f*; Druckgasflasche *f*, Bombe *f*
cylinder oxygen Bombensauerstoff *m*
cylindrical zylindrisch, zylinderförmig, Zylinder...
cylindrical batch mixer Kreiselmischer *m*
cylpebs Cylpebs(e) *pl (zylindrische Mahlkörper in Kugelmühlen)*
***p*-cymene** *p*-Cymen *n*, *p*-Cymol *n*, *1-Methyl-4-isopropylbenzen *n*
cysteic acid Cysteinsäure *f*, 2-Amino-3-sulfopropionsäure *f*
cysteine Cystein *n*, Cys, Thioserin *n*, α-Amino-β-mercaptopropionsäure *f*
cystine Cystin *n*, Dicystein *n*, *3,3′-Dithiobis-(2-aminopropionsäure) *f*
cystine link *(Bioch)* Cystin-Bindeglied *n*, Cystin-Brücke *f*, Disulfidbrücke *f*
cytidylic acid *(Bioch)* Cytidylsäure *f*, Cytidin-5′-monophosphat *n*
cytochemistry Zytochemie *f*
cytochrome oxidase *(Bioch)* Cytochromoxidase *f*, warburgsches Atmungsferment *n*, Cytochrom-aa_3-Komplex *m*
cytochrome reductase *(Bioch)* Cytochromreductase *f*
cytostatic zytostatisch
cytostatic agent Zytostatikum *n*
cytotoxic zytotoxisch, zellschädigend
cytotoxin Zytotoxin *n*, Zellgift *n*

Czerny-Turner arrangement *(Spektr)* Czerny-Turner-Aufstellung *f*

D

D *s.* 1. difunctional; 2. Debye; 3. density
D configuration D-Konfiguration *f*
2,4-D 2,4-D *n*, 2,4-Dichlorphenoxyessigsäure *f (Herbizid)*
d-form (+)-Form *f*, rechtsdrehende Form *f*, Rechtsform *f*
2D-NMR *s.* two-dimensional NMR spectroscopy
D unit difunktionelle Einheit *f*, D-Einheit *f*, Grundbaustein *m* D *(zum Aufbau von Polymeren)*
DADI 1. *s.* direct analysis of daughter ions; 2. = dianisidine diisocyanate
d.a.f. *s.* dry ash-free
daily intake (dose) *(Tox)* Tagesdosis *f*, *(i. e. S.)* duldbare [zulässige] Tagesdosis *f (in mg/kg Körpermasse)*
dairy chemistry Milchchemie *f*, Chemie *f* der Milch (und Milchprodukte), milchwirtschaftliche Chemie *f*
dairying Milchwirtschaft *f*, Molkereiwesen *n*
daltonian [daltonide] compounds *s.* daltonides
daltonides Daltonide *npl*, daltonide [stöchiometrisch zusammengesetzte] Verbindungen *fpl*
Dalton's law of partial pressures daltonsches Partialdruckgesetz *n*, daltonsches Gesetz *n* der Partialdrücke [Teildrücke]
damage analysis Schadensanalyse *f*
damage to the atmosphere Schädigung *f* der Atmosphäre
damar *s.* dammar
dammar (resin) Dammar(harz) *n*
damp *v* 1. anfeuchten, befeuchten, feucht [nass] machen, benetzen; 2. dämpfen *(Schwingungen)*
damp feucht
damp Feuchtigkeit *f*; Dunst *m*
dAMP *s.* deoxyadenylic acid
dampen *v s.* damp
damping 1. Anfeuchten *n*, Befeuchten *n*, Feuchtung *f*, Benetzen *n*, Netzen *n*; 2. Dämpfen *n (von Schwingungen)*
damping coefficient [factor] Dämpfungsfaktor *m*
dampness Feuchte *f*, Feuchtigkeit *f*, Nässe *f*
danger class Gefahrenklasse *f*
dangerous to health gesundheitsgefährdend, gesundheitsschädlich, gesundheitsschädigend
dangerously explosive hochexplosiv
dangerously poisonous hochgiftig, hochtoxisch, stark giftig
Daniell cell Daniell-Element *n*, Daniell-Kette *f*
Daniell trap *(Lab)* daniellscher Hahn *m (Gasbrenner)*
dansyl chloride Dansylchlorid *n*, 1-Dimethylaminonaphthalen-5-sulfonylchlorid *n*
dansylation Dansylierung *f*
DAP *s.* 1. diammonium hydrogen phosphate; 2. diallyl phthalate resin
dark current Dunkelstrom *m*
dark phase *(Bot, Bioch)* Dunkelphase *f*, Dunkelperiode *f (der Photosynthese)*
dark reaction Dunkelreaktion *f (Photochemie)*
darken *v* dunkel färben; sich dunkel [dunkler] färben, dunkel [dunkler] werden, nachdunkeln
darkening Dunkelfärbung *f*, Dunkelwerden *n*, Nachdunkeln *n*
darkness reaction Dunkelreaktion *f (Photochemie)*
Darzens (glycidic ester) condensation Darzens-Erlenmeyer-Claisen-Kondensation *f*, Darzens-Reaktion *f (Glycidesterkondensation)*
dative bond dative (kovalente) Bindung *f*, koordinative Bindung *f*, Donator-Akzeptor-Bindung *f*
dative covalence [covalency] 1. koordinative Wertigkeit *f*, Koordinationszahl *f*, Zähligkeit *f*; 2. *s.* dative bond
daughter strand *(Bioch)* Tochterstrang *m (der DNA)*
DBP *s.* dibutyl phthalate
DBPC *s.* 2,6-di-*tert*-butyl-*p*-cresol
d.c. *(direct current)* Gleichstrom *m*
DC *(direct current)* Gleichstrom *m*
d.c. polarography Gleichstrompolarographie *f* • **by d.c. polarography** gleichstrompolarographisch

DCB 1. = *o*-dichlorobenzene; 2. = 1,4-dichlorobutane
dCMP *s.* deoxycytidylic acid
DCP *s.* d.c. polarography
D.D.T. DDT *n*, Dichlordiphenyltrichlorethan *n (Kontaktinsektizid)*
DDT DDT *n*, Dichlordiphenyltrichlorethan *n (Kontaktinsektizid)*
de Broglie relation(ship) De-Broglie-Beziehung *f*
de Broglie wavelength De-Broglie-Wellenlänge *f*, Materiewellenlänge *f*
deacetoxylation Desacetoxylierung *f*
deacetylate *v* entacetylieren
deacetylation Entacetylieren *n*, Entacetylierung *f*
deacidification Entsäuern *n*, Entsäuerung *f*; Abstumpfung *f*, Neutralisation *f (saurer Lösungen)*
deacidify *v* entsäuern; abstumpfen, neutralisieren *(saure Lösungen)*
Deacon process Deacon-Verfahren *n (zur Herstellung von Chlor)*
deactivate *v* desaktivieren, unwirksam machen *(Katalysatoren)*; *(bei Enzymen meist)* inaktivieren
deactivating collision desaktivierender Stoß *m*, Stoß *m* zweiter Art
deactivating group inaktivierende Gruppe *f*
deactivation Desaktivierung *f*, Unwirksammachen *n (von Katalysatoren)*; *(bei Enzymen meist)* Inaktivierung *f*
deacylate *v* desacylieren
deacylation Desacylierung *f*
dead 1. abgestanden, schal, fad(e) *(Getränke)*; 2. matt, glanzlos, stumpf *(Farben)*; 3. *(Ech)* stromlos; spannungslos; 4. *(Min)* taub
dead polymer totes Polymer(e) *n*
dead space Totraum *m*
dead spot Totzone *f*
dead-stop titration Dead-stop-Titration *f*, Stillstandstitration *f*, biamperometrische Titration *f*, Polarisationsstromtitration *f*
dead-time *(Anal)* Totzeit *f*
dead volume Totvolumen *n*
dead zone Totzone *f*
deadenylylation *(Bioch)* Desadenylylierung *f*
deadly poisonous tödlich (giftig), letal
deaerate *v* entlüften; entgasen
deaeration Entlüften *n*, Entlüftung *f*; Entgasen *n*, Entgasung *f*
dealdolization Desaldolierung *f*
dealkalization 1. Abstumpfung *f*, Neutralisation *f (basischer Lösungen)*; 2. *(Hyd)* Entkarbonisierung *f (mittels Ionenaustauschs)*
dealkalize *v* 1. abstumpfen, neutralisieren *(basische Lösungen)*; 2. *(Hyd)* entkarbonisieren *(mittels Ionenaustauschs)*
dealkalizer *(Hyd)* Ionenaustauscheranlage *f* zur Entkarbonisierung, Teilentsalzungsanlage *f*
dealkoxylation Desalkoxylierung *f*
dealkylate *v* dealkylieren, entalkylieren
dealkylation Dealkylieren *n*, Dealkylierung *f*, Entalkylierung *f*
dealuminate *v* dealuminieren
dealumination Dealuminierung *f*
deamidate *v* desamidieren, entamidieren
deamidation Desamidierung *f*, Entamidierung *f*
deaminase Desaminase *f (hydrolytische Desaminierung bewirkendes Enzym)*
deaminate *v* desaminieren
deamination Desaminierung *f*
dearomatization Desaromatisierung *f*
dearylation Desarylierung *f*, Entarylierung *f*
deasphalt *v* entasphaltieren, asphaltartige Anteile ausfällen
deasphaltation Entasphaltieren *n*, Entasphaltierung *f*, Ausfällen *n* asphaltartiger Anteile
deasphalting Entasphaltieren *n*, Entasphaltierung *f*, Ausfällen *n* asphaltartiger Anteile
death curve *(Biot)* Abtötungskurve *f (Sterilisation)*
death rate *(Biot)* Abtötungsrate *f (Sterilisation)*
debenzolization Entbenzol(ier)en *n*, Entbenzolierung *f*
debenzolize *v* entbenzol(ier)en
debrominate *v* debromieren
debromination Debromierung *f*
debutanization Entbutanisieren *n*, Entbutanisierung *f*, Debutanisierung *f*, Butanabtrennung *f*
debutanize *v* entbutanisieren, debutanisieren

debutylation Debutylierung *f*
Debye Debye-Einheit *f*, Debye *n (SI-fremde Einheit des Dipolmoments von Molekülen; 1 D = 3,33 · 10^{-30} Cm)*
Debye force Induktionskraft *f*
Debye-Hückel theory Debye-Hückel-Theorie *f (der starken Elektrolyte)*
Debye-Scherrer diagram *(Krist)* Debye-Scherrer-Diagramm *n*, Debyeogramm *n*
Debye-Scherrer(-Hull) method Debye-Scherrer-Verfahren *n*, Pulververfahren *n (der Kristallstrukturanalyse)*
Debye T^3 law debyesches T^3-Gesetz *n (der Molwärmen)*
Debye temperature Debye-Temperatur *f*, charakteristische Temperatur *f*
decadentate zehnzähnig, zehnzählig *(Komplexchemie)*
decaffeinate *v* coffeinfrei [coffeinarm] machen, entcoffeinieren
decaffeinize *v s.* decaffeinate
decahydrate Decahydrat *n*
decalcification 1. Entkalken *n*; 2. *(Bioch)* Dekalzifikation *f*, Dekalzifizierung *f (z. B. der Knochen)*
decalcify *v* 1. entkalken; 2. *(Bioch)* dekalzifizieren *(z. B. Knochen)*
decane *(org Ch)* Decan *n*
decanedioic acid *Decandisäure *f*, *Octan-1,8-dicarbonsäure *f*, Sebacinsäure *f*
decanoic acid Decansäure *f*
decant *v* **(off)** (ab)dekantieren, (vorsichtig) abgießen; abfüllen, umfüllen
decantate dekantierte Flüssigkeit *f*
decantation Dekantieren *n*, Dekantierung *f*, Dekantation *f*, (vorsichtiges) Abgießen *n*, (vorsichtiges) Abfließenlassen *n*; Abfüllen *n*, Umfüllen *n*
decanter 1. *(Lab)* Dekantiergefäß *n*, Dekantiertopf *m*; 2. *(Tech)* Dekantierapparat *m*, Dekanter *m*, Dekanteur *m*
decapeptide Decapeptid *n (Peptid mit 10 Aminosäureeinheiten)*
decarbalkoxylation Decarbalkoxylierung *f*
decarbobenzoxylation Decarbobenzoxylierung *f*
decarbonate *v* entkarbonisieren
decarbonation Entkarbonisierung *f*
decarbonylation *(org Ch)* Decarbonylierung *f*
decarboxylate *v* decarboxylieren
decarboxylation Decarboxylieren *n*, Decarboxylierung *f*
decarburization *(Met)* Entkohlen *n*, Entkohlung *f*
decarburize *v (Met)* entkohlen
decationize *v* von Kationen befreien
decay *v* 1. abnehmen, geringer werden, schwinden; abklingen; 2. absterben, verfallen, zugrunde gehen, zerstört werden; verwesen, sich zersetzen; 3. *(Lebm)* verderben; 4. *(Kern)* zerfallen; 5. *(Geol)* verwittern
decay 1. Abnahme *f*; Abklingen *n*; 2. Verfall *m*; Verwesung *f*, Zersetzung *f*; 3. *(Lebm)* Verderben *n*; 4. *(Kern)* Zerfallen *n*, Zerfall *m*; 5. *(Geol)* Verwittern *n*, Verwitterung *f*
decay chain *s.* decay series
decay coefficient [constant] *(Kern)* Zerfallskonstante *f*
decay curve Abklingkurve *f*
decay energy *(Kern)* Zerfallsenergie *f*
decay family *s.* decay series
decay heat *(Kern)* Zerfallswärme *f*
decay product *(Kern)* Zerfallsprodukt *n*
decay rate *(Kern)* Zerfallsgeschwindigkeit *f*
decay series (radioaktive) Zerfallsreihe *f*, radioaktive Familie *f*
decay time Abklingdauer *f*
decelerate *v* verzögern, verlangsamen, bremsen; sich verzögern [verlangsamen]
deceleration Verzögerung *f*, Verlangsamung *f*, Abbremsung *f*, negative Beschleunigung *f*
1-decene Dec-1-en *n*
decenoic acid Decensäure *f*
dechlorinate *v* 1. dechlorieren; 2. *(Hyd)* entchloren
dechlorination 1. Dechlorierung *f*; 2. *(Hyd)* Entchlorung *f*
decigram range *(Anal)* Dezigrammbereich *m (0,1 bis 1 g)*
decimal balance Dezimalwaage *f*
decimol Dezimol *n (0,1 Mol)*
decimolar solution 0,1 m-Lösung *f*, m/10-Lösung *f*
decinormal solution 0,1 n-Lösung *f*, n/10-Lösung *f*
decline curve *(Tox)* Abbaukurve *f*
declined [declining] growth phase

(Hyd, Biot) abnehmende Wachstumsphase *f (Mikroorganismenwachstum)*
decoct *v* 1. abkochen, auskochen *(z. B. Drogen)*; 2. einengen *(Lösungen)*
decoction 1. Abkochen *n*, Auskochen *n (z. B. von Drogen)*; 2. *(Pharm)* Dekokt *n*, Abkochung *f*, Absud *m*
decoctum *(Pharm)* Dekokt *n*, Abkochung *f*, Absud *m*
decolor *v s.* decolorize
decolorant Entfärber *m*, Entfärbungsmittel *n*, Bleichmittel *n*
decolor(iz)ation Entfärbung *f*, Bleichung *f*
decolorize *v* entfärben, bleichen; sich entfärben
decolorizer Entfärber *m*, Entfärbungsmittel *n*, Bleichmittel *n*
decolour... *siehe: decolor...*
decomposability Zersetzlichkeit *f*, Zersetzbarkeit *f*, Abbaubarkeit *f*
decomposable zersetzlich, zersetzbar, abbaubar
decompose *v* 1. zersetzen, abbauen; zerlegen, (auf) spalten; sich zersetzen, zerfallen; 2. *(Bio)* verwesen
decomposer 1. Zersetzerzelle *f*, Zersetzer *m*, Pile *f (Elektrolyse)*; 2. abbauender [zersetzender, zerstörender] Mikroorganismus *m*
decomposition 1. Zersetzung *f*, Abbau *m*; Zerlegung *f*, Spaltung *f*, Aufspaltung *f*; Zerfall *m*; 2. *(Bio)* Verwesung *f* • **undergo decomposition** sich zersetzen
decomposition potential Zersetzungsspannung *f*, Zersetzungspotenzial *n*
decomposition reaction Zersetzungsreaktion *f*; Zerfallsreaktion *f*
decontaminate *v (Kern)* dekontaminieren, entaktivieren, entgiften, *(i. e. S.)* entgasen, entseuchen, entstrahlen; reinigen, säubern
decontamination *(Kern)* Dekontaminierung *f*, Entaktivierung *f*, Entgiftung *f*, *(i. e. S.)* Entgasung *f*, Entseuchung *f*, Entstrahlung *f*; Reinigung *f*, Säuberung *f*
decopperize *v* entkupfern
decorative coating Dekorschicht *f*, dekorative Schutzschicht *f*
decouple *v* entkoppeln
decrease in solubility Löslichkeitserniedrigung *f*
decrease in volume Volumenabnahme *f*, Volumenverminderung *f*
decyanation Decyanierung *f*
decyanoethylation Decyanethylierung *f*
decyclization Decyclisierung *f*, Ringöffnung *f*, Ringspaltung *f*, Ring(auf)sprengung *f*
dedimerization Dedimerisierung *f*, Dedimerisation *f*
de-electronation Oxidation *f*
de-emulsification *s.* demulsification
de-energization Desaktivierung *f*
de-energize *v* desaktivieren
deep bed filtration *(Hyd)* Tiefenfiltration *f*
deep mining *(Bgb)* Untertagebau *m*, Abbau *m* unter Tage, Tiefbau(betrieb) *m*, Förderung *f* [Abbau *m*] im Tiefbau, untertägiger Bergbau *m*
de-ethanization Entethanisierung *f*, Deethanisierung *f*
de-ethanize *v* entethanisieren, deethanisieren
de-ethoxylation Desethoxylierung *f*
de-ethylation Desethylierung *f*
defat *v* entfetten
defect 1. *(Krist)* Fehler *m*, Baufehler *m*, Defekt *m*; 2. *(Krist)* Fehlstelle *f*
defect conduction *(phys Ch)* Mangelleitung *f*, Löcherleitung *f*, p-Leitung *f*
defect density Defektdichte *f*
defect electron Defektelektron *n*, Loch *n*
defect-free *(Krist)* fehlerfrei, streng geordnet, ideal
defective fehlerhaft
deferrization *(Hyd)* Enteisenung *f*
deferrize *v (Hyd)* enteisenen
deferrize-deficiency 1. Mangel *m*, Fehlen *n*; 2. Fehler *m*, Defekt *m*
deferrize disease *(Landw, Med)* Mangelkrankheit *f*
deferrize of oxygen Sauerstoffmangel *m*, Sauerstoffdefizit *n*
deflagrate *v* deflagrieren, (explosionsfrei) abbrennen, verpuffen
deflagration Deflagration *f*, (explosionsfreies) Abbrennen *n*, Verpuffung *f*
deflect *v* 1. ablenken, auslenken; ausschlagen *(Zeiger)*; 2. sich durchbiegen *(Flächen bei Belastung)*
deflection 1. Ablenken *n*, Ablenkung *f*, Auslenken *n*, Auslenkung *f*; Ausschlag *m (eines Zeigers)*; 2. Durchbiegung *f (von Flächen bei Belastung)*

deflector Ablenkplatte *f*, Ablenkwand *f*, Ablenkblech *n*, Prallplatte *f*, Prallwand *f*, Prallblech *n*, Leitblech *n*, Lenkblech *n*, Verteilerplatte *f*, Verteiler *m*
deflexion *s.* deflection
deflocculant 1. Dispergens *n*, Dispersionsmittel *n*, Dispergiermittel *n*; 2. *(Koll)* Peptisator *m*, Peptisationsmittel *n*
deflocculate *v* 1. entflocken, zerteilen, dispergieren; 2. *(Koll)* peptisieren; sich entflocken
deflocculation 1. Entflockung *f*, Zerteilung *f*, Dispergierung *f*; 2. *(Koll)* Peptisation *f*; 3. *(Ker)* Verflüssigen *n*, Verdünnung *f (eines Glasurschlickers)*
defluorinate *v* entfluorieren
defluorination Entfluorierung *f*
defoam *v* entschäumen
defoamer Schaumdämpfungsmittel *n*, Schaumdämpfer *m*, Schaumzerstörungsmittel *n*, Entschäumungsmittel *n*, Entschäumer *m*, Antischaummittel *n*
defoaming agent Schaumdämpfungsmittel *n*, Schaumdämpfer *m*, Schaumzerstörungsmittel *n*, Entschäumungsmittel *n*, Entschäumer *m*, Antischaummittel *n*
defoliant Entlaubungsmittel *n*, Entblätterungsmittel *n*, Defoliationsmittel *n*
deform *v* deformieren, verformen *(unbeabsichtigt)*; umformen *(beabsichtigt)*; sich verformen
deformability Verformbarkeit *f*
deformability behaviour *(Tech)* Verformungsverhalten *n*
deformation Deformation *f*, Verformung *f (unbeabsichtigt)*; Umformen *n (beabsichtigt)*; Form(ver)änderung *f*, Gestalts(ver)änderung *f*
deformation vibration Deformationsschwingung *f*, δ-Schwingung *f*, Knickschwingung *f*
deformylation Deformylierung *f*
defrost *v* abtauen, enteisen; entfrosten, auftauen
DEG *s.* diethylene glycol
deg C *s.* degree Celsius
deg F *s.* degree Fahrenheit
degas *v* entgasen
degasification Entgasen *n*, Entgasung *f*
degasify *v* entgasen
degaussing Entmagnetisierung *f*
degeneracy 1. *(phys Ch)* Entartung *f*; 2. *(Bioch)* Degeneriertheit *f (des genetischen Codes)*
degenerate *v (phys Ch)* entarten, degenerieren
degenerate 1. *(phys Ch)* entartet, degeneriert; 2. *(Bioch)* degeneriert *(genetischer Code)*
degradability Abbaubarkeit *f*
degradable abbaubar
degradation 1. Abbau *m*, Degradation *f*, Zerlegung *f (Tätigkeit)*; Abbau *m*, Zersetzung *f (Vorgang)*; 2. *(Boden)* Degradation *f*
degradation curve *(Tox)* Abbaukurve *f*
degradation pathway *(Bioch)* Abbauweg *m*
degradation reaction Abbaureaktion *f*
degrade *v* 1. abbauen, degradieren, zerlegen; sich abbauen [zersetzen]; degradieren *(Boden)*; 2. *(phys Ch)* degradieren, Energie zerstreuen
degrease *v* entfetten
degreaser Entfettungsmittel *n*
degreasing agent Entfettungsmittel *n*
degree Grad *m*
degree British *s.* degree Clark
degree Celsius [centigrade] Grad *m* Celsius, °C
degree Clark [English] *(Hyd)* Grad *m* englischer Härte, englischer Härtegrad *m (1°eH = 8,0 mg CaO/l H_2O)*
degree Fahrenheit Grad *m* Fahrenheit, °F
degree French *(Hyd)* Grad *m* französischer Härte, französischer Härtegrad *m (1°fH = 5,6 mg CaO/l H_2O)*
degree German *(Hyd)* Grad *m* deutscher Härte, deutscher Härtegrad *m (1°dH = 10,0 mg CaO/l H_2O)*
degree Kelvin *s.* kelvin
degree of acidity Säuregrad *m*, Aziditätsgrad *m*
degree of branching Verzweigungsgrad *m*
degree of clarification *(Hyd)* Reinigungsgrad *m*
degree of conversion Umwandlungsgrad *m*, Umsetzungsgrad *m*, Konvertierungsgrad *m*, Umsatz *m*
degree of cross-linking Vernetzungsgrad *m*, Vernetzungsdichte *f*
degree of crystallinity Kristallinitätsgrad *m*

degree of deprotonation Deprotonierungsgrad *m*
degree of dilution Verdünnungsgrad *m*
degree of disorder *(phys Ch)* Grad *m* der Unordnung
degree of dissociation Dissoziationsgrad *m*
degree of freedom *(phys Ch)* Freiheitsgrad *m*, Freiheit *f*, *(i. e. S.)* Anzahl *f* der Freiheitsgrade [Freiheiten]
degree of hardness *(Hyd)* Härtegrad *m*, Härtestufe *f*, Härtewert *m*
degree of ionization Dissoziationsgrad *m (von Salzen in Lösungen)*; Ionisationsgrad *m*, Ionisierungsgrad *m (von Gasen)*
degree of order [orientation] Orientierungsgrad *m*, Ordnungsgrad *m*
degree of pollution Verschmutzungsgrad *m*, Verunreinigungsgrad *m*
degree of polymerization Polymerisationsgrad *m*
degree of purity Reinheitsgrad *m*
degree of rotational freedom *(phys Ch)* Rotationsfreiheitsgrad *m*
degree of separation Trenngrad *m*, Trennschärfe *f*
degree of substitution Substitutionsgrad *m*
degree of swelling *s.* swelling degree
degree of translational freedom *(phys Ch)* Translationsfreiheitsgrad *m*
degree of vibrational freedom *(phys Ch)* Schwingungsfreiheitsgrad *m*
dehalogenation De(s)halogenierung *f*, Halogenentzug *m*
dehydrase Dehydratase *f*
dehydrate *v* 1. dehydratisieren *(aus Verbindungen H und OH als Wasser abspalten)*; 2. entwässern, dehydratisieren *(z. B. Hydrate)*; 3. trocknen, entfeuchten, *(Feststoffe auch)* entwässern; 4. *(Lebm)* trocknen, dehydratisieren
dehydrated alcohol absoluter [wasserfreier] Alkohol *m*, Absolutalkohol *m*, wasserfreier Ethylalkohol *m*
dehydrating agent 1. wasserabspaltendes Mittel *n*, Dehydratisierungsmittel *n*; 2. wasserentziehendes Mittel *n*, Entwässerungsmittel *n*, Trockenmittel *n*; 3. *(Lebm)* Dehydratisierungsmittel *n*
dehydration 1. Dehydratisierung *f*, Dehydratation *f (Abspaltung von H und OH als Wasser aus Verbindungen)*; 2. Entwässerung *f*, Dehydratisierung *f*, Dehydratation *f (z. B. von Hydraten)*; 3. Trocknung *f*, Entfeuchtung *f*, *(bei Feststoffen auch)* Entwässerung *f*; 4. *(Lebm)* Trocknung *f*, Dehydratisierung *f*
dehydroarene Dehydroaromat *m*
dehydroaromatization Dehydroaromatisierung *f*
dehydrobenzene Dehydrobenzen *n*
dehydrocyclization Dehydrocyclisierung *f*, Cyclodehydrierung *f*
dehydrocyclodimerization Dehydrocyclodimerisierung *f*, Dehydrocyclodimerisation *f*
dehydrodimerization Dehydrodimerisierung *f*, oxidative Dimerisierung *f* [Dimerisation *f*]
dehydrofreeze *v (Lebm)* gefriertrocknen
dehydrofreezing *(Lebm)* Gefriertrocknung *f*
dehydrogenase Dehydrogenase *f*
dehydrogenate *v* dehydrieren, Wasserstoff abspalten [entziehen]
dehydrogenating agent Dehydrierungsmittel *n*
dehydrogenation Dehydrieren *n*, Dehydrierung *f*, Dehydration *f*, Wasserstoffabspaltung *f*, Wasserstoffentzug *m*
dehydrogenize *v s.* dehydrogenate
dehydrohalogenate *v* dehydrohalogenieren, Halogenwasserstoff abspalten [entziehen]
dehydrohalogenation Dehydrohalogenierung *f*, Halogenwasserstoffabspaltung *f*, Halogenwasserstoffentzug *m*
dehydroisomerization Dehydroisomerisierung *f*
dehydrotosylation *indirekte Wasserabspaltung über den Tosylester eines Alkohols zwecks Einführung von Doppelbindungen*
dehydroxyalkylation Dehydroxyalkylierung *f*
dehydroxylation Dehydroxylierung *f*
deice *v* enteisen, abtauen
de-icing fluid Enteisungsflüssigkeit *f*
deink *v (Pap)* entfärben, deinken
deintercalation Deinterkalation *f*, Disinterkalation *f*
deionization 1. Deionisation *f*, Deioni-

sierung *f*, Entionisation *f*, Entionisierung *f*; 2. *(Hyd)* Vollentsalzung *f (mittels Ionenaustauschs)*
deionize *v* 1. deionisieren, entionisieren; 2. *(Hyd)* vollentsalzen *(mittels Ionenaustauschs)*
deionized water vollentsalztes Wasser *n*, Deionat *n*
deisobutanization Entisobutanisierung *f*, Deisobutanisierung *f*, Isobutanentfernung *f*
deisobutanize *v* entisobutanisieren, deisobutanisieren
delaminate *v* (auf)spalten, in Schichten zerlegen [trennen, spalten], zerspalten; sich aufspalten, aufblättern, in Schichten zerfallen
delamination Aufspaltung *f*, Aufblättern *n (von Schichtstoffen)*; Schichtentrennung *f*, Schicht(en)spaltung *f*
delay in boiling Siedeverzug *m*, Siedeverzögerung *f*
delayed precipitation Nachfällung *f*
deleterious schädlich, *(i. e. S.)* gesundheitsschädlich
deleteriousness Schädlichkeit *f*
deletion Deletion *f (Genetik, Verlust von Chromosom- oder Genabschnitt)*
deliberate poisoning absichtliche Vergiftung *f (eines Katalysators)*
deliquesce *v* zerfließen
deliquescence 1. Zerfließen *n*; Zerfließlichkeit *f*; 2. Schmelzflüssigkeit *f*
deliquescent zerfließlich, zerfließend
deliver *v* fördern, liefern
delivery 1. Förderung *f*, Lieferung *f*; 2. Fördermenge *f*, Förderleistung *f*; Förderstrom *m*, Lieferstrom *m*
delocalization energy Mesomerieenergie *f*, Delokalisierungsenergie *f*, Resonanzenergie *f*, Sonderenergie *f*
delocalize *v* delokalisieren
delocalized bond delokalisierte Bindung *f*
delq. *s.* deliquescent
delta bond δ-Bindung *f*
demagnetization Entmagnetisierung *f*
demagnetize *v* entmagnetisieren
demanganization Entmanganung *f*
demanganize *v* entmanganen
demask *v* demaskieren *(Komplexchemie)*
demercuration *(org Ch)* Demercurierung *f*
demetalation Demetallierung *f*
demethanization Entmethanisieren *n*, Entmethanisierung *f*, Demethanisierung *f*
demethanize *v* entmethanisieren, demethanisieren
demethoxylation Demethoxylierung *f*
demethylate *v* entmethylieren, demethylieren
demethylation Entmethylieren *n*, Entmethylierung *f*, Demethylierung *f*, Demethylation *f*
demineralization 1. Demineralisation *f*, Entmineralisierung *f*; 2. *(Hyd)* Entsalzung *f*, *(i. e. S.)* Vollentsalzung *f* mittels Ionenaustauschs
demineralize *v* 1. entmineralisieren, demineralisieren; 2. *(Hyd)* entsalzen, *(i. e. S.)* vollentsalzen mittels Ionenaustauschs
demineralized water vollentsalztes Wasser *n*, Deionat *n*
demineralizer *(Hyd)* Entsalzungsanlage *f*, *(i. e. S.)* Vollentsalzungsanlage *f*
demix *v* sich entmischen
Demjanov rearrangement *(org Ch)* Demjanov-Umlagerung *f*, demjanovsche Methode *f (der Ringerweiterung)*
demulsification Brechen *n* [Spalten *n*] von Emulsionen, Trennen *n* [Zerstören *n*] von Emulsionen, Dismulgieren *n*, Entemulsionieren *n*
demulsifier Demulgator *m*, Emulsionsentmischer *m*, Emulsionsspalter *m*, Emulsionsbrecher *m*
demulsify *v* dismulgieren, entemulsionieren, eine Emulsion brechen [spalten], eine Emulsion trennen [zerstören]
denaturability *(Bioch)* Denaturierbarkeit *f*
denaturable *(Bioch)* denaturierbar
denaturant 1. *(Lebm)* Vergällungsmittel *n*, Denaturierungsmittel *n*; 2. *(Bioch)* Denaturierung auslösender Stoff *m*
denaturation 1. *(Lebm)* Vergällen *n*, Denaturieren *n*; 2. *(Bioch)* Denaturierung *f*, Denaturation *f*, Entfaltung *f (von Proteinen)*
denature *v* 1. *(Lebm)* vergällen, denaturieren; 2. *(Bioch)* denaturieren

denatured alcohol vergällter [denaturierter] Alkohol *m*
denatured DNA denaturierte [einsträngige] DNA *f* [Desoxyribonucleinsäure *f*]
denaturize *v s.* denature
dendrimer Dendrimer *n (Polymerarchitektur)*
dendrite Dendrit *m*, Baumkristall *m*
dendritic(al) *(Krist)* dendritisch, verzweigt, verästelt
denitrate *v* denitrieren
denitration Denitrieren *n*, Denitrierung *f*
denitrification 1. Denitrifikation *f*, Denitrifizierung *f (Reduktion von Nitraten durch Bodenbakterien)*; 2. *s.* denitration
denitrifiers *s.* denitrifying bacteria
denitrify *v* denitrifizieren
denitrifying bacteria denitrifizierende Bakterien *npl*, Denitrifikationsbakterien *npl*, Denitrifikanten *mpl*, Denitrifikatoren *mpl*
denitrosation Denitrosierung *f*
densification Verdichtung *f*
densify *v* verdichten
densimeter Densimeter *n*, Dichtemesser *m*
densimetry Densimetrie *f*, Dichtemessung *f*, Dichtebestimmung *f*
densitometer 1. *(Foto)* Dens(it)ometer *n*, Densograph *m*, Schwärzungsmesser *m*; 2. *s.* densimeter
densitometry Densitometrie *f*, Schwärzungsmessung *f*
density 1. Dichte *f (Masse je Volumeneinheit)*; 2. Schwärzungsdichte *f*, (optische) Dichte *f*, Schwärzung *f*, Extinktion *f*; 3. *(i. w. S.)* Dichte *f*, Dichtheit *f*
density balance Dichtewaage *f*
density-gradient centrifugation *(Anal)* Dichtegradientenzentrifugation *f*
density separation Trennung *f* nach der Dichte, Dichtesortierung *f*
dental alloy Dentallegierung *f*
dental gold (alloy) Zahngold *n*, Dentalgold *n*, Dentalgoldlegierung *f*
deodorant desodor(is)ierend, geruchsbeseitigend, geruchszerstörend
deodorant *s.* deodorizer
deodorization Desodor(is)ieren *n*, Desodor(is)ierung *f*, Desodor(is)ation *f*, Geruchlosmachen *n*, Geruchfreimachen *n*
deodorize *v* desodor(is)ieren, geruchlos [geruchfrei] machen, von üblem Geruch befreien
deodorizer Desodor(is)ierungsmittel *n*, Desodorans *n*, Deodorant *n*, desodor(is)ierendes [geruchsbeseitigendes, geruchszerstörendes] Mittel *n*, Geruchsverbesserer *m*
deoil *v* entölen
deoxidant Desoxidationsmittel *n*
deoxidate *v s.* deoxidize
deoxid(iz)ation Desoxidieren *n*, Desoxidation *f*
deoxidize *v* desoxidieren, von Sauerstoff befreien, *(Stahl auch)* beruhigen
deoxidizer agent *s.* deoxidant
deoxidizing agent *s.* deoxidant
deoxy sugar Desoxyzucker *m*
deoxyadenylic acid *(Bioch)* Desoxyadenosin-5′-phosphat *n*, dAMP
deoxycorticosterone 11-Desoxycorticosteron *n*, Cortexon *n*, 21-Hydroxypregn-4-en-3,20-dion *n*
deoxycortone 11-Desoxycorticosteron *n*, Cortexon *n*, 21-Hydroxypregn-4-en-3,20-dion *n*
deoxycytidylic acid *(Bioch)* Desoxycytidin-5′-phosphat *n*, dCMP
deoxygenate *v* desoxidieren, von Sauerstoff befreien
deoxygenated blood venöses Blut *n*
deoxygenated haemoglobin reduziertes [sauerstoffarmes, desoxygeniertes] Hämoglobin *n*
deoxygenation Desoxidation *f*
deoxyguanylic acid *(Bioch)* Desoxyguanosin-5′-phosphat *n*, dGMP
deoxypentose Desoxypentose *f*
deoxyribonuclease *(Bioch)* Desoxyribonuclease *f*, DNase *f (zu den Phosphatasen gehörendes Enzym)*
deoxyribonucleic acid *(Bioch)* Desoxyribonucleinsäure *f*, DNS *f*, *(jetzt meist)* DNA *f*
deoxyribonucleoprotein Desoxyribonucleoprotein *n*
deoxyribose Desoxyribose *f*, 2-Ribodesose *f (ein Monosaccharid)*
deoxythymidine 5′-phosphate *(Bioch)* Desoxy-thymidin-5′-phosphat *n*, dTMP

deoxyxylose Desoxyxylose *f*, Xylodesose *f (ein Monosaccharid)*
deozonation *(Hyd)* Entozon(isier)ung *f*
deozonize *v* desozon(is)ieren, entozonisieren
deparaffin *v s.* deparaffinize
deparaffinization Entparaffinieren *n*, Entparaffinierung *f*
deparaffinize *v* entparaffinieren
dependent joint action *(Tox)* Abhängigkeitsverbundwirkung *f (Wirkung zweier physiologisch verschieden angreifender Gifte, wobei ein Gift den Effekt des anderen verstärkt)*
depentanization Entpentanisieren *n*, Entpentanisierung *f*, Depentanisierung *f*
depentanize *v* entpentanisieren, depentanisieren
dephenolate *v* entphenol(ier)en
dephenol(iz)ation Entphenol(ier)ung *f*
dephenolize *v* entphenol(ier)en
dephlegmate *v (Dest)* dephlegmieren, rektifizieren, aufstärken
dephlegmation *(Dest)* Dephlegmierung *f*, Dephlegmation *f*, Teilkondensation *f*, Aufstärkung *f*
dephlegmator *(Dest)* Dephlegmator *m*, Teilkondensator *m*
dephosphorization Entphosphoren *n*, Entphosphorung *f*
dephosphorize *v* entphosphoren
dephosphorylate *v* dephosphorylieren
dephosphorylation Dephosphorylierung *f*
deplete *v* erschöpfen, ausbeuten *(z. B. Rohstoffe)*
depletion Erschöpfung *f*, Ausbeutung *f (z. B. von Rohstoffen)*; Verarmung *f (z. B. an Nährstoffen)*
depletion of dissolved oxygen *(Hyd)* Sauerstoffzehrung *f*, Sauerstoffschwund *m*
deploy *v* ausbreiten; sich ausbreiten
depolarization Depolarisation *f*
depolarize *v* depolarisieren
depolarizer Depolarisator *m*
depolymerization Depolymerisation *f*, Depolymerisierung *f*
depolymerize *v* depolymerisieren
deposit *v* abscheiden, ausscheiden, ablagern, sedimentieren, niederschlagen; sich abscheiden [ausscheiden], sich (ab)setzen, sich ablagern [niederschlagen], sedimentieren, einen Niederschlag [Bodensatz] bilden, zur Ausscheidung gelangen
deposit *v* **on a tip** auf Halde ablagern, aufhalden, verkippen *(Abprodukte)*
deposit 1. Abscheidung *f*, Niederschlag *m*, Ablagerung *f*; aufgelagertes Pflanzenschutzmittel *n (z. B. auf Blättern)*; 2. *(Geol)* Lager *n*, Lagerstätte *f*, Vorkommen *n*, Depot *n*
deposition 1. Abscheidung *f*, Ablagerung *f*; 2. *s.* deposit
deposition potential Abscheidungspotenzial *n*
deposition rate Aufdampfrate *f*
depress *v* erniedrigen, herabsetzen, senken
depression 1. Depression *f*, Erniedrigung *f*, Senkung *f*; 2. Vertiefung *f*
deprive *v* **of** entziehen
depropanization Entpropanisieren *n*, Entpropanisierung *f*, Depropanisierung *f*, Propanabtrennung *f*
depropanize *v* entpropanisieren, depropanisieren
deprotect *v (org Ch)* entschützen
deprotection *(org Ch)* Entschützen *n*
deproteinization Deproteinisieren *n*, Enteiweißen *n*
deproteinize *v* deproteinisieren, enteiweißen
deprotonation Deprotonierung *f*
depside Depsid *n (Ester aus zwei oder mehreren Phenolcarbonsäuren)*
depsipeptide *(Bioch)* Depsipeptid *n*
depth Tiefe *f*; Höhe *f (einer Schüttung)*
depth filter *(Hyd)* Tiefenfilter *n*
deriv. *s.* derivative
derivable ableitbar
derivation Ableitung *f*, Herleitung *f*
derivative Derivat *n*, Abkömmling *m*
derivative polarography *(Anal)* Derivativpolarographie *f*
derivative spectroscopy Derivativspektroskopie *f*, Ableitungsspektroskopie *f*
derivative thermogravimetry derivative Thermogravimetrie *f*, Differenzial-Thermogravimetrie *f*
derivatization *(Anal)* Derivatisierung *f*, Überführung *f* [Umwandlung *f*] in Derivate, Derivatbildung *f*

derivatize *v (Anal)* derivatisieren, in ein Derivat überführen [umwandeln]
desalinate *v (Hyd)* entsalzen *(salzhaltige Wässer)*
desalination *(Hyd)* Entsalzung *f (salzhaltiger Wässer)*
desalinator *(Hyd)* Entsalzungsaggregat *n*
desalinization *s.* desalination
desalt *v s.* desalinate
desaminase *s.* deaminase
descale *v* 1. entkrusten, *(meist)* von Kesselstein befreien; 2. *(Met)* entzundern
descaling agent Mittel *n* gegen Kesselstein, Kesselsteingegenmittel *n*, Kesselsteinlösemittel *n*
descending chromatography absteigende Chromatographie *f*
descriptive chemistry deskriptive [beschreibende] Chemie *f*
desensitization *(Bioch)* Desensibilisierung *f*
desensitize *v (Bioch)* desensibilisieren
desiccant Trockenmittel *n*, Trockenmedium *n*, Trocknungsmittel *n*, Entfeuchtungsmittel *n*
desiccate *v* 1. trocknen, entfeuchten, *(Feststoffe auch)* entwässern; 2. *(Lebm)* trocknen, dehydratisieren, *(Obst auch)* dörren
desiccation 1. Trocknung *f*, Entfeuchtung *f*, *(bei Feststoffen auch)* Entwässerung *f*; 2. *(Lebm)* Trocknung *f*, Dehydratisierung *f*, *(bei Obst auch)* Dörren *n*
desiccator 1. Exsikkator *m (Gerät)*; 2. *s.* desiccant
desilicification *(Hyd)* Entkieselung *f*
desilicify *v (Hyd)* entkieseln
desilylation Desilylierung *f*
desludge *v* entschlammen
desludging *(Hyd)* Entschlammung *f*
desmo-enzyme Desmoenzym *n (ein extrazelluläres Enzym)*
desmolysis Desmolyse *f (Sprengung einer C–C-Bindung)*
desmotropic desmotrop, tautomer
desmotropism Desmotropie *f*, Tautomerie *f*
desmotropy Desmotropie *f*, Tautomerie *f*
desorb *v* desorbieren
desorption Desorption *f (Entweichen oder Austreiben sorbierter Gase aus Sorptionsmitteln)*
desoxy... *siehe: deoxy...*
destruction Zerstörung *f*, Vernichtung *f*, Abbau *m*, Destruktion *f*
destructive destruktiv, zersetzend, abbauend, *(besonders Werkstoffprüfung)* zerstörend
destructive distillation trockene [destruktive] Destillation *f*, Trockendestillation *f*, Zersetzungsdestillation *f*
desublimate *v* solidensieren
desublimation Solidensieren *n*
desugar *v* entzuckern
desugarization Entzuckern *n*, Entzuckerung *f*
desugarize *v s.* desugar
desulfur(ize) *v s.* desulphurize
desulphation Desulfatierung *f*
desulphonate *v* desulfonieren
desulphonation Desulfonierung *f*
desulphur(iz)ation Desulfurieren *n*, Desulfurierung *f*, Entschwefeln *n*, Entschwefelung *f*
desulphurize *v* desulfurieren, entschwefeln
detach *v* ablösen
detachment Ablösung *f*
detar *v* entteeren
detarrer Teer(ab)scheider *m*
detarring Entteeren *n*, Entteerung *f*
detect *v (Anal)* nachweisen, feststellen; auffinden *(Naturstoffe)*
detectability 1. *(Anal)* Nachweisbarkeit *f*; 2. *s.* detection limit
detectable nachweisbar
detectable by odour geruchlich wahrnehmbar
detectable by taste geschmacklich wahrnehmbar
detection *(Anal)* Nachweis *m*, Feststellung *f*; Auffindung *f (von Naturstoffen)*
detection agent Nachweismittel *n*, Nachweisreagens *n*
detection limit Nachweisgrenze *f*, Erfassungsgrenze *f*
detection reaction Nachweisreaktion *f*
detection sensitivity Nachweisempfindlichkeit *f*
detectivity *s.* detection limit
detector 1. Messfühler *m*, Fühler *m*; Messeinrichtung *f*; 2. *(Anal)* Detektor

m; Strahlungsdetektor m, Strahlungsempfänger m
detention time Verweilzeit f, Aufenthaltszeit f, Rückhaltezeit f, Haltezeit f, Durchlaufzeit f
detergency builder Aufbaustoff m, Builder m, Gerüstsubstanz f, Gerüststoff m *(zum Aufbau synthetischer Waschmittel)*
detergent reinigend, Reinigungs...; Wasch...
detergent 1. Reinigungsmittel n; Waschmittel n; 2. s. synthetic detergent; 3. Detergent m, Schlamminhibitor m, absetzverhinderndes Mittel n *(für Schmieröle)*
detergent additive Detergent-Additiv(e) n, Detergentzusatz m, Reinigungszusatz m, reinigender Zusatz m
detergent-sanitizer Desinfektionsmittel n mit Reinigungswirkung
detergent surfactant waschaktive Substanz f, waschaktiver Stoff m, WAS
deteriorate v schlechter werden, sich verschlechtern, an Wert verlieren; sich entmischen *(Emulsionen)*; verschlechtern, im Wert herabsetzen [mindern]
deterioration 1. Verschlechterung f, Herabminderung f, *(i. e. S.)* Wertminderung f; 2. Entmischung f *(von Emulsionen)*
determinable bestimmbar
determination Bestimmung f, Feststellung f, Ermittlung f
determine v bestimmen, feststellen, ermitteln
detoluate v enttoluenen
detoluation Enttoluenen n
detonate v detonieren, *(i. w. S.)* explodieren; detonieren lassen, zur Detonation bringen
detonating gas Knallgas n
detonation Detonation f, *(i. w. S.)* heftige Explosion f; Sprengung f
detosylation *(org Ch)* Detosylierung f
detoxicate v entgiften
detoxi(fi)cation Entgiften n, Entgiftung f
detoxify v entgiften
detrimental schädlich
detritiation Detritierung f
detritylation Detritylierung f
deuridylic *(Bioch)* Desuridyl...
deuterate v deuterieren, mit Deuterium [schwerem Wasserstoff] markieren
deuteration Deuterierung f, Markierung f mit Deuterium [schwerem Wasserstoff]
deuteride Deuterid n
deuterio group Deuteriogruppe f
deuterium D Deuterium n, schwerer Wasserstoff m *(Wasserstoffisotop $^{2}_{1}H$)*
deuterium exchange Deuteriumaustausch m
deuterium oxide Deuteriumoxid n, schweres Wasser n
deuterize v s. deuterate
deut(er)on *(Kern)* Deuteron n
Devarda's alloy devardasche Legierung f *(Reduktionsmittel)*
develop v 1. entwickeln; 2. *(Farb)* erzeugen; sich entwickeln; sich zeigen
developed dye Entwicklungsfarbstoff m
developer 1. *(Farb, Foto)* Entwickler m; 2. *(Chromat)* Elutionsmittel n, *Eluent m
developing bath *(Foto)* Entwicklerbad n, Entwicklungsbad n
developing chamber *(Chromat)* Entwicklungskammer f, Steiggefäß n
development Entwicklung f
development factor *(Foto)* Entwicklungsfaktor m, Gammawert m, Gamma n, Kontrastfaktor m, Gradation f
development of (antibiotic) resistance *(Biot)* Resistenzentwicklung f
device Einrichtung f, Vorrichtung f, Gerät n
devitrification Entglasung f
devitrified glass Glaskeramik f, glaskeramischer Stoff m, Vitrokeram n
devitrify v entglasen
devoid of structure strukturlos, gefügelos, unstrukturiert
devolatilization 1. Abnahme f der flüchtigen Bestandteile; 2. Entfernen n der flüchtigen Bestandteile
devolatilize v von flüchtigen Bestandteilen befreien
devolution Umwandlung f *(eines Elements durch radioaktiven Zerfall)*
dew-point Taupunkt m
dew-point curve *(Dest)* Taulinie f, Kondensationslinie f
Dewar flask (gläsernes) Dewar-Gefäß n, Weinhold-Gefäß n, weinhold-dewarsches Gefäß n
Dewar vessel Dewar-Gefäß n, Metall-

Dewar-Gefäß *n (zum Transport flüssiger Gase)*
dewater *v* entwässern, entfeuchten, *(manchmal auch)* eindicken, verdicken; sich entwässern lassen
dewax *v* entwachsen, entparaffinieren
dewaxing Entwachsen *n*, Entparaffinieren *n*, Entparaffinierung *f*
dextran *(Bioch)* Dextran *n*
dextrin Dextrin *n*, Stärkegummi *n*
dextrinate *v* 1. dextrinieren, in Dextrin umwandeln [überführen]; 2. dextrinieren, mit Dextrin versetzen
dextrine *s.* dextrin
dextrinization Umwandlung *f* [Überführung *f*] in Dextrin, Dextrinierung *f*; Dextrinbildung *f*, Dextrinierung *f*
dextrinize *v* dextrinieren, in Dextrin umwandeln [überführen]; sich in Dextrin umwandeln, dextrinieren
dextro *s.* dextrorotatory
dextrogyr(at)e *s.* dextrorotatory
dextrogyratory *s.* dextrorotatory
dextrogyrous *s.* dextrorotatory
dextrolactic acid L(+)-Milchsäure *f*, Rechtsmilchsäure *f*, Fleischmilchsäure *f*
dextrorotary *s.* dextrorotatory
dextrorotation Rechtsdrehung *f*
dextrorotatory rechtsdrehend, (+)-drehend, (+)-... *(Stereochemie)*
dextrose Dextrose *f*, Glucose *f*, D-Glucose *f*, Traubenzucker *m*
dextrotartaric acid (+)-Weinsäure *f*, L-Threarsäure *f*, Rechtsweinsäure *f*
DGA = diglycolamine
dGMP *s.* deoxyguanylic acid
DI *s.* deionization
diabatic photoreaction diabatische [nicht adiabatische] Photoreaktion *f*
diacetic acid *s.* acetoacetic acid
diacetyl Biacetyl *n*, *Butan-2,3-dion *n*
diacetylation Diacetylierung *f*
diacid 1. zweisäurig, zweibasig, zweiwertig *(Base)*; 2. Dihydrogen... *(Salz)*
diacid zweibasige [zweiprotonige, zweiwertige] Säure *f*
diacid salt Dihydrogensalz *n (einer dreibasigen Säure)*
diacylamine Diacylamin *n*
diad 1. zweiwertig; 2. *(Krist)* zweizählig
diad zweiwertiges Element *n*; zweiwertige Atomgruppe *f*
diad axis *(Krist)* zweizählige Achse *f* [Drehachse *f*]
diagnostic reagent Diagnostikum *n*
diagonal relationship [similarity] Schrägbeziehung *f (im Periodensystem)*
dialcohol zweiwertiger Alkohol *m*, Diol *n*
dialdehyde Dialdehyd *m*
dialkene Dien *n*, Diolefin *n*
dialkoxychlorophosphine $(RO)_2PCl$ Dialkylphosphorigsäurechlorid *n*
dialkyl ether Dialkylether *m*
dialkylate *v* dialkylieren
dialkylation Dialkylierung *f*
dialkylbenzene Dialkylbenzen *n*
diallyl phthalate resin *(Kst)* Diallylphthalatharz *n*, DAP
***N,N*-diallyl-2-chloroacetamide** 2-Chlor-*N,N*-diallylacetamid *n*, CDAA *n*
dialuramide Uramil *n*, Dialuramid *n*, 5-Aminobarbitursäure *f*
dialysable dialysierbar
dialysate Dialysat *n*; *(manchmal)* Dialysiergut *n*, zu dialysierende Flüssigkeit *f*
dialyse *v* dialysieren
dialyser Dialysator *m*, Dialysierapparat *m*
dialysing membrane Dialysiermembran *f*
dialysis Dialyse *f*
dialysis cell Dialysierzelle *f*
dialytic cell Dialysierzelle *f*
dialyze *v s.* dialyse
diamagnet Diamagnetikum *n*, diamagnetischer Stoff *m*
diamagnetic susceptibility diamagnetische Suszeptibilität *f*
diamagnetism Diamagnetismus *m*
diamide Diamid *n*, *(i. e. S.)* $H_2N–NH_2$ Hydrazin *n*
diamine Diamin *n*
1,4-diaminobenzene 1,4-Diaminobenzen *n*, *p*-Phenylendiamin *n*
4,4′-diaminobiphenyl 4,4′-Diaminobiphenyl *n*, Benzidin *n*
diammine mercuric chloride Diamminquecksilber(II)-chlorid *n*, schmelzbares weißes Präzipitat *n*
diammonium hydrogen phosphate Diammoniumphosphat *n*, Ammoniumhydrogenphosphat *n*
diamond Diamant *m*
diamond lattice *(Krist)* Diamantgitter *n*

diamond-like 1. diamantartig; 2. *s.* diamondoid
diamond pyramid hardness Vickershärte *f*, HV
diamondoid diamantoid, adamantoid *(Ringsystem)*
dianion Dianion *n*, Doppelanion *n*
diaphragm Scheidewand *f*, Trennungswand *f*, Zwischenwand *f*, Membran *f*; Zwischenboden *m*; Diaphragma *n*, Membran *f*, Scheidewand *f (Elektrolyse, Dialyse)*
diaphragm cell Diaphragmazelle *f*, Diaphragmenzelle *f (Elektrolyse)*
diaphragm pump Membranpumpe *f*, Diaphragmapumpe *f*
1,2-diaryldiazane 1,2-Diarylhydrazin *n*, Hydrazoaren *n*
1,2-diarylhydrazine 1,2-Diarylhydrazin *n*, Hydrazoaren *n*
diastereogenic diastereogen *(Stereochemie)*
diastereo(iso)mer Diastereomer(e) *n*, diastereomere Verbindung *f*
diastereo(iso)meric diastereomer
diastereoselectivity *(org Ch)* Diastereoselektivität *f*
diastereotopic diastereotop *(Stereochemie)*
diatom(aceous) earth Diatomeenerde *f*, Infusorienerde *f*, Kieselgur *f*
diatomic zweiatomig
diatomic zweiatomige Verbindung *f*
diatomic alcohol $C_nH_{2n}(OH)_2$ zweiwertiger Alkohol *m*, Diol *n*, *(i. w. S.)* Glykol *n*
diauxic growth diauxisches Wachstum *n (von Mikroorganismen)*
diauxy *(Bioch)* Diauxie *f (zwei- oder mehrphasiger Wachstumsverlauf)*
diaxial diaxial *(Stereochemie)*
diazene Diazen *n*, Diimid *n*
diazine Diazin *n*
diazo compound Diazoverbindung *f*
diazo dye Diazofarbstoff *m*
diazo group –N=N– Diazogruppe *f*, Diazorest *m*
diazoamino compound RN=NNHR Diazoaminoverbindung *f*
diazoate RN=NOM Diazotat *n*
diazocoupling Diazokupplung *f*, Azokupplung *f*
diazomethane Diazomethan *n*
diazonium group $[ArN\equiv N]^+$ Diazoniumgruppe *f*
diazonium salt Diazoniumsalz *n*, Diazoniumverbindung *f*
diazooxide Diazooxid *n*, Diazophenol *n*, Diazocyclohexadienon *n*
diazophenol Diazooxid *n*, Diazophenol *n*, Diazocyclohexadienon *n*
diazotate *s.* diazoate
diazotizable diazotierbar
diazotization Diazotieren *n*, Diazotierung *f*
diazotize *v* diazotieren
dibasic 1. zweibasig, zweiprotonig, zweiwertig *(Säure)*; 2. zweisäurig, zweibasig, zweiwertig *(Base oder basisches Salz)*
dibasic phosphate $M^I_3HPO_4$ sekundäres Phosphat *n*, Hydrogenphosphat *n*
dibasicity 1. Dibasizität *f (von Säuren)*; 2. Zweiwertigkeit *f (von Basen oder basischen Salzen)*
dibenzo-1,4-dioxin Dibenzo-1,4-dioxin *n*, Dibenzo-*p*-dioxin *n*, Dioxin *n*
dibenzo-*p*-dioxin Dibenzo-1,4-dioxin *n*, Dibenzo-*p*-dioxin *n*, Dioxin *n*
dibenzoylation Dibenzoylieren *n*, Dibenzoylierung *f*, zweifache Benzoylierung *f*
dibenzyl Bibenzyl *n*, *1,2-Diphenylethan *n*
dibenzyl ether Dibenzylether *m*
dibenzylation Dibenzylierung *f*
diborane Diboran *n*
dibromide Dibromid *n*
dibrominate *v* dibromieren, zweifach bromieren
dibromination Dibromieren *n*, Dibromierung *f*, zweifache Bromierung *f*
dibromo compound Dibromverbindung *f*
dibromomethane Dibrommethan *n*
dibromothymolsulphonephthalein Dibromthymolsulfophthalein *n*, Bromthymolblau *n (ein pH-Indikator)*
dibutyl phthalate Dibutylphthalat *n*
2,6-di-*tert*-butyl-*p*-cresol 2,6-Di-*tert*-butyl-*p*-cresol *n*, 2,6-Di-*tert*-butyl-4-methylphenol *n*, DBPC *(Antioxidans)*
dicarboxylic acid Dicarbonsäure *f*, zweibasige Carbonsäure *f*
dichloride Dichlorid *n*
dichlorination Dichlorierung *f*

dichloro derivative Dichlorderivat *n*
dichloroacetic acid Dichloressigsäure *f*, Dichlorethansäure *f*
dichloroacetylation Dichloracetylierung *f*
dichlorodiethyl sulphide Dichlordiethylsulfid *n*, Senfgas *n*
dichlorodifluoromethane Dichlordifluormethan *n*
dichlorodiphenyltrichloroethane Dichlordiphenyltrichlorethan *n*, DDT, *1,1,1-Trichlor-2-2-bis(*p*-chlorphenyl)-ethan *n*
1,2-dichloroethane 1,2-Dichlorethan *n*, Ethylendichlorid *n*
dichloroethanoic acid *s.* dichloroacetic acid
dichlorofluoromethane Dichlorfluormethan *n*
dichloromethane Dichlormethan *n*
2,4-dichlorophenoxyacetic acid 2,4-Dichlorphenoxyessigsäure *f*, 2,4-D *n* *(Herbizid)*
4-(2,4-dichlorophenoxy)butyric acid 4-(2,4-Dichlorphenoxy)buttersäure *f*, 2,4-DB *n* *(Herbizid)*
dichlorprop Dichlorprop *n*, 2,4-DP *n*, 2-(2′,4′-Dichlorphenoxy)propionsäure *f* *(Herbizid)*
dichroic *(Krist, Koll)* dichroitisch, doppelfarbig
dichroism *(Krist, Koll)* Dichroismus *m*, Doppelfarbigkeit *f*
dichromate $M^{I}_{2}Cr_2O_7$ Dichromat *n*
dichromate titration Chromatometrie *f*
dicyanogen Dicyan *n*
dicyclic *(org Ch)* bicyclisch
1,2-didehydroarene Arin *n*, Aryn *n*, 1,2-Dehydroaromat *n*
1,2-didehydrobenzene 1,2-Dehydrobenzen *n*, Benzin *n*, Benzyn *n*
didepside Didepsid *n* *(Ester aus zwei Phenolcarbonsäuren)*
di-derivative Diderivat *n*
dideuterio... zweifach deuteriert
die *v* *(Kern)* zugrunde gehen
die *v* **away** *s.* die down 1.
die *v* **down** 1. nachlassen, abklingen, zum Stillstand kommen, erlöschen *(Reaktion)*; 2. erlöschen *(Flamme)*
die 1. Matrize *f*, *(meist)* Unterwerkzeug *n*, Form *f*, Pressform *f*, Stanzform *f*, Schnittplatte *f*, Schneidplatte *f*, Stanzwerkzeug *n*, Schnittwerkzeug *n*; 2. Hohlform *f*, Form *f*, *(Metallverarbeitung auch)* Gesenk *n*, (metallische) Dauergießform *f*, Metallform *f*, Kokille *f*; 3. *(Kst)* Düse *f*
Dieckmann condensation [reaction] Dieckmann-Reaktion *f*, dieckmannsche (intramolekulare) Esterkondensation *f*
dielectric dielektrisch, Dielektrizitäts...
dielectric Dielektrikum *n*, Nichtleiter *m*
dielectric constant Dielektrizitätskonstante *f*, DK, *(i. e. S.)* relative Dielektrizitätskonstante *f*
dielectric loss dielektrischer Durchschlag *m*
dielectric resistance Durchschlagswiderstand *m*
dielectric strength Durchschlagfestigkeit *f*, dielektrische [elektrische] Festigkeit *f*
dielectrometer DK-Meter *n*, Dekameter *n*, Dielektrizitätskonstante-Messer *m*
dielectrometry *(Anal)* *Dielektrometrie *f*, Dekametrie *f*, DK-Metrie *f* *(Bestimmung der Dielektrizitätskonstante)*
dielectrophoresis Dielektrophorese *f* *(Trennung von Zellen und Nanoteilchen)*
Diels-Alder addition [condensation] *s.* Diels-Alder reaction
Diels-Alder reaction Diels-Alder-Reaktion *f*, Diels-Alder-Synthese *f*, diels-aldersche Diensynthese *f*, Dien-1,4-Addition *f*
diene *(org Ch)* Dien *n*
dienoic zweifach ungesättigt
dienophile Dienophil *n*, Philodien *n*, dienophile Komponente *f*, philodiener Partner *m*
dienophilic dienophil, philodien
diequatorial diäquatorial, diäquatorisch *(Stereochemie)*
diesel fuel (oil) Dieselkraftstoff *m*, DK
diester *(org Ch)* Diester *m*
dietary fiber Ballaststofffaser *f*
dietary lipid Nahrungslipid *n*
diethyl ether Diethylether *m*, Ethoxyethan *n*, *(i. e. S.)* Ether *m*
diethyl sulphate Schwefelsäurediethylester *m*, Diethylsulfat *n*
diethyl sulphide Diethylsulfid *n*, Ethylthioethan *n*

diethylamine Diethylamin *n*
diethylene glycol Diethylenglykol *n*, 2-(2-Hydroxyethoxy)ethanol *n*
diethylenetriamine pentaacetic acid Diethylentriaminpentaessigsäure *f*, DPE
diethylmagnesium Diethylmagnesium *n*
diethylstilboestrol Diethylstilböstrol *n*, *trans-P,P'*-Dihydroxydiethylstilben *n*
difference in solubility Löslichkeitsunterschied *m*
difference spectrum Differenzspektrum *n*
differential aeration cell Belüftungselement *n*, Belüftungszelle *f*, Sauerstoffkonzentrationszelle *f*, Evans-Element *n*, Evans-Korrosionselement *n*
differential centrifugation *(Bioch)* Differenzialzentrifugation *f*
differential distillation Differenzialdestillation *f*, offene [differenzielle] Destillation *f*
differential head Differenzdruckhöhe *f*
differential heat differenzielle [intermediäre] Wärme *f*
differential polarography *(Anal)* *Differenzpolarographie *f*, Differenzialpolarographie *f*
differential pressure Differenzdruck *m*, Wirkdruck *m*
differential pulse polarography *(Anal)* Differenz-Pulspolarographie *f*
differential reactor Differenzialreaktor *m*
differential scanning calorimetry Kalorimetrie *f* mit Differenzialabtastung, Differenzialscanningkalorimetrie *f*, DSK
differential thermal analysis Differenzialthermoanalyse *f*, DTA • **by differential thermal analysis** differenzialthermoanalytisch
differential thermogravimetry derivative Thermogravimetrie *f*, Differenzial-Thermogravimetrie *f*
differential titration Differenzialtitration *f*
difficultly soluble schwerlöslich, schwer [wenig] löslich, slö, wl.
diffraction Beugung *f*, Diffraktion *f*
diffraction angle Beugungswinkel *m*
diffraction grating Beugungsgitter *n*
diffraction pattern Beugungsbild *n*
diffraction spectrum Beugungsspektrum *n*, Gitterspektrum *n*, Normalspektrum *n*
diffusate Diffusat *n*
diffuse *v* diffundieren, eindringen *(in feiner Verteilung)*; sich ausbreiten; eindringen lassen
diffuse *v* **away** wegdiffundieren, abdiffundieren
diffuse *v* **in** eindiffundieren
diffuse *v* **out** herausdiffundieren
diffuse *v* **through** hindurchdiffundieren
diffuse double layer *(phys Ch)* diffuse Doppelschicht *f*
diffuse series *(Spektr)* d-Serie *f*, erste Nebenserie *f*
diffusible diffusionsfähig
diffusion Diffusion *f*; Transfusion *f (von Gasen durch eine poröse Trennwand)*
diffusion barrier Diffusionsmembran *f (elektrochem. Sensor)*
diffusion cell Diffuseur *m*, Diffusionsapparat *m*
diffusion coefficient Diffusionskoeffizient *m*, Diffusionskonstante *f*, Diffusionszahl *f*
diffusion-controlled *(phys Ch)* diffusionsgesteuert, diffusionsbedingt, diffusionsbestimmt, diffusionskontrolliert
diffusion(-controlled) current Diffusionsstrom *m*
diffusion furnace Diffusionsofen *m*
diffusion-limited *s.* diffusion-controlled
diffusion membrane *s.* diffusion barrier
diffusion potential Diffusionspotenzial *n*, Flüssigkeits(diffusions)potenzial *n*, Diffusions(galvani)spannung *f*
diffusion pump Diffusionspumpe *f*
diffusion tube Diffusionsrohr *n*
diffusional diffusorisch, Diffusions...
diffusionless transition diffusionslose Umwandlung *f*
diffusivity 1. Diffusionskoeffizient *m*, Diffusionskonstante *f*; 2. Temperaturleitfähigkeit *f*, Temperaturleitkoeffizient *m*, Temperaturleitzahl *f*
difluoride Difluorid *n*
difluorination Difluorierung *f*
difluorodichloromethane Dichlordifluormethan *n*
diformate Diformiat *n*
difunctional difunktionell, bifunktionell
difunctional unit difunktionelle Einheit *f*,

D-Einheit *f*, Grundbaustein *m* D *(zum Aufbau von Polymeren)*
digest *v* 1. aufschließen *(durch Hitze und/oder Lösungsmittel)*; digerieren; 2. ausziehen, auslaugen, digerieren; 3. *(Bio)* verdauen; 4. sich zersetzen; 5. altern *(Niederschläge)*; 6. *(Hyd)* (biologisch, biochemisch, mikrobiell) abbauen *(organische Abwasserinhaltsstoffe)*; (aus)faulen *(Schlamm)*
digested sludge *(Hyd)* ausgefaulter Schlamm *m*, Faulschlamm *m*
digester *(Hyd)* biologische Reinigungsanlage *f (für aerobe und anaerobe Abwasser- und Schlammbehandlung)*; *(i. e. S.)* Faulbehälter *m*, Schlammfaulbehälter *m*
digester gas Faulgas *n*, Biogas *n*
digestibility 1. Aufschließbarkeit *f*; 2. Ausziehbarkeit *f*; 3. Verdaulichkeit *f*, Bekömmlichkeit *f (von Lebensmitteln)*; 4. *(Hyd)* Faulfähigkeit *f*, Fäulnisfähigkeit *f*
digestible 1. aufschließbar; 2. ausziehbar, auslaugbar; 3. verdaulich, bekömmlich *(Lebensmittel)*; 4. *(Hyd)* faulfähig, fäulnisfähig
digesting agent Aufschlussmittel *n*, Aufschlusschemikalie *f*, Aufschließungsreagens *n*
digestion 1. Aufschließen *n*, Aufschluss *m (durch Hitze und/oder Lösungsmittel)*; Digerieren *n*; 2. Ausziehen *n*, Auslaugen *n*, Digerieren *n*; 3. *(Bio)* Verdauung *f*, Digestion *f*; 4. Alterung *f (eines Niederschlags)*; 5. *(Hyd)* (biologischer, biochemischer, mikrobieller) Abbau *m (organischer Abwasserinhaltsstoffe)*; Faulung *f*, Ausfaulen *n (des Schlamms)*
digestion mix Aufschlussmittelgemisch *n*, Reaktionsgemisch *n*, *(beim Kjeldahl-Aufschluss auch)* Katalysatorgemisch *n*
digestion tank *(Hyd)* Faulbehälter *m*, Schlammfaulbehälter *m*
digestive enzyme Verdauungsenzym *n*
diglyceride Diglycerid *n*
diglycol(l)ic acid Diglykolsäure *f*, *2,2′-Oxydiethansäure *f*
digonal hybridization digonale Hybridisierung *f*, *sp*-Hybridisierung *f*
dihalide Dihalogenid *n*
dihalogenation Dihalogenierung *f*
dihydrate Dihydrat *n*
dihydric 1. Dihydrogen... *(Salze)*; 2. zweiwertig, mit zwei Hydroxylgruppen [OH-Gruppen] *(Alkohole)*
dihydric alcohol *Diol *n*, zweiwertiger Alkohol *m*
dihydrogen Dihydrogen...
dihydrogen H_2 Diwasserstoff *m*
dihydroxy compound Dihydroxyverbindung *f*
2,5-dihydroxybenzoic acid 2,5-Dihydroxybenzoesäure *f*, Gentisinsäure *f*
dihydroxyphenylalanine Dihydroxyphenylalanin *n*, Dopa *n*, *2-Amino-3-(3,4-dihydroxyphenyl)propansäure *f*
dihydroxysuccinic acid Dihydroxybernsteinsäure *f*, *2,3-Dihydroxybutandisäure *f*, Weinsäure *f*
diimide Diimid *n*, Diazen *n*
diiodide Diiodid *n*
diisocyanate Diisocyanat *n (Verbindung mit zwei –N=C=O-Gruppen)*
diketo ester Diketoester *m*
diketone Diketon *n*
dil. *s.* dilute
dilatancy *(Koll)* Dilatanz *f*
dilatant *(Koll)* dilatant
dilatation Dilatation *f*, Ausdehnung *f*, Expansion *f*
dilatometer Dilatometer *n (Gerät zur Messung der Wärmeausdehnung)*
dilatometric dilatometrisch
dilatometry Dilatometrie *f*, Messung *f* der Wärmeausdehnung, dilatometrische Untersuchung *f*
diluent *(Anstr)* Verschneidmittel *n*, Verschnittmittel *n*, Streckmittel *n*, Verschnittlöser *m*, Verdünnungsmittel *n*, Verdünner *m (für Lösungsmittel)*; Streckmittel *n*, Streckstoff *m (für Feststoffe)*
dilute *v* verdünnen; strecken
dilute verdünnt, verd.
dilute sludge *(Hyd)* Dünnschlamm *m*
diluted waste acid Dünnsäure *f*
diluting agent 1. Verdünnungsmittel *n*; 2. *s.* diluent
dilution 1. Verdünnen *n*, Verdünnung *f*; Strecken *n*, Streckung *f*; 2. (verdünnte) Lösung *f*
dilution law ostwaldsches Verdünnungsgesetz *n*

dilution series Verdünnungsreihe *f*
dimedon(e) *(Anal)* Dimedon *n*, 5,5-Dimethyl-cyclohexan-1,3-dion *n*
dimer Dimer(e) *n*
dimeric dimer
dimerization Dimerisation *f*, Dimerisierung *f*
dimerize *v* dimerisieren
di-π-methane rearrangement Di-π-methan-Umlagerung *f (Photochemie)*
dimethoxymethane *Dimethoxymethan *n*, Formaldehyddimethylacetal *n*, Methylal *n*
dimethoxymethyl group $-CH(OCH_3)_2$ Dimethoxymethylgruppe *f*
dimethyl arsinic acid Dimethylarsinsäure *f*, Kakodylsäure *f*
dimethyl ether Dimethylether *m*, Methoxymethan *n*
dimethyl formamide Dimethylformamid *n*
dimethyl glyoxal Dimethylglyoxal *n*, *Butan-2,3-dion *n*, Biacetyl *n*
dimethyl sulphate Schwefelsäuredimethylester *m*, Dimethylsulfat *n*
dimethyl sulphoxide Dimethylsulfoxid *n*, DMSO
dimethylamine Dimethylamin *n*
dimethylbenzene Dimethylbenzen *n*, Xylen *n*, Xylol *n*
dimethyldichlorosilane Dimethyldichlorsilan *n*
dimethylformamide Dimethylformamid *n*, DMF
dimethylketone Dimethylketon *n*, Aceton *n*, *Propan-2-on *n*
dimethylphenol Dimethylphenol *n*, Xylenol *n*
dimorphic dimorph
dimorphous dimorph
Dimroth condenser Dimroth-Kühler *m*
dimsyl ion Methylsulfinylcarbanion *n*
dinegative zweifach negativ (geladen)
dinitration Dinitrieren *n*, Dinitrierung *f*
dinitride Dinitrid *n*
dinitrile Dinitril *n*
dinitro compound Dinitroverbindung *f*
dinitrobenzoic acid Dinitrobenzoesäure *f*, Dinitrobenzencarbonsäure *f*
dinitrocresol Dinitrocresol *n*
dinitrogen monoxide Distickstoffoxid *n*, Stickstoff(I)-oxid *n*
dinitrophenol Dinitrophenol *n*, DNP, Hydroxydinitrobenzen *n*
dinitrophenyl derivative Dinitrophenylderivat *n*, DNP-Derivat *n*
dinoseb Dinoseb *n*, Dinitro-*o*-*sec*-butylphenol *n*, DNBP
diode laser Diodenlaser *m*, Halbleiterlaser *m*
diol *Diol *n*, zweiwertiger Alkohol *m*
diolefin(e) Diolefin *n*
diolefinic diolefinisch, Diolefin...
dioxan(e) Dioxan *n (Diether des Glykols)*
dioxide Dioxid *n*
dioxin Dioxin *n*, Dibenzo-1,4-dioxin *n*, Dibenzo-*p*-dioxin *n*, *(i. e. S.)* 2,3,7,8-Tetrachlordibenzo-1,4-dioxin *n*, 2,3,7,8-TCDD, TCDD, Tetrachlorodioxin *n*
dioxygen O_2 Disauerstoff *m*
dioxygenase *(Bioch)* Dioxygenase *f*, Sauerstofftransferase *f*
dip *v* 1. (ein)tauchen; 2. *(Anstr)* tauchlackieren
dip-coat *v* 1. durch Tauchen beschichten, tauchen; 2. *(Kst)* Heißtauchen
dip 1. Tauchen *n*, Eintauchen *n*; 2. *(Landw)* Tauchbad *n*; Bademittel *n*
dip coating 1. Tauchbeschichten *n*; Tauchlackieren *n*; 2. *(Kst)* Heißtauchen *n*; 3. Tauchanstrichstoff *m*; 4. Tauchschicht *f*
dipeptide Dipeptid *n (Peptid mit 2 Aminosäureeinheiten)*
diphenyl *s.* biphenyl
diphenyl dye Diphenylfarbstoff *m*
diphenyl ketone *Diphenylketon *n*, Benzoylbenzen *n*, Benzophenon *n*
diphenylamine Diphenylamin *n*
diphenyldiazene Azobenzen *n*, Diphenyldiimid *n*
diphenyldiimide Azobenzen *n*, Diphenyldiimid *n*
1,2-diphenylethane *1,2-Diphenylethan *n*, Bibenzyl *n*
diphenylmethane Diphenylmethan *n*, Benzylbenzen *n*
1,3-diphenylpropenone 1,3-Diphenylpropenon *n*, Benzylidenacetophenon *n*, Benzalacetophenon *n*, Chalkon *n*
diphenylthiocarbazone Diphenylthiocarbazon *n*, Dithizon *n*
diphosgene Diphosgen *n*, Perchlorameisensäuremethylester *m*, Chlor-

ameisensäuretrichlormethylester *m*, Perchlormethylformiat *n*
diphosphate Diphosphat *n*, Pyrophosphat *n*
diphosphoglyceric acid *(Bioch)* Diphosphoglycerolsäure *f*
diphosphoric acid Diphosphorsäure *f*, Pyrophosphorsäure *f*
diphosphorous acid diphosphorige Säure *f*
dipolar Dipol..., polar
dipolar bond koordinative [dative] Bindung *f*, Koordinationsbindung *f*, Donator-Akzeptor-Bindung *f*
dipolar molecule Dipolmolekül *n*, polares Molekül *n*
dipole Dipol *m*; Dipolmolekül *n*
dipole-dipole attraction Dipol-Dipol-Anziehung *f*
dipole force Dipolkraft *f*
dipole moment Dipolmoment *n*
dipositive zweifach positiv (geladen)
dipotassium hydrogen phosphate Dikaliumphosphat *n*, Kaliumhydrogenphosphat *n*
dipping 1. Tauchen *n*, Eintauchen *n*; 2. *(Anstr)* Tauchen *n*, Tauchlackieren *n*
dipping bath Tauchbad *n*
dipping electrode Tauchelektrode *f*, Eintauchelektrode *f*
dipping refractometer Eintauchrefraktometer *n*
dipping tank Tauchbehälter *m*, Tauchtank *m*, Tauchgefäß *n*, Tauchwanne *f*, Tauchbottich *m*
dipping varnish Tauchlack *m*
diprimary *(org Ch)* diprimär
diprotic acid zweibasige [zweiprotonige, zweiwertige] Säure *f*
diradical Diradikal *n*, Biradikal *n*
direct *v* dirigieren, lenken (*einen zweiten Substituenten nach der m- bzw. o-, p-Stellung*)
direct analysis direkte Analyse *f*
direct analysis of daughter ions Ionenenergie-Spektroskopie *f* zum Nachweis metastabiler Zerfälle
direct-contact poison Kontaktgift *n*, Berührungsgift *n*
direct coupling galvanische Kopplung *f*
direct current Gleichstrom *m*
direct distillation Gleichstromdestillation *f*
direct dye(stuff) Direktfarbstoff *m*, direktziehender [substantiver] Farbstoff *m*, Substantivfarbstoff *m*
direct fermentation *(Biot)* direkte Fermentation *f*
direct fungicide direktes Fungizid *n* *(durch Kontakt mit dem Pilzmyzel wirkend)*; Fungizid *n* mit kurativer Wirkung
direct-light microscopy Auflichtmikroskopie *f*
direct methanol fuel cell Direkt-Methanol-Brennstoffzelle *f*
direct potentiometry *(Anal)* Direktpotenziometrie *f*
direct reaction Hinreaktion *f*
direct-reduction process *(Met)* Direktreduktionsverfahren *n* *(zur Eisengewinnung)*
direct smelting *(Met)* direktes [unmittelbares] Schmelzen *n*
direct synthesis direkte Synthese *f*, Direktsynthese *f* *(z. B. zur Herstellung von Chlorsilanen)*
direct titration direkte Titration *f*
direct weighing einfache [direkte] Wägung *f*, Proportional(itäts)wägung *f*, Kompensationswägung *f*
directed biosynthesis *(Biot)* gerichtete Biosynthese *f*
directed interesterification [transesterification] gelenkte [gerichtete] Umesterung *f*
direction of flow Strömungsrichtung *f*, Stromrichtung *f*, Fließrichtung *f*
direction of propagation Ausbreitungsrichtung *f*, Fortpflanzungsrichtung *f*
direction of vibration Schwingungsrichtung *f*
directional distribution (räumliche) Ausrichtung *f*, Orientierung *f*
directional preference Richtungsbevorzugung *f*
directive influence dirigierender Einfluss *m*, dirigierende Wirkung *f* *(von Substituenten)*
dirt Schmutz *m*
dirt particle Schmutzteilchen *n*, Schmutzpartikel *n*, Verschmutzungsteilchen *n*, Verunreinigungspartikel *n*
dirtiness Verschmutzung *f*, verschmutzter Zustand *m*
dirty schmutzig, verschmutzt

dirty water Schmutzwasser *n*
disaccharide Disaccharid *n*, Zweifachzucker *m*
disagreeable unangenehm, schlecht, widerlich, übel *(Geruch)*
disappearance curve *(Tox)* Abbaukurve *f*
disazo dye Disazofarbstoff *m*
disc *s.* disk
disc electrophoresis *(Anal)* Disk-Elektrophorese *f (mit diskontinuierlichem Gelsystem)* • **by disc electrophoresis** diskelektrophoretisch
discard *v* 1. *(Lab)* (als unbrauchbar) verwerfen; 2. *(Tech)* abstoßen; 3. *(Aufb)* ausscheiden, abscheiden, aussortieren, aussondern, austragen
discard *(Aufb)* Ausscheiden *n*, Abscheiden *n*, Aussortieren *n*, Aussondern *n*, Austrag *m*
discarded verbraucht *(z. B. Lösung)*
discharge *v* 1. austragen, ausbringen, entnehmen; ablassen, ausströmen lassen *(Fluide)*; entladen, entleeren; abstechen *(Hochofen)*; ablaufen, auslaufen, abfließen, ausströmen; 2. fördern *(Pumpe, Kompressor)*; 3. entladen *(Physik)*
discharge *v* **into a receiving stream** in den Vorfluter einleiten [ableiten] *(Abwasser)*
discharge *v* **into the air** emittieren, in die Atmosphäre abgeben *(Luftschadstoffe)*
discharge 1. Austragen *n*, Austrag *m*, Ausbringen *n*, Entnahme *f*; Ablassen *n*; Entladen *n*, Entleeren *n*; Ablaufen *n*, Auslaufen *n*, Abfließen *n*, Ausströmen *n*; 2. Förderung *f (von Pumpen, Kompressoren)*; 3. Entladung *f (Physik)*; 4. Fördermenge *f (von Pumpen, Kompressoren)*; 5. *s.* discharge point; 6. *s.* waste discharge
discharge cycle Entladezyklus *m*
discharge electrode Sprühelektrode *f*, Entladungselektrode *f*
discharge head Förderhöhe *f (z. B. einer Pumpe)*
discharge limits *(Hyd)* Grenzwerte *mpl* für die Einleitung von Abwasserinhaltsstoffen in Vorfluter, Abwassergrenzwerte *mpl*
discharge pipe Austrag(s)rohr *n*, Abflussrohr *n*, Ausflussrohr *n*, Abzugsrohr *n*, Auslaufrohr *n*, Ablaufrohr *n*, Austrittsrohr *n*, Abflussstutzen *m*
discharge plasma Gasentladungsplasma *n*
discharge point 1. Austragstelle *f*, Entnahmestelle *f*; Austrittsstelle *f*; 2. *(Hyd)* Einleitungsstelle *f (für Abwasser)*
discharge potential Entladungspotenzial *n*
discharge pressure Förderdruck *m*, Lieferdruck *m*
discharge regulations [standards] *(Hyd)* Einleitungsbedingungen *fpl (für Abwasser)*
discharge stroke Druckhub *m*, Förderhub *m*
discharge tube 1. *s.* discharge pipe; 2. Entladungsröhre *f*, Gasentladungsröhre *f*, geißlersche Röhre *f*
discharge valve Austragventil *n*, Auslassventil *n*, Abflussventil *n*, Ablassventil *n*, Leerventil *n*
discolor *v s.* discolour
discoloration Verfärbung *f*, Farb(ver)änderung *f*; Verbleichen *n*, Verblassen *n*, Verschießen *n*; Entfärbung *f*
discolour *v* verfärben; entfärben; sich verfärben; verbleichen, verblassen, verschießen; sich entfärben, die Farbe verlieren
discolouration *s.* discoloration
discontinuous diskontinuierlich, unstetig, unterbrochen
discontinuous phase *s.* disperse phase
discontinuous transition Umwandlung *f* [Phasenumwandlung *f*] 1. Ordnung [Art], Übergang *m* [Phasenübergang *m*] 1. Ordnung [Art], diskontinuierliche Umwandlung *f* [Phasenumwandlung *f*]
discrete spectrum diskretes Spektrum *n*, Linienspektrum *n*
discrimination *(Anal)* Diskriminierung *f*
disecondary disekundär
disgusting unangenehm, schlecht, widerlich, übel *(Geruch)*
dish Schale *f*
dished nach innen gewölbt, konkav
dishwasher *(Lab)* Spülmaschine *f*
disilane Disilan *n*
disilicate Disilicat *n*
disilicide Disilicid *n*

disinfect *v* desinfizieren, entkeimen, entseuchen
disinfectant desinfizierend, keimtötend
disinfectant Desinfektionsmittel *n*, Desinfiziens *n*, desinfizierendes [keimtötendes] Mittel *n*, Entkeimungsmittel *n*
disinfectant solution Desinfektionslösung *f*
disinfection Desinfektion *f*, Entkeimung *f*, Entseuchung *f*
disintegrate *v* 1. auflösen *(in die Bestandteile)*; desintegrieren; (ver)mahlen *(weiche bis mittelharte Stoffe)*; sich auflösen *(in die Bestandteile)*; 2. *(Kern)* zerfallen; verwittern, zerfallen
disintegration 1. Zerfall *m*, Auflösung *f* *(in die Bestandteile)*; Desintegration *f*; Mahlen *n*, Vermahlen *n* *(weicher bis mittelharter Stoffe)*; Verwitterung *f*, Zerfall *m*; 2. *(Kern)* Zerfall *m*
disintegration chain *s.* disintegration series
disintegration constant Zerfallskonstante *f*
disintegration energy Zerfallsenergie *f*
disintegration family *s.* disintegration series
disintegration rate Aktivität *f*, Umwandlungsrate *f* *(eines radioaktiven Stoffes)*
disintegration series (radioaktive) Zerfallsreihe *f*, radioaktive Familie *f*
disintegrator Desintegrator *m*, Desintegratormühle *f* *(Mühle für weiche bis mittelharte Stoffe)*
β-disintegrator *(Kern)* β-Strahler *m*, Betastrahler *m*
disk Scheibe *f*; Platte *f*
disk electrode Scheibenelektrode *f*
disk electrophoresis *s.* disc electrophoresis
disk filter Scheibenfilter *n*
disk-like 1. scheibenförmig; 2. *s.* diskotic
disk mill Scheibenmühle *f*, Mahlscheibenmühle *f*, Tellermühle *f*; Scheibenkolloidmühle *f*
disk-type agitator Scheibenrührer *m*
diskotic diskotisch *(Flüssigkristall)*
dislocation *(Krist)* Versetzung *f*, Dislokation *f*
dismutate *v* dismutieren
dismutation Dismutation *f*
dismutation reaction Dismutationsreaktion *f*
dismutative dismutativ
disodium phosphate Dinatriumphosphat *n*, Natriumhydrogenphosphat *n*
disorder 1. *(phys Ch)* Unordnung *f*; 2. *(Krist)* Fehlordnung *f*, Unordnung *f*
disordered 1. *(phys Ch)* ungeordnet; 2. *(Krist)* fehlgeordnet, ungeordnet
disorient *v* entorientieren
dispel *v* verjagen, vertreiben, austreiben *(flüchtige Stoffe)*
dispense *v* abziehen *(Flüssigkeiten)*
dispenser 1. Verteiler *m*, Probengeber *m*; 2. Spender *m* *(Gerät zur bequemen Entnahme von Materialien)*
dispensing buret(te) Vorratsbürette *f*
dispersal *s.* dispersion
dispersant *s.* dispersing agent
disperse *v* dispergieren, fein verteilen [zerteilen]
disperse dispers, fein verteilt [zerteilt]
disperse medium *s.* dispersion medium
disperse particles disperse [dispergierte] Teilchen *npl*
disperse phase disperse Phase *f*, dispergierte [zerteilte, innere, offene] Phase *f*, Dispersum *n*
disperse-phase rule *(Koll)* Bodenkörperregel *f*
disperse system disperses System *n*, Dispersion *f*
dispersibility Dispergierbarkeit *f*
dispersible dispergierbar
dispersing additive Dispersionszusatz *m*, Dispersantadditiv *n*, Dispersantzusatz *m*
dispersing agent Dispergier(hilfs)mittel *n*, Dispersionsmittel *n*, Dispersant *m*
dispersion 1. Dispersion *f*, Feinverteilung *f*, feine Verteilung *f* [Zerteilung *f*] *(Zustand)*; Dispergieren *n*, Dispergierung *f*, Feinzerteilen *n*, Feinverteilen *n* *(Herstellen einer Dispersion)*; Dispersion *f* *(von Wellen)*; 2. *s.* disperse system
dispersion adhesive 1. *(Kst)* Dispersionsklebstoff *m*; 2. *(Kst)* Klebstoffdispersion
dispersion colloid Dispersoidkolloid *n*, Dispersionskolloid *n*, Phasenkolloid *n*
dispersion force Dispersionskraft *f*, London-(van-der-Waals-)Kraft *f*
dispersion medium Dispersionsmittel *n*, Dispersionsmedium *n*, Dispergens *n*
dispersity Dispersität *f*, Dispersions-

zustand *m*, disperser Zustand *m*; Dispersionsgrad *m*, Dispersitätsgrad *m*
dispersive dispergierend, Dispersions..., dispersiv
dispersoid 1. Dispersoid *n*; 2. *(Koll)* Dispersoidkolloid *n*, Dispersionskolloid *n*, Phasenkolloid *n*
dispiro compound [hydrocarbon] Dispiro-Verbindung *f*
displace *v* verlagern, verschieben; verdrängen; ersetzen
displaceable verschiebbar; verdrängbar; ersetzbar
displacement 1. Verlagerung *f*, Verschiebung *f*; Verdrängung *f*; Ersetzung *f*; 2. Fördermenge *f* je Umdrehung, Fördervolumen *n*
displacement analysis [chromatography] Verdrängungsanalyse *f*, Verdrängungschromatographie *f*
displacement titration Verdrängungstitration *f*
displacive transition displazive Umwandlung *f* [Phasenumwandlung *f*]
disposal Beseitigung *f*, Entsorgung *f*; Deponie *f*, Ablagerung *f*
disposal site Deponie *f*; Entsorgungsstandort *m*
disposal to land Deponie *f*, Ablagerung *f* *(von Abprodukten)*
disposal to sea Beseitigung *f* durch Einleitung ins Meer; Beseitigung *f* durch Abklappen [Verkippen ins Meer]
disposal to tip Ablagerung *f* auf Halde, Aufhalden *n*, Verkippen *n*
dispose *v* **of** beseitigen; deponieren, ablagern *(Abprodukte)*
dispose *v* **to sea** ins Meer einleiten; abklappen, ins Meer verkippen
dispose *v* **to tip** auf Halde ablagern, aufhalden, verkippen *(Abprodukte)*
disproportionate *v* disproportionieren; sich disproportionieren
disproportionation Disproportionierung *f*
disproportionation reaction Disproportionierungsreaktion *f*
disrotatory *(org Ch)* disrotatorisch
disrupt *v* lösen, (auf)spalten, sprengen *(eine chemische Bindung)*
dissd. = dissolved
dissimilation Dissimilation *f*, Katabolismus *m*
dissipate *v* ableiten, abführen *(Wärme)*
dissipation 1. Dissipation *f*, *(besonders quantitativ auch)* Energiedissipation *f* *(Übergang einer beliebigen Energieform in Wärme)*; 2. *s.* dissipation of heat
dissipation of heat Wärmeableitung *f*
dissipative dissipativ
dissociable dissoziabel, dissoziationsfähig
dissociate *v* dissoziieren, sich aufspalten, zerfallen; dissoziieren, aufspalten, trennen
dissociation Dissoziation *f*, Aufspaltung *f*, *(Tätigkeit auch)* Trennung *f*, *(Vorgang auch)* Zerfall *m*
dissociation constant Dissoziationskonstante *f*
dissociation equilibrium Dissoziationsgleichgewicht *n*
dissoluble 1. löslich, (auf)lösbar; 2. lösbar *(mechanische Verbindung)*
dissolute (auf)gelöst
dissolution Lösen *n*, Auflösen *n*; Inlösunggehen *n*, Auflösung *f*
dissolution rate Auflösungsgeschwindigkeit *f*
dissolvability Löslichkeit *f*, Lösbarkeit *f*
dissolvable löslich, (auf)lösbar
dissolve *v* (auf)lösen; sich (auf)lösen, in Lösung gehen
dissolve *v* **away [out]** herauslösen
dissolve *v* **with difficulty** schwerlöslich sein, sich schwer lösen
dissolved constituent *(Hyd)* gelöster Inhaltsstoff *m*
dissolved organic carbon *(Hyd)* gelöster organischer [organisch gebundener] Kohlenstoff *m*
dissolved organic chlorine *(Hyd)* gelöstes organisches [organisch gebundenes] Chlor *n*
dissolved oxygen *(Hyd)* gelöster Sauerstoff *m*, Gelöstsauerstoff *m*
dissolved oxygen deficit *(Hyd)* Sauerstoffdefizit *n*, Sättigungsdefizit *n* *(Differenz zwischen Sättigungswert und tatsächlichem Sauerstoffgehalt)*
dissolved oxygen sag *(Hyd)* Sauerstoffschwund *m*, Sauerstoffzehrung *f*
dissolved solid pollutants *(Hyd)* gelöste Schmutzstoffe *mpl* [Verunreinigungen *fpl*]
dissolved solids 1. *(Anal)* Gelöstes *n*,

gelöste Stoffe *mpl*; 2. *(Hyd)* (echt) gelöste Stoffe *mpl* [Wasserinhaltsstoffe *mpl*]
dissolvent Lösungsmittel *n*, Lösemittel *n*, Lsgm., Löser *m*, Solvens *n*
dissymmetric(al) asymmetrisch, unsymmetrisch
dissymmetry Asymmetrie *f*, Unsymmetrie *f*
dist. *s.* distillable
distasteful unschmackhaft, von unangenehmem Geschmack, widerlich schmeckend
distd. = distilled
distg. = distilling
distil *v* destillieren, *(bei der Alkoholherstellung auch)* brennen; überdestillieren, übergehen
distil *v* **along with** gemeinsam destillieren mit
distil *v* **off** abdestillieren, abtreiben; abdestillieren, entweichen
distil *v* **out** (her)ausdestillieren
distil *v* **over** überdestillieren, übertreiben; überdestillieren, übergehen
distil *v* **together** gemeinsam übergehen
distill *v s.* distil
distillability Destillierbarkeit *f*
distillable destillierbar
distilland Destillationsgut *n*, Destilliergut *n*, *(während der Destillation auch)* Destillans *n*
distillate Destillat *n*
distillate drain Destillatabzug *m*
distillate fraction Destillat(ions)fraktion *f*
distillate receiver Destillat(ions)vorlage *f*, Destilliervorlage *f*, Destillatsammler *m*, Vorlage *f*
distillation Destillieren *n*, Destillation *f*, *(bei der Alkoholherstellung auch)* Brennen *n*; Überdestillieren *n*, Übergehen *n* • **by (means of) distillation** durch Destillation, destillativ
distillation analysis Siedeanalyse *f*
distillation apparatus Destilliergerät *n*, Destillationsgerät *n*, Destillierapparat *m*, Destillationsapparat *m*, *(bei der Alkoholherstellung auch)* Brenngerät *n*, Brennapparat *m*
distillation at constant pressure isobare Destillation *f*
distillation boiler Destillierblase *f*, Destillationsblase *f*, Blase *f*
distillation column Destillierkolonne *f*, Destillationskolonne *f*, Destilliersäule *f*
distillation curve Destillationskurve *f*, Destillationslinie *f*
distillation flask Destillierkolben *m*, Destillationskolben *m*
distillation range Destillationsbereich *m*, Siedebereich *m*, Siedegrenzen *fpl*
distillation residue Destillationsrückstand *m*
distillation stillhead 1. Kolonnenkopf *m*; 2. *(Lab)* Destillieraufsatz *m*
distillation tower Destillationsturm *m*
distillation under vacuum Vakuumdestillation *f*, Unterdruckdestillation *f*, Destillation *f* im Vakuum, Destillation *f* unter vermindertem Druck
distilled in vacuo vakuumdestilliert, im [unter] Vakuum destilliert
distilled spirit(s) Branntwein *m*
distilled water destilliertes Wasser *n*, Aqua destillata
distillery Brennerei *f*, Branntweinbrennerei *f*, Spiritusbrennerei *f*
distilling ... *siehe: distillation ...*
distn. *s.* distillation
distort *v* verzerren; verdrehen
distortion Verzerrung *f*; Verdrehung *f*
distortion polarization Verschiebungspolarisation *f*
distribute *v* verteilen
distribution Verteilung *f*
distribution coefficient Verteilungskoeffizient *m*, *Verteilungskonstante *f*
distribution curve Verteilungskurve *f*
distribution equilibrium Verteilungsgleichgewicht *n*
distribution law Verteilungsgesetz *n*; nernstsches Verteilungsgesetz *n*, nernstscher Verteilungssatz *m*
distribution of particle size Korn(größen)verteilung *f*; Teilchengrößenverteilung *f*, Partikelgrößenverteilung *f*
distribution ratio Verteilungsverhältnis *n*, Distributionsverhältnis *n*
distributor Verteiler *m*, Verteilvorrichtung *f*
disubstituted disubstituiert, zweifach substituiert
disubstitution Disubstitution *f*
disulphate $M^I_2S_2O_7$ Disulfat *n*
disulphide MS_2 Disulfid *n*
disulphide bond [bridge] 1. Disulfid-

brücke *f*, Disulfidbindung *f*, –S–S-Brücke *f*, –S–S-Bindung *f*; 2. *(Bioch)* Cystin-Brücke *f*
disulphide (cross-)link *s.* disulphide bond
disulphide linkage *s.* disulphide bond
disulphonate Disulfonat *n*
disulphonation Disulfonierung *f*
disulphonic acid Disulfonsäure *f*
disulphuric acid Dischwefelsäure *f*
disulphurous acid dischweflige Säure *f*
diterpene Diterpen *n*
diterpenoid Diterpenoid *n*
diterpenoid alkaloid Diterpenalkaloid *n*
dithio acid $C_nH2_{n+1}CSSH$ Dithiosäure *f*
dithio compound Dithioverbindung *f*
dithioacetal Dithioacetal *n*, Mercaptal *n*
dithiocarbamate NH_2CSSM^I Dithiocarbamat *n*
dithiocarbamic acid Dithiocarbamidsäure *f*
dithiocarbonic acid Dithiokohlensäure *f*, *(in der Form HOCSSH auch)* Thiolthionkohlensäure *f*, *(in der Form HSCOSH auch)* Dithiolkohlensäure *f*
dithiocarboxylation Dithiocarboxylierung *f*
dithiocarboxylic acid Thionthiolsäure *f*, Carbodithiosäure *f*
dithioketal Dithioketal *n*, Mercaptal *n*
dithionate $M^I_2S_2O_6$ Dithionat *n*
dithionic acid Dithionsäure *f*
dithionite 1. $M^I_2S_2O_4$ Dithionit *n*; 2. *(Farb)* $Na_2S_2O_4$ Dithionit *n*, Natriumdithionit *n*
dithionous acid dithionige Säure *f*
dithiooxamide Dithiooxamid *n*, Rubeanwasserstoff *m*
dithiophosphate $M^I_3PS_2O_2$ Dithiophosphat *n*
dithiophosphoric acid Dithiophosphorsäure *f*
dithizone Dithizon *n*, Diphenylthiocarbazon *n*
ditrigonal-scalenohedral *(Krist)* ditrigonal-skalenoedrisch
diunsaturated doppelt [zweifach] ungesättigt
diurea Diharnstoff *m*, 4-Urazin *n*
diuretic diuretisch, harntreibend
diuretic (agent) Diuretikum *n*, harntreibendes [diuretisches] Mittel *n*
divacancy Doppelleerstelle *f*
divalence Zweiwertigkeit *f*, Bivalenz *f*
divalency Zweiwertigkeit *f*, Bivalenz *f*
divalent zweiwertig, bivalent
divariant divariant, bivariant, zweifachfrei
diverse ion effect Wirkung *f* fremdioniger Zusätze, Einsalzeffekt *m*
divisible teilbar; spaltbar *(Racemat)*
division 1. Teilung *f*, Strichteilung *f (eines Messinstruments)*; 2. Skalenteil *m*, Teilstrichabstand *m*; 3. Teilstrich *m*
DLS *s.* dynamic light scattering
DMA *s.* dynamic mechanical analysis
DME *s.* dropping-mercury electrode
DMF *s.* dimethyl formamide
D.M.F. *s.* dry mineral-matter-free
DMFC *s.* direct methanol fuel cell
d.m.m.f. *s.* dry mineral-matter-free
DMSO *s.* dimethyl sulphoxide
DNA *(Bioch)* DNA *f*, Desoxyribonucleinsäure *f*
DNA ligase DNA-Ligase *f*
DNA polymerase DNA-Polymerase *f*
DNA replication DNA-Replikation *f*
DNA sequence DNA-Sequenz *f*
DNA strand DNA-Strang *m*
DNase *s.* deoxyribonuclease
DNBP *s.* dinoseb
DO *s.* dissolved oxygen
Döbereiner's law of triads döbereinersche [Döbereiners] Triadenregel *f*
Dobson unit Dobson-Unit *f*, Dobson-Einheit *f (Ozongehalt in der Luft)*
DOC *s.* dissolved organic carbon
DOCl *s.* dissolved organic chlorine
docosane Docosan *n*
docosanoic acid Docosansäure *f*
docosenoic acid Docosensäure *f*
dodecahedron *(Krist)* Dodekaeder *n*, Zwölfflächner *m*, Zwölfflach *n*
dodecahydrate Dodekahydrat *n*
dodecane *(org Ch)* Dodecan *n*
dodecanedioic acid Dodecandisäure *f*, Decan-1,10-dicarbonsäure *f*
dodecanoic acid *Dodecansäure *f*, Laurinsäure *f*
1-dodecene Dodec-1-en *n*
dodecenoic acid Dodecensäure *f*
dodecyl alcohol Dodecylalkohol *m*, *Dodecan-1-ol *n*, Laurylalkohol *m*
dodecyne *(org Ch)* Dodecin *n*
DoE = Department of Environment
Doebner-Miller reaction Doebner-Miller-Reaktion *f*, Doebner-Miller-Synthe-

se *f (zur Darstellung von Chinolin und dessen Derivaten)*

dolomite 1. *(Min)* Dolomit *m (Calciummagnesiumcarbonat)*; 2. *(Geol)* Dolomit *m*, Dolomitgestein *n*

dolomitic dolomitisch, Dolomit...

dolomitic lime Dolomitkalk *m*, Magnesiumbranntkalk *m*, Graukalk *m (Calcium-Magnesium-Oxid)*

domain 1. Bereich *m*, Bezirk *m*; 2. *(Krist)* Domäne *f*, *(bei Ferromagnetika auch)* weißscher Bezirk *m*, Weiß-Bezirk *m*, ferromagnetische Domäne *f*

domain wall *(Krist)* Bloch-Wand *f*, blochsche Wand *f*, Domänenwand *f*, Antiphasengrenze *f*

dome 1. *(Tech)* Haube *f*; 2. *(Dest)* Glocke(nkappe) *f*, Austauschglocke *f*

domed konvex gewölbt

domestic chemistry Haushaltchemie *f*

domestic refuse [rubbish] Haus(halt)-müll *m*, häuslicher Müll *m*

domestic sewage [waste water] Haushaltabwasser *n*, häusliches Abwasser *n*

donate *v* abgeben, liefern, spenden

donation Abgabe *f*

donation of electrons Elektronenabgabe *f*

Donnan (membrane) equilibrium donnansches Membrangleichgewicht *n*, Donnan-Gleichgewicht *n*

donor Donator *m*

donor-acceptor bond Donator-Akzeptor-Bindung *f*, koordinative [dative] Bindung *f*

donor atom Donatoratom *n*

donor site *(Bioch)* Don(at)orort *m*, Peptidbindungsort *m*, Peptidylort *m*, P-Ort *m*, Ausgang *m*

DOPA Dopa *n*, *2-Amino-3-(3,4-dihydroxyphenyl)propensäure *f*, Dihydroxyphenylalanin *n*

dopa Dopa *n*, *2-Amino-3-(3,4-dihydroxyphenyl)propensäure *f*, Dihydroxyphenylalanin *n*

dopant Dotierstoff *m*, Dotierungsmaterial *n*

dopant level Dotierungsgrad *m*

dope *v* dopen *(Mineralölprodukte)*

dope 1. Zusatzstoff *m*; Dope *m(n)*, Dope-Stoff *m*, Dope-Mittel *n (Zusatzstoff in Mineralölprodukten, < 1 %)*; 2. *(Tox)* Dopingmittel *n*

doping Dotieren *n*, Dotierung *f (Halbleiter)*

doping process Dotierverfahren *n*

Doppler broadening *(Spektr)* Doppler-Verbreiterung *f*

Doppler effect Doppler-Effekt *m*

Dorr thickener Dorr-Eindicker *m*

dosage 1. Dosieren *n*, Dosierung *f*, Zumessen *n*, Zuteilen *n*; Verabreichung *f*, Applikation *f*; 2. *s.* dosage figure

dosage figure 1. Dosis *f*; 2. *(Tech)* Einsatzmenge *f*

dosage-response curve Dosis-Wirkungs-Kurve *f*, Dosis-Effekt-Kurve *f*

dose *v* dosieren, zumessen, zuteilen

dose 1. Dosis *f*; 2. *(Tech)* Einsatzmenge *f*

dose of radiation Strahlungsdosis *f*, Strahlendosis *f*

dose rate Dosisrate *f*, Dosisleistung *f*

dosimeter Dosimeter *n*, Dosismesser *m*, Dosismessgerät *n*, Strahlungsdosimeter *n*

dosimetry Dosimetrie *f*, Dosismessung *f*

dosing apparatus Dosiergerät *n*, Dosierapparat *m*, Dosiereinrichtung *f*, Zuteileinrichtung *f*, Zuteiler *m*

dot formula [symbol] Elektronenformel *f*

double-beam spectro(photo)meter Zweistrahlspektrometer *n*, Zweistrahlspektralphotometer *n*, Zweistrahlgerät *n*, Doppelstrahlspektrometer *n*, Doppelstrahlgerät *n*

double bond Doppelbindung *f*, Zweifachbindung *f*

double-chained polymer Leiterpolymer(e) *n*

double-charge layer *(Koll)* elektrische Doppelschicht *f*

double decomposition doppelte Umsetzung *f*, Metathese *f*, doppelter Umsatz *m*, Wechselumsetzung *f*, Wechselzersetzung *f*

double displacement reaction doppelte Verdrängungsreaktion *f*, Ping-pong-Reaktion *f (Enzymkinetik)*

double-distilled water doppelt destilliertes Wasser *n*, Aqua redestillata [bidestillata]

double-effect evaporator Zweikörperverdampfer *m*, Zweistufenverdampfer *m*

double-focus(s)ing mass spectrograph doppel(t)fokussierender Massenspektrograph *m*
double-helical *(Bioch)* doppelhelikal, doppelsträngig, Doppelstrang..., zweisträngig, Zweistrang...
double helix *s.* double-strand(ed) helix
double layer Doppelschicht *f*
double-layer potential elektrokinetisches Potenzial *n*, Zeta-Potenzial *n*, ζ-Potenzial *n*
double oxide Doppeloxid *n*
double-peaked distribution *(Anal)* bimodale [zweigipflige] Verteilung *f*
double-reciprocal plot *(Bioch)* Lineweaver-Burk-Auftragung *f*, grafische Darstellung *f* nach Lineweaver und Burk
double refraction Doppelbrechung *f*
double refraction of flow Strömungsdoppelbrechung *f*
double replacement *s.* double decomposition
double resonance technique *(Spektr)* Doppelresonanzmethode *f*
double ring closure doppelter Ringschluss *m*
double salt Doppelsalz *n*
double-stage zweistufig, Zweistufen...
double-strand DNA Doppelstrang-DNA *f*, doppelsträngige DNA *f*
double-strand(ed) helix *(Bioch)* Doppelhelix *f*, doppelsträngige Helix *f*
double-strength *(Ch)* doppelt konzentriert, von doppelter Konzentration
double weighing Vertauschungswägung *f* (nach Gauß)
doublet 1. Dublett *n*, Elektronendublett *n*, gemeinsames Elektronenpaar *n* zweier Atome; 2. Dublett *n*, Spektralliniendublett *n*, Doppellinie *f*; 3. Dublett *n (Benennung für die Duplizität)*
doublet splitting Dublettaufspaltung *f*
doublet term Dublett-Term *m*
doubling time *(Biot)* Verdopplungszeit *f (Mikroorganismenwachstum)*
doubly bonded [bound] doppelt [zweifach] gebunden
doubly charged zweifach geladen
doubly unsaturated doppelt ungesättigt
dough *v* teigig werden
dough 1. teigartige Masse *f*, Teig *m*; 2. *(Lebm)* Teig *m (Backmasse)* • **make into a dough** anteigen
dough mixer [mixing machine] 1. Mischmaschine *f*, Mischer *m*, Knetmaschine *f*, Knetwerk *n*, Knetapparat *m*, Kneter *m*; 2. *(Lebm)* Teigmischmaschine *f*, Teigmischer *m*, Teigknetmaschine *f*, Teigkneter *m*
doughiness 1. teigige [pastöse] Konsistenz *f*; 2. *(Lebm)* Teigkonsistenz *f*
doughy teigartig, teigig
downcomer *(Dest)* Ablaufrohr *n*, Rücklaufrohr *n*, Fallrohr *n*, Rückflussrohr *n*
downdraft *s.* downdraught
downdraught Unterwind *m*, Saugzug *m*
downflow Abwärtsstrom *m* • **operate downflow** im Abwärtsstrom arbeiten; im Abwärtsstrom fahren
downflow evaporator Fallfilmverdampfer *m*, Fallstromverdampfer *m*
downflow filter Abwärtsfilter *n*
downhill reaction Bergabwärtsreaktion *f*
downpipe *(Dest)* Ablaufrohr *n*, Rücklaufrohr *n*, Fallrohr *n*, Rückflussrohr *n*
downspout *s.* downpipe
downstream processing *(Biot)* Produktabtrennung *f* und -aufarbeitung *f*
downtake *s.* downpipe
downtime Betriebsunterbrechung *f*; Maschinenausfallzeit *f*, Stillstandszeit *f*; Wartezeit *f*
downward current abwärts gerichtete Strömung *f*, abwärts gerichteter Strom *m*
downward distillation absteigende Destillation *f*
DP *s.* degree of polymerization
2,4-DP 2,4-DP *n*, 2-(2,4-Dichlorphenoxy)propionsäure *f*, Dichlorprop *n (Herbizid)*
DPA *s.* diphenylamine
DPH *s.* diamond pyramid hardness
DPN *(diphosphopyridine nucleotide) s.* nicotinamide-adenine dinucleotide
DPP *s.* differential pulse polarography
drag *s.* drag force
drag force hydrodynamischer Widerstand *m*; aerodynamischer Widerstand *m*
drag-in Eintrag *m*, eingeschleppte Lösung *f*
drag-out Austrag *m*, herausgeschleppte Lösung *f*

drag-out losses Ausschleppverluste *mpl*
dragée Dragée *n*
Dragendorff alkaloid reagent *s.* Dragendorff's solution
Dragendorff's solution Dragendorffs Reagens *n (Alkaloidnachweis)*
drain *v* ablassen, entleeren, leerlaufen lassen *(z. B. ein Becken)*; ablaufen lassen *(Flüssigkeiten)*; abtropfen lassen; leerlaufen *(z. B. ein Becken)*; ablaufen; abtropfen *(Flüssigkeiten)*
drain *v* **away** ableiten, abführen, abziehen, abgießen; ablaufen lassen
drain *v* **down** ablassen, entleeren, leerlaufen lassen *(z. B. ein Becken)*
drain-dry *v* durch Abtropfenlassen trocknen
drain *v* **off** *s.* drain away
drain *v* **through** durchfließen, abfließen; durchfließen lassen, abfließen lassen
drain *v* **to the sewer** in die Kanalisation einleiten [ableiten] *(Abwasser)*
drain Abfluss *m*, Ablauf *m*, Ablass *m*, Auslass *m*, Ausfluss *m*, Abflussöffnung *f*; Ausguss *m*; Drän *m*; Drän(rohr)strang *m*, Rohrstrang *m*; Dränrohr *n*, Dränröhre *f*
drain cock Ablasshahn *m*
drain line Abflussleitung *f*, Ablassleitung *f*; Abwasserleitung *f*
drain of nutrients *(Boden)* Nährstoffentzug *m*
drain pipe Ablaufrohr *n*
drain plug Ablassstopfen *m*
drain tap *s.* drain cock
drain valve Ablassventil *n*
drainage Dränung *f*, Entwässerung *f*; Entleerung *f*; Ablassen *n*; Abtropfenlassen *n*; Ablaufen *n*; Abtropfen *n*; Leerlaufen *n*
drainage pipe Ablaufrohr *n*, Ablaufleitung *f*, Abflussrohr *n*, Abflussleitung *f*; Dränrohr *n*, Dränröhre *f*
drainage system Entwässerungsanlage *f*
drainage water Sickerwasser *n*
drastic energisch, scharf, drastisch *(z. B. Reaktionsbedingungen)*
drastic stark wirkendes Arzneimittel *n*, *(i. e. S.)* starkes Abführmittel *n*, Drastikum *n*
draught Zug *m*, Schornsteinzug *m*, Kaminzug *m*
draught of air Luftstrom *m*
draw *v* 1. ziehen *(z. B. Glas, Kunststoffe)*; recken, strecken *(Fasern, Folien)*; 2. (an)saugen, anziehen *(z. B. Flüssigkeiten)*
draw *v* **down** abziehen *(Flüssigkeiten)*; laufen lassen
draw *v* **in** (an)saugen, anziehen *(z. B. Flüssigkeiten)*
draw *v* **off** abziehen *(Flüssigkeiten)*; abstechen *(flüssige Metalle)*
draw *v* **samples** Proben entnehmen
draw Abziehen *n*, Abzug *m (einer Flüssigkeit aus einem Tank, einer Leitung)*
drawing 1. Ziehen *n (z. B. von Glas, Kunststoffen)*; 2. Saugen *n*, Ansaugen *n*, Anziehen *n (z. B. von Flüssigkeiten)*; 3. *s.* draw
drawn yarn gereckte Seide *f*
dregs Bodensatz *m*, Satz *m*, Sediment *n*
drench *v* tränken
drench shower Notbrause *f*, Löschbrause *f*, Feuerlöschbrause *f*
dress *v* 1. *(Met)* aufbereiten; 2. *(Landw)* düngen; 3. *(Text)* appretieren, schlichten
dressing 1. *(Met)* Aufbereitung *f*; 2. *(Landw)* Düngung *f*; 3. *(Text)* Appretur *f*, Appretieren *n*, Schlichten *n*; 4. *(Landw)* Düngemittel *n*; 5. *(Text)* Appreturmittel *n*, Appret *n*, Schlichte *f*
dried milk Trockenmilch *f*, Milchpulver *n*
dried sludge *(Hyd)* Trockenschlamm *m*, getrockneter Schlamm *m*
drier 1. *(Anstr)* Trockenstoff *m*, Trocknungsstoff *m*, Trockner *m*, Sikkativ *n*; 2. *s.* dryer 1.
drift *v (Anal)* abwandern, driften *(z. B. Basislinie)*
drift Drift *f (z. B. des Nullpunkts)*; Abtrift *f*, Abdrift *f (von Pflanzenschutzmitteln)*
drinking-water disinfection Trinkwasserentkeimung *f*
drinking-water regulations [standards] Trinkwasser-Standard *m*, Gütebedingungen *fpl* für Trinkwasser, Güteanforderungen *fpl* [Beschaffenheitsanforderungen *fpl*] an Trinkwasser
drinking-water treatment Trinkwasseraufbereitung *f*
drip *v* tropfen, tröpfeln

drip-proof tropfwassergeschützt
drive *v* antreiben, betreiben
drive *v* **off [out]** austreiben, heraustreiben, vertreiben, verjagen
drive Antrieb *m*
driving force [potential] 1. Triebkraft *f*, treibende Kraft *f*; 2. *(phys Ch)* (chemische) Triebkraft *f*, (chemische) Affinität *f*, Reaktionsaffinität *f*
drop *v* 1. tropfen; 2. sinken, abfallen *(Messwerte)*; (ab)senken, vermindern, verringern *(Messwerte)*
drop 1. Tropfen *m*; 2. Sinken *n*, Abfallen *n*, Fallen *n*, Abfall *m*, Fall *m*, Sturz *m* *(z. B. der Temperatur, Spannung)*
drop by drop tropfenweise
drop in head Druckabfall *m*
drop in voltage Spannungsabfall *m*, Spannungsabnahme *f*, Spannungsrückgang *m*
drop of temperature Temperaturabfall *m*
drop reaction Tüpfelreaktion *f*
drop shape analysis Tropfenformanalyse *f*
droplet Tröpfchen *n*, kleiner Tropfen *m*
dropper Tropfer *m*, *(i. w. S.)* Tropfenzähler *m*
dropping-ball viscometer Kugelfallviskosimeter *n*
dropping bottle Tropffläschchen *n*, Tropfflasche *f*, Tropfglas *n*, *(i. w. S.)* Tropfenzähler *m*
dropping electrode Tropfelektrode *f*
dropping funnel Tropftrichter *m*
dropping-mercury electrode Quecksilbertropfelektrode *f*
dropping pipet(te) Tropfpipette *f*, Stechpipette *f*, *(i. w. S.)* Tropfenzähler *m*
dropping point Tropfpunkt *m*
dropwise addition tropfenweises Zugeben *n* [Zusetzen *n*], Zutropfen *n*
dross Schaum *m*, Metallschaum *m* *(auf geschmolzenem Metall)*
drug Arzneimittel *n*; Arzneidroge *f*; Narkotikum *n*, *(i. e. S.)* Rauschgift *n*
drug delivery Wirkstoffabgabe *f*
drug toxicosis Arzneimittelvergiftung *f*
drum Trommel *f*, Walze *f*
drum dryer Walzentrockner *m*
drum filter Trommelfilter *n*, Drehfilter *n*
drum mill Trommelmühle *f*
drum mixer Trommelmischer *m*, Mischtrommel *f*
drum screen Trommelsieb *n*, Rundsieb *n*, Wälzsieb *n*
drum-type vacuum filter Vakuumtrommelfilter *n*, Trommelsaugfilter *n*, *(meist)* Vakuumtrommelzellenfilter *n*, Saugzellenfilter *n*
dry *v* 1. trocknen, entfeuchten, *(Feststoffe auch)* entwässern; 2. *(Lebm)* trocknen, dehydratisieren, *(Obst auch)* dörren; trocknen, trocken werden; austrocknen; abtrocknen; eintrocknen
dry *v* **off** (ab)trocknen
dry *v* **out** austrocknen
dry *v* **to constant weight** bis zur Massekonstanz trocknen
dry analysis trockene Analyse *f*, Analyse *f* auf trockenem Wege, *(besonders Metalle)* trockene Vorprüfung *f*
dry ash-free wasser- und aschefrei, waf
dry ashing trockenes Veraschen *n*, Veraschung *f* auf trockenem Wege
dry battery 1. Trockenbatterie *f*; 2. *s.* dry cell
dry blending Mischen *n* [Vermischen *n*, Vermengen *n*] von Feststoffkomponenten
dry-bulb temperature Temperatur *f* des trockenen Thermometers, Trockenkugeltemperatur *f*
dry cell Trockenelement *n*
dry chemical (fire) extinguisher Pulverlöscher *m*, Trockenlöscher *m*
dry deposition trockene Deposition *f*, Trockenablagerung *f* *(von Schadstoffen aus der Atmosphäre)*
dry desulphuration Trockenentschwefelung *f* *(von Gasen)*
dry distillation trockene Destillation *f*, Trockendestillation *f*, Entgasung *f*
dry filter Trockenfilter *n*
dry gas cleaning Trockengasreinigung *f*, Trockenreinigung *f* eines Gases
dry grinding Trocken(ver)mahlen *n*, Trocken(ver)mahlung *f*
dry hydrogen chloride trockener Chlorwasserstoff *m*
dry ice Trockeneis *n*, Kohlensäureschnee *m*, festes Kohlendioxid *n*
dry matter Trockensubstanz *f*
dry milk Trockenmilch *f*, Milchpulver *n*

dry mineral-matter-free trocken [lufttrocken] und mineral(stoff)frei
dry mixing 1. Mischen *n* [Vermischen *n*, Vermengen *n*] von Feststoffkomponenten; 2. *(Ker)* Trockenaufbereitung *f*
dry natural gas trockenes Erdgas *n* [Naturgas *n*], Trockengas *n* *(Hauptanteil Methan)*
dry product 1. getrocknetes [trockenes] Gut *n*, Trockengut *n*; 2. *(Lebm)* Trockenware *f*
dry run Probelauf *m* *(einer Apparatur)*
dry saturated steam trockengesättigter Dampf *m*, Trockendampf *m*
dry screening Trockensieben *n*
dry sludge *(Hyd)* Trockenschlamm *m*
dry sludge solids *(Hyd)* Schlammfeststoffgehalt *m*, Schlammtrockensubstanz *f*
dry steam Trockendampf *m*, trockengesättigter Dampf *m*
dry volume Schüttvolumen *n*
dry weight Trocken(substanz)masse *f*, Trockenstoffmasse *f*
dry weight of biomass Trockenbiomasse *f*, Biotrockenmasse *f*, BTM
dry weight of cells *(Biot)* Zelltrockenmasse *f*, Zelltrockensubstanz *f*, Zell-TS *f*
dry yeast Trockenhefe *f*
dryer 1. Trockner *m*, Trockenapparat *m*, Trockenmaschine *f*; 2. *s.* drier 1.
dryer-pulverizer Mahltrocknungsanlage *f*
drying 1. Trocknung *f*, Entfeuchtung *f*, *(bei Feststoffen auch)* Entwässerung *f*; 2. *(Lebm)* Trocknung *f*, Dehydratisierung *f*, *(bei Obst auch)* Dörren *n*; Trocknung *f*; Austrocknung *f*; Abtrocknung *f*; Eintrocknung *f*
drying agent Trockenmittel *n*, Trockenmedium *n*, Trocknungsmittel *n*, Entfeuchtungsmittel *n*
drying by evaporation Verdampfungstrocknung *f*; Verdunstungstrocknung *f*
drying cabinet Trockenschrank *m*
drying curve Trocknungskurve *f*
drying oil trocknendes Öl *n*
drying oven Trockenofen *m*; Trockenschrank *m*
drying pistol Trockenpistole *f*, Röhrentrockner *m*
drying plant Trockenanlage *f*, Trocknungsanlage *f*
drying rate Trockengeschwindigkeit *f*, Trocknungsgeschwindigkeit *f*, spezifische Austauschfeuchtemenge *f*
drying temperature Trocknungstemperatur *f*, Trockentemperatur *f*
drying to constant weight Trocknen *n* bis zur Massekonstanz
dryness Trockenheit *f*; Trockne *f*
d.s. *s.* difficultly soluble
DS *s.* 1. dissolved solids; 2. degree of substitution
DSA *s.* drop shape analysis
DSC = differential scanning calorimetry
DTA *s.* differential thermal analysis
DTG *s.* differential thermogravimetry
dTMP *s.* deoxythymidine 5′-phosphate
DTPA *s.* diethylenetriamine pentaacetic acid
dual collision *(phys Ch)* Zweierstoß *m*
dual filter *s.* dual media filter
dual-function catalyst bifunktioneller Katalysator *m*
dual media filter Zweischichtenfilter *n*, Zweifachfilter *n*
dubnium Db Dubnium *n* *(Element 105)*
duct Leitung *f*, Leitungsrohr *n*; Kanal *m*
ductile duktil, dehnbar, streckbar, ziehbar, formbar
ductile fracture Zähbruch *m*
ductility Duktilität *f*, Dehnbarkeit *f*, Streckbarkeit *f*, Ziehbarkeit *f*, Formbarkeit *f*
dull *v* mattieren, den Glanz verringern; abstumpfen; trüben; matt werden; stumpf werden; trüb werden
dull matt; stumpf; trüb(e)
dull red heat dunkle Rotglut *f*, Dunkelrotglut *f*
dull redness dunkle Rotglut *f*, Dunkelrotglut *f*
dulling Mattieren *n*; Abstumpfen *n*; Trübung *f*, Abtrübung *f*; Mattwerden *n*; Stumpfwerden *n*; Trübwerden *n*
dullness Mattheit *f*, mattes Aussehen *n*, Glanzlosigkeit *f*; Stumpfheit *f*; Trübheit *f*
Dulong-Petit's law *(phys Ch)* Dulong-Petit-Regel *f*, dulong-petitsche Regel *f*
Dumas' method for nitrogen Stickstoffbestimmung *f* nach Dumas
dummy tube blindes Rohr *n*, Blindrohr *n*
dump *v* ausschütten, auskippen, aus-

stoßen, auswerfen, abwerfen; verkippen *(Abprodukte)*
dump leaching *(Aufb)* Haldenlaugung *f*
dumping Ausschütten *n*, Auskippen *n*, Ausstoßen *n*, Auswerfen *n*, Abwerfen *n*; Verkippen *n (von Abprodukten)*
dumping at sea Einbringen *n* [Verkippen *n*, Versenken *n*] von Abprodukten ins Meer, Verkippen *n* auf hoher See, Abklappen *n*
duplet Dublett *n*, Elektronendublett *n*, gemeinsames Elektronenpaar *n* zweier Atome
duplex *(Bioch)* Doppelstrang *m (der DNA)*
duplex helix *(Bioch)* Doppelhelix *f*
durability Haltbarkeit *f*, Dauerhaftigkeit *f*, Lebensdauer *f*; Standfestigkeit *f (von Werkzeugen)*
durable haltbar, dauerhaft, beständig
duration of exposure Einwirkungsdauer *f*
durene Duren *n*, 1,2,4,5,-Tetramethylbenzen *n*
duromer Duromer(e) *n (hartelastischer Kunststoff)*
durometer 1. Härteprüfer *m*, Härtemesser *m*, Durometer *n*; 2. *(Gum)* Weichheitsprüfer *m*, Weichheitsprüfgerät *n*
durometer hardness Durometerhärte *f (Shore-Härte)*
dust *v* 1. stauben; verstauben; 2. bestäuben, einstäuben, (ein)pudern; stäuben *(mit Stäubemitteln)*; verstäuben *(Stäubemittel)*
dust 1. Staub *m*; 2. *(Landw)* Stäubemittel *n (z. B. Insektizid)*
dust catcher Staubsammler *m*, Staubfänger *m*, Staubabscheider *m*, Entstauber *m*, Staubsack *m*
dust collection Staubabscheidung *f*
dust collector *s.* dust catcher
dust concentration Staubkonzentration *f*
dust control Staubbekämpfung *f*
dust-control filter Filtrationsentstauber *m*, Filtrationsabscheider *m*, Staubfilter *n*
dust deposition Staubablagerung *f*
dust explosion Staubexplosion *f*
dust fall Staubablagerung *f*, Staubsedimentation *f*
dust filter *s.* dust-control filter
dust formation Staubbildung *f*; Staubanfall *m*
dust-free staubfrei; nicht staubend
dust-laden staubbeladen, mit Staub beladen, staubführend, staubhaltig
dust loading Staubbeladung *f*, Staubgehalt *m*
dust particle Staubteilchen *n*, Staubpartikel *n*, Staubkorn *n*
dust separator *s.* dust catcher
dusting 1. Stauben *n*; Verstauben *n*, Verstaubung *f*; 2. Bestäuben *n*, Einstäuben *n*, Einpudern *n*, Pudern *n*; Stäuben *n (mit Stäubemitteln)*; Verstäuben *n*, Verstäubung *f (von Stäubemitteln)*
dusting powder *(Landw)* Stäubemittel *n*
dustless staubfrei; nicht staubend
dustproof staubdicht, staubundurchlässig
dyad zweiwertiges Element *n*; zweiwertige Atomgruppe *f*
dyadic zweiwertig
dye *v* (an)färben
dye *v* **directly** direkt [unmittelbar] färben [anfärben]
dye Farbstoff *m*
dye absorption [acceptance] Farbstoffaufnahme *f*
dye bath Färbebad *n*, Färbeflotte *f*, Farbbad *n*, Farbflotte *f*, Flotte *f*
dye chemist Farbstoffchemiker *m*
dye chemistry Farbstoffchemie *f*
dye fastness Farbstoffechtheit *f*
dye intermediate Farbstoffzwischenprodukt *n*, Farbenzwischenstoff *m*
dye laser Farbstofflaser *m*
dye liquor *s.* dye bath
dye salt Farbsalz *n*
dye uptake Farbstoffaufnahme *f*
dyeing Färben *n*, Färberei *f*
dyeing assistant Färbe(rei)hilfsmittel *n*
dyeing power Farbkraft *f*, Färbekraft *f*, Färbefähigkeit *f*
dyestuff Farbstoff *m*
dyestuff chemistry Farbstoffchemie *f*
dyewood Farbholz *n*
dyke Gesteinsgang *m*, Gang *m*
dynamic equilibrium dynamisches Gleichgewicht *n*, Fließgleichgewicht *n*
dynamic isomer Tautomer(e) *n*, tautomere Form *f*

dynamic isomerism Tautomerie *f*
dynamic light scattering dynamische Lichtstreuung *f*
dynamic mechanical analysis dynamisch mechanische Analyse *f*
dynamic quenching dynamische Löschung *f (Photochemie)*
dynamic range *(Anal)* dynamischer Bereich *m*
dynamic steady state *s.* dynamic equilibrium
dynamic stress dynamische Beanspruchung *f*
dynamic viscosity dynamische [absolute] Viskosität *f* [Zähigkeit *f*]
dynamics Dynamik *f*
dynamite Dynamit *n*
dysprosia Dysprosium(III)-oxid *n*
dysprosium Dy Dysprosium *n*
dystectic mixture dystektisches Gemisch *n (Stoffgemisch mit höchstem Schmelzpunkt)*

E

e-bond e-Bindung *f*, äquatoriale Bindung *f*
E_T-value E_T-Wert *m (zur Charakterisierung der Lösungsmittelpolarität)*
EA-MS *s.* electron attachment mass spectroscopy
E.A.N. *s.* effective atomic number
earth *v* erden *(Geräte, Anlagen)*
earth colour [pigment] Erdpigment *n*, natürliches anorganisches Pigment *n*
earth pitch Asphalt *m*, Erdpech *n*
earth wax Erdwachs *n*, Bergwachs *n*, Bergtalg *m*, Ozokerit *m*
earthenware Irdengut *n*, Irdenware *f*, Tongut *n*; Steingut *n*
earthnut oil Erdnussöl *n*, Arachisöl *n (von Arachis hypogaea L.)*
earth's core *(Geol)* Erdkern *m*, Barysphäre *f*
earth's crust Erdrinde *f*, Erdkruste *f*
earthy erdig, Erd..., erd(e)haltig; erdartig
ease of serviceability [servicing] *s.* ease of use
ease of use leichte Handhabbarkeit *f* [Bedienbarkeit *f*]
easily soluble leichtlöslich
easy to ignite leicht entzündbar [zündend]
easy to operate leicht bedienbar
eat *v* zerfressen, anfressen, angreifen, (oberflächlich) zerstören, korrodieren; sich hineinfressen
eat *v* **away** abtragen *(durch Korrosion)*
eat *v* **through** durchfressen
ebonite Hartgummi *m*, Hartkautschuk *m*
ebullated [ebullating] bed wallendes Bett *n (spezielle Form einer Wirbelschicht)*
ebullience Kochen *n*, Sieden *n*, Aufwallen *n*
ebulliency Kochen *n*, Sieden *n*, Aufwallen *n*
ebullient kochend, siedend, aufwallend
ebullioscopic constant ebullioskopische Konstante *f*, molale Siedepunktserhöhung *f*
ebullioscopy Ebullioskopie *f*
ebullition Kochen *n*, Sieden *n*, Aufwallen *n*
EC = Enzyme Commission
EC number *(Bioch)* Nummer *f* der EC-Nomenklatur *(Enzymklassifizierung)*
ECD *s.* electron capture detector
echelette grating *(Spektr)* Echelettegitter *n*
echelon grating *(Spektr)* Echelongitter *n*, Stufengitter *n*
ECL *s.* electrochemiluminescence
eclipsed synperiplanar, ekliptisch, verdeckt *(Stereochemie)*
eclipsed conformation [form] synperiplanare Konformation *f (Torsionswinkel 0°)*
economic poison ökonomisches Gift *n*, *(i. e. S.)* Pflanzenschutzmittel *n*
economizer Ekonomiser *m*, Speisewasservorwärmer *m*, Abgasvorwärmer *m*, Rauchgasvorwärmer *m*
ecosystem Ökosystem *n*
ecotrophology Ökotrophologie *f (Haushalts- und Ernährungswissenschaft)*
ectoenzyme Ektoenzym *n (in der Zellmembran lokalisiertes Enzym)*
ectohormone Ektohormon *n*, Pheromon *n*
ED *s.* electrodialysis
EDC *s.* ethylene dichloride
eddy Wirbel *m*, Strudel *m*

eddy current 1. Wirbelströmung *f*, turbulente Strömung *f*; 2. (elektrischer) Wirbelstrom *m*
eddy-diffusion Wirbeldiffusion *f*, Wirbeldispersion *f*, Streudiffusion *f*, Kanaldispersion *f*
edge Kante *f*, Rand *m*
edge dislocation *(Krist)* Kantenversetzung *f*, Stufenversetzung *f*, Taylor-Orowan-Versetzung *f*
edge filter Oberflächenfilter *n*; Spaltfilter *n*
edge filtration Oberflächenfiltration *f*
edibility Essbarkeit *f*, Genießbarkeit *f*
edible essbar, genießbar
edible acid Genusssäure *f*
edible oil Speiseöl *n*
edibles Esswaren *fpl*, Lebensmittel *npl*, Nahrungsmittel *npl*
Edison accumulator [cell] Edison-Akkumulator *m*, NiFe-Akkumulator *m*, Stahlakkumulator *m*
EDL *s.* electric double-layer
Edman degradation *(Bioch)* Edman-Abbau *m (von Proteinen)*
EDTA *s.* ethylenediamine tetraacetic acid
eductor Ejektor *m*, Saugstrahlpumpe *f*
EDX *s.* energy-dispersive X-ray spectroscopy
EFA *s.* essential fatty acid
effect 1. Effekt *m*, Auswirkung *f*, Wirkung *f*; 2. Verdampferstufe *f*, Stufe *f*, Körper *m*
α-effect α-Effekt *m (Verstärkung der Nukleophilie)*
effective wirksam, effektiv, nutzbar, Wirk..., Nutz...
effective atomic number effektive Kernladungszahl *f* [Ordnungszahl *f*]
effective chimney height effektive Schornsteinhöhe *f*, Effektivhöhe *f* [effektive Höhe *f*] des Schornsteins
effective dose *(Tox)* Wirk(ungs)dosis *f*
effective stack height *s.* effective chimney height
effectiveness Wirksamkeit *f*, Effektivität *f*
effector *(Bioch)* Effektor *m*, Steuermetabolit *m*
effervesce *v* 1. (auf)brausen, (auf-)schäumen; 2. *(Lebm)* moussieren
effervescence 1. Aufbrausen *n*, Brausen *n*, Aufschäumen *n*, Schäumen *n*; 2. *(Lebm)* Moussieren *n*
effervescency 1. Aufbrausen *n*, Brausen *n*, Aufschäumen *n*, Schäumen *n*; 2. *(Lebm)* Moussieren *n*
effervescent wine Schaumwein *m*, Sekt *m*
efficiency 1. Nutzleistung *f*; Wirkungsgrad *m*; Ausnutzungsgrad *m*; 2. Wirksamkeit *f*
efficiency factor 1. Wirkungsgrad *m*; 2. Ausbeutefaktor *m (bei der Radikalkettenpolymerisation)*
efficiency of biosynthesis *(Biot)* Biosyntheseleistung *f*
efficiency of luminous source Lichtausbeute *f (in lm/W)*
efficiency of separation Trennerfolg *m*; Abscheidegrad *m*, Ausscheidungsgrad *m*; Entstaubungsgrad *m*
efficient wirksam, wirkungsvoll, leistungsfähig; mit hohem Wirkungsgrad; brauchbar
effloresce *v* 1. *(Min)* effloreszieren, ausblühen; 2. *(Krist)* unter Kristallwasserverlust verwittern
efflorescence 1. *(Min)* Effloreszenz *f*, Ausblühung *f*; 2. *(Krist)* Verwitterung *f* unter Kristallwasserverlust
effluence 1. Ausfließen *n*, Ausfluss *m*, Abfließen *n*, Abfluss *m*, Ablaufen *n*, Ablauf *m*, Ausströmen *n*; Ausstrahlung *f*; 2. *s.* effluent
effluent ausfließend, ausströmend; ausstrahlend
effluent 1. ablaufende [abfließende] Flüssigkeit *f*, Ablauf *m*, Abfluss *m*; 2. *(Hyd)* in Vorfluter [Kanalisation] einzuleitendes [abzuleitendes] Abwasser *n*, Abwasserablauf *m (eines Betriebs)*; Abfluss *m*, Ablauf *m (einer Reinigungsanlage)*; (gereinigtes) abfließendes Abwasser *n*
effluent disposal Abwasserbeseitigung *f*, Abwasserentsorgung *f*
effluent flow Abwasseranfall *m*, Abwasserabfluss *m*
effluent limitations *(Hyd)* Einleitungsbeschränkungen *fpl (für Abwässer in Vorfluter)*
effluent loading *(Hyd)* Schmutz(stoff)-belastung *f*, *(quantitativ:)* Schmutzstofflast *f (z. B. des Betriebsabwassers)*

effluent pipe Ausflussrohr *n*, Ablaufrohr *n*, Abflussrohr *n*, Austrittsrohr *n*
effluent regulations *s.* effluent standards
effluent restrictions *s.* effluent limitations
effluent sample Abwasserprobe *f*
effluent standards *(Hyd)* Abwassereinleitungsbedingungen *fpl*
effluent treatment plant Abwasserbehandlungsanlage *f*, ABA, Abwasserreinigungsanlage *f*, Kläranlage *f*, KA, Klärwerk *n (für Betriebsabwasser)*
effluent water *(Hyd)* ablaufendes Wasser *n*, Ablaufwasser *n*, Ablauf *m*; Reinwasser *n*, Klarwasser *n (bei der Wasseraufbereitung)*
effluent water quality Ablaufqualität *f*, Ablaufgüte *f*; Reinwassergüte *f*, Klarwassergüte *f (bei der Wasseraufbereitung)*; Filtratqualität *f*, Filtratgüte *f (bei der Filtration)*; Weichwassergüte *f*, Reinwassergüte *f*, Deionatgüte *f (beim Ionenaustausch)*
efflux Ausfluss *m*, Abfluss *m*
effuse *v* ausströmen, ausfließen
effusion Effusion *f*, Ausströmen *n*, Ausfließen *n*, Ausfluss *m*, Erguss *m*
EG *s.* ethylene glycol
egg albumin 1. Ei(er)albumin *n*, Ovalbumin *n*; 2. *s.* egg white
egg white Weißei *n*, Eiklar *n*
egg yolk Eigelb *n*, Eidotter *n*, Dotter *n*
EHO = Environmental Health Officer
eicosane Eicosan *n*
eicosanoic acid Eicosansäure *f*
eigenfunction Eigenfunktion *f*
eigenstate Eigenzustand *m*
eigenvalue Eigenwert *m*
eight-electron configuration Elektronenoktett-Anordnung *f*, Edelgaskonfiguration *f*
eight-membered ring achtgliedriger Ring *m*, Acht(er)ring *m*
Einhorn reaction *(org Ch)* Einhorn-Reaktion *f*, Haloform-Reaktion *f*
Einstein Einstein *n*, E *(1 Mol Photonen)*
Einstein equation Einstein-Gleichung *f*, einsteinsche (Masse-Energie-)Gleichung *f*
einsteinium Es Einsteinium *n*
eject *v* 1. (hin)ausdrücken, auswerfen, ausstoßen; 2. *(Kern)* emittieren, ausstoßen, (her)ausschleudern *(Teilchen)*
ejection 1. Ausdrücken *n*, Auswerfen *n*, Ausstoßen *n*; 2. *(Kern)* Emission *f*, Emittierung *f*, Ausstoß *m*, Herausschleudern *n (von Teilchen)*
ejector 1. Ausdrücker *m*, Ausstoßer *m*, Auswerfer *m*, Auswurfvorrichtung *f*; 2. Ejektor *m*, Saugstrahlpumpe *f*
elaeostearic acid Eläostearinsäure *f*, *Octadeca-9,11,13-triensäure *f*
elaidic acid Elaidinsäure *f*, *trans-Octadec-9-ensäure *f*
elaidinization Elaidin(is)ierung *f*
elastic collision elastischer Stoß *m*
elastic deformation elastische Verformung *f* [Deformation *f*]
elastic limit Elastizitätsgrenze *f*
elastic modulus Elastizitätsmodul *m*, E-Modul *m*
elasticity Elastizität *f*, Federkraft *f*
elastin Elastin *n (ein Protein des Bindegewebes)*
elastomer Elastomer(e) *n*
elastomeric elastomer
elastomeric fibre Elastomerfaser *f*
elbow (fitting) Krümmer *m*, Rohrkrümmer *m*, Knie(stück) *n*, Winkelstück *n*
ELDOR *s.* electron double resonance
electric elektrisch
electric arc (elektrischer) Lichtbogen *m*
electric-arc furnace Lichtbogen(elektro)ofen *m*, Elektrolichtbogenofen *m*
electric charge elektrische Ladung *f*
electric circuit Stromkreis *m*
electric current (elektrischer) Strom *m*
electric-dipole moment (elektrisches) Dipolmoment *n*
electric double-layer elektrische [elektrochemische] Doppelschicht *f*, Ladungsdoppelschicht *f*
electric field effect *(Spektr)* Stark-Effekt *m*
electric furnace Elektroofen *m*, E-Ofen *m*
electric-furnace steel Elektrostahl *m*, E-Stahl *m*
electric-resistance furnace (elektrischer) Widerstandsofen *m*, widerstandsbeheizter Ofen *m*, Ofen *m* mit Widerstandsheizung
electrical elektrisch

electrical conductance elektrischer Leitwert *m*, Wirkleitwert *m*
electrical conductivity elektrische Leitfähigkeit *f*, elektrisches Leitvermögen *n*
electrical double refraction elektrische Doppelbrechung *f*
electrical insulator (elektrischer) Isolator *m*, Nichtleiter *m*, Dielektrikum *n*
electrical precipitation elektrisches [elektrostatisches] Abscheiden *n* [Reinigen *n*], Elektroabscheiden *n*, Elektroreinigen *n*, *(i. e. S.)* Elektroentstaubung *f*
electrical precipitator Elektroabscheider *m*, Elektrofilter *n*, *(i. e. S.)* Elektroentstauber *m*
electrical resistance (elektrischer) Widerstand *m*
electrically neutral elektroneutral
electrification 1. (elektrisches) Aufladen *n*; 2. Elektrifizieren *n*, Elektrifizierung *f*
electrify *v* 1. (elektrisch) aufladen, (elektrisch) laden; 2. elektrifizieren
electroactive elektroaktiv, elektrochemisch aktiv
electroanalysis Elektroanalyse *f*, elektrochemische Analyse *f*, elektroanalytische Bestimmung *f*
electroanalytical elektroanalytisch
electrocatalysis Elektrokatalyse *f*, elektrochemische Katalyse *f*
electrocatalytic elektrokatalytisch
electroceramics Elektrokeramik *f*
electrochemical elektrochemisch • **by an electrochemical route** auf elektrochemischem Wege
electrochemical cell galvanische [elektrochemische] Zelle *f* [Kette *f*]
electrochemical fluorination elektrochemische Fluorierung *f*, Elektrofluorierung *f*
electrochemical method of analysis [detection] elektrochemische Analysenmethode *f* [Nachweismethode *f*]
electrochemical series elektrochemische Spannungsreihe *f*
electrochemiluminescence Elektrolumineszenz *f*
electrochemist Elektrochemiker *m*
electrochemistry Elektrochemie *f*
electrochromic effect Stark-Effekt *m* *(Aufspaltung von Spektrallinien)*
electrocoating Galvanisierung *f*, Metallbeschichtung *f*
electrocratic *(Koll)* elektrokratisch *(durch elektrische Ladung beständig)*
electrocyclic reaction elektrocyclische Reaktion *f*
electrode Elektrode *f*
electrode potential Elektrodenpotenzial *n*
electrode reaction Elektrodenreaktion *f*
electrodeposit *v* 1. elektrochemisch [galvanisch] abscheiden; 2. elektrophoretisch abscheiden
electrodeposit elektrochemisch [galvanisch] hergestellte Schutzschicht *f*
electrodeposition 1. elektrochemisches [galvanisches] Abscheiden *n*, elektrolytische Abscheidung *f*, galvanisches Auftragen *n*; 2. elektrophoretisches Abscheiden *n*
electrodialysis Elektrodialyse *f*
electroendosmosis Elektro(end)osmose *f*
electroerosion Elektroerosion *f*
electroetching galvanisches Ätzen *n*
electrofiltration Elektrofiltration *f*, Elektroentstaubung *f* *(von Gasen)*
electroforming *(Ech)* Galvanoplastik *f*, Galvanoformung *f*, Elektroformung *f*
electrofuge elektrofuge Gruppe *f*
electrogalvanize *v* elektrochemisch [galvanisch] verzinken
electrogalvanizing elektrochemisches [galvanisches] Verzinken *n*
electrographite Elektrographit *m*
electrogravimetry Elektrogravimetrie *f*
electroinactive elektrochemisch inaktiv
electrokinetic potential elektrokinetisches Potenzial *n*, Zeta-Potenzial *n*, ζ-Potenzial *n*
electroluminescence Elektrolumineszenz *f*
electrolyse *v s.* electrolyze
electrolysis Elektrolyse *f*
electrolyte Elektrolyt *m*
electrolyte acid Akkumulatorensäure *f*, Akkusäure *f*
1:1 electrolyte 1-1-wertiger Elektrolyt *m*
electrolytic elektrolytisch, Elektrolyt...; elektrolytisch, Elektrolyse...
electrolytic capacitor Elektrolytkondensator *m*
electrolytic cell 1. Elektrolyse(n)zelle *f*,

elektrolytische Zelle *f*, Elektrolysierzelle *f*; 2. *s.* electrochemical cell
electrolytic conductivity (spezifische) elektrolytische Leitfähigkeit *f*, (spezifisches) elektrolytisches Leitvermögen *n*
electrolytic conductor elektrolytischer Leiter *m*, Elektrolyt *m*, Leiter *m* zweiter Ordnung [Klasse], Ionenleiter *m*
electrolytic copper Elektrolytkupfer *n*, E-Kupfer *n*, E-Cu
electrolytic dissociation elektrolytische Dissoziation *f* [Zersetzung *f*]
electrolytic nickel Elektrolytnickel *n*, E-Nickel *n*, Katodennickel *n*
electrolytic oxidation elektrolytische [elektrochemische] Oxidation *f*
electrolytic reduction elektrolytische Reduktion *f*
electrolytic refining elektrochemische [elektrolytische] Raffination *f*, Elektroraffination *f*
electrolytic separation elektrolytische Trennung *f*
electrolyze *v* elektrolysieren, elektrolytisch zersetzen [zerlegen]
electrolyzer 1. *(Lab)* Elektrolysegerät *n*, Elektrolyseapparatur *f*; 2. *s.* electrolytic cell
electromagnetic spectrum elektromagnetisches Spektrum *n*
electromeric effect Mesomerie-Effekt *m*, M-Effekt *m*, elektromerer Effekt *m*, E-Effekt *m*
electrometer Elektrometer *n*
electrometric titration elektrometrische [potenziometrische] Titration *f* [Maßanalyse *f*]
electromotive force (elektrische) Urspannung *f*, *(veraltet:)* elektromotorische Kraft *f*, EMK *f*
electromotive (force) series elektrochemische Spannungsreihe *f*
electron Elektron *n*, Negatron *n*
electron-accepting elektronenaufnehmend, elektronenziehend, elektronenverbrauchend
electron acceptor Elektronenakzeptor *m*, Elektronen(auf)nehmer *m*, Elektronenverbraucher *m*
electron affinity Elektro(nen)affinität *f*, E. A.
electron attachment mass spectroscopy Elektronenanlagerungs-Massenspektroskopie *f*
electron-attracting elektronenanziehend, elektronensuchend, elektronenfreundlich, elektrophil, kationoid
electron attraction Elektronenzug *m*
electron beam Elektronenstrahl *m*
electron beam lithography Elektronenstrahllithographie *f*
electron bombardment Elektronenbeschuss *m*, Elektronenbeschießung *f*, Elektronenbombardement *n*
electron capture detector Elektroneneinfangdetektor *m*, ECD
electron carrier Elektronen(über)träger *m*
electron-carrying elektronenübertragend
electron-carrying protein *(Bioch)* Elektronenüberträger-Protein *n*, Elektronen-Carrier-Protein *n*
electron cloud Elektronenwolke *f*
electron configuration Elektronenkonfiguration *f*, Elektronenanordnung *f*
electron deficiency Elektronenmangel *m*, Elektronenunterschuss *m*
electron-deficient elektronenarm, Elektronenmangel...
electron-deficient compound Elektronenmangelverbindung *f*
electron density Elektronendichte *f*, Elektronenbelegung *f*, Elektronenkonzentration *f*
electron diffraction Elektronenbeugung *f*
electron distribution Elektronenverteilung *f*
electron-donating elektronenabgebend, elektronenliefernd, elektronenspendend
electron donor Elektronendonator *m*, Elektronenspender *m*, Elektronengeber *m*
electron-donor-acceptor complex Elektronen-Donator-Akzeptor-Komplex *m*, EDA-Komplex *m*, Ladungsübertragungskomplex *m*, Charge-transfer-Komplex *m*, CT-Komplex *m*
electron dot formula [structure] Elektronenformel *f*
electron double resonance *(Anal)* Elektronendoppelresonanz *f*
electron energy Elektronenenergie *f*

electron energy level Elektronenniveau *n*
electron energy loss spectroscopy Elektronenenergieverlust-Spektroskopie *f*, Energieverlustspektroskopie *f*
electron energy state *s.* electron energy level
electron excess Elektronenüberschuss *m*
electron exchange Elektronenaustausch *m*
electron-exchange resin Elektronenaustauscherharz *n*
electron exchanger Elektronenaustauscher *m*
electron excitation Elektronenanregung *f*
electron flow Elektronenfluss *m*
electron gap Elektronenlücke *f*
electron gas Elektronengas *n*
electron hole Defektelektron *n*, Elektronenloch *n*, Loch *n*, Mangelelektron *n*
electron-hole pair Elektron-Defektelektron-Paar *n*, Elektron-Loch-Paar *n*, Exziton *n*
electron impact ionization Elektronenstoßionisation *f*
electron jump Elektronensprung *m*
electron lattice Elektronengitter *n*
electron mass Elektronen(ruhe)masse *f*, Ruhemasse *f* des Elektrons
electron micrograph elektronenmikroskopische Abbildung *f*
electron microprobe Elektronen-Mikrosonde *f*
electron microscope Elektronenmikroskop *n*
electron microscopy Elektronenmikroskopie *f*
electron migration Elektronenwanderung *f*
electron-multiplier phototube *s.* multiplier phototube
electron nuclear double resonance Elektron-Kern-Doppelresonanz *f*, ENDOR
electron octet Elektronenoktett *n*
electron pair Elektronenpaar *n*
electron-pair acceptor Elektronenpaarakzeptor *m*
electron-pair bond Elektronenpaarbindung *f*, Atombindung *f*, homöopolare [kovalente, unpolare] Bindung *f*, Kovalenz *f*
electron-pair donor Elektronenpaardonator *m*
electron-pair method Elektronenpaarmethode *f*, Valenzbindungsmethode *f*, VB-Methode *f*, Valenzstrukturmethode *f*, Heitler-London-Slater-Pauling-Methode *f*, HLSP-Methode *f*, Spinmethode *f*
electron paramagnetic resonance *s.* electron spin resonance
electron polarization Elektronenpolarisation *f*
electron population Elektronenbesetzung *f*
electron probe *(Anal)* Elektronenstrahlmikrosonde *f*
electron-probe microanalysis Elektronenstrahlmikroanalyse *f*
electron-pulling *s.* electron-withdrawing
electron release Elektronenabgabe *f*
electron-releasing elektronenabgebend, elektronenliefernd, elektronenspendend
electron resonance *s.* electron paramagnetic resonance
electron rest mass *s.* electron mass
electron-rich elektronenreich
electron shell Elektronenschale *f*
electron spectroscopy Elektronenspektroskopie *f*
electron spectroscopy for chemical analysis röntgenstrahlangeregte Photoelektronenspektroskopie *f*, Elektronenspektroskopie *f* für die chemische Analyse
electron spectrum Elektronenspektrum *n*
electron spin Elektronenspin *m*, Spin *m*
electron spin quantum number Elektronenspinquantenzahl *f*
electron spin resonance Elektronen(spin)resonanz *f*, ESR, (elektronen)paramagnetische Resonanz *f*, EPR
electron state Elektronenzustand *m*
electron structure Elektronenstruktur *f*, Elektronenaufbau *m*
electron transfer Elektronenübertragung *f*; Elektronenübergang *m*, Elektronentransfer *m (Redoxreaktion)*
electron transfer reaction Elektronen-

übertragungsreaktion *f*, Elektronentransferreaktion *f* *(Redoxreaktion)*
electron-transferring protein *(Bioch)* Elektronenüberträgerprotein *n*
electron transport Elektronentransport *m*
electron-transport chain *(Bioch)* Elektronentransportkette *f*, Atmungskette *f*
electron-volt Elektronenvolt *n*, eV
electron withdrawal Elektronenentzug *m*
electron-withdrawing elektronenziehend
π **electron** π-Elektron *n*
σ **electron** σ-Elektron *n*
electronegative elektronegativ
electronegativity Elektronegativität *f*
electroneutrality Elektroneutralität *f*
electronic elektronisch, Elektronen...
electronic charge Elektronenladung *f*
electronic conduction Elektronenleitung *f*
electronic conductivity Elektronenleitfähigkeit *f*, elektronische Leitfähigkeit *f*
electronic conductor Elektronenleiter *m*, Leiter *m* erster Ordnung
electronic emission Elektronenemission *f*, Elektronenaustritt *m*
electronic excitation energy Elektronenanregungsenergie *f*
electronic formula Elektronenformel *f*
electronic ground state Elektronengrundzustand *m*, elektronischer Grundzustand *m*
electronic level Elektronenniveau *n*
electronic mass Elektronenmasse *f*
electronic region Elektronenhülle *f*, Atomhülle *f*
electronic semiconductor (elektronischer) Halbleiter *m*, Elektronenhalbleiter *m*
electronic spectroscopy Elektronenanregungsspektroskopie *f*
electronic spectrum Elektronenanregungsspektrum *n*, Elektronensprungspektrum *n*
electronic subshell Unterschale *f*
electronic symbol *s.* electronic formula
electronic transition Elektronenübergang *m*
electronic work function Elektronenaustrittsarbeit *f*, Austrittsarbeit *f*, Abtrennungsarbeit *f*, Ablösearbeit *f*
electronically excited elektronisch angeregt
electroosmosis Elektro(end)osmose *f*
electrophile Elektrophil *n*, elektrophiles Reagens *n* [Agens *n*]
electrophilic elektrophil, elektronensuchend, elektronenfreundlich, elektronenanziehend, kationoid
electrophilic addition elektrophile Addition *f*
electrophilic substitution elektrophile Substitution *f*
electrophilicity Elektrophilie *f*, elektrophile Reaktivität *f*
electrophoresis Elektrophorese *f*
electrophoretic mobility elektrophoretische Beweglichkeit *f*
electroplate *v* galvanisieren, elektrochemisch [galvanisch] beschichten, elektroplattieren
electroplating Galvanostegie *f*, Elektroplattierung *f*; Galvanisieren *n*, Elektroplattieren *n* *(Tätigkeit)*
electroplating bath 1. Galvanisierbad *n* *(flüssigkeitsgefüllter Behälter)*; 2. *s.* electroplating solution; 3. *s.* electroplating tank
electroplating of plastics Kunststoffgalvanisierung *f*
electroplating solution Elektrolyt *m*, Elektrolytflüssigkeit *f*, Galvanisierelektrolyt *m*
electroplating tank Galvanisierbehälter *m*, Elektrolytbehälter *m*
electropolymerisation *(Kst)* elektrochemische Polymerisation *f*
electropositive elektropositiv
electrorefining elektrochemische [elektrolytische] Raffination *f*, Elektroraffination *f*
electrospray *(Spektr)* Elektrospray *n*
electrospray ionization *(Spektr)* Elektrospray-Ionisierung *f*
electrostatic attraction elektrostatische [elektrische] Anziehung *f*, coulombsche Anziehung *f*
electrostatic charge elektrostatische Ladung *f* [Aufladung *f*]
electrostatic force elektrostatische [coulombsche] Kraft *f*
electrostatic powder coating elektrostatische Pulverbeschichtung *f*, elektrostatische Pulverlackierung *f*

electrostatic precipitation elektrisches [elektrostatisches] Abscheiden *n* [Reinigen *n*], Elektroabscheidung *f*, Elektroreinigung *f*, *(i. e. S.)* Elektroentstaubung *f*
electrostatic precipitator Elektroabscheider *m*, Elektrofilter *n*, *(i. e. S.)* Elektroentstauber *m*
electrostatic repulsion elektrostatische Abstoßung *f*
electrosteel Elektrostahl *m*, E-Stahl *m*
electrosynthesis Elektrosynthese *f*
electrothermal elektrothermisch
electrovalence 1. elektrochemische Wertigkeit *f* [Valenz *f*], Elektrovalenz *f*; 2. Ionenbeziehung *f*, Ionenbindung *f*, heteropolare Bindung *f*
electrovalency 1. elektrochemische Wertigkeit *f* [Valenz *f*], Elektrovalenz *f*; 2. Ionenbeziehung *f*, Ionenbindung *f*, heteropolare Bindung *f*
electrovalent elektrovalent
electrovalent bond *s.* electrovalence 2.
element 1. (chemisches) Element *n*, (chemischer) Grundstoff *m*; 2. Element *n*, Grundbestandteil *m*, wesentlicher Bestandteil *m*; 3. Bauelement *n*, Bauteil *n*
elemental 1. elementar, ungebunden; 2. *(Min)* gediegen
elemental analysis Elementaranalyse *f*
elemental copper elementares [gediegenes] Kupfer *n*
elemental sulphur elementarer Schwefel *m*, Elementarschwefel *m*
elementary 1. elementar, grundlegend; 2. *s.* elemental
elementary analysis Elementaranalyse *f*
elementary charge Elementarladung *f*, elektrisches Elementarquantum *n*, e
elementary chemical reaction chemische Elementarreaktion *f*
elementary particle Elementarteilchen *n*
elementary reaction Elementarreaktion *f*, Urreaktion *f*
eleostearic acid *s.* elaeostearic acid
elevate *v* 1. (in die Höhe) heben, hochfördern, nach oben tragen; 2. erhöhen, steigern
elevation 1. Anhebung *f*, Hebung *f*; 2. Erhöhung *f*, Steigerung *f*; 3. Aufriss *m*
elevator Stellförderer *m*, Senkrechtförderer *m*, Elevator *m*
eliminability Eliminierbarkeit *f*
eliminable eliminierbar
eliminate *v* 1. eliminieren, beseitigen, entfernen; 2. *(org Ch)* eliminieren, abspalten
elimination 1. Eliminierung *f*, Beseitigung *f*, Entfernung *f*; 2. *(org Ch)* Eliminierung *f*, Abspaltung *f*
elimination-addition mechanism Eliminierungs-Additions-Mechanismus *m*
elimination of water Wasserabspaltung *f*, Wasserentzug *m*, Entwässern *n*, Entwässerung *f*
elimination reaction Eliminierung(sreaktion) *f*, Abspaltungsreaktion *f*
ellagic acid Ellagsäure *f*, 4,5,6,4′,5′,6′-Hexahydroxydiphensäuredilacton *n*
ellipsometry *(Spektr)* Ellipsometrie *f*
elliptical polarization elliptische Polarisation *f*
elongate-prismatic *(Krist)* langprismatisch
elongation 1. Verlängerung *f*, Längenänderung *f*; 2. *(Bioch)* Elongation *f*, Kettenverlängerung *f*
ELS *s.* energy-loss spectroscopy
eluant Elutionsmittel *n*, *Eluent *m*
eluate Eluat *n* *(durch Herauslösen adsorbierter Stoffe gewonnene Flüssigkeit)*
elucidate *v* aufklären, ermitteln, erforschen
elucidation Aufklärung *f*, Ermittlung *f*, Erforschung *f*
eluent *s.* eluant
eluotropic series *(Chromat)* eluotrope Reihe *f*
elute *v* eluieren, herausspülen, herauslösen, auswaschen *(adsorbierte Stoffe aus festen Adsorptionsmitteln)*
eluting agent *s.* eluant
elution Elution *f*, Eluieren *n* *(adsorbierter Stoffe aus festen Adsorptionsmitteln)*
elution chromatogram Elutionschromatogramm *n*, Durchlaufchromatogramm *n*
elution chromatography *Elutionschromatographie *f*, Durchlaufchromatographie *f*
elution curve *(Chromat)* Elutionskurve *f*, Konzentrationsprofil *n*

elutriate *v* (ab)schlämmen, ausschlämmen, abschwemmen
elutriation Abschlämmen *n*, Ausschlämmen *n*, Schlämmen *n*, Abschwemmen *n*
emanate *v* emanieren, ausströmen, ausfließen; emanieren, ausstrahlen, aussenden *(z. B. radioaktive Strahlen)*
emanation Emanation *f*, Ausströmen *n*, Ausfließen *n*; Emanation *f*, Ausstrahlen *n*, Aussenden *n (z. B. radioaktiver Strahlen)*
embarrass *v* behindern *(z. B. eine Reaktion)*
Embden-Meyerhof-Parnas scheme (of glycolysis) *(Bioch)* Embden-Meyerhof-Parnas-Weg *m*, Embden-Meyerhof-Parnas-Abbauweg *m*, EMP-Weg *m*, glykolytischer Abbau *m*, Glykolyse *f*, Fructose-1,6-diphospat-Weg *m*, FDP-Weg *m*
embrittle *v* spröd(e) machen; brüchig machen; spröd(e) werden, verspröden; brüchig werden
embrittlement Versprödung *f*, Sprödwerden *n*
emerald *(Min)* Smaragd *m*
emergency shower Notbrause *f*, Löschbrause *f*, Feuerlöschbrause *f*
emetic Emetikum *n*, Brechmittel *n*
emf, e.m.f., EMF, E.M.F. *s.* electromotive force
emission Emission *f*, *(von Strahlung auch)* Aussendung *f*, Ausstrahlung *f*, Abstrahlung *f*, *(von Schadstoffen, Teilchen auch)* Ausstoß *m*, Ausstoßung *f*
emission control Emissionsminderung *f*
emission inventory Emissionskataster *m*
emission sink Emissionssenke *f*, Einwirkungsstelle *f (von Luftschadstoffen)*
emission source Emissionsquelle *f*
emission spectroscopy Emissionsspektroskopie *f*, Emissionsspektralanalyse *f*, ES
emission spectrum Emissionsspektrum *n*
emissive emittierend, Emissions..., ausstrahlend, aussendend
emissivity *(Spektr)* Emissionskoeffizient *m*, Emissionsvermögen *n*; Emissionsgrad *m*, Schwärzegrad *m*, Emissionszahl *f (für Wärmestrahlung aussendende Körper)*
emit *v* emittieren, aussenden, ausstrahlen, abstrahlen; emittieren, ausstoßen *(Teilchen)*
emittance 1. spezifische Ausstrahlung *f*; 2. *s.* emissivity
emitter Emitter *m*, Strahler *m*, Strahlungsquelle *f*; Emitterelektrode *f*, Emissionselektrode *f*
α-emitter *(Kern)* α-Strahler *m*, Alphastrahler *m*
emmenagogue *(Pharm)* Emmenagogum *n*, Menagogum *n*, menstruationsförderndes Mittel *n*
EMP *s.* electron microprobe
EMP pathway *s.* Embden-Meyerhof-Parnas scheme
empirical empirisch (abgeleitet), erfahrungsgemäß, Erfahrungs...
empirical formula 1. *(Ch)* empirische Formel *f*, Summenformel *f*, Bruttoformel *f*; 2. empirisch abgeleitete Formel *f (Mathematik, Physik)*
empty *v* (ent)leeren
empty band leeres [unbesetztes] Energieband *n* [Leitungsband *n*, Leitfähigkeitsband *n*]
emulsifiability Emulgierbarkeit *f*
emulsifiable emulgierbar
emulsification Emulgieren *n*, Emulgierung *f*, Emulsionieren *n*, Emulsionsbildung *f*
emulsifier 1. Emulgator *m*, Emulgier(ungs)mittel *n*, Emulsionsbildner *m*, Emulgens *n*, Emulsionsvermittler *m*; 2. Emulgiermaschine *f*, Emulgator *m*
emulsify *v* emulgieren, emulsionieren
emulsifying agent *s.* emulsifier 1.
emulsion 1. Emulsion *f*; 2. *s.* photographic emulsion
emulsion breakdown [breaking] Brechen *n* [Spalten *n*, Trennen *n*, Zerstören *n*] von Emulsionen, Dismulgieren *n*, Entemulsionieren *n*; Brechen *n* [Zerfallen *n*] einer Emulsion, Entmischung *f* [Spaltung *f*] einer Emulsion, Emulsionsentmischung *f*, Emulsionsspaltung *f*
emulsion layer *(Foto)* Emulsionsschicht *f*
emulsion polymer Emulsionspolymerisat *n*, Emulsionspolymer(e) *n*
emulsion polymerization Emulsions-

polymerisation *f*, Polymerisation *f* in Emulsion
emulsion sensitivity *(Foto)* Schichtempfindlichkeit *f*
emulsion stabilizer Emulsionsstabilisator *m*
emulsive Emulsions...; emulsionsartig
emulsoid Emulsoid *n*, Emulsionskolloid *n*
enamel *v* 1. emaillieren; 2. mit Emaillelackfarbe lackieren
enamel 1. Email *n*; 2. emailähnliche Schutzschicht *f*, Lackschutzschicht *f*; 3. emaillierter Gegenstand *m*; 4. Emaillelack *m*; 5. *(Bio)* Zahnschmelz *m*, Schmelz *m*
enamine *(org Ch)* Enamin *n*
enantiomer Enantiomer(e) *n*, Spiegelbildisomer(e) *n*, optisches Isomer(e) *n*
enantiomer separation Enantiomerentrennung *f* *(Stereochemie)*
enantiomeric enantiomer, spiegelbildisomer, optisch isomer
enantiomeric excess Enantiomerenüberschuss *m* *(Stereochemie)*
enantiomeric purity Enantiomerenreinheit *f* *(Stereochemie)*
enantiomeric separation *s.* enantiomer separation
enantiomerically pure enantiomerenrein *(Stereochemie)*
enantiomerism Enantiomerie *f*, Spiegelbildisomerie *f*, optische Isomerie *f*
enantiomorph *(Krist)* enantiomorphe Form *f*
enantiomorphic *s.* enantiomorphous
enantiomorphism *(Krist)* Enantiomorphie *f*
enantiomorphous *(Krist)* enantiomorph
enantioselective *(org Ch)* enantioselektiv
enantioselectivity *(org Ch)* Enantioselektivität *f*
enantiotropic enantiotrop, wechselseitig [ineinander] umwandelbar *(Modifikationen)*
enantiotropic transition enantiotrope Umwandlung *f* [Phasenumwandlung *f*]
enantiotropy Enantiotropie *f*, wechselseitige Umwandelbarkeit *f* *(von Modifikationen)*
encapsulate *v* 1. *(Kst)* einbetten; 2. *s.* microencapsulate
encapsulation 1. *(Kst)* Einbetten *n*, Einbettung *f*, Vergießen *n*; 2. *s.* microencapsulation; 3. *(Kern)* Einkapselung *f*; 4. Kapselung *f* *(Umweltschutz)*
enclose *v* einschließen
enclosed molecule Gastmolekül *n*
enclosure Einschluss *m*
enclosure compound Einschlussverbindung *f*
encounter Begegnung *f*, Zusammentreffen *n*, Zusammenstoß *m* *(von Molekülen in Lösung)*
encrust *v* inkrustieren, eine Kruste bilden, verkrusten; inkrustieren, mit einer Kruste überziehen [bedecken], überkrusten
encrustation 1. Inkrustierung *f*, Inkrustation *f*, Überkrustung *f*; 2. Inkrustation *f*, Kruste *f*
end group analysis Endgruppenbestimmung *f*, Endgruppenanalyse *f*
end-point Endpunkt *m* *(z. B. der Titration)*; Endsiedepunkt *m*, Siedeendpunkt *m*, Siedeende *n*, SE, Endkochpunkt *m*
end-point detection [determination] Endpunktsbestimmung *f*
end-product inhibition *(Bioch)* Endprodukthemmung *f*, Rückkopplungshemmung *f*, negative Rückkopplung *f*, negatives Feed-back *n*
end-to-end distance Endpunktabstand *m*, Kettenendenabstand *m*, Fadenendenabstand *m*
endergonic endergon(isch), energieverbrauchend *(Reaktion)*
endo product endo-Produkt *n*
endocellular enzyme intrazelluläres [sessiles] Enzym *n*, Zellenzym *n*
endocyclic endocyclisch *(z. B. Doppelbindung)*
endoenzyme *s.* endocellular enzyme
endogenous growth phase *(Hyd)* endogene Wachstumsphase *f*, letale Phase *f*, Absterbephase *f*
endomethylene bridge *(org Ch)* Endomethylenbrücke *f*
ENDOR *s.* electron nuclear double resonance
endosmosis Endosmose *f*
endosmotic endosmotisch
endothermic endotherm, wärmeaufnehmend, wärmeverbrauchend, wärmebindend

endotoxin Endotoxin *n*
ene *(org Ch)* En-Komponente *f*
ene reaction En-Reaktion *f*
enediol *(org Ch)* Endiol *n*
energetic coupling energetische Kopplung *f*
energetically favourable [favoured] energetisch begünstigt [bevorzugt]
energization Aktivierung *f*; Anhebung *f* *(von Elektronen auf eine energiereichere Bahn)*
energize *v* aktivieren; (an)heben, emporheben *(Elektronen auf eine energiereichere Bahn)*
energizer Aktivierungsmittel *n*, aktivierendes Mittel *n*, Aktivierungszusatz *m*, Verstärker *m*, Aktivator *m*
energy Energie *f*, Arbeitsvermögen *n* *(eines physikalischen Systems)*
energy balance Energiebilanz *f*
energy band Energieband *n*, Energiebereich *m*
energy barrier Energiebarriere *f*, Energiewall *m*, Energieberg *m*, Energiegebirge *n*, Energieschwelle *f*
energy carrier Energieträger *m*
energy consumption Energieverbrauch *m*, Energieaufnahme *f*
energy content Energieinhalt *m*
energy conversion Energie(um)wandlung *f*, Energieumformung *f*, Energieumsetzung *f*
energy density Energiedichte *f*
energy-dispersive X-ray spectroscopy energiedispersive Röntgenspektroskopie *f*
energy gap Energielücke *f*, verbotenes Band *n*, verbotene Zone *f*, verbotener [nicht zugelassener] Energiebereich *m*
energy level Energieniveau *n*, Energiezustand *m*, Energieterm *m*, Term *m*, Elektronenniveau *n*
energy-level diagram Energieniveaudiagramm *n*, Energieniveauschema *n*, Termdiagramm *n*, Termschema *n*
energy-loss spectroscopy Energieverlust-Spektroskopie *f*
energy of activation Aktivierungsenergie *f*
energy of binding Bindungsenergie *f*
energy of formation Bildungsenergie *f*
energy of radiation Strahlungsenergie *f*, Strahlenenergie *f*
energy of rotation [rotational motion] *(Spektr)* Rotationsenergie *f*
energy of translational motion *(Spektr)* Translationsenergie *f*
energy of vibrational motion *(Spektr)* Schwingungsenergie *f*
energy quantum Energiequant *n*
energy-requiring 1. energieverbrauchend; 2. *(Bioch)* endergon(isch)
energy state Energiezustand *m*
energy storage *(Bioch)* Energiespeicherung *f* *(Kohlenhydrate)*
energy transformation *(Bioch)* Energieumsatz *m*
energy-yielding 1. energieliefernd, energieabgebend; 2. *(Bioch)* exergon(isch)
engineering chemistry Ingenieurchemie *f*
engineering plastics Konstruktionskunststoff *m*
English degree of hardness *(Hyd)* Grad *m* englischer Härte, englischer Härtegrad *m* *(1 °eH = 8,0 mg CaO/l* H_2O*)*
enlarging adapter Übergangsstück *n* (Form) B *(mit großer Hülse auf kleinem Kern)*
enmesh *v* 1. räumlich vernetzen *(z. B. Stärkemoleküle)*; 2. einhüllen, einschließen *(z. B. Schmutzteilchen)*
enmeshment 1. räumliche Vernetzung *f* *(z. B. von Stärkemolekülen)*; 2. Einschluss *m* *(z. B. von Schmutzteilchen)*
enol *(org Ch)* Enol *n*
enol form [structure] Enolform *f*
enolate *(org Ch)* Enolat *n*
enolic *(org Ch)* enolisch, Enol...
enolic form [structure] Enolform *f*
enolization *(org Ch)* Enolisierung *f*
enolize *v* *(org Ch)* enolisieren
enophile *(org Ch)* Enophil *n*
enrich *v* anreichern
enriching section *(Dest)* Rektifizierteil *m*, Rektifikationsteil *m*, Rektifizierzone *f*, Rektifikationszone *f*, Verstärkungsteil *m*, Obersäule *f*
enrichment factor *(Kern)* Anreicherungsgrad *m*, Anreicherungsfaktor *m*
enrichment line *(Dest)* Verstärkungsgerade *f*, Verstärkungslinie *f*
entanglement Verhakung *f* *(von Polymerketten)*

entanglement network Verhakungsnetzwerk *n*
enter *v* eintreten
enter *v* **into chemical union** *s.* enter into combination
enter *v* **into combination** eine Verbindung eingehen
enter *v* **into reaction** in Reaktion treten, eine Reaktion eingehen
entering group eintretende Gruppe *f*
enthalpimetric enthalpiemetrisch
enthalpimetric analysis Enthalpiemetrie *f*
enthalpy Enthalpie *f*, Wärmeinhalt *m*, gibbssche Wärmefunktion *f*, H
enthalpy change Enthalpieänderung *f*
enthalpy-entropy chart Enthalpie-Entropie-Diagramm *n*, Mollier-Diagramm *n*, i-s-Diagramm *n*
enthalpy of formation Bildungsenthalpie *f*
enthalpy of melting Schmelzenthalpie *f*
enthalpy of reaction Reaktionsenthalpie *f*
entire lot [mass] *(Anal)* Gesamtmenge *f*, Grundgesamtheit *f*
entity Objekt *n*, Gebilde *n*
Entner-Doudoroff pathway *(Bioch)* Entner-Doudoroff-Weg *m*, Entner-Doudoroff-Abbauweg *m*, ED-Weg *m*, 2-Keto-3-desoxy-6-phosphogluconat-Weg *m*, KDPG-Weg *m (des Glucoseabbaus)*
entrain *v* mitreißen, mitführen
entrain *v* **out** *(Dest)* herausschleppen
entrained-bed reactor Flugstromreaktor *m*, Flugwolkereaktor *m*, Flugstaubwolkenreaktor *m*, Transportreaktor *m*
entrained dust Flugstaub *m (bei Wirbelschichtverfahren)*
entrainer *(Dest)* Schleppmittel *n*, Schlepper *m*, Mitnehmer *m*, Azeotropbildner *m*; Trägergas *n (bei der Sublimation)*
entraining agent *s.* entrainer
entrainment *(Dest)* Mitreißen *n*, Mitführen *n*
entrainment distillation *(Dest)* azeotrope Destillation *f*, Azeotropdestillation *f*
entrainment loss Verlust *m* durch Mitreißen
entrap *v* einschließen *(z. B. Kristallwasser)*
entrapment Einschluss *m*
entrapped enzyme *(Biot)* eingeschlossenes Enzym *n*
entropy Entropie *f*
entropy at absolute zero Nullpunktsentropie *f*, S_0
entropy chart 1. T-s-Diagramm *n*, Temperatur-Entropie-Diagramm *n*; 2. i-s-Diagramm *n*, Enthalpie-Entropie-Diagramm *n*, Mollier-Diagramm *n*
entropy decrease Entropieabnahme *f*
entropy increase Entropiezunahme *f*
entropy of reaction Reaktionsentropie *f*
entry site *(Bioch)* Eingang *m*, Akzeptorort *m*, A-Ort *m*, Rezeptorort *m*, Decodierungsort *m*, Aminoacylort *m*, Erkennungsregion *f*
envelope conformation *(Bioch)* Briefumschlagkonformation *f*, Kuvertkonformation *f*
environment Umgebung *f*, Milieu *n*; Umwelt *f*
environmental analysis Umweltanalytik *f*
environmental chemical Umweltchemikalie *f*
environmental chemist Umweltchemiker *m*
environmental chemistry Umweltchemie *f*
environmental contaminant Umweltschadstoff *m*
environmental impact Umweltbelastung *f*
environmental lead Umweltblei *n*
environmental pollutant Umweltschmutzstoff *m*, Umweltschadstoff *m*
environmental pollution Umweltverschmutzung *f* (durch Schadstoffe)
environmental protection Umweltschutz *m*
environmental radioactivity Umweltradioaktivität *f*
environmental sample *(Anal)* Umweltprobe *f*
environmental technology Umwelt(schutz)technik *f*
environmental temperature Umgebungstemperatur *f*
environmental toxicant Umweltgift *n*, toxischer Umweltstoff *m*

environmental toxicity Umwelttoxizität *f*
environmental toxicology Umwelttoxikologie *f*, Ökotoxikologie *f*
environmentally acceptable umweltverträglich, umweltfreundlich, umweltgünstig, umweltkompatibel
environmentally beneficial [benign, compatible] *s.* environmentally acceptable
environmentally damaging umweltschädigend
environmentally dangerous *s.* environmentally hazardous
environmentally friendly *s.* environmentally acceptable
environmentally hazardous umweltgefährlich, umweltgefährdend
environmentally persistent in der Umwelt persistent
environmentally safe umweltsicher
enzymatic enzymatisch, fermentativ, Enzym..., Ferment...
enzymatic activity *s.* enzyme activity
enzymatic catalysis Enzymkatalyse *f*
enzymatic conversion enzymatische Umwandlung *f* [Umsetzung *f*], Enzymumwandlung *f*, *(i. e. S.)* enzymatische Verzuckerung *f (Stärkekonversion)*
enzymatic decomposition [degradation] enzymatischer Abbau *m*
enzymatic hydrolysis enzymatische Hydrolyse *f*
enzymatic oxidation enzymatische Oxidation *f*
enzymatic reaction enzymatische [enzymkatalysierte] Reaktion *f*, Enzymreaktion *f*
enzymatic splitting enzymatische Spaltung *f*
enzymatic synthesis enzymatische Synthese *f*
enzyme Enzym *n*, Ferment *n*
enzyme activity Enzymaktivität *f*, enzymatische Aktivität *f*
enzyme assay *(Bioch)* Enzym-Assay *n*
enzyme biosynthesis Enzymbiosynthese *f*
enzyme biotechnology *s.* enzyme technology
enzyme bonding to a carrier Trägerbindung *f* eines Enzyms
enzyme-bound enzymgebunden
enzyme carrier Enzymträger *m*
enzyme catalysis Enzymkatalyse *f*
enzyme catalyst enzymatischer Katalysator *m*
enzyme-catalyzed enzymkatalysiert
enzyme chemistry Enzymchemie *f*
enzyme classification number *s.* EC number
enzyme complex Enzymkomplex *m*
enzyme conjugate *(Bioch)* Enzymkonjugat *n*
enzyme electrode Enzymelektrode *f*
enzyme fixation Enzymfixierung *f*
enzyme immobilization Enzymimmobilisierung *f*
enzyme inactivation Enzyminaktivierung *f*
enzyme induction Enzyminduktion *f*
enzyme inhibitor Enzyminhibitor *m*
enzyme kinetics Enzymkinetik *f*
enzyme-membrane reactor *(Biot)* Enzymmembranreaktor *m*
enzyme protein Enzymprotein *n*
enzyme reactor *(Biot)* Enzymreaktor *m*
enzyme recovery Enzymaufarbeitung *f*; Enzymrückgewinnung *f*
enzyme specificity Enzymspezifität *f*
enzyme-substrate complex Enzym-Substrat-Komplex *m*, ES-Komplex *m*, Michaelis-Komplex *m*
enzyme technology Enzymtechnologie *f*, Enzymtechnik *f*
enzyme unit *(Bioch)* Katal *n*, kat *(Einheit der Enzymaktivität)*
enzyme yield Enzymausbeute *f*
enzymic *s.* enzymatic
enzymologist Enzymologe *m*, Enzymforscher *m*
enzymology Enzymologie *f*
EO *s.* ethylene oxide
EOCl *s.* extractable organic chlorine compounds
eosin(e) Eosin *n*, 2,4,5,7-Tetrabromfluorescein *n*
EP lubricant *s.* extreme-pressure lubricant
EPA = Environmental Protection Agency *(U.S.)*
ephedrine hydrochloride Ephedrinhydrochlorid *n*
epichlorohydrin Epichlorhydrin *n*, *1-Chlor-2,3-epoxypropan *n*, Chlormethyloxiran *n*
epihalohydrin Epihalogenhydrin *n*

epimer Epimer(e) *n*
epimeric epimer
epimerism Epimerie *f* *(Stereochemie)*
epimerization Epimerisierung *f*
epimerize *v* epimerisieren
epitaxial *(Krist)* epitaktisch, Epitaxie…
epitaxic *(Krist)* epitaktisch, Epitaxie…
epitaxy *(Krist)* Epitaxie *f*, Orientierungsbeziehung *f*, Topotaxie *f*
EPMA *s.* electron-probe microanalysis
epoxidation Epoxidation *f*, Epoxidierung *f*
epoxide Epoxid *n*, Oxiran *n*
epoxidize *v* epoxidieren
epoxies Epoxidharze *npl*, Epoxyharze *npl*
epoxy group Epoxidgruppe *f*, Epoxygruppe *f*
epoxy resin Epoxidharz *n*, EP
1,2-epoxypropane 1,2-Epoxypropan *n*, Methyloxiran *n*, Propylenoxid *n*
EPR *(electron paramagnetic resonance)* *s.* ESR
EPT *s.* ethylene-propylene terpolymer
eq *s.* gram equivalent
eqn. = equation
equation of state Zustandsgleichung *f*
ρσ-equation *(org Ch)* Hammett-Gleichung *f*
equatorial/be äquatorial stehen
equatorial bond äquatoriale Bindung *f*, e-Bindung *f*
equatorial conformation äquatoriale Konformation *f*, e
equilibrate *v* ins Gleichgewicht bringen [setzen], äquilibrieren, ausbalancieren, ausgleichen; auswuchten; im Gleichgewicht halten
equilibration Herstellung *f* des Gleichgewichts, Äquilibrierung *f*, Ausbalancieren *n*; Auswuchten *n*; Aufrechterhaltung *f* des Gleichgewichts
equilibrium Gleichgewicht *n*; Ruhelage *f* • **be at [in] equilibrium** im Gleichgewicht sein • **go toward equilibrium** dem Gleichgewicht zustreben • **re-establish equilibrium** das Gleichgewicht wiederherstellen
equilibrium constant Gleichgewichtskonstante *f*, Massenwirkungskonstante *f*
equilibrium diagram Zustandsdiagramm *n*, Phasendiagramm *n*
equilibrium distillation Gleichgewichtsdestillation *f*, geschlossene [integrale] Destillation *f*, (kontinuierliche) Entspannungsdestillation *f*
equilibrium distribution Gleichgewichtsverteilung *f*
equilibrium position Gleichgewichtslage *f*
equilibrium potential Gleichgewichtspotenzial *n*, Gleichgewichtsgalvanispannung *f*, Nernst-Potenzial *n*
equilibrium state Gleichgewichtszustand *m*
equimolar äquimolar
equimolecular 1. äquimolekular; 2. *s.* equimolar
equipartition principle [theorem] Gleichverteilungssatz *m*, Äquipartitionstheorem *n*, Äquipartitionsprinzip *n*
equipment Ausrüstung *f*, Ausstattung *f*, Einrichtung *f*, Gerätschaften *pl*, Gerätepark *m*
equipotential surface Äquipotenzialfläche *f*, Potenzialfläche *f*, Niveaufläche *f*
equivalence point Äquivalenzpunkt *m*
equivalent äquivalent, gleichwertig, Äquivalent…; gleich stark *(Bindung)*
equivalent Äquivalent *n*, äquivalente [äquimolekulare] Menge *f*
equivalent conductance Äquivalentleitfähigkeit *f*
equivalent point Äquivalenzpunkt *m*
equivalent weight (relative) Äquivalentmasse *f*
erbia Erbiumoxid *n*
erbium Er Erbium *n*
ERDA = Energy Research and Development Administration
erfc *s.* error function
ergone 1. Biokatalysator *m*, Wirkstoff *m*, Ergin *n*; 2. Ergon *n*, Wirkungsgruppe *f* *(eines Enzyms)*
ergot alkaloid Mutterkornalkaloid *n*, Ergotalkaloid *n*
Erlenmeyer flask Erlenmeyer-Kolben *m*
erosion corrosion Erosionskorrosion *f*, Strömungskorrosion *f*
error function Fehlerfunktion *f*, Fehlerintegral *n*
error in measuring Messfehler *m*, wahrer Fehler *m*
erucic acid Erucasäure *f*, **cis*-Docos-13-ensäure *f*

erythrina alkaloid Erythrinaalkaloid *n*
erythro form erythro-Form *f (Stereochemie)*
ES 1. *s.* emission spectroscopy; 2. = Electrochemical Society
ES complex *s.* enzyme-substrate complex
e.s. unit elektrostatische Einheit *f* [Ladungseinheit *f*]
ESCA *s.* electron spectroscopy for chemical analysis
escape *v* 1. entweichen, austreten, ausströmen, abströmen, abfließen; 2. entgehen *(z. B. der Oxidation)*
escape *v* **notice [observation]** der Aufmerksamkeit entgehen, sich der Beobachtung entziehen
escape peak *(Anal)* Ausreißer *m*
esculent essbar, genießbar
ESI *s.* electrospray ionization
ESR *(electron spin resonance)* Elektronen(spin)resonanz *f*, ESR, (elektronen)paramagnetische Resonanz *f*, EPR
esr *(electron spin resonance)* Elektronen(spin)resonanz *f*, ESR, (elektronen)paramagnetische Resonanz *f*, EPR
ESR spectroscopy Elektronen(spin)resonanzspektroskopie *f*, ESR-Spektroskopie *f*, paramagnetische Resonanzspektroskopie *f*, EPR-Spektroskopie *f*
ESR spectrum Elektronen(spin)resonanzspektrum *n*, ESR-Spektrum *n*, paramagnetisches Resonanzspektrum *n*, EPR-Spektrum *n*
essence 1. Essenz *f*, Auszug *m*, Extrakt *m*; 2. flüchtiger Stoff *m*; 3. Wohlgeruch *m*
essential 1. *(Bioch)* essenziell, unentbehrlich, lebensnotwendig, unersetzlich; 2. ätherisch, flüchtig
essential amino acid essenzielle Aminosäure *f*
essential fatty acid essenzielle Fettsäure *f*
essential oil ätherisches Öl *n*
essentiality *(Bioch)* Unentbehrlichkeit *f*, Unersetzlichkeit *f*
establish *v* herstellen *(z. B. Gleichgewicht, Kontakt)*
establish *v* **the structure** die Konstitution feststellen [aufklären, ermitteln, sichern]
establishment Herstellung *f (z. B. des Gleichgewichts, Kontakts)*
ester Ester *m*
ester condensation Esterkondensation *f*
ester exchange Esteraustausch *m*
ester interchange Umesterung *f*
ester saponification Esterverseifung *f*, *(i. e. S.)* Verseifung *f*, Esterspaltung *f*
esterification Verestern *n*, Veresterung *f*
esterify *v* verestern
estimate *v* bestimmen; schätzen
estimate *(Anal)* Schätzwert *m*
estimation Bestimmung *f*; Schätzung *f*
estrogen *s.* oestrogen
et. *s.* ethyl ether
Étard reaction étardsche Reaktion *f* *(Überführung aromatisch oder heterocyclisch gebundener Methylgruppen in Aldehyd- oder Ketogruppen)*
etch *v* ätzen
etch resist Ätzschutz *m*
etchant Ätzmittel *n*
etching Ätzen *n*, Ätzung *f*, Anätzen *n*; Beizen *n*
etching mask Ätzmaske *f*
ethanal *Ethanal *n*, Acetaldehyd *m*
ethane Ethan *n*
1,2-ethane diamine Ethan-1,2-diamin *n*, Ethylendiamin *n*
ethanedioic acid *Ethandisäure *f*, Oxalsäure *f*
1,2-ethanediol *Ethan-1,2-diol *n*, Ethylenglykol *n*
ethanethiol Ethanthiol *n*, Ethylhydrosulfid *n*
ethanethiolic acid Thioessigsäure *f*
ethanoic acid *Ethansäure *f*, Essigsäure *f*
ethanoic anhydride Ethansäureanhydrid *n*, Acetanhydrid *n*, Essigsäureanhydrid *n*
ethanol *Ethanol *n*, Ethylalkohol *m*
ethanol fermentation *(Biot)* Ethanolfermentation *f*, Ethanolgärung *f*
ethanolamine Ethanolamin *n*, *(i. e. S.)* Monoethanolamin *n*, *2-Aminoethan-1-ol *n*, Colamin *n*
ethanolic ethanolisch
ethanolysis Ethanolyse *f*

ethanoyloxy group $-CH_3CO_2$ Acetoxygruppe *f*
ethene *Ethen *n*, Ethylen *n*
ether 1. R–O–R′ Ether *m*, *(i. e. S.)* C_2H_5–O–C_2H_5 Diethylether *m*, Ether *m*, *Ethoxyethan *n*; 2. *s.* ester
ether cleavage Etherspaltung *f*
etherate Etherat *n*
ethereal etherisch, Ether...
ethereal oil *s.* essential oil
etherification Verethern *n*, Veretherung *f*
etherify *v* verethern
ethine *Ethin *n*, Acetylen *n*
ethinyl group –C≡CH Ethinylgruppe *f*, Ethinylrest *m*
ethoxide Ethylat *n*
ethoxyaniline Ethoxyanilin *n*, Phenetidin *n*, *Aminophenylethylether *m*
ethoxybenzene *Ethoxybenzen *n*, Ethylphenylether *m*, Phenetol *n*
ethoxyethane *Ethoxyethan *n*, Diethylether *m*, *(i. e. S.)* Ether *m*
ethyl acetate Essigsäureethylester *m*, Essigester *m*, Ethylacetat *n*
ethyl acetoacetate Acetessigsäureethylester *m*, Acetessigester *m*
ethyl alcohol Ethylalkohol *m*, *Ethanol *n*
ethyl bromide Ethylbromid *n*, Bromethan *n*
ethyl butyrate Buttersäureethylester *m*, Ethylbutyrat *n*
ethyl carbamate Carbamidsäureethylester *m*, Ethylcarbamat *n*, Ethylurethan *n*
ethyl chloride Ethylchlorid *n*, Chlorethan *n*
ethyl ester Ethylester *m*
ethyl ether Diethylether *m*, Ethoxyethan *n*, *(i. e. S.)* Ether *m*
ethyl formate Ameisensäureethylester *m*, Ethylformiat *n*
ethyl gasoline verbleites [gebleites, mit Tetraethylblei versetztes] Benzin *n*, Ethylbenzin *n*, Bleibenzin *n*
ethyl group C_2H_5– Ethylgruppe *f*, Ethylrest *m*
ethyl halide Ethylhalogenid *n*, Halogenethan *n*
ethyl malonate Malonsäurediethylester *m*, Malonester *m*, Diethylmalonat *n*
ethyl propionate Propionsäureethylester *m*, Ethylpropionat *n*
ethyl silicone Ethylpolysiloxan *n*, Ethylsilicon *n*
ethyl sulphate Schwefelsäurediethylester *m*, Diethylsulfat *n*
ethylal Ethylal *n*, *Diethoxymethan *n*
ethylamine Ethylamin *n*, Monoethylamin *n*
ethylate *v* ethylieren
ethylate Ethylat *n*
ethylating reagent Ethylierungsmittel *n*
ethylation Ethylierung *f*
ethylbenzene Ethylbenzen *n*
ethylcellulose Ethylcellulose *f*, EC, Triethylcellulose *f*, Celluloseethylether *m*
ethylene 1. $H_2C{=}CH_2$ Ethen *n*, Ethylen *n*; 2. *s.* ethylene group
ethylene chloride *s.* ethylene dichloride
ethylene dichloride Ethylendichlorid *n*, *1,2-Dichlorethan *n*
ethylene glycol Ethylenglykol *n*, *(i. e. S.)* Glykol *n*, *Ethan-1,2-diol *n*
ethylene group $-CH_2CH_2-$ Ethylengruppe *f*
ethylene oxide Ethylenoxid *n*, Oxiran *n*, 1,2-Epoxyethan *n*
ethylene-propylene terpolymer Ethylen-Propylen-Terpolymer(e) *n*
ethylene-vinylacetate copolymer Ethylen-Vinylacetat-Copolymerisat *n*
ethylenediamine Ethylendiamin *n*, 1,2-Ethandiamin *n*
ethylenediamine tetraacetic acid Ethylendiamintetraessigsäure *f*, EDTE
ethylenedicarboxylic acid *Ethan-1,2-dicarbonsäure *f*, *Butandisäure *f*, Bernsteinsäure *f*
ethyleneimine Ethylenimin *n*, Dimethylenimin *n*
ethylenic hydrocarbon Ethylenkohlenwasserstoff *m*, *Alken *n*, Olefin *n*
ethylidene group $=CHCH_3$ Ethylidengruppe *f*
ethylsulphuric acid Ethylschwefelsäure *f*, Ethylhydrogensulfat *n*, Schwefelsäuremonoethylester *m*
ethylthioethane Ethylthioethan *n*, Diethylsulfid *n*
ethyne *Ethin *n*, Acetylen *n*
ethynyl group –C≡CH Ethinylgruppe *f*, Ethinylrest *m*
ethynylation Ethinylierung *f*
EtOH *s.* ethyl alcohol

ETS *(electron-transport system) s.* electron-transport chain
eucalyptol(e) Eucalyptol *n*, 1,8-Cineol *n*, 1,8-Epoxy-*p*-menthan *n*
eucolloid Eukolloid *n*
eugenol Eugenol *n*, *1-Allyl-4-hydroxy-3-methoxybenzen *n*
europia Europiumoxid *n*
europium Eu Europium *n*
eutactic polymer eutaktisches [sterisch regelmäßiges] Polymer(e) *n*
eutectic eutektisch
eutectic *s.* 1. eutectic mixture; 2. eutectic point
eutectic mixture eutektische Mischung *f*, eutektisches Gemisch *n*, Eutektikum *n*
eutectic point eutektischer Punkt *m*
eutectoid eutektoid
eutectoid Eutektoid *n*
eutrophic *(Hyd)* eutroph *(reich an gelösten Pflanzennährstoffen)*
eutrophication *(Hyd)* Eutrophierung *f*
eV *s.* electron-volt
EVA *s.* ethylene-vinylacetate copolymer
evacuate *v* evakuieren, auspumpen, (ent)leeren, *(i. e. S.)* luftleer machen; abpumpen *(Gase, Flüssigkeiten)*
evacuation Evakuieren *n*, Evakuierung *f*
evanescent vergänglich, schwindend *(z. B. Färbung)*; unbeständig, instabil *(chemische Verbindung)*
evaporability Verdampfbarkeit *f*
evaporable verdampfbar
evaporate *v* verdampfen; verdunsten *(unterhalb des normalen Siedepunkts)*; eindampfen *(Lösungen)*; abdampfen, verdampfen (lassen) *(eine Flüssigkeit)*; ausdampfen *(Komponenten eines Flüssigkeitsgemischs)*
evaporate *v* **off** abdampfen
evaporate *v* **to dryness** zur Trockne eindampfen
evaporate *v* **to low [small] bulk [volume]** einengen
evaporating basin [dish] Abdampfschale *f*
evaporation Verdampfung *f*, Verdampfen *n*; Verdunstung *f*, Verdunsten *n* *(unterhalb des normalen Siedepunkts)*; Eindampfen *n (von Lösungen)*; Abdampfen *n*, Verdampfen(lassen) *n (einer Flüssigkeit)*; Ausdampfen *n (von Komponenten eines Flüssigkeitsgemischs)*
evaporative Verdampfungs...; Verdunstungs... *(unterhalb des normalen Siedepunkts)*
evaporator Verdampf(ungs)apparat *m*, Eindampfapparat *m*, Verdampfer *m*, Eindampfer *m*
even out ausgleichen *(z. B. die Konzentration)*
even-membered geradzahlig *(Ringverbindung)*
even-number(ed) geradzahlig *(Fettsäure)*
evil smelling übel riechend
evolution of heat Wärmeentwicklung *f*, Wärmeabgabe *f*
evolution of hydrogen Wasserstoffentwicklung *f*, Wasserstoffabscheidung *f*
evolve *v* entwickeln, erzeugen, abgeben; sich entwickeln, entstehen
evolved/be sich entwickeln, entstehen
Ewens-Bassett number Ewens-Bassett-Zahl *f*, Ladungszahl *f*
EXAFS = extended X-ray absorption fine structure
examination Prüfung *f*, Überprüfung *f*, Untersuchung *f*
exceedingly sensitive *(Anal)* höchst empfindlich
excess überschüssig, Überschuss...; übermäßig, Über...
excess Überschuss *m*; Übermaß *n* • **in slight excess of** mit einem geringen Überschuss an
excess charge Überschussladung *f*
excess electron Überschusselektron *n*
excess energy Überschussenergie *f*, Energieüberschuss *m*
excess of acid Säureüberschuss *m*
excess of electrons Elektronenüberschuss *m*
excess pressure Überdruck *m*
excess solid [solute] Bodenkörper *m* *(im Gleichgewicht mit seiner Lösung)*
excessive überschüssig, Überschuss...; übermäßig, Über...
exchange *v* austauschen
exchange Austausch *m*
exchange chromatography Austauschchromatographie *f*, *(i. e. S.)* Ionenaustauschchromatographie *f*

exchange current density *(Ech)* Austauschstromdichte *f*
exchange energy Austauschenergie *f*
exchange of heat Wärmeaustausch *m*, Wärmeübertragung *f*, Wärmetransport *m*
exchange of materials Stoffaustausch *m*
exchange reaction Austauschreaktion *f*, *(i. e. S.)* Ionenaustauschreaktion *f*
exchange site Austauschplatz *m (eines Ionenaustauschers)*
exchangeability 1. *(Ch)* Austauschbarkeit *f*; 2. *(Tech)* Auswechselbarkeit *f*
exchangeable 1. *(Ch)* austauschbar; 2. *(Tech)* auswechselbar
exchanger Austauscher *m*
excimer Excimer *n (Komplex aus angeregtem und nicht angeregtem Molekül gleicher Sorte)*
excimer laser Excimerlaser *m*
exciplex Exciplex *m (Komplex aus angeregtem Molekül A und nicht angeregtem Molekül B)*
excision repair *(Bioch)* Exzisionsreparatur *f (der DNA)*
excitability Anregbarkeit *f*
excitable anregbar
excitation Anregung *f*
excitation energy Anregungsenergie *f*
excitation spectrum Anregungsspektrum *n*
excite *v* anregen
excited atom angeregtes Atom *n*
excited complex *s.* exciplex
excited state angeregter Zustand *m*, Anregungszustand *m*
exciton Exziton *n*, Exciton *n*, Elektron-Loch-Paar *n*, Elektron-Defektelektron-Paar *n*
excluded volume ausgeschlossenes Volumen *n*, Ausschlussvolumen *n (eines Makromolekülknäuels)*
exclusion chromatography *s.* gel chromatography
exclusion of oxygen Sauerstoffausschluss *m*
exclusion principle (paulisches) Ausschließungsprinzip *n*, Pauli-Prinzip *n*
exergonic exergon(isch), energieabgebend, energieliefernd *(Reaktion)*
exfoliate *v* abblättern, abschiefern, abschilfern, abplatzen, abspringen
exfoliation Abblättern *n*, Abschiefern *n*, Abschilfern *n*, Abplatzen *n*, Abspringen *n*
exhaust *v* 1. auspumpen, luftleer machen, evakuieren; 2. erschöpfen, vollständig ausnutzen *(Vorräte oder Inhaltsstoffe)*; 3. erschöpfend extrahieren
exhaust gas Abgas *n*; Auspuffgas *n*
exhaust hood Abzug(sschrank) *m*, Digestorium *n*
exhaust-steam-pressure regulator Abdampfdruckregler *m*
exhaust tube Abgasschlauch *m*
exhaust-valve box Abblaseventilkammer *f*
exhausted ion exchanger *(Hyd)* erschöpfter Ionenaustauscher *m*
exhauster Saugzuglüfter *m*, Exhaustor *m*
exhaustion 1. Auspumpen *n*, Evakuieren *n*; 2. Erschöpfung *f*, vollständige Ausnutzung *f (von Vorräten oder Inhaltsstoffen)*; 3. erschöpfende Extraktion *f*
exhaustive methylation erschöpfende Methylierung *f*
exhibit *v* zeigen, aufweisen
exit Austritt *m*, Auslass *m*, Auslauf *m*
exit air Abluft *f*
exocyclic exocyclisch
exoelectron Exoelektron *n*
exoenzyme *s.* extracellular enzyme
exoergic 1. exergon(isch), energieabgebend, energieliefernd; 2. *s.* exothermic
exosmosis Exosmose *f*
exothermic exotherm(isch), wärmeabgebend, wärmeliefernd
exotoxin Exotoxin *n*, Ektotoxin *n*
expand *v* ausdehnen, expandieren; entspannen *(Gase)*; (ver)schäumen, aufschäumen *(Kunststoffe)*; sich ausdehnen
expandability 1. *(Kst)* Verschäumbarkeit *f*, Aufschäumbarkeit *f*; 2. *s.* expansibility
expandable 1. *(Kst)* verschäumbar, aufschäumbar; 2. *s.* expansible
expandable polystyrene aufschäumbares Polystyrol *n*
expanded plastic Schaum(kunst)stoff *m*
expanded polystyrene Schaumpolystyrol *n*, geschäumtes Polystyrol *n*

expanding agent Treibmittel *n*, Blähmittel *n*
expansibility Ausdehnbarkeit *f*, Dehnbarkeit *f*
expansible (aus)dehnbar, expansibel
expansion Ausdehnung *f*, Dehnung *f*, Expansion *f*; Entspannung *f (von Gasen)*; Aufweitung *f (des Kristallgitters)*; Schäumen *n*, Aufschäumen *n*, Verschäumen *n (von Kunststoffen)*
expansion behaviour Dehnungsverhalten *n*
expansion coefficient Ausdehnungskoeffizient *m*, Ausdehnungszahl *f*
expansion effect Expansionseffekt *m*
expectation Erwartungswert *m*, Mittelwert *m (Statistik)*
expected value Erwartungswert *m*, Mittelwert *m (Statistik)*
expectorant *(Pharm)* Expektorans *n*, auswurfförderndes Mittel *n*
expel *v* (her)austreiben, vertreiben, verdrängen, verjagen *(Gase)*; abpressen, herauspressen *(Öl)*; ausstoßen, emittieren, aussenden *(radioaktive Teilchen)*; ausstoßen, abspalten *(Moleküle bei Eliminierungsreaktionen)*
experiment *v* experimentieren, Versuche anstellen
experiment Experiment *n*, Versuch *m*
experimental experimentell, experimental, Experimental..., Versuchs...
experimental Experimentelles *n*, Methodik *f (in Abhandlungen)*
experimental arrangement Versuchsanordnung *f*
experimental chemistry Experimentalchemie *f*, experimentelle Chemie *f*
experimental error Versuchsfehler *m*
experimental procedure Versuchsdurchführung *f*
experimental technique Experimentiertechnik *f*, Experimentierkunst *f*
experimentalize *v* experimentieren, Versuche anstellen
expl. = explodes
explode *v* explodieren, *(Behälter auch)* (zer)platzen, bersten, zerknallen; explodieren lassen, zur Explosion bringen
explosibility Explosivität *f*, Explosibilität *f*, Explosionsfähigkeit *f*
explosible explosiv, explosibel, leicht explodierend, explosionsgefährlich
explosion Explosion *f*
explosion hazard Explosionsgefahr *f*
explosion limits Explosionsgrenzen *fpl*
explosion-proof explosionsgeschützt, explosionssicher
explosive explosiv, explosibel, leicht explodierend, explosionsgefährlich; Explosions..., Spreng...; explosionsartig
explosive Explosivstoff *m*, *(i. e. S.)* Sprengstoff *m*
explosive limits [range] Explosionsgrenzen *fpl*
explosively explosionsartig, unter Explosion
explosiveness *s.* explosibility
explosives industry Explosivstoffindustrie *f*, *(i. e. S.)* Sprengstoffindustrie *f*
expose *v* 1. exponieren, aussetzen *(z. B. einer Strahlung oder korrosiven Einflüssen)*; *(Werkstoffprüfung auch)* auslagern; 2. *(Foto)* belichten, exponieren
expose *v* **to air** der Luft aussetzen
expose *v* **to light** 1. dem Licht aussetzen; 2. *s.* expose 2.
exposure 1. Exponierung *f*, Exposition *f*, Aussetzen *n*, *(Werkstoffprüfung auch)* Auslagerung *f*; Beanspruchung *f*; 2. *(Foto)* Belichten *n*, Belichtung *f*, Exposition *f* • **on exposure to air** an der Luft, unter Lufteinwirkung, bei Luftzutritt • **on exposure to light** unter Lichteinwirkung [Lichteinfluss], am [im] Licht, bei Belichtung
express *v* auspressen, abpressen, ausdrücken
expression Auspressen *n*, Abpressen *n*, Ausdrücken *n*
expulsion Austreiben *n*, Heraustreiben *n*, Vertreiben *n*, Verdrängen *n*, Verjagen *n (von Gasen)*; Abpressen *n*, Herauspressen *n (von Öl)*; Ausstoßen *n*, Emission *f*, Aussenden *n (von radioaktiven Teilchen)*; Ausstoßen *n*, Ausstoßung *f*, Abspaltung *f (bei Eliminierungsreaktionen)*
exsiccate *v* (aus)trocknen
exsiccation Trocknung *f*, Austrocknung *f*
extend *v* 1. (aus)dehnen; 2. strecken *(Chemikalien durch Zusatzstoffe)*

extended-chain gestrecktkettig *(Polymer)*
extended length Konturlänge *f*, Gesamtlänge *f (einer Molekülkette)*
extender 1. Streckmittel *n*, Verschnittmittel *n*, Extender *m*; 2. *(Kst)* Füllmittel *n*, Füllstoff *m*, Beschwerungsmittel *n*
extensibility Dehnbarkeit *f*, Ausdehnbarkeit *f*, Streckbarkeit *f*
extension Dehnung *f*, Ausdehnung *f*, Längenzunahme *f*
extensive property [quantity] extensive [additive] Größe *f*, Quantitätsgröße *f*
extent of dilution Verdünnungsgrad *m*
extent of reaction Reaktionslaufzahl *f*
exterior durability Außenbeständigkeit *f*, Witterungsbeständigkeit *f*, Wetterbeständigkeit *f*, Wetterfestigkeit *f*
exterior paint Außenanstrichstoff *m*, Außenanstrichfarbe *f*
external 1. äußerer, außen befindlich, Außen...; 2. *(Pharm)* extern, äußerlich, zum äußerlichen Gebrauch
external electron Außenelektron *n*, äußeres Elektron *n*
external phase geschlossene [zusammenhängende] Phase *f*, Dispersionsphase *f*
external plasticization äußere Weichmachung *f*
external pressure Außendruck *m*
external shell äußere Schale *f*, Außenschale *f*
external surface (area) äußere Oberfläche *f*, Außen(ober)fläche *f*
extinction 1. Extinktion *f (Strahlungsschwächung durch Absorption und Streuung)*; 2. *(Spektr)* Extinktion *f*, Absorbanz *f*; Extinktion *f*, Auslöschung *f (Polarimetrie, Röntgenbeugung)*; 3. *(Foto)* Schwärzung *f*, Deckung *f*, Extinktion *f*, logarithmische Opazität *f*
extinction coefficient *(Spektr)* Extinktionskoeffizient *m*, Extinktionskonstante *f*
extinguish *v* löschen, zum Erlöschen bringen; auslöschen *(optisch)*
extracellular enzyme extrazelluläres Enzym *n*, Exoenzym *n*
extract *v* 1. extrahieren, ausziehen; 2. gewinnen *(Erze, Minerale, Metalle)*
extract with ether ausethern, mit Ether ausschütteln
extract Extrakt *m*, Auszug *m*
extract phase Extraktphase *f*
extract yield Extraktausbeute *f*
extractability Extrahierbarkeit *f*
extractable 1. extrahierbar, ausziehbar; 2. gewinnbar *(Erze, Minerale, Metalle)*
extractable (matter) extrahierbarer Stoff *m*, extrahierbare Substanz *f*
extractable organic chlorine compounds *(Hyd)* extrahierbare organische Chlorverbindungen *fpl*
extractant Extraktionsmittel *n*
extracting liquid [solvent] Extraktionsmittel *n*
extraction 1. Extrahieren *n*, Extraktion *f*, Ausziehen *n*; 2. Gewinnung *f (von Erzen, Mineralen, Metallen)*
extraction apparatus Extraktionsapparat *m*, Extrakteur *m*, Extraktor *m*
extraction battery Extraktionsbatterie *f*
extraction column Extraktionssäule *f*, Extraktionskolonne *f*
extraction efficiency Extraktionseffizienz *f*
extraction thimble Extraktionshülse *f*
extractive extraktiv, extrahierend, durch Extraktion erfolgend
extractive Extrakt(iv)stoff *m*
extractive distillation extraktive [extrahierende] Destillation *f*, Extraktivdestillation *f*, Distex-Prozess *m*
extractive esterification extraktive Veresterung *f*
extractive metallurgy erzeugende Metallurgie *f*
extractor Extraktionsapparat *m*, Extraktor *m*, Extrakteur *m*
extraneous material [substance] Fremdstoff *m*, fremder Stoff *m*, fremde Substanz *f*
extranuclear electron Hüllenelektron *n*
extranuclear region Atomhülle *f*, Elektronenhülle *f*
extreme-pressure lubricant Höchstdruckschmiermittel *n*, Höchstdruckschmierstoff *m*, Extreme-Pressure-Schmiermittel *n*, EP-Schmiermittel *n*
extreme value Extremwert *m*, Spitzenwert *m*, Extremum *n*
extremely slightly soluble sehr schwer [wenig] löslich, ssl., swl.
extremely toxic äußerst giftig
extremum *s.* extreme value

extrinsic factor Extrinsic-Faktor *m*, tierischer Eiweißfaktor *m*, Hämogen *n*, Vitamin *n* B_{12}
extrudate Extrudat *n*, Extrudiererzeugnis *n*, Strangpresserzeugnis *n*
extrude *v* extrudieren, strangpressen, spritzen
extruder Extruder *m*, Strangpresse *f*, Spritzmaschine *f*
extrusion 1. Extrudieren *n*, Strangpressen *n*, Spritzen *n*; Spinnen *n* [Erspinnen *n*] aus der Schmelze; 2. *s.* extrudate
extrusion compound Extrudiermasse *f*, Strangpressmischung *f*
extrusion moulding Extrudieren *n*, Strangpressen *n*, Spritzen *n*
exudate Exsudat *n*, Ausscheidungsprodukt *n*, Ausscheidung *f*, Ausschwitzungsprodukt *n*
exudation 1. Ausscheiden *n*, Absondern *n*, Ausschwitzen *n*; Austreten *n*; 2. *s.* exudate
exude *v* ausscheiden, absondern; ausschwitzen *(z. B. Harze)*; sich ausscheiden [absondern], austreten
eye irritant Augenreizstoff *m*
eye protection Augenschutz *m*

F

f. *s.* face-centred
F strain effect *(org Ch)* sterische Verzögerung *f*, F-strain-Effekt *m*
FAB = fast atom bombardment
fabric (cloth) Textilgewebe *n*, Fasergewebe *n*, (textiler) Stoff *m*, Gewebe *n*
fabric detergent Waschmittel *n*
fabric filter Tuchfilter *n*, Stofffilter *n*, Gewebefilter *n*; Gewebeabscheider *m*
face 1. Fläche *f*, Oberfläche *f*; Kristallfläche *f*; 2. *(org Ch, Nom)* Seite *f (eines Rings)*
face-centred *(Krist)* flächenzentriert, fl.z.
face-centred cubic *(Krist)* kubisch-flächenzentriert, kfz
face-centred lattice *(Krist)* flächenzentriertes Gitter *n*
face protector Gesichtsschutzschale *f*, Gesichtsschutz *m*
facilitated diffusion *(Bioch)* passiv katalysierter Transport *m*, erleichterte Diffusion *f (durch Membranen hindurch)*
factitious künstlich, nachgemacht, unecht
factor Faktor *m*, Koeffizient *m*, Beiwert *m*
factory effluent Betriebsabwasser *n*, betriebliches Abwasser *n*, Werksabwasser *n*
FAD *s.* flavin adenine dinucleotide
fade *v* verschießen, ausbleichen, verbleichen, verblassen, sich verfärben
fade *v* **away** verschwinden *(z. B. Färbung)*; abklingen *(Reaktion)*
fading Verschießen *n*, Ausbleichen *n*, Verbleichen *n*, Verblassen *n*
Fah *s.* Fahrenheit
Fahr *s.* Fahrenheit
Fahrenheit 1. Fahrenheitgrad *m*, Grad *m* Fahrenheit, °F; 2. Fahrenheitskale *f*
Fahrenheit scale *s.* Fahrenheit 2.
faintly acid schwach sauer
faintly alkaline schwach alkalisch
faintly defined unscharf [ungenau] definiert
fair mäßig *(Bewertung von Eigenschaften)*
fair yield brauchbare Ausbeute *f*
Fajans (adsorption indicator) method Endpunktsbestimmung *f* nach Fajans *(mit Adsorptionsindikatoren)*; Adsorptionstitration *f* nach Fajans *(Argentometrie)*
fall *v* 1. (ab)fallen, (ab)sinken, abnehmen, sich vermindern [verringern], zurückgehen; 2. *(in den Grundzustand)* zurückspringen *(Elektronen)*
fall *v* **off** *s.* fall 1.
fall *v* **to powder** zu Pulver zerfallen
fall Fallen *n*, Abfallen *n*, Abfall *m*, Sinken *n*, Absinken *n*, Abnehmen *n*, Abnahme *f*, Rückgang *m*, Minderung *f*, Verminderung *f*
fall of potential Potenzialabfall *m*
fall-off curve Abklingkurve *f*
falling-ball impact test Kugelfallversuch *m*
falling-ball visco(si)meter Kugelfallviskosimeter *n*
falling-film distillation Destillation *f* mit fallendem Film
falling-film evaporator Fallfilmverdampfer *m*, Fallstromverdampfer *m*
fallout 1. Fallout *n (Absetzen von Luft-*

schadstoffen); 2. Fallout *m (abgelagerte Luftschadstoffe)*; *(i. e. S.)* radioaktiver Fallout *m*
family 1. Familie *f*, Gruppe *f*, Verwandtschaftsgruppe *f (im Periodensystem)*; 2. *s.* family of compounds
family of compounds Stoffklasse *f*, Verbindungsklasse *f*
family relationship Vertikalbeziehung *f (im Periodensystem)*
fan Ventilator *m*, Lüfter *m*, Gebläse *n*
fanning plume fächerförmige Rauchfahne *f*
far infrared ferner Infrarotbereich *m*, fernes Infrarot(gebiet) *n*, FIR
far-infrared spectral region *s.* far infrared
far-infrared spectroscopy Spektroskopie *f* im fernen Infrarot
faradaic current faradayscher Strom *m*, Faraday-Strom *m*
Faraday constant Faraday-Konstante *f*, faradaysche Konstante *f*, F *(= 96 486 Coulomb)*
Faraday effect Faraday-Effekt *m*, Magnetorotation *f*, magnetische Drehung *f* (der Polarisationsebene), magnetisches Dreh(ungs)vermögen *n*
Faraday-Tyndall effect [phenomenon] *(Koll)* Faraday-Tyndall-Effekt *m*, Tyndall-Effekt *m*, Tyndall-Phänomen *n*
farina feines Mehl *n*; Kartoffelstärke *f*, Kartoffelmehl *n*, Stärkemehl *n*
farinaceous 1. mehlig, mehlartig; bemehlt *(Oberfläche)*; 2. stärkehaltig; stärkereich
fast base Echt(färbe)base *f*, Echtfarbbase *f*
fast breeder reactor *(Kern)* schneller Brutreaktor *m* [Brüter *m*]
fast color *(Am) s.* fast dye
fast dye Echtfarbstoff *m*, echter Farbstoff *m*
fast emulsion *(Foto)* hochempfindliche Emulsion *f*
fast-evaporating leichtflüchtig
fast neutron *(Kern)* schnelles Neutron *n*
fast reaction schnelle Reaktion *f*
fast red Echtrot *n*
fast salt *(Farb)* Echt(färbe)salz *n*
fast solvent schnellflüchtiges [leichtflüchtiges] Lösungsmittel *n*
fast to acid 1. säurefest, säurebeständig, beständig gegen Säuren; 2. *(Farb)* säureecht
fast to alkali 1. alkalifest, alkalibeständig, beständig gegen Alkalien; 2. *(Farb)* alkaliecht
fast to light lichtecht, lichtbeständig
fast-weighing balance Schnellwaage *f*
fastness 1. Beständigkeit *f*, Widerstandsfähigkeit *f*; 2. *(Farb)* Echtheit *f*
fastness of colour Farbechtheit *f*
fastness to light Lichtechtheit *f*, Lichtbeständigkeit *f*
fat 1. fett(ig), fetthaltig, ölhaltig, Fett..., Öl...; 2. fett, fruchtbar *(Boden)*
fat Fett *n*
fat acid höhere Fettsäure *f (mit 12 bis 24 C-Atomen im Molekül)*
fat chemistry Fettchemie *f*
fat deterioration Fettverderb *m*, Fettverderben *n*
fat globule Fettkügelchen *n*, Fetttröpfchen *n*
fat hardening [hydrogenation] Fetthärtung *f*, Fetthydrierung *f*
fat hydrolysis Fetthydrolyse *f*, Hydrolyse *f* der Fette
fat metabolism Fettstoffwechsel *m*
fat oxidation Fettoxidation *f*
fat solubility Fettlöslichkeit *f*
fat-soluble fettlöslich
fat splitting Fettspaltung *f*, Lipolyse *f*
fat spoilage *s.* fat deterioration
fatal dose *(Tox)* letale [tödliche] Dosis *f*, LD, Tödlichkeitsdosis *f*, dosis letalis
fatal poisoning tödliche Vergiftung *f*
fatigue *v (Tech)* ermüden
fatigue *(Tech)* Ermüdung(serscheinung) *f*
fatigue behaviour Ermüdungsverhalten *n*
fatigue crack Ermüdungsriss *m*
fatigue life Grenzlastspielzahl *f*, Grenzschwingspielzahl *f (Werkstoffprüfung)*
fatigue limit Ermüdungsgrenze *f*, Dauer(schwing)festigkeit *f*
fatigue proof ermüdungssicher, ermüdungsfrei
fatlike fettartig, fettähnlich
fattiness Fettigkeit *f*, Fetthaltigkeit *f*
fatty fett(ig), fetthaltig; fettartig; Fett...
fatty acid 1. Fettsäure *f*; 2. *s.* fat acid
fatty alcohol Fettalkohol *m*, Alkanol *n*
fatty oil fettes Öl *n*, Fettöl *n*
fault Mangel *m*, Fehler *m*, Defekt *m*

fault finding Fehlersuche *f*
faulty fehlerhaft *(Material)*; defekt *(Mechanismus)*
Favorskii rearrangement Favorskij-Umlagerung *f (von α-Halogenketonen zu Carbonsäuren oder deren Estern)*
F.B.P. *s.* final boiling point
FC *s.* 1. fixed carbon; 2. forced circulation
fcc *s.* face-centred cubic
FCC *s.* fluid catalytic cracking
FD *s.* field desorption
F.D.A. = Food and Drug Administration *(U.S.)*
feasible (technisch) durchführbar
feathery fed(e)rig
febrifuge *(Pharm)* Fiebermittel *n*, Antipyretikum *n*
feculence Trübung *f*, Trübe *f*; Bodensatz *m*, Satz *m*
feculent trüb(e), getrübt, unrein, undurchsichtig, verunreinigt, schlammig
feebly acid(ic) schwach sauer
feebly basic schwach basisch
feed *v* 1. einspeisen, zuspeisen, aufgeben, zuführen, eintragen, einfüllen *(Material)*; speisen, beschicken, füllen *(z. B. Reaktoren, Öfen)*; 2. *(Biot)* füttern *(Mikroorganismen)*; zufüttern, verfüttern *(Nährstoffe)*
feed *v* **in** einspeisen, eintragen, einfüllen
feed 1. Einspeisen *n*, Zuspeisen *n*, Aufgeben *n*, Aufgabe *f*, Zuführen *n*, Zufuhr *f*, Eingeben *n*, Eingabe *f*, Eintragen *n*, Eintrag *m*, Einfüllen *n (von Material)*; Speisen *n*, Beschicken *n*, Füllen *n (z. B. von Reaktoren, Öfen)*; 2. Beschickungsmaterial *n*, Beschickung *f*, Aufgabegut *n*, Füllung *f*, Füllmasse *f*, Schüttung *f*, Einsatzgut *n*, Einsatzstoff *m*, Einsatzmaterial *n*, Einsatz *m*, Eintrag *m*; Charge *f*, Ansatz *m*, Satz *m (bei diskontinuierlicher Arbeitsweise)*; 3. *(Landw)* Futter(mittel) *n*; 4. *(Biot)* Futter *n (für Mikroorganismen)*
feed additive Futtermittelzusatzstoff *m*, Futterzusatz *m*, Futtersupplement *n*
feed-additive antibiotic Fütterungsantibiotikum *n*, nutritives Antibiotikum *n*, Antibiotikum *n* als Futtermittelzusatz
feed-back ... *siehe: feedback ...*
feed-forward activation *(Bioch)* Aktivitätsförderung *f*, positive Rückkopplung *f*
feed-forward regulation *(Bioch)* Vorwärtskontrolle *f*
feed gas Einsatzgas *n*
feed inlet Beschickungsöffnung *f*, *(für Feststoffe auch)* Eintragöffnung *f*, Aufgabeöffnung *f*, Beladeöffnung *f*, Füllöffnung *f*, Füllloch *n*, *(für Flüssigkeiten auch)* Zulauf *m*
feed material *s.* feed 2.
feed opening *s.* feed inlet
feed pipe Beschickungsrohr *n*, Zuführungsrohr *n*, *(für Feststoffe auch)* Eintragrohr *n*, *(für Flüssigkeiten auch)* Zulaufrohr *n*, Einlaufrohr *n*
feed pump Dosierpumpe *f*
feed rate Eintragmenge *f* je Zeiteinheit
feed slurry *(Filtr)* (zulaufende) Trübe *f*
feed solids (concentration) *(Hyd)* Feststoffgehalt *m* [Feststoffkonzentration *f*] im Zulauf
feed stock *s.* feed 2. und 3.
feed valve Zulaufventil *n*
feed water Speisewasser *n*
feed-water make-up Zusatz(kesselspeise)wasser *n*, Zusatzspeisewasser *n*
feed zone 1. Beschickungszone *f*, Speisezone *f*, *(bei Flüssigkeiten auch)* Zulaufzone *f*; 2. *(Kst)* Einzugszone *f (im Extruder)*
feedback Rückführung *f*
feedback control 1. Regeln *n*, Regelung *f*; 2. *(Bioch)* Feed-back-Regulation *f*
feedback control system Regelkreis *m*
feedback inhibition *(Bioch)* Produktenhemmung *f*, Rückkopplungshemmung *f*, Feed-back-Hemmung *f*, Feed-back-Inhibition *f*, negatives Feed-back *n*
feeder Beschickungsvorrichtung *f*, Aufgabevorrichtung *f*, Aufgeber *m*, Eintragvorrichtung *f*, Einspeisevorrichtung *f*, Speiser *m*, Feeder *m*
feeding 1. *s.* feed 1.; 2. *(Biot)* Füttern *n (von Mikroorganismen)*; Zufütterung *f*, Verfütterung *f*, Feeding *n (von Nährstoffen)*
feeding device *s.* feeder
feedstock *s.* feed 2. und 3.
feedstuff *s.* feed 3.
Fehling's (reagent, solution) fehling-

sche Lösung *f (zum Nachweis reduzierender Zucker)*
Feigl micro-reaction feiglsche Reaktion *f*
feldspar *(Min)* Feldspat *m*
feldspathic *(Min)* feldspathaltig; Feldspat...
felspar *s.* feldspar
felt *v* 1. filzen, zu Filz machen; filzen, sich verfilzen; 2. mit Filz überziehen [auskleiden]
felt Filz *m*
felt-fabric filter Faser(vlies)filter *n*, Faserschichtfilter *n*
felting Filzen *n*, Filzbildung *f*, Verfilzen *n*, Verfilzung *f*
female ground glass joint Schliffhülse *f*; Schliffpfanne *f*, Schliffschale *f*
female mould *(Kst)* Gesenk *n*, Matrize *f*
femtogram range *(Anal)* Femtogrammbereich *m* *(10^{-15} bis 10^{-12} g)*
femtomolar range femtomolarer Bereich *m* *(10^{-15} bis 10^{-12} Mol)*
ferment *v* 1. gären, vergären *(einem Gärungsvorgang unterliegen)*; vergären *(Stoffe einem Gärungsverfahren unterziehen)*; fermentieren *(Tabak, Tee)*; 2. *(Biot)* fermentieren
ferment Enzym *n*, Ferment *n*
fermentability Vergärbarkeit *f*, Gärungsfähigkeit *f*; Fermentierbarkeit *f*
fermentable vergärbar, gär(ungs)fähig; fermentierbar
fermentables 1. vergärbare Stoffe *mpl*; 2. *(Biot)* fermentierbare Anteile *mpl (des Substrats)*
fermentation Gärung *f (Vorgang)*; Vergärung *f (Tätigkeit)*; Fermentation *f (von Tabak, Tee)*; Reifung *f (von Käse)*; (biotechnologische) Fermentation *f*
fermentation alcohol Gärungsalkohol *m*
fermentation broth 1. Gärungsflüssigkeit *f*; 2. *(Biot)* Fermentationsbrühe *f*, Fermenterbrühe *f*, Fermentationslösung *f*, Fermentationsflüssigkeit *f*, Fermenterflüssigkeit *f*
fermentation chemistry Gärungschemie *f*
fermentation culture Impfkultur *f* für Fermenter
fermentation enzyme Gärungsenzym *n*, Gärungsferment *n*
fermentation ethanol [ethyl alcohol] Gärungsethanol *n*, Gärungsethylalkohol *m*
fermentation industry Gärungsindustrie *f*, Gärungsgewerbe *n*
fermentation lactic acid Gärungsmilchsäure *f*, DL-Milchsäure *f*
fermentation liquor *s.* fermentation broth
fermentation pathway Gärungsablauf *m*; Fermentationsweg *m*
fermentation period *s.* fermentation time
fermentation product Gärungsprodukt *n*; Fermentationsprodukt *n*
fermentation rate Fermentationsrate *f*, Produktbildungsrate *f*, Umsatzrate *f*
fermentation route Fermentationsweg *m*
fermentation slurry [solution] *s.* fermentation broth
fermentation substrate Fermentationssubstrat *n*
fermentation tank Gärbottich *m*, Gärtank *m*
fermentation temperature Gärtemperatur *f*; Fermentationstemperatur *f*, Bebrütungstemperatur *f*
fermentation time Gärdauer *f*; Fermentationszeit *f*, Fermentationsdauer *f*
fermentation vinegar Gärungsessig *m*
fermentation yield Fermentationsausbeute *f*
fermentative fermentativ, enzymatisch, Ferment..., Enzym...; gärend, Gärung erregend, gärungserregend, Gär(ungs)...
fermenter 1. *(Biot)* Fermenter *m*, Fermentor *m*, Fermentationsreaktor *m*; 2. Gärbottich *m*, Gärtank *m*
fermenter throughput Fermenterdurchsatz *m*
fermenter with recycle (loop) Rezirkulationsfermenter *m*
fermentor *s.* fermenter
Fermi-Dirac statistics Fermi-(Dirac-) Statistik *f*
Fermi distribution Fermi-Verteilung *f (der Elektronen in Metallen)*
Fermi energy Fermi-Energie *f*
Fermi level Fermi-Niveau *n*, Fermi-Kante *f*, fermische Grenzenergie *f*
Fermi resonance Fermi-Resonanz *f*
fermion *(Kern)* Fermion *n*

fermium Fm Fermium *n*
Fernbach flask Fernbach-Kolben *m*, Kulturkolben *m* nach Fernbach
ferrate Ferrat *n*
ferrian Eisen(III)-haltig
ferric Eisen..., *(i. e. S.)* Eisen(III)-..., *(manchmal auch)* Eisen(VI)...; eisenhaltig
ferric chelate Eisen(III)-chelat *n*
ferric chloride Eisen(III)-chlorid *n*, Eisentrichlorid *n*
ferric compound Eisen(III)-Verbindung *f*
ferric ferrocyanide Eisen(III)-hexacyanoferrat(II) *n*
ferric hydroxide Eisen(III)-oxidhydrat *n*, *(veraltet:)* Eisen(III)-hydroxid *n*
ferric ion Eisen(III)-Ion *n*
ferric iron dreiwertiges Eisen *n*
ferric orthophosphate Eisen(III)-orthophosphat *n*, Eisen(III)-phosphat *n*
ferric oxide Eisen(III)-oxid *n*, Eisentrioxid *n*, Dieisentrioxid *n*
ferric potassium sulphate Kaliumeisen(III)-sulfat-12-Wasser *n*, Kaliumeisenalaun *m*
ferric salt Eisen(III)-Salz *n*
ferric sulphate Eisen(III)-sulfat *n*, Eisentrisulfat *n*, Dieisentrisulfat *n*
ferric sulphide Eisen(II)-sulfid *n*, Dieisentrisulfid *n*
ferricyanide Cyanoferrat(III) *n*, Hexacyanoferrat(III) *n*
ferrihaem Hämin *n*, Eisen(III)-protoporphyrin *n*
ferrihaemoglobin Eisen(III)-hämoglobin *n*
ferrimagnetic transition ferrimagnetische Umwandlung *f* [Phasenumwandlung *f*]
ferrimagnetism Ferrimagnetismus *m*
ferrite 1. Ferrit *m*, α-Eisen *n (Eisen-Kohlenstoff-Mischkristall im α- und δ-Bereich)*; 2. $M^{I}FeO_2$ Ferrat(III) *n*; 3. Ferrit *m (keramischer Magnetwerkstoff aus Fe_2O_3 und anderen Metalloxiden)*
ferritic ferritisch
ferroalloy Ferrolegierung *f*
ferroan Eisen(II)-haltig
ferroboron Ferrobor *n*
ferrocarbon *s.* pig iron
ferrocene Ferrocen *n*, Eisen-bis-cyclopentadienyl *n*
ferrochromium Ferrochrom *n (Chrom-Eisen-Legierung)*
ferrocyanide Cyanoferrat(II) *n*, Hexacyanoferrat(II) *n*
ferroelastic transition ferroelastische Umwandlung *f* [Phasenumwandlung *f*]
ferroelectric ferroelektrisch
ferroelectric (material, substance) ferroelektrisches Material *n*, ferroelektrischer Stoff *m*, Ferroelektrikum *n*
ferroelectric transition ferroelektrische Umwandlung *f* [Phasenumwandlung *f*]
ferroferric oxide Eisen(II,III)-oxid *n*
ferrohaem Häm *n*, Eisenporphyrin *n* IX, Ferroporphyrin *n* IX, Eisen(II)-protoporphyrin *n*
ferrohaemoglobin Eisen(II)-hämoglobin *n*
ferroheme *s.* ferrohaem
ferromagnetic ferromagnetisch
ferromagnetic *s.* ferromagnetic material [substance]
ferromagnetic domain *(Krist)* weißscher Bezirk *m*, Weiß-Bezirk *m*, ferromagnetische Domäne *f*
ferromagnetic material [substance] ferromagnetisches Material *n*, ferromagnetischer Stoff *m*, Ferromagnetikum *n*
ferromagnetic susceptibility ferromagnetische Suszeptibilität *f*
ferromagnetic transition ferromagnetische Umwandlung *f* [Phasenumwandlung *f*]
ferromagnetism Ferromagnetismus *m*
ferromanganese Ferromangan *n*
ferromolybdenum Ferromolybdän *n*
ferronickel Ferronickel *n*
ferrophosphorus Ferrophosphor *m*
ferrosilicon Ferrosilicium *n*
ferroso-ferric compound Eisen(II,III)-Verbindung *f*
ferrotitanium Ferrotitan *n*
ferrotungsten Ferrowolfram *n*
ferrous Eisen..., *(i. e. S.)* Eisen(II)-...; eisenhaltig
ferrous ammonium sulphate Ammoniumeisen(II)-sulfat-6-Wasser *n*, mohrsches Salz *n*
ferrous chelate Eisen(II)-chelat *n*
ferrous chloride Eisen(II)-chlorid *n*, Eisendichlorid *n*

ferrous ferricyanide Eisen(II)-hexacyanoferrat(III) *n*
ferrous ferrocyanide Eisen(II)-hexacyanoferrat(II) *n*
ferrous hydroxide Eisen(II)-hydroxid *n*, Eisendihydroxid *n*
ferrous iron zweiwertiges Eisen *n*
ferrous metal Eisenmetall *n*, Eisenwerkstoff *m*
ferrous metallurgy Eisenmetallurgie *f*, Eisenhüttenwesen *n*, Schwarzmetallurgie *f*
ferrous orthophosphate Eisen(II)-orthophosphat *n*, Eisen(II)-phosphat *n*
ferrous oxide Eisen(II)-oxid *n*, Eisenmonoxid *n*
ferrous salt Eisen(II)-Salz *n*
ferrous scrap Eisenschrott *m*, Alteisen *n*
ferrous sulphate Eisen(II)-sulfat *n*
ferrous sulphide Eisen(II)-sulfid *n*, Eisenmonosulfid *n*
ferrovanadium Ferrovanadium *n*
ferrozirconium Ferrozirconium *n*
fertile 1. fruchtbar, fertil; 2. *(Kern)* brütbar, brutfähig, in Spaltstoff umwandelbar
fertile material *(Kern)* Brutmaterial *n*, Brutstoff *m*
fertility 1. Fruchtbarkeit *f*, Fertilität *f*; 2. *(Kern)* Brütbarkeit *f*, Brutfähigkeit *f*, Vermehrungsfähigkeit *f*
fertilization 1. *(Landw)* Düngung *f*; 2. *(Kern)* Brüten *n*, Brutvorgang *m*, Vermehrung *f*
fertilize *v (Landw)* düngen
fertilizer *(Landw)* Düngemittel *n*, Dünger *m*
ferulic acid Ferulasäure *f*, Kaffeesäure-3-methylether *m*, 3-(4-Hydroxy-3--methoxyphenyl)propensäure *f*
FESEM *s.* 1. field emission scanning electron microscope; 2. field emission scanning electron microscopy
Feulgen reaction feulgensche Reaktion *f* [Nuclealreaktion *f*] *(Methode zur Bestimmung von Desoxyribonucleinsäure)*
FFA = free fatty acids
FFDCA = Federal Food, Drug, and Cosmetic Act *(U.S.)*
FFF *s.* field flow fractionation
FGAM = formylglycinamidine ribonucleotide
FGAmR = formylglycinamidine ribonucleotide
FGAR = formylglycinamide ribonucleotide
FGD *s.* flue-gas desulphurization
FI *s.* field ionization
FIA *s.* 1. flow injection analysis; 2. fluorescence indicator analysis
FIB *s.* focus ion beam
fiber *s.* fibre
fiberization *(Am)* Defibrierung *f*, Zerfaserung *f*, Zerlegung *f* in Einzelfasern
fiberize *v (Am)* defibrieren, zerfasern, in Einzelfasern zerlegen
fibre 1. *(Bio)* Faser *f*, Fiber *f*; 2. *(Text)* Faser *f*; Faserstoff *m*
fibre blend Fasermischung *f*
fibre drawing process Fadenziehverfahren *n (Herstellung von Chemiefaserstoffen)*
fibre-entrapped enzyme *(Biot)* in Fasern [Hohlfasern] eingeschlossenes Enzym *n*
fibre-filled moulding material *(Kst)* Formmasse *f* mit Faserstoffverstärkung [Faserstofffüllung]
fibre glass *s.* fibreglass
fibre-reinforced plastic faserverstärkter Kunststoff *m*
fibre staple 1. Faser *f*, Spinnfaser *f*; Stapelfaser *f (bei Chemiefaserstoffen)*; 2. Faserlänge *f*, Stapellänge *f*, Stapel *m*
fibreboard Faserplatte *f*, *(meist)* Holzfaserplatte *f*
fibreglass Glasfaserstoff *m*, Faserglas *n*; Glasfaser *f*
fibril(la) Fibrille *f*, Fäserchen *n*, Teilfäserchen *n*
fibrillar protein *s.* fibrous protein
fibrillate *v* fibrillieren, in Teilfäserchen [Fibrillen] aufspalten, defibrillieren
fibrillation Fibrillieren *n*, Fibrillierung *f*, Fibrillenbildung *f*, Defibrillierung *f*
fibrin *(Bioch)* Fibrin *n*
fibrinogen *(Bioch)* Fibrinogen *n*
fibrinolytic fibrinolytisch
fibroin Fibroin *n*, Seidenstoff *m (Eiweißkörper der Naturseide)*
fibrous fas(e)rig, faserartig, Faser..., fibrös
fibrous colloid Linearkolloid *n*
fibrous polymer (elementar)fadenbildendes Polymer(e) *n*

fibrous protein fibrilläres [faserartiges] Protein *n*, Fibrillärprotein *n*, Strukturprotein *n*, Faserprotein *n*, Gerüsteiweiß *n*
Fick's law (of diffusion) ficksches Diffusionsgesetz *n* [Gesetz *n*]
FID *s.* flame ionization detector
field desorption Felddesorption *f*, FD
field direction Feldrichtung *f*
field effect Feldeffekt *m*
field emission microscope Feldemissionsmikroskop *n*, Feldelektronenmikroskop *n*
field emission scanning electron microscope Feldemissionsrasterelektronenmikroskop *n*, FESEM
field emission scanning electron microscopy Feldemissionsrasterelektronenmikroskopie *f*, FESEM
field experiment Feldversuch *m*, Freilandversuch *m*, Feldtest *m*
field flow fractionation *(Anal)* Feld-Fluss-Fraktionierung *f*
field ionization Feldionisation *f*, FI
field-ionization mass spectrometry Feldionisations-Massenspektrometrie *f*, FIMS
field strength Feldstärke *f*
field test [trial] *s.* field experiment
FIFRA = Federal Insecticide, Fungicide, and Rodenticide Act *(U.S.)*
filament 1. *(Text)* Elementarfaden *m*, Filament *n*, Endlosfaser *f*; 2. Glühfaden *m*, Leuchtfaden *m (einer Glühlampe)*; Heizfaden *m (einer Elektronenröhre)*
filament yarn Seide *f*, Endlosgarn *n*
fill *v* 1. füllen; 2. *(Tech)* beschicken; füllen, mit Füllstoffen versetzen
fill *v* **out** auffüllen, ausbauen *(Elektronenschalen)*
fill *v* **up** auffüllen, nachfüllen
fill Füllung *f*; Einbauten *mpl*, Einbau *m (z. B. in einem Kühlturm)*
filler 1. Füllstoff *m*, Füllmittel *n*, Füllmaterial *n*, Füller *m*, Füllmasse *f*; 2. *(Kst)* Streckmittel *n*, Harzträger *m*; 3. *(Anstr)* Spachtelmasse *f*, Spachtel *m(f)*; 4. *(Bau)* Zuschlag(stoff) *m*
filling 1. Füllen *n*, Füllung *f*; 2. *(Tech)* Beschicken *n*; Füllen *n*, Versetzen *n* mit Füllstoffen; 3. *s.* filler 1.
filling agent *s.* filler 1.
filling funnel *(Lab)* Pulvertrichter *m*
filling level Füllhöhe *f*, Füllstand *m*
filling material *s.* filler 1.
fillings Füllkörper *mpl*
film 1. Film *m*, (dünne) Schicht *f*, Häutchen *n*; (dünner) Überzug *m (vorgefertigt)*; 2. *(Kst)* Folie *f*, *(i. e. S.)* Feinfolie *f (Dicke unter 0.01 inch = 0,25 mm)*; 3. *(Foto)* Film *m*
film blowing machine Folienblasmaschine *f*
film boiling Filmverdampfung *f*
film casting machine Feinfoliengießmaschine *f*, Filmgießmaschine *f*
film condensation Filmkondensation *f*
film diffusion Filmdiffusion *f*
film evaporator Dünnschichtverdampfer *m*
film fault *(Anstr)* Anstrichfehler *m*, Anstrichmangel *m*
film fermenter *(Biot)* Filmfermenter *m*, Filmreaktor *m*
film former Filmbildner *m*, filmbildender Stoff *m*, filmbildende Substanz *f*, filmbildendes Material *n*
film thickness Filmdicke *f*, Schichtdicke *f*
filter *v* filtrieren *(bei Trennung Flüssigkeit - Feststoff)*; filtern *(bei Trennung Gas - Feststoff)*; sich filtrieren lassen, filtrierbar sein
filter *v* **aseptically** sterilfiltrieren, durch Filtration keimfrei machen
filter *v* **by suction** (ab)nutschen, absaugen
filter *v* **off** abfiltrieren
filter *v* **out** abfiltrieren, durch Filtration entfernen; sich beim Filtrieren absetzen
filter Filter *n*
filter aid Filterhilfsmittel *n*, Filterhilfsstoff *m*, Filterhilfe *f*, Filtrationshilfe *f*
filter bed Filterbett *n*, Filter(schütt)-schicht *f*, Filterschüttung *f*
filter blinding Filterverstopfung *f*, Filterversatz *m*
filter cake Filterkuchen *m*; Zentrifugenkuchen *m*
filter candle [cartridge] Filterkerze *f*, Filterpatrone *f*
filter cloth Filtergewebe *n*, Filtertuch *n*, Filterstoff *m*
filter crucible Filtertiegel *m*, Filtriertiegel *m*

filter efficiency Filterwirkungsgrad *m*
filter effluent Filterablauf *m*, Filtrat *n*
filter fabric *s.* filter cloth
filter flask Saugflasche *f*
filter leaf Filterblatt *n (des Blattfilters)*
filter mass Filtermasse *f*, Filtriermasse *f*, Filtrierstoff *m*
filter material [medium] Filtermedium *n*, Filtermittel *n*, Filtermaterial *n*
filter-medium filtration Tiefenfiltration *f*
filter paper Filterpapier *n*, Filtrierpapier *n*
filter precoat Anschwemm(filter)schicht *f*, Precoatschicht *f*, Filterhilfsschicht *f*
filter press Filterpresse *f*
filter respirator Atemschutzfiltergerät *n*
filter sterilization *(Biot)* Sterilfiltration *f*
filter stick *(Lab)* Filterstäbchen *n*, Filterstab *m*, Filtrierstäbchen *n*
filter tube 1. *(Tech)* Filterkerze *f*; Filterröhre *f*; 2. *(Lab)* Filterröhrchen *n*, Filtrierröhrchen *n*
filter-type dust collector Filtrationsentstauber *m*, Filtrationsabscheider *m*, Staubfilter *n*
filter yield Feststoffleistung *f*, Feststoffausbeute *f (in kg/m^2 · h; Vakuumfiltration)*
filterability Filtrierbarkeit *f*
filterable filtrierbar
filtering Filtrieren *n*, Filtration *f (bei Trennung Flüssigkeit - Feststoff)*; Filtern *n*, Filterung *f (bei Trennung Gas - Feststoff)*
filtering centrifugal [centrifuge] Filterzentrifuge *f*, Filtrierzentrifuge *f*, Siebzentrifuge *f*, Zentrifugal(kraft)filter *n*
filtering effect Filterwirkung *f*
filtering funnel Filtertrichter *m*
filtering pressure Filtrationsdruck *m*
filtermasse Filtermasse *f* aus faserigen Stoffen
filtrate *v s.* filter
filtrate Filtrat *n*
filtrate jar Filtrierstutzen *m*
filtration Filtration *f*, Filtrieren *n (bei Trennung Flüssigkeit - Feststoff)*; Filterung *f*, Filtern *n (bei Trennung Gas - Feststoff)*
filtration accelerator Filtrationshilfe *f*, Filterhilfe *f*, Filterhilfsmittel *n*, Filterhilfsstoff *m*
filtration by gravity Filtration *f* unter Wirkung der Schwerkraft
filtration curve Filtrationskurve *f*, Durchflusskurve *f*
filtration enrichment *(Biot)* Filtrationsanreicherung *f*
filtration with activated carbon *(Hyd)* Aktivkohlefiltration *f*, Filtration *f* über A-Kohle
60° filtration funnel Analysentrichter *m*
FIMS *s.* field-ionization mass spectrometry
fin 1. *(Tech)* Rippe *f*; Kühlrippe *f*; 2. Grat *m*
fin-tube heat exchanger Rippenrohrwärmeübertrager *m*
final BOD *(Hyd)* Rest-BSB *m*, BSB-Endwert *m*, BSB-Restkonzentration *f* im Ablauf
final boiling point Siedeendpunkt *m*, Siedeende *n*, SE, Endsiedepunkt *m*, Endkochpunkt *m*
final chlorination *(Hyd)* Nachchlorung *f*
final concentration Endkonzentration *f*
final content Restgehalt *m*
final disposal Endlagerung *f*, endgültige [dauerhafte] Ablagerung *f (von Abprodukten)*
final hardness *(Hyd)* Resthärte *f*, RH
final moisture content Endfeuchte(beladung) *f*
final period Nachperiode *f*, Nachversuch *m (einer kalorimetrischen Messung)*
final processing step Nachbehandlungsstufe *f*
final product 1. Endprodukt *n*; 2. *(Tech)* Finalprodukt *n*, Enderzeugnis *n*; Fertigprodukt *n*, Fertigerzeugnis *n*
final reaction product Endprodukt *n*, Endstoff *m*
final treatment 1. Endbehandlung *f*; 2. *(Hyd)* Endreinigung *f*, Schlussreinigung *f*, Nachreinigung *f*
fine 1. fein(körnig) *(z. B. Schüttgut)*; 2. fein, rein *(Edelmetall)*; 3. fein, dünn *(Folie)*; 4. fein, genau, präzis *(Messgeräteeinstellung)* • **in a fine state** in fein verteiltem Zustand
fine ceramics Feinkeramik *f*, feinkeramische Erzeugnisse *npl*
fine chemical Feinchemikalie *f*
fine dust Feinstaub *m*
fine-fabric detergent Feinwaschmittel *n*, Leichtwaschmittel *n*
fine filtration Feinfiltration *f*

fine-grain feinkörnig, Feinkorn..., dicht *(Gefüge)*
fine-grained 1. feinkristallin(isch), kleinkristallin; 2. *s.* fine-grain
fine grinding Fein(ver)mahlen *n*, Fein(aus)mahlung *f*, feine Mahlung *f* [Ausmahlung *f*]; Feinschleifen *n*, Feinschliff *m*
fine material Feingut *n*
fine-pore(d) feinporig, mit feinen Poren, feinporös, engporig, kleinlückig
fine sieve Feinsieb *n*
fine splitting *(Spektr)* Feinaufspaltung *f*
fine stoneware *(Ker)* Feinsteinzeug *n*
fine structure Feinstruktur *f (von Atomspektren)*
fine-structure rule *(Spektr)* 3. [dritte] hundsche Regel *f*
fine weigh Feinwägung *f*
finely dispersed feindispers, fein dispergiert
finely divided fein zerteilt, fein zerkleinert; fein verteilt, in feiner Verteilung
fineness 1. Feinheit *f*, *(i. e. S.)* Mahlfeinheit *f*; 2. Feingehalt *m (an Edelmetall in tausend Teilen einer Legierung)*
fines 1. fein(st)e Kornfraktionen *fpl*, Fein(st)gut *n*, Fein(st)korn *n*, Feines *n*, Feinanteile *mpl*, feinerer Teil *m*, Unterkorn *n*, *(Siebklassierung auch)* Siebfeines *n*; 2. *(Bgb)* Feinkohle *f*; Feinerz *n*; 3. *(Aufb)* Feingut *n*, Schlamm *m*
finger-type condenser Kühlfinger *m*, kalter Finger *m*, Einhängekühler *m*
fingerprint *(Anal)* Fingerabdruck *m*, Fingerprint *m*
finish *v* 1. beend(ig)en, vollenden, zu Ende führen; enden, aufhören; 2. *(Tech)* fertig bearbeiten, fertig stellen, nachbearbeiten; veredeln, die Oberflächenbeschaffenheit [Oberflächengüte] verbessern, die Oberfläche behandeln; 3. *(Text)* appretieren, ausrüsten, veredeln; 4. *(Anstr)* mit einem Schlussanstrich [Deckanstrich] versehen
finish 1. Oberflächenzustand *m*, Oberflächenbeschaffenheit *f*, Oberflächengüte *f*, Finish *n*; 2. *(Text)* Appretur *f*, Ausrüstung *f*; 3. *(Anstr)* Schlussanstrich *m*, Deckanstrich *m*; 4. *(Text)* Appret *n*, Appreturmittel *n*; 5. *(Anstr)* Decklack *m*, Überzugslack *m*; 6. *s.* finishing
finished product Endprodukt *n*, Finalprodukt *n*, Enderzeugnis *n*; Fertigprodukt *n*, Fertigerzeugnis *n*
finished water *(Hyd)* aufbereitetes Wasser *n*, Reinwasser *n*
finishing 1. Beend(ig)en *n*, Vollenden *n*, Vollendung *f*; 2. *(Tech)* Fertigbearbeitung *f*, Fertigstellung *f*, Nachbearbeitung *f*; Veredelung *f*, Verbesserung *f* der Oberflächenbeschaffenheit [Oberflächengüte], Oberflächenbehandlung *f*; 3. *(Text)* Appretieren *n*, Appretur *f*, Ausrüsten *n*, Ausrüstung *f*, Veredelung *f*; 4. *(Anstr)* Schlussanstrich *m*, Deckanstrich *m (Tätigkeit)*
Finkelstein exchange [reaction] *(org Ch)* Finkelstein-Austausch *m*, Finkelstein-Reaktion *f*
finned tube Rippenrohr *n*
FIR *s.* far infrared
fire *v* 1. anzünden, entzünden; feuern, (be)heizen *(Kessel)*; zünden *(Sprengstoff)*; 2. *(Ker)* brennen
fire-refine *v* feuerraffinieren, im Schmelzfluss raffinieren
fire Feuer *n* • **catch fire** Feuer fangen
fire blanket Feuerlöschdecke *f*
fire extinguisher 1. Feuerlöscher *m*, Feuerlöschapparat *m*, Feuerlöschgerät *n*; 2. Feuerlöschmittel *n*
fire-extinguishing agent Feuerlöschmittel *n*
fire hazard Brandgefahr *f*
fire point Brennpunkt *m*, BP
fire protection Brandschutz *m*
fire refining Feuerraffination *f*, Raffination *f* im Schmelzfluss
fire resistance Feuerbeständigkeit *f*, Feuerwiderstandsfähigkeit *f*
fire-resistant feuerbeständig, feuerwiderstandsfähig
fire-retardant feuerhemmend
fire retardant Feuerschutzmittel *n*, feuerhemmendes Mittel *n*, Flamm(en)-schutzmittel *n*
firebrick Schamottestein *m*, Schamotteziegel *m*, feuerfester Stein *m*
fireclay Schamotteton *m*, Feuer(fest)ton *m*, feuerfester Ton *m*
firedamp 1. Grubengas *n (Hauptbestandteil Methan)*; 2. schlagende Wet-

ter *pl*, Schlagwetter *pl* *(explosives Gemisch aus Grubengas und Luft)*

fireproof *v* feuerfest machen, feuersicher [feuerbeständig, brandsicher, unbrennbar] machen

fireproof feuerfest, feuersicher, feuerbeständig, unbrennbar

fireproofing agent Flamm(en)schutzmittel *n*, Feuerschutzmittel *n*

first ionization potential erstes Ionisierungspotenzial *n*, erste Ionisierungsspannung *f*, erste Ionisierungsarbeit *f*, *(in eV:)* erste Ionisierungsenergie *f*

first law of thermodynamics Erster Hauptsatz *m* der Thermodynamik

first-order nucleophilic substitution (reaction) monomolekulare nucleophile Substitution *f*, S_N1-Reaktion *f*

first-order rate constant Geschwindigkeitskonstante *f* erster Ordnung

first-order rate equation Geschwindigkeitsgleichung *f* erster Ordnung, Zeitgesetz *n* [Geschwindigkeitsgesetz *n*] erster Ordnung

first-order reaction Reaktion *f* erster Ordnung

first-order transition Umwandlung *f* [Phasenumwandlung *f*] erster Ordnung [Art], Übergang *m* [Phasenübergang *m*] erster Ordnung [Art], diskontinuierliche Umwandlung *f* [Phasenumwandlung *f*]

first run(nings) *(Dest)* Vorlauf *m*

first waste treatment stage *(Hyd)* erste Reinigungsstufe *f*, mechanische Reinigung *f*, Vorreinigung *f*

Fischer-Hepp rearrangement Fischer-Hepp-Umlagerung *f* *(aromatischer Nitrosamine zu p-Nitrosaminen in Gegenwart von Säuren)*

Fischer indol synthesis fischersche Indolsynthese *f* *(aus Phenylhydrazonen)*

Fischer projection Fischer-Projektion *f* *(zur Abbildung von Molekülen)*

Fischer-Tropsch process [synthesis] Fischer-Tropsch-Verfahren *n*, Fischer-Tropsch-Synthese *f*, Kogasinverfahren *n* *(zur Herstellung von Kohlenwasserstoffen)*

fish poison 1. Fischgift *n* *(für Fische toxische Substanz)*; 2. Ichthyotoxin *n*, Fischgift *n* *(fischeigenes Gift)*

fish-tail burner Fischschwanzbrenner *m*, Breitbrenner *m*, Schlitzbrenner *m*, Schnittbrenner *m*

fishiness *(Lebm)* Fischigkeit *f*, fischiger Geschmack *m*, Fischgeschmack *m* *(verdorbener Fette)*

fishy fischig, fischartig *(Geruch, Geschmack)*

fissile 1. *(Min, Krist)* spaltbar; 2. *s.* fissionable

fissility 1. *(Min, Krist)* Spaltbarkeit *f*; 2. *s.* fissionability

fission 1. Spalten *n*, Spaltung *f*, Aufspalten *n*, Aufspaltung *f*; 2. *(Kern)* Spaltung *f*, Kernspaltung *f*; Zerfall *m*

fission chain *(Kern)* Spaltkette *f*, Spaltproduktreihe *f*

fission energy Spalt(ungs)energie *f*, Kernspaltungsenergie *f*

fission fragment Spaltstück *n*, Bruchstück *n*

fission neutron Spaltneutron *n*

fission product Spalt(ungs)produkt *n*

fission reaction Spalt(ungs)reaktion *f*

fission yield Spalt(produkt)ausbeute *f*

fissionability *(Kern)* Spaltbarkeit *f*

fissionable *(Kern)* spaltbar

fissionable (material) spaltbares Material *n*, Spaltmaterial *n*, Spaltstoff *m*

Fittig reaction [synthesis] Fittig-Reaktion *f*, fittigsche Synthese *f* *(Darstellung von Bi- und Polyarylen)*

five-carbon chain C_5-Kette *f*

five-carbon sugar Pentose *f*

five-coordinate(d) pentakoordiniert, fünffach koordiniert; fünffach koordinativ gebunden, mit fünf koordinativen Bindungen

five days BOD [biochemical oxygen demand] fünftägiger biochemischer Sauerstoffbedarf *m*, BSB_5(-Wert) *m*

five-fold degenerate *(phys Ch)* fünffach entartet [degeneriert]

five-membered ring fünfgliedriger Ring *m*, Fünf(er)ring *m*

fix *v* 1. binden, fixieren; 2. *(Foto)* fixieren; 3. *(Landw)* festlegen *(Nährstoffe im Boden)*

fixation 1. Bindung *f*, Fixierung *f*; 2. *(Foto)* Fixieren *n*, Fixierung *f*, Fixage *f*; 3. *(Landw)* Festlegung *f* *(von Nährstoffen im Boden)*

fixation of nitrogen Stickstoffbindung *f*, Stickstoff-Fixierung *f*

fixed bed Festbett *n*, ruhendes [statisches] Bett *n* [Feststoffbett *n*], ruhende [statische] Schüttung *f*
fixed-bed fermenter *(Biot)* Festbettreaktor *m*
fixed-bed ion exchanger Festbett(ionen)austauscher *m*
fixed-bed reactor Festbettreaktor *m*
fixed-bed synthesis Festbettsynthese *f*
fixed carbon fixer [fester] Kohlenstoff *m*
fixed nitrogen gebundener [fixierter] Stickstoff *m*
fixed point Fixpunkt *m*, Festpunkt *m*, Fundamentalpunkt *m*
fixer *(Foto)* Fixiermittel *n*, Fixiersalz *n*; Fixierbad *n*
fixing 1. Fixieren *n*, Fixierung *f*; 2. *(Foto)* Fixage *f*
fixing agent 1. *(Foto, Text)* Fixiermittel *n*; 2. *(Text, Anstr)* Nachbehandlungsmittel *n (zur Echtheitserhöhung von Färbungen)*
fixing bath Fixierbad *n*
fixing salt Fixiersalz *n*
flake *v* 1. in Schichten zerlegen, Schichten abspalten; 2. *s.* flake off
flake *v* **off** abblättern, abschuppen, abplatzen, abspringen
flake Flocke *f*, Schuppe *f*
flaking Flocken *n*, Flockung *f*; Abblättern *n*, Abschuppen *n*, Schuppenbildung *f*, Abplatzen *n*, Abspringen *n*; Klumpen *n*
flame *v (Bioch)* abflammen *(zum Abtöten von Mikroorganismen)*
flame Flamme *f*
flame atomic absorption spectrometry Flammenatomabsorptionsspektralphotometrie *f*, Flammen-AAS *f*
flame coal Flammkohle *f*
flame colo(u)ration Flammenfärbung *f*
flame-emission spectroscopy Atomemissionsspektroskopie *f*
flame ionization detector Flammenionisationsdetektor *m*, FID
flame photometric detector Flammenphotometerdetektor *m*, flammenphotometrischer Detektor *m*, FPD
flame photometry Flammenphotometrie *f*
flame resistance Flammbeständigkeit *f*, Flammwidrigkeit *f*
flame-resistant flammbeständig, flammwidrig
flame-retardant flammenhemmend
flame retardant [retarder] Flammverzögerungsmittel *n*
flame spectroscopy Flammenspektroskopie *f*
flame spraying 1. Flammspritzen *n*; 2. *(Kst)* Wärmespritzen *n*, *(bei Metallen auch)* Metallspritzen *n*, Spritzmetallisieren *n*, Schoopisieren *n*
flame spreader Breitbrenneraufsatz *m*, Schnittbrenneraufsatz *m*, Schlitzaufsatz *m*
flameproof *v* flammsicher [flammfest, unentflammbar] ausrüsten
flameproof 1. flammsicher, flammfest, nicht [schwer] entflammbar, unentflammbar; 2. explosionsgeschützt, explosionssicher
flameproofing agent Flamm(en)schutzmittel *n*, Feuerschutzmittel *n*
flameproofness Flammsicherheit *f*, Flammfestigkeit *f*, Nichtentflammbarkeit *f*, Unentflammbarkeit *f*
flammability Entflammbarkeit *f*, Entzündbarkeit *f*, Entzündlichkeit *f*; Brennbarkeit *f*
flammable entflammbar, entzündbar, entzündlich; brennbar
flammable limits Entflammbarkeitsbereich *m*, Zündbereich *m*
flange Flansch *m*
flanged angeflanscht, Flansch...
flareback Flammenrückschlag *m*, Zurückschlagen *n (einer Flamme)*
flash *v* 1. entflammen; aufflammen, aufblitzen; 2. sehr rasch verdampfen *(z. B. durch Entspannung)*
flash *v* **back** zurückschlagen *(Flammen in Brennern)*
flash Aufflammen *n*, Aufblitzen *n*
flash curve Gleichgewichtssiedekurve *f*, Siedekurve *f* bei geschlossener Verdampfung, Flashkurve *f*
flash distillation (kontinuierliche) Entspannungsdestillation *f*, geschlossene [integrale] Destillation *f*
flash evaporation Entspannungsverdampfung *f*, Stoßverdampfung *f*, Blitzverdampfung *f*, Flashverdampfung *f*
flash lamp Blitzlampe *f*
flash photolysis Blitzlichtphotolyse *f*
flash point Flammpunkt *m*, FP.

flash-point apparatus [tester] Flammpunktprüfer *m*, Flammpunkt(prüf)gerät *n*, Flammpunktapparat *m*
flash pyrolysis *(org Ch)* Blitzpyrolyse *f*
flash vaporization *s.* flash evaporation
flashback Zurückschlagen *n (einer Flamme)*; Flammenrückschlag *m*
flask Kolben *m*; Flasche *f*
flat 1. flach, eben; 2. fahl, matt, stumpf *(Farbton)*; 3. schal, fad(e), abgestanden *(Geschmack)*
flat-bottom(ed) flask Stehkolben *m*
flat-bottom(ed) long-neck flask Langhalsstehkolben *m*
flat-bottom(ed) short-neck flask Kurzhalsstehkolben *m*
flat-flame burner Breitbrenner *m*, Schnittbrenner *m*, Schlitzbrenner *m*
flat flange joint Planschliffverbindung *f*
flat varnish Mattlack *m*
flatness 1. Flachheit *f*, Ebenheit *f*; 2. Fahlheit *f*, Mattheit *f*, Stumpfheit *f (von Farbtönen)*; 3. Schalheit *f*, Fadheit *f (Geschmack)*
flatten *v (Farb)* abstumpfen
flatten *v* **out** abklingen *(Reaktion)*
flavan *(org Ch)* Flavan *n*
flavanoid Flavanoid *n*, Flavan-Derivat *n*
flavanone *(org Ch)* Flavanon *n*
flavin 1. Flavin *n*, Isoalloxazin *n*; 2. Quercetin *n*, 5,7,3′,4′-Tetrahydroxyflavonol *n*
flavin adenine dinucleotide Flavinadenin-dinucleotid *n*, FAD
flavin enzyme Flavinenzym *n*, Gelbes Enzym *n*, Oxidationsenzym *n*, Flavoprotein *n*
flavin mononucleotide Flavinmononucleotid *n*, FMN
flavone Flavon *n*, 2-Phenylchromon *n*, 2-Phenyl-1,4-chromenon *n*
flavonoid Flavonoid *n*
flavoprotein *s.* flavin enzyme
flavor *s.* flavour
flavour *v* würzen *(Speisen)*; aromatisieren, mit Aromastoffen versetzen, parfümieren *(z. B. Tabak)*
flavour Aroma *n*, Wohlgeschmack *m* und Wohlgeruch *m*; Würze *f*, Blume *f (des Weins)*
flavour enhancer Geschmacksverstärker *m*, geschmacksverstärkender Stoff *m*
flavourful aromatisch, wohlschmeckend und wohlriechend
flavouring (material, matter, substance) Aromastoff *m*, Geschmacks- und Geruchsstoff *m*
flavourless geschmacklos und geruchlos
flaw 1. Materialfehler *m*, Fehler *m*, Defekt *m*; 2. *(Krist)* Lockerstelle *f*; 3. *(Ker)* Riss *m*, Sprung *m*
flax-seed oil Leinöl *n*
flex cracking Biegerissbildung *f*, Biegeermüdung *f*
flex-cracking resistance Biegerisswiderstand *m*, Biegerissfestigkeit *f*
flex life Dauerbiegefestigkeit *f*, Ermüdungsbeständigkeit *f*
flexibility Biegsamkeit *f*, Flexibilität *f*
flexibilize *v (Kst)* weich machen, flexibilisieren
flexibilizer Weichmacher *m (für Epoxidharze)*
flexible biegsam, flexibel, elastisch, federnd; nicht starr *(z. B. Kupplung)*
flexible plastic weicher [weich gestellter] Kunststoff *m*
flexible PVC Weich-PVC *n*, PVC *n* weich
flexing life *s.* flex life
flexural impact strength test Schlagbiegeversuch *m*
flexural modulus Biegemodul *m (aus dem Biegeversuch ermittelter E-Modul)*
flexural strength Biegefestigkeit *f*
FLIC *s.* 1. fluorescence interface contrast microscope; 2. fluorescence interface contrast microscopy
flint glass Flintglas *n*
flint pebble Flintstein *m*; Flintsteinmahlkörper *m*
flip-flop enzyme Flipflop-Enzym *n*
float *v* flotieren *(durch Aufschwemmen aufbereiten)*; aufschwimmen; zum Aufschwimmen bringen
float *v* **off** ausschwimmen; fortschwemmen, abschwemmen, forttragen
float-and-sink method Schwimm- und Sinkverfahren *n*, Schwimm-Sink-Verfahren *n*, Schwerflüssigkeitsverfahren *n*, *(meist)* Schwertrübeverfahren *n*
float material [product] *(Aufb)*

Schwimmgut *n*, aufschwimmendes Gut *n*
float zone process Zonenschmelzverfahren *n (Halbleiterherstellung)*
floatability Schwimmfähigkeit *f*; Flotierbarkeit *f*
floatable schwimmfähig; flotierbar
floatables schwimmfähige Stoffe *mpl*
floatation Flotieren *n*, Flotation *f*, Schwimmaufbereitung *f*
floatation agent *s.* floatation reagent
floatation beneficiation flotative Aufbereitung *f*, Aufbereitung *f* durch Flotation
floatation cell Flotationszelle *f*
floatation-clarification *(Hyd)* Abwasserflotation *f*
floatation reagent Flotationsmittel *n*, Flotationsreagens *n*, Schwimmmittel *n*
floatation separation flotative Trennung *f*
floatation tailings Flotationsberge *pl*, Flotationsabgänge *mpl*, Abgänge *mpl*
floating *(Anstr)* vertikales Ausschwimmen *n (Entmischen der Pigmente)*
floating fraction Schwimmanteil *m*, Schwimmgut *n*, aufschwimmendes Gut *n*
floating solids *(Hyd)* Schwimmstoffe *mpl*
floats *(Aufb)* Schwimmgut *n*, aufschwimmendes Gut *n*
floc Flocke *f*; Flocken *fpl*
flocculant 1. Flockungsmittel *n*, Flockungschemikalie *f*, Ausflockungsmittel *n*, Flocker *m*; 2. *(Koll)* Koagulationsmittel *n*, Koagulator *m*
flocculate *v* 1. (aus)flocken, durch Flockung eliminieren, zur Ausflockung bringen; 2. *(Koll)* koagulieren, pektisieren; (aus)flocken, sich zusammenballen, koagulieren, pektisieren; 3. *(Hyd)* flocken *(Roh- und Abwasser)*; 4. *(Hyd)* zu Flocken verbinden, absetzfähige Flocken bilden *(nach der Entstabilisierung)*
flocculate 1. Flocke *f (eines Niederschlags)*; 2. flockiger Niederschlag *m*; 3. *(Koll)* Koagulat *n*
flocculating agent *s.* flocculant
flocculation 1. Ausflockung *f*, Flockung *f*; 2. *(Koll)* Koagulation *f*, Pektisation *f*; 3. *(Hyd)* Flockung *f*, Flockungsbehandlung *f (von Roh- und Abwasser)*; 4. *(Hyd)* Flockenbildung *f*, Bildung *f* absetzfähiger Flocken *(nach der Entstabilisierung)*; Flockulation *f (Flockungsbehandlung unter Einsatz von Polymeren als Flockungsmittel)*
flocculation aid *(Hyd)* Flockungshilfsmittel *n*, Flockungsbeschleuniger *m*
flocculation by electrolytes *(Koll)* Elektrolytkoagulation *f*
flocculation chemical Flockungschemikalie *f*
flocculation value 1. Flockungswert *m*; 2. *(Koll)* Koagulationswert *m*
floccule Flocke *f*, Flöckchen *n*
flocculence Flockigkeit *f*
flocculent flockig, flockenartig, flockenförmig
flocculent *s.* flocculant
flock 1. Flocke *f*; 2. *(Text)* Kurzfaser *f*, Flocke *f*
flood *v* 1. fluten *(eine Kolonne)*; *(unbeabsichtigt)* überfluten *(eine Kolonne)*; 2. *(Anstr)* (in horizontaler Richtung) ausschwimmen
flooding 1. Fluten *n*, Flutung *f (einer Kolonne)*; Überfluten *n*, Überflutung *f (einer Kolonne durch unbeabsichtigten Flüssigkeitsstau)*; 2. *(Anstr)* horizontales Ausschwimmen *n (Entmischen der Pigmente)*
flooding point Flutpunkt *m*, Überflutungsgrenze *f*, Spuckgrenze *f (einer Kolonne)*
floor 1. Boden *m*; Fußboden *m*; 2. *(Lebm)* Horde *f*, Rösthorde *f*, Darrhorde *f*; 3. *(Gär)* Malztenne *f*
Florence flask Kochkolben *m*, Kochflasche *f*, Langhalskolben *m*, *(i. e. S.)* Langhals-Stehkolben *m*
Florentine flask [receiver] Florentiner Flasche *f*, Scheideflasche *f (eine Vorlage)*
Flory theta temperature Flory-Temperatur *f*, Θ-Temperatur *f*
flotation *s.* floatation
flour 1. Mehl *n*, Staub *m*, feines Pulver *n*; 2. *(Lebm)* (feines) Mehl *n*, *(i. e. S.)* feines Weizenmehl *n*
flour improver Mehlverbesserungsmittel *n (zur Verbesserung der backtechnischen Eigenschaften)*; Mehlbleich(ungs)mittel *n*

floury 1. mehlig, mehlartig; 2. bemehlt
flow *v* 1. fließen, strömen; 2. *(Anstr)* verlaufen, fließen; fließen lassen
flow *v* **back** zurückfließen
flow *v* **out** ausfließen, ausströmen
flow *v* **round** umspülen
flow *v* **through** durchströmen
flow 1. Fließen *n*, Fluss *m*, Strömen *n*, Strömung *f*; Strom *m*; 2. *(Anstr)* Verlauf *m*
flow behaviour Fließverhalten *n*
flow birefringence *(Anal)* Strömungsdoppelbrechung *f*
flow channel Strömungskanal *m*
flow chart *s.* flow diagram
flow curve Fließkurve *f*
flow diagram Fließdiagramm *n*, Fließschema *n*, Fließbild *n*
flow direction Fließrichtung *f*, Strömungsrichtung *f*
flow injection analysis Fließinjektionsanalyse *f*, FIA
flow measurement Durchfluss(mengen)messung *f*, Mengenstrommessung *f*
flow method Strömungsmethode *f (zur Untersuchung schneller Reaktionen)*
flow pattern 1. Strömungsform *f (im Reaktor)*; 2. *s.* flow diagram
flow point Fließpunkt *m*, Fließgrenze *f*, Fließtemperatur *f (einer Schmelze)*
flow rate 1. Fließgeschwindigkeit *f*, Strömungsgeschwindigkeit *f*, Durchflussgeschwindigkeit *f*, Durchlaufgeschwindigkeit *f (in* $m \cdot s^{-1}$*)*; 2. Durchfluss *m (in* m^3/h*)*
flow rate indicator Durchflussanzeiger *m*
flow reactor Durchflussreaktor *m*, kontinuierlicher [kontinuierlich durchflossener] Reaktor *m*
flow sheet *s.* flow diagram
flow technique *s.* flow method
flow temperature Fließtemperatur *f*
flow time Durchlaufzeit *f*
flowability Fließvermögen *n*, Fließfähigkeit *f*; Rieselvermögen *n*, Rieselfähigkeit *f (von Schüttgut)*
flowable fließfähig; rieselfähig *(Schüttgut)*
flowers of sulphur Schwefelblüte *f*, Schwefelblume *f*
flowing fluid strömendes Fluid *n*
flowing properties Fließeigenschaften *fpl*
flowmeter Durchfluss(mengen)messer *m*, Mengenstrommesser *m*, Strömungsmesser *m*
flowsheet *s.* flow diagram
flowstream reactor stationär arbeitender Reaktor *m*
fluctuation in concentration Konzentrationsschwankung *f*
flue Zug *m*, Heizzug *m*, Feuerzug *m*, Ofenzug *m*
flue dust Hüttenstaub *m*, *(i. w. S.)* Hüttenrauch *m*
flue gas Rauchgas *n*
flue-gas analysis Rauchgasanalyse *f*
flue-gas desulphurization Rauchgasentschwefelung *f*
flue-gas scrubber Rauchgaswäscher *m*
fluff *v* flockig machen, auflockern; flockig werden
fluffy flaumig; locker *(Schüttgut, Niederschlag)*
fluid fluid *(gasförmig, dampfförmig oder flüssig)*
fluid Fluid *n (Gas, Dampf oder Flüssigkeit)*
fluid bed Fließbett *n*, Fließschicht *f*, Wirbelbett *n*, Wirbelschicht *f*
fluid-bed reactor Wirbelschichtreaktor *m*, Wirbelbettreaktor *m*
fluid-bed technique Fließbett-Technik *f*, Wirbelschichttechnik *f*, Staubfließtechnik *f*, Fluid-Technik *f*
fluid catalyst Fließbettkatalysator *m*, Wirbelbettkatalysator *m*, Staubfließkatalysator *m*, Fließstaubkontakt *m*
fluid catalytic cracking *(Erdöl)* katalytisches Wirbelschichtkracken *n*, katalytisches Kracken *n* in (der) Wirbelschicht, Kracken *n* in Wirbelschicht mit Fließbettkatalysator
fluid flow 1. Flüssigkeitsströmung *f*; Gasströmung *f*; 2. Flüssigkeitsstrom *m*; Gasstrom *m*
fluid friction hydrodynamischer Widerstand *m*
fluid fuel 1. flüssiger Brennstoff *m*; 2. flüssiger Kraftstoff *m*; 3. flüssiger Treibstoff *m*, *(i. e. S.)* flüssiger Raketentreibstoff *m*
fluid medium strömendes Medium *n*, Trägerstrom *m (Wirbelbettverfahren)*

fluid phase fluide Phase *f*
fluid process Fluidverfahren *n*, Wirbelschichtverfahren *n*, Staubfließverfahren *n*, Fließ(bett)verfahren *n*
fluid-solid reaction Fluid-Feststoff-Reaktion *f*, Feststoff-Fluid-Reaktion *f*
fluid-solid reactor Fluid-Feststoff-Reaktor *m*, Reaktor *m* für Fluid-Feststoff-Reaktionen
fluid-solid transition Übergang *m* vom flüssigen in den festen Zustand
fluidify *v* 1. verflüssigen; sich verflüssigen; 2. *s.* fluidize
fluidity 1. Fließvermögen *n*, Fließfähigkeit *f*; 2. *(phys Ch)* Fluidität *f (Kehrwert der Viskosität)*
fluidization Fluidisieren *n*, Fluidisation *f*, Überführen *n* in den Fließbettzustand
fluidize *v* fluidisieren, in den Fließbettzustand überführen; in den Fließbettzustand übergehen
fluidized bed *s.* fluid bed
fluidized-bed bioreactor *(Biot)* Wirbelschicht(bio)reaktor *m*, Wirbelbettreaktor *m*
fluidized-bed combustion Wirbelschichtverbrennung *f*
fluidized-bed fermenter *s.* fluidized-bed bioreactor
fluidized-bed gasification *(Carb)* Wirbelschichtvergasung *f*, Wirbelbettvergasung *f*
fluidized-bed incinerator *(Hyd)* Wirbelschichtofen *m (Schlammverbrennung)*
fluidized-bed reactor Wirbelschichtreaktor *m*, Wirbelbettreaktor *m*
fluidized process *s.* fluid process
fluidizing gas Wirbelgas *n*
flume Kanal *m*, *(i. e. S.)* Venturi-Kanal *m*, Messgerinne *n (zur Mengenstrommessung)*
fluo... *(in chemischen Benennungen) siehe: fluoro...*
fluoracetic acid Fluoressigsäure *f*
fluoranthene *(org Ch)* Fluoranthen *n*
fluorapatite *(Min)* Fluorapatit *m*
fluorene *(org Ch)* Fluoren *n*
fluoresce *v* fluoreszieren
fluorescein Fluorescein *n*, Resorcinolphthalein *n*
fluorescence Fluoreszenz *f*; Fluoreszenzstrahlung *f*
fluorescence dye Fluoreszenzfarbstoff *m*
fluorescence indicator analysis Fluoreszenzindikatoranalyse *f*
fluorescence interface contrast microscope Fluoreszenz-Interface-Kontrastmikroskop *n*
fluorescence interface contrast microscopy Fluoreszenz-Interface-Kontrastmikroskopie *f*
fluorescence lifetime *(Spektr)* Fluoreszenzlebenszeit *f*, Fluoreszenzlebensdauer *f*
fluorescence marker Fluoreszenzmarker *m*
fluorescence quantum yield *(Spektr)* Fluoreszenzquantenausbeute *f*
fluorescence quenching Fluoreszenzlöschung *f*
fluorescence spectroscopy Fluoreszenzspektroskopie *f*
fluorescence spectrum Fluoreszenzspektrum *n*
fluorescent fluoreszierend, Fluoreszenz..., Leucht...
fluorescent agent Fluoreszenzstoff *m*, Leuchtstoff *m*
fluorescent brightener [brightening agent] optischer Aufheller *m*, optisches Aufhellungsmittel *n* [Bleichmittel *n*], Weißtöner *m*, aufhellender Leuchtstoff *m*
fluorescent indicator Fluoreszenzindikator *m*
fluorescent probe Fluoreszenzsonde *f*
fluoridate *v (Hyd)* fluoridieren, fluorisieren
fluoridation *(Hyd)* Fluoridierung *f*, Fluorisierung *f*
fluoride Fluorid *n*
fluoridize *v s.* fluoridate
fluorimetry *s.* fluorometry
fluorinate *v* fluorieren
fluorinated alkane Fluoralkan *n*
fluorinated rubber Fluorkautschuk *m*, Fluorelastomer(e) *n*
fluorinating agent Fluorierungsmittel *n*, Fluorüberträger *m*
fluorination Fluorieren *n*, Fluorierung *f*
fluorine F Fluor *n*
fluorite *(Min)* Fluorit *m*, Flussspat *m (Calciumfluorid)*
fluoroalkane Fluoralkan *n*

fluoroalkylation Fluoralkylierung *f*
fluoroamination Fluoraminierung *f*
fluorobenzene Fluorbenzen *n*
fluoroborate Fluoroborat *n*
fluoroboric acid Fluoroborsäure *f*, Borfluorwasserstoffsäure *f*
fluorocarbon Fluorkohlenwasserstoff *m*, fluorierter Kohlenwasserstoff *m*
fluorocarbon elastomer Fluorelastomer(e) *n*, Fluorkautschuk *m*, *(i. e. S.)* Fluorcarbonkautschuk *m*
fluoroelastomer *s.* fluorocarbon elastomer
fluorogen *s.* fluorophore
fluorohydric Fluorwasserstoff...
fluorohydric acid *s.* hydrofluoric acid
fluorometer Fluorimeter *n*, Fluorophotometer *n*
fluorometric fluorometrisch
fluorometry Fluorometrie *f*, Fluorimetrie *f*, Fluoreszenzmessung *f*
fluoro-olefin Fluorolefin *n*, Fluoralken *n*, fluoriertes Alken *n*, fluorierter ungesättigter Kohlenwasserstoff *m*
fluorophore Fluorophor *m*, fluoreszierendes Prinzip *n (Fluoreszenz verursachender Molekülanteil)*
fluoroplastic Fluorcarbonkunststoff *m*
fluorosilicate Fluorosilicat *n*, Hexafluorosilicat *n*
fluorosilicic acid Fluorokieselsäure *f*, Hexafluorokieselsäure *f*
fluorosilicone rubber Fluorsiliconkautschuk *m*
fluorosulphuric acid Fluoroschwefelsäure *f*
5-fluorouracil 5-Fluoruracil *n*
fluorspar *(Min)* Flussspat *m*, Fluorit *m* *(Calciumfluorid)*
flush *v* 1. (aus)spülen, auswaschen; 2. strömen, sich ergießen; strömen lassen
flush *v* **away** fortspülen
flush *v* **out** (aus)spülen, auswaschen
fluted filter (paper) Faltenfilter *n*
fluted funnel (Jenaer) Analysentrichter *m*, Rippentrichter *m*
flux *v* 1. schmelzen, flüssig machen; 2. aufschließen *(durch Schmelzen)*; fluxen, mit Flussmittel behandeln; 3. *(Kst)* plastifizieren, weich machen, erweichen
flux *v* **on** aufschmelzen
flux 1. Fließen *n*, Fluss *m*; Massenfluss *m*, Teilchenfluss *m*; 2. Lötmittel *n*; 3. *s.* fluxing agent
flux of material Materialfluss *m*
fluxing agent 1. Flussmittel *n*, Fluss *m*, Schmelzmittel *n*; 2. *(Met)* Zuschlag *m*
fluxional molecule Molekül *n* mit fluktuierenden Bindungen
fly ash Flugasche *f*
FM = *N*-formyl morpholine
FMC *s.* final moisture content
FMN *s.* flavin mononucleotide
foam *v* 1. (auf)schäumen, Schaum bilden; zum Schäumen bringen, schaumig [schäumend] machen; 2. *(Kst)* (ver)schäumen, aufschäumen
foam *v* **over** überschäumen
foam 1. Schaum *m*; 2. *s.* foamed plastic
foam breaking Schaumbrechen *n*, Schaumbrechung *f*, Schaumzerstörung *f (mechanisch)*
foam-control agent Schaumregulierer *m*, Schaumregulator *m*
foam destruction Schaumzerstörung *f*, Schaumbrechen *n*, Schaumbrechung *f* *(chemisch)*
foam extinguisher Schaumlöscher *m*
foam formation Schaumbildung *f*, Schaumentwicklung *f*
foam glass Schaumglas *n*
foam inhibitor Mittel *n* gegen Schaumbildung, Schaumverhinderungsmittel *n*, Schaumverhütungsmittel *n*
foam prevention Schaumverhütung *f*
foam rubber Schaumgummi *m*, Latexschaum(gummi) *m*
foam stability Schaumbeständigkeit *f*
foam stabilizer [stabilizing compound] Schaumstabilisator *m*
foamable *(Kst)* verschäumbar, aufschäumbar
foamed plastic Schaum(kunst)stoff *m*
foaminess Schaumigkeit *f*; Schaumkraft *f*
foaming agent 1. Schaum(erzeugungs)mittel *n*, Schaumerzeuger *m*, Schaumbildner *m*, Schäummittel *n*, Schäumer *m*; 2. *(Kst)* Treibmittel *n*
foamy schaum(art)ig; mit Schaum bedeckt
focus *v* fokussieren
focus ion beam *(Spektr)* fokussierter Ionenstrahl *m*

fog Nebel *m*
fog application *(Landw)* Nebeln *n*
fog droplet Nebeltröpfchen *n*
fog generator *(Landw)* Nebelgerät *n*, Nebelblaser *m*, Atomiseur *m*
fog track *(phys Ch)* Nebelspur *f*
foil Folie *f*, *(i. e. S.)* Metallfolie *f*, *(manchmal auch)* Kunststofffolie *f*
folate *(Bioch)* Folat *n*, Pteroylglutamat *n* *(Salz oder Ester der Folsäure)*
fold height [period] Faltungslänge *f*, Faltungshöhe *f*, Faltungsdicke *f* *(von Polymerketten)*
folded filter Faltenfilter *n*
foliage application *(Landw)* Blattapplikation *f*
foliage burn Blattverbrennungen *fpl* *(z. B. durch Pflanzenschutzmittel)*
foliated blätt(e)rig, geblättert, schieferig, schuppig
folic acid Folsäure *f*, Pteroylglutaminsäure *f*
folinic acid Folinsäure *f*, Leukovorin *n*, N^5-Formyltetrahydrofolsäure *f*, Citrovorumfaktor *m*
Folin's reagent Folins Reagens *n* [Aminosäurereagens *n*] *(β-Naphthochinon und Schwefelsäure zum Aminosäurenachweis)*
follicle-stimulating hormone *Follitropin *n*, follikelstimulierendes Hormon *n*, Follikelstimulierungshormon *n*, FSH
follicle-stimulating-hormone releasing hormone *Gonadoliberin *n*, GnRH
follicular hormone Follikelhormon *n*
following strand *(Bioch)* Folgestrang *m* *(der DNA)*
food acid Genusssäure *f*
food additive Lebensmittelzusatz(stoff) *m*, Fremdstoff *m* [fremder Stoff *m*] in Lebensmitteln
food chain Nahrungskette *f*, *(bei Tieren auch)* Freßkette *f*
food chemist Lebensmittelchemiker *m*
food chemistry Lebensmittelchemie *f*, Nahrungsmittelchemie *f*
food color *(Am)* *s.* food dye
food colorant [coloring] *s.* food dye
food contaminant Lebensmittelkontaminant *m*
food contamination *(Lebm)* Lebensmittelvergiftung *f*, Nahrungsmittelvergiftung *f*
food control Lebensmittelüberwachung *f*
food deterioration Qualitätsbeeinträchtigung *f* [Qualitätsminderung *f*] von Lebensmitteln
food dye Lebensmittelfarbstoff *m*, Lebensmittelfarbe *f*
food fat Nahrungsfett *n*, Speisefett *n*, Genussfett *n*
food fortification Lebensmittelanreicherung *f* *(z. B. mit Vitaminen)*
food industry Lebensmittelindustrie *f*, Nahrungsmittelindustrie *f*
food intolerance Lebensmittelunverträglichkeit *f*, Nahrungsmittelunverträglichkeit *f*
food intoxication *s.* food contamination
food irradiation Lebensmittelbestrahlung *f*
food law Lebensmittelgesetz *n*, Lebensmittelrecht *n*
food legislation *s.* food law
food microbiology Lebensmittelmikrobiologie *f*
food oil Speiseöl *n*
food packaging Lebensmittelverpackung *f*
food poisoning *s.* food contamination
food preservative Konservierungsmittel *n* [Konservierungsstoff *m*] für Lebensmittel
food processing Lebensmittelverarbeitung *f*
food protein Nahrungsprotein *n*, Nahrungseiweiß *n*
food science Ernährungswissenschaft *f*
food spoilage Lebensmittelverderb *m*
food technology Lebensmitteltechnologie *f*
food value Nährwert *m*
foodstuff Nährstoff *m*; Nahrungsmittel *n*
foots 1. Niederschlag *m*, Bodensatz *m*, Satz *m*, Sediment *n*, Ausscheidung *f*, Ausscheidungsprodukt *n*, Ablagerung *f*; 2. *(Dest)* Nachlauf *m*
forbidden band [gap] verbotenes Band *n*, verbotene Zone *f*, verbotener [nicht zugelassener] Energiebereich *m*, Energielücke *f*
forbidden transition *(phys Ch)* verbotener Übergang *m*
force *v* **out** auspressen, abpressen

force *v* **through** hindurchpressen, (hin-)durchdrücken
force *v* **through the filter** unter Druck filtrieren; abnutschen, absaugen
force Kraft *f*
force constant Kraftkonstante *f*
force field Kraftfeld *n*
force of attraction Anziehungskraft *f*, Attraktionskraft *f*
force of interaction Wechselwirkungskraft *f*
force of repulsion Abstoßungskraft *f*
forced circulation Zwang(s)umlauf *m*, Zwanglauf *m*, Zwangskreislauf *m*, Zwangszirkulation *f*, Umpump *m*
forced convection erzwungene Konvektion *f* [Wärmekonvektion *f*], Wärmekonvektion *f* bei erzwungener Strömung
forced draught künstlicher [künstlich erzeugter] Zug *m*, Saugzug *m* *(z. B. eines Ofens)*
forceps Zange *f*; Pinzette *f*
foreign atom Fremdatom *n*, *(im Kristallgitter auch)* Stör(stellen)atom *n*, Fremdstörstelle *f*, (chemische) Störstelle *f*
foreign element Fremdelement *n*
foreign ion Fremdion *n*
foreign material [matter] Fremdstoff *m*, fremder Stoff *m*, Fremdsubstanz *f*; Verunreinigung *f*
foreign molecule Fremdmolekül *n*
foreign protein *(Bioch)* Fremdprotein *n*, fremdes Eiweiß *n*
foreign substance *s.* foreign material
forensic chemist Gerichtschemiker *m*
forensic chemistry forensische [gerichtliche] Chemie *f*, Gerichtschemie *f*
forepump Vor(vakuum)pumpe *f*
forerun Vorlauf *m*
forerunning(s) Vorlauf *m*
foreruns Vorlauf *m*
forevacuum Vorvakuum *n*
forewarming Vorwärmung *f*
forge pig iron Puddelroheisen *n*
forked chain verzweigte Kette *f*
form *v* bilden, darstellen, formen; sich bilden [ausbilden, entwickeln], entstehen
form 1. Form *f*, Gestalt *f*; 2. *(Kst)* Werkzeug *n*, Form *f* • **in combined form** gebunden, in gebundener Form
***d*-form** (+)-Form *f*, rechtsdrehende Form *f*, Rechtsform *f*
***l*-form** (-)-Form *f*, linksdrehende Form *f*, Linksform *f*
formability Formbarkeit *f*
formal Formal *n*, Formaldehydacetal *n*, *(i. e. S.)* $CH_2(OCH_3)_2$ Formaldehyddimethylacetal *n*, *Dimethoxymethan *n*, Formal *n*
formal charge formale Ladung *f*, Formalladung *f*
formaldehyde Formaldehyd *m*, *Methanal *n*
formaldoxime Formaldoxim *n*, Formaldehydoxim *n*
formalinize *v* mit Formaldehyd behandeln
formamide Formamid *n*, Ameisensäureamid *n*, Methanamid *n*
formate Formiat *n* *(Salz oder Ester der Ameisensäure)*
formation Bildung *f*, Ausbildung *f*, Entwicklung *f*, Entstehung *f*
formation constant Komplexbildungskonstante *f*, Komplexstabilitätskonstante *f*, Bildungskonstante *f*, Stabilitätskonstante *f*
formation of bonds Bindungsbildung *f*, Ausbildung *f* von Bindungen; Knüpfen *n* von Bindungen
formation reaction Bildungsreaktion *f*
formazan *(org Ch)* Formazan *n*
formiate *s.* formate
formic acid Ameisensäure *f*, Methansäure *f*
formimidation Formimidierung *f*
forming 1. Formen *n*, Formung *f*, Formgebung *f*; 2. *(Kst)* Umformen *n*, spanloses Formen *n*
formolization Formolisierung *f*, Behandlung *f* mit Formaldehyd [Methanal]
formolize *v* formolisieren, mit Formaldehyd [Methanal] behandeln
formolysis Formolyse *f*
formula 1. Formel *f*; 2. Rezept *n*, Vorschrift *f*
formula conversion Formelumsatz *m*
formula weight (relative) Formelmasse *f*
formulate *v* 1. formulieren, durch eine Formel charakterisieren [zum Ausdruck bringen, symbolisieren]; 2. formulieren, nach Rezept aufbauen,

sachgemäß zubereiten, konfektionieren, finalisieren
formulation 1. Formulierung *f*, Charakterisierung *f* [Symbolisierung *f*] durch eine Formel; 2. Formulierung *f*, Aufbau *m* nach Rezept, sachgemäße Zubereitung *f*, Konfektionierung *f*, Finalisierung *f*; 3. formuliertes Handelsprodukt *n*, Formulierung *f*, Zubereitung *f (nach festgelegter Rezeptur gefertigt)*; 4. *(Kst)* Ansatz *m (Ausgangsgemisch)*
formyl group Formylgruppe *f*, Formylrest *m*
formylate *v* formylieren
formylation Formylierung *f*
fortification Verstärkung *f*, Stärkung *f*; Anreicherung *f*; Wirkungssteigerung *f (z. B. eines Insektizids durch Kombination mit einem anderen Mittel)*
fortify *v* verstärken, stärken, aufstärken; anreichern; die Wirkung steigern *(z. B. eines Insektizids durch Kombination mit einem anderen Mittel)*
forward rate constant Geschwindigkeitskonstante *f* der Hinreaktion
forward reaction Hinreaktion *f*
forward reaction rate Geschwindigkeit *f* der Hinreaktion
fossil fuel fossiler [mineralischer] Brennstoff *m*, Fossilbrennstoff *m*
foul smelling übel riechend, unangenehm [widerlich] riechend
fouling 1. Verschmutzung *f*; 2. Bewuchs *m*, Anwuchs *m (z. B. an Schiffen)*
found in nature natürlich vorkommend
foundry Gießerei *f*
foundry (pig) iron Gießerei(roh)eisen *n*
four-carbon compound *(Bioch)* Vierkohlenstoffverbindung *f*, C-4-Verbindung *f*, C-4-Körper *m*
four-component system Vierkomponentensystem *n*, Vierstoffsystem *n*, quaternäres System *n*
four-coordinate(d) vierfach koordinativ gebunden, mit vier koordinativen Bindungen; vierfach koordiniert *(Komplex)*
four-fold degenerate *(phys Ch)* vierfach entartet [degeneriert]
four-membered ring viergliedriger Ring *m*, Vier(er)ring *m*
four-neck flask Vierhalskolben *m*
four-step reaction Vierstufenreaktion *f*
four-way valve Vierwegeventil *n*, Vierweg(e)hahn *m*
Fourier transform *s.* Fourier transformation
Fourier transform infrared spectrometer Fourier-Transform-IR-Spektrometer *n*, FT-IR-Spektrometer *n*
Fourier transform mass spectrometry Fourier-Transform-Massenspektrometrie *f*
Fourier transform NMR spectroscopy Fourier-Transform-NMR-Spektroskopie *f*, FTNMR
Fourier transformation *(Anal)* Fourier-Transformation *f*
F-6-P = fructose-6-phosphate
fraction *v s.* fractionate
fraction 1. *(Dest, Anal)* Fraktion *f*; 2. Fraktion *f*, Kornklasse *f*, Klasse *f*; 3. Bruchteil *m*
fraction collector *(Dest, Anal)* Fraktionssammler *m*
fractional charge Teilladung *f*, Partialladung *f*, partielle Ladung *f*
fractional crystallization 1. fraktionierte Kristallisation *f*; 2. *(Geol)* Kristallisationsdifferentiation *f*, fraktionierte Kristallisation *f*
fractional distillation fraktionierte [fraktionierende] Destillation *f*, Fraktionieren *n*, Fraktionierung *f*
fractional-order rate equation Geschwindigkeitsgleichung *f* gebrochener Ordnung *(Kinetik)*
fractional precipitation fraktionierte [stufenweise] Fällung *f*
fractionate *v* 1. fraktionieren; 2. *(Dest)* fraktioniert [stufenweise] destillieren; 3. *(Aufb)* klassieren
fractionating column *(Dest)* Fraktionierkolonne *f*, Fraktioniersäule *f*, Fraktionator *m*
fractionating flask *(Dest)* Fraktionierkolben *m*, Fraktionskolben *m*
fractionation 1. Fraktionieren *n*, Fraktionierung *f*; 2. *(Dest)* fraktionierte [fraktionierende] Destillation *f*
fracture Bruch *m*
fracture plane Bruchebene *f*
fracture surface Bruchfläche *f*
fragile zerbrechlich; brüchig
fragility Zerbrechlichkeit *f*; Brüchigkeit *f*

fragment *v* zerlegen, abbauen, *(i. e. S.)* fragmentieren *(Moleküle)*; zerfallen
fragment Bruchstück *n*, Spaltstück *n*, Rest *m*
fragment ion Fragmention *n*, Bruchstückion *n*
fragmentation Zerlegung *f*, Abbau *m*, *(i. e. S.)* Fragmentierung *f (von Molekülen)*; Zerfall *m*
fragmentation path Zerfallsweg *m*
fragrance Wohlgeruch *m*, Duft *m*
fragrant wohlriechend, duftend, aromatisch riechend
framework Gerüst *n*; Rahmen *m*
francium Fr Francium *n*
Fraunhofer lines *(Spektr)* Fraunhofer-Linien *fpl*, fraunhofersche Linien *fpl*
free *v* befreien
free frei, ungebunden, unverbunden
free carbon dioxide *(Hyd)* freie Kohlensäure *f*
free chlorine *(Hyd)* freies (wirksames) Chlor *n*
free electron freies [frei bewegliches] Elektron *n*
free-electron theory of metals Elektronentheorie *f* der Metalle
free electrophoresis *(Anal)* (träger)freie Elektrophorese *f*, Tiselius-Elektrophorese *f*, Methode *f* der wandernden Grenzfläche(n)
free energy 1. (gibbssche) freie Enthalpie *f*, gibbssche Funktion *f*, G; 2. freie Energie *f*, F
free energy G *s.* free energy 1.
free energy of activation freie Aktivierungsenthalpie *f*
free enthalpy *s.* free energy 1.
free-filtering leicht filtrierend [filtrierbar]
free from aromatic hydrocarbons aromatenfrei
free from odour geruchlos, geruchfrei
free from polluting substances schmutzstofffrei, unverschmutzt, nicht verschmutzt [belastet] *(Wasser, Luft)*
free from taste geschmack(s)frei, geschmacklos
free nitrogen fixation *(Landw)* nicht symbio(n)tische Stickstoffbindung *f* [Fixierung *f* von Luftstickstoff], Stickstoffbindung *f* durch freilebende Mikroorganismen
free of ... *siehe: free from ...*
free path freie Weglänge *f*
free radical (freies) Radikal *n*
free-radical chain reaction Radikalkettenreaktion *f*, radikalische Kettenreaktion *f*
free-radical mechanism Radikalmechanismus *m*, radikalischer Mechanismus *m*
free-radical polymerization Radikalkettenpolymerisation *f*, radikalische Polymerisation [Kettenpolymerisation] *f*
free-radical reaction radikalische Reaktion *f*, Radikalreaktion *f*
free-residual chlorination *(Hyd)* Brechpunktchlorung *f*, Knickpunktchlorung *f*
free residual chlorine *(Hyd)* freies wirksames Restchlor *n*, Restgehalt *m* an freiem (wirksamen) Chlor, Restchlorgehalt *m*, Chlorüberschuss *m*
free rotation freie Rotation [Drehbarkeit] *f*
free settling freies Absetzen [Sedimentieren] *n*
freedom from colour Farblosigkeit *f*, Farbfreiheit *f*
freedom from odour Geruchlosigkeit *f*, Geruch(s)freiheit *f*
freedom from taste Geschmack(s)freiheit *f*
freely filterable leicht filtrierend [filtrierbar]
freely soluble leicht löslich, leichtlöslich, ll
freezable gefrierbar
freeze *v* 1. gefrieren; erstarren *(Schmelze)*; 2. *(Lebm)* gefrieren, einfrieren; 3. einfrieren, verzögern, bremsen *(Reaktionen)*; 4. sich festfressen
freeze-dry *v* gefriertrocknen, lyophil trocknen, lyophilisieren
freeze *v* **out** ausfrieren
freeze dehydration *s.* freeze drying
freeze-dried gefriergetrocknet
freeze drying Gefriertrocknung *f*, Sublimationstrocknung *f*, Tiefkühltrocknung *f*, Lyophilisation *f*
freeze-thaw Gefrier-Tau(-Prozess) *m*
freeze-out Ausfrieren *n*
freezing 1. Gefrieren *n*; Erstarren *n (einer Schmelze)*; 2. *(Lebm)* Gefrieren *n*, Einfrieren *n*; 3. Einfrieren *n*, Verzögerung *f*,

Bremsung *f (von Reaktionen)*; 4. *(Hyd)* Ausfrieren *n*

freezing curve *s.* freezing-point curve

freezing mixture Kältemischung *f*

freezing point Gefrierpunkt *m*, Gefriertemperatur *f*; Erstarrungspunkt *m*, Erstarrungstemperatur *f (einer Schmelze)*

freezing-point curve Gefrierkurve *f*; Erstarrungskurve *f (einer Schmelze)*

freezing-point depression [lowering] Gefrierpunktserniedrigung *f*, Gefrierpunktsdepression *f*

French degree of hardness *(Hyd)* Grad *m* französischer Härte, französischer Härtegrad *m (1 °fH = 5,6 mg CaO/l H_2O)*

Frenkel defect [disorder] *(Krist)* Frenkel-Defekt *m*, Frenkel-Fehlordnung *f (aus einem Frenkel-Paar bestehend)*

frequency 1. Frequenz *f*, Schwingungszahl *f*, Schwing(ungs)frequenz *f*; 2. Frequenz *f*, Häufigkeit *f*

frequency distribution Häufigkeitsverteilung *f (Statistik)*

frequency doubling *(Spektr)* Frequenzverdopplung *f*

frequency factor Frequenzfaktor *m*, Aktionskonstante *f*, Häufigkeitsfaktor *m*, präexponentieller [vorexponentieller] Faktor *m*, Präexponentialfaktor *m (Kinetik)*

frequency of neighbouring *(Bioch)* Nachbarschaftshäufigkeit *f (von Purin- und Pyrimidinbasen)*

frequency range [region] *(Anal)* Frequenzgebiet *n*, Frequenzbereich *m*

frequency shift Frequenzverschiebung *f*

fresh acid Frischsäure *f*, frische Säure *f*

fresh catalyst Frischkatalysator *m*, frischer Katalysator *m*

fresh water Frischwasser *n*; Süßwasser *n*

fresh-water treatment *(Hyd)* Wasseraufbereitung *f*, Wasserbehandlung *f*

freshly prepared frisch bereitet [hergestellt]

fretting corrosion Reibkorrosion *f*, Reiboxidation *f*, Tribokorrosion *f*

Freundlich adsorption isotherm freundlichsche Adsorptionsisotherme *f*, Freundlich-Isotherme *f*

FRH *s.* follicle-stimulating-hormone releasing hormone

FRI = Fermentation Research Institute *(Japan)*

friability Bröckligkeit *f*; Zerreiblichkeit *f (z. B. des Kokses, der Kohle)*

friable bröck(e)lig, krüm(e)lig, mürbe; zerreiblich, zerreibbar *(z. B. Koks, Kohle)*

friction Reibung *f*, Friktion *f*

friction force Reib(ungs)kraft *f*, Reibung *f*

friction loss Reibungsverlust *m*

friction resistance Reibungswiderstand *m*

frictional constant Reib(ungs)koeffizient *m*, Reib(ungs)faktor *m*, Reibwert *m*

frictional heat Reibungswärme *f*

frictionless flow reibungsfreie Strömung *f*

Friedel-Crafts acylation Friedel-Crafts-Acylierung *f*

Friedel-Crafts alkylation Friedel-Crafts-Alkylierung *f*

Friedel-Crafts condensation Friedel-Crafts-Kondensation *f*

Friedel-Crafts reaction [synthesis] Friedel-Crafts-Reaktion *f*, Friedel-Crafts-Synthese *f*, friedel-craftssche Reaktion *f* [Synthese *f*]

Friedländer synthesis Friedländer-Synthese *f*, friedländersche Chinolinsynthese *f*

Friedrichs (reflux) condenser Kühler *m* nach Friedrichs, Schraubenkühler *m*, Spiralkühler *m* (nach Friedrichs)

Fries migration *s.* Fries reaction

Fries reaction [rearrangement] Fries-Reaktion *f*, friessche Umlagerung *f* [Verschiebung *f*] *(von Phenolestern in Phenolketone)*

frigorific kälteerzeugend

fringed micelle Fransenmizelle *f*

Frings acetator *(Biot)* Frings-Generator *m*

frit *v (Ker)* fritten

frit *(Ker)* Fritte *f*

fritted glass filter Glasfritte *f*, Glasfilter *n*

Fröhde reagent Fröhdes Reagens *n (Natriummolybdat in konzentrierter Schwefelsäure)*

front 1. *(Anal)* Vorderflanke *f*, Vorderfront *f (eines Peaks)*; 2. *(Chromat)* Front *f (des Fließmittels)*

front end *(Dest)* Vorlauf *m*

frontal chromatography Frontalchromatographie *f*
frontier orbital Grenzorbital *n*
fronting Bartbildung *f*, *Fronting *n* *(in Chromatogrammen)*
frost point Frostpunkt *m*, Gefrierpunkt *m*
frost resistance Frostbeständigkeit *f*
frostproof frostbeständig
froth *v* 1. zum Schäumen bringen, schaumig [schäumend] machen; 2. *(Kst)* verschäumen; (auf)schäumen; sich mit Schaum bedecken
froth *v* **over** überschäumen
froth *v* **up** aufschäumen
froth Schaum *m*
froth floatation Schaumflotieren *n*, Schaumflotation *f*, Schaumschwimmaufbereitung *f*
froth-preventing agent Mittel *n* gegen Schaumbildung, Schaumverhinderungsmittel *n*, Schaumverhütungsmittel *n*
froth suppression Schaumverhinderung *f*, Schaumverhütung *f*
frothiness Schaumigkeit *f*
frothing 1. Schäumen *n*, Aufschäumen *n*; 2. *(Kst)* Verschäumen *n*; Schaumbildung *f*
frothing agent Schaum(erzeugungs)mittel *n*, Schaumerzeuger *m*, Schaumbildner *m*, Schäummittel *n*, Schäumer *m*
frothy schaum(art)ig; mit Schaum bedeckt
FRP *s.* fibre-reinforced plastic
Fru *s.* fructose
fructose Fructose *f*, Fruchtzucker *m*, Lävulose *f*, *(i. e. S.)* D-Fructose *f*
fruit acid Fruchtsäure *f*
fruit essence Fruchtessenz *f*, Fruchtäther *m*, Fruchtaroma *n*
fruit sugar Fruchtzucker *m*, Fructose *f*, Lävulose *f*
fruit vinegar Obstessig *m*, Fruchtessig *m*
fruity frucht(art)ig, obstartig, Frucht...
FS = Federal Specifications
FSB = Federal Specifications Board
FSH *s.* follicle-stimulating hormone
FT *s.* Fourier transformation
FTMS *s.* Fourier transform mass spectrometry
FTNMR (spectroscopy) *s.* Fourier transform NMR spectroscopy
fuchsin(e) Fuchsin *n*, Magenta *n*, Rosanilin(hydrochlorid) *n*
fuchsinesulphurous acid fuchsinschweflige Säure *f*
fuchsonimine *(Farb)* Fuchsonimin *n*
fuel *v* (be)heizen, mit Brennstoff beschicken; (be) tanken, auftanken *(mit Kraftstoff oder Treibstoff)*; mit Kernbrennstoff [Brennstoff] beschicken
fuel Brennstoff *m*, Brennmaterial *n*, Heiz(brenn)stoff *m*, Feuerungsmaterial *n*; Kraftstoff *m*; Treibstoff *m (für Flugtriebwerke)*; Kernbrennstoff *m*, Brennstoff *m*
fuel additive Kraftstoffadditiv *n*
fuel alcohol Kraftsprit *m*, Kraftspiritus *m*, Treibspiritus *m (für Raketen)*
fuel alcohol from biomass *(Biot)* Biosprit *m*
fuel cell *(Ech)* Brennstoffzelle *f*, Brennstoffelement *n*
fuel chemist Brennstoffchemiker *m*
fuel chemistry Brennstoffchemie *f*
fuel element 1. *(Kern)* Brenn(stoff)element *n*; 2. *s.* fuel cell
fuel for jet planes Düsentreibstoff *m*
fuel gas Brenngas *n*, Heizgas *n*, Beheizungsgas *n*; Treibgas *n (für Motoren und Turbinen)*
fuel oil Heizöl *n*
fuel reprocessing *(Kern)* Brennstoffwiederaufbereitung *f*, Brennstoffwiederaufarbeitung *f*
fuel-reprocessing plant *(Kern)* Brennstoffwiederaufbereitungsanlage *f*, Brennstoffwiederaufarbeitungsanlage *f*
fuel rod *(Kern)* Brenn(stoff)stab *m*, Brennstoffstange *f*
fuel value Heizwert *m*
fugacity *(phys Ch)* Fugazität *f*
fugacity coefficient Fugazitätskoeffizient *m*
fugitive flüchtig
full-scale vollständig; großtechnisch, in großtechnischem Maßstab
full-scale factory production großtechnische Herstellung *f*, Großproduktion *f*
full-scale plant Produktionsbetrieb *m (im Vergleich zur Pilotanlage)*
fullerene *(org Ch)* Fulleren *n*
fully miscible unbegrenzt mischbar

fulminate Fulminat *n (Salz der Knallsäure)*
fulminic acid Knallsäure *f*, Blausäureoxid *n*, Formonitriloxid *n*
fulvalen *(org Ch)* Fulvalen *n*
fulvene Fulven *n*, 5-Methylencyclopenta-1,3-dien *n*, *(i. w. S.)* Fulven *n*, Fulvenkohlenwasserstoff *m*
fumarate Fumarat *n (Salz oder Ester der Fumarsäure)*
fumaric acid Fumarsäure *f*, *trans-Butendisäure *f*, *trans-Ethen-1,2-dicarbonsäure *f*
fume *v* rauchen; räuchern
fume *v* **off** *(Lab)* abrauchen
fume 1. Dampf *m (meist geruchsintensiv)*; 2. Rauch *m (Suspension von Schwebeteilchen in Gas)*; 3. *(Met) direkt von den Abgasen eines Bleischachtofens mitgeführter Flugstaub*
fume chamber *s.* fume hood
fume closet *(Lab)* Abzugskasten *m*, Tischabzugsschrank *m*
fume cupboard *s.* fume hood
fume hood Abzug(s)schrank *m*, Abzug *m*, Digestorium *n*, Kapelle *f*
fumigant Räuchermittel *n*; Vergasungsmittel *n*, Begasungsmittel *n*, Durchgasungsmittel *n*
fumigate *v* ausräuchern, durchräuchern *(Räume)*; beräuchern, dem Rauch aussetzen *(Gegenstände, Bäume)*; begasen, durchgasen *(den Boden)*
fumigating plume Verrauchungsfahne *f*
fumigation Ausräuchern *n*, Durchräuchern *n (von Räumen)*; Beräuchern *n (von Gegenständen, Bäumen)*; Begasen *n*, Durchgasen *n (des Bodens)*
fuming sulphuric acid rauchende Schwefelsäure *f*, Oleum *n*
function *v* wirken *(z. B. als Promotor)*
function plotter x-y-Schreiber *m*
functional group funktionelle [charakteristische] Gruppe *f*
functional name *(Nom)* funktioneller Name *m*, funktionelle Benennung *f*
functionality Funktionalität *f*
fundamental *s.* fundamental vibration
fundamental band *(Spektr)* Grundschwingungsbande *f*
fundamental building block Grundbaustein *m*, grundlegender [fundamentaler] Baustein *m*, Fundamentalbaustein *m*
fundamental chemistry Grundlagenchemie *f*
fundamental constituent Grundbestandteil *m*
fundamental equation *(phys Ch)* Fundamentalgleichung *f*
fundamental frequency *(Spektr)* Grundschwingungsfrequenz *f*
fundamental molecule Grundmolekül *n*
fundamental operation Grundoperation *f*, grundlegende Operation *f*
fundamental particle Elementarteilchen *n*
fundamental process Grundprozess *m*, Grundvorgang *m*; Grundverfahren *n*
fundamental reaction Grundreaktion *f*, grundlegende [fundamentale] Reaktion *f*
fundamental series *(Spektr)* f-Serie *f*
fundamental substance Grundstoff *m*, Grundsubstanz *f*
fundamental unit 1. Grundeinheit *f (eines Moleküls)*; 2. Basiseinheit *f*, Grundeinheit *f (eines Einheitensystems)*
fundamental vibration Grundschwingung *f*, Normalschwingung *f*, Eigenschwingung *f*
fungal pilzlich, Pilz...; pilzartig
fungal amylase Pilzamylase *f*
fungal biomass Pilzbiomasse *f*
fungal enzyme Pilzenzym *n*
fungal fermentation *(Biot)* Pilzfermentation *f*
fungal lipase Pilzlipase *f*
fungal mycelium Pilzmyzel *n*
fungal protease Pilzprotease *f*
fungal strain *(Biot)* Pilzstamm *m*
fungicidal 1. fungizid (wirksam), antifungal, pilztötend, pilzwirksam; 2. *(Pharm)* antimykotisch
fungicidal activity [**effectiveness**, **efficiency**] fungizide Wirksamkeit *f*
fungicide Fungizid *n*, pilztötendes Mittel *n*
fungicide-resistant fungizidresistent
fungistat Fungistatikum *n*
fungistatic fungistatisch, das Pilzwachstum hemmend
fungitoxic *s.* fungicidal
fungitoxicity *s.* fungicidal activity

funnel Trichter *m*
funnel heater Heißwassertrichter *m*, Warmwassertrichter *m*
funnel rack [stand] Filtrierstativ *n*
fur *v* Kesselstein ansetzen
fur 1. Kesselstein *m*; 2. Weinstein *m*
fural *s.* furfural
furaldehyde Furaldehyd *m*, *(i. e. S.)* 2-Furaldehyd *m*, Furfural *n*
furan *(org Ch)* Furan *n*
furan resin Furanharz *n*
furancarboxylic acid Furancarbonsäure *f*, *(i. e. S.)* Furan-2-carbonsäure *f*, Brenzschleimsäure *f*
2,5-furandione Furan-2,5-dion *n*, *cis*-Butendisäureanhydrid *n*, Maleinsäureanhydrid *n*
furanose Furanose *f (ein Disaccharid)*
furanose form Furanoseform *f*, furanoide Form *f (Zucker)*
furanoside Furanosid *n*
furazan Furazan *n*, 1,2,5-Oxadiazol *n*
furfural 1. Furfural *n*, 2-Furaldehyd *m*; 2. *s.* furfurylidene
furfural resin Furfuralharz *n*
furfuryl alcohol Furfurylalkohol *m*, 2-Hydroxymethylfuran *n*
furfurylidene Furfuryliden *n*, Fur-2-ylmethylen *n (Molekülrest)*
furil Furil *n*, Di-α-furyldiketon *n*
furnace (technischer) Ofen *m*, Industrieofen *m*
furnace atmosphere Ofenatmosphäre *f*
furnace coke Hochofenkoks *m*
furnace gas Verbrennungs(ab)gas *n*, Ofengas *n*; Gichtgas *n*
furnace lining Ofenzustellung *f*, Ofenauskleidung *f*, Ofenfutter *n*, Ofenausmauerung *f*; Hochofenfutter *n*
furnish *v* 1. beschicken, füllen *(z. B. einen Kocher)*; eintragen *(z. B. Kochgut)*; 2. liefern *(z. B. Elektronen)*
furoate Furoat *n (Salz oder Ester der Furancarbonsäure)*
furocoumarin *(org Ch)* Furocumarin *n*
furoin *(org Ch)* Furoin *n*
further reaction Weiterreaktion *f*, weitere Reaktion *f*
furylacrylic acid Furylacrylsäure *f*, Furalessigsäure *f*
furylidene *(org Ch)* Furyliden *n*
fuse *v* 1. schmelzen; (miteinander) verschmelzen; 2. *(org Ch)* kondensieren, anellieren *(Ringe)*
fuse *v* **in** einschmelzen *(in ein Glasgerät)*
fuse *v* **to** ankondensieren *(Ringe)*
fuse *v* **together** 1. verschmelzen, zusammenschmelzen; 2. *(org Ch)* (miteinander) kondensieren *(Ringe)*
fused catalyst Schmelzkatalysator *m*
fused electrolyte Schmelze *f*
fused nucleus *(org Ch)* kondensierter [anellierter] Kern *m*
fused polycyclic aromatic hydrocarbons kondensierte Arene *npl*
fused quartz Quarzglas *n*, durchsichtiges [klares] Kieselglas *n*
fused ring *(org Ch)* kondensierter [anellierter] Ring *m*
fused-ring hydrocarbon kondensierter [anellierter] Ringkohlenwasserstoff *m*
fused salt Salzschmelze *f*
fused-salt electrolysis Schmelzflusselektrolyse *f*
fused silica Quarzgut *n*, durchscheinendes [undurchsichtiges] Kieselglas *n*
fusel oil Fuselöl *n*
fusibility Schmelzbarkeit *f*
fusible schmelzbar
fusible alloy leichtschmelzende [leichtschmelzbare, niedrigschmelzende] Legierung *f*
fusidic acid Fusidinsäure *f (Antibiotikum)*
fusion 1. Schmelzen *n*, Schmelze *f*; Verschmelzen *n*; 2. *(Kern)* Fusion *f*, Verschmelzung *f*; 3. *(org Ch)* Kondensation *f*, Ringkondensation *f*, Anellierung *f*; 4. Schmelze *f (Zustand)*; 5. Schmelzaufschluss *m (schwerlöslicher Substanzen)*
fusion cone *(Ker)* Schmelzkegel *m*, Schmelzkörper *m*, Brennkegel *m*, *(meist)* Seger-Kegel *m*
fusion crucible Schmelztiegel *m*
fusion name *(Nom)* Fusionsname *m*, Verschmelzungsname *m*, Anellierungsname *m*
fusion position Kondens(ations)stelle *f*, Anellierungsstelle *f (Bindungslehre)*
fusion prefix *(Nom)* Anellierungspräfix *n*
fusion reagent *(Lab)* Flussmittel *n*
fusion with alkali Alkalischmelze *f*
FWHM = full width (at) half maximum

G

G *s.* gauche conformation
g. equiv. *s.* gram equivalent
g-m counter *s.* Geiger(-Müller) counter
G-M counter *s.* Geiger(-Müller) counter
G-6-P = glucose-6 phosphate
GABA *s.* gamma aminobutyric acid
Gabriel (phthalimide) synthesis (of primary amines) Gabriel-Synthese *f*, gabrielsche Aminsynthese *f (durch Phthalimidspaltung)*
GAC *(granular activated carbon)* gekörnte A-Kohle *f*, Kornkohle *f*
GAC filtration Kornkohlefiltration *f*, Filtration *f* über gekörnte Aktivkohle
gadoleic acid Gadoleinsäure *f*, *Eicos-9-ensäure *f*
gadolinia Gadoliniumoxid *n*
gadolinium Gd Gadolinium *n*
gage *s.* gauge
gain *v* anlagern, aufnehmen
gain *v* **electrons** Elektronen aufnehmen
gain *v* **weight** an Gewicht zunehmen; an Masse zunehmen
gain in energy Energiegewinn *m*
gain in weight Gewichtszunahme *f*; Masse(n)zunahme *f*
gain of electrons Elektronenaufnahme *f*
Gal = galactose
galactaric acid Schleimsäure *f*, Galactarsäure *f*, Tetrahydroxyadipinsäure *f*, *2,3,4,5-Tetrahydroxyhexandisäure *f*
galactosaemia *(Med)* Galaktosämie *f*
***d*-galacturonic acid** D-Galacturonsäure *f*
gall 1. *(Gerb, Bot)* Galle *f*, Pflanzengalle *f*; 2. *(Med)* Galle *f*, Gallenflüssigkeit *f*
gallate 1. Gallat(III) *n (Galliumverbindung)*; 2. 3,4,5-Trihydroxybenzoat *n*, Gallat *n (Salz oder Ester der Gallussäure)*
gallic 1. Gallium(III)-...; 2. Gallussäure...
gallic acid Gallussäure *f*, 3,4,5-Trihydroxybenzoesäure *f*
gallium Ga Gallium *n*
gallium hydroxide Gallium(III)-hydroxid *n*
gallium oxychloride Galliumoxidchlorid *n*
gallium trichloride Galliumtrichlorid *n*, Gallium(III)-chlorid *n*
gallotannic acid Gallotannin *n*, Tanningerbstoff *m*, Gallusgerbsäure *f*, *(i. e. S.)* Tannin *n (Estergemisch aus Glucose und Gallussäure und/oder deren Depsiden)*
gallotannin Gallotannin *n*, Tanningerbstoff *m*, Gallusgerbsäure *f*, *(i. e. S.)* Tannin *n (Estergemisch aus Glucose und Gallussäure und/oder deren Depsiden)*
gallous Gallium(I,III)-...
gallstone *(Med)* Gallenstein *m*
galmei 1. *(Min)* Smithsonit *m*, (Edler) Galmei *m*, Zinkspat *m (Zinkcarbonat)*; 2. *(Min)* Hemimorphit *m*, (Gemeiner) Galmei *m*, Kieselgalmei *m*, Kieselzinkerz *n (Zinkhydroxysilicat)*
galmey 1. *(Min)* Smithsonit *m*, (Edler) Galmei *m*, Zinkspat *m (Zinkcarbonat)*; 2. *(Min)* Hemimorphit *m*, (Gemeiner) Galmei *m*, Kieselgalmei *m*, Kieselzinkerz *n (Zinkhydroxysilicat)*
GalNAc = *N*-acetylgalactosamine
Galvani potential Galvani-Potenzial *n*, inneres elektrisches Potenzial *n*
galvanic galvanisch; anodisch wirksam *(Schutzschicht)*
galvanic cell galvanische Zelle *f* [Kette *f*], galvanisches Element *n*
galvanic connection galvanische Kopplung *f*
galvanic corrosion galvanische Korrosion *f*, Kontaktkorrosion *f*, Berührungskorrosion *f*
galvanic couple [element] *s.* galvanic cell
galvanization Galvanisierung *f*, Verzinken *n*, *(i. e. S.)* Feuerverzinken *n*
galvanize *v* verzinken, *(i. e. S.)* feuerverzinken
galvanizing *s.* galvanization
galvanometer Galvanometer *n*
gamma aminobutyric acid γ-Aminobuttersäure *f*
gamma globulin *(Bioch)* γ-Globulin *n*
gamma helix γ-Helix *f (Strukturvariante von Polypeptidketten)*
gamma hydrogen Wasserstoffatom *n* in γ-Stellung [γ-Position]
gamma iron γ-Eisen *n*
gamma radiation Gammastrahlung *f*
gamma ray Gammastrahl *m*, γ-Strahl *m*
gamma transition *s.* glass transition

ganglion blocking agent *(Pharm)* Ganglienblocker *m*, Ganglioplegikum *n*
gangue (mineral) Gangart *f*, taubes Gestein *n* [Mittel *n*], Bergemittel *n*, Nichterz *n*
gap 1. Hohlraum *m*, Lücke *f*, Spalt *m*, Spalte *f*; 2. *s.* forbidden band
garbage Abfall *m*, Müll *m*
garbage composting Müllkompostierung *f*
garbage disposal Müllbeseitigung *f*, Müllentsorgung *f*
garbage disposal plant Müllaufbereitungsanlage *f*, Müllbehandlungsanlage *f*
garbage dump Müllkippe *f*
garbage incineration Müllverbrennung *f*, Müllveraschung *f*
garbage incinerator Müllverbrennungsanlage *f*, MVA, *(i. e. S.)* Müllverbrennungsofen *m*
garnet *(Min)* Granat *m*
gas. *s.* gaseous
gas *v* 1. gasen, Gas abgeben; 2. *(Landw, Lebm)* begasen; mit Gas vergiften
gas 1. Gas *n*; 2. *s.* gasoline
gas absorption Gasabsorption *f*, Gasaufnahme *f*
gas analysis Gasanalyse *f*
gas-analytical gasanalytisch
gas balance Gas(dichte)waage *f*
gas blowtorch Gebläsebrenner *m*
gas bubble Gasblase *f*, Gasbläschen *n*
gas burette Gasbürette *f*
gas burner Gasbrenner *m*
gas calorimeter Gaskalorimeter *n*, Kalorimeter *n* für Gase
gas cell *(Spektr)* Gasküvette *f*; (elektrisches) Gaselement *n*, (elektrische) Gaszelle *f*, Gaskette *f*
gas centrifuge Gaszentrifuge *f*
gas centrifuge process *(Kern)* Gaszentrifugenverfahren *n*, Zentrifugenverfahren *n*, Schleuderverfahren *n (zur Isotopentrennung)*
gas chromatograph Gaschromatograph *m*
gas-chromatographic gaschromatographisch
gas chromatography Gaschromatographie *f*
gas chromatography-mass spectrometry (coupling) Gaschromatographie-Massenspektrometrie-Kopplung *f*
gas clathrate compound *s.* gas hydrate
gas cleaning [clean-up] Gasreinigung *f*
gas coal Gaskohle *f*
gas constant (universelle, allgemeine) Gaskonstante *f*
gas coulometer Gascoulometer *n*, *(i. e. S.)* Knallgascoulometer *n*
gas cylinder Gasflasche *f (Druckflasche)*
gas density balance Gas(dichte)waage *f*
gas desulphurization Gasentschwefelung *f*
gas detector Gasspürgerät *n*
gas discharge tube Gasentladungsröhre *f*, Entladungsröhre *f*, geißlersche Röhre *f*
gas dispersion [distribution] tube Begasungsfilter *n*, Begasungsröhrchen *n*, *(mit zylindrischem Frittenteil auch)* Begasungsfilterkerze *f*
gas drying jar *(Lab)* Trockenturm *m*, Chlorcalciumzylinder *m*
gas electrode Gaselektrode *f*
gas enrichment Gasanreicherung *f*
gas-entry tube Gaseinleitungsrohr *n*
gas equation Gasgleichung *f*
gas evolution Gasentwicklung *f*, Gasabscheidung *f*
gas filter [filtering tube] Gasfilter *n*
gas filtration Filtern *n* (von Gasen), Gasfiltration *f*
gas flame coal Gasflammkohle *f*
gas flow Gasströmung *f*
gas-flow counter tube *(Kern)* Gasdurchflusszählrohr *n*
gas flowmeter Gasdurchflussmesser *m*, Strömungsmesser *m* für Gase
gas generator 1. *(Lab)* Gasentwicklungsapparat *m*, Gasentwickler *m*; 2. *s.* gas producer
gas-holder Gasbehälter *m*, *(besonders für Stadtgas:)* Gasometer *m*
gas hydrate Gashydrat *n*, Eishydrat *n (Käfigeinschlussverbindung)*
gas inlet pipe Gaszuführungsrohr *n*
gas jar Standzylinder *m*
gas laser Gaslaser *m*
gas law Gasgesetz *n*
gas-level(l)ing bulb Niveaukugel *f*, Niveaubirne *f*
gas lift 1. Gaslift *m*, Gasheber *m*; 2. *s.* gas lifting

gas lifting Gasliften *n*, Gas(lift)förderung *f*, Förderung *f* mittels Gaslift [Druckgas]
gas liquefaction Gasverflüssigung *f*
gas-liquid (partition) chromatography Gas-flüssig-Chromatographie *f*, Gas-Liquidus-Chromatographie *f*, GLC, Gas-Verteilungschromatographie *f*
gas-liquid reaction Gas-flüssig-Reaktion *f*
gas-liquid reactor Gas-flüssig-Reaktor *m*, Reaktor *m* für Gas-flüssig-Reaktionen
gas main Gasleitung *f*
gas making Gaserzeugung *f*, Gasherstellung *f*
gas manufacture Gaserzeugung *f*, Gasherstellung *f*
gas mask Gasmaske *f*, Atemschutzmaske *f*
gas meter Gasmesser *m*, Gaszähler *m*
gas mixture Gasgemisch *n*
gas oil Gasöl *n*, Dieselöl *n*, Treiböl *n*
gas-oil fermentation [process] *(Biot)* Gasöl-Fermentation *f*, Gasölverfahren *n (SCP-Herstellung)*
gas outlet Gasaustritt *m*, Gasaustrittsöffnung *f*, Gasausgang *m*, Gasabgang *m*, Gasabzug *m*
gas-permeable gasdurchlässig, gaspermeabel
gas phase gasförmige Phase *f*, Gasphase *f*
gas-phase reaction Gasphasenreaktion *f*
gas pipette Gaspipette *f*
gas plating Gasplattieren *n*, Reaktionsbeschichten *n* aus der Gasphase, Beschichten *n* nach dem CVD-Verfahren, CVD-Beschichten *n*; Abscheiden *n* aus der Gasphase
gas plume Rauchfahne *f*, Abgasfahne *f*
gas poisoning Gasvergiftung *f*
gas pressure Gasdruck *m*
gas processing Gasaufbereitung *f*, Gasreinigung *f*
gas producer Gaserzeuger *m*, Gasgenerator *m*
gas purification Gasreinigung *f*
gas reaction Gasreaktion *f*
gas scrubber *s.* gas washer
gas-solid chromatography Gas-fest-Chromatographie *f*, Gas-Solidus-Chromatographie *f*, GSC, Gas-Adsorptionschromatographie *f*
gas-solid reaction Gas-Feststoff-Reaktion *f*
gas-solid reactor Gas-Feststoff-Reaktor *m*, Reaktor *m* für Gas-Feststoff-Reaktionen
gas soot Gasruß *m*
gas spirit Gasbenzin *n*
gas stream Gasstrom *m*
gas stripping Gasstrippen *n*
gas thermometer Gasthermometer *n*
gas-tight gasdicht, gasundurchlässig
gas transfer coefficient *(Hyd)* Gasaustauschkoeffizient *m*, Gasübergangskoeffizient *m*, Gasübergangsbeiwert *m*
gas treating [treatment] Gasaufbereitung *f*, Gasnachbehandlung *f*
gas valve Gasventil *n*, Gasschieber *m*
gas washer Gaswäscher *m*, Wäscher *m*, Wascher *m*, Skrubber *m*, *(für Stäube und Dämpfe in Gasen auch)* Nass(staub)abscheider *m*, Nassentstauber *m*
gas washing Gaswaschen *n*, Gaswäsche *f*, Waschen *n*, *(zur Entfernung von Stäuben und Dämpfen auch)* Nassentstauben *n*
gas-washing bottle Gaswaschflasche *f*
gas welding Gas(schmelz)schweißen *n*, autogenes Schweißen *n*, Autogenschweißen *n*
gaseous gasförmig, gasartig
gaseous diffusion process *(Kern)* Gasdiffusionsverfahren *n*, Diffusionsverfahren *n (zur Isotopentrennung)*
gaseous fuel gasförmiger Brennstoff *m*; gasförmiger Kraftstoff *m*
gaseous hydrogen chloride gasförmiger Chlorwasserstoff *m*, Chlorwasserstoffgas *n*
gaseous reaction Gasreaktion *f*
gasholder *s.* gas-holder
gasification Vergasen *n*, Vergasung *f*
gasifier Vergaser *m*, Vergasungsapparat *m*
gasify *v* vergasen
gasket Dichtungsmanschette *f*, Dichtung *f (für nicht gegeneinander bewegte Teile)*
gaslifting *(Erdöl)* Gasliften *n*, Gas(lift)förderung *f*, Förderung *f* mittels Gaslifts [Druckgases]

gasohol *Benzin-Ethylalkohol-Gemisch (90 : 10 %) als Kraftstoff*
gasoline Benzin *n (besonders als Vergaserkraftstoff)*
gasoline additive Benzinadditiv(e) *n*
gasoline-soluble benzinlöslich
gasometer *s.* gas-holder
gassing 1. Gasen *n*, Gasentwicklung *f*; 2. *(Landw, Lebm)* Begasung *f*; Gasvergiftung *f*
gastight gasdicht, gasundurchlässig
gastric acid Magensäure *f*
gastric inhibitory polypeptide *(Bioch)* gastrisches Inhibitor-Polypeptid *n*
gate valve Schieber *m*, Absperrschieber *m*, Schieberventil *n*
gathering *quantitative Mitfällung einer nur schwach vertretenen Substanz*
Gattermann-Koch reaction [synthesis] Gattermann-Koch-Synthese *f*, gattermann-kochsche Synthese *f (von Phenolaldehyden)*
gauche conformation gauche-Konformation *f (Torsionswinkel ±120°)*
gauche form schiefe [windschiefe, synclinale] Form *f*, syn-Form *f*
gauge *v* kalibrieren, *(wenn amtlich:)* eichen *(Gefäße)*
gauge 1. Messgerät *n*, Messapparat *m*, Messinstrument *n*, Messer *m*; 2. *(Text)* Warendichte *f*
Gaussian *(Anal)* gaußförmig
Gaussian curve gaußsche Verteilungskurve *f (Statistik)*
Gaussian distribution function gaußsche [normale] Verteilungsfunktion *f*
Gauss's method of double weighing gaußsche Wägung *f*, Vertauschungswägung *f*
gauze Netz *n*, Siebgewebe *n*; Mull *m*, Gaze *f*
gauze catalyst Netzkontakt *m*
gauze electrode Netzelektrode *f*
Gay-Lussac law gay-lussacsches Gesetz *n*, Gay-Lussac-Gesetz *n*
GBL = γ-butyrolactone
GC *s.* gas chromatography
GC-MS *s.* gas chromatography-mass spectrometry
GDH = glutamate dehydrogenase
GDP *s.* guanosine diphosphate
gear oil Getriebeöl *n*
gear pump Zahnradpumpe *f*
gegenion Gegenion *n*
Geiger(-Müller) counter [tube] Geiger-Müller-Zählrohr *n*, Geiger-Zähler *m*
Geissler tube Geißler-Röhre *f*, geißlersche Röhre *f*
gel. *s.* gelatinous
gel *v* 1. erstarren, fest werden, steif werden, *(i. e. S.)* gel(atin)ieren, zu Gelee erstarren; 2. *(Koll)* in den Gelzustand übergehen, ausflocken, koagulieren
gel *(Koll)* Gel *n*
gel chromatography Gelchromatographie *f*, Ausschlusschromatographie *f*
• **by gel chromatography** gelchromatographisch
gel effect *(Kst)* Geleffekt *m*, Trommsdorff-(Norrish-)Effekt *m*
gel electrophoresis Gelelektrophorese *f*
gel-entrapped enzyme *(Biot)* in Gel(-matrix) eingeschlossenes Enzym *n*
gel filtration (chromatography) Gelfiltrationschromatographie *f*, GFC, Gelfiltration *f*
gel permeation chromatography Gelpermeationschromatographie *f*, GPC, Gelpermeation *f (Ausschlusschromatographie in nicht wässrigen Systemen)*
• **by gel permeation chromatography** gelpermeationschromatographisch
gel point 1. Gel(atin)ierungstemperatur *f*; 2. *(Kst)* Gelpunkt *m*
gel-sol transformation Gel-Sol-Umwandlung *f*
gel-sol transition Gel-Sol-Übergang *m*
gelate *v s.* gel
gelatin Gelatine *f*; Gallerte *f*, Gallert *n*
gelatin dynamite Sprenggelatine *f*
gelatin sol Gelatinesol *n*
gelatinate *v s.* gelatinize
gelatinization Gelatinieren *n*, Gelatinierung *f*
gelatinize *v* gelatinieren, in Gelee überführen; gel(atin)ieren, zu Gelee erstarren
gelatinizer Geliermittel *n*, Gelatinierungsmittel *n*, Gelierstoff *m*
gelatinizing agent Geliermittel *n*, Gelatinierungsmittel *n*, Gelierstoff *m*
gelatinous gelatineartig, gelatinös, gallertartig, gallertig; gelatinehaltig
gelatinous mass Gallerte *f*, Gallert *n*
gelation 1. Erstarren *n*, Festwerden *n*,

Steifwerden *n*; Gel(atin)ieren *n*, Gel(atin)ierung *f*, Gallertbildung *f*; 2. *(Koll)* Gelbildung *f*

gem Schmuckstein *m (bearbeiteter Edel- oder Halbedelstein)*

gem-... *siehe: geminal*

geminal geminal, gem-... *(zwei identische Substituenten am gleichen Kohlenstoffatom)*

gemstone Edelstein *m (unbearbeitet)*

gene cloning *(Biot)* Genklonierung *f*

gene manipulation *(Biot)* Genmanipulation *f*, genetische [gentechnologische] Manipulation *f*

gene technology *(Biot)* Gentechnologie *f*, Gentechnik *f*

gene transfer *(Biot)* Gentransfer *m*, Genübertragung *f*, Genverpflanzung *f*

general acid-base catalysis allgemeine Säure-Base-Katalyse *f*

general chemistry allgemeine Chemie *f*

general corrosion flächenhafte [abtragende] Korrosion *f*, Flächenkorrosion *f*, Flächenfraß *m*

general formula allgemeine Formel *f*, Allgemeinformel *f*

general gas constant (allgemeine, universelle) Gaskonstante *f*

general herbicide *(Landw)* Totalherbizid *n*

general-purpose allgemein [universell] verwendbar, Allzweck..., Universal...

generate *v* erzeugen, generieren

generation Erzeugung *f*, Entwicklung *f*; Entstehung *f*

generation of gases Gasentwicklung *f*

generator 1. Generator *m (Schachtofen zur Gaserzeugung)*; *(i. e. S.)* Generatorschacht *m*; 2. Dampferzeuger *m*

generic name [term] 1. Sammelname *m*, allgemeiner [generischer] Name *m*, Gattungsname *m*; 2. nicht wortgeschützter Name *m*, freier Warenname *m*, Freiname *m*

genetic code *(Bioch)* genetischer Code *m*

genetic engineering *s.* gene technology

genetically engineered microorganisms *(Biot)* genetisch [gentechnologisch] veränderte Mikroorganismen *mpl*

Geneva system of naming [nomenclature] Genfer Nomenklatursystem *n*

gentisic acid Gentisinsäure *f*, 2,5-Dihydroxybenzoesäure *f*

gentle gelinde, mäßig, schwach, vorsichtig

gentle distillation schonende Destillation *f*

gentle heating (gelindes) Erwärmen *n*

geochemical geochemisch

geochemist Geochemiker *m*

geochemistry Geochemie *f*, geologische Chemie *f*

geometric isomer geometrisches Isomer(e) *n*, *cis-trans*-Isomer(e) *n*

geometric isomerism geometrische Isomerie *f*, *cis-trans*-Isomerie *f*

geometric mean *(Anal)* geometrisches Mittel *n*

geosphere Geosphäre *f*

geraniol Geraniol *n*, *3,7-Dimethyl-octa-2,6(2,7)-dien-1-ol *n*

germ oil Keimöl *n*

German degree of hardness *(Hyd)* Grad *m* deutscher Härte, deutscher Härtegrad *m (1 °dH = 10,0 mg CaO/l H_2O)*

germanate Germanat(IV) *n*

germane Monogerman *n*, Germaniumtetrahydrid *n*

germanic Germanium..., *(i. e. S.)* Germanium(IV)-...

germanite Germanat(II) *n*

germanium Ge Germanium *n*

germanium dichloride Germaniumdichlorid *n*, Germanium(II)-chlorid *n*

germanium dioxide Germaniumdioxid *n*, Germanium(IV)-oxid *n*

germanium tetrachloride Germaniumtetrachlorid *n*, Germanium(IV)-chlorid *n*

germanium tetrahydride Germaniumtetrahydrid *n*, *Monogerman *n*

germanous Germanium..., *(i. e. S.)* Germanium(II)-...

germicidal keimtötend, entkeimend

germicide Germizid *n*, keimtötendes [keimfreimachendes] Mittel *n*

getter Getter(stoff) *m*, Fangstoff *m*, Gettermetall *n (Vakuumtechnik)*

GFC *s.* gel filtration (chromatography)

gg. *s.* gauge

GG *s.* gamma globulin

GH *s.* growth hormone

ghost peak *(Anal)* Geisterpeak *m*, Störpeak *m*, Fremdpeak *m*

giant molecule Makromolekül *n*, Riesenmolekül *n*
gibberellic acid *(Bioch)* Gibberellinsäure *f*
gibberellin *(Bioch)* Gibberellin *n*
Gibbs adsorption equation gibbssche Adsorptionsgleichung *f*
Gibbs adsorption isotherm gibbssche Adsorptionsisotherme *f*
Gibbs-Duhem equation (gibbs-)duhemsche Gleichung *f*
Gibbs free energy (gibbssche) freie Enthalpie *f*, thermodynamisches Potenzial *n*
Gibbs free energy of activation freie Aktivierungsenthalpie *f*
Gibbs function *s.* Gibbs free energy
Gibbs-Helmholtz equation Gibbs-Helmholtz-Gleichung *f*
Gibbs phase rule (gibbssches) Phasengesetz *n*, (gibbssche) Phasenregel *f*
GIC *s.* graphite intercalation compound
gild *v* vergolden
gin Gin *m (Wacholderbranntwein)*
GIP *s.* gastric inhibitory polypeptide
give *v* **a swirl to** in einen Wirbel versetzen
give *v* **off [up]** abgeben
glacial acetic acid Eisessig *m*
glance mineralischer Glanz *m (Mineralsulfide)*
glance pitch reiner Asphalt *m*
glancing angle *(Krist)* Glanzwinkel *m*, braggscher Winkel *m*
gland 1. *(Bio, Med)* Drüse *f*; 2. *(Tech)* Stopfbuchse *f*, Stopfbuchsenbrille *f*
glass 1. Glas *n*; 2. *(Geol)* Glas *n*, Gesteinsglas *n*
glass bar Glasstab *m*
glass beaker Becherglas *n*
glass bell Glasglocke *f*
glass blower's lamp Glasbläserlampe *f*, Gebläselampe *f*
glass blowing Glasblasen *n*
glass bottle Glasflasche *f*
glass bulb Glaskolben *m*
glass ceramic Glaskeramik *f*, glaskeramischer Stoff *m*, Vitrokeram *n*, Vitrokeramik *f*
glass cloth *s.* glass fabric
glass cover Deckglas *n (Sonnenenergie)*
glass cutter Glasschneider *m*
glass cuvette Glasküvette *f*
glass dish Glasschale *f*
glass electrode Glaselektrode *f*
glass fabric Glas(faser)gewebe *n*
glass fibre Glasfaser *f*; Glasfaserstoff *m*, GL
glass-fibre fabric *s.* glass fabric
glass-fibre laminate Glasfaserschichtstoff *m*, GFS, Glasfaserlaminat *n*
glass-fibre reinforced plastic glasfaserverstärkter Kunststoff *m*, GFK
glass-fibre roving [strand] Glasseidenroving *m*, Roving *m*, Glasseidenstrang *m*
glass-fibre yarn Glas(stapel)fasergarn *n*
glass filament Glasfaden *m*
glass filter Glasfilter *n*, Glasfritte *f*
glass filtering crucible Glasfiltertiegel *m*, Glasfrittentiegel *m*
glass flask Glaskolben *m*
glass-forming glasbildend
glass frit 1. Glasfritte *f*; 2. Glasurmasse *f (Glasieren)*
glass funnel Glastrichter *m*
glass furnace Glas(schmelz)ofen *m*
glass grinding Glasschliff *m*
glass industry Glasindustrie *f*
glass insulator Glasisolator *m*
glass insurance Glasversicherung *f*
glass-kiln Glasschmelzofen *m*
glass-like carbon glasartiger Kohlenstoff *m*, Glaskohlenstoff *m*
glass-like state glasiger [glasartiger] Zustand *m*, Glaszustand *m*
glass-lined mit Glas ausgekleidet, mit Glasauskleidung
glass machining *(Tech)* Glasbearbeitung *f*
glass making Glasblasen *n*
glass-melting pot Glashafen *m*
glass-melting tank Glasschmelzwanne *f*
glass membrane Glasmembran *f*
glass membrane electrode Glaselektrode *f*
glass panel Glasscheibe *f*
glass pipe Glasrohr *n*, Glasröhre *f*
glass piping Glasrohr(material) *n*, Röhrenglas *n*
glass plate *s.* glass panel
glass powder Glaspulver *n*
glass rod Glasstab *m*
glass sand *(Geol)* Quarzsand *m*
glass silk Glasseide *f*

glass stopper Glasstopfen *m*, Glasstöpsel *m*
glass-stoppered bottle Glasstopfenflasche *f*, Glasstöpselflasche *f*
glass strand Glasseidenspinnfaden *m*
glass suction filter Glas(filter)nutsche *f*
glass temperature *s.* glass-transition temperature
glass transition Glasübergang *m*, Übergang *m* in den Glaszustand, Glasumwandlung *f (einer unterkühlten Schmelze)*
glass-transition temperature 1. Glas(übergangs)temperatur *f*, Glasumwandlungstemperatur *f (von Polymeren)*; 2. *(Glas)* Transformationstemperatur *f*, Glasbildungstemperatur *f*, Einfriertemperatur *f*, Einfrierpunkt *m (Viskosität* 10^{13} *dPa · s)*
glass tube Glasrohr *n*, Glasröhre *f*
glass vessel Glasgefäß *n*
glass wool Glaswolle *f*
glass working Glasbearbeitung *f*
glassware Glasware *f*, Glaswaren *fpl*, Glasartikel *mpl*, Glas *n*; Glasgeräte *npl*
glassy glasartig, glasähnlich, glasig
glassy carbon glasartiger Kohlenstoff *m*, Glaskohlenstoff *m*
glassy material glasiger Stoff *m*, Glas *n*; glasiger Werkstoff *m*
glassy phase Glasphase *f*, glasige Phase *f*
glassy state glasiger [glasartiger] Zustand *m*, Glaszustand *m*
Glauber salt Glaubersalz *n*, Natriumsulfat-10-Wasser *n*
glaze *v (Ker)* glasieren
glaze *(Ker)* Glasur *f*
GLC *s.* gas-liquid (partition) chromatography
GlcN *s. D*-glucosamine
GlDH = glutamate dehydrogenase
GLDH = glutamate dehydrogenase
glide plane *(Krist)* Gleitebene *f*
global rate of reaction Gesamtreaktionsgeschwindigkeit *f*
globin portion Globinanteil *m (des Hämoglobins)*
globoside *(Bioch)* Globosid *n (neutrales Ceramid-Oligosaccharid)*
globular colloid Sphärokolloid *n*, globuläres Kolloid *n*
globular protein globuläres [kugelförmiges] Protein *n*, Globulärprotein *n*, Sphäroprotein *n*, Kugelprotein *n*, globulärer Eiweißkörper *m*
globulin Globulin *n (ein Globulärprotein)*
gloss Glanz *m*, glänzendes Aussehen *n*, Oberflächenglanz *m*
glossy glänzend, Glanz...
glove box Glovebox *f*, Handschuhbox *f*, Manipulationskammer *f*
glow *v* glimmen, glühen
glow Glimmen *n*, Glühen *n*; Glut *f*
glow discharge Glimmentladung *f*
glucagon Glucagon *n (ein Bauchspeicheldrüsenhormon)*
glucan Glucan *n*
glucaric acid Zuckersäure *f*, Glucozuckersäure *f (eine Form der 2,3,4,5-Tetrahydroxyhexandisäure)*
glucitol D-Sorbitol *m (ein Zuckeralkohol)*
glucocorticoid Glucocorticoid *n (ein Nebennierenrindenhormon)*
gluconeogenesis *(Bioch)* Gluconeogenese *f*, Glucoseneubildung *f (aus nicht kohlenhydratartigen Stoffen)*
***d*-gluconic acid** D-Gluconsäure *f*
***D*-glucosamine** D-Glucosamin *n*, GlcN, Chitosamin *n*, 2-Amino-2-deoxy-α-D-glucopyranose *f*
glucose Glucose *f*, *(i. e. S.)* D-Glucose *f*, Traubenzucker *m*, Dextrose *f*
glucose phosphate Glucosephosphat *n*
glucosic Glucose...
glucoside Glucosid *n (ein Glykosid mit Glucose als Zuckerkomponente)*
glucosidic glucosidisch, Glucosid...
glucuronate pathway *(Bioch)* Glucuronatweg *m*, Glucuronat-Xylulose-Zyklus *m*, D-Glucuronat-L-Gulonat-Weg *m*
glucuronic acid Glucuronsäure *f*
glucuronide *(org Ch)* Glucuronid *n*
glue *v* (ver)leimen, (ver)kleben; mit Leim bestreichen
glue Leim *m*
glutamate Glutamat *n (Salz der Glutaminsäure)*
glutamic acid Glutaminsäure *f*, Glu, 2-Aminoglutarsäure *f*, *2-Aminopentandisäure *f*
glutamine Glutamin *n*, Glu-NH_2
glutar(di)aldehyde Glutaraldehyd *m*, Glutar(säure)dialdehyd *m*, *Pentandial *n*

glutaric acid Glutarsäure *f*, *Pentandisäure *f*
glutathione Glutathion *n*, *N*-(*N*-γ-Glutamylcysteinyl)glycin *n*
gluten Gluten *n*, Kleber *m*, Klebereiweiß *n*
glutinous klebrig
gly. *s.* glycerol
glycan Glykan *n*, Polysaccharid *n*
glycaric acid Aldarsäure *f*, Zuckerdicarbonsäure *f*
glyceraldehyde Glyceraldehyd *m*, *2,3-Dihydroxypropanal *n*
glycerate Glycerat *n* *(Salz der Glycersäure)*
glyceric acid Glycersäure *f*, *2,3-Dihydroxypropansäure *f*
glyceride Glycerid *n*, Acylglycerol *n* *(Ester des Glycerols mit Fettsäuren)*
glycerol Glycerol *n*, Glycerin *n*, *Propan-1,2,3-triol *n*
glycerol fermentation Glycerolfermentation *f*
glycerol trinitrate Glyceroltrinitrat *n*
glycidic ester Glycidester *m*
glycine Glycin *n*, Gly, Aminoethansäure *f*, Aminoessigsäure *f*, Glykokoll *n*
glycocholic acid Glykocholsäure *f* *(eine Gallensäure)*
glycocoll *s.* glycine
glycogen Glykogen *n*, Leberstärke *f*, Tierstärke *f*, tierische Stärke *f*
glycogenesis *(Bioch)* Glykogenese *f*
glycogenolysis *(Bioch)* Glykogenolyse *f*, Glykogenabbau *m*
glycol $C_nH_{2n}(OH)_2$ Glykol *n*, *Diol *n*, zweiwertiger Alkohol *m*, *(i. e. S.)* $HOCH_2CH_2OH$ Glykol *n*, 1,2-Glykol *n*, Ethylenglykol *n*, *Ethan-1,2-diol *n*
glycolaldehyde Glykolaldehyd *m*, Hydroxyacetaldehyd *m*, *Hydroxyethanal *n*
glycolipid(e) *(Bioch)* Glykolipid *n*
glycollic acid Glykolsäure *f*, Hydroxyessigsäure *f*
glycolysis Glykolyse *f*
glycolytic pathway [scheme] *(Bioch)* glykolytischer Abbau *m*, Glycolyse *f*, Embden-Meyerhof-Parnas-Weg *m*, Embden-Meyerhof-Parnas-Abbauweg *m*, EMP-Weg *m*, FDP-Weg *m*
glyconic acid Aldohexonsäure *f* *(allg. für ein Aldose-Zuckerderivat der Formel $HOCH_2(CHOH)_4COOH$)*
glycoprotein Glykoprotein *n*, Glykoproteid *n*, Eiweißzucker *m*
glycosamine Aminozucker *m*
glycoside Glykosid *n*
glycosidic bond [linkage] glykosidische Bindung *f* [Verknüpfung *f*], Glykosidbindung *f*
glycosphingolipid Glycosphingolipid *n*
glycosuria Glykosurie *f* *(Ausscheidung von Zuckern im Harn)*; *(i. e. S.)* Glucosurie *f*
glycosylation Glykosylierung *f*, Glykosidierung *f*, *(i. e. S.)* Glucosylierung *f*
glycuronic acid Uronsäure *f* *(allg. für eine Aldehydcarbonsäure der Zuckerreihe)*; *(i. e. S.)* Glucuronsäure *f*
glyoxal Glyoxal *n*, Oxalaldehyd *m*, *Ethandial *n*
glyoxylate cycle *(Bioch)* Glyoxylat-Zyklus *m*, Glyoxylsäurezyklus *m*, Krebs-Kornberg-Zyklus *m*
glyoxylic acid Glyoxylsäure *f*, Oxoessigsäure *f*, Oxoethansäure *f*, Oxalaldehydsäure *f*
glyptal (resin) Glyptal(harz) *n*, Glycerol-Phthalsäure-Harz *n*
GMP = guanosine-5-monophosphate
GnRH *s.* gonadotropin releasing hormone
go *v* verlaufen *(Reaktion)*
go *v* **into solution** in Lösung gehen
go *v* **toward equilibrium** dem Gleichgewicht zustreben
goggles Schutzbrille *f*
Golay cell [detector] *(Spektr)* Golay-Zelle *f*, Golay-Detektor *m*
gold Au Gold *n*
gold amalgam 1. Goldamalgam *n*; 2. *(Min)* Goldamalgam *m*
gold-bearing goldhaltig, goldführend
gold leaf Blattgold *n*
gold monochloride Gold(mono)chlorid *n*, Gold(I)-chlorid *n*
gold nanoparticles *s.* colloidal gold
gold number *(Koll)* Goldzahl *f*
gold sol Goldsol *n*
gold trichloride Goldtrichlorid *n*, Gold(III)-chlorid *n*
Goldschmidt radius goldschmidtscher Radius *m* [Ionenradius *m*], Goldschmidt-Radius *m*

Goldschmidt's process Goldschmidt-Verfahren *n*, Thermitverfahren *n*, Aluminothermie *f*
gonadotropic hormone gonadotropes Hormon *n*, Gonadotropin *n*
gonadotropin releasing hormone *Gonadoliberin *n*
Gooch crucible [filter] Gooch-Tiegel *m*, Filtertiegel *m* nach Gooch
GOT = glutamate-oxaloacetate transaminase
Gouy layer *(phys Ch)* Gouy-Chapman-Schicht *f*, gouy-chapmansche Doppelschicht *f*
GPC *s.* gel permeation chromatography
G6PD = glucose-6-phosphate dehydrogenase
G6PDH = glucose-6-phosphate dehydrogenase
GPT = glutamate-pyruvate transaminase
grade *v* sortieren *(besonders nach Qualitätsmerkmalen)*
grade Grad *m*, Stufe *f*; Sorte *f*, Qualität *f*, Güteklasse *f*, Klasse *f*, Gütegrad *m*; Korngröße *f*, Körnung *f*, *(beim Sieben auch)* Siebfeinheit *f*
graded-density filter Mehrschicht(en)-filter *n*
gradient Gradient *m*, Gefälle *n*
gradient elution Gradient(en)elution *f*
gradient layer Gradientschicht *f*
grading into size Klassieren *n*, Klassierung *f*, Trennen *n* [Trennung *f*] nach Größenklassen [Korngrößen]
graduate *v* 1. graduieren, mit genauer Einteilung versehen, in Grade (ein)teilen, unterteilen; 2. gradieren, verstärken, konzentrieren *(Lösungen durch Abdunsten)*
graduate Messgefäß *n*, Mensur *f*
graduated cylinder Messzylinder *m*, Maßzylinder *m*, Mensur *f*
graduated pipette Messpipette *f*, Teilpipette *f*
graduated to contain auf Einguss geteilt [graduiert]
graduated to deliver auf Ablauf [Auslauf] geteilt [graduiert]
graduation 1. Graduierung *f*, Gradeinteilung *f*; Teilstrich *m*, *(i. w. S.)* Teilungsbild *n (z. B. an Messgefäßen)*; 2. Gradieren *n*, Verstärken *n*, Konzentrieren *n (von Lösungen durch Abdunsten)*
graft *v* (auf)pfropfen, aufpolymerisieren, anpolymerisieren
graft *s.* graft polymer
graft copolymer Pfropfcopolymer(e) *n*, Kammpolymer *n*, Pfropfcopolymerisat *n*
graft polymer Pfropfpolymer(e) *n*, Graft-polymer(e) *n*, Pfropfpolymerisat *n*
grafting Pfropfen *n*, Pfropfung *f*, Aufpfropfen *n*, Aufpolymerisation *f*, Anpolymerisation *f*
grafting reaction Pfropfreaktion *f*
Graham's salt grahamsches Salz *n (vorwiegend Natriumpolyphosphate)*
grain *v* granulieren, körnen, körnig machen
grain 1. Korn *n*, Körnchen; 2. *(Foto)* Korn *n*; 3. *(Zucker)* Korn *n*, Kristallkörnchen *n*; 4. *(Krist)* Korn *n*; 5. *(Landw)* Samenkorn *n*, *(i. e. S.)* Getreidekorn *n*; Getreide *n*; 6. körnige Beschaffenheit *f*; 7. *(Geol)* Feinkorn *n*, Gefüge *n*
grain alcohol Kornbranntwein *m*, Getreidebranntwein *m*, *(i. w. S.)* Ethanol *n*
grain boundary Korngrenze *f*
grain-boundary diffusion Korngrenzendiffusion *f*
grain size Körnung *f*, mittlere Korngröße *f*; Korndurchmesser *m*
grain size distribution Korngrößenverteilung *f*
grain structure Gefüge *n*
grained granuliert, gekörnt; gemasert; genarbt
graininess *(Tech, Foto)* Körnigkeit *f*
grainy körnig, granulös
gram atom Grammatom *n*
gram equivalent Grammäquivalent *n*, Grammval *n*, Val *n*
gram-molecular weight *s.* gram molecule
gram molecule Mol *n*, Grammmolekül *n*
gram range *(Anal)* Grammbereich *m (1 bis 10 g)*
Gram-negative gramnegativ *(Bakterien)*
Gram-positive grampositiv *(Bakterien)*
gran. *s.* granular
granular körnig, granulös
granular activated carbon gekörnte A-Kohle *f*, Kornkohle *f*

granular bed körnige Schüttschicht *f*, Körnerhaufwerk *n*, Körnerschüttung *f*
granular material Granulat *n*
granularity körnige Struktur *f*, Körnigkeit *f*
granulate *v* granulieren, körnen, körnigmachen
granulation 1. Granulieren *n*, Granulierung *f*, Körnigmachen *n*, Körnen *n*, *(i. e. S.)* Granulatformen *n*, Aufbaugranulieren *n*, Krümeln *n*, Abbaugranulieren *n*, Schneiden *n*; 2. Körnung *f*
granulator Granulierapparat *m*, Granuliermaschine *f*, Granuliereinrichtung *f*, Granulator *m*, Granulatformer *m*
granule Granalie *f*, Körnchen *n*, Korn *n*
granule density Korndichte *f (von Schüttgütern)*
grape sugar Traubenzucker *m*, Glucose *f*, Dextrose *f*
graph 1. Diagramm *n*, Schaubild *n*, grafische Darstellung *f*; 2. Kurve *f*
graphic formula Formelbild *n*, Strukturformel *f*, Konstitutionsformel *f*
graphical presentation [representation] *s.* graph 1.
graphite Graphit *m*
graphite crucible Graphittiegel *m*
graphite electrode Graphitelektrode *f*
graphite fibre Graphitfaser *f*
graphite filament Graphitfaden *m*
graphite furnace *(Spektr)* Graphit(rohr)ofen *m*, *(erhitzte)* Graphitrohrküvette *f*
graphite intercalation compound Graphit-Interkalationsverbindung *f*, Graphit-Einlagerungsverbindung *f*
graphite lattice Graphitgitter *n*
graphite material Graphitwerkstoff *m*, Graphitmaterial *n*
graphite-moderated reactor *(Kern)* graphitmoderierter Reaktor *m*, Graphitreaktor *m*
graphite rod Graphitstab *m*
graphite whisker Graphitwhisker *m*
graphitic graphitisch, Graphit...; graphithaltig
graphitic carbon graphitischer Kohlenstoff *m*
graphitizable carbon graphitierbarer Kohlenstoff *m*
graphitization 1. Graphitieren *n*, Einstäuben *n* [Beschichten *n*] mit Graphit; 2. Graphitieren *n*, Übergang *m* in Graphit *(von Kohlenstoff)*; 3. Graphit(is)ieren *n*, graphitisierendes Glühen *n (von Stahl)*; 4. Graphitisierung *f*, Spongiose *f (eine Korrosionsform von Gusseisen)*
graphitization heat treatment thermische Graphitierung *f*
graphitize *v* 1. graphitieren, mit Graphit einstäuben [beschichten]; 2. graphitieren, in Graphit übergehen *(Kohlenstoff)*; 3. graphitisierend glühen *(Stahl)*; 4. Spongiose erleiden, der Graphitisierung unterliegen *(Gusseisen)*
graphitizing furnace Graphitier(ungs)ofen *n*
GRAS (substance) *(Am: generally recognized as safe) durch eine Expertenkommission für unschädlich erklärter Lebensmittelzusatzstoff*
grate Rost *m*; Gitter *n*
grating 1. Rost *m*; Gitter *n*; 2. *(phys Ch)* Beugungsgitter *n*, Gitter *n*
grating spectrograph Gitterspektrograph *m*
grating spectrum Gitterspektrum *n*
gravel 1. Kies *m*; 2. *(Med)* Harngrieß *m*
gravel filter Kiesfilter *n*
gravimetric gravimetrisch
gravimetric analysis Gewichtsanalyse *f*, gravimetrische Analyse *f*
gravimetric factor Faktor *m (Gewichtsanalyse)*
gravimetry Gravimetrie *f*, Gewichtsanalyse *f*
gravitational field Gravitationsfeld *n*, Schwerefeld *n*
gravitational force [pull] Schwerkraft *f*, Erdanziehungskraft *f*, Gravitationskraft *f*
gravity Gravitation *f*, Schwerkraft *f*, Massenanziehung *f*; Schwere *f*
gravity clarifier *(Hyd)* Klärbecken *n*, Absetzbecken *n*
gravity collector Schwerkraftabscheider *m*
gravity concentration Schwerkraftaufbereitung *f*, Dichtesortierung *f*, Dichteanreicherung *f*
gravity die metallische Dauer(gieß)form *f*, Metallform *f*, Kokille *f*
gravity die casting 1. Dauerformguss *m*, (normales) Kokillengießen *n*, (normaler) Kokillenguss *m*; 2. Dauerformguss-

stück *n*, Kokillengussstück *n*, Kokillengussteil *n*
gravity feed Schwerkraftzuführung *f*, Gefällezuführung *f (von Material)*; Schwerkraftbeschickung *f*, Schwerkraftspeisung *f*, *(mit Flüssigkeiten auch)* Fließspeisung *f (einer Anlage)*
gravity filter hydrostatisches Filter *n*, Schwerkraftfilter *n*
gravity filtration Schwerkraftfiltration *f*
gravity return natürlicher Umlauf *m*, Naturumlauf *m*
gravity sedimentation Schwerkraftsedimentation *f*, Schwerkraftabsetzung *f*
gravity separation Schwerkraftabscheidung *f*, Schwerkrafttrennung *f*, Schweretrennung *f*
gravity settling *s.* 1. gravity sedimentation; 2. gravity separation
gravity thickener Schwerkraft-Eindicker *m*
grease *v* (ein)fetten; *(i. w. S.)* (ein-)schmieren
grease ausgelassenes [ausgeschmolzenes, zerlassenes] tierisches Fett *n*; Schmierfett *n*, Schmiere *f*
grease removal *(Hyd)* Fettabscheidung *f*
grease solvent Fettlösungsmittel *n*, Fettlöser *m*
greasiness Fettigkeit *f*; Schmierigkeit *f*
greasy fettig; schmierig
green 1. Grün *n (Farbempfindung)*; 2. grüner [grünfärbender] Farbstoff *m*, Grün *n*
green biotechnology grüne Biotechnologie *f (Landwirtschaft)*
green compact grünes Teil *n*, Grünling *m (ungesinterter Pressling in der Pulvermetallurgie)*
green electrode grüne Elektrode *f (vor dem Brennen)*
green glass Grünglas *n*
green paste grüne [rohe] Elektroden(stampf)masse *f*, grüne [rohe] Söderberg-Masse *f*
green pellet Grünpellet *n*
green state *(Ker)* grüner [roher] Zustand *m*
greenhouse effect Treibhauseffekt *m*
greenhouse gas Treibhausgas *n*
grey 1. Grau *n (Farbempfindung)*; 2. grauer [graufärbender] Farbstoff *m*, Grau *n*
grey biotechnology graue Biotechnologie *f (Abfallwirtschaft)*
grey body *(phys Ch)* grauer Körper *m* [Strahler *m*], Graukörper *m*
grey iron Grauguss *m*
grey tin graues Zinn *n (eine Modifikation)*
greying *(Text)* Vergrauen *n*
grid Gitter *n*, Rost *m*
Grignard compound Grignard-Verbindung *f*
Grignard reaction *s.* Grignard synthesis
Grignard reagent Grignard-Reagens *n (Lösung eines Organomagnesiumhalogenids)*
Grignard synthesis Grignard-Synthese *f*, Grignard-Reaktion *f*
grignardization Grignardierung *f*
grime Schmutz *m*
grind *v* 1. zerkleinern; (zer)mahlen, vermahlen; schroten *(z. B. Malz, Getreide)*; 2. schleifen *(Metall, Glas, Holz)*; 3. anreiben *(Pigmente)*
grind *v* **finely** fein mahlen
grind *v* **in** einschleifen
grindability 1. Mahlbarkeit *f*, Vermahlbarkeit *f*; 2. Schleifbarkeit *f*
grindable 1. mahlbar; 2. schleifbar
grinder 1. Mühle *f*; 2. Schleifmaschine *f*
grinding aid Mahlhilfsstoff *m*, Mahlhilfe *f*
grinding ball Mahlkugel *f*
grinding machine 1. Mahlmaschine *f*; 2. Schleifmaschine *f*
grinding mill Mühle *f*
grinding roll Mahlwalze *f*
grinding wheel Schleifscheibe *f*
grindstone 1. Mahlstein *m*, Mühlstein *m*; 2. Schleifstein *m*
grit 1. *(Geol)* Grus *m*; 2. *(Bau)* Splitt *m*; Grobsand *m*; 3. *(Lebm)* Grieß *m*; Schrot *n(m)*; 4. Grobstaub *m (in der Atmosphäre)*
grittiness grießartige [körnige] Beschaffenheit *f*
gritty grießig, grießartig, grießförmig, körnig
grizzly (screen) 1. Stangen(sieb)rost *m*, Stab(sieb)rost *m*, Siebrost *m*; 2. Stabrostsiebmaschine *f*, Rostsiebmaschine *f*
groove Nut *f*; Furche *f*, Rinne *f*, Rille *f*
grooved genutet; geriffelt, gerillt
gross composition Bruttozusammensetzung *f*

gross heat of combustion spezifischer Brennwert *m*, *(veraltet:)* Verbrennungswärme *f*, oberer Heizwert *m*, H_o
gross heating value *s.* gross heat of combustion
gross sample *(Anal)* Bruttoprobe *f*, Ausgangsprobe *f*
ground *v* erden *(Geräte, Anlagen)*
ground 1. zerkleinert; gemahlen; 2. geschliffen
ground 1. Grund *m*, Boden *m*; 2. *(Bgb)* Gestein *n*, Gebirge *n*
ground coat Untergrundanstrich *m*, Grund(ier)anstrich *m*, Grundierung *f*
ground electronic state Elektronengrundzustand *m*, elektronischer Grundzustand *m*
ground-glass flask Schliffkolben *m*
ground-glass joint Schliffverbindung *f*
ground-glass stopper Schliffstopfen *m*, eingeschliffener Glasstopfen *m*
ground-in eingeschliffen
ground-level concentration bodennahe Konzentration *f (insbesondere von Schadstoffen)*
ground-mass Grundmasse *f*, Einbettungsmasse *f*, Matrix *f*
ground state Grundzustand *m*, Normalzustand *m*, Grundniveau *n*, Grundterm *m (eines Atoms)*
ground-state (electronic) configuration Elektronenkonfiguration *f* im Grundzustand
ground substance Grundsubstanz *f*
ground surface Schlifffläche *f*
ground term *s.* ground state
ground-water pollutants Schmutzstoffe *mpl* [Verunreinigungen *fpl*] im Grundwasser
ground-water pollution Grundwasserverschmutzung *f*, Grundwasserverunreinigung *f*
groundnut oil Erdnussöl *n*, Arachisöl *n*
groundnut protein Erdnussprotein *n*, Erdnusseiweiß *n*
groundwood (pulp) *(Pap)* (mechanischer) Holzschliff *m*, (mechanischer) Holzstoff *m*, Holzmasse *f*, Schleifmasse *f*
group *v* in eine Gruppe zusammenfassen, in Gruppen anordnen, gruppieren; zu Gruppen zusammenbauen
group Gruppe *f (z. B. des Periodensystems)*; Gruppe *f*, Rest *m (eines Moleküls)*
group analysis Gruppenanalyse *f*
group frequency *(Spektr)* Gruppenfrequenz *f*, Schlüsselfrequenz *f*
group of compounds Stoffgruppe *f*
group reagent Gruppenreagens *n*
group separation *(Anal)* Gruppentrennung *f*, Trennung *f* in Gruppen
group-specific gruppenspezifisch *(Enzym)*
group transfer *(Bioch, org Ch)* Gruppenübertragung *f*, Gruppentransfer *m*
group transfer reaction *(org Ch)* Gruppentransferreaktion *f*
grow *v* 1. *(Bio)* wachsen; 2. zunehmen, ansteigen, sich vergrößern, (an)wachsen; 3. züchten *(Kristalle)*
growth 1. *(Bio)* Wachstum *n*, Wuchs *m*; 2. Wachstum *n*, Wachsen *n (einer Polymerkette)*; 3. Zunahme *f*, Anstieg *m*, Vergrößerung *f*
growth hormone 1. *(Med)* Wachstumshormon *n*, somatotropes Hormon *n*, STH, Somatotropin *n*; 2. *(Bot)* Wuchshormon *n*
growth inhibitor Wachstumsinhibitor *m*, Hemmstoff *m*
growth mechanism Wachstumsmechanismus *m*
growth medium Nährmedium *n*, Nährsubstrat *n*, Wachstumsmedium *n*
growth phase *(Biot)* Wachstumsphase *f (Mikroorganismenwachstum)*
growth rate 1. *(Krist, Kst)* Wachstumsgeschwindigkeit *f*; 2. *(Biot)* Wachstumsrate *f (von Mikroorganismen)*
growth reaction Wachstumsreaktion *f (bei der Polymerisation)*
growth regulator Wachstumsregulator *m*, Wuchsstoffmittel *n*, *(mit hemmender Wirkung auch)* Wuchsstoffherbizid *n*
growth step Wachstumsschritt *m (bei der Polymerisation)*
growth substance Wuchsstoff *m*, wachstumsfördernder Stoff *m*
g.r.p. *s.* glass-fibre reinforced plastic
G.R.P. *s.* glass-fibre reinforced plastic
Grubbs catalysts *(org Ch)* Grubbs-Katalysatoren *mpl (Metathese)*
G.S.C. *s.* gas-solid chromatography
GTP *s.* guanosine triphosphate

guaiacol Guajacol *n*, 2-Methoxyphenol *n*, Brenzcatechinmonomethylether *m*
guaiol Guajol *n (Sesquiterpen-Alkohol)*
guanidine Guanidin *n*, Iminoharnstoff *m*, *Aminomethanamidin *n*
guanine Guanin *n*, 2-Amino-6-hydroxypurin *n*, 2-Aminohypoxanthin *n*
guanosine Guanosin *n (ein Nucleosid)*
guanosine diphosphate Guanosindiphosphat *n*, GDP
guanosine triphosphate Guanosintriphosphat *n*, GTP
guanylic acid *(Bioch)* Guanylsäure *f*
guest Gast *m (bei Einschlussverbindungen)*; Gastelement *n*
guest atom Gastatom *n*
guest molecule Gastmolekül *n*
Guldberg and Waage law Massenwirkungsgesetz *n*, MWG
gum 1. Pflanzengummi *n*, Gummi *n*; 2. *s.* caoutchouc; 3. *s.* resin; 4. *(Erdöl)* Gum *m*, Harz *n*
gum arabic Arabisches Gummi *n*, Gummiarabikum *n*, Gummi arabicum *n*, Akaziengummi *n (von Acacia-Arten, besonders von A. senegal (L.) Willd.)*
gum benzoin Benzoe *f*, Benzoeharz *n (von Styrax-Arten)*
gum butea Buteakino *n*, Bengalkino *n*, Bengalisches Kino *n (von Butea superba Roxb.)*
gum content *(Erdöl)* Gumgehalt *m*, Harzgehalt *m*
gum copal Kopal *m (Sammelname für hochschmelzende pflanzliche Harze)*
gum formation *(Erdöl)* Gumbildung *f*, Harzbildung *f*, Verharzung *f*
gum guaiac Guajakharz *n (von Guajacum officinale L. und G. sanctum L.)*
gum inhibitor *(Erdöl)* Oxidationsinhibitor *m*, Oxidationsverhinderer *m*, Antioxidans *n*
gum mastic Mastix *m*, Mastixharz *n (von Pistacia lentiscus L.)*
gum resin Gummiharz *n*
gum rosin Balsamkolophonium *n*, Balsamharz *n*, Harz *n* aus Rohterpentin
gum tragacanth Tragantgummi *n*, Tragant *m (von Astragalus-Arten)*
guncotton Schieß(baum)wolle *f*
gunpowder Schwarzpulver *n*, Sprengpulver *n*, *(i. w. S.)* Schießpulver *n*
gutta-percha Guttapercha *f(n)*
Gutzeit test gutzeitsche Probe *f* [Arsenprobe *f*], Arsenprobe *f* nach Gutzeit
Guyard reaction Guyard-Reaktion *f (Komproportionierung von MnO_4^- und Mn^{++} zu MnO_2)*
gypsum *(Min)* Gips(stein) *m*, Selenit *m* *($CaSO_4 \cdot 2 H_2O$)*
gypsum cement gebrannter Gips *m (im Wesentlichen $CaSO_4 \cdot 0,5 H_2O$)*
gyromagnetic ratio *(Spektr)* gyromagnetisches Verhältnis *n*

H

H-bonding *(phys Ch)* Wasserstoff(brücken)bindung *f*
H-form exchanger *(Hyd)* H-Austauscher *m*, H_2-Austauscher *m*, Ionenaustauscher *m* in der Wasserstoffform [H-Form, H_2-Form], Wasserstoff(ionen)-austauscher *m*
Haber(-Bosch) process Haber-Bosch-Verfahren *n (Ammoniaksynthese)*
habit Habitus *m*, Kristallhabitus *m*
habit-forming *(Pharm)* suchterzeugend
hadron *(Kern)* Hadron *n*
haem Häm *n*, Eisenporphyrin *n* IX, Ferroporphyrin *n* IX, Eisen(II)-protoporphyrin *n*
haem enzyme Hämenzym *n*, Eisenporphyrinenzym *n*
haem protein Hämoprotein *n*
haematin Hämatin *n (oxidierte Form des Häms)*
haematite *(Min)* Hämatit *m*, Haematit *m* *(Eisen(III)-oxid)*
haemin Hämin *n*, Eisen(III)-protoporphyrin *n*, *(i. e. S.)* Hämin(chlorid) *n*, Chlorhämin *n*, Protohämin *n*
haemochrome Hämochrom(ogen) *n*
haemochromogen Hämochrom(ogen) *n*
haemoglobin Hämoglobin *n*, Hb, roter Blutfarbstoff *m*
haemolysin Hämolysin *n*
haemolysis Hämolyse *f*
haemolyze *v* hämolysieren; sich auflösen, zerfallen *(rote Blutkörperchen)*
haemoprotein Hämoprotein *n*
haemostatic Hämostatikum *n*, blutstillendes Mittel *n*
hafnium Hf Hafnium *n*
hafnium carbide Hafniumcarbid *n*

hafnium oxide Hafniumdioxid *n*, Hafnium(IV)-oxid *n*
hair dye Haarfärbemittel *n*, Haarfärbepräparat *n*, Haarfarbe *f*
hair shampoo Haarwaschmittel *n*, Haarreinigungsmittel *n*, Kopfwaschmittel *n*, Haarwäsche *f*
hairpin coil [tube] Haarnadelrohr *n*
half-cell Halbzelle *f*, Halbelement *n*
half-cell (electrode) potential Halbzellenpotenzial *n*
half-ester Halbester *m*
half-intensity breadth [width] *s.* half-width
half-life (period) 1. *(Kern, phys Ch)* Halbwertszeit *f*; 2. *(Biot)* Halblebenszeit *f*, Halbwertszeit *f (der Enzymaktivität)*
half-mask facepiece Halbmaske *f (Atemschutz)*
half-reaction Teilreaktion *f*
half silk *(Text)* Halbseide *f*
half-time Halbwertszeit *f (einer Reaktion)*
half-wave potential Halbstufenpotenzial *n*, Halbwellenpotenzial *n*
half-width (at half maximum) *(Anal)* Halbwertsbreite *f*, Halbhöhenbreite *f*
halide Halogenid *n*
hallucinogen *(Tox)* Halluzinogen *n*, Psychodysleptikum *n*, Phantastikum *n*, Psychotikum *n*, Psychotomimetikum *n*, Psychosomimetikum *n*
hallucinogenic *(Tox)* halluzinogen
halo... Halogen...
haloacid Halogencarbonsäure *f*
haloacylation Halogenacylierung *f*
haloalkane Halogenalkan *n*, Alkylhalogenid *n*
haloalkylation Halogenalkylation *f*
haloamination Halogenaminierung *f*
haloarylation Halogenarylierung *f*
halocarbon Halogenkohlenwasserstoff *m*
halodecarboxylation Halodecarboxylierung *f*, halogenierende Decarboxylierung *f*
haloethane Halogenethan *n*, Ethylhalogenid *n*
haloform Halogenoform *n*, Trihalogenmethan *n*
haloform reaction *(org Ch)* Haloformreaktion *f*, Einhorn-Reaktion *f*
halogen Halogen *n*
halogen compound Halogenverbindung *f*
halogen derivative Halogenderivat *n*, Halogenabkömmling *m*
halogen-substituted halogensubstituiert
halogenate *v* halogenieren
halogenated hydrocarbon Halogenkohlenwasserstoff *m*
halogenated organic compounds organische Halogenverbindungen *fpl*, Organohalogenverbindungen *fpl*
α-halogenated α-halogeniert, in α-Stellung halogeniert
halogenation Halogenierung *f*
halogenide Halogenid *n*
halogenolysis Halogenolyse *f*
halohydrin Halogenhydrin *n (Glycerolderivat)*
haloketone Halogenketon *n*
halomethane Halogenmethan *n*, Methylhalogenid *n*
halonium ion Halonium-Ion *n*
haloorganic compounds *s.* halogenated organic compounds
haloorganics *s.* halogenated organic compounds
halt (eutektischer) Haltepunkt *m*
Hamiltonian (operator) *(phys Ch)* Hamilton-Operator *m*
hammer mill Hammermühle *f*
hammer scale Hammerschlag *m (Eisen(II,III)-oxid)*
Hammett equation [ρσ-equation] *(org Ch)* Hammett-Gleichung *f*
Hammond postulate [principle] Hammond-Postulat *n*, Hammond-Prinzip *n (Reaktionskinetik)*
hand-operated handbetrieben, Hand...
hand operation Handbetrieb *m*
handblown glass mundgeblasenes Glas *n*, Mundblaseglas *n*
handle *v* 1. handhaben, gebrauchen, bedienen; umgehen mit *(z. B. Gasen)*; sich handhaben lassen, funktionieren; 2. behandeln *(z. B. Abwasser, Schlamm)*; 3. (be)fördern, transportieren, weiterleiten *(Stoffe)*; (über)leiten, überführen *(Abgase)*; 4. durchsetzen *(eine Stoffmenge)*
handling 1. Handhabung *f*, Gebrauch *m*, Bedienung *f*; Umgang *m (z. B. mit Gasen)*; 2. Behandlung *f (z. B. von

Abwasser, Schlamm); 3. Beförderung *f*, Förderung *f*, Transport *m*, Weiterleitung *f (von Stoffen)*; Leitung *f*, Überleitung *f*, Überführung *f (von Abgasen)*
handwheel Handrad *n*
hang *v* **up** sich ansetzen, hängen bleiben
hank *(Text)* Strang *m*, Strähn *m*
hank dyeing Strangfärben *n*, Färben *n* im Strang, Färben *n* in Strangform
Hantzsch pyridine synthesis hantzschsche Pyridinsynthese *f*
Hantzsch-Widman name *(Nom)* Hantzsch-Widman-Name *m*
hapten *(Bioch)* Hapten *n*, Halbantigen *n*
haptophore haptophore Gruppe *f*, Haftgruppe *f (eines Toxinmoleküls nach Ehrlichs Seitenkettentheorie)*
haptophoric group haptophore Gruppe *f*, Haftgruppe *f (eines Toxinmoleküls nach Ehrlichs Seitenkettentheorie)*
hard 1. hart *(1. feste Stoffe; 2. Wasser)*; 2. hart, durchdringend *(Strahlung)*; 3. *(Lebm)* sauer, herb, streng *(Geschmack)*; 4. *(Am)* hochprozentig, stark *(Getränke)*
hard acid harte Säure *f (nach der Einteilung von Pearson)*
hard base harte Base *f (nach der Einteilung von Pearson)*
hard carbide *(Met)* hartes Carbid *n* [Metallkarbid *n*], Hartcarbid *n*
hard fat Hartfett *n*
hard glass Hartglas *n*
hard material Hartstoff *m*, harter Stoff *m*
hard metal (alloy) Hartmetall *n*, *(i. e. S.)* Sinterhartmetall *n*, Carbidhartmetall *n*, Sinterkarbid(metall) *n*
hard metal carbide *s.* hard carbide
hard porcelain Hartporzellan *n*
hard radiation harte [energiereiche] Strahlung *f*
hard resin hartes Naturharz *n*, Hartharz *n*
hard rubber Hartgummi *m*, Hartkautschuk *m*
hard settling hartes Absetzen *n*, Bildung *f* eines harten [festen] Bodensatzes
hard water hartes Wasser *n*, Hartwasser *n*
hard wax Hartwachs *n*
hardboard Hartfaserplatte *f*
harden *v* 1. härten, hart machen; 2. *(Anstr, Kst)* aushärten; 3. *(Hyd)* aufhärten *(Wasser)*; (er)härten, hart [fest] werden; 4. *(Anstr, Kst)* aushärten; abbinden *(Mörtel)*
hardenable härtbar
hardened case *(Met)* gehärtete Randschicht *f* [Randzone *f*, Einsatzschicht *f*], Härteschicht *f*
hardened fat gehärtetes [hydriertes] Fett *n*, Hartfett *n*
hardener Härter *m*, Härtungsmittel *n*, Härtemittel *n*, Härtezusatz *m*; Härtebad *n*
hardening 1. Härten *n*, Härtung *f*; 2. *(Anstr, Kst)* Aushärten *n*, Aushärtung *f*; 3. *(Hyd)* Aufhärtung *f (des Wassers)*; Härten *n*, Härtung *f*, Erhärten *n*, Erhärtung *f*, Hartwerden *n*, Festwerden *n*; 4. *(Anstr, Kst)* Aushärten *n*, Aushärtung *f*; Abbinden *n (von Mörtel)*
hardening agent Härter *m*, Härtungsmittel *n*, Härtemittel *n*, Härtezusatz *m*
hardness 1. Härte *f (1. fester Stoffe; 2. des Wassers)*; 2. Härte *f*, Stärke *f (von Strahlung)*; 3. *(Lebm)* Herbe *f*, Herbheit *f*, Herbigkeit *f (im Geschmack)* • **add hardness to the water** das Wasser aufhärten • **remove hardness from water** Wasser enthärten
hardness constituent [element] *(Hyd)* Härtebildner *m*
hardness-forming *(Hyd)* härtebildend
hardness level *(Hyd)* Härtegehalt *m*
hardness removal *(Hyd)* Enthärtung *f*
hardness scale Härteabscheidungen *fpl*, Inkrustationen *fpl (durch Härtebildner)*; Wasserstein *m*, *(i. e. S.)* Kesselstein *m*
hardness testing Härteprüfung *f*, Härtebestimmung *f*
harmal(a) alkaloid Harmalaalkaloid *n*
harmful schädlich; gefährlich
harmful material Schadstoff *m*
harmfulness Schädlichkeit *f*; Gefährlichkeit *f*
harmless unschädlich; ungefährlich
harmlessness Unschädlichkeit *f*; Ungefährlichkeit *f*
harmonic vibration harmonische Schwingung *f*
harsh scharf, herb, ätzend, beißend
harshness Schärfe *f*, Herbe *f*, Herbheit *f*, Herbigkeit *f*
hartshorn Hirschhornsalz *n*

harvest *v (Biot)* ernten *(Fermentationsprodukte)*
harvest *(Biot)* Ernte *f (des Fermentationsprodukts)*
haschisch [**hasheesh**, **hashish**] Haschisch *m(n)*, Marihuana *n (von Cannabis indica Lam.)*
hassium Hs Hassium *n (Element 108)*
Hatch-Slack (CO_2 incorporation) pathway *(Bioch)* Hatch-Slack-Kortschak-Zyklus *m*, C_4-Säurenzyklus *m*, C_4-Dicarbonsäure-Weg *m*, C_4-Zyklus *m*
Haworth formula [projection formula] sterische Projektionsformel *f* von Haworth, Haworth-Formel *f (für Zucker)*
hazard of explosion Explosionsgefahr *f*
hazardous material gefährlicher Stoff *m*
hazardous substance Gefahrstoff *m*
hazardous to human health gesundheitsgefährdend, gesundheitsschädlich, gesundheitsschädigend
Hb *s.* haemoglobin
HC *s.* hydrocracking
HCB *s.* hexachlorobenzene
hcp *s.* hexagonal close-packed
HD ... *siehe: high-density ...*
HDPE *s.* high-density polyethylene
HDS *s.* hydrodesulphurization
H.E. *s.* high explosive
head 1. (statische) Druckhöhe *f*, *(i. w. S.)* Druck *m*; 2. *(Tech)* Haube *f*, Helm *m*; 3. Spritzkopf *m*, Kopf *m (einer Spritzmaschine)*
head of water 1. *(Hyd)* Wasserüberstauhöhe *f*, Stauhöhe *f*, Wasserüberstau *m*, Überstau *m*, Stau *m (über dem Filtermittel)*; 2. *(Hyd)* Strömungsdruck *m*
head-to-head polymerization Kopf-Kopf-Polymerisation *f*
head-to-tail polymerization Kopf-Schwanz-Polymerisation *f*
heads *(Dest)* Vorlauf *m*
headspace Dampfraum *m*, Kopfraum *m*, Headspace *m (dampfgefüllter Raum über der flüssigen Phase in geschlossenen Gefäßen)*
headspace (gas) analysis Dampfraumanalyse *f*, Kopfraumanalyse *f*, Headspace-Analyse *f*, Head-Space-Technik *f*, Dampfraumtechnik *f (Gaschromatographie)*
health hazard Gesundheitsgefährdung *f*
heart poison Herzgift *n*
hearth Herd *m*, Arbeitsherd *m (eines Flammofens)*; Herd *m*, Etage *f*, Arbeitsetage *f (eines Etagenröstofens)*; Gestell *n (eines Hochofens)*
hearth furnace Herdofen *m*
heat *v* erwärmen, erhitzen; (be)heizen *(Räume, Kessel)*; sich erwärmen [erhitzen], warm [heiß] werden
heat *v* **gently [slightly]** gelinde erhitzen, gelinde [mäßig] erwärmen, vorsichtig erwärmen
heat *v* **to boiling** zum Sieden [Kochen] bringen
heat *v* **to redness** auf [bis zur] Rotglut erhitzen, rotglühen
heat *v* **to the boil** *s.* heat to boiling
heat-treat *v* wärmebehandeln, warmbehandeln, hitzebehandeln
heat *v* **up** aufheizen, hochheizen
heat 1. Wärme *f (Energieform)*; 2. Hitze *f*, Wärme *f (Empfindung)*; 3. Dauer *f* der Wärmeeinwirkung • **in intense heat** bei hoher Temperatur
heat balance Wärmebilanz *f*, Wärmehaushalt *m*
heat capacity Wärmekapazität *f*
heat carrier Wärme(über)träger *m*, Wärmeübertragungsmittel *n*
heat change Wärmeänderung *f*, *(i. e. S.)* Wärmetönung *f*
heat conduction Wärmeleitung *f*
heat conductivity Wärmeleitfähigkeit *f*, (spezifisches) Wärmeleitvermögen *n*, *(quantitativ auch)* Wärmeleitzahl *f*
heat content Enthalpie *f*, *(veraltet:)* Wärmeinhalt *m*, gibbssche Wärmefunktion *f*
heat damage Hitzebeschädigung *f*
heat-denaturable hitzedenaturierbar
heat denaturation Hitzedenaturierung *f*
heat energy Wärmeenergie *f*
heat exchange Wärmeaustausch *m*, Wärmeübertragung *f*, Wärmetransport *m*
heat exchange coefficient Wärmeaustauschkoeffizient *m*
heat exchange medium Wärmeübertragungsmittel *n*, Wärmeübertragungsmedium *n*, Wärme(über)träger *m*
heat exchanger Wärme(aus)tauscher *m*,

Wärmeaustauschapparat *m*, Wärmeübertrager *m*
heat flow Wärmefluss *m*; *(quantitativ)* Wärmestrom *m*, Wärmeleistung *f*
heat flux Wärmestromdichte *f*, Wärmestromintensität *f*, *(bei Dampfkesseln:)* Heizflächenbelastung *f*
heat generation Wärmeerzeugung *f*; Wärmeentwicklung *f*
heat-giving wärmeliefernd
heat inactivation thermische Inaktivierung *f*, Hitzeinaktivierung *f*
heat input 1. Wärmezufuhr *f*; 2. *(Hyd)* Wärmeeintrag *m (in den Vorfluter)*
heat-insulating material Wärmedämmstoff *m*
heat insulation Wärmedämmung *f*
heat-labile thermisch labil [instabil], thermolabil, wärmeunbeständig, hitzelabil
heat lability Thermolabilität *f*, Wärmeunbeständigkeit *f*, Hitzelabilität *f*
heat loss Wärmeverlust *m*
heat of adsorption Adsorptionswärme *f*
heat of combustion 1. Verbrennungswärme *f*; 2. *(phys Ch)* stoffmengenbezogener [molarer] Heizwert *m*
heat of combustion per mole stoffmengenbezogener [molarer] Brennwert *m*, *(veraltet:)* molare Verbrennungswärme *f*
heat of condensation Kondensationswärme *f*
heat of formation Bildungswärme *f*, *(i. e. S.)* molare Bildungswärme *f*
heat of fusion Schmelzwärme *f*
heat of reaction Reaktionswärme *f*
heat of reaction at constant pressure Reaktionsenthalpie *f*
heat of reaction at constant volume Reaktionsenergie *f*
heat of solution Lösungswärme *f*, Lösewärme *f*
heat of transformation [transition] Umwandlungswärme *f*
heat of vaporization Verdampfungswärme *f*; Verdunstungswärme *f*
heat processing *s.* heat treatment
heat pump Wärmepumpe *f*
heat radiation Wärmestrahlung *f*, thermische Strahlung *f*
heat recovery Wärmerückgewinnung *f*
heat release Wärmefreisetzung *f*
heat resistance Wärmebeständigkeit *f*, Wärmefestigkeit *f*, Wärmestabilität *f*, Hitzebeständigkeit *f*, thermische Beständigkeit *f*, Temperaturbeständigkeit *f (von Werkstoffen)*; Warmfestigkeit *f (von Stahl bis 580 °C)*; Hitzebeständigkeit *f (von Stahl über 600 °C)*; Hitzeresistenz *f*, Thermoresistenz *f*, Hitzebeständigkeit *f (von Mikroorganismen)*
heat-resistant wärmebeständig, wärmefest, wärmestabil, hitzebeständig, thermisch beständig, temperaturbeständig *(Werkstoffe)*; warmfest *(Stahl bis 580 °C)*; hitzebeständig *(Stahl über 600 °C)*; hitzeresistent, thermoresistent, hitzebeständig *(Mikroorganismen)*
heat-resisting wärmebeständig, wärmefest, wärmestabil, hitzebeständig, thermisch beständig, temperaturbeständig *(Werkstoffe)*; warmfest *(Stahl bis 580 °C)*; hitzebeständig *(Stahl über 600 °C)*; hitzeresistent, thermoresistent, hitzebeständig *(Mikroorganismen)*
heat-sensitive wärmeempfindlich, hitzeempfindlich
heat sensitivity Wärmeempfindlichkeit *f*, Hitzeempfindlichkeit *f*
heat sink 1. *(Tech)* Kühlblech *n*; 2. *(Tech)* Kühlkörper *m*; 3. *(Tech)* Lamellenkühler *m*
heat source Wärmequelle *f*
heat-stable *s.* heat-resistant
heat sterilization Hitzesterilisation *f*, thermische Sterilisation *f*
heat theorem Wärmetheorem *n*
heat tonality Wärmetönung *f*
heat transfer Wärmeübertragung *f*, Wärmeaustausch *m*, Wärmetransport *m*; Wärmeübergang *m*
heat-transfer coefficient Wärmeübergangskoeffizient *m*
heat-transfer fluid [medium] Wärmeübertragungsmittel *n*, Wärmeübertragungsmedium *n*, Wärme(über)träger *m*
heat transmission 1. Wärmeübertragung *f*, Wärmeaustausch *m*, Wärmetransport *m*; 2. *(quantitativ:)* Wärmestrom *m*
heat treatment Wärmebehandlung *f*, Warmbehandlung *f*, Hitzebehandlung *f*, thermische Behandlung *f*
heater Heizelement *n*, Heizkörper *m*;

Heizeinrichtung *f*, Heizvorrichtung *f*, Heizapparat *m*, Heizgerät *n*, Heizer *m*, Erhitzer *m*
heating bath Heizbad *n*
heating coil Heizschlange *f*
heating element Heizkörper *m*, Heizelement *n*
heating funnel Heißwassertrichter *m*, Warmwassertrichter *m*
heating jacket [mantle] Heizmantel *m*
heating oil Heizöl *n*
heating rate Aufheiz(ungs)geschwindigkeit *f*, Erwärmungsgeschwindigkeit *f*, Erhitzungsgeschwindigkeit *f*
heating steam Heizdampf *m*
heating-up Aufheizen *n*
heating value *s.* heat of combustion
heavier-than-heavy component am schwersten siedende Komponente *f* *(eines Mehrstoffgemischs)*
heavily loaded stark belastet, hochbelastet
heavily polluted water stark verschmutztes [verunreinigtes, belastetes] Wasser *n*
heavy 1. schwer; 2. zäh(flüssig), schwerflüssig; steif *(z. B. Paste)*; hart, fest *(z. B. Teig)*; 3. *(Dest)* hochsiedend, höhersiedend, schwer(er) flüchtig *(Fraktion)*
heavy ceramics Grobkeramik *f*
heavy chemical porcelain chemisch-technisches Porzellan *n*
heavy chemicals Schwerchemikalien *fpl* *(in großem Umfang hergestellte Säuren, Salze und Alkalien)*
heavy ends *(Dest)* Nachlauf *m*; Siedeschwanz *m (bei Siedeanalysen)*
heavy gasoline Schwerbenzin *n*, schweres Benzin *n*
heavy hydrogen ^2_1H, D schwerer Wasserstoff *m*, Deuterium *n*
heavy-ion linear accelerator Linearbeschleuniger *m* für Schwerionen
heavy isotope schweres Isotop *n*
heavy-medium separation Schwerflüssigkeitsaufbereitung *f*, Schwerflüssigkeitssortieren *n*, Schwimm-(und-)Sink-Aufbereitung *f*, *(i. e. S.)* Trennung *f* in Schwertrüben, Schwertrübeaufbereitung *f*, Schwertrübescheidung *f*
heavy metal Schwermetall *n*
heavy-metal contamination *(Hyd)* Belastung *f* durch Schwermetalle
heavy-metal poisoning Schwermetallvergiftung *f*
heavy-metal salt Schwermetallsalz *n*
heavy mineral schweres Mineral *n*, Schwermineral *n*
heavy naphtha Schwerbenzin *n*, schweres Benzin *n*, Naphtha *n(f)*
heavy oil Schweröl *n*, schweres Öl *n*
heavy spar *(Min)* Schwerspat *m*, Baryt *m* *(Bariumsulfat)*
heavy water D_2O schweres Wasser *n*, Deuteriumoxid *n*
HEED *s.* high-energy electron diffraction
HEETP *s.* height equivalent to an effective theoretical plate
height equivalent to a theoretical plate [stage] 1. Höhe *f* einer theoretischen Trennstufe, theoretische Trennstufenhöhe *f*; 2. *(Dest)* äquivalente Füllkörperhöhe *f*; 3. *(Chromat)* Schichtlänge *f* eines theoretischen Bodens
height equivalent to an effective theoretical plate *(Chromat)* effektive Trennstufenhöhe *f* [Bodenhöhe *f*]
height of one transfer unit *(Dest)* Höhe *f* einer Übertragungseinheit, HTU
Heisenberg uncertainty relation (heisenbergsche) Unbestimmtheitsrelation *f*, Unschärferelation *f*, Ungenauigkeitsrelation *f*, Unbestimmtheitsbeziehung *f*, Unschärfebeziehung *f*, Ungenauigkeitsbeziehung *f*
Heitler-London-Slater-Pauling method Heitler-London-Slater-Pauling-Methode *f*, HLSP-Methode *f*, Valenzbindungsmethode *f*, Valenzstrukturmethode *f*, VB-Methode *f*, Spinmethode *f*, Elektronenpaarmethode *f*
helical helikal, schraubenförmig, helixförmig; Helix...
helical conformation Helixkonformation *f*
helical conveyor Schneckenförderer *m*
helical structure Helixstruktur *f*, Schraubenstruktur *f (von Makromolekülen)*
helicity 1. Helizität *f (von masselosen Teilchen in der Quantenmechanik)*; 2. Helizität *f (von Molekülen in der Stereochemie)*
helium He Helium *n*

helium-cadmium laser Helium-Cadmium-Laser *m*, He-Cd$^+$-Laser *m*
helium ionization detector Helium(ionisations)detektor *m*, HID
helium nucleus Heliumkern *m*, α-Teilchen *n*
helix 1. Helix *f (eines Makromoleküls)*; 2. Wendel *f (Füllkörper)*
helix-breaking helixbrechend
helix-coil transition *(Bioch)* Helix-Knäuel-Übergang *m*, Helix-Knäuel-Umwandlung *f*
helix-forming helixbildend
helix sense Gängigkeit *f*
Hell-Volhard-Zelinsky reaction Hell-Volhard-Zelinsky-Reaktion *f (α-Halogenierung aliphatischer Carbonsäuren)*
Helmholtz double layer *(phys Ch)* helmholtzsche Doppelschicht *f*
Helmholtz free energy freie Energie *f*, F
HEMA *s.* 2-hydroxyethyl methacrylate
hematite *s.* haematite
heme *s.* haem
hemiacetal Halbacetal *n*, Hemiacetal *n*
hemiaminal Halbaminal *n*, α-Aminoalkohol *m*
hemicellulose Hemicellulose *f*
hemicolloid Hemikolloid *n*
hemicrystalline halbkristallin
hemiformal $HOCH_2OR$ Halbformal *n (Halbacetal des Formaldehyds)*
hemiketal Hemiketal *n (Halbacetal eines Ketons)*
hemin *s.* haemin
hemiterpene Hemiterpen *n*, *(i. e. S.)* Isopren *n*
hemo... *siehe: haemo...*
hemp 1. Hanf *m*, Cannabis L.; 2. Hanffaser *f*; 3. Haschisch *m(n)*, Marihuana *n (von Cannabis indica Lam.)*
Hempel gas pipette hempelsche Gaspipette *f*, Hempel-Pipette *f*
Henry's law henrysches Gesetz *n* [Absorptionsgesetz *n*]
Henry's law constant Henry-Konstante *f*, Gas-Löslichkeitskonstante *f*
hepar test Heparprobe *f (zum Schwefelnachweis)*
hepatotoxic hepatotoxisch, leberschädigend
heptachlor Heptachlor *n*, Heptachlorendomethylentetrahydroinden *n (Insektizid)*
***n*-heptadecanoic acid** *Heptadecansäure *f*, Margarinsäure *f*
heptane Heptan *n*
1,7-heptanedicarboxylic acid *1,7-Heptandicarbonsäure *f*, *Nonandisäure *f*, Azelainsäure *f*
heptanedioic acid *Heptandisäure *f*, Pimelinsäure *f*
heptanoic acid Heptansäure *f*
1-heptanol Heptan-1-ol *n*
heptavalent siebenwertig, heptavalent
heptose Heptose *f (Monosaccharid mit sieben Kohlenstoffatomen)*
heptoxide Heptoxid *n*
***n*-heptyl alcohol** Heptylalkohol *m*, *Heptan-1-ol *n*
heptylic aldehyde Heptylaldehyd *m*, *(i. e. S.)* *Heptanal *n*
1-heptyne Hept-1-in *n*
herb Kraut *n*
herbicidal herbizid, Herbizid..., pflanzentötend, *(i. e. S.)* unkrautbekämpfend, unkrautvernichtend
herbicide Herbizid *n*, pflanzentötendes Mittel *n*, *(i. e. S.)* Unkrautbekämpfungsmittel *n*, Unkrautvertilgungsmittel *n*
hermetic(al) hermetisch (dicht, verschlossen), hermetisch [gas- und wasserdicht, luft- und feuchtigkeitsdicht] abgeschlossen
heroin Heroin *n*, Diacetylmorphin *n*, Diamorphin *n (Rauschgift)*
Herz compound Herz-Verbindung *f*, herzsche Verbindung *f* [Zwischenverbindung *f*] *(ein Benzodithiazol)*
Hess's law (of heat summation) hessscher Satz *m*, hesssches Gesetz *n*, Gesetz *n* der konstanten Wärmesummen
hetarene *s.* heteroarene
hetaryl group *s.* heteroaryl group
hetero-analogous heteroanalog
heteroalkene Heteroalken *n*
heteroallene Heteroallen *n*
heteroarene Heteroaren *n*, Hetaren *n*, Heteroaromat *m*
heteroaryl group Heteroarylgruppe *f*, Hetarylgruppe *f*, Heteroarylrest *m*
heteroatom Heteroatom *n*
heteroconjugated system heterocyclisches konjugiertes System *n*, heteroaromatisches System *n*

heterocumulene Heterokumulen *n*
heterocycle *s.* heterocyclic (compound)
heterocyclic heterocyclisch
heterocyclic (compound) Heterocyclus *m*, heterocyclische Verbindung *f*
heterocyclic ring Heteroring *m*
heterocyclyl group Heterocyclylgruppe *f*, Heterocyclylrest *m*
heterodiene Heterodien *n*
heteroexcimer *s.* exciplex
heterofermentative *(Biot)* heterofermentativ *(Mikroorganismen)*
heterogeneity Heterogenität *f*, Uneinheitlichkeit *f*, Ungleichartigkeit *f*
heterogeneous heterogen, uneinheitlich, ungleichartig (zusammengesetzt)
heterogeneous catalysis heterogene Katalyse *f*, Heterogenkatalyse *f*, Kontaktkatalyse *f*, Oberflächenkatalyse *f*
heterogeneous catalyst heterogener Katalysator *m*, Katalysator *m* der heterogenen Katalyse, Kontakt(katalysator) *m*, Oberflächenkatalysator *m*
heterogeneous equilibrium heterogenes Gleichgewicht *n*
heterogeneous nuclear RNA *(Bioch)* heterogene RNA *f* des Kerns, hnRNA *(Sammelname)*
heterogeneous nucleation *(Krist)* heterogene [simultane, athermische] Keimbildung *f*
heterogeneous reaction heterogene Reaktion *f*
heterolactic fermentation *(Bioch)* heterofermentative Gärung *f*, gemischte Säuregärung *f*
heterolysis *s.* heterolytic cleavage
heterolytic cleavage heterolytische Bindungsspaltung *f*, Heterolyse *f*
heterolytic reaction heterolytische Reaktion *f*
heteronuclear molecule heteronukleares Molekül *n*, Molekül *n* aus verschiedenartigen Atomen
heteropolar bond heteropolare Bindung *f*, Ionenbindung *f*, Ionenbeziehung *f*
heteropolar compound (hetero)polare Verbindung *f*, Ionenverbindung *f*
heteropoly acid Heteropolysäure *f*
heteropolymer Heteropolymer(e) *n*, Heteropolymerisat *n*
heteropolymeric heteropolymer
heteropolymerization Heteropolymerisation *f*
heteropolysaccharide Heteropolysaccharid *n*
heterotacticity Heterotaktizität *f (von Polymeren)*
heterotrophic *(Bioch)* heterotroph
heterotrophs heterotrophe Mikroorganismen *mpl*
heterotrophy *(Bioch)* Heterotrophie *f*, heterotrophe Ernährung *f*
HETP *s.* height equivalent to a theoretical plate
hevea rubber Hevea-Kautschuk *m*
hex. *s.* hexagonal
hexa *s.* hexamethylenetetramine
hexaaqua... Hexaquo...
hexachloride Hexachlorid *n*
hexachlorobenzene Hexachlorbenzen *n*, Perchlorbenzen *n*
hexachloroethane *Hexachlorethan *n*, Perchlorethan *n*
hexachlorophene Hexachlorophen *n*, 3,3′,5,5′,6,6′-Hexachlor-2,2′-dihydroxydiphenylmethan *n*
hexachloroplatinate Hexachloroplatinat(IV) *n*
hexacoordinate(d) sechsfach koordinativ gebunden, mit sechs koordinativen Bindungen; sechsfach koordiniert *(Komplex)*
***n*-hexacosanoic acid** *Hexacosansäure *f*, Cerotinsäure *f*
hexacovalent koordinativ sechswertig
hexacyanoferrate 1. $M^I_4[Fe(CN)_6]$ Hexacyanoferrat(II) *n*; 2. $M^I_3[Fe(CN)_6]$ Hexacyanoferrat(III) *n*
hexad *(Krist)* sechszählig, 6-zählig *(Symmetrieachse)*
hexadecanoic acid *Hexadecansäure *f*, Palmitinsäure *f*
hexadentate sechszähnig, sechszählig *(Komplexchemie)*
2,4-hexadienoic acid *Hexa-2,4-diensäure *f*, Sorbinsäure *f*
hexafluoride Hexafluorid *n*
hexagonal 1. *(Krist)* hexagonal, hex., sechseckig; 2. *(Krist)* sechszählig, 6-zählig *(Symmetrieachse)*
hexagonal close-packed *(Krist)* hexagonal dicht gepackt
hexagonal close(st) packing *(Krist)*

hexagonal(e) dichteste Kugelpackung *f* [Packung *f*]
hexagonal lattice *(Krist)* hexagonales Gitter *n*
hexahelicene Hexahelicen *n*
hexahydrate Hexahydrat *n*
hexamethylenediamine Hexamethylendiamin *n*, *Hexan-1,6-diamin *n*
hexamethylenetetramine Hexamethylentetramin *n*, Hexamin *n*, Methenamin *n*, 1,3,5,7-Tetraazaadamantan *n*
hexamine *s.* hexamethylenetetramine
hexane Hexan *n*
hexanedioic acid *Hexandisäure *f*, *Butan-1,4-dicarbonsäure *f*, Adipinsäure *f*
hexanitrocobaltate 1. $M^I_4[Co(NO_2)_6]$ Hexanitrocobaltat(II) *n*; 2. $M^I_3[Co(NO_2)_6]$ Hexanitrocobaltat(III) *n*
hexanoic acid Hexansäure *f*
1-hexanol *Hexan-1-ol *n*, Hexylalkohol *m*
hexapositive sechsfach positiv (geladen)
hexasubstituted hexasubstituiert
hexatomic sechsatomig
hexavalency Sechswertigkeit *f*, Hexavalenz *f*
hexavalent sechswertig, hexavalent
2-hexene Hex-2-en *n*
hexitol Hexitol *n*, Hexanhexol *n (sechswertiger Alkohol)*
hexose Hexose *f (Monosaccharid mit 6 Kohlenstoffatomen)*
hexuronic acid 1. Hexuronsäure *f*; 2. *s.* ascorbic acid
1-hexyne Hex-1-in *n*
h.f.-... *siehe: high-frequency ...*
HF-... *siehe: high-frequency ...*
HF alkylation *s.* hydrofluoric-acid alkylation
HF laser Wasserstofffluoridlaser *m*, HF-Laser *m*
HIC *s.* hydrophobic interaction chromatography
high-activity waste *(Kern)* hochaktiver [heißer] Abfall *m*
high-added-value product hochveredeltes Produkt *n*
high-alloy steel hochlegierter Stahl *m*
high-analysis hochprozentig
high-boiling hochsiedend, schwersiedend
high-boiling azeotrope Maximumazeotrop *n*, Azeotrop *n* mit Maximumsiedepunkt, Siedepunktsmaximumazeotrop *n*
high-boiling component *(Dest)* Schwerersiedendes *n*, schwerersiedende [höhersiedende] Komponente *f*, schwererflüchtiger Anteil *m*, Phlegma *n*
high-capacity hochkapazitiv, hochleistungsfähig, Hochleistungs...
high-carbon hochgekohlt, hochkohlenstoffhaltig, kohlenstoffreich, mit hohem Kohlenstoffgehalt *(z. B. Stahl)*
high-concentration hochkonzentriert
high-density polyethylene Polyethylen *n* hoher Dichte, Niederdruckpolyethylen *n*, Linearpolyethylen *n*
high dose tolerance Hochdosistoleranz *f (Immunologie)*
high-duty Hochleistungs...
high-efficiency Hochleistungs...
high-energy energiereich, hochenergetisch
high-energy chemistry Hochenergiechemie *f*, Chemie *f* hoher Energien
high-energy electron diffraction Beugung *f* schneller [hochenergetischer] Elektronen, Hochenergieelektronenbeugung *f*
high-energy radiation energiereiche Strahlung *f*, hochenergetische [harte] Strahlung *f*
high explosive brisanter Explosivstoff *m* [Sprengstoff *m*]
high-frequency conductometry Hochfrequenzkonduktometrie *f*
high-frequency heating Hochfrequenz(be)heizung *f*, Hochfrequenzerhitzung *f*, Hochfrequenzerwärmung *f*
high-frequency titration Hochfrequenztitration *f*, oszillometrische Titration *f*
high-fusion(-point) *s.* high-melting(-point)
high-grade ore hochwertiges [reichhaltiges, reiches] Erz *n*
high-impact polystyrene hochschlagfestes Polystyren *n*
high-intensity mixing Schnellmischen *n*, Schnellrühren *n*
high-level active *(Kern)* hochaktiv
high-melting(-point) hochschmelzend, schwerschmelzbar

high-molecular hochmolekular
high-molecular-weight von hoher relativer Molekülmasse
high-performance Hochleistungs...
high-performance liquid affinity chromatography Hochleistungs-Flüssigkeits-Affinitätschromatographie *f*
high-performance liquid chromatography *s.* high-pressure liquid chromatography
high-performance polymer Hochleistungskunststoff *m*
high-performance thin-layer chromatography Hochleistungs-Dünnschichtchromatographie *f*
high polymer Hochpolymer(e) *n*
high-polymeric hochpolymer
high-polymeric chemistry Chemie *f* der Hochpolymeren
high-precision weighing Feinwägung *f*
high-pressure chemistry Hochdruckchemie *f*
high-pressure filtration Hochdruckfiltration *f*
high-pressure gasification *(Carb)* Hochdruckvergasung *f*
high-pressure hydrogenation Hochdruckhydrierung *f*
high-pressure liquid chromatography Hochdruckflüssig(keits)chromatographie *f*, Hochleistungsflüssigchromatographie *f*, schnelle Flüssig(keits)chromatographie *f*
high-pressure modification Hochdruckmodifikation *f*
high-pressure polyethylene Hochdruckpolyethylen *n*, verzweigtes Polyethylen *n*, Polyethylen *n* niedriger Dichte
high-pressure press *(Lab)* Hochdruckpresse *f*
high-pressure steam Hochdruckdampf *m*
high-pressure synthesis Hochdrucksynthese *f*
high-pressure vessel Hochdruckgefäß *n*, Hochdruckbehälter *m*
high-producing strain *(Biot)* Hochleistungsstamm *m (von Mikroorganismen)*
high-purity water hochreines Wasser *n*
high-rate filter Schnellfilter *n*
high-resolution nmr [nuclear magnetic resonance] hochauflösende magnetische Kernresonanz *f*
high-speed centrifuge Ultrazentrifuge *f*, Hochleistungszentrifuge *f*
high-speed drying Schnelltrocknung *f*, Schnelltrocknen *n*
high-spin complex Anlagerungskomplex *m*, Normalkomplex *m*, High-spin-Komplex *m*
high-strength acid hochkonzentrierte [hochprozentige] Säure *f*
high-strength waste (water) *(Hyd)* hochkonzentriertes Abwasser *n*, hochbelastetes [stark verschmutztes, starkes, „dickes"] Abwasser *n*
high-temperature alloy Hochtemperaturlegierung *f*
high-temperature behaviour Hochtemperaturverhalten *n*, Wärmeverhalten *n*, Verhalten *n* bei hohen Temperaturen
high-temperature carbonization Hochtemperaturverkokung *f*, HT-Verkokung *f*, Normalverkokung *f*, *(i. e. S.)* Verkokung *f*, Hochtemperaturentgasung *f*, HT-Entgasung *f*
high-temperature chemistry Hochtemperaturchemie *f*
high-temperature corrosion Hochtemperaturkorrosion *f*
high-temperature furnace Hochtemperaturofen *m*
high-temperature material Hochtemperaturwerkstoff *m*, hochtemperaturbeständiger Werkstoff *m*
high-temperature pyrolysis Hochtemperaturpyrolyse *f*
high-temperature reactor Hochtemperaturreaktor *m*
high-temperature resistance Hochtemperaturbeständigkeit *f*, Beständigkeit *f* [Widerstandsfähigkeit *f*] gegen hohe Temperaturen
high-temperature-resistant hochtemperaturbeständig, beständig [widerstandsfähig] gegen hohe Temperaturen
high-temperature short-time pasteurization *(Lebm)* Kurzzeitpasteurisation *f*, Hochkurzpasteurisation *f*, HTST-Pasteurisation *f*
high-temperature stability *s.* high-temperature resistance

high-temperature treatment Hochtemperaturbehandlung *f*
high throughput Hochdurchsatz *m*
high throughput screening Hochdurchsatztesten *n*
high-vacuum distillation Hochvakuumdestillation *f*
high-vacuum technology Hochvakuumtechnik *f*; Hochvakuumtechnologie *f*
high-viscosity hochviskos, von hoher Viskosität, zähflüssig
high-voltage electrophoresis Hochspannungselektrophorese *f*
high-yield mit hoher Ausbeute, Hochausbeute..., ergiebig
higher-boiling höhersiedend, schwerersiedend
higher heating value *s.* gross heat of combustion
higher hydrocarbon höherer Kohlenwasserstoff *m*
higher-melting höherschmelzend, schwererschmelzbar
higher-membered höhergliedrig
higher-order reaction Reaktion *f* höherer Ordnung
higher-order transition Umwandlung *f* [Phasenumwandlung *f*] höherer Ordnung, Übergang *m* [Phasenübergang *m*] höherer Ordnung
higher-valency höherwertig
higher-valent höherwertig
highest(-energy) occupied molecular orbital höchstes besetztes Molekülorbital *n*
highly active 1. hochaktiv, hochwirksam; 2. *s.* highly radioactive
highly alloyed hochlegiert
highly aromatic hocharomatisch
highly branched stark verzweigt
highly coloured intensiv farbig, farbkräftig, farbstark
highly concentrated hochkonzentriert
highly cross-linked stark vernetzt, hochvernetzt
highly disperse hochdispers
highly loaded stark belastet, hochbelastet
highly ordered structure *(Bioch, Krist)* hochgeordnete Struktur *f*
highly oriented hochorientiert
highly poisonous *s.* highly toxic
highly polymerized hochpolymer
highly pure hochrein
highly radioactive stark radioaktiv, hochaktiv, heiß
highly reactive hochreaktiv, hochreaktionsfähig
highly refractory hochfeuerfest
highly selective hochselektiv
highly sensitive hochempfindlich
highly toxic hochgiftig, hochtoxisch, stark giftig
highly unsaturated hochungesättigt
highly viscous hochviskos, von hoher Viskosität
highly volatile leichtflüchtig, hochflüchtig
Hill reaction *(Bioch)* hillsche Reaktion *f*, Hill-Reaktion *f (lichtabhängige Reduktion bei der Photosynthese)*
hinder *v* behindern
hindered rotation *(org Ch)* Hinderung *f* [Aufhebung *f*] der freien Drehbarkeit
hindrance (sterische) Hinderung *f*
Hinsberg reaction [test] Hinsberg-Reaktion *f (zur Trennung primärer, sekundärer und tertiärer Amine)*
hippuric acid Hippursäure *f*, Benzoylaminoessigsäure *f*
Hirsch funnel Hirsch-Trichter *m*, Trichter *m* nach Hirsch
histamine Histamin *n*, β-Imidazol-4(5)-ethylamin *n*
histidine Histidin *n*, His, α-Amino-β-imidazolylpropionsäure *f*
histochemistry Histochemie *f*
histone Histon *n (ein Protein)*
Hittorf number hittorfsche Überführungszahl *f*
HMDA *s.* hexamethylenediamine
HMO *s.* Hückel molecular orbital
hn RNA = heterogeneous nuclear ribonucleic acid
HOC *s.* halogenated organic compounds
Hoffmann clamp Quetschhahn *m* [Schlauchklemme *f*] nach Hoffmann, Schraubenquetschhahn *m*, Schraubklemme *f*
Hofmann degradation of amides hofmannscher Säureamidabbau *m*, Hofmann-Abbau *m* von Säureamiden
Hofmann elimination Hofmann-Eliminierung *f*, hofmannscher Abbau *m (quartärer Ammoniumhydroxide)*

Hofmann-Martius rearrangement Hofmann-Martius-Umlagerung *f (von N-alkylierten Anilinhydrochloriden oder -bromiden in Alkylaniline)*
Hofmann mustard-oil reaction hofmannsche Senfölreaktion *f*
Hofmann orientation *(org Ch)* Hofmann-Orientierung *f*
Hofmann reaction *s.* Hofmann degradation of amides
Hofmeister series hofmeistersche [lyotrope] Reihe *f (der Ionen im Verhalten gegenüber Eiweiß)*
holarrhena alkaloid Holarrhenaalkaloid *n*, Kurchialkaloid *n*
hold *v* 1. enthalten, fassen; 2. binden *(z. B. Moleküle)*
hold-up *(Dest, Tech)* Betriebsinhalt *m*
hold-up time Verweilzeit *f*, Aufenthaltszeit *f*, Retentionszeit *f*, Rückhaltezeit *f*, Haltezeit *f*, Stehzeit *f*, Standzeit *f*, Durchgangszeit *f*, Durchlaufzeit *f*
holder 1. Halter *m*, Haltevorrichtung *f*; 2. Behälter *m*
holding temperature Aufbewahrungstemperatur *f*, Lager(ungs)temperatur *f*
holdup *s.* hold-up
hole 1. Loch *n*, Öffnung *f*; Bohrung *f*; 2. *(Krist)* Leerstelle *f*, Fehlstelle *f*, Gitterleerstelle *f*, Gitterfehlstelle *f*, Gitterlücke *f*; 3. *(phys Ch)* Loch *n*, Elektronenloch *n*, Defektelektron *n*, Mangelelektron *n*
hole conduction Lochleitung *f*, Löcherleitung *f*, Mangelleitung *f*, Defekt(elektronen)leitung *f*
hole-electron pair Loch-Elektron-Paar *n*, Elektron-Defektelektron-Paar *n*
hole transport layer Lochleitungsschicht *f (organische Leuchtdioden)*
holiday Fehlstelle *f (in einer Schutzschicht)*
hollow-cathode lamp *(Spektr)* Hohlkatodenlampe *f*, HKL
hollow fibre membrane *(Hyd)* Hohlfasermembran *f (Umkehrosmose)*
holmia Holmiumoxid *n*
holmium Ho Holmium *n*
holocellulose Holocellulose *f*
holocrystalline holokristallin, vollkristallin
holoenzyme Holoenzym *n*
holoside *(org Ch)* Holosid *n*
HOMO *s.* highest(-energy) occupied molecular orbital
homoaromatic *(org Ch)* homoaromatisch
homoaromaticity *(org Ch)* Homoaromatizität *f*
homoconjugation *(org Ch)* Homokonjugation *f*
homocycle homocyclische [isocyclische, carbocyclische] Verbindung *f*, Homocyclus *m*, Isocyclus *m*, Carbocyclus *m*
homocyclic compound homocyclische [isocyclische, carbocyclische] Verbindung *f*, Homocyclus *m*, Isocyclus *m*, Carbocyclus *m*
homofermentative *(Biot)* homofermentativ *(Mikroorganismen)*
homogeneity Homogenität *f*, Einheitlichkeit *f*, Gleichartigkeit *f*
homogeneous homogen, einheitlich, gleichartig (zusammengesetzt)
homogeneous catalysis homogene Katalyse *f*, Homogenkatalyse *f*
homogeneous kinetics Homogenkinetik *f*
homogeneous nucleation *(Krist)* homogene Keimbildung *f*, spontane [sporadische, thermische] Keimbildung *f*
homogeneous reaction homogene Reaktion *f*
homogeneous reactor 1. homogenphasiger Reaktor *m*, Reaktor *m* für homogene Reaktionssysteme; 2. *(Kern)* homogener Reaktor *m*, Homogenreaktor *m*
homogeneous throughout vollständig [vollkommen, durch und durch] homogen
homogenization 1. Homogenisieren *n*, Homogenisierung *f*, Vergleichmäßigen *n*, gleichmäßiges Durcharbeiten *n* [Vermischen *n*, Verteilen *n*]; 2. Homogenisierungsglühen *n*, homogenisierendes Glühen *n*, Homogenisieren *n*, Diffusionsglühen *n (von Legierungen)*
homogenize *v* 1. homogenisieren, vergleichmäßigen, gleichmäßig durcharbeiten [vermischen, verteilen]; 2. homogenisierend glühen, homogenisieren, diffusionsglühen *(Legierungen)*
homogenizer Homogenisator *m*, Ho-

mogenisiermaschine *f*, Homogenisiervorrichtung *f*, Homogenisierungsapparat *m*
homolactic fermentation *(Bioch)* glykolytischer Abbau *m*, Glykolyse *f*, Embden-Meyerhof-Parnas-Weg *m*, EMP-Weg *m*, FDP-Weg *m*
homolog *s.* homologue
homologization Homologisierung *f*
homologous series *(org Ch)* homologe Reihe *f*
homologue Homolog *n*, homologe Verbindung *f*
homolysis homolytische Bindungsspaltung *f*, Homolyse *f*
homolytic (bond) cleavage homolytische Bindungsspaltung *f*, Homolyse *f*
homonuclear molecule homonukleares Molekül *n*, Molekül *n* aus gleichen Atomen
homopolar bond [linkage] homöopolare [kovalente, unpolare] Bindung *f*, Atombindung *f*, Elektronenpaarbindung *f*, Kovalenz *f*
homopolymer Homopolymer(e) *n*, Homopolymerisat *n*, Unipolymer(e) *n*
homopolymeric homopolymer
homopolymerization Homopolymerisation *f*
homopolysaccharide Homopolysaccharid *n*
homotropic effector *(Bioch)* homotroper Effektor *m* [Steuermetabolit *m*]
homotropylium cation Homotropylium-Kation *n*
honeycomb structure wabenförmige [wabenartige] Struktur *f*, Wabenstruktur *f*
hood *(Lab)* Abzug *m*, Abzug(s)schrank *m*, Digestorium *n*; Rauchfangdach *n*; Dunsthaube *f*, Schwadenhaube *f*
Hookean behaviour hookesches [idealelastisches] Verhalten *n (Werkstoffkunde)*
Hooker (diaphragm) cell Hooker-Zelle *f*, Hooker-Diaphragmazelle *f (zur Alkalichloridelektrolyse)*
Hooke's law lineares Elastizitätsgesetz *n*
hop *v* hopfen, mit Hopfen versetzen *(Würze)*
hop bitters Hopfenbitterstoffe *mpl*
hop extract(ive) Hopfenextrakt *m*
hop resin Hopfenharz *n*
hopped wort gehopfte Würze *f*
hopper 1. Trichter *m*, Fülltrichter *m*, Einfülltrichter *m*, Zuführungstrichter *m*, Schütttrichter *m*, Beschickungstrichter *m*; 2. Silo *n(m)*, Bunker *m*, Behälter *m (mit Untenentleerung)*
Höppler (falling-ball) viscometer Höppler-Viskosimeter *n*, Höppler-Kugelfallviskosimeter *n*
horizontal autoclave liegender [horizontaler] Autoklav *m*
horizontal centrifuge Horizontalzentrifuge *f*
horizontal chromatography Horizontalchromatographie *f*
horizontal development *(Chromat)* Horizontalentwicklung *f*, horizontale Entwicklung *f*
horizontal relationship Horizontalbeziehung *f (im Periodensystem)*
hormonal hormonal, hormonell
hormonal control hormonale [humorale] Regulation *f*, hormonale Steuerung *f* • **under hormonal control** hormonal gesteuert
hormone Hormon *n*
hormone weed killer Wuchsstoffherbizid *n*, Unkrautbekämpfungsmittel *n* auf Wuchsstoffbasis, Wuchsstoffpräparat *n*, Wuchsstoffmittel *n*
hose Schlauch *m*
hose cock Schlauchklemme *f*, Quetschhahn *m*
hose coupling Schlauchverbindung *f*
host *s.* host component
host cell *(Bioch)* Wirtszelle *f*
host component Wirtskomponente *f*, Wirtssubstanz *f*, Wirt *m*
host-guest complex Einschlussverbindung *f*, Inklusionsverbindung *f*
host lattice *(Krist)* Wirtsgitter *n*
host molecule Wirtsmolekül *n*
hot-galvanize *v* feuerverzinken, schmelzflüssig verzinken
hot-press *v* heißpressen, warmpressen, heiß pressen [verpressen] *(Metallpulver, Ölsaaten)*; unter Druck sintern *(Metallpulver)*
hot 1. heiß, warm; 2. heiß, stark radioaktiv
hot-air oven *(Lab)* Heißluftschrank *m*
hot-air sterilizer Heißluftsterilisator *m*, Heißluftsterilisierschrank *m*

hot atom heißes [hochenergiereiches, hoch angeregtes] Atom *n*
hot-atom chemistry Heißatomchemie *f*, Chemie *f* heißer Atome, Rückstoßchemie *f*
hot chlorination Heißchlorierung *f*
hot-dip aluminizing Feueraluminieren *n*, Schmelztauchaluminieren *n*, Heißaluminieren *n*, Tauchalitieren *n*
hot-dip galvanizing Feuerverzinken *n*, Schmelztauchverzinken *n*, Heißverzinken *n*
hot embossing 1. *(Kst)* Farbprägen *n*; 2. *(Kst)* Heißprägen *n*
hot filtration Heißfiltration *f*
hot forming Warmformen *n*, Warmformgebung *f*
hot gas Heißgas *n*, heißes Gas *n*
hot hardness Warmhärte *f*
hot-melt coating 1. Heißtauchen *n*, Schmelztauchen *n*; 2. Heißtauchmasse *f*, Schmelztauchmasse *f*; 3. Heißtauchschutzschicht *f*
hot pressing Heißpressen *n*, Warmpressen *n (von Metallpulver, Ölsaaten)*; Drucksintern *n*, Presssintern *n*, Heißsintern *n (von Metallpulver)*
hot-saturated solution heiß gesättigte Lösung *f*
hot-setting *(Anstr, Kst)* heißhärtend, heißabbindend
hot spot Überwärmungszone *f*, Überhitzungszone *f (in einem Reaktor)*
hot-stage microscope Heiztischmikroskop *n*
hot-state reaction heiße Reaktion *f*, Heißatomreaktion *f*
hot strength Warmfestigkeit *f*
hot waste *(Kern)* hochaktiver [heißer] Abfall *m*
hot-water funnel Heißwassertrichter *m*, Warmwassertrichter *m*
hot working 1. Warmverarbeitung *f*; 2. *s.* hot forming
hotmelt adhesive *(Kst)* Schmelzkleber *m*
hotness 1. Wärme *f*; 2. Temperatur *f*
Houben-Hoesch synthesis Hoesch-Ketonsynthese *f*, Houben-Hoesch-Synthese *f*
Houdry fixed-bed process Houdry-Festbettverfahren *n*, katalytisches Houdry-Krackverfahren *n*
house *v* umhausen *(Anlagen)*
household ammonia Salmiakgeist *m*, Ammoniakwasser *n*
household laundry detergent Haushaltwaschmittel *n*
household refuse Haus(halt)müll *m*, häuslicher Müll *m*
household waste Haushaltabfall *m*, häuslicher Abfall *m*
housing Gehäuse *n*
HP *s.* high-density polyethylene
HPLAC *s.* high-performance liquid affinity chromatography
HPLC *s.* high-pressure liquid chromatography
HPTLC *s.* high-performance thin-layer chromatography
HRF *s.* high-rate filter
HSA *s.* human serum albumin
HSAB principle HSAB-Konzept *n*, Prinzip *n* der harten und weichen Säuren und Basen
HT ... *siehe: high-temperature ...*
HTL *s.* hole transport layer
HTST pasteurization *s.* high-temperature short-time pasteurization
HTU *s.* height of one transfer unit
Huang-Minlon reaction [reduction] Huang-Minlon-Reduktion *f*, Huang-Minlon-Reaktion *f (von Aldehyden und Ketonen)*
Hückel molecular orbital Hückel-Molekülorbital *n*, hückelsches Molekülorbital *n*
Hückel rule Hückel-Regel *f (für konjugierte Kohlenwasserstoffe mit 4n + 2 π-Elektronen)*
Hückel [4n + 2] rule Hückel-Regel *f (für konjugierte Kohlenwasserstoffe mit 4n + 2 π-Elektronen)*
hue Ton *m*, Farbton *m*, Farbtönung *f*, Farbschattierung *f*, Nuance *f*, Stich *m*
human pollutants *(Hyd)* anthropogene Schmutzstoffe *mpl*, *(i. e. S.)* anthropogene Schadstoffe *mpl*
human pollution *(Hyd)* anthropogene Verschmutzung *f* [Verunreinigung *f*]
human serum albumin *(Bioch)* Humanserumalbumin *n*
human toxicity Humantoxizität *f*
humate *(Boden)* Humat *n*
Hume-Rothery phase *(Krist)* Hume-Rothery-Phase *f*

humectant Feuchthaltemittel *n*; Benetzungsmittel *n*, Netzmittel *n*, Netzer *m*
humic acid Huminsäure *f*, Humussäure *f*
humic substance Huminstoff *m*, Huminsubstanz *f*
humid feucht, humid
humidification Befeuchtung *f*, Anfeuchtung *f*
humidifier Befeuchter *m*, Befeuchtungsapparat *m*
humidify *v* befeuchten, anfeuchten
humidity Feuchtigkeit *f*, Feuchte *f (in Gasen, auf Oberflächen)*; Feuchtebeladung *f (bezogen auf eine Masseneinheit des Gases)*
humidity chart Feuchtigkeitsdiagramm *n*, Feuchtigkeitstafel *f*
humulon(e) Humulon *n*, α-Hopfenbittersäure *f*, α-Bittersäure *f*, α-Lupulinsäure *f*
humus Humus *m*, organische Substanz *f (im Boden)*; Huminstoff *m*, Huminsubstanz *f*
Hund-Mulliken-Lennard-Jones-Hückel method Hund-Mulliken-Lennard-Jones-Hückel-Methode *f*, Molekülorbitalmethode *f*, MO-Methode *f*, Methode *f* der Molekülorbitale
Hund rule hundsche Regel *f (der Energie von Atomzuständen)*
Hunsdiecker cleavage [reaction] Esterspaltung *f* nach Hunsdiecker, Hunsdiecker-Spaltung *f*, Hunsdiecker-Reaktion *f (Silbersalz-Decarboxylierung)*
HWHM *s.* half-width (at half maximum)
HWI = Hazardous Waste Inspectorate
hyaluronic acid *(Bioch)* Hyaluronsäure *f*
hybrid hybrid
hybrid Hybrid *n (Bindungslehre)*
hybrid material Hybridmaterial *n*
hybrid orbital Hybridorbital *n*
***sp*-hybrid** *sp*-Hybrid *n*, digonales Hybrid *n*
hybridization *(Bioch)* Hybridisierung *f*
***sp*^{2}-hybridization** *sp*^{2}-Hybridisierung *f*, trigonale Hybridisierung *f*
hybridize *v* hybridisieren
hybridoma (cell) *(Biot)* Hybridomzelle *f*, Hybridom(a) *n (antikörperproduzierende Zellhybride)*
hydantoin Hydantoin *n*, Glykolharnstoff *m*
hydnocarpic acid Hydnocarpussäure *f*, 11-(2-Cyclopentenyl)undecansäure *f*
hydracid sauerstofffreie Säure *f*
hydracrylic acid Hydracrylsäure *f*, *3-Hydroxypropionsäure *f*, Propan-3-olsäure *f*, Ethylenmilchsäure *f*
hydrargyrism Quecksilbervergiftung *f*
hydrate *v* hydratisieren
hydrate Hydrat *n*
hydrate water Hydratwasser *n*
hydrated 1. hydratisiert, Hydrat...; 2. gelöscht *(Kalk)*
hydration Hydratation *f*, Hydratisierung *f*, Wasseranlagerung *f*
hydration number (of ions) Hydratationszahl *f*
hydration sheath Hydrathülle *f*, Hydratationssphäre *f*
hydraulic cement hydraulischer [wasserbindender] Zement *m*, Wasserzement *m*
hydraulic classification Aufstromklassieren *n*, Gegenstromklassieren *n*
hydraulic cyclone separator Hydrozyklon *m*, Nasszyklon *m*
hydraulic fluid Hydraulikflüssigkeit *f*
hydraulic lime *(Bau)* hydraulischer Kalk *m*, *(bei nur 10 bis 15 % löslichen sauren Bestandteilen auch)* Wasserkalk *m*
hydraulic-operated hydraulisch betätigt [bewegt, betrieben, angetrieben]
hydrazide Hydrazid *n*
hydrazide imide Hydrazidimid *n*
hydrazine Hydrazin *n*, Diamid *n*
hydrazinium sulphate Hydraziniumsulfat *n*
hydrazinolysis Hydrazinolyse *f*
hydrazo compound Hydrazoverbindung *f*
hydrazoarene Hydrazoaren *n*, 1,2-Diarylhydrazin *n*
hydrazoate M^IN_3 Azid *n*
hydrazobenzene Hydrazobenzen *n*, 1,2-Diphenylhydrazin *n*
hydrazoic acid Stickstoffwasserstoffsäure *f*, Azoimid *n*
hydrazone Hydrazon *n*
hydride Hydrid *n*
hydride ion H^- Hydridion *n*
hydroalcoholic alkoholisch-wässrig
hydroalkylation hydrierende Alkylierung *f*, reduktive [reduzierende] Alkylierung *f*
hydroaromatic compounds Hydroaro-

maten *pl*, hydroaromatische Verbindungen *fpl*
hydroboration Hydroborierung *f (Addition von Diboran an Olefine)*
hydrobromic acid Bromwasserstoffsäure *f*
hydrobromination Hydrobromierung *f*
hydrocarbon Kohlenwasserstoff *m*, KW-Stoff *m*
hydrocarbon chain Kohlenwasserstoffkette *f*
hydrocarbon fermentation *(Biot)* Kohlenwasserstoff-Fermentation *f*
hydrocarbon polymer polymerer [polymerisierter] Kohlenwasserstoff *m*
hydrocarbon residue Kohlenwasserstoffrest *m*
hydrocarbon resin Kohlenwasserstoffharz *n*
hydrocarbylsulphanyl group Sulfenylgruppe *n*
hydrocarbylsulphenylium ion Sulfenylium-Ion *n*
hydrocellulose Hydrocellulose *f (hydrolytisch abgebaute Cellulose)*
hydrochemistry Hydrochemie *f*, Wasserchemie *f*
hydrochloric acid Chlorwasserstoffsäure *f*, Salzsäure *f*
hydrochloride *(org Ch)* Hydrochlorid *n*
hydrochlorination Behandlung *f* mit Chlorwasserstoff; Hydrochlorierung *f*, Chlorwasserstoffanlagerung *f*
hydroclone Hydrozyklon *m*, Nasszyklon *m*
hydrocondensation Hydrokondensation *f*, reduktive [reduzierende] Kondensation *f*
hydrocrackate *(Erdöl)* Hydrokrackprodukt *n*
hydrocracking *(Erdöl)* Hydrokracken *n*, Hydrokrackung *f*, hydrierendes Kracken *n* [Spalten *n*], hydrierende Krackung *f* [Spaltung *f*], spaltende Druckhydrierung *f*
hydrocyanation Hydrocyanierung *f*
hydrocyanic acid Cyanwasserstoffsäure *f*, Blausäure *f*
hydrocyclone (separator) *s.* hydroclone
hydrodealkylation Hydrodesalkylierung *f*, reduktive [reduzierende] Desalkylierung *f*
hydrodesulphurization Hydrodesulfurierung *f*, hydrierende Entschwefelung *f*, Hydroentschwefelung *f*, Wasserstoffentschwefelung *f*, HDS
hydrodimerization reduktive Dimerisierung *f*
hydrofining Hydrofining *n (katalytische Entschwefelung und Produktverbesserung von Erdölfraktionen bei Anwesenheit von Wasserstoff)*
hydrofluoric acid Fluorwasserstoffsäure *f*, Flusssäure *f*
hydrofluoric-acid alkylation Alkylierung *f* mit Fluorwasserstoffsäure [Flusssäure, Fluorwasserstoff], Flusssäurealkylierung *f*, Fluorwasserstoffalkylierung *f*, HF-Alkylierung *f*
hydrofluoride *(org Ch)* Hydrofluorid *n*
hydrofluorination Hydrofluorierung *f*
hydroform *v* hydroformieren
hydroformate Hydroformat *n*, Hydroforming-Produkt *n*
hydroforming Hydroform(ier)en *n*, Hydroforming *n (ein katalytisches Reformieren mit Wasserstoffumlauf)*
hydroformylation Hydroformylierung *f*, Oxosynthese *f*
hydroformylation reaction Oxo(synthese)reaktion *f*, Roelen-Reaktion *f*
hydrogasification *(Carb)* hydrierende Vergasung *f*, Wasserstoffvergasung *f*
hydrogasify *v (Carb)* hydrierend vergasen
hydrogel Hydrogel *n*
hydrogen H Wasserstoff *m*, Hydrogen *n*
hydrogen abstraction reaction Wasserstoffabspaltungsreaktion *f*, H-Abspaltungsreaktion *f*
hydrogen acceptor Wasserstoffakzeptor *m*
hydrogen acid Wasserstoffionen abspaltende Säure *f*, *(i. w. S.)* Proton(en)säure *f*, Brönsted-Säure *f*, Protonendonator *m*, Dysprotid *n*
hydrogen atom Wasserstoffatom *n*
hydrogen bond Wasserstoff(brücken)-bindung *f*, H-Bindung *f*, Wasserstoffbrücke *f*
hydrogen-bonded über Wasserstoff verbunden, durch eine Wasserstoffbrücke verbunden
hydrogen bonding Wasserstoff(brücken)bindung *f*, Bindung *f* über Was-

serstoffbrücken, Bildung *f* [Ausbildung *f*] von Wasserstoffbrücken
hydrogen bromide Bromwasserstoff *m*, Hydrogenbromid *n*
hydrogen carbonate Hydrogencarbonat *n*
hydrogen-cation exchanger Wasserstoff(ionen)austauscher *m*, H-Austauscher *m*, H_2-Austauscher *m*, Ionenaustauscher *m* in der Wasserstoffform [H-Form, H_2-Form]
hydrogen chloride Chlorwasserstoff *m*, Hydrogenchlorid *n*
hydrogen compound Wasserstoffverbindung *f*
hydrogen cracking 1. Wasserstoffrisskorrosion *f*, Wasserstoffkrankheit *f*; 2. *s.* hydrogen embrittlement
hydrogen cyanide Cyanwasserstoffsäure *f*, Blausäure *f*, Hydrogencyanid *n*
hydrogen-donating wasserstoffliefernd
hydrogen donor Wasserstoffdonator *m*
hydrogen electrode Wasserstoffelektrode *f*
hydrogen embrittlement 1. Wasserstoffversprödung *f*, H-Versprödung *f*, katodische Spannungsrisskorrosion *f* *(Vorgang)*; 2. Wasserstoffbrüchigkeit *f* *(Ergebnis)*
hydrogen evolution Wasserstoffentwicklung *f*, Wasserstoffabscheidung *f*
hydrogen fluoride Fluorwasserstoff *m*, Hydrogenfluorid *n*
hydrogen-form exchanger *s.* hydrogen-cation exchanger
hydrogen gas Wasserstoffgas *n*
hydrogen half-cell Wasserstoffhalbelement *n*, Wasserstoffelektrode *f*
hydrogen halide Halogenwasserstoff *m*
hydrogen ion Wasserstoffion *n*, H^+-Ion *n*
hydrogen-ion activity Wasserstoffionenaktivität *f*
hydrogen-ion concentration Wasserstoffionenkonzentration *f*
hydrogen-ion exponent Wasserstoffionenexponent *m*, pH-Wert *m*
hydrogen nucleus Wasserstoffkern *m*, Proton *n*
hydrogen overvoltage *(Ech)* Wasserstoffüberspannung *f*
hydrogen-oxygen fuel cell Wasserstoff-Sauerstoff-Brennstoffzelle *f*, Knallgaszelle *f*, Knallgaselement *n*
hydrogen-oxygen reaction Knallgasreaktion *f*
hydrogen peroxide Wasserstoffperoxid *n*, Hydrogenperoxid *n*
hydrogen phosphate Hydrogenphosphat *n*
hydrogen scale elektrochemische Spannungsreihe *f* *(mit Normalwasserstoffelektrode als Bezugselektrode; besonders in älterer angelsächsischer Literatur mit umgekehrten Vorzeichen)*
hydrogen spectrum Wasserstoff(atom)spektrum *n*
hydrogen sulphate M^IHSO_4 Hydrogensulfat *n*
hydrogen sulphide 1. H_2S Schwefelwasserstoff *m*, Wasserstoffsulfid *n*, Monosulfan *n*; 2. *s.* hydrogensulphide
hydrogen sulphite M^IHSO_3 Hydrogensulfit *n*
hydrogen transfer Wasserstoffübertragung *f*
hydrogenability Hydrierbarkeit *f*
hydrogenable hydrierbar
hydrogenate *v* hydrieren, *(bei Ölen und Fetten auch)* härten
hydrogenate *v* **catalytically** katalytisch hydrieren
hydrogenated oil hydriertes [gehärtetes] Öl *n* [Fett *n*], Hartfett *n*
hydrogenation Hydrieren *n*, Hydrierung *f*, *(bei Ölen und Fetten auch)* Härten *n*, Härtung *f*
hydrogenation catalyst Hydrier(ungs)-katalysator *m*
hydrogenation reaction Hydrierungsreaktion *f*
hydrogenic wasserstoffähnlich *(im Kernaufbau)*
hydrogenize *v* hydrieren, *(Öle und Fette auch)* härten
hydrogenolysis Hydrogenolyse *f*, hydrogenolytische Spaltung *f*
hydrogenous wasserstoffhaltig; Wasserstoff...
hydrogensulphide M^IHS Hydrogensulfid *n*, primäres [saures] Sulfid *n*
hydrohalic acid Halogenwasserstoffsäure *f*
hydrohalide Hydrohalogenid *n* *(halogenwasserstoffsaures Salz einer basischen organischen Verbindung)*
hydrohalogen Halogenwasserstoff *m*

hydroisomerization Hydroisomerisierung *f*, reduktive [reduzierende] Isomerisierung *f*
hydrolase Hydrolase *f (hydrolytische Spaltungen katalysierendes Enzym)*
hydrolysability Hydrolysierbarkeit *f*
hydrolysable hydrolysierbar
hydrolysate Hydrolysat *n*
hydrolyse *v* hydrolysieren; Hydrolyse erleiden, der Hydrolyse unterliegen, hydrolysiert werden, hydrolysieren
hydrolysis 1. Hydrolyse *f*; 2. Aquotisierung *f (Komplexchemie)*
hydrolysis constant Hydrolysenkonstante *f*
hydrolyte hydrolysierter [zu hydrolysierender] Stoff *m*
hydrolytic degradation hydrolytischer Abbau *m*, Abbau *m* durch Hydrolyse
hydrolytic stability Hydrolysebeständigkeit *f*, Hydrolysestabilität *f*, Hydrolyseresistenz *f*
hydrolyze *v s.* hydrolyse
hydrometallurgical hydrometallurgisch, nassmetallurgisch
hydrometallurgy Hydrometallurgie *f*, Nassmetallurgie *f*
hydrometer Aräometer *n*, Senkwaage *f*, Senkspindel *f*, Spindel *f*, Densimeter *n*
hydrometric(al) aräometrisch, Aräometer...
hydrometry Dichtebestimmung *f* von Flüssigkeiten (mit Aräometern)
hydron Hydron *n (allg. für ein Kation eines Wasserstoffisotops)*
hydronium ion Hydroniumion *n (hydratisiertes Wasserstoffion)*
hydroperoxide Hydroperoxid *n (allg. für eine organische Verbindung mit der Gruppe –O–OH)*
hydroperoxyl radical Hydroperoxyradikal *n*, HO_2-Radikal *n*
hydrophilic hydrophil *(Ion, Atomgruppe, Kolloid)*; hydrophil, wasseraufnehmend, mit Wasser benetzbar *(Werkstoffe, Textilien)*
hydrophilicity Hydrophilie *f*
hydrophobic hydrophob *(Ion, Atomgruppe, Kolloid)*; hydrophob, wasserabstoßend, wasserabweisend, mit Wasser nicht benetzbar *(Werkstoffe, Textilien)*
hydrophobic bond *s.* hydrophobic interaction
hydrophobic interaction 1. hydrophobe Wechselwirkung *f*; 2. *(Bioch)* hydrophobe Bindung *f*, hydrophobe Stapelungskräfte *fpl*, Basenaufstockungskräfte *fpl (zur Stabilisierung der DNA-Doppelhelix)*
hydrophobic interaction chromatography hydrophobe Interaktionschromatographie *f*, HIC, hydrophobe Chromatographie *f*, Hydrophobie-Chromatographie *f*
hydrophobicity Hydrophobie *f*
hydrophobing agent Hydrophobier(ungs)mittel *n*, hydrophobierendes Mittel *n*
hydropolymerization Hydropolymerisation *f*, Hydropolymerisierung *f*
hydroquinone Hydrochinon *n*, *Benzen-1,4-diol *n*
hydrorefining *(Erdöl)* hydrierende Raffination *f*, Druckwasserstoffraffination *f*
hydrosol Hydrosol *n*
hydrostatic head hydrostatischer Druck *m*
hydrostatic head filter hydrostatisches Filter *n*, Schwerkraftfilter *n*
hydrosulphide Hydrogensulfid *n*
hydrosulphite M^IHSO_3 Hydrogensulfit *n*
hydrotreating Hydrotreating *n (katalytische Raffination von Erdölfraktionen mit Wasserstoff zur Entfernung von Schwefel- und Stickstoffverbindungen)*
hydrotreatment Hydrotreating *n (katalytische Raffination von Erdölfraktionen mit Wasserstoff zur Entfernung von Schwefel- und Stickstoffverbindungen)*
hydrotrope hydrotrope Verbindung *f*
hydrotropic hydrotrop *(die Löslichkeit organischer Verbindung in Wasser erhöhend)*
hydrotropy Hydrotropie *f (Erhöhung der Löslichkeit organischer Verbindungen in Wasser)*
hydrous wasserhaltig
hydroxamic acid Hydroxamsäure *f*
hydroxide Hydroxid *n*
hydroxide ion OH^- Hydroxidion *n*
hydroxo salt Hydroxosalz *n*
hydroxoaluminate Hydroxoaluminat *n*
hydroxonium ion Hydroxoniumion *n*, Oxoniumion *n*

hydroxy group *s.* hydroxyl group
hydroxy ketone Hydroxyketon *n*, Keto(n)alkohol *m*
hydroxy-substituted hydroxysubstituiert
α-hydroxy acid 2-Hydroxy(carbon)säure *f*, α-Hydroxysäure *f*
β-hydroxy acid 3-Hydroxy(carbon)säure *f*, β-Hydroxysäure *f*
hydroxyacetic acid Hydroxyessigsäure *f*, *Hydroxyethansäure *f*, Glykolsäure *f*
hydroxyaldehyde Hydroxyaldehyd *m*
hydroxyalkylation Hydroxyalkylierung *f*, Oxyalkylierung *f*
hydroxyanthraquinone Hydroxyanthrachinon *n*
hydroxybenzaldehyde Hydroxybenzaldehyd *m*
hydroxybenzene Phenol *n*, Hydroxybenzen *n*
***o*-hydroxybenzoic acid** *o*-Hydroxybenzoesäure *f*, Salicylsäure *f*
hydroxybutanedioic acid *Hydroxybutandisäure *f*, Äpfelsäure *f*
β-hydroxybutyraldehyde 3-Hydroxybutyraldehyd *m*, *3-Hydroxybutanal *n*, Acetaldol *n*
hydroxycarboxylic acid Hydroxycarbonsäure *f*
hydroxycinnamic acid Hydroxyzimtsäure *f*, Cumarsäure *f*, *Hydroxyphenylpropensäure *f*
2-hydroxyethyl methacrylate 2-Hydroxyethylmethacrylat *n*
hydroxyketone Hydroxyketon *n*, Keto(n)alkohol *m*
hydroxyl-form exchanger *(Hyd)* OH-Austauscher *m*, $(OH)_2$-Austauscher *m*, Ionenaustauscher *m* in der Hydroxylform [OH-Form, $(OH)_2$-Form], Anionenaustauscher *m*
hydroxyl group Hydroxylgruppe *f*, Hydroxylrest *m*, OH-Gruppe *f*
hydroxyl number *s.* hydroxyl value
hydroxyl radical Hydroxylradikal *n*
hydroxyl-terminated mit endständiger Hydroxylgruppe
hydroxyl value Hydroxylzahl *f*, OHZ *(Kennzahl der Fette und fetten Öle)*
hydroxylamine Hydroxylamin *n*
hydroxylammonium chloride Hydroxylammoniumchlorid *n*
hydroxylate *v* hydroxylieren
hydroxylation Hydroxylieren *n*, Hydroxylierung *f*
hydroxylic oxygen Hydroxylsauerstoff *m*
hydroxymethyl group $-CH_2OH$ Hydroxymethylgruppe *f*
hydroxymethylation Hydroxymethylierung *f*, Oxymethylierung *f*
2-hydroxypropionic acid 2-Hydroxypropionsäure *f*, *2-Hydroxypropansäure *f*, Ethylidenmilchsäure *f*, *(i. e. S.)* Milchsäure *f*
hydroxyquinoilne Hydroxychinolin *n*
hydroxysalt Hydroxysalz *n*
hydroxytoluene Hydroxytoluen *n*, Cresol *n*
hyg. *s.* hygroscopic
hygrometer Hygrometer *n*, Feuchtemesser *m*, Luftfeuchtigkeitsmesser *m*
hygrometry Hygrometrie *f*, Luftfeuchtigkeitsmessung *f*
hygroscopic hygroskopisch, hygr., wasseranziehend
hygroscopic moisture [water] hygroskopisches [hygroskopisch gebundenes] Wasser *n*
hygroscopicity Hygroskopizität *f*
***dl*-hyoscyamine** DL-Hyoscyamin *n*, Atropin *n*
hyperacidity *(Med)* Superazidität *f*, Hyperazidität *f (des Magensafts)*
hyperbranched *(Kst)* hochverzeigt *(in Polymerstrukturen)*
hypercholester(ol)aemia *(Med)* Hypercholesterolämie *f (erhöhter Cholesterolgehalt des Bluts)*
hyperchromic effect 1. *(Bioch)* Hyperchromie-Effekt *m (bei thermischer Denaturierung der DNA)*; 2. *(Spektr)* hyperchromer Effekt *m*
hyperchromicity *(Bioch)* Hyperchromie *f (der DNA bei thermischer Denaturierung)*
hyperconjugation Hyperkonjugation *f*, Baker-Nathan-Effekt *m*
hypereutectic übereutektisch
hypereutectoid übereutektoid(isch)
hyperfiltration *(Hyd)* Hyperfiltration *f*
hyperfine structure *(phys Ch)* Hyperfeinstruktur *f*, HFS
hyperglyc(a)emia *(Med)* Hyperglykämie *f (erhöhter Blutzuckergehalt)*

hyperglyc(a)emic *(Med)* hyperglykämisch
hypergolic fuel [rocket propellant] hypergoler [selbstzündender] Treibstoff *m* [Raketentreibstoff *m*], Hypergol *m*
hyperligated complex stark gebundener Komplex *m*
hyperlip(a)emia *(Med)* Hyperlipämie *f*, Lipämie *f*, *(i. w. S.)* Hyperlipidämie *f* *(erhöhter Fett- bzw. Fett- und Lipoidgehalt des Bluts)*
hyperoxide M^IO_2 Hyperoxid *n*
hypersensitive überempfindlich, hypersensibel
hypersensitivity Überempfindlichkeit *f*, Hypersensibilität *f*
hypersorption Hypersorption *f*
hyperstoichiometric überstöchiometrisch
hyperstructure Überstruktur *f*
hypervitaminosis *(Med)* Hypervitaminose *f*
hypnotic (drug) *(Pharm)* Hypnotikum *n*, Schlafmittel *n*
hypoacidity *(Med)* Subazidität *f*, Hypoazidität *f (des Magensafts)*
hypochlorite M^IOCl Hypochlorit *n*
hypochlorite bleach (liquor) Hypochloritbleichlauge *f*
hypochlorous acid hypochlorige Säure *f*
hypochromic effect *(Spektr)* hypochromer Effekt *m*
hypochromicity Hypochromie *f (von Nucleinsäuren)*
hypochromism Hypochromie *f (verminderte Extinktion)*
hypocrystalline hypokristallin, halbkristallin
hypoeutectic untereutektisch
hypoeutectoid untereutektoid(isch)
hypoglyc(a)emia *(Med)* Hypoglykämie *f (verminderter Blutzuckergehalt)*
hypoglyc(a)emic *(Med)* hypoglykämisch
hypohalogenation Hypohalogenierung *f*
hypohalous acid unterhalogenige Säure *f*
hypoligated complex schwach gebundener Komplex *m*
hyponitrite $M^I_2N_2O_2$ Hyponitrit *n*
hyponitrous acid hyposalpetrige Säure *f*
hypophosphate $M^I_4P_2O_6$ Hypophosphat *n*
hypophosphite $M^IH_2PO_2$ Hypophosphit *n*
hypophosphoric acid Hypophosphorsäure *f*
hypophosphorous acid hypophosphorige Säure *f*
hypophyseal hormone Hypophysenhormon *n*
hypostoichiometric unterstöchiometrisch
hyposulphate $M^I_2S_2O_6$ Dithionat *n*
hyposulphite 1. $M^I_2S_2O_4$ Dithionit *n*; 2. $M^I_2S_2O_3$ Thiosulfat *n*; 3. *(Foto)* Fixiersalz *n*, Natriumthiosulfat *n*
hyposulphurous acid dithionige Säure *f*
hypotensive blutdrucksenkend, antihypertonisch
hypotensive drug *(Pharm)* blutdrucksenkendes Mittel *n*, Hypotensivum *n*, Antihypertonikum *n*, Mittel *n* gegen Bluthochdruck
hypsochrome *s.* hypsochromic group
hypsochromic hypsochrom, farberhöhend, farbaufhellend
hypsochromic group hypsochrome [farberhöhende] Gruppe *f*
hypsochromic shift hypsochrome Verschiebung *f* [Farbtonverschiebung *f*], Blauverschiebung *f*, Hypsochromie *f*, Farberhöhung *f*, Farbaufhellung *f*, negative Farbänderung *f (Verschiebung der Absorption in den kurzwelligen Spektralbereich)*
hysteresis Hysterese *f*, Hysteresis *f*

I

i. *s.* insoluble
I effect *s.* induction effect
IBAD *s.* ion beam assisted deposition
IBID *s.* ion beam induced deposition
iboga alkaloid Ibogaalkaloid *n*
I.B.P. *s.* initial boiling point
ICDH = isocitrate dehydrogenase
ice Eis *n*, Natureis *n*
ice bath Eisbad *n*
ice calorimeter Eiskalorimeter *n*
ice crusher Eismühle *f*
ice water Eiswasser *n*
ichthyotoxin Ichthyotoxin *n*, Fischgift *n* *(fischeigenes Gift)*
ICP *s.* inductively coupled plasma

ICR *s.* ion cyclotron resonance
I.C.R.P. = International Commission on Radiological Protection
ICSH *s.* interstitial-cell-stimulating hormone
i.d. *s.* inside diameter
I.D. *s.* inside diameter
ideal gas ideales [vollkommenes] Gas *n*
ideal-gas equation [law] (thermische) Zustandsgleichung *f* idealer Gase, ideales Gasgesetz *n*, ideale Gasgleichung *f*
ideal reactor (strömungstechnisch) idealer Reaktor *m*
ideality Idealität *f*, Idealzustand *m*
identically charged gleichsinnig geladen
identifiable identifizierbar; nachweisbar
identification Identifizierung *f*; Nachweis *m*
identification limit Nachweisgrenze *f*, Erfassungsgrenze *f*
identify *v* identifizieren; nachweisen
identity period Identitätsperiode *f (bei Polymeren)*
idiophase *(Biot)* Idiophase *f*, Produktbildungsphase *f*
idle außer Betrieb (befindlich), stillstehend
idose Idose *f (ein Monosaccharid)*
iduronic acid *(Bioch)* Iduronsäure *f*
IEE *s.* induced electron emission
Ifn *s.* interferon
IFN *s.* interferon
Ig *s.* immunoglobulin
ignitability Entzündbarkeit *f*, Entzündlichkeit *f*
ignitable entzündbar, entzündlich
ignite *v* entzünden, anzünden *(Brennstoff)*; zünden *(Gasgemisch, Sprengstoff)*; glühen; sich entzünden, zünden
ignite *v* **to constant weight** bis zur Massekonstanz glühen
ignition 1. Entzünden *n*, Anzünden *n (eines Brennstoffs)*; Zünden *n*, Zündung *f (eines Gasgemischs, eines Sprengstoffs)*; Entzündung *f*, Zündung *f (Vorgang)*; 2. *(Anal)* Glühen *n*
ignition crucible Schmelztiegel *m*
ignition point *s.* ignition temperature
ignition source Zündquelle *f*
ignition temperature Zünd(ungs)temperatur *f*, Entzündungstemperatur *f*, *(i. e. S.)* Selbstentzündungstemperatur *f*, Selbstzündtemperatur *f*, SZT, Zündpunkt *m*
ignition test tube *(Lab)* Glühröhrchen *n*
IIR *s.* isobutylene-isoprene rubber
IKES *s.* ion kinetic energy spectroscopy
Ileu = isoleucine
ill-defined schlecht [mangelhaft] definiert
ill-smelling übel riechend, unangenehm riechend
illuminance Beleuchtungsstärke *f*
illumination/under unter Lichteinwirkung [Lichteinfluss, Einstrahlung von Licht], am [im] Licht, bei Belichtung
ilmenite *(Min)* Ilmenit *m (Eisen(II)-metatitanat)*
IMA *s.* ion-probe microanalysis
imbalance Ungleichgewicht *n*
imbibe *v* aufnehmen, aufsaugen, einsaugen, absorbieren, imbibieren *(Flüssigkeiten)*; (durch)tränken, imprägnieren, imbibieren *(mit Flüssigkeiten)*
imbibition Flüssigkeitsaufnahme *f*, Imbibition *f*; Tränkung *f*, Durchtränkung *f*, Imprägnierung *f*, Imbibition *f*, Durchdringung *f (mit Flüssigkeiten)*
Imhoff tank *(Hyd)* Emscherbrunnen *m*, Imhoff-Tank *m*
imidazole Imidazol *n*
imidazole alkaloid Imidazolalkaloid *n*
imide Imid *n*
imidic acid Imidsäure *f*
imido ester Imidoester *m*, Iminoether *m*
imido group Imidogruppe *f (=NH-Gruppe als Ersatz für O=)*
imine Imin *n*, Iminoverbindung *f*
iminium compound Iminiumverbindung *f (Salz mit dem Kation* $R_2C{=}N^+R_2$*)*
imino acid Iminosäure *f*
imino group Iminogruppe *f (=NH-Gruppe als Ersatz für 2* H^-*)*
imino urea Iminoharnstoff *m*, Guanidin *n*, *Aminomethanamidin *n*
immerse *v* (ein)tauchen, untertauchen, versenken
immersion Tauchen *n*, Eintauchen *n*, Untertauchen *n*, Versenken *n*, Immersion *f*
immersion heater Tauchsieder *m*
immersion objective Immersionsobjektiv *n*, Immersionssystem *n*

immersion refractometer Eintauchrefraktometer *n*
immiscible un(ver)mischbar, nicht mischbar
immobile phase *(Chromat)* stationäre [unbewegliche] Phase *f*
immobilization 1. *(Landw)* Immobilisierung *f*, biologische Festlegung *f (von Nährstoffen in Bodenmikroorganismen)*; 2. *(Biol)* Immobilisierung *f (von Enzymen und Zellen)*
immobilize *v* 1. *(Landw)* immobilisieren, biologisch festlegen *(Nährstoffe)*; 2. *(Biot)* immobilisieren *(Enzyme und Zellen)*
immonium compound *s.* iminium compound
immune globulin Immunglobulin *n*, Ig
immunity Immunität *f*
immunization Immunisierung *f*
immunize *v* immunisieren, immun machen
immunochemical immun(o)chemisch
immunochemistry Immun(o)chemie *f*
immunoelectrophoresis *(Anal)* Immun(o)elektrophorese *f*
immunoglobulin Immunglobulin *n*, Ig
immunology Immunologie *f*
immunostimulant Immunstimulans *n (Stoff, der bestimmte Immunreaktionen fördert)*
immunosuppressant Immunsuppressor *m (Stoff, der bestimmte Immunreaktionen hemmt)*
immunosuppressor Immunsuppressor *m (Stoff, der bestimmte Immunreaktionen hemmt)*
imonium compound *s.* iminium compound
IMP = inosine 5′-monophosphate
impact *v* zusammenstoßen, zusammenprallen; aufprallen, anprallen, anstoßen, aufschlagen
impact Stoß *m*; Zusammenstoß *m*, Zusammenprall *m*; Aufprall *m*, Anprall *m*, Anstoß *m*; Schlag *m*; Aufschlag *m*
impact area Aufschlagsbereich *m*
impact electron Stoßelektron *n*, Anstoßelektron *n*
impact energy Aufschlagsenergie *f*
impact hardness Schlaghärte *f*
impact ionization Stoßionisation *f*, Stoßionisierung *f*
impact resistance Schlagfestigkeit *f*, Schlagzähigkeit *f*, Stoßfestigkeit *f*
impact-resistant schlagfest, stoßfest
impact test Schlagprüfung *f*
impact velocity Aufschlagskraft *f*
impart *v* **a colour** eine Färbung geben [erteilen] *(z. B. einer Flamme)*
impedance *(Anal)* Impedanz *f*, Wechselstromwiderstand *m*, Scheinwiderstand *m*, komplexer Widerstand *m*
impedance measurement *(Ech)* Impedanzmessung *f*
impedance spectroscopy *(Ech)* Impedanzspektroskopie *f*
impedance spectrum *(Ech)* Impedanzspektrum *n*
impedimetric titration *(Anal)* Hochfrequenztitration *f*, oszillometrische Titration *f*
impedimetry *(Anal)* Hochfrequenztitration *f*, oszillometrische Titration *f*
impenetrable *s.* impermeable
imperfect crystal gestörter [realer] Kristall *m*, Realkristall *m*
imperfect gas reales Gas *n*
imperfection *(Krist)* Störung *f*, *(i. w. S.)* Fehler *m*
impermeability Undurchlässigkeit *f*, Impermeabilität *f*
impermeable undurchlässig, impermeabel
impermeable to water wasserdicht, wasserundurchlässig
impervious *s.* impermeable
impinge *v* 1. aufprallen, anprallen, auftreffen, anstoßen; zusammenprallen, zusammenstoßen; 2. in scharfem Strahl lenken, auftreffen lassen *(Gas, Dampf auf eine Oberfläche)*
impingement Aufprall *m*, Anprall *m*, Anstoß *m*; Zusammenprall *m*, Zusammenstoß *m*; Stoß *m*
impoverish *v* verarmen lassen
impoverishment Verarmung *f*
impregnant *s.* impregnation material
impregnate *v* 1. imprägnieren, (durch-)tränken; 2. *(Lebm)* mit Kohlensäure imprägnieren [versetzen], karbonisieren *(Getränke)*
impregnation 1. Imprägnieren *n*, Imprägnierung *f*, Tränken *n*, Durchtränken *n*, Tränkung *f*, Durchtränkung *f*; 2. *(Lebm)* Imprägnieren *n* [Versetzen *n*]

mit Kohlensäure, Karbonisieren *n (von Getränken)*; 3. *s.* impregnation material
impregnation material Imprägnier(ungs)mittel *n*, Tränkmittel *n*, Tränkstoff *m*
impressed-current cathodic protection katodischer Schutz *m* durch Fremdstrom
impure unrein, unsauber, nicht rein, verunreinigt, verschmutzt, mit Beimischungen
impurity 1. Unreinheit *f*; 2. Verschmutzung *f*, Verunreinigung *f*, Fremdbestandteil *m*, Fremdstoff *m*, Beimengung *f*, Beimischung *f*
impurity atom Fremdatom *n*, *(im Kristallgitter auch)* Stör(stellen)atom *n*, Fremdstörstelle *f*, (chemische) Störstelle *f*
impurity semiconductor Störstellen(halb)leiter *m*, Fremdstoffhalbleiter *m*, Extrinsic-Halbleiter *m*
in-cloud scavenging Rain-out *n*, Ausregnen *n*
in-plane vibration δ-Schwingung *f*, Deformationsschwingung *f* in der Ebene
in situ in situ, an Ort und Stelle, vor Ort *(z. B. Untersuchung, Probennahme)*
in situ analysis In-situ-Analyse *f*, Analyse *f* an Ort und Stelle
in vitro in vitro, unter künstlichen Bedingungen
inactivate *v* desaktivieren, unwirksam machen *(Katalysatoren)*; *(bei Enzymen meist)* inaktivieren
inactivation Desaktivierung *f*, Unwirksammachen *n (von Katalysatoren)*; *(bei Enzymen meist)* Inaktivierung *f*
inactive 1. inaktiv, reaktionslos, nicht reaktionsfähig, reaktionsträge, inert; 2. optisch inaktiv, racemisch; 3. nicht radioaktiv • **render inactive** *s.* inactivate
inactive gas inaktives [inertes, indifferentes] Gas *n*, (reaktions)träges Gas *n*, Inertgas *n*
inactivity Inaktivität *f*, Trägheit *f*
INAH *(isonicotinic acid hydrazide) s.* isoniazid
incandesce *v* weißglühen, auf [bis zur] Weißglut erhitzen
incandescence Weißglühen *n*, Weißglut *f*
incandescency Weißglühen *n*, Weißglut *f*
incandescent weißglühend; glühend, Glüh...
incidence Einfall *m (z. B. eines Strahls)*
incidence angle Einfall(s)winkel *m*
incident einfallend, auffallend, auftreffend, Einfalls...
incident light einfallendes [auffallendes] Licht *n*; Primärlicht *n*
incidental food additive Lebensmittelkontaminant *m*, Kontaminant *m* in Lebensmitteln
incinerate *v* veraschen; verbrennen
incinerating dish Veraschungsschälchen *n*
incineration Veraschen *n*, Veraschung *f*; Verbrennen *n*, Verbrennung *f*
incineration of refuse Müllverbrennung *f*, Müllveraschung *f*
incineration of sludge *(Hyd)* Schlammverbrennung *f*, Schlammveraschung *f*
incineration plant Müllverbrennungsanlage *f*, MVA
incineration test Abbrandversuch *m*
incinerator Müllverbrennungsanlage *f*, MVA, *(i. e. S.)* Müllverbrennungsofen *m*
inclusion Einschluss *m*, Einlagerung *f*, Inklusion *f*
inclusion compound Einschlussverbindung *f*
incoherence Inkohärenz *f (von Strahlung)*
incoherent radiation inkohärente Strahlung *f*
incombustible un(ver)brennbar, nicht brennbar
incombustible Unverbrennbares *n*, Unverbrennliches *n*
incoming air Zuluft *f*
incompatibility Unverträglichkeit *f*, Inkompatibilität *f*
incompatible unverträglich, inkompatibel
incomplete combustion unvollständige [unvollkommene] Verbrennung *f*
incomplete precipitation unvollständige Fällung *f*
incompletely miscible begrenzt [teilweise] mischbar
incongruent melting *(phys Ch)* inkongruentes Schmelzen *n*
incongruent melting point peritektischer Punkt *m*

incongruent reaction peritektische Reaktion *f* [Übergangsreaktion *f*]
incorporate *v* einlagern, einschließen; einbauen, inkorporieren *(Atome oder Atomgruppen in Moleküle)*
incorporation Einlagerung *f*, Einschluss *m*; Einbau *m*, Inkorporation *f (von Atomen oder Atomgruppen in Moleküle)*
increase in pressure Druckerhöhung *f*, Druckanstieg *m*
increase in volume Volumenzunahme *f*
incrust *v* inkrustieren, eine Kruste bilden, verkrusten; inkrustieren, mit einer Kruste überziehen, überkrusten
incrustants Inkrusten *pl*, inkrustierende Substanzen *fpl*, Inkrustsubstanzen *fpl*
incrustation 1. Inkrustierung *f*, Inkrustation *f*, Überkrustung *f*; 2. Inkrustation *f*, Kruste *f*
incubate *v (Biot)* bebrüten, inkubieren *(Fermentationslösung)*
incubation *(Biot)* Bebrütung *f*, Inkubation *f (einer Fermentationslösung)*
incubator Brutschrank *m*
indanthrone vat dye Indanthron-Küpenfarbstoff *m*
indent *v* eindrücken, einprägen; (ein-)kerben
indentation Eindruck *m*, Einprägung *f*; Kerbe *f*
indentation hardness Eindruckhärte *f*, Eindringhärte *f*, Druckhärte *f*
independent joint action *(Tox)* Unabhängigkeitsverbundwirkung *f (ohne gegenseitige Beeinflussung der Komponenten)*
indeterminacy principle (heisenbergsches) Unbestimmtheitsprinzip *n*, (heisenbergsches) Unschärfeprinzip *n*
indeterminate error *(Anal)* zufälliger Fehler *m*
indicate *v* 1. anzeigen; 2. angeben, kennzeichnen; 3. indizieren, beziffern
indication 1. Anzeige *f (von Messgeräten)*; 2. Angabe *f*, Kennzeichnung *f*; 3. Indizierung *f*, Bezifferung *f*
indicator 1. *(Ch)* Indikator *m*; 2. Anzeigevorrichtung *f*, Anzeiger *m*, Zeiger *m*
indicator change Indikatorumschlag *m*
indicator electrode Indikatorelektrode *f*, Messelektrode *f*
indicator paper Indikatorpapier *n*, Reagenzpapier *n*, Prüfpapier *n*
indicator range Umschlagbereich *m*, Umschlaggebiet *n (eines Indikators)*
indifferent electrolyte Grundelektrolyt *m*, Leitelektrolyt *m*, Zusatzelektrolyt *m*, indifferenter Elektrolyt *m*
indifferent gas indifferentes [inertes, inaktives] Gas *n*, (reaktions)träges Gas *n*, Inertgas *n*
indigestibility 1. Unverdaulichkeit *f*; 2. *(Hyd)* Fäulnisunfähigkeit *f*
indigestible 1. *(Lebm)* unverdaulich; 2. *(Hyd)* nicht faulfähig, fäulnisunfähig
indigo (blue) Indig(o)blau *n*, Indigo *m(n)*, Indigotin *n*
indigoid dye Indigofarbstoff *m*, indigoider Farbstoff *m*
indirect arc(-heated) furnace indirekter Lichtbogenofen *m*, Lichtbogenofen *m* mit indirekter Beheizung, Lichtbogenofen *m* mit reiner Strahlungsbeheizung, Lichtbogenstrahlungsofen *m*
indirect solvent latentes Lösungsmittel *n*, mittelbarer [latenter] Löser *m*, Hilfslöser *m*, Hilfslösungsmittel *n*
indissoluble unlöslich, nicht löslich, unl., unlö, nl
indium In Indium *n*
indium hydroxide Indium(III)-hydroxid *n*
indium-tin-oxide Indium-Zink-Oxid *n (organische Leuchtdioden)*
indium trichloride Indiumtrichlorid *n*, Indium(III)-chlorid *n*
indium trioxide Indiumtrioxid *n*, Indium(III)-oxid *n*
individual Individuum *n*
individual cell *(Biot)* Einzelzelle *f*
individual chemical entity [substance] chemisches Individuum *n*
individual order Teilordnung *f (einer Reaktion)*
individual ring *(Nom)* Einzelring *m*
individual step Teilschritt *m (einer Reaktion)*
indivisible unteilbar; nicht spaltbar *(Racemat)*
indole Indol *n*, 2,3-Benzopyrrol *n*
indole alkaloid Indolalkaloid *n*
indoleacetic acid Indolyl-3-essigsäure *f*, β-Indolylessigsäure *f*, IES *f*
indophenol Indophenol *n*, Chinonphenolimin *n*

indoxyl Indoxyl *n*, 3-Hydroxyindol *n*
indoxylcarboxylic acid Indoxylcarbonsäure *f*
induce *v* **crystallization** die Kristallisation anregen [auslösen]
induced dipole moment induziertes Dipolmoment *n*
induced electron emission induzierte Elektronenemission *f*, IEE
induced enzyme induziertes Enzym *n*
induced nucleation *(Krist)* heterogene Keimbildung *f*
induced radioactivity induzierte [künstliche] Radioaktivität *f*
induced reaction induzierte Reaktion *f*
inducer *s.* inductor 2.
induction Induktion *f*
induction effect *(phys Ch)* Induktionseffekt *m*, induktiver Effekt *m*, I-Effekt *m*
induction furnace Induktionsofen *m*, induktionsbeheizter Ofen *m*
induction period *(phys Ch)* Induktionsperiode *f*, Anlaufperiode *f*, Startperiode *f*
inductive effect 1. *(phys Ch)* induktiver Effekt *m*, Induktionseffekt *m*, I-Effekt *m*; 2. *(Bioch)* Induktionswirkung *f*
inductively coupled plasma *(Anal)* induktiv gekoppeltes Plasma *n*, ICP
inductor 1. Induktor *m (bei induzierten Reaktionen)*; 2. *(Bioch)* Induktor *m*
industrial alcohol technischer Alkohol *m*, Industriealkohol *m*
industrial chemical Industriechemikalie *f*, industrielle Chemikalie *f*
industrial chemist industrieller [technischer] Chemiker *m*, Industriechemiker *m*, Werkschemiker *m*
industrial chemistry industrielle [technische] Chemie *f*, Industriechemie *f*
industrial enzyme *(Biot)* Industrie-Enzym *n*, industrielles [technisches] Enzym *n*
industrial fermentation *(Biot)* industrielle Fermentation *f*, Fermentation *f* in technischem Maßstab
industrial fertilizer Handelsdünger *m*
industrial gas technisches Gas *n (in technischem Ausmaß erzeugt, brennbar oder nicht brennbar)*; Industriegas *n*
industrial hygiene Arbeitshygiene *f*
industrial microbiology industrielle [technische] Mikrobiologie *f*
industrial off-gas Industrieabgas *n*
industrial waste 1. industrielle Abprodukte *npl*, Industrieabfälle *mpl*, Fabrikationsabfälle *mpl*; 2. *s.* industrial waste water
industrial waste water Industrieabwasser *n*, industrielles Abwasser *n*, Abwasser *n* von Industriebetrieben
industrial yeast *(Biot)* industriell genutzte Hefe *f*, Kulturhefe *f*, Zuchthefe *f*
ineffective unwirksam, wirkungslos
inelastic collision unelastischer Stoß *m*
inelastic neutron scattering *(Kern, Anal)* unelastische Neutronenstreuung *f*
inert inert, inaktiv, reaktionsträge, reaktionslos
inert atmosphere inerte [indifferente] Atmosphäre *f*
inert gas 1. Inertgas *n*, inertes [indifferentes, inaktives] Gas *n*, (reaktions-)träges Gas *n*; 2. *s.* noble gas
inert substance inerter [indifferenter] Stoff *m*, Inertstoff *m*, Inertsubstanz *f*
inertia Trägheit *f*, Beharrungsvermögen *n*
inertness Trägheit *f*, Reaktionsträgheit *f*, Inaktivität *f*
infinite dilution unendliche Verdünnung *f*
infinitely dilute unendlich verdünnt
infl *s.* 1. inflammable; 2. influence
inflame *v* entflammen, (ent)zünden; entflammen, sich entzünden
inflammability Entflammbarkeit *f*, Entzündbarkeit *f*, Entzündlichkeit *f*, Brennbarkeit *f*
inflammable entflammbar, entzündbar, entzündlich, brennbar, inflammabel
inflammation Entflammung *f*, Entzündung *f*, Zündung *f*
inflection point Wendepunkt *m (einer Kurve)*
inflow 1. Zufließen *n*, Zufluss *m*, Zulaufen *n*, Einfließen *n*, Einströmen *n*; 2. *s.* influent
influence *v* beeinflussen, einwirken auf
influence Einfluss *m*, Einwirkung *f*, Wirkung *f*
influent zufließende [zulaufende] Flüssigkeit *f*, Zufluss *m*, Zulauf *m*, Einlauf *m*
information transfer *(Bioch)* Informationsübertragung *f*

infrared infrarot, Infrarot..., IR-...
infrared Infrarot *n*
infrared absorption spectroscopy IR-Absorptionsspektroskopie *f*
infrared-active *(Anal)* infrarot-aktiv
infrared detector Infrarotdetektor *m*
infrared-inactive *(Anal)* infrarot-inaktiv
infrared lamp Infrarotlampe *f*, Infrarotstrahler *m*
infrared microscope Infrarotmikroskop *n*
infrared radiation Infrarotstrahlung *f*, infrarote Strahlung *f*
infrared (spectral) region Infrarotbereich *m*, Infrarotgebiet *n*, IR-Spektralbereich *m*
infrared spectroscopy Infrarotspektroskopie *f*
infrared spectrum Infrarotspektrum *n*
infrared-transmitting infrarotdurchlässig
infrared-transparent infrarotdurchlässig
infusibility Unschmelzbarkeit *f*
infusible unschmelzbar, *(i. w. S.)* schwer schmelzbar
infusion 1. *(Pharm)* Aufguss *m*, Infus(um) *n*, Infusionslösung *f*; 2. *(Pharm)* Infusion *f*, Einfließenlassen *n*
infusorial earth Infusorienerde *f*, Diatomeenerde *f*, Kieselgur *f*
ingest *v* mit der Nahrung aufnehmen
ingestible *(Pharm)* zum Einnehmen
ingestion Aufnahme *f* mit der Nahrung
ingot *(Met)* Block *m*, Gussblock *m*, Rohblock *m*, Ingot *m*, Barren *m*
ingot iron [steel] Flussstahl *m*
ingrain dye Entwicklungsfarbstoff *m*, *(i. e. S.)* (auf der Faser erzeugter) unlöslicher Azofarbstoff *m*
ingredient 1. Bestandteil *m*, Inhaltsstoff *m*; 2. *(Lebm, Pharm)* Ingrediens *n*, Ingredienz *f*, Zusatzstoff *m*
inherent toxicity innewohnende Toxizität *f*
inhibit *v* inhibieren, hemmen, verzögern, bremsen, behindern
inhibition Inhibition *f*, Inhibitorwirkung *f*, Hemmung *f*, Verzögerung *f*, *(i. e. S.)* negative Katalyse *f*, Antikatalyse *f*
inhibitor Inhibitor *m*, Hemmstoff *m*, Verzögerer *m*, Passivator *m*, *(i. e. S.)* negativer Katalysator *m*, Antikatalysator *m*
inhibitory inhibierend, hemmend, verzögernd, bremsend, hindernd
inhibitory effect Inhibitorwirkung *f*, Hemmwirkung *f*
inhomogeneity Inhomogenität *f*, innere Uneinheitlichkeit *f*
inhomogeneous inhomogen, innerlich uneinheitlich
initial boiling point Siedebeginn *m*, SB, Anfangssiedepunkt *m*, Beginn-Siedepunkt *m*, Beginn-Kochpunkt *m*
initial concentration Anfangskonzentration *f*, Ausgangskonzentration *f*
initial material Ausgangsstoff *m*, Ausgangssubstanz *f*
initial period Vorperiode *f*, Vorversuch *m* *(einer kalorimetrischen Messung)*
initial reaction rate Anfangsgeschwindigkeit *f* (der Reaktion)
initial state Ausgangszustand *m*, Anfangszustand *m*
initiate *v* initiieren, anregen, einleiten, in Gang bringen, starten, auslösen
initiated by irradiation strahleninitiiert *(z. B. Polymerisation)*
initiation 1. Initiierung *f*, Anregung *f*, Einleitung *f*, Start *m (einer Reaktion)*; 2. *(Kst)* Kettenstart *m*
initiation codon *(Bioch)* Startcodon *n*, Initiationscodon *n*, Initiatorcodon *n*, Initiatortriplett *n*
initiation reaction Startreaktion *f*, Primärreaktion *f*
initiation step Startschritt *m*, Primärschritt *m*, Primärstufe *f (einer Reaktion)*
initiator Initiator *m*, Aktivator *m*
inject *v* 1. injizieren, (ein)spritzen; 2. freisetzen *(z. B. Schadstoffe in die Luft)*
injection 1. Injektion *f*, Einspritzung *f*; 2. Freisetzung *f (z. B. von Schadstoffen in die Luft)*
injection block *(Chromat)* Einspritzblock *m*
injection embossing *(Kst)* Spritzprägen *n*
injection-moulded part spritzgegossenes Formteil *n*, Spritzgussteil *n*, Spritzling *m*
injection moulding 1. Spritzgießverfahren *n*, Spritzgussverfahren, Spritzgie-

ßen *n*, Spritzguss *m*; 2. *s.* injection-moulded part
injection-moulding compound Spritzgussmasse *f*
injection port *(Chromat)* Einspritzstelle *f*, *(am Gaschromatographen)* Einspritzblock *m*
injection pressure Spritzdruck *m*, Einspritzdruck *m*
injection valve *(Anal)* Einspritzventil *n*, Injektionsventil *n*
injector 1. Injektor *m*, Druckstrahlpumpe *f*; 2. *(Anal)* Einspritzvorrichtung *f*
ink 1. Druckfarbe *f*, (grafische) Farbe *f*; 2. Tinte *f*
inky tintig *(Geschmack, Aussehen)*
inlet 1. Einlass *m*, Einlauf *m (Öffnung)*; 2. zufließende [zulaufende] Flüssigkeit *f*, Zufluss *m*, Zulauf *m*
inlet air Zuluft *f*; Frischluft *f*
inlet port Eintrittsöffnung *f*, Einlauföffnung *f*
inlet system *(Anal)* Probeneinlasssystem *n*
inlet temperature Eingangstemperatur *f*, Eintrittstemperatur *f*
inlet valve Einlassventil *n*
INN *s.* international non-proprietary name
inner anhydride *s.* intramolecular anhydride
inner complex Innerkomplex *m*, Innenkomplex *m*, Durchdringungskomplex *m*
inner electron inneres Elektron *n*, Elektron *n* der inneren Schalen
inner orbital complex *s.* inner complex
inner potential *(Ech)* Galvani-Potenzial *n*, inneres elektrisches Potenzial *n*
inner quantum number innere Quantenzahl *f*, Gesamtdrehimpulsquantenzahl *f*
inner salt inneres [intramolekulares] Salz *n*, Zwitterion *n*, Ampho-Ion *n*
inner shell innere Schale *f*, Innenschale *f*
inner-sphere complex Innensphärenkomplex *m*, Innersphäre-Komplex *m* *(Koordinationslehre)*
inner surface innere Oberfläche *f*, Innen(ober)fläche *f*
innocuous unschädlich, harmlos
inoculate *v* 1. *(Biot)* (be)impfen, animpfen *(ein Nährmedium)*; 2. *(Krist)* impfen
inoculation 1. *(Biot)* Beimpfen *n*, Beimpfung *f*, Animpfen *n*, Animpfung *f*, Impfung *f (eines Nährmediums)*; 2. *(Krist)* Impfen *n*
inoculation medium *(Biot)* Impfmedium *n*, *(i. e. S.)* Impflösung *f*
inoculum *(Biot)* Impfgut *n*, Impfmaterial *n*, Impfkultur *f*, Inokulum *n*
inoculum cultivation Impfgutanzucht *f*
inodorous geruchlos, geruchfrei
inoffensive nicht aggressiv
inorganic anorganisch
inorganic acid anorganische Säure *f*
inorganic benzene anorganisches Benzen *n*, Borazin *n*, *Cyclotriborazan *n*
inorganic-chemical anorganisch-chemisch
inorganic-chemical technology anorganisch-chemische Technik *f*; anorganisch-chemische Technologie *f*
inorganic chemist Anorganiker *m*
inorganic chemistry anorganische Chemie *f*
inorganic coating 1. anorganischer Beschichtungsstoff *m*; anorganischer Anstrichstoff *m*, Anstrichstoff *m* mit anorganischem Bindemittel; 2. anorganische nicht metallische Schutzschicht *f*
inorganic fertilizer Mineraldüngemittel *n*, mineralisches Düngemittel *n*
inorganic polymer anorganisches Polymer(e) *n*
inorganic salt anorganisches Salz *n*
inorganics anorganische Verbindungen *fpl*
inosilicate *(Min)* Inosilicat *n*, Kettensilicat *n*
inosine Inosin *n*, Ino, Hypoxanthosin *n*
inosinic acid Inosinsäure *f*, Inosin-5′-monophosphat *n (Nucleotid)*
inosite Inositol *n*, *Cyclohexan-1,2,3,4,5,6-hexol *n*
inositol Inositol *n*, *Cyclohexan-1,2,3,4,5,6-hexol *n*
input Aufwand *m*, Einsatzmenge *f*
input of energy Energiezufuhr *f*; Energieeintrag *m*, zugeführte Energie *f*
INS *s.* inelastic neutron scattering
insect attractant Insektenanlockmittel *n*, Insektenlockstoff *m*
insect repellent Insektenabwehrmittel *n*,

Insektenschutzmittel *n*, insektenvertreibendes [insektenabschreckendes, insektenabstoßendes] Mittel *n*
insecticidal insektizid, insektentötend, Insekten vernichtend
insecticide Insektizid *n*, Insektenbekämpfungsmittel *n*, Insektenvertilger *m*
insecticide resistance Insektizidresistenz *f*
insensitive unempfindlich
insensitivity Unempfindlichkeit *f*
inseparable un(ab)trennbar
insert *v* einführen, einsetzen, einbringen; einlagern, einbauen, einfügen *(z. B. Atome in Zwischengitterplätze)*; einschieben; sich einschieben *(Atome und Atomgruppen in Bindungen)*
insertion Einführen *n*, Einsetzen *n*, Einbringen *n*; Einlagerung *f*, Einbau *m*, Einfügen *n (z. B. von Atomen in Zwischengitterplätze)*; Insertion *f*, Einschub *m (von Atomen und Atomgruppen in Bindungen)*
insertion polymerization Insertionspolymerisation *f*, Polyinsertion *f*
insertion reaction 1. Einschubreaktion *f*, Einschiebungsreaktion *f*; 2. *(Krist)* topotaktische Reaktion *f*
inside diameter Innendurchmesser *m*
inside surface innere Oberfläche *f*, Innen(ober)fläche *f*
insol *s.* insoluble
insolubility Unlöslichkeit *f*, Nichtlöslichkeit *f*
insolubilize *v* unlöslich machen
insoluble unlöslich, nicht löslich, unl., unlö, nl
insoluble unlösliche [nicht lösliche] Substanz *f*, Unlösliches *n*
insoluble chloride group *(Anal)* Salzsäure-Gruppe *f*, erste Gruppe *f*
insoluble in water wasserunlöslich
insoluble sulphide group *(Anal)* Schwefelwasserstoff-Gruppe *f*, zweite Gruppe *f*
inspissate *v* einengen, eindampfen, eindicken, verdicken; dickflüssig [dick, zähflüssig, zäh] werden
inspissation Einengen *n*, Eindampfen *n*, Eindicken *n*, Verdicken *n*
instability Instabilität *f*, Unbeständigkeit *f*, Labilität *f*
instability constant (of complex ions) Instabilitätskonstante *f*, Komplexdissoziationskonstante *f*, Dissoziationskonstante *f*
instable unstabil, instabil, unbeständig, labil
instrument reading gemessener Wert *m*, Ablesewert *m*
instrumental analysis Instrumentenanalyse *f*
instrumental error Instrumentenfehler *m*, Gerät(e)fehler *m*
insulate *v* isolieren, mit Isoliermaterial einhüllen, *(gegen Schall oder Wärme besser:)* dämmen
insulation 1. Isolieren *n*, Isolierung *f*, Isolation *f*, *(gegen Schall oder Wärme besser:)* Dämmung *f*; 2. *s.* insulation material
insulation material Isolierstoff *m*, Isoliermaterial *n*, *(gegen Schall oder Wärme besser:)* Dämmstoff *m*
insulator 1. Isolator *m*, Isolationskörper *m*; 2. Isolierstoff *m*, Isoliermaterial *n*, *(gegen Schall oder Wärme besser:)* Dämmstoff *m*
intake 1. Einlass *m*, Eintritt *m*, Einlauf *m*, Eintrittsstelle *f*, Einlaufstelle *f*; 2. Aufnahme *f (z. B. von Stoffen durch den Körper)*; 3. aufgenommene Menge *f*
integral chromatogram Integralchromatogramm *n*
integral detector Integraldetektor *m* *(Gaschromatographie)*
integral heat integrale Wärme *f*
integral reactor Integralreaktor *m*
intensely coloured intensiv farbig, farbkräftig, farbstark
intensification Verstärkung *f*, Intensivierung *f*
intensification of colour Farbvertiefung *f*
intensify *v* verstärken, intensivieren; vertiefen *(Farbe)*
intensity Intensität *f*, Stärke *f*
intensity distribution *(Spektr)* Intensitätsverteilung *f*
intensity of radiation 1. Strahlstärke *f*; 2. Strahlungsflussdichte *f*
intensive property [quantity] intensive Größe *f*, Intensitätsgröße *f*, Qualitätsgröße *f*
intentional food additive Lebensmittelzusatzstoff *m*, Zusatzstoff *m*, Additiv *n*

inter-ester innerer Ester *m*
interact *v* wechselwirken, in Wechselwirkung treten; wechselwirken, in Wechselwirkung stehen
interaction Wechselwirkung *f*
interaction force Wechselwirkungskraft *f*
interatomic interatomar
interatomic distance [separation, spacing] Atom(kern)abstand *m*, Kernabstand *m*
interband transition Band-Band-Übergang *m (bei Festkörpern)*
interbonding angle Bindungswinkel *m*, Valenzwinkel *m*
intercalate *v* interkalieren, einlagern *(Gastmoleküle zwischen Schichten)*
intercalate Interkalat *n*
intercalation Interkalation *f*, Einlagerung *f (von Gastmolekülen zwischen Schichten)*; Interkalation *f*, Einschub *m (planarer Ringsysteme zwischen die gestapelten Basen von Nucleinsäuren)*
intercalation compound Interkalationsverbindung *f*, *(schichtförmige)* Einlagerungsverbindung *f*
intercept *v* abfangen *(reaktive Zwischenprodukte)*
intercept *(Krist)* Achsenabschnitt *m*, Parameter *m*
interception 1. Abfangen *n (reaktiver Zwischenprodukte)*; 2. *s.* intercept
interchange gegenseitiger [wechselseitiger] Austausch *m*
interchange esterification Umesterung *f*
interchangeability gegenseitige [wechselseitige] Austauschbarkeit *f*
interchangeable gegenseitig [wechselseitig] austauschbar
intercombination *(Spektr)* Interkombination *f*
interconversion (reaction) gegenseitige [wechselseitige] Umwandlung *f* [Umsetzung *f*], Interkonversion *f*
interconvert *v* gegenseitig [wechselseitig] umwandeln [umsetzen]; sich gegenseitig umwandeln [umsetzen]
interconvertible gegenseitig [wechselseitig] umwandelbar interkonvertierbar
intercrystalline interkristallin, zwischenkristallin
intercrystalline corrosion *s.* intergranular corrosion
interelectronic repulsion interelektronische Abstoßung *f*, Elektronenabstoßung *f*
interesterification Umesterung *f*
interesterify *v* umestern
interface Grenzfläche *f*, Phasengrenzfläche *f*
interface-active grenzflächenaktiv
interfacial activity Grenzflächenaktivität *f*
interfacial polycondensation Grenzflächen(poly)kondensation *f*
interfacial polymerization Grenzflächenpolymerisation *f*
interfacial tension Grenzflächenspannung *f*
interfere *v* **with** eingreifen in, beeinflussen, interferieren mit; beeinträchtigen, behindern, stören
interference gegenseitige Beeinflussung *f (von Substituenten)*; Interferenz *f (von Wellen)*
interferent *(Anal)* Störstoff *m*, Störsubstanz *f*
interfering element *(Anal)* Störelement *n*
interfering ion Störion *n*
interferogram *(Anal)* Interferogramm *n*
interferometer Interferometer *n*
interferometry Interferometrie *f*
interferon *(Bioch)* Interferon *n*
interfuse *v* verschmelzen
interfusion Verschmelzung *f*
intergranular corrosion interkristalline Korrosion *f*, Korngrenzenkorrosion *f*
interhalogen (compound) Interhalogenverbindung *f*, Interhalogen *n (Verbindung von Halogenen untereinander)*
interionic interionisch
interionic force interionische Wechselwirkungskraft *f*
interior paint Innenanstrichfarbe *f*
interlaboratory test *(Anal)* Ringversuch *m*
interlayer Zwischenschicht *f*, Trennschicht *f*; Zwischenlage *f*
interleave *v* schichtförmig einlagern; zwischen Schichten einlagern
intermediary metabolism *(Bioch)* Intermediärstoffwechsel *m*, Zwischenstoffwechsel *m*, Intermediärmetabolismus *m*

intermediary product *s.* intermediate
intermediate intermediär, Zwischen...
intermediate Intermediat *n*, Transient *m*, Zwischenstoff *m*, Zwischenprodukt *n* *(einer chemischen Reaktion)*; Zwischenprodukt *n*, Zwischengut *n*, Zwischenstoff *m*, Zwischenerzeugnis *n*, Intermediärprodukt *n*, Vorprodukt *n* *(in der Verfahrenstechnik)*
intermediate compound Zwischenverbindung *f*, Intermediärverbindung *f*, intermediäre Verbindung *f*
intermediate cut *(Dest)* Zwischenfraktion *f*, Zwischenlauf *m*, Übergangsfraktion *f*
intermediate product Zwischenprodukt *n*, Zwischengut *n*, Zwischenstoff *m*, Zwischenerzeugnis *n*, Intermediärprodukt *n*, Vorprodukt *n*
intermediate stage Zwischenstufe *f*, Zwischenstadium *n*
intermediate state Zwischenzustand *m*
intermetallic (compound, phase) intermetallische Phase *f* [Verbindung *f*]
intermittent diskontinuierlich, periodisch, unstetig
intermix *v* vermischen; sich vermischen
intermolecular attraction intermolekulare [zwischenmolekulare] Anziehung *f*
intermolecular (bonding) force (zwischen)molekulare Kraft *f*, Molekularkraft *f*
internal 1. innerer, innen befindlich, Innen...; 2. *(Pharm)* intern, innerlich, zum inneren Gebrauch, zum Einnehmen
internal anhydride inneres [intramolekulares] Anhydrid *n*
internal energy innere Energie *f*
internal phase innere Phase *f*, disperse [dispergierte, zerteilte, offene] Phase *f*, Dispersum *n*
internal plasticizer innerer Weichmacher *m*
internal pressure 1. Innendruck *m*, innerer Druck *m* *(in einem Behälter)*; 2. *(phys Ch)* Binnendruck *m*, Kohäsionsdruck *m*
internal salt inneres [intramolekulares] Salz *n*, Zwitterion *n*, Ampho-Ion *n*
internal standard *(Anal)* innerer [interner] Standard *m*
internal surface (area) innere Oberfläche *f*, Innen(ober)fläche *f*
internals Einbauten *mpl*
international non-proprietary name WHO-Name *m* *(nicht schützbarer Arzneimittelname)*
international unit internationale Einheit *f*, I.E. *(von Wirkstoffen)*
internuclear distance [separation] Kernabstand *m*, Atom(kern)abstand *m*
interpenetrate *v* einander [sich] durchdringen
interpenetrating network Durchdringungsnetzwerk *n*, interpenetrierendes Netzwerk *n*
intersolubility gegenseitige [wechselseitige] Löslichkeit *f*
interstice 1. Zwischenraum *m*, Lücke *f*; 2. *s.* interstitial lattice site
interstitial interstitiell, Interstitial..., Einlagerungs...
interstitial atom Zwischengitteratom *n*, Einlagerungsatom *n*, Atom *n* auf Zwischengitterplatz
interstitial carbide Einlagerungscarbid *n*, interstitielles Carbid *n*
interstitial-cell-stimulating hormone *Lutropin *n*, zwischenzellenstimulierendes [interstitielle Zellen stimulierendes, luteinisierendes] Hormon *n*, ICSH, Luteinisierungshormon *n*, LH, Gelbkörperreifungshormon *n*
interstitial compound Einlagerungsverbindung *f*, interstitielle Verbindung *f*
interstitial ion Zwischengitterion *n*, Ion *n* auf Zwischengitterplatz
interstitial lattice site Zwischengitterplatz *m*, Zwischengitterstelle *f*
interstitial (solid) solution Einlagerungsmischkristall *m*, Interstitialmischkristall *m*, interstitielle feste Lösung *f*, Interstitiallösung *f*
intertwined verdrillt, ineinandergewunden *(Makromolekülketten)*
interwound verdrillt, ineinandergewunden *(Makromolekülketten)*
intimate innig *(Durchmischung)*; eng, innig *(Kontakt)*
intoxicant berauschendes Getränk *n*; Rauschgift *n*
intoxicate *v* 1. berauschen, *(i. e. S.)* trunken machen; 2. *(Med)* vergiften
intoxication 1. Trunkenmachen *n*; Berauschteit *f*, *(i. e. S.)* Trunkenheit *f*; 2. *(Med)* Intoxikation *f*, Vergiftung *f*

intra-ester innerer Ester *m*
intracellular enzyme intrazellulares [sessiles] Enzym *n*, Zellenzym *n*
intrachain disulphide bond *(Bioch)* Disulfidspange *f*, –S–S-Spange *f*
intramolecular intramolekular, innermolekular
intramolecular anhydride intramolekulares [inneres] Anhydrid *n*
intramolecular rearrangement intramolekulare [innermolekulare] Umlagerung *f*
intramolecular salt intramolekulares [inneres] Salz *n*, Zwitterion *n*, Ampholon *n*
intrastrand linkage *s.* intrachain disulphide bond
intrinsic angular momentum innerer Drehimpuls *m*, Eigendrehimpuls *m*, Spin *m*
intrinsic (electrical) conductivity Eigen(halb)leitfähigkeit *f*, Intrinsic-Leitfähigkeit *f*, Eigenleitung *f*
intrinsic energy innere Energie *f*
intrinsic factor *(Bioch)* Intrinsic factor *m*, innerer Faktor *m*
intrinsic pressure Kohäsionsdruck *m*, Binnendruck *m*
intrinsic property *(phys Ch)* innere Eigenschaft *f*
intrinsic semiconductor Eigenhalbleiter *m*, Halbleiter *m* mit Eigenleitfähigkeit
intrinsic viscosity Grenzviskositätszahl *f*, Staudinger-Index *m*
introduce *v* einbringen, eintragen; einleiten, einführen *(Substituenten)*
introduction Einbringen *n*, Eintragen *n*; Einleiten *n*; Einführung *f (von Substituenten)*
invariable unveränderlich, invariabel
invariant invariant, non-variant, ohne Freiheit(sgrad)
inverse Raman spectroscopy inverse Raman-Spektroskopie *f*, IRS
inversion 1. Inversion *f*, Invertierung *f*, Umkehr(ung) *f (der Konfiguration)*; 2. *(Krist)* Umwandlung *f*; Inversion *f (Sonderfall der atmosphärischen Schichtung)*
inversion centre Inversionszentrum *n*, Symmetriezentrum *n (eines Moleküls)*
inversion height Inversionshöhe *f*, Sperrschichthöhe *f (in der Atmosphäre)*
inversion of configuration Konfigurationsumkehr *f*, Walden-Umkehr *f*, Konfigurationswechsel *m*, Inversion *f* (der Konfiguration)
inversion of phases Phasenumkehr *f*
inversion point Umwandlungspunkt *m (von Modifikationen)*
invert *v* 1. umkehren *(Vorgänge)*; invertieren *(Konfiguration)*; umdrehen, umkehren; 2. *(Krist)* sich umwandeln
invert sugar Invertzucker *m*
inverted micelle inverse Mizelle *f*
inverted population *(phys Ch)* inverse Besetzung *f*
investigate *v* untersuchen
investigation of structure Strukturuntersuchung *f*
inviscid flow reibungsfreie Strömung *f*
inviscid fluid ideales [reibungsfreies, pascalsches] Fluid *n*
iodate $M^{I}IO_3$ Iodat *n*
iodic Iod..., *(i. e. S.)* Iod(V)-...
iodic acid Iodsäure *f*
iodide $M^{I}I$ Iodid *n*
iodide of starch Iodstärke *f*
iodimetry indirektes iodometrisches Titrationsverfahren *n (Titration mit Iod)*
iodinate *v* iodieren
iodination Iodieren *n*, Iodierung *f*
iodine I Iod *n*
iodine colour reaction *(Bioch, Anal)* Iodstärkereaktion *f*
iodine dioxide Ioddioxid *n*
iodine monochloride Iod(mono)chlorid *n*, Iod(I)-chlorid *n*
iodine number *s.* iodine value
iodine pentoxide Iodpentoxid *n*, Diiodpentoxid *n*, Iod(V)-oxid *n*
iodine trichloride Iodtrichlorid *n*, Iod(III)-chlorid *n*
iodine value Iodzahl *f*, IZ
iodine vapour Ioddampf *m*
iodization Iodieren *n*, Iodierung *f*
iodize *v* iodieren
iodized (table) salt Iodspeisesalz *n*
iodobenzene Iodbenzen *n*
iodoform Iodoform *n*, Triiodmethan *n*
iodometric iodometrisch
iodometry Iodometrie *f*, iodometrisches Titrationsverfahren *n (Titration mit oder von Iod)*; *(i. e. S.)* indirektes iodometrisches Titrationsverfahren *n (Titration von Iod)*

iodonium compound Iodoniumverbindung *f*
iodonium ion Iodonium-Ion *n*
iodous Iod..., *(i. e. S.)* Iod(III)-...
ion Ion *n*
ion activity Ionenaktivität *f*
ion association Ionenassoziation *f*
ion atmosphere Ionenwolke *f*, Ionenatmosphäre *f (in Elektrolytlösungen)*
ion beam Ionenstrahl *m*
ion beam assisted deposition ionenstrahlgestütztes Abscheiden *n*
ion beam induced deposition ionenstrahlinduziertes Abscheiden *n*
ion bombardment Ionenbeschuss *m*, Ionenbombardement *n*
ion carrier *(Bioch)* Ionophor *m*, Ionencarrier *m*
ion chromatography Ionenchromatographie *f*
ion cloud *s.* ion atmosphere
ion cyclotron resonance Ionenzyklotron-Resonanz *f*
ion detector Ionendetektor *m*
ion exchange 1. Ionenaustausch *m*; 2. *(Boden)* Ionenumtausch *m*
ion-exchange chromatography Ionenaustauschchromatographie *f*
ion-exchange column Ionenaustauschersäule *f*
ion-exchange equilibrium Ionenaustauschgleichgewicht *n*
ion-exchange membrane Ionenaustausch(er)membran *f*
ion-exchange resin Ionenaustausch(er)harz *n*, Austausch(er)harz *n*, Ionenaustauscher *m* auf Kunstharzbasis, Kunstharz(ionen)austauscher *m*
ion-exchange site Austauschplatz *m (eines Ionenaustauschers)*
ion exchanger Ionenaustauscher *m*
ion exclusion Ionenausschluss *m*
ion-exclusion chromatography Ionenausschlusschromatographie *f*
ion imaging Ionenabbildungsspektroskopie *f*
ion implantation Ionenimplantation *f (Dotierungstechnik)*
ion-ion recombination Ionenrekombination *f*, Ion-Ion-Rekombination *f*
ion kinetic energy spectroscopy Ionenenergie-Spektroskopie *f* zur Analyse metastabiler Zerfälle
ion lattice *(Krist)* Ionengitter *n*
ion microprobe Ionenmikrosonde *f*
ion migration Ionenwanderung *f*, Migration *f*
ion mobility Ionenbeweglichkeit *f*
ion pair Ionenpaar *n*
ion-pair chromatography Ionenpaarchromatographie *f*
ion plating Ionenplattieren *n (Metallbeschichtungstechnik)*
ion probe *s.* ion microprobe
ion-probe microanalysis Ionenstrahl-Mikroanalyse *f*
ion radical Radikalion *n*
ion-scattering spectroscopy Ionenstreuungsspektroskopie *f*
ion-selective ionenselektiv, *(bei Elektroden auch)* ionensensitiv
ion-sensitive electrode *(Anal)* ionensensitive [ionenselektive] Elektrode *f*
ion-sensitive field-effect transistor *(Anal)* ionensensitiver Feldeffekttransistor *m*
ion separation Ionentrennung *f*
ion source Ionenquelle *f*
ionic ionisch, ional, Ionen...; ionoid *(Addition)*; ionisiert
ionic attraction Anziehung *f* zwischen Ionen (entgegengesetzter Ladung)
ionic balance *s.* ionic equilibrium
ionic bond Ionenbeziehung *f*, Ionenbindung *f*, heteropolare Bindung *f*
ionic character Ionencharakter *m*
ionic charge Ionenladung *f*
ionic charge number Ionenladungszahl *f*
ionic compound Ionenverbindung *f*, heteropolare Verbindung *f*
ionic conductance (spezifische) Ionenleitfähigkeit *f*
ionic conduction Ionenleitung *f*
ionic conductivity Ionenleitfähigkeit *f*
ionic conductor Ionenleiter *m*, Leiter *m* zweiter Ordnung
ionic crystal Ionenkristall *m*
ionic crystal lattice *(Krist)* Ionengitter *n*
ionic elimination ionische Eliminierung *f*
ionic equilibrium Ionengleichgewicht *n*, Elektrolytgleichgewicht *n*
ionic equivalent conductance Ionenäquivalentleitfähigkeit *f*
ionic formula Ionenformel *f*
ionic interaction interionische Wech-

selwirkung *f*, Ionenwechselwirkung *f*, Ion-Ion-Wechselwirkung *f*
ionic liquid ionische Flüssigkeit *f (Solarzellen)*
ionic plasma Ionenplasma *n*
ionic polymerization ionische Polymerisation *f*, Ionenpolymerisation *f*
ionic product Ionenprodukt *n*
ionic radius Ionenradius *m*
ionic reaction Ionenreaktion *f*, ionische Reaktion *f*
ionic repulsion Abstoßung *f* zwischen Ionen (gleicher Ladung)
ionic solid ionischer Festkörper *m*
ionic solution Elektrolytlösung *f*
ionic strength Ionenstärke *f*
ionic surfactant ionogener grenzflächenaktiver Stoff *m*
ionic susceptibility Ionensuszeptibilität *f*
ionic valence [valency] Ionenwertigkeit *f*
ionizable dissoziabel *(in Lösungen)*; ionisierbar *(Gase)*
ionization elektrolytische Dissoziation *f (in Lösungen)*; Ionisation *f*, Ionisierung *(von Gasen)* • **capable of ionization** *s.* ionizable
ionization constant Dissoziationskonstante *f*
ionization detector *(Anal)* Ionisationsdetektor *m*
ionization energy Ionisierungsenergie *f*
ionization equilibrium Dissoziationsgleichgewicht *n*
ionization gauge Ionisationsvakuummeter *n*
ionization potential Ionisationspotenzial *n*, Ionisierungsspannung *f*
ionization reaction Dissoziationsreaktion *f*; Ionisierungsreaktion *f*
ionize *v* dissoziieren, in Ionen zerfallen *(in Lösungen)*; ionisieren *(Gase)*
ionizing radiation ionisierende Strahlung *f*
ionogenic ionogen, ionenerzeugend; ionisierbar
ionomer Ionomer(e) *n*
ionophilic ionophil
ionophore *(Bioch)* Ionophor *m*, ionophore Substanz *f*, Ionencarrier *m*
ionophoresis Ionophorese *f*
ionophoric *(Bioch)* ionophor
ionotropic ionotrop
ionotropy Ionotropie *f*
IOS = International Organization for Standardization
ipecacuanha alkaloid Ipecacuanha-alkaloid *n*
IPN *s.* interpenetrating network
ipso-attack *(org Ch)* ipso-Angriff *m*, ipso-Substitution *f*
IR *s.* 1. infrared; 2. isoprene rubber
IR drop IR-Abfall *m*, iR-Abfall *m*, (ohmscher) Spannungsabfall *m*, Widerstandspolarisation *f*, Widerstandsüberspannung *f*
iR drop IR-Abfall *m*, iR-Abfall *m*, (ohmscher) Spannungsabfall *m*, Widerstandspolarisation *f*, Widerstandsüberspannung *f*
iridescence Irisieren *n*, Schillern *n*
iridescent irisierend
iridium Ir Iridium *n*
iron *v* **out** ausgleichen *(z. B. die Konzentration)*
iron Fe Eisen *n*
iron alloy Eisenlegierung *f*
iron-ammonium sulphate Ammoniumeisen(III)-sulfat-12-Wasser *n*, Ammoniumeisenalaun *m*
iron-bearing eisenhaltig
iron carbide Eisencarbid *n*, Zementit *m*
iron carbonyl Eisenpentacarbonyl *n*
iron catalyst Eisenkatalysator *m*, Eisenkontakt *m*
iron dichloride Eisendichlorid *n*, Eisen(II)-chlorid *n*
iron disulphide Eisen(II)-disulfid *n*
iron hydroxide floc *(Hyd)* Eisenhydroxidflocken *fpl*
iron metallurgy Eisenmetallurgie *f*, Eisenhüttenwesen *n*, Schwarzmetallurgie *f*
iron monosulphide Eisenmonosulfid *n*, Eisen(II)-sulfid *n*
iron monoxide Eisenmonoxid *n*, Eisen(II)-oxid *n*
iron nitride Eisennitrid *n*
iron ore Eisenerz *n*
iron oxide red Eisenoxidrot *n*
iron pentacarbonyl Eisenpentacarbonyl *n*
iron porphyrin *(Bioch)* Eisenporphyrin *n*
iron porphyrin enzyme Eisenporphyrinenzym *n*, Hämenzym *n*

iron-potassium sulphate Kaliumeisen(III)-sulfat-12-Wasser *n*, Kaliumeisenalaun *m*
iron pyrite(s) *(Min)* Pyrit *m*, Eisenkies *m*, Schwefelkies *m (Eisen(II)-sulfid)*
iron removal *(Hyd)* Enteisenung *f*
iron silicide Eisensilicid *n*
iron sponge Eisenschwamm *m*
iron-sulphur protein Eisen-Schwefel-Protein *n*, Fe-S-Protein *n*
iron tetracarbonyl Eisentetracarbonyl *n*
iron triad Eisengruppe *f*, Gruppe *f* der Eisenmetalle *(im Periodensystem)*
iron trichloride Eisentrichlorid *n*, Eisen(III)-chlorid *n*
irradiance 1. Strahlungsflussdichte *f*; 2. *(Spektr)* Bestrahlungsstärke *f*
irradiate *v* 1. bestrahlen; 2. ausstrahlen
irradiation 1. Bestrahlen *n*, Bestrahlung *f*; 2. Ausstrahlen *n*
irradiation cross-linking Strahlenvernetzung *f*
irreversible irreversibel, nicht umkehrbar
irreversible colloid irreversibles [irresolubles] Kolloid *n*
irreversible thermodynamics irreversible Thermodynamik *f*, Thermodynamik *f* irreversibler Prozesse, Thermodynamik *f* der Nichtgleichgewichtsprozesse
irreversible transition irreversible Umwandlung *f* [Phasenumwandlung *f*]
irritant Reizmittel *n*, Reizstoff *m*
irritate *v* reizen
IRS *s.* inverse Raman spectroscopy
i.s. *s.* insoluble
isenthalpic is(o)enthalpisch
isentrope *(phys Ch)* Isentrope *f (Diagrammkurve einer reversiblen adiabatischen Zustandsänderung)*
isentropic isentrop(isch) *(z. B. Reaktionsreihe)*
ISFET *s.* ion-sensitive field-effect transistor
island growth Inselwachstum *m (auf Oberflächen)*
ISO = International Organization for Standardization
iso acid isomere Säure *f*
isoalkane Isoalkan *n*
isoamyl alcohol Isoamylalkohol *m*, *3-Methylbutan-1-ol *n*
isoascorbic acid Isoascorbinsäure *f*, Erythorbinsäure *f*, 3-Oxo-D-glucofuranolacton *n*
isobar *(phys Ch)* Isobare *f*
isobaric coefficient of thermal expansion Ausdehnungskoeffizient *m*, thermischer Volumenausdehnungskoeffizient *m*, Volumen-Temperatur-Koeffizient *m*, Raumausdehnungskoeffizient *m*
isobaric distillation isobare Destillation *f*
isobaric heat capacity Wärmekapazität *f* bei konstantem Druck
isobutane Isobutan *n*, *2-Methylpropan *n*
isobutyl alcohol Isobutylalkohol *m*, *2-Methylpropan-1-ol *n*
isobutylene-isoprene rubber Isobutylen-Isopren-Kautschuk *m (Butylkautschuk)*
isobutyric acid Isobuttersäure *f*, *2-Methylpropansäure *f*
isochore *(phys Ch)* Isochore *f*
isochoric heat capacity Wärmekapazität *f* bei konstantem Volumen
isocitrate Isocitrat *n (Zwischenprodukt des Citrat-Zyklus)*
isocitric acid Isocitronensäure *f*, *1-Hydroxypropan-1,2,3-tricarbonsäure *f*
isocitric acid dehydrogenase Isocitratdehydrogenase *f*, Isocitronensäuredehydrogenase *f*
isoclined chain isoclinale Kette *f*
isocyanate R–N=C=O Isocyanat *n*
isocyanate resin Isocyanatkunststoff *m*, Polyurethan *n*, PUR
isocyanate rubber Urethankautschuk *m*, Polyurethankautschuk *m*
isocyanide Isocyanid *n*, Isonitril *n*
isocyclic compound isocyclische [carbocyclische, homocyclische] Verbindung *f*, Isocyclus *m*, Carbocyclus *m*
isodisperse isodispers
isodispersion *(Koll)* isodisperses System *n*
isodynamic *(Bioch)* isodynam *(Enzyme)*
isoelectric focus(s)ing *(Anal)* isoelektrische Fokussierung *f*
isoelectric point isoelektrischer [elektrisch neutraler] Punkt *m*, I. P.
isoelectronic isoelektronisch, isoster
isoenzyme Isoenzym *n*
isoeugenol Isoeugenol *n*, *4-Hydroxy-3-methoxy-1-propenylbenzen *n*

isoflavone Isoflavon *n*
isogel *(Koll)* Isogel *n*
isohydric von gleichem pH-Wert, mit gleicher Hydroniumionenkonzentration
isoionic point isoionischer Punkt *m*
isokinetic line Isotache *f*
isokinetic temperature isokinetische Temperatur *f*
isolable isolierbar, rein darstellbar
isolate *v* 1. isolieren, rein darstellen, abtrennen; 2. *s.* insulate
isolated double bond isolierte Doppelbindung *f*
isolated system abgeschlossenes [isoliertes] System *n*
isolation 1. Isolierung *f*, Reindarstellung *f*, Abtrennung *f*; 2. *s.* insulation 1.
isoleucine Isoleucin *n*, Ile, 2-Amino-3-methyl-valeriansäure *f*
isologous series *(org Ch)* isologe Reihe *f*
isologue *(org Ch)* Isolog *n*
isomer Isomer(e) *n*
isomer distribution Isomerenverteilung *f*
isomer shift *(Spektr)* Isomerieverschiebung *f*
isomeric compound isomere Verbindung *f*, Isomer(e) *n*
isomerism Isomerie *f*
isomerization Isomerisierung *f*, Isomerisation *f* • **undergo isomerization** sich isomerisieren, isomerisiert werden
isomerization catalyst Isomerisierungskatalysator *m*
isomerize *v* sich isomerisieren, isomerisiert werden; isomerisieren
isomorphic *s.* isomorphous
isomorphism *(Krist)* Isomorphie *f*, Isomorphismus *m*, Gleichgestaltigkeit *f*
isomorphous isomorph, gleichgestaltig
isoniazid Isonicotinsäurehydrazid *n*, INH, Pyridin-4-carbonsäurehydrazid *n*
isonicotinic acid Isonicotinsäure *f*, Pyridin-4-carbonsäure *f*
isonicotinic acid hydrazide *s.* isoniazid
isonitrile Isonitril *n*, Isocyanid *n*
isonitroso compound Isonitrosoverbindung *f*, Oxim *n*
isooctane Isooctan *n*, *(i. e. S.)* $(CH_3)_2CHCH_2C(CH_3)_3$ Isooctan *n*, 2,2,4-Trimethylpentan *n*
isoosmotic isosmotisch, isotonisch
isopentane Isopentan *n*, *2-Methylbutan *n*
isophthalic acid Isophthalsäure *f*, Benzen-*m*-dicarbonsäure *f*, Benzen-1,3-dicarbonsäure *f*
isopiestic solutions Lösungen *fpl* gleichen Dampfdrucks
isopleth Isoplethe *f*
isopoly acid Isopolysäure *f*
isopolymorphism Isopolymorphie *f*
isoprene Isopren *n*, *2-Methylbuta-1,3-dien *n*
isoprene rubber Isoprenkautschuk *m* *(synthetisches Polyisopren)*
isoprenoid aus Isoprenmolekülen aufgebaut
isoprenoids Isoprenoide *npl (Sammelname für Terpene und Steroide)*
isopropyl alcohol Isopropylalkohol *m*, *Propan-2-ol *n*
isopropyl group $(CH_3)_2CH$– Isopropylgruppe *f*, Methylethylgruppe *f*
isopropylamine Isopropylamin *n*, 2-Aminopropan *n*, *1-Methylethylamin *n*
isopropylbenzene *Isopropylbenzen *n*, Cumen *n*, Cumol *n*
isoquinoline Isochinolin *n*, 3,4-Benzopyridin *n*, Benzo[c]pyridin *n*, 2-Benzazin *n*
isoquinoline(-type) alkaloid Isochinolin-Alkaloid *n*, Alkaloid *n* mit Isochinolinring
isorubber Iso-Kautschuk *m*
isostere 1. isostere Verbindung *f*; 2. *(phys Ch)* Isostere *f (eine Kurve)*
isosteric isoster(isch), isoelektronisch
isosterism Isosterie *f*
isosynthesis Isosynthese *f*
isotach Isotache *f*
isotachophoresis *(Anal)* Isotachophorese *f (Sonderform der freien Elektrophorese)*
isotactic polymer isotaktisches Polymer(e) *n*
isotacticity Isotaktizität *f*
isotherm Isotherme *f*
isothermal distillation isotherme Destillation *f*
isothermal operation isotherme Fahrweise *f*
isothermal reactor isothermer Reaktor *m*
isothermal transformation isotherme Umwandlung *f*, Umwandlung *f* bei gleich bleibender Temperatur

isothiocyanate Isothiocyanat *n*, Isothiocyansäureester *m*, Senföl *n*
isotonic isotonisch, isosmotisch
isotonic sodium chloride solution *(Med)* physiologische Kochsalzlösung *f*
isotope Isotop *n*
isotope dilution analysis Isotopenverdünnungsanalyse *f*
isotope effect Isotopieeffekt *m*, Isotopeneffekt *m*
isotope exchange Isotopenaustausch *m*
isotope exchange equilibrium *(Kern)* Isotopenaustauschgleichgewicht *n*
isotope separation Isotopentrennung *f*
isotopic isotop; Isotopen..., Isotopie...
isotopic abundance Isotopenhäufigkeit *f*
isotopic atom isotopes Atom *n*
isotopic composition Isotopenzusammensetzung *f*
isotopic indicator *s.* isotopic tracer
isotopic purity Isotopenreinheit *f*
isotopic ratio Isotopenverhältnis *n*
isotopic tracer Isotopentracer *m*, Isotopenindikator *m*, Indikatorisotop *n*, Leitisotop *n*, Zusatzisotop *n*
isotopism Isotopie *f*
isotopy Isotopie *f*
isotropic *(Krist)* isotrop
isotropy *(Krist)* Isotropie *f*
isotypic *(Krist)* isotyp
isotypy *(Krist)* Isotypie *f*
isovaleral [isovaleraldehyde, isovalerial] Isovaleraldehyd *m*, *3-Methylbutanal *n*
isovaleric acid Isovaleriansäure *f*, *3-Methylbutansäure *f*
isovel Isotache *f*
isovol Isovole *f*, Linie *f* [Kurve *f*] gleichen Gehalts an flüchtiger Substanz
isozyme *s.* isoenzyme
ISS *s.* ion-scattering spectroscopy
issue *v* aussenden, emittieren, ausströmen lassen; entweichen, ausströmen, abziehen
It diagram It-Diagramm *n*, I-t-Diagramm *n*, Wärmeinhalt-Temperatur-Diagramm *n*
itaconic acid Itaconsäure *f*, Methylenbernsteinsäure *f*, *Prop-2-en-2,3-dicarbonsäure *f*
ITO *s.* indium-tin-oxide
I.U. *s.* international unit
IUB = International Union of Biochemistry
IUC 1. = International Union of Chemistry; 2. = International Union of Crystallography
IUCr = International Union of Crystallography
IUPAC = International Union of Pure and Applied Chemistry
IUPAC name nach den IUPAC-Regeln gebildete Benennung *f*
IUPAC rule *(Nom)* IUPAC-Regel *f*, Regel *f* der IUPAC
I.V. *s.* iodine value

J

j-j coupling *(phys Ch)* jj-Kopplung *f*
j-value Photodissoziationsgeschwindigkeit *f*, Photodissoziationskonstante *f*
Jablonski diagram *(phys Ch)* Jablonski-Termschema *n*, Jablonski-Diagramm *n*
jacket *v* ummanteln, verkleiden, umhüllen
jacket Mantel *m*, Ummantelung *f*, Verkleidung *f*, Umhüllung *f*, Hülle *f*, Kapsel *f*
jacketed coil condenser Intensivkühler *m*
jacketed kettle Mantelkessel *m*, Mantelgefäß *n*, Mantelbehälter *m*
jacketed pipe [tube] Mantelrohr *n*
Jahn-Teller transition Jahn-Teller-Phasenübergang *m*
jar 1. Gefäß *n (aus Glas oder Keramik)*; 2. *(Lab)* Glaszylinder *m*
jar brush *(Lab)* Gläserbürste *f*
jar mill Kugelmühle *f (mit Mahltopf)*
jarosite *(Min)* Gelbeisenerz *n*
jasper *(Min)* Jaspis *m (Silicium(IV)-oxid)*
Javel(le) water javellesche Lauge *f*, Eau de javelle *n (wässrige Kaliumhypochloritlösung)*
jaw breaker [crusher] Backenbrecher *m*
jellification 1. Gel(atin)ieren *n*, Gel(atin)ierung *f*, Gallertbildung *f*; 2. *(Koll)* Gelbildung *f*
jellify *v* gelatinieren, in Gelee überführen; gel(atin)ieren, zu Gelee erstarren
jelly *v* gelatinieren, in Gelee überführen; gel(atin)ieren, zu Gelee erstarren

jelly 1. Gelee *n*, Gallerte *f*, Gallert *n*; 2. *(Koll)* Gel *n*; 3. *(Lebm)* Götterspeise *f*
jelly-like 1. gallertartig, gallertig; 2. *(Koll)* gelartig; salbenartig
jet *v* 1. ausstoßen, ausschleudern; 2. druckspülen; 3. ausströmen, herausschießen
jet 1. Strahl *m*; 2. Düse *f*
jet fuel Düsentreibstoff *m*
jet mill Strahlmühle *f*
jet of steam Dampfstrahl *m*
jet pump Treibmittelpumpe *f*, Strahlpumpe *f*
jig *v* 1. (ständig ruckartig) auf- und abbewegen, hin- und herbewegen; 2. *(Aufb)* setzen, in der Setzmaschine aufbereiten
jig *(Aufb)* Setzmaschine *f*, Setzapparat *m*
jig washer *(Aufb)* Nasssetzmaschine *f*
jigging *(Aufb)* Setzen *n*, Setzarbeit *f*
jj-coupling *(phys Ch)* jj-Kopplung *f*
Johnson noise *(Anal)* thermisches Rauschen *n*, Wärmerauschen *n*
join *v* **together** miteinander verbinden
joining peptide *(Bioch)* J-Peptid *n*
joint Verbindung *f*; Anschluss *m*
joint action Verbundwirkung *f*
joint grease Dichtungsfett *n*
joule Joule *n*, J *(SI-Einheit für Arbeit, Energie und Wärmemenge)*
Joule-Thomson coefficient Joule-Thomson-Koeffizient *m*
Joule-Thomson effect Joule-Thomson-Effekt *m (isenthalpischer Drosseleffekt)*
journal Zapfen *m (z. B. einer Welle)*
journal bearing Zapfenlager *n*, Radialgleitlager *n*
juice Saft *m (von Früchten, Gemüse, Fleisch)*
junction 1. Grenzfläche *f*, Berührungsfläche *f*, Berührungszone *f*; 2. *(phys Ch)* Phasengrenzfläche *f*; Übergang *m (z. B. im Halbleiter)*; 3. *(Kst)* Vernetzungspunkt *m (Polymernetzwerke)*
junction potential Grenzflächenpotenzial *n*
junk Altmaterial *n*, Altwaren *pl*; Abfall *m*
jute packing Jutedichtung *f*, Jutepackung *f*
juxtaposed benachbart

K

K capture K-Elektroneneinfang *m*, K-Einfang *m*, inverse Betastrahlung *f*
K-electron capture K-Elektroneneinfang *m*, K-Einfang *m*, inverse Betastrahlung *f*
K meson *s.* kaon
K shell K-Schale *f (eines Atoms)*
kallidin Kallidin *n (ein Decapeptid)*
kallikrein Kallikrein *n (hormonartiger Wirkstoff der Bauchspeicheldrüse)*
kaolin 1. *(Min)* Kaolin *m*, Porzellanerde *f*, weißer [reiner] Ton *m*, China Clay *m(n)*; 2. *(Tech)* Kaolin *m*, Schlämmkaolin *m*, geschlämmte Porzellanerde *f*
kaon (particle) *(Kern)* Kaon *n*
Karl Fischer reagent Karl-Fischer-Reagens *n*, Karl-Fischer-Lösung *f*
Karl Fischer titration Karl-Fischer-Titration *f*, Karl-Fischer-Wasserbestimmung *f*
kat *s.* katal
katabolic *s.* catabolic
katal *(Bioch)* Katal *n*, kat *(Einheit der Enzymaktivität)*
katharometer *(Anal)* Katharometer *n*, Wärmeleitfähigkeits(mess)zelle *f*, Wärmeleitfähigkeitsdetektor *m*, WLD
kaurene Kauren *n (ein Diterpen)*
kauri (copal, gum) Kaurikopal *m*, Kauriharz *n*, Kaurigum *m*, Cowrikopal *m*
KBr pellet *(Spektr)* KBr-Pressling *m*
KBr-pellet technique *(Spektr)* KBr-Presstechnik *f*
keep *v* **at the boil** am Sieden [Kochen] halten
keep *v* **for an indefinite period** unbegrenzt haltbar sein
keep *v* **in progress** in Gang halten *(Reaktion)*
keep *v* **in solution** in Lösung halten
keep *v* **in suspension** in Suspension halten, in der Schwebe halten
keep *v* **near the boil** in Nähe des Siedepunkts halten
keep *v* **screened from the light** lichtgeschützt aufbewahren
keeping quality *(Lebm)* Haltbarkeit *f*
keeping time Haltbarkeitsdauer *f*, Haltbarkeitserwartung *f*
Kekulé formula *(org Ch)* Kekulé-Formel *f*

Kekulé(-like) structures Kekulé-Strukturen *fpl*, Kekulé-Grenzstrukturen *fpl*
kelvin Kelvin *n (SI-Einheit der thermodynamischen Temperatur und Temperaturdifferenz)*
Kelvin (temperature) scale Kelvin-Skale *f (der absoluten Temperatur)*
kephalin Kephalin *n*, Cephalin *n (ein Glycerophospholipid)*
kernel 1. Atomrumpf *m*, Rumpf *m*; 2. Kern *m*; Korn *n*
kernel oil Kernöl *n*
kerosene *s.* kerosine
kerosine Petroleum *n*, Leuchtöl *n*, Kerosin *n*
kerosine fraction Petrol(eum)fraktion *f*, Kerosinfraktion *f*
Kerr cell *(Spektr)* Kerr-Zelle *f*
Kerr effect Kerr-Effekt *m (Doppelbrechung)*
ketal *s.* ketone acetal
ketalization Ketalisierung *f*
ketazine Ketazin *n (Kondensationsprodukt von Ketonen mit Hydrazin)*
ketene Keten *n (organische Verbindung mit der Gruppe =C=C=O)*; *(i. e. S.)* $CH_2C{=}O$ Keten *n*
ketide Polyketid *n*, Acetogenin *n*
ketimine Ketimin *n*, Ketonimin *n (Imin der allgemeinen Formel $R_2C{=}NR'$)*
keto acid Keto(carbon)säure *f*
keto alcohol Keto(n)alkohol *m*, Hydroxyketon *n*
keto aldehyde Keto(n)aldehyd *m*
keto carbene Ketocarben *n*
keto carboxylic acid *s.* keto acid
keto compound Keto(n)verbindung *f*
keto-enol tautomerism Keto-Enol-Tautomerie *f*
keto ester Keto(säure)ester *m*, Ketocarbon(säure)ester *m*
keto form Ketoform *f*
keto group Ketogruppe *f*, ketonartig gebundene Carbonylgruppe *f*, Ketongruppe *f*
ketoamine Ketoamin *n*
ketogenesis *(Med, Bioch)* Ketogenese *f*, Ketonkörperbildung *f*, Acetonkörperentstehung *f*
ketogenic *(Med, Bioch)* ketogen
ketoglutaric acid Ketoglutarsäure *f*, Oxoglutarsäure *f*, *Oxopentandisäure *f*, *(i. e. S.)* α-Ketoglutarsäure *f*, 2-Oxoglutarsäure *f*, *2-Oxopentandisäure *f*
ketoketene $(R)_2C{=}C{=}O$ Ketoketen *n*
ketol Ketol *n*, Monohydroxyketon *n*, *(i. w. S.)* Keto(n)alkohol *m*, Hydroxyketon *n*, *(ungenau:)* Ketol *n*
ketolysis Ketolyse *f*
ketolytic ketolytisch
ketonaemia *(Med)* Ketonämie *f (abnorme Anhäufung von Ketonkörpern im Blut)*
ketone *(org Ch)* Keton *n*
ketone acetal Ketonacetal *n*
ketone body *(Bioch)* Ketonkörper *m*, Acetonkörper *m*
ketone-body formation *s.* ketogenesis
ketone (carbonyl) group *s.* keto group
ketone resin Ketonharz *n*
ketonic ketonig, ketonhaltig; Keton…
ketonic carbonyl group *s.* keto group
ketonic cleavage [fission] Ketonspaltung *f*
ketonic function Keto(n)funktion *f*
ketonic group *s.* keto group
ketonic rancidity Ketonranzigkeit *f*, Parfümranzigkeit *f*
ketonic sugar *s.* ketose
ketonimine *s.* ketimine
ketonimine dye (stuff) Ketoniminfarbstoff *m*, Ketonimidfarbstoff *m*
ketonization Ketonisierung *f*
ketonize *v* 1. ketonisieren; in ein Keton übergehen; 2. *(Lebm)* ketonig werden
ketose Ketose *f*, Keto(n)zucker *m*
ketoside Ketosid *n (Glykosid, das bei der Hydrolyse eine Ketose ergibt)*
ketosidic ketosidisch
ketosteroid *(Bioch)* Ketosteroid *n*
ketoxime *(org Ch)* Ketoxim *n*
ketyl Ketyl *n*, Metallketyl *n*
key chemical Grundchemikalie *f*, Chemiegrundstoff *m*, (chemischer) Grundstoff *m*, Schlüsselstoff *m*, Schlüsselsubstanz *f*
key enzyme Schlüsselenzym *n*
key intermediate 1. wichtiges Zwischenprodukt *n*; 2. *(Bioch)* Schlüsselintermediäres *n*, Schlüsselverbindung *f*, Schlüsselsubstanz *f*
key reaction Schlüsselreaktion *f*
key substance Schlüsselsubstanz *f*
kieselgu(h)r Kieselgur *f*, Diatomeenerde *f*, Infusorienerde *f*

kill *v* 1. *(Met)* beruhigen *(eine Schmelze)*; 2. *(Hyd)* (ab)töten *(pathogene Keime)*
kill of microbes *(Hyd)* Keim(ab)tötung *f*
killed steel beruhigter [beruhigt vergossener, ruhig vergossener] Stahl *m*
kiln *v* 1. *(Ker)* (im Ofen) brennen; 2. *(Ch)* kalzinieren *(Rohstoffe zum Austreiben von Wasser oder Kohlendioxid erhitzen)*; 3. *s.* kiln-dry
kiln 1. *(Ker)* Brennofen *m*; 2. *(Ch)* Kalzinierofen *m (zum Austreiben von Wasser oder Kohlendioxid)*; 3. *(Lebm)* Trockenofen *m*
kiln-dry *v* 1. im Ofen trocknen; 2. *(Lebm)* (im Ofen) darren, dörren
kiln-dried *(Lebm)* ofengetrocknet, ofentrocken; gedarrt, gedörrt
kiln dryer *(Lebm)* Trockenofen *m*; Trockenkammer *f*; Darre *f*, Darrofen *m*
kindle *v* anzünden, entzünden; sich entzünden
kinematic viscosity kinematische Viskosität *f* [Zähigkeit *f*], Viskositäts-Dichte-Verhältnis *n*
kinetic control kinetische Steuerung *f* [Kontrolle *f*]
kinetic energy kinetische Energie *f*, Bewegungsenergie *f*
kinetic friction kinetische Reibung *f*
kinetic hindrance kinetische Hinderung *f (einer Reaktion)*
kinetic theory of gases kinetische Gastheorie *f*
kineticist Kinetiker *m*
kinetics Kinetik *f*
kinin Kinin *n (im tierischen Organismus erzeugtes, pharmakologisch hochaktives Polypeptid)*; Kinin *n*, Phytokinin *n*, Cytokinin *n (pflanzlicher Wirkstoff)*
kink *(Krist)* Kerbstelle *f (Baufehler)*
kino (gum) Kino *n*, Kinoharz *n*, Kinogummi *n (eingetrockneter gerbstoffreicher Saft aus den Rinden verschiedener tropischer Bäume)*
Kipp (gas) generator kippscher Apparat *m* [Gasentwickler *m*], Gasentwicklungsapparat *m* nach Kipp
Kirchhoff's law kirchhoffsches Gesetz *n (der Temperaturabhängigkeit von Reaktionsenthalpien)*
Kjeldahl determination Kjeldahl-Bestimmung *f*, kjeldahlsche Stickstoffbestimmung *f*, Stickstoffbestimmung *f* nach Kjeldahl
Kjeldahl digestion apparatus Kjeldahl-Apparat *m*, Stickstoffbestimmungsapparat *m* nach Kjeldahl
Kjeldahl flask Kjeldahl-Kolben *m*
kjeldahlization Aufschluss *m* im Kjeldahl-Apparat
kjeldahlize *v* im Kjeldahl-Apparat aufschließen
klystron valve *(Anal)* Klystron *n*, Triftröhre *f*
knead *v* (durch)kneten, plasti(fi)zieren
kneader *s.* kneading machine
kneader mixer Knetmischer *m*, Mischkneter *m*
kneading machine Knetmaschine *f*, Knetwerk *n*, Knetapparat *m*, Kneter *m*, *(oft auch)* Mischmaschine *f*, Mischer *m*, *(i. e. S.)* Schaufelkneter *m*, Doppelarmkneter *m*, Doppelpaddelmischer *m*
knife Messer *n*; Schaber *m*, Schabemesser *n*, Abstreifer *m*, Abstreifermesser *n*, Rakel *f*
knife-discharge centrifuge Schälzentrifuge *f*, Schälschleuder *f*
knife edge Schneide *f (z. B. an einer Analysenwaage)*
knock Klopfen *n (im Vergasermotor)*
knock-down agent [poison] Knock-down-Mittel *n (sofort wirkendes Insektizid)*
knock-free klopffrei, nicht klopfend
knock-prone klopffreudig, klopfempfindlich, klopfstark
knock-proof klopffest
knock rating 1. Klopfwertbestimmung *f*, Klopf(wert)prüfung *f*, Ermittlung *f* der Klopffestigkeit; 2. Klopfwert *m*, Klopffestigkeit *f*
knock-rating engine Klopfprüfmotor *m*
knock resistance Klopffestigkeit *f*
knock suppressor Antiklopfmittel *n*, Klopfbremse *f*
knock value Klopfwert *m*, Klopffestigkeit *f*
knocker Klopfwerk *n*, Klopfvorrichtung *f*
knocking Klopfen *n (im Vergasermotor)*
knocking properties Klopfeigenschaften *fpl*
Knoevenagel condensation [reaction] Knoevenagel-Reaktion *f*, Knoevena-

gel-Kondensation *f*, knoevenagelsche Synthese *f*
Knoop hardness test Härteprüfung *f* nach Knoop
Knorr pyrrole synthesis knorrsche Pyrrolsynthese *f*
Knudsen diffusion Knudsen-Diffusion *f*
Knudsen diffusivity Knudsen-Diffusionskoeffizient *m*
Kofler (hot) bench *(Lab)* koflersche Heizbank *f*
Kohlrausch bridge *(phys Ch)* Kohlrausch-Brücke *f*
Kohlrausch law kohlrauschsches Gesetz *n (der unabhängigen Ionenwanderung)*
kola (nut) Kolanuß *f (von verschiedenen Kolaarten)*
Kolbe electrolysis reaction kolbesche Kohlenwasserstoffsynthese *f* [Paraffinsynthese *f*], Kolbe-Synthese *f* von Kohlenwasserstoffen
Kolbe hydrocarbon synthesis kolbesche Kohlenwasserstoffsynthese *f* [Paraffinsynthese *f*], Kolbe-Synthese *f* von Kohlenwasserstoffen
Kolbe reaction of phenols Kolbe-Synthese *f* von Phenolcarbonsäuren, kolbesche Salicylsäuresynthese *f*
Kolbe Schmitt reaction [synthesis] Kolbe-Schmitt-Reaktion *f*, Kolbe-Schmitt-Synthese *f (von Phenolcarbonsäuren unter Druck)*
kollergang Kollergang *m*, Kollermühle *f*
Kosower Z-value Z-Wert *m (zur Charakterisierung der Lösungsmittelpolarität)*
Kovats index *(Chromat)* Kováts-Index *m*, Retentionsindex *m* nach Kováts
kraft paper Kraft(pack)papier *n*
kraft process Kraftzellstoffverfahren *n*
kraft pulp Kraftzellstoff *m (hochfester Sulfatzellstoff)*
Krebs (citric acid) cycle *(Bioch)* (Szent-Györgyi-)Krebs-Zyklus *m*, Citronensäurezyklus *m*, Citratzyklus *m*, Tricarbonsäurezyklus *m*
Krebs-Henseleit cycle *(Bioch)* Krebs-Henseleit-Zyklus *m*, Ornithinzyklus *m*, (Arginin-)Harnstoff-Zyklus *m*
Kröcker bomb Berthelot-Mahler-Kröcker-Bombe *f*, berthelotsche Bombe *f*
Kroll process *(Met)* Kroll-Verfahren *n (ein Reduktionsverfahren)*
Krupp iron sponge process Krupp-Eisenschwammverfahren *n*
Krupp melting process Krupp-Einschmelzverfahren *n*
Krupp stacking cutter Krupp-Stapelschneide *f*
krypton Kr Krypton *n*
krypton(-ion) laser Krypton(ionen)-Laser *m*
kynurenic acid Kynurensäure *f*, 4-Hydroxy-2-chinolincarbonsäure *f*
kynurenine Kynurenin *n*, *o*-Aminophenazylaminoessigsäure *f*
kynuric acid Kynursäure *f*, Oxanilsäure-2-carbonsäure *f*
kynurine Kynurin *n*, 4-Hydroxychinolin *n*

L

l-form (-)-Form *f*, linksdrehende Form *f*, Linksform *f*
L configuration L-Konfiguration *f*
L shell L-Schale *f (eines Atoms)*
lab *s.* laboratory
lab. *s.* laboratory
label *v* 1. kennzeichnen, angeben; etikettieren, beschriften; 2. *(Anal)* markieren *(Substanzen durch Einbau gut nachweisbarer meist radioaktiver Atome kenntlich machen)*; 3. *(Nom)* indizieren, beziffern
label 1. Etikett *n*, Aufschrift *f*; Markierung *f*; 2. *(Anal)* markiertes Atom *n*
labelled compound *(Anal)* markierte Verbindung *f*
labelling 1. Kennzeichnen *n*, Kennzeichnung *f*; Etikettieren *n*, Beschriftung *f*; 2. *(Anal)* Markieren *n*, Markierung *f (Kenntlichmachung einer Substanz durch den Einbau gut nachweisbarer meist radioaktiver Atome)*; 3. *(Nom)* Indizierung *f*, Bezifferung *f*
labile labil, instabil, unbeständig
lability Labilität *f*, Instabilität *f*, Unbeständigkeit *f*
laboratorial Labor(atoriums)...
laboratory Labor(atorium) *n* • **on the laboratory scale** im Labormaßstab, labortechnisch, labormäßig

laboratory apparatus Laborgerät *n*, Laborapparat *m*
laboratory assistant Laborant *m*, Chemielaborant *m*; Laborantin *f*, Chemielaborantin *f*
laboratory chemical Laborchemikalie *f*
laboratory conditions Laborbedingungen *fpl*
laboratory determination Laborbestimmung *f*
laboratory examination *s.* laboratory investigation
laboratory glassware Laborgeräte *f* aus Glas
laboratory instrument *s.* laboratory apparatus
laboratory investigation Laboruntersuchung *f*, Laborprüfung *f*
laboratory method Labormethode *f*
laboratory preparation präparative Darstellung *f*, Darstellung *f* im Labor
laboratory procedure Laborverfahren *n*
laboratory-scale labortechnisch, im Labormaßstab, labormäßig
laboratory test Laborversuch *m*, Labortest *m*
laboratory testing Laborprüfung *f*
labware Laborgeräte *npl*
lachrymator Augenreizstoff *m*, Tränenreizstoff *m*, Tränengas *n*
lachrymatory augenreizend, tränenreizend, tränenerregend, tränenerzeugend, lakrimogen
lack of reaction Reaktionslosigkeit *f*
lacquer physikalisch trocknender Lack *m* [Klarlack *m*], *(i. e. S.)* Nitrolack *m*; physikalisch trocknende Lackfarbe *f*, *(i. e. S.)* Nitrolackfarbe *f*
lacquer remover Abbeizmittel *n*
lacrimator *s.* lachrymator
lactalbumin Lactalbumin *n*, Milchalbumin *n*
lactam *(org Ch)* Lactam *n*
β-lactam antibiotic *(Biot)* β-Lactamantibiotikum *n*, Beta-Lactam-Antibiotikum *n*
lactase *(Bioch)* Laktase *f*
lactase deficiency *s.* lactose intolerance
lactate Lactat *n*
lactate fermentation *(Bioch)* Lactatgärung *f*
lactic acid Milchsäure *f*, *(i. e. S.)* *2-Hydroxypropansäure *f*
lactic-acid bacteria Milchsäurebakterien *npl*
lactic-acid fermentation Milchsäuregärung *f*
lactic anhydride Milchsäureanhydrid *n*, *2-Hydroxypropansäureanhydrid *n*
lactic dehydrogenase Lactatdehydrogenase *f*, Milchsäuredehydrogenase *f*, LDH
lactide Lactid *n*
lactiferous latexführend, milchsaftführend
lactim *(org Ch)* Lactim *n*
lactogen *s.* lactogenic hormone
lactogenic hormone lactogenes [luteotropes] Hormon *n*, LTH, *Prolactin *n*, PRL
lactoglobulin Lactoglobulin *n*
lactone Lacton *n (innerer Ester von Hydroxysäuren)*
lactone rule Lactonregel *f* (der optischen Drehung), hudsonsche Lactonregel *f*
lactonic acid Lactonsäure *f (allg. für ein Lactonderivat mit Carboxylgruppe am Lactonring)*
lactonitrile Lactonitril *n*, Milchsäurenitril *n*, *2-Hydroxypropannitril *n*
lactonization Lactonisierung *f*, Lactonbildung *f*
lactonize *v* lactonisieren
lactose Lactose *f*, Milchzucker *m*
lactose fermentation Milchzuckergärung *f*
lactose intolerance *(Lebm)* Laktoseintoleranz *f*
ladder conformation *(Bioch)* *cis*-Leiter-Konformation *f (der DNA)*
ladder polymer Leiterpolymer(e) *n*
ladle *(Met)* Pfanne *f*
laevo *s.* laevorotatory
laevolactic acid D(-)-Milchsäure *f*, Linksmilchsäure *f*
laevorotary *s.* laevorotatory
laevorotation Linksdrehung *f*
laevorotatory linksdrehend, (-)-drehend, (-)-...
laevorotatory sugar linksdrehender Zucker *m*
laevotartaric acid (-)-Weinsäure *f*, D-Threarsäure *f*, Linksweinsäure *f*
laevul(in)ic acid Lävulinsäure *f*, *4-Oxopentansäure *f*, 4-Oxovaleriansäure *f*

laevulose Lävulose *f*, D-Fructose *f*, Fruchtzucker *m*
lag phase [time] *(Hyd, Biot)* Verzögerungsphase *f*, lag-Phase *f*
lagoon *(Hyd)* Abwasserteich *m*; durchflossenes Erdbecken *n*; Schlamm(entwässerungs)teich *m*
lagooning *(Hyd)* Teichreinigung *f*, Behandlung *f* in Abwasserteichen, Teichverfahren *n*
lake *(Anal, Text)* Farblack *m*
LALLS *s.* low-angle laser light scattering
Lambert-Beer law lambert-beersches Gesetz *n (der Lichtabsorption)*
Lambert(-Bouguer) law (of absorption) lambertsches Gesetz *n* [Absorptionsgesetz *n*], bouguer-lambertsches Gesetz *n* [Absorptionsgesetz *n*]
lamella Lamelle *f*
lamella evaporator Lamellenverdampfer *m*
lamella heater Lamellenwärmeaustauscher *m*
lamellar lamellar, schichtförmig, geschichtet
lamellar compound lamellare Verbindung *f*, schichtförmig ausgebildete Einlagerungsverbindung *f*
laminar flow 1. laminare Strömung *f*, Laminarströmung *f*, Schichtenströmung *f*; 2. *s.* clean bench
laminate *v* 1. beschichten, laminieren; 2. *(Pap)* bekleben, kaschieren; 3. zu einem Schichtpaket zusammenpressen; 4. in Schichten aufspalten; sich in Schichten aufspalten; 5. (zu einer Schicht) auswalzen
laminate 1. Schicht(press)stoff *m*, Laminat *n*, geschichteter Werkstoff *m*; 2. *s.* laminated fabric
laminated fabric Hartgewebe *n*, Hgw
laminated material [plastic] *s.* laminate 1.
laminated safety (sheet) glass geschichtetes Sicherheitsglas *n*, Verbund(sicherheits)glas *n*, Mehrscheiben(sicherheits)glas *n*, Mehrschichten(sicherheits)glas *n*
lamination 1. Beschichten *n*, Laminieren *n*; 2. *(Pap)* Bekleben *n*, Kaschieren *n*; 3. Zusammenpressen *n* zu einem Schichtpaket; 4. Aufspalten *n* in Schichten; 5. Auswalzen *n* zu einer Schicht
LAMMA *s.* laser microprobe mass analysis
LAMS *s.* laser mass spectrometry
land disposal 1. Deponie *f*, Ablagerung *f (von Abprodukten)*; 2. *s.* land treatment
land treatment (of waste water) *(Hyd)* Abwasserlandbehandlung *f*, Abwasserbodenbehandlung *f*, landwirtschaftliche Abwasserbehandlung *f*
Landé factor g Landé-Faktor *m* g, g-Faktor *m*, spektroskopischer Aufspaltungsfaktor *m*
Landé g factor Landé-Faktor *m* g, g-Faktor *m*, spektroskopischer Aufspaltungsfaktor *m*
Landé splitting factor *s.* Landé g factor
landfill disposal (of wastes) Abproduktbeseitigung *f* durch Geländeauffüllung
landfill gas extraction Gasgewinnung *f* aus Mülldeponien
Langmuir adsorption isotherm langmuirsche Adsorptionsisotherme *f*, Langmuir-Isotherme *f*
lanolin(e) Lanolin *n*, gereinigtes Wollfett *n*
lanthana *s.* lanthanum trioxide
lanthanide *s.* lanthanoid
lanthanoid Lanthanoid *n*, Lanthanoidenelement *n*, Element *n* der Lanthanreihe
lanthanoid contraction Lanthanoidenkontraktion *f*
lanthanoid series Lanthanoidenreihe *f*, Lanthanoidengruppe *f*
lanthanum La Lanthan *n*
lanthanum trioxide Lanthan(tri)oxid *n*, Lanthan(III)-oxid *n*
lapis (lazuli) *(Min)* Lapislazuli *m*, Lasurit *m*, Lasurstein *m (ein Tektosilicat)*
Laplace [Laplacian] operator Laplace-Operator *m*, Delta-Operator *m*
lard Schweineschmalz *n*, Schmalz *n*, Schweinefett *n*
lard oil Lardöl *n*, Specköl *n*, Fettöl *n*, Schmalzöl *n*
large-capacity hochkapazitiv, hochleistungsfähig, Hochleistungs...
large-pore(-size) großporig
large-pored großporig
large ring großer Ring *m*, Makroring *m* *(aus 13 oder mehr Gliedern bestehend)*

large-scale in großem Umfang [Maße, Maßstab], ausgedehnt, Groß..., Massen..., *(oft auch)* großtechnisch, in großtechnischem Maßstab, großindustriell
large-scale chemistry Großchemie *f*
large-scale process großtechnisches Verfahren *n*, (technischer) Großprozess *m*
large-scale unit Großanlage *f*
lariat ether Lassoether *m*, Lariatether *m* *(ein Kronenether)*
Larmor frequency *(Spektr)* Larmor-Frequenz *f*, Präzessionsfrequenz *f*
Larmor precession Larmor-Präzession *f*
larvacide Larvizid *n*, Larvengift *n*
larvicide Larvizid *n*, Larvengift *n*
LAS = linear alkyl sulphonate
laser *(light amplification by stimulated emission of radiation)* Laser *m*, *(i. e. S.)* Laser-Oszillator *m*
laser beam Laserstrahl *m*
laser chemistry Laserchemie *f*
laser desorption mass spectrometry Laser-Desorptions-Massenspektrometrie *f*, LD
laser-excited atomic fluorescence spectrometry laserangeregte Atomfluoreszenzspektrometrie *f*
laser-induced fluorescence laserinduzierte Fluoreszenz *f*
laser mass spectrometry Laser-Massenspektrometrie *f*, LAMS
laser microprobe mass analysis Lasermikrosonden-Massenspektrometrie *f*
laser Raman spectroscopy Laser-Raman-Spektroskopie *f*
Lassaigne test lassaignesche Probe *f*, Lassaigne-Probe *f*, Stickstoffprobe *f* nach Lassaigne
last *v* **well** gut beständig sein
latent heat latente Wärme *f*, Umwandlungswärme *f*
latent solvent latenter [mittelbarer] Löser *m*, latentes Lösungsmittel *n*, Hilfslöser *m*, Hilfslösungsmittel *n*
lateral chain Seitenkette *f*
lateral diffusion Querdiffusion *f*
latex 1. Latex *m*, Milchsaft *m*; 2. Latex *m*, Kunstharzdispersion *f*
latex adhesive Latexklebstoff *m*
latex foam (rubber) Latexschaum(-gummi) *m*, Schaumgummi *m*
latex paint Latex(anstrich)farbe *f*
lather *v* schäumen
lather Schaum *m*, *(i. e. S.)* Seifenschaum *m*
lattice *(Krist)* Gitter *n*
lattice constant Gitterkonstante *f*
lattice defect Gitter(bau)fehler *m*, Kristall(bau)fehler *m*, Kristallstörung *f*, Gitterstörung *f*, Gitterdefekt *m*, Störstelle *f*
lattice deformation Gitterverformung *f*
lattice disorder Gitterfehlordnung *f*
lattice distance Gitterabstand *m*, Gitterperiode *f*, Netzebenenabstand *m*
lattice energy Gitterenergie *f*
lattice imperfection *s.* lattice defect
lattice-inclusion compound Gittereinschlussverbindung *f*
lattice ion Gitterion *n*, gittereigenes Ion *n*
lattice mismatch Gitterfehlanpassung *f*
lattice plane Gitterebene *f*, Netzebene *f*
lattice site Gitterplatz *m*, Gitterstelle *f*
lattice vacancy Gitterleerstelle *f*, Gitterfehlstelle *f*, Gitterlücke *f*, Leerstelle *f*, Fehlstelle *f*
lattice vibration Gitterschwingung *f*
lattice window Gitterfenster *n*
Laue method *(Krist)* Laue-Verfahren *n*, Laue-Methode *f*
Laue pattern [photograph] Laue-Aufnahme *f*, Laue-Bild *n*, Laue-Fotogramm *n*, Laue-Diagramm *n*
laughing gas Lachgas *n*, Lustgas *n* *(Stickstoff(I)-oxid)*
launder *v* waschen *(Textilien)*
laundering fastness Waschechtheit *f*, Waschbeständigkeit *f*
laundry detergent Waschmittel *n*
lauraldehyde Lauraldehyd *m*, Laurinaldehyd *m*, *Dodecanal *n*
laurate Laurat *n* *(Salz oder Ester der Laurinsäure)*
lauric acid Laurinsäure *f*, *Dodecansäure *f*
lauryl alcohol Laurylalkohol *m*, *Dodecan-1-ol *n*
laurylamine Laurylamin *n*, *Dodecylamin *n*
Laves phase *(Krist)* Laves-Phase *f* *(bei intermetallischen Verbindungen)*
law of conservation of energy Gesetz *n* von der Erhaltung der Energie, Ener-

gieerhaltungsgesetz *n*, Energieerhaltungssatz *m*, Energieprinzip *n*

law of constant proportions Gesetz *n* der konstanten Proportionen [Massenverhältnisse]

law of equipartition (of energy) Gesetz *n* der Gleichverteilung der Energie, Gleichverteilungssatz *m*, Äquipartitionstheorem *n*

law of equivalent proportions Gesetz *n* der äquivalenten Proportionen

law of independent migration of ions Gesetz *n* der unabhängigen Ionenwanderung, kohlrauschsches Gesetz *n*

law of inertia Trägheitsgesetz *n*

law of mass action Massenwirkungsgesetz *n*, MWG

law of multiple proportions Gesetz *n* der multiplen Proportionen

lawrencium Lr Lawrencium *n*

laxative *(Pharm)* Laxans *n*, Laxativum *n*, mildes Abführmittel *n*

lay *v* **down** abscheiden, ausfällen, niederschlagen *(Schutzschicht)*

layer 1. Schicht *f*, Lage *f*, *(manchmal auch)* Film *m*; 2. *(Foto)* Emulsionsschicht *f*; 3. *(Geol)* Schicht *f*, Bank *f*; 4. *(Boden)* Horizont *m*

layer growth Schichtwachstum *m*

layer lattice *(Krist)* Schicht(en)gitter *n*

layer sequence Schichtfolge *f*

layer structure Schichtstruktur *f*

layered geschichtet

layered structure Schichtstruktur *f*

layering Schichtenbildung *f*

LC *s.* liquid chromatography

LC-MS *s.* liquid chromatography-mass spectroscopy coupling

LCAO *(linear combination of atomic orbitals)* LCAO, Linearkombination *f* von Atomorbitalen

LCAO approximation LCAO-Näherung *f*

LCO_2 = liquid carbon dioxide

LD *s.* 1. laser desorption mass spectrometry; 2. lethal dose

L.D. polythene *s.* low-density polyethylene

LDPE *s.* low-density polyethylene

Le Châtelier(-Braun) principle le-châtelier-braunsches Prinzip *n*, Prinzip *n* des kleinsten Zwangs, Prinzip *n* des beweglichen Gleichgewichts

leach *v* 1. auslaugen, auswaschen, herauslösen; ausziehen, *(i. e. S.)* (Feststoffe) extrahieren; 2. *(Aufb)* laugen *(Erze durch saure oder alkalische Laugung anreichern)*; 3. *(Biot)* mikrobiell (aus)laugen, leachen

leach *v* **out** 1. austreten; 2. *s.* leach 1.

leach 1. Auslaugbehälter *m*, Auslaugbottich *m*, Extraktionsgefäß *n*; 2. Lauge *f*; 3. *s.* leaching

leach solution *(Aufb)* (metallhaltige) Lauge *f*

leachate Auszug *m (aus Drogen oder Böden)*; Perkolat *n*

leaching 1. Auslaugung *f*, Auswaschung *f*, Herauslösen *n*, Ausziehen *n*, *(i. e. S.)* Feststoffextraktion *f*, Extraktion *f* von Feststoffen, Extraktion *f* fest-flüssig, Fest-flüssig-Extraktion *f*; 2. *(Aufb)* Laugen *n*, Laugung *f (von Erzen)*; 3. *(Biot)* mikrobielle [biologische] Laugung *f* [Erzaufbereitung *f*], mikrobielles Leaching *n*, Bioleaching *n*, Biobergbau *m*

leaching agent Laugungsmittel *n*, Laugemittel *n*

leaching solution *(Aufb)* Lauge *f*, Löseflüssigkeit *f*, Laugungsflüssigkeit *f*, Leaching-Flüssigkeit *f (Chemikalien oder Bakterien enthaltend)*

leaching tank [**trough**, **vat**, **vessel**] *s.* leach 1.

lead *v* 1. (ver)bleien, mit Tetraethylblei versetzen *(Kraftstoffe)*; 2. *s.* lead-coat; 3. sich mit Blei überziehen; 4. leiten, führen

lead *v* **away** ableiten, abführen

lead-coat *v* verbleien *(Werkstoffe)*

lead-line *v* mit Blei auskleiden, ausbleien

lead 1. Pb Blei *n*; 2. Zuleitung *f*; 3. Steigung *f (z. B. einer Schnecke)*

lead acetate Blei(II)-acetat *n*, Bleidiacetat *n*

lead-acid accumulator [battery] Bleiakkumulator *m*, Bleisammler *m*

lead-alkali silicate glass Bleialkalisilicatglas *n*

lead alloy Bleilegierung *f*

lead arsenate Bleiarsenat *n*, Blei(II)-tetroxoarsenat(V) *n*; *(als Insektizid vorwiegend:)* saures Bleiarsenat *n*, Blei(II)-hydrogenarsenat(V) *n*

lead azide [azoimide] Blei(II)-azid *n*

lead blast furnace Bleischachtofen *m*, Blei(schmelz)ofen *m*
lead carbonate Blei(II)-carbonat *n*
lead-chamber process Bleikammerverfahren *n (zur Schwefelsäuregewinnung)*
lead chromate Blei(II)-chromat *n*
lead cladding 1. Plattieren *n* mit Blei, *(i. w. S.)* Verkleiden *n* [Überziehen *n*] mit Blei; 2. Bleiplattierüberzug *m*, *(i. w. S.)* Bleiverkleidung *f*, Bleiüberzug *m*
lead coating 1. Verbleien *n*; 2. Blei(schutz)schicht *f*
lead contamination *(Tox)* Bleikontamination *f*
lead crown glass Kronflintglas *n*
lead crystal glass Bleikristall(glas) *n*
lead dichloride Bleidichlorid *n*, Blei(II)-chlorid *n*
lead dioxide Bleidioxid *n*, Blei(IV)-oxid *n*
lead-free bleifrei
lead glance *(Min)* Bleiglanz *m*, Galenit *m (Blei(II)-sulfid)*
lead glass Bleiglas *n*
lead line Anschlussleitung *f*, Zuführungsleitung *f*
lead lining 1. Auskleiden *n* mit Blei, Ausbleien *n*; 2. Bleiauskleidung *f*, *(i. w. S.)* (dicke) Bleiauflage *f*, (dicke) Bleischutzschicht *f*
lead monoxide Blei(mon)oxid *n*, Blei(II)-oxid *n*
lead nitrate Blei(II)-nitrat *n*
lead paint bleihaltiger Anstrichstoff *m*, Bleifarbe *f*
lead pipe Bleirohr *n*
lead poisoning Bleivergiftung *f*
lead refining Bleiraffination *f*
lead storage battery [cell] Bleiakkumulator *m*, Bleisammler *m*
lead sulphate Blei(II)-sulfat *n*
lead sulphide Blei(II)-sulfid *n*
lead-sulphide cell *(Spektr)* Bleisulfidzelle *f*, PbS-Zelle *f*
lead tetraethyl Tetraethylblei *n*
lead tetramethyl Tetramethylblei *n*
lead tetraoxide [tetroxide] Pb_3O_4 Blei(II,IV)-oxid *n*, Bleimennige *f*, Mennige *f*
lead tree Bleibaum *m*
leaded gasoline verbleites [gebleites, mit Tetraethylblei versetztes] Benzin *n*, Ethylbenzin *n*, Bleibenzin *n*
leading reactant [substrate] führendes [obligatorisches] Substrat *n (Enzymkinetik)*
leadless bleifrei
leaf filter Blattfilter *n*, Scheibenfilter *n*
leaflet *(Krist)* Blättchen *n*
LEAFS *s.* laser-excited atomic fluorescence spectrometry
leak *v* lecken, leck [undicht, defekt] sein; durchsickern, ausströmen, entweichen
leak 1. Leck *n*; 2. Lecken *n*; Durchsickern *n*, Ausströmen *n*, Entweichen *n*
• **spring a leak** undicht werden
leakage 1. Lecken *n*; 2. Leckage *f*, Leckverlust *m*, Sickerverlust *m*; Leckflüssigkeit *f*; Leckgas *n*; Leckluft *f*; 3. Durchbruch *m (Ionenaustausch)*
leakage curve Durchbruchkurve *f (Adsorption, Ionenaustausch)*
leakage test Dichtigkeitstest *m*
leakiness Undichtheit *f*, Undichtigkeit *f*
leakproof leckfrei, lecksicher, dicht
leaky undicht, leck
lean mager, arm
lean gas Schwachgas *n*, Armgas *n*
lean ore geringwertiges [geringhaltiges, armes] Erz *n*, Armerz *n*, Magererz *n*
leather Leder *n*
leather chemist Gerbereichemiker *m*, Lederindustrie-Chemiker *m*
leathery lederartig, lederähnlich
leave *v* austreten, abgehen, entweichen *(z. B. Dampf)*; verlassen
leave *v* **at the top** *(Dest)* über Kopf abgehen
leaven *v (Lebm)* (auf)gehen lassen *(Teig)*; (an)säuern, einsäuern, zur Gärung bringen *(Teig)*
leaven *(Lebm)* Treibmittel *n*, Triebmittel *n*, Teiglockerungsmittel *n*; Sauer(teig) *m*
leavening 1. *(Lebm)* Gehenlassen *n (von Teig)*; Säuern *n*, Ansäuern *n*, Einsäuern *n (von Teig)*; 2. *s.* leaven
leavening agent *s.* leaven
leaving group Austrittsgruppe *f*, Abgangsgruppe *f*, austretende Gruppe *f*
Leblanc process Leblanc-Verfahren *n (zur Sodagewinnung)*
Leblanc soda Leblanc-Soda *f*
lecithin *(Bioch)* Lecithin *n (ein Phospholipid)*
Leclanché cell Leclanché-Element *n*
lectin *(Bioch)* Lektin *n (ein Glykoprotein)*

LEED *s.* low-energy electron diffraction
lees Niederschlag *m*, Bodensatz *m*, Satz *m*, Trub *m*
left(-hand) circularly polarized linkszirkular polarisiert
left-handed 1. *(Krist)* Links...; 2. linksdrehend, (-)-drehend *(optische Aktivität)*
left-handed helix linksgängige [linksdrehende] Helix *f*
left-handed sense Linksgängigkeit *f* *(einer Helix)*
leg Schenkel *m* *(z. B. eines U-Rohrs)*
legal chemist Gerichtschemiker *m*
legal chemistry Gerichtschemie *f*, forensische [gerichtliche] Chemie *f*
LEL *s.* lower explosive limit
length of exposure 1. Expositionsdauer *f*, Einwirkungsdauer *f*; 2. *(Foto)* Belichtungsdauer *f*, Belichtungszeit *f*
length of path *(Anal)* Weglänge *f*, Schichtdicke *f*
lengthen *v (org Ch)* verlängern *(eine Kette)*
Lennard-Jones potential Lennard-Jones-Potenzial *n*
lens 1. Linse *f*; 2. *(Foto)* Objektiv *n*
lens aperture Linsenapertur *f*, Linsenöffnung *f*
lenticular linsenförmig
lepton *(Kern)* Lepton *n*
less volatile component *(Dest)* Schwerersiedende *n*, schwerersiedende [höhersiedende] Komponente *f*, schwererflüchtiger Anteil *m*, Phlegma *n*
lethal dose letale [tödliche] Dosis *f*, LD, Tödlichkeitsdosis *f*, dosis latalis
lethal dose 50 mittlere letale Dosis, LD 50 *(letale Dosis für 50 % der Versuchstiere)*
lethal temperature *(Biot)* Abtötungstemperatur *f*
leucine Leucin *n*, Leu, *2-Amino-4-methylpentansäure *f*
leuco base Leukobase *f*
leuco compound Leukoverbindung *f*
leucoindigo Leuk(o)indigo *m*
leucopararosaniline Leukopararosanilin *n*, Paraleukorosanilin *n*
leucoquinizarine Leukochinizarin *n*, *Anthracen-1,4,9,10-tetrol *n*
leucovorin Leucovorin *n*, Folinsäure *f*, N^5-Formyltetrahydrofolsäure *f*, Citrovorumfaktor *m*
leukotriene *(org Ch)* Leukotrien *n*
level *v* 1. (ein)ebnen, planieren; 2. ausgleichen; nivellieren *(Titration)*; 3. *(Anstr, Text)* egalisieren; 4. *(Text)* egalfärben
level *v* **off** sich einspielen, sich einpegeln *(z. B. pH-Wert)*
level *v* **out** ausgleichen
level *(Farb)* egal, gleichmäßig
level 1. Grad *m*, Stufe *f*, Niveau *n*; 2. Gehalt *m*, Anteil *m*; Konzentration *f*; 3. Füllstand *m*, Füllhöhe *f*, Niveau *n*, Spiegel *m*, Pegel *m*; 4. *(Boden)* Horizont *m*
level dyeing Egalfärben *n*, Egalisieren *n*, Egalisierung *f*
level measurement Füllhöhenmessung *f*, Füllstand(s)messung *f*, Niveaumessung *f*
level of pollutants *(Hyd)* Schmutzstoffgehalt *m*, Schmutzstoffanteil *m*
level of purity Reinheitsgrad *m*
levelling 1. Nivellierung *f (Titration)*; 2. *(Anstr, Text)* Egalisieren *n*, Egalisierung *f*; 3. *(Text)* Egalfärbung *f*
levelling agent *(Farb)* Egalisier(hilfs)mittel *n*, Egalisierungsmittel *n*, Egalisierer *m*
levelling bottle Niveauflasche *f*, Niveaugefäß *n*, Ausgleich(s)gefäß *n* *(Gasanalyse)*
levelling effect *(phys Ch)* nivellierender Effekt *m*, Nivellierung *f*
lever Hebel *m*, Hebestange *f*
levo... *siehe: 1. laevo...; 2. laevu...*
levu... *siehe: 1. laevo...; 2. laevu...*
Lewis acid Lewis-Säure *f (allg. für eine Verbindung mit Säurecharakter im Sinne von Lewis)*
Lewis base Lewis-Base *f (allg. für eine Verbindung mit Basecharakter im Sinne von Lewis)*
Lewis formula Elektronenformel *f*
Lewis theory of acids and bases *(phys Ch)* Säure-Base-Theorie *f* von Lewis
LFE relationship *s.* linear free energy relation(ship)
LH *s.* luteinizing hormone
liberate *v* freisetzen, entwickeln *(Gase)*; entwickeln, entbinden, freisetzen *(Wärme)*; freisetzen, in Freiheit set-

zen *(chemische Verbindungen)*; freisetzen, freilegen *(Nährstoffe im Boden)*; befreien *(von Verunreinigungen)*
liberation Freisetzung *f*, Entwicklung *f* *(von Gasen)*; Entwicklung *f*, Entbindung *f*, Freisetzung *f (von Wärme)*; Freisetzung *f (chemischer Verbindungen)*; Freisetzung *f*, Freilegung *f (von Nährstoffen im Boden)*; Befreiung *f (von Verunreinigungen)*
library (reference) spectrum Literaturspektrum *n*, Vergleichsspektrum *n*, Musterspektrum *n*
licanic acid *(org Ch)* Licansäure *f*, Couepinsäure *f*
lichen acid Flechtensäure *f*
lid Deckel *m*
Liebig condenser Liebig-Kühler *m*, liebigscher Kühler *m*
Liesegang rings Liesegang-Ringe *mpl*, liesegangsche Ringe *mpl*
LIF *s.* laser-induced fluorescence
life Lebensdauer *f*, Dauerhaftigkeit *f*, Haltbarkeit *f*, Laufzeit *f*
life hazard Lebensgefahr *f*
lifetime *(phys Ch)* (mittlere) Lebensdauer *f*; Halbwertszeit *f (Kinetik, Kernphysik)*
lift *v* liften, (hinauf)heben, nach oben befördern
lift Lift *m*, Heber *m*, Hebevorrichtung *f*
lift air Förderluft *f*, Luft *f* für Airlift
lifting 1. Liften *n*, Heben *n*, Hebung *f*, Förderung *f (z. B. des Katalysators beim katalytischen Kracken)*; 2. Hochziehen *n*, Hochgehen *n (Anquellen und Loslösen eines Anstrichs durch Lösungsmittel)*
ligancy Koordinationszahl *f*, Zähligkeit *f*, koordinative Wertigkeit *f*
ligand Ligand *m*
ligand exchange chromatography Liganden(austausch)chromatographie *f*
ligand field theory Ligandenfeldtheorie *f*, LFT
ligate *v (Bioch)* ligieren *(DNA-Enden verknüpfen)*
ligating atom Ligatoratom *n*, Donoratom *n*
ligation *(Bioch)* Ligation *f*, Ligierung *f* *(Verknüpfen komplementärer Enden von DNA)*
light *v* 1. anzünden, anbrennen; 2. beleuchten
light 1. hell, licht; 2. leicht; 3. *(Dest)* leicht(er)siedend, niedrigsiedend, tiefsiedend; 4. *(Lebm)* leicht, mit geringem Alkoholgehalt, alkoholarm, schwach alkoholhaltig [alkoholisch]
light Licht *n*
light-absorbing lichtabsorbierend
light absorption Lichtabsorption *f*
light-catalyzed lichtkatalysiert, photokatalysiert
light chain *(Bioch)* leichte Kette *f*, λ-Kette *f*
light component *s.* low-boiling component
light-conducting fiber Lichtleitfaser *f*
light degradation Lichtabbau *m*, Abbau *m* durch Licht, photochemischer Abbau *m*
light efficiency *s.* efficiency of luminous source
light emission Lichtemission *f*
light emitting diode Leuchtdiode *f*
light emitting material lichtemittierendes Material *n*
light-fast lichtecht, lichtbeständig
light fastness Lichtechtheit *f*, Lichtbeständigkeit *f*
light filter Lichtfilter *n*
light fraction 1. *(Dest)* leichte [niedrigsiedende, tiefsiedende] Fraktion *f*; 2. *(Aufb)* Leichtgut *n*
light gasoline Leichtbenzin *n*, leichtes Benzin *n*
light hydrogen ^{1}H leichter Wasserstoff *m*, Protium *n*
light metal Leichtmetall *n*
light microscope Lichtmikroskop *n*
light naphtha *s.* light gasoline
light oil Leichtöl *n*, leichtes Öl *n*
light path *(Spektr)* optische Weglänge *f*
light petroleum Petrol(eum)äther *m* *(Siedebereich 40 bis 70 °C)*
light quant(um) Lichtquant *n*, Photon *n*
light reaction Lichtreaktion *f*, Hellreaktion *f (Photochemie)*
light resistance *s.* light fastness
light scattering Lichtstreuung *f*
light-scattering photometer Streulichtphotometer *n*
light-sensitive lichtempfindlich
light sensitivity Lichtempfindlichkeit *f*
light source Lichtquelle *f*
light stabilizer Lichtstabilisator *m*

light transmission Lichtdurchlässigkeit *f*
light-water reactor *(Kern)* Leichtwasserreaktor *m*
light wave Lichtwelle *f*
light yield *s.* efficiency of luminous source
lighter component *s.* low-boiling component
lighter-than-light component *(Dest)* leichtest siedende Komponente *f*
lightly branched schwach verzweigt
lightproof lichtundurchlässig, lichtdicht
lignan Lignan *n*
ligneous substance Holzsubstanz *f*
lignin Lignin *n*, Holzstoff *m*
ligninsulphonic acid Ligninsulfonsäure *f*, Lignosulfonsäure *f*
lignite Braunkohle *f*; verfestigte [erhärtete] Braunkohle *f (nach der Kohlenklassifikation der ASTM)*; Xylit *m(n)*, Braunkohlenxylit *m(n)*, Braunkohlenholz *n*, Hylit *m(n)*, xylitische [holzartige] Braunkohle *f*
lignitic coal Braunkohle *f (nach der Kohlenklassifikation der ASTM)*
lignocellulose Lignocellulose *f (mit Lignin inkrustierte Cellulose)*
ligroin(e) Ligroin *n (Siedebereich 90 bis 120 °C)*; Petrol(eum)äther *m (Siedebereich 40 bis 70 °C)*; Leichtbenzin *n*, leichtes Benzin *n (Siedebereich 20 bis 135 °C)*
like charge gleichnamige Ladung *f*
like-charged (mit) gleicher Ladung
likelihood of collision Stoßwahrscheinlichkeit *f*
likely structure wahrscheinliche Struktur *f*
lime *v* 1. *(Landw)* kalken, mit Kalk düngen; 2. *(Zucker)* scheiden
lime *v* **out** (ab)kalken, auskalken, mit Kalk behandeln *(Zwischenprodukte)*
lime-soften *v (Hyd)* durch Kalkfällung enthärten
lime 1. Kalk *m*; 2. *(Landw)* Kalkdünger *m*, Kalkdüngemittel *n*; 3. *s.* liming 2.
lime burning Kalkbrennen *n*
lime feldspar *(Min)* Kalkfeldspat *m*, Anorthit *m (Calciumalumodisilicat)*
lime fertilizer Kalkdüngemittel *n*, Kalkdünger *m*
lime glass Kalkglas *n*
lime hydrate Kalkhydrat *n*, Calciumhydroxid *n*, Löschkalk *m*, gelöschter Kalk *m*
lime hydration *s.* lime slaking
lime hypophosphite Calciumhypophosphit *n*
lime kiln Kalk(brenn)ofen *m*
lime milk Kalkmilch *f (Suspension von Calciumhydroxid in Wasser)*
lime mortar Kalkmörtel *m*
lime nitrogen Kalkstickstoff *m*, Calciumcyanamid *n*
lime precipitation *(Hyd)* Kalkfällung *f*
lime refractory *(Ker)* Calciumoxiderzeugnis *n*
lime saccharate Kalksaccharat *n*, Calciumsaccharat *n*
lime shaft kiln Kalk-Schachtofen *m*
lime slaking Kalklöschen *n*
lime soap Kalkseife *f*
lime-soda softening *(Hyd)* Enthärtung *f* nach dem Kalk-Soda-Verfahren
lime softening *(Hyd)* Enthärtung *f* durch Kalkfällung
lime sulphur Schwefelkalk *m*, *(i. e. S.)* Schwefelkalkbrühe *f (Pflanzenschutzmittel)*
lime tunnel kiln Kalk-Ringofen *m*
lime wash *(Landw)* Kalkbrühe *f*, Kalkmilch *f*
lime-zeolite (softening) process *(Hyd)* Kalk-Zeolith-Verfahren *n*
limestone Kalkstein *m*
liming 1. *(Landw)* Kalken *n*, Kalkung *f*; 2. *(Zucker)* Scheidung *f*, Kalkung *f*
limit Grenze *f*; Grenzwert *m*
limit of detection Nachweisgrenze *f*, Erfassungsgrenze *f*
limiting concentration Grenzkonzentration *f*
limiting current *(phys Ch)* Grenzstrom *m*
limiting frequency Grenzfrequenz *f*
limiting structure Grenzstruktur *f*
limiting value Grenzwert *m*
limiting viscosity number Grenzviskositätszahl *f*, Staudinger-Index *m*
limonene Limonen *n*, *p*-Mentha-1,8(9)-dien *n*, *1-Methyl-4-isopropenylcyclohex-1-en *n*
limonite *(Min)* Limonit *m*, Brauneisenerz *n*, Brauner Glaskopf *m (wasserhaltiges Eisenoxidhydroxid)*
limy kalkhaltig, kalkig; kalkartig
limy soil Kalkboden *m*

Linde cycle [process] Linde-Verfahren *n* *(zur Verflüssigung von Gasen)*
line *v* 1. auskleiden; ausmauern; 2. *(Met)* zustellen, (aus)füttern; 3. *(Pap)* bekleben, kaschieren
line 1. Linie *f*; 2. Leitung *f*
line defect *(Krist)* Liniendefekt *m*, Linienfehler *m*, eindimensionaler [linienhafter] Gitterfehler *m* [Defekt *m*]
line formula Konstitutionsformel *f*, Strukturformel *f*, Valenzstrichformel *f*
line grating *(Spektr)* Liniengitter *n*, Strichgitter *n*
line spectrum Linienspektrum *n*, Atomspektrum *n*
linear acceleration Linearbeschleunigung *f*
linear accelerator linearer Beschleuniger *m* [Teilchenbeschleuniger *m*], Linearbeschleuniger *m*
linear accelerator for electrons Linearbeschleuniger *m* für Elektronen
linear combination of atomic orbitals Linearkombination *f* von Atomorbitalen, LCAO
linear expansity linearer thermischer Ausdehnungskoeffizient *m*
linear free energy relation(ship) lineare Beziehung *f* der freien Enthalpie, LFE-Beziehung *f*
linear frequency Frequenz *f*, Periodenfrequenz *f*, Schwingungszahl *f*
linear molecule lineares Molekül *n*; lineares [fadenförmiges] Makromolekül *n*, Fadenmolekül *n*
linear polarization lineare Polarisation *f*
linear polymer Linearpolymer(e) *n*, lineares [eindimensionales] Polymer(e) *n*
linearly polarized linear polarisiert
linen fabric Leinen(gewebe) *n*, Leinenstoff *m*, leinener Stoff *m*, Leinwand *f*
linen fibre Flachsfaser *f*
Lineweaver-Burk method Lineweaver-Burk-Methode *f* *(zur Auswertung kinetischer Daten)*
Lineweaver-Burk plot Lineweaver-Burk-Diagramm *n*
linin 1. Linin *n* *(Linum cartharticum principle)*; Bitterstoff *m*, Abführlein *n* *(Heilkraut)*; 2. Oxychromatin *n* *(Zytologie, Material im Zellkern)*
lining 1. Auskleiden *n*; Ausmauern *n*; 2. *(Met)* Zustellen *n*, Füttern *n*; 3. *(Pap)* Bekleben *n*, Kaschieren *n*; 4. Auskleidung *f*; Ausmauerung *f*; 5. *(Met)* Zustellung *f*, Futter *n*
link *v* verbinden, verketten, *(Farbstoff an die Unterlage)* binden, *(auf der Unterlage)* fixieren; sich verbinden
link *v* **together** miteinander verbinden [verknüpfen]; sich (miteinander) verbinden
linkage 1. Verbindung *f*, Verkettung *f*, Verknüpfung *f*; 2. (chemische) Bindung *f*
linolate Linoleat *n* *(Salz oder Ester der Linolsäure)*
linoleate Linoleat *n* *(Salz oder Ester der Linolsäure)*
linoleic acid Linolsäure *f*, Leinölsäure *f*, *Octadeca-9,12-diensäure *f*
linolenic acid Linolensäure *f*, *Octadeca-9,12,15-triensäure *f*
linseed oil Leinöl *n*
linseed-oil varnish Leinöllack *m*; Leinölfirnis *m*
lip Ausguss *m* *(z. B. eines Becherglases)*
lipid Lipid *n* *(allg. für ein Fett oder Lipoid)*
lipid bilayer *(Bio)* Lipiddoppelschicht *f*, Doppellipidschicht *f*
lipid metabolism Lipidstoffwechsel *m*
lipid-soluble lipidlöslich
lipoid Lipoid *n*, komplexes [polares] Lipid *n*
lipolysis Lipolyse *f*, Fettspaltung *f*
lipolytic lipolytisch, fettspaltend
lipolytic enzyme fettspaltendes Enzym *n*
lipophile lipophil, fettfreundlich, sich mit Fett mischend, in Fett löslich
lipophilic lipophil, fettfreundlich, sich mit Fett mischend, in Fett löslich
lipopolymer Lipopolymer *n* *(fettlösliches Polymer)*
lipopolysaccharide Lipopolysaccharid *n*, LPS
lipoproteid(e) Lipoproteid *n*
lipoprotein Lipoproteid *n*
liposoluble fettlöslich
liq. *s.* liquid
liquate *v* **(out)** *(Met)* (ab)seigern *(Komponenten aus Rohmetall unter Ausnutzung unterschiedlicher Schmelzpunkte abtrennen)*; seigern, sich ausscheiden *(aus Legierungen)*
liquation 1. *(Met)* Seigern *n*, Abseigern *n*;

2. *(Met)* Seigerung *f (als unerwünschte Erscheinung)*
liquefaction 1. Verflüssigen *n*, Verflüssigung *f*; 2. *(Met)* Schmelzen *n*; Flüssigwerden *n*, Verflüssigung *f*
liquefaction of air Luftverflüssigung *f*
liquefiable verflüssigbar
liquefied gas 1. verflüssigtes Gas *n*, Flüssiggas *n*; 2. *s.* liquefied petroleum gas
liquefied natural gas verflüssigtes Erdgas *n*, LNG
liquefied petroleum gas verflüssigtes Erdölgas *n*, Flüssiggas *n*, LPG
liquefy *v* 1. verflüssigen; sich verflüssigen, flüssig werden; 2. *(Met)* schmelzen
liquid 1. flüssig, tropfbar; 2. klar, durchsichtig
liquid Flüssigkeit *f*
liquid air flüssige Luft *f*
liquid ammonia flüssiges [verflüssigtes] Ammoniak *n*
liquid body Flüssigkeitskörper *m*, Flüssigkeitsvolumen *n*, Flüssigkeitsmasse *f*
liquid carbon dioxide extraction *(Biot)* Extraktion *f* mit flüssigem CO_2
liquid chromatography *Flüssigchromatographie *f*, Flüssigkeitschromatographie *f*, LC • **by liquid chromatography** flüssigkeitschromatographisch
liquid chromatography-mass spectroscopy coupling Flüssigchromatographie-Massenspektroskopie-Kopplung *f*
liquid crystal Flüssigkristall *m*, flüssiger Kristall *m*, kristalline [anisotrope, mesomorphe] Flüssigkeit *f*, Mesophase *f*
liquid cyclone (separator) Hydrozyklon *m*, Nasszyklon *m*
liquid extraction *s.* liquid-liquid extraction
liquid fertilizer Flüssigdünger *m*
liquid film Flüssigkeitsfilm *m*
liquid fuel flüssiger Brennstoff *m*; flüssiger Kraftstoff *m*; Flüssigtreibstoff *m*, *(i. e. S.)* flüssiger Raketentreibstoff *m*
liquid gas Flüssiggas *n (verflüssigte Kohlenwasserstoffe)*
liquid glass Wasserglas *n*
liquid hold-up *(Dest)* Flüssigkeitsinhalt *m*
liquid holding volume Flüssigkeitsfassungsvermögen *n*
liquid-junction potential Flüssigkeits(diffusions)potenzial *n*, Diffusionspotenzial *n*, Diffusions(galvani)spannung *f*
liquid laser Flüssigkeitslaser *m*
liquid-liquid chromatography *Flüssig-flüssig-Chromatographie *f*, Liquidus-Liquidus-Chromatographie *f*, LLC, Flüssig-Verteilungschromatographie *f*
liquid-liquid extraction Solventextraktion *f*, Flüssig-flüssig-Extraktion *f*, Extraktion *f* flüssig-flüssig, Extraktion *f* in flüssigen Systemen
liquid-liquid interface Flüssig-flüssig-Grenzfläche *f*, Grenzfläche *f* [Phasengrenze *f*] flüssig-flüssig
liquid-liquid partition Flüssig-flüssig-Verteilung *f*
liquid membrane *(Koll)* Flüssigmembran *f*
liquid-motivated ejector Flüssigkeitsstrahlpumpe *f*; Flüssigkeitsstrahlgebläse *n*, Flüssigkeitsstrahlverdichter *m*
liquid nitrogen flüssiger [verflüssigter] Stickstoff *m*, Flüssigstickstoff *m*
liquid operating hold-up Betriebsinhalt *m*, Arbeitsinhalt *m (z. B. einer Kolonne)*
liquid oxygen flüssiger [verflüssigter] Sauerstoff *m*, Flüssigsauerstoff *m*
liquid paraffin 1. Paraffinöl *n*; 2. *(Pharm)* flüssiges Paraffin *n*
liquid phase flüssige Phase *f*, Flüssig(keits)phase *f*, *(bei der Hochdruckhydrierung meist)* Sumpfphase *f*
liquid-phase process Flüssigphaseverfahren *n*
liquid-phase reaction Flüssigphasereaktion *f*, Reaktion *f* in (homogener) flüssiger Phase
liquid propellant flüssiger Raketentreibstoff *m*, Flüssigtreibstoff *m*
liquid-receiver *(Lab)* Flüssigkeitsvorlage *f*
liquid resin Flüssigharz *n*
liquid-solid chromatography *Flüssig-fest-Chromatographie *f*, Liquidus-Solidus-Chromatographie *f*, LSC, Flüssig-Adsorptionschromatographie *f*
liquid-solid extraction Feststoffextraktion *f*, Extraktion *f* von Feststoffen, Extraktion *f* fest-flüssig, Fest-flüssig-Extraktion *f*

liquid-solid interface Flüssig-fest-Grenzfläche *f*, Grenzfläche *f* flüssig-fest
liquid-solid(-phase heterogeneous) reaction Flüssig-fest-Reaktion *f*, Reaktion *f* im Phasensystem fest-flüssig
liquid sugar Flüssigzucker *m*, flüssiger Zucker *m*
liquid throughput Flüssigkeitsdurchsatz *m*
liquid to be pumped Förderflüssigkeit *f*, zu fördernde Flüssigkeit *f*
liquid under examination [investigation] Untersuchungsflüssigkeit *f*
liquid-vapour interface Flüssigkeit-Dampf-Grenzfläche *f*, Grenzfläche *f* flüssig-gasförmig
liquidus (curve, line) Liquiduskurve *f*, Liquiduslinie *f*, Flüssiglinie *f*, Löslichkeitskurve *f*
liquor 1. Flüssigkeit *f*; 2. *(Tech)* Lauge *f*; Brühe *f*; 3. *(Text)* Flotte *f*, Bad *n*; 4. *(Lebm)* alkoholisches [geistiges] Getränk *n*
litharge Bleiglätte *f*, Bleimonoxid *n*, Blei(II)-oxid *n*
lithia Lithiumoxid *n*
lithiate *v* mit Lithium [Lithiumverbindungen] versetzen [verbinden]
lithic acid Harnsäure *f*, 2,6,8-Trihydroxypurin *n*
lithium Li Lithium *n*
lithium aluminium hydride Lithiumalanat *n*, Lithiumaluminiumhydrid *n*
lithium amide Lithiumamid *n*
lithium aryl Aryllithium *n*
lithium chloride Lithiumchlorid *n*
lithium hydride Lithiumhydrid *n*
lithium hydroxide Lithiumhydroxid *n*
lithium molybdate Lithiummolybdat(VI) *n*
lithium perchlorate Lithiumperchlorat *n*
lithocholic acid Lithocholsäure *f*, 3-Hydroxycholansäure *f*
lithography Lithographie *f*
lithop(h)one Lithopone *f*, Lithopon *n* *(Weißpigment aus ZnS und $BaSO_4$)*
lithosphere *(Geol)* Lithosphäre *f*, Gesteinsmantel *m*, Gesteinshülle *f*, Gesteinskruste *f*
litmus Lackmus *n(m)* *(Flechtenfarbstoff)*
litmus paper [test paper] Lackmuspapier *n*
Littrow mounting *(Spektr)* Littrow-Aufstellung *f*, Littrow-Anordnung *f*
Littrow spectrograph Littrow-Spektrograph *m*
live steam Frischdampf *m*, Direktdampf *m*, direkter [gespannter] Dampf *m*
liver of sulphur Schwefelleber *f*, Hepar sulfuris *(technisches Kaliumsulfid)*
living polymer lebendes Polymer(e) *n*
living polymerization lebende Polymerisation *f*
lixiviate *v* (aus)laugen, auswaschen, herauslösen, ausziehen, extrahieren
lixiviation Auslaugen *n*, Auslaugung *f*, Laugung *f*, Auswaschen *n*, Herauslösen *n*, Ausziehen *n*, Extrahieren *n*, Extraktion *f*
LLC *s.* liquid-liquid chromatography
LN_2 = liquid nitrogen
LNG *s.* liquefied natural gas
load *v* 1. beladen, beschicken, füllen, chargieren *(z. B. Gefäße)*; 2. (auf)laden, aufgeben, einfüllen, einschütten, einlegen, eintragen, einwerfen *(Füllgut)*; 3. beladen *(Ionenaustauscher, Adsorbentien, Gase)*; 4. beschweren, füllen *(mit Füllstoffen)*; 5. belasten *(mechanisch)*
load 1. Ladung *f*, Füllung *f*, Charge *f*; 2. Beladung *f*, Belastung *f*, Last *f* *(Produkt aus Konzentration der Inhaltsstoffe und Flüssigkeitsmenge bzw. Gasvolumen je Zeiteinheit)*; 3. Last *f*, Belastung *f* *(mechanisch)*
load-elongation curve Kraft-Längenänderungs-Kurve *f*, Kraft-Verlängerungs-Kurve *f*
load-extension curve Kraft-Längenänderungs-Kurve *f*, Kraft-Verlängerungs-Kurve *f*
loadability *(Tech, Anal)* Belastbarkeit *f*
loaded gefüllt, füllstoffhaltig, mit Füllstoffen
loading 1. Beladen *n*, Beschicken *n*, Füllen *n*, Chargieren *n* *(z. B. von Gefäßen)*; 2. Aufladen *n*, Laden *n*, Aufgeben *n*, Einfüllen *n*, Einschütten *n*, Einlegen *n*, Eintragen *n*, Einwerfen *n* *(von Füllgut)*; 3. Beladen *n*, Beladung *f* *(von Ionenaustauschern, Adsorbentien, Gasen)*; 4. Beschweren *n*, Füllen *n* *(mit Füllstoffen)*; 5. Füllstoff *m*, Beschwerungsmaterial *n*; 6. *s.* load 2.

loading agent *s.* loading material
loading capacity *(Dest)* Belastbarkeit *f*
loading material Füllstoff *m*, Zuschlag *m*, Zusatz *m*, Beschwerungsmaterial *n*
loading point 1. Belastungsgrenze *f*; 2. *(Dest)* Staupunkt *m*
loading rate per unit area *(Hyd)* Flächenbelastung *f*, Flächenbeschickung *f (in kg/m^2 · a bei Trockenbeeten; in m^3/m^2 · h bei Tropfkörpern)*; Oberflächenbelastung *f*, Oberflächenbeschickung *f (in m^3/m^2 · h bei Absetzbecken)*
loading rate per unit volume *(Hyd)* Raumbelastung *f (in kg/m^3 · d)*
loam Lehm *m*
loamy lehmhaltig, lehmig
lobaric acid Lobarsäure *f*, Stereokaulsäure *f*
lobe pump Drehkolbenpumpe *f*, Kapselpumpe *f*
lobelia alkaloid Lobeliaalkaloid *n*
local anaesthetic Lokalanästhetikum *n*, Mittel *n* zur örtlichen Schmerzausschaltung
local cell [galvanic element] Lokalelement *n (ein Korrosionselement)*
local mixing Mikrovermischung *f*
localization signal *s.* signal peptide
localized bond lokalisierte Bindung *f*
localized corrosion örtliche Korrosion *f*, Lokalkorrosion *f*
locant *(Nom)* Stellungsziffer *f*
locate *v* unterbringen *(Atome auf Zwischengitterplätzen)*; bringen *(auf Zwischengitterplätze)*
location 1. Stelle *f*; 2. Unterbringung *f (von Atomen auf Zwischengitterplätzen)*
lock Schleuse *f*, Schleuseneinrichtung *f*
lock-and-key relationship *(Bioch)* Schlüssel-Schloss-Beziehung *f*
lock hopper Beschickungsschleuse *f*
locking device Verriegelung *f*; Feststelleinrichtung *f*, Arretierung *f*, Arretiervorrichtung *f*; Hahnsicherung *f*
lode *(Geol)* Gang *m*, Ader *f*
loess Löss *m*
log-log plot doppelt logarithmisches Diagramm *n*
log phase *s.* logarithmic growth phase
logarithmic growth phase *(Hyd, Biot)* logarithmische [exponentielle] Wachstumsphase *f*, log-Phase *f (Mikroorganismenwachstum)*
logarithmic normal distribution logarithmische Normalverteilung *f*
loglog paper doppelt logarithmisch geteiltes Papier *n*
LOI *s.* loss on ignition
London force London-(van-der-Waals-) Kraft *f*, Dispersionskraft *f*
lone electron pair einsames [freies] Elektronenpaar *n*
long-chain langkettig
long-chain fatty acid höhere Fettsäure *f*
long-chain polymer langkettiges Polymer(e) *n*
long-fibre langfas(e)rig
long-lived radical persistentes [langlebiges] Radikal *n*
long-neck flask Langhalskolben *m*
long-oil alkyd langöliges Alkyd(harz) *n*
long-pathlength cell *(Spektr)* Langweg(gas)küvette *f*, Langweggaszelle *f*
long period lange [große] Periode *f*, Langperiode *f (im Periodensystem)*
long periodic table Langperiodensystem *n*
long-range coupling [interaction] Fernkopplung *f*, Weitwegkopplung *f (Bindungstheorie)*
long-range order *(Krist)* Fernordnung *f*
long residue [residuum] langer Destillationsrückstand *m*, atmosphärischer Rückstand *m*, Topprückstand *m*
long sifter Langsiebmaschine *f*
long-term experiment Langzeitversuch *m*
long-term fermentation *(Biot)* Langzeitfermentation *f*
long-term stability Langzeitstabilität *f*
long-tube evaporator Langrohrverdampfer *m*
long wavelength region Langwellenbereich *m*; langwelliger Spektralbereich *m*, langwelliges Spektralgebiet *n*
longer-chained längerkettig
longest-lived isotope langlebigstes Isotop *n*
longitudinal diffusion Längsdiffusion *f*, axiale Diffusion *f*, Longitudinaldiffusion *f*
longitudinal expansion coefficient linearer thermischer Ausdehnungskoeffizient *m*

longitudinal mixing Längsvermischung *f*, Axialvermischung *f*, axiale Rückvermischung *f*
longitudinal reactor ideales Strömungsrohr *n*
loop reactor Schlaufenreaktor *m*, Kreislaufreaktor *m*
loop-type fermenter *(Biot)* Schlaufenfermenter *m*
looping plume schleifenförmige Rauchfahne *f*
loose locker, lose
loose ion pair solvensgetrenntes [äußeres] Ionenpaar *n (durch ein oder mehr als ein Lösungsmittelmolekül getrennt)*
loosen *v* **(up)** (auf)lockern *(eine chemische Bindung)*; auflockern *(ein Filterbett)*
Lorentzian profile Lorentz-Profil *n*, lorentzförmiges Profil *n (einer Spektrallinie)*
Loschmidt number *veraltete, früher als avogadrosche Zahl bezeichnete Maßeinheit für die Anzahl der Moleküle in 1 ml eines idealen Gases bei 0 °C und 1 atm (s. dagegen Avogadro number)*
lose *v* abgeben, verlieren
loss Verlust *m*, Abnahme *f*, Abgang *m*, Schwund *m*, Verarmung *f*
loss angle Verlustwinkel *m*
loss by radiation Strahlungsverlust *m*, Abstrahlungsverlust *m*
loss in strength Festigkeitsverlust *m*, Festigkeitsrückgang *m*, Festigkeitseinbuße *f*
loss of energy Energieverlust *m*
loss of material Substanzverlust *m*
loss on ignition Glühverlust *m*
loss tangent 1. (dielektrischer) Verlustfaktor *m*; 2. Verlustfaktor *m (mechanisch)*
Lossen rearrangement lossensche Umlagerung *f*, Lossen-Abbau *m (von Hydroxamsäuren zu primären Aminen)*
Lothar Meyer periodic table Tabelle *f* des Periodensystems nach Lothar Meyer
lotus effect Lotuseffekt *m (Selbstreinigung von Oberflächen)*
lotus plant *(Bio)* Lotuspflanze *f (Lotuseffekt)*
low-added-value product geringveredeltes Produkt *n*
low-alkali glass alkaliarmes Glas *n*
low-alloy steel niedriglegierter Stahl *m*
low-analysis niederprozentig
low-angle laser light scattering Laser(licht)-Kleinwinkelstreuung *f*
low boiler Niedrigsieder *m*, niedrigsiedende Komponente *f (Lösungsmittel)*
low-boiling tiefsiedend, niedrigsiedend, leichtsiedend
low-boiling azeotrope Minimumazeotrop *n*, Azeotrop *n* mit Minimumsiedepunkt, Siedepunktsminimumazeotrop *n*
low-boiling component *(Dest)* Leichtersiedendes *n*, leichtersiedende Komponente *f*, leichterflüchtiger Anteil *m*
low-carbon niedriggekohlt, kohlenstoffarm, mit niedrigem Kohlenstoffgehalt *(z. B. Stahl)*
low-dead-volume totvolumenarm
low-density polyethylene Polyethylen *n* niedriger Dichte, Hochdruckpolyethylen *n*, verzweigtes Polyethylen *n*
low-energy energiearm, niederenergetisch
low-energy electron diffraction Beugung *f* langsamer [niederenergetischer] Elektronen
low explosive Treibstoff *m*, Schießstoff *m*, Schießmittel *n*, Pulversprengstoff *m*
low-fat fettarm
low-frequency niederfrequent
low-fusion(-point) *s.* low-melting
low-grade minderwertig, geringwertig, wertarm, von minderer Qualität
low-grade ore geringwertiges [geringhaltiges, armes] Erz *n*, Armerz *n*, Magererz *n*
low-intensity intensitätsschwach
low-level active *(Kern)* niedrigaktiv
low-melting niedrigschmelzend, leichtschmelzend, tiefschmelzend, mit niedrigem Schmelzpunkt, leichtschmelzbar
low-melting-point *s.* low-melting
low-molar-mass von geringer Molmasse
low-molecular niedrigmolekular, niedermolekular
low-noise *(Anal)* rauscharm

low-octane niedrigoctan(zahl)ig, mit niedriger [geringer] Octanzahl
low-oxygen sauerstoffarm, mit niedrigem [geringem] Sauerstoffgehalt
low-pressure boiler Niederdruckkessel *m*
low-pressure mercury (vapour) lamp Quecksilberniederdrucklampe *f*
low-pressure (process) polyethylene Niederdruckpolyethylen *n*, Linearpolyethylen *n*, Polyethylen *n* hoher Dichte
low-pressure steam Niederdruckdampf *m*
low-quality minderwertig
low-rate filter Langsamfilter *n*
low redness schwache Rotglut *f*
low-shaft furnace Niederschachtofen *m*
low-solubility schwachlöslich
low-spin complex Durchdringungskomplex *m*, Innerkomplex *m*, Innenkomplex *m*, Low-spin-Komplex *m*
low-strain spannungsarm *(cyclische Verbindung)*
low-sulphur schwefelarm, S-arm, mit niedrigem [geringem] Schwefelgehalt
low-temperature behaviour Tieftemperaturverhalten *n*, Kälteverhalten *n*, Verhalten *n* bei Kälte [tiefen Temperaturen]
low-temperature carbonization Tieftemperaturverkokung *f*, Schwelung *f*, Schwelen *n*, Verschwelung *f*, Verschwelen *n*, Tieftemperaturentgasung *f*
low-temperature engineering Tieftemperaturtechnik *f*, Kryotechnik *f*
low-temperature hydrogenation Tieftemperaturhydrierung *f*
low-temperature material Tieftemperaturwerkstoff *m*
low-temperature polymerization Tieftemperaturpolymerisation *f*
low-temperature properties Tieftemperatureigenschaften *fpl*, Kälteeigenschaften *fpl*
low-temperature reaction Tieftemperaturreaktion *f*
low-temperature resistance Tieftemperaturbeständigkeit *f*, Kältebeständigkeit *f*, Kältefestigkeit *f*
low-temperature technology Tieftemperaturtechnik *f*, Kryotechnik *f*
low vacuum Grobvakuum *n*
low-viscosity niedrigviskos, leichtviskos, wenig viskos, von niedriger Viskosität, dünnflüssig
low-voltage electrophoresis Niederspannungselektrophorese *f*
lower bainite *(Met)* unterer Bainit *m*, Unterbainit *m*, Gefüge *n* der unteren Zwischenstufe
lower-boiling niedriger [leichter] siedend, leichtersiedend
lower consolute [critical solution] temperature untere kritische Lösungstemperatur *f* [Mischungstemperatur *f*], unterer kritischer Lösungspunkt *m* [Mischungspunkt *m*]
lower detection limit *(Anal)* (untere) Nachweisgrenze *f*, (untere) Erfassungsgrenze *f*
lower explosive limit untere Explosionsgrenze *f*
lower fatty acid niedere Fettsäure *f*
lower-frequency niedrigerfrequent
lower heating value spezifischer Heizwert *m*, H_u, *(veraltet:)* unterer Heizwert *m*
lower observed effect level niedrigste Dosis *f* mit beobachtbarer toxischer Wirkung
lower-valency niederwertig
lower-valent niederwertig
lowering of the freezing point Gefrierpunktserniedrigung *f*, Gefrierpunktsdepression *f*
lowering of the melting point Schmelzpunktserniedrigung *f*, Schmelzpunktsdepression *f*
lowering of vapour pressure Dampfdruckerniedrigung *f*
lowest(-energy) unoccupied molecular orbital niedrigstes unbesetztes Molekülorbital *n*
Lowry method (for protein determination) Lowry-Methode *f*, Proteinbestimmungsmethode *f* nach Lowry
LP steam *s.* low-pressure steam
L.P. gas *s.* liquefied petroleum gas
L.P.G. *s.* liquefied petroleum gas
lq. *s.* liquid
LRH *(lutropin releasing hormone)* *Gonadoliberin *n*, GnRH
LS coupling *(Spektr)* Spin-Bahn-Kopplung *f*, Spin-Bahn-Wechselwirkung *f*,

LS-Kopplung *f*, Russel-Saunders-Kopplung *f*
LSC *s.* liquid-solid chromatography
LSD *s.* lysergic acid diethylamide
lube *s.* lubricant
lubricant Schmierstoff *m*, Schmiermittel *n*, *(i. e. S.)* Schmierflüssigkeit *f*; Gleitmittel *n*
lubricate *v* (ab)schmieren
lubricating additive Schmiermitteladditiv *n*
lubricating grease Schmierfett *n*
lubricating oil Schmieröl *n*
lubrication Schmieren *n*, Schmierung *f*
Luggin capillary *(phys Ch)* lugginsche Kapillare *f*
Lugol's solution lugolsche Lösung *f* [Iodlösung *f*] *(wässrige Lösung von Iod und Kaliumiodid)*
luminesce *v* lumineszieren
luminescence Lumineszenz *f*, Leuchtanregung *f*, kaltes Leuchten *n*
luminescence analysis Lumineszenzanalyse *f*
luminescence behaviour Lumineszenzverhalten *n*
luminescence indicator Lumineszenzindikator *m*
luminescence spectroscopy Lumineszenzspektroskopie *f*, Lumineszenzspektralanalyse *f*
luminescence spectrum Lumineszenzspektrum *n*
luminescent lumineszent, lumineszierend, Lumineszenz...
luminophor(e) Luminophor *m*, Lumineszenzstrahler *m*, *(i. e. S.)* Leuchtstoff *m (im sichtbaren Spektralbereich strahlend)*
luminosity factor *s.* efficiency of luminous source
luminous leuchtend; glänzend; Leucht..., Licht...
luminous arc process Lichtbogenverfahren *n*
luminous efficiency *s.* efficiency of luminous source
luminous flame leuchtende Flamme *f*
luminous flux Lichtstrom *m*
LUMO *s.* lowest(-energy) unoccupied molecular orbital
lump *v* klumpen, Klumpen bilden, klumpig werden, sich zusammenballen
lump Klumpen *m*, Brocken *m*, Stück *n*
lump ore Stückerz *n*
lumpy klumpig, klumpenförmig, Klumpen..., stückig, Stück...
lunar rock Mondgestein *n*
lung injurant Lungengift *n*, *(Militärchemie auch)* lungenschädigender Kampfstoff *m*
lung irritant Lungenreizstoff *m*
lupin(e) alkaloid Lupinenalkaloid *n*, Chinolizidinalkaloid *n*
Lurgi high-pressure process *(Carb)* Lurgi-Druckvergasungsverfahren *n*, Lurgi-Druckgasverfahren *n*
Lurgi pressure-gasification process *(Carb)* Lurgi-Druckvergasungsverfahren *n*, Lurgi-Druckgasverfahren *n*
Lurgi pressure gasifier [generator] *(Carb)* Lurgi-Druckvergaser *m*, Lurgi-Druckgasgenerator *m*, Lurgi-Druckgaserzeuger *m*
Lurgi Spülgas low-temperature carbonization process *(Carb)* Lurgi-Spülgas(schwel)verfahren *n*
luster *s.* lustre
lustre Glanz *m*, Schein *m*, Schimmer *m*
lustreless glanzlos, matt
lustrous glänzend, Glanz...
lute *v* (ver)kitten, *(i. w. S.)* verschmieren, dichten
lute Kitt *m*, Dichtungsmasse *f*; Gummi-(dichtungs)ring *m*
luteinize *v* luteinisieren
luteinizing hormone *Lutropin *n*, luteinisierendes [zwischenzellenstimulierendes] Hormon *n*, LH, Gelbkörperreifungshormon *n*, interstitielle Zellen stimulierendes Hormon *n*, ICSH
luteotropin *Prolactin *n*, PRL, luteotropes [lactogenes] Hormon *n*, LTH
lutetia Lutetiumoxid *n*
lutetium Lu Lutetium *n*
lutidine Lutidin *n*, Dimethylpyridin *n*
lyate ion Lyat-Ion *n*
lycine Betain *n*, Lycin *n*, Oxyneurin *n*, Carboxymethyltrimethylammonium *n*, Trimethylammonioacetat *n*, Trimethylglycin *n*
lycopene Lycopin *n*, Dicaroten *n (ein Naturfarbstoff)*
lye 1. Lauge *f*, *(i. e. S.)* Alkalilauge *f*; 2. Extrakt *m*, Auszug *m*
Lyman series *(Spektr)* Lyman-Serie *f*

lymph *(Med)* Lymphe *f*
lyo-enzyme Lyoenzym *n (im Zellinhalt gelöstes Enzym)*
lyogel Lyogel *n*
lyolysis Lyolysis *f*, Solvolyse *f*
lyonium ion Lyonium-Ion *n*
lyophilic lyophil, lösungsmittelanziehend
lyophilization Lyophilisation *f*, Lyophilisierung *f*, Gefriertrocknung *f*, Sublimationstrocknung *f*, Tiefkühltrocknung *f*
lyophilize *v* lyophilisieren, gefriertrocknen
lyophilized culture *(Biot)* gefriergetrocknete Kultur *f*, Lyophilisat *n*
lyophobic lyophob, lösungsmittelabstoßend
lyotropic liquid crystal lyotroper Flüssigkristall *m* [flüssiger Kristall *m*]
lyotropic order [series] lyotrope Reihe *f*, hofmeistersche Ionenreihe *f* [Reihe *f*]
lysergic acid Lysergsäure *f*
lysergic acid alkaloid Lysergsäurealkaloid *n*
lysergic acid diethylamide Lysergsäurediethylamid *n*, LSD *n*
lysine *(Bioch)* Lysin *n*, Lys, *2,6-Diaminohexansäure *f*
lysis *(Bio)* Lyse *f (Zerfall einer Zelle)*
lysozyme Lysozym *n (bakteriolytisches Enzym)*
lyzed cells *(Biot)* lysierte Zellen *fpl*

M

m. al. *s.* methyl alcohol
M shell M-Schale *f (eines Atoms)*
M unit M-Einheit *f*, monofunktionelle Einheit *f*, Grundbaustein *m* M *(zum Aufbau von Polymeren)*
ma = myria, 10^4
MAA *s.* methacrylic acid
MAb *s.* monoclonal antibody
MAC *s.* maximum allowable concentration
maceral Mazeral *n*, Gefügebestandteil *m*, Aufbauelement *n*, Komponente *f (der Kohle)*
macerate *v* mazerieren, aufschließen *(besonders Pflanzenteile)*
maceration Mazerieren *n*, Mazeration *f*, Aufschließen *n (besonders von Pflanzenteilen)*
machinability maschinelle Bearbeitbarkeit *f*
machine oil Maschinen(schmier)öl *n*
macro sample Makroprobe *f (> 0,1 g)*
macroanalysis Makroanalyse *f*
macroanalytical makroanalytisch
macrochemical makrochemisch
macrochemistry Makrochemie *f*
macrocrystalline makrokristallin
macrocycle Makrocyclus *m*, Großring *m*
macrocyclic (compound) makrocyclische Verbindung *f*, Verbindung *f* mit großem Ring, Großringverbindung *f*
macroelement *s.* macronutrient
macrofibril Makrofibrille *f*
macrofibrillar makrofibrillär
macroinitiator Makroinitiator *m*
macroion Makroion *n*
macrolide antibiotic *(Biot)* Makrolid-Antibiotikum *n*, makrolides Antibiotikum *n*
macromer Makromer *n*
macromethod Makromethode *f*
macromolecular makromolekular
macromolecular monomer *s.* macromer
macromolecule Makromolekül *n*, Riesenmolekül *n*
macronutrient *(Landw)* Makronährstoff *m*, Kernnährstoff *m*
macropore Makropore *f*, *(bei Katalysatoren auch)* Transportpore *f*
macroporous makroporös
macroradical Makroradikal *n*
macroreticular makroretikulär
macroring Makroring *m*, großer Ring *m (aus 13 oder mehr Gliedern)*
macrostructure Makrogefüge *n*, Makrostruktur *f*
made up of/be bestehen aus, sich zusammensetzen aus
Madelung constant Madelung Konstante *f*, madelungsche Konstante *f (der Gitterenergie)*
magenta Magenta *n*, Fuchsin *n*, Rosanilin(hydrochlorid) *n*
magic acid magische Säure *f (Gemisch aus Fluorschwefelsäure und Antimonpentafluorid)*
magma 1. *(Geol)* Magma *n*; 2. *(Tech)* Masse *f*, Brei *m*

magmatic *(Geol)* magmatisch, Magma...
magnaflux Magnetpulver *n*
magnesia Magnesia *f*, Magnesiumoxid *n*
magnesia alba Magnesiumhydroxidcarbonat-5-Wasser *n*
magnesia usta Magnesiumoxid *n*
magnesian 1. magnesiahaltig, Magnesia..., magnesiumoxidhaltig, Magnesiumoxid...; 2. Magnesium...
magnesite brick (feuerfester) Magnesitstein *m*
magnesium Mg Magnesium *n*
magnesium ammonium phosphate Ammoniummagnesiumphosphat *n*
magnesium carbonate Magnesiumcarbonat *n*
magnesium carbonate hardness *(Hyd)* Magnesium-Karbonathärte *f*, Mg-KH
magnesium chloride Magnesiumchlorid *n*
magnesium hardness *(Hyd)* Magnesiahärte *f*, MgH
magnesium hydride Magnesiumhydrid *n*
magnesium hydroxide Magnesiumhydroxid *n*
magnesium non-carbonate hardness *(Hyd)* Magnesium-Nichtkarbonathärte *f*, Mg-NKH
magnesium oxide Magnesiumoxid *n*
magnesium sulphate Magnesiumsulfat *n*
magnesium sulphate hardness *(Hyd)* Magnesiumsulfathärte *f*
magnetic magnetisch, Magnet...
magnetic dipole moment magnetisches Dipolmoment *n* [Moment *n*]
magnetic field magnetisches Feld *n*, Magnetfeld *n*
magnetic moment *s.* magnetic dipole moment
magnetic quantum number magnetische [bahnmagnetische, räumliche] Quantenzahl *f*, magnetische Bahn(drehimpuls)quantenzahl *f*, Magnetquantenzahl *f*, Orientierungsquantenzahl *f*, Richtungsquantenzahl *f*
magnetic resonance spectroscopy magnetische Resonanzspektroskopie *f*
magnetic rotation Magnetorotation *f*, magnetische Drehung *f* (der Polarisationsebene), Faraday-Effekt *m*
magnetic stirrer [stirring apparatus] magnetisches Rührwerk *n*, Magnetrührer *m*
magnetic susceptibility magnetische Suszeptibilität *f*
magnetic transition magnetischer Übergang *m*
magnetite *(Min)* Magnetit *m*, Magneteisenerz *n* *(Eisen(II,III)-oxid)*
magnetizable magnetisierbar
magnetization Magnetisierung *f*
magnetize *v* magnetisieren
magneto-optical rotatory dispersion *(Anal)* magnetooptische Rotationsdispersion *f*
magnetochemistry Magnetochemie *f*
magneton Magneton *n* *(Maß für magnetische Momente von Atomen, Atomkernen und Molekülen)*
Maillard reaction *(Lebm)* Maillard-Reaktion *f*, nicht enzymatische Bräunungsreaktion *f*, Carbonylaminoreaktion *f*
main Hauptleitung *f*, Hauptstrang *m*, Hauptrohr *n*
main alkaloid Hauptalkaloid *n*
main chain Hauptkette *f* *(eines verzweigten Moleküls)*
main component Hauptkomponente *f*, Grundkomponente *f*, Hauptbestandteil *m*
main fraction *(Dest)* Hauptfraktion *f*, Hauptlauf *m*
main group Hauptgruppe *f* *(des Periodensystems)*
main group element Hauptgruppenelement *n*
main reaction Hauptreaktion *f*
mains voltage Netzspannung *f*
maintain *v* 1. unterhalten, warten, instand halten; 2. (aufrecht)erhalten, halten *(z. B. eine Temperatur)*
maintain *v* **constant** konstant halten
maintain *v* **in suspension** in Suspension halten, in der Schwebe halten
maintain *v* **sterility** *(Biot)* steril halten
maintenance 1. Wartung *f*, Instandhaltung *f*, Erhaltung *f*; 2. Aufrechterhaltung *f* *(z. B. einer Temperatur)*
maize oil Mais(keim)öl *n*
maize starch Maisstärke *f*

major constituent 1. *(Anal)* Hauptbestandteil *m (*100 bis 1 %)*; 2. *(Hyd)* Hauptinhaltsstoff *m*
major element 1. *(Met)* Grundmetall *n*, Hauptbestandteil *m (einer Legierung)*; 2. *(Landw)* Kernnährstoff *m*, Makronährstoff *m*
major tranquilizer *(Pharm)* Neuroleptikum *n*, Antipsychotikum *n*, Psycholeptikum *n*, Antihalluzinogen *n*
make *v* 1. herstellen, erzeugen; 2. *(Lab)* darstellen
make *v* **a bond** eine Bindung herstellen [bilden]
make *v* **acidic** (an)säuern
make *v* **alkaline [basic]** 1. alkalisch machen, alkalisieren; 2. *(Farb)* alkalisch stellen
make *v* **good** ausgleichen *(ein Defizit)*
make *v* **into a dough [paste]** anteigen, anpasten
make *v* **into a pulp** *(Pap)* zu Halbstoff aufschließen, *(i. e. S.)* zu Holzschliff verschleifen, chemisch aufschließen
make *v* **neutral** neutralisieren
make *v* **red hot** auf Rotglut erhitzen
make *v* **to measure** gezielt [nach Maß] aufbauen [herstellen], „maßschneidern" *(z. B. Polymere)*
make *v* **up** 1. ausmachen, bilden *(z. B. den Hauptbestandteil)*; zusammensetzen, aufbauen *(ein Ganzes, z. B. ein Molekül)*; 2. ausgleichen; auffüllen, Verluste decken, nachdosieren
make *v* **up to volume** auffüllen
make 1. Ausführung *f*, Bauart *f*, Typ *m*; Fabrikat *n*, Erzeugnis *n*, Produkt *n (einer bestimmten Firma)*; 2. Herstellung *f*, Erzeugung *f*, Fabrikation *f*; 3. Produktionsmenge *f*, Produktion *f*, Ausstoß *m*; 4. Beschaffenheit *f*, Zustand *m*, Verfassung *f*
make-up 1. Zusammensetzung *f*, Aufbau *m*; 2. Ausgleich *m*; Auffüllung *f*; Deckung *f* von Verlusten, Nachdosierung *f*; 3. Zusatzchemikalie *f (zur Deckung von Verlusten)*
makeup *s.* make-up
malachite *(Min)* Malachit *m*, Kupferspat *m (Kupfer(II)-dihydroxidcarbonat)*
malate Malat *n (Salz oder Ester der Äpfelsäure)*
MALDI-TOF *s.* matrix-assisted laser desorption/ionization-time of flight mass spectroscopy
male mould *(Kst)* Stempel *m*, Stempelprofil *n*, Pressstempel *m*, Patrize *f*, *(beim Streck- und Vakuumformen:)* Positivform *f*
male spherical joint Schliffkugel *f*, Schliffball *m*
male tapered joint Schliffkern *m*
maleate Male(in)at *n (Salz oder Ester der Maleinsäure)*
maleic acid Maleinsäure *f*, *cis*-Ethylen-1,2-dicarbonsäure *f*
malic acid Äpfelsäure *f*, *Hydroxybutandisäure *f*
malic enzyme Malatenzym *n*, Malathydrogenase *f*
malleable (cast) iron Temperguss *m*
MALLS *s.* multi-angle laser light scattering
malo-lactic fermentation Äpfelsäure-Milchsäuregärung *f*
malodorous übel riechend
malonate Malonat *n (Salz oder Ester der Malonsäure)*
malonic acid Malonsäure *f*, *Propandisäure *f*
malonic ester Malonester *m*, Malonsäurediethylester *m*, Diethylmalonat *n*
malonitrile Malonsäuredinitril *n*, Methylendicyanid *n*
malonyl urea Malonylharnstoff *m*, Barbitursäure *f*
malt *v* (ver)mälzen; mit Malz versetzen
malt Malz *n*
malt enzyme Malzenzym *n*
malt extract Malzextrakt *m*, Maltosesirup *m*
malt sugar Malzzucker *m*, Maltose *f*
malted barley Gerstenmalz *n*
malting Mälzen *n*, Vermälzung *f*, Malzbereitung *f*, Mälzerei *f*
maltose Maltose *f*, Malzzucker *m*
malty malzartig, Malz...
mammalian toxicity Giftigkeit *f* [Toxizität *f*] für Säugetiere
man-made fibre Chemiefaser *f*; Chemiefaserstoff *m*
man-made rubber synthetischer [künstlicher] Kautschuk *m*, Synthesekautschuk *m*, Kunstkautschuk *m*, SK
ManA *s.* mannuronic acid

mandelate Mandelat *n* *(Salz oder Ester der Mandelsäure)*
mandelic acid Mandelsäure *f*, α-Hydroxyphenylessigsäure *f*, Phenylglycolsäure *f*
manganate Manganat *n*
manganese Mn Mangan *n*
manganese carbide Mangancarbid *n*
manganese dioxide Mangandioxid *n*, Mangan(IV)-oxid *n*, Braunstein *m*
manganese green Mangangrün *n*, Kasseler Grün *n* *(Bariummanganat)*
manganese heptoxide Manganheptoxid *n*, Mangan(VII)-oxid *n*
manganese hydroxide Mangan(IV)-oxiddihydroxid *n*
manganese monoxide Mangan(mon)-oxid *n*, Mangan(II)-oxid *n*
manganese nodule *(Geol)* Manganknolle *f*
manganese removal 1. Manganentfernung *f*; 2. *(Hyd)* Entmanganung *f*
manganese soap Manganseife *f*
manganese steel Mangan(hart)stahl *m*
manganese trioxide Mangantrioxid *n*, Mangan(VI)-oxid *n*
manganic Mangan..., *(i. e. S.)* Mangan(III)-...
manganic acid Mangansäure *f*
manganic oxide Mangan(III)-oxid *n*, Dimangantrioxid *n*
manganite 1. Manganat(IV) *n*, Oxomanganat(IV) *n*; 2. *(Min)* Manganit *m* *(Mangan(III)-hydroxid)*
mangano-manganic oxide Mangan-(II,IV)-oxid *n*, rotes Manganoxid *n*
manganous Mangan..., *(i. e. S.)* Mangan(II)-...
manganous chloride Mangan(II)-chlorid *n*, Mangandichlorid *n*
manganous hydroxide Mangan(II)-hydroxid *n*
manganous oxide Mangan(II)-oxid *n*, Manganmonoxid *n*
manganous sulphate Mangan(II)-sulfat *n*
manhole 1. Mannloch *n*, Einsteigöffnung *f*, Einsteigluke *f*; 2. *(Hyd)* Einsteigschacht *m*
manifold Sammelleitung *f*, Sammelrohr *n*; Verteiler *m*, Verteilungsrohr *n*, Rohrverteiler *m*
mannan Mannan *n* *(ein Polysaccharid)*
Mannich reaction Mannich-Reaktion *f* *(Aminoalkylierung zur Darstellung tertiärer Basen)*
mannitol Mannitol *n* *(ein Zuckeralkohol)*
mannose Mannose *f* *(ein Monosaccharid)*
mannuronic acid Mannuronsäure *f*
manometer Druckmesser *m*, Manometer *n*
manometric manometrisch, Manometer..., Druck(messer)...
mantle *(Tech)* Mantel *m*
manually operated handbetätigt, manuell betätigt
manufacture *v* erzeugen, herstellen
manufacture Erzeugung *f*, Herstellung *f*, Fabrikation *f*
manufactured gas technisches [künstlich hergestelltes] Gas *n* *(brennbar)*
manufacturer Hersteller *m*, Herstellerbetrieb *m*, Herstellerfirma *f*, Produzent *m*
manufacturing process 1. Produktionsprozess *m*; Produktionsverfahren *n*, Herstellungsverfahren *n*; 2. Verarbeitungsprozess *n*; Verarbeitungsverfahren *n*
many-membered vielgliedrig, hochgliedrig *(Ringverbindung)*
many-step synthesis Mehrstufensynthese *f*
MAO = monoaminooxidase
maple sugar Ahornzucker *m* *(von Acer saccharum Marsh.)*
maraging *(Met)* Martensitaushärtung *f*
marble Marmor *m*
marc 1. Trester *pl*; 2. Extraktionsrückstand *m*
margaric acid Margarinsäure *f*, *Heptadecansäure *f*
margarine Margarine *f*
marihuana Marihuana *n*, Haschisch *m(n)* *(von Cannabis indica Lam.)*
marijuana Marihuana *n*, Haschisch *m(n)* *(von Cannabis indica Lam.)*
marine animal oil Seetieröl *n*, Marineöl *n*
marine disposal Beseitigung *f* durch Einleiten ins Meer; Beseitigung *f* durch Abklappen [Verkippen ins Meer]
marine oil *s.* marine animal oil
marine paint meerwasserbeständige Anstrichfarbe *f*

marine pollution Meerwasserverunreinigung *f*
marine salt Meersalz *n*, Seesalz *n*
Mariotte law boyle-mariottesches Gesetz *n*
mark *v* markieren, kennzeichnen, (be-)zeichnen
mark *v* **off** abflecken, abfärben, ausbluten
mark 1. Marke *f*, Strichmarke *f*, Eichmarke *f*; 2. *s.* marc 2.
marker Markierungssubstanz *f*, Markersubstanz *f*
marker enzyme Leitenzym *n*, Marker-Enzym *n*, Indikatorenzym *n*
marker protein Marker-Protein *n*
Markovnikov addition Markovnikov-Addition *f*, ionoide Addition *f* *(von Halogenwasserstoffen an aliphatische Doppelbindungen)*
marsh gas Sumpfgas *n* *(Methan-Kohlendioxid-Gemisch)*
Marsh's test (for arsenic) marshsche Arsenprobe *f*, marshsche Probe *f* (auf Arsen), Arsenprobe *f* nach Marsh
martensite *(Met)* Martensit *m*
martensitic *(Met)* martensitisch
martensitic transition martensitische Umwandlung *f*
maser *(microwave amplification by stimulated emission of radiation)* Maser *m*
mash *v* *(Gär)* (ein)maischen
mash *(Gär)* Maische *f*
masher Maischapparat *m*
mask *v* 1. maskieren *(Kationen)*; 2. blockieren, inaktivieren, unwirksam machen *(reaktionsfähige Gruppe oder Position)*
masking 1. Maskierung *f* *(von Kationen)*; 2. Blockierung *f*, Inaktivierung *f* *(einer reaktionsfähigen Gruppe oder Position)*
masking agent Maskierungsmittel *n*, Maskierungsreagens *n*, maskierendes Mittel *n*, maskierender Stoff *m*
mass Masse *f*
mass-action expression [law] Massenwirkungsgesetz *n*, MWG *n*
mass-analyzed ion kinetics spectrometry Ionenenergie-Spektroskopie *f* zum Nachweis metastabiler Zerfälle
mass balance Masse(n)bilanz *f*, Stoffbilanz *f*
mass defect Masse(n)defekt *m*, Massenschwund *m*
mass degree of swelling *s.* mass swelling degree
mass density Massendichte *f*, *(i. e. S.)* Dichte *f*, Raumdichte *f* der Masse
mass-dyed *(Text)* (er)spinngefärbt, düsengefärbt
mass equivalent Masse(n)äquivalent *n*
mass flux Stoffstrom *m*, Massenstrom *m*, Mengenstrom *m*
mass number Masse(n)zahl *f*, Kernmasse(n)zahl *f*
mass per unit volume Masse(n)konzentration *f* *(z. B. g/l bei Lösungen)*
mass percentage Masseprozent *n*, Masse%
mass polymerization Substanzpolymerisation *f*, Massepolymerisation *f*, Polymerisation *f* in Masse [Substanz]
mass ratio Massenverhältnis *n*
mass scale Massenskala *f*, Atommassenskala *f*
mass spectrometer Massenspektrometer *n*
mass-spectrometric massenspektrometrisch
mass spectrometry Massenspektrometrie *f* • **by mass spectrometry** massenspektrometrisch
mass spectrum Massenspektrum *n*
mass swelling degree Massenquellungsgrad *m* *(Polymernetzwerke)*
mass transfer Stoffübertragung *f*; Stoffübergang *m* *(innerhalb einer Phase)*
mass transfer between phases Stoffdurchgang *m* *(durch eine Phasengrenze)*
mass-transfer coefficient Stoffübergangskoeffizient *m*
mass-transfer resistance Stoffübergangswiderstand *m*
mass transport Stofftransport *m*
massive 1. massiv, wuchtig; 2. gediegen *(Edelmetall)*; 3. *(Kern)* massereich
masterbatch *v* *(Gum)* Masterbatches [Vormischungen] herstellen; mit Masterbatches mischen
masterbatch *(Gum)* Masterbatch *m*, Vormischung *f*, Grundmischung *f*
mastic (gum) Mastix *m*, Mastixharz *n* *(von Pistacia lentiscus L.)*

masticate *v (Gum)* mastizieren *(durch Kneten bei normaler Temperatur unter Lufteinwirkung erweichen)*
mastication *(Gum)* Mastikation *f*, Mastizieren *n*
masticator *(Gum)* Mastikator *m*, Mastiziermaschine *f*
mastix *s.* mastic (gum)
mat matt, glanzlos
mat 1. Matte *f*, Vlies *n (z. B. aus Glasfasern)*; 2. *(Pap)* Stoffbahn *f*, Papierbahn *f*, Papiervlies *n*, Faserfilz *m*; 3. *(Hyd)* Filterhaut *f*
material Material *n*, Werkstoff *m*; Material *n*, Stoff *m*, Substanz *f*; Gut *n*
material balance Stoffbilanz *f*
material being distilled Destillationsgut *n*, Destilliergut *n*, Destillationsmaterial *n*, Destillans *n*
material being dried trocknendes Gut *n*, Trockengut *n*
material change Stoff(ver)änderung *f*, stoffliche Veränderung *f* [Änderung *f*], Material(ver)änderung *f*
material of construction Konstruktionswerkstoff *m*
material to be adsorbed Adsorbend *m*, zu adsorbierender Stoff *m*
material to be distilled Destillationsgut *n*, Destilliergut *n*, Destillationsmaterial *n*
material to be dried Trockengut *n*, zu trocknendes Gut *n*, Trockneraufgabegut *n*
material to be tested Untersuchungsmaterial *n*, zu untersuchendes Material *n*
material under examination [investigation] Untersuchungsmaterial *n*, Untersuchungssubstanz *f*
materials conversion Stoff(um)wandlung *f*, Stoffumsetzung *f*, stoffliche Umwandlung *f* [Umsetzung *f*]
materials science Werkstoffkunde *f*, Werkstoffwissenschaft *f*
materials testing Werkstoffprüfung *f*
matrix 1. *(phys Ch)* Matrix *f*; 2. Grundmasse *f (einer Legierung)*; Bindemittel *n*; 3. *(Geol)* Grundmasse *f*, Matrix *f*; 4. *(Min)* Muttergestein *n*; taubes Gestein *n* [Mittel *n*], Bergemittel *n*, Nichterz *n*; 5. *(Ker)* Grundmasse *f*, Einbettungsmasse *f*, Matrix *f*; 6. *(Biot)* Matrix *f*, Trägermatrix *f (Zell- und Enzymfixierung)*
matrix-assisted laser desorption/ionization-time of flight mass spectroscopy matrixunterstützte Laserdesorptions-/Ionisations-Flugzeit-Massenspektrometrie *f*
matrix effect [interference effect] *(phys Ch, Anal)* Matrixeffekt *m*, Dritter-Partner-Effekt *m*
matt matt, glanzlos
matte Stein *m (künstlich erschmolzenes Gemisch von Metallsulfiden)*
matte-smelt *v* auf Stein (ver)schmelzen
matte smelting Steinschmelzen *n*, Schmelzen *n* [Verschmelzen *n*] auf Stein
matter Materie *f*, Stoff *m*, Substanz *f*; Material *n*, Werkstoff *m*, Stoff *m*
maturation Reifen *n*, Reifung *f*
mature *v* reifen, reif werden; reifen (lassen) *(z. B. Wein)*; ablagern
mature reif, ausgereift; abgelagert *(z. B. Wein)*
maturity Reife *f (Zustand)*
mauve Mauvein *n*, Mauve *n (basischer Azinfarbstoff)*
mauvein(e) Mauvein *n*, Mauve *n (basischer Azinfarbstoff)*
maverick *(Anal)* Ausreißer *m*, abweichender Wert *m*
maximum allowable concentration höchstzulässige Konzentration *f*, *(i. e. S.)* maximale Arbeitsplatzkonzentration *f*, MAK-Wert *m*, MAK
maximum and minimum thermometer Maximum-Minimum-Thermometer *n*
maximum-boiling azeotrope Maximumazeotrop *n*, Azeotrop *n* mit Maximumsiedepunkt, Siedepunktsmaximumazeotrop *n*
maximum dose 1. Maximaldosis *f*; 2. *(Pharm, Landw)* größte Gabe *f*
maximum emission allowance Abgas-Grenzwerte *mpl*
maximum emission concentration Emissionsgrenzwert *m*
maximum-multiplicity principle [rule] (hundsches) Prinzip *n* der größten Multiplizität, (hundsche) Regel *f* der größten Multiplizität, hundsches Multiplizitätsprinzip *n* [Prinzip *n*], (erste) hundsche Regel *f*

maximum permissible concentration höchstzulässige Konzentration *f*, Grenzkonzentration *f*, zulässige Höchstkonzentration *f*, Toleranzgrenze *f*
maximum tolerance *(Tox)* duldbare [zulässige, zugelassene] Höchstmenge *f*, Toleranz(dosis) *f*, Toleranzwert *m*
maximum useful work *(phys Ch)* maximale Nutzarbeit *f* (der Reaktion)
maximum work funktion *(phys Ch)* freie Energie *f*, F
Maxwell-Boltzmann distribution Maxwell-Boltzmann-Verteilung *f (der Geschwindigkeit von Teilchen)*
MB/D = thousands of barrels per day
MBE *s.* molecular beam epitaxy
McCabe-Thiele construction [method] *(Dest)* McCabe-Thiele-Verfahren *n*, grafisches Verfahren *n* nach Thiele und McCabe *(zur Trennstufenermittlung)*
MCFC *s.* molten carbonate fuel cell
McLafferty rearrangement McLafferty-Umlagerung *f (bei der massenspektrometrischen Fragmentierung)*
McLeod gauge McLeod-Manometer *n*, McLeod-Vakuummeter *n*, Kompressionsmanometer *n* nach McLeod
MDC *s.* multidimensional chromatography
ME ... *siehe: multiple-effect ...*
MEA *s.* monoethanolamine
meal 1. *(Lebm)* (grobes) Mehl *n*, Schrotmehl *n*; 2. Mehl *n*, Pulver *n (z. B. aus Mineralen, Gesteinen)*
mealy mehlig, mehlartig
mean *s.* mean value
mean free path *(phys Ch)* mittlere freie Weglänge *f*
mean life(time) *(phys Ch)* (mittlere) Lebensdauer *f*
mean residence time mittlere Verweilzeit *f*
mean value Mittelwert *m*, Mittel *n*, Durchschnitt(swert) *m*
measurable messbar
measurable range Messbereich *m*
measure *v* 1. messen; 2. *s.* measure off [out]
measure *v* **off [out]** abmessen
measure 1. Maß *n*, Ausmaß *n*; 2. Maßnahme *f*; 3. Maß *n (z. B. Messgefäß)*; 4. Maß *n*, Maßeinheit *f* • **make to measure** gezielt [nach Maß] aufbauen [herstellen], „maßschneidern" *(z. B. Polymere)*
measured quantity Messgröße *f*
measured value Messwert *m*
measurement Messung *f*
measurement error Messfehler *m*
measurement of pH pH-(Wert-)Messung *f*
measurement technique Messmethode *f*
measuring accuracy Messgenauigkeit *f*
measuring cell Messzelle *f*
measuring cylinder Messzylinder *m*, Maßzylinder *m*, Mensur *f*
measuring electrode Messelektrode *f*, Indikatorelektrode *f*
measuring flask Messkolben *m*, Maßkolben *m*
measuring glass Messglas *n*
measuring pipette Messpipette *f*, Teilpipette *f*
measuring range Messbereich *m*
measuring van Messwagen *m*
measuring vessel Messgefäß *n*, Messbehälter *m*
meat curing Fleischpökeln *n*, Pökeln *n*, Einpökeln *n*
meat extract Fleischextrakt *m*
meat tenderizer Zartmachersalz *n*, Fleischmürbesalz *n*, Tenderizer *m*
MEC enzyme *s.* microencapsulated enzyme
mechanical equivalent of heat mechanisches Wärmeäquivalent *n*
mechanical filter respirator Schwebstofffiltergerät *n*, Staubschutzmaske *f*
mechanical pulping mechanischer Aufschluss *m* des Holzes, Holzschlifferzeugung *f*, Holzschliffgewinnung *f*
mechanical wood pulp (mechanischer) Holzschliff *m*, (mechanischer) Holzstoff *m*, Holzmasse *f*, Schleifmasse *f*
mechanism of action Wirkungsmechanismus *m*, Wirkungsweise *f*
mechanism of reaction Reaktionsmechanismus *m*
mechanochemical mechanochemisch
mechanochemistry Mechanochemie *f*
media depth *(Filtr)* Schichthöhe *f*, Schichtdicke *f*, Betthöhe *f*, Bettiefe *f*
median effective dose *(Tox)* mittlere effektive Dosis *f*, ED 50

median lethal dose *(Tox)* mittlere letale Dosis *f*, LD 50 *(letale Dosis für 50 % der Versuchstiere)*
mediated transport *(Bioch)* katalysierter [vermittelter] Transport *m (durch Membranen hindurch)*
mediator *(Bioch)* Mediator *m (Botenstoff)*
medical chemistry medizinische Chemie *f*
medicament Medikament *n*, Arzneimittel *n*
medicinal medizinisch, arzneilich
medicinal chemist Pharmakochemiker *m*, Arzneimittelchemiker *m*
medicinal chemistry pharmazeutische Chemie *f*, Pharmakochemie *f*, Arzneimittelchemie *f*
medicinal oil medizinisches Öl *n*, Medizin(al)öl *n*, Öl *n* für medizinische Zwecke
medicinal substance Arzneistoff *m*
medicine 1. Medizin *f (Wissenschaft)*; 2. Arzneimittel *n* für inneren Gebrauch
medicine dropper Tropfpipette *f*, Stechpipette *f*
medium 1. Medium *n*, Mittel *n*, Stoff *m*, Träger *m*; 2. Lösungsmittel *n*; 3. *(Biot)* Medium *n*, Nährmedium *n*, Nährsubstrat *n*; 4. *(Anstr)* Bindemittel *n*; Bindemittellösung *f*
medium boiler *(Dest)* Mittelsieder *m*, mittelsiedendes Lösungsmittel *n*
medium-chain fatty acid mittlere Fettsäure *f*
medium-density polyethylene Polyethylen *n* mittlerer Dichte
medium gasoline Mittel(schwer)benzin *n*
medium-grained mittel(fein)körnig
medium-level active *(Kern)* mittelaktiv
medium naphtha *s.* medium gasoline
medium-oil alkyd mittelöliges Alkyd(-harz) *n*
medium-pressure synthesis Mitteldrucksynthese *f*
medium ring mittlerer [mittelgroßer] Ring *m (aus 8 bis 12 Gliedern bestehend)*
medium-sulphur mit mittlerem Schwefelgehalt
medium-temperature carbonization Mitteltemperaturverkokung *f*, Mitteltemperaturentgasung *f*
medium-viscosity mittelviskos, von mittlerer Viskosität
Meerwein-Ponndorf-Verley reduction Meerwein-Ponndorf-Verley-Reduktion *f*, Meerwein-Ponndorf-Verley-Carbonylreduktion *f*, meerwein-ponndorfsche Reaktion *f*
Meisenheimer adduct [complex, compound] *(org Ch)* (Jackson-)Meisenheimer-Komplex *m*
meitnerium Mt Meitnerium *n (Element 109)*
MEK *s.* methyl ethyl ketone
Meker burner Méker-Brenner *m*
melamine Melamin *n*, Triamino-*sym*-triazin *n*
melamine(-formaldehyde) resin Melamin(-Formaldehyd)harz *n*
melanocyte-stimulating hormone *Melanotropin *n*, melanocytenstimulierendes Hormon *n*
melanocyte-stimulating hormone releasing hormone *Melanoliberin *n*, MRH
melanotropin (release) inhibitory hormone *Melanostatin *n*, Melanotropin-release-inhibierendes Hormon *n*, MRIH, MIH
melanotropin releasing hormone *Melanoliberin *n*, MRH
mellitic acid Mellithsäure *f*, Honigsteinsäure *f*, *Benzenhexacarbonsäure *f*
melt *v* 1. (zer)schmelzen, flüssig werden; schmelzen; 2. *(Met)* einschmelzen *(z. B. Schrott)*
melt *v* **down** niederschmelzen, einschmelzen, zusammenschmelzen
melt *v* **out** *(Bioch)* schmelzen *(Desoxyribonucleinsäure)*
melt *v* **together** zusammenschmelzen, verschmelzen
melt 1. Schmelze *f*, Schmelzfluss *m*; Schmelze *f*, Charge *f*; 2. Schmelzen *n*; 3. *(Met)* Einschmelzen *n (z. B. von Schrott)*
melt spinning Schmelzspinnen *n*, Spinnen *n* [Erspinnen *n*] aus der Schmelze
meltability Schmelzbarkeit *f*
meltable schmelzbar
melter 1. Schmelzvorrichtung *f*; 2. *(Glas)* Schmelzwanne *f*, Schmelzraum *m*, Schmelzteil *m*
melting crucible Schmelztiegel *m*

melting curve Schmelzkurve *f*
melting diagram Schmelzdiagramm *n*
melting enthalpy Schmelzenthalpie *f*
melting furnace Schmelzofen *m*
melting point Schmelzpunkt *m*, Fließpunkt *m*
melting-point depression [lowering] Schmelzpunkterniedrigung *f*, Schmelzpunktdepression *f*
melting range Schmelzbereich *m*, Schmelzintervall *n*
membrane Membran *f*, Wand *f*, Scheidewand *f*, Diaphragma *n*
membrane-bound *(Bioch)* membrangebunden
membrane equilibrium Membrangleichgewicht *n*
membrane filtration Membranfiltration *f*
membrane lipid *(Bioch)* Membranlipid *n*
membrane osmometer Membranosmometer *n*
membrane osmometry Membranosmose *f (Molmassenbestimmung)*
membrane potential Membranpotenzial *n*
membrane protein Membranprotein *n*
membrane separation *(Hyd)* Trennung *f* mittels semipermeabler Membranen
membrane transport *(Bioch)* Membrantransport *m*, Transport *m* durch eine Membran
Mendeléeff periodic system mendeleevsches [mendelejewsches] Periodensystem *n*, Mendeleev-System *n*
mendelevium Md Mendelevium *n*
meniscus Meniskus *m*
menstruum Lösungsmittel *n*, Extraktionsmittel *n (zur Gewinnung von Drogenauszügen)*
***p*-menthane** *p*-Menthan *n*, 1-Methyl-4-isopropyl-cyclohexan *n*
menthene Menthen *n*, *(i. e. S.)* Δ^3-Menthen *n*
menthone Menthon *n*, *p*-Menthan-3-on *n*
meq *s.* milliequivalent
mequiv. *s.* milliequivalent
mercaptal Dithioacetal *n*, Mercaptal *n*
mercaptan 1. *(org Ch)* Thiol *n*, Mercaptan *n*; 2. *(Erdöl)* Mercaptan *n*
mercaptide Thiolat *n*, Mercaptid *n (allg. für Salze der Thioalkohole)*
mercapto group HS– Thiolgruppe *f*, Mercaptogruppe *f*
mercaptoacetic acid Mercaptoessigsäure *f*, Thioglykolsäure *f*
mercaptole Dithioketal *n*, Mercaptol *n*
mercerization *(Text)* Merzerisieren *n*, Merzerisierung *f*, Merzerisation *f*, Laugen *n (kurzzeitige Behandlung von Baumwollgarnen und -geweben mit konzentrierter kalter Natronlauge)*
mercerize *v (Text)* merzerisieren, laugen
mercerizing assistant *(Text)* Merzerisierhilfsmittel *n*
mercurial quecksilberhaltig; Quecksilber...
mercurial Quecksilberverbindung *f*
mercurialism Quecksilbervergiftung *f*
mercuric Quecksilber..., *(i. e. S.)* Quecksilber(II)-...
mercuric acetate Quecksilber(II)-acetat *n*
mercuric arsenate Quecksilber(II)-arsenat(V) *n*
mercuric bromide Quecksilber(II)-bromid *n*, Quecksilberdibromid *n*
mercuric chloride Quecksilber(II)-chlorid *n*, Quecksilberdichlorid *n*, Sublimat *n*
mercuric cyanide Quecksilber(II)-cyanid *n*
mercuric fluoride Quecksilber(II)-fluorid *n*, Quecksilberdifluorid *n*
mercuric fulminate Quecksilber(II)-fulminat *n*, Knallquecksilber *n*
mercuric iodide Quecksilber(II)-iodid *n*, Quecksilberdiiodid *n*
mercuric nitrate Quecksilber(II)-nitrat *n*
mercuric oxide Quecksilber(II)-oxid *n*
mercuric sulphate Quecksilber(II)-sulfat *n*
mercuric sulphide Quecksilber(II)-sulfid *n*
mercuric thiocyanate Quecksilber(II)-thiocyanat *n*
mercurimetric mercurimetrisch
mercurimetry Mercurimetrie *f (Titration mit Quecksilber(II)-nitratlösung)*
mercurization Mercurierung *f (Einführung von Quecksilber in organische Verbindungen)*
mercurize *v* mercurieren *(Quecksilber in organische Verbindungen einführen)*
mercurometric mercurometrisch

mercurometry Mercurometrie *f (Titration mit Quecksilber(I)-nitratlösung)*
mercurous Quecksilber..., *(i. e. S.)* Quecksilber(I)-...
mercurous bromide Quecksilber(I)-bromid *n*
mercurous chloride Quecksilber(I)-chlorid *n*, Kalomel *n*
mercurous fluoride Quecksilber(I)-fluorid *n*
mercurous iodide Quecksilber(I)-iodid *n*
mercurous nitrate Quecksilber(I)-nitrat *n*
mercurous oxide Quecksilber(I)-oxid *n*
mercurous sulphate Quecksilber(I)-sulfat *n*
mercury Hg Quecksilber *n*
mercury barometer Quecksilberbarometer *n*
mercury cell Quecksilberzelle *f*, Amalgamzelle *f (Elektrolyse)*
mercury-cell process Quecksilberverfahren *n*, Amalgamverfahren *n (Elektrolyse)*
mercury column Quecksilbersäule *f*
mercury dichloride Quecksilberdichlorid *n*, Quecksilber(II)-Chlorid *n*, Sublimat *n*
mercury diffusion pump *s.* mercury-vapour pump
mercury dropping cathode Quecksilbertropfkatode *f*
mercury electrode Quecksilberelektrode *f*
mercury-film electrode *(Ech)* Quecksilberfilmelektrode *f*
mercury fulminate Quecksilber(II)-fulminat *n*, Knallquecksilber *n*
mercury lamp *s.* mercury-vapour lamp
mercury manometer Quecksilbermanometer *n*
mercury poisoning Quecksilbervergiftung *f*
mercury porosimetry Quecksilberporosimetrie *f (Porositätsprüfung)*
mercury pump *s.* mercury-vapour pump
mercury thermometer Quecksilberthermometer *n*
mercury tong(s) Quecksilberzange *f*
mercury-vapour lamp Quecksilber(dampf)lampe *f*, Quecksilberdampfbrenner *m*
mercury-vapour pump Quecksilber-Diffusionspumpe *f*, Diffusionspumpe *f* mit Quecksilberfüllung, Quecksilberpumpe *f*
merge *v (Kern)* verschmelzen
Merrifield (peptide) synthesis Merrifield-Synthese *f*, Festphasen-Peptidsynthese *f*
MES *s.* Mössbauer(-effect) spectroscopy
mescaline Mescalin *n*, Mezcalin *n*, 3,4,5-Trimeth-oxy-β-phenethylamin *n*
mesh 1. Masche *f (Sieböffnung)*; 2. Siebnummer *f*, Maschenzahl *f (Anzahl der Maschen je Zoll linear)*
mesh size 1. Maschenweite *f*; 2. *Korngröße, ausgedrückt durch die Maschenzahl des Siebs*
mesityl oxide Mesityloxid *n*, *2-Methylpent-2-en-4-on *n*
mesitylene Mesitylen *n*, *1,3,5-Trimethylbenzen *n*
meso compound Mesoverbindung *f*
meso form *meso*-Form *f*, Mesoform *f*
meso sample *Mesoprobe *f*, Halbmikroprobe *f (0,1 bis 0,01 g)*
meso-trace analysis Meso-Spuren-Analyse *f (*Probenmasse 10^{-1} bis 10^{-2} g)*
mesocolloid Mesokolloid *n*
mesoionic compound mesoionische Verbindung *f*
mesomeric mesomer, Mesomerie...
mesomeric effect Mesomerieeffekt *m*, M-Effekt *m*
mesomeric energy Mesomerieenergie *f*, Resonanzenergie *f*, Delokalisierungsenergie *f*, Sonderenergie *f*
mesomerism Mesomerie *f*, Resonanz *f*, Strukturresonanz *f*
mesomorphic state mesomorpher [flüssigkristalliner] Zustand *m*
mesomorphism *s.* mesomorphic state
meson *(Kern)* Meson *n*
mesophase Mesophase *f*, mesomorphe [kristalline, anisotrope] Flüssigkeit *f*, flüssiger Kristall *m*, Flüssigkristall *m*
mesophiles *(Biot)* mesophile Mikroorganismen *mpl (Wachstumsoptimum 20 bis 45 °C)*
mesophilic microorganisms *(Biot)* mesophile Mikroorganismen *mpl (Wachstumsoptimum 20 bis 45 °C)*
mesoxalic acid Mesoxalsäure *f*, Oxomalonsäure *f*, *Oxopropandisäure *f*

messenger *(Bio)* Botenstoff *m*
messenger RNA Messenger-RNA *f*, Boten-RNA *f*, mRNA *f*, Messenger-Ribonucleinsäure *f*
mesylate Mesylat *n*, Mesylester *m*, Methansulfonsäureester *m*, Methansulfonat *n*
mesylation Mesylierung *f*
meta metaständig, *m*-ständig, in *meta*-Stellung [*m*-Stellung, 1,3-Stellung] • **in meta position** metaständig, *m*-ständig, in *meta*-Stellung [*m*-Stellung, 1,3-Stellung] • **be (located, situated) meta** sich in *meta*-Stellung [*m*-Stellung, 1,3-Stellung] befinden, in *meta*-Stellung stehen, metaständig [*m*-ständig] sein
meta attack Angriff *m* an der *meta*-Position *(einer aromatischen Verbindung)*
meta compound *meta*-Verbindung *f*, *m*-Verbindung *f*
meta-directing *meta*-dirigierend, *m*-dirigierend, nach der *meta*-Stellung dirigierend
meta substituent *meta*-Substituent *m*, *m*-Substituent *m*, Substituent *m* in *meta*-Stellung [*m*-Stellung], metaständiger Substituent *m*
meta-substituted *meta*-substituiert, *m*-substituiert
meta substitution *meta*-Substitution *f*, *m*-Substitution *f*
metabisulphite $M^I_2S_2O_3$ Disulfit *n*
metabolic *(Bioch)* metabolisch, Stoffwechsel...
metabolic antagonist Antimetabolit *m* *(Enzyminhibitor, meist bei kompetitiver Hemmung wirkend)*
metabolic chain Stoffwechselkette *f*
metabolic pathway Stoffwechselweg *m*, Stoffwechselablauf *m*, Stoffwechselbahn *f*
metabolic pool Stoffwechselpool *m*, Stoffwechselreservoir *n*, Sammelbecken *n* des Stoffwechsels, labile Mischphase *f*
metabolic product *s.* metabolite
metabolic route [sequence] *s.* metabolic pathway
metabolism *(Bioch)* Stoffwechsel *m*, Metabolismus *m*
metabolite *(Bioch)* Metabolit *m*, Stoffwechselprodukt *n*
metabolizable *(Bioch)* metabolisierbar
metabolize *v* *(Bioch)* metabolisieren, verstoffwechseln, (im Zuge des Stoffwechsels) umsetzen; metabolisieren, stoffwechseln
metaformaldehyde α-Trioxymethylen *n*, *1,3,5-Trioxan *n*
metal Metall *n*
metal alkyl Alkylmetallverbindung *f*
metal atom Metallatom *n*
metal bath Metallbad *n*
metal carbide Metallcarbid *n*
metal carbonyl Metallcarbonyl *n*, Carbonyl *n*
metal chelate (complex, compound) Metallchelatkomplex *m*, Metallchelatverbindung *f*, Metallchelat *n*
metal chemistry Metallchemie *f*
metal complex Metallkomplex *m*
metal deposition Metallabscheidung *f*
metal extraction 1. Metallgewinnung *f*; 2. *(Biot)* Metallextraktion *f (mit Hilfe von Mikroorganismen)*
metal hydride Metallhydrid *n*
metal-insulator transition Metall-Isolator-Übergang *m*
metal ion Metallion *n*
metal mask Metallmaske *f*
metal-metal bond Metallbindung *f*
metal mordant dye Metallkomplex-Säurefarbstoff *m*
metal-nonmetal transition Metall-Isolator-Übergang *m*
metal-organic chemical vapour deposition metallorganische Gasphasenabscheidung *f*
metal-organic framework metallorganisches Gerüst *n*
metal phthalocyanine Metallphthalocyanin *n*
metal pigment Metallpigment *n*, metallisches Pigment *n*
metal stearate Metallstearat *n*
metal sulphide Metallsulfid *n*
metalation Metallierung *f*
metaldehyde Metaldehyd *m* *(Polymer des Acetaldehyds)*
metallic metallisch, metallen, Metall...
metallic alloy Metalllegierung *f*
metallic bond metallische Bindung *f*, Metallbindung *f*
metallic catalyst Metallkatalysator *m*

metallic coating metallische Schutzschicht *f*
metallic fibre Metallfaser *f*; Metallfaserstoff *m*, metallischer Faserstoff *m*, MT
metallic filament Metallfaden *m*, (dünner) Metalldraht *m*
metallic lattice metallisches Gitter *n*, Metallgitter *n*
metallic lustre metallischer Glanz *m*, Metallglanz *m*
metallic soap Metallseife *f*
metallic state metallischer Zustand *m*, Metallzustand *m*
metallicity Metallizität *f*, metallischer Charakter *m*
metalliferous metallhaltig; metallführend, erzführend
metallization 1. Metallspritzen *n*, Spritzmetallisieren *n*, *(i. w. S.)* Metallisieren *n*; 2. Umwandlung *f* in die metallische Form [Modifikation]
metallize *v* 1. spritzmetallisieren, *(i. w. S.)* metallisieren; 2. sich in die metallische Form [Modifikation] umwandeln
metallo-organic compound metallorganische Verbindung *f*, Organometallverbindung *f*, Metallorganyl *n*
metallocarbene *(org Ch)* Metallcarben *n* *(Metathese)*
metallocarbene complex *(org Ch)* Metallcarbenkomplex *m (Metathese)*
metallocene *(org Ch)* Metallocen *n* *(Metall-π-Komplex)*
metalloenzyme Metall(o)enzym *n*
metallographer Metallograph *m*
metallographic metallographisch
metallography Metallographie *f*, Metallbeschreibung *f*; Metallkunde *f*, Metallogie *f*, *(i. w. S.)* Metallographie *f*
metalloid 1. Halbmetall *n*; 2. *(Met) Legierungsbildner für Eisen und Stahl*; 3. Nichtmetall *n*
metalloprotein Metallprotein *n*
metallothionein *(Biot)* Metallothionein *n* *(metallbindendes mikrobielles Protein)*
metallurgical metallurgisch, hüttenmännisch
metallurgical chemist Hüttenchemiker *m*
metallurgical chemistry metallurgische Chemie *f*, Hüttenchemie *f*
metallurgical coke Hüttenkoks *m*, metallurgischer [hüttenfähiger, verhüttungsfähiger] Koks *m*, Koks *m* für metallurgische Zwecke
metallurgist Metallurg(e) *m*
metallurgy Metallurgie *f*, Hüttenkunde *f*, Hüttenwesen *n*
metamagnetic transition metamagnetische Umwandlung *f*
metamer Metamer(e) *n*
metameric metamer
metamerism Metamerie *f (eine Form der Strukturisomerie)*
metametal Metametall *n*
metanilic acid Metanilsäure *f*, **m*-Aminobenzensulfonsäure *f*
metaperiodate M^IIO_4 *Tetroxoperiodat *n*, Metaperiodat *n*
metaphosphate M^IPO_3 Metaphosphat *n*
metaphosphoric acid Metaphosphorsäure *f*
metasilicate $M^I_2SiO_3$ Metasilicat *n*, *Trioxosilicat *n*
metastability Metastabilität *f*
metastable state metastabiler Zustand *m*
metathesis *(org Ch)* Metathese *f*, doppelte Umsetzung *f*, doppelter Umsatz *m*, Wechselumsetzung *f*, Wechselzersetzung *f*
metathesis catalyst *(org Ch)* Metathesekatalysator *m*
metathetic(al) metathetisch, Metathese...
metathetic(al) reaction Metathesereaktion *f*, doppelte Umsatzreaktion *f*
meter *v* dosieren, zumessen
meter Messgerät *n*, Messer *m*
metering Dosieren *n*, Dosierung *f*, Zumessen *n*
metering pump 1. Dosier(ungs)pumpe *f*, Zumesspumpe *f*; 2. *(Kst)* Titerpumpe *f*, Spinnpumpe *f*
methacrylate Methacrylat *n*
methacrylate polymer [resin] Polymethacrylat *n*, Polymethacrylsäureester *m*
methacrylic acid Methacrylsäure *f*, *2-Methylpropensäure *f*
methaemoglobin Methämoglobin *n*, Met-Hb, Hämiglobin *n*
methanal *Methanal *n*, Formaldehyd *m*
methanamide Methanamid *n*, Ameisensäureamid *n*, Formamid *n*

methanation Methanisierung *f (von Kohlenmonoxid durch katalytisches Hydrieren)*
methane Methan *n*
methane bacteria *(Hyd)* Methanbakterien *npl*, methanogene Bakterien *npl*
methane-dicarboxylic acid *Methandicarbonsäure *f*, *Propandisäure *f*, Malonsäure *f*
methane fermentation 1. *(Hyd, Biot)* alkalische Gärung *f* [Methangärung *f*] *(Methangewinnung aus Abprodukten)*; 2. *(Biot)* Methanfermentation *f (Biomasseherstellung mit Methan als C-Quelle)*
methane-oxidizing bacteria *(Biot)* Methanoxidierer *mpl*, methanoxidierende Bakterien *npl*
methanediyl group $=CH_2$ Methylengruppe *f*
methanization *s.* methanation
methanize *v* methanisieren *(Kohlenmonoxid durch katalytisches Hydrieren in Methan überführen)*
methanoic acid *Methansäure *f*, Ameisensäure *f*
methanol *Methanol *n*, Methylalkohol *m*
methanol fermentation *(Biot)* Methanolfermentation *f*
methanolic methanolisch, methylalkoholisch
methanolysis Methanolyse *f*
methemoglobin *s.* methaemoglobin
methenamine Methenamin *n*, Hexamethylentetramin *n*, Hexamin *n*
methide *(org Ch)* Methid *n*
methine dye Methinfarbstoff *m*
methine group =CH-Methingruppe *f*
methionine Methionin *n*, Met, *2-Amino-4-methyl-mercaptobuttersäure *f*, *2-Amino-4-methylthiobuttersäure *f*
method Methode *f*, Verfahren *n*, Technik *f*, Arbeitsweise *f*
method of application Anwendungsweise *f*
method of half-times Halbwertszeitmethode *f (zur Bestimmung der Reaktionsordnung)*
method of measurement Messmethode *f*, Messverfahren *n*
method of separation Trennmethode *f*, Trennverfahren *n*
method of working Arbeitsweise *f*
methoxide CH_3OM^I Methoxid *n*, Methylat *n*
methoxy group Methoxygruppe *f*
methoxybenzene *Methoxybenzen *n*, *Methylphenylether *m*, Anisol *n*
***p*-methoxybenzoic acid** *p*-Methoxybenzoesäure *f*, *4-Methoxybenzencarbonsäure *f*, Anissäure *f*
methoxyl group $-OCH_3$ Methoxylgruppe *f*, Methoxylrest *m*
methoxylation Methoxylierung *f*
methoxymethane Methoxymethan *n*, Dimethylether *m*
methyl acetate Essigsäuremethylester *m*, Methylacetat *n*
methyl alcohol Methylalkohol *m*, *Methanol *n*
methyl bromide *Brommethan *n*, Methylbromid *n*
methyl cellulose Methylcellulose *f*
methyl chloride *Chlormethan *n*, Methylchlorid *n*
methyl ester Methylester *m*
methyl ether Dimethylether *m*, *Methoxymethan *n*
methyl ethyl ketone Methylethylketon *n*, *Butan-2-on *n*
methyl group CH_3- Methylgruppe *f*, Methylrest *m*
methyl halide *Halogenmethan *n*, Methylhalogenid *n*
methyl iodide *Iodmethan *n*, Methyliodid *n*
methyl isobutyl ketone Methylisobutylketon *n*, *4-Methylpentan-2-on *n*
methyl methacrylate Methacrylsäuremethylester *m*, Methylmethacrylat *n*
methyl nitrene Methylnitren *n*
methyl orange Methylorange *n*, Goldorange *n*, Orange *n* III, Helianthin *n*
methyl phenyl ether *s.* methoxybenzene
methyl phenyl ketone *Methylphenylketon *n*, Acetylbenzen *n*, Acetophenon *n*
methyl propyl ketone Methylpropylketon *n*, *Pentan-2-on *n*
methyl radical 1. (freies) Methylradikal *n*; 2. *s.* methyl group
methyl red Methylrot *n*, **p*-Dimethylaminoazobenzen-*o*-carbonsäure *f*
methyl salicylate Salicylsäuremethylester *m*, Methylsalicylat *n*

methyl silicone Methylpolysiloxan *n*, Methylsilicon *n*
methyl sulphate Schwefelsäuredimethylester *m*, Dimethylsulfat *n*
methyl sulphoxide Dimethylsulfoxid *n*, DMSO
methyl violet Methylviolett *n*
methyl yellow Dimethylgelb *n*, Buttergelb *n*, **p*-Dimethylaminoazobenzen *n*
methylal Methylal *n*, Formaldehyddimethylacetal *n*, *Dimethoxymethan *n*
methylamine Methylamin *n*, *(i. e. S.)* CH_3NH_2 Methylamin *n*, Monomethylamin *n*
methylate *v* 1. *(org Ch)* methylieren; 2. vergällen, denaturieren *(mit Methanol oder auch anderen Mitteln)*
methylate CH_3OM^I Methylat *n*, Methoxid *n*
methylated spirit vergällter [denaturierter] Alkohol *m* [Spiritus *m*]
methylation 1. *(org Ch)* Methylierung *f*; 2. Vergällung *f*, Denaturierung *f*, Denaturation *f (mit Methanol oder auch anderen Mitteln)*
methylbenzene Methylbenzen *n*, Toluen *n*
methylbenzoic acid Methylbenzoesäure *f*, Methylbenzencarbonsäure *f*, Toluylsäure *f*
3-methylbutanal *3-Methylbutanal *n*, Methylbutyraldehyd *m*
2-methylbutane *2-Methylbutan *n*, Isopentan *n*
3-methylbutanoic acid *3-Methylbutansäure *f*, Isovaleriansäure *f*
methylcellosolve Methylcellosolve *n (Methylglycolether)*
2-methyl-4-chlorophenoxyacetic acid 2-Methyl-4-chlorphenoxyessigsäure *f*, MCPA *n (Herbizid)*
methylchlorosilane Methylchlorsilan *n*
methylcyclohexane Methylcyclohexan *n*, Hexahydrotoluen *n*
methylenation Methylenierung *f*
methylene 1. CH_2 Carben *n*, Methylen *n (instabiles freies Radikal)*; 2. *s.* methylene group
methylene blue Methylenblau *n*
methylene bromide *Dibrommethan *n*, Methylenbromid *n*
methylene chloride *Dichlormethan *n*, Methylenchlorid *n*
methylene group $=CH_2$ Methylengruppe *f*
methylene halide *Dihalogenmethan *n*, Methylenhalogenid *n*
methylene iodide *Diiodmethan *n*, Methyleniodid *n*
methylenediphenol Bisphenol *n*, Methylendiphenol *n*
methylethylketone Ethylmethylketon *n*, *Butan-2-on *n*
methylglyoxal Methylglyoxal *n*, *2-Oxopropanal *n*
methylnaphthalene Methylnaphthalen *n*
2-methyl-1,4-naphthoquinone 2-Methyl-1,4-naphthochinon *n*, Menadion *n*, Vitamin *n* K_3
methylotrophic bacteria *(Biot)* methylotrophe Bakterien *npl*
methylotrophs *(Biot)* methylotrophe Bakterien *npl*
2-methyloxirane Methyloxiran *n*, 1,2-Epoxypropan *n*, Propylenoxid *n*
2-methylpropanoic acid *2-Methylpropansäure *f*, Isobuttersäure *f*
2-methylpropenoic acid *2-Methylpropensäure *f*, 2-Methacrylsäure *f*
methylsulphanyl group $CH_3S–$ Methylthiogruppe *f*
methylsulphonyl group $CH_3SO_2–$ Methylsulfonylgruppe *f*
methylthio group $CH_3S–$ Methylthiogruppe *f*
methyltrichlorosilane Methyltrichlorsilan *n*
metric tons annually Tonnen *fpl* je Jahr
mevalonic acid Mevalonsäure *f*, 3,5-Dihydroxy-3-methylvaleriansäure *f*
mf *s.* moisture-free
MIBK *s.* methyl isobutyl ketone
mica Glimmermineral *n*, Glimmer *m*
micaceous glimmerig, Glimmer...; glimmerhaltig
micellar mizellar, Mizell(ar)...
micellar catalysis mizellare Katalyse *f*
micellar colloid Mizellkolloid *n*, Assoziationskolloid *n*
micelle *(Koll)* Mizelle *f*
micellize *v* Mizellen bilden
Michael condensation [reaction] Michael-Addition *f (nucleophile Methylenaddition)*
Michaelis-Arbuzov reaction Michaelis-

Arbuzov-Reaktion *f (zur Knüpfung einer Phosphor-Kohlenstoff-Bindung)*
Michaelis-Menten complex *(Bioch)* Michaelis-Menten-Komplex *m*, Enzym-Substrat-Komplex *m*
Michaelis-Menten constant *(Bioch)* Michaelis-Menten-Konstante *f*
Michler's ketone Michlers Keton *n*, Bis(*p*-dimethylaminophenyl)-keton *n*
micro-alloyed mikrolegiert
micro-assembly Mikrobaugruppe *f*
micro-crack Mikroriss *m*
micro-current Mikrostrom *m*
micro grits Mikrokörnungen *fpl*
micro-perforated mikroperforiert
micro sample Mikroprobe *f (< 0,01 g)*
micro-scale procedure Arbeitstechnik *f* der Mikroanalyse, Mikroarbeitsweise *f*, Mikromethode *f*, Milligramm-Methode *f*
micro-trace *(Anal)* Mikrospur *f (*10^{-7} *bis* 10^{-4} *ppm)*
micro-trace analysis Mikro-Spuren-Analyse *f (*Probenmasse* 10^{-3} *bis* 10^{-2} *g)*
microanalysis Mikroanalyse *f*
microanalytic(al) mikroanalytisch
microarray Mikroarray *n (Molekularbiologie)*
microbalance *leichte Feinwaage für Lasten bis zu 0,1 g*
microbes *s.* microorganisms
microbial mikrobiell
microbial activity *(Biot, Hyd)* Mikrobenaktivität *f*, Bioaktivität *f* [biochemische Aktivität *f*] der Mikroorganismen
microbial breakdown mikrobieller Abbau *m*
microbial concentration Mikroorganismenkonzentration *f*, Biomassekonzentration *f*
microbial contaminants *(Biot)* Infektionsmikroorganismen *mpl*, Kontaminanten *mpl*
microbial count 1. *(Biot)* Mikroorganismenzahl *f*; 2. *(Hyd)* Bakterienzahl *f*
microbial culture *(Biot)* Mikroorganismenkultur *f*
microbial decomposition [degradation] mikrobieller Abbau *m*
microbial dry weight *(Biot)* Trockensubstanz *f* der Mikroorganismenzellen, Zelltrockenmasse *f*, Zelltrockensubstanz *f*, Zell-TS *f*
microbial fermentation 1. *(Biot)* Mikroorganismenfermentation *f*, mikrobielle Fermentation *f*; 2. *(Lebm)* mikrobielle Gärung *f*
microbial insecticide mikrobielles Insektizid *n*, Bioinsektizid *n*
microbial leaching *(Biot)* mikrobielle [biologische] Laugung *f* [Erzaufbereitung *f*], mikrobielles Leaching *n*, Bioleaching *n*, Biobergbau *m*
microbial mass *(Biot, Hyd)* Mikroorganismenmasse *f*, (mikrobielle) Biomasse *f*
microbial polysaccharide mikrobielles Polysaccharid *n*
microbial product mikrobielle Substanz *f*, Naturstoff *m* mikrobiellen Ursprungs
microbial protein mikrobielles Protein *n*, Mikroorganismeneiweiß *n*
microbial strain *(Biot)* Mikroorganismenstamm *m*
microbicidal mikrobizid, Mikroorganismen abtötend
microbicide mikrobizides [Mikroorganismen abtötendes] Mittel *n*
microbiological mikrobiologisch
microbiology Mikrobiologie *f*
microbioreactor Mikrobioreaktor *m*
microbomb *(Lab)* Mikrobombe *f*
microbore column Mikropacksäule *f (Gaschromatographie)*
microbubbles feinste [sehr kleine] Bläschen *npl*
microburet(te) Mikrobürette *f*, Feinbürette *f*, Bankbürette *f*
microburner Mikrobrenner *m*
microcalorimeter Mikrokalorimeter *n*
microcalorimetric mikrokalorimetrisch
microcalorimetry Mikrokalorimetermethode *f*
microcapsule *(Biot)* Mikrokapsel *f*
microcarrier *(Biot)* Mikroträger *m (für Zellkulturen)*
microcell *(Spektr)* Mikroküvette *f*
microcellular rubber mikroporöser Gummi *m*, Moosgummi *m*, Porengummi *m*
microchannel Mikrokanal *m*
microchemical mikrochemisch
microchemistry Mikrochemie *f*

microchromatography Mikrochromatographie *f*
microcomponent Mikrokomponente *f*, Mikrobestandteil *m*
microconstituent Mikrokomponente *f*, Mikrobestandteil *m*
microconsumers abbauende [zersetzende, zerstörende] Mikroorganismen *mpl*, Zersetzer *mpl*, Zerstörer *mpl*
microcrystalline mikrokristallin, feinkristallin
microcrystalline wax mikrokristallines Wachs *n* [Paraffin *n*, Paraffinwachs *n*], Mikrowachs *n*, Mikroparaffin *n*
microdetermination Mikrobestimmung *f*
microdistillation Mikrodestillation *f*
microdroplet Mikrotröpfchen *n*
microelectrode Mikroelektrode *f*
microelectrophoresis *(Anal)* Mikro(elektro)phorese *f*
microelement *s.* micronutrient
microemulsion Mikroemulsion *f*
microencapsulate *v* 1. in Mikrokapseln einschließen, einkapseln; 2. *(Biot)* verprillen
microencapsulated enzyme *(Biot)* eingekapseltes [gekapseltes, mikroenkapsuliertes] Enzym *n*, Enzym *n* in Mikrokapseln
microencapsulated form Prillform *f* • **in microencapsulated form** *(Biot)* in Prillform
microencapsulation 1. Einschluss *m* in Mikrokapseln, Verkapselung *f*, Mikroeinkapselung *f*, Mikroverkapselung *f*; 2. *(Biot)* Verprillung *f*
microexamination Mikrountersuchung *f*, mikroskopische Untersuchung *f*
microfibril Mikrofibrille *f*
microfibrillar mikrofibrillär
microfluidic Mikrofluidik *f*
microgel Mikrogel *n*
microgram range *(Anal)* Mikrogrammbereich *m* *(10^{-6} bis 10^{-3} g)*
microhardness Mikrohärte *f*
microheterogeneity Mikroheterogenität *f*
microkatal *(Bioch)* Mikrokatal *n*, µkat *(Einheit der Enzymaktivität)*
micromanipulator Mikromanipulator *m*, Feinmanipulator *m*
micromethod Mikromethode *f*
micromixing Mikrovermischung *f*
micromolecular mikromolekular
micron Mikron *n (mikroskopisch sichtbares Schwebeteilchen)*
micronutrient *(Landw)* Spurenelement *n*, Mikronährstoff *m*, Mikroelement *n*, Hochleistungselement *n*, Spurennährstoff *m*
microorganisms Mikroorganismen *mpl*, Mikroben *fpl*
micropipet(te) Mikropipette *f*
micropollutant *(Hyd)* Mikroverunreinigung *f*
micropore Mikropore *f*
microporous mikroporös
microprobe *(Anal)* Mikrosonde *f*
microscience Mikrowissenschaft *f*
microscope Mikroskop *n*
microscope slide Objektträger *m*
microscopic reversibility mikroskopische Reversibilität *f*
microscopic stain Färbemittel *n* für die Mikroskopie
microscopy Mikroskopie *f*
microsomal *(Bioch)* mikrosomal
microspray *(Spektr)* Mikrospray *n (Elektrospraytechnik)*
microstate *(phys Ch)* Mikrozustand *m*
microstructure Mikrogefüge *n*, Mikrostruktur *f*
microsyringe *(Anal)* Mikrodosierspritze *f*, Mikroliterspritze *f*
microtitration Mikrotitration *f*
microvalve Mikroventil *n*
microwave *(Lab)* Mikrowelle *f*
microwave detector *(Spektr)* Mikrowellendetektor *m*
microwave furnace *(Lab)* Mikrowellenofen *m*
microwave-induced plasma mikrowelleninduziertes Plasma *n*, MIP
microwave reactor *(Lab)* Mikrowellenreaktor *m*
microwave spectroscopy Mikrowellenspektroskopie *f*
mid-boiling point mittlerer Siedepunkt *m*
mid infrared mittleres Infrarot(gebiet) *n*, MIR
middle fraction *(Dest)* Mittelfraktion *f*, mittlere Fraktion *f*
middling Mittelgut *n*, Mittelprodukt *n*, Zwischenfraktion *f*

middlings product Mittelgut *n*, Mittelprodukt *n*, Zwischenfraktion *f*
migrate *v* wandern, migrieren *(z. B. Ionen, Erdöl)*
migrate *v* **in** (hin)zuwandern, (hin)einwandern
migrate *v* **out** fortwandern, (hin)auswandern, wegwandern, abwandern
migration Migration *f*, Wandern *n*, Wanderung *f (z. B. von Ionen, Erdöl)*
migration current Wanderungsstrom *m*, Migrationsstrom *m*
migration rate [velocity] Wanderungsgeschwindigkeit *f (z. B. der Ionen)*
MIH *s.* melanotropin (release) inhibitory hormone
MIKES *s.* mass-analyzed ion kinetics spectrometry
mil = millilitre
mild mild, schonend, gelinde, leicht
mild conditions milde [schonende] Bedingungen *fpl*
mild oxidation schonende Oxidation *f*
mild steel Weichstahl *m*, weicher Stahl *m*
mildew resistance Schimmelbeständigkeit *f*, Widerstandsfähigkeit *f* gegen Schimmel
mildew-resistant schimmelbeständig, schimmelfest
mildly exothermic schwach exotherm
mildly toxic schwach giftig [toxisch]
milk 1. Milch *f (von Säugetieren)*; 2. Milchsaft *m*, Latex *m (von Pflanzen)*; Pflanzenmilch *f (aus Samen)*; Kokosmilch *f*; 3. Milch *f (Suspension von Stärke, verschiedener Oxide oder anderer weißer Pulver)*
milk albumin Lactalbumin *n*, Milchalbumin *n*
milk fat Milchfett *n*
milk glass Milchglas *n*
milk globulin Lactoglobulin *n*, Milchglobulin *n*
milk of lime Kalkmilch *f*
milk of sulphur Schwefelmilch *f*, gefällter Schwefel *m*
milk protein Milchprotein *n*, Milcheiweiß *n*
milk serum protein Molkenprotein *n*, Molkeneiweiß *n*
milk solids Milchtrockenmasse *f*
milk sugar Milchzucker *m*, Lactose *f*
milkiness milchige Trübung *f*
milky milchig, milchartig; milchhaltig
milky sap Milchsaft *m*, Latex *m (von Pflanzen)*
mill *v* 1. mahlen, vermahlen, zermahlen *(grob bis fein)*; in einer Mühle zerkleinern *(Erz)*; 2. schälen, enthülsen *(Körner)*
mill 1. Mühle *f (Betrieb)*; 2. Mühle *f (Maschine zum Zerkleinern fester Stoffe)*; Aufbereitungsanlage *f (für Erze)*
mill drying Mahltrocknen *n*, Mahltrocknung *f*
mill runner Läufer *m*, Laufstein *m*, Laufrolle *f*
mill scale Walzzunder *m*
mill water Betriebswasser *n*, Brauchwasser *n*, Nutzwasser *n*
Miller (crystal) indices *(Krist)* Miller-Indizes *mpl*, millersche Indizes *mpl*
milliequivalent Milli(gramm)äquivalent *n*, Millival *n*, mVal, mval
milligram analysis Mikroanalyse *f*
milligram range *(Anal)* Milligrammbereich *m (0,001 bis 0,01 g)*
millimetres of mercury Millimeter *mpl* Quecksilbersäule *(Einheit des Drucks von Fluiden; 1 mm Hg = 1 Torr = 133,3 Pa)*
millimolar solution 0,001 m-Lösung *f*, m/1000-Lösung *f*
millinormal solution 0,001 n-Lösung *f*, n/1000-Lösung *f*
Millon's base millonsche Base *f*, Dihydroxoquecksilber(II)-ammoniumhydroxid *n*
Millon's reagent Millons Reagens *n (zum Nachweis von Eiweißstoffen)*
mine *v (Bgb)* abbauen
mine Grube *f*, Bergwerk *n*, Zeche *f*; Mine *f*
mine gas Grubengas *n*
mine-run salt Rohsalz *n*, *(Düngemittelindustrie i. e. S.)* Kalirohsalz *n*
mineral mineralisch, Mineral...; anorganisch
mineral Mineral *n*; Erz *n*
mineral acid Mineralsäure *f*, anorganische Säure *f*
mineral chemistry Mineralchemie *f*, mineralogische Chemie *f*
mineral content 1. Mineral(stoff)gehalt *m*; 2. *(Hyd)* Mineralsalzgehalt *m*
mineral dressing Mineralaufbereitung *f*,

bergmännische [bergbauliche] Aufbereitung *f*
mineral fertilizer Mineraldünger *m*, mineralischer Dünger *m*
mineral fibre mineralische Faser *f*, Mineralfaser *f*; Mineralfaserstoff *m*
mineral leaching *(Biot)* Minerallaugung *f*
mineral matter mineralischer Stoff *m*, mineralische Substanz *f*, Mineralstoff *m*, Mineralsubstanz *f*, mineralisches Material *n*
mineral oil Mineralöl *n*, mineralisches Öl *n*
mineral pigment natürliches anorganisches Pigment *n*, Erdpigment *n*
mineral pitch Asphalt *m*, Erdpech *n*, Erdharz *n*
mineral processing *s.* mineral dressing
mineral salt Mineralsalz *n*, *(i. e. S.)* Steinsalz *n*
mineral spirit(s) Lösungsbenzin *n*
mineral substance *s.* mineral matter
mineral tanning Mineralgerbung *f*
mineral water Mineralwasser *n*
mineral wax mineralisches Wachs *n*
mineral wool Mineralwolle *f*
mineralization *(Boden, Hyd, Geol)* Mineralisation *f*, Mineralisierung *f*
mineralize *v (Boden, Hyd, Geol)* mineralisieren
mineralocorticoid *(Bioch)* Mineralcorticoid *n*
mineralogical mineralogisch
mineralogist Mineralog(e) *m*
mineralogy Mineralogie *f*, Mineralkunde *f*
mingle *v* (ver)mischen, zusammenmischen, *(Feststoffe auch)* (ver)mengen; sich vermischen
minimum-boiling azeotrope Minimumazeotrop *n*, Azeotrop *n* mit Minimumsiedepunkt, Siedepunktsminimumazeotrop *n*
minimum detection limit *(Anal)* (untere) Nachweisgrenze *f*, (untere) Erfassungsgrenze *f*
minimum dosage [dose] Mindestdosis *f*
minimum reflux ratio *(Dest)* Mindestrücklaufverhältnis *n*
minimum temperature Mindesttemperatur *f*, Minimaltemperatur *f*
mining Bergbau *m*
minium Mennige *f*, Bleimennige *f*, Blei(II)-orthoplumbat *n*
minor base *(Bioch)* seltene Base *f*, seltener Nucleinsäurebaustein *m (Purin- oder Pyrimidinderivat)*
minor constituent 1. *(Anal)* Nebenbestandteil *m (*1 bis 0,01 %)*; 2. *(Hyd)* Begleitstoff *m*
minor element 1. Legierungszusatz *m*, Nebenbestandteil *m*, Minorelement *n*; 2. *s.* micronutrient
minor tranquilizer *(Pharm)* Tranquilizer *m*, Ataraktikum *n*, Anxiolytikum *n*, Psychosedativum *n*
minority carrier Minoritäts(ladungs)träger *m*, Minderheitsträger *m*
minus material Unterkorn *n*, Feinkorn *n*, Feingut *n (beim Klassieren)*; *(beim Siebklassieren auch)* Siebfeines *n*, Siebfeine *f*, *(beim Windsichten auch)* Sichtfeines *n*
MIP *s.* microwave-induced plasma
mirror-image isomer Spiegelbildisomer(e) *n*, Enantiomer(e) *n*, optisches Isomer(e)
mirror-image isomerism Enantiomerie *f*, Spiegelbildisomerie *f*, optische Isomerie *f*
miscella *(Lebm)* Miszella *f (mit Fett beladenes Extraktionsmittel)*
misch metal Mischmetall *n*, Cer-Mischmetall *n (Legierung von Seltenerdmetallen)*
miscibility Mischbarkeit *f*
miscibility gap Mischungslücke *f*
miscible mischbar
miscible in all proportions beliebig [in jedem Verhältnis] mischbar
misnomer *(Nom)* Fehlbenennung *f*, falscher Name *m*
miticidal akarizid, milbentötend
miticide Akarizid *n*, Milbengift *n*
mitochondrial deoxyribonucleic acid mitochondriale Desoxyribonucleinsäure *f*, Mitochondrien-DNA *f*, mtDNA *f*
mitochondrial ribonucleic acid mitochondriale Ribonucleinsäure *f*, mtRNA *f*
mitogenic *(Bioch)* Mitose auslösend
mitotic poison *(Bio)* Mitosegift *n*
mix *v* (ver)mischen, zusammenmischen, *(Feststoffe auch)* (ver)mengen, *(Flüssigkeiten auch)* verrühren, *(Feststoffe*

mit Flüssigkeiten auch) anrühren; sich (ver)mischen
mix *v* **intimately** innig vermischen
mix *s.* mixture
mixed acid Mischsäure *f*, *(meist:)* Salpeterschwefelsäure *f*, Nitriersäure *f*
mixed anhydride gemischtes Anhydrid *n*
mixed-bed column *(Hyd, Chromat)* Mischbettsäule *f*
mixed catalyst Mischkatalysator *m*, Mehrstoffkatalysator *m*
mixed crystal Mischkristall *m*
mixed fermentation *(Bioch)* gemischte Säuregärung *f*, heterofermentative Gärung *f*
mixed fertilizer Mehrnährstoffdünger *m*, Kombinationsdünger *m*
mixed-function oxygenase mischfunktionelle Oxygenase *f*, Monooxygenase *f*, Hydroxylase *f*
mixed gas Mischgas *n*
mixed glyceride gemischtes [gemischtsäuriges] Glycerid *n*, gemischtes Acylglycerol *n*
mixed indicator Mischindikator *m*, Indikatorgemisch *n*
mixed-media filter Mehrschicht(en)filter *n*
mixed melting point Mischschmelzpunkt *m*
mixed oxide Mischoxid *n*
mixed phase Mischphase *f*, Gemischtphase *f*, gemischte Phase *f*
mixed potential Mischpotenzial *n*, Misch(galvani)spannung *f*
mixed triglyceride gemischtes [gemischtsäuriges, heterogenes] Triglycerid *n*
mixer Mischer *m*, Mischapparat *m*, Mischmaschine *f*, Rührer *m*, Rührwerk *n*, Rührmaschine *f*; Kneter *m*, Knetwerk *n*, Knetmaschine *f*; Mischer *m*, Mixer *m*, Mischgefäß *n*, Mischbehälter *m*
mixer-settler Mischer-Scheider-Extrakteur *m*, Mischer-Settler-Extraktor *m*, Misch-Trenn-Behälter *m*
mixing arm Knetarm *m*; Rührarm *m*
mixing blade Rühr(er)schaufel *f*, Rühr(er)flügel *m*; Mischflügel *m*; Knetschaufel *f*
mixing element Rührorgan *n*; Knetorgan *n*
mixing height Mischungshöhe *f (von Stoffen in der Atmosphäre)*
mixing machine Mischmaschine *f*; Rührmaschine *f*; Knetmaschine *f*
mixing nozzle Mischdüse *f*
mixing ratio Mischungsverhältnis *n*
mixing vessel Mischgefäß *n*, Mischkessel *m*, Misch(er)behälter *m*; Rühr(werk)kessel *m*, Rühr(werk)behälter *m*
mixture 1. Mischung *f*, Gemisch *n*, *(bei Feststoffen auch)* Gemenge *n*; 2. *(Pharm)* Mixtur *f*; 3. Mischen *n*, Vermischen *n*, *(bei Feststoffen besser)* Vermengen *n*
mixture of solids Feststoffgemisch *n*, Gemenge *n*
mixture of substances Stoffgemisch *n*
ml RNA = giant messenger-like ribonucleic acid
MMA *s.* methyl methacrylate
MMCF/D = millions of cubic feet per day
MMLB/Y = millions of pounds per year
MMT/Y = millions of long tons per year
MO *s.* molecular orbital
mobile hydrogen atom bewegliches Wasserstoffatom *n*
mobile phase *(Chromat)* mobile [bewegliche] Phase *f*
mobility Beweglichkeit *f*
mobility of ions Ionenbeweglichkeit *f*
mobilization *(Bioch)* Mobilisierung *f*
mobilize *v (Bioch)* mobilisieren
Möbius system Möbius-System *n*, Anti-Hückel-System *n (cyclisch konjugierter Kohlenwasserstoff mit Phasensprung)*
MOCVD *s.* metal-organic chemical vapour deposition
modacrylic fibre Modacrylfaser *f*; Modacrylfaserstoff *m*
mode *(Anal)* Gipfel *m (einer Verteilung)*
mode-locked laser phasengekoppelter [modengekoppelter] Laser *m*
mode of action Wirkungsweise *f*
mode of operation 1. Betriebsweise *f*, Arbeitsweise *f*; 2. *(Tech)* Fahrweise *f*
mode of vibration Schwingungsform *f*, Mode *f*
model compound Modellverbindung *f*
model reaction Modellreaktion *f*
moderate *v (Kern)* moderieren, (ab-) bremsen, verlangsamen *(Neutronen)*
moderate heat mäßige Wärme *f*

moderately concentrated mäßig konzentriert
moderately soluble mäßig löslich
moderately strong [weak] mittelstark *(z. B. Säure)*
moderation *(Kern)* Bremsen *n*, Bremsung *f*, Abbremsen *n*, Abbremsung *f* *(von Neutronen)*
moderator *(Kern)* Moderator *m*, Moderatorsubstanz *f*, Bremssubstanz *f*, Bremsstoff *m*
modification 1. Modifizierung *f*, Abänderung *f*, Abwandlung *f (z. B. eines Verfahrens)*; 2. Modifikation *f (Zustandsform eines Elements oder einer Verbindung)*
modified modifiziert, abgeändert, abgewandelt
modified acrylic fibre Modacrylfaser *f*; Modacrylfaserstoff *m*
modified starch modifizierte Stärke *f*
modifier 1. Modifikationsmittel *n*; 2. *(Aufb)* Regler *m*, regelndes Reagens *n* [Mittel *n*, Schwimmmittel *n*]; 3. *(Bioch)* allosterischer Effektor *m*
modify *v* modifizieren, abändern, abwandeln *(z. B. ein Verfahren)*
modifying agent *s.* modifier
modulation frequency *(Spektr)* Modulationsfrequenz *f*
modulator *(Bioch)* modulierender Effektor *m*, Modulator *m*
modulus of elasticity Elastizitätsmodul *m*, E-Modul *m*
MOF *s.* metal organic framework
Mohr method (titration) Titration *f* [Bestimmung *f*] nach Mohr *(Argentometrie)*
Mohr's salt mohrsches Salz *n* [Doppelsalz *n*], Ammoniumeisen(II)-sulfat-6-Wasser *n*
Mohs' (hardness) scale Härteskala *f* nach Mohs, mohssche Härteskala *f*, Mohs-Skala *f*
Mohs' scale of hardness Härteskala *f* nach Mohs, mohssche Härteskala *f*, Mohs-Skala *f*
moiety Teil *m*, Anteil *m*, Komponente *f*
moist feucht; nass
moist dryness beginnende Trockne *f*
moist pellet Grünpellet *n*
moisten *v* anfeuchten, befeuchten; feucht werden
moisture Feuchte *f*, Feuchtigkeit *f*; Nässe *f*; Feuchtegehalt *m*, Feuchtigkeitsgehalt *m*
moisture-carrying capacity Feuchteaufnahmevermögen *n*, Feuchtigkeitsaufnahmevermögen *n*
moisture content Feuchtegehalt *m*, Feuchtigkeitsgehalt *m*; Feuchtebeladung *f*
moisture-free wasserfrei, wf, trocken
moisture resistance Feuchtigkeitsbeständigkeit *f*, Feuchtebeständigkeit *f*, Feuchtigkeitsfestigkeit *f*, Feuchtefestigkeit *f*
moisture-resistant feuchtigkeitsbeständig, feuchtigkeitsfest
moisture-vapour resistance Wasserdampfundurchlässigkeit *f*
moisture-vapour transmission Wasserdampfdurchlässigkeit *f*
moisturize *v* feucht halten
moisturizer Feuchthaltemittel *n*
mol *s.* mole
molal molal, gewichtsmolar
molal boiling-point(-elevation) constant molale [molare, molekulare] Siedepunktserhöhung *f*, ebullioskopische Konstante *f*
molal concentration *s.* molality
molal freezing-point(-depression) constant molale [molare, molekulare] Gefrierpunktserniedrigung *f*, kryoskopische Konstante *f*
molality Molalität *f*, Kilogramm-Molarität *f*, kg-Molarität *f*
molar absorption coefficient *(Spektr)* molarer Absorptionskoeffizient *m*, molarer (dekadischer) Extinktionskoeffizient *m*
molar absorptivity *s.* molar absorption coefficient
molar activity molare Aktivität *f (Enzymkinetik)*
molar concentration *s.* molarity
molar conductance molare Leitfähigkeit *f*, molares Leitvermögen *n*
molar decadic absorption coefficient *s.* molar absorption coefficient
molar extinction coefficient molarer Extinktionskoeffizient *m*
molar heat (capacity) molare Wärmekapazität *f*, Molwärme *f*
molar mass Molmasse *f*, molare Masse *f*

molar mass distribution Molmassenverteilung *f (Polymere)*
molar polarization Mol(ekular)polarisation *f*
molar rotation molare Drehung *f*
molar susceptibility molare Suszeptibilität *f*, Molsuszeptibilität *f*
molar volume Molvolumen *n*, stoffmengenbezogenes Volumen *n*
molarity Molarität *f*, Liter-Molarität *f*, Volumenmolarität *f*
molasses 1. *(Zucker)* Melasse *f*, Ablauf *m* II; 2. Sirup *m*
mold *v s.* mould
mole Mol *n*, Grammmol(ekül) *n*
mole fraction Molenbruch *m*
molecular molekular, Molekular..., Molekül...
molecular acid Neutralsäure *f*, Molekülsäure *f*
molecular aggregate Molekülaggregat *n*, molekulares Aggregat *n*
molecular association Molekülassoziation *f*, Assoziation *f*
molecular attraction intermolekulare [zwischenmolekulare] Anziehung *f*
molecular band *(Spektr)* Molekülbande *f*
molecular beam Molekularstrahl *m*, Molekülstrahl *m*
molecular beam epitaxy Molekularstrahlepitaxie *f*
molecular colloid Molekülkolloid *n*
molecular complex Molekülkomplex *m*
molecular compound Molekülverbindung *f*, Molekularverbindung *f*
molecular conformation Molekülkonformation *f*
molecular crystal Molekülkristall *m*
molecular diffusion molekulare Diffusion *f*
molecular distillation Molekulardestillation *f*
molecular eigenfunction Moleküleigenfunktion *f*
molecular electronic spectrum Molekülelektronenspektrum *n*
molecular exclusion chromatography Ausschlusschromatographie *f*, Größenausschlusschromatographie *f*, Gelchromatographie *f*
molecular force Molekularkraft *f*, (zwischen)molekulare Kraft *f*
molecular formula Molekularformel *f*, Molekülformel *f*
molecular fragment Molekülfragment *n*, Molekülbruchstück *n*
molecular geometry Molekülgeometrie *f*
molecular inclusion compound Moleküleinschlussverbindung *f*
molecular interaction intermolekulare [zwischenmolekulare] Wechselwirkung *f*
molecular ion Molekülion *n*
molecular lattice molekulares Gitter *n*, Molekülgitter *n*
molecular manufacturing molekulare Fertigung *f*
molecular mass Molekülmasse *f*, Molekularmasse *f*, molekulare Masse *f*
molecular motion [movement] Molekularbewegung *f*
molecular orbital Molekülorbital *n*, MO *n*
molecular-orbital method Molekülorbitalmethode *f*, MO-Methode *f*, Methode *f* der Molekülorbitale, Hund-Mulliken-Lennard-Jones-Hückel-Methode *f*
molecular oxygen O_2 molekularer Sauerstoff *m*
molecular peak *(Spektr)* Molekülpeak *m*, Ausgangslinie *f* [Massenpeak *m*] im Massenspektrum
molecular polarizability Molekülpolarisierbarkeit *f*, molekulare Polarisierbarkeit *f*, Polarisierbarkeit *f* des Moleküls
molecular rotation molekulare Drehung *f*, Molekulardrehung *f*, Molekularrotation *f*
molecular sieve Molekularsieb *n*, Molekülsieb *n*
molecular spectroscopist Molekülspektroskopiker *m*
molecular spectroscopy Molekülspektroskopie *f*
molecular spectrum Molekülspektrum *n*, Bandenspektrum *n*
molecular spin orbital Molekülspinorbital *n*, molekulares Spinorbital *n*
molecular still Molekulardestillierapparat *m*
molecular structure Molekülstruktur *f*
molecular vibration Molekülschwingung *f*
molecular weight relative Molekülmasse *f*, *(veraltet:)* Molekulargewicht *n*

molecular-weight average Mittelwert *m* der relativen Molekülmasse
molecular-weight determination Bestimmung *f* der relativen Molekülmasse
molecular-weight distribution Verteilung *f* der relativen Molekülmassen
molecularity (of reaction) Molekularität *f* (einer Reaktion), Reaktionsmolekularität *f*
molecularly disperse molekulardispers
molecule Molekül *n*, Molekel *f*
Molisch reaction Molisch-Reaktion *f*, molischsche Reaktion *f (Nachweis von Kohlenhydraten)*
Mollier chart Mollier-Diagramm *n*, i-s-Diagramm *n*, Enthalpie-Entropie-Diagramm *n*
mollus(ca)cide Molluskizid *n*, Mittel *n* gegen Schnecken, Schneckenbekämpfungsmittel *n*, Schneckenvertilgungsmittel *n*
molozonide Molozonid *n*, 1,2,3-Trioxolan *n*
molten geschmolzen, schmelzflüssig
molten alkali (bath) Alkalischmelze *f*, alkalische Schmelze *f*
molten carbonate fuel cell Schmelzkarbonat-Brennstoffzelle *f*
molten metal schmelzflüssiges Metall *n*, Metallschmelze *f*
molten salt Salzschmelze *f*, geschmolzenes Salz *n*
molybdate $M^{I}_{2}MoO_4$ Molybdat(VI) *n*
molybdenum Mo Molybdän *n*
molybdenum blue Molybdänblau *n*, Molybdän(V,VI)-oxid *n*
molybdenum carbide Molybdäncarbid *n*
molybdenum pentoxide Molybdänpentoxid *n*, Molybdän(V)-oxid *n*
molybdenum steel Molybdänstahl *m*
molybdenum trioxide Molybdäntrioxid *n*, Molybdän(VI)-oxid *n*
molybdic Molybdän...; höherwertigem Molybdän entsprechend
molybdic acid Molybdänsäure *f*
molybdic sulphide Molybdän(VI)-sulfid *n*, Molybdändisulfid *n*
molybdoiron protein *(Bioch)* Molybdän-Eisen-Protein *n*
molybdophosphate Molybdatophosphat *n*, Dodecamolybdatophosphat *n*
molybdous Molybdän...; niederwertigem Molybdän entsprechend
moment of inertia Trägheitsmoment *n*
moment of momentum Drehimpuls *m*, Impulsmoment *n*
momentum 1. Impuls *m*, Bewegungsgröße *f*; 2. *(Tech)* bewegende Kraft *f*, Triebkraft *f*
momentum (balance) equation Impulsbilanzgleichung *f*, Stoßgleichung *f*
momentum flux Impulsstrom *m (je Flächen- und Zeiteinheit)*
momentum transport Impulstransport *m*
mon. *s.* monoclinic
MON *s.* motor octane (number)
monatomic einatomig, monoatomar
monatomicity Einatomigkeit *f*
monazite *(Min)* Monazit *m (Cerphosphat)*
Mond carbonyl process *(Met)* Mond-Verfahren *n*, Mond-Niederdruckcarbonylverfahren *n*
Mond process *(Tech)* Mond-Verfahren *n (Reinigung von Nickel)*
monitor *v* überwachen *(durch wiederholtes Messen)*
monitoring Überwachung *f (durch wiederholtes Messen)*; Monitoring *n*; Verfolgung *f (von chemischen Reaktionen)*
mono derivative Monoderivat *n*
monoacid 1. einsäurig, einbasig, einwertig *(Base)*; 2. Monohydrogen..., Hydrogen... *(Salz)*
monoacid einbasige [einwertige, einprotonige] Säure *f*
monoalkylate *v* monoalkylieren
monoalkylation Monoalkylierung *f*
monoalkylbenzene Monoalkylbenzen *n*
monoamide Monamid *n*
monoamine Monoamin *n*
monoazo dye Monoazofarbstoff *m*
monobasic 1. einbasig, einwertig, einprotonig *(Säure)*; 2. einsäurig, einbasig, einwertig *(Base oder basisches Salz)*
monobasic sodium (ortho)phosphate Natriumdihydrogenphosphat *n*, Mononatriumphosphat *n*
monobasicity 1. Monobasizität *f (von Säuren)*; 2. Einwertigkeit *f (von Basen)*
monobromide Monobromid *n*
monobromobenzene Monobrombenzen *n*, Brombenzen *n*

monocalcium phosphate Calciumdihydrogenphosphat *n*
monocarboxylic acid Monocarbonsäure *f*, einbasige Carbonsäure *f*
monochloride Monochlorid *n*
monochlorinate *v* monochlorieren
monochlorination Monochlorierung *f*
monochloro derviative Monochlorderivat *n*
monochlorobenzene Monochlorbenzen *n*, Chlorbenzen *n*
monochloromethane Monochlormethan *n*, *Chlormethan *n*, Methylchlorid *n*
monochloroparaffin Monochloralkan *n*, monochlorierter Paraffinkohlenwasserstoff *m*
monochlorosilane Monochlorsilan *n*, Chlorsilan *n*
monochromatic monochromatisch, einfarbig
monochromatism Monochromasie *f* *(einer Strahlung)*
monochromatize *v* monochromatisieren *(Licht)*
monochromator *(Spektr)* Monochromator *m*, Strahlungsfilter *n*
monoclinic *(Krist)* monoklin, monokl.
monoclinic-prismatic monoklin-prismatisch
monoclinic sulphur monokliner Schwefel *m*, β-Schwefel *m*
monoclonal antibody *(Biot)* monoklonaler Antikörper *m*, MAk, monAk
monoculture *(Biot)* Reinkultur *f*, Monokultur *f* *(von Mikroorganismen)*
monocyclic monocyclisch *(Kohlenwasserstoff)*
monodentate einzähnig, einzählig *(Komplexchemie)*
monodisperse(d) monodispers, homodispers, isodispers
monoenoic *(org Ch)* einfach ungesättigt
monoester Monoester *m*
monoethanolamine Monoethanolamin *n*, *2-Aminoethan-1-ol *n*
monofil Monofil(garn) *n*, monofile Seide *f*
monofilament (yarn) Monofil(garn) *n*, monofile Seide *f*
monofluoride Monofluorid *n*
monofunctional unit monofunktionelle Einheit *f*, M-Einheit *f*, Grundbaustein *m* M *(zum Aufbau von Polymeren)*
monoglyceride Monoglycerid *n*
monohalide Monohalogenid *n*
monohalogenate *v* monohalogenieren
monohalogenation Monohalogenierung *f*
monohydrate Monohydrat *n*
monohydric 1. einwertig, mit einer Hydroxylgruppe [OH-Gruppe] *(Alkohol, Phenol)*; 2. Monohydrogen..., Hydrogen... *(Salz)*
monohydride Monohydrid *n*
monohydrogen Monohydrogen..., Hydrogen...
monohydrogen Monowasserstoff *m*, atomarer [einatomiger] Wasserstoff *m*
monohydrogen phosphate Monohydrogenphosphat *n*, Hydrogenphosphat *n*
monohydroxy(lic) Monohydroxy...
monoisotopic monoisotop
monolayer monomolekulare Schicht *f*, monomolekularer Film *m* [Oberflächenfilm *m*], Monolage *f*, Monoschicht *f*, Monomolekularfilm *m*
monomer Monomer(e) *n*, monomere Substanz *f*
monomer reactivity ratio Copolymerisationsparameter *m*
monomeric monomer
monomolecular monomolekular, unimolekular, 1-molekular
monomolecular film [layer] *s.* monolayer
monomolecular reaction monomolekulare [unimolekulare] Reaktion *f*
mononitrate *v* mononitrieren
mononitration Mononitrieren *n*, Mononitrierung *f*
mononitro compound Mononitroverbindung *f*
mononuclear einkernig
mononucleotide Mononucleotid *n*
monoolefin Monoolefin *n*, Alken *n*
monooxide Monoxid *n*
monophosphate Monophosphat *n*, Orthophosphat *n*, Phosphat *n*
monopolar cell *(Ech)* Unipolarzelle *f*, unipolare Zelle *f*
monopotassium phosphate Monokaliumphosphat *n*, Kaliumdihydrogenphosphat *n*
monoprotic acid einprotonige [einbasige, einwertige] Säure *f*

monosaccharide Monosaccharid *n*
monosilane Monosilan *n*, Silan *n*
monosodium glutamate Mononatriumglutamat *n*, Natriumglutamat *n*
monospiro compound [hydrocarbon] Monospiro-Verbindung *f*
monosubstituted monosubstituiert, einfach substituiert
monosubstitution Monosubstitution *f*, Einfachsubstitution *f*
monosulphide Monosulfid *n*
monosulphonic acid Monosulfonsäure *f*
monotectic reaction monotektische Reaktion *f* [Übergangsreaktion *f*]
monotectoid reaction monotektoide Reaktion *f* [Übergangsreaktion *f*]
monoterpene *(org Ch)* Monoterpen *n*
monoterpene alkaloid Monoterpenalkaloid *n*
monothioacetal Monothioacetal *n*
monothiocarboxylic acid Thiosäure *f*, Carbothiosäure *f*
monothioketal Monothioketal *n*
monotropic monotrop, einseitig umwandelbar *(Modifikationen)*
monotropic transition monotrope Umwandlung *f*
monotropy Monotropie *f*, einseitige Umwandelbarkeit *f*
monounsaturated einfach ungesättigt
monovalence Einwertigkeit *f*
monovalency Einwertigkeit *f*
monovalent einwertig, monovalent
monovariant monovariant, univariant, einfachfrei
monoxide Monoxid *n*
montan wax Montanwachs *n*
Monte-Carlo simulation Monte-Carlo-Simulation *f (Mathematik)*
moor peat Hochmoortorf *m*
MORD *s.* magneto-optical rotatory dispersion
mordant *v* beizen
mordant Beizmittel *n*, Beizstoff *m*, Beize *f*
mordant color *(Am) s.* mordant dye
mordant dye Beizenfarbstoff *m*, adjektiver [beizenfärbender] Farbstoff *m*
mordanting Beizen *n*, Beizung *f*
morphine *(org Ch)* Morphin *n*
morphine alkaloid Morphinalkaloid *n*
morpholine Morpholin *n*, Tetrahydro-1,4-oxazin *n*
morphological *(Krist)* morphologisch
morphology *(Krist)* Morphologie *f*, Gestalt *f*
mortar 1. *(Lab)* Mörser *m*; Reibschale *f*; 2. *(Bau)* Mörtel *m*
mosaic bloc *(Krist)* Mosaikblock *m*, Mosaikblöckchen *n*
mosaic structure [texture] *(Krist)* Mosaikstruktur *f*, Mosaiktextur *f*, Mosaikbau *m*
Moseley law *(Spektr)* moseleysches Gesetz *n*
Mössbauer effect Mößbauer-Effekt *m*
Mössbauer(-effect) spectroscopy Mößbauer-Spektroskopie *f*
Mössbauer spectrum Mößbauer-Spektrum *n*
mossy zinc granuliertes Zink *n*, Zinkgranalien *fpl*
mother liquid [liquor] Mutterlauge *f*, Stammlauge *f*, Urlauge *f*
mother of pearl Perlmutter *f*, Perlmutt *n*
mother-of-pearl sulphur perlmuttartiger Schwefel *m (eine Schwefelmodifikation)*
mother rock Muttergestein *n*, Ursprungsgestein *n*, Ausgangsgestein *n*
mother substance Muttersubstanz *f*, Grundsubstanz *f*, Stammkörper *m*, Grundkörper *m*
mothproof *v* mottenfest machen, mottenecht [mottensicher, mottenbeständig] machen
mothproof mottenfest, mottenecht, mottensicher, mottenbeständig
mothproofing agent Motten(schutz)-mittel *n*
motivating fluid Treibmittel *n (bei Strahlpumpen)*
motive steam Treibdampf *m (z. B. bei Dampfstrahlpumpen)*
motor fuel Motor(en)kraftstoff *m*
motor gasoline Motor(en)benzin *n*, Fahrbenzin *n*
motor method Motormethode *f*, Motorverfahren *n*, F-2-Methode *f (zur Octanzahlbestimmung)*
motor octane (number) Motor-Octanzahl *f*, F 2
motor oil Motorenöl *n*
motor-operated durch Elektromotor betätigt, elektromechanisch betätigt

mould *v* 1. formpressen, formen; 2. (ver)schimmeln, schimm(e)lig werden
mould 1. Form *f*, Gießform *f*, Gussform *f*; 2. *(Kst)* Form *f*, Werkzeug *n*, Pressform *f*, Presswerkzeug *n*; 3. Schimmel *m*; Schimmelpilz *m*
mould cure [curing] *(Gum)* Vulkanisation *f* [Heizung *f*] in Formen, Formheizung *f*
mould fermentation *durch Schimmelpilze bewirkte Gärung*
moulded articles [goods] Formartikel *mpl*, Formteile *npl*
moulding 1. Formpressen *n*, Herstellung *f* von Formartikeln [Formteilen], Formen *n*; 2. Formartikel *m*, Formteil *n*; 3. Schimmeln *n*, Verschimmeln *n*
moulding compound Formmasse *f*
moulding press *(Kst)* Formpresse *f*, Presse *f*
moulding resin Pressharz *n*
moulding sand Formsand *m*
mount *v* anordnen, aufstellen; anbringen, befestigen; montieren
mounting 1. Anordnung *f*, Aufstellung *f*, *(Tätigkeit auch)* Montage *f*; 2. Halterung *f*
move *v* wandern *(z. B. im Chromatogramm)*
moving-bed reactor Bewegtbettreaktor *m*, Wanderbettreaktor *m*
moving-boundary electrophoresis [method] (träger)freie Elektrophorese *f*, Tiselius-Elektrophorese *f*, Methode *f* der wandernden Grenzfläche(n)
moving-coil galvanometer Drehspulgalvanometer *n*
moving phase *(Chromat)* mobile [bewegliche] Phase *f*
mp *s.* melting point
m.p. *s.* melting point
6-MP = 6-mercaptopurine
MPD *s.* multiphoton dissociation
MPI *s.* multiphoton ionization
MRH *s.* melanotropin releasing hormone
mRNA *s.* messenger RNA
MS *s.* mass spectrometry
MSF distillation [evaporation] *s.* multistage flash evaporation
MSG *s.* monosodium glutamate
MSH *s.* melanocyte-stimulating hormone
MSO *s.* molecular spin orbital
mta *s.* metric tons annually
MT/D = thousands of long tons per day
mtDNA *s.* mitochondrial deoxyribonucleic acid
mtRNA *s.* mitochondrial ribonucleic acid
mu meson *(Kern)* Myon *n*
mucic acid Schleimsäure *f*, Mucinsäure *f*, Tetrahydroxyadipinsäure *f*, *2,3,4,5-Tetrahydroxyhexandisäure *f*
mucilage Schleim *m*, *(i. e. S.)* Pflanzenschleim *m*
mucilaginous schleimig, schleimhaltig
mucoid schleimartig
mucopolysaccharide Mucopolysaccharid *n*
mucoprotein Mucoproteid *n*, Mucoprotein *n*
mucus Schleim *m*
mud Schlamm *m*, Schlick *m*
muffle furnace *(Lab)* Muffelofen *m*
mull technique Suspensionstechnik *f* *(IR-Spektroskopie)*
muller Laufstein *m*, Läufer *m*, Mahlwalze *f*
Müller-Rochow process Müller-Rochow-Verfahren *n*, Rochow-Verfahren *n*, direktes Verfahren *n*, Direktverfahren *n*, direkte Methode *f* *(zur Herstellung von Chlorsilanen)*
mullite *(Min)* Mullit *m* *(Aluminiumsilicat)*
multi-angle laser light scattering Mehrwinkel-Laserlichtstreuung *f*
multiacidic mehrsäurig, mehrbasig, mehrwertig *(Base)*
multibasic mehrbasig, mehrprotonig *(Säure)*
multicell Mehrzellen…, Vielzellen…
multicellular foam offenzelliger Schaum(kunst)stoff *m*
multicentre Mehrzentren…, polyzentrisch
multicentre bond Mehrzentrenbindung *f*, Vielzentrenbindung *f*
multicentre reaction Mehrzentrenreaktion *f*
multichamber Mehrkammer…
multichannel analyser Mehrkanalspektrometer *n*
multicompartment Mehrkammer…; Mehrzellen…, Vielzellen…
multicomponent system Mehrkomponentensystem *n*, Mehrstoffsystem *n*, polynäres System *n*

multidentate mehrzähnig, vielzähnig, mehrzählig *(Komplexchemie)*
multidimensional chromatography mehrdimensionale Chromatographie *f*
multienzyme complex Multienzymkomplex *m*
multifil Multifil(garn) *n*, multifile [polyfile] Seide *f*
multifilament (yarn) Multifil(garn) *n*, multifile [polyfile] Seide *f*
multifunctional polyfunktionell, mehrfunktionell
multihearth furnace mehrherdiger Ofen *m*, *(meist:)* Mehretagenofen *m*, mehretagiger Ofen *m*
multilayer Multilagen...
multilayer Mehrfachschicht *f*, multimolekulare Schicht *f*
multilayer adsorption Mehrschichtenadsorption *f*
multilayer filter Mehrschicht(en)filter *n*
multimembered vielgliedrig, hochgliedrig *(Ringverbindung)*
multimolecular layer *s.* multilayer
multinecked flask Mehrhalskolben *m*
multinutrient fertilizer Mehrnährstoffdünger *m*, Kombinationsdünger *m*
multipass mehrgängig *(z. B. Wärmeübertrager)*
multiphase system Mehrphasensystem *n*, mehrphasiges [heterogenes] System *n*
multiphoton absorption Multiphotonenabsorption *f*
multiphoton dissociation Multiphotonendissoziation *f*, MPD
multiphoton ionization Multiphotonenionisation *f*, MPI
multiple vielfach, mehrfach, multipel, Viel(fach)..., Mehr(fach)...
multiple 1. Vielfache *n*, Multiple *n*; 2. *s.* unit
multiple bond Mehrfachbindung *f*, mehrfache Bindung *f*
multiple decay *s.* multiple disintegration
multiple development *(Chromat)* Mehrfachentwicklung *f*
multiple disintegration Mehrfachzerfall *m*, Dualzerfall *m*, dualer [verzweigter] Zerfall *m*, (radioaktive) Verzweigung *f*
multiple-effect evaporator Mehrstufenverdampfer *m*, Mehrkörperverdampfer *m*, Mehrfachverdampfer *m*, Kaskadenverdampfer *m*
multiple-hearth incinerator *(Hyd)* Drehetagenofen *m (Schlammverbrennung)*
multiple-layer insulation Vakuum-Vielschichtisolierung *f*, Superisolierung *f* *(Kältetechnik)*
multiple proportions multiple Proportionen *fpl*
multiple resistance *(Biot)* Mehrfachresistenz *f*
multiple scattering Mehrfachstreuung *f*
multiple unit Mehrfachapparat *m*, Vielfachapparat *m*
multiplet *(phys Ch)* Multiplett *n*
multiplet splitting *(Spektr)* Multiplettaufspaltung *f*
multiplication *(Biot)* Vermehrung *f*, Propagierung *f (von Mikroorganismen)*
multiplicity Multiplizität *f (eines elektronischen Zustands)*
multiplicity rule *(Spektr)* Erste [1.] hundsche Regel *f*, hundsche *f* Regel der größten Multiplizität
multiplier phototube Sekundärelektronenvervielfacher *m*, SEV, Photo(elektronen)vervielfacher *m*, Photomultiplier *m*
multi-ply mehrschichtig, mehrlagig, Mehrschichten..., Mehrlagen...
multiport plug valve Mehrweg(e)hahn *m*
multiproduct unit 1. Mehrproduktenapparat *m*; 2. *(Aufb)* Mehrgutapparat *m*
multipurpose fermenter *(Biot)* Universalfermenter *m*
multirotation Multirotation *f*, Mutarotation *f*
multistage mehrstufig, Mehrstufen...
multistage evaporator mehrstufiger Verdampfer *m*, Mehrstufenverdampfer *m*, Mehrkörperverdampfer *m*
multistage flash evaporation mehrstufige Flash-Verdampfung *f*, Multiflash-Verdampfung *f*
multistep mehrstufig, vielstufig, Mehrstufen...
multistranded mehrsträngig *(z. B. Makromolekülketten)*
multisweep polarography *(Anal)* *Multisweep-Polarographie *f*, oszillographische Polarographie *f* mit Wechselstrom
multitube fixed-bed reactor Rohrbündelreaktor *m*, Röhrenreaktor *m*

multitube-flow reactor *s.* multitube fixed-bed reactor
multivalence Mehrwertigkeit *f*
multivalency Mehrwertigkeit *f*
multivalent mehrwertig
multiwall carbon nanotube mehrwandiges Kohlenstoff-Nanoröhrchen *n*
mundic *(Min)* Eisenkies *m*, Schwefelkies *m*, Pyrit *m (Eisendisulfid)*
municipal refuse *s.* municipal solid waste
municipal sewage kommunales Abwasser *n*, Kommunalabwasser *n*, *(i. e. S.)* städtisches Abwasser *n*
municipal sewage treatment plant kommunale Abwasserbehandlungsanlage *f* [Abwasserreinigungsanlage *f*, Kläranlage *f*], *(i. e. S.)* städtische Abwasserbehandlungsanlage *f* [Abwasserreinigungsanlage *f*, Kläranlage *f*]
municipal solid waste (feste) kommunale Abprodukte *npl*, kommunaler Müll *m*
municipal waste 1. kommunale Abprodukte *npl*; 2. *s.* municipal sewage
Muntz metal Muntz-Metall *n (60 % Cu, 40 % Zn und geringer Zusatz an Pb)*
muon *(Kern)* Myon *n*
muon chemistry Myonchemie *f*
muon spin rotation Myonspinrotation *f*, µSR
muonium chemistry Myoniumchemie *f*
muramic acid Muraminsäure *f*, 2-Amino-2-desoxy-3-O-(α-D-carboxyethyl)-D-glucose *f*
murein Murein *n*, Peptidoglykan *n (ein Polysaccharid-Peptid-Komplex der Bakterienzellwand)*
murexide Murexid *n*, Ammoniumpurpurat *n*
murexide test Murexidprobe *f (Harnsäurenachweis mittels Murexidreaktion)*
muscarine Muscarin *n*, 2-Methyl-3-hydroxy-5-dimethylaminomethyl-tetrahydrofuran-hydroxymethylat *n*
muscle adenylic acid Muskeladenylsäure *f*, 5′-Adenylsäure *f*, Adenosin-5′-monophosphorsäure *f*, Adenosin-5′-monophosphat *n*
muscovite *(Min)* Muskovit *m*, Kaliglimmer *m*
muscovy glass *(Min)* Muskovit *m*, Kaliglimmer *m*
museum jar Präparatenglas *n*, (Ausstellungs-)Schauglas *n*
musk Moschus *m*
musk-like moschusartig, Moschus...
must 1. Most *m*; 2. Moder *m*, Schimmel *m*
mustard gas Senfgas *n*, Mustardgas *n*, Lost *m*, Yperit *n*
mustard oil Isothiocyanat *n*, Isothiocyansäureester *m*, Senföl *n*
mutagen *(Bioch)* Mutagen *n*, mutationsauslösendes Agens *n*
mutagenesis *(Bioch)* Mutagenese *f*
mutagenic *(Bioch)* mutationsauslösend, mutagen (wirkend)
mutagenic agent *s.* mutagen
mutagenicity *(Bioch)* Mutagenität *f*, mutationsauslösende Wirkung *f*
mutarotate *v* Mutarotation zeigen *(Zucker)*
mutarotation Mutarotation *f*, Multirotation *f (Veränderung der optischen Drehung bei Zuckern)*
mutasynthesis *(Biot)* Muta(tions)synthese *f*, Mutations-Biosynthese *f*
mutation-inducing *s.* mutagenic
mutual gegenseitig, wechselseitig, beiderseitig, Wechsel...
mutual decomposition Wechselzersetzung *f*, Wechselumsetzung *f*, doppelte Umsetzung *f*, doppelter Umsatz *m*, Metathese *f*
mutual solubility gegenseitige Löslichkeit *f*
mutually interconvertible gegenseitig [wechselseitig] ineinander umwandelbar
mutually soluble ineinander löslich
MVA *s.* mevalonic acid
MVT *s.* moisture-vapour transmission
m.w. *s.* molecular weight
M.W. *s.* molecular weight
MWCNT *s.* multiwall carbon nanotube
MWD *s.* molecular-weight distribution
MWNT *s.* multiwall carbon nanotube
M.W.P. *s.* mechanical wood pulp
mycelial dry weight Myzeltrockenmasse *f*, Myzeltrockensubstanz *f*
mycelial suspension Myzelsuspension *f*
mycelium-bound myzelgebunden

mycelium pellets Myzelpellets *npl*, Myzelkugeln *fpl*
mycolic acid Mycolsäure *f*
myoglobin Myo(hämo)globin *n*
myosin Myosin *n (zu den Globulinen gehörendes Muskelprotein)*
myristaldehyde Myristaldehyd *m*, Myristinaldehyd *m*, *Tetradecanal *n*
myristate Myristat *n (Salz oder Ester der Myristinsäure)*
myristic acid Myristinsäure *f*, *Tetradecansäure *f*
myristyl alcohol Myristylalkohol *m*, *Tetradecan-1-ol *n*
myrrh gum Myrrhenharz *n*, Myrrhe *f (von Commiphora-Arten)*

N

N-alkylation N-Alkylierung *f*
N-shell *(Kern)* N-Schale *f*
N solution Normallösung *f*, normale Lösung *f*, 1n-Lösung *f*
N-terminal N-terminal, N-endständig
n-doping n-Dotierung *f*
n-type conducting [semiconducting] n-leitend, überschussleitend
n-type semiconductor Halbleiter *m* vom n-Typ, n-(Typ-)Halbleiter *m*, Elektronenüberschusshalbleiter *m*, Überschusshalbleiter *m*
N_A *s.* Avogadro constant
Na-form exchanger *(Hyd)* Na-Austauscher *m*, Na_2-Austauscher *m*, Natriumaustauscher *m*, Ionenaustauscher *m* in der Natriumform [Na-Form], Neutralaustauscher *m*
NAA *s.* neutron-activation analysis
NAD *s.* nicotinamide-adenine dinucleotide
NADP *s.* nicotinamide-adenine dinucleotide phosphate
nail polish Nagellack *m*
nail polish remover Nagellackentferner *m*
naked flame offene [freibrennende] Flamme *f*
nalixidic acid *(Bioch)* Nalixidinsäure *f*
name *v* benennen
name reaction Namenreaktion *f*
Nametkin rearrangement Nametkin-Umlagerung *f (eines Carbeniumions mit Wanderung einer Methylgruppe)*
naming Benennung *f*, Namengebung *f*
naming system Benennungssystem *n*
NANA = *N*-acetyl neuraminic acid
nano range Nanobereich *m*
nanoarchitecture Nanoarchitektur *f*
nanobead Nanoperle *f*
nanobiotechnological nanobiotechnologisch
nanobiotechnology Nanobiotechnologie *f*
nanocapsule Nanokapsel *f*
nanochannel Nanokanal *m*
nanochemical nanochemisch
nanochemistry Nanochemie *f*
nanocluster Nanocluster *m*
nanocolloid Nanokolloid *n*
nanocomposite *(Kst)* Nanoverbundmaterial *n*, Nanokomposit *m*
nanocrystal Nanokristall *m*
nanodomain Nanodomäne *f*
nanoelectrode Nanoelektrode *f*
nanoelectronics Nanoelektronik *f*
nanoengineering Nanotechnik *f*, Nanometertechnik *f*
nanofibre Nanofaser *f*
nanofiller Nanofüllstoff *m*
nanofilm Nanofilm *m*
nanogram range *(Anal)* Nanogrammbereich *m (10^{-9} bis 10^{-6} g)*
nanoimprint lithography Nanoabdrucklithographie *f*
nanokatal *(Bioch)* Nanokatal *n*, nkat *(Einheit der Enzymaktivität)*
nanolayer Nanoschicht *f*
nanolithography Nanolithographie *f*
nanoparticle Nanoteilchen *n*, Nanopartikel *n*
nanopore Nanopore *f*
nanoporous nanoporös
nanorod Nanostäbchen *n*, Nanostab *m*
nanoscale engineering Nanotechnik *f*, Nanometertechnik *f*
nanoscale probe Nanosonde *f*
nanosciences Nanowissenschaften *fpl*
nanoscopic scale nanoskopische Skala *f*
nanosensor Nanosensor *m*
nanosphere Nanokugel *f*
nanospray *(Spektr)* Nanospray *n (Elektrospraytechnik)*

nanostructure Nanostruktur *f*, nanodimensionierte Struktur *f*
nanostructured nanostrukturiert, im Nanometerbereich strukturiert
nanotechnology Nanotechnologie *f*
nanotrace *(Anal)* Nanospur *(10^{-7} bis 10^{-10} ppm)*
nanotube Nanoröhrchen *n*, Nanoröhre *f*
naphtha Benzin *n (besonders für technische Zwecke oder als Reformingstock)*; Schwerbenzin *n*, schweres Benzin *n*, Naphtha *n(f)*; Ligroin *n (Siedebereich 90-120 °C)*
naphthalene Naphthalen *n*, Naphthalin *n*
naphthalene carboxylic acid *s.* naphthoic acid
naphthalene sulphonyl chloride Naphthalensulfonylchlorid *n*
naphthalene 1,2,3,4-tetrahydride Tetrahydronaphthalen *n*
1,8-naphthalene dicarboxylic acid Naphthalen-1,8-dicarbonsäure *f*, Naphthalsäure *f*
α-naphthaleneacetic acid α-Naphthylessigsäure *f*
naphthalenesulphonic acid Naphthalensulfonsäure *f*
naphthalic acid Naphthalsäure *f*, Naphthalen-1,8-dicarbonsäure *f*
naphthalol Naphthalol *n*, Betol *n*, Salicylsäure-β-naphthylester *m*
naphthenate Naphthenat *n*
naphthene Naphthen *n*, Cycloalkan *n*
naphthene-base petroleum naphthen(bas)isches Erdöl *n* [Rohöl *n*], Naphthenerdöl *n*, Naphthen(basis)öl *n*, Erdöl *n* auf Naphthenbasis
naphthenic naphthenartig, naphthenisch, gesättigt alicyclisch
naphthenic acid Naphthensäure *f*
naphthenic petroleum *s.* naphthene-base petroleum
naphthindigo Naphth(alen)indigo *m*, Naphth(alen)indolindigo *m*
naphthionic acid Naphthionsäure *f*, Naphth-1-ylamin-4-sulfonsäure *f*
naphthoic acid Naphthoesäure *f*, Naphthalencarbonsäure *f*
naphthol AS Naphthol *n* AS, 2-Hydroxy-3-naphthoesäureanilid *n*, Anilidsäure *f*
1-naphthol *s.* α-naphthol
2-naphthol *s.* β-naphthol
α-naphthol α-Naphthol *n*, Naphth-1-ol *n*
β-naphthol β-Naphthol *n*, Naphth-2-ol *n*
naphtholsulphonic acid Naphtholsulfonsäure *f*
naphthoquinone Naphthochinon *n*
naphthyl group $C_{10}H_7$– Naphthylgruppe *f*, Naphthylrest *m*
naphthylamine Naphthylamin *n*
naphthylaminesulphonic acid Naphthylaminsulfonsäure *f*, Aminonaphthalensulfonsäure *f*
narcotic narkotisch (wirkend)
narcotic Narkotikum *n*, Narkosemittel *n*
narrow-bore column *(Chromat)* enge [englumige] Säule *f*
narrow-neck(ed) bottle Enghalsflasche *f*
narrow-neck(ed) flask Enghalskolben *m*
NAS = National Academy of Sciences *(U.S.)*
nasal nuisance Geruchsbelästigung *f*
nascent naszierend, in statu nascendi
native nativ, natürlich; gediegen *(z. B. Schwefel, Gold)*
native asphalt Naturasphalt *m*, natürlicher Asphalt *m*
native conformation *(Bioch)* native Konfiguration *f (der Desoxyribonucleinsäure)*
native metal gediegenes Metall *n*
native protein natives Protein *n* [Eiweiß *n*]
natron 1. Kristallsoda *f*, Natriumcarbonat-10-Wasser *n*; 2. *(Min)* Natron *n*, Soda *f*
natron cellulose Natroncellulose *f*
natural abundance natürliche Häufigkeit *f* [Isotopenhäufigkeit *f*], natürliches Vorkommen *n (eines Elements)*
natural asphalt *s.* native asphalt
natural cellulose natürliche [native] Cellulose *f*
natural color *(Am) s.* natural dye
natural compound *(Biot)* Naturstoff *m*, *(i. e. S.)* Pflanzenstoff *m*, pflanzlicher Inhaltsstoff *m*
natural convection freie Konvektion *f* [Wärmekonvektion *f*], Eigenkonvektion *f*, Wärmekonvektion *f* bei freier Strömung
natural dye Naturfarbstoff *m*, natürlicher Farbstoff *m*
natural enzyme originäres Enzym *n*

natural fibre natürliche [native] Faser *f*, Naturfaser *f*; Naturfaserstoff *m*
natural filtration Schwerkraftfiltration *f*
natural frequency Eigenfrequenz *f*, charakteristische Frequenz *f*
natural gas Naturgas *n*, Erdgas *n*
natural gasoline Natur(gas)benzin *n*, Erdgasbenzin *n*, Rohrkopfbenzin *n*, Casinghead-Benzin *n*
natural glass Naturglas *n*, Gesteinsglas *n*
natural gum Naturgummi *n*
natural hydrocarbon Naturkohlenwasserstoff *m*
natural methanol roher Methylalkohol *m* [Holzgeist *m*], Rohholzgeist *m*
natural penicillin *(Biot)* natürliches Penicillin *n*
natural pigment natürliches Pigment *n*
natural product Naturprodukt *n*, Naturstoff *m*
natural product synthesis Naturstoffsynthese *f*
natural resin Naturharz *n*, natürliches Harz *n*
natural rubber Naturkautschuk *m*, NK
natural silk Naturseide *f*, reine [echte] Seide *f*
naturally occurring natürlich (vorkommend), Natur...
naval brass [bronze] Marinemessing *n* *(etwa 60 % Cu, 1 % Sn, Rest Zn)*
NBR *s.* 1. nitrile-butadiene rubber; 2. acrylonitrile-butadiene rubber
NBS 1. = National Bureau of Standards; 2. = New British Standard; 3. *s.* *N*-bromosuccinimide
NCAR = National Center for Atmospheric Research
NCR *s.* nitrile-chloroprene rubber
nd = needles
NDP = nucleoside diphosphate
NDT *s.* non-destructive testing
near-equilibrium annäherndes Gleichgewicht *n*, Prägleichgewicht *n*
near infrared naher Infrarotbereich *m*, nahes Infrarot(gebiet) *n*, NIR
near-infrared range naher Infrarotbereich *m*, nahes Infrarot(gebiet) *n*, NIR
near ultraviolet naher Ultraviolettbereich *m*, nahes Ultraviolett(gebiet) *n*, Quarz-UV *n*, QUV, Quarz-UV-Bereich *m*
near-ultraviolet range naher Ultraviolettbereich *m*, nahes Ultraviolett(gebiet) *n*, Quarz-UV *n*, QUV, Quarz-UV-Bereich *m*
nearest-neighbour (sequence) frequency *(Bioch)* Nachbarschaftshäufigkeit *f (von Purin- und Pyrimidinbasen)*
nebulization Vernebeln *n*, Verneb(e)lung *f*
nebulize *v* vernebeln, aerosolieren, aerolisieren
nebulizer 1. *(Landw)* Nebelblaser *m*, Nebelgerät *n*, Atomiseur *m*; 2. *(Anal)* Zerstäuber *m*
neck Hals *m (z. B. einer Flasche)*; Zapfen *m (einer Walze)*
needle *(Krist)* Nadel *f*
needle-like nadelförmig, nadelartig, nadelig, Nadel...
needle-shaped nadelförmig, nadelartig, nadelig, Nadel...
needle valve Nadelventil *n*
Neel point [temperature] *(Krist)* Neel-Temperatur *f*
negative *(Foto)* Negativ *n*
negative catalysis negative Katalyse *f*, Antikatalyse *f*, Inhibition *f*, Hemmung *f*, Verzögerung *f*
negative catalyst negativer Katalysator *m*, Antikatalysator *m*, Inhibitor *m*, Hemmstoff *m*, Verzögerer *m*, Passivator *m*
negative electron *s.* negatron
negative proton Antiproton *n*, ^{-}p
negatively charged negativ geladen
negativity Negativität *f*
negatron Negatron *n*, *(normales, negativ geladenes)* Elektron *n*
negligible zu vernachlässigen(d), vernachlässigbar
neighbouring benachbart, nachbarständig, vicinal, Nachbar...
neighbouring atom Nachbaratom *n*
neighbouring group effect Nachbargruppeneffekt *m*
NEL *s.* no-effect-level
nematic liquid crystal nematischer Flüssigkristall *m* [flüssiger Kristall *m*]
nematic phase *(Koll)* nematische Phase *f*, pl-Phase *f*
nematocidal nematizid, nematozid

nematocide Nematizid *n*, Nematozid *n* *(Mittel gegen Fadenwürmer)*
neodymia Neodymoxid *n*
neodymium Nd Neodym *n*
neodymium laser Neodymlaser *m*
neodymium oxide Neodymoxid *n*
neon Ne Neon *n*
neon discharge tube Neonröhre *f*
neoprene Neopren *n*, Poly-2-chlorbuta-1,3-dien *n*
neoprene rubber Neoprenkautschuk *m*
nephelometer Nephelometer *n*, Trübungsmesser *m*
nephelometric titration nephelometrische Titration *f*
nephelometry Nephelometrie *f*, Tyndallometrie *f*, Streulichtmessung *f*
nephrotoxic nephrotoxisch, nierenschädigend
neptunium Np Neptunium *n*
neral Neral *n*, Citral *n* B, *cis*-Citral *n* *(ein Terpenaldehyd)*
Nernst distribution law nernstsches Verteilungsgesetz *n*, nernstscher Verteilungssatz *m*
Nernst equation nernstsche Gleichung *f*, Nernst-Gleichung *f*, Nernst-Beziehung *f*
Nernst heat theorem nernstscher Wärmesatz *m*, nernstsches Wärmetheorem *n*
Nernst layer nernstsche Diffusionsschicht *f*
nerol Nerol *n*, *3,7-Dimethylocta-2,6-dien-1-ol *n*
nerve gas Nervengas *n*, *(Militärchemie i. w. S.)* Nervengift *n*
nerve poison Nervengift *n*
nervonic acid Nervonsäure *f*, Selachensäure *f*, *cis*-Tetracos-15-ensäure *f*
nesosilicate *(Min)* Nesosilicat *n*, Inselsilicat *n*
Nessler's reagent [solution] Neßlers Reagens *n* *(zum Nachweis von Ammoniak)*
net calorific value spezifischer Heizwert *m*, H_u, *(veraltet:)* unterer Heizwert *m*
net charge Nettoladung *f*
net equation Nettogleichung *f*
net reaction Nettoreaktion *f*
net work *(phys Ch)* Nutzarbeit *f*
network chain *(Kst)* Netzkette *f* *(Polymernetzwerke)*
network former *(Glas)* Netzwerkbildner *m*
network-forming ion *(Glas)* Netzwerkbildner *m*
network modifier *(Glas)* Netzwerkwandler *m*, Netzwerkänderer *m*, Modifizierer *m*
network-modifying ion *(Glas)* Netzwerkwandler *m*, Netzwerkänderer *m*, Modifizierer *m*
neuraminic acid Neuraminsäure *f* *(ein Aminozucker)*
neurohypophyseal hormone neurohypophysäres Hormon *n*, Hormon *n* der Neurohypophyse, Hypophysenhinterlappenhormon *n*
neuroleptic *(Pharm)* neuroleptisch
neuroleptic (drug) *(Pharm)* Neuroleptikum *n*, Antipsychotikum *n*, Psycholeptikum *n*, Antihalluzinogen *n*
neurotoxic neurotoxisch, nervenschädigend
neurotoxicity Neurotoxizität *f*
neurotoxin Neurotoxin *n*, Nervengift *n*
neutral neutral • **make neutral** neutralisieren
neutral equilibrium indifferentes Gleichgewicht *n*
neutral fat Neutralfett *n*
neutral oil Neutralöl *n*, neutrales Öl *n*
neutral range Neutralbereich *m*
neutral-reacting neutral reagierend
neutral red Neutralrot *n*, Toluylenrot *n* *(ein Redoxindikator)*
neutral-salt error Neutralsalzfehler *m*, Salzfehler *m* *(Messfehler bei pH-Bestimmungen)*
neutral sodium sulphite process *(Pap)* Neutralsulfitverfahren *n*, NSSC-Verfahren *n*
neutrality Neutralität *f* • **around neutrality** um den Neutralpunkt
neutralization Neutralisieren *n*, Neutralisation *f*
neutralization curve Neutralisationskurve *f*
neutralization indicator Neutralisationsindikator *m*
neutralization number Neutralisationszahl *f*
neutralization reaction Neutralisationsreaktion *f*
neutralization titration Neutralisations-

analyse *f*, Neutralisationstitration *f*, Säure-Base(n)-Titration *f*, Protolysetitration *f*
neutralize *v* neutralisieren, abstumpfen
neutralizer 1. Neutralisationsmittel *n*, Neutralisierungsmittel *n*; 2. Neutralisator *m*, Neutralisationsbehälter *m*; Neutralisationsanlage *f*
neutralizing agent *s.* neutralizer 1.
neutrino *(Kern)* Neutrino *n*, ν
neutron *(Kern)* Neutron *n*
neutron absorber Neutronenabsorber *m*, Neutronenfänger *m*
neutron-activation analysis Neutronenaktivierungsanalyse *f*, NAA
neutron beam Neutronenstrahl *m (gebündelt)*
neutron bombardment Neutronenbeschuss *m*, Neutronenbestrahlung *f*
neutron-capture cross section Neutroneneinfangquerschnitt *m*
neutron diffraction Neutronenbeugung *f*
neutron emission Neutronenemission *f*
neutron irradiation Neutronenbestrahlung *f*
neutron mass [rest mass] Neutronen(ruhe)masse *f*, Ruhemasse *f* des Neutrons
neutron source Neutronenquelle *f*
newton Newton *n*, N *(SI-Einheit der Kraft)*
Newtonian behaviour newtonsches Verhalten *n (Rheologie)*
Newtonian fluid [liquid] newtonsche Flüssigkeit *f*
NHE *s.* normal hydrogen electrode
N.H.E. *s.* normal hydrogen electrode
NHI protein *s.* non-haem iron protein
niacin Niacin *n*, Nicotinsäure *f*, Pyridin-3-carbonsäure *f*
Nichols furnace Nichols-Ofen *m*, Mehretagenröstofen *m* nach Nichols
nickel *v* vernickeln
nickel-plate *v* vernickeln, *(meist:)* elektrochemisch [galvanisch] vernickeln
nickel Ni Nickel *n*
nickel alloy Nickellegierung *f*
nickel-cadmium battery [cell] Nickel-Cadmium-Akkumulator *m*
nickel carbonyl *s.* nickel tetracarbonyl
nickel catalyst Nickelkatalysator *m*
nickel crucible Nickeltiegel *m*
nickel dimethylglyoxime Nickeldimethylglyoxim *n*, Nickeldiacetyldioxim *n*
nickel-iron (alkaline) battery Nickel-Eisen-Akkumulator *m*, NiFe-Akkumulator *m*
nickel matte *(Met)* Kupfer-Nickel-Rohstein *m*, Nickelstein *m*
nickel monoxide Nickel(mon)oxid *n*, Nickel(II)-oxid *n*
nickel nitrate Nickel(II)-nitrat *n*
nickel phthalocyanine Nickelphthalocyanin *n*
nickel plating Vernickeln *n*, *(meist:)* elektrochemisches [galvanisches] Vernickeln *n*
nickel-plating bath Nickelelektrolyt *m*, Vernickelungselektrolyt *m*
nickel steel Nickelstahl *m*
nickel sulphate Nickelsulfat *n*
nickel tetracarbonyl Nickeltetracarbonyl *n*
nickelic Nickel..., *(i. e. S.)* Nickel(III)-...
nickeliferous nickelhaltig
nickelize *v* vernickeln
nickelous Nickel..., *(i. e. S.)* Nickel(II)-...
nickelous chloride Nickel(II)-chlorid *n*, Nickeldichlorid *n*
nickelous oxide Nickel(II)-oxid *n*, Nickeloxid *n*
Nicol prism Nicol-Prisma *n*, nicolsches Prisma *n*, Nicol *n*
nicotinamide Nicotin(säure)amid *n*, Pyridin-3-carbonsäureamid *n*
nicotinamide-adenine dinucleotide Nicotinamid-adenin-dinucleotid *n*, NAD, *(bisher:)* Diphosphopyridinnucleotid *n*, DPN^+
nicotinamide-adenine dinucleotide phosphate *(Bioch)* Nicotinsäureamid-adenin-dinucleotidphosphat *n*, NADP
nicotinate Nicotinat *n (Salz oder Ester der Nicotinsäure)*
nicotine poisoning Nicotinvergiftung *f*
nicotinic acid Nicotinsäure *f*, Pyridin-3-carbonsäure *f*, Niacin *n*
ninhydrin reaction [test] Ninhydrinreaktion *f (zum Nachweis von Proteinen und Aminosäuren)*
niobic Niob..., *(i. e. S.)* Niob(V)-...
niobic acid Niobpentoxidhydrat *n*, Niobsäure *f*
niobium Nb Niob *n*
niobium pentoxide Niobpentoxid *n*, Niob(V)-oxid *n*
niobous Niob..., *(i. e. S.)* Niob(III)-...

nioxime Nioxim *n*, 1,2-Cyclohexandiondioxim *n*
nip (of the rolls) Walzenspalt *m*
NIR *s.* near infrared
niter *s.* nitre
nitramide Nitramid *n*, Imidosalpetersäure *f*
nitramine 1. *s.* nitramide; 2. Nitramin *n* *(Nitramid-Derivat der allgemeinen Formel RHN–NO_2 oder R_1R_2N–NO_2)*
nitrate *v* nitrieren
nitrate Nitrat *n*
nitrate bacteria Nitratbakterien *npl*, Nitratbildner *mpl (Gattung Nitrobacter)*
nitrate nitrogen Nitratstickstoff *m*
nitrate removal *(Hyd)* Nitrateliminierung *f*, NO_3-Elimination *f*, Nitratentfernung *f*
nitrating acid 1. Nitriersäure *f*; 2. *(Tech)* Mischsäure *f (Gemisch aus konzentrierter Salpeter- und Schwefelsäure)*
nitrating agent Nitrier(ungs)mittel *n*, Nitrierreagens *n*
nitration Nitrieren *n*, Nitrierung *f*
nitrator Nitrierer *m*, Nitrator *m*, Nitrierapparat *m*, Nitriergefäß *n*
nitre 1. Kalisalpeter *m*, Kaliumnitrat *n*; 2. Chilesalpeter *m*, Natronsalpeter *m*, Natriumnitrat *n*; 3. Salpetrigsäureethylester *m*, Ethylnitrit *n*; 4. *s.* natron
nitrene :NR Nitren *n*
nitric Stickstoff...; höherwertigem Stickstoff entsprechend, *(meist)* Stickstoff(V)-...
nitric acid Salpetersäure *f*
nitride *v* 1. *(Met)* nitrierhärten, nitrieren, durch Nitrierung härten; 2. *(Met)* aufsticken, versticken *(unerwünschter Vorgang)*
nitride Nitrid *n*
nitride hardening *(Met)* Nitrierhärten *n*, Nitrieren *n*, Nitridhärten *n*, Stickstoffhärten *n*
nitriding *(Met)* Nitrierhärten *n*, Nitrieren *n*, Nitridhärten *n*, Stickstoffhärten *n*
nitrification *(Landw)* Nitrifikation *f*, Nitrifizierung *f*
nitrifying bacteria nitrifizierende Bakterien *npl*, Nitrifikationsbakterien *npl*, Nitrobakterien *npl*, Nitrifikanten *mpl (allg. für Nitrit- und Nitratbakterien)*
nitrile RC≡N Nitril *n (Derivat der Blausäure)*
nitrile-butadiene rubber Nitrilkautschuk *m*, Butadien-Acrylnitrilkautschuk *m (Copolymerisat)*
nitrile-chloroprene rubber Nitril-Chloroprenkautschuk *m (Chlorbutadien-Acrylnitril-Copolymerisat)*
nitrile group ≡N Nitrilgruppe *f*
nitrile rubber *s.* nitrile-butadiene rubber
nitrilium ion Nitrilium-Ion *n*
nitrilotriacetic acid Nitrilotriessigsäure *f*, NTE
nitrite Nitrit *n*
nitrite-forming bacteria Nitritbakterien *npl*, Nitritbildner *mpl*
nitrite nitrogen Nitritstickstoff *m*
nitro compound Nitroverbindung *f*
nitro dye Nitrofarbstoff *m*
nitro group –NO_2 Nitrogruppe *f*
***aci*-nitro compound** *aci*-Nitroverbindung *f*, Nitronsäure *f*
nitroalkane *Nitroalkan *n*, Nitroparaffin *n*
nitroaniline Nitranilin *n*, *Aminonitrobenzen *n*
nitrobacteria *s.* 1. nitrifying bacteria; 2. nitrate bacteria
nitrobenzene Nitrobenzen *n*
nitrocellulose Cellulosenitrat *n*, Nitratcellulose *f*, Cellulosesalpetersäureester *m*, *(veraltet:)* Nitrozellulose *f*
nitrocellulose lacquer Cellulosenitratlack *m*, Nitrolack *m*
nitrocobaltate Nitrocobaltat *n*
nitrocotton Schieß(baum)wolle *f*
nitroethane Nitroethan *n*
nitrogen N Stickstoff *m*, Nitrogen *n*
nitrogen alkylation N-Alkylierung *f*
nitrogen balance *(Bioch)* Stickstoffbilanz *f*
nitrogen-centred radical Stickstoffradikal *n*
nitrogen dioxide Stickstoffdioxid *n*, Stickstoff(IV)-oxid *n*
nitrogen family Stickstoffgruppe *f*, Stickstoff-Phosphor-Gruppe *f (im Periodensystem)*
nitrogen fertilizer Stickstoffdüngemittel *n*, Stickstoffdünger *m*
nitrogen fixation Stickstoffbindung *f*, Stickstoff-Fixierung *f (chemisch oder biologisch)*
nitrogen-fixing bacteria stickstoffbindende [nitrogene] Bakterien *npl*, Nitrogenbakterien *npl (freien Luftstickstoff bindend)*

nitrogen laser Stickstoff-Laser *m*, N_2-Laser *m*
nitrogen metabolism *(Bioch)* Stickstoff-Stoffwechsel *m*
nitrogen monoxide 1. N_2O Distickstoff(mon)oxid *n*, Stickstoff(I)-oxid *n*; 2. NO Stickstoff(mon)oxid *n*, Stickstoff(II)-oxid *n*, Stickoxid *n*
nitrogen mustard Stickstoffsenfgas *n*, Stickstofflost *m*, Stickstoffyperit *n*, N-Lost *m*, N-Yperit *n*
nitrogen oxide Stickstoffoxid *n (i. w. S.)*
nitrogen oxides Nitrosegase *npl*
nitrogen pentoxide Stickstoffpentoxid *n*, Distickstoffpentoxid *n*, Stickstoff(V)-oxid *n*
nitrogen removal Abgasentstickung *f*
nitrogen tetraoxide [tetroxide] Distickstofftetroxid *n*, Stickstofftetroxid *n*
nitrogen trioxide Stickstofftrioxid *n*, Distickstofftrioxid *n*, Stickstoff(III)-oxid *n*
nitrogenous stickstoffhaltig, Stickstoff enthaltend, Stickstoff...
nitroglycerine Glyceroltrinitrat *n*, *(veraltet:)* Nitroglycerin *n*
nitromethane Nitromethan *n*
nitronaphthalene Nitronaphthalen *n*
nitrone Nitron *n (N-Oxid einer schiffschen Base)*
nitronic acid Nitronsäure *f*, *aci*-Nitroverbindung *f*
nitronium ion NO_2^+ Nitroniumion *n*, Nitrylion *n*
nitroparaffin *Nitroalkan *n*, Nitroparaffin *n*
nitroprusside Nitroprussid *n*, Pentacyanonitrosylferrat *n*
nitrosamide Nitrosamid *n*
nitrosamine Nitrosamin *n*
nitrosate *v* nitrosieren, die Nitrosogruppe einführen *(in eine organische Verbindung)*
nitrosate *(org Ch)* Nitrosat *n*
nitrosating agent Nitrosier(ungs)mittel *n*, nitrosierendes Agens *n*
nitrosation Nitrosierung *f*
nitroso compound Nitrosoverbindung *f*
nitroso dye Nitrosofarbstoff *m*
nitroso group –N=O Nitrosogruppe *f*
nitrosobenzene Nitrosobenzen *n*
***p*-nitrosophenol** *p*-Nitrosophenol *n*
nitrostarch Nitrostärke *f*, Stärkenitrat *n (Sprengstoff)*
nitrosyl ion NO^+ Nitrosylion *n*
nitrous 1. nitros, nitrosehaltig *(Stickoxid enthaltend)*; 2. salpeterhaltig; salpetrig, salpeterartig, Salpeter...; 3. Stickstoff...; niederwertigem Stickstoff entsprechend, *(meist)* Stickstoff(III)-...
nitrous acid salpetrige Säure *f*
nitrous gases nitrose Gase *npl (Gemisch aus Luft und Stickstoffoxiden)*
nitrous oxide Stickstoff(I)-oxid *n*, Distickstoffoxid *n*
NMR *(nuclear magnetic resonance)* NMR, kernmagnetische Resonanz *f*, magnetische Kernresonanz *f*
nmr *(nuclear magnetic resonance)* NMR, kernmagnetische Resonanz *f*, magnetische Kernresonanz *f*
NMR spectrometer NMR-Spektrometer *n*, Kernresonanzspektrometer *n*
NMR spectroscopy NMR-Spektroskopie *f*, kernmagnetische Resonanzspektroskopie *f*, (magnetische) Kernresonanzspektroskopie *f*
NMR spectrum NMR-Spektrum *n*, kernmagnetisches Resonanzspektrum *n*, magnetisches Kernresonanzspektrum *n*
no-effect-level Dosis *f* ohne toxische Wirkung
no observed effect level nachweisbare Dosis *f* ohne toxische Wirkung
nobelium No Nobelium *n*
nobility *s.* noble character
noble character *(Ech)* edler [stark elektropositiver] Charakter *m*
noble gas Edelgas *n*
noble-gas clathrate Edelgasclathrat *n*, Edelgaseinschlussverbindung *f*
noble-gas configuration Edelgaskonfiguration *f*
noble metal edles Metall *n*, Edelmetall *n*
nodal plane Knotenfläche *f*, Knotenebene *f (Orbitaltheorie)*
nodular *(Min)* knollig
nodular cast iron Kugelgraphit(grau)-guss *m*, sphärolithischer [globularer] Grauguss *m*, GGG
nodule-forming bacteria Knöllchenbakterien *npl (Rhizobium-Arten)*
nodulize *v* pellet(is)ieren, zu Pellets [Kügelchen] formen

NOE *s.* nuclear Overhauser effect
NOEL *s.* no observed effect level
noise *(Anal)* Rauschen *n*
noise level Rauschpegel *m*
nomenclature Nomenklatur *f*
nomenclature rule Nomenklaturregel *f*
nomenclature system Nomenklatursystem *n*
non-adiabatic reactor nicht adiabater Reaktor *m*
non-ageing nicht alternd, alterungsbeständig, alterungssicher
non-aqueous chemistry Chemie *f* in nicht wässrigen Systemen
non-aqueous solvent nicht wässriges Lösungsmittel *n*
non-aqueous titration Titration *f* in nicht wässriger Lösung, Titration *f* in wasserfreiem Medium
non-aromatic nicht aromatisch
non-benzenoid nicht benzoid
non-biodegradable compounds [organics] biologisch [biochemisch] nicht [schwer] abbaubare organische Verbindungen *fpl* [Inhaltsstoffe *mpl*], resistente [abbauresistente, persistente, biochemisch stabile] organische Verbindungen *fpl* [Inhaltsstoffe *mpl*]
non-bonding electron einsames [nicht bindendes] Elektron *n*
non-bonding orbital nicht bindendes Orbital *n*
non-carbonate hardness *(Hyd)* Nichtkarbonathärte *f*, NKH, permanente [bleibende] Härte *f*
non-catalytic reaction unkatalysierte [nicht katalytische, nicht katalysierte] Reaktion *f*
non-combustible nicht brennbar, un(ver)brennbar
non-combustible (matter) Unverbrennbares *n*, Unverbrennliches *n*
non-competitive inhibition nicht kompetitive Hemmung *f (Enzymkinetik)*
non-conducting nicht leitend, dielektrisch
non-conductor Nichtleiter *m*, Dielektrikum *n*
non-conjugated nicht konjugiert
non-contaminated unverschmutzt, nicht verschmutzt [belastet] *(Wasser, Luft)*
non-corroding nicht korrodierend, korrosionsbeständig, korrosionsfest
non-corrosive nicht korrosiv [aggressiv], korrosionsinaktiv
non-crystalline nicht kristallin, amorph, gestaltlos, formlos, ohne Kristallform
non-cumulative nicht kumuliert
non-cyclic(al) nicht zyklisch *(z. B. Reaktion)*; acyclisch *(chemische Verbindung)*
non-Daltonian compounds *s.* non-stoichiometric compounds
non-daltonide compounds *s.* non-stoichiometric compounds
non-degenerate nicht entartet
non-degradable nicht abbaubar, abbauresistent
non-destructive testing zerstörungsfreie Prüfung *f*
non-discolouring nicht verfärbend, farbbeständig
non-drying oil nicht trocknendes Öl *n*
non-eclipsed conformation gestaffelte [± antiperiplanare] Konformation *f*
non-electroactive elektrochemisch inaktiv
non-electrolyte Nichtelektrolyt *m*
non-enzym(at)ic browning (reaction) *(Lebm)* nicht enzymatische Bräunungsreaktion *f*, Maillard-Reaktion *f*, Amin-Zucker-Bräunung *f*, Carbonylaminoreaktion *f*
non-equilibrium Ungleichgewicht *n*
non-equilibrium thermodynamics Thermodynamik *f* der Nichtgleichgewichtsprozesse, irreversible Thermodynamik *f*, Thermodynamik *f* irreversibler Prozesse
non-essential *(Bioch)* nicht essenziell, entbehrlich
non-exchangeable nicht austauschbar
non-explosive nicht explosiv [explosibel]
non-faradaic admittance nicht faradaysche Admittanz *f*
non-ferrous alloy NE-Legierung *f*
non-ferrous metal Nichteisenmetall *n*, NE-Metall *n*
non-ferrous metallurgy Nichteisenmetallurgie *f*, Nichteisen(metall)hüttenwesen *n*
non-fissile nicht spaltbar
non-fissionable nicht spaltbar

non-flammability Nichtentflammbarkeit *f*, Unentflammbarkeit *f*, Flammwidrigkeit *f*, Flammbeständigkeit *f*
non-flammable nicht entflammbar [brennbar], unentflammbar flammwidrig, flammbeständig
non-graphitic carbon nicht graphitischer Kohlenstoff *m*
non-haem iron *(Bioch)* Nichthämeisen *n*, Nicht-Hämineisen *n*, nicht-hämgebundenes Eisen *n*
non-haem iron protein Nichthämeisenprotein *n*, Nicht-Hämin-Eisenprotein *n*, NHI-Protein *n*
non-harmonic vibration *(Spektr)* anharmonische Schwingung *f*
non-heme ... *siehe: non-haem ...*
non-homogeneity Inhomogenität *f*
non-hygroscopic nicht hygroskopisch
non-ideal flow nicht ideale [reale] Strömung *f*
non-ideal solution reale Lösung *f*
non-insecticidal insektenverträglich, insektenschonend, unschädlich für Insekten *(z. B. Fungizide)*
non-ionic nicht ionogen [ionisch], ioneninaktiv, nicht ionisierend
non-ionic nicht ionogener (grenzflächenaktiver) Stoff *m*, nicht ionogenes Tensid *n*
non-isothermal reactor nicht isothermer Reaktor *m*
non-isotropic anisotrop
non-knocking nicht klopfend, klopffrei
non-luminous flame Heizflamme *f*, nicht leuchtende Flamme *f*
non-magnetic nicht magnetisch, unmagnetisch
non-mediated process *(Bioch)* nicht katalysierter [vermittelter] Transportprozess *m (durch Membranen hindurch)*
non-metal Nichtmetall *n*
non-metallic nicht metallisch, Nichtmetall...
non-miscible nicht mischbar
non-natural künstlich, synthetisch
non-Newtonian behaviour nicht-newtonsches Verhalten *n (Rheologie)*
non-Newtonian fluid [liquid] nichtnewtonsche Flüssigkeit *f*
non-noble metal Unedelmetall *n*, unedles Metall *n*
non-nutritional sweetener Süßstoff *m*
non-nutritive sweetener Süßstoff *m*
non-odorous geruchlos
non-oxidative nicht oxidierend
non-oxidizing nicht oxidierend
non-pathogenic nicht pathogen, apathogen
non-persistent pesticide nicht persistentes Pflanzenschutzmittel *n*
non-phenolic nicht phenolisch
non-phytotoxic pflanzenverträglich
non-pigmented *(Anstr)* unpigmentiert, nicht pigmentiert
non-poisonous ungiftig, nicht giftig [toxisch], atoxisch
non-polar 1. nicht polar, unpolar, apolar; 2. unpolar, homöopolar *(chemische Bindung)*
non-polar bond homöopolare [kovalente] Bindung *f*, Elektronenpaarbindung *f*, Atombindung *f*
non-polarizable unpolarisierbar, nicht polarisierbar
non-proprietary name nicht wortgeschützter Name *m*, Freiname *m*, freier Warenname *m*
non-protein fraction [moiety] Nichtprotein(an)teil *m*, Nichteiweißanteil *m*, Nichteiweißkomponente *f (z. B. von Enzymen)*
non-protein nitrogen Nichtproteinstickstoff *m*, Nichteiweißstickstoff *m*
non-purgeable organic halides *(Hyd)* nicht austreibbare Menge *f* an Organohalogenverbindungen
non-quantitative unvollständig, unvollkommen *(z. B. Fällung)*
non-quinonoid nicht chinoid
non-radiative strahlungslos, strahlungsfrei
non-radioactive isotope stabiles [nicht radioaktives] Isotop *n*
non-reactive reaktionsunfähig, nicht reaktionsfähig
non-reactivity Reaktionsunfähigkeit *f*
non-reducing nicht reduzierend
non-residue cracking rückstandsloses Kracken *n*, Kracken *n* auf Koksrückstand, Kracken *n* mit Kokungsarbeitsweise, Kracken *n* nach der Verkokungsfahrweise
non-returnable Einweg...

non-reversible irreversibel, nicht umkehrbar [rückläufig]
non-rigid plastic weicher [weich gestellter] Kunststoff *m*
non-saponifiable unverseifbar, nicht verseifbar
non-scaling 1. zunderbeständig, zunderfrei; 2. weich *(Wasser)*
non-selective herbicide nicht selektives Herbizid *n*, allgemein [total] wirkendes Herbizid *n*, Totalherbizid *n*
non-sludging oil „nicht verharzendes" Öl *n (bei Luftzutritt keinen Schlamm bildend)*
non-solvent Nichtlöser *m*, inaktiver Löser *m*, inaktives Lösungsmittel *n*, Nichtlösungsmittel *n*
non-specific unspezifisch
non-staining nicht verfärbend
non-steady-flow reactor nicht stationär arbeitender Reaktor *m*
non-steroid compound nicht-steroide Verbindung *f*
non-stoichiometric compounds nicht stöchiometrische [daltonide] Verbindungen *fpl*, berthollide Verbindungen *fpl*, Berthollide *npl*
non-stoichiometry Nichtstöchiometrie *f*, Unstöchiometrie *f*
non-sugar nicht zuckerartig, zuckerfrei
non-sugar portion Nichtzuckeranteil *m*
non-sugar substance Nichtzuckerstoff *m*
non-superimposable nicht deckungsgleich *(Stereochemie)*
non-symbiotic nitrogen fixation *(Landw)* nicht symbio(n)tische Stickstoffbindung *f* [Fixierung *f* von Luftstickstoff], Stickstoffbindung *f* durch freilebende Mikroorganismen
non-systemic *(Landw)* nicht systemisch *(nicht über das Gefäßsystem der Pflanze wirkend)*
non-toxic nicht toxisch, atoxisch, ungiftig, nicht giftig
non-toxicity Ungiftigkeit *f*
non-uniform uneinheitlich, nicht einheitlich
non-uniformity Uneinheitlichkeit *f*, Nichteinheitlichkeit *f*
non-variant nonvariant, invariant, ohne Freiheit(sgrad)
non-viscous flow reibungsfreie Strömung *f*
non-viscous fluid ideales [reibungsfreies, pascalsches] Fluid *n*
non-volatile matter nicht flüchtiger Stoff *m*, nicht flüchtige Bestandteile *mpl*, Nichtflüchtiges *n*
non-volatile total organic carbon *(Hyd)* nicht flüchtiger organisch gebundener Kohlenstoff *m*
non-weathering nicht verwitternd
non-wettable unbenetzbar, nicht benetzbar
non-wovens Textilverbundstoffe *mpl*, Vliesstoffe *mpl*, Vlieswaren *fpl*, ungewebte Textilien *pl*
non-yellowing vergilbungsbeständig, nicht (ver)gilbend, nicht gelb werdend
nonacosane Nonacosan *n*
nonanedioic acid *Nonandisäure *f*, *Heptan-1,7-dicarbonsäure *f*, Azelainsäure *f*
nonanoic acid Nonansäure *f*
***n*-nonyl alcohol** Nonylalkohol *m*, *Nonan-1-ol *n*
1-nonyne Non-1-in *n*
norbornene Norbornen *n*, *Bicyclo[2,2,1]hept-2-en *n*
normal alkane unverzweigtes [geradkettiges] Alkan *n*
normal alkyl group *n*-Alkylgruppe *f*
normal calomel electrode Normalkalomelelektrode *f*, Kalomelnormalelektrode *f*
normal distribution *(Anal)* Gauß-Verteilung *f*, (gaußsche) Normalverteilung *f*
normal distribution function *(Anal)* gaußsche [normale] Verteilungsfunktion *f*
normal electrode Normalelektrode *f*
normal hydrogen electrode Normalwasserstoffelektrode *f*, Wasserstoffnormalelektrode *f*
normal mode of vibration Normalschwingung *f*, Grundschwingung *f*, Eigenschwingung *f*
normal potential Normalpotenzial *n*
normal pressure Norm(al)druck *m*, Standarddruck *m*, Normalluftdruck *m*
normal-pressure synthesis Normaldrucksynthese *f*
normal salt normales [neutrales] Salz *n*, Neutralsalz *n*

normal solution 1. Normallösung *f*, normale Lösung *f*, n-Lösung *f*; 2. *(Med, Bio)* isotonische Lösung *f*
normal state Normalzustand *m*, Grundzustand *m*, Grundniveau *n*, Grundterm *m (eines Atoms)*
normal stress Normalspannung *f (senkrecht zum Querschnitt)*
normal temperature Norm(al)temperatur *f*, *(i. e. S.)* Zimmertemperatur *f*, Raumtemperatur *f*, gewöhnliche Temperatur *f*
normal temperature and pressure Norm(al)temperatur *f* und -druck *m*, physikalischer Normzustand *m (Standardbedingungen von 0 °C und 1,01325 10^5 Pa)*
normal volume Norm(al)volumen *n*
normality Normalität *f (ein Konzentrationsmaß)*
normalize *v (Met)* normalglühen, normalisierend glühen, normalisieren
normalizing *(Met)* Normal(isierungs)glühen *n*, normalisierendes Glühen *n*, Normalisieren *n*
Norrish type I photoreaction Norrish-Typ-I-Reaktion *f (α-Spaltung einer Carbonylverbindung)*
Norrish type II photoreaction Norrish-Typ-II-Reaktion *f (Verschiebung eines γ-Wasserstoffatoms einer Carbonylverbindung)*
nose irritant Nasen-Rachen-Reizstoff *m*
notate *v* bezeichnen
notation Bezeichnung *f*; Bezeichnungsweise *f*; Notation *f (Kurzdarstellung chemischer Strukturen, besonders zur Speicherung in Datenverarbeitungsanlagen)*
notation system Bezeichnungssystem *n*, Bezeichnungsweise *f*; Notationssystem *n*
notch Kerbe *f*, Kerb *m*
notch impact resistance [strength] Kerbschlagzähigkeit *f*
notch-sensitive kerbempfindlich
novolak (resin) Novolak *m*, Novolakharz *n*
NOx *s.* nitrogen oxides
noxious schädlich
noxious-smelling übel riechend
nozzle *v* 1. bedüsen *(mit Flüssigkeiten)*; 2. mit Düsen versehen
nozzle Düse *f*
NPOX *s.* non-purgeable organic halides
NR *s.* natural rubber
NSSC = neutral sodium sulphite semichemical
NSSC process *(Pap)* Neutralsulfitverfahren *n*, NSSC-Verfahren *n*
NTA *s.* nitrilotriacetic acid
NTP *s.* normal temperature and pressure
N.T.P. *s.* normal temperature and pressure
nuclear nuklear, Kern...
nuclear chain reaction nukleare Kettenreaktion *f*, Kernkettenreaktion *f*
nuclear charge Kernladung *f*
nuclear chemistry Kernchemie *f*
nuclear collision Kern(zusammen)stoß *m*
nuclear constituent Kernbaustein *m*, Kernteilchen *n*, Nukleon *n*
nuclear cross section Wirkungsquerschnitt *m* des Atomkerns [Kerns], nuklearer Wirkungsquerschnitt *m*
nuclear decay Kernzerfall *m*
nuclear disintegration *s.* 1. nuclear decay; 2. nuclear fission
nuclear energy Kernenergie *f*, Atomenergie *f*
nuclear energy level Kern(energie)niveau *n*, Kernterm *m*
nuclear excitation Kernanregung *f*
nuclear fission (gesteuerte) Kernspaltung *f*, Atomkernspaltung *f*
nuclear fragments Kernbruchstücke *npl*
nuclear fuel Kernbrennstoff *m*
nuclear-fuel reprocessing Brennstoffwiederaufbereitung *f*, Brennstoffwiederaufarbeitung *f*
nuclear fusion Kernfusion *f*, Kernverschmelzung *f*
nuclear isobar Kernisobar *n*, Isobar *n*, isobare Atomart *f*
nuclear isomer 1. *(Kern)* Kernisomer(e) *n*; 2. *(org Ch)* Skelettisomer(e) *n*, Kettenisomer(e) *n*
nuclear isomerism 1. *(Kern)* Kernisomerie *f*; 2. *(org Ch)* Skelettisomerie *f*, Kettenisomerie *f*
nuclear magnetic moment magnetisches Kernmoment *n* [Moment *n* des Kerns], kernmagnetisches Moment *n*
nuclear magnetic resonance kernma-

gnetische Resonanz *f*, magnetische Kernresonanz *f*, NMR
nuclear magneton Kernmagneton *n*, KM
nuclear mass Kernmasse *f*
nuclear mass number Masse(n)zahl *f*, Kernmasse(n)zahl *f*
nuclear Overhauser effect Kern-Overhauser-Effekt *m*, NOE
nuclear particle Kernbaustein *m*, Kernteilchen *n*, Nukleon *n*
nuclear physics Kernphysik *f*
nuclear quadrupole moment Kernquadrupolmoment *n*, Quadrupolmoment *n* des Kerns
nuclear quadrupole resonance spectroscopy Kernquadrupolresonanzspektroskopie *f*
nuclear radiation Kernstrahlung *f*
nuclear reaction Kernreaktion *f*
nuclear reactor Kernreaktor *m*
nuclear resonance Kernresonanz *f*
nuclear spin (angular momentum) Kernspin *m*, Kerndrehimpuls *m*
nuclear spin quantum number Kernspinquantenzahl *f*
nuclear transformation [transmutation] Kernumwandlung *f*
nuclear waste radioaktiver Abfall *m*, *(i. e. S.)* Atommüll *m*
nuclear waste disposal Beseitigung *f* des radioaktiven Abfalls, *(i. e. S.)* Atommüllbeseitigung *f*
nucleate *v* 1. als Keim wirken; 2. *(Krist)* als Kristall(isations)keim wirken; Keime bilden, Kristall(isations)keime bilden *(aus eigenem Material)*; als Kern wirken, als Kristallisationskern wirken; Kerne bilden, Kristall(isations)kerne bilden *(aus fremdem Material)*; 3. einleiten, auslösen, anregen, initiieren, starten, in Gang bringen *(eine Reaktion)*
nucleation 1. Keimbildung *f*; 2. *(Krist)* Kristall(isations)keimbildung *f (aus eigenem Material)*; Kernbildung *f*, Kristall(isations)kernbildung *f (aus fremdem Material)*; 3. Einleitung *f*, Auslösung *f*, Anregung *f*, Initiierung *f (einer Reaktion)*
nucleation agent Keimbildner *m*, keimbildendes Mittel *n*, keimbildender Zusatz *m*
nucleation and growth Keimbildung *f* und -wachstum *n*; Bergabdiffusion *f (in einer Legierung)*
nucleation centre [site] Kristallisationszentrum *n*; Kondensationszentrum *n (beim Phasenübergang dampfförmig - flüssig)*
nucleator *s.* nucleation agent
nucleic acid *(Bioch)* Nucleinsäure *f*
nucleofuge nucleofuge Gruppe *f*
nucleohistone *(Bioch)* Nucleohiston *n*
nucleon Nukleon *n*, Kernteilchen *n*, Kernbaustein *m*
nucleonics Nukleonik *f*
nucleophile Nucleophil *n*, nucleophiles [anionisches] Reagens *n*
nucleophilic nucleophil
nucleophilic substitution nucleophile [anionoide] Substitution *f*, SN-Reaktion *f*, nucleophiler Austausch *m*
nucleophilicity Nucleophilie *f*, *(quantitativ auch:)* nucleophile Kraft *f* [Potenz *f*]
nucleoprotein Nucleoprotein *n*, Nucleoproteid *n*
nucleoside Nucleosid *n*
nucleoside antibiotic *(Biot)* Nucleosid-Antibiotikum *n*
nucleotide Nucleotid *n*, Nucleosidphosphat *n*
nucleotide-excision repair Nucleotid-Exzisionsreparatur *f*
nucleotide sequence Nucleotidsequenz *f*, Basensequenz *f*
nucleus 1. Nukleus *m*, Atomkern *m*, Kern *m*; 2. *(org Ch)* Kern *m*, Ring *m*; 3. *(Bio)* Nucleus *m*, Zellkern *m*, Kern *m*; 4. *s.* nucleus of crystallization
nucleus of crystallization Kristall(isations)keim *m*, Eigenkeim *m*; Kristall(isations)kern *m*, Fremdkeim *m*
nuclide Nuklid *n*
nuclidic mass Nuklid(en)masse *f*
nuisance threshold Geruchsschwelle *f*
null-balance instrument Nullinstrument *n*, Ausgleich(mess)instrument *n*
null method (of measurement) Nullmethode *f*, Kompensationsmethode *f*, Abgleichmethode *f*
number *v* nummerieren, beziffern, durchnummerieren
number 1. Zahl *f*; Ziffer *f*; Nummer *f*; 2. Anzahl *f*, Menge *f*
number average Zahlenmittel *n*

number average molecular weight zahlenmittlere Molmasse *f*
number of actual plates [trays] *(Dest)* praktische [wirkliche] Bodenzahl *f*
number of theoretical plates [trays] *(Dest)* theoretische Trennstufenzahl *f* [Bodenzahl *f*]
Nusselt number Nusselt-Zahl *f*, Nusselt-Kennzahl *f (des Wärmeübergangs)*
nutrient nährend, nahrhaft, Nähr...; Nährstoff...
nutrient Nährstoff *m*, Nahrungsstoff *m*
nutrient agar Nähragar *m(n)*, Agarnährboden *m*
nutrient balance Nährstoffbilanz *f*
nutrient broth Nährbouillon *f*, Nährflüssigkeit *f*, Nährlösung *f*, Kulturflüssigkeit *f*, Kulturlösung *f (für Mikroorganismen)*
nutrient deficiency Nährstoffmangel *m*, Nährstoffarmut *f*
nutrient demand *s.* nutrient requirements
nutrient element Nähr(stoff)element *n*
nutrient enrichment Nährstoffanreicherung *f*
nutrient line *(Bioch)* Nährstofflinie *f (im Periodensystem)*
nutrient medium Nährmedium *n*, Nährsubstrat *n*
nutrient ratio Nährstoffverhältnis *n*
nutrient requirements Nährstoffbedarf *m*, Nährstoffansprüche *mpl*
nutrient salt Nährsalz *n*
nutrient solution 1. Nährlösung *f (Pflanzenernährung)*; 2. *s.* nutrient broth
nutrient supply Nährstoffversorgung *f*, Nährstoffzufuhr *f*, Nährstoffnachlieferung *f*
nutrient value Nährwert *m*
nutrition science Ernährungswissenschaft *f*
nutrition scientist Ernährungswissenschaftler *m*
nutritional die Ernährung betreffend, Ernährungs..., Nahrungs..., Nähr...
nutritional disease ernährungsabhängige Krankheit *f*
nutritional quality Nährwert *m*
nutritional supplement Futtermittelzusatz(stoff) *m*, Futterzusatz *m*, Futtersupplement *n*
nutritional value Nährwert *m*
nutritionally indispensable amino acid essenzielle Aminosäure *f*
nutritionist *s.* nutrition scientist
nutritious 1. *(Lebm)* nahrhaft, nährend; 2. *(Landw)* nährstoffreich
nutritive 1. Ernährungs..., Nahrungs..., Nähr...; 2. *s.* nutritious
nutritive Nährstoff *m*, Nahrungsstoff *m*
nutritive value Nährwert *m*
nutsch (filter) Filternutsche *f*, Nutsche *f*
nutsche Filternutsche *f*, Nutsche *f*
nux vomica Brechnuß *f*, Krähenauge *n (von Strychnos nuxvomica L.)*
NVTOC *s.* non-volatile total organic carbon

O

O-P toxicant toxisches Phosphororganikum *n*, toxisches organisches Phosphorderivat *n*
o-rh. *s.* orthorhombic
obey *v* **a law** ein Gesetz befolgen, einem Gesetz gehorchen [folgen]
object to be weighed Wägegut *n*, abzuwägende Substanz *f*
objectionable 1. zu beanstanden(d), störend, schädlich; 2. unangenehm, schlecht, aufdringlich, widerlich, übel *(Geruch)*
obligatory reactant [substrate] obligatorisches [führendes] Substrat *n (Enzymkinetik)*
obnoxious übel riechend, unangenehm [widerlich] riechend
observational error Beobachtungsfehler *m*, Ablesefehler *m*
observed effect level/no nachweisbare Dosis *f* ohne toxische Wirkung
obtain *v* erhalten, gewinnen
obtain *v* **pure** rein erhalten, in reinem Zustand gewinnen
obtainable commercially im Handel erhältlich
occlude *v* okkludieren, einschließen
occlusion Okklusion *f*, Einschluss *m*
occupation number Besetzungszahl *f*
occupational disease Berufskrankheit *f*
occupational hygiene Arbeitshygiene *f*
occupational hygienist Arbeitshygieniker *m*
occupy *v* besetzen, einnehmen *(ein*

Energieniveau, einen Gitterplatz); einnehmen *(ein bestimmtes Volumen)*
occur *v* vorkommen, auftreten, sich finden
occurrence Vorkommen *n*, Auftreten *n*
ocean disposal Beseitigung *f* durch Einleiten ins Meer; Beseitigung *f* durch Abklappen [Verklappung, Verkippen ins Meer]
ocher *(Min)* Ocker *m*
ochre *(Min)* Ocker *m*
ochreous ockerhaltig
oct. *s.* octahedral
9,12-octadecadienoic acid *Octadeca-9,12-diensäure *f*, Linolsäure *f*
***n*-octadecanoic acid** *Octadecansäure *f*, Stearinsäure *f*
1-octadecanol *Octadecan-1-ol *n*, Stearylalkohol *m*
9,12,15-octadecatrienoic acid *Octadeca-9,12,15-triensäure *f*, Linolensäure *f*
***cis*-9-octadecenoic acid** **cis*-Octadec-9-ensäure *f*, Ölsäure *f*
9-octadecynoic acid *Octadec-9-in-säure *f*, Stearolsäure *f*
octafluoride Octafluorid *n*
octahedral *(Krist)* octaedrisch, achtflächig
octahedral orientation octaedrische Anordnung *f (von Liganden)*
octahedron *(Krist)* Octaeder *n*, (regelmäßiger, regulärer) Achtflächner *m*
octahydrate Octahydrat *n*
octahydride Octahydrid *n*
octamer Octamer(e) *n*
octane Octan *n*
octane number Octanzahl *f*, OZ *f*
octane rating 1. Octanzahlbestimmung *f*, OZ-Bestimmung *f*; 2. *s.* octane number
octanoic acid *Octansäure *f*, Heptan-1-carbonsäure *f*
1-octanol *Octan-1-ol *n*, Octylalkohol *m*
octasubstituted octasubstituiert
octavalency Achtwertigkeit *f*, Octavalenz *f*
octavalent achtwertig, octavalent
octet Oktett *n*, Achtergruppe *f*, Achtergruppierung *f*, Achterschale *f*
octet expansion Oktettaufweitung *f*
octet rule Oktettregel *f*, Oktettprinzip *n*
octet structure Elektronenoktett-Anordnung *f*, Edelgaskonfiguration *f*
octose Octose *f (Monosaccharid mit 8 Kohlenstoffatomen)*
octoxide Octoxid *n*
1-octyne *Oct-1-in *n*
OD *s.* oven-dry
O.D. = outer diameter
odd electron ungepaartes Elektron *n*
odd-membered ring ungeradzahliger Ring *m*
odd-number(ed) ungeradzahlig *(z. B. Fettsäure)*
odorant duftend, wohlriechend, Duft..., Riech...
odorant Odorans *n*, Odor(is)ierungsmittel *n*, Odoriermittel *n*
odorization Odor(is)ierung *f*, Odor(is)ieren *n (von Stadtgas, Erdgas)*
odorize *v* odor(is)ieren *(Stadtgas, Erdgas)*
odorous *s.* odorant
odour 1. Geruch *m (neutral oder unangenehm)*; 2. *s.* odour-causing substance
odour-causing substance geruchsbildender [geruchsaktiver] Stoff *m*, Geruchsbildner *m*, Geruchsstoff *m*, Riechstoff *m*, Osmogen *n*
odour removal Geruchsbeseitigung *f*, Entfernung *f* von Geruchsbeeinträchtigungen
odour threshold *(Hyd)* Geruchsschwellenkonzentration *f*, RSK, Geruchsschwelle *f*
odourless geruchslos, geruch(s)frei, nicht riechend
OEM = original equipment manufacturer
oenological önologisch, weinfachkundlich
oenology Önologie *f*, Weinkunde *f*
OES *s.* optical emission spectroscopy
oestradiol Östradiol *n (ein Sexualhormon)*
oestrogen Östrogen *n*, östrogenes Hormon *n*, Follikelhormon *n*
off-flavour Geschmacks- und Geruchsfehler *m*, Geschmacks- und Geruchsabweichung *f*, Offflavour *m*
off-gas Abgas *n*
offensive aggressiv
offset *v* ausgleichen, kompensieren

offset bleibende [dauernde] Dehnung *f* *(Werkstoffprüfung)*
0.2 % offset yield strength [stress] 0,2-Dehngrenze *f*, 0,2-%-Grenze *f*, 0,2-Grenze *f*
offshore disposal Hochseeverklappung *f (von Abfällen)*
offtake Entnahme *f*
offtake pipe Austragrohr *n*, Ableitungsrohr *n*, Austrittsrohr *n*, Abzugsrohr *n*
OH-form exchanger *(Hyd)* OH-Austauscher *m*, $(OH)_2$-Austauscher *m*, Ionenaustauscher *m* in der Hydroxylform [OH-Form, $(OH)_2$-Form], Anionenaustauscher *m*
OH-furnace *s.* open-hearth furnace
OH group OH-Gruppe *f*, OH-Rest *m*, Hydroxylgruppe *f*
OH radical 1. (freies) OH-Radikal *n*, (freies) Hydroxylradikal *n*; 2. *s.* OH group
ohmic continuity galvanischer Durchgang *m*
ohmic drop [overpotential, polarization] ohmscher Spannungsabfall *m*, IR-Abfall *m*, iR-Abfall *m*, Widerstandspolarisation *f*, Widerstandsüberspannung *f*
oil *v* ölen; schmieren
oil-harden *v* ölhärten, in Öl härten
oil-quench *v* in Öl abschrecken [ablöschen]
oil Öl *n*
oil-base(d) paint Ölfarbe *f*
oil bath Ölbad *n*
oil-bearing 1. *(Geol)* (erd)ölführend, (erd)ölhaltig; 2. ölhaltig *(pflanzliches Gewebe)*
oil-bearing seed Ölsaat *f*, ölhaltige Samen *mpl*
oil cake Ölkuchen *m*, Presskuchen *m*
oil cracking Ölkracken *n*, Ölspaltung *f*
oil deposit Erdöllagerstätte *f*, Erdölvorkommen *n*
oil-extended rubber ölgestreckter [ölplastizierter, ölhaltiger] Kautschuk *m*
oil field Ölfeld *n*, Erdölfeld *n*
oil gas Ölgas *n*
oil gasification Ölvergasung *f*
oil genesis Erdölgenesis *f*, Erdölentstehung *f*, Erdölbildung *f*
oil green Chromoxidgrün *n*, Ölgrün *n*, grüner Zinnober *m*, Laubgrün *n* *(Chrom(III)-oxid)*
oil hardening Ölhärten *n*, Ölhärtung *f*, Härten *n* in Öl
oil-in-water emulsion Öl-in-Wasser-Emulsion *f*, Ö/W
oil length *(Anstr)* Ölgehalt *m (bezogen auf Harz)*; Verhältnis *n* Öl/Harz
oil-modified resin ölmodifiziertes Harz *n*
oil of bay Bayöl *n (von Pimenta racemosa (Mill.) J. W. Moore)*
oil of bitter almond(s) Bittermandelöl *n*
oil of cloves Nelkenöl *n*, Gewürznelkenöl *n (von Syzygium aromaticum (L.) Merr. et L. M. Perry)*
oil of hartshorn ätherisches [dippelsches] Tieröl *n (aus Knochen oder Knochenteer)*
oil of juniper Wacholder(beer)öl *n (von Juniperus communis L.)*
oil of lavender Lavendelöl *n (von Lavandula-Arten)*
oil of mirbane *(Kosmet)* Mirbanöl *n*, Mirbanessenz *f (Nitrobenzen)*
oil of mustard Senföl *n*
oil of nutmeg ätherisches Muskatnußöl *n (von Myristica fragrans Houtt.)*
oil of roses Rosenöl *n*
oil of sassafras Sassafrasöl *n (von Sassafras albidum (Nutt.) Nees)*
oil of thyme Thymianöl *n (von Thymus vulgaris L. und T. zygis L.)*
oil of turpentine Terpentinöl *n*
oil of wintergreen Wintergrünöl *n (von Gaultheria procumbens L.)*
oil paint Ölfarbe *f*
oil plant Ölpflanze *f*, Ölfrucht *f*
oil quenching Ölabschrecken *n*, Abschrecken *n* in Öl, Ölablöschen *n*
oil-reactive resin ölreaktives Harz *n*
oil refinery Erdölraffinerie *f*
oil removal *(Hyd)* Ölabscheidung *f*, Ölabtrennung *f*
oil resistance Ölfestigkeit *f*, Ölbeständigkeit *f*
oil-resistant ölfest, ölbeständig
oil-resisting ölfest, ölbeständig
oil scrubbing *s.* oil washing
oil separator Ölabscheider *m*, Ölseparator *m*, Ölfänger *m*
oil-soluble öllöslich
oil-source rock Erdölmuttergestein *n*
oil spill Ölhavarie *f*

oil tanning Sämischgerbung *f*, *(i. w. S.)* Fettgerbung *f*
oil-to-resin ratio Öl-Harz-Verhältnis *n*, Verhältnis *n* von Öl zu Harz
oil varnish Öllack *m*; Ölfirnis *m* *(in der grafischen Technik)*
oil washing Ölwäsche *f*, Auswaschen *n* mit Waschöl
oil-water interface Grenzfläche *f* Öl gegen Wasser
oil well Ölbohrloch *n*, Ölbohrung *f*, Ölbrunnen *m*, Erdölbohrloch *n*, Erdölbohrung *f*; Ölquelle *f*, Erdölquelle *f*
oil yellow 1. *s.* oil yellow II; 2. *allgemein für einen synthetischen, fettlöslichen, gelben Farbstoff*
oil yellow II Buttergelb *n*, Dimethylgelb *n*, *p*-Dimethylaminoazobenzen *n*
oilcloth Wachstuch *n*
oiliness 1. Öligkeit *f*, Fettigkeit *f*; 2. *(Lebm)* altöliger Geschmack *m* *(verdorbener fetter Öle)*; Peroxidranzigkeit *f* *(von fetten Ölen)*; 3. Schmierfähigkeit *f*, Schmiergüte *f*, Schmierwert *m*, Öligkeit *f*, Schmierergiebigkeit *f* *(eines Schmieröls)*
oilproof öldicht, ölundurchlässig *(Papier)*
oiltight öldicht *(mechanische Verbindungsstellen)*
oily öl(art)ig; ölhaltig, Öl...
ointment *(Pharm)* Salbe *f*
ol bridge *s.* olation bridge
olate *v* verolen
olation Verolung *f* *(Bildung von Nebenvalenzbrücken zwischen Metallatomen und OH-Gruppen)*
olation bridge Olbrücke *f*
oleaginous öl(art)ig; ölhaltig, Öl...
oleate Oleat *n* *(Salz oder Ester der Ölsäure)*
OLED *s.* organic light emitting diode
olefin (hydrocarbon) Olefin *n*, *Alken *n*, Ethylenkohlenwasserstoff *m*
olefin metathesis *(org Ch)* Olefinmetathese *f*
olefin series Olefinreihe *f*, Alkenreihe *f*, Klasse *f* der Alkene
olefination Olefinierung *f*
olefinic olefinisch, Olefin..., *Alken...
olefinic linkage olefinische Doppelbindung *f*, C=C-Doppelbindung *f*
oleic acid Ölsäure *f*, **cis*-Octadec-9-ensäure *f*
olein 1. Olein *n*, Glyceroloieat *n*, *(i. e. S.)* Triolein *n*, Glyceroltrioleat *n*; 2. Olein *n*, Elain *n*, Stearinöl *n*, technische [rohe] Ölsäure *f*; 3. sulfatiertes Öl *n*
oleophobic oleophob, ölabweisend
oleoresin Oleoresin *n*, Ölharz *n* *(natürlich vorkommende Lösung von Harz in flüchtigen Ölen)*
oleoresin acid Harzsäure *f*
oleoresinous varnish Ölharzlack *m*, Harz-Öl-Lack *m*, Öl-Naturharz-Lack *m*
oleum Oleum *n*, rauchende Schwefelsäure *f*
oligofunctional oligofunktionell
oligomer Oligomer(e) *n* *(Molekül mit 3 bis 10 Struktureinheiten)*
oligomeric oligomer
oligomerization Oligomerisation *f*
oligonucleotide *(Bioch)* Oligonucleotid *n*
oligopeptide Oligopeptid *n* *(Peptid mit maximal 10 Aminosäuren)*
oligosaccharide Oligosaccharid *n*
olivine *(Min)* Olivin *m* *(Magnesiumeisen(II)-orthosilicat)*
OMA *s.* optical multichannel analyser
ommochrome Ommochrom *n* *(Naturfarbstoff)*
OMP = orotidine 5'-monophosphate
on-line analysis Prozessanalyse *f*, dynamische Analyse *f*
on-site in situ, an Ort und Stelle, vor Ort
on-site analysis Analyse *f* an Ort und Stelle, Vor-Ort-Analyse *f*
once-through cooling Durchflusskühlung *f*, Durchlaufkühlung *f*
once-through evaporator Durchlaufverdampfer *m*
oncogenic tumorerzeugend, tumorinduzierend, onkogen
oncogenous tumorerzeugend, tumorinduzierend, onkogen
one-carbon compound Einkohlenstoffverbindung *f*, C-1-Verbindung *f*, C-1-Körper *m*
one-carbon metabolism C_1-Stoffwechsel *m*
one-component system Einkomponentensystem *n*, Einstoffsystem *n*, unitäres System *n*
one-electron atom Einelektronenatom *n*
one-pack system *(Anstr)* Einkomponentensystem *n*

one-pass evaporator Durchlaufverdampfer *m*
one-phase system Einphasensystem *n*, einphasiges [homogenes] System *n*
one-pot reaction *(org Ch)* Eintopfreaktion *f*
one-stage process *s.* one-step process
one-step process Einstufenverfahren *n*, einstufiges Verfahren *n*
one-step reaction einstufige Reaktion *f*, Einstufenreaktion *f*, Einschrittreaktion *f*
one-substrate reaction Ein-Substrat-Reaktion *f (Enzymkinetik)*
one-way reaction irreversible [nicht umkehrbare] Reaktion *f*
onium compound Oniumverbindung *f*
Onsager (reciprocal) relations onsagersche Reziprozitätsbeziehungen *fpl*
onset Einsetzen *n (einer Reaktion)*
onyx *(Min)* Onyx *m (Silicium(IV)-oxid)*
ooze Schlamm *m*, Schlick *m*
OP *s.* osmotic pressure
OPA *s. o*-phthalic aldehyde
opacify *v* trüben *(Glas, Kunststoffe)*; trüb werden, sich trüben
opacity Undurchsichtigkeit *f*, Opazität *f*, Lichtundurchlässigkeit *f*; Trübung *f*
opal *(Min)* Opal *m (Silicium(IV)-oxid)*
opalesce *v* opaleszieren, opalisieren
opalescence Opaleszenz *f*, Opaleszieren *n*, Opalisieren *n*
opalescent opaleszierend, opalisierend
opaque 1. undurchsichtig, nicht durchscheinend, opak, lichtundurchlässig; milchig, trüb; 2. *(Anstr)* gut deckend
opaque glass Trübglas *n*, opakes Glas *n*, *(i. e. S.)* Milchglas *n*
open-cast mining [working] Tagebauförderung *f*, Tagebaubetrieb *m*, Förderung *f* im Tagebau
open-cell foam offenzelliger Schaum(kunst)stoff *m*
open-chain hydrocarbon offenkettiger [kettenförmiger, acyclischer, aliphatischer] Kohlenwasserstoff *m*
open-cycle system plant Anlage *f* mit offenem Kreislauf
open-hearth furnace Siemens-Martin-Ofen *m*, SM-Ofen *m*; Herdofen *m*
open-hearth process Herdfrischverfahren *n*, Siemens-Martin-Verfahren *n*, SM-Verfahren *n*
open-hearth steel Herdfrischstahl *m*, herdgefrischter Stahl *m*, Siemens-Martin-Stahl *m*, SM-Stahl *m*
open pit mine Tagebau *m*
open shell offene Schale *f (Bindungslehre)*
open steam direkter Dampf *m*, Direktdampf *m*, Frischdampf *m*
open system *(phys Ch, Bioch)* offenes System *n*
open tubular (capillary) column *(Chromat)* Kapillarsäule *f*, Trennkapillare *f*
opencast [opencut] mining *s.* opencast mining
operate *v* betreiben, fahren *(z. B. eine Anlage)*; bedienen, betätigen; steuern; antreiben; arbeiten, in Betrieb sein
operate *v* **adiabatically** adiabat betreiben; adiabat arbeiten
operate *v* **batchwise [intermittently]** diskontinuierlich fahren
operate *v* **isothermally** isotherm betreiben; isotherm arbeiten
operating characteristics Betriebsdaten *pl*, Betriebsparameter *mpl*
operating hold-up *(Dest)* Arbeitsinhalt *m*, dynamischer Kolonneninhalt *m*
operating life 1. Nutzungsdauer *f*, Gebrauchswertdauer *f*, Lebensdauer *f*, Betriebsdauer *f*, Standzeit *f*; 2. *(Hyd)* Laufzeit *f (z. B. eines A-Kohle-Filters)*
operating line Arbeitslinie *f*, Betriebslinie *f (Destillation, Extraktion)*
operating parameters *s.* operating characteristics
operating pressure Betriebsdruck *m*, Arbeitsdruck *m*; Steuerdruck *m*
operating temperature Betriebstemperatur *f*, Arbeitstemperatur *f*
operation Betrieb *m*, Arbeitsweise *f*, Fahrweise *f*, Betriebsmethode *f*; Arbeitsgang *m*, Operation *f*
operon *(Bioch)* Operon *n (gemeinsam wirkende Gengruppe)*
opium Opium *n*, Rohopium *n*
opium alkaloid Opiumalkaloid *n*, Mohnalkaloid *n*
OPN = *o*-phthalonitrile
Oppenauer oxidation [reaction] oppenauersche Oxidation *f*, Oppenauer-Reaktion *f (Alkoholdehydrierung)*
opposed ekliptisch, verdeckt, ± synperiplanar *(Stereochemie)*

opposing reaction Gegenreaktion *f*, gegenläufige Reaktion *f*
opposite in sign mit umgekehrtem Vorzeichen
opposite spins entgegengesetzte [antiparallele] Spins *mpl*
oppositely charged entgegengesetzt geladen
optic axis *(Krist)* optische Achse *f*
optical activity optische Aktivität *f*, optisches Dreh(ungs)vermögen *n*, Rotationspolarisation *f*
optical antipode *s.* optical isomer
optical birefringence optische Doppelbrechung *f*
optical bleaching agent optischer Aufheller *m*, optisches Aufhellungsmittel *n* [Bleichmittel *n*], Weißtöner *m*
optical brightener optischer Aufheller *m*, optisches Aufhellungsmittel *n* [Bleichmittel *n*], Weißtöner *m*
optical density 1. *(Foto)* (optische) Dichte *f*, Schwärzung *f*, Deckung *f*, Extinktion *f*, logarithmische Opazität *f*; 2. *(Spektr)* Extinktion *f*, Absorbanz *f*
optical emission spectroscopy optische Emissionsspektroskopie *f*
optical inversion waldensche Umkehrung *f*, Inversion *f*
optical isomer optisches Isomer(e) *n*, Enantiomer(e) *n*, Spiegelbildisomer(e) *n*
optical isomerism optische Isomerie *f*, Enantiomerie *f*, Spiegelbildisomerie *f*
optical multichannel analyser *(Spektr)* optischer Mehrkanalanalysator *m* [Vielkanalanalysator *m*], OMA
optical pathlength *(Spektr)* optische Weglänge *f*
optical rotation optische Drehung *f* [Rotation *f*], Drehung *f* der Polarisationsebene
optical rotatory dispersion optische Rotationsdispersion *f*, ORD
optical specificity *(Bioch)* optische Spezifität *f*
optical spectroscopy optische Spektroskopie *f*
optical spectrum optisches Spektrum *n*
optically active optisch aktiv
optically anisotropic optisch anisotrop
optically inactive optisch inaktiv
optically isotropic optisch isotrop
optoelectric optoelektrisch
optoelectronics Optoelektronik *f*
oral *(Pharm)* oral (verabreichbar, wirksam)
oral toxicity orale Toxizität *f*
orbit Bahn *f*, Umlaufbahn *f*
orbital Orbital *n(m)*
orbital angular momentum Bahndrehimpuls *m*, Orbitaldrehimpuls *m*
orbital angular momentum quantum number *s.* orbital quantum number
orbital electron Orbitalelektron *n*, Bahnelektron *n*
orbital magnetic moment bahnmagnetisches [orbitalmagnetisches] Moment *n*, magnetisches Bahnmoment *n*
orbital moment(um) *s.* orbital angular momentum
orbital quantum number Bahndrehimpulsquantenzahl *f*, Drehimpulsquantenzahl *f*, Orbitaldrehimpulsquantenzahl *f*, Nebenquantenzahl *f*, azimutale Quantenzahl *f*, Azimut(al)quantenzahl *f*
***p* orbital** *p*-Orbital *n*
***s* orbital** *s*-Orbital *n*
***sp* orbital** *sp*-Orbital *n*
π-orbital π-Orbital *n*, π-Molekülorbital *n*, π-MO *n*
σ-orbital σ-Orbital *n*, σ-Molekülorbital *n*, σ-MO *n*
ORD *s.* optical rotatory dispersion
order Ordnung *f*
order-disorder transformation [transition] Ordnung-Unordnung-Umwandlung *f*, Ordnung-Unordnung-Übergang *m*, Ordnungsumwandlung *f*
order of magnitude Größenordnung *f*
order of precedence *(Nom)* Prioritätenfolge *f*, Rangfolge *f*
order of reaction Reaktionsordnung *f*
order of seniority *s.* order of precedence
ordinal number Ordnungszahl *f*, OZ, Kernladungszahl *f*, Z
ore Erz *n*
ore-bearing erzhaltig, erzführend
ore deposit Erzlagerstätte *f*, Erzlager *n*, Erzvorkommen *n*
ore dressing Erzaufbereitung *f*
ore mineral Erzmineral *n*
organ culture *(Biot)* Organkultur *f*
organic organisch
organic acid organische Säure *f*
organic-chemical organisch-chemisch

organic chemical organische Chemikalie *f*, Organochemikalie *f*
organic-chemical technology 1. organisch-chemische Technik *f*; 2. organisch-chemische Technologie *f*
organic chemist Organochemiker *m*, Organiker *m*
organic chemistry organische Chemie *f*
organic-chemistry nomenclature Nomenklatur *f* der organischen Chemie
organic compound organische Verbindung *f*
organic contaminants *(Hyd)* organische Schmutzstoffe *mpl*
organic electrosynthesis organische Elektrosynthese *f*
organic ester lubricant [oil] Esteröl *n*
organic gel Organogel *n*
organic halide organische Halogenverbindung *f*, Organohalogenverbindung *f*
organic light emitting diode organische Leuchtdiode *f*
organic load(ing) *(Hyd)* organische Belastung *f*, Belastung *f* mit organischen Inhaltsstoffen [Substanzen]
organic pigment organisches Pigment *n*
organic pollutants *s.* organic contaminants
organic polymer organisches Polymer(e) *n*
organic salt organisches Salz *n*
organic slime Faulschlamm *m*
organic solid organischer Festkörper *m*
organic solvent organisches Lösungsmittel *n*
organic sulphur organischer [organisch gebundener] Schwefel *m*
organics 1. organische Verbindungen *fpl*; 2. *(Hyd)* organische Abwasserinhaltsstoffe *mpl*; 3. *s.* organic contaminants
organoboron compound borhaltige organische Verbindung *f*, *(i. e. S.)* bororganische Verbindung *f*, Organoborverbindung *f (mit direkter B-C-Bindung)*
organocell process *(Pap)* Organocell-Verfahren *n*
organochlorine compound organische Chlorverbindung *f*, *(i. e. S.)* chlorierter Kohlenwasserstoff *m*
organochlorosilane Organochlorsilan *n*, organisches Chlorsilan *n*
organoderivative organisches Derivat *n*, *(i. e. S.)* Organoderivat *n*
organofunctional organofunktionell, carbofunktionell
organohalide Organohalogenverbindung *f*, organische Halogenverbindung *f*
organoleptic examination [testing] *(Lebm, Pharm)* organoleptische [sinnesphysiologische] Prüfung *f* [Untersuchung *f*], Organoleptik *f*
organomagnesium halide Organomagnesiumhalogenid *n*
organomercury compound quecksilberhaltige organische Verbindung *f*, *(i. e. S.)* quecksilberorganische Verbindung *f*, Organoquecksilberverbindung *f (mit direkter Hg-C-Bindung)*
organometallic chemistry Organometallchemie *f*, Metallorganik *f*
organometallic compound metallhaltige organische Verbindung *f*, *(i. e. S.)* metallorganische [organometallische] Verbindung *f*, Organometallverbindung *f (mit direkter Metall-Kohlenstoff-Bindung)*
organophosphorus compound phosphorhaltige organische Verbindung *f*, *(i. e. S.)* phosphororganische Verbindung *f*, Organophosphorverbindung *f (mit direkter P-C-Bindung)*
organosilicon compound siliciumhaltige organische Verbindung *f*, *(i. e. S.)* Organosiliciumverbindung *f*, siliciumorganische Verbindung *f (mit direkter Si-C-Bindung)*
organosiloxane polymer Organosiloxanpolymer(e) *n*, Silikonpolymer(e) *n*
organotin compound zinnhaltige organische Verbindung *f*, *(i. e. S.)* zinnorganische Verbindung *f*, Organozinnverbindung *f (mit direkter Sn-C-Bindung)*
orient(ate) *v* orientieren
orientation 1. Orientierung *f*, (räumliche) Ausrichtung *f (von Molekülen)*; Ortsbestimmung *f (für Substituenten)*; 2. *(Krist)* Orientierung *f*
orientation polarization Orientierungspolarisation *f*
orifice kleine [enge] Öffnung *f*, Mündung *f*, Austritt *m*, Auslauföffnung *f*; Düse *f*; Mundstück *n*; Blende *f*; Sieblochung *f*,

Bohrung *f*, Öffnung *f*, Lochweite *f (eines Siebs)*
original state Ausgangszustand *m*, Anfangszustand *m*
ornithine Ornithin *n (Zwischenprodukt des Harnstoff-Zyklus)*
ornithine cycle *(Bioch)* Ornithinzyklus *m*, (Arginin-)Harnstoff-Zyklus *m*, Krebs-Henseleit-Zyklus *m*
orotic acid Orotsäure *f*, Uracil-4-carbonsäure *f*
ORP *(oxidation-reduction potential) s.* redox potential
orpiment (yellow) Auripigment *n*, Operment *n*, Königsgelb *n*, Rauschgelb *n*, gelbes Arsenik *n* [Schwefelarsen *n*] *(Arsen(III)-sulfid)*
Orsat analyser [apparatus] Orsat-Apparat *m*, Orsat-Gerät *n*, Gasanalysenapparat *m* nach Orsat
ortho orthoständig, *o*-ständig, in *ortho*-Stellung [*o*-Stellung, 1,2-Stellung] • **in ortho position** orthoständig, *o*-ständig, in *ortho*-Stellung [*o*-Stellung, 1,2-Stellung] • **be (located, situated) ortho** sich in *ortho*-Stellung [*o*-Stellung, 1,2-Stellung] befinden, in *ortho*-Stellung stehen, orthoständig [*o*-ständig] sein
ortho acid Orthocarbonsäure *n*
ortho compound *ortho*-Verbindung *f*, *o*-Verbindung *f*
ortho-directing *ortho*-dirigierend, nach der *ortho*-Stellung dirigierend
ortho-fused orthokondensiert, orthoanelliert
ortho hydrogen *ortho*-Wasserstoff *m*, Orthowasserstoff *m*
ortho isomer *ortho*-Isomer(e) *n*, *o*-Isomer(e) *n*
ortho orientation *ortho*-Orientierung *f*
ortho-para-directing group *ortho-para*-dirigierende Gruppe *f*, *ortho-para*-dirigierender Substituent *m*, Substituent *m* erster Ordnung
ortho-peri-fused *ortho*-peri-kondensiert, *ortho*-peri-anelliert
ortho substituent *ortho*-Substituent *m*, *o*-Substituent *m*, Substituent *m* in *ortho*-Stellung [*o*-Stellung], *ortho*-ständiger Substituent *m*
ortho-substituted *ortho*-substituiert, *o*-substituiert
ortho substitution *ortho*-Substitution *f*, *o*-Substitution *f*
orthoborate $M^I_3BO_3$ Orthoborat *n*, Trioxoborat *n*
orthoboric acid Orthoborsäure *f*, Monoborsäure *f*, Trioxoborsäure *f*, Borsäure *f*
orthoclase *(Min)* Orthoklas *m (Kaliumalumotrisilicat)*
orthohydrogen Orthowasserstoff *m*, *ortho*-Wasserstoff *m*
orthophosphate Orthophosphat *n*
orthophosphoric acid Orthophosphorsäure *f*, (gewöhnliche) Phosphorsäure *f*, Monophosphorsäure *f*
orthopositronium Ortho-Positronium *n (Positron und Elektron mit entgegengesetzten Spins)*
orthorhombic *(Krist)* (ortho)rhombisch
orthorhombic (crystal) system rhombisches System *n*
orthosilicate $M^I_4SiO_4$ Orthosilicat *n*, *Tetroxosilicat *n*
orthosilicic acid Orthokieselsäure *f*, *Tetroxokieselsäure *f*
orthotellurate Orthotellurat *n*
orthotelluric acid Orthotellursäure *f*, Hexoxotellursäure *f*
Orton reaction [rearrangement] Orton-Umlagerung *f (von N-Halogenarylaminen zu 1,2- und 1,4-Halogen-arylaminen)*
osazonation Osazonierung *f*
osazone Osazon *n*
oscillate *v* oszillieren, schwingen, pendeln, vibrieren; schwingen [pendeln] lassen, hin und her schwenken [bewegen]
oscillating reaction oszillierende Reaktion *f*
oscillation Oszillation *f*, Schwingung *f*
oscillator Oszillator *m*
oscillographic polarography *s.* oscillopolarography
oscillometry Hochfrequenzkonduktometrie *f*, Oszillometrie *f*
oscillopolarographic oszillopolarographisch
oscillopolarography *Oszillopolarographie *f*, oszillographische Polarographie *f*, Oszillationspolarographie *f*
OSHA = Occupational Safety and Health Administration

osmate $M^{I}_{2}OsO_4$ Osmat(VI) *n*, Tetroxoosmat(VI) *n*
osmic Osmium...; höherwertigem Osmium entsprechend
osmic acid Tetroxoosmium(VI)-säure *f*, Osmiumsäure *f*
osmium Os Osmium *n*
osmium tetroxide Osmiumtetroxid *n*, Osmium(VIII)-oxid *n*
osmolality Osmolalität *f (molare Gesamtkonzentration gelöster Stoffe je kg Lösungsmittel)*
osmolarity Osmolarität *f (molare Gesamtkonzentration gelöster Stoffe je l Lösung)*
osmole Osmol *n (1 osmol =1 mol gelöster Substanz je kg Lösungsmittel beziehungsweise je l Lösung)*
osmometer Osmometer *n*
osmometry Osmometrie *f*
osmosis Osmose *f*
osmotic pressure osmotischer Druck *m*
osmous Osmium...; niederwertigem Osmium entsprechend
osone *(org Ch)* Oson *n*
Ostwald dilution law ostwaldsches Verdünnungsgesetz *n*
Ostwald process Ostwald-Verfahren *n (zur Gewinnung von NO_2)*
Ostwald rule ostwaldsche Stufenregel *f*
Ostwald viscometer Ostwald-Viskosimeter *n*, Kapillarviskosimeter *n* nach Ostwald
Ostwald viscosity pipette Ostwald-Viskosimeter *n*, Kapillarviskosimeter *n* nach Ostwald
OSW *s.* overall organic substances in water
ouabain Ouabain *n*, g-Strophantin *n*, Strophantin *n* G, Gratus-Strophantin *n (Glykosid)*
out-of-plane vibration γ-Schwingung *f*, Deformationsschwingung *f* aus der Molekülebene heraus
outage time Ausfallzeit *f*, Stillstandszeit *f*
outbreathing Entweichen *n* von Dämpfen
outcrop *v (Geol)* anstehen, ausgehen, ausstreichen, zutage treten
outcrop *(Geol)* Anstehendes *n*, anstehendes Gestein *n*
outdoor durability Außenbeständigkeit *f*, Witterungsbeständigkeit *f*, Wetterbeständigkeit *f*, Wetterfestigkeit *f*
outdoor weathering test Frei(luft)bewitterungsversuch *m*, Naturbewitterungsversuch *m*, Bewitterungsversuch *m*
outer complex Anlagerungskomplex *m*, Außenkomplex *m*, Normalkomplex *m*, Highspin-Komplex *m*
outer Helmholtz plane *(Ech)* äußere Helmholtz-Fläche *f*
outer potential äußeres elektrisches Potenzial *n*, Volta-Potenzial *n*
outer shell äußere Schale *f*, Außenschale *f*, Valenzschale *f*
outer-shell electron Außenelektron *n*, Valenzelektron *n*
outer-sphere complex Außensphärenkomplex *m*, Outer-sphere-Komplex *m*
outermost electron Außenelektron *n*, Valenzelektron *n*
outermost shell äußere Schale *f*, Außenschale *f*, Valenzschale *f*
outflow 1. Ausfließen *n*, Ausströmen *n*, Abfließen *n*, Ablaufen *n*; 2. ablaufende [abfließende] Flüssigkeit *f*, Ablauf *m*, Abfluss *m*; 3. *s.* outlet 1.
outgas *v* entgasen
outgassing Entgasen *n*, Entgasung *f*; Gasaushauchung *f*, Aushauchung *f*, Exhalation *f (von Vulkanen)*
outlet 1. Ablass *m*, Auslass *m*, Ablauf *m*, Ausfluss *m*, Austrag *m (Öffnung)*; 2. ablaufende [abfließende] Flüssigkeit *f*, Ablauf *m*, Abfluss *m*; 3. *(Hyd)* Abwassereinleitungsstelle *f (eines Betriebs)*; 4. Absatzmarkt *m*, Absatzgebiet *n*, Absatzmöglichkeit *f*; Verwendungsmöglichkeit *f*
outlet air Abluft *f*
outlet stream 1. Austragsmassestrom *m*, Austragsgutstrom *m*; 2. Auslauf *m*
outlet tube Ablassrohr *n*, Ablaufrohr *n*, Abflussrohr *n*
outlet valve Ablassventil *n*, Auslassventil *n*, Abflussventil *n*, Austragventil *n*, Leerventil *n*
outlier *(Anal)* Ausreißer *m*, abweichender Wert *m*
output Ausstoß *m*, Leistung *f*
output stream Austragsmassestrom *m*, Austragsgutstrom *m*
outside air Außenluft *f*

oven 1. *(Tech)* Ofen *m*; 2. *(Ker)* Brennofen *m*; 3. *(Lab)* Trockenschrank *m*
oven-dry *v* 1. *(Tech)* im Ofen trocknen; 2. *(Lab)* im Trockenschrank trocknen
oven-dry 1. *(Tech)* im Ofen getrocknet, ofentrocken; 2. *(Lab)* im Trockenschrank getrocknet *(bei 100 bis 110 °C)*; absolut trocken
over-develop *v (Foto)* überentwickeln
over-expose *v (Foto)* überbelichten
overall column efficiency *(Dest)* mittleres Verstärkungsverhältnis *n (Quotient aus theoretischer und praktischer Trennstufenzahl)*
overall conversion Gesamtumsatz *m*
overall course Gesamtablauf *m (einer Reaktion)*
overall efficiency Gesamtwirkungsgrad *m*
overall equation Bruttogleichung *f*
overall formation constant Brutto-Komplexbildungskonstante *f*, Bruttostabilitätskonstante *f*, kumulative Stabilitätskonstante *f*
overall heat-transfer coefficient Wärmedurchgangskoeffizient *m*
overall mass-transfer coefficient Stoffdurchgangskoeffizient *m*
overall order (of reaction) Gesamtreaktionsordnung *f*
overall organic substances in water *(Hyd)* Gesamtgehalt *m* an organischen Wasserinhaltsstoffen
overall rate Gesamtgeschwindigkeit *f*
overall reaction Gesamtreaktion *f*, Bruttoreaktion *f*
overall yield Gesamtausbeute *f*
overbleach *v* überbleichen
overbleaching Überbleichung *f*, Überbleiche *f*
overburden *(Bgb)* Deckgebirge *n*, Deckschicht *f*, Deckschichten *fpl*
overcure *v* 1. *(Kst, Anstr)* überhärten; 2. *(Gum)* übervulkanisieren, übervernetzen
overcure 1. *(Kst, Anstr)* Überhärten *n*, Überhärtung *f*; 2. *(Gum)* Übervulkanisation *f*, Übervernetzung *f*
overcuring 1. *(Kst, Anstr)* Überhärten *n*, Überhärtung *f*; 2. *(Gum)* Übervulkanisation *f*, Übervernetzung *f*
overdosage Überdosierung *f*
overdose *v* überdosieren
overdose Überdosis *f*
overdry *v* übertrocknen, zu lange trocknen
overdrying Übertrocknen *n*, Übertrocknung *f*
overdye *v* überfärben *(1. durch Verwendung von zuviel Farbstoff; 2. absichtlich unter Einsatz eines zweiten Farbstoffs)*
overfeed *v* überdosieren
overfeed(ing) Überdosierung *f*
overfertilization Überdüngung *f*
overflow *v* überlaufen, überfließen, überströmen; überfluten
overflow 1. Überlauf *m*, Überlaufvorrichtung *f*; 2. *s.* overflow product
overflow product Überlaufprodukt *n*, Überlaufgut *n*, Überlauf *m (beim Klassieren)*; *(beim Siebklassieren auch)* Siebüberlauf *m*; Klarflüssigkeit *f (beim Hydroklassieren)*
overglaze colour Aufglasurfarbe *f*; Schmelzfarbe *f*; Emailfarbe *f*
overgrind *v* übermahlen, totmahlen
overgrowth *(Krist)* Überwachsung *f*
Overhauser effect *(Spektr)* Overhauser-Effekt *m*
overhead (product) 1. *(Dest)* Kopfprodukt *n*, Überkopfprodukt *n*; 2. *(Erdöl)* Toppprodukt *n*
overheads *s.* overhead (product)
overheat *v* überhitzen
overlap *v* überlappen; sich überlappen *(Orbitale, genetischer Code)*
overlap Überlappen *n*, Überlappung *f (von Orbitalen, des genetischen Codes)*
overlapping concentration Überlappungskonzentration *f (in polymeren Lösungen)*
overload *v* über(be)lasten
overpotential *s.* overvoltage
overpressure Überdruck *m*
overproduction 1. *(Tech, Biot)* Überproduktion *f*; 2. *(Bioch)* vermehrte Bildung *f*
overrun *v* übersteigen, überschreiten *(einen Richtwert)*
overrun *v* **the end point** *(Anal)* übertitrieren
overshoot *v s.* overrun
oversize (material, product) Überkorn *n*, Grobkorn *n*, Grobgut *n (beim Klas-*

sieren); *(beim Siebklassieren auch)* Siebgrobes *n*, Siebgröbe *f*, *(beim Windsichten auch)* Sichtgrobes *n*, Grieß *m*
overstoving *(Anstr)* Überbrennen *n* *(Überschreiten der Einbrennbedingungen)*
overtannage Übergerbung *f*, Totgerbung *f*
overtitrate *v* übertitrieren
overtone *(Spektr)* Ober(ton)schwingung *f*, Oberton *m*
overtreat *v* übermäßig behandeln
overuse *v* übermäßig gebrauchen, zu reichlich anwenden *(z. B. Biozide)*
overuse übermäßiger Gebrauch *m*, zu reichliche Anwendung *f (z. B. von Bioziden)*
overvoltage (elektrochemische, elektrolytische) Überspannung *f*
ovicidal ovizid, eiabtötend
ovicide Ovizid *n*, Eiergift *n (Ei abtötendes Schädlingsbekämpfungsmittel)*
O/W emulsion Öl-in-Wasser-Emulsion *f*
oxalacetic acid Oxalessigsäure *f*, Ketobernsteinsäure *f*
oxalaldehyde Oxalaldehyd *m*, Glyoxal *n*, *Ethandial *n*
oxalate *v* mit Oxalaten behandeln, *(i. e. S.)* durch Zusatz von Oxalaten ungerinnbar [gerinnungsunfähig] machen *(Blut)*
oxalate Oxalat *n (Salz oder Ester der Oxalsäure)*
oxalic acid Oxalsäure *f*, Kleesäure *f*, *Ethandisäure *f*
oxalic ester Oxalsäurediethylester *m*
oxaloacetate Oxaloacetat *n (Zwischenprodukt des Citrat-Zyklus)*
oxaloacetate pathway *(Bioch)* Oxalacetatweg *m (des Asparagins)*
oxalonitrile Dicyan *n*, Oxalsäurenitril *n*
oxalosuccinate Oxalosuccinat *n (Zwischenprodukt des Citrat-Zyklus)*
oxalosuccinic acid Oxal(o)bernsteinsäure *f*, 1-Oxopropan-1,2,3-tricarbonsäure *f*
oxalyl urea Oxalylharnstoff *m*, Parabansäure *f*, Imidazoltrion *n*
oxamic acid Oxamidsäure *f*, Oxalsäuremonamid *n*
oxamide Oxamid *n*, *Ethandiamid *n*
oxazine dye Oxazinfarbstoff *m*
oxazole Oxazol *n*
oxazolone Oxazolon *n*, Oxazolinon *n*, Azlacton *n*
oxethane Oxethan *n*, Oxacyclobutan *n*
oxidant Oxidans *n*, Oxidationsmittel *n*
oxidation Oxidation *f* • **undergo oxidation** oxidieren
oxidation-inhibiting oxidationshemmend
oxidation inhibitor Oxidationsinhibitor *m*, Oxidationsverhinderer *m*
oxidation number *s.* oxidation state
oxidation potential Oxidationspotenzial *n*
oxidation product Oxidationsprodukt *n*
oxidation reaction Oxidationsreaktion *f*
oxidation-reduction Oxidoreduktion *f*
oxidation resistance Oxidationsbeständigkeit *f*, Beständigkeit *f* gegen oxidative Einflüsse
oxidation-resistant oxidationsbeständig, beständig gegen oxidative Einflüsse
oxidation-sensitive oxidationsempfindlich
oxidation state Oxidationszahl *f*, Oxidationsstufe *f*
oxidation-susceptible oxidationsempfindlich
oxidative addition oxidative Addition *f*
oxidative coupling oxidative Kupplung *f*
oxidative deamination *(Bioch)* oxidative Desaminierung *f*
oxidative decarboxylation oxidative Decarboxylierung *f*
oxidative degradation oxidativer Abbau *m*, oxidative Degradation *f*, Abbau *m* durch Oxidation
oxidative deterioration *(Lebm)* oxidativer Verderb *m*
oxidative enzyme Oxidationsenzym *n*
oxidative fermentation oxidative [aerobe] Gärung *f*
oxidative phosphorylation *(Bioch)* oxidative Phosphorylierung *f*, Atmungskettenphosphorylierung *f*
oxidative photodegradation photooxidativer Abbau *m*
oxidative rancidity Peroxidranzigkeit *f*, Peroxidigkeit *f*
oxide Oxid *n*
oxide barrier Oxidsperrschicht *f*
oxide catalyst oxidischer Katalysator *m*

oxide ceramics 1. oxidkeramische Erzeugnisse *npl*, Oxidkeramik *f*; 2. oxidkeramische Stoffe *mpl*; 3. Oxidkeramik *f (als Zweig)*
oxide-coating 1. *(Met)* oxidische Deckschicht *f*, Oxidschicht *f (spontan entstanden)*; 2. *(Met)* Oxidschutzschicht *f*, oxidische Schutzschicht *f (künstlich erzeugt oder verstärkt)*
oxide film Oxidfilm *m*, Oxidhaut *f*, Oxidbelag *m*, dünne Oxidschicht *f*
oxidic oxidisch
oxidimetric oxidimetrisch
oxidimetry Oxidimetrie *f*, Redoxanalyse *f*
oxidizable oxidierbar, oxidabel
oxidize *v* oxidieren
oxidize *v* **back** zurückoxidieren
oxidize *v* **to higher valency** aufoxidieren
oxidizer 1. Sauerstoffträger *m (in Explosivstoffen)*; 2. *s.* oxidant
oxidizing agent *s.* oxidant
oxidizing atmosphere oxidierende Atmosphäre *f*
oxidizing flame Oxidationsflamme *f*
oxido-reductase Oxidoreductase *f*
oxidoreduction Oxidoreduktion *f*, Redoxreaktion *f*, Oxidations-Reduktions-Reaktion *f*
oxime Oxim *n*, Isonitrosoverbindung *f*
oxinate Oxinat *n (Komplexverbindung des Oxins)*
oxindole Oxindol *n*, 2-Oxoindolin *n*
oxindole alkaloid Oxindolalkaloid *n*
oxine Oxin *n*, 8-Hydroxychinolin *n*
oxirane Oxiran *n*, Epoxid *n*
oxo acid Oxosäure *f*, Sauerstoffsäure *f*
oxo compound Oxoverbindung *f*
oxo-cyclo tautomerism Oxo-Cyclo-Tautomerie *f*
oxo group =O Oxogruppe *f*
oxo process *s.* 1. oxo reaction; 2. oxo synthesis
oxo reaction Oxo(synthese)reaktion *f*, Roelen-Reaktion *f*
oxo salt Oxosalz *n*
oxo synthesis Oxosynthese *f*, Hydroformylierung *f*
oxoacid *s.* oxo acid
oxoethanoic acid Oxoethansäure *f*, Glyoxylsäure *f*
oxonium ion 1. Oxoniumion *n*; 2. Hydroniumion *n (hydratisiertes Wasserstoffion)*
oxonium salt Oxoniumsalz *n*
3-oxopentanedioic acid *3-Oxopentandisäure *f*, Acetondicarbonsäure *f*, 3-Ketoglutarsäure *f*
oxy 1. sauerstoffhaltig, Oxy...; 2. *(org Ch)* hydroxylhaltig, Hydroxy...
oxy acid *s.* oxo acid
oxyacetylene welding *s.* oxygen-acetylene welding
oxyacid *s.* oxo acid
oxybiotic oxybiontisch, aerob
oxychloride Oxidchlorid *n*
oxychlorination Oxychlorierung *f*
oxydation *s.* oxidation
oxyde *s.* oxide
oxygen O Sauerstoff *m*, Oxygen *n*
oxygen-acetylene welding Acetylen-Sauerstoff-Schweißen *n*, Autogenschweißen *n*
oxygen acid *s.* oxo acid
oxygen affinity Sauerstoffaffinität *f*, Affinität *f* zu(m) Sauerstoff
oxygen alkylation O-Alkylierung *f*
oxygen bomb calorimeter *(Lab)* Bombenkalorimeter *n*, Kalorimeterbombe *f*, kalorimetrische Bombe *f*, Verbrennungsbombe *f*
oxygen carrier Sauerstoff(über)träger *m*
oxygen-carrying sauerstoff(über)tragend
oxygen-centred radical Sauerstoffradikal *n*
oxygen compound Sauerstoffverbindung *f*
oxygen concentration cell Sauerstoffkonzentrationszelle *f*, Belüftungselement *n*, Belüftungszelle *f*, Evans-Element *n*, Evans-Korrosionselement *n*
oxygen-consuming sauerstoffverbrauchend, sauerstoffzehrend
oxygen deficiency 1. Sauerstoffmangel *m*; 2. Sauerstoffdefizit *n (quantitativ)*; 3. *(Hyd)* Sättigungsdefizit *n (Differenz zwischen Sättigungswert und tatsächlichem Sauerstoffgehalt)*
oxygen-deficient sauerstoffarm
oxygen demand Sauerstoffbedarf *m*, *(quantitativ auch)* Sauerstoffbedarfswert *m*
oxygen-depleted an Sauerstoff verarmt
oxygen-depleting sauerstoffzehrend
oxygen depletion Sauerstoffverarmung *f*, Verarmung *f* an Sauerstoff

oxygen evolution Sauerstoffentwicklung *f*, Sauerstoffabscheidung *f*
oxygen exclusion Ausschluss *m* von Sauerstoff
oxygen generation Sauerstofferzeugung *f*
oxygen-hydrogen coulometer Knallgascoulometer *n*
oxygen overvoltage *(Ech)* Sauerstoffüberspannung *f*
oxygen partial pressure Sauerstoffpartialdruck *m*
oxygen process of steelmaking Sauerstoff(auf)blasverfahren *n*, Sauerstofffrischverfahren *n*, (Sauerstoff-) Blasstahlverfahren *n*, Oberwindfrischverfahren *n*
oxygen supply Sauerstoffzufuhr *f*, Sauerstoffantransport *m*
oxygen transfer 1. Sauerstoffübergang *m*; Sauerstoffübertragung *f*; 2. *(Hyd)* Sauerstoffeintrag *m*; 3. *(Hyd)* Sauerstoffeintrag *m*, Sauerstoffeintragswert *m (in g* $O_2/m^3 \cdot h$*)*
oxygen transfer agent Sauerstoffüberträger *m*
oxygen transferase Sauerstofftransferase *f*, Dioxygenase *f*
oxygen transport 1. Sauerstofftransport *m*; 2. *(Hyd)* Sauerstoffübertragung *f*, Sauerstoffeintrag *m*, Sauerstoffzufuhr *f*
oxygen uptake Sauerstoffaufnahme *f*
oxygenate *v* mit Sauerstoff sättigen; mit Sauerstoff anreichern; oxidieren
oxygenated sauerstoffgesättigt; sauerstoffbeladen, mit Sauerstoff angereichert; oxidiert
oxygenation Sättigung *f* mit Sauerstoff; Anreicherung *f* mit Sauerstoff; Oxidation *f*
oxygenic reactor *(Biot)* Reaktor *m* mit Sauerstoffeintrag
oxyhaemocyanin Oxyhämocyanin *n* *(kupferhaltiger Blutfarbstoff)*
oxyhaemoglobin Oxyhämoglobin *n*
oxyhalide Oxidhalogenid *n*
oxyhydrogen blowpipe [burner] Knallgasgebläse *n*
oxyhydrogen gas Knallgas *n (i. e. S.)*
oxyneurine Oxyneurin *n*, Betain *n*, Lycin *n*, Carboxymethyltrimethylammonium *n*, Trimethylammonioacetat *n*, Trimethylglycin *n*
oxysalt 1. Oxosalz *n (Salz einer Oxosäure)*; 2. Oxidsalz *n*, *(i. w. S.)* basisches Salz *n*
oxytocic (agent) *(Pharm)* Uterotonikum *n*, Oxytozikum *n*, Wehenmittel *n*, oxytozisches [wehenerregendes] Mittel *n*
ozokerite *(Min)* Ozokerit *m*, Bergwachs *n*, Erdwachs *n*, Bergtalg *m*
ozonate *v s.* ozonize
ozonation Ozoni(si)erung *f*
ozonator *s.* ozone generator
ozone Ozon *n*, Trisauerstoff *m*
ozone-destroying ozonzerstörend
ozone destruction Ozonzerstörung *f*
ozone generator Ozongenerator *m*, Ozonisator *m*
ozone hole Ozonloch *n*
ozone layer Ozonsphäre *f*, Ozonschicht *f* *(der Stratosphäre)*
ozone photolysis Photolyse *f* von Ozon
ozone resistance Ozonbeständigkeit *f*, Ozonfestigkeit *f*, Ozonresistenz *f*, Ozonwiderstand *m*
ozone-resistant ozonbeständig, ozonfest, ozonresistent
ozone-resisting ozonbeständig, ozonfest, ozonresistent
ozonic ozonartig; ozonhaltig; Ozon...
ozonide Ozonid *n*
ozonify *v s.* ozonize
ozonization 1. Ozon(is)ierung *f*; 2. *(Hyd)* Ozonung *f*, Ozonbehandlung *f*
ozonize *v* 1. ozon(is)ieren *(z. B. Luft, Wasser, chemische Verbindungen)*; 2. ozonisieren, in Ozon verwandeln; sich in Ozon verwandeln
ozonizer *s.* ozone generator
ozonolysis Ozonolyse *f*, Ozon(id)spaltung *f*, Ozonabbau *m*
ozonosphere Ozonosphäre *f*, Ozonschicht *(der Stratosphäre)*

P

p-doping p-Dotierung *f*
p-n junction pn-Übergang *m*
P site *(Bioch)* P-Ort *m*, Peptidbindungsort *m*, Peptidylort *m*, Donorort *m*, Ausgang *m*
p-type conducting [semiconducting] p-leitend, defektleitend
p-type semiconductor Halbleiter *m* vom

p-Typ, p-(Typ-)Halbleiter *m*, Defekt(elektronen)halbleiter *m*, Mangelhalbleiter *m*
PA *s.* phthalic anhydride
PABA = *p*-aminobenzoic acid
pacemaker enzyme Schrittmacherenzym *n*
pack *v* 1. packen *(Chromatographiesäulen)*; (mit Füllstoffen) füllen *(Kolonnen, Säulen)*; 2. *(Met)* einpacken, verpacken, einbetten, umgeben *(z. B. mit einem Zementationspulver)*; 3. einpacken, verpacken, paketieren *(Produkte)*; 4. (ab)dichten
pack 1. Packen *m*, Ballen *m*; Packung *f* *(abgepackte Menge)*; 2. *(Med)* Packung *f*; 3. *(Tech)* Einbau *m*, Einbauten *mpl*
pack carburizing Aufkohlen *n* in festen Kohlungsmitteln, Pulveraufkohlen *n*, Zementieren *n* in festen Einsatzmitteln, Zementieren *n* in festem Einsatz, Pulverzementieren *n*
package *v* verpacken, einpacken, paketieren
package 1. Einpacken *n*, Verpacken *n*; 2. Packung *f (abgepackte Menge)*; 3. Verpackung *f*, Emballage *f*, Verpackungsmaterial *n*
packed bed bioreactor [fermenter] *(Biot)* Festbettreaktor *m*
packed capillary column Mikrosäule *f*, gepackte Kapillarsäule *f (Gaschromatographie)*
packed column 1. *(Dest)* Füllkörperkolonne *f*; 2. *(Chromat)* gepackte Säule *f*
packing 1. Packen *n (von Chromatographiesäulen)*; Füllen *n* (mit Füllstoffen); 2. *(Met)* Einpacken *n*, Verpacken *n*, Einbetten *n*, Umgeben *n (z. B. mit einem Zementationspulver)*; Einpacken *n*, Verpacken *n (von Produkten)*; Dichten *n*, Abdichten *n*; 3. Packung *f*, Füllung *f*, Schüttung *f*; Füllung *f*, Füllmaterial *n (für Kolonnen, Säulen)*; Verpackung *f*, Verpackungsmaterial *n*, Emballage *f*; Dichtung *f (Bauelement)*; 4. *(Krist)* Packung *f*, Kugelpackung *f*
packing density Packungsdichte *f*
packing material 1. Verpackungsmaterial *n*, Verpackung *f*, Emballage *f*; 2. *(Met)* Einpackmittel *n*, Einbettungsmittel *n*, Einbettungswerkstoff *m*, Einbettungsmasse *f*; 3. Dicht(ungs)werkstoff *m*, Dichtungsstoff *m*, Dicht(ungs)material *n*, Dicht(ungs)mittel *n*; Packungs(werk)stoff *m*, Packungsmaterial *n*; Füllmaterial *n*, Füllung *f*, Füllstoff *m*
packings Füllkörper *mpl*
PACN *s.* polyacrylonitrile
pad Unterlage *f*, Kissen *n*, Polster *n*
pad dyeing *(Text)* Klotzfärben *n*, Klotzfärbung *f*, Foulardfärbung *f*
paddle Schaufelblatt *n*, Schaufel *f*, Blatt *n*, Flügel *m*, Paddel *n*
paddle agitator (mixer) Blattrührer *m*, Paddelrührer *m*
paddle granulator Schaufelgranulator *m*
PAFC *s.* phosphoric-acid fuel cell
PAGE = polyacrylamide gel electrophoresis
PAH *s.* polycyclic aromatic (hydrocarbon)
pain killer [reliever] *(Pharm)* schmerzstillendes [schmerzlinderndes] Mittel *n*, Schmerzlinderungsmittel *n*, Analgetikum *n*
pain-relieving schmerzstillend, schmerzlindernd, analgetisch
paint *v (Anstr)* (an)streichen
paint 1. pigmentierter Anstrichstoff *m*, Anstrichfarbe *f*; 2. *s.* paint coat
paint binder Anstrichstoffbindemittel *n*, Farb(en)bindemittel *n*
paint chemist Lack- und Farbenchemiker *m*
paint chemistry Lack- und Farbenchemie *f*
paint coat Anstrich *m*, Anstrichschicht *f*
paint coating *s.* 1. paint coat; 2. paint 1.
paint extender Extender *m*, Füllstoff *m*, Verschneidmittel *n*, Verschnittmittel *n*, Streckmittel *n*
paint film Anstrichfilm *m*
paint-film defect Anstrichfehler *m*, Anstrichmangel *m*
paint pigment Anstrichpigment *n*, Farbpigment *n*, Pigment *n*
paint polish Lackglanzmittel *n*
paint remover Abbeizmittel *n*, Abbeizstoff *m*, Entlackungsmittel *n*, Lackentferner *m*
paint spraying Farbspritzen *n*
paint system Anstrichsystem *n*, Anstrichaufbau *m*

paint vehicle 1. Bindemittellösung *f*; 2. *s.* paint binder
painter's naphtha Lackbenzin *n*
painter's putty Glaserkitt *m*, Fensterkitt *m*
pair *v* sich paaren, ein Paar bilden, paarweise [zu einem Paar] zusammentreten, sich paarweise vereinigen; paaren *(Elektronen, Nukleonen)*
pair of electrons Elektronenpaar *n*
pairing Paarung *f*, Paarbildung *f (von Elektronen, Nukleonen)*
pale oil helles Öl *n*, Pale Oil *n (leichtgefärbtes raffiniertes Schmieröldestillat)*
palladate Palladat *n (Komplexsalz mit Pd als Zentralatom)*
palladic Palladium..., *(i. e. S.)* Palladium(IV)-...
palladic oxide Palladium(IV)-oxid *n*, Palladiumdioxid *n*
palladium Pd Palladium *n*
palladium black Palladiummohr *n*, Palladiumschwarz *n*
palladium diiodide Palladiumdiiodid *n*, Palladium(II)-iodid *n*
palladium hydride Palladiumhydrid *n*
palladium monoxide Palladiummonoxid *n*, Palladium(II)-oxid *n*
palladous Palladium..., *(i. e. S.)* Palladium(II)-...
palladous potassium chloride Kaliumtetrachloropalladat(II) *n*
palm kernel (nut) oil Palmkernöl *n*, *(i. e. S.)* Palmkernfett *n (raffiniertes Palmkernöl)*
palmitate Palmitat *n (Salz oder Ester der Palmitinsäure)*
palmitic acid Palmitinsäure *f*, *Hexadecansäure *f*
palmitin Palmitinsäureglycerylester *m*, *(i. e. S.)* Tripalmitin *n*, Glyceryltripalmitat *n*
pan 1. Pfanne *f*, Mulde *f*, Trog *m*, Schüssel *f*, Schale *f*, Wanne *f*; 2. Waagschale *f*, Schale *f*
PAN = peroxy acetyl nitrate
panchromatize *v (Foto)* panchromatisch machen [sensibilisieren]
pantoic acid *(Bioch)* Pantoinsäure *f*
pantothenic acid Pantothensäure *f*, *N*-(2,4-Dihydroxy-3,3-dimethylbutyryl)-β-alanin *n*
paper Papier *n* • **make into paper** zu [auf] Papier verarbeiten
paper chromatogram Papierchromatogramm *n*
paper-chromatographic papierchromatographisch
paper chromatography Papierchromatographie *f*
paper colouring Papierfärbung *f*, Papierfärben *n*, Papierfärberei *f (mit löslichen oder unlöslichen Farbstoffen)*
paper converting *(Pap)* Papierverarbeitung *f*
paper dyeing Papierfärbung *f*, Papierfärben *n*, Papierfärberei *f (mit löslichen Farbstoffen)*
paper electrophoresis Papierelektrophorese *f*
paper filler Papierfüllstoff *m*
paper filter Papierfilter *n*
paper finishing Papierausrüstung *f*, Fertigstellung *f* des Papiers
paper machine Papier(herstellungs)maschine *f*, *(i. e. S.)* Langsieb(papier)maschine *f*
paper machine wire Papiermaschinensieb *n*
paper mill waste (water) Papierfabrikabwasser *n*
paper print *(Foto)* Papierabzug *m*, Papierbild *n*
paper pulp 1. Papierzellstoff *m*, Zellstoff *m* für die Papierindustrie; 2. *s.* paper stock
paper sizing Papierleimung *f*
paper stock Papierrohstoff *m*, Rohstoff *m* für die Papiererzeugung, Faserrohstoff *m*, Papierfaserstoff *m*; (fertiger) Papierstoff *m*, Papiermasse *f*, Papierbrei *m*, Ganzstoff *m*, Ganzzeug *n*, Stoff *m*
paper web Papierbahn *f*, Stoffbahn *f*, Papiervlies *n*, Faserfilz *m*
paperboard Pappe *f*; Karton *m*
papermaking Papierherstellung *f*, Papiererzeugung *f*, Papierfabrikation *f*, Papiermachen *n*, Papiermacherei *f*
PAPS = 3-phosphoadenosine-5-phosphosulphate
para paraständig, *p*-ständig, in *para*-Stellung [*p*-Stellung, 1,4-Stellung] • **in para position** paraständig, *p*-ständig, in *para*-Stellung [*p*-Stellung, 1,4-Stel-

lung] • **be (located, situated) para** sich in *para*-Stellung [*p*-Stellung, 1,4-Stellung] befinden, in *para*-Stellung stehen, paraständig [*p*-ständig] sein
para hydrogen *para*-Wasserstoff *m*, Parawasserstoff *m*
para isomer *para*-Isomer(e) *n*, *p*-Isomer(e) *n*
para red 1. *s.* paranitraniline red; 2. *Benennung für mehrere rote Azofarbstoffe*
para substituent *para*-Substituent *m*, *p*-Substituent *m*, Substituent *m* in *para*-Stellung [*p*-Stellung], *para*-ständiger Substituent *m*
para-substituted *para*-substituiert, *p*-substituiert
para-substitution *para*-Substitution *f*, *p*-Substitution *f*
Para (rubber) Parakautschuk *m*, Para(-gummi) *m (von Hevea brasiliensis (H. B. K.) Muell Arg.)*
parabanic acid Parabansäure *f*, Oxalylharnstoff *m*, Imidazoltrion *n*
parabens *(Lebm)* 4-Hydroxybenzoesäureester *mpl*, *p*-Hydroxybenzoesäureester *mpl*, PHB-Ester *mpl (Konservierungsmittel)*
parachor *(phys Ch)* Parachor *m(n)*, P
paracrystal 1. Parakristall *m (kristallähnlicher Körper mit nur annähernder Gitterperiodizität)*; 2. Metakristall *m (Gitterordnung in weniger als 3 Dimensionen)*; 3. *s.* liquid crystal
paracrystalline 1. parakristallin; 2. metakristallin
paraffin 1. *s.* paraffin wax; 2. *s.* paraffin hydrocarbon; 3. Petroleum *n*, Kerosin *n*, Leuchtöl *n*
paraffin-base petroleum paraffinbasisches [paraffinisches, paraffinöses] Erdöl *n* [Rohöl *n*], Paraffin(basis)öl *n*, Erdöl *n* auf Paraffinbasis
paraffin hydrocarbon Paraffinkohlenwasserstoff *m*, gesättigter Kohlenwasserstoff *m*, *Alkan *n*
paraffin oil 1. Paraffinöl *n*; 2. *(Pharm)* flüssiges Paraffin *n*; 3. Petroleum *n*, Kerosin *n*, Leuchtöl *n*
paraffin series Paraffinreihe *f*, Alkanreihe *f*
paraffin slack wax Paraffingatsch *m*, Gatsch *m*
paraffin wax (festes) Paraffin *n*, Festparaffin *n*, Paraffinwachs *n*
paraffinic paraffinisch; Paraffin...
paraform(aldehyde) Paraformaldehyd *m*
parafuchsine Parafuchsin *n*
parahydrogen Parawasserstoff *m*, para-Wasserstoff *m*
paraldehyde Paraldehyd *m*, Paracetaldehyd *m*, *2,4,6-Trimethyl-1,3,5-trioxan *n*
parallax error *(Anal)* Parallaxenfehler *m*, parallaktischer Fehler *m* [Ablesefehler *m*], Parallaxe *f*
parallel-disk viscosimeter Parallelplattenviskosimeter *n*
parallel flow Parallelstrom *m*, Gleichstrom *m*
parallel reaction Parallelreaktion *f*, Nebenreaktion *f*
parallel synthesis Parallelsynthese *f*
paramagnetic paramagnetisch
paramagnetic (material) paramagnetischer Stoff *m*, paramagnetische Substanz *f*, Paramagnetikum *n*
paramagnetic resonance paramagnetische Resonanz *f*, EPR, PMR, Elektronenspinresonanz *f*, ESR, Elektronenresonanz *f*
paramagnetic resonance spectroscopy paramagnetische Resonanzspektroskopie *f*, EPR-Spektroskopie *f*, Elektronen(spin)resonanzspektroskopie *f*, ESR-Spektroskopie *f*
paramagnetic susceptibility paramagnetische Suszeptibilität *f*
paramagnetism Paramagnetismus *m*
paranitraniline red Paranitranilinrot *n*, Nitranilinrot *n*, Pigmentrot *n*, Pararot *n*, 1-(4-Nitrophenyl)azonaphth-2-ol *n*
parapositronium Para-Positronium *n (Positron und Elektron mit gleich gerichteten Spins)*
pararosaniline Pararosanilin *n*
pararosolic acid Pararosolsäure *f*, *p*-Rosolsäure *f*, Aurin *n*
parasympathomimetic *(Pharm)* parasympath(ik)omimetisch
parasympathomimetic agent Parasympath(ik)omimetikum *n*
parathion Parathion *n (0,0-Diethyl-0-p-nitrophenylester der Thiophosphorsäure, ein Insektizid)*

parathyroid hormone Parathormon *n*, *Parathyrin *n*, PTH
parent 1. Stamm *m*; 2. *s.* parent element; 3. *s.* parent substance
parent compound Stammverbindung *f*, Ausgangsverbindung *f*
parent element *(Kern)* Anfangsglied *n*, Mutterelement *n*, Muttersubstanz *f*, Ausgangselement *n (einer Zerfallsreihe)*
parent hydrocarbon Grundkohlenwasserstoff *m*, Stammkohlenwasserstoff *m*
parent lot *(Anal)* Gesamtmenge *f*, Grundgesamtheit *f*, Prüfgut *n*
parent name Stammname *m*
parent structure Grundgerüst *n (eines Moleküls)*
parent substance Muttersubstanz *f*, Grundsubstanz *f*, Grundstoff *m*, Ausgangssubstanz *f*, Ausgangsstoff *m*
parison *(Kst)* Rohling *m*, Vorformling *m*
Parr bomb *(Lab)* Parr-Bombe *f*
part by volume Raumteil *m*, Volumenteil *m*
part by weight Masseteil *m*
partial partial, partiell, Partial..., teilweise, Teil...
partial analysis Teilanalyse *f*
partial combustion partielle Verbrennung *f*, Teilverbrennung *f*
partial condensation partielle Kondensation *f*, Teilkondensation *f*, Dephlegmation *f*, Dephlegmierung *f*
partial condenser Teilkondensator *m*, Dephlegmator *m*
partial heat of solution differenzielle Lösungswärme *f* [Lösungsenthalpie *f*]
partial miscibility begrenzte [teilweise] Mischbarkeit *f*
partial molar quantity partielle molare Größe *f*
partial molar volume partielles molares Volumen *n*, partielles Molvolumen *n*
partial order (of reaction) Reaktionsordnung *f* bezüglich eines Reaktanten, Teilordnung *f*
partial oxidation partielle [teilweise] Oxidation *f*, Partialoxidation *f*
partial pressure Partialdruck *m*, Teildruck *m*
partial reaction Teilreaktion *f*
partial treatment *(Hyd)* Teilreinigung *f*
partially miscible teilweise [begrenzt] mischbar
partially soluble teilweise löslich
participate *v* teilnehmen *(z. B. an einer Reaktion)*
particle 1. Teilchen *n*, Partikel *n*; 2. *(Kern)* Korpuskel *n*
particle accelerator Teilchenbeschleuniger *m*
particle collision Teilchenzusammenstoß *m*, Teilchenkollision *f*
particle emission Partikelemission *f*
particle growth Teilchenwachstum *n*
particle-induced X-ray emission *(Anal)* partikelinduzierte Röntgenemission *f*
particle radiation Teilchenstrahlung *f*, Korpuskularstrahlung *f*
particle shape Teilchenform *f*; Kornform *f*, Korngestalt *f*
particle size 1. Teilchengröße *f*, Partikelgröße *f*; 2. *(Tech)* Korngröße *f*, *(bei gröberem Material auch)* Stückgröße *f*; Tröpfchengröße *f*
particle-size distribution Korn(größen)verteilung *f*; Teilchengrößenverteilung *f*, Partikelgrößenverteilung *f*
particle-size range Kornspanne *f*, Korngrößenbereich *m*, Körnungsbereich *m*, Teilchengrößenbereich *m*
α-particle *(Kern)* α-Teilchen *n*, Alphateilchen *n (Heliumkern)*
β-particle *(Kern)* β-Teilchen *n*, Betateilchen *n (Elektron)*
particulate matter 1. partikulärer Stoff *m (fest oder flüssig)*; 2. *s.* particulates
particulates 1. suspendierte Teilchen *npl* [Partikel *npl*, Feststoffteilchen *npl*], Schwebeteilchen *npl*, Suspensionspartikel *npl*, Trübeteilchen *npl*; 2. Staubteilchen *npl*; Stäube *mpl*
parting 1. Lösen *n*, Ablösen *n*, Trennen *n*, Trennung *f*; 2. *(Aufb)* Scheiden *n*, Scheidung *f (von Gold und Silber)*
parting agent [compound] Trennmittel *n*
partition *v* 1. teilen, trennen; abteilen; 2. verteilen
partition 1. Teilen *n*, Trennen *n*; Abteilen *n*; Verteilen *n*; 2. Fach *n*, Abteilung *f*; 3. Trennwand *f*, Scheidewand *f*, Zwischenwand *f*, Membran *f*
partition chromatography Verteilungschromatographie *f* • **by partition**

chromatography verteilungschromatographisch
partition coefficient 1. Verteilungskoeffizient *m*, *Verteilungskonstante *f*; 2. Segregationskonstante *f*, Abscheidungskonstante *f*, Verteilungskoeffizient *m* *(in der Zonenschmelztheorie)*
partition function *(phys Ch)* Verteilungsfunktion *f*, Zustandssumme *f*
partition law Verteilungsgesetz *n*; nernstsches Verteilungsgesetz *n*, nernstscher Verteilungssatz *m*
partition ratio Verteilungsverhältnis *n*
partly saponified partiell verseift
parts per billion Teile *mpl* je Billion Teile; *(Am)* Teile *mpl* je Milliarde Teile
parts per hundred million Teile *mpl* je hundert Millionen Teile, pphm
parts per million Teile *mpl* je Million Teile, ppm
parts per million by mass Teile *mpl* je Million Teile, ppmm *(Masseanteil)*
parts per million by volume Teile *mpl* je Million Teile, ppmv *(Volumenanteil)*
parts per thousand pro mille, Teile *mpl* je tausend Teile, ‰
parts per trillion by volume Teile *mpl* je Trillion Teile, pptv *(Volumenanteil)*
PAS *s.* photoacoustic spectroscopy
Paschen-Runge mounting *(Spektr)* Paschen-Runge-Aufstellung *f*, Paschen-Runge-Anordnung *f*
Paschen series *(Spektr)* Paschen-Serie *f*
Pase = phosphatase
pass *v* 1. einleiten; (hindurch)leiten *(Gase, Flüssigkeiten)*; 2. durchfließen, durchströmen *(z. B. eine Leitung)*; passieren *(einen Anlagenteil)*; durchlaufen, hindurchgehen *(z. B. Papierbahn durch die Trockenpartie)*; 3. übergehen *(in einen anderen Zustand)*; 4. *s.* pass over
pass *v* **in** einleiten
pass *v* **over** überleiten; überdestillieren, übergehen
pass *v* **through** 1. (hin)durchleiten; 2. passieren, hindurchgehen, hindurchtreten, *(durch ein Sieb auch)* hindurchfallen, *(durch ein Filter auch)* hindurchlaufen
pass *v* **up(wards)** nach oben steigen, aufsteigen, emporsteigen; aufwärts [nach oben] führen [transportieren], hochfördern
pass Durchgang *m*, Passage *f*
passage Durchgang *m*, Durchlauf *m*, Durchfluss *m*, Durchtritt *m*
passivate *v* passivieren *(ein unedles Metall)*
passivation Passivierung *f (eines unedlen Metalls)*
passivator Passivierungsmittel *n*, Passivator *m*
passive passiv, inaktiv
passivity Passivität *f*
paste *v* anpasten, anteigen, anreiben, anrühren
paste 1. Paste *f*, Brei *m*, Teig *m*; Kleister *m*; 2. *(Ker)* Masse *f* • **make into a paste** *s.* paste
paste-like pastenartig, pastenförmig, pastös, pastig
paste mill [mixer] Kneter *m*
pasteurization Pasteurisieren *n*, Pasteurisierung *f*, Pasteurisation *f*
pasteurize *v* pasteurisieren
pasteurizer Pasteuriseur *m*, Pasteurisierapparat *m*, *(i. e. S.)* Milcherhitzer *m*, Milcherhitzungsapparat *m*
pastille *(Pharm)* Pastille *f*
pasting Anpasten *n*, Anteigen *n*, Anreiben *n*, Anreibung *f*, Anrühren *n*
pasty pastenartig, pastenförmig, pastös, pastig
Paterno-Büchi reaction Paterno-Büchi-Reaktion *f (zu einem Oxethanderivat führende Photocycloaddition)*
path 1. Weg *m (einer Reaktion)*; 2. *(phys Ch)* Weg *m*, Bahn *f*, Flugbahn *f*; Weglänge *f*
path dependence Wegabhängigkeit *f*
path independence Wegunabhängigkeit *f*
pathogen pathogener [krankheitserregender] Keim *m*, Krankheitserreger *m*
pathogenic pathogen, krankheitserregend, krankheitserzeugend
pathway Weg *m (einer Reaktion)*
pathway of metabolism Stoffwechselweg *m*, Stoffwechselablauf *m*, Stoffwechselbahn *f*
patina Patina *f*, Edelrost *m*
patrix Patrize *f*, Stempel *m*, Stempelprofil *n*, Pressstempel *m*

pattern Muster *n*, Probe *f*; Schablone *f*, Lehre *f*; Modell *n*
Pauli (exclusion) principle (paulisches) Ausschließungsprinzip *n*, Pauli-Prinzip *n*
Pauly (protein) reaction paulysche Diazoreaktion *f* [Reaktion *f*]
pay ore abbauwürdiges Erz *n*
PBB *s.* polybrominated biphenyl
PBI *s.* polybenzimidazole
PC *s.* paper chromatography
PCB *s.* polychlorinated biphenyls
PCDD *s.* polychlorinated dibenzo-*p*-dioxin
PCDF *s.* polychlorinated dibenzofuran
PCP 1. Pentachlorphenol *n*, PCP *n* *(Herbizid)*; 2. *s.* polychloroprene
PCR *s.* polymerase chain reaction
pct = per cent
PCTFE *s.* polychlorotrifluoroethylene
pcu *(pound centigrade unit) s.* centigrade heat unit
pd. *s.* powder
PD *s.* potential difference
p. d. *s.* potential difference
P.D. *s.* pressure distillate
PDMS *s.* polydimethylsiloxane
PE *s.* 1. pentaerythritol; 2. polyethylene
peak 1. Höchstwert *m*, Maximum *n*, Scheitelwert *m*; Scheitelpunkt *m*, Gipfelpunkt *m (z. B. einer Kurve)*; 2. *(Anal)* Peak *m*, Spitze *f*, spitzes Signal *n*, Peaksignal *n*, Zacke *f*
peak area Peakfläche *f*, Fläche *f* unter dem Peak, Signalfläche *f*
peak broadening Signalverbreiterung *f*, Bandenverbreiterung *f*, Peakaufweitung *f*
peak elution volume *(Chromat)* Peakelutionsvolumen *n*
peak height Peakhöhe *f*, Signalhöhe *f*
peak integration Peakintegration *f*
peak intensity Peakintensität *f*, Signalintensität *f*
peak position Peaklage *f*, Signallage *f*, Signalposition *f*
peak shape Peakform *f*, Signalgestalt *f*, Signalform *f*
peak tailing *(Chromat)* Schwanzbildung *f*, Schweifbildung *f*, *Tailing *n*
peak temperature Spitzentemperatur *f*
peak width at half height *Peakbreite *f* in halber Höhe, Halbhöhenbreite *f*
pearl polymerization Perlpolymerisation *f*, Suspensionspolymerisation *f*
pearlite *(Met)* Perlit *m (lamellares Aggregat aus Ferrit und Zementit)*
pearlitic perlitisch
peat Torf *m*
peat coke Torfkoks *m*
peat gas Gas *n* aus Torf, Torfgas *n*
peat tar Torfteer *m*
peaty torfig
pebble 1. Kiesel(stein) *m*; 2. Pebble *n*, Wärmestein *m*, Steinkugel *f (Kugel aus hitzebeständigem Material oder Koks im Wärmesteinerhitzer)*
pebble heater Pebble-Heater *m*, Wärmesteinerhitzer *m*, Steinerhitzer *m*
pebble mill Kugelmühle *f (mit Flintstein- oder Hartporzellankugelfüllung)*
pectate Pektat *n (Salz oder Ester der Pektinsäure)*
pectic acid Pektinsäure *f*, Poly-D-galacturonsäure *f*
pectic enzyme pektisches Enzym *n*, Pektin(glykosid)ase *f*, pektinabbauendes [pektinspaltendes] Enzym *n*
pectin Pektin *n (i. w. S.)*; Pektinstoff *m*; Pektin *n*, pektinige Säure *f (teilweise oder voll methoxylierte Poly-D-galacturonsäure)*
pectinate Pektinat *n (Salz oder Ester der pektinigen Säure)*
pectinic acid pektinige Säure *f*, Pektin *n (teilweise oder voll methoxylierte Poly-D-galacturonsäure)*
pectization Pektisation *f*, Koagulation *f*, Flockung *f*, Ausflockung *f*
pectize *v* pektisieren, koagulieren, (aus)flocken
pectolytic pekt(in)olytisch, pektinabbauend, pektinspaltend
pedologic(al) bodenkundlich, pedologisch
pedologist Bodenkundler *m*
pedology Bodenkunde *f*, Pedologie *f*
PEDOT *s.* poly(3,4-ethylenedioxythiophene)
peel *v* schälen, enthülsen; abschälen, ablösen *(z. B. eine Folie)*; sich ablösen, abplatzen, abspringen, sich abschälen, abblättern *(Schicht)*
peel *v* **off** sich ablösen, abplatzen, abspringen, sich abschälen, abblättern *(Schicht)*

PEG *s.* polyethylene glycol
Peierls transition Peierls-Übergang *m*, Peierls-Phasenübergang *m (ein Metall-Isolator-Übergang)*
pelargonic acid Pelargonsäure *f*, *Nonansäure *f*
pelargonic aldehyde Pelargonaldehyd *m*, *Nonanal *n*
pellagra-preventive factor Antipellagravitamin *n*, Vitamin *n* PP, PP-Faktor *m (Nicotinsäureamid oder Nicotinsäure)*
pellet 1. Pellet *n*, Kugel *f*, Kügelchen *n*, Granalie *f*, Körnchen *n*, kug(e)liges Granulatkorn *n*; Pille *f*; Plätzchen *n*; 2. *(Spektr)* Pressling *m*, Tablette *f*
pelletize *v* 1. pellet(is)ieren, zu Pellets [Kügelchen] formen, rollgranulieren; 2. *(Aufb)* pellet(is)ieren, kugelsintern *(Erz)*
pelletized catalyst pillenförmiger [in Pillenform gepresster] Katalysator *m*, Katalysator *m* in Pillenform
pelletizer Pellet(is)iermaschine *f*, Granulierapparat *m*, Granulator *m*
pelletizing conveyor Granulierband *n*
pelletizing machine Pellet(is)iermaschine *f*, Granulierapparat *m*, Granulator *m*
pellicular packing [support] *(Chromat)* schalenporöser Füllkörper *m* [Trennkörper *m*], PLB-Teilchen *n*, Dünnschichtteilchen *n*
PEMFC *s.* proton exchange membrane fuel cell
pen recorder *(Anal)* Schreiber *m*
penam Penam *n (Vertreter einer Gruppe von Antibiotika)*
pendant drop hängender Tropfen *m*
penem Penem *n (Vertreter einer Gruppe von Antibiotika)*
penetrate *v* (durch)tränken, imprägnieren; penetrieren, durchdringen, durchwandern, passieren; eindringen
penetrating odour durchdringender [penetranter] Geruch *m*
penetrating radiation durchdringende [harte] Strahlung *f*
penetration 1. Tränkung *f*, Durchtränkung *f*, Imprägnierung *f*; Penetration *f*, Durchdringung *f*, Durchwandern *n*; Eindringen *n*; 2. *s.* penetration depth
penetration complex Durchdringungskomplex *m*, Innerkomplex *m*, Innenkomplex *m*
penetration depth Eindringtiefe *f*
penetration twins *(Krist)* Durchwachsungszwillinge *mpl*, Penetrationszwillinge *mpl*, Durchdringungszwillinge *mpl*
penicillamine Penicillamin *n*, β,β-Dimethylcystein *n*, β-Thiolvalin *n*
penicillanic acid Penicillansäure *f (Baustein der Penicilline)*
penicillic acid Penicill(in)säure *f*, 3-Methoxy-5-methyl-4-oxo-hexa-2,5-diensäure *f*
penicillin Penicillin *n (Vertreter einer Gruppe von Antibiotika)*
penicilloate Penicilloat *n (bei der Alkali-Inaktivierung entstehendes Abbauprodukt eines Penicillins)*
penicilloic acid Penicillo(in)säure *f (allg. für eine Amidodicarbonsäure der Formel* $RCONH(C_6H_{10}NS)(COOH)_2$*)*
penillic acid Penillsäure *f (ein Inaktivierungsprodukt eines Penicillins in saurer Lösung)*
penilloic acid Penillosäure *f*
Penning gauge Kaltkatoden-Ionisationsvakuummeter *n*, Penning-Vakuummeter *n*, Philips-Vakuummeter *n*
Pensky-Martens flash-point apparatus [tester] Flammpunktprüfer *m* nach Pensky-Martens, Pensky-Martens-Gerät *n*, Pensky-Martens-Apparat *m*
pentaaqua... Pentaquo...
pentaaquo... Pentaquo...
pentabasic 1. fünfbasig, fünfwertig, fünfprotonig *(Säure)*; 2. fünfwertig, fünfsäurig *(Base oder basisches Salz)*
pentacene Pentacen *n*, Dibenzo[b,i]anthracen *n*
pentachloride Pentachlorid *n*
pentachlorophenol Pentachlorphenol *n*, PCP *(Herbizid)*
pentacid fünfsäurig *(Base)*
pentacoordinate(d) pentakoordiniert, fünffach koordiniert
pentacovalent koordinativ fünfwertig
pentadecane Pentadecan *n*
pentadecanoic acid Pentadecansäure *f*
pentadecanol Pentadecanol *n*, *(i. e. S.)* *Pentadecan-1-ol *n*, Pentadecylalkohol *m*
pentadecyl alcohol Pentadecylalkohol *m*, *Pentadecan-1-ol *n*
pentadentate fünfzähnig, fünfzählig *(Komplexchemie)*

pentaerythritol Pentaerythritol *n*, *2,2-Bis(hydroxymethyl)propan-1,3-diol *n*
pentahydrate Pentahydrat *n*
pentahydric fünfwertig, mit 5 Hydroxylgruppen [OH-Gruppen] *(Alkohol, Phenol)*
pentahydroxy 1. Pentahydroxy...; 2. *s.* pentahydric
pentamethylenediamine *Pentamethylendiamin *n*, *Pentan-1,5-diamin *n*, Cadaverin *n*
pentammine Pentammin *n (ein Ammin mit 5 Ammoniakmolekülen)*
pentanal *Pentanal *n*, Valeraldehyd *m*
pentane Pentan *n*
pentanecarboxylic acid Pentancarbonsäure *f*
pentanedial *Pentandial *n*, Glutar(säure)dialdehyd *m*, Glutaraldehyd *m*
pentanedicarboxylic acid Pentandicarbonsäure *f*
pentanedioic acid *Pentandisäure *f*, Glutarsäure *f*
pentanediol Pentandiol *n (zweiwertiger Alkohol des Pentans)*
pentanedione Pentandion *n (Diketon des Pentans)*
pentanegative fünffach negativ (geladen)
pentanoic acid *Pentansäure *f*, Valeriansäure *f*
pentanol Pentanol *n*, *(i. e. S.)* Pentan-1-ol *n*, Pentylalkohol *m*
pentanone *(org Ch)* Pentanon *n*
pentapeptide Pentapeptid *n (Peptid mit 5 Aminosäureeinheiten)*
pentapositive fünffach positiv (geladen)
pentasubstituted pentasubstituiert
pentasulphide Pentasulfid *n*
pentavalence Fünfwertigkeit *f*, Pentavalenz *f*
pentavalency Fünfwertigkeit *f*, Pentavalenz *f*
pentavalent fünfwertig, pentavalent
1-pentene Pent-1-en *n*
pentite Pentitol *n (fünfwertiger Alkohol)*
pentitol Pentitol *n (fünfwertiger Alkohol)*
pentosan Pentosan *n (aus Pentosen aufgebautes Polysaccharid)*
pentose Pentose *f (Monosaccharid mit 5 Kohlenstoffatomen)*
pentose phosphate pathway *(Bioch)* Pentosephoshatweg *m*, PP-Weg *m*, oxidativer Pentosephosphat-Zyklus *m*, oxidativer Pentoseweg *m*, direkter Oxidationsweg *m*, Warburg-Dickens-Horecker-Schema *n*, Warburg-Dickens-Horecker-Abbauweg *m*, Phosphogluconat-Weg *m*
pentoxide Pentoxid *n*
pentyne *(org Ch)* Pentin *n*
peptide Peptid *n*
peptide alkaloid Peptidalkaloid *n*, Alkaloid *n* der Peptidgruppe
peptide antibiotic Peptid-Antibiotikum *n*
peptide bond Peptidbindung *f*, peptidische Verknüpfung *f*, peptidartige Bindung *f*
peptide moiety [portion] Peptid(an)teil *m*, Peptidkomponente *f*
peptidyl-tRNA site *(Bioch)* Peptidbindungsort *m*, Peptidylort *m*, P-Ort *m*, Donorort *m*, Ausgang *m*
peptization 1. *(Koll)* Peptisierung *f*, Peptisation *f*, Gel-Sol-Umwandlung *f*, Gel-Sol-Übergang *m*; 2. *(Gum)* Peptisierung *f*, Peptisation *f*, Plastizierung *f* [Abbau *m*, Erweichung *f*] mit Peptisiermitteln
peptize *v* 1. *(Koll)* peptisieren; 2. *(Gum)* peptisieren, mit Peptisiermitteln erweichen [abbauen, plastizieren]
peptizer 1. *(Koll)* Peptisier(ungs)mittel *n*, Peptisationsmittel *n*, Peptisator *m*; 2. *(Gum)* Peptisier(ungs)mittel *n*, chemisches Plastizier(ungs)mittel *n* [Abbaumittel *n*]
per-acid 1. (anorganische) Persäure *f*, Übersäure *f*; 2. *s.* peroxy acid 1.; 3. *s.* peroxo acid
per cent composition prozentuale Zusammensetzung *f*
per-salt Persalz *n*
per substitution Persubstitution *f*
peracetic acid *s.* peroxyacetic acid
peracid *s.* 1. per-acid 1.; 2. peroxy acid 1.; 3. peroxo acid
perbenzoic acid *s.* peroxybenzoic acid
percentage prozentualer Anteil *m*; Prozentgehalt *m*, prozentualer Gehalt *m*, %-Gehalt *m*
percentage by mass Masseprozent *n*, Masse%
percentage by volume Volumenprozent *n*, Vol.-%

percentage error *(Anal)* *prozentualer Fehler *m*
percentage recovery *(Anal)* *Wiederfindungsrate *f*
perchlorate Perchlorat *n*
perchloric acid Perchlorsäure *f*
perchloromethane Kohlenstofftetrachlorid *n*, Tetrachlorkohlenstoff *m*, *Tetrachlormethan *n*, Perchlormethan *n*
percolate *v* durchsickern, durchlaufen; durchsickern [durchlaufen] lassen; perkolieren, durchseihen
percolate Perkolat *n*
percolation Durchsickern *n*, Durchlaufen *n*; Durchsickernlassen *n*; Perkolieren *n*, Perkolation *f*
percolation leaching *(Aufb)* Sickerlaugung *f*, Perkolation *f*
percolation water Sickerwasser *n*
percolator Perkolator *m*
perfect fluid ideales [reibungsfreies, pascalsches] Fluid *n*
perfect gas ideales [vollkommenes] Gas *n*
perfect gas law ideales Gasgesetz *n*, ideale Gasgleichung *f*, (thermische) Zustandsgleichung *f* idealer Gase
perfect-gas temperature scale ideale gasthermometrische Temperaturskala *f*
perfect lattice *(Krist)* ideales Gitter *n*, Idealgitter *n*
perfect mixing ideale Durchmischung *f*
perfect plate *(Dest)* idealer [theoretischer] Boden *m*, *(i. w. S.)* theoretische Trennstufe *f*
perfectly homogeneous vollkommen [vollständig, durch und durch] homogen
perfluorinate *v* perfluorieren
perfluorination Perfluorierung *f*
perfluoroethylene Perfluorethen *n*, *Tetrafluorethen *n*
perforated bowl centrifuge Filterzentrifuge *f*, Filtrierzentrifuge *f*, Siebzentrifuge *f*, Zentrifugal(kraft)filter *n*
perforated-plate column [tower] *(Dest)* Siebbodenkolonne *f*
perforation Perforation *f*, Perforieren *n*, Perforierung *f*; Sieblochung *f*, Bohrung *f*, Lochweite *f*, Öffnung *f (eines Siebs)*
perforator *(Lab)* Perforator *m (zur kontinuierlichen Extraktion)*
performance Leistungsverhalten *n*, Betriebsverhalten *n*, Leistung *f*
perfume 1. Wohlgeruch *m*, Duft *m*; 2. Parfüm *n*, Riechstoff *m*, Geruchsstoff *m*
perhydrogenate *v* perhydrieren
perhydrogenation Perhydrierung *f*
peri-fused perikondensiert, perianelliert
peri position peri-Stellung *f*, Peristellung *f*
pericyclic reaction *(org Ch)* pericyclische Reaktion *f*
period 1. Periode *f (z. B. im Periodensystem)*; 2. Periode *f*, Dauer *f*, Zeitabschnitt *m*, Zeit *f*
period of half-change *(phys Ch)* Halbwertszeit *f*
period of induction 1. *(phys Ch)* Induktionsperiode *f*, Anlaufperiode *f*, Startperiode *f*; 2. *s.* period of latency
period of latency *(Tox, Med)* Latenzzeit *f (Dauer bis zum Sichtbarwerden einer Reizwirkung)*
periodate Periodat *n*
periodic acid Periodsäure *f*, Überiodsäure *f*
periodic property periodische [sich periodisch wiederholende] Eigenschaft *f (der Elemente)*
periodic system (of the elements) Periodensystem *n* (der Elemente), PSE *n*, natürliches System *n* (der Elemente)
periodic table (of the elements) 1. Tabelle *f* [Tafel *f*] des Periodensystems (der Elemente); 2. *s.* periodic system
periodicity Periodizität *f*, periodischer Charakter *m*, periodische Wiederholung *f*
peripheral speed Umfangsgeschwindigkeit *f*
periplasmic binding protein periplasmatisches Bindungsprotein *n*
peristaltic pump *(Lab)* peristaltische Pumpe *f*
peritectic point *(phys Ch)* peritektischer Punkt *m*
peritectic reaction peritektische Reaktion *f* [Übergangsreaktion *f*]
peritectoid reaction peritektoide Reaktion *f* [Übergangsreaktion *f*]
Perkin condensation [reaction] Perkin-

Reaktion *f*, perkinsche Reaktion *f* [Synthese *f*], Perkin-Kondensation *f*
perlite *s.* pearlite
permanent gas permanentes Gas *n*
permanent hardness *(Hyd)* permanente [bleibende] Härte *f*, Nichtkarbonathärte *f*, NKH, Permanenthärte *f*
permanent magnet Dauermagnet *m*, Permanentmagnet *m*
permanent-mould casting Dauerformguss *m*, (normales) Kokillengießen *n*, (normaler) Kokillenguss *m*
permanganate Permanganat *n*, Manganat(VII) *n*
permanganate number Permanganatzahl *f*
permanganate test Baeyer-Probe *f*, baeyersche Probe *f*, Permanganatprobe *f (zum Nachweis von Alkenen)*
permanganate titration Titration *f* mit Permanganat, manganometrische Titration *f*
permanganic acid Permangansäure *f*
permeability Permeabilität *f*, Durchlässigkeit *f*, Durchdringbarkeit *f*
permeability coefficient Durchlässigkeitskoeffizient *m*
permeability to gas(es) Gasdurchlässigkeit *f*
permeable permeabel, durchlässig, durchdringbar
permeate *v* permeieren, durchdringen
permeate *(Hyd)* Permeat *n*, Dünnwasser *n*, Reinwasser *n (Membranfiltration)*
permeation Permeation *f*, Durchdringung *f*
permeation chromatography Gel(permeations)chromatographie *f*, GPC • **by permeation chromatography** gelpermeationschromatographisch
permeativity Permeiervermögen *n*, Eindringvermögen *n*, Durchdringungsvermögen *n*; Eindringfähigkeit *f*, Durchdringungsfähigkeit *f (z. B. von Pflanzenschutzmitteln)*
permissible daily body intake *(Tox)* duldbare [zulässige] Tagesdosis *f (in mg/kg Körpermasse)*
permitted discharge values *(Hyd)* Richtwerte *mpl* für Abwasserinhaltsstoffe
permittivity Dielektrizitätskonstante *f*, DK, *(i. e. S.)* absolute Dielektrizitätskonstante *f*
permselective membrane *(Hyd)* selektiv-permeable Membran *f*
peroxidase test Peroxidasetest *m*, Peroxidaseprobe *f*
peroxidation Peroxidieren *n*, Peroxidierung *f*, Peroxidation *f*; Epoxidieren *n*, Epoxidierung *f*, Epoxidation *f*
peroxide Peroxid *n*; H_2O_2 Wasserstoffperoxid *n*, Hydrogenperoxid *n*
peroxide bleaching Peroxidbleiche *f*
peroxide number Peroxidzahl *f*, Peroxidwert *m*
peroxide rancidity Peroxid(ranz)igkeit *f*
peroxide value *s.* peroxide number
peroxidic peroxidisch, Peroxid...
peroxidize *v* peroxidieren; epoxidieren
peroxo acid Peroxosäure *f*
peroxoborate Peroxoborat *n*
peroxoboric acid Peroxoborsäure *f*
peroxocarbonate Peroxocarbonat *n*
peroxocarbonic acid Peroxokohlensäure *f*
peroxochromate Peroxochromat *n*
peroxochromic acid Peroxochromsäure *f*
peroxodisulphate Peroxodisulfat *n*
peroxodisulphuric acid Peroxodischwefelsäure *f*
peroxophosphate Peroxophosphat *n*, *(i. e. S.)* $M^{I}_{3}[PO_5]$ Peroxomonophosphat *n*
peroxophosphoric acid Peroxophosphorsäure *f*, *(i. e. S.)* H_3PO_5 Peroxomonophosphorsäure *f*
peroxosulphate Peroxosulfat *n*, *(i. e. S.)* $M^{I}_{2}SO_5$ Peroxomonosulfat *n*
peroxosulphuric acid Peroxoschwefelsäure *f*, *(i. e. S.)* H_2SO_5 Peroxomonoschwefelsäure *f*
peroxy acid 1. Peroxysäure *f*, Persäure *f*; 2. *s.* peroxo acid
peroxy... Peroxy...; Peroxo... *(bei koordinativ gebundener –O–O-Gruppe)*
peroxyacetic acid Peressigsäure *f*, Peroxyessigsäure *f*, Peroxyethansäure *f*
peroxybenzoic acid Perbenzoesäure *f*, Peroxybenzoesäure *f*, Benzopersäure *f*
peroxycarboxylic acid Peroxycarbonsäure *f*
peroxyformic acid Perameisensäure *f*,

Peroxyameisensäure *f*, Permethansäure *f*
persist *v* beständig [stabil, persistent] sein *(z. B. Biozide)*; bestehen bleiben
persistence *(phys Ch)* Persistenz *f*; Beständigkeit *f*, Stabilität *f*, Persistenz *f* *(z. B. von Bioziden)*
persistence curve Abbaukurve *f (von Bioziden)*
persistent beständig, stabil, persistent *(z. B. Biozide)*; anhaltend, nachhaltig
persistent radical persistentes [langlebiges] Radikal *n*
personal error persönlicher [subjektiver] Fehler *m*
Peru balsam Perubalsam *m*, Peruanischer [Indischer] Balsam *m (von Myroxylon balsamum (L.) Harms var. pereirae)*
perylene Perylen *n*, peri-Dinaphthylen *n*
PES *s.* photoelectron spectroscopy
PESIS *s.* photoelectron spectroscopy of inner-shell electrons
pest control Schädlingsbekämpfung *f*
pesticide Pestizid *n*, Schädlingsbekämpfungsmittel *n*
pesticide residue Pestizidrückstand *m*, Schädlingsbekämpfungsmittelrückstand *m*
pestle *(Lab)* Stößel *m*, Pistill *n*, Mörserkeule *f*
PET *s.* polyethylene terephthalate
Petri (culture) dish Petrischale *f*
petrochemical 1. petrochemisch, gesteinschemisch; 2. petrolchemisch
petrochemical Petrolchemikalie *f*, Erdölchemikalie *f*, chemisches Produkt *n* auf Erdölbasis
petrochemistry 1. Petrochemie *f*, Gesteinschemie *f*; 2. Petrolchemie *f*, Erdölchemie *f*
petrographic(al) petrographisch
petrography Petrographie *f*, Gesteinskunde *f*
petrol Benzin *n (besonders als Vergaserkraftstoff)*
petrolatum Petrolat(um) *n*, Rohvaseline *f*, Rohvaselin *n*
petroleum Erdöl *n*
petroleum asphalt Erdölbitumen *n*, Erdölasphalt *m*, Petrolasphalt *m*
petroleum-based auf Erdölbasis, erdölstämmig
petroleum benzin Petroleumbenzin *n*
petroleum-derived aus Erdöl erzeugt
petroleum distillate Erdöldestillat *n*
petroleum distillation Erdöldestillation *f*
petroleum ether Petrol(eum)äther *m (Siedebereich 40 bis 70 °C)*; Leichtbenzin *n*, leichtes Benzin *n (Siedebereich 20 bis 35 °C)*
petroleum refinery Erdölraffinerie *f*
petroleum refining Erdölverarbeitung *f*, Erdölraffination *f*
petroleum residue Erdölrückstand *m*, Erdölresiduum *n*
petroleum spirit(s) Lösungsbenzin *n*
petroleum tar Erdölteer *m*
petroleum wax Erdölwachs *n*, Erdölparaffin *n*, Paraffinwachs *n* aus Erdöl
petrologic(al) petrologisch
petrology Petrologie *f*
PFIMS *s.* pyrolysis field ionization mass spectrometry
PFT *s.* pulse Fourier transform
Pfund series *(Spektr)* Pfund-Serie *f*
PGA *s.* pteroylglutamic acid
PGC *s.* pyrolysis gas chromatography
pH adjustment Einstellung *f* des pH-Werts
pH control pH-(Wert-)Regelung *f*; pH-Kontrolle *f*, Kontrolle *f* [Überwachung *f*] des pH-Werts
pH-dependent pH-(Wert-)abhängig
pH electrode pH-Elektrode *f*
pH-independent pH-(Wert-)unabhängig
pH indicator 1. pH-Indikator *m*; 2. pH-Anzeigegerät *n*
pH meter pH-Messgerät *n*, pH-Meter *n*
pH paper pH-Papier *n*
pH range pH-Bereich *m*, pH-Gebiet *n*
pH scale pH-Skala *f*
pH standard pH-Standard *m*
pH value pH-Wert *m*, pH *m(n)*, pH-Zahl *f*, Wasserstoff(ionen)exponent *m*
phage-resistant *(Biot)* phagenresistent
PhAR *s.* photosynthetically active radiation
pharmaceutic(al) pharmazeutisch
pharmaceutic(al) Pharmazeutikum *n*, pharmazeutisches Mittel *n*, Arzneimittel *n*
pharmaceutic(al) chemist Pharmakochemiker *m*; Arzneimittelchemiker *m*; Apotheker *m*
pharmaceutic(al) chemistry pharma-

zeutische Chemie *f*, Pharmakochemie *f*, Arzneimittelchemie *f*
pharmaceutic(al) dosage form Arzneiform *f*
pharmaceutic(al) preparation pharmazeutisches Präparat *n*, Arzneipräparat *n*, Arzneizubereitung *f*
pharmaceutics Pharmazeutik *f*
pharmac(eut)ist Pharmazeut *m*; Apotheker *m*
pharmacodynamic pharmakodynamisch
pharmacodynamics Pharmakodynamik *f*, Pharmakodynamie *f*
pharmacognostic pharmakognostisch
pharmacognosy Pharmakognosie *f*, Drogenkunde *f*
pharmacokinetics Pharmakokinetik *f*
pharmacologic(al) pharmakologisch
pharmacology Pharmakologie *f*
pharmacon Pharmakon *n (1. Arzneimittel; 2. Gift)*
pharmacop(o)eia Pharmakopöe *f*, Arzneibuch *n*
pharmacopoeial Pharmakopöe..., Arzneibuch..., offizinell; den Anforderungen des Arzneibuchs entsprechend
pharmacy Pharmazie *f*
phase Phase *f*
phase boundary Phasengrenze *f*
phase change *s.* phase transition
phase-contrast microscope Phasenkontrastmikroskop *n*
phase diagram Phasendiagramm *n*, Zustandsdiagramm *n*
phase difference Phasendifferenz *f*, Phasenunterschied *m*
phase equilibrium Phasengleichgewicht *n*, Gleichgewicht *n* der Phasen
phase inversion Phasenumkehr *f*
phase rule (of Gibbs) (gibbssches) Phasengesetz *n*, (gibbssche) Phasenregel *f*
phase separation Phasentrennung *f*
phase shift Phasenverschiebung *f*, Phasendrehung *f*
phase-transfer catalysis Phasentransferkatalyse *f*
phase-transfer catalyst Phasentransferkatalysator *m*
phase transition Phasenumwandlung *f*, Phasenübergang *m*, Phasentransformation *f*, Phasenänderung *f*
phase transition temperature Phasenübergangstemperatur *f*
Phe = phenylalanine
phellandrene Phellandren *n (ein monocyclisches Terpen)*
phenacetin Phenacetin *n*, 4-Ethoxyacetanilid *n*
phenacyl chloride Phenacylchlorid *n*, α-Chloracetophenon *n*, ω-Chloracetophenon *n*
phenanthraquinone Phenanthrenchinon *n*, *(i. e. S.)* 9,10-Phenanthrenchinon *n*, 9,10-Dihydro-9,10-dioxophenanthren *n*
phenanthrene Phenanthren *n (ein aromatischer Kohlenwasserstoff)*
phenanthroline *(org Ch)* Phenanthrolin *n*
phenarsazine chloride Phenarsazinchlorid *n*, Diphenylaminchlorarsin *n*, 10-Chlor-5,10-dihydrophenarsazin *n*
phenazine Phenazin *n*, Dibenzopyrazin *n*, Azophenylen *n*
phenazine-based dye Phenazinfarbstoff *m*, Azinfarbstoff *m*
phenazone Phenazon *n*, 1-Phenyl-2,3-dimethyl-5-pyrazolon *n*
phenetidine Phenetidin *n*, Ethoxyanilin *n*, Aminophenylethylether *m*
phenetole Phenetol *n*, Ethoxybenzen *n*, Ethylphenylether *m*
phenol *(org Ch)* Phenol *n*
phenol-formaldehyde resin Phenol-Formaldehyd-Harz *n*
phenol-furfural resin Phenolfurfuralharz *n*
phenol removal *(Hyd)* Entphenol(ier)ung *f*
phenolate *v* phenolisieren
phenolate Phenolat *n*
phenolic phenolisch, Phenol...; phenolhaltig
phenolic 1. Phenol *n (i. w. S.)*; 2. Phenolharz *n*; Phenoplast *m*, Phenolharzkunststoff *m*
phenolic acid Phenolcarbonsäure *f*, Hydroxybenzoesäure *f*
phenolic aldehyde Phenolaldehyd *m*
phenolic ester Phenolester *m*
phenolic hydroxyl phenolisches Hydroxyl [OH] *n*, phenolische Hydroxylgruppe *f*, Phenolhydroxyl *n*
phenolic moulding compound Phenolharzpressmasse *f*

phenolic resin Phenolharz *n*
phenolic varnish Phenolharzlack *m*
phenolize *v s.* phenolate
phenolphthalein Phenolphthalein *n*, 3,3-Bis-(*p*-hydroxyphenyl)phthalid *n*
phenolsulphonephthalein Phenolsulfonphthalein *n*, Phenolrot *n*
phenolsulphonic acid Phenolsulfonsäure *f*, Hydroxybenzensulfonsäure *f*
phenonium ion Phenonium-Ion *n*
phenoplast Phenoplast *m*, Phenolharzkunststoff *m*
phenothiazine Phenothiazin *n*
phenoxide Phenolat *n*
phenoxy group Phenoxygruppe *f*, Phenoxyrest *m*
phenoxyacetic acid Phenoxyessigsäure *f*, *O*-Phenylglykolsäure *f*
phenoxymethylpenicillin Phenoxymethylpenicillin *n*, Penicillin *n* V
phenyl acetate Essigsäurephenylester *m*, Phenylacetat *n*
phenyl benzoate Benzoesäurephenylester *m*, Phenylbenzoat *n*
phenyl cation Phenylkation *n*
phenyl ester Phenylester *m*
phenyl group Phenylgruppe *f*, Phenylrest *m*
***N*-phenyl amide** Anilid *n*, *N*-Phenylamid *n*
***N*-phenyl imine** Anil *n (eine sich vom Anilin oder einem anderen primären aromatischen Amin ableitende schiffsche Base)*
***N*-phenylacetamide** *N*-Phenylacetamid *n*, Acetanilid *n*
phenylacetic acid Phenylessigsäure *f*, 1′-Tolylsäure *f*, 1′-Toluencarbonsäure *f*
α-**phenylacrylic acid** α-Phenylacrylsäure *f*, 2-Phenylacrylsäure *f*, Atropasäure *f*
β-**phenylacrylic acid** β-Phenylacrylsäure *f*, *trans*-β-Phenylacrylsäure *f*, **trans*-3-Phenylpropensäure *f*, (gewöhnliche) Zimtsäure *f*, *trans*-Zimtsäure *f*, Cinnamsäure *f*
β-**phenylalanine** β-Phenylalanin *n*, Phe, *2-Amino-3-phenylpropionsäure *f*
phenylalkane Phenylalkan *n*
phenylamine Phenylamin *n*, Aminobenzen *n*, Anilin *n*
phenylamino group C_6H_5NH- Anilinogruppe *f*
phenylate *v* phenylieren
phenylation Phenylierung *f*
phenylchloroform Phenylchloroform *n*, Benzotrichlorid *n*, α-Trichlortoluen *n*, ω-Trichlormethylbenzen *n*
2-phenylchromenylium salt 2-Phenylchromenyliumsalz *n*, 2-Phenylchromyliumsalz *n*, Flavyliumsalz *n*
phenylene blue Phenylenblau *n (ein Indaminfarbstoff)*
phenylene group Phenylengruppe *f*, Phenylenrest *m*
phenylenediamine *Phenylendiamin *n*, Diaminobenzen *n*
phenylethyl alcohol Phenylethylalkohol *m*, *(i. e. S.)* Phenethylalkohol *m*, *2-Phenylethanol *n*, (2-Hydroxyethyl)-benzen *n*
phenylethylamine *Phenethylamin *n*, Aminoethylbenzen *n*
phenylethylene Phenylethen *n*, Vinylbenzen *n*, Styren *n*
phenylglycine Phenylglycin *n*, Phenylglykokoll *n*, *(i. e. S.)* *N*-Phenylglycin *n*, Anilinoessigsäure *f*
phenylhydrazine Phenylhydrazin *n*
phenylhydrazone Phenylhydrazon *n*
phenylhydroxyacetic acid 2-Hydroxyphenylessigsäure *f*, Phenylglykolsäure *f*, Mandelsäure *f*
phenylium Phenylkation *n*
phenylmercuric acetate Phenylquecksilberacetat *n*, PMAS *n*, PMA *n*
2-phenyl-2-propanol 2-Phenylpropan-2-ol *n*, α,α-Dimethylbenzylalkohol *m*
phenylpropionic acid Phenylpropionsäure *f*, *Phenylpropansäure *f*
phenylpropynoic acid *Phenylpropinsäure *f*, Phenylpropiolsäure *f*
phenylpyruvic acid Phenylbrenztraubensäure *f*, Benzylglyoxylsäure *f*
pheromone Pheromon *n*, Ektohormon *n*
pheron Pheron *n*, Träger *m (eines Enzyms)*
Philips gauge Kaltkatoden-Ionisationsvakuummeter *n*, Philips-Vakuummeter *n*, Penning-Vakuummeter *n*
philodiene *(org Ch)* Philodien *n*
phlobaphene Phlobaphen *n*, Gerbstoffrot *n*
phlori(d)zin Phloridzin *n*, Phlorrhizin *n* *(Glucosid)*

phloroglucin(ol) Phloroglucinol *n*, 1,3,5-Trihydroxybenzen *n*
phonon Phonon *n*, Schallquant *n*
phorone Phoron *n*, 2,6-Dimethylhepta-2,5-dien-4-on *n*
phosgenation Phosgenieren *n*, Phosgenierung *f*
phosgene Phosgen *n*, Kohlenoxidchlorid *n*, Carbo-nylchlorid *n*, Kohlensäuredichlorid *n*
phosphamic acid Phosphamsäure *f*, Amidophosphorsäure *f*
phosphane Phosphan *n (Phosphorwasserstoff der allgemeinen Formel P_nH_{n+})*; Phosphorhydrid *n*, Phosphorwasserstoff *m*, *(i. e. S.)* PH_3 Phosphan *n*, Monophosphan *n*, *(unsubstituiert auch:)* Phosphin *n*
λ^5-phosphane Phosphoran *n*, λ^5-Phosphan *n*
phosphate *v* phosphatieren *(eine Phosphatschicht als Rostschutz aufbringen)*
phosphate Phosphat *n*, *(i. e. S.)* $M^I_3PO_4$ Orthophosphat *n*
phosphate fertilizer Phosphatdüngemittel *n*, Phosphatdünger *m*
phosphate treatment *(Hyd)* Phosphatreinigung *f*, Enthärtung *f* mit Phosphat, Trinatriumphosphatverfahren *n*
phosphatide Phosphatid *n*, Phospholipid *n*
phosphatidic acid Phosphatidsäure *f*, Diglyceridphosphat *n*
3-phosphatidyl ethanolamine Phosphatidylethanolamin *n*, Ethanolkephalin *n*, Colaminkephalin *n*, *(i. e. S.)* Kephalin *n*
3-phosphatidyl serine Phosphatidylserin *n*, Serinkephalin *n*
phosphation Phosphatieren *n*, Phosphatierung *f (Aufbringen einer Phosphatschicht als Rostschutz)*
phosphene Diphosphen *n*
phosphenic acid $HOPO_2$ Phosphensäure *f*
phosphide Phosphid *n*
phosphine Phosphan *n*, Monophosphan *n*, *(als unsubstituiertes PH_3 auch:)* Phosphin *n*
phosphine oxide Phosphanoxid *n*, Phosphinoxid *n*, Oxophosphoran *n*
phosphinic acid RPH(=O)(OH) *Phosphinsäure *f*, Phosphinigsäure *f (der Beilstein-Nomenklatur)*
phosphinous acid $R_1R_2P(OH)$ Phosphinigsäure *f*
phosphite Phosphit *n*
phosphodiester Phosphodiester *m*
6-phosphogluconate *(Bioch)* 6-Phosphogluconat *n*
phosphoglyceride *(Bioch)* Phosphoglycerid *n*
phosphoketolase pathway *(Bioch)* Phospho(hexo)ketolaseweg *m*
phospholipid Phospholipid *n*, Phosphatid *n*
phosphomolybdate Tetrakistrimolybdatophosphat *n*, Dodecamolybdatophosphat *n*, Molybdatophosphat *n*
phosphomolybdic acid Tetrakistrimolybdatophosphorsäure *f*, Dodecamolybdatophosphorsäure *f*, Molybdatophosphorsäure *f*
phosphonate Phosphonat *n*; Phosphonsäureester *m*
phosphonic acid $RP(=O)(OH)_2$ Phosphonsäure *f*, Alkylphosphonsäure *f*
phosphonium Phosphonium *n*
phosphonium compound Phosphoniumverbindung *f*
phosphonium ylide Phosphoniumylid *n*, Phosphorylid *n*, P-Ylid *n*, Alkylidenphosphoran *n*, Phosphinalkylen *n*
phosphonous acid $RP(OH)_2$ Phosphonigsäure *f*, phosphonige Säure *f*
phosphonylation Phosphonylierung *f*
phosphoproteide Phosphoprotein *n*, Phospho(r)proteid *n*
phosphoprotein Phosphoprotein *n*, Phospho(r)proteid *n*
phosphor Phosphor *m (phosphoreszenzfähiger Stoff)*
phosphorane Phosphoran *n*, λ^5-Phosphan *n*
phosphoresce *v* phosphoreszieren
phosphorescence Phosphoreszenz *f*
phosphorescence spectroscopy Phosphoreszenzspektroskopie *f*
phosphorescence spectrum Phosphoreszenzspektrum *n*
phosphorescent phosphoreszierend
phosphoric 1. Phosphor…; höherwertigem Phosphor entsprechend, *(meist)* Phosphor(V)-…; 2. phosphoreszierend
phosphoric acid Phosphorsäure *f*, *(i. e.*

S.) H_3PO_4 Orthophosphorsäure *f*, Monophosphorsäure *f*
phosphoric-acid fuel cell phosphorsaure Brennstoffzelle *f*
phosphoric oxide Phosphor(V)-oxid *n*, Phosphorpentoxid *n*
phosphorolysis *(Bioch)* Phosphorolyse *f*
phosphorolytic *(Bioch)* phosphorolytisch
phosphorous 1. Phosphor...; niederwertigem Phosphor entsprechend, *(meist)* Phosphor(III)-...; 2. phosphoreszierend
phosphorous acid H_2PHO_3 *Phosphonsäure *f*, Phosphorigsäure *f*
phosphorus P Phosphor *m*
phosphorus halide Phosphorhalogenid *n*
phosphorus hydride Phosphor(III)-hydrid *n*, Phosphin *n*, Phosphorwasserstoff *m*, *(i. e. S.)* PH_3 Monophosphin *n*
phosphorus metabolism Phosphorstoffwechsel *m*, *(i. e. S.)* Phosphatstoffwechsel *m*
phosphorus nitride Phosphornitrid *n*, Triphosphorpentanitrid *n*, Phosphorstickstoff *m*
phosphorus oxychloride Phosphor(V)-oxidchlorid *n*, Phosphoryl(V)-chlorid *n*
phosphorus pentachloride Phosphorpentachlorid *n*, Phosphor(V)-chlorid *n*
phosphorus pentoxide Phosphorpentoxid *n*, Phosphor(V)-oxid *n*
phosphorus salt Phosphorsalz *n*, Natriumammoniumhydrogenphosphat-4-Wasser *n*
phosphorus tetroxide P_2O_4 Phosphortetroxid *n*
phosphorus thiochloride Thiophosphoryl(V)-chlorid *n*
phosphorus trichloride Phosphortrichlorid *n*, Phosphor(III)-chlorid *n*
phosphorus trioxide Phosphortrioxid *n*, Phosphor(III)-oxid *n*
phosphorus ylide *s.* phosphonium ylide
phosphorylate *v (Bioch)* phosphorylieren
phosphorylating agent *(Bioch)* Phosphorylierungsmittel *n*
phosphorylation *(Bioch)* Phosphorylierung *f*
phosphotungstate Tetrakistriwolframatophosphat *n*, Dodecawolframatophosphat *n*, Wolframatophosphat *n*
phosphotungstic acid Tetrakistriwolframatophosphorsäure *f*, Dodecawolframatophosphorsäure *f*, Wolframatophosphorsäure *f*
phosphyne Diphosphin *n*
photo... 1. *(phys Ch)* Photo..., Licht...; 2. Foto... *(die Fotografie betreffend)*
photo-Fries rearrangement Photo-Fries-Umlagerung *f (von Estern)*
photo resist Photolack *m*
photoacoustic photoakustisch
photoacoustic spectroscopy photoakustische Spektroskopie *f*, Photoakustikspektroskopie *f*, PA-Spektroskopie *f*, PAS, Optoakustikspektroskopie *f*, OA-Spektroskopie *f*
photoacoustic spectrum Photoakustikspektrum *n*
photoactivation *(Bioch)* Photoaktivierung *f*, Aktivierung *f* durch Licht
photoaddition Photoaddition *f*, photochemische Addition *f*
photocatalysis Photokatalyse *f*
photocatalyst Photokatalysator *m*
photocell Photoelement *n*
photochemical excitation photochemische Anregung *f*
photochemical reaction photochemische Reaktion *f* [Umsetzung *f*], Photoreaktion *f*, Lichtreaktion *f*
photochemical reactor photochemischer Reaktor *m*, Photoreaktor *m*
photochemistry Photochemie *f*
photochromic glass photochromatisches [photochromes, phototropes] Glas *n*
photochromism Photochromie *f*, Phototropie *f (Farbänderung beim Einwirken von Strahlungsenergie)*
photoconductance Photoleitwert *m*
photoconductive photoleitend
photoconductive cell Photowiderstandszelle *f*, Photowiderstand *m*
photoconductivity Photoleitfähigkeit *f*, lichtelektrische [photoelektrische] Leitfähigkeit *f*
photocrosslinking photochemische Vernetzung *f*
photocuring photochemische Härtung *f* [Aushärtung *f*] *(z. B. von Monomeren in dünner Schicht)*

photocurrent Photo(elektronen)strom *m*
photocyclic photocyclisch
photocyclization Photocyclisierung *f*
photocycloaddition Photocycloaddition *f*
photodecomposition *s.* photolysis
photodegradation photochemischer Abbau *m*, Abbau *m* durch Licht, Lichtabbau *m*
photodimerization Photodimerisierung *f*
photodiode Photodiode *f*, Leuchtdiode *f*
photodissociation Photodissoziation *f*
photoelectric cell Photoelement *n*
photoelectric effect photoelektrischer [lichtelektrischer] Effekt *m*, Photoeffekt *m*
photoelectric electron-multiplier tube *s.* photomultiplier (tube)
photoelectric scanning photoelektrische Abtastung *f*
photoelectrochemical *(Ech)* photoelektrochemisch
photoelectrochemistry *(Ech)* Photoelektrochemie *f*
photoelectron Photoelektron *n*
photoelectron spectroscopy Photoelektronenspektroskopie *f*, PES
photoelectron spectroscopy of inner-shell electrons Photoelektronenspektroskopie *f* innerer Elektronen
photoelimination Photoeliminierung *f*
photoemission Photoemission *f*, lichtelektrische Emission *f*
photoevolution of oxyen *(Bioch)* Sauerstoffentwicklung *f* bei Belichtung *(von Pflanzen)*
photoexcitation photochemische Anregung *f*
photoflash Lichtblitz *m*, Photolyseblitz *m* *(Kinetik)*
photogalvanic cell photogalvanische Zelle *f*
photografting Photografting *n (zur Modifizierung biomaterieller Oberflächen)*
photographic chemistry fotografische Chemie *f*, Chemie *f* der Fotografie [fotografischen Prozesse]
photographic emulsion fotografische [lichtempfindliche] Emulsion *f*, Fotoemulsion *f*; fotografische [lichtempfindliche] Schicht *f*, Emulsion *f*
photographic film (fotografischer) Film *m*
photoinactivation Photoinaktivierung *f*
photoinduced polymerization photoaktivierte Polymerisation *f*
photoinitiation Photoinitiation *f*, Photoinitiierung *f*
photoionization Photoionisierung *f*
photoionization detector Photoionisationsdetektor *m*, PID
photolithography Photolithographie *f*
photolithotrophs photolithotrophe Organismen *mpl (organische Kohlenstoffverbindungen mittels Lichtenergie aus anorganischen Kohlenstoffverbindungen aufbauend)*
photoluminescence Photolumineszenz *f*
photolysable durch Photolyse zersetzbar [zerlegbar]
photolysis Photolyse *f*, Zersetzung *f* durch Licht, Photozersetzung *f*
photolytic photolytisch
photomacrograph Makrofotografie *f*, Makrofoto *n*, Makroaufnahme *f*, makrofotografische Aufnahme *f*
photomask Photomaske *f*
photometer Photometer *n*
photometric titration photometrische Titration *f*
photometry Photometrie *f*
photomicrograph Mikrofotografie *f*, Mikrofoto *n*, Mikroaufnahme *f*, mikrofotografische Aufnahme *f*
photomultiplier (tube) Photo(elektronen)vervielfacher *m*, Sekundärelektronenvervielfacher *m*, SEV *m*, Photomultiplier *m*
photon Photon *n*, Lichtquant *n*
photoorganotrophs photoorganotrophe Organismen *mpl (organische Kohlenstoffverbindungen teils mittels Lichtenergie, teils unter Nutzung fremder Kohlenhydrate aufbauend)*
photooxidation Photooxidation *f*, photochemische Oxidation *f*
photophosphorylate *v (Bioch)* photophosphorylieren
photophosphorylation *(Bioch)* Photophosphorylierung *f*
photopigment Augenpigment *n*, Sehstoff *m*
photopolymer Photopolymer(e) *n*
photopolymerization Photopolymerisation *f*

photoreaction photochemische Reaktion *f* [Umsetzung *f*], Photoreaktion *f*, Lichtreaktion *f*
photoreactivation *(Bioch)* Photoreaktivierung *f*
photoreactor Photoreaktor *m*
photorearrangement photochemische Umlagerung *f*
photoreceptor Photorezeptor *m*
photoreduction Photoreduktion *f*, photochemische Reduktion *f*
photoresist Photoresist *n (lichtempfindliches Polymer)*
photoresist dye Photoresistfarbstoff *m*
photoresist resin Photoresistlack *m*
photoresistance lichtelektrischer Widerstand *m*, Photowiderstand *m*
photoresistor Photowiderstand *m*, Photowiderstandszelle *f*
photorespiration *(Bioch)* Lichtatmung *f*
photosensitive lichtempfindlich; photosensibel, photosensitiv
photosensitive surface Photoschicht *f*, photosensibler Film *m*
photosensitivity Lichtempfindlichkeit *f*; Photosensibilität *f*
photosensitization Photosensibilisation *f*, Photosensibilisierung *f*, Sensibilisierung *f*
photosensitize *v* (photo)sensibilisieren
photosensitizer 1. Photosensibilisator *m*; 2. *(Foto)* Sensibilisator *m*
photostability Lichtbeständigkeit *f*, Lichtunempfindlichkeit *f*
photostabilization Lichtunempfindlichmachen *n*
photostabilizer Lichtstabilisator *m*, photostabilisierendes Mittel *n (Gummi)*
photostable lichtbeständig, lichtunempfindlich
photoswitchable dye photoschaltbarer Farbstoff *m*
photosynthesis Photosynthese *f*
photosynthetic photosynthetisch
photosynthetically active radiation photosynthetisch ausnutzbare Strahlung *f*
phototrophic bacteria lichtnutzende Bakterien *npl*
phototrophs phototrophe Organismen *mpl (organische Kohlenstoffverbindungen mit Hilfe von Lichtenergie aufbauend)*
phototube Photozelle *f*
photovoltaic photovoltaisch
photovoltaic cell Sperrschicht(photo)zelle *f*, Sperrschicht(photo)element *n*, Photoelement *n*
photovoltaic effect Sperrschicht(photo)effekt *m*, Photo-Volta-Effekt *m*, Photovolteffekt *m*
photovoltaics Photovoltaik *f (Solarenergietechnik)*
phthalamic acid Phthalamidsäure *f*, Phthalsäuremonamid *n*
phthalamide Phthalamid *n*, Phthalsäurediamid *n*
phthalate Phthalat *n (Salz oder Ester der Phthalsäure)*
phthalein *(org Ch)* Phthalein *n*
phthalic anhydride Phthalsäureanhydrid *n*, *Benzen-1,2-dicarbonsäureanhydrid *n*
***o*-phthalic acid** Phthalsäure *f*, *Benzen-*o*-dicarbonsäure *f*, *Benzen-1,2-dicarbonsäure *f*
***o*-phthalic aldehyde** *o*-Phthalaldehyd *m*, OPA, *o*-Phthaldialdehyd *m*, 1,2-Phthalaldehyd *m*
phthalide Phthalid *n*
phthalimide Phthalimid *n*, Phthalsäureimid *n*, Isoindol-1,3-dion *n*
phthalocyanine dye Phthalocyaninfarbstoff *m*
phthalonitrile Phthalodinitril *n*, Phthalsäuredinitril *n*, 1,2-Dicyanobenzen *n*
phycocyan(in) Phykocyan *n (Farbstoff der blaugrünen Algen)*
phycoerythrin Phykoerythrin *n (Farbstoff der Rotalgen)*
phylloquinone Phyllochinon *n*, Vitamin *n* K, Koagulationsvitamin *n*, antihämorrhagisches Vitamin *n*
phyllosilicate *(Min)* Phyllosilicat *n*, Schichtsilicat *n*, Blattsilicat *n*
physical adsorption physikalische [van-der-waalssche] Adsorption *f*, Physisorption *f*
physical-chemical *s.* physicochemical
physical chemist Physikochemiker *m*
physical chemistry physikalische Chemie *f*, Physikochemie *f*
physical constant physikalische Konstante *f*, Naturkonstante *f*
physical network Verhakungsnetzwerk *n*

physical stability mechanische Festigkeit *f*
physical state Aggregatzustand *m*
physical vapour deposition *(Tech)* physikalische Gasphasenabscheidung *f (vakuumbasiertes Beschichtungsverfahren)*
physicochemical physikalisch-chemisch, physikochemisch
physiologic(al) chemistry physiologische Chemie *f*
physiologic(al) saline [salt solution] *(Med)* physiologische Kochsalzlösung *f*
physisorption *s.* physical adsorption
phytic acid Phytinsäure *f*
phytin Phytin *n (Ca-Mg-Salz der Phytinsäure)*
phytochemical pflanzenchemisch
phytochemical defense pflanzenchemische Abwehr *f*
phytochemistry Phytochemie *f*, Pflanzenchemie *f*
phytocidal pflanzentötend
phytohormone Phytohormon *n*, Pflanzenhormon *n*
phytology Pflanzenkunde *f*
phytoprotein Pflanzeneiweiß *n*
phytosterol Phytosterol *n*, pflanzliches Sterol *n*, Pflanzensterol *n*
phytotoxic phytotoxisch, pflanzenschädigend
phytotoxicant pflanzenschädigender Stoff *m*
phytotoxicity Phytotoxizität *f*, Giftigkeit *f* für Pflanzen
phytotoxin pflanzliches Toxin *n*
PI *s.* polyimide
pi-adduct *(org Ch)* π-Komplex *m*
pi bond π-Bindung *f*, π-Elektronenpaarbindung *f*
pi meson *s.* pion
pi orbital π-Orbital *n*, π-Molekülorbital *n*, π-MO *n*
pick *v* (aus)wählen, sortieren, (aus)lesen, (aus)klauben
pick *v* **up** aufnehmen *(Flüssigkeiten)*
picked ore Scheideerz *n*
pickle *v* 1. *(Lebm)* nasspökeln; (ein)säuern, *(bei Fisch auch)* marinieren; 2. (ab)beizen *(Saatgut, Metall)*
pickle 1. *(Lebm)* Pökellake *f*, Pökelsalzlösung *f*; Marinade *f*, Aufguss *m (z. B. gewürzte Salz-Essig-Lösung)*; 2. Beize *f*, Beizmittellösung *f*
pickled sheet Mattblech *n*
pickling 1. *(Lebm)* Nasspökeln *n*, Nasspökelung *f*; Säuern *n*, Säuerung *f*, Einsäuern *n*, *(bei Fisch auch)* Marinieren *n*; 2. Beizen *n*, Abbeizen *n*, Beize *f*
pickling acid *(Met)* Beizablauge *f*, Brennsäure *f*
pickling agent Abbeizmittel *n*, Beizmittel *n*
pickling bath *(Met)* Ätzbad *n*, Beizbad *n*, Dekapierbad *n*, Vorbrenne *f*, Gelbbrenne *f*
pickling inhibitor Beizinhibitor *m*
pickling liquor Beizlauge *f*
pickling plant *(Met)* Beizanlage *f*, Abbrenneinrichtung *f*
pickling process 1. *(Gerb)* Pickeln *n*, Pickelverfahren *n*; 2. *(Met)* Beizen *n*, Ätzen *n*
pickling solution *s.* pickle
pickling time 1. *(Met)* Beizdauer *f*; 2. *(Lebm)* Pökeldauer *f*
pickling vat *(Met)* Abbrennkessel *m*, Beizbehälter *m*
pickling water Beizwasser *n*
pickup 1. *(Text)* Flüssigkeitsaufnahme *f*, Flottenaufnahme *f*; 2. *(Tech)* Laden *n*, Verladen *n*
pickup velocity Abtastgeschwindigkeit *f*
picogram range *(Anal)* Pikogrammbereich *m* (10^{-12} *bis* 10^{-9} *g)*
picolinic acid Pikolinsäure *f*, Pyridin-2-carbonsäure *f*
picotrace *(Anal)* Pikospur *f*, Pikomenge *f* (*10^{-10} *bis* 10^{-13} *ppm)*
picram(in)ic acid Pikraminsäure *f*, 2-Amino-4,6-dinitrophenol *n*
picrate Pikrat *n (Salz oder Ester der Pikrinsäure)*
picric acid Pikrinsäure *f*, *2,4,6-Trinitrophenol *n*
Pictet-Gams reaction [synthesis] Pictet-Gams-Synthese *f (von Isochinolinen)*
PID *s.* photoionization detector
piezochemistry Piezochemie *f*
piezoelectric crystal piezoelektrischer Kristall *m*, Piezokristall *m*
piezoelectric resistor piezoelektrischer Widerstand *m*
piezoelectricity Piezoelektrizität *f*

PIF *s.* prolactin release-inhibiting factor
pig 1. *(Met)* Massel *f*; 2. *(Dest)* Spinne *f*
pig iron Roheisen *n*, Masseleisen *n*
pig lead Werkblei *n (Rohblei)*
pigment *v* 1. pigmentieren, färben; 2. *(Gum)* füllen, Füllstoffe zusetzen
pigment 1. Pigment *n*, Körperfarbe *f*; Pigmentfarbstoff *m*; 2. *(Gum)* Füllstoff *m*
pigment assembly Pigmentkollektiv *n (von Photosynthesefarbstoffen)*
pigment-binder ratio Pigment-Bindemittel-Verhältnis *n*
pigment dye(stuff) Pigmentfarbstoff *m*
pigment grinding Pigmentanreibung *f*
pigment volume concentration Pigmentvolumenkonzentration *f*, PVK
pigmentation 1. Pigmentation *f*, Färbung *f*; 2. *(Gum)* Füllung *f*, Zusatz *m* [Zuschlag *m*] von Füllstoffen, Füllstoffdosierung *f*
pile 1. *(Tech)* Haufen(speicher) *m*; 2. *(Kern)* Reaktor *m*, Kernreaktor *m*
pill *v* pillieren, zu Kügelchen formen
pill 1. *(Kst)* Vorformling *m*; 2. *(Pharm)* Pille *f*
pilot plant Pilotanlage *f*, Halbbetriebsanlage *f*, halbtechnische Versuchsanlage *f*; Versuchsbetrieb *m*
pilot-plant(-scale) production Produktion *f* in halbtechnischem Maßstab
pimelic acid Pimelinsäure *f*, *Pentan-1,5-dicarbonsäure *f*, *Heptandisäure *f*
pin *(Tech)* Stift *m*, Bolzen *m*, Dorn *m*
pin-disk mill *s.* pinned-disk disintegrator
pin(-type) mill *s.* pinned-disk disintegrator
pinacoid *(Krist)* Pinakoid *n*, Zweiflächner *m*
pinacoidal *(Krist)* pinakoidal
pinacol Pinakol *n*, Tetramethylethylenglykol *n*
pinacol rearrangement Pinakol-Pinakolon-Umlagerung *f*
pinacolone Pinakolon *n*, *3,3-Dimethylbutan-2-on *n*, *tert*-Butylmethylketon *n*
pinchcock Quetschhahn *m*, Quetschklemme *f*, Schlauchklemme *f*
pine oil Pine Oil *n*, Kienöl *n (von Pinus-Arten)*; Kiefernnadelöl *n*; Kiefernzapfenöl *n*
pine resin [rosin] Kiefernharz *n*, *(i. w. S.)* Terpentinharz *n*, Balsamharz *n*, Harz *n* aus Rohterpentin, Kolophonium *n*, Balsamkolophonium *n*, Geigenharz *n*
pine tar Kiefern(holz)teer *m*
pinene Pinen *n (bicyclisches Monoterpen)*
ping-pong reaction Ping-pong-Reaktion *f*, doppelte Verdrängungsreaktion *f*
pinhole Stiftloch *n*, Pore *f*, Nadelstich *m (in einer Schutzschicht)*
pink salt Pinksalz *n*, Ammoniumchlorostannat(IV) *n*
pinned-disk disintegrator [mill] Stift-(scheiben)mühle *f*, Schlagstiftmühle *f*, Dismembrator *m*
pion *(Kern)* Pion *n*, π-Meson *n*
pipe *v* 1. verrohren, mit Rohren [einer Rohrleitung] versehen; durch eine Rohrleitung verbinden; 2. in Rohrleitungen führen, durch Rohre leiten
pipe 1. Rohr *n*, Röhre *f*, Leitung *f*; 2. Lunker *m (im Gussblock)*
pipe clamp Rohrschelle *f*
pipe coil Rohrschlange *f*, Rohrspirale *f*
pipe friction Rohrreibung *f*
pipe joint Rohrverbindung *f*, Rohrverbinder *m*
pipe still *(Erdöl, Dest)* Röhrenofen *m*, Röhrenerhitzer *m*, Röhren-Chemieofen *m*
pipe wall Rohrwand(ung) *f*
pipeline Pipeline *f*, Überlandrohrleitung *f*, Fernleitung *f*, Rohrleitung *f*, Leitung *f*
pipeline(-quality) gas Ferngas *n*, Pipeline-Gas *n*
piperazine Piperazin *n*, Diethylendiamin *n*
piperidine Piperidin *n*
piperidine alkaloid Piperidinalkaloid *n*
piperonal Piperonal *n*, Heliotropin *n*
pipet *s.* pipette
pipette *v* pipettieren
pipette *v* **into** einpipettieren
pipette *v* **off** abpipettieren
pipette Pipette *f*
pipetting Pipettieren *n*, Pipettierung *f*
piping 1. Verrohren *n*, Verrohrung *f*; Verbinden *n* durch eine Rohrleitung; 2. Führen *n* in Rohrleitungen, Leiten *n* durch Rohre; 3. Rohrmaterial *n*; 4. Rohrleitungssystem *n*, Rohrnetz *n*; 5. Lunkern *n*, Lunkerung *f*, Lunker(aus)bildung *f (im Gussblock)*

piranha solution Piranha-Lösung *f* *(Substratreiningung)*
Pirani gauge Pirani-Manometer *n*, Pirani-Vakuummeter *n*, Wärmeleitvakuummeter *n*
piscicide Fischbekämpfungsmittel *n*
piston Kolben *m*, Stempel *m*, Schieber *m*
piston flow Pfropfenströmung *f*
piston pump Kolbenpumpe *f (i. e. S.)*
pit 1. Grube *f*; 2. *(Kst)* Grübchen *n*, Loch *n* *(Pressfehler)*
pit tannage Grubengerbung *f*
PITC = phenylisothiocyanate
pitch *v* pichen, teeren
pitch 1. Pech *n*; 2. *(Pap)* schädliches Harz *n*; 3. Pichen *n*, Teeren *n*; 4. Abstand *m (regelmäßig)*; Teilung *f (z. B. von Ketten, Zahnrädern)*; Steigung *f*, Ganghöhe *f (z. B. von Förderschnecken, Schraubenrührern, Gewinde)*; 5. *(Bioch)* Ganghöhe *f*; Identitätsperiode *f (einer Helix)*
pitch coal Pechkohle *f*, Glanz(braun)-kohle *f*
pitch coke Pechkoks *m*
pitchblende *(Min)* Pechblende *f*, Uraninit *m*, Uranpecherz *n (Uran(IV)-oxid)*
pitched-blade turbine Schrägblattrührer *m*
pitchy pechartig
Pitot tube Pitot-Rohr *n*, Staurohr *n*, pitotsche Röhre *f*
pitting 1. Lochfraß *m*, *(im Anfangsstadium auch)* punktförmige Anfressung *f*, Grübchenbildung *f*; 2. Grübchenbildung *f (in elektrochemisch abgeschiedenen Schichten)*
pitting corrosion Loch(fraß)korrosion *f*, lochförmige Korrosion *f*, *(im Anfangsstadium auch)* punktförmige [grubenförmige] Korrosion *f*, Grübchenkorrosion *f*
pivalic acid Pivalinsäure *f*, *Dimethylpropansäure *f*, Trimethylessigsäure *f*
pivoted-bucket conveyor Pendelbecherwerk *n*, Schaukelbecherwerk *n*
PIXE *s.* particle-induced X-ray emission
place *v* 1. eintragen, (ein)bringen *(z. B. Chemikalien in ein Gefäß)*; 2. *(Landw)* ausbringen, applizieren, placieren *(z. B. Düngemittel)*
place *v* **in** einsetzen
place *v* **on a tip** auf Halde ablagern, aufhalden, verkippen *(Abprodukte)*
place isomerism Stellungsisomerie *f*, Substitutionsisomerie *f*
placebo *(Pharm)* Placebo *n*, Leermedikament *n*, Scheinmedikament *n*, indifferentes [unwirksames] Mittel *n*
placement *(Landw)* Ausbringung *f*, Applikation *f*, Placierung *f (z. B. von Düngemitteln)*
plagioclase *(Min)* Plagioklas *m (ein Feldspat)*
plain 1. eben, flach, plan; glatt; 2. zusatzfrei; rein, unlegiert *(Stahl)*
plain carbon steel Kohlenstoffstahl *m*, unlegierter Stahl *m (mit bis zu 2 % Kohlenstoff)*
plain clarifier *(Hyd)* flaches Absetzbecken *n*, Flachbecken *n*
plain cylinder *(Lab)* Standzylinder *m*
plait point kritischer Punkt *m* [Mischungspunkt *m*] bei gleich bleibender Temperatur
Planck quantum of action *s.* Planck's constant
Planck's hypothesis plancksche Quantenhypothese *f*, plancksche Hypothese *f* der Energiequanten
Planck's constant Planck-Konstante *f*, plancksche Konstante *f*, plancksches Wirkungsquantum *n*
plane of polarization Polarisationsebene *f*
plane of symmetry 1. Symmetrieebene *f*; 2. *(Krist)* Spiegelebene *f*
plant 1. Anlage *f*, Betriebsanlage *f*; Betrieb *m*, Werk *n*, Fabrik *f*; 2. Pflanze *f*
plant biochemistry Pflanzenbiochemie *f*
plant biomass pflanzliche Biomasse *f*, photosynthetisch entstandene Pflanzenmasse *f*
plant cell culture *(Biot)* pflanzliche Zellkultur *f*, Pflanzenzellkultur *f*
plant colouring matter pflanzlicher Farbstoff *m*, Pflanzenfarbstoff *m*
plant-derived pflanzlicher Herkunft, pflanzlichen Ursprungs
plant fat Pflanzenfett *n*, pflanzliches Fett *n*
plant-growth regulator Wachstumsregulator *m*, Wuchsstoffmittel *n*, *(mit hemmender Wirkung auch)* Wuchsstoffherbizid *n*

plant gum Pflanzengummi *n*
plant material pflanzlicher Rohstoff *m*
plant performance Anlagenleistung *f*
plant poison pflanzliches Gift *n*, Pflanzengift *n*
plant-produced phytogen
plant protective protektives [vorbeugend wirkendes] Pflanzenschutzmittel *n*
plant protein pflanzliches Protein *n*, Pflanzenprotein *n*
plant sewage *s.* plant waste water
plant source pflanzliche Rohstoffquelle *f*
plant sterol pflanzliches Sterol *n*, Pflanzensterol *n*, Phytosterol *n*
plant technology Anlagentechnik *f*
plant tissue culture *(Biot)* pflanzliche Gewebekultur *f*
plant waste treatment *(Hyd)* betriebliche Abwasserbehandlung *f*
plant waste water Betriebsabwasser *n*, betriebliches Abwasser *n*, Werksabwasser *n*
plant water Betriebswasser *n*, Brauchwasser *n*, Nutzwasser *n*
plasma 1. *(phys Ch)* Plasma *n*; 2. *(Med)* Blutplasma *n*, Plasma *n*; 3. *(Bio)* Protoplasma *n*, Plasma *n*
plasma arc torch Lichtbogenplasmabrenner *m*
plasma burner Plasmabrenner *m*
plasma chemistry Plasmachemie *f*
plasma cutting Plasmaschneiden *n*
plasma deposition Abscheidung *f* aus dem Plasma
plasma gun Plasmabbrenner *m*
plasma polymerization Plasmapolymerisation *f*
plasma protein Plasmaprotein *n*
plasma separation *(Bioch)* Plasmaseparation *f*
plasma spray Plasmaspray *n (Beschichtungstechnik)*
plasma torch Plasmabrenner *m*
plasmalogen Plasmalogen *n (Vertreter einer Gruppe von Glycerophospholipiden)*
plasmid *(Bioch)* Plasmid *n*
plaster (mortar) Putzmörtel *m*; Gipsmörtel *m*
plaster of Paris gebrannter Gips *m (im Wesentlichen $CaSO_4 \cdot 0,5\ H_2O$)*; (kristalliner) Gips *m (Calciumsulfat-2-Wasser)*
plastic plastisch, verformbar, bildsam
plastic (organischer) Kunststoff *m*
plastic ageing Kunststoffalterung *f*
plastic composite material Kunststoff-Verbundmaterial *n*
plastic concrete Kunststoffbeton *m*
plastic deformation plastische Verformung *f* [Deformation *f*], bleibende [irreversible, nicht elastische] Verformung *f*
plastic film Kunststofffolie *f*
plastic flow plastisches Fließen *n* *(Werkstoffkunde)*
plastic foam Schaum(kunst)stoff *m*
plastic laminate Schichtpressstoff *m*
plastic pipe Kunststoffrohr *n*
plastic resin Kunststoffharz *n*
plastic sulphur plastischer Schwefel *m*, γ-Schwefel *m*
plastic waste Kunststoffabfall *m*
plastic yield *s.* plastic deformation
plastical optical fibre polymerer Lichtwellenleiter *m*
plasticate *v* 1. *(Gum, Kst)* plastizieren; 2. *(Gum)* mastizieren *(bei Raumtemperatur)*; thermisch abbauen *(bei erhöhter Temperatur)*
plastication 1. *(Gum, Kst)* Plastizieren *n*, Plastizierung *f*; 2. *(Gum)* Mastikation *f* *(bei Raumtemperatur)*; thermischer Abbau *m (bei erhöhter Temperatur)*
plasticator 1. *(Kst)* Plastikator *m*, Plastiziermaschine *f*; 2. *(Gum)* Mastikator *m*, Mastiziermaschine *f*, Plastikator *m*
plasticity Plastizität *f*, Verformbarkeit *f*, Bildsamkeit *f*
plasticization 1. *(Kst, Gum)* Plastifizierung *f*, Plastifikation *f*, Weichmachung *f*; 2. *(Gum)* Peptisierung *f*, Peptisation *f*; 3. *s.* plastication
plasticize *v* 1. *(Kst, Gum)* plastifizieren, weich machen; 2. *(Gum)* peptisieren; 3. *s.* plasticate
plasticized PVC Weich-PVC *n*, PVC-Weich *n*, PVC-W
plasticizer 1. *(Kst, Gum)* Weichmacher *m*, Weichmachungsmittel *n*, Plastifizier(ungs)mittel *n*, Plastifikationsmittel *n*, weich [plastisch] machender Zusatz *m*; 2. *(Gum)* (chemisches) Abbaumittel *n*, Peptisier(ungs)mittel *n*

plastigel Plastigel *n*
plastisol Plastisol *n*
plastomer Plastomer(e) *n*
plastoquinone *(Bioch)* Plastochinon *n*, PQ
plate *v* 1. beschichten; 2. *(Ech)* elektrochemisch [galvanisch] beschichten, galvanisieren; 3. aufbringen *(Metalle)*; *(i. e. S.)* elektrochemisch [galvanisch] abscheiden; sich abscheiden lassen
plate *v* **out** sich elektrochemisch [galvanisch] abscheiden [niederschlagen]; elektrochemisch [galvanisch] abscheiden
plate 1. Platte *f*, Tafel *f*; Blech *n*; 2. *(Dest)* Boden *m*, Austauschboden *m*; 3. dünne elektrochemisch [galvanisch] hergestellte Schutzschicht *f*; 4. *(Krist)* Tafel *f*; 5. *(Ech)* (flache) Elektrode *f*, Platte *f*
plate-and-frame (filter) press Rahmen(filter)presse *f*
plate column Bodenkolonne *f*, Bodensäule *f*
plate count *(Chromat)* Trennstufenzahl *f*
plate efficiency *(Dest)* Bodenwirkungsgrad *m*
plate heat exchanger Plattenwärmeaustauscher *m*, Plattenwärmeübertrager *m*
plate height *(Dest, Anal)* Bodenhöhe *f*, Trennstufenhöhe *f*, äquivalente Füllkörperhöhe *f*, HETP-Wert *m*
plate number *(Dest)* Bodenzahl *f*
plate precipitator Plattenelektrofilter *n*, Platten(elektro)abscheider *m*
plate tower Bodenkolonne *f*, Bodensäule *f*
platen press Etagenpresse *f*, Packpresse *f*, Plattenpresse *f*
platformate *(Erdöl)* Platformat *n*, Platformingprodukt *n*, Platformerprodukt *n*
platforming *(Erdöl)* Platforming *n*, Platformen *n*, Platin-Reforming *n*, Reformieren *n* an Platinkatalysatoren [Platinkontakten]
platinate *v s.* platinize
platinate Platinat *n (Komplexsalz mit Pt als Zentralatom)*
plating 1. Beschichten *n*; 2. *(Ech)* elektrochemisches [galvanisches] Beschichten *n*, Galvanisieren *n*; 3. Aufbringen *n (von Metallen)*; *(i. e. S.)* elektrochemisches [galvanisches] Abscheiden *n*; 4. elektrochemisch [galvanisch] hergestellte Schutzschicht *f*
plating bath 1. Elektrolyt *m*, Galvanisierelektrolyt *m*, Elektrolytflüssigkeit *f*; 2. Galvanisierbehälter *m*, Elektrolytbehälter *m*
platinic Platin..., *(i. e. S.)* Platin(IV)-...
platinic sodium chloride Natriumhexachloroplatinat(IV) *n*
platinic sulphide Platin(IV)-sulfid *n*, Platindisulfid *n*
platiniferous platinhaltig
platinization Platinieren *n*
platinize *v* platinieren
platinized platinum electrode platinierte Platinelektrode *f*
platinous Platin..., *(i. e. S.)* Platin(II)-...
platinous sodium chloride Natriumchloroplatinat(II) *n*
platinous sulphide Platin(II)-sulfid *n*, Platinmonosulfid *n*
platinum Pt Platin *n*
platinum black Platinmohr *n*, Platinschwarz *n*
platinum catalyst Platinkatalysator *m*, platinhaltiger Katalysator *m*, Platinkontakt *m*
platinum-coated platiniert
platinum crucible Platintiegel *m*
platinum dichloride Platindichlorid *n*, Platin(II)-chlorid *n*
platinum dioxide Platindioxid *n*, Platin(IV)-oxid *n*
platinum electrode Platinelektrode *f*
platinum mohr *s.* platinum black
platinum monosulphide Platinmonosulfid *n*, Platin(II)-sulfid *n*
platinum monoxide Platinmonoxid *n*, Platin(II)-oxid *n*
platinum plating *(Ech)* Platinieren *n*
platinum sponge Platinschwamm *m*
platinum tetrachloride Platintetrachlorid *n*, Platin(IV)-chlorid *n*
platy *(Krist)* tafelförmig, tafelig
PLB *s.* porous layer bead
pleasant-smelling angenehm riechend, von angenehmem Geruch
pleated filter Faltenfilter *n*
pleated-sheet conformation [structure] *(Bioch)* Faltblattstruktur *f*, β-Struktur *f (von Polypeptidketten)*
PLED *s.* polymer light emitting diode

pleochro(mat)ic *(Krist)* pleochroitisch *(je nach Betrachtungsrichtung verschiedene Farben zeigend)*
pleochro(mat)ism *(Krist)* Pleochroismus *m*
pliability Biegsamkeit *f*, Geschmeidigkeit *f*
pliable biegsam, geschmeidig
pliant biegsam, geschmeidig
plot *v* eintragen *(in ein Koordinatensystem)*
PLOT column *(porous layer open tubular column)* PLOT-Säule *f*, Dickschichtsäule *f (Gaschromatographie)*
plug *v* verstopfen, verlegen, versetzen, zusetzen, verschließen; zustöpseln; verstopfen, sich zusetzen
plug Stopfen *m*, Pfropfen *m*, Stöpsel *m*; Zapfen *m*, Spund *m (an Fässern)*; Küken *n*, Hahnküken *n*
plug cock Hahn *m*, Hahnventil *n*; Zapfen *m*, Spund *m (an Fässern)*
plug-flow reactor Pfropfenströmungsreaktor *m*, (strömungstechnisch) idealer Rohrreaktor *m*
plug valve Hahn *m*, Hahnventil *n*
pluggage Verstopfen *n*, Verlegen *n*, Zusetzen *n*, Versetzen *n*, Versatz *m*, Verschließen *n*
plugging Verstopfen *n*, Verlegen *n*, Zusetzen *n*, Versetzen *n*, Versatz *m*, Verschließen *n*
plumbago Graphit *m*
plumbane Plumban *n*, Bleiwasserstoff *m*
plumbate Plumbat *n*, *(i. e. S.)* Plumbat(IV) *n*
plumber's solder Blei-Zinn-Weichlot *n*, Zinnlot *n*, Lötzinn *n (mit 50 bis 75 % Pb)*
plumbic Blei…, *(i. e. S.)* Blei(IV)…
plumbiferous bleihaltig
plumbite $M^{I}_{2}PbO_2$ Plumbat(II) *n*
plumbous Blei…, *(i. e. S.)* Blei(II)-…
plumbous oxide Blei(II)-oxid *n*, Bleimonoxid *n*
plumbous sulphide Blei(II)-sulfid *n*
plume Rauchfahne *f*, Abgasfahne *f*
plunge *v* (ein)tauchen
plunger Tauchkolben *m*, Plunger(kolben) *m (einer Pumpe)*; Stempel *m*, Pressstempel *m (eines Presswerkzeugs)*
plunger pump Tauchkolbenpumpe *f*, Plungerpumpe *f*
plunger-type injection machine *(Kst)* Kolbenspritzgussmaschine *f*
plus material Überkorn *n*, Grobkorn *n*, Grobgut *n (beim Klassieren)*; *(beim Siebklassieren auch)* Siebgrobes *n*, Siebgröbe *f*, *(beim Windsichten auch)* Sichtgrobes *n*, Grieß *m*
plutonium Pu Plutonium *n*
ply 1. Lage *f*, Schicht *f*; Zwischenlage *f*; 2. Furnierplatte *f*
PMA *s.* phosphomolybdic acid
PMDA *s.* pyromellitic dianhydride
pmr *s.* 1. proton magnetic resonance; 2. paramagnetic resonance
PMR *s.* 1. proton magnetic resonance; 2. paramagnetic resonance
PMT *s.* photomultiplier (tube)
pneumatic pneumatisch; Druckluft…
pneumatic atomizer Druckluftversprüher *m*; Druckluftzerstäuber *m (für Feststoffe)*
pneumatic cell *(Spektr)* Golay-Zelle *f*, pneumatische Zelle *f*, Golay-Detektor *m*
pneumatic classification Sichten *n*, Windsichten *n*, Aeroklassieren *n*
pneumatic sampler *(Anal)* pneumatischer [pneumatisch arbeitender] Probenstecher *m*
Podbielniak contactor [extractor] Podbielniak-Zentrifugalextraktor *m*, Podbielniak-Extraktor *m*, Podbielniak-Zentrifuge *f*
Poggendorff compensation method poggendorffsche Kompensationsmethode *f* [Kompensationsschaltung *f*]
point charge Punktladung *f*
point defect punktförmiger Gitterfehler *m*, Punktstörung *f*, Punktdefekt *m*, punktförmige Fehlstelle *f*, Punktfehlstelle *f*
point group 1. *(Krist)* Punkt(symmetrie)-gruppe *f*, Kristallklasse *f*, Symmetrieklasse *f*; 2. *(Spektr)* Punktgruppe *f*
point mutation *(Biot)* Punktmutation *f*
point of attack *(Bioch)* Angriffsort *m*
point of change Umschlagspunkt *m* *(Titration)*
point of entry Eintrittsstelle *f*, Einlaufstelle *f*, Einspritzstelle *f*, Eintritt *m*, Einlauf *m*
point of fusion Kondensationsstelle *f*,

Anellierungsstelle *f*, gemeinsame Seite *f*
point of neutrality Neutralpunkt *m (Titration)*
point of zero charge *(phys Ch)* Nullladungspunkt *m*
point source (of emission) Punktquelle *f*
pointed flask Spitzkolben *m*
pointer Zeiger *m*, Zunge *f*, Nadel *f (z. B. eines Messgeräts)*
Poiseuille flow (Hagen-)Poiseuille-Strömung *f*, laminare Rohrströmung *f*
Poiseuille's equation [formula] (Hagen-)Poiseuille-Gleichung *f (der laminaren Rohrströmung)*
poison *v* vergiften
poison Gift *n*
poison gas Giftgas *n*
poisoning Vergiftung *f*, Intoxikation *f*
poisoning hazard Vergiftungsgefahr *f*
poisonous giftig, toxisch
poisonousness Giftigkeit *f*, Toxizität *f*
Poisson distribution Poisson-Verteilung *f*
polar-aprotic solvent polares aprot(on)isches Lösungsmittel *n*
polar bond 1. heteropolare Bindung *f*, Ionenbindung *f*, Ionenbeziehung *f*; 2. *s.* polar covalent bond
polar compound (hetero)polare Verbindung *f*, Ionenverbindung *f*
polar covalent bond polare (kovalente) Bindung *f*
polarimeter Polarimeter *n*
polarimetry Polarimetrie *f*
polarity Polarität *f*
polarizability Polarisierbarkeit *f*
polarizable polarisierbar
polarization Polarisation *f*, Polarisierung *f*
polarization current Polarisationsstrom *m*
polarization plane Polarisationsebene *f*
polarization potential Polarisationspotenzial *n*
polarization resistance Polarisationswiderstand *m*
polarize *v* polarisieren
polarizer 1. Polarisator *m (Gerät)*; 2. polarisierende Substanz *f*
polarizing angle Polarisationswinkel *m*
polarizing microscope Polarisationsmikroskop *n*
polarogram Polarogramm *n*, polarographische Kurve *f*
polarograph 1. Polarograph *m*, Polarographiegerät *n*; 2. *s.* polarogram
polarographic analysis polarographische Analyse *f*
polarographic wave polarographische Stufe *f* [Welle *f*]
polarography Polarographie *f*
Polenske number [value] Polenske-Zahl *f*, Po-Z, PZ *(Kennzahl der Fette und fetten Öle)*
policeman Gummiwischer *m (für Rührstäbe)*
polish *v* 1. polieren, glätten; schleifen *(Edelsteine)*; 2. blankfiltrieren, polierfiltrieren; 3. *(Hyd)* schönen *(gereinigte Abwässer)*
polish Politur *f*
polished section Anschliff *m (in der Auflichtmikroskopie)*
polished thin section polierter Dünnschliff *m*
polishing 1. Polieren *n*, Glätten *n*; Schleifen *n (von Edelsteinen)*; 2. *s.* polishing filtration; 3. *(Hyd)* (ionogene) Feinstreinigung *f (Ionenaustausch)*; Schönung *f (gereinigter Abwässer)*
polishing agent Poliermittel *n*
polishing filter Klärfilter *n*, Blankfilter *n*, Polierfilter *n*, Feinfilter *n*
polishing filtration Klärfiltration *f*, Blankfiltration *f*, Polierfiltration *f*, Feinfiltration *f*
polishing rouge Polierrot *n*, Englischrot *n (Eisen(III)-oxid)*
polishing wheel Polierscheibe *f*
pollutant Schmutzstoff *m*, Verschmutzung *f*, Verunreinigung *f*, *(i. e. S.)* Schadstoff *m*
pollutant gas Schadgas *n*
pollutant limit value Schadstoffgrenzwert *m*
pollutant load(ing) 1. Schmutz(stoff)last *f*, Schmutz(stoff)belastung *f*, *(i. e. S.)* Schadstoffbelastung *f*; 2. *(Hyd)* Abwasserlast *f*, Abwasserbelastung *f*, Belastung *f* mit Abwasserinhaltsstoffen
polluted verschmutzt, verunreinigt, (mit Schmutzstoffen) belastet *(Wasser, Luft)*; abwasserbelastet *(Gewässer)*
pollution Verschmutzung *f*, Verunreinigung *f*, *(i. e. S.)* Verschmutzung *f* durch

Schadstoffe, schädliche Verschmutzung *f*
pollution control Schmutzstoffbekämpfung *f*, *(i. e. S.)* Schadstoffbekämpfung *f*, Schadstoffverminderung *f*, Schadstoffkontrolle *f*
pollution emission Schadstoffemission *f*
pollution load(ing) *s.* pollutant load(ing)
pollution of the environment Umweltverschmutzung *f*
polonium Po Polonium *n*
polyacetal *s.* polyoxymethylene
polyacetylene *s.* polyyne
polyacid mehrsäurig, mehrwertig *(Base)*
polyacid mehrbasige [mehrwertige] Säure *f*, Polysäure *f*
polyacrylamide Polyacrylamid *n*, Polyacrylsäureamid *n*
polyacrylate Polyacrylat *n*, Polyacrylsäureester *m*
polyacrylic acid Polyacrylsäure *f*
polyacrylonitrile Polyacrylnitril *n*, PAN
polyacrylonitrile fibre Polyacrylnitrilfaser *f*, Acrylfaser *f*; Polyacrylnitrilfaserstoff *m*, PVY, Acrylfaserstoff *m*
polyad mehrwertig, vielwertig, polyad
polyad mehrwertiges Element *n*; mehrwertige Atomgruppe *f*
polyaddition Polyaddition *f*
polyadenylation *(Bioch)* Polyadenylierung *f*
polyalcohol Polyalkohol *m*, Polyol *n*, mehrwertiger Alkohol *m*
polyallomer *(Kst)* Polyallomer(e) *n*
polyamide Polyamid *n*, polymeres Amid *n*, PA
polyamide fibre Polyamidfaser *f*; Polyamidfaserstoff *m*, PA
polyampholyte Polyampholyt *m*
polyanion Polyanion *n*
polyaryl ether Polyarylether *m*
polyatomic mehratomig, vielatomig, polyatomar
polyatomic mehratomige [vielatomige, polyatomare] Verbindung *f*
polybasic 1. mehrbasig, mehrwertig, mehrprotonig *(Säure)*; 2. mehrsäurig, mehrbasig, mehrwertig *(Base oder basisches Salz)*
polybasicity 1. Mehrbasigkeit *f*, Mehrwertigkeit *f (einer Säure)*; 2. Mehrsäurigkeit *f*, Mehrwertigkeit *f (einer Base oder eines basischen Salzes)*
polybenzimidazole Polybenzimidazol *n*
polyblend Polyblend *n*, Polymermischung *f (Gemisch mehrerer Thermoplaste)*
polybrominated biphenyl Polybrombiphenyl *n*
polybutadiene Polybutadien *n*
polybutene Polybuten *n*, Polybutylen *n*
polybutylene Polybuten *n*, Polybutylen *n*
polycaprolactam Polycaprolactam *n*
polycarbazole Polycarbazol *n*
polycarbodiimide Polycarbodiimid *n*
polycarbonate Polycarbonat *n*, PC
polycarboxylic acid Polycarbonsäure *f*
polycation Polykation *n*
polychlorinated biphenyls polychlorierte Biphenyle *npl*, PCB
polychlorinated dibenzo-*p*-dioxin polychloriertes Dibenzo-*p*-dioxin *n*, Polychlordibenzo-1,4-dioxin *n*, PCDD
polychlorinated dibenzofuran polychloriertes Dibenzofuran *n*, PCDF
polychloroprene Polychloropren *n*
polychlorotrifluoroethylene Polychlortrifluorethylen *n*, PCTFE
polychromasia Polychromasie *f (einer Strahlung)*
polycondensate Polykondensat *n*
polycondensation Polykondensation *f*
polycrystal Vielkristall *m*, Polykristall *m*
polycrystalline polykristallin
polycyclic aromatic (hydrocarbon) polycyclischer aromatischer Kohlenwasserstoff *m*, PAK, mehrkerniger Aromat *m*
polycyclic compound polycyclische Verbindung *f*, Polycyclus *m*
polycyclic hydrocarbon polycyclischer Kohlenwasserstoff *m*
polydentate mehrzähnig, vielzähnig, mehrzählig *(Komplexchemie)*
polydimethylsiloxane Poly(dimethylsiloxan) *n*
polydisperse polydispers
polydisperse system polydisperses System *n*
polydispersity Polydispersität *f*
polyelectrolyte Polyelektrolyt *m*
polyelectrolyte complex Polyelektrolytkomplex *m*
polyene Polyen *n (Verbindung mit meh-*

reren konjugierten Kohlenstoffdoppelbindungen)
polyene antibiotic polyenes Antibiotikum *n*, Polyen-Antibiotikum *n*
polyene macrolide antibiotic polyenes Makrolid-Antibiotikum *n*
polyenoic *(org Ch)* mehrfach [stark] ungesättigt
polyepoxide Polyepoxid *n*
polyester Polyester *m*
polyester fibre Polyesterfaser *f*; Polyesterfaserstoff *m*, PE
polyester resin Polyesterharz *n*, Polyester *m*
polyether Polyether *m*, Polyalkylenoxid *n*
polyethylene Polyethylen *n*, PE
polyethylene fibre Polyethylenfaser *f*; Polyethylenfaserstoff *m*, PT
polyethylene glycol Polyethylenglykol *n*, PEG
polyethylene imine Polyethylenimin *n*
polyethylene oxide Polyethylenoxid *n*
polyethylene terephthalate Polyethylenterephthalat *n*, PETP
poly(3,4-ethylenedioxythiophene) Poly(3,4-ethylendioxythiophen) *n (Material in organischen Leuchtdioden)*
polyformaldehyde *s.* polyoxymethylene
polyforming *(Erdöl)* Polyformen *n*, Polyforming *n (thermisches Reformieren mit Gasrückführung)*
polyfunctional polyfunktionell, mehrfunktionell
polyglycol Polyglykol *n*
polyhalide Polyhalogenid *n*
polyhalogenate *v* mehrfach halogenieren
polyhalogenation Mehrfachhalogenierung *f*
polyhedral *(Krist)* polyedrisch, vielflächig
polyhedron *(Krist)* Polyeder *n*, Vielflach *n*, Vielflächner *m*
polyhydrate Polyhydrat *n*
polyhydric mehrere Hydroxylgruppen enthaltend
polyhydric alcohol *s.* polyol
polyhydric phenol mehrwertiges Phenol *n*
polyhydroxy acid Polyhydroxysäure *f*
polyhydroxy alcohol *s.* polyol
polyhydroxy compound Polyhydroxyverbindung *f*
polyimidazopyrrolone Polyimidazopyrrolon *n*, Pyrron *n*
polyimide Polyimid *n*
polyion Polyion *n*, Mehrfachion *n*
polyisobutene Polyisobuten *n*, Polyisobutylen *n*, PIB
polyisobutylene Polyisobuten *n*, Polyisobutylen *n*, PIB
polyisoprene Polyisopren *n*
polyisotopic polyisotop
polyisotopic element Mischelement *n*
polyketide Polyketid *n*, Acetogenin *n*
polyketone Polyketon *n*
polymer Polymer(e) *n*, Polymerisat *n*
polymer backbone Hauptkette *f (eines Polymers)*
polymer blend *s.* polyblend
polymer chain Polymerkette *f*
polymer chemistry Polymer(en)chemie *f*, makromolekulare Chemie *f*
polymer coating Abdeckschicht *f*
polymer concrete Polymerbeton *m (Kunstharzbeton)*
polymer for medical application medizinischer Kunststoff *m*
polymer gasoline Polymer(isations)-benzin *n*
polymer latex Kunststofflatex *m*
polymer light emitting diode polymere Leuchtdiode *f*
polymer network *(Kst)* Polymernetzwerk *n*
polymer processing Kunststoffverarbeitung *f*
polymerase chain reaction Polymerase-Kettenreaktion *f*
polymeric polymer
polymeric nanocomposite Polymer-Nanokomposit *m*
polymerizability Polymerisationsfähigkeit *f*; Polymerisierbarkeit *f*
polymerizable polymerisationsfähig; polymerisierbar
polymerizate *s.* polymer
polymerization Polymerisation *f*, Polymerisieren *n*
polymerization initiator Polymerisationsinitiator *m*, Polymerisationserreger *m*, Starter *m*, Aktivator *m*
polymerize *v* polymerisieren
polymethacrylate Polymethacrylat *n*, Polymethacrylsäureester *m*

polymethacrylic acid Polymethacrylsäure *f*
polymethine dye Polymethinfarbstoff *m*
polymethyl methacrylate Polymethylmethacrylat *n*, PMMA
poly-4-methyl-1-pentene Poly-4-methylpent-1-en *n*, PMP
polymethylene Polymethylen *n*
polymolecular polymolekular
polymolecularity Polymolekularität *f*
polymorph 1. polymorpher [in mehreren Modifikationen auftretender] Stoff *m*; 2. polymorphe Form *f*
polymorphic polymorph, vielgestaltig
polymorphic transition polymorphe Umwandlung *f* [Phasenumwandlung *f*]
polymorphism Polymorphie *f*, Polymorphismus *m*, Vielgestaltigkeit *f*
polymyxin Polymyxin *n* *(Eiweißchemie)*
polynosic fibre polynosische Faser *f*, Polynosefaser *f*; polynosischer Faserstoff *m*, Polynosefaserstoff *m*
polynuclear complex mehrkerniger [polynuklearer] Komplex *m*, Mehrkernkomplex *m*
polyol Polyol *n*, Polyalkohol *m*, mehrwertiger Alkohol *m*
polyolefin Polyolefin *n*, PO
polyorganosiloxane *Polyorganosiloxan *n*, Organopolysiloxan *n*, organisches Polysiloxan *n*, Silicon *n*
polyose *s.* polysaccharide
polyoxymethylene Polyoxymethylen *n*, POM, Polyformaldehyd *m*, Polyacetal *n*
polyparabanic acid Polyparabansäure *f*
polypeptide antibiotic Polypeptidantibiotikum *n*
polypeptide chain Polypeptidkette *f*
polyphenylene Polyphenylen *n*
polyphenylene oxide Polyphenylenoxid *n*, PPO
polyphenylene sulphide Polyphenylensulfid *n*, PPS, Polythio-1,4-phenylen *n*
polyphenylene sulphone Polyphenylensulfon *n*, Polysulfon *n*
polyphosphate Polyphosphat *n*
polyphosphazene Polyphosphazen *n*
polyphosphoric acid Polyphosphorsäure *f* *(allg. für eine Phosphorsäure der Formel H_{n+2} $[P_nO_{3n+1}]$)*
polypropene *s.* polypropylene
polypropylene Polypropylen *n*, PP
polyprotic acid mehrbasige [mehrprotonige, mehrwertige] Säure *f*
polypyrrole Polypyrrol *n* *(elektrisch-leitfähiges Polymer)*
polyquinane Polyquinan *n* *(polycyclischer Kohlenwasserstoff)*
polyquinene Polyquinen *n* *(ungesättigter polycyclischer Kohlenwasserstoff)*
polyreaction Polyreaktion *f* *(Polymerisations-, Polykondensations- oder Polyadditionsreaktion)*
polysaccharide Polysaccharid *n*
polysilicic acid Polykieselsäure *f*
polysiloxane Polysiloxan *n*, polymeres Siloxan *n*
polyspiro compound Polyspiroverbindung *f*
polystyrene Polystyrol *n*, *Polystyren *n*, PS
poly(styrenesulfonate) Polystyrolsulfonat *n* *(Material in organischen Leuchtdioden)*
polysubstituted polysubstituiert, mehrfach substituiert
polysubstitution Polysubstitution *f*, Mehrfachsubstitution *f*
polysulphide Polysulfid *n*
polysulphide rubber Polysulfidkautschuk *m*, Thioplast *m*
polysulphone Polysulfon *n*, Polyphenylsulfon *n*
polysynthetic twinning *(Krist)* polysynthetische Zwillingsbildung *f* [Verzwillingung *f*]
polyterpene *(org Ch)* Polyterpen *n*
polytetrafluoroethylene Polytetrafluorethylen *n*, PTFE
polythene *s.* polyethylene
polythionate $M^I_2[S_xO_6]$ Polythionat *n*
polythionic acid Polythionsäure *f*
polythiophene Polythiophen *n* *(elektrisch-leitfähiges Polymer)*
polythiourea Thioharnstoffharz *n*, Polythiocarbamid *n*
polytriazine Polytriazin *n*, Triazinpolymer(e) *n*
polyunsaturated fatty acid mehrfach ungesättigte Fettsäure *f*
polyurea Harnstoffharz *n*, Polyharnstoff *m*
polyurethane Polyurethan *n*, PUR
polyurethane elastomer Polyurethanelastomer(e) *n*, Urethanelastomer(e) *n*

polyurethane foam Polyurethanschaum(stoff) *m*
polyurethane rubber Polyurethankautschuk *m*, Urethankautschuk *m*
polyuronic acid Polyuronsäure *f*
polyuronid(e) Polyuronid *n*
polyvalence Mehrwertigkeit *f*
polyvalency Mehrwertigkeit *f*
polyvalent mehrwertig
polyvinyl acetal Polyvinylacetal *n*
polyvinyl acetate Polyvinylacetat *n*, PVAC
polyvinyl alcohol Polyvinylalkohol *m*, PVA
polyvinyl butyral Polyvinylbutyral *n*, PVB
polyvinyl carbazole Polyvinylcarbazol *n*, PCV
polyvinyl chloride Polyvinylchlorid *n*, PVC
polyvinyl-chloride acetate Polyvinylchloridacetat *n*, PVCA
polyvinyl ester Polyvinylester *m*
polyvinyl ether Polyvinylether *m*
polyvinyl fibre Polyvinylfaser *f*; Polyvinylfaserstoff *m*, PV
polyvinyl fluoride Polyvinylfluorid *n*
polyvinyl formal Polyvinylformal *n*, Polyvinyl-Formaldehydacetal *n*
polyvinylidene chloride Polyvinylidenchlorid *n*, PVDC
polyvinylidene cyanide Polyvinylidencyanid *n*
polyvinylidene fluoride Polyvinylidenfluorid *n*, PVDF
poly-*p*-xylene Poly-*p*-xylylen *n*
poly-*p*-xylylene Poly-*p*-xylylen *n*
polyyne Polyin *n*, Polyacetylen *n*, Polyacetylenkohlenwasserstoff *m*
pomace Brei *m*, (breiige) Masse *f*; Mus *n*, Pulpe *f*, Pülpe *f*, Pulp *m*, Fruchtmasse *f*, Fruchtbrei *m*; Maische *f*; Trester *pl*; Pressrückstand *m*, Presskuchen *m*
Pomeranz-Fritsch reaction [synthesis] Pomeranz-Fritsch-Synthese *f (von Isochinolinen)*
pond 1. *(Tech)* Sumpf *m*; 2. Teich *m*, Becken *n*; 3. *(Hyd)* aerober [aerob arbeitender] Abwasserteich *m*, *(i. e. S.)* Oxidationsteich *m*
pool 1. *(Geol)* erdölführendes [erdgashaltiges] Speichergestein *n*; 2. *(Tech)* Becken *n*; 3. *(Bioch)* Pool *m*, Sammelbecken *n*
poor arm; schlecht, gering
poor in hydrogen wasserstoffarm
poorly absorbing schwach absorbierend
poorly conducting schlecht leitend
poorly defined mangelhaft [schlecht] definiert
poorly reproducible schlecht [schwer] reproduzierbar
poorly soluble wenig [schwer] löslich, schwerlöslich
popcorn-polymer Popcorn-Polymer *n (schaumiges Polymerisatkorn)*
populate *v* besetzen *(z. B. ein Energieband mit Elektronen)*
population 1. Besetzung *f (z. B. eines Energiebands mit Elektronen)*; 2. Grundgesamtheit *f (Statistik)*
population diagram Besetzungsschema *n (von Molekülorbitalen)*
population inversion Besetzungsinversion *f (des Energieniveaus von Atomen)*
porcelain Porzellan *n*
porcelain clay Porzellanerde *f*, weißer [reiner] Ton *m*, Kaolin *m*, China Clay *m(n)*
porcelain crucible Porzellantiegel *m*
porcelain enamel Email *n*
porcelain packing Füllkörper *mpl* aus Porzellan
porcelainization 1. Brennen *n* zu Porzellan; 2. Emaillieren *n*
porcelainize *v* 1. zu Porzellan brennen; 2. emaillieren
porcelainous 1. aus Porzellan, porzellanen; 2. porzellanartig, porzellanähnlich
porcel(l)aneous 1. aus Porzellan, porzellanen; 2. porzellanartig, porzellanähnlich
porcel(l)anous 1. aus Porzellan, porzellanen; 2. porzellanartig, porzellanähnlich
pore Pore *f*
pore diffusion Porendiffusion *f*, innere Diffusion *f*
pore size Porenweite *f*, Porengröße *f*
pore-size distribution Porengrößenverteilung *f*
pore-volume distribution Porenvolumenverteilung *f*
poriferous porös, porig, mit Poren versehen

porosimeter Porosimeter *n (Gerät zur Messung der Porosität)*
porosity Porosität *f*, Porigkeit *f*; Porosität *f*, Durchlässigkeit *f*; relatives Porenvolumen *n*, relativer Porenraum *m*
porous porig, porös, mit Poren versehen; porös, durchlässig; schwammig
porous layer bead *(Chromat)* Festschichtkügelchen *n*, schalenporöser Füllkörper *m* [Trennkörper *m*], Dünnschichtteilchen *n*, PLB-Teilchen *n*
porphine Porphin *n*, *Porphyrin *n (cyclisches Tetrapyrrol)*
porphobilinogen *(Bioch)* Porphobilinogen *n*
porphyrin Porphyrin *n, (unsubstituiert als* $C_20H_{14}N_4$ *auch:)* Porphin *n*
porphyry *(Geol)* Porphyr *m*
port Öffnung *f*, Stutzen *m*, Mund *m*, Anschluss *m*
portable tragbar; transportabel, ortsbeweglich
portion Teil *m*, Anteil *m*; Portion *f*, (abgemessene) Menge *f*, Quantum *n* • **in portions** portionsweise
portland cement Portlandzement *m*, PZ
position *v* einstellen, in die richtige Lage [Stellung] bringen
position Stellung *f*, Position *f*, Lage *f (am Molekül)*; Ort *m (eines Elementarteilchens)* • **in meta position** in *meta*-Stellung [*m*-Stellung, 1,3-Stellung], metaständig, *m*-ständig • **in ortho position** in *ortho*-Stellung [*o*-Stellung, 1,2-Stellung], orthoständig, *o*-ständig • **in para position** in *para*-Stellung [*p*-Stellung, 1,4-Stellung], paraständig, *p*-ständig
position isomer Stellungsisomer(e) *n*
position isomerism Stellungsisomerie *f*, Substitutionsisomerie *f*
position selectivity *(org Ch)* Positionsselektivität *f*
positive *(Foto)* Positiv *n*
positive beta decay *(Kern)* β^+-Zerfall *m*, Positronenzerfall *m*
positive displacement pump Verdrängerpumpe *f*, Verdrängungspumpe *f*
positive electrode positive [positiv geladene] Elektrode *f*, Anode *f*
positive electron positives Elektron *n*, Positron *n*
positive ion positives [positiv geladenes] Ion *n*, Kation *n*
positive ray positiver Strahl *m*, Kanalstrahl *m*
positive valency positive (elektrochemische) Wertigkeit *f*
positively charged positiv geladen
positron Positron *n*, positives Elektron *n*
positron chemistry Positronchemie *f*
positronium Positronium *n (aus Positron und Elektron bestehendes System)*
positronium chemistry Positroniumchemie *f*
postchlorinate *v* 1. nachchlorieren; 2. *(Hyd)* nachchloren
postchlorination 1. Nachchlorieren *n*, Nachchlorierung *f*; 2. *(Hyd)* Nachchloren *n*, Nachchlorung *f*
postcolumn derivatization *(Chromat)* Derivatisierung *f* nach der Trennsäule, Nachsäulenderivatisierung *f*
postcure *(Kst)* Nachhärten *n*, Nachhärtung *f*
postcuring *(Kst)* Nachhärten *n*, Nachhärtung *f*
postemergence herbicide Nachauflaufherbizid *n*
posterior-lobe hormone Hypophysenhinterlappenhormon *n*, neurohypophysäres Hormon *n*, Hormon *n* der Neurohypophyse
postexposure *(Foto)* Nachbelichtung *f*
postperiod Nachperiode *f*, Nachversuch *m (einer kalorimetrischen Messung)*
postpolymerization Nachpolymerisation *f*
postprecipitation Nachfällung *f*
posttreatment Nachbehandlung *f*
pot 1. Topf *m*; 2. *(Glas)* Hafen *m*; 3. *(Dest)* Blase *f*; 4. *(Kst)* Spritztopf *m*, Füllzylinder *m*, Füllraum *m*; 5. *(Landw)* Kulturgefäß *n*
pot life Gebrauchsdauer *f*, Topfzeit *f*, Verarbeitungszeit *f (von Reaktionsanstrichstoffen)*
pot press Kachelpresse *f*, Trogpresse *f*, Ringpresse *f*, Schachtelpresse *f*
pot still Blasendestillierapparat *m*
potable trinkbar
potable Getränk *n*
potable water standards Trinkwasser-Standard *m*, Gütebedingungen *fpl* für Trinkwasser, Güteanforderungen *fpl*

[Beschaffenheitsanforderungen *fpl*] an Trinkwasser
potable water treatment Trinkwasseraufbereitung *f*
potash 1. Pottasche *f*, Kaliumcarbonat *n*; 2. Kaliumhydroxid *n*, Ätzkali *n*; 3. Kaliumoxid *n*, Kali *n*; 4. Kali(salz) *n*
potash feldspar *(Min)* Kalifeldspat *m* *(Kaliumalumotrisilicat)*
potash fertilizer Kalidüngemittel *n*, Kalidünger *m*
potash mica *(Min)* Kaliglimmer *m*, Muskovit *m*
potash salt Kali(salz) *n*
potassic kaliumhaltig, Kalium...; kalihaltig, Kali...
potassiferous kalihaltig, Kali...
potassium K Kalium *n*
potassium acetate Kaliumacetat *n*
potassium alum Kalium(aluminium)-alaun *m*, Kalialaun *m*, Alaun *m*, Kaliumaluminiumsulfat-12-Wasser *n*
potassium aluminate Kaliumaluminat *n*
potassium amalgam Kaliumamalgam *n*
potassium antimonyl tartrate Kaliumantimonotartrat *n*, Kaliumantimonyltartrat *n*
potassium bromate Kaliumbromat *n*
potassium bromide pellet *(Spektr)* KBr-Scheibe *f*, KBr-Tablette *f*
potassium carbonate Kaliumcarbonat *n*, Pottasche *f*
potassium chlorate Kaliumchlorat *n*
potassium chloride Kaliumchlorid *n*
potassium chromate Kaliumchromat *n*
potassium-chrome alum Kaliumchromalaun *m*, Chromalaun *m*, Kaliumchrom(III)-sulfat-12-Wasser *n*
potassium cyanate Kaliumcyanat *n*
potassium cyanide Kaliumcyanid *n*, Cyankali *n*
potassium dichromate Kaliumdichromat(VI) *n*
potassium fluoride Kaliumfluorid *n*
potassium fluoroborate Kaliumtetrafluoroborat *n*
potassium fluorosilicate Kaliumhexafluorosilicat *n*
potassium hexacyanoferrate(II) Kaliumhexacyanoferrat(II) *n*, gelbes Blutlaugensalz *n*
potassium hexacyanoferrate(III) Kaliumhexacyanoferrat(III) *n*, rotes Blutlaugensalz *n*
potassium hexanitrocobaltate(III) Kaliumhexanitrocobaltat(III) *n*
potassium hydrogencarbonate Kaliumhydrogencarbonat *n*
potassium hydrogenfluoride Kaliumhydrogenfluorid *n*
potassium hydrogensulphate Kaliumhydrogensulfat *n*
potassium hydrogensulphide Kaliumhydrogensulfid *n*
potassium hydrogensulphite Kaliumhydrogensulfit *n*
potassium hydrogentartrate Kaliumhydrogentartrat *n*
potassium hydroxide Kaliumhydroxid *n*, Ätzkali *n*
potassium hydroxide fusion Kaliumhydroxidschmelze *f*
potassium hydroxide solution Kaliumhydroxidlösung *f*, Kalilauge *f*
potassium hypochlorite Kaliumhypochlorit *n*
potassium iodate Kaliumiodat *n*
potassium iodide Kaliumiodid *n*
potassium-iodide starch paper Kaliumiodidstärkepapier *n*, Iodkaliumstärkepapier *n*, Ozonpapier *n*
potassium-magnesium sulphate Kaliummagnesiumsulfat *n*
potassium manganate Kaliummanganat *n*
potassium nitrate Kaliumnitrat *n*
potassium nitrite Kaliumnitrit *n*
potassium oleate Kaliumoleat *n*
potassium orthophosphate Kaliumorthophosphat *n*, Trikaliumphosphat *n*
potassium oxalate Kaliumoxalat *n*
potassium oxide Kaliumoxid *n*
potassium perchlorate Kaliumperchlorat *n*
potassium permanganate Kaliumpermanganat *n*, Kaliummanganat(VII) *n*
potassium peroxide 1. Kaliumperoxid *n*; 2. *s.* potassium superoxide
potassium peroxydisulphate Kaliumperoxodisulfat *n*
potassium pyrosulphate Kaliumdisulfat *n*, Dikaliumdisulfat *n*
potassium-selective K^+-sensitiv *(Elektrode)*
potassium soap Kali(um)seife *f*

potassium-sodium tartrate Kaliumnatriumtartrat *n*, Seignettesalz *n*
potassium sulphate Kaliumsulfat *n*
potassium sulphite Kaliumsulfit *n*
potassium superoxide Kaliumhyperoxid *n*
potassium tartrate Kaliumtartrat *n*
potassium tetrathionate Kaliumtetrathionat *n*
potassium thiocyanate Kaliumthiocyanat *n*
potassium xanthate [xanthogenate] Kaliumxanthogenat *n*, *(i. e. S.)* Kaliumethylxanthogenat *n*
potato alcohol Kartoffelspiritus *m*, Kartoffelsprit *m*
potato starch Kartoffelstärke *f*
potency 1. Wirkkraft *f*, Wirkungsstärke *f*, Wirksamkeit *f*, Potenz *f*; 2. Potenz *f*, Verdünnung *f (Homöopathie)*
potent stark wirkend, wirksam
potential 1. potenziell *(z. B. Energie)*; 2. *(Ech)* Potenzial...
potential *(Ech)* Potenzial *n*
potential barrier Potenzialwall *m*, Potenzialschwelle *f*, Potenzialberg *m*, Potenzialbarriere *f*
potential-determining potenzialbestimmend
potential difference Potenzialdifferenz *f*, Potenzialunterschied *m*
potential drop Potenzialabfall *m*, Potenzialrückgang *m*
potential energy potenzielle Energie *f*, Lageenergie *f*, Zustandsenergie *f*
potential-energy contour map Potenzialflächendiagramm *n*, Schichtliniendiagramm *n* [Höhenliniendiagramm *n*] der Potenzialfläche, Niveauliniendarstellung *f* der Potenzialfläche
potential-energy profile Energieprofil *n*
potential-energy (reaction) surface Potenzial(energie)fläche *f*
potential gradient Potenzialgradient *m*, Potenzialgefälle *n*
potential head Lagehöhe *f*, Ortshöhe *f*, geodätische Höhe *f*
potential jump Potenzialsprung *m*
potential-scanning coulometry Coulometrie *f* mit kontinuierlich geändertem Potenzial
potentiate *v* 1. potenzieren, steigern, verstärken; 2. potenzieren, verdünnen *(homöopathische Mittel)*
potentiation 1. Potenzierung *f*, Steigerung *f*, Verstärkung *f*; 2. Potenzieren *n*, Verdünnen *n (von homöopathischen Mitteln)*
potentiometer Potenziometer *n*
potentiometric analysis potenziometrische Analyse *f*, *(i. e. S.)* potenziometrische [elektrometrische] Maßanalyse *f* [Titration *f*]
potentiometric titration potenziometrische [elektrometrische] Titration *f*
potentiometry Potenziometrie *f*
potter's clay [earth] Töpferton *m*
pottery Töpferei *f*; Töpferware *f*, *(i. e. S.)* Steingut *n*
pour *v* gießen, schütten; fließen, strömen, rinnen
pour *v* **off** abgießen, dekantieren
pour point Fließpunkt *m*, Pourpoint *m* *(niedrigste Temperatur, bei der ein Kraftstoff beim Abkühlen gerade noch fließt)*
pourability Gießbarkeit *f*, Vergießbarkeit *f*
pourable (ver)gießbar
powder *v* 1. pulverisieren, (zer)pulvern; auf Staubfeinheit mahlen; zu Pulver zerfallen; 2. (ein)pudern, bestäuben
powder *v* **together** gemeinsam verreiben
powder 1. Pulver *n*, Mehl *n*; 2. *(Kosmet)* Puder *m* • **in fine powder** fein gepulvert, fein pulverisiert, feinpulverig
powder coating Pulverbeschichtung *f*
powder diffraction camera Pulverdiffraktometer *n*
powder extinguisher Pulverlöscher *m*, Trockenlöscher *m*
powder funnel Pulvertrichter *m*
powder insulation Pulverisolierung *f* *(Tieftemperaturtechnik)*
powder metal gepulvertes Metall *n*, Metallpulver *n*, metallisches Pulver *n*
powder-metallurgical pulvermetallurgisch
powder metallurgy Pulvermetallurgie *f*
powder method (of analysis) *(Krist)* Pulververfahren *n*, Pulvermethode *f*
powder pattern [photograph, picture] *(Krist)* Pulverdiagramm *n*, Pulveraufnahme *f*

powdery pulv(e)rig, pulverförmig; zerreiblich, zerreibbar
power *v* speisen *(z. B. einen Stromkreis)*
power 1. Kraft *f*, Stärke *f*; Energie *f*; 2. Leistung *f (im physikalischen Sinne)*; 3. Vermögen *n*, Fähigkeit *f*; Leistungsvermögen *n*, Leistungsfähigkeit *f*
power alcohol from biomass Biosprit *m*
power connection Stromanschluss *m*
power consumption Leistungsaufnahme *f*
power gas Kraftgas *n*
power plant Kraftwerk *n*, Elektrizitätswerk *n*
power requirements Kraftbedarf *m*, Kraftaufwand *m*; Energiebedarf *m*, Elektroenergiebedarf *m*
power supply Stromzufuhr *f*, Stromversorgung *f*; Stromquelle *f*
POX *s.* purgeable organic halogens
PP *s.* polypropylene
PP factor *s.* pellagra-preventive factor
ppb *s.* parts per billion
ppbv = parts per billion by volume
pphm *s.* parts per hundred million
ppm *s.* parts per million
ppmm *s.* parts per million by mass
ppmv *s.* parts per million by volume
PPO *s.* polyphenylene oxide
PPS *s.* polyphenylene sulphide
ppt *s.* 1. precipitate; 2. parts per thousand
pptd. = precipitated
ppt'n *s.* precipitation
pptv *s.* parts per trillion by volume
PRA = phosphoribosylamine
practicality Anwendbarkeit *f*, Durchführbarkeit *f*; Brauchbarkeit *f (einer Methode)*
Prandtl boundary layer prandtlsche Strömungsgrenzschicht *f*, Grenzschicht *f* der turbulenten Strömung
praseo salt Praseosalz *n (Komplexchemie)*
praseodymia Praseodym(III)-oxid *n*
praseodymium Pr Praseodym *n*
praseodymium chloride Praseodym(III)-chlorid *n*
praseodymium dioxide Praseodymdioxid *n*, Praseodym(IV)-oxid *n*
praseodymium trioxide Praseodymtrioxid *n*, Praseodym(III)-oxid *n*
preamplifier *(Anal)* Vorverstärker *m*
preblend *v* vormischen
preblend Vormischung *f*
prebloom spray *(Landw)* Vorblütenspritzung *f*
precaution Vorsichtsmaßnahme *f* • **take precautions** Vorsichtsmaßnahmen treffen
precedence *(Nom)* Priorität *f*, Vorrang *m*
precession *(phys Ch)* Präzession *f*
prechlorinate *v* 1. *(Pap)* vorchlorieren, vorbleichen; 2. *(Hyd)* vorchloren
prechlorination 1. *(Pap)* Vorchlorierung *f*, Vorbleiche *f*; 2. *(Hyd)* Vorchloren *n*, Vorchlorung *f*
precious metal Edelmetall *n*
precious stone Edelstein *m*
precip. = precipitated
precipitability Fällbarkeit *f*; Abscheidbarkeit *f*
precipitable (aus)fällbar; abscheidbar
precipitant Fällungsmittel *n*, Fällungs(re)agens *n*, Ausfällungsmittel *n*
precipitate *v* 1. (aus)fällen, präzipitieren; abscheiden; 2. *(Met)* niederschlagen, präzipitieren; kondensieren; ausfallen; sich abscheiden; 3. *(Met)* sich ausscheiden; sich kondensieren
precipitate (chemischer) Niederschlag *m*, Fällung *f*
precipitating agent *s.* precipitant
precipitating electrode Niederschlagselektrode *f*
precipitation 1. Fällen *n*, Ausfällen *n*, Präzipitieren *n*; Abscheiden *n*, *(i. e. S.)* Elektroabscheiden *n*; 2. *(Met)* Niederschlagen *n*, Präzipitieren *n*; Kondensieren *n*; Ausfallen *n*; Abscheidung *f*; 3. *(Met)* Ausscheidung *f*; Kondensation *f*; 4. (atmosphärischer) Niederschlag *m*; 5. *s.* precipitate
precipitation analysis Fällungs(maß)analyse *f*, Fällungsmethode *f*
precipitation chromatography Fällungschromatographie *f*, FC
precipitation hardening Ausscheidungshärten *n*, Aushärten *n*
precipitation reaction Fällungsreaktion *f*
precipitation softening *(Hyd)* Enthärtung *f* durch Fällverfahren, *(i. e. S.)* Enthärtung *f* durch Kalkfällung
precipitation tank Absetzbecken *n*, Absetzbehälter *m*, Absitzbehälter *m*,

Abscheider *m*, Fällkasten *m*, Klärtank *m*, Klärbehälter *m*
precipitation titration Fällungstitration *f*
precision *(Anal)* *Genauigkeit *f*, Präzision *f*
precision balance Präzisionswaage *f*
preclarification *(Hyd)* Vorklärung *f*, Vorreinigung *f*
preclarify *v (Hyd)* vorklären, vorreinigen
preclean *v* vorreinigen
precleaner 1. Vorreiniger *m*; 2. Vorabscheider *m*, Grobabscheider *m*
precoat filter Anschwemmfilter *n*, Precoatfilter *n*
precolumn *(Chromat)* Vorsäule *f*
precolumn derivatization Derivatisierung *f* vor der Trennsäule, Vorsäulenderivatisierung *f*
preconcentrate *v* vorkonzentrieren
preconcentration Vorkonzentrierung *f*, Vorkonzentration *f*
precondition *v* 1. vorbehandeln; 2. *(Hyd)* vorreinigen
preconditioning 1. Vorbehandlung *f*; 2. *(Hyd)* Vorreinigung *f*
precultivate *v (Biot)* vorkultivieren
preculture *(Biot)* Vorkultur *f*
precure *v (Kst)* vorhärten
precursor Vorläufer *m*, Vorstufe *f*, Präkursor *m*; Vorläufersubstanz *f (insbesondere eines Luftschadstoffes)*; *(i. e. S.)* Primärschadstoff *m*
predioxin *(Tox)* Prädioxin *n*
predissociation *(phys Ch)* Prädissoziation *f*
predry *v* vortrocknen
predryer Vortrockner *m*
predrying Vortrocknen *n*, Vortrocknung *f*
pre-electrolysis Vorelektrolyse *f*
pre-emergence herbicide Vor(auf)laufherbizid *n*
pre-enzyme *s.* proenzyme
pre-equilibrium vorgelagertes Gleichgewicht *n*
pre-evaporator Vorverdampfer *m*
pre-expand *v (Kst)* vorschäumen
pre-exponential factor präexponentieller Faktor *m*, Präexponentialfaktor *m* *(der Arrhenius-Gleichung)*
preferential adsorption bevorzugte [selektive] Adsorption *f*
prefermentation *(Biot)* Vorfermentation *f*; Vorgärung *f*, Angärung *f*
prefermenter *(Biot)* Vorfermenter *m*
preferred name *(Nom)* Vorzugsbenennung *f*
prefilt (feed, slurry) *(Filtr)* Aufgabegut *n*, Filtergut *n*, Trübe *f*
prefilter Vorfilter *n*
prefiltration Vorfiltration *f*
prefoam *v (Kst)* vorschäumen
prefractionate *v (Dest, Chromat)* vorfraktionieren
prefractionation *(Dest, Chromat)* Vorfraktionierung *f*
preharvest interval Karenzzeit *f*, Wartezeit *f*, Karenzfrist *f*, Wartefrist *f (nach dem Einsatz von Pflanzenschutzmitteln)*
preheat *v* vorwärmen, anwärmen, vorerhitzen; vorheizen
preheater Vorerhitzer *m*, Vorheizer *m*, Vorheizofen *m*, Vorwärmer *m*, Vorwärmgerät *n*
prehydrolysis Vorhydrolyse *f*
preimpregnate *v* 1. vorimprägnieren; 2. *(Kst)* vorbeharzen
preimpregnation Vorimprägnierung *f*
preionization Präionisation *f*, Autoionisation *f*, Selbstionisation *f*
preirradiation *(Kst)* Vorbestrahlung *f*
preliminary cooling Vorkühlung *f*
preliminary examination [investigation] Voruntersuchung *f*
preliminary separation *(Chromat)* Vortrennung *f*
preliminary test Vorprüfung *f*, Vorversuch *f*, Vorprobe *f*; orientierende Prüfung *f (von Pflanzenschutzmitteln)*
preliminary treatment Vorbehandlung *f*
premetallize *v* vormetallisieren
premix *v* vormischen
premix 1. Vormischung *f*; 2. *(Kst)* Premix-Masse *f (vorgemischtes harzgetränktes Glasfasermaterial)*
premix burner Brenner *m* mit Vormischung
pre-oxidation vorherige [vorhergehende] Oxidation *f*
pre-oxidize *v* vorher oxidieren
pre-ozonation *(Hyd)* Vorozon(is)ierung *f*
preparation 1. *(Lab)* Darstellung *f*, Herstellung *f*; Ansetzen *n*, Bereitung *f*, Zubereitung *f (nach Vorschrift)*; 2. Vorbereiten *n*, Vorbereitung *f*; Aufbereiten *n*, Aufbereitung *f*; Vorbehandlung *f*,

Präparieren *n*; 3. Haltbarmachen *n*, Präparieren *n*; 4. Präparat *n*; 5. *(Med)* Mittel *n*, Arzneimittel *n*, Arzneizubereitung *f*, Fertigpräparat *n*
preparative chemistry präparative Chemie *f*, Präparatenchemie *f*
preparative method Darstellungsmethode *f*, Herstellungsmethode *f*
preparative-scale präparativ, in präparativem Maßstab
preparatory treatment Vorbehandlung *f*
prepare *v* 1. *(Lab)* darstellen, herstellen; ansetzen, (zu)bereiten *(nach Vorschrift)*; 2. vorbereiten; aufbereiten; vorbehandeln, präparieren; 3. haltbar machen, präparieren
prepare *v* **in pure form** rein [in reinem Zustand] darstellen
preparing salt Präpariersalz *n (Natriumhexahydroxostannat(IV))*
pre-period Vorperiode *f (einer kalorimetrischen Messung)*
prephenic acid Prephensäure *f (ein Cyclohexadienderivat)*
preplanting application *(Landw)* Applikation *f* [Ausbringen *n*] vor dem Auspflanzen
preplastication *(Kst)* Vorplastizieren *n*, Vorplastizierung *f*
prepn. *s.* preparation
prepolymer Vorpolymer(e) *n*, Vorpolymerisat *n*
prepreg *(Kst)* Prepreg *n*, vorimprägniertes Glasfasermaterial *n*
preprocessing Vorbehandlung *f*
preproduction stage Stadium *n* der großtechnischen Erprobung
preroast *v* vorrösten
prescribe *v* 1. vorschreiben; 2. *(Med)* verschreiben
prescription 1. Vorschrift *f*; 2. *(Med)* Rezept *n*
prescription drug [pharmaceutical] rezeptpflichtiges Arzneimittel *n*
presedimentation basin *(Hyd)* Vorklärbecken *n*
presence of air Luftanwesenheit *f*, Luftgegenwart *f*
preservation Konservieren *n*, Konservierung *f*, Haltbarmachung *f*
preservation jar Präparatenglas *n*, Schauglas *n*
preservative (agent) Konservierungsmittel *n*, Konservierungsstoff *m*
preserve *v* konservieren, haltbar machen
preserve 1. Konserve *f*, *(i. e. S.)* Obstkonserve *f*; Marmelade *f*; 2. Arbeitsschutzmittel *n*, *(i. e. S.)* Schutzbrille *f*
preset value vorgegebener Wert *m*
presinter *v* vorsintern
presintering Vorsinterung *f*
presowing application *(Landw)* Vorsaatanwendung *f*
presowing herbicide Vorsaatherbizid *n*
press *v* pressen; zusammenpressen; abpressen; auspressen; (nieder)drücken *(z. B. Hebel)*
press *v* **off** abpressen
press Presse *f*
press cake Presskuchen *m*
press cloth Presstuch *n*
press-juice Presssaft *m*
press liquor Presslauge *f*
pressed disk *(Spektr)* Pressling *m*, Tablette *f (Probesubstanz vorwiegend in Kaliumbromid)*
pressed disk technique *(Spektr)* Presstechnik *f*, *(i. e. S.)* Kaliumbromid-Presstechnik *f*
pressing 1. Pressen *n*, Pressung *f*, Abpressen *n*; Auspressen *n*; 2. *(Met)* Pressteil *n*
pressure-diecast *v* druckgießen, unter Druck gießen, durch Druck vergießen, auf Druckgießmaschinen vergießen, im Druckguss herstellen
pressure Druck *m*
pressure atomizer Druckversprüher *m*; Druckzerstäuber *m (für Feststoffe)*
pressure broadening Druckverbreiterung *f*, Stoßverbreiterung *f*, Lorentz-Verbreiterung *f (von Spektrallinien)*
pressure casting Druckgießen *n*, Druckguss *m*, Gießen *n* unter Druck
pressure chamber Druckkammer *f*
pressure-compensating vessel Druckausgleichsgefäß *n*
pressure cycle reactor *(Biot)* Druckumlauffermenter *m*
pressure cylinder 1. Druck(gas)flasche *f*, Bombe *f*; 2. Druckkörper *m (eines Filters)*; 3. *(Kst)* Druckzylinder *m*
pressure-dependent druckabhängig
pressure distillate Druckdestillat *n*
pressure distillation Druckdestillation *f*

pressure drop Druckabfall *m*, Druckverlust *m*
pressure filter Druckfilter *n*, Überdruckfilter *n*
pressure filtration Druckfiltrieren *n*, Druckfiltration *f*; Druckfiltern *n (von Gasen)*
pressure gasification *(Carb)* Druckvergasung *f*
pressure gasifier *(Carb)* Druckvergaser *m*, Druckgasgenerator *m*, Druckgaserzeuger *m*
pressure gauge Manometer *n*, Druckmesser *m*
pressure head (statische) Druckhöhe *f*, *(i. w. S.)* Druck *m*
pressure hydrogenation Druckhydrierung *f*
pressure-independent druckunabhängig
pressure-induced transition druckinduzierte Umwandlung *f*
pressure jump Drucksprung *m*
pressure leaching *(Aufb)* Drucklaugung *f*
pressure nutsche Druck(filter)nutsche *f*
pressure relief valve Überdruckventil *n*, Druckbegrenzungsventil *n*, Sicherheitsventil *n*
pressure sensor Drucksensor *m*
pressure tubing Druckschlauchmaterial *n*, Druckschlauch *m*
pressure vessel Druckgefäß *n*, Druckbehälter *m*, Autoklav *m*
pressurize *v* unter Druck setzen
pressurized-air floatation *(Hyd)* Druckflotation *f*
pre-steady-state kinetics Prästationärenzymkinetik *f*
prestressed (safety) glass vorgespanntes Glas *n*
pretan *v* angerben, vorgerben
pretannage Vorgerbung *f*
pretreat *v* vorbehandeln
pretreatment 1. Vorbehandlung *f*; 2. *(Hyd)* Voraufbereitung *f (von Trinkwasser)*; Vorreinigung *f (von Abwasser)*
previtamin Provitamin *n*
prewashing Vorwäsche *f*
preweigh *v* vorher wägen
PRH *s.* prolactin-releasing hormone
prill *v* 1. verperlen, sprühkristallisieren, prillen *(eine Schmelze oder konzentrierte Lösung in Granulat verwandeln)*; 2. *s.* microencapsulate
prill 1. Metallkönig *m*, Regulus *m*; 2. *durch Sprühkristallisation erzeugte Granalie*
prilling 1. Verperlen *n*, Sprühkristallisieren *n*, Prillen *n*; 2. *s.* microencapsulation
primaquine Primaquin *n*, 8-(4′-Amino-1′-methyl-butylamino-)6-methoxychinolin *n*
primary alcohol primärer Alkohol *m*
primary amine primäres Amin *n*
primary battery *(Ech)* Primärbatterie *f*
primary beam Primärstrahl *m*
primary carbon atom primäres Kohlenstoffatom *n*
primary cell *(Ech)* Primärelement *n*, Primärzelle *f (nicht regenerierbar)*
primary clarifier *(Hyd)* Vorklärbecken *n*
primary colour Grundfarbe *f (Farbentheorie)*
primary column Vordestillationskolonne *f*
primary component *(Farb)* Erstkomponente *f*, aktive Komponente *f*, Diazo(tierungs)komponente *f*
primary crystallization primäre Kristallisation *f*, Hauptkristallisation *f*
primary distillation Primärdestillation *f*, Erstdestillation *f*, primäre [erste] Destillation *f*
primary electron primäres Elektron *n*, Primärelektron *n*
primary element *s.* primary nutrient
primary energy Primärenergie *f*
primary explosive Initialsprengstoff *m*, Zünd(spreng)stoff *m*, Initialexplosivstoff *m*
primary fermentation Hauptgärung *f*
primary fraction Grundfraktion *f*
primary metabolism *(Bioch)* Primärmetabolismus *m*, Primärstoffwechsel *m*, Grundstoffwechsel *m*, Hauptstoffwechsel *m*
primary metabolite *(Bioch)* primärer Metabolit *m*, Primärmetabolit *m*, primäres Stoffwechselprodukt *n*
primary mineral primäres Mineral *n*
primary nucleation *(Krist)* primäre Keimbildung *f*, Primärkeimbildung *f*
primary nutrient Kernnährstoff *m*
primary photochemical reaction pho-

tochemische Primärreaktion *f*, primäre photochemische Reaktion *f*
primary photoreaction photochemische Primärreaktion *f*, primäre photochemische Reaktion *f*
primary plasticizer primärer [lösender] Weichmacher *m*, Primärweichmacher *m*
primary pollutant Primärschadstoff *m*
primary ray Primärstrahl *m*
primary reaction Primärreaktion *f*
primary salt primäres Salz *n (einer mehrbasigen Säure)*
primary standard 1. *(Anal)* Urtitersubstanz *f*; 2. Primärstandard *m (pH-Messung)*
primary structure *(Bioch)* Primärstruktur *f*
primary tar Primärteer *m*, Urteer *m*
primary valence (bond) Hauptvalenz(-bindung) *f*
primary waste treatment (stage) *(Hyd)* erste Reinigungsstufe *f*, mechanische Reinigung *f*, Vorreinigung *f*
prime *v* 1. füllen; 2. zünden, als Zünder dienen *(Initialsprengstoffe)*; 3. *(Anstr)* grundieren; 4. spucken *(Kessel, Destillationskolonnen)*
prime steam Frischdampf *m*, Direktdampf *m*, direkter [gespannter] Dampf *m*
primer 1. Sprengkapsel *f*, Zündhütchen *n*, Zünder *m*; 2. Grundiermittel *n*, Grund(ier)anstrichmittel *n*; 3. *(Bioch)* Starter *m*, Startermolekül *n*, Primer *m*; 4. *s.* primary explosive
priming 1. Füllen *n*; 2. Stoßzündung *f*; 3. *s.* primary explosive; 4. *s.* primer 1.; 5. *(Anstr)* Grundieren *n*, Grundierung *f*; 6. *s.* priming coat; 7. Spucken *n (von Kesseln und Destillationskolonnen)*
priming coat Grund(ier)anstrich *m*, Untergrundanstrich *m*, Grundierung *f*
principal alkaloid Hauptalkaloid *n*
principal colouring material Hauptfarbstoff *m*
principal metal Grundmetall *n*
principal quantum number Hauptquantenzahl *f*
principal reaction Hauptreaktion *f*
principal series *(Spektr)* p-Serie *f*
principle 1. Prinzip *n*, Grundsatz *m*, Lehrsatz *m*, Satz *m*, Gesetz *n*, Regel *f*; 2. Grundbestandteil *m*, Prinzip *n*
principle of corresponding states *(phys Ch)* Prinzip *n* der korrespondierenden Zustände
principle of detailed balancing *(org Ch)* Prinzip *n* des detaillierten Gleichgewichts
principle of equipartition (of energy) Äquipartitionsprinzip *n* (der Energie), Äquipartitionstheorem *n*, Gleichverteilungssatz *m*
principle of mass action Massenwirkungsgesetz *n*, MWG *n*
principle of maximum multiplicity (hundsches) Prinzip *n* der größten Multiplizität, (hundsche) Regel *f* der größten Multiplizität, hundsches Multiplizitätsprinzip *n* [Prinzip *n*], (erste) hundsche Regel *f*
principle of microscopic reversibility *(phys Ch)* Prinzip *n* der mikroskopischen Reversibilität
print *v* 1. drucken; bedrucken; 2. *(Foto)* kopieren, abziehen
print 1. Druck *m*; Abdruck *m*; 2. *(Foto)* Kopie *f*, Abzug *m*
printing ink Druckfarbe *f*, grafische Farbe *f*, *(i. e. S.)* Druckerschwärze *f*
prior equilibrium vorgelagertes Gleichgewicht *n*
priority *(Nom)* Priorität *f*, Vorrang *m*
prism *(Krist)* Prisma *n*, Säule *f*
prism spectrograph Prismenspektrograph *m*
prismatic *(Krist)* prismatisch
probability density Aufenthaltswahrscheinlichkeitsdichte *f*, Wahrscheinlichkeitsdichte *f*
probability distribution Wahrscheinlichkeitsverteilung *f*
probability of finding a particle Aufenthaltswahrscheinlichkeit *f* eines Teilchens
probe 1. *(Anal)* Sonde *f*; 2. *(Chromat)* Probe *f (mit definierten Eigenschaften)*
procaryote *s.* prokaryote
procedure Arbeitsweise *f*, Methode *f*, Verfahren *n*, Technik *f*
proceed *v* 1. fortschreiten; vonstatten gehen, verlaufen *(Reaktionen)*; 2. verfahren, vorgehen

proceed *v* **poorly [sluggishly]** träge [schleppend] verlaufen *(Reaktion)*
proceed *v* **smoothly** glatt verlaufen *(Reaktion)*
proceed *v* **vigorously** stürmisch verlaufen *(Reaktion)*
process *v* behandeln; bearbeiten; verarbeiten
process *v* **wastes** *(Hyd)* Abwässer behandeln
process *v* **water** *(Hyd)* Wasser aufbereiten
process 1. Verfahren *n*, Arbeitsmethode *f*, Prozess *m*; 2. Vorgang *m*, Ablauf *m*, Verlauf *m*, Prozess *m*; 3. Steuerstrecke *f*, Regelstrecke *f*, Strecke *f*
process annealing *(Met)* Zwischenglühen *n*, Zwischenglühung *f*
process chemical Prozesschemikalie *f*
process contamination *(Hyd)* produktionsbedingte Belastung *f*, Verschmutzung *f* durch Produktaustritt
process control Prozesssteuerung *f*, Verfahrenssteuerung *f*
process design Prozessgestaltung *f*
process development Verfahrensentwicklung *f*
process engineer Verfahrenstechniker *m*, Verfahrensingenieur *m*
process engineering Verfahrenstechnik *f*
process gas Prozessgas *n*
process goal Verfahrensziel *n*, Prozessziel *n*
process heat Prozesswärme *f*
process industry Verfahrensindustrie *f*
process quantity Prozessgröße *f*
process stage Verfahrensstufe *f*, Prozessstufe *f*
process steam Prozessdampf *m*
process step Verfahrensschritt *m*, Prozessschritt *m*
process variable Prozessvariable *f*
process waste (water) Prozessabwasser *n*, Produktionsabwasser *n*, Fabrikationsabwasser *n*
process water Prozesswasser *n*, Produktionswasser *n*, Fabrikationswasser *n*, Wasser *n* im Fabrikationsbetrieb
process yield Produkt(ions)ausbeute *f*
process yield coefficient *(Biot)* Ausbeutekoeffizient *m*, Ertragskoeffizient *m*, ökonomischer Koeffizient *m*
processability Verarbeitbarkeit *f*
processable verarbeitbar
processing 1. Behandlung *f*; Bearbeitung *f*; Verarbeitung *f*; 2. Arbeitsweise *f*, Betrieb *m*, Fahrweise *f*, Betriebsmethode *f*; 3. *(Bioch)* Reifung *f*, Processing *n (neugebildeter Ribonucleinsäure)*
processing agent [aid] Verarbeitungshilfsmittel *n*
processing characteristics verarbeitungstechnische Kennwerte *mpl*
processing chemical 1. Behandlungschemikalie *f*; 2. *(Hyd)* Aufbereitungschemikalie *f*
processing industry (weiter)verarbeitende Industrie *f*, Verarbeitungsindustrie *f*
processing properties Verarbeitungseigenschaften *fpl*
processing specification Verarbeitungsvorschrift *f*, Verarbeitungsrichtlinie *f*
processing technique Bearbeitungstechnik *f*
processing technology Verarbeitungstechnik *f*; Verarbeitungstechnologie *f*
prochiral prochiral *(Stereochemie)*
prochirality Prochiralität *f (Stereochemie)*
producer 1. Gasgenerator *m*, Generator *m*; 2. *(Biot)* Produktbildner *m*, Produzent *m*, Metabolitbildner *m*, Stoffproduzent *m*; 3. Hersteller *m*, Produzent *m*
producer gas Generatorgas *n*
producer-gas equilibrium Generatorgasgleichgewicht *n*, Boudouard-Gleichgewicht *n*, boudouardsches Gleichgewicht *n*
product 1. Produkt *n*, Erzeugnis *n*; Gut *n*; 2. *(Biot)* (gebildetes) Produkt *n*, Bildungsprodukt *n*, Fermentationsprodukt *n*
product cooler *(Dest)* Produktkühler *m*
product formation 1. Produktbildung *f*; 2. *(Biot)* Stoff(wechselprodukt)bildung *f*, Metabolitproduktion *f*, Stoffproduktion *f*
product formation phase *(Biot)* Produktbildungsphase *f*, Idiophase *f*
product gas Produktgas *n*
product inhibition *(Bioch)* Produkt(en)-hemmung *f*

product of metabolism *(Bioch)* Stoffwechselprodukt *n*, Metabolit *m*
product of value Wertstoff *m*
product phase Produktphase *f*
product recovery *(Biot)* Produktaufarbeitung *f*
product take-off *(Dest)* Destillatabnahme *f*
product yield Produkt(ions)ausbeute *f*
production 1. Herstellung *f*, Erzeugung *f*, Produktion *f*; 2. Bildung *f*
production culture *(Biot)* Produktionskultur *f*, Hauptkultur *f*, Mastkultur *f*
production fermentation *(Biot)* Produktionsfermentation *f*, Hauptfermentation *f*
production fermenter *(Biot)* Produktionsfermenter *m*, Hauptfermenter *m*
production-line status Produktionsreife *f*
production metallurgy erzeugende Metallurgie *f*
production on a large scale großtechnische Herstellung *f*
production output Ausstoß *m*, Produktionsleistung *f*, Fertigungsmenge *f*, Produktionszahl *f*
production plant Produktionsanlage *f*, Betrieb *m*, großtechnische Anlage *f (im Vergleich zur Pilotanlage)*
production process Herstellungsverfahren *n*, Produktionsverfahren *n*
production strain *(Biot)* Produktionsstamm *m*, Arbeitsstamm *m*, Leistungsstamm *m (von Mikroorganismen)*
proenzyme Proenzym *n*, Enzymogen *n*, Zymogen *n*
proferment Proenzym *n*, Enzymogen *n*, Zymogen *n*
progenitor *s.* precursor
progestational hormone Gestagen *n*, Corpusluteum-Hormon *n*, Gelbkörperhormon *n (besonders Progesteron)*
progesterone Progesteron *n (Hormon)*
progestin Progesteron *n (Hormon)*
programmed cell death *(Bio)* programmierter Zelltod *m*, Apoptose *f*
programmed-temperature chromatography temperaturprogrammierte Chromatographie *f*
progress *v* fortschreiten *(Reaktionen)*
progress-of-reaction diagram Reaktionsablaufdiagramm *n*
projection formula Projektionsformel *f*
prokaryote *(Bioch)* Pro(to)karyont *m*, Protokaryot *m*
prokaryotic cell *(Bioch)* pro(to)karyotische Zelle *f*, Protokaryotenzelle *f*
prolactin *(Bioch)* *Prolactin *n*, PRL, lactogenes [luteotropes] Hormon *n*, LTH
prolactin release-inhibiting factor Prolactostatin *n*
prolactin-releasing hormone *Prolactoliberin *n*, Prolactin-Releasinghormon *n*, PRH
prolamin(e) Prolamin *n (Eiweißbestandteil des Getreidemehls)*
proline Prolin *n*, Pro, Pyrrolidin-2-carbonsäure *f*
promethium Pm Promethium *n*
promote *v* 1. fördern, beschleunigen *(z. B. eine Reaktion)*; 2. *(phys Ch)* heben *(ein Elektron von einem Orbital in ein anderes)*
promoter 1. Promotor *m*, Aktivator *m*, Beschleuniger *m*, synergetischer Verstärker *m (Katalyse)*; 2. *(Aufb)* Sammler *m*, Kollektor *m*
promotion 1. Förderung *f*, Beschleunigung *f (z. B. einer Reaktion)*; 2. *(phys Ch)* Promovierung *f*, Promotion *f*
promotion energy *(phys Ch)* Promotionsenergie *f*
prone to oxidation oxidationsanfällig, oxidationsfreudig
proneness to oxidation Oxidationsanfälligkeit *f*, Oxidationsfreudigkeit *f*
proof *v* 1. beweisen; 2. dicht [undurchlässig, undurchdringlich] machen; imprägnieren; gummieren
proof 1. dicht, undurchlässig, undurchdringlich; 2. *(Gär) 57,1 Vol.-% (U.K.) bzw. 50,0 Vol.-% (U.S.) Alkohol enthaltend*
proof 1. Beweis *m*; 2. *(Gär) Alkoholgehalt bezogen auf 57,1 Vol.-% (U.K.) bzw. 50,0 Vol.-% (U.S.)*; 3. Reagenzglas *n*
proof of structure Strukturbeweis *m*, Konstitutionsbeweis *m*
0.2 % proof stress 0,2-%-Dehngrenze *f*
proofing 1. Dichtmachen *n*, Undurchlässigmachen *n*, Undurchdringlichmachen *n*; Imprägnieren *n*, Imprägnierung *f*; Gummieren *n*, Gummierung *f*; 2. Imprägnierungsmittel *n*; Dichtungs-

mittel *n*; 3. gummierter Stoff *m*, gummiertes Gewebe *n*
proofness Dichtheit *f*, Dichtigkeit *f*, Undurchlässigkeit *f*, Undurchdringlichkeit *f*
prooxidant (agent) Prooxidans *n (Stoff, der schon in Spuren Oxidationsvorgänge fördert)*
propadiene *Propadien *n*, Allen *n*
propagate *v* 1. sich ausbreiten [fortpflanzen]; weitertragen; 2. *(Biot)* vermehren; sich vermehren *(Mikroorganismen)*
propagating flask *(Lab)* Kulturkolben *m*
propagation 1. Ausbreitung *f*, Fortpflanzen *n*, Fortpflanzung *f*; Weitertragen *n*; 2. Kettenwachstum *n*, Kettenfortpflanzung *f (z. B. bei der Polymerisation)*; 3. *(Biot)* Vermehrung *f*, Propagierung *f (von Mikroorganismen)*
propagation factor *(Bioch)* Elongationsfaktor *m*, Transferfaktor *m*
propagation reaction Wachstumsreaktion *f*, Fortpflanzungsreaktion *f (z. B. bei der Polymerisation)*
propanal *Propanal *n*, Propionaldehyd *m*
propane Propan *n*
propane dewaxing Propanentparaffinierung *f*, Entparaffinierung *f* [Entparaffinieren *n*] mit Propan
propanedioic acid *Propandisäure *f*, Malonsäure *f*
1,2,3-propanetriol *Propan-1,2,3-triol *n*, Glycerol *n*
2-propanone *Propan-2-on *n*, Dimethylketon *n*, Aceton *n*
propargyl alcohol Propargylalkohol *m*, *Prop-2-in-1-ol *n*
propellane *(org Ch)* Propellan *n*
propellant Treibstoff *m*, Treibmittel *n*, *(für Aerosole auch)* Treibgas *n*
propellant explosive Schießstoff *m*, Schießmittel *n*, Treibstoff *m*, Treibmittel *n*
propellant gas Treibgas *n*
propeller agitator [mixer] Propellerrührer *m*, Propellerrührwerk *n*, Schraubenrührer *m*, Propellermischer *m*, Propellermischgerät *n*
propenal *Propenal *n*, Acrylaldehyd *m*, Acrolein *n*
propene *(org Ch)* Propen *n*
1,2,3-propenetricarboxylic acid *Propen-1,2,3-tricarbonsäure *f*, Aconitsäure *f*
propenoic acid *Propensäure *f*, Acrylsäure *f*
2-propen-1-ol *Prop-2-en-1-ol *n*, Allylalkohol *m*
propensity Neigung *f*
property of state Zustandsgröße *f*
β-propiolactone β-Propiolacton *n*, Propan-3-olid *n*
propiolic acid Propiolsäure *f*, *Propinsäure *f*
propionaldehyde *Propanal *n*, Propionaldehyd *m*
propionate Propionat *n (Salz oder Ester der Propionsäure)*
propionic acid Propionsäure *f*, *Propansäure *f*
propionitrile Propionitril *n*, *Propannitril *n*, Ethylcyanid *n*
propionyl group $C_2H_5CO–$ Propionylgruppe *f*, Propionylrest *m*
propiophenone Propiophenon *n*, *Ethylphenylketon *n*, Propionylbenzen *n*
proportion *v* dosieren, zuteilen, zumessen
proportion Proportion *f*, Verhältnis *n*; Anteil *m* • **in molecular proportions** in molekularem Verhältnis • **miscible in all proportions** in jedem Verhältnis mischbar
proportion by volume Volumenverhältnis *n*, Raumverhältnis *n*
proportionality constant [factor] Proportionalitätsfaktor *m*
proportioning Dosierung *f*, Zuteilung *f*, Zumessung *f*
proportioning pump Dosier(ungs)pumpe *f*, Zumesspumpe *f*
proprietary name geschützter Name *m*
propyl acetate Essigsäurepropylester *m*, Propylacetat *n*
propyl aldehyde *Propanal *n*, Propionaldehyd *m*
***n*-propyl alcohol** Propylalkohol *m*, *Propan-1-ol *n*
propylamine Propylamin *n*
propylation Propylierung *f*
propylene Propen *n*
propylene oxide Propylenoxid *n*, 1,2-Epoxypropan *n*, Methyloxiran *n*

1,2-propylene glycol *Propan-1,2-diol, Propylenglykol *n*
propyne *(org Ch)* Propin *n*
2-propyn-1-ol *Prop-2-in-1-ol *n*, Propargylalkohol *m*
prospect *v* prospektieren, schürfen, aufspüren, suchen, *(nach Erz)* graben, *(nach Öl)* bohren
prospecting Prospektierung *f*, Prospektion *f*, Schürfung *f*, Aufspürung *f*, Suche *f*
prostaglandin Prostaglandin *n (Vertreter einer Gruppe tierischer Hormone)*
prostanoic acid *(Bioch)* Prostansäure *f*
prosthetic group prosthetische Gruppe *f (bei Proteinen)*; *(bei Enzymen auch)* aktive Gruppe *f*, Wirk(ungs)gruppe *f*, Coenzym *n*
protactinium Pa Protactinium *n*
protamine Protamin *n (eine Eiweißart)*
protease Protease *f*, proteolytisches [proteinspaltendes, eiweißspaltendes] Enzym *n*
protectant *s.* 1. protective agent; 2. protective toxicant
protected from light lichtgeschützt, vor Licht geschützt
protecting group Schutzgruppe *f*
protection from light Lichtschutz *m*
protective schützend, Schutz...
protective Schutzmittel *n*, Schutzstoff *m*; protektives [protektiv wirkendes, vorbeugend wirkendes] Mittel *n (zum Pflanzenschutz)*
protective agent 1. Schutzmittel *n*, Schutzstoff *m*; 2. *(Gum)* Alterungsschutzmittel *n*
protective atmosphere Schutz(gas)atmosphäre *f*
protective clothing Schutz(be)kleidung *f*
protective coating 1. Schutzschichtstoff *m*, Beschichtungsstoff *m*, Beschichtungsmaterial *n*; Anstrichstoff *m* für Schutzanstriche; 2. Schutzschicht *f*; Schutzanstrich *m*
protective colloid Schutzkolloid *n*
protective film Schutzfilm *m*, Schutzhaut *f*, dünne Schutzschicht *f*
protective fungicide Residualfungizid *n*, protektives Fungizid *n*
protective gloves Schutzhandschuhe *mpl*
protective goggles Schutzbrille *f*
protective insecticide protektives [vorbeugend wirkendes] Insektizid *n*
protective measure Schutzmaßnahme *f*
protective shell Schutzhülle *f*
protective toxicant protektives [vorbeugend wirkendes] Gift *n* [Pflanzenschutzmittel *n*]
proteic *s.* proteinaceous
proteid(e) 1. Proteid *n*, zusammengesetzter [konjugierter] Eiweißstoff *m* [Eiweißkörper *m*]; 2. *s.* protein
protein Protein *n*, Eiweiß *n*, Eiweißstoff *m*, Eiweißkörper *m*
protein biosynthesis *(Bio)* Proteinbiosynthese *f*, Eiweißsynthese *f*
protein-bound proteingebunden
protein breakdown [catabolism] Proteinabbau *m*, Eiweißabbau *m*, Proteinkatabolismus *m*
protein chemist Proteinchemiker *m*, Eiweißchemiker *m*
protein chemistry Proteinchemie *f*, Eiweißchemie *f*
protein cleavage *s.* proteolysis
protein component *s.* protein moiety
protein-containing fiber proteinhaltige Faser *f*
protein content Eiweißgehalt *m*
protein deficiency Eiweißmangel *m*
protein degradation *s.* protein breakdown
protein denaturation Proteindenaturierung *f*, Proteindenaturation *f*
protein determination Eiweißbestimmung *f*
protein fibre Eiweißfaser *f*, Proteinfaser *f*; Eiweiß(chemie)faserstoff *m*, CE, Proteinfaserstoff *m*
protein folding Proteinfaltung *f*
protein food proteinreiche [eiweißreiche] Lebensmittel *npl*
protein hydrolyzate Proteinhydrolysat *n*
protein isolate Proteinisolat *n*
protein metabolism Proteinstoffwechsel *m*, Eiweißstoffwechsel *m*
protein moiety Proteinanteil *m*, Proteinkomponente *f*
protein nitrogen Eiweißstickstoff *m*
protein plastic Proteinkunststoff *m*, Kunsthorn *n*
protein synthesis Proteinsynthese *f*, Eiweißsynthese *f*

proteinaceous proteinartig, eiweißartig; proteinisch, Protein…, Eiweiß…
proteinase inhibitor Proteinaseinhibitor *m*
proteohormone Proteohormon *n*
proteolipid *(Bioch)* Proteolipid *n*
proteolysis Proteolyse *f*, Eiweißspaltung *f*
proteolytic enzyme proteolytisches [proteinspaltendes, eiweißspaltendes] Enzym *n*, Protease *f*
proteose Proteose *f (Spaltprodukt des Proteins)*
protic prot(on)isch
protide $^1H^-$ Protid *n*
protium ^1H Protium *n*, leichter Wasserstoff *m*
protocatechualdehyde Protocatechualdehyd *m*, 3,4-Dihydroxybenzaldehyd *m*
protocatechuic acid Protocatechusäure *f*, 3,4-Di-hydroxybenzoesäure *f*
protogenic protogen, protonenliefernd, protonenabgebend, H^+-abgebend, protonenabspaltend
protohaem Häm *n*, Eisenporphyrin *n* IX, Ferroporphyrin *n* IX, Eisen(II)-protoporphyrin *n*
protolysis Protolyse *f*, protolytische Reaktion *f (Brönsteds Säure-Base-Theorie)*
protolyte Protolyt *m (in Brönsteds Säure-Base-Theorie)*
protolytic protolytisch *(Brönsteds Säure-Base-Theorie)*
protomer(ic unit) Protomer(e) *n (kleinste identische Untereinheit der Eiweiße)*
proton Proton *n*, p
proton acceptor Protonenakzeptor *m*, Protonennehmer *m*, Emprotid *m*
proton acid Proton(en)säure *f*, Brönsted-Säure *f*
proton affinity Protonenaffinität *f*
proton donor Protonengeber *m*, Protonendonator *m*, Dysprotid *n*
proton exchange Protonenaustausch *m*
proton exchange membrane Protonenaustauschmembran *f*
proton exchange membrane fuel cell Niedertemperatur-Protonenaustauschmembran-Brennstoffzelle *f*
proton exchange reaction Protonenaustauschreaktion *f (Säure-Base-Reaktion)*
proton magnetic resonance Protonen-Kernresonanz *f*, Protonenresonanz *f*
proton mass Protonen(ruhe)masse *f*, Ruhemasse *f* des Protons
proton pump *(Bioch)* Protonenpumpe *f*
proton shift Protonenverschiebung *f*
proton transfer reaction Protonenübertragungsreaktion *f*
protonate *v* protonieren
protonation Protonierung *f*
protonic acid *s.* proton acid
protophilic protonophil, Protonen aufnehmend [anlagernd], H^+-aufnehmend
prototropic prototrop
prototropic rearrangement *s.* prototropy
prototropy *(org Ch)* Protomerie *f*, Prototropie *f (Positionsänderung eines Protons)*
provitamin Provitamin *n*
proximate analysis 1. Pauschalanalyse *f (zur Bestimmung der Hauptbestandteile von Lebensmitteln)*; 2. Immediatanalyse *f*, Kurzanalyse *f (zur Bestimmung des Gehalts an Wasser, Asche und flüchtigen Bestandteilen in Kohle)*
PRPP = 5-phosphoribosyl-1-pyrophosphate
Prussian blue Berliner Blau *n*, Preußischblau *n*
prussiate 1. M^ICN Cyanid *n*; 2. $M^I_4[Fe(CN_6)]$ Hexacyanoferrat(III) *n*, Cyanoferrat(III) *n*; 3. Prussid *n*, Prussiat *n*, Pentacyanoferrat *n*
PSD *s.* pore-size distribution
pseudo acid Pseudosäure *f*
pseudo base Pseudobase *f*
pseudo-first-order reaction Reaktion *f* pseudo-erster Ordnung
pseudo-zero-order reaction Reaktion *f* pseudo-nullter Ordnung
pseudoaromatics *(org Ch)* Pseudoaromaten *pl*
pseudoasymmetric(al) pseudoasymmetrisch
pseudoasymmetry Pseudoasymmetrie *f*
pseudocatalysis Pseudokatalyse *f*
pseudocrystalline pseudokristallin
pseudocumene Pseudocumen *n*, *1,2,4-Trimethylbenzen *n*

pseudohalide Pseudohalogenid *n*
pseudohalogen Pseudohalogen *n*
pseudoideal solution pseudoideale Lösung *f*, Theta-Lösung *f*
pseudomorph pseudomorpher Kristall *m*, Afterkristall *m*
pseudomorphic *(Krist)* pseudomorph
pseudomorphism *(Krist)* Pseudomorphie *f*
pseudopeptide Pseudopeptid *n*
pseudoplastic (fluid) strukturviskose [pseudoplastische] Flüssigkeit *f*
pseudoplasticity Strukturviskosität *f*, strukturviskoses Verhalten *n*, Pseudoplastizität *f*
pseudoracemic pseudoracemisch
pseudorotation Pseudorotation *f*
pseudosalt Pseudosalz *n*
pseudosolution Pseudolösung *f (Sol oder Suspension)*
pseudosymmetric(al) pseudosymmetrisch
pseudosymmetry Pseudosymmetrie *f*
pseudounimolecular reaction Reaktion *f* pseudo-erster Ordnung, pseudomonomolekulare Reaktion *f*
pseudouridylic acid *(Bioch)* Pseudouridylsäure *f*
PSS *s.* poly(styrenesulfonate)
psychedelic drug *s.* psychotomimetic (drug)
psychochemical Psychochemikalie *f*; psychotoxischer Kampfstoff *m*, Psychogift *n*
psychodysleptic *s.* psychotomimetic (drug)
psychomimetic *s.* psychotomimetic
psychopharmacologic agent [drug] Psychopharmakon *n*, Psychotropikon *n*, psychotropes Pharmakon *n*
psychostimulant *(Pharm)* Psychostimulans *n*, Psychotonikum *n*
psychotomimetic *(Pharm)* psychotomimetisch, halluzinogen
psychotomimetic (drug) *(Pharm)* Psychodysleptikum *n*, Halluzinogen *n*, Phantastikum *n*, Psychotikum *n*, Psychotomimetikum *n*, Psychosomimetikum *n*
psychotropic agent *s.* psychopharmacologic agent
psychrometer Psychrometer *n (ein Luftfeuchtemesser)*
psychrometry Luftfeuchtemessung *f*
psychrophiles *(Biot)* psychrophile Mikroorganismen *mpl (Wachstumsoptimum 5 bis 20 °C)*
psychrophilic microorganisms *(Biot)* psychrophile Mikroorganismen *mpl (Wachstumsoptimum 5 bis 20 °C)*
PTC *s.* phase-transfer catalysis
pteridine Pteridin *n (unsubstituierter Grundkörper der Pterine)*
pterin *(org Ch)* Pterin *n*
pteroic acid *(Bioch)* Pteroinsäure *f*
pteroylglutamate Pteroylglutamat *n*, Folat *n (Salz oder Ester der Pteroylglutaminsäure)*
pteroylglutamic acid Pteroylglutaminsäure *f*, Folsäure *f*
PTFCE *(polytrifluorochloroethylene) s.* polychlorotrifluoroethylene
PTFE *s.* polytetrafluoroethylene
puckered conformation gewinkelte Konformation *f (von Kohlenstoffringen)*
puddle *v* 1. umrühren; 2. *(Met)* puddeln
puddling process Puddelverfahren *n*, Flammofenfrischverfahren *n (zur Herstellung von Schweißstahl)*
pug mill Knetwerk *n*, Knetmaschine *f*, Kneter *m*
puke *v (Dest)* spucken
pulegone Pulegon *n (ein Terpenketon)*
Pulfrich refractometer Pulfrich-Refraktometer *n*
pull *v* 1. ziehen *(z. B. eine Reaktion)*; 2. *(Glas)* ziehen
pull *v* **out** ausziehen *(Glas zu einer Kapillare)*
pull 1. Anziehung *f (durch elektrostatische Kräfte)*; 2. Zug *m (von Feuerstätten)*
pulp *v* 1. *(Aufb)* aufschwemmen; 2. *(Pap)* zu Halbstoff aufschließen, *(i. e. S.)* zu Holzschliff verschleifen, chemisch aufschließen
pulp 1. Brei *m*, breiige Masse *f*, Pulpe *f*, Pülpe *f*, Pulp *m*; 2. *(Lebm)* Pülpe *f*, *(i. e. S.)* Fruchtfleisch *n*, Fruchtmark *n*, Obstpülpe *f*; 3. *(Aufb)* Trübe *f*, Suspension *f*; 4. *(Pap)* Faser(stoff)brei *m*, Stoffbrei *m*, Fasermasse *f*, Stoffmasse *f*, Stoff *m*; Faserhalbstoff *m*, Halbstoff *m*, Stoff *m*, Halbzeug *n*; 5. *s.* mechanical wood pulp • **make into (a) pulp** *s.* pulp 2.

pulp bleaching Zellstoffbleiche *f*
pulp board *(Pap)* Halbstoff *m* in Bogenform [trockenen Bogen], *(i. e. S.)* Holzschliff *m* in Bogenform [Pappenform], Holzschliffpappe *f*, Holzschliffblätter *npl* [Zellstoff *m*] in Bogenform [Pappenform], Zellstoffpappe *f*, Zellstoffblätter *npl*
pulp brightness Stoffweiße *f*, Weißgehalt *m* [Weißgrad *m*] des Zellstoffs
pulp industry Zellstoffindustrie *f*
pulp of ore Erztrübe *f*, Stofftrübe *f*, Trübe *f*
pulp wood *(Pap)* Papierholz *n*, Faserholz *n*, *(i. e. S.)* Schleifholz *n*, Zellstoffholz *n*, Chemieholz *n*
pulper *(Pap)* Pulper *m*, Stoff(auf)löser *m*
pulping *(Pap)* Aufschließen *n*, Aufschließung *f*, Aufschluss *m*
pulping agent *(Pap)* Aufschlussmittel *n*, Aufschlusschemikalie *f*, Aufschließungsreagens *n*
pulping liquor *(Pap)* Aufschlusslösung *f*, Kochflüssigkeit *f*, Kochlösung *f*
pulping of wood Holzaufschluss *m*
pulpy breiig, breiartig; fleischig
pulsate *v* pulsieren
pulsation Pulsation *f*, pulsierende Bewegung *f*, Pulsieren *n*
pulse Fourier transform Puls-Fourier-Transformation *f*, PFT
pulse polarography *(Anal)* Pulspolarographie *f*
pulse radiolysis Pulsradiolyse *f (zur Untersuchung schneller Reaktionen)*
pulsed column Pulsationskolonne *f*, pulsierte Kolonne *f*
pulsed laser *(Spektr)* gepulster Laser *m*
pulsed laser deposition Laserstrahlverdampfen *n (PVD-Verfahren)*
pultrusion Pultrusion *f (Strangziehen bes. von glasfaserverstärkten Harzen)*
pulverization Pulverisieren *n*, Pulvern *n*, Zerpulvern *n*; Mahlen *n* auf Staubfeinheit; Mahlen *n*, Vermahlen *n (harter bis mittelharter Stoffe)*
pulverize *v* pulverisieren, (zer)pulvern; auf Staubfeinheit mahlen; (ver)mahlen *(harte bis mittelharte Stoffe)*; zu Pulver zerfallen
pulverized-coal burner Kohlenstaubbrenner *m*, Staubbrenner *m*
pulverized-coal gasification *(Carb)* Staubvergasung *f*, Schwebevergasung *f*, Flugstromvergasung *f*, Vergasung *f* in der Flugstaubwolke [Staubwolke]
pulverized fuel Brenn(stoff)staub *m*
pulverizer *s.* pulverizing mill
pulverizing mill Mühle *f (für harte bis mittelharte Stoffe)*; Staubmühle *f*, Pulverisiermühle *f*, Fein(st)mühle *f*
pulverulent 1. pulv(e)rig, pulverförmig; 2. brüchig, (leicht) zerbröckelnd, zerkrümelnd; 3. staubig, staubbedeckt
pumice Bimsstein *m*
pumiceous bimssteinartig; Bimsstein...; bimssteinähnlich
Pummerer rearrangement Pummerer-Umlagerung *f (von Sulfoxiden zu α-Acyloxysulfiden)*
pump *v* pumpen
pump *v* **off** abpumpen
pump Pumpe *f*
pump laser Pumplaser *m*
pumpable pumpfähig
pumping fluid Treibmittel *n (Strahlpumpe)*
punch *v* lochen; stanzen; prägen
punch Stempel *m*, Prägestempel *m*, Pfaff *m*
punched-plate screen Lochblechsieb *n*
punching machine *(Met)* Stanzmaschine *f*
puncture-proof durchschlagfest
puncture strength Durchschlagfestigkeit *f*, dielektrische [elektrische] Festigkeit *f*
pungent stechend *(Geruch)*; brennend, scharf *(Geschmack)*
pungent smelling stechend riechend
pure rein
pure chemical Reinchemikalie *f*
pure chemistry reine Chemie *f*
pure culture Reinkultur *f*, Reinzucht *f*
pure metal reines Metall *n*, Reinmetall *n*
pure product Reinprodukt *n*, Reinerzeugnis *n*
pure research Grundlagenforschung *f*
pure substance reine Substanz *f*, Reinsubstanz *f*, reiner Stoff *m*, Reinstoff *m*
purgative Purgans *n*, Purgativum *n*, Abführmittel *n (mäßig stark)*
purge *v* reinigen, säubern; entlüften; spülen *(mit Inertgas)*
purge gas Spülgas *n (Gaschromatographie)*

purgeable organic halogens *(Hyd)* austreibbarer [flüchtiger] Anteil *m* an organischen Halogenverbindungen, flüchtige halogenorganische Verbindungen *fpl*
purification 1. Reinigung *f*, *(besonders von Ölen und Metallen auch)* Raffination *f*; 2. Reindarstellung *f*
purifier 1. Reiniger *m*, Reinigungsapparat *m*; 2. Reinigungsmittel *n*
purify *v* 1. reinigen, *(besonders Öle, Metalle auch)* raffinieren; 2. reindarstellen
purine alkaloid Purinalkaloid *n*
purine base Purinbase *f*
purine metabolism Purinstoffwechsel *m*
purity Reinheit *f*
purity level Reinheitsgrad *m*
purity specification Reinheitsanforderung *f*, Anforderung *f* an den Reinheitsgrad
purpose-built Spezial...
purpuric acid Purpursäure *f*
purpurin Purpurin *n*, 1,2,4-Trihydroxy-9,10-anthra-chinon *n*
put *v* **into solution** in Lösung bringen, (auf)lösen
put *v* **through** durchsetzen *(eine Stoffmenge)*
put *v* **up** 1. zusammenstellen; 2. errichten; 3. einbringen *(in ein Behältnis)*
putrefacient *s.* putrefactive
putrefaction Faulen *n*, Fäule *f*, Fäulnis *f* *(Zersetzung durch Mikroorganismen bei gehemmtem Sauerstoffzutritt)*
putrefactive fäulniserregend, Fäulnis...
putrefactive bacteria Fäulnisbakterien *npl*, Putride *npl*
putrefactive fermentation Fäulnisgärung *f*
putrefiable *s.* putrescible
putrefy *v* (ver)faulen
putrescent (ver)faulend; Fäulnis...
putrescibility Faulfähigkeit *f*, Fäulnisfähigkeit *f*
putrescible faul(ungs)fähig, fäulnisfähig, ausfaulbar
putrid faul(ig), angefault
putrid slime Faulschlamm *m*
putridity Fäule *f*, Fäulnis *f*
putty *v* (ver)kitten
putty Kitt *m*, Klebkitt *m*, *(i. e. S.)* Glaserkitt *m*, Fensterkitt *m*
PV *(permanganate value)* *s.* permanganate number
p,V-work Volumen(änderungs)arbeit *f*
PVA *s.* polyvinyl acetate
PVC *s.* 1. polyvinyl chloride; 2. pigment volume concentration
PVD *s.* physical vapour deposition
PVF *s.* polyvinyl fluoride
PVP = polyvinylpyrrolidone
pycnometer Pyknometer *n*, Wägefläschchen *n* (für Dichtemessungen)
pycnometric pyknometrisch, Pyknometer...
pyran *(org Ch)* Pyran *n*
pyranose Pyranose *f* *(ringförmiges Monosaccharid)*
pyranoside *(Bioch)* Pyranosid *n*
pyrazine Pyrazin *n*, 1,4-Diazin *n*, Piazin *n*
pyrazole Pyrazol *n*, 1,2-Diazol *n*
pyrazolone *(org Ch)* Pyrazolon *n*
pyrazolone dye Pyrazolonfarbstoff *m*
pyrene *(org Ch)* Pyren *n*
pyrethrine Pyrethrin *n* *(Insektizid)*
pyrethroid Pyrethroid *n* *(Insektizid mit pyrethrinartiger Wirkung)*
pyridazine Pyridazin *n*, 1,2-Diazin *n*, Oiazin *n*
pyridine *(org Ch)* Pyridin *n*
pyridine alkaloid Pyridinalkaloid *n*
pyridine base Pyridinbase *f*
3-pyridine carboxylic acid Pyridin-3-carbonsäure *f*, Nicotinsäure *f*, Niacin *n*
pyridoxal phosphate Pyridoxalphosphat *n* *(Coenzym in der Decarboxylase)*
pyrimidine Pyrimidin *n*, 1,3 Diazin *n*, Miazin *n*
pyrimidine base *(Bioch)* Pyrimidinbase *f*
pyrite *(Min)* Pyrit *m*, Eisenkies *m*, Schwefelkies *m* *(Eisendisulfid)*
pyrites 1. Kies *m* *(sulfidisches Erz)*; 2. *s.* pyrite
pyroacid Pyrosäure *f*
pyroborate Tetraborat *n*, Heptoxotetraborat *n*
pyrocatechol Brenzcatechin *n*, *Benzen-1,2-diol *n*, 1,2-Dihydroxybenzen *n*
pyrocellulose Kollodiumwolle *f*, Kolloxylin *n* *(niedrignitrierte Cellulose)*
pyrogallol Pyrogallol *n*, 1,2,3-Trihydroxybenzen *n*
pyrolysis Pyrolyse *f*
pyrolysis field ionization mass spec-

trometry Pyrolyse-Feldionisations-Massenspektrometrie *f*, PFIMS

pyrolysis gas chromatography Pyrolyse-Gaschromatographie *f*

pyrolysis reaction Pyrolysereaktion *f*

pyrolytic pyrolytisch

pyrolyzate Pyrolyseprodukt *n*

pyrolyze *v* pyrolysieren

pyromellitic acid Pyromellit(h)säure *f*, Benzen-1,2,4,5-tetracarbonsäure *f*

pyromellitic dianhydride Pyromellit(h)-säuredianhydrid *n*

pyrometallurgical pyrometallurgisch

pyrometallurgy Pyrometallurgie *f*

pyrometer Pyrometer *n*, Strahlungstemperaturmesser *m*, Strahlungsthermometer *n*

pyrometric pyrometrisch

pyrometric cone *(Ker)* Schmelzkegel *m*, Schmelzkörper *m*, Brennkegel *m*, *(meist)* Seger-Kegel *m*

pyrometry Pyrometrie *f*, Messung *f* hoher Temperaturen

pyromucic acid Brenzschleimsäure *f*, Furan-2-carbonsäure *f*

pyrone Pyron *n (allg. für eine Gruppe ketonartiger heterocyclischer Verbindungen)*

pyrophoric pyrophor, selbstentzündlich

pyrophoric iron pyrophores [reduziertes] Eisen *n (durch Reduktion mit Wasserstoff hergestelltes Eisenpulver)*

pyrophosphate Diphosphat *n*, Pyrophosphat *n*

pyrophosphoric acid Diphosphorsäure *f*, Pyrophosphorsäure *f*

pyrophosphorylation *(Bioch)* Pyrophosphorylierung *f*

pyrosulphate $M^{I}_{2}S_{2}O_{7}$ Disulfat *n*

pyrosulphite $M^{I}_{4}S_{2}O_{5}$ Disulfit *n*

pyrosulphuric acid Dischwefelsäure *f*

pyrosulphurous acid dischweflige Säure *f*

pyrotartaric acid Brenzweinsäure *f*, Methylbernsteinsäure *f*, *Methylbutandisäure *f*

pyrotechnic pyrotechnisch

pyrotechnics 1. Pyrotechnik *f*, Feuerwerkstechnik *f*, Feuerwerkerei *f*; 2. pyrotechnische Erzeugnisse *npl*

pyroxylin(e) Kollodiumwolle *f*, Kolloxylin *n (niedrignitrierte Cellulose)*

pyrrole Pyrrol *n*, Imidol *n*, Azol *n*

pyrrolidine Pyrrolidin *n*, Tetrahydropyrrol *n*, Tetramethylenimin *n*

pyrrolidine alkaloid Pyrrolidinalkaloid *n*

pyrrone Pyrron *n*, Polyimidazopyrrolon *n*

pyruvate Pyruvat *n (Salz oder Ester der Brenztraubensäure)*

pyruvic acid Pyruvinsäure *f*, Brenztraubensäure *f*, *Oxopropansäure *f*

PZC *s.* point of zero charge

Q

Q electron Q-Elektron *n*

Q shell Q-Schale *f (der Elektronenhülle)*

Q unit Q-Einheit *f*, tetrafunktionelle Einheit *f (Struktureinheit)*

QCM *s.* quartz crystal microbalance

quadrivalence Vierwertigkeit *f*, Tetravalenz *f*

quadrivalency Vierwertigkeit *f*, Tetravalenz *f*

quadrivalent vierwertig, tetravalent

quadruple-effect evaporator Quadrupeleffektverdampfer *m*, Vierkörperverdampfer *m*, Vierstufenverdampfer *m*

quadruple ion Quadrupolion *n*

quadruple point *(phys Ch)* Quadrupelpunkt *m*

quadruply bonded [bound, connected] vierfach gebunden

quadruply degenerate *(phys Ch)* vierfach entartet [degeneriert]

quadrupole Quadrupol *m*

quadrupole mass spectrometer Quadrupolmassenspektrometer *n*

quadrupole moment Quadrupolmoment *n*, *(i. e. S.)* Kernquadrupolmoment *n*

qualitative analysis qualitative Analyse *f*

qualitative analysis scheme Trennungsschema *n*, Trennungsgang *m*

quality 1. Qualität *f*, Beschaffenheit *f*, Eigenschaft; 2. Qualität *f*, Güte *f*

quality assurance Qualitätssicherung *f*, Gütesicherung *f*

quality control Qualitätskontrolle *f*, Gütekontrolle *f*

quality test Qualitätsprüfung *f*

quant *(phys Ch)* Quant *n*

quantitate *v* quantitativ bestimmen

quantitation quantitative Bestimmung *f*,

Mengenbestimmung *f*, Bestimmung *f* des Mengenanteils
quantitative analysis quantitative Analyse *f*
quantity Quantität *f*, Menge *f*, Anzahl *f*; Größe *f* • **liberal quantities of** reichlich
quantity being measured Messgröße *f*
quantity of electricity Elektrizitätsmenge *f*
quantity of heat Wärmemenge *f*
quantity of substance Stoffmenge *f*
quantity to be measured Messgröße *f*
quantization Quanteln *n*, Quantelung *f*, Quantisieren *n*, Quantisierung *f*
quantization of energy Energiequantelung *f*, Quantisierung *f* der Energie
quantize *v* quanteln, quantisieren
quantum 1. Quantum *n*, Menge *f*, Anzahl *f*; 2. *(phys Ch)* Quant *n*
quantum chemistry Quantenchemie *f*
quantum dot Quantenpunkt *m (Halbleiter-Nanostruktur)*
quantum-mechanical quantenmechanisch
quantum mechanics Quantenmechanik *f*
quantum number Quantenzahl *f*
quantum of action Planck-Konstante *f*, plancksche Konstante *f*, plancksches Wirkungsquantum *n*
quantum of light Lichtquant *n*, Photon *n*
quantum theory Quantentheorie *f*
quantum yield Quantenausbeute *f (bei der Photosynthese)*
quarrying Tagebauförderung *f*, Tagebaubetrieb *m*, Förderung *f* im Tagebau
quarter *v (Anal)* vierteln, quartieren, kreuzteilen
quarter-transition-time potential *(Ech)* Viertelstufenpotenzial *n*
quartet splitting *(phys Ch)* Quartettaufspaltung *f*
quartz *(Min)* Quarz *m (Siliciumdioxid)*
quartz crystal microbalance Quarzmikrowaage *f*
quartz crystal microgravimetry Quarzkristallmikrogravimetrie *f*
quartz glass Quarzglas *n*, durchsichtiges [klares] Kieselglas *n*
quartz lamp Quarzlampe *f*
quartzite *(Min)* Quarzit *m*
quartzitic quarzitisch
quartzose [quartzous, quartzy] 1. quarzhaltig, Quarz ...; 2. *(Geol)* quarzführend; aus Quarz (bestehend); 3. quarzähnlich, quarzartig
quasi-crystalline quasikristallin
quasi-equilibrium Quasigleichgewicht *n*
quasi-stationary quasistationär
quaternary alloy quaternäre Legierung *f*
quaternary ammonium compound quartäre Ammoniumverbindung *f*
quaternary salt Quartärsalz *n*, quartäres Salz *n*
quaternary structure *(Bioch)* Quartärstruktur *f*
quaternary system quaternäres System *n*, Vierkomponentensystem *n*, Vierstoffsystem *n*
quaternization Quaternisierung *f (z. B. von Aminen)*
quaternize *v* quaternisieren *(z. B. Amine)*
quench *v* 1. abschrecken, (rasch) abkühlen; 2. *(Met)* abschreckhärten, (durch Abschrecken) härten; 3. (aus-)löschen *(Lichtbogen)*; löschen *(Fluoreszenz)*; 4. löschen, quenchen *(Reaktionen)*
quench-harden *v* abschreckhärten, durch Abschrecken härten
quench hardening Abschreckhärten *n*
quencher 1. Löscher *m (Kinetik)*; 2. *s.* quenching agent 1.
quenching agent 1. Löscher *m*, Quencher *m (Photochemie)*; 2. *s.* quenching medium
quenching bath Abschreckbad *n*
quenching medium Abschreckmittel *n*; Ablöschmittel *n*
quenching of fluorescence Fluoreszenzlöschung *f*
quick-acting schnellwirkend, raschwirkend
quick assay Schnelltest *m*
quick-curing *(Kst)* schnellhärtend
quick-drying schnelltrocknend, schnell trocknend
quick test Schnelltest *m*
quicklime gebrannter [ungelöschter] Kalk *m*, Branntkalk *m (Calciumoxid)*
quicksilver Quecksilber *n*
quiescent tank Absetztank *m*, Absetzbehälter *m*
quinaldine Chinaldin *n*, 2-Methylchinolin *n*

quinaldinic acid Chinaldinsäure *f*, Chinolin-2-carbonsäure *f*
quinalizarin(e) Chinalizarin *n*, Alizarinbordeaux *n*, 1,2,5,8-Tetrahydroxyanthrachinon *n*
quinazoline Chinazolin *n*, 1,3-Benzodiazin *n*, Benzopyrimidin *n*
quinhydrone Chinhydron *n*
quinhydrone electrode Chinhydronelektrode *f*
quinic acid Chinasäure *f*, 1,3,4,5-Tetrahydroxycyclohexancarbonsäure *f*
quinidine Chinidin *n*, β-Chinin *n*, Conchinin *n*
quinine *(org Ch)* Chinin *n*
quinizarin condensation Chinizarinkondensation *f*
quinoid chinoid
quinol Hydrochinon *n*, *Benzen-1,4-diol *n*
quinolate Chinolat *n (Salz oder Ester der Chinolinsäure)*
quinoline Chinolin *n*, Benzo[b]pyridin *n*
quinoline alkaloid Chinolinalkaloid *n*, Chinaalkaloid *n*, Alkaloid *n* mit Chinolinring
quinoline dye Chinolinfarbstoff *m*
quinolinic acid Chinolinsäure *f*, Pyridin-2,3-dicarbonsäure *f*
quinolizidine ring Chinolizidinring *m*
quinone Chinon *n*, *(i. e. S.)* $O{=}C_2H_4{=}O$ Chinon *n*, *p*-Chinon *n*, *p*-Benzochinon *n*, Cyclohexa-2,5-dien-1,4-dion *n*
quinone imine dye Chinoniminfarbstoff *m*
quinone oxime Chinonoxim *n*
quinonoid dye chinoider Farbstoff *m*, Chinonfarbstoff *m*
quinonoid ring chinoider Ring *m*, Chinonring *m*
quinoxaline *(org Ch)* Chinoxalin *n*
quinque-positive fünffach positiv (geladen)
quinquevalence Fünfwertigkeit *f*, Pentavalenz *f*
quinquevalency Fünfwertigkeit *f*, Pentavalenz *f*
quinquevalent fünfwertig, pentavalent
quintuple-effect evaporator Fünfkörperverdampfer *m*, Fünfstufenverdampfer *m*
quintuply bonded [bound, connected] fünffach gebunden
quintuply degenerate *(phys Ch)* fünffach entartet [degeneriert]
quite poisonous [toxic] hochgiftig, hochtoxisch, stark giftig

R

R-M number *s.* Reichert-Meissl number
rabble *v* krählen, (um)rühren
rabble arm Krähl(werks)arm *m*, Rührarm *m*, Krähler *m (eines Etagenofens)*
race *v* zu hoch steigen *(Temperatur)*
racemate 1. Racemat *n (optisch inaktives 1 : 1-Gemisch oder optisch inaktive Verbindung zweier optischer Antipoden)*; 2. Racemat *n*, DL-Tartrat *n (Salz oder Ester der Traubensäure)*
racemic acid racemische Weinsäure *f* [Säure *f*], DL-Weinsäure *f*, Traubensäure *f*
racemic mixture racemisches Gemisch *n*
racemic pair optisches Antipodenpaar *n*
racemization Racemisierung *f*
racemize *v* racemisieren
rack *v* abfüllen *(Bier auf Fässer)*; abziehen *(Wein auf Fässer)*; füllen *(Fässer)*
rack 1. Gestell *n*, Ständer *m*, Rahmen *m*; 2. *(Hyd)* Rechen *m*, Rechenanlage *f*, Rechenvorrichtung *f*, *(meist)* Grobrechen *m*
rad Rad *n*, rd *(SI-fremde Einheit der Energiedosis; 1 rd = 10^{-2} J/kg)*
radial chromatography Zirkularchromatographie *f*, Ringchromatographie *f*
radial diffusion radiale Diffusion *f*
radial distribution function radiale Verteilungsfunktion *f*
radial flow radiale Strömung *f*
radial-flow (catalytic) reactor Radialstromreaktor *m*
radial mixing Quer(ver)mischung *f*
radial quantum number radiale Quantenzahl *f*
radiant strahlend, Strahlungs…
radiant energy Strahlungsenergie *f*
radiant excitance spezifische Ausstrahlung *f*
radiant flux Strahlungsfluss *m*, Strahlungsleistung *f*
radiant heat Strahlungswärme *f*
radiant power *s.* radiant flux

radiate *v* strahlen; (aus)strahlen, abstrahlen, emittieren; bestrahlen
radiation Strahlung *f*; Ausstrahlen *n*, Ausstrahlung *f*, Abstrahlung *f*, Emission *f*; Bestrahlung *f*
radiation absorption Strahlungsabsorption *f*
radiation-chemical strahlenchemisch, strahlungschemisch
radiation chemistry Strahlenchemie *f*, Strahlungschemie *f*
radiation counter Strahlenzähler *m*, Strahlenzählrohr *n*, Strahlungszähler *m*
radiation damage Strahlungsschaden *m*, Strahlenschaden *m*
radiation detector Strahlendetektor *m*
radiation dosage [dose] Strahlungsdosis *f*, Strahlendosis *f*
radiation dosimetry Strahlendosimetrie *f*
radiation field Strahlungsfeld *n*
radiation hazard Strahlengefährdung *f*, Strahlengefahr *f*
radiation-induced polymerization strahleninduzierte [strahlenchemische] Polymerisation *f*
radiation protection Strahlungsschutz *m*, Strahlenschutz *m*
radiation resistance Strahlungsbeständigkeit *f*, Strahlenbeständigkeit *f*, Strahlenresistenz *f*
radiation-resistant strahlungsbeständig, strahlenbeständig, strahlenresistent
radiation sickness Strahlenkrankheit *f*, Strahlensyndrom *n*
radiation source Strahlungsquelle *f*, Strahlenquelle *f*
radiation spectrum Strahlungsspektrum *n*
radiationless transition strahlungsloser [strahlungsfreier] Übergang *m*
radiative capture Strahlungseinfang *m*
radiative deactivation Strahlungsdesaktivierung *f*, strahlende Desaktivierung *f*
radiative lifetime mittlere [natürliche] Strahlungslebensdauer *f*
radiative transition strahlender Übergang *m*, Strahlungsübergang *m*
radiator Strahler *m*
radical radikalisch
radical 1. *(org Ch)* Gruppe *f*, Rest *m*; 2. *(org Ch)* (freies) Radikal *n*
radical-chain reaction Radikalkettenreaktion *f*
radical former Radikalbildner *m*
radical ion Radikalion *n*
radical polymerization Radikalkettenpolymerisation *f*, radikalische Polymerisation *f* [Kettenpolymerisation *f*]
radical trap Radikalfänger *m*, Radikalfalle *f*
radicofunctional name radikofunktioneller Name *m (aus den Benennungen von Molekülrest und funktioneller Gruppe zusammengesetzt)*
radio-frequency polarography *(Anal)* Hochfrequenzpolarographie *f*, HFP
radio labelling Radiomarkierung *f*
radioactive chain *s.* radioactive series
radioactive contamination radioaktive Kontamination *f* [Verunreinigung *f*, Verseuchung *f*]
radioactive dating *s.* radiometric dating
radioactive decay radioaktiver Zerfall *m*, Atomzerfall *m*, Kernzerfall *m*
radioactive disintegration *s.* radioactive decay
radioactive displacement law Verschiebungssatz *m*, Verschiebungsgesetz *n*
radioactive element radioaktives Element *n*, Radioelement *n*
radioactive fallout (radioaktiver) Fallout *m*
radioactive isotope radioaktives [instabiles] Isotop *n*, Radioisotop *n*
radioactive nuclide radioaktives Nuklid *n*, Radionuklid *n*
radioactive series radioaktive Familie *f* [Zerfallsreihe *f*]
radioactive tracer *s.* radiotracer
radioactive transformation radioaktive Umwandlung *f*, radioaktiver Übergang *m*
radioactive waste radioaktiver Abfall *m*, *(i. e. S.)* Atommüll *m*
radioactively labelled radioaktiv markiert, radioindiziert
radioactivity Radioaktivität *f*
radioautograph Autoradiogramm *n*, autoradiographische Aufnahme *f*
radioautographic autoradiographisch

radioautography Autoradiographie *f*, Radioautographie *f*
radiocarbon dating Radiokohlenstoffdatierung *f*, Altersbestimmung *f* [Datierung *f*] mit Hilfe der Radiocarbonmethode [Kohlenstoff-14-Methode, ^{14}C-Methode]
radiochemical radiochemisch, strahlenchemisch
radiochemist Radiochemiker *m*
radiochemistry 1. Radiochemie *f*; 2. *s.* radiation chemistry
radioelement Radioelement *n*, radioaktives Element *n*
radiograph Röntgenaufnahme *f*, Röntgenbild *n*, Röntgenogramm *n*
radiographic radiographisch, Radiographie...; röntgenographisch, Röntgen...
radiography Radiographie *f (Verfahren zur Werkstoffprüfung)*; Röntgenographie *f*, Röntgenfotografie *f*
radioimmunoassay Radioimmunoassay *n*, radioimmunologischer Versuch *m (besonders zur Hormonbestimmung)*
radioisotope radioaktives [instabiles] Isotop *n*, Radioisotop *n*
radiology Radiologie *f*, Strahlenkunde *f*
radioluminescence Radiolumineszenz *f*
radiolysis Radiolyse *f*, strahlenchemische Zersetzung *f*
radiometer Radiometer *n*
radiometric dating radiometrische [physikalische, absolute] Altersbestimmung *f*, Altersbestimmung *f* durch Radionuklide
radiometry Radiometrie *f*
radiomimetic (agent, compound) *(Bio)* Radiomimetikum *n*, Ruhekerngift *n*
radionuclide Radionuklid *n*, radioaktives Nuklid *n*
radiopasteurization Strahlenpasteurisierung *f*
radiosensitive strahlungsempfindlich, strahlenempfindlich
radiosensitivity Strahlungsempfindlichkeit *f*, Strahlenempfindlichkeit *f*
radiotracer Radiotracer *m*, Radioindikator *m*, radioaktiver Tracer *m* [Indikator *m*]
radium Ra Radium *n*
radium series Radiumreihe *f*, Uran-Radium-Reihe *f*, Uran-Radium-Zerfallsreihe *f*
radius of gyration Trägheitsradius *m*
radon Rn Radon *n*
raffinate Raffinat *n*, Solvat *n (im Lösungsmittel unlösliche Anteile bei der Solventextraktion)*
RAFT *s.* reversible addition fragmentation chain transfer polymerization
rag pulp *(Pap)* Hadern(halb)stoff *m*, Lumpenhalbstoff *m*, Hadernhalbzeug *n*
rags *(Pap)* Hadern *mpl*, Lumpen *mpl*
raies ultimes *(Spektr)* Grundlinien *fpl*, Nachweislinien *fpl*, *(veraltet:)* letzte Linien *fpl*
rain-out 1. Rain-out *n*, Ausregnen *n (von Schadstoffen durch Bildung von Regentropfen)*; 2. Rain-out *m (ausgefallene Schadstoffe)*
raining bed Regenbett *n (Fließbettsystem mit fallenden Feststoffteilchen in aufströmendem Fluid)*
rainout *s.* rain-out
raise *v* 1. steigern, erhöhen *(z. B. Temperatur)*; 2. *(Lebm)* (auf)gehen lassen *(Teig)*
raise *v* **to the boil** zum Kochen bringen
rake *v* 1. durch Kratzen fördern; krählen *(Schlamm)*; 2. auskratzen
rake 1. Rechen *m*; 2. *s.* raking arm
rake classifier Rechenklassierer *m*
raking arm Krählarm *m*, Rechenarm *m (eines Eindickers)*
rakings *(Hyd)* Rechengut *n*
RAM *s.* relative atomic mass
ram Kolben *m*, Stempel *m*
Raman-active *(Spektr)* ramanaktiv
Raman effect Raman-Effekt *m*, Smekal-Raman-Effekt *m*
Raman line Raman-Linie *f*
Raman scattering Raman-Streuung *f*
Raman shift Raman-Verschiebung *f*
Raman spectroscopy Raman-Spektroskopie *f*
Raman spectrum Raman-Spektrum *n*
ramming Rammen *n*, Stampfen *n*
rancid ranzig
rancidification Ranzigwerden *n*
rancidify *v* ranzig werden; ranzig machen [werden lassen]
rancidity Ranzidität *f*, Ranzigkeit *f*, ranziger Geruch *m*
random regellos, zufällig, ungeordnet

random coil statistisches Knäuel *n*, Zufallsknäuel *n (Proteinchemie)*
random copolymer statistisches Copolymer(e) *n*
random distribution *(Krist)* zufällige [regellose, statistische] Verteilung *f*
random error Zufallsfehler *m*, zufälliger [subjektiver] Fehler *m*
random sample *(Anal)* Zufalls(stich)probe *f*
random sampling *(Anal)* Zufallsprobe(nent)nahme *f*
random system *(phys Ch)* ungeordnetes System *n*
randomly distributed *(Krist)* zufällig [regellos, statistisch] verteilt
randomness *(Krist)* Unordnung *f*
Raney catalyst Raney-Katalysator *m*, Skelettkatalysator *m*, Skelettkontakt *m*
Raney nickel (catalyst) Raney-Nickel *n*
range Bereich *m*, Gebiet *n*; Abstand *m*, Entfernung *f*; Reichweite *f*; Umfang *m*, Spanne *f*
range of measurement Messbereich *m*
range of pH pH-Bereich *m*, pH-Gebiet *n*
rank Rang *m*, Rangstufe *f*, *(bei Kohle auch)* Inkohlungsgrad *m* • **by rank** *(Carb)* nach dem Inkohlungsgrad
Rankine cycle Clausius-Rankine-Prozess *m (Thermodynamik)*
Rankine scale [temperature scale] Rankine-Skala *f*
Raoult's law *(phys Ch)* raoultsches Gesetz *n*
rap *v* (be)klopfen
rape(-seed) oil Kolzaöl *n*, Kohlsaatöl *n*, *(i. e. S.)* Rapsöl *n (aus Brassica napus L. em. Metzg.)*; Rüb(sen)öl *n (aus Brassica rapa L. em. Metzg.)*
rapid analysis Schnellanalyse *f*
rapid filter Schnellfilter *n*
rapid-hardening schnellbindend; schnellhärtend
rapid mixing *(Hyd)* Schnellrühren *n*, Schnellmischen *n*
rapid reaction schnelle Reaktion *f*
rapid tannage Schnellgerbung *f*
rapid test Schnellversuch *m*
rapper Klopfvorrichtung *f*, Klopfer *m*
rare base *(Bioch)* seltene Base *f*, seltener Nucleinsäurebaustein *m (Purin- oder Pyrimidinderivat)*
rare-earth elements [metals] Seltenerdmetalle *npl*, Metalle *npl* der Seltenerden
rare earths 1. seltene Erden *pl*, Seltenerden *pl*; 2. *s.* rare-earth elements
rare gas Edelgas *n*
rarefaction 1. Verdünnen *n*; 2. verdünnter Zustand *m*; 3. Porosität *f*
rarefiable verdünnbar
rarefy *v* 1. verdünnen; 2. porös werden
rarify *v* 1. verdünnen; 2. porös werden
Raschig process Raschig-Verfahren *n (1. katalytische Herstellung von Phenol aus Benzen; 2. Herstellung von Hydrazin)*
Raschig ring Raschig-Ring *m (ein Füllkörper)*
Rast method [micromethod] Mikromethode *f* von Rast, Rast-Methode *f*, Camphermethode *f* (nach Rast) *(zur Bestimmung der relativen Molekülmasse)*
rat insulin *(Biot)* Ratteninsulin *n*
rate 1. Geschwindigkeit *f*; 2. Verhältnis *n*, Rate *f*; 3. Menge *f* je Zeiteinheit; Anzahl *f* je Zeiteinheit
rate coefficient Reaktionsgeschwindigkeitskonstante *f*, RGK, Geschwindigkeitskonstante *f (einer komplexen Reaktion)*
rate constant Reaktionsgeschwindigkeitskonstante *f*, RGK, Geschwindigkeitskonstante *f (einer Elementarreaktion)*
rate-controlling *s.* rate-limiting
rate-determining *s.* rate-limiting
rate equation [law] Reaktionsgeschwindigkeitsgleichung *f*, Geschwindigkeitsgleichung *f*, Zeitgesetz *n*, Geschwindigkeitsgesetz *n*
rate-limiting geschwindigkeitsbestimmend
rate-limiting reaction Schrittmacherreaktion *f*
rate-limiting step geschwindigkeitsbestimmender Schritt *m (einer Reaktion)*
rate meter Durchfluss(mengen)messer *m*, Mengenstrommesser *m*, Strömungsmesser *m*
rate of appearance Bildungsgeschwindigkeit *f*
rate of combustion Verbrennungsge-

schwindigkeit *f*, Brenngeschwindigkeit *f*
rate of consumption Verbrauchsgeschwindigkeit *f*
rate of conversion Umwandlungsgeschwindigkeit *f*, Umsetzungsgeschwindigkeit *f*, Umsatzgeschwindigkeit *f*
rate of corrosion Korrosionsgeschwindigkeit *f*
rate of crystallization Kristallisationsgeschwindigkeit *f*
rate of death *(Biot)* Abtötungsrate *f* *(Sterilisation)*
rate of decay Zerfallsgeschwindigkeit *f*
rate of deposition Aufdampfrate *f*
rate of detonation Detonationsgeschwindigkeit *f*
rate of diffusion Diffusionsgeschwindigkeit *f*; Transfusionsgeschwindigkeit *f (von Gasen durch eine poröse Trennwand)*
rate of disappearance Verbrauchsgeschwindigkeit *f*
rate of discharge Lieferstrom *m*, Förderstrom *m*
rate of disintegration Zerfallsgeschwindigkeit *f*; Zerfallsrate *f*
rate of dissolution Lösegeschwindigkeit *f*, Lösungsgeschwindigkeit *f*, Auflösungsgeschwindigkeit *f*
rate of drying Trockengeschwindigkeit *f*, Trocknungsgeschwindigkeit *f*, spezifische Austauschfeuchtemenge *f*
rate of elution Elutionsgeschwindigkeit *f*
rate of evaporation Verdampfungsgeschwindigkeit *f*; Verdunstungsgeschwindigkeit *f*
rate of filtration Filtriergeschwindigkeit *f (bei Flüssigkeiten)*; Filtergeschwindigkeit *f (bei Gasen)*
rate of flow 1. Fließgeschwindigkeit *f*, Strömungsgeschwindigkeit *f*, Durchflussgeschwindigkeit *f*, Durchlaufgeschwindigkeit *f (in* $m \cdot s^{-1}$*)*; 2. Durchfluss *m (in* $m^3 \cdot h^{-1}$*)*
rate of formation Bildungsgeschwindigkeit *f*
rate of generation Bildungsgeschwindigkeit *f*
rate of growth 1. *(Krist, Kst)* Wachstumsgeschwindigkeit *f*; 2. *(Biot)* Wachstumsrate *f (von Mikroorganismen)*
rate of heat transfer Wärmestrom *m (bei Wärmedurchgang)*
rate of heating Erwärmungsgeschwindigkeit *f*, Aufheiz(ungs)geschwindigkeit *f*, Erhitzungsgeschwindigkeit *f*
rate of migration [movement] Wanderungsgeschwindigkeit *f*
rate of nucleation Keimbildungsgeschwindigkeit *f*
rate of product formation *(Biot)* Produktbildungsrate *f*, Fermentationsrate *f*, Umsatzrate *f*
rate of reaction Reaktionsgeschwindigkeit *f*
rate of rotation Winkelgeschwindigkeit *f*; Drehzahl *f*
rate of sedimentation [settling] Absetzgeschwindigkeit *f*, Sedimentationsgeschwindigkeit *f*, Absinkgeschwindigkeit *f*, Sinkgeschwindigkeit *f*
rate of shear Scherrate *f*, Schergeschwindigkeit *f*
rate of transformation *s.* rate of conversion
rate of vaporization Verdampfungsgeschwindigkeit *f*; Verdunstungsgeschwindigkeit *f*
rate of withdrawal *(Dest)* Entnahmeverhältnis *n*
raticide Rattenbekämpfungsmittel *n*
rating Bewertung *f (von Eigenschaften)*
ratio of concentrations Konzentrationsverhältnis *n*
ratio of retention *(Chromat)* Retentionsrate *f*, Retentionsverhältnis *n*
ratio of the intercepts (kristallographisches) Achsenverhältnis *n*
rational index law *(Krist)* Rationalitätsgesetz *n*, Gesetz *n* der rationalen Indizes [Parameterverhältnisse], haüysches Gesetz *n*
rauwolfia alkaloid Rauwolfiaalkaloid *n*
raw acid Rohsäure *f*
raw clay Rohton *m*
raw coal Roh(förder)kohle *f*
raw gas Rohgas *n*
raw juice 1. Rohsaft *m*; 2. *(Zucker)* Diffusionssaft *m*, Rohsaft *m*
raw material Rohmaterial *n*, Rohstoff *m*, Ausgangsmaterial *n*, Ausgangsstoff *m*
raw ore Roherz *n*, Fördererz *n*

raw protein Rohprotein *n*
raw rubber Rohkautschuk *m*
raw sewage Rohabwasser *n*
raw silk Rohseide *f*, Bastseide *f*
raw starch Rohstärke *f*, Grünstärke *f*
raw stock Rohcharge *f*
raw sugar Rohzucker *m*
raw waste water Rohabwasser *n*
raw water Rohwasser *n*
ray of light Lichtstrahl *m*
α-ray α-Strahl *m*, Alphastrahl *m*
β-ray β-Strahl *m*, Betastrahl *m*
β-ray disintegration Beta-Zerfall *m*, β-Zerfall *m*, Beta-Umwandlung *f*
γ-ray γ-Strahl *m*, Gammastrahl *m*
Rayleigh-Jeans law *(phys Ch)* rayleigh-jeanssches Strahlungsgesetz *n*
Rayleigh scattering *(Spektr)* Rayleigh-Streuung *f*
rayon Regeneratcellulosefaserstoff *m*, RZ, Celluloseregeneratseide *f*, Reyon *n(m)*; *(früher i. w. S.)* Chemieseide *f*, Kunstseide *f*
RBE *s.* relative biological effectiveness
RCM *s.* ring closing metathesis
RDA *s.* recommended daily allowance
rDNA *s.* recombinant DNA
Re. *s.* Reynolds number
re-absorb *v* 1. wieder absorbieren [aufsaugen, einsaugen, aufnehmen]; 2. *s.* resorb
reacidify *v* erneut ansäuern
react *v* 1. reagieren; zur Reaktion bringen; 2. rückwirken
react *v* **further** weiterreagieren
react *v* **sluggishly** träge [schleppend] reagieren
react *v* **stormily** stürmisch reagieren
react *v* **to form ...** unter Bildung von ... reagieren
react *v* **vigorously [violently]** heftig reagieren
reactant 1. Reaktionspartner *m*, Reaktionsteilnehmer *m*, Reaktant *m*, reagierender Stoff *m*; 2. Substrat *n (Enzymchemie)*
reaction Reaktion *f*
reaction adhesive Reaktionsklebstoff *m*
reaction chain Reaktionskette *f*
reaction cross section Reaktionsquerschnitt *m*
reaction energy Reaktionsenergie *f*
reaction enthalpy Reaktionsenthalpie *f*
reaction equation Reaktionsgleichung *f*, Umsatzgleichung *f*
reaction Gibbs function freie Reaktionsenthalpie *f*
reaction half-life Halbwertszeit *f* (der Reaktion)
reaction half-time Halbwertszeit *f* (der Reaktion)
reaction heat Reaktionswärme *f*
reaction intermediate Intermediat *n*, Transient *m*, Zwischenstoff *m*, Zwischenprodukt *n (einer chemischen Reaktion)*
reaction isotherm Reaktionsisotherme *f*
reaction kinetics Reaktionskinetik *f*, chemische Kinetik *f*
reaction mechanism Reaktionsmechanismus *m*
reaction molecularity Reaktionsmolekularität *f*, Molekularität *f* (der Reaktion)
reaction order Reaktionsordnung *f*
reaction path(way) 1. Reaktionsweg *m*; 2. *s.* reaction mechanism
reaction period Hauptperiode *f*, Hauptversuch *m (einer kalorimetrischen Messung)*
reaction product Reaktionsprodukt *n*
reaction rate Reaktionsgeschwindigkeit *f*, Umsetzungsgeschwindigkeit *f*
reaction-rate constant Reaktionsgeschwindigkeitskonstante *f*, RGK, Geschwindigkeitskonstante *f (einer Elementarreaktion)*
reaction sequence *(org Ch)* Reaktionsablauf *m*, Reaktionsfolge *f*, Reaktionssequenz *f*
reaction-specific *(Bioch)* reaktionsspezifisch, wirkungsspezifisch
reaction specificity *(Bioch)* Reaktionsspezifität *f*, Wirkungsspezifität *f (eines Enzyms)*
reaction vessel Reaktionsgefäß *n*, Reaktionskessel *m*, Reaktionsbehälter *m*
reactionless reaktionslos
reactivate *v* reaktivieren
reactivation Reaktivierung *f*
reactive reaktiv, reaktionsfähig, reaktionsfreudig, reaktionsbereit; unedel *(Metall)*
reactive adhesive *(Kst)* Reaktionsklebstoff *m*
reactive dye Reaktivfarbstoff *m*

reactive site aktiviertes Zentrum *n (an einer Polymerkette)*
reactivity Reaktivität *f*, Reaktionsfähigkeit *f*, Reaktionsvermögen *n*, Reaktionsfreudigkeit *f*, Reaktionsbereitschaft *f*
reactivity ratio Copolymerisationsparameter *m*
reactor 1. Reaktor *m*, Reaktionsapparat *m*, Stoffumsetzer *m*, Umsetzer *m*; Reaktionsofen *m*; Reaktionsraum *m*, Reaktionskammer *f*; 2. *s.* fermenter; 3. Kernreaktor *m*, Reaktor *m*
reactors in series Reaktorkaskade *f*
read *v* **(off)** ablesen
readily oxidizable leicht oxidierbar
readily soluble leichtlöslich, ll, leicht [gut] löslich
readily volatile leichtflüchtig
reading 1. Ablesen *n*, Ablesung *f (z. B. eines Skalenwerts)*; 2. Skalenwert *m*, Stand *m*, Anzeige *f* • **show a constant reading** einen konstanten Wert aufweisen, konstant sein • **take a reading of** ablesen
reading error Ablesefehler *m*
readjust *v* wieder einstellen *(auf einen Wert)*
readout Ablesen *n*, Ablesung *f (z. B. eines Skalenwerts)*
reaffirm *v* **the presence of** nachweisen
reagent Reagens *n*
reagent blank *(Anal)* Reagenzienblindwert *m*
reagent-grade analysenrein, zur Analyse, p. a.
reagent purity Analysenreinheit *f*
real crystal realer [gestörter] Kristall *m*, Realkristall *m*
real gas reales Gas *n*
realgar *(Min)* Realgar *m (Tetrarsentetrasulfid)*
rear attack *(org Ch)* Angriff *m* von der Rückseite her *(z. B. bei Umlagerungen)*
rearrange *v* umlagern, umgruppieren; sich umlagern [umgruppieren]
rearrangement Umlagerung *f*, Umgruppierung *f*
reasonably soluble mäßig [einigermaßen] löslich
reassay Neubestimmung *f*, nochmalige Bestimmung *f*
reboil *v* nochmals kochen [sieden]; (wieder) aufkochen
reboiler Destillierblase *f*, Blase *f*, Verdampfungsofen *m (einer Rektifizierkolonne)*
rebound Rückprall *m*, Rückstoß *m*, Zurückspringen *n*
recapture *v* wieder auffangen; wieder einfangen *(z. B. Ionen)*
recapture Wiederauffangen *n*; Wiedereinfangen *n (z. B. von Ionen)*
recarburization *(Met)* Wiederaufkohlen *n*
recarburize *v (Met)* wiederaufkohlen
recarburizer *(Met)* Aufkohlungsmittel *n*
recarburizing agent *(Met)* Aufkohlungsmittel *n*
receding contact angle Rückzugsrandwinkel *m*, Kontraktionsberührungswinkel *m (Prüfung grenzflächenaktiver Stoffe)*
receiver 1. Sammelbehälter *m*, Sammelgefäß *n*, Auffangbehälter *m*, Auffanggefäß *n*, Auffänger *m*; Auffangschale *f*; 2. *(Dest)* Vorlage *f*, Destilliervorlage *f*, Destillat(ions)vorlage *f*, Destillatsammler *m*
receiver adapter *(Dest)* Vorstoß *m*
receiving water *(Hyd)* Vorfluter *m*
recent frisch, neu, rezent
receptacle Behälter *m*, Sammelbehälter *m*, Gefäß *n*, Sammelgefäß *n*
receptivity to dyestuffs Farbstoffaufnahmevermögen *n*, Anfärbbarkeit *f*
receptor area Emissionssenke *f*, Einwirkungsstelle *f (von Luftschadstoffen)*
recharge Regenerierung *f*, Regeneration *f (eines erschöpften Ionenaustauschers)*
recheck *v* nachprüfen, nochmals (über)prüfen
recheck *(Tox)* Nachprüfung *f (zur Kontrolle der Hauptprüfung)*
recipe Rezept *n*; Rezeptur *f*
reciprocate *v* hin- und herbewegen; hin- und hergehen, sich hin- und herbewegen
reciprocating compressor Kolbenverdichter *m*, Kolbenkompressor *m*
reciprocating pump Hubkolbenpumpe *f*, Kolbenpumpe *f* mit hin- und hergehenden Kolben

reciprocating screen Vibrationssieb *n*, Vibratorsieb *n*, Vibrosieb *n*, Zittersieb *n*
recirculate *v s.* recycle
recirculation 1. Umlauf *m*, Zirkulation *f*; 2. *s.* recycling 1.
reclaim *v* 1. wiedergewinnen, (zu)rückgewinnen, rückführen; wiederaufbereiten; 2. *(Gum, Kst)* regenerieren
reclaim 1. *(Gum)* Regenerat *n*, regenerierter Kautschuk *m*, Regenerativgummi *m*; 2. *(Kst)* Regenerat *n*
reclaiming agent Regeneriermittel *n*
reclamation Regenerierung *f*
recognition sequence [site] Erkennungsregion *f (in der DNA)*
recoil atom Rückstoßatom *n*
recoil chemistry Heißatomchemie *f*, Chemie *f* heißer Atome, Rückstoßchemie *f*
recoil electron Rückstoßelektron *n*, Compton-Elektron *n*
recoil nucleus Rückstoßkern *m*
recombinant DNA *(Bioch)* rekombinierte DNA *f* [Desoxyribonucleinsäure *f*], rDNA *f*, rDNS *f*
recombinant DNA technique (DNA-)Rekombinationstechnik *f (Methode)*
recombination Rekombination *f*, Wiedervereinigung *f*
recombination reaction Rekombinationsreaktion *f*
recombine *v* wiedervereinigen, rekombinieren; sich wiedervereinigen, sich rekombinieren
recommended daily allowance *(Lebm)* empfohlene Tagesaufnahme *f* [tägliche Aufnahme *f*] *(von Nährstoffen)*
recommended limit Richtwert *m*
reconcentrate *v* aufsättigen
reconstructive transition rekonstruktive Umwandlung *f* [Phasenumwandlung *f*]
reconversion Rückumwandlung *f*, Rück(ver)wandlung *f*
reconvert *v* rückumwandeln, zurückverwandeln, rück(ver)wandeln; sich rückumwandeln [zurückverwandeln], sich rück(ver)wandeln
record *v* aufzeichnen, registrieren *(Messwerte)*
recorder Schreiber *m*, Schreibwerk *n*
recover *v* 1. wiedergewinnen, (zu)rückgewinnen, rückführen; (wieder)aufbereiten; 2. *(Gum, Kst)* regenerieren; 3. gewinnen; 4. *(Biot)* aufarbeiten *(das Produkt)*; 5. sich nachbilden *(z. B. Isotope)*; 6. *(Gum)* rückfedern, sich erholen, zurückspringen
recoverable wiedergewinnbar
recovery 1. Wiedergewinnung *f*, Rückgewinnung *f*, Rückführung *f*; Aufbereitung *f*; 2. *(Gum, Kst)* Regenerierung *f*, Regeneration *f*; 3. Gewinnung *f*; 4. *(Biot)* Aufarbeitung *f (des Produkts)*; 5. Ausbeute *f*; 6. Nachbildung *f (z. B. von Isotopen)*; 7. *(Gum)* (elastische) Erholung *f*, Rückverformung *f*, Rückfederung *f*, Zurückspringen *n*, elastischer Anteil *m*
recovery loss *(Biot)* Aufarbeitungsverlust *m*
recovery of chemicals Chemikalienrückführung *f*, Chemikalienrückgewinnung *f*
recovery of heat Wärmerückgewinnung *f*
recovery of values Wertstoff(rück)gewinnung *f*
recrystallization 1. Umkristallisieren *n (zum Reinigen)*; 2. *(Met)* (primäre, verformungsbedingte) Rekristallisation *f (nach plastischer Verformung)*
recrystallization annealing Rekristallisationsglühen *n*, rekristallisierendes Glühen *n*
recrystallize *v* 1. umkristallisieren *(zum Reinigen)*; wieder auskristallisieren; 2. *(Met)* rekristallisieren *(nach plastischer Verformung)*
rectification 1. Rektifikation *f*, Rektifizierung *f*, Gegenstromdestillation *f*; 2. Justieren *n*, Justierung *f*, Justage *f*, (richtige) Einstellung *f (eines Instruments)*
rectification still *(Dest)* Rektifizierapparat *m*, Rektifikationsapparat *m*, Rektifizieranlage *f*
rectify *v* 1. *(Dest)* rektifizieren; 2. justieren (richtig) einstellen *(Instrumente)*
rectifying column *(Dest)* Rektifiziersäule *f*, Rektifizierkolonne *f*, Rektifikationskolonne *f*, Austauschsäule *f*, Austauschkolonne *f*, Trennsäule *f*, Trennkolonne *f*
recuperative furnace Rekuperativofen *m*

recuperator Rekuperator *m*, rekuperativer Vorwärmer *m*
recycle *v* 1. im Kreislauf führen [fahren], rundführen, umpumpen; 2. (in den Kreislauf) zurückführen, dem Kreislauf wieder zuführen [zusetzen], rezirkulieren
recycle 1. Kreislauf *m*; 2. *s.* recycling 1.
recycle gas Kreislaufgas *n*, Umlaufgas *n*, Zirkulationsgas *n*
recycle pump Umlaufpumpe *f*, Umwälzpumpe *f*, Kreislaufpumpe *f*, Zirkulationspumpe *f*
recycle reactor Kreislaufreaktor *m*; Schlaufenreaktor *m*
recycling 1. Kreislaufführung *f*, Kreislauffahrweise *f*, zirkulierende Fahrweise *f*, Kreislauffahren *n*; 2. Recycling *n*, Rückführung *f* von Abprodukten
recycling of water Kreislaufwasserführung *f*, Wasserkreislaufführung *f*, Kreislaufnutzung *f* des Wassers
recycling water Kreislaufwasser *n*
red 1. Rot *n (Farbempfindung)*; 2. roter [rotfärbender] Farbstoff *m*, Rot *n*
red biotechnology rote Biotechnologie *f (Medizin/Pharmazie)*
red brass Rotguss *m*, Rotmetall *n*, Rotmessing *n (85 % Cu, 15 % Zn)*
red copper ore *(Min)* Rotkupfererz *n*, Cuprit *m (Kupfer(I)-oxid)*
red heat Rotglut *f*
red hot rotglühend
red lead Mennige *f*, Bleimennige *f*, rotes Bleioxid *n*, Bleioxidrot *n (Blei(II)-orthoplumbat)*
red mercuric sulphide Zinnober *m*, rotes Quecksilbersulfid *n*, *Quecksilber(II)-sulfid *n*
red phosphorus roter Phosphor *m (des Handels)*
red precipitate rotes Präzipitat *n*, rotes Quecksilber(II)-oxid *n*
red prussiate of potash rotes Blutlaugensalz *n*, Kaliumhexacyanoferrat(III) *n*
red-shift Rotverschiebung *f*, bathochrome Verschiebung *f* [Farbverschiebung *f*]
redeposit *v* sich wieder absetzen [ablagern, niederschlagen]
redetermination nochmalige Bestimmung *f*
redissolution Wiederauflösung *f*, Rücklösung *f*
redissolve *v* wiederauflösen, wiederlösen; sich wieder (auf)lösen
redistil *v* nochmals [erneut, wiederholt] destillieren, umdestillieren, redestillieren
redistillation nochmalige [erneute] Destillation *f*, Zweitdestillation *f*, Redestillation *f*, Redestillieren *n*
redness Rotglut *f*
redox catalyst Redoxkatalysator *m*
redox couple Redoxpaar *n*
redox equilibrium Redoxgleichgewicht *n*
redox potential Redoxpotenzial *n*, Reduktions-Oxidations-Potenzial *n*
redox reaction Redoxreaktion *f*, Reduktions-Oxidations-Reaktion *f*, Oxidoreduktion *f*
redox titration Redoxtitration *f*, Reduktions-Oxidations-Titration *f*
redoxase Oxidoreductase *f*, Redoxase *f*
reduce *v* 1. reduzieren, (ver)mindern, verringern, herabsetzen, senken; zerkleinern; sich vermindern [verkleinern, verringern], (ab)sinken, abnehmen, zurückgehen, nachlassen, (ab)fallen; 2. *(Ch)* reduzieren; reduziert werden, sich reduzieren lassen
reduce *v* **to ashes** veraschen
reduce *v* **to fibres** defibrieren, zerfasern, in Einzelfasern zerlegen
reduce *v* **to powder** pulverisieren, (zer)pulvern; auf Staubfeinheit mahlen
reduced equation of state reduzierte Zustandsgleichung *f*
reduced property reduzierte Größe *f* [Zustandsgröße *f*]
reduced variable reduzierte Zustandsvariable *f*
reducer 1. Übergangsstück *n*, Reduzierstück *n*; 2. Reduzent *m*, organische Stoffe abbauender [zersetzender, zerstörender] Organismus *m*; 3. *s.* reductant
reducibility Reduzierbarkeit *f*
reducible reduzierbar
reducing adapter *(Lab)* Übergangsstück *n* (Form) A, Reduzierstück *n (mit kleiner Hülse auf großem Kern)*
reducing agent *s.* reductant

reducing atmosphere reduzierende Atmosphäre *f*
reducing flame Reduktionsflamme *f*
reducing gas Reduktionsgas *n*
reducing sugar reduzierender Zucker *m*
reducing valve Reduzierventil *n*, Druckminder(ungs)ventil *n*, Druckminderer *m*
reductant Reduktionsmittel *n*, reduzierendes Mittel *n*, Reduktor *m*, Desoxidationsmittel *n*
reduction 1. Reduktion *f*, Reduzierung *f*, Verminderung *f*, Minderung *f*, Verringerung *f*, Herabsetzung *f*, Senkung *f*; Zerkleinerung *f*; Sinken *n*, Absinken *n*, Abnahme *f*, Rückgang *m*, Nachlassen *n*, Abfallen *n*, Abfall *m*, Fallen *n*; 2. *(Ch)* Reduktion *f*
reduction oxidation potential *s.* redox potential
reduction potential Reduktionspotenzial *n*
reductive reduzierend, reduktiv
reductive coupling reduktive Kupplung *f*
reductive elimination *(org Ch)* reduktive Eliminierung *f*
Redwood visco(si)meter Redwood-Viskosimeter *n*
reel *v* **(up)** 1. *(Pap, Text)* aufwickeln, aufrollen; 2. *(Text)* haspeln, winden
re-emission Wiederausstrahlung *f*
re-equilibration Wiederherstellung *f* des Gleichgewichts
re-esterification Umesterung *f*, Reesterifizierung *f*
re-esterify *v* umestern
re-evaporation Rückverdampfung *f*, erneute Verdampfung *f*
re-examination nochmalige Prüfung *f* [Überprüfung *f*, Untersuchung *f*]
referee check Schiedsanalyse *f (forensische Chemie)*
reference *s.* 1. reference compound; 2. reference sample
reference beam *(Spektr)* Vergleichsstrahl *m*, Referenzstrahl *m*
reference cell *(Spektr)* Vergleichsküvette *f*, Referenzküvette *f*
reference compound Vergleichsverbindung *f*, Referenzverbindung *f*
reference electrode [element] Bezugselektrode *f*, Referenzelektrode *f*
reference peak Bezugssignal *n*
reference sample Vergleichsprobe *f*
reference solution Vergleichslösung *f*
reference spectrum Vergleichsspektrum *n*
reference standard Bezugsstandard *m*, Vergleichsstandard *m*
reference substance Bezugssubstanz *f*, Vergleichssubstanz *f*
reference voltage Bezugsspannung *f*, Vergleichsspannung *f*, Referenzspannung *f*
refilter *v* erneut [nochmals] filtrieren
refine *v* 1. reinigen, veredeln, *(besonders Erdöl, Zucker, Metalle)* raffinieren; 2. *(Glas)* läutern, lauter schmelzen; 3. *(Erdöl)* verarbeiten, *(i. w. S.)* raffinieren
refinery 1. *(Erdöl, Zucker)* Raffinerie *f*; 2. *(Met)* Hütte *f*
refinery gas Raffineriegas *n*, Raffiniergas *n*
refining 1. Reinigung *f*, Veredelung *f*, *(besonders Erdöl, Zucker, Metalle)* Raffination *f*; 2. *(Erdöl)* Verarbeitung *f*, *(i. w. S.)* Raffination *f*
reflect *v* **(back)** reflektieren, zurückwerfen, zurückstrahlen, spiegeln
reflectance Reflexionsgrad *m*, Reflexionsvermögen *n*
reflection Reflexion *f*, Reflektierung *f*, Rückstrahlung *f*, Spiegelung *f*
reflection coefficient *s.* reflectance
reflection electron microscopy Reflexionselektronenmikroskopie *f*
reflection factor *s.* reflectance
reflection grating *(Spektr)* Reflexionsgitter *n*
reflection high-energy electron diffraction *(Anal)* Reflexionsbeugung *f* schneller Elektronen, RHEED
reflectivity 1. Reflexionsvermögen *n*, Reflexionsfähigkeit *f*, Reflexionskraft *f*, Rückstrahlungsvermögen *n*; 2. Reflexionszahl *f (Wärmestrahlung)*
reflux *v* am Rückflusskühler kochen, unter Rückfluss(kühlung) erhitzen [kochen]
reflux 1. *(Dest)* Rückfluss *m*, Rücklauf *m*, Reflux *m*; 2. *s.* reflux condenser
reflux condenser 1. *(Tech)* Rücklaufkondensator *m*; 2. *(Lab)* Rückflusskühler *m*
reflux ratio Rücklaufverhältnis *n*, Rückflussverhältnis *n*

refluxing Rückflusskochen *n*, Kochen *n* am Rückflusskühler
refolding *(Bioch)* Rückfaltung *f (der Polypeptidkette)*
re-form *v* 1. zurückbilden; sich zurückbilden; 2. sich erneut bilden
reform *v* reformieren *(Benzinkohlenwasserstoffe)*
reformate Reformat *n*, reformiertes Produkt *n*
re-formation 1. Rückbildung *f*; 2. erneute Bildung *f*
Reformatsky reaction Reformatsky-Reaktion *f*, Reformatsky-Synthese *f (von β-Hydroxycarbonsäureestern)*
reforming Reform(ier)en *n*, Reformierung *f*, Reforming *n (Änderung der Molekülstruktur und -größe von Benzinkohlenwasserstoffen)*
reforming catalyst Reformingkatalysator *m*
reforming plant Refom(ier)anlage *f*, Reformierungsanlage *f*, Reforming-Anlage *f*, Reformer *m*
refraction Brechung *f*, Refraktion *f*
refractive lichtbrechend
refractive index Brechzahl *f*, Brechungsindex *m*
refractive index detector *(Anal)* Brechungsindexdetektor *m*, RI-Detektor *m*
refractivity Brechungsvermögen *n*, Refraktionsvermögen *n*
refractometer Refraktometer *n*
refractometric refraktometrisch, Refraktometer...
refractometry Refraktometrie *f*
refractoriness Feuerfestigkeit *f*, Feuerbeständigkeit *f*
refractory feuerfest, ff., feuerbeständig; Schamotte...
refractory feuerfestes Material *n*; feuerfestes Erzeugnis *n* [Produkt *n*], *(i. e. S.)* feuerfester Stein *m*
refractory clay feuerfester Ton *m*, Feuerfestton *m*, Schamotteton *m*
refractory lining feuerfeste Auskleidung *f*
refractory organics *(Hyd)* biologisch [biochemisch] nicht abbaubare organische Inhaltsstoffe *npl*, resistente [abbauresistente, persistente, biochemisch stabile] organische Inhaltsstoffe *npl*
refrigerant kühlend
refrigerant Kältemittel *n*
refrigerate *v* kühlen; sich abkühlen
refrigerating brine Kälteträger *m*, *(i. e. S.)* Kältesole *f*, Kühlsole *f*, Sole *f*
refrigerating machine Kältemaschine *f*
refrigeration Kühlen *n*, Kühlung *f (insbesondere auf Temperaturen unterhalb Atmosphärentemperatur)*; Kälteerzeugung *f*
refrigeration performance Kälteleistung *f*, Kühlvermögen *n*
refrigerator Kältemaschine *f*; Kühlschrank *m*
re-fuse *v (Met)* nochmals [wieder] schmelzen, umschmelzen
refuse 1. Müll *m*; 2. *(Aufb)* Berge *pl*, Abgänge *mpl*
refuse disposal Müllbeseitigung *f*, Müllentsorgung *f*
refuse incineration Müllverbrennung *f*, Müllveraschung *f*
refuse incinerator Müllverbrennungsanlage *f*, MVA, *(i. e. S.)* Müllverbrennungsofen *m*
refuse tip Müllkippe *f*
regenerable regenerierbar *(z. B. Aktivkohle)*
regenerant 1. Wiederbelebungsmittel *n (Adsorption)*; 2. *(Hyd)* Regenerier(ungs)mittel *n*, Regenerationsmittel *n*, Regenerierchemikalie *f (Ionenaustausch)*
regenerant acid Regeneriersäure *f*
regenerant brine Natriumchloridlösung *f* [Kochsalzlösung *f*, Kochsalzsole *f*] zum Regenerieren *(Ionenaustausch)*
regenerant caustic Regenerierlauge *f*
regenerant solution Regenerier(mittel)-lösung *f*
regenerate *v* 1. regenerieren, wiederaufbereiten, wiedergewinnen; 2. regenerieren, wieder beleben *(z. B. Sorbens, Ionenaustauscher, Katalysator, Fixierbad)*
regenerated cellulose regenerierte Cellulose *f*, Regeneratcellulose *f*
regenerated cellulose fibre Regeneratcellulosefaser *f*, Celluloseregeneratfaser *f*; Regeneratcellulosefaserstoff *m*, Celluloseregeneratfaserstoff *m*

regeneration 1. *(Text)* Regenerierung *f*, Wiederaufbereitung *f*, Wiedergewinnung *f*; 2. Regenerierung *f*, Regeneration *f*, Wiederbelebung *f (z. B. eines Sorbens, Ionenaustauschers, Katalysators, Fixierbads)*; 3. Regeneration *f*, Wärmeregeneration *f (z. B. beim Kokereibetrieb)*
regenerative furnace Regenerativofen *m*
regenerator 1. Regenerator *m*, Wärmeregenerator *m*, regenerativer Vorwärmer *m*; 2. Regenerator *m*, Kiln *m*, Katalysatorregenerator *m*, Regenerierofen *m (beim katalytischen Kracken)*
region Bereich *m*, Gebiet *n*, Zone *f*
regioselective *(org Ch)* regioselektiv
regioselectivity *(org Ch)* Regioselektivität *f*
regiospecific *(Biot)* regiospezifisch
regiospecificity *(Biot)* Regiospezifität *f*
regrind *v* wieder mahlen [aufarbeiten] *(z. B. Kunststoffabfälle)*
regular regulär, regelmäßig, normal
regular boiling konstantes Sieden *n*, Siedekonstanz *f*
regulation Vorschrift *f*
regulator Regler *m*, Reglersubstanz *f (bei der Polymerisation)*
regulator gene *(Bioch)* Regulator-Gen *n*
regulatory enzyme Kontrollenzym *n*
regulatory protein Regulatorprotein *n*, Regulationsprotein *n*
regulus *(Met, Lab)* Regulus *m*, Metallkönig *m*
rehydrogenate *v* aufhydrieren
rehydrogenation Aufhydrierung *f*
Reichert-Meissl number [value] Reichert-Meissl-Zahl *f*, reichert-meisslsche Zahl *f*, R-M-Z, RMZ *(von Fetten und Ölen)*
re-ignite *v* nochmals glühen
Reimer-Tiemann reaction [synthesis] Reimer-Tiemann-Synthese *f*, reimer-tiemannsche Synthese *f* [Phenolaldehydsynthese *f*]
reineckate Reineck(e)at *n (vom Reinecke-Salz sich ableitende Verbindung)*
Reinecke salt Reinecke-Salz *n*, Ammoniumtetrathiocyanatodiammin-chromat(III)-1-Wasser *n*
reinforce *v* verstärken; armieren, bewehren
reinforced concrete armierter [bewehrter] Beton *m*, Stahlbeton *m*
reinforced plastic verstärkter Kunststoff *m*
reinforcement 1. Verstärkung *f*; Armierung *f*, Bewehrung *f*; 2. *s.* reinforcing agent
reinforcing agent Verstärkungsmittel *n*, Verstärkungsmaterial *n*, Verstärker *m*
reinforcing filler [ingredient] Verstärkerfüllstoff *m*, verstärkender [verstärkend wirkender aktiver] Füllstoff *m*
re-isolate *v (Anal)* wieder [erneut] isolieren
Reissert compound *(org Ch)* Reissert-Verbindung *f (durch Addition von Cyanid-Ionen an N-Benzoylchinoliniumchlorid erhältliche Verbindung)*
reject *v* 1. verwerfen *(Substanzen)*; 2. *(Nom)* ablehnen, verwerfen
reject Ausschuss *m*
relative atomic mass relative Atommasse *f*
relative biological effectiveness relative biologische Wirksamkeit *f*, RBW *f*
relative configuration relative Konfiguration *f*
relative density relative Dichte *f*, Dichtezahl *f (Verhältnis der Dichte eines Stoffs zur Dichte eines Bezugsstoffs)*; *(bei gleichem Aggregatzustand auch)* Dichteverhältnis *n*
relative group specificity Bindungsspezifität *f*, Reaktionsspezifität *f (eines Enzyms)*
relative humidity relative Feuchte *f* [Feuchtigkeit *f*]
relative retention *(Chromat)* relative Retention *f*
relative volatility relative Flüchtigkeit *f*
relaxation Relaxation *f*
relaxation method Relaxationsmethode *f*, Relaxationstechnik *f*
relaxation time Relaxationszeit *f*, Relaxationsperiode *f*
release *v* 1. freisetzen, entwickeln *(Gase)*; entwickeln, entbinden, freisetzen *(Wärme)*; freisetzen, in Freiheit setzen *(chemische Verbindungen)*; 2. entspannen, die Spannung aufheben; 3. (ab)trennen, ablösen, herausnehmen *(aus Formen)*
release *v* **electrons** Elektronen abgeben

release 1. Freisetzung *f*, Entwicklung *f* *(von Gasen)*; Entwicklung *f*, Entbindung *f*, Freisetzung *f* *(von Wärme)*; Freisetzung *f* *(chemischer Verbindungen)*; 2. Entspannen *n*, Aufheben *n* der Spannung; 3. Trennen *n*, Abtrennen *n*, Ablösen *n*, Herausnehmen *n* *(aus Formen)*; 4. Trennvermögen *n*
releasing hormone Releasinghormon *n*, Freisetzungshormon *n*, *Liberin *n*
relief Ablass *m*, Ablassvorrichtung *f* *(z. B. für Luft)*
relief valve Überdruckventil *n*, Druckbegrenzungsventil *n*, Sicherheitsventil *n*
relieve *v* entlasten
relieve *v* **down** vermindern, herabsetzen *(Druck)*
rem *s.* roentgen equivalent man
REM *s.* reflection electron microscopy
remainder Rest *m*, restliches Material *n*
remaining activity *(Bioch)* Restaktivität *f* *(eines Enzyms)*
remanufacture Wiederaufbereitung *f*, Regeneration *f* *(z. B. von Altpapier)*
remedial measure Gegenmaßnahme *f*, Abhilfemaßnahme *f*
remelt *v* wieder [nochmals] schmelzen, umschmelzen
remnant 1. Rest *m*; 2. *(Bioch)* Restkörper *m* *(aus Chylomikronen hervorgehend)*
remote control Fernsteuerung *f*
remote indication Fernanzeige *f*
remote operation Fernbetätigung *f*, Fernbedienung *f*
removable entfernbar, eliminierbar
removable by filtration abfiltrierbar
removal Entfernen *n*, Entfernung *f*, Beseitigung *f*, Eliminierung *f*, Elimination *f*; Entziehung *f*, Entzug *m* *(z. B. von Wasser)*; Ablassen *n* *(z. B. von Luft)*; Abtrennung *f* *(z. B. einer Säure)*; Abscheidung *f* *(z. B. von Ammoniak)*; Austragen *n*, Austrag *m* *(eines Guts)*
removal efficiency *(Hyd)* Reinigungsleistung *f*, Eliminationsleistung *f*
removal of odour *(Hyd)* Geruchsbeseitigung *f*, Entfernung *f* von Geruchsbeeinträchtigungen
removal of taste *(Hyd)* Geschmacksbeseitigung *f*, Entfernung *f* von Geschmacksbeeinträchtigungen
removal of turbidity *(Hyd)* Trübstoffentfernung *f*, Trübstoffelimination *f*
remove *v* entfernen, beseitigen, eliminieren; entziehen *(z. B. Wasser)*; ablassen *(z. B. Luft)*; abtrennen *(z. B. eine Säure)*; abscheiden *(z. B. Ammoniak)*; austragen *(ein Gut)*
remove *v* **hardness** *(Hyd)* enthärten
remover Entferner *m*
renaturation *(Bioch)* Renaturierung *f* *(denaturierter Proteine)*
renature *v* renaturieren *(Proteine)*
render *v* *(Lebm)* auslassen, (aus-)schmelzen *(zur Fettgewinnung)*
render *v* **passive** *(phys Ch)* passivieren
render *v* **soluble** löslich machen
rendered fat ausgelassenes Fett *n*, Schmalz *n*
rennet *v* (ver)laben, mit Lab versetzen
rennet 1. Lab *n* *(aus zerkleinerten Labmägen gewonnenes Produkt)*; 2. *s.* rennin
renneting Labung *f*, Verlabung *f* *(der Milch)*
rennin Rennin *n*, Lab(ferment) *n*, Chymosin *n*
reoxidation Reoxidation *f*, Wiederoxidation *f*, Rückoxidation *f*, Zurückoxidieren *n*
rep (unit) *s.* roentgen equivalent physical
repair enzyme *(Bioch)* Reparaturenzym *n*
repeat unit 1. *(Krist)* Elementarzelle *f*; 2. *s.* repeating unit
repeatability *(Anal)* Wiederholbarkeit *f*, Wiederholstreubereich *m*
repeated flexural strength Dauerbiegefestigkeit *f*
repeating unit Wiederholungseinheit *f*, sich wiederholende Einheit *f*, (ständig) wiederkehrende Einheit *f*, Repetiereinheit *f*
repel *v* 1. abweisen, abstoßen; 2. vertreiben *(Tiere durch Abschreckmittel)*
repellency 1. Abweisungsvermögen *n*, abweisende Kraft *f* [Eigenschaft *f*], abweisendes Verhalten *n*; 2. Abschreckwirkung *f* *(gegen schädliche Tiere)*
repellent 1. abweisend *(Oberfläche)*; 2. *(Landw)* abschreckend
repellent 1. *(Text)* Abstoßungsmittel *n*, Schutzmittel *n* mit abstoßender Wir-

kung; 2. *(Landw)* Abschreckmittel *n*, Abschreckstoff *m*, Abwehrmittel *n*, Abwehrstoff *m*, Repellentstoff *m*, Repellent *n*
replace *v* ersetzen, substituieren, austauschen
replaceable ersetzbar, substituierbar, austauschbar
replacement Ersetzen *n*, Ersatz *m*, Substitution *f*, Austausch *m*
replacement reaction Substitutionsreaktion *f*, Austauschreaktion *f*
replacement titration Verdrängungstitration *f*
replenish *v* nachfüllen, auffüllen, ergänzen
replenishment Nachfüllung *f*, Auffüllung *f*, Ergänzung *f*
replica Oberflächenabdruck *m*, Abdruck *m (für elektronenmikroskopische Untersuchungen)*
replicate *v* replizieren *(Nucleinsäuren)*
replicate Wiederholung *f (einer Versuchsvariante zur gleichen Zeit)*
replication 1. Re(du)plikation *f (von Nucleinsäuren)*; 2. *s.* replicate
replicative *(Bioch)* replikativ *(DNA)*
Reppe chemistry Reppe-Chemie *f (Zweig der Acetylenchemie)*
Reppe synthesis Reppe-Synthese *f (verschiedener Verbindungen aus Acetylen)*
reprecipitate *v* umfällen; erneut [wieder] ausfallen
reprecipitation Umfällung *f*; erneutes Ausfallen *n*
represent *v* **spatially** räumlich darstellen
repress *v* 1. unterdrücken *(eine Reaktion)*; 2. *(Bioch)* reprimieren
repression *(Bioch)* Repression *f (Rückkopplungshemmung der Enzymbildung)*
repressor-corepressor complex *(Bioch)* Repressor-Corepressor-Komplex *m*
repressor-inducer complex *(Bioch)* Repressor-Induktor-Komplex *m*
reprocess *v* wiederverarbeiten, nachverarbeiten, wiederaufarbeiten
reprocessing Wiederverarbeitung *f*, Nachverarbeitung *f*, Wiederaufarbeitung *f*
reprocessing plant *(Kern)* Wiederaufbereitungsanlage *f*
reproducibility *(Anal)* Reproduzierbarkeit *f*, Vergleichstreubereich *m*
reproducible *(Anal)* reproduzierbar • **poorly reproducible** schlecht [schwer] reproduzierbar
repulp *v* wiederaufschlämmen, aufrühren
repulsion Abstoßung *f*, Repulsion *f*
repulsive 1. *(phys Ch)* abstoßend, Abstoß...; 2. widerlich, abstoßend *(Geruch)*
repulsive force abstoßende Kraft *f*, Abstoßungskraft *f*
requirements Bedarf *m* • **meet the requirements** den Bedarf decken
RER *s.* rough endoplasmatic reticulum
rerinse *v* nachspülen
rerun *v* redestillieren, nochmals [erneut, wiederholt] destillieren, umdestillieren
rerunning Redestillieren *n*, Redestillation *f*, Zweitdestillation *f*, nochmalige [erneute] Destillation *f*
resaturate *v* aufsättigen
resaturation Aufsättigung *f*
rescan *v* 1. nochmals abtasten; 2. *(Anal)* nochmals durchfahren [abfahren] *(einen Messbereich)*
research laboratory Forschungslabor(atorium) *n*
research octane number Research-Octanzahl *f*, ROZ
reservoir Reservoir *n*, Speicher *m*; (offener) Behälter *m*; Tank *m*
reservoir rock Speichergestein *n*, Erdölspeichergestein *n*
resid *(Erdöl)* Rückstand *m*, Residuum *n* • **run resid** Rückstände [Topprückstände] aufarbeiten
residence time Verweilzeit *f*, Aufenthaltszeit *f*, Retentionszeit *f*, Rückhaltezeit *f*, Haltezelt *f*, Standzeit *f*, Durchlaufzeit *f*
residence-time distribution Verweilzeitverteilung *f*
residual biomass Abfallbiomasse *f*
residual chlorine *(Hyd)* Restchlor *n*, *(qualitativ auch)* Restchlorgehalt *m*, Restgehalt *m* an freiem (wirksamem) Chlor, Chlorüberschuss *m*
residual content Restgehalt *m*

residual current *(phys Ch)* Reststrom *m*, Grundstrom *m*
residual deformation bleibende Verformung *f* [Deformation *f*, Gestaltsänderung *f*], Verformungsrest *m*, Formänderungsrest *m*
residual fuel (oil) *(Erdöl)* Rückstandsheizöl *n*
residual hardness *(Hyd)* Resthärte *f*, RH
residual hydrocarbons *(Hyd)* Restkohlenwasserstoffe *mpl*
residual moisture content [level] Restfeuchtegehalt *m*; Restfeuchtebeladung *f*
residual organic content *(Hyd)* Restlast *f* [Restteil *m*] an organischen Inhaltsstoffen, restliche [im Ablauf verbleibende] organische Schmutzstoffe *mpl*, organische Restverschmutzung *f*
residual radiation Reststrahlung *f*
residual solid matter fester Rückstand *m*
residual stress Restspannung *f*
residual sulphur content Restschwefelgehalt *m*
residual toxicity Rückstandstoxizität *f* *(z. B. von Insektiziden)*
residual valence [valency] Restvalenz *f*
residual water Restwasser *n*
residual water content Restwassergehalt *m*
residue 1. Rückstand *m*, Rest *m*; 2. *(Erdöl)* Residuum *n*; 3. Filterkuchen *m*; 4. Rest *m*, Gruppe *f*, Molekülrest *m*
residue cracking Kracken *n* auf flüssigen Rückstand, Kracken *n* mit Rückstandsarbeitsweise, Kracken *n* nach der Entspannungsfahrweise
residue-life 50 per cent *(Tox)* Rückstands-Halbwertszeit *f*, RL_{50}
residue on ignition Glührückstand *m*
residue-persistence curve *(Tox)* Abbaukurve *f*
residue tolerance *(Tox)* duldbare Rückstandsmenge *f* [Restmenge *f*]
residuum *(Erdöl)* Rückstand *m*, Residuum *n*
resilience Zurückschnellen *n*, Zurückfedern *n*, Zurückspringen *n*; *(quantitativ:)* Verformungsarbeit *f*, elastischer Wirkungsgrad *m*, *(i. e. S.)* elastischer Wirkungsgrad *m* bei Stoß-Druck-Beanspruchung, Rückprallelastizität *f*, Stoßelastizität *f*
resin *v (Kst)* beharzen, mit Harz tränken
resin Harz *n (natürlich oder synthetisch)*
resin acid Harzsäure *f*, Resinosäure *f*
resin bead Harzkorn *n*, Harzkügelchen *n (Ionenaustausch)*
resin soap Harzseife *f (Salz einer Harzsäure)*
resin-to-oil ratio Harz-Öl-Verhältnis *n*, Verhältnis *n* von Harz zu Öl
resinate Resinat *n (Harzseife oder Harzester)*
resinify *v* verharzen
resinoid harzartig
resinoid 1. harzartiger Bestandteil *m*, Gummiharz *n*; 2. *(Kosmet)* Resinoid *n (alkoholischer Extrakt aus Harzen und anderen Drogen)*; 3. Duroplast *m*, hitzehärtbarer [wärmehärtbarer] Kunststoff *m*
resinous harz(halt)ig, harzreich *(z. B. Holz)*; harzartig
resist 1. Abdeckung *f*, (isolierende) Schutzabdeckung *f*, Schutzschicht *f*, Isolation *f (Galvanotechnik)*; 2. Fotoresist *m*, Resist *m*, Fotolack *m*, Kopierlack *m (Fotolithographie)*
resistance 1. Widerstand *m*, Beständigkeit *f*, Stabilität *f*, Festigkeit *f*, Widerstandsfähigkeit *f*, Resistenz *f*; 2. (elektrischer) Widerstand *m*, Widerstandswert *m*, Ohmwert *m*
resistance bridge elektrische Widerstandsbrücke *f*
resistance furnace (elektrischer) Widerstandsofen *m*, widerstandsbeheizter Ofen *m*, Ofen *m* mit Widerstandserhitzung
resistance glass Hartglas *n*
resistance thermometer Widerstandsthermometer *n*
resistance to abrasion 1. Abriebwiderstand *m*, Abriebbeständigkeit *f*, Abriebfestigkeit *f*; 2. *(Text)* Scheuerbeständigkeit *f*, Scheuerfestigkeit *f*
resistance to acid(s) Säurebeständigkeit *f*, Säurefestigkeit *f*, Säureresistenz *f*
resistance to alkali(es) Alkalibeständigkeit *f*, Laugenbeständigkeit *f*, Laugenfestigkeit *f*

resistance to antibiotics *(Biot)* Antibiotikaresistenz *f*
resistance to chemicals Chemikalienbeständigkeit *f*, Chemikalienfestigkeit *f*, Widerstandsfähigkeit *f* [Resistenz *f*] gegen Chemikalien
resistance to corrosion Korrosionsbeständigkeit *f*, Korrosionsfestigkeit *f*; *(quantitativ:)* Korrosionswiderstand *m*, Korrosionsbeständigkeit *f*
resistance to fluid flow Strömungswiderstand *m*
resistance to friction Reibungswiderstand *m*
resistance to grease Fettbeständigkeit *f*
resistance to high temperature(s) Hochtemperaturbeständigkeit *f*, Hochhitzebeständigkeit *f*, Beständigkeit *f* [Widerstandsfähigkeit *f*] gegen hohe Temperaturen
resistance to hydrolysis Hydrolysebeständigkeit *f*
resistance to light Lichtbeständigkeit *f*, Lichtechtheit *f*
resistance to oil Ölfestigkeit *f*, Ölbeständigkeit *f*
resistance to oxidation Oxidationsbeständigkeit *f*, Beständigkeit *f* gegen oxidative Einflüsse
resistance to shock Stoßfestigkeit *f*
resistance to swelling Quellbeständigkeit *f*
resistance to wear Verschleißfestigkeit *f*, Abnutzungsbeständigkeit *f*
resistance to weathering (agencies) Witterungsbeständigkeit *f*, Wetterbeständigkeit *f*, Wetterfestigkeit *f*
resistant beständig, fest, widerstandsfähig, resistent
resistant mutant *(Biot)* Resistenzmutante *f*
resistant to abrasion 1. abriebbeständig, abriebfest; 2. *(Text)* scheuerbeständig, scheuerfest
resistant to acid(s) säurebeständig, beständig gegen Säuren, säurefest
resistant to alkali(es) alkalibeständig, laugenbeständig, laugenfest
resistant to biological degradation *(Hyd)* abbauresistent, biologisch nicht abbaubar, biologisch stabil
resistant to chemicals chemikalienbeständig, chemikalienfest, chemikalienresistent
resistant to corrosion korrosionsbeständig, korrosionsfest
resistant to grease fettbeständig
resistant to high temperature(s) hochtemperaturbeständig, hochhitzebeständig
resistant to hydrolysis hydrolysebeständig, hydrolyseresistent
resistant to oil ölfest, ölbeständig
resistant to oxidation oxidationsbeständig, beständig gegen oxidative Einflüsse
resistant to wear 1. verschleißfest, abnutzungsbeständig; 2. *(Text)* scheuerfest
resistant to weathering witterungsbeständig, wetterbeständig, wetterfest
resistivity (spezifischer) Widerstand *m*
resistor Widerstand *m (als Bauelement)*; Widerstandskörper *m*, Heizwiderstand *m*
resite Resit *n*, Phenolharz *n* im C-Zustand, *(i. e. S.)* C-Harz *n*
resitol Resitol *n*, Phenolharz *n* im B-Zustand, B-Harz *n (i. e. S.)*
reslurry *v* wiederaufschlämmen, aufrühren
resol Resol *n*, Phenolharz *n* im A-Zustand, *(i. e. S.)* A-Harz *n*
resolidification Wiederverfestigung *f*
resoluble colloid reversibles [wiederauflösbares] Kolloid *n*
resolution 1. *(org Ch)* Trennung *f*, Spaltung *f*; Racematspaltung *f*, Aufspaltung *f (von Racematen)*; 2. Auflösung *f*, Auflösungsvermögen *n (Optik, Analytik)*
resolvability 1. Trennbarkeit *f*, Spaltbarkeit *f (von Racematen)*; 2. Auflösbarkeit *f (Optik, Analytik)*
resolvable 1. trennbar, (auf)spaltbar *(Racemate)*; 2. auflösbar *(Optik, Analytik)*
resolve *v* 1. trennen, (auf)spalten *(Racemate)*; 2. auflösen *(Optik, Analytik)*
resolving power Auflösungsvermögen *n*, Auflösung *f (Optik, Analytik)*
resonance 1. Resonanz *f*; 2. *(org Ch)* Mesomerie *f*, Resonanz *f*, Strukturresonanz *f*

resonance absorption *(Kern)* Resonanzabsorption *f*
resonance effect *(org Ch)* Mesomerieeffekt *m*, M-Effekt *m*, mesomerer Effekt *m*
resonance energy *(org Ch)* Mesomerieenergie *f*, Resonanzenergie *f*, Delokalisierungsenergie *f*, Sonderenergie *f*
resonance fluorescence Resonanzfluoreszenz *f*, Resonanzstrahlung *f*
resonance frequency Resonanzfrequenz *f*
resonance hybrid *(org Ch)* mesomere Verbindung *f*, Resonanzhybrid *n*
resonance-ionization spectroscopy Resonanzionisationsspektroskopie *f*
resonance line *(Spektr)* Resonanzlinie *f*
resonance neutron Resonanzneutron *n*
resonance peak *(Spektr)* Resonanzsignal *n (im Diagramm)*
resonance radiation Resonanzstrahlung *f*
resonance Raman effect *(Spektr)* Resonanz-Raman-Effekt *m*
resonance Raman scattering Resonanz-Raman-Streuung *f*
resonance Raman spectroscopy Resonanz-Raman-Spektroskopie *f*
resonance scattering Resonanzstreuung *f*, anomale elastische Streuung *f*
resonance signal *(Spektr)* Resonanzsignal *n*
resonance spectrum Resonanzspektrum *n*
resonance stabilization *(org Ch)* Mesomeriestabilisierung *f*, Resonanzstabilisierung *f*
resonance-stabilized *(org Ch)* mesomeriestabilisiert, resonanzstabilisiert
resonance state mesomerer Zustand *m*, Resonanzzustand *m*
resonance structure *(org Ch)* mesomere Grenzstruktur *f*, Resonanzstruktur *f*
resonance transition *(Spektr)* Resonanzübergang *m*
resonant *(Spektr)* resonanzaktiv
resorb *v* resorbieren, aufnehmen, einsaugen, aufsaugen; resorbiert werden
resorcin(ol) Resorcinol *n*, *Benzen-1,3-diol *n*
resorcin(ol)-formaldehyde resin Resorcinharz *n*
resorption Resorption *f*, Aufnahme *f*, Einsaugung *f*, Aufsaugung *f*
respiration Respiration *f*, Atmung *f*
respiration inhibitor Atmungsinhibitor *m*, Atmungshemmstoff *m*
respirator Atemschutzmaske *f*
respiratory respiratorisch, Atmungs..., Atem...
respiratory chain *(Bioch)* Atmungskette *f*
respiratory-chain phosphorylation *(Bioch)* Atmungskettenphosphorylierung *f*, oxidative Phosphorylierung *f*
respiratory enzyme Atmungsferment *n*
respiratory irritant Lungenreizstoff *m*
respiratory pigment Atmungspigment *n*, Sauerstofftransportpigment *n*, respiratorisches Pigment *n*
respiratory poison Atemgift *n*
respiratory protection Atemschutz *m*
respiratory quotient [ratio] *(Bioch)* respiratorischer Quotient *m* [Koeffizient *m*], Atmungsquotient *m*, Atmungskoeffizient *m*, Respirationsquotient *m*, RQ
respond *v* 1. reagieren *(auf äußere Einflüsse)*; 2. *(Anal)* ansprechen *(Messgerät)*
response 1. *(Bioch, Pharm)* Wirkung *f (Auswirkung)*; Reaktion *f*; 2. *(Anal)* Ansprechen *n (eines Geräts)*; 3. *s.* response time
response time *(Anal)* Ansprechdauer *f*, Ansprechzeit *f (von Messgeräten)*
rest mass Ruh(e)masse *f*
resting phase *(Biot)* stationäre Phase *f (Mikroorganismenwachstum)*
restore *v* wiederherstellen
restoring force rücktreibende Kraft *f*, Rückstellkraft *f*
restricted diffusion behinderte Diffusion *f*
restricted rotation *(Spektr)* behinderte Rotation *f*
restriction 1. Drosselung *f*; 2. Drossel(-stelle) *f*, Verengung *f*, Drosselwiderstand *m*, Widerstand *m (gegen Strömung)*
restriction endonuclease [enzyme] *(Bioch)* Restriktionsendo(desoxy)ribonuclease *f*, Restriktionsenzym *n*

resultant conductance Gesamtleitfähigkeit *f*, Gesamtleitvermögen *n*
resuspend *v* wieder suspendieren *(bereits abgesetzte Feststoffteilchen)*
resuspension erneute Suspension *f* *(bereits abgesetzter Feststoffteilchen)*
ret *v* rösten *(z. B. Flachs)*
ret Röste *f*, *(i. e. S.)* Flachsröste *f*
retard *v* verzögern, verlangsamen
retardation Verzögerung *f*, Verlangsamung *f*
retarder 1. Verzögerungsmittel *n*, Verzögerer *m*, Hemmstoff *m*, Inhibitor *m*, Bremsmittel *n*, Retardiermittel *n*, Retarder *m*; 2. *(Bioch)* Retardans *n*, synthetischer Hemmstoff *m*; 3. negativer Katalysator *m*, Antikatalysator *m*, Passivator *m*
retention Beibehaltung *f*, Erhaltung *f*; Zurückhaltung *f*, Retention *f*
retention index *(Chromat)* Retentionsindex *m*
retention period (time) Verweilzeit *f*, Aufenthaltszeit *f*, Retentionszeit *f*, Rückhaltezeit *f*, Haltezeit *f*, Standzeit *f*, Durchlaufzeit *f*
retention volume *(Chromat)* *Retentionsvolumen *n*, Rückhaltevolumen *n*
retinoid *(org Ch)* Retinoid *n*
retort *v* in der Retorte erhitzen
retort Retorte *f*
retort furnace Retortenofen *m*, Destillierofen *m*
retort stand *(Lab)* Stativ *n*
re-treat *v* nochmals [wiederholt] behandeln, nachbehandeln
re-treatment nochmalige [wiederholte] Behandlung *f*, Nachbehandeln *n*, Nachbehandlung *f*
retroaddition Cycloeliminierung *f*
retrocycloaddition Cycloeliminierung *f*
retro-ene reaction *(org Ch)* Retro-En-Reaktion *f*
retroinhibition *(Bioch)* Endprodukthemmung *f*, Rückkopplungshemmung *f*, negative Rückkopplung *f*, negatives Feed-back *n*
retropinacolin rearrangement Retropinacolinumlagerung *f*
retrosynthesis *(org Ch)* Retrosynthese *f*
retting Rösten *n*, Röste *f*, *(i. e. S.)* Flachsröste *f*, Flachsrotte *f*
return *v* zurückführen, wieder zuführen; zurückfüllen; zurückfließen
return *v* **to the circuit** in den Kreislauf zurückführen, dem Kreislauf wieder zuführen [zusetzen], rezirkulieren
return 1. Rückführung *f*, Rückleitung *f*; 2. Rücklauf *m (Material)*; 3. *s.* return pipe
return flow Rückfluss *m*
return pipe Rücklaufrohr *n*, Rückleitung *f*
return water Rücklaufwasser *n*
reusable wiederverwendbar
reuse *v* wiederverwenden, von neuem verwenden
reuse Wiederverwendung *f*
reverberatory furnace [kiln] Flammofen *m*, Flammenofen *m*
reversal Umstellen *n*, Umstellung *f*, Umkehr(ung) *f*, Wechsel *m (z. B. der Arbeitsrichtung)*
reversal of phases Phasenumkehr *f*
reverse *v* umstellen, umkehren, wechseln *(z. B. die Arbeitsrichtung)*
reverse micelle inverse Mizelle *f*
reverse osmosis Umkehrosmose *f*, Reversosmose *f*, umgekehrte Osmose *f*
reverse-phase ... *siehe: reversed-phase ...*
reverse rate constant Geschwindigkeitskonstante *f* der Rückreaktion
reverse reaction Rückreaktion *f*
reverse reaction rate Geschwindigkeit *f* der Rückreaktion
reverse transcriptase *(Bioch)* Revertase *f*, RNA-abhängige DNA-Polymerase *f*
reversed-phase chromatography *Umkehrphasenchromatographie *f*, Chromatographie *f* an Umkehrphasen
reversed phases Umkehrphasen *fpl*, umgekehrte [vertauschte] Phasen *fpl*, unpolare stationäre Phasen *fpl*
reversibility Reversibilität *f*, Umkehrbarkeit *f*
reversible reversibel, umkehrbar, in beiden Richtungen verlaufend
reversible addition fragmentation chain transfer polymerization Polymerisation *f* durch degenerativen Kettentransfer
reversible colloid reversibles [wiederauflösbares] Kolloid *n*
reversible denaturation *(Bioch)* reversible Denaturierung *f*

reversible hydrogen electrode reversible Wasserstoffelektrode *f*
reversible isothermal work of expansion reversible isotherme [technische] Volumenarbeit *f*
reversible reaction reversible [umkehrbare] Reaktion *f*
reversible transition reversible Umwandlung *f* [Phasenumwandlung *f*]
reversibly convertible ineinander [wechselseitig] umwandelbar
reversion 1. Umwandlung *f*; 2. Umschlagen *n (einer Emulsion)*; 3. *(Lebm, Gum)* Reversion *f*; 4. *(Lebm)* Fischigwerden *n (von Fetten)*
revert *v* 1. sich umwandeln; 2. umschlagen *(Emulsionen)*; 3. *(Lebm, Gum)* der Reversion unterliegen, Reversion erleiden; 4. *(Lebm)* fischig werden *(Fette)*
revivification Regenerierung *f*; Wiederbelebung *f*, Reaktivierung *f (von Aktivkohle)*
revivify *v* regenerieren; wiederbeleben, reaktivieren *(Aktivkohle)*
revolving filter Drehfilter *n*, Trommelfilter *n*
revolving grate Drehrost *m*
revolving screen Wälzsieb *n*, Trommelsieb *n*, Rundsieb *n*
reweigh *v* nochmals [erneut] wägen
rework *v* (wieder) aufarbeiten
Reynolds number *(phys Ch)* Reynolds-Zahl *f*, reynoldssche Zahl *f*
RF *(releasing factor) s.* releasing hormone
Rf value Rf-Wert *m (Papier- und Dünnschichtchromatographie)*
R_F value *(Chromat)* R_F-Wert *m*, Retentionsfaktor *m*
R.H. *s.* 1. red heat; 2. relative humidity; 3. Rockwell hardness
rh *s.* rh factor
rh factor *(Bioch)* Rhesus-Faktor *m*, Rh-Faktor *m*
rh-negative Rh-negativ
rh-positive Rh-positiv
RHE *s.* reversible hydrogen electrode
RHEED *s.* reflection high-energy electron diffraction
rhenate $M^I_2ReO_4$ Rhenat(VI) *n*
rhenic acid Perrheniumsäure *f*, Rhenium(VII)-säure *f*
rhenite $M^I_2ReO_3$ Rhenat(IV) *n*
rhenium Re Rhenium *n*
rhenium heptoxide Rheniumheptoxid *n*, Dirheniumheptoxid *n*, Rhenium(VII)-oxid *n*
rhenium hexachloride Rheniumhexachlorid *n*, Rhenium(VI)-chlorid *n*
rheological behaviour Fließverhalten *n*, rheologisches Verhalten *n*
rheology Rheologie *f*, Fließkunde *f*
rheopectic behaviour rheopexes Verhalten *n*
rheopexy *(Koll)* Rheopexie *f*
rhesus factor *(Bioch)* Rhesus-Faktor *m*, Rh-Faktor *m*
rho-sigma correlation *(org Ch)* Hammett-Beziehung *f*, Hammett-Korrelation *f*
rhodamine *(org Ch)* Rhodamin *n*
rhodanate M^ISCN Thiocyanat *n*; *(veraltet:)* Rhodanid *n*
rhodanic acid Rhodanwasserstoffsäure *f*, Thiocyansäure *f*
rhodanide *s.* rhodanate
rhodanize *v* rhodinieren, mit Rhodium beschichten
rhodate Rhodinat *n*
rhodium Rh Rhodium *n*
rhodium black Rhodiummohr *n*, Rhodiumschwarz *n (fein verteiltes Rhodium)*
rhodium plating Rhodinieren *n (Galvanotechnik, Vakuumbedampfung)*
rhodium sesquioxide Dirhodiumtrioxid *n*, Rhodium(III)-oxid *n*
rhodium sulphate Rhodium(III)-sulfat *n*
rhodium trichloride Rhodiumtrichlorid *n*, Rhodium(III)-chlorid *n*
rhodopsin Rhodopsin *n*, Sehpurpur *m*
rhomb. *s.* rhombic
rhombic *(Krist)* (ortho)rhombisch
rhombic crystal system rhombisches System *n*
rhombic sulphur rhombischer Schwefel *m*, α-Schwefel *m*
rhombohedral *(Krist)* rhomboedrisch
rhombohedron *(Krist)* Rhomboeder *n*
rhythmic precipitation rhythmische Fällung *f*
RI detector *(Anal)* Brechungsindexdetektor *m*, RI-Detektor *m*
RIA *s.* radioimmunoassay
Rib *s.* ribose
ribbed gerippt, gerillt, geriffelt
riboflavin(e) Riboflavin *n*, Vitamin *n* B_2

ribonucleate Ribonucleat *n (Salz der Ribonucleinsäure)*
ribonucleic acid *(Bioch)* Ribonucleinsäure *f*, RNS *f*, *(jetzt meist)* RNA *f*
ribonucleoprotein Ribonucleoproteid *n*
ribonucleotide *(Bioch)* Ribonucleotid *n*
ribose Ribose *f (zu den Pentosen gehörendes Monosaccharid)*
ribosomal RNA ribosomale RNA *f* [Ribonucleinsäure *f*], rRNA *f*
ribosome *(Bioch)* Ribosom *n*
ribosyl group Ribosylgruppe *f*, Ribosylrest *m*
ribulose-1,5-bisphosphate carboxylase *(Bioch)* Ribulose-1,5-diphosphatcarboxylase *f*
ribulose diphosphate cycle *(Bioch)* Ribulose-Diphosphat-Zyklus *m*, Calvin-Zyklus *m*, Photosynthesezyklus *m*, reduktiver Pentosephosphatzyklus *m*
ribulose monophosphate cycle *(Bioch)* Ribulose-Monophosphat-Zyklus *m*, Quayle-Zyklus *m*
RIC = Royal Institute of Chemistry
rice starch Reisstärke *f*
rich concrete fetter Beton *m*
rich gas Reichgas *n*, reiches Gas *n (spezifischer Brennwert 37465 bis 35617 kJ/m^3)*; Starkgas *n (spezifischer Brennwert 15085 bis 23045 kJ/m^3)*
rich in carbon kohlenstoffreich
rich in oxygen sauerstoffreich
rich oil reiches [beladenes] Waschöl *n*, fettes Öl *n (beim Absorptionsverfahren)*
ricinoleate Ricinoleat *n*
ricinol(e)ic acid Ricinolsäure *f*, Ricinusölsäure *f*, *cis*(-)-12-Hydroxyoctadec-9-ensäure *f*
rider *(Lab)* Reiter *m*, Reiterwägestück *n*
riffle Rille *f*, Riffel *f*, Leiste *f*
right(-hand) circularly polarized rechtszirkular polarisiert
right-handed 1. *(Krist)* Rechts...; 2. rechtsdrehend, (+)-drehend *(optische Aktivität)*
right-handed helix rechtsgängige [rechtsdrehende] Helix *f*
right-handed sense Rechtsgängigkeit *f* *(einer Helix)*
rigid starr, steif; fest, stabil; hart
rigid foam 1. fester Schaum *m*; 2. harter Schaumstoff *m*, Hartschaum(stoff) *m*
rigid PVC Hart-PVC *n*, PVC-hart *n*
rigidity Starrheit *f*, Steifigkeit *f*, Steifheit *f*; Festigkeit *f*, Stabilität *f*; Härte *f*
rimmed [rimming] steel unberuhigter [unberuhigt vergossener, unruhig vergossener] Stahl *m*
rind Rinde *f*, Schale *f*, Kruste *f*
ring Ring *m*
ring burner *(Lab)* Ringbrenner *m*, Gasheizkranz *m*
ring-carbon atom Ringkohlenstoffatom *n*
ring-chain tautomerism Ring-Ketten-Tautomerie *f*, cyclisch-offene Tautomerie *f*
ring cleavage *s.* ring opening
ring closing metathesis Ringschlussmetathese *f*
ring closure Ringschluss *m*, Cyclisierung *f*
ring compound Ringverbindung *f*, ringförmige [cyclische] Verbindung *f*
ring contraction Ringverengung *f*
ring enlargement [expansion] Ringerweiterung *f*
ring fission *s.* ring opening
ring-halogenated kernhalogeniert
ring member Ringglied *n*
ring-methyl group ringständige Methylgruppe *f*
ring opening Ringöffnung *f*, Ringspaltung *f*, Ring(auf)sprengung *f*, Decyclisierung *f*
ring-opening metathesis polymerization Polymerisation *f* durch Ringöffnung, Ringöffnungsmetathese
ring-opening polymerization Ringöffnungspolymerisation *f*, ringöffnende Polymerisation *f*
ring-opening reaction Ringöffnungsreaktion *f*
ring-roll mill Ringrollenmühle *f (i. w. S.)*; Ringwalzenmühle *(i. w. S.)*; Fremdkraftwälzmühle *f*, Fremdkraftrollenmühle *f*, Fremdkraftringmühle *f*
ring silicate *(Min)* Cyclosilicat *n*
ring stand *(Lab)* Stativ *n*
ring strain Ringspannung *f*
ring synthesis Ringsynthese *f*
Rinman(n)'s green Rinmans Grün *n*, Kobaltgrün *n*
rinse *v* (ab)spülen; ausspülen *(Gefäße)*
rinse *v* **away** *s.* rinse off

rinse *v* **downflow** *(Hyd)* im Abwärtsstrom auswaschen *(Ionenaustausch)*
rinse *v* **off** abspülen, wegspülen
rinse *v* **out** ausspülen
rinse 1. Spülen *n*, Abspülen *n*; Ausspülen *n (von Gefäßen)*; 2. Spülmittel *n*, Nachspülmittel *n*
rinse water Spülwasser *n*
rinsing 1. Spülen *n*, Abspülen *n*; Ausspülen *n (von Gefäßen)*; 2. Spülflüssigkeit *f*
rinsing agent Spülmittel *n*
rinsings Spülflüssigkeit *f*
riot control agent Polizeikampfmittel *n*
ripen *v* reifen (lassen); reifen, reif werden
ripening agent Reifebeschleuniger *m*, Reifungsbeschleuniger *m (meist Ethylen)*
RIS *s.* resonance-ionization spectroscopy
rise *v* sich erhöhen, (an)steigen, zunehmen; aufsteigen *(Gase)*
rise in [of] temperature Temperaturanstieg *m*, Temperaturerhöhung *f*
riser (pipe, tube) 1. Steigrohr *n*; 2. *(Dest)* Kamin *m*, Dampfkamin *m*, Kaminstummel *m*, Dämpfestutzen *m*, Dampfhals *m (einer Bodenkolonne)*
rising-film evaporator Kletterfilmverdampfer *m*
risk of contamination 1. Kontaminationsgefahr *f*; 2. *(Biot)* Infektionsgefahr *f*
Ritz-Rydberg combination principle *(phys Ch)* ritz-rydbergsches Kombinationsprinzip *n*
river pollution Flussverschmutzung *f*, Flussverunreinigung *f*
RL_{50} *s.* residue-life 50 per cent
RMS speed *s.* root-mean-square speed
RNA *(Bioch)* RNA *f*, RNS *f*, Ribonucleinsäure *f*
RNA-directed DNA polymerase *(Bioch)* RNA-abhängige DNA-Polymerase *f*, Revertase *f*, Umkehrtranskriptase *f*
RNA-directed RNA polymerase RNA-abhängige RNA-Polymerase *f*, RNA-Synthetase *f*
RNA replicase *s.* RNA-directed RNA polymerase
RNA splicing *(Bioch)* Spleißen *n* der RNA
RO *s.* reverse osmosis
road octane number Straßen-Octanzahl *f*, SOZ
road salt Tausalz *n*, Auftausalz *n*
roaring flame rauschende Flamme *f*
roast *v* 1. *(Met)* (ab)rösten *(Erze)*; 2. *(Lebm)* rösten
roast-reaction process *(Met)* Röstreaktionsverfahren *n*
roast-reduction process Röstreduktionsverfahren *n*
roasted product Röstprodukt *n*
roasted pyrites Kiesabbrand *m*
roaster Röstofen *m*
roasting furnace Röstofen *m*
roasting hearth Röstherd *m*
roasting of pyrites Kiesabröstung *f*, *(i. e. S.)* Pyritröstung *f*
roborant *(Pharm)* Kräftigungsmittel *n*, stärkendes Mittel *n*, Roborans *n*
Rochelle salt Seignettesalz *n*, Rochellesalz *n*, Kaliumnatriumtartrat *n*
rock *v* schütteln; wiegen; schaukeln
rock Gestein *n*
rock asphalt Naturasphalt *m*
rock crystal 1. *(Min)* Bergkristall *m (Quarzvarietät)*; 2. Kristallglas *n*
rock fabric *(Geol)* Gefüge *n*
rock-forming mineral gesteinsbildendes Mineral *n*
rock salt Steinsalz *n*
rock-salt lattice *(Krist)* Steinsalzgitter *n*, Natriumchloridgitter *n*
rock wool Steinwolle *f*, Gesteinswolle *f*, Mineralwolle *f*
rocket fuel [propellant] Raketentreibstoff *m*
rocking vibration *(Spektr)* Schaukelschwingung *f*, Rocking-Schwingung *f*, ρ-Schwingung *f*
Rockwell hardness Rockwellhärte *f*, HR
rod Stange *f*, Stab *m*
rod-like molecule stabförmiges [stäbchenförmiges] Molekül *n*
rod mill Stabmühle *f*
rod screen Stangenrostsieb *n*, Stabrostsieb *n*, Rostsieb *n*
rodenticidal rodentizid; Rodentizid...
rodenticide Rodentizid *n*, Nagetiergift *n*
roentgen Röntgen-Einheit *f*, Röntgen *n*, *R (SI-fremde Einheit der Exposition oder der Ionendosis einer ionisierenden Strahlung)*
roentgen equivalent man Roentgen-equivalent-man *n*, Rem-Einheit *f*,

Rem *n*, rem *(biologisches Röntgenäquivalent)*
roentgen equivalent physical Roentgen-equivalent-physical *n*, Rep-Einheit *f*, Rep *n*, rep *(physikalisches Röntgenäquivalent)*
roentgen rays Röntgenstrahlen *mpl*
roll *v* rollen; walzen
roll *v* **into a plate** auswalzen
roll-mill *v* in der Walzenmühle mahlen
roll Walze *f*, Rolle *f*; Walzenpresse *f*, Walzenmaschine *f*
roll crusher Walzenbrecher *m*
roll doctor Schaber *m*, Abnahmeschaber *m (an einer Walze)*
roll nip Walzenspalt *m*
roll scale Walz(en)zunder *m*, Walzsinter *m*
roller Walze *f*, Rolle *f*; Zylinder *m*
roller bearing Rollenlager *n*, Walzenlager *n*
roller dryer Walzentrockner *m*; Kontaktbahnentrockner *m*, Zylindertrockner *m*, Kurzschleifentrockner *m*
roller mill 1. Walzenmühle *f*, Walzenstuhl *m*; 2. Wälzmühle *f*, Rollmühle *f*, *(i. w. S.)* Ringmühle *f*; 3. *s.* ring-roll mill
roller press Walzenpresse *f*
rolling mill Walzwerk *n*
ROMP *s.* ring-opening metathesis polymerization
RON *s.* research octane number
röntgen *s.* roentgen
room temperature Zimmertemperatur *f*, ZT, Raumtemperatur *f*
root-mean-square error mittlere quadratische Abweichung *f*, Standardabweichung *f*, mittlerer quadratischer Fehler *m*
root-mean-square speed [velocity] mittlere quadratische Geschwindigkeit *f*, Wurzel *f* aus dem mittleren Geschwindigkeitsquadrat
rope dyeing *(Text)* Strangfärben *n*, Färben *n* im Strang, Färben *n* in Strangform
ropiness Zähflüssigkeit *f*, Dickflüssigkeit *f*
ropy zähflüssig, dickflüssig, fadenziehend
rosaniline dye Rosanilinfarbstoff *m*, Fuchsinfarbstoff *m*
rose oil Rosenöl *n*
Rose crucible Schmelztiegel *m* nach Rose, Rose-Tiegel *m*
Rosenmund reaction [reduction] Rosenmund-Reduktion *f*, Rosenmund-Zajcev-Reduktion *f*, Rosenmund-Zajcevs Säurechloridreduktion *f*
Rose's alloy [metal] Roses [rosesches] Metall *n*, Legierung *f* nach Rose
rosin Kolophonium *n*, Terpentinharz *n*, Geigenharz *n*
rosin acid Harzsäure *f*
rosin oil Terpentinharzöl *n*, Harzöl *n (aus trockener Destillation von Kolophonium)*
rosolic acid 1. Rosolsäure *f (Methylderivat des Aurins)*; 2. Aurin *n*
rot *v* 1. verrotten; 2. *(Geol)* verwittern; 3. rösten *(z. B. Flachs)*
rot resistance Verrottungsbeständigkeit *f*, Verrottungsfestigkeit *f*, Fäulnisbeständigkeit *f*, Fäulnisfestigkeit *f*
rot-resistant verrottungsbeständig, verrottungsfest; fäulnisbeständig, fäulnisfest
rotameter Schwebekörper(-Durchfluss)messer *m*, Rotameter *n*
rotary barrel mixer *(Tech)* Mischtrommel *f*
rotary blower Rotationsgebläse *n*, Umlaufkolbengebläse *n*
rotary column *s.* rotary still
rotary compressor Rotationsverdichter *m*, Rotationskompressor *m*, rotierender Verdichter *m*
rotary crusher Glockenmühle *f*, Kegelmühle *f*
rotary-cup atomizer Becherversprüher *m*
rotary-cup (oil) burner Dreh(zerstäuber)brenner *m*, Drehölbrenner *m*, Rotations(zerstäuber)brenner *m*, Rotationsölbrenner *m (einer Ölfeuerungsanlage)*
rotary-disk atomizer Scheibenversprüher *m*; Scheibenzerstäuber *m (für Feststoffe)*
rotary-disk contactor [extractor] Drehscheibenextraktor *m*, Drehscheibenextrakteur *m*, Drehscheibenkolonne *f*
rotary-disk filter Scheibenfilter *n*
rotary-drum filter Trommelfilter *n*
rotary-drum vacuum filter Vakuum-

trommelfilter *n*, Trommelsaugfilter *n*, *(meist)* Vakuumtrommelzellenfilter *n*, Saugzellenfilter *n*
rotary dryer Drehtrommeltrockner *m*, Trommeltrockner *m*
rotary (film) evaporator Rotationsverdampfer *m*
rotary filter Drehfilter *n*, Rotationsfilter *n*
rotary kiln Drehofen *m*, rotierender Ofen *m*, Reaktionsdrehofen *m*
rotary mixer *(Tech)* Drehmischer *m*
rotary muffle kiln Muffeldrehrohrofen *m*
rotary polarization Rotationspolarisation *f*, optische Aktivität *f*, optisches Dreh(ungs)vermögen *n*
rotary pump Rotationspumpe *f*, rotierende Pumpe *f*, Umlaufkolbenpumpe *f*, Kolbenpumpe *f* mit rotierendem Kolben
rotary still Rotationskolonne *f*, Destillierapparat *m* mit rotierender Verdampferfläche
rotary vacuum filter Vakuumdrehfilter *n*, Vakuumrotationsfilter *n*
rotary washer Rotationswäscher *m*, Rotationsnassabscheider *m*, rotierender [umlaufender] Wäscher *m*
rotatability Drehbarkeit *f*
rotatable drehbar
rotate *v* rotieren, sich drehen, umlaufen; drehen
rotated Fischer projection gedrehte Fischer-Projektion *f (zur Abbildung von Kettenmolekülen)*
rotating-crystal method Drehkristallverfahren *n*, Drehkristallmethode *f*
rotating-crystal photograph Drehkristallaufnahme *f*, Drehkristalldiagramm *n*
rotating-cylinder viscometer of Couette Couette-Viskosimeter *n*, Couette-Apparat *m*, Rotationsviskosimeter *n* nach Couette
rotating-disk electrode rotierende Scheibenelektrode *f*
rotating kiln *s.* rotary kiln
rotation Rotation *f*, Drehung *f*, Umlaufen *n*; Umdrehung *f*
rotation spectrum Rotationsspektrum *n*
rotation-vibration spectrum Rotations-Schwingungs-Spektrum *n*
rotational angle Rotationswinkel *m*, Torsionswinkel *m*, Konformationswinkel *m*, Diederwinkel *m (Stereochemie)*
rotational degree of freedom Rotationsfreiheitsgrad *m*
rotational energy Rotationsenergie *f*
rotational energy level *(Spektr)* Rotationsniveau *n*
rotational excitation *(Spektr)* Rotationsanregung *f*
rotational isomer Rotationsisomer(e) *n*
rotational isomerism Rotationsisomerie *f*, Konformationsisomerie *f*, Konstellationsisomerie *f*
rotational partition function Rotationsverteilungsfunktion *f*, Verteilungsfunktion *f* der Rotationsenergie
rotational quantum number Drehimpulsquantenzahl *f*, Rotationsquantenzahl *f*
rotational spectrum Rotationsspektrum *n*
rotational state Rotationszustand *m*
rotational transition *(Spektr)* Rotationsübergang *m*
rotational viscometer Rotationsviskosimeter *n*
rotationally excited *(Spektr)* rotationsangeregt
rotatory dispersion Rotationsdispersion *f*, R. D.
rotatory power Dreh(ungs)vermögen *n*
rotaxane *(org Ch)* Rotaxan *n*
rotenoid Rotenoid *n (dem Rotenon verwandte Verbindung)*
rotenone Rotenon *n (Insektizid)*
rotproof verrottungsbeständig, verrottungsfest, fäulnisbeständig, fäulnisfest
rotproofing agent Verrottungsschutzmittel *n*, Fäulnisverhütungsmittel *n*
rough rau *(Oberfläche)*
rough endoplasmatic reticulum *(Bioch)* raues [granuläres] endoplasmatisches Reticulum *n*, RER
rough screening Grobsiebung *f*
rough wood Rohholz *n*
roughing filter Grobfilter *n*, Vorfilter *n*
roughness Rauigkeit *f*, Rauheit *f*
round-bottom (boiling) flask Rundkolben *m*, Kochkolben *m*
round-bottom long-neck flask Langhalsrundkolben *m*
round-bottom short-neck flask Kurzhalsrundkolben *m*

round-bottom two-neck flask Zweihalsrundkolben *m*
round-bottom wide-neck flask Weithalsrundkolben *m*
round filter Rundfilter *n*
routine analysis Routineanalyse *f*, Serienanalyse *f*
roving Roving *m*, Glasseidenstrang *m*; Roving *m*, Vorgarn *n*
Rowland grating mount Rowland-Aufstellung *f*, Rowland-Anordnung *f*
royal jelly Weisel(zellen)futtersaft *m*, Gelée royale *n*
RPC *s.* reversed-phase chromatography
RQ *s.* respiratory quotient
RRE *s.* resonance Raman effect
rRNA *s.* ribosomal RNA
RRS *s.* resonance Raman scattering
RS system R-S-System *n*, R,S-System *n* *(zur Konfigurationsangabe von Stereoisomeren)*
RTD *s.* residence-time distribution
rubber(-coat) *v* gummieren, mit Gummi beschichten
rubber Kautschuk *m(n)*; Gummi *m*
rubber cement *s.* rubber solution
rubber chemist Kautschukchemiker *m*; Gummichemiker *m*
rubber chemistry Kautschukchemie *f*; Gummichemie *f*
rubber-coated fabric gummiertes Gewebe *n*, gummierter Stoff *m*
rubber compound *s.* rubber stock
rubber derivative Kautschukderivat *n*
rubber elasticity Gummielastizität *f*, Kautschukelastizität *f*, Entropieelastizität *f*
rubber-elasticity behaviour gummielastisches [kautschukelastisches, entropieelastisches] Verhalten *n*
rubber gasket Gummidichtung *f*
rubber gloves Gummihandschuhe *mpl*
rubber goods Gummiwaren *fpl*, Gummiartikel *mpl*, Gummifabrikate *npl*
rubber hose Gummischlauch *m*
rubber hydrocarbon Kautschukkohlenwasserstoff *m*, Kautschuk-KW *m*
rubber hydrochloride Kautschukhydrochlorid *n*, Hydrochlorkautschuk *m*
rubber latex Kautschuklatex *m*, Kautschukmilchsaft *m*, Kautschukmilch *f*
rubber policeman *(Lab)* Gummiwischer *m*
rubber scrap Abfallgummi *m*, Altgummi *m*, Gummiabfälle *mpl*, Vulkanisatabfälle *mpl*; unvulkanisierte Fabrikationsabfälle *mpl*
rubber solution Kautschuklösung *f*; Gummilösung *f*
rubber solvent Kautschuklösungsmittel *n*; Gummilösungsmittel *n*
rubber stock Kautschukmischung *f*
rubber stopper Gummistopfen *m*, Gummistöpsel *m*
rubber-tipped glass rod *(Lab)* Glasstab *m* mit Gummiwischer
rubber-tipped policeman *s.* rubber-tipped glass rod
rubber tree Kautschukbaum *m*, *(i. e. S.)* Parakautschukbaum *m*, Hevea brasiliensis [H.B.K.] Muell. Arg.
rubber tube Gummischlauch *m*
rubber vulcanizate Kautschukvulkanisat *n*
rubber waste *s.* rubber scrap
rubber-yielding plant kautschukliefernde [kautschukbildende, kautschukführende, kautschukhaltige] Pflanze *f*, Kautschukpflanze *f*, Kautschukgewächs *n*, Kautschukträger *m*
rubberize *v* gummieren *(mit Gummi beschichten oder imprägnieren)*
rubbery gummiartig
rubeanic acid Rubeanwasserstoffsäure *f*, Rubeanwasserstoff *m*, Dithiooxalsäurediamid *n*, Dithiooxamid *n*
rubidic Rubidium...
rubidium Rb Rubidium *n*
rubidium carbonate Rubidiumcarbonat *n*
rubidium chloride Rubidiumchlorid *n*
rubidium hydroxide Rubidiumhydroxid *n*
rubidium peroxide Rubidiumperoxid *n*
rubin(e) number Rubinzahl *f* *(zur Messung der Schutzkolloidwirkung)*
Rubisco *s.* ribulose-1,5-bisphosphate carboxylase
ruby 1. *(Min)* Rubin *m*; 2. Rubin(rot) *n*
ruby laser Rubinlaser *m*
Ruff degradation ruffscher Abbau *m* *(der Zucker)*
Ruggli high-dilution principle ruggli-zieglersches Verdünnungsprinzip *n*
rule of maximum multiplicity (hund-

sche) Regel *f* der größten Multiplizität, (erste) hundsche Regel *f*
rule of thumb Faustregel *f*
ruled grating *(Spektr)* Strichgitter *n*, Liniengitter *n*
run *v* 1. fließen, strömen, laufen, rinnen; auslaufen *(Farbe)*; laufen, undicht [leck] sein *(Gefäß)*; laufen [fließen, ausströmen] lassen *(eine Flüssigkeit)*; 2. verlaufen, ablaufen, laufen *(chemische Reaktion)*; laufen, in Gang [Betrieb] sein, arbeiten *(Maschine)*; gehen, funktionieren *(Gerät)*; in Gang setzen, fahren *(z. B. eine Anlage)*; durchführen, ausführen, *(Versuche auch)* anstellen, laufen lassen; führen *(Flüssigkeiten)*; 3. schmelzen, sich verflüssigen *(z. B. Metalle)*; schmelzen, tauen *(Eis)*; 4. destillieren
run *v* **a blank (determination)** eine Blindprobe [Blindreaktion] durchführen
run *v* **adiabatically** adiabat betreiben; adiabat arbeiten
run *v* **an analysis** eine Analyse durchführen
run *v* **back** zurücklaufen
run *v* **down** herunterlaufen
run *v* **in** zulaufen lassen, einlaufen [zufließen, einfließen] lassen, zulassen; zutropfen lassen
run *v* **isothermally** isotherm betreiben; isotherm arbeiten
run *v* **off** ablaufen lassen, ablassen, abziehen; abgießen
run *v* **out** auslaufen *(z. B. Farbe)*
run *v* **to the sewer** in die Kanalisation einleiten [ableiten] *(Abwasser)*
run *v* **wild** außer Kontrolle geraten
run 1. Versuch *m*; 2. *(Dest)* Fraktion *f*
run-of-mine *(Bgb)* Rohfördergut *n*
run-of-mine coal Grubenkohle *f*, Förderkohle *f*, Roh(förder)kohle *f*
run-of-mine ore Fördererz *n*, (aus dem Bergwerk kommendes) Roherz *n*
running *(Dest)* Fraktion *f*
running time Laufzeit *f (z. B. eines Chromatogramms, eines Filters)*
running water fließendes Wasser *n*
runoff Ablauf *m*, Auslauf *m*, Abfluss *m*
runoff cock Ablasshahn *m*
runoff pipe Ablaufrohr *n*; Ablaufstutzen *m*, Entleerungsstutzen *m*
rupture *v* (auf)sprengen, spalten
rupture Sprengung *f*, Aufsprengung *f*, Spaltung *f*, Bruch *m*
rupture disk Berstscheibe *f*, Reißscheibe *f*, Platzscheibe *f*
Russel-Saunders coupling *(Spektr)* Russel-Saunders-Kopplung *f*, Spin-Bahn-Kopplung *f*, Spin-Bahn-Wechselwirkung *f*, LS-Kopplung *f*
rust *v* rosten
rust Rost *m*
rust converter Rostumwandler *m (Korrosion)*
rust film Rostfilm *m*, dünne Rostschicht *f*
rust-inhibiting rostinhibierend, rostschützend, rostverhindernd, rostverhütend
rust-inhibiting grease Rostschutzfett *n*
rust-inhibiting oil Rostschutzöl *n*
rust inhibition Rostschutz *m*, Rostverhütung *f*
rust inhibitor Rostinhibitor *m*, Rostschutzmittel *n*, Rostverhütungsmittel *n*
rust-preventing additive Rostschutzadditiv *n*, rostverhinderndes [rostverhütendes] Additiv *n*, Antirostadditiv *n*
rust prevention *s.* rust inhibition
rust preventive temporäres Rostschutzmittel *n (Öl, Fett oder Wachs)*
rust protection Rostschutz *m*
rust-protective rostschützend, Rostschutz...
rust-protective paint Rostschutzfarbe *f*, rostschützende Anstrichfarbe *f*
rust-removing agent Rostentfernungsmittel *n*, Entrostungsmittel *n*, Rostentferner *m*
rust resistance Rostbeständigkeit *f*
rust-resistant rostbeständig, nicht rostend
rust-resisting rostbeständig, nicht rostend
rustless rostfrei, nicht rostend, rostbeständig, rostsicher *(z. B. Stahl)*
ruthenate Ruthenat *n*
ruthenic Ruthenium...; höherwertigem Ruthenium entsprechend
ruthenic chloride Ruthenium(IV)-chlorid *n*, Rutheniumtetrachlorid *n*
ruthenious Ruthenium...; niederwertigem Ruthenium entsprechend
ruthenious chloride Ruthenium(II)-chlorid *n*, Rutheniumdichlorid *n*
ruthenium Ru Ruthenium *n*

ruthenium dioxide Rutheniumdioxid *n*, Ruthenium(IV)-oxid *n*
ruthenium hydroxide Ruthenium(III)-hydroxid *n*
ruthenium sesquioxide Ruthenium(III)-oxid *n*
Rutherford atom (model) rutherfordsches Atommodell *n*
rutherfordium Rf Rutherfordium *n (Element 104)*
rutile *(Min)* Rutil *m (Titan(IV)-oxid)*
Rydberg atom Rydberg-Atom *n*
Rydberg constant [number] Rydberg-Konstante *f*, rydbergsche Konstante *f*, Rydberg-Zahl *f*
Rydberg series *(Spektr)* Rydberg-Serie *f*
Rydberg state *(Spektr)* Rydberg-Zustand *m*
Rydberg transition *(Spektr)* Rydberg-Übergang *m*
rye starch Roggenstärke *f*

S

s. *s.* 1. soluble; 2. solubility
S protein S-Protein *n (Spaltprodukt der Ribonuclease)*
Sabatier-Senderens reaction [reduction] sabatier-senderenssche Reaktion *f* [Reduktion *f*] *(von organischen Verbindungen mit Wasserstoff)*
SAC = Society for Analytical Chemistry
saccharate Saccharat *n (Salz oder Ester der Zuckersäure)*
saccharic acid Zuckersäure *f*, *(i. e. S.)* D-Glucozuckersäure *f (eine Tetrahydroxyadipinsäure)*
saccharide Saccharid *n*, Kohlenhydrat *n*
saccharification Verzuckerung *f*, Saccharifikation *f*
saccharification of wood Holzverzuckerung *f*
saccharify *v* 1. in Zucker verwandeln, verzuckern, saccharifizieren; 2. zuckern, süßen
saccharimeter Saccharimeter *n (ein Polarimeter zur Bestimmung des Zuckergehalts)*
saccharin Saccharin *n*, *o*-Benzoesäuresulfimid *n*
saccharine zuckerhaltig; zuckerartig, Zucker…; zuck(e)rig, zuckersüß
saccharinic acid Saccharinsäure *f*, Tetrahydroxyhexandisäure *f*, *(i. e. S.)* Saccharinsäure *f*, 2,3,4,5-Tetrahydroxyhexandisäure *f*
saccharolactic acid Galactozuckersäure *f*, Schleimsäure *f*, Tetrahydroxyadipinsäure *f*, *2,3,4,5-Tetrahydroxyhexandisäure *f*
saccharometer Saccharometer *n (eine Senkspindel zur Bestimmung der Dichte einer Zuckerlösung)*
saccharomycete Saccharomyzet *m*, Hefepilz *m*
saccharose 1. *s.* sucrose; 2. *allgemein für ein Di- oder Trisaccharid*
Sachse-Mohr concept of strainless rings Sachse-Mohr-Theorie *f* (spannungsfreier Ringe), sachse-mohrsche Spannungstheorie *f*
sacrificial anode Opferanode *f*, Aktivanode *f*
sacrificial protection katodischer Schutz *m* [Korrosionsschutz *m*] durch [mit] Opferanoden
saddle Sattel(füll)körper *m*
safe for factory processing verarbeitungssicher
safeguard Sicherung *f*, Schutz *m*, Wächter *m*
safety 1. Sicherheit *f*; 2. Unschädlichkeit *f*
safety clothing Schutz(be)kleidung *f*, Arbeitsschutzkleidung *f*
safety explosive 1. Sicherheitssprengstoff *m*; 2. *(Bgb)* Wettersprengstoff *m*
safety glass Sicherheitsglas *n*, Schutzglas *n*
safety glasses [goggles] Schutzbrille *f*
safety instruction Sicherheitsvorschrift *f*
safety measure Sicherheitsmaßnahme *f*
safety of operation Betriebssicherheit *f*
safety precautions Sicherheitsvorkehrungen *fpl*
safety principle Sicherheitsmaßregel *f*
safety regulation Sicherheitsbestimmung *f*, Arbeitsschutzordnung *f*
safety rule Sicherheitsmaßregel *f*
safety shower Notbrause *f*, Löschbrause *f*, Feuerlöschbrause *f*
safety valve Sicherheitsventil *n*
safety wash bottle Sicherheitswaschflasche *f*
safranin(e) Safranin *n (ein Azinfarbstoff)*

safrol(e) Safrol *n*, *1,2-Methylendioxy-4-(prop-2-enyl)benzen *n*, 4-Allyl-1,2-methylendioxybenzen *n*
sag *v* 1. durchhängen, sich durchbiegen, durchsacken, absacken; 2. *(Anstr)* (ab)laufen, Gardinen [Vorhänge] bilden
sag Durchhang *m*, Durchbiegung *f*
sagging 1. Durchhängen *n*, Durchhang *m*, Durchbiegen *n*, Durchbiegung *f*, Durchsacken *n*, Absacken *n*; 2. *(Anstr)* Ablaufen *n*, Laufen *n*, Läuferbildung *f*, Gardinenbildung *f*, Vorhangbildung *f*
sago Sagostärke *f*, Palm(en)sago *m*, Palmenstärke *f (von Metroxylon-Arten)*
sal ammoniac *s.* salmiac
sal soda *s.* salt of soda
salicyl alcohol Salicylalkohol *m*, Saligenin *n*, 2-Hydroxybenzylalkohol *m*
salicylaldehyde Salicylaldehyd *m*, *o*-Hydroxybenzaldehyd *m*
salicylamide Salicyl(säure)amid *n*, *o*-Hydroxybenzamid *n*, *o*-Hydroxybenzencarbonsäureamid *n*
salicylanilide Salicyl(säure)anilid *n*, *N*-Phenylsalicylamid *n*, *o*-Hydroxy-*N*-phenylbenzencarbonsäureamid *n*
salicylate Salicylat *n (Salz oder Ester der Salicylsäure)*
salicylic acid Salicylsäure *f*, 2-Hydroxybenzoesäure *f*
saliferous salzhaltig, salzführend
salifiable salzbildungsfähig
salification Salzbildung *f*
salify *v* ein Salz bilden, Salze bilden; in ein Salz überführen *(eine Säure oder Base)*
salimeter *s.* salinimeter
saline 1. salzhaltig, salzig, Salz...; 2. salzartig
saline 1. Salzquelle *f*; 2. Salzsee *m*; 3. Salzlagerstätte *f*, Salzlager *n*; 4. Saline *f*, Salzsiederei *f*, Salzwerk *n*; 5. Siedepfanne *f*; 6. Sole *f*, Salzlake *f*, Salzlösung *f*; 7. *(Pharm)* physiologische Kochsalzlösung *f*; 8. *(Pharm)* salinisches Abführmittel *n*
saline solution Salzlösung *f*; Kochsalzlösung *f*
saline water Salzwasser *n*, Sole *f*, salzhaltiges Wasser *n*
saline water conversion Meerwasserentsalzung *f*
salinimeter Salz(gehalt)messer *m*, Salzwaage *f*, Halometer *n (Aräometer für Salzlösungen)*
salinity 1. Salzhaltigkeit *f*, Salzigkeit *f*; 2. Salzartigkeit *f*
saliva Speichel *m*
SALLS *s.* small-angle light scattering
salmiac 1. Salmiak *m*, Salmiaksalz *n*, Ammoniumchlorid *n*; 2. *(Min)* (natürlicher) Salmiak *m*, Salammoniak *m (Ammoniumchlorid)*
salmonella test Ames-Test *m*, Rückmutationstest *m (zur Prüfung der Mutagenität von Chemikalien)*
salt *v* salzen
salt *v* **out** aussalzen
salt Salz *n*, *(i. e. S.)* Kochsalz *n*
salt bath *(Met)* Salzbad *n*, Salzschmelze *f*
salt bridge Salzbrücke *f*, Elektrolytbrücke, Elektrolytschlüssel *m*, (elektrolytischer) Stromschlüssel *m*
salt brine Salzlake *f*, Salzlauge *f*, Salzlösung *f*, Sole *f*, Lake *f*
salt error Salzfehler *m*, Neutralsalzfehler *m (Messfehler bei pH-Bestimmungen)*
salt formation Salzbildung *f*
salt-forming salzbildend
salt-free 1. salzfrei; 2. *(Lebm)* salzlos
salt-ice mixture Kältemischung *f* aus Salz und Eis
salt-like salzartig, salzähnlich
salt of hartshorn Hirschhornsalz *n*, (gewöhnliches) Ammoniumcarbonat *n (Gemisch aus Ammoniumhydrogencarbonat und Ammoniumcarbamat)*
salt of phosphorus Phosphorsalz *n*, Natriumammoniumhydrogenphosphat-4-Wasser *n*
salt of soda Kristallsoda *f*, Waschsoda *f*, Soda *f*, Natriumcarbonat-10-Wasser *n*
salt of sorrel Kleesalz *n*, Sauerkleesalz *n*, Bitterkleesalz *n (Kaliumtetraoxalat oder ein Gemisch aus diesem und Kaliumhydrogenoxalat)*
salt of tartar Weinsteinsalz *n*, Kaliumcarbonat *n*
salt pair Salzpaar *n*
salt solution Salzlösung *f*; Kochsalzlösung *f*
salt spray test Salznebelversuch *m*, Salzsprühversuch *m*, Salzsprühtest *m*, Aerosolversuch *m (zur Korrosionsprüfung)*
salt taste Salzgeschmack *m*

salt water Salzwasser *n*, Sole *f*, salzhaltiges Wasser *n*
salt works Salzwerk *n*, Saline *f*, Salzsiederei *f*
saltern Saline *f*, Salzwerk *n*, Salzsiederei *f*; Salzgarten *m*, Saline *f*
saltiness Salzigkeit *f (Geschmacksempfindung)*
salting 1. Salzen *n*, Salzung *f*; 2. Salzablagerung *f*
salting-in effect Einsalzeffekt *m*
salting-out chromatography Aussalzchromatographie *f*
salting-out effect Aussalzeffekt *m*
saltish etwas salzig
saltpetre 1. Kalisalpeter *m*, Kaliumnitrat *n*; 2. Chilesalpeter *m*, Natronsalpeter *m*, Natriumnitrat *n*
salty salzig; salzhaltig
SAM *s.* scanning Auger microscopy
samaria Samarium(III)-oxid *n*
samaric Samarium..., *(i. e. S.)* Samarium(III)-...
samarium Sm Samarium *n*
samarium dichloride Samariumdichlorid *n*, Samarium(II)-chlorid *n*
samarium oxide Samarium(III)-oxid *n*
samarium trichloride Samariumtrichlorid *n*, Samarium(III)-chlorid *n*
samarous Samarium..., *(i. e. S.)* Samarium(II)-...
sample *v* Proben entnehmen
sample 1. *(Anal)* Probe *f*, Substanzprobe *f*, Stichprobe *f*, *(i. w. S.)* Prüfgut *n*, (gesamte verfügbare) Probe *f*; 2. *(Werkstoffprüfung)* Probe *f (i. w. S.)*; Muster *n*; Probestück *n (zur Herstellung von Prüflingen)*; 3. *s.* specimen • **collect [draw] samples** Proben entnehmen
sample analysed Analysenprobe *f*, analysierte Probe *f*
sample application 1. Probeauftragen *n*, Probenaufgabe *f (auf eine Schicht)*; 2. *s.* sample introduction
sample cell *(Spektr)* Probenküvette *f*
sample changer Probenwechselvorrichtung *f*, Probenwechsler *m*
sample collection Proben(ent)nahme *f*, Probenahme *f*, Probenehmen *n*
sample collector *s.* sampler 2.
sample discrimination Probendiskriminierung *f*
sample dispenser Probengeber *m*
sample divider Probenteiler *m*
sample division Teilen *n* von Proben, *(i. w. S.)* Probenverjüngung *f (Teilen und Verwerfen von Teilmengen)*
sample injection port *(Chromat)* Einspritzstelle *f*, *(am Gaschromatographen)* Einspritzblock *m*
sample injector *(Chromat)* Probengeber *m*
sample inlet Probeneinlass *m*
sample introduction Probenaufgabe *f*, Probeneingabe *f*, Probenzufuhr *f (zu einer Säule)*
sample loop *(Chromat)* Probenschleife *f*, Speicherschleife *f*
sample port Probeneinlass *m*
sample preparation Probenvorbereitung *f*, Probenaufbereitung *f*, Probenpräparation *f*
sample preservation Probenkonservierung *f*
sample quantity Probenmenge *f*
sample reducer Probenteiler *m*
sample reduction 1. Probenzerkleinerung *f*; 2. Probenverjüngung *f (Teilen und Verwerfen von Teilmengen)*; *(i. w. S.)* Probenverarbeitung *f (einschließlich der Zerkleinerung)*
sample size 1. Probenumfang *m*, Probengröße *f (im statistischen Sinne)*; 2. *s.* sample quantity
sample splitter *(Chromat)* Gasstromteiler *m*, Strömungsteiler *m*, Splitter *m*
sample storage Probenaufbewahrung *f*
sample throughput Probendurchsatz *m*
sample transfer Probenüberführung *f*, Probentransfer *m*
sample under analysis Analysenprobe *f (während der Untersuchung)*
sampler 1. Probenehmer *m (Person)*; 2. Probenehmer *m*, Probenahmegerät *n*, Probenahmevorrichtung *f*
sampling 1. Proben(ent)nahme *f*, Probenahme *f*, Probenehmen *n*; 2. Proben(auf)gabe *f*, Dosierung *f*
sampling error Probenahmefehler *m*, Probenfehler *m*
sampling location Probenahmeort *m*
sampling site Probenahmestelle *f*, Entnahmestelle *f*
sampling specification Probe(nent)nahmevorschrift *f*

sampling technique Probenahmetechnik *f*, Probenahmemethode *f*
sampling thief Probenstecher *m*
sand bath Sandbad *n*
sand filter Sandfilter *n*
sand product Sand *m*, Grobgut *n*, Rückstand *m (beim Klassieren)*
sandalwood oil Sandel(holz)öl *n (von Santalum album L.)*
sandblast *v* sandstrahlen
sandblasting Sandstrahlen *n*, Sandstrahlreinigung *f*
Sandmeyer (diazo) reaction Sandmeyer-Reaktion *f*, sandmeyersche Reaktion *f (Ersatz der Diazogruppe durch andere Reste)*
sandwich bond Sandwichbindung *f*, Doppelkegelbindung *f*
sandwich complex Sandwichkomplex *m*
sandwich compound Sandwichverbindung *f*
Sanger's reagent sangersches Reagens *n (2,4-Dinitrofluorbenzen zum Nachweis von Aminosäuren und Proteinen)*
sanitary land fill geordnete Deponie *f*
sanitary sewage [waste, waste water] *(Hyd)* häusliches Abwasser *n*; Sanitärabwasser *n*, Sozialabwasser *n*, Abwasser *n* aus sanitären und sozialen Einrichtungen, Wirtschaftsabwasser *n (eines Betriebs)*
sanitaryware Sanitärkeramik *f*
sanitization Hygienisierung *f*
SANS *s.* small-angle neutron scattering
santonin Santonin(lacton) *n*
santoninic acid Santoninsäure *f*
sap *v* entsaften
sap Saft *m*
SAP *s.* superabsorbent polymer
saponifiable verseifbar
saponifiable matter Verseifbares *n*
saponification Verseifung *f*, Saponifikation *f*
saponification number [value] Verseifungszahl *f*, VZ
saponify *v* verseifen
sapphire *(Min)* Saphir *m (Aluminiumoxid)*
saprogenic saprogen, fäulniserregend
saprogenous saprogen, fäulniserregend
sapropel Sapropel *n(m)*, Faulschlamm *m*
sapropelic coal Sapropel(it)kohle *f*, Faulschlammkohle *f*, sapropelitische Kohle *f*
saprophyte Saprophyt *m*, Fäulnisbewohner *m*
saprophytic saprophytisch, fäulnisbewohnend
sarcolactic acid Fleischmilchsäure *f*, Rechtsmilchsäure *f*, L(+)-Milchsäure *f*
sarcosine Sarkosin *n*, Sar, Methylaminoessigsäure *f*, *N*-Methylglykokoll *n*
sarin Sarin *n (Methylfluorphosphonsäureisopropylester, ein Kampfstoff)*
SAS *s.* sodium-aluminium sulphate
sassafras oil Sassafrasöl *n (von Sassafras albidum (Nutt.) Nees)*
satd. = saturated
satellite DNA *(Bioch)* Satelliten-DNA *f*
satellite line *(Spektr)* Satellitenlinie *f*
satn. *s.* saturation
SATP = standard ambient temperature and pressure *(298,15 K, 1 bar)*
saturate *v* 1. sättigen *(Lösungen)*; 2. absättigen *(Valenzen)*; 3. imprägnieren, tränken
saturated calomel electrode *(Ech)* gesättigte Kalomelelektrode *f*, GKE
saturated fatty acid gesättigte Fettsäure *f*
saturated hydrocarbon gesättigter Kohlenwasserstoff *m*, *(i. e. S.)* Paraffinkohlenwasserstoff *m*, *Alkan *n*
saturated solution gesättigte Lösung *f*
saturated steam gesättigter Wasserdampf *m* [Dampf *m*], Sattdampf *m*
saturation 1. Sättigung *f (von Lösungen)*; 2. Absättigung *f*, Sättigung *f (von Valenzen)*; 3. Imprägnierung *f*, Tränkung *f*
saturation capacity Sättigungsbeladung *f (Adsorption)*
saturation concentration Sättigungskonzentration *f*
saturation humidity Sättigungsfeuchte *f*, Sättigungsfeuchtigkeit *f*; Sättigungswassergehalt *m*
saturation of valence Valenzabsättigung *f*
saturation temperature Taupunkt *m*
saturation vapour pressure Sättigungs(dampf)druck *m*
sausage flask Säbelkolben *m* (nach Anschütz), Schwertkolben *m*
Sauter mean diameter Sauter-Durch-

messer *m (zur Kennzeichnung von Teilchengrößen)*
Saybolt viscometer *(Erdöl)* sayboltsches Viskosimeter *n*, Saybolt-Universalviskosimeter *n*
Saytzeff rule *(org Ch)* Zajcev-Regel *f*, Saizew-Regel *f*
SBA *s. sec*-butyl alcohol
SBF *s.* simulated body fluid
SBP spirit *s.* special boiling-point spirit
SBR *s.* styrene-butadiene rubber
SC *s.* spreading coefficient
scaffold Gerüst *n*
scale *v* 1. verkrusten, *(i. e. S.)* mit Kesselstein bedecken; verkrusten, *(i. e. S.)* Kesselstein ansetzen; 2. *(Met)* (ver-)zundern; 3. entkrusten, *(i. e. S.)* von Kesselstein befreien; 4. *(Met)* entzundern; 5. *s.* scale off
scale *v* **down** senken, drosseln *(Produktion)*
scale *v* **off** abblättern, abschuppen, abplatzen, abspringen
scale *v* **up** (im Maßstab) vergrößern *(eine Anlage)*; steigern, erhöhen *(Produktion)*
scale *v* **up to industrial use** in den industriellen Maßstab überführen
scale 1. Skala *f*, Stufenfolge *f*; Skale *f*, Maßeinteilung *f*, Gradeinteilung *f (an Messgeräten)*; 2. dünne Schicht *f*, Belag *m*, Ablagerung *f*, Niederschlag *m*; Inkrustation *f*; 3. *(Hyd)* Kesselstein *m*; Wasserstein *m (in Rohren)*; 4. *(Met)* Zunder *m*, Sinter *m*; 5. Schuppe *f*; 6. Waagschale *f*; Waage *f* • **on a [the] laboratory scale** im Labormaßstab • **on a [the] large scale** in großem Maßstab, großtechnisch
scale deposit Kesselsteinablagerung *f*
scale division Skalenteil *m*, Teilstrichabstand *m*; Skalen(teil)strich *m*
scale formation 1. Steinbildung *f*, *(i. e. S.)* Kesselsteinbildung *f*; Wassersteinbildung *f (in Rohren)*; 2. *(Met)* Zunderbildung *f*
scale-forming hardness temporäre [vorübergehende] Härte *f*, Karbonathärte *f*, KH
scale inhibitor Kesselsteinverhütungsmittel *n*
scale of electronegativity Elektronegativitätsskala *f*
scale-up Maßstabsvergrößerung *f*, maßstäbliche Vergrößerung *f (einer Anlage, eines Verfahrens)*; Steigerung *f*, Erhöhung *f (der Produktion)*
scales *(einfache)* Waage *f*
scaling 1. Messung *f* [Einstellung *f*] nach einer Skale; 2. Verkrusten *n*, Verkrustung *f*; 3. *(Hyd)* Wassersteinbildung *f*, Steinbildung *f*, *(i. e. S.)* Kesselsteinbildung *f*; 4. *(Met)* Zundern *n*, Verzunderung *f*; 5. Entfernen *n* von Kesselstein; 6. *(Met)* Entzundern *n*; 7. Abschuppen *n*, Abblättern *n*, Abplatzen *n*, Abspringen *n*
scaly schuppig
scan *v* 1. abtasten; 2. *(Anal)* scannen, durchfahren, abfahren *(einen Messbereich)*
scan *(Anal)* Durchfahren *n*, Abfahren *n (eines Messbereichs)*
scandia Scandiumoxid *n*
scandium Sc Scandium *n*
scandium oxide Scandiumoxid *n*
scanning Auger microscopy Scanning-Auger-Mikroskopie *f*
scanning calorimeter Scanningkalorimeter *n*
scanning electron microscopy Rasterelektronenmikroskopie *f*, REM
scatter *v* (zer)streuen
scatter Streuen *n*, Zerstreuen *n*, Streuung *f*, Zerstreuung *f*
scattered electron Streuelektron *n*, gestreutes Elektron *n*
scattered light Streulicht *n*, gestreutes [zerstreutes, diffuses] Licht *n*, Beugungslicht *n*
scattering angle Streu(ungs)winkel *m*
scattering cross section Streu(ungs)-querschnitt *m*
scavenge *v* reinigen, säubern, spülen
scavenger 1. Reinigungsmittel *n*, Spülmittel *n*; 2. Fänger *m*, Fängersubstanz *f (z. B. für Radikale, Elektronen)*; 3. Kollektor *m*, Sammler *m (zum Entfernen von Verunreinigungen)*
S.C.E. *s.* saturated calomel electrode
SCF *s.* self-consistent field
scf = standard cubic feet *(1 scf = 0,02832 m^3)*
scfd = standard cubic feet per day *(Maß für Durchsatz)*
Schardinger dextrin Schardinger-Dextrin *n*, Cyclodextrin *n*

Scheele's green Scheeles [scheelesches] Grün *n* *(Kupferarsenit)*
Schellbach burette Schellbach-Bürette *f*, Bürette *f* mit Schellbachstreifen
scheme of analysis Analysengang *m*, Analysenweg *m*
scheme of analysis for the anions Anionentrennungsgang *m*
Schiff base RHC=NR′ schiffsche Base *f*, Azomethin *n*
Schiff reagent schiffsches [Schiffs] Reagens *n* *(meist fuchsinschweflige Säure, zum Aldehydnachweis)*
schist *(Geol)* Schiefer *m*
schistose *(Geol)* schiefrig, geschiefert; schieferartig
Schlenk technique *(Lab)* Schlenk-Technik *f* *(für den Umgang mit luftempfindlichen Substanzen)*
Schlenk tube *(Lab)* Schlenk-Rohr *n* *(für den Umgang mit luftempfindlichen Substanzen)*
schlieren method Schlierenverfahren *n* *(zur Untersuchung von Dichtegradienten)*
schlieren optics Schlierenoptik *f*
Schlippe's salt schlippesches Salz *n*, Natriumthioantimonat(V)-9-Wasser *n*
Schmidt reaction Schmidt-Reaktion *f*, schmidtsche Reaktion *f* *(Carbonylabbau mit Stickstoffwasserstoffsäure)*
Schotten-Baumann reaction Schotten-Baumann-Reaktion *f*, schotten-baumannsche Reaktion *f* *(von Säurechloriden mit Alkoholen oder Aminen)*
Schrock catalysts *(org Ch)* Schrock-Katalysatoren *mpl* *(Metathese)*
Schrödinger (wave) equation Schrödinger-Gleichung *f*, schrödingersche Wellengleichung *f* [Gleichung *f*]
Schweinfurth green Schweinfurter Grün *n* *(Kupferarsenitacetat)*
Schweizer's reagent Schweizer-Reagens *n*, Schweizers Reagens *n* *(Kupferoxidammoniak-Lösung)*
SCI = Society of Chemical Industry
scintillate *v* szintillieren, aufblitzen, aufleuchten, flimmern
scintillation Szintillieren *n*, Szintillation *f*, Aufblitzen *n*, Aufleuchten *n*, Flimmern *n*
scintillation counter Szintillationszähler *m*
scintillation spectrometer Szintillationsspektrometer *n*
scintillator Szintillator *m*
scission Spaltung *f*
scission reaction Spalt(ungs)reaktion *f*
scissor vibration Scherenschwingung *f*, Deformationsschwingung *f*, Spreizschwingung *f*
scleroprotein Skleroprotein *n*, Gerüsteiweiß *n*, Strukturprotein *n*, Faserprotein *n*, fibrilläres Protein *n*
scleroscope Skleroskop *n*, Härtemesser *m*, Fallhärteprüfer *m*, Rücksprunghärteprüfer *m*
scopolamine Scopolamin *n* *(Alkaloid)*
scopoletin Scopoletin *n*, Chrysatropasäure *f*, Gelseminsäure *f*, 6-Methoxy-7-hydroxycumarin *n*
scorch *v* versengen, verbrennen; verätzen, verbrennen *(Pflanzenteile durch Chemikalien)*
scorch Brandfleck *m*, Sengfleck *m*; Verätzung *f*, Verbrennung *f* *(an Pflanzen durch Chemikalien)*
scoria *(Met, Geol)* Schlacke *f*
scoriaceous schlackenartig, schlackenähnlich, schlackig
SCOT column *(solid coated open tubular column)* SCOT-Säule *f*, *(i. e. S.)* Dünnschicht-Kapillarsäule *f* *(Gaschromatographie)*
scour *v* 1. reinigen; scheuern, blank putzen [reiben]; auswaschen, (aus-)spülen; abbeizen; 2. *(Met)* entzundern, beizen; 3. *(Text)* entfetten, entschweißen *(Wolle)*; 4. *(Text)* entbasten, degummieren, entschälen *(Seide)*; 5. *(Met)* ausfressen *(das Ofenfutter)*
scouring 1. Reinigen *n*, Reinigung *f*; Scheuern *n*; Auswaschen *n*, Spülen *n*, Ausspülen *n*; Abbeizen *n*; 2. *(Met)* Entzundern *n*, Beizen *n*; 3. *(Text)* Entfetten *n*, Entschweißen *n* *(von Wolle)*; 4. *(Text)* Entbasten *n*, Degummieren *n*, Entschälen *n* *(von Seide)*; 5. *(Met)* Ausfressung *f* *(im Ofenfutter)*
scouring agent 1. Reinigungsmittel *n*, Reiniger *m*; Scheuermittel *n*, Putzmittel *n*; Waschmittel *n*, Spülmittel *n*; Abbeizmittel *n*; 2. *(Met)* Beize *f*; 3. *(Text)* Entschweißungsmittel *n* *(für Wolle)*; 4. *(Text)* Entbastungsmittel *n* *(für Seide)*
scouring bath 1. Reinigungsbad *n*,

Waschbad *n*; Spülbad *n*; 2. *(Met)* Beizbad *n*; 3. *(Text)* Entschweißbad *n* *(für Wolle)*; 4. *(Text)* Entbastungsbad *(für Seide)*
SCP *s.* single-cell protein
SCR *s.* styrene-chloroprene rubber
scrap 1. Abfall *m*; Ausschuss *m*; 2. *(Met)* Schrott *m*, Altmetall *n*
scrap reclaiming Wiederaufarbeiten *n* von Abfällen
scrape *v* kratzen; reiben
scrape *v* **off** abkratzen, abschaben
scraper Abstreifer *m*, Schaber *m*; Ausräumer *m*
scraper(-chain) conveyor Kratzerförderer *m (i. w. S.)*; Stegkettenförderer *m (i. w. S.)*
scratch *v* ritzen, kratzen
scratch cooler Kratzkühler *m*
scratch hardness Ritzhärte *f*, Ritzfestigkeit *f*, Kratzfestigkeit *f*
scratch heater Kratzwärmeaustauscher *m*
screen *v* 1. (durch)sieben, absieben, aussieben, siebklassieren; 2. (ab-)schirmen; abblenden *(Licht)*; 3. durchmustern, screenen, eine Vorauswahl treffen
screen 1. Sieb *n*, Klassiersieb *n*; Siebapparat *m*, Siebmaschine *f*, Siebvorrichtung *f*; 2. Schirm *m*; Röntgenschirm *m*; Bildschirm *m*, Projektionswand *f*; 3. Filter *n*, Blende *f*
screen analysis Siebanalyse *f*
screen aperture Sieb(öffnungs)weite *f*, Sieböffnung *f*, Maschenweite *f*
screen centrifuge Filterzentrifuge *f*, Filtrierzentrifuge *f*, Siebzentrifuge *f*, Zentrifugal(kraft)filter *n*
screen classification Siebklassieren *n*
screen classifier Siebapparat *m*, Siebmaschine *f*, Siebvorrichtung *f*
screen filter Siebfilter *n*; Rahmenfilter *n*, Flächenabscheider *m*
screen fines Siebfeines *n*, Siebfeine *f*, Feinkorn *n*, Feingut *n*, Unterkorn *n*
screen oversize Siebgrobes *n*, Siebgröbe *f*, Grobkorn *n*, Grobgut *n*, Überkorn *n*
screen size 1. Korngröße *f*, Körnung *f*, Teilchengröße *f (beim Siebklassieren)*; 2. *s.* screen aperture
screen sizing Sieben *n*, Siebklassieren *n*
screen-type mill Siebmühle *f*
screen undersize *s.* screen fines
screening 1. Sieben *n*, Absieben *n*, Durchsieben *n*, Aussieben *n*, Siebklassieren *n*; 2. Abschirmen *n*, Abschirmung *f*; 3. Durchmusterung *f*, Screening *n*, Vorauswahl *f*
screening constant *(Kern)* Abschirm(ungs)konstante *f*, Abschirmungszahl *f*
screening machine Siebapparat *m*, Siebmaschine *f*, Sieb *n*
screw Schraube *f*, Schraubenspindel *f*; Schnecke *f (Fördereinrichtung)*
screw axis *(Krist)* Schraubenachse *f*
screw compressor clamp *(Lab)* Schraubenquetschhahn *m*, Schraubklemme *f* (nach Hoffmann)
screw conveyor Schneckenförderer *m*
screw dislocation *(Krist)* Schraubenversetzung *f*, Burgers-Versetzung *f*
screw injection (moulding) machine *(Kst)* Schneckenspritzgießmaschine *f*, Spritzgießmaschine *f* mit Schneckenkolben
screw press Schneckenpresse *f*, Extruder *m*
screw-type compressor Schraubenverdichter *m*, Schraubenkompressor *m*, Schneckenverdichter *m*, Schneckenkompressor *m*
screw-type mixer Schneckenmischer *m*, Schneckenrührer *m*, Schneckenkneter *m*, Schraubenkneter *m*
scrub *v* 1. schrubben, scheuern, reinigen *(mit einer Bürste)*; 2. waschen *(Gase)*
scrubber 1. Wäscher *m*, Wascher *m*, Gaswäscher *m*, Skrubber *m*, *(für Stäube und Dämpfe auch)* Nass(staub)abscheider *m*, Nassentstauber *m*; 2. *s.* scrubbing tower
scrubbing 1. Schrubben *n*, Scheuern *n*, Reinigen *n (mit einer Bürste)*; 2. Waschen *n*, Gaswaschen *n*, Gaswäsche *f*, *(zur Entfernung von Stäuben und Dämpfen auch)* Nassentstaubung *f*
scrubbing liquid Waschflüssigkeit *f (für Gase beim Nassabscheiden)*
scrubbing tower Turmwäscher *m*, Waschturm *m*, Gaswaschturm *m*, Skrubber *m*
scrupulously clean peinlich sauber
scuff resistance Abriebfestigkeit *f*, Ab-

riebbeständigkeit *f (bei gleitender oder rollender Reibung und Sand oder Staub)*
scum 1. Schaum *m*, Abschaum *m*; 2. *(Met)* Schlacke *f*; 3. *s.* scum layer
scum layer [mat] *(Hyd)* schaumartige Schwimmschicht *f*, Schaumschicht *f*, Schwimmschlamm *m (z. B. bei der Flotation)*
scum skimmer *(Hyd)* Schaumabstreifer *m*, Schwimmschlammabstreifer *m*, Schwimmschlamm(räum)schild *m*
SCWO *s.* supercritical water oxidation
SDH = succinate dehydrogenase
SDS = sodium dodecylsulphate
sea disposal Beseitigung *f* durch Einleiten ins Meer; Beseitigung *f* durch Abklappen [Verkippen ins Meer]
sea dumping Einbringen *n* [Verkippen *n*, Versenken *n*] von Abprodukten ins Meer, Verkippen *n* auf hoher See, Abklappen *n*
sea salt Seesalz *n*, Meersalz *n*
seaborgium Sg Seaborgium *n (Element 106)*
seal *v* 1. (ver)schließen, abschließen; (ab)dichten; zuschmelzen, verschmelzen, (ver)siegeln; 2. *(Anstr)* (ab)sperren *(den Untergrund)*; 3. *(Met)* nachverdichten *(bei der anodischen Oxidation)*; 4. *(Kst)* siegeln, (ver)schweißen *(Folien)*
seal *v* **up** zuschmelzen, abschmelzen
seal Verschluss *m*, Abschluss *m*, Dichtung *f*, Abdichtung *f*, Sperre *f*, Packung *f*
sealant *s.* sealing material
sealed-tube decomposition Aufschluss *m* im Bombenrohr [Einschmelzrohr, Schießrohr]
sealing cup Abdichtmanschette *f*
sealing fluid Sperrflüssigkeit *f*, Dichtungsflüssigkeit *f*
sealing material Dicht(ungs)werkstoff *m*, Dicht(ungs)stoff *m*, Dicht(ungs)material *n*, Dicht(ungs)mittel *n*
seam 1. Saum *m*, Naht *f*; 2. *(Geol)* Flöz *n*, Lager *n*
season *v* 1. altern; trocknen, ablagern *(Holz)*; 2. *(Lebm)* würzen
seasoning 1. Altern *n*, Alterung *f*; Trocknen *n*, Trocknung *f*, Ablagern *n (von Holz)*; 2. *(Lebm)* Würzen *n*; 3. *(Lebm)* Würzmittel *n*
sebaceous talgig, talgartig, Talg...
sebacic acid Sebacinsäure *f*, *Decandisäure *f*, *Octan-1,8-dicarbonsäure *f*
SEC *s.* size-exclusion chromatography
second law of thermodynamics zweiter Hauptsatz *m* der Thermodynamik, Entropiesatz *m*
second-order nucleophilic substitution (reaction) bimolekulare nucleophile Substitution *f*, S_N2-Reaktion *f*
second-order rate constant Geschwindigkeitskonstante *f* zweiter Ordnung
second-order rate equation [law] Geschwindigkeitsgleichung *f* zweiter Ordnung, Zeitgesetz *n* [Geschwindigkeitsgesetz *n*] zweiter Ordnung
second-order reaction Reaktion *f* zweiter Ordnung
second-order transition Umwandlung *f* [Phasenumwandlung *f*, Phasenübergang *m*] 2. Ordnung, kontinuierliche Umwandlung *f*, kontinuierlicher Übergang *m*, *(bei Polymeren auch)* Glasumwandlung *f*, Glasübergang *m*, Gamma-Umwandlung *f*, α-Anomalie *f*
second-order transition point [temperature] Umwandlungstemperatur *f* [Umwandlungspunkt *m*] 2. Ordnung, *(bei Polymeren auch)* Glas(umwandlungs)temperatur *f*, Glasumwandlungspunkt *m*, Glasübergangstemperatur *f*, Einfriertemperatur *f*, Einfrierpunkt *m*
second-stage intercalation compound Interkalationsverbindung *f* [Einlagerungsverbindung *f*] 2. Stufe
second substituent Zweitsubstituent *m*
secondary alcohol sekundärer Alkohol *m*
secondary amine RC–NH–CR sekundäres Amin *n*
secondary battery *(Ech)* Sekundärbatterie *f*, Akkumulator *m*
secondary cell *(Ech)* Sekundärelement *n*, Sekundärzelle *f (regenerierbar)*
secondary clarifier *(Hyd)* Nachklärbecken *n*
secondary component *(Farb)* Kupplungskomponente *f*, passive Komponente *f*, Zweitkomponente *f*, Kupplungskörper *m*

secondary crystallization sekundäre Kristallisation *f*, Nachkristallisation *f*
secondary digestion tank *(Hyd)* Nachfaulraum *m*, Nachfaulbecken *n*, Nachfaulbehälter *m*, zweiter Faulbehälter *m (bei zweistufiger Faulung)*
secondary electron sekundäres Elektron *n*, Sekundärelektron *n*, SE *n*
secondary emission Sekundäremission *f*
secondary-emission electron multiplier Sekundärelektronenvervielfacher *m*, SEV, Photo(elektronen)vervielfacher *m*, Photomultiplier *m*
secondary fermentation Nachgärung *f*, *(Bierherstellung auch)* Lagerung *f*
secondary ion mass spectrometry Sekundärionenmassenspektrometrie *f*, SIMS
secondary kinetic isotope effect sekundärer kinetischer Isotopieeffekt *m*
secondary metabolism *(Bioch, Biot)* Sekundärmetabolismus *m*, Sekundärstoffwechsel *m*, Nebenstoffwechsel *m*
secondary metabolite *(Bioch, Biot)* sekundärer Metabolit *m*, Sekundärmetabolit *m*, sekundäres Stoffwechselprodukt *n*
secondary mineral sekundäres Mineral *n*; allothigenes [nicht am Fundort entstandenes] Mineral *n*
secondary nucleation *(Krist)* sekundäre Keimbildung *f*, Sekundärkeimbildung *f*
secondary phosphate $M^I_2HPO_4$ Hydrogenphosphat *n*
secondary photochemical reaction photochemische Sekundärreaktion *f*
secondary plasticizer sekundärer [nicht lösender] Weichmacher *m*, Sekundärweichmacher *m*
secondary pollutant Sekundärschadstoff *m*
secondary product Nebenprodukt *n*
secondary reaction Nebenreaktion *f*, Sekundärreaktion *f*
secondary salt sekundäres Salz *n*, *(besser:)* Monohydrogensalz *n*, Hydrogensalz *n (einer dreibasigen Säure)*
secondary screening Nachklassierung *f*, Nachsortierung *f*; Feinsortierung *f*
secondary settling *(Hyd)* Nachklärung *f*
secondary settling tank *(Hyd)* Nachklärbecken *n*
secondary standard Sekundärstandard *m*, sekundärer Standard *m*
secondary structure 1. *(Bioch)* Sekundärstruktur *f*; 2. *(Krist)* Mosaikblöckchen *n*
secondary treatment 1. Nachbehandeln *n*, Nachbehandlung *f*; 2. *s.* secondary waste treatment
secondary valence Nebenvalenz *f*
secondary valence bond Nebenvalenzbindung *f*
secondary waste treatment (stage) *(Hyd)* zweite [biologische] Reinigungsstufe *f*
secrete *v (Bio)* sezernieren, absondern
secretion Sekretion *f*, Absonderung *f*
secretory *(Bio)* sekretorisch, den Sekretionsvorgang betreffend, Sekretions...
sectile schneidbar
section 1. Abschnitt *m*, Teil *m*; 2. Schnitt *m (Mikroskopie)*; 3. *(Geol)* Schliff *m*; 4. *(Kst)* Profil *n*
sedative *(Pharm)* Sedativum *n*, Beruhigungsmittel *n*
sediment *v* abscheiden, absitzen lassen *(Feststoffe aus Flüssigkeiten)*; einen Niederschlag bilden *(Flüssigkeit)*; sich (ab)setzen, sich abscheiden [ausscheiden, ablagern, niederschlagen], ausfallen, (ab)sinken, absitzen, sedimentieren *(Feststoffe)*
sediment *v* **out** abscheiden, absitzen lassen *(Feststoffe aus Flüssigkeiten)*; einen Niederschlag bilden *(Flüssigkeit)*
sediment 1. Sediment *n*, Niederschlag *m*, Bodensatz *m*, Satz *m*, Ausscheidung *f*, Ausscheidungsprodukt *n*, Ablagerung *f*; 2. *(Lebm)* Trub *m*
sedimentary sedimentär, Sediment(ations)..., Niederschlags..., Absatz..., Absetz...
sedimentate *v* abscheiden, niederschlagen, ausfällen, sedimentieren *(aus Flüssigkeiten)*
sedimentation 1. Abscheiden *n*, Absitzenlassen *n (von Feststoffen aus Flüssigkeiten)*; 2. *(Hyd)* mechanische Klärung *f*; Bildung *f* eines Niederschlags *(aus Flüssigkeiten)*; Absitzen *n*, Ausfallen *n*, Sinken *n*, Absinken *n*, Sedimentation *f (von Feststoffen)*

sedimentation basin *(Hyd)* Absetzbecken *n*, Klärbecken *n*
sedimentation centrifuge Sedimentierzentrifuge *f*, Absetzzentrifuge *f*, Vollmantelzentrifuge *f*, Dekantierzentrifuge *f*
sedimentation coefficient [constant] Sedimentationskonstante *f*, Sedimentationskoeffizient *m* *(Molekülmassenbestimmung)*
sedimentation equilibrium Sedimentationsgleichgewicht *n*
sedimentation rate Absetzgeschwindigkeit *f*, Sedimentationsgeschwindigkeit *f*, Absinkgeschwindigkeit *f*, Sinkgeschwindigkeit *f*
sedimentation sludge *(Hyd)* Sedimentationsschlamm *m*, Absetzschlamm *m*, Klärschlamm *m*
sedimentation tank 1. Sedimentationsbehälter *m*, Sedimentationsgefäß *n*, Absetzbehälter *m*, Absetzgefäß *n*, Absetzer *m*; 2. *(Hyd)* Absetzbecken *n*, Klärbecken *n*
seed *v* 1. *(Krist)* impfen, animpfen; 2. *(Biot)* (be)impfen, animpfen *(ein Nährmedium)*
seed 1. Samen *m*; 2. *s.* seed culture
seed crystal Impfkristall *m*, Keimkristall *m*, Anregekristall *m*
seed culture *(Biot)* Impfkultur *f*, Einsaatkultur *f*, frisch eingesäte Kultur *f*
seed oil Samenöl *n*
seed protectant Saat(gut)beizmittel *n*
seed technique Saattechnik *f*
seeded crystallization *(Krist)* Kristallisation *f* durch Impfen [Animpfen]
seep *v* **away** versickern
seep water *(Hyd)* Sickerwasser *n*, Gravitationswasser *n*
seepage Versickern *n*; Einsickern *n*, Durchsickern *n*
Seger cone *(Ker)* Seger-Kegel *m*, Brennkegel *m* [Schmelzkegel *m*] nach Seger
segregate *v* 1. absondern, entmischen, trennen, (ab)scheiden, isolieren; 2. *(Met)* seigern; sich absondern [entmischen]; 3. *(Met)* seigern
segregated flow Segregation *f*, Getrenntfluss *m*
segregation 1. Absonderung *f*, Entmischung *f*, Trennung *f*, Scheidung *f*, Abscheidung *f*, Segregation *f*; 2. *(Met)* Seigern *n*, Seigerung *f* *(1. Trennen von Metallschmelzen; 2. unerwünschte Entmischung von Legierungsbestandteilen)*
Seignette salt Seignettesalz *n*, Kaliumnatriumtartrat *n*
seize *v* sich festfressen [festklemmen]
seizing Festfressen *n*
seizure Festfressen *n*
selacholeic acid Selachensäure *f*, Nervonsäure *f*, *cis*-Tetracos-15-ensäure *f*
select *v* (aus)wählen; auslesen, selektieren
selection Auswählen *n*, Auswahl *f*, Wahl *f*; Auslese *f*, Selektion *f*
selection principle [rule] *(Spektr)* Auswahlregel *f*
selective selektiv (wirkend), auswählend
selective adsorption selektive Adsorption *f*
selective herbicide Selektivherbizid *n*, selektives [selektiv wirkendes] Herbizid *n*
selective hydrogenation selektive Hydrierung *f*, *(bei Fetten auch)* selektive Härtung *f*
selective solvent Selektivlösungsmittel *n*, selektives [selektiv wirkendes] Lösungsmittel *n*
selectivity 1. Selektivität *f*, selektive Wirkung *f*, Selektivwirkung *f*; 2. *(Anal)* Trennschärfe *f*
selectivity coefficient Selektivitätskoeffizient *m* *(Ionenaustausch)*
selenate $M^{I}_{2}SeO_4$ Selenat *n*
selenic Selen...; höherwertigem Selen entsprechend, *(meist)* Selen(VI)-...
selenic acid Selensäure *f*
selenide $M^{I}_{2}Se$ Selenid *n*
selenious Selen...; niederwertigem Selen entsprechend, *(meist)* Selen(IV)-...
selenious acid selenige Säure *f*
selenite $M^{I}_{2}SeO_3$ Selenit *n*
selenium Se Selen *n*
selenium cell Selenzelle *f*
selenium dioxide Selen(IV)-oxid *n*, Selendioxid *n*
selenium oxychloride Selenoxidchlorid *n*
selenone Selenon *n*
selenonium compound Selenoniumverbindung *f*

selenonium ion Selenonium-Ion *n*
self-absorption Selbstabsorption *f*, Eigenabsorption *f (Photochemie)*
self-aggregation Selbstaggregation *f*
self-assembled monolayer selbstangeordnete Monolage *f*
self-assembling system *(Bioch)* sich selbst organisierendes System *n*
self-assembly Selbstanordnung *f*
self-balancing recorder Kompensationsschreiber *m*
self-catalysis Autokatalyse *f*
self-cleaning Selbstreinigung *f (von Oberflächen)*
self-condensation Selbstkondensation *f*, extramolekulare Kondensation *f*
self-consistent field *(phys Ch)* selbstkonsistentes Feld *n*
self-curing *(Kst)* selbsthärtend, eigenhärtend
self-diffusion Selbstdiffusion *f*, Eigendiffusion *f*
self-dissociation *s.* self-ionization
self-extinguishing selbst(ver)löschend, selbstauslöschend
self-ignition temperature Selbstzünd(ungs)temperatur *f*, Selbstentzündungstemperatur *f*, SZT, Selbstzündpunkt *m*, Zündpunkt *m*
self-ionization Autoprotolyse *f*, Eigendissoziation *f*
self-ionization constant Autoprotolysekonstante *f*
self-ionization reaction Autoprotolysereaktion *f*
self-organisation Selbstorganisation *f*
self-purification Selbstreinigung *f*
self-quenching Selbstlöschung *f (Photochemie)*
self-quenching counter selbstlöschendes Zählrohr *n*, selbstlöschender Zähler *m*, Auslösezählrohr *n*, Auslösezähler *m*
self-reinforcing *(Kst)* selbstverstärkend
self-supporting *s.* self-sustained
self-sustained sich selbst unterhaltend *(Reaktion)*
self-sustaining sich selbst unterhaltend *(Reaktion)*
SEM *s.* scanning electron microscopy
semibatch reactor halbkontinuierlicher [halbkontinuierlich arbeitender] Reaktor *m*, Semibatch-Reaktor *m*
semicarbazide Semicarbazid *n*
semicarbazone Semicarbazon *n (Reaktionsprodukt von Aldehyden oder Ketonen mit Semicarbazid)*
semichemical pulping Halbzellstoffaufschluss *m*, halbchemischer Aufschluss *m*; Neutralsulfitverfahren *n*
semicolloid Semikolloid *n*, Halbkolloid *n*
semicommercial(-scale) halbtechnisch, in halbtechnischem Maßstab
semiconducting material Halbleitermaterial *n*, halbleitender Werkstoff *m*, Halbleiter *m*
semiconductivity Halbleitung *f*
semiconductor Halbleiter *m*
semiconductor laser Diodenlaser *m*, Halbleiterlaser *m*
semiconservation *(Bioch)* semikonservative Replikation *f (von DNA)*
semiconservative replication *(Bioch)* semikonservative Replikation *f (von DNA)*
semicrystalline halbkristallin
semidine rearrangement [transformation] Semidinumlagerung *f*, halbseitige [halbe] Benzidinumlagerung *f*
semidrying oil halbtrocknendes [mäßig trocknendes, schwach trocknendes] Öl *n*
semi-ester Halbester *m*
semihard rubber halbharter Gummi *m*, Halbhartgummi *m*, Semiebonit *n*
semikilled steel halbberuhigter Stahl *m*
semiladder polymer Halb-Leiterpolymer(e) *n*
semiliquid halbflüssig, dickflüssig, zähflüssig
semilog(arithmic) paper halblogarithmisches Papier *n*, einfach logarithmisch geteiltes Papier *n*, Exponentialpapier *n*
semimetal Halbmetall *n*
semimetallic halbmetallisch
semimicro analysis Halbmikroanalyse *f*, Semimikroanalyse *f*
semimicro balance Halbmikro(analysen)waage *f*
semimicro sample Halbmikroprobe *f*, *Mesoprobe *f (0,1 bis 0,01 g)*
semimicro-scale im Halbmikromaßstab
semiordered *(phys Ch)* halbgeordnet
semipermeability Semipermeabilität *f*

semipermeable semipermeabel, halbdurchlässig, einseitig durchlässig
semipermeable membrane semipermeable [halbdurchlässige] Membran *f*, semipermeable Scheidewand *f* [Trennwand *f*, Wand *f*, Haut *f*]
semipolar (double) bond semipolare [halbpolare] Bindung *f*
semiprecious stone Halbedelstein *m*
semiquantitative halbquantitativ, semiquantitativ
semiquinone *(org Ch)* Semichinon *n*, Halbchinon *n*, Merichinon *n*
semirefractory halbfeuerfest
semirigid plastic halbharter Kunststoff *m*
semisolid halbfest
semisynthesis Semisynthese *f*
semisynthetic fibre Regeneratfaser *f*; Regeneratfaserstoff *m*
semisynthetical semisynthetisch, halbsynthetisch
semisynthetics semisynthetisch [halbsynthetisch] hergestellte Stoffe *mpl*
semisystematic [semitrivial] name halbsystematischer [halbtrivialer] Name *m*, Halbtrivialname *m*, Semitrivialname *m*
semivitreous *(Ker)* halbverglast, halbglasartig
semiworks Halbbetriebsanlage *f*, Pilotanlage *f*, halbtechnische Versuchsanlage *f*, Versuchsbetrieb *m*
senecio alkaloid Senecioalkaloid *n*, Pyrrolizidinalkaloid *n*
seniority *(Nom)* Priorität *f*, Vorrang *m*
sense of rotation Dreh(ungs)sinn *m*
sensitive to acid säureempfindlich
sensitive to light lichtempfindlich
sensitiveness *s.* sensitivity
sensitivity *(Anal, Foto)* Empfindlichkeit *f*
sensitivity of detection *s.* sensitivity
sensitivity to acid Säureempfindlichkeit *f*
sensitivity to light Lichtempfindlichkeit *f*
sensitization 1. Sensibilisierung *f*; 2. *(Foto)* Empfindlichmachen *n*, Lichtempfindlichmachen *n*
sensitize *v* 1. sensibilisieren; 2. *(Foto)* (licht)empfindlich machen
sensitizer Sensibilisator *m*
sensitizing dye Sensibilisierungsfarbstoff *m*, sensibilisierender Farbstoff *m*
sensor Sensor *m*, Messfühler *m*
sensory examination [testing] organoleptische [sinnesphysiologische] Prüfung *f* [Untersuchung *f*], Organoleptik *f*
S.E.P. *s.* standard electrode potential
separability Trennbarkeit *f*, Abtrennbarkeit *f*, Abscheidbarkeit *f*
separable (ab)trennbar, abscheidbar
separate *v* 1. (auf)trennen *(in Komponenten)*; 2. *(Met)* scheiden; (ab)trennen, separieren, abscheiden *(Komponenten)*; sich scheiden [entmischen], sich (voneinander) trennen *(Gemische, Phasen)*; sich absondern [abscheiden] *(Komponenten)*
separate *v* **away [off]** (ab)trennen, separieren, abscheiden
separate *v* **out** (ab)trennen, separieren, abscheiden; sich ausscheiden [abscheiden], ausfallen
separating agent 1. Trennmittel *n*, Scheidemittel *n*; 2. *(Dest)* Trennmittel *n*, Zusatzmittel *n*, Zusatzstoff *m*, Hilfsstoff *m*, *(bei der Azeotropdestillation auch)* Schleppmittel *n*, Mitnehmer *m*, *(bei der Extraktivdestillation auch)* Extraktionsstoff *m*
separation 1. Trennen *n*, Auftrennen *n* *(in Komponenten)*; 2. *(Met)* Scheiden *n*; Abtrennen *n*, Trennen *n*, Abscheiden *n* *(von Komponenten)*; Entmischung *f*, Trennung *f* *(von Gemischen, Phasen)*; Absonderung *f*, Abscheidung *f* *(von Komponenten)*; 3. Abstand *m* *(z. B. im Kristallgitter)*
separation column *(Chromat)* Trennsäule *f*
separation efficiency Trennwirksamkeit *f*, Trennerfolg *m*; Trennleistung *f*
separation factor *(Anal)* Trennfaktor *m*, Separationsfaktor *m*
separation into groups *(Anal)* Gruppentrennung *f*, Trennung *f* in Gruppen
separation method Trennmethode *f*, Trennverfahren *n*, Trenntechnik *f*
separation of enantiomers *s.* enantiomer separation
separation scheme Trennungsgang *m*
separation task Trennaufgabe *f*, Trennproblem *n*
6σ separation *(Chromat)* 6σ-Trennung *f*, Basislinientrennung *f*
separator 1. Abscheider *m*, Scheider *m*,

Separator *m*, *(i. e. S.)* Zentrifuge *f*, Trennschleuder *f*; 2. Entstauber *m*, *(für hohe Rohgasstaubgehalte)* Abscheider *m*; 3. Sichter *m*
separator tank Absetztank *m*, Absetzbehälter *m*, Absetzgefäß *n*, Abscheider *m*, Scheider *m* *(Solventextraktion)*
separatory funnel Scheidetrichter *m*, Schütteltrichter *m*, Trenntrichter *m*
septavalent *s.* heptavalent
septic tank *(Hyd)* Faulgrube *f*
septum 1. Scheidewand *f*; 2. *(Hyd)* Stützgewebe *n* *(eines Anschwemmfilters)*
septum filtration *(Hyd)* Filtration *f* mittels Anschwemmfilters, Anschwemmfiltration *f*
sequence Sequenz *f*, Reihenfolge *f*, Folge *f*
sequence analysis [determination] *(Bioch)* Sequenzanalyse *f*, Sequenzermittlung *f* *(von Proteinen)*
sequence homology *(Bioch)* Sequenzhomologie *f* *(bei Proteinen)*
sequence of reactions Reaktionsfolge *f*
sequential analysis 1. Sequenzanalyse *f* *(Gegensatz: Simultananalyse)*; 2. *s.* sequence analysis
sequential mechanism *(Bioch)* sequenzieller Mechanismus *m* [Reaktionsmechanismus *m*]
sequester *v* maskieren
sequestering agent Maskierungsmittel *n*, Maskierungsreagens *n*, maskierender Stoff *m*, Komplexbildner *m*, Sequestriermittel *n*
sequestrant *s.* sequestering agent
sequestration Maskierung *f*, Komplexbildung *f*
SER *s.* smooth endoplasmatic reticulum
sericin Sericin *n*, Seidenleim *m*, Seidenbast *m* *(Eiweißkörper der Naturseide)*
series Serie *f*, Reihe *f*, Folge *f*; Gruppe *f* *(z. B. von Elementen, Verbindungen)*
series of reactions Reaktionskette *f*
series of stirred tanks Rührkesselkaskade *f*, Mischreaktorkaskade *f*, Kaskadenreaktor *m*, mehrstufiger Rührreaktor *m*
serine Serin *n*, Ser, 2-Amino-3-hydroxypropionsäure *f*
serine pathway *(Biot)* Serin-Weg *m* *(Methanolfermentation)*
serine protease *(Bioch)* Serinprotease *f*
serotonin Serotonin *n*, Enteramin *n*, 5-Hydroxy-tryptamin *n*, 5-HT
SERS *s.* surface-enhanced Raman spectroscopy
serum albumin Serumalbumin *n*, Blut(serum)albumin *n*, Plasmaalbumin *n*
serum globulin Serumglobulin *n*
serum protein Serumprotein *n*
service cycle 1. *(Filtr)* Arbeitszyklus *m*, Arbeitstakt *m*, Arbeitsphase *f*; 2. *(Hyd)* Beladungsvorgang *m*, Zyklus *m* Beladen, Austauschzyklus *m*, Austauschvorgang *m* *(Ionenaustausch)*
service life Gebrauchswertdauer *f*, Nutzungsdauer *f*
service temperature Gebrauchstemperatur *f*
sesquioxide Sesquioxid *n* *(veraltet, besser durch stocksche Benennung zu ersetzen)*
sesquiterpene *(org Ch)* Sesquiterpen *n*
sesquiterpenoid *(org Ch)* Sesquiterpenoid *n*
ses(ter)terpenoid *(org Ch)* Sesterterpenoid *n*
set *v* 1. aufstellen, aufbauen, montieren *(eine Apparatur)*; 2. hart [fest, starr] werden, erstarren *(z. B. Kolloide)*; erhärten, aushärten, sich verfestigen, abbinden *(Bindemittel)*; hart [fest, starr] werden lassen; 3. *(Gär)* anstellen *(Maische)*; 4. *(Text)* fixieren *(Gewebe, Farbstoffe)*
set *v* **free** freisetzen
set *v* **to a gel** *(Lebm)* gelieren, (zu Gelee) erstarren
set *v* **to *N*** auf *N* einstellen *(eine Lösung)*
set *v* **up** 1. aufstellen, aufbauen, montieren *(eine Apparatur)*; 2. erhärten; zusammenbacken, verbacken
set 1. Satz *m*; Block *m*, Einheit *f*, Gruppe *f*; 2. bleibende Verformung *f* [Deformation *f*, Gestaltsänderung *f*], Formänderungsrest *m*, Verformungsrest *m*; 3. *s.* setting 2.
set of conditions Bedingungsgefüge *n*, Faktorengefüge *n*, Faktorenkonstellation *f*
set of weights Wägesatz *m*
set point Sollwert *m*

set-up 1. Aufbau(satz) *m*, Apparatur *f*; 2. *s.* setting-up

setting 1. Aufstellen *n*, Aufbau *m*, Montage *f (einer Apparatur)*; 2. Hartwerden *n*, Festwerden *n*, Erstarren *n*, Erstarrung *f (z. B. von Kolloiden)*; Erhärtung *f*, Aushärtung *f*, Verfestigung *f*, Abbinden *n (von Bindemitteln)*; 3. *(Gär)* Anstellen *n (der Maische)*; 4. *(Text)* Fixierung *f (von Gewebe, Farbstoffen)*; 5. *s.* set 1.

setting accelerator *(Bau)* Abbindebeschleuniger *m*

setting point Erstarrungspunkt *m*, EP *m*; Stockpunkt *m*, Setting Point *m (bei Ölen)*

setting retarder *(Bau)* Abbindeverzögerer *m*

setting-up Erhärten *n*; Zusammenbacken *n*, Verbacken *n*

settle *v* 1. abscheiden, absitzen lassen *(Feststoffe aus Flüssigkeiten)*; einen Niederschlag bilden *(Flüssigkeit)*; 2. sich (ab)setzen, sich abscheiden [ausscheiden, ablagern, niederschlagen], ausfallen, (ab)sinken, absitzen, sedimentieren *(Feststoffe)* • **allow to settle** absitzen lassen • **let settle** absitzen lassen

settle *v* **down [out]** *s.* settle 2.

settleability *(Hyd)* Absetzfähigkeit *f*, Sedimentationsfähigkeit *f*, Absetzbarkeit *f*, Abscheidbarkeit *f*, Sedimentierbarkeit *f (z. B. von Flocken)*

settleable *(Hyd)* absetzfähig, sedimentationsfähig, sedimentierfähig, absetzbar, abscheidbar, sedimentierbar *(z. B. Flocken)*

settleable solids *(Hyd)* absetzbare Stoffe *mpl*, Absetzstoffe *mpl*, Sinkstoffe *mpl*

settled material Sinkgut *n*

settler 1. Sedimentationsbehälter *m*, Sedimentationsgefäß *n*, Absetzbehälter *m*, Absetzgefäß *n*, Absetzer *m*; 2. *(Hyd)* Absetzbecken *n*, Klärbecken *n*; Abscheider *m*, Scheider *m*, Absetzbehälter *m*, Absetzgefäß *n*, Absetztank *m (Solventextraktion)*

settling 1. Abscheiden *n*, Absitzenlassen *n (von Feststoffen aus Flüssigkeiten)*; Bildung *f* eines Niederschlags *(von Flüssigkeiten)*; Absitzen *n*, Ausfallen *n*, Sinken *n*, Absinken *n*, Sedimentation *f (von Feststoffen)*; 2. Niederschlag *m*, Bodensatz *m*, Satz *m*, Ausscheidung *f*, Ausscheidungsprodukt *n*, Ablagerung *f*, Sediment *n*; 3. *(Lebm)* Trub *m*

settling basin *(Hyd)* Absetzbecken *n*, Klärbecken *n*

settling chamber Absetzkammer *f*, Abscheidekammer *f*, Staubkammer *f*; Kammerabscheider *m*

settling cone Absetzkonus *m*

settling curve Absetzkurve *f*, Sedimentationskurve *f*

settling cylinder Absetzglas *n*, Standzylinder *m*, Standglas *n*

settling tank *s.* settler

settlings *s.* settling 2.

seven-membered ring siebengliedriger Ring *m*, Sieben(er)ring *m*

severe conditions of cracking scharfe Krackbedingungen *fpl*

sewage *(Hyd)* häusliches Abwasser *n*; kommunales Abwasser *n*, Schmutzwasser *n*

sewage-contaminated abwasserbelastet, durch Abwässer verschmutzt

sewage disposal Abwasserbeseitigung *f*, Abwasserentsorgung *f*

sewage farm Rieselfeld *n*

sewage gas Faulgas *n*, Biogas *n*

sewage pollution Verschmutzung *f* [Verunreinigung *f*] durch Abwässer, Abwasserbelastung *f*

sewage pump Abwasserpumpe *f*, Schmutzwasserpumpe *f*

sewage purification Abwasserreinigung *f*

sewage sample Abwasserprobe *f*

sewage sludge Abwasserschlamm *m*

sewage treatment Abwasserbehandlung *f*

sewage-treatment lagoon Abwasserteich *m*

sewage-treatment plant Abwasserbehandlungsanlage *f*, ABA, Abwasserreinigungsanlage *f*, Kläranlage *f*, KA, Klärwerk *n*

sewage water *s.* sewage

sewer *v* ablassen *(Flüssigkeiten)*

sewer Abwasserkanal *m*, Ablaufkanal *m*, Ablasskanal *m*

sewer system Kanalisationssystem *n*, Kanal(isations)netz *n*, Kanalisation *f*, Entwässerungssystem *n*, Entwässe-

rungsnetz *n*, Abwasser(ableitungs)netz *n*
sewerage 1. *(Hyd)* Abwasserableitung *f* und -beseitigung *f*; 2. *(Hyd)* Stadtentwässerung *f*; Ortsentwässerung *f*; 3. *s.* sewer system; 4. *s.* sewage
sex attractant Sexuallockstoff *m*
sex hormone Sexualhormon *n*, Geschlechtshormon *n*, Sexagen *n*
sexadentate sechszähnig, sechszählig *(Komplexchemie)*
sexavalent *s.* hexavalent
sextuple evaporator Sechsstufenverdampfer *m*, Sechskörperverdampfer *m*
sexual attractant *s.* sex attractant
S.G. *(specific gravity) s.* relative density
shade *v* schattieren, nuancieren, (ab)tönen
shade Schattierung *f*, Nuancierung *f*, Nuance *f*, Farbnuance *f*, Farbton *m*, Farbtönung *f*, Stich *m*
shaft 1. Welle *f*; Stange *f*; 2. Schacht *m* *(eines Schachtofens)*
shaft furnace *(Met)* Schachtofen *m*
shaft kiln Schachtofen *m* *(besonders zum Kalzinieren und Sintern)*
shake *v* (durch)schütteln
shake *v* **out** ausschütteln *(mit Extraktionsmittel)*
shake Schütteln *n*, Durchschütteln *n*; Ausschütteln *n*
shaker *s.* shaking machine
shaker screen Schüttelsieb *n*, Schüttelsortierer *m*
shaking Schütteln *n*, Durchschütteln *n* • **with shaking** unter Schütteln [Umschütteln] • **with constant shaking** unter ständigem Schütteln • **with occasional shaking** unter gelegentlichem Schütteln
shaking bottle Schüttelflasche *f*
shaking machine Schüttelmaschine *f*, Schüttelapparat *m*
shaking screen *s.* shaker screen
shaking table Schüttelherd *m*; Stoßherd *m*, Wurfherd *m*
shale (clay) *(Geol)* Schiefer(ton) *m*
shale oil Schieferöl *n*
shaly schief(e)rig, schieferartig, geschiefert; schieferhaltig
shape *v* formen, gestalten; profilieren
shape 1. Form *f*, Gestalt *f*; Profil *n*; 2. Formteil *n*
shape memory effect Formgedächtniseffekt *m*
shape memory polymers Formgedächtnispolymere *npl*
sheet metal forming *(Tech)* Umformverfahren *n*
shape resonance *(phys Ch)* Gestaltresonanz *f*
shaping Formen *n*, Gestalten *n*, Formgebung *f*
share *v* (auf)teilen; teilhaben [beteiligt sein] an
shared electron gemeinsames Elektron *n*
shared pair of electrons gemeinsames Elektronenpaar *n*, Bindungselektronenpaar *n*
sharp-melting scharf schmelzend, mit scharfem Schmelzpunkt
sharpness of separation Trennschärfe *f*, Trenngrad *m*
shatter Zertrümmern *n*, Zertrümmerung *f*, Zerspringen *n*, Zersplittern *n*, Splittern *n*
shattering power Sprengkraft *f*, Brisanz *f*
shatterproof splitterfest, splittersicher, splitterfrei, bruchsicher
shavings Späne *mpl*; Metallspäne *mpl*
SHE *s.* 1. superheavy element; 2. standard hydrogen electrode
shear *v* 1. (ab)scheren, (ab)schneiden; scherzerkleinern; 2. verschieben, einer Schubwirkung [Scherung] aussetzen
shear 1. Scherung *f*, Schub *m*; 2. (große) Schere *f*
shear force Scherkraft *f*
shear modulus (of elasticity) Schermodul *m*, Schubmodul *m*, Gleitmodul *m*
shear rate Scherrate *f*, Schergeschwindigkeit *f*
shear strength Scherfestigkeit *f*
shear stress Scherspannung *f*, Schubspannung *f*, Scherbeanspruchung *f*, Schubbeanspruchung *f*
shear thickening Fließverfestigung *f*
shear thinning Strukturviskosität *f*, strukturviskoses Verhalten *n*
sheath Hülle *f*, Mantel *m*, Umhüllung *f*, Ummantelung *f*
sheathe *v* umhüllen, ummanteln
sheathing 1. Umhüllen *n*, Umhüllung *f*,

Ummanteln *n*, Ummantelung *f*; 2. *s.* sheath
sheen Glanz *m*, Schimmer *m*
sheet 1. Schicht *f*; 2. *(Met)* Blech *n*; 3. *(Pap)* Bogen *m*; 4. *(Kst)* Folie *f (als Stück; Dicke > 0.01 inch = 0,25 mm)*
sheet aluminium Aluminiumblech *n*
sheet-drawing process Tafelglasziehverfahren *n*
sheet extrusion *(Kst)* Folienstrangpressen *n*
sheet filter Schichten(platten)filter *n*
sheet glass Tafelglas *n*
sheet silicate *(Min)* Phyllosilicat *n*, Blattsilicat *n*, Schichtsilicat *n*
sheet structure Faltblattstruktur *f*, Beta-Struktur *f (von Eiweißen)*
shelf dryer Plattentrockner *m*
shelf life 1. Lager- und Verarbeitbarkeitsdauer *f*, Gebrauchsfähigkeitsdauer *f*; 2. *(Lebm)* Lagerdauer *f*
shell Schale *f*, Hülle *f*, Mantel *m*; Gehäuse *n*
shell-and-tube heat exchanger Rohrbündelwärmeübertrager *m*
shell model Schalenmodell *n (der Elektronenbahnen und Atomkerne)*
shell structure Schalenstruktur *f*, Schalenaufbau *m*
shell-type boiler Mantelkessel *m*, Kessel *m* in Mantelform
shellac Schellack *m*
shellfish poison Muscheltoxin *n*
sherardize *v* sherardisieren, in Pulver verzinken *(nach dem Diffusionsverfahren)*
sherardizing Sherardisieren *n*, Verzinken *n* im Pulver *(Diffusionsverzinken nach Sherard)*
shielding constant *(phys Ch)* Abschirmkonstante *f*
shielding gas Schutzgas *n*
shift *v* 1. verschieben; sich verschieben, wandern; 2. konvertieren *(Synthesegase)*
shift *v* **back** zurückwandern
shift 1. Verschiebung *f*, *(bei Atomen und Atomgruppen auch)* Platzwechsel *m*, Wanderung *f*; 2. Konvertierung *f*, Konversion *f (von Synthesegas)*
shift conversion *(Carb)* CO-Konvertierung *f*, Kohlenmonoxid-Konversion *f*
shift to the red Rotverschiebung *f*, Verschiebung *f* nach Rot
1,2-shift 1,2-Umlagerung *f*, Ortho-Umlagerung *f*
shikimic acid Shikimisäure *f*, 3,4,5-Trihydroxycyclohexen-1-carbonsäure *f*
shikimic acid pathway *(Bioch)* Shikimisäureweg *m*
Shine-Delgarno sequence *(Bioch)* Shine-Delgarno-Sequenz *f (purinreiche Sequenz der Messenger-RNA)*
shiny glänzend
shipment Versand *m*
shipping regulation Versandvorschrift *f*
shock-absorbing stoßdämpfend
shock front Stoßfront *f (Kinetik)*
shock tube technique Stoßrohrverfahren *n*, Stoßwellenverfahren *n (zur Untersuchung schneller Reaktionen)*
shock wave Stoßwelle *f (Kinetik, Sprengtechnik)*
shockproof stoßfest, stoßsicher
Shore hardness Shore-Härte *f*
short-circuit *v* kurzschließen
short-acting *(Pharm)* raschwirkend, Kurzzeit...
short-chain kurzkettig
short-chain branching Kurzkettenverzweigung *f*
short-circuiting 1. Kurzschließen *n*; 2. Ausbildung *f* [Auftreten *n*] von Kurzschlussströmungen, Auftreten *n* eines Kurzschlusses *(im Reaktor)*
short-lived *(phys Ch)* kurzlebig
short-neck flat-bottom flask Kurzhalsstehkolben *m*
short-oil alkyd kurzöliges Alkyd(harz) *n*
short-path cracking Kurzwegkrackung *f*
short-path distillation Kurzwegdestillation *f*
short-path evaporator Kurzwegverdampfer *m*
short-path high-vacuum distillation Kurzwegdestillation *f*
short period kurze [kleine] Periode *f*, Kurzperiode *f (im Periodensystem)*
short periodic table Kurzperiodensystem *n*
short-range order *(Krist)* Nahordnung *f*
short residue [residuum] kurzer Destillationsrückstand *m*, Vakuumrück-

stand *m*, Rückstand *m* einer Vakuumkolonne, Short Residuum *n*
short-term ... *siehe: short-time ...*
short-time fermentation *(Biot)* Kurzzeitfermentation *f*
short-time high-temperature pasteurization *(Lebm)* Kurzzeitpasteurisation *f*, Hochkurzpasteurisation *f*, HTST-Pasteurisation *f*
short-time test Kurzzeitversuch *m*, Kurz(zeit)test *m*
short-tube evaparator Kurzrohrverdampfer *m*
short-wavelength region Kurzwellenbereich *m*; kurzwelliger Spektralbereich *m*, kurzwelliges Spek-tralgebiet *n*
shortstop *v* abstoppen, inhibieren, abbrechen *(die Polymerisation)*
shortstop agent Abstoppmittel *n*, Polymerisationsabstoppmittel *n*, Polymerisationsabstopper *m*, Inhibitor *m*
shortstopping agent Abstoppmittel *n*, Polymerisationsabstoppmittel *n*, Polymerisationsabstopper *m*, Inhibitor *m*
shot *v* verperlen; granulieren
shot 1. Schuss *m*, Sprengung *f*, Explosion *f*; 2. *(Kst)* Schuss *m*; 3. verperltes Produkt *n*; Granulat *n*
shot noise *(Anal)* Schrotrauschen *n*
shovel Schaufelblatt *n*, Schaufel *f*, Blatt *n*, Flügel *m*
show bottle Schauglas *n*, Präparatenglas *n*
shower *v* rieseln lassen
shower of electrons Elektronenschauer *m*
shower-proof regendicht, wasserdicht
shred *v* zerfasern; zerkleinern *(Müll)*
shredder Zerfaserer *m*
shredding machine *s.* shredder
shreds Schnitzel *npl*
shrink *v* 1. schrumpfen, schwinden; 2. *(Text)* krumpfen, einlaufen, eingehen
shrink resistance *(Text)* Schrumpffestigkeit *f*, Schrumpfbeständigkeit *f*, Krumpffestigkeit *f*, Krumpfbeständigkeit *f*
shrink-resistant *(Text)* schrumpffest, schrumpfbeständig, krumpffest, krumpfbeständig, nicht krumpfend [schrumpfend, einlaufend]
shrinkage 1. Schrumpfen *n*, Schrumpfung *f*, Schwinden *n*, Schwindung *f*, Schwund *m*; 2. *(Text)* Krumpfen *n*, Krumpfung *f*, Einlaufen *n*, Eingehen *n*
shrinkage crack Schrumpfriss *m*, Schwind(ungs)riss *m*, Schwundriss *m*
shrinking effect Schrumpfungseffekt *m* *(Kunststoffverarbeitung)*
shrinking factor Kontraktionsfaktor *m* *(verzweigte Polymere)*
shut *v* **down** außer Betrieb setzen, stilllegen
shut *v* **off** absperren; schließen *(ein Hahnventil)*
shut *s.* shutdown
shutdown Außerbetriebsetzung *f*, Stilllegung *f*
shuttle (system) *(Bioch)* Shuttle *m*, einschleusendes Stoffsystem *n*, Schleuse *f*
shuttle vector bifunktioneller Vektor *m*, Shuttle-Vektor *m*
sial *(Geoch)* Sial *n*, Sal *n*, Granitschale *f*
sialic acid Sial(in)säure *f (allg. für eine acylierte Neuraminsäure)*
SIC *s.* specific inductive capacity
siccative Sikkativ *n*, Trockenstoff *m*, Trockner *m*
sickle-cell haemoglobin *(Bioch)* Sichelzellenhämoglobin *n*, Hämoglobin *n* S, HbS
side/on mit der Breitseite dem Betrachter zugekehrt *(Atomgruppe)*
side-band *(Spektr)* Nebenbande *f*
side chain Seitenkette *f*
side-chain degradation Seitenkettenabbau *m*
side-chain halogenation Seitenkettenhalogenierung *f*, Halogenierung *f* in der Seitenkette
side-chain substitution Seitenkettensubstitution *f*
side effect *(Pharm)* Nebenwirkung *f*
side-group Seitengruppe *f*
side of fusion *(Nom)* Anellierungsseite *f*, Anellierungsstelle *f*, Kondensstelle *f*, gemeinsame Seite *f*
side product 1. Nebenprodukt *n (einer Reaktion)*; 2. *(Dest)* Seitenprodukt *n*
side reaction Nebenreaktion *f*
side stream Seitenstrom *m*
side stripper *(Erdöl)* Seitenturm *m*, Seitenkolonne *f*, Abstreiferkolonne *f*, Abtreiberkolonne *f*, Strippkolonne *f*, Ab-

streifer *m*, Nebenturm *m*, Nebenkolonne *f*, Stripper *m*
siderite 1. Siderit *m*, Eisenmeteorit *m*, Meteoreisen *n*; 2. *(Min)* Siderit *m*, Eisenspat *m (Eisen(II)-carbonat)*
siderophile element *(Geoch, Met)* siderophiles [eisenfreundliches] Element *n (mit geringer Affinität zu Sauerstoff und Schwefel)*
Siemens-Martin ... *siehe: open-hearth ...*
sienna Sienaerde *f (Eisenoxidhydrat)*
sieve *v* sieben
sieve Sieb *n*
sieve analysis Siebanalyse *f*
sieve area Siebfläche *f*
sieve-plate cascade reactor *(Biot)* Siebbodenreaktor *m*
sieve-plate column [tower] Siebbodenkolonne *f*
sieve size Sieböffnungsgröße *f*, Öffnungsweite *f* (des Siebs)
sift *v* (durch)sieben, (durch)beuteln
sifter Siebapparat *m*, Siebmaschine *f*, Siebvorrichtung *f*, Sieb *n*
sigma-adduct *(org Ch)* σ-Komplex *m*
sigma bond Sigmabindung *f*, σ-Bindung *f*, σ-Elektronenpaarbindung *f*
sigma-constant *(org Ch)* Substituentenkonstante *f*
sigma-orbital σ-Orbital *n*, σ-Molekülorbital *n*, σ-MO *n*
sigmatropic rearrangement *(org Ch)* sigmatrope Umlagerung *f*
1,5 sigmatropic shift of hydrogen 1,5-Wasserstoff-Verschiebung *f*
signal peptide *(Bioch)* Signalsequenz *f*
signal recognition particle *(Bioch)* Signalerkennungspartikel *n*
signal sequence *s.* signal peptide
signal splitting *(Spektr)* Signalaufspaltung *f*
signal-to-noise (ratio) *(Anal)* Signal-(zu-)Rausch-Verhältnis *n*, Signal-Untergrund-Verhältnis *n*
silane Si_nH_{2n+2} Silan *n*, Siliciumwasserstoff *m*, Siliciumhydrid *n*, *(i. e. S.)* SiH_4 Silan *n*, Monosilan *n*
silanization *(Chromat)* Silanisierung *f*
silanize *v (Chromat)* silanisieren
silanol Silanol *n (eine Siliciumverbindung mit OH-Gruppen)*; *(i. e. S.)* H_3SiOH Silanol *n*
silazane Silazan *n (siliciumorganische Verbindung)*
silica Siliciumdioxid *n*, Silicium(IV)-oxid *n*
silica gel Kiesel(säure)gel *n*
silica glass Kieselglas *n*
silica refractory SiO_2-reiches Schamotteerzeugnis *n (mit > 92 % Siliciumdioxid)*
silica sol Kiesel(säure)sol *n*
silica surface Siliciumoberfläche *f*
silicate 1. Silicat *n*; 2. *(Tech)* Silikat *n*
silicate glass Silikatglas *n*
siliceous siliciumdioxidhaltig, kieselsäurehaltig, Kiesel(säure)...; silicatisch, Silicat...
silicic acid Kieselsäure *f*, *(i. e. S.)* H_4SiO_4 Orthokieselsäure *f*, Tetroxokieselsäure *f*, Kieselsäure *f*
silicic-acid ester Kieselsäureester *m*, Siliciumester *m*
silicide Silicid *n*
silicify *v (Geoch)* verkieseln, sili(zi)fizieren
silicomanganese steel Mangansiliciumstahl *m*
silicon Si Silicium *n*
silicon bronze Siliciumbronze *f*
silicon carbide Siliciumcarbid *n*
silicon-carbon bond Silicium-Kohlenstoff-Bindung *f*, Kohlenstoff-Silicium-Bindung *f*
silicon cast iron siliciumlegiertes Gusseisen *n*, Siliciumguss *m*
silicon dioxide Siliciumdioxid *n*, Silicium(IV)-oxid *n*
silicon-functional siliciumfunktionell
silicon hydride *s.* silane
silicon iron Siliciumeisen *n*
silicon rectifier Siliciumgleichrichter *m*
silicon tetrachloride Siliciumtetrachlorid *n*, Silicium(IV)-chlorid *n*, Tetrachlorsilan *n*
silicon tetrafluoride Siliciumtetrafluorid *n*, Silicium(IV)-fluorid *n*, Tetrafluorsilan *n*
silicon wafer Siliciumscheibe *f*
silicone-treat *v* mit Silikonen behandeln, silikonisieren
silicone Silikon *n*, *Polyorganosiloxan *n*, Organopolysiloxan *n*
silicone-coated silikonbeschichtet
silicone elastomer Silikonelastomer(e) *n*, elastomeres Silikon *n*

silicone film Silikon(schutz)film *m*, silikonhaltiger Film *m*, Silikonhaut *f*
silicone fluid Silikonflüssigkeit *f*, Silikonfluid *n*, Silikonöl *n*
silicone grease Silikonfett *n*
silicone lubricant Silikonschmiermittel *n*
silicone oil *s.* silicone fluid
silicone polymer Organopolysiloxanpolymer(e) *n*, Silikonpolymer(e) *n*
silicone resin Organopolysiloxanharz *n*, Silikon(kunst)harz *n*
silicone rubber Silikongummi *m (Silikonkautschukvulkanisat)*
silicone-treated silikonbehandelt, silikonisiert
silicone treatment Silikonbehandlung *f*, Behandeln *n* mit Silikonen, Silikonisieren *n*, Silikonisierung *f*
silicone varnish Silikonlack *m*
siliconization 1. Silikonisieren *n*, Silikonisierung *f*, Silikonbehandlung *f*, Behandeln *n* mit Silikonen; 2. *(Met)* Insilizieren *n*, Silizieren *n (Zementation mit Silicium)*
siliconize *v* 1. silikonisieren, mit Silikonen behandeln; 2. *(Met)* (in)silizieren *(mit Silicium zementieren)*
silicono... (HO)OSi– Silicono...
silicosis *(Med)* Silikose *f*
silicospiegel Silikospiegel *m (eine Eisen-Silicium-Mangan-Legierung)*
silicothermic silikothermisch
silk Seide *f*
silky seidenartig, seidig
sillimanite refractory *(Ker)* Sillimaniterzeugnis *n*
silo Silo *n(m)*, Bunker *m*, Behälter *m*, Zellenspeicher *m*
siloxane Siloxan *n*
siloxane polymer Siloxanpolymer(e) *n*
siloxen(e) Siloxen *n*
silt *v* **(up)** verschlammen
silt 1. Schluff *m (Korngröße 0,026 bis 0,002 mm)*; 2. *(Geol)* Silt *m*, Feinsand *m*
silver *v* versilbern
silver-plate *v* elektrochemisch [galvanisch] versilbern
silver Ag Silber *n*
silver azide Silberazid *n*
silver bromide 1. Silberbromid *n*; 2. *(Foto)* Bromsilber *n*
silver chloride 1. Silberchlorid *n*; 2. *(Foto)* Chlorsilber *n*
silver-chloride electrode Silberchloridelektrode *f*
silver coulometer Silbercoulometer *n*
silver cyanide Silbercyanid *n*
silver electrode Silberelektrode *f*
silver fluoride Silberfluorid *n*, *(i. e. S.)* AgF Silber(I)-fluorid *n*, Silber(mono)-fluorid *n*
silver fulminate AgONC Silberfulminat *n*
silver halide 1. Silberhalogenid *n*; 2. *(Foto)* Halogensilber *n*
silver iodide 1. Silberiodid *n*; 2. *(Foto)* Iodsilber *n*
silver mirror Silberspiegel *m (durch Reduktion erzeugter Silberniederschlag)*
silver nitrate Silbernitrat *n*
silver-nitrate paper Silbernitratpapier *n*
silver oxide Silberoxid *n*, *(i. e. S.)* Ag_2O Silber(I)-oxid *n*, Silberoxid *n*
silver plating Versilbern *n*
silver-plating bath Versilberungselektrolyt *m*, Silberelektrolyt *m*
silver ruby *(Min)* dunkles Rotgültigerz *n*, Pyrargyrit *m (Antimon(III)-silbersulfid)*
silver-silver chloride electrode Silber-Silberchlorid-Elektrode *f*
silver sulphate Silbersulfat *n*
silver sulphide Silbersulfid *n*
silver thiocyanate Silberthiocyanat *n*
silvichemical Silvichemikalie *f*
silvicide Gehölzvernichtungsmittel *n*, Arborizid *n*
silyl group H_3Si– Silylgruppe *f*
silylate *v* silylieren
silylation Silylierung *f*
silylene H_2Si= Silylen *n*
silylidyne HSi≡ Silylidin *n*
sima *(Geoch)* Sima *n*, Simazone *f*, Gabbroschale *f*
simazine Simazin *n*, 6-Chlor-2,4-bisethylamino-1,3,5-triazin *n (Herbizid)*
similar joint action *(Tox)* Ähnlichkeitsverbundwirkung *f*
similarity Ähnlichkeit *f*
similitude Ähnlichkeit *f*
Simmons-Smith reaction Simmons-Smith-Reaktion *f (zur Darstellung von Cyclopropanen)*
simple distillation einfache Destillation *f*, Gleichstromdestillation *f*, Abtrieb *m*, Abtreiben *n*
simple lipid einfaches Lipid *n*

simple protein einfacher Eiweißstoff *m* [Eiweißkörper *m*], einfaches [reines] Protein *n*
simple reaction einfache Reaktion *f*
simple replacement Substitution *f*
simple salt einfaches Salz *n*, Einfachsalz *n*
simple step Elementarschritt *m (einer Reaktion)*
simple sugar einfacher Zucker *m*, Einfachzucker *m*, Monosaccharid *n*
simple triglyceride gleichsäuriges [einsäuriges, homogenes] Triglycerid *n*
simplest (possible) formula einfachst mögliche Formel *f*, einfachste Formel *f* [Summenformel *f*, Bruttoformel *f*], *(i. e. S.)* empirische Formel *f*, Verhältnisformel *f*, stöchiometrische Formel *f* [Grundformel *f*], Substanzformel *f*
SIMS *s.* secondary ion mass spectrometry
simulated body fluid künstliche Körperflüssigkeit *f*
simulation Simulation *f (Mathematik)*
simultaneous gleichzeitig
simultaneous analysis Simultanbestimmung *f*
simultaneous precipitation Mitfällung *f*
simultaneous reaction Simultanreaktion *f*
sinapic acid Sinapinsäure *f*, Hydroxydimethoxyzimtsäure *f*
sinapic alcohol Sinapinalkohol *m*
single-acting einfachwirkend
single-beam instrument *(Spektr)* Einstrahlgerät *n*
single bond Einfachbindung *f*
single cell *(Bio, Biot)* Einzelzelle *f*
single-cell protein Einzellerprotein *n*, Single-cell-Protein *n*, SCP *n*
single-compartment thickener Einkammereindicker *m*
single covalent bond Einfachbindung *f*
single crystal Ein(zel)kristall *m*, Einling *m*
single-crystal fibre Einkristallfaser *f*
single-crystal structure determination *(Krist)* Einkristallstrukturaufklärung *f*
single daylight press Einetagenpresse *f*
single-displacement mechanism sequenzieller Mechanismus *m (Enzymkinetik)*
single dose *(Pharm)* Einzeldosis *f*
single-effect evaporator Einstufenverdampfer *m*, Einkörperverdampfer *m*
single-electrode potential Einzel(elektroden)potenzial *n*, E
single-electron bond Einzelelektron(en)bindung *f*, Singulettbindung *f*
single-flash curve Gleichgewichtssiedekurve *f*, Siedekurve *f* bei geschlossener Verdampfung, Flashkurve *f*
single-function catalyst monofunktioneller Katalysator *m*
single-ion activity Einzelionenaktivität *f*
single line *(Spektr)* Einzellinie *f*
single medium filter Einschicht(en)filter *n*, Einfachfilter *n*
single-pan analytical balance Einschalen-Analysenwaage *f*
single-pass evaporator Durchlaufverdampfer *m*
single-purpose fermenter *(Biot)* Spezialfermenter *m*
single-scattering *(phys Ch)* Einfachstreuung *f*
single-stage compressor einstufiger Verdichter *m* [Kompressor *m*]
single-stage fermentation *(Biot)* einstufige Fermentation *f*, Einzelfermentation *f*
single-step reaction Einstufenreaktion *f*
single strand Einzelstrang *m (z. B. eines Makromoleküls)*
single-strand DNA *(Bioch)* Ein(zel)-strang-DNA *f*, ein(zel)strängige DNA *f*
single-strand DNA-binding protein *(Bioch)* ssb-Protein *n*, HD-Protein *n*, helixdestabilisierendes Protein *n*
single-strand helix *(Bioch)* Einzelhelix *f*, einsträngige Helix *f*
single-strand polymer einsträngiges Polymer *n*
single-sweep polarography *(Anal)* Katodenstrahlpolarographie *f*, *Single-sweep-Polarographie *f*
single-trip container Einwegbehälter *m*
single-wall carbon nanotube einwandiges Kohlenstoff-Nanoröhrchen *n*
single withdrawal einphasige Entnahme *f (beim Extrahieren)*
singlet molecular oxygen Singulett-Sauerstoff *m*
singlet-singlet transition Singulett-(Singulett-)Übergang *m*

singlet state Singulettzustand *m*, S-Zustand *m*
singlet-triplet transition Singulett-Triplett-Übergang *m*
singly bonded [bound] einfach gebunden
sinigrin Sinigrin *n* *(ein Thioglykosid)*
sink *v* (ab)sinken; untersinken; versenken
sink 1. Abwaschbecken *n*, Abflussbecken *n*, Spülbecken *n*, Ausgussbecken *n*, Ausguss *m*; 2. Senke *f* *(von Schadstoffen, als Gegensatz zur Schadstoffquelle)*; 3. *s.* sewer
sink-float separation Schwimm-(und-) Sink-Aufbereitung *f*, Schwerflüssigkeitsaufbereitung *f*, *(i. e. S.)* Schwertrübeaufbereitung *f*
sinter *v* sintern
sinter Sinter *m*, Sintergut *n*
sinter roasting Sinterrösten *n*, Sinterröstung *f*
sintered carbide Sinterhartmetall *n*, (gesintertes) Hartmetall *n*, HM
sintered-glass crucible Glasfrittentiegel *m*, Glasfiltertiegel *m*
sintered-glass filter Glasfritte *f*, Glasfilter *n*
sintered material Sinterwerkstoff *m*
sintered metal Sintermetall *n*
sintered steel Sinterstahl *m*
sintering Sintern *n*, Sinterung *f*
sintering furnace Sinterofen *m*
sintering machine Sinterapparat *m*, Sintermaschine *f*
siphon *v* **(off)** abhebern, *(i. w. S.)* absaugen
siphon Heber *m*, Saugheber *m*; Siphon *m*, Siphonflasche *f*; Saugstrahlpumpe *f*, Ejektor *m*
SIR *s.* styrene-isoprene rubber
sirup *s.* syrup
S.I.T. *s.* spontaneous-ignition temperature
site of attack Angriffsstelle *f*, Angriffsort *m*
site of synthesis *(Bioch)* Syntheseort *m*
site-specific *(Biot)* regiospezifisch
site specificity *(Biot)* Regiospezifität *f*
six-carbon chain C_6-Kette *f*
six-coordinate(d) sechsfach koordinativ gebunden, mit sechs koordinativen Bindungen; sechsfach koordiniert *(Komplex)*
six-membered ring sechsgliedriger Ring *m*, Sechs(er)ring *m*
six-pi-electron system 6π-Elektronensystem *n*
six-valent sechswertig
sixfold axis of symmetry sechszählige Symmetrieachse *f*
size *v* 1. klassieren, nach Korn(größen)-klassen trennen [einteilen, zerlegen]; 2. *(Text)* schlichten; 3. *(Pap)* leimen; 4. bemessen, dimensionieren, auslegen *(Anlagen)*
size-separate *v* nach der Größe trennen
size 1. Größe *f*, Umfang *m*, Format *n*, Weite *f*, Abmessung *f*; 2. *s.* sizing material
size analysis Siebanalyse *f*, *(i. w. S.)* Körnungsanalyse *f*
size classification *s.* sizing 1.
size determination Größenbestimmung *f*
size distribution Größenverteilung *f* *(z. B. von Körnern)*; Weitenverteilung *f* *(z. B. von Poren)*
size-exclusion chromatography *(Chromat)* Größenausschlusschromatographie *f*; Ausschlusschromatographie *f*
size fraction Fraktion *f*, Korn(größen)-klasse *f*, Kornfraktion *f*, Korngruppe *f*
size of separation Trennkorngröße *f*, Kornscheide *f*
size range Kornspanne *f*, Korngrößenbereich *m*, Körnungsbereich *m*
size reduction Zerteilen *n*, Zerteilung *f*, *(bei festen Stoffen:)* Zerkleinern *n*, Zerkleinerung *f*
size separation *s.* sizing 1.
sizing 1. Klassieren *n*, Trennen *n* nach Korn(größen)klassen; 2. *(Text)* Schlichten *n*; 3. *(Pap)* Leimung *f*, Leimen *n*, Papierleimung *f*; 4. *s.* sizing material; 5. Bemessung *f*, Auslegen *n* *(z. B. von Anlagen)*
sizing bath 1. *(Text)* Schlichtebad *n*, Schlichteflotte *f*; 2. *(Pap)* Leimbad *n*
sizing material 1. *(Text)* Schlichte *f*, Schlichtemittel *n*, Appretur *f*; 2. *(Pap)* Leimungsmaterial *n*, Leim(stoff) *m*, Leim(ungs)mittel *n*
skatole Skatol *n*, 3-Methylindol *n*
skeletal isomer Skelettisomer(e) *n*, Kettenisomer(e) *n*

skeletal isomerism Skelettisomerie *f*, Kettenisomerie *f*
skeletal polysaccharide Gerüstpolysaccharid *n*
skeletal protein Gerüsteiweiß *n*, Skleroprotein *n*, Strukturprotein *n*, Faserprotein *n*
skeletal rearrangement Gerüstumlagerung *f*
skeletal vibration Gerüstschwingung *f*
skeleton Skelett *n*, Gerüst *n*; Hauptkette *f*, Rückgratkette *f*, Verbindungsstamm *m (eines verzweigten Moleküls)*
skeleton catalyst Skelettkatalysator *m*, Skelettkontakt *m*, Raney-Katalysator *m*
skew *s.* skewness
skew form schiefe [windschiefe, synclinale] Form *f*, *syn*-Form *f (Stereochemie)*
skewness Schiefe *f (z. B. einer Kurve)*
skim *v* 1. abstreifen, abstreichen *(z. B. Schaum)*; abschöpfen *(z. B. Verunreinigungen)*; entschäumen *(Flüssigkeiten)*; 2. *(Lebm)* entrahmen, abrahmen *(Milch)*; 3. *(Erdöl)* (ab)toppen, skimmen; 4. (ab)schälen *(z. B. in der Schälzentrifuge)*
skim *v* **off** *s.* skim 1.
skim milk entrahmte Milch *f*, Magermilch *f*
skimmer 1. Abstreifer *m*, Abstreicher *m*; Abschäumer *m*, Abschäumlöffel *m*, Schaumlöffel *m*, Schaumkelle *f*; 2. Schäler *m*; Schällöffel *m*; Schälmesser *n*; Schälrohr *n*; Schälteller *m*
skimmer centrifuge Schälzentrifuge *f*, Schälschleuder *f*; Schälrohrzentrifuge *f*, Schälrohrschleuder *f*
skimmings 1. Abgeschäumtes *n*, Schaum *m*, Abgeschöpftes *n*; 2. *(Met)* Schlacke *f*; 3. *(Hyd)* Schwimmstoff *m*, Schwimmstoffe *mpl*, Schwimmschlamm *m*
skin 1. Haut *f*; 2. *(Met)* Gusshaut *f*; Walzhaut *f*
skin absorption *(Pharm, Tox)* Aufnahme *f* [Resorption *f*] über die Haut
skin-irritant hautreizend
skin irritant hautreizender Stoff *m*, Hautreizstoff *m*
skip 1. Förderkübel *m*, Transportkübel *m*, Kippkübel *m*, Fördergefäß *n*, Skip *m*, Skipgefäß *n*; 2. Förderkorb *m*
Skraup quinoline synthesis skraupsche Chinolinsynthese *f* [Synthese *f*], Skraup-Synthese *f*
slab 1. Platte *f*, Tafel *f*, Fliese *f*; 2. *(Met)* Bramme *f*; 3. *(Gum)* Platte *f*, Kautschukplatte *f*, Kautschukkuchen *m*; Slab *m (Rohkautschukhandelssorte)*
slab gel Plattengel *n (Elektrophorese)*
slack wax Gatsch *m*, Paraffingatsch *m*
slacklime *s.* slaked lime
slag *v* verschlacken, Schlacke bilden, sich in Schlacke verwandeln; in Schlacke verwandeln; ausschlacken, entschlacken
slag Schlacke *f*
slag cement Hüttenzement *m*
slag-forming schlackenbildend
slag wool Schlackenwolle *f*, Schlackenfaser *f*
slagging Verschlackung *f*, Schlackenbildung *f*; Entschlackung *f*
slaggy 1. schlackenähnlich, schlackenartig; 2. schlackenreich, schlackig
slake *v* 1. (ab)löschen *(Kalk)*; 2. zerbröckeln, zerfallen *(z. B. Kohle)*
slaked lime Löschkalk *m*, gelöschter Kalk *m (Calciumhydroxid)*
slate Schiefer *m*, Tonschiefer *m*; Schieferplatte *f*
slaty schief(e)rig, schieferartig, schieferähnlich; schieferhaltig
sleeping drug *(Pharm)* Schlafmittel *n*, Hypnotikum *n*
sliced film [sheet] Schälfolie *f*, geschälte Folie *f*
slide 1. *(Lab)* Reiter *m*, Reiterwägestück *n*; 2. *(Foto)* Diapositiv *n*, Dia *n*; 3. Objektträger *m (Mikroskopie)*
slide cover glass Deckglas *n (für Objektträger)*
slight gelinde, mäßig, schwach, vorsichtig
slightly acidic schwach sauer (reagierend)
slightly soluble schwer [wenig] löslich, schwerlöslich, slö, wl.
slime 1. Schleim *m*; 2. Schlamm *m*, Schlick *m*; 3. *(Aufb)* Feingut *n*
sliminess 1. Schleimigkeit *f*; 2. Schlammigkeit *f*

slimy 1. schleimig, Schleim...; 2. schlammig
slip *v* gleiten, rutschen
slip 1. Gleiten *n*, Rutschen *n*; Schlupf *m*; Gleitfähigkeit *f*; 2. *(Ker)* Schlicker *m*, Schlempe *f*; 3. *s.* slippage loss
slip plane *(Krist)* Gleitebene *f*
slippage Schlupf *m*
slippage loss Leckverlust *m*, Sickerverlust *m*
slipperiness Schlüpfrigkeit *f*, Glätte *f*
slippery schlüpfrig, glatt, glitschig, schmierig
slit *v* aufschlitzen, aufschneiden; schneiden, längsschneiden, längstrennen, einen Riss bekommen, reißen
slit Schlitz *m*, Spalt *m*
slit film *(Text)* Folienfaden *m*, Folienbändchen *n*
slit sieve funnel *(Lab)* Schlitzsiebnutsche *f*, Glas-Büchner-Trichter *m*
slit width Spaltbreite *f (eines Spektrographen)*
slop 1. Schmutzwasser *n*; 2. *(Lebm)* Schlempe *f*, Brennereischlempe *f*
slope Neigung *f*, Gefälle *n*; schiefe Ebene *f*, Schräge *f*
slot die *(Kst)* Breitschlitzdüse *f (zur Folienherstellung)*
slot nozzle Schlitzdüse *f*, Flachstrahldüse *f*
slow *v* **(down)** 1. verlangsamen, (ab-)bremsen, drosseln, herabsetzen, vermindern *(die Geschwindigkeit)*; 2. *(Kern)* moderieren; sich verlangsamen
slow-acting 1. langsam wirkend; 2. *(Tox)* schleichend wirkend
slow emulsion *(Foto)* geringempfindliche [niedrigempfindliche] Emulsion *f*, Emulsion *f* niedriger Empfindlichkeit
slow-evaporating langsamflüchtig, schwerflüchtig
slow mixing Langsamrühren *n*, Langsammischung *f*
slow neutron 1. langsames Neutron *n*; 2. *s.* thermal neutron
slow sand filter Langsamfilter *n*
slow sand filtration Langsamfiltration *f*
slow stirring *s.* slow mixing
slow-striking *(Text)* langsamziehend *(Farbstoffe)*
slowness Trägheit *f*
sludge Schlamm *m*
sludge activation Schlammbelebung *f*
sludge cake Schlammkuchen *m*; Filterkuchen *m*; Zentrifugenkuchen *m*
sludge composting *(Hyd)* Schlammkompostierung *f*
sludge concentration 1. Schlammeindickung *f*, Schlammanreicherung *f*; 2. Schlammkonzentration *f*, Schlammeindickungsgrad *m (in kg TS/m^3)*
sludge concentrator Schlammeindicker *m*
sludge conditioning Schlammkonditionierung *f*, Schlammvorbehandlung *f*, Schlammaufbereitung *f*
sludge dewatering Schlammentwässerung *f*, Feststoffentwässerung *f*
sludge digester Schlammfaulbehälter *m*, Faulbehälter *m*
sludge digestion Schlamm(aus)faulung *f*
sludge discharge Schlammaustrag *m*
sludge disposal Schlammbeseitigung *f*
sludge-drying bed Schlammtrockenbeet *n*, Trockenbeet *n*
sludge fertilizer Klärschlamm *m (als Düngemittel)*
sludge floc Schlammflocken *fpl*
sludge flocculation Schlammflockung *f*
sludge furnace Schlammverbrennungsofen *m*
sludge gas Faulgas *n*, Biogas *n*
sludge incineration Schlammverbrennung *f*, Schlammveraschung *f*
sludge lagoon Schlammentwässerungsbecken *n*, Schlamm(entwässerungs)teich *m*
sludge layer Schlammschicht *f*, Schlammablagerung *f*
sludge loading Schlammbelastung *f*, Schlammlast *f*
sludge processing Schlammbehandlung *f*
sludge production 1. Schlammproduktion *f*; 2. Schlammanfall *m*
sludge pump Schlammpumpe *f*, Dickstoffpumpe *f*
sludge-raking mechanism Krählwerk *n*, Rührwerk *n*
sludge scraper Schlammräumer *m*, Schlammkratzer *m*
sludge separation Schlammabtrennung *f*, Schlammabscheidung *f*
sludge settling Schlammabsetzung *f*

sludge solids 1. Schlammfeststoffe *mpl*; 2. Schlammfeststoffgehalt *m*, Schlammtrockensubstanz *f*
sludge thickener Schlammeindicker *m*
sludge thickening Schlammeindickung *f*
sludge treatment Schlammbehandlung *f*, Abwasserschlammbehandlung *f*
sludge zone Schlammzone *f*, Eindick(ungs)zone *f*, Dickschlammzone *f*
slugging bed stoßende Wirbelschicht *f* *(mit Gasblasen von der Größe des Reaktordurchmessers)*
sluggish träge, schleppend *(Reaktion)*
slurry *v* aufschlämmen
slurry 1. Schlamm *m*, Schlämme *f*, Aufschlämmung *f*, (dünner) Brei *m*; 2. *(Filtr)* Trübe *f*; 3. *(Hyd)* Schlammtrübe *f*, Abwasserschlammsuspension *f*; Schlammwasser *n* *(beim Schlammkontaktverfahren)*
slurry pump Schlammpumpe *f*
slurry reactor Suspensionsreaktor *m*
slush 1. Schlamm *m*, Schlicker *m*; 2. *s.* slushing grease
slushing grease Korrosionsschutzfett *n*, Konservierungsfett *n*, *(bei Eisen und Stahl auch)* Rostschutzfett *n*
slushing oil Korrosionsschutzöl *n*, Konservierungsöl *n*, *(bei Eisen und Stahl auch)* Rostschutzöl *n*
small-angle light scattering Kleinwinkellichtstreuung *f*
small-angle neutron scattering Kleinwinkelneutronenstreuung *f*
small-angle X-ray scattering Röntgenkleinwinkelstreuung *f*
small-bore column *(Chromat)* Mikrosäule *f*
small fermenter *(Biot)* Kleinfermenter *m*
small-pore(-size) engporig
small-pored engporig
small ring kleiner Ring *m* *(3- und 4-gliedrig)*
small-ring compound Verbindung *f* mit kleinem Ring, Kleinringverbindung *f*
small-scale experiment Kleinversuch *m*
small-scale test Vorversuch *m* mit kleinen Substanzmengen
small waste-water treatment plant *(Hyd)* Kleinkläranlage *f*
smalt Smalte *f*, Schmalte *f*, Blaufarbenglas *n* *(Cobalt(II)-kaliumsilicat)*
smear *v* 1. einschmieren, bestreichen; aufschmieren, aufstreichen *(Schmiermittel)*; einsalben *(Haut)*; 2. beschmieren, beschmutzen; 3. schmieren, schmierig sein
smear 1. Schmiere *f*; klebrige Substanz *f*; 2. *s.* ointment; 3. Fettfleck *m*, Schmutzfleck *m*
smectic liquid crystal smektischer Flüssigkristall *m* [flüssiger Kristall *m*]
smectic phase smektische Phase *f*, bz-Phase *f*
smectic state smektischer Aggregatzustand *m* *(kristalliner Flüssigkeiten)*
smell *v* riechen
smell Geruch *m*, Duft *m*; übler [schlechter] Geruch *m*, Gestank *m*
smelly nuisance Geruchsbelästigung *f*
smelt *v* verhütten *(Erze)*; erschmelzen *(Metalle aus Erz)*
smelt Schmelze *f*, Schmelzfluss *m*
smelter 1. Schmelzofen *m*; 2. Schmelzer *m* *(Arbeiter)*
smelting Verhütten *n*, Verhüttung *f* *(von Erzen)*; Erschmelzen *n* *(von Metallen aus Erz)*
smelting flux Schmelzfluss *m*
smelting furnace Schmelzofen *m*
smog Smog *m*, Stadtnebel *m* *(Luftverunreinigungstyp)*
smog emergency Smogalarm *m*
smoke *v* 1. *(Lebm, Gum)* räuchern; 2. ausräuchern *(Räume)*; 3. rauchen
smoke Rauch *m*
smoke monitor Rauchmeldeanlage *f*
smoke point 1. *(Erdöl)* Rußpunkt *m*; 2. *(Lebm)* Rauchpunkt *m*
smokeless fuel rauchloser [rauchfreier, rußfreier] Brennstoff *m*
smokestack Schornstein *m*, Esse *f*, Schlot *m*, Kamin *m*
smoky flame rußende [leuchtende] Flamme *f*
smooth *v* glätten
smooth *v* **out** ausgleichen *(z. B. die Konzentration)*
smooth 1. glatt, eben *(Oberfläche)*; 2. glatt, gleichmäßig *(Reaktionsablauf)*
smooth endoplasmatic reticulum *(Bioch)* glattes [agranuläres] endoplasmatisches Reticulum *n*
smooth transition Umwandlung *f* [Phasenumwandlung *f*] höherer Ordnung,

Übergang *m* [Phasenübergang *m*] höherer Ordnung
smoothing Glätten *n*, Glättung *f*
smoothing-out of concentration *(Hyd)* Konzentrationsausgleich *m*, Vergleichmäßigung *f* der Konzentration, Abfangen *n* der Konzentrationsschwankungen
smoothness Glätte *f*
smoulder *v* schwelen, glimmen
smoulder-proof nicht nachglimmend
smudge *v* 1. beschmutzen; verschmieren; 2. schwelen
smudge Schmutz *m*; Schmutzfleck *m*
smut Rußflocke *f*
S_N reaction S_N-Reaktion *f*, nucleophile [anionoide] Substitution *f*
S_N1 reaction S_N1-Reaktion *f*, monomolekulare nucleophile Substitution *f*
S_N2 reaction S_N2-Reaktion *f*, bimolekulare nucleophile Substitution *f*
S/N ratio *s.* signal-to-noise (ratio)
snake venom Schlangengift *n*
sneeze gas Nasen-Rachen-Reizstoff *m*
S.N.F. *s.* solids-not-fat
SNG *s.* substitute natural gas
snowout 1. Snowout *n (Abscheidung von Luftschadstoffen durch Schnee)*; 2. Snowout *m (mit Schnee abgelagerte Luftschadstoffe)*; *(i. e. S.)* radioaktiver Snowout *m*
SO *s.* spin orbital
soak *v* 1. sich vollsaugen; 2. einsickern, versickern; (hindurch)sickern; durchdringen, durchtränken *(Flüssigkeit)*; 3. tränken, einweichen; 4. *(Ker)* (ein-)sumpfen; 5. *(Met)* ausgleichglühen; 6. *s.* soak up
soak *v* **up** aufsaugen
soap Seife *f*
soap builder Builder *m*, Gerüststoff *m*, Aufbaustoff *m (in Waschmitteln)*
soap chromatography Seifenchromatographie *f*
soap film flowmeter Seifenfilmströmungsmesser *m*, Seifenblasenfilmmesser *m*
soap micelle Seifenmizelle *f*
soapless soap synthetisches Waschmittel *n* [Reinigungsmittel *n*], Detergens *n*, Detergent *n*
socket 1. Hülse *f*, *(bei Kegelschliffen auch)* Schliffhülse *f*; Schliffpfanne *f*, Schliffschale *f (einer Kugelschliffverbindung)*; 2. Muffe *f*; Flansch *m*; Rohrstutzen *m*
SOCMA = Synthetic Organic Chemical Manufacturers Association
soda 1. *(Min)* Soda *f*, Natron *n*, Natrit *m (Natriumcarbonat-10-Wasser)*; 2. *s.* sodium carbonate; 3. *s.* soda ash; 4. *s.* sodium carbonate decahydrate; 5. *s.* sodium hydrogencarbonate; 6. *s.* sodium hydroxide
soda alum Sodaalaun *m*, Natronalaun *m*, Natriumalaun *m (Natriumaluminiumsulfat-12-Wasser)*
soda ash (kristall)wasserfreie Soda *f*, kalzinierte Soda *f*, wasserfreies Natriumcarbonat *n*
soda bordeaux Burgunder Brühe *f*, Kupfersodabrühe *f (Pflanzenschutzmittel)*
soda crystals Kristallsoda *f*, Natriumcarbonat-10-Wasser *n*
soda glass Natronglas *n*
soda lime Natronkalk *m (Ätznatron-Ätzkalk-Gemisch)*
soda-lime glass Sodakalkglas *n*, Kalknatronglas *n*, Natronkalkglas *n*
soda liquor [lye] *(Pap)* Natronkochlauge *f*
soda nitre 1. Natronsalpeter *m*, Natriumnitrat *n*; 2. *(Min)* Nitronatrit *m*
soda pulp Natron(zell)stoff *m*, Sodazellstoff *m*
soda pulping *(Pap)* Natronaufschluss *m*, Sodaaufschluss *m*
soda-silica glass Natriumsilikatglas *n*
soda soap Natronseife *f*, Natriumseife *f*
soda water glass Natronwasserglas *n*
Söderberg (continuous) electrode Söderberg-Elektrode *f*, selbstbrennende Elektrode *f*, Dauerelektrode *f*
sodium Na Natrium *n*
sodium acetate Natriumacetat *n*
sodium acetylide Natriumacetylid *n*
sodium acid carbonate *s.* sodium hydrogencarbonate
sodium alginate Natriumalginat *n*
sodium alkoxide Natriumalkoholat *n*
sodium alkyl Natriumalkyl *n*
sodium aluminate Natriumaluminat *n*
sodium-aluminium sulphate Natriumaluminiumsulfat *n*
sodium amalgam Natriumamalgam *n*

sodium amide Natriumamid *n*
sodium *para*-aminobenzoate Natrium-*p*-aminobenzoat *n*
sodium-ammonium hydrogenphosphate Natriumammoniumhydrogenphosphat-4-Wasser *n*, Phosphorsalz *n*
sodium arsenite Natriumarsenit *n*, Natriumarsenat(III) *n*
sodium ascorbate Natriumascorbat *n*
sodium azide Natriumazid *n*
sodium benzoate Natriumbenzoat *n*
sodium bicarbonate *s.* sodium hydrogencarbonate
sodium bichromate *s.* sodium dichromate
sodium bisulphite liquor Natriumhydrogensulfitlösung *f*, Natriumbisulfitlösung *f*
sodium (1:2) borate *s.* borax
sodium borohydride *Natriumtetrahydridoborat *n*, *(veraltet:)* Natriumborhydrid *n*
sodium bromate Natriumbromat *n*
sodium bromide Natriumbromid *n*
sodium carbide *s.* sodium acetylide
sodium carbonate Natriumcarbonat *n*
sodium carbonate decahydrate Natriumcarbonat-10-Wasser *n*
sodium carbonate extract *(Anal)* Sodaauszug *m*
sodium carboxymethylcellulose Natriumcarboxymethylcellulose *f*
sodium caseinate Natriumcaseinat *n*
sodium cellulose xanthate Natriumcellulosexanthogenat *n*
sodium chlorate Natriumchlorat *n*
sodium chloride Natriumchlorid *n*
sodium chloride brine *(Hyd)* Kochsalzsole *f (Ionenaustausch)*
sodium chloride lattice Natriumchloridgitter *n*, Steinsalzgitter *n*
sodium chlorite Natriumchlorit *n*
sodium chloroplatinate Natriumhexachloroplatinat(IV) *n*
sodium chromate Natriumchromat *n*
sodium citrate Natriumcitrat *n*
sodium cobaltinitrite Natriumhexanitrocobaltat(III) *n*
sodium cyanate Natriumcyanat *n*
sodium cyanide Natriumcyanid *n*
sodium cycle exchange *(Hyd)* Ionenaustausch *m* mit Natriumaustauscher [Na-Austauscher, Neutralaustauscher], Kationen-Neutralaustauschverfahren *n*, Neutralaustausch *m*
sodium D line Natrium-D-Linie *f*, Na-D-Linie *f*
sodium diacetate Natriumdiacetat *n (mit angelagerter Essigsäure)*
sodium dichromate Natriumdichromat(VI) *n*
sodium diethyldithiocarbamate Natrium-*N,N*-diethyldithiocarbamat *n*
sodium dihydrogen phosphate Natriumdihydrogenphosphat *n*, Mononatriumphosphat *n*
sodium dithionite 1. Natriumdithionit *n*; 2. *(Farb)* Dithionit *n*
sodium ethylate Natriumethylat *n*
sodium exchanger *(Hyd)* Natriumaustauscher *m*, Ionenaustauscher *m* in der Natriumform [Na-Form], Na-Austauscher *m*, Na_2-Austauscher *m*, Neutralaustauscher *m*
sodium ferricyanide Natriumhexacyanoferrat(III) *n*
sodium ferrocyanide Natriumhexacyanoferrat(II) *n*
sodium fluoride Natriumfluorid *n*
sodium fluorosilicate Natriumhexafluorosilicat *n*
sodium folate Natriumfolat *n*, Natriumpteroylglutamat *n*
sodium formate Natriumformiat *n*
sodium gluconate Natriumgluconat *n*
sodium glutamate Natriumglutamat *n*
sodium hydride Natriumhydrid *n*
sodium hydrogencarbonate Natriumhydrogencarbonat *n*
sodium hydrogensulphate Natriumhydrogensulfat *n*, Mononatriumsulfat *n*
sodium hydrogensulphite 1. Natriumhydrogensulfit *n*; 2. *(Tech)* Natriumbisulfit *n*
sodium hydroxide Natriumhydroxid *n*, Ätznatron *n*
sodium hypochlorite Natriumhypochlorit *n*
sodium iodate Natriumiodat *n*
sodium iodide Natriumiodid *n*
sodium lactate Natriumlactat *n*
sodium lauryl sulphate Natriumlaurylsulfat *n*, *Natriumdodecylsulfat *n*
sodium metaborate Natriummetaborat *n*

sodium metasilicate Natriummetasilicat *n*
sodium methylate Natriummethylat *n*, Natriummethoxid *n*
sodium molybdate Natriummolybdat(VI) *n*
sodium 12-molybdophosphate Natriumdodecamolybdatophosphat *n*
sodium nitrate Natriumnitrat *n*
sodium nitrite Natriumnitrit *n*
sodium nitroprusside Nitroprussidnatrium *n*, Natriumnitroprussid *n*, Natriumnitroprussiat *n*, Dinatriumpentacyanonitrosylferrat *n*
sodium oxalate Natriumoxalat *n*
sodium palmitate Natriumpalmitat *n*
sodium perborate Natriumperoxoborat *n*
sodium perchlorate Natriumperchlorat *n*
sodium peroxide Natriumperoxid *n*
sodium peroxydisulphate Natriumperoxodisulfat *n*
sodium persulphate *s.* sodium peroxydisulphate
sodium phosphate Natriumphosphat *n*
sodium phosphate bead Natriummetaphosphatperle *f*, Phosphorsalzperle *f*
sodium polysulphide Natriumpolysulfid *n*
sodium propionate Natriumpropionat *n*
sodium pteroylglutamate *s.* sodium folate
sodium pyrophosphate Natriumdiphosphat *n*, Natriumpyrophosphat *n*
sodium saccharin Kristallose *f*, Kristallsaccharin *n* *(Natriumsalz des Saccharins)*
sodium salicylate Natriumsalicylat *n*
sodium selenate Natriumselenat *n*
sodium selenite Natriumselenit *n*
sodium silicate Natriumsilicat *n*, Natronwasserglas *n*
sodium soap Natriumseife *f*, Natronseife *f*
sodium stearate Natriumstearat *n*
sodium sulphate Natriumsulfat *n*
sodium sulphide Natriumsulfid *n*
sodium sulphite Natriumsulfit *n*
sodium tartrate Natriumtartrat *n*
sodium tetraborate Natriumtetraborat *n*
sodium tetrasulphide Natriumtetrasulfid *n*
sodium thiocyanate Natriumthiocyanat *n*
sodium thiosulphate Natriumthiosulfat *n*
sodium uranyl acetate Natriumuranylacetat *n*
sodium-vapour (discharge) lamp Natrium(dampf)lampe *f*
soft 1. weich; 2. *(Am)* alkoholfrei *(Getränke)*
soft acid weiche Säure *f* *(nach der Einteilung von Pearson)*
soft base weiche Base *f* *(nach der Einteilung von Pearson)*
soft glass weiches [leicht schmelzbares] Glas *n*, Weichglas *n*
soft ionization *(Anal)* schonende Ionisierung *f*
soft-magnetic material weichmagnetischer Werkstoff *m*
soft porcelain Weichporzellan *n*
soft radiation weiche [energiearme] Strahlung *f*
soft resin Weichharz *n*
soft rubber Weichgummi *m*
soft steel Weichstahl *m*, weicher Stahl *m*
soft water weiches Wasser *n*, Weichwasser *n*
soft wax Weichwachs *n*
soften *v* 1. weich [geschmeidig, biegsam, plastisch] machen, erweichen, aufweichen; weich werden, erweichen, aufweichen; 2. *(Hyd)* enthärten
soften *v* **completely** *(Hyd)* vollenthärten
softened water *(Hyd)* Weichwasser *n*, enthärtetes Wasser *n*
softener *s.* 1. softening agent; 2. softening plant
softening agent 1. Weichmacher *m*, erweichendes Mittel *n*, Erweichungsmittel *n*; 2. *(Hyd)* Enthärter *m*, Enthärtungsmittel *n*
softening plant *(Hyd)* Enthärtungsanlage *f*, Enthärter *m*
softening point Erweichungspunkt *m*, Erweichungstemperatur *f*
softening range Erweichungsbereich *m*, Erweichungsintervall *n*, Erweichungszone *f*
softening temperature *s.* softening point
soil *v* anschmutzen, beschmutzen; schmutzig werden

soil 1. Boden *m*, Grund *m*, Erdreich *n*, Erde *f*; 2. Verschmutzen *n*, Beschmutzen *n*; 3. Schmutz *m*, Verschmutzung *f*; Schmutzfleck *m*
soil acidification Bodenversauerung *f*
soil acidity Bodenazidität *f*
soil amendment 1. Bodendesinfektionsmittel *n (nicht total wirkend)*; 2. *s.* soil conditioner
soil chemistry Bodenchemie *f*, Chemie *f* des Bodens
soil colloid Bodenkolloid *n*
soil compaction Bodenverdichtung *f*
soil conditioner Boden(struktur)verbesserungsmittel *n*
soil contamination Bodenverschmutzung *f*
soil corrosion Bodenkorrosion *f*, Erdbodenkorrosion *f*
soil fertility Bodenfruchtbarkeit *f*
soil formation Bodenbildung *f*
soil fumigant Bodenbegasungsmittel *n (zur Entseuchung)*
soil fungicide Bodenfungizid *n*
soil horizon Bodenhorizont *m*
soil microorganisms Bodenmikroorganismen *mpl*
soil moisture Bodenfeuchte *f*, Bodenfeuchtigkeit *f*
soil nutrient Bodennährstoff *m*
soil pH Boden-pH-Wert *m*
soil reaction Bodenreaktion *f*
soil sample Bodenprobe *f*
soil science Bodenkunde *f*, Pedologie *f*
soil sterilant 1. Totalherbizid *n*; 2. Bodendesinfektionsmittel *n*
soil structure Bodenstruktur *f*, Bodengefüge *n*
soiling Verschmutzung *f (an Sachgütern durch Luftschadstoffe)*
sol. *s.* 1. soluble; 2. solution
sol *(Koll)* Sol *n*
sol-gel change [transformation, transition] Sol-Gel-Umwandlung *f*, Sol-Gel-Transformation *f*
solanin(e) Solanin *n (Steroidalkaloid verschiedener Solanum-Arten)*
solanum alkaloid Solanumalkaloid *n*
solar distillation *s.* solar evaporation 2.
solar evaporation 1. solare Eindunstung *f (zur Salzgewinnung)*; 2. *(Hyd)* Verdampfung *f* [Destillation *f*] durch Sonnenbestrahlung *(zur Meerwasserentsalzung)*
solar furnace Sonnenofen *m*
solar radiation solare Strahlung *f*, Sonnenstrahlung *f*
solar salt Seesalz *n*, Meersalz *n (durch solare Eindunstung gewonnen)*
solar still *(Hyd)* Sonnendestillationsanlage *f (zur Meerwasserentsalzung)*
solate *v* in den Solzustand übergehen, ein Sol bilden
solation *(Koll)* Gel-Sol-Umwandlung *f*, Gel-Sol-Übergang *m*
solder *v* löten; sich löten lassen
solder Lot *n*, *(i. e. S.)* Weichlot *n*
soldering 1. Löten *n*, Lötung *f*; 2. Lötverbindung *f*, Lötstelle *f*
soldering flux Flussmittel *n*
solenoid valve Magnetventil *n*
solid fest; dicht, kompakt; massiv
solid 1. Festkörper *m*, Feststoff *m*, fester Körper *m* [Stoff *m*]; 2. *(Anal)* Trockenmasse *f*, Trockensubstanz *f*, TS; 3. Festkörper *m*, feste Phase *f*, Kristallisat *n (beim Zonenschmelzen)*
solid-bowl centrifuge Sedimentierzentrifuge *f*, Absetzzentrifuge *f*, Vollmantelzentrifuge *f*, Dekantierzentrifuge *f*
solid carbon dioxide Trockeneis *n*, festes Kohlendioxid *n*
solid constituent 1. Festbestandteil *m*; 2. *(Lebm)* Trockenbestandteil *m*
solid diffusion Festkörperdiffusion *f*, Diffusion *f* von Festkörpern
solid electrolyte fester Elektrolyt *m*, Festelektrolyt *m*
solid extraction *s.* solid-liquid extraction
solid fat festes Fett *n*, Talg *m*
solid foam fester Schaum *m*
solid fuel 1. fester Brennstoff *m*, Festbrennstoff *m*; fester Kraftstoff *m*, Festkraftstoff *m*; 2. *s.* solid rocket fuel
solid-liquid extraction Extraktion *f* fest-flüssig; Fest-flüssig-Extraktion *f*, Feststoffextraktion *f*, Extraktion *f* von Feststoffen
solid-liquid interface Grenzfläche *f* fest-flüssig, Fest-flüssig-Grenzfläche *f*
solid material *s.* solid 1. und 2.
solid matter *s.* solid 1.
solid paraffin Festparaffin *n*, festes Paraffin *n*, Hartparaffin *n*

solid phase feste Phase *f*; Bodenkörper *m*
solid-phase organic synthesis organische Festphasensynthese *f*
solid-phase protein synthesis Synthese *f* an polymeren Trägern, Merrifield-Synthese *f* *(zur Darstellung von Polypeptiden)*
solid-phase synthesis Festphasensynthese *f*
solid point Stockpunkt *m* *(bei Ölen)*
solid propellant 1. fester Kraftstoff *m*, Festkraftstoff *m*; 2. *s.* solid rocket fuel
solid rocket fuel [propellant] fester Raketentreibstoff *m*, Festtreibstoff *m*
solid-solid interface Grenzfläche *f* fest-fest, Fest-fest-Grenzfläche *f*
solid-solid mixing Vermengen *n*, Mischen *n* von Feststoffkomponenten
solid-solid reaction Feststoff-Feststoff-Reaktion *f*, Feststoffreaktion *f*
solid solubility Festkörperlöslichkeit *f*, Löslichkeit *f* in festem Zustand
solid solute Bodenkörper *m* *(im Gleichgewicht mit seiner Lösung)*
solid solution feste Lösung *f*, Mischkristall *m*, *(i. e. S.)* Einlagerungsmischkristall *m*
solid state fester Zustand *m*, Festkörperzustand *m*
solid-state chemistry Festkörperchemie *f*
solid-state diffusion Festkörperdiffusion *f*, Diffusion *f* von [in] festen Stoffen
solid-state extraction Festphasenextraktion *f*
solid-state laser Festkörperlaser *m*
solid-state micro extraction Festphasenmikroextraktion *f*
solid-state physics Festkörperphysik *f*
solid-state reaction Festkörperreaktion *f*
solid substance feste Substanz *f*, Festsubstanz *f*
solid waste 1. feste Abprodukte *npl* [Abfallstoffe *mpl*], Müll *m*; 2. *(Hyd)* Abwasserschlamm *m*
solid waste disposal 1. Beseitigung *f* fester Abprodukte, Abproduktbeseitigung *f*, Abfallbeseitigung *f*, Müllbeseitigung *f*, Müllentsorgung *f*; 2. *(Hyd)* Schlammbeseitigung *f*
solidensing Solidensieren *n* *(direkter Übergang eines Gases in den festen Aggregatzustand)*
solidification Festwerden *n*, Erstarren *n*, Erstarrung *f*, Verfestigung *f*
solidification point Erstarrungspunkt *m*, Erstarrungstemperatur *f*; Stockpunkt *m* *(bei Ölen)*
solidification range Erstarrungsintervall *n*
solidify *v* verfestigen, erstarren lassen, fest [hart, starr] werden lassen; sich verfestigen, erstarren, fest [hart, starr] werden, verhärten, erhärten
solidifying agent Verfestigungsmittel *n*, Festiger *m*
solids 1. Feststoffe *mpl*; 2. *(Anal)* Trockenmasse *f*, Trockensubstanz *f*, TS
solids contact clarifier [unit] *(Hyd)* Schlammkontaktanlage *f*
solids content 1. Fest(stoff)gehalt *m*, Feststoffanteil *m*; 2. *(Pap, Hyd)* Gehalt *m* an Trockensubstanz, TS-Gehalt *m*, Trockengehalt *m*
solids-liquid separation Trennen *n* von festen Stoffen und Flüssigkeiten, Trennung *f* flüssig-fest
solids mixture Feststoffgemisch *n*, Gemenge *n*
solids-not-fat *(Lebm)* fettfreie Trockenmasse *f*
solids particle Fest(stoff)teilchen *n*
solids recovery Feststoffabscheidung *f* *(Filtration, Zentrifugation)*
solids recovery filtration Trennfiltration *f*
solids removal Feststoffentfernung *f*
solids separation Feststoffabscheidung *f*
solidus (curve, line) Erstarrungskurve *f*, Soliduskurve *f*, Soliduslinie *f*
solidus point Soliduspunkt *m*
soliquid *disperses System aus flüssigem Dispersionsmittel und festem dispersem Anteil, z. B. Suspension*
soln. *s.* solution
solubility Löslichkeit *f* • **give solubility to** löslich machen
solubility chart Löslichkeitsdiagramm *n*
solubility coefficient Löslichkeitskoeffizient *m*, Löslichkeitskonstante *f*
solubility curve Löslichkeitskurve *f*
solubility equilibrium Löslichkeitsgleichgewicht *n*, Lösungsgleichgewicht *n*

solubility product Löslichkeitsprodukt *n*
solubility promotor *s.* solutizer
solubilization Solubilisierung *f*, Solubilisation *f*, Löslichmachung *f*
solubilize *v* löslich machen
solubilizer *s.* solutizer
solubilizing effect lösungsvermittelnde Wirkung *f*
solubilizing power *s.* solvency
soluble löslich; dispergierbar, emulgierbar *(Öle)* • **easily soluble** leichtlöslich, ll., leicht [gut] löslich • **mutually soluble** ineinander löslich • **sparingly soluble** schwerlöslich, schwer [wenig] löslich, slö, wl.
soluble glass Wasserglas *n*
soluble in all proportions unbegrenzt löslich
soluble in water wasserlöslich
soluble nitrocellulose Kolloxylin *n*, Kollodiumwolle *f (niedrignitrierte Cellulose)*
soluble oil lösliches Öl *n*; emulgierbares [wasserlösliches] Öl *n (durch Emulgatorzusatz mit Wasser mischbar)*
soluble ribonucleic acid *s.* transfer ribonucleic acid
soluble saccharin Kristallose *f*, Kristallsaccharin *n (Natriumsalz des Saccharins)*
soluble starch lösliche [dünnkochende] Stärke *f*
solute 1. gelöster Stoff *m*, Gelöstes *n*; 2. gelöster Stoff *m*, Beimengung *f*, Verunreinigung *f (beim Zonenschmelzen)*
solute molecule gelöstes Molekül *n*
solution Lösung *f*, Lsg. • **bring into solution** in Lösung bringen, lösen • **go into solution** in Lösung gehen, sich lösen
solution assistant *s.* solutizer
solution-dyed *(Text)* (er)spinngefärbt, düsengefärbt
solution-grown in Nährlösung kultiviert
solution of caustic soda Natronlauge *f*, Natriumhydroxidlösung *f*
solution of potassium hydroxide Kalilauge *f*, Kaliumhydroxidlösung *f*
solution polymerization Lösungspolymerisation *f*
solution pressure Lösungsdruck *m*, Lösungstension *f*
solution reaction Reaktion *f* in Lösung, Lösungsreaktion *f*
solution spinning Erspinnen *n* [Spinnen *n*] aus Lösungen, Lösungsspinnverfahren *n (Herstellung von Chemiefaserstoffen)*
solution strength Lösungskonzentration *f*
solution tested Probelösung *f*, Untersuchungslösung *f*, geprüfte [untersuchte] Lösung *f*, Prüf(lings)lösung *f*
solution to be analyzed *s.* solution to be tested
solution to be tested Probelösung *f*, Untersuchungslösung *f*, zu prüfende [untersuchende] Lösung *f*, Prüf(lings)-lösung *f*
solutizer Lösungsvermittler *m*, Lösungsverbesserer *m*, Lösungsbeschleuniger *m*, Lösungshilfsmittel *n*, Löslichkeitsverbesserer *m*, Löslichkeitsvermittler *m*, Solutizer *m*
solutizing agent Lösungsvermittler *m*, Lösungsverbesserer *m*, Lösungsbeschleuniger *m*, Lösungshilfsmittel *n*, Löslichkeitsverbesserer *m*, Löslichkeitsvermittler *m*, Solutizer *m*
solvate *v* solvatisieren
solvate Solvat *n*
solvating power Solvatationskraft *f*, Solvatisierungsvermögen *n*
solvation Solvatation *f*, Solvatisierung *f*, Solvatisieren *n*
solvation energy Solvatationsenergie *f*
solvation reaction Solvatationsreaktion *f*
solvation sheath Solvathülle *f*
Solvay process Solvay-Verfahren *n*, Ammoniak-Soda-Verfahren *n*
solve *v* (auf)lösen
solvency Lösungsvermögen *n*, Lösevermögen *n*, Lösefähigkeit *f*, Lösekraft *f*
solvent (auf)lösend
solvent Lösungsmittel *n*, Lsgm., Lösemittel *n*, Löser *m*, Solvens *n*
solvent-based adhesive Lösungsmittelklebstoff *m*, Lösungsmittelkleber *m*
solvent cage Lösungsmittelkäfig *m*
solvent dewaxing Lösungsmittelentparaffinierung *f*, Solvententparaffinierung *f*, Solvententwachsung *f*, Entparaffi-

nierung *f* [Entwachsung *f*] mit Lösungsmitteln
solvent drag *(Bioch) Mitnahme von Substanzen durch Membranporen beim Wassertransport*
solvent extraction Solventextraktion *f*, Extraktion *f* flüssig-flüssig, Flüssig-flüssig-Extraktion *f*, Extraktion *f* in flüssigen Systemen
solvent front *(Chromat)* Lösungsmittelfront *f*, Laufmittelfront *f*, Fließmittelfront *f*
solvent isotope effect Lösungsmittel-Isotopieeffekt *m*
solvent migration-distance *(Chromat)* Fließmittelwanderungsstrecke *f*
solvent mixture Lösungsmittelgemisch *n*
solvent naphtha Solventnaphtha *n(f)* *(Lösungsbenzol oder Schwerbenzin als Steinkohlenteer- bzw. Erdölfraktion)*
solvent partition Ausschütteln *n*
solvent phase Lösungsmittelphase *f*, Lösemittelphase *f*
solvent polymerization Lösungspolymerisation *f*
solvent power *s.* solvency
solvent recovery Lösungsmittelrückgewinnung *f*, Lösungsmittelwiedergewinnung *f*
solvent refining Solventraffination *f*, Lösungs(mittel)raffination *f*
solvent resistance Lösungsmittelbeständigkeit *f*, Widerstandsfähigkeit *f* [Beständigkeit *f*] gegen Lösungsmittel
solvent-resisting lösungsmittelbeständig, lösungsmittelfest
solvent-separated ion pair solvensgetrenntes [äußeres] Ionenpaar *n (durch mehrere Lösungsmittelmoleküle getrennt)*
solvent-shared ion pair solvensgetrenntes [äußeres] Ionenpaar *n (durch nur ein Lösungsmittelmolekül getrennt)*
solvent shift *(Spektr)* Lösungsmittelverschiebung *f*
solvent spinning Erspinnen *n* [Spinnen *n*] aus Lösungen, Lösungsspinnverfahren *n (Herstellung von Chemiefasern)*
solvent tannage Lösungsmittelgerbung *f*
solvent-thinned paint lösungsmittelverdünnbare [mit Lösungsmitteln verdünnbare] Farbe *f (als Gegensatz zur wasserverdünnbaren Farbe)*
solvent vapour Lösungsmitteldampf *m*
solventless lösungsmittelfrei
solvolysis Solvolyse *f*, Lyolysis *f*
solvolysis reaction Solvolysereaktion *f*
solvolytic solvolytisch
solvolyze *v* durch Solvolyse aufspalten [zersetzen] *(gelöste Stoffe)*
solvolyze *v* **off** durch Solvolyse abspalten
solvus (curve, line) *(phys Ch)* Solvuskurve *f*
soman Soman *n (Methylfluorphosphonsäurepinacolylester, ein Kampfstoff)*
somatotropic hormone somatotropes Hormon *n*, Somatotropin *n*, STH, Wachstumshormon *n*
somatotropin *s.* somatotropic hormone
somatotropin inhibitory hormone *Somatostatin *n*
somatotropin releasing hormone *Somatoliberin *n*
Sommelet reaction Sommelet-Reaktion *f (zur Darstellung von Aldehyden)*
Sommelet rearrangement Sommelet-Umlagerung *f (Isomerisierung quartärer Benzylammoniumverbindungen)*
SOMO = singly occupied molecular orbital
sonic agglomeration *s.* sonic precipitation
sonic cyclone Schallaerozyklon *m*
sonic precipitation akustisches Abscheiden *n (von Staub aus Gasen)*
sonicate *v* beschallen
sonication Beschallung *f*
sonoluminescence Sonolumineszenz *f*
soot Ruß *m*
sooty 1. rußend; 2. rußig, *(i. w. S.)* geschwärzt
soporific Schlafmittel *n*, Hypnotikum *n*
sorb *v* sorbieren
sorbate Sorptiv *n*, sorbierter [aufgenommener] Stoff *m*, *(während der Aufnahme)* Sorbend *m*
sorbent Sorbent *m*, Sorbens *n*, Sorptionsmittel *n*, aufnehmender Stoff *m*
sorbic acid Sorbinsäure *f*, *Hexa-2,4-diensäure *f*
sorbite *(Met)* Sorbit *m (feinlamellarer Perlit)*

sorbitic *(Met)* sorbitisch
sorbitol D-Sorbitol *n*, *D-Glucitol *n (ein Zuckeralkohol)*
***d*-sorbitol** D-Sorbitol *n*, *D-Glucitol *n (ein Zuckeralkohol)*
Sorel cement Sorel-Zement *m*, Magnesiazement *m*
sorosilicate *(Min)* Sorosilicat *n*, Gruppensilicat *n*
sorption Sorption *f*
sorption isotherm Sorptionsisotherme *f*
sorptive sorptiv
sorrel salt Kleesalz *n*, Bitterkleesalz *n*, Sauerkleesalz *n (Kaliumtetraoxalat oder ein Gemisch aus diesem und Kaliumhydrogenoxalat)*
sound-absorbing material Schalldämmstoff *m*
sound absorption Schallabsorption *f*
sound energy Schallenergie *f*
sound field Schallfeld *n*
sound frequency Schallfrequenz *f*
sound power Schalleistung *f*
sound pressure Schalldruck *m*
sound source Schallquelle *f*
sound wave Schallwelle *f*
sour *v* 1. *(Lebm)* (an)säuern; sauer [stichig] werden; 2. *(Boden)* versauern
sour 1. *(Lebm)* sauer, herb, scharf; gegoren; ranzig; 2. *(Boden)* sauer, kalkarm; 3. *(Erdöl)* sauer, doktor-positiv, mit positivem Doktortest
sour gas saures Erdgas *n (enthält Schwefelwasserstoff und Thiole)*; saures Gas *n*, Sauergas *n (enthält saure Bestandteile wie Schwefelwasserstoff, Kohlendioxid und Cyanwasserstoff)*
source 1. Ausgangsstoff *m*; 2. Quelle *f*; Herkunft *f*; Hersteller *m* • **from natural sources** natürlich (vorkommend), Natur...
source of electrons Elektronendonator *m*, Elektronenspender *m*
source of energy Energiequelle *f*, Energielieferant *m*, Energiespender *m*
source of raw material Rohstoffquelle *f*
sourdough Sauer(teig) *m*
souring *(Lebm)* Säuerung *f*, Säuern *n*, Ansäuern *n*; Sauerwerden *n*, Stichigwerden *n*
sourish *(Lebm)* säuerlich, angesäuert
sourness *(Lebm)* Säure *f*, Herbheit *f*
Southern blot *(Bioch) Identifizierung von DNA-Abschnitten durch Elektrophorese nach Hybridisierung*
Soxhlet (extractor) Soxhlet-Extraktor *m*, Soxhlet-Apparat *m*, soxhletscher Extraktionsapparat *m*, Soxhlet *m*
Soxhlet thimble Extraktionshülse *f* nach Soxhlet
soybean oil Soja(bohnen)öl *n*
soybean protein Sojaprotein *n*, Sojaeiweiß *n*
s.p. *s.* setting point
sp^3 centre sp^3-Zentrum *n (Bindungstheorie)*
sp. gr. *(specific gravity) s.* relative density
sp hybrid sp-Hybrid *n (Bindungstheorie)*
sp hybridization sp-Hybridisierung *f (Bindungstheorie)*
space Raum *m*; Zwischenraum *m*, Intervall *n*, Abstand *m*; Entfernung *f*
space-centred *(Krist)* raumzentriert, innenzentriert
space charge Raumladung *f*
space formula Raumformel *f*, Stereoformel *f*, Konfigurationsformel *f*
space group *(Krist)* Raum(symmetrie)-gruppe *f*
space lattice *(Krist)* Raumgitter *n*, *(i. e. S.)* Translationsgitter *n*, Elementargitter *n*
space quantization Richtungsquantelung *f*, räumliche Quantelung *f*, Raumquantelung *f*
space-time yield Raum-Zeit-Ausbeute *f*
spacer 1. Zwischenstück *n*, Distanzstück *n*; 2. Seitenkette *f*, Kohlenwasserstoffarm *m (in der Gelchromatographie)*; 3. *(Biot)* Spacer *m (zwischengeschaltete Seitenkette; Enzymfixierung)*; 4. Spacer *m (zur Strukturaufklärung verwendetes bifunktionelles Reagens, das mit anderen Gruppen nur bei bestimmtem Abstand zwischen diesen reagiert)*
spacing 1. Abstand *m*, Zwischenraum *m*, Intervall *n*; Entfernung *f*; 2. Einteilen *n (in Abstände)*; 3. *s.* spacing of the planes
spacing of the planes *(Krist)* Netzebenenabstand *m*, Gitterabstand *m*, Gitterperiode *f*
spall *v* abspalten; zertrümmern; abplatzen, abblättern; (zer)splittern
spallation 1. Abspaltung *f*; Abplatzen *n*,

Abblättern *n*; 2. Spallation *f*, Kernzersplitterung *f*, Kernzertrümmerung *f*
sparger 1. Begasungsrohr *n*; 2. *(Hyd)* Verteilereinrichtung *f*, Luftverteiler *m*
sparingly soluble schwerlöslich, schwer [wenig] löslich, slö., wl.
spark *v* Funken sprühen, funken; zünden
spark Funke *m*
spark excitation *(Spektr)* Funkenanregung *f*
spark spectrum Funkenspektrum *n*
sparkle *v* 1. schäumen, perlen, moussieren *(Wein)*; 2. *s.* spark
sparkling water Sprudel *m (Tafelwasser natürlicher Herkunft)*
sparkling wine Schaumwein *m*, Sekt *m*
spasmolytic *(Pharm)* spasmolytisch, krampflösend
spasmolytic *(Pharm)* Spasmolytikum *n*, krampflösendes Mittel *n*
spathic *(Min)* spatig, spatartig, blätterig
spathic iron (ore) *(Min)* Eisenspat *m*, Siderit *m (Eisen(II)-carbonat)*
spatial arrangement [orientation] räumliche [geometrische] Anordnung *f*
spatial representation räumliche Darstellung *f*
spatter *v* 1. spritzen *(Flüssigkeitsteilchen herausschleudern)*; 2. *(Met)* spratzen *(Metalltröpfchen herausschleudern)*; 3. verspritzen; 4. bespritzen
spatula Spatel *m* • **a spatula-point** eine Spatelspitze • **a spatula-tipful** eine Spatelspitze
special boiling-point spirit Siedegrenzenbenzin *n*
special-purpose für Spezialzwecke, Spezial..., Sonder...
special reagent spezifisches Reagens *n*
special solid waste Sondermüll *m*
specialty alloy Sonderlegierung *f*
specialty chemicals *in geringem Umfang hergestellte Chemikalien*; *etwa: Feinchemikalien*
speciation *(Anal)* Speziesbestimmung *f*
species-specific *(Bioch)* speziesspezifisch, artspezifisch
species specificity *(Bioch)* Speziesspezifität *f*, Artspezifität *f*
specific acid catalysis spezifische Säurekatalyse *f*
specific activity spezifische Aktivität *f (Enzymkinetik)*
specific adsorption *(phys Ch)* Chemisorption *f*
specific base catalysis spezifische Basekatalyse *f*
specific conductance (spezifische elektrische) Leitfähigkeit *f (Materialkenngröße)*
specific gravity *s.* relative density
specific-gravity balance mohr-westphalsche Waage *f (zur Dichtebestimmung)*
specific-gravity bottle [flask] Wägefläschchen *n* für Dichtemessungen, Pyknometer *n*
specific heat (capacity) spezifische Wärmekapazität *f*
specific inductive capacity Dielektrizitätszahl *f*, Permittivitätszahl *f*, relative Dielektrizitätskonstante *f* [Permittivität *f*]
specific production efficiency [rate] *(Biot)* spezifische Fermentationsrate *f* [Bildungsrate *f*] *(mg Produkt/mg Zell-TS · h)*
specific reaction rate [velocity] spezifische Reaktionsgeschwindigkeit *f*, Reaktionsgeschwindigkeitskonstante *f*, Geschwindigkeitskonstante *f*
specific refraction [refractive index, refractivity] spezifische Refraktion *f*
specific resistance spezifischer Widerstand *m*
specific rotation *s.* specific rotatory power
specific rotatory power spezifisches Dreh(ungs)vermögen *n*, spezifische Drehung *f*
specific surface spezifische Oberfläche *f*
specific susceptibility spezifische Suszeptibilität *f*, Massensuszeptibilität *f*, Grammsuszeptibilität *f*
specific thermal capacity *s.* specific heat
specific thermal conductivity (spezifisches) Wärmeleitvermögen *n*, Wärmeleitfähigkeit *f*, *(quantitativ auch)* Wärmeleitzahl *f*
specific volume spezifisches Volumen *n*
specific weight Wichte *f (Gewichtskraft durch Volumen)*
specification Einsatzrichtlinie *f*, Spezifikation *f (für ein bestimmtes Produkt)*;

Verarbeitungsrichtlinie *f*, Verarbeitungsvorschrift *f*; Prüfungsvorschrift *f* • **meet the specifications** den Vorschriften entsprechen
specificity Spezifität *f*
specify *v* vorschreiben, festlegen
specimen Probe *f*, *(Werkstoffprüfung auch)* Prüfling *m*, Prüfkörper *m*, Probekörper *m*
specimen jar Schauglas *n*, Präparatenglas *n*
speck Fleck(en) *m*, Unreinheit *f*
specking Fleckenbildung *f*
specpure = spectroscopically pure
spectral spektral, Spektral...
spectral analysis Spektralanalyse *f*, *(i. e. S.)* optische Emissionsspektralanalyse *f*
spectral band *(Anal)* Bande *f*, Spektralbande *f*
spectral colour Spektralfarbe *f*
spectral dispersion spektrale Zerlegung *f*
spectral irradiance spektrale Strahlungsflussdichte *f*
spectral line Spektrallinie *f*
spectral radiance Strahl(ungs)dichte *f*
spectral radiant intensity spektrale Strahlungsintensität *f*
spectral range [region] Spektralbereich *m*, Spektralgebiet *n*
spectral response [responsivity] spektrale Empfindlichkeit *f*
spectral series Spektralserie *f*, Serie *f*
spectral term Spektralterm *m*
spectro(-quality) grade spektralrein
spectrochemical spektrochemisch
spectrogram Spektrogramm *n*
spectrograph Spektrograph *m*
spectrographic spektrographisch
spectrometer Spektrometer *n*
spectrometric spektrometrisch
spectrometry Spektrometrie *f*
spectrophotometer Spektralphotometer *n*, Spektrophotometer *n*
spectrophotometric spektralphotometrisch, spektrophotometrisch
spectrophotometry Spektralphotometrie *f*, Spektrophotometrie *f*
spectroscope Spektroskop *n*
spectroscopic spektroskopisch
spectroscopic displacement law of Kossel and Sommerfeld spektroskopischer Verschiebungssatz *m* von Sommerfeld und Kossel, sommerfeldkosselscher Verschiebungssatz *m*
spectroscopic splitting factor *(Spektr)* Aufspaltungsfaktor *m*, g-Faktor *m*, gyromagnetische Konstante *f*, gyromagnetischer Faktor *m*
spectroscopist Spektroskopiker *m*
spectroscopy Spektroskopie *f* • **by molecular spectroscopy** molekülspektroskopisch
spectrum Spektrum *n*
spectrum of activity Wirkungsspektrum *n*, Wirkungsbereich *m*, Wirkungsbreite *f (z. B. von Pflanzenschutzmitteln)*
specular iron (ore) *(Min)* Eisenglanz *m*, Spekularit *m (gut kristallisierter Hämatit)*
specularite *(Min)* Eisenglanz *m*, Spekularit *m (gut kristallisierter Hämatit)*
speed *v* **(up)** beschleunigen
speed 1. Geschwindigkeit *f*; Drehzahl *f*; 2. *(Foto)* Empfindlichkeit *f*
speed of absorption Absorptionsgeschwindigkeit *f*, Aufziehgeschwindigkeit *f (von Farbstoffen)*
speed of cure 1. *(Gum)* Vulkanisationsgeschwindigkeit *f*, Heizgeschwindigkeit *f*; 2. *(Anstr, Kst)* Härtungsgeschwindigkeit *f*, Aushärtungsgeschwindigkeit *f*
speed of propagation Ausbreitungsgeschwindigkeit *f*
speed of reaction Reaktionsgeschwindigkeit *f*
speed of response *(Anal)* Ansprechgeschwindigkeit *f*
speise Speise *f (Zwischenerzeugnis beim Verschmelzen von schwefel-, arsen- und antimonhaltigen Erzen)*
speiss Speise *f (Zwischenerzeugnis beim Verschmelzen von schwefel-, arsen- und antimonhaltigen Erzen)*
spelter Rohzink *n*
spent verbraucht, (aus)gebraucht, erschöpft *(z. B. Lösung)*
spent brine ablaufende [verbrauchte] Sole *f (Chloralkalielektrolyse)*; verbrauchte Regenerier(mittel)lösung *f (bei der Regenerierung von Austauscherharzen)*
spent catalyst verbrauchter [erschöpfter] Katalysator *m*

spent fermentation broth *(Biot)* verbrauchte Nährlösung *f* [Fermentationslösung *f*]
spent-liquor recovery *(Pap)* Ablaugengewinnung *f*, Ablaugenregeneration *f*, Laugenregeneration *f*
spent regenerant *(Hyd)* verbrauchte Regenerier(mittel)lösung *f*, Eluat *n*
spent soda Sodaablauge *f*
spermatocidal *(Pharm)* spermizid, spermienabtötend
spermatocide *(Pharm)* spermizides [spermienabtötendes] Mittel *n*
spermicidal s. spermatocidal
sphalerite *(Min)* Sphalerit *m*, Zinkblende *f*, Blende *f* *(Zinksulfid)*
sphene *(Min)* Sphen *m* *(Titanit-Varietät)*
sphere 1. Sphäre *f*, Bereich *m*; 2. Kugel *f*
spherical sphärisch, kugelförmig, Kugel...
spherical aberration sphärische Aberration *f*, Öffnungsfehler *m*
spherical-conical adapter Übergangsstück *n* mit Normschliffhülse [NS-Hülse] und Kugelschliffball [KS-Ball]; Übergangsstück *n* mit Kugelschliffschale [KS-Schale] und Normschliffkern [NS-Kern]
spherical joint Kugelschliffverbindung *f*
spherical shell Kugelschale *f*
spherical symmetry Kugelsymmetrie *f*
spherical-top molecule Kugelkreisel *m*
spherically symmetric(al) kugelsymmetrisch
sphericity Sphärizität *f*, Kugelförmigkeit *f*
spherocolloid Sphärokolloid *n*
spheroidal sphäroidal
spheroidal graphite Kugelgraphit *m*
spheroidizing (anneal) *(Met)* Weichglühen *n*, Weichglühung *f*
sphingolipid(e) *(Bioch)* Sphingolipoid *n*
sphingomyelin Sphingomyelin *n* *(ein glycerolfreies Phosphatid)*
sphingosine *(Bioch)* Sphingosin *n*
spice *v* würzen
spice Gewürz *n*
spicy gewürzt, würzig, aromatisch
spiegel (iron) Spiegeleisen *n* *(manganhaltige Roheisensorte)*
spigot 1. Zapfen *m*, Pflock *m* *(zum Verschließen von Fässern)*; 2. Hahnventil *n*
spike *(Spektr, Chromat)* Spike *m* *(steiler, kurzer Schreiberausschlag)*
spill *v* verschütten, verspritzen
spill *v* **out** *(Bioch)* ausscheiden; ausschütten *(Hormone)*
spillage *(Hyd)* Spritzwässer *npl*
spills *(Hyd)* Spritzwässer *npl*
spin *v* 1. spinnen *(Garn)*; verspinnen *(Spinnlösungen)*; erspinnen *(Chemiefaserstoffe)*; 2. schnell [mit hoher Drehzahl] rotieren [umlaufen]; 3. (ab-) schleudern
spin coat *v* aufschleudern *(auf Wafer)*
spin *v* **down** zentrifugieren
spin Spin *m*, Eigendrehimpuls *m*, innerer Drehimpuls *m*
spin adduct Spin-Addukt *n*
spin-allowed electronic transition spinerlaubter Elektronenübergang *m*
spin angular momentum Spindrehimpuls *m*
spin coater Schleuderbeschichter *m*
spin coating Aufschleudern *n* *(auf Wafer)*
spin dyeing Spinnfärbung *f*, Erspinnfärbung *f*, Düsenfärbung *f*
spin-forbidden transition spinverbotener Übergang *m*
spin-glass transition Übergang *m* in den Spinglaszustand
spin labelling *(Spektr)* Spinmarkierung *f*
spin-lattice interaction [relaxation] *(Spektr)* Spin-Gitter-Wechselwirkung *f*, Spin-Gitter-Relaxation *f*
spin mapping dreidimensionale Spektroskopie *f*, Zeugmatographie *f*
spin multiplicity Spinmultiplizität *f*
spin-non-paired electrons Elektronen *npl* mit parallelen Spinmomenten
spin-orbit coupling [interaction] *(Spektr)* Spin-Bahn-Kopplung *f*, Spin-Bahn-Wechselwirkung *f*, Russel-Saunders-Kopplung *f*, LS-Kopplung *f*
spin orbital Spinorbital *n*
spin-paired electrons Elektronen *npl* mit antiparallelen Spinmomenten
spin population Spinpopulation *f*
spin quantum number Spinquantenzahl *f*
spin-spin coupling [interaction] Spin-Spin-Kopplung *f*, Spin-Spin-Wechselwirkung *f*, Dipol-Dipol-Wechselwirkung *f*
spin-spin relaxation Spin-Spin-Relaxation *f*

spin-spin splitting Spin-(Spin-)Kopplungsaufspaltung *f*
spin state Spinzustand *m*
spinel *(Min)* Spinell *m*, *(i. e. S.)* Magnesiospinell *m (Magnesiumaluminat)*
spinnability Verspinnbarkeit *f*, *(quantitativ auch)* technischer Spinnwert *m*
spinneret *(Text)* Spinndüse *f*, Spinnbrause *f*
spinning schnellrotierend, schnellumlaufend, mit hoher Drehzahl rotierend [umlaufend]
spinning 1. Spinnen *n (von Garn)*; Verspinnen *n (von Spinnlösungen)*; Erspinnen *n (von Chemiefaserstoffen)*; 2. schnelle Rotation *f*; 3. Schleudern *n*, Abschleudern *n*
spinning atomizer rotierender Versprüher *m*; rotierender Zerstäuber *m (für Feststoffe)*
spinning band column *(Dest)* Drehbandkolonne *f*
spinning bath Spinnbad *n*, Fällbad *n*
spinning bulb Spinnzwiebel *f*
spinning-cup atomizer Becherversprüher *m*
spinning-disk atomizer Scheibenversprüher *m*
spinning dope *s.* spinning solution
spinning head Spinnkopf *m*
spinning pot Spinntopf *m*, Spinnzentrifuge *f*
spinning pump Spinnpumpe *f*, Dosierpumpe *f*, Titerpumpe *f*
spinning solution Spinnlösung *f*, Erspinnlösung *f*
spinning-top atomizer [sprayer] Kreiselversprüher *m*
spinodal curve *(phys Ch)* Spinodale *f*
spinodal decomposition spinodale Entmischung *f*, Bergaufdiffusion *f (in einer übersättigten Legierung)*
spiral coil Spiralrohrschlange *f*
spiral conveyor Schneckenförderer *m*
spiral-plate heat exchanger Spiralwärmeübertrager *m*
spiral-tube heat exchanger Spiralrohrbündel-Wärmeübertrager *m*, Kreuzgegenstrom-Wärmeübertrager *m*
spiran(e) Spiran *n*, Spirokohlenwasserstoff *m*, spirocyclische Verbindung *f*, Spiroverbindung *f*
spirit 1. Spiritus *m*, Sprit *m (gewerblich hergestelltes Ethanol)*; Branntwein *m*; 2. *(Pharm)* Spiritus *m*, arzneiliche Spirituose *f*; 3. Destillat *n (besonders von Lösungsmitteln und Kraftstoffen)*; 4. Benzin *n (besonders als Vergaserkraftstoff)*
spirit of turpentine Terpentinöl *n*
spirit-soluble spritlöslich, alkohollöslich, *(i. w. S.)* löslich in organischen Lösungsmitteln
spirit varnish Spirituslack *m*
spirits of wine Weingeist *m*, Ethylalkohol *m*
spiro compound [hydrocarbon] *s.* spiran(e)
spiro polymer Spiropolymer(e) *n*
spiro ring system Spirosystem *n*, spirocyclisches System *n*
spit *v (Met)* spratzen
spitzkasten *(Aufb)* Spitzkasten *m*
splash *v* spritzen; verspritzen
splash Spritzer *m*
splashing loss Spritzverlust *m*
splice *v* 1. (zusammen)spleißen; 2. *(Bioch)* spleißen *(RNA)*
splicing *(Bioch)* Spleißen *n*, Splicing *n (der RNA)*
split *v* 1. (auf)spalten; sich spalten; 2. *(Chromat)* splitten, teilen *(den Gasstrom)*
split *v* **off [out]** abspalten; sich abspalten
split *v* **up** *s.* split
split fibre Foliefaden *m*, Foliefaser *f*, Spaltfaser *f*
split flow *(Chromat)* (ungleich) geteilter Gasstrom *m*
split injection [sampling] *(Chromat)* Dosierung *f* [Probenaufgabe *f*, Probengabe *f*] mit Gasstromteilung, Splitdosierung *f*
split-stream ion exchanger *(Hyd)* Ionenaustauscheranlage *f* in Teilstromschaltung
splitless injection [sampling] *(Chromat)* splitlose Injektion *f* [Dosierung *f*], Proben(auf)gabe *f* ohne Gasstromteilung
splitter *(Chromat)* Gasstromteiler *m*, Strömungsteiler *m*, Splitter *m*
splitting 1. Spaltung *f*, Aufspaltung *f*, Spalten *n*, Aufspalten *n*; 2. *(Chromat)* Strömungsteilung *f*, Splitten *n (Teilen des Gasgemischs in zwei ungleiche Ströme)*

splitting factor g Landé-Faktor *m* g, g-Faktor *m*, spektroskopischer Aufspaltungsfaktor *m*
splitting-off Abspalten *n*, Abspaltung *f*
splitting-out Abspalten *n*, Abspaltung *f*
splitting pattern *(phys Ch)* Aufspaltungsbild *n*
splitting-up *s.* splitting
spodium Spodium *n*, Knochenkohle *f*
spoil *v* verderben, unbrauchbar machen; verderben, schlecht [unbrauchbar] werden *(Lebensmittel)*
spoil pile *(Bgb)* Abraumhalde *f*
spoilage 1. Verderb *m*, Verderben *n*, Schlechtwerden *n (von Lebensmitteln)*; 2. verdorbenes Material *n*
spoilage microorganisms mikrobielle Verderbniserreger *mpl*
sponge glass Schaumglas *n*
sponge iron Schwammeisen *n*, Eisenschwamm *m*
sponge plastic Schwammkunststoff *m*
sponge rubber Schwammgummi *m*
spongelike *s.* spongy
sponging agent *(Gum, Kst)* Treibmittel *n*, Blähmittel *n*
spongy schwamm(art)ig, locker, porös
spongy-nickel catalyst Nickelschwammkatalysator *m*
spongy platinum Platinschwamm *m*
spontaneous spontan, freiwillig, von selbst ablaufend
spontaneous emission spontane Emission *f (von Strahlung)*
spontaneous-ignition temperature Selbstzünd(ungs)temperatur *f*, SZT, Selbstentzündungstemperatur *f*, Selbstzündpunkt *m*, Zündpunkt *m*
spontaneous mutation *(Biot)* Spontanmutation *f*, spontane Mutation *f*
spontaneous nucleation *(Krist)* homogene Keimbildung *f*, spontane [sporadische, thermische] Keimbildung *f*
spore germinating medium *(Biot)* Keimungsmedium *n*
spore suspension *(Biot)* Sporensuspension *f*
sporulation medium *(Biot)* Sporulationsmedium *n*
SPOS *s.* solid-phase organic synthesis
spot *v* tüpfeln
spot Fleck(en) *m*, Unreinheit *f*, Klecks *m*, Tupfen *m*
spot plate Tüpfelplatte *f*
spot remover Fleck(en)entfernungsmittel *n*, Fleckenentferner *m*, Fleckenreinigungsmittel *n*, Detachiermittel *n*
spot test Tüpfelprobe *f*
spot-test analysis Tüpfelanalyse *f*
spout Ausguss *m (eines Laborgeräts)*; Abflussrohr *n*, Speirohr *n*
spouted bed Sprudelschicht *f (bei Wirbelschichtverfahren)*
SPR *s.* surface plasmon resonance spectroscopy
spray *v* 1. (ver)spritzen *(Flüssigkeiten)*; versprühen, verdüsen *(in feine Tröpfchen)*; bespritzen *(mit Flüssigkeit)*; besprühen, übersprühen, bedüsen *(mit feinen Tröpfchen)*; 2. *(Aufb)* bebrausen, abbrausen, (mit einer Brause) abspülen; 3. *(Anstr)* spritzen
spray-dry *v* im Sprühverfahren trocknen
spray 1. Sprühmittel *n*, Sprühflüssigkeit *f*, Nebelmittel *n*; Spritzmittel *n*, Spritzflüssigkeit *f*; 2. *(Kosmet)* Spray *m(n)*, Sprühmittel *n*; 3. Sprühnebel *m*; 4. *s.* spray can
spray absorber Sprühabsorber *m*
spray can Sprühdose *f*
spray chamber Spritzkammer *f*; Sprühkammer *f*
spray coat Spritzschicht *f*
spray coating 1. Spritzen *n (von Oberflächen)*; 2. Spritzschicht *f*
spray column Sprühkolonne *f (Gasreinigung)*
spray cone Spritzkegel *m*; Sprühkegel *m*
spray cooler Rieselkühler *m*, Berieselungskühler *m*
spray dryer Sprühtrockner *m*
spray drying Sprühtrocknung *f*
spray gun 1. Spritzpistole *f*, Sprühpistole *f*; 2. *(Anstr)* Farbspritzpistole *f*
spray injury Spritzschaden *m (durch Pflanzenschutzmittel)*
spray load *s.* spray residue
spray mist *(Landw)* Sprühnebel *m*
spray nozzle Sprühdüse *f*
spray painting Farbspritzen *n*, Spritzen *n*
spray reagent Sprühreagens *n*
spray residue Spritzrückstand *m*, Spritzbelag *m*, Spritzniederschlag *m (von Pflanzenschutzmitteln)*
spray scrubber Sprühwascher *m (zur Gasreinigung)*

spray swathe Sprühschleier *m*; Spritzschleier *m*
spray tower Sprühturm *m*
spray washer *s.* spray scrubber
sprayer Versprüher *m*, Sprüheinrichtung *f*
spread *v* **(out)** ausbreiten, verteilen; ablegen, auslegen; spreiten *(unlösliche oder schwerlösliche Stoffe auf der Oberfläche von Flüssigkeiten zu einer monomolekularen Schicht)*; sich ausbreiten [ausdehnen]
spreader *s.* spreading agent
spreading agent Netzmittel *n*
spreading coefficient Ausbreitungskoeffizient *m*, Spreitungskoeffizient *m*; Benetzungskoeffizient *m*
spreading efficiency Verteilerwirksamkeit *f (von Netzmitteln)*
spreading factor *(Bioch)* Ausbreitungsfaktor *m*
spreading pressure Spreitungsdruck *m*
spring 1. Quelle *f*; 2. Feder *f*
spring-actuated relief valve *s.* spring safety valve
spring balance Federwaage *f*
spring-loaded mit Federdruck, federbelastet
spring safety valve federbelastetes Sicherheitsventil *n*, Federsicherheitsventil *n*
sprinkle *v* 1. streuen; 2. spritzen; sprühen; bespritzen, besprengen, berieseln
sprinkler *(Hyd)* Regner *m*
sprinkler system Sprinkler-Feuerlöschanlage *f*
spruce turpentine Fichtenöl *n*
sprue *(Kst)* Anguss(kegel) *m*
SPT 1. = selective population transfer; 2. = soft pulse transfer
spume *v* schäumen, sich mit Schaum bedecken
spume Schaum *m*
spumescent [spumous, spumy] schaum(art)ig; mit Schaum bedeckt
spun-bonded product *(Text)* nach dem Schmelzspinnverfahren hergestellter Vliesstoff *m*, Schmelzspinnverbundstoff *m*
spun-dyed (er)spinngefärbt, düsengefärbt
spun fibre Spinnfaser *f*, gesponnene Faser *f*
spurious peak *(Anal)* Geisterpeak *m*, Störpeak *m*, Fremdpeak *m*
sputter *v* sputtern *(im Vakuum zer- oder aufstäuben)*
sputter-ion pump Ionen(getter)pumpe *f (Vakuumtechnik)*
sputtering Vakuumzerstäuben *n*, Ionen-Plasma-Zerstäuben *n*, Sputtern *n*, Katodenzerstäuben *n*
spy hole Schauloch *n*
squalene Squalen *n (aliphatisches Triterpen)*
square planar orientation quadratisch ebene Anordnung *f (von Liganden)*
square-wave polarography Rechteckwellenpolarographie *f*, *Square-wave-Polarographie *f*
squeegee Quetschwalze *f*, Gummiquetscher *m*; Rakel *f (für Siebdruck)*
squeeze *v* 1. zusammenpressen, zusammenquetschen, zusammendrücken; auspressen, ausquetschen, ausdrücken *(z. B. Früchte)*; herauspressen, abpressen, abquetschen *(z. B. Saft)*; 2. einzwängen *(z. B. Atome in eine Gitterlücke)*; 3. sich erschöpfen, verknappen, knapp werden *(natürliche Vorkommen)*
squeeze *v* **through** durchpressen
squeeze 1. Zusammenpressen *n*, Zusammenquetschen *n*, Zusammendrücken *n*; Auspressen *n*, Ausquetschen *n*, Ausdrücken *n (z. B. von Früchten)*; Herauspressen *n*, Abpressen *n*, Abquetschen *n (z. B. von Saft)*; 2. Einzwängung *f (z. B. von Atomen in Gitterlücken)*; 3. Erschöpfung *f*, Verknappung *f (natürlicher Vorkommen)*
squeeze bottle Spritzflasche *f* aus weichem Kunststoff
squirrel-cage disintegrator [mill] Schlagkorbmühle *f*, *(i. e. S.)* Desintegrator *m*, *(i. e. S.)* Desintegratormühle *f*, Schleudermühle *f*
μSR *s.* muon spin rotation
S.R.B. *(straight-run benzine) s.* straight-run gasoline
SRH *s.* somatotropin releasing hormone
s.s. *s.* slightly soluble
SS *s.* suspended solids
SSB protein *s.* single-strand DNA-binding protein
SSD *s.* steady-state distribution

stab culture Stichkultur *f (Mikrobiologie)*
stability 1. Beständigkeit *f*, Widerstandsfähigkeit *f*, Stabilität *f*, Dauerhaftigkeit *f (chemisch, thermisch)*; Stabilität *f*, Festigkeit *f (mechanisch)*; 2. *(Lebm)* Haltbarkeit *f* • **come into stability** Stabilität erlangen, stabil werden *(von Phasen, Modifikationen)*
stability at high temperatures Hochtemperaturbeständigkeit *f*, Beständigkeit *f* [Widerstandsfähigkeit *f*] gegen hohe Temperaturen
stability constant Komplexstabilitätskonstante *f*, Komplexbildungskonstante *f*, Stabilitätskonstante *f*, Bildungskonstante *f*
stability in storage Lagerbeständigkeit *f*, Lager(halt)barkeit *f*, Lagerstabilität *f*
stability of colour Farb(ton)beständigkeit *f*, Farbenbeständigkeit *f*, Farb(ton)echtheit *f*
stability to acid(s) Säurebeständigkeit *f*, Säurefestigkeit *f*, Säurewiderstandsfähigkeit *f*, Säureresistenz *f*
stability to alkali(es) Alkalibeständigkeit *f*, Laugenbeständigkeit *f*, Laugenfestigkeit *f*
stability to chemical attack chemische Unangreifbarkeit *f* [Beständigkeit *f*], Beständigkeit *f* gegen chemische Einwirkungen; Chemikalienbeständigkeit *f*, Widerstandsfähigkeit *f* gegen Chemikalien
stabilization lagoon [pond] *(Hyd)* Stabilisierungsteich *m*
stabilization tank *(Hyd)* Regenerationsbecken *n*, Regenerierungsbecken *n*, Stabilisierungsbecken *n (Belebtschlammregenerierung)*
stabilize *v* 1. stabilisieren; sich stabilisieren; 2. *(Met)* stabilglühen, stabilisierend glühen
stabilizer 1. Stabilisier(ungs)mittel *n*, Stabilisator *m*; 2. *(Erdöl)* Stabilisator *m*, Stabilisierkolonne *f*, Stabilisationskolonne *f*, Stabilizer *m*
stabilizing 1. Stabilisieren *n*, Stabilisierung *f*, Stabilisation *f*; 2. *s.* stabilizing anneal
stabilizing agent Stabilisier(ungs)mittel *n*, Stabilisator *m*
stabilizing anneal *(Met)* Stabil(isierungs)glühen *n*, Stabil(isierungs)glühung *f*, stabilisierendes Glühen *n*
stabilizing energy Stabilisierungsenergie *f*, dewarsche Resonanzenergie *f*
stable 1. beständig, widerstandsfähig, stabil, dauerhaft *(chemisch, thermisch)*; stabil, fest *(mechanisch)*; 2. *(Lebm)* haltbar
stable in storage lagerbeständig, lagerbar
stable isotope stabiles [nicht radioaktives] Isotop *n*
stable orbit stabile [stationäre] Bahn *f*, Sollbahn *f*
stable to acid(s) säurebeständig, säurestabil, säurefest, beständig gegen Säurewirkung
stable to alkali(es) alkalibeständig, laugenbeständig, laugenfest
stable to oxidation oxidationsbeständig, beständig gegen oxidative Einflüsse
stable towards ... *siehe: stable to ...*
stack *v* 1. (auf)stapeln; 2. *s.* stack up
stack *v* **up** aufeinander schichten
stack 1. Kamin *m*, Schornstein *m*, Esse *f*; Schacht *m (eines Hochofens)*; 2. Stapel *m*; Satz *m (z. B. von Walzen)* • **be lost at the stack** durch den Schornstein abgehen
stack gas Rauchgas *n*
stack loss Flugstaubverlust *m*, Rauchgasverlust *m*
stacked packing Packung *f*, geordnete Kolonnenfüllung *f*
stacking 1. Stapeln *n*, Aufstapeln *n*; 2. *(Krist)* Stapelung *f*
stacking fault *(Krist)* Stapelfehler *m*
stacking interaction *(Bioch)* Wechselwirkung *f* durch Stapelkräfte *(in der DNA-Doppelhelix)*
stage Stufe *f*, Stadium *n*, Phase *f*, Schritt *m*
stage 1 intercalation compound Interkalationsverbindung *f* [Einlagerungsverbindung *f*] 1. Stufe
stage of plasticity plastischer Zustand *m*, Erweichungszustand *m*
stage of treatment 1. Behandlungsstufe *f*, Verfahrensstufe *f*, Prozessstufe *f*; 2. *(Hyd)* Aufbereitungsstufe *f*
stagger *v* gegeneinander versetzen, versetzt anordnen

staggered gegeneinander versetzt, versetzt angeordnet; gestaffelt, auf Lücke, ±anti-periplanar *(Stereochemie)*
staggered configuration [conformation] gestaffelte [±anti-periplanare] Konformation *f*, Atom-Lücke-Konformation *f*
stagnant zone stagnierende Zone *f*
stagnation point Staupunkt *m (Strömungslehre)*
stagnation pressure Staudruck *m*, *(i. e. S.)* Gesamtdruck *m*, Ruhedruck *m*
stain *v* beflecken, anflecken, beschmutzen, anschmutzen; (an)färben, beizen *(Holz, Glas, Textilien)*
stain 1. Fleck *m*; 2. Färbemittel *n*; Beize *f*, Holzbeize *f*; 3. *(Glas)* Beize *f*, Farbbeize *f*
stain remover Fleck(en)entfernungsmittel *n*, Fleckenentferner *m*, Detachiermittel *n*
stainless korrosionsbeständig, *(Eisen und Stahl auch)* nicht rostend, rostbeständig, rostfrei, rostsicher
staircase conformation Sesselkonformation *f*
stalagmometer Stalagmometer *n (zur Bestimmung der Oberflächenspannung von Flüssigkeiten)*
stalagmometry Stalagmometrie *f (Messung der Oberflächenspannung von Flüssigkeiten)*
stamp Stempel *m*
stand Stativ *n*
stand clamp Stativklemme *f*
stand oil Standöl *n*
standard 1. Standard *m*; Normal *n*; 2. *s.* standard substance • **bring to standard strength** auf Typ bringen *(einen Farbstoff)*
standard acid Säure-Maßlösung *f*, eingestellte Säure *f*
standard atmosphere Normalatmosphäre *f*, physikalische Atmosphäre *f*, atm *(SI-fremde Einheit des Drucks*; *1 atm = 107325 Pa)*
standard base Lauge-Maßlösung *f*, eingestellte Lauge *f*
standard buffer (solution) Standardpufferlösung *f*, standardisierte Pufferlösung *f*, Standardpuffer *m*
standard calibration Eichung *f (von Gefäßen)*
standard calomel electrode Standardkalomelelektrode *f*
standard cell Standard(normal)element *n*, Standardzelle *f*, Normalelement *n*
standard chemical potential chemisches Standardpotenzial *n*
standard conditions 1. Standardbedingungen *fpl*, Norm(al)bedingungen *fpl*; 2. *(Text)* Norm(al)klima *n*
standard deviation Standardabweichung *f*, mittlere quadratische Abweichung *f*, mittlerer quadratischer Fehler *m (Statistik)*
standard electrode potential Standardelektrodenpotenzial *n*, Standardpotenzial *n* (der Elektrode)
standard emf Standard-(Bezugs-)EMK *f*
standard enthalpy Standardenthalpie *f*
standard enthalpy of formation Standardbildungsenthalpie *f*
standard entropy Standardentropie *f*
standard entropy of activation Standardaktivierungsentropie *f*
standard free energy freie Standardenthalpie *f*
standard free energy of formation freie Standardbildungsenthalpie *f*
standard Gibbs energy freie Standardenthalpie *f*
standard Gibbs function of formation freie Standardbildungsenthalpie *f*
standard ground glass joint Standardschliffverbindung *f*, Standardschliff *m*, Normschliff *m*
standard half-cell potential Standardpotenzial *n* des Halbelements
standard heat of formation Standardbildungswärme *f*; Standardbildungsenthalpie *f*
standard hydrogen electrode Standardwasserstoffelektrode *f*
standard method Standardmethode *f*, Standardverfahren *n*, standardisierte Methode *f*, standardisiertes Verfahren *n*
standard molar Gibbs function molare freie Standardenthalpie *f*
standard of emf EMK-Normal *n*, Urspannungsnormal *n*
standard oxidation potential Standardoxidationspotenzial *n*
standard plane *(Krist)* Hauptsymmetrieebene *f*

standard potential 1. Standard(normal)potenzial *n*; 2. *s.* standard electrode potential
standard pressure Standarddruck *m*, Norm(al)druck *m*, Normalluftdruck *m*
standard procedure *s.* standard method
standard reaction enthalpy Standardreaktionsenthalpie *f*
standard reaction entropy Standardreaktionsentropie *f*
standard reagent Titersubstanz *f*; Standardreagens *n*
standard redox potential Standardredoxpotenzial *n*, Standardpotenzial *n* des Redoxsystems
standard reduction potential Standardreduktionspotenzial *n*
standard reference state Standardzustand *m*
standard sieve scale [series] Standardsiebskala *f*, Normalsiebskala *f*, Standardsiebreihe *f*, Normalsiebreihe *f*
standard solution Maßlösung *f*, Standardlösung *f*, standardisierte [eingestellte] Lösung *f*; Vergleichslösung *f*
standard state Standardzustand *m*
standard substance Standard(bezugs)substanz *f*, standardisierte Substanz *f*
standard technique *s.* standard method
standard temperature and pressure Norm(al)temperatur *f* und -druck *m*, physikalischer Normzustand *m* *(Standardbedingungen von 0 °C und 1,01325 105 Pa)*
standard test(ing) method standardisierte Prüfmethode *f* [Untersuchungsmethode *f*]
standard third law entropy molare Standardentropie *f*
standard titrant [titrimetric substance] Titersubstanz *f*
standard unit Standardeinheit *f*, Normaleinheit *f*
standard value Standardwert *m*, Norm(al)wert *m*
standard voltage Standard(bezugs)spannung *f*
standard volume Norm(al)volumen *n*
standard Weston normal cadmium cell Weston-Normalelement *n*, Weston-Element *n*
standardization 1. Standardisierung *f*; 2. Einstellen *n*, Einstellung *f (auf einen vorgegebenen Richtwert)*
standardize *v* 1. standardisieren; 2. einstellen *(auf einen vorgegebenen Richtwert)*
standards for discharge *(Hyd)* Einleitungsbedingungen *fpl (für Abwasser)*
standpipe Standrohr *n*
stannane *s.* stannic hydride
stannate Stannat *n*
stannic Zinn..., *(i. e. S.)* Zinn(IV)-...
stannic acid Zinnsäure *f*
stannic ammonium chloride Ammoniumhexachlorostannat(IV) *n*, Pinksalz *n*
stannic chloride Zinn(IV)-chlorid *n*, Zinntetrachlorid *n*
stannic hydride Zinnhydrid *n*, Zinn(IV)-hydrid *n*, Zinnwasserstoff *m*, Stannan *n*
stannic oxide Zinn(IV)-oxid *n*, Zinndioxid *n*
stannic sulphide Zinn(IV)-sulfid *n*, Zinndisulfid *n*
stannous Zinn..., *(i. e. S.)* Zinn(II)-...
stannous chloride Zinn(II)-chlorid *n*, Zinndichlorid *n*
stannous sulphide Zinn(II)-sulfid *n*, Zinnmonosulfid *n*
Stanton number Stanton-Zahl *f (Verhältnis Stoffübergangsstrom zu Konvektionsstrom)*
staple fibre Faser *f*, Spinnfaser *f*, *(bei Chemiefaserstoffen)* Stapelfaser *f*
staple length Faserlänge *f*, Stapellänge *f*, Stapel *m*
star feeder Zellenradaufgeber *m*, Zell(en)rad *n*
star polymer *(Kst)* Sternpolymer *n (Polymerarchitektur)*
star valve Zellradschleuse *f*, Zellenschleuse *f*
starch Stärke *f*
starch block electrophoresis Stärkeblockelektrophorese *f*
starch breakdown *s.* starch degradation
starch chain Stärkekette *f*
starch conversion 1. Stärkeumwandlung *f*; 2. *(Lebm)* Stärkeverzuckerung *f*, Stärkekonversion *f (mit Säuren und Enzymen)*
starch-converting enzyme *s.* starch-splitting enzyme

starch degradation Stärkeabbau *m*, Stärkezerlegung *f*
starch-digesting enzyme *s.* starch-splitting enzyme
starch flour Stärkemehl *n*
starch gel electrophoresis Stärkegelelektrophorese *f*
starch glue Stärkeleim *m*, Stärkekleister *m*
starch grain [granule] Stärkekorn *n*, Stärkekörnchen *n*
starch gum *s.* dextrin
starch hydrolysis Stärkehydrolyse *f*, hydrolytischer Stärkeabbau *m*
starch hydrolyzate Stärkehydrolysat *n*
starch iodide paper Iod(id)stärkepapier *n*
starch liquefaction Stärkeverflüssigung *f*
starch nitrate Stärkenitrat *n*, Nitrostärke *f (Sprengstoff)*
starch plant Stärkepflanze *f (überschüssigen Assimilatzucker in Stärke umwandelnd)*
starch retrogradation Stärkeretrogradation *f*
starch saccharification *(Lebm)* Stärkeverzuckerung *f*, Stärkekonversion *f (mit Säuren und Enzymen)*
starch-splitting enzyme stärkespaltendes [stärkeabbauendes] Enzym *n*, Amylase *f*
starchy stärkehaltig, stärkeführend; Stärke...
Stark effect Stark-Effekt *m (Aufspaltung von Spektrallinien)*
Stark-splitting Stark-Aufspaltung *f*
start *v* 1. initiieren, anregen, einleiten, in Gang bringen *(Reaktion)*; einsetzen, in Gang kommen, beginnen, anfangen *(Reaktion)*; 2. *s.* start up
start *v* **up** in Betrieb nehmen [setzen], in Gang setzen, anfahren, anlassen, *(bei Öfen auch)* anheizen
start 1. Beginn *m*, Anfang *m*, Einsetzen *n (einer Reaktion)*; 2. *s.* start-up; 3. *(Chromat)* Startfleck *m*
start-up Inbetriebnahme *f*, Ingangsetzen *n*, Anfahren *n*, Anlassen *n*, *(bei Öfen auch)* Anheizen *n*
starter culture 1. *(Biot)* Starterkultur *f*; 2. *(Lebm)* Säureweckerkultur *f*
starter fertilizer Startdünger *m*
starting concentration Anfangskonzentration *f*
starting material 1. Ausgangsmaterial *n*, Ausgangsstoff *m*, Ausgangssubstanz *f*, Ausgangsgut *n*; 2. *(Biot)* Ausgangssubstrat *n*
starting value Anfangswert *m*
starvation *(Landw)* Nährstoffmangel *m*
state Zustand *m*
state diagram Termschema *n*, Energieniveauschema *n*, Schema *n* der Energiezustände
state function Zustandsfunktion *f*
state of aggregation 1. Aggregatzustand *m*; 2. Anordnung *f (von Atomen, Atomgruppen oder Molekülen)*
state of equilibrium Gleichgewichtszustand *m*
state of matter Aggregatzustand *m*, Formart *f*
static bed Festbett *n*, ruhendes [statisches] Bett *n* [Feststoffbett *n*], ruhende [statische] Schüttung *f*
static catalyst Festbettkatalysator *m*, Festbettkontakt *m*, festliegender [fester, ruhender] Katalysator *m* [Kontakt *m*]
static electrification statische Aufladung *f*
static friction Haftreibung *f*
static head (statische) Druckhöhe *f*
static light scattering statische Lichtstreuung *f*
static loading statische Belastung *f*
static method statische Methode *f (zur Messung von Dampfdrücken)*
static modulus of elasticity statischer Elastizitätsmodul *m (Zugversuch)*
static pressure statischer Druck *m*
static quenching statische Löschung *f (Photochemie)*
static seal Dichtung *f (für nicht gegeneinander bewegte Teile)*
static stress statische Beanspruchung *f*
stationary phase 1. *(Chromat)* stationäre [unbewegliche] Phase *f*; 2. *(Biot)* stationäre Phase *f (Mikroorganismenwachstum)*
stationary screen Sieb *n* mit unbewegter Siebfläche
stationary state *s.* steady state
statistical mechanics *s.* statistical thermodynamics

statistical thermodynamics statistische Thermodynamik *f*
statistical weight *(phys Ch)* statistisches Gewicht *n*
statistically valid statistisch gesichert
steady state stationärer [stabiler] Zustand *m*; (quasi)stationärer Zustand *m*, Stationärzustand *m*, Fließgleichgewicht *n*, Fließgleichgewichtszustand *m* *(Kinetik)*
steady-state approximation method *s.* steady-state method
steady-state concentration (quasi)stationäre Konzentration *f*, Gleichgewichtskonzentration *f*
steady-state culture *(Biot)* Kultur *f* in statischem Zustand, statische Kultur *f*
steady-state distribution Gleichgewichtsverteilung *f*
steady-state flow stationäre Strömung *f*
steady-state kinetics *(Bioch)* Stationärenzymkinetik *f*
steady-state method Methode *f* [Prinzip *n*] des quasistationären Zustands, Quasistationaritätsprinzip *n*, (bodensteinsches) Stationaritätsprinzip *n*, Bodenstein-Prinzip *n* *(Kinetik)*
steady-state potential Gleichgewichtspotenzial *n*, Gleichgewichtsgalvanispannung *f*, Nernst-Potenzial *n*
steady-state reactor stationär arbeitender Reaktor *m*
steady-state value (quasi)stationärer Wert *m*, Gleichgewichtswert *m*
steal *v* **away** abfangen
steam *v* 1. dämpfen, mit Dampf [Wasserdampf] behandeln; ausdämpfen, mit Wasserdampf desorbieren; 2. *(Carb)* gasen
steam-distil *v* mit Wasserdampf destillieren
steam Dampf *m*, *(i. e. S.)* Wasserdampf *m*
steam autoclave Dampfautoklav *m*, Dampfgefäß *n*
steam bath Dampfbad *n*
steam blanching *(Lebm)* Dampfblanchieren *n*
steam boiler Dampfkessel *m*
steam calorimeter Dampfkalorimeter *n*, Kondensationskalorimeter *n*
steam can Dampfkanne *f*
steam-carbon reaction Reaktion *f* des Kohlenstoffs mit Wasserdampf, Wassergasreaktion *f*
steam coal Dampf(kessel)kohle *f*, Kesselkohle *f*
steam coil Dampf(heiz)schlange *f*
steam condenser Kühler *m* *(für Dämpfe)*; Kondensator *m*
steam decomposition 1. Dampfzersetzung *f*; 2. Dampfzersetzungsgrad *m*
steam-distillable mit Wasserdampf destillierbar
steam distillation Wasserdampfdestillation *f*
steam-driven dampfgetrieben, durch Dampfdruck betrieben, mit Dampfantrieb
steam dryer Dampftrockner *m*, Dampftrockenapparat *m*
steam economy *Kehrwert des spezifischen Heizdampfbedarfs des Verdampfers*
steam ejector *s.* steam-jet ejector
steam generator Dampferzeuger *m*, Dampfentwickler *m*
steam-heated dampfbeheizt, mit Dampf beheizt, Dampf...
steam-iron method [process] Dampf-Eisen-Verfahren *n* *(zur Wasserstofferzeugung)*
steam-jacketed mit Dampfmantel
steam-jet ejector Dampfstrahlejektor *m*, Dampfstrahlsauger *m*
steam line Dampfleitung *f*
steam main Dampf(haupt)leitung *f*; Dampfleitungsnetz *n*
steam-oxygen gasification *(Carb)* Dampf-Sauerstoff-Vergasung *f*, Vergasung *f* mit Wasserdampf und Sauerstoff
steam pipe Dampf(leitungs)rohr *n*; Dampfheizungsrohr *n*
steam raising Dampferzeugung *f*
steam stripping *s.* steam distillation
steam superheater Dampfüberhitzer *m*
steam trap Kondens(at)wasserableiter *m*, Kondenstopf *m*, Spritzwasserfalle *f* *(am Verdampfer)*
steam tube *s.* steam pipe
steam-turbine oil Dampfturbinenöl *n*
steam valve Dampfventil *n*
steam-volatile wasserdampfflüchtig, mit Wasserdampf flüchtig

steam vulcanization *(Gum)* Dampfvulkanisation *f*, Vulkanisation *f* in Dampf

steaming 1. Dämpfen *n*, Dampfbehandlung *f*; 2. *(Farb)* Dämpfen *n*, Dämpferpassage *f*; 3. *(Carb)* Dampfen *n*, Dampfung *f*, Nassbetrieb *m (bei der Gaserzeugung)*; Gasen *n*, Gasung *f*, Kaltblasen *n*

stearamide Stearamid *n*, Octadecansäureamid *n*

stearate Stearat *n (Salz oder Ester der Stearinsäure)*

stearic acid Stearinsäure *f*, *Octadecansäure *f*

stearin 1. Stearin *n*, *(i. e. S.)* Tristearin *n*, Glyceroltristearat *n*, Glyceryloctadecanoat *n*; 2. Stearin *n (technische Stearinsäure)*

stearolic acid Stearolsäure *f*, *Octadec-9-insäure *f*

stearyl alcohol Stearylalkohol *m*, *Octadecan-1-ol *n*

steatite *(Min)* Steatit *m*, Speckstein *m*

steel Stahl *m*

steel casting 1. Stahlgussstück *n*; 2. Stahlguss *m*

steel cylinder Stahlflasche *f*, Stahlbombe *f*

steel ingot Stahlblock *m*, Rohstahlblock *m*

steel jacket Stahlmantel *m*

steel scrap Stahlschrott *m*

steel shell Stahlpanzer *m*

steelmaking process Stahlerzeugungsverfahren *n*, Stahlherstellungsverfahren *n*, Stahlgewinnungsverfahren *n*

steep *v* (ein)weichen, (ein)wässern, tränken, quellen

steep *s.* 1. steeping; 2. steeping liquor

steepage *s.* steeping

steeping Weichen *n*, Einweichen *n*, Wässern *n*, Einwässern *n*, Tränken *n*, Quellen *n*; Trogtränkung *f*, Tauchtränkung *f (Holzkonservierung)*

steeping liquor Einweichflüssigkeit *f*, Tränkflüssigkeit *f*

steepness Steilheit *f (einer Kurve oder Elektrode)*

Stefan-Boltzmann law Stefan-Boltzmann-Gesetz *n (des Emissionsvermögens schwarzer Körper)*

Steffen process Steffen-Verfahren *n*, Kalkentzuckerungsverfahren *n*, Kalksaccharatverfahren *n*, (steffensches) Ausscheidungsverfahren *n (zur Entzuckerung von Melasse)*

Steinbühl yellow *s.* baryta yellow

stem 1. Ansatzrohr *n (einer Gasbürette)*; Stiel *m (eines Trichters)*; Messkapillare *f*, Messkapillarrohr *n (eines Thermometers)*; 2. Spindel *f (eines Ventils)*

stem correction Fadenkorrektur *f (Thermometrie)*

Stengel process Stengel-Verfahren *n (zur Herstellung von Ammoniumnitrat)*

step Schritt *m*, Stufe *f (eines Verfahrens, einer Reaktion)*

step-growth copolymerization Co-Stufenpolymerisation *f*

step-growth polymerization Stufen(wachstums)polymerisation *f*

step height *(Chromat)* Stufenhöhe *f*

step reaction Stufenreaktion *f*

step-reaction degradation Kettenspaltung *f*

Stephen reaction [reduction] Stephen-Reduktion *f*, Stephen-Reaktion *f (zur Darstellung von Aldehyden aus Nitrilen)*

stepwise degradation stufenweiser [schrittweiser] Abbau *m*

stepwise elution *(Chromat)* Stufeneluierung *f*, stufenweise Elution *f*

stepwise formation constant *s.* successive formation constant

stepwise process stufenweise ablaufender Vorgang *m*; stufenweise geführtes Verfahren *n*

stepwise reaction Folgereaktion *f*, *(besonders bei Polymerisationen:)* Stufenreaktion *f*

stereoblock polymer Stereoblockpolymer(e) *n*

stereochemical stereochemisch

stereochemical formula Stereoformel *f*

stereochemistry Stereochemie *f*

stereoelectronic stereoelektronisch

stereoisomer Stereoisomer(e) *n*, Stereomer(e) *n*

stereoisomeric stereoisomer

stereoisomerism Stereoisomerie *f*, stereochemische [räumliche] Isomerie *f*

stereomer *s.* stereoisomer

stereoregular stereoregulär, stereoreguliert, sterisch geordnet, stereogeordnet

stereoregularity Stereoregularität *f*
stereoselective polymerization stereoselektive Polymerisation *f*
stereoselectivity Stereoselektivität *f*
stereospecific catalysis stereospezifische Katalyse *f*
stereospecific polymerization stereospezifische Polymerisation *f*
stereospecificity Stereospezifität *f*, stereochemische Spezifität *f*
steric acceleration sterische Beschleunigung *f*, B-strain-Effekt *m*
steric factor sterischer Faktor *m*, Wahrscheinlichkeitsfaktor *m*
steric hindrance [**inhibition, limitation**] sterische Hinderung *f*
steric retardation sterische Verzögerung *f*, F-strain-Effekt *m*
sterically feasible sterisch möglich
sterically hindered sterisch gehindert
sterile steril, keimfrei
sterile filtration *(Hyd, Biot)* Sterilfiltration *f*, Entkeimungsfiltration *f*
sterility Sterilität *f*, Keimfreiheit *f*
sterilization Sterilisation *f*, Sterilisierung *f*, Entkeimung *f*, Keimfreimachung *f*
sterilize *v* sterilisieren, entkeimen, keimfrei machen
sterilize *v* **by filtration** *(Hyd, Biot)* sterilfiltrieren, durch Filtration keimfrei machen
sterilizer Sterilisator *m*, Sterilisationsapparat *m*, Entkeimungsapparat *m*
Stern-Gerlach experiment Stern-Gerlach-Versuch *m*, stern-gerlachscher Versuch *m (zum Nachweis des Elektronenspins)*
Stern-Volmer kinetic relationship Stern-Volmer-Beziehung *f*, Stern-Volmer-Gleichung *f (der Fluoreszenzlöschung)*
sternutative *(Pharm)* Niesmittel *n*, Sternutatorium *n*; Nasen-Rachen-Reizstoff *m (Militärchemie)*
sternutator *(Pharm)* Niesmittel *n*, Sternutatorium *n*; Nasen-Rachen-Reizstoff *m (Militärchemie)*
sternutatory Niesen erregend
steroid Steroid *n*
steroid alkaloid Steroidalkaloid *n*, steroides Alkaloid *n*
steroid chemistry Steroidchemie *f*
steroid hormone Steroidhormon *n*
steroidal alkaloid *s.* steroid alkaloid
sterol *(Bioch)* Sterol *n*
Stevens rearrangement Stevens-Umlagerung *f*, stevenssche Umlagerung *f (quartärer Ammoniumverbindungen)*
STH *s.* somatotropic hormone
stibane Stiban *n*, Antimonwasserstoff *m*, Antimonhydrid *n (allgemeine Formel Sb_nH_{n+2})*
stibial antimonhaltig; Antimon...
stibic Antimon..., *(i. e. S.)* Antimon(V)-...
stibine Stiban *n*, Monostiban *n*, Antimonwasserstoff *m*, *(unsubstituiertes SbH_3 auch:)* Stibin *n*
stibinic acid RR′SbOOH Stibinsäure *f*
stibious Antimon..., *(i. e. S.)* Antimon(III)-...
stibnite *(Min)* Stibnit *m*, Antimonit *m*, Antimonglanz *m*, Grauspießglanz *m (Antimon(III)-sulfid)*
stibonium compound Antimoniumverbindung *f*
stick *v* 1. durchstechen, durchstoßen; stecken; 2. ankleben; aufkleben, (ver)kleben; 3. haften, (fest)kleben; klemmen, sich verklemmen, hängen bleiben, stecken (bleiben), festsitzen
stick *v* **together** zusammenkleben, *(stückiges Gut)* zusammenbacken
stick spectrum Strichspektrum *n*
sticker Haftmittel *n*, Haftstoff *m*
stickiness Klebrigkeit *f*
sticking agent *s.* sticker
sticky klebrig, klebend
sticky end *(Bioch)* klebriges [kohäsives] Ende *n*, sticky end *(an Spaltstellen der DNA nach Einwirkung von Restriktionsenzymen)*
stiction Haftreibung *f*
stiff steif, starr
stiffen *v* (ver)steifen
stiffening Versteifen *n*, Versteifung *f*
stiffness Steifigkeit *f*
stiffness in bend [flexure] Biegesteifigkeit *f*
stiffness in torsion Verdrehungssteifigkeit *f*, Torsionssteifigkeit *f*
stilbene Stilben *n*, *trans*-1,2-Diphenylethen *n*
stilbene dye Stilbenfarbstoff *m*
still Destillationsapparat *m*, Destillierapparat *m*; Destillationsanlage *f*, Destillieranlage *f*

still head 1. Kolonnenkopf *m*; 2. *(Lab)* Destillieraufsatz *m*
still pot 1. Destillationsblase *f*, Destillierblase *f*, Blase *f*, Destillationsgefäß *n*; 2. *(Tech)* Verdampfungsofen *m*
still receiver Destilliervorlage *f*, Destillat(ions)vorlage *f*, Destillatsammler *m*, Vorlage *f*
still residue Destillationsrückstand *m*
stillage 1. *(Gär)* Schlempe *f*, Brennereischlempe *f*; 2. Gestell *n*, Stellage *f*
stimulant Stimulans *n*, Anregungsmittel *n*, Reizmittel *n*
stimulated emission stimulierte Emission *f*
stimulatory drug *s.* stimulant
stimulatory modulator *(Bioch)* fördernder [positiver] Modulator *m*
stir *v* rühren
stir *v* **in** einrühren
stir *v* **together** zusammenrühren
stir *v* **up** aufrühren; aufwirbeln *(durch Gas)*
stirrable rührbar
stirred autoclave Rühr(werks)autoklav *m*
stirred bioreactor (fermenter) *(Biot)* gerührter Bioreaktor *m*, Rühr(kessel)-reaktor *m*, Rühr(kessel)fermenter *m*
stirred-tank reactor Rühr(kessel)reaktor *m*, Rührkessel *m*, Rührtankreaktor *m*, Mischreaktor *m*
stirred-tank reactors in series Rührkesselkaskade *f*, Kaskadenreaktor *m*, Mischreaktorkaskade *f*, mehrstufiger Rührreaktor *m*
stirrer 1. Rührer *m*; Rührwerk *n (Rührer mit Motor)*; Rührmaschine *f*, Rührapparat *m*; 2. Rührarm *m*
stirrer blade Rühr(er)blatt *n*, Rühr(er)-flügel *m*, Rühr(er)schaufel *f*, Mischflügel *m*
stirrer speed Rührerdrehzahl *f*
stirring arm Rührarm *m*
stirring bar Rühranker *m (eines magnetischen Rührwerks)*
stirring rod Rührstab *m*
stirrup Waagschalenaufhängung *f*, Schalengehänge *n*, Gehänge *n*
Stobbe condensation [reaction] Stobbe-Reaktion *f*, Stobbe-Kondensation *f* *(von Bernsteinsäureestern)*
stock 1. Grundwerkstoff *m*, Ausgangsmaterial *n*, zu verarbeitendes Material *n*; 2. *(Gum)* Mischung *f*; 3. *(Met)* Gicht *f*; 4. *(Pap)* Papierrohstoff *m*, Rohstoff *m* für die Papiererzeugung, Faserrohstoff *m*, Papierfaserstoff *m*; Faser(stoff)brei *m*, Stoffbrei *m*, Fasermasse *f*, Stoffmasse *f*, Stoff *m*; Faserhalbstoff *m*, Halbstoff *m*, Stoff *m*, Halbzeug *n*; (fertiger) Papierstoff *m*, Ganzstoff *m*, Ganzzeug *n*, Stoff *m*; 5. *(Geol)* Stock *m*; 6. Lagervorrat *m*, Vorrat *m*
stock bin Vorratsbehälter *m*, Vorratsbunker *m*, Vorratssilo *n(m)*
stock culture *(Biot)* Stammkultur *f*
stock density *(Pap)* Stoffdichte *f*, Stoffkonzentration *f*, Stoffkonsistenz *f*
stock disintegration *(Pap)* Stoffmahlung *f*
stock dyeing *(Text)* Färben *n* in der Flocke [Wolle]
stock emulsion Stammemulsion *f*
stock level Gichthöhe *f*, Beschickungshöhe *f (eines Hochofens)*
stock liquor *(Text)* Stammflotte *f*, Stammansatz *m*
stock room Lagerraum *m*, Vorratsraum *m*, Lager *n*
stock solution Vorratslösung *f*, *(besonders bei Titrationen)* Stammlösung *f*
stock tank Speichertank *m*
stock vat *(Farb)* Stammküpe *f*
Stock naming *(Nom)* Benennung *f* nach den Regeln von Stock
Stock nomenclature stocksche Nomenklatur *f*
Stock notation stocksche Bezeichnungsweise *f (der Wertigkeit)*
Stock number Stock-Zahl *f*, Oxidationszahl *f*
stockpiling Lagern *n*, Lagerung *f*, Vorratshaltung *f*
stoichiometric coefficient stöchiometrischer Koeffizient *m* [Faktor *m*]
stoichiometric equation stöchiometrische Gleichung *f*, (chemische) Reaktionsgleichung *f*, Stoffgleichung *f*
stoichiometric factor *s.* stoichiometric coefficient
stoichiometric formula stöchiometrische Formel *f* [Grundformel *f*], *(i. e. S.)* empirische Formel *f*, einfachst mögliche Formel *f*, einfachste Formel *f* [Summenformel *f*, Bruttoformel *f*],

Verhältnisformel *f*, Substanzformel *f*
stoichiometric number Stöchiometriezahl *f*
stoichiometry 1. Stöchiometrie *f*; 2. stöchiometrische Verhältnisse *npl*, Stöchiometrie *f*
stoke Stokes *n* *(SI-fremde Einheit der kinematischen Viskosität)*
Stokes law 1. stokessches Gesetz *n* [Reibungsgesetz *n*], Stokes-Gesetz *n*; 2. *s.* Stokes rule
Stokes line *(Spektr)* Stokes-Linie *f*, stokessche Linie *f*
Stokes rule stokessche Regel *f*, stokessches Fluoreszenzgesetz *n*
Stokes shift *(Spektr)* Stokes-Verschiebung *f*
stoneware *(Ker)* Steinzeug *n*
stop valve Absperrventil *n*
stopcock Absperrhahn *m*, Sperrhahn *m*, *(i. w. S.)* Hahn *m*
stopoff *s.* resist 1.
stopper *v* (mit einem Stopfen) verschließen, zustöpseln, verstöpseln
stopper 1. Stopfen *m*, Stöpsel *m*, Pfropfen *m*; Spund *m* *(am Fass)*; *(Am)* Hahnküken *n*; 2. Abstoppmittel *n*, Polymerisationsabstoppmittel *n*, Polymerisationsstopper *m*, Inhibitor *m*
stopping-off lokales Abschirmen *n* [Abdecken *n*] *(von Oberflächenteilen zum Schutz vor chemischem Angriff)*
storability Lagerbarkeit *f*, Lagerfähigkeit *f*
storable lagerbar, lagerfähig
storage 1. Lagerung *f*, Einlagerung *f*, Aufbewahrung *f*; Speicherung *f* *(von Gasen)*; 2. *s.* storage tank; 3. *s.* storeroom; 4. *s.* storehouse
storage basin *(Hyd)* Speicherbecken *n*, Stapelbecken *n*
storage battery Akku(mulator) *m*, Sammler *m*, Batterie *f*
storage bin Vorratsbehälter *m*, Vorratsbunker *m*, Vorratssilo *n(m)*
storage bottle 1. Vorratsflasche *f*; 2. *(Lab)* Standflasche *f*
storage cell *(Ech)* Sekundärelement *n*, Sekundärzelle *f* *(regenerierbar)*
storage-hopper Bunker *m*, Silo *n(m)*, Behälter *m* *(mit Untenentleerung)*
storage life 1. Lager- und Verarbeitbarkeitsdauer *f*, Gebrauchsfähigkeitsdauer *f*; 2. *(Lebm)* Lagerdauer *f*
storage lipid *(Bioch)* Reservelipid *n*
storage modulus Speichermodul *m*
storage protein *(Bioch)* Speicherprotein *n*
storage resistance [stability] Lagerbeständigkeit *f*, Lager(halt)barkeit *f*, Lagerstabilität *f*
storage ring Speicherring *m* *(zur Speicherung hochenergetischer, elektrisch geladener Teilchen)*
storage tank 1. Lagertank *m*, Lagerbehälter *m*, Vorratstank *m*, Vorratsbehälter *m*, Speicherbehälter *m*; 2. *(Pap)* Vorratsbütte *f*; 3. *(Hyd)* Stapeltank *m*
store *v* lagern, aufbewahren
store *v* **in cool place** kühl lagern
store *v* **in dry place** trocken lagern
store 1. Bestand *m*, Vorrat *m*, Reserve *f*; 2. *s.* storeroom; 3. *s.* storehouse
storehouse Lagerhalle *f*, Lager(haus) *n*, Speicher *m*, Magazin *n*, Depot *n*
storeroom Lagerraum *m*, Vorratsraum *m*, Lager *n*
storing Lagerung *f*, Einlagerung *f*, Aufbewahrung *f*; Vorratshaltung *f*
storm-water retention basin Regen(wasser)rückhaltebecken *n*, Regenwasseraufhaltebecken *n*
stove *v* 1. erwärmen, erhitzen; 2. im Ofen trocknen, *(i. w. S.)* thermisch [durch Wärme] trocknen; 3. *(Anstr)* einbrennen, im Ofen trocknen [härten]; härten *(Gießharz)*
stove 1. Ofen *m*; 2. *(Tech)* Brennofen *m*; 3. *(Anstr)* Trockenofen *m*, Einbrennofen *m*
stoving enamel Einbrennemaillelackfarbe *f*, ofentrocknende Emaillelackfarbe *f*, *(i. w. S.)* ofentrocknende Lackfarbe *f*
stoving lacquer Einbrennlack *m*, ofentrocknender Lack *m* *(physikalisch trocknend)*; ofentrocknende Lackfarbe *f*
stoving paint Einbrennanstrichstoff *m*, ofentrocknender Anstrichstoff *m*
stoving synthetic Alkyd-Aminharz-Einbrennlack *m*, ofentrocknender Alkyd-Aminharz-Lack *m*
stoving time Härtedauer *f* *(bei Gießharzen)*; Einbrenndauer *f* *(bei Lacken)*

stoving varnish Einbrennlack *m*, ofentrocknender Lack *m*
S.T.P. *s.* standard temperature and pressure
STP *s.* standard temperature and pressure
straight carbon steel reiner [unlegierter] Kohlenstoffstahl *m*
straight-chain geradkettig, mit gerader Kette
straight-chain hydrocarbon geradkettiger Kohlenwasserstoff *m*
straight distillation *s.* straight-run distillation
straight-lobe compressor Roots-Gebläse *n*, Wälzkolbenverdichter *m*
straight-run distillation direkte [einfache] Destillation *f*, Straight-run-Destillation *f*
straight-run gasoline [naphtha, spirit] Straight-run-Benzin *n*, SR-Benzin *n*, Destillat(ions)benzin *n*, direkt herausdestilliertes Benzin *n*
strain *v* 1. (durch)seihen, (durch)sieben, *(i. w. S.)* filtrieren; durchsickern, durchlaufen *(Filtrat)*; 2. verformen, *(i. e. S.)* dehnen
strain 1. Formänderung *f*, Verformung *f*, *(i. e. S.)* Längenänderung *f (in Prozent der ursprünglichen Messlänge)*; Spannung *f (in Molekülen)*; 2. *s.* stress; 3. *(Bioch, Biot)* Stamm *m*, Rasse *f (von Mikroorganismen)*
strain at break Bruchdehnung *f*
strain development *(Biot)* Stammentwicklung *f*, Stammoptimierung *f*
strain-free spannungsfrei
strain point [temperature] *(Glas)* unterer Kühlpunkt *m*, untere Kühltemperatur *f*, 15-Stunden-Entspannungstemperatur *f*
strain theory Spannungstheorie *f*
(+) strain *(Biot)* plus-Stamm *m*
(-) strain *(Biot)* minus-Stamm *m*
strainer 1. Seiher *m*, (feines) Sieb *n*, *(i. w. S.)* Filter *n*; Siebeinsatz *m*, Siebeinlage *f*, Lochplatte *f*; 2. *(Gum)* Siebkopf-Spritzmaschine *f*, Siebpresse *f*, Strainer *m*; 3. *s.* strainer bottom
strainer bottom *(Gär)* Läuter(bottich)-boden *m*, Siebboden *m*
straining 1. Durchseihen *n*, Sieben *n*, *(i. w. S.)* Filtrieren *n*; 2. *s.* strain 1.
straining cloth Seihtuch *n*
strainless spannungsfrei
strand 1. Strang *m*; 2. *(Text)* Strahn *m*, Faserbündel *n*, Spinnfaden *m*, Elementarfadenbündel *n*; 3. Trum *m(n) (eines Stetigförderers)*
strand break *(Bioch)* Strangbruch *m*
strange particle *(Kern)* seltsames Teilchen *n*, „strange particle"
strangeness *(Kern)* Strangeness *f*, Seltsamkeit *f*
strass Strass *m (Glassorte mit hohem Bleigehalt)*
stratified sampling *(Anal)* geschichtete Probe(nent)nahme *f*
stratify *v* schichten, in Schichten auseinander ziehen, auffächern *(z. B. das Aufgabegut bei der Herdaufbereitung)*; Schichten bilden
stratum 1. Schicht *f*; 2. *(Geol)* Schicht *f*, Bank *f*
straw cellulose (vollaufgeschlossener) Stroh(zell)stoff *m*
straw pulp (gelber) Strohstoff *m*, Gelbstrohstoff *m*; (vollaufgeschlossener) Stroh(zell)stoff *m*
stray light Streulicht *n*
stray radiation Streustrahlung *f*
streak 1. Streifen *m*; 2. *(Glas, Geol)* Schliere *f*; 3. *s.* streak colour
streak colour *(Min)* Strichfarbe *f*, Strich *m*
stream Strom *m (von Fluiden)*; Strömung *f* • **be on stream** in Betrieb sein
stream filament Stromfaden *m (Rheologie)*
stream gold Waschgold *n*, Seifengold *n*
stream meter Durchfluss(mengen)messer *m*, Mengenstrommesser *m*
stream of gas Gasstrom *m*
stream of water Wasserstrom *m*
stream pollution *(Hyd)* Flussverschmutzung *f*, Flussverunreinigung *f*
stream splitter *(Chromat)* Strömungsteiler *m*
stream tube Flüssigkeitszylinder *m (Rheologie)*
streaming birefringence [double refraction] Strömungsdoppelbrechung *f*
streaming potential Strömungspotenzial *n*
streamline flow laminare Strömung *f*,

Laminarströmung *f*, Schichtenströmung *f*
Strecker degradation Strecker-Abbau *m (von Aminosäuren zu Aldehyden)*
Strecker reaction Strecker-Synthese *f*, streckersche Synthese *f (von Aminosäuren)*
strength 1. Festigkeit *f*, Dauerhaftigkeit *f*, Haltbarkeit *f*; 2. Stärke *f*, Wirkungskraft *f*, Wirkungsgrad *m*; 3. *(Pharm)* Wirkungsstärke *f*; 4. Stärke *f*, Intensität *f*, Echtheit *f (z. B. einer Farbe)*; 5. Dichte *f*, Konzentration *f (einer Lösung)*
strength in compression Druckfestigkeit *f*
strength in tension Zugfestigkeit *f*
strength of sewage *(Hyd)* Abwasserkonzentration *f*, Stärke *f* [Schmutzstoffgehalt *m*] des Abwassers
strengthen *v* 1. verstärken *(Lösungen)*; 2. *(Glas)* verfestigen
stress *v* (mechanisch) beanspruchen, spannen
stress (mechanische) Beanspruchung *f*, Spannung *f*
stress corrosion Spannungskorrosion *f*, *(i. e. S.)* Spannungsrisskorrosion *f*
stress cracking Spannungsrissbildung *f*
stress-optical spannungsoptisch
stress relaxation Spannungsrelaxation *f*
stress relief anneal(ing) *(Met)* Spannungsfreiglühen *n*, Spannungsarmglühen *n*, entspannendes Glühen *n*
stress-strain diagram Beanspruchungs-Dehnungs-Diagramm *n*, Spannungs-Dehnungs-Diagramm *n*
stress-strain properties Zugdehnungseigenschaften *fpl*, Zugverformungseigenschaften *fpl*, Zug-Dehnungs-Verhalten *n*
stretch *v* strecken, dehnen, längen, ausziehen, spannen; recken *(Fasern, Folien)*; (aus)weiten
stretch 1. (elastische) Dehnung *f*; 2. Dehnbarkeit *f*; 3. *(Text)* Stretch *m*; 4. Strecken *n*, Dehnen *n*; Recken *n (von Fasern, Folien)*
stretch forming *(Kst)* Streckformen *n*
stretch ratio *(Text)* Gesamtlängenverhältnis *n (Verhältnis der Feinheit des ungereckten zu der des gereckten Fadens)*
stretch spinning *(Text)* Streckspinnen *n*
stretching vibration Valenzschwingung *f*, ν-Schwingung *f*, Streckschwingung *f*
strike *v* **back** zurückschlagen *(Flamme)*
strike *v* **through** durchschlagen *(Farben)*
strike bath (solution) *stark verdünnte Elektrolytlösung zum Abscheiden dünner Zwischenschichten, z. B. Vorverkupferungselektrolyt*
strike-through Durchschlagen *n (von Farben)*
striking-back Zurückschlagen *n (der Flamme)*
string galvanometer Saitengalvanometer *n*, Fadengalvanometer *n*
strip *v* 1. austreiben, desorbieren; 2. *(Dest)* abstreifen, abtreiben, (ab)strippen, ausdämpfen; 3. *s.* strip away
strip *v* **away** ablösen *(chemisch oder mechanisch)*; abbeizen *(Altanstriche)*; abziehen *(als Ganzes)*
strip *v* **off [out]** *s.* strip 1.
strip-chart recorder Bandschreiber *m*
strip heater Bandheizkörper *m*, Heizband *n*
strip mining Tagebauförderung *f*, Tagebaubetrieb *m*, Förderung *f* im Tagebau *(meist von Kohle im Parallelabbau in langen Streifen)*
stripped atom hochionisiertes Atom *n*
stripped gas ausgetriebenes Gas *n*
stripper 1. *(Dest)* Abtreiber *m*; 2. *s.* stripping section; 3. *s.* stripper column
stripper column 1. *(Dest)* Abtriebssäule *f*, Abtreibkolonne *f*; 2. *(Erdöl)* Abstreiferkolonne *f*, Seitenkolonne *f*, Strippkolonne *f*, Abstreifer *m*, Seitenturm *m*, Nebenturm *m*, Stripper *m*
stripping column *s.* stripper column
stripping gas Strippgas *n*
stripping oil Abstreiferöl *n*, Waschöl *n*
stripping operating line *(Dest)* Abtriebsgerade *f*
stripping reaction *(Kern)* Abstreifreaktion *f*, Stripping-Reaktion *f*
stripping section 1. *(Dest)* Abtreibeteil *m*, Abtriebsteil *m*, Untersäule *f*; 2. *(Erdöl)* Abstreiferzone *f*, Abstrippzone *f*, Ausdämpfsektion *f*, Ausdämpfungsteil *m*
stripping still *s.* stripper column
stroke 1. Hub *m*; 2. *s.* stroke length
stroke length Hublänge *f*, Hub *m*
strong acid 1. starke Säure *f (nach dem*

Dissoziationsgrad); 2. konzentrierte Säure *f*
strong-acid cation exchanger [resin] *(Hyd)* stark saurer Kationenaustauscher *m*
strong base starke Base *f (nach dem Dissoziationsgrad)*
strong-base anion exchanger [resin] *(Hyd)* stark basischer Anionenaustauscher *m*
strong electrolyte starker Elektrolyt *m*
strong liquor Dicklauge *f*
strong sewage *(Hyd)* hochkonzentriertes [hochbelastetes] Abwasser *n*, starkes [stark verschmutztes, „dickes"] Abwasser *n*
strong-smelling stark duftend; von durchdringendem [intensivem] Geruch
strong waste *s.* strong sewage
strongly acid(ic) stark sauer
strongly basic stark basisch [alkalisch]
strontia Strontiumoxid *n*
strontic Strontium...
strontium Sr Strontium *n*
strontium carbonate Strontiumcarbonat *n*
strontium chloride Strontiumchlorid *n*
strontium hydroxide Strontiumhydroxid *n*
strontium nitrate Strontiumnitrat *n*
strontium oxide Strontiumoxid *n*
strontium sulphate Strontiumsulfat *n*
strophanthin Strophanthin *n (Alkaloid)*
strophanthin-G Strophanthin *n* G, g-Strophanthin *n*, Ouabain *n (Glykosid)*
structural 1. strukturell, Struktur..., Bau..., Gefüge...; 2. konstruktiv, Konstruktions...
structural analysis Strukturanalyse *f*
structural chemistry Strukturchemie *f*
structural constituent Gefügebestandteil *m*
structural formula Strukturformel *f*, Konstitutionsformel *f*
structural isomer Strukturisomer(e) *n*
structural isomerism Strukturisomerie *f*
structural parachor Strukturparachor *m*, Bindungsparachor *m*
structural protein Strukturprotein *n*, Gerüsteiweiß *n*, Faserprotein *n*, fibrilläres Protein *n*, Skleroprotein *n*
structural study Strukturuntersuchung *f*
structural substance *(Bioch)* Gerüstsubstanz *f*
structural transition strukturelle Umwandlung *f* [Phasenumwandlung *f*]
structural unit strukturelle Einheit *f*, Struktureinheit *f*
structural viscosity Strukturviskosität *f*, strukturviskoses Verhalten *n*
structurally identical strukturidentisch
structure Struktur *f*, Gefüge *n*, Aufbau *m*, Bau *m*, Konstitution *f*
structure cell *(Krist)* Elementarzelle *f*
structure determination [elucidation] Strukturaufklärung *f*, Konstitutionsaufklärung *f*, Strukturermittlung *f*, Konstitutionsermittlung *f*, Konstitutionserforschung *f*
structure type Strukturtyp *m*, Kristallgittertyp *m*
structureless strukturlos, gefügelos, unstrukturiert
strychnos alkaloid Strychnosalkaloid *n*
S.T.T. *s.* short-time test
Stuart model Stuart-Modell *n*, Stuart-Kalotte *f (zur Darstellung von Molekülstrukturen)*
stucco Stuckgips *m*
stuff *(Pap)* (fertiger) Papierstoff *m*, Ganzstoff *m*, Ganzzeug *n*, Stoff *m*; (mechanischer) Holzschliff *m*, (mechanischer) Holzstoff *m*, Holzmasse *f*, Schleifmasse *f*
stuff engine Holländer *m*, Mahlholländer *m*, Messerholländer *m*, Ganzzeugholländer *m*
stuffing box Stopfbuchse *f*, Stopfbuchspackung *f*
styphnate Styphnat *n (Salz oder Komplex der Styphninsäure)*
styphnic acid Styphninsäure *f*, 2,4,6-Trinitroresorcinol *n*, 1,3-Dihydroxy-2,4,6-trinitrobenzen *n*
styptic styptisch, adstringierend, zusammenziehend, *(i. e. S.)* hämostyptisch, blutstillend
styptic Styptikum *n*, Adstringens *n*, adstringierendes [zusammenziehendes] Mittel *n*, adstringierende Substanz *f*, *(i. e. S.)* Hämostyptikum *n*, blutstillendes Mittel *n*, Blutstillungsmittel *n*
stypticity Adstringenz *f*, adstringierende Wirkung *f*
styrenate *v* styrolisieren

styrenated alkyd styrolisiertes Alkydharz *n*, Styrolalkydharz *n*
styrene Styren *n*, Vinylbenzen *n*, Styrol *n*
styrene-butadiene rubber Styren-Butadien-Kautschuk *m*, Butadien-Styren-Kautschuk *m*, Styrenkautschuk *m*, SBK *(Butadien-Styren-Copolymerisat)*
styrene-chloroprene rubber Styren-Chloropren-Kautschuk *m (Chlorbutadien-Styren-Copolymerisat)*
styrene-isoprene rubber Styren-Isopren-Kautschuk *m (Isopren-Styren-Copolymerisat)*
styrone Zimtaldehyd *m*, *3-Phenylpropenal *n*
styryl alcohol [carbinol] Zimtalkohol *m*, Cinnamylalkohol *m*, *3-Phenylprop-2-en-1-ol *n*
subacid 1. schwach sauer; 2. *(Lebm)* säuerlich; 3. *(Med)* subazid
subacidity *(Med)* Subazidität *f*, Hyp(o)azidität *f*
subalkaline schwach alkalisch
subatomic subatomar
subbituminous coal subbituminöse Kohle *f*, Glanz(braun)kohle *f*, bituminöse [steinkohlenähnliche] Braunkohle *f*
subclass Gruppe *f*, Enzymgruppe *f*, Unterklasse *f (Enzymsystematik)*
subcloning *(Bioch)* Umklonieren *n*, Subklonieren *n (von DNA-Abschnitten)*
subcool *v* unterkühlen
subcooled boiling örtliches Sieden *n*, Oberflächensieden *n*, Verdampfungskühlung *f*
subcritical *(Kern, Krist)* unterkritisch
subcuticular residue subkutikulärer Rückstand *m (von Pflanzenschutzmitteln)*
suberane Suberan *n*, *Cycloheptan *n*
subereous 1. korken, aus Kork, Kork...; 2. korkartig, korkähnlich
suberic acid Korksäure *f*, *Hexan-1,6-dicarbonsäure *f*, *Octandisäure *f*
suberin Suberin *n (hochmolekulare Korksubstanz)*
suberone Suberon *n*, *Cycloheptanon *n*
suberyl alcohol Suberylalkohol *m*, Suberol *n*, Cycloheptanol *n*
subgroup Untergruppe *f*; Nebengruppe *f (des Periodensystems)*
subl = sublimes
sublethal dose subletale Dosis *f*
sublevel Unterniveau *n (von Elektronen)*
sublimable sublimierbar
sublimand Sublimand *m*, Sublimiergut *n*, zu sublimierender Stoff *m*
sublimate *v* sublimieren
sublimate 1. Sublimat *n (Endprodukt der Sublimation)*; 2. Sublimat *n*, Quecksilber(II)-chlorid *n*
sublimation Sublimation *f*
sublimation curve Sublimations(druck)kurve *f*
sublimation point [temperature] Sublimationspunkt *m*, Sbp., Sublimationstemperatur *f*
sublimed sulphur sublimierter Schwefel *m*, Schwefelblüte *f*, Schwefelblume *f*
submerge *v* (ein)tauchen, untertauchen
submerged submers, untergetaucht
submerged burner Tauchbrenner *m*
submerged culture *(Biot)* Submerskultur *f*, submerse Kultur *f*, Suspensionskultur *f*, Submerssuspensionskultur *f*, Tiefkultur *f*
submerged fermentation *(Biot)* Submersfermentation *f*
submerged fermenter *(Biot)* Submersfermenter *m*, Submersreaktor *m*, Reaktor *m* für Submersverfahren, Submerstank *m*
submergence 1. Eintauchen *n*; 2. Eintauchtiefe *f*, Tauchtiefe *f*; 3. *(Dest)* Dampfdurchdringtiefe *f*
submersed *s.* submerged
submetallic lustre *(Min)* Halbmetallglanz *m*
submicro sample Submikroprobe *f (*10^{-3} bis 10^{-4} g)*
submicron Submikron *n*
submicroscopic submikroskopisch
submicrostructure Submikrostruktur *f*, Submikrogefüge *n*
submolecule Submolekül *n*
subnuclear particle *s.* nucleon
suboxide Suboxid *n*
subsample Teilprobe *f*
subsaturated untersättigt
subsaturation Untersättigung *f*
subshell Unterschale *f*
subside *v* 1. (ab)sinken, sich (ab)setzen, sich abscheiden [ausscheiden], sich ablagern [niederschlagen], absitzen, ausfallen, zur Abscheidung gelangen,

sedimentieren; 2. abklingen, nachlassen, zum Stillstand kommen, erlöschen *(Reaktion)*
subsidence 1. Sinken *n*, Setzen *n*, Setzung *f*; 2. Bodensatz *m*, Satz *m*, Niederschlag *m*, Sediment *n*; 3. Abklingen *n*, Nachlassen *n*, Erlöschen *n (einer Reaktion)*
subsidiary band Nebenbande *f*
subsidiary quantum number Nebenquantenzahl *f*, Bahndrehimpulsquantenzahl *f*, Drehimpulsquantenzahl *f*, Orbitaldrehimpulsquantenzahl *f*, Azimut(al)quantenzahl *f*, azimutale Quantenzahl *f*
substance Substanz *f*, Stoff *m*
substance to be ... *siehe auch: material to be ...*
substance to be analysed Analysensubstanz *f*, zu analysierende Substanz *f*
substance under analysis Analysensubstanz *f (während der Analyse)*
substance under investigation Versuchssubstanz *f*, Untersuchungssubstanz *f*, Probesubstanz *f*
substantially free praktisch frei
substantially soluble größtenteils löslich
substantive dye Direktfarbstoff *m*, direktziehender [substantiver] Farbstoff *m*, Substantivfarbstoff *m*
substituent Substituent *m*
substituent constant Substituentenkonstante *f*
substituent effect Substituenteneffekt *m*
substitutability 1. Substituierbarkeit *f*, Austauschbarkeit *f*; Austauschfähigkeit *f*, Austauschaktivität *f (von Atomgruppen)*; 2. Einsetzbarkeit *f (als Ersatzstoff)*
substitutable 1. substituierbar, austauschbar; austauschfähig, austauschaktiv *(Atomgruppe)*; 2. (als Ersatzstoff) einsetzbar
substitute *v* 1. substituieren, ersetzen, austauschen, auswechseln; 2. als Ersatzstoff dienen für, ersetzen
substitute 1. Ersatz(stoff) *m*, Ersatzmittel *n*, Surrogat *n*, Austauschstoff *m*; 2. *(Gum)* Faktis *m*
substitute natural gas Erdgasaustauschgas *n*, synthetisches Erdgas *n*, Erdgasersatzgas *n*, Ersatzerdgas *n*
***m*-substituted** *m*-substituiert, metasubstituiert, 1,3-substituiert
***N*-substituted** *N*-substituiert *(z. B. Carbamidsäure)*
***o*-substituted** *o*-substituiert, orthosubstituiert, 1,2-substituiert
***p*-substituted** *p*-substituiert, parasubstituiert, 1,4-substituiert
substitution Substitution *f*, Ersatz *m*, Austausch *m*
substitution isomerism Substitutionsisomerie *f*, Stellungsisomerie *f*
substitution reaction Substitutionsreaktion *f*, Austauschreaktion *f*
substitution tautomerism Substitutionstautomerie *f*
substitution weighing Substitutionswägung *f*
substitutional alloy Substitutionslegierung *f*
substitutional solid solution Substitutionsmischkristall *m*
substitutive name Substitutionsname *m*
substrate 1. *(Ch)* Substrat *n*; 2. *(Bioch, Biot, Hyd)* Substrat *n*, Nährsubstrat *n*; 3. *(Anstr)* Substrat *n*, Träger *m*, Farbträger *m*
substrate binding *(Bioch)* Substratbindung *f*
substrate conversion efficiency *(Biot)* Stoffwandlungsleistung *f*
substrate-induced *(Bioch)* substratvermittelt
substrate inhibition *(Bioch)* Substrat(überschuss)hemmung *f*
substrate-level phosphorylation *(Bioch)* Substrat(ketten)phosphorylierung *f*
substrate saturation Substratsättigung *f (Enzymkinetik)*
substrate-specific *(Bioch)* substratspezifisch
substrate specificity *(Bioch)* Substratspezifität *f*
substructure 1. *(Tech)* Unterbau *m*; 2. *(Krist)* Unterstruktur *f*
sub-subclass Untergruppe *f*, Unter-Unterklasse *f (Enzymsystematik)*
subsubmicron Amikron *n (mikroskopisch unsichtbares Schwebeteilchen)*
subsurface mining Untertagebau *m*,

Abbau *m* unter Tage, Tiefbau(betrieb) *m*, Förderung *f* [Abbau *m*] im Tiefbau, untertägiger Bergbau *m*
sub-trace analysis Sub-Spuren-Analyse *f (*Probenmasse* 10^{-3} *bis* 10^{-4} *g)*
subtraction solid solution Subtraktionsmischkristall *m*
subtractive name Subtraktionsname *m*, Subtraktivname *m*
subunit Untereinheit *f*, *(Eiweißchemie auch)* SU
successive formation constant konsekutive [individuelle, sukzessive] Stabilitätskonstante *f* [Komplexbildungskonstante *f*]
successive reaction Folgereaktion *f*
successive reactions law ostwaldsche Stufenregel *f*, Stufenregel *f* von Ostwald
succinate Succinat *n (Salz oder Ester der Bernsteinsäure)*
succinate dehydrogenase Succinatdehydrogenase *f*
succindialdehyde Succin(di)aldehyd *m*, Bernsteinsäuredialdehyd *m*, *Butan-1,4-dial *n*
succinic acid Bernsteinsäure *f*, *Butandisäure *f*, Ethan-1,2-dicarbonsäure *f*
succinimide Succinimid *n*, Bernsteinsäureimid *n*
succinite Succinit *m*, Bernstein *m*
succinoxidase Succinatoxidase *f*
succinylate *v* sukzinylieren
succinylation Sukzinylierung *f*
succus Succus *m*, Saft *m*, *(i. e. S.)* Pflanzensaft *m*
suck *v* (an)saugen, einsaugen
suck *v* **off** absaugen
sucrase Saccharase *f*, Invertase *f*, β-Fructosidase *f (rohrzuckerspaltendes Enzym)*
sucrate Saccharat *n (Alkali- oder Erdalkalisalz des Rohrzuckers)*
sucrochemistry Sucrochemie *f*
sucrose Saccharose *f*, Rohrzucker *m*
sucrose ester Saccharoseester *m*
sucrose ether Saccharoseether *m*
sucrose phosphate Saccharosephosphat *n*
suction-filter *v* 1. im Vakuum filtrieren, mit Unterdruck filtrieren; 2. *(Lab)* (ab-)nutschen
suction Saugen *n*, Saugung *f*, Ansaugen *n*, Ansaugung *f*; Absaugen *n*; Einsaugen *n*; Saugzug *m*, Zug *m*
suction conveyor Saugförderer *m*
suction filter 1. Saugfilter *n*, Vakuumfilter *n*, Unterdruckfilter *n*; 2. *(Lab)* Filternutsche *f*, Nutsche *f*
suction filtration 1. Saugfiltration *f*, Vakuumfiltration *f*, Unterdruckfiltration *f*; 2. *(Lab)* Abnutschen *n*, Nutschen *n*
suction flask Saugflasche *f*
suction friction head Saugreibungshöhe *f*
suction head Saughöhe *f*
suction line [pipe] Saugrohr *n*, Saugleitung *f*, Ansaugrohr *n*, Ansaugleitung *f*
suction pressure Saugdruck *m*, Ansaugdruck *m*
suction pump Saugpumpe *f*
suction side Saugseite *f*
suction stroke Saughub *m*
sudorific diaphoretisch, schweißtreibend
sudorific *(Pharm)* Diaphoretikum *n*, schweißtreibendes Mittel *n*
sugar Zucker *m*
sugar acid Zuckersäure *f*
sugar alcohol Zuckeralkohol *m*
sugar anhydride Zuckeranhydrid *n*
sugar-bearing zuckerhaltig
sugar beet Zuckerrübe *f*, Beta vulgaris L. var. altissima Doell
sugar-containing zuckerhaltig
sugar conversion Zuckerumwandlung *f*
sugar derivative Zuckerderivat *n*, Zuckerabkömmling *m*
sugar ester Zuckerester *m*
sugar moiety Zuckerkomponente *f*, Zuckeranteil *m (z. B. eines Glykosids)*
sugar of lead Bleizucker *m*, Blei(II)-acetat *n*
sugar of milk Milchzucker *m*, Lactose *f*
sugar plant Zuckerpflanze *f (ohne Umwandlung des Assimilatzuckers in Stärke)*
sugar refining Raffinieren *n* [Raffination *f*] des Zuckers, Zuckerraffination *f*
sugar series Zuckerreihe *f*
sugar substitute Zuckerersatzstoff *m*
sugar wort Süßwurzel *f*
sugary zuck(e)rig, zuckerhaltig, (zucker)süß
suggestive of/be schließen lassen auf
sulf... *siehe: sulph...*

sulpha drug Sulfonamidpräparat *n*, Arzneimittel *n* auf Sulfonamidbasis
sulphacetamide Sulfacetamid *n (ein Sulfonamid)*
sulphadiazine Sulfadiazin *n*, 2-(*p*-Aminobenzensulfonamido)-pyrimidin *n*
sulphamate Sulfamat *n*, Sulfamidat *n*, Amidosulfat *n*
sulphamic acid Amidoschwefelsäure *f*, Sulfamidsäure *f*
sulphane Sulfan *n (allg. für einen kettenförmigen Schwefelwasserstoff)*
sulphanilamide Sulfanilamid *n*, *p*-Aminobenzensulfonamid *n*, *p*-Aminophenylsulfonamid *n*
sulphanilic acid Sulfanilsäure *f*, *p*-Aminobenzensulfonsäure *f*, Anilin-*p*-sulfonsäure *f*
sulphanyl group HS– Thiolgruppe *f*, Mercaptogruppe *f*
sulphapyridine Sulfapyridin *n*, 2-(*p*-Aminobenzensulfonamido)-pyridin *n*
sulphate *v* 1. sulfatieren *(Alkohole mit Schwefelsäure verestern)*; 2. sulfatisieren *(einen kristallinen Blei(II)-sulfat-Niederschlag im Bleiakku bilden)*
sulphate Sulfat *n*
sulphate hardness *(Hyd)* Sulfathärte *f*, *(i. e. S.)* Gipshärte *f*
sulphate ion SO_4^{-2} Sulfat-Ion *n*, Sulfatanion *n*
sulphate of potash 1. schwefelsaures Kali *n (Düngemittel mit Kaliumsulfat als Hauptbestandteil)*; 2. *s.* potassium sulphate
sulphate pulp Sulfatzellstoff *m*
sulphate pulping *(Pap)* Sulfataufschluss *m*
sulphate reducers desulfurierende Bakterien *npl*, Desulfurikanten *mpl*
sulphate-reducing bacteria *s.* sulphate reducers
sulphate scale *(Hyd)* Sulfatstein *m*
sulphated castor oil Sulforicinoleat *n*, Türkischrotöl *n*, sulfatiertes [sulfoniertes] Ricinusöl *n*
sulphathiazol(e) Sulfathiazol *n*, 2-(4-Aminobenzensulfonamido)-thiazol *n*
sulphation 1. Sulfatieren *n*, Sulfatierung *f (Veresterung von Alkoholen mit Schwefelsäure)*; 2. Sulfatisieren *n*, Sulfatisierung *f (Bildung von kristallinem Blei(II)-sulfat im Bleiakku)*
sulphatize *v* sulfatisieren *(Metallsulfide z. B. durch Rösten in Sulfate umwandeln)*
sulphenamide Sulfenamid *n*
sulphene Sulfen *n*
sulphenic acid Sulfensäure *f*
sulphenium ion Sulfenylium-Ion *n*
sulphenyl group Sulfenylgruppe *n*
sulphenylium ion Sulfenylium-Ion *n*
sulphide *v s.* sulphidize
sulphide Sulfid *n*
sulphide dye(stuff) Schwefelfarbstoff *m*, Schwefelfarbe *f*, Sulfinfarbe *f*
sulphide ore sulfidisches schwefelhaltiges Erz *n*, Sulfiderz *n*, Schwefelerz *n*
sulphide stress corrosion Sulfid-Spannungsrisskorrosion *f*
sulphide sulphur Sulfidschwefel *n*
sulphidic sulfidisch, schwefelhaltig, schwefelführend, Schwefel...
sulphidize *v* 1. sulfidieren; 2. *(Text)* xanthogenieren
sulphine Sulfin *n*
sulphinic acid Sulfinsäure *f*
sulphitation Sulfitieren *n*, Schwefeln *n*, Schwefelung *f*, *(Zucker auch)* Sulfitation *f*, Schwefelsaturation *f (Behandlung mit Schwefeldioxid oder schwefliger Säure)*
sulphite *v* sulfitieren, schwefeln *(mit Schwefeldioxid oder schwefliger Säure behandeln)*
sulphite Sulfit *n*
sulphite cooking liquor *(Pap)* Sulfit(koch)säure *f*, Kochsäure *f*
sulphite kraft paper Sulfitkraftpapier *n*
sulphite pulp Sulfit(zell)stoff *m*, Bisulfitzellstoff *m*
sulphite pulping *(Pap)* Sulfitaufschluss *m*
sulphite spent liquor *(Pap)* Sulfitablauge *f*, Ablauge *f*, Urlauge *f*
sulphocarbonate $M^I_2CS_3$ Thiocarbonat *n*
sulphochlorinate *v* sulfochlorieren
sulphochlorination Sulfochlorierung *f*
sulphocyanate M^ISCN Thiocyanat *n*
sulphocyanide M^ISCN Thiocyanat *n*
sulphofication *(Boden)* Sulfurikation *f (mikrobielle Oxidation von organisch gebundenem Schwefel zu Sulfat)*
sulphonamide Sulfonsäureamid *n*, Sulf(on)amid *n*

sulphonate *v* sulfonieren, sulfurieren, *(i. w. S.)* sulfieren; sich sulfonieren [sulfurieren] lassen, *(i. w. S.)* sich sulfieren lassen
sulphonate Sulfonat *n*
sulphonating agent Sulfonierungsmittel *n*, Sulfonierungsagens *n*, Sulfurierungsmittel *n*, sulfonierendes Agens *n* [Mittel *n*]
sulphonation Sulfonieren *n*, Sulfonierung *f*, Sulfurieren *n*, Sulfurierung *f*, *(i. w. S.)* Sulfieren *n*, Sulfierung *f*
***N*-sulphonation** Sulfaminierung *f*, *N*-Sulfonierung *f*
sulphone Sulfon *n*
sulphonephthalein Sulfonphthalein *n*
sulphonic acid Sulfonsäure *f*
sulphonic acid group $-SO_3H$ Sulfogruppe *f*, Sulfonsäuregruppe *f*
sulphonium ion Sulfonium-Ion *n*
sulphonyl chloride Sulfochlorid *n*, Sulfonsäurechlorid *n*
sulphonyl group $=SO_2$ Sulfonylgruppe *f*
sulphonylation Sulfonylierung *f*
sulphonyloxy group OSO_2R Sulfonyloxygruppe *f*
sulphosalicylic acid Sulfosalicylsäure *f*, Salicylsäure-5-sulfonsäure *f*
sulphourea Thioharnstoff *m*, Thiocarbamid *n*
sulphoxidation Sulfoxidation *f*
sulphoxide R_2SO Sulfoxid *n*
sulphoxylate M^ISO_2 Sulfoxylat *n*
sulphoxylic acid Sulfoxylsäure *f*
sulphur *v* schwefeln *(mit Schwefel, Schwefeldioxid oder Sulfiten behandeln)*; *(i. e. S.)* mit Schwefel ausräuchern
sulphur S Schwefel *m*, Sulfur *m*
sulphur bacteria Schwefelbakterien *npl*, Thiobakterien *npl*
sulphur black Schwefelschwarz *n*, schwarzer Schwefelfarbstoff *m*
sulphur blue Schwefelblau *n*, blauer Schwefelfarbstoff *m*
sulphur bridge Schwefel(vernetzungs)-brücke *f*, Schwefelvernetzungsstelle *f*
sulphur compound Schwefelverbindung *f*
sulphur cross-link(age) *s.* sulphur bridge
sulphur-cured mit Schwefel vulkanisiert [vernetzt]
sulphur dichloride Schwefeldichlorid *n*, Schwefel(II)-chlorid *n*
sulphur dioxide Schwefeldioxid *n*, Schwefel(IV)-oxid *n*
sulphur dye Schwefelfarbstoff *m*, Schwefelfarbe *f*, Sulfinfarbe *f*
sulphur flowers Schwefelblüte *f*, Schwefelblume *f*, sublimierter Schwefel *m*
sulphur heptoxide Schwefelheptoxid *n*, Dischwefelheptoxid *n*
sulphur hexafluoride Schwefelhexafluorid *n*, Schwefel(VI)-fluorid *n*
sulphur indigo blue Schwefelindigoblau *n*
sulphur link *s.* sulphur bridge
sulphur monochloride Dischwefeldichlorid *n*, Dischwefelchlorid *n*
sulphur monofluoride Dischwefeldifluorid *n*, Dischwefelfluorid *n*
sulphur monoxide Schwefelmonoxid *n*, Schwefel(II)-oxid *n*
sulphur nitride Schwefelnitrid *n*, *(i. e. S.)* Tetraschwefeltetranitrid *n*
sulphur-reducing bacteria desulfurierende Bakterien *npl*, Desulfurikanten *mpl*
sulphur sesquioxide Dischwefeltrioxid *n*
sulphur sol Schwefelsol *n*
sulphur tetrachloride Schwefeltetrachlorid *n*, Schwefel(IV)-chlorid *n*
sulphur tetrafluoride Schwefeltetrafluorid *n*, Schwefel(IV)-fluorid *n*
sulphur tetraoxide [tetroxide] Schwefeltetroxid *n*
sulphur trioxide Schwefeltrioxid *n*, Schwefel(VI)-oxid *n*
sulphur vulcanization Schwefelvulkanisation *f*, Vernetzung *f* mit Schwefel
sulphur yellow Schwefelgelb *n*, gelber Schwefelfarbstoff *m*
α-sulphur α-Schwefel *m*, rhombischer Schwefel *m*
β-sulphur β-Schwefel *m*, monokliner Schwefel *m*
γ-sulphur γ-Schwefel *m*, plastischer Schwefel *m*
λ-sulphur λ-Schwefel *m*, cyclo-Octaschwefel *m*
μ-sulphur μ-Schwefel *m*, catena-Polyschwefel *m*
sulphurate *v s.* sulphurize

sulphurated potash Schwefelleber *f*, Hepar sulfuris *(technisches Kaliumsulfid)*
sulphuration Sulfurierung *f*
sulphureous 1. aus Schwefel bestehend; schwefelhaltig; 2. schwefelartig; schwefelfarben
sulphuric Schwefel...; höherwertigem Schwefel entsprechend, *(meist)* Schwefel(VI)-...
sulphuric acid Schwefelsäure *f*
sulphuric-acid alkylation *(Erdöl)* Schwefelsäurealkylierung *f*, Alkylierung *f* mit Schwefelsäure
sulphurization Schwefeln *n*, Schwefelung *f*
sulphurize *v* schwefeln, mit Schwefel [Schwefelverbindungen] verbinden [versetzen]
sulphurous Schwefel...; niederwertigem Schwefel entsprechend, *(meist)* Schwefel(IV)-...
sulphurous acid schweflige Säure *f*
sulphuryl chloride Sulfurylchlorid *n*
sulphydrate $M^{I}HS$ Hydrogensulfid *n*
sulphydryl (group) HS– Sulfhydrylgruppe *f*, *Mercaptogruppe *f*, HS-Gruppe *f*
sum band *(Spektr)* Summenbande *f*
sump Sumpf *m*; Sammelbehälter *m*
sunflower (seed) oil Sonnenblumenöl *n*
sunproofing agent Ozonschutzmittel *n*, Antiozonans *n*, Antiozonisator *m*
sunscreen (agent) *(Kosmet)* Sonnenschutzmittel *n*
sunscreen oil Sonnenschutzöl *n*
suntan make-up *(Kosmet)* Sonnenbräunungsmittel *n*, Hautbräunungsmittel *n*
suntan preparation *(Kosmet)* Sonnenbräunungsmittel *n*, Hautbräunungsmittel *n*
superabsorbent superabsorbierend
superabsorbent polymer Superabsorber *m*
superabsorber Superabsorber *m*
superacid supersauer, superazid, sehr stark sauer
superacid Supersäure *f*
superacidity Superazidität *f*
superalkaline hochalkalisch, sehr stark alkalisch
superalloy Superlegierung *f*, Superalloy *n*
supercarburize *v* *(Met)* überkohlen
superchlorinate *v* *(Hyd)* hochchloren
superchlorination *(Hyd)* Hochchlorung *f*, Überschusschlorung *f*
supercoiled helix *s.* superhelix
supercoiling *(Bioch)* Superknäuelung *f* *(von Desoxyribonucleinsäure)*
superconduct *v* Supraleitung [Supraleitfähigkeit] zeigen
superconducting supraleitend, supraleitfähig
superconducting transition Umwandlung *f* [Phasenumwandlung *f*] normal-supraleitend, Übergang *m* Normalleitung-Supraleitung
superconductive *s.* superconducting
superconductive temperature Sprungtemperatur *f*, Sprungpunkt *m* in Supraleitern
superconductivity Supraleitfähigkeit *f*, Supraleitung *f*
superconductor Supraleiter *m*
supercool *v* unterkühlen
supercritical *(Kern, phys Ch)* überkritisch, superkritisch
supercritical carbon dioxide überkritisches Kohlendioxid *n*
supercritical gas extraction *(Carb)* Extraktion *f* mit Gas im superkritischen Zustand, Extraktion *f* mit überkritischen Gasen
supercritical water oxidation überkritische Wasseroxidation *f*
superexcited state superangeregter [hochangeregter] Zustand *m*
superfluid supraflüssig, superflüssig, superfluid, suprafluid
superfluid Supraflüssigkeit *f*
superfluid state supraflüssiger [superfluider] Zustand *m*
superfluidity Supraflüssigkeit *f*, Suprafluidität *f*
superheat *v* überhitzen
superheated steam überhitzter Wasserdampf *m*, Heißdampf *m*
superheavy element *(Kern)* superschweres Element *n*
superhelical *(Bioch)* superhelikal, Superhelix...
superhelix *(Bioch)* Superhelix *f*, Tripelhelix *f*, dreisträngige Helix *f*
superimposable deckungsgleich *(Stereochemie)*

superimpose *v* übereinanderlagern, übereinanderlegen; überlagern
superinsulation Superisolierung *f*, Vakuum-Vielschichtisolierung *f* *(Kältetechnik)*
superlattice *(Krist)* Übergitter *n*, Überstruktur *f*
supermolecular structure übermolekulare Struktur *f*
supernatant liquid [liquor] überstehende Flüssigkeit *f*, Überstand *m*, Flüssigkeitsüberstand *m*
superoxide M^IO_2 Hyperoxid *n*
superphosphate Superphosphat *n* *(Düngemittel)*
superposition of peaks Signalüberlagerung *f*, Signalüberlappung *f*, *(im Diagramm:)* Peaküberlagerung *f*, Peaküberlappung *f*
superpure reinst, von höchster Reinheit
superpure aluminium Reinstaluminium *n*
superrefining extreme Reinigung *f*, Hochreinigung *f*
superrefractory hochfeuerfest
supersaturate *v* übersättigen
supersaturated solution übersättigte Lösung *f*
supersaturation Übersättigung *f*
supersecondary structure *(Bioch)* Supersekundärstruktur *f* *(der Proteine)*
superstructure *(Krist)* Überstruktur *f*, Übergitter *n*
supplementary component [material] Hilfsstoff *m*
supplementation Zufügung *f*, Zugabe *f*, Zusatz *m*
support *v* 1. halten; 2. unterhalten *(z. B. die Verbrennung)*
support Träger *m*, Trägermaterial *n*, Trägersubstanz *f* *(z. B. für Katalysatoren)*
support-coated open tubular column SCOT-Säule *f*, *(i. e. S.)* Dünnschicht-Kapillarsäule *f* *(Gaschromatographie)*
supported catalyst Trägerkatalysator *m*
supporting electrolyte *(Ech)* Grundelektrolyt *m*, Leitelektrolyt *m*, Zusatzelektrolyt *m*, indifferenter Elektrolyt *m*
suppository *(Pharm)* Suppositorium *n*, Stuhlzäpfchen *n*
suppress *v* unterdrücken
suppressor mutation *(Biot)* Suppressormutation *f*
suprafacial *(org Ch)* suprafacial, s
supramolecular assembly [system] Assoziat *n*
surface *v* oberflächenbehandeln
surface-harden *v* oberflächenhärten, oberflächlich härten
surface Oberfläche *f*; Fläche *f*
surface-active grenzflächenaktiv, *(an der Grenzfläche Flüssigkeit-Luft auch)* oberflächenaktiv, oberflächenwirksam
surface-active agent [**compound, substance**] grenzflächenaktiver Stoff *m*, *(an der Grenzfläche Flüssigkeit-Luft auch)* oberflächenaktiver Stoff *m*, oberflächenaktive Substanz *f* [Verbindung *f*], oberflächenaktives Mittel *n*, *(als synthetischer Stoff auch)* Tensid *n*
surface activity Grenzflächenaktivität *f*, *(an der Grenzfläche Flüssigkeit-Luft auch)* Oberflächenaktivität *f*
surface adhesion *(Kst)* Oberflächenhaftung *f*
surface aeration *(Hyd)* Oberflächenbelüftung *f*
surface anaesthetic *(Med)* Oberflächenanästhetikum *n*
surface antigen *(Biot)* Oberflächenantigen *n*
surface area Oberfläche *f*, Oberflächengröße *f*, Flächengröße *f*
surface catalysis Oberflächenkatalyse *f*, Kontaktkatalyse *f*, heterogene Katalyse *f*
surface charge *(phys Ch)* Oberflächenladung *f*
surface charge density Oberflächenladungsdichte *f*
surface chemistry Oberflächenchemie *f*
surface coating 1. Anstrich *m*; Oberflächen(schutz)schicht *f*; 2. Anstrichstoff *m*, Anstrichmittel *n*; Beschichtungs(werk)stoff *m*, Beschichtungsmaterial *n*
surface conductivity Oberflächenleitfähigkeit *f*, Oberflächenleitvermögen *n*
surface coverage Oberflächenbedeckung *f*
surface crack Oberflächenriss *m*
surface culture *(Biot)* Oberflächenkultur *f*
surface defect 1. *(Krist)* flächenhafter

Gitterfehler *m*; 2. *(Krist)* flächenhafte Gitterfehlstelle *f*
surface denaturation *(Bioch)* Oberflächendenaturierung *f*
surface diffusion Oberflächendiffusion *f*
surface diffusivity Oberflächendiffusionskoeffizient *m*
surface effect Oberflächeneffekt *m*, Oberflächenwirkung *f*
surface energy Oberflächenenergie *f*
surface-enhanced Raman spectroscopy oberflächenverstärkte Laser-Raman-Spektroskopie *f*
surface fermentation *(Biot)* Oberflächenfermentation *f*
surface film Oberflächenfilm *m*, Oberflächenhaut *f*
surface film reactor *(Biot)* Oberflächenreaktor *m*, Reaktor *m* für Oberflächenverfahren
surface filter Oberflächenfilter *n*
surface filtration Oberflächenfiltration *f*
surface force Oberflächenkraft *f*
surface friction *(Kst)* Oberflächenreibung *f*
surface hardening Oberflächenhärten *n*, *(i. e. S.)* Oberflächenhärten *n* mit Diffusion
surface heterogeneity Oberflächenheterogenität *f*
surface-inactive oberflächeninaktiv
surface layer 1. Oberflächenschicht *f*; 2. *(Met)* Oberflächenzone *f*, Randschicht *f*, Randzone *f*, Rand *m*
surface mining Tagebauförderung *f*, Förderung *f* im Tagebau, Tagebau(-betrieb) *m*
surface modification *(Kst)* Oberflächenmodifizierung *f*
surface nucleation *(Krist)* sekundäre Keimbildung *f*, Sekundärkeimbildung *f*
surface oxidation Oberflächenoxidation *f*, oberflächliche Oxidation *f*
surface patterning Oberflächenmusterung *f*
surface phenomenon Oberflächenerscheinung *f*
surface plasmon resonance spectroscopy *(Spektr)* Oberflächenplasmonresonanzspektroskopie *f*
surface potential Oberflächenpotenzial *n*, Flächenpotenzial *n*
surface process *(Biot)* Oberflächen(kultur)verfahren *n*, Emersverfahren *n* *(Fermentation)*
surface properties Oberflächeneigenschaften *fpl*
surface protection Oberflächenschutz *m*
surface quarrying *s.* surface mining
surface reaction (chemische) Oberflächenreaktion *f*, Grenzflächenreaktion *f*
surface-renewal theory Oberflächenerneuerungstheorie *f* *(Stoffübertragung)*
surface resistance Oberflächenwiderstand *m*
surface resistivity spezifischer Oberflächenwiderstand *m*
surface roughness Oberflächenrauigkeit *f*, Oberflächenrauheit *f*
surface temperature Oberflächentemperatur *f*
surface tension Oberflächenspannung *f*
surface texture Oberflächentextur *f*
surface treating [treatment] Oberflächenbehandlung *f*
surfacer Spachtelmasse *f*, Spachtel *m(f)*
surfacing Oberflächenbehandlung *f*
surfactant *s.* surface-active agent
surplus of electrons Elektronenüberschuss *m*
surroundings temperature Umgebungstemperatur *f*
survival factor *(Biot)* Überlebensfaktor *m*
survival time *(Tox)* Überlebensdauer *f*
susceptibility Empfindlichkeit *f*, Suszeptibilität *f*
susceptibility to failure Störanfälligkeit *f*
susceptibility to oxidation Oxidationsempfindlichkeit *f*
susceptibility to substitution Substitutionsbereitschaft *f*
susceptible anfällig, empfindlich
susceptible to failure störanfällig
susceptible to oxidation oxidationsempfindlich
suspend *v* 1. suspendieren, aufschwemmen, aufschlämmen; 2. aufhängen
suspended centrifuge Hängezentrifuge *f*, hängende Zentrifuge *f*
suspended material [matter] suspendierter Stoff *m*, Schweb(e)stoff *m*, Trübungsstoff *m*

suspended particles suspendierte Teilchen *npl* [Partikel *npl*, Feststoffteilchen *npl*], Schwebeteilchen *npl*, Suspensionspartikel *npl*, Trübeteilchen *npl*
suspended solid pollutants *(Hyd)* suspendierte Schmutzstoffe *mpl* [Verunreinigungen *fpl*]
suspended solids 1. *(Hyd)* suspendierte Feststoffe *mpl* [Feststoffteilchen *npl*], feste Schwebstoffe *mpl*, grobdisperse Wasserinhaltsstoffe *mpl* (> 5 · 10^{-4} *mm)*; 2. *s.* suspended-solids content
suspended-solids content [level] *(Hyd)* Gehalt *m* [Anteil *m*] an suspendierten Feststoffen, Schwebstoffgehalt *m*
suspending agent 1. Suspendiermittel *n*, Suspensionsmittel *n*, Suspensionsmedium *n*, Stabilisator *m (für Suspensionen)*; 2. *(Anstr)* Absetzverhinderungsmittel *n*, Schwebemittel *n*
suspending capacity Suspendiervermögen *n*
suspensibility Suspendierbarkeit *f*
suspensible suspendierbar
suspension Suspension *f*, Aufschwemmung *f*, Aufschlämmung *f* • **keep [maintain] in suspension** in Suspension halten, in der Schwebe halten
suspension firing (of coal) Kohlenstaubfeuerung *f*, Staubfeuerung *f*
suspension gasification *(Carb)* Staubvergasung *f*, Schwebevergasung *f*, Flugstromvergasung *f*, Vergasung *f* in der Flugstaubwolke [Staubwolke]
suspension polymerization Suspensionspolymerisation *f*, Perlpolymerisation *f*, Kornpolymerisation *f*
suspension polyvinyl chloride Suspensionspolyvinylchlorid *n*, S-PVC *n*
suspension roasting *(Met)* Suspensionsröstung *f*, *(i. w. S.)* Schweberöstung *f*
suspensoid Suspensoid *n*, Suspensionskolloid *n*
suspensoid catalytic cracking Suspensoid-Kracken *n*, Kracken *n* mit suspendiertem Katalysator
sustain *v* unterhalten *(z. B. die Verbrennung)*
S.V. *(saponification value) s.* saponification number
Svedberg unit Svedberg-Einheit *f (Molmassenbestimmung durch Sedimentation)*
swab *v* 1. abwischen, aufwischen; 2. auftragen
swarm of molecules Molekülschwarm *m*
Swarts reaction Swarts-Reaktion *f (zur Darstellung von Fluoralkanen)*
swathe Schwaden *m*
SWCNT *s.* single-wall carbon nanotube
sweat *v* schwitzen; ausschwitzen
sweat *v* **off** *(Dest)* langsam übertreiben
sweating Schwitzen *n*, Schwitzung *f (von Paraffin zum Entfernen von Ölresten)*
sweep *v* **forward** mitreißen, mitführen
sweep *v* **out** 1. säubern; 2. austreiben, verdrängen *(Gase)*
sweep *v* **through** hindurchtransportieren *(Gaschromatographie)*
sweep coil *(Spektr)* Sweep-Spule *f*
sweep generator *(Spektr)* Sweep-Generator *m*
sweepings *(Met)* Gekrätz *n*
sweeps *(Met)* Gekrätz *n*
sweet 1. süß; 2. *(Erdöl)* süß, gesüßt, doktor-negativ, mit negativem Doktortest
sweet gasoline süßes [gesüßtes] Benzin *n*
sweet oil süßes [gesüßtes] Erdölprodukt *n* [Erdöldestillat *n*], doktor-negatives Öl *n*, Öl *n* mit negativem Doktortest
sweet-smelling süßlich riechend, mit süßlichem Geruch
sweet water *(Hyd)* Süßwasser *n*
sweeten *v* 1. (ab)süßen, versüßen; 2. *(Erdöl)* süßen *(um Schwefelverbindungen zu entfernen bzw. in Disulfide zu überführen)*
sweetened 1. (ab)gesüßt, versüßt; 2. *(Erdöl)* gesüßt, süß, doktor-negativ, mit negativem Doktortest
sweetener *(Lebm)* Süß(ungs)mittel *n*
sweetening agent 1. *(Erdöl)* Süßungsmittel *n*; 2. *s.* sweetener
sweetening process *(Erdöl)* Süß(ungs)verfahren *n*, Sweetening-Prozess *m*
sweetish süßlich
Sweetland filter Sweetland-Filter *n*, Sweetland-Filterpresse *f*, Druckfilter *n* S, Sweetland-Presse *f*
Sweetland (filter) press Sweetland-Fil-

ter *n*, Sweetland-Filterpresse *f*, Druckfilter *n* S, Sweetland-Presse *f*
sweetness Süße *f (Geschmacksempfindung)*
swell *v* **(up)** (auf)quellen
swell ratio *(Kst)* Strangaufweitungsverhältnis *n (beim Extrudieren)*
swelling 1. Quellen *n*, Quellung *f*, Aufquellen *n*, Aufquellung *f*; 2. Blähen *n*, Blähung *f*, Aufblähen *n*, Aufblähung *f (der Kohle)*
swelling agent Quellungsmittel *n*
swelling degree Quellungsgrad *m (Polymernetzwerke)*
swelling equilibrium Quellungsgleichgewicht *n*
swelling from the die *(Kst)* Strangaufweitung *f (beim Extrudieren)*
swelling kinetic Quellkinetik *f*
swelling resistance Quellbeständigkeit *f*, Quellfestigkeit *f*
swept through/be passieren, durchströmen *(Gaschromatographie)*
swill *v* (ab)spülen
swill *v* **out** ausspülen
swill water Spülwasser *n*
swills Spülwässer *npl*
swing open 1. aufklappen; 2. ausschwingen
swing schwenkbar, Schwenk...; Schwing...
swinging-bucket rotor ausschwingender Rotor *m (einer Zentrifuge)*; Rotor *m* mit frei ausschwingenden Bechern
swingout centrifuge head *s.* swinging-bucket rotor
swings in load *(Hyd)* Belastungsschwankungen *fpl*, Lastschwankungen *fpl*
swirl *v* (um)schwenken
swivel clamp holder Universaldoppelmuffe *f (drehbare Doppelmuffe für Stativklemmen)*
SWNT *s.* single-wall carbon nanotube
sydnone *(org Ch)* Sydnon *n*
sydnone imine Sydnonimin *n*
sylvin(e) *(Min)* Sylvin *n (Kaliumchlorid)*
sylvite *(Min)* Sylvin *n (Kaliumchlorid)*
symbiotic nitrogen fixation Stickstoff-Fixierung *f* durch symbiontische Mikroorganismen, symbio(n)tische Stickstoffbindung *f* [Stickstoff-Fixierung *f*] *(besonders durch Knöllchenbakterien)*
symmetric-top molecule symmetrisches Kreiselmolekül *n*, symmetrischer Kreisel *m*
symmetrical symmetrisch
symmetrical anhydride *(org Ch)* symmetrisches Anhydrid *n*
symmetrical breathing vibration Pulsationsschwingung *f*, Atemschwingung *f*, Breathing-Schwingung *f*
symmetrical ether R–O–R einfacher [symmetrischer] Ether *m*
symmetry class Symmetrieklasse *f*, Kristallklasse *f*, Punkt(symmetrie)-gruppe *f*
symmetry element *(Krist, Spektr)* Symmetrieelement *n*
symmetry factor *(Ech)* Durchtrittsfaktor *m*
symmetry group *s.* symmetry class
symmetry operation Symmetrieoperation *f*
symmetry plane 1. Symmetrieebene *f*; 2. *(Krist)* Spiegelebene *f*
symmetry transformation *s.* symmetry operation
sympathetic reaction induzierte Reaktion *f*
sympatholytic sympath(ik)olytisch
sympatholytic (agent) *(Pharm)* Sympath(ik)olytikum *n*
sympathomimetic sympath(ik)omimetisch
sympathomimetic (agent) *(Pharm)* Sympath(ik)omimetikum *n*
symport *(Bioch)* Symport *m*, Co-Transport *m (Form des Carrier-Transports)*
syn syn-..., in syn-Stellung befindlich
• **be syn** in syn-Stellung stehen
syn-anti isomerism syn-anti-Isomerie *f*
syn position syn-Stellung *f*
synante(c)tic *(Geoch)* synantetisch *(durch Reaktion zweier Minerale miteinander entstanden)*
synchrocyclotron Synchrozyklotron *n*, frequenzmoduliertes Zyklotron *n*, FM-Zyklotron *n*
synchronous reaction synchrone Reaktion *f (unter gleichzeitigem Lösen und Bilden von Bindungen)*
synchrotron *(Kern)* Synchrotron *n*
synchrotron radiation Synchrotronstrahlung *f*

synclinal conformation synclinale Konformation *f (Torsionswinkel ±60°)*
syncrude (oil) *s.* synthetic crude oil
syndet *s.* synthetic detergent
syndiotactic polymer syndiotaktisches Polymer(e) *n*
syndiotacticity Syndiotaktizität *f*
syneresis Synärese *f*, Synaerese *f (spontaner Flüssigkeitsaustritt aus Gelen)*
synergism Synergismus *m*, synergistische Wirkung *f*
synergist Synergist *m*
synergistic effect synergistischer Effekt *m*
synerize *v* Synärese zeigen *(Gele)*
synperiplanar conformation synperiplanare Konformation *f (Torsionswinkel 0°)*
syntan Syntan *n*, synthetischer [künstlicher] Gerbstoff *m*
synthesis Synthese *f*, Aufbau *m*, Vereinigung *f*
synthesis gas Synthesegas *n (Gasgemisch als Ausgangsprodukt für Synthesen)*
synthesis path(way) *(Bioch)* Syntheseweg *m*
synthesis step Syntheseschritt *m*
synthesize *v* synthetisieren, synthetisch [durch Synthese] herstellen
synthesize *v* **from scratch** voll synthetisieren
synthetic synthetisch, Synthese...
synthetic 1. synthetisch gewonnene Substanz *f*, Syntheseprodukt *n*; 2. *s.* synthetic fibre; 3. *s.* synthetic rubber
synthetic ammonia synthetisches Ammoniak *n*, Syntheseammoniak *n*
synthetic chemist Synthesechemiker *m*
synthetic chemistry Synthesechemie *f*, synthetische Chemie *f*
synthetic crude oil *(Carb)* synthetisches Roh(erd)öl *n*
synthetic detergent Detergens *n*, Detergent *n*, Syndet *n*, synthetisches Reinigungsmittel *n* [Waschmittel *n*]
synthetic fibre Synthesefaser *f*, synthetische Faser *f*; Synthesefaserstoff *m*, synthetischer Faserstoff *m*, CS
synthetic graphite künstlicher Graphit *m*
synthetic medium *(Biot)* synthetisches [künstliches] Nährmedium *n*
synthetic natural gas synthetisches Erdgas *n*, Erdgasaustauschgas *n*, Erdgasersatzgas *n*, Ersatzerdgas *n*
synthetic organic chemistry organische Synthesechemie *f*, synthetische organische Chemie *f*
synthetic resin Kunstharz *n*
synthetic-resin foam Kunstharzschaum *m*
synthetic-resin varnish Kunstharzlack *m*
synthetic rubber synthetischer [künstlicher] Kautschuk *m*, Synthesekautschuk *m*, SK, Kunstkautschuk *m*
synthetic tannin [tanning agent] *s.* syntan
synthetical *s.* synthetic
synthetize *v s.* synthesize
synthol Synthol *n (Gemisch von Alkoholen, Ketonen, Säuren und etwa 2 % Kohlenwasserstoffen)*
syphon *s.* siphon
syrup 1. *(Ch)* Sirup *m*, sirupdicke Masse *f*; 2. *(Lebm)* Sirup *m (dickflüssiger Obst- oder Zuckersaft)*; *(i. e. S.)* Speisesirup *m*; 3. *(Pharm)* Sirup *m*
syruplike sirupartig, sirupös, dickflüssig, klebrig
syrupy sirupartig, sirupös, dickflüssig, klebrig
system of two components Zweikomponentensystem *n*, Zweistoffsystem *n*, binäres System *n*
systematic error systematischer Fehler *m*
systematic name systematischer [rationeller, systematisch gebildeter] Name *m*, systematische Benennung *f*
systemic (chemical) systemisches [innertherapeutisches] Mittel *n*, Innertherapeutikum *n (Pflanzenschutz)*
Szilard-Chalmers effect Szilard-Chalmers-Effekt *m*
Szilard-Chalmers method Szilard-Chalmers-Verfahren *n*, Szilard-Chalmers-Methode *f (zur Abtrennung von Radionukliden von ihrer isotopen Ausgangssubstanz)*

T

T *s.* 1. transmission ratio; 2. transport number; 3. trans conformation
2,4,5-T *s.* 2,4,5-trichlorophenoxyacetic acid
T piece T-Stück *n*, T-förmiges Zwischenabzweigstück *n*
T-shape connecting tube T-Stück *n*, T-förmiges Zwischenabzweigstück *n*
T unit *s.* trifunctional unit
TA *s.* 1. terephthalic acid; 2. thermal analysis
table 1. *(Aufb)* Herd *m*, Setzherd *m*; 2. *(Aufb)* Herdtafel *f*, Herdplatte *f (eines Setzherds)*
table salt Tafelsalz *n*, Speisesalz *n*
table vinegar Speiseessig *m*, Tafelessig *m*
tablet *v* tablettieren
tablet Tablette *f*
tabletting machine [press] Tablettiermaschine *f*, Tablettenpresse *f*
tabling *(Aufb)* Herdsortieren *n*, Herdaufbereitung *f*, Herdarbeit *f*
tabular tafel(förm)ig
tack Klebrigkeit *f*
tackiness Klebrigkeit *f*
tacky klebrig
tacticity Taktizität *f (von Polymeren)*
Tafel diagram [plot] *(Ech)* Tafel-Diagramm *n*
tag *v* markieren *(eine Substanz durch den Einbau gut nachweisbarer, meist radioaktiver Atome kenntlich machen)*
tail 1. Schwanz *m (eines Monomeren)*; 2. *(Chromat)* Schwanz *m*, Schweif *m*; 3. *(Aufb)* Abgänge *mpl*, Berge *pl*; 4. *(Dest)* Nachlauf *m*
tail-to-tail polymerization Schwanz-Schwanz-Polymerisation *f*
tailing 1. *(Chromat)* Schwanzbildung *f*, Schweifbildung *f*, *Tailing *n*; 2. *s.* tailings
tailings 1. *(Aufb)* Abgänge *mpl*, Berge *pl*, *(Flotation auch)* Flotationsberge *pl*; 2. Überlaufprodukt *n*, Überlaufgut *n*, Überlauf *m (Klassierung)*; *(bei körnigem Gut auch)* Überkorn *n*, Grobkorn *n*, Grobgut *n*, *(beim Siebklassieren auch)* Siebgrobes *n*, *(beim Windsichten auch)* Sichtgrobes *n*; 3. *(Dest)* Nachlauf *m*
tailor(-make) *v* gezielt [nach Maß] aufbauen [herstellen], „maßschneidern" *(z. B. Polymere)*
tails *s.* tailings
take *v* fassen *(z. B. eine Flüssigkeitsmenge)*
take *v* **off** *(Dest)* abnehmen, entnehmen
take *v* **out** entnehmen; entziehen *(z. B. Wasser)*
take *v* **overhead** *(Dest)* über Kopf abnehmen
take *v* **samples** Proben entnehmen
take *v* **up** aufnehmen *(einen Stoff)*; abziehen, abnehmen *(z. B. ein Extrudat)*
take-off *(Dest)* Abnahme *f*, Entnahme *f*
take-up Aufnahme *f (von Stoffen)*; Abziehen *n*, Abzug *m*, Abnahme *f (z. B. des Extrudats)*
talc 1. *(Min)* Talk *m*, Talkum *n*, Talkstein *m (Magnesiumdihydroxytetrasilicat)*; synthetischer Talk *m*; Talkum *n*, Talk *m (pulvrige Form des Talks)*; 2. *(Min)* Muskovit *m*, Kaliglimmer *m (ein Phyllosilicat)*
talcose talkig, Talk...
talcum Talkum *n*, Talk *m (pulvrige Form des Talks)*
tall oil Tallöl *n*, Kiefernöl *n*, Harzöl *n*
tallow Talg *m*
tallowiness Talgigkeit *f*, Talggeschmack *m*, talgiger Geschmack *m (verdorbener Fette)*; Peroxidranzigkeit *f (von Fetten)*
tallowy talg(art)ig
TAN *s.* total acid number
tan *v* gerben
tan *s.* 1. tanning material; 2. tanning agent
tandem cuvette *(Spektr)* Tandemküvette *f*, geteilte Küvette *f*
tank *v* im Tank (ab)lagern
tank 1. Tank *m*, (geschlossener) Behälter *m*; Becken *n*; Wanne *f*, Trog *m*; Pfanne *f*; 2. *(Gum)* Kessel *m*, Vulkanisierkessel *m*, Vulkanisationskessel *m*; 3. *(Glas)* Wanne *f*, Schmelzwanne *f*
tank car Eisenbahnkesselwagen *m*, Eisenbahntankwagen *m*
tank crystallizer offener feststehender Kristallisator *m*, Kastenkristallisator *m*, Pfannenkristallisator *m*, Kristallisierpfanne *f*
tank farm Tanklager *n*
tank furnace *(Glas)* Wannenofen *m*

tank glass Wannenglas *n*, in der Wanne geschmolzenes Glas *n*
tank reactor Rühr(kessel)reaktor *m*, Rührkessel *m*, Mischreaktor *m*
tank-type flow reactor kontinuierlicher Rührkesselreaktor *m*, Durchflussmischreaktor *m*
tanker Tanker *m*, Tankschiff *n*; Straßentankwagen *m*, Tankauto *n*, Autokesselwagen *m*; Tankflugzeug *n*
tannage Gerben *n*, Gerbung *f*, Gerberei *f*
tannate Tannat *n (Salz oder Ester des Tannins)*
tanner's red Gerbstoffrot *n*, Gerberrot *n*, Phlobaphen *n*
tannery 1. Gerberei *f*, Gerbanlage *f*; 2. Gerberei *f*, Gerben *n*, Gerbung *f*
tannic acid *s.* tannin
tannide *mikrochemisch eine Gerbstoffreaktion zeigender Stoff*
tanniferous plant Gerbstoffpflanze *f*, gerbstoffhaltige Pflanze *f*
tannin 1. Gallotannin *n*, Tanningerbstoff *m*, Gallusgerbsäure *f*, Tannin *n*; 2. pflanzlicher Gerbstoff *m*, *(i. w. S.)* Tannin *n*; 3. *m*-Digallussäure *f*; 4. *Estergemisch aus Glucose und Gallussäure und/oder deren Depsiden, Hauptanteil Penta-m-digalloyl-β-glucose*
tanning 1. Gerben *n*, Gerbung *f*, Gerberei *f*; 2. *(Kosmet)* Bräunung *f (der Haut)*
tanning agent Gerbstoff *m*, gerbender Stoff *m*, gerbende Substanz *f (gerbender Anteil des Gerbmittels)*
tanning bark Gerbrinde *f*
tanning index Durchgerbungszahl *f*
tanning material Gerbmittel *n*, Gerbmaterial *n*
tanning oil Gerböl *n*
tanning pit [vat] Gerbgrube *f*
tantalate Tantalat *n*
tantalic Tantal..., *(i. e. S.)* Tantal(V)-...
tantalic acid Tantalsäure *f*, *(besser:)* Tantalpentoxidhydrat *n*
tantalous Tantal..., *(i. e. S.)* Tantal(III)-...
tantalum Ta Tantal *n*
tantalum carbide Tantalcarbid *n*
tantalum nitride Tantalnitrid *n*
tantalum pentachloride Tantalpentachlorid *n*, Tantal(V)-chlorid *n*
tantalum pentoxide Tantal(pent)oxid *n*, Tantal(V)-oxid *n*
tanwood Gerbholz *n*, gerbstoffhaltiges Holz *n*
tanyard Gerberei *f*, Gerbanlage *f*
tap *v* 1. zapfen *(Flüssigkeiten, Latex)*; anzapfen *(Fässer, Kautschukbäume)*; 2. *(Met)* abstechen, ablassen; abziehen *(Schlacke)*; 3. *(Glas)* ablassen, abstechen, anstechen *(eine Wanne)*; ausschöpfen *(einen Hafen)*
tap *v* **out** abstechen, ablassen *(schmelzflüssiges Metall)*
tap 1. Hahn *m*; 2. Zapfen *m*, Spund *m*
tap funnel Tropftrichter *m*
tap water Leitungswasser *n*
tapered joint Kegelschliffverbindung *f*
taphole *(Met)* Abstichloch *n*, Stichloch *n*, Abstichöffnung *f*, Abstich *m*, Stich *m*
tapioca (starch) Tapioka(stärke) *f (von Manihot utilissima Pohl)*
tapping channel Abstichrinne *f*, Abflussrinne *f*
tar *v* teeren
tar Teer *m*
tar acid Teersäure *f*
tar base Teerbase *f*
tar distillation Teerdestillation *f*
tar hydrogenation Teerhydrierung *f*
tar oil Teeröl *n*
tar pitch Teerpech *n*
tar sand Teersand *m*, Asphaltsand *m*
tar separator Teer(ab)scheider *m*
tare *v* (aus)tarieren
target 1. Target *n*; 2. *(Kern)* Targetkern *m*; 3. *(Bioch)* Empfänger *m*, Erfolgsorgan *n*
targeting signal *s.* signal peptide
tarnish *v* anlaufen, blind werden
tarnish 1. Anlaufen *n*, Blindwerden *n*; 2. *s.* tarnish film
tarnish film Anlauffilm *m*, Anlaufschicht *f*, *(bei Wärmeeinwirkung auch)* Anlassfilm *m*
tarry 1. teer(art)ig; 2. geteert
tartar Weinstein *m*, Kaliumhydrogentartrat *n*
tartaric acid Weinsäure *f*, *2,3-Dihydroxybutandisäure *f*
***d*-tartaric acid** (+)-Weinsäure *f*, Rechtsweinsäure *f*, rechtsdrehende Weinsäure *f*
***dl*-tartaric acid** (±)-Weinsäure *f*, racemische Weinsäure *f* [Säure *f*], Traubensäure *f*

l-tartaric acid (-)-Weinsäure *f*, Linksweinsäure *f*, linksdrehende Weinsäure *f*
tartrate Tartrat *n (Salz oder Ester der Weinsäure)*
tartrazine Tartrazin *n*, Hydrazingelb *n* O, Echtwollgelb *n*, Säuregelb *n*, Echtlichtgelb *n*, Flavazin *n* T
tartronic acid Tartronsäure *f*, *Hydroxypropandisäure *f*
tast polarography Tastpolarographie *f*
taste *v* 1. auf Geschmack prüfen; 2. schmecken, einen bestimmten Geschmack haben
taste Geschmack *m* • **have an acid taste** sauer schmecken
taste-causing substance geschmacksverursachender [geschmacksbildender] Stoff *m*, Geschmacksstoff *m*
taste-impairing geschmacksbeeinträchtigend
taste-modifying geschmacksverändernd
taste threshold *(Hyd)* Geschmacksschwellenkonzentration *f*, GSK, Geschmacksschwelle *f*
taurine Taurin *n*, 2-Aminoethan-1-sulfonsäure *f*
taurocholic acid *(Bioch)* Taurocholsäure *f*
tautomer Tautomer(e) *n*
tautomeric tautomer
tautomerism Tautomerie *f*
tautomerization Tautomerisierung *f*
tautomerize *v* tautomerisieren
TBA *s.* thiobarbituric acid
2,3,6-TBA *s.* 2,3,6-trichlorobenzoic acid
TBN *s.* total base number
TBT (compound) *(Tox)* Tributylzinnverbindung *f*
TC *s.* total carbon
TCA *s.* 1. tricarboxylic acid; 2. trichloroacetic acid
TCB *s.* trichlorobenzene
TCD *(thermal conductivity detector) s.* thermal-conductivity cell
TCDD *s.* 2,3,7,8-tetrachlorodibenzo-1,4-dioxin
2,3,7,8-TCDD *s.* 2,3,7,8-tetrachlorodibenzo-1,4-dioxin
TCE *s.* trichloroethane
TCP *s.* tricresyl phosphate
T/D = long tons per day *(U.K. 1 long ton = 1 016,05 kg; U.S. 1 long ton = 907,187 kg)*
TDI *s.* toluene diisocyanate
TDS *s.* total dissolved solids
tds *s.* total dissolved solids
TEA *s.* triethanolamine
tear *v* (zer)reißen
tear 1. Träne *f*, Tropfen *m*, Lacktropfen *m (beim Tauchlackieren)*; 2. Reißen *n*, Zerreißen *n*; Einriss *m*
tear gas Tränengas *n*
tear initiation strength Einreißfestigkeit *f*, Einreißwiderstand *m*, Widerstand *m* gegen Einreißen
tear resistance [strength] Zerreißfestigkeit *f*, Reißfestigkeit *f*
teaspoon/a level einen gestrichenen Teelöffel voll
teaspoonful/a einen Teelöffel voll *(U.K. etwa 4,93 cm^3; U.S. etwa 4,74 cm^3)*
teat pipette Tropfpipette *f*, Stechpipette *f*, Tropfer *m*
technate Didym *n (Nd-Pr-Gemisch)*
technetium Tc Technetium *n*
technical ceramics technische Keramik *f*
technical sulphuric acid technische [technisch reine] Schwefelsäure *f*
technically pure technisch (rein), techn.
technician 1. Techniker *m*; 2. *Facharbeiter in einem technischen Beruf*
technique Arbeitstechnik *f*, Arbeitsverfahren *n*, Technik *f*, Verfahren *n*, Methode *f*
technological chemist technischer Chemiker *m*
technological chemistry technische Chemie *f*
technologist Techniker *m*
technology 1. Technik *f (Gesamtheit der Tätigkeiten und Arbeitsmittel in der Produktion)*; 2. (allgemeine, vergleichende) Technologie *f (Wissenschaftszweig)*; (spezielle) Technologie *f (der Herstellung bestimmter Produkte)*
tectosilicate *(Min)* Tektosilicat *n*, Gerüstsilicat *n*
tee *v* mit einem T-Stück verbinden
tee (connector) T-Stück *n*, T-förmiges Zwischenabzweigstück *n*
TEELS *s.* transmission electron energy loss spectrometry

teem *v (Met)* gießen *(flüssige Metalle aus einer Gießpfanne in Formen)*
TEG *s.* triethylene glycol
teichoic acid *(Bioch)* Teichonsäure *f*
TEL *s.* tetraethyllead
tele-substitution *(org Ch)* tele-Substitution *f*
tellurate Tellurat *n*
telluric Tellur...; höherwertigem Tellur entsprechend, *(meist)* Tellur(VI)-...
telluric acid Tellursäure *f*, Orthotellursäure *f*, Hexoxotellursäure *f*
telluric chloride Tellur(IV)-chlorid *n*, Tellurtetrachlorid *n*
telluric oxide Tellur(VI)-oxid *n*, Tellurtrioxid *n*
telluride M^I_2Te Tellurid *n*
tellurite $M^I_2TeO_3$ Tellurit *n*
tellurium Te Tellur *n*
tellurium hexafluoride Tellurhexafluorid *n*, Tellur(VI)-fluorid *n*
tellurium hydride Tellurwasserstoff *m*, Tellurhydrid *n*
tellurium monoxide Tellur(mon)oxid *n*, Tellur(II)-oxid *n*
telluronium compound Telluroniumverbindung *f*
tellurous Tellur...; niederwertigem Tellur entsprechend, *(meist)* Tellur(IV)-...
tellurous acid tellurige Säure *f*
tellurous chloride Tellur(II)-chlorid *n*, Tellurdichlorid *n*
tellurous oxide Tellur(IV)-oxid *n*, Tellurdioxid *n*
telogen Telogen *n (Mittel zur Regelung des Polymerisationsgrads)*
telomer Telomer(e) *n*, Telomerisat *n*
telomerization reaction Telomerisationsreaktion *f*
TEM *s.* 1. transmission electron microscope; 2. transmission electron microscopy
temper *v* 1. *(Met)* anlassen; 2. *(Kst)* tempern; 3. *(Glas)* verspannen, vorspannen, härten
temper colour Anlassfarbe *f*, Anlauffarbe *f*
temperature bath Temperierbad *n*, Thermostat *m*
temperature coefficient Temperaturkoeffizient *m*, Temperaturbeiwert *m*
temperature coefficient of viscosity Viskositäts-Temperatur-Koeffizient *m*, Temperaturkoeffizient *m* der Viskosität [Zähigkeit], VTC
temperature control Temperaturregelung *f*
temperature-dependent temperaturabhängig
temperature drop Temperaturabfall *m*
temperature-entropy chart Temperatur-Entropie-Diagramm *n*, T-s-Diagramm *n*
temperature equilibrium Temperaturgleichgewicht *n*
temperature fluctuation Temperaturschwankung *f*
temperature gradient Temperaturgradient *m*, Temperaturgefälle *n*
temperature-independent temperaturunabhängig
temperature-insensitive temperaturunempfindlich, temperaturbeständig
temperature jump Temperatursprung *m*
temperature limit Temperaturgrenze *f*
temperature range Temperaturbereich *m*
temperature resistance Temperaturbeständigkeit *f*
temperature-resistant temperaturbeständig
temperature-resisting temperaturbeständig
temperature scale Temperaturskale *f*
temperature-sensitive temperaturempfindlich
temperature-stable *s.* temperature-resistant
tempered (safety) glass Einscheibensicherheitsglas *n*, ESG
tempering oil Anlassöl *n*
template *(Bioch)* Matrize *f*
temporary hardness *(Hyd)* temporäre [vorübergehende] Härte *f*, Karbonathärte *f*, KH
tenacious zäh; festhaftend; zugfest
tenacity Zähigkeit *f*; Haftfestigkeit *f*; Zugfestigkeit *f*
tendency to corrode Korrosionsneigung *f*, Korrosionsbestreben *n*, Korrosionstendenz *f*
tendency to react Reaktionsbestreben *n*
tensammetric *(Ech)* tensammetrisch
tensammetry *(Ech)* Tensammetrie *f*
tensile failure Bruch *m* durch Zugbeanspruchung

tensile force Zugkraft *f*
tensile modulus Dehngrenze *f (Beanspruchung bei vorgeschriebener Verlängerung)*
tensile strength Zugfestigkeit *f*
tensile stress Zugspannung *f*
tensile stress-strain characteristics [properties] Zugdehnungseigenschaften *fpl*, Zugverformungseigenschaften *fpl*, Zug-Dehnungs-Verhalten *n*
tensile test Zugversuch *m*
tension Spannung *f*, Zugspannung *f*
tenth molar solution 0,1 m-Lösung *f*, m/10-Lösung *f*
terbia Terbium(III)-oxid *n*
terbium Tb Terbium *n*
terbium chloride Terbium(III)-chlorid *n*
terbium oxide Terbium(III)-oxid *n*
terdentate dreizähnig, dreizählig *(Ligand)*
terephthalic acid Terephthalsäure *f*, Benzen-1,4-dicarbonsäure *f*
term 1. *(phys Ch)* Term *m*, Energieniveau *n*, Energieterm *m*, Energiezustand *m*; 2. Term *m (Mathematik)*; 3. Fachwort *n*, Fachausdruck *m*, Terminus *m* (technicus)
terminal 1. End...; 2. *(org Ch)* endständig, terminal
terminal group endständige [terminale] Gruppe *f*, Endgruppe *f*
terminal knife edge Endschneide *f*, Seitenschneide *f (an Analysenwaagen)*
terminal position Endstellung *f*
terminate *v* abbrechen, beenden *(Reaktionskette)*
termination 1. Abbruch *m*, Beendigung *f*; 2. *(Nom)* Endung *f*; 3. *(Bioch)* Termination *f*; 4. *s.* termination of chains
termination of chains Kettenabbruch *m*, Beendigung *f* des Kettenwachstums
termination reaction Abbruchreaktion *f*
terminator codon *(Bioch)* Terminationscodon *n*, Stop(p)codon *n*, Punktcodon *n*
termolecular reaction trimolekulare [termolekulare] Reaktion *f*, Dreierreaktion *f*
ternary alloy ternäre Legierung *f*, Dreistofflegierung *f*
ternary compound ternäre Verbindung *f*
ternary system ternäres System *n*, Dreistoffsystem *n*, Dreikomponentensystem *n*
terpene Terpen *n*
terpene alcohol Terpenalkohol *m*
terpene hydrocarbon Terpenkohlenwasserstoff *m*
terpenoid Terpenoid *n*, Terpenverbindung *f*, *(i. w. S.)* Terpen *n*
terpenoid alkaloid Terpenalkaloid *n*
terpin Terpin *n*, 1,8-Terpin *n*
terpinene Terpinen *n*
terpineol Terpineol *n (ungesättigter Terpenalkohol)*
terpinolene Terpinolen *n*, Mentha-1,4(8)-dien *n*
terpolymer Terpolymer(e) *n*
terra-cotta *(Ker)* Terrakotta *f*, Terrakotte *f*
terracotta *(Ker)* Terrakotta *f*, Terrakotte *f*
tert. *s.* tertiary
tertiary tertiär
tertiary alcohol tertiärer Alkohol *m*
tertiary amine tertiäres Amin *n*
tertiary-butylation *tert.*-Butylierung *f*
tertiary salt tertiäres [neutrales] Salz *n (einer dreibasigen Säure)*
tertiary structure *(Bioch)* Tertiärstruktur *f*
tertiary waste treatment (stage) *(Hyd)* dritte Reinigungsstufe *f*, weitergehende Abwasserbehandlung *f*, Nachreinigung *f*
tervalency Dreiwertigkeit *f*, Trivalenz *f*
tervalent dreiwertig, trivalent
test *v* testen, prüfen, untersuchen
test *v* **by spotting** tüpfeln
test *v* **for** prüfen auf, nachweisen
test Test *m*, Prüfung *f*, Versuch *m*, Prüfversuch *m*, Untersuchung *f*; Nachweis *m* • **conduct a test** einen Versuch durchführen • **under test** zu prüfen(d)
test arrangement Versuchsanordnung *f*
test material Probematerial *n*, Probengut *n*, Testmaterial *n*, Untersuchungsmaterial *n*
test method Testverfahren *n*, Prüfverfahren *n*, Prüfmethode *f*, Untersuchungsverfahren *n*, Untersuchungsmethode *f*
test paper Reagenzpapier *n*, Prüfpapier *n*, Indikatorpapier *n*
test piece Prüfkörper *m*, Probekörper *m*, Probe *f*, Prüfling *m*
test reaction Nachweisreaktion *f*
test sample Untersuchungsprobe *f*,

Endprobe *f (beim Verjüngen von Sammelproben)*
test solution 1. Prüflösung *f (zum Prüfen verwendet)*; 2. Probelösung *f*, Prüf(lings)lösung *f*, Untersuchungslösung *f*, zu prüfende [untersuchende] Lösung *f*
test specimen *s.* test piece
test substance Versuchssubstanz *f*, Probesubstanz *f*, Untersuchungssubstanz *f*
test tube Reagenzglas *n*, Probierglas *n*
test-tube centrifuge Flaschenzentrifuge *f*, Flaschenschleuder *f*
test-tube experiment Reagenzglasversuch *m*
tester 1. Prüfer *m*, Prüfender *m*; 2. Prüfgerät *n*, Prüfvorrichtung *f*, Prüfer *m*, Tester *m*
testing Prüfung *f*, Untersuchung *f*, Testung *f*
testosterone Testosteron *n (Sexualhormon)*
tetanic poison Krampfgift *n*
tetr. = tetragonal
tetraacetate Tetraacetat *n*
tetraatomic vieratomig
tetrabasic 1. vierbasig, vierwertig, vierprotonig *(Säure)*; 2. viersäurig, vierbasig, vierwertig *(Base oder basisches Salz)*
tetrabasicity 1. Tetrabasizität *f (von Säuren)*; 2. Vierwertigkeit *f (von Basen)*
tetraborate $M^I_2B_4O_7$ Tetraborat *n*, Heptoxotetraborat *n*
tetraboric acid Tetraborsäure *f*, Heptoxotetraborsäure *f*
tetrabromofluorescein Tetrabromfluorescein *n*
tetrachloride Tetrachlorid *n*
2,3,7,8-tetrachlorodibenzo-1,4-dioxin *(Tox)* 2,3,7,8-Tetrachlordibenzo-1,4-dioxin *n*, 2,3,7,8-Tetrachloridbenzo-*p*-dioxin *n*, 2,3,7,8-TCDD, TCDD, Tetrachlorodioxin *n*, Dioxin *n*
2,3,7,8-tetrachlorodibenzo-*p*-dioxin *(Tox)* 2,3,7,8-Tetrachlordibenzo-1,4-dioxin *n*, 2,3,7,8-Tetrachloridbenzo-*p*-dioxin *n*, 2,3,7,8-TCDD, TCDD, Tetrachlorodioxin *n*, Dioxin *n*
tetrachloroethylene *Tetrachlorethen *n*, Perchlorethen *n*
tetrachloromethane *Tetrachlormethan *n*, Tetrachlorkohlenstoff *m*, Kohlenstofftetrachlorid *n*, Tetra *m*
tetrachlorosil(ic)ane Tetrachlorsilan *n*
tetracid viersäurig, vierbasig, vierwertig *(Base)*
tetracid vierbasige [vierprotonige, vierwertige] Säure *f*
tetracoordinate(d) vierfach koordinativ gebunden, mit vier koordinativen Bindungen; vierfach koordiniert *(Komplex)*
tetracovalent koordinativ vierwertig
tetracycline *(Biot)* Tetracyclin *n (Antibiotikum und allg. für einen Vertreter einer Antibiotikagruppe)*
tetrad *(Krist)* vierzählig, 4-zählig
tetrad vierwertiges Element *n*; vierwertige Atomgruppe *f*
tetradecanoic acid *Tetradecansäure *f*, Myristinsäure *f*
tetradentate vierzähnig, vierzählig *(Komplexchemie)*
tetraethyllead Tetraethylblei *n*
tetrafluoroethylene Tetrafluorethen *n*, Perfluorethen *n*
tetrafunctional unit tetrafunktionelle Einheit *f*, Q-Einheit *f (Struktureinheit)*
tetragonal system tetragonales Kristallsystem *n*
tetrahalide Tetrahalogenid *n*
tetrahedral arrangement tetraedrische Anordnung *f*, Tetraederanordnung *f*
tetrahedral hybridization tetraedrische Hybridisierung *f*, sp^3-Hybridisierung *f*
tetrahydrate Tetrahydrat *n*, 4-Hydrat *n*
tetrahydrofuran Tetrahydrofuran *n*, Oxolan *n (ein cyclischer Ether)*
tetrakisazo dye Tetrakisazofarbstoff *m*
tetramer Tetramer(e) *n*
tetramethyl biarsine [diarsyl] *Tetramethyldiarsan *n*, Kakodyl *n*
tetramethylen(e)imine Tetramethylenimin *n*, Tetrahydropyrrol *n*, Pyrrolidin *n*
tetramethylethylene glycol Tetramethylethylenglykol *n*, *2,3-Dimethylbutan-2,2-diol *n*, Pinakol *n*
tetramethylsilane Tetramethylsilan *n*, TMS
tetrammine Tetrammin *n*, Tetramminkomplex *m*
tetraoxide Tetroxid *n*
tetrapeptide Tetrapeptid *n (Peptid mit vier Aminosäureeinheiten)*
tetrapositive vierfach positiv (geladen)

tetrapyrrole Tetrapyrrol *n*
tetrasaccharide Tetrasaccharid *n*
tetrasil(ic)ane Tetrasilan *n*
tetrasodium salt Tetranatriumsalz *n*
tetrasubstituted tetrasubstituiert, vierfach substituiert
tetrasubstitution Tetrasubstitution *f*
tetrasulphide Tetrasulfid *n*
tetraterpene *(Bioch)* Tetraterpen *n*
tetraterpenoid *(Bioch)* Tetraterpenoid *n*
tetratomic 1. vieratomig; 2. *s.* tetravalent
tetravalency Tetravalenz *f*, Vierwertigkeit *f*
tetravalent tetravalent, vierwertig
tetrazine Tetrazin *n (heterocyclischer Sechsring mit vier Stickstoffatomen)*
tetrazo compound Tetrazoverbindung *f*
tetritol Tetritol *n (vierwertiger Zuckeralkohol)*
tetrose Tetrose *f (Monosaccharid mit vier Kohlenstoffatomen)*
tetroxide *s.* tetraoxide
textile Gewebe *n*, Stoff *m*
textile auxiliary Textilhilfsmittel *n*, Textilveredlungsmittel *n*
textile chemist Textilchemiker *m*
textile chemistry Textilchemie *f*
textile fibre Textilfaser *f*; Textilfaserstoff *m*
textile finishing Textilveredlung *f*, Textilausrüstung *f*, Appretieren *n*
textiles Textilien *pl*, Textilerzeugnisse *npl*
texture Textur *f*
textured texturiert
TFE *s.* tetrafluoroethylene
Tg (point) *s.* glass-transition temperature
TGA *s.* thermogravimetric analysis
TH *s.* thyrotrop(h)ic hormone
thallic Thallium..., *(i. e. S.)* Thallium(III)-...
thallic oxide Thallium(III)-oxid *n*
thallic sulphate Thallium(III)-sulfat *n*
thallium Tl Thallium *n*
thallium monochloride Thalliummonochlorid *n*, Thallium(I)-chlorid *n*
thallium monoxide Thalliummonoxid *n*, Dithalliummonoxid *n*, Thallium(I)-oxid *n*
thallium trichloride Thalliumtrichlorid *n*, Thallium(III)-chlorid *n*
thallium trioxide Thalliumtrioxid *n*, Thallium(III)-oxid *n*
thallous Thallium..., *(i. e. S.)* Thallium(I)-...
thallous carbonate Thallium(I)-carbonat *n*
thallous chloride Thallium(I)-chlorid *n*, Thalliummonochlorid *n*
thallous hydroxide Thallium(I)-hydroxid *n*
thallous oxide Thallium(I)-oxid *n*, Dithalliummonoxid *n*, Thalliummonoxid *n*
thallous sulphate Thallium(I)-sulfat *n*
theine Coffein *n*, Thein *n*, 1,3,7-Trimethylxanthin *n*
Theisen disintegrator Theisen-Desintegrator *m*, Desintegratorgaswäscher *m* [Tellerwäscher *m*] nach Theisen, Theisen-Wäscher *m*
Thénard blue Thénards Blau *n*, Kobaltblau *n* I *(Cobaltaluminat)*
theobromine Theobromin *n*, 3,7-Dimethylxanthin *n (Alkaloid)*
theophylline Theophyllin *n*, 1,3-Dimethylxanthin *n (Alkaloid)*
theoretical/5 % in excess of the 5 % über dem theoretischen Wert, 5 % mehr als der Theorie entspricht
theoretical/5 % over the 5 % über dem theoretischen Wert, 5 % mehr als der Theorie entspricht
theoretical chemistry theoretische Chemie *f*
theoretical plate [tray] *(Dest)* theoretischer [idealer] Boden *m*, *(i. w. S.)* theoretische Trennstufe *f*
theoretical yield theoretische Ausbeute *f*
theory/in slight excess of über dem theoretischen Wert, etwas mehr als der Theorie entspricht
theory of absolute reaction rate Theorie *f* der absoluten Reaktionsgeschwindigkeit
theory of valency Valenztheorie *f*, Valenzlehre *f*, Wertigkeitstheorie *f*
therapeutic therapeutisch (wirksam)
therapeutic (agent) Therapeutikum *n*, therapeutischer Wirkstoff *m*, therapeutisch wirksamer Stoff *m*, Heilmittel *n*
therapeutic dose therapeutische Dosis *f*, Heildosis *f*
thermal activation thermische Aktivierung *f (einer Reaktion)*
thermal analysis thermische Analyse *f*, Thermoanalyse *f*

thermal capacity Wärmekapazität *f*
thermal conduction Wärmeleitung *f*
thermal conductivity Wärmeleitfähigkeit *f*, (spezifisches) Wärmeleitvermögen *n*, *(quantitativ auch)* Wärmeleitzahl *f*
thermal-conductivity cell [detector] Wärmeleitfähigkeits(mess)zelle *f*, Wärmeleitfähigkeitsdetektor *m*, WLD
thermal cracking thermisches [radikalisches] Kracken *n* [Spalten *n*], Kracken *n* über Radikale
thermal decomposition thermische Zersetzung *f*
thermal degradation thermischer Abbau *m*, Abbau *m* durch Wärme, Wärmeabbau *m*
thermal denaturation *(Bioch)* Hitzedenaturierung *f*
thermal-diffusion method Thermodiffusionsverfahren *n*, (clusiussches) Trennrohrverfahren *n*, Clusius-Dickel-Verfahren *n* *(zur Trennung gasförmiger Isotope)*
thermal diffusivity Temperaturleitfähigkeit *f*, Temperaturleitkoeffizient *m*, Temperaturleitzahl *f*
thermal efficiency thermischer Wirkungsgrad *m*, Wärmewirkungsgrad *m*
thermal energy thermische Energie *f*, Wärmeenergie *f*
thermal equilibrium thermisches [thermodynamisches] Gleichgewicht *n*
thermal excitation thermische Anregung *f*, Temperaturanregung *f*
thermal expansion thermische Ausdehnung *f*, Wärmeausdehnung *f*
thermal flux Wärmestromdichte *f*, Wärmestromintensität *f*
thermal insulation Wärmedämmung *f*
thermal insulator Wärmedämmstoff *m*
thermal ionization thermische Ionisierung *f* [Ionisation *f*]
thermal ionization detector thermoionischer Detektor *m*, TID
thermal lability Thermolabilität *f*, thermische Unbeständigkeit *f*
thermal motion Wärmebewegung *f*
thermal neutron thermisches Neutron *n*
thermal noise *(Anal)* thermisches Rauschen *n*, Wärmerauschen *n*, Johnson-Rauschen *n*
thermal plasma chemistry thermische Plasmachemie *f*
thermal process technology thermische Verfahrenstechnik *f*
thermal radiation Wärmestrahlung *f*, thermische Strahlung *f*
thermal reactor thermischer Reaktor *m*
thermal resistance thermischer Widerstand *m*, Wärme(leit)widerstand *m*
thermal-shock resistance Temperaturwechselbeständigkeit *f*, Widerstand *m* gegen Temperaturwechsel
thermal-shock-resistant temperaturwechselbeständig
thermal spray Thermospray *n* *(Beschichtungstechnik)*
thermal stability thermische Beständigkeit *f* [Stabilität *f*, Festigkeit *f*], Wärmebeständigkeit *f*, Wärmestabilität *f*, Wärmefestigkeit *f* *(von Werkstoffen)*
thermal transfer Wärmeübertragung *f*, Wärmeaustausch *m*, Wärmetransport *m*; Wärmeübergang *m*
thermally stable thermisch beständig [stabil], wärmebeständig, wärmestabil, wärmefest *(Werkstoffe)*
thermionic detector *(Anal)* Thermoionisationsdetektor *m*
thermite process Thermitverfahren *n*, aluminothermisches Verfahren *n*
thermoanalysis *s.* thermal analysis
thermochemical calorie thermochemische Kalorie *f* *(SI-fremde Einheit der Wärmemenge; 1 cal_{th} = 4,184 J)*
thermochemist Thermochemiker *m*
thermochemistry Thermochemie *f*
thermochromic thermochrom
thermochromism Thermochromie *f* *(reversible Farbänderung bei Erwärmung)*
thermocouple Thermopaar *n* *(aus zwei Metallen)*; *(i. w. S.)* Thermoelement *n* *(Thermopaar mit Messgerät)*
thermodiffusion Thermodiffusion *f*
thermoduric thermoresistent, hitzeresistent, hitzebeständig *(Mikroorganismen)*
thermodynamic cycle thermodynamischer Kreisprozess *m*
thermodynamic efficiency thermodynamischer Wirkungsgrad *m*
thermodynamic equation of state

thermodynamische Zustandsgleichung *f*
thermodynamic equilibrium constant (thermodynamische) Gleichgewichtskonstante *f*
thermodynamic potential thermodynamisches Potenzial *n*, (gibbssche) freie Enthalpie *f*, G
thermodynamic temperature thermodynamische [absolute] Temperatur *f*, Kelvin-Temperatur *f*
thermodynamics Thermodynamik *f*
thermoelectron Thermoelektron *n*, Glühelektron *n*
thermogram Thermogramm *n*
thermogravimetric analysis thermogravimetrische Analyse *f*
thermogravimetry Thermogravimetrie *f*
thermolabile thermolabil, thermisch labil [unbeständig], wärmeunbeständig
thermolability Thermolabilität *f*, thermische Unbeständigkeit *f*
thermoluminescence Thermolumineszenz *f*
thermolysis Thermolyse *f*
thermolytic thermolytisch
thermomechanical analysis thermomechanische Analyse *f*, TMA
thermometer Thermometer *n*
thermometric analysis thermometrische Analyse *f*
thermometric conductivity Temperaturleitfähigkeit *f*, Temperaturleitkoeffizient *m*, Temperaturleitzahl *f*
thermometry Thermometrie *f*, Temperaturmessung *f* (mit Hilfe des Thermometers)
thermooxidative degradation thermooxidativer Abbau *m*
thermophiles *(Biot)* thermophile Mikroorganismen *mpl (Wachstumsoptimum 45 bis 55 °C)*
thermophilic microorganisms *(Biot)* thermophile Mikroorganismen *mpl (Wachstumsoptimum 45 bis 55 °C)*
thermoplastic thermoplastisch
thermoplastic Thermoplast *m*, Plastomer(e) *n*, thermoplastischer Kunststoff *m*
thermosetting adhesive *(Kst)* Duroplastklebstoff *m*
thermosetting plastic [resin] Duroplast *m*, Duromer(e) *n*, hitzehärtbarer [wärmehärtbarer] Kunststoff *m*
thermostability Thermostabilität *f*, Hitzestabilität *f (besonders von Enzymen, Vitaminen)*
thermostable thermostabil, hitzestabil *(besonders Enzyme, Vitamine)*
thermostat *v* thermostat(is)ieren, temperieren
thermostat Thermostat *m*
thermostatic bath Badthermostat *m*
thermostatting Thermostatierung *f*
thermotropic liquid crystal thermotroper Flüssigkristall *m* [flüssiger Kristall *m*]
theta solvent Theta-Lösungsmittel *n*
theta temperature Theta-Temperatur *f*
THF *s.* tetrahydrofuran
thiamin(e) Thiamin *n*, Vitamin *n* B_1
thiazine dye Thiazinfarbstoff *m*
thiazole dye Thiazolfarbstoff *m*
thiazolidine Thiazolidin *n*
thick 1. dick; dicht; dickflüssig, zähflüssig, strengflüssig, viskos; trüb(e); 2. *(Geol)* mächtig
thick-tabular *(Krist)* dicktaflig
thicken *v* eindicken, dickflüssig [zähflüssig, viskos] machen *(Flüssigkeiten)*; verdichten; entwässern; eindicken, dickflüssig [zähflüssig, viskos] werden
thickener 1. Eindicker *m*, Eindickmittel *n*, Dickungsmittel *n*, Verdicker *m*, Verdickungsmittel *n*, Verdickungszusatz *m*; 2. Eindicker *m*, Eindickapparat *m*
thickener-clarifier *(Hyd)* Kläreindicker *m*
thickening agent *s.* thickener 1.
thickness 1. Dicke *f*; Dichte *f*; Dickflüssigkeit *f*, Zähflüssigkeit *f*, Viskosität *f*; 2. *(Geol)* Mächtigkeit *f*
thief sampling Probenahme *f* mittels Probenstechers
Thiele melting point tube Schmelzpunkt(bestimmungs)apparat *m* nach Thiele
thin *v* verdünnen
thin dünn; dünnflüssig
thin film cells Dünnschichtzellen *fpl*
thin film deposition Abscheiden *n* dünner Filme
thin film solar cells Dünnschichtzellen *fpl*
thin film technology Dünnfilmtechnik *f*

thin-layer chromatography Dünnschichtchromatographie *f*, DC
thin mercury film electrode *(Ech)* Quecksilberfilmelektrode *f (gebräuchlich ist auch mercury-film electrode)*
thin section *(Min)* Dünnschliff *m*
thin-tabular *(Krist)* dünntaflig
thinner *(Anstr)* Verdünner *m*, Verdünnungsmittel *n*, Verdünnung *f (zur Einstellung auf die gewünschte Verarbeitungsviskosität)*
thinning agent *(Anstr)* Verdünner *m*, Verdünnungsmittel *n*, Verdünnung *f (zur Einstellung auf die gewünschte Verarbeitungsviskosität)*
thio derivative Thioderivat *n*
thioacetal Thioacetal *n*
thioacetic acid Thioessigsäure *f*
thioacid Thiosäure *f*
thioalcohol *Thiol *n*, Thioalkohol *m*
thioaldehyde Thioaldehyd *m*
thioamide Thioamid *n*
thiobarbituric acid Thiobarbitursäure *f*, TBA
thiocarbamide *s.* thiourea
thiocarbanilide Thiocarbanilid *n*, *N,N′*-Diphenylthioharnstoff *m*
thiocarbonate $M^I_2[CS_3]$ Thiocarbonat *n*
thiocarbonic acid Thiokohlensäure *f*
thiocarboxylic acid Thiocarbonsäure *f*
thiocyanate M^ISCN Thiocyanat *n*
thiocyanic acid Thiocyansäure *f*
thiocyanogen number [value] *(Lebm)* Rhodanzahl *f*, rhodanometrische Iodzahl *f*, RhZ *(Kennzahl der Fette und fetten Öle)*
thioglycollic acid Thioglykolsäure *f*, Mercaptoessigsäure *f*
thioketal Thioketal *n*
thiol *(org Ch)* R–SH *Thiol *n*
thiolysis Thiolyse *f*, thiolytische [thioklastische] Spaltung *f (Fettsäureabbau)*
thion-thiol rearrangement Thion-Thiol-Umlagerung *f*
thionation 1. Schwefeln *n*, Schwefelung *f*; 2. *(Farb)* Verschwefeln *n*, Verschwefelung *f*
thionyl chloride Thionylchlorid *n*
thiophene Thiophen *n*
thiophenol Thiophenol *n*
thiophosgene Thiophosgen *n*, Thiocarbonyldichlorid *n*
thiophosphate Thiophosphat *n*
thiosemicarbazide Thiosemicarbazid *n*, Hydrazinthiocarbonsäureamid *n*
thiosulphate $M^I_2S_2O_3$ Thiosulfat *n*
thiosulphite $M^I_2S_2O_2$ Thiosulfit *n*
thiourea Thioharnstoff *m*, Thiocarbamid *n*
third-law entropy absolute molare Entropie *f*
third law of thermodynamics dritter Hauptsatz *m* der Thermodynamik, nernstscher Wärmesatz *m*, nernstsches Wärmetheorem *n*
third-order reaction Reaktion *f* dritter Ordnung
third-stage intercalation compound Interkalationsverbindung *f* [Einlagerungsverbindung *f*] 3. Stufe
third-stage treatment *(Hyd)* dritte Reinigungsstufe *f*, weitergehende Abwasserbehandlung *f*, Nachreinigung *f*
thixotrope *(Koll)* thixotropes Gel *n*
thixotropic agent Thixotropier(ungs)-mittel *n*, thixotroper [Thixotropie erzeugender] Stoff *m*
thixotropy Thixotropie *f*
THM *s.* trihalomethane
Thomas process Thomas-Verfahren *n*, Thomas-Konverterverfahren *n*, basisches Windfrischverfahren *n* [Bessemer-Verfahren *n*]
Thomas steel Thomas-Stahl *m*, Thomas-Konverterstahl *m*
thorate $M^I_2ThO_3$ Thorat *n*
thoria Thoriumoxid *n*, Thoriumdioxid *n*
thoriate *v* thorieren, mit einer Thoriumschicht versehen
thoric Thorium…
thorium Th Thorium *n*
thorium chloride Thoriumchlorid *n*
thorium decay series *s.* thorium series
thorium dioxide Thoriumdioxid *n*, Thoriumoxid *n*
thorium oxalate Thoriumoxalat *n*
thorium series Thorium(zerfalls)reihe *f*
thorium sulphate Thoriumsulfat *n*
thorough mixing Durchmischen *n*
Thorpe reaction Thorpe-Reaktion *f (zur Darstellung von Cycloalkanonen)*
thousandth normal solution 0,001 n-Lösung *f*, n/1000-Lösung *f*
thread 1. Faden *m*; 2. Gewinde *n*; Gewindegang *m*
threadlike molecule Fadenmolekül *n*

three-bladed dreischauflig, Dreiblatt... *(Rührer)*
three-body collision Dreierstoß *m*
three-carbon acid C_3-Carbonsäure *f*, Fettsäure *f* mit drei C-Atomen
three-carbon compound C_3-Verbindung *f*, C_3-Körper *m*
three-carbon tautomerism Dreikohlenstofftautomerie *f*
three-centre bond Dreizentrenbindung *f*
three-component system Dreikomponentensystem *n*, Dreistoffsystem *n*, ternäres System *n*
three-coordinate(d) dreifach koordinativ gebunden, mit drei koordinativen Bindungen; dreifach koordiniert *(Komplex)*
three-dimensional network dreidimensionales [räumliches] Netzwerk *n*
three-fold axis of symmetry *(Krist)* dreizählige Achse *f* [Drehachse *f*]
three-fold degenerate *(phys Ch)* dreifach entartet [degeneriert]
three-membered ring dreigliedriger Ring *m*, Drei(er)ring *m*
three-neck(ed) flask Dreihalskolben *m*
three-roll(er) mill Dreiwalzenmühle *f*, Dreiwalzenstuhl *m*, Dreiwalze *f*
three-stage pump dreistufige Pumpe *f*
three-step reaction Dreistufenreaktion *f*, dreistufige Reaktion *f*
three-way cock [stopcock, tap] *(Lab)* Dreiwegehahn *m*
threonine Threonin *n*, Thr, α-Amino-β-hydroxybuttersäure *f*
threshold 1. Schwelle *f* *(Physik)*; 2. Schwellenwert *m*, Grenzwert *m*
threshold limit value maximale Arbeitsplatzkonzentration *f*, MAK
threshold of odour *(Hyd)* Geruchsschwellenkonzentration *f*, RSK, Geruchsschwelle *f*
threshold of taste *(Hyd)* Geschmacksschwellenkonzentration *f*, GSK, Geschmacksschwelle *f*
throttle *v* drosseln
throttle Drossel(ung) *f*, Drosselpfropfen *m*
throughput 1. Durchsatz *m*, Durchsatzmenge *f*; 2. *(Dest)* Belastung *f*
throw *v* **down** (aus)fällen, niederschlagen, abscheiden
thrust bearing Axiallager *n*
thulium Tm Thulium *n*
thyme camphor *s.* thymol
thyme oil Thymianöl *n (von Thymus vulgaris L. und T. zygis L.)*
thymidylic acid *(Bioch)* Thymidylsäure *f*
thymol Thymol *n*, Thymiancampher *m*, Thymiansäure *f*, 2-Hydroxy-4-isopropyltoluen *n*
thymol blue Thymolblau *n*, Thymolsulfophthalein *n (pH-Indikator)*
thymolphthalein Thymolphthalein *n (pH-Indikator)*
thymolsulphon(e)phthalein *s.* thymol blue
thyroglobulin Thyreoglobulin *n*
thyroid hormone Schilddrüsenhormon *n*
thyrotrop(h)ic hormone thyreotropes Hormon *n*, Thyreotropin *n*, TSH
thyrotrop(h)in *s.* thyrotrop(h)ic hormone
thyrotrop(h)in-releasing factor [hormone] *Thyroliberin *n*, Thyreotropin-Releasinghormon *n*, TRH
thyroxin(e) Thyroxin *n*, Tetraiodthyronin *n*
TID *s.* thermal ionization detector
tie line Gleichgewichtslinie *f*, Kon(n)ode *f (im Gleichgewichtsdiagramm)*
Tiffeneau reaction [rearrangement] Tiffeneau-Reaktion *f*, Tiffeneau-Umlagerung *f (von α-Aminoalkoholen in Aldehyde und Ketone)*
tight ion pair Kontaktionenpaar *n*, inneres Ionenpaar *n*
tiglic acid Tiglinsäure *f*, **trans*-2-Methylbut-2-ensäure *f*
tile 1. Fliese *f*, Platte *f*; Dachziegel *m*; Hohlziegel *m*; 2. *(Dest)* Halbzylinder *m (Füllkörper)*
timber preservation [proofing] Holzschutz *m*, Holzkonservierung *f*
time constant Zeitkonstante *f*
time course zeitlicher Verlauf *m* [Ablauf *m*]
time-dependent (non-newtonian) fluid zeitabhängige (nicht-newtonsche) Flüssigkeit *f*, rheoinstabile Flüssigkeit *f*
time domain *(Spektr)* Zeitdomäne *f*
time-independent (non-newtonian) fluid zeitunabhängige (nicht-newtonsche) Flüssigkeit *f*, rheostabile Flüssigkeit *f*

time-of-flight mass spectrometer Flugzeit-Massenspektrometer *n*, dynamisches Spektrometer *n*
time-of-flight secondary ion mass spectroscopy Flugzeit-Sekundärionen-Massenspektrometrie *f*
time of residence Verweilzeit *f*, Aufenthaltszeit *f*, Retentionszeit *f*, Rückhaltezeit *f*, Haltezeit *f*, Standzeit *f*, Durchlaufzeit *f*
time of settling Absetzdauer *f*, Absetzzeit *f*
time reaction Zeitreaktion *f*
time-resolved spectroscopy zeitaufgelöste Spektroskopie *f*
time-temperature-transformation diagram Zeit-Temperatur-Umwandlungsdiagramm *n*, Zeit-Temperatur-Umwandlungsschaubild *n*, ZTU-Diagramm *n*, ZTU-Schaubild *n*
tin *v* 1. verzinnen; 2. *(Lebm)* eindosen, in Dosen [Büchsen] konservieren; in Dosen verpacken
tin-plate *v* verzinnen, *(i. e. S.)* elektrochemisch [galvanisch] verzinnen
tin Sn Zinn *n*
tin alloy Zinnlegierung *f*
tin bath 1. Zinnschmelze *f (beim Feuerverzinnen)*; 2. Zinnelektrolyt *m*, Verzinnungselektrolyt *m (Galvanotechnik)*; 3. Zinnbad *n (Gefäß mit Schmelze oder Elektrolyt)*
tin bronze Zinnbronze *f (1. Kupfer-Zinn-Legierung; 2. Zinn(IV)-sulfid)*
tin cry Zinngeschrei *n*, Zinnschrei *m*
tin dichloride Zinndichlorid *n*, Zinn(II)-chlorid *n*
tin dioxide Zinndioxid *n*, Zinn(IV)-oxid *n*
tin disease *s.* tin pest
tin disulphide Zinndisulfid *n*, Zinn(IV)-sulfid *n*
tin hydride Zinnwasserstoff *m*, Stannan *n*, Zinn(IV)-hydrid *n*
tin monosulphide Zinnmonosulfid *n*, Zinn(II)-sulfid *n*
tin pest [plague] Zinnpest *f*
tin plating Verzinnen *n*, *(i. e. S.)* elektrochemisches [galvanisches] Verzinnen *n*
tin stone *(Min)* Zinnstein *m*, Cassiterit *m (Zinn(IV)-oxid)*
tin tetrachloride Zinntetrachlorid *n*, Zinn(IV)-chlorid *n*
tincal *(Min)* Tinkal *m*, Borax *m (Natriumtetraborat-10-Wasser)*
tincture *(Pharm)* Tinktur *f*
tinge *v* (an)färben, tönen
tinge Anflug *m*, Stich *m*, Tönung *f*
tint *v* färben, *(i. e. S.)* anfärben, (ab)tönen, nuancieren
tint Färbung *f*, *(i. e. S.)* Farbton *m*, Farbtönung *f*, Farbschattierung *f*, Nuance *f*, Stich *m*
tip *v* abkippen, verkippen *(auf Halde)*
tip *(Aufb)* Halde *f*
Tiselius method Tiselius-Elektrophorese *f*, (träger)freie Elektrophorese *f*, Methode *f* der wandernden Grenzflächen
Tishchenko reaction Tischtschenko-Reaktion *f (zur Darstellung von Estern aus Aldehyden)*
tissue 1. Gewebe *n*, (textiler) Stoff *m*, Textilgewebe *n*; 2. *(Bio)* Gewebe *n*
tissue culture *(Bio)* Gewebekultur *f*, Gewebezüchtung *f*
tissue engineering Gewebeentwicklung *f*, Gewebezüchtung *f (synthetisch)*
tissue fluid Gewebeflüssigkeit *f*
tissue sample *(Bioch)* Gewebeprobe *f*
titanate Titanat *n*
titania Titan(IV)-oxid *n*, Titandioxid *n*
titanic Titan..., *(i. e. S.)* Titan(IV)-...
titanic acid Titansäure *f*
titanic chloride Titan(IV)-chlorid *n*, Titantetrachlorid *n*
titanic oxide Titan(IV)-oxid *n*, Titandioxid *n*
titanic sulphate Titan(IV)-sulfat *n*
titaniferous titanhaltig
titanium Ti Titan *n*
titanium carbide Titancarbid *n*
titanium diboride Titandiborid *n*
titanium dichloride Titandichlorid *n*, Titan(II)-chlorid *n*
titanium dioxide Titandioxid *n*, Titan(IV)-oxid *n*
titanium disulphide Titandisulfid *n*, Titan(IV)-sulfid *n*
titanium hydride Titanhydrid *n*
titanium monoxide Titanmonoxid *n*, Titan(II)-oxid *n*
titanium nitride Titannitrid *n*, Titan(III)-nitrid *n*
titanium tetrachloride Titantetrachlorid *n*, Titan(IV)-chlorid *n*

titanium trichloride Titantrichlorid *n*, Titan(III)-chlorid *n*
titanium white Titanweiß *n*
titanometry Titanometrie *f*
titanous Titan..., *(i. e. S.)* Titan(III)-...
titanous chloride Titan(III)-chlorid *n*, Titantrichlorid *n*
titer *s.* titre
titrable *s.* titratable
titrant Titrationsmittel *n*, Titrator *m*, Titrant *m*, Titrans *n*
titratable titrierbar
titrate *v* titrieren
titration Titration *f*, Titrieren *n*
titration curve Titrationskurve *f*
titration error Titrierfehler *m*, Titrationsfehler *m*
titration flask Titrierkolben *m*
titrator Titriergerät *n*, Titriervorrichtung *f*, Titrator *m*
titre 1. Titer *m (Gehalt einer Maßlösung)*; 2. *(Text)* Titer *m (Kennzeichnung der Feinheit von Garnen)*
titrimeter Titrimeter *n*
titrimetric titrimetrisch, volumetrisch, maßanalytisch
titrimetric analysis 1. (einzelne) titrimetrische [volumetrische] Analyse *f*, Titrieranalyse *f*, Maßanalyse *f*; 2. *s.* titrimetry
titrimetric factor *(Anal)* Faktor *m*, Korrekturfaktor *m*, Normalitätsfaktor *m*
titrimetric method *(Anal)* Titrationsmethode *f*, titrimetrische Methode *f*
titrimetry Titrimetrie *f*, Volumetrie *f*; Maßanalyse *f*, Titrieranalyse *f (Zweig der Analytik)*
TLC *s.* thin-layer chromatography
TLV *s.* threshold limit value
TMA *s.* thermomechanical analysis
TMFE *s.* thin mercury film electrode
TMS *s.* tetramethylsilane
T.N.T. *s.* trinitrotoluene
tobacco alkaloid Tabakalkaloid *n*
TOC *s.* total organic carbon
TOC content *(Hyd)* Gehalt *m* an organisch gebundenem Kohlenstoff, TOC-Gehalt *m*
TOD *s.* total oxygen demand
TOF mass spectrometer Flugzeit-Massenspektrometer *n*
TOF-SIMS *s.* time-of-flight secondary ion mass spectroscopy
TOF spectrometer Flugzeit-Massenspektrometer *n*
tol. *s.* toluene
tolan Tolan *n*, *Diphenylethin *n*
tolerance 1. Toleranz *f*, zulässige Abweichung *f*; 2. *(Tox)* Toleranz(dosis) *f*, Toleranzwert *m*, duldbare [zulässige, zugelassene] Höchstmenge *f* [Restmenge *f*]
tolerance limit *(Hyd)* Grenzwert *m*
Tollens' reagent Tollens-Reagens *n (ammoniakalische Silbernitratlösung)*
toluene Toluen *n*, Toluol *n*, *Methylbenzen *n*
toluene diisocyanate Toluendiisocyanat *n*, TDI
***p*-toluenesulphonate** *p*-Toluensulfonat *n*, *p*-Toluensulfonsäureester *m*, Tosylester *m*, Tosylat *n*
***p*-toluenesulphonic acid** *p*-Toluensulfonsäure *f*, 1-Methylbenzensulfonsäure *f*
***p*-toluenesulphonyl chloride** *p*-Toluensulfonylchlorid *n*, *p*-Toluensulfo(nsäure)chlorid *n*, Tosylchlorid *n*
toluidine Toluidin *n*, Aminotoluen *n*
toluylene red Toluylenrot *n*, Neutralrot *n (ein Redoxindikator)*
tombac Tombak *m(n) (Cu-Zn-Legierung mit 30 % oder mehr Zn)*
tombak Tombak *m(n) (Cu-Zn-Legierung mit 30 % oder mehr Zn)*
tonka-bean camphor Tonkabohnencampher *m*, Cumarin *n*, 1,2-Benzopyron *n*, *2-Oxo-1,2-chromen *n*
tons per annum *s.* tons per year
tons per day Tonnen *fpl* je Tag, t/d, *(inkorrekt:)* Tagestonnen, tato, Tato
tons per year Tonnen *fpl* je Jahr, t/Jhr, *(inkorrekt:)* Jahrestonnen, jato, Jato
tool steel Werkzeugstahl *m*
top *v* 1. herausdestillieren; 2. *(Erdöl)* toppen; 3. *(Text)* überfärben, nachfärben, nachdecken, schönen
top 1. Oberteil *n*; Spitze *f*; Deckel *m*; 2. Gicht *f (des Hochofens)*
top-blown oxygen converter process Sauerstoff(auf)blasverfahren *n*, Sauerstoff-Frischverfahren *n*, Aufblas(e)verfahren *n*, Blasstahlverfahren *n*, Oberwindfrischverfahren *n*
top-feed(ing) filter Filter *n* mit Obenaufgabe

top fermentation Obergärung *f*
top-fermented obergärig
top-fermenting obergärig
top fraction *(Erdöl)* Kopffraktion *f*
top of the column *(Dest)* Kolonnenkopf *m*
top product 1. *(Dest)* Kopfprodukt *n*, Überkopfprodukt *n*; 2. *(Erdöl)* Toppprodukt *n*
top scum *(Hyd)* schaumartige Schwimmschicht *f*
top strand Obertrum *m(n)*, Volltrum *m(n)*, Lasttrum *m(n) (eines Stetigförderers)*
top-suspended centrifuge Hängezentrifuge *f*, hängende Zentrifuge *f*
topcoat 1. Deckschicht *f*; 2. *(Anstr)* Deckanstrich *m*, Schlussanstrich *m*
topical anaesthetic Lokalanästhetikum *n*, Mittel *n* zur örtlichen Schmerzausschaltung
topochemical reaction topochemische Reaktion *f*
topochemistry Topochemie *f*
topotactic reaction topotaktische Reaktion *f*
topotaxy Topotaxie *f*
topped crude (petroleum) getopptes [abgetopptes, reduziertes] Rohöl *n* [Öl *n*], Topped Crude *n*, Reduced Oil *n*
topping 1. Toppen *n*, Toppung *f*, Toppdestillation *f*, atmosphärische [unter atmosphärischem Druck arbeitende] Destillation *f*, Normaldruckdestillation *f*; 2. *(Text)* Überfärben *n*, Überfärbung *f*, Nachfärben *n*, Nachfärbung *f*
tops *(Erdöl)* Toppprodukt *n*
torch Brenner *m*; Lötlampe *f*; Schweißbrenner *m*
torque Drehmoment *n*, Moment *n*
torrefaction *(Lebm, Aufb)* Rösten *n*, Röstung *f*
torrefy *v (Lebm, Aufb)* rösten
torsion Torsion *f*, Verdrehung *f*, Drillung *f*, Verwindung *f*
torsion angle Torsionswinkel *m*, Konformationswinkel *m*, Rotationswinkel *m*, Diederwinkel *m*
torsion balance Torsionswaage *f*, Drehwaage *f*
torsional mode [oscillation] *(Spektr)* Torsionsschwingung *f*, Twisting-Schwingung *f*, τ-Schwingung *f*
torsional strain Torsionsspannung *f*, Pitzer-Spannung *f (in Molekülen)*
tortuosity factor Krümmungsfaktor *m*, Labyrinthfaktor *m (Diffusionstheorie)*
tosyl chloride Tosylchlorid *n*, *p*-Toluensulfonylchlorid *n*, *p*-Toluensulfo(nsäure)chlorid *n*
tosylate Tosylat *n*, Tosylester *m*, *p*-Toluensulfonsäureester *m*, *p*-Toluensulfonat *n*
total acid number *Maßzahl für die freie Gesamtsäure in mg KOH je Gramm Probesubstanz*
total anion (content) *(Hyd)* Gesamtanionenwert *m*, GAW, Gesamtanionen *npl*
total base number *Maßzahl für den Alkaligehalt, ausgedrückt in mg KOH als Äquivalent für die zur Neutralisation aufgewandte Säuremenge je Gramm Probesubstanz*
total carbon *(Hyd)* Gesamtkohlenstoff *m*
total cation (content) *(Hyd)* Gesamtkationenwert *m*, GKW, Gesamtkationen *npl*
total concentration Gesamtkonzentration *f*
total condensation *(Dest)* Totalkondensation *f*
total dissolved solids 1. *(Hyd)* Gehalt *m* an echt gelösten Stoffen [Wasserinhaltsstoffen], Gesamtmenge *f* an (echt) gelösten Stoffen; 2. *(Hyd)* Gesamtsalzgehalt *m*
total dry solids Gesamttrockensubstanz *f*, Gesamttrockenmasse *f*
total electron spin angular momentum Gesamtspindrehimpuls *m*, Gesamteigendrehimpuls *m*
total-energy equation Energiebilanzgleichung *f*
total hardness (of water) *(Hyd)* Gesamthärte *f*, GH
total heat of solution ganze [totale] Lösungswärme *f* [Lösungsenthalpie *f*]
total ionic content *(Hyd)* Gesamtionengehalt *m*
total order (of reaction) Gesamtreaktionsordnung *f*
total organic carbon *(Hyd)* Gesamtmenge *f* an organisch gebundenem Kohlenstoff, TOC
total organic halides *(Hyd)* Gesamtgehalt *m* an Organohalogenverbindungen

total oxygen demand *(Hyd)* Gesamtsauerstoffbedarf *m*
total quantum number Hauptquantenzahl *f*
total rate of reaction Gesamtreaktionsgeschwindigkeit *f*
total reaction Gesamtreaktion *f*
total reflection Totalreflexion *f*
total reflux operation Fahrweise *f* [Arbeitsweise *f*] mit totalem Rücklauf, Fahrweise *f* ohne Destillatabnahme
total solids 1. *(Lebm)* Gesamttrockensubstanz *f*, Gesamttrockenmasse *f*; 2. *(Hyd)* Gesamtfeststoffgehalt *m*, Gesamtfeststoffmenge *f*
total suspended solids (level) *(Hyd)* Gesamtgehalt *m* an suspendierten Feststoffen, Gesamtschwebstoffgehalt *m*, Gesamtschwebstoffe *mpl*
total synthesis Totalsynthese *f*
total throughput Gesamtdurchsatz *m*
total trihalomethanes *(Hyd)* Gesamtgehalt *m* an Trihalo(gen)methanen
total water hardness *(Hyd)* Gesamthärte *f*, GH
total yield Gesamtausbeute *f*
tough fracture Zähbruch *m*
toughen *v* 1. zäh machen; 2. *(Glas)* verfestigen; zäh werden
toughened polystyrene schlagfestes Polystyrol *n*
tower 1. Turm *m*; 2. *(Dest)* Säule *f*, Kolonne *f*
tower fermentation *(Biot)* Turmfermentation *f*
tower fermenter [reactor] *(Biot)* Turmreaktor *m*, Turmfermenter *m*
tower scrubber Turmwäscher *m*, Waschturm *m*, Gaswaschturm *m*, Skrubber *m*
town gas Stadtgas *n*
town refuse Stadtmüll *m*
TOX *s.* total organic halides
toxic toxisch, giftig • **be toxic to** giftig sein für
toxic *s.* toxicant
toxic action toxische Wirkung *f*, Gift(ein)wirkung *f*
toxic agent *s.* toxicant
toxic dose toxische [giftig wirkende] Dosis *f*, Giftdosis *f*
toxic principle toxisches Prinzip *n*, giftiger Bestandteil *m*
toxic substance *s.* toxicant
toxic waste(s) Giftmüll *m*
toxicant Giftstoff *m*, Gift *n*, giftiger Wirkstoff *m*, giftige Substanz *f*, Toxikum *n*
toxicity Toxizität *f*, Giftigkeit *f*
toxicologic(al) toxikologisch
toxicologic(al) chemistry toxikologische Chemie *f*
toxicologist Toxikologe *m*, Giftkundiger *m*, Giftsachverständiger *m*
toxicology Toxikologie *f*, Giftkunde *f*
toxification Vergiftung *f*
toxify *v* vergiften
toxin Toxin *n*
toxophore toxophore Gruppe *f* *(die Giftwirkung auslösender Anteil eines Toxins nach Ehrlichs Seitenkettentheorie)*
toxophoric group toxophore Gruppe *f* *(die Giftwirkung auslösender Anteil eines Toxins nach Ehrlichs Seitenkettentheorie)*
2,4,5-TP *s.* 2,4,5-trichlorophenoxypropionic acid
tpa *(tons per annum) s.* tons per year
TPA *s.* 1. terephthalic acid; 2. two-photon absorption
tpd *s.* tons per day
TPP = thiamine pyrophosphate
tpy *s.* tons per year
trace 1. *(Anal)* Spur *f* *(*10^2 *bis* 10^{-4} *ppm)*; 2. Schreibspur *f*, Spur *f*, Kurve *f*
trace analysis Spurenanalyse *f*
trace component [constituent] Spurenbestandteil *m*
trace element *(Landw)* Spurenelement *n*, Mikronährstoff *m*, Mikroelement *n*, Hochleistungselement *n*, Spurennährstoff *m*
trace metal Spurenmetall *n*
tracer Tracer *m*, Indikator *m*, Tracersubstanz *f*, Indikatorsubstanz *f*, Markierungssubstanz *f*, *(i. e. S.)* Indikatorisotop *n*, Leitisotop *n*, Zusatzisotop *n*
tracer method Tracermethode *f*, Indikatormethode *f*, *(i. e. S.)* Leitisotopenmethode *f*, Isotopenmethode *f*
trade effluent *(Hyd)* Industrieabwasser *n* *(in Kanalisation oder Vorfluter einzuleiten)*
trade name Handelsname *m*, Handelsbezeichnung *f*, kommerzieller Name *m*

trade waste 1. Industriemüll *m*, Industrieabfälle *mpl*, (feste) industrielle Abprodukte *npl*; 2. Industrieabwasser *n*, industrielles Abwasser *n*
tragacanth gum Tragantgummi *n*, Tragant *m (von Astragalus-Arten)*
tranquil(l)izer *(Pharm)* Tranquilizer *m*, Ataraktikum *n*, Anxiolytikum *n*, Psychosedativum *n*
tranquillizing drug *(Pharm)* Tranquilizer *m*, Ataraktikum *n*, Anxiolytikum *n*, Psychosedativum *n*
trans *trans-...*, *trans*-ständig, in *trans*-Stellung befindlich • **be trans** trans [in *trans*-Stellung] stehen, *trans*-ständig (angeordnet) sein
trans conformation *trans*-Konformation *f (Torsionswinkel 0°)*
transamination Transaminierung *f*
transcription *(Bioch)* Transkription *f*, Umschreibung *f (Übertragung genetischer Informationen durch Enzyme)*
transduction *(Bioch)* Transduktion *f (Übertragung genetischer Informationen durch Bakteriophagen)*
transesterification Umesterung *f*
transesterify *v* umestern
transfection Transfektion *f (Einschleusen fremder Nucleinsäure in Zellen)*
transfer *v* 1. übertragen, abgeben, transportieren *(Elektronen, Ionen)*; überführen, übertragen, bringen *(Substanzen in andere Gefäße)*; transportieren *(Güter)*; übertragen, transportieren *(Wärme)*; 2. *(Hyd)* eintragen *(Sauerstoff)*; übergehen *(Elektronen, Substanzen, Wärme)*
transfer 1. Übertragung *f*, Abgabe *f*, Transport *m (von Elektronen, Ionen)*; Überführen *n*, Übertragen *n (von Substanzen)*; Transport *m (von Gütern)*; Übertragung *f*, Transport *m (von Wärme)*; 2. *(Hyd)* Eintrag *m (von Sauerstoff)*; Übergang *m (von Elektronen, Substanzen, Wärme)*
transfer coefficient 1. Austauschkoeffizient *m (Stoffaustausch)*; 2. *(Ech)* Durchtrittsfaktor *m*
transfer of heat Wärmeübertragung *f*
transfer pipette Vollpipette *f*
transfer rate Austauschgeschwindigkeit *f*, Austauschrate *f (Stoffaustausch)*
transfer reaction Übertragungsreaktion *f (Polymerisation)*
transfer resistance Übergangswiderstand *m (Stoffübertragung)*
transfer ribonucleic acid Transfer-Ribonucleinsäure *f*, Transfer-RNA *f*, t-RNA *f*, lösliche [aktivierende] Ribonucleinsäure *f*, l-RNA *f*
transference Überführung *f*, Übertragung *f*; Übergang *m*
transference number *(Ech)* Überführungszahl *f*
transform *v* umwandeln, überführen; sich umwandeln, übergehen
transformation 1. Umwandlung *f*, Überführung *f*; Übergehen *n*, Übergang *m*, Verwandlung *f*, Umwandlung *f*; 2. *(Kern)* Umwandlung *f*, Kernumwandlung *f*; 3. *(phys Ch)* Transformation *f (von Energieformen)*; 4. *(Bioch)* Transformation *f (Genübertragung durch Einfügen isolierter DNA in Zellen)*; 5. Umlagerung *f (von Atomen oder Atomgruppen)* • **undergo transformation** sich umwandeln, übergehen
transformation constant *(Kern)* Zerfallskonstante *f*
transformation of materials Stoff(um)wandlung *f*, Stoffumsetzung *f*, stoffliche Umwandlung *f* [Umsetzung *f*]
transformation range *(Glas)* Transformationsintervall *n*
transglycosidation Transglykosylierung *f*, Transglykosidierung *f (enzymatische Übertragung einer Glykosylgruppe)*
transglycosylation Transglykosylierung *f*, Transglykosidierung *f (enzymatische Übertragung einer Glykosylgruppe)*
transient unbeständig, instabil *(z. B. Radikale)*
transient state *(Bioch)* Übergangsphase *f*, Übergangszustand *m*
transit peptide *s.* signal peptide
transit time Durchgangszeit *f*, Durchlaufzeit *f (durch eine Behandlungsanlage)*
transition Übergehen *n*, Übergang *m*, Umwandlung *f*, Metamorphose *f*; Umschlag *m*, Umschlagen *n (eines Indikators)*; Transition *f (Austausch von DNA-Bausteinen)*
transition element Übergangselement *n*, Übergangsmetall *n*

transition entropy Phasenumwandlungsentropie *f*
transition interval 1. *(phys Ch)* Transformationsintervall *n*; 2. *(Kst)* Einfriergebiet *n*; 3. Umschlagsintervall *n*, Umschlagsgebiet *n*, Übergangsbereich *m (eines Farbindikators)*; Übergangsintervall *n*, Umwandlungsintervall *n (eines Doppelsalzes)*
transition metal *s.* transition element
transition point *(Krist, phys Ch)* Umwandlungspunkt *m*; *(bei inkongruent schmelzenden Verbindungen)* peritektischer Punkt *m*
transition series Übergangsreihe *f (im Periodensystem)*
transition state Übergangszustand *m*, aktivierter Komplex *m (Kinetik)*
transition temperature 1. Umwandlungstemperatur *f*, Übergangstemperatur *f*; 2. *(Kst)* Einfriertemperatur *f*; 3. Sprungtemperatur *f*, Sprungpunkt *m (eines Supraleiters)*
transition-type system *(phys Ch)* Peritektikum *n*
transitory state Zwischenzustand *m*
translation 1. *(Bioch)* Translation *f (Umsetzung genetischer Informationen in eine Aminosäuresequenz)*; 2. *(Krist)* Translation *f*, Parallelverschiebung *f*
translational degree of freedom Translationsfreiheitsgrad *m*
translational diffusion coefficient translatorischer Diffusionskoeffizient *m*
translational partition function Translationsverteilungsfunktion *f*, Verteilungsfunktion *f* der Translationsenergie
translucence 1. Lichtdurchlässigkeit *f*, *(unvollkommene)* Transparenz *f*; 2. *(Ker)* Durchscheinbarkeit *f*
translucency 1. Lichtdurchlässigkeit *f*, *(unvollkommene)* Transparenz *f*; 2. *(Ker)* Durchscheinbarkeit *f*
translucent lichtdurchlässig, *(unvollkommen)* transparent, durchscheinend
transmembrane protein *(Bioch)* Transmembranprotein *n*
transmetalation *(org Ch)* Transmetallierung *f*
transmission 1. Übersendung *f*, Versand *m*, Beförderung *f*; 2. *(Tech)* Getriebe *n*; Transmission *f (zur Drehmomentübertragung)*; 3. *s.* transmittance 1.; 4. *s.* transmission ratio
transmission electron energy loss spectrometry Transmissions-Elektronenenergieverlust-Spektrometrie *f*
transmission electron microscope Transmissionselektronmikroskop *n*
transmission electron microscopy Transmissionselektronmikroskopie *f*
transmission ratio Durchlassgrad *m*, Transmissionsgrad *m (Optik)*
transmissivity Durchlasszahl *f (Wärmestrahlung)*
transmittance 1. Durchlässigkeit *f (Optik)*; 2. *s.* transmission ratio
transmittancy 1. Intensitätsverhältnis *n (Kalorimetrie)*; 2. *s.* transmittance 1.
transmitted-light technique Durchlichtmethode *f (Mikroskopie)*
transmitted radiation *(Spektr)* (hin-) durchgelassene Strahlung *f*
transmutation Umwandlung *f*, Stoffumwandlung *f*, Elementumwandlung *f*, Transmutation *f*
transmute *v* umwandeln *(Elemente)*; sich umwandeln
transparency Transparenz *f*, Durchsichtigkeit *f*, Lichtdurchlässigkeit *f*; Strahlendurchlässigkeit *f*
transparent transparent, durchsichtig, lichtdurchlässig
transport number *(Ech)* Überführungszahl *f*
transport reaction *(phys Ch, Bioch)* Transportreaktion *f*
transport reactor Transportreaktor *m*, Flugstromreaktor *m*, Flugwolkenreaktor *m*, Flugstaubwolkenreaktor *m*
transposition Umlagerung *f*
transtactic trans-taktisch, tt *(Polymer)*
transtacticity trans-Taktizität *f (von Polymeren)*
transuranian [transuranic, transuranium] element Transuran *n*
trap *v* 1. auffangen, abfangen, festhalten; einschließen *(z. B. Gase in Kristallen)*; 2. *(Hyd)* zurückhalten *(Schmutzteilchen in der Filterschicht)*
trap 1. Auffanggefäß *n*, Sammelgefäß *n*; Geruchsverschluss *m*, Wasserverschluss *m*, Siphon *m*; Abscheider *m*, Falle *f*; 2. Trap *m*, Haftstelle *f*, Fang-

stelle *f*, Rekombinationszentrum *n*, Trapniveau *n* *(für Elektronen)*
trapping reaction Abfangreaktion *f*
Traube reaction Traube-Reaktion *f* *(zur Darstellung von Purinen)*
travel *v* wandern
travel 1. Lauf *m*, Wanderung *f*; 2. *s.* travel rate
travel rate Laufgeschwindigkeit *f*, Wanderungsgeschwindigkeit *f*
tray 1. Pfanne *f*; Trog *m*; Schale *f*; 2. Trockenschale *f* *(eines Trockners)*; Trockenblech *n*, Siebblech *n*, Trockenhorde *f*, Horde *f* *(eines Trockners)*; 3. *(Dest)* Boden *m*
tray column (tower) *(Dest)* Bodenkolonne *f*, Bodensäule *f*
treat *v* behandeln, aufbereiten, *(Erdölprodukte, Gase auch)* nachbehandeln; versetzen; bearbeiten *(mechanisch)*
treated water aufbereitetes Wasser *n*, Reinwasser *n*
treating process 1. Behandlungsverfahren *n*; 2. *(Erdöl)* Nachbehandlungsverfahren *n*, Raffinationsverfahren *n*
treatment Behandlung *f*, Aufbereitung *f*, *(bei Erdölprodukten, Gasen auch)* Nachbehandlung *f*, *(bei Erdölprodukten auch)* Raffination *f* *(i. e. S.)*; Bearbeitung *(mechanisch)* • **give a mild cracking treatment** mild [gelinde] kracken
treatment chemical 1. Behandlungschemikalie *f*; 2. *(Hyd)* Aufbereitungschemikalie *f*
treatment plant 1. Behandlungsanlage *f*; 2. *(Hyd)* Reinigungsanlage *f*
treatment stage 1. Behandlungsstufe *f*, Verfahrensstufe *f*, Prozessstufe *f*; 2. *(Hyd)* Aufbereitungsstufe *f*, Reinigungsstufe *f*
tree *v* *(Krist)* Dendriten bilden, sich baumartig [als verästelte kristalline Masse] abscheiden, *(i. e. S.)* einen Bleibaum bilden
tree *(Krist)* Dendrit *m*, *(i. e. S.)* Bleibaum *m*
TRF *s.* thyrotrop(h)in-releasing factor [hormone]
TRH *s.* thyrotrop(h)in-releasing factor [hormone]
tri. *s.* triclinic
triacetate Triacetat *n*
triacid 1. dreisäurig, dreibasig, dreiwertig *(Base)*; 2. Trihydrogen... *(Salz)*
triacid dreibasige [dreiprotonige, dreiwertige] Säure *f*
triacidic *s.* triacid
triad 1. dreiwertig; 2. *(Krist)* dreizählig, 3-zählig
triad 1. dreiwertiges Element *n*; dreiwertige Atomgruppe *f*; 2. Triade *f* *(Dreiergruppe von Elementen mit ähnlichen Eigenschaften)*
triads of Döbereiner Döbereiner-Triaden *fpl*
trial Versuch *m* • **through trial and error** empirisch
trial-and-error procedure (empirisches) Näherungsverfahren *n*
triangular-wave voltammetry *(Anal)* Dreieckwellenvoltammetrie *f*
triarylmethane dye Triarylmethanfarbstoff *m*, Triphenylmethanfarbstoff *m*, Tritylfarbstoff *m*
triatomic dreiatomig
triazine Triazin *n* *(ein heterocyclischer Sechsring mit drei N-Atomen)*
triazine polymer Triazinpolymer(e) *n*, Polytriazin *n*
triazole Triazol *n* *(ein heterocyclischer Fünfring mit drei N-Atomen)*
tribasic 1. dreibasig, dreiwertig, dreiprotonig *(Säure)*; 2. dreisäurig, dreibasig, dreiwertig *(Base oder basisches Salz)*
tribasic sodium (ortho)phosphate Trinatriumphosphat *n*, Natrium(ortho)-phosphat *n*
tribasicity 1. Tribasizität *f* *(von Säuren)*; 2. Dreiwertigkeit *f* *(von Basen)*
triblock copolymer *(Kst)* Triblockcopolymer *n*
tribochemistry Tribochemie *f*
triboluminescence Tribolumineszenz *f*, Reibungslumineszenz *f*, Trennungsleuchten *n*
triboluminescent tribolumineszierend
tribromomethane *Tribrommethan *n*, Bromoform *n*
tributyl tin (compound) *(Tox)* Tributylzinnverbindung *f*
tric. *s.* triclinic
tricarboxylic acid Tricarbonsäure *f*, dreibasige Carbonsäure *f*

tricarboxylic-acid cycle Tricarbonsäurezyklus *m*, Citronensäurezyklus *m*, Citratzyklus *m*, (Szent-Györgyi-)Krebs-Zyklus *m*
trichloride Trichlorid *n*
trichloro derivative Trichlorderivat *n*
trichloroacetic acid Trichloressigsäure *f*, *Trichlorethansäure *f*, TCA
trichloroacetonitrile Trichloracetonitril *n*, Chlorcyanhydrin *n*
trichlorobenzene *(org Ch)* Trichlorbenzol *n*
2,3,6-trichlorobenzoic acid 2,3,6-Trichlorbenzoesäure *f*, *(als Herbizid auch)* 2,3,6-TBA
trichloroethane Trichlorethan *n*
trichloromethane Trichlormethan *n*, Chloroform *n*
2,4,5-trichlorophenoxyacetic acid 2,4,5-Trichlorphenoxyessigsäure *f*, *(als Herbizid auch)* 2,4,5-T
2,4,5-trichlorophenoxypropionic acid 2,4,5-Trichlorphenoxypropionsäure *f*, *(als Herbizid auch)* Fenoprop *n*, 2,4,5-TP
2,3,6-trichlorophenylacetic acid 2,3,6-Trichlorphenylessigsäure *f*, *(als Herbizid auch)* Fenac *n*
trichlorosil(ic)ane Trichlorsilan *n*
α-trichlorotoluene α-Trichlortoluen *n*, ω-Trichlormethylbenzen *n*, *Benzotrichlorid *n*
trichloro-*sym*-triazine Trichlor-*sym*-triazin *n*, Cyanursäurechlorid *n*
trichroism *(Krist)* Trichroismus *m*, Dreifarbigkeit *f*
trickle *v* tröpfeln; rieseln, rinnen; sickern; tröpfeln (lassen), träufeln; rieseln lassen; versickern lassen
trickle *v* **down** herabtröpfeln; herabrieseln, herabrinnen
trickle-bed reactor Rieselreaktor *m*
trickling film reactor *(Biot)* Tropfkörperreaktor *m*, Rieselfilmreaktor *m*
trickling filter *(Hyd)* Tropfkörper *m*
triclinic *(Krist)* triklin, trikl.
tricresyl phosphate Phosphorsäuretricresylester *m*, Tricresylphosphat *n*
tridentate dreizähnig, dreizählig *(Komplexchemie)*
triene *(org Ch)* Trien *n*
triester *(org Ch)* Triester *m*
triethanolamine Triethanolamin *n*, Tri(2-hydroxyethyl)amin *n*
triethyl aluminium Triethylaluminium *n*
triethyl phosphate Phosphorsäuretriethylester *m*, Triethylphosphat *n*
triethylamine Triethylamin *n*
triethylene glycol Triethylenglykol *n*, Triglykol *n*, *2,2′-Ethylendioxydiethanol *n*
triflic acid *s.* trifluoromethane sulphonic acid
trifluoroacetic acid Trifluoressigsäure *f*, *Trifluorethansäure *f*
trifluoroethanoic acid *s.* trifluoroacetic acid
trifluoromethane sulphonic acid Trifluormethansulfonsäure *f*
trifunctional unit trifunktionelle Einheit *f*, T-Einheit *f*, Grundbaustein *m* T *(zum Aufbau von Polymeren)*
trig. *s.* trigonal
trigger *v* 1. auslösen *(z. B. eine Kettenreaktion)*; 2. *(Bioch)* triggern
triglyceride Triglycerid *n*, Neutralglycerid *n*
triglycol *s.* triethylene glycol
trigonal *(Krist)* trigonal, trig.
trigonal hybridization trigonale Hybridisierung *f*, sp^2-Hybridisierung *f*
trihalide Trihalogenid *n*
trihalomethane Trihalo(gen)methan *n*
trihydrate Trihydrat *n*
trihydric alcohol dreiwertiger Alkohol *m*
trihydroxide Trihydroxid *n*
1,2,3-trihydroxybenzene 1,2,3-Trihydroxybenzen *n*, Pyrogallol *n*
3,4,5-trihydroxybenzoic acid 3,4,5-Trihydroxybenzoesäure *f*, Gallussäure *f*
trihydroxylic alcohol dreiwertiger Alkohol *m*
1,2,3-trihydroxypropane *Propan-1,2,3-triol *n*, Glycerol *n*
triiodomethane Triiodmethan *n*, Iodoform *n*
trimellitic acid Trimellithsäure *f*, *1,2,4-Benzentricarbonsäure *f*
trimer Trimer(e) *n*, trimere Verbindung *f*
trimeric trimer
trimerization Trimerisierung *f*, Trimerisation *f*
trimerize *v* trimerisieren
trimethylaluminium Trimethylaluminium *n*

trimethylamine Trimethylamin *n*
***N,N,N*-trimethylammonioacetate** Trimethylammonioacetat *n*, Betain *n*, Carboxymethyl-trimethylammonium *n*, Trimethylglycin *n*, Oxyneurin *n*, Lycin *n*
trimethylsilyl group $-SiMe_3$ Trimethylsilylgruppe *f*
trimethylsilylation Trimethylsilylierung *f*
trimolecular trimolekular
trinegative dreifach negativ (geladen)
trinitrate Trinitrat *n*
trinitration Trinitrierung *f*
trinitride Trinitrid *n*
2,4,6-trinitrophenol *2,4,6-Trinitrophenol *n*, Pikrinsäure *f*
trinitrotoluene Trinitrotoluen *n*, TNT *n*, Tritol *n*
triol dreiwertiger Alkohol *m*
triose Triose *f (Monosaccharid mit drei Kohlenstoffatomen)*
trioxide Trioxid *n*
tripeptide Tripeptid *n (Peptid mit drei Aminosäureeinheiten)*
triphenyl phosphate Triphenylphosphat *n*, Phosphorsäuretriphenylester *m*
triphenylamine dye Triphenylaminfarbstoff *m*
triphenylmethane dye Triphenylmethanfarbstoff *m*
triphenylmethyl *s.* trityl
triphosphate $M^I_5P_3O_{10}$ Triphosphat *n*
triple bond dreifache Bindung *f*, Dreifachbindung *f*
triple-effect evaporator Tripeleffektverdampfer *m*, Dreikörperverdampfer *m*, Dreistufenverdampfer *m*
triple helix *(Bioch)* Tripelhelix *f*, Superhelix *f*, dreisträngige Helix *f*
triple neck adapter Dreifachaufsatz *m*
triple point Tripelpunkt *m*
triple salt Tripelsalz *n*
triplet 1. Triplett *n*, Triplettspektrallinie *f*; 2. *(Bioch)* Triplett *n*, Dreierkombination *f (genetischer Code)*
triplet-singlet transition Triplett-Singulett-Übergang *m*
triplet state Triplettzustand *m*, T Zustand *m*
triplet-triplet transition Triplett-Triplett-Übergang *m*
triply bonded [bound, connected] dreifach gebunden
triply unsaturated dreifach ungesättigt
tripositive dreifach positiv (geladen)
triprotic acid dreiprotonige [dreibasige, dreiwertige] Säure *f*
tris buffer *(Bioch)* Trispuffer *m*
trisaccharide Trisaccharid *n*
trisilane $H_3Si-SiH_2-SiH_3$ Trisilan *n*
trisilicate Trisilicat *n*
trisodium salt Trinatriumsalz *n*
trispiro compound [hydrocarbon] Trispiro-Verbindung *f*
trisubstituted trisubstituiert, dreifach substituiert
trisubstitution product Trisubstitutionsprodukt *n*
trisulphate $M^I_2S_3O_{10}$ Trisulfat *n*
trisulphide Trisulfid *n*
trisulphonic acid Trisulfonsäure *f*
triterpene *(org Ch)* Triterpen *n*
triterpenoid *(org Ch)* Triterpenoid *n (mit 30 C-Atomen)*
trithionate $M^I_2S_3O_6$ Trithionat *n*
tritiate *v* tritiieren, mit Tritium markieren
tritiation Tritiierung *f*
tritium T Tritium *n (Wasserstoffisotop* ${}^3_1H$*)*
tritium-labelled mit Tritium markiert
triton 1. *(Kern)* Triton *n (Kern des Tritiumatoms)*; 2. *s.* trinitrotoluene
triturate *v* pulverisieren, (zer)pulvern, zerreiben, verreiben, zermahlen, zerstoßen
trituration 1. Pulverisieren *n*, Pulvern *n*, Zerpulvern *n*, Zerreiben *n*, Verreiben *n*, Zermahlen *n*, Zerstoßen *n*; 2. *(Pharm)* Verreibung *f*, Trituration *f*
trityl Trityl *n*, Triphenylmethyl *n (beständiges freies Radikal)*
trivalency Dreiwertigkeit *f*, Trivalenz *f*
trivalent dreiwertig, trivalent
trivial name Trivialname *m*, unsystematischer [nicht systematisch gebildeter] Name *m*
tRNA *s.* transfer ribonucleic acid
tropane alkaloid Tropa(n)alkaloid *n*, Tropinalkaloid *n*, Tropein *n*
trophophase *(Biot)* Trophophase *f*, Ernährungsphase *f*, Phase *f* des aktiven Wachstums
tropilidene Tropiliden *n*, Cyclohepta-1,3,5-trien *n*
tropolone Tropolon *n (Cycloheptanderivat)*
tropone Tropon *n*, Cyclohepta-2,4,6-trienon *n*

tropylium ion Tropylium-Ion *n*, Cycloheptatrienylium-Ion *n*
trough 1. Mulde *f*, Trog *m*, Wanne *f*; Bottich *m*; Schiff *n*; 2. Rinne *f*; 3. *(Bioch)* Spalt *m (in Hydrolasemolekülen)*
trough mixer Trogmischer *m*, Muldenmischer *m*; Trogkneter *m*
Trouton's law [rule] *(phys Ch)* troutonsche Regel *f*
Trp *s.* tryptophan(e)
true formula wahre Formel *f* [Summenformel *f*, Bruttoformel *f*]
true solution echte [molekolare] Lösung *f*
true solvent echtes [aktives] Lösungsmittel *n*
true transference number wahre Überführungszahl *f*
trypan red Trypanrot *n*
trypanocidal *(Pharm)* trypanozid (wirkend), gegen Trypanosomen wirksam
trypsin Trypsin *n (zu den Proteinasen gehörendes Enzym)*
tryptophan(e) Tryptophan *n*, Trp, α-Amino-β-indolpropionsäure *f*
ts *s.* temperature-sensitive
T.S. *s.* total solids
Tschitschibabin ... *siehe: Chichibabin ...*
Tschugaeff reaction Čugaev-Reaktion *f*, Tschugajew-Reaktion *f (zur Darstellung von Alkenen)*
TSH *(thyroid-stimulating hormone) s.* thyrotrop(h)ic hormone
tsp *s.* teaspoonful/a
tspn *s.* teaspoonful/a
t. t. experiment *s.* test-tube experiment
TTHM level *s.* total trihalomethanes
TTHMs *s.* total trihalomethanes
T.T.T. diagram *s.* time-temperature-transformation diagram
tub Bottich *m*, Bütte *f*, Kübel *m*, Fass *n*
tub conformation Bootkonformation *f*, Wannenkonformation *f (Stereochemie)*
tube *v* 1. verrohren, mit Rohren [einer Rohrleitung] versehen; durch eine Rohrleitung verbinden; 2. in Rohrleitungen führen, durch Rohre leiten
tube bank Rohrregister *n*, Rohrreihe *f (eines Wärmeübertragers)*
tube fermenter *s.* tubular-flow reactor
tube filter Kerzenfilter *n*
tube furnace 1. *(Erdöl, Dest)* Röhrenofen *m*, Röhrenerhitzer *m*; 2. *(Ker, Met)* Röhrenofen *m*, Rohrofen *m*
tube mill Rohrmühle *f*
tube rotary dryer Röhrentrockner *m*
tube-side liquid [medium] Rohrmedium *n*
tubing 1. Rohre *npl*, Rohrmaterial *n*; 2. Rohrleitung *f*; 3. Schlauchmaterial *n (ohne Gewebeeinlage)*; 4. Spritzen *n* von Schläuchen; 5. Verrohren *n*, Verrohrung *f*
tubocurare *(Tox)* Tubocurare *n*, Topfcurare *n (von Chondrodendron-Arten)*
tubular röhrenförmig, röhrenartig, Röhren..., Rohr...
tubular (bowl) centrifuge Rohrzentrifuge *f*
tubular condenser *(Dest)* Röhrenkühler *m*
tubular evaporator Rohrverdampfer *m*
tubular fermenter *s.* tubular-flow reactor
tubular-flow reactor Rohrreaktor *m*, Strömungsrohrreaktor *m*, Röhren(strömungs)reaktor *m*, Strömungsrohr *n*, Tubularreaktor *m*
tubular heat exchanger Röhrenwärmeübertrager *m*, Röhrenwärmeaustauscher *m*
tubular loop reactor 1. Schlaufenreaktor *m*, Kreislaufreaktor *m*; 2. *(Biot)* Schlaufenfermenter *m*
tumble *v* trommeln, rommeln *(Kleinteile)*
tumbler 1. Drehtrommel *f*; 2. *s.* tumbler mixer
tumbler mixer Fallmischer *m*, Freifallmischer *m*
tumbling barrel Trommel *f*, Rommelfass *n (zur Behandlung von Kleinteilen)*
tumbling mill Sturzmühle *f*
tumorigenic tumorerzeugend, tumorerregend, tumorinduzierend, onkogen
tumorigenicity tumorerzeugende Wirkung *f*
tumour-causing *s.* tumorigenic
tun Tonne *f*, Fass *n*
tunable laser *(Spektr)* abstimmbarer [durchstimmbarer] Laser *m*
tungstate Wolframat *n*
tungsten W Wolfram *n*
tungsten bronze Wolframbronze *f*
tungsten carbide Wolframcarbid *n*
tungsten dichloride Wolframdichlorid *n*, Wolfram(II)-chlorid *n*

tungsten dioxide Wolframdioxid *n*, Wolfram(IV)-oxid *n*
tungsten disulphide Wolframdisulfid *n*, Wolfram(IV)-sulfid *n*
tungsten hexachloride Wolframhexachlorid *n*, Wolfram(VI)-chlorid *n*
tungsten pentachloride Wolframpentachlorid *n*, Wolfram(V)-chlorid *n*
tungsten source *(Spektr)* Wolframbandlampe *f*
tungsten tetrachloride Wolframtetrachlorid *n*, Wolfram(IV)-chlorid *n*
tungsten trioxide Wolframtrioxid *n*, Wolfram(VI)-oxid *n*
tungstic Wolfram..., *(i. e. S.)* Wolfram(VI)-...
tungstic acid 1. H_2WO_4 (gelbe) Wolframsäure *f*; 2. $H_2WO_4 \cdot H_2O$ (weiße) Wolframsäure *f*; 3. *s.* tungsten trioxide
tungstic oxide Wolfram(VI)-oxid *n*, Wolframtrioxid *n*
tunnel effect *(phys Ch)* Tunneleffekt *m*
tunnel furnace [kiln] Tunnelofen *m*
tunnel(l)ing *(phys Ch)* Tunnelung *f* *(Durchdringen der Energiebarriere)*
turbid trüb(e)
turbidimeter Trübungsmesser *m*, Turbidimeter *n*
turbidimetric titration turbidimetrische Titration *f*, Trübungstitration *f*
turbidimetry Turbidimetrie *f*, Trübungsanalyse *f*
turbidity 1. Trübe *f*, Trübung *f*, Trübheit *f*; 2. Trübe *f (Suspension feiner Teilchen)*; 3. *(Hyd)* Trübstoffe *mpl*, trübende Stoffe *mpl*, Schwebstoffe *mpl*, Schweb *m (z. B. im Rohwasser)*
turbidity point Trübungspunkt *m*
turbine fuel Turbinentreibstoff *m*
turbine mixer Turbinenrührer *m*, Turbomischer *m*, Turborührer *m*
turbulence Turbulenz *f*
turbulent diffusion Wirbeldiffusion *f*, Streudiffusion *f*, Kanaldispersion *f*
turbulent flow turbulente Strömung *f*, Wirbelströmung *f*
Turkey-red oil Türkischrotöl *n (sulfatiertes Ricinusöl)*
turmeric paper Kurkumapapier *n (ein Reagenzpapier)*
turn *v* 1. drehen; sich drehen; 2. (ver-)färben; sich (ver)färben; umschlagen *(Farbe)*; 3. umwandeln, verwandeln; sich umwandeln [verwandeln]
turn 1. Drehung *f*, Umdrehung *f*; 2. Umwandlung *f*, Verwandlung *f*; 3. *(Bioch)* Windung *f (einer α-Helix)*
turnover *(Bioch)* Stoffumsatz *m*, Turnover *n*
turnover number *(Bioch)* Wechselzahl *f*, katalytische Zentrumsaktivität *f*
turnover time 1. *(Bioch)* Verweildauer *f*; 2. *(Bioch)* Neubildungsdauer *f*
turpentine Terpentin *n(m) (Balsam von Nadelhölzern)*
tuyere Windform *f*, Winddüse *f*, Blasform *f (des Hochofens)*
tweezers Pinzette *f*
twice-distilled doppelt [zweifach] destilliert
twin (crystal) Zwilling(skristall) *m*, Kristallzwilling *m*
twin-screw extruder Doppelschneckenextruder *m*, Zweischneckenextruder *m*
twin-shell blender [mixer] V-Mischer *m*, Hosenmischer *m*
twinning *(Krist)* Zwillingsbildung *f*, Verzwillingung *f*
twist conformation Twistbootkonformation *f*
twisted verdrillt, ineinandergewunden *(Makromolekülketten)*
twisting vibration *(Spektr)* Torsionsschwingung *f*, Twisting-Schwingung *f*, τ-Schwingung *f*
two-body collision Zweierstoß *m*
two-carbon acid C_2-Carbonsäure *f*, Fettsäure *f* mit zwei C Atomen
two-carbon fragment *(Bioch)* C_2-Bruchstück *n*, C_2-Einheit *f*
two-centre bond Zweizentrenbindung *f*
two-centre (molecular) orbital Zweizentren(molekül)orbital *n*
two-colour indicator zweifarbiger Indikator *m*
two-component system Zweikomponentensystem *n*, Zweistoffsystem *n*, binäres System *n*
two-dimensional chromatography zweidimensionale Chromatographie *f*
two-dimensional NMR spectroscopy zweidimensionale NMR-Spektroskopie *f*

two-electron bond Zweielektronenbindung *f*
two-fold axis of symmetry zweizählige Symmetrieachse *f* [Drehachse *f*]
two-fold degenerate *(phys Ch)* zweifach entartet
two-neck(ed) flask Zweihalskolben *m*
two-pass zweigängig *(z. B. Wärmeübertrager)*
two-phase system Zweiphasensystem *n*, zweiphasiges System *n*
two-photon absorption *(Anal)* Zweiphotonenabsorption *f*
two-photon excitation Zweiphotonenanregung *f*
two-position control Zweipunktregelung *f*
two-product unit 1. Zweiproduktenapparat *m*; 2. *(Aufb)* Zweigutapparat *m*
two-roll mill Zweiwalzen-Walzwerk *n*; Zweiwalzenmühle *f*
two-stage process Zweistufenverfahren *n*, zweistufiges Verfahren *n*, zweistufiger Vorgang *m*
two-stage resin Novolakharz *n*, Novolak *m*
two-step process *s.* two-stage process
two-step reaction zweistufige Reaktion *f*, Zweistufenreaktion *f*, Zweischrittreaktion *f*
two-stranded *(Bioch)* zweisträngig, Zweistrang..., doppelsträngig, Doppelstrang..., doppelhelikal
two-substrate reaction Zwei-Substrat-Reaktion *f* *(Enzymkinetik)*
two-way stopcock [tap] *(Lab)* Zweiwegehahn *m*
two-zone gasifier *(Carb)* Schwelgenerator *m*, Doppelgasgenerator *m*, Doppelgaserzeuger *m*
twyer *s.* tuyere
Tyndall cone *(Koll)* Tyndall-Kegel *m*
Tyndall effect [phenomenon] Tyndall-Effekt *m*, Tyndall-Phänomen *n*, Faraday-Tyndall-Effekt *m*
tyndallometer Tyndall(o)meter *n*
typical equipment [inventory] Grundausstattung *f* *(z. B. eines Labors)*
tyramine Tyramin *n*, *p*-Hydroxyphenylethylamin *n*
tyrosine Tyrosin *n*, Tyr, L-α-Amino-β-(*p*-hydroxyphenyl)propionsäure *f*

U

U-tube heat exchanger U-Rohr-Wärmeaustauscher *m*, U Rohr-Wärmeübertrager *m*
U-tube manometer U-Rohr-Manometer *n*
Ubbelohde (suspended-level) viscometer Ubbelohde-Viskosimeter *n* (mit hängendem Kugelniveau)
UDP *s.* uridine diphosphate
UEL *s.* upper explosive limit
UF *s.* ultrafiltration
Ugi reaction Ugi-Reaktion *f* *(zur Darstellung von α-Aminocarbonsäureamiden oder Peptiden)*
UHT (processed) milk ultrahocherhitzte Milch *f*, UHT Milch *f*
Ullmann reaction Ullmann-Reaktion *f*, ullmannsche Reaktion *f* *(Diarylsynthese)*
ultimate analysis Elementaranalyse *f*
ultimate disposal Endlagerung *f*, endgültige [dauerhafte] Ablagerung *f* *(von Abprodukten)*
ultimate organic analysis organische Elementaranalyse *f*
ultimate tensile strength Zugfestigkeit *f*
ultimate vacuum Endvakuum *n*, Enddruck *m*
ultra-high performance concrete ultrahochfester Beton *m*
ultra-high temperature process *(Lebm)* Ultrahocherhitzungsverfahren *n*, UHT-Verfahren *n* *(zur Sterilisation)*
ultra-high vacuum Ultrahochvakuum *n*
ultra-trace analysis Ultra-Spuren-Analyse *f* *(*Probenmasse $\leq 10^{-4}$ g)*
ultracentrifugation Ultrazentrifugieren *n*, Ultrazentrifugation *f*
ultracentrifuge Ultrazentrifuge *f*
ultrafilter *v* ultrafiltrieren, durch Ultrafilter filtrieren
ultrafilter Ultrafilter *n*
ultrafiltration Ultrafiltration *f*, Ultrafiltrieren *n*
ultrafiltration reactor *(Biot)* Membranreaktor *m*
ultramicro sample Ultramikroprobe *f* *(* $<10^{-4}$ g)*
ultramicroanalysis Ultramikroanalyse *f*

ultramicrochemistry Ultramikrochemie *f*
ultramicroscope Ultramikroskop *n*
ultramicroscopic ultramikroskopisch
ultrapure ultrarein, höchstrein, extrem rein
ultrapurification extreme [sehr hohe] Reinigung *f*, Hochreinigung *f*, Feinreinigung *f*
ultrared infrarot, Infrarot…, IR-…
ultrasonic bath Ultraschallbad *n*
ultrasonic cleaning Ultraschallreinigung *f*
ultrasonic testing Ultraschallprüfung *f* *(Werkstoffprüfung)*
ultraviolet ultraviolett, Ultraviolett…, UV-…
ultraviolet photoelectron spectroscopy UV-Photoelektronen-Spektroskopie *f*
ultraviolet radiation UV-Strahlung *f*
ultraviolet region UV-Bereich *m*, UV-Gebiet *n*
ultraviolet spectroscopy UV-Spektroskopie *f*
ultraviolet spectrum UV-Spektrum *n*
ultraviolet-visible … *siehe: UV-VIS …*
UMP = uridine 5′-monophosphate
unaerated bioreactor *(Biot)* unbelüfteter [unbegaster] Bioreaktor *m*
unalloyed unlegiert, rein, Rein…
unary system unitäres System *n*, Einstoffsystem *n*, Einkomponentensystem *n*
unattackable unangreifbar
unattended operation wartungsfreier Betrieb *m*
unbalance Ungleichgewicht *n*
unbleached ungebleicht
unbonded ungebunden
unbound ungebunden
unbranched unverzweigt, linear
unbranched-chain geradkettig, unverzweigt
unbuffered ungepuffert
unburned unverbrannt; ungebrannt
unburnt unverbrannt; ungebrannt
uncatalyzed nicht katalysiert, unkatalysiert
uncertainty principle (heisenbergsches) Unbestimmtheitsprinzip *n*, (heisenbergsches) Unschärfeprinzip *n*
unchanged unverändert; unangegriffen, nicht angegriffen; nicht umgesetzt [umgewandelt], unumgesetzt
uncharged ungeladen, ladungslos, (elektrisch) neutral
uncoiling Entknäuelung *f (von Molekülen)*
uncombined frei, ungebunden, unverbunden
uncompetitive (enzyme) inhibition unkompetitive Hemmung *f (Enzymkinetik)*
unconditional acceptable daily intake *(Tox)* zulässige [duldbare] Tagesdosis *f* *(ohne Einschränkungen, in mg je kg Körpergewicht)*
uncontaminated unverschmutzt, nicht verschmutzt [belastet] *(Wasser, Luft)*
unconverted unumgesetzt, nicht umgesetzt [umgewandelt], nicht in Reaktion getreten, unverbraucht
uncouple *v (Bioch)* entkoppeln
uncoupling *(Bioch)* Entkopplung *f*
unctious 1. salbenartig; fettig, ölig; schmierig; 2. *(Boden)* gar
unctuous 1. salbenartig; fettig, ölig; schmierig; 2. *(Boden)* gar
unctuousness 1. Salbenartigkeit *f*; Fettigkeit *f*, Öligkeit *f*; Schmierigkeit *f*; 2. *(Boden)* Gare *f*
undecanal *(org Ch)* Undecanal *n*
undecane *(org Ch)* Undecan *n*
undecanoic acid Undecansäure *f*
undecanone *(org Ch)* Undecanon *n*
undecene *(org Ch)* Undecen *n*
undecenoic acid Undecensäure *f*
undecyl alcohol Undecylalkohol *m*, Alkohol *m* C 11, *Undecan-1-ol *n*
undecylic acid *s.* undecanoic acid
undecyne *(org Ch)* Undecin *n*
undecynoic acid Undecinsäure *f*
underbleach *v* unterbleichen, unvollständig bleichen
underbleaching Unterbleichung *f*, Unterbleiche *f*, unvollständige Bleichung *f*
undercoat Zwischenanstrich *m*, Voranstrich *m*
undercoat paint Zwischenanstrichstoff *m*, Voranstrichstoff *m*, Vorstreichfarbe *f*
undercure 1. *(Kst, Anstr)* Unterhärtung *f*; 2. *(Gum)* Untervulkanisation *f*, Untervernetzung *f*
undercured 1. *(Kst, Anstr)* nicht ausgehärtet, ungenügend gehärtet; 2. *(Gum)* untervulkanisiert, untervernetzt

underdosage Unterdosierung *f*
underdose *v* unterdosieren
underdry *v* unvollständig [unzureichend] trocknen
underdrying unvollständige [unzureichende] Trocknung *f*
underfeed *v* unterdosieren
underfeed(ing) Unterdosierung *f*
underflow 1. Unterlauf *m*, Unterlaufprodukt *n*, *(Flotation auch)* Sinkgut *n*, Rückstand *m*, *(Siebklassieren auch)* Siebunterlauf *m*, Siebdurchlauf *m*, Siebdurchgang *m*; 2. *(Hyd)* geklärter Ablauf *m*, Klarwasser *n*, ablaufendes [entöltes] Wasser *n* *(Ölabscheidung)*
underglaze colour Unterglasurfarbe *f*
undergo *v* **chemical change** sich chemisch verändern, eine chemische Veränderung erleiden, chemisch verändert werden
undergo *v* **conversion** eine Verwandlung [Umsetzung] erleiden, umgewandelt [umgesetzt] werden
undergo *v* **decomposition** 1. sich zersetzen, zersetzt werden, zerfallen; 2. *(Bio)* verwesen
undergo *v* **fission** *(Kern)* zerfallen, sich spalten
undergo *v* **further reaction** weiterreagieren
undergo *v* **reaction** reagieren, eine Reaktion erleiden
underground burial untertägige Deponie *f (von Abprodukten)*
underground mining Untertagebau *m*, Abbau *m* unter Tage, Tiefbau(betrieb) *m*, Förderung *f* [Abbau *m*] im Tiefbau, untertägiger Bergbau *m*
underload *v* unterbelasten *(eine Anlage)*
undersize (material, product) Unterkorn *n*, Feinkorn *n*, Feingut *n (beim Klassieren)*; *(beim Siebklassieren auch)* Siebfeines *n*, Siebfeine *f*, *(beim Windsichten auch)* Sichtfeines *n*
undervulcanization Untervulkanisation *f*, Untervernetzung *f*
undervulcanized untervulkanisiert, untervernetzt
undetermined component nicht bestimmte Komponente *f (einer zu analysierenden Probe)*
undigestible 1. unverdaulich; 2. *(Anal)* nicht aufschließbar; 3. *(Hyd)* nicht faulfähig [fäulnisfähig] *(Schlamm)*
undiluted unverdünnt
undissociated undissoziiert
undissolved solid [solute] Bodenkörper *m (im Gleichgewicht mit seiner Lösung)*
unequal-arm balance ungleicharmige Balkenwaage *f*
unfermentable unvergärbar, nicht vergärbar
unfilled ungefüllt, nicht gefüllt, füllstofffrei, ohne Füllstoff
unfilt(e)rable nicht filtrierbar [filtrierend]
unfolding Entfaltung *f (z. B. von Makromolekülen)*
uniaxial *(Krist)* (optisch) einachsig
unibivalent electrolyte ein-zwei-wertiger [1-2-wertiger] Elektrolyt *m*
unicellular foam geschlossenzelliger Schaum(kunst)stoff *m*
unidentate einzähnig, einzählig *(Komplexchemie)*
uniform substance einheitlicher Stoff *m*, einheitliche Substanz *f*
uniform throughout vollständig [vollkommen, durch und durch] einheitlich
unimolecular film [layer] monomolekulare Schicht *f*, Monoschicht *f*, monomolekularer Film *m* [Oberflächenfilm *m*]
unimolecular reaction monomolekulare [unimolekulare] Reaktion *f*
uninegative einfachnegativ
uninflammability Flammwidrigkeit *f*, Nichtentflammbarkeit *f*, Unentflammbarkeit *f*, Flammbeständigkeit *f*
uninflammable flammwidrig, nicht entflammbar [brennbar], unentflammbar, flammbeständig
union 1. Vereinigung *f*, Verbindung *f*; 2. Verband *m*; 3. Rohrverbindung *f*, Rohrverbindungsstück *n*; Schlauchverbindung *f*, Schlauchverbindungsstück *n*
union of atoms 1. Vereinigung *f* von Atomen; 2. Atomverband *m*
unionized undissoziiert; unionisiert, nicht ionisiert
unipositive einfachpositiv
unit 1. Einheit *f*, Bezugseinheit *f*, Maßeinheit *f*; 2. *(Ch, Krist)* Einheit *f*, Struktureinheit *f*, Baueinheit *f*; 3. *(Tech)* Einheit *f*, Apparateeinheit *f*, Apparate-

gruppe *f*, Anlage *f (als Teil eines Betriebs)*
unit area 1. Flächeneinheit *f*; 2. *(Aufb)* Einheitsfläche *f*, durchsatzbezogene Klärfläche *f (je Tonne Trockensubstanz je Tag, bei Eindickern)*
unit cell *(Krist)* Elementarzelle *f*
unit charge Elementarladung *f*, elektrisches Elementarquantum *n*, e
unit chemical process *s.* unit process
unit concentration Konzentration *f* 1
unit height *Länge h der konformativen Repetiereinheit einer Helix*
unit of amount of substance Stoffmengeneinheit *f*
unit operation (physikalische) Grundoperation *f*, (physikalische) Einheitsoperation *f*
unit physical operation *s.* unit operation
unit plane *(Krist)* Hauptsymmetrieebene *f*
unit process (verfahrenstechnischer, stofflicher) Grundprozess *m*, (chemischer) Einheitsprozess *m*, (chemische) Grundreaktion *f*
unit time Zeiteinheit *f* • **per unit time** je [in der] Zeiteinheit
unit volume Volumeneinheit *f* • **per unit volume** je Volumeneinheit
unitary rate constant Reaktionsgeschwindigkeitskonstante *f*, RGK, Geschwindigkeitskonstante *f*
uniunivalent electrolyte ein-ein-wertiger [1-1-wertiger] Elektrolyt *m*
univalence Einwertigkeit *f*
univalency Einwertigkeit *f*
univalent einwertig, monovalent
univariant univariant, monovariant, einfachfrei
universal gas constant (universelle, allgemeine) Gaskonstante *f*
universal indicator Universalindikator *m*
universe Grundgesamtheit *f (Statistik)*
unknown unbekannte Substanz *f*, unbekanntes Material *n*
unleaded fuel bleifreier [ungebleiter, unverbleiter] Kraftstoff *m*
unlike charge ungleichnamige Ladung *f*
unmix *v* sich entmischen
unmixing Entmischung *f*
unmodified unmodifiziert, nicht modifiziert, unverändert
unoccupied unbesetzt *(z. B. Orbital)*
unordered ungeordnet, wirr
unpaired electron ungepaartes Elektron *n*
unpigmented 1. unpigmentiert, nicht pigmentiert; 2. *s.* unfilled
unplasticized PVC unplastifiziertes [weichmacherfreies] PVC *n*; Hart-PVC *n*, PVC hart *n*, PVC-H *n*
unpolluted unverschmutzt, nicht verschmutzt [belastet] *(Wasser, Luft)*
unprocessed nicht verarbeitet, roh
unreacted unumgesetzt, nicht umgesetzt [umgewandelt], nicht in Reaktion getreten, unverbraucht
unreactive reaktionsunfähig, nicht reaktionsfähig
unreinforced unverstärkt *(z. B. Kunststoff)*
unsaponifiable matter [residue] unverseifbarer Anteil *m*, Unverseifbares *n*, UV
unsaturated un(ab)gesättigt; nicht völlig abgesättigt • **highly unsaturated** hochungesättigt
unsaturated fatty acid ungesättigte Fettsäure *f*
unsaturated polyester ungesättigter Polyester *m*, UP
unsaturation Ungesättigtheit *f*, ungesättigter Charakter *m* [Zustand *m*]
unseeded crystallization spontane Kristallisation *f*
unseeded solution *(Krist)* ungeimpfte Lösung *f*
unshared electron pair einsames [freies] Elektronenpaar *n*
unslaked lime ungelöschter [gebrannter] Kalk *m*, Branntkalk *m*, Ätzkalk *m (Calciumoxid)*
unspent unverbraucht *(Chemikalien)*
unstable instabil, unstabil, unbeständig, labil • **make unstable** destabilisieren
unsteady-state conduction nicht stationärer Wärmetransport *m*
unsupported catalyst Katalysator *m* ohne Träger [Stützmaterial]
unsymmetrical compound unsymmetrische [asymmetrische] Verbindung *f*
unsymmetrical ether gemischter [unsymmetrischer] Ether *m*
untreated unbehandelt, nicht behandelt, roh

unvulcanized unvulkanisiert, unvernetzt, nicht vernetzt
unwind *v* 1. abrollen, abwickeln; 2. *(Bioch)* entspiralisieren, aufzwirbeln, entdrillen, entzwirnen *(DNA-Doppelstrang)*; entspiralisiert werden
unwinding 1. Abrollen *n*, Abwickeln *n*; 2. *(Bioch)* Entspiralisierung *f*, Aufzwirbelung *f*, Entdrillen *n*, Entzwirnen *n (des DNA-Doppelstrangs)*
upflow Auf(wärts)strom *m*, aufsteigender Strom *m* • **operate upflow** im Aufwärtsstrom arbeiten; im Aufwärtsstrom fahren
upflow filter *(Hyd)* Aufwärtsfilter *n*
upflow filtration *(Hyd)* Filtration *f* im Aufwärtsstrom, Aufwärtsfiltration *f*
upper consolute [critical solution] temperature obere kritische Lösungstemperatur *f* [Mischungstemperatur *f*], oberer kritischer Lösungspunkt *m* [Mischungspunkt *m*]
upper explosive limit obere Explosionsgrenze *f*
UPS *s.* ultraviolet photoelectron spectroscopy
upset *v* stören *(z. B. den Betriebsablauf)*
upset Havarie *f*, Betriebsstörung *f*
uptake of nutrients Nährstoffaufnahme *f*
uptake of oxygen Sauerstoffaufnahme *f*
upward current aufwärts gerichtete Strömung *f*, Auf(wärts)strom *m*, aufsteigender Strom *m*
UQ = ubiquinone
uracil Uracil *n*, 2,4-Dioxotetrahydropyrimidin *n*
uramil Uramil *n*, Dialuramid *n*, 5-Aminobarbitursäure *f*
uranate Uranat *n*
urania *s.* uranium dioxide
uranic Uran...; höherwertigem Uran entsprechend, *(meist)* Uran(VI)-...
uranic acid Uransäure *f*, *(i. e. S.)* H_2UO_4 oder $UO_2(OH)_2$ (normale) Uransäure *f*, Uranylhydroxid *n*
uranide *s.* uranoid
uranin(e) Uranin *n*, Fluoresceinnatrium *n* *(Dinatriumsalz des Fluoresceins)*
uranium U Uran *n*
uranium decay series *s.* uranium series
uranium dioxide Urandioxid *n*, Uran(IV)-oxid *n*
uranium(-235) enrichment Anreicherung *f* von Uran-235
uranium fission *(Kern)* Uranspaltung *f*
uranium hexafluoride Uranhexafluorid *n*, Uran(VI)-fluorid *n*
uranium hydride Uranhydrid *n*
uranium oxychloride *s.* uranyl chloride
uranium-radium (disintegration) series *s.* uranium series
uranium series Uranzerfallsreihe *f*, Uran-Radium-Zerfallsreihe *f*, Uran-Radium-Reihe *f*
uranium tetrafluoride Urantetrafluorid *n*, Uran(IV)-fluorid *n*
uranium trioxide Urantrioxid *n*, Uran(VI)-oxid *n*
uranoid Transuran *n (Element einer Actinoidenteilreihe)*
uranous Uran...; niederwertigem Uran entsprechend, *(meist)* Uran(IV)-...
uranyl acetate Uranylacetat *n*
uranyl chloride Uranylchlorid *n*
uranyl nitrate Uranylnitrat *n*
uranyl salt Uranylsalz *n*, Uranylverbindung *f*, Dioxouran(VI)-salz *n*
urate Urat *n (Salz der Harnsäure)*
urban sewage städtisches Abwasser *n*
urea Harnstoff *m*, Carbamid *n*, Kohlensäurediamid *n*; Harnstoff *m*, Harnstoffderivat *n*
urea adduct Harnstoffaddukt *n*, Harnstoffadditionsverbindung *f*
urea clathrate Harnstoffeinschlussverbindung *f*
urea complex Harnstoffkomplex *m*, Harnstoffkomplexverbindung *f*
urea cycle *(Bioch)* (Arginin-)Harnstoff-Zyklus *m*, Ornithinzyklus *m*, Krebs-Henseleit-Zyklus *m*
urea-formaldehyde resin Harnstoff-Formaldehyd-Harz *n*
urea inclusion compound Harnstoffeinschlussverbindung *f*
urea resin 1. Harnstoffharz *n*, Carbamidharz *n*, Harnstoff-Aldehyd-Harz *n*; 2. *s.* urea-formaldehyde resin
ureide *(org Ch)* Ureid *n*
ureido group NH2CONH– Ureidogruppe *f*, Carbamidogruppe *f*
ureolytic harnstoffspaltend
ureotelic *(Bioch)* harnstoffausscheidend, ureotel
urethan(e) Urethan *n*, Carbamidsäu-

reester *m*, *(i. e. S.)* Ethylurethan *n*, Carbamidsäureethylester *m*; Polyurethan *n*, (polymeres) Urethan *n*
urethan(e) elastomer Polyurethanelastomer(e) *n*, Urethanelastomer(e) *n*
urethan(e) foam Polyurethanschaum(-stoff) *m*, Urethanschaumstoff *m*
uric acid Harnsäure *f*, 2,6,8-Trihydroxypurin *n*, Purin-2,6,8-triol *n*
uricolysis Uricolyse *f (Oxidation der Harnsäure durch Uricase zu Allantoin)*
uricolytic uricolytisch
uricotelic *(Bioch)* harnsäureausscheidend, urikotel
uridine Uridin *n*, Uracil-*D*-ribosid *n*
uridine diphosphate Uridindiphosphat *n*, UDP
uridine monophosphate Uridinmonophosphat *n*, Uridin(mono)phosphorsäure *f*, UMP, *(i. w. S.)* Uridylsäure *f*
uridine triphosphate Uridintriphosphat *n*, Uridintriphosphorsäure *f*, UTP
uronic acid Uronsäure *f (allg. für eine Aldehydcarbonsäure der Zuckerreihe)*
use *v* **topically [in topical applications]** *(Pharm)* lokal anwenden, zur Lokalbehandlung verwenden
use *v* **up** aufbrauchen
use temperature Anwendungstemperatur *f*
used catalyst gebrauchter Katalysator *m*, Altkatalysator *m*
used oil Altöl *n*
useful life 1. Nutzungsdauer *f*, Gebrauchswertdauer *f*, Lebensdauer *f*, Standzeit *f*; 2. *(Hyd)* Laufzeit *f (z. B. der A-Kohle)*
useful material Wertstoff *m*
useful maximum work maximale Nutzarbeit *f*
USEPA = US Environmental Protection Agency
user-friendly nutzerfreundlich
USP 1. = United States Pharmacopeia; 2. = United States Patent
USP grade *den Forderungen des Arzneibuchs der USA entsprechend*
USS = United States Standard
uterotonic *(Pharm)* Uterotonikum *n*, Oxytozikum *n*, Wehenmittel *n*, oxytozisches [wehenerregendes] Mittel *n*
UTP *s.* uridine triphosphate
UV ... *siehe: ultraviolet ...*
UV-VIS = ultraviolet and visible
UV-VIS absorption spectroscopy UVS-Absorptionsspektroskopie *f*
UV-VIS range UVS-Bereich *m*
UV-VIS spectroscopy UVS-Spektroskopie *f*, Spektroskopie *f* im sichtbaren und UV-Gebiet, Spektroskopie *f* im Sichtbaren und Ultraviolett
UV-VIS spectrum UVS-Spektrum *n*

V

V-phase Phase *f* geringerer Dichte *(bei Stofftrennverfahren)*
vacancy *(Krist)* Leerstelle *f*, Fehlstelle *f*, Gitterleerstelle *f*, Gitterfehlstelle *f*, Gitterlücke *f*
vacancy pair Leerstellenpaar *n*
vacant frei, leer, unbenutzt, vakant
vacant lattice position *s.* vacancy
vacant (lattice) site *s.* vacancy
vaccenic acid Vaccensäure *f*, Octadec-11-ensäure *f*
vacuo/in im Vakuum
vacuum Vakuum *n* • **create a vacuum** ein Vakuum herstellen • **operate under a vacuum** unter Vakuum arbeiten
vacuum column *(Dest)* Vakuumkolonne *f*
vacuum concentration Vakuumeindampfung *f*
vacuum dehydration *s.* vacuum drying
vacuum deposition *(Kst)* Vakuumaufdampfung *f*
vacuum desiccator Vakuumexsikkator *m*
vacuum distillate Vakuumdestillat *n*
vacuum distillation Vakuumdestillation *f*, Destillation *f* im Vakuum, Destillation *f* unter vermindertem Druck
vacuum-distilled vakuumdestilliert, im [unter] Vakuum destilliert
vacuum-dry *v* unter [im] Vakuum trocknen
vacuum dryer Vakuumtrockenapparat *m*, Vakuumtrockner *m*
vacuum drying Vakuumtrocknen *n*, Vakuumtrocknung *f*
vacuum evaporation Vakuumverdampfung *f*
vacuum evaporator Vakuumverdampfer *m*, Vakuumverdampfapparat *m*

vacuum filter Vakuumfilter *n*, Saugfilter *n*, Unterdruckfilter *n*
vacuum filtration Vakuumfiltration *f*, Saugfiltration *f*, Unterdruckfiltration *f*
vacuum gauge Vakuummessgerät *n*, Vakuummeter *n*
vacuum pump Vakuumpumpe *f*
vacuum still Vakuumdestillierapparat *m*; Vakuumkolonne *f*
vacuum tube Elektronenröhre *f*
VAE copolymer Vinylacetat-Ethylen-Copolymer(e) *n*
vaguely defined unscharf [ungenau] definiert
valence 1. Wertigkeit *f*, Valenz(zahl) *f*; 2. *(besonders Am)* Valenz *f (als Einheit der Bindekraft)*
valence angle Valenz(bindungs)winkel *m*, Bindungswinkel *m*
valence band Valenzband *n*
valence bond Valenzbindung *f*
valence-bond theory Valenzbindungstheorie *f*, Valenzstrukturtheorie *f*, VB-Theorie *f*, Theorie *f* [Bindungstheorie *f*] der Valenzstrukturen [Elektronenpaarbindungen], Heitler-London-Slater-Pauling-Theorie *f*, HLSP-Theorie *f*
valence electron Valenzelektron *n*, Außenelektron *n*
valence electron pair Valenzelektronenpaar *n*
valence orbital Valenzorbital *n*
valence shell Valenzschale *f*, äußere Schale *f*, Außenschale *f*
valence structural formula Valenzstrichformel *f*, Strukturformel *f*, Konstitutionsformel *f*
valence vibration Valenzschwingung *f*, ν-Schwingung *f*, Streckschwingung *f*
valency 1. Valenz *f (als Einheit der Bindekraft)*; 2. *s.* valence 1.; 3. *s.* combining power
valeraldehyde Valeraldehyd *m*, *(i. e. S.)* $CH_3[CH_2]_3CHO$ *Pentanal *n*
valerate Valerat *n (Salz oder Ester einer Valeriansäure)*
valeric acid Valeriansäure *f*, *(i. e. S.)* $CH_3[CH_2]3COOH$ *Pentansäure *f*
valine Valin *n*, Val, α-Aminoisovaleriansäure *f*, *2-Amino-3-methylbutansäure *f*
value 1. Wert *m*; 2. Kennwert *m*, Wert *m*, Index *m*, Kennziffer *f*, Kennzahl *f*
γ-value γ-Wert *m*, abbesche Zahl *f*, reziproke relative Dispersion *f*
values Wertstoffe *mpl*
valve Ventil *n*, Absperrvorrichtung *f*
valve tray Ventilboden *m*, Klappenboden *m*
van der Waals bond Van-der-Waals-Bindung *f*, van-der-waalssche Bindung *f*
van der Waals equation (of state) van-der-waalssche Zustandsgleichung *f* [Gleichung *f*]
van der Waals forces (of attraction) Van-der-Waals-Kräfte *fpl*, van-der-waalssche Kräfte *fpl* [Anziehungskräfte *fpl*]
vanadate Vanadat *n*
vanadic Vanadium...; höherwertigem Vanadium entsprechend, *(meist)* Vanadium(V)-...
vanadic acid Vanadiumsäure *f*
vanadic fluoride Vanadium(V)-fluorid *n*, Vanadiumpentafluorid *n*
vanadic oxide Vanadium(V)-oxid *n*, Divanadiumpentoxid *n*, Vanadiumpentoxid *n*
vanadium V Vanadium *n*
vanadium dichloride Vanadiumdichlorid *n*, Vanadium(II)-chlorid *n*
vanadium dioxide *s.* vanadium tetroxide
vanadium pentafluoride Vanadiumpentafluorid *n*, Vanadium(V)-fluorid *n*
vanadium pentoxide Vanadiumpentoxid *n*, Divanadiumpentoxid *n*, Vanadium(V)-oxid *n*
vanadium tetroxide Vanadiumtetroxid *n*, Vanadiumdioxid *n*, Vanadium(IV)-oxid *n*
vanadium trichloride Vanadiumtrichlorid *n*, Vanadium(III)-chlorid *n*
vanadium trifluoride Vanadiumtrifluorid *n*, Vanadium(III)-fluorid *n*
vanadium trioxide Vanadiumtrioxid *n*, Divanadiumtrioxid *n*, Vanadium(III)-oxid *n*
vanadous Vanadium...; niederwertigem Vanadium entsprechend, *(meist)* Vanadium(III)-...
vanadous chloride *s.* 1. vanadium dichloride; 2. vanadium trichloride
vanadous oxide Vanadium(III)-oxid *n*, Divanadiumtrioxid *n*, Vanadiumtrioxid *n*

vanadyl dichloride Vanadyldichlorid *n*, Vanadiumoxiddichlorid *n*, Vanadium-(IV)-oxidchlorid *n*
vanadyl monochloride Vanadylmonochlorid *n*, Vanadiumoxidmonochlorid *n*, Vanadium(III)-oxidchlorid *n*
vanadyl salt Vanadylsalz *n*, Vanadiumoxidsalz *n*
vanadyl trichloride Vanadyltrichlorid *n*, Vanadiumoxidtrichlorid *n*, Vanadium(V)-oxidchlorid *n*
vanadylic die Vanadylgruppe VO^{3+} enthaltend
vanadylous die Vanadylgruppe VO^{+} enthaltend
vane Schaufel *f*, Flügel *m*
vane stirrer Blattrührer *m*
vanillal Vanillal *n*, Bourbonal *n*, 4-Hydroxy-3-ethoxybenzaldehyd *m*
vanillaldehyde *s.* vanillin
vanillic acid Vanillinsäure *f*, 4-Hydroxy-3-methoxybenzoesäure *f*
vanillic aldehyde *s.* vanillin
vanillin Vanillin *n*, 4-Hydroxy-3-methoxybenzaldehyd *m*
van't Hoff equation Van't-Hoff-Gleichung *f*, van't-hoffsche Gleichung *f*
van't Hoff intermediate van't-hoffscher Zwischenstoff *m*
van't Hoff reaction isobar van't-hoffsche Reaktionsisobare *f*
van't Hoff reaction isochore van't-hoffsche Reaktionsisochore *f*
van't Hoff reaction isotherm van't-hoffsche Reaktionsisotherme *f*
vapor *s.* vapour
vaporization Verdampfen *n*, Verdampfung *f*; *(unterhalb des normalen Siedepunkts)* Verdunsten *n*, Verdunstung *f*
vaporize *v* verdampfen, in Dampf überführen; verdunsten lassen; verdampfen; *(unterhalb des normalen Siedepunkts)* verdunsten
vaporizer Verdampfer *m*
vaporous dampfförmig
vapour Dampf *m*; Brüden(dampf) *m*, Abdampf *m*
vapour-compression machine Kompressionskältemaschine *f*
vapour density Dampfdichte *f*, Gasdichte *f*
vapour deposition Aufdampfung *f*
vapour diffusion pump Diffusionspumpe *f*
vapour phase Dampfphase *f*, dampfförmige Phase *f*, *(bei der Hochdruckhydrierung meist)* Gasphase *f*
vapour-phase cracking Dampfphasekracken *n*, Gasphasekracken *n*, Kracken *n* in der Dampfphase [Gasphase]
vapour-phase oxidation Dampfphaseoxidation *f*, Gasphaseoxidation *f*
vapour-phase polymerization Gasphasepolymerisation *f*
vapour pressure Dampfdruck *m*
vapour-pressure curve Dampfdruckkurve *f*
vapour-pressure depression [lowering] Dampfdruckerniedrigung *f*
vapour-pressure osmometry Dampfdruckosmose *f (Molmassenbestimmung)*
var. *s.* variety
variable of state Zustandsvariable *f*
variance 1. *(phys Ch)* Freiheit *f*, Freiheitsgrad *m*, *(i. e. S.)* Anzahl *f* der Freiheiten [Freiheitsgrade]; 2. Varianz *f (Streuungsquadrat)*
variate Zufallsvariable *f (Statistik)*
variation Abweichung *f*, Veränderung *f*, Schwankung *f*
variation in concentration Konzentrationsschwankung *f*
variety *(Min)* Varietät *f*, Var., V., Abart *f*
varnish *v* lackieren
varnish Lack *m*, Klarlack *m*, *(i. e. S.)* oxidativ trocknender Lack *m* [Klarlack *m*]
varnish additive Lackadditiv *n*
varnish black Lackschwarz *n*
varnish-makers' (and painters') naphtha Lackbenzin *n*
varnish red Lackrot *n*
varnish remover Lackentferner *m*, Abbeizmittel *n*
varnish wool Lackwolle *f*
vasoconstrictive vasokonstriktorisch, gefäßkontrahierend, gefäßvereng(e(r)nd
vasoconstrictor Vasokonstriktor *m*, gefäßverengendes [gefäßkontrahierendes] Mittel *n*
vasodepressor *s.* vasodilator
vasodilating vasodilatatorisch, gefäßerweiternd

vasodilator Vasodilatans *n*, Vasodilator *m*, Gefäßerweiterungsmittel *n*, gefäßerweiterndes Mittel *n*
vasoexcitor *s.* vasoconstrictor
vat *v* 1. *(Farb)* in der Küpe behandeln; 2. *(Farb)* (ver)küpen
vat 1. Bottich *m*, Bütte *f*, Trog *m*, Wanne *f*, großes Gefäß *n*; 2. *(Farb)* Küpe *f*; 3. *(Farb)* Küpe *f (Lösung der Küpenfarbstoffe)*
vat dye Küpenfarbstoff *m*
vat dyeing Küpenfärberei *f*
vat sizing *(Pap)* Oberflächenleimung *f* im Leimbadtauchverfahren
vatting *(Farb)* Verküpen *n*, Verküpung *f*
VB ... *siehe: valence-bond ...*
VC *s.* 1. vinyl chloride; 2. vinylidene chloride
VC ... *siehe: vapour-compression ...*
v.d. *s.* vapour density
VD *s.* vacuum distillate
vee-type mixer V-Mischer *m*, Hosenmischer *m*
Vegard's law [rule] *(Krist)* vegardsche Regel *f*
vegetable pflanzlich, vegetabilisch, Pflanzen...
vegetable drug pflanzliche Droge *f*
vegetable fat pflanzliches Fett *n*, Pflanzenfett *n*
vegetable fibre pflanzliche Faser *f*, Pflanzenfaser *f*; pflanzlicher Faserstoff *m*
vegetable material [matter] Pflanzensubstanz *f*, Pflanzenstoff *m*, Pflanzenmaterial *n*, Pflanzenmasse *f*, pflanzliche [vegetabilische] Substanz *f*
vegetable oil pflanzliches Öl *n*, Pflanzenöl *n*
vegetable protein pflanzliches Protein *n*, Pflanzenprotein *n*
vegetable substance *s.* vegetable material
vegetable tannage pflanzliche [vegetabilische] Gerbung *f*, Gerbung *f* mit pflanzlichen Gerbstoffen
vehicle 1. Bindemittel *n*; 2. Bindemittellösung *f*
velocity Geschwindigkeit *f*
velocity constant Geschwindigkeitskonstante *f*, Reaktionsgeschwindigkeitskonstante *f*, spezifische Reaktionsgeschwindigkeit *f*
velocity distribution Geschwindigkeitsverteilung *f*
velocity mapping *(phys Ch)* Geschwindigkeitskartographie *f*
velocity of flow Fließgeschwindigkeit *f*, Strömungsgeschwindigkeit *f*, Durchflussgeschwindigkeit *f*, Durchlaufgeschwindigkeit *f*
velocity of impact Aufschlaggeschwindigkeit *f*
velocity of light Lichtgeschwindigkeit *f*
velocity of migration Wanderungsgeschwindigkeit *f*
velocity of reaction Reaktionsgeschwindigkeit *f*
velocity of transformation Umwandlungsgeschwindigkeit *f*, Umsetzungsgeschwindigkeit *f*
vena contracta Einschnürungsstelle *f*, Kontraktionsstelle *f*, Stelle *f* des kleinsten Strömungsquerschnitts
venom tierisches Gift *n*
vent Austritt *m*, Austrittsöffnung *f*; Ablass *m*, Ablassöffnung *f (zum Druckausgleich)*; Entlüftung *f*, Entlüftungsöffnung *f*, Lüftungsöffnung *f*
vent gas Abgas *n*
vent pipe Entlüftungsrohr *n*, Abzugsrohr *n*, Abzugslutte *f*
vent valve Entlüftungsventil *n*
ventilating hood *(Lab)* Abzug *m*
venting Lüften *n*, Be- und Entlüften *n*
venting device Be- und Entlüftungseinrichtung *f (in Tankanlagen)*
Venturi scrubber Venturi-Wäscher *m*, Venturi-Wascher *m*, Venturi-Nassabscheider *m*, Venturi-Skrubber *m*
Venturi tube Venturi-Rohr *n*, Venturi-Düse *f*
Venturi washer *s.* Venturi scrubber
veratraldehyde Veratrumaldehyd *m*, Protocatechualdehyddimethylether *m*
veratric acid Veratrumsäure *f*, 3,4-Dimethoxybenzoesäure *f*
veratrole Veratrol *n*, Brenzcatechindimethylether *m*, *1,2-Dimethoxybenzen *n*
verdigris Grünspan *m*, Spanisches Grün *n (Gemisch aus basischen Kupfer(II)-acetaten)*
vermicide Vermizid *n*, wurmtötendes Mittel *n*

vermifuge Helminthagogum *n*, wurmabtreibendes Mittel *n*, Wurmmittel *n*
vertex Spitze *f*, Ecke *f (z. B. eines Tetraedermoleküls)*
vertical column Vertikalspalte *f*, Vertikalreihe *f (des Periodensystems)*
vertical-flow clarifier *(Hyd)* vertikal durchströmte Absetzanlage *f*, vertikal durchflossenes Absetzbecken *n*, Klärbecken *n* [Absetzbecken *n*, Absetzanlage *f*] mit vertikaler Durchströmung
vertical-flow sedimentation tank *s.* vertical-flow clarifier
vertical kiln Schachtofen *m*
vertical oven Vertikalofen *m*, Senkrechtofen *m*, vertikaler [senkrechter] Ofen *m (z. B. zur Gaserzeugung)*
vertical relationship Vertikalbeziehung *f (im Periodensystem)*
vertical-tube evaporator Vertikalrohrverdampfer *m*
very-high-density lipoprotein Lipoprotein *n* sehr hoher Dichte, VHDL
very-high-pressure boiler Höchstdruckkessel *m*
very-high-purity argon Reinargon *n*
very-high-resolution electron energy loss spectrometry hochauflösende Schwingungs-Elektronenenergieverlust-Spektrometrie *f*
very little soluble *s.* very slightly soluble
very-low-density lipoprotein Lipoprotein *n* sehr geringer Dichte, VLDL
very poisonous *s.* very toxic
very slightly soluble sehr wenig löslich
very soluble sehr leicht löslich
very toxic hochgiftig, hochtoxisch, stark giftig
vesicant blasenziehend
vesicant 1. blasenziehendes Mittel *n*; 2. *s.* vesicatory gas
vesicant war gas *s.* vesicatory gas
vesicate *v* zur Blasenbildung reizen, Blasen ziehen *(auf der Haut)*; Blasen bilden, sich mit Blasen bedecken *(Haut)*
vesication 1. Blasenbildung *f (auf der Haut)*; 2. Blase *f*
vesicatory *s.* vesicant
vesicatory gas blasenziehender Kampfstoff *m*, *(i. e. S.)* hautschädigender Kampfstoff *m*
vesicular 1. blasig; 2. *(Geol)* zellig
vessel Gefäß *n*, Behälter *m*
vetiver oil Vetiveröl *n (von Vetiveria zizanioides (L.) Nash)*
VHDL *s.* very-high-density lipoprotein
VHRELS *s.* very-high-resolution electron energy loss spectrometry
VI *s.* viscosity index
vial Glasfläschchen *n*, Phiole *f*
vibrate *v* in Schwingungen versetzen, schwingen [vibrieren, oszillieren] lassen, schütteln, rütteln; schwingen, vibrieren, oszillieren, zittern
vibrating (ball) mill schwingende Kugelmühle *f*, Schwingmühle *f*
vibrating screen Vibrationssieb *n*, Vibratorsieb *n*, Vibrosieb *n*, Zittersieb *n*
vibrating stress schwingende Beanspruchung *f*, Vibrationsbeanspruchung *f*
vibration Schütteln *n*, Rütteln *n*; Schwingung *f*, Vibration *f*, Oszillation *f*
vibration amplitude Schwing(ungs)amplitude *f*; Schwingweite *f*, Schwingungsweite *f*, Schwingweg *m*; Hublänge *f*, Hub *m*
vibration frequency Schwingungsfrequenz *f*
vibration-rotation spectrum Rotations-Schwingungs-Spektrum *n*
vibration spectrum Schwingungsspektrum *n*
vibrational Schwing(ungs)..., Vibrations..., Oszillations..., oszillatorisch
vibrational band *(Spektr)* Schwingungsbande *f*
vibrational degree of freedom Schwingungsfreiheitsgrad *m*
vibrational energy Schwingungsenergie *f*
vibrational energy level Schwingungsniveau *n*
vibrational frequency Schwingungsfrequenz *f*
vibrational mode Schwingungsform *f*, Mode *f*
vibrational partition function Schwingungsverteilungsfunktion *f*, Verteilungsfunktion *f* der Schwingungsenergie
vibrational quantum number Schwingungsquantenzahl *f*, Vibrationsquantenzahl *f*

vibrational spectroscopy Schwingungsspektroskopie *f*
vibrational spectrum Schwingungsspektrum *n*
vibrational state Schwingungszustand *m*
vibrational term Schwingungsterm *m*
vibrational transition Schwingungsübergang *m*
vibrator Rüttler *m*, Vibrator *m*, Schwingungs(ein)rüttler *m*
vic-... *(Nom)* vic-..., 1,2,3,-... oder 1,2,3,4-... *(identische nachbarständige Substituenten)*
Vicat softening point [temperature] *(Kst)* Erweichungspunkt *m* nach Vicat, Formbeständigkeit *f* (in der Wärme) nach Vicat, Vicat-Zahl *f*
vicinal nachbarständig *(Atome im Molekül)*; *(i. e. S.)* vicinal, in 1,2,3-Stellung, in 1,2,3,4-Stellung
Vickers (diamond pyramid) hardness Vickershärte *f*, HV
Victor Meyer method (of determining molecular weights) Bestimmung *f* der relativen Molekülmasse nach Victor Meyer
viewing aperture Schauöffnung *f*, Schauloch *n*
viewing window Sichtfenster *n*, Sichtglas *n*, Schauglas *n*
vigorous kräftig, stark, energisch; lebhaft, heftig, stürmisch *(Reaktion)*
Vigreux column *(Dest)* Vigreux-Kolonne *f*
vile abstoßend, widerlich *(Geruch)*
Vilsmeier-Haack reaction [synthesis] Vilsmeier-Haack-Reaktion *f*, Vilsmeier-Haack-Synthese *f (von Aldehyden durch Formylierung aromatischer Verbindungen)*
vinegar Essig *m*
vinegar bacteria Essig(säure)bakterien *npl*
vinegar fermentation Essig(säure)gärung *f*
vinegar generator Essig(säure)generator *m*, Essigbildner *m*
vinyl acetate Vinylacetat *n*
vinyl alcohol Vinylalkohol *m*, *Ethenol *n*
vinyl benzene Vinylbenzen *n*, Styren *n*, Styrol *n*
vinyl cation Vinylkation *n*
vinyl chloride Vinylchlorid *n*, *Chlorethen *n*
vinyl compound Vinylverbindung *f*
vinyl ester Vinylester *m*
vinyl ether CH_2=CH–OR Vinylether *m*
vinyl group –CH=CH_2 Vinylgruppe *f*, Vinylrest *m*, Ethenylgruppe *f*
vinyl halide Vinylhalogenid *n*
vinyl homologue *s.* vinylog
vinyl plastic Vinyl(harz)kunststoff *m*, Vinoplast *m*
vinyl polymerization Vinylpolymerisation *f*
vinyl radical 1. (freies) Vinylradikal *n*; 2. *s.* vinyl group
vinyl resin Vinylharz *n*
vinyl sulphone dye Vinylsulfonfarbstoff *m*
vinylene group Vinylengruppe *f*, Vinylenrest *m*
vinylic Vinyl...
vinylidene chloride Vinylidenchlorid *n*, *1,1-Dichlorethen *n*
vinylidene cyanide Vinylidencyanid *n*, *1,1-Dicyanethen *n*
vinylog Vinylhomologe *n*, Vinyloge *n*
vinylogous vinylhomolog, vinylog
vinylogy Vinylhomologie *f*, Vinylogie *f (Übertragung der Eigenschaften des C-Atoms einer Carbonylgruppe auf ein anderes C-Atom im Molekül)*
violently explosive hochexplosiv
violet phosphorus violetter [hittorfscher] Phosphor *m*
violuric acid Violursäure *f*, 5-Isonitrosobarbitursäure *f*, Isonitrosomalonylharnstoff *m*, Alloxan-5-oxim *n*
viral protein Virusprotein *n*, Viruseiweiß *n*
virgin 1. rein, unvermischt; roh, unbehandelt; 2. gediegen *(z. B. Schwefel, Gold)*
virgin metal Primärmetall *n*
virial coefficient *(phys Ch)* Virialkoeffizient *m*
virial equation (of state) Virialgleichung *f*, Virialsatz *m*
virucidal virentötend, virustötend, viruzid
virucide virentötendes [virustötendes, viruzides] Mittel *n* [Präparat *n*]
visbreaking *s.* viscosity breaking
viscid 1. klebrig; 2. *s.* viscous 1.
viscoelastic viskoelastisch

viscoelasticity *(Kst)* Viskoelastizität *f*
viscometer Viskosimeter *n*
viscometric viscosimetrisch
viscometry Viskosimetrie *f*, Viskositätsmessung *f*
viscose 1. Viskose...; 2. *s.* viscous 1.
viscose Viskose *f*
viscose fibre Viskosefaser *f*; Viskosefaserstoff *m*, VI
viscose rayon Viskoseseide *f*, Viskosereyon *n*
viscose silk *(Text)* Kunstseide *f*, Viskoseseide *f*
viscose (spinning) solution Viskose(er)spinnlösung *f*
viscose staple fibre Viskosefaser *f*, Zellwolle *f*
viscosimetric *s.* viscometric
viscosity Viskosität *f*, Zähigkeit *f*, Zähflüssigkeit *f*
viscosity average Viskositätsmittel *n*
viscosity average molecular weight viskositätsmittlere Molmasse *f*
viscosity breaking Viskositätsbrechen *n*, Viskositätsbrechung *f*, Visbreaking *n*
viscosity coefficient Viskositätskoeffizient *m*, Viskositätskonstante *f*, dynamische Viskosität *f*, Konstante *f* der inneren Reibung
viscosity index Viskositätsindex *m*, VI *(Kennzahl für Schmieröle)*
viscosity-temperature coefficient Viskositäts-Temperatur-Koeffizient *m*, VTC, Temperaturkoeffizient *m* der Viskosität [Zähigkeit]
viscous 1. viskos, zähflüssig, dickflüssig; 2. *s.* viscid 1.
viscous elasticity Viskoelastizität *f*
viscous flow viskoses Fließen *n*
viscous force innere Reibung *f*, Flüssigkeitsreibung *f*
visible range [region] *(Spektr)* sichtbarer Bereich *m*, Sichtbares *n*
visible spectroscopy Spektroskopie *f* im sichtbaren Bereich
visible to the naked eye mit unbewaffnetem Auge sichtbar
visual colorimeter visuelles Kolorimeter *n*
visual colorimetry visuelle Kolorimetrie *f*
visual examination Sichtprüfung *f*, visuelle Beurteilung *f*
visual pigment *(Bioch)* Augenpigment *n*, Augenfarbstoff *m*, Seh(farb)stoff *m*
visual purple Sehpurpur *m*, Rhodopsin *n*
vitamer Vitamere *n*, Vitaminanaloge *n*
vitamin Vitamin *n*
vitamin B12 compound Corrinoid *n*, Corrin *n*, Cobalamin *n*, Verbindung *f* der Vitamin-B12-Gruppe
vitamin concentrate Vitaminkonzentrat *n*
vitamin deficiency Vitaminmangel *m*
vitamin intake Vitaminaufnahme *f*
vitamin precursor Provitamin *n*
vitamined vitaminreich
vitaminization Vitamin(is)ieren *n*, Vitamin(is)ierung *f*
vitaminize *v* vitamin(is)ieren, mit Vitaminen anreichern
vitellus Dotter *m(n)*, Eidotter *m(n)*, Eigelb *n*, Vitellus *m*
vitreous 1. gläsern, aus Glas; 2. glasartig, glasähnlich, glasig
vitreous carbon glasartiger Kohlenstoff *m*, Glaskohlenstoff *m*
vitreous china *(Ker)* Vitreous China *n*, Halbporzellan *n*, Porzellangut *n*
vitreous enamel Email *n*, Emaille *f*
vitreous phase glasige Phase *f*, Glasphase *f*
vitreous silica durchsichtiges [klares] Kieselglas *n*, Quarzglas *n*
vitrification 1. *(Ker)* Verglasung *f*; 2. *(Geol)* Frittung *f*, Verglasung *f*, oberflächliche Anschmelzung *f*; 3. *(Gum)* Glasumwandlung *f*, Umwandlung *f* zweiter Ordnung, Gamma-Umwandlung *f*, α-Anomalie *f*
vitrify *v (Ker)* verglasen, dicht [glasig] sintern
vitriol *v* 1. mit Schwefelsäure behandeln; 2. *(Met)* mit Schwefelsäure beizen
vitriol 1. Vitriol *n (kristallwasserhaltiges Sulfat eines zweiwertigen Metalls)*; 2. *s.* vitriol oil
vitriol oil Schwefelsäure *f*, *(veraltet)* Vitriolöl *n*
vitroceramic Vitrokeram *n*, Glaskeramik *f*, Keramik *f* aus Glas, glaskeramischer Stoff *m*
vivianite *(Min)* Blaueisenerz *n*
vivid lebhaft
VLDL *s.* very-low-density lipoprotein
v.m. *s.* volatile matter

VM *s.* volatile matter
V.M.(&)P. naphtha *s.* varnish-makers' (and painters') naphtha
VOCs *s.* volatile organic compounds
void Hohlraum *m*; Pore *f*; Fehlstelle *f (in einer Schutzschicht)*
void fraction *s.* voidage
void space Hohlraum *m*
void volume Zwischenkornvolumen *n*, Kornzwischenraumvolumen *n*, Lückenvolumen *n (einer Schüttschicht)*; Hohlraumvolumen *n*, Porenvolumen *n (z. B. eines Katalysators)*
voidage relatives Porenvolumen *n*, relativer Porenraum *m*, Porosität *f*
volat. 1. *s.* volatile; 2. = volatilizes
volatile (leicht) flüchtig
volatile constituent flüchtiger Bestandteil *m*, flüchtige Komponente *f*
volatile matter flüchtige Substanz *f*, flüchtiger Stoff *m*, flüchtige Bestandteile *mpl*, Flüchtiges *n*
volatile oil ätherisches Öl *n*
volatile organic compounds *(Hyd)* flüchtige organische Wasserinhaltsstoffe *mpl*
volatiles *s.* volatile matter
volatility Flüchtigkeit *f*, Fugazität *f*
volatilizable leicht zu verflüchtigen(d), leicht verdampfbar
volatilization Verdampfung *f*, Verflüchtigung *f*; Verdunstung *f (unterhalb des normalen Siedepunkts)*; Vergasung *f (z. B. von Pflanzenschutzmitteln)*
volatilize *v* verdampfen, verflüchtigen; verdampfen, sich verflüchtigen; verdunsten *(unterhalb des normalen Siedepunkts)*; vergasen *(z. B. Pflanzenschutzmittel)*
Volhard titration Titration *f* [Bestimmung *f*] nach Volhard *(Argentometrie)*
Volta potential Volta-Potenzial *n*, äußeres elektrisches Potenzial *n*
voltage (elektrische) Spannung *f*
voltage drop Spannungsabfall *m*
voltage proof durchschlagsfest
voltage rise Spannungsanstieg *m*
voltaic cell [element] galvanisches [elektrochemisches] Element *n*, galvanische [elektrochemische] Zelle *f* [Kette *f*]
voltameter Coulometer *n*, Voltameter *n*
voltammeter Spannungs-und-Strom-Messer *m*, Volt-und-Ampere-Meter *n*
voltammetric voltammetrisch
voltammetry *(Anal)* Voltammetrie *f*
voltammogram voltammetrische Kurve *f*
voltmeter Spannungsmesser *m*, Voltmeter *n*
volume Volumen *n*, Rauminhalt *m*, Inhalt *m*
volume degree of swelling *s.* volume swelling degree
volume element Volumenelement *n*, Raumelement *n*
volume flux Volumendurchsatz *m*
volume of activation Aktivierungsvolumen *n*
volume of pore space Porenvolumen *n*, Hohlraumvolumen *n*
volume of the molecules Molekülvolumen *n*, Eigenvolumen *n* der Moleküle
volume ratio Volumenverhältnis *n*
volume swelling degree Volumenquellungsgrad *m (Polymernetzwerke)*
volumetric volumetrisch, *(bei Lösungen auch)* titrimetrisch, maßanalytisch, Titrations…
volumetric analysis 1. *(einzelne)* volumetrische Analyse *f*, *(bei Lösungen auch)* titrimetrische Analyse *f*, Titrieranalyse *f*, Maßanalyse *f*; 2. Volumetrie *f*, *(bei Lösungen auch)* Titrimetrie *f*, Maßanalyse *f*, Titrieranalyse *f (Zweig der Labortechnik)*
volumetric factor *(Anal)* Faktor *m*, Korrekturfaktor *m*, Normalitätsfaktor *m*
volumetric flask Messkolben *m*, Maßkolben *m*, Messflasche *f*
volumetric flow rate 1. Volumenstrom *m*; 2. *(Chromat)* *Volumenfluss *m*
volumetric pipette Vollpipette *f*
volumetric production efficiency [rate] *(Biot)* volumetrische Fermentationsrate *f (mg Produkt/l · h)*
volumetric rate of flow *s.* volumetric flow rate
volumetric rate of reaction *s.* volumetric production efficiency
vortex *v* verwirbeln
vortex *(großer)* Wirbel *m*, Trombe *f*, Strudel *m*
vortex dryer Wirbeltrockner *m*
vortex formation Trombenbil-

dung *f (unerwünschte Erscheinung z. B. in einem Reaktor)*

vortexing 1. Verwirbeln *n*, Verwirbelung *f*; 2. *s.* vortex formation

v.p. *s.* vapour pressure

VP *s.* vapour pressure

v.s. *s.* very soluble

V.S.P. *s.* Vicat softening point

v.s.s. *s.* very slightly soluble

vug(g) *(Min)* Druse *f*

vugh *(Min)* Druse *f*

vulcanite Hartgummi *m*, Hartkautschuk *m*

vulcanizable vulkanisierbar, vernetzbar

vulcanizate Vulkanisat *n*

vulcanization Vulkanisation *f*, Vernetzung *f*, Heizung *f*

vulcanization accelerator Vulkanisationsbeschleuniger *m*

vulcanization assistant Vulkanisationshilfsmittel *n*

vulcanization rate Vulkanisationsgeschwindigkeit *f*, Heizgeschwindigkeit *f*

vulcanization reaction Vulkanisationsreaktion *f*

vulcanize *v* vulkanisieren, vernetzen, heizen

vulcanize *v* **at room temperature** kaltvulkanisieren, in der Kälte vulkanisieren, bei Raumtemperatur vulkanisieren

vulcanized rubber vulkanisierter Kautschuk *m*, Kautschukvulkanisat *n*, Gummi *m*

vulcanizer Vulkanisierapparat *m*, Vulkanisiergerät *n*, Vulkanisationsapparat *m*, Heizapparat *m*, Heizer *m*, Vulkanisator *m*

vulcanizing agent Vulkanisationsmittel *n*, Vulkanisationsagens *n*, Vulkanisiermittel *n*, Vernetzungsmittel *n*, Vernetzer *m*

vulcanizing autoclave Vulkanisierkessel *m*, Vulkanisationskessel *m*; Autoklav *m*, Autoklav(en)presse *f*, Autoklavheizpresse *f*

vulnerable angreifbar *(z. B. durch Substituenten)*

W

Wackenroder's solution wackenrodersche Flüssigkeit *f (Gemisch von Polythionsäuren)*

Wadsworth mounting *(Spektr)* Wadsworth-Aufstellung *f*, Wadsworth-Anordnung *f*

waft *v* fächeln *(zur Geruchsprüfung)*

wagging vibration *(Spektr)* Nickschwingung *f*, Wagging-Schwingung *f*, ω-Schwingung *f*

Wagner-Meerwein rearrangement *(org Ch)* Wagner-Meerwein-Umlagerung *f*, Retropinacolinumlagerung *f*

Walden inversion Walden-Umkehr *f*, Konfigurationsumkehr *f*, Konfigurationswechsel *m*, Inversion *f* (der Konfiguration)

Walden law [rule] *(phys Ch)* waldensche Regel *f*

wall Wand(ung) *f*, Mantel *m*

wall-coated open tubular capillary column Dünnschicht-Kapillarsäule *f*, Dünnfilm-Kapillarsäule *f (Gaschromatographie)*

wall collision Stoß *m* mit der Gefäßwand *(bei Kettenreaktionen)*

wall effect Wandeinfluss *m (auf den Reaktionsverlauf)*

wall-reaction Wandreaktion *f*

Wallach rearrangement Wallach-Umlagerung *f (von Azoxybenzen zu 4-Hydroxyazobenzen)*

war gas chemischer Kampfstoff *m*

Warburg's yellow enzyme warburgsches (gelbes) Atmungsferment *n*, Cytochromoxidase *f*

warm *v* erwärmen; vorwärmen; (be)heizen

warm *v* **gently [moderately, slightly]** gelinde erhitzen, gelinde [mäßig, vorsichtig] erwärmen

warm *v* **up** *s.* warm

warming(-up) Erwärmen *n*, Erwärmung *f*; Vorwärmen *n*; Heizen *n*, Beheizen *n*, Beheizung *f*

warming-up period Aufheizperiode *f*, Aufheizabschnitt *m*

wash *v* 1. waschen, spülen, reinigen; durchwaschen; auswaschen; 2. *(Aufb)* läutern; 3. *(Zucker)* decken, abdecken

wash *v* **countercurrently** im Gegenstrom waschen
wash *v* **out** auswaschen
wash *v* **through** durchspülen, durchwaschen
wash 1. Waschen *n*, Waschung *f*, Wäsche *f*, Spülen *n*, Reinigen *n*; Durchwaschen *n*; Auswaschen *n*; 2. Gärlösung *f*, Gärflüssigkeit *f*, Gärgut *n*, Maische *f*; 3. *(Landw)* Brühe *f*, Spritzbrühe *f (zur Schädlingsbekämpfung)*; 4. *(Kosmet)* Wasser *n*, Lotion *f*; 5. *s.* wash liquid 1.; 6. *s.* wash water
wash bottle *s.* washing bottle
wash column Waschkolonne *f*
wash liquid [liquor] 1. Waschflüssigkeit *f*, Waschlauge *f*, Reinigungsflüssigkeit *f*; 2. *(Text)* Waschflotte *f*; 3. *s.* wash water
wash-out *s.* washout
wash solution [solvent] *s.* wash liquid 1.
wash water Waschwasser *n*; Spülwasser *n*
washer 1. Waschmaschine *f*, Wäscher *m*; 2. Scheibe *f*, Ring *m*
washer-cooler Waschkühler *m*, Kühl(er)wascher *m*
washing agent 1. Waschmittel *n*; 2. *(Text)* Waschrohstoff *m*
washing bath *(Text)* Waschflotte *f*, Waschlauge *f*
washing bottle 1. Waschflasche *f*, Gaswaschflasche *f*; 2. Spritzflasche *f*
washing composition Waschmittel *n*
washing device Waschapparat *m*
washing efficiency Wascheffekt *m*, Waschwirkung *f*, Reinigungswirkung *f*
washing liquid [liquor] *s.* 1. wash liquid 1.; 2. wash water
washing-off Auswaschen *n*, Spülen *n*
washing powder Waschpulver *n*
washing soda Kristallsoda *f*, Waschsoda *f*, Soda *f*, Natriumcarbonat-10-Wasser *n*
washing water Waschwasser *n*; Spülwasser *n*
washings Waschwasser *n*; (verbrauchte) Waschflüssigkeit *f*
washout 1. Washout *n (Auswaschen von Luftschadstoffen durch atmosphärische Niederschläge)*; 2. Washout *m (ausgewaschene Luftschadstoffe)*
wastage 1. Verlust *m*; Verschleiß *m*, Abnutzung *f*; Schwinden *n*, Schwund *m*; 2. Verschwendung *f*
waste 1. Ab(fall)produkt *n*, Ab(fall)stoff *m*, Abfall *m*, Abfälle *mpl*, Abgänge *mpl*; Müll *m*; Ausschuss *m*; 2. *(Hyd)* Abwasser *n*; 3. *s.* waste rock; 4. *s.* radioactive waste
waste acid Abfallsäure *f*
waste crock *(Lab)* irdener Abfallkübel *m*
waste discharge 1. Abwassereinleitung *f*, Abwasserableitung *f (in Vorfluter oder Kanalisation)*; 2. in Vorfluter oder Kanalisation einzuleitendes Abwasser *n*, Abwasserablauf *m*
waste discharge standards Abwassereinleitungsbedingungen *fpl*
waste disposal 1. Abproduktbeseitigung *f*; 2. Abwasserentsorgung *f*
waste disposal site Deponie *f*
waste-dumping at sea Einbringen *n* [Verkippen *n*, Versenken *n*] von Abprodukten ins Meer, Verkippen *n* auf hoher See, Abklappen *n*
waste effluent 1. in Vorfluter oder Kanalisation einzuleitendes Abwasser *n*, Abwasserablauf *m (eines Betriebs)*; 2. *s.* waste treatment effluent
waste elimination *s.* waste disposal
waste from chemical plants Abwasser *n* der chemischen Industrie, Chemieabwasser *n*
waste gas Abgas *n*
waste heat Ab(fall)wärme *f*, Abhitze *f*
waste-heat boiler Abhitze(dampf)kessel *m*
waste-heat recovery Abwärmerückgewinnung *f*, Abhitzerückgewinnung *f*
waste jar *(Lab)* Abfallkübel *m*, Abfallbehälter *m*
waste liquor Ablauge *f*
waste material 1. *s.* waste 1.; 2. Altmaterial *n*, Alt(roh)stoff *m*, Sekundärrohstoff *m*
waste paper Altpapier *n*, Abfallpapier *n*, Papierabfälle *mpl*, Makulatur *f*; Ausschusspapier *n*, Makulatur *f*
waste product Ab(fall)produkt *n*
waste production 1. Entstehen *n* [Anfall *m*] von Abprodukten; 2. Abwasseranfall *m*
waste recovery *s.* by-product recovery
waste rock taubes Gestein *n*, Gangart *f*, Nichterz *n*

waste rubber Altgummi *m*; Abfallgummi *m*, Gummiabfälle *mpl*, Vulkanisatabfälle *mpl*; (unvulkanisierte) Fabrikationsabfälle *mpl*
waste sludge Abwasserschlamm *m*
waste solids feste Abprodukte *npl* [Abfallstoffe *mpl*], Müll *m*
waste tip Müllkippe *f*; Abraumhalde *f*, Halde *f*
waste treatment 1. Abproduktbehandlung *f*, Abfallbehandlung *f*; 2. Abwasserbehandlung *f*, Abwasserreinigung *f*
waste treatment effluent (gereinigtes) abfließendes Abwasser *n*, Endablauf *m*, Abwasserablauf *m*, Kläranlagenablauf *m*, Ablauf *m* der Abwasserbehandlungsanlage
waste treatment lagoon Abwasserteich *m*
waste treatment plant Abwasserbehandlungsanlage *f*, ABA, Abwasserreinigungsanlage *f*, Kläranlage *f*, KA, Klärwerk *n*
waste treatment process Abwasserbehandlungsverfahren *n*, Abwasserreinigungsverfahren *n*
waste water Abwasser *n*
waste-water analysis Abwasseranalyse *f*
waste-water clarification Abwasserklärung *f*, mechanische Abwasserreinigung *f*
waste-water component Abwasserinhaltsstoff *m*
waste-water contaminant *s.* waste-water pollutant
waste-water control Abwasserüberwachung *f*
waste-water disposal Abwasserbeseitigung *f*, Abwasserentsorgung *f*
waste-water load(ing) Abwasserlast *f*, Abwasserbelastung *f*, Belastung *f* mit [durch] Abwasser
waste-water pollutant Abwasserschmutzstoff *m*, Abwasserverschmutzung *f*, Verunreinigung *f* im Abwasser, *(i. e. S.)* Abwasserschadstoff *m*
waste-water pond Abwasserteich *m*
waste-water purification Abwasserreinigung *f*
waste-water sample Abwasserprobe *f*
waste-water treatment Abwasserbehandlung *f*, Abwasserreinigung *f*
waste wood Abfallholz *n*
wasteful unvorteilhaft; ohne Nebenprodukt(en)gewinnung arbeitend, ohne [unter Verzicht auf] Nebenproduktgewinnung
wastewater *s.* waste water
watch glass Uhrglas *n*, Uhrglasschale *f*
water *v* befeuchten, bewässern, (be-)sprengen, (be)gießen; tränken; mit Wasser versorgen
water-cool *v* mit Wasser kühlen
water-quench *v* 1. durch [mit] Wasser ablöschen *(z. B. Koks)*; 2. *(Met)* in Wasser abschrecken [ablöschen]
water-absorbing wasseraufnehmend, wasserabsorbierend, wasseranziehend, hygroskopisch
water absorption Wasserabsorption *f*, Wasseraufnahme *f*
water-abstracting agent wasserentziehendes Mittel *n*
water addition 1. Wasserzusatz *m*; 2. Wasseranlagerung *f*
water affinity Affinität *f* zu(m) Wasser
water analysis Wasseranalyse *f*, wasseranalytische Untersuchung *f*
water aspirator *(Lab)* Wasserstrahlpumpe *f*
water-attracting *s.* water-absorbing
water-base paint *s.* water paint
water bath Wasserbad *n*
water calorimeter Wasserkalorimeter *n*, Flüssigkeitskalorimeter *n*, Mischungskalorimeter *n*
water characteristics Wassergütemerkmale *npl*, Kennzeichen *npl* [Kennwerte *mpl*] der Wasserbeschaffenheit
water chemist Wasserchemiker *m*, Hydrochemiker *m*
water chemistry Wasserchemie *f*, Hydrochemie *f*
water circuit Wasserkreislauf *m*, Wasserumlauf *m*
water clarification *s.* water treatment
water clean-up Wasserreinigung *f*
water-clear wasserklar, wasserhell
water conditioning Wasseraufbereitung *f*
water constituent Wasserinhaltsstoff *m*

water contaminant s. 1. water pollutant; 2. water constituent
water contamination s. water pollution
water content Wassergehalt *m*
water cooling Wasserkühlung *f*, Kühlung *f* mit Wasser
water cycle Wasserkreislauf *m*, Wasserumlauf *m*
water demineralizer Entsalzungsanlage *f*, *(i. e. S.)* Vollentsalzungsanlage *f*
water desalination Wasserentsalzung *f*
water-dilutable wasserverdünnbar, mit Wasser verdünnbar
water droplet Wassertröpfchen *n*
water ejector s. water-operated ejector
water electrolysis Wasserelektrolyse *f*, Wasserzersetzung *f*
water electrolyzer Wasserzersetzungszelle *f*, Wasserzersetzer *m*
water entrainer *(Dest)* Wasserschlepper *m*
water envelope Hydrathülle *f*
water equivalent Wasserwert *m (z. B. eines Kalorimeters)*
water filter Wasserfilter *n*
water filtration Wasserfiltration *f*
water for industrial purposes [use] Industriewasser *n*, Betriebswasser *n (für Industriebetriebe)*; (industrielles) Brauchwasser *n*
water gas Wassergas *n*
water-gas generator Wassergasgenerator *m*, Wassergaserzeuger *m*
water-gas reaction 1. (heterogene) Wassergasreaktion *f (Reduktion von Wasser mittels Kohlenstoffs)*; 2. s. water-gas-shift reaction
water-gas shift CO-Konvertierung *f*, Kohlenmonoxid-Konversion *f*
water-gas-shift equilibrium Wassergasgleichgewicht *n*
water-gas-shift reaction Konvertierungsreaktion *f* (des Kohlenmonoxids), homogene Wassergasreaktion *f (Reduktion von Kohlendioxid mittels Wasserstoffs)*
water glass Wasserglas *n (Lösung von Alkalisilicaten in Wasser)*
water hardness Wasserhärte *f*
water-immiscible nicht mit Wasser mischbar
water-in-fat emulsion Wasser-in-Fett-Emulsion *f*
water-in-oil emulsion Wasser-in-Öl-Emulsion *f*
water-insolubility Wasserunlöslichkeit *f*
water-insoluble wasserunlöslich, nicht wasserlöslich, in Wasser unlöslich, nicht in Wasser löslich
water-jacketed wassergekühlt, mit Wassermantel, Wassermantel...
water jet Wasserstrahl *m*
water leachate wässriger Auszug *m*
water metabolism *(Bioch)* Wasserstoffwechsel *m*, Wasserhaushalt *m*
water-miscible mit Wasser mischbar
water molecule Wassermolekül *n*
water of capillarity Kapillarwasser *n*, Porensaugwasser *n*
water of constitution Konstitutionswasser *n*, konstitutiv gebundenes Wasser *n*
water of crystallization Kristallwasser *n*
water of deliquescence von hygroskopischen Stoffen aufgenommenes Wasser *n*
water of dilution Verdünnungswasser *n*
water of hydration Hydratwasser *n*
water-operated ejector Wasserstrahlpumpe *f*
water paint Wasserfarbe *f*, wässriger Anstrichstoff *m*
water phase wässrige Phase *f*
water pollutant Schmutzstoff *m* im Wasser, Wasserverunreinigung *f*, *(i. e. S.)* schädlicher Wasserinhaltsstoff *m*, Wasserschadstoff *m*
water pollution Gewässerverschmutzung *f*, Gewässerverunreinigung *f*, Gewässerbelastung *f*, Wasserverschmutzung *f*, Wasserverunreinigung *f*
water pollution control Gewässerschutz *m*, Schutz *m* der Gewässer vor Verunreinigungen, Gewässerreinhaltung *f*, Wasserreinhaltung *f*, Kontrolle *f* der Reinhaltung der Gewässer
water properties s. water quality
water protection s. water pollution control
water pump *(Lab)* Wasserstrahlpumpe *f*
water purification s. water treatment
water quality Wassergüte *f*, Wasserqualität *f*, Wasserbeschaffenheit *f*
water-quality characteristics s. water characteristics
water-quality criteria 1. Wassergüte-

bedingungen *fpl*; 2. *s.* water characteristics
water-quality parameters *s.* water characteristics
water-quality standards Wassergütebedingungen *fpl*
water quenching 1. Löschen *n*, Ablöschen *n*, Ablöschung *f (z. B. des Kokses)*; 2. *(Met)* Wasserabschrecken *n*, Wasserabschreckung *f*, Wasserablöschung *f*, Abschrecken *n* in Wasser
water recirculation Kreislaufwasserführung *f*, Wasserkreislaufführung *f*, Kreislaufnutzung *f* des Wassers
water recovery Wasserrückgewinnung *f*
water recycle Wasserkreislauf *m*, Wasserumlauf *m*
water removal Wasserentzug *m*, Wasserabtrennung *f*
water repellency Wasserabweisungsvermögen *n*, wasserabweisende Kraft *f* [Eigenschaft *f*], wasserabweisendes Verhalten *n*, wasserabweisender Charakter *m*, Wasserabweisung *f*, Wasserabstoßung *f*, Abperleffekt *m*
water-repellent wasserabweisend, wasserabstoßend, hydrophob
water repellent Hydrophobier(ungs)mittel *n*, wasserabweisendes [wasserabstoßendes, hydrophobierendes] Mittel *n*
water resistance Wasserbeständigkeit *f*, Wasserfestigkeit *f*
water-resistant wasserbeständig, wasserfest
water reuse Wasserwiederverwendung *f*, Wasserwiederverbrauch *m*, Wassermehrfachnutzung *f*
water sample Wasserprobe *f*
water sampler Wasserschöpfer *m*, Schöpfgefäß *n* für Probenentnahme, Probenschöpfer *m*, Entnahmegerät *n*
water sampling Entnahme *f* von Wasserproben, Probenentnahme *f*, Probenahme *f*
water scrubbing Wasserwäsche *f*, Druckwasserwäsche *f (von Synthesegas)*
water separator Wasserabscheider *m*
water sheath Wasserhülle *f (z. B. von Ionen)*
water sizing Hydroklassieren *n*, Nassklassieren *n*
water-slaked lime Löschkalk *m*, gelöschter Kalk *m*
water softener 1. Wasserenthärter *m*, Wasserenthärtungsmittel *n*; 2. *s.* water-softener plant
water-softener plant Wasserenthärtungsanlage *f*
water softening Wasserenthärtung *f*
water solubility Wasserlöslichkeit *f*, Löslichkeit *f* in Wasser
water-soluble wasserlöslich
water-soluble paint wasserlöslicher Lack *m*
water solution wässrige [wässerige] Lösung *f*
water suction pump *(Lab)* Wasserstrahlpumpe *f*
water-swellable in Wasser quellbar
water technology 1. Wasser(aufbereitungs)technik *f*, Wasserreinigungstechnik *f*; 2. Wasser(reinigungs)technologie *f*
water-thinned paint wasserverdünnbarer [mit Wasser verdünnbarer] Anstrichstoff *m*
water-tight wasserdicht, wasserundurchlässig
water treatment Wasseraufbereitung *f*, Wasserbehandlung *f*, Wasserreinigung *f*
water treatment chemistry Verfahrenschemie *f* der Wasseraufbereitung
water treatment plant Wasseraufbereitungsanlage *f*, Wasserbehandlungsanlage *f*, Wasserwerk *n*, WW
water treatment process Wasseraufbereitungsverfahren *n*
water treatment sludge Wasserwerksschlamm *m*
water treatment technology *s.* water technology
water treatment works *s.* water treatment plant
water-tube boiler Wasserrohrdampferzeuger *m*, Wasserrohrkessel *m*, Siederohrdampferzeuger *m*, Siederohrkessel *m*
water vapour Wasserdampf *m*
water vapour absorption Wasserdampfaufnahme *f*, Feuchteaufnahme *f*
water vapour permeability [transmission] Wasserdampfdurchlässigkeit *f*
water-white wasserhell, wasserklar

water works *s.* water treatment plant
waterborne coating *(Anstr)* Wasserlack *m*, Lack *m* auf Wasserbasis
waterproof *v* 1. *(Text)* wasserdicht [wasserundurchlässig] machen; 2. wasserdicht abschließen, kapseln *(Geräte)*
waterproof 1. *(Text)* wasserdicht, wasserundurchlässig; 2. wassergeschützt, wasserdicht abgeschlossen, gekapselt *(Geräte)*
waterworks *s.* water treatment plant
watery wässrig, wässerig
Watson-Crick model Watson-Crick-Modell *n (der Desoxyribonucleinsäure)*
wave Welle *f*
wave equation Wellengleichung *f*
wave function Wellenfunktion *f*
wave height Stufenhöhe *f (Polarographie)*
wave mechanics Wellenmechanik *f*
wave number Wellenzahl *f*
wavelength Wellenlänge *f*
wavenumber Wellenzahl *f*
wax *v* wachsen
wax Wachs *n (i. e. S., einschließlich Kohlenwasserstoffwachs)*
wax-like wachsartig, wachsähnlich
waxlike wachsartig, wachsähnlich
waxy 1. wachsartig, wachsähnlich; 2. wächsern, aus Wachs
3-way cock *(Lab)* Dreiweg(e)hahn *m*
WCOT = wall-coated open tubular capillary column
weak acid schwache Säure *f (nach dem Dissoziationsgrad)*
weak-acid cation exchanger [resin] *(Hyd)* schwach saurer Kationenaustauscher *m*
weak base schwache Base *f (nach dem Dissoziationsgrad)*
weak-base anion exchanger [resin] *(Hyd)* schwach basischer Anionenaustauscher *m*
weak bond schwache Bindung *f*
weak electrolyte schwacher Elektrolyt *m*
weak liquor Dünnlauge *f*
weak sewage *(Hyd)* gering konzentriertes Abwasser *n*, schwach belastetes [„dünnes"] Abwasser *n*, Dünnwasser *n*
weaken *v* schwächen *(z. B. eine Bindung)*
weakly acid(ic) schwach sauer
weakly basic schwach basisch [alkalisch]
weakly coloured schwach (an)gefärbt, farbschwach, schwachfarbig
wear Verschleiß *m*, Abnutzung *f*
wear debris Abriebmaterial *n*
wear resistance Abriebwiderstand *m*, Abriebbeständigkeit *f*, Abriebfestigkeit *f*, Abnutzungsbeständigkeit *f*, Verschleißwiderstand *m*, Verschleißfestigkeit *f*
wear-resistant abriebbeständig, abriebfest, abnutzungsbeständig, verschleißbeständig, verschleißfest
wear-resisting abriebbeständig, abriebfest, abnutzungsbeständig, verschleißbeständig, verschleißfest
weather *v* 1. *(Geoch)* verwittern; 2. bewittern *(der Witterung aussetzen)*
weather resistance Wetterbeständigkeit *f*, Wetterfestigkeit *f*, Witterungsbeständigkeit *f*
weather-resistant wetterbeständig, wetterfest, witterungsbeständig
weathering device Bewitterungsgerät *n*
weathering product Verwitterungsprodukt *n*
weathering resistance *s.* weather resistance
weathering test Bewitterungsversuch *m*
web 1. Gewebe *n*; 2. *(Kst)* endlose Bahn *f*; 3. *s.* web of paper
web coating *(Kst)* Beschichten *n* endloser Bahnen
web of paper Papierbahn *f*, Stoffbahn *f*, Papiervlies *n*, Faserfilz *m*
web of pulp Zellstoffbahn *f*
Wedge roaster [roasting furnace] Wedge-Ofen *m*, Mehretagenröstofen *m* nach Wedge
weed control agent *s.* weed killer
weed killer Unkrautbekämpfungsmittel *n*, Unkrautvertilgungsmittel *n*, Herbizid *n*
weedicide *s.* weed killer
weigh *v* 1. wiegen *(eine Masse feststellen)*; wiegen *(eine Masse haben)*; 2. *s.* weigh out
weigh *v* **in** einwägen, einwiegen
weigh *v* **out** abwiegen
weigh *v* **to 1 mg accuracy** auf ein Milligramm genau wägen

weigh *v* **to the nearest mg** *s.* weigh to 1 mg accuracy
weigh *v* **up** abwiegen
weighing bottle Wägegläschen *n*, Wägeglas *n*
weighing error Wägefehler *m*, Wägungsfehler *m*
weighing form Wägeform *f*, Wägungsform *f*
weighing piggy Wägeröhrchen *n*, Wägeschweinchen *n*, Schweinchen *n*
weighing scoop Wägeschiffchen *n*
weighing tube *s.* weighing piggy
weight *v* belasten, beschweren
weight 1. Masse *f (Eigenschaft eines Körpers, in einem Schwerefeld ein Gewicht anzunehmen; Einheit: kg)*; 2. Gewicht *n (die von einer Masse ausgeübte Kraft; Einheit: kp)*; 3. Wägestück *n (zum Wägen benutzter Körper von bestimmter Masse)*; Gewichtsstück *n (Körper nicht festgelegter Masse zum Lastausgleich)*
weight average Gewichtsmittel *n*
weight average molecular weight gewichtsmittlere Molmasse *f*
weight gain Masse(n)zunahme *f*; Gewichtszunahme *f*
weight loss Masse(n)verlust *m*; Gewichtsverlust *m*
weighting agent Streckmittel *n*, Verschnittmittel *n*
Weiss (molecular magnetic) field *(Krist)* weißscher Bezirk *m*
Weissenberg effect Weissenberg-Effekt *m (bei viskoelastischen Flüssigkeiten)*
weld *v* schweißen; sich schweißen lassen, schweißbar sein
weld 1. Schweißstelle *f*; Schweißnaht *f*, (geschweißte) Naht *f*; 2. Schweißung *f*
welding Schweißen *n*, Schweißung *f*
welding flux Schweiß(fluss)mittel *n*
welding torch Schweißbrenner *m*
well-defined wohldefiniert, gut definiert
well-developed *(Krist)* gut ausgebildet
well-mixed (flow) reactor 1. Rühr(kessel)reaktor *m*, Rührkessel *m*, Mischreaktor *m*; 2. *(Hyd)* vollständiges Mischbecken *n*; 3. *(Biot)* vollständig gemischter Fermenter *m*
well-mixed reactors in series Rührkesselkaskade *f*, Mischreaktorkaskade *f*, Kaskadenreaktor *m*, mehrstufiger Rührreaktor *m*
well-ordered *(phys Ch)* wohlgeordnet
Werner complex Koordinationsverbindung *f*, Komplexverbindung *f*
Weston normal cell *(Ech)* Weston-Normalelement *n*, Weston-Element *n*
Westphal balance westphalsche Waage *f (zur Dichtebestimmung)*
wet *v* (be)netzen, befeuchten, anfeuchten
wet-clean *v* 1. nassreinigen, waschen; 2. nassaufbereiten, auf nassem Wege aufbereiten
wet air oxidation Nassoxidation *f (als natürlicher Vorgang)*; Nassverbrennung *f*, Nass-Luft-Oxidation *f (von Abprodukten)*
wet analysis Analyse *f* auf nassem Wege
wet-and-dry-bulb hygrometer [thermometer] Psychrometer *n (ein Luftfeuchtemesser)*
wet ashing *(Lab)* feuchtes [nasses] Veraschen *n*
wet assay *(Met)* nasse Probe *f*, Nassprobe *f*
wet basis *s.* wet-weight basis
wet-bulb temperature Feuchtkugeltemperatur *f*, Temperatur *f* des feuchten Thermometers, Kühlgrenztemperatur *f*
wet-bulb thermometer feuchtes Thermometer *n*, Feuchtthermometer *n*
wet classification Nassklassieren *n*, Hydroklassieren *n*
wet classifier Nassklassierer *m*, Hydroklassierer *m*
wet cleaner Nassreiniger *m*
wet cleaning 1. Nassreinigung *f*, Waschen *n*, Wäsche *f*; 2. Nassaufbereitung *f*, Aufbereitung *f* auf nassem Wege
wet collection Nass(staub)abscheidung *f*; nassmechanische Abscheidung *f*
wet collector Nass(staub)abscheider *m*, Nassentstauber *m*, Waschentstauber *m*, Skrubber *m*, Wäscher *m*, Wascher *m*
wet combustion Nassverbrennung *f*, Nass-Luft-Oxidation *f (von Abprodukten)*
wet condenser nasser Mischkondensator *m*, Einspritzkondensator *m*

wet-cyclone classifier Hydrozyklon *m*, Nasszyklon *m*
wet deposition nasse Deposition *f*, Nassablagerung *f (von Säuren aus der Atmosphäre)*
wet galvanizing galvanisches Verzinken *n*
wet gas 1. nasses Gas *n*; 2. *s.* wet natural gas
wet grinding *s.* wet milling
wet jig Nasssetzmaschine *f*, hydraulische Setzmaschine *f*, hydraulischer Setzapparat *m*
wet metallurgy Nassmetallurgie *f*, Hydrometallurgie *f*
wet milling (process) Nass(ver)mahlen *n*, Nass(ver)mahlung *f*
wet natural gas nasses [feuchtes] Erdgas *n* [Naturgas *n*], Nassgas *n*
wet oxidation *s.* wet air oxidation
wet precipitator Nasselektroabscheider *m*, Nasselektrofilter *n*
wet process 1. nasses Verfahren *n*, Nassverfahren *n*; 2. Nassaufbereitung *f*
wet-process plant Nassaufbereitungsanlage *f*
wet product Nassgut *n*, Feuchtgut *n (bei der Trocknung)*
wet screening Nasssieben *n*
wet scrubbing *s.* wet collection
wet spinning Nassspinnen *n*, Nassspinnverfahren *n*
wet steam Nassdampf *m*
wet strength Nassfestigkeit *f*
wet table *(Aufb)* Nassherd *m*
wet tabling *(Aufb)* nasses Herdsortieren *n*
wet washing *s.* wet cleaning 2.
wet-weight basis Bezugsbasis *f* Masse des feuchten Stoffs
wettability Benetzbarkeit *f*, Netzbarkeit *f*
wettable (be)netzbar
wetted-wall absorber Rieselabsorber *m*, Absorber *m* mit berieselter Wand, Dünnschichtabsorber *m*, Fallstromabsorber *m*
wetted-wall column Rieselfilmkolonne *f*, leere Kolonne *f*
wetting ability Benetzungsfähigkeit *f*, Netzfähigkeit *f*, Benetzungsvermögen *n*, Netzvermögen *n*
wetting agent [aid] Benetzungsmittel *n*, Netzmittel *n*, Netzer *m*
wetting angle Benetzungswinkel *m*, Randwinkel *m*
wetting tension Benetzungsspannung *f*
W.H. *s.* white heat
wheat starch Weizenstärke *f*
Wheatstone bridge Wheatstone-Brücke *f*
Wheland intermediate *(org Ch)* Wheland-Intermediat *m*
whipping tube Siedekapillare *f*
whirl *v* **up** aufwirbeln
whisker Whisker *m*, Haarkristall *m*, Nadelkristall *m*, Faserkristall *m*, Einkristallfaser *f*
white 1. Weiß *n (Farbempfindung)*; 2. weißer [weißfärbender] Farbstoff *m*, Weiß *n*
white arsenic 1. (weißes) Arsenik *n*, Arsentrioxid *n*, Arsen(III)-oxid *n*; 2. *(Min)* Arsenolith *m (Arsen(III)-oxid)*
white biotechnology weiße Biotechnologie *f (Industrieprozesse)*
white bole *(Min)* weißer [reiner] Ton *m*, Porzellanerde *f*, Kaolin *m*, China Clay *m(n)*
white cast iron weißes Gusseisen *n*, Weißguss *m*, Hartguss *m*
white heat Weißglut *f*
white-hot weißglühend
white lead Bleiweiß *n*, Karbonatbleiweiß *n (Bleihydroxidcarbonat)*
white metal Weißmetall *n*, Lagerweißmetall *n (Legierung zum Ausgießen von Lagern)*; Spurstein *m*, Weißmetall *n (praktisch eisenfreier Kupferstein mit 78 bis 80 % Kupfer)*
white (mineral) oil Weißöl *n*
white phosphorus weißer Phosphor *m*
white precipitate 1. $Hg(NH_3)_2C_2$ schmelzbares weißes Präzipitat *n*; 2. $Hg(NH_2)Cl$ unschmelzbares weißes Präzipitat *n*
white rust weißer Rost *m*, Zinkrost *m*
white tin β-Zinn *n (metallische Modifikation des Zinns)*
white vitriol *(Min)* Zinkvitriol *m*, Goslarit *m (Zinksulfat-7-Wasser)*
α-**white phosphorus** kubisch kristallisierter weißer Phosphor *m*
β-**white phosphorus** hexagonal kristallisierter weißer Phosphor *m*
whiten *v (Pap)* bleichen, entfärben, den Weißgehalt erhöhen, auf eine höhere

Weiße bleichen, einen hohen Weißgehalt schaffen [herstellen]; weiß werden
whiteness Weißgehalt *m*, Weißgrad *m*, Weiße *f*
whitening *(Pap)* Bleichen *n*, Bleichung *f*, Entfärben *n*, Entfärbung *f*, Schaffung *f* [Herstellung *f*] eines hohen Weißgehalts, Erhöhung *f* des Weißgehalts; Weißwerden *n*
whitening agent Weißtöner *m*, (optischer) Aufheller *m*, (optisches) Aufhellungsmittel *n*, optisches Bleichmittel *n*, *(in Waschmitteln auch)* Weißmacher *m*
whiteware *(Ker) dichte Ware mit weißem Scherben*
whiting (fein) gemahlene Kreide *f*; (fein) gemahlener Kalkstein *m*
whiting putty Glaserkitt *m*
whizzer Schleuder *f*, Zentrifuge *f*; Schleudertrockner *m*, Zentrifugaltrockenmaschine *f*, Trockenzentrifuge *f*
wide-angle X-ray pattern Weitwinkelaufnahme *f (kristallographische Röntgenaufnahme)*
wide-band decoupling Breitbandentkopplung *f*
wide-line NMR [nuclear magnetic resonance] Breitlinienkernresonanz *f*
wide-mouth bottle Weithalsflasche *f*, Pulverflasche *f*
wide-mouth flask Weithalskolben *m*
wide-stemmed funnel Pulvertrichter *m*
Widmannstätten figures [lines, patterns] widmannstättensche Figuren *fpl (Metallographie)*
Wien displacement law wiensches Verschiebungsgesetz *n*
Wien effect Wien-Effekt *m*, Feldstärkeeffekt *m*
Wijs (iodine monochloride) solution Wijs-Lösung *f*, wijssche Lösung *f (Lösung von Iodmonochlorid in Eisessig zur Bestimmung der Iodzahl von Ölen und Fetten)*
Wijs special solution Wijs-Lösung *f*, wijssche Lösung *f (Lösung von Iodmonochlorid in Eisessig zur Bestimmung der Iodzahl von Ölen und Fetten)*
wildsteel unberuhigter Stahl *m*
Willgerodt reaction Willgerodt-Reaktion *f (Säureamidsynthese mittels Redoxamidierung von Alkyl-Aryl-Ketonen)*
Williamson (ether) synthesis Williamson-Synthese *f (von Ethern durch Alkoholatalkylierung)*
Wilson (cloud) chamber (wilsonsche) Nebelkammer *f*, Wilson-Kammer *f*
wine stone (roher) Weinstein *m*, Rohweinstein *m*, Kaliumhydrogentartrat *n*
wing shape policeman Spatenwischer *m (ein Gummiwischer)*
wing tip [top] *(Lab)* Breitbrenneraufsatz *m*, Schnittbrenneraufsatz *m*, Schlitzaufsatz *m*
Winkler generator Winkler-Generator *m*, Winkler-Gaserzeuger *m*
Winkler process Winkler-Verfahren *n*, Winkler-Vergasungsverfahren *n*, Winkler-Generatorverfahren *n*
wipe *v* **dry** trockenwischen
wiper Abstreifer *m*, Schaber *m*; Ausräumer *m*
wire 1. Draht *m*; 2. *(Pap)* Sieb *n (der Papiermaschine)*
wire cloth Drahtgewebe *n*, Metallgewebe *n*
wire gauze 1. Drahtnetz *n*; 2. *s.* wire cloth
wire-gauze packing Drahtnetzfüllkörper *m*
wire screen [sieve] Drahtsieb *n*, Metallsieb *n*
withdraw *v* abziehen, ableiten, abführen, absaugen, entnehmen, abnehmen *(Flüssigkeiten oder Gase)*
withdraw *v* **by pipette** abpipettieren
withdraw *v* **electrons** Elektronen (ent)ziehen
withdraw *v* **samples** Proben entnehmen
withdrawal Abziehen *n*, Ableiten *n*, Abführen *n*, Absaugen *n*, Entnahme *f*, Abnahme *f (von Flüssigkeiten oder Gasen)*
withdrawal of samples Proben(ent-)nahme *f*, Probenahme *f*
witherite *(Min)* Witherit *m (Bariumcarbonat)*
Witt filter plate *(Lab)* Filterscheibe *f* nach Witt
Witt jar wittscher Topf *m*
Wittig ether rearrangement Wittig-Umlagerung *f (von Ethern zu Alkoholen)*
Wittig reaction wittigsche Phosphin-Methylen-Reaktion *f*, Wittig-Reaktion *f (aufbauende Carbonylolefinierung)*
Wittig reagent Phosphoniumylid *n*,

Phosphorylid *n*, *P*-Ylid *n*, Alkylidenphosphoran *n*, Phosphinalkylen *n*
w/o [W/O] emulsion *s.* water-in-oil emulsion
woad Blaufarbstoff *m*
wobble base *(Bioch)* Wobblebase *f*, 3'-Codon-Base *f*
Wohl degradation [reaction] Wohl-Abbau *m (von Aldosen zum nächstniederen Zucker)*
Wolff-Kishner reaction [reduction] Wolff-Kishner-Reaktion *f*, Wolff-Kishner-Reduktion *f*, Reduktion *f* nach Wolff-Kishner
wolfram W Wolfram *n*
wolframate Wolframat *n (Salz der Wolframsäure)*
wolframic Wolfram..., *(i. e. S.)* Wolfram(VI)-...
wolframic acid 1. H_2WO_4 (gelbe) Wolframsäure *f*; $H_2WO_4 \cdot H_2O$ (weiße) Wolframsäure *f*; 2. WO_3 Wolfram(VI)-oxid *n*
wolframite *(Min)* Wolframit *m (Eisenmanganwolframat)*
wollastonite *(Min)* Wollastonit *m (Calciummetasilicat)*
woman chemist Chemikerin *f*
wood alcohol Holzgeist *m*, Methylalkohol *m*, *Methanol *n*
wood carbonization Holzverkohlung *f*, Holzentgasung *f*
wood cellulose Holzcellulose *f*, Holzzellstoff *m*
wood chemistry Holzchemie *f*
wood distillation Holzdestillation *f*, Trockendestillation *f* des Holzes
wood fibre Holzfaser *f*
wood gas Holzgas *n*
wood hydrolysis Holzhydrolyse *f*
wood preservation Holzschutz *m*, Holzkonservierung *f*
wood preservative Holzschutzmittel *n*
wood pulp *(Pap)* (mechanischer) Holzschliff *m*, (mechanischer) Holzstoff *m*, Holzmasse *f*, Schleifmasse *f*; Holzzellstoff *m*, Zellstoff *m*
wood rosin Wurzelharz *n*, Wurzelkolophonium *n*, Extraktionskolophonium *n*, Extraktionsharz *n*
wood saccharification Holzverzuckerung *f*
wood spirit Holzgeist *m*, Methylalkohol *m*, *Methanol *n*, *(i. e. S.)* roher Holzgeist *m* [Methylalkohol *m*], Rohholzgeist *m*
wood sugar Holzzucker *m (Produkt der Holzverzuckerung)*; Holzzucker *m*, Xylose *f*
wood tar Holzteer *m*
wood vinegar Holzessig *m*
wood wastes Holzabfälle *mpl*
Wood's metal woodsches Metall *n*, woodsche Legierung *f*
Woodward rule *(Spektr)* Woodward-Regel *f*, woodwardsche Absorptionsregel *f*
wool alcohol Wollfettalkohol *m*
wool dye Wollfarbstoff *m*
wool fat Wollfett *n*, *(i. e. S.)* Lanolin *n*, gereinigtes Wollfett *n*
wool grease Wollfett *n*
wool wax Wollwachs *n*, Rohwollfett *n*
work *v* 1. arbeiten; 2. arbeiten lassen *(eine Apparatur)*; 3. kneten, durcharbeiten *(viskose Massen)*; 4. *(Bgb)* abbauen; 5. *s.* work in
work *v* **for** verarbeiten auf
work *v* **in** 1. *(Ech)* durcharbeiten, einarbeiten *(Elektrolyte im Gleichstrom)*; 2. *(Tech, Kosmet)* einarbeiten *(Mischungsbestandteile, Salbe)*
work *v* **opencast** im Tagebau arbeiten
work *v* **up** aufarbeiten
work done geleistete [umgesetzte, gewonnene] Arbeit *f*
work function 1. Austrittsarbeit *f*, Ablösearbeit *f*, Abtrennungsarbeit *f (von Elektronen)*; 2. freie Energie *f*, F
work of adhesion Adhäsionsarbeit *f*, Haftarbeit *f*
work of cohesion Kohäsionsarbeit *f*
work of compression Verdichtungsarbeit *f*, Kompressionsarbeit *f*
work of expansion Volumen(änderungs)arbeit *f*
workability 1. Bearbeitbarkeit *f*; Verarbeitbarkeit *f*; 2. *(Ker)* Formbarkeit *f*; 3. *(Bgb)* Abbaubarkeit *f*; Abbauwürdigkeit *f*
workable 1. bearbeitbar; verarbeitbar; 2. *(Ker)* formbar; 3. *(Bgb)* abbaubar; abbauwürdig
working 1. Arbeiten *n*; 2. Kneten *n*, Durcharbeiten *n (von viskosen Massen)*; 3. *(Bgb)* Abbau *m*, Verhieb *m*

working electrode Arbeitselektrode *f*, Versuchselektrode *f*, Messelektrode *f*
working level *(Tox)* konventioneller Richtwert *m (für Toleranzdosen)*
working life Lebensdauer *f*, Nutzungsdauer *f*, Betriebsdauer *f*, Standzeit *f*, Gebrauchswertdauer *f*
working line *(Dest)* Arbeitslinie *f*
working point *(Tech)* Betriebspunkt *m*
working pressure Arbeitsdruck *m*, Betriebsdruck *m*
working range Verarbeitungsbereich *m*
working temperature Arbeitstemperatur *f*, Betriebstemperatur *f*
working-up Aufarbeitung *f*
working volume Arbeitsvolumen *n*, Betriebsvolumen *n*, Füllvolumen *n (eines Reaktors)*
workplace maximum maximale Arbeitsplatzkonzentration *f*, MAK-Wert *m*
works chemist Betriebschemiker *m*
works effluent Betriebsabwasser *n*, betriebliches Abwasser *n*, Werksabwasser *n*
workup Aufarbeitung *f*
worm 1. *(Tech)* Schnecke *f*; 2. *(Dest)* Kühlschlange *f*
worm conveyor Schneckenförderer *m*
wort Würze *f*, Bierwürze *f*
wort kettle Würze(koch)kessel *m*, Würzepfanne *f*, Braukessel *m*, Braupfanne *f*
Woulfe [Woulff] bottle woulfesche [woulffsche] Flasche *f*
woven(-fabric) filter Tuchfilter *n*, Stofffilter *n*, Gewebefilter *n*; Gewebeabscheider *m*
woven-glass-fibre cloth Glasfasergewebe *n*
wrought alloy Knetlegierung *f*
wrought iron Schweißstahl *m*
wt *s.* weight
Wurtz-Fittig reaction [synthesis] Wurtz-Fittig-Reaktion *f*, Wurtz-Fittig-Synthese *f (zur Darstellung von alkylierten Aromaten)*
Wurtz reaction [synthesis] wurtzsche Reaktion *f* [Synthese *f*], Wurtz-Reaktion *f*, Wurtz-Synthese *f (zur Darstellung von Alkanen)*
w/v = weight to volume
WVT *s.* water vapour permeability [transmission]
WWB *s.* wet-weight basis

X

X-amorphous röntgenamorph
X radiation Röntgenstrahlung *f*
x-ray *v* (mit Röntgenlicht) durchstrahlen; eine Röntgenaufnahme anfertigen von
X ray Röntgenstrahl *m*
X-ray absorption Röntgen(strahlen)absorption *f*, Absorption *f* der Röntgenstrahlen [Röntgenstrahlung]
X-ray analysis Röntgen(strahl)analyse *f*, *(i. e. S.)* Röntgen(fein)strukturanalyse *f*, Röntgenfeinstrukturuntersuchung *f*, Strukturanalyse *f* mit Röntgenstrahlen
X-ray-analytical röntgenanalytisch
X-ray back reflection Laue technique Laue-Verfahren *n*, Laue-Aufnahmetechnik *f*
X-ray crystal(-structure) analysis Röntgenkristallstrukturanalyse *f*, röntgenographische Kristallstrukturanalyse *f*, Kristallstrukturanalyse *f* mit Röntgenstrahlen, Röntgen(struktur)-analyse *f* von Kristallen
X-ray crystallographic analysis *s.* X-ray crystal(-structure) analysis
X-ray diagram *s.* X-ray pattern
X-ray diffraction Röntgen(strahl)beugung *f*, Röntgen(strahl)diffraktion *f*
X-ray diffraction analysis *s.* X-ray crystal(-structure) analysis
X-ray diffraction diagram *s.* X-ray diffractogram
X-ray diffraction method Röntgenbeugungsmethode *f*, Röntgendiffraktionsmethode *f*
X-ray diffraction pattern [photograph, picture] *s.* X-ray diffractogram
X-ray diffractogram Röntgenbeugungsdiagramm *n*, Röntgendiffraktionsaufnahme *f*
X-ray diffractometer Röntgendiffraktometer *n*
X-ray emission Röntgenemission *f*
X-ray emission spectrum *s.* X-ray spectrum
X-ray examination *s.* X-ray investigation
X-ray fluorescence [fluorescent] analysis Röntgenfluoreszenzanalyse *f*
X-ray image *s.* X-ray pattern
X-ray-induced Auger electron spec-

troscopy Auger-Elektronen-Spektroskopie *f* mit Röntgenstrahlanregung
X-ray-induced photoelectron spectroscopy röntgenstrahlangeregte Photoelektronenspektroskopie *f*, Elektronenspektroskopie *f* für die chemische Analyse
X-ray investigation Röntgenuntersuchung *f*, röntgenographische Untersuchung *f*
X-ray irradiation Röntgenbestrahlung *f*
X-ray level Röntgenterm *m*, Röntgenniveau *n*
X-ray method Röntgen(strahl)methode *f*, Röntgenverfahren *m*, röntgenographische Methode *f*, röntgenographisches Verfahren *n*
X-ray microanalysis Röntgen(strahl)-mikroanalyse *f*
X-ray microscope Röntgen(strahl)mikroskop *n*
X-ray microscopy Röntgenmikroskopie *f*, Mikroskopie *f* mit Röntgenstrahlen
X-ray pattern 1. Röntgenaufnahme *f*, Röntgenbild *n*; 2. *(Krist)* Röntgendiagramm *n*, Röntgen(fot)ogramm *n*
X-ray photoelectron spectroscopy *s.* X-ray-induced photoelectron spectroscopy
X-ray photograph [picture] *s.* X-ray pattern
X-ray powder camera Röntgenpulverkamera *f*, Pulverkamera *f*
X-ray powder diffraction *s.* X-ray powder method
X-ray powder diffractometer Pulverdiffraktometer *n*
X-ray powder method Pulververfahren *n*, Pulvermethode *f (Kristallstrukturuntersuchung)*
X-ray powder pattern [photograph] Pulverdiagramm *n*, Pulveraufnahme *f*
X-ray powder spectroscopy Röntgenspektroskopie *f* nach dem Pulververfahren [Debye-Scherrer-Verfahren]
X-ray radiation Röntgenstrahlung *f*
X-ray range [region] *(Anal)* Röntgen(wellen)gebiet *n*, Gebiet *n* der Röntgenstrahlen
X-ray rotation photograph Drehkristallaufnahme *f*
X-ray scattering Röntgen(strahl-)streuung *f*, Streuung *f* der Röntgenstrahlen
X-ray small-angle scattering Röntgenkleinwinkelstreuung *f*, RKS
X-ray spectrometer Röntgenspektrometer *n*
X-ray spectrometry Röntgenspektrometrie *f*
X-ray spectroscopy Röntgenspektroskopie *f*
X-ray spectrum Röntgenspektrum *n*, Röntgenstrahlen(emissions)spektrum *n*
X-ray structure analysis Röntgen(fein)-strukturanalyse *f*, Röntgenfeinstrukturuntersuchung *f*, Strukturanalyse *f* mit Röntgenstrahlen
X-ray unit Röntgeneinheit *f*, Einheit *f* der Röntgendosis
X-Y recorder x-y-Schreiber *m*, Kurvenschreiber *m*
XAES *s.* X-ray-induced Auger electron spectroscopy
xanthan gum *(Biot)* Xanthan *n (ein mikrobielles Polysaccharid)*
xanthate *v (Text)* xanthogenieren, sulfidieren
xanthate Xanthogenat *n*, Xanthat *n (Salz oder Ester einer Xanthogensäure)*
xanthation *(Text)* Xanthogenieren *n*, Sulfidieren *n*
xanthene Xanthen *n*, Dibenzopyran *n*, *(i. e. S.)* Xanthenderivat *n*, Xanthen *n*
xanthene dye(stuff) Xanthenfarbstoff *m*
xanthic acid 1. *s.* xanthogenic acid; 2. HOCSSH Dithiokohlensäure *f*
xanthine Xanthin *n*, Xan, 2,6-Dihydroxypurin *n*, *(i. w. S.)* Xanthinderivat *n*, Xanthin *n*
xanthogen Xanthogen *n*
xanthogenate *s.* xanthate
xanthogenic acid Xanthogensäure *f*, *(i. e. S.)* C_2H_5OCSSH Ethylxanthogensäure *f*, Dithiokohlensäure-O-ethylester *m*
xanthone Xanthon *n*, Dibenzo-γ-pyron *n*, 9-Oxoxanthen *n*
xanthophyll Xanthophyll *n*
xanthoproteic acid Xanthoproteinsäure *f*
xanthoproteic reaction Xanthoproteinreaktion *f*
xanthoprotein Xanthoprotein *n*

xanthoprotein reaction *s.* xanthoproteic reaction
xanthosine Xanthosin *n*, Xanthin-9-ribosid *n*
xanthurenic acid Xanthurensäure *f*, 4,8-Dihydroxychinaldinsäure *f*, 4,8-Dihydroxychinolin-2-carbonsäure *f*
xanthylic acid *(Bioch)* Xanthylsäure *f*, Xanthosin 5′-monophosphat *n*, XMP
xanthylium salt Xanthyliumsalz *n*
XDA = xylylene diamine
XDI = xylylene diisocyanate
xenate Xenat *n (Salz der Xenonsäure)*
xenic acid Xenonsäure *f*
xenobiotic Xenobiotikum *n (vom Menschen in die Natur gebrachter Fremdstoff)*
xenomorphic *(Krist)* xenomorph, allotriomorph, fremdgestaltig
xenon Xe Xenon *n*
xenon lamp Xenonlampe *f*
xenylamine Xenylamin *n*, 4-Aminobiphenyl *n*
xerogel Xerogel *n*, Trockengel *n*
ximenynic acid Ximeninsäure *f*, Santalbsäure *f*, Santalbinsäure *f*, *Octadec-11-en-9-insäure *f*
XMP = xanthosine 5′-monophosphate
XPS *s.* X-ray-induced photoelectron spectroscopy
XRD *s.* 1. X-ray diffraction; 2. X-ray powder method
XRS *s.* X-ray spectroscopy
xylan Xylan *n*, Holzgummi *n*, Hemicellulose *f* A
xylene Xylen *n*, Xylol *n*, *Dimethylbenzen *n*
xylenol Xylenol *n*, *Dimethylhydroxybenzen *n*
xylenol resin Xylenolharz *n*
xylidine Xylidin *n*, Aminodimethylbenzen *n*
xylite Xylit *m(n)*, Braunkohlenholz *n*, xylitische [holzartige] Braunkohle *f*
xylitol Xylitol *n (ein Zuckeralkohol)*
xylonic acid Xylonsäure *f (eine 2,3,4,5-Tetrahydroxypentansäure)*
xylose Xylose *f*, Holzzucker *m*
xylulose Xylulose *f*, Xyloketose *f*
xylyl bromide *α-Bromxylen *n*, Xylylbromid *n*
xylylene bromide *α,α′-Dibromxylen *n*, *Bis(brommethyl)benzen *n*, Xylylenbromid *n*
xylylic acid Xylylsäure *f*, Dimethylbenzoesäure *f*, Dimethylbenzencarbonsäure *f*

Y

Y-shape connecting tube Y-Stück *n*, Ypsilonstück *n*, Y-förmiges Zwischenabzweigstück *n* [Verbindungsstück *n*]
YAG *s.* yttrium-aluminium garnet
yarn Garn *n*, Faden *m*
yarn-dye *v* im Garn färben
yarn dyeing Garnfärben *n*, Färben *n* im [von] Garn
yeast *v* mit Hefe versetzen; gären
yeast Hefe *f*
yeast culture Hefekultur *f*
yeast fermentation 1. *(Biot)* Hefefermentation *f*; 2. *(Gär)* Hefegärung *f*
yeast-like hefig, hefeartig
yeast nucleic acid Hefenucleinsäure *f*
yeast protein Hefeprotein *n*, Hefeeiweiß *n*
yeast RNA Heferibonucleinsäure *f*, Hefe-RNA *f*
yellow *v* vergilben
yellow 1. Gelb *n (Farbempfindung)*; 2. gelber [gelbfärbender] Farbstoff *m*, Gelb *n*
yellow brass Gelbmessing *n*, Gelbguss *m*
yellow enzyme Flavinenzym *n*, Gelbes Enzym *n*, Flavoprotein *n*
yellow heat Gelbglut *f*
yellow lead oxide Bleiglätte *f*, Blei(II)-oxid *n*
yellow mercuric oxide 1. gelbes [gefälltes] Quecksilber(II)-oxid *n*; 2. *(Pharm)* gelbes Präzipitat *n*
yellow ochre *(Min)* Limonit *m*, gelber Ocker *m*, Brauneisenerz *n*, Brauner Glaskopf *m (wasserhaltiges Eisenoxidhydroxid)*
yellow phosphorus gelber [weißer, farbloser] Phosphor *m*
yellow precipitate *s.* yellow mercuric oxide
yellow prussiate of potash gelbes Blutlaugensalz *n*, Kaliumhexacyanoferrat(II) *n*

yellowing Vergilben *n*, Vergilbung *f*, Gelbwerden *n*
yield *v* 1. (er)geben, liefern; abgeben; 2. fließen *(Werkstoffe)*
yield 1. Ausbeute *f (bei chemischen Reaktionen)*; 2. *(Biot)* Ausbeute *f (an Stoffwechselprodukten)*; 3. *(Hyd, Biot)* Ertrag *m*, Ausbeute *f (an Mikroorganismenmasse)*; 4. *(Met)* Ausbeute *f*, Ausbringen *n*; 5. *(Landw)* Ertrag *m*; 6. Fließen *n (von Werkstoffen)*
yield coefficient *(Biot)* Ertragskoeffizient *m*, Ausbeutekoeffizient *m*, ökonomischer Koeffizient *m (Mikroorganismenwachstum)*
yield of product Produkt(ions)ausbeute *f*
yield phenomenon Fließerscheinung *f*
yield point Fließgrenze *f*, Streckgrenze *f (im Spannungs-Dehnungs-Diagramm)*
yield strain Fließdehnung *f*
yield strength Dehngrenze *f*, technische Elastizitätsgrenze *f*, praktische Fließgrenze *f (mit vorgegebener bleibender Dehnung in Prozent)*
yield stress Fließspannung *f*, Streckspannung *f*; (binghamsche) Fließgrenze *f (Rheologie)*
yield value untere Fließgrenze *f* [Streckgrenze *f*]
ylide *(org Ch)* Ylid *n*
yohimbé alkaloid Yohimbealkaloid *n*
yolk 1. Eidotter *m(n)*, Dotter *m(n)*, Eigelb *n*; 2. *(Text)* Wollfett *n*, Wollschweiß *m*
young fustic *(Farb)* Ungarisches Gelbholz *n*, Junger Fustik *m*, Jungfustik *m*, Fiset(te)holz *n (von Cotinus coggygria Scop.)*
Young's modulus (of elasticity) Elastizitätsmodul *m*, E-Modul *m*, youngscher Modul *m*
yperite Yperit *n*, Schwefelyperit *n*, S-Yperit *n (Dichlordiethylsulfid, Kampfstoff)*
ytterbia Ytterbium(III)-oxid *n*
ytterbic Ytterbium..., *(i. e. S.)* Ytterbium(III)-...
ytterbic chloride Ytterbiumtrichlorid *n*, Ytterbium(III)-chlorid *n*
ytterbic sulphate Ytterbiumsulfat *n*, Ytterbium(III)-sulfat *n*
ytterbium Yb Ytterbium *n*
ytterbium dichloride Ytterbiumdichlorid *n*, Ytterbium(II)-chlorid *n*
ytterbium oxide Ytterbium(III)-oxid *n*
ytterbium trichloride Ytterbiumtrichlorid *n*, Ytterbium(III)-chlorid *n*
ytterbous Ytterbium..., *(i. e. S.)* Ytterbium(II)-...
ytterbous chloride Ytterbiumdichlorid *n*, Ytterbium(II)-chlorid *n*
yttria Yttriumoxid *n*
yttric Yttrium...
yttrium Y Yttrium *n*
yttrium-aluminium garnet Yttrium-Aluminium-Granat *m*
yttrium carbonate Yttriumcarbonat *n*
yttrium chloride Yttriumchlorid *n*
yttrium earth Yttererde *f (als Sammelname)*
yttrium hydroxide Yttriumhydroxid *n*
yttrium oxide Yttriumoxid *n*

Z

Z average Zentrifugenmittel *n*, z-Mittel *n (von Makromolekülen)*
Z-value Z-Wert *m (zur Charakterisierung der Lösungsmittelpolarität)*
ZAAS *s.* Zeeman atomic absorption spectrophotometry
Zaitsev rule Zajcev-Regel *f (der Orientierung von β-Eliminierungen)*
Zanzibar copal Sansibarkopal *m (von Trachylobium verrucosum Oliv. und T. hornemannianum Heyne)*
Zeeman atomic absorption spectrophotometry Zeeman-Atomabsorptionsspektrophotometrie *f*
Zeeman broadening Zeeman-Verbreiterung *f (von Spektrallinien bei Einwirkung magnetischer Felder)*
Zeeman effect *(phys Ch)* Zeeman-Effekt *m*, Zeeman-Aufspaltung *f* (der Energieniveaus)
Zeeman energy level *s.* Zeeman term
Zeeman term *(phys Ch)* Zeeman-Term *m*, Zeeman-Niveau *n*
zein Zein *n (Prolamin des Maises)*
zein plastic Zeinkunststoff *m*
zein staple Zeinfaser *f*
zeolite *(Min)* Zeolith *m (allg. für eine Gruppe von Tektosilicaten)*
zeolite exchange resin künstlicher Mineralaustauscher *m* [Zeolith *m*]
zeolitic water zeolithisches [zeolithisch

gebundenes] Wasser *n*, Zeolithwasser *n*
Zerewitinoff determination Cerevitinov-Bestimmung *f (aktiver H-Atome)*
zero *v* auf den Nullpunkt einstellen
zero *s.* zero point
zero adjustment Nullpunkteinstellung *f*
zero concentration Konzentration *f* Null
zero-field splitting *(Spektr)* Nullfeldaufspaltung *f*, Kristallfeldaufspaltung *f*
zero group nullte Gruppe *f*, Gruppe *f* 0 *(im Periodensystem)*
zero method Nullmethode *f*, Kompensationsmethode *f*, Abgleichmethode *f* *(Messtechnik)*
zero-order reaction Reaktion *f* nullter Ordnung
zero point Nullpunkt *m*, Ausgangspunkt *m (einer Skale)*
zero-point energy Nullpunkt(s)energie *f*
zero point of charge 1. Ladungsnullpunkt *m*; 2. *(phys Ch)* isoelektrischer Punkt *m*
zero-point vibration Nullpunkt(s)schwingung *f*
zero tolerance *(Tox)* Nulltoleranz *f*, Toleranzwert *m* 0
zero-valent nullwertig
zeroing Nullpunkteinstellung *f*
zeroth electronic state Elektronengrundzustand *m*, elektronischer Grundzustand *m*
zeroth law of thermodynamics nullter Hauptsatz *m* der Thermodynamik
zeta potential Zeta-Potenzial *n*, ζ-Potenzial *n*, elektrokinetisches Potenzial *n*
Ziegler-Natta catalyst Ziegler-(Natta-)Katalysator *m*, Ziegler-Kontakt *m*
Ziegler process Ziegler-Verfahren *n*, Niederdruckverfahren *n* nach Ziegler *(zur Olefinpolymerisation)*
zigzag chain Zickzackkette *f*
Zimm plot Zimm-Diagramm *n (zur Molmassenbestimmung bei Polymeren)*
Zimmermann-Reinhardt (preventive) solution Reinhardt-Zimmermann-Lösung *f (Phosphorsäure, Schwefelsäure, Mangan(II)-sulfat)*
zinc *v* verzinken
zinc Zn Zink *n*
zinc amalgam Zinkamalgam *n*
zinc baryta white Barytzinkweiß *n*, Lithopone *f*, Lithopon *n*
zinc blende *(Min)* Zinkblende *f*, Sphalerit *m (Zinksulfid)*
zinc-blende lattice *(Krist)* Zinkblendegitter *n*
zinc chloride Zinkchlorid *n*
zinc coating Verzinken *n*, Verzinkung *f*
zinc-dust distillation Zinkstaubdestillation *f*
zinc fluorosilicate Zinkhexafluorosilicat *n*
zinc grey Zinkgrau *n*
zinc oxide Zinkoxid *n*
zinc-oxide catalyst Zinkoxidkatalysator *m*
zinc plating elektrochemisches [galvanisches] Verzinken *n*
zinc spar *(Min)* Zinkspat *m*, Smithsonit *m* *(Zinkcarbonat)*
zinc vitriol 1. Zinkvitriol *n*, Zinksulfat-7-Wasser *n*; 2. *(Min)* Goslarit *m*, natürlicher Zinkvitriol *m*, weißer Vitriol *m* *(Zinksulfat-7-Wasser)*
zinc white Zinkweiß *n (rohes Zinkoxid)*
zinc yellow Zink(chrom)gelb *n*, Zitronengelb *n (Gemisch aus Kalium-, Zink- und Chromoxid)*
zincate Zinkat *n*
zincic Zink...
zincous Zink...
zineb Ẓineb *n*, Zinkethylenbisdithiocarbamat *n*
zingerone Zingeron *n*, 3-Methoxy-4-hydroxybenzylaceton *n*
zingiberene Zingiberen *n (ein monocyclischer Sesquiterpenkohlenwasserstoff)*
ziram Ziram *n*, Zinkdimethyldithiocarbamat *n*
zircon *(Min)* Zirkon *m (Zirconiumorthosilicat)*
zircon porcelain Zirkonporzellan *n*
zirconate Zirconat *n*
zirconic Zirconium...
zirconic acid Zirconium(IV)-hydroxid *n*, Zirconiumoxidaquat *n*, *(inkorrekt:)* Zirconsäure *f*
zirconium Zr Zirconium *n*
zirconium carbide Zirconiumcarbid *n*
zirconium dichloride Zirconiumdichlorid *n*, Zirconium(II)-chlorid *n*
zirconium dioxide Zirconiumdioxid *n*, Zirconium(IV)-oxid *n*
zirconium hydride Zirconiumhydrid *n*

zirconium oxide ZrO_2 Zirconium(IV)-oxid *n*, Zirconiumdioxid *n*
zirconium oxybromide Zirconiumoxid-bromid *n*
zirconium oxychloride Zirconiumoxid-chlorid *n*
zirconium silicate Zirconiumsilicat *n*
zirconium sulphate Zirconium(IV)-sulfat *n*
zirconium tetrachloride Zirconium-tetrachlorid *n*, Zirconium(IV)-chlorid *n*
zirconium trichloride Zirconiumtrichlorid *n*, Zirconium(III)-chlorid *n*
zirconyl chloride *s.* zirconium oxy-chloride
zonal centrifugation *(Anal)* Zonenzentrifugation *f*
zone-melt *v* zonenschmelzen
zone-purify *v s.* zone-refine
zone-refine *v* zonenreinigen, durch Zonenschmelzen reinigen
zone Zone *f*
zone axis *(Krist)* Zonenachse *f*
zone electrophoresis *(Anal)* Zonen-elektrophorese *f*
zone melting Zonenschmelzen *n*
zone of combustion Verbrennungszone *f*
zone of oxidation Oxidationszone *f*
zone of preheat Vorwärm(e)zone *f*, Vorheizzone *f*
zone purification *s.* zone refining
zone refining Zonenreinigen *n*, Zonenreinigung *f*
zone settling *(Hyd)* Zonensedimentation *f*
zone spreading *(Chromat)* Bandenverbreiterung *f*
zoosterol *(org Ch)* Zoosterin *n*, tierisches Sterin *n*
ZP *s.* zeta potential
ZPC *s.* zero point of charge
zwitterion Zwitterion *n*, inneres [intramolekulares] Salz *n*, Ampho-Ion *n*
zwitterionic zwitterionisch
zwitterionic compound *s.* zwitterion
zymogen Zymogen *n*, Proenzym *n*, Enzymogen *n*
zymogenic gärungserregend, zymogen, zymotisch; Zymogen...
zymogenous gärungserregend, zymogen, zymotisch; Zymogen...

Deutsch-Englisch

A

a 1. = absolut; 2. *s.* antarafacial
A *s.* Anode
ABA *s.* Abwasserbehandlungsanlage
Abart *f* variety, modification
Abbau *m* 1. degradation, decomposition, breakdown, fragmentation; 2. *(bioch)* katabolism, catabolism; 3. mining, working, extracting, winning *(as of coal)* • **Abbau erleiden** undergo degradation
Abbau *m***/aerober biologischer** aerobic degradation [decomposition, digestion]
Abbau *m***/anaerober biologischer** anaerobic degradation [decomposition, digestion]
Abbau *m***/bakterieller** bacterial degradation
Abbau *m***/barbier-wielandscher** Barbier-Wieland degradation *(for eliminating a carbon atom from a side chain)*
Abbau *m***/biochemischer** biochemical degradation [decomposition, digestion]
Abbau *m***/biologischer** biodegradation, biological digestion *(of organics)*
Abbau *m***/braunscher** Braun degradation *(of tertiary amines)*
Abbau *m***/chemischer** chemical degradation
Abbau *m***/curtiusscher** Curtius rearrangement [transformation] *(the decomposition of acid azides to give isocyanates and nitrogen)*
Abbau *m***/enzymatischer [fermentativer]** enzymatic degradation
Abbau *m***/glykolytischer** *(bioch)* glycolytic pathway, Embden-Meyerhof-Parnas scheme (of glycolysis), EMP scheme, homolactic fermentation
Abbau *m***/hofmannscher** Hofmann degradation *(of quaternary ammonium hydroxides)*
Abbau *m***/hydrolytischer** hydrolytic degradation
Abbau *m***/mikrobieller** microbial degradation [decomposition]
Abbau *m***/oxidativer** oxidative degradation [cleavage]
Abbau *m***/photochemischer** photochemical [photolytic, light] degradation, photodegradation, photolysis
Abbau *m***/ruffscher** Ruff degradation *(of sugars)*
Abbau *m***/schrittweiser [stufenweiser]** stepwise degradation
Abbau *m***/thermischer** thermal degradation, degradation by heat
Abbau *m***/thermooxidativer** thermooxidative degradation
Abbau *m* **unter Tage** underground [deep] mining
abbaubar 1. *(org ch)* degradable, decomposable; 2. *(mine)* workable
abbaubar/biologisch biodegradable
abbaubar/nicht non-degradable
Abbaubarkeit *f* 1. *(org ch)* degradability, decomposability; 2. *(mine)* workability
Abbaubarkeit *f***/biologische** biodegradability
abbauen *v* 1. degrade, decompose, disintegrate, break down, fragment; 2. *(bioch)* catabolize; 3. *(mine)* mine, work, extract, win
abbauen *v***/biochemisch** degrade biochemically
abbauen *v***/biologisch** digest, biodegrade *(organics)*
abbauen *v***/sich** degrade
Abbaukurve *f (tox)* degradation curve, decline [disappearance, loss, decay] curve, (residue-)persistence curve
Abbauprodukt *n* degradation [decomposition, breakdown] product
Abbaureaktion *f* degradation [degradative] reaction, decomposition [breakdown] reaction
abbauresistent non-degradable, *(hyd specif)* non-biodegradable, resistant to biological degradation, biologically inert
Abbauweg *m* 1. degradation [degradative] pathway; 2. *(bioch)* catabolic pathway [route]
Abbauwürdigkeit *f (mine)* workability
abbeizen *v* strip, pickle, remove old paint
Abbeizmittel *n* paint stripper [remover], organic coating stripper, lacquer remover, pickling agent
Abbe-Refraktometer *n* Abbe refractometer
Abbindebeschleuniger *m (build)* (setting, cementing) accelerator

abbinden *v (build)* set, harden
Abbindeverzögerer *m (build)* (setting) retarder
abblasen *v* 1. blow off *(e.g. steam)*; 2. *(pap)* blow (off) *(a digester)*; 3. *(met)* blow out *(a blast furnace)*
Abblaserohr *n (tech)* blow off pipe
Abblaseventilkammer *f* exhaust-valve box
Abblasleitung *f* blow pipe, blowpipe
Abblasstellen *fpl (tech)* blow off points
abblättern *v* peel (off), flake (off), scale (off), spall (off), exfoliate
Abbrand *m* 1. *(met)* calcine *(product of roasting)*; 2. *(nucl)* burn-up
Abbrandverhältnis *n* burning rate
Abbrandversuch *m* incineration test
abbrausen *v (min tech)* spray *(classified material)*
abbrechen *v* arrest *(a reaction)*; terminate, break *(a chain)*; terminate, shortstop *(polymerization)*
abbremsen *v* 1. slow down, decelerate, retard; 2. *(nucl)* moderate
Abbremsung *f* 1. slowing-down, deceleration, retardation; 2. *(nucl)* moderation
Abbrenneinrichtung *f s.* Beizanlage
Abbrennkessel *m (met)* pickling vat
Abbruch *m* arrest *(of a reaction)*; termination, breaking *(of a chain)*; termination, shortstopping *(of polymerization)*
Abdampf *m* exhaust [dead] steam
Abdampfdruckregler *m* exhaust-steam-pressure regulator
abdampfen *v* evaporate (off); bleed *(gas chromatography)*
Abdampfen *n* evaporation; bleeding *(gas chromatography)*
Abdampfschale *f (lab)* evaporating dish, *(UK)* basin
abdecken *v* 1. cover (over) *(e.g. a vessel)*; 2. *s.* abdecken/lokal
abdecken v/lokal stop off *(protect desired areas from chemical action)*
Abdeckschicht *f* polymer coating
Abdeckung *f* resist, stop-off (coating) *(electroplating)*
abdekantieren *v* decant off
abdestillieren *v* distil off, strip (off, out), remove by distillation; distil off *(of a liquid)*
abdichten *v* make tight, seal, pack, ca(u)lk; waterproof; gasproof
Abdichtmanschette *f* sealing cup
abdiffundieren *v* diffuse away
Abdrift *f (agric)* blow-off, drift *(as of pesticides)*
Abdruck *m* impression, mark, stamp; print; replica *(plastic film produced on surface structures for electron microscopy)*
abdunsten *v* evaporate
abfahren *v* scan *(a measuring range)*
abfahren v/nochmals rescan
Abfall *m* 1. waste (material), refuse, scrap; rubbish, *(Am)* garbage; 2. *(plast, glass)* cull; 3. *s.* Abfallen
Abfall m/häuslicher domestic [household] waste
Abfall m/radioaktiver nuclear [radioactive] waste
Abfallbehälter *m (lab)* waste jar
Abfallbehandlung *f s.* Abproduktbehandlung
Abfallbiomasse *f* residual biomass
abfallen *v* drop, decrease, fall (off) *(as of measuring values)*
Abfallen *n* drop, decrease, fall *(as of measuring values)*
Abfallen n des Drucks pressure drop
Abfallgummi *m* rubber scrap [waste], scrap [waste] rubber
Abfallholz *n (pap)* waste wood
Abfallkübel *m (lab)* waste jar, *(earthenware also)* waste [disposal] crock
Abfallpapier *n* waste [old] paper
Abfallsäure *f* waste [residuary] acid
Abfallstoff *m s.* Abprodukt
abfangen *v* catch, trap, intercept, steal away *(reactive intermediates)*
Abfangen *n* catching, trapping, interception *(of reactive intermediates)*
Abfangreaktion *f* trapping reaction
abfärben *v* lose colour, bleed, mark off
abfiltern *v s.* abfiltrieren
abfiltrieren *v* filter off [out], remove by filtration
abfließen *v* flow (off), run (off), drain (off)
abfließen lassen *v* drain (off), run (off), discharge
Abfließen *n* flowing-off, draining-off, drainage, effluence
Abfluss *m* 1. *s.* Abfließen; 2. discharge

(point), outlet, outflow, drain; 3. *s.* Ablauf 3.

Abflussbecken *n (lab)* (bench) sink

Abflussleitung *f* discharge [delivery] line, drain [drainage, runoff, outlet] line; waste line

Abflussöffnung *f* discharge opening, discharge outlet [door, aperture, port]

Abflussrohr *n* discharge tube [pipe], drainage [delivery] tube, runoff [effluent, outlet] tube; waste tube [pipe]

Abflussventil *n* discharge [outlet] valve

abführen *v* 1. lead off [away], carry off, drain off [away], draw off, withdraw *(liquids or gases)*; exhaust *(gases)*; dissipate, eliminate *(heat)*; 2. *(pharm)* purge

Abführlein *n* linin *(herb)*

Abführmittel *n (pharm)* cathartic

abfüllen *v*/**auf Fässer** barrel, rack

abfüllen *v*/**auf Flaschen** bottle (in, up), *(Am)* can

Abgabe *f* release, liberation, evolution *(as of gases, heat)*; release, donation *(of electrons, ions)*

Abgang *m* 1. loss; 2. waste (material, product); 3. *s.* Abgänge

Abgänge *mpl (min tech)* (waste) tailings, refuse

Abgangsgruppe *f s.* Austrittsgruppe

Abgas *n* waste [exit] gas, off-gas; exhaust gas *(from combustion engines)*

Abgasentstickung *f* nitrogen removal

Abgasfahne *f* (gas) plume

Abgas-Grenzwerte *mpl* maximum emission allowance

Abgasreiniger *m* collector

Abgasschlauch *m* exhaust tube

Abgasvorwärmer *m* economizer

abgebaut werden *v* undergo degradation

abgeben *v* release, liberate, give off, evolve *(e.g. gas, heat)*; release, donate, lose *(electrons, ions)*; lose *(hydration water)*

abgeben *v*/**in die Atmosphäre** discharge into the air *(air pollutants)*

abgehen *v* leave *(as of vapours)*

abgehen *v*/**durch den Schornstein** be lost at the stack

abgehen *v*/**über Kopf** *(distil)* leave at the top

abgeschlossen/hermetisch hermetically sealed

abgestanden flat, stale, insipid *(beverages)*

abgetragen werden *v* ablate *(by melting, vaporization, or decomposition at high temperatures)*

abgießen *v* pour off, drain (off), run (off)

abgießen *v*/**vorsichtig** decant

Abgleichmethode *f* null(-balance) method, zero method *(of measurement)*

Abgleiten *n* crystal [translation] gliding *(along a crystal plane)*

Abhängigkeitsverbundwirkung *f (tox)* dependent joint action

Abhäsion *f* abhesion *(loss of adhesion)*

abhäsiv abhesive

abhebern *v* siphon (off)

Abhilfemaßnahme *f* remedial measure

Abhitze *f s.* Abwärme

Abietat *n* abietate *(salt or ester of abietic acid)*

Abietinsäure *f* abietic acid *(a resin acid)*

a-Bindung *f* a-bond, axial bond

abkippen *v*/**auf Halde** tip, dispose of

abklappen *v* dump at sea, dispose of to sea, dispose of in the ocean *(waste products)*

Abklappen *n* ocean [sea] dumping, sea burial, waste dumping at sea

Abklärflasche *f (lab)* aspirator

Abklingdauer *f* decay time

abklingen *v* die [quieten] down, flatten out *(of a reaction)*; fade away, subside *(of vibrations)*

Abklingkurve *f* decay curve, fall-off curve *(kinetics)*

abkochen *v* 1. boil; 2. *(pharm)* decoct; 3. *(text)* bowk, buck, kier-boil

Abkochung *f (pharm)* decoction

abkratzen *v* scrape off, scratch off

abkühlen *v* cool (down), *(if rapidly:)* chill, *(relating to metal:)* quench

abkühlen lassen *v* allow to cool, let cool

abkühlen *v*/**sich** cool (down), *(if rapidly:)* chill

Abkühlungskurve *f* cooling curve

Abkühlzyklus *m* cooling cycle

ablagern *v* 1. sediment, deposit *(of liquids)*; dispose of *(waste products)*; 2. *(food)* age, mature; season *(timber)*

ablagern *v*/**auf Halde** dispose of tip,

dispose of by tipping, deposit on a tip *(waste products)*

ablagern v/sich deposit, settle (out), *(esp from liquids:)* sediment, subside

ablagern v/sich wieder redeposit

Ablagerung *f* 1. deposit, *(esp from liquids:)* sediment, (bottom) settlings, BS, bottoms, dregs; 2. *(food, ferm)* lees, foots, *(esp of biological matter:)* fouling; scale *(on the inside of a boiler)*; 3. deposition, sedimentation; disposal (land) *(of waste products)*; 4. *(food)* ageing, maturing; seasoning *(of timber)*

Ablagerung *f* **auf Halde** disposal to tip

Ablagerung f/dauerhafte [endgültige] final [ultimate] disposal *(of waste products)*

Ablagerung f/saure acid deposition

Ablass *m* discharge, outlet, drain; vent, relief *(for lessening excessive pressure)*

ablassen *v* let off [out], discharge, exhaust, *(relating to liquid also)* run (off), drain (off), sewer, tap, *(relating to gas also)* blow off; discharge, drain (down), exhaust, tap, empty *(a vessel)*

Ablasshahn *m* discharge cock, outlet [delivery, runoff] cock, drain cock, drain tap, dispensing spigot

Ablassleitung *f* discharge [delivery] line, drainage [drain, runoff, outlet] line; waste line

Ablassöffnung *f* discharge opening [outlet, door, aperture, port]; vent, relief *(for gases and vapours)*

Ablassrohr *n* discharge tube [pipe], drainage [delivery] tube, runoff [effluent, outlet] tube; waste tube [pipe]

Ablassstopfen *m* drain plug

Ablassventil *n* discharge valve, drain [outlet] valve

Ablassvorrichtung *f* discharging device, discharger

Ablation *f* ablation *(wearing away of material by melting, vaporization, or decomposition at high temperatures)*

Ablationsschicht *f* ablative coating

ablativ ablative *(removable by melting, vaporization, or decomposition at high temperatures)*

Ablauf *m* 1. *s.* Ablaufen 1.; 2. discharge (point), outlet, outflow, drain; 3. runoff, effluent, efflux *(material)*; 4. *(hyd)* effluent [outgoing] water, outflow, outlet; runoff water *(land treatment of waste water)*; discharge *(from a waste-water treatment plant)*; 5. course *(as of a reaction or process)* • **auf Ablauf geteilt [graduiert]** calibrated [graduated] deliver, calibrated for delivery

Ablauf m/zeitlicher time course *(of a reaction)*

Ablaufeigenschaft *f* drainage behaviour

ablaufen *v* 1. drain (off), run (off), flow (off) *(of liquids)*; curtain, sag *(of surface coatings)*; 2. run, proceed, take its course *(of a reaction)*

ablaufen lassen *v* 1. drain (off), let drain, run (off), siphon; 2. *(hyd)* send to the drain

Ablaufen *n* 1. draining-off, drainage, flowing-off, effluence, outflow *(of liquids)*; curtaining, sagging *(of surface coatings)*; 2. running, proceeding, taking its course *(of a reaction)*

Ablaufgüte *f (hyd)* effluent (water) quality

Ablaufkanal *m* effluent [drainage] channel, *(esp for waste water:)* sewer

Ablaufrohr *n* 1. *(distil)* downpipe, downspout, downtake, downcomer, rundown pipe, drain pipe; 2. *s.* Abflussrohr

Ablaufwasser *n (hyd)* effluent water

Ablauge *f* 1. waste [spent] lye, waste [spent] liquor; 2. *(pap)* sulphite spent [waste] liquor, spent (pulping) liquor

Ablaugeregeneration *f (pap)* waste-liquor [spent-liquor] recovery

ableiten *v* 1. lead away [off], carry off, drain away [off], draw off, withdraw *(liquids or gases)*; dissipate, eliminate *(heat)*; 2. derive *(one chemical compound from another)*

ableiten v/in den Vorfluter *(hyd)* discharge into a receiving stream

ableiten v/in die Kanalisation *(hyd)* discharge [drain, run] to the sewer

Ableitung *f* 1. leading-away, leading-off, carrying-off, drainage, withdrawal *(of liquids or gases)*; dissipation, elimination *(of heat)*; 2. derivation *(of one chemical compound from another)*

Ablenkblech *n* deflector (plate)

ablenken *v* deflect, turn off

Ablenkung *f* deflection

Ablenkwinkel *m* angle of deflection

Ablesefehler *m (anal)* reading [observational] error
Ablesefehler *m*/**parallaktischer** parallax error
ablesen *v* read off, take a reading
Ablesewert *m (lab)* instrument reading
Ablesung *f* reading, readout
ablöschen *v* quench *(coke, metal)*; slake *(esp lime)*
ablöschen *v*/**in [mit] Öl** oil-quench *(metal)*
ablöschen *v*/**in [mit] Wasser** water-quench *(metal)*
Ablöschmittel *n* quenching medium [agent] *(for metals)*
Ablösearbeit *f (phys ch)* (electronic) work function
ablösen *v* strip, detach, peel *(e.g. a film)*; release *(as from a mould)*; dissolve *(e.g. a soluble substance from a vessel wall)*; detach *(electrons)*
ablösen *v*/**sich** peel (off), scale (off), flake (off), shell (off)
Ablösung *f* **eines Elektrons** electron detachment
Abluft *f* exhaust [exit] air, outlet [outgoing, leaving] air
Abluftreiniger *m* collector
abmessen *v* measure (out, off), meter, batch *(liquids, bulk material)*
Abmessung *f* size, dimension
Abnahme *f* 1. take-off, withdrawal, discharge *(of liquids or gases)*; 2. take-off; 3. *(plast)* take-up, haul-off *(spot or device)*; 4. decrease, loss, drop, fall, reduction *(as of measuring values)*
abnehmen *v* 1. take off; 2. *(plast)* take up, haul off *(e.g. extrudate)*; withdraw, discharge *(liquids or gases)*; 3. decrease, drop, fall, reduce *(as of measuring values)*
abnehmen *v*/**über Kopf** *(distil)* take overhead
abnutschen filter by [with, under] suction, *(Am)* suction-filter
Abnutschen *n* suction filtration
abnutzen *v*/**sich** wear (out); abrade *(by friction)*
Abnutzung *f* wear
Abnutzung *f* **durch Abrieb** abrasion, abrasive wear, attrition
abnutzungsbeständig wear-resistant; abrasion-resistant *(to friction)*
Abnutzungsbeständigkeit *f* wear resistance; abrasion resistance *(to friction)*
abpipettieren *v* pipette off, withdraw by pipette
abplatzen *v* exfoliate, flake (off), scale (off), chip (off), peel (off), spall
abpressen *v* press (off), squeeze (off), force out, expel, express
Abprodukt *n* waste (product)
Abprodukt *n*/**festes** solid waste (material), waste solid
Abprodukt *n*/**gewerbliches** commercial waste
Abprodukt *n*/**industrielles** industrial waste
Abprodukt *n*/**kommunales** municipal waste
Abproduktbehandlung *f* waste treatment
Abproduktbeseitigung *f* waste disposal
Abproduktbeseitigung *f* **durch Einleitung ins Meer** ocean disposal
Abproduktbeseitigung *f* **durch Geländeauffüllung** landfill disposal (of wastes)
Abproduktnutzung *f* waste utilization
Abproduktverwertung *f* waste utilization
abpuffern *v* buffer (off)
abpumpen *v* pump off, evacuate
abrahmen *v* cream (off), skim *(milk)*
abrauchen *v* fume off
Abraumhalde *f (mine)* spoil pile
abreiben *v* abrade, scour
Abrieb *m* 1. abrasion, abrasive wear, attrition; 2. rubbings, dust
abriebbeständig *s.* abriebfest
abriebfest abrasion-resistant
Abriebfestigkeit *f* abrasion resistance; scuff resistance *(with sliding or rolling friction)*
Abriebmaterial *n* wear debris
Abriebprüfung *f* abrasion test
Abriebversuch *m* abrasion test
Abriebwiderstand *m* abrasion resistance
abrollen *v* unreel, reel off, unwind
abrösten *v* roast, burn *(e.g. ores)*
abs. = absolut
ABS 1. *s.* Abscisinsäure; 2. = Alkylbenzensulfonat

absacken *v* 1. bag (up) *(bulk material)*; 2. sag *(of coatings)*
absättigen *v* saturate *(valencies)*
Absättigung *f* saturation *(of valencies)*
absatzweise batch(wise) *(operation)*
absaugen *v* 1. suck off, siphon (off), withdraw, exhaust; 2. *(filtr)* filter by [with, under] suction, *(Am)* suction-filter
Absaugen *n* 1. suction, siphoning, withdrawal, exhaustion; 2. *(filtr)* suction filtration
abschaben *v* scrape off, doctor off
abschälen *v* peel (off) *(e.g. a foil)*; skim *(as in the knife-discharge centrifuge)*
abschälen *v***/sich** peel (off)
abschäumen *v* skim (off), scum
Abschäumer *m* skimmer
abscheidbar separable, *(from liquids also)* precipitable, settleable, *(from gases also)* settleable, *(esp electrostatically:)* precipitable
Abscheidbarkeit *f* separability, *(from liquids also)* precipitability, settleability, *(from gases also)* settleability, *(esp electrostatically:)* precipitability
Abscheidegrad *m* 1. separation [collection] efficiency *(gas cleaning)*; 2. *(hyd)* solids removal efficiency [performance] *(with vacuum filtration)*; degree of solids recovery, recovery *(with centrifugation)*
Abscheidekammer *f* separating chamber, settling [settlement, fall-out] chamber, expansion [drop-out] box *(gas cleaning)*
abscheiden *v* 1. separate, eliminate, remove, segregate; sediment(ate), precipitate, throw [lay] down *(bottoms)*; 2. *(min tech)* discard *(tailings)*; deposit *(coating material)*; evolve, set free, liberate *(gases)*; deposit, sediment *(of a liquid)*
abscheiden *v***/elektrochemisch [elektrolytisch, galvanisch]** electrodeposit
abscheiden *v***/gemeinsam** co-deposit
abscheiden *v***/sich** separate (out), *(of precipitates also)* settle (down, out), sediment, precipitate (out), deposit, subside
abscheiden *v***/sich baumartig** *(cryst)* tree
abscheiden *v***/sich elektrochemisch [elektrolytisch]** plate out
abscheiden *v***/sich gemeinsam** co-deposit
Abscheiden *n* 1. separation, elimination, removal, segregation; sedimentation, precipitation *(of bottoms)*; 2. *(min tech)* discard *(of tailings)*; deposition *(of coating material)*; evolution, liberation *(of gases)*; deposition, sedimentation *(of bottoms by a liquid)*
Abscheiden *n***/akustisches** sonic precipitation [agglomeration]
Abscheiden *n* **aus der Gasphase/chemisches** chemical vapour deposition, CVD *(via decomposable gaseous intermediates, mostly halides)*
Abscheiden *n* **dünner Filme** thin film deposition
Abscheiden *n***/elektrisches** electrostatic [electrical] precipitation *(gas cleaning)*
Abscheiden *n***/elektrochemisches [elektrolytisches]** electrodeposition *(of coatings)*
Abscheiden *n***/elektrostatisches** *s.* Abscheiden/elektrisches
Abscheiden *n***/galvanisches** *s.* Abscheiden/elektrochemisches
Abscheiden *n***/ionenstrahlgestütztes** ion beam assisted deposition, IBAD
Abscheiden *n***/ionenstrahlinduziertes** ion beam induced deposition, IBID
Abscheider *m* 1. separator, precipitator, catcher, trap; settler, separator [settling] tank *(fluid extraction)*; 2. *(min tech)* collector; 3. *(nucl)* trap
Abscheidezyklon *m* centrifugal cyclone separator, cyclone
Abscheidung *f* 1. *s.* Abscheiden; 2. deposit, sediment, settling(s), bottom sediments [settlings], B.S., bottoms
• **zur Abscheidung gelangen** *s.* abscheiden/sich
Abscheidung *f* **aus dem Plasma** plasma deposition
Abscheidung *f***/elektrolytische** electrodeposition
Abscheidungspotenzial *n* deposition potential
abscheren *v* shear (off)
abschirmen *v* shield, screen off, blanket; protect *(atomic groups in a molecule)*
Abschirmkonstante *f* 1. *(phys ch)*

shielding constant; 2. *(nucl)* screening constant
Abschirmungszahl *f (nucl)* screening constant
abschlämmen *v* elutriate
Abschlämmen *n* elutriation
abschleudern *v* centrifuge (off), spin
Abschleudern *n* centrifugation, spinning
abschließen *v* 1. terminate *(a process)*; 2. seal *(e.g. a vessel)*
abschließen *v***/wasserdicht** waterproof
Abschluss *m* 1. termination *(of a process)*; 2. seal *(as of a vessel)*; 3. exclusion *(as of air during a reaction)*
Abschlusskappe *f* closing cap
Abschlussventil *n* closing valve
abschmutzen *v (dye, coat, tann)* bleed (off)
abschneiden *v* cut off [away], shear (off)
Abschnitt *m* 1. region, section *(as in a molecule)*; 2. section, part *(of equipment)*; 3. stage *(of a process)*; 4. period *(of time)*; 5. cutting, chip *(of material)*
abschöpfen *v* 1. skim (off); 2. *(met)* rabble
Abschreckbad *n* quenching bath
abschrecken *v* 1. quench, cool rapidly; 2. *(met)* chill; 3. deter *(animals)*
abschrecken *v***/in Öl** oil-quench *(metal)*
abschrecken *v***/in Wasser** water-quench *(metal)*
abschreckhärten *v* quench-harden
Abschreckmittel *n* 1. quenching medium [agent]; 2. repellent, deterrent *(against animals)*
Abschreckwirkung *f* repellency, deterrent action *(on animals)*
abschuppen *v***/sich** scale (off), flake (off), chip
abschwächen *v* 1. attenuate *(an action)*; 2. weaken, thin *(e.g. a solution)*
Abschwächung *f* 1. attenuation *(of an action)*; 2. weakening, thinning *(as of a solution)*
Abscisinsäure *f* abscisic acid, ABA *(a plant hormone)*
abseigern *v (met)* liquate, melt down, smelt
abseihen *v* strain
absenken *v* lower *(a level, a temperature)*
Absetzanlage *f* 1. settling unit [plant], sedimentation [precipitation] unit, settler, precipitator; 2. *(hyd)* clarification unit [plant]
Absetzapparat *m* settler, precipitator
absetzbar settleable *(e.g. contaminants, floc)*
absetzbar/nicht non-settleable
Absetzbarkeit *f* settleability *(as of contaminants, floc)*
Absetzbecken *n* 1. settling tank [basin], sedimentation [precipitation] tank; 2. *(hyd)* gravity clarifier [settler], quiescent tank [basin]
Absetzbecken *n* **mit horizontaler Durchströmung** *(hyd)* horizontal-flow (sedimentation) tank
Absetzbecken *n* **mit vertikaler Durchströmung** *(hyd)* vertical-flow (sedimentation) tank, continuous vertical tank
Absetzbecken *n* **nach Imhoff** *(hyd)* Imhoff tank
Absetzbehälter *m* settling [sedimentation] tank, settler; settling [separator] tank, settler *(fluid extraction)*
Absetzdauer *f (hyd)* settling period [time], clarification time
absetzen *v* sediment, deposit *(of a liquid)*
absetzen *v***/sich** settle (down, out), sediment, precipitate, deposit, subside
absetzen *v***/sich wieder** redeposit
Absetzen *n* settling, sedimentation, precipitation, deposition, subsiding
Absetzen *n***/behindertes** *(min tech)* hindered settling
Absetzen *n***/freies** *(min tech)* free settling
Absetzen *n***/hartes** *(coat)* hard settling
Absetzer *m* settler, precipitator; coalescer *(for collecting liquid droplets)*
Absetzgeschwindigkeit *f* settling [sedimentation] rate, rate of gravity settling, subsidence rate
Absetzglas *n* settling cylinder
Absetzkammer *f* settling [sedimentation] chamber, expansion chamber
Absetzkonus *m* settling cone
Absetzkurve *f* settling curve
Absetzschlamm *m (hyd)* clarification [clarifier] sludge, sedimentation (tank) sludge
Absetzstoffe *mpl (hyd)* settleable solids
Absetztank *m s.* Absetzbehälter
Absetzverhinderungsmittel *n* antisettling [suspending] agent

Absetzzentrifuge *f* sedimentation(-type) centrifuge, sedimenting [solid-bowl, solid-wall, full-bowl] centrifuge, centrifugal decanter
absieben *v* screen (out), sieve (out); scalp (out) *(coarser grades)*
absinken *v* 1. sediment, settle (out), sink, subside *(of particles)*; 2. decrease, drop, fall *(as of measuring values)*
Absinken *n* 1. sedimentation, settling *(of particles)*; 2. decrease, drop, fall, decline *(as of measuring values)*
Absinkgeschwindigkeit *f* settling [sedimentation] rate [velocity]
Absitz... *siehe: Absetz...*
absitzen *v* settle (down, out), sediment, precipitate, deposit, subside
absitzen lassen *v* allow to deposit [settle], let settle *(from liquids)*
Absitzen *n* settling, sedimentation, precipitation, deposition, subsiding
absondern *v* 1. separate, eliminate, isolate, segregate; 2. *(biol)* secrete, exude *(e.g. resin)*
absondern *v***/sich** 1. separate, segregate *(from a mixture)*; 2. *(biol)* exude
Absonderung *f* 1. separation, elimination, isolation, segregation; 2. *(biol)* secretion; 3. *(biol)* secretion, *(relating to useless substances also)* excretum
Absorbanz *f (spectr)* absorbance *(preferred term)*; absorbancy, absorbency, extinction, optical density
Absorbat *n* 1. absorbate, absorbed material [substance]; 2. *absorbed material together with the absorbent*
Absorbend *m* material to be absorbed
Absorbens *n* absorbent, absorber, absorbing agent [material, substance]
Absorber *m* absorber, absorption apparatus [unit]
Absorber *m* **mit berieselter Wand** wetted-wall(-column) absorber
absorbierbar absorbable
Absorbierbarkeit *f* absorbability
absorbieren *v* absorb, imbibe
absorbieren *v***/wieder** re-absorb
absorbierend/schwach poorly absorbing
Absorpt *n* absorbate, absorbed material [substance]
Absorptiometer *n* absorptiometer
Absorptiometrie *f* absorptiometry
absorptiometrisch absorptiometric
Absorption *f* absorption
Absorption *f* **der Röntgenstrahlen [Röntgenstrahlung]** X-ray absorption
Absorptionsanalyse *f* absorptiometry
Absorptionsapparat *m s.* Absorber
Absorptionsbande *f* absorption band, spectral absorbance band
Absorptionsbatterie *f* absorption train *(for gases)*
Absorptionsbereich *m* absorbing region
absorptionsfähig absorptive, absorbent, absorbing
Absorptionsfähigkeit *f* absorbency
Absorptionsflasche *f* absorption bottle
Absorptionsflüssigkeit *f* absorption [absorbing] liquid
Absorptionsgebiet *n* absorbing region
Absorptionsgesetz *n***/bouguer-lambertsches** Bouguer-Lambert law (of absorption), Lambert's absorption law
Absorptionsgesetz *n***/henrysches** Henry's law (of absorption)
Absorptionsgesetz *n***/lambert-beersches** Beer-Lambert(-Bouguer) law
Absorptionsgesetz *n***/lambertsches** *s.* Absorptionsgesetz/bouguer-lambertsches
Absorptionskälteanlage *f* absorption refrigeration system [machine]
Absorptionskältemaschine *f* absorption refrigeration system [machine]
Absorptionskoeffizient *m* 1. *(phys ch, spectr)* absorption coefficient; 2. *(spectr)* absorptivity
Absorptionskoeffizient *m***/bunsenscher** *(phys ch)* Bunsen (absorption) coefficient
Absorptionskoeffizient *m***/molarer** *(spectr)* molar (decadic) absorption coefficient, molar absorptivity
Absorptionskolben *m* absorption flask
Absorptionskolonne *f* 1. absorption [absorbing] column; 2. *(tech)* absorption tower
Absorptionsküvette *f (anal)* absorption cell
Absorptionslinie *f* absorption line
Absorptionsmittel *n* absorbing agent [material, substance], absorbent, absorber
Absorptionspipette *f* absorption (gas) pipette

Absorptionsquerschnitt *m* 1. absorption cross section *(photochemistry)*; 2. *(nucl)* capture cross section
Absorptionsregel f/woodwardsche *(spectr)* Woodward's rule
Absorptionsrohr *n* absorption tube
Absorptionssäule *f* 1. absorption [absorbing] column; 2. *(tech)* absorption tower
Absorptionsspektrometer *n* absorption spectrometer
Absorptionsspektroskopie *f* absorption spectroscopy
Absorptionsspektrum *n* absorption spectrum
Absorptionsvermögen *n* absorbency; *(relating to radiation:)* absorptivity, absorptive power, *(quantitatively also)* absorption coefficient [factor]
Absorptionszahl *f* absorption coefficient [factor] *(for radiation)*; *(specif relating to heat radiation)* absorptivity
abspalten *v* split [cleave] off, eliminate, abstract *(atoms or atomic groups from molecules)*; expel *(molecules in elimination reactions)*; release *(gases)*; detach *(electrons)*
abspalten *v***/durch Solvolyse** *(bioch)* solvolyze off
abspalten *v***/Schichten** flake
abspalten *v***/sich** split [cleave] off
Abspaltung *f* splitting-off, cleaving-off, elimination *(of atomic groups from molecules)*; expulsion *(of molecules in elimination reactions)*; release *(of gases)*
Abspaltung *f* **eines Elektrons** electron detachment
Abspaltungsreaktion *f* elimination [abstraction] reaction
absperren *v* 1. shut off; 2. *(coat)* seal
Absperrhahn *m* stopcock
Absperrschieber *m* gate valve
Absperrventil *n* cut-off [shut-off] valve, stop [block] valve
absplittern *v* splinter off, chip (off)
abspringen *v* scale (off), flake (off), exfoliate, peel (off), shell (off)
abspülen *v* 1. rinse, swill *(e.g. a vessel)*; rinse off [down, away], swill off *(e.g. impurities)*; 2. *(min tech)* spray
Abstand *m* 1. distance, space, spacing; 2. interval *(period)*
abstechen *v* tap (out), draw off *(liquid metals)*
abstellen *v* turn off, shut off *(water, gas)*; stop *(a machine)*
Absterbephase *f (biot, hyd)* death [endogenous growth] phase
Abstich *m* 1. tapping *(of liquid metals)*; 2. *s.* Abstichloch
Abstichloch *n (met)* taphole, tapping hole
Abstichöffnung *f (met)* taphole, tapping hole
Abstichrinne *f (met)* tapping channel, runner
abstoppen *v* 1. stop (off, up); 2. *(rubber)* shortstop *(polymerization)*
Abstoppmittel *n (rubber)* shortstopping agent, shortstop, stopper
abstoßen *v* 1. repel *(as by electric charge)*; 2. discard, reject *(useless material)*
abstoßend 1. repellent, repulsive, vile *(esp smell)*; 2. repulsive *(force)*
Abstoßung *f* repulsion
Abstoßung *f***/coulombsche** Coulomb repulsion
Abstoßung *f***/elektrostatische** electrostatic repulsion
Abstoßung *f***/interelektronische** interelectronic [electron-electron] repulsion
Abstoßung *f* **zwischen Ionen (gleicher Ladung)** ionic repulsion
Abstoßungskraft *f* repulsive force, force of repulsion
abstrahlen *v* radiate, emit
Abstrahlung *f* radiation, emission
Abstrahlungsverlust *m* radiation loss, loss by radiation
abstreifen *v* 1. doctor off *(as from a roll)*; skim (off) *(scum)*; 2. *(distil)* strip (off, out)
Abstreifer *m* 1. *(distil)* stripper; 2. *s.* Abstreifermesser
Abstreiferkolonne *f (distil)* stripping column [still], stripper
Abstreifermesser *n* scraper [doctor] knife [blade], scraper
Abstreiferöl *n* stripping oil
Abstreiferteil *m s.* Abtriebsteil
Abstreiferzone *f s.* Abtriebsteil
Abstreifreaktion *f* stripping reaction
Abstrippzone *f s.* Abtriebsteil
abströmen *v* flow off [away], run off *(of liquids)*; escape *(of gases)*

abstumpfen *v* 1. *(ch)* blunt, neutralize, *(relating to acids also)* deacidify, *(relating to alkalies also)* dealkalize; 2. blunt *(glassware after cutting)*
Absud *m* decoction, decoctum
absüßen *v* sweeten, dulcify
abtasten *v* scan *(a measuring range)*
abtasten *v***/nochmals** rescan
Abtastgeschwindigkeit *f* pickup velocity
Abtastung *f***/photoelektrische** photoelectric scanning
abtauen *v* defrost *(e.g. a surface)*; thaw (off) *(ice)*
abtönen *v* tint, shade, tone
abtoppen *v* skim *(petroleum)*
Abtötungskurve *f* death curve *(sterilization)*
Abtötungsrate *f* death rate, rate of death *(sterilization)*
Abtötungstemperatur *f* lethal temperature *(sterilization)*
Abtrag *m* (surface) removal, *(esp by corrosion:)* eating away, *(esp by mechanical forces:)* wearing (away), erosion
abtragen *v* remove, *(esp by corrosion:)* eat away, *(esp by mechanical forces:)* wear (away), erode
abtreiben *v* 1. distil off, strip (off, out); 2. *(met)* cupel
Abtreiben *n* 1. simple distillation, stripping; 2. *(met)* cupellation
Abtreiber *m s.* Abtriebskolonne
Abtreiberkolonne *f s.* Abtriebskolonne
abtrennbar separable
Abtrennbarkeit *f* separability
abtrennen *v* separate (off), separate out [away], isolate *(something from a mixture)*; abstract, separate by distillation; detach *(an electron)*
Abtrennung *f* separation, isolation *(from a mixture)*; abstraction, separation by distillation; detachment *(of an electron)*
Abtrennungsarbeit *f* work function (of electrons), electronic work function
Abtrieb *m* simple distillation, stripping
Abtriebsgerade *f (distil)* stripping operating line
Abtriebskolonne *f (distil)* stripping column [still], stripper
Abtriebssäule *f (distil)* stripping column [still], stripper
Abtriebsteil *m* stripping [exhausting] section *(of a fractionating column)*
Abtrift *f* blow-off, drift *(of pesticides)*
abtrocknen *v* dry (off, up)
abtropfen *v* drain, drip [trickle] off
abtropfen lassen *v* drain
Abtropfen *n* drainage
Abtropfenlassen *n* drainage
abtupfen *v***/mit Fließpapier** blot
abwandern *v* 1. migrate out *(as of ions)*; 2. *(anal)* drift *(as of a base-line)*
Abwärme *f* waste heat
Abwärmerückgewinnung *f* waste-heat recovery
Abwärtsfilter *n (hyd)* downflow filter
Abwärtsfiltration *f (hyd)* downflow filtration
Abwärtsstrom *m* downflow, downward flow [current], descending current • **im Abwärtsstrom arbeiten** operate downflow *(of a filter)* • **im Abwärtsstrom auswaschen** rinse downflow *(ion exchange)* • **im Abwärtsstrom fahren** operate downflow *(a filter)*
abwaschen *v* wash (up), rinse
Abwasser *n* waste [discharge] water, waste, effluent, *(esp of domestic origin:)* sewage
Abwasser *n***/abfließendes (gereinigtes)** discharge from a waste-water treatment plant, waste treatment effluent, sewage plant effluent
Abwasser *n***/abzuleitendes** waste discharge, *(from an industrial plant:)* effluent discharge, waste effluent, outfall
Abwasser *n* **aus sanitären und sozialen Einrichtungen** sanitary waste water *(from industrial plants)*
Abwasser *n***/betriebliches** plant waste water, plant [factory, works] effluent, plant discharge
Abwasser *n* **der chemischen Industrie** chemical plant's waste, waste from chemical plants
Abwasser *n***/einzuleitendes** waste discharge; *(from an industrial plant:)* effluent discharge, waste effluent, outfall
Abwasser *n***/gereinigtes** discharge from a waste-water treatment plant, waste treatment effluent, treated effluent, sewage plant effluent
Abwasser *n***/gering konzentriertes**

low-contaminated [low-strength] waste, weak sewage

Abwasser *n*/**gewerbliches** commercial waste

Abwasser *n*/**häusliches** domestic sewage, household [sanitary] sewage, sewage water

Abwasser *n*/**hochkonzentriertes** high-contaminated [high-strength] waste, strong sewage

Abwasser *n*/**industrielles** industrial waste water, industrial discharge [effluent], trade effluent [waste]

Abwasser *n*/**kommunales** municipal [community] sewage [waste water]

Abwasser *n*/**städtisches** town [urban, municipal] sewage

Abwasserablauf *m* 1. waste-water effluent [discharge]; 2. effluent flow

Abwasserablauf *m* **eines Betriebs** plant effluent, waste effluent [discharge], outfall

Abwasserableitung *f* waste discharge

Abwasserableitung *f* **und -beseitigung** *f* sewerage

Abwasseranfall *m* waste(-water) flow, effluent flow

Abwasseraufbereitung *f s.* Abwasserbehandlung

Abwasserbehandlung *f* waste(-water) treatment, effluent treatment, *(esp municipally:)* sewage treatment

Abwasserbehandlung *f*/**betriebliche** plant waste(-water) treatment

Abwasserbehandlung *f*/**biologische** biological waste(-water) treatment

Abwasserbehandlung *f*/**kommunale** municipal sewage treatment

Abwasserbehandlung *f*/**landwirtschaftliche** land treatment of waste water, land disposal

Abwasserbehandlung *f* **nach dem Belebtschlammverfahren** activated-sludge waste [sewage] treatment, treatment by activated-sludge technique

Abwasserbehandlung *f*/**städtische** urban [municipal] sewage treatment

Abwasserbehandlung *f*/**weitergehende** third-stage treatment, advanced [tertiary] waste treatment

Abwasserbehandlungsanlage *f* waste-treatment unit, effluent-treatment station *(of an industrial plant)*; (municipal) sewage-treatment plant

Abwasserbehandlungsanlage *f* **mit biologischer Reinigungsstufe** secondary sewage-treatment plant

Abwasserbehandlungsverfahren *n* waste(-water)treatment process, *(esp municipally:)* sewage-treatment process

abwasserbelastet sewage-contaminated, polluted *(e.g. river)*

Abwasserbelastung *f* 1. waste-water loading; 2. *s.* Abwasserlast

Abwasserbeseitigung *f* waste-water disposal, waste disposal [elimination]

Abwassereinleitung *f* waste discharge

Abwassereinleitungsbedingungen *fpl* waste discharge standards, effluent standards [quality requirements, regulations]

Abwassereinleitungsstelle *f* waste-water discharge site, point of waste-water discharge, plant outfall [outlet]

Abwasserentsorgung *f s.* Abwasserbeseitigung

Abwassergrenzwerte *mpl* discharge limits

Abwasserinhaltsstoffe *mpl* waste-water components

Abwasserkanal *m* sewer

Abwasserklärung *f*/**mechanische** waste-water clarification

Abwasserkonzentration *f* waste concentration, strength of sewage

Abwasserlandbehandlung *f* land treatment of waste water, land disposal

Abwasserlast *f* waste-water pollution load, waste-water load(ing), waste load, pollutant [contaminant] load [loading], load of pollution [contamination]

Abwasserleitung *f* sewer [waste] line

Abwasserprobe *f* waste-water sample

Abwasserreinigung *f s.* Abwasserbehandlung

Abwasserschadstoff *m* waste-water pollutant [contaminant]

Abwasserschlamm *m* waste(-treatment) sludge, sludge from waste-water treatment

Abwasserschlammbehandlung *f* waste-water sludge treatment

Abwasserschlammsuspension *f* 1.

water slurry; 2. *(filtr)* feed sludge, filter feed slurry
Abwasserschmutzstoff *m* waste-water pollutant [contaminant]
Abwasserteich *m* (waste-treatment, sewage-treatment) lagoon, waste(-water) pond
Abwasserüberwachung *f* waste-water control
Abwehr *f*/**pflanzenchemische** phytochemical defense
Abwehrmittel *n s.* Abschreckmittel 2.
Abwehrprotein *n (med)* immune [antibody] protein
abweichen *v* deviate, be different, differ, depart *(from a standard or expected value)*
Abweichung *f* deviation, departure
Abweichung *f*/**mittlere quadratische** standard deviation, root mean square error, *(in a graph also)* half-peak width *(width at 0.607 h)*
Abweichung *f*/**zulässige** tolerance, allowance
abweisen *v* repel
abweisend repellent
Abweisungsvermögen *n* repellency
abwerfen *v* discharge, dump *(as from a conveyor)*
abwickeln *v* unwind, unreel, reel off
abwiegen *v* weigh (out, up)
abzentrifugieren *v* centrifuge off
abziehen *v* 1. draw off [down], withdraw, dispense, drain off *(liquids)*; tap *(slag)*; escape, issue *(as of vapours)*; 2. strip *(a dye)*; peel off *(coatings)*; 3. *(phot)* print; 4. *(plast)* take up *(extrudate)*
abziehen *v*/**auf Fässer** barrel, rack
abziehen *v*/**auf Flaschen** bottle (in, up)
Abzug *m* 1. *(lab)* (exhaust) hood, fume hood [chamber, closet]; 2. discharge (point), outlet, drain; 3. *(phot)* (contact) print; 4. *(plast)* take-up *(of extrudate)*
Abzugskasten *m (lab)* local exhaust hood, fume closet
Abzugsrohr *n* discharge pipe [tube], drainage [outlet, offtake, effluent] pipe; waste pipe [tube]; vent [fume] pipe
Abzugsschrank *m (lab)* fume cupboard [chamber]
Abzweig *m* branch
abzweigen *v* branch off *(of pipelines)*
Abzweigung *f* 1. branching-off *(of pipelines)*; 2. branch
Acen *n (org ch)* acene
Acenaphthen *n (org ch)* acenaphthene
Acetacetat *n (bioch)* acetoacetate
Acetal *n* $R{-}CH(OR')_2$ acetal, *(specif)* $CH_3CH(OC_2H_5)_2$ acetal, *1,1-diethoxyethane
Acetaldehyd *m* acetaldehyde, *ethanal
Acetaldol *n* acetaldol, aldol, 3-hydroxybutyraldehyde, *3-hydroxybutanal
Acetalharz *n* acetal resin
acetalisieren *v* acetalate
Acetalisierung *f* acetalation
Acetamid *n* acetamide
Acetamidin *n (org ch)* acetamidine
Acetamidogruppe *f* $-NHCOCH_3$ acetamido [acetylamino] group
Acetanhydrid *n* acetic anhydride
Acetanilid *n* acetanilide, *N*-phenylacetamide
Acetat *n* acetate
Acetat *n*/**aktiviertes** *s.* Acetyl-Coenzym A
Acetatfaser *f* acetate fibre
Acetatfaserstoff *m* acetate fibre
Acetatseide *f* cellulose acetate rayon, acetate fila-ment yarn
Acetessigester *m s.* Acetessigsäureethylester
Acetessigsäure *f* acetoacetic acid, 3-oxobutyric acid
Acetessigsäureethylester *m* ethyl acetoacetate, acetoacetic ester
Acetoacetylierung *f* acetoacetylation
Acetoethylierung *f* acetoethylation
Acetogenin *n* acetogenin, polyketide, ketide
Acetoin *n* acetoin, dimethylketol, *3-hydroxybutan-2-one
Acetolyse *f* acetolysis
Aceton *n* acetone, *propanone, dimethylketone
Aceton-Butanol-Gärung *f (biot)* acetone-butanol fermentation
Acetondicarbonsäure *f* acetonedicarboxylic acid, ADA, 3-oxoglutaric acid, *3-oxopentanedioic acid
Acetonharz *n* acetone resin
Acetonierung *f* acetonation
Acetonitril *n* acetonitrile, AN
Acetonkörper *m (bioch)* acetone [ketone] body

Acetophenon *n* acetophenone, acetylbenzene, *methyl phenyl ketone
Acetoxygruppe *f* CH_3CO_2– acetoxy [ethanoyloxy] group
Acetoxylierung *f* acetoxylation
Acetoxymethylierung *f* acetoxymethylation
Acetylaminogruppe *f* acetylamino group, acetamido group
Acetylbenzen *n* acetylbenzene, *methyl phenyl ketone, acetophenone
Acetylbutyrylcellulose *f* cellulose acetate butyrate, CAB
Acetylcellulose *f* acetylated cellulose, cellulose acetate, CA
Acetylcholin *n* acetylcholine
Acetyl-Coenzym *n* **A** acetyl coenzyme A, acetyl Co A, active acetate
Acetylen *n* 1. acetylene, *ethyne; 2. *s.* Acetylenkohlenwasserstoff
Acetylencarbonsäure *f* acetylenic acid *(any carboxylic acid having a triple bond)*
Acetylendicarbonsäure *f* acetylenedicarboxylic acid, *butynedioic acid
Acetylenentwickler *m* acetylene gas generator
acetylenisch acetylenic
Acetylenkohlenwasserstoff *m* acetylenic hydrocarbon, *alkyne
Acetylgruppe *f* CH_3CO– acetyl group [residue]
Acetylharnstoff *m* acetylurea, *N*-monoacetylurea
Acetylid *n* $M^I_2C_2$ acetylide
acetylierbar acetyl(at)able
acetylieren *v* acetylate, acetylize
Acetylierung *f* acetyl(iz)ation
Acetylierungsmittel *n* acetylation [acetylating] agent
Acetylrest *m s.* Acetylgruppe
Acetylsalicylsäure *f O*-acetylsalicylic acid, *o*-acetoxybenzoic acid
Acetylureid *n s.* Acetylharnstoff
Acetylzahl *f* acetyl value [number] *(of fats and oils)*
Achat *m (min)* agate
Acheson-Verfahren *n* Acheson process *(for producing artificial graphite)*
achiral achiral *(stereochemistry)*
Achiralität *f* achirality *(stereochemistry)*
Achse *f***/dreizählige** axis of threefold symmetry, three-fold axis of symmetry
Achse *f***/kristallographische** crystal(lographic) axis
Achse *f***/optische** optic(al) axis
Achsenabschnitt *m (cryst)* intercept, parameter
Achsensymmetrie *f (cryst)* axial symmetry
achsensymmetrisch *(cryst)* axisymmetric
Achsenverhältnis *n* axial ratio, ratio of the intercepts
Achsenverhältnis *n***/kristallographisches** crystallographic axial ratio, ratio of the intercepts
Achsenwinkel *m (cryst)* axial angle
Achtergruppe *f* octet
Achterring *m s.* Achtring
Achterschale *f* octet
achtflächig *(cryst)* octahedral, oct.
Achtflächner *m (cryst)* octahedron
Achtring *m* eight-membered ring
achtwertig octavalent
Achtwertigkeit *f* octavalency
Acidoligand *m* anion(ic) ligand
aci-Form *f* aci form *(of nitro compounds)*
Aconitsäure *f* aconitic acid, *propene-1,2,3-tricarboxylic acid
Aconit(um)alkaloid *n* aconitum [aconite] alkaloid
Acoxylierung *f* ac(yl)oxylation
ACP = acyl-carrier protein
Acridinfarbstoff *m* acridine dye
Acrolein *n* acrolein, acraldehyde, *propenal
Acrylaldehyd *m s.* Acrolein
Acrylamid *n* acrylamide, *propenamide
Acrylat *n* acrylate
Acrylelastomer(e) *n* acrylate [acrylic, polyacrylate] elastomer
Acrylfaser *f* acrylic [polyacrylonitrile] fibre
Acrylfaserstoff *m* acrylic [polyacrylonitrile] fibre
Acrylglas *n* acrylic glass
Acrylharz *n* acrylic resin, acrylate [acrylic-acid] resin
Acryllack *m* acrylic lacquer
Acrylnitril *n* acrylonitrile, cyanoethylene, vinyl cyanide, *propene nitrile
Acrylnitril-Butadien-Kautschuk *m* acrylonitrile-butadiene rubber
Acrylnitril-Butadien-Styrol-Kunststoff

m acrylonitrile-butadiene-styrene plastic
Acrylsäure *f* acrylic acid, *propenoic acid
Acrylsäureamid *n s.* Acrylamid
Acrylsäureester *m* acrylic-acid ester, acrylic ester
Acrylsäurenitril *n s.* Acrylnitril
ACTH *s.* Hormon/adrenocorticotropes
Actinid(enelement) *n s.* Actinoidenelement
Actinium *n* Ac actinium
Actiniumzerfallsreihe *f* actinium (decay) series
Actinoid *n s.* Actinoidenelement
Actinoidenelement *n* actinoid (element)
Actinoidengruppe *f s.* Actinoidenreihe
Actinoidenkontraktion *f* actinoid contraction
Actinoidenreihe *f* actinoid group [series]
acyclisch acyclic, non-cyclic(al)
Acylaminoalkylierung *f* acylaminoalkylation, amidoalkylation
***N*-Acylaminocarbonsäure** *f N*-acyl amino carboxylic acid
Acylglycerin *n s.* Glycerid
Acylglycerol *n s.* Glycerid
Acylgruppe *f* (carboxylic) acyl group, acyl residue
Acylhalogenid *n* acyl halide *(acid halide of a carboxylic acid)*
acylierbar acylable
acylieren *v* acylate
Acylierung *f* acylation
Acylierungsmittel *n* acylating agent
Acylium-Ion *n* acylium ion, acyl cation
Acylkation *n* acylium ion, acyl cation
Acyloin *n* acyloin *(a keto alcohol of the general formula R–CO–CH–R′–OH)*
Acyloinaddition *f* acyloin condensation
Acyloinkondensation *f* acyloin condensation
Acyloinumlagerung *f* acyloin rearrangement
Acyloxylierung *f* ac(yl)oxylation
Acylradikal *n* (free) acyl radical
Acylrest *m s.* Acylgruppe
Adamantan *n (org ch)* adamantane
adamantoid diamondoid, diamondlike *(ring system)*
Addend *m* addend
addieren *v* add *(an atom or atomic group to a molecule)*
addieren *v***/sich** add
Addition *f* addition
Addition *f***/durch Peroxide ausgelöste [initiierte]** peroxide-initiated addition, anti-Markovnikov addition
Addition *f***/elektrophile** electrophilic addition
Addition *f***/ionoide** ionic [Markovnikov] addition
Addition *f***/oxidative** oxidative addition
Addition *f***/photochemische** photoaddition
α-Addition *f* α-addition, alpha-addition (reaction)
Additionsmischkristall *m* addition solid solution
Additionspolymer(e) *n* addition polymer
Additionspolymerisation *f* addition polymerization
Additionsreaktion *f* addition reaction
Additionsverbindung *f* addition [additive] compound
additiv additive
Additiv *n* additive *(a substance added to another one in small amounts; specif a substance added to mineral-oil products in a quantity from 1 to 10 %)*
Additiv *n***/rostverhinderndes [rostverhütendes]** rust-preventing additive
Additivitätsprinzip *n* additivity principle *(of the contribution of substituent groups to the properties of the molecule)*
Addukt *n* adduct
Adduktkautschuk *m* adduct rubber
Adenin *n* adenine, 6-aminopurine
Adenosindiphosphat *n s.* Adenosindiphosphorsäure
Adenosindiphosphorsäure *f* adenosine diphosphoric acid, adenosine diphosphate, ADP
Adenosinmonophosphat *n s.* Adenosinmonophosphorsäure
Adenosin-5′-monophosphat *n s.* Adenosin-5′-phosphorsäure
Adenosinmonophosphorsäure *f* adenosine monophosphoric acid, adenosine monophosphate, AMP, adenylic acid, AA
Adenosin-5′-phosphorsäure *f* adenosine 5′-phosphoric acid, adenosine 5′-monophosphate, AMP, muscle adenylic acid

Adenosintriphosphat *n s.* Adenosintriphosphorsäure
Adenosintriphosphorsäure *f* adenosine triphosphoric acid, adenosine triphosphate, ATP
Adenylbernsteinsäure *f* adenylosuccinic acid
adenylieren *v* adenylate
Adenylierung *f* adenylation
Adenylierungsmittel *n* adenylating agent
Adenylosuccinat *n* adenylosuccinate
Adenylpyrophosphat *n s.* Adenosintriphosphorsäure
Adenylsäure *f s.* Adenosinmonophosphorsäure
adenylylieren *v (bioch)* adenylylate
Adenylylierung *f (bioch)* adenylylation
Ader *f (geol)* lode, vein
Adermin *n* adermin(e), 3-hydroxy-4,5-di-(hydroxymethyl)-2-methylpyridine *(vitamin B_6)*
ADH *s.* Hormon/antidiuretisches
Adhärend *m* adherend *(a body to be attached to another one by an adhesive)*
Adhärens *n* adhesive (agent, substance), adhesion agent
adhärieren *v* adhere
adhärierend adherent, adhesive
Adhäsion *f* adhesion, adherence
Adhäsionsarbeit *f* adhesional work, work of adhesion
adhäsionsfähig adherent, adhesive
Adhäsionsfestigkeit *f* adhesive [adhesion] strength
Adhäsionskraft *f* adhesive force
Adhäsionsvermögen *n* adhesiveness, adhesion [adhesive] power, adherence
adhäsiv adhesive, adherent
Adiabasie *f* adiabaticity
adiabat adiabatic
Adiabate *f* adiabat(ic), adiabatic curve [line]
adiabatisch adiabatic
adiatherman atherm(an)ous
Adion *n* adion *(an ion adsorbed on a surface)*
Adipamid *n* adipamide
Adipat *n* adipate *(salt or ester of adipic acid)*
Adipinsäure *f* adipic acid, *hexanedioic acid
Adipinsäurediamid *n* adipamide
Adipinsäuredinitril *n* adiponitrile
Adiponitril *n* adiponitrile
Adjuvans *n (pharm)* adjuvant
Admittanz *f (anal)* admittance
ADP *s.* Adenosindiphosphorsäure
Adrenalin *n* adrenaline, *1-(3,4-dihydroxyphenyl)-2-methylaminoethanol
Adrenocorticotropin *n* adrenocorticotropic hormone, ACTH, corticotropin
Adsorbat *n* adsorbed material [substance], adsorbate; adsorption complex, adsorbent-adsorbate complex [system], adsorbate
Adsorbend *m* material to be adsorbed
Adsorbens *n* adsorbent (material), adsorbing agent [material, substance]
Adsorber *m* 1. adsorber *(apparatus)*; 2. *s.* Adsorbens
adsorbierbar adsorbable
Adsorbierbarkeit *f* adsorbability
adsorbieren *v* adsorb
adsorbieren *v*/**chemisch** chemisorb
adsorbieren *v*/**physikalisch** adsorb physically
Adsorpt *n* adsorbed material [substance], adsorbate
Adsorption *f* adsorption
Adsorption *f*/**aktivierte** *s.* Adsorption/chemische
Adsorption *f*/**bevorzugte** *s.* Adsorption/selektive
Adsorption *f*/**chemische** chemical [activated] adsorption, chemisorption, chemosorption
Adsorption *f*/**physikalische** physical [van der Waals] adsorption, physisorption
Adsorption *f*/**selektive** selective [preferential] adsorption
Adsorption *f*/**van-der-waalssche** *s.* Adsorption/physikalische
Adsorptionsanalyse *f* adsorption analysis
Adsorptionsapparat *m* adsorber
Adsorptionschromatographie *f* adsorption chromatography, chromatographic adsorption analysis
adsorptionschromatographisch by adsorption chromatography
adsorptionsfähig adsorbent, adsorptive, adsorbing

Adsorptionsfähigkeit *f* adsorption ability, adsorptive capacity, adsorptiveness
Adsorptionsfiltration *f* adsorption filtration
Adsorptionsgleichgewicht *n* adsorption equilibrium
Adsorptionsgleichung *f*/**gibbssche** Gibbs adsorption equation
Adsorptionsindikator *m* adsorption indicator
Adsorptionsisotherme *f* adsorption isotherm
Adsorptionsisotherme *f*/**freundlichsche** Freundlich (adsorption) isotherm
Adsorptionsisotherme *f*/**gibbssche** Gibbs (adsorption) isotherm
Adsorptionsisotherme *f*/**langmuirsche** Langmuir (adsorption) isotherm
Adsorptionsisotherme *f* **nach Brunauer, Emmett und Teller** BET isotherm
Adsorptionsmittel *n* adsorbent (material), adsorbing agent [material, substance]; sweetener *(for dry-cleaning)*
Adsorptionsschicht *f* adsorbed layer
Adsorptionstitration *f* **nach Fajans** Fajans (adsorption indicator) method *(argentometry)*
Adsorptionswärme *f* heat of adsorption
Adsorptionszentrum *n* adsorption centre [site]
adsorptiv *s.* adsorptionsfähig
Adsorptiv *n* adsorptive, material [substance] to be adsorbed
Adstringens *n* *(pharm)* astringent, styptic
Adstringenz *f* astringency, stypticity
adstringierend astringent, styptic
aerob *(biol)* aerobic, oxybiotic
Aerobier *m* *(biol)* aerobe
Aerobiont *m* *(biol)* aerobe
Aerogel *n* aerogel
Aeroklassieren *n* air classifying [sizing], air separation [sweeping, elutriation], pneumatic classification
Aerosol *n* aerosol
aerosolieren *v* aerosolize, nebulize
Aerosol(is)ierung *f* aerosolization
Aerosolnebel *m* aerosol mist
Aerosoltreibmittel *n* aerosol propellant
AES *s.* Auger-Elektronenspektroskopie
Affinität *f* affinity
Affinität *f*/**chemische** chemical affinity
Affinität *f* **zu Wasser** water affinity
Affinitätschromatographie *f* affinity chromatography
affinitätschromatographisch by affinity chromatography
AFID = Alkaliflammenionisationsdetektor
AFS *s.* Atomfluoreszenzspektrometrie
Afterkristall *m* pseudomorph
Agar(-Agar) *m(n)* agar(-agar), agar gel, Japan agar [isinglass], Chinese gelatin
Agargel *n s.* Agar
Agargel-Elektrophorese *f* *(anal)* agar gel electrophoresis
Agarnährboden *m* nutrient agar
Agens *n* agent
Agens *n*/**angreifendes** attacking agent
Agens *n*/**chemisches [chemisch wirksames]** chemical agent
Agens *n*/**elektrophiles** electrophilic agent, electrophile
Agens *n*/**mutagenes [mutationsauslösendes]** mutagenic agent, mutagen
Agens *n*/**nucleophiles** nucleophilic agent, nucleophile
Agglomerat *n* agglomerate
Agglomeration *f* agglomeration
agglomerieren *v* agglomerate
Agglutination *f* agglutination
agglutinieren *v* agglutinate
agglutinierend agglutinant
Agglutinogen *n* agglutinogen
Aggregat *n* aggregate
Aggregat *n*/**molekulares** molecular [molecule] aggregate
Aggregation *f* aggregation
Aggregation *f*/**molekulare** molecular aggregation
Aggregatzustand *m* state of aggregation, physical state
Aggregatzustandsänderung *f* change of state
aggregieren *v* aggregate
aggregieren *v*/**sich** aggregate
aggressiv aggressive, offensive *(chemicals)*
aggressiv/nicht non-corrosive
Aggressivität *f* aggressivity, aggressiveness, offensiveness *(of chemicals)*
Aglucon *n* aglucon(e) *(non-sugar portion of a glucoside)*
Aglykon *n* aglycon(e) *(non-sugar portion of a glycoside)*

Agon *n* agon, prosthetic [active] group, coenzyme
Agonist *m (pharm)* agonist *(any of a class of remedies which alter cell properties after joining with a receptor)*
Agrarchemie *f* agrochemistry, agricultural chemistry
Agrarchemiker *m* agricultural chemist
Agrikulturchemie *f s.* Agrarchemie
Agrochemie *f s.* Agrarchemie
Agrochemikalie *f* agricultural chemical, agrochemical
A-Harz *n* A-stage [one-stage] resin, resol
AHG *s.* Globulin (A)/antihämophiles
Ähnlichkeit *f* 1. similarity; 2. *(phys ch)* similitude *(collectively for geometrical, dynamical and kinematic similarity)*
Ähnlichkeitsverbundwirkung *f (tox)* similar joint action
Ahornzucker *m* maple sugar
AIBN *s.* Azobisisobutyronitril
AICAR = 5(4)-Aminoimidazol-4(5)-carboxamidribotid
Airlift *m* 1. air lift, mammoth [air-lift] pump; 2. *(petrol)* air lift, air-lift pump, riser *(catalyst-handling system in catalytic cracking)*
Aitken-Partikel *n* Aitken particle *(aerosol particle below 0.1 mm in diameter)*
Aitken-Teilchen *n* Aitken particle *(aerosol particle below 0.1 mm in diameter)*
akarizid acaricidal, miticidal
Akarizid *n* acaricide, miticide
A-Kation *n* class [a] metal ion
Akaziengummi *n* acacia [Arabic] gum, gum arabic *(from Acacia specc., esp from A. senegal (L.) Willd.)*
Akkretion *f* accretion, enlargement *(of aerosols)*
Akku *m s.* Akkumulator
Akkumulation *f* accumulation
Akkumulator *m* accumulator, (storage) battery
Akkumulator(en)säure *f* battery [electrolyte] acid
A-Kohle *f* activated carbon
Aktionskonstante *f* frequency factor, (Arrhenius) pre-exponential factor, Arrhenius factor, A factor *(kinetics)*
aktiv/elektrochemisch electroactive
aktiv/optisch optically active
Aktivanode *f* sacrificial [galvanic, expendable] anode
Aktivator *m* 1. activator, activating agent [substance]; promoter, promoting agent *(catalysis)*; 2. *(rubber)* (polymerization) initiator, initiating agent, reaction catalyst; 3. *(met)* energizer *(additive to a carburizer)*
Aktivatoreffekt *m (tox)* activation
Aktivchlor *n (pap)* available [active] chlorine
Aktiverde *f* active [activated] earth
aktivierbar activable, capable of being activated
aktivieren *v* 1. activate; energize *(molecules)*; 2. *(rubber)* boost; radioactivate
Aktivierung *f* 1. activation; energization *(of molecules)*; 2. *(rubber)* boosting; radioactivation
Aktivierung f/thermische thermal activation *(of a reaction)*
Aktivierungsanalyse *f* activation analysis
Aktivierungsenergie *f* activation energy
Aktivierungsenergie f/arrheniussche Arrhenius energy of activation, Arrhenius activation energy
Aktivierungsenthalpie *f* activation enthalpy, enthalpy of activation
Aktivierungsenthalpie f/freie Gibbs (free) energy of activation, free energy of activation, activation Gibbs function
Aktivierungsquerschnitt *m* activation cross section
Aktivierungsvolumen *n* activation volume
Aktivität *f* 1. activity *(of a catalyst)*; 2. activity *(corrected mole fraction of a substance in a mixed phase)*; 3. activity, disintegration rate *(of a radioactive substance)*
Aktivität f/biologische biological activity
Aktivität f/katalytische catalytic activity
Aktivität f/molare molar activity *(enzyme kinetics)*
Aktivität f/optische optical activity [rotatory power], rotary polarization
Aktivität f/spezifische specific activity *(enzyme kinetics)*
Aktivitätsförderung *f (bioch)* feed-forward activation [regulation], positive feedback
Aktivitätskoeffizient *m* activity coefficient [factor]

Aktivkohle *f* activated carbon
Aktivkohle ***f*/gekörnte** granulated [granular] activated carbon, GAC, granular carbon, GC
Aktivkohlefilter *n* (activated-)carbon filter, carbon canister
Aktivkohlefiltration *f* (activated-)carbon filtration, CF, filtration with activated carbon
Aktivsauerstoff *m* active [available] oxygen *(as in bleaching)*
Aktivstelle *f* active site [centre]
Aktivzentrum *n* active site [centre]
Aktor *m* actor *(in induced reactions)*
Aktuator *m (anal)* actuator
Akzeptor *m* acceptor; electron acceptor
Akzeptorort *m (bioch)* acceptor site, A site, entry [recognition, decoding, aminoacyl] site, aminoacyl-tRNA site
Akzeptorstelle *f (bioch)* acceptor site, A site, entry [recognition, decoding, aminoacyl] site, aminoacyl-tRNA site
Akzeptorverbindung *f* acceptor compound
Akzessibilität *f* accessibility
AL *s.* Alginatfaserstoff
Alan *n* alane, aluminium hydride
Alanat *n* $M^I[AlH_4]$ tetrahydridoaluminate, alanate *(a complex hydride)*
Alanin *n* alanine, 2-aminopropionic acid
ALAT = L-Alanin-2-ketoglutarat-Aminotransferase
Alaun *m* $M^IM^{III}(SO_4)_2 \cdot 12\ H_2O$ alum, *(specif)* $KAl(SO_4)_2 \cdot 12\ H_2O$ potassium [potash] alum
alaunhaltig containing alum, aluminous
Albumin *n* albumin, *(sometimes)* albumen *(one of a class of simple proteins)*
Albuminoid *n* albuminoid
Aldarsäure *f (org ch)* aldaric acid
Aldazin *n (org ch)* aldazine
Aldehyd *m* aldehyde
Aldehyd ***m*/aliphatischer** aliphatic aldehyde, alkanal
Aldehydcarbonsäure *f s.* Aldehydsäure
Aldehydgruppe *f* –CHO aldehyde [aldehydic] group
aldehydig aldehydic
Aldehyd-Keton-Umlagerung *f* aldehyde-ketone-rearrangement
Aldehydkondensation *f* aldehyde condensation
Aldehydsäure *f* aldehydic acid
Aldehydzucker *m* aldehyde sugar
Aldimin *n* aldimine
Aldohexonsäure *f* aldonic [glyconic] acid *(any of several acids derived from an aldose; general formula $HOCH_2(CHOH)_4COOH$)*
Aldohexose *f* aldohexose *(a hexose containing an aldehyde group)*
Aldoketen *n* RCH=C=O aldoketene
Aldol *n* aldol *(any of a class of 3-hydroxyaldehydes)*; *(specif)* $CH_3CH(OH)CH_2CHO$ acetaldol, aldol, 3-hydroxybutyraldehyde
Aldoladdition *f s.* Aldolkondensation
Aldolkondensation *f* aldol condensation, aldolization
Aldolkondensation ***f*/gekreuzte** crossed aldol condensation
Aldonsäure *f* $HOCH_2(CHOH)_nCOOH$ aldonic acid *(any of a class of acids derived from aldoses)*
Aldose *f* aldose *(any of a class of sugars containing one –CHO group per molecule)*
Aldoxim *n* R–CH=NOH aldoxime
Algenbekämpfungsmittel *n s.* Algizid
Algenrohmasse *f (biot)* algal biomass
Algenrohsubstanz *f (biot)* algal biomass
Algensäure *f s.* Alginsäure
Alginat *n* alginate *(salt or ester of alginic acid)*
Alginatfaser *f (text)* alginate fibre
Alginatfaserstoff *m (text)* alginate fibre
Alginatseide *f (text)* alginate yarn
Alginsäure *f* alginic acid *(a polymer of D-mannuronic acid)*
Algizid *n* algicide, algaecide
Alicyclen *pl* alicyclics, alicyclic compounds
alicyclisch alicyclic, cycloaliphatic
Aliphaten *pl* aliphatics, aliphatic compounds
Aliphatenchemie *f* aliphatic chemistry
aliphatisch aliphatic
aliquot *(anal)* aliquot
aliquotieren *v (anal)* aliquot
alitieren *v* aluminize *(make metals oxidation-resistant by heating them with powdered aluminium)*
Alizarinfarbstoff *m* alizarin dye
Alk. *s.* Alkohol
Alkadiin *n (org ch)* alkadiyne
alkal. *s.* alkalisch

Alkali *n* alkali
Alkalialkoholat *n* alkali alkoxide [alcoholate]
Alkalibatterie *f (el ch)* alkaline cell
alkalibeständig resistant to alkali(es), alkali-stable, alkali-resistant, alkali-resisting, alkali-proof
alkalibeständig/nicht alkali-unstable, alkali-labile
Alkalibeständigkeit *f* alkali resistance, resistance [stability] to alkali(es)
Alkaliboratglas *n* alkali borate glass
Alkalicellulose *f* alkali cellulose, *(Am)* soda cellulose
Alkalichloridelektrolyse *f* chloralkali electrolysis
alkaliecht *(dye)* alkali-fast, fast to alkali(es)
alkaliempfindlich alkali-sensitive, sensitive to alkalies
Alkaliempfindlichkeit *f* alkali sensitivity, sensitivity to alkalies
Alkalien *npl* alkalies, alkalis
Alkalien... *siehe: Alkali...*
Alkalifehler *m* alkali(ne) error *(of a glass electrode)*
Alkalifeldspat *m* alkali feldspar
Alkaliglas *n* alkali glass
Alkalihalogenid *n* alkali halide [halogenide]
Alkalihydroxid *n* alkali hydroxide
Alkaliion *n* alkali ion
Alkalilauge *f* lye
alkalilöslich alkali-soluble, soluble in alkali(es)
Alkalimetall *n* alkali(ne) metal
Alkalimetallhalogenid *n* alkali halide [halogenide]
Alkalimetallhydroxid *n* alkali hydroxide
Alkalimetallion *n* alkali metal ion
Alkalimetalloxid *n* alkali oxide
Alkalimeter *n* alkalimeter *(for measuring the proportion of alkali in a solution)*
Alkalimetrie *f* alkalimetry
alkalimetrisch alkalimetric
Alkalinität *f* alkalinity *(referring to hydrogen carbonates dissolved in water)*
Alkaliphosphatglas *n* alkali phosphate glass
Alkalisalz *n* alkali(-metal) salt
Alkalisilikatglas *n* alkali silicate glass
alkalisch alkaline, basic • **alkalisch machen** alkalify, alkali(ni)ze, make alkaline [basic] • **alkalisch reagieren** react alkaline • **alkalisch stellen** alkalify, alkali(ni)ze, make alkaline [basic] • **alkalisch werden** alkalify
alkalisch/schwach mildly alkaline, alkalescent, subalkaline
alkalisch/stark highly alkaline
Alkalischmachen *n* alkali(ni)zation
Alkalischmelze *f* 1. fusion with alkali, alkali [caustic] fusion; 2. alkali [caustic] fusion, molten alkali (bath)
alkalisieren *v* alkalify, alkali(ni)ze, make alkaline [basic]
Alkalisierung *f* alkali(ni)zation
Alkalispaltung *f* alkali cleavage
Alkalität *f* alkalinity, basicity, *(quantitatively also)* basic strength
alkaliunbeständig alkali-unstable, alkali-labile
alkaliunlöslich alkali-insoluble, insoluble in alkali(es), base-insoluble
Alkaloid *n* alkaloid
alkaloidisch alkaloidal
Alkamin *n* alkamine, amino alcohol
Alkan *n* *alkane, paraffin (hydrocarbon), saturated hydrocarbon
Alkan *n*/**geradkettiges [unverzweigtes]** normal [straight-chain] alkane
Alkan *n*/**verzweigtes [verzweigtkettiges]** branched-chain alkane
Alkanal *n* alkanal
Alkandiylgruppe *f* alkanediyl group, alkylene
Alkandiylrest *n* alkanediyl group, alkylene
Alkanfermentation *f (biot)* alkane fermentation
Alkanol *n* alkanol
Alkanolamin *n* alkanolamine
Alkanon *n* alkanone
Alkanphosphonigsäure *f* $RP(OH)_2$ alkylphosphonous [alkylphosphinic] acid, *(Am)* alkanephosphonous acid
Alkanreihe *f* alkane family
Alkansäure *f* alkanoic acid
Alkatrien *n (org ch)* alkatriene
Alkatriin *n (org ch)* alkatriyne
Alken *n* *alkene, olefin (hydrocarbon)
Alken *n*/**fluoriertes** *fluoroalkene, fluoroolefin
Alkenreihe *f* alkene series [family], olefin series

Alkensäure *f* alkenoic acid *(any of the monounsaturated fatty acids)*
Alkenylbenzen *n* alkenylbenzene, aromatic-alkene compound
Alkenylierung *f* alkenylation
Alkin *n* *alkyne, acetylenic hydrocarbon
Alkinmethatese *f (org ch)* alkyne metathesis
Alkinol *n* *alkynol, acetylenic alcohol
Alkinsäure *f* alkynoic acid
Alkohol *m* alcohol, *(specif)* C_2H_5OH ethyl alcohol, *ethanol
Alkohol *m*/**absoluter** absolute [dehydrated, anhydrous] alcohol
Alkohol *m*/**denaturierter** denatured alcohol, methylated spirit
Alkohol *m*/**dreiwertiger** trihydroxylic [trihydric] alcohol
Alkohol *m*/**einwertiger** monohydroxylic [monohydric] alcohol
Alkohol *m*/**mehrwertiger** polyhydroxylic alcohol, polyhydric [polyhydroxy, polyfunctional] alcohol, polyalcohol, polyol
Alkohol *m*/**primärer** primary alcohol
Alkohol *m*/**sekundärer** secondary alcohol
Alkohol *m*/**technischer** commercial [industrial, non-beverage] alcohol, commercial spirit
Alkohol *m*/**tertiärer** tertiary alcohol
Alkohol *m*/**vergällter** *s.* Alkohol/denaturierter
Alkohol *m*/**vierwertiger** tetrahydroxylic [tetrahydric] alcohol
Alkohol *m*/**wasserfreier** *s.* Alkohol/absoluter
Alkohol *m*/**zweiwertiger** dihydroxylic [dihydric] alcohol, dialcohol, diol, *(relating to aliphatic compounds also)* glycol, aliphatic diol
Alkoholat *n* *alkoxide, alcoholate
Alkoholauszug *m* alcoholic extract
alkoholfrei alcohol-free, non-alcoholic, soft
Alkoholgärung *f* alcoholic fermentation
alkoholisch alcoholic, spirituous
alkoholisch-wässrig hydroalcoholic
alkoholisieren *v* alcoholize
Alkoholisierung *f* alcoholization
Alkoholkraftstoff *m* alcohol fuel
alkohollöslich alcohol-soluble, spirit-soluble, soluble in alcohol
Alkoholthermometer *n* alcohol-in-glass thermometer
alkoholunlöslich insoluble in alcohol
Alkoholyse *f* alcoholysis • **einer Alkoholyse unterwerfen** alcoholyze
Alkoxyborierung *f* alkoxyboration
Alkoxycarbonylierung *f* alkoxycarbonylation, carbalkoxylation
Alkoxydealkylierung *f* alkoxydealkylation
Alkoxygruppe *f s.* Alkoxylgruppe
Alkoxyhalogenierung *f* alkoxyhalogenation
Alkoxylalkylierung *f* alkoxylalkylation
Alkoxylgruppe *f* $C_nH_{2n+1}O$– alkoxyl [alkoxy] group, alkoxyl [alkoxy] residue
Alkoxylierung *f* alkoxylation
Alkyd *n s.* Alkydharz
Alkydharz *n* alkyd (resin)
Alkydharz *n*/**kurzöliges** short-oil alkyd
Alkydharz *n*/**langöliges** long-oil alkyd
Alkydharz *n*/**mittelöliges** medium-oil alkyd
Alkydharz *n*/**styrolisiertes** styrenated alkyd
Alkydharzanstrichstoff *m* alkyd paint [coating]
Alkyl *n* alkyl
Alkylamin *n* alkylamine
Alkylans *n* alkylating agent
Alkylarensulfonat *n* alkylarene sulphonate
Alkylarylsilikon *n s.* Alkylarylsiloxan
Alkylarylsiloxan *n* alkyl aryl siloxane
Alkylarylsulfonat *n s.* Alkylarensulfonat
Alkylat *n* alkylate
Alkylbenzen *n* alkylbenzene
Alkylbenzensulfonat *n* alkyl benzene sulphonate, ABS
Alkylbenzensulfonat *n*/**lineares** linear alkyl benzene sulphonate, LAS, α-benzenesulphonate
Alkylbenzol *n s.* Alkylbenzen
Alkylbrücke *f* alkyl bridge
Alkylcyanid *n* alkyl cyanide
Alkylderivat *n* alkyl derivative
Alkylgruppe *f* alkyl group [residue]
Alkylhalogenid *n* alkyl halide [halogenide], *haloalkane
Alkylidengruppe *f* alkylidene group
Alkylidenphosphoran *n* phosphonium [phosphorus] ylide, Wittig reagent
Alkylidenrest *m* alkylidene group

alkylieren *v* alkylate
Alkylierung *f* alkylation, alkanation, alkyl substitution
Alkylierung *f***/hydrierende** hydroalkylation, reductive alkylation
Alkylierung *f* **mit Fluorwasserstoffsäure [Flusssäure]** hydrofluoric-acid alkylation, HF alkylation
Alkylierung *f* **mit Schwefelsäure** sulphuric-acid alkylation
Alkylierung *f***/reduktive [reduzierende]** *s.* Alkylierung/hydrierende
***N*-Alkylierung** *f N*-alkylation, nitrogen alkylation
***O*-Alkylierung** *f O*-alkylation, oxygen alkylation
Alkylierungsmittel *n* alkylating agent
Alkylmetallverbindung *f* alkylmetal compound
Alkylperoxid *n* alkyl peroxide
Alkylphenolharz *n* alkyl phenol resin
Alkylphosphonsäure *f* $RP=O(OH)_2$ alkylphosphonic acid
Alkylradikal *n* alkyl radical, *(specif)* free alkyl radical
Alkylrest *m* alkyl residue [group]
Alkylschwefelsäure *f* alkylsulphuric acid
Alkylsilikon *n* *alkyl siloxane, alkyl silicone
Alkylsulfonat *n* alkyl sulphonate
Alkylsulfonsäure *f* alkylsulphonic acid
Alkylverbindung *f* alkyl compound
Allantoin *n* allantoin, glyoxylic diureide
Allantoinsäure *f* allantoic acid, diureidoacetic acid
Allelochemikalie *f* allelochemical *(a substance produced by organisms and acting on other organisms)*
allelotrop allelotropic
Allelotropie *f* allelotropism *(coexistence of tautomeric forms)*
Allen *n* allene, *(specif)* $CH_2=C=CH_2$ *propadiene, allene
allenisch *(org ch)* allenic
Allenisomerie *f* allene isomerism
Allen-Kegel *m (min tech)* Allen cone classifier
Allen-Konus *m (min tech)* Allen cone classifier
Allergen *n (med)* allergen
Alles-oder-Nichts-Übergang *m* concerted transition [transfer] *(enzyme kinetics)*
Allgemeinformel *f* general formula
Allihn-Kühler *m* Allihn [bulb] condenser
all-*cis*-Isomer(e) *n* all-*cis* isomer
all-*trans*-Isomer(e) *n* all-*trans* isomer
Allokatalyse *f* allocatalysis
Allomerisierung *f (bioch)* allomerization
Allophanamid *n* allophanamide, carbamyl urea, biuret
Allophansäureamid *n s.* Allophanamid
Allose *f* allose *(an aldohexose)*
Allosterie *f (bioch)* allosterism
allosterisch allosteric
allotriomorph *(cryst)* allotriomorphic, anhedral, xenomorphic
allotrop allotropic
Allotrop *n* allotrope, allotropic modification [form] *(of an element)*
Allotropie *f* allotropism, allotropy *(existence of an element in two or more different modifications)*
Allylalkohol *m* allyl alcohol, AA, *prop-2-en-1-ol
Allylanion *n* allyl anion
Allylchlorid *n* allyl chloride, *3-chloropropene
Allylchlorierung *f* allylic chlorination
Allylester-Kunststoff *m* allyl plastic
Allylgruppe *f* $CH_2=CH-CH_2-$ allyl [allylic] group [residue]
Allylharz *n* allyl(ic) resin
Allylharz-Kunststoff *m* allyl plastic
Allylierung *f* allylation
allylisch allylic
Allylkation *n* allyl cation
Allylkondensation *f* allylic condensation
Allylrest *m s.* Allylgruppe
allylständig allylic
Allylsubstitution *f* allylic substitution
Allylumlagerung *f* allylic rearrangement
Alphacellulose *f* alpha [chemical] cellulose, α-cellulose
Alphastrahler *m* alpha emitter
Alphastrahlung *f* alpha radiation
Alphateilchen *n* alpha particle
Alphazerfall *m* alpha decay [disintegration]
ALS *s.* Alginatseide
Alteisen *n* scrap iron, ferrous scrap
altern *v* age *(e.g. a rubber product)*; undergo ageing, age, *(of precipitates also)* digest
Altersbestimmung *f* 1. age determination; 2. *(geol)* dating

Altersbestimmung f/radiometrische radiometric dating
Alterung *f* ageing, aging, *(of precipitates also)* digestion
Alterung f/beschleunigte [künstliche] artificial [accelerated] ageing
alterungsbeständig *v* resistant ageing, non-ageing
Alterungsbeständigkeit *f* resistance to ageing, ageing resistance
Alterungsschutzmittel *n (rubber, plast)* antioxidant, age-resister
Altgummi *m* rubber scrap [waste], scrap [waste] rubber
Altkatalysator *m* used catalyst
Altmaterial *n* waste (material)
Altmetall *n* scrap (metal)
Altöl *n* used oil
Altpapier *n* waste [old] paper
Alt(roh)stoff *m* waste material
Alu *n s.* Aluminium
alumetieren *v* alumetize
Aluminat *n* aluminate
Aluminid *n* aluminide
aluminieren *v* aluminize *(without formation of alloy layers)*
Aluminium *n* Al aluminium, *(Am)* aluminum
Aluminiumacetat *n* aluminium acetate
Aluminiumblech *n* sheet aluminium
Aluminiumbronze *f* aluminium bronze
Aluminiumchlorid *n* aluminium chloride
Aluminiumfluorid *n* aluminium fluoride
Aluminiumhexafluorosilicat *n* aluminium hexafluorosilicate
Aluminiumlegierung *f* aluminium alloy
Aluminiumnitrat *n* aluminium nitrate
Aluminiumoxid *n* aluminium oxide
Aluminiumseife *f* aluminium soap
Aluminiumsulfat *n* aluminium sulphate
Aluminiumtriäthyl *n s.* Triethylaluminium
Aluminiumtrimethyl *n s.* Trimethylaluminium
Aluminiumwasserstoff *m* aluminium hydride, alane
Aluminogel *n* alumina gel, gelatinous aluminium hydroxide
Aluminothermie *f* aluminothermics, aluminothermy, Goldschmidt's process
aluminothermisch aluminothermic
Alumosilicat *n* aluminosilicate
Amadori-Umlagerung *f* Amadori rearrangement *(of glycosylamines into 1-aminoketoses)*
Amalgam *n* amalgam *(an alloy of mercury)*
Amalgamation *f* amalgamation
Amalgamator *m s.* Amalgamierapparat
Amalgamierapparat *m* amalgamation apparatus, amalgamator
amalgamieren *v* amalgam(ate), amalgamize
Amalgamierung *f* amalgamation
Amalgamverfahren *n* (intermediate) mercury electrode process, mercury-cell process *(electrolysis)*
Amalgamzelle *f* mercury (cathode) cell *(electrolysis)*
ambident *(org ch)* ambident *(possessing two reactive centres)*
ambient ambient, surrounding
ambifunktionell *s.* ambident
ambivalent *s.* ambident
Ameisensäure *f* formic acid, methanoic acid
Ameisensäureamid *n* formamide, *methanamide
Ameisensäureethylester *m* ethyl formate
Americium *n* Am americium
Ames-Test *m* Ames [salmonella] test *(for determining the mutagenicity of chemicals)*
A-Metallion *n* class [a] metal ion
Amid *n* $RCONH_2$ amide, azanide; M^INH_2 (metal) amide
Amidhydrazon *n* amide hydrazone
amidieren *v* amidate
Amidierung *f* amidation
Amidin *n (org ch)* amidine, *(specif)* $RC(=NH)NH_2$ carboxamidine
Amidinocarbonylgruppe *f s.* Amidinogruppe
Amidinogruppe *f* $-C(=NH)NH_2$ amidine group
amidisch amidic
Amidoalkylierung *f* amidoalkylation, acylaminoalkylation
Amidogruppe *f* $-CONH_2$ amido group
Amidokohlensäure *f* carbamic acid
Amidophosphorsäure *f* amidophosphoric acid, phosphamic acid
Amidoschwefelsäure *f* amidosulphuric acid, sulphamic acid

Amidosulfat *n* $NH_2SO_3M^I$ amidosulphate, sulphamate
Amidoxim *n* amide oxime
Amidrazon *n* amidrazone
Amidstickstoff *m* amide nitrogen
Amikron *n* amicron, subsubmicron *(a disperse particle invisible under the microscope)*
Amin *n* amine
Amin *n*/**biogenes** biogenic amine
Amin *n*/**primäres** RNH_2 primary amine
Amin *n*/**sekundäres** R_1R_2NH secondary amine
Amin *n*/**tertiäres** $R_1R_2R_3N$ tertiary amine
Aminal *n* aminal, geminal diamine
aminieren *v* aminate
Aminierung *f* amination
Aminoacetal *n* aminoacetal
Aminoacetanilid *n* aminoacetanilide
Aminoacylierung *f* aminoacylation
Aminoacylort *m s.* Akzeptorort
α-**Aminoadipinsäure-Weg** *m (bioch)* α-aminoadipic-acid pathway
Aminoalkohol *m* amino alcohol, alkamine
Aminoanteil *m* amino moiety
Aminoazoverbindung *f* aminoazo compound
Aminobenzen *n* aminobenzene, phenylamine, aniline
***p*-Aminobenzensulfonamid** *n p*-aminobenzene-sulphonamide, sulphanilamide
Aminobenzensulfonsäure *f* aminobenzenesulphonic acid, anilinesulphonic acid
Aminobenzoesäure *f* aminobenzoic acid
Aminobenzol *n s.* Aminobenzen
Aminobenzylierung *f* aminobenzylation
Aminobernsteinsäure *f* aminosuccinic acid, aspartic acid, asparaginic acid, *aminobutanedioic acid
Aminobuttersäure *f (bioch)* aminobutyric acid
γ-**Aminobuttersäure** *f (bioch)* gamma aminobutyric acid, GABA
Aminocarbonsäure *f s.* Aminosäure
2-Amino-2-deoxy-α-D-glucopyranose *f* D-glucosamine, 2-amino-2-deoxy-α-D-glucopyranose
Aminodicarbonsäure *f* amino dicarboxylic acid
Aminoessigsäure *f* aminoacetic acid, glycine
2-Aminoethan-1-ol *n* *2-aminoethan-1-ol, 2-aminoethyl alcohol, monoethanolamine, MEA
Aminoethansäure *f s.* Aminoessigsäure
Aminogruppe *f* NH_2– amino group [residue]
Aminokomponente *f* amino moiety
Aminolyse *f* aminolysis
Aminomethan *n* methylamine
aminomethylieren *v* aminomethylate
Aminomethylierung *f* aminomethylation
Aminonaphthalensulfonsäure *f* naphthylaminesulphonic acid, aminonaphthalenesulphonic acid
Aminophenol *n* aminophenol, hydroxyaniline
Aminoplast *m* aminoplastic
Aminopolycarbonsäure *f* amino polycarboxylic acid
2-Aminopropionsäure *f* 2-aminopropionic acid, alanine
Aminosäure *f* amino acid, amino carboxylic acid
Aminosäure *f*/**essenzielle [unentbehrliche]** *(bioch)* essential amino acid
α-**Aminosäure** *f* α-amino acid, α-amino carboxylic acid
Aminosäureanalysator *m* amino-acid analyser
Aminosäureeinheit *f* amino-acid unit
Aminosäure(reihen)folge *f s.* Aminosäuresequenz
Aminosäurerest *m* amino-acid residue
Aminosäuresequenz *f* amino-acid sequence, sequence of amino-acid residues
Aminotoluen *n* aminotoluene, toluidine
Aminotoluol *n* aminotoluene, toluidine
Aminoverbindung *f* amino compound
Aminoxid *n* amine oxide
Aminozucker *m* amino sugar, glycosamine
Aminsynthese *f*/**gabrielsche** Gabriel phthalimide synthesis, Gabriel synthesis of primary amines
Amin-Zucker-Bräunung *f (food)* amino-sugar browning, Maillard-type browning, non-enzymatic browning

Ammin *n* ammine *(coordination chemistry)*
Ammoniak *n* ammonia
Ammoniak *n*/**flüssiges** liquid ammonia
Ammoniak *n*/**gasförmiges** gaseous ammonia, ammonia gas
Ammoniakalaun *m* ammonia [ammonium] alum
ammoniakalisch ammoniac(al)
Ammoniakat *n* ammoniate, ammonate
Ammoniakbildner *mpl* ammonifiers, ammonifying bacteria
Ammoniakgas *n* ammonia gas, gaseous ammonia
Ammoniaklösung *f* ammonia, aqua ammonia
Ammoniaklösung *f*/**wässrige** ammonia water, aqua ammonia
Ammoniakoxidation *f* ammonia oxidation, oxidation of ammonia
Ammoniak-Soda-Verfahren *n* (Solvay's) ammonia soda process, Solvay process
Ammoniakstickstoff *m* ammonia [ammoniacal] nitrogen
Ammoniaksynthese *f* ammonia synthesis
Ammoniakverbrennung *f* ammonia oxidation, combustion of ammonia
Ammoniakwäscher *m* ammonia scrubber
Ammonifizierung *f (agric)* ammonification
ammonisieren *v* ammoniate *(fertilizers)*
Ammonisierung *f* 1. ammoniation *(of fertilizers)*; 2. *s.* Ammonifizierung
Ammonium *n* ammonium
Ammoniumacetat *n* ammonium acetate
Ammoniumalaun *m s.* Ammoniakalaun
Ammoniumbenzoat *n* ammonium benzoate
Ammoniumcarbamat *n* ammonium carbamate, ammonium aminoformate
Ammoniumcarbonat *n* ammonium carbonate
Ammoniumchlorid *n* ammonium chloride, salmiac
Ammoniumdichromat *n* ammonium dichromate
Ammoniumeisenalaun *m* *ammonium iron alum, ferric ammonium alum
Ammoniumformiat *n* ammonium formate
Ammoniumgruppe *f* NH_2– ammonium group [residue]
Ammoniumhydrogencarbonat *n* ammonium hydrogencarbonate
Ammoniumhydrogensulfid *n* ammonium hydrogensulphide
Ammoniumhydroxid *n* ammonium hydroxide
Ammoniummolybdat *n* ammonium molybdate
Ammoniumnitrat *n* ammonium nitrate
Ammoniumperoxodisulfat *n* ammonium peroxodisulphate
Ammoniumpikrat *n* ammonium picrate
Ammoniumradikal *n* ammonium (free) radical
Ammoniumrest *m* ammonium residue [group]
Ammoniumrhodanid *n s.* Ammoniumthiocyanat
Ammoniumsalz *n* ammonium salt
Ammoniumstickstoff *m s.* Ammoniakstickstoff
Ammoniumsulfamat *n s.* Ammoniumsulfamidat
Ammoniumsulfamidat *n* ammonium sulphamate, AMS
Ammoniumsulfaminat *n* ammonium sulphamate, AMS
Ammoniumsulfat *n* ammonium sulphate
Ammoniumsulfid *n* ammonium sulphide
Ammoniumthiocyanat *n* ammonium thiocyanate
Ammoniumthiosulfat *n* ammonium thiosulphate
Ammoniumwolframat *n* *ammonium wolframate, ammonium tungstate
Ammoniumylid *n* ammonium ylide, amine ylide
Ammonolyse *f* ammonolysis
ammonolytisch ammonolytic
Ammonotelie *f* ammonotelism *(excretion of nitrogen as ammonia)*
Ammonoxidation *f* ammoxidation
Ammonsalpetersprengstoff *m* ammonium nitrate explosive
amöbizid amoebicidal
Amöbizid *n* amoebicide
amorph amorphous
Amorphie *f* amorphousness, amorphicity
AMP *s.* Adenosinmonophosphorsäure
Amperemeter *n* amperemeter, ammeter

Amperometrie *f* amperometry
amperometrisch amperometric
Amphetamin *n* amphetamine, 1-phenyl-2-propylamine
amphibol(isch) *(bioch)* amphibolic
Amphibolismus *m (bioch)* amphibolism
Amphipathisch *(bioch)* amphipathic
amphi-Stellung *f* amphi-position
Ampho-Ion *n* amphion, amphoteric [ampholyte] ion, dipole [dipolar, dual, hybrid] ion, zwitterion, zwitterionic compound, inner [internal, intramolecular] salt
Ampholyt *m* ampholyte, amphoteric electrolyte, amphiprotic substance
amphoter amphoteric, amphiprotic
Amphoterie *f* amphoterism *(phenomenon)*; amphotericity *(of a substance)*
Ampulle *f* ampoule
amu *s.* Masseeinheit/atomare
Amylacetat *n* amyl acetate
Amylalkohol *m* $C_5H_{11}OH$ amyl alcohol, *(specif)* $CH_3(CH_2)_3CH_2OH$ *pentan-1-ol, *(deprecated:)* *n*-amyl alcohol
Amylgruppe *f* $CH_3(CH_2)_3CH_2$– amyl group [residue]
Amylnitrit *n* $C_5H_{11}ONO$ amyl nitrite, *(specif)* $(CH_3)_2CHCH_2CH_2ONO$ (ordinary) amyl nitrite, *3-methyl-1-butyl nitrite
amyloid amyloid(al)
Amylopektin *n* amylopectin *(main component of starch)*
anabol *(bioch)* anabolic
anabolisch *(bioch)* anabolic
Anabolismus *m (bioch)* anabolism
anaerob anaerobic
Anaerobier *m* anaerobe
Analeptikum *n (pharm)* analeptic, central nervous system stimulant
Analgetikum *n (pharm)* analgesic, pain-reliever, pain-killer
analgetisch analgesic
Analog(e) *n* analogue
Analysator *m s.* Analysengerät
Analyse *f* analysis • **eine Analyse durchführen** perform [run, do] an analysis • **zur Analyse** *s.* analysenrein
Analyse *f* **an Ort und Stelle** on-site analysis
Analyse *f* **auf trockenem Wege** dry analysis
Analyse *f***/chemische** chemical analysis
Analyse *f***/direkte** direct analysis
Analyse *f***/diskontinuierliche** discontinuous [batch] analysis
Analyse *f***/dynamisch mechanische** dynamic mechanical analysis, DMA
Analyse *f***/dynamische** on-line analysis
Analyse *f***/elektrochemische** electroanalysis
Analyse *f***/gravimetrische** gravimetric analysis, analysis by weight
Analyse *f***/kolorimetrische** colorimetric analysis
Analyse *f***/polarographische** polarographic analysis
Analyse *f***/potenziometrische** potentiometric analysis
Analyse *f***/qualitative** qualitative [compositional] analysis
Analyse *f***/quantitative** quantitative analysis
Analyse *f***/technische** commercial analysis
Analyse *f***/thermische** thermal analysis, thermoanalysis
Analyse *f***/thermogravimetrische** thermogravimetric analysis, TGA
Analyse *f***/thermomechanische** thermomechanical analysis, TMA
Analyse *f***/thermometrische** thermometric analysis
Analyse *f***/trockene** dry analysis
Analysenbefund *m* analytical result
Analysendurchführung *f* analytical procedure
Analyseneichfunktion *f* calibration function
Analysenergebnis *n* analytical result
Analysenfehler *m* analytical error
Analysengang *m* scheme [course] of analysis
Analysengerät *n* analyser, analysis apparatus, analytical instrument [unit]
Analysenmethode *f* analytical [analysis] method, method of analysis
Analysenmethodik *f* analytical technique [procedure, operation]
Analysenprobe *f* sample analysed [under analysis]
Analysenprotokoll *n* analysis report
analysenrein reagent-grade, analytical-reagent-quality
Analysenreinheit *f* analytical-reagent quality

Analysensubstanz *f* analysand, substance to be analysed; substance under analysis
Analysentrichter *m* fluted funnel, 60° filtration funnel
Analysenverfahren *n s.* Analysenmethodik
Analysenwaage *f* analytical balance
Analysenweg *m* analytical approach, scheme of analysis
Analysenwert *m* analytical value
analysierbar analysable
analysieren *v* analyse
Analyt *m* analyte
Analytik *f* analytical chemistry
Analytiker *m* analytical chemist [scientist], (chemist-)analyst
analytisch analytic(al)
Anaphorese *f (phys ch)* anaphoresis
anästhesierend anaesthetic
Anästhesierungsmittel *n* anaesthetic
Anästhetikum *n* anaesthetic
Anatoxin *n* anatoxin *(detoxicated toxin)*
anätzen *v* 1. etch; 2. *(med)* cauterize *(by chemicals)*
anbacken *v* bake on, stick on
anblasen *v* blow in *(a blast furnace)*
anbrennen *v* 1. ignite, catch fire; 2. *s.* anzünden
anbringen *v* attach, mount
Änderung *f* change, alteration
Änderung *f***/stoffliche** change of material, material change
Androgen *n* androgen *(any of a class of sex hormones)*
andrücken *v* press, contact
aneinanderhaften *v* stick together
anellieren *v (org ch)* anellate, fuse
Anellierung *f (org ch)* anellation, fusion
Anellierungsseite *f* side [point, position] of fusion
Anellierungsstelle *f* side [point, position] of fusion
anerkennen *v* approve, schedule, register *(e.g. a pesticide)*
Anerkennung *f***/amtliche** (government) approval, registration, *(Am)* label clearance *(as of a pesticide)*
Anerkennungsverfahren *n* approval scheme *(as for pesticides)*
anfahren *v* start up *(e.g. a reactor)*
Anfall *m* 1. accumulation; 2. *s.* Anfallmenge
anfällig susceptible
Anfallmenge *f* amount *(as of product or waste)*
Anfangsfeuchte(beladung) *f* initial moisture content, IMC
Anfangsgeschwindigkeit *f* **der Reaktion** initial reaction rate
Anfangsglied *n* 1. initial member; 2. *(nucl)* parent element *(of a decay series)*
Anfangskonzentration *f* initial [starting] concentration
Anfangssiedepunkt *m* initial boiling point, I.B.P.
Anfangswert *m* initial [starting] value
Anfangszustand *m* initial [original] state
anfärbbar 1. dyeable; 2. *(biol)* stainable
Anfärbbarkeit *f* 1. dyeability, dye receptivity, receptivity to dyestuffs; 2. *(biol)* stainability
anfärben *v* 1. tint, tinge, *(with soluble colouring matter:)* dye, *(Am)* color, *(with suspended colouring matter:)* paint; 2. *(biol)* stain *(for microscopical investigation)*
anfärben *v***/direkt** *(text)* dye directly
anfeuchten *v* moisten, wet, damp(en), humidify
anflecken *v* stain
anfressen *v* corrode, eat, attack
anfressen *v***/punktförmig** pit
Angärung *f* pre-fermentation
angegriffen werden *v* undergo attack
angerben *v* pretan, colour
angreifbar attackable, vulnerable *(chemically)*; corrodible *(metal)*; affectable, affectible *(physically as by heat)*
Angreifbarkeit *f* vulnerability, capability of being attacked, liability to attack *(chemically)*; affectability, affectibility *(physically as by heat)*
angreifen *v* attack *(chemically)*; corrode, eat *(metals)*; affect *(physically as by heat)*
Angriff *m* attack
Angriff *m* **an der *meta*-Position** meta attack
Angriff *m***/bakterieller** bacterial attack
Angriff *m* **von der Rückseite (her)** rear [back-side] attack *(as in Walden inversion)*
Angriffsort *m* site [point] of attack
Angriffsstelle *f* site [point] of attack
Anguss(kegel) *m (plast)* gate, sprue

anhaften *v* adhere
Anhaften *n* adherence, adhesion
anhaftend adherent, adhesive
anharmonisch anharmonic
Anharmonizität *f* anharmonicity
anhäufen *v* accumulate
anhäufen *v*/**sich** accumulate, collect
Anhäufung *f* accumulation
anheben *v* elevate, raise *(e.g. temperature, boiling point)*; raise, energize *(electrons to an excited state)*
Anhebung *f* elevation, raising *(as of temperature, boiling point)*; raising, energization *(of electrons to an excited state)*
anheizen *v* heat up
Anhydrid *n* (acid) anhydride
Anhydrid *n*/**gemischtes** mixed anhydride
Anhydrid *n*/**inneres [intramolekulares]** inner [internal, intramolecular] anhydride
Anhydrid *n*/**symmetrisches** symmetric anhydride
anhydri(di)sieren *v* anhydr(id)ize
Anhydrisierung *f* anhydr(id)ization
Anhydrit *m (min)* anhydrite *(anhydrous calcium sulphate)*
Anhydrobase *f* anhydrobase
Anhydrozucker *m* anhydrosugar
Anil *n* anil, *N*-phenyl imine
Anilid *n* 1. anilide, *N*-phenyl amide *(an N-acyl derivative of aniline)*; 2. anilide *(a salt of aniline)*
Anilin *n* aniline, aminobenzene
Anilinfarbstoff *m* aniline dye
Anilingelb *n* aniline yellow, *p*-aminoazobenzene, solvent yellow 1
Anilinharz *n* aniline(-formaldehyde) resin
Anilinhydrochlorid *n* aniline hydrochloride, aniline salt
Anilinogruppe *f* C_6H_5NH- anilino [phenylamino] group
Anilin-*p*-sulfonsäure *f* aniline-*p*-sulphonic acid, sulphanilic acid
Anilinvergiftung *f* aniline poisoning
animpfen *v* 1. *(cryst)* seed; 2. *(biot)* inoculate, seed *(a nutrient solution)*
Animpfen *n* 1. *(cryst)* seeding; 2. *(biot)* inoculation, seeding *(of a nutrient solution)*
Anion *n* anion
Anionbase *f* anion base
anionenaktiv anion-active, anionic
Anionenaustausch *m* anion exchange
Anionenaustauscher *m* 1. anion [hydroxyl-form] exchanger, OH-form exchanger, anionite; 2. *(hyd)* anion unit *(of a demineralizing system)*
Anionenaustauscher *m*/**schwach basischer** weak-base [weakly basic] anion exchanger, weak-base deionizer [resin]
Anionenaustauscher *m*/**stark basischer** strong-base [strongly basic] anion exchanger, strong-base deionizer [resin]
Anionenelektrode *f* anodized [anodic] electrode
Anionenfehlstelle *f* anion vacancy
Anionenleerstelle *f* anion vacancy
Anionentrennungsgang *m* anion scheme, scheme of analysis for the anions
anionisch anionic
anionoid anionoid, nucleophilic
anionotrop anionotropic
Anionotropie *f* anionotropy
Anionsäure *f* anion acid
Aniontensid *n* anionic surfactant
anisodesmisch *(cryst)* anisodesmic
Anisol *n* anisole, *methoxybenzene, methyl phenyl ether
anisotrop anisotropic, non-isotropic, aeolotropic
anisotrop/optisch optically anisotropic
Anisotropie *f* anisotropy, aeolotropism
Anissäure *f* anisic acid, *p*-methoxybenzoic acid
Ankergruppe *f* active group *(ion exchange)*
ankohlen *v* char
ankondensieren *v* fuse *(a ring to another one)*
ankuppeln *v (dye)* couple
Anlage *f* 1. plant; unit; 2. arrangement *(as of a trial)*; layout *(of a plant)*
Anlage *f* **mit geschlossenem Kreislauf** closed-cycle system plant
Anlage *f* **mit offenem Kreislauf** open-cycle system plant
Anlagenleistung *f* plant performance
Anlagentechnik *f* plant technology
anlagern *v* add *(a chemical compound for preparing an addition compound)*;

attach, gain *(e.g. a proton or electron - said of an atom or compound)*
anlagern v/koordinativ coordinate *(e.g. molecules)*
anlagern v/sich 1. become attached, attach oneself, *(esp of atoms or atomic groups:)* undergo addition, add; 2. *(physically:)* adsorb, undergo adsorption
anlagern v/sich koordinativ coordinate
Anlagerung *f* 1. attachment, *(esp of atoms or atomic groups:)* addition; 2. *(physically:)* adsorption
Anlagerung f/koordinative coordination
Anlagerungskomplex *m* outer complex, high-spin [spin-free] complex
Anlagerungsreaktion *f* addition reaction
Anlagerungsverbindung *f* addition compound
anlassen *v* 1. start (up) *(e.g. a machine)*; 2. temper *(metals)*
Anlassfarbe *f (coat)* temper colour, annealing colour
Anlassfilm *m* tarnish (film)
Anlassöl *n* tempering oil
anlaufen *v* 1. tarnish; 2. start (up) *(as of a machine)*; get under way *(of production)*
Anlauffarbe *f* temper colour
Anlaufperiode *f (phys ch)* induction period
Anlaufschicht *f* tarnish (film)
Anlockmittel *n* attractant *(as for insects)*
Anlockstoff *m* attractant *(as for insects)*
Anlösung *f* partial solution
annehmen *v* 1. assume, suppose, suggest; 2. accept *(e.g. a theory)*; 3. acquire *(a configuration)*; 4. take on *(a colour)*; 5. *(text)* take, accept *(a dye)*
Annihilation *f (phys ch)* annihilation, energy pooling
Annulen *n* annulene
[8]Annulen *n* [8]annulene, cycloocta-1,3,5,7-tetraene
Anode *f* anode, anelectrode, positive electrode
Anodenpotenzial *n* anode potential
Anodenraum *m* anode compartment
Anodenreaktion *f* anode reaction
Anodenschlamm *m* anode slime [sludge, mud]
Anodenspannung *f* anode voltage
Anodenstrom *m* anode current
anodisch anodic
anodisch wirksam galvanic *(coating)*
anodisieren *v (met)* anodize
Anodisierung *f (met)* anodizing, anodic oxidation [coating]
Anolyt *m* anolyte *(electrolyte in the anode compartment)*
anomal anomalous, abnormal
Anomalie *f* anomaly
α-Anomalie *f* glass transition, gamma [second-order] transition, vitrification *(of polymers)*
anomer anomeric *(stereochemistry)*
Anomer(e) *n* anomer *(stereochemistry)*
Anomerisierung *f* anomerization *(stereochemistry)*
anordnen *v* arrange, mount, align *(e.g. parts of an apparatus)*
anordnen v/sich align *(as of chain molecules)*
Anordnung *f* 1. arrangement, array, alignment *(as of chain molecules)*; 2. *(spectr)* mounting *(of a monochromator)*
Anordnung f/geometrische *s.* Anordnung/räumliche
Anordnung f/octaedrische octahedral orientation *(of ligands)*
Anordnung f/quadratisch ebene square planar orientation *(of ligands)*
Anordnung f/räumliche spatial arrangement [orientation]
Anordnung f/tetraedrische tetrahedral arrangement [orientation]
Anordnung f/v-förmige angular [bent] orientation *(of ligands)*
Anorganiker *m* inorganic chemist
anorganisch inorganic
anorganisch-chemisch inorganic-chemical
Anorthit *m (min)* anorthite *(calcium aluminium silicate)*
Anpassstück *n* adapter, adaptor
anpasten *v* make into a paste, paste
Anprall *m* impact, impingement
anprallen *v* impact, impinge
anrauen *v* roughen
anregbar excitable
Anregbarkeit *f* excitability
anregen *v* excite *(an atom)*; activate, initiate, start *(a reaction)*; induce *(crystallization)*
Anregung *f* excitation *(of an atom)*; ac-

tivation, initiation *(of a reaction)*; induction *(of crystallization)*
Anregung f/photochemische photochemical excitation, photoexcitation
Anregung f/thermische thermal excitation
Anregung f von Atomen atomic excitation
Anregungsenergie *f* excitation energy
Anregungsmittel *n (pharm)* stimulant, stimulus, stimulatory drug
Anregungsspektrum *n* excitation spectrum
Anregungszustand *m* excited [energized] state
anreiben *v* grind *(pigments)*
Anreibung *f (coat)* grinding
anreichern *v* enrich; concentrate, beneficiate *(ore)*
anreichern v/mit Sauerstoff enrich with oxygen, oxygenate, oxygenize
anreichern v/mit Vitaminen vitaminize, enrich [fortify] with vitamins
anreichern v/sich accumulate
Anreicherung *f* 1. enrichment; concentration, beneficiation *(of ore)*; 2. accumulation
Anreicherung f von Uran-235 uranium(-235) enrichment
Anreicherungsanlage *f (min tech)* concentration plant, concentrator
Anreicherungsfaktor *m (nucl)* enrichment factor
Anreicherungsgrad *m (nucl)* enrichment factor
anrühren *v* temper, mix *(for preparing a slurry)*; paste, make into a paste
ansammeln v/sich accumulate, collect
Ansammlung *f* accumulation, collection
Ansatz *m* 1. batch, charge, charge [charging] stock, feed material, feed(-stock); 2. *(plast)* formulation; 3. crust *(of undesired solid matter)*
ansäuern *v* 1. acidify, make acidic, acidulate; 2. *(food)* sour; leaven *(dough)*
ansäuern v/erneut reacidify
Ansäuern *n* 1. acidification, acidulation; 2. *(food)* souring; leavening *(of dough)*
Ansäuerungsmittel *n* acidulant, acidulent
Ansaugdruck *m* suction pressure
ansaugen *v* draw (in), suck, aspirate
Ansaugen *n* drawing, suction, aspiration
Ansaugrohr *n* suction pipe [tube]
Ansa-Verbindung *f (org ch)* ansa compound
anschließen *v* attach, connect
Anschliff *m (geol)* ground [polished] face, ground surface [section] *(for direct-light microscopy)*
Anschluss *m* 1. attachment, connection *(act)*; 2. *s.* Anschlussstelle; 3. *s.* Anschlussstück
Anschlussrohr *n* connecting tube
Anschlussstelle *f* joint, junction
Anschlussstück *n* connecting piece, connection, joint
anschmutzen *v* 1. stain; 2. *(text)* soil
Anschmutzung *f* 1. staining; 2. *(text)* soiling
Anschütz-Aufsatz *m (distil)* Anschütz head
anschwellen *v* swell (up)
Anschwemmfilter *n* 1. precoat(ed) filter; 2. *(hyd)* septum filter
Anschwemmfiltration *f* 1. precoat filtration; 2. *(hyd)* septum filtration
ansetzen *v* 1. prepare *(e.g. a reaction mixture)*; 2. charge *(a furnace)*; 3. scale, become encrusted *(with a hard deposit)*; become covered *(e.g. with rust)*
ansetzen v/Kesselstein scale, fur
ansetzen v/Rost rust
ansetzen v/sich deposit
ANSI = American National Standards Institute
Ansprechdauer *f (anal)* response time
ansprechen *v* respond *(as to a test or agent)*
Ansprechgeschwindigkeit *f (anal)* speed of response
anstehen *v (geol)* crop out, outcrop
Anstehendes *n (geol)* bedrock
ansteigen *v* increase, rise, grow
anstellen *v* 1. turn on, start, set in motion; 2. carry out, run *(an experiment)*; 3. *(ferm)* pitch *(wort)*; set *(mash)*
anstellen v/Versuche experiment(alize), run [carry out] experiments
Anstieg *m* increase, rise, growth
Anstoß *m* impact, impingement
Anstoßelektron *n* impact electron
anstoßen *v* impact, impinge
anstreichen *v* paint, brush, coat
Anstrich *m* paint, coat (of paint)

Anstrichaufbau *m s.* Anstrichsystem
Anstrichbindemittel *n s.* Anstrichstoffbindemittel
Anstrichfarbe *f* paint, pigmented coating
Anstrichfehler *m* paint-film defect [fault]
Anstrichfilm *m* paint film
Anstrichschaden *m s.* Anstrichfehler
Anstrichstoff *m* paint, coating (material)
Anstrichstoff *m*/**anorganischer** inorganic paint [coating]
Anstrichstoff *m*/**ofentrocknender** stoving [baking] paint
Anstrichstoff *m*/**wässriger** water(-base) paint
Anstrichstoffbindemittel *n* paint [coating] binder
Anstrichsystem *n* painting [coating] system
Antagonismus *m* antagonism
Antagonist *m* 1. antagonist; 2. *(bioch)* competitive inhibitor, antimetabolite, metabolic antagonist
antarafacial *(org ch)* antarafacial, a
Antazidum *n (pharm)* (gastric) antacid *(an agent which counteracts superacidity)*
anteigen *v* 1. make into a paste, paste; 2. *(rubber)* make into a dough
Anteil *m* portion, proportion, constituent (part), component, *(esp if one of two approximately equal portions:)* moiety
Anteil *m* **an suspendierten Feststoffen** *(hyd)* suspended solids content [concentration, level]
Anteil *m*/**elastischer** rebound, recovery *(of strain)*
Anteil *m*/**hochsiedender [höhersiedender]** *(distil)* less volatile component, high-boiling component
Anteil *m*/**leichterflüchtiger [leichtersiedender]** *(distil)* more volatile component, M.V.C., low-boiling component, light(er) component
Anteil *m*/**prozentualer** percentage
Anteil *m*/**schwererflüchtiger [schwerersiedender]** *s.* Anteil/hochsiedender [höhersiedender]
Anteil *m*/**unverseifbarer** unsaponifiable matter [residue]
Anteile *mpl*/**fermentierbare** *(biot)* fermentables *(of the substrate)*
Anthanthronfarbstoff *m* anthanthrone dye
Anthelminthikum *n (pharm)* anthelmint(h)ic
anthelminthisch *(pharm)* anthelmint(h)ic
Anthocyan(in) *n* anthocyan(in) *(any of a class of plant pigments)*
Anthracen *n* anthracene
anthrachinoid anthraquinonoid
Anthrachinon *n* anthraquinone
Anthrachinonfarbstoff *m* anthraquinone [anthraquinonoid] dye
Anthranilsäureester *m* anthranilate
Anthrazit *m (coal)* anthracite
Anthrimid *n (dye)* anthrimide
Anthron *n* anthrone, 9*H*-anthracen-9-one
anthropogen anthropogenic
Antiabsetzmittel *n* antisettling [suspending] agent
antiadhäsiv abhesive
antiaromatisch *(org ch)* antiaromatic
Antiarthritikum *n (pharm)* antiarthritic
antibakteriell antibacterial
antibindend antibonding
Antibindung *f* 1. antibond; 2. antibonding
Antibiose *f* antibiosis
antibiotikaresistent resistant to antibiotics, antibiotic-resistant
Antibiotikaresistenz *f* resistance to antibiotics, antibiotic resistance
Antibiotikarückstand *m* antibiotic residue
Antibiotikum *n (pharm)* antibiotic (agent, substance)
Antibiotikum *n*/**makrolides** macrolide antibiotic
Antibiotikum *n*/**nutritives** feed-additive antibiotic
Antibiotikum *n*/**polyenes** polyene antibiotic
antibiotisch *(pharm)* antibiotic
Anticodon *n (bioch)* anticodon *(a sequence of three nucleotides)*
Antidepressivum *n (pharm)* antidepressant
Antidiabetikum *n (pharm)* antidiabetic
antidiabetisch antidiabetic
Antidiarrhoikum *n (pharm)* antidiarrhoeic, styptic
antidiarrhöisch antidiarrhoeal
Antidiuretikum *n (pharm)* antidiuretic

antidiuretisch antidiuretic
Antidot *n (tox)* antidote
Antiemetikum *n (pharm)* antiemetic
Antienzym *n* antienzyme
Antiepileptikum *n (pharm)* antiepileptic
antiferromagnetisch antiferromagnetic
Antiferromagnetismus *m* antiferromagnetism
Antifiebermittel *n s.* Antipyretikum
Antifoulingfarbe *f* anti-fouling (marine) paint, marine anti-fouling paint
antifungal antifungal, fungicidal
Antigen *n (bioch)* antigen
Antigen *n***/agglutinierbares** agglutinogen
Antigenität *f (bioch)* antigenicity
Antigenwirkung *f (bioch)* antigenicity
Antihalluzinogen *n (pharm)* neuroleptic (drug), antipsychotic, major tranquilizer
Antihidrotikum *n* antihidrotic, antiperspirant, perspiration check
Antihistaminikum *n (pharm)* antihistamine
Antihistaminpräparat *n (pharm)* antihistamine
Antihormon *n* antihormone
Anti-Hückel-System *n* Möbius system *(a cyclic conjugated hydrocarbon)*
Antihypertonikum *n (pharm)* antihypertensive [hypotensive] drug, blood pressure depressant
antihypertonisch antihypertensive, hypotensive
anti-Isomer(e) *n* anti isomer
Antikatalysator *m* anticatalyst, negative catalyst, inhibitor, retarder
Antikatalyse *f* anticatalysis, negative catalysis, inhibition
antikatalytisch anticatalytic
Antiklopfmittel *n* antiknock (agent, additive), knock suppressor, octane improver
Antikoagulans *n (pharm)* anticoagulant
Antikoaguliermittel *n (pharm)* anticoagulant
anti-Konformation *f* 1. anti conformation, A *(of macromolecules, torsion angle ±60°)*; 2. *(org ch)* *anticlinal conformation, ac *(torsion angle ±120°)*
anti-Konstellation *f* 1. anti conformation, A *(of macromolecules, torsion angle ±60°)*; 2. *(org ch)* *anticlinal conformation, ac *(torsion angle ±120°)*
Antikonvulsans *n (pharm)* anticonvulsive
Antikonvulsivum *n (pharm)* anticonvulsive
antikonzeptionell contraceptive
Antikonzeptivum *n (pharm)* contraceptive
Antikonzipiens *n (pharm)* contraceptive
Antikörper *m* antibody, immune body
Antikörper *m***/monoklonaler** *(biot)* monoclonal antibody, MAb
antileukozytär antileucocytic
Antimalariamittel *n (pharm)* antimalarial
Antimalariawirksamkeit *f* antimalarial activity
Anti-Markovnikov-Addition *f* anti-Markovnikov addition, peroxide-initiated addition
Antimaterie *f* antimatter
Antimetabolit *m (bioch)* antimetabolite, metabolite [metabolic] antagonist *(enzyme inhibitor acting mostly in competitive inhibition)*
antimikrobiell antimicrobial
Antimikrobiotikum *n (pharm)* antimicrobial agent
Antimon *n* Sb antimony
Antimonat *n* antimonate
Antimonbutter *f* butter of antimony, mineral butter *(antimony(III) chloride)*
Antimon(III)-chlorid *n* antimony(III) chloride, antimony trichloride
Antimon(V)-chlorid *n* antimony(V) chloride, antimony pentachloride
Antimongelb *n* antimony yellow, Naples yellow *(lead antimonate)*
Antimonglanz *m s.* Antimonit
Antimonhydrid *n s.* Antimonwasserstoff
Antimonid *n* antimonide
Antimonit *m (min)* antimonite, stibnite, antimony glance, grey antimony *(antimony(III) sulphide)*
Antimon(III)-oxid *n* antimony(III) oxide, antimony trioxide
Antimon(III,V)-oxid *n* antimony(III,V) oxide, antimony tetraoxide
Antimon(V)-oxid *n* antimony(V) oxide, antimony pentaoxide
Antimon(III)-oxidchlorid *n* antimony(III) chloride oxide
Antimonspeise *f (met)* antimonial speiss
Antimonwasserstoff *m* antimony hy-

dride, stibane, *(if unsubstituted also)* antimony trihydride, stibine
Antimonweiß *n* antimony white *(antimony(III) oxide)*
Antimutagen *n (bioch)* antimutagen *(a substance which lowers the ratio of mutations)*
Antimykotikum *n (pharm)* antifungal agent [drug], fungicide
antimykotisch antifungal, fungicidal • **antimykotisch wirken** have antifungal activity
Antineuralgikum *n (pharm)* antineuralgic
antineuralgisch antineuralgic
Antineutrino *n* antineutrino
Antineutron *n* antineutron
Antinukleon *n* antinucleon
Antioxidans *n* antioxidant (agent), antioxidizing agent, antioxygen
Antioxidationsmittel *n* antioxidant (agent), antioxidizing agent, antioxygen
Antiozonant *m* antiozonant, antiozidant, sunproofing agent
Antiozonisator *m* antiozonant, antiozidant, sunproofing agent
Antipellagravitamin *n* pellagra-preventive factor, pp factor *(either nicotinic acid or nicotinamide)*
antiperiplanar non-eclipsed, staggered *(stereochemistry)*
Antiperspirant *n* antiperspirant, antihidrotic, perspiration check
Antiphasengrenze *f (cryst)* Bloch [domain] wall, boundary of antiphases
Antiphlogistikum *n (pharm)* antiphlogistic (agent)
antiphlogistisch antiphlogistic
Antipode *m* antipode, optical antipode [opposite], mirror-image isomer
Antipode *m*/**optischer** antipode, optical antipode [opposite], mirror-image isomer
Antipodenpaar *n*/**optisches** racemic pair
Antiproton *n* antiproton, negative proton
Antipsychotikum *n (pharm)* antipsychotic, neuroleptic (drug), major tranquilizer
Antipyretikum *n (pharm)* antipyretic, febrifuge
antipyretisch antipyretic
Antirachitikum *n (pharm)* antirachitic (agent)
antirachitisch antirachitic
Antirheumatikum *n (pharm)* antirheumatic
antirheumatisch antirheumatic
Antischaummittel *n* antifoaming [antifrothing] agent, antifoam (agent, compound)
Antiseptikum *n (pharm)* antiseptic (agent)
antiseptisch antiseptic
antiskorbutisch antiscorbutic
Antispasmodikum *n (pharm)* antispasmodic
Antistatikmittel *n* antistatic (agent, additive), static eliminator
Antistatikum *n* antistatic (agent, additive), static eliminator
antistatisch antistatic
anti-Stellung *f* anti-position • **in anti-Stellung stehen** be anti
Antisterilitätsvitamin *n* antisterility vitamin
anti-Stokes-Raman-Spektroskopie *f*/**kohärente** coherent anti-Stokes Raman spectroscopy, CARS
antisymmetrisch antisymmetric *(orbital)*
anti-syn-Isomerie *f* anti-syn isomerism
Antiteilchen *n* antiparticle
Antitoxin *n* antitoxin
antitoxisch antitoxic
Antitranspirationsmittel *n* antiperspirant, antihidrotic, perspiration check
antitumoral antitumour
Antitumormittel *n (pharm)* antitumour agent
Antitussivum *n (pharm)* antitussive
antiviral antiviral
Antivitamin *n* antivitamin
antreiben *v* drive
Antrieb *m* drive
anwärmen *v* preheat
anwendbar applicable
Anwendbarkeit *f* applicability
anwenden *v* apply, use
anwenden *v*/**lokal** *(pharm)* use topically [in topical applications]
anwenden *v*/**zu reichlich** overuse *(e.g. pesticides)*
Anwendung *f* application, use
anwendungsspezifisch application-specified

Anwendungsweise *f* mode [method] of application
Anxiolytikum *n (pharm)* anxiolytic (sedative), tranquillizing drug, (minor) tranquil(l)izer
Anzahl *f* **der Freiheiten [Freiheitsgrade]** *(phys ch)* variance, number of degrees of freedom
Anzahl *f* **je Zeiteinheit** rate
anzapfen *v* tap
Anzeige *f* indication, reading *(of a measuring instrument)*
Anzeigegerät *n* indicating instrument, indicator
Anzeigeinstrument *n* indicating instrument, indicator
anzeigen *v* indicate; register *(measuring values automatically)*
anziehen *v* attract; draw (in) *(e.g. a liquid)*
Anziehung *f* attraction
Anziehung *f***/coulombsche** *s.* Anziehung/elektrische
Anziehung *f***/elektrische [elektrostatische]** electrostatic [electrical] attraction, Coulomb [coulombic] attraction
Anziehung *f***/intermolekulare** (inter-) molecular attraction
Anziehung *f* **zwischen Ionen (entgegengesetzter Ladung)** ionic attraction
Anziehung *f***/zwischenmolekulare** (inter)molecular attraction
Anziehungskraft *f* attractive force
Anziehungskräfte *fpl***/van-der-waalssche** van der Waals forces (of attraction)
Anzucht *f s.* Kultivierung
anzünden *v* light *(a burner)*; ignite, light, kindle *(fuels)*; light, kindle *(a fire)*
AO *s.* Atomorbital
A-Ort *m s.* Akzeptorort
Apatit *m (min)* apatite *(calcium fluorophosphate)*
Apertur *f* aperture *(measure for the quality of an optical system)*
Apex *m* apex *(of a liquid cyclone)*
Äpfelsäure *f* malic acid, *hydroxybutanedioic acid
Äpfelsäure-Milchsäure-Gärung *f* malolactic fermentation
Aphrodisiakum *n* aphrodisiac
Apodisation *f (spectr)* apodization
Apoenzym *n* apoenzyme, protector
Apoferment *n* apoenzyme, protector
apolar apolar, non-polar
Apoprotein *n (bioch)* apoprotein
Apoptose *f s.* Zelltod/programmierter
Aporphinalkaloid *n* aporphine alkaloid
Apotheker *m* pharmacist, pharmaceutic(al) chemist
Apo-Umlagerung *f* apo rearrangement
Apparat *m* apparatus
Apparat *m***/kippscher** Kipp (gas) generator
Apparatekonstante *f* apparatus constant
Apparatur *f* apparatus, equipment
Applikation *f* 1. *(agric)* application, placement *(of fertilizers or pesticides)*; 2. *(pharm)* application, administration, dosage *(of a medicine)*
Applikation *f* **vor dem Auspflanzen** preplanting application
applizieren *v* 1. *(agric)* apply, place *(fertilizers or pesticides)*; 2. *(pharm)* apply, administer *(a medicine)*
appretieren *v (text, tann)* finish
Appretur *f* 1. *(text, tann)* finishing; 2. *(text, tann)* finish; 3. *s.* Appreturmittel
Appreturmittel *n (text, tann)* finish, finishing agent [compound]
aprot(on)isch aprotic
Aquakomplex *m* aquo complex
Aquametrie *f* aquametry
Aquat *n* aquate *(any of various salts containing hydration water esp in nonstoichiometric amounts)*
Aquation *f* aquation, aquatization, aquotization *(coordination chemistry)*
äquatorial equatorial *(stereochemistry)*
• **äquatorial stehen** be equatorial
äquilibrieren *v* equilibrate
Äquilibrierung *f* equilibration
äquimolar equimolar
äquimolekular equimolecular
Äquipartitionsprinzip *n* equipartition principle [theorem], law of equipartition (of energy)
Äquipartitionstheorem *n* equipartition principle [theorem], law of equipartition (of energy)
Äquipotenzialfläche *f* equipotential surface, potential energy surface
äquivalent equivalent
Äquivalent *n* equivalent
Äquivalentleitfähigkeit *f* equivalent conductance

Äquivalentmasse *f* equivalent [combining] weight
Äquivalentmasse *f*/relative equivalent [combining] weight
Äquivalenzpunkt *m* equivalence [equivalent] point
Aquoion *n* aquo-ion, aquated ion
Aquokomplex *m* aquo complex
Aquotisierung *f s.* Aquation
Aquoverbindung *f* aquo compound
Arachidonsäure *f* arachidonic acid
Arachisöl *n* arachis [peanut] oil
Araliphat *m* araliphatic compound
araliphatisch araliphatic
Aralkylsilicon *n s.* Alkylarylsiloxan
Aralkylsiloxan *n s.* Alkylarylsiloxan
Aralkylsulfonat *n s.* Alkylarensulfonat
Aräometer *n* araeometer, hydrometer
Aräometrie *f* araeometry, hydrometry
aräometrisch araeometric(al), hydrometric(al)
Arbeit *f*/geleistete [umgesetzte] work done
arbeiten *v*/adiabat operate [run] adiabatically
arbeiten *v*/isotherm operate [run] isothermally
arbeitend/absatzweise [diskontinuierlich] batch *(apparatus)*
Arbeitsdruck *m* operating [working] pressure
Arbeitselektrode *f* working [measuring] electrode
Arbeitsgang *m* operation
Arbeitshygiene *f* industrial [occupational] hygiene
Arbeitshygieniker *m* industrial [occupational] hygienist
Arbeitsinhalt *m (distil)* operating hold-up
Arbeitslinie *f (distil)* operating [working] line, material-balance line
Arbeitsmethode *f* 1. *(tech)* procedure, operating method; 2. *(lab)* technique
Arbeitsplatz *m*/reiner *(anal)* clean box *(provided with filtered air)*
Arbeitsplatzgrenzwert *m* **für kurzzeitige Belastung** short-time exposure limit, STEL
Arbeitsplatzkonzentration *f*/maximale maximum allowable concentration, MAC
Arbeitsplatztoleranzwert *m*/biologischer biological exposure index, BAT
Arbeitsschutz *m* industrial safety
Arbeitsschutzanordnung *f* safety regulation
Arbeitsschutzkleidung *f* safety clothing
Arbeitsschutzsalbe *f* barrier cream
Arbeitsspannung *f* operating [working] voltage
Arbeitsstamm *m (biot)* working strain, producing [production] strain *(of microorganisms)*
Arbeitstakt *m* service step [run]
Arbeitstechnik *f* technique
Arbeitstemperatur *f* operating [working] temperature
Arbeitsverfahren *n s.* Arbeitstechnik
Arbeitsvermögen *n* energy *(of a physical system)*
Arbeitsvolumen *n* 1. working volume; 2. *(biot)* fermenter [fermentation] volume
Arbeitsweise *f* 1. (mode of) operation *(of an apparatus)*; 2. *s.* Arbeitsmethode
Arbeitsweise *f*/diskontinuierliche batch operation
Arbeitsweise *f* mit totalem Rücklauf total-reflux operation
Arbeitszyklus *m* operating [working, service] cycle, *(ion exchange also)* exhaust-regenerate cycle
Arborizid *n* brushkiller, silvicide
Aren *n* arene, aromatic hydrocarbon
Arenenium-Ion *n* arenium [arenonium] ion
Arenepoxid *n* arene epoxide [oxide]
Arenoxid *n* arene epoxide [oxide]
Arg *s.* Arginin
Argentometrie *f* argentometry
argentometrisch argentometric
Arginin *n* 2-amino-5-guanidinovaleric acid
Arginin-Harnstoff-Zyklus *m s.* Harnstoffzyklus
Argon *n* Ar argon
Argon(ionen)-Laser *m (spectr)* argon (ion) laser
Arin *n* aryne, 1,2-didehydroarene *(any of a class of transient dehydrogenated derivatives of aromatic compounds)*; *(specif)* benzyne
arm poor *(as in content)*; lean *(gas)*; lean, low-grade *(ores)*; barren, weak *(solution)*; poor, infertile, barren, thin *(soil)*

Armerz *n* lean [low-grade] ore
Armgas *n* lean gas
armieren *v* reinforce
Armierung *f* reinforcement
Armlauge *f* barren liquor [solution] *(in cyaniding)*
Aroma *n* 1. flavour, aroma; 2. *s.* Aromastoff
Aromabildung *f* aroma development
Aromastoff *m* aroma [aromatic] substance [body] *(occurring in food)*; congener(ic) *(occurring in distilled beverage spirits)*; aroma ingredient, aromatizing product, flavouring (material, matter, substance) *(for food)*
Aromat *m* aromatic (hydrocarbon, compound)
Aromat *m*/**mehrkerniger** polynuclear [polycyclic] aromatic hydrocarbon, PAH
aromatenfrei free from aromatic hydrocarbons
aromatisch 1. *(org ch)* aromatic; 2. *(food)* aromatic, flavourful
aromatisch/nicht non-aromatic
aromatisieren *v* 1. *(org ch)* aromatize; 2. *(food)* aromatize, flavour
Aromatisierung *f* 1. *(org ch)* aromatization; 2. *(food)* aromatization, flavouring
Aromatizität *f* 1. *(org ch)* aromaticity, aromatic character; 2. *(coal)* (carbon) aromaticity *(fraction of carbon in aromatic form)*; 3. *(food)* aromaticity
Aroylbenzoesäure *f* aroylbenzoic acid
Aroylgruppe *f* aroyl group [residue]
arretieren *v* arrest
Arretierung *f* 1. arrestment, arresting; 2. *s.* Arretiervorrichtung
Arretiervorrichtung *f* arresting mechanism
Arrhenius-Aktivierungsenergie *f* Arrhenius energy of activation, Arrhenius activation energy
Arrhenius-Gleichung *f* Arrhenius equation
Arsan *n* As_nH_{n+2} arsane, arsenic hydride, *(specif)* AsH_3 arsane, arsine, arsenic(III) hydride, arsenic trihydride
Arsen *n* As arsenic
Arsen *n*/**gelbes** yellow arsenic, α-arsenic
Arsen *n*/**graues [metallisches]** grey [metallic] arsenic, γ-arsenic
Arsen *n*/**schwarzes** black arsenic, β-arsenic
Arsenat(III) *n* arsenite
Arsenat(V) *n* arsenate
Arsen(III)-chlorid *n* arsenic(III) chloride, arsenic trichloride
arsenhaltig containing arsenic, arsenical, arsenian, *(esp ores:)* arseniferous, arsenic-bearing
Arsenhydrid *n s.* Arsan
Arsenid *n* arsenide
Arsenik *n* white arsenic, arsenic(III) oxide
Arsenik *n*/**gelbes** yellow arsenic (sulphide), arsenic yellow, king's yellow [gold], royal yellow, orpiment (yellow) *(technically pure arsenic(III) sulphide)*
Arsenik *n*/**weißes** *s.* Arsenik
Arsenikalie *f* arsenical
Arsenikvergiftung *f* arsenic poisoning
Arsenolith *m (min)* arsenolite *(arsenic(III) oxide)*
Arsen(III)-oxid *n* arsenic(III) oxide, arsenic trioxide, white arsenic
Arsen(V)-oxid *n* arsenic(V) oxide, arsenic pentaoxide
Arsenprobe *f* arsenic test
Arsenprobe *f*/**bettendorfsche** Bettendorf's test (for arsenic)
Arsenprobe *f*/**gutzeitsche** Gutzeit's test (for arsenic)
Arsenprobe *f*/**marshsche** Marsh's test (for arsenic)
Arsen(III)-säure *f* arsenious acid
Arsen(V)-säure *f* arsenic acid
Arsenspeise *f (met)* arsenical speiss
Arsen(III)-sulfid *n* arsenic(III) sulphide, arsenic trisulphide
Arsenvergiftung *f* arsenic poisoning
Arsenvergiftung *f*/**chronische [gewerbliche]** arsenicalism
Arsenwasserstoff *m s.* Arsan
Arsin *n* 1. AsR_3 arsine *(any of several organic compounds)*; 2. AsH_3 arsine, arsane, arsenic(III) hydride, arsenic trihydride
Arsinsäure *f* arsinic acid *(any of several organic acids R_2–As(O)OH)*
Arsoniumgruppe *f* arsonium group [residue]
Arsoniumverbindung *f* arsonium compound
Arsonsäure *f* arsonic acid *(any of several organic acids R–As(O)$(OH)_2$)*

Arsonylierung *f* arsonation
artspezifisch *(bioch)* species-specific
Arylalkylierung *f* ar(yl)alkylation
Arylazogruppe *f* arylazo group [residue]
Arylazorest *m* arylazo group [residue]
Arylen *n* arylene (group), arenediyl group *(any of a class of bivalent radicals derived from an aromatic hydrocarbon)*
Arylengruppe *f s.* Arylen
Arylenrest *m s.* Arylen
Arylether *m* aryl [aromatic] ether
Arylgruppe *f* aryl group [residue], Ar
Arylhalogenid *n* aryl halide
arylieren *v* arylate
Arylierung *f* arylation
Arylkation *n* aryl cation
Arylmethylgruppe *f* $ArCH_2$– arylmethyl group
Arylmethylrest *m* $ArCH_2$– arylmethyl group
Arylsiloxan *n* aryl siloxane
Arylsulfonsäure *f* arylsulphonic acid
Arzneibuch *n* pharmacopoeia • **den Anforderungen des Arzneibuchs entsprechend** pharmacopoeial, *(UK)* B.P. grade, *(US)* U.S.P. grade
Arzneidroge *f* drug
Arzneiform *f* (pharmaceutical) dosage form
arzneilich medicinal
Arzneimittel *n* pharmaceutical (preparation), medicinal drug, medicament, remedy, *(if for internal use also)* medicine
Arzneimittel *n* **auf Sulfonamidbasis** sulphonamide drug, *(if antibacterial:)* sulpha drug
Arzneimittel *n* **für inneren Gebrauch** medicine
Arzneimittel *n***/rezeptpflichtiges** prescription pharmaceutical
Arzneimittelchemie *f* pharmaceutical [medicinal] chemistry
Arzneimittelchemiker *m* pharmaceutical [medicinal] chemist
Arzneimittelvergiftung *f* drug intoxication
Arzneistoff *m* medicinal substance
ASAT = Aspartat-Aminotransferase
Asbest *m* asbestos
Asbestdiaphragma *n* asbestos diaphragm
Asbestdrahtnetz *n* asbestos(-covered wire) gauze
Asbestfaser *f* asbestos fibre
Asbestfilter *n* asbestos filter
Asbestmembran *f s.* Asbestdiaphragma
Asbeststaub *m* asbestos dust
Asbestzement *m* asbestos cement, cement-asbestos
Asche *f* ash(es)
aschefrei ash-free, free from ash, *(esp filter paper:)* ashless
Aschegehalt *m* ash content, *(in analysing coal also)* ash yield • **mit hohem Aschegehalt** high-ash • **mit niedrigem Aschegehalt** low-ash
Ascheschmelzpunkt *m* ash-fusion temperature
Ascorbigen *n* ascorbigen *(composed of ascorbic acid and protein)*
Ascorbinsäure *f* ascorbic acid
Asepsis *f* asepsis
aseptisch aseptic
Asp *s.* Asparaginsäure
Asparagin *n* asparagine, *(specif)* $HOOC–CH(NH_2)CH_2CONH_2$ β-asparagine
Asparaginsäure *f* aspartic acid, Asp, asparaginic acid, aminosuccinic acid
Asphalt *m* 1. *(petrol)* petroleum asphalt, (artificial) asphalt; 2. *(build)* asphalt *(asphaltic bitumen mixed with mineral matter)*; 3. *s.* Asphalt/natürlicher
Asphalt *m***/natürlicher** *(min)* (natural, native) asphalt, mineral [earth] pitch
Asphalt *m***/reiner** glance pitch
Asphaltbasisöl *n* asphalt-base petroleum [crude oil, crude], asphaltic petroleum
Asphaltbitumen *n* asphaltic bitumen
Asphalten *n* asphaltene *(high-molecular-weight hydrocarbon)*
asphaltisch asphaltic
Asphaltit *m* asphaltite *(natural asphalt)*
Asp-NH$_2$ *s.* Asparagin
Assimilat *n* assimilate, photosynthate
Assimilation *f* assimilation
assimilierbar assimilable
Assimilierbarkeit *f* assimilability
assimilieren *v* assimilate
Assoziat *n* supramolecular assembly, complex
Assoziation *f* 1. *(ch)* (molecular) association; 2. *(min)* association

Assoziationskolloid *n* association [micellar] colloid
Assoziationsreaktion *f* association reaction
assoziieren *v* associate
assoziieren *v***/sich** associate
Astat *n* At astatine
ASTM = American Society for Testing and Materials
Aston-Massenspektrograph *m* Aston mass spectrograph
asymm. *s.* asymmetrisch
Asymmetrie *f* asymmetry, dissymmetry
Asymmetrieparameter *m (glass)* asymmetry parameter
Asymmetriepotenzial *n* asymmetry potential
Asymmetriezentrum *n* asymmetric centre, centre of asymmetry
asymmetrisch asymmetric(al), dissymmetric, unsymmetrical, *(nomencl as a prefix:) asym-*
ataktisch *(plast)* atactic
Ataraktikum *n (pharm)* (minor) tranquillizer, tranquillizing drug, anxiolytic (sedative)
Atemgift *n* respiratory poison
Atemschutz *m* respiratory protection
Atemschutzmaske *f* respirator, breathing mask
Atemschwingung *f* (symmetrical) breathing vibration
ätherisch essential, volatile, ethereal *(oils)*
atherman atherm(an)ous
atmophil atmophil(e) *(found in the atmosphere)*
Atmosphäre *f* atmosphere
Atmosphäre *f***/indifferente [inerte]** inert atmosphere
Atmosphäre *f***/kontrollierte** controlled atmosphere *(of a furnace)*
Atmosphäre *f***/oxidierende** oxidizing atmosphere
Atmosphäre *f***/physikalische** physical atmosphere, atm *(1 atm = 101325 Pa)*
Atmosphäre *f***/reduzierende** reducing atmosphere
Atmosphäre *f***/technische** technical atmosphere, at *(1 at = 98066.5 Pa)*
Atmosphärenchemie *f* chemistry of atmosphere, atmospheric chemistry
Atmosphärendruck *m* atmospheric [air] pressure
atmosphärisch atmospheric(al)
Atmung *f* breathing, respiration
Atmungsferment *n* respiratory enzyme
Atmungsferment *n***/warburgsches** Warburg's enzyme, *(specif)* cytochrome oxidase
Atmungsferment *n***/warburgsches gelbes** Warburg's yellow enzyme, old yellow enzyme
Atmungshemmstoff *m* respiration inhibitor
Atmungsinhibitor *m* respiration inhibitor
Atmungskette *f (bioch)* respiratory [oxidative] chain, electron-transport chain [sequence, system]
Atmungskettenphosphorylierung *f (bioch)* respiratory-chain phosphorylation, oxidative phosphorylation
Atmungspigment *n (bioch)* respiratory pigment
Atmungsquotient *m (bioch)* respiratory quotient [ratio], RQ
Atom *n* atom
Atom *n***/adsorbiertes** adatom, adsorbed atom
Atom *n***/angeregtes** excited [activated] atom
Atom *n* **auf Zwischengitterplatz** interstitial (atom)
Atom *n***/heißes [hoch angeregtes]** hot atom
Atom *n***/hochionisiertes** stripped atom
Atom *n***/isotopes** isotopic atom
Atom *n***/markiertes** labelled [tagged] atom, label
Atomabsorptionsspektrometrie *f* atomic absorption spectrometry, AAS
Atomabsorptionsspektroskopie *f* atomic absorption spectroscopy
Atomabstand *m s.* Kernabstand
Atomaggregat *n* atomic aggregate, aggregate of atoms
Atomanordnung *f s.* Atomkonfiguration
atomar atomic(al)
Atombau *m* atomic structure
Atombindung *f* atomic [homopolar] bond, covalent [non-polar] bond, (shared-)electron-pair bond
Atomemissionsspektrometrie *f* atomic emission spectrometry, AES
Atomenergie *f s.* Kernenergie

Atomfluoreszenzspektrometrie *f* atomic fluorescence spectrometry, AFS
Atomfluoreszenzspektrometrie *f***/laserangeregte** laser-excited atomic fluorescence spectrometry, LEAFS
Atomgewicht *n s.* Atommasse/relative
Atomgitter *n* atom(ic) lattice
Atomgruppe *f* group of atoms
Atomgruppe *f***/dreiwertige** triad
Atomgruppe *f***/mehrwertige** polyad
Atomgruppe *f***/vierwertige** tetrad
Atomgruppe *f***/zweiwertige** diad, dyad
Atomhülle *f* extranuclear [electronic] region
Atomigkeit *f s.* Atomizität
Atomisator *m* atomizer
Atomiseur *m (agric)* low-volume mist blower, air-blast sprayer, fog generator [appliance], nebulizer
atomisieren *v* atomize
Atomisierung *f* atomization
atomistisch atomistic
Atomizität *f* atomicity *(number of atoms contained in a molecule)*
Atomkern *m* (atomic) nucleus
Atomkette *f* atomic chain, chain of atoms
Atomkonfiguration *f* (atom, atomic) configuration
Atomkristall *m* covalent crystal
Atomlagenabscheidung *f* atom layer deposition, ALD
Atom-Lücke-Konformation *f* staggered conformation
Atommasse *f s.* Atommasse/absolute
Atommasse *f***/absolute** atomic mass, mass of atom
Atommasse *f***/relative** relative atomic mass, RAM
Atommasseeinheit *f* atomic mass unit
Atommodell *n* atom(ic) model
Atommodell *n***/bohrsches [bohr-rutherfordsches]** Bohr atom (model), Bohr-Rutherford atom (model)
Atommodell *n***/bohr-sommerfeldsches** Bohr-Sommerfeld atom (model)
Atommodell *n***/rutherfordsches** Rutherford atom (model), nuclear atom (model), nuclear model of the atom
Atommüll *m* radioactive [nuclear] waste
Atommüllbeseitigung *f* radioactive [nuclear] waste disposal
Atomniveau *n* atomic (energy) level
Atomnummer *f* atomic [ordinal] number, A.N., *(symbol)* Z
Atomorbital *n* atomic orbital, AO
Atompolarisation *f* atom(ic) polarization
Atomprozent *n* atomic percent(age), atom %
Atomradius *m***/kovalenter** covalent radius (of atoms)
Atomreaktor *m s.* Kernreaktor
Atomrumpf *m* (atomic) core, (atomic) kernel
Atomspektroskopie *f* atomic spectroscopy
atomspektroskopisch by atomic spectroscopy
Atomspektrum *n* atomic [line] spectrum
Atomspin *m* atomic spin
Atomsuszeptibilität *f* atomic susceptibility, susceptibility per gram atom
Atomübertragungsreaktion *f* atom transfer radical polymerization, ATRP
Atomumwandlung *f* atomic transmutation
Atomverband *m* union of atoms
Atomverbindung *f* atomic compound
Atomverhältnis *n (anal)* atomic (combining) ratio, atomic proportion, proportion by atoms
Atomverschiebung *f* atomic shift
Atomvolumen *n* atomic volume
Atomzerfall *m* atomic [radioactive] disintegration
Atomzustand *m* atomic state
atoxisch non-toxic, non-poisonous
ATP *s.* Adenosintriphosphorsäure
ATP-Citratlyase *f* ATP-citrate lyase, citrate cleavage enzyme
ATR *s.* Totalreflexion/abgeschwächte
atro absolutely dry, bone-dry, B.D.
Atropin *n* atropine, DL-hyoscyamine *(alkaloid)*
Atropisomer(e) *n* atropisomer, atropoisomer
Atropisomerie *f* atropisomerism, atropo-isomerism
Attenuator *m (bioch)* attenuator *(DNA segment acting as a controlling element)*
Attraktans *n* attractant *(as for insects)*
Attraktion *f* attraction
Attraktionskraft *f* attractive force [power]
Ätzalkalien *npl* caustic alkalies

Ätzbad *n s.* Beizbad

ätzen *v* 1. etch; 2. *(tox)* cause burns *(on the skin)*; corrode *(the skin)*; 3. *(med)* cauterize *(e.g. by chemicals)*

Ätzen *n* 1. etching; 2. *(med)* cauterization *(e.g. by chemicals)*; 3. *(met)* pickling process

Ätzen *n***/galvanisches** electroetching

ätzend 1. caustic; 2. *(med)* cauterant, cauterizing

Ätzkali *n* caustic potash *(potassium hydroxide)*

Ätzkalk *m* 1. caustic [burnt] lime, quicklime *(calcium oxide)*; 2. slaked [hydrated] lime, slacklime *(calcium hydroxide)*

Ätzlösung *f* etching solution

Ätzmaske *f* etching mask

Ätzmittel *n* 1. etching reagent, etchant; engraver's acid *(usually nitric acid)*; 2. *(med)* caustic (agent), cauterant, cautery

Ätznatron *n* caustic (soda) *(sodium hydroxide)*

Ätznatronschmelze *f* caustic-soda fusion

Ätzresist *m s.* Ätzschutz

Ätzschutz *m* etch resist

Ätzung *f s.* Ätzen

Ätzwirkung *f* causticity

aufarbeiten *v* process *(e.g. ores)*; work up, recover, prepare *(products of value)*; reprocess, rework *(used or discarded material)*

aufarbeiten *v***/Topprückstände** *(petrol)* run resid

Aufarbeitung *f* processing *(e.g. of ores)*; working-up, workup, recovery, preparation *(of products of value)*; reprocessing, reworking *(of used or discarded material)*

Aufarbeitungsverlust *m* recovery loss *(as in recovering a product of value)*

Aufbau *m* 1. structure, constitution, build-up, make-up *(of a molecule)*; composition, make-up *(of a chemical compound or mixture)*; set-up, structure, build *(of an apparatus)*; build, construction *(of a plant)*; 2. build-up, formation *(of an electrical field)*; 3. *(bioch)* anabolism; 4. *s.* Aufbauen

aufbauen *v* build up, synthesize *(a chemical compound)*; set (up), erect *(an apparatus)*; make up *(a whole)*

aufbauen *v***/ein Mischungsrezept** compound

aufbauen *v***/gezielt** make measure, tailor(-make) *(e.g. polymers)*

aufbauen *v***/nach Rezept** formulate

aufbauen *v***/sich** build up *(of an electrical field)*

Aufbauen *n* building-up, build-up, synthesis *(of a chemical compound)*; setting(-up), erection *(of an apparatus)*

Aufbaugranulieren *n* granulation, pelletizing

Aufbauprinzip *n (phys ch)* aufbau [building-up] principle

Aufbaureaktion *f* build-up reaction; aufbau [chain-extension] reaction *(polymerization)*

Aufbaustoff *m* (detergency, soap) builder

Aufbauweg *m (bioch)* anabolic pathway [route]

aufbereiten *v* prepare; treat, process *(water)*; condition *(boiler feed water)*; dress, beneficiate *(ore)*

aufbereiten *v***/in der Setzmaschine** *(min tech)* jig

aufbereiten *v***/nass** *(min tech)* wet-clean

aufbereiten *v***/trocken** *(min tech)* dry-clean

Aufbereitung *f* preparation; treatment *(of water)*; conditioning *(of boiler feed water)*; dressing, beneficiation *(of ore)*

Aufbereitung *f***/bergbauliche** mineral dressing, minerals beneficiation

Aufbereitung *f***/flotative** floatation beneficiation

Aufbereitung *f***/nasse** *(min tech)* wet cleaning [washing]

Aufbereitung *f***/trockene** *(min tech)* dry [pneumatic] cleaning

Aufbereitungsanlage 1. *(hyd)* treatment plant [works]; 2. *(hyd)* treatment unit *(of an industrial plant)*; 3. *(min tech)* ore dressing plant

Aufbereitungsstufe *f (hyd)* treatment stage

Aufbereitungstechnik *f* mineral technology

aufbewahren *v***/kühl** store in a cool place

aufbewahren *v*/**lichtgeschützt** keep screened from the light
Aufbewahrung *f* storing, storage, keeping
Aufbewahrungstemperatur *f* storage [holding] temperature
Aufblas(e)verfahren *n* top-blown oxygen converter process, basic oxygen converter [furnace, steel] process, oxygen process of steelmaking, oxygen lance process
aufblättern *v* exfoliate, delaminate, cleave *(of laminated material)*
Aufblättern *n* exfoliation, delamination, cleavage *(of laminated material)*
aufbrauchen *v* use up, consume *(as in a reaction)*
aufbrausen *v* effervesce
Aufbrausen *n* effervescence, effervescency
aufbrechen *v* break up *(molecules, bonds)*
Aufbrechen *n* breaking-up, breakup *(of molecules, bonds)*
aufbringen *v* apply *(coatings)*
Aufbringen *n* application *(of coatings)*
Aufdampfen *n* vapour deposition, deposition from vapour, vapour condensation plating
Aufdampfen *n* **im Vakuum** vacuum deposition, deposition in vacuo
Aufdampfrate *f* deposition rate, rate of deposition
Aufdampfung *f* vapour deposition
aufdringlich objectionable *(smell)*
Aufenthaltswahrscheinlichkeit *f* probability of finding *(a particle in a specified location)*
Aufenthaltswahrscheinlichkeitsdichte *f* probability density
Auffangbehälter *m* receiving tank, receiver
Auffangelektrode *f* collector [collecting] electrode, collector
auffangen *v* collect, catch, trap, receive
Auffänger *m* receiver, catcher; electron acceptor; target *(as of an X-ray tube)*; ion collector *(of a mass spectrometer)*
Auffanggefäß *n* 1. collection vessel, receiver, catcher, catch pot, trap; 2. *(lab)* catch container
aufflammen *v* flash, burst into flame
Aufflammen *n* flash(ing)
Aufflammen *n*/**verzögertes** afterflaming *(fireproofing)*
auffüllen *v* 1. fill up *(liquid or bulk material)*; 2. *(anal)* make up volume; fill out *(electron shells)*; replenish *(a store)*
Aufgabe *f* charging, feeding, filling, loading, furnishing
Aufgabegut *n* 1. feed(stock), feed material, charge (stock), batch; feed slurry [pulp] *(as in centrifugation)*; 2. *(filtr)* prefilt (feed, slurry)
Aufgabeöffnung *f* feed [charging] hole, feed inlet [opening]
Aufgabevorrichtung *f* feeder, loader
aufgeben *v* feed, *(esp if discontinuously or relating to a definite quantity:)* charge
Aufgeber *m* feeder, loader
aufgehen *v* rise *(of yeast, dough)*
aufgehen lassen *v* raise, leaven *(dough)*
Aufglasurfarbe *f* *(ceram)* on-glaze colour, overglaze [enamel] colour
Aufguss *m* infusion, *(by boiling also)* decoction
aufhalden *v* dispose of to tip, dispose of by tipping, deposit on a tip *(waste products)*
Aufhalden *n* disposal to tip *(of waste products)*
aufhärten *v* add hardness *(to the water)*
Aufhärtung *f* adding of hardness *(to the water)*; hardening
aufheizen *v* heat up
Aufheizung *f* heating(-up)
Aufheizzeit *f* heating-up time [period]
aufhellen *v* brighten *(esp by adding optically active agents)*; bleach, clear
Aufheller *m* brightener, brightening agent
Aufheller *m*/**optischer** optical bleaching agent, (optical) brightening agent, (optical) whitening agent, optical brightener [bleach], fluorescent brightener [bleach]
Aufhellungsmittel *n s.* Aufheller
aufhydrieren *v* rehydrogenate
Aufhydrierung *f* rehydrogenation
Aufkegeln *n* **und Kreuzteilen** *n* **[Vierteln]** *n* *(anal)* coning and quartering
aufklären *v* elucidate *(e.g. constitution)*
Aufklärung *f* elucidation *(as of constitution)*
Aufkleben *n* *(plast)* adhesive bonding
aufkochen *v* boil up

aufkochen lassen *v* boil up, bring to the boil
aufkochen v/wieder reboil
Aufkochen *n* boiling-up, *(process also)* ebullition
aufkohlen *v (met)* carburize
aufkohlen v/im Salzbad bath-carburize, liquid-carburize
aufkohlen v/in der Randschicht [Randzone] case-carburize
aufkohlen v/in festen Kohlungsmitteln pack-carburize
aufkohlen v/in flüssigen Mitteln *s.* aufkohlen/im Salzbad
aufkohlen v/in gasförmigen Mitteln gas-carburize
Aufkohlen *n (met)* carburizing, carburization
Aufkohlen *n* **im Salzbad** bath carburizing, liquid [liquid-salt, molten-salt] carburizing
Aufkohlen *n* **in der Randschicht [Randzone]** case carburizing
Aufkohlen *n* **in festen Kohlungsmitteln** (solid-)pack carburizing
Aufkohlen *n* **in flüssigen Mitteln** *s.* Aufkohlen im Salzbad
Aufkohlen *n* **in gasförmigen Mitteln** gas carburizing
Aufkohlungsbad *n (met)* carburizing bath
Aufkohlungsmittel *n (met)* carburizing agent [compound], (case-hardening) carburizer
Aufkohlungsofen *m (met)* carburizing furnace [oven]
Aufkohlungsschicht *f (met)* (carburized) case
Aufkohlungstiefe *f (met)* carburizing [case] depth
Aufladung f/(elektro)statische electrostatic charging, static electrification
Auflichtmikroskopie *f* direct-light microscopy
auflösen *v* 1. dissolve; 2. disintegrate *(into constituent elements)*
auflösen v/sich 1. dissolve, undergo dissolution; 2. disintegrate *(into constituent elements)*
auflösen v/sich wieder redissolve
Auflösung *f* 1. dissolution; 2. disintegration *(into constituent elements)*; 3. *s.* Auflösungsvermögen
Auflösungsgeschwindigkeit *f* dissolution rate
Auflösungsvermögen *n (anal, phot)* resolving power, resolution
Aufnahme *f* 1. uptake, take-up *(of substances)*; *(by the human body:)* intake; absorption, take-up, pick-up *(of liquids)*; absorption *(of gases)*; acceptance, acquisition *(of electrons)*; 2. *(phot)* taking; 3. photograph, picture
Aufnahme f/autoradiographische autoradiograph, radioautograph
Aufnahme f/makrofotografische photomacrograph
Aufnahme f/mikrofotografische photomicrograph
aufnahmefähig absorptive, absorbent
Aufnahmefähigkeit *f* (absorbing, absorption) capacity, absorbency
Aufnahmevermögen *n* capacity *(of containing)*; (absorbing, absorption) capacity, absorbency
aufnehmbar 1. absorbable; 2. *(agric)* available *(nutrients)*
Aufnehmbarkeit *f* 1. absorbability; 2. *(agric)* availability *(of nutrients)*
aufnehmen *v* 1. take up *(substances)*; 2. *(tox)* take in; absorb, take [pick] up *(liquids)*; absorb *(gases)*; gain, accept, acquire *(electrons)*; 3. take a photograph
Aufoxidation *f* oxidation to higher valency, further oxidation
aufoxidieren *v* oxidize to higher valency
aufpfropfen *v* graft *(polymers)*
aufpolymerisieren *v s.* aufpfropfen
Aufprall *m* impingement, impact(ion)
aufprallen *v* impinge, impact
Aufprallerosion *f* impingement attack
aufquellen *v* swell (up)
aufrahmen *v (rubber, plast, food)* cream
aufrauen *v* 1. roughen; 2. *(text)* raise (a nap), nap
aufrollen *v (pap, text)* reel (up)
aufrühren *v* agitate, stir up; repulp, reslurry
aufsättigen *v* resaturate, reconcentrate
Aufsättigung *f* resaturation, reconcentration
aufsaugen *v* suck [soak] up, imbibe
Aufsaugen *n* suction, imbibition
Aufsaugung *f* suction, imbibition

aufschäumbar *(plast)* expandable, foamable
Aufschäumbarkeit *f (plast)* expandability, foamability
aufschäumen *v* 1. foam, froth *(a substance)*; 2. *(plast)* expand, foam; foam (up), froth (up), effervesce *(of a substance)*
Aufschlaggeschwindigkeit *f* velocity of impact, impact velocity
Aufschlagsbereich *m* impact area
Aufschlagsenergie *f* impact energy
Aufschlagskraft *f* impact velocity
aufschlämmen *v* suspend, slurry
Aufschlämmung *f* suspension, slurry
aufschleudern *v* spin coat
Aufschleudern *n* spin coating *(on wafer)*
aufschließbar digestible
Aufschließbarkeit *f* digestibility, digestibleness
aufschließen *v* 1. digest, decompose, open up *(by heat or solvents)*; 2. *(biol)* macerate; 3. *(pap)* cook, pulp; repulp *(waste paper)*
Aufschluss *m* 1. digestion, decomposition, opening-up *(by heat or solvents)*; 2. *(biol)* maceration; 3. *(pap)* cooking, pulping
Aufschluss *m*/**alkalischer** *(pap)* alkaline pulping
Aufschluss *m*/**chemischer** *(pap)* (full) chemical pulping
Aufschluss *m* **des Holzes/mechanischer** *(pap)* mechanical [groundwood] pulping
Aufschluss *m*/**halbchemischer** *(pap)* semichemical pulping
Aufschluss *m* **im Bombenrohr** [**Einschmelzrohr, Schießrohr**] sealed-tube decomposition
Aufschluss *m*/**saurer** *(pap)* acid pulping
Aufschlusslösung *f (pap)* pulping [cooking, digestion] liquor
Aufschlussmittel *n* 1. digesting [decomposing] agent; 2. *(pap)* pulping [cooking] agent [chemical]
aufschmelzen *v* burn on, weld [flux] on *(e.g. lead to form a coating)*
aufschmieren *v* smear *(e.g. a lubricant)*
aufschwemmen *v* 1. suspend; 2. *(min tech)* pulp
Aufschwemmung *f* suspension
aufschwimmen *v* float, rise
Aufschwimmen *n* floating, rising, rise • **zum Aufschwimmen bringen** float
aufspalten *v* 1. cleave, crack (up), split *(chemical compounds)*; *(relating to ring molecules also)* open; cleave, break, crack (up), split, disrupt *(chemical bonds)*; resolve *(racemic mixtures)*; 2. split (up), cleave, delaminate *(mechanically)*; 3. *s.* aufspalten/sich
aufspalten *v*/**durch Solvolyse** solvolyze
aufspalten *v*/**in Fibrillen [Teilfäserchen]** fibrillate *(fibres)*
aufspalten *v*/**sich** 1. crack, decompose, split up *(of chemical compounds)*; cleave, crack *(chemical bonds)*; dissociate *(into ions)*; 2. split, cleave, delaminate *(mechanically)*
Aufspaltung *f* 1. cleavage, cracking, decomposition *(of chemical compounds)*; cleavage, breaking, cracking, splitting, disruption, fission, scission *(of chemical bonds)*; resolution *(of racemic mixtures)*; 2. splitting(-up), cleavage, delamination *(mechanically)*
Aufspaltungsbild *n (phys ch)* splitting pattern
Aufspaltungsfaktor *m* spectroscopic splitting factor, Landé splitting factor g, g factor
Aufspaltungsfaktor *m*/**spektroskopischer** spectroscopic splitting factor, Landé splitting factor g, g factor
aufsprengen *v (org ch)* rupture *(a ring)*
Aufsprengung *f (org ch)* rupture *(of a ring)*
aufstärken *v* 1. fortify; 2. *(distil)* dephlegmate
Aufstärkung *f* 1. fortification; 2. *(distil)* dephlegmation
aufsteigen *v* rise, ascend, pass up(-wards)
aufstellen *v* 1. erect, set (up), mount *(e.g. an apparatus)*; 2. determine, calculate *(a formula)*; formulate *(an equation)*
Aufstellung *f* 1. erection, setting(-up), mounting *(as of an apparatus)*; 2. determination, calculation *(of a formula)*; formulation *(of an equation)*
aufsticken *v (met)* nitride
Aufsticken *n (met)* nitride hardening, nitrogen (case-)hardening, nitriding, nitridation

Aufstrom *m* ascending [upward] current, upflow, upward flow
Aufstromklassieren *n* hydraulic classification [separation]
auftauen *v* thaw, defrost
aufteilen *v*/**in gleiche Anteile** *(anal)* aliquot
auftragen *v* apply, coat onto a support *(e.g. coating material)*
Auftragen *n* application *(as of coating material)*
Auftragmaschine *f* coating machine, coater
Auftreten *n* **von Kurzschlussströmungen** by-passing, short circuiting *(in a reactor)*
Auftrieb *m* buoyancy
Auftriebskraft *f* buoyancy [buoyant] force
Auftrittspotenzial *n (spectr)* appearance potential
aufwallen *v* boil (up), bubble (up)
Aufwallen *n* boiling, bubbling, ebullience, ebullition
aufwallend ebullient
Aufwärtsfilter *n (hyd)* upflow [upward-flow] filter
Aufwärtsfiltration *f (hyd)* upflow filtration
Aufwärtsstrom *m* upflow, upward flow [current], ascending current • **im Aufwärtsstrom arbeiten** operate upflow *(of a filter)* • **im Aufwärtsstrom fahren** operate upflow *(a filter)*
aufweisen *v*/**einen konstanten Wert** show a constant reading
Aufweitung *f* expansion *(of the crystal lattice)*
aufwickeln *v* reel (up), wind (up)
aufwirbeln *v* stir up, whirl up *(e.g. a suspension by means of a gas stream)*
aufzehren *v* consume *(e.g. a reactant)*
Aufzehrung *f* consumption *(as of a reactant)*
aufziehen *v* attach, key *(dyes to fibres)*
Aufziehen *n (dye)* attachment, keying, strike
Aufziehgeschwindigkeit *f (dye)* rate [speed] of absorption
Aufziehvermögen *n (dye)* absorptive [absorbing] capacity, absorptive [absorbing] power, *(Am)* pile-on property
aufzwirbeln *v (bioch)* unwind, untwist *(the DNA double strand)*
Aufzwirbelung *f (bioch)* unwinding, untwisting *(of the DNA double strand)*
Augenfarbstoff *m* eye [visual] pigment, photopigment
Augenpigment *n* eye [visual] pigment, photopigment
augenreizend lachrymatory, irritating the eye
Augenreizstoff *m* lachrymator, eye irritant
Augenreizung *f* eye irritation
Augenschutz *m* eye protection
Auger-Effekt *m (phys ch)* Auger effect
Auger-Elektron *n* Auger electron
Auger-Elektronenspektroskopie *f* Auger (electron) spectroscopy, AES
Auger-Elektronenspektroskopie *f* **mit Röntgenstrahlanregung** X-ray-induced Auger electron spectroscopy, XAES
Aurat *n* aurate *(a salt of auric acid)*
Aureolin *n* aureolin, cobalt [Indian] yellow *(potassium hexanitrocobaltate)*
Aurin *n* aurin(e) dyestuff, *(specif)* $(HOC_6H_4)_2{=}C_6H_4{=}O$ aurin(e)
Auripigment *n (min)* auripigment *(arsenic(III) sulphide)*; orpiment (yellow), yellow arsenic (sulphide), arsenic yellow, king's yellow [gold], royal yellow *(arsenic(III) sulphide)*
Ausbeute *f* yield, recovery, gain ratio • **mit hoher Ausbeute** high-yield
Ausbeute *f*/**brauchbare** fair yield
Ausbeute *f*/**theoretische** theoretical yield
Ausbeutekoeffizient *m* 1. *(biot)* process yield coefficient *(fermentation process)*; 2. *s.* Ertragskoeffizient
ausbeuten *v* exploit *(mineral resources)*
Ausbeuteverlust *m* yield loss
Ausbeutung *f* exploitation *(of mineral resources)*
ausbilden *v*/**sich** form
Ausbildung *f* **von Bindungen** bond formation
Ausbildung *f* **von Kurzschlussströmungen** by-passing, short-circuiting *(in a reactor)*
ausblasen *v* blow out
ausbleichen *v* bleach out *(something)*; fade
ausbleien *v* lead-line, lead-clad
Ausbleiung *f* lead lining [cladding]

ausblühen *v* bloom (out), effloresce *(of crystals)*
Ausblühen *n* blooming, efflorescence *(of crystals)*
Ausblühung *f (cryst, min)* bloom, efflorescence
ausbluten *v (dye, coat, tann)* bleed (off, through), mark off
Ausbrand *m* burn-out
ausbreiten *v* spread (out), diffuse, propagate
ausbreiten *v***/sich** spread (out), diffuse, propagate
Ausbreitung *f* spreading, diffusion, propagation
Ausbreitungsfaktor *m (bioch)* spreading factor, hyaluronidase *(a family of enzymes)*
Ausbreitungsgeschwindigkeit *f* speed of propagation
Ausbreitungskoeffizient *m* spreading coefficient, SC
ausbrennen *v* 1. *(text)* burn out; 2. *(coal, petrol)* burn out [off]; deflagrate *(of explosives)*
ausbringen *v* 1. discharge; 2. *(agric)* apply, place *(fertilizers or pesticides)*; 3. yield, produce
Ausbringen *n* 1. discharge, discharging; 2. *(agric)* application, placement, placing *(of fertilizers or pesticides)*; 3. yield, recovery
Ausbringen *n* **vom Flugzeug aus** aeroplane application
Ausbringen *n* **vor dem Auspflanzen** preplanting application
ausdampfen *v* evaporate
Ausdampfen *n* evaporation
ausdämpfen *v* 1. steam; 2. *(distil)* strip (off, out)
Ausdämpfsektion *f* stripping section *(of a fractionating column)*
ausdehnbar expansible, expandable, extensible, extendible
Ausdehnbarkeit *f* expansibility, expandability, extensibility, extendibility
ausdehnen *v* expand, extend
ausdehnen *v***/sich** expand, increase in volume *(as with heat)*; spread *(over an area)*
Ausdehnung *f* 1. expansion, extension *(act or process)*; 2. extension *(range)*
Ausdehnung *f***/adiabatische** adiabatic expansion
Ausdehnung *f***/kubische (räumliche)** cubic(al) expansion
Ausdehnung *f***/thermische** thermal expansion
Ausdehnungskoeffizient *m* coefficient of expansion, expansion coefficient, isobaric coefficient of thermal expansion, isobaric thermal expansivity
Ausdehnungskoeffizient *m***/kubischer** coefficient of cubic(al) expansion, coefficient of volume expansion
Ausdehnungskoeffizient *m***/linearer** coefficient of linear expansion
Ausdehnungskoeffizient *m***/linearer thermischer** coefficient of elongation, linear expansity, longitudinal expansion coefficient
Ausdehnungskoeffizient *m***/mittlerer linearer** average linear expansion coefficient
Ausdehnungskoeffizient *m***/räumlicher** *s.* Ausdehnungskoeffizient/kubischer
ausdrücken *v* 1. press out, squeeze (out), express; 2. push out, blow out, discharge *(solid material)*; 3. *(plast)* eject
Ausdrücken *n* 1. squeeze, expression; 2. discharge *(of solid material)*; 3. *(plast)* ejection
Ausdrücker *m (plast)* ejector, knockout
ausethern *v* extract [shake out] with ether, *(broadly)* extract, shake out
Ausethern *n* ether extraction, *(broadly)* extraction, shaking-out
Ausfall *m* 1. loss *(of material)*; 2. failure, break-down, outage *(of an apparatus)*; 3. discharge (point)
Ausfall *m* **durch Ermüdung** fatigue failure
ausfällbar precipitable
Ausfällbarkeit *f* precipitability
ausfallen *v* 1. precipitate (out), settle (out), come down, separate out, sediment, deposit, subside *(from a solution)*; 2. fail, break down *(of an apparatus)*
ausfällen *v* precipitate, throw down; cement *(a metal by a more active one)*
Ausfällung *f* precipitation; cementation *(of a metal by a more active one)*

Ausfällungsmittel *n* precipitant, precipitating agent, precipitator
ausfaulen *v (hyd)* digest *(of sludge)*
Ausfaulen *n (hyd)* digestion *(of sludge)*
Ausfaulung *f (hyd)* digestion *(of sludge)*
ausfließen *v* flow out, discharge, effuse
Ausfließen *n* outflow, discharge, effusion, effluence
ausfließend effluent
ausflocken *v* 1. flocculate, coagulate, clot, curd(le); 2. *(coll)* pectize
Ausflockung *f* 1. flocculation, coagulation, clotting, curdling; 2. *(coll)* pectization • **zur Ausflockung bringen** flocculate, coagulate
Ausflockungsmittel *n* flocculating [coagulating] agent, flocculant, coagulant, coagulator
Ausfluss *m* 1. effluent, efflux, outflow *(material)*; 2. effluence, efflux, outflow *(process)*; 3. drain, outlet *(site)*
Ausflussrohr *n* effluent [discharge] tube [pipe]
ausfrieren *v* freeze out
Ausfrieren *n* freeze-out, freezing-out
ausführen *v*/**einen Versuch** run [carry out] an experiment
ausfüttern *v (met)* line
Ausgang *m (bioch)* donor [condensing] site, peptidyl-tRNA site, P site
Ausgangselement *n* parent element *(of a radioactive series)*
Ausgangsgestein *n* parent [mother, source] rock
Ausgangskonzentration *f* initial concentration
Ausgangslinie *f* **im Massenspektrum** parent (mass) peak
Ausgangsmaterial *n* starting [parent, raw] material, parent substance, stock, source, base material
Ausgangsprobe *f (anal)* gross sample
Ausgangspunkt *m* 1. starting point; 2. *(bioch)* initiation site; 3. zero (point) *(of a scale)*
Ausgangsstoff *m s.* 1. Ausgangsmaterial; 2. Ausgangselement
Ausgangssubstanz *f s.* 1. Ausgangsmaterial; 2. Ausgangselement
Ausgangssubstrat *n (biot)* basal [main growth] medium
Ausgangsverbindung *f* parent compound
Ausgangszustand *m* original [initial] state
ausgebildet/gut *(cryst)* well-developed
ausgebraucht spent *(solution)*
ausgehärtet/nicht *(plast)* undercured
ausgekleidet/basisch *(met)* basic-lined
ausgekleidet/mit Glas glass-lined
ausgekleidet/sauer *(met)* acid-lined
ausgereift mature
ausgießen *v* pour out
Ausgleich *m* compensation; counterbalance *(esp of a force)*; make-up *(esp for material lost or used up)*; equilibration, levelling *(esp of weights)*; smoothing-out, balancing-out *(esp of concentrations)*
ausgleichen *v* compensate; (counter-) balance *(esp a force)*; make up *(esp losses of material)*; equilibrate, level (out) *(esp weights)*; smooth [balance] out *(esp concentrations)*
Ausgleichgefäß *n* 1. *(tech)* expansion tank [vessel]; 2. *(lab)* levelling bottle
Ausgleich(mess)instrument *n* null-balance instrument
Ausguss *m* drain, sink; (pour-out) lip, spout *(as of a beaker)*
Ausgussbecken *n* (bench) sink
aushärten *v* 1. *(met)* precipitation-harden *(relating to steel)*; age(-harden) *(relating to light metal)*; 2. *(coat, plast)* cure, harden; 3. harden *(relating to fats)*; 4. set *(of concrete)*
Aushärtung *f* 1. *(met)* precipitation hardening *(of steel)*; ageing, age-hardening *(of light metal)*; 2. *(coat, plast)* curing, cure; 3. hardening *(of fats)*; 4. setting, set *(of concrete)*
Aushärtung *f*/**photochemische** photocuring *(as of polymers in the form of a film)*
Aushärtung *f*/**vorzeitige** *(plast)* premature curing *(moulding defect)*
Aushärtungsgeschwindigkeit *f (coat, plast)* speed of cure
Aushärtungszeit *f (coat, plast)* cure [curing] time
Aushauchung *f* exhalation *(as from volcanoes)*
aushebern *v* siphon (off)
auskalken *v* lime out *(in manufacturing organic intermediates)*

ausklauben *v (min tech)* remove, pick (off)
auskleiden *v* line
auskleiden *v***/mit Blei** lead-line
Auskleidung *f* lining *(act or material)*
Auskleidung *f***/basische** *(met)* basic lining
Auskleidung *f***/feuerfeste** refractory lining
Auskleidung *f* **mit Blei** lead lining
Auskleidung *f***/saure** *(met)* acid lining
auskochen *v* boil out *(a vessel)*; decoct *(material)*
Auskochen *n* boiling-out *(of vessels)*; decoction *(of material)*
auskristallisieren *v* crystallize (out)
auskristallisieren *v***/wieder** recrystallize
Auslass *m* 1. discharge *(act)*; 2. discharge (point), outlet, exit, drain, outflow, runoff
auslassen *v (food)* render, try (out) *(fat)*
Auslassventil *n* outlet valve, delivery [discharge] valve
Auslauf *m* discharge (point), outlet, exit, drain, outflow, runoff • **auf Auslauf geteilt [graduiert]** calibrated [graduated] to deliver, calibrated for delivery
auslaufen *v* 1. run out, flow out, *(unintentionally)* leak out *(of liquids)*; 2. discharge, *(unintentionally:)* leak *(of vessels)*
Auslauföffnung *f* discharge opening [outlet, door, aperture, port]
Auslaufrohr *n* discharge [outlet] tube [pipe]
auslaugbar leachable, extractable, extractible
Auslaugbarkeit *f* leachability, extractability
Auslaugbehälter *m* leaching tank [vat, vessel], leach
Auslaugbottich *m* leaching tank [vat, vessel], leach
auslaugen *v* 1. leach (out), lixiviate, extract; 2. *(soil)* eluviate, dilute *(nutrients from the eluvial horizon)*
Auslaugung *f* 1. leach(ing), lixiviation, extraction; 2. *(soil)* eluviation, chemical denudation
Auslegen *n* sizing *(of a plant)*
auslenken *v* deflect
Auslenkung *f* deflection
Auslese *f* selection
auslesen *v* 1. select, pick out; 2. *(min tech)* remove, pick (off)
Auslesen *n* 1. selection; 2. *(min tech)* removal, picking
auslöschen *v* extinguish, quench *(optically)*; quench *(an electric arc)*; extinguish *(a flame)*
Auslöschung *f* extinction *(polarimetry, X-ray diffraction)*
auslösen *v* 1. initiate, bring about *(a reaction)*; trigger *(a chain reaction)*; 2. dissolve out *(substances by solvents)*
Auslösezähler *m (nucl)* self-quenched [self-quenching] counter
Auslösezählrohr *n (nucl)* self-quenched [self-quenching] counter
Auslösung *f* 1. initiation *(of a reaction)*; triggering *(of a chain reaction)*; 2. dissolving-out *(of substances by solvents)*
ausmahlen *v* grind (thoroughly)
Ausmahlung *f***/feine** fine grinding
Ausmahlung *f***/grobe** coarse grinding
ausmauern *v* brick-line
Ausmauerung *f* brick lining *(act or material)*
Ausnutzungsgrad *m* 1. efficiency; 2. *(agric)* recovery *(of fertilizer by a crop)*
auspressen *v* 1. press out, express, force out, squeeze (out); 2. *(plast)* extrude
Auspressen *n* 1. pressing, expression, squeeze; 2. *(plast)* extrusion, extruding
Auspuffgas *n* exhaust gas
auspumpen *v* evacuate, exhaust, pump out
Auspumpen *n* evacuation, exhaustion
ausquetschen *v* squeeze (out), press out, express, force out
Ausquetschen *n* squeeze, pressing, expression
ausräuchern *v* fumigate, smoke (out)
Ausräuchern *n* fumigation, smoking
Ausregnen *n* rain out, in-cloud scavenging *(of pollutants by raindrops)*
Ausreißer *m (anal)* outlier, maverick; escape peak
ausrichten *v* align
ausrichten *v***/sich** align, orient
Ausrichtung *f* alignment, directional distribution, orientation
ausrüsten *v* 1. equip; 2. *(pap, text)* finish
ausrüsten *v***/knirschend** *(text)* scroop

Ausrüstung *f* 1. equipment; 2. *(pap, text)* finish(ing)
Aussalzchromatographie *f* salting-out chromatography, hydrophobic interaction chromatography, HIC
Aussalzeffekt *m* salting-out effect
aussalzen *v* salt out, *(esp soap:)* grain, cut
Aussalzung *f* salting out, *(esp of soap:)* graining, cutting
ausscheiden *v* 1. separate, eliminate; 2. *(min tech)* screen out; discard *(tailings)*; deposit, sediment *(of a liquid)*; evolve, set free, liberate *(gases)*; 3. *(biol)* excrete, *(esp resins:)* exude
ausscheiden *v*/**sich** 1. separate (out), *(of precipitates also)* settle (out), sediment, deposit, precipitate (out), subside; 2. *(biol)* exude *(esp of resins)*
Ausscheiden *n* 1. separation, elimination; 2. *(min tech)* discard *(of tailings)*; deposition, sedimentation *(of bottoms by a liquid)*; evolution, liberation *(of gases)*; 3. *(biol)* excretion, *(esp of resins:)* exudation
Ausscheidung *f* 1. *s.* Ausscheiden; 2. sediment, settling(s), bottom sediment [settlings], B.S., bottoms, deposit; 3. *(ferm)* lees, dregs; 4. *(biol)* excretion product, excretum, *(esp of resins:)* exudate • **zur Ausscheidung gelangen** *s.* ausscheiden/sich
Ausscheidungshärten *n* age [precipitation] hardening
Ausscheidungsprodukt *n s.* Ausscheidung 2.
ausschlacken *v* slag
Ausschlag *m* deflection *(of a pointer)*
ausschlagen *v* deflect
ausschlämmen *v* elutriate
Ausschlämmen *n* elutriation
Ausschleppverluste *mpl* drag-out losses
Ausschließungsprinzip *n*/**paulisches** (Pauli) exclusion principle
Ausschluss *m* **von Luft** exclusion of air
Ausschlusschromatographie *f* size-exclusion chromatography, SEC, gel filtration
ausschlusschromatographisch by exclusion chromatography, by gel filtration
Ausschlussvolumen *n* excluded volume *(of a coiled macromolecule)*
Ausschuss *m* 1. reject, waste (material); 2. *(plast)* cull
ausschütteln *v* shake out, extract by shaking
ausschütteln *v*/**mit Ether** extract [shake out] with ether
Ausschütteln *n* shake, shaking(-out), extraction by shaking, solvent partition
ausschütten *v* 1. dump (out), pour out; 2. *(bioch)* spill out *(hormones)*
ausschwimmen *v* *(coat)* flood *(segregate horizontally)*
ausschwitzen *v* 1. *(bot, text)* exude *(e.g. resin or lubricant)*; 2. *(petrol)* sweat
Ausschwitzungsprodukt *n* exudate, exudation
Außenanstrichfarbe *f* exterior paint
Außenbeständigkeit *f* outdoor [exterior] durability
aussenden *v* emit, radiate *(light)*; emanate, emit, issue *(e.g. radioactive particles)*
Außendruck *m* external pressure
Aussendung *f* emission, radiation *(of light)*; emanation, emission *(as of radioactive particles)*
Außenelektron *n* outer [outside, external] electron, valence(-shell) electron
Außenluft *f* external air, outdoor [outside] air
Außenoberfläche *f* external [outer] surface
Außenschale *f* external shell, outer(-most) shell
Außenschleifen *n* *(ceram)* external grinding
Außensphärenkomplex *m* outer-sphere complex *(coordination chemistry)*
Außerbetriebsetzung *f* shut(down)
aussetzen *v* 1. expose *(as to radiation or air)*; 2. fail *(of an engine)*
Aussetzen *n* 1. exposure *(as to radiation or air)*; 2. failure *(of an engine)*
aussieben *v* screen (out)
aussondern *v* 1. separate (out), eliminate; 2. *(min tech)* discard *(tailings)*
Aussondern *n* 1. separation, elimination; 2. *(min tech)* discard
aussortieren *v* 1. cull, eliminate; 2. *(min tech)* discard *(tailings)*

Aussortieren *n* 1. culling, elimination; 2. *(min tech)* discard
ausspülen *v* rinse (out), scour, flush (out), swill out
aussteinen *v* brick-line
Ausstellungs-Schauglas *n* museum jar
Ausstoß *m* 1. (production) output, make; 2. discharge *(of solid material)*; 3. emission *(of pollutants)*
ausstoßen *v* 1. eject, expel *(e.g. radioactive particles)*; 2. expel *(molecules in elimination reactions)*; 3. push out, discharge *(solid material)*; 4. emit *(pollutants)*
Ausstoßen *n* 1. ejection, expulsion *(as of radioactive particles)*; 2. expulsion *(of molecules in elimination reactions)*; 3. pushing-out, discharge *(of solid material)*
ausstrahlen *v* radiate, emit *(light)*; emit, emanate, issue *(e.g. radioactive particles)*
ausstrahlen *v*/**wieder** reradiate, re-emit
Ausstrahlung *f* radiation, emission *(of light)*; emission, emanation *(as of radioactive particles)*
Ausstrahlung *f*/**spezifische** emittance, radiant excitance
ausstreichen *v* 1. spread out *(on a surface)*; 2. *(geol)* outcrop, crop out
ausströmen *v* 1. flow out, discharge, effuse, *(unintentionally:)* leak out *(of liquids)*; escape *(of gases, steam)*; 2. emanate, emit *(heat)*
ausströmen lassen *v* discharge, run
Ausströmen *n* 1. outflow, discharge, effluence, effusion, *(unintentionally:)* leak(age) *(of liquids)*; escape *(of gases, steam)*; 2. emanation, emission *(of heat)*
austarieren *v* tare, (counter)balance
Austausch *m* exchange *(of ions, heat)*; substitution, placement *(of groups of atoms)*; substitution *(of material)*
Austausch *m*/**wechselseitiger** interchange
austauschaktiv capable of substituting, substitutable *(atomic group)*
Austauschaktivität *f* substitutability
austauschbar exchangeable *(ions)*; substitutable, replaceable *(groups of atoms)*; substitutable *(material)*
austauschbar/nicht non-exchangeable
austauschbar/wechselseitig interchangeable
Austauschbarkeit *f* exchangeability *(of ions)*; substitutability, replaceability *(of groups of atoms)*; substitutability *(of material)*
Austauschbarkeit *f*/**wechselseitige** interchangeability
Austauschboden *m (distil)* plate, tray
Austauschchromatographie *f* (ion-)exchange chromatography
austauschen *v* exchange *(ions)*; substitute, replace *(groups of atoms)*; substitute *(material)*
austauschen *v*/**wechselseitig** interchange
Austauschenergie *f* exchange (binding) energy
Austauscher *m* exchanger *(apparatus or material)*
Austauscherharz *n* ion-exchange resin
Austauscherharz *n*/**chelatbildendes** chelate (ion-exchange) resin
Austauschfeuchtemenge *f*/**spezifische** drying rate
Austauschglocke *f (distil)* bubble cap, dome
Austauschkapazität *f* (ion-)exchange capacity
Austauschkoeffizient *m* transfer coefficient *(mass transfer)*
Austauschkolonne *f* rectifying [rectification] column
Austauschplatz *m* (ion-)exchange site *(of an ion exchanger)*
Austauschrate *f* transfer rate *(mass transfer)*
Austauschreaktion *f* substitution [replacement] reaction, *(esp relating to ions:)* exchange reaction
Austauschsäule *f s.* Austauschkolonne
Austauschstoff *m* substitute
Austauschstromdichte *f (el ch)* exchange current density
Austauschzyklus *m* service [loading] cycle, service [operating] run, exhaustion cycle *(ion exchange)*
Austenit *m (met)* austenite *(a solid solution of carbon in gamma iron)*
austenitisch *(met)* austenitic
austenitisieren *v (met)* austen(it)ize
Austenitisierung *f (met)* austenitization

Austenitisierungstemperatur *f (met)* austenitizing temperature
Austrag *m* 1. *s.* Austragen; 2. discharge (point), outlet, exit; 3. *(el ch)* drag-out
austragen *v* 1. discharge; 2. *(min tech)* discard *(tailings)*
Austragen *n* 1. discharge; 2. *(min tech)* discard
Austragmassestrom *m* output [outlet] stream
Austragrohr *n* discharge [outlet] tube [pipe]
Austragstelle *f* discharge (point), outlet, exit
Austragventil *n* discharge [outlet] valve
austreiben *v* 1. expel, dispel, drive off [out], sweep out *(gases from liquids)*; 2. *(distil)* strip (off, out), distil off
Austreiben *n* 1. expulsion *(of gases from liquids)*; stripping, desorption *(of sorbed gases from sorbent material)*; 2. *(distil)* stripping
austreten *v* escape, leave, pass out, issue *(of gases, liquids)*; exude *(esp of resins)*; leave *(of electrons, groups of atoms)*
Austreten *n* escape *(of gases, liquids)*; exudation *(esp of resins)*
Austritt *m* 1. *(phys ch)* emission *(as of heat, light, or electrons)*; 2. *s.* Austreten; 3. discharge (point) exit, outlet; outflow
Austrittsarbeit *f* work function *(of electrons or ions)*
Austrittsgruppe *f (org ch)* leaving group
Austrittsöffnung *f* discharge opening [outlet, door, aperture, port]
Austrittsrohr *n* discharge [outlet] pipe [tube], off-take (pipe); outflow [effluent] pipe
Austrittsstelle *f* discharge (point), exit, outlet; outflow
austrocknen *v* dry (out, up), exsiccate
Austrocknung *f* drying, exsiccation
Auswahlregel *f (phys ch)* selection principle [rule]
auswalzen *v* 1. roll into a plate, laminate; 2. *(plast)* calender
auswandern *v* migrate out *(as of ions)*
auswaschen *v* 1. wash (out), rinse (out), flush (out), scour *(e.g. a vessel)*; scrub *(gases)*; wash *(a filter cake)*; leach (out), wash out *(soluble substances)*; 2. *(chromat)* elute; 3. *(soil)* eluviate, leach (out) *(nutrients from the eluvial horizon)*
Auswaschen *n* wash out, below-cloud scavenging *(of pollutants by atmospheric precipitation)*
auszementieren *v* cement *(a metal by a more active one)*
Auszementierung *f* cementation *(of a metal by a more active one)*
ausziehbar extractable, extractible
ausziehen *v* extract, leach, lixiviate *(material containing valuable components)*; extract, leach out *(soluble substances)*; exhaust *(e.g. a dye liquor)*
Ausziehen *n* extraction, leach(ing), lixiviation *(of material containing valuable components)*; extraction, leaching-out *(of soluble substances)*; exhaustion *(as of a dye liquor)*
Auszug *m* extract, essence; leachate
Auszug *m*/**alkoholischer** alcoholic extract
Auszug *m*/**wässriger** water leachate
Autogenmühle *f* autogenous mill
Autogenschmelzen *n* autogenous smelting
Autogenschweißen *n* autogenous (gas) welding, oxyacetylene [oxygen-acetylene] welding, oxywelding
Autoionisation *f* autoionization, preionization
Autokatalysator *m* autocatalytic agent
Autokatalyse *f* autocatalysis, self-catalysis
autokatalytisch autocatalytic
Autoklav *m* 1. autoclave, (large-)pressure cooker; 2. *s.* Autoklavheizpresse • **im Autoklaven behandeln [kochen]** autoclave
Autoklavenbehandlung *f* autoclaving
Autoklavheizpresse *f (rubber)* autoclave press, vulcanizer autoclave, pot heater (vulcanizer)
autoklavieren *v* autoclave
Autokorrelationsfunktion *f* autocorrelation function
Autolysat *n (biol)* autolysate
Autolyse *f (biol)* autolysis, autolytic decomposition • **Autolyse auslösen** autolyze, induce autolysis • **Autolyse erleiden** autolyze, undergo autolysis • **der Autolyse unterliegen** autolyze, undergo autolysis

autolytisch *(biol)* autolytic
automorph *(min)* automorphic, idiomorphic, idiomorphous, euhedral
Autoprotolyse *f* autoprotolysis, self-ionization
Autoprotolysegleichgewicht *n* autoprotolysis equilibrium
Autoprotolysekonstante *f* autoprotolysis [self-ionization] constant
Autoprotolysereaktion *f* autoprotolysis [self-ionization] reaction
Autoracemisation *f* autoracemization
Autoracemisierung *f* autoracemization
Autoradiogramm *n* autoradiograph, radioautograph
Autoradiographie *f* autoradiography, radioautography
autoradiographisch autoradiographic, radioautographic
autotroph *(biol)* autotrophic
Autotrophie *f (biol)* autotrophy
Autoxidation *f* autoxidation, autooxidation, *(specif)* spontaneous oxidation
Autoxidator *m* autoxidator
auxochrom auxochromic
Auxochrom *n* auxochrome, auxochromic group
Auxochrom *n*/**negatives** negative [acidic] auxochromic group
Auxochrom *n*/**positives** positive [basic] auxochromic group
auxotroph *(biol)* auxotrophic
Avogadro-Konstante *f* Avogadro constant, N_A *(number of particles in 1 mol of a substance)*
Avogadro-Zahl *f* Avogadro number, N_O *(dimensionless Avogadro constant)*
axial axial
axialsymmetrisch axially symmetric(al), axisymmetric
Axialvermischung *f* axial [longitudinal] mixing
AZ *s.* 1. Acetatfaserstoff; 2. Acetylzahl
Azan *n* N_nH_{n+2} azane
Azelainaldehyd *m* azelaic aldehyde
Azelainsäure *f* azelaic acid, *nonanedioic acid
Azelat *n* azelate *(salt or ester of azelaic acid)*
azeotrop azeotropic
Azeotrop *n* azeotrope, azeotropic mixture
Azeotrop *n* **mit Maximumsiedepunkt** maximum-boiling azeotrope, high-boiling azeotrope
Azeotrop *n* **mit Minimumsiedepunkt** minimum-boiling azeotrope, low-boiling azeotrope
Azeotropbildner *m (distil)* azeotroping [entraining] agent, azeotrope-former, entrainer
Azeotropdestillation *f* azeotropic [entrainment] distillation
Azeotropie *f* azeotropy
azeotropisch azeotropic
Azeotroppunkt *m* azeotropic point
azid acidic *(having the character of an acid)*
Azid *n* 1. M^IN_3 azide, hydrazoate; 2. *(org ch)* RN_3 azide
azidifizieren *v* acidify *(an H atom)*
Azidifizierung *f* acidification *(of an H atom)*
Azidimeter *n* acidimeter, acidometer
Azidimetrie *f* acidimetry
azidimetrisch acidimetric
Azidität *f* acidity, acid strength
Azidität *f*/**aktuelle** active acidity
Aziditätsfunktion *f* acidity function
Aziditätsgrad *m* degree of acidity
Aziditätskonstante *f* acidity constant
Azidogruppe *f* $-N_3$ azido group
Azidokomplex *m* acido complex
Azidolyse *f* acidolysis
azidolytisch acidolytic
Azimut(al)quantenzahl *f* azimuthal [subsidiary] quantum number, orbital (angular momentum) quantum number
Azin *n* azine *(any of two classes of organic compounds containing two or more N atoms)*
Azinfarbstoff *m* azine [phenazine-based] dye
Azlacton *n* azlactone, oxazolone
Azobenzen *n* azobenzene, diphenyldiimide, diphenyldiazene
Azobindung *f* azo link
Azobisisobutyronitril *n* azobisisobutyronitrile, AIBN
Azobrücke *f* azo link
Azodicarbonamid *n* azodicarbonamide, azoformamide
Azofarbstoff *m* azo dye
Azofarbstoff *m*/**auf der Faser erzeugter** *s.* Azofarbstoff/unlöslicher

Azofarbstoff *m*/**unlöslicher** azoic (dye), insoluble azo dye, *(Am)* ice color
Azogruppe *f* –N=N– azo group
Azoimid *n* azoimide, hydrazoic acid
Azokörper *m* R–N=N–R′ azo compound
Azokupplung *f* azocoupling, diazocoupling
Azol *n* azole *(any of a class of heterocyclic compounds containing N)*
Azomethin *n* R_2C=NR′ azomethine, *(if RHC=NR′ also)* Schiff base, Schiff's base
Azomethinfarbstoff *m* azomethine dye
Azophenylen *n* azophenylene, phenazine
Azotometer *n* azotometer, nitrometer
Azoverbindung *f* R–N=N–R′ azo compound
Azoxygruppe *f* azoxy group
Azoxyverbindung *f* azoxy compound
Azulen *n* azulene *(bicyclic hydrocarbon possessing 10 p-electrons)*

B

Bachbildung *f* channel formation, channelling *(as in column packings or reactors)*
backen *v* 1. *(food)* bake; 2. *(ch)* bake *(amines for the purpose of sulphonation)*; 3. *(coal)* cake
Backenbrecher *m* jaw breaker [crusher]
Backgroundkonzentration *f* background concentration [background] level, baseline concentration
Backpulver *n* baking powder
Backtriebmittel *n* baking agent
Bad *n* 1. bath; 2. *(tech)* dip, steep; 3. *(text)* bath, liquor
badaufkohlen *v (met)* liquid-carburize, bath-carburize
Badaufkohlen *n (met)* liquid [bath] carburizing, liquid-salt [molten-salt] carburizing
Badthermostat *m* thermostatic bath
badzementieren *s.* badaufkohlen
Baeyer-Probe *f* Baeyer test, permanganate test *(for detecting alkenes)*
Baeyer-Spannung *f (org ch)* (Baeyer's) angle strain
Baeyer-Villiger-Oxidation *f* Baeyer-Villiger oxidation [reaction] *(of ketones into esters)*
Baeyer-Villiger-Reaktion *f* Baeyer-Villiger oxidation [reaction] *(of ketones into esters)*
Bagasse *f* (sugar cane) bagasse, begass(e), megass(e) *(remains of sugar cane)*
Bahn *f* 1. orbit *(of electrons surrounding a nucleus)*; path *(of particles leaving an atom)*; 2. *(bioch)* route, pathway; 3. *(pap, plast, text)* web
Bahn *f*/**anabolische** *(bioch)* anabolic route [pathway]
Bahn *f*/**endlose** *(plast)* web
Bahn *f*/**katabolische** *(bioch)* catabolic route [pathway]
Bahndrehimpuls *m* orbital (angular) momentum
Bahndrehimpulsquantenzahl *f* azimuthal [subsidiary] quantum number, orbital (angular momentum) quantum number
Bahndrehimpulsquantenzahl *f*/**magnetische** magnetic quantum number
Bahnelektron *n* orbital electron
Bahnmoment *n*/**magnetisches** orbital magnetic moment
Bainit *m*/**oberer** *(met)* upper bainite
Bainit *m*/**unterer** lower bainite
Bainitumwandlung *f (phys ch)* bainite transition
Bajonettkupplung *f* bayonet coupling [joint]
Baker-Venkataraman-Umlagerung *f* Baker-Venkataraman rearrangement *(of 2-acyloxy acetophenones into 1,3-diketones)*
bakteriell bacterial
Bakterien *npl*/**denitrifizierende** denitrifying bacteria, denitrifiers
Bakterien *npl*/**desulfurierende** sulphate-reducing [sulphur-reducing] bacteria, sulphate reducers
Bakterien *npl*/**lichtnutzende** phototrophic bacteria *(e.g. purple bacteria, green sulphur bacteria)*
Bakterien *npl*/**methanogene** methanogenic [methane-forming] bacteria, methanogens
Bakterien *npl*/**methanoxidierende** methane-oxidizing bacteria

Bakterien *npl*/**methylotrophe** methylotrophic bacteria, methylotrophs
Bakterien *npl*/**nitrifizierende** nitrifying bacteria, nitrobacteria *(collectively for nitrite and nitrate bacteria)*
Bakterien *npl*/**nitrogene [stickstoffbindende]** nitrogen-fixing bacteria
bakteriendicht bacteria-tight
Bakterieneiweiß *n s.* Protein/bakterielles
Bakterienfilter *n* bacteriological filter, bacteria-retaining filter
bakterientötend bacteri(o)cidal
bakterienwachstumshemmend bacteriostatic
Bakterienzahl *f (hyd)* bacterial [bacteriological] count, bacterial number
Bakteriochlorophyll *n (bioch)* bacteriochlorophyll
Bakteriostatikum *n* bacteriostat(ic)
bakteriostatisch bacteriostatic
bakterizid bacteri(o)cidal
Bakterizid *n* bactericide
Balkenwaage *f* beam balance
Balkenwaage *f*/**ungleicharmige** unequal-arm balance
Ballast(stoff) *m* 1. impurity, diluent *(as of fuel)*; 2. *(food)* bulk
Ballaststofffaser *f* dietary fiber
Ballon *m* 1. balloon flask; demijohn; *(esp for acids:)* carboy; 2. *(rubber)* bulb
Balmer-Serie *f* Balmer series *(of the hydrogen spectrum)*
Balsam *m* balsam, balm
Balsam *m*/**Peruanischer** Peru [black] balsam *(from Myroxylon balsamum (L.) Harms var. pereirae)*
Balsamharz *n* gum [common] rosin
Balsamkolophonium *n* gum [common] rosin
Bamford-Stevens-Reaktion *f* Bamford-Stevens reaction *(for preparing diazoalkanes)*
Bananenbindung *f* banana [bent] bond, banana-like [banana-shaped] bond
Banbury-Innenmischer *m* Banbury (mixer), intensive mixer
Band *n* 1. band, tape; apron, belt, band, strand *(of a conveyor)*; strip chart *(of a recorder)*; 2. *(phys ch)* band
Band *n*/**erlaubtes** *(phys ch)* allowed band
Band *n*/**verbotenes** *(phys ch)* forbidden band, energy gap
Band-Band-Übergang *m* interband transition *(solids)*
Bande *f* 1. (spectral) band; 2. *(chromat)* band
Bandenaufspaltung *f (spectr)* band splitting
Bandenbreite *f (spectr, chromat)* bandwidth
Bandenkante *f (spectr)* band edge [head]
Bandenkopf *m (spectr)* band edge [head]
Bandenspektrum *n* band(ed) spectrum, molecular spectrum
Bandenverbreiterung *f* 1. *(spectr, chromat)* band broadening; 2. *(chromat)* column band broadening, zone spreading
Bändermodell *n* band theory
Bandfilter *n* belt [band] filter
Bandfilterpresse *f (hyd)* belt (filter) press, belt pressure filter
Bandförderer *m* belt [band] conveyor
Bandheizkörper *m* band [strip] heater, heater band
Bandlücke *f* band gap
Bandpassfilter *m (spectr)* bandpass filter
Band-pass-Filter *m (spectr)* bandpass filter
Bandsaugfilter *n* (rotary-drum) belt-type vacuum filter
Bandschreiber *m* strip-chart recorder
Bandstruktur *f* band structure
Bandtheorie *f* **der Festkörper** band theory of solids
Bandtrockner *m* belt (tunnel, trough) dryer, conveyor dryer, moving-band dryer
Bandwaage *f* conveyor scale, (feed-)belt weigher, weigh(ing) belt
Bank *f (geol)* bed, stratum
Bankbürette *f* microburet(te)
Baratte *f (text)* baratte, xanthator, (xanthating) churn
Barbier-Wieland-Abbau *m* Barbier-Wieland degradation [reaction] *(of carboxylic acids)*
Barbier-Wieland-Reaktion *f* Barbier-Wieland degradation [reaction] *(of carboxylic acids)*

Barbiturat *n (org ch)* barbiturate
Barbitursäure *f* barbituric acid, pyrimidinetrione
Baritflint(glas) *n* barium flint (glass)
Baritkron(glas) *n* barium crown (glass)
Barium *n* Ba barium
Bariumdichromat *n* barium dichromate
Bariumhydrid *n* barium hydride
Bariumhydrogensulfid *n* barium hydrogensulphide
Bariumhydroxid *n* barium hydroxide
Bariumlack *m* barium lake
Bariumoxid *n* barium oxide
Bariumperoxid *n* barium peroxide
Bariumperoxodisulfat *n* barium peroxodisulphate
Bariumsulfat *n* barium sulphate
Bariumsulfid *n* barium sulphide, barium monosulphide
Barn *n (nucl)* barn *(a unit of area for measuring cross section)*
Barometerdruck *m* barometric pressure
Barren *m (met)* ingot, billet
Barrierekunststoff *m* barrier polymer *(packing technology)*
Bartbildung *f* fronting, leading, bearding *(in a chromatogram)*
Baryon *n (nucl)* baryon
Barysphäre *f (geol)* barysphere, centrosphere, earth's core [nucleus]
Baryt *m (min)* barite, baryte(s), heavy spar *(barium sulphate)*
Baryterde *f s.* Bariumoxid
Barytgelb *n* baryta yellow, ultramarine [Steinbühl] yellow, yellow ultramarine, lemon chrome, gelbin *(barium chromate)*
Barytwasser *n (anal)* baryta water *(aqueous solution of barium hydroxide)*
Barytweiß *n* fixed [permanent] white, blanc fixe *(precipitated barium sulphate)*
Barytzinkweiß *n* zinc baryta white, Orr's white
Basalt *m* basalt
basaltisch basaltic
Base *f* base
Base *f***/harte** hard base *(according to Pearson's classification)*
Base *f***/korrespondierende** conjugate base
Base *f***/millonsche** Millon's base *(a mercurammine compound)*
Base *f***/schiffsche** R'N=CHR Schiff [Schiff's] base, azomethine (compound)
Base *f***/schwache** weak base
Base *f***/seltene** *(bioch)* rare [minor] base *(purine or pyrimidine derivative)*
Base *f***/starke** strong base
Base *f***/weiche** soft base *(according to Pearson's classification)*
Basekatalyse *f* base [basic, hydroxide-ion] catalysis
Basekatalyse *f***/allgemeine** general base catalysis
Basekatalyse *f***/spezifische** specific base catalysis
basekatalysiert base-catalyzed
Basenanhydrid *n* basic anhydride
Basenäquivalenz *f (bioch)* base equivalence *(within DNA)*
Basenaustausch *m* base exchange
Basenbindungsvermögen *n* base-binding capacity [power]
Basen-Excisionsreparatur *f (bioch)* base excision repair
Basengehalt *m* 1. basicity *(of a solution)*; 2. *(soil)* alkalinity
Basenpaar *n (bioch)* base pair
Basenpaarung *f (bioch)* base pairing
Basensequenz *f (bioch)* base [nucleotide] sequence
Basenstapelung *f (bioch)* base stacking
Basenstärke *f* base [basic] strength
Basentriplett *n (bioch)* codon, coding [nucleotide] triplet
Basenumtausch *m (soil)* base exchange
basiphil *(bot)* basophilous *(growing preferably in alkaline soil)*
basisch 1. basic, alkaline; 2. *(met)* basic • **basisch ausgekleidet** *(met)* basic-lined • **basisch stellen** *(dye)* basify, make alkaline, alkal(in)ize
basisch/schwach weakly [feebly] basic, low-alkalinity
basisch/stark strongly basic, high-alkalinity
Basischstellen *n* basification
Basiseinheit *f* base [fundamental] unit *(of a system of units)*
Basislinie *f (anal)* base-line
Basislinientrennung *f (chromat)* 6σ separation
Basizität *f* 1. alkalinity, basicity *(of a solution)*; 2. basicity, basic capacity *(of*

an acid); 3. basicity *(of atoms, ions, or molecules)* • **die Basizität schwächend** base-weakening
Basizität f/aktuelle active alkalinity
Basoid *n (soil)* basoid *(colloidal substance saturated with OH ions)*
basophil 1. basophile, basophilic, basophilous *(having an affinity for basic dyes)*; 2. *(bot)* basophilous *(growing preferably in alkaline soils)*
Bassin *n* basin, tank
Bastseide *f* raw [gum] silk, grege, greige
BAT *s.* Arbeitsplatztoleranzwert/biologischer
Batch-Fermentation *f (biot)* batch(-process) fermentation, batch-fed fermentation
Batch-Fermenter *m (biot)* batch fermenter
bathochrom bathochromic
Bathochromie *f* bathochromic shift [displacement]
Batterie *f* 1. battery *(a group of uniform devices)*; bench *(a group of retorts in a coke oven)*; 2. *(el ch)* battery
Bau *m* 1. structure, constitution *(of a molecule)*; set-up, structure, build *(of an apparatus)*; 2. building, build(ing)-up, construction, structure, setting
Bauart *f* build, make, design
Baubestandteil *m* building constituent
Baueinheit *f* building unit
Bauelement *n* structural element [entity, unit]
Baufehler *m (cryst)* structural defect
Bauformel *f* constitutional formula *(of molecules)*
Baukalk *m* building [mason's] lime *(chiefly calcium hydroxide)*
Baukeramik *f* building [structural] ceramics
Baumé-Skale *f* Baumé scale
Baumkristall *m* dendrite
Baumwolle *f* cotton
Baumwollfarbstoff *m* cotton dye, *(Am)* cotton color
Baumwollfaser *f* cotton fibre
Baumwollkernöl *n* cotton(seed) oil
Baumwollsamenöl *n* cotton(seed) oil
Baumwollwachs *n* cotton wax
Baustein *m* 1. building block [brick, unit], structural element [entity, unit] *(as of a chemical compound)*; 2. *(lab)* module
Baustoff *m* building [construction] material
Bauxit *m* bauxite, beauxite
bauxitisch bauxitic
Bauzement *m* building cement
Bayer-Verfahren *n* Bayer process *(digestion of bauxite in sodium hydroxide solution)*
Bayöl *n* bay [myrcia] oil *(from Pimenta racemosa (Mill.) I. W. Moore)*
beanspruchen *v* 1. stress; 2. expose *(as to aggressive agents)*
Beanspruchung *f* 1. stressing, stress application; 2. stress *(force acting across a unit area)*; 3. exposure *(as to aggressive agents)*
Beanspruchung f/chemische chemical exposure
Beanspruchung f/dynamische dynamic stress
Beanspruchung f/schwingende vibrating stress
Beanspruchung f/statische static stress
Beanspruchungs-Dehnungs-Diagramm *n* stress-strain diagram
bearbeitbar workable
Bearbeitbarkeit *f* workability
Bearbeitbarkeit f/maschinelle machinability
bearbeiten *v* work, treat, process, machine
Bearbeitung *f* working, treatment, processing, machining
Bearbeitungstechnik *f* processing technique
Beattie-Bridgeman-Zustandsgleichung *f (phys ch)* Beattie-Bridgeman equation (of state)
bebrüten *v (biot)* incubate *(the fermentation broth)*
Bebrütung *f (biot)* incubation *(of the fermentation broth)*
Béchamp-Reaktion *f* Béchamp reaction [reduction] *(of aromatic nitro compounds)*
Béchamp-Reduktion *f* Béchamp reaction [reduction] *(of aromatic nitro compounds)*
Béchamp-Verfahren *n* Béchamp process *(for producing primary aromatic amines)*

Becher *m* 1. cup; 2. *(lab)* beaker; 3. *(tech)* bucket, scoop *(as of an elevator)*
Becheraufzug *m s.* Becherwerk
Becherelevator *m s.* Becherwerk
Becherglas *n* beaker
Becherversprüher *m* spinning-cup atomizer [sprayer], rotary-cup atomizer [sprayer]
Becherwerk *n* bucket elevator
Becherzerstäuber *m s.* Becherversprüher
Becken *n* 1. basin, tank; 2. *s.* Ausgussbecken
Beckmann-Thermometer *n* Beckmann thermometer
Beckmann-Umlagerung *f* Beckmann (oxime-amide) rearrangement
Bedarf *m* requirements, demands • **den Bedarf decken an** meet the requirements [demands] of
bedienbar operable
bedienbar/leicht easy-to-operate
Bedienbarkeit *f* operability
Bedienbarkeit *f***/leichte** *s.* Bedienungskomfort
bedienen *v* operate, run, handle *(e.g. an apparatus)*
Bedienung *f* operation, running, handling *(as of an apparatus)*
Bedienungskomfort *m* operational convenience, ease [simplicity] of servicing, ease of serviceability
Bedingungen *fpl***/milde [schonende]** mild conditions
Bedingungsgefüge *n* set of conditions
bedüsen *v* nozzle *(with a liquid)*
beeinflussen *v* influence, affect, *(esp negatively)* interfere with
Beeinflussung *f* affection, *(esp negatively:)* interference
beeinträchtigen *v* interfere with
beenden *v* finish, complete; terminate, break *(the growth of a chain molecule)*
Beendigung *f* finishing, completion; termination *(of chain growth)*
befestigen *v* attach, mount, anchor
befeuchten *v* moisten, wet, damp(en), humidify, water, dew
Befeuchtung *f* moistening, wetting, damp(en)ing, humidification, watering, dewing
Befeuchtungsapparat *m* humidifier
befeuern *v* fire, heat
beflecken *v* stain
befreien *v* liberate *(from contaminants)*
befreien *v***/von flüchtigen Bestandteilen** devolatilize
befreien *v***/von Kationen** decationize
befreien *v***/von Kesselstein** (de)scale
Befreiung *f* liberation *(from contaminants)*
begasen *v* 1. *(agric, food)* fumigate, gas; 2. *(hyd, biot)* aerate
Begaser *m (hyd, biot)* aerator
Begasung *f* 1. *(agric, food)* fumigation, gassing, exposure to gas; 2. *(hyd, biot)* aeration
Begasungsfilter *n* air stone, fritted gas bubbler; gas dispersion [distribution] tube
Begasungsmittel *n (agric, food)* fumigant
Begasungsrohr *n* gas dispersion [distribution] tube, sparger
Begegnung *f* encounter *(of molecules in solutions)*
begichten *v* burden, charge *(a blast furnace)*
Beginn-Kochpunkt *m* initial boiling point, I.B.P.
Beginn-Siedepunkt *m* initial boiling point, I.B.P.
Begleitalkaloid *n* companion alkaloid
Begleitreaktion *f* concurrent reaction
Begleitstoff *m* 1. companion (substance), accompanying substance, admixture; 2. *(hyd)* minor constituent
Begleitsubstanz *f* 1. companion (substance), accompanying substance, admixture; 2. *(hyd)* minor constituent
begünstigt/energetisch energetically favoured [favourable]
Behälter *m* vessel, receptacle, container, *(for liquids esp if large:)* tank, *(esp if circular and open:)* basin; *(for gases:)* holder; *(if funnel-shaped, for bulk material:)* hopper
behandeln *v* treat, process
behandeln *v***/anodisch** *(met)* anodize
behandeln *v***/im Autoklaven** autoclave
behandeln *v***/in der Küpe** *(dye)* vat
behandeln *v***/mit Arsen** arsenicate
behandeln *v***/mit Bleicherde** clay
behandeln *v***/mit Brom** brominate
behandeln *v***/mit Chlor** chlorinate, chlorinize

behandeln *v*/**mit Chlorwasserstoff** hydrochlorinate
behandeln *v*/**mit Dampf** steam
behandeln *v*/**mit Formaldehyd** 1. treat with formaldehyde; 2. *(pharm)* formalinize, formolize
behandeln *v*/**mit Kalk** lime out *(as in manufacturing organic intermediates)*
behandeln *v*/**mit Salzlake [Salzlösung]** *(food, tann)* brine
behandeln *v*/**mit Schwefel** sulphur
behandeln *v*/**nochmals** re-treat
behandeln *v*/**übermäßig** overtreat
Behandlung *f* treatment, processing
Behandlung *f*/**anodische** *(met)* anodizing, anodization, anodic treatment [oxidation, coating]
Behandlung *f* **in Abwasserteichen** *(hyd)* pond treatment, ponding, lagooning
Behandlung *f* **mit Bleicherde** clay treatment
Behandlung *f* **mit Chlorwasserstoff** hydrochlorination
Behandlung *f* **mit Salzlake [Salzlösung]** *(food, tann)* brining
Behandlung *f*/**nochmalige** re-treatment
Behandlung *f*/**thermische** heat treatment
Behandlung *f* **vor dem Auspflanzen** preplanting treatment *(of a soil with pesticides)*
Behandlung *f*/**wiederholte** re-treatment
Behandlungsanlage *f* treatment plant; treatment unit *(of an industrial plant)*
Behandlungschemikalie *f* 1. treatment [processing] chemical; 2. *(hyd)* water-treating chemical
Behandlungsstufe *f* treatment stage
Behandlungsverfahren *n* treating process
Beharrungsvermögen *n* inertia
beharzen *v (plast)* resin
beheizen *v* heat, fire, warm (up)
beheizt/mit Dampf steam-heated
Beheizung *f* heating, firing
Beheizung *f*/**elektrische** electric heating
Beheizung *f*/**induktive** induction heating
Beheizung *f* **mit Gas** gas-fired heating
Behensäure *f* *docosanoic acid, *(deprecated:)* behenic acid
behindern *v* hinder, inhibit, embarrass
Behinderung *f* hindrance, inhibition, embarrassment
Beilstein-Probe *f* Beilstein's test *(for detecting halogens in organic compounds)*
beimengen *v* admix
beimischen *v* admix
Beimischung *f* 1. admixture *(act)*; 2. admixture, impurity • **mit Beimischungen** impure
beimpfen *v (biot)* seed, inoculate
Beimpfung *f (biot)* seeding, inoculation
Beiprodukt *n s.* Nebenprodukt
beißend acrid, pungent *(taste, smell)*
Beistoff *m* inert ingredient; corrective *(in building up active-substance mixtures)*
Beiwert *m* index; factor; coefficient
Beizablauge *f (met)* pickling acid
Beizahl *f* index; factor; coefficient
Beizanlage *f (met)* pickling plant
Beizbad *n (met)* scouring bath, pickling bath
Beizbehälter *m (met)* pickling vat
Beizdauer *f (met)* pickling time
Beize *f* 1. mordant *(for treating textiles or microscopic preparations)*; stain *(for treating rubber, glass, or wood)*; 2. *(agric)* pickle; 3. *(met)* scouring agent, pickle; 4. *s.* Beizen
beizen *v* 1. mordant *(textiles or microscopic preparations)*; stain *(rubber, glass, or wood)*; 2. *(tann)* bate; 3. *(agric)* pickle; 4. *(met)* pickle, scour *(for removing scale)*
Beizen *n* 1. mordanting *(of textiles or microscopic preparations)*; staining *(of rubber, glass, or wood)*; 2. *(agric)* pickling; 3. *(met)* pickling, pickling process, scouring *(for removing scale)*
Beizenfarbstoff *m (text)* mordant dye(-stuff), adjective dye
Beizinhibitor *m* pickling inhibitor
Beizkasten *m s.* Beizbehälter
Beizlauge *f* pickling liquor
Beizmittel *n s.* Beize 1.
Beizsäure *f s.* Beizablauge
Beizstoff *m s.* Beize 1.
Beizwasser *n* pickling water
bekämpfen *v* control
Bekämpfung *f* control
beladen *v* charge, feed, load
Beladung *f* 1. charging, feeding, loading; 2. *(hyd)* loading *(ion exchange)*; 3. load; *(of liquids:)* concentration
Belag *m* 1. coat(ing), cover(ing), layer, *(if*

thin:) film; scale, encrustation *(esp of alien substance)*; overlay *(for wood)*; 2. *(food)* bloom

Belastbarkeit *f* 1. *(tech)* loadability, load-bearing capacity; 2. *(distil)* loading capacity; 3. *(anal)* loadability, maximum possible loading

belasten *v* 1. load, weight; 2. stress; 3. *(distil, hyd)* load

belastet/mit Schmutzstoffen polluted, contaminated *(water, air)*

belastet/nicht unpolluted, uncontaminated, non-contaminated *(water, air)*

Belastung *f* 1. *(distil, hyd)* loading, *(hyd quantitatively:)* load, loading (rate); 2. *(tox)* exposure

Belastung *f* **durch Schwermetalle** *(hyd)* heavy-metal contamination

Belastung *f* **mit organischen Inhaltsstoffen** *(hyd)* organic load(ing)

Belastung *f***/produktionsbedingte** *(hyd)* process contamination

Belastung *f***/statische** static loading

Belastung *f***/zeitgewichtete durchschnittliche** *(tox)* time-weighted average, TWA

Belastungsgrenzwert *m***/kurzzeitiger** *(tox)* short-time exposure limit, STEL

Belastungsschwankung *f (hyd)* load fluctuation, variation in loadings

beleben *v* activate *(e.g. floatation)*

Belebtschlamm *m (hyd)* activated sludge, active [biological] sludge

Belebtschlammbecken *n s.* Belebungsbecken

Belebtschlammverfahren *n s.* Belebungsverfahren

Belebungsbecken *n (hyd)* aeration tank, aeration [aerated] basin, activated-sludge tank

Belebungsverfahren *n (hyd)* activated-sludge process [method]

belegen *v* cover, coat

Belegung *f***/monomolekulare** *(phys ch)* monolayer coverage

Beleuchtungsstärke *f* illuminance

belichten *v* 1. expose to light; 2. *(phot)* expose

Belichtung *f* 1. exposure to light; 2. *(phot)* exposure • **bei Belichtung** on exposure to light

Belichtungsdauer *f (phot)* duration of exposure, exposure time

Belladonnaalkaloid *n* belladonna alkaloid

Belousov-Zabotinskij-Reaktion *f* Belousov-Zhabotinskii reaction, BZ reaction *(an oscillating reaction)*

belüften *v* aerate

Belüfter *m (hyd)* aerator

Belüftung *f* aeration

Belüftungsbecken *n s.* Belebungsbecken

Belüftungselement *n* (differential) aeration cell, oxygen (concentration) cell *(corrosion)*

Belüftungszelle *f* (differential) aeration cell, oxygen (concentration) cell *(corrosion)*

bemessen *v* size *(a treatment unit)*

Bemessung *f* sizing *(of a treatment unit)*

benachbart adjacent, neighbouring, vicinal, juxtaposed *(substituents)*

benennen *v (nomencl)* name, denominate

Benennung *f* 1. *(nomencl)* naming, denomination; 2. name, term

Benennung *f***/funktionelle** functional name

Benennung *f***/nach den IUPAC-Regeln gebildete** IUPAC name

Benennung *f* **nach den Regeln von Stock** Stock naming

Benennung *f***/systematische** systematic name

Benennungssystem *n (nomencl)* naming system

benetzbar wettable, hydrophilic, hydrophile

benetzbar/nicht non-wettable, hydrophobic, hydrophobe

Benetzbarkeit *f* wettability, ability of being wetted

benetzen *v* perfuse, *(with water also)* wet, moisten, humidify, dew, water, damp(en)

Benetzung *f* perfusion, *(with water also)* wetting, moistening, humidification, dewing, watering, damp(en)ing

Benetzungskoeffizient *m* spreading coefficient, SC

Benetzungsmittel *n* wetting agent [aid], wetter

Benetzungsspannung *f* wetting tension

Benetzungsvermögen *n* wetting power [ability]

Benetzungswinkel *m* wetting [contact] angle
Bengalkino *n* Bengal [butea] gum *(from Butea superba Roxb.)*
Bengough-Stuart-Verfahren *n* Bengough-Stuart process, chromic-acid (anodizing) process
Bentonit(ton) *m* bentonite (clay)
Benzalaceton *n* benzalacetone,*4-phenyl-but-3-en-2-one
Benzalchlorid *n* benzal chloride, α,α dichlorotoluene, benzylidene chloride
Benzaldehyd *m* benzaldehyde
Benzalgrün *n* malachite [benzal] green
Benzamid *n* benzamide
Benzanthronfarbstoff *m* benzanthrone dye
Benzen *n* benzene, *(commercial product:)* benzole, benzol
Benzen *n***/anorganisches** inorganic benzene, borazole, triborine triamine
Benzenabkömmling *m* benzene derivative
Benzenazimid *n* benzeneazimide, benzotriazole, aziminobenzene
Benzencarbonsäure *f* benzene carboxylic acid, *(specif)* C_6H_5COOH benzoic acid
Benzenderivat *n* benzene derivative
Benzendicarbonsäure *f* benzenedicarboxylic acid
Benzenenium-Ion *n* s. Benzenium-Ion
Benzenhexacarbonsäure *f* benzenehexacarboxylic acid, mellitic acid
Benzenhexachlorid *n* benzene hexachloride, BHC, hexachlorocyclohexane
Benzenium-Ion *n* $C_6H_7^+$ benzenium [benzenonium, phenonium] ion, benzene carbonium ion
Benzenkern *m* benzene ring [nucleus]
Benzenkohlenwasserstoffe *mpl* benzene [aromatic] hydrocarbons, aromatics
Benzenmonosulfonsäure *f* benzenesulphonic acid
Benzenreihe *f* benzene series
Benzenring *m* benzene ring [nucleus]
Benzensulfonamid *n* benzene sulphonamide
Benzensulfonsäure *f* benzenesulphonic acid
Benzentetracarbonsäure *f* benzenetetracarboxylic acid
Benzentricarbonsäure *f* benzenetricarboxylic acid
Benzidin *n* benzidine, 4,4′-diaminobiphenyl
Benzidinumlagerung *f* benzidine rearrangement [transformation, conversion]
Benzil *n* benzil, bibenzoyl, diphenyl diketone
Benzilsäure *f* benzilic acid
Benzilsäureumlagerung *f* benzilic-acid rearrangement
Benz-in *n* benzyne, 1,2-didehydrobenzene
Benzin *n (ch)* benzin(e); gasoline, petrol, (motor) spirit *(as a motor fuel)*; (petroleum) naphtha *(esp for technical purposes or as a reformer feedstock)*
Benzin *n***/bleifreies** unleaded [lead-free] gasoline
Benzin *n***/direkt herausdestilliertes** straight-run gasoline, distillate gasoline, straight-run benzine, S.R.B.
Benzin *n***/leichtes** gasoline *(boiling range 30 to 100 °C)*; light gasoline [benzine, spirit, naphtha]
Benzin *n***/schweres** heavy gasoline *(boiling range 150 to 210 °C)*
Benzin *n***/süßes** sweet gasoline
Benzin *n***/verbleites** leaded [ethyl] gasoline
Benzinadditiv *n* gasoline additive
Benzoat *n* benzoate
***p*-Benzochinon** *n p*-benzoquinone, *(proper)* quinone, *cyclohexadiene-1,4-dione
Benzoe *f* benzoin, benzoin [Benjamin] gum [resin] *(from Styrax specc.)*
Benzoeharz *n* benzoin, benzoin [Benjamin] gum [resin] *(from Styrax specc.)*
Benzoesäure *f* benzoic acid, benzenecarboxylic acid
Benzoesäureanhydrid *n* benzoic anhydride
Benzoesäurephenylester *m* phenyl benzoate
***o*-Benzoesäuresulfimid** *n o*-sulphobenzoic imide, saccharin
Benzofuran *n* benzofuran, coumarone, cumarone
benzoid benzenoid

benzoid/nicht non-benzenoid
Benzoin *n* benzoin, α-hydroxybenzyl phenyl ketone
Benzoinkondensation *f* benzoin condensation
Benzol *n* 1. benzole, benzol *(commercial product)*; 2. *s.* Benzen • **mit Benzol anreichern [beladen, sättigen]** benzolize
Benzol... *siehe: Benzen... (for chemical compounds)*
Benzopersäure *f* perbenzoic acid
Benzophenanthren *n* benzophenanthrene
Benzophenon *n* benzophenone, *benzoyl benzene, *diphenyl ketone
2,3-Benzopyridin *n* 2,3-benzpyridine, benzo[*b*]pyridine, quinoline, 1-benzazine
3,4-Benzopyridin *n* 3,4-benzpyridine, benzo[*c*]pyridine, isoquinoline, 2-benzazine
Benzopyrimidin *n* benzpyrimidine
1,2-Benzopyron *n* 1,2-benzopyrone, coumarin, 2H-chromen-2-one, α-chromone
2,3-Benzopyrrol *n* 2,3-benzpyrrole, indole
Benzotriazol *n* benzotriazole
Benzotrichlorid *n* benzotrichloride, α,α,α-trichlorotoluene
Benzoylaminoessigsäure *f s.* Benzoylglycin
Benzoylglycin *n* benzoylglycine, benzoylaminoacetic acid, hippuric acid
Benzoylglykokoll *n* benzoylglycine, benzoylaminoacetic acid, hippuric acid
Benzoylgrün *n s.* Benzalgrün
Benzoylgruppe *f* C_6H_5CO– benzoyl group [residue]
benzoylieren *v* benzoylate
Benzoylierung *f* benzoylation
Benzoyloxylierung *f* benzoyloxylation
Benzoylperoxid *n* benzoyl peroxide, dibenzoyl peroxide
Benzpyren *n* benzopyrene
Benzylacetat *n* benzyl acetate
Benzylalkohol *m* benzyl alcohol
Benzylbenzen *n* benzylbenzene, diphenylmethane
Benzylcellulose *f* benzyl cellulose
Benzylchlorid *n* benzyl chloride, α-chlorotoluene
Benzylcyanid *n* benzyl cyanide, ω-cyanotoluene, phenylacetonitrile
Benzylgruppe *f* $C_6H_5CR_2$– benzylic group [residue], *(specif)* $C_6H_5CH_2$– benzyl group [residue]
Benzylidenchlorid *n* benzylidene chloride, α,α-dichlorotoluene, benzal chloride
Benzylierung *f* benzylation
Benzylpenicillin *n* benzylpenicillin
Benzylradikal *n* benzyl radical, *(specif)* free benzyl radical
Benzylrest *m s.* Benzylgruppe
Benzyn *n* benzyne, 1,2-didehydrobenzene
beobachten *v* observe, watch, study
Beobachtung *f* observation • **sich der Beobachtung entziehen** escape observation
Beobachtungsfehler *m* observational error
Bereich *m* 1. region, range *(as of measurement or of state)*; 2. area, region *(locally)*; region, section *(as of a molecule)*; 3. sphere *(as of a science)*
Bereich *m***/dynamischer** dynamic range *(of an analyser)*
Bereich *m***/femtomolarer** femtomolar range *(from 10^{-12} to 10^{-15})*
Bereich *m***/kristalliner** crystalline region
Bereich *m***/sichtbarer** *(spectr)* visible range [region]
bereiten *v* prepare, make (ready)
bereitet/frisch freshly prepared
Bereitung *f* preparation, making
Bergabdiffusion *f* nucleation and growth *(in an alloy)*
Bergabwärtsreaktion *f* downhill reaction
Bergaufdiffusion *f* spinodal decomposition *(in a supersaturated alloy)*
Bergbau *m* mining
Berge *pl (min tech)* tailings, tails, waste tailing, refuse
Bergemittel *n* gangue (mineral), matrix
Bergius-Hochdruckverfahren *n* Bergius (hydrogenation) process
Bergius-Hydrierverfahren *n* Bergius (hydrogenation) process
Bergkristall *m (min)* rock crystal *(a variety of quartz)*

Bergtalg *m (min)* ozokerite, earth [ader] wax, native paraffin
Bergwachs *n (min)* ozokerite, earth [ader] wax, native paraffin
berieseln *v* sprinkle, spray; scrub *(gases)*
Berieselung *f* sprinkling, spraying; scrubbing *(of gases)*
Berieselungskühler *m* spray cooler
Berkelium *n* Bk berkelium
Bernstein *m* amber, succinite
Bernsteinsäure *f* succinic acid, *butanedioic acid
Bernsteinsäuredialdehyd *m* succindialdehyde, *butane-1,4-dial
Bernsteinsäureimid *n* succinimide, 2,5-dioxopyrrolidine
bersten *v* crack, break, *(esp of surfaces)* burst
Berstscheibe *f* rupture [bursting] disk *(in pressure relief devices)*
Berstwiderstand *m (plast, pap)* bursting strength
Berthelot-Kalorimeter *n* Berthelot calorimeter
Berthollide *npl* berthollid(e)s, berthollide [non-Daltonian, non-daltonide, non-stoichiometric] compounds
Berthollidverbindungen *fpl* berthollid(e)s, berthollide [non-Daltonian, non-daltonide, non-stoichiometric] compounds
Berufskrankheit *f* occupational disease
beruhigen *v* kill *(a smelt)*; deoxidize *(steel)*
Beruhigungsmittel *n* 1. *(pharm)* sedative; 2. *(met)* killing agent
berühren contact
berühren *v*/**sich** contact
Berührungsfläche *f* surface of contact, *(quantitatively:)* area of contact, contact area
Berührungsgift *n* (direct) contact poison, contact toxicant
Berührungskorrosion *f* galvanic [contact] corrosion, bimetallic [two-metal] corrosion
Berührungstrocknen *n* contact [conduction, indirect] drying
Berührungswinkel *m* contact angle *(in testing surface-active substances)*
Beryll *m (min)* beryl *(beryllium aluminium silicate)*
Beryllat *n* beryllate
Beryllid *n* beryllide
Beryllіose *f s.* Berylliumkrankheit
Beryllium *n* Be beryllium
Berylliumhalogenid *n* beryllium halide [halogenide]
Berylliumhydrid *n* beryllium hydride
Berylliumhydroxid *n* beryllium hydroxide
Berylliumkrankheit *f (med)* beryllium disease, beryll(i)osis *(lung damage caused by inhalation of beryllium dust)*
Berylliumoxid *n* beryllium oxide
Beryllose *f s.* Berylliumkrankheit
Beschaffenheit *f* quality, constitution, nature, *(of man-made products also)* make
beschallen *v* sonicate
Beschallung *f* sonication
beschichten *v* coat, *(esp with metal:)* plate; laminate *(as with a plastic film)*; overlay *(as with veneer)*
beschichten *v*/**durch Tauchen** dip-coat
beschichten *v*/**mit Blei** lead-coat
beschichten *v*/**mit Graphit** coat with graphite, graphitize, *(Am)* graphite
beschichten *v*/**mit Platin** platinize, platinate
Beschichten *n* coating, *(esp with metal:)* plating; laminating *(as with a plastic film)*; overlaying *(as with veneer)*
Beschichten *n* **durch Tauchen** dip coating
Beschichten *n* **endloser Bahnen** web coating
Beschichtung *f s.* 1. Beschichten; 2. Schicht
Beschichtungsstoff *m*/**anorganischer** inorganic coating (material)
Beschichtungsstoff *m*/**organischer** organic coating (material)
beschicken *v* 1. feed, charge, fill, load, furnish; 2. *(nucl)* fuel *(a reactor)*; 3. *(met)* burden
Beschickung *f* 1. feeding, charging, filling, loading, furnishing; 2. *(nucl)* fuelling; 3. *(met)* burdening; 4. *s.* Beschickungsmaterial
Beschickungsmaterial *n* 1. feed(stock), feed material, *(with discontinuous operation also)* charge (stock), load, batch; 2. *(met)* burden
Beschickungsöffnung *f* feed inlet [opening, hole], charging hole

Beschickungsrohr *n* feed pipe [tube]
Beschickungsschleuse *f* entry lock, inlet sluice, *(if conical:)* lock hopper
Beschichtungstechnik *f (coat)* coating technique
Beschichtungstechnologie *f (coat)* coating technology
Beschickungstrichter *m* feed(ing) hopper, feed(ing) funnel, charging [loading] hopper [funnel]
Beschickungsvorrichtung *f* feeder, loader, *(for solid fuel also)* stoker
beschießen *v (nucl)* bombard
Beschießen *n (nucl)* bombardment
Beschlag *m* 1. bloom; 2. *(glass)* tarnish *(defect)*
beschlagen *v* bloom
beschleunigen *v* accelerate, promote, speed (up)
Beschleuniger *m* 1. accelerator, promoter, accelerating [promoting] agent; 2. *(nucl)* (particle) accelerator
Beschleunigung *f* acceleration, promotion
Beschleunigung *f***/anchimere** anchimeric assistance *(of the reaction rate by neighbouring groups)*
Beschleunigung *f***/negative** deceleration
Beschleunigung *f***/sterische** steric acceleration, B strain effect
Beschleunigungsspannung *f* accelerating potential
beschmieren *v* smear
beschmutzen *v* soil, pollute, stain, *(esp with greasy matter:)* smear
Beschuss *m* **mit Elektronen** *(nucl)* electron bombardment
beschweren *v (pap, text)* weight, load *(with fillers or sizing material)*
Beschwerungsmaterial *n* weighting [loading] material [agent], load(ing)
beseitigen *v* eliminate, remove; dispose of *(waste products)*
beseitigen *v***/durch Geländeauffüllung** dispose of by landfill
Beseitigung *f* elimination, removal; disposal *(of waste products)*
Beseitigung *f* **des radioaktiven Abfalls** radioactive-waste [nuclear-waste] disposal
Beseitigung *f* **durch Abklappen** disposal to [at] sea, ocean [marine] disposal
Beseitigung *f* **durch Einleitung ins Meer** disposal to [at] sea, ocean [marine] disposal
Beseitigung *f* **durch Geländeauffüllung** landfill disposal
Beseitigung *f* **durch Verkippen ins Meer** *s.* Beseitigung durch Abklappen
Beseitigung *f* **fester Abprodukte** solid-waste disposal
besetzen *v* populate, occupy, fill *(an energy level)*; occupy *(a lattice vacancy)*; charge *(e.g. a furnace)*
Besetzung *f* population, occupation, filling *(of an energy level)*; occupation *(of a lattice vacancy)*; charging *(as of a furnace)*
Besetzung *f***/inverse** inverted population
Besetzungsgrad *m* degree of population *(of an energy level)*
Besetzungsinversion *f* population inversion *(of an energy level)*
Besetzungsschema *n* population [filling] diagram *(of molecular orbitals)*
Besetzungszahl *f* occupation number *(number of electrons in a shell)*
besprengen *v* sprinkle, water
bespritzen *v* spray, sprinkle
besprühen *v* spray, dew
Bessemer-Konverter *m (met)* Bessemer converter
bessemern *v (met)* bessemerize, convert
Bessemer-Stahl *m* Bessemer [acid] steel
Bessemer-Verfahren *n (met)* Bessemer [converter] process, *(specif)* acid (Bessemer) process
Bessemer-Verfahren *n***/basisches** basic Bessemer [converter] process, basic process, Thomas(-Gilchrist) process
Bessemer-Verfahren *n***/saures** acid Bessemer [converter] process, acid process
beständig 1. resistant, resisting, stable *(to an agent)*; 2. *(dye, text)* fast, proof; persistent *(biocide)*; durable, stable *(material)* • **gut beständig sein** last well
beständig/chemisch chemically resistant [stable], resistant [stable] to chemical attack

beständig gegen Alkalien resistant to alkali(es), alkali-resistant, alkali-stable
beständig gegen oxidative Einflüsse resistant [stable] to oxidation, oxidatively stable
beständig gegen Säuren resistant to acid(s), acid-resistant, acid-stable
beständig/thermisch heat-resistant, thermally stable *(e.g. plastics)*
Beständigkeit *f* 1. resistance, stability *(to an agent)*; 2. *(dye, text)* fastness, proofness; persistence *(of a biocide)*; durability, stability *(of a material)*
Beständigkeit *f***/chemische** chemical resistance [stability], resistance [stability] to chemical attack
Beständigkeit *f* **gegen Alkalien** resistance [stability] to alkali(es), alkali resistance
Beständigkeit *f* **gegen Chemikalien** resistance to chemicals
Beständigkeit *f* **gegen hohe Temperaturen** resistance to high temperature(s), high-temperature resistance [stability]
Beständigkeit *f* **gegen Lösungsmittel** resistance to solvents, solvent resistance
Beständigkeit *f* **gegen oxidative Einflüsse** resistance to oxidation, oxidation [oxidative] resistance
Beständigkeit *f* **gegen Säuren** resistance [stability] to acid(s), acid resistance
Beständigkeit *f***/thermische** heat resistance, thermal stability *(as of plastics)*
Bestandteil *m* constituent, component, ingredient
Bestandteil *m***/flüchtiger** volatile [fugitive] constituent
Bestandteil *m***/giftiger** toxic principle
Bestandteil *m***/nicht zuckerartiger** nonsugar, *(esp)* aglycon *(of a glycoside)*
Bestandteil *m***/wesentlicher** integral [essential] constituent
Bestandteil *m***/wirksamer** active ingredient [principle]
Bestandteile *mpl***/flüchtige** volatile matter, v.m., VM, volatiles
Bestandteile *mpl***/nicht flüchtige** nonvolatile matter
bestäuben *v* dust, powder
bestehen *v* **aus** be made up of, consist of
bestimmbar determinable, *(qualitatively also)* identifiable
Bestimmbarkeit *f* determinability, *(qualitatively also)* identifiability
bestimmen *v* determine
bestimmen *v***/qualitativ** determine qualitatively, identify
bestimmen *v***/quantitativ** determine quantitatively, quantitate, estimate
Bestimmung *f* determination
Bestimmung *f* **der relativen Molekülmasse** molecular-weight determination
Bestimmung *f* **des organisch gebundenen Kohlenstoffs** *(hyd)* TOC determination
Bestimmung *f* **nach Mohr** Mohr titration [method] *(argentometry)*
Bestimmung *f* **nach Volhard** Volhard titration [method] *(argentometry)*
Bestimmung *f***/nochmalige** redetermination
Bestimmung *f***/qualitative** qualitative determination, identification
Bestimmung *f***/quantitative** quantitative determination, quantitation, estimation
bestrahlen *v* 1. (ir)radiate; 2. *(nucl)* bombard
bestrahlen *v***/mit Röntgenstrahlen** X-ray
Bestrahlung *f* 1. (ir)radiation; 2. *(nucl)* bombardment
Bestrahlung *f* **mit energiereicher [harter] Strahlung** high-energy irradiation
Bestrahlungsstärke *f (spectr)* irradiance
Betacellulose *f* beta cellulose
BET-Adsorptionsisotherme *f* BET isotherm
Betain *n* betaine *(any of a class of zwitterionic compounds)*; *(specif)* betaine, *NNN*-trimethylammonioacetate, oxyneurine, lycine
Beta-Lactam-Antibiotikum *m (biot)* β-lactam antibiotic
Betarückstreuung *f* beta backscatter(-ing)
Betaspektrum *n* beta-ray spectrum
Betastrahlen *mpl* beta rays
Betastrahler *m* beta emitter
Betastrahlung *f* beta radiation

Betastruktur *f* sheet structure *(of proteins)*
Betateilchen *n* beta particle
betätigt/elektromechanisch electromechanically operated
betätigt/hydraulisch hydraulic-operated
betätigt/manuell manually operated
Betätigungsorgan *n* actuator
betäubend anaesthetic
Betazerfall *m* beta decay [disintegration], beta-ray decay [disintegration]
BET-Gleichung *f (phys ch)* BET equation, Brunauer-Emmett-Teller relationship
BET-Isotherme *f* BET isotherm
Beton *m* concrete
Beton *m***/armierter [bewehrter]** reinforced concrete
Beton *m***/fetter** rich [good] concrete
Beton *m***/magerer** lean(-mixed) concrete, poor concrete
Beton *m***/ultrahochfester** ultra-high performance concrete
Betonzerfall *m* concrete disintegration
Betonzuschlag(stoff) *m* concrete aggregate
Betrag *m* amount, quantum, value
betreiben *v***/adiabat** operate [run] adiabatically
betreiben *v***/isotherm** operate [run] isothermally
Betrieb *m* 1. plant, works, factory; 2. operation • **außer Betrieb (befindlich)** out of operation, idle • **außer Betrieb setzen** shut (down), close down *(a factory)*; cut out of service, put out of operation, stop *(a machine)* • **in Betrieb (befindlich)** in operation, at [in] work, working, on-stream • **in Betrieb nehmen** put [set] in operation, set in action, start (up), prime *(a machine)* • **in Betrieb sein** be in operation, run, be running [working, operating] *(of a machine)* • **in Betrieb setzen** put [set] in operation, set in action, start (up), prime *(a machine)*
Betrieb *m***/chargenweiser** *s.* Betrieb/diskontinuierlicher
Betrieb *m***/chemischer** chemical plant [works]
Betrieb *m***/diskontinuierlicher** batch operation
Betrieb *m***/kontinuierlicher** continuous operation
Betrieb *m***/wartungsfreier** unattended operation
betrieben/diskontinuierlich discontinuous, *(Am)* batch
betrieben/kontinuierlich continuous
Betriebsabwasser *n (hyd)* plant waste water, plant [factory, works] effluent, plant discharge
Betriebsanlage *f* plant, works
Betriebschemiker *m* industrial [works] chemist
Betriebsdaten *pl* operating parameters [characteristics]
Betriebsdauer *f* operating [working] life
Betriebsdruck *m* operating [working] pressure
Betriebsinhalt *m* 1. *(tech)* hold-up; 2. *(distil)* column hold-up
Betriebsparameter *mpl s.* Betriebsdaten
Betriebspunkt *m (tech)* working point
Betriebsreaktor *m* commercial(-scale) reactor
Betriebssicherheit *f* safety of operation
Betriebsstörung *f* upset, stoppage, breakdown
Betriebstemperatur *f* operating [working] temperature
Betriebsunterbrechung *f* downtime, down period
Betriebsverhalten *n* performance
Betriebsvolumen *n (biot)* working [fermenter, fermentation] volume
Betriebswasser *n s.* Brauchwasser
Betriebsweise *f* mode of operation
Bett *n (tech, geol)* bed
Bett *n***/ruhendes [statisches]** *(tech)* fixed [static] bed
Bett *n***/wallendes** *(tech)* ebullating [ebullated] bed *(special form of a fluidized bed)*
Betthöhe *f* 1. *(tech)* bed depth [height, level]; 2. *(hyd)* media depth
Bettiefe *f* 1. *(tech)* bed depth [height, level]; 2. *(hyd)* media depth
Betweenanen *n* betweenanene *(any of a class of bicyclic alkenes)*
beuchen *v (text)* kier-boil, bowk, buck
Beugung *f* diffraction
Beugung *f* **am Kristall(gitter)** crystal diffraction

Beugung *f* **hochenergetischer Elektronen** *s.* Beugung schneller Elektronen
Beugung *f* **langsamer [niederenergetischer] Elektronen** low-energy electron diffraction, LEED
Beugung *f* **schneller Elektronen** high-energy electron diffraction, HEED
Beugungsbild *n* diffraction pattern
Beugungsgitter *n* diffraction grating
Beugungsspektrum *n* diffraction [normal] spectrum
Beugungswinkel *m* diffraction angle
Be- und Entlüftungseinrichtung *f* venting device *(for tanks)*
Beurteilung *f***/visuelle** visual examination
Beutel *m* bag
Beutelfilter *n* bag filter
beuteln *v* bolt, sift *(e.g. flour)*
bevorzugt/energetisch energetically favourable [favoured]
bewässern *v* water, irrigate
Bewässerung *f* watering, irrigation
bewegen *v* move; agitate *(e.g. a reaction mixture)*; *(rapidly up and down or to and fro:)* jig
beweglich mobile, movable
Beweglichkeit *f* mobility
Beweglichkeit *f***/elektrophoretische** electrophoretic mobility
Bewegtbettreaktor *m* moving-bed reactor
Bewegung *f* motion; agitation *(as of a reaction mixture)*
Bewegung *f***/brownsche** Brownian motion [movement]
Bewegung *f***/pulsierende** pulsation
bewehren *v* reinforce
Bewehrung *f* reinforcement
Beweis *m* proof
beweisen *v* proof
bewerten *v* evaluate, rate *(properties)*
Bewertung *f* evaluation, rating *(of properties)*
bewittern *v* weather
Bewitterung *f* 1. weathering; 2. *s.* Bewitterungsbeanspruchung
Bewitterungsbeanspruchung *f* outdoor weathering [exposure], atmospheric exposure
Bewitterungsgerät *n* weathering device
Bewitterungsversuch *m* weathering test, *(esp)* outdoor weathering [exposure] test
Bewuchs *m* fouling, marine growth
bezeichnen *v* 1. *(nomencl)* notate, designate; 2. *s.* benennen
Bezeichnung *f* 1. *(nomencl)* notation, designation; 2. *s.* Benennung
Bezeichnungssystem *n (nomencl)* system of notation, notation system
Bezeichnungsweise *f***/stocksche** Stock notation [scheme], Stock's system *(of indicating valencies)*
Beziehung *f* **der freien Enthalpie/lineare** linear free-energy relation(ship), LFE relationship, linear Gibbs energy relation
beziffern *v* number, index, indicate, label
Bezifferung *f* numbering, indexing, indication, labelling
Bezirk *m* 1. region, range *(as of measurement or state)*; 2. *(cryst)* domain *(in ferromagnetic substances)*
Bezirk *m***/kristalliner** crystalline region
Bezirk *m***/weißscher** *(cryst)* Weiss (molecular magnetic) field, ferromagnetic domain
Bezugsbasis *f* **Masse des feuchten Stoffs** wet(-weight) basis, WWB
Bezugsbasis *f* **Trockenmasse** bone-dry-weight basis, BDWB, dry(-weight) basis
Bezugselektrode *f* reference [comparison] electrode
Bezugslinie *f (spectr)* standard [comparison] line
Bezugssignal *n (anal)* reference peak
Bezugsspannung *f* reference voltage
Bezugsstandard *m* reference standard
Bezugssubstanz *f* reference [standard] substance [compound]
BHA *s.* Butylhydroxyanisol
B-Harz *n* B-stage resin, resitol
Biacetyl *n* biacetyl, *butanedione
Biaryl *n (org ch)* biaryl
biaxial *(cryst)* biaxial
Bibenzyl *n* bibenzyl, *sym*-diphenylethane
bicyclisch bicyclic, dicyclic
Bicyclus *m* bicyclic compound
Biegefestigkeit *f* bending strength, transverse [flexural] strength
Biegemodul *m* flexural modulus
biegen *v* bend, flex

Biegerissbildung *f* flex cracking
Biegerissfestigkeit *f* flex(-cracking) resistance, resistance to flex cracking
Biegerisswiderstand *m* flex(-cracking) resistance, resistance to flex cracking
Biegesteifigkeit *f* flexural rigidity, stiffness in bend [flexure]
Biegeversuch *m (plast)* bending test
biegsam flexible, pliable, pliant
Biegsamkeit *f* flexibility, pliability
Bienengift *n* bee venom
Bienenharz *n* bee glue, propolis, balm
Bienenwachs *n* beeswax
Bier *n* beer, *(collectively also)* malt beverages; *(if top-fermented and strongly hopped:)* ale
Bier *n***/obergäriges** top-fermented [top-fermentation] beer
Bier *n***/untergäriges** bottom-fermented [bottom-fermentation] beer
Bierbrauen *n* brewing
Bierbrauerei *f* brewery
Bierhefe *f* beer yeast, brewing [brewer's, brewery] yeast
Bierwürze *f* brewer's wort
bifunktionell bifunctional, difunctional
Biguanid *n* biguanide, diguanide
Biharnstoff *m* biurea, dicarbamylhydrazine
Bikomponentenfaser *f (text)* bicomponent [conjugate] fibre
Bilanz *f* balance
Bilanzgleichung *f* balanced equation
bilanzieren *v* balance *(a reaction equation)*
Bilanzraum *m* control volume *(energy balance)*
bilden *v* form *(e.g. crystals or a precipitate)*; make up *(e.g. the main component)*
bilden *v***/Blasen** bubble *(of gas or water)*; blister *(of metal or paint)*; vesicate, blister *(of the skin)*
bilden *v***/Chelate** chelate
bilden *v***/ein Gel** gel(ate)
bilden *v***/ein Sol** solate
bilden *v***/eine Kruste** encrust, incrust
bilden *v***/einen Bodenkörper [Bodensatz]** *s.* bilden/einen Niederschlag
bilden *v***/einen Komplex** complex
bilden *v***/einen Niederschlag** (form a) precipitate, sediment, settle (down, out), deposit, subside; sediment, deposit *(of a solution)*
bilden *v***/Klumpen** clot, clog
bilden *v***/Kristalle** crystallize (out); form crystals
bilden *v***/Mizellen** micellize
bilden *v***/Schlacke** 1. slag; 2. *(coal)* clinker
bilden *v***/sich** form, *(esp of crusts or layers also)* build up
bildsam plastic, *(esp relating to metal:)* ductile
Bildsamkeit *f* plasticity, *(esp relating to metal:)* ductility
Bildung *f* formation *(as of crystals or a precipitate)*; *(esp of crusts or layers also)* build-up
Bildung *f* **eines Bodenkörpers [Niederschlags]** sedimentation, deposition
Bildung *f***/erneute** reformation
Bildungsenergie *f* energy of formation
Bildungsenthalpie *f* enthalpy of formation, heat of formation at constant pressure
Bildungsgeschwindigkeit *f* rate of formation [generation], rate of appearance
Bildungskonstante *f* formation constant
Bildungsrate *f***/spezifische** *(biot)* specific production efficiency [rate], specific rate of product formation *(mg product/mg cell dry weight · h)*
Bildungsreaktion *f* formation reaction
Bildungswärme *f* heat of formation
Bildungsweg *m (bioch)* anabolic route
Bilirubin *n (bioch)* bilirubin
Biliverdin *n (bioch)* biliverdin
Billiter-Zelle *f* Billiter cell *(electrolysis)*
Bimetall *n* bimetal
bimetallisch bimetallic
Bimetallthermometer *n* bimetallic thermometer
bimolekular bimolecular
Bimsstein *m* pumice (stone)
bimssteinartig pumiceous
Bindemittel *n* 1. binding [bonding] agent [material], binder; adhesive agent [substance], adhesive; 2. *(pharm)* excipient *(as for pills)*; 3. *(geol)* cement, cementing agent, agglutinant
Bindemittellösung *f* (coating, paint) vehicle, carrier, medium

binden *v* 1. bond, link, bind *(atoms)*; 2. adsorb *(dust particles)*; absorb *(gases)*; 3. *(bot, soil)* fix *(atmospheric nitrogen)*; 4. *(geol)* cement, agglutinate *(particles)*; 5. *s.* abbinden

binden *v*/**komplex** complex

binden *v*/**koordinativ** coordinate *(atoms or molecules)*

binden *v*/**sich** bind *(to a molecule)*

Binder *m (build)* binder, binding agent [material]

Bindigkeit *f* covalence *(chemical-bond theory)*

Bindung *f* 1. bonding, linkage, binding *(of atoms)*; adsorption *(of dust particles)*; absorption *(of gases)*; 2. *(bot, soil)* fixation *(of atmospheric nitrogen)*; 3. bond *(between atoms)* • **eine Bindung eingehen** bond • **eine Bindung herstellen** make a bond

Bindung *f*/**äquatoriale** equatorial bond, e-bond

Bindung *f*/**axiale** axial bond, a-bond

Bindung *f*/**chemische** 1. chemical bonding; 2. chemical bond

Bindung *f*/**dative (koordinative)** *s.* Bindung/koordinative 2.

Bindung *f*/**delokalisierte** delocalized bond

Bindung *f*/**gebogene** bent [banana] bond, banana-like [banana-shaped] bond

Bindung *f*/**glykosidische** glycosidic bond

Bindung *f*/**halbpolare** *s.* Bindung/koordinative 2.

Bindung *f*/**heteropolare** (hetero)polar bond, ionic [electrostatic, electrovalent] bond, electrovalence

Bindung *f*/**homöopolare** homopolar [atomic] bond, covalent [non-polar] bond, (shared-)electron-pair bond

Bindung *f*/**hydrophobe** *(bioch)* hydrophobic interaction, *(deprecated:)* hydrophobic bond

Bindung *f*/**koordinative** 1. coordination; 2. dipolar [coordinate, dative] bond, donor-acceptor bond

Bindung *f*/**kovalente** *s.* Bindung/homöopolare

Bindung *f*/**lokalisierte** localized bond

Bindung *f*/**metallische** metal(lic) bond

Bindung *f*/**peptidartige** peptide bond

Bindung *f*/**polare** *s.* 1. Bindung/heteropolare; 2. Bindung/polare kovalente

Bindung *f*/**polare kovalente** polar (covalent) bond

Bindung *f*/**schwache** weak bond

Bindung *f*/**semipolare** *s.* Bindung/koordinative

Bindung *f*/**unpolare (kovalente)** *s.* Bindung/homöopolare

Bindung *f*/**van-der-waalssche** van der Waals bond

δ-Bindung *f* δ bond, delta bond

π-Bindung *f* π bond, pi bond

σ-Bindung *f* σ bond, sigma bond

Bindungsabstand *m s.* Bindungslänge

Bindungsbildung *f* bond formation

Bindungsbruch *m* bond breakage [breaking]

Bindungsdipolmoment *n* bond dipole moment

Bindungsdissoziationsenergie *f* bond-dissociation energy

Bindungselektron *n* bonding electron, valency [outermost, optical] electron

Bindungselektronenpaar *n* bonding [sharing] electron pair, bonding pair of electrons, shared pair of electrons

Bindungsenergie *f* bond energy, bonding energy

Bindungsenergie *f*/**mittlere** mean bond energy, bonding energy

Bindungsfestigkeit *f* bond(ing) strength

Bindungsgrad *m* bond number [order]

Bindungskraft *f* bond(ing) force, combining force

Bindungslänge *f* bond length [distance]

Bindungslockerung *f* antibonding

Bindungsorbital *n* bond orbital

Bindungsordnung *f s.* Bindungsgrad

Bindungsort *m* binding site

Bindungsparachor *m* structural parachor *(chemical-bond theory)*

Bindungsprotein *n (bioch)* binding protein

Bindungsprotein *n*/**periplasmatisches** periplasmic binding protein

Bindungsschwingung *f* bond vibration

Bindungsspaltung *f* bond cleavage [fission, scission, dissociation]

Bindungsspaltung *f*/**heterolytische** heterolysis, heterolytic (bond) cleavage, heterolytic (bond) fission

Bindungsspaltung *f*/**homolytische** ho-

molysis, homolytic (bond) cleavage, homolytic (bond) fission, bond dissociation
Bindungsspaltungsenergie *f* bond dissociation energy
Bindungsspezifität *f* bond [relative group] specificity *(of an enzyme)*
Bindungssphäre *f* boundary [bounding] surface *(chemical-bond theory)*
Bindungsstärke *f* bond(ing) strength
Bindungsstelle *f s.* Bindungsort
Bindungsumgruppierung *f* bond rearrangement
Bindungswertigkeit *f s.* Bindigkeit
Bindungswinkel *m* bond [valence] angle, interbonding angle
Bindungszahl *f* bond number
Bingham-Körper *m* Bingham body [plastic]
Binnendruck *m* cohesion [internal, intrinsic] pressure
Binodal-Kurve *f (phys ch)* binodal curve, conodal [coexistence] curve
bioabbaubar biodegradable
Bioaffinitätschromatographie *f* affinity chromatography
Bioaktivität *f* biological activity
bioanorganisch bioinorganic
Biobergbau *m (biot)* microbial leaching, microbiological [bacterial] leaching
Biochemie *f* biochemistry, biological chemistry
Biochemikalie *f* biochemical (product)
Biochemiker *m* biochemist
biochemisch biochemical
Biochip *m (biot)* biochip, biologically based microchip
Biodiesel *m* biodiesel *(biofuel)*
Bioelektrizität *f* bioelectricity
Bioenergie *f* bioenergy
Bioethanol *m* bioethanol *(biofuel)*
Biofilm *m* biofilm
Biofilter *n* biofilter, biological filter
Biogas *n* 1. biogas; 2. *(hyd)* digester [sludge] gas, sewage [sewer] gas
Biogasreaktor *m* biogas [anaerobic] digester
biogen biogenic, biogenous
biogeochemisch biogeochemical
Bioindikation *f* bioindication
Bioindikator *m* bioindicator
Bioingenieurtechnik *f s.* Bioverfahrenstechnik
Bioingenieurwesen *n s.* Bioverfahrenstechnik
Bioinsektizid *n (biot)* bioinsecticide, microbial insecticide
Biokatalysator *m* biocatalyst, biochemical catalyst, ergone
Biokatalyse *f* biocatalysis
biokatalytisch biocatalytic
Biokinetik *f* biokinetic
Biokompatibilität *f* biocompatibility
Biokonversion *f (biot)* bioconversion, biochemical [microbial] conversion, biotransformation, biochemical [microbial] transformation
Biokorrosion *f* biocorrosion
Biokraftstoff *m* biofuel, biomass fuel
Bioleaching *n s.* Biobergbau
biologisch verträglich biocompatible
biolumineszent bioluminescent
Biolumineszenz *f* bioluminescence
Biomakromolekül *n s.* Biopolymer(e)
Biomasse *f* biomass
Biomasse *f***/mikrobielle** microbial mass
Biomasse *f***/pflanzliche** plant biomass
Biomassekonzentration *f* biomass concentration; microbial concentration, concentration of microorganisms [suspended biomass]
Biomembran *f* biological membrane
Biomimetik *f (org ch)* biomimetics
biomimetisch *(org ch)* biomimetic
Biomineralisation *f (bioch)* biomineralization
Biomolekül *n (bioch)* biomolecule
bioorganisch bioorganic
Biooxidation *f* bio-oxidation, biological oxidation
Biopolymer(e) *n* biopolymer, biological polymer, biomacromolecule
Bioprozesstechnik *f* bioprocess technology
Bioreaktor *m (biot)* bioreactor, biological [biochemical microbiological] reactor
Biosäule *f (biot)* column reactor
Bioschlamm *m (hyd)* biological sludge, activated [active] sludge
Biose *f* biose *(monosaccharide containing two carbon atoms)*
Biosensor *m (biot)* biosensor, biologically based sensor
Biosignal *n* biological signal
Biosorption *f* biosorption
Biosynthese *f* biosynthesis

Biosynthese f/gerichtete *(biot)* directed biosynthesis
Biosynthese f/gesteuerte [gelenkte] *(biot)* controlled biosynthesis
Biosyntheseleistung *f (biot)* efficiency of biosynthesis
Biosyntheseweg *m* biosynthetic [biosynthesis] pathway [route]
Biotechnik *f* biotechnology
biotechnisch biotechnological
Biotechnologie *f* biotechnology
Biotechnologie f/blaue blue biotechnology *(sea products)*
Biotechnologie f/graue grey biotechnology *(waste control)*
Biotechnologie f/grüne green biotechnology *(agriculture)*
Biotechnologie f/rote red biotechnology *(medical processes)*
Biotechnologie f/weiße white biotechnology *(industrial processes)*
biotechnologisch biotechnological
Biotest *m (tox)* bioassay (test), biological assay [test]
biotisch biotic
Biotransformation *f s.* Biokonversion
Biotreibstoff *m (biot)* biofuel
Biotrockenmasse *f (biot)* dry weight of biomass
Bioverfahrenstechnik *f* bioengineering
Bioverfügbarkeit *f (pharm)* bioavailability
biozid biocidal *(killing organisms)*
Biozid *n* biocide *(chemical which kills organisms)*; *(in a narrower sense)* pesticide, (pest) control agent
Biphenyl *n* biphenyl, phenylbenzene
Biphenyle *npl*/**polychlorierte** polychlorinated biphenyls, PCB's
Bipolarzelle *f* bipolar cell *(electrolysis)*
Biradikal *n* biradical, diradical
Birch-Hückel-Reaktion *f* Birch reduction *(of organic compounds by metallic sodium dissolved in liquid ammonia)*
Birkeland-Eyde-Verfahren *n* Birkeland-Eyde process *(for manufacturing nitric acid)*
Birkenöl *n* (sweet-)birch oil
Birkenteeröl *n* birch tar oil
Bischler-Napieralski-Reaktion *f* Bischler-Napieralski reaction *(for synthesizing isoquinoline)*
Bis-harnstoff *m s.* Biharnstoff
Bismut *n* Bi bismuth
Bismutan *n* Bi_nH_{n+2} bismutane, bismuth hydride, *(specif)* BiH_3 bismuthane, bismuthine, bismuth hydride
Bismutat *n* bismuthate
bismuthaltig bismuth-containing, *(esp relating to ores:)* bismuthiferous, bismuth-bearing
Bismuthydrid *n s.* Bismutan
Bismut(III)-hydroxid *n* bismuth(III) hydroxide, bismuth trihydroxide
Bismutin *n* 1. BiR_3 bismuthine *(any of several organic compounds)*; 2. BiH_3 bismuthane, bismuthine, bismuth hydride
Bismutit *m (min)* bismutite *(bismuth carbonate oxide)*
Bismut(III)-oxid *n* bismuth(III) oxide, bismuth trioxide
Bismut(V)-oxid *n* bismuth(V) oxide, bismuth pentaoxide
Bismutoxidchlorid *n* bismuth chloride oxide
Bismutpentoxid *n s.* Bismut(V)-oxid
Bismutsäure *f* bismuthic acid
Bismut(III)-sulfid bismuth(III) sulphide, bismuth trisulphide
Bismuttri... *siehe: Bismut(III)-...*
Bismutwasserstoff *m s.* Bismutan
Bisphenol *n* bisphenol, methylenediphenol
bistabil bistable *(reacting system)*
Bistabilität *f* bistability *(of a reacting system)*
Bisulfit-Additionsverbindung *f* bisulphite addition compound
Bisulfitaddukt *n* bisulphite addition compound
Bisulfitzellstoff *m (pap)* sulphite pulp
Bittermandelöl *n* bitter almond oil, amygdala amara oil
Bittersalz *n* 1. bitter salt, Epsom salt(s); 2. *(min)* epsomite *(magnesium sulphate 7-water)*
Bitterstoff *m* 1. bitter principle [substance], linin; 2. *(ferm)* bittern
Bitumen *n* bitumen, *(specif)* asphaltic bitumen
Bitumen n/geblasenes *s.* Blasbitumen
Bitumenanstrichstoff *m* bituminous paint [coating]
bitumenhaltig bituminous
Bitumensand *m* bituminous sand

Bitumenschiefer *m* bituminous [oil] shale
bituminös bituminous
bitumisch bituminous
Biuret *n* biuret, ureidoformamide
Biuretreaktion *f* biuret reaction *(for determining proteins)*
bivalent bivalent, divalent
Bivalenz *f* bivalence, divalence
bivariant bivariant, divariant
Bixin *n (food)* bixin *(dye from Bixa orellana L.)*
B-Kation *n* class (b) metal ion
Blanc fixe *n* blanc fixe, permanent white *(precipitated barium sulphate)*
Blanchierapparat *m (food)* blancher
blanchieren *v (food)* blanch
Blankfilter *n* polishing [clarifying] filter
Blankfiltration *f* polishing (filtration), clarification
blankfiltrieren *v* polish
blankglühen *v* bright-anneal
Blasbitumen *n* blown bitumen, (air-) blown asphalt, mineral rubber
Bläschen *n* bubble
Bläschenbildung *f* bubbling
Blase *f* 1. bubble; blister *(a defect in material)*; 2. *(met)* blow-hole, gas cavity; 3. *(plast)* void; vesication, blister *(on the skin)*; 4. distillation boiler, still pot, *(with rectifying apparatus:)* reboiler • **Blasen bilden** bubble *(of gas or water)*; blister *(of metal or paint)*; vesicate, blister *(of the skin)* • **in Blasen aufsteigen** bubble (up) • **in Blasen aufsteigen lassen** bubble
Blasegas *n* blow [blast] gas *(in manufacturing producer gas)*
blasen *v* blow
Blasenbildung *f* 1. bubbling, formation of bubbles; 2. *(tech)* blistering, blister formation; 3. *(med)* vesication, blistering • **Blasen ziehend** *(med)* vesicant, vesicatory, blistering • **zur Blasenbildung reizen** *(med)* vesicate, blister
Blasendestillierapparat *m* pot still
Blasenkammer *f (nucl)* bubble chamber
Blasensäule *f* bubble column
Blasensäulenfermenter *m (biot)* bubble column [tower] fermenter
Blasensäulenreaktor *m* bubble column
Blasenzähler *m (lab)* bubble counter
Blasfolie *f (plast)* blown film
Blasformen *n (plast)* blow moulding
Blasformteil *n (plast)* blow moulding
Blashochofen *m (met)* blast furnace
Blaskopf *m* blow head
Blasluft *f* blow air
Blasrohr *n* blow-pipe
Blasstahl *m* basic oxygen (furnace) steel
Blasstahlverfahren *n* basic oxygen (converter) process, basic oxygen furnace [steel] process, top-blown oxygen converter process, oxygen process of steelmaking, oxygen-lance process
Blatt *n (tech)* leaf *(of a filter or of metal)*; paddle, blade, shovel *(as of an agitator)*
Blattapplikation *f (agric)* foliage application *(as of pesticides)*
Blättchen *n (cryst)* leaflet
blättchenförmig foliate, leaf-like
Blattfilter *n* leaf filter
blattförmig foliate, leaf-like
Blattgold *n* gold leaf
Blattgrün *n* leaf green, chlorophyll
blättrig 1. lamellar, foliated; 2. *(min)* spathic, spathose
Blattrührer *m* vane stirrer, paddle [leaf] agitator
Blattsilicat *n (min)* phyllosilicate
Blattverbrennungen *fpl (agric)* foliage burn *(as by pesticides)*
Blau *n*/**Berliner** *(ch)* Prussian [Berlin] blue, *(commercially)* iron [cyanide] blue *(a complex iron cyanide)*
Blau *n*/**Bremer** Bremen blue, blue verditer, copper blue *(a basic copper carbonate)*
Blau *n*/**Pariser** *s.* Blau/Berliner
Blau *n*/**Thénards** Thénard blue, cobalt blue [ultramarine], king's blue *(cobalt aluminate)*
Blauanlaufen *n* blueing *(of tools)*; blooming *(of oil varnishes)*
Blaueisenerz *n (min)* vivianite, blue iron ore
blauempfindlich *s.* blausensibilisiert
bläuen *v* blue
Blaufarbenglas *n* smalt, powder blue *(cobalt(II) potassium silicate)*
Blaufarbstoff *m* woad
Blaumasse *f (text)* cuprammonium cellulose
Blausäure *f* hydrogen cyanide, hydrocyanic acid

Blausäureglykosid *n* cyanogenetic [cyanophoric] glycoside
Blausäureoxid *n* formonitrile oxide, fulminic acid
blausensibilisiert blue-sensitive, blue-sensitized
blaustichig bluish
Blauton *m* blue shade
Bläuungsmittel *n* blueing (agent) *(for improving the degree of whiteness)*
Blauverschiebung *f* blue-shift, hypsochromic effect
Blaze-Gitter *n (spectr)* blazed grating
Blech *n (met)* sheet, plate
Blei *n* Pb lead • **mit Blei auskleiden** lead-line, lead-clad • **mit Blei überziehen** lead-coat
Blei(II)-acetat *n* lead(II) acetate, lead diacetate
Bleiakkumulator *m* lead [lead-acid] accumulator [battery], lead storage battery
Bleialkalisilicatglas *n* lead-alkali silicate glass
Bleiauskleidung *f* lead lining, internal lead cladding
Bleibaum *m (el ch)* lead tree • **einen Bleibaum bilden** tree
Bleibenzin *n* leaded [ethyl] gasoline
Bleichapparat *m* bleaching apparatus; bleacher
Bleichbad *n s.* Bleichflotte
bleichbar bleachable
Bleichbottich *m* bleaching vat [chest]
Bleichchlor *n***/aktives [wirksames]** *(pap)* available chlorine
Bleiche *f* bleach(ing)
bleichecht fast to bleach(ing), resistant to bleaching
Bleichechtheit *f* fastness to bleach(ing), resistance to bleaching, bleach-fastness
bleichen *v* 1. *(of human agent:)* bleach, whiten, brighten, decolorize; 2. *(of substances:)* bleach, fade
Bleicherde *f* bleaching [decolorizing] earth [clay], active earth
Bleicherde *f***/aktivierte [künstlich aktive]** *(petrol)* activated clay
Bleicherdebehandlung *f* clay treating [treatment] *(of oils)*
Bleicherdekontakt *m* clay catalyst [contact]
Bleicherderaffination *f s.* Bleicherdebehandlung
Bleichflotte *f (text, pap)* bleaching liquor [lye, solution]
Bleichflüssigkeit *f (text, pap)* bleaching liquor [lye, solution]
Bleichhilfsmittel *n* bleaching assistant
Bleichkalk *m* chlorinated lime, chloride of lime, bleaching powder
Bleichkessel *m* bleaching vat [chest]
Bleichlauge *f* bleaching liquor [solution]
Bleichlauge *f***/javellesche** Javel(le) water *(aqueous solution of potassium hypochlorite)*
Blei(II)-chlorid *n* lead(II) chloride, lead dichloride
Bleichmittel *n* bleaching agent [material], bleach; decolorizing agent, decolorizer, decolorant
Bleichmittel *n***/optisches** optical bleaching [brightening] agent, optical [fluorescent] brightener [bleach]
Blei(II)-chromat *n* lead(II) chromate
Bleichton *m* bleaching clay
Bleichturm *m (pap)* bleaching tower, bleacher
Bleidi... *siehe: Blei(II)-...*
bleien *v* lead *(fuels)*
Bleifarbe *f* lead paint
bleifrei lead-free, leadless
Bleiglanz *m (min)* lead glance, galena, galenite *(lead(II) sulphide)*
Bleiglas *n* lead glass
Bleiglätte *f* litharge, yellow lead oxide *(lead(II) oxide)*
bleihaltig lead-containing, *(esp ores:)* plumbiferous
Bleikammerverfahren *n* (lead-)chamber process *(for producing sulphuric acid)*
Bleikontamination *f (tox)* lead contamination
Bleikrankheit *f s.* Bleivergiftung
Bleikristallglas *n* lead crystal glass
Bleilegierung *f* lead alloy
Bleimennige *f* red lead (oxide), minium *(lead(II) tetraoxoplumbate(IV))*
Bleimonoxid *n s.* Blei(II)-oxid
Blei(II)-nitrat *n* lead(II) nitrate
Bleioxid *n***/rotes** *s.* Bleimennige
Blei(II)-oxid *n* lead(II) oxide, lead monooxide
Blei(II,IV)-oxid *n* 1. trilead tetraoxide,

lead(II) tetraoxoplumbate(IV); 2. dilead trioxide, lead(II) trioxoplumbate(IV)
Bleioxidrot *n s.* Bleimennige
Bleiraffination *f* lead refining
Bleirohr *n* lead pipe
Bleisammler *m s.* Bleiakkumulator
Bleischachtofen *m* lead blast furnace
Bleischutzschicht *f* lead coating
Blei(II)-sulfat *n* lead(II) sulphate
Blei(II)-sulfid *n* lead(II) sulphide
Bleiüberzug *m s.* Bleischutzschicht
Bleivergiftung *f* lead poisoning, saturnism, plumbism
Bleiwasserstoff *m* lead hydride, plumbane
Bleiweiß *n* white lead, ceruse *(lead carbonate hydroxide)*
Bleizucker *m* sugar of lead, salt of Saturn *(lead(II) acetate)*
Blende *f* 1. shield, screen; orifice *(of a flowmeter)*; 2. *(spectr)* attenuator; 3. *(min)* blende *(any of several sulphidic minerals)*; 4. *s.* Zinkblende
blenden *v (petrol)* blend
blind dull, tarnished *(surface)* • **blind werden** tarnish
Blindbestimmung *f* blank determination
Blindflansch *m* blind flange, blank
Blindlösung *f* blank solution
Blindprobe *f s.* Blindversuch
Blindrohr *n* dummy tube
Blindversuch *m* blank (test), negative control • **einen Blindversuch anstellen [durchführen]** run a blank, perform a blank test
Blindwerden *n* tarnish(ing) *(of surfaces)*
Blindwert *m* 1. blank reading [value, measure]; 2. *(tox)* predosage level
Blitzlampe *f* flash lamp
Blitzlichtphotolyse *f* flash photolysis
Blitzpyrolyse *f (org ch)* flash pyrolysis
Blitzverdampfung *f* instantaneous vaporization
Bloch-Wand *f (cryst)* Bloch wall, domain wall
Block *m* 1. block *(e.g. a piece of material)*; 2. *(met)* ingot *(of iron)*; pig *(of nonferrous metal)*; 3. block *(comprising equal or similar elements)*; 4. *(bioch)* block *(site, state, or process)*
Blockcopolymer(e) *n* block (co)polymer
Blockcopolymerisat *n* block (co)polymer
Blocker *m (bioch)* blocking agent
blockieren *v* block, mask *(reactive groups or sites)*; block *(reactions)*
Blockierung *f* blocking, masking *(of reactive groups or sites)*; blocking *(of reactions)*
blockig blocky
Blockpolymer(e) *n s.* Blockcopolymer(e)
Blockschaltbild *n* block diagram
Blume *f* bouquet, aroma, flavour *(as of wine)*
Blutalbumin *n s.* Blutserumalbumin
blutdrucksenkend hypotensive, antihypertensive
blutdrucksteigernd raising blood pressure, pressor
bluten *v (dye, coat, tann)* bleed (off, through)
Blutfarbstoff *m* blood pigment
Blutgerinnung *f* blood coagulation [clotting]
Blutgift *n* blood poison, *(as war gas also)* systemic poison
Blutlaugensalz *n***/gelbes** yellow prussiate of potash, yellow potassium prussiate, potassium hexacyanoferrate(II)
Blutlaugensalz *n***/rotes** red prussiate of potash, red potassium prussiate, potassium hexacyanoferrate(III)
Blutplasmaersatz *m* blood plasma extender
Blutserumalbumin *n* blood (serum) albumin, serum albumin, seralbumin
blutstillend styptic
Blutstillungsmittel *n* styptic
Blutzucker *m* blood sugar [glucose]
B-Metallion *n* class [b] metal ion
Boden *m* 1. *(tech)* bottom, base *(as of a vessel)*; 2. *(distil)* plate, tray; deck *(as of flat screens)*; 3. soil • **mit halbkugeligem [rundem] Boden** round-bottomed
Boden *m***/idealer** *s.* Boden/theoretischer
Boden *m***/praktischer** *(distil)* actual [practice] plate
Boden *m***/theoretischer** *(distil)* theoretical [perfect] plate
Bodenazidität *f* soil acidity
Bodenbegasungsmittel *n* soil fumigant
Bodenbildung *f (soil)* pedogenesis
Bodenchemie *f* soil chemistry
Bodendesinfektionsmittel *n* soil sterilant, *(specif)* soil fumigant

Bodenentleerung *f* bottom discharge
Bodenfeuchte *f* soil moisture
Bodenfeuchtigkeit *f* soil moisture
Bodenfraktion *f (distil)* bottom fraction
Bodenfruchtbarkeit *f* soil fertility
Bodenfungizid *n* soil fungicide
Bodengefüge *n* soil structure
Bodenhöhe *f (anal, distil)* plate height
Bodenhorizont *m* soil horizon
Bodenkolloid *n* soil colloid
Bodenkolonne *f (distil)* plate [tray] column
Bodenkörper *m* 1. excess solute, solid [undissolved] solute, excess [undissolved] solid *(in equilibrium with its solution)*; 2. *s.* Bodenprodukt
Bodenkörperregel *f (coll)* disperse-phase rule
Bodenkorrosion *f* soil corrosion
Bodenkunde *f* soil science, pedology
Bodenkundler *m* soil scientist, pedologist
bodenkundlich pedologic(al)
Bodenmikroorganismen *mpl* soil microorganisms
Bodennährstoff *m* soil nutrient
Boden-pH-Wert *m* soil pH
Bodenprobe *f* sample of soil
Bodenprodukt *n (distil)* bottom product, bottoms
Bodenreaktion *f* soil reaction
Bodensatz *m* 1. sediment, subsidence, (bottom) settlings, B.S., dregs, bottoms; 2. *(food, ferm)* lees, foots
Bodensäule *f s.* Bodenkolonne
Bodenschicht *f* bottom layer
Bodenstein-Prinzip *n* Bodenstein (steady-state) approximation, steady-state (approximation) method, stationary-state method
Bodenstruktur *f* soil structure
Bodenverbesserungsmittel *n* soil conditioner [ameliorant], (soil) amendment
Bodenverdichtung *f* soil compaction
Bodenversauerung *f* 1. soil souring [acidification]; 2. soil sourness
Bodenverschmutzung *f* soil contamination
Bodenwirkungsgrad *m (distil)* plate efficiency (factor), tray efficiency (factor)
Bodenzahl *f* 1. *(distil, chromat)* number of plates, plate number; 2. *(chromat)* plate count
Bodenzahl *f*/praktische [tatsächliche] number of actual plates [trays]
Bodenzahl *f*/theoretische theoretical plate number, number of theoretical plates [trays]
Bogen *m* 1. sheet *(of paper or pulp)*; 2. (electric) arc
Bogen... *siehe auch: Lichtbogen...*
Bogen-Anregung *f* arc excitation
Bogenentladung *f* arc discharge
Bogenspektrum *n* arc spectrum
Bohn-Schmidt-Reaktion *f* Bohn-Schmidt reaction *(for preparing polyhydroxyanthraquinones)*
bohren *v*/nach Erdöl bore [drill, prospect] for oil
Bohren *n (petrol)* drilling
Bohrium *n* Bh bohrium *(element 107)*
Bohrlochkopfgas *n* casing-head gas
Bohrung *f* 1. boring, bore, (bore)hole; perforation, orifice *(of a screen)*; 2. *(petrol)* (oil) well; 3. *s.* Bohren
Bolometer *n* bolometer *(a resistance thermometer)*
bolometrisch bolometric
Boltzmann-Konstante *f* Boltzmann constant
Boltzmann-Statistik *f* (Maxwell-)Boltzmann statistics
Boltzmann-Verteilung *f (phys ch)* Boltzmann distribution
Bolus *m*/weißer *(min)* white bole, bolus alba, china [porcelain] clay, kaolin(e)
bombardieren *v (nucl)* bombard
Bombardierung *f (nucl)* bombardment
Bombe *f* bomb; (pressure) cylinder *(for liquefied gas)*
Bombe *f*/berthelotsche *s.* Bombe nach Berthelot/kalorimetrische
Bombe *f*/kalorimetrische calorimeter [calorimetric, explosion] bomb, (oxygen) bomb calorimeter
Bombe *f* nach Berthelot/kalorimetrische Berthelot [Kröcker, Mahler] bomb
Bombenkalorimeter *n s.* Bombe/kalorimetrische
Bombenrohr *n* **(nach Carius)** *(lab)* Carius [bomb] tube
Bombensauerstoff *m* cylinder oxygen
Boot *n (lab)* boat
Bootform *f* boat [tub] form *(stereochemistry)*

Bootkonformation *f* boat [tub] conformation
Bor *n* B boron
Boran *n* borane, boron hydride, *(specif)* BH_3 borane(3)
Borat *n* borate, *(specif)* $M^I_3BO_3$ orthoborate, trioxoborate
Borax *m* 1. borax, sodium tetraborate 10-water; 2. *(min)* tincal
Borax *m***/gebrannter [kalzinierter]** calcined borax, burnt [anhydrous, dehydrated] borax, borax usta, sodium tetraborate
Borax *m* **usta** *s.* Borax/gebrannter
Boraxperle *f* borax bead
Borazin *n* borazine, *(specif)* $B_3N_3H_6$ borazine, borazole
Borazol *n s.* Borazin
Borcarbid *n* boron carbide
Borhydrid *n* boron hydride, borane
Borid *n* boride
Borin *n* borine *(any of a class of compounds B_nH_{n+2})*; *(specif)* BH_3 borane(3)
Bornan *n (org ch)* bornane
Borneol *n* borneol, bornyl alcohol, 2-hydroxybornane
Born-Haber-Kreisprozess *m* Born-Haber (thermochemical) cycle
Bornitrid *n* boron nitride
Born-Oppenheimer-Näherung *f* Born-Oppenheimer approximation *(for estimating electronic transitions)*
Bornylalkohol *m s.* Borneol
Borosilikatglas *n* borosilicate [hard] glass
Bor(III)-oxid *n* boron(III) oxide, boron trioxide, boric oxide
Borsäure *f* boric acid, *(specif)* $B(OH)_3$ orthoboric acid, trioxoboric acid
Borsäureester *m* boric-acid ester, borate ester
Borstahl *m* boron steel
Borverbindung *f* boron compound • **mit Borverbindungen versetzen** boronate *(fertilizers)*
Borwasserstoff *m* boron hydride, borane
Bose-Einstein-Statistik *f* Bose-Einstein statistics
Boson *n* boson
Boten-Ribonucleinsäure *f* messenger ribonucleic acid, messenger RNA
Boten-RNA *f* messenger ribonucleic acid, messenger RNA
Botenstoff *m (biol)* messenger
Bottich *m* tub, vat
Boudouard-Gleichgewicht *n* producer-gas equilibrium
Boudouard-Reaktion *f* air-carbon reaction
Bouillon *f (pharm)* broth
Bouillonkultur *f (pharm)* broth culture
Bourbonal *n* bourbonal, 3-ethoxy-4-hydroxybenzaldehyde
Bouveault-Blanc-Reduktion *f* Bouveault-Blanc reduction *(of esters to alcohols)*
BP *s.* Brennpunkt
Brackett-Serie *f (spectr)* Brackett series *(of the hydrogen atom)*
brackig brackish
Brackwasser *n* brackish water
Bradford-Bestimmung *f* Bradford assay *(protein chemistry)*
Bragg-Methode *f* Bragg method (of crystal analysis), Bragg rotating crystal method
Bramme *f (met)* slab
Brand *m* 1. burning; 2. *(ceram)* firing, burning, baking; 3. fire; 4. *(ceram)* burn *(batch of ceramic ware)*
Brandfleck *m* scorch, burn, burned spot
Brandgefahr *f* fire hazard
Brandschutz *m* fire protection [prevention]
brandsicher fireproof • **brandsicher machen** fireproof
Branntkalk *m* burnt [burned] lime, caustic lime, quicklime *(calcium oxide)*
Branntwein *m* (distilled) spirit(s), *(loosely:)* alcohol
Brauchbarkeit *f* practicality *(of a method)*
Brauchwasser *n* plant [industrial] water, mill [commercial] water
brauen *v* brew
Brauerei *f* 1. brewery, brewhouse; 2. brewing
Brauereichemie *f* brewing chemistry
Braugerste *f* brewing [brewer's, malting] barley
Braukessel *m s.* Braupfanne
Braumalz *n* brewer's malt, malt for brewing

Brauneisenerz *n (min)* brown iron ore [stone], limonite *(a variety of goethite)*
Braunerde *f* brown soil
Braunhuminsäure *f (soil)* brown humic acid
Braunkohle *f* lignite, *(esp if low-quality:)* brown coal, *(according to the ASTM coal classification:)* lignitic coal
Braunkohle f/bituminöse bituminous lignite, lignitous [subbituminous] coal
Braunkohle f/holzartige [lignitische] *s.* Braunkohle/xylitische
Braunkohle f/verfestigte lignite *(according to the ASTM coal classification)*
Braunkohle f/xylitische xylite, woody lignite [brown coal]
Braunkohlenkoks *m* lignitic [brown-coal] coke
Braunkohlenteer *m* lignite [brown-coal] tar
Braunkohlenxylit *m(n)* xylite, woody lignite [brown coal]
Braunspat *m (min)* ankerite
Braunstein *m* manganese dioxide
Bräunung f/nicht enzymatische *(food)* non-enzymatic browning, Maillard-type [amino-sugar] browning
Bräunung *f* **vom Maillard-Typus** *s.* Bräunung/nicht enzymatische
Bräunungsreaktion f/nicht enzymatische *(food)* non-enzymatic browning reaction, Maillard [carbonyl-amine] reaction
Braupfanne *f* brew [wort] kettle, (wort) copper
brausen *v* spray, sprinkle; effervesce
Brausen *n* spraying, sprinkling; effervescence, effervescency
Brauwasser *n* brewing water [liquor]
Bravais-Gitter *n (cryst)* Bravais lattice
Breathing-Schwingung *f* (symmetrical) breathing vibration
brechen *v* 1. break; mill, crush *(rock, ore)*; chop *(bark)*; roll *(flax)*; break, crack *(of emulsions)*; 2. refract, break *(light)*
brechen v/eine Emulsion break [crack] an emulsion, demulsify, de-emulsify
Brechen *n* 1. breaking, breakage; milling, crush(ing) *(of rock, ore)*; chopping *(of bark)*; rolling *(of flax)*; 2. refraction, breaking *(of light)*
Brechen *n* **einer Emulsion** breaking [cracking] of an emulsion, demulsification, de-emulsification; emulsion breakdown [breaking] *(process)*
Brecher *m* crusher
Brecherwalzwerk *n* breaker, breaking [break-down] mill
Brechgut *n* material being [to be] crushed
Brechmittel *n (pharm)* emetic
Brechplatte *f* crushing [breaker] plate
Brechpunkt *m* 1. *(petrol)* breaking point; 2. *(hyd)* breakpoint
Brechpunktchlorung *f (hyd)* breakpoint [free-residual] chlorination
Brechung *f (phys ch)* refraction
Brechungsindex *m s.* Brechzahl
Brechungskoeffizient *m s.* Brechzahl
Brechungsvermögen *n* refractivity
Brechungswinkel *m* angle of refraction
Brechwalze *f* crushing [crusher] roll
Brechzahl *f* refraction [refractive] index
Brei *m* pulp, paste
Brei m/dünner slurry
brei(art)ig pulpy
Breitbandantibiotikum *n* broad-spectrum antibiotic
Breitbandentkopplung *f* broad-band [wide-band] decoupling
Breitbandspektrum *n* broad-band spectrum
Breitbrenner *m (lab)* flat-flame [fish-tail] burner
Breitbrenneraufsatz *m* burner wing top [tip], flat burner head, (burner) flame spreader
Breitlinienkernresonanz *f* broad-line [wide-line] NMR
Breitschlitzdüse *f (plast)* slot die *(for sheet forming)*
Breitspektrumantibiotikum *n s.* Breitbandantibiotikum
bremsen *v* 1. slow down, decelerate, retard; inhibit *(a reaction)*; 2. *(nucl)* moderate
Bremsen *n* 1. slowing down, deceleration, retardation, retarding; inhibition *(of a reaction)*; 2. *(nucl)* moderation
Bremsstrahlung *f (nucl)* bremsstrahlung
Bremssubstanz *f (nucl)* moderator, slowing-down agent
Bremsung *f (nucl)* moderation
brennbar combustible, burnable, *(esp relating to liquids:)* (in)flammable

brennbar/nicht incombustible, noncombustible, *(esp relating to liquids:)* non-flammable
Brennbares *n* combustible(s)
Brennbarkeit *f* combustibility, *(esp relating to liquids:)* (in)flammability
Brennelement *n s.* Brennstoffelement 1.
brennen *v* 1. burn; 2. *(ceram)* fire, burn, bake; calcine *(e.g. limestone)*; bake *(shaped green carbon bodies)*; 3. distil *(alcohol)*
brennen *v*/**zu Porzellan** porcelainize
Brennen *n* 1. burning; 2. *(ceram)* firing, burning, baking; calcining, calcination *(as of limestone)*; baking *(of shaped green carbon bodies)*; 3. distillation *(of alcohol)*
Brenner *m* 1. *(lab, tech)* burner; (welding) gun, torch; 2. distiller *(profession)*
Brenner *m* **mit Vormischung** premix burner
Brennerei *f (ferm)* distillery
Brennereischlempe *f* stillage, (distillery) slop, vinasse, spent wash
Brenngas *n* fuel [combustible] gas
Brenngerät *n* distillation apparatus *(in alcohol distillation)*
Brenngeschwindigkeit *f* rate of combustion
Brennkammer *f* combustion chamber [space]
Brennkegel *m (ceram)* pyrometric cone
Brennkegel *m* **nach Seger** Seger cone
Brennmaterial *n* fuel
Brennofen *m (ceram)* kiln
Brennöl *n* burning oil
Brennpunkt *m* 1. burning [fire] point *(of liquid fuel)*; 2. focus *(optics)*
Brennraum *m* furnace chamber, combustion chamber [space]
Brennsäure *f (met)* pickling acid
Brennspiritus *m* mineralized methylated spirit
Brennstab *m s.* Brennstoffstab
Brennstaub *m* pulverized [powdered] fuel
Brennstoff *m* (combustion) fuel; (nuclear) fuel
Brennstoff *m*/**fester** solid fuel
Brennstoff *m*/**flüssiger** liquid fuel
Brennstoff *m*/**fossiler** fossil fuel
Brennstoff *m*/**gasförmiger** gaseous fuel
Brennstoff *m*/**rauchfreier [rauchloser]** smokeless fuel
Brennstoff *m*/**umweltfreundlicher** clean fuel
Brennstoffchemie *f* fuel chemistry
Brennstoffchemiker *m* fuel chemist
Brennstoffelement *n* 1. *(nucl)* fuel element; 2. *(el ch)* fuel cell
Brennstoffstab *m (nucl)* fuel rod
Brennstoffstange *f (nucl)* fuel rod
Brennstoffstaub *m s.* Brennstaub
Brennstoffwiederaufarbeitung *f* (nuclear) fuel reprocessing, reactor fuel reprocessing
Brennstoffwiederaufbereitung *f* (nuclear) fuel reprocessing, reactor fuel reprocessing
Brennstoffwiederaufbereitungsanlage *f (nucl)* fuel-reprocessing plant
Brennstoffzelle *f (el ch)* fuel cell
Brennstoffzelle *f*/**alkalische** alkaline fuel cell, AFC
Brennstoffzelle *f*/**phosphorsaure** phosphoric-acid fuel cell, PAFC
Brennwert *m* calorific value
Brennwert *m*/**molarer** *s.* Brennwert/stoffmengenbezogener
Brennwert *m*/**spezifischer** gross heat of combustion, *(deprecated terms:)* gross calorific value, higher heating value
Brennwert *m*/**stoffmengenbezogener** heat of combustion per mole
Brenzcatechin *n* catechol, pyrocatechol, **o*-dihydroxybenzene
Brenzcatechindimethylether *m* catechol dimethyl ether, veratrol, *1,2-dimethoxybenzene
Brenzcatechinmonomethylether *m* catechol methyl ether, *o*-methoxyphenol, guaiacol
Brenzschleimsäure *f* pyromucic acid, furan-2-carboxylic acid
Brenztraubensäure *f* pyruvic acid, *2-oxopropionic acid
Brenzweinsäure *f* pyrotartaric acid, methylsuccinic acid, *2-methylbutane-1,4-dioic acid
Brewster-Winkel *m* Brewster angle
Briefumschlag-Konformation *f* envelope conformation
Brightstock *m (petrol)* bright stock
Brightstock-Öl *n (petrol)* bright stock
Brikett *n* briquet(te)

brikettierbar briquettable
Brikettierbarkeit *f* briquettability, briquetting properties [qualities]
Brikettiereigenschaften *fpl* briquettability, briquetting properties [qualities]
brikettieren *v* briquette
Brikettierpresse *f* briquetting [briquette] press
brillant brilliant
Brillantfarbstoff *m* brilliant dye
Brillantgrün *n* brilliant [emerald] green *(a basic triphenylmethane dye)*; emerald green, chrome green, Mittler's green, Guignet's green *(hydrated chromium oxide)*
Brillanz *f* brilliance, brilliancy
Brillouin-Polyeder *n (cryst)* Brillouin polyhedron
Brinellhärte *f* Brinell hardness
bringen *v*/**auf Typ** *(dye)* bring to standard strength
bringen *v*/**in Kontakt** contact
bringen *v*/**in Lösung** bring [put] into solution
bringen *v*/**ins Gleichgewicht** bring into equilibrium, equilibrate
bringen *v*/**zum Erlöschen** extinguish
bringen *v*/**zum Kochen** bring [raise] to the boil
bringen *v*/**zum Schäumen** foam, froth
bringen *v*/**zum Stehen** arrest *(a reaction)*
bringen *v*/**zur Ausflockung** coagulate, curdle
bringen *v*/**zur Detonation** detonate
bringen *v*/**zur Explosion** explode
bringen *v*/**zur Kristallisation** crystallize out
bringen *v*/**zur Reaktion** react
Brisanz *f* brisance, shattering power
Brocken *m* lump *(as of coal, sugar)*
bröcklig friable, crumbly
Bröckligkeit *f* friability, crumbliness
Brom *n* Br bromine
Bromaceton *n* bromoacetone, *1-bromopropanone
Bromacetoxylierung *f* bromoacetoxylation
Bromacetylierung *f* bromoacetylation
Bromacylierung *f* bromoacylation
Bromalid *n (org ch)* bromalide
Bromalkan *n* bromoalkane
Bromalkylierung *f* bromoalkylation
Bromat *n* $M^{I}BrO_3$ bromate • **mit Bromat behandeln** bromate
Bromatometrie *f* bromatometry
bromatometrisch bromatometric
Brombenzen *n* bromobenzene
Brombenzoesäure *f* bromobenzoic acid
Brombenzol *n s.* Brombenzen
***N*-Brombernsteinsäureimid** *n N*-bromosuccinimide, NBS
Bromcresolgrün *m* bromocresol green *(a pH indicator)*
Bromcresolpurpur *m* bromocresol purple *(a pH indicator)*
Bromdesulfonierung *f* bromodesulphonation
Bromessigsäure *f* bromoacetic acid, *bromoethanoic acid
Bromethan *n* bromoethane
Bromethansäure *f s.* Bromessigsäure
Bromethen *n* bromoethylene, *bromoethene, vinyl bromide
Brom(III)-fluorid *n* bromine(III) fluoride, bromine trifluoride
Brom(V)-fluorid *n* bromine(V) fluoride, bromine pentafluoride
Bromid *n* $M^{I}Br$ bromide
bromieren *v* brominate
bromieren *v*/**zweifach** dibrominate
Bromierung *f* bromination
Bromierung *f*/**zweifache** dibromination
Bromierungsmittel *n* brominating agent
Bromismus *m (tox)* bromism
Bromlactonisierung *f* bromolactonization
Brommethan *n* bromomethane
Brommethylierung *f* bromomethylation
Bromoform *n* bromoform, *tribromomethane
Bromolyse *f* brominolysis
Bromometrie *f* bromometry
bromometrisch bromometric
Bromoniumverbindung *f* bromonium compound
Brompentafluorid *n* bromine pentafluoride
Bromphenolblau *n* bromophenol blue, tetrabromophenolsulphonephthalein *(a pH indicator)*
3-Brompropen *n* *3-bromopropene, allyl bromide
Bromsäure *f* bromic acid
Bromsilber *n (phot)* silver bromide

***N*-Bromsuccinimid** *n* *N*-bromosuccinimide
Bromthymolblau *n* bromothymol blue, dibromothymolsulphonephthalein *(a pH indicator)*
Bromtrifluorid *n* bromine trifluoride
Bromwasser *n* bromine water
Bromwasserstoff *m* hydrogen bromide
Bromwasserstoffsäure *f* hydrobromic acid
α-**Bromxylen** *n* α-bromoxylene
α-**Bromxylol** *n* α-bromoxylene
Bromzahl *f (anal)* bromine number *(for characterizing the degree of unsaturation of fatty oils)*
Brönsted-Base *f* Brønsted(-Lowry) base
Brönsted-Katalysegesetz *n* (Brønsted) catalysis law
Brönsted-Säure *f* Brønsted(-Lowry) acid, proton(ic) acid, hydrogen acid
Bronze *f* bronze, *(specif)* tin bronze
bronzieren *v* bronze
Bruch *m* 1. breaking, breakage, fracture *(process)*; 2. *(min)* cleavage *(the manner in which a mineral may be cleft or split)*; 3. break(age), fracture *(result)*; 4. *(min)* fracture *(texture of a broken surface)*; 5. scrap *(discarded metal collected for melting down)*; 6. *(ceram, glass)* breakage *(broken ware)*; cullet *(broken glass for remelting)*; curd *(the coagulated part of milk used for cheese-making)*
Bruchdehnung *f* strain after fracture, strain at break
Bruchebene *f* fracture plane
Bruchfestigkeit *f* fracture strength
Bruchfläche *f* fracture facet [surface]
brüchig brittle • **brüchig machen** make brittle, embrittle • **brüchig werden** become brittle, embrittle
Brüchigkeit *f* brittleness
Bruchkraft *f* breaking load *(bending test)*
bruchsicher shatterproof
Bruchstück *n* 1. fragment; 2. *(nucl)* fission fragment
Bruchstückion *n* fragment ion
Brucinpapier *n* brucine paper *(for detecting nitrous acid)*
Brücke *f* 1. bridge *(chemical-bond theory)*; 2. *(plast, rubber)* cross-link(age)
Brückenatom *n* bridge atom
brückenbildend bridging
Brückenbildung *f* bridging
Brückenbindung *f* bridge-type bond
Brücken-Kohlenwasserstoff *m* bridged hydrocarbon
Brückenkopf *m* bridgehead *(in bridgering systems)*
Brückenkopfatom *n* bridgehead atom
Brückenringverbindung *f* bridged-ring compound
Brückensauerstoff *m* bridging oxygen
Brüden(dampf) *m* exhaust [dead] steam, vapour
Brühe *f* 1. broth *(a culture medium)*; 2. *(tann)* liquor; 3. *(agric)* wash *(pest control)*
Brühe *f*/**Bordelaiser** Bordeaux mixture *(fungicide)*
Brühe *f*/**Burgunder** Burgundy mixture, soda bordeaux *(fungicide)*
Brunauer-Emmett-Teller-Adsorptionsisotherme *f* Brunauer Emmett-Teller isotherm, BET isotherm
brünieren *v* brown *(metal surfaces)*
Brunnen *m* 1. well; 2. *(hyd)* water well; *(natural fountain:)* spring, *(specif)* mineral spring; 3. *(petrol)* (oil) well; 4. mineral waters
brütbar *(nucl)* fertile
Brütbarkeit *f (nucl)* fertility
Brüten *n (nucl)* breeding, fertilization
Brüter *m s.* Brutreaktor
Brutmaterial *n (nucl)* fertile material
Brutreaktor *m (nucl)* breeder reactor
Brutreaktor *m*/**schneller** fast breeder reactor
Brutschrank *m* incubator • **im Brutschrank aufbewahren** incubate
Brutstoff *m s.* Brutmaterial
Bruttoformel *f* empirical (molecular) formula
Bruttoformel *f*/**einfachste** simplest (possible) formula, stoichiometric formula
Bruttoformel *f*/**wahre** true [empirical molecular] formula
Bruttogleichung *f* overall equation *(of a reaction)*
Brutto-Komplexbildungskonstante *f* overal formation constant
Brutto-Komplexstabilitätskonstante *f* overal formation constant
Bruttoprobe *f (anal)* gross sample
Bruttoreaktion *f* overall reaction

Bruttozusammensetzung *f* gross composition
BSB *m (biochemischer Sauerstoffbedarf)* BOD, biochemical oxygen demand
BSB-Belastung *f (hyd)* BOD loading
BSB-Endwert *m (hyd)* final [residual] BOD
BSB-Last *f (hyd)* BOD loading
BSB_5-Wert *m* five-day BOD, five-days biochemical oxygen demand, BOD_5
B-strain-Effekt *m s.* Beschleunigung/sterische
BTM *s.* Biotrockenmasse
BTX-Aromaten *mpl* benzene-toluene-xylenes, BTX
Bucherer-(Lepetit-)Reaktion *f* Bucherer reaction *(conversion of a naphthylamine to a naphthol or vice versa)*
Büchner-Nutsche *f* Büchner filter [funnel]
Büchner-Trichter *m* Büchner filter [funnel]
Buckminster-Fullerene *npl* buckytube
Buna *m(n) s.* Bunakautschuk
Bunakautschuk *m* buna (rubber)
bündeln *v* concentrate *(rays)*
Bündelung *f* concentration *(of rays)*
Bunker *m* bunker, bin, silo, (storage) tank, *(if funnel-shaped:)* (storage) hopper
Bunkerkohle *f* bunker coal
Bunkeröl *n* bunker fuel (oil)
Bunsenbrenner *m* Bunsen burner
Bunsentrichter *m* Bunsen [long-stemmed] funnel
Bunte-Bürette *f* Bunte (gas) burette
Bunte-Salz *n* Bunte salt *(salts, usually sodium salts, of S-alkylthiosulphuric acid)*
Buntglas *n* coloured glass
Bürette *f* burette
Bürette *f* **mit automatischer Nullpunkteinstellung** automatic zero burette
Bürettenhahn *m* burette stopcock [valve]
Bürettenhalter *m* burette clamp [holder]
Bürettenklemme *f* burette clamp [holder]
Burgers-Versetzung *f (cryst)* screw [Burgers] dislocation
Bürste *f* brush
bürsten *v* brush
Bürstenbelüfter *m (hyd)* brush aerator
Butadien *n* butadiene, *(specif)* buta-1,3-diene
Butadien-Acrylnitril-Copolymerisat *n* butadiene-acrylonitrile copolymer
Butadien-Acrylnitril-Kautschuk *m* butadiene-acrylonitrile rubber, nitrile(-butadiene) rubber, NBR
Butadienkautschuk *m* butadiene rubber, BR
Butadien-Styren-Copolymerisat *n* butadiene-styrene copolymer
Butadien-Styren-Kautschuk *m* butadiene-styrene rubber, styrene-butadiene rubber, SBR
Butadiin *n* butadiyne
Butan *n* butane
Butanal *n* *butanal, butyric aldehyde
Butan-1,4-dial *n* *butane-1,4-dial, succindialdehyde
Butandial-(1,4) *n* *butane-1,4-dial, succindialdehyde
Butan-1,4-dicarbonsäure *f* *butane-1,4-dicarboxylic acid, *hexanedioic acid, adipic acid
Butandicarbonsäure-(1,4) *f* *butane-1,4-dicarboxylic acid, *hexanedioic acid, adipic acid
Butan-2,3-dion *n* butane-2,3-dione
Butandion-(2,3) *n* butane-2,3-dione
Butandisäure *f* *butanedioic acid, *ethane-1,2-dicarboxylic acid, succinic acid
Butan-1-ol *n* butan-1-ol
Butan-2-ol *n* butan-2-ol
Butanol-(1) *n* butan-1-ol
Butanol-(2) *n* butan-2-ol
***sec*-Butanol** *n s.* Butan-2-ol
***tert*-Butanol** *n* *2-methylpropan-2-ol, *tert*-butyl alcohol
Butanol-(3)-al-(1) *n s.* 3-Hydroxybutanal
Butan-2-on *n* *butan-2-one, 2-oxobutane
Butanon-(2) *n* *butan-2-one, 2-oxobutane
Butansäure *f s.* Buttersäure
Buteakino *n* butea gum, gum butea, Bengal kino *(from Butea superba Roxb.)*
But-1-en *n* but-1-ene
But-2-en *n* but-2-ene
Buten-(1) *n* but-1-ene
Buten-(2) *n* but-2-ene

But-2-enal *n* *but-2-enal, crotonic aldehyde
Buten-(2)-al-(1) *n* *but-2-enal, crotonic aldehyde
Butendisäure *f* butenedioic acid, ethylene-1,2-dicarboxylic acid
***cis*-Butendisäureanhydrid** *n* **cis*-butenedioic anhydride, maleic anhydride, furan-2,5-dione
Butenpolymer(e) *n* butylene polymer
***cis*-But-2-ensäure** *f* **cis*-but-2-enoic acid, isocrotonic acid
***cis*-Buten-(2)-säure** *f* **cis*-but-2-enoic acid, isocrotonic acid
***trans*-But-2-ensäure** *f* **trans*-but-2-enoic acid, crotonic acid
***trans*-Buten-(2)-säure** *f* **trans*-but-2-enoic acid, crotonic acid
Butin *n* 1. butin, 7,3′,4′-trihydroxyflavanone; 2. butyne *(either of two isomeric alkynes)*
But-1-in *n* but-1-yne
But-2-in *n* but-2-yne
Butin-(1) *n* but-1-yne
Butin-(2) *n* but-2-yne
Butindisäure *f* butynedioic acid
Bütte *f* vat, tub
butterartig buttery
Butterfarbe *f* butter dye, *(Am)* butter color
Butterfett *n* butter fat
Buttergelb *n* butter yellow, *p*-dimethylaminoazobenzene
Buttersäure *f* butyric acid, *butanoic acid
***i*-Buttersäure** *f* isobutyric acid, *2-methylpropanoic acid
Buttersäureethylester *m* ethyl butyrate
Buttersäuregärung *f* butyric(-acid) fermentation
Butylacetat *n* butyl acetate
***n*-Butylacetat** *n* butyl acetate
***n*-Butylalkohol** *m s.* Butan-1-ol
***sec*-Butylalkohol** *m s.* Butan-2-ol
***tert*-Butylalkohol** *m s. tert*-Butanol
Butylgruppe *f* C_4H_9– butyl group [residue]
Butylhalogenid *n* butyl halide
Butylhydroxyanisol *n (food)* butylated hydroxyanisole, BHA *(antioxidant)*
Butylierung *f* butylation
Butylkautschuk *m* butyl rubber, BR
Butylrest *m s.* Butylgruppe
Butyraldehyd *m* butyric aldehyde, butanal
Butyrat *n* butyrate
bz-Phase *f (coll)* smectic phase
B-Zustand *m (plast)* B stage

C

C *s.* Chemiefaserstoff
CA 1. *s.* Celluloseacetat; 2. = anorganische Chemiefaserstoffe
CAB *s.* Celluloseacetatbutyrat
Cadmium *n* Cd cadmium
Cadmiumgelb *n* cadmium yellow, aurora [orient] yellow *(cadmium sulphide)*
Cadmiumhydroxid *n* cadmium hydroxide
Cadmium(II)-oxid *n* cadmium oxide
Cadmiumrot *n* cadmium red
Cadmiumsulfat *n* cadmium sulphate
Cadmiumsulfid *n* cadmium sulphide
Cadmiumvergiftung *f* cadmium poisoning
Caesium *n* Cs caesium
Caesiumchlorid *n* caesium chloride
Caesiumhydrid *n* caesium hydride
Caesiumhydroxid *n* caesium hydroxide
Caesiumiodid *n* caesium iodide
Caesiumoxid *n* caesium oxide
Caesiumsulfat *n* caesium sulphate
Caesiumsulfid *n* caesium sulphide
Cahn-Ingold-Prelog-System *n* Cahn-Ingold-Prelog system, CIP system *(for designating absolute configuration)*
Calcit *m (min)* calcite, calc-spar, lime spar *(calcium carbonate)*
Calcium *n* Ca calcium
Calciumcarbid *n* calcium carbide, calcium acetylide, carbide *(proper)*
Calciumcarbonat *n* calcium carbonate
Calciumcarbonathärte *f (hyd)* calcium carbonate hardness
Calciumcarbonatstein *m (hyd)* calcium carbonate scale
Calciumchlorid *n* calcium chloride
Calciumcyanamid *n* calcium cyanamide
Calciumfluorid *n* calcium fluoride
Calciumhärte *f (hyd)* calcium hardness
Calciumhydrid *n* calcium hydride
Calciumhydrogencarbonat *n* calcium hydrogencarbonate
Calciumhydroxid *n* calcium hydroxide

Calciumhypochlorit *n* calcium hypochlorite
Calciumhypochlorit-Bleichlauge *f* (calcium hypochlorite) bleach liquor
Calciumnichtcarbonathärte *f (hyd)* calcium non-carbonate hardness
Calciumoxid *n* calcium oxide
Calciumoxiderzeugnis *n (ceram)* lime refractory
Calciumperoxid *n* calcium peroxide
Calciumseife *f* calcium soap
Calciumspiegel *m* calcium level *(of blood)*
Calciumstearat *n* calcium stearate
Calciumsulfat *n* calcium sulphate
Calciumsulfid *n* calcium sulphide
Calebassenalkaloid *n* calabash(-curare) alkaloid, C-alkaloid
Calebassencurare-Alkaloid *n* calabash(-curare) alkaloid, C-alkaloid
Californium *n* Cf californium
Calixaren *n (org ch)* calixarene
C-Alkaloid *n s.* Calebassencurare-Alkaloid
Calvin-Bassham-Zyklus *m s.* Calvin-Zyklus
Calvin-Pflanze *f s.* C_3-Pflanze
Calvin-Zyklus *m* Calvin(-Bassham) cycle, Calvin pathway, C_3 pentose phosphate cycle, C_3 pathway, ribulose diphosphate cycle *(of CO_2 assimilation)*
Camphen *n* camphene *(a bicyclic terpene hydrocarbon)*
Campher *m* camphor, camphan-2-one *(a bicyclic terpene ketone)*
campherartig camphoraceous
Camphermethode *f* **nach Rast** Rast camphor method, Rast (molecular weight) method, Rast micromethod
Campheröl *n* camphor oil *(from Cinnamomum camphora (L.) Sieb.)*
Camphersäure *f* camphoric acid, 1,2,2-trimethylcyclopentane-1,3-dicarboxylic acid
Candelkohle *f s.* Cannelkohle
Cannelkohle *f* cannel [candle, jet] coal
Cannizzaro-Reaktion *f* Cannizzaro reaction *(aldehyde dismutation)*
Cantilever *m* cantilever
CAP *s.* Celluloseacetatpropionat
Caprinat *n* *decanoate *(salt or ester of decanoic acid)*; *(deprecated:)* caprate
***n*-Caprinsäure** *f* *decanoic acid, *(deprecated:)* capric acid
Caprolactam *n* caprolactam
Capronat *n* *hexanoate *(a salt or ester of hexanoic acid)*; *(deprecated:)* capronate, caproate
***n*-Capronsäure** *f* *hexanoic acid, *(deprecated:)* caproic acid
Caprylat *n* *octanoate *(a salt or ester of octanoic acid)*; *(deprecated:)* caprylate
***n*-Caprylsäure** *f* *octanoic acid, *(deprecated:)* *n*-caprylic acid
Capsomer *n (bioch)* capsomere
Carbalkoxylierung *f* alkoxycarbonylation, carbalkoxylation
Carbaluminierung *f* carblumination
Carbamat *n* NH_2COOR carbamate, aminoformate *(salt or ester of carbamic acid)*
Carbamid *n* carbamide, urea
Carbamidharz *n* urea resin
Carbamidsäure *f* carbamic acid, aminoformic acid
Carbamidsäureethylester *m* ethyl carbamate, ethyl aminoformate
Carbamoylcarbonsäure *f* carbamoyl carboxylic acid, amic acid
Carbamoylethylierung *f* carbam(o)ylethylation
Carbam(o)ylierung *f* carbam(o)ylation, carboxamidation
Carbanion *n* carbanion
Carbazol *n* carbazole, dibenzopyrrole
Carbazon *n* carbazone, *(specif)* $H_2NNHCON{=}NH$ carbazone
Carben *n* $R_1–C–R_2$ carbene, *(specif)* $:CH_2$ carbene, methylene
Carbeniat-Anion *n s.* Carbanion
Carbeniat-Ion *n s.* Carbanion
Carbeniumion *n* carbenium ion, carbocation *(having three-coordinate carbon)*
Carbeniumkation *n* carbenium ion, carbocation *(having three-coordinate carbon)*
Carbenoid *n* carbenoid
Carbethoxylierung *f* carb(o)ethoxylation, ethoxycarbonylation
Carbethoxymethylierung *f* carbethoxymethylation, ethoxycarbonylmethylation
Carbid *n* carbide, *(specif)* CaC_2 calcium carbide, calcium acetylide

Carbid *n***/gesintertes** sintered (hard, metal) carbide
Carbid *n***/hartes** hard (metal) carbide
Carbid *n***/interstitielles** interstitial carbide
Carbidhartmetall *n* cemented (hard) carbide, cemented hard metal
carbidisch carbidic
Carbidkalk *m* carbide [acetylene] lime *(impure calcium hydroxide)*
Carbmetallierung *f* carbometallation
Carbobenzoxylierung *f* carbobenzoxylation
Carbobromierung *f* bromoformylation, bromocarbonylation
Carbochemie *f* coal chemistry
Carbocuprierung *f* carbocupration
carbocyclisch carbocyclic, homocyclic, isocyclic
Carbocyclus *m* carbocyclic compound, homocyclic [isocyclic] compound
Carbodiimid *n* carbodiimide
Carbodithiosäure *f* dithiocarboxylic acid
carbofunktionell carbon-functional, organofunctional
Carbokation *n* carbocation *(positively charged carbon atom)*
Carbolithiierung *f* carbolithiation
Carbolsäure *f* carbolic acid, phenol, hydroxybenzene
Carbolsäure *f***/rohe** cresylic acid *(a crude mixture of phenols, cresols, and xylenols)*
Carbomethoxylierung *f* carbomethoxylation, methoxycarbonylation
Carbon *n* C carbon
Carbonat *n* $M^I_2CO_3$ carbonate
Carbonatbleiweiß *n* white lead, ceruse *(lead carbonate hydroxide)*
Carbonathärte *f (hyd)* carbonate hardness [alkalinity], temporary hardness
Carbonatisation *f (geoch)* carbonatization
Carbonatisierung *f (geoch)* carbonatization
Carbonatstein *m (hyd)* carbonate scale
Carbonisation *f* 1. *(food)* carbonation, aeration, impregnation with carbon dioxide; 2. carbonation *(sodium carbonate manufacture)*
carbonisieren *v* 1. *(food)* carbonate, aerate, impregnate with carbon dioxide; 2. carbonate *(in sodium-carbonate manufacture)*
carbonitrieren *v* carbonitride *(steel)*
Carbonitrierung *f* carbonitriding, nicarbing *(of steel)*
Carbonitril *n* carbonitrile
Carboniumion *n* 1. carbonium ion *(having five-coordinate carbon)*; 2. *s.* Carbeniumion
Carbonohydrazid *n* carbonohydrazide, carbazide, *(specif)* $H_2NNHCONHNH_2$ carbonohydrazide, carbazide
Carbonsäure *f* carboxylic acid
Carbonsäure *f***/dreibasige** tricarboxylic acid
Carbonsäure *f***/einbasige** monocarboxylic acid
Carbonsäure *f***/mehrbasige** polycarboxylic acid
Carbonsäure *f***/zweibasige** dicarboxylic acid
Carbonsäureabbau *m***/barbier-wielandscher** Barbier-Wieland degradation [reaction]
Carbonsäureamid *n* carboxylic acid amide, carboxamide, carboxazylic acid
Carbonyl *n* carbonyl (compound), metal carbonyl
Carbonylaminoreaktion *f (food)* carbonyl-amine reaction, Maillard reaction, non-enzymatic browning reaction
Carbonylchlorid *n* carbonyl chloride, phosgene
Carbonylgruppe *f* >C=O carbonyl group
Carbonylgruppe *f***/ketonartig gebundene** ketonic [ketone] carbonyl group, ketonic [ketone, keto] group
carbonylieren *v* carbonylate
Carbonylierung *f* carbonylation
Carbonyloxid *n* carbonyl oxide
Carbonylsulfid *n* carbonyl sulphide
Carboran *n* carborane, *carbaborane
Carbothiosäure *f* monothiocarboxylic acid
Carboxamid *n* carboxamide, carboxylic acid amide
Carboxamidin *n* $RC(=NH)NH_2$ carboxamidine, amidine
Carboxyalkylierung *f* carboxyalkylation
Carboxyethylierung *f* carboxyethylation
Carboxylat *n* carboxylate *(a salt or ester of a carboxylic acid)*

Carboxylende *n* C-terminal group [residue] *(in proteins)*
Carboxylgruppe *f* –COOH carboxyl group
carboxylieren *v* carboxylate
Carboxylierung *f* carboxylation
Carboxymethylcellulose *f* carboxymethylcellulose, CM cellulose, CMC
Carboxymethylierung *f* carboxymethylation
Carboxymethyl-trimethylammonium *n* betaine, *NNN*-trimethylammonioacetate, oxyneurine, lycine
Carius-Rohr *n (lab)* Carius [bomb] tube
Carminativum *n (pharm)* carminative
Carnot-Prozess *m (phys ch)* Carnot cycle
carosche Säure *f* Caro's acid
Caroten *n* carotene
Carotenoid *n (org ch)* carotenoid
Carotin *n s.* Caroten
Casein *n* casein
Caseinat *n* caseinate
Caseinfaser *f* casein fibre [staple]
Caseinfaserstoff *m* casein fibre
Casinghead-Benzin *n* casing-head gasoline
Casinghead-Gas *n* casing-head gas
Cassiterit *m (min)* cassiterite, tin stone *(tin(IV) oxide)*
Castner-Zelle *f (el ch)* Castner cell
Cat-Benzin *n* cat-cracked gasoline
Catcracken *n* cat cracking, catalytic [catalyst] cracking
Catecholamin *n* catecholamine
Catenan *n (org ch)* catenane
Catena-Verbindung *f (org ch)* catena [catenation] compound
Catformen *n (petrol)* catforming
Catforming-Verfahren *n (petrol)* catforming process
Cavitand *m (org ch)* cavitand
C_2-Bruchstück *n (bioch)* two-carbon fragment
C–C-Bindung *f s.* C–C-Einfachbindung
C–C-Bindung *f*/**einfache** *s.* C–C-Einfachbindung
C=C-Bindung *f s.* C–C-Doppelbindung
C≡C-Bindung *f s.* C–C-Dreifachbindung
C–C-Doppelbindung *f* carbon(-carbon) double bond, C=C bond, olefinic linkage
C–C-Dreifachbindung *f* carbon(-carbon) triple bond, C≡C bond, acetylenic linkage
C–C-Einfachbindung *f* carbon(-carbon) single bond, carbon-carbon bond, C–C bond
CCP = Carboxyl-Carrier-Protein
C/C-Verbundwerkstoff *m* carbon-carbon composite, carbon fibre-reinforced carbon, CFC
CDAA *s.* 2-Chlor-*N,N*-diallylacetamid
C_4-Dicarbonsäureweg *m (bioch)* C_4 cycle [pathway], C_4 dicarboxylic acid cycle, Hatch-Slack (CO_2 incorporation) pathway
cDNS *s.* DNA/komplementäre
CE *s.* Eiweißchemiefaserstoff
CED *(cohesive energy density) s.* Energiedichte/kohäsive
C_2-Einheit *f (bioch)* two-carbon fragment
Celluloid *n* celluloid
Cellulose *f* cellulose
Cellulose *f*/**native [natürliche]** natural cellulose
Cellulose *f*/**regenerierte** regenerated cellulose
α-Cellulose *f* alpha [chemical] cellulose, α-cellulose
β-Cellulose *f* beta cellulose, β-cellulose
γ-Cellulose *f* gamma cellulose, γ-cellulose
Celluloseacetat *n* cellulose acetate, CA, acetylated cellulose
Celluloseacetatbutyrat *n* cellulose acetate butyrate
Celluloseacetatfaser *f* cellulose acetate fibre
Celluloseacetatfaserstoff *m* cellulose acetate fibre
Celluloseacetatpropionat *n* cellulose acetate propionate
Celluloseacetatseide *f* cellulose acetate rayon, acetate filament yarn
Cellulosechemiefaser *f* cellulosic fibre
Cellulosechemiefaserstoff *m* cellulosic fibre
Celluloseerzeugnis *n* cellulosic
Celluloseester *m* cellulose ester
Celluloseether *m* cellulose ether
Celluloseethylether *m* ethylcellulose
Cellulosefaser *f* cellulose fibre
Cellulosefaserstoff *m* cellulose fibre
Cellulosenitrat *n* cellulose nitrate

Cellulosenitratlack *m* cellulose nitrate lacquer
Cellulosepropionat *n* cellulose propionate
Celluloseregeneratfaser *f* regenerated cellulose fibre
Celluloseregeneratfaserstoff *m* regenerated cellulose fibre
Celluloseregeneratseide *f* regenerated cellulose fibre
Cellulosexanthat *n s.* Cellulosexanthogenat
Cellulosexanthogenat *n* cellulose xanthate
cellulosisch cellulosic
Celsius-Skala *f* centigrade [Celsius] scale
Celsius-Temperaturskala *f* centigrade [Celsius] scale
Celsius-Thermometerskala *f s.* Celsius-Skala
C-endständig C-terminal *(amino acids in proteins)*
Cephalin *n (bioch)* cephalin, kephalin *(any of various glycerophospholipids)*; *(specif)* 3-phosphatidyl ethanolamine, 3-phosphatidyl serine
Cephalosporin *n* cephalosporin *(antibiotic)*
Cer *n* Ce cerium
Ceramid *n (bioch)* ceramide, *N*-acylsphingosine
Cercarbid *n* cerium carbide
Cerdioxid *n s.* Cerium(IV)-oxid
Cerebrosid *n* cerebroside *(a lipoid)*
Ceresin *n* ceresin (wax), ceresine
Cerevitinov-Bestimmung *f* Zerewitinoff determination *(of active H atoms)*
Cer(III)-fluorid *n* cerium(III) fluoride, cerium trifluoride
Cer(IV)-fluorid *n* cerium(IV) fluoride, cerium tetrafluoride
Cerimetrie *f* ceri(o)metry, cerate oxidimetry
Ceriterde *f* cerite earth *(any of one group of rare-earth metal oxides)*
Cermet *n* cer(a)met, ceramal, ceramel
Cer-Mischmetall *n* misch metal
Cerotinsäure *f* hexacosanoic acid, *(deprecated:)* cerotic acid
Cer(III)-oxid *n* cerium(III) oxide, cerium trioxide
Cer(IV)-oxid *n* cerium(IV) oxide, cerium dioxide
Certetrafluorid *n s.* Cer(IV)-fluorid
Cetan *n* *hexadecane, *(deprecated:)* cetane
Cetanzahl *f* cetane number [rating]
Cetanziffer *f* cetane number [rating]
CFC-Verbundwerkstoff *m* carbon-carbon composite, carbon fibre-reinforced carbon, CFC
CFR-Motormethode *f* CFR motor method *(for measuring the antiknock qualities of fuels)*
Chalkogen *n* chalcogen *(any of the elements oxygen, sulphur, selenium, tellurium)*
Chalkogenid *n* chalcogenide *(a binary compound of a chalcogen)*
Chalkon *n* chalcone, chalkone, *(specif)* $C_6H_5COCH{=}CHC_6H_5$ chalcone, chalkone, 1,3-diphenylpropenone, benzylideneacetophenone, benzalacetophenone
Chalkopyrit *m (min)* chalcopyrite, chalkopyrite, copper pyrites *(copper(I) iron(III) sulphide)*
chaotrop *(bioch)* chaotropic
Charakter *m* 1. *(ch)* character; 2. base *(of crude petroleum)*
Charakter *m*/**amphoterer** amphotericity
Charakter *m*/**aromatischer** aromatic character, aromaticity
Charakter *m*/**edler** *(el ch)* noble character [nature], nobility
Charakter *m*/**ungesättigter** unsaturation
Charakter *m*/**wasserabweisender** water repellency
Charakterisierung *f* characterization *(as of chemical compounds)*
Charakterisierung *f* **durch eine Formel** formulation
Charakteristik *f* characteristics; characteristic curve
Charge *f* charge (stock), batch, feed material, feed (stock), load; batch *(product)*
Chargenbetrieb *m* batch operation
Chargendestillation *f* batch distillation
Chargenmischer *m* batch mixer
Chargennitrierung *f* batch nitration
Chargentrockner *m* batch dryer
Chargenverfahren *n* batch process

chargenweise batchwise
chargieren *v* 1. charge, load, feed, fill; 2. *(text)* weight *(add sizing material)*
C-Harz *n* C-stage resin, resite
C–H-Bindung *f* carbon-hydrogen bond
Chelat *n s.* Chelatverbindung
Chelataustauscher *m s.* Chelatharz
Chelatbildner *m* chelating agent
Chelatbildung *f* chelation, chelate formation
Chelatbildungskonstante *f* chelate formation constant
Chelateffekt *m* effect of chelation, chelate effect
chelatgebunden chelated
Chelatharz *n* chelate (ion-exchange) resin, chelating (ion exchange) resin
chelatisieren *v* chelate
Chelatkomplex *m s.* Chelatverbindung
Chelatligand *m* chelating ligand
Chelatometrie *f* chelatometry, chelatometric titration
chelatometrisch chel(at)ometric
Chelator *m s.* Chelatbildner
Chelatring *m* chelate ring
Chelatstabilitätskonstante *f* chelate stability constant
Chelatverbindung *f* chelate (compound, complex), crab's-claw complex
Chemie *f* chemistry
Chemie *f***/aliphatische** aliphatic chemistry
Chemie *f***/allgemeine** general chemistry
Chemie *f***/analytische** analytical chemistry
Chemie *f***/angewandte** applied chemistry
Chemie *f***/anorganische** inorganic chemistry
Chemie *f***/beschreibende** descriptive chemistry
Chemie *f***/biologische** biological chemistry, biochemistry
Chemie *f* **der Atmosphäre** chemistry of atmosphere, atmospheric chemistry
Chemie *f* **der Hochpolymeren** polymer [high polymeric] chemistry
Chemie *f* **der Kohlenstoffverbindungen** chemistry of the carbon compounds
Chemie *f* **der Koordinationsverbindungen** chemistry of coordination compounds, coordination chemistry
Chemie *f* **der Milch (und Milchprodukte)** dairy chemistry
Chemie *f* **des Wassers** water [aquatic] chemistry, hydrochemistry
Chemie *f***/deskriptive** descriptive chemistry
Chemie *f* **exotischer Atome** chemistry of exotic atoms
Chemie *f***/experimentelle** experimental chemistry
Chemie *f***/forensische** forensic [legal] chemistry
Chemie *f***/fotografische** photographic chemistry, chemistry of photography
Chemie *f***/geologische** geological chemistry, geochemistry
Chemie *f***/gerichtliche** *s.* Chemie/forensische
Chemie *f***/heiße** *s.* Chemie heißer Atome
Chemie *f* **heißer Atome** hot-atom chemistry, recoil chemistry
Chemie *f* **hoher Energien** high-energy chemistry
Chemie *f***/industrielle** industrial chemistry
Chemie *f***/klinische** clinical chemistry
Chemie *f***/kosmetische** cosmetic chemistry
Chemie *f***/landwirtschaftliche** agricultural chemistry, agrochemistry
Chemie *f***/makromolekulare** polymer [high-polymeric] chemistry
Chemie *f***/medizinische** medical chemistry
Chemie *f***/metallurgische** metallurgical chemistry
Chemie *f***/mineralogische** mineral(ogical) chemistry
Chemie *f* **neuer Atome** chemistry of exotic atoms
Chemie *f***/organische** organic chemistry
Chemie *f***/pharmazeutische** pharmaceutic(al) chemistry, medicinal chemistry
Chemie *f***/physikalische** physical chemistry
Chemie *f***/physiologische** physiologic(al) chemistry
Chemie *f***/präparative** preparative chemistry
Chemie *f***/reine** pure chemistry
Chemie *f***/synthetische** synthetic chemistry

Chemie *f*/synthetische organische synthetic organic chemistry
Chemie *f*/technische technological chemistry
Chemie *f*/theoretische theoretical chemistry
Chemie *f*/toxikologische toxicological chemistry
Chemieabwasser *n (hyd)* waste (water) from chemical plants, chemical plant's waste
Chemieanlage *f* chemical plant
Chemieanlagenbau *m* chemical-plant construction
Chemiebetrieb *m* chemical works [plant]
Chemiefaser *f (text)* man-made fibre, staple (fibre) *(a fibre of relatively short length cut from continuous filaments)*
Chemiefaserstoff *m (text)* man-made fibre
Chemiefeindlichkeit *f* chemophobia
Chemiegrundstoff *m* basic chemical (material)
Chemieindustrie *f* chemical (processing) industry
Chemieingenieur *m* chemical engineer
Chemieingenieurtechnik *f* chemical engineering technology
Chemieingenieurwesen *n* chemical engineering
Chemieingenieurwissenschaft *f* chemical engineering science
Chemielabor *n* chemistry laboratory
Chemielaborant *m* laboratory assistant
Chemiemüll *m* chemical waste(s)
Chemiepumpe *f* chemical [process] pump
Chemierohstoff *m* chemical raw material
Chemieseide *f* man-made continuous filament yarn
Chemietechnik *f* chemical technology
chemietechnisch chemical-technological, chemico-technological
Chemietechnologe *m* chemical technologist
Chemiewerk *n* chemical works [plant]
Chemikalie *f* chemical
Chemikalie *f*/industrielle industrial chemical
Chemikalien *fpl*/dosierte [zugesetzte] chemical feed
chemikalienbeständig resistant to chemicals, chemical-resistant
Chemikalienbeständigkeit *f* resistance to chemicals
Chemikaliendosierung *f* chemical feeding
chemikalienfest *s.* chemikalienbeständig
chemikalienresistent *s.* chemikalienbeständig
Chemikalienrückführung *f* chemical recovery, recovery [conservation] of chemicals
Chemikalienverlust *m* loss of chemicals, chemical loss
Chemikalienzugabe *f* chemical addition [feeding, introduction]
Chemikalienzusatz *m* 1. *s.* Chemikalienzugabe; 2. *(quantitatively:)* chemical feed
Chemiker *m* chemist
Chemiker *m*/industrieller industrial chemist
Chemiker *m*/organischer *s.* Organochemiker
Chemiker *m*/technischer technical [technological] chemist
Chemikerin *f* woman chemist
Chemilumineszenz *f* chemiluminescence
chemilumineszierend chemiluminescent
chemisch chemical • **auf chemischem Wege** by a chemical route, by chemical means, chemically
chemisch-technisch chemical-technological, chemico-technological
chemisch-technologisch chemical-technological, chemico-technological
Chemismus *m* chemism *(of a reaction)*
chemisorbieren *v* chemisorb, chemosorb
Chemisorption *f* chemisorption, chemosorption, chemical [activated] adsorption
Chemophobie *f* chemophobia
chemoselektiv chemoselective *(reagent)*
Chemoselektivität *f* chemoselectivity *(of a reagent)*
Chemostat *m (biot)* chemostat
Chemosteril(is)ans *n* chemosterilant

Chemosterilisierungsmittel *n* chemosterilant
Chemosynthese *f* chemosynthesis *(microbiology)*
chemosynthetisch chemosynthetic *(microbiology)*
Chemotaxis *f (biol)* chemotaxis
Chemotechniker *m* chemical technician
Chemotherapeutikum *n* chemotherapeutant, chemotherapeutic agent
chemotherapeutisch chemotherapeutic(al)
Chemotherapie *f* chemotherapy, chemotherapeutics
chemotroph chemo(auto)trophic
Chemotrophie *f (biol)* chemoautotrophy *(gain of assimilation energy from oxidation processes)*
Chemurgie *f* chemurgy *(industrial utilization of organic raw materials)*
Chilesalpeter *m* Chile saltpetre [nitre, nitrate], Chilean [Chilian] nitrate, soda nitre *(sodium nitrate)*
Chinaalkaloid *n s.* Chinarindenalkaloid
Chinaldin *n* quinaldine, 2-methylquinoline
Chinaldinsäure *f* quinaldic acid, quinoline-2-carboxylic acid
Chinalizarin *n* quinalizarin, 1,2,5,8-tetrahydroxyanthraquinone
Chinarindenalkaloid *n* cinchona [quinoline-type] alkaloid
Chinasäure *f* quinic acid, *1,3,4,5-tetrahydroxycyclohexane-1-carboxylic acid
Chinazolin *n* quinazoline, 5,6-benzpyrimidine
Chinhydron *n (org ch)* quinhydrone
Chinhydronelektrode *f* quinhydrone electrode
Chinidin *n* quinidine *(a cinchona alkaloid)*
Chinin *n* quinine *(a cinchona alkaloid)*
Chinizarinkondensation *f* quinizarin condensation
chinoid quin(on)oid
chinoid/nicht non-quinonoid
Chinolat *n* quinolate *(salt or ester of quinolinic acid)*
Chinolin *n* quinoline, chinoline, 2,3-benzpyridine
Chinolinalkaloid *n* quinoline alkaloid
Chinolinblau *n* cyanine blue, pigment blue 15
Chinolinfarbstoff *m* quinoline dye
Chinolinsäure *f* quinolinic acid, pyridine-2,3-dicarboxylic acid
Chinolinsynthese *f***/friedländersche** Friedländer quinoline synthesis
Chinolinsynthese *f***/skraupsche** Skraup quinoline synthesis
Chinolizidin *n* quinolizidine
Chinolizidinalkaloid *n* quinolizidine [lupin] alkaloid
Chinolizidinring *m* quinolizidine ring
Chinon *n* quinone, *(specif) p*-benzoquinone, cyclohexadiene-1,4-dione
Chinonfarbstoff *m* quin(on)oid dye
Chinoniminfarbstoff *m* quinone imine dye
Chinonoxim *n* quinonoxime
Chinonring *m* quinone ring
Chinoxalin *n* quinoxaline, benzo-1,4-diazine
chiral chiral *(stereochemistry)*
Chiralität *f* chirality *(stereochemistry)*
Chiralitätszentrum *n* chiral centre *(stereochemistry)*
Chitin *n* chitin *(a polysaccharide)*
chitinig chitinous
chitinös chitinous
Chitosamin *n* chitosamine, D-glucosamine, GlcN, 2-amino-2-deoxy-D-glucopyranose
Chlor *n* Cl chlorine • **mit Chlor behandeln** chlorinate, chlorinize
Chlor *n***/aktives** *s.* Chlor/wirksames
Chlor *n***/freies (wirksames)** *(hyd)* free (available) chlorine
Chlor *n***/gebundenes** *(hyd)* combined chlorine
Chlor *n***/gelöstes organisches [organisch gebundenes]** *(hyd)* dissolved organic chlorine, DOCl
Chlor *n***/wirksames** active [available] chlorine
Chloracetal *n* chloroacetal, diethyl chloroacetal
Chloraceton *n* chloroacetone, *chloropropanone
Chloracetophenon *n* chloroacetophenone, *(specif)* α-chloroacetophenone
Chloracetylierung *f* chloroacetylation
Chlorakne *f (med)* chloracne
Chloral *n* chloral, trichloroacetaldehyde
Chloralhydrat *n* chloral hydrate, trichloroacetaldehyde hydrate

Chloralkalielektrolyse *f* chlor-alkali electrolysis [operation]
Chloralkali-Elektrolysezelle *f* chlor-alkali [alkali-chlorine] cell
Chlorameisensäure *f* chloroformic acid
Chlorameisensäureester *m* chloroformic acid ester, chloroformate, chlorocarbonate
Chlorameisensäuretrichlormethylester *m* trichloromethyl chloroformate, diphosgene
Chloramin *n* chloramine, *N*-chloroamine, *(specif)* NH_2Cl chloramine
***N*-Chloramin** *n* chloramine, *N*-chloroamine, *(specif)* NH_2Cl chloramine
Chloraminierung *f* chloroamination
Chloranil *n* chloranil, tetrachlor-*p*-benzoquinone
Chloranilin *n* chloroaniline
Chlorarylierung *f* chloroarylation
Chlorat *n* M^IClO_3 chlorate
Chlorbenzen *n* chlorobenzene
Chlorbenzol *n* chlorobenzene
Chlorbleiche *f* 1. chlorine bleaching; 2. *(text)* chemic(k)
Chlorbutylkautschuk *m* chlorobutyl rubber, chlorinated butyl rubber
Chlorcalciumröhrchen *n (lab)* calcium chloride tube
Chlorcalciumzylinder *m* (gas) drying jar
Chlorcyan *n* cyanogen chloride, chlorine cyanide
Chlorcyanhydrin *n* chlorocyanohydrin
Chlordehydroxylierung *f* chlorodehydroxylation
2-Chlor-*N,N*-diallylacetamid *n* 2-chloro-*N,N*-diallylacetamide, CDAA *(a herbicide)*
Chlordioxid *n* chlorine dioxide
Chlordioxidbleiche *f (pap)* chlorine dioxide bleaching
chloren *v* chlorinate
Chloressigsäure *f* chloroacetic acid
Chlorethan *n* chloroethane
2-Chlorethanol *n* 2-chloroethanol, 2-chloroethyl alcohol
Chlorethansäure *f* chloroethanoic acid, chloroacetic acid
Chlorethen *n* *chloroethene, vinyl chloride
Chlorethylgruppe *f* $ClCH_2CH_2-$ chloroethyl group
Chlorethylierung *f* chloroethylation
Chlorfluorierung *f* chlorofluorination
Chlorgas *n* chlorine gas
Chlorgeruch *m (hyd)* chlorine odour
Chlorgeschmack *m (hyd)* chlorinous taste
Chlorheptoxid *n s.* Chlor(VII)-oxid
Chlorhydrin *n* chlorohydrin
Chlorid *n* M^ICl chloride
Chloridhärte *f (hyd)* chloride hardness
Chloridschmelze *f* chloride melt
Chloridsole *f* chloride brine
chlorieren *v* chlorinate
Chlorierung *f* chlorination
Chlorierungskessel *m* chlorinating vessel, chlorinator
Chlorierungsmittel *n* chlorinating agent
Chlorierungsprodukt *n* chlorinated product, chlorinate
Chlorinität *f* chlorinity, Cl%
Chloriodierung *f* chloroiodination, iodochlorination
Chlorit *n* M^IClO_2 chlorite
Chloritbleiche *f (text)* chlorite bleaching
Chlorkalk *m* chlorinated lime, chloride of lime
Chlorkautschuk *m* chlorinated rubber
Chlorkohlensäure *f s.* Chlorameisensäure
Chlorkohlenwasserstoff *m* chlorinated hydrocarbon
Chlormethan *n* *chloromethane, methyl chloride
Chlormethoxylierung *f* chloromethoxylation
Chlormethylierung *f* chloromethylation
Chloroform *n* chloroform, trichloromethane
Chlorokomplex *m* chlorocomplex
Chlorolyse *f* chlorinolysis
Chloroniumverbindung *f* chloronium compound
Chlorophyll *n (bioch)* chlorophyll, leaf green
Chlorophyllid *n (bioch)* chlorophyllide
Chloroplatinat(II) *n* $M^I_2[PtCl_4]$ chloroplatinate(II), tetrachloroplatinate(II)
Chloroplatin(IV)-säure *f* chloroplatinic(IV) acid, hexachloroplatinic(IV) acid
Chloropren *n* chloroprene, *2-chlorobuta-1,3-diene
Chloroprenkautschuk *m* chloroprene rubber, CR

Chlorosäure *f* chloro acid *(coordination chemistry)*
Chloroschwefelsäure *f* chlorosulphuric acid
Chlorosulfat *n* chlorosulphate
Chlor(I)oxid *n* chlorine(I) oxide, dichlorine monooxide
Chlor(IV)-oxid *n* chlorine(IV) oxide, chlorine dioxide
Chlor(VII)-oxid *n* chlorine(VII) oxide, dichlorine heptaoxide
Chlorphenol *n* chlorophenol, chlorinated phenol
Chlorphenylierung *f* chlorophenylation
Chlorphosphonierung *f* chlorophosphonation, chlorophosphonylation
Chlorphosphonylierung *f* chlorophosphonation, chlorophosphonylation
3-Chlorpropan-1,2-diol *n* *3-chloropropane-1,2-diol, (glycerol) α-monochlorohydrin
Chlorpropanon *n* *chloropropanone, chloroacetone
Chlorsäure *f* chloric acid
Chlorsilan *n* chlorosilane, *(specif)* SiH_3Cl chloromonosilane
Chlorsilan *n***/organisches** organochlorosilane
chlorsulfonieren *v* chlorosulphonate
Chlorsulfonierung *f* chlorosulphonation
Chlortetroxid *n* chlorine tetraoxide
α-Chlortoluen *n* α-chlorotoluene
α-Chlortoluol *n* α-chlorotoluene
Chlortrifluorethylen *n* chlorotrifluoroethylene, CFE
Chlorüberschuss *m (hyd)* (free) residual chlorine, (amount of) chlorine residual, free (available) chlorine residual, excess (of) chlorine
Chlorung *f (hyd)* chlorination
Chlorungsmittel *n (hyd)* chlorinating agent, chlorination chemical
Chlorverbindungen *fpl***/extrahierbare organische** *(hyd)* extractable organic chlorine compounds, EOCl
Chlorwasser *n* 1. chlorine water; 2. *(hyd)* chlorine solution
Chlorwasserstoff *m* hydrogen chloride
Chlorwasserstoff *m***/gasförmiger** *s.* Chlorwasserstoffgas
Chlorwasserstoff *m***/trockener [wasserfreier]** anhydrous hydrogen chloride
Chlorwasserstoffgas *n* hydrogen chloride gas, gaseous hydrogen chloride
Chlorwasserstoffsäure *f* hydrochloric acid
Cholesterin *n s.* Cholesterol
cholester(in)isch cholesteric
Cholesterol *n (bioch)* cholesterol
Cholsäure *f* cholic acid, 3,7,12-trihydroxy-5b-cholan-24-oic acid
Chrom *n* Cr Chrom
Chromalaun *m* chrome alum, *(specif)* $KCr(SO_4)_2 \cdot 12H_2O$ potassium [potash] chrome alum, chrome potash alum
Chromat *n* $M^I_2CrO_4$ chromate
chromat(is)ieren *v* chromate, chromatize
Chromatogramm *n* chromatogram
Chromatographie *f* chromatography
Chromatographie *f***/absteigende** descending chromatography
Chromatographie *f***/hydrophobe** hydrophobic (interaction) chromatography
Chromatographie *f***/mehrdimensionale** multidimensional chromatography, MDC
Chromatographie *f***/temperaturprogrammierte** programmed-temperature (gas) chromatography
Chromatographie *f***/zweidimensionale** two-dimensional chromatography
Chromatographiekammer *f* chromatographic [chromatography] chamber, chromatographic cabinet
Chromatographiepapier *n* chromatographic paper
chromatographieren *v* chromatograph
Chromatographiesäule *f* chromatographic column
chromatographisch chromatographic
Chromatometrie *f* chromatometry, dichromate titration
chromatometrisch chromatometric
Chrombeize *f (text)* chrome [chromium] mordant
Chromcarbonyl *n* chromium carbonyl, chromium hexacarbonyl
Chrom(II)-chlorid *n* chromium(II) chloride, chromium dichloride
Chrom(III)-chlorid *n* chromium(III) chloride, chromium trichloride
Chromdi... *siehe auch: Chrom(II)-...*
chromdiffundieren *v* chromize

Chromdioxid *n s.* Chrom(IV)-oxid
Chromeisenerz *n s.* Chromit
2*H*-Chromen-2-on *n s.* α-Chromon
Chromentwicklungsfarbstoff *m* after-chrome [chrome-developed] dye
Chromfarbstoff *m s.* Chromierungsfarbstoff
Chromgelb *n* chrome yellow, lemon chrome *(mixture or mixed crystals of lead chromate and lead sulphate)*
Chromgerbbrühe *f (tann)* chrome liquor
chromgerben *v* chrome
Chromgerbung *f* chrome [chromium] tannage, chroming
Chrom(II)-hydroxid *n* chromium(II) hydroxide
Chrom(III)-hydroxid *n* chromium(III) hydroxide
chromieren *v (dye)* chrome; chromize *(metals)*
Chromierung *f (dye)* chroming; chromizing *(of metals)*
Chromierungsfarbstoff *m* chrome (mordant) dye
Chromit *m (min)* chromite, chrome iron ore, chromic iron *(iron(II) chromium(III) oxide)*
Chromleder *n* chrome leather
Chrommolybdänstahl *m* chrome molybdenum steel
Chrommonoxid *n s.* Chrom(II)-oxid
Chromnickelstahl *m* chrome nickel steel
Chromogen *n* chromogen *(a compound containing a chromophore)*
Chromoisomer(e) *n* chromoisomer
Chromoisomerie *f* chromoisomerism
α-Chromon *n* α-chromone, 1,2-benzopyrone, coumarin, 2*H*-chromen-2-one
chromophor chromophoric
Chromophor *m* chromophore, *(specif)* chromophoric [colour-bearing] group
Chrom(II)-oxid *n* chromium(II) oxide, chromium monooxide
Chrom(III)-oxid *n* chromium(III) oxide, dichromium trioxide
Chrom(IV)-oxid *n* chromium(IV) oxide, chromium dioxide
Chrom(VI)-oxid *n* chromium(VI) oxide, chromium trioxide
Chrom(VI)-oxidchlorid *n s.* Chromylchlorid
Chromoxidgrün *n* chrome (oxide) green, green cinnabar, oil green *(chromium(III) oxide)*
Chromoxidhydratgrün *n* chrome [emerald] green, Mittler's [Guignet's] green, transparent chromium oxide *(hydrated chromium oxide)*
Chromrot *n* chromate [chrome, Persian] red, American vermilion *(a basic lead chromate)*
Chromsäure *f* chromic acid
Chromsäureverfahren *n* chromic-acid process, Bengough-Stuart process *(anodic oxidation)*
Chromschwefelsäure *f (lab)* (chromic acid) cleaning mixture
Chromstahl *m* chromium [chrome] steel
Chrom(II)-sulfat *n* chromium(II) sulphate
Chrom(III)-sulfat *n* chromium(III) sulphate
Chromtri... *siehe auch: Chrom(III)-...*
Chromtrioxid *n s.* Chrom(VI)-oxid
Chromylchlorid *n* chromyl chloride, chromium(VI) dichloride dioxide
Chromzinnober *m s.* Chromrot
Chronopotenziometrie *f (anal)* chronopotentiometry
Chrysatropasäure *f* chrysatropic acid, scopoletin, 7-hydroxy-6-methoxycoumarin
Chrysen *n* chrysene, 1,2-benzophenanthrene
CI = Colour Index
Čičibabin-Kohlenwasserstoff *m* Chichibabin hydrocarbon
CIDEP *s.* Elektronenpolarisation/chemisch induzierte dynamische
CI-Massenspektrometer *n* chemical ionization mass spectrometer
Cinchonaalkaloid *n s.* Chinarindenalkaloid
cine-Substitution *f (org ch)* cine-substitution
Cinnamat *n* cinnamate *(salt or ester of cinnamic acid)*
Cinnamsäure *f s.* Zimtsäure
CIP-System *n s.* Cahn-Ingold-Prelog-System
cis-Konformation *f* cis conformation, C *(torsion angle ± 180°)*
cis-Leiter-Konformation *f (bioch)* ladder conformation *(of deoxyribonucleic acid)*
cisoid cisoid

cis-ständig cis • **cis-ständig (angeordnet) sein** be cis
cis-Stellung *f* cis position
cis-taktisch *(plast)* cistactic
cis-Taktizität *f (plast)* cistacticity
cis-trans-Isomer(e) *n* cis-trans isomer, geometric isomer
cis-trans-Isomerie *f* cis-trans isomerism, geometric isomerism
cis-trans-Umwandlung *f* cis-trans conversion
Citral *n* **A** citral a, geranial, **trans*-3,7-dimethylocta-2,6-dienal
Citral *n* **B** citral b, neral, **cis*-3,7-dimethylocta-2,6-dienal
***cis*-Citral** *n s.* Citral B
***trans*-Citral** *n s.* Citral A
Citrat *n* citrate *(salt or ester of citric acid)*
Citrat(*si*&)-Synthase *f (bioch)* citrate-condensing enzyme
Citratzyklus *m (bioch)* citric-acid [tricarboxylic-acid] cycle, TCA cycle, Krebs cycle
Citronensäure *f* citric acid, *2-hydroxypropane-1,2,3-tricarboxylic acid
Citronensäuregärung *f* citric-acid fermentation
Citronensäurezyklus *m s.* Citratzyklus
Citrovorumfaktor *m* citrovorum factor, leucovorin, N^5-formyltetrahydrofolic acid, folinic acid
C_6-Kette *f* six-carbon chain
C_3-Körper *m (bioch)* three-carbon compound
Claisen-Kolben *m* Claisen (distilling) flask
Claisen-Kondensation *f* Claisen condensation *(between esters or between esters and ketones)*
Claisen-Umlagerung *f* Claisen rearrangement *(O-allyl C-allyl ortho rearrangement)*
Clathrat *n* clathrate (inclusion compound), cage compound
Clathratverbindung *f* clathrate (inclusion compound), cage compound
Claude-Verfahren *n* Claude process *(synthesis of ammonia from nitrogen and hydrogen)*
Claus-Anlage *f* Claus (sulphur) plant
Clemmensen-Reduktion *f* Clemmensen reduction *(of aldehydes or ketones to hydrocarbons)*
Click-Chemie *f (org ch)* click chemistry
Click-Reaktion *f (org ch)* click reaction
Clusius-Dickel-Verfahren *n* Clusius-Dickel method, thermal-diffusion method *(for separating isotopes)*
Clusius-Trennrohr *n* Clusius column *(for separating isotopes)*
Cluster *m* cluster
CM-Cellulose *f* CM cellulose, carboxymethylcellulose, CMC
CMP = kapazitiv gekoppeltes Mikrowellenplasma
CoA *s.* Coenzym A
Cobalamin *n (bioch)* cobalamine, corrinoid, corrin, corphyrin, vitamin B12 compound
Cobalt *n* Co cobalt
Cobalt... *siehe auch: Kobalt... (for technical terms)*
Cobalt(II)-acetat *n* cobalt(II) acetate
Cobaltammin *n* cobalt ammine
Cobaltat *n* cobaltate
Cobaltchelat *n* cobalt chelate
Cobalt(II)-chlorid *n* cobalt(II) chloride, cobalt dichloride
Cobaltdi... *siehe: Cobalt(II)-...*
Cobalt(II)-hexafluorosilicat *n* cobalt(II) hexafluorosilicate, cobalt fluorosilicate
Cobaltiak *n* cobalt(III) ammine
Cobaltkatalysator *m* cobalt catalyst
Cobaltkontakt *m* cobalt catalyst
Cobaltmonoxid *n s.* Cobalt(II)-oxid
Cobalt(II)-nitrat *n* cobalt(II) nitrate
Cobalt(II)-oxid *n* cobalt(II) oxide, cobalt monooxide
Cobalt(III)-oxid *n* cobalt(III) oxide, dicobalt trioxide
Cobalt(II)-sulfat *n* cobalt(II) sulphate
Cobalt(III)-sulfat *n* cobalt(III) sulphate
Cobalttrioxid *n s.* Cobalt(III)-oxid
Code *m*/**genetischer** *(bioch)* genetic code
Codeeinheit *f s.* Codon
Codestillation *f* codistillation
Codetriplett *n s.* Codon
codieren *v (bioch)* code *(the sequence of amino acids)*
Codon *n (bioch)* codon, coding [nucleotide] triplet
C–O-Doppelbindung *f* carbonyl double bond

CO_2-Entfernung *f* carbon-dioxide removal
Coenzym *n* coenzyme, prosthetic [active] group
Coenzym *n* **A** coenzyme A, HSCoA
Coenzymatisch *(bioch)* coenzymatic
Coesit *n* coesite *(high-pressure modification of SiO_2)*
Cofaktor *m (biot)* cofactor *(enzyme technology)*
Coffein *n* caffeine, 1,3,7-trimethylxanthine
coffeinfrei caffeine-free, decaffeinated
coffeinhaltig caffeinic, containing caffeine
CO_2-Fixierung *f* CO_2 fixation
C=O-Gruppe *f* C=O group, carbonyl group
CO-Hämoglobin *n* carboxyhaemoglobin, carbon monoxy haemoglobin
Coion *n (phys ch)* co-ion
Cokatalysator *m* co-catalyst
Cokatalyse *f* co-catalysis
CO-Konvertierung *f (coal)* carbon monoxide conversion [shift], CO conversion, water-gas shift, shift conversion
Colaminkephalin *n* 3-phosphatidyl ethanolamine
Comonomer(e) *n* comonomer
compoundieren *v* compound *(oils)*
Compton-Effekt *m* Compton effect, Compton scattering
Compton-Elektron *n* Compton (recoil) electron
Compton-Streuung *f s.* Compton-Effekt
Conchinin *n* quinidine, conquinine, conchinine *(a cinchona alkaloid)*
Coniin *n* coniine, 2-propylpiperidine *(alkaloid)*
COOH-Gruppe *f* –COOH carboxyl group
Cope-Eliminierung *f* Cope elimination *(pyrolysis of amine oxides)*
Cope-Umlagerung *f (org ch)* Cope rearrangement
Copigment *n (bioch)* copigment
Copolyaddition *f* copolyaddition
Copolykondensation *f* copolycondensation
Copolymer(e) *n* copolymer
Copolymer(e) *n***/alternierendes** alternating copolymer
Copolymer(e) *n***/statistisches** random copolymer
Copolymerisat *n s.* Copolymer(e)
Copolymerisation *f* copolymerization
Copolymerisation *f***/alternierende** alternating copolymerization
Copolymerisation *f***/azeotrope** azeotropic copolymerization
Copolymerisation *f***/statistische** random copolymerization
Copolymerisationsparameter *m* (copolymerization, monomer) reactivity ratio
copolymerisieren *v* copolymerize
Cordieritkeramik *f* cordierite ceramics
Corepressor *m* corepressor *(enzyme kinetics)*
Cornu-Prisma *n (spectr)* Cornu prism
Coronand *m* coronand *(monocyclic crown compound)*
Coronat *n* coronate *(chelate complex of a coronand)*
Corrin *n (bioch)* corrin, corrinoid, cobalamine, corphyrin, vitamin B12 compound
Corrinoid *n (bioch)* corrin, corrinoid, cobalamine, corphyrin, vitamin B12 compound
Corticoid *n (bioch)* corticoid
Corticosteroid *n (bioch)* corticosteroid
Corticosteron *n* corticosterone *(an adrenocortical hormone)*
Corticotropin *n* corticotropin, adrenocorticotropic hormone, ACTH
Cortisol *n* cortisol *(an adrenocortical hormone)*
Cortison *n* cortison *(an adrenocortical hormone)*
Co-Stufenpolymerisation *f* step-growth [step-reaction] copolymerization
Cosubstrat *n (bioch)* cosubstrate
Cosynthese *f (biot)* cosynthesis
Co-Transport *m (bioch)* symport *(one form of carrier transport)*
Cottrell-Abscheider *m* Cottrell precipitator
Cottrell-Staubfilter *n* Cottrell precipitator
Couepinsäure *f* couepic acid, licanic acid *(either of two isomeric oxoalkenoic acids)*
Couette-Strömung *f* Couette flow *(rheology)*

Couette-Viskosimeter *n* Couette viscometer, rotating-cylinder viscometer of Couette
Coulomb-Abstoßung *f* Coulomb repulsion
Coulomb-Anziehung *f* Coulomb attraction
Coulomb-Gesetz *n* Coulomb law
Coulomb-Wechselwirkung *f* coulombic interaction
Coulometer *n* coulo(mb)meter, voltameter
Coulometrie *f* coulometry
Coulometrie *f***/amperostatische** amperostatic [galvanostatic] coulometry, coulometry at constant current
Coulometrie *f* **bei konstantem Potenzial** *s.* Coulometrie/potenzialkontrollierte
Coulometrie *f* **bei konstanter Stromstärke** *s.* Coulometrie/amperostatische
Coulometrie *f***/galvanostatische** *s.* Coulometrie/amperostatische
Coulometrie *f* **mit kontinuierlich geändertem Potenzial** potential-scanning coulometry
Coulometrie *f***/potenzialkontrollierte [potenziostatische]** potentiostatic [controlled-potential] coulometry
coulometrisch coulometric
Coulter-Verfahren *n (anal)* Coulter-method *(particle size determination)*
Covolumen *n (phys ch)* covolume
CP *s.* Cellulosepropionat
C-Peptid *n (bioch)* connecting peptide
C_3-Pflanze *f* C_3 plant *(producing three-carbon compounds as first intermediates of photosynthesis)*
C_4-Pflanze *f* C_4 plant *(producing four-carbon compounds as first intermediates of photosynthesis)*
C-Quelle *f (biot)* carbon source
Crack... *siehe: Krack...*
Cresol *n* cresol, hydroxytoluene, methylphenol
Cresolharz *n* cresol [cresylic] resin
Cresolrot *n* cresol red, *o*-cresolsulphonephthalein
Cross-Flow-Filtration *f (biot)* cross-flow filtration
Crotonaldehyd *m* crotonaldehyde, *but-2-enal
Crotonisierung *f* crotonization
Crotonsäure *f* crotonic acid, *but-2-enoic acid, *(specif) trans*-but-2-enoic acid
Cryptand *m* cryptand *(spherical crown compound)*
Cryptat *n* cryptate *(chelate complex of a cryptand)*
CS *s.* 1. Synthesefaserstoff; 2. Citronensäure
C_4-Säurenzyklus *m (bioch)* C_4 cycle [pathway], C_4 dicarboxylic acid cycle, Hatch-Slack (CO_2 incorporation) pathway
CSB *m (chemischer Sauerstoffbedarf)* COD, chemical oxygen demand
C_1-Stoffwechsel *m* one-carbon metabolism
CSV *s.* Sauerstoffverbrauch/chemischer
ct *s.* cis-taktisch
C-terminal C-terminal *(of amino acids in proteins)*
CT-Komplex *m* charge-transfer complex, electron-donor-acceptor complex
CTP = Cytidin-5′-triphosphat
Cuen *n s.* Kupferethylendiamin
Čugaev-Reaktion *f* Chugaev reaction *(for obtaining alkenes)*
Cumarin *n* coumarin, cumarin, 1,2-benzopyrone, 2*H*-chromen-2-one, α-chromone
Cumaron *n* benzofuran, *(deprecated:)* coumarone, cumarone
Cumaronharz *n* coumaron(-indene) resin
Cumaron-Inden-Harz *n* coumaron(-indene) resin
Cumarsäure *f* coumaric [cumaric] acid, hydroxycinnamic acid, *(specif) o*-coumaric acid, *trans-o*-hydroxycinnamic acid
Cumen *n* cumene, 2-phenylpropane, isopropylbenzene
Cumenhydroperoxid *n* cumene hydroperoxide, CHP
Cumol *n s.* Cumen
Cuprat *n* cuprate
Cuprit *m (min)* cuprite, red [ruby] copper ore *(copper(I) oxide)*
Curarealkaloid *n* curare alkaloid
Curie-Punkt *m* Curie point [temperature]

(marking the transition between ferromagnetism and paramagnetism)
Curie-Temperatur *f* Curie point [temperature] *(marking the transition between ferromagnetism and paramagnetism)*
Curium *n* Cm curium
Curtius-Abbau *m* Curtius degradation, Curtius reaction [rearrangement, transformation] *(of acid azides to give primary amines)*
Curtius-Reaktion *f* Curtius degradation, Curtius reaction [rearrangement, transformation] *(of acid azides to give primary amines)*
CVD-Beschichtungstechnik *f* chemical vapour deposition, CVD *(via decomposable gaseous intermediates, mostly halides)*
CVD-Verfahren *n* chemical vapour deposition, CVD *(via decomposable gaseous intermediates, mostly halides)*
C_3-Verbindung *f (bioch)* three-carbon compound
C_4-Verbindung *f (bioch)* four-carbon compound
CW-Technik *f (anal)* continuous-wave technique
Cyan *n* cyanogen, oxalonitrile
Cyanalkylierung *f* cyanoalkylation
Cyanamid *n* cyanamide, carbodiimide
Cyanat *n* ROC≡N cyanate *(salt or ester of cyanic acid)*
Cyanchlorid *n* chlorine cyanide, cyanogen chloride
Cyanethylierung *f* cyanoethylation
Cyangruppe *f* CN– cyano group
Cyanhydrin *n* cyanohydrin *(any of a class of compounds R′–C(OH)(CN)–R″)*
Cyanid *n* RCN cyanide *(salt or derivative of hydrogen cyanide)*
Cyanidlaugerei *f (met)* cyanidation, cyaniding
Cyanidlaugung *f (met)* cyanidation, cyaniding
cyanieren *v* cyanide *(steel)*
Cyanieren *n* cyaniding, cyanide (case-) hardening *(of steel)*
Cyanierung *f (org ch)* cyanation, cyanogenation, cyanylation
Cyanin *n* 1. cyanin *(an anthocyanin)*; 2. *s.* Cyaninfarbstoff
Cyaninfarbstoff *m* cyanine dye
Cyankali *n s.* Kaliumcyanid
Cyanmethylierung *f* cyanomethylation
Cyanoaurat(I) *n* $M^I[Au(CN)_2]$ cyanoaurate(I), dicyanoaurate(I)
Cyanoaurat(III) *n* $M^I[Au(CN)_4]$ cyanoaurate(III), tetracyanoaurate(II)
Cyanoferrat(II) *n* $M^I_4[Fe(CN)_6]$ cyanoferrate(II), hexacyanoferrate(II)
Cyanoferrat(III) *n* $M^I_3[Fe(CN)_6]$ cyanoferrate(III), hexacyanoferrate(III)
Cyanoniccolat *n* cyanoniccolate, cyanonickelate
Cyanoplatinat(II) *n* $M^I_2[Pt(CN)_4]$ cyanoplatinate(II), tetracyanoplatinate(II)
Cyanoplatinat(IV) *n* $M^I_4[Pt(CN)_6]$ cyanoplatinate(IV), hexacyanoplatinate(IV)
Cyanrest *m s.* Cyangruppe
Cyansäure *f* cyanic acid
Cyansäureester *m* ROC≡N cyanate
Cyanurchlorid *n* cyanuric chloride
Cyanursäure *f* cyanuric acid
Cyanursäurechlorid *n s.* Cyanurchlorid
Cyanwasserstoff *m* hydrogen cyanide
Cyanwasserstoffsäure *f* hydrogen cyanide
cyclisch cyclic
cyclisch/nicht acyclic, non-cyclic(al)
cyclisieren *v (org ch)* cyclize
Cyclisierung *f (org ch)* cyclization
Cyclisierungsreaktion *f* cyclization reaction
Cycloaddition *f (org ch)* cycloaddition
cycloaliphatisch cycloaliphatic, alicyclic
Cycloalkadien *n* cycloalkadiene
Cycloalkan *n* *cycloalkane, cyclane, naphthene
Cycloalkatrien *n* cycloalkatriene
Cycloalken *n* cycloalkene, cyclic olefin
Cycloalkenylierung *f* cycloalkenylation
Cycloalkin *n* cycloalkyne
Cycloalkylgruppe *f* cycloalkyl group
Cycloalkylierung *f* cycloalkylation
Cyclodehydratisierung *f* cyclodehydration
cyclodehydrieren *v* cyclodehydrogenate, dehydrocyclize
Cyclodehydrierung *f* cyclodehydrogenation, dehydrocyclization
Cyclodextrin *n* cyclodextrin, Schardinger dextrin
Cyclodimerisierung *f* cyclodimerization
Cycloeliminierung *f* cycloelimination,

cycloreversion, retroaddition, retrocycloaddition
Cycloheptatrienylium-Ion *n* cycloheptatrienylium ion, tropylium ion
Cyclohexan *n* *cyclohexane, hexahydrobenzene
Cyclohexanol *n* *cyclohexanol, hexahydrophenol
Cyclohexen *n* cyclohexene, tetrahydrobenzene
Cyclohexylierung *f* cyclohexylation
Cyclokautschuk *m* cyclorubber, cyclized rubber
Cyclokohlenwasserstoff *m* cyclic hydrocarbon
Cyclokondensation *f* cyclocondensation
Cycloocta-1,3,5,7-tetraen *n* cycloocta-1,3,5,7-tetraene, [8]annulene
Cycloolefin *n s.* Cycloalken
Cyclooligomerisation *f* cyclooligomerization
Cycloparaffin *n s.* Cycloalkan
Cyclopentadien *n* cyclopentadiene
Cyclopentadienyl-Anion *n* cyclopentadienyl anion
Cyclopentan *n* cyclopentane
Cyclopentylierung *f* cyclopentylation
Cyclopeptid *n* cyclic peptide
Cyclophan *n* cyclophane *(bridged aromatic system)*
Cyclo-phenylalkan *n* cyclophane *(bridged aromatic system)*
Cyclopropan *n* cyclopropane
Cyclopropenylkation *f* cyclopropenyl cation
Cyclosilicat *n (min)* cyclosilicate, ring silicate
Cyclotrimerisation *f* cyclotrimerization
Cyclotronstrahlung *f* cyclotron radiation
Cyclovoltammetrie *f (el ch)* cyclovoltammetry
cyclovoltammetrisch *(el ch)* cyclovoltammetric
Cymen *n* cymene, isopropyltoluene, isopropylmethylbenzene
Cymol *n* cymene, isopropyltoluene, isopropylmethylbenzene
Cystein *n* cysteine, *2-amino-3-mercaptopropionic acid
Cysteinsäure *f* cysteic acid, *2-amino-3-sulphopropionic acid
Cystin *n* cystine, dicysteine, *3,3′-dithiobis[2-aminopropanoic acid]
Cystin-Bindeglied *n* cystine link
Cystin-Brücke *f* cystine link
Cytidylsäure *f (bioch)* cytidylic acid
Cytochemie *f* cytochemistry
Cytochromoxidase *f* cytochrome [indophenol] oxidase, Warburg's respiratory enzyme
Cytochromreductase *f* cytochrome reductase
Cytokinin *n* cytokinin *(any of various plant growth factors)*
Cytotoxin *n* cytotoxin
cytotoxisch cytotoxic
CZ *s.* Cellulosechemiefaserstoff
Czerny-Turner-Aufstellung *f (spectr)* Czerny-Turner arrangement
C_3-Zyklus *m (bioch)* C_3 cycle [pathway], C_3 pentose phosphate cycle, Calvin(-Bassham) cycle, Calvin pathway, ribulose diphosphate cycle *(of CO_2 assimilation)*
C_4-Zyklus *m (bioch)* C_4 cycle [pathway], C_4 dicarboxylic acid cycle, Hatch-Slack (CO_2 incorporation) pathway

D

d- = dextrogyr
D *s.* Dichte
2,4-D *s.* 2,4-Dichlorphenoxyessigsäure
Daltonide *npl* daltonides, daltonian [daltonide] compounds
Dammar(harz) *n* dammar (resin), gum dammar *(esp from several species of the family Dipterocarpaceae)*
dAMP *s.* Desoxyadenosin-5′-phosphat
Dampf *m* vapour, *(specif)* water vapour, steam; fume *(visible volatile chemicals)* • **in Dampf überführen** vaporize • **mit Dampf behandeln** treat with steam, steam • **mit Dampf beheizt** steam-heated
Dampf *m*/**direkter** live steam, prime [direct, open] steam
Dampf *m*/**gesättigter** saturated vapour; saturated steam
Dampf *m*/**überhitzter** superheated steam
Dampfautoklav *m* steam autoclave
Dampfbad *n* steam bath

Dampfbehandlung *f* steam treatment, steaming
Dampfbeheizung *f* steam heating
Dampfblanchieren *n (food)* steam blanching
Dampfdestillation *f* steam distillation
Dampfdichte *f* vapour density, v.d.
Dampfdruck *m* vapour pressure [tension], v.p., VP
Dampfdruckerniedrigung *f* vapour-pressure lowering [depression]
Dampfdruckkurve *f* vapour-pressure curve
Dampfdruckosmose *f* vapour-pressure osmometry *(molar mass determination)*
Dampf-Eisen-Verfahren *n* steam-iron process *(for producing hydrogen)*
dampfen *v* steam
dämpfen *v* 1. steam; 2. damp(en) *(e.g. a motion)*; damp(en), attenuate *(e.g. the violence of a reaction)*
Dämpfen *n* 1. steaming; 2. damp(en)ing *(as of a motion)*; damp(en)ing, attenuation *(as of a violent reaction)*
Dampferzeuger *m* steam generator, boiler
Dampferzeugung *f* steam generation [raising]
dampfförmig vaporous
dampfgetrieben steam-driven
Dampfheizschlange *f* steam coil
Dampfkalorimeter *n* steam calorimeter
Dampfkanne *f (lab)* steam can
Dampfkessel *m* (steam) boiler
Dampfkesselkohle *f* steam(-raising) coal
Dampfleitung *f* steam line, *(if large:)* steam main
Dampfleitungsrohr *n* steam pipe [tube]
Dampfmantel *m* steam jacket • **mit Dampfmantel** steam-jacketed
Dampfphase *f* vapour phase
Dampfphasekracken *n (petrol)* vapour-phase cracking
Dampfphasenoxidation *f* vapour-phase oxidation
Dampfraum *m* 1. vapour head [chamber, space], flash chamber, body *(of an evaporator)*; 2. headspace *(vapour phase above liquid phase in a closed vessel)*
Dampfraumanalyse *f (chromat)* headspace (gas) analysis
Dampf-Sauerstoff-Vergasung *f (coal)* steam-oxygen gasification
Dampfstrahlejektor *m* steam-jet ejector, steam-motivated [steam-operated] ejector
Dampfstrahlpumpe *f* steam injector
Dampftrockner *m* steam dryer; vapour dryer *(using a vaporizable liquid drying agent)*
Dampfturbinenöl *n* steam-turbine oil
Dampfüberhitzer *m* steam superheater
Dämpfung *f* 1. steaming; 2. damp(en)ing *(as of a motion)*; 3. *(rubber)* hysteresis; damp(en)ing, attenuation *(as of a violent reaction)*
Dämpfungsfaktor *m* damping factor [coefficient]
Dampfventil *n* steam valve
Dampfvulkanisation *f (rubber)* steam curing [cure, vulcanization]
Dampfzersetzung *f* steam decomposition
Dampfzersetzungsgrad *m* (degree of) steam decomposition
Daniell-Element *n* Daniell cell
Daniell-Kette *f* Daniell cell
daniellscher Hahn *m (lab)* Daniell trap *(gas burner)*
Dansylierung *f (org ch)* dansylation
DAP *s.* Diallylphthalatharz
Darre *f* drying kiln, kiln dryer
darren *v* kiln-dry, kiln *(e.g. malt)*
Darrhorde *f* kiln floor
Darrofen *m* drying kiln, kiln dryer
darstellbar/rein isolable *(natural product)*
darstellen *v* prepare, make; isolate *(natural products)*
darstellen *v***/räumlich** represent spatially
darstellen *v***/rein** prepare in pure form, prepare in a pure condition [state], isolate *(natural products)*
Darstellung *f* 1. preparation, making; isolation *(of natural products)*; 2. representation *(as of measuring values)*
Darstellung *f***/grafische** graphical [diagrammatic] representation, graph, chart, diagram
Darstellung *f***/präparative** laboratory preparation
Darstellung *f***/räumliche** spatial representation

Darstellungsmethode *f* method of preparation, preparative method
Darzens-Erlenmeyer-Claisen-Kondensation *f s.* Darzens-Reaktion
Darzens-Reaktion *f* Darzens (glycidic ester) condensation
Daten *pl***/kritische** *(phys ch)* critical constants [data]
Dauer *f* **eines Arbeitszyklus** cycle time *(of a treatment unit)*
Dauerbetrieb *m* continuous operation [working]
Dauerbiegefestigkeit *f* repeated flexural strength, flex(ing) life
Dauerformguss *m* permanent-mould casting, gravity die-casting
dauerhaft durable, permanent, stable
Dauerkultur *f* continuous culture *(of microorganisms)*
Dauerschwingkorrosion *f* corrosion fatigue
Dauerstrichbetrieb *m (spectr)* continuous-wave technique
Dauerwärmebeständigkeit *f* continuous heat resistance
2,4-DB *s.* 4-(2′,4′-Dichlorphenoxy)buttersäure
DBPC *s.* 2,6-Di-*tert*-butyl-*p*-cresol
DC *s.* Dünnschichtchromatographie
dCMP *s.* Desoxycytidin-5′-phosphat
DDNP *s.* Diazodinitrophenol
DDT *s.* Dichlordiphenyltrichlorethan
Dead-stop-Titration *f* dead-stop titration
dealkylieren *s.* desalkylieren
dealuminieren *v* dealuminate
Dealuminierung *f* dealumination
de-Broglie-Beziehung *f* de Broglie relation(ship)
de-Broglie-Wellenlänge *f* de Broglie wavelength
debromieren *v* debrominate
Debromierung *f* debromination
Debutanisierung *f* debutanization
Debutylierung *f* debutylation
Debye *n s.* Debye-Einheit
Debye-Einheit *f* Debye (unit), D *(non-SI unit of dipole moment of molecules)*
Debye-Hückel-Theorie *f (phys ch)* Debye-Hückel theory
Debyeogramm *n s.* Debye-Scherrer Diagramm
Debye-Scherrer-Aufnahme *f (cryst)* Debye-Scherrer diagram [pattern, photograph]
Debye-Scherrer-Diagramm *n (cryst)* Debye-Scherrer diagram [pattern, photograph]
Debye-Scherrer-Methode *f (cryst)* Debye-Scherrer(-Hull) method
Debye-Temperatur *f* (Debye) characteristic temperature
Decahydrat *n* decahydrate
Decan *n* decane
Decandisäure *f* *decanedioic acid, sebacic acid
Decansäure *f* decanoic acid
Decapeptid *n* decapeptide
Decarbalkoxylierung *f* decarbalkoxylation
Decarbobenzoxylierung *f* decarbobenzoxylation
decarbonylieren *v* decarbonylate
Decarbonylierung *f* decarbonylation
decarboxylieren *v* decarboxylate
Decarboxylierung *f* decarboxylation
Decarboxylierung *f***/halogenierende** halodecarboxylation
Decarboxylierung *f***/oxidative** oxidative decarboxylation
Dec-1-en *n* dec-1-ene
Decensäure *f* decenoic acid
dechlorieren *v* dechlorinate
Dechlorierung *f* dechlorination
Dec-1-in *n* dec-1-yne
Deckanstrich *m* 1. *(coat)* finish(ing) *(act)*; 2. *(coat)* finish(ing) coat, finish, top coat(ing), topcoat, cover coat(ing)
Decke *f* 1. cover; 2. *s.* Deckgebirge
Deckel *m* lid, cap, top, cover, hood • **mit einem Deckel verschlossen [versehen]** lidded
decken *v* cover, top *(with a finish or another dye)*; cover *(as of a pigment)*
deckend/einander coincident *(stereochemistry)*
Deckfähigkeit *f s.* Deckvermögen
Deckgebirge *n (mine)* overburden, roof rock, rock cover
Deckglas *n* cover glass, *(microscopy also)* slide cover glass, glass cover *(solar energy)*
Deckgrün *n* chrome green *(a mixture of iron blue and chrome yellow)*
Deckkraft *f s.* Deckvermögen
Deckschicht *f* 1. top [final, cover] coat-

ing *(as of a protective coating system)*; coating *(spontaneously formed)*; *(if thin:)* film *(as of oxides)*; 2. *s.* Deckgebirge
Deckschicht f/oxidische oxide coating *(on metals)*
Deckung *f* coincidence *(in space or time)*
• **zur Deckung bringen** make (to) coincide
deckungsgleich superimposable
deckungsgleich/nicht non-superimposable
Deckvermögen *n* covering power, coverage, *(of paints also)* hiding [obliterating, opacifying] power
Decodierungsort *m s.* Akzeptorort
Decyanethylierung *f* decyanoethylation
Decyanierung *f* decyanation
Decyclisierung *f* decyclization, ring cleavage [opening]
Dedimerisierung *f* dedimerization
deethanisieren *v* de-ethanize
Deethanisierung *f* de-ethanization
defekt defective, faulty
Defekt *m* 1. defect, fault, *(in material also)* flaw; 2. *(cryst)* defect, imperfection
Defekt *m*/**eindimensionaler [linienhafter]** *(cryst)* line(ar) defect
Defektdichte *f* defect density
Defektelektron *n* defect electron, (electron) hole
Defektelektronenleitung *f s.* Defektleitung
Defekt(halb)leiter *m (phys ch)* defect (semi)conductor, p-type (semi)conductor
Defektleitung *f (phys ch)* hole conduction
defibrieren *v (pap)* defibre, defibrate, reduce to fibres, *(Am)* (de)fiberize
Defibrierung *f (pap)* defib(e)ring, defibration, *(Am)* (de)fiberization
definiert/gut well-defined *(e.g. compound)*
definiert/mangelhaft [schlecht] ill-defined, poorly defined
definiert/ungenau [unscharf] vaguely [faintly] defined
definiert/wohl well-defined *(compound)*
Deflagration *f* deflagration
deflagrieren *v* deflagrate
Defoliationsmittel *n (agric)* defoliant
Deformation f/bleibende residual [plastic, permanent] deformation, residual set
Deformation f/elastische elastic deformation
Deformation f/irreversible [plastische] *s.* Deformation/bleibende
Deformationsschwingung *f* deformation [scissor] vibration, deformation mode *(of molecules)*
deformieren *v* deform
Deformylierung *f* deformylation
degenerieren *v (phys ch)* degenerate
degeneriert *(phys ch, bioch)* degenerate
Degeneriertheit *f (bioch)* degeneracy *(of the genetic code)*
dehalogenieren *v* dehalogenate
Dehalogenierung *f* dehalogenation
dehnbar extensible, expansible, expandable, expandible, *(esp of metal:)* ductile
Dehnbarkeit *f* extensibility, expansibility, expandability, *(esp of metal:)* ductility
dehnen *v* extend, expand, stretch
Dehngrenze *f* tensile stress at a given elongation, yield strength, proof stress, *(deprecated:)* (tensile) modulus
0,2 %-Dehngrenze *f* 0.2 % offset yield strength [stress], 0.2 % proof stress, yield strength 0.2 % offset
Dehnung *f* extension, expansion, stretch, *(esp relating to materials testing:)* strain
Dehnung f/bleibende plastic [permanent] strain, offset
Dehnung f/elastische 1. elastic strain; 2. *(text)* stretch
Dehnungsverhalten *n* expansion behaviour
Dehydrase *f s.* Dehydrogenase
Dehydratase *f* dehydrase
Dehydratation *f s.* 1. Dehydratisierung 1.; 2. Dehydratisierung 2.
dehydratisieren *v* 1. dehydrate *(to remove H and OH as water from compounds)*; 2. dehydrate *(e.g. hydrates)*; 3. *(food)* dehydrate, desiccate, dry
Dehydratisierung *f* 1. dehydration *(removal of H and OH as water from compounds)*; 2. dehydration *(as of hydrates)*; 3. *(food)* dehydration, desiccation, drying
Dehydratisierungsmittel *n* dehydrating agent, dehydrator

dehydrieren *v* dehydrogenate, dehydrogenize
Dehydrierung *f* dehydrogenation
Dehydrierungsmittel *n* dehydrogenating agent
Dehydroaromat *m* dehydroarene
1,2-Dehydroaromat *n* aryne, 1,2-didehydroarene
Dehydroaromatisierung *f* dehydroaromatization
Dehydrobenzen *n* dehydrobenzene
Dehydrobenzol *n* dehydrobenzene
1,2-Dehydrobenzen *n* benzyne, 1,2-didehydrobenzene
Dehydrocyclisierung *f* dehydrocyclization
Dehydrocyclodimerisierung *f* dehydrocyclodimerization
Dehydrodimerisierung *f* dehydrodimerization, oxidative dimerization
Dehydrogenase *f* dehydrogenase
dehydrohalogenieren *v* dehydrohalogenate
Dehydrohalogenierung *f* dehydrohalogenation
Dehydroisomerisierung *f* dehydroisomerization
Dehydroxyalkylierung *f* dehydroxyalkylation
Dehydroxylierung *f* dehydroxylation
D-Einheit *f* difunctional [bifunctional] unit, D unit *(structural element of macromolecules)*
Deinterkalation *f* deintercalation, disintercalation
Deionat *n (hyd)* deionized [demineralized] water *(ion exchange)*
Deionisation *f* deionization, DI
deionisieren *v* deionize
deisobutanisieren *v* deisobutanize
Deisobutanisierung *f* deisobutanization
Dekalzifikation *f (bioch)* decalcification
dekalzifizieren *v (bioch)* decalcify
Dekameter *n* dielectrometer, dielectric constant meter
Dekametrie *f* dielectrometry, dielectric constant measurement
Dekantation *f* decantation, decanting, pouring-off
dekantieren *v* decant, pour off
Dekantiergefäß *n* decanter, decanting jar
Dekantiertopf *m* decanter, decanting jar
Dekantierzentrifuge *f* centrifugal decanter, sedimentation [sedimenting] centrifuge, solid-bowl [solid-wall, full-bowl] centrifuge
Dekapierbad *n s.* Ätzbad
Dekokt *n* decoctum, decoction
dekontaminieren *v (nucl)* decontaminate
Dekontaminierung *f (nucl)* decontamination
Dekorschicht *f* decorative coating
Deletion *f* deletion *(genetics, missing of chromosome)*
delokalisieren *v (phys ch)* delocalize
Delokalisierung *f (phys ch)* delocalization
Delokalisierungsenergie *f* delocalization [resonance, mesomeric] energy
demaskieren *v* demask *(coordination chemistry)*
Demaskierung *f* demasking *(coordination chemistry)*
Demercurierung *f (org ch)* demercuration
Demetallierung *f* demetalation
demethanisieren *v* demethanize
Demethanisierung *f* demethanization
Demethoxylierung *f* demethoxylation
demethylieren *v* demethylate
Demethylierung *f* demethylation
Demineralisation *f* demineralization
demineralisieren *v* demineralize
Demjanov-Umlagerung *f* Demjanov rearrangement *(of primary cycloaliphatic amines)*
Demulgator *m* demulsifier, emulsion breaker
demulgieren *v* demulsify, de-emulsify, break, crack
Demulgieren *n* demulsification, de-emulsification, breaking, cracking
Demulgierung *f* demulsification, de-emulsification, breaking, cracking
denaturierbar *(bioch)* denaturable
Denaturierbarkeit *f (bioch)* denaturability
denaturieren *v* 1. *(food)* denature, denaturize, *(ethanol also)* methylate; 2. *(bioch)* denature
Denaturierung *f* 1. *(food)* denaturation, *(of ethanol also)* methylation; 2. *(bioch)* denaturation

Denaturierung f/reversible *(bioch)* reversible denaturation
Denaturierungsmittel *n (food)* denaturant, denaturing agent
Dendrimer *n* dendrimer *(polymeric architecture)*
Dendrit *m (cryst)* dendrite
dendritisch *(cryst)* dendritic(al)
denitrieren *v* denitrate
Denitrierung *f* denitration
Denitrifikanten *mpl* denitrifying bacteria, denitrifiers
Denitrifikation *f* denitrification *(reduction of nitrates brought about by denitrifying bacteria)*
Denitrifikationsbakterien *npl* denitrifying bacteria, denitrifiers
denitrifizieren *v* denitrify
Denitrosierung *f* denitrosation
Densimeter *n* densimeter, densitometer, *(for liquids also)* hydrometer
Densimetrie *f* densimetry
Densitometer *n (phot)* densitometer
Densitometrie *f (phot)* densitometry
Dentalgold *n* dental gold (alloy)
Dentalgoldlegierung *f* dental gold (alloy)
Dentallegierung *f* dental alloy
depentanisieren *v* depentanize
Depentanisierung *f* depentanization
Dephlegmator *m (distil)* dephlegmator, (counter-current) partial condenser, partial-condensation head
dephlegmieren *v (distil)* dephlegmate
Dephlegmierung *f (distil)* dephlegmation, partial condensation
dephosphorylieren *v* dephosphorylate
Dephosphorylierung *f* dephosphorylation
Depolarisation *f* depolarization
Depolarisator *m* depolarizer
depolarisieren *v* depolarize
Depolymerisation *f* depolymerization
depolymerisieren *v* depolymerize
Deponie *f* 1. disposal (to land) *(of waste products)*; 2. (waste) disposal site, *(in hollows)* landfill
Deponie f/geordnete sanitary landfill
Deponie f/untertägige underground burial, deep burial on land *(of waste products)*
deponieren *v* dispose of *(waste products)*; *(if underground also)* bury
Deposition f/nasse wet deposition *(of pollutants from the atmosphere)*
Deposition f/saure acid deposition
Deposition f/trockene dry deposition
depropanisieren *v* depropanize
Depropanisierung *f* depropanization
deproteinisieren *v* deproteinize
Deproteinisierung *f* deproteinization
Deprotonierung *f* deprotonation
Deprotonierungsgrad *m* degree of deprotonation
Depsid *n* depside *(ester consisting of two or more phenolic benzoic acids)*
Depsipeptid *n (bioch)* depsipeptide
Derivat *n* derivative • **in ein Derivat überführen** *(anal)* derivatize
Derivat n/organisches organoderivative
Derivatbildung *f* 1. formation of derivatives; 2. *s.* Derivatisierung
Derivatisierung *f (anal)* derivatization
Derivatisierung *f* **nach der Trennsäule** *(chromat)* post-column derivatization
Derivatisierung *f* **vor der Trennsäule** *(chromat)* pre-column derivatization
Derivativpolarographie *f (anal)* derivative polarography
Derivativspektroskopie *f* derivative spectroscopy
DES *s.* Diethylstilböstrol
Desacetoxylierung *f* deacetoxylation
desacylieren *v* deacylate
Desacylierung *f* deacylation
Desadenylylierung *f (bioch)* deadenylylation
desaktivieren *v* deactivate, inactivate *(catalysts)*; deactivate, de-energize *(molecules)*
Desaktivierung *f* deactivation, inactivation *(of catalysts)*; deactivation, de-energization *(of molecules)*
Desaktivierung f/strahlende radiative deactivation
Desaktivierung f/strahlungslose radiationless deactivation
Desaldolierung *f* dealdolization
Desalkoxylierung *f* dealkoxylation
desalkylieren *v* dealkylate
Desalkylierung *f* dealkylation
Desalkylierung f/reduktive [reduzierende] hydrodealkylation, reductive dealkylation
desamidieren *v* deamidate
Desamidierung *f* deamidation

Desaminase *f* deaminase
desaminieren *v* deaminate
Desaminierung *f* deamination
Desaromatisierung *f* dearomatization
Desarylierung *f* dearylation
desensibilisieren *v (phot, bioch)* desensitize
Desensibilisierung *f (phot, bioch)* desensitization
Desethoxylierung *f* deethoxylation
Desethylierung *f* deethylation
Desilylierung *f* desilylation
Desinfektion *f* disinfection
Desinfektionslösung *f* disinfectant solution
Desinfektionsmittel *n* disinfectant
Desinfektionsmittel *n* **mit Reinigungswirkung** detergent-sanitizer
Desinfiziens *n* disinfectant
desinfizieren *v* disinfect
desinfizierend disinfectant
Desintegration *f* disintegration
Desintegrator *m s.* 1. Desintegratorgaswäscher; 2. Desintegratormühle
Desintegratorgaswäscher *m* disintegrator (gas) washer, disintegrator
Desintegratorgaswäscher *m* **nach Theisen** Theisen disintegrator
Desintegratormühle *f* disintegrator, *(specif)* cage [squirrel-cage] disintegrator [mill], bar mill
desintegrieren *v* disintegrate
Desmoenzym *n* desmo-enzyme *(any of a group of extracellular enzymes)*
Desmoferment *n* desmo-enzyme *(any of a group of extracellular enzymes)*
Desmolyse *f* desmolysis
desmotrop desmotropic
Desmotropie *f* desmotropy, desmotropism, dynamic isomerism
Desodorans *n* deodorant, deodorizer
desodorieren *v* deodorize
desodorierend deodorant
Desodorierung *f* deodorization
Desodorierungsmittel *n* deodorant, deodorizer
desorbieren *v* desorb, strip (off, out)
desorbieren *v*/**mit Wasserdampf** steam
Desorption *f* desorption, stripping
Desoxidation *f* deoxid(iz)ation, deoxygenation
Desoxidationsmittel *n* 1. deoxidant, deoxidizer, deoxidizing agent; 2. *(met)* scavenger
desoxidieren *v* 1. deoxidize, deoxidate, deoxygenate; 2. *(met)* scavenge
Desoxyadenosin-5′-phosphat *n (bioch)* deoxyadenylic acid
Desoxycorticosteron *n (bioch)* deoxycorticosterone
Desoxycytidin-5′-phosphat *n (bioch)* deoxycytidylic acid
Desoxyguanosin-5′-phosphat *n (bioch)* deoxyguanylic acid
Desoxypentose *f* deoxypentose *(a monosaccharide)*
Desoxyribonuclease *f (bioch)* deoxyribonuclease, DNase
Desoxyribonucleinsäure *f (bioch)* deoxyribonucleic acid, DNA
Desoxyribonucleoprotein *n (bioch)* deoxyribonucleoprotein
Desoxyribose *f* deoxyribose *(a monosaccharide)*
Desoxythymidin-5′-phosphat *n (bioch)* deoxythymidine 5′-phosphate
Desoxyzucker *m (bioch)* deoxy sugar
desozon(is)ieren *v* deozonize
destabilisieren *v* destabilize, make unstable
Destillans *n* material being distilled
Destillat *n* distillate
Destillatabnahme *f* product take-off, distillate drain
Destillatbenzin *n* straight-run gasoline, distillate gasoline, straight-run benzine, S.R.B.
Destillation *f* distillation
Destillation *f*/**absteigende** downward distillation
Destillation *f*/**azeotrope** azeotropic [entrainment] distillation
Destillation *f*/**direkte** straight(-run) distillation, simple distillation
Destillation *f*/**diskontinuierliche** batch distillation
Destillation *f*/**einfache** *s.* Destillation/direkte
Destillation *f*/**extrahierende [extraktive]** extractive distillation
Destillation *f*/**fraktionierende [fraktionierte]** fractional distillation, fractionation
Destillation *f* **im Vakuum** vacuum dis-

tillation, distillation under vacuum [reduced pressure]
Destillation *f*/integrale equilibrium distillation
Destillation *f*/isobare isobaric distillation, distillation at constant pressure
Destillation *f*/isotherme isothermal distillation, distillation at constant temperature
Destillation *f*/kontinuierliche continuous distillation
Destillation *f* mit fallendem Film falling-film distillation
Destillation *f* mit Zusatzstoff(en) co-distillation
Destillation *f* nach ASTM ASTM distillation
Destillation *f*/nochmalige redistillation, rerun(ning)
Destillation *f*/offene differential distillation
Destillation *f*/schonende gentle distillation
Destillation *f*/trockene dry [destructive] distillation
Destillations... *siehe auch: Destillier...*
Destillationsanlage *f* 1. *(tech)* distillation [distilling] plant, distillery, still; 2. *(lab)* distillation unit, still
Destillationsapparat *m* distillation apparatus, still
Destillationsbereich *m* distillation range
Destillationsgut *n* distilland, material to be distilled; material being distilled
Destillationskurve *f* distillation curve
Destillationslinie *f* distillation curve
Destillationsrückstand *m* distillation [still] residue
Destillationsrückstand *m*/kurzer short residue [residuum]
Destillationsrückstand *m*/langer long residue [residuum]
Destillationsturm *m* distillation tower
destillativ by (means of) distillation
Destillatsammler *m s.* Destilliervorlage
Destillatvorlage *f s.* Destilliervorlage
Destillier... *siehe auch: Destillations...*
Destillieraufsatz *m* distillation head, stillhead, distillation connecting tube
destillierbar/mit Dampf [Trägerdampf, Wasserdampf] steam-distillable
Destillierbarkeit *f* distillability
Destillierblase *f* distillation boiler, still pot, *(with rectifying apparatus:)* reboiler
destillieren *v* distil
destillieren *v*/fraktioniert fractionate
destillieren *v*/mit Dampf [Trägerdampf, Wasserdampf] steam-distil
Destillierkolben *m* distillation flask
Destillierkolonne *f* distillation column
Destillierofen *m* distillation furnace, retort furnace [oven]
Destilliersäule *f* distillation column
destilliert/doppelt twice-distilled
destilliert/unter Vakuum vacuum-distilled, distilled in vacuo
Destilliervorlage *f* (distillate, distillation) receiver
Destilliervorstoß *m* adapter
Desulfatierung *f* desulphation
desulfonieren *v* desulphonate
Desulfonierung *f* desulphonation
desulfurieren *v* desulphurize, desulphur
Desulfurierung *f* desulphurization, desulphuration
Desulfurikanten *mpl* sulphate reducers, sulphate-reducing bacteria, sulphur-reducing bacteria
Desuridylderivat *n (bioch)* deuridylic derivative
Detektion *f*/amperometrische *(anal)* amperometric detection
Detektor *m (anal)* detector
Detektor *m*/flammenphotometrischer flame photometric detector
Detektor *m*/thermoionischer thermal ionization detector, TID
Detergens *n* (synthetic) detergent, syndet, soapless soap
Detergent *m* 1. detergent *(for holding in suspension insoluble matter)*; 2. *s.* Detergens
Detonation *f* detonation
Detonationsgeschwindigkeit *f* detonation rate
detonieren *v* detonate
detosylieren *v* detosylate
Detosylierung *f* detosylation
Detritierung *f* detritiation
Detritylierung *f* detritylation
Deuterid *n* deuteride
deuterieren *v* deuterate, deuterize
Deuterierung *f* deuteration
Deuteriogruppe *f* deuterio group

Deuterium *n* ^2_1H D deuterium, heavy hydrogen
Deuteriumaustausch *m* deuterium exchange
Deuteriumoxid *n* deuterium oxide, heavy water
Deuteron *n (nucl)* deuteron
Dewar-Gefäß *n* Dewar (flask, vessel) *(for holding liquid gases)*
Dextran *n (bioch)* dextran(e)
Dextrin *n* dextrin(e), British [starch] gum
dextrinieren *v* dextrinize
Dextrinierung *f* dextrinization
dextrogyr *s.* rechtsdrehend
dextrorotatorisch dextrorotatory *(stereo chemistry)*
Dextrose *f* dextrose, D-glucose *(a monosaccharide)*
Dezigrammbereich *m (anal)* decigram range *(0.1 to 1 g)*
Dezimalwaage *f* decimal balance
Dezimol *n* decimol
dGMP *s.* Desoxyguanosin-5′-phosphat
Diacetyl *n s.* Biacetyl
Diacetylierung *f* diacetylation
Diacetylmorphin *n* diacetylmorphine, diamorphine, heroin *(a narcotic)*
Diacylamin *n* diacylamine
Diagnostikum *n* diagnostic reagent
Diagramm *n* graph, diagram, chart, graphical [diagrammatic] representation
Diagramm *n***/doppeltlogarithmisches** log-log plot, Ellingham diagram
Dialdehyd *m* dialdehyde
Dialkylbenzen *n* dialkylbenzene
Dialkylbenzol *n* dialkylbenzene
Dialkylether *m* dialkyl ether
dialkylieren *v* dialkylate
Dialkylierung *f* dialkylation
Dialkylphosphorigsäurechlorid *n* $(RO)_2PCl$ dialkoxychlorophosphine, dialkylphosphorochloridite
Diallylphthalatharz *n* diallyl phthalate resin, DAP
Dialysat *n* dialysate
Dialysator *m* dialyser
Dialyse *f* dialysis
dialysierbar dialysable
dialysieren *v* dialyse
Dialysiermembran *f* dialysing membrane
Dialysierzelle *f* dialysis [dialytic] cell
Diamagnetikum *n* diamagnet, diamagnetic (substance)
diamagnetisch diamagnetic
Diamagnetismus *m* diamagnetism
Diamant *m* diamond
diamantartig diamond-like
diamanten adamantine
Diamantgitter *n (cryst)* diamond lattice
Diamantglanz *m* brilliant lustre
diamantoid diamondoid, diamond-like *(ring system)*
Diamid *n* diamide, hydrazine
Diamin *n* diamine
Diaminobenzen *n* diaminobenzene, phenylenediamine
4,4′-Diaminobiphenyl *n* 4,4′-diaminobiphenyl, benzidine
2,6-Diaminohexansäure *f* 2,6-diaminohexanoic acid, lysine
Diamminquecksilber(II)-chlorid *n* diamminemercury(II) chloride, diammine mercuric chloride, fusible white precipitate
Diammoniumhydrogenphosphat *n* ammonium hydrogenphosphate
Diamorphin *n* diamorphine, diacetylmorphine, heroin *(a narcotic)*
Dianion *n* dianion
Diantimonpentoxid *n* diantimony pentaoxide, antimony(V) oxide
Diaphoretikum *n (pharm)* diaphoretic, sudorific
diaphoretisch diaphoretic, sudorific
Diaphragma *n* diaphragm, membrane
Diaphragmazelle *f* diaphragm cell *(electrolysis)*
diäquatorial diequatorial *(stereochemistry)*
diäquatorisch diequatorial *(stereochemistry)*
Diarsenpentasulfid *n* diarsenic pentasulphide, arsenic(V) sulphide
Diarsenpentoxid *n* diarsenic pentaoxide, arsenic(V) oxide
Diarsentrioxid *n* diarsenic trioxide, arsenic(III) oxide, arsenic
Diarsentrisulfid *n* diarsenic trisulphide, arsenic(III) sulphide
1,2-Diarylhydrazin *n* 1,2-diarylhydrazine, hydrazoarene, 1,2-diaryldiazane
diastereogen diastereogenic *(stereo chemistry)*
diastereomer diastereo(iso)meric

Diastereomer(e) *n* diastereo(iso)mer
Diastereoselektivität *f* *(org ch)* diastereoselectivity
diastereotop diastereotopic *(stereo chemistry)*
diatherm(an) diathermanous, diathermic
Diathermansie *f* diatherma(n)cy
Diatomeenerde *f* diatomaceous [infusorial] earth, kieselguhr
Diauxie *f* diauxy *(diphasic or polyphasic course of growth with microorganisms)*
diauxisch diauxic *(said of the growth of microorganisms which shows a diphasic or polyphasic course)*
diaxial diaxial
Diazen *n* diazene, diimide
1,2-Diazin *n* 1,2-diazine, pyridazine
1,3-Diazin *n* 1,3-diazine, pyrimidine
1,4-Diazin *n* 1,4-diazine, pyrazine
Diazoaminoverbindung *f* R–N=N–NHR diazoamino compound
Diazodinitrophenol *n* diazodinitrophenol, DDNP
Diazofarbstoff *m* diazo dye
Diazogruppe *f* –N=N– diazo group
Diazokupplung *f* diazo [azo] coupling
Diazomethan *n* diazomethane
Diazoniumgruppe *f* $[Ar–N\equiv N]^+$ diazonium group
Diazoniumsalz *n* diazonium salt
Diazooxid *n* diazooxide, diazophenol, diazocyclohexadienone
Diazophenol *n s.* Diazooxid
Diazorest *m* –N=N– diazo group
Diazotat *n* R–N=N–OM diazoate, diazotate
diazotierbar diazotizable
diazotieren *v* diazotize
Diazotierung *f* diazotization
Diazoverbindung *f* diazo compound
Dibasizität *f* dibasicity
Dibenzo-1,4-dioxin *n* dibenzo-1,4-dioxin, dibenzo-*p*-dioxin, dioxin
Dibenzo-1,4-dioxin *n***/polychloriertes** polychlorinated dibenzo-1,4-dioxin, PCDD
Dibenzo-*p*-dioxin *n* dibenzo-1,4-dioxin, dibenzo-*p*-dioxin, dioxin
Dibenzofuran *n* dibenzofuran
Dibenzofuran *n***/polychloriertes** polychlorinated dibenzofuran, PCDF
dibenzoylieren *v* dibenzoylate
Dibenzoylierung *f* dibenzoylation
Dibenzoylperoxid *n* dibenzoyl peroxide, benzoyl peroxide
Dibenzylether *m* dibenzyl ether
Dibenzylierung *f* dibenzylation
Diboran *n* diborane
Dibromid *n* dibromide
dibromieren *v* dibrominate
Dibromierung *f* dibromination
Dibrommethan *n* dibromomethane
Dibromthymolsulfophthalein *n* dibromothymolsulphonephthalein, bromothymol blue *(a pH indicator)*
Dibromverbindung *f* dibromo compound
2,6-Di-*tert*-butyl-*p*-cresol *n* 2,6-di-*tert*-butyl-*p*-cresol, 4-methyl-2,6-di-*tert*-butylphenol *(an antioxidant)*
Dibutylphthalat *n* dibutyl phthalate, DBP
Dicarbonsäure *f* dicarboxylic acid
Dichloracetylierung *f* dichloroacetylation
Dichlorderivat *n* dichloro derivative
Dichlordiethylsulfid *n* dichlorodiethyl sulphide, di-2-chloroethyl sulphide
Dichlordifluormethan *n* dichlorodifluoromethane
Dichlordiphenyltrichlorethan *n* dichlorodiphenyltrichloroethane, DDT *(a contact insecticide)*
Dichloressigsäure *f* dichloroacetic acid, dichloroethanoic acid
Dichlorethan *n* dichloroethane
Dichlorethansäure *f* dichloroethanoic acid, dichloroacetic acid
Dichlorfluormethan *n* dichloromonofluoromethane
Dichlorid *n* dichloride
Dichlorierung *f* dichlorination
Dichlormethan *n* dichloromethane
4-(2′,4′-Dichlorphenoxy)buttersäure *f* 4-(2′,4′-dichlorophenoxy)butyric acid, 2,4-DB *(a herbicide)*
2,4-Dichlorphenoxyessigsäure *f* 2,4-dichlorophenoxyacetic acid, 2,4-D *(a herbicide)*
Dichlorprop *n* dichlorprop, 2,4-DP, 2-(2′,4′-dichlorophenoxy)propionic acid *(a herbicide)*
Dichroismus *m* *(cryst, coll)* dichroism
dichroitisch *(cryst, coll)* dichroic
Dichromat *n* $M^I_2Cr_2O_7$ dichromate
dicht tight, proof, impermeable, impenetrable, impervious; leaktight, leak-

proof; dense, close-grained, fine-grained, compact *(structure)*
Dichte *f* 1. density, *(specif)* mass density, D *(mass of a substance per unit volume)*; strength *(of a solution)*; 2. *s.* Dichte/optische
Dichte *f*/**optische** optical density
Dichte *f*/**relative** relative density *(ratio of the density of a material to the density of some standard material)*
Dichte *f*/**scheinbare** apparent density
Dichtebestimmung *f* densimetry
Dichtebestimmung *f* **von Flüssigkeiten** hydrometry
Dichtegradientenzentrifugation *f (anal)* density-gradient centrifugation
Dichtemesser *m* densimeter
dichten *v* make tight, seal, pack, ca(u)lk
Dichtesortierung *f* density separation [cut], gravity concentration
Dichteverhältnis *n* relative density *(ratio of the density of a material to the density of some standard material at the same state of matter)*
Dichtewaage *f* density balance
dichtgepackt *(cryst)* close-packed
Dichtheit *f* tightness, proofness; leak tightness, leakproofness
Dichtigkeit *f* tightness, proofness; leak tightness, leakproofness
Dichtigkeitstest *m* leakage test
Dichtmaterial *n* sealant
Dichtmittel *n* sealant
Dichtung *f* packing, seal, *(for parts without relative motion also)* static seal, gasket, *(for moving parts also)* dynamic seal
Dichtungsfett *n* joint grease
Dichtungsflüssigkeit *f* sealing liquid [fluid]
Dichtungsmanschette *f* gasket
Dichtungsmasse *f* lute, luting, ca(u)lking compound
Dichtungsmaterial *n* packing [sealing, gasketing] material
Dicke *f* thickness
dickflüssig viscous, viscose, thick, syrupy, ropy • **dickflüssig werden** thicken, inspissate
Dickflüssigkeit *f* viscosity, thickness, ropiness
Dicklauge *f* concentrated [strong] liquor
Dickschlamm *m* thickened [concentrated] sludge, *(as discharge of a thickener also)* thickened liquor, thick slime, thickener pulp
Dickschlammzone *f* sludge zone
Dickstoffpumpe *f* sludge [thick-liquor] pump
dicktaflig *(cryst)* thick-tabular
Dickungsmittel *n* thickening agent, thickener, viscosifier
Dicobalttrioxid *n* dicobalt trioxide, cobalt(III) oxide
Dicyan *n* N≡C–C≡N dicyanogen, cyanogen (gas), oxalonitrile
1,1-Dicyanethen *n* 1,1-dicyanoethylene
Dicyanoaurat(I) *n* $M^I[Au(CN)_2]$ dicyanoaurate(I), cyanoaurate(I)
Dicystein *n* cystine, dicysteine, *3,3′-dithio-*bis*(2-aminopropanoic acid)
Didepsid *n* didepside *(ester consisting of two phenolic benzoic acids)*
Diderivat *n* diderivative
Didym(metall) *n* didymium *(an alloy consisting of neodymium and praseodymium)*
Dieckmann-Reaktion *f* Dieckmann reaction *(intramolecular condensation of esters)*
Dieisentrioxid *n* diiron trioxide, iron(III) oxide, ferric oxide
Dielektrikum *n* dielectric (material), non-conductor
dielektrisch dielectric, non-conducting
Dielektrizitätskonstante *f* (absolute) permittivity
Dielektrizitätskonstante *f*/**absolute** (absolute) permittivity
Dielektrizitätskonstante *f*/**relative** *s.* Dielektrizitätszahl
Dielektrizitätskonstante-Messer *m* dielectrometer, dielectric constant meter
Dielektrizitätszahl *f* relative dielectric constant, relative permittivity, specific inductive capacity, SIC
Dielektrometrie *f* dielectrometry, dielectric constant measurement
Dielektrophorese *f* dielectrophoresis *(separation of cells or nanoparticles)*
Diels-Alder-Reaktion *f* Diels-Alder reaction [addition, condensation], diene reaction [synthesis]
Diels-Alder-Synthese *f* Diels-Alder reaction [addition, condensation], diene reaction [synthesis]

Dien *n* diene *(a compound with two esp conjugated C–C double bonds)*
Dienmetathese f/acyclische acyclic diene metathesis, ADMET
dienophil dienophilic
Dienophil *n* dienophile
Dieselkraftstoff *m* diesel fuel [oil]
Diester *m* diester
1,1-Diethoxyethan *n* *1,1-diethoxyethane, acetal
Diethoxymethan *n* *diethoxymethane, diethylformal, formaldehyde diethyl acetal
Diethylamin *n* diethylamine
Diethylendiamin *n* diethylenediamine, piperazine
Diethylenglykol *n* diethylene glycol, DEG, *di(-2-hydroxyethyl) ether
Diethylentriaminpentaessigsäure *f* diethylenetriamine pentaacetic acid, DTPA
Diethylether *m* diethyl ether
Diethylmagnesium *n* diethylmagnesium
Diethylmalonat *n* diethyl malonate, malonic ester
Diethylstilböstrol *n* diethylstilboestrol
Diethylsulfat *n* diethyl sulphate
Diethylsulfid *n* diethyl sulphide, ethylthioethane
Differenzdruck *m* differential pressure
Differenzdruckhöhe *f* differential head
Differenzialdestillation *f* differential distillation
Differenzialreaktor *m* differential reactor
Differenzial-Thermoanalyse *f* differential thermal analysis, DTA
differenzialthermoanalytisch by differential thermal analysis
Differenzial-Thermogravimetrie *f* derivative [differential] thermogravimetry
Differenzialtitration *f* differential titration
Differenzialzentrifugation *f (bioch)* differential centrifugation
Differenzpolarographie *f (anal)* differential polarography
Differenz-Pulspolarographie *f (anal)* differential pulse polarography
Differenzspektrum *n* difference spectrum
Diffraktion *f* diffraction
diffundieren *v* diffuse
Diffusat *n* diffusate
Diffuseur *m* diffuser, diffusor, diffusing tank, diffusion cell
Diffusion *f* diffusion
Diffusion f/axiale axial [longitudinal] diffusion
Diffusion f/behinderte restricted diffusion
Diffusion f/erleichterte *(bioch)* facilitated diffusion
Diffusion f/innere internal [pore] diffusion
Diffusion f/molekulare molecular diffusion
Diffusion f/radiale radial diffusion
Diffusion *f* **von Festkörpern** solid(-state) diffusion
Diffusionsapparat *m s.* Diffuseur
diffusionschromieren *v* chromize
diffusionsfähig diffusible
Diffusionsfähigkeit *f* diffusibility
Diffusionsgeschwindigkeit *f* diffusion rate
Diffusionsgesetz n/ficksches Fick's law (of diffusion)
diffusionsgesteuert diffusion-controlled, under diffusion control, diffusion-limited
diffusionsglühen *v* homogenize *(alloys)*
Diffusionskoeffizient *m* diffusion coefficient, diffusivity
Diffusionskoeffizient m/kooperativer cooperative diffusion coefficient
Diffusionskoeffizient m/translatorischer translational diffusion coefficient
Diffusionskonstante *f* diffusion coefficient, diffusivity
Diffusionskontrolle *f s.* Diffusionssteuerung
diffusionskontrolliert *s.* diffusionsgesteuert
Diffusionsmembran *f* diffusion barrier, diffusion membrane *(electrochemical sensor)*
Diffusionsofen *m (tech)* barrel furnace, diffusion furnace
Diffusionspotenzial *n* diffusion [liquid-junction] potential
Diffusionspumpe *f* (vapour) diffusion pump
Diffusionsrohr *n (tech)* diffusion tube
Diffusionsschicht f/nernstsche Nernst (diffusion) layer

Diffusionssteuerung *f (phys ch)* diffusion control
Diffusionsstrom *m* 1. *(phys ch)* diffusion(-controlled) current; 2. *(phys ch)* diffusive flux
diffusionsverchromen *v* chromize
diffusorisch diffusional
Difluorid *n* difluoride
Difluorierung *f* difluorination
Diformiat *n* diformate
difunktionell difunctional, D, bifunctional
***m*-Digallussäure** *f* digallic acid, *m*-digallic acid, gallic acid 3-monogallate, 5,6-dihydroxy-3-carboxyphenyl ester of gallic acid
digerieren *v* digest *(by heat or solvents)*
Digerieren *n* digestion *(by heat or solvents)*
Digestorium *n (lab)* (fume) hood, fume chamber [cupboard, closet]
Diglycerid *n* diglyceride
Diglykolsäure *f* diglycolic [diglycollic] acid, *2,2′-oxydiethanoic acid
Dihalogenalkan *n* alkyl dihalide
Dihalogenid *n* dihalogenide, dihalide
Dihalogenierung *f* dihalogenation
Diharnstoff *m* diurea, *p*-urazine
Dihydrat *n* dihydrate
Dihydrogenphosphat *n* $M^IH_2PO_4$ dihydrogen phosphate
Dihydrogensalz *n* dihydrogen salt *(of a tribasic acid)*
2,5-Dihydroxybenzoesäure *f* 2,5-dihydroxybenzoic acid, gentisic acid
Dihydroxybernsteinsäure *f* dihydroxysuccinic acid, tartaric acid, *2,3-dihydroxybutanedioic acid
Dihydroxyphenylalanin *n* dihydroxyphenylalanine, dopa, DOPA
2,3-Dihydroxypropanal *n* *2,3-dihydroxypropanal, glyceraldehyde
2,3-Dihydroxypropansäure *f* *2,3-dihydroxypropanoic acid, glyceric acid
Dihydroxyverbindung *f* dihydroxy compound
Diimid *n* diimide, diazene
Diiodid *n* diiodide
Diiodmethan *n* diiodomethane
Diisocyanat *n* diisocyanate *(a compound having two –N=C=O groups)*
Dikaliumdisulfat *n* dipotassium disulphate, potassium disulphate
Diketoester *m* diketo ester
Diketon *n* diketone
dilatant *(coll)* dilatant
Dilatanz *f (coll)* dilatancy, shear thickening, inverse plasticity
Dilatation *f* dilatation
Dilatometer *n* dilatometer
Dilatometrie *f* dilatometry
dilatometrisch dilatometric
Dimangantrioxid *n* dimanganese trioxide, manganese(III) oxide
Dimedon *n (anal)* dimedon, 5,5-dimethylcyclohexane-1,3-dione
dimensionieren *v* size
Dimensionierung *f* sizing
dimer dimeric
Dimer(e) *n* dimer
dimerisieren *v* dimerize, *(process also)* undergo dimerization
Dimerisierung *f* dimerization
Dimerisierung *f***/oxidative** oxidative dimerization, dehydrodimerization
Dimerisierung *f***/reduktive [reduzierende]** reductive dimerization, hydrodimerization
Di-π-methan-Umlagerung *f* di-π-methane rearrangement *(photochemistry)*
Dimethoxymethan *n* *dimethoxymethane, methylal, formaldehyde dimethyl acetal
Dimethoxymethylgruppe *f* $-CH(OCH_3)_2$ dimethoxymethyl group
Dimethylamin *n* dimethylamine
Dimethylarsinsäure *f* dimethylarsinic acid, cacodylic acid
Dimethylbenzen *n* dimethylbenzene, xylene
2,5-Dimethylbenzoesäure *f* 2,5-dimethylbenzoic acid, *p*-xylylic acid
α,α-Dimethylbenzylalkohol *m* 2-phenyl-2-propanol, dimethylphenyl carbinol, DMPC
5,5-Dimethylcyclohexan-1,3-dion *n* s. Dimedon
Dimethyldichlorsilan *n* dimethyldichlorosilane
Dimethylenimin *n* dimethyleneimine, aziridine
Dimethylether *m* dimethyl ether
Dimethylformamid *n* dimethyl formamide, DMF
Dimethylglyoxal *n* dimethylglyoxal, biacetyl, *butane-2,3-dione

Dimethylketon *n* dimethyl ketone, acetone, *propanone
Dimethylphenol *n* dimethylphenol, xylenol
Dimethylsulfat *n* dimethyl sulphate
Dimethylsulfoxid *n* dimethyl sulphoxide, DMSO
dimolekular bimolecular
dimorph dimorphic, dimorphous
Dimroth-Kühler *m (lab)* Dimroth condenser
Dinatriumsalz *n* disodium salt
Dinitrid *n* dinitride
dinitrieren *v* dinitrate
Dinitrierung *f* dinitration
Dinitril *n* dinitrile
Dinitrobenzoesäure *f* dinitrobenzoic acid
Dinitrocresol *n* dinitrocresol
Dinitrophenol *n* dinitrophenol
Dinitrophenylderivat *n* dinitrophenyl derivative, DNP derivative
Dinitroverbindung *f* dinitro compound
Dinoseb *n* Dinoseb, DNBP, 2-*sec*-butyl-4,6-dinitrophenol *(a pesticide)*
Diodenlaser *m* diode [semiconductor] laser
Diol *n* diol, dialcohol, dihydroxylic [dihydric] alcohol
Diolefin *n* diolefin *(a hydrocarbon with two double bonds)*
diolefinisch diolefinic
Dioxan *n* dioxan
Dioxid *n* dioxide
Dioxin *n* dibenzo-1,4-dioxin, dibenzo-*p*-dioxin, dioxin, *(specif)* 2,3,7,8-tetrachlorodibenzo-1,4-dioxin, 2,3,7,8-tetrachlorodibenzo-*p*-dioxin, 2,3,7,8-TCDD, TCDD, dioxin
Dipeptid *n* dipeptide
Diphenylamin *n* diphenylamine, DPA, phenylaniline
Diphenyldiimid *n* diphenyldiimide, diphenyldiazene, *azobenzene
Diphenyldiketon *n* *diphenyl diketone, diphenylglyoxal, benzil, bibenzoyl
Diphenylenimin *n* diphenyleneimine, carbazole, dibenzopyrrole
1,2-Diphenylethan *n* 1,2-diphenylethane, bibenzyl
Diphenylethen *n* diphenylethylene, stilbene
Diphenylethin *n* diphenylethyne, tolane
Diphenylfarbstoff *m* diphenyl dye
Diphenylglykolsäure *f* diphenylglycollic acid, benzilic acid
Diphenylimid *n s.* Diphenylenimin
Diphenylketon *n* *diphenyl ketone, benzophenone, benzoyl benzene
Diphenylmethanfarbstoff *m* diphenylmethane dye
1,3-Diphenylpropenon *n* 1,3-diphenylpropenone, chalcone, benzylideneacetophenone
Diphenylthiocarbazon *n* diphenylthiocarbazone, dithizone
Diphosgen *n* diphosgene, trichloromethyl chloroformate
Diphosphat *n* $M^{I}_{4}P_2O_7$ diphosphate, pyrophosphate
Diphosphen *n* phosphene
Diphosphin *n* phosphyne
Diphosphorsäure *f* diphosphoric acid, pyrophosphoric acid
Dipol *m* dipole
dipolar dipolar
Dipol-Dipol-Anziehung *f* dipole-dipole attraction
Dipol-Dipol-Wechselwirkung *f* dipolar [dipole-dipole] interaction, dipolar [spin-spin] coupling
Dipolkraft *f* dipole force
Dipolmolekül *n* dipolar [dipole] molecule
Dipolmoment *n* dipole moment
Dipolmoment *n***/elektrisches** electric (dipole) moment
Dipolmoment *n***/induziertes** induced dipole moment
Dipolmoment *n***/magnetisches** magnetic (dipole) moment
diprimär *(org ch)* diprimary
Diradikal *n* *biradical, diradical
Direktdampf *m* live [open] steam, prime [direct] steam
Direktfarbstoff *m* direct [substantive] dye [dyestuff], *(Am)* direct color
Direkt-Methanol-Brennstoffzelle *f* direct methanol fuel cell, DMFC
Direktpotenziometrie *f* direct potentiometry
Direktreduktionsverfahren *n (met)* direct reduction process *(for producing iron)*
Direktsynthese *f* direct synthesis *(as for obtaining chlorosilanes)*
direktziehend *(dye)* direct, substantive

dirigieren *v* direct *(a substituent into a position)*
dirigierend/nach der *meta*-Stellung meta-directing
dirigierend/nach der *ortho*-Stellung ortho-directing
dirigierend/nach der *ortho*- und *para*-Stellung ortho-para-directing
Disaccharid *n* disaccharide
Disauerstoff *m* dioxygen
Disazofarbstoff *m* bisazo [disazo] dye
Dischwefeldichlorid *n* disulphur dichloride
Dischwefelheptoxid *n* disulphur heptaoxide
Dischwefelsäure *f* disulphuric acid
Dischwefeltrioxid *n* disulphur trioxide, sulphur(III) oxide
disekundär disecondary
Disilan *n* disilane
Disilicat *n* disilicate
Disilicid *n* disilicide
Disinterkalation *f* deintercalation, disintercalation
Disk-Elektrophorese *f (anal)* disc electrophoresis *(having a discontinuous gel system)*
diskelektrophoretisch *(anal)* by disc electrophoresis
diskontinuierlich discontinuous, batch(wise), intermittent *(operation)*
diskotisch diskotic, disk-like *(liquid crystals)*
Diskriminierung *f (anal)* discrimination
Dislokation *f (cryst)* dislocation
dismulgieren *v* demulsify, de-emulsify, break, crack *(an emulsion)*
Dismulgieren *n* demulsification, de-emulsification, emulsion breakdown [breaking], breaking [cracking] of emulsions
Dismutation *f* dismutation
dismutativ dismutative
dismutieren *v* dismutate
Dispergens *n s.* Dispersionsmittel 1.
dispergierbar dispersible
Dispergierbarkeit *f* dispersibility
dispergieren *v* disperse; peptize, deflocculate *(colloids)*
dispergierend dispersive
Dispergierung *f* dispersion, dispersal; peptization, deflocculation *(of colloids)*
dispers dispersed, *(Am)* disperse
Dispersant *m s.* Dispersionsmittel 1.
Dispersantadditiv *n s.* Dispersionszusatz
Dispersion *f* 1. dispersion, dispersal *(of particles)*; 2. dispersion, disperse system; 3. dispersion *(of waves)*; *(quantitatively also)* dispersivity, differential refractivity
Dispersion f/axiale axial dispersion *(in a tubular reactor)*
Dispersion f/kolloide colloidal dispersion
Dispersionsgrad *m* 1. degree of dispersion; 2. *(coll)* dispersity
Dispersionsklebstoff *m (plast)* dispersion adhesive
Dispersionskolloid *n* dispersion colloid, dispersoid
Dispersionskraft *f* dispersion force, London [transient polarization] force *(acting between molecules)*
Dispersionsmedium *n s.* Dispersionsphase
Dispersionsmittel *n* 1. dispersing [deflocculating] agent, dispersant, deflocculant, deflocculator; 2. *s.* Dispersionsphase
Dispersionsphase *f* continuous [external] phase, disperse [dispersive, dispersion] medium *(of a disperse system)*
Dispersionszusatz *m* dispersing [dispersant] additive
Dispersität *f* dispersity
dispersiv dispersive
Dispersoid *n* dispersoid
Dispersum *n* internal [discontinuous] phase, disperse(d) phase *(of a disperse system)*
Dispiro-Verbindung *f* dispiro compound [hydrocarbon]
disproportionieren *v* disproportionate
Disproportionierung *f* disproportionation, self-oxidation-reduction, auto-oxidation-reduction
Disproportionierungsreaktion *f* disproportionation reaction
disrotatorisch *(org ch)* disrotatory
Dissimilation *f (bioch)* dissimilation, catabolism
dissimilatorisch *(bioch)* dissimilative, catabolic
Dissipation *f* (energy) dissipation
dissipativ dissipative

dissoziabel dissociable, *(esp if into ions)* ionizable
Dissoziation *f* dissociation, splitting-up, *(esp if into ions)* ionization
Dissoziation *f*/**elektrolytische** electrolytic ionization [dissociation]
Dissoziationsenergie *f* (bond-)dissociation energy
Dissoziationsgleichgewicht *n* dissociation equilibrium
Dissoziationsgrad *m* degree of dissociation
Dissoziationskonstante *f* dissociation constant, *(with acids and bases also)* affinity constant; instability constant *(of complex ions)*
Dissoziationsreaktion *f* dissociation reaction
dissoziieren *v* dissociate, split up, *(esp if into ions)* ionize
Distanzstück *n* spacer
Distickstoffmonoxid *n* dinitrogen monooxide, nitrogen(I) oxide
Distickstoffpentoxid *n* dinitrogen pentaoxide, nitrogen(V) oxide
Distickstofftetroxid *n* dinitrogen tetraoxide
Distickstofftrioxid *n* dinitrogen trioxide, nitrogen(III) oxide
disubstituiert disubstituted, bis-substituted
Disubstitution *f* disubstitution
Disulfat *n* $M^I_2S_2O_7$ disulphate
Disulfid *n* disulphide
Disulfidbrücke *f* 1. disulphide bridge [bond, cross-link], –S–S-linkage; 2. *(bioch)* interchain disulphide bond, cystine link
Disulfidspange *f* *(bioch)* intrachain disulphide bond, intrastrand linkage
Disulfit *n* $M^I_2S_2O_5$ disulphite
Disulfonat *n* disulphonate
disulfonieren *v* disulphonate
Disulfonierung *f* disulphonation
Disulfonsäure *f* disulphonic acid
Disulfosäure *f* disulphonic acid
Diterpen *n* diterpene
Diterpenalkaloid *n* diterpenoid alkaloid
Diterpenoid *n* diterpenoid
Dithioacetal *n* dithioacetal, mercaptal
3,3′-Dithiobis(2-aminopropionsäure) *f* *3,3′-dithio*bis*(2-aminopropanoic acid), cystine, dicysteine
Dithiocarbamat *n* NH_2CSSM^I dithiocarbamate
Dithiocarbamidsäure *f* dithiocarbamic acid
Dithiocarboxylierung *f* dithiocarboxylation
Dithioketal *n* dithioketal, mercaptole
Dithiokohlensäure *f* dithiocarbonic acid
Dithiolkohlensäure *f* dithiocarbonic acid
Dithionat *n* dithionate
Dithionit *n* 1. $M^I_2S_2O_4$ dithionite; 2. *(dye)* $Na_2S_2O_4$ (sodium) dithionite, (sodium) hydrosulphite
Dithionsäure *f* dithionic acid
Dithiophosphat *n* $M^I_3PS_2O_2$ dithiophosphate
Dithiophosphorsäure *f* dithiophosphoric acid
Dithiosäure *f* $C_nH_{2n+1}CSSH$ dithio acid
Dithioverbindung *f* dithio compound
Dithizon *n* dithizone, diphenylthiocarbazone
ditrigonal-skalenoedrisch *(cryst)* ditrigonal-scalenohedral
Diuretikum *n* *(pharm)* diuretic
diuretisch diuretic
Divanadiumpentoxid *n* divanadium pentaoxide, vanadium(V) oxide
Divanadiumtrioxid *n* divanadium trioxide, vanadium(III) oxide
divariant bivariant, divariant
Diwasserstoff *m* H_2 dihydrogen
DK *s.* 1. Dielektrizitätskonstante; 2. Dieselkraftstoff
DK-Meter *n s.* Dekameter
D-Konfiguration *f* D configuration
DMF *s.* Dimethylformamid
DMSO *s.* Dimethylsulfoxid
DNA *f* DNA, deoxyribonucleic acid
DNA *f*/**denaturierte** denatured DNA
DNA *f*/**doppelsträngige** double-strand(ed) DNA
DNA *f*/**komplementäre** complementary DNA, cDNA
DNA *f*/**mitochondriale** mitochondrial DNA
DNA *f*/**neukombinierte [rekombinierte]** recombinant DNA, rDNA
DNA-Kopie *f* copy DNA, cDNA
DNA-Ligase *f* DNA ligase
DNA-Polymerase *f* DNA polymerase
DNA-Polymerase *f*/**RNA-abhängige** RNA-directed DNA polymerase, reverse transcriptase

DNA-Rekombinationstechnik *f (biot)* recombinant DNA technology; recombinant DNA technique
DNA-Replikation *f* DNA replication
DNase *s.* Desoxyribonuclease
DNA-Sequenz *f* DNA sequence
DNA-Strang *m* DNA strand
DNBP *s.* Dinoseb
DNP *s.* Dinitrophenol
DNP-Derivat *n s.* Dinitrophenylderivat
DNS *(Desoxyribonucleinsäure) s.* DNA
Döbereiner-Triaden *fpl* triads of Döbereiner
Dobson-Einheit *f* Dobson unit *(ozone content of air)*
Docosan *n* docosane
Docosansäure *f* docosanoic acid
Docosensäure *f* docosenoic acid
Dodecahydrat *n* dodecahydrate
Dodecamolybdatophosphat *n* $M^I_3[PMo_{12}O_{40}]$ dodecamolybdophosphate
Dodecan *n* dodecane
Dodecanal *n* *dodecanal, dodecylaldehyde
Dodecandisäure *f* dodecanedioic acid
Dodecan-1-ol *n* *dodecan-1-ol, alcohol C-12
Dodecansäure *f* *dodecanoic acid, lauric acid
Dodecawolframatophosphat *n* $M^I_3[PW_{12}O_{40}]$ dodecawolframophosphate, dodecatungstophosphate
Dodec-1-en *n* dodec-1-ene
Dodecensäure *f* dodecenoic acid
Dodec-2-in *n* dodec-2-yne
Dodecylamin *n* dodecylamine
Dodekaeder *n (cryst)* dodecahedron
Doebner-Miller-Reaktion *f* Doebner-Miller reaction [synthesis] *(for obtaining quinoline and its derivatives)*
doktor-negativ *(petrol)* sweet(ened), doctor-sweet
doktor-positiv *(petrol)* sour
Dolomit *m (min)* dolomite, magnesian limestone *(calcium magnesium carbonate)*
dolomitisch dolomitic
Dolomitkalk *m* dolomitic [dolomite] lime *(calcium magnesium oxide)*
Domäne *f (cryst)* domain
Domäne *f***/ferromagnetische** ferromagnetic domain, Weiss (molecular magnetic) field
Domänenwand *f (cryst)* domain [Bloch] wall
Donator *m* donor, donator
Donator-Akzeptor-Bindung *f* donor-acceptor bond, dipolar [coordinate] bond, coordination (bond)
Donatoratom *n* donor [ligating] atom *(coordination chemistry)*
Donatorort *m (bioch)* donor [condensing] site, peptidyl-tRNA site, P site
Donnan-Gleichgewicht *n* Donnan (membrane) equilibrium
Donor *m s.* Donator
Dopa *n s.* Dihydroxyphenylalanin
Dope *m(n)* dope *(a substance added to mineral-oil products in quantities below 1 %)*
Dope-Mittel *n* dope *(a substance added to mineral-oil products in quantities below 1 %)*
dopen *v* dope *(to treat mineral-oil products with a dope)*
Dope-Stoff *m s.* Dope
Dopingmittel *n (pharm)* dope
Doppelarmkneter *m* double-arm mixer
Doppelbindung *f* double (covalent) bond
Doppelbindung *f***/gekreuzte** crossed double bond
Doppelbindung *f***/isolierte** isolated double bond
Doppelbindung *f***/konjugierte** conjugated double bond
Doppelbindung *f***/kumulierte** cumulated double bond
Doppelbindung *f***/olefinische** olefinic double bond
doppelbrechend doubly refracting [refractive], birefringent
Doppelbrechung *f* double refraction, birefringence
Doppelbrechung *f***/elektrische** electrical double refraction
Doppelbrechung *f***/optische** optical birefringence
doppelfarbig *(cryst, coll)* dichroic
Doppelfarbigkeit *f (cryst, coll)* dichroism
Doppelgaserzeuger *m (coal)* predistillation (gas) producer, carbonizing [double gas, two-stage] generator, two-zone gasifier
Doppelgasgenerator *m (coal)* predistil-

lation (gas) producer, carbonizing [double gas, two-stage] generator, two-zone gasifier
doppelhelikal *(bioch)* double-helical
Doppelhelix *f (bioch)* double [duplex] helix, double-strand(ed) helix
Doppelkegelbindung *f* sandwich bond *(coordination chemistry)*
Doppelleerstelle *f* divacancy *(a crystal defect)*
Doppellinie *f (spectr)* doublet
Doppellipidschicht *f s.* Lipiddoppelschicht
Doppelmembran *f (bioch)* bilayer membrane
Doppelmuffe *f (lab)* clamp holder, bosshead
Doppeloxid *n* double oxide
Doppelresonanzmethode *f (spectr)* double-resonance method [technique]
Doppelrohrwärmeaustauscher *m* double-pipe heat exchanger, concentric-tube heat exchanger
Doppelrohrwärmeübertrager *m* double-pipe heat exchanger, concentric-tube heat exchanger
Doppelsalz *n* double salt
Doppelschicht *f* 1. *(phys ch)* double layer; 2. *(bioch)* bilayer
Doppelschicht f/diffuse diffuse double layer
Doppelschicht f/elektrische [elektrochemische] electric(al) double layer, EDL, double-charge layer
Doppelschicht f/gouy-chapmansche Gouy layer
Doppelschneckenextruder *m* twin-screw extruder
Doppelstrahlgerät *n* double-beam spectro(photo)meter
Doppelstrahlspektrometer *n* double-beam spectro(photo)meter
Doppelstrang *m (bioch)* double strand, duplex *(of DNA)*
Doppelstrang-DNA *f (bioch)* double-strand DNA
doppelsträngig 1. *(bioch)* double-strand(ed), double-helical, bihelical; 2. *(tech)* two-stranded *(e.g. conveyor)*
Doppler-Effekt *m (spectr)* Doppler effect, Doppler (frequency) shift
Doppler-Verbreiterung *f (spectr)* Doppler broadening
Dorr-Eindicker *m* Dorr thickener [agitator], Dorr settling tank
dörren *v* 1. dry, desiccate, dehydrate *(fruit)*; dry-cure *(meat)*; 2. *s.* darren
Dosierapparat *m* dosing apparatus, metering [proportioning] apparatus
Dosiereinrichtung *f* dosing mechanism, metering [proportioning] mechanism, feeder
Dosiereinrichtung *f* **für Chemikalien** chemical feeder
dosieren *v* dose, meter, proportion, batch
Dosierpumpe *f* 1. dosing pump, metering [proportioning, controlled-volume, feed] pump; 2. *(plast)* spinning pump
Dosierung *f* 1. dosage, dosing, metering, proportioning; 2. *(chromat)* sampling, injection
Dosierung *f* **mit Gasstromteilung** *(chromat)* split sampling [injection]
Dosierung *f* **ohne Gasstromteilung** *(chromat)* splitless sampling [injection]
Dosimeter *n* dosimeter, dosemeter, dosage meter
Dosimetrie *f* dosimetry, dosage measurement
Dosis *f* dose, dosage
Dosis f/giftig wirkende *s.* Dosis/toxische
Dosis f/kleinste wirksame minimum effective dose
Dosis f/letale lethal [fatal] dose
Dosis *f* **mit beobachtbarer toxischer Wirkung/niedrigste** lower observed effect level, LOEL
Dosis f/mittlere effektive median effective dose
Dosis f/mittlere letale median lethal dose, lethal dose 50
Dosis *f* **ohne toxische Wirkung** no-effect-level, NEL
Dosis *f* **ohne toxische Wirkung/nachweisbare** no observed effect level, NOEL
Dosis f/subletale sublethal dose
Dosis f/therapeutische therapeutic dose
Dosis f/tödliche *s.* Dosis/letale
Dosis f/toxische toxic dose
Dosisleistung *f* dose rate
Dosis-Wirkungs-Kurve *f* dosage-response curve

Dotieren *n* doping *(semiconductors)*
Dotierstoff *m* dopant
Dotierung *f s.* Dotieren
Dotierungsgrad *m* dopant level
Dotierungsmaterial *n s.* Dotierstoff
Dotierverfahren *n* doping process
Dotter *m(n)* (egg) yolk, vitellus
d.P. *s.* Packung/dichteste
2,4-DP *s.* Dichlorprop
DPE *s.* Diethylentriaminpentaessigsäure
DPP *s.* Differenz-Pulspolarographie
Dragée *n* dragée
Draht *m* 1. wire; 2. *(text)* twist
Drahtgewebe *n* wire cloth [gauze], metal fabric
Drahtnetzfüllkörper *m* wire-gauze packing
Drahtsieb *n* wire screen [sieve], metal screen
dränieren *v* drain
Dränrohr *n* 1. field-drain pipe, drainage pipe [tube]; 2. *(hyd)* underdrain, *(Am)* drain tile
Dränröhre *f* 1. field-drain pipe, drainage pipe [tube]; 2. *(hyd)* underdrain, *(Am)* drain tile
Drän(rohr)strang *m* drain
Dränung *f* drainage
Drastikum *n (pharm)* drastic, strong purgative
Drehachse *f* 1. axis of rotation; 2. *(cryst)* axis of symmetry
Drehachse *f*/**dreizählige** *(cryst)* threefold [triad] axis of symmetry
Drehachse *f*/**zweizählige** *(cryst)* twofold [diad] axis of symmetry
Drehbandkolonne *f (distil)* spinning band column, rotating-strip column
drehbar rotatable
Drehbarkeit *f* rotatability
Drehbarkeit *f*/**freie** free rotation
***d*-drehend** *s.* (+)-drehend
***l*-drehend** *s.* (-)-drehend
(+)-drehend dextrorotatory, dextrogyrate
(-)-drehend laevorotatory, laevogyrate
Drehetagenofen *m (hyd)* multiple-hearth incinerator *(sludge incineration)*
Drehfilter *n* rotary [rotating] filter, drum filter
Drehimpuls *m* angular momentum, moment of momentum, spin
Drehimpuls *m*/**innerer** intrinsic angular momentum *(of elementary particles)*
Drehimpulsquantenzahl *f* azimuthal [subsidiary] quantum number, orbital (angular momentum) quantum number
Drehkolbenpumpe *f* lobe pump
Drehkristallaufnahme *f* rotating-crystal photograph [diagram], X-ray rotation photograph
Drehkristalldiagramm *n* rotating-crystal photograph [diagram], X-ray rotation photograph
Drehkristallmethode *f* rotating-crystal method *(of crystal analysis)*
Drehkristallverfahren *n* rotating-crystal method *(of crystal analysis)*
Drehmischer *m (tech)* rotary mixer
Drehmoment *n* torque
Drehofen *m* 1. rotary kiln; 2. *(met)* rotary furnace
Drehölbrenner *m* rotary-cup (oil) burner, rotary [spinning-cup] burner
Drehrohrofen *m* rotary kiln
Drehrost *m* rotating [revolving] grate
Drehscheibenextrakteur *m* rotary-disk contactor [extractor, tower]
Drehscheibenextraktor *m* rotary-disk contactor [extractor, tower]
Drehscheibenfilter *n* disk-type rotary vacuum filter, rotary disk filter
Drehscheibenkolonne *f s.* Drehscheibenextrakteur
Drehsinn *m* sense of rotation
Drehspulgalvanometer *n* moving-coil galvanometer
Drehtrommel *f* rotary [rotating] drum [cylinder], tumbler, tumbling barrel
Drehtrommeltrockner *m* rotary (drum) dryer, rotatory dryer
Drehung *f* **der Polarisationsebene** *s.* Drehung/optische
Drehung *f* **(der Polarisationsebene)/ magnetische** magnetic rotation, Faraday effect
Drehung *f*/**molare** molar rotation
Drehung *f*/**optische** optical rotation
Drehung *f*/**spezifische** specific rotation
Drehungsvermögen *n*/**magnetisches** magnetic rotatory power
Drehungsvermögen *n*/**molares** molar rotatory power
Drehungsvermögen *n*/**optisches** optical activity [rotatory power]

Drehungsvermögen *n***/spezifisches** specific rotatory power
Drehwaage *f* torsion balance
Drehwert *m* amount [angle] of rotation
Drehwinkel *m s.* Drehwert
Drehwinkel *m***/optischer** *s.* Drehwert
Drehzahl *f* number of revolutions, rate of rotation, speed
Drehzerstäuberbrenner *m s.* Drehölbrenner
dreiatomig triatomic
dreibasig tribasic, triprotic *(acid)*; tribasic, triacid *(base)*
dreibindig trivalent, tervalent *(relating to homopolar bonds)*
Dreieckwellenvoltammetrie *f (anal)* triangular wave voltammetry
Dreierkombination *f (bioch)* triplet (codon) *(in the genetic code)*
Dreierreaktion *f* termolecular reaction
Dreierstoß *m* ternary collision, triple [three-body] collision
dreifach negativ (geladen) trinegative
dreifach positiv (geladen) tripositive
Dreifachaufsatz *m* triple-neck adapter, three-neck(ed) adapter
Dreifachbindung *f* triple (covalent) bond
Dreifarbigkeit *f (cryst)* trichroism
dreifunktionell trifunctional
dreiglied(e)rig three-membered
dreihalsig three-neck(ed)
Dreihalskolben *m* three-neck(ed) flask
Dreikohlenstoffkörper *m* three-carbon compound
Drei-Kohlenstoff-Tautomerie *f* three-carbon tautomerism
Dreikomponentensystem *n* ternary [tertiary, three-component] system
Dreikörperverdampfer *m* triple-effect evaporator [evaporating unit]
dreiprotonig triprotic, tribasic *(acid)*
Dreiring *m* three-membered (carbon) ring
dreisäurig triacid, tribasic *(base)*
dreischauflig three-bladed *(agitator)*
Dreistofflegierung *f* ternary [three-component] alloy
Dreistoffsystem *n* ternary [three-component] system
Dreistufenreaktion *f* three-step reaction
Dreistufenverdampfer *m s.* Dreikörperverdampfer
Dreiwalzenmühle *f* three-roll mill
Dreiwalzenstuhl *m* three-roll mill
Dreiweg(e)hahn *m (lab)* three-way stopcock [tap]
dreiwertig trivalent, tervalent *(element)*; tribasic, triprotic *(acid)*; tribasic, triacid *(base)*; trihydric *(alcohol)*
Dreiwertigkeit *f* trivalency, tervalency *(of an element)*; tribasicity *(of an acid or base)*
dreizählig 1. *(cryst)* triad, threefold; 2. *s.* dreizähnig
dreizähnig tridentate, terdentate *(ligand)*
Dreizentrenbindung *f* three-centre bond
Drift *f (anal)* drift *(as of a base line)*
driften *v (anal)* drift *(as of a base line)*
Dritte-Partner-Effekt *m (phys ch)* matrix (interference) effect
Droge *f***/pflanzliche** vegetable drug
Drogenkunde *f* pharmacognosy, pharmacogn(os)ia
Drosselklappe *f* butterfly valve
Drosselklappenventil *n* butterfly valve
drosseln *v* slow down, choke, throttle
Drosselstelle *f* constriction, restriction, throttle
Drosselventil *n* butterfly valve
Druck *m* 1. pressure; 2. printing *(process)*; 3. print *(printed material)* • **unter Druck setzen** pressurize
Druck *m***/barometrischer** barometric pressure
Druck *m***/hydrostatischer** hydrostatic head
Druck *m***/kritischer** critical pressure
Druck *m***/osmotischer** osmotic pressure, OP
Druck *m***/statischer** static pressure *(of a fluid)*
Druckabfall *m* pressure drop [loss]
druckabhängig pressure-dependent
Druckanstieg *m* pressure increase, increase in pressure
Druckausgleichsgefäß *n* pressure-compensating vessel
Druckbelüftung *f* compressed air aeration *(wastewater)*
Druckdestillat *n* pressure distillate, P.D.
Druckdestillation *f* pressure distillation
drücken *v* press, push *(e.g. a lever or button)*; blow *(a liquid into a reservoir)*
Druckerhöhung *f* pressure increase, increase in pressure

Druckerschwärze *f* printing [printer's] ink
Druckfarbe *f* printing ink
Druckfass *n* blowcase, acid egg
Druckfestigkeit *f* compressive [compression] strength
Druckfilter *n* pressure filter, pressure-filtration funnel
Druckfiltration *f* pressure filtration
Druckfiltrieren *n* pressure filtration
Druckflotation *f (hyd)* (pressurized-)air floatation, dissolved-air pressure floatation, dissolved-air floatation without recycle, DAF
Druckfügen *n (tech)* clinching
Druckgaserzeuger *m (coal)* pressurized-gas producer, pressure gasifier
Druckgasgenerator *m (coal)* pressurized-gas producer, pressure gasifier
Druckgefäß *n* pressure vessel [tank]
druckgießen *v* (pressure-)diecast
Druckguss *m* (pressure) diecasting, pressure casting • **im Druckguss herstellen** *s.* druckgießen
Druckhöhe *f* (pressure) head
Druckhöhe *f***/statische** (pressure) static head
Druckhydrierung *f* hydrogenation under pressure, pressure hydrogenation
Druckhydrierung *f***/spaltende** *(petrol)* hydrocracking
Druckkammer *f* pressure chamber
Drucklaugung *f (min tech)* pressure leaching
Druckluft *f* compression [compressed] air
druckluftbetätigt air-operated, air-driven
Drucklufttheber *m* air lift, mammoth [air-lift] pump
Druckluftversorgung *f* compressed-air supply
Druckluftversprüher *m* pneumatic [auxiliary-fluid] atomizer
Druckluftzerstäuber *m* 1. pneumatic atomizer *(for solids)*; 2. *s.* Druckluftversprüher
Druckmesser *m* pressure gauge, manometer
Druckminder(ungs)ventil *n* (pressure-) reducing valve
Druckmischer *m* bubbler
Drucksensor *m* pressure sensor
Drucksintern *n* sintering under pressure, hot pressing
Druckspeicher *m***/hydraulischer** hydraulic accumulator
Drucksprung *m* pressure jump
druckspülen *v* jet
Drucksterilisator *m* autoclave sterilizer
Drucksterilisierapparat *m* autoclave sterilizer
Druckstrahlpumpe *f* injector
Druckumlauffermenter *m (biot)* pressure cycle reactor
druckunabhängig pressure-independent
Druckverbreiterung *f* pressure [collision] broadening *(of spectral lines)*
Druckvergaser *m (coal)* pressure gasifier, pressurized gas producer
Druckvergasung *f (coal)* (elevated-) pressure gasification
Druckverlust *m* 1. pressure loss; 2. pressure drop
Druckversprüher *m* pressure atomizer
Druckwasserstoffraffination *f (petrol)* hydrorefining
Druckwasserwäsche *f* water scrubbing
Druckwelle *f* blast
Druckzerstäuber *m s.* Druckversprüher
DS *s.* Diethylstilböstrol
dTA = derivative Thermoanalyse
DTA *s.* Differenzial-Thermoanalyse
DTG *s.* Differenzial-Thermogravimetrie
dTMP *s.* Desoxythymidin-5′-phosphat
Dualzerfall *m* branched [multiple] disintegration [decay], branching
Dublett *n* 1. *(spectr)* doublet; 2. duplet, doublet *(the structure in which two atoms share a pair of electrons)*
Dublettaufspaltung *f (spectr)* doublet splitting
Dublettspektrum *n* doublet spectrum
Dublett-Term *m (spectr)* doublet term
Dubnium *n* Db dubnium *(element 104)*
duktil ductile
Duktilität *f* ductility
Dulong-Petit-Regel *f (phys ch)* Dulong and Petit's law
Düngemittel *n* fertilizer
Düngemittel *n***/anorganisches [mineralisches]** mineral fertilizer
düngen *v* fertilize, dress
Dünger *m* fertilizer, *(esp of animal excreta also)* (farm) manure

Düngung *f* fertilization, (fertilizer) dressing • **auf Düngung ansprechen [reagieren]** respond to fertilizing
Dunkelfärbung *f* darkening
Dunkelperiode *f* dark phase *(of photosynthesis)*
Dunkelphase *f* dark phase *(of photosynthesis)*
Dunkelreaktion *f* dark(ness) reaction *(photochemistry)*
Dunkelrotglut *f* dull redness [red heat]
Dunkelstrom *m (spectr)* dark current
Dünnfilmtechnik *f* thin film technology
dünnflüssig thin, highly liquid [fluid], low-viscosity
Dünnflüssigkeit *f* thinness, low viscosity
Dünnlauge *f* weak [dilute] liquor
Dünnsäure *f* diluted waste acid
Dünnschicht *f* thin layer, (thin) film
Dünnschichtabsorber *m* wetted-wall(-column) absorber
Dünnschichtchromatographie *f* thin-layer chromatography, TLC
dünnschichtchromatographisch by thin-layer chromatography
Dünnschicht-Kapillarsäule *f* wall-coated open tubular capillary column, WCOT column *(gas chromatography)*
Dünnschichtteilchen *n (chromat)* porous layer bead, PLB, pellicular packing [support], solid core support
Dünnschichtverdampfer *m* film evaporator
Dünnschichtzellen *fpl* thin film (solar) cells
Dünnschlamm *m (hyd)* dilute sludge
Dünnschliff *m (min)* thin section
Dünnschliff *m*/**polierter** polished thin section
dünntaflig *(cryst)* thin-tabular
Dunst *m* damp, haze *(consisting of droplets)*; fume, smoke *(consisting of solid particles)*
Dunsthaube *f* hood, air dome
durcharbeiten *v* 1. work [knead] thoroughly *(e.g. a dough)*; 2. homogenize *(an emulsion)*; 3. *(el ch)* work (in), deal with, dummy *(a plating bath)*
durchbiegen *v*/**sich** sag
Durchbiegung *f* sag(ging)
durchbluten *v (dye, coat, tann)* bleed (through), strike through
durchbrechen *v* break through *(ion exchange, filtration)*
Durchbruch *m* breakthrough, leakage *(ion exchange, filtration)*
Durchbruchkurve *f* breakthrough [leakage] curve
Durchbruchpunkt *m* break(through) point, point of breakthrough
durchdringen *v* penetrate, permeate
durchdringen *v*/**sich** interpenetrate
durchdringend penetrating *(e.g. odour)*
Durchdringung *f* penetration, permeation
Durchdringungskomplex *m* penetration complex, inner [low-spin, inner-orbital, spin-paired] complex
Durchdringungsnetzwerk *n (plast)* interpenetrating network, IPN
Durchdringungsvermögen *n* penetrating power, permeativity
Durchdringungszwillinge *mpl (cryst)* penetration twins
durchdrücken *v* force through
durchfahren *v (anal)* scan *(a measuring range)*
durchfahren *v*/**nochmals** rescan
durchfließen *v* flow [pass, run] through
Durchfluss *m* 1. flow, flowing-through, passage; 2. *(glass)* throat, flow hole
Durchflussanzeiger *m* flow rate indicator
Durchflussfermenter *m (biot)* continuous fermenter
Durchflussgeschwindigkeit *f* flow rate
Durchflusskühlung *f* once-through cooling
Durchfluss(mengen)messer *m* flowmeter, rate [fluid] meter, rate-of-flow volume meter
Durchflussmischreaktor *m* tank-type flow reactor, continuous-flow stirred-tank reactor
Durchflussreaktor *m* (continuous-)flow reactor, continuous reactor
Durchflussrührfermenter *m (biot)* continuous stirred-tank fermenter
Durchflussrührkessel *m* continuous stirred tank
Durchflussrührkesselfermenter *m s.* Durchflussrührfermenter
durchfressen *v* eat through, corrode
durchführbar/technisch practicable, feasible

Durchführbarkeit *f* practicality
durchführen *v* carry out, run, perform, conduct
Durchgang *m* passage
Durchgang *m*/**galvanischer** ohmic continuity
durchgasen *v (agric, food)* fumigate
Durchgasung *f (agric, food)* fumigation
Durchgasungsmittel *n (agric, food)* fumigant, fumigator
Durchhang *m* sag(ging)
durchhängen *v* sag
durchkochen *v* cook thoroughly
Durchlassgrad *m (phot, anal)* transmission ratio, transmittance, transmittancy
durchlässig permeable
Durchlässigkeit *f* 1. permeability *(relating to particles)*; 2. transmittance *(relating to radiation)*
Durchlässigkeitskoeffizient *m* permeability coefficient
Durchlasszahl *f* transmissivity *(radiation of heat)*
Durchlauf *m* passage
durchlaufen *v* 1. run [pass] through *(e.g. stages)*; 2. *s.* hindurchlaufen; 3. *s.* durchströmen
Durchlaufentwicklung *f* overrun development *(paper chromatography)*; continuous development *(thin-layer chromatography)*
Durchlaufkühlung *f* 1. once-through cooling; 2. *(glass)* continuous annealing
Durchlaufofen *m (met)* continuous furnace
Durchlaufverdampfer *m* once-through evaporator, single-pass [one-pass] evaporator
Durchlaufverfahren *n* continuous process
Durchlaufzeit *f* retention time, detention [transit, hold-up] time, flow time, holding time [period]
durchleiten *v* pass (through)
Durchlichtmethode *f* transmitted-light technique *(microscopy)*
durchmischen *v* intermix, mix [blend] together; mix thoroughly
Durchmischung *f* intermixture; thorough mixing
Durchmischung *f*/**ideale** perfect mixing
Durchmischung *f*/**vollständige** complete mixing
durchmustern *v (pharm, biot)* screen *(for valuable compounds)*
durchnummerieren *v (nomencl)* number
durchpressen *v* squeeze through
durchrühren *v* stir, agitate
durchsacken *v* sag
Durchsatz *m* throughput, *(relating to a liquid also)* flow rate
durchscheinend translucent, translucid
durchscheinend/nicht opaque
Durchschlag *m*/**dielektrischer** dielectric loss
durchschlagen *v* 1. *(dye, coat, tann)* bleed (through), strike through; 2. puncture *(a dielectric)*
Durchschlagfestigkeit *f*/**dielektrische [elektrische]** dielectric strength, breakdown [puncture] strength, electric strength
durchschlagsfest voltage proof
Durchschlag(s)spannung *f* breakdown voltage
Durchschlagswiderstand *m* dielectric resistance
Durchschnittsausbeute *f* average yield
Durchschnittsprobe *f* average sample
durchschütteln *v* shake
durchseihen *v* strain, percolate
durchsetzen *v* put through, batch, handle *(a definite quantity of material)*
Durchsetzfügen *n s.* Druckfügen
durchsichtig transparent
Durchsichtigkeit *f* transparency
durchsickern *v* 1. *(filtr)* trickle through, seep through, percolate, strain; 2. leak
durchsickern lassen *v* pass (through), percolate, strain *(a liquid)*
durchsieben *v* screen *(e.g. coal, gravel)*; sieve, sift *(e.g. flour)*
durchspülen *v* wash thoroughly [through]
durchstrahlen *v*/**mit Röntgenstrahlen** X-ray
durchströmen *v* flow (through), pass, *(gas chromatography also)* be swept through
durchtränken *v* soak, impregnate, imbibe, penetrate, saturate
Durchtritt *m* passage
Durchtrittsfaktor *m (el ch)* transfer coefficient, symmetry factor

Durchwachsungszwillinge *mpl (cryst)* penetration twins
Duromer(e) *n* duromer
Durometer *n* durometer, hardness tester [meter]
Durometerhärte *f* durometer [Shore] hardness
Duroplast *m* thermosetting plastic [resin], thermoset (resin)
Duroplastklebstoff *m (plast)* thermosetting adhesive
Düse *f* 1. nozzle; 2. *(plast)* die
Düsenfärbung *f (text)* dope [spin] dyeing
düsengefärbt *(text)* dope-dyed, spin-dyed, spun-dyed, mass-dyed, solution-dyed
Düsentreibstoff *m* fuel for jet planes, (turbo)jet fuel
Dynamik *f* dynamics
Dynamit *n* dynamite
Dysprosium *n* Dy dysprosium
Dysprotid *n* proton don(at)or, protonic acid, Brønsted-Lowry acid

E

EA-MS *s.* Elektronenanlagerungs-Massenspektroskopie
e-Bindung *f* e-bond, equatorial bond
Ebullioskopie *f* ebullioscopy
ebullioskopisch ebullioscopic
EC *s.* Ethylcellulose
Echelettegitter *n (spectr)* echelette [blazed] grating
Echelongitter *n (spectr)* echelon grating
echt 1. fast *(dye)*; 2. genuine *(noble metal, leather, gem)*; real, pure *(silk)*
Echt(farb)base *f* fast base, *(Am)* fast color base
Echtfärbebase *f* fast base, *(Am)* fast color base
Echtfärbesalz *n* fast salt, *(Am)* fast color salt
Echtfarbstoff *m* fast dye
Echtheit *f* fastness *(of dyes)*; genuineness *(of noble metals or gems)*
Echtrot *n* fast red
Eckventil *n* angle valve
ED_{50} *s.* Dosis/mittlere effektive
EDA-Komplex *m s.* Elektronen-Donor-Akzeptor-Komplex
edel noble, unreactive, non-reactive
Edelgas *n* noble [inert, rare] gas
Edelgasclathrat *n* noble-gas [inert-gas] clathrate
Edelgaskonfiguration *f* noble-gas (electronic) configuration, inert-gas (electronic) structure, octet structure
Edelmetall *n* noble [precious] metal
Edelstein *m* precious stone, gemstone
Edison-Akkumulator *m (el ch)* Edison accumulator [cell]
Edman-Abbau *m (bioch)* Edman degradation *(of proteins)*
ED-Weg *m (bioch)* Entner-Doudoroff pathway
E-Effekt *m s.* Effekt/elektromerer
Effekt *m***/elektromerer** *s.* Effekt/mesomerer
Effekt *m***/hyperchromer** *(bioch, spectr)* hyperchromic effect
Effekt *m***/hypochromer** *(spectr)* hypochromic effect
Effekt *m***/induktiver** *(phys ch)* inductive [induction] effect, I effect
Effekt *m***/lichtelektrischer** *s.* Effekt/photoelektrischer
Effekt *m***/mesomerer** mesomeric [electromeric, resonance] effect
Effekt *m***/nivellierender** *(phys ch)* levelling effect
Effekt *m***/photoelektrischer** photoelectric effect
Effekt *m***/synergistischer** synergistic effect, synergism
α-Effekt *m (org ch)* α-effect, alpha-effect *(enhancement of nucleophilicity)*
Effektor *m (bioch)* effector
Effektor *m***/allosterischer** allosteric effector, modifier, determinant *(enzyme kinetics)*
Effektor *m***/heterotroper** heterotropic effector
Effektor *m***/homotroper** homotropic effector
Effloreszenz *f (min)* efflorescence
effloreszieren *v (min)* effloresce
egalfärben *v (text)* level, dye level
egalisieren *v (text)* level, dye level
Egalisierungsmittel *n (dye)* levelling agent
Eialbumin *n* egg albumin
eichen *v* calibrate *(e.g. measuring apparatus)*; adjust *(balance weights or measures)*; gauge *(vessels)*

Eichfunktion *f* calibration function
Eichkurve *f* calibration curve [graph]
Eichlösung *f* calibrating solution
Eichstandard *m* calibration standard
Eichung *f* calibration *(as of measuring apparatus)*; adjustment *(of balance weights or measures)*; standard calibration, gauging *(of vessels)*
Eicosan *n* eicosane
Eicosansäure *f* *eicosanoic acid, arachidic acid
Eigelb *n* (egg) yolk, vitellus
Eigenabsorption *f* self-absorption
Eigendiffusion *f* self-diffusion
Eigendissoziation *f* self-dissociation, self-ionization, autoprotolysis
Eigendrehimpuls *m* intrinsic angular momentum, spin *(of elementary particles)*
Eigenfrequenz *f* natural frequency *(of vibration)*
Eigenfunktion *f (phys ch)* eigenfunction
Eigenhalbleiter *m* intrinsic semiconductor
Eigenionisation *f* autoionization
Eigenkeim *m* crystal nucleus, nucleus of crystallization, nucleation centre
Eigenkonvektion *f* natural convection
Eigenschaft *f* property
Eigenschaft f/additive additive property
Eigenschaft f/innere intrinsic property
Eigenschaft f/kolligative colligative property *(depending only on the number of particles)*
Eigenschaft f/konstitutive constitutive property
Eigenschaft f/periodische periodic property *(of the elements)*
Eigenschwingung *f* normal [fundamental] vibration, normal mode of vibration *(of molecules)*
Eigenwert *m* eigenvalue
Eigenzustand *m* eigenstate
Eiklar *n* (egg) albumen, egg white, glair(e)
Eimer *m* bucket *(as of an elevator)*
einachsig uniaxial
einachsig/optisch optically uniaxial
einarbeiten *v* 1. work in, intermingle *(constituents of a mix)*; 2. work (in), deal with, dummy *(a plating bath)*
einatomig monoatomic
Einatomigkeit *f* monoatomicity
einbasig monobasic, monoprotic *(acid)*; monobasic, monoacid *(base)*
Einbau *m* 1. insertion *(as of atoms into a lattice)*; incorporation *(as of nutrients into organic substance)*; 2. *s.* Einbauten
einbauen *v* insert *(e.g. atoms into a lattice)*; incorporate *(e.g. nutrients into organic substance)*
Einbauten *mpl* internals, *(in a cooling tower also)* pack, fill, *(in a rotary dryer also)* (internal) flights; baffles *(for directing a fluid stream)*
einbetten *v* 1. embed, imbed; 2. *(met)* pack *(as with a carburizing powder)*; embed *(completely in a medium)*; encapsulate *(by dip coating)*; pot *(in a container)*
Einbetten *n (plast)* encapsulation
einblasen *v* blow in
Einbrennemaillack *m* stoving [baking] enamel
Einbrennemaille *f* stoving [baking] enamel
einbrennen *v* 1. burn in *(e.g. pigments)*; 2. *(coat)* stove, bake
Einbrennlack *m* stoving [baking] varnish; *(if pigmented:)* stoving [baking] lacquer
Einbrennofen *m (coat)* stoving [baking] oven, stove
einbringen *v* introduce, place
Einbringen *n* introduction, placing
eindampfen *v* evaporate, concentrate by evaporation, boil down, inspissate
Eindampfen *n* evaporation, concentration by evaporation, boildown, boiling-down, inspissation
eindicken *v* 1. thicken, concentrate, *(esp by evaporation:)* boil down, inspissate; 2. *(food)* condense; 3. *(coat)* body
Eindicker *m* 1. thickener, concentrator; 2. *s.* Eindickmittel
Eindickmittel *n* thickening agent, thickener, viscosifier
Eindickung *f* 1. thickening, concentration, *(esp by evaporation)* boildown, inspissation; 2. *(food)* condensation; 3. *(coat)* bodying
Eindick(ungs)zone *f* 1. thickening region [section, zone]; 2. *(hyd)* sludge zone
eindiffundieren *v* diffuse in(wards)
eindosen *v (food)* can, tin
eindringen *v* penetrate, permeate, *(of liquids or gases also)* diffuse

Eindringen *n* penetrating, penetration, permeation, *(of liquids or gases also)* diffusion
Eindringhärte *f s.* Eindruckhärte
Eindringtiefe *f* (depth of) penetration; depth of indentation *(hardness testing)*
eindrücken *v* 1. indent; 2. blow in *(e.g. a gas into a vessel)*
Eindruckhärte *f* indentation hardness *(materials testing)*
eindunsten *v* evaporate down
Eindunstung *f***/solare** solar evaporation
einebnen *v* level
ein-ein-wertig uniunivalent
Einelektronenatom *n* one-electron atom
einengen *v* evaporate to low [small] bulk, concentrate to small volume
Einetagenpresse *f* single-daylight [one-daylight] press
einfach negativ (geladen) uninegative
einfach positiv (geladen) unipositive
Einfachbindung *f* single (covalent) bond
Einfachfilter *n* single-medium filter
Einfachstreuung *f (phys ch)* single scattering
Einfachsubstitution *f* monosubstitution
Einfachzucker *m* simple sugar, monosaccharide
Einfall(s)winkel *m* angle of incidence
Einfang *m (nucl)* capture
einfangen *v (nucl)* capture
Einfangquerschnitt *m (nucl)* capture cross section
einfarbig *(phot)* monochromatic, monochrome
einfetten *v* grease
Einfluss *m***/dirigierender** directive influence
einfrieren *v* 1. *(food)* freeze; 2. freeze *(chemical reactions)*; 3. *(glass, plast, rubber)* exhibit transition
Einfrieren *n* 1. *(food)* freezing; 2. freezing *(of chemical reactions)*; 3. *(glass, plast, rubber)* transition
Einfriergebiet *n (plast)* transition interval
Einfriertemperatur *f (glass, plast, rubber)* glass (transition) temperature, transition temperature, Tg point, Tg
einfügen *v* insert *(e.g. atoms into interstitial lattice sites)*
Einfügen *n* insertion *(as of atoms into interstitial lattice sites)*
einführen *v* introduce
Einführung *f* introduction
einfüllen *v* fill (in), feed (in), charge, load
Einfülltrichter *m* 1. *(lab)* chemical funnel; 2. *(tech)* (feed, charge) hopper, loading [charging] hopper
Eingang *m (bioch)* entry site, acceptor [recognition, decoding, aminoacyl-tRNA] site
eingehen *v* 1. go *(into a product)*; 2. *(text)* shrink, contract
eingehen *v***/eine Bindung** bond
eingehen *v***/eine chemische Reaktion** enter into (chemical) reaction
eingehen *v***/eine (chemische) Verbindung** enter into chemical combination [union], combine
eingelagert intercalated, intercalary *(molecules, ions)*
eingeschliffen ground-in
Einheit *f* 1. unit *(as of measuring values)*; 2. *(tech)* unit, set
Einheit *f***/difunktionelle** difunctional [bifunctional] unit, D unit *(a structural element of macromolecules)*
Einheit *f***/internationale** international unit, I.U. *(of biochemically active substances)*
Einheit *f***/monofunktionelle** monofunctional unit, M unit *(of macromolecules)*
Einheit *f***/strukturelle** structural unit
Einheit *f***/tetrafunktionelle** tetrafunctional unit, Q unit *(of macromolecules)*
Einheit *f***/trifunktionelle** trifunctional unit, T unit *(of macromolecules)*
einheitlich uniform, *(relating to composition also)* homogeneous
einheitlich/chemisch chemically uniform
einheitlich/nicht non-uniform
Einheitlichkeit *f* uniformity, *(relating to composition also)* homogeneity
Einheitlichkeit *f***/chemische** chemical uniformity
Einheitsfläche *f* unit area
Einhorn-Reaktion *f (org ch)* Einhorn [haloform] reaction
einhüllen *v* 1. envelop, enwrap; 2. *(hyd)* enmesh *(e.g. dirt particles)*; encapsulate *(liquid drops)*
Einkammereindicker *m* single-compartment thickener, unit thickener

einkapseln *v* 1. encapsulate; 2. *(biot)* microencapsulate, prill *(an enzyme)*
Einkapselung *f (nucl)* encapsulation
einkernig mononuclear
Einkohlenstoffverbindung *f* one-carbon compound
Einkomponentensystem *n* 1. *(phys ch)* unary [one-component] system; 2. *(plast, coat)* one-component system, one-pack(age) system, single-pack(-age) system
Einkörperverdampfer *m* single-effect evaporator
Einkristall *m* single crystal, monocrystal
Einkristallaufnahme *f* crystallogram
Einkristallfaser *f* single-crystal fibre, (crystal) whisker
Einkristallstrukturaufklärung *f (cryst)* single-crystal structure determination
einlagern *v* 1. store; 2. *(cryst)* insert, include; intercalate *(guest molecules between layers)*
Einlagerung *f* 1. storing, storage; 2. *(cryst)* insertion, inclusion; intercalation *(of guest molecules between layers)*
Einlagerungsatom *n* interstitial (atom)
Einlagerungscarbid *n* interstitial carbide
Einlagerungsmischkristall *m* interstitial (solid) solution
Einlagerungsverbindung *f* interstitial compound; intercalation compound
Einlagerungsverbindung *f***/schichtförmig ausgebildete** lamellar compound
Einlagerungsverbindung *f* **1. Stufe** first-stage intercalation compound, stage 1 intercalation compound
Einlass *m* inlet, intake
Einlassventil *n* inlet valve
Einlauf *m* 1. inlet, point of entry; 2. influent *(material)*
einlaufen *v* 1. flow in; 2. *(text)* shrink, contract
einlaufen lassen *v* run in, infuse
einleiten *v* 1. introduce, pass in *(material)*; 2. *(hyd)* discharge, run, drain *(as into a sewer system)*; 3. initiate, start *(a reaction)*
Einleitung *f* 1. introduction, passing-in *(of material)*; 2. initiation, start *(of a reaction)*
Einleitungsbedingungen *fpl (hyd)* discharge standards [regulations], effluent limitations [restrictions]
Einleitungsstelle *f (hyd)* discharge point, point of waste-water discharge
einmaischen *v (ferm)* mash
einnehmen *v* 1. *(pharm)* take; 2. occupy *(e.g. the interstices in a lattice)*; cover *(an area)*; occupy *(a volume)*
einpegeln *v***/sich** level off *(as of pH values)*
Einphasensystem *n* one-phase system, homogeneous system
einpipettieren *v* pipette *(a liquid into a vessel)*
einpökeln *v* cure
einprotonig monoprotic, monobasic *(acid)*
einpudern *v* powder, dust
Einrichtung *f* installation, facility; arrangement; equipment
einrühren *v* stir in
Einsalzeffekt *m* salting-in effect, diverse-ion effect
Einsatz *m* 1. application *(as of a chemical)*; 2. *(tech)* charging, feeding, batching; 3. *(met)* (carburized) case; 4. tray *(as of a desiccator or column)*; 5. *s.* Einsatzgut
Einsatzgas *n* feed gas
Einsatzgut *n* feed(stock), feed material, charge (stock), charging stock
Einsatzhärte *f (met)* case hardness
einsatzhärten *v (met)* case-harden
Einsatzhärtungstiefe *f (met)* case depth
Einsatzmenge *f* amount [quantity] required, dose, dosage, input
Einsatzrichtlinie *f* specification
Einsatzschicht *f***/gehärtete** *(met)* (hardened) case
Einsatzstahl *m* case-hardening steel; case-hardened steel
Einsatztulpe *f (lab)* crucible adapter
Einsatzverchromen *n* chromizing, chromium cementation
einsaugen *v* suck (in, up), imbibe, absorb
Einsaugen *n* suction, imbibition, absorption
einsäurig monoacid, monobasic *(base)*
Einschalenanalysenwaage *f* single-pan analytical balance
Einscheibensicherheitsglas *n* tempered safety glass
Einschichtenfilter *n* single-medium filter
einschieben *v* insert *(atoms or groups)*

einschieben v/sich insert *(of atoms or groups)*
einschleifen *v* grind in
einschließen *v* enclose, (en)trap, include, occlude; enmesh *(particles as in a network)*
einschließen v/in Mikrokapseln *(biot)* microencapsulate, prill *(an enzyme)*
Einschluss *m* enclosure, entrapment, inclusion, occlusion; enmeshment *(of particles as in a network)*
Einschluss *m* **in Mikrokapseln** *(biot)* (micro)encapsulation, prilling *(of an enzyme)*
Einschlussverbindung *f* inclusion compound [complex], enclosure compound
einschmelzen *v* 1. fuse [seal] in *(e.g. in a glass tube)*; 2. *(met)* melt (down)
Einschmelzrohr *n* **nach Carius** Carius [bomb] tube
Einschnürung(sstelle) *f* constriction, waist, throat *(as of a tube)*; *(esp in flow measurement:)* vena contracta
Einschrittreaktion *f* one-step reaction
einschrumpfen *v* shrink, contract
Einschub *m* 1. insertion *(of atoms or groups)*; 2. *(bioch)* intercalation *(of planar ring systems between the stacked bases of nucleic acids)*
Einschubreaktion *f* insertion reaction
einschütten *v* 1. fill in; 2. *(tech)* charge, load, feed (in)
einsetzen *v* 1. apply *(e.g. a certain chemical)*; 2. charge, feed, batch *(a certain quantity)*; 3. insert, place in *(mechanically)*; 4. *(met)* carburize; 5. start *(of a reaction)*
Einsetzen *n* 1. application *(as of a certain chemical)*; charging, feeding, batching *(of a certain quantity)*; 2. insertion *(mechanically)*; 3. *(met)* carburizing, carburization; 4. start, onset *(of a reaction)*
einsickern *v* seep in, trickle in, soak in, infiltrate
Einsickern *n* seepage, trickling, soaking, infiltration
einspannen *v* clamp, fix, attach
einspeisen *v* charge, feed (in)
Einspeisung *f* 1. charging, feeding; 2. charge, charging stock, feed(stock)
einspielen v/sich level off *(as of pH value)*
Einspritzblock *m* (sample) injection port, injection block *(gas chromatography)*
Einspritzdruck *m* injection pressure
einspritzen *v* 1. inject; 2. *(plast)* mould in
Einspritzkondensator *m* jet [wet] condenser
Einspritzventil *n* *(chromat)* injection valve
Einspritzvorrichtung *f* *(chromat)* injector
Einsteigluke *f* manhole, manway
Einsteigöffnung *f* manhole, manway
Einstein *n* Einstein, E *(1 mole of photons)*
Einsteinium *n* Es einsteinium
einstellen *v* adjust *(e.g. instruments)*; standardize *(chemicals)*
einstellen v/sich establish *(as of equilibrium)*
Einstellung *f* adjustment *(as of instruments)*; standardization *(of chemicals)*; establishment *(of chemical equilibrium)*
Einstellung *f* **des pH-Werts** pH adjustment
Einstrahlgerät *n* *(spectr)* single-beam instrument
einsträngig single-strand(ed)
Einstufenreaktion *f* single-step [one-step] reaction
Einstufenverdampfer *m* single-effect evaporator
Einstufenverfahren *n* one-stage [one-step] process
Ein-Substrat-Reaktion *f* *(bioch)* one-substrate reaction
eintauchen *v* dip, plunge, immerse, immerge
Eintauchrefraktometer *n* immersion [dipping] refractometer
einteilen *v* classify; divide *(e.g. a scale)*
Einteilung *f* classification; division *(as of a scale)* • **mit (genauer) Einteilung versehen** graduate, scale
Eintopfreaktion *f* *(org ch)* one-pot reaction
Eintrag *m* 1. charge, charging stock, feed(stock), feed material, load; 2. charging, feeding, loading, introduction *(of material into a reactor)*; 3. *(lab)* placing *(in a vessel)*; 4. *(hyd)* transfer *(of oxygen into water)*; input *(of matter or energy into a system)*

Eintrag *m*/**anthropogener** anthropogenic input *(into the environment)*
eintragen *v* 1. charge, feed (in), load, introduce *(material into a reactor)*; 2. *(lab)* place *(in a vessel)*; 3. *(hyd)* transfer *(oxygen into water)*; 4. plot *(in a coordinate system)*; register *(a trademark)*
Eintragmenge *f* **in der Zeiteinheit** rate of feeding, feed rate
eintreten *v* 1. enter *(of material being charged)*; 2. occur *(of an event)*
Eintritt *m* 1. entry, entrance *(of material)*; 2. *s.* Eintrittsstelle; 3. occurrence *(of an event)*
Eintrittsöffnung *f* inlet (port), intake
Eintrittsstelle *f* point of entry, inlet, intake
Eintrittstemperatur *f* inlet temperature
eintrocknen *v* dry (up)
Einwaage *f* weighed portion
einwägen *v* weigh in
einwandern *v* migrate in *(as of ions)*
Einwegbehälter *m* non-returnable container, single-trip [one-trip] container
einweichen *v* steep, soak
Einweichflüssigkeit *f* 1. steeping liquor; 2. *(text)* steep
einwertig monovalent, univalent *(element)*; monobasic, monoprotic *(acid)*; monobasic, monoacid *(base)*; monohydric *(alcohol)*
Einwertigkeit *f* monovalence, monovalency, univalence, univalency *(of an element)*; monobasicity *(of an acid or base)*
einwirken *v* act
einwirken *v* **auf/störend** interfere with, affect
einwirken lassen *v*/**aufeinander** react
Einwirkung *f* action
Einwirkungsdauer *f* exposure [contact] time, duration of exposure
Einwirkungsstelle *f* emission sink, receptor area *(for pollutants)*
einzählig *s.* einzähnig
einzähnig monodentate, unidentate *(coordination chemistry)*
Einzeldosis *f (pharm)* single dose
Einzelelektrode *f* single electrode, half-cell, half-element
Einzelfaser *f* single [individual] fibre • **in Einzelfasern zerlegen** defibre, defibrate, reduce to fibres, shred, *(Am)* (de)fiberize
Einzelfermentation *f (biot)* single-stage fermentation
Einzelhelix *f (bioch)* single-strand(ed) helix
Einzelionenaktivität *f* single-ion activity
Einzellerprotein *n (biot)* single-cell protein, SCP
Einzellinie *f (spectr)* single line
Einzelpotenzial *n* single(-electrode) potential
Einzelring *m (nomencl)* single [individual] ring
Einzelstrang-DNA *f* single-strand(ed) DNA
Einzelzelle *f (biot)* single [individual] cell
ein-zwei-wertig unibivalent
Eisbad *n* ice bath
Eisen *n* Fe iron
Eisen *n*/**dreiwertiges** trivalent [ferric] iron
Eisen *n*/**nicht hämgebundenes** non-haem iron
Eisen *n*/**pyrophores [reduziertes]** pyrophoric iron *(prepared by hydrogen reduction)*
Eisen *n*/**zweiwertiges** bivalent [divalent, ferrous] iron
Eisenbahnkesselwagen *m* rail tank (car), tank car [wagon]
Eisenbahntankwagen *m* rail tank (car), tank car [wagon]
Eisencarbid *n* 1. iron carbide; 2. *(met)* Fe_3C cementite, cemented carbide, iron carbide
Eisen(II)-chelat *n* iron(II) chelate, ferrous chelate
Eisen(III)-chelat *n* iron(III) chelate, ferric chelate
Eisen(II)-chlorid *n* iron(II) chloride, iron dichloride, ferrous chloride
Eisen(III)-chlorid *n* iron(III) chloride, iron trichloride, ferric chloride
Eisen(II)-disulfid *n* iron(II) disulphide
Eisenerz *n* iron ore
Eisenglanz *m (min)* specular iron [ore], specularite *(a variety of haematite)*
Eisengruppe *f* iron group
eisenhaltig iron-containing, *(esp ores:)* iron-bearing
eisen(II)-haltig ferroan
eisen(III)-haltig ferrian

Eisen(II)-hämoglobin *n* ferrohaemoglobin
Eisen(III)-hämoglobin *n* ferrihaemoglobin
Eisen(II)-hexacyanoferrat(III) *n* iron(II) hexacyanoferrate(II), iron(II) cyanoferrate(III)
Eisenhydroxidflocken *fpl (hyd)* iron (hydroxide) floc
Eisenkatalysator *m* iron catalyst
Eisenkies *m (min)* pyrite, iron pyrite(s), mundic *(iron(II) disulphide)*
Eisenkontakt *m s.* Eisenkatalysator
Eisenlegierung *f* iron alloy
Eisenmetall *n* ferrous metal
Eisenmetallurgie *f* ferrous [iron] metallurgy
Eisenmeteorit *m* iron meteorite, meteoric iron, (holo)siderite
Eisenmonoxid *n s.* Eisen(II)-oxid
Eisennitrid *n* iron nitride
Eisen(II)-orthophosphat *n* iron(II) orthophosphate, ferrous orthophosphate, ferrous phosphate
Eisen(II)-oxid *n* iron(II) oxide, iron monooxide, ferrous oxide
Eisen(II,III)-oxid *n* iron(II,III) oxide, triiron tetraoxide, ferrosoferric oxide
Eisen(III)-oxid *n* iron(III) oxide, diiron trioxide, ferric oxide, ferric trioxide
Eisenoxidrot *n* iron oxide red, red oxide, chemical red
Eisenpentacarbonyl *n* iron pentacarbonyl
Eisenporphyrinenzym *n (bioch)* haem [iron porphyrin] enzyme
Eisen(II)-salz *n* iron(II) salt, ferrous salt
Eisen(III)-salz *n* iron(III) salt, ferric salt
Eisenschrott *m* scrap iron, ferrous scrap
Eisenschwamm *m* iron sponge, sponge iron
Eisensilicid *n* iron silicide
Eisenspat *m (min)* spathic iron (ore), siderite *(iron(II) carbonate)*
Eisen(II)-sulfat *n* iron(II) sulphate, ferrous sulphate
Eisen(III)-sulfid *n* iron(III) sulphide, ferric sulphide
Eisentetracarbonyl *n* iron tetracarbonyl
Eisen(II)-Verbindung *f* iron(II) compound, ferrous compound
Eisen(III)-Verbindung *f* iron(III) compound, ferric compound
Eisessig *m* glacial acetic acid
Eisfarbe *f* azoic dye, insoluble azo dye, *(Am)* ice color
Eisfarbstoff *m* azoic dye, insoluble azo dye, *(Am)* ice color
Eiskalorimeter *n* ice calorimeter
Eismühle *f* ice crusher
Eiswasser *n* ice water
Eiweiß *n* 1. *(bioch)* protein; 2. *(food)* egg white, albumen, glair
Eiweiß... *siehe auch: Protein...*
Eiweißbestimmung *f (bioch)* protein determination
Eiweißchemiefaserstoff *m (text)* protein man-made fibre
Eiweißgehalt *m (bioch)* protein content
Eiweißkörper *m (bioch)* protein
Eiweißmangel *m* protein deficiency
Eiweißstickstoff *m* protein nitrogen
Eiweißstoff *m s.* Protein
Eiweißsynthese *f s.* Proteinbiosynthese
Eiweißzucker *m* glycoprotein, glycoproteid
Ejektor *m* ejector, eductor
ekliptisch eclipsed, opposed *(stereochemistry)*
Ekonomiser *m* economizer, boiler feed preheater
Ektoenzym *n* ectoenzyme *(localized on the outer surface of the cell)*
Ektohormon *n* ectohormone, pheromone
Ektotoxin *n* exotoxin
E-Kupfer *n* electrolytic copper
Elaidin(is)ierung *f* elaidinization *(conversion of oleic acid into its trans isomer)*
Elaidinsäure *f* elaidic acid, **trans*-octadec-9-enoic acid
Elain *n* olein, commercial oleic acid
Elainsäure *f* oleic acid, **cis*-octadec-9-enoic acid
Eläostearinsäure *f* elaeostearic acid, *octadeca-9,11,13-trienoic acid
Elastin *n* elastin *(a scleroprotein)*
elastisch elastic
Elastizität *f* elasticity
Elastizitätsgesetz *n*/**lineares** Hooke's law
Elastizitätsgrenze *f* elastic limit
Elastizitätsgrenze *f*/**technische** yield strength
Elastizitätsmodul *m* modulus of elas-

ticity, Young's modulus (of elasticity), elastic modulus
Elastizitätsmodul *m*/**statischer** static modulus of elasticity *(tensile test)*
elastomer elastomeric
Elastomer(e) *n* elastomer
Elastomerfaser *f* elastomeric fibre, *(Am)* snap-back fiber
elektrisch electric(al)
elektrisch neutral uncharged
Elektrizität *f* electricity
Elektrizitätsleiter *m* conductor of electricity, electric conductor
Elektrizitätsmenge *f* quantity of electricity
Elektroabscheider *m s.* Elektrofilter
elektroaktiv electroactive
Elektroanalyse *f* electroanalysis
elektroanalytisch electroanalytical
Elektrochemie *f* electrochemistry
Elektrochemiker *m* electrochemist
elektrochemisch electrochemical • **auf elektrochemischem Wege** by an electrochemical route
Elektrode *f* electrode
Elektrode *f*/**ionenselektive [ionensensitive]** ion-selective [ion-sensitive] electrode
Elektrode *f*/**ringförmige** annular electrode
Elektrode *f*/**selbst(ein)brennende** Soderberg (continuous) electrode, (Soderberg) self-baking electrode
Elektrodenmasse *f* electrode paste
Elektrodenmasse *f*/**grüne [rohe]** green paste
Elektrodenpotenzial *n* electrode potential
Elektrodenreaktion *f* electrode reaction
Elektrodenspannung *f* electrode voltage
Elektrodialysator *m* electrodialyser
Elektrodialyse *f* electrodialysis, ED
Elektroendosmose *f* electro(end)osmosis
Elektroentstauber *m s.* Elektrofilter
Elektroentstaubung *f* electrical [electrostatic] precipitation
Elektroerosion *f* electroerosion
Elektrofilter *n* electrical [electrostatic] precipitator
Elektrofiltration *f* electrical [electrostatic] precipitation, electrofiltration
Elektrofluorierung *f s.* Fluorierung/elektrochemische
Elektroformung *f (el ch)* electroforming
Elektrographit *m* electrographite
Elektrogravimetrie *f* electrogravimetric [electrodeposition] analysis, electrolytic (deposition) analysis, analytical electrodeposition
elektrogravimetrisch electrogravimetric
elektroinaktiv electroinactive, nonelectroactive
Elektrokatalyse *f* electrocatalysis
elektrokatalytisch electrocatalytic
Elektrokeramik *f* electroceramics
elektrokinetisch electrokinetic
elektrokratisch *(coll)* electrocratic *(stabilized by electric charge)*
Elektrolichtbogenofen *m* electric-arc furnace
Elektrolumineszenz *f* electroluminescence, electrogenerated chemiluminescence, electrochemiluminescence, ECL
Elektrolyse *f* electrolysis
Elektrolyse *f* **bei kontrolliertem Potenzial** controlled-potential electrolysis
Elektrolyseapparatur *f (lab)* electrolyzer
Elektrolysegerät *n (lab)* electrolyzer
Elektrolysezelle *f* electrolysis [electrolytic] cell
elektrolysieren *v* electrolyse
Elektrolyt *m* electrolyte; electroplating solution [bath]
Elektrolyt *m*/**amphoterer** amphoteric electrolyte, ampholyte
Elektrolyt *m*/**ein-ein-wertiger** uniunivalent electrolyte, 1:1 electrolyte
Elektrolyt *m*/**ein-zwei-wertiger** unibivalent electrolyte
Elektrolyt *m*/**fester** solid electrolyte
Elektrolyt *m*/**schwacher** weak electrolyte
Elektrolyt *m*/**starker** strong electrolyte
Elektrolytbrücke *f* salt bridge
Elektrolytgleichgewicht *n* ionic equilibrium
elektrolytisch electrolytic
Elektrolytkoagulation *f (coll)* flocculation by electrolytes
Elektrolytkondensator *m* electrolytic capacitor

Elektrolytkupfer *n* electrolytic copper
Elektrolytlösung *f* electrolytic [ionic] solution, solution of electrolytes; electroplating solution [bath]
Elektrolytnickel *n* electrolytic [cathode] nickel
Elektrolytschlüssel *m s.* Elektrolytbrücke
elektromer electromeric
Elektron *n* electron
Elektron *n*/**antibindendes** antibonding electron
Elektron *n*/**äußeres** outer [outside, external] electron
Elektron *n*/**einsames** unshared [nonbonding] electron
Elektron *n*/**freies [frei bewegliches]** free electron
Elektron *n*/**gemeinsames** shared electron
Elektron *n*/**gepaartes** paired electron
Elektron *n*/**gestreutes** scattered electron
Elektron *n*/**inneres** inner(-shell) electron
Elektron *n*/**positives** positive electron, posit(r)on, antielectron
Elektron *n*/**primäres** primary [initiating] electron
Elektron *n*/**sekundäres** secondary electron • **Elektronen abgeben** release [lose] electrons • **Elektronen aufnehmen** accept electrons, gain [acquire] electrons • **Elektronen (ent)ziehen** withdraw electrons
π-**Elektron** *n* π electron, pi electron, unsaturation electron
σ-**Elektron** *n* σ electron, sigma electron
Elektron-Defektelektron-Paar *n* electron-hole [hole-electron] pair
elektronegativ electronegative
Elektronegativität *f* electronegativity
Elektronegativitätsskala *f* scale of electronegativity
Elektronen *npl* **mit antiparallelen Spinmomenten** spin-paired electrons
Elektronen *npl* **mit parallelen Spinmomenten** spin-non-paired electrons
Elektronenabgabe *f* electron release [donation]
elektronenabgebend electron-releasing, electron-donating, electron-providing
elektronenabstoßend electron-repelling
Elektronenabstoßung *f* electron-electron repulsion, interelectronic repulsion
elektronenaffin electron-attracting
Elektronenaffinität *f* electron affinity
Elektronenakzeptor *m* electron acceptor
Elektronenanlagerungs-Massenspektroskopie *f* electron-attachment mass spectroscopy, EA-MS
Elektronenanordnung *f s.* Elektronenkonfiguration
Elektronenanregung *f* electron excitation
Elektronenanregungsenergie *f* electronic excitation energy
Elektronenanregungsspektroskopie *f* electronic spectroscopy
Elektronenanregungsspektrum *n* electronic spectrum
elektronenanziehend electron-attracting
elektronenarm electron-deficient
Elektronenaufbau *m s.* Elektronenstruktur
Elektronenaufnahme *f* electron acceptance, gain of electrons
elektronenaufnehmend electron-accepting
Elektronenaustausch *m* electron exchange
Elektronenaustauscher *m* electron exchanger
Elektronenaustauscherharz *n* electron-exchange resin
Elektronenaustrittsarbeit *f* electronic work function
Elektronenbeschuss *m* electron bombardment
Elektronenbesetzung *f* electron population
Elektronenbeugung *f* electron diffraction
Elektronendichte *f* electron [electronic charge] density; electron probability density
Elektronendon(at)or *m* electron donor
Elektronen-Donor-Akzeptor-Komplex *m* charge transfer complex, electron-donor-acceptor complex
Elektronendoppelresonanz *f (anal)* electron double resonance, ELDOR

Elektronendublett *n* doublet, duplet
Elektroneneinfangdetektor *m* electron capture detector
Elektronenemission *f* electron(ic) emission
Elektronenemission *f***/induzierte** induced electron emission, IEE
Elektronenenergie *f* electron(ic) energy
Elektronenentzug *m* electron removal
Elektronenfluss *m* electron flow
Elektronenformel *f* electron-dot formula [structure], electronic [dot] formula, Lewis formula [structure]
Elektronengas *n* electron gas
Elektronengitter *n* electron lattice
Elektronengrundzustand *m* electronic ground state, ground [zeroth] electronic state
Elektronenhalbleiter *m* electronic semiconductor
Elektronenhülle *f* electronic [extranuclear] region
Elektronenkonfiguration *f* electron(ic) configuration, electron(ic) arrangement, orbital electron arrangement
Elektronenkonfiguration *f* **im Grundzustand** ground-state (electronic) configuration
Elektronenladung *f* electron(ic) charge
Elektronenlawine *f* (electron) avalanche, Townsend avalanche
Elektronenleiter *m* electronic conductor
Elektronenleitfähigkeit *f* electronic conductivity
Elektronenleitung *f* electronic conduction
elektronenliefernd *s.* elektronenspendend
Elektronenloch *n* electron hole
Elektronenlücke *f* electron gap
Elektronenmangel *m* electron deficiency, shortage of electrons
Elektronenmangelverbindung *f* electron-deficient compound
Elektronenmikroskop *n* electron microscope
Elektronenmikroskopie *f* electron microscopy
Elektronenmikrosonde *f* electron microprobe, EMP
Elektronenniveau *n* electronic (energy) level, term
Elektronenoktett *n* electron octet
Elektronenoktett-Anordnung *f* octet structure, eight-electron configuration, noble-gas (electronic) configuration
Elektronenpaar *n* electron pair
Elektronenpaar *n***/bindendes** bonding pair of electrons
Elektronenpaar *n***/einsames [freies]** lone (electron) pair, lone [unshared] pair of electrons
Elektronenpaar *n***/gemeinsames** shared pair (of electrons), sharing electron pair
Elektronenpaarakzeptor *m* electron-pair acceptor
Elektronenpaarbindung *f* 1. (shared-) electron-pair bond, covalent [non-polar] bond, homopolar [atomic] bond *(state)*; 2. covalent bonding *(process)*
Elektronenpaardonator *m* electron-pair donor
Elektronenpolarisation *f* electron polarization
Elektronenpolarisation *f***/chemisch induzierte dynamische** chemically induced dynamic electron polarization
Elektronenröhre *f* vacuum tube
Elektronenruh(e)masse *f* electron (rest) mass
Elektronenschale *f* electron shell
Elektronenspektroskopie *f* electron spectroscopy
Elektronenspektroskopie *f* **für die chemische Analyse** X-ray-induced photoelectron spectroscopy, electron spectroscopy for chemical analysis
Elektronenspektrum *n* 1. electron spectrum; 2. *s.* Elektronenanregungsspektrum
elektronenspendend electron-donating, electron-releasing, electron-providing
Elektronenspender *m* electron donor
Elektronenspin *m* electron spin
Elektronenspinquantenzahl *f* electron spin quantum number
Elektronenspinresonanz *f* electron paramagnetic [spin] resonance, EPR, ESR, paramagnetic (electronic) resonance, PMR
Elektronenspinresonanzspektroskopie *f* electron paramagnetic [spin] resonance spectroscopy, EPR spectroscopy, ESR spectroscopy

Elektronensprung *m* electron jump
Elektronenstoßionisation *f* electron impact ionization
Elektronenstrahl *m* electron beam *(if bundled)*; electron ray *(if single)*
Elektronenstrahllithographie *f* electron beam lithography
Elektronenstrahlmikroanalyse *f* electron-probe microanalysis, EPMA
Elektronenstrahlmikrosonde *f* electron (micro)probe, EMP
Elektronenstruktur *f* electron(ic) structure
6π-Elektronensystem *n* six-pi-electron system
Elektronentheorie *f* **der Metalle** free-electron theory of metals
Elektronentransfer *m* electron transfer *(redox reactions)*
Elektronentransferreaktion *f* electron transfer reaction *(redox reaction)*
Elektronentransportkette *f (bioch)* electron-transport chain [sequence, system], ETS, respiratory chain
Elektronenübergang *m* electronic transition
Elektronenübergang *m***/spinerlaubter** spin-allowed electronic transition
Elektronenüberschuss *m* excess [surplus] of electrons
elektronenübertragend electron-transferring, electron-carrying
Elektronenüberträger *m* electron carrier
Elektronenüberträgerprotein *n* electron-transferring [electron-carrier] protein
Elektronenübertragung *f* electron transfer
Elektronenübertragungsreaktion *f* electron-transfer reaction
Elektronenunterschuss *m* electron deficiency, shortage of electrons
elektronenverbrauchend electron-accepting
Elektronenverbraucher *m* electron acceptor
Elektronenverteilung *f* electron distribution
Elektronenvolt *n* electron-volt, eV, E.V.
Elektronenwanderung *f* electron migration
Elektronenwolke *f* electron(ic) cloud
elektronenziehend electron-attracting, electron-pulling, electron-withdrawing, electrophilic
Elektronenzug *m* electron attraction
Elektronenzustand *m* electron(ic) state
elektroneutral electrically neutral
Elektroneutralität *f* electroneutrality
elektronisch electronic
Elektron-Kern-Doppelresonanz *f (anal)* electron nuclear double resonance, ENDOR
Elektroofen *m* electric furnace
Elektroosmose *f* electro(end)osmosis
elektrophil electrophilic, electron-attracting, electron-withdrawing
Elektrophil *n* electrophile, electrophilic reagent
Elektrophilie *f* electrophilicity, *(quantitatively also)* electrophilic power
Elektrophorese *f* electrophoresis
Elektrophorese *f***/freie [trägerfreie]** free [moving-boundary] electrophoresis, Tiselius method
elektrophoretisch electrophoretic
elektroplattieren *v* electroplate, *(specif)* coat [plate] continuously *(steel strip)*
Elektroplattieren *n* electroplating, *(specif)* continuous strip coating *(of steel strip)*
elektropositiv electropositive
Elektroraffination *f* electrorefining, electrolytic refining
Elektroreinigen *n* electrical [electrostatic] precipitation, electrofiltration
Elektrospray *n (spectr)* electrospray
Elektrospray-Ionisierung *f (spectr)* electrospray ionization, ESI
Elektrostahl *m* electrosteel, electric(-furnace) steel
elektrostatisch electrostatic
Elektrosynthese *f* electrosynthesis
Elektrosynthese *f***/organische** organic electrosynthesis
elektrothermisch electrothermal, electrothermic
elektrovalent electrovalent
Elektrovalenz *f* 1. electrovalence *(of an atom in an ionic bond)*; 2. *s.* Ionenbeziehung
Element *n s.* 1. Element/chemisches; 2. Element/elektrochemisches; 3. Grundbestandteil
Element *n***/atmophiles** atmophile ele-

ment *(an element comparatively concentrated in the atmosphere)*
Element *n*/**biophiles** biophile element
Element *n*/**chalkophiles** chalcophile element
Element *n*/**chemisches** (chemical) element
Element *n*/**elektrochemisches** electrochemical cell, galvanic [voltaic] cell [element]
Element *n*/**galvanisches** *s.* Element/elektrochemisches
Element *n*/**mehrwertiges** polyvalent [multivalent] element
Element *n*/**radioaktives** radioactive element, radioelement
Element *n*/**siderophiles** *(geoch, met)* siderophile element *(having little affinity to oxygen and sulphur)*
Element *n*/**superschweres** superheavy element, SHE
elementar *(ch)* elemental
Elementaranalyse *f* ultimate [elemental, elementary] analysis
Elementaranalysenapparat *m* combustion train
Elementarfaden *m (text)* filament, continuous fibre [filament] *(a natural or man-made fibre of great or indefinite length)*
Elementarfadenbündel *n (text)* strand
Elementarladung *f* elementary [unit] charge
Elementarreaktion *f* elementary reaction
Elementarreaktion *f*/**chemische** elementary chemical reaction
Elementarschritt *m* elementary (reaction) step, simple step
Elementarteilchen *n* elementary [fundamental] particle
Elementarwürfel *m (cryst)* cube
Elementarzelle *f (cryst)* unit [structure] cell
Elementsymbol *n* chemical sign [symbol]
Elementumwandlung *f* transmutation
Elevator *m* elevator, *(of a rotary-drilling installation also)* drill pipe elevator
eliminierbar eliminable
Eliminierbarkeit *f* eliminability
eliminieren *v* 1. eliminate; 2. *(hyd)* remove *(water constituents)*
Eliminierung *f* 1. elimination; 2. *(hyd)* removal *(of water constituents)*
Eliminierung *f*/**reduktive** *(org ch)* reductive elimination
α-**Eliminierung** *f (org ch)* α-elimination, alpha-elimination
Eliminierungs-Additions-Mechanismus *m* elimination-addition mechanism
Eliminierungsreaktion *f* elimination reaction
Ellagsäure *f* ellagic acid *(a phenolic dilactone)*
Ellipsometrie *f (spectr)* ellipsometry
Elongation *f (bioch)* elongation *(of polypeptide chains)*
Elongationsfaktor *m (bioch)* elongation [propagation, transfer] factor
Eluat *n* eluate *(liquid obtained by washing out adsorbed substances)*
Eluent *m* eluent, eluant, elutant, eluting agent [solvent]
Eluentenstrom *m* eluent stream
eluieren *v* elute
Elution *f* elution
Elution *f*/**stufenweise** stepwise elution
Elutionschromatogramm *n* elution chromatogram
Elutionschromatographie *f* elution chromatography
elutionschromatographisch by elution chromatography
Elutionsgeschwindigkeit *f* rate of elution
Elutionskurve *f* elution curve, concentration profile
Elutionsmittel *n s.* Eluent
Email *n* (vitreous) enamel, *(Am)* porcelain enamel
Emaillackfarbe *f* enamel [hard-gloss] paint, *(loosely:)* enamel
Emaillackfarbe *f*/**ofentrocknende** stoving enamel
Emaille *f s.* Email
emaillieren *v* enamel
Emanation *f (nucl)* (radioactive) emanation
emanieren *v* emanate
Emballage *f* packing [packaging] material, packing, package
Embden-Meyerhof-Parnas-Abbauweg *m (bioch)* Embden-Meyerhof-Parnas pathway [scheme], EMP pathway

[scheme], glycolytic pathway, homolactic fermentation
Embden-Meyerhof-Parnas-Weg *m (bioch)* Embden-Meyerhof-Parnas pathway [scheme], EMP pathway [scheme], glycolytic pathway, homolactic fermentation
Emersverfahren *n (biot)* surface(-culture) process
Emetikum *n (pharm)* emetic
Emission *f* 1. emission, issue; 2. *(nucl)* ejection, expulsion
Emission f/lichtelektrische photoemission
Emission f/spontane spontaneous emission
Emission f/stimulierte stimulated emission
Emission f/thermische thermionic emission
Emissionsgrad *m* (thermal) emissivity *(of bodies radiating heat)*
Emissionsgrenzwert *m* **für Luftverunreinigungen** maximum emission concentration, air emission standard
Emissionskataster *m* emission inventory
Emissionskoeffizient *m (spectr)* emissivity, emission coefficient
Emissionsminderung *f* emission control
Emissionsquelle *f* emission source
Emissionssenke *f* emission sink, receptor area *(for pollutants)*
Emissionsspektralanalyse *f* emission spectroscopy
emissionsspektralanalytisch by emission spectroscopy
Emissionsspektroskopie *f* emission spectroscopy
Emissionsspektroskopie f/optische optical emission spectroscopy, OES
Emissionsspektrum *n* emission spectrum
Emissionsvermögen *n* 1. emissive power; 2. *s.* Emissionskoeffizient
Emissionszahl *f s.* Emissionsgrad
Emitter *m* emitter
emittieren *v* emit, radiate *(light)*; emit, issue, emanate *(e.g. radioactive particles)*
emittierend emissive
EMK electromototive force, e.m.f., EMF
EMK-Normal *n* standard of e.m.f.
Emmenagogum *n (pharm)* emmenagogue
E-Modul *m* elastic modulus, Young's modulus (of elasticity), modulus of elasticity
empfängnisverhütend contraceptive
empfindlich susceptible, sensitive *(material)*; labile, sensitive *(chemicals)*; sensitive, delicate *(instrument, test)*
empfindlich/höchst exceedingly sensitive *(test)*
Empfindlichkeit *f* 1. susceptibility, sensitivity, sensitiveness *(of material)*; 2. *(phot)* speed; lability, sensitivity *(of chemicals)*; sensitivity *(of instruments, tests)*
Empfindlichkeit f/spektrale spectral responsivity [response, sensitivity]
empirisch empirical, through trial and error
emporheben 1. lift; buoy up *(in a liquid)*; 2. energize, raise *(electrons into an excited state)*
emporkriechen *v* creep up *(the sides of a vessel)*
emporsteigen *v* rise, ascend, pass up(-wards)
Emprotid *n* proton acceptor, Brønstedt-Lowry base
EMP-Weg *m s.* Embden-Meyerhof-Parnas-Weg
Emscherbrunnen *m* Imhoff tank *(water purification)*
Emulgator *m* 1. emulsification machine, emulsifier; 2. emulsifier, emulsifying agent
Emulgens *n s.* Emulgator 2.
emulgierbar emulsifiable, emulsible
Emulgierbarkeit *f* emulsifiability, emulsibility
emulgieren *v* emulsify
Emulgiermittel *n s.* Emulgator 2.
Emulgierung *f* emulsification
Emulsion *f* emulsion
Emulsion f/fotografische photographic [sensitive] emulsion
Emulsion f/geringempfindliche *(phot)* slow emulsion
Emulsion f/hochempfindliche *(phot)* fast [high-speed] emulsion
Emulsion f/wässrige aqueous emulsion
emulsionsartig emulsive

Emulsionsbildner *m* emulsifying agent, emulsifier
Emulsionsbildung *f* emulsification
Emulsionsbrecher *m s.* Emulsionsspalter
Emulsionsentmischer *m s.* Emulsionsspalter
Emulsionskolloid *n s.* Emulsoid
Emulsionspolymer(e) *n* emulsion polymer
Emulsionspolymerisat *n* emulsion polymer
Emulsionspolymerisation *f* emulsion polymerization
Emulsionsschicht *f (phot)* emulsion layer
Emulsionsspalter *m* emulsion breaker, demulsifier
Emulsionsspaltung *f* breaking [cracking] of emulsions, de-emulsification, demulsification *(act)*; emulsion breakdown [breaking] *(process)*
Emulsionsstabilisator *m* emulsion stabilizer
Emulsoid *n* emulsoid (colloid)
Enamin *n (org ch)* enamine
enantiomer enantiomeric, enantiomorphic, enantiomorphous
Enantiomer(e) *n* enantiomer, enantiomorph, enantiomorphous [mirror-image] isomer, optical isomer [antipode], antimer
enantiomerenrein enantiomerically pure *(stereo chemistry)*
Enantiomerenreinheit *f* enantiomeric purity *(stereo chemistry)*
Enantiomerentrennung *f* enantiomer separation, enantiomeric separation, separation of enantiomers *(stereo chemistry)*
Enantiomerenüberschuss *m* enantiomeric excess *(stereo chemistry)*
Enantiomerie *f* enantiomerism, enantiomorphism, optical [mirror-image] isomerism
enantiomorph *s.* enantiomer
enantioselektiv *(org ch)* enantioselective
Enantioselektivität *f (org ch)* enantioselectivity
enantiotrop enantiotropic
Enantiotropie *f* enantiotropy
Enddruck *m* final pressure, *(vacuum technology also)* ultimate [maximum] vacuum
Ende *n***/klebriges [kohäsives]** *(bioch)* sticky end
endergon(isch) endergonic, energy-demanding
Endfeuchte(beladung) *f* final moisture content, FMC
Endgruppe *f* terminal [end] group *(of a molecule)*
Endgruppenanalyse *f* end group analysis
Endgruppenbestimmung *f* end group analysis
Endiol *n (org ch)* enediol
Endkochpunkt *m* final boiling point, F.B.P.
Endkonzentration *f* final concentration [strength]
Endlagerung *f* final [ultimate] disposal *(of waste products)*
Endlosfaser *f s.* Elementarfaden
Endlosgarn *n (text)* continuous-filament yarn
endocyclisch endocyclic *(e.g. double bond)*
endogen endogenous
Endomethylenbrücke *f* endomethylene bridge • **mit Endomethylenbrücke** endomethylene-bridged
endo-Produkt *n* endo product
Endosmose *f* endosmosis
endosmotisch endosmotic
endotherm endothermic
Endotoxin *n* endotoxin
Endprodukt *n* 1. *(phys ch)* end product, final (reaction) product; 2. *(tech)* final product, finished [consumer] product
Endprodukthemmung *f (bioch)* end-product inhibition
Endprodukthemmung *f***/konzertierte [multivalente]** concerted inhibition
Endpunkt *m* end point *(as of a titration)*
Endpunktabstand *m* end-to-end distance, chain-end distance
Endpunktsbestimmung *f* end-point determination
Endreinigung *f (hyd)* final (effluent) treatment
Endschneide *f* end [terminal] knife edge *(of a balance)*
Endsiedepunkt *m* final boiling point, F.B.P.

endständig terminal • **mit endständiger Hydroxylgruppe** hydroxyl-terminated
Endstellung *f* terminal position
Endung *f (nomencl)* ending, termination
Endvakuum *n* ultimate [maximum] vacuum
energetisch begünstigt [bevorzugt] energetically favourable [favoured]
Energie *f* 1. *(phys ch)* energy; 2. *(tech)* power, *(esp relating to electricity:)* energy • **von geringer Energie** low-energy, poor in energy
Energie *f*/freie free energy, *(specif)* Helmholtz free energy, maximum work function
Energie *f*/innere intrinsic [internal] energy
Energie *f*/kinetische kinetic energy
Energie *f*/potenzielle potential energy
Energie *f*/thermische thermal energy
Energieabgabe *f* energy output, release of energy
energieabgebend *s.* energieliefernd
energiearm low-energy, poor in energy
Energieaufnahme *f* energy consumption
Energieband *n s.* Energiebereich
Energiebarriere *f s.* Energieschwelle
Energiebedarf *m* 1. *(phys ch)* energy demand; 2. *(tech)* power demand [requirements, needs], *(esp relating to electricity:)* energy demand [needs]
Energiebeitrag *m* energy contribution
Energiebereich *m* energy range [band], band
Energiebereich *m*/erlaubter allowed band
Energiebereich *m*/verbotener forbidden band, energy gap
Energieberg *m s.* Energieschwelle
Energiebilanz *f* energy balance
Energiebilanzgleichung *f* total-energy equation
Energiedichte *f* energy density
Energiedichte *f*/kohäsive cohesive energy density, C.E.D.
energiedispersiv energy-dispersive, energy-dispersion
Energiedissipation *f* dissipation of energy
Energieeintrag *m* energy input
energieerhaltend energy-conserving
Energieerhaltung *f* energy conservation
Energieerhaltungssatz *m* energy principle, law of conservation of energy
Energiegewinn *m* gain in energy
Energieinhalt *m* energy content
Energielieferant *m* source of energy
energieliefernd 1. energy-releasing, energy-yielding, exoergic; 2. *(bioch)* exergonic
Energielücke *f* energy gap, forbidden band
Energieniveau *n* energy level, term
Energieniveaudiagramm *n* energy(-level) diagram, level scheme, state diagram
Energieniveauschema *n* energy(-level) diagram, level scheme, state diagram
Energieprinzip *n* energy principle, law of conservation of energy
Energieprofil *n* (potential-)energy profile
Energiequant *n* energy quantum
Energiequantelung *f* quantization of energy
Energiequelle *f* source of energy
energiereich high-energy, rich in energy, energy-rich
Energieschwelle *f* energy barrier [hump]
Energiespeicherung *f (bioch)* energy storage *(carbohydrates)*
Energiestufe *f* energy level, term
Energieterm *m* energy level, term
Energieträger *m* energy carrier
Energieüberschuss *m* excess energy
Energieumsatz *m (bioch)* energy transformation
Energieumwandlung *f* 1. energy change; 2. *(intentionally:)* transformation of energy, energy conversion
Energieverbrauch *m* 1. energy consumption; 2. *(tech)* power consumption
energieverbrauchend 1. energy-requiring, energy-demanding, endoergic; 2. *(bioch)* endergonic
Energieverlust *m* loss of energy
Energieverlust-Spektroskopie *f* energy-loss spectroscopy, ELS
Energiezufuhr *f* energy input
Energiezustand *m* energy state
energisch vigorous, drastic *(treatment)*
Enghalsflasche *f* narrow-mouth(ed) bottle
Enghalskolben *m* narrow-mouth(ed) flask, narrow-neck(ed) flask
engklassiert closely graded

engporig fine-pore(d), finely pored, small-pore(-size)
engsiedend *(distil)* close-boiling
E-Nickel *n* electrolytic [cathode] nickel
En-Komponente *f (org ch)* ene
Enol *n (org ch)* enol
Enolat *n (org ch)* enolate
Enolform *f* enol(ic) form
enolisch enolic
enolisieren *v* enolize
Enolisierung *f* enolization
Enophil *n (org ch)* enophile
En-Reaktion *f (org ch)* ene reaction
entacetylieren *v* deacetylate
Entacetylierung *f* deacetylation
entaktivieren *v* 1. deactivate; 2. *(nucl)* decontaminate
Entaktivierung *f* 1. deactivation; 2. *(nucl)* decontamination
entalkylieren *v* dealkylate
Entalkylierung *f* dealkylation
entamidieren *v* deamidate
Entamidierung *f* deamidation
entarten *v (phys ch)* degenerate
entartet/dreifach *(phys ch)* triply [three-fold] degenerate
entartet/nicht non-degenerate
Entartung *f (phys ch)* degeneracy
Entartung *f***/zufällige** *(phys ch)* accidental degeneracy
Entarylierung *f* dearylation
entasphaltieren *v* deasphalt
Entasphaltierung *f* deasphalting, deasphaltation
entbenzol(ier)en *v* debenzolize
Entbenzol(ier)ung *f* debenzolization
entbutanisieren *v* debutanize
Entbutanisierung *f* debutanization
entchloren *v* dechlorinate
Entchlorung *f* dechlorination
enteisen *v* de-ice, defrost
enteisenen *v* deferrize *(e.g. water)*
Enteisenung *f* deferrization *(as of water)*
Enteisung *f* de-icing, defrosting
Enteisungsflüssigkeit *f* de-icing fluid
enteiweißen *v* deproteinize
Enteiweißung *f* deproteinization
entemulsionieren *v* demulsify, de-emulsify, break, crack *(an emulsion)*
Entemulsionieren *n* demulsification, de-emulsification, emulsion breakdown [breaking], cracking of emulsions
entethanisieren *v* de-ethanize
Entethanisierung *f* de-ethanization
Entfaltung *f* 1. unfolding *(as of macromolecules)*; 2. *(bioch)* denaturation *(of proteins)*
entfärben *v* 1. decolour, decolorize, discolour, bleach; 2. *(pap)* whiten, brighten; deink *(waste paper)*
entfärben *v***/sich** decolour, decolorize, discolour, bleach out
Entfärbung *f* 1. decolorization, discoloration, bleaching; 2. *(pap)* whitening, brightening; deinking *(of waste paper)*
Entfärbungsmittel *n* decolorizing [stripping] agent [assistant], decolorizer, decolorant
entfernen *v* eliminate, remove, discharge, abstract
Entfernen *n* elimination, removal, discharge, abstraction
Entferner *m* remover
Entfernung *f s.* Entfernen
entfetten *v* degrease, defat; scour *(wool)*
Entfettung *f* degreasing, defatting; scouring *(of wool)*
Entfettungsmittel *n* degreasing agent, degreaser
entfeuchten *v* dehumidify, dehydrate, desiccate, dry
Entfeuchtung *f* dehumidification, dehydration, desiccation, drying
Entfeuchtungsmittel *n* desiccating [drying] agent, desiccant
entflammbar (in)flammable
entflammbar/nicht uninflammable, non-(in)flammable, flame-proof
Entflammbarkeit *f* (in)flammability
Entflammbarkeitsbereich *m* flammable limits
entflammen *v* flash, inflame, burst into flame; inflame *(something)*
Entflammung *f* inflammation
entflocken *v (coll)* deflocculate
entflocken *v***/sich** deflocculate
Entflockung *f* deflocculation
entfluorieren *v* defluorinate
Entfluorierung *f* defluorination
entgasen *v* 1. degas, degasify, outgas; 2. *(coal)* coke, carbonize
Entgasung *f* 1. degassing, degasification, outgassing; 2. *(coal)* coking, carbonization
entgegenwirken *v* counteract

entgehen *v*/**der Aufmerksamkeit** escape notice [observation]
entgiften *v* 1. detoxicate, detoxify; 2. *(nucl)* decontaminate
Entgiftung *f* 1. detoxication, detoxification; 2. *(nucl)* decontamination
entglasen *v* devitrify
Entglasung *f* devitrification
Enthalpie *f* enthalpy, heat content
Enthalpie *f*/**(gibbssche) freie** free enthalpy, Gibbs (free) energy, Gibbs function, free energy G
Enthalpieänderung *f* enthalpy change
Enthalpie-Entropie-Diagramm *n* enthalpy-entropy chart [diagram], Mollier chart
Enthalpiemetrie *f* enthalpimetric analysis
enthalpiemetrisch enthalpimetric
enthärten *v (hyd)* soften, remove hardness from *(water)*
enthärten *v*/**durch Kalkfällung** lime-soften, lime-treat
Enthärter *m* 1. softener, softening agent [chemical]; 2. *s.* Enthärtungsanlage
Enthärtung *f (hyd)* softening
Enthärtung *f* **durch Fällverfahren** precipitation softening, softening by precipitation
Enthärtung *f* **durch Kalkfällung** lime softening, softening by lime treatment
Enthärtung *f* **mittels Ionenaustauschs** ion-exchange softening
Enthärtungsanlage *f (hyd)* softening unit [plant], (water) softener
Enthärtungsmittel *n (hyd)* softener, softening agent [chemical]; alkali builder *(in soaps)*
Entionisation *f* deionization
entionisieren *v* deionize
Entionisierstäbe *mpl* anti-static bars
entisobutanisieren *v* deisobutanize
Entisobutanisierung *f* deisobutanization
entkalken *v (bioch, soil)* decalcify
Entkalkung *f (bioch, soil)* decalcification
entkarbonisieren *v (hyd)* decarbonate, decarbonize, *(by ion exchange also)* dealkalize
Entkarbonisierung *f (hyd)* decarbon(iz)ation, *(by ion exchange also)* dealkalization, dealkalizing
entkeimen *v* sterilize, *(relating to pathogenic organisms:)* disinfect
entkeimend germicidal
Entkeimung *f* sterilization, *(relating to pathogenic organisms:)* disinfection
Entkeimungsfiltration *f* sterile filtration
Entkeimungsmittel *n* germicide, *(for killing pathogenic organisms:)* disinfectant
entknäueln *v*/**sich** uncoil *(of molecules)*
Entknäuelung *f* uncoiling *(of molecules)*
entkohlen *v (met)* decarburize
Entkohlung *f (met)* decarburization
entkoppeln *v* 1. *(bioch)* uncouple, decouple; 2. *(spectr)* decouple
Entkopplung *f* 1. *(bioch)* uncoupling, decoupling; 2. *(spectr)* decoupling
entkrusten *v* (de)scale
entkupfern *v* decopperize
entladen *v* 1. *(tech)* discharge, unload; 2. discharge *(physics)*
Entladezyklus *m* discharge cycle
Entladung *f* 1. *(tech)* discharge, unloading; 2. discharge *(physics)*
Entladungselektrode *f* discharge electrode
Entladungspotenzial *n* discharge potential
Entladungsröhre *f* (gas) discharge tube
Entlaubungsmittel *n* defoliant
entleeren *v* discharge, empty, drain (down), evacuate *(a vessel)*; siphon, drain *(a liquid)*; dump *(bulk material)*
Entleerung *f* discharge, emptying, drainage, evacuation *(of a vessel)*; siphoning, drainage *(of a liquid)*; dumping *(of bulk material)*
Entleerungsstutzen *m* discharge [run-off] pipe
entlüften *v* deaerate, degas, bleed *(e.g. a vessel)*
Entlüftung *f* 1. deaeration, degassing, bleeding *(as of a vessel)*; 2. *s.* Entlüftungseinrichtung
Entlüftungseinrichtung *f* air relief [vent]
Entlüftungsrohr *n* vent [blow] pipe
Entlüftungsventil *n* vent valve
entmagnetisieren *v* demagnetize
Entmagnetisierung *f* demagnetization, degaussing
entmanganen *v* demanganize
Entmanganung *f* demanganization
entmethanisieren *v* demethanize

Entmethanisierung *f* demethanization
entmethylieren *v* demethylate
Entmethylierung *f* demethylation
entmineralisieren *v* demineralize
Entmineralisierung *f* demineralization
entmischen *v/sich* unmix, separate, segregate; break, crack, deteriorate *(of emulsions)*
Entmischung *f* unmixing, separation, segregation
Entmischung *f* **einer Emulsion** emulsion breaking [breakdown], breaking [cracking] of emulsions
Entmischung *f***/spinodale** spinodal decomposition *(of a supersaturated liquid or solid solution)*
Entnahme *f* 1. take-off, offtake, withdrawal, discharge; 2. *s.* Entnahmevorrichtung
Entnahme *f***/einphasige** single withdrawal *(on extracting)*
Entnahmeverhältnis *n (distil)* rate of withdrawal
Entnahmevorrichtung *f* discharging device, discharger, discharge, take-off, offtake
entnehmen *v* take off, withdraw, discharge
Entner-Doudoroff-Abbauweg *m (bioch)* Entner-Doudoroff pathway
Entner-Doudoroff-Weg *m (bioch)* Entner-Doudoroff pathway
entölen *v* deoil
Entölung *f* deoiling
entozonisieren *v* deozonize
Entozonisierung *f* deozonation
entparaffinieren *v* deparaffin(ize), dewax
Entparaffinierung *f* deparaffinization, dewaxing
Entparaffinierung *f* **mit Lösungsmitteln** solvent dewaxing
Entparaffinierung *f* **mit Propan** propane dewaxing
entpentanisieren *v* depentanize
Entpentanisierung *f* depentanization
entphenol(ier)en *v* dephenolize, dephenolate
Entphenol(ier)ung *f* dephenol(iz)ation
entphosphoren *v* dephosphorize
Entphosphorung *f* dephosphorization
entpropanisieren *v* depropanize
Entpropanisierung *f* depropanization
entrahmen *v* cream (off), skim *(milk)*
entrinden *v (pap)* (de)bark, peel, *(Am)* ross
Entropie *f* entropy
Entropie *f***/absolute molare** third-law entropy
Entropieabnahme *f* entropy decrease
Entropieelastizität *f* rubber [long-range] elasticity
Entropiezunahme *f* entropy increase, positive entropy change, gain of entropy
Entropiezuwachs *m* entropy increase, positive entropy change, gain of entropy
Entrostungsmittel *n* rust-removing agent, rust remover
entsalzen *v* free from salt; desalinate, desalt *(sea water)*; deionize, demineralize *(fresh water)*
Entsalzung *f* desalin(iz)ation, desalting *(of sea water)*; deionization, demineralization *(of fresh water)*
Entsalzungsanlage *f* desalinating [desalination, desalting] plant *(for seawater treatment)*; deionization plant *(for fresh-water treatment)*
entsäuern *v* deacidify, neutralize
Entsäuerung *f* deacidification, neutralization
entschäumen *v* defoam, *(mechanically:)* skim (off)
Entschäumer *m* 1. *(biot)* foam separator; 2. *s.* Entschäumungsmittel
Entschäumung *f* defoaming, *(mechanically:)* skimming(-off)
Entschäumungsmittel *n* defoaming agent, defoamer, foam breaker [killer]
entschlacken *v* slag, free from slag
Entschlackung *f* 1. slagging; 2. *(coal)* clinker discharge
entschlammen *v* desludge, *(relating to very fine particles:)* deslime
entschützen *v (org ch)* deprotect
Entschützen *n (org ch)* deprotection
entschwefeln *v* desulphur(ize)
Entschwefelung *f* desulphur(iz)ation
Entschwefelung *f***/hydrierende** hydrodesulphurization, HDS
entseuchen *v* 1. *(med)* disinfect; 2. *(nucl)* decontaminate
Entseuchung *f* 1. *(med)* disinfection; 2. *(nucl)* decontamination

entsorgen *v* dispose
Entsorgung *f* disposal
entspannen *v* expand, release
Entspannung *f* expansion, release, stress relief
Entspannungsdestillation *f* flash distillation, continuous equilibrium vaporization
Entspannungsdestillation *f***/kontinuierliche** flash distillation, continuous equilibrium vaporization
Entspannungsverdampfung *f* flash [instantaneous] vaporization, flash evaporation
entspiralisieren *v (bioch)* untwist, unwind *(the DNA double strand)*
Entspiralisierung *f (bioch)* untwisting, unwinding *(of the DNA double strand)*
entstauben *v* (de)dust
Entstauber *m* dust catcher [collector, separator, settler]
Entstaubung *f* dedusting
Entstaubung *f***/elektrische [elektrostatische]** electrical [electrostatic] precipitation
Entstaubungsgrad *m* collection efficiency
entstehen *v* originate, be formed, form
Entstehung *f* 1. origination, formation, generation, nascency; 2. *(geoch)* genesis *(as of coal or petroleum)*
entteeren *v* detar
Entteerung *f* detarring, tar separation
enttoluolen *v* detoluate
entwachsen *v* dewax
entwässern *v* dehydrate, dewater, desiccate, dry
Entwässerung *f* dehydration, dewatering, desiccation, drying
Entwässerungsanlage *f* drainage system
Entwässerungsmittel *n* dehydrating agent, dehydrator
Entwässerungsnetz *n* sewer [sewage] system, sewerage (system) • **mit einem Entwässerungsnetz versehen** sewer
entweichen *v* escape, leak (out), issue, leave
entwickeln *v* 1. evolve, generate, liberate, release *(e.g. gas or heat)*; 2. develop, evolve *(e.g. a method)*; 3. develop *(a photograph or chromatogram)*
entwickeln *v***/sich** evolve, be evolved, form, be formed
Entwickler *m* 1. *(phot)* developing agent, developer; 2. *(dye)* developing agent, developer, coupling component
Entwicklung *f* 1. evolution, generation, liberation, release *(as of gas or heat)*; development, evolution *(as of a method)*; 2. development *(of a photograph or a chromatogram)*
Entwicklung *f***/absteigende** *(chromat)* descending development
Entwicklung *f***/aufsteigende** *(chromat)* ascending development
Entwicklung *f***/horizontale** *(chromat)* horizontal development
Entwicklungsbad *n (phot)* developing bath
Entwicklungsfaktor *m (phot)* development factor, gamma value
Entwicklungsfarbstoff *m* developed [ingrain] dye
Entwicklungskammer *f* developing [chromatography] chamber
entziehen *v* abstract, extract, withdraw
entziehen *v***/sich der Beobachtung** escape notice
Entziehung *f s.* Entzug
entzuckern *v* desugar(ize)
Entzuckerung *f* desugarization
Entzug *m* abstraction, extraction, removal, withdrawal
entzündbar (in)flammable, ignitable, ignitible
entzündbar/leicht readily flammable [ignitible], easy to ignite
entzündbar/nicht non-flammable
Entzündbarkeit *f* (in)flammability, ignitability, ignitibility
entzünden *v* ignite, kindle, light, inflame
entzünden *v***/sich** ignite, kindle, light, inflame
entzundern *v (met)* (de)scale, scour
entzündlich *s.* entzündbar
entzündungshemmend *(pharm)* antiphlogistic, anti-inflammatory
Enzym *n* enzyme • **mit Enzymen anreichern** enzymize
Enzym *n***/acylübertragendes** acyl(-carrier) enzyme
Enzym *n***/analytisches** analytical enzyme
Enzym *n***/citratkondensierendes**

(citrate-)condensing enzyme, citrate synthase, citrogenase
Enzym *n*/**citratspaltendes** citrate cleavage enzyme, ATP-citrate lyase
Enzym *n*/**eingeschlossenes** *(biot)* entrapped enzyme
Enzym *n*/**eiweißabbauendes [eiweißspaltendes]** proteolytic [protein-digesting] enzyme, protease
Enzym *n*/**extrazelluläres** extracellular [exocellular] enzyme, exoenzyme
Enzym *n*/**fettspaltendes** lipolytic enzyme
Enzym *n*/**gekapseltes** *(biot)* microencapsulated [MEC] enzyme
Enzym *n*/**immobilisiertes** 1. bound enzyme; 2. *(biot)* immobilized enzyme
Enzym *n*/**industrielles** *(biot)* industrial enzyme, enzyme of industrial importance
Enzym *n*/**induziertes** induced enzyme
Enzym *n*/**intrazelluläres** intracellular [endocellular] enzyme, endoenzyme
Enzym *n*/**pektinabbauendes [pektinspaltendes]** pectolytic [pectic] enzyme, pectinase
Enzym *n*/**proteinspaltendes [proteolytisches]** *s.* Enzym/eiweißabbauendes
Enzym *n*/**stärkeabbauendes [stärkespaltendes]** starch-splitting enzyme, starch-reducing [starch-converting, amylolytic] enzyme, amylase
Enzym *n*/**technisches** *s.* Enzym/industrielles
Enzym *n*/**trägerfixiertes [trägergebundenes]** *(biot)* (carrier-)bound enzyme
Enzym-Assay *n (bioch)* enzyme assay
enzymatisch enzym(at)ic
Enzymaufarbeitung *f* enzyme preparation [recovery]
Enzymausbeute *f (biot)* enzyme yield
Enzymbiosynthese *f* enzyme biosynthesis
Enzymchemie *f* enzyme chemistry
Enzymelektrode *f* enzyme electrode
Enzymfixierung *f (biot)* enzyme fixation
enzymgebunden enzyme-bound
Enzymgruppe *f* subclass *(classification of enzymes)*
Enzymimmobilisierung *f (biot)* enzyme immobilization
Enzyminaktivierung *f* enzyme inactivation
Enzyminduktion *f (biot)* enzyme induction
Enzyminhibitor *m* enzyme inhibitor
Enzymkatalyse *f* enzyme [enzymatic] catalysis
enzymkatalysiert enzyme-catalyzed
Enzymkinetik *f* enzyme kinetics
Enzymkomplex *m* enzyme complex
Enzymkonjugat *n (bioch)* enzyme conjugate
Enzymmembranreaktor *m (biot)* enzyme-membrane reactor
Enzymologe *m* enzymologist
Enzymologie *f* enzymology
Enzymprotein *n* enzyme protein, apoenzyme, colloid carrier
Enzymreaktion *f* enzyme [enzymatic] reaction
Enzymreaktor *m (biot)* enzyme reactor
Enzymrückgewinnung *f (biot)* enzyme recovery
Enzymspezifität *f* enzyme specificity
Enzym-Substrat-Komplex *m* enzyme-substrate complex, ES complex, Michaelis(-Menten) complex
Enzymtechnik *f* enzyme technology
Enzymtechnologie *f* enzyme technology
Enzymträger *m (biot)* enzyme carrier, binding support
Enzymumwandlung *f* enzym(at)ic conversion
E-Ofen *m s.* Elektroofen
Eosin *n* eosin, tetrabromofluorescein
EP *s.* 1. Epoxidharz; 2. Erstarrungspunkt
Ephedrinhydrochlorid *n* ephedrine hydrochloride
Epichlorhydrin *n* epichlorohydrin, α-epichlorohydrin, chloropropylene oxide, chloromethyloxirane
Epihalogenhydrin *n* epihalohydrin
epimer epimeric
Epimer(e) *n* epimer(ide)
Epimerie *f* epimerism
epimerisieren *v* epimerize
Epimerisierung *f* epimerization
epitaktisch *(cryst)* epitaxial *(growing in oriented manner on a different crystalline substrate)*
Epitaxie *f (cryst)* epitaxy *(oriented growth on a different crystalline substrate)*
Epithelschutzvitamin *n* antiinfective [antixerophthalmic] vitamin, vitamin A

Epoxid *n* epoxide, oxirane
Epoxidation *f* epoxidation
Epoxidharz *n* epoxide [epoxy, ethoxylene] resin
epoxidieren *v* epoxidize
Epoxidring *m* epoxide [epoxy] ring
Epoxyethan *n* *epoxyethane, oxirane, ethylene oxide, EO
1,2-Epoxypropan *n* 1,2-epoxypropane, 2-methyloxirane, propylene oxide
EP-Schmiermittel *n* EP [extreme-pressure] lubricant
Erbium *n* Er erbium
Erdalkali *n* alkaline earth
Erdalkalimetall *n* alkaline-earth metal
erdartig earthy
Erdbodenkorrosion *f* soil corrosion
Erde *f* 1. *(ch)* earth; 2. *(agric)* soil, earth
erden *v* earth
Erden *pl***/seltene** rare earths
Erdgas *n* natural gas
Erdgas *n***/feuchtes [nasses]** wet (natural) gas
Erdgas *n***/saures** sour gas *(containing hydrogen sulphide and thiols)*
Erdgas *n***/synthetisches** synthetic [substitute] natural gas, SNG
Erdgas *n***/trockenes** dry (natural) gas
Erdgas *n***/verflüssigtes** liquefied natural gas, LNG
Erdgasbenzin *n* natural gasoline
erdig earthy
Erdkern *m* earth's core [nucleus], barysphere, centrosphere
Erdkruste *f* earth's crust
Erdnusseiweiß *n* peanut protein
Erdnussöl *n* peanut oil
Erdöl *n* petroleum, rock oil
Erdöl *n***/asphaltbasisches [asphaltisches]** asphalt-base petroleum [crude oil, crude], asphaltic petroleum
Erdöl *n***/naphthenbasisches [naphthenisches]** naphthene-base petroleum [crude oil, crude], naphthenic petroleum
Erdöl *n***/paraffinbasisches [paraffinisches]** paraffin-base petroleum [crude oil, crude], paraffinic petroleum
Erdöl *n***/rohes** crude petroleum [oil], crude
Erdölasphalt *m* petroleum [artificial] asphalt, asphaltic residue
Erdölbildung *f* petroleum [oil] genesis
Erdölbitumen *n* s. Erdölasphalt
Erdölbohrloch *n* petroleum [oil] well
Erdölchemie *f* petroleum chemistry, petrochemistry
Erdölchemikalie *f* petrochemical
Erdöldestillat *n* petroleum distillate
Erdöldestillat *n***/saures** sour oil
Erdöldestillat *n***/süßes** sweet oil
Erdöldestillation *f* petroleum distillation
Erdölentstehung *f* petroleum [oil] genesis
Erdölfeld *n* oil field [pool, reservoir]
erdölführend petroliferous, petroleum-bearing, oil-bearing
Erdölgas *n***/verflüssigtes** liquefied petroleum gas, L.P. gas, L.P.G.
erdölhaltig *s.* erdölführend
Erdöllagerstätte *f* oil deposit, oil-field
Erdölmuttergestein *n* oil(-source) rock, mother rock
Erdölparaffin *n* petroleum wax
Erdölprodukt *n* petroleum product
Erdölquelle *f* petroleum [oil] well
Erdölraffination *f* petroleum refining
Erdölraffinerie *f* petroleum [oil] refinery
Erdölresiduum *n* petroleum residue
Erdölrückstand *m* petroleum residue
Erdölspeichergestein *n* reservoir rock
erdölstämmig petroleum-based
Erdölteer *m* petroleum tar
Erdölverarbeitung *f* petroleum refining
Erdölvorkommen *n* oil occurrence; oil deposit
Erdölwachs *n* petroleum wax
Erdpech *n* 1. mineral pitch, asphalt(e); 2. *(min)* elastic bitumen, mineral caoutchouc, elaterite
Erdpigment *n* earth [mineral, natural] pigment, earth colour
erdverlegt buried
Erdwachs *n (min)* earth [ader] wax, native paraffin, ozokerite
erfassen *v* detect, identify *(as in analyses)*
Erfassungsgrenze *f (anal)* detection [identification] limit
Erfassungsgrenze *f***/absolute** absolute detection limit
Erfassungsgrenze *f***/untere** lower [minimum] detection limit
Ergebnis *n***/experimentelles** experimental result
ergiebig high-yield

ergießen *v*/**sich** flush, pour, spill
Ergin *n* biocatalyst, biochemical catalyst, ergone
Ergotalkaloid *n* ergot alkaloid
Erhalt *m* obtaining *(as by synthesis)*
erhalten *v* 1. obtain *(as by synthesis)*; 2. retain *(e.g. flavour)*
erhalten *v*/**rein** obtain pure [in pure form], obtain in a pure condition [state]
erhältlich/im Handel commercially available [obtainable]
Erhaltung *f* conservation *(as of mass or energy)*
Erhaltungssatz *m* conservation law
erhärten *v* harden, set; *(if too rapidly or unintentionally:)* set up
Erhärtung *f* hardening, set(ting); *(if too rapidly or unintentionally:)* set-up
erhitzen *v* heat
erhitzen *v*/**auf [bis zur] Rotglut** heat to redness, make red-hot
erhitzen *v*/**auf [bis zur] Weißglut** incandesce
erhitzen *v*/**gelinde** heat gently [moderately, slightly]
erhitzen *v*/**in der Retorte** retort
erhitzen *v*/**mit fächelnder Flamme** brush with the free flame
erhitzen *v*/**sich** heat
erhitzen *v*/**zum Sieden** heat to boiling
erhöhen *v* increase *(e.g. number, quantity)*; raise, elevate *(e.g. temperature, boiling point)*
erhöhen *v*/**sich** rise *(as of temperature, boiling point)*; increase *(of number, quantity)*
Erhöhung *f* *(act:)* raising, raise, elevation; *(process:)* rise, elevation *(as of temperature, boiling point)*; increase *(of number or quantity)*
erholen *v*/**sich** *(tech)* recover
Erholung *f* *(tech)* recovery
Erholung *f*/**elastische** *(rubber)* elastic recovery, rebound
erkalten *v* cool (down), chill
erkalten lassen *v* allow to chill, chill
Erkennungsregion *f* *(bioch)* recognition site [sequence], acceptor [entry, aminoacyl tRNA, decoding] site, A site
erkochen *v* *(pap)* cook
Erlenmeyer-Kolben *m* Erlenmeyer flask
Erlenmeyer-Kolben *m*/**weithalsiger** wide-mouthed Erlenmeyer flask, beaker flask
erlöschen *v* 1. go out, *(gradually:)* die out [down] *(of a flame)*; 2. die down, subside *(of a reaction)*
ermitteln *v* elucidate, establish, determine, find out
Ermittlung *f* elucidation, establishment, determination
ermüden *v* *(tech)* fatigue
Ermüdung *f* *(tech)* fatigue
Ermüdungsbeständigkeit *f* fatigue resistance [strength], resistance to fatigue
Ermüdungsbruch *m* fatigue failure [fracture]
ermüdungsfrei *s.* ermüdungssicher
Ermüdungsgrenze *f* fatigue limit
Ermüdungsriss *m* fatigue crack
ermüdungssicher fatigue proof
Ermüdungsverhalten *n* fatigue behaviour
Ernährung *f* 1. nourishment; 2. *(biol, med)* nutrition
Ernährungsphase *f* *(biot)* trophophase *(growth of microorganisms)*
Ernährungswissenschaft *f* nutrition science
Ernährungswissenschaftler *m* nutrition scientist, nutritionist
erniedrigen *v*/**sich** decrease *(as of boiling point)*
Erniedrigung *f* *(act:)* depression, lowering; *(process:)* decrease *(as of boiling point)*
Ernte *f* *(biot)* harvest *(of the fermentation product)*
ernten *v* *(biot)* harvest *(the fermentation product)*
Erosionskorrosion *f* erosion corrosion
erregen *v* excite, activate
Erregung *f* excitation, activation
Ersatz *m* 1. replacement, substitution; 2. substitute
Ersatzstoff *m* substitute
erschmelzen *v* smelt *(metals)*
erschöpfen *v*/**sich** squeeze *(of resources)*
erschöpfend exhaustive
Erschöpfung *f* exhaustion *(as of a dye bath)*; exhaustion, squeeze *(of natural resources)*

erschütterungsfrei vibration-free, vibrationless
ersetzbar replaceable, displaceable, substitutable
ersetzen *v* replace, displace, substitute
Ersetzen *n* replacement, displacement, substitution
erspinnen *v* spin *(chemical fibres)*
Erspinnen *n* **aus der Schmelze** melt spinning, (melt) extrusion
Erspinnen *n* **aus Lösungen** solution [solvent] spinning
Erspinnfärbung *f* spin [dope] dyeing
erspinngefärbt spin-dyed, spun-dyed, dope-dyed, *(in a melt also)* mass-dyed, *(in solution also)* solution-dyed
Erspinnlösung *f* spinning solution, (spinning) dope
erstarren *v* 1. solidify, harden, set, freeze, congeal; 2. *(coll)* gel(ate)
erstarren lassen *v* solidify, congeal, set
Erstarren *n* 1. solidification, hardening, set(ting), freezing, congealing, congelation; 2. *(coll)* gelation
Erstarrung *f* 1. solidification, hardening, set(ting), freezing, congealing, congelation; 2. *(coll)* gelation
Erstarrungsintervall *n* solidification range
Erstarrungskurve *f* freezing(-point) curve, solidus curve *(of a melt)*
Erstarrungspunkt *m* solidification [setting] point, s.p., freezing [congealing] point [temperature]
Erstarrungstemperatur *f* solidification [setting] point, s.p., freezing [congealing] point [temperature]
Erstdestillation *f* primary distillation
ersticken *v* blanket, choke *(fire)*
Erstkomponente *f (dye)* primary component, diazo [diazonium] component
Ertrag *m* 1. *(agric)* yield; 2. *(biot)* growth yield *(of microorganisms)*
Ertragskoeffizient *m (biot)* growth yield coefficient *(of microorganisms)*
Erucasäure *f* erucic acid, **cis*-docos-13-enoic acid
erwärmen *v* heat, *(specif)* heat gently [moderately, slightly], warm
erwärmen *v*/**gelinde [mäßig]** warm gently [moderately, slightly]
erwärmen *v*/**sich** heat, warm (up)
Erwärmung *f* heating, *(specif)* gentle heating, moderate [slight] heating, warming
Erwärmungsgeschwindigkeit *f* heating rate
Erwartungswert *m* expected value, expectation *(statistics)*
erweichen *v* soften
erweichen *v*/**mit Peptisiermitteln** *(rubber)* peptize
Erweichung *f* softening
Erweichung *f* **mit Peptisiermitteln** *(rubber)* peptization
Erweichungsbereich *m* softening range
Erweichungsintervall *n* softening range
Erweichungspunkt *m* softening point [temperature]
Erweichungspunkt *m* **nach Vicat** *(plast)* Vicat softening point [temperature], V.S.P., Vicat needle point
Erythorbinsäure *f* erythorbic acid, isoascorbic acid
Erythrinaalkaloid *n* erythrina alkaloid
Erythroform *f* erythro form *(stereochemistry)*
Erz *n* ore
Erz *n*/**abbauwürdiges** pay ore
Erz *n*/**armes [geringhaltiges]** lean [low-grade] ore
Erz *n*/**reiches [reichhaltiges]** high-grade ore
Erz *n*/**sulfidisches** sulphide ore
Erzaufbereitung *f* ore dressing [beneficiation]
Erzaufbereitung *f*/**mikrobielle** *(biot)* microbial leaching, microbiological [bacterial] leaching
erzeugen *v* manufacture, make, produce *(e.g. chemicals)*; generate *(e.g. steam)*
Erzeugnis *n* 1. product, *(esp if relating to its origin)* make; 2. *(distil)* liquid product
Erzeugnis *n*/**feuerfestes** refractory (product)
Erzeugnisse *npl*/**feinkeramische** fine ceramics [ceramic ware]
Erzeugnisse *npl*/**oxidkeramische** oxide-ceramic products, oxide ceramics
Erzeugnisse *npl*/**pyrotechnische** pyrotechnics
Erzeugung *f* manufacture, making, make, *(esp over a specified period)* production; generation *(as of steam)*
erzhaltig ore-bearing, metalliferous
Erzlager *n* ore deposit, orebody

Erzlagerstätte *f* ore deposit, orebody
Erzmineral *n* ore mineral
Erzvorkommen *n* ore deposit, orebody
ES *s.* Emissionsspektroskopie
ESG *s.* Einscheibensicherheitsglas
ESMA *s.* Elektronenstrahlmikroanalyse
ESR *s.* Elektronenspinresonanz
ESR-Spektroskopie *f* ESR [EPR] spectroscopy
Esse *f* chimney, (smoke)stack
Essenz *f* essence
essenziell essential
essenziell/nicht *(bioch)* non-essential
Essig *m* vinegar
Essigester *m s.* Essigsäureethylester
Essiggenerator *m (biot)* vinegar [trickling] generator
Essigsäure *f* acetic acid, *ethanoic acid
Essigsäure *f***/aktivierte** active acetate, acetyl coenzyme A, acetyl CoA
Essigsäureanhydrid *n* acetic anhydride, *ethanoic anhydride
Essigsäurebakterien *npl* acetic-acid bacteria, vinegar bacteria
Essigsäurebutylester *m* butyl acetate
Essigsäureethylester *m* ethyl acetate, acetic ester
Essigsäuregärung *f* acetic(-acid) fermentation, vinegar fermentation
Essigsäuremethylester *m* methyl acetate
Essigsäurephenylester *m* phenyl acetate
Essigsäurepropylester *m* propyl acetate
Essigsäureureid *n* acetylurea, *N*-monoacetylurea
E-Stahl *m s.* Elektrostahl
Ester *m* ester
Ester *m***/innerer** intra-ester, inter-ester
Esteraustausch *m* ester exchange
Esterkondensation *f* ester condensation
Esterkondensation *f***/claisensche** Claisen condensation
Esterkondensation *f***/dieckmannsche (intramolekulare)** Dieckmann condensation [reaction]
Esteröl *n* (organic) ester oil, (organic) ester lubricant
Esterspaltung *f* ester cleavage
Esterspaltung *f* **nach Hunsdiecker** Hunsdiecker cleavage [reaction]
Esterverseifung *f* ester saponification
Etagenpresse *f* multidaylight press, (multiple-)daylight press, (multi)platen press
Ethan *n* ethane
Ethanal *n* *ethanal, acetaldehyde
Ethandial *n* *ethanedial, oxaldehyde, glyoxal
Ethan-1,2-dicarbonsäure *f* *ethane-1,2-dicarboxylic acid, *butanedioic acid, succinic acid
Ethan-1,2-diol *n* *ethane-1,2-diol, ethylene glycol, EG
Ethandisäure *f* *ethanedioic acid, oxalic acid
Ethanol *n* *ethanol, ethyl alcohol
Ethanolamin *n* ethanolamine, *(specif)* *2-aminoethane-1-ol, colamine
Ethanolgärung *f* ethanol(ic) fermentation
ethanolisch ethanolic
Ethanolyse *f* ethanolysis
Ethansäure *f* *ethanoic acid, acetic acid
Ethansäureanhydrid *n* *ethanoic anhydride, acetic anhydride
Ethanthiol *n* *ethanethiol, ethyl hydrosulphide
Ethen *n* *ethene, ethylene
Ethencarbonsäure *f* ethylenecarboxylic acid, *propenoic acid, acrylic acid
Ethenol *n* *ethenol, vinyl alcohol
Ether *m* R–O–R′ ether, *(specif)* $C_2H_5OC_2H_5$ diethyl ether, ordinary ether
Ether *m***/einfacher** R–O–R symmetrical ether
Ether *m***/gemischter** R′–O–R″ mixed [unsymmetrical] ether
Ether *m***/wasserfreier** absolute ether
Etherat *n* etherate
etherisch ethereal, etherial
Etherspaltung *f* ether cleavage
Ethin *n* acetylene, *ethyne
Ethindicarbonsäure *f* acetylenedicarboxylic acid, *butynedioic acid
Ethinylgruppe *f* –C≡CH ethynyl group
ethinylieren *v* ethynylate
Ethinylierung *f* ethynylation
Ethoxyanilin *n* ethoxyaniline, phenetidine, aminophenol ethyl ether
Ethoxybenzen *n* ethoxybenzene, ethyl phenyl ether, phenetole
Ethoxyethan *n* *ethoxyethane, diethyl ether

Ethylacetat *n* ethyl acetate, acetic ester
Ethylal *n* ethylal, *diethoxymethane, formaldehyde diethyl acetal
Ethylalkohol *m* ethyl alcohol, *ethanol
Ethylalkohol *m*/**wasserfreier** absolute (ethyl) alcohol
Ethylamin *n* ethylamine
Ethylat *n* $C_2H_5OM^I$ ethylate, ethoxide
Ethylbenzen *n* ethylbenzene
Ethylbenzin *n* ethyl gasoline
Ethylbromid *n* *bromoethane, ethyl bromide
Ethylbutyrat *n* ethyl butyrate, *ethyl butanoate
Ethylcarbamat *n* ethyl aminoformate, ethyl carbamate
Ethylcellulose *f* ethylcellulose
Ethylchlorid *n* *chloroethane, ethyl chloride
Ethylcyanid *n* ethyl cyanide, propionitrile
Ethylen *n* 1. ethylene, *ethene; 2. *s.* Ethylenkohlenwasserstoff
Ethylendiamin *n* ethylenediamine, 1,2-ethanediamine
Ethylendiamintetraessigsäure *f* ethylenediamine tetraacetic acid, EDTA
Ethylen-1,2-dicarbonsäure *f* ethylene-1,2-dicarboxylic acid, butenedioic acid
Ethylendichlorid *n* *1,2-dichloroethane, ethylene dichloride, EDC
Ethylenglykol *n* ethylene glycol, EG, *ethane-1,2-diol
Ethylengruppe *f* $-CH_2CH_2$ ethylene group
Ethylenimin *n* ethyleneimine, aziridine
Ethylenkohlenwasserstoff *m* ethylenic hydrocarbon, *alkene, olefin(e)
Ethylenmilchsäure *f* ethylenelactic acid, *3-hydroxypropionic acid
Ethylenoxid *n* ethylene oxide, EO, oxirane, *epoxyethane
Ethylen-Propylen-Terpolymerisat *n* ethylene-propylene terpolymer, EPT
Ethylen-Vinylacetat-Copolymerisat *n* ethylene-vinylacetate copolymer, EVA
Ethylester *m* ethyl ester
Ethylformiat *n* ethyl formate
Ethylgruppe *f* C_2H_5- ethyl group [residue]
Ethylhalogenid *n* ethyl halide [halogenide]
Ethylidengruppe *f* $=CHCH_3$ ethylidene group
ethylieren *v* ethylate
Ethylierung *f* ethylation
Ethylierungsmittel *n* ethylating reagent
Ethylphenylether *m* ethyl phenyl ether, ethoxybenzene, phenetole
Ethylphenylketon *n* ethyl phenyl ketone, propiophenone
Ethylpolysiloxan *n* *polyethylsiloxane, ethyl silicone
Ethylpropionat *n* ethyl propionate
Ethylschwefelsäure *f* ethyl sulphuric acid, ethyl hydrogensulphate
Ethylthioethan *n* ethylthioethane, diethyl sulphide
Ethylurethan *n* ethyl aminoformate, ethyl carbamate
Ethylxanthogensäure *f* ethylxanthogenic acid, xanthogenic acid, dithiocarbonic acid *O*-ethyl ester
Etikett *n* label; tag
etikettieren *v* label; tag
E_T-Wert *m* E_T-value *(for characterizing the polarity of solvents)*
Eugenol *n* eugenol, 1-allyl-4-hydroxy-3-methoxybenzene
Eukalyptol *n* eucalyptol, cineole, cineole-1,8
Eukolloid *n* eucolloid
Europium *n* Eu europium
Eutektikum *n* eutectic (mixture)
eutektisch eutectic
eutektoid eutectoid
Eutektoid *n* eutectoid
eutroph *(hyd)* eutrophic *(rich in dissolved plant nutrients)*
Eutrophierung *f* *(hyd)* eutrophication
evakuieren *v* evacuate, exhaust
Evakuierung *f* evacuation, exhaustion
Evans-Element *n* (differential) aeration cell, oxygen (concentration) cell
Evans-Korrosionselement *n* (differential) aeration cell, oxygen (concentration) cell
Ewens-Bassett-Zahl *f* *(nomencl)* Ewens-Bassett number, charge number *(of ionic charge)*
Excimer *n* *(spectr)* excimer, excited dimer *(consisting of an excited and a non-excited molecule of the same kind)*
Excimerlaser *m* excimer laser
Exciplex *m* exciplex, excited complex *(consisting of an excited molecule A and a non-excited molecule B)*

exergon(isch) 1. exoergic, energy-releasing, energy-yielding; 2. *(bioch)* exergonic
Exhaustor *m* exhauster
exocyclisch exocyclic
Exoelektron *n* exoelectron
Exoenzym *n* 1. *s.* Enzym/extrazelluläres; 2. *enzyme acting on the terminal parts of a chain molecule*
Exosmose *f* exosmosis
exosmotisch exosmotic
exotherm(isch) exothermic, heat-liberating
Exotoxin *n* exotoxin
expandieren *v* expand, dilate
Expansion *f* expansion, dilatation
Expansion *f***/adiabatische** adiabatic expansion
Expansionseffekt *m* expansion effect
Expektorans *n (pharm)* expectorant
Expektorantium *n (pharm)* expectorant
Experiment *n* experiment
Experimentalchemie *f* experimental chemistry
experimentell experimental
Experimentelles *n* experimental *(in treatises)*
experimentieren *v* experiment(alize)
Experimentiertechnik *f* experimental technique
explodieren *v* explode
explodieren lassen *v* explode, blow up
Explosion *f* explosion, blast, shot • **zur Explosion bringen** set off
explosionsartig explosive
Explosionsgefahr *f* explosion hazard, danger of explosion
explosionsgefährlich explosive, explosible
explosionsgeschützt explosion-proof
Explosionsgeschwindigkeit *f* explosion velocity
Explosionsgrenze *f* explosive [explosion, explosivity] limit
Explosionsgrenze *f***/obere** upper explosive limit, UEL
Explosionsgrenze *f***/untere** lower explosive limit, LEL
explosionssicher explosion-proof
explosiv explosive, explosible
explosiv/nicht non-explosive
Explosivität *f* explosiveness, explosivity, explosibility
Explosivstoff *m* explosive
Explosivstoff *m***/brisanter** high explosive, H.E.
Exponent *m* exponent, superscript numeral, numeric superscript
exponieren *v* 1. expose *(as to radiation or air)*; 2. *(phot)* expose (to light)
Exposition *f* 1. exposure *(as to radiation or air)*; 2. *(phot)* exposure (to light)
Expositionsdauer *f (tox)* exposure duration [time], length of exposure *(as of animals to pesticides)*
Expositionszeit *f (tox)* exposure duration [time], length of exposure *(as of animals to pesticides)*
Exsikkator *m (lab)* desiccator
Exsudat *n* exudate, exudation
Extender *m* extender, extending filler
Extinktion *f (spectr)* absorbance, absorbancy, extinction, optical density
Extinktionskoeffizient *m (spectr)* absorbance [extinction] coefficient
Extinktionskoeffizient *m***/molarer (dekadischer)** molar (decadic) absorption coefficient, molar absorptivity
Extinktionskoeffizient *m***/spezieller [spezifischer]** absorbency index, absorptivity *(proper)*
extrahierbar extractable, extractible
Extrahierbarkeit *f* extractability, extractibility
extrahieren *v* extract, leach, lixiviate *(material containing valuable components)*; extract, leach out *(soluble substances)*
extrahierend extractive
Extrakt *m* extract, essence
Extraktausbeute *f* extract yield
Extrakteur *m* extractor, extraction apparatus; contactor *(for solvent extraction)*
Extraktion *f* extraction, leach(ing), lixiviation *(of material containing valuable components)*; extraction, leaching-out *(of soluble substances)*
Extraktion *f* **fest-flüssig** solid-liquid extraction, leaching
Extraktion *f* **flüssig-flüssig** liquid(-liquid) extraction, solvent extraction
Extraktion *f* **mit flüssigem** CO_2 *(biot)* liquid carbon dioxide extraction
Extraktion *f* **mit überkritischen Gasen** *(coal)* supercritical gas extraction

Extraktionsapparat *m* extraction apparatus, extractor; contactor *(for solvent extraction)*
Extraktionsbatterie *f* extraction battery
Extraktionseffizienz *f* extraction efficiency
Extraktionsgefäß *n* extraction [extracting] vessel, *(esp with water also)* leaching tank [trough, vat, vessel], leach
Extraktionshülse *f* extraction thimble
Extraktionshülse *f* **nach Soxhlet** Soxhlet thimble
Extraktionskolonne *f* extraction column; contactor *(for solvent extraction)*
Extraktionsmaschine *f* contactor *(for solvent extraction)*
Extraktionsmittel *n* extracting agent [solvent], extractant, *(esp for extracting soluble principles from drugs:)* menstruum
Extraktionsrückstand *m* 1. extraction residue; 2. *(pharm, food)* marc, mark
Extraktionssäule *f* extraction column
extraktiv extractive
Extraktivdestillation *f* extractive distillation
Extraktphase *f* extract phase
Extraktstoff *m* extractive material [matter, substance], extractive
Extremwert *m* extreme (value), extremum
Extrudat *n* extrudate
Extruder *m* extruder, extruding machine, extrusion press
extrudieren *v (plast)* extrude
Extrudiererzeugnis *n* extrudate, extruded article
Extrudiermasse *f* extrusion compound
Exzisionsreparatur *f (bioch)* excision repair
Exziton *n (phys ch)* exciton

F

Fabrik *f*/**chemische** chemical works [plant]
Fabrikat *n* product, article; make *(as of a specific plant or country)*
Fabrikation *f* manufacture, make, making, *(esp over a specified period:)* production
Fabrikationsabfälle *mpl* industrial waste(s)
Fabrikationsabwasser *n* process waste water
Fabrikationswasser *n* process(ing) water, water for manufacturing use
fächeln *v* fan; waft *(for testing odours)*
FAD *s.* Flavin-adenin-dinucleotid
fad(e) stale, insipid, flat, tasteless, dead • **fad(e) werden** stale
Faden *m* 1. filament *(as of carbon or metal)*; thread *(as of glass, plastic, rubber)*; 2. *(text)* yarn *(for knitting or weaving fabrics)*
Fadenendenabstand *m* end-to-end distance, chain-end distance *(of macromolecules)*
Fadenkorrektur *f* (emergent) stem correction, exposed thread correction *(with liquid thermometers)*
Fadenmolekül *n* thread(like) molecule, filamentary [linear] molecule
Fadenziehen *n* 1. *(coat)* cobwebbing, stringing *(a defect)*; 2. *(text)* fibre drawing
fadenziehend ropy, stringy
Fadheit *f* staleness, insipidness, flatness
fahl flat *(shade)*
fahren *v* operate, run *(e.g. an apparatus)*
fahren *v*/**diskontinuierlich** operate intermittently [batch-wise]
Fahrweise *f (ch)* (mode of) operation
Fahrweise *f*/**adiabat(isch)e** adiabatic operation
Fahrweise *f*/**diskontinuierliche** batch operation
Fahrweise *f*/**isotherme** isothermal operation
Fahrweise *f* **mit totalem Rücklauf** *(distil)* total-reflux operation
Faktor *m* factor, coefficient; volumetric [titrimetric] factor *(titrimetry)*; gravimetric [analytical] factor *(gravimetry)*
Faktor *m*/**gyromagnetischer** gyromagnetic factor, g-factor
Faktor *m*/**innerer** *(bioch)* intrinsic factor *(a substance produced by stomach and intestinal mucosa)*
Faktor *m*/**präexponentieller** *(phys ch)* (Arrhenius) pre-exponential factor, Arrhenius [A] factor, frequency factor
Faktor *m*/**sterischer** *(phys ch)* steric [probability] factor

Faktor *m***/stöchiometrischer** stoichiometric factor [coefficient]
g*-Faktor** *m/landéscher** Landé g-factor, (Landé) splitting factor g *(gyromagnetic ratio of electrons)*
fällbar precipitable
Fällbarkeit *f* precipitability
Falle *f* trap
fallen *v* 1. fall, drop *(as of bulk material)*; 2. decrease, fall, drop *(as of measuring values)*
fällen *v* precipitate, throw down
Fallen *n* drop *(as of bulk material)*; decrease, fall, drop *(as of measuring values)*; fall *(as of a liquid level)*
Fallfilmverdampfer *m* falling-film evaporator, downflow evaporator
Fallkörperviskosimeter *n* falling-body viscometer
Fallmischer *m* tumbler [tumbling] mixer, tumbler
Fallout *n* fallout *(the settling of particles from the atmosphere)*
Fallout *m* 1. fallout *(particles which settle from the atmosphere)*; 2. *s.* Fallout/radioaktiver
Fallout *m***/radioaktiver** radioactive [atomic] fallout
Fallstromabsorber *m* wetted-wall(-column) absorber
Fallstromverdampfer *m* falling-film evaporator, downflow evaporator
Fällung *f* 1. precipitation; 2. precipitate
Fällung *f***/fraktionierte [gebrochene]** fractional precipitation
Fällung *f***/rhythmische** rhythmic precipitation
Fällung *f***/stufenweise** fractional precipitation
Fällung *f***/unvollständige** incomplete precipitation
Fällung *f***/vollständige** complete precipitation
Fällungsagens *n s.* Fällungsmittel
Fällungsanalyse *f* (volumetric) precipitation analysis, precipitation titration
Fällungschromatographie *f* precipitation chromatography
Fällungsmittel *n* 1. precipitating agent, precipitant; 2. *(hyd)* softening agent
Fällungsreaktion *f* precipitation reaction
Fällungstitration *f s.* Fällungsanalyse
Faltblattstruktur *f (bioch)* (pleated-) sheet structure, pleated-sheet conformation *(of polypeptide chains)*
Faltenbalg *m* corrugated bellows
Faltenfilter *n* pleated [folded, fluted] filter
Faltungslänge *f* fold period [height] *(of macromolecules)*
Familie *f* family *(as in the periodic system)*
Familie *f***/radioaktive** radioactive [decay] family [chain], radioactive (decay) series, decay series
Fänger *m* 1. *(tech)* catcher, trap; 2. *s.* Fängersubstanz
Fängersubstanz *f* scavenger *(e.g. for radicals, electrons)*
Faraday *n s.* Faraday-Konstante
Faraday-Effekt *m* Faraday effect, magnetic rotation
Faraday-Konstante *f* Faraday constant, faraday, F *(equivalent to 96486 coulomb)*
Faraday-Strom *m* faradaic current
Farbänderung *f* change in colour, colour change
Farbbad *n* dye bath [liquor]
Farbbase *f* dye [colour] base
Farbbeize *f* stain *(for glass, wood)*
farbbeständig colour-fast, non-discolouring, fadeless
Farbbeständigkeit *f* colour fastness [stability]
Farbe *f* 1. colour *(sensation)*; 2. colouring matter, *(Am)* color; *(in or for suspension:)* paint, pigment; *(for walls and ceilings:)* distemper; *(for artists:)* paint, colour; *(in or for solution:)* dye; *(for typography:)* (printing) ink; *(for glass:)* stain • **Farbe annehmen** colour • **die Farbe verlieren** discolour, fade
Farbe *f***/bewuchsverhindernde** antifouling (marine) paint
Farbe *f***/grafische** (printing) ink
Farbe *f***/lösungsmittelverdünnbare** solvent-thinned paint
Farbe *f***/wasserverdünnbare** water-thinned paint
Färbebad *n* dye bath [liquor]
Färbebase *f* dye [colour] base
farbecht colour-fast, non-discolouring, fadeless
Farbechtheit *f* colour fastness [stability]
Färbeflotte *f (text)* dye bath [liquor]
Färbehilfsmittel *n* dyeing assistant [aid]

Färbekraft *f* 1. colouring [tinctorial] power; 2. *(text)* dyeing power
Färbekufe *f (text)* dye back, dye(ing) vessel, dye(ing) vat
Färbemittel *n* colouring matter [agent]
Färbemittel *n* **für die Mikroskopie** microscopic stain
färben *v* 1. colour, *(using suspensions:)* paint, *(using solutions:)* dye, *(slightly:)* tint, tinge, *(using dry colouring matter:)* pigment; stain *(esp wood, glass, or tissues for microscopy)*; 2. *(food)* add colorants, colour
färben *v*/**direkt** *(text)* dye directly
färben *v*/**im Garn** *(text)* yarn-dye
färben *v*/**sich** colour; stain *(locally)*
färben *v*/**sich dunkel [dunkler]** darken
Färben *n* 1. colouring, *(using suspensions:)* painting, *(using solutions:)* dyeing, *(slight:)* tinting, tinging, *(using dry colouring matter:)* pigmenting; staining *(esp of wood, glass, or of tissues for microscopy)*; 2. *(food)* colouring
Färben *n* **im Garn** *(text)* yarn dyeing
Färben *n* **im Strang** *(text)* rope [hank] dyeing
Färben *n* **in der Flocke** *(text)* (loose) stock dyeing
Färben *n* **in der Wolle** *(text)* stock dyeing
Farbenchemie *f s.* 1. Farbstoffchemie; 2. Lack- und Farbenchemie
Färberei *f* 1. dyeing; 2. dye-house, dyeworks
Färbereihilfsmittel *n* dyeing assistant [aid]
Farbfehler *m* 1. colour defect, off-colour; 2. chromatic aberration *(optics)*
Farbfilter *n* colour filter
farbgebend 1. colour-causing *(e.g. water constituents)*; 2. chromophoric *(atomic group)*
Farbglas *n* stained [coloured] glass
Farbholz *n* dyewood
Farbindikator *m* colour indicator
Farbintensität *f* colour intensity, colouring strength
Farbkraft *f* 1. colouring [tinctorial] power; 2. *(text)* dyeing power
farbkräftig of strong colouring [tinctorial] power; highly [intensely] coloured, brilliant
Farblack *m (text, anal)* lake, *(Am)* color lake
farblos colourless, free from colour
Farblosigkeit *f* colourlessness, freedom from colour
Farbnuance *f* 1. shade, tint, tinge, tone, hue; 2. *(text)* cast
Farbpigment *n* coloured [paint] pigment
Farbprägen *n (plast)* hot embossing
Farbreaktion *f* colour reaction
Farbsalz *n* dye salt
Farbspritzen *n* spray painting, paint spraying
Farbspritzpistole *f* paint spray(ing) gun, paint sprayer
Farbstoff *m* 1. colouring matter [substance], colorant; dye(stuff) *(soluble organic compound)*; pigment *(insoluble matter of organic or inorganic origin)*; stain *(for wood, glass, microscopical investigation)*; 2. *(biol)* pigment
Farbstoff *m*/**adjektiver** adjective [mordant] dye
Farbstoff *m*/**anionischer** acid dye
Farbstoff *m*/**anthrachinoider** anthraquinone [anthraquinonoid] dye
Farbstoff *m*/**basischer** basic dye; basic stain *(for microscopy)*
Farbstoff *m*/**direktziehender** direct [substantive] dye
Farbstoff *m*/**echter** fast dye
Farbstoff *m*/**indigoider** indigoid dye
Farbstoff *m*/**kationischer** cationoid dye
Farbstoff *m*/**natürlicher** natural dye [colouring matter], biochrome
Farbstoff *m*/**pflanzlicher** plant pigment, *(for use:)* plant colouring matter, vegetable dye
Farbstoff *m*/**photoschaltbarer** photoswitchable dye
Farbstoff *m*/**saurer** acid(ic) dye
Farbstoff *m*/**sensibilisierender** sensitizing dye
Farbstoff *m*/**substantiver** substantive [direct] dye
Farbstoffaufnahme *f* dye uptake [absorption, acceptance]
Farbstoffaufnahmevermögen *n* dye receptivity, receptivity to dyestuffs
Farbstoffbase *f* dye [colour] base
Farbstoffchemie *f* dye(stuff) chemistry
Farbstoffchemiker *m* dye chemist
Farbstoffechtheit *f* dye fastness

Farbstofflaser *m (spectr)* dye laser
Farbstoffreagens *n (bioch)* colour(-producing) reagent
Farbstoffzwischenprodukt *n* dye intermediate
Farbton *m* 1. shade, tint, tone, hue; 2. *(text)* cast • **einen Farbton treffen** match a shade • **im Farbton abfallen** be off-shade
Farbtonbeständigkeit *f* colour fastness [stability]
Farbtonechtheit *f* colour fastness [stability]
Farbtonverschiebung *f* shift [displacement] of shade
Farbtonverschiebung *f*/**bathochrome** bathochromic shift [displacement], red-shift
Farbtonverschiebung *f*/**hypsochrome** hypsochromic shift [displacement], blue-shift
farbtragend chromophoric
Farbträger *m* 1. chromophore, chromophoric group; 2. *(coat)* substrate
Farbumschlag *m* change in colour, (sharp) colour change, *(titrimetry also)* indicator transition
Färbung *f* 1. *s.* Färben; 2. coloration, colour; 3. *(biol)* pigmentation; 4. *(result of treatment with suspensions:)* paint, *(with solutions:)* dye; *(of wood, glass, tissues for microscopy:)* stain • **eine Färbung erteilen** impart a colour *(as to a flame)*
farbverstärkend auxochromic
farbvertiefend bathochromic
Farin(zucker) *m* brown sugar, muscovado
Faser *f* 1. *(biol)* fibre; 2. *(text)* (staple) fibre *(a natural or man-made fibre of comparatively short length)*
Faser *f*/**gesponnene** spun fibre
Faser *f*/**keramische** ceramic fibre
Faser *f*/**mineralische** mineral fibre
Faser *f*/**native [natürliche]** natural fibre
Faser *f*/**pflanzliche** vegetable fibre
Faser *f*/**polynosische** polynosic fibre
Faser *f*/**proteinhaltige** protein-containing fibre
Faser *f*/**synthetische** (completely, fully) synthetic fibre, synthetic polymer fibre, synthetic
Faser *f*/**tierische** animal fibre
faserartig fibrous
Faserbündel *n (text)* strand
Fäserchen *n* fibril(la)
Faserfilter *n* felt(-fabric) filter
Fasergewebe *n* fabric (cloth)
Faserhalbstoff *m (pap)* half-stuff, half-stock
faserig fibrous
Faserkristall *m* (crystal) whisker
Faserlänge *f* fibre length [staple], staple (length)
Fasermischung *f* fibre blend
Faserplatte *f* fibre board
Faserprotein *n* fibrous protein, structural [skeletal] protein, scleroprotein
Faserschichtfilter *n* felt-fabric filter
Faserstoff *m (text)* fibre, fibrous material
Faserstoff *m*/**metallischer** metallic (fibre)
Faserstoffchemie *f* chemistry of fibres
Faservliesfilter *n* felt(-fabric) filter
fasrig fibrous
Fass *n* drum, vat, *(if wooden:)* barrel, cask, *(if small:)* keg, *(if very large:)* tun • **im Fass behandeln [durcharbeiten]** *(tann)* drum
Fassamalgamation *f* barrel amalgamation
fassen *v* hold *(a certain quantity)*; take *(a load)*
Fassungsvermögen *n* capacity
Faulbehälter *m (hyd)* digestion tank, (anaerobic) digester
faulen *v* 1. *(biol)* putrefy; 2. *(hyd)* digest
Faulen *n* 1. *(biol)* putrefaction; 2. *(hyd)* digestion
faulend putrescent
faulfähig 1. putrefiable, putrescible; 2. *(hyd)* digestible
faulfähig/nicht 1. unputrefiable, imputrescible; 2. *(hyd)* indigestible
Faulfähigkeit *f* 1. putrescibility; 2. *(hyd)* digestibility
Faulgas *n (hyd)* digester [sludge] gas, sewage [sewer] gas
faulig putrid
Fäulnis *f* putrefaction
Fäulnisbakterien *npl* putrefactive bacteria
fäulnisbeständig rotproof, rot-resistant
Fäulnisbeständigkeit *f* rotproofness, rot resistance

fäulniserregend putrefactive, putrefacient, saprogenic, saprogenous
Fäulnisgärung *f* putrefactive fermentation
Faulraum *m (hyd)* digester chamber [compartment]
Faulschlamm *m (hyd)* organic slime, putrid slime, digesting sludge, (anaerobic) digested sludge; sapropel *(hydrobiology)*
Faulung *f (hyd)* (anaerobic) digestion, anaerobic decomposition *(of sludge)*
faulungsunfähig *(hyd)* indigestible
Faustregel *f* rule of thumb
Favorskij-Umlagerung *f* Favorskii rearrangement *(of α-haloketones to acids or esters)*
Fayence *f (ceram)* faience
FCKW *s.* Fluorchlorkohlenwasserstoff
FDP-Weg *m s.* Embden-Meyerhof-Parnas-Weg
Feder *f (tech)* spring
federbelastet spring-loaded
Federsicherheitsventil *n* spring safety valve, spring-actuated relief valve
Federwaage *f* spring balance
Feedback *n***/negatives** *(bioch)* feedback inhibition
Feedback *n***/positives** positive feedback, feed-forward activation
Feedback-Hemmung *f (bioch)* feedback inhibition
Feedback-Inhibition *f (bioch)* feedback inhibition
Feedback-Regulation *f (bioch)* feedback control [regulation]
Fehlbenennung *f* misnomer
Fehler *m* 1. defect, flaw, fault *(in material)*; 2. *(cryst)* defect, imperfection; 3. error *(statistics)*
Fehler *m***/methodischer** error of method
Fehler *m***/mittlerer quadratischer** root mean square error, standard deviation, *(in a graph also)* half-peak width *(width at 0.607 h)*
Fehler *m***/parallaktischer** parallax error
Fehler *m***/prozentualer** percentage error
Fehler *m***/subjektiver** personal error
Fehler *m***/systematischer** systematic [constant] error, *bias (error)
Fehler *m***/wahrer** error of measurement, error in measuring
Fehler *m***/zufälliger** accidental [random, indeterminate] error
fehlerfrei *(cryst)* defect-free
Fehlerfunktion *f* error function, erfc
Fehlergrenze *f* limit of error
fehlerhaft defective, faulty, imperfect
Fehlerintegral *n* error function, erfc
Fehlerkompensation *f* compensation of errors
Fehlersuche *f* fault finding
fehlgeordnet *(cryst)* disordered
Fehlordnung *f (cryst)* disorder
Fehlordnung *f***/frenkelsche** Frenkel disorder
Fehlstelle *f (cryst)* (lattice) defect; holiday, void *(in a coating)*
Fehlstelle *f***/frenkelsche** Frenkel defect
Fehlstelle *f***/punktförmige** point defect
Fehlstellenpaar *n***/frenkelsches** *(cryst)* Frenkel pair, vacancy-interstitial [interstitial-vacancy] pair
Feinaufspaltung *f (spectr)* fine splitting
Feinausmahlung *f* fine grinding
Feinbürette *f* microburette
Feinchemikalie *f* fine chemical
feindispers finely dispersed
Feinerz *n* fine ore
Feinfilter *n* polishing [clarifying] filter
Feinfiltration *f* polishing (filtration), clarification
Feinfolie *f (plast)* film *(thickness < 0.01 inch)*
Feingehalt *m* fineness *(of a gold or silver alloy in parts per thousand)*
Feingut *n* 1. *(min tech)* fine material, fines, slime; 2. *s.* Feinkorn 1.
Feinheit *f* fineness *(as of particles or fibres)*
Feinkeramik *f* fine ceramics [ceramic ware]
Feinkohle *f* fine coal, (coal) fines
Feinkorn *n* 1. undersize (material, product), fines, minus material *(classifying)*; *(screening also)* screen undersize [fines]; 2. *(geol)* grain; 3. *(phot)* fine grain
feinkörnig 1. fine-grain(ed), fine; 2. *(min)* close-grained *(texture)*
feinkristallin(isch) finely crystalline, fine-grained
Feinmahlung *f* fine grinding
Feinmühle *f* fine-grinding mill, fine grinder, pulverizing mill, pulverizer

feinporig fine-pore(d)
feinpulverig finely powdered
Feinreinigung *f* ultrapurification
Feinschliff *m* fine grinding
Feinsieb *n* fine sieve
Feinsortierung *f* fine [secondary] screening
Feinstaub *m* fine dust
Feinsteinzeug *n (ceram)* fine stoneware
Feinstmühle *f* pulverizing mill, pulverizer
Feinstreinigung *f*/**ionogene** *(hyd)* polishing *(ion exchange)*
Feinstruktur *f* fine structure *(of atomic spectra)*
Feinstrukturanalyse *f s.* Kristallstrukturanalyse
Feinwägung *f* high-precision weighing, fine weigh
Feinwaschmittel *n* fine-fabric detergent, light-duty detergent
Feinzerteilung *f* dispersion, dispersal
FEL *s.* Freielektronenlaser
Feld *n (phys ch)* field • **nach tieferen Feldern verschoben** *(spectr)* downfield
Feld *n*/**atomares** atomic field
Feld *n*/**elektrostatisches** electrostatic field
Feld *n*/**entmagnetisierendes** demagnetizing field
Feld *n*/**magnetisches** magnetic field
Feld *n*/**selbstkonsistentes** self-consistent field, SCF
Felddesorption *f* field desorption, FD
Feldeffekt *m (phys ch)* field [direct] effect
Feldeffekttransistor *m*/**chemisch sensitiver** chemically sensitive field effect transistor, CHEMFET, ChemFET
Feldeffekttransistor *m*/**ionensensitiver** ion-sensitive field effect transistor, ISFET
Feldelektronenmikroskop *n* field emission microscope
Feldemissions(elektronen)mikroskop *n* field emission microscope
Feldemissionsrasterelektronenmikroskop *n* field emission scanning electron microscope, FESEM
Feldemissionsrasterelektronenmikroskopie *f* field emission scanning electron microscopy, FESEM
Feld-Fluss-Fraktionierung *f (anal)* field flow fractionation, FFF
Feldionisation *f* field ionization, FI
Feldionisations-Massenspektrometrie *f* field-ionization mass spectrometry, FIMS
Feldrichtung *f (phys ch)* field direction
Feldspat *m (min)* fel(d)spar, feldspath
feldspathaltig *(min)* fel(d)spathic
Feldstärke *f* field strength
Feldstärkeeffekt *m* Wien effect
Feldtest *m* field experiment [test, trial]
Feldversuch *m* field experiment [test, trial]
Femtogrammbereich *m (anal)* femtogram range *(10^{-15} to 10^{-12} g)*
Fenac *n* fenac, 2,3,6-trichlorophenylacetic acid *(a herbicide)*
Fensterkitt *m* (glazier's, painter's) putty
Ferment *n* enzyme, ferment
Ferment... *siehe: Enzym...*
Fermentation *f* fermentation
Fermentationsausbeute *f* fermentation yield
Fermentationsbrühe *f* fermentation broth [liquor, slurry, solution], fermenter broth
Fermentationsdauer *f* length of fermentation, fermentation [incubation] period
Fermentationsflüssigkeit *f s.* Fermentationsbrühe
Fermentationslösung *f s.* Fermentationsbrühe
Fermentationsrate *f* fermentation [conversion] rate, rate of product formation
Fermentationsrate *f*/**spezifische** specific production efficiency [rate], specific rate of product formation *(mg product/mg cell dry weight · h)*
Fermentationsrate *f*/**volumetrische** volumetric production efficiency [rate], volumetric rate of reaction *(mg product/l · h)*
Fermentationsreaktor *m s.* Fermenter
Fermentationssubstrat *n* fermentation substrate
Fermentationsweg *m* fermentation [fermentative] pathway, fermentation route
fermentativ enzym(at)ic, ferment(at)ive
Fermenter *m* fermenter, reactor
Fermenter *m*/**biotechnologischer** fermenter, bioreactor

Fermenter *m* **für diskontinuierliches Verfahren** batch fermenter
Fermenter *m* **für kontinuierliches Verfahren** continuous fermenter, through-flow bioreactor
Fermenter *m*/**vollständig gemischter** completely mixed fermenter, homogeneously mixed bioreactor, well mixed (flow) reactor
Fermenterbrühe *f s.* Fermentationsbrühe
Fermenterdurchsatz *m* fermenter throughput
fermentierbar fermentable
Fermentierbarkeit *f* fermentability
fermentieren *v* ferment
Fermi-Dirac-Statistik *f* Fermi(-Dirac) statistics
Fermi-Energie *f* Fermi energy
Fermi-Niveau *n* Fermi (characteristic energy) level
Fermion *n* *(nucl)* fermion
Fermi-Resonanz *f* Fermi resonance
Fermi-Statistik *f s.* Fermi-Dirac-Statistik
Fermium *n* Fm fermium
Fermi-Verteilung *f* Fermi distribution
Fernanzeige *f* remote indication
Fernbachkolben *m* Fernbach flask *(for propagating microorganisms)*
Fernbedienung *f* remote operation
Fernbetätigung *f* remote operation
Ferngas *n* pipeline (quality) gas
Fernkopplung *f* *(phys ch)* long-range coupling [interaction]
Fernleitung *f* (long-distance) pipeline
Fernordnung *f* *(cryst)* long-range (crystalline) order
Fernsteuerung *f* remote control
Ferrat(III) *n* $M^{I}FeO_2$ ferrate(III)
ferrimagnetisch ferrimagnetic
Ferrimagnetismus *m* ferrimagnetism
Ferrit *m* 1. *(met)* ferrite *(a solid solution of carbon in alpha or delta iron)*; 2. ferrite *(a magnetic material of the formula* $M^{II}Fe_2O_4$*)*
ferritisch ferritic
Ferrobor *n* ferroboron
Ferrocen *n* ferrocene, dicyclopentadienyl iron
Ferrochrom *n* ferrochromium
Ferroelektrikum *n* ferroelectric (material, substance)
ferroelektrisch ferroelectric
Ferrolegierung *f* ferro-alloy
Ferromagnetikum *n* ferromagnetic (material, substance)
ferromagnetisch ferromagnetic
Ferromagnetismus *m* ferromagnetism
Ferromangan *n* ferromanganese
Ferromolybdän *n* ferromolybdenum
Ferronickel *n* ferronickel
Ferrophosphor *m* ferrophosphorus
Ferrosilicium *n* ferrosilicon
Ferrotitan *n* ferrotitanium
Ferrovanadium *n* ferrovanadium
Ferrowolfram *n* ferrotungsten
Ferrozirconium *n* ferrozirconium
Fertigbearbeitung *f* finish(ing)
fertigen *v* produce, manufacture, make *(technical articles)*
Fertigerzeugnis *n s.* Fertigprodukt
Fertigpräparat *n* *(pharm)* preparation
Fertigprodukt *n* final [finished] product [stock]
Fertigung *f* production, manufacture, fabrication *(of technical articles)*
Fertigung *f*/**molekulare** molecular manufacturing
fertil fertile
Fertilität *f* fertility
Fertilitätsvitamin *n* antisterility vitamin
Ferulasäure *f* ferulic acid, 4-hydroxy-3-methoxycinnamic acid
FESEM *s.* 1. Feldemissionsrasterelektronenmikroskopie; 2. Feldemissionsrasterelektronenmikroskop
fest 1. solid *(as opposed to liquid)*; strong, firm, rigid *(gel)*; 2. strong, stable *(chemical bond)*; strong *(material or joint)*; resistant, stable *(to destructive influences)* • **fest werden** 1. solidify, harden, set, congeal; 2. *(coll)* gel(ate) • **fest werden lassen** solidify, set, congeal
Festbestandteil *m* solid
Festbett *n* fixed bed, static [dense] bed
Festbettionenaustauscher *m* fixed-bed (ion) exchanger
Festbettkatalysator *m* fixed-bed catalyst, static catalyst
Festbettreaktor *m* 1. fixed-bed reactor; 2. *(biot)* fixed-bed [packed-bed] bioreactor [fermenter]
Festbettreaktor *m*/**katalytischer** catalytic fixed-bed reactor, fixed-bed catalytic reactor

Festbettsynthese *f* fixed-bed synthesis
Festbrennstoff *m* solid fuel
Festelektrolyt *m* solid electrolyte
Fest-fest-Grenzfläche *f* solid-solid interface
Fest-flüssig-Extraktion *f* solid(-liquid) extraction, leaching
Fest-flüssig-Grenzfläche *f* solid-liquid interface
festfressen *v*/**sich** seize, freeze *(as of joints)*
Festfressen *n* seizing, seizure, freezing *(as of joints)*
festhaftend tenacious, adherent
Festigkeit *f* strength, stability *(of a chemical bond)*; strength, firmness, rigidity *(of a gel)*; strength *(of a material or joint)*; resistance, strength, stability *(to destructive influences)*
Festigkeit *f*/**dielektrische** dielectric strength, breakdown [puncture] strength
Festigkeit *f*/**mechanische** 1. physical stability *(e.g. gelmatrix)*; 2. physical strength *(e.g. coal)*
Festigkeitsverlust *m* loss in strength *(if partially)*; loss of strength *(if totally)*
festklemmen *v* clamp
Festkörper *m* solid
Festkörper *m*/**ionischer** ionic solid
Festkörper *m*/**organischer** organic solid
Festkörperchemie *f* solid-state chemistry
Festkörperdiffusion *f* solid(-state) diffusion
Festkörperlaser *m* solid-state laser
Festkörperlöslichkeit *f* solid solubility
Festkörperphysik *f* solid-state physics
Festkörperreaktion *f* solid-state reaction
festlegen *v* fix *(nutrients in a soil)*
festlegen *v*/**biologisch** immobilize *(nutrients by the action of soil microorganisms)*
Festlegung *f* fixation *(of nutrients in a soil)*
Festlegung *f*/**biologische** immobilization *(of nutrients by the action of soil microorganisms)*
Festparaffin *n* solid paraffin, paraffin wax
Festphasenextraktion *f* solid-state extraction
Festphasenmikroextraktion *f* solid-state micro extraction
Festphasen-Peptidsynthese *f* Merrifield (peptide) synthesis, solid-phase protein synthesis
Festphasensynthese *f* solid-phase synthesis
Festphasensynthese *f*/**organische** solid-phase organic synthesis, SPOS
Festphasen-Verfahren *n* Merrifield (peptide) synthesis, solid-phase protein synthesis
Festpunkt *m* fixed point
Festschichtkügelchen *n (chromat)* porous layer bead
feststellen *v* 1. *(anal)* determine, identify, detect; establish *(e.g. the structure of molecules)*; 2. arrest, fasten, fix *(mechanically)*
Feststellung *f (anal)* determination, identification, detection; establishment *(as of the structure of molecules)*
Feststoff *m* solid (matter, substance)
Feststoffabscheidung *f* solids separation [capture, recovery], separation of solid particles
Feststoffanteil *m s.* Feststoffgehalt
Feststoffbett *n* solid bed
Feststoffbett *n*/**bewegtes** moving bed
Feststoffbett *n*/**ruhendes [statisches]** fixed [static] bed
Feststoffe *mpl*/**suspendierte** *(hyd)* suspended solids, SS, suspended solid matter *(> $5 \cdot 10^{-4}$ mm)*
Feststoffentfernung *f* solids removal
Feststoffextraktion *f* (liquid-)solid extraction, solid-phase extraction, leaching
Feststoff-Feststoff-Reaktion *f* solid-solid reaction
Feststoffgehalt *m* solids (content), solids concentration [level]
Feststoffgemisch *n* solids mixture, bulk blend
Feststoffkonzentration *f s.* Feststoffgehalt
Feststoffreaktion *f* solid-solid reaction
Feststoffteilchen *n* solids particle
Feststoffteilchen *npl*/**suspendierte** 1. particulates, suspended particles; 2.

(hyd) suspended solids, SS, suspended solid matter *(> 5 · 10^{-4} mm)*
Festtreibstoff *m* solid (rocket) fuel, solid (rocket) propellant
Festwerden *n* 1. solidification, hardening, set(ting), freezing, congelation, congealing; 2. *(coll)* gelation
fett *(ch)* fatty *(e.g. oil)*
Fett *n* 1. *(ch, food)* fat; 2. *(tech)* grease
Fett *n*/**gehärtetes [hydriertes]** hardened [hydrogenated] fat
Fett *n*/**pflanzliches** vegetable [plant] fat
Fett *n*/**tierisches** animal fat
Fettabscheidung *f (hyd)* grease removal
fettähnlich fatlike
Fettalkohol *m* fatty alcohol
fettartig fatlike
fettbeständig resistant to grease, grease-resistant
Fettbeständigkeit *f* resistance to grease, grease resistance
Fettchemie *f* fat chemistry
fetten *v (tech)* grease, lubricate; compound *(oils)*
fettfreundlich lipophilic, fat-liking
Fettgerbung *f* chamois [oil] tannage
fetthaltig fat-containing, fatty, adipose
Fetthärtung *f* fat hardening [hydrogenation]
Fetthydrierung *f* fat hardening [hydrogenation]
Fetthydrolyse *f* fat hydrolysis, saponification of fat
fettig fatty, oily, greasy, unctuous *(consistency or substance)*
Fettigkeit *f* fattiness, oiliness, greasiness, unctuousness
Fettkügelchen *n* fat globule
Fettlöser *m* grease [fat] solvent
fettlöslich fat-soluble, soluble in fat, liposoluble
Fettlöslichkeit *f* fat solubility, solubility in fat, liposolubility
Fettlösungsmittel *n* grease [fat] solvent
Fettöl *n (ch)* fat(ty) oil, fixed oil *(as opposed to volatile oil)*
Fettoxidation *f* fat oxidation
Fettsäure *f* fatty acid
Fettsäure *f*/**essenzielle** essential fatty acid, EFA
Fettsäure *f*/**gesättigte** saturated fatty acid
Fettsäure *f*/**höhere** long-chain fatty acid, fat acid *(containing 12 to 24 carbon atoms)*
Fettsäure *f*/**mehrfach ungesättigte** polyunsaturated [polyethenoid] fatty acid
Fettsäure *f*/**mittlere** medium-chain fatty acid
Fettsäure *f*/**niedere** lower fatty acid
Fettsäure *f*/**ungesättigte** unsaturated fatty acid
Fettsäureamid *n* fatty (acid) amide
Fettsäureester *m* fatty acid ester
Fettsäureranzidität *f* hydrolytic rancidity
Fettsäureranzigkeit *f* hydrolytic rancidity
fettspaltend lipolytic, fat-splitting
Fettspaltung *f* lipolysis, fat splitting, cleavage of fats
Fettstoffwechsel *m* fat [triglyceride] metabolism
Fettverderb *m* fat deterioration
Fettverseifung *f* saponification of fat, fat hydrolysis
feucht moist, damp, wet, *(relating to air also)* humid • **feucht werden** become moist [wet], moisten
Feuchte *f* moisture, dampness, wetness, *(relating to air also)* humidity
Feuchte *f*/**absolute** absolute humidity *(of air)*
Feuchte *f* **der lufttrockenen Probe** *s.* Feuchte/hygroskopische
Feuchte *f*/**gebundene** bound moisture
Feuchte *f*/**hygroskopische** air-dried moisture
Feuchte *f*/**kritische** critical humidity *(of air)*; critical moisture content *(of solids)*
Feuchte *f*/**relative** relative humidity, R.H., percentage humidity [saturation] *(of air)*
Feuchteanteil *m* moisture content wet weight basis
Feuchteaufnahme *f* moisture absorption [pickup]
Feuchteaufnahmevermögen *n* moisture-carrying capacity *(of air)*
Feuchtebeladung *f* humidity *(of a gas)*; moisture content *(of a solid)*
Feuchtebeladung *f*/**kritische** critical humidity *(of a gas)*; critical moisture content *(of a solid)*
feuchtebeständig moisture-resistant, moisture-proof

Feuchtebeständigkeit *f* moisture resistance
Feuchtegehalt *m* moisture content, *(specif)* moisture content dry weight basis
Feuchtegehalt *m* **bezogen auf Feuchtmasse** *s.* Feuchteanteil
Feuchtegehalt *m* **bezogen auf Trockenmasse** *s.* Feuchtesatz
Feuchtemesser *m* hygrometer; moisture meter [tester] *(for determining the percentage of moisture in a material)*
Feuchtesatz *m* moisture content dry weight basis
Feuchtgut *n* wet [damp] product [feed] *(on drying)*
Feuchthaltemittel *n* moisturizer, humectant
Feuchtigkeit *f s.* Feuchte
Feuchtigkeits... *siehe auch: Feuchte...*
feuchtigkeitsabweisend moisture-repellent
Feuchtkugeltemperatur *f* wet bulb [wet-surface] temperature
Feuchtluft *f* humified [moisture-laden] air
Feuchtthermometer *n* wet-bulb thermometer
Feuchtung *f* damp(en)ing
Feuer *n* fire
Feueraluminieren *n* hot-dip aluminizing
feuerbeständig 1. fire-resistant; 2. *(ceram)* refractory; *(by treatment:)* fireproof • **feuerbeständig machen** fireproof
Feuerbeständigkeit *f* 1. fire resistance; 2. *(ceram)* refractoriness; *(by treatment:)* fireproofness
feuerfest *s.* feuerbeständig
Feuerfestmaterial *n* refractory (material)
feuerhemmend fire-retardant, fire-retarding
Feuerlöschbrause *f* drench [safety, emergency] shower
Feuerlöschdecke *f* fire blanket
Feuerlöscher *m* (fire) extinguisher
Feuerlöschgerät *n* (fire) extinguisher
Feuerlöschmittel *n* fire-extinguishing agent
feuern *v* fire, fuel
Feuerraffination *f* fire refining
feuerraffinieren *v* fire-refine
Feuerraum *m* combustion chamber [space], furnace chamber, firebox *(of an industrial furnace)*
Feuerschutzmittel *n* flameproofing [fireproofing] agent, fire retardant
Feuerung *f* 1. firing; 2. *s.* Feuerraum; 3. *s.* Feuerungsmaterial
Feuerungsmaterial *n* fuel
Feuerverzinken *n* (hot-dip) galvanizing, zinc dipping
ff. = feuerfest
FGAR = *N*-Formylglycinamidribotid
FIA *s.* Fließinjektionsanalyse
Fibrillärprotein *n* fibrillar [fibrous] protein
Fibrille *f (biol, text)* fibril(la)
Fibrillenbildung *f* fibrillation
fibrillieren *v* fibrillate
Fibrillierung *f* fibrillation
Fibrin *n (bioch)* fibrin
Fibrinogen *n (bioch)* fibrinogen
fibrinolytisch fibrinolytic
Fibroin *n* fibroin *(the insoluble protein of silk)*
fibrös fibrous
Fichtenöl *n* spruce turpentine
Fiebermittel *n* antipyretic, febrifuge
fiebersenkend antipyretic, febrifuge, antifebrile
Filament *n (text)* filament, continuous fibre [filament] *(a natural or man-made fibre of great or indefinite length)*
Film *m* film
Film *m***/fotografischer** photographic film
Film *m***/monomolekularer** monomolecular [unimolecular] film [layer], monolayer
Film *m***/photosensibler** *s.* Photoschicht
filmbildend film-forming, filmogenic
Filmbildner *m* film former, film-forming component [material, substance], filmogen
Filmdicke *f* film thickness
Filmdiffusion *f* film diffusion
Filmfermenter *m (biot)* (biological, microbial) film fermenter
Filmgießmaschine *f (plast)* film casting machine
Filmkondensation *f* film(-type) condensation
Filmreaktor *m s.* Filmfermenter
Filmverdampfung *f* film boiling
Filter *n(m)* filter, *(if working without pressure also)* strainer

Filter *n(m)***/bakteriendichtes** bacteriological [bacteria-retaining] filter
Filter *n(m)***/biologisch arbeitendes** *(hyd)* biological filter, biofilter
Filter *n(m)***/hydrostatisches** gravity [hydrostatic head] filter
Filter *n(m)* **mit Obenaufgabe** top-feeding filter
Filter *n(m)* **mit Untenaufgabe** bottom-feeding filter
Filterablauf *m* 1. filter effluent, filtrate; 2. *(hyd)* filtered [effluent] water
Filterbett *n* filter [filtration] bed
Filterblatt *n* filter leaf
Filterfläche *f* filter surface (area), filtering surface, filter (bed) area, area of filtration
Filtergewebe *n* filter [filtration] cloth [fabric]
Filtergut *n* prefilt (slurry, feed), material being [to be] filtered
Filterhilfsmittel *n* filter aid, filtration accelerator
Filterhilfsschicht *f* filter precoat, precoat filter cake
Filterkerze *f* filter candle [cartridge]
Filterkuchen *m* filter cake, (filter)cake
Filtermasse *f* filter mass
Filtermedium *n* filter medium [material], filtering [filtration] medium
filtern *v* filter *(esp gases)*
Filtern *n* filtration *(esp of gases)*
Filternutsche *f* nutsch(e), nutsch filter
Filternutsche *f* **nach Büchner** Büchner funnel [filter]
Filterpapier *n* filter paper
Filterpatrone *f* filter cartridge [candle]
Filterpresse *f* filter press
Filterröhrchen *n (lab)* filter tube
Filterscheibe *f* filter disk
Filterscheibe *f* **nach Witt** *(lab)* Witt (filter) plate
Filterschicht *f* 1. layer of filtering material *(of a multilayer filter)*; 2. *s.* Filterschüttschicht
Filterschüttschicht *f* filter [filtration] bed
Filterstab *m (lab)* filter stick
Filterstäbchen *n (lab)* filter stick
Filtertiegel *m* filtering crucible
Filtertiegel *m* **nach Gooch** Gooch crucible [filter]
Filtertrichter *m* filtering [fritted-disk] funnel
Filtertuch *n* filter [filtration] cloth [fabric]
Filterung *f* filtration *(of gases)*
Filterungs... *siehe: 1. Filter...; 2. Filtrations...*
Filterverstopfung *f* filter blinding
Filterwirkung *f* filtering action; filtering effect
Filterwirkungsgrad *m* filter efficiency
Filterzentrifuge *f s.* Filtrierzentrifuge
Filtrat *n* 1. filtrate, filter effluent; 2. *(hyd)* filtered [effluent] water
Filtratgüte *f* 1. filtrate quality; 2. *(hyd)* effluent (water) quality
Filtration *f* filtration, *(without pressure also)* straining • **durch Filtration entfernen** remove by filtration, filter out • **durch Filtration keimfrei machen** *(biot)* filter aseptically, sterilize by filtration
Filtration *f* **im Abwärtsstrom** downflow filtration
Filtration *f* **im Aufwärtsstrom** upflow filtration
Filtration *f* **mittels Anschwemmfilters** septum filtration
Filtration *f* **über A-Kohle [Aktivkohle]** *(hyd)* activated-carbon filtration, filtration with activated carbon, carbon filtration, CF
Filtration *f* **über körniges Filtermaterial** granular-medium filtration, (granular) bed filtration
Filtration *f* **unter Ausnutzung der Schwerkraft** gravity [natural] filtration
Filtrations... *siehe auch: Filtrier...*
Filtrationsanreicherung *f (biot)* filtration enrichment
Filtrationsdruck *m* filtering [filtration] pressure
Filtrationsentstauber *m* filter-type dust collector, dust(-control) filter, dust collection filter
Filtrationshilfe *f* filter aid
Filtrationskurve *f* filtration curve
Filtrier... *siehe auch: 1. Filtrations...; 2. Filter...*
filtrierbar filt(e)rable
filtrierbar/leicht freely filt(e)rable, free-filtering
filtrierbar/nicht unfilt(e)rable
filtrierbar/schwer poorly [difficultly] filterable
Filtrierbarkeit *f* filterability, filtering ability

filtrieren *v* filter, filtrate, *(without pressure also)* strain
filtrieren *v*/**durch Ultrafilter** ultrafilter
filtrieren lassen *v*/**sich** filter *(well or poorly)*
filtrieren *v*/**nochmals** refilter
filtrieren *v*/**unter Druck** filter with pressure, force through the filter
Filtrierstativ *n (lab)* filter [funnel] rack [stand]
Filtrierstutzen *m* filtrate jar
Filtrierzentrifuge *f* filtering centrifuge [centrifugal], centrifugal (filter), screen [perforate bowl] centrifuge
Filz *m* felt • **mit Filz auskleiden** line with felt, felt • **mit Filz überziehen** cover with felt, felt
Filzbildung *f* felting
filzen *v (text)* felt
FI-Massenspektroskopie *f* field-ionization mass spectrometry, FIMS
FIMS field-ionization mass spectrometry, FIMS
Finger *m*/**kalter** *(lab)* cold finger (condenser, trap), finger-type condenser, acorn condenser
Fingerabdruckbereich *m (spectr)* finger-print region
Finkelstein-Austausch *m (org ch)* Finkelstein exchange [reaction]
Finkelstein-Reaktion *f (org ch)* Finkelstein exchange [reaction]
FIR *s.* Infrarotgebiet/fernes
Firnis *m* boiled oil
fischartig fishy *(smell, taste)*
Fischbekämpfungsmittel *n* piscicide
Fischer-Hepp-Umlagerung *f* Fischer-Hepp rearrangement *(of aromatic nitrosamines to p-nitrosoarylamines in the presence of acids)*
Fischer-Projektion *f* Fischer projection *(for displaying molecules)*
Fischer-Projektion *f*/**gedrehte** rotated Fischer projection
Fischer-Tropsch-Synthese *f* Fischer-Tropsch synthesis *(of hydrocarbons)*
Fischgeschmack *m* reversion flavour, fishy taste *(of spoiled fats)*
Fischgift *n* 1. *(med)* ichthyotoxin; 2. *(agric)* fish poison *(e.g. several pesticides)*
fischig fishy *(smell, taste)* • **fischig werden** revert *(of fats)*
Fischigkeit *f* fishiness *(of spoiled fats)*
Fischigwerden *n* reversion *(of fats)*
Fischschwanzbrenner *m (lab)* fish-tail [bats-wing] burner
Fiset(te)holz *n (dye)* fustet, young fustic *(from Cotinus coggygria Scop.)*
Fixierbad *n (phot)* fixing [hypo] bath, fixer
fixieren *v* fix
Fixiermittel *n* 1. *(phot)* fixing agent, fixer; 2. *(text)* fixing agent
Fixiersalz *n (phot)* fixing salt, fixer
Fixierung *f* fixation
Fixierung *f*/**nicht symbiotische** *(agric)* non-symbiotic fixation, free fixation, azofication *(of atmospheric nitrogen)*
Fixierung *f*/**symbiotische** *(agric)* symbiotic fixation *(of atmospheric nitrogen)*
Fixpunkt *m* fixed point
Flachbecken *n (hyd)* plain clarifier
Fläche *f* 1. surface; 2. *(cryst)* face; 3. (surface) area *(as of an industrial plant)*
Flächenbelastung *f (hyd)* surface loading (rate), loading rate per unit area *(drying beds: $kg/m^2 \cdot a$; sedimentation tanks, trickling filters: $m^3/m^2 \cdot h$)*
Flächeneinheit *f* unit area
Flächengröße *f* (surface) area
Flächenkorrosion *f* general corrosion
Flächenpotenzial *n* surface potential
flächenzentriert *(cryst)* face-centred
Flachstrahldüse *f* slot [flat-spray] nozzle
flammbeständig flame-resistant, uninflammable, non-(in)flammable, flameproof
Flammbeständigkeit *f* flame resistance [resistivity], uninflammability, nonflammability
Flamme *f* flame • **mit fächelnder Flamme erhitzen [erwärmen]** brush with the free flame
Flamme *f*/**leuchtende** luminous [surface] flame
Flamme *f*/**nicht leuchtende** non-luminous flame, volume flame
Flamme *f*/**offene** free [naked] flame
Flamme *f*/**rauschende** roaring flame
Flamme *f*/**rußende** smoky flame
Flammen-AAS *f* flame AAS, flame atomic absorption spectrometry
Flammenatomabsorptionsspektr(alphot)ometrie *f* flame AAS, flame atomic absorption spectrometry
Flammenfärbung *f* flame coloration

flammenhemmend flame-retardant
Flammenionisationsdetektor *m* flame ionization detector, FID
Flammenphotometerdetektor *m (anal)* flame photometric detector
Flammenphotometrie *f* flame photometry
flammenphotometrisch flame-photometric
Flammenrückschlag *m* flareback, flashback
Flammenspektroskopie *f* flame spectroscopy
flammfest flameproof
Flammfestausrüstung *f (text)* flameproof finish [impregnation], fireproof [flame-resistant] finish [impregnation]
Flammfestigkeit *f* flameproofness
Flammkohle *f* flame coal
Flammofen *m* 1. reverberatory [air] furnace; 2. *(ceram)* reverberatory kiln
Flammpunkt *m* flash point
Flammpunktprüfer *m* flash-point apparatus [tester], flash tester
Flammpunktprüfer *m* **nach Pensky-Martens** Pensky-Martens (flash-point) tester
Flammschutzmittel *n* flameproofing [fireproofing] agent
Flansch *m* flange, socket
Flasche *f* bottle, flask
Flasche *f***/Florentiner** *(distil)* Florentine flask [receiver]
Flasche *f***/mariottesche** *(lab)* Mariotte bottle [flask], aspirator
Flasche *f***/woulfesche** *(lab)* Woulfe bottle
Flaschenabfüllmaschine *f* bottling machine, bottler
Flaschengärung *f* bottle fermentation
Flaschengas *n* cylinder gas, *(propane or butane or mixture of both:)* bottle(d) gas
Flaschenglas *n* bottle glass
Flaschenzentrifuge *f* bottle centrifuge
Flash-Kurve *f (distil)* (single-)flash curve
Flash-Verdampfung *f* flash [instantaneous] evaporation
Flash-Verdampfung *f***/mehrstufige** multiflash evaporation, MSF [multistage flash] evaporation
Flavan *n (org ch)* flavan
Flavanoid *n (org ch)* flavanoid
Flavanon *n (org ch)* flavanone
Flavin *n* flavin(e); *(specif)* isoalloxazine
Flavin-adenin-dinucleotid *n* flavin(e) adenine dinucleotide, FAD
Flavinenzym *n* flavin(e) enzyme, flavoenzyme, flavoprotein, yellow enzyme
Flavinferment *n* flavin(e) enzyme, flavoenzyme, flavoprotein, yellow enzyme
Flavinmononucleotid *n* flavin(e) mononucleotide, FMN, riboflavin-5′-phosphate
Flavon *n (org ch)* flavone
Flavoproteid *n s.* Flavinenzym
Flavoprotein *n s.* Flavinenzym
Flavyliumsalz *n* flavylium salt, 2-phenylchromenylium salt
Flechtensäure *f* lichen acid
Fleck *m* 1. spot, *(if undesirable also)* speck, blotch, stain; 2. *(chromat)* spot
Fleckenbildung *f* spotting, specking
Fleckenentferner *m* stain [spot] remover
Fleckenentfernungsmittel *n* stain [spot] remover
Fleischbrühe *f* broth
Fleischextrakt *m* meat extract
Fleischmilchsäure *f* sarcolactic acid, L(+)-lactic acid, dextrorotatory lactic acid
Fleischmürbesalz *n (food)* meat tenderizer
flexibel flexible
flexibilisieren *v (plast)* flexibilize
Flexibilität *f* flexibility
Fliehkraft *f* centrifugal force
Fliehkraftabscheider *m* centrifugal separator [collector]
Fliehkraftabscheidung *f* centrifugal separation
Fliehkraftmühle *f* centrifugal mill
Fliehkraftsichter *m* centrifugal classifier
Fliehkraftversprüher *m* centrifugal atomizer, spinning disk (atomizer)
Fliese *f* tile, slab
Fließbetrieb *m* continuous operation [working]
Fließbett *n* fluid(ized) bed, boiling bed
Fließbett... *siehe auch: Wirbelschicht...*
Fließbettkatalysator *m* fluidized catalyst, fluid(-bed) catalyst
Fließdehnung *f* yield strain
Fließdiagramm *n* flow diagram [chart, sheet]
Fließdialyse *f* continuous dialysis

Fließeigenschaften *fpl s.* Fließverhalten
fließen *v* 1. flow, run; 2. *(coat)* flow; 3. yield *(materials science)*
fließen lassen *v* run *(a liquid)*
Fließen *n* 1. flow, flux; plastic flow *(one kind of deformation)*; 2. yield *(materials science)*
Fließen *n***/kaltes** cold flow *(of thermoplastics)*
Fließen *n***/plastisches** plastic flow
Fließen *n***/viskoses** viscous flow
Fließerscheinung *f* yield phenomenon *(as in metals under tension)*
fließfähig flowable; fusible *(melt)*
Fließfähigkeit *f* flowability, fluidity; fusibility *(of a melt)*
Fließgeschwindigkeit *f* flow rate [velocity], liquid flow velocity
Fließgleichgewicht *n* dynamic equilibrium [steady state], steady [stationary] state
Fließgrenze *f* 1. yield point *(materials testing)*; 2. *(coat)* yield value; 3. *(met)* flow point
Fließgrenze *f***/praktische** yield strength *(materials testing)*
Fließgrenze *f***/untere** yield value *(materials testing)*
Fließinjektionsanalyse *f* flow-injection analysis, FIA
Fließkurve *f* flow curve
Fließmittelfront *f (chromat)* solvent front
Fließmittelwanderungsstrecke *f (chromat)* solvent migration distance
Fließpapier *n* absorbent paper
Fließpunkt *m* 1. melting point, m.p.; pour point *(of oils)*; 2. *(met)* flow point
Fließrichtung *f* flow direction
Fließschicht *f s.* Fließbett
Fließspannung *f* yield stress
Fließspeisung *f* gravity feed *(of an apparatus with liquids)*
Fließstaubkontakt *m* fluidized catalyst, fluid(-bed) catalyst
Fließtemperatur *f* 1. flow temperature; 2. *(met)* flow point
Fließverfahren *n s.* Wirbelschichtverfahren
Fließverfestigung *f* shear thickening
Fließverhalten *n* flow [rheological] behaviour, flow(ing) properties
Fließvermögen *n* flowability, fluidity; fusibility *(of a melt)*
Flintglas *n* flint glass
Flintsteinmahlkörper *m* flint pebble
Flip-flop-Enzym *n* flip-flop enzyme
Flöckchen *n* floccule
Flocke *f* 1. flake, *(esp in suspensions:)* floc; 2. *(text)* flock
flocken *v* 1. flocculate, coagulate, clot, curdle; 2. *(coll)* pectize; 3. *(hyd)* coagulate, flocculate
Flocken *fpl (hyd)* floc(s) • **Flocken bilden** agglomerate into floc(s)
flockenartig flake-like, flocculent
Flockenbildung *f* 1. flocculation, coagulation, clotting; 2. *(hyd)* floc formation [building]
flockig flocculent
Flockung *f* 1. flocculation, coagulation, clotting, curdling; 2. *(coll)* pectization; 3. *(hyd)* coagulation *(esp using metal coagulants)*; flocculation *(esp using polymer flocculants)*
Flockung *f* **und Sedimentation** *f (hyd)* clarification-flocculation, flocculation with sedimentation
Flockungsbehandlung *f (hyd)* coagulation [coagulant] treatment *(esp using metal coagulants)*; flocculation [flocculant] treatment *(esp using polymer flocculants)*
Flockungsfiltration *f (hyd)* coagulation-filtration
Flockungshilfsmittel *n (hyd)* coagulation [flocculation] aid *(as ground clay or activated silica)*
Flockungsmittel *n (hyd)* coagulant *(esp metal salts)*; flocculant *(esp polymers)*
Flockungswert *m* flocculation value *(of an electrolyte)*
Flory-Temperatur *f* Flory theta temperature
Flotation *f* flo(a)tation
Flotation *f***/differenzielle** differential [selective] floatation
Flotation *f***/kollektive** collective [bulk] floatation
Flotationsabgänge *mpl* floatation tailings
Flotationsberge *pl* floatation tailings
Flotationsmittel *n* floatation (re)agent
Flotationszelle *f* floatation cell [unit]
flotierbar floatable
Flotierbarkeit *f* floatability
flotieren *v* float *(e.g. ore)*

Flotte *f* liquor
Flottenaufnahme *f (text)* pick-up (of liquor)
Flottenverhältnis *n* 1. *(text)* liquor [bath] ratio, bath length; 2. *(pap)* liquor(-to-wood) ratio, liquid-to-solid ratio
Flöz *n (mine)* stratum, layer, *(if thin also)* seam
flüchtig volatile, fugitive
flüchtig/leicht highly [readily] volatile, high-volatile
flüchtig/schwer difficultly volatile, slow-evaporating, heavy
flüchtig/schwerer less volatile
Flüchtige *n* volatile matter, v.m.
Flüchtigkeit *f* volatility
Flugasche *f* fly ash
Flugbahn *f (nucl)* path
Flügel *m* blade, shovel, vane, paddle *(as of an agitator)*
Flugkraftstoff *m* aviation fuel
Flugstaub *m* entrained dust *(in fluidized-bed processes)*
Flugstaubverlust *m* stack loss
Flugstromreaktor *m* transport [entrained-bed] reactor
Flugstromvergasung *f (coal)* entrained-bed gasification, suspension [dilute-phase] gasification, entrainment [pulverized-coal] gasification
Flugzeit-Massenspektrometer *n* TOF [time-of-flight] mass spectrometer
Flugzeit-Sekundärionen-Massenspektrometrie *f* time-of-flight secondary ion mass spectroscopy, TOF-SIMS
fluid fluid
Fluid *n* fluid
Fluid *n***/ideales** ideal fluid, perfect [non-viscous] fluid
Fluid *n***/reales** actual fluid
Fluid *n***/strömendes** flowing fluid
Fluid-Feststoff-Reaktion *f* fluid-solid (heterogeneous) reaction
Fluidisation *f* fluidization
fluidisieren *v* fluidize
Fluidität *f (phys ch)* fluidity *(reciprocal of viscosity)*
Fluidtechnik *f* fluid-bed technique, fluidization [fluidized-bed, boiling-bed] technique
Fluidverfahren *n* fluid-bed process, fluid(ized) process
Fluo... *siehe: Fluoro... (for chemical compounds)*
Fluor *n* F fluorine
Fluoralkan *n* fluoroalkane, fluorinated alkane, aliphatic fluorocarbon
Fluoralken *n* *fluoroalkene, fluoro-olefin
Fluoralkylierung *f* fluoroalkylation
Fluoraminierung *f* fluoroamination
Fluoranthen *n (org ch)* fluoranthene
Fluorcarbonkunststoff *m* fluoroplastic
Fluorchlorkohlenwasserstoff *m* chlorofluorocarbon, CFC, freon
Fluorelastomer(e) *n* fluorocarbon elastomer, fluoroelastomer
Fluoren *n* fluorene, 2,3-benzindene, diphenylenmethane
Fluorescein *n* fluoresc(e)in
Fluoressigsäure *f* fluoroacetic acid
Fluoreszenz *f* fluorescence
Fluoreszenz *f***/laserinduzierte** laser-induced fluorescence, LIF
Fluoreszenzfarbstoff *m* fluorescence dye
Fluoreszenzindikator *m* fluorescent indicator
Fluoreszenzindikatoranalyse *f* fluorescence indicator analysis, FIA
Fluoreszenz-Interface-Kontrastmikroskop *n* fluorescence interface contrast microscope, FLIC
Fluoreszenz-Interface-Kontrastmikroskopie *f* fluorescence interface contrast microscopy, FLIC
Fluoreszenzlebensdauer *f s.* Fluoreszenzlebenszeit
Fluoreszenzlebenszeit *f (spectr)* fluorescence lifetime
Fluoreszenzlöschung *f* fluorescence quenching
Fluoreszenzmarker *m* fluorescence marker
Fluoreszenzquantenausbeute *f (spectr)* fluorescence quantum yield
Fluoreszenzsonde *f* fluorescent probe
Fluoreszenzspektroskopie *f* fluorescence spectroscopy
Fluoreszenzspektrum *n* fluorescence spectrum
Fluoreszenzstoff *m* fluorescent agent [substance]
Fluoreszenzstrahlung *f* fluorescence radiation
fluoreszieren *v* fluoresce

fluoreszierend fluorescent
Fluorid *n* M^IF fluoride
fluoridieren *v* fluoridize, fluoridate *(drinking water)*
Fluoridierung *f* fluoridation *(of drinking water)*
fluorieren *v* fluorinate
Fluorierung *f* fluorination
Fluorierung *f***/elektrochemische** electrochemical fluorination
Fluorierungsmittel *n* fluorinating agent
Fluorit *m (min)* fluorite, fluor-spar *(calcium fluoride)*
Fluorkautschuk *m* fluorinated [fluorine] rubber
Fluorkohlenwasserstoff *m* fluorocarbon
Fluoroborat *n* fluoroborate, tetrafluoroborate
Fluoroborsäure *f* fluoroboric acid, tetrafluoroboric acid
Fluorokieselsäure *f* fluorosilicic acid, hexafluorosilicic acid, sand acid
Fluorolefin *n* *fluoroalkene, fluoro-olefin
Fluorometrie *f* fluorometry, fluorimetry
fluorometrisch fluorometric, fluorimetric
Fluorophor *m* fluorophore, fluorogen *(a radical which causes fluorescence)*
Fluoroschwefelsäure *f* fluorosulphuric acid
Fluorosilicat *n* fluorosilicate
Fluorsiliconkautschuk *m* fluorosilicone rubber
Fluorwasserstoff *m* hydrogen fluoride
Fluorwasserstoffalkylierung *f* hydrofluoric-acid alkylation, HF alkylation
Fluorwasserstoffsäure *f* hydrofluoric acid
Fluss *m* 1. flow, flux; 2. *s.* Flussmittel
flüssig liquid, fluid • **flüssig werden** liquefy, melt, fuse, flux
Flüssigdünger *m* liquid fertilizer
Flüssig-fest-Chromatographie *f* liquid-solid chromatography, LSC
Flüssig-fest-Grenzfläche *f* liquid-solid interface
Flüssig-fest-Reaktion *f* liquid-solid(-phase heterogeneous) reaction
Flüssig-flüssig-Chromatographie *f* liquid-liquid chromatography, LLC
Flüssig-flüssig-Extraktion *f* liquid(-liquid) extraction, solvent extraction
Flüssig-flüssig-Verteilung *f* liquid-liquid partition
Flüssig-flüssig-Verteilungschromatographie *f* liquid-liquid partition chromatography
Flüssiggas *n* liquefied (petroleum) gas, L.P. gas, L.P.G., liquid gas *(liquefied hydrocarbons)*
Flüssigharz *n* liquid resin
Flüssigkeit *f* 1. liquid *(as opposed to solid or gas)*; 2. *(tech, food, biol)* liquor; 3. liquidity, fluidity *(state)*
Flüssigkeit *f***/anisotrope** anisotropic [crystalline] liquid
Flüssigkeit *f***/dekantierte** decantate
Flüssigkeit *f***/ionische** ionic liquid *(solar cells)*
Flüssigkeit *f***/kristalline [mesomorphe]** *s.* Flüssigkeit/anisotrope
Flüssigkeit *f***/pseudoplastische** pseudoplastic fluid
Flüssigkeit *f***/rheopexe** rheopectic fluid
Flüssigkeit *f***/strukturviskose** pseudoplastic fluid
Flüssigkeit *f***/thixotrope** thixotropic fluid
Flüssigkeit *f***/überstehende** supernatant liquid [liquor]
Flüssigkeit-Dampf-Grenzfläche *f* liquid-vapour interface
Flüssigkeit-Flüssigkeit-Grenzfläche *f* liquid-liquid interface, liquid junction, dineric interface
Flüssigkeitsabscheider *m* coalescer *(for collecting liquid droplets)*
Flüssigkeitsaufnahme *f* 1. uptake of liquid, imbibition; 2. *(text)* pick-up
Flüssigkeitschromatographie *f* liquid chromatography, LC
Flüssigkeitschromatographie *f***/schnelle** high-pressure liquid chromatography, HPLC
Flüssigkeitschromatographie *f***/überkritische** supercritical fluid chromatography
flüssigkeitschromatographisch by liquid chromatography
Flüssigkeitsdiffusionspotenzial *n* liquid-junction potential, diffusion potential
Flüssigkeitsdurchsatz *m* liquid throughput
Flüssigkeitsfilm *m* liquid film

Flüssigkeitsinhalt *m (distil)* liquid hold-up
Flüssigkeitskörper *m* fluid [liquid] body
Flüssigkeitslaser *m (anal)* liquid laser
Flüssigkeitsreibung *f* viscous force, internal friction
Flüssigkeitssäule *f* column of liquid
Flüssigkeitsstrahl *m* jet of liquid
Flüssigkeitsstrom *m* 1. liquid flow; 2. *(tech)* liquor flow
Flüssigkeitsströmung *f* 1. liquid flow; 2. *(tech)* liquor flow
Flüssigkeitsthermometer *n* liquid expansion thermometer
Flüssigkeitsüberstand *m* supernatant liquid [liquor]
Flüssigkeitsvolumen *n* liquid body
Flüssigkeitsvorlage *f (lab)* (liquid-)receiver
Flüssigkeitszylinder *m* stream tube *(rheology)*
Flüssigkristall *m* liquid crystal, mesophase
Flüssigkristall *m*/**cholester(in)ischer** cholesteric liquid crystal
Flüssigkristall *m*/**lyotroper** lyotropic liquid crystal
Flüssigkristall *m*/**nematischer** nematic liquid crystal
Flüssigkristall *m*/**smektischer** smectic liquid crystal
Flüssigkristall *m*/**thermotroper** thermotropic liquid crystal
Flüssigmembran *f (coll)* liquid membrane
Flüssigphase *f* liquid phase
Flüssigphasereaktion *f* liquid-phase reaction, liquid-liquid reaction
Flüssigphaseverfahren *n* liquid-phase process
Flüssigsauerstoff *m* liquid oxygen
Flüssigstickstoff *m* liquid nitrogen
Flüssigtreibstoff *m* liquid fuel *(propellant)*
Flüssigwerden *n* liquefaction, melting, fusing, fluxing
Flüssigzucker *m* liquid sugar
Flussmittel *n* 1. flux(ing agent); 2. *(lab)* fusion reagent; soldering flux; brazing flux
Flusssäure *f* hydrofluoric acid
Flusssäurealkylierung *f* hydrofluoric-acid alkylation, HF alkylation
Flussspat *m (min)* fluor-spar, fluorite *(calcium fluoride)*
Flussstahl *m* ingot iron [steel]
Flussverschmutzung *f* river [stream] pollution
Flussverunreinigung *f* river [stream] pollution
fluten *v (distil)* flood *(e.g. a packed column)*
Fluten *n (distil)* flooding *(as of a packed column)*
Flutpunkt *m (distil)* flooding point
Flutung *f (distil)* flooding
fluxen *v* 1. *(met, petrol)* flux; 2. *(petrol)* cut back
fl. z. *s.* flächenzentriert
F-1-Methode *f* F 1 [research] method *(for octane rating)*
F-2-Methode *f* F 2 [motor] method *(for octane rating)*
FMN *s.* Flavinmononucleotid
FM-Zyklotron *n* frequency-modulated cyclotron, synchrocyclotron
fokussieren *v* focus
Fokussierung *f*/**isoelektrische** *(anal)* isoelectric focussing, electrofocussing
Folat *n* folate, pteroylglutamate *(salt or ester of folic acid)*
Folgereaktion *f* consecutive reaction, subsequent [successive, stepwise] reaction
Folgestrang *m (bioch)* following strand *(of DNA)*
Folie *f (plast)* film, *(if thickness > 0.01 inch:)* sheeting *(as a web)*; sheet *(as a piece)*; *(esp relating to metal:)* foil
Folie *f*/**biaxial gereckte [verstreckte]** biaxially oriented film
Folie *f*/**extrudierte** extruded film [sheet]
Folie *f*/**gegossene** cast film [sheet]
Folie *f*/**geschälte** sliced film [sheet]
Folie *f*/**geschäumte** expanded sheet
Foliefaser *f* split fibre
Folienblaskopf *m (plast)* blow head
Folienblasmaschine *f (plast)* film blowing machine
Folienstrangpressen *n (plast)* film extrusion, *(relating to products with thickness > 0.01 inch:)* sheet extrusion
Folinsäure *f* folinic acid, N^5-formyltetrahydrofolic acid, leucovorin, citrovorum factor

Follikelhormon *n* follicular [oestrus-producing] hormone, oestrogen
Follikelreifungshormon *n* follicle-stimulating hormone, FSH
Follikelstimulierungshormon *n* follicle-stimulating hormone, FSH
Folsäure *f* folic acid, pteroylglutamic acid, PGA
Förderband *n* 1. conveyor [conveying] belt, apron; 2. *s.* Gurtbandförderer
Förderbandtrockner *m* conveyor dryer, belt (tunnel) dryer, moving-band dryer
Förderdruck *m* discharge [delivery] pressure
Förderer *m* conveyor
Förderer *m*/**pneumatischer** air conveyor
Fördererz *n* run-of-mine ore, raw [as-mined] ore
Förderflüssigkeit *f* liquid being [to be] pumped
Fördergurt *m* conveyor [conveying] belt
Fördergut *n* material being [to be] conveyed
Förderhöhe *f* discharge [delivery] head *(as of a pump)*
Förderhub *m* discharge [delivery] stroke *(as of a pump)*
Förderkohle *f* run-of-mine coal, raw coal
Förderkübel *m* skip (car)
Förderleistung *f* delivery (volume), discharge, *(of pumps also)* displacement
Förderluft *f* conveying air, *(directed upwards also)* lift air
Fördermenge *f s.* Förderleistung
fördern *v* 1. mine, extract, win *(e.g. coal)*; 2. convey *(bulk material)*; deliver, discharge *(of pumps, compressors)*; 3. promote *(e.g. a reaction)*; stimulate *(e.g. growth)*
Förderschnecke *f* 1. conveying [conveyor] screw [worm]; 2. *s.* Schneckenförderer
Förderstrom *m* rate of delivery [discharge, flow], delivery
Förderung *f* 1. mining, extracting, winning *(as of coal)*; 2. conveying *(of bulk material)*; delivery, discharge *(of pumps, compressors)*; 3. promotion *(as of reaction)*; stimulation *(as of growth)*
Förderung *f* **im Tagebau** surface [open-cut, open-cast] mining, open pit method, *(esp relating to ores:)* (surface) quarrying
Förderung *f* **im Tiefbau [Untertagebau]** deep mining, underground mining [working]
Förderung *f* **mittels Airlifts** air lifting, air-lift transport
Förderung *f* **mittels Gaslifts** gas lifting
Förderung *f*/**pneumatische** air conveying
Form *f* 1. form, shape; *(isomerism:)* form; 2. *(tech)* mould • **in gebundener Form** in combined form
Form *f*/**enantiomorphe** enantiomorph, enantiomer, enantiomorphous form [isomer], optical antipode [isomer], antimer
Form *f*/**furanoide** furanose form *(of sugars)*
Form *f*/**linksdrehende** laevo(rotatory) form, (-) form *(of an optically active compound)*
Form *f*/**polymorphe** polymorphic form, polymorph
Form *f*/**pyranoide** pyranose form *(of sugars)*
Form *f*/**rechtsdrehende** dextro(rotatory) form, (+) form *(of an optically active compound)*
Form *f*/**tautomere** tautomeric form, dynamic isomer
(+)-Form *f s.* Form/rechtsdrehende
(-)-Form *f s.* Form/linksdrehende
***aci*-Form** *f* aci form *(of a nitro compound)*
***d*-Form** *f s.* Form/rechtsdrehende
***l*-Form** *f s.* Form/linksdrehende
Formal *n* formal, formaldehyde acetal *(any acetal derived from formaldehyde and an alcohol)*
Formaldehyd *m* formaldehyde, *methanal
Formaldehydacetal *n s.* Formal
Formaldehyddimethylacetal *n* formaldehyde dimethyl acetal, methylal, *dimethoxymethane
Formaldoxim *n* formaldoxime, formaldehyde oxime
Formalladung *f (phys ch)* formal charge
Formamid *n (org ch)* formamide
Formänderung *f* deformation, strain
Formänderungsrest *m* residual [permanent] deformation [set]
Formazan *n (org ch)* formazan

formbar formable, shap(e)able, ductile; mouldable
Formbarkeit *f* formability, shap(e)ability, ductility; mouldability
Formbeständigkeit *f (plast, text)* dimensional stability
Formbeständigkeit *f* **in der Wärme** *(plast)* heat deflection [distortion] point [temperature]
Formbeständigkeit *f* **in der Wärme nach Vicat** *(plast)* Vicat softening point [temperature], V.S.P., Vicat needle point
Formel *f* formula
Formel *f***/allgemeine** general formula
Formel *f***/angenäherte** approximate formula
Formel *f***/stöchiometrische** stoichiometric [empirical] formula, simplest (possible) formula
Formel *f***/wahre** true formula
Formelbild *n* graphic formula
Formelgewicht *n* formula weight
Formelmasse *f* formula weight
Formelsprache *f* chemical shorthand
Formelumsatz *m* formula conversion
formen *v* form, shape; mould
Formgebung *f* forming, shaping, profiling
Formgedächtniseffekt *m* shape memory effect
Formgedächtnispolymere *npl* shape memory polymers
Formgussstück *n* casting
Formhöhlung *f (plast)* mould cavity
Formiat *n* $HCOOM^{I}$ *formate, formiate
Formimidierung *f* formimidation
Formmasse *f (plast)* moulding material [compound]
Formmasse *f***/wärmehärtbare** thermosetting moulding material
Formolyse *f* formolysis
Formonitriloxid *n* formonitrile oxide, fulminic acid
Formpresse *f (plast)* moulding [compression] press
Formpressen *n (plast)* (compression) moulding
Formpressstoff *m (plast)* compression-moulding material
Formsand *m* moulding sand
Formteil *n* moulding, blank, shape
Formteil *n***/gegossenes** cast moulding
Formteile *npl* moulded articles [goods, parts, products]
Formteile *npl***/spritzgegossene** injection-moulded articles
formulieren *v* formulate
Formulierung *f* formulation *(1. expressing with a formula*; *2. compounding in accordance with a recipe)*
Formveränderung *f* deformation, strain
Formylgruppe *f* –C(=O)H formyl group [residue]
formylieren *v* formylate
Formylierung *f* formylation
Forschungslabor(atorium) *n* research laboratory
fortpflanzen *v* propagate *(e.g. a reaction chain)*
fortpflanzen *v***/sich** propagate *(as of a reaction)*
Fortpflanzung *f* propagation
Fortpflanzungsreaktion *f* propagation reaction
Fortpflanzungsrichtung *f* direction of propagation
fortschreiten *v* proceed, advance, progress
fortschwemmen *v* float off
fortspülen *v* wash away, rinse [flush] away
fortwandern *v* migrate out *(as of ions)*
Fossilbrennstoff *m* fossil fuel
F_0-Teil *n (bioch)* F_0 factor *(of the ATP synthetase)*
Foto... *siehe auch: Photo...*
Fotoemulsion *f* photographic [sensitive] emulsion
Fotopapier *n* photo(graphic) paper
Foulardfärbung *f (text)* pad(ded) dyeing
Fourier-Transformation *f (anal)* Fourier transform(ation), FT
Fourier-Transform-IR-Spektrometrie *f* Fourier transform infrared spectrometry, FT-IR spectrometry
Fourier-Transform-Massenspektrometrie *f* Fourier transform mass spectrometry, FTMS
Fourier-Transform-NMR-Spektroskopie *f* Fourier transform NMR spectroscopy, FT-NMR spectroscopy, FTNMR
FP. *s.* Flammpunkt
FPD *s.* Flammenphotometerdetektor
fragmentieren *v* fragment *(e.g. molecules)*

Fragmentierung *f* fragmentation *(as of molecules)*
Fragment-Ion *n* fragment ion
Fraktion *f* 1. *(distil)* fraction, cut; 2. size fraction *(in classifying)*
Fraktion *f*/hochsiedende [höhersiedende] high-boiling [higher-boiling] fraction, heavy fraction
Fraktion *f*/niedrigsiedende low-boiling fraction, light fraction
fraktionieren *v* fractionate, fraction
Fraktionierkolben *m* fractionating flask
Fraktionierkolonne *f* fractionating column, fractionator
Fraktioniersäule *f* fractionating column, fractionator
Fraktionierung *f* 1. fractionation; 2. *(distil)* fractional distillation
Fraktionssammler *m* fraction collector
Francium *n* Fr francium
Fransenmizelle *f* fringed micelle
Fraunhofer-Linien *fpl* Fraunhofer lines
frei free; vacant *(orbitals)*
frei/praktisch substantially free
Freie-Enthalpie-Beziehung *f*/lineare linear free energy relation(ship), LFE relationship, linear Gibbs energy relation
Freielektronenlaser *m* free-electron laser, FEL
Freifallmischer *m* tumbling mixer, tumbler
Freiheit *f (phys ch)* variance, degree of freedom • **in Freiheit setzen** liberate, release • **ohne Freiheit** non-variant, invariant
Freiheitsgrad *m (phys ch)* degree of freedom, variance
Freilandversuch *m* field experiment [test, trial]
Freiluftbewitterungsversuch *m* outdoor weathering test
Freilufttrocknung *f* 1. air drying; 2. *(ceram)* hack drying
Freiname *m* non-proprietary name, generic name [term]
freisetzen *v* release, liberate, set free
Freisetzung *f* release, liberation
Freisetzungshormon *n* releasing hormone
freiwillig spontaneous
Fremdatom *n* foreign [impurity] atom
Fremdelement *n* foreign element
fremdgestaltig *(cryst)* xenomorphic, allotriomorphic, anhedral
Fremdion *n* 1. foreign ion; 2. *(cryst)* nonlattice ion
fremdionig foreign-ion
Fremdkeim *m* crystal nucleus, nucleus of crystallization *(consisting of foreign material)*
Fremdkraftringmühle *f* ring(-roll) mill, centrifugal attrition mill, centrifugal grinder
Fremdkraftrollmühle *f* ring(-roll) mill, centrifugal attrition mill, centrifugal grinder
Fremdkraftwälzmühle *f* ring(-roll) mill, centrifugal attrition mill, centrifugal grinder
Fremdmolekül *n* foreign molecule
Fremdpeak *m* spurious [ghost] peak
Fremdprotein *n (biot)* foreign protein
Fremdstoff *m* admixture, impurity, foreign [extraneous] matter [material, substance]
Fremdstoff *m* **in Futtermitteln** feed additive
Fremdstoff *m* **in Lebensmitteln** food additive
Fremdstoffhalbleiter *m* impurity [extrinsic] semiconductor
Frenkel-Defekt *m (cryst)* Frenkel defect [disorder] *(lattice vacancy plus interstitial atom)*
Frenkel-Fehlordnung *f (cryst)* Frenkel defect [disorder] *(lattice vacancy plus interstitial atom)*
Frequenz *f* frequency, *(specif)* linear frequency, ν *(Hz sec^{-1})*; angular frequency, ω *(rad sec^{-1})*
Frequenz *f*/charakteristische characteristic frequency, proper [natural] frequency
Frequenzbereich *m* frequency range [region]
Frequenzfaktor *m* frequency factor, (Arrhenius) pre-exponential factor, Arrhenius factor, A factor *(kinetics)*
Frequenzmethode *f*/variable *(spectr)* continuous-wave technique, CW technique
Frequenzverdopplung *f* frequency doubling
Frequenzverschiebung *f* frequency shift

Freundlich-Isotherme *f* Freundlich (adsorption) isotherm
Friedel-Crafts-Acylierung *f* Friedel-Crafts acylation
Friedel-Crafts-Alkylierung *f* Friedel-Crafts alkylation
Friedel-Crafts-Kondensation *f* Friedel-Crafts condensation
Friedel-Crafts-Reaktion *f* Friedel-Crafts reaction [synthesis] *(for preparing hydrocarbons or ketones)*
Friedländer-Synthese *f* Friedländer synthesis *(for preparing quinoline derivatives)*
Fries-Reaktion *f* Fries reaction [rearrangement, migration] *(conversion of phenolic esters into hydroxyketones)*
Frings-Generator *m (biot)* Frings acetator
Frischdampf *m* live steam, direct [prime, open] steam
frischen *v (met)* blow
Frischluft *f* fresh air
Frischsäure *f* fresh acid
Frischwasser *n* fresh water
Fritte *f* 1. *(ceram)* frit, agglomerate; 2. *(tech)* diffuser [air] stone *(for dissolving gas in a liquid)*; 3. *(agric)* frit *(containing micronutrients)*; 4. *s.* Glasfritte
fritten *v (ceram)* frit, sinter, agglomerate *(by heat)*; sinter, agglomerate *(under the influence of heat)*
Frittenwaschflasche *f (lab)* sintered-plate wash-bottle
Front *f (chromat)* front, boundary *(of the mobile solvent)*
Frontalchromatographie *f* frontal chromatography
frostbeständig frost-resistant, frost-proof
Frostbeständigkeit *f* frost resistance, resistance to frost [freezing]
Frostpunkt *m* frost [freezing] point
Frostschutzmittel *n* antifreeze (agent)
Fruchtaroma *n* fruit essence
Fruchtäther *m* fruit essence
fruchtbar fertile *(soil)*
Fruchtbarkeit *f* fertility *(of soil)*
Fruchtbrei *m* pomace, squash
Fruchtessenz *f* fruit essence
Fruchtessig *m* fruit vinegar
Fruchtsäure *f* fruit acid
Fruchtzucker *m s.* Fructose
Fructose *f* fructose, fruit sugar, Fru
f-Serie *f (spectr)* fundamental series
FSH *s.* Follikelstimulierungshormon
F-strain-Effekt *m s.* Verzögerung/sterische
FT-IR-Spektrometrie *f s.* Fourier-Transform-IR-Spektrometrie
Fuchsin *n* fuchsin(e), magenta, rosaniline
Fuchsinfarbstoff *m* fuchsine [rosaniline] dye
Fuchsonimin *n (dye)* fuchsonimine
Fugat *n* centrifugate
Fugazität *f (phys ch)* fugacity
Fugazitätskoeffizient *m (phys ch)* fugacity coefficient
Fühler *m* sensing device, sensor, detector
führen *v* 1. conduct *(a process)*; run *(a factory)*; 2. lead, conduct, pipe *(e.g. vapour in piping)*
führen *v*/**im Kreislauf** recycle, (re)circulate
führen *v*/**in Rohrleitungen** pipe
füllen *v* 1. fill; 2. *(tech)* feed, charge, load, furnish *(e.g. a furnace or reactor)*; 3. *(distil)* pack *(a column with packing material)*; 4. load, fill *(a product with fillers)*
Fulleren *n (org ch)* fullerene
Füllkörper *m* 1. filler (material), filling (agent, material); 2. *(distil)* packing body, piece of packing
Füllkörper *mpl* packing *(of or for a column or tower)*
Füllkörper *mpl* **aus Porzellan** porcelain packing
Füllkörper *mpl*/**schalenporöse** *(chromat)* porous layer beads, pellicular packing [support], solid core support
Füllkörperhöhe *f*/**äquivalente** height equivalent to a theoretical plate [stage], HETP, HETS
Füllkörperkolonne *f* packed column
Füllkörpersäule *f* packed column
Füllloch *n* charging [filling] door, feed inlet [hole]
Füllmaterial *n* 1. load, batch, feed(-stock), feed material, charge, charging stock *(as for a reactor)*; packing (material) *(as for a column or tower)*; packing medium *(of a trickling filter)*; 2. *s.* Füllstoff 1.

Füllöffnung *f* charging [filling] door, feed inlet [hole]
Füllrohr *n* charging pipe
Füllstand *m* filling level, fill height
Füllstandsmessung *f* level measurement
Füllstoff *m* 1. filler, filling [loading] material, load(er); 2. *s.* Füllmaterial 1.
füllstofffrei unfilled, unloaded, filler-free
füllstoffhaltig filled, loaded
Füllstutzen *m* charging [filling, feeding] pipe
Fülltrichter *m* (charge, feed) hopper, feed(ing) funnel
Füllung *f* 1. filling; 2. *(tech)* feeding, charging, loading, furnishing *(as of a furnace or reactor)*; 3. filling, loading *(of a product for conditioning)*; 4. *s.* Füllmaterial 1.; 5. *s.* Füllstoff 1.
Füllvolumen *n* 1. filling volume; 2. *(tech)* feeding [charging, loading] volume; 3. *(biot)* working [fermenter, fermentation] volume
Fulminat *n* $CNOM^{I}$ fulminate
Fulvalen *n (org ch)* fulvalen
Fulven *n* fulvene *(5-methylenecyclopenta-1,3-diene or any of its derivatives)*
Fumarat *n* fumarate *(salt or ester of fumaric acid)*
Fumarsäure *f* fumaric acid, **trans*-butenedioic acid
Fundamentalbaustein *m* fundamental [basic] building block
Fundamentalgleichung *f (phys ch)* fundamental equation
Fundamentalpunkt *m* fixed point
fünfbasig pentabasic, pentaprotic *(acid)*; pentabasic, pentaacid *(base)*
Fünfkörperverdampfer *m* quintuple-effect evaporator
Fünfring *m* five-membered ring
fünfsäurig pentaacid *(base)*
fünfwertig pentavalent, quinquevalent *(element)*; pentabasic, pentaprotic *(acid)*; pentabasic, pentaacid *(base)*; pentahydric *(alcohol)*
Fünfwertigkeit *f* pentavalence, pentavalency, quinquevalence *(of elements)*
fünfzählig *s.* fünfzähnig
fünfzähnig pentadentate *(ligand)*
Fungistatikum *n* fungistat
fungistatisch fungistatic
fungitoxisch fungitoxic
fungizid fungicidal, antifungal
Fungizid *n* 1. fungicide; 2. *(pharm)* antifungal drug
Fungizid *n***/direktes** direct [eradicant] fungicide
fungizidresistent fungicide-resistant
Funkenanregung *f (spectr)* spark excitation
Funkenspektrum *n* spark [flash] spectrum
Funktionalität *f* functionality
Fural *n s.* Furfural
2-Furaldehyd *m s.* Furfural
Furan *n (org ch)* furan
Furancarbonsäure *f* furancarboxylic acid
Furandion *n* furandione
Furanharz *n* furan resin
Furanoseform *f* furanose form *(of sugars)*
Furanosid *n* furanoside
Furazan *n (org ch)* furazan
Furfural *n* furfural, 2-furylaldehyde
Furfuralharz *n* furfural resin
Furfurylalkohol *m* furfuryl alcohol, 2-hydroxymethylfuran
α-Furfurylalkohol *m* furfuryl alcohol, 2-hydroxymethylfuran
Furil *n* furil, αα-furil, di-2-furylglyoxal, di-α-furyl diketone
Furocumarin *n (org ch)* furocoumarin
Furoin *n* furoin, αα-furoin, 1,2-difuryl-2-oxoethanol
Furylacrylsäure *f* furylacrylic acid, 3-α-furylacrylic acid, furfuralacetic acid
Fusidinsäure *f* fusidic acid *(antibiotic)*
Fusionsname *m (nomencl)* fusion name
Futter *n* 1. *(met)* lining, refractory; 2. *(agric)* feed(stuff), feeding stuff; 3. *(biot)* feed
Futter *n***/basisches** *(met)* basic lining [refractory]
Futter *n***/saures** *(met)* acid lining [refractory]
Futtermittel *n (agric)* feed(stuff), feeding stuff, animal food [feeding material]
Futtermittelzusatz(stoff) *m* feed additive [supplement], nutritional supplement
füttern *v* 1. *(met)* line; 2. *(biot)* feed *(microorganisms)*
Fütterungsantibiotikum *n (biot)* feed-

additive antibiotic, antibiotic used in animal feed
FUV = ferner Ultraviolettbereich

G

Gabe *f*/**größte** *(pharm)* maximum dose
Gabriel-Synthese *f* Gabriel phthalimide synthesis (of primary amines), Gabriel synthesis
G-Actin *n* G actin *(a globular muscle protein)*
Gadoleinsäure *f* gadoleic acid, eicos-9-enoic acid
Gadolinium *n* Gd gadolinium
Galactarsäure *f* galactosaccharic acid, mucic acid *(a tetrahydroxydicarboxylic acid)*
Galactozuckersäure *f s.* Galactarsäure
Galacturonsäure *f* galacturonic acid *(a pentahydroxycarboxylic acid)*
Galaktosämie *f (med)* galactosaemia
Galenit *m (min)* galena, galenite *(lead(II) sulphide)*
Gallat *n* 1. gallate(III) *(gallium compound)*; 2. 3,4,5-trihydroxybenzoate, gallate *(salt or ester of gallic acid)*
Gallenflüssigkeit *f (med)* bile, gall
Gallensäure *f* bile acid
Gallenstein *m (med)* gallstone, biliary calculus
Gallert *n* gelatin(e), jelly, gelatinous mass [substance]
gallertartig gelatinous, gelatiniform, jelly-like
Gallertbildung *f* jellification
Gallerte *f s.* Gallert
Gallium *n* Ga gallium
Gallium(III)-chlorid *n* gallium(III) chloride, gallium trichloride
Gallium(III)-hydroxid *n* gallium(III) hydroxide
Gallotannin *n s.* Gallusgerbsäure
Gallusgerbsäure *f* gallotannic acid, gallotannin, tannic acid *(glucose esterified with gallic acid or depsides of it)*
Gallussäure *f* gallic acid, 3,4,5-trihydroxybenzoic acid
Galmei *m* 1. *(min)* calamine, galmei, galmey; 2. *(pharm)* calamine *(zinc oxide with a small amount of ferric oxide)*
Galvani-Potenzial *n (el ch)* Galvani [inner] potential
galvanisch 1. galvanic *(cell, current, anode)*; 2. electroplated, electrodeposited *(coating)*
Galvanisierbad *n s.* 1. Galvanisierelektrolyt; 2. Galvanisierbehälter
Galvanisierbehälter *m* (electro)plating tank
Galvanisierelektrolyt *m* electroplating solution
galvanisieren *v* electroplate, plate
Galvanisierung *f* galvanization, galvanizing, electrocoating
Galvanometer *n*/**ballistisches** ballistic galvanometer
Galvanoplastik *f* electroforming, galvanoplastics, galvanoplasty
galvanoplastisch galvanoplastic
Galvanostegie *f* electroplating
Gammastrahlen *mpl* gamma rays
Gammastrahlung *f* gamma radiation
Gammaumwandlung *f (rubber)* glass transition, gamma [second-order] transition, vitrification
Gammawert *m (phot)* gamma value, development factor
Gang *m* 1. *(geol)* vein, dike, *(of metal ore also)* lode; 2. *(ch)* course *(of a reaction)*; 3. *(tech)* flight *(as of a worm shaft)* • **in Gang halten** keep in progress *(a reaction)* • **in Gang setzen** initiate, start up *(a reaction)*
Gangart *f (mine)* gangue, gang, waste rock, matrix
Ganghöhe *f (tech, bioch)* pitch *(of a screw or an alpha helix)*
Gängigkeit *f (bioch)* helix sense
Ganglienblocker *m (pharm)* ganglion blocking agent, ganglionic blockader
Ganglioplegikum *n (pharm)* ganglion blocking agent, ganglionic blockader
Gär... *siehe auch: Gärungs...*
Gärbottich *m* fermentation vat [vessel], fermenter
Gärdauer *f* fermentation period, length of fermentation
gären *v* ferment, yeast
Gärflüssigkeit *f* fermenting liquor, wash
Garn *n (text)* spun(-staple) yarn, yarn
Garnfärben *n* yarn dyeing
Gärtank *m* fermentation [fermenting] tank, fermenter

Gärtemperatur *f* fermentation temperature, fermenter set temperature
Gärung *f* fermentation
Gärung *f***/aerobe** aerobic [oxidative] fermentation
Gärung *f***/alkalische** *(hyd)* methane fermentation
Gärung *f***/alkoholische** alcohol(ic) fermentation
Gärung *f***/anaerobe** anaerobic fermentation
Gärung *f***/bakterielle** bacterial fermentation
Gärung *f***/heterofermentative** heterolactic [mixed] fermentation
Gärung *f***/mikrobielle** microbial fermentation
Gärung *f***/oxidative** oxidative [aerobic] fermentation
Gärung *f***/saure** acid fermentation
Gärungs... *siehe auch: Gär...*
Gärungsablauf *m* fermentation pathway [route], course of fermentation
Gärungsalkohol *m* fermentation alcohol
Gärungschemie *f* fermentation chemistry, zymurgy
Gärungsenzym *n* fermentation enzyme
gärungserregend fermentative, zymogenic, zymogenous
Gärungsessig *m* fermentation vinegar
Gärungsethanol *n* fermentation ethanol
gärungshemmend antifermentative
Gärungsindustrie *f* fermentation industry
Gärungsmilchsäure *f* lactic acid of fermentation, DL-lactic acid, ordinary lactic acid, i-lactic acid *(optically inactive)*
Gärungsprodukt *n* fermentation product
Gas *n* gas
Gas *n***/ausgetriebenes** stripped gas
Gas *n***/ideales** ideal [perfect] gas
Gas *n***/inertes** inert [inactive, indifferent] gas
Gas *n***/permanentes** permanent gas
Gas *n***/reales** imperfect [real, actual] gas
Gas *n***/reiches** rich gas *(gross calorific value 7500 to 8500 kcal/m^3, equalling 31401 to 35588 kJ/m^3)*
Gas *n***/saures** 1. sour gas *(containing acid components as hydrogen sulphide, carbon dioxide, and hydrogen cyanide)*; 2. acid gas *(carbon dioxide, hydrogen sulphide)*
Gas *n***/technisches** industrial gas, *(esp if combustible:)* manufactured gas
Gas *n***/verflüssigtes** liquefied gas
Gasabscheidung *f* 1. gas separation; 2. *s.* Gasentwicklung
Gasanalyse *f* gas analysis
Gasanalysenapparat *m* **nach Orsat** Orsat (gas analysis) apparatus, Orsat analyser
gasanalytisch gas-analytical
Gasanreicherung *f* gas enrichment
gasartig gaseous
Gasaufbereitung *f* gas processing [treating, treatment]
Gasaufnahme *f* gas absorption [uptake]
Gasaushauchung *f* outgassing *(of a volcano)*
Gasaustauschkoeffizient *m (hyd)* gas transfer coefficient
Gasaustritt *m* 1. escape of gas; 2. gas outlet
Gasbehälter *m* gas tank [container], *(esp for town gas:)* gasholder, gasometer
Gasbeton *m* gas [gassy] concrete
Gasbläschen *n* gas bubble
Gasbrenner *m* gas burner
Gasbürette *f* gas(-measuring) burette
gascarbonitrieren *v* gas-cyanide, carbonitride, dry-cyanide, dry-nitride *(steel)*
Gaschromatograph *m* gas chromatograph
Gaschromatographie *f* gas chromatography, GC
Gaschromatographie-Massenspektrometrie-Kopplung *f* gas chromatography-mass spectrometry coupling, GC-MS
gaschromatographisch by gas chromatography; gas-chromatographic *(technique, system)*
Gascoulometer *n* gas coulometer
gasdicht gastight, gasproof
Gasdichtewaage *f* gas (density) balance
Gasdiffusionsverfahren *n* gaseous diffusion process *(for separating isotopes)*
Gasdruck *m* gas pressure [drive]
Gasdurchflussmesser *m* gas flowmeter
Gasdurchflusszählrohr *n (nucl)* gas-

flow counter tube, gas-flow radiation counter, flow counter
gasdurchlässig permeable to gas, gas-permeable
Gasdurchlässigkeit *f* permeability to gas, gas permeability
Gase *npl***/nitrose** nitrous gases
Gaseinleitungsrohr *n* gas-entry tube
Gaselektrode *f* gas electrode
Gaselement *n (el ch)* gas cell
gasen *v* 1. gas; 2. steam *(in producing water gas)*
Gasentladungsröhre *f* (gas) discharge tube
Gasentschwefelung *f* gas desulphurization
Gasentwickler *m s.* Gasentwicklungsapparat
Gasentwicklung *f* generation [evolution] of gas, gassing
Gasentwicklungsapparat *m* gas generator
Gasentwicklungsapparat *m***/kippscher** Kipp (gas) generator, Kipp's apparatus
Gaserzeuger *m (tech)* gas producer [generator]
Gaserzeugung *f* gas manufacture [making, production]
Gas-fest-Chromatographie *f* gas-solid chromatography, GSC, gas adsorption chromatography
Gas-Feststoff-Reaktion *f* gas-solid (heterogeneous) reaction
Gasfilter *n* 1. gas filter; 2. *(lab)* gas filtering [filtration] tube; 3. chemical filtering element *(respiratory protection)*
Gasfiltration *f* gas filtration
Gasflammkohle *f* gas flame coal
Gasflasche *f* gas cylinder
Gas-flüssig-Chromatographie *f* gas-liquid (partition) chromatography, GLC
Gas-flüssig-fest-Chromatographie *f* gas-liquid-solid chromatography, GLSC
Gas-flüssig-Reaktion *f* gas-liquid reaction
Gas-flüssig-Reaktor *m* gas-liquid reactor
gasförmig gaseous
Gasgemisch *n* gas mixture
Gasgenerator *m* gas producer [generator]
Gasgewinnung *f* **aus Mülldeponien** landfill gas extraction
Gasgleichung *f* gas equation
Gasgleichung *f***/ideale** *s.* Zustandsgleichung idealer Gase
Gasheber *m* gas lift
Gashydrat *m* gas hydrate *(any of a group of clathrate compounds)*
Gaskalorimeter *n* gas calorimeter
Gaskohle *f* gas(-making) coal
Gaskonstante *f* gas(-law) constant
Gaskonstante *f***/allgemeine [universelle]** (general) gas constant
Gasküvette *f (spectr)* gas cell
Gaslaser *m* gas laser
Gasleitung *f* gas main
Gasliftförderung *f* gas lift(ing), gas-lift transport
Gasmaske *f* gas mask
Gasmesser *m* gas meter
Gasöl *n (petrol)* gas oil
Gasöl-Fermentation *f (biot)* gas-oil fermentation [process]
Gasometer *m* 1. *(lab)* gasometer; 2. *s.* Gasbehälter
Gasphase *f* gas phase; vapour phase *(in high-pressure hydrogenation)*
Gasphasekracken *n* vapour-phase cracking
Gasphasenabscheidung *f***/metallorganische** metal-organic chemical vapour deposition, MOCVD
Gasphasenabscheidung *f***/physikalische** *(tech)* physical vapour deposition, PVD *(vacuum based coating process)*
Gasphasenoxidation *f* vapour-phase oxidation
Gasphasepolymerisation *f* vapour-phase polymerization, gaseous [gas-phase] polymerization
Gasphasereaktion *f* gas-phase reaction
Gaspipette *f* gas pipette
Gasreaktion *f* gas(eous) reaction
Gasreinigung *f* gas cleaning [clarification] *(esp removing solid and liquid components)*; gas purification *(esp removing gaseous components)*
Gasruß *m* gas black [soot]
Gasschieber *m* gas valve
Gas(schmelz)schweißen *n* gas [autogenous] welding
Gasspürgerät *n* gas detector

Gasstrippen *n (hyd)* gas stripping
Gasstrom *m* gas stream [flow]
Gasstrom *m***/(ungleich) geteilter** *(chromat)* split flow
Gasstromteiler *m (chromat)* (sample) splitter
Gasströmung *f* gas flow
Gast *m* guest *(within a host molecular entity)*
Gastatom *n* guest atom
Gastelement *n* guest element
Gastheorie *f***/kinetische** kinetic theory of gases
Gasthermometer *n* gas thermometer
Gastmolekül *n* guest [enclosed] molecule
Gasübergangskoeffizient *m (hyd)* gas transfer coefficient
gasundurchlässig impervious [impermeable] to gas, gastight, gasproof
Gasventil *n* gas valve
Gasverflüssigung *f* gas liquefaction
Gasvergiftung *f* gas poisoning, gassing
Gaswaage *f* gas (density) balance, dasymeter
Gaswäsche *f* 1. gas washing [scrubbing], wet gas cleaning; 2. gas-washing system
Gaswäscher *m* gas washer, scrubber
Gaswaschflasche *f (lab)* gas-washing bottle
Gaswaschturm *m* gas-washing tower, scrubbing tower, tower scrubber
Gaszähler *m* gas meter
Gaszelle *f (el ch)* gas cell
Gaszentrifugenverfahren *n* gas centrifuge process *(for separating isotopes)*
Gaszuführungsrohr *n* gas feed pipe [tube], gas inlet pipe, gas conductor
Gaszustand *m* gaseous state
Gatsch *m* (paraffin) slack wax
Gattermann-Koch-Synthese *f* Gattermann-Koch reaction *(for preparing phenolic aldehydes)*
Gattungsname *m* generic name [term]
gauche-Konformation *f* gauche conformation, G *(torsion angle ± 120°)*
gaußförmig *(anal)* Gaussian
GAW *s.* Gesamtanionenwert
Gay-Lussac-Gesetz *n* Gay-Lussac law
Gaze *f* gauze
GDP *s.* Guanosindiphosphat
Gebiet *n* region, range *(as of a diagram or scale)*
Gebilde *n***/molekulares** molecular entity
Gebirge *n (mine)* ground
Gebläse *n* fan, (air) blower, air-blast system
Gebläsebrenner *m* blowlamp, blowtorch, bench blowpipe, blast [nozzle-mix] burner, blast [glass blower's] lamp
Gebläselampe *f* blowlamp, blowtorch, bench blowpipe, blast [nozzle-mix] burner, blast [glass blower's] lamp
Gebläse(schacht)ofen *m* blast furnace *(for obtaining iron)*
gebrauchen *v***/übermäßig** overuse *(e.g. pesticides)*
Gebrauchsdauer *f* (liquid) pot life *(of reaction coatings)*
Gebrauchsfähigkeitsdauer *f* shelf [storage] life
Gebrauchswertdauer *f* working [operating, useful, service] life
gebraucht spent *(e.g. solution)*
gebunden combined; bonded, bound, linked, connected *(chemical-bond theory)* • **in gebundener Form** in combined form, in the combined state
gebunden/chemisch chemically bonded
gebunden/doppelt doubly bonded
gebunden/dreifach triply bonded
gebunden/einfach singly bonded
gebunden/koordinativ coordinate, complexed
gebunden/kovalent covalently bonded
gebunden/vierfach quadruply bonded
gebunden/vierfach koordinativ four-coordinate(d)
gediegen *(min, met)* virgin, elemental, native, massive
geeicht/auf Ablauf [Auslauf] calibrated to deliver, calibrated for delivery *(by an authority)*
geeicht/auf Einguss calibrated to contain, calibrated for content *(by an authority)*
Gefahrenklasse *f* danger class
gefährlich dangerous, harmful
Gefährlichkeit *f* dangerousness, harmfulness
Gefahrstoff *m* hazardous substance
Gefälle *n* drop, slope; *(quantitatively)*

gradient • **mit Gefälle** slanted *(e.g. pipe)*
gefällt/frisch freshly precipitated
gefärbt coloured *(naturally)*; *(by human agent:)* dyed, *(Am)* colored
gefärbt/deutlich appreciably coloured
gefärbt/schwach feebly [weakly] coloured
Gefäß *n* vessel, receptacle, container, *(made of glass or earthenware also)* jar, *(if spherical also)* bulb, *(if large also)* vat
gefäßerweiternd vasodilating
Gefäßerweiterungsmittel *n (pharm)* vasodilator, vasodepressor
gefäßverengend vasoconstrictive
Gefäßverengungsmittel *n (pharm)* vasoconstrictor
gefrieren *v* freeze, congeal
gefrieren lassen *v* freeze, congeal
gefriergetrocknet freeze-dried
Gefrierkurve *f* freezing(-point) curve
Gefrierpunkt *m* freezing point [temperature], frost point
Gefrierpunktsdepression *f s.* Gefrierpunktserniedrigung
Gefrierpunktserniedrigung *f* freezing-point depression [lowering]
Gefrierpunktserniedrigung *f***/molale** [**molare**, **molekulare**] molal freezing-point(-depression) constant, molal [molar, molecular] depression constant, cryoscopic constant
Gefrierschutzmittel *n* antifreeze (agent)
Gefrier-Tau(-Prozess) *m* freeze-thaw
gefriertrocknen *v* 1. freeze-dry, lyophilize; 2. *(food)* dehydrofreeze
Gefriertrocknung *f* 1. freeze drying, lyophilization; 2. *(food)* dehydrofreezing
Gefüge *n* 1. *(met)* (grain) structure, grain; 2. *(geol)* rock fabric, *(relating to the larger features:)* structure, *(relating to the smaller features:)* texture
Gefüge *n***/bainitisches** *s.* Gefüge der Zwischenstufe
Gefüge *n* **der oberen Zwischenstufe** *(met)* upper bainite
Gefüge *n* **der unteren Zwischenstufe** *(met)* lower bainite
Gefüge *n* **der Zwischenstufe** *(met)* bainite (structure), bainitic structure
Gefügebestandteil *m* 1. structural constituent; 2. *(coal)* maceral
gefügelos *(met, geol)* structureless, devoid of structure, *(relating to the smaller features:)* textureless, devoid of texture
gefüllt/mit Füllstoffen 1. *(pap, tann)* filled, loaded; 2. *(rubber)* loaded, pigmented
Gegendruck *m* counterpressure, back pressure
Gegenelektrode *f* counterelectrode
Gegen-EMK *f* counter [back] electromotive force, counter [back] e.m.f.
Gegengift *n s.* Gegenmittel
Gegenion *n* counterion, gegenion
gegenläufig counterrotating, contrarotating
Gegenmasse *f* counterweight, counterpoise
Gegenmaßnahme *f* remedial measure
Gegenmittel *n (tox)* antidote
Gegenreaktion *f* opposing reaction, opposed [oppositely directed] reaction
Gegensinn-RNA *f (bioch)* antisense RNA
Gegenstrom *m* countercurrent (flow), counterflow • **im Gegenstrom** countercurrent(ly), in countercurrent, in counterflow
Gegenstromchromatographie *f* countercurrent chromatography
Gegenstromdestillation *f* rectification
Gegenstromdestillation *f***/diskontinuierliche** batch rectification
Gegenstromdestillation *f***/kontinuierliche [stetige]** continuous rectification
Gegenstromextraktion *f* countercurrent extraction [separation]
Gegenstromfahrweise *f* countercurrent operation
Gegenstromklassieren *n* countercurrent [hydraulic] classification
Gegenstromprinzip *n* countercurrent principle • **nach dem Gegenstromprinzip** countercurrent(ly), in countercurrent, in counterflow
Gegenstromverteilung *f* countercurrent distribution
Gegenstromwärmeaustauscher *m* countercurrent heat exchanger
Gegenuhrzeigersinn/im anticlockwise
Gehalt *m* content, concentration; load(ing) *(of a fluid with undesired matter)* • **auf Gehalt prüfen** *(pharm, met)* assay
Gehalt *m* **an gelösten Stoffen [Was-**

serinhaltsstoffen] *(hyd)* total dissolved solids (level), TDS, dissolved solids (concentration, content, level), dissolved-solute load
Gehalt *m* **an organisch gebundenem Kohlenstoff** *(hyd)* TOC [total organic carbon] content
Gehalt *m* **an suspendierten Feststoffen** *(hyd)* suspended solids content [concentration, level]
Gehalt *m* **an Trockenmasse [Trockensubstanz]** solid(s) content
Gehalt *m***/prozentualer** percentage
Gehäuse *n* casing, case, *(esp of pumps, motors, bearings)* housing, *(if rectangular also)* box, *(if domed, hemispherical or spherical also)* shell
gehen *v***/in Lösung** go into solution, dissolve
gehen lassen *v* raise, leaven *(dough)*
gehen *v***/zugrunde** *(nucl)* decay, die
gehindert/sterisch sterically hindered
Gehölzvernichtungsmittel *n* brushkiller, silvicide
Geiger-Müller-Zählrohr *n (nucl)* Geiger(-Müller) counter, Geiger(-Müller) tube, G-M counter [tube]
Geisterpeak *m (anal)* ghost [spurious] peak
gekörnt granular, granulate(d), grained, grainy
Gekrätz *n (met)* sweepings, sweeps
Gel *n (coll)* gel
Gel *n***/thixotropes** thixotrope
geladen/dreifach triply charged
geladen/dreifach negativ trinegative
geladen/dreifach positiv tripositive
geladen/einfach singly charged
geladen/einfach negativ uninegative
geladen/einfach positiv unipositive
geladen/entgegengesetzt oppositely charged
geladen/fünffach quintuply charged
geladen/fünffach negativ pentanegative, quinque-negative
geladen/fünffach positiv pentapositive, quinque-positive
geladen/gleichsinnig identically charged
geladen/negativ negatively charged, negative
geladen/positiv positively charged, positive
geladen/vierfach quadruply charged
geladen/vierfach negativ tetranegative
geladen/vierfach positiv tetrapositive
geladen/zweifach doubly charged
geladen/zweifach negativ dinegative
geladen/zweifach positiv dipositive
gelartig *(coll)* gel-like, gelatinous
Gelatine *f* gelatin(e)
gelatineartig gelatinous, gelatiniform
Gelatinesol *n* gelatin sol
gelatinieren *v (food)* gelatinize, gelatinate, jellify, jelly
Gelatiniermittel *n* gelatinizing [gelling] agent, gelatinizer
Gelatinierung *f* gelatin(iz)ation, gelatification, gelation, jellification
Gelatinierungstemperatur *f* gel point, gelation temperature
gelatinös gelatinous, gelatiniform
Gelbbrenne *f s.* Ätzbad
Gelbeisenerz *n (min)* jarosite
Gelbglut *f* yellow heat
Gelbildung *f* gel formation, gelling, gelation, jellification
Gelchromatographie *f* gel (permeation) chromatography, GPC
Gelee *n* jelly • **zu Gelee erstarren** gelatinize, gelatinate, *(Am)* jellify
Gelée royale *f (pharm)* royal jelly, queen-bee's nutrient jelly
Geleffekt *m (plast)* gel effect, Trommsdorff effect
Gelelektrophorese *f (anal)* gel electrophoresis
Gelfiltration *f* gel filtration, molecular exclusion [sieve] chromatography
gelieren *v (coll, food)* gel(ate), jelly
Geliermittel *n* gelatinizing [gelling] agent, gelatinizer
Gelierstoff *m* gelatinizing [gelling] agent, gelatinizer
Gelierung *f (coll, food)* gelation, jellification
Gelierungstemperatur *f* gel point, gelation temperature
gelinde gentle, slight *(e.g. heating)*
Gelose *f* gelose, *(broadly)* agar (gel), agar-agar, Japan agar [isinglass], Chinese gelatin
Gelöste *n (anal)* dissolved solids, DS, solute (material)
Gelöstsauerstoff *m* dissolved oxygen, DO

Gelpermeationschromatographie *f* gel (permeation) chromatography, GPC
Gelpunkt *m* gel point *(polycondensation)*
Gelseminsäure *f* gelseminic acid, scopoletin, 7-hydroxy-6-methoxycoumarin
Gel-Sol-Übergang *m* gel-sol transition
Gel-Sol-Umwandlung *f* gel-sol transformation, peptization, solation
Gelzustand *m (coll)* gel state [condition] • **in den Gelzustand übergehen** gel
gem- *s.* geminal
Gemenge *n* 1. solids mixture, bulk blend; 2. *(glass)* batch
geminal geminal, gem- *(two identical substituents on the same carbon atom)*
Gemisch *n* mixture, mix
Gemisch *n***/azeotropes** azeotropic mixture, azeotrope, constant-boiling-point mixture
Gemisch *n***/binäres** binary mixture
Gemisch *n***/dystektisches** dystectic mixture *(a mixture having a maximum melting point)*
Gemisch *n***/eutektisches** eutectic (mixture)
Gemisch *n***/inniges** intimate mixture, blend
Gemisch *n***/racemisches** racemic mixture
Gen *n* gene, (DNA) cistron
Genauigkeit *f* 1. accuracy; 2. *(anal)* precision
Generator *m* producer, generator *(for manufacturing gas or steam)*
Generatorgas *n* producer gas
Generatorgasgleichgewicht *n* producer-gas equilibrium
generieren *v* generate *(e.g. reactive species)*
Generierung *f* generation *(as of reactive species)*
Genklonierung *f (biot)* gene cloning
Genmanipulation *f (biot)* gene [genetic] manipulation
Gentechnik *f (biot)* genetic technique
Gentechnologie *f (biot)* gene [genetic] technology, genetic engineering
Gentisinsäure *f* gentisic acid, 2,5-dihydroxybenzoic acid
Gentransfer *m (biot)* gene transfer
Genübertragung *f (biot)* gene transfer
Genusssäure *f* culinary [food] acid, edible (organic) acid
Geochemie *f* geochemistry
Geochemiker *m* geochemist
geochemisch geochemical
geordnet/sterisch stereoregular
geordnet/wohl well-ordered *(structure)*
Geosphäre *f* geosphere
gepackt/dicht *(cryst)* close-packed
gepackt/hexagonal dicht hexagonal close-packed, hcp
gepackt/kubisch dicht cubic close-packed, ccp
gepulvert/fein finely powdered
gequantelt quantized
geradkettig straight-chain, unbranched-chain
geradzahlig even-number(ed) *(e.g. fatty acid)*
Geranial *n* geranial, citral a, *trans*-3,7-dimethylocta-2,6-dienal
Geraniol *n* geraniol *(either of two dimethyloctadienol isomers)*
Gerät *n* 1. *(ch)* apparatus, *(if no specific term otherwise implicit:)* device; instrument *(esp for measuring)*; 2. *(collectively:)* apparatus(es), *(Am)* equipment
Geräte *npl***/wissenschaftliche** scientific apparatus, research equipment
Gerätefehler *m* instrumental error
Gerätekonstante *f* apparatus constant
gerben *v* tan
gerben *v***/mit Chromsalzen** chrome
Gerberei *f* 1. tannage, tannery, tanning; 2. tanning plant, tanyard, tannery
Gerbereichemie *f* chemistry of tanning [leather manufacture]
Gerbereichemiker *m* leather chemist
Gerbgrube *f* tanning [tan, suspender] pit, tannin vat
Gerbholz *n* tanwood
Gerbmittel *n* tanning material, tan
Gerböl *n* tanning oil
Gerbrinde *f* tan [tanner's, tanning] bark
Gerbstoff *m* 1. tanning agent, *(if of vegetable origin also)* tannin; 2. *s.* Gerbmittel
Gerbstoff *m***/pflanzlicher** (vegetable) tannin
Gerbstoff *m***/synthetischer** syntan, synthetic tannin [tanning agent]

Gerbstoffpflanze *f* tanniferous [tanning] plant
Gerbstoffrot *n* tanner's red, phlobaphene
Gerbung *f* tannage, tanning, tannery
Gerbung *f*/**pflanzliche** vegetable tannage
Gerbung *f*/**synthetische** syntan tannage
Gerichtschemie *f* forensic [legal] chemistry
Gerichtschemiker *m* forensic [legal] chemist
geriffelt ribbed, grooved
gerillt ribbed, grooved
geringwertig low-grade, poor
gerinnbar coagulable, congealable
Gerinnbarkeit *f* coagulability
gerinnen *v* coagulate, congeal, curd(le), clot, *(relating to milk also)* sour
gerinnen lassen *v* coagulate, congeal, curd(le)
Gerinnen *n* coagulation, congelation, curdling, clotting, *(relating to milk also)* souring
Gerinnsel *n* clot, curd
gerinnungshemmend anticoagulant
Gerinnungspunkt *m* coagulation [curdling] point, *(milk also)* setting point
gerippt ribbed
German *n* germane, *(specif)* GeH_4 monogermane, germane, germanium tetrahydride
Germanat *n* germanate *(salt or ester of germanic acid)*
Germanium *n* Ge germanium
Germanium(II)-chlorid *n* germanium(II) chloride, germanium dichloride, germanous chloride
Germanium(IV)-chlorid *n* germanium(IV) chloride, germanium tetrachloride, germanic chloride
Germaniumdioxid *n s.* Germanium(IV)-oxid
Germanium(IV)-oxid germanium(IV) oxide, germanium dioxide, germanic oxide
Germaniumtetrahydrid *n* germanium tetrahydride, monogermane, germane
Germizid *n* germicide
Gerstenmalz *n* barley malt, malted barley
Gerstenstärke *f* barley starch
Geruch *m* smell, *(esp if unpleasant:)* odour, *(if pleasant:)* scent, *(Am)* perfume, *(if pleasant but penetrating:)* aroma
Geruch *m*/**angenehmer** scent, pleasant smell
Geruch *m*/**durchdringender** penetrating odour
Geruch *m*/**frucht(art)iger** fruity odour [fragrance]
Geruch *m* **nach faulen Eiern** rotten-egg odour
Geruch *m*/**ranziger** rancid odour
Geruch *m*/**stechender** pungent smell
Geruch *m*/**übler** bad smell, stench, stink
Geruch *m*/**unangenehmer** (unpleasant) odour
geruchlos inodorous, odourless, odour-free • **geruchlos machen** deodorize
Geruchlosigkeit *f* inodorousness
Geruchsbelästigung *f* nasal nuisance
geruchsbeseitigend deodorant
Geruchsbeseitigung *f* 1. deodorization; 2. *(hyd)* odour removal
Geruchsbildner *m* odour-causing [odour-producing] substance
Geruchsentfernung *f* 1. deodorization; 2. *(hyd)* odour removal
Geruchsschwelle *f s.* Geruchsschwellenkonzentration
Geruchsschwellenkonzentration *f* threshold-odour concentration, threshold of odour, odour [nuisance] threshold
Geruchsstoff *m* 1. odorous substance, odourcausing, odour-producing (odour-bearing) substance; 2. *(food)* flavouring matter; 3. *(cosmet)* perfume
Geruchsverschluss *m* water seal, (siphon) trap
Gerüst *n* 1. skeleton *(as of a molecule)*; 2. *(tech)* scaffold, framework
Gerüst *n*/**metallorganisches** metal-organic framework, MOF
Gerüsteiweiß *n* structural [skeletal, fibrous] protein, scleroprotein
Gerüsteiweißstoff *m* structural [skeletal, fibrous] protein, scleroprotein
Gerüstpolysaccharid *n* skeletal polysaccharide
Gerüstschwingung *f* skeletal vibration
Gerüstsilicat *n (min)* tectosilicate
Gerüststoff *m* 1. *(bioch)* skeletal substance; 2. (detergency, soap) builder *(a*

substance added to synthetic detergents)
Gerüstsubstanz *f* 1. *(bioch)* skeletal substance; 2. (detergency, soap) builder *(a substance added to synthetic detergents)*
Gerüstumlagerung *f* skeletal rearrangement
Gesamtablauf *m* overall course *(of a reaction)*
Gesamtanionenwert *m (hyd)* total anion content, total anion(s)
Gesamtausbeute *f* total [overall] yield
Gesamtdrehimpulsquantenzahl *f* total angular momentum quantum number, inner quantum number
Gesamtdruck *m* total pressure
Gesamtdurchsatz *m* total throughput
Gesamteigendrehimpuls *m* total electron spin angular momentum
Gesamtfeststoffgehalt *m (hyd)* total solids, T.S.
Gesamtfeststoffmenge *f (hyd)* total solids, T.S.
Gesamtgehalt *m* **an organisch gebundenem Kohlenstoff** *(hyd)* total organic carbon, TOC
Gesamtgehalt *m* **an organischen Inhaltsstoffen** *(hyd)* overall organic substances
Gesamtgehalt *m* **an Organohalogenverbindungen** *(hyd)* total organic halides, TOX
Gesamtgehalt *m* **an suspendierten Feststoffen** *(hyd)* total suspended solids, total solids in suspension
Gesamtgehalt *m* **an Trihalo(gen)methanen** *(hyd)* total trihalomethanes, TTHM level, TTHMs
Gesamtgeschwindigkeit *f* overall rate [velocity]
Gesamthärte *f (hyd)* total (water) hardness
Gesamtionengehalt *m (hyd)* total ionic content
Gesamtkationenwert *m (hyd)* total cation content, total cation(s)
Gesamtkohlenstoff *m (hyd)* total carbon, TC
Gesamtkonzentration *f* total concentration
Gesamtlänge *f* contour [extended] length *(of a molecular chain)*
Gesamtlängenverhältnis *n (text)* stretch ratio
Gesamtleitfähigkeit *f* total [resultant] conductance
Gesamtleitvermögen *n* total [resultant] conductance
Gesamtlösung *f* bulk solution
Gesamtmenge *f (anal)* entire [parent] lot, entire mass
Gesamtreaktion *f* overall [total] reaction
Gesamtreaktionsgeschwindigkeit *f* overall [total] rate of reaction
Gesamtreaktionsordnung *f* overall [total] order of reaction
Gesamtsalzgehalt *m (hyd)* total mineral level, total salts
Gesamtsauerstoffbedarf *m (hyd)* total oxygen demand, TOD
Gesamtschwebstoffe *mpl (hyd)* total suspended solids (level), total solids in suspension
Gesamtschwebstoffgehalt *m (hyd)* total suspended solids (level), total solids in suspension
Gesamtspindrehimpuls *m* total electron spin angular momentum, resultant spin angular momentum
Gesamttrockenmasse *f* 1. *(hyd)* total dry solids; 2. *(food)* total solids, T.S.
Gesamttrockensubstanz *f* 1. *(hyd)* total dry solids; 2. *(food)* total solids, T.S.
Gesamtumsatz *m* overall conversion
Gesamtwirkungsgrad *m* overall efficiency *(as of a pump)*
gesättigt saturated *(solution or compound)*
geschichtet lamellar
Geschlechtshormon *n* sex hormone
geschlossenkettig closed-chain
geschlossenzellig closed-cell *(foamed plastic)*
Geschmack *m* taste, *(if pleasant:)* flavour • **von angenehmem Geschmack** tasteful • **von unangenehmem Geschmack** distasteful
Geschmack *m*/**altöliger** oiliness *(as of spoiled fats)*
Geschmack *m*/**fischiger** fishy taste, fishiness
Geschmack *m*/**ranziger** rancid taste
Geschmack *m*/**talgiger** tallowy taste
geschmacklos 1. *(ch)* tasteless, free from taste; 2. *(food)* tasteless, insipid

geschmacksbeeinträchtigend impairing the flavour [taste]
Geschmacksfehler *m* flavour defect, off-flavour
Geschmacksschwelle(nkonzentration) *f (hyd)* taste threshold, threshold of taste
Geschmacksstoff *m* 1. *(food)* flavouring matter [substance]; 2. *(hyd)* taste-causing [taste-bearing] substance
geschmacksverändernd taste-modifying
Geschmacksverstärker *m (food)* flavour enhancer
Geschwindigkeit *f* velocity, speed *(as of particles)*; rate *(as of reactions)*
Geschwindigkeit *f* **der Hinreaktion** forward reaction rate, forward rate (of reaction)
Geschwindigkeit *f* **der Rückreaktion** reverse reaction rate, reverse rate (of reaction), back rate
Geschwindigkeit *f***/mittlere quadratische** root-mean-square speed [velocity], RMS speed
geschwindigkeitsbestimmend rate-limiting, rate-controlling, rate-determining
Geschwindigkeitsgleichung *f* rate(-of-reaction) equation, empirical differential rate equation, rate law
Geschwindigkeitsgleichung *f* **erster Ordnung** first-order rate equation [law]
Geschwindigkeitsgleichung *f* **gebrochener Ordnung** fractional-order rate equation [law]
Geschwindigkeitsgleichung *f* **pseudoerster Ordnung** pseudo-first-order rate equation [law]
Geschwindigkeitsgleichung *f* **zweiter Ordnung** second-order rate equation [law]
Geschwindigkeitskartographie *f (phys ch)* velocity mapping
Geschwindigkeitskonstante *f* rate coefficient, *(relating to elementary reactions:)* (reaction-)rate constant, unitary rate constant
Geschwindigkeitskonstante *f* **der bimolekularen Reaktion** bimolecular rate constant
Geschwindigkeitskonstante *f* **der Hinreaktion** forward rate constant
Geschwindigkeitskonstante *f* **der Rückreaktion** reverse rate constant
Geschwindigkeitskonstante *f* **erster Ordnung** first-order rate constant
Geschwindigkeitskonstante *f* **zweiter Ordnung** second-order rate constant
Geschwindigkeitsverteilung *f* distribution of velocities [speeds], velocity distribution
Gesenk *n* 1. *(plast)* female form [mould], force; 2. *(met)* die
Gesetz *n* law, principle • **ein Gesetz befolgen** obey a law • **einem Gesetz folgen [gehorchen]** obey a law
Gesetz *n***/avogadrosches** Avogadro law [hypothesis] *(of the number of molecules in gases)*
Gesetz *n***/bouguer-lambertsches** Bouguer-Lambert law (of absorption), Lambert's absorption law
Gesetz *n***/boyle-mariottesches** law of Boyle-Mariotte, Boyle [Mariotte] law *(of gas pressure and volume)*
Gesetz *n***/daltonsches** Dalton law *(of partial pressures)*
Gesetz *n* **der äquivalenten Proportionen** law of equivalent proportions
Gesetz *n* **der Gleichverteilung der Energie** law of equipartition of energy, equipartition principle
Gesetz *n* **der konstanten Proportionen** law of constant [definite] proportions, Prout law
Gesetz *n* **der konstanten Wärmesummen** *s.* Gesetz/hesssches
Gesetz *n* **der multiplen Proportionen** law of multiple proportions
Gesetz *n* **der rationalen Indizes [Parameterverhältnisse]** *s.* Gesetz/haüysches
Gesetz *n* **der unabhängigen Ionenwanderung** law of independent migration of ions
Gesetz *n***/gay-lussacsches** Gay-Lussac law *(of gas volume and temperature)*
Gesetz *n***/haüysches** *(cryst)* Haüy law, rational index law
Gesetz *n***/hesssches** Hess's law (of heat summation), law of constant heat summation
Gesetz *n***/kirchhoffsches** Kirchhoff's law *(of the temperature dependence of reaction enthalpies)*

Gesetz *n*/**kohlrauschsches** Kohlrausch law *(of independent migration of ions)*
Gesetz *n*/**lambert-beersches** Beer-Lambert(-Bouguer) law, (Bouguer-)Lambert-Beer law *(of light absorption)*
Gesetz *n*/**moseleysches** Moseley law *(of the wave numbers in X-ray spectra)*
Gesetz *n* **von der Erhaltung der Energie** law of conservation of energy, energy principle
Gesetz *n* **von der Erhaltung der Masse** law of conservation of mass [matter]
gesichert/statistisch statistically valid
Gesichtsschutz *m* face protector
gestaffelt staggered, non-eclipsed *(stereochemistry)*
Gestagen *n* progestational hormone
Gestalt *f* 1. shape, form; 2. *(cryst)* morphology
gestaltlos amorphous
Gestaltresonanz *f (phys ch)* shape resonance
Gestaltsänderung *f* deformation, strain
Gestaltsänderung *f*/**bleibende** permanent deformation [set], residual deformation [set], set
Gestank *m* bad smell, stench, stink
Gestein *n* 1. rock; 2. *(mine)* ground
Gestein *n*/**anstehendes** bedrock
Gestein *n*/**taubes** waste rock, gangue, matrix
Gesteinschemie *f* petrochemistry
gesteinschemisch petrochemical
Gesteinsgang *m* vein, dike
Gesteinskunde *f* petrography
Gesteinsmantel *m* lithosphere
Gesteinswolle *f* rock wool
Gestell *n* 1. rack; 2. *(met)* hearth, crucible *(of a blast furnace)*
gestrecktkettig extended-chain
gesundheitsgefährdend hazardous to health
Gesundheitsgefährdung *f* health hazard
gesundheitsschädlich dangerous [injurious] to health, harmful, deleterious
gesüßt *(petrol)* sweet(ened)
geteilt/auf Ablauf [Auslauf] graduated [calibrated] to deliver, graduated for delivery
geteilt/auf Einguss graduated [calibrated] to contain, graduated for content
Getränk *n* beverage, drink, potable
Getränk *n*/**alkoholisches** alcoholic [spirituous] beverage, spirit
Getreidebranntwein *m* grain alcohol
Getreidestärke *f* cereal starch
Getriebeöl *n* gear oil
getrocknet/im Trockenschrank oven-dried, oven-dry, OD
Getter(stoff) *m* getter *(vacuum technology)*
Gewässerbelastung *f* water pollution
Gewässerreinhaltung *f* water pollution control, water protection
Gewässerschutz *m* water pollution control, water protection
Gewässerverschmutzung *f* water pollution
Gewässerverunreinigung *f* water pollution
Gewebe *n* 1. fabric (cloth), cloth, web; 2. *(text)* textile (fabric); 3. (biological) tissue
Gewebe *n*/**beschichtetes** coated fabric
Gewebe *n*/**gummiertes** rubberized fabric, rubbered [rubber-coated, proofed] fabric, proofing
Gewebe *n*/**kaschiertes** coated [combined] fabric
Gewebeentwicklung *f* tissue engineering *(synthetic)*
Gewebefilter *n* fabric filter, woven(-fabric) filter
Gewebekultur *f (biot)* tissue culture
Gewebeprobe *f (bioch)* tissue sample
Gewebezüchtung *f s.* Gewebeentwicklung
Gewerbeschutzsalbe *f* barrier cream
Gewicht *n* 1. weight *(the force by which the mass of a substance is attracted by gravity; unit: N)*; 2. *s.* Gewichtsstück; 3. *s.* Wägestück
Gewicht *n*/**spezifisches** *s.* Wichte
Gewicht *n*/**statistisches** *(phys ch)* statistical weight
Gewichtsanalyse *f* gravimetric analysis, analysis by weight
Gewichtskonstanz *f* constancy of weight, constant weight • **bis zur Gewichtskonstanz glühen** ignite to constant weight
Gewichtsmittel *n* weight average
Gewichtsstück *n* weight *(a heavy object*

of indefinite mass for counterbalancing)
Gewichtsverlust *m* loss in weight, weight loss
Gewichtszunahme *f* gain in weight, weight increase
gewinnen *v* 1. obtain, recover *(a reaction product)*; isolate *(from natural products)*; 2. *(mine)* mine, extract, win
gewinnen *v*/**in reinem Zustand** obtain pure [in pure form], obtain in a pure condition [state]
Gewinnung *f* 1. recovery *(of a reaction product)*; isolation *(from natural products)*; 2. *(mine)* mining, extraction, winning
Gewürz *n* seasoning, *(of vegetable origin:)* spice, *(esp salt and pepper:)* condiment
g-Faktor *m* g-factor, gyromagnetic factor
g-Faktor *m*/**landéscher** Landé g-factor, (Landé) splitting factor g *(gyromagnetic ratio of electrons)*
GFK *s.* Kunststoff/glasfaserverstärkter
GFS *s.* Glasfaserschichtstoff
GGG *s.* Grauguss/globularer
GH *s.* Gesamthärte
Gibberellin *n (bioch)* gibberellin
Gibberellinsäure *f (bioch)* gibberellic acid
Gibbs-Helmholtz-Gleichung *f* Gibbs-Helmholtz equation
Gicht *f* 1. *(met)* throat, top *(part of a blast furnace)*; 2. *(met)* stock, burden, charge(stock), feed
Gichtgas *n* furnace [blast-furnace] gas
Gichthöhe *f (met)* stock level
Gieß... *siehe auch: Guss...*
gießbar 1. capable of being poured, pourable; 2. castable *(molten material)*
Gießbarkeit *f* 1. pourability; 2. castability *(of molten material)*
gießen *v* 1. pour; 2. cast *(molten material into moulds)*; teem *(molten ferrous metal into ingot moulds)*
gießen *v*/**unter Druck** pressure-diecast
Gießen *n* 1. pouring; 2. casting *(of molten material into moulds)*; teeming *(of molten ferrous metal into ingot moulds)*
Gießen *n* **unter Druck** pressure diecasting
Gießerei *f* foundry
Gießerei(roh)eisen *n* foundry (pig) iron
Gießfolie *f (plast)* cast film, *(if thickness > 0.01 inch:)* cast sheet
Gießform *f* (casting) mould
Gießharz *n* cast(ing) resin
Gießling *m* casting
Gift *n* poison, toxicant, toxic (substance)
Gift *n*/**ökonomisches** *s.* Pflanzenschutzmittel
Gift *n*/**pflanzliches** plant poison
Gift *n*/**protektives [protektiv wirkendes]** *(agric)* protective toxicant
Gift *n*/**systemisches** *(agric)* systemic poison
Gift *n*/**tierisches** venom
Gifteinwirkung *f* toxic action
Giftgas *n* poison(ous) gas
giftig poisonous, toxic(al) • **giftig sein für** be toxic to
giftig/äußerst extremely poisonous [toxic]
giftig/schwach mildly poisonous [toxic]
giftig/stark highly [very, quite] poisonous [toxic], dangerously poisonous
Giftigkeit *f* toxicity, poisonousness
Giftigkeit *f*/**akute** acute toxicity
Giftigkeit *f*/**chronische** chronic toxicity
Giftigkeit *f* **für Pflanzen** phytotoxicity
Giftigkeit *f* **für Säugetiere** mammalian toxicity
Giftmüll *m* toxic [hazardous] waste(s)
Giftstoff *m* toxicant, toxic (substance)
Giftwirkung *f* poisoning [poisonous, toxic] action; poisoning effect
Gipfel *m (anal)* mode *(of a distribution)*
Gipfelpunkt *m* peak *(of a curve)*
Gips *m (min)* gypsum *(calcium sulphate-2-water)*
Gips *m*/**gebrannter** calcined [anhydrous] gypsum, gypsum cement, plaster of Paris *(essentially calcium sulphate 0.5-water)*
Gipshärte *f (hyd)* sulphate hardness
Gipsmörtel *m* gypsum mortar, plaster (mortar)
Gipsstein *m (hyd)* calcium sulphate scale
Gitter *n* 1. *(cryst)* lattice; 2. grating *(optics)*; 3. *(tech)* grating, grid; checker chamber, *(Am)* chequer chamber *(of a blast-furnace stove)*
Gitter *n*/**flächenzentriertes** face-centred lattice
Gitter *n*/**geblaztes** *(spectr)* blazed grat-

ing *(grooves shaped in a sawtooth pattern)*
Gitter *n*/**hexagonales** hexagonal lattice
Gitter *n*/**ideales** perfect lattice
Gitter *n*/**innenzentriertes** body-centred lattice
Gitter *n*/**kubisches** cubic lattice
Gitter *n*/**metallisches** metallic lattice
Gitter *n*/**molekulares** molecular lattice
Gitter *n*/**raumzentriertes** *s.* Gitter/innenzentriertes
Gitterabstand *m (cryst)* lattice distance [spacing], spacing of the planes
Gitteraufbau *m (cryst)* lattice structure
Gitterebene *f (cryst)* lattice [atomic, net] plane
Gittereinschlussverbindung *f* lattice inclusion compound
Gitterenergie *f (cryst)* lattice energy
Gitterfehlanpassung *f* lattice mismatch
Gitterfehler *m (cryst)* lattice defect [imperfection, irregularity]
Gitterfehler *m*/**eindimensionaler** line [linear] defect
Gitterfehler *m*/**flächenhafter** surface defect
Gitterfehler *m*/**linienhafter** line(ar) defect
Gitterfehler *m*/**punktförmiger** point defect
Gitterfehlordnung *f* lattice disorder
Gitterfehlstelle *f s.* Gitterleerstelle
Gitterfenster *n* lattice window
Gitterion *n (cryst)* lattice ion
Gitterkonstante *f (cryst)* lattice constant
Gitterleerstelle *f (cryst)* lattice vacancy, hole (position), vacant site, vacant lattice site [position]
Gittermauerwerk *n* chequerwork, chequer brickwork, *(Am)* checkerwork, checker
Gitterplatz *m (cryst)* lattice site [position]
Gitterschwingung *f* 1. lattice vibration; 2. *(spectr)* lattice mode
Gitterspektrograph *m* (diffraction) grating spectrograph
Gitterspektrum *n* grating [diffraction, normal] spectrum
Gitterstelle *f (cryst)* lattice site [position]
Gitterverformung *f* lattice deformation
GKE *s.* Kalomelelektrode/gesättigte
GKW *s.* Gesamtkationenwert
GL *s.* Glasfaserstoff
Glanz *m* lustre, gloss, brightness
Glanz *m*/**metallischer** metallic lustre
Glanz *m*/**mineralischer** glance
Glanzbad *n* bright plating solution [bath]
Glanzbildner *m* brightener, brightening agent *(electroplating)*
Glanzbraunkohle *f* bright brown coal, black [bituminous] lignite, pitch coal
glänzend lustrous, glossy, bright
Glanzkohle *f* 1. bright coal; 2. *s.* Glanzbraunkohle
glanzlos lustreless, dull, mat(t), nonbright, *(of colours also)* dead
Glanzlosigkeit *f* lustrelessness, dullness, mattness
Glanznickel *m (min)* brigth [brilliant] nickel
Glanzschutzlack *m (coat)* clearcoat
Glanzsilber *n* bright silver
Glanzwinkel *m (cryst)* glancing [Bragg] angle
Glas *n* 1. glass *(material)*; *(broadly also)* glassy material; 2. glassware *(equipment)* • **mit Glas ausgekleidet** glass-lined
Glas *n*/**alkaliarmes** low-alkali glass
Glas *n*/**alkalihaltiges** alkali glass
Glas *n*/**farbiges** coloured glass
Glas *n*/**leicht schmelzbares** soft (sealing) glass
Glas *n*/**mundgeblasenes** handblown [hand-made] glass
Glas *n*/**opakes** opaque [opal] glass
Glas *n*/**photochrom(atisch)es** photochromic glass
Glas *n*/**vorgespanntes** prestressed glass
glasartig 1. glassy, glass-like, vitreous; 2. *(ceram)* vitrified, vitreous
Glasartikel *mpl* glassware
Glasauskleidung *f* glass lining • **mit Glasauskleidung** glass-lined
Glasbearbeitung *f (tech)* glass working [manipulation], glass machining
Glasbildner *m* glass-forming substance, glass former
Glasbildungstemperatur *f* glass-transition temperature *(at which the viscosity of the glass is 10^{13} dPa s)*
Glasblasen *n* glass blowing, glass making
Glasbläserlampe *f* glass-blower's lamp

Glas-Büchner-Trichter *m (lab)* slit sieve funnel
Glaselektrode *f* glass (membrane) electrode
Gläserbürste *f (lab)* beaker [jar] brush
Glaserkitt *m* (glazier's) putty, painter's [whiting] putty
gläsern vitreous, glassy
Glasfaden *m* glass filament
Glasfaser *f* glass fibre
Glasfasergarn *n* glass-fibre yarn, staple-fibre glass yarn
Glasfasergewebe *n* glass(-fibre) fabric, woven-glass-fibre cloth, glass cloth
Glasfaserschichtstoff *m* glass-fibre laminate
Glasfaserstoff *m* fibre [fibrous, spun] glass
glasfaserverstärkt glass(-fibre) reinforced
Glasfilter *n* glass filter; sintered-glass [fritted-glass] filter
Glasfilternutsche *f* (all-)glass suction filter
Glasfiltertiegel *m* sintered-glass (filtering) crucible, glass filtering crucible
Glasfläschchen *n* (glass) vial
Glasflasche *f* glass bottle
Glasfritte *f* sintered-glass [fritted-glass] filter, glass frit
Glasgefäß *n* glass vessel
Glasgeräte *npl* glassware
Glasglocke *f* glass bell, bell-jar
Glashafen *m* glass(-melting) pot
glasieren *v (ceram)* glaze
glasig 1. vitreous, glassy, glass-like; 2. *(geol)* vitrophyric, vitreous
Glasindustrie *f* glass industry
Glasisolator *m* glass insulator
Glaskeramik *f* glass ceramic(s), vitroceramic, devitrified [neo-ceramic] glass
Glaskohlenstoff *m* glass-like carbon, glassy [vitreous] carbon
Glaskolben *m* glass flask [bulb]
Glasküvette *f* glass cuvette, cell
Glasmacherpfeife *f* blow pipe, blow(ing) iron
Glasmembran *f* glass membrane *(as of a glass electrode)*
Glasnutsche *f s.* Glasfilternutsche
Glasofen *m* glass(-melting) furnace
Glasphase *f* vitreous [glassy] phase [state]
Glasrohr *n* glass pipe [tube]; *(collectively:)* glass piping [tubing]
Glasschale *f* glass dish
Glasscheibe *f* glass panel, glass plate
Glasschliff *m* glass grinding
Glasschmelzofen *m* glass(-melting) furnace, glass-kiln
Glasschmelzwanne *f* (glass-)melting tank
Glasschneider *m* glass cutter *(tool or worker)*
Glasseide *f* glass silk
Glasseidenspinnfaden *m* continuous glass filament
Glasseidenstrang *m* (glass-fibre) roving, glass-fibre strand
Glasstab *m* glass bar, glass rod, cane
Glasstapelfasergarn *n* staple-fibre glass yarn, glass-fibre yarn
Glasstopfen *m* glass stopper • **mit Glasstopfen** glass-stoppered
Glasstopfen *m***/eingeschliffener** ground(-glass) stopper
Glastemperatur *f s.* Glasumwandlungspunkt
Glastrichter *m* glass funnel
Glasübergang *m s.* Glasumwandlung
Glasumwandlung *f* 1. glass transition; 2. *(rubber)* gamma [second-order] transition, vitrification
Glasumwandlungspunkt *m* 1. glass(-transition) temperature, Tg (point); 2. *(rubber)* second-order transition temperature [point]
Glasumwandlungstemperatur *f* 1. glass(-transition) temperature, Tg (point); 2. *(rubber)* second-order transition temperature [point]
Glasur *f (ceram)* glaze
Glasurmasse *f* glass frit *(glazing)*
Glasversicherung *f* glass insurance
Glaswaren *fpl* glassware
Glaswolle *f* glass wool
Glaszustand *m* vitreous [glassy, glass-like] state
Glaszylinder *m* jar
glatt 1. smooth *(surface)*; 2. plain *(shape)*; 3. smooth *(e.g. course of reaction)*
Glätte *f* smoothness *(of a surface)*
glätten *v* smooth, polish

Glaubersalz *n* 1. Glauber salt; 2. *(min)* mirabilite *(sodium sulphate 10-water)*

GLC *(Gas-Liquidus-Chromatographie) s.* Gas-flüssig-Chromatographie

gleichgestaltig *(cryst)* isomorphic, isomorphous

Gleichgestaltigkeit *f (cryst)* isomorphism

Gleichgewicht *n* equilibrium, balance • **das Gleichgewicht wiederherstellen** re-establish equilibrium • **dem Gleichgewicht zustreben** go toward equilibrium • **Gleichgewicht herstellen** establish equilibrium • **im Gleichgewicht halten** keep in equilibrium, equilibrate • **im Gleichgewicht sein** be at [in] equilibrium • **ins Gleichgewicht bringen [setzen]** equilibrate, bring into equilibrium

Gleichgewicht *n*/**annäherndes** near-equilibrium

Gleichgewicht *n*/**biologisches** biological [biotic] equilibrium, biotic balance

Gleichgewicht *n*/**boudouardsches** producer-gas equilibrium

Gleichgewicht *n*/**dynamisches** dynamic equilibrium, steady state

Gleichgewicht *n*/**heterogenes** heterogeneous equilibrium

Gleichgewicht *n*/**indifferentes** neutral equilibrium

Gleichgewicht *n*/**thermisches** thermal equilibrium

Gleichgewicht *n*/**thermodynamisches** thermodynamic equilibrium

Gleichgewicht *n*/**vorgelagertes** pre-equilibrium, prior equilibrium

Gleichgewichtsdestillation *f* equilibrium distillation

Gleichgewichtskonstante *f* equilibrium constant

Gleichgewichtskonstante *f*/**thermodynamische** thermodynamic equilibrium constant

Gleichgewichtskonzentration *f* 1. equilibrium concentration; 2. *(bioch)* steady-state concentration

Gleichgewichtslage *f* position of equilibrium

Gleichgewichtslinie *f* tie line, conode *(in the equilibrium diagram)*

Gleichgewichtspotenzial *n* equilibrium [steady-state] potential

Gleichgewichtsreaktion *f* balanced reaction

Gleichgewichtssiedekurve *f* (single-) flash curve

Gleichgewichtsverteilung *f* equilibrium distribution, steady-state distribution, SSD

Gleichgewichtswert *m* equilibrium value

Gleichgewichtszustand *m* equilibrium state

gleichmäßig 1. uniform, even *(also of a process)*; smooth *(e.g. course of a reaction)*; 2. *(dye)* level

Gleichmäßigkeit *f* 1. uniformness, evenness *(also of a process)*; smoothness *(as of a reaction)*; 2. *(dye)* levelness

Gleichstrom *m* 1. cocurrent [concurrent, parallel] flow *(of two fluids)*; 2. direct current, d.c., D.C. *(of electricity)* • **im Gleichstrom (geführt)** cocurrent *(fluids)*

Gleichstromdestillation *f* simple [direct] distillation

Gleichstromdestillation *f*/**einfache** simple [direct] distillation

Gleichstromfahrweise *f* cocurrent operation

Gleichstrompolarographie *f* d.c. polarography

gleichstrompolarographisch by d.c. polarography

Gleichung *f* equation

Gleichung *f*/**arrheniussche** Arrhenius equation

Gleichung *f*/**chemische** chemical(-reaction) equation, reaction equation

Gleichung *f*/**clausius-clapeyronsche** Clapeyron-Clausius equation

Gleichung *f*/**stöchiometrische** stoichiometric [balanced] equation

Gleichung *f*/**van-der-waalssche** van der Waals equation (of state)

Gleichung *f*/**van't-hoffsche** van't Hoff equation

Gleichung *f* **von Brunauer, Emmett und Teller** Brunauer-Emmett-Teller equation [relationship], BET equation

Gleichverteilungssatz *m* **(der Energie)** equipartition principle [theorem], law of equipartition (of energy)

gleichzeitig simultaneous, coincident

Gleichzeitigkeit *f* simultaneity, simultaneousness, coincidence
Gleitebene *f (cryst)* slip(ping) plane, glide [gliding] plane
gleiten *v* slide, slip
Gleiten *n* slide, slip
Gleitmittel *n* 1. lubricant, lubricating agent, lube, slip additive [agent]; 2. *(pharm)* intestinal lubricant; 3. *(plast)* external lubricant; mould lubricant *(foundry)*
Gleitmodul *m* shear modulus (of elasticity), modulus of rigidity [elasticity in shear]
Gleitung *f* crystal [translation] gliding *(along a crystal plane)*
Gliederbandförderer *m* apron conveyor
glimmen *v* glow (feebly); smoulder, burn faintly
Glimmen *n* (feeble) glow; smouldering
Glimmentladung *f* glow discharge
Glimmer *m* mica
glimmerartig micaceous
Globosid *n (bioch)* globoside
Globulärprotein *n* globular protein
Globulin *n* globulin *(any of a class of simple proteins)*
Globulin *n* **(A)/antihämophiles** antihaemophilic factor, AHF
γ-**Globulin** *n* gamma globulin, GG
Glocke *f* 1. *(distil)* (bubble) cap, bubbler; crown *(of a melting furnace)*; bell, cone *(of a blast furnace)*; 2. *(lab)* bell jar *(for protecting objects esp under vacuum)*
Glockenboden *m (distil)* bubble(-cap) tray, bubble(-cap) plate
Glockenbodenkolonne *f (distil)* bubble-cap column, bubble-tray [bubble-plate] column
Glockenbronze *f* bell bronze
Glockenmühle *f* cone mill, conical grinder, rotary crusher
Glover-Säure *f* Glover (tower) acid, brown oil of vitriol, B.O.V. *(chamber process)*
Glu *s.* Glutaminsäure
Glucagon *n* glucagon *(a protein produced by the pancreas)*
Glucan *n (org ch)* glucan
***D*-Glucitol** *n* D-glucitol, D-sorbitol, D-sorbite, sorbol
Glucocorticoid *n* glucocorticoid *(an adrenal cortex hormone)*
Gluconeogenese *f (bioch)* gluconeogenesis *(formation of sugars from non-carbohydrate precursors)*
***D*-Gluconsäure** *f* D-gluconic acid
***D*-Glucosamin** *n* D-glucosamine, GlcN, chitosamine, 2-amino-2-deoxy-*D*-glucopyranose
Glucose *f* glucose, *(specif) D*-glucose, dextrose
Glucosephosphat *n* glucose phosphate
Glucosid *n* glucoside *(a glucoside that yields glucose on hydrolysis)*
glucosidisch glucosidic, glucosidal
Glucozuckersäure *f* glucosaccharic acid, saccharic acid, glucaric acid *(one form of 2,3,4,5-tetrahydroxyhexanedioic acid)*
Glucuronatweg *m (bioch)* glucuronate pathway
Glucuronid *n* glucuronide, glucuronoside
Glucuronsäure *f* glucuronic [glycuronic] acid
glühen *v* 1. ignite, *(esp limestone, ores:)* calcine; 2. *(ceram)* bake; 3. *(met)* anneal; glow *(of a hot substance)*
glühen *v***/bis zur Gewichtskonstanz [Massekonstanz]** ignite to constant weight
glühen *v***/nochmals** re-ignite
Glühen *n* 1. ignition, *(esp of limestone, ores:)* calcination; 2. *(ceram)* baking; 3. *(met)* anneal(ing); glow *(of a hot substance)*
Glühen *n***/entspannendes** *(met)* stress-relieving anneal, stress relief (anneal)
Glühen *n***/graphitisierendes** *(met)* graphitizing
Glühen *n***/homogenisierendes** *(met)* homogenizing
Glühen *n* **in Schutzgas** *(met)* bright annealing
Glühen *n***/normalisierendes** *(met)* normalizing
Glühen *n***/rekristallisierendes** *(met)* recrystallization annealing
Glühen *n***/stabilisierendes** *(met)* stabilizing
glühend 1. glowing; red-hot *(metal)*; incandescent *(hot body)*; 2. *(geol)* igneous *(magma)*
Glühfaden *m* incandescent filament
Glühröhrchen *n* ignition (test) tube

Glührückstand *m* residue on ignition
Glühschale *f (lab)* ignition dish
Glühschiffchen *n* combustion boat
Glühverlust *m* loss on ignition, LOI, ignition loss
Glu-NH$_2$ *s.* Glutamin
Glutamat *n* glutamate *(salt or ester of glutamic acid)*
Glutamin *n* glutamine
Glutaminsäure *f* glutamic acid, 1-aminopropane-1,3-dicarboxylic acid
Glutaraldehyd *m* glutaraldehyde, glutaric dialdehyde, 1,3-pentanedial
Glutarsäure *f* glutaric acid, propane-1,3-dicarboxylic acid
Glutathion *n* glutathione, glutamylcysteinylglycine
Gluten *n (bioch)* gluten
Glv. *s.* Glühverlust
Gly *s.* Glykokoll
Glyceraldehyd *m* glyceraldehyde, *2,3-dihydroxypropanal
Glycerat *n* glycerate *(salt or ester of glyceric acid)*
Glycerid *n* glyceride, acylglycerol *(ester of glycerol with fatty acids)*
Glycerid *n*/**gemischtes [gemischtsäuriges]** mixed [component] glyceride
Glycerin *n s.* Glycerol
Glycerol *n* glycerol, *propane-1,2,3-triol • **mit Glycerol behandeln** glycerolize, glycerolate
Glycerol-α-chlorhydrin *n* glycerol α-chlorohydrin, *3-chloropropane-1,2-diol
Glycerolgärung *f* glycerol fermentation
Glyceroltrinitrat *n* glycerol trinitrate, glyceryl trinitrate
Glycersäure *f* glyceric acid, *2,3-dihydroxypropanoic acid
Glycin *n* glycine, aminoacetic acid
Glykan *n* glycan, polysaccharide
Glykocholsäure *f* glycocholic acid *(a bile acid)*
Glykogen *n* glycogen, animal [liver] starch
Glykogenabbau *m (bioch)* glycogenolysis
Glykogenese *f (bioch)* glycogenesis
Glykogenolyse *f (bioch)* glycogenolysis
Glykokoll *n* glycine, glycocoll, aminoacetic acid
Glykol *n* $C_nH_{2n}(OH)_2$ glycol, (aliphatic) diol, dihydric alcohol, *(specif)* ethane-1,2-diol
1,2-Glykol *n* ethylene glycol, *1,2-ethane-1,2-diol
Glykolaldehyd *m* glycol(l)aldehyde, hydroxyacetaldehyde
Glykolharnstoff *m* glycollylurea, hydantoin, imidazolidine-2,4-dione
Glykolipid *n (bioch)* glycolipid(e)
Glykolsäure *f* glycol(l)ic acid, hydroxyacetic acid
Glykolyse *f (bioch)* glycolysis, homolactic fermentation, Embden-Meyerhof-Parnas scheme [pathway]
Glykoproteid *n* glycoprotein, glycopeptide
Glykoprotein *n* glycoprotein, glycopeptide
Glykosid *n* glycoside, oside
Glykosid *n*/**herzaktives** cardiac glycoside
Glykosidbindung *f* glycosidic bond
Glykosidierung *f* glycosylation
glykosidisch glycosidic
Glykosphingolipid *n (bioch)* glycosphingolipid
Glykosurie *f* glycosuria *(excretion of sugars in urine)*
Glykosylierung *f* glycosylation
Glyoxal *n* glyoxal, oxalaldehyde, *ethanedial
Glyoxylatzyklus *m (bioch)* glyoxylate cycle
Glyoxylsäure *f* glyoxylic acid
Glyptal(harz) *n* glyptal (resin), glycerol phthalic resin, phthalic glyceride resin
GMP = Guanosin-5′-monophosphat
GnRH *(Gonadotropin-Releasing-Hormon) s.* Gonadoliberin
Golay-Detektor *m (spectr)* detector Golay, Golay cell
Gold *n* Au gold
Gold *n*/**kolloidales** colloidal gold, gold nanoparticles
Goldamalgam *m (min)* gold amalgam
Gold(I)-chlorid *n* gold(I) chloride, gold monochloride, aurous chloride
Gold(III)-chlorid *n* gold(III) chloride, gold trichloride, auric chloride
Gold(I,III)-chlorid *n* gold(I,III) chloride, digold tetrachloride, auroso-auric chloride

Gold(I)-cyanid *n* gold(I) cyanide, gold monocyanide, aurous cyanide
Gold(III)-cyanid *n* gold(III) cyanide, gold tricyanide, auric cyanide
goldhaltig 1. gold-bearing; 2. *(geoch)* auriferous
Gold(I)-hydroxid *n* gold(I) hydroxide, aurous hydroxide
Gold(III)-hydroxid *n* gold(III) hydroxide, auric hydroxide, auric acid
Gold(I)-oxid *n* gold(I) oxide, digold monooxide, aurous oxide
Gold(III)-oxid *n* gold(III) oxide, digold trioxide, auric oxide
Goldschmidt-Radius *m* Goldschmidt radius *(of ions)*
Goldschmidt-Verfahren *n* Goldschmidt process, aluminothermics, aluminothermy
Goldsol *n* gold sol
Goldzahl *f (coll)* gold number
Gonadoliberin *n* gonadoliberin, GnRH
Goochtiegel *m* Gooch crucible [filter]
Goochtiegelhalter *m* crucible adapter
Goslarit *m (min)* goslarite, zinc vitriol *(hydrous zinc sulphate)*
GOT = Glutamat-Oxalacetat-Transaminase
Götterspeise *f (food)* jelly
Gouy-Chapman-Schicht *f (hyd)* diffused layer *(coagulation)*
G-6-P = Glucose-6-Phosphat
G6PDH = Glucose-6-phosphatdehydrogenase
GPT = Glutamat-Pyruvat-Transaminase
Grad *m* degree; grade *(of purity of chemicals)* • **in Grade (ein)teilen** graduate
Grad *m* **Celsius** degree centigrade [Celsius], deg C, °C
Grad *m* **der Unordnung** *(phys ch)* degree of disorder
Grad *m* **deutscher Härte** *(hyd)* German degree of hardness, degree German *(1 °dH = 10.0 mg CaO/l H_2O)*
Grad *m* **englischer Härte** *(hyd)* English degree of hardness, degree Clark [English, British] *(1 °eH = 8.0 mg CaO/l H_2O)*
Grad *m* **Fahrenheit** degree Fahrenheit, deg F, °F
Grad *m* **französischer Härte** *(hyd)* French degree of hardness, degree French *(1 °fH = 5.6 mg CaO/l H_2O)*
Gradation *f (phot)* gradation, development factor, gamma value
Gradeinteilung *f* graduation, division
Gradient *m* gradient
Gradient(en)elution *f* gradient elution
Gradientschicht *f* gradient layer
gradieren *v* graduate *(solutions by evaporation)*
Gradieren *n* graduation *(of solutions by evaporation)*
graduieren *v* graduate, calibrate
graduiert/auf Ablauf [Auslauf] graduated [calibrated] to deliver, graduated for delivery
graduiert/auf Einguss graduated [calibrated] to contain, graduated for content
Graduierung *f* graduation, calibration
Graftpolymer(e) *n* graft (polymer)
Graham-Salz *n* Graham's salt *(a sodium metaphosphate glass)*
Grammäquivalent *n (phys ch)* gram equivalent, eq
Grammatom *n* gram atom, gram-atomic weight
Grammbereich *m (anal)* gram range *(1 to 10 g)*
Grammmol(ekül) *n* gram molecule [mole], gram-molecular weight, mole
Grammsuszeptibilität *f* susceptibility per gram, specific [mass] susceptibility
gramnegativ Gram-negative *(bacteriology)*
grampositiv Gram-positive *(bacteriology)*
Granalie *f* granule, pellet, agglomerate
Granat *m (min)* garnet
Granulat *n* granulate, granular material, *(molten droplets which have solidified:)* shot
Granulatformen *n* agglomerating, agglomeration *(of powder by means of a liquid)*
Granulation *f* granulation, granulating, graining, *(solidification of molten droplets also)* shotting, *(of powder by means of a liquid also)* agglomeration
Granulatkorn *n* pellet, granule, *(formed from powder also)* agglomerate
Granulator *m* granulator, pelletizer, granulating [pelletizing] machine

Granulierband *n (tech)* pelletizing conveyor
granulieren *v* granulate, grain
granuliert granulate(d), grained
granulös granular, grainy, granulate
Graphit *m* graphite • **mit Graphit auskleiden** line with graphite, graphitize, *(Am)* graphite • **mit Graphit beschichten** coat with graphite, graphitize, *(Am)* graphite
Graphit *m*/**künstlicher** synthetic [artificial] graphite
Graphit-Einlagerungsverbindung *f* graphite intercalation compound, GIC
Graphitelektrode *f* graphite electrode
Graphitfaden *m* graphite filament
Graphitfaser *f* graphite fibre
Graphitgitter *n* graphite lattice
graphitieren *v* graphitize, *(Am)* graphite
Graphitierofen *m* graphitizing furnace
Graphitierung *f*/**thermische** graphitization heat treatment
Graphit-Interkalationsverbindung *f* graphite intercalation compound, GIC
graphitisch graphitic
Graphitreaktor *m (nucl)* graphite-moderated reactor
Graphitrohrküvette *f (spectr)* graphite (resistance) furnace
Graphitrohrofen *m (spectr)* graphite (resistance) furnace
Graphitstab *m* graphite rod
Graphittiegel *m* graphite [plumbago] crucible
Graphitwerkstoff *m* graphite material
Graphitwhisker *m* graphite whisker
Grat *m* 1. burr *(produced in cutting metal)*; flash, fin *(on castings)*; 2. *(plast)* burr, flash
Grauguss *m* grey (cast) iron; *(from a mould:)* grey-iron casting
Grauguss *m*/**globularer [sphärolithischer]** nodular [spheroidal graphite] iron, ductile iron
Graukörper *m* grey body
Grauspießglanz *m (min)* grey antimony, antimony glance, antimonite, stibnite *(antimony(III) sulphide)*
Gravimetrie *f* gravimetry
gravimetrisch gravimetric(al)
Gravitationsfeld *n* gravitational field
Gravitationskraft *f* force of gravitation [gravity], gravitational force [pull]

Grenze *f* 1. limit; 2. boundary *(as of an atom or phase)*
0,2 %-Grenze *f* 0.2 % offset yield strength [stress], 0.2 % proof stress, yield strength 0.2 % offset *(materials testing)*
Grenzfläche *f* boundary [bounding] surface, interface, junction
Grenzfläche *f* **fest-fest** solid-solid interface
Grenzfläche *f* **fest-flüssig** solid-liquid interface
Grenzfläche *f* **flüssig-flüssig** liquid-liquid interface, liquid junction, dineric interface
Grenzfläche *f* **flüssig-gasförmig** liquid-vapour interface
Grenzfläche *f* **Luft gegen Flüssigkeit** air-liquid interface
Grenzfläche *f* **Öl gegen Wasser** oil-water interface
grenzflächenaktiv interface-active, interfacially active, surface-active, surface-tension-lowering
Grenzflächenaktivität *f* interfacial [surface] activity
Grenzflächen(poly)kondensation *f* interfacial polycondensation
Grenzflächenpolymerisation *f* interfacial polymerization
Grenzflächenpotenzial *n* junction potential
Grenzflächenspannung *f* interfacial tension
Grenzfrequenz *f* limiting frequency
Grenzkonzentration *f* 1. limiting concentration; 2. *(tox)* maximum permissible [admissible] concentration
Grenzlastspielzahl *f* fatigue life *(materials testing)*
Grenzorbital *n* frontier orbital
Grenzschicht *f* boundary layer
Grenzschicht *f*/**laminare** laminar boundary layer
Grenzschicht *f*/**turbulente** turbulent [Prandtl] boundary layer
Grenzschwingspielzahl *f* fatigue life *(materials testing)*
Grenzstrom *m* limiting current
Grenzstruktur *f* limiting structure
Grenzstruktur *f*/**mesomere** contributing structure, resonance [resonating] structure

Grenzviskosität *f* intrinsic viscosity, limiting viscosity number
Grenzviskositätszahl *f* intrinsic viscosity, limiting viscosity number
Grenzwert *m* limiting value, limit, *(relating to toxic materials also)* boundary [tolerance] limit
Grieß *m* oversize (material, product) *(air classifying)*
grießartig gritty
Grignardierung *f (org ch)* grignardization
Grignard-Reagens *n (org ch)* Grignard reagent *(solution of an organometallic halide)*
Grignard-Reaktion *f (org ch)* Grignard reaction *(for obtaining organometallic halides)*
Grignard-Synthese *f (org ch)* Grignard synthesis *(using a Grignard reagent)*
Grignard-Verbindung *f (org ch)* Grignard compound
grob 1. coarse *(screen)*; 2. coarse, rough *(e.g. bulk material)*; 3. rough *(method)*
Grobabscheider *m* precleaner *(gas cleaning)*
Grobausmahlung *f* coarse grinding
grobdispers coarsely dispersed
Grobeinstellung *f* coarse adjustment
grobfaserig coarse-fibred
Grobfilter *n* coarse [roughing] filter
Grobgut *n* 1. coarse material; 2. *(hyd)* coarse solids; 3. *s.* Grobkorn
Grobkeramik *f* 1. heavy ceramics *(branch)*; 2. heavy clay product [ware]
Grobkorn *n* oversize (material, product), tailings, tails, plus material *(classifying)*; *(screening also)* screen oversize, plus mesh
grobkörnig coarse-grained
Grobkörnigkeit *f* coarseness
grobkristallin coarse-crystalline, coarsely crystalline
grobmahlen *v* crush, grind coarsely
Grobmahlung *f* crushing, coarse grinding
grobmaschig coarse-mesh(ed)
Grobsand *m* coarse sand, grit
Grobsiebung *f* coarse [rough] screening
Grobstaub *m* grit *(in the atmosphere)*
Grobvakuum *n* low vacuum
grobzerkleinern *v* crush, grind coarsely
Grobzerkleinerung *f* crushing, coarse grinding
Großanlage *f* large-scale unit
Großchemie *f* large-scale chemistry
Größe *f (phys ch)* quantity
Größe *f*/**additive [extensive]** extensive quantity
Größe *f*/**intensive** intensive quantity
Größe *f*/**kritische** critical constant
Größe *f*/**partielle molare** partial molar [molal] quantity
Größe *f*/**reduzierte** reduced variable [property]
Größenausschlusschromatographie *f* size-exclusion chromatography, GPC, SEC
Größenbestimmung *f* size determination *(classifying)*
Größenordnung *f* order of magnitude
Größenverteilung *f* size distribution *(classifying)*
großindustriell large-scale
großporig large-pored
Großproduktion *f* large-scale production [manufacture, fabrication]; *(as opposed to pilot-plant production:)* full-scale factory production
Großprozess *m* large-scale process
Großring *m* large ring, macrocycle
Großringverbindung *f* macrocyclic compound, macroring [large-ring] compound
großstückig blocky, lumpy
großtechnisch large-scale; *(as opposed to pilot-scale:)* full-scale
Grubbs-Katalysatoren *mpl (org ch)* Grubbs catalysts *(metathesis)*
Grübchenkorrosion *f* pitting corrosion
Grube *f* 1. pit; 2. mine
Grubengas *n* mine gas, filty, fire-damp
Grubenkohle *f* run-of-mine coal
Grün *n*/**Kasseler** Cassel [manganese] green *(barium manganate)*
Grün *n*/**Rinmans** Rinman's [cobalt] green *(consisting essentially of cobalt and zinc oxides)*
Grün *n*/**scheeles** Scheele's green *(copper arsenite)*
Grün *n*/**Schweinfurter** Paris [Schweinfurth] green *(copper acetate arsenite)*
Grundanstrich *m* 1. priming *(act)*; 2. base coat, primer (coat), priming [prime] coat, primary [ground] coat
Grundanstrichmittel *n* primer (coating), priming coat material

Grundanstrichstoff *m* primer (coating), priming coat material
Grundausstattung *f* typical [small-scale] equipment, typical inventory
Grundbaustein *m* basic [fundamental] building block; base unit, (basic) repeating unit *(of polymers)*
Grundbaustein *m* **D** difunctional [bifunctional] unit, D unit *(for building up polymers)*
Grundbaustein *m* **M** monofunctional unit, M unit *(for building up polymers)*
Grundbaustein *m* **T** trifunctional unit, T unit *(for building up polymers)*
Grundbestandteil *m* main [principal, chief] component, main constituent [ingredient], base, basis
Grundchemikalie *f* basic [key] chemical
Grundeinheit *f* 1. basic [fundamental] unit, *(esp of polymers)* base unit, (basic) repeating unit; 2. base [fundamental] unit *(of a system of units)*
Grundeinheit *f*/**strukturelle** (basic) structural unit, structural element [entity]
Grundelektrolyt *m* support(ing) electrolyte, base [indifferent] electrolyte *(polarography)*
Grundformel *f*/**stöchiometrische** stoichiometric [simplest] formula
Grundfraktion *f (petrol)* primary fraction
Grundgerüst *n* parent structure, basic framework *(of a molecule)*; backbone, skeleton *(of a chain molecule)*
Grundgesamtheit *f* 1. universe, population *(statistics)*; 2. *(anal)* entire [parent] lot, entire mass
grundieren *v (coat)* prime
Grundiermittel *n* primer
Grundierung *f s.* 1. Grundanstrich; 2. Grundanstrichmittel
Grundkohlenwasserstoff *m* parent hydrocarbon
Grundkomponente *f s.* Grundbestandteil
Grundkörper *m* parent [mother] substance
Grundlagenchemie *f* fundamental chemistry
Grundlagenforschung *f* basic (science) research, pure research
Grundlinie *f (anal)* base line
Grundlinien *fpl (spectr)* persistent [ultimate] lines
Grundlösung *f* basis solution
Grundmasse *f (geol, ceram, met)* matrix, ground-mass
Grundmetall *n* principal [base] metal *(of an alloy)*
Grundmischung *f (rubber)* masterbatch, base stock [compound, mix], mother [blank] stock
Grundmolekül *n* basic [fundamental] molecule, *(esp of polymers:)* base molecule, monomer
Grundoperation *f* 1. basic [fundamental] operation; 2. *s.* Grundoperation/physikalische
Grundoperation *f* **der chemischen Verfahrenstechnik** chemical engineering unit operation
Grundoperation *f*/**physikalische** basic physical operation, *(esp chemical engineering:)* unit operation
Grundprozess *m* fundamental [basic] process
Grundprozess *m*/**stofflicher [verfahrenstechnischer]** unit process
Grundreaktion *f* basic [fundamental] reaction
Grundschicht *f (coat)* base coat
Grundschwingung *f* fundamental [normal] vibration, normal mode of vibration
Grundschwingungsbande *f (spectr)* fundamental band
Grundschwingungsfrequenz *f (spectr)* fundamental (vibration) frequency
Grundstoff *m* basic [fundamental] substance
Grundstoff *m*/**chemischer** basic chemical (material), key chemical
Grundstoffwechsel *m (bioch)* primary metabolism
Grundstrom *m (phys ch)* residual current
Grundstruktur *f* basic structure
Grundsubstanz *f* 1. parent [mother] substance *(of chemical compounds)*; 2. *(bioch)* ground substance *(as of tissues)*
Grundsubstrat *n (biot)* basal [main growth] medium
Grundumsatz *m (bioch)* basal metabolism, basal metabolic rate

Grundverfahren *n* basic [fundamental] method [process]
Grundvorgang *m* basic [fundamental] process
Grundwasserverschmutzung *f* groundwater pollution [contamination]
Grundwasserverunreinigung *f* groundwater pollution [contamination]
Grundzustand *m (phys ch)* ground state [term], atomic ground [unexcited] state, normal state
Grundzustand *m*/**elektronischer** electronic ground state, ground [zeroth] electronic state
Grünglas *n* green glass
Grünling *m* green compact *(in powder metallurgy)*
Grünpellet *n* green [moist] pellet
Grünspan *m* verdigris, aerugo *(a mixture of basic copper(II) acetates)*
Gruppe *f* group *(of atoms)*; family, group *(in the periodic system)*; subclass *(of the enzyme classifying system)*; set, battery, bank *(of equipment)*
Gruppe *f*/**aktive** 1. active [prosthetic] group, coenzyme, agon *(enzymology)*; 2. *s.* Gruppe/austauschaktive
Gruppe *f*/**aktivierende** activating group
Gruppe *f*/**austauschaktive** active group *(ion exchange)*
Gruppe *f*/**austretende** leaving group
Gruppe *f*/**auxochrome** *(dye)* auxochromic group, auxochrome
Gruppe *f*/**bathochrome** *(dye)* bathochromic group, bathochrome, bathychrome
Gruppe *f*/**chromophore** *(dye)* chromophoric [colour-bearing] group, chromophore
Gruppe *f* **der Eisenmetalle** iron group
Gruppe *f*/**eintretende** entering group
Gruppe *f*/**elektrofuge** electrofuge
Gruppe *f*/**endständige** terminal group
Gruppe *f*/**erste** *(anal)* insoluble chloride group
Gruppe *f*/**farberhöhende** hypsochromic group
Gruppe *f*/**farbgebende [farbtragende]** *s.* Gruppe/chromophore
Gruppe *f*/**farbvermehrende [farbverstärkende]** *s.* Gruppe/auxochrome
Gruppe *f*/**farbvertiefende** *s.* Gruppe/bathochrome
Gruppe *f*/**funktionelle** functional group
Gruppe *f*/**haptophore** haptophoric group, haptophore *(portion of a toxin molecule which binds it to a body cell)*
Gruppe *f*/**hypsochrome** *(dye)* hypsochromic group, hypsochrome
Gruppe *f*/**inaktivierende** deactivating group
Gruppe *f*/**nucleofuge** nucleofuge
Gruppe *f*/**nullte** zero group *(in the periodic system)*
Gruppe *f*/**ortho-para-dirigierende** ortho-para-directing group, ortho-para director
Gruppe *f*/**prosthetische** prosthetic [conjugate] group *(protein chemistry)*; *(relating to enzymes also)* active group, coenzyme
Gruppe *f*/**terminale** terminal group
Gruppe *f*/**toxophore** toxophoric group, toxophore
Gruppe *f*/**zweite** *(anal)* insoluble sulphide group
Gruppenanalyse *f* group analysis
Gruppenfrequenz *f (spectr)* group frequency
Gruppenreagens *n* group reagent
Gruppensilicat *n (min)* sorosilicate
gruppenspezifisch group-specific *(enzyme)*
Gruppenspezifität *f* (absolute) group specificity *(of enzymes)*
Gruppentransfer *m (bioch, org ch) s.* Gruppenübertragung
Gruppentransferreaktion *f (org ch)* group transfer reaction
Gruppenübertragung *f (bioch)* group transfer
Grus *m* 1. *(soil)* grit; 2. *(coal)* breeze
GSC *(Gas-Solidus-Chromatographie) s.* Gas-fest-Chromatographie
GSK *s.* Geschmacksschwelle(nkonzentration)
g-Strophantin *n* G-strophantin *(a glycoside)*
GTP *s.* Guanosintriphosphat
Guajacol *n* guaiacol, *o*-methoxyphenol
Guajakharz *n* guaiac (resin), gum guaiac *(from Guajacum officinale L. and G. sanctum L.)*
Guajol *n* guaiol *(a sesquiterpene alcohol)*
Guanidin *n* guanidine, iminourea

Guanin *n* guanine, 6-hydroxy-2-aminopurine
Guanosin *n* guanosine, guanine riboside *(a nucleoside)*
Guanosindiphosphat *n (bioch)* guanosine diphosphate
Guanosintriphosphat *n (bioch)* guanosine triphosphate
Guanylsäure *(bioch)* guanylic acid
Gumbildung *f (petrol)* gum formation, gumming
Gumgehalt *m (petrol)* gum content
Gummi *m* rubber • **mit Gummi ausgekleidet** rubber-lined • **mit Gummi beschichtet** rubber-coated
Gummi *m*/**halbharter** semihard rubber, semiebonite
Gummi *m*/**mikroporöser** microporous rubber
Gummi *n* gum
Gummi *n*/**arabisches** gum arabic, acacia [Arabic] gum *(from Acacia specc., esp from A. senegal (L.) Willd.)*
Gummiabfälle *mpl* rubber scrap [waste], scrap [waste] rubber
Gummiarabikum *n s.* Gummi/arabisches
gummiartig rubber-like, rubbery
Gummiartikel *m* rubber article [product]
Gummibelag *m* rubber covering
Gummichemie *f* rubber chemistry
Gummichemiker *m* rubber chemist
Gummidichtung *f* rubber packing [seal] *(on moving parts)*; rubber gasket *(with static application)*
Gummielastizität *f* rubber [long-range] elasticity
gummieren *v* rubber(ize)
Gummierung *f* 1. rubberizing, proofing; 2. rubber coat(ing); *(inner surfaces:)* rubber lining
Gummihandschuhe *mpl* rubber gloves
Gummiharz *n* gum resin
Gummilösung *f* rubber solution; rubber cement
Gummilösungsmittel *n* rubber solvent
Gummiquetscher *m* squeegee
Gummiring *m* rubber ring [annulus]
Gummischlauch *m* rubber hose [tubing]
Gummistopfen *m* rubber stopper [bung]
Gummiwaren *fpl* rubber goods [products, articles]
Gummiwischer *m (lab)* rubber-tipped glass rod, (rubber) policeman, bobby
Gurt *m* belt
Gurtbandförderer *m* belt [band] conveyor
Guss *m* 1. casting *(product)*; 2. *s.* Gießen
Guss... *siehe auch: Gieß...*
Gussblase *f* blow hole, blister *(a material fault)*
Gussblock *m* ingot
Gussbronze *f* cast bronze
Gusseisen *n* cast iron
Gusseisen *n*/**siliciumlegiertes** silicon cast iron
Gusseisen *n*/**weißes** white cast iron
Gusslegierung *f* cast(ing) alloy
Gussstahl *m* cast steel
Gussstück *n* casting
Gussteil *n* casting
Gussverfahren *n* casting *(manufacturing process)*
Gut *n* material, stuff, *(if already treated also)* product
Gut *n*/**abgeröstetes** *(met)* calcine
Gut *n*/**aufschwimmendes** *(min tech)* floating material [fraction], floats
Gut *n*/**zu behandelndes** material to be treated
Gutaufgabe *f s.* Guteintrag
Gutaustrag *m* 1. discharge *(act or process)*; 2. discharge (point)
Gutaustritt *m* 1. discharge *(act or process)*; 2. discharge (point)
Gütebedingungen *fpl* quality standards
Gütebedingungen *fpl* **für Trinkwasser** potable [drinking] water standards [regulations]
Guteintrag *m* 1. charging *(act or process)*; 2. charging point
Güteklasse *f* grade
Gütekontrolle *f* quality control
Gütesicherung *f (anal)* quality assurance
Guttapercha *f(n)* gutta-percha *(from Payena and Palaquium specc.)*
Guyard-Reaktion *f* Guyard reaction *(oxidation of Mn^{++} by MnO_4^-)*
GV *s.* Glühverlust
GW *s.* Wasser/gebundenes

H

H *s.* Heizwert
Haarfärbemittel *n* hair dye
Haarkristall *m* (crystal) whisker
Haarnadelrohr *n* hairpin tube
Haarriss *m* hair crack, craze *(surface defect)*
Haarrissbildung *f* hair-(line) cracking, (micro)crazing *(on surfaces)*
Haarröhrchen *n* capillary (tube)
Haarwäsche *f* hair wash [shampoo]
Haarwaschmittel *n* hair wash [shampoo]
Haber-Bosch-Verfahren *n* Haber(-Bosch) process *(for synthesizing ammonia from H_2 and N_2)*
Habitus *m* habit *(as of crystals)*
hacken *v (pap)* chip, chop
Hackmaschine *f (pap)* chipper, chopper, chipping [chopping] machine
Häckselmaschine *f (pap)* chopping machine, chopper, cutter *(for straw)*
häckseln *v (pap)* chop, cut *(straw)*
Hadern *pl (pap)* rags
Hadernhalbstoff *m (pap)* all-rag furnish, rag pulp [stuff, stock], non-woody pulp
Hadron *n (nucl)* hadron
Hafen *m (glass, ceram)* pot
Haferstärke *f* oat starch
Hafnat *n* $M^I_2HfO_3$ hafnate
Hafnium *n* Hf hafnium
Hafniumcarbid *n* hafnium carbide
Hafnium(IV)-oxid *n* hafnium(IV) oxide, hafnium dioxide
haften *v* adhere, stick
Haften *n* adhesion, adherence, sticking
haftend adhesive, adherent
haftfähig adhesive, adherent
Haftfähigkeit *f* adhesiveness, adherence, adhesive power [capacity], sticking power [capacity]
Haftfestigkeit *f* tenacity, adhesive strength
Haftmittel *n* 1. adhesive, sticking agent; 2. *(rubber)* bonding agent; 3. *(text)* coupling agent *(for laminates)*; 4. *(agric)* deposit builder *(in pesticide formulations)*
Haftreibung *f* static friction, stiction
Haftstelle *f (phys ch)* trap *(as for recombination of electrons and defect electrons)*
Haftvermittler *m* 1. adhesion promoter; 2. *(plast)* coupling [anchoring] agent
Hagen-Poiseuille-Gleichung *f* Poiseuille's equation [formula] *(of laminar flow in pipes)*
Hagen-Poiseuille-Strömung *f* Poiseuille flow
Hahn *m* cock, tap, plug valve [cock, bib]
Hahnhülse *f* socket, cock shell [barrel]
Hahnküken *n* (cock) plug, *(Am)* stopper
Hahnventil *n* plug valve [cock, bib]
Halbacetal *n* hemiacetal
halbacetalartig hemiacetal-like
halbacetalisch hemiacetal-like
Halbaminal *n* hemiaminal, α-amino alcohol
Halbantigen *n (bioch)* hapten, haptene
Halbchinon *n* semiquinone
halbdurchlässig semipermeable
Halbedelstein *m* semiprecious stone
Halbelement *n (phys ch)* half-element, half-cell
Halbester *m* semi-ester, half-ester
halbfest semisolid
halbflüssig semiliquid, semifluid
Halbformal *n* hemiformal *(any of the hemiacetals of formaldehyde)*
halbgeordnet *(phys ch)* semi-ordered
halbhart half-hard *(cold-rolled metal)*; medium-hard *(pitch)*; semihard *(rubber)*; semirigid *(plastic)*
Halbhöhenbreite *f (anal)* (peak) width at half height, (peak) half width
Halbkoks *m* coal char
Halbkolloid *n* semicolloid
halbkontinuierlich semicontinuous, semibatch
halbkristallin(isch) semicrystalline, hemicrystalline, hypocrystalline
Halblebenszeit *f (biot)* half life *(of enzyme activity)*
halbleitend semiconducting, semiconductive
Halbleiter *m* semiconductor
Halbleiterlaser *m* semiconductor [diode] laser
Halbleitermaterial *n* semiconducting material
Halb-Leiterpolymer(e) *n* semiladder polymer
Halbleitung *f* semiconductivity

Halbmaske *f* half-mask facepiece *(respiratory protection)*
Halbmetall *n* semimetal, crossroads element
Halbmetallglanz *m (min)* submetallic lustre
halbmetallisch semimetallic
Halbmikroanalyse *f* semimicro [centigram] analysis
Halbmikromaßstab *m*/**im** semimicroscale
Halbmikromethode *f* semimicro method, centigram procedure
Halbmikroprobe *f* *meso sample, semimicro sample *(0.1 to 0.01 g)*
Halbmikrowaage *f* semimicro balance
Halbporzellan *n* semiporcelain, vitreous china
halbquantitativ semiquantitative
Halbseide *f (text)* half silk
Halbstoff *m (pap)* half-stuff, half-stock
• **zu Halbstoff aufschließen** *(pap)* make into (a) pulp, reduce to pulp, pulp
Halbstoff *m* **in Bogenform [Pappenform]** half-stuff board, (solid) pulp board, sheets [laps] of pulp, lap(ped) pulp
Halbstufenpotenzial *n* half-wave potential
halbsynthetisch semisynthetical
halbtechnisch pilot-(plant-)scale
Halbtrivialname *m* semitrivial [semisystematic] name
halbverglast *(ceram)* semivitreous, semivitrified
Halbwellenpotenzial *n* half-wave potential
Halbwert(s)breite *f (anal)* half-width (at half maximum), HWHM, half-intensity width *(of a signal)*
Halbwertszeit *f* 1. half-life (period), half-time, half-value period, time [period] of half-change *(of reactants)*; 2. *(nucl)* (radioactive) half-life; 3. *(biot)* half-life *(of enzyme activity)*; 4. *s.* Halbwertszeit der Reaktion; 5. *s.* Rückstands-Halbwertszeit
Halbwertszeit *f* **der Reaktion** (reaction) half-life, (reaction) half-time, half reaction time
Halbwertszeitmethode *f* method of half-times *(for determining the reaction order)*
Halbzelle *f (phys ch)* half-cell, half-element
Halbzellenpotenzial *n* half-cell potential
Halbzellstoff *m* semichemical pulp
Halbzellstoffaufschluss *m* semichemical pulping
Halde *f (min tech)* tip, (waste) dump
Haldenlaugung *f (min tech)* (waste-) dump leaching, slope leaching
Haldenleaching *n (min tech)* (waste-) dump leaching, slope leaching
halluzinogen *(tox)* hallucinogenic, psycho(to)mimetic
Halluzinogen *n (tox)* hallucinogen, psychodysleptic, psychotomimetic (drug), psychedelic drug
Haloform-Reaktion *f (org ch)* haloform [Einhorn] reaction
Halogen *n* halogen
Halogenacylierung *f* haloacylation
Halogenalkan *n* haloalkane, alkyl halide
Halogenalkylierung *f* haloalkylation
Halogenaminierung *f* haloamination
Halogenarylierung *f* haloarylation
Halogencarbonsäure *f* haloacid
Halogenderivat *n* halo(gen) derivative
Halogenentzug *m* removal of halogen
Halogenethan *n* haloethane, ethyl halide
Halogenhydrin *n* halohydrin *(any of a class of glycerol derivatives)*
Halogenid *n* halide, halogenide
halogenieren *v* halogenate
α-halogeniert α-halogenated, alpha-halogenated
halogeniert/mehrfach polyhalogenated
Halogenierung *f* halogenation
α-Halogenierung *f* alpha-halogenation
Halogenketon *n* haloketone
Halogenkohlenwasserstoff *m* halogenated hydrocarbon, halocarbon
Halogenmethan *n* halomethane, methyl halide
Halogenoform *n* haloform *(any of the triholomethanes)*
Halogenolyse *f* halogenolysis
halogensubstituiert halogen-substituted
Halogenverbindung *f* halo(gen) compound
Halogenverbindungen *fpl*/**organische** haloorganic [halogenated organic] compounds, HOC, haloorganics, organic halides, organohalides

Halogenwasserstoff *m* hydrogen halide [halogenide], hydrohalogen
Halogenwasserstoffabspaltung *f* removal of hydrogen halide
Halogenwasserstoffsäure *f* hydrohalic acid
Halometer *n* sali(ni)meter *(a hydrometer for salt solutions)*
Halonium-Ion *n* halonium ion
Hals *m* neck *(as of a bottle or shaft)*
haltbar durable, stable • **haltbar machen** preserve, conserve • **unbegrenzt haltbar sein** keep indefinitely • **haltbar verpackt** packed for prolonged storage
Haltbarkeit *f* 1. durability, stability; 2. *(food)* storage [keeping] quality
Haltbarkeitsdauer *f* 1. life(time), service [length of] life; 2. *(food)* keeping time
Haltbarmachung *f* preservation, conservation
halten *v* support *(mechanically)*; keep, maintain *(e.g. a definite temperature)*
halten *v***/am Kochen [Sieden]** keep at the boil
halten *v***/im Gleichgewicht** equilibrate, keep in equilibrium
halten *v***/in Lösung** keep in solution
halten *v***/in Suspension** keep in suspension
halten *v***/instand** maintain
halten *v***/konstant** maintain constant
halten *v***/nahe am Sieden** keep near the boil
Haltepunkt *m* halt
Haltepunkt *m***/eutektischer** eutectic halt
Halter *m* holder, clamp, clip, gripping mechanism
Halterung *f* holder, clamp, clip, gripping mechanism
Haltezeit *f* residence time, retention [hold-up, holding, detention] time
Häm *n (bioch)* haem, protohaem, *(Am)* heme
Hämatin *n* haematin *(oxidized form of haem)*
Hämatit *m (min)* (red) haematite *(iron(III) oxide)*
Hämenzym *n* haem [iron porphyrine] enzyme
Hämiglobin *n* haemiglobin, methaemoglobin
Hamilton-Operator *m (phys ch)* Hamiltonian (operator), energy operator
Hämin *n* haemin, ferrihaem
Hammermühle *f* hammer mill [disintegrator]
Hammerschlag *m* hammer scale *(iron(II,III) oxide)*
Hammett-Beziehung *f (phys ch)* Hammett relation [correlation], rho-sigma correlation, rs-correlation
Hammett-Gleichung *f* Hammett equation, (Hammett) rs equation, rho-sigma equation
Hammond-Postulat *n* Hammond postulate [principle] *(kinetics)*
Hammond-Prinzip *n* Hammond postulate [principle] *(kinetics)*
Hämochrom(ogen) *n* haemochromogen, haemochrome
Hämocyanin *n* haemocyanin *(a respiratory pigment of numerous invertebrate animals)*
Hämogen *n (bioch)* extrinsic factor, vitamin B_{12}
Hämogenase *f (bioch)* intrinsic factor
Hämoglobin *n* haemoglobin, Hb
Hämoglobin *n***/desoxygeniertes [reduziertes]** *s.* Hämoglobin/sauerstoffarmes
Hämoglobin *n* **S** haemoglobin S, sickle-cell haemoglobin
Hämoglobin *n***/sauerstoffarmes** deoxygenated haemoglobin
Hämolyse *f* haemolysis
hämolysieren *v* haemolyze
Hämolysin *n* haemolysin *(toxic substance produced by certain bacteria)*
Hämoprotein *n* haemoprotein, haem [iron porphyrine] protein
Hämostyptikum *n (pharm)* haemostatic, styptic
hämostyptisch haemostatic, styptic
handbetätigt manually operated, hand-operated
Handbetrieb *m* manual [hand] operation
handbetrieben manually operated, hand-operated
Handelschemiker *m* commercial chemist
Handelsdünger *m* commercial fertilizer
Handelsname *m* trade name [term], commercial name

Handelsqualität *f* commercial [market] grade
handelsüblich commercial, commercially available
Handfeuerlöscher *m* fire extinguisher, portable (fire) extinguisher
Handhabbarkeit f/leichte ease of use, ease [simplicity] of servicing
handhaben *v* handle
handhaben lassen v/sich handle
Handhabung *f* handling
Handrad *n* handwheel
Handschuhbox *f (nucl)* glove box
Hanffaser *f* hemp fibre *(from Cannabis sativa L.)*
Hängezentrifuge *f* (top-)suspended centrifuge, overdriven centrifuge
Hanglaugung *f (min tech)* heap leaching
Hangleaching *n (min tech)* heap leaching
Hapten *n (bioch)* hapten, haptene
Harmalaalkaloid *n* harmal(a) alkaloid
harmlos harmless, innocuous, benign
Harmlosigkeit *f* harmlessness, innocuousness, benignity
Harnsäure *f* uric acid, 2,6,8-trihydroxypurine
harnsäureausscheidend *(bioch)* uricotelic
Harnstoff *m* urea, carbamide
Harnstoffaddukt *n* urea adduct
harnstoffausscheidend *(bioch)* ureotelic
Harnstoffeinschlussverbindung *f* urea clathrate [inclusion compound]
Harnstoff-Formaldehyd-Harz *n* urea(-formaldehyde) resin, polyurea
Harnstoffkomplex *m* urea complex
Harnstoffkomplexverbindung *f* urea complex
harnstoffspaltend ureolytic
Harnstoffzyklus *m (bioch)* urea [ornithine] cycle, Krebs-Henseleit cycle
harntreibend diuretic
hart hard *(as of metals, water, radiation)* • **hart werden** harden, solidify, chill, set
härtbar hardenable
Härtbarkeit *f* hardenability
Hartblei *n* hard(ened) lead, antimonial lead, regulus metal *(a lead alloy containing up to 15 % antimony)*
Hartcarbid *n* hard (metal) carbide
Härte *f* hardness *(as of metals, water, radiation)*
Härte f/permanente *(hyd)* permanent [non-carbonate] hardness
Härte f/temporäre *(hyd)* temporary [carbonate] hardness, carbonate alkalinity
Härteabscheidungen *fpl (hyd)* hardness [mineral] scale, hard-water depositions [residue]
Härtebestimmung *f* determination of hardness, hardness testing
härtebildend *(hyd)* hardness-forming, hardness-causing, hardness-producing
Härtebildner *m (hyd)* hardness constituent, hardness-producing substance
Härtegehalt *m (hyd)* hardness level
Härtegrad *m* degree of hardness
Härtemesser *m s.* Härteprüfer
härten *v* 1. *(met)* harden; 2. *(plast, coat)* cure; 3. *(glass)* temper, strengthen; hydrogenize, hydrogenate, harden *(fats and oils)*
härten v/im Ofen *(plast, coat)* stove, *(Am)* bake
Härten *n* 1. *(met)* hardening; 2. *(plast, coat)* curing, cure; 3. *(glass)* tempering, strengthening; hydrogenation, hardening *(of fats and oils)*
Härten n im Ofen *(plast, coat)* stoving, *(Am)* baking
Härteprüfer *m* hardness tester [meter]; *(using a drill:)* durometer; *(using the rebound of a ball:)* scleroscope; *(using a stylus:)* sclerometer
Härteprüfung *f* hardness test(ing)
Härter *m (plast, coat)* curing [hardening] agent, hardener
Härteskala f nach Mohs Mohs' scale (of hardness)
Härtezeit *f* 1. *(met)* hardening time; 2. *(plast, coat)* curing [cure] time; 3. *(plast)* stoving time *(of cast resins)*
Hartfaserplatte *f* hard-board
Hartfett *n* solid [hard] fat; hydrogenated [hardened] fat
Hartgewebe *n* laminated fabric, synthetic-resin-bonded fabric sheet
Hartglas *n* resistance [hard] glass
Hartgummi *m* hard rubber, ebonite, vulcanite

Hartguss *m* chilled [white] cast iron, chill-cast iron
Hartharz *n* hard resin
Hartlot *n* brazing [hard] solder, brazing alloy [metal]
hartlöten *v* braze, hard-solder
Hartmetall *n* 1. hard metal *(composite consisting of hard materials and metals)*; 2. *s.* Hartmetall/gesintertes
Hartmetall *n***/gesintertes** cemented (hard) carbide, sintered carbide, cemented hard metal
Hartparaffin *n* hard [solid] paraffin, ceresin (wax)
Hartporzellan *n* hard(-paste) porcelain
Hart-PVC *n* unplasticized [rigid] PVC
Hartschaum(stoff) *m (plast)* rigid foam
Härtung *f s.* Härten
Härtungskatalysator *m (plast)* curing catalyst
Härtungsmittel *n* 1. hardening agent, hardener; 2. *s.* Härter
Hartwachs *n* hard wax
Hartwasser *n* hard water
Hartwerden *n* hardening, set(ting)
Harz *n* 1. resin; 2. *(petrol)* gum
Harz *n* **im A-Zustand** A-stage resin
Harz *n***/ölmodifiziertes** oil-modified resin
harzartig resinous, resinoid, resiny
Harzbildung *f* 1. *(bot)* resin formation; 2. *(petrol)* gum formation, gumming
Harzgehalt *m* 1. resin content; 2. *(petrol)* gum content
harzhaltig resiniferous, resinous
Harzkorn *n* resin bead *(ion exchange)*
Harzöl *n* resin oil, liquid resin
Harz-Öl-Verhältnis *n* resin-to-oil ratio, resin/oil ratio
Harzsäure *f* oleoresin [resin, rosin] acid
Harzseife *f* resin soap *(salt of a resin acid)*
Haschisch *n(m)* hashish, hasheesh, haschisch, marihuana, marijuana *(from Cannabis indica Lam.)*
haspeln *v (text)* reel (up)
Hatch-Slack-Kortschak-Zyklus *m (bioch)* Hatch-Slack (CO_2 incorporation) pathway, C_4 (dicarboxylic acid) cycle, C_4 pathway
Haube *f (tech)* hood, head, cap, *(esp of a furnace)* dome, crown
Haufen *m* heap, pile
Haufenspeicher *m* pile
Häufigkeit *f* frequency; abundance *(as of an element or isotope)*
Häufigkeitsfaktor *m* frequency factor, (Arrhenius) pre-exponential factor, Arrhenius factor, A factor *(kinetics)*
Häufigkeitsverteilung *f* frequency distribution
Haufwerkfilter *n* bed filter
Hauptalkaloid *n* main alkaloid, major [principal, chief] alkaloid
Hauptbande *f (spectr)* centre band
Hauptbestandteil *m* main [principal, chief] component, main constituent [ingredient], base, basis; *major constituent *(100 to 1 %)*; major element *(of an alloy)*
Hauptfarbstoff *m* principal colouring material
Hauptfermentation *f (biot)* production [trade] fermentation
Hauptfermenter *m (biot)* main [production] fermenter
Hauptfraktion *f* main fraction
Hauptgärung *f* main fermentation
Hauptgruppe *f* 1. main group *(of the periodic system)*; 2. *(nomencl)* principal function; 3. major class *(of the enzyme classifying system)*
Hauptinhaltsstoff *m (hyd)* major constituent
Hauptkette *f* main chain, parent [fundamental] chain, backbone (chain), skeleton *(of a branched molecule)*
Hauptkomponente *f s.* Hauptbestandteil
Hauptkristallisation *f* primary cystallization
Hauptkultur *f (biot)* main [production] culture
Hauptkulturmedium *n (biot)* basal [main growth] medium
Hauptlauf *m (distil)* main fraction
Hauptleitung *f* main
Hauptmasse *f* bulk
Hauptmenge *f* bulk
Hauptperiode *f* reaction period *(of calorimetric measurements)*
Hauptprodukt *n* main product, major [principal, chief] product
Hauptquantenzahl *f* principal [total] quantum number
Hauptreaktion *f* main [major, principal] reaction

Hauptsatz *m* **der Thermodynamik/ nullter** zeroth law of thermodynamics
Hauptstoffwechsel *m (bioch)* primary metabolism
Hauptstrang *m* main *(as of a pipe system)*
Hauptsymmetrieebene *f (cryst)* unit [standard] plane
Hauptvalenz *f* primary [principal] valency
Hauptvalenzbindung *f* primary [major] valency bond
Haushaltabfall *m* domestic [household] waste
Haushaltabwasser *n* domestic sewage
Haushaltchemie *f* domestic chemistry
Haushaltmüll *m* domestic [household] refuse, domestic [private] rubbish
Haushaltwaschmittel *n* household (laundry) detergent, household laundering formulation, consumer detergent
Haut *f* 1. *(coat, met)* skin; 2. *(tann)* hide
Haut *f***/künstliche** artificial skin
Häutchen *n* film, membrane
hautreizend skin-irritant
Hautreizstoff *m* skin irritant
Havarie *f* upset
Haworth-Formel *f* Haworth (projection) formula *(for sugars)*
HB *s.* Brinellhärte
Hb *s.* Hämoglobin
Hb-CO *s.* Kohlenmonoxidhämoglobin
H-Bindung *f* 1. hydrogen bonding [linkage]; 2. hydrogen bridge bond
HbS *s.* Sichelzellenhämoglobin
HCH *s.* Hexachlorcyclohexan
Hebel *m* lever
heben *v* lift, elevate; promote, energize *(electrons into an excited state)*
Heber *m* 1. *(lab)* siphon, syphon; 2. *(tech)* lift
hebern *v* siphon, syphon
Hebevorrichtung *f* lift(ing device), elevator
He-Cd$^+$-Laser *m* helium-cadmium laser, cadmium-helium laser
Hefe *f* yeast
Hefe *f***/industriell genutzte** industrial yeast
hefeartig yeast-like
Hefeeiweiß *n* yeast protein
Hefefermentation *f* yeast fermentation
Hefegärung *f* yeast fermentation
Hefekultur *f* yeast culture
Hefenucleinsäure *f* yeast nucleic acid
Hefepilz *m* yeast (plant)
Hefe-RNA *f* yeast RNA
heftig vigorous *(e.g. reaction)*
Heildosis *f* therapeutic dose
Heilkraut *n* herb
Heilmittel *n* therapeutic agent, curative drug
Heilwirkung *f* curative action; curative effect
heiß 1. hot; 2. *(nucl)* highly (radio)active, hot • **heiß werden** become hot, heat
Heißaluminieren *n* hot-dip aluminizing
Heißatomchemie *f* hot-atom chemistry, recoil chemistry
Heißatomreaktion *f* hot-state reaction
Heißchlorierung *f* hot chlorination
Heißdampf *m* superheated steam
Heißfiltration *f* hot filtration
Heißgas *n* hot gas
Heißluftschrank *m* hot-air oven, air-circulating oven
Heißluftsterilisator *m* hot-air sterilizer
Heißluftsterilisierschrank *m* hot-air sterilizer
Heißprägen *n (plast)* hot embossing
heißpressen *v* hot-press, *(relating to powders also)* sinter under pressure
heißsintern *v* hot-press, sinter under pressure
heißtauchen *v* 1. *(coat)* hot-dip *(for applying organic coatings for temporary protection)*; 2. *(plast)* dip-mould *(using external moulds for producing gloves etc.)*
Heißtauchmasse *f (coat)* hot-melt [hot-dip] coating
Heißtauchschutzschicht *f (coat)* hot-melt [hot-dip] coating
Heißverzinken *n* (hot-dip) galvanizing
Heißwassertrichter *m (lab)* hot-water funnel, heating funnel, funnel heater
Heitler-London-Slater-Pauling-Theorie *f* Heitler-London-Slater-Pauling theory, HLSP theory, electron-pair [valence-bond] theory, VB theory *(quantum chemistry)*
Heizbad *n* heating bath
Heizband *n* heating tape [band], strip [band] heater
Heizbrennstoff *m* fuel
Heizdampf *m* heating steam

Heizelement *n* heating element [unit], heater
heizen *v* 1. heat, fire, fuel; 2. *(rubber)* cure, vulcanize
Heizer *m* 1. heater *(apparatus)*; 2. *(rubber)* vulcanizer, vulcanizing apparatus
Heizflamme *f* volume [non-luminous] flame *(of a Bunsen burner)*
Heizgas *n* fuel [heating] gas
Heizkammer *f* heating chamber; calandria *(of an evaporator)*
Heizkörper *m* heating unit [element], heater
Heizmantel *m* heating jacket [blanket, mantle]
Heizöl *n* fuel [heating] oil
Heizschlange *f* heating coil
Heiztischmikroskop *n* hot-stage microscope
Heizung *f* 1. heating; 2. *(rubber)* cure, vulcanization
Heizvorrichtung *f* heating device, heater
Heizwert *m* 1. *(phys ch)* calorific value *(per unit weight or unit volume)*; *(specif)* net calorific value; 2. *(broadly:)* fuel value
Heizwert *m***/molarer** *s.* Heizwert/stoffmengenbezogener
Heizwert *m***/oberer** *s.* Brennwert/spezifischer
Heizwert *m***/spezifischer** net calorific value, lower heating value
Heizwert *m***/stoffmengenbezogener** heat of combustion, heat(ing) value
Heizwert *m***/unterer** *s.* Heizwert/spezifischer
heizwertarm of low calorific value
heizwertreich of high calorific value
Heizwiderstand *m* heating resistor
Heizzug *m* (heating) flue
helikal *(bioch)* helical
Helium *n* He helium
Helium-Cadmium-Laser *m* helium-cadmium laser, cadmium-helium laser
Heliumdetektor *m (anal)* helium ionization detector
Heliumionisationsdetektor *m (anal)* helium ionization detector
Heliumkern *m* helium nucleus, α-particle
Helix *f (bioch)* helix
Helix *f***/doppelsträngige** double-strand(ed) helix, double helix
Helix *f***/dreisträngige** superhelix, triple(-stranded) helix
Helix *f***/einsträngige** single-strand(ed) helix
Helix *f***/linksdrehende [linksgängige]** left-handed helix
Helix *f***/rechtsdrehende [rechtsgängige]** right-handed helix
α-Helix *f (bioch)* alpha helix
γ-Helix *f (bioch)* gamma helix
helixbildend helix-forming
helixbrechend helix-breaking
helixförmig *(bioch)* helical
Helix-Knäuel-Übergang *m (bioch)* helix-coil transition
Helix-Knäuel-Umwandlung *f (bioch)* helix-coil transition
Helixkonformation *f* helical conformation
Helixstruktur *f* helix [helical, helicoidal] structure *(of protein molecules)*
Helizität *f* helicity *(1. quantum mechanics; 2. bonding theory)*
Helligkeit *f* brightness, luminosity, *(esp quantitatively)* luminous intensity, intensity of light
Hellreaktion *f* light reaction *(photochemistry)*
Hell-Volhard-Zelinsky-Reaktion *f* Hell-Volhard-Zelinsky reaction *(a-halogenation of aliphatic carboxylic acids)*
Helmholtz-Fläche *f (el ch)* Helmholtz plane
Helminthagogum *n (pharm)* helminthagogue, vermifuge
Hemiacetal *n (org ch)* hemiacetal
Hemicellulose *f* hemicellulose, pseudocellulose
Hemiketal *n* hemiketal *(hemiacetal of a ketone)*
Hemikolloid *n* hemicolloid
Hemimellit(h)säure *f* hemimellitic acid, benzene-1,2,3-tricarboxylic acid
Hemiterpen *n (org ch)* hemiterpene
hemmen *v* inhibit, retard
hemmend inhibitory, inhibitive, retardant
Hemmstoff *m* 1. inhibiting [retarding] substance [agent], inhibitor, retarder, anticatalyst, negative catalyst; 2. *(biol)* growth inhibitor
Hemmstoff *m***/kompetitiver** *(bioch)* competitive inhibitor

Hemmung *f* inhibition, retardation, anticatalysis, negative catalysis
Hemmung f/kompetitive competitive inhibition *(enzyme kinetics)*
Hemmung f/konzertierte *(bioch)* concerted inhibition
Hemmung f/multivalente *s.* Hemmung/konzertierte
Hemmung f/nicht kompetitive noncompetitive inhibition *(enzyme kinetics)*
Hemmung f/unkompetitive uncompetitive [anticompetitive] inhibition *(enzyme kinetics)*
Hemmwirkung *f (bioch)* inhibitory action; inhibitory effect
Hempel-Pipette *f* Hempel gas pipette
Hendec... *siehe: Undec...*
Henry-Konstante *f (phys ch)* Henry's law constant
Heparprobe *f* hepar test *(for detecting sulphur)*
hepatotoxisch *(med)* hepatotoxic
Heptachlor *n* heptachlor *(an insecticide)*
Heptadecansäure *f* heptadecanoic acid
Heptan *n* heptane
Heptanal *n* heptanal
Heptan-1-carbonsäure *f* heptane-1-carboxylic acid, octanoic acid
Heptan-1,7-dicarbonsäure *f* *heptane-1,7-dicarboxylic acid, *nonanedioic acid, azelaic acid
Heptandisäure *f* *heptanedioic acid, *pentane-1,5-dicarboxylic acid, pimelic acid
Heptansäure *f* heptanoic acid
heptavalent heptavalent, septivalent
Heptose *f* heptose *(monosaccharide containing 7 carbon atoms per molecule)*
Heptoxid *n* heptaoxide
Heptoxotetraborat *n* $M^I_2B_4O_7$ heptaoxotetraborate, tetraborate
Heptoxotetraborsäure *f* heptaoxotetraboric acid, tetraboric acid
***n*-Heptylaldehyd** *m s.* Heptanal
***n*-Heptylalkohol** *m* heptan-1-ol
herabrieseln *v* trickle down
herabrinnen *v* trickle down
herabsetzen *v* lower, reduce, decrease; slow down *(the velocity)*; relieve (down) *(the pressure)*
Herabsetzung *f* lowering, reduction, decrease; slowing-down *(of velocity)*; relief *(of pressure)*
herabtröpfeln *v* trickle down
herausdestillieren *v* distil out, top
herausdiffundieren *v* diffuse out
herausdrücken *v* push [blow] out *(as from a reactor)*; squirt (out) *(as from a nozzle)*
herauslösen *v* dissolve out, lixiviate, leach (out); *(esp relating to adsorbed substances:)* eluate, elute
herauspressen *v* press [squeeze] out, expel *(e.g. oil)*
herausschleppen *v (distil)* entrain out
herausschleudern *v (nucl)* eject
herausspülen *v* rinse out; eluate, elute *(adsorbed substances from a solid adsorbent)*
heraustreiben *v* expel, drive out [off] *(e.g. gases)*
herb harsh, hard, sour; rough, dry *(wine)*
Herbe *f (food)* harshness, hardness, tartness, sourness; roughness, dryness *(of wine)*
Herbheit *f (food)* harshness, hardness, tartness, sourness; roughness, dryness *(of wine)*
herbizid herbicidal(ly active)
Herbizid *n* herbicide, weed-killer, weed control agent
Herbizid n/nicht selektives non-selective herbicide
Herbizid n/selektives [selektiv wirkendes] selective herbicide
Herbizid n/systemisches systemic [translocated] herbicide, translocation weed-killer
Herd *m* 1. *(min tech)* (concentrating) table, concentrate [concentrator] table; 2. *(met)* hearth
Herdarbeit *f s.* Herdsortieren
Herdaufbereitung *f s.* Herdsortieren
Herdfrischstahl *m* open-hearth steel
Herdfrischverfahren *n* open-hearth process
Herdfrischverfahren n/basisches basic open-hearth process
Herdfrischverfahren n/saures acid open-hearth process
Herdofen *m* hearth furnace
Herdplatte *f* deck, table *(of a concentrating table)*
Herdsortieren *n (min tech)* tabling

Herdsortieren *n*/**nasses** wet tabling
Herdsortieren *n*/**trockenes** dry tabling
hergestellt/großtechnisch produced on the large scale
Herkunft *f* origin, source • **pflanzlicher Herkunft** of vegetable [plant] origin, plant-derived • **tierischer Herkunft** of animal origin
hermetisch (abgeschlossen, dicht, verschlossen) hermetic(al)
Heroin *n* heroin, diacetylmorphine *(a narcotic)*
herstellen *v* 1. *(tech)* manufacture, make, produce; 2. *(lab)* prepare; 3. establish *(e.g. equilibrium or contact)*
herstellen *v*/**gezielt** make to measure, tailor(-make) *(e.g. polymers)*
herstellen *v*/**großtechnisch** produce on the large scale
herstellen *v*/**Masterbatches [Vormischungen]** *(rubber)* masterbatch, mix into a masterbatch
Hersteller(betrieb) *m* manufacturer, maker, producer
Herstellerfirma *f* manufacturer, maker, producer
Herstellung *f* 1. *(esp relating to know-how:)* manufacture, make, *(esp relating to economical aspects:)* production; 2. *(lab)* preparation; 3. establishment *(as of equilibrium or contact)*
Herstellung *f*/**großtechnische** large-scale production; *(as opposed to pilot-plant production:)* full-scale factory production
Herstellung *f* **in halbtechnischem Maßstab** pilot(-plant-scale) production
Herstellung *f* **von Vormischungen** *(rubber)* masterbatching
Herstellungsmethode *f* 1. manufacturing method, method of production; 2. *(lab)* method of preparation, preparative method
Herstellungsverfahren *n* manufacturing process
herunterkühlen *v* cool down
herunterrieseln *v* trickle down
Herzgift *n* heart poison
Herzglykosid *n* cardiac glycoside
Herz-Verbindung *f (dye)* (intermediate) Herz compound *(a benzodithiazole)*
herzwirksam cardioactive, cardiac-active
Heteroalken *n* heteroalkene
Heteroallen *n* heteroallene
heteroanalog hetero-analogous
Heteroaren *n* heteroarene, hetarene
Heteroaromat *m* heteroarene, hetarene
Heteroarylgruppe *f* heteroaryl [hetaryl] group
Heteroarylrest *m* heteroaryl [hetaryl] group
Heteroatom *n* heteroatom
heterocyclisch *(org ch)* heterocyclic
Heterocyclus *m (org ch)* heterocycle, heterocyclic (compound)
Heterocyclylgruppe *f* heterocyclyl group
Heterocyclylrest *m* heterocyclyl group
Heterodien *n* heterodiene
heterofermentativ *(bioch)* heterofermentative
heterogen heterogeneous
Heterogenität *f* heterogeneity
Heterogenkatalyse *f* heterogeneous catalysis
Heterokumulen *n* heterocumulene
Heterolyse *f* heterolysis, heterolytic cleavage [fission, bond fission]
heterolytisch heterolytic
heteropolymer heteropolymeric
Heteropolymer(e) *n* heteropolymer, heterogeneous polymer
Heteropolymerisat *n* heteropolymer, heterogeneous polymer
Heteropolymerisation *f* heteropolymerization
Heteropolysaccharid *n* heteropolysaccharide
Heteropolysäure *f* heteropoly acid
Heteroring *m* heterocyclic ring
heterotaktisch *(plast)* heterotactic
Heterotaktizität *f (plast)* heterotacticity
heterotroph *(biol)* heterotrophic
Heterotrophie *f (biol)* heterotrophy
HETP-Wert *m (distil)* height equivalent to a theoretical plate, HETP, plate height
Hevea-Kautschuk *m* hevea rubber *(from Hevea brasiliensis (H.B.K.) Muell. Arg.)*
Hexachlorbenzen *n* hexachlorobenzene, HCB, perchlorobenzene
Hexachlorbenzol *n* hexachlorobenzene, HCB, perchlorobenzene

Hexachlorcyclohexan *n* hexachlorocyclohexane
Hexachlorethan *n* hexachloroethane, perchloroethane
Hexachlorid *n* hexachloride
Hexachlorophen *n* hexachlorophene, *3,3′,5,5′,6,6′-hexachloro-2,2′-dihydroxydiphenylmethane
Hexachloroplatinat(IV) *n* $M^I_2[PtCl_6]$ hexachloroplatinate(IV), chloroplatinate(IV)
Hexachloroplatin(IV)-säure *f* hexachloroplatinic(IV) acid, chloroplatinic(IV) acid
Hexacosansäure *f* *hexacosanoic acid, cerotic acid, cerinic acid
Hexacyanocobaltat(III) *n* $M^I_3[Co(CN)_6]$ hexacyanocobaltate(III), cyanocobaltate(III)
Hexacyanoferrat(III) *n* $M^I_3[Fe(CN)_6]$ hexacyanoferrate(III), cyanoferrate(III)
Hexacyanoplatinat(IV) *n* $M^I_4[Pt(CN)_6]$ hexacyanoplatinate(IV)
Hexadecan *n* hexadecane
Hexadecansäure *f* hexadecanoic acid
Hexa-2,4-diensäure *f* *hexa-2,4-dienoic acid, sorbic acid
Hexadien-(2,4)-säure *f* *hexa-2,4-dienoic acid, sorbic acid
Hexafluorid *n* hexafluoride
Hexafluorokieselsäure *f* hexafluorosilicic acid, fluorosilicic acid
Hexafluorosilicat *n* $M^I_2[SiF_6]$ hexafluorosilicate, fluorosilicate
hexagonal *(cryst)* hexagonal
Hexahelicen *n (org ch)* hexahelicene
Hexahydrat *n* hexahydrate
Hexahydrophenol *n* hexahydrophenol, cyclohexanol
Hexahydrotoluen *n* hexahydrotoluene, *methylcyclohexane
Hexahydrotoluol *n* hexahydrotoluene, *methylcyclohexane
Hexamethylendiamin *n* hexamethylene diamine, HMDA
Hexamethylentetramin *n* hexamethylenetetramine, hexamine, methenamine
Hexamin *n* 1. hexamine, methenamine, hexamethylenetetramine; 2. di-2,4,6-trinitrophenylamine, hexanitrodiphenylamine, hexite
Hexan *n* hexane
Hexan-1,6-dicarbonsäure *f* hexane-1,6-dicarboxylic acid, octanedioic acid
Hexandicarbonsäure-(1,6) *f* hexane-1,6-dicarboxylic acid, octanedioic acid
Hexandisäure *f* *hexanedioic acid, adipic acid
Hexanitrocobaltat(III) *n* $M^I_3[Co(NO_2)_6]$ hexanitrocobaltate(III), nitrocobaltate(III)
Hexan-1-ol *n* hexan-1-ol
Hexanol-(1) *n* hexan-1-ol
Hexansäure *f* hexanoic acid
1,4,7,10,13,16-Hexaoxacyclooctadecan *n* 1,4,7,10,13,16-hexaoxacyclooctadecane, 18-crown 6
Hexaquo... hexaaqua..., hexaaquo...
hexasubstituiert hexasubstituted
hexavalent hexavalent
Hexavalenz *f* hexavalence
Hex-1-en *n* hex-1-ene
Hexen-(1) *n* hex-1-ene
Hex-1-in *n* hex-1-yne
Hexin-(1) *n* hex-1-yne
Hexit *m* hexitol *(any of the hexahydroxy alcohols $HOCH_2[CHOH]_4CH_2OH$)*
Hexitol *n* hexitol *(any of the hexahydroxy alcohols $HOCH_2[CHOH]_4CH_2OH$)*
Hexose *f* hexose *(monosaccharide containing 6 oxygen atoms per molecule)*
Hexoxotellursäure *f* hexaoxotelluric acid, orthotelluric acid, telluric acid
***n*-Hexylalkohol** *m s.* Hexan-1-ol
HF-... *siehe: 1. Hochfrequenz...; 2. Fluorwasserstoff...*
HFS *s.* Hyperfeinstruktur
Hgw *s.* Hartgewebe
HID *s.* Heliumionisationsdetektor
Hilfsausrüstung *f* ancillary equipment
Hilfselektrode *f* auxiliary electrode
Hilfslöser *m* 1. cosolvent, indirect [latent] solvent; 2. *(distil)* solvent
Hilfslösungsmittel *n* 1. cosolvent, indirect [latent] solvent; 2. *(distil)* solvent
Hilfsmittel *n* 1. auxiliary contrivance [device]; 2. *s.* Hilfsstoff
Hilfsstoff *m* 1. auxiliary [supplementary] agent, aid; 2. *(pharm)* adjuvant; corrective *(in building-up active-substance mixtures)*; 3. *(distil)* separating agent
Hill-Reaktion *f (bioch)* Hill reaction
hinauswandern *v* migrate out *(as of ions)*

Hinderung *f* hindrance
Hinderung *f* **der freien Drehbarkeit** hindered rotation
Hinderung *f***/kinetische** kinetic hindrance
Hinderung *f***/sterische** steric hindrance [inhibition, limitation]
hindurchdiffundieren *v* diffuse through
hindurchdrücken *v* press [force] through
hindurchlaufen *v* 1. pass (through), percolate *(as through a medium)*; 2. *(filtr)* strain, filter; pass (through) *(as through a sieve)*
hindurchlaufen lassen *v* 1. pass (through), percolate *(e.g. through a medium)*; 2. *(filtr)* strain, filter
Hindurchlaufen *n* 1. passage, percolation *(as through a medium)*; 2. *(filtr)* straining, filtering; passage *(as through a sieve)*
hindurchleiten *v* pass
hindurchperlen *v* bubble *(of a gas passing a liquid)*
hindurchperlen lassen *v* bubble *(a gas through a liquid)*
hindurchpressen *v* force through
hindurchsickern *v* percolate
hineinfressen *v***/sich** eat *(as of an acid into metal)*
hineinwandern *v* migrate in *(as of ions)*
Hinreaktion *f* direct [forward] reaction
Hinsberg-Reaktion *f* Hinsberg reaction *(for separating primary, secondary, and tertiary amines)*
Hintergrund *m (nucl, anal)* background
Hintergrundkonzentration *f* background [baseline] concentration, background level
hin- und herbewegen *v* 1. reciprocate; 2. *(tann)* rock *(pelts in a rocker frame)*
hin- und herbewegen *v***/sich** reciprocate
hinzufügen *v* add
hinzugeben *v* add
H-Ionen... *siehe: Wasserstoffionen...*
Hippursäure *f* hippuric acid, benzoylaminoacetic acid
Hirschhornsalz *n* (salt of) hartshorn, commercial ammonium carbonate, sal volatile *(a mixture consisting of ammonium hydrogencarbonate and ammonium carbamate)*
Hirsch-Trichter *m* Hirsch funnel
His *s.* Histidin
Histamin *n* histamine, 4-(ω-aminoethyl)-glyoxaline
Histidin *n* histidine, 2-amino-3-imidazolylpropionic acid
Histochemie *f* histochemistry
Hitze *f* heat
Hitze... *siehe auch: Wärme...*
hitzebeständig heat-resistant, thermoduric *(microorganisms)*; heat-resistant *(steel, more than 600 °C)*
Hitzebeständigkeit *f* heat resistance *(1. of microorganisms; 2. of steel, more than 600 °C)*
hitzedenaturierbar heat-denaturable
Hitzedenaturierung *f* heat denaturation
Hitzelabilität *f* heat lability
hitzeresistent heat-resistant, thermoduric *(microorganisms)*
Hitzeresistenz *f* heat resistance *(of microorganisms)*
Hitzesterilisation *f* heat sterilization
HKL *s.* Hohlkatodenlampe
HLSP-Theorie *f s.* Heitler-London-Slater-Pauling-Theorie
HM *s.* Hartmetall
HM-... *siehe: Halbmikro...*
hnRNA heterogeneous nuclear RNA
H_o *(oberer Heizwert) s.* Brennwert/spezifischer
hochaktiv 1. highly active; 2. *(nucl)* high-level active, highly radioactive
hochalkalisch highly alkaline, superalkaline
hocharomatisch highly aromatic
hochbelastet highly [heavily] loaded
hochchloren *v* superchlorinate *(water)*
hochchlorieren *v* superchlorinate *(water)*
Hochchlorung *f (hyd)* superchlorination
hochdispers highly disperse
Hochdosistoleranz *f* high dose tolerance *(immunology)*
Hochdruckbehälter *m* high-pressure vessel
Hochdruckchemie *f* high-pressure chemistry
Hochdruckdampf *m* high-pressure steam, HP steam
Hochdruckfiltration *f* high-pressure filtration
Hochdruck-Flüssigkeitschromatogra-

phie *f* high-performance [high-pressure] liquid chromatography, HPLC
Hochdruck-flüssigkeitschromatographisch by high-pressure liquid chromatography
Hochdruckhydrierung *f* high-pressure hydrogenation
Hochdruckhydrier(ungs)verfahren *n* **nach Bergius** Bergius process, berginization
Hochdruckmodifikation *f* high-pressure modification
Hochdruckpolyethylen *n* high-pressure [low-density, L.D.] polyethylene, LDPE, branched polyethylene
Hochdruckpresse *f (lab)* high-pressure press
Hochdrucksynthese *f* high-pressure synthesis
Hochdruckvergasung *f (coal)* high-pressure gasification
Hochdurchsatz *m* high throughput
Hochdurchsatztesten *n* high throughput screening
hochempfindlich highly sensitive
Hochenergiechemie *f* high-energy chemistry
hochexplosiv highly explosive, violently [dangerously] explosive
Hochfest-Kohlenstofffaser *f* carbon fibre type HT *(Young's modulus between 150 and 300 GPa)*
hochfeuerfest highly refractory, super-refractory
hochflüchtig highly volatile
hochfördern *v* pass up(wards), elevate
Hochfrequenzerhitzung *f* high-frequency [radio-frequency] heating, electronic heating
Hochfrequenzerwärmung *f* high-frequency [radio-frequency] heating, electronic heating
Hochfrequenzkonduktometrie *f (anal)* high-frequency conductometry
Hochfrequenzpolarographie *f (anal)* radio-frequency polarography
Hochfrequenztitration *f* high-frequency titration, impedimetric titration
hochgekohlt high-carbon *(e.g. steel)*
hochgiftig highly [very] poisonous [toxic], quite poisonous [toxic], dangerously poisonous
hochgliedrig many-membered, multimembered
hochheizen *v* heat up
hochhitzebeständig resistant to high temperature(s)
Hochhitzebeständigkeit *f* resistance to high temperature(s)
hochkohlenstoffhaltig high-carbon *(e.g. steel)*
hochkonzentriert highly concentrated, high-concentration
hochkriechen *v* creep up *(the sides of a vessel)*
Hochkurzerhitzung *f (food)* short-time heat processing, high-temperature short-time pasteurization [heat treatment], HTST pasteurization
Hochkurzpasteurisation *f (food)* short-time heat processing, high-temperature short-time pasteurization [heat treatment], HTST pasteurization
hochlegiert highly alloyed
Hochleistungs-Dünnschichtchromatographie *f* high-performance thin-layer chromatography, HPTLC
Hochleistungselement *n (agric)* microelement, micronutrient, minor (nutrient) element
hochleistungsfähig high-capacity, large-capacity
Hochleistungsflüssigchromatographie *f* high-performance [high-pressure] liquid chromatography, HPLC
Hochleistungs-Flüssigkeits-Affinitätschromatographie *f* high-performance liquid affinity chromatography, HPLAC
Hochleistungskunststoff *m (plast)* high-performance polymer
Hochleistungsstamm *m (biot)* high-producing [high-yield] strain *(of microorganisms)*
Hochleistungszentrifuge *f* high-speed [high-capacity] centrifuge
Hochmodul-Kohlenstofffaser *f* carbon fibre type HM [high modulus] *(Young's modulus larger than 300 GPa)*
hochmolekular high-molecular
Hochofen *m* blast furnace
Hochofenfutter *n* blast-furnace lining
Hochofen(gicht)gas *n* blast-furnace gas
Hochofenkoks *m* (blast-)furnace coke

Hochofenmöller *m* charge (stock), burden, stock, feed
hochorientiert highly oriented *(bond)*
hochpolymer high-polymeric, highly polymerized
Hochpolymer(e) *n* high polymer
hochprozentig high-percent, high-percentage, high-analysis; *(relating to spirits:)* strong, *(Am)* hard
hochreaktionsfähig highly reactive
hochreaktiv highly reactive
hochrein highly pure [purified], high-purity
Hochreinigung *f* ultrapurification *(zone melting)*
hochschmelzend high-melting(-point), high-fusion
Hochseeverklappung *f* offshore disposal *(of waste)*
hochselektiv highly selective
hochsiedend high-boiling, heavy
Hochspannungselektrophorese *f* high-voltage electrophoresis
höchstauflösend *(spectr)* very-high-resolution
Höchstdruckkessel *m* very-high-pressure boiler
Höchstdruckschmierstoff *m* extreme-pressure lubricant, EP lubricant
Höchstkonzentration *f***/zulässige** *(tox)* maximum permissible [admissible, allowable] concentration, MAC
Höchstmenge *f***/duldbare [zulässige]** *(tox)* (maximum) tolerance
höchstrein ultrapure
Hochtemperaturbehandlung *f* high-temperature treatment
hochtemperaturbeständig resistant to high temperature(s), stable at high temperature(s)
Hochtemperaturbeständigkeit *f* high-temperature resistance [stability, durability]
Hochtemperaturchemie *f* high-temperature chemistry
Hochtemperaturkoks *m* high-temperature coke
Hochtemperaturkorrosion *f* high-temperature corrosion, dry [hot] corrosion
Hochtemperaturlegierung *f* high-temperature alloy
Hochtemperaturofen *m* 1. high-temperature furnace; 2. *(ceram)* high-temperature kiln
Hochtemperaturpyrolyse *f* high-temperature py-rolysis, HTP
Hochtemperaturreaktor *m* high-temperature reactor
Hochtemperaturverhalten *n* high-temperature behaviour
Hochtemperaturverkokung *f* high-temperature carbonization [coking]
Hochtemperaturwerkstoff *m* high-temperature material
hochtoxisch *s.* hochgiftig
hochungesättigt highly unsaturated
Hochvakuumdestillation *f* high-vacuum distillation
Hochvakuumtechnik *f* high-vacuum technique
hochveredelt highly refined, high-added-value *(product)*; highly finished *(surface)*
hochverzeigt *(plast)* hyperbranched *(in polymer structures)*
hochviskos highly viscous, high-viscosity
hochwirksam highly active, potent
Hof *m (phot, spectr, chromat)* halo
Hofmann-Abbau *m* Hofmann degradation *(of amides)*
Hofmann-Eliminierung *f* Hofmann elimination *(of quaternary ammonium hydroxides)*
Hofmann-Martius-Umlagerung *f* Hofmann-Martius rearrangement *(of hydrohalides of N-alkylanilines into alkylanilines)*
Hofmann-Orientierung *f (org ch)* Hofmann orientation
Höhe *f* **des Schornsteins/effektive** effective chimney [stack] height
Höhe *f* **einer theoretischen Trennstufe** height equivalent to a theoretical stage [plate], HETS, HETP
Höhe *f* **einer Übertragungseinheit** *(distil)* height of one transfer unit, HTU
Höhe *f***/geodätische** potential head
Höhenformel *f***/barometrische** barometric formula
Höhenliniendiagramm *n* **der Potenzialfläche** potential-energy contour map
höhergliedrig higher-membered
höherschmelzend higher-melting

höherwertig of higher valence, higher-valent, higher-valency *(chemical-bond theory)*; polyhydric *(alcohol)*; higher-analysis *(as of a commercial product)*

Hohlfasermembran *f (hyd)* hollow fibre membrane *(reverse osmosis)*

Hohlform *f (plast)* die

Hohlkatodenlampe *f* hollow-cathode lamp

Hohlkörperblasen *n (plast)* blow moulding

Hohlraum *m* 1. cavity, hollow [void] space; 2. *(plast)* void *(a defect)*; 3. *(chromat)* interstice

Hohlraumstrahler *m* black-body radiator

Hohlraumstrahlung *f* black-body radiation

Hohlraumvolumen *n* 1. void [pore] volume *(of a catalyst)*; 2. *(chromat)* interstitial volume; 3. *(soil)* volume of pore space

Holarrhenaalkaloid *n* holarrhena alkaloid

Holländer *m (pap)* Hollander, Hollander beater [beating engine], pulp [stuff] engine, pulp grinder

Holmium *n* Ho holmium

Holocellulose *f* holocellulose

holoedrisch *(cryst)* holohedral

Holoenzym *n* holoenzyme

holokristallin holocrystalline

Holosid *n (org ch)* holoside

Holzaufschluss *m (pap)* pulping of wood

Holzbeize *f* stain

Holzcellulose *f* wood cellulose

Holzchemie *f* wood chemistry

Holzdestillation *f* wood distillation

Holzessig *m* wood vinegar, pyroligneous acid *(crude acetic acid obtained by wood distillation)*

Holzfaser *f* wood fibre

Holzgas *n* wood(-distillation) gas; wood producer gas

Holzgeist *m* wood [pyroligneous] spirit, natural methanol

Holzgeist *m***/roher** wood [pyroligneous] spirit, natural methanol

Holzgummi *n* xylan *(a pentosan)*

Holzhydrolyse *f* wood hydrolysis

Holzkohle *f* (wood) charcoal

Holzkonservierung *f* 1. wood preservation; 2. *(build)* timber proofing

Holzschliff *m (pap)* wood pulp, *(specif)* mechanical pulp • **zu Holzschliff verschleifen** reduce to pulp, make into (a) pulp, pulp

Holzschliff *m***/chemischer** chemigroundwood

Holzschliff *m***/mechanischer** mechanical pulp [wood pulp], MWP, groundwood (pulp)

Holzschliffpappe *f (pap)* wood-pulp board, board [sheets, laps] of mechanical wood pulp

Holzschutz *m* 1. wood preservation; 2. *(build)* timber proofing

Holzschutzmittel *n* wood preservative

Holzstoff *m* 1. lignin; 2. *s.* Holzschliff

Holzsubstanz *f* ligneous substance

Holzteer *m* wood tar

Holzverkohlung *f* wood carbonization, charcoal burning

Holzverzuckerung *f* saccharification of wood

Holzzellstoff *m* 1. *(pap)* (chemical) wood pulp, CWP; 2. *s.* Holzcellulose

Holzzucker *m* wood sugar, xylose

HOMO *s.* Molekülorbital/höchstes besetztes

homoaromatisch *(org ch)* homoaromatic

Homoaromatizität *f (org ch)* homoaromaticity

homocyclisch *(org ch)* homocyclic, isocyclic, carbocyclic

Homocyclus *m (org ch)* homocycle, homocyclic (compound)

homodispers monodisperse

homofermentativ *(bioch)* homofermentative

homogen homogeneous

homogen/vollkommen [vollständig] perfectly homogeneous, homogeneous throughout

Homogenisator *m* homogenizer

Homogenisierapparat *m* homogenizer

homogenisieren *v* homogenize

Homogenisierung *f* homogenization, homogenizing *(as of emulsions)*

Homogenisierungsglühen *n* homogenization, homogenizing *(of alloys)*

Homogenität *f* homogeneity

Homogenkatalyse *f* homogeneous catalysis

Homogenkinetik *f* homogeneous kinetics
Homogenreaktor *m (nucl)* homogeneous reactor
Homokonjugation *f (org ch)* homoconjugation
homolog homologous
Homolog(e) *n* homologue, homolog
Homologisierung *f* homologization
Homolyse *f* homolysis, homolytic (bond) cleavage, homolytic (bond) fission, bond dissociation
homolytisch homolytic
homöopolar covalent, homopolar, nonpolar *(chemical-bond theory)*
homopolymer homopolymeric
Homopolymer(e) *n* homopolymer
Homopolymerisat *n* homopolymer
Homopolymerisation *f* homopolymerization
Homopolysaccharid *n* homopolysaccharide
Homotropyliumkation *n* homotropylium cation
Hooker-Diaphragmazelle *f* Hooker (diaphragm) cell *(for electrolyzing brine)*
Hooker-Zelle *f* Hooker (diaphragm) cell *(for electrolyzing brine)*
hopfen *v* hop *(the wort)*
Hopfenbittersäure *f* hop bitter acid
Hopfenextrakt *m* hop extract(ive)
Hopfenharz *n* hop resin
Höppler-Kugelfallviskosimeter *n* Höppler (falling-ball) viscometer
Höppler-Viskosimeter *n* Höppler (falling-ball) viscometer
HO_2-Radikal *n* HO_2-radical, hydroperoxyl radical
Horde *f* (kiln) floor, tray *(as of a dryer)*
Horizontalbeziehung *f* horizontal relationship *(in the periodic system)*
Horizontalchromatographie *f* horizontal chromatography
Horizontalentwicklung *f (chromat)* horizontal development
Horizontalzelle *f* horizontal (diaphragm) cell *(electrolysis)*
Horizontalzentrifuge *f* horizontal centrifuge
Hormon *n* hormone
Hormon *n***/adrenocorticotropes** adrenocorticotropic hormone, corticotropin
Hormon *n***/antidiuretisches** antidiuretic hormone, ADH, vasopressin
Hormon *n***/follikelstimulierendes** follicle-stimulating hor-mone, FSH
Hormon *n***/luteinisierendes** luteinizing hormone, LH, interstitial-cell-stimulating hormone, ICSH, prolan B
Hormon *n***/luteotrop(h)es** *s.* Prolactin
Hormon *n***/melanozytenstimulierendes** melanocyte-stimulating hormone, *melanotropin
Hormon *n***/neurohypophysäres** neurohypophyseal [posterior-lobe] hormone
Hormon *n***/östrogenes** oestrus-producing hormone
Hormon *n***/somatotropes** somatotropic hormone, somatropin, STH
Hormon *n***/thyreotropes** thyrotrop(h)ic [thyroid-stimulating] hormone, TH, thyrotrop(h)in
hormonal hormonal • **hormonal gesteuert** under hormonal control
Hosenmischer *m* twin-shell [vee-type] mixer [blender]
Houben-Hoesch-Synthese *f* Houben-Hoesch synthesis *(of ketones)*
Houdry-Festbettverfahren *n (petrol)* Houdry fixed-bed process
Houdry-Krackverfahren *n***/katalytisches** *(petrol)* Houdry catalytic cracking process
HR *s.* Rockwellhärte
HSAB-Konzept *n* HSAB principle *(acid-base theory)*
HSK-Zyklus *m s.* Hatch-Slack-Kortschak-Zyklus
HT-... *siehe: Hochtemperatur...*
HTU *s.* Höhe einer Übertragungseinheit
H_u *(unterer Heizwert) s.* Heizwert/spezifischer
Huang-Minlon-Reaktion *f* Huang-Minlon reaction [reduction] *(of aldehydes and ketones)*
Huang-Minlon-Reduktion *f* Huang-Minlon reaction [reduction] *(of aldehydes and ketones)*
Hub *m* stroke (length) *(of a pump)*
Hubkolbenpumpe *f* reciprocating pump
Hückel-Molekülorbital *n* Hückel molecular orbital
Hückel-Regel *f (org ch)* Hückel (4n + 2) rule, Hückel rule *(relating to conjugated*

hydrocarbons having 4n + 2π-electrons)
Hülle *f* 1. shell, envelope, sheath *(of an atom)*; 2. *(tech)* jacket, sheath(ing), cover, shell
Hülle-Kern-Teilchen *npl s.* Kern-Schale-Teilchen
Hüllenelektron *n* extranuclear electron
Hüllprotein *n* coat [capsid] protein
Hülse *f* 1. *(tech)* socket, jacket; socket *(of a ground joint)*; 2. *(lab)* thimble *(for extracting)*
Humanserumalbumin *n (pharm, bioch)* human serum albumin, HSA
Humantoxizität *f* human toxicity
Humat *n (soil)* humate
Hume-Rothery-Phase *f (cryst)* Hume-Rothery phase
Huminsäure *f* humic acid
Huminstoff *m* humic substance [matter]
Huminsubstanz *f* humic substance [matter]
Humulon *n* humulone, α-lupulinic acid
Humus *m* humus
Humussäure *f* humic acid
Hund-Mulliken-Lennard-Jones-Hückel-Theorie Hund-Mulliken-Lennard-Jones-Hückel theory, molecular-orbital theory
Hunsdiecker-Reaktion *f* 1. Hunsdiecker reaction *(decarboxylation of the silver salt of an organic acid)*; 2. *s.* Hunsdiecker-Spaltung
Hunsdiecker-Spaltung *f* Hunsdiecker cleavage *(of esters)*
Hustenmittel *n* antitussive
Hut *m (geol)* cap
Hütte *f (met)* refinery, smelting plant
Hüttenchemie *f* metallurgical chemistry
Hüttenchemiker *m* metallurgical chemist
Hüttenkoks *m* metallurgical [blast-furnace] coke
Hüttenkunde *f* metallurgy
hüttenmännisch metallurgical
Hüttenrauch *m* flue dust
Hüttenstaub *m* flue dust
Hüttenwesen *n* metallurgy
Hüttenzement *m* blast-furnace (slag) cement, slag cement
HV *s.* Vickers-Härte
Hyaluronsäure *f* hyaluronic acid *(a mucopolysaccharide)*
hybrid hybrid
***sp*-Hybrid** *n sp*-hybrid, digonal hybrid
hybridisieren *v* hybridize
Hybridisierung *f (bioch)* hybridization
***sp*-Hybridisierung** *f sp* hybridization, digonal hybridization
sp^2-Hybridisierung *f* sp^2-hybridization, trigonal hybridization
sp^3-Hybridisierung *f* sp^3 hybridization, tetrahedral hybridization
Hybridmaterial *n* hybrid material
Hybridom *n (biot)* hybridoma (cell)
Hybridorbital *n* hybrid (bond) orbital
Hydantoinsäure *f* hydantoic acid, ureidoacetic acid
Hydnocarpussäure *f* hydnocarpic acid, *11-(2-cyclopentenyl) undecanoic acid
Hydracrylsäure *f* hydracrylic acid, *3-hydroxypropionic acid
Hydrat *n* hydrate
Hydratation *f* hydration, aqua(tiza)tion
Hydratationssphäre *f* hydration sphere
Hydratationszahl *f* hydration number
Hydrathülle *f* hydration sheath, water envelope
hydratisieren *v* hydrate, aquate
Hydratisierung *f* hydration, aqua(tiza)-tion
Hydratwasser *n* water of hydration
Hydraulikflüssigkeit *f* hydraulic fluid [medium]
Hydraulikspeicher *m* hydraulic accumulator
Hydrazid *n* hydrazide
Hydrazidimid *n* hydrazide imide
Hydrazin *n* hydrazine
Hydrazinolyse *f* hydrazinolysis
Hydrazinspaltung *f* hydrazinolysis
Hydrazoaren *n* hydrazoarene, 1,2-diarylhydrazine, 1,2-diaryldiazane
Hydrazobenzen *n* hydrazobenzene
Hydrazobenzol *n* hydrazobenzene
Hydrazon *n* hydrazone
Hydrazoverbindung *f* hydrazo compound
Hydrid *n* hydride
hydrierbar hydrogenable
hydrieren *v* hydrogenate, hydrogenize
Hydrierkatalysator *m* hydrogenating [hydrogenation] catalyst
Hydrierofen *m* converter
Hydrierung *f* hydrogenation

Hydrierung f/katalytische catalytic hydrogenation
Hydrierung f/selektive selective hydrogenation
Hydrierungs... *siehe: Hydrier...*
Hydroaromaten *pl* hydroaromatic compounds
hydroaromatisch hydroaromatic
Hydroborierung *f* hydroboration *(addition of diborane to alkenes)*
Hydrocellulose *f* hydrocellulose
Hydrochemie *f* hydrochemistry, water [aquatic] chemistry
Hydrochinon *n* quinol, hydroquinone, *p*-dihydroxybenzene
Hydrochlorid *n* hydrochloride
Hydrochlorierung *f* hydrochlorination
Hydrocyanierung *f* hydrocyanation
Hydrodesalkylierung *f* hydrodealkylation, reductive dealkylation
Hydrodesulfurierung *f* hydrodesulphurization, HDS
Hydrofluorid *n* hydrofluoride
Hydrofluorierung *f* hydrofluorination
Hydroformat *n (petrol)* hydroformate
hydroformieren *v (petrol)* hydroform *(to reform by catalytic dehydrogenation and cyclization)*
Hydroforming-Produkt *n (petrol)* hydroformate
Hydroformylierung *f* hydroformylation, oxo synthesis
Hydrogel *n* hydrogel
Hydrogencarbonat *n* M^IHCO_3 hydrogencarbonate
Hydrogenfluorid *n* M^IHF_2 hydrogen difluoride
Hydrogenolyse *f* hydrogenolysis
Hydrogenperoxid *n* hydrogen peroxide, dioxidane
Hydrogenphosphat *n* $M^I_2HPO_3$ hydrogenphosphate
Hydrogensalz *n* acid salt
Hydrogensulfat *n* M^IHSO_4 hydrogensulphate
Hydrogensulfid *n* M^IHS hydrogensulphide
Hydrogensulfit *n* M^IHSO_3 hydrogensulphite
Hydroisomerisierung *f* hydroisomerization, reductive isomerization
Hydroklassieren *n* wet classification
Hydrokondensation *f* hydrocondensation, reductive condensation
Hydrokracken *n (petrol)* hydrocracking, hydrogenation cracking, HC
Hydrokrackprodukt *n (petrol)* hydrocrackate
Hydrolase *f* hydrolase, hydrolytic enzyme
Hydrolysat *n* hydrolyzate
Hydrolyse *f* hydrolysis • **Hydrolyse erleiden** undergo hydrolysis, hydrolyze
Hydrolyse f/alkalische alkaline [base] hydrolysis
Hydrolyse f/enzymatische enzymatic hydrolysis
Hydrolyse f/saure acid hydrolysis
hydrolysebeständig resistant to hydrolysis
Hydrolysebeständigkeit *f* resistance to hydrolysis, hydrolytic stability
Hydrolysenkonstante *f* hydrolysis constant
hydrolysierbar hydrolyzable
hydrolysieren *v* hydrolyze; undergo hydrolysis, hydrolyze
hydrolytisch hydrolytic
Hydrometallurgie *f* hydrometallurgy, wet metallurgy
hydrometallurgisch hydrometallurgical
Hydron *n* hydron *(any of the cations of a hydrogen isotope)*
Hydroniumion *n* hydronium [oxonium] ion, H_3O^+ ion
Hydroniumionenkonzentration *f* hydronium-ion concentration • **mit gleicher Hydroniumionenkonzentration** isohydric
Hydroperoxid *n* hydroperoxide
Hydroperoxyradikal *n* hydroperoxyl radical, HO_2-radical
hydrophil hydrophilic, hydrophile
Hydrophilie *f* hydrophilicity
hydrophob hydrophobic, hydrophobe, water-repellent
Hydrophobchromatographie *f* hydrophobic (interaction) chromatography
Hydrophobie *f* hydrophobicity
Hydrophobier(ungs)mittel *n* hydrophobing agent, water repellent
Hydropolymerisation *f* hydropolymerization
Hydrosol *n* hydrosol, aquasol
Hydrotropie *f* hydrotropy

Hydroxamsäure *f (org ch)* hydroxamic acid
Hydroxid *n* hydroxide
Hydroxidsalz *n* hydroxide salt
Hydroxoaluminat *n* hydroxoaluminate
Hydroxoniumion *n s.* Hydroniumion
Hydroxosalz *n* hydroxo salt
Hydroxyacetaldehyd *m* hydroxyacetaldehyde, glycollaldehyde
Hydroxyaldehyd *m* hydroxyaldehyde
Hydroxyalkylierung *f* hydroxyalkylation
Hydroxybenzen *n* hydroxybenzene, phenol
Hydroxybenzoesäure *f* hydroxybenzoic acid, phenolic benzoic acid, phenolic acid
***o*-Hydroxybenzoesäure** *f* *o*-hydroxybenzoic acid, salicylic acid
Hydroxybenzol *n* hydroxybenzene, phenol
Hydroxybernsteinsäure *f* hydroxysuccinic acid, malic acid, *hydroxybutanedioic acid
3-Hydroxybutanal *n* *3-hydroxybutanal, 3-hydroxybutyraldehyde, acetaldol, aldol
Hydroxycarbonsäure *f* hydroxycarboxylic acid, hydroxy acid
Hydroxychinolin *n* hydroxyquinoline
Hydroxydinitrobenzen *n (bioch)* dinitrophenol
Hydroxyessigsäure *f* hydroxyacetic acid, glycollic acid, *hydroxyethanoic acid
2-Hydroxyethylmethacrylat *n* 2-hydroxyethyl methacrylate, HEMA
Hydroxyketon *n* hydroxy ketone, keto alcohol
Hydroxylamin *n* hydroxylamine
Hydroxylammoniumchlorid *n* hydroxylammonium chloride
Hydroxylgruppe *f* hydroxyl [hydroxy] group, OH group • **mit einer Hydroxylgruppe** monohydric *(alcohol, phenol)* • **mit endständiger Hydroxylgruppe** hydroxyl-terminated • **mit mehreren Hydroxylgruppe** polyhydric *(alcohol, phenol)*
Hydroxylgruppe *f*/phenolische phenolic hydroxyl group
hydroxyl(gruppen)haltig containing hydroxyl (groups), hydroxy
hydroxylieren *v* hydroxylate
Hydroxylierung *f* hydroxylation
Hydroxylradikal *n* OH radical, HO radical, hydroxyl radical
Hydroxylsauerstoff *m* hydroxylic oxygen
Hydroxylzahl *f* hydroxyl value [number] *(of fats and fatty oils)*
Hydroxymethylgruppe *f* $-CH_2OH$ hydroxymethyl group
Hydroxymethylierung *f* hydroxymethylation
2-Hydroxyphenylessigsäure *f* phenylhydroxyacetic acid
2-Hydroxypropannitril *n* 2-hydroxypropane nitrile, lactonitrile, acetaldehyde cyanohydrin
2-Hydroxypropionsäure *f* *2-hydroxypropionic acid, lactic acid
3-Hydroxypropionsäure *f* *3-hydroxypropionic acid, ethylenelactic acid
Hydroxysäure *f* hydroxycarboxylic acid, hydroxy acid
2-Hydroxysäure *f* 2-hydroxycarboxylic acid, α-hydroxy acid
α-Hydroxysäure *f* 2-hydroxycarboxylic acid, α-hydroxy acid
3-Hydroxysäure *f* 3-hydroxycarboxylic acid, β-hydroxy acid
β-Hydroxysäure *f* 3-hydroxycarboxylic acid, β-hydroxy acid
hydroxysubstituiert hydroxy-substituted
Hydroxytoluen *n* hydroxytoluene, methylphenol, cresol
Hydroxyzimtsäure *f* hydroxycinnamic acid, coumaric acid, *hydroxyphenylpropenoic acid
2-Hydroxyzimtsäurelacton *n* coumarin, 2H-chromen-2-one, α-chromone, 1,2-benzopyrone
Hydrozyklon *m* hydroclone, hydrocyclone, liquid cyclone (separator), hydraulic cyclone separator, wet cyclone classifier
Hygienisierung *f* sanitization
hygr. *s.* hygroskopisch
Hygrometer *n* hygrometer
Hygrometrie *f* hygrometry
hygroskopisch hygroscopic(al), water-absorbing, water-attracting
hygroskopisch/nicht non-hygroscopic
Hygroskopizität *f* hygroscopicity

Hylit *m(n)* xylite, (woody) lignite, woody brown coal
***DL*-Hyoscyamin** *n* DL-hyoscyamine, atropine *(alkaloid)*
Hyperazidität *f* hyperacidity, superacidity *(of gastric juice)*
Hypercholesterolämie *f (med)* hypercholesterolaemia, hypercholesteraemia
hyperchrom hyperchromic *(exhibiting enlarged extinction)*
Hyperchromie *f* 1. hyperchromism *(enlarged extinction)*; 2. *(bioch)* hyperchromicity *(of the DNA with thermal denaturation)*
Hyperchromie-Effekt *m* hyperchromic effect *(with thermal denaturation of DNA)*
Hyperfeinstruktur *f* hyperfine structure
Hyperfiltration *f* hyperfiltration *(a branch of reverse osmosis)*
Hyperglykämie *f (med)* hyperglycaemia
hyperglykämisch *(med)* hyperglycaemic
Hypergol *m* hypergol, hypergolic fuel [rocket propellant]
Hyperkonjugation *f* hyperconjugation, no-bond resonance
Hyperlipidämie *f (med)* hyperlipaemia *(in a larger sense; enlarged fat and lipoid content in blood)*
Hypersensibilität *f* hypersensitivity
Hypersorption *f* hypersorption
hyperverzweigt *s.* hochverzweigt
Hypervitaminose *f* hypervitaminosis
Hypnotikum *n* hypnotic, soporific, somnifacient, somnificant, sleeping drug
Hypoazidität *f* subacidity, hypoacidity *(of gastric juice)*
Hypochlorit *n* M^IOCl hypochlorite
Hypochloritbleichlauge *f* hypochlorite bleach (liquor)
hypochrom hypochromic *(exhibiting reduced extinction)*
Hypochromie *f* hypochromism *(reduced extinction)*; hypochromicity *(of nucleic acids)*
Hypoglykämie *f (med)* hypoglycaemia
hypoglykämisch *(med)* hypoglycaemic
Hypohalogenierung *f* hypohalogenation
hypokristallin hypocrystalline
Hyponitrit *n* $M^I_2N2O_2$ hyponitrite
Hypophosphat *n* $M^I_4P_2O_6$ hypophosphate
Hypophosphit *n* $M^IH_2PO_2$ *phosphinate, hypophosphite
Hypophosphorsäure *f* hypophosphoric acid
Hypophysenhinterlappenhormon *n* neurohypophyseal [posterior-lobe] hormone
Hypophysenhormon *n* hypophyseal hormone
Hypophysenvorderlappenhormon *n* adenohypophyseal [anterior-pituitary] hormone
Hypotensivum *n* antihypertensive drug, blood-pressure depressant
Hypothese *f***/avogadrosche** Avogadro hypothesis [law]
Hypothese *f* **der Energiequanten/ plancksche** Planck's hypothesis *(of quantized energy)*
hypsochrom hypsochromic
Hypsochromie *f* hypsochromic shift
Hysterese *f* hysteresis

I

Ibogaalkaloid *n* iboga alkaloid
ICDH = Isocitratdehydrogenase
ICP *s.* Plasma/induktiv gekoppeltes
ICS *s.* Isocitronensäure
ICSH *(interstitial-cell-stimulating hormone) s.* Hormon/luteinisierendes
ideal 1. ideal, perfect; 2. *(cryst)* defect-free • **sich ideal verhalten** behave perfectly
Idealzustand *m* ideality
Identitätsperiode *f* 1. identity period *(of a macromolecule)*; 2. *(bioch)* pitch *(of a helix)*
idiomorph *(min)* idiomorphic, idiomorphous, euhedral, automorphic *(having the proper crystal form)*
Idiophase *f (biot)* idiophase, production [product formation] phase
Idose *f* idose *(a monosaccharide)*
Iduronsäure *f* iduronic acid
I.E. *s.* Einheit/internationale
I-Effekt *m s.* Induktionseffekt
IES *s.* Indolyl-3-essigsäure
Ile *s.* Isoleucin

Ilmenit *m (min)* ilmenite *(iron(II) metatitanate)*
Imhoff-Brunnen *m (hyd)* Imhoff tank
Imid *n* imide
Imidazolalkaloid *n* imidazole alkaloid
Imidoester *m* imido ester, imino ether
Imidogruppe *f* imido group [residue]
Imidsäure *f* imidic acid
Imin *n* $R^1R^2C{=}NR^3$ imine
Iminiumverbindung *f* iminium compound
Iminogruppe *f* –NH– imino group [residue], aminediyl group
Iminoharnstoff *m* iminourea, carbamidine, guanidine
Iminosäure *f* imino acid
Immission *f* air-borne pollution, pollutant input, immission
immobilisieren *v* immobilize *(e.g. enzymes or nutrients)*
Immobilisierung *f* immobilization *(as of enzymes or nutrients)*
Immunelektrophorese *f (anal)* immunoelectrophoresis
Immunglobulin *n* immunoglobulin, Ig, immune (serum) globulin, antibody globulin
immunisieren *v* immunize
Immunisierung *f* immunization
Immunochemie *f* immunochemistry
immunochemisch immunochemical
Immunstimulans *n* immunostimulant *(agent which enhances certain immune reactions)*
Immunsuppressor *m* immunosuppressor, immunosuppressant *(agent which suppresses certain immune reactions)*
IMP *s.* Inosin-5′-monophosphat
Impedanz *f (anal)* impedance
Impedanzmessung *f (el ch)* impedance measurement
Impedanzspektroskopie *f (el ch)* impedance spectroscopy
Impedanzspektrum *n (el ch)* impedance spectrum
impfen *v (cryst, biot)* seed, inoculate
Impffermenter *m (biot)* inoculum [culture] fermenter [tank], seed (culture) fermenter, seed tank
Impfgut *n (biot)* inoculum
Impfgutanzucht *f (biot)* inoculum cultivation [build-up], cultivation of inoculum
Impfkristall *m* seed crystal
Impfkultur *f (biot)* seed culture
Impfkultur *f* **für Fermenter** fermentation culture
Impflösung *f (biot)* inoculation [inoculum] medium
Impfung *f (cryst, biot)* seeding, inoculation
imprägnieren *v* 1. impregnate, imbibe; 2. *(text)* (water)proof, impregnate
imprägnieren *v***/mit** $\mathbf{CO_2}$ carbonate, aerate, impregnate *(beverages)*
Imprägniermittel *n* impregnant, impregnation (material), impregnating material, proofing
Imprägnierung *f* 1. impregnation, imbibition *(act)*; 2. *(text)* (water)proofing impregnation; 3. impregnation, finish *(state)*
Impulsbilanzgleichung *f* momentum balance equation
Impulsmoment *n* moment of momentum, angular momentum
Impulsstrom *m* momentum flux *(of liquids per unit time and unit volume)*
Impulstransport *m* momentum transport
inaktiv 1. inactive, inert, passive; 2. *(rubber)* non-reinforcing
inaktiv/elektrochemisch electroinactive, non-electroactive
inaktiv/optisch optically inactive
inaktivieren *v* inactivate, deactivate, block, *(reactive groups or sites also)* mask
Inaktivierung *f* inactivation, deactivation, blocking, *(of reactive groups or sites also)* masking
Inaktivierung *f***/thermische** thermal [heat] inactivation
Inaktivität *f* inactivity
Inbetriebnahme *f* start-up, starting(-up)
inchromieren *v* chromize
Inchromierstahl *m* chromized steel
Inchromierung *f* chromizing
Indanthron-Küpenfarbstoff *m* indanthrone vat dye
indifferent inert, indifferent, inactive, passive
Indifferenz *f* inertness, indifference
Indifferenz *f***/chemische** chemical inertness
Indigo *m(n)* indigo (blue)

Indigofarbstoff *m* indigoid (dye)
Indikator *m* 1. *(ch)* indicator, *(relating to isotopes preferably:)* tracer; 2. *(food)* indicator ingredient [substance]
Indikator *m*/**basischer** basic indicator
Indikator *m*/**radioaktiver** radioactive tracer [indicator], radiotracer
Indikator *m*/**saurer** acid indicator
Indikator *m*/**zweifarbiger** two-colour indicator
Indikatorbase *f* basic indicator
Indikatorelektrode *f* indicator electrode
Indikatorenzym *n (bioch)* marker enzyme
Indikatorisotop *n* (isotopic) tracer
Indikatormethode *f* tracer method
Indikatororganismus *m s.* Bioindikator
Indikatorpapier *n* (colour-)indicator paper, test paper, *(Am)* reaction paper
Indikatorsäure *f* acid indicator
Indikatorumschlag *m* indicator change
Indium *n* In indium
Indium(III)-chlorid *n* indium(III) chloride, indium trichloride
Indium(III)-hydroxid *n* indium(III) hydroxide
Indium(III)-oxid *n* indium(III) oxide, diindium trioxide
Indium-Zink-Oxid *n* indium-tin-oxide, ITO *(organic light emitting diodes)*
Individuum *n*/**chemisches** chemical individual, individual chemical entity [substance], chemically individual substance
Indizes *mpl*/**millersche** Miller (crystal) indices
indizieren *v (nomencl)* index, label, indicate
Indizierung *f (nomencl)* indexing, labelling, indication
Indol *n* indole, 2,3-benzpyrrole
Indolalkaloid *n* indole alkaloid
Indolsynthese *f*/**fischersche** Fischer indole synthesis
Indolyl-3-essigsäure *f* 3-indolylacetic acid
Indophenol *n* indophenol
Indoxyl *n* indoxyl, 3-hydroxyindole
Indoxylcarbonsäure *f* indoxylcarboxylic acid
Induktionseffekt *m* induction [inductive] effect, I effect, field effect
Induktionskraft *f* induction [Debye] force, induced dipole force
Induktionsofen *m* induction [inductance, induction-heated] furnace
Induktionsperiode *f* induction period
Induktor *m* 1. inductor *(in chemical reactions)*; 2. *(bioch)* inducer
Industrie *f*/**biotechnologische** bioprocess industry
Industrie *f*/**chemische** chemical industry
Industrie *f*/**(weiter)verarbeitende** processing industry
Industrieabfälle *mpl* industrial [trade] waste
Industrieabgas *n* industrial off-gas
Industrieabwasser *n* industrial waste water, industrial discharge [effluent], trade effluent [waste]
Industriechemikalie *f* industrial chemical
Industriechemiker *m* industrial chemist
Industrie-Enzym *n (biot)* industrial enzyme, enzyme of industrial importance
Industriefermenter *m (biot)* commercial(-scale) fermenter
Industriemüll *m* industrial solid waste(s)
Industriewasser *n* industrial water, water for industrial purpose [use]
ineinandergewunden interwound, twisted *(macromolecules)*
inert inert, inactive, passive, indifferent
Inertgas *n* inert gas, indifferent [inactive] gas
Inertstoff *m* inert substance
Inertsubstanz *f* inert substance
Infektion *f (biot)* contamination *(with foreign microorganisms)*
infektionsgefährdet sein *v (biot)* have a high risk of contamination
Infektionsmikroorganismen *mpl (biot)* contaminant [contaminating] microorganisms, contaminants
Informationsübertragung *f (bioch)* information transfer
infrarot infrared, ultrared
Infrarot *n* 1. infrared (radiation); 2. *s.* Infrarotgebiet
Infrarotabsorptionsspektroskopie *f* infrared absorption spectroscopy
infrarot-aktiv infrared-active
Infrarotdetektor *m (chromat)* infrared detector

infrarotdurchlässig infrared-transmitting, infrared-transparent
Infrarotdurchlässigkeit *f* infrared transmittance [transmission]
Infrarotgebiet *n* infrared (spectral) region
Infrarotgebiet *n*/**fernes** far infrared (spectral region), FIR
Infrarotgebiet *n*/**mittleres** mid infrared (spectral region)
Infrarotgebiet *n*/**nahes** near infrared (spectral region), NIR
infrarot-inaktiv infrared-inactive
Infrarotmikroskop *n* infrared microscope
Infrarotspektroskopie *f* infrared spectroscopy
Infrarotspektrum *n* infrared spectrum
Infrarotstrahler *m* infrared radiator [lamp]
Infrarotstrahlung *f* infrared radiation
Infus *n (pharm)* infusion *(solution obtained by steeping vegetable matter in a hot solvent)*
Infusorienerde *f* diatom(aceous) earth, infusorial earth, kieselguhr
Infusum *n s.* Infus
Ingenieurchemie *f* engineering chemistry
Ingenieurtechnik *f*/**chemische** chemical engineering technology
Ingenieurwesen *n*/**chemisches** chemical engineering
Ingrediens *n* ingredient
Ingredienz *f* ingredient
Inhalt *m* content *(relating to energy, heat)*; contents *(something contained in a vessel)*; content(s), capacity, volume *(of a vessel)*
Inhaltsstoff *m* ingredient, constituent
Inhaltsstoffe *mpl (hyd)* (water) constituents • **frei von organischen Inhaltsstoffen** organic-free
Inhaltsstoffe *mpl*/**abbauresistente [biologisch nicht abbaubare] organische** refractory [non-biodegradable] organics, biologically inert refractory matter
Inhaltsstoffe *mpl*/**gelöste** dissolved constituents
inhibieren *v* inhibit, *(polymerization also)* shortstop
Inhibitor *m* inhibitor, inhibiting substance, retarder, retarding agent, *(relating to polymerization also)* stopper, shortstop, shortstopping agent
Inhibitor *m*/**kompetitiver** *(bioch)* competitive inhibitor, antimetabolite, metabolic antagonist
Inhibitor-Polypeptid *n*/**gastrisches** *(bioch)* gastric inhibitory polypeptide, GIP
Inhibitorwirkung *f* inhibitory [inhibiting] effect
inhomogen inhomogeneous, non-homogeneous
Inhomogenität *f* inhomogeneity, non-homogeneity
Initialsprengstoff *m* initiator, initiating [primary] explosive, primer, detonator, initial detonating agent
Initiationscodon *n (bioch)* (chain) initiation codon, initiator codon
Initiator *m* initiator, initiating agent
Initiatorcodon *n s.* Initiationscodon
Initiatortriplett *n s.* Initiationscodon
initiieren *v* initiate, start
Injektionsspritze *f* hypodermic syringe, injection [sample-charging] syringe *(gas chromatography)*
Injektionsventil *n (anal)* injection valve
Inklusionsverbindung *f* inclusion compound
inkohärent incoherent *(radiation)*
Inkohärenz *f* incoherence *(of radiation)*
Inkohlungsgrad *m* degree of coalification, rank • **von hohem Inkohlungsgrad** high-rank • **von mittlerem Inkohlungsgrad** medium-rank • **von niedrigem Inkohlungsgrad** low-rank
Inkrustation *f* incrustation, encrustation
Inkrusten *pl* incrustants, encrustants, incrusting material [matter, substance]
inkrustieren *v* incrust, encrust
Inkubation *f (biot)* incubation
inkubieren *v (biot)* incubate
Inlösunggehen *n* dissolution
Innenanstrichfarbe *f* interior [indoor] paint
Innendruck *m* internal pressure
Innenfläche *f* inner surface
Innenmischer *m* closed mixer
Innenoberfläche *f* internal surface (area), inner surface (area)
Innenschale *f* inner shell *(of an atom)*
Innensphärenkomplex *m* inner-sphere complex

innenzentriert *(cryst)* space-centred, body-centred
Innere *n* **der flüssigen Phase** bulk liquid, bulk of the liquid phase
Innerkomplex *m* inner(-orbital) complex, penetration [low-spin, spin-paired] complex
innerlich *(pharm)* internal
innermolekular intramolecular
Innertherapeutikum *n* systemic (chemical) *(pest control)*
innig intimate *(e.g. contact or mixture)*
Ino *s.* Inosin
Inosilicat *n (min)* inosilicate, chain silicate *(any of a class of polymeric silicates)*
Inosin *n* inosine *(a riboside)*
Inosin-5′-monophosphat *n* inosine-5′-monophosphate, IMP, inosinic acid *(a nucleotide)*
Inosinsäure *f* inosine-5′-monophosphate, IMP, inosinic acid *(a nucleotide)*
Inositol *m* inositol, hexahydroxycyclohexane
Insektenabwehrmittel *n* insect repellent, insectifuge
Insektenlockstoff *m* insect attractant
insektentötend insecticidal
insektenverträglich non-insecticidal *(e.g. fungicides)*
insektizid insecticidal
Insektizid *n* insecticide
Insektizid *n***/mikrobielles** microbial insecticide, bioinsecticide
Insektizid *n***/pflanzliches** insecticide of plant origin, botanical
Insektizidresistenz *f* insecticide resistance *(of animals)*
Inselsilicat *n (min)* nesosilicate
Inselwachstum *m* island growth *(on surfaces)*
Insertion *f* insertion *(of atoms or groups into bonds)*
Insertionspolymerisation *f* insertion polymerization
insilizieren *v* siliconize *(metals for protection)*
Insilizierung *f* siliconization *(of metals for protection)*
In-situ-Analyse *f* in situ analysis
instabil unstable, instable, labile, transient, *(relating to isotopes also)* evanescent
Instabilität *f* instability, lability, transience
Instabilitätskonstante *f* instability constant *(as of complex ions)*
Instandhaltung *f* maintenance, upkeep
Instrumentalanalyse *f* instrumental analysis
Instrumentenanalyse *f* instrumental analysis
Instrumentenfehler *m* instrumental error
Integralchromatogramm *n* integral chromatogram
Integraldetektor *m* integral detector *(gas chromatography)*
Integralreaktor *m* integral (tubular-flow) reactor
Intensitätsgröße *f* intensive quantity *(being independent of the mass of the system concerned)*
intensitätsschwach low-intensity
intensitätsstark high-intensity
Intensitätsverhältnis *n (spectr)* intensity ratio; transmittancy *(colorimetry)*
Intensitätsverteilung *f (spectr)* intensity distribution
Intensivkühler *m* jacketed coil condenser
interatomar interatomic
Interferenz *f* interference *(physics)*
interferieren *v* interfere
Interferogramm *n (spectr)* interferogram
Interferometer *n (anal)* interferometer
Interferometrie *f (anal)* interferometry
Interferon *n (biot)* interferon, Ifn, IFN
Interhalogen *n* interhalogen (compound)
Interhalogenverbindung *f* interhalogen (compound)
interionisch interionic
Interkalat *n* intercalate
Interkalation *f* intercalation *(1. introduction of guest molecules between layers; 2. insertion of planar ring systems between the stacked bases of nucleic acids)*
Interkalationsverbindung *f* intercalation compound
Interkalationsverbindung *f* **1. Stufe** first-stage intercalation compound, stage 1 intercalation compound
interkalieren *v* intercalate *(to introduce guest molecules between layers)*

Interkombination *f* *(spectr)* intercombination
Interkonversion *f* interconversion (reaction)
interkristallin intercrystalline
Intermediärprodukt *n* intermediate product [substance], (reaction) intermediate
Intermediärstoffwechsel *m* *(bioch)* intermediary metabolism
Intermediärverbindung *f* intermediate (compound)
Intermediat *n* *s.* Zwischenprodukt
intermolekular intermolecular
Interstitialmischkristall *m* interstitial (solid) solution
interstitiell interstitial
Intoxikation *f* 1. poisoning; 2. *(med)* intoxication
intramolekular intramolecular
intrazellulär intracellular, endocellular
invariant invariant, non-variant
Inversion *f* **der Konfiguration** *s.* Inversion/waldensche
Inversion *f***/waldensche** Walden inversion, inversion of configuration
Inversionshöhe *f* inversion height *(in the atmosphere)*
Inversionszentrum *n* inversion centre *(of a molecule)*
Inversvoltammetrie *f* **(an der Anode)** *(anal)* anodic stripping analysis [voltammetry], ASV
inversvoltammetrisch by anodic stripping analysis
invertieren *v* invert *(the configuration of a molecule)*
Invertierung *f* inversion *(of the configuration of a molecule)*
Invertzucker *m* invert(ed) sugar
Iod *n* I iodine
Iodat *n* $M^{I}IO_3$ iodate
Iodbenzen *n* iodobenzene
Iodbenzol *n* iodobenzene
Iod(I)-chlorid *n* iodine(I) chloride, iodine monochloride
Iodchlorierung *f* iodochlorination, chloroiodination
Ioddampf *m* iodine vapour
Iodid *n* $M^{I}I$ iodide
Iodidstärkepapier *n* starch iodide paper
iodieren *v* iodize, iodinate
Iodierung *f* iodization, iodination
Iodkaliumstärkepapier *n* potassium-iodide starch paper
Iodmethan *n* *iodomethane, methyl iodide
Iodoform *n* iodoform, tri-iodomethane
Iodometrie *f* iodometry, iodimetry
iodometrisch iodometric, iodimetric
Iodonium-Ion *n* iodonium ion
Iodoniumverbindung *f* iodonium compound
Iod(V)-oxid *n* iodine(V) oxide, diiodine pentaoxide
Iodsäure *f* iodic acid
Iodspeisesalz *n* iodized (table) salt
Iodstärke *f* iodide of starch, starch-iodine complex
Iodstärkepapier *n* starch iodide paper
Iodstärkereaktion *f* *(bioch, anal)* iodine colour reaction
Iodwasserstoffsäure *f* hydroiodic acid
Iodzahl *f* iodine (absorption) value, I.V., iodine [Huebl] number *(characterizing the unsaturation of a compound or mixture)*
Ion *n* ion
Ion *n***/komplexes** complex ion
Ion *n***/negatives** negative ion, anion
Ion *n***/positives** positive ion, cation
ional ionic
Ionenabbildungsspektroskopie *f* ion imaging
Ionenaktivität *f* ion(ic) activity
Ionenäquivalentleitfähigkeit *f* ionic equivalent conductance, equivalent ion(ic) conductance
Ionenassoziation *f* ion association
Ionenausschluss *m* ion exclusion
Ionenausschlusschromatographie *f* ion exclusion chromatography
Ionenaustausch *m* ion(ic) exchange
Ionenaustausch *m* **mit H-Austauscher** hydrogen cycle (cation) exchange, H-form exchange, hydrogen-cation exchange, ion exchange in [on] the hydrogen cycle
Ionenaustausch *m* **mit Natriumaustauscher** sodium-cycle exchange, Na-form exchange, ion exchange in [on] the sodium cycle
Ionenaustauschchromatographie *f* ion-exchange chromatography
Ionenaustauscher *m* ion exchanger, ion-exchange material

Ionenaustauscher *m*/**erschöpfter** exhausted ion exchanger
Ionenaustauscher *m* **in der H-Form [H_2-Form]** hydrogen(-form) exchanger, H-form exchanger, hydrogen-cation exchanger
Ionenaustauscher *m* **in der Na-Form [Natriumform]** sodium [Na-form] exchanger
Ionenaustauscher *m* **in der OH-Form** hydroxyl-form [OH-form] exchanger
Ionenaustauscheranlage *f* ion-exchange unit [system]
Ionenaustauscherharz *n* ion-exchange resin
Ionenaustauschermembran *f* ion-exchange membrane
Ionenaustauschersäule *f* ion-exchange column
Ionenaustauschgleichgewicht *n* ion-exchange equilibrium
Ionenaustauschreaktion *f* ion-exchange reaction
Ionenaustauschvollentsalzungsanlage *f* ion-exchange demineralizer
Ionenbeschuss *m* ion(ic) bombardment
Ionenbeweglichkeit *f* ion(ic) mobility
Ionenbeziehung *f* ionic bond, heteropolar [polar, electrostatic, electrovalent] bond, electrovalence
Ionenbindung *f* 1. electrovalent linkage, ionic [polar, heteropolar, electrostatic] linkage *(process)*; 2. *s.* Ionenbeziehung
Ionencharakter *m* ionic character
Ionenchromatographie *f* ion chromatography
ionenchromatographisch ion-chromatographic, by ion chromatography
Ionendetektor *m* ion detector
Ionenenergie-Spektroskopie *f* **zum Nachweis metastabiler Zerfälle** direct analysis of daughter ions, DADI, mass-analysed ion kinetics spectrometry, ion kinetic energy spectroscopy, IKES
Ionenformel *f* ionic formula
Ionengetterpumpe *f* getter-ion [sputter-ion] pump
Ionengitter *n (cryst)* ionic (crystal) lattice
Ionengleichgewicht *n* ionic equilibrium [balance]
Ionenimplantation *f* ion implantation *(doping process)*
Ionenkristall *m* ionic crystal
Ionenladung *f* ionic charge
Ionenladungszahl *f* ionic charge number
Ionenleiter *m* ionic [electrolytic] conductor
Ionenleitfähigkeit *f* ion(ic) conductivity
Ionenleitfähigkeit *f*/**spezifische** ion(ic) conductance
Ionenleitung *f* ionic conduction
Ionenmikrosonde *f* ion (micro)probe
Ionenpaar *n* ion pair
Ionenpaar *n*/**äußeres** loose ion pair, *(esp if separated by only a single solvent molecule:)* solvent-shared ion pair, *(esp if separated by more than one solvent molecule:)* solvent-separated ion pair
Ionenpaar *n*/**inneres** tight ion pair, intimate [contact] ion pair
Ionenpaarchromatographie *f* ion-pair chromatography
Ionenplasma *n* ionic plasma
Ionenplattieren *n* ion plating *(technique to form metal coatings)*
Ionenprodukt *n* ion(ic) product
Ionenquelle *f* ion source
Ionenradius *m* ionic radius
Ionenradius *m*/**goldschmidtscher** Goldschmidt radius
Ionenreaktion *f* ion(ic) reaction, ion-ion reaction
Ionenreihe *f*/**hofmeistersche** lyotropic order [series]
Ionenrekombination *f* ion(-ion) recombination
ionenselektiv ion-selective
ionensensitiv ion-sensitive, *(relating to electrodes also:)* ion-selective
ionenspezifisch ion-specific
Ionenstärke *f* ionic strength
Ionenstrahl *m* ion beam
Ionenstrahl *m*/**fokussierter** *(spectr)* focus ion beam, FIB
Ionenstrahl-Mikroanalyse *f* ion-probe microanalysis, IMA
Ionenstreuungsspektroskopie *f* ion-scattering spectroscopy, ISS
Ionensuszeptibilität *f* ionic susceptibility *(susceptibility per gram ion)*
Ionentrennung *f* ion separation
Ionenumtausch *m* ion exchange
Ionenverbindung *f* ionic [polar, heteropolar] compound

Ionenwanderung *f* ion(ic) migration, migration of ions
Ionenwechselwirkung *f* ionic [ion-ion] interaction, interionic action
Ionenwertigkeit *f* ionic valence
Ionenwolke *f (phys ch)* ion cloud, ion(ic) atmosphere
Ionenzyklotron-Resonanz *f* ion-cyclotron resonance, ICR
Ionisation *f* ionization
Ionisation *f* **bei Atmosphärendruck** *(spectr)* atmospheric pressure ionization, API
Ionisation *f* **bei Atmosphärendruck/chemische** *(spectr)* atmospheric pressure chemical ionization, APCI
Ionisation *f***/chemische** chemical ionization
Ionisation *f***/lawinenartige** cumulative ionization
Ionisation *f***/thermische** thermal ionization
Ionisationsdetektor *m* ionization detector
Ionisationsenergie *f* ionization energy
Ionisationsenergie *f***/erste** first ionization energy
Ionisationsgrad *m* degree of ionization
Ionisationspotenzial *n* ionization [ionizing] potential
Ionisationspotenzial *n***/erstes** first ionization potential
Ionisationsreaktion *f* ionization reaction
Ionisationsvakuummeter *n* ionization (vacuum) gauge
ionisch ionic
ionisch/nicht non-ionic
ionisierbar ionizable, capable of ionization
ionisieren *v* ionize
ionisiert/nicht unionized
Ionisierung *f s.* Ionisation
Ionisierungsquerschnittdetektor *m* *(chromat)* cross-section detector
ionogen ionogenic
ionogen/nicht non-ionogenic, non-ionic
ionoid *(org ch)* ionic *(addition)*
Ionomer(e) *n* ionomer
ionophil ionophilic
ionophor *(bioch)* ionophoric
Ionophor *n (bioch)* ionophore, ionophoric agent, ion carrier
Ionophorese *f* ionophoresis
ionotrop ionotropic
Ionotropie *f* ionotropy
I.P. *s.* Punkt/isoelektrischer
Ipecacuanhaalkaloid *n* ipecacuanha alkaloid
ipso-Angriff *m (org ch)* ipso-attack
ipso-Substitution *f (org ch)* ipso-attack
IQD *s.* Ionisierungsquerschnittdetektor
IR *s.* Infrarot
IR-Abfall *m s.* Spannungsabfall/ohmscher
iR-Abfall *m s.* Spannungsabfall/ohmscher
Irdengut *n* earthenware
Irdenware *f* earthenware
Iridium *n* Ir iridium
irisieren *v* be iridescent, iridesce
Irisieren *n* iridescence
irreversibel irreversible, non-reversible
IR-Spektralbereich *m s.* Infrarotgebiet
i-s-Diagramm *n* enthalpy-entropy chart [diagram], Mollier chart
isenthalpisch *(phys ch)* isenthalpic *(e.g. expansion)*
isentrop *(phys ch)* is(o)entropic *(e.g. series of reactions)*
Isentrope *f (phys ch)* isentrope *(the representation of an isentropic process in a thermodynamic diagram)*
Isoalkan *n* isoalkane
Isoamylalkohol *m* isoamyl alcohol, *3-methylbutan-1-ol
Isoascorbinsäure *f* *D*-araboascorbic acid, isoascorbic acid, erythorbic acid
isobar isobaric
Isobare *f (phys ch)* isobar
Isobutan *n* isobutane, *2-methylpropane
Isobuttersäure *f* isobutyric acid, *2-methylpropionic acid
Isobutylalkohol *m* isobutyl alcohol, *2-methylpropan-1-ol
Isobutylen-Isopren-Kautschuk *m* isobutylene-isoprene rubber, IIR
Isochinolin *n* isoquinoline, benzo[c]pyridine, 2-benzazine
Isochinolinalkaloid *n* isoquinoline alkaloid
Isochore *f (phys ch)* isochore
Isocitrat *n* isocitrate *(intermediate in citric-acid cycle)*
Isocitratdehydrogenase *f* isocitric acid dehydrogenase

Isocitronensäure *f* isocitric acid, *1-hydroxypropane-1,2,3-tricarboxylic acid
Isocrotonsäure *f* isocrotonic acid, **cis*-but-2-enoic acid
Isocyanat *n* R–N=C=O isocyanate
Isocyanid *n* R–N≡C isocyanide, isonitrile, carbylamine
isocyclisch isocyclic, homocyclic, carbocyclic
Isocyclus *m* isocyclic [homocyclic, carbocyclic] compound
isodispers isodisperse, monodisperse
isodynam isodynamic
isoelektrisch isoelectric
isoelektronisch isoelectronic, isosteric
Isoenzym *n* isoenzyme, isozyme
Isoeugenol *n* isoeugenol, 2-methoxy-4-propenylphenol
Isoflavon *n* isoflavone
Isogel *n* *(coll)* isogel
Iso-Kautschuk *m* isorubber
Isolation *f* 1. insulation (material); insulation; resist, stop-off (coating) *(electroplating)*; 2. insulation *(against something)*
Isoleucin *n* isoleucine, ileu *(an amino acid)*
isolierbar isolable
isolieren *v* 1. *(ch)* isolate, separate, segregate; 2. insulate *(against something)*
Isoliermaterial *n* insulation (material)
Isolierstoff *m* insulation (material)
Isolierung *f* 1. *(ch)* isolation, separation, segregation; 2. insulation *(against something)*
Isologe *n* isologue
isomer isomeric
isomer/optisch enantiomeric
Isomer *n* isomer, isomeric compound
Isomer *n***/geometrisches** geometric(al) isomer, cis-trans isomer
Isomer *n***/optisches** optical isomer [antipode], enantiomer, antimer, mirror-image isomer
***o*-Isomer** *n* *o*-isomer, ortho isomer
***p*-Isomer** *n* *p*-isomer, para isomer
Isomerenverteilung *f* isomer distribution
Isomerie *f* isomerism
Isomerie *f***/geometrische** geometrical [cis-trans] isomerism
Isomerie *f***/optische** optical isomerism, enantiomerism, mirror-image isomerism
Isomerie *f***/räumliche [stereochemische]** stereoisomerism, space isomerism
***anti-syn*-Isomerie** *f* anti-syn isomerism
Isomerieverschiebung *f* *(spectr)* isomer shift
isomerisieren *v* isomerize
isomerisieren *v***/sich** isomerize, undergo isomerization
Isomerisierung *f* isomerization, molecular rearrangement
Isomerisierungskatalysator *m* isomerization catalyst
isomorph *(cryst)* isomorphous
Isomorphie *f* *(cryst)* isomorphism
Isonicotinsäure *f* isonicotinic acid, pyridine-4-carboxylic acid
Isonitril *n* isonitrile, isocyanide, carbylamine
Isonitrosoverbindung *f* oxime, isonitroso compound
Isooctan *n* isooctane, *(specif)* $(CH_3)_2CHCH_2C(CH_3)_3$ 2,2,4-trimethylpentane
Isopentan *n* *2-methylbutane, isopentane
Isophthalsäure *f* isophthalic acid, *m*-phthalic acid, benzene-*m*-dicarboxylic acid
Isoplethe *f* isopleth
Isopolymorphie *f* isopolymorphism
Isopolysäure *f* isopoly acid
Isopren *n* isoprene, *2-methylbuta-1,3-diene
Isoprenkautschuk *m* isoprene rubber, IR
Isoprenoid *n* isoprenoid *(general term for terpenes and steroids)*
Isopropylalkohol *m* propan-2-ol
Isopropylamin *n* isopropylamine, 2-aminopropane, *1-methylethylamine
Isopropylbenzen *n* isopropylbenzene, cumene, 2-phenylpropane
Isopropylgruppe *f* $(CH_3)_2CH–$ isopropyl group [residue]
isosmotisch is(o)osmotic, isotonic
isoster isosteric, isoelectronic
Isostere *f* *(phys ch)* isostere
Isosterie *f* isosterism
Isosynthese *f* isosynthesis
Isotache *f* isotach, isokinetic line, isovel

Isotachophorese *f (anal)* isotachophoresis *(special type of electrophoresis)*
isotaktisch isotactic
Isotaktizität *f* isotacticity
isotherm isothermal
Isotherme *f* isotherm
Isothiocyanat *n* isothiocyanate, mustard oil
Isothiocyansäureester *m* isothiocyanate, mustard oil
isotonisch isotonic, is(o)osmotic
isotop isotopic
Isotop *n* isotope
Isotop *n***/künstlich erzeugtes** artificial isotope
Isotop *n***/langlebiges** long-lived isotope
Isotop *n***/radioaktives** radioactive [unstable] isotope, radioisotope
Isotop *n***/schweres** heavy isotope
Isotop *n***/stabiles** stable [non-radioactive] isotope
Isotopenaustausch *m* isotope [isotopic] exchange
Isotopenaustauschgleichgewicht *n (nucl)* isotope exchange equilibrium
Isotopenhäufigkeit *f* isotopic abundance
Isotopenindikator *m* isotopic tracer [indicator]
isotopenmarkiert (isotopically) labelled
Isotopenmethode *f* tracer [atom tagging] method
isotopenrein isotopically pure
Isotopenreinheit *f* isotopic purity
Isotopentrennung *f* isotope [isotopic] separation
Isotopenverdünnungsanalyse *f* isotopic dilution analysis
Isotopenverhältnis *n* abundance [isotopic] ratio
Isotopenzusammensetzung *f* isotopic composition
Isotopie *f* isotopism, isotopy
Isotopieeffekt *m* isotope effect
isotrop *(cryst)* isotropic
Isotropie *f (cryst)* isotropism, isotropy
isotyp *(cryst)* isotypic
Isotypie *f (cryst)* isotypy
Isovaleraldehyd *m* isovaleraldehyde, *3-methyl-butanal
Isovaleriansäure *f* isovaleric acid, *3-methylbutanoic acid
Isovalersäure *f* isovaleric acid, *3-methylbutanoic acid
Isovole *f* isovol *(line of equal volatile matter)*
Itaconsäure *f* itaconic acid, *prop-2-ene-1,2-dicarboxylic acid
IT-Diagramm *n* IT diagram, heat-content/temperature diagram
I-U-Kennlinie *f s.* Strom-Spannungs-Kurve
I-U-Kurve *f s.* Strom-Spannungs-Kurve
IUPAC IUPAC, International Union of Pure and Applied Chemistry
IUPAC-Regel *f (nomencl)* IUPAC rule

J

Jablonski-Diagramm *n (phys ch)* Jablonski diagram
Jablonski-Termschema *n (phys ch)* Jablonski diagram
Jackson-Meisenheimer-Komplex *m (org ch)* Meisenheimer complex [compound, adduct]
Jahn-Teller-Phasenübergang *m* Jahn-Teller transition
Jahrestonnen *fpl s.* Tonnen je Jahr
Jantzen-Verteilung *f* countercurrent extraction [separation]
Jaspis *m (min)* jasper *(a variety of quartz)*
jato *(Jahrestonnen) s.* Tonnen je Jahr
Jato *(Jahrestonnen) s.* Tonnen je Jahr
jj-Kopplung *f (phys ch)* jj coupling
Jod *n s.* Iod
Johnson-Rauschen *n (anal)* Johnson [thermal] noise
Joule *n* joule, J *(SI unit of work, energy, and heat)*
Joule-Thomson-Effekt *m* Joule-Thomson effect
Joule-Thomson-Koeffizient *m* Joule-Thomson coefficient
J-Peptid *n (bioch)* joining peptide
Jungfustik *m (dye)* young fustic, fustet *(from Cotinus coggygria Scop.)*
Justage *f s.* Justierung
justieren *v* adjust, rectify *(instruments)*
Justierschraube *f* adjusting screw *(as of a balance)*
justiert/auf Auslauf calibrated [graduated] to deliver, calibrated for delivery

justiert/auf Einguss calibrated to contain, calibrated for content
Justierung *f* adjustment, rectification *(of an instrument)*
Jutedichtung *f* jute packing
Jutepackung *f* jute packing
JZ *(Jodzahl) s.* Iodzahl

K

KA *s.* 1. Caseinfaserstoff; 2. Kläranlage
Kachelpresse *f* pot press *(as for expression of oil-seeds)*
Käfig *m* cage *(consisting of molecules in liquids and solids)*
Käfig(einschluss)verbindung *f* cage [clathrate] compound, clathrate (inclusion compound)
Kakaobutter *f* cocoa butter [oil, fat], cacao butter
Kakaofett *n* cocoa butter [oil, fat], cacao butter
Kakaoöl *n* cocoa butter [oil, fat], cacao butter
Kakodyl *n* cacodyl, tetramethyldiarsine
Kakodylsäure *f* cacodylic acid, dimethylarsinic acid
Kalander *m* calender (machine)
Kalander *m* **zum Belegen von Geweben** *(rubber)* coating calender
kalandern *v* calender
kalandrieren *v* calender
Kali *n s.* 1. Kalisalz; 2. Kalidüngemittel; 3. Kalium
Kali *n***/schwefelsaures** sulphate of potash *(a fertilizer chiefly consisting of K_2SO_4)*
kalibrieren *v* calibrate *(measuring apparatus)*
Kalibrierfunktion *f (anal)* calibration function
Kalibriergerade *f (anal)* calibration curve, calibration line
Kalibrierkurve *f (anal)* calibration curve
Kalibrierprobe *f (anal)* calibration sample
Kalibrierung *f* calibration *(of measuring apparatus)*
Kalidüngemittel *n* potash [potassic] fertilizer
Kalidünger *m* potash [potassic] fertilizer
Kaliglimmer *m (min)* potash [potassium] mica, muscovite *(a phyllosilicate)*
kalihaltig potassiferous, potassic
Kalilauge *f* potash lye, caustic potash solution, potassium hydroxide solution
Kalirohsalz *n* potash ore, mine-run salt
Kalisalpeter *m* saltpetre, nitre, nitrate of potash *(potassium nitrate)*
Kalisalz *n* potash (salt)
Kaliseife *f* potassium soap, potash (soft) soap
Kalium *n* K potassium
Kaliumacetat *n* potassium acetate
Kaliumalaun *m* potassium [potash] alum, aluminium potassium sulphate 12-water
Kaliumaluminat *n* potassium aluminate
Kaliumaluminiumalaun *m s.* Kaliumalaun
Kaliumaluminiumsulfat *n* aluminium potassium sulphate
Kaliumamalgam *n* potassium amalgam
Kaliumantimonotartrat *n* potassium antimonyl tartrate
Kaliumantimonyltartrat *n s.* Kaliumantimonotartrat
Kaliumbromat *n* potassium bromate
Kaliumbromid *n (spectr)* potassium bromide
Kaliumcarbonat *n* potassium carbonate, potash, salt of tartar
Kaliumchlorat *n* potassium chlorate
Kaliumchlorid *n* potassium chloride
Kaliumchromalaun *m* chrome [potash] alum, potassium chrome alum
Kaliumchromat *n* potassium chromate
Kaliumcyanat *n* potassium cyanate
Kaliumcyanid *n* potassium cyanide
Kaliumdichromat *n* potassium dichromate
Kaliumdihydrogenphosphat *n* potassium dihydrogen phosphate
Kaliumdisulfat *n* potassium disulphate
Kaliumeisenalaun *m* iron potassium alum, iron potassium sulphate 12-water
Kaliumethylxanthogenat *n* potassium ethyl xanthate
Kaliumfluorid *n* potassium fluoride
kaliumhaltig potassium-containing, potassic
Kaliumhexacyanoferrat(II) *n* potassium hexacyanoferrate(II), yellow prussiate of potash

Kaliumhexacyanoferrat(III) *n* potassium hexacyanoferrate(III), red prussiate of potash
Kaliumhexafluorosilicat *n* potassium hexafluorosilicate
Kaliumhexanitrocobaltat(III) *n* potassium hexanitrocobaltate(III)
Kaliumhydrogencarbonat *n* potassium hydrogencarbonate
Kaliumhydrogenfluorid *n* potassium hydrogenfluoride
Kaliumhydrogenphosphat *n* dipotassium hydrogenphosphate
Kaliumhydrogensulfat *n* potassium hydrogensulphate
Kaliumhydrogensulfid *n* potassium hydrogensulphide
Kaliumhydrogensulfit *n* potassium hydrogensulphite
Kaliumhydrogentartrat *n* potassium hydrogentartrate
Kaliumhydroxid *n* potassium hydroxide
Kaliumhydroxidschmelze *f* 1. potassium hydroxide fusion, molten potassium hydroxide; 2. potassium hydroxide fusion *(act)*
Kaliumhyperoxid *n* potassium hyperoxide
Kaliumhypochlorit *n* potassium hypochlorite
Kaliumiodat *n* potassium iodate
Kaliumiodid *n* potassium iodide
Kaliumiodidstärkepapier *n* potassium-iodide starch paper
Kaliummagnesiumsulfat *n* magnesium potassium sulphate
Kaliummanganat *n* potassium manganate
Kaliummanganat(VII) *n s.* Kaliumpermanganat
Kaliumnatriumtartrat *n* potassium sodium tartrate
Kaliumnitrat *n* potassium nitrate
Kaliumnitrit *n* potassium nitrite
Kaliumoleat *n* potassium oleate, *(specif)* potassium octadec-9-enoate
Kaliumorthophosphat *n* potassium orthophosphate
Kaliumoxalat *n* potassium oxalate
Kaliumoxid *n* potassium oxide
Kaliumperchlorat *n* potassium perchlorate
Kaliumpermanganat *n* potassium permanganate
Kaliumperoxid *n* potassium peroxide
Kaliumperoxodisulfat *n* potassium peroxodisulphate
Kaliumsulfat *n* potassium sulphate
Kaliumsulfit *n* potassium sulphite
Kaliumtartrat *n* potassium tartrate
Kaliumtetrachloropalladat(II) *n* potassium tetrachloropalladate(II)
Kaliumtetrafluoroborat *n* potassium tetrafluoroborate
Kaliumtetrathionat *n* potassium tetrathionate
Kaliumthiocyanat *n* potassium thiocyanate
Kaliumxanthogenat *n* potassium xanthate, *(specif)* C_2H_5OCSSK potassium ethylxanthate
Kalk *m* lime • **mit Kalk behandeln** lime out *(manufacture of organic intermediates)* • **mit Kalk düngen** *(agric)* lime
Kalk *m*/**gebrannter** burnt [burned, caustic] lime, quicklime *(calcium oxide)*
Kalk *m*/**gelöschter** hydrated [slaked, water-slaked] lime, lime hydrate *(calcium hydroxide)*
Kalk *m*/**hydraulischer** hydraulic [water] lime *(a lime which will harden under water)*
Kalk *m*/**ungelöschter** unslaked lime, quicklime *(calcium oxide)*
kalkartig limy, calcareous
Kalkboden *m* limy [calcareous] soil
Kalkbrennen *n* lime burning
Kalkbrennofen *m* lime kiln
Kalkbrühe *f (agric)* lime wash, whitewash
Kalkdüngemittel *n* lime (fertilizer)
Kalkdünger *m* lime (fertilizer)
kalken *v* 1. *(agric, sugar)* lime; 2. lime out *(manufacture of organic intermediates)*
Kalkentzuckerungsverfahren *n* Steffen process *(for desugarizing molasses)*
Kalkfällung *f (hyd)* lime precipitation • **durch Kalkfällung enthärten** *(hyd)* lime-soften
Kalkfeldspat *m (min)* lime feldspar
Kalkgestein *n* calcareous rock
Kalkglas *n* lime glass
kalkhaltig limy, calcareous
Kalkhärte *f (hyd)* calcium hardness
Kalkhydrat *n s.* Kalk/gelöschter

kalkig limy, calcareous
Kalklöschen *n* lime slaking [hydration]
Kalkmilch *f* 1. milk [slurry] of lime, lime milk; 2. *(agric)* lime wash, whitewash *(a suspension of calcium hydroxide or hydrated lime in water)*
Kalkmörtel *m* lime mortar
Kalknatronglas *n* soda-lime glass
Kalk-Nichtkarbonathärte *f (hyd)* calcium non-carbonate hardness
Kalkofen *m* lime kiln
kalkreich rich in lime, high-lime
Kalk-Ringofen *m* lime ring furnace
Kalksaccharat *n* lime saccharate
Kalksaccharatverfahren *n* **nach Steffen** Steffen process *(for desugarizing molasses)*
Kalk-Schachtofen *m* lime shaft kiln
Kalkseife *f* lime soap
Kalkspat *m (min)* calc-spar, calcite *(calcium carbonate)*
Kalkstein *m* limestone
Kalkstickstoff *m* nitrolime, lime nitrogen *(calcium cyanamide)*
Kalktuff *m* calcareous tufa, tufaceous limestone
Kalkung *f (agric, sugar)* liming
Kalk-Zeolith-Verfahren *n (hyd)* lime-zeolite (softening) process
Kallidin *n* kallidin *(a decapeptide)*
Kallikrein *n* kallikrein *(a pancreatic enzyme)*
Kalomel *n* calomel, mercury(I) chloride
Kalomelelektrode *f* calomel electrode
Kalomelelektrode *f***/gesättigte** saturated calomel electrode, S.C.E.
Kalomelnormalelektrode *f* normal calomel electrode
Kalorie *f* calorie *(non-SI unit of heat energy; 1 cal = 4.1868 J)*
Kalorie *f***/thermochemische** thermochemical calorie *(non-SI unit of heat energy;* $1\ cal_{th} = 4.184\ J$*)*
Kalorimeter *n* calorimeter
Kalorimeter *n* **für Gase** gas calorimeter
Kalorimeterbombe *f (lab)* calorimeter [calorimetric, explosion] bomb, (oxygen) bomb calorimeter
Kalorimetrie *f* calorimetry
kalorimetrisch calorimetric(al)
kalorisch caloric
kalorisieren *v (met)* calorize
kaltabbindend *(plast)* cold-curing, cold-setting
kaltblasen *v* steam *(in producing water gas)*
Kältebad *n* cold bath
kältebeständig cold-resistant
Kältebeständigkeit *f* cold [low-temperature] resistance
Kältebeständigkeitsprüfung *f* cold [low-temperature] test
Kältebruchtemperatur *f* brittle(-point) temperature
kälteerzeugend frigorific, refrigerant
Kälteerzeugung *f* cold production
Kälteleistung *f* refrigeration performance
Kältemaschine *f* refrigerating machine, refrigerator
Kältemischung *f* freezing mixture
Kältemittel *n* refrigerant
Kältesole *f* refrigerating [cooling] brine
Kälteträger *m* secondary refrigerant, *(esp)* refrigerating [cooling] brine
Kälteverhalten *n* low-temperature behaviour [characteristics]
Kaltformteil *n (plast)* cold moulding
kalthärtend *(plast)* cold-curing, cold-setting
Kalthärtung *f (plast)* cold curing
Kaltkatoden-Ionisationsvakuummeter *n* cold-cathode ionization gauge, Penning [Philips] gauge
Kaltkautschuk *m* cold [polymerized] rubber
Kaltpolymerisation *f* cold polymerization
kaltpressen *v* 1. *(met)* cold-press; 2. *(plast)* cold-mould; 3. *(food)* cold-draw, cold-press *(oils)*
Kaltpressmasse *f (plast)* cold-moulding compound [material]
kaltrecken *v (plast)* cold-draw
kaltstrecken *v (plast)* cold-draw
Kaltumformen *n* cold forming [working]
kaltverarbeiten *v* cold-work
kaltverstrecken *s.* kaltrecken
Kaltverzinken *n* cold galvanizing *(using zinc-pigmented coatings)*
Kaltvulkanisat *n* cold vulcanizate
Kaltvulkanisation *f* cold vulcanization [curing]
kaltvulkanisieren *v* vulcanize [cure] at room temperature

kaltvulkanisierend room-temperature-vulcanizing, RTV
Kaltwalzen *n* cold rolling
Kaltwerden *n* cooling(-down)
kaltziehen *v (met)* cold-draw
kalzinieren *v* calcine, burn
Kalzinierofen *m* calcining furnace, calciner
Kalzinierung *f* calcination, calcining, burning
Kalzinierungsprodukt *n* calcine
Kamin *m* 1. (smoke)stack, chimney; 2. *(distil)* riser (pipe, tube), chimney *(of a cap)*
Kaminzug *m* draught, *(Am)* draft
Kammblende *f (spectr)* comb-shaped shutter, comb attenuator
Kammer *f* chamber, cabinet, box, *(if one of several:)* compartment
Kammerabscheider *m* (gravity) settling chamber *(gas cleaning)*
Kammerfilterpresse *f* chamber (filter) press
Kammerofen *m* 1. *(met)* chamber furnace; 2. *(ceram)* chamber [box] kiln
Kammer(schwefel)säure *f* chamber (sulphuric) acid
Kammertrockner *m* 1. cabinet dryer *(with only one chamber)*; compartment dryer *(with multiple chambers)*; 2. *(ceram)* chamber dryer
Kammerverfahren *n* chamber process *(for manufacturing sulphuric acid)*
Kammpolymer *n s.* Pfropfcopolymer
Kampfer *m s.* Campher
Kampfstoff *m* warfare agent
Kampfstoff *m***/chemischer** chemical weapon [warfare agent], war gas
Kampfstoff *m***/hautschädigender** blister gas [agent], vesicant
Kampfstoff *m***/lungenschädigender** lung irritant [injurant]
Kampfstoff *m***/psychotoxischer** psychochemical
Kanadabalsam *m* Canada turpentine, balsam of fir *(from Abies balsamea (L.) Mill.)*
Kanadaterpentin *n(m)* Canada turpentine, balsam of fir *(from Abies balsamea (L.) Mill.)*
Kanal *m (tech)* canal, *(esp if tubular:)* channel, duct
Kanalbildung *f* channel formation, channel(l)ing *(as in column packings or reactors)*
Kanaldispersion *f* eddy [turbulent] dispersion [diffusion]
Kanaleinschlussverbindung *f* channel inclusion compound
Kanalisation *f* sewer [sewage] system
• **in die Kanalisation ableiten [einleiten]** drain [run] to the sewer *(waste water)*
Kanalisationsnetz *n s.* Kanalisation
Kanalisationssystem *n s.* Kanalisation
Kanalisierung *f* channel(l)ing *(with ion diffraction)*
Kanalstrahl *m* canal [positive] ray
Kanister *m* can
Kännelkohle *f* cannel coal
Kante *f* edge
Kantenfilter *n (spectr)* cut-off filter
Kantenversetzung *f (cryst)* edge dislocation
kanzerogen carcinogenic, cancerogenic, cancer-producing
Kanzerogen *n* carcinogen
Kaolin *m* kaolin(e), china [porcelain] clay, white bole, bolus alba
Kaon *n (nucl)* kaon (particle), K meson
Kapazitätsstrom *m* capacity current *(polarography)*
Kapelle *f* 1. *(met)* cupel; 2. fume hood
kapillar capillary
kapillaraktiv capillary-active
Kapillaraszension *f* capillary rise
Kapillarattraktion *f* capillarity
Kapillardepression *f* capillary depression
Kapillardruck *m* capillary pressure
Kapillare *f* capillary (tube)
Kapillare *f***/lugginsche** *(phys ch)* Luggin capillary, capillary salt bridge
Kapillareffekt *m* capillary action, capillarity
Kapillarelektrophorese *f (anal)* capillary electrophoresis
Kapillar-Gaschromatographie *f (anal)* capillary gas chromatography
kapillarinaktiv capillary-inactive
Kapillarität *f* capillarity
Kapillarkondensation *f* capillary condensation
Kapillarkraft *f* capillary force
Kapillarrohr *n* capillary (tube)
Kapillarröhrchen *n* capillary (tube)

Kapillarsäule *f (chromat)* capillary column, *(specif)* open tubular (capillary) column
Kapillarsäule *f*/**gepackte** packed capillary column
Kapillarviskosimeter *n* capillary viscometer, viscosity pipette
Kapillarviskosimeter *n* **nach Ostwald** Ostwald (capillary) viscometer, Ostwald viscosity pipette
Kapillarwasser *n (soil)* capillary water
Kapillarwirkung *f* capillary action [attraction]
Kapsel *f* capsule
kapseln *v* can, encase, enclose *(e.g. a pump or motor)*; waterproof
Kapselmembran *f* capsule membrane
Kapseloberfläche *f* capsule surface
Kapselpumpe *f* lobe pump
Kapselung *f* encapsulation *(environmentalism)*
Kapsid *n (bioch)* capsid protein, (viral) coat protein
Karamell *m* caramel, caramelized sugar
karamellisieren *v* caramelize
Karamellisierung *f* caramelization
Karbo... *siehe auch: Carbo...*
Karbonisation *f* 1. *(text)* carbonization *(of wool)*; 2. *(food)* carbonation, impregnation (with carbon dioxide)
karbonisieren *v* 1. *(text)* carbonize *(wool)*; 2. *(food)* carbonate, impregnate (with carbon dioxide)
karbothermisch *(met)* carbothermic, carbothermal
Karburator *m* carburet(t)or *(for carburetting gases)*
karburieren *v* carburet *(gases)*
Kardiotonikum *n (pharm)* cardiotonic, cardiac tonic *(an agent strengthening the cardiac musculature)*
Karenzfrist *f* preharvest interval *(after application of pesticides)*
Karkassengummi *m* carcass [casing] rubber
Karl-Fischer-Reagens *n* Karl Fischer reagent
Karl-Fischer-Titration *f* Karl Fischer titration
Karl-Fischer-Wasserbestimmung *f* Karl Fischer titration
Karte *f*/**topographische** biochemical map *(as of RNA)*
Kartoffelmehl *n s.* Kartoffelstärke
Kartoffelspiritus *m* potato alcohol
Kartoffelsprit *m* potato alcohol
Kartoffelstärke *f* potato starch [flour], farina
Karton *m* cardboard, (paper)board
karzinogen carcinogenic, cancerogenic, cancer causing
Karzinogen *n* carcinogen
kaschieren *v (pap)* laminate, line
Kasein *n s.* Casein
käsig curd(l)y
Kaskade *f (tech)* cascade *(kind of multistage systems)*
Kaskadenboden *m* cascade tray *(rectification)*
Kaskadenbodenkolonne *f* baffle(-plate) column, baffle tower (extractor)
Kaskadenreaktor *m* series of stirred-tank reactors, stirred-tank reactors in series
Kaskadentrockner *m* cascade dryer
Kaskadenverdampfer *m* cascade [multiple-effect] evaporator
kastenaufkohlen *v* box-carburize *(steel)*
kasteneinsetzen *v* box-carburize *(steel)*
kastenglühen *v* box-anneal *(castings)*
Kastenkristallisator *m* tank crystallizer
kastenzementieren *v* box-carburize *(steel)*
Kastoröl *n* castor oil *(from Ricinus communis L.)*
katabol(isch) *(bioch)* catabolic
katabolisieren *v (bioch)* catabolize
• **katabolisiert werden** catabolize
Katabolismus *m (bioch)* catabolism
Katabolit *m (bioch)* catabolite
Katal *n (bioch)* katal, enzyme unit *(of enzyme activity)*
Katalaseaktivität *f (bioch)* catalase activity
Katalysator *m* (reaction) catalyst, catalyzer, cat
Katalysator *m*/**bifunktioneller** bifunctional [dual-function] catalyst
Katalysator *m*/**enzymatischer** *(bioch)* enzyme catalyst
Katalysator *m*/**fester** 1. solid catalyst; 2. *s.* Katalysator/ruhender
Katalysator *m*/**heterogener** heterogeneous [contact] catalyst
Katalysator *m*/**komplexkoordinativer** coordination [complexing] catalyst

Katalysator *m*/**monofunktioneller** single-function catalyst
Katalysator *m*/**negativer** negative catalyst, anticatalyst, retarder, inhibitor
Katalysator *m* **ohne Stützmaterial [Träger]** unsupported catalyst
Katalysator *m*/**oxidischer** oxide catalyst
Katalysator *m*/**perlförmiger** bead catalyst
Katalysator *m*/**pillenförmiger** pelletized catalyst
Katalysator *m*/**ruhender** static catalyst
Katalysator *m*/**technischer** commercial catalyst
Katalysatoraktivität *f* catalyst activity
Katalysatorbett *n* catalyst bed
Katalysatordesaktivierung *f* catalyst deactivation
Katalysatorgemisch *n* catalyst mixture
Katalysatorgift *n* catalyst poison
Katalysatorkomplex *m* catalyst-reactant complex
Katalysatoroberfläche *f* catalyst surface
Katalysatorpellet *n* catalyst pellet
Katalysatorpille *f* catalyst pellet
Katalysatorregenerierung *f* catalyst regeneration
Katalysatorrückgewinnung *f* catalyst recycling
Katalysatorschicht *f* catalyst bed
Katalysatorselektivität *f* catalyst selectivity
Katalysatorträger *m* catalyst support [carrier]
Katalyse *f* catalysis
Katalyse *f*/**bifunktionelle** bifunctional catalysis
Katalyse *f*/**elektrochemische** *s.* Elektrokatalyse
Katalyse *f*/**heterogene** heterogeneous catalysis, contact [surface] catalysis
Katalyse *f*/**homogene** homogeneous catalysis
Katalyse *f*/**kovalente** covalent catalysis
Katalyse *f*/**mizellare** micellar catalysis
Katalyse *f*/**negative** inhibition
Katalyse *f*/**stereospezifische** stereospecific catalysis
Katalysegesetz *n*/**brönstedsches** (Brønsted) catalysis law
Katalysekonstante *f* catalytic coefficient
Katalyseofen *m* catalytic reactor
katalysieren *v* catalyse
katalysiert/durch Basen base-catalysed
katalysiert/nicht uncatalyzed
katalytisch catalytic
Kataphorese *f* cataphoresis
kataphoretisch cataphoretic
Katharometer *n (anal)* katharometer
Kathartikum *n (pharm)* cathartic
Kation *n* cation
kation(en)aktiv cation-active, cationic
Kationenaustausch *m* cation exchange
Kationenaustauscher *m* cation(ic) exchanger
Kationenaustauscher *m*/**schwach saurer** *(hyd)* weak-acid cation exchanger
Kationenaustauscher *m*/**stark saurer** *(hyd)* strong-acid cation exchanger
Kationenfehlstelle *f* cation vacancy
Kationenleerstelle *f* cation vacancy
Kationen-Neutralaustauschverfahren *n (hyd)* sodium cycle exchange
Kationentrennungsgang *m (anal)* scheme of analysis for the cations, cation scheme
Kationenüberführungszahl *f* cation transference number
Kationenumtausch *m* cation exchange
kationisch cationic
Kationit *m* cation exchanger
kationkapillaraktiv cation-active
kationoid cationoid, electrophilic, electron-attracting
kationotrop cationotropic
Kationotropie *f* cationotropy
Kationsäure *f (phys ch)* cation(ic) acid
Katkracken *n (petrol)* catalytic [cat] cracking
Katode *f* cathode, negative electrode
Katodenpotenzial *n* cathode potential
Katodenraum *m* cathode compartment
Katodenreaktion *f* cathode reaction
Katodenstrahl *m* cathode ray
Katodenstrahlpolarographie *f (anal)* single-sweep polarography
Katodenstrom *m* cathode current
Katodenzerstäuben *n* cathode sputtering
katodisch cathodic

Katolyt *m* catholyte *(electrolyte surrounding a cathode)*
Kauren *n* kaurene *(a terpene)*
Kauriharz *n* kauri copal [gum, resin] *(from Agathis australis Salisb.)*
Kaurikopal *m* kauri copal [gum, resin] *(from Agathis australis Salisb.)*
kaustifizieren *v* causticize
Kaustifizierung *f* causticization
Kaustikum *n (med)* caustic agent
kaustisch caustic
Kauterisation *f (med)* cauterization *(burning of tissue with a caustic or heat)*
Kautschuk *m(n)* rubber, caoutchouc
Kautschuk *m*/**cyclisierter** cyclized rubber, cyclorubber
Kautschuk *m*/**kalt polymerisierter** cold (polymerized) rubber
Kautschuk *m*/**ölgestreckter** [**ölhaltiger**, **ölplastizierter**] oil-extended rubber
Kautschuk *m*/**regenerierter** reclaimed rubber, reclaim
Kautschuk *m*/**synthetischer** synthetic rubber, artificial [man-made, chemical] rubber
Kautschuk *m*/**vulkanisierter** vulcanized [cured] rubber
Kautschukbaum *m* rubber tree
Kautschukchemie *f* rubber chemistry
Kautschukchemiker *m* rubber chemist
Kautschukderivat *n* rubber derivative
Kautschukelastizität *f* rubber elasticity
Kautschukhydrochlorid *n* rubber hydrochloride
Kautschukkohlenwasserstoff *m* rubber hydrocarbon
Kautschuklatex *m* rubber latex
Kautschuklösung *f* rubber solution
Kautschuklösungsmittel *n* rubber solvent
Kautschukmilch *f* rubber latex
Kautschukmilchsaft *m* rubber latex
Kautschukmischung *f* rubber stock [mixture, compound]
Kautschukpflanze *f* rubber(-yielding) plant
Kautschukvulkanisat *n* rubber vulcanizate, vulcanized rubber
Kavitation *f* cavitation
KBr-Tablette *f (spectr)* KBr [potassium bromide] pellet
KDPG-Weg *m (bioch)* Entner-Doudoroff pathway *(of glucose degradation via 2-keto-3-deoxy-6-phosphogluconate)*
Kegelmischer *m* conical mixer
Kegelmühle *f* conical grinder, cone mill, rotary crusher
Kegelschliffverbindung *f* conical (ground-glass) joint, tapered joint
Keim *m* 1. *(cryst)* nucleus; 2. *(med, food, hyd)* germ, microbe • **als Keim wirken** nucleate • **Keime bilden** nucleate
Keim *m*/**krankheitserregender** [**pathogener**] pathogen
Keimabtötung *f s.* Keimtötung
Keimbildner *m (cryst)* nucleation [nucleating] agent
Keimbildung *f (cryst)* nucleation, nucleus formation
Keimbildung *f*/**heterogene** heterogeneous [induced] nucleation
Keimbildung *f*/**homogene** homogeneous [spontaneous] nucleation
Keimbildung *f*/**primäre** primary nucleation
Keimbildung *f*/**sekundäre** secondary [surface] nucleation
Keimbildung *f* **und Keimwachstum** *n* nucleation and growth
Keimbildungsgeschwindigkeit *f* rate of nucleation
keimfrei *(med)* sterile, aseptic • **keimfrei machen** sterilize
Keimfreiheit *f (med)* sterility, asepsis
Keimfreimachung *f (med)* sterilization
Keimkristall *m* seed crystal
Keimöl *n (food)* germ oil
keimtötend sterilizing, germicidal, disinfectant
Keimtötung *f* kill(ing) of microbes
Keimungsmedium *n (biot)* germination [spore-germinating] medium
Keimzahl *f (med, food, hyd)* bacterial count *(per unit amount of sample)*
K-Einfang *m* K(-electron) capture
Kekulé-Formel *f (org ch)* Kekulé formula
Kekulé-Grenzstrukturen *fpl* Kekulé(-like) structures
Kekulé-Strukturen *fpl* Kekulé(-like) structures
Kelvin-Skale *f* Kelvin (temperature) scale
Kelvin-Temperatur *f* thermodynamic [absolute] temperature
Kennlinie *f* characteristic curve
Kennwert *m* characteristic (value)

Kennwert *m*/**verarbeitungstechnischer** processing characteristic
Kennzahl *f* characteristic, coefficient, number, value, index
Kennziffer *f* characteristic, coefficient, number, value, index
Kerametall *n* cermet, ceramal
Keramik *f* ceramics
Keramik *f*/**chemische** chemical ceramics
Keramik *f*/**technische** technical ceramics
Keramikfaser *f* ceramic fibre
Keramikfilter *n* ceramic filter
keramisch ceramic
kerbempfindlich notch-sensitive
kerben *v* notch, indent
Kerbschlagzähigkeit *f* notch impact resistance [strength]
Kern *m* 1. nucleus *(of an atom, of a cell, of crystallization)*; 2. *(lab)* cone *(of a ground-glass joint)* • **als Kern wirken** nucleate • **Kerne bilden** nucleate *(of foreign material)*
Kern *m*/**aromatischer** aromatic [benzene] nucleus
Kern *m*/**kondensierter** *(org ch)* condensed [fused] nucleus
Kernabstand *m* internuclear [interatomic] distance, bond length
Kernaktivität *f* core activity
Kernanregung *f* nuclear excitation
Kernbaustein *m* nuclear constituent [particle], nucleon
Kernbildung *f* nucleation, nucleus formation *(crystallization, condensation)*
Kernbrennstoff *m* nuclear [fission] fuel
Kernbruchstück *n* nuclear fragment
Kernchemie *f* nuclear chemistry
Kerndrehimpuls *m* nuclear spin [angular momentum]
Kernenergie *f* nuclear [atomic] energy
Kernenergieniveau *n* nuclear energy level
Kernenzym *n (bioch)* core enzyme
Kernfusion *f* nuclear fusion
kernhalogeniert ring-halogenated
Kernisobar *n* nuclear isobar
Kernisomer(e) *n* nuclear isomer
Kernisomerie *f* nuclear isomerism
Kernkettenreaktion *f* nuclear chain reaction
Kernladung *f* nuclear charge
Kernladungszahl *f* atomic [ordinal] number, A.N.
Kernladungszahl *f*/**effektive** effective atomic number, E.A.N.
Kernmagneton *n* nuclear magneton
Kernmasse *f* nuclear mass
Kernmasse(n)zahl *f* (nuclear) mass number, nuclear number
Kernmoment *n*/**magnetisches** nuclear magnetic moment
Kernnährstoff *m (agric)* primary nutrient [element]
Kernöl *n (food)* kernel oil
Kern-Overhauser-Effekt *m* nuclear Overhauser effect, NOE
Kernphysik *f* nuclear physics
Kernpolarisation *f*/**chemisch induzierte dynamische** chemically induced dynamic nuclear polarization, CIDNP
Kernquadrupolmoment *n* nuclear quadrupole moment
Kernquadrupolresonanzspektroskopie *f* nuclear quadrupole resonance spectroscopy
Kernreaktion *f* nuclear reaction
Kernreaktor *m* (nuclear) reactor
Kernresonanz *f* nuclear resonance
Kernresonanz *f*/**hochauflösende magnetische** high-resolution NMR [nuclear magnetic resonance]
Kernresonanz *f*/**magnetische** nuclear magnetic resonance, NMR
Kernresonanzspektrometer *n* NMR [nuclear magnetic resonance] spectrometer
Kernresonanzspektroskopie *f* NMR [nuclear magnetic resonance] spectroscopy
Kernresonanzspektroskopie *f*/**magnetische** NMR [nuclear magnetic resonance] spectroscopy
Kernresonanzspektrum *n* NMR [nuclear magnetic resonance] spectrum
Kernresonanzspektrum *n*/**magnetisches** NMR [nuclear magnetic resonance] spectrum
Kern-Schale-Teilchen *n* core-shell particle
Kernspaltung *f* nuclear fission
Kernspaltung *f*/**gesteuerte** nuclear fission
Kernspaltungsenergie *f* fission energy

Kernspin *m* nuclear spin [angular momentum]
Kernspinquantenzahl *f* nuclear spin quantum number
Kernstrahlung *f* nuclear radiation
Kernteilchen *n* nuclear particle [constituent], nucleon
Kernumwandlung *f* transmutation, nuclear transformation
Kernumwandlung *f***/künstliche** artificial transmutation [transformation]
Kernverschmelzung *f* (nuclear) fusion
Kernzerfall *m* nuclear decay [disintegration]
Kernzersplitterung *f* spallation
Kernzertrümmerung *f* spallation
Kernzusammenstoß *m* nuclear collision
Kerosin *n* kerosine, paraffin (oil)
Kerr-Effekt *m* Kerr effect *(double refraction)*
Kerr-Zelle *f (spectr)* Kerr cell
Kerzenfilter *n* tube filter
Kessel *m* boiler
Kesselkohle *f* steam coal
Kesselspeisewasser *n* boiler feed(ing) water
Kesselstein *m* (boiler) scale, hardness scale, fur • **Kesselstein ansetzen** scale, fur • **von Kesselstein befreien** (de)scale
Kesselsteinablagerung *f* scale deposit
Kesselsteinbildung *f (hyd)* scale formation, scaling
Kesselsteingegenmittel *n* descaling agent
Kesselsteinlösemittel *n* descaling agent
Kesselsteinverhütungsmittel *n* scale inhibitor
Ketalisierung *f* ketalization
Ketazin *n* ketazine *(an azine formed from a ketone)*
Keten *n* keten(e) *(an organic compound having a =C=C=O group)*; *(specif)* $CH_2C=O$ ketene
Ketimin *n* ketimine *(an imine having the general structure $R_2C=NR'$)*
Ketoamin *n* ketoamine
Ketobernsteinsäure *f* oxosuccinic acid, oxalacetic acid
Ketocarben *n* keto carbene
Ketocarbonsäure *f* keto carboxylic acid, keto acid
Ketocarbonsäureester *m* keto ester
Keto-Enol-Tautomerie *f* keto-enol tautomerism
Ketoester *m* keto(nic) ester
Ketoform *f* keto form
Ketofunktion *f* ketonic function
ketogen *(bioch)* ketogenic
Ketoglutarsäure *f* oxoglutaric acid, ketoglutaric acid, *(specif)* 2-oxoglutaric acid
3-Ketoglutarsäure *f* 3-oxoglutaric acid, acetonedicarboxylic acid
Ketogruppe *f* keto group, ketonic (carbonyl) group
Ketoketen *n* $(R)_2C=C=O$ ketoketene
Ketol *n* ketol, keto alcohol, hydroxy ketone
Ketolyse *f* ketolysis
ketolytisch ketolytic
Keton *n (org ch)* ketone • **in ein Keton übergehen** ketonize
Keton *n***/Michlers** Michler's ketone, di-*p*-dimethylaminophenyl ketone
Ketonacetal *n* ketone acetal, ketal
Ketonaldehyd *m* keto aldehyde
Ketonalkohol *m* keto alcohol, ketol, hydroxy ketone
Ketonämie *f (med)* ketonaemia
ketonhaltig ketonic
Ketonharz *n* ketone resin
ketonig ketonic • **ketonig werden** *(food)* ketonize
Ketonimid *n s.* Ketimid
Ketonimin *n s.* Ketimid
ketonisieren *v* ketonize
Ketonisierung *f* ketonization
Ketonkörper *m (bioch)* ketone [acetone] body
Ketonkörperbildung *f (bioch, med)* ketogenesis, ketone-body formation
Ketonranzigkeit *f* ketonic rancidity
Ketonspaltung *f* ketonic cleavage [fission]
Ketosäure *f s.* Ketocarbonsäure
Ketose *f s.* Ketozucker
Ketosid *n* ketoside *(a glycoside which yields a ketose on hydrolysis)*
ketosidisch ketosidic
Ketosteroid *n (bioch)* ketosteroid
Ketoverbindung *f* keto (ketonic) compound
Ketoxim *n (org ch)* ketoxime
Ketozucker *m* ketose, ketomic sugar

Kette *f* chain
Kette f/anticlinale anticlined chain *(of a polymer)*
Kette f/elektrochemische [galvanische] galvanic [voltaic] cell
Kette f/gerade straight [unbranched] chain
Kette f/geschlossene *(org ch)* closed chain
Kette f/isoclinale isoclined chain *(of a polymer)*
Kette f/verzweigte *(org ch)* branched [forked] chain
Kettenabbrecher *m* chain stopper [terminator]
Kettenabbruch *m* chain termination [breakage]
Kettenbildung *f (plast)* chain formation
Kettenendenabstand *m* end-to-end distance, chain-end distance
Kettenförderer *m* chain conveyor
Kettenfortpflanzung *f* chain propagation
Kettenfortpflanzungsreaktion *f* chain-propagating reaction
Kettenisomer(e) *n* chain [skeletal] isomer
Kettenisomerie *f* chain [skeletal] isomerism
Kettenkonformation *f* chain conformation *(of a polymer)*
Kettenmechanismus *m* chain(-reaction) mechanism
Kettenmechanismus m/verzweigter branched-chain mechanism
Kettenmolekül *n* chain molecule
Kettenpolymerisation *f* chain(-growth) polymerization
Kettenpolymerisation f/radikalische (free-)radical polymerization
Kettenreaktion *f* chain reaction
Kettenreaktion f/nukleare nuclear chain reaction
Kettenreaktion f/radikalische free-radical chain reaction
Kettensegment *n* chain segment
Kettensilicat *n (min)* inosilicate, chain silicate
Kettenspaltung *f* chain scission
Kettenstart *m* chain initiation
Kettenträger *m* chain carrier
Kettenübertragung *f* chain transfer
Kettenverbindung *f* chain compound
Kettenverlängerung *f* 1. chain extension; 2. *(bioch)* elongation *(of polypeptide chains)*
Kettenverzweigung *f* chain branching
Kettenverzweigungsmechanismus *m* branched-chain mechanism
Kettenwachstum *n* chain growth [propagation]
Kettenwachstumspolymerisation *f s.* Kettenpolymerisation
Ketyl *n* ketyl
kfz *(kubisch-flächenzentriert) (cryst)* face-centred cubic
Kiefernharz *n* pine resin
Kiefernnadelöl *n* pine-needle oil, Scotch fir oil *(from Pinus sylvestris L.)*
Kiefernöl *n* tall oil
Kiefernteer *m* pine tar
Kienöl *n* pine oil *(from Pinus specc.)*
Kies *m* 1. gravel; 2. *(mine, met)* pyrites
Kiesabbrand *m (met)* roasted pyrites
Kiesabröstung *f* roasting [burning] of pyrites
Kiesel *m* pebble
Kieselgel *n* silica gel
Kieselglas *n* vitreous silica, silica glass
Kieselglas n/durchscheinendes fused [translucend vitreous] silica
Kieselglas n/durchsichtiges [klares] transparent (vitreous) silica, fused quartz, quartz glass
Kieselglas n/undurchsichtiges non-transparent vitreous silica
Kieselgur *f* kieselguhr, infusorial [diatomaceous] earth
Kieselsäure *f* silicic acid
Kieselsäureester *m* silicic-acid ester
kieselsäurehaltig siliceous
Kiesel(säure)sol *n* silica sol
Kieselstein *m* pebble
Kiesfilter *n* gravel filter
Kilogramm-Molarität *f* molality, molal concentration
Kinetik *f* kinetics
Kinetik f/chemische chemical [reaction] kinetics
Kinetiker *m* kineticist
Kinin *n (bioch)* kinin
Kino *n* kino (gum)
Kino n/Bengalisches Bengal kino, butea gum *(from Butea superba Roxb.)*
Kinogummi *n s.* Kino
Kinoharz *n s.* Kino

Kippkübel *m* skip (car)
Kirschrotglut *f* cherry-red heat
kistenglühen *v* box-anneal *(metals)*
Kitt *m* cement, *(esp for sealing)* lute; putty *(for glass)*
kitten *v* cement; putty *(glass)*
Kjeldahl-Apparat *m* Kjeldahl digestion apparatus • **im Kjeldahl Apparat aufschließen** kjeldahlize
Kjeldahl-Bestimmung *f* Kjeldahl determination
Kjeldahl-Kolben *m* Kjeldahl flask
KK *s.* Kohlenstoff/kohlenstofffaserverstärkter
Klammer *f* 1. *(lab)* clamp, clip; 2. *(nomencl)* bracket
Klappenboden *m* valve tray
klar clear *(e.g. liquid)*; bright • **klar werden** clarify, become clear
Kläranlage *f (hyd)* waste [sewage, effluent] treatment plant
Kläranlage *f*/**kommunale [städtische]** (municipal) sewage-treatment plant
Kläranlagenablauf *m* waste treatment effluent
Klärapparat *m* clarifying apparatus, clarifier
Klärbecken *n (hyd)* (gravity) clarifier, settler, settling basin, sedimentation basin [tank]
Klärbecken *n* **mit vertikaler Durchströmung** vertical-flow clarifier
Kläreindicker *m (hyd)* clarifier-thickener, thickener-clarifier
klären *v* clarify, clear
klären *v*/**sich** clarify, (become) clear
Klärfilter *n* clarifying [polishing] filter
Klarfiltrat *n* clear filtrate
Klärfiltration *f* clarification, polishing (filtration)
Klarflüssigkeit *f* clarified liquid; overflow product *(wet classification)*
Klärgefäß *n* clarifying [settling] tank, clarifier
Klarheit *f* clarity *(as of a filtrate or crystal)*
Klärhilfsmittel *n* clarifying agent, clarifier, clarificant
Klarlack *m (coat)* varnish, clearcoat
Klärschlamm *m* 1. *(hyd)* clarification [sedimentation] sludge; 2. *(agric)* sludge fertilizer
Klärung *f* clarification
Klärung *f*/**mechanische** sedimentation
Klärvorrichtung *f* clarifier
Klarwasser *n* 1. clear water, effluent [clean] water *(water treatment)*; 2. *(hyd)* underflow *(oil removal)*
Klarwassergüte *f* effluent water quality *(water treatment)*
Klärwerk *n s.* Kläranlage
Klärzentrifuge *f* clarifying centrifuge, centrifugal clarifier
Klasse *f* class *(as of compounds or substances)*
Klassierapparat *m s.* Klassierer
klassieren *v (tech)* classify, size, grade (into size)
Klassierer *m* classifier
Klassierkegel *m* cone classifier
Klassiersieb *n* sizing screen
Klassierung *f* (size) classification, sizing, grading into size
klassifizieren *v (nomencl)* classify
Klassifizierung *f (nomencl)* classification
klauben *v (min, tech)* pick, sort
Klebelösung *f (rubber)* cement
kleben *v* bond, stick, glue, cement; adhere, stick
Kleben *n* (adhesive) bonding, sticking, gluing, cementation; adherence, adhesion
klebend sticky, adhesive, adherent
Kleber *m s.* 1. Klebereiweiß; 2. Klebstoff
Klebereiweiß *n (food, bioch)* gluten
Klebflüssigkeit *f (plast)* adhesive liquid
Klebkraft *f* adhesiveness
klebrig sticky, tacky
Klebrigkeit *f* stickiness, tack, tackiness
Klebstoff *m* adhesive (agent), cement
Klebstoffbestandteil *m (plast)* adhesive constituent
Klebstoffdispersion *f (plast)* dispersion adhesive
Klebstofflöser *m (plast)* adhesive dissolver
Klebstofftechnologie *f (plast)* adhesive technology
Klebvermögen *n* adhesiveness
Kleeblattstruktur *f (bioch)* cloverleaf arrangement [conformation] *(of transfer RNA)*
Kleesalz *n* salt of sorrel, sorrel salt *(potassium tetraoxalate pure or mixed with potassium hydrogenoxalate)*
Kleesäure *f* oxalic acid
Kleinfermenter *m (biot)* small fermenter

Kleinkläranlage *f (hyd)* small wastewater treatment plant
kleinkristallin finely crystalline
kleinlückig fine-pore(d)
Kleinringverbindung *f (org ch)* small-ring compound
Kleinversuch *m* small-scale experiment
Kleinwinkellichtstreuung *f* small-angle light scattering, SALLS
Kleinwinkelneutronenstreuung *f* small-angle neutron scattering, SANS
Kleister *m* paste
Klemme *f* clamp, clip
Kletterfilmverdampfer *m* rising-film [climbing-film] evaporator
Klimagerät *n* (air) conditioner
Klimakammer *f* climatic chamber
klimatisieren *v* (air-)condition
Klinker *m* clinker (brick)
Klonierung *f (biot)* cloning
Klopfeigenschaften *fpl* knocking properties
klopfempfindlich knock-prone
klopfen *v* knock *(of a carburetting fuel)*
Klopfen *n* knock(ing) *(of a carburetting fuel)*
klopfend/nicht knock-free, non-knocking *(carburetting fuel)*
Klopfer *m* rapper *(gas cleaning)*
klopffest knock-proof *(carburetting fuel)*
Klopffestigkeit *f* 1. knock resistance; 2. *s.* Klopfwert
klopffrei knock-free, non-knocking *(ignition of carburetting fuel)*
Klopfprüfmotor *m* knock-rating engine
Klopfvorrichtung *f (filtr)* knocker; rapper *(gas cleaning)*
Klopfwert *m* antiknock value, knock value [rating]
Klopfwertbestimmung *f* (anti)knock rating
Klopfwertprüfung *f* (anti)knock rating
Klotzfärben *n (text)* pad dyeing
klumpen *v* agglomerate, clot, clog
Klumpen *m* lump, clot • **Klumpen bilden** *s.* klumpen
klumpig lumpy • **klumpig werden** *s.* klumpen
Klystron *n (anal)* klystron (valve)
KM *s.* Kernmagneton
KMK *s.* Mizellkonzentration/kritische
Knallgas *n* detonating gas, *(specif)* oxyhydrogen gas
Knallgascoulometer *n* oxygen-hydrogen coulometer
Knallgasgebläse *n* oxyhydrogen blowpipe [burner]
Knallgasreaktion *f* hydrogen-oxygen reaction
Knallquecksilber *n* fulminating mercury, mercury [mercuric] fulminate
Knallsäure *f* fulminic acid, formonitrile oxide
Knäuel *n(m)* coil *(as of a macromolecule)*
Knäuel *n***/statistisches** random [statistical] coil
Knäuelmolekül *n* coiled molecule
Knäuelung *f* coiling *(as of a macromolecule)*
Knetarm *m* mixing arm
kneten *v* knead
Kneter *m s.* Knetwerk
Knetlegierung *f* wrought alloy
Knetmaschine *f s.* Knetwerk
Knetmischer *m* kneader mixer
Knetorgan *n* kneading [mixing] element
Knetschaufel *f* mixing blade
Knetwerk *n* kneading machine, kneander, (dough) mixer, pug mill
Knickpunkt *m* 1. break *(as of a curve)*; 2. *(hyd)* breakpoint
Knickpunktchlorung *f (hyd)* breakpoint [free-residual] chlorination
Knickpunktfeuchte *f* critical humidity [moisture content]
Knickpunkt-Feuchtebeladung *f* critical humidity [moisture content]
Knickschwingung *f (spectr)* bending vibration
Knie *n* elbow (fitting)
Knochenkohle *f* bone char(coal), spodium
knochentrocken *(ceram)* bone-dry
Knockdown-Mittel *n* knock-down agent [poison] *(any of a class of immediately acting insecticides)*
Knoevenagel-Kondensation *f (org ch)* Knoevenagel condensation *(of aldehydes and ketones with compounds having an activated methylene group)*
Knöllchenbakterien *npl* nodule-forming bacteria *(Rhizobium specc.)*
knollig *(min)* nodular
Knotenfläche *f* nodal plane *(wave mechanics)*

Knotenpunkt *m (bioch)* branch point *(of metabolism)*
Knudsen-Diffusion *f* Knudsen diffusion
Knudsen-Diffusionskoeffizient *m* Knudsen diffusivity
Koagel *n* coagel
Koagulans *n s.* Koagulationsmittel
Koagulant *m s.* Koagulationsmittel
Koagulat *n* coagulate, coagulum
Koagulation *f* 1. coagulation, clotting, curdling; 2. *(coll)* pectization
Koagulationsfiltration *f (hyd)* coagulation-filtration
Koagulationsmittel *n* coagulant, coagulator
Koagulationsvitamin *n* antihaemorrhagic vitamin, phylloquinone
Koagulationswert *m (coll)* coagulation value
Koagulator *m s.* Koagulationsmittel
koagulierbar coagulable
Koagulierbarkeit *f* coagulability
koagulieren *v* 1. coagulate, curd, clot; 2. *(coll)* pectize
Koagulum *n s.* Koagulat
Koaleszenz *f* coalescence
koaleszieren *v* coalesce
koal(is)ieren *v* coalesce
Koaxialzylinderviskosimeter *n* coaxial cylinder viscometer
Koazervat *n* coacervate
Koazervation *f* coacervation
Koazervierung *f* coacervation
Kobalt *n s.* Cobalt
Kobaltblau *n* cobalt [ultramarine] blue *(cobalt aluminate)*
Kobaltgelb *n* cobalt yellow *(potassium hexanitrocobaltate)*
Kobaltgrün *n* cobalt [Rinman's] green, Rinma(n)n's green
Kobaltsikkativ *n (coat)* cobalt drier
Kobalttrockner *m (coat)* cobalt drier
Kobaltultramarin *n s.* Kobaltblau
kochbeständig resistant [fast] to boiling
kochecht resistant [fast] to boiling
kochen *v* 1. boil; 2. *(food, pap)* cook
• **kochen lassen** boil
kochen *v*/**nochmals** reboil
kochen *v*/**unter Rückfluss** reflux
Kochen *n* 1. boil(ing); 2. *(food, pap)* cooking; 3. *(met)* boil *(of a blown smelt)*
• **am Kochen halten** keep at the boil
• **zum Kochen bringen** bring [raise] to the boil
Kocher *m (food, pap)* cooker
Kocherfüllung *f (pap)* cook
Kocherinhalt *m (pap)* cook
kochfest *s.* kochbeständig
Kochkessel *m* cooking kettle
Kochkolben *m* boiling [Florence] flask, round-bottom (boiling) flask
Kochpunkt *m* boiling point, bp, boiling temperature
Kochsalz *n* common salt
Kochsalzlösung *f*/**physiologische** *(med, pharm)* physiological salt solution, isotonic sodium chloride solution, (physiologic) saline
Kochsalzsole *f* sodium chloride brine
Kochsalzsole *f* **zum Regenerieren** regenerant brine, brine regenerant *(ion exchange)*
Kochsäure *f (pap)* cooking acid [liquor]
Kochung *f (pap)* cook(ing), boil
Köder *m* bait, baiting agent
Ködermittel *n* bait, baiting agent
Koeffizient *m*/**ökonomischer** *(biot)* growth yield coefficient *(growth of microorganisms)*
Koeffizient *m*/**respiratorischer** respiratory quotient [ratio]
Koeffizient *m*/**stöchiometrischer** stoichiometric coefficient [factor]
Koexistenz *f* coexistence *(as of two phases)*
koexistieren *v* coexist
Kogasinverfahren *n (coal)* Kogasin [Fischer-Tropsch] process [synthesis]
Kohärenz *f* coherence
kohärieren *v* cohere
Kohäsion *f* cohesion
Kohäsionsarbeit *f* work of cohesion
Kohäsionsdruck *m* cohesion pressure, intrinsic [internal] pressure
Kohäsionsenergiedichte *f* cohesive energy density
Kohäsionskraft *f* cohesive [cohesion] force
kohäsiv cohesive
Kohle *f* 1. coal; 2. *(el ch)* carbon (electrode)
Kohle *f*/**anthrazitische** anthracitic coal
Kohle *f*/**bituminöse** bituminous coal
Kohle *f*/**sapropelitische** sapropelic coal

Kohle *f***/subbituminöse** subbituminous coal
Kohle... *siehe auch: Kohlen...*
kohleähnlich coal-like
Kohleaufbereitung *f* coal preparation [dressing]
Kohlebogen *m* carbon arc
Kohlebrei *m* coal slurry
Kohlechemie *f* coal chemistry
Kohleelektrode *f* carbon (electrode)
Kohleentgasung *f* coal carbonization
Kohleentschwefelung *f* coal desulphurization
Kohle-Erz-Brikett *n* coal-ore briquette
Kohleextraktion *f* coal extraction
Kohlefilter *n* carbon [char] filter
Kohlefolgeprodukt *n* coal product
Kohlehydrierung *f* coal hydrogenation
kohlen *v* 1. carburize *(steel)*; 2. char *(of wood)*
Kohlen... *siehe auch: Kohle...*
Kohlendioxid *n* carbon dioxide
Kohlendioxid *n***/festes** solid carbon dioxide, dry ice
Kohlendioxid *n***/überkritisches** supercritical carbon dioxide
Kohlendioxidassimilation *f* carbon dioxide assimilation
Kohlendioxidentfernung *f* carbon dioxide removal
Kohlendioxidschnee *m* carbon dioxide ice [snow]
Kohlendisulfid *n* carbon disulphide
Kohlengas *n* coal gas
Kohlenhydrat *n* carbohydrate, saccharide
Kohlenmonoxid *n* carbon monoxide
Kohlenmonoxidhämoglobin *n* carbon monoxy haemoglobin, carboxyhaemoglobin
Kohlenmonoxid-Konversion *f* carbon monoxide conversion [shift], CO conversion, water-gas shift, shift conversion
Kohlenoxid *n s.* Kohlenmonoxid
Kohlenoxidchlorid *n* carbon dichloride oxide, carbonyl chloride, phosgene
Kohlenoxidsulfid *n* carbon oxide sulphide, carbonyl sulphide
Kohlensäure *f* 1. carbonic acid; 2. *(tech)* carbon dioxide • **mit Kohlensäure imprägnieren [versetzen]** *(food)* impregnate *(beverages)*
Kohlensäure *f***/freie** *(hyd)* free carbon dioxide
Kohlensäure *f***/gebundene** *(hyd)* bound carbon dioxide
Kohlenschwelung *f* coal carbonization
Kohlenstaub *m* coal dust, breeze; pulverized [powdered] coal
Kohlenstaubbrenner *m* pulverized-coal burner
Kohlenstaubfeuerung *f* pulverized-coal firing, suspension firing (of coal)
Kohlenstoff *m* C carbon
Kohlenstoff *m***/fester [fixer]** fixed carbon, FC
Kohlenstoff *m***/gelöster organischer [organisch gebundener]** *(hyd)* dissolved organic carbon, DOC
Kohlenstoff *m***/glasartiger** glass-like carbon, glassy [vitreous] carbon
Kohlenstoff *m***/graphitierbarer** graphitizable carbon
Kohlenstoff *m***/graphitischer** graphitic carbon
Kohlenstoff *m***/kohlenstofffaserverstärkter** carbon fibre-reinforced carbon, CFC, carbon-carbon composite
Kohlenstoff *m***/nicht flüchtiger organisch gebundener** *(hyd)* non-volatile total organic carbon
Kohlenstoff *m***/nicht graphitischer** non-graphitic carbon
kohlenstoffarm poor in carbon, low-carbon *(e.g. steel)*
Kohlenstoffatom *n***/asymmetrisches** asymmetric carbon atom
Kohlenstoffatom *n***/primäres** primary carbon atom
Kohlenstoffaufnahme *f (met)* carbon restoration
Kohlenstoffausbrand *m (tech)* carbon combustion
Kohlenstoffbilanz *f* carbon balance
Kohlenstoffchemie *f* chemistry of the carbon compounds
Kohlenstoffdioxid *n s.* Kohlendioxid
Kohlenstoffdisulfid *n* carbon disulphide
Kohlenstoffeinsatzstahl *m* carbon case-hardening steel, carbon-carburizing steel
Kohlenstofffaden *m* carbon filament
Kohlenstofffaser *f* carbon fibre
Kohlenstofffaser-Kohlenstoff-Verbundwerkstoff *m* carbon fibre-rein-

forced carbon, CFC, carbon-carbon composite

Kohlenstoffgarn *n* carbon yarn

Kohlenstoffgehalt *m* carbon content • **mit hohem Kohlenstoffgehalt** high-carbon • **mit niedrigem Kohlenstoffgehalt** low-carbon

Kohlenstoffgerüst *n* carbon skeleton

kohlenstoffhaltig containing carbon, carbonaceous

Kohlenstoffkette *f* carbon chain

Kohlenstoff-Kohlenstoff-Bindung *f* carbon-carbon bond, C–C bond

Kohlenstoff-Kohlenstoff-Doppelbindung *f* carbon-carbon double bond, C=C bond

Kohlenstoff-Kohlenstoff-Dreifachbindung *f* carbon-carbon triple bond, C≡C bond

Kohlenstoff-Kohlenstoff-Einfachbindung *f* carbon-carbon single bond, C–C bond

Kohlenstoffmaterial *n* carbon material

Kohlenstoff-Nanofaser *f* carbon nanofibre, CNF

Kohlenstoff-Nanoröhrchen *n* carbon nanotube, CNT

Kohlenstoff-Nanoröhrchen *n***/einwandiges** single-wall carbon nanotube, SWCNT, SWNT

Kohlenstoff-Nanoröhrchen *n***/mehrwandiges** multiwall carbon nanotube, MWCNT, MWNT

Kohlenstoffquelle *f (biot)* carbon source

Kohlenstoffradikal *n* carbon-centred radical

kohlenstoffreich rich in carbon, high-carbon

Kohlenstoffring *m* carbon ring

Kohlenstoff-Silicium-Bindung *f* carbon-silicon bond

Kohlenstoff-12-Skala *f* carbon-12 scale *(of atomic weights)*

Kohlenstoffstahl *m* carbon steel

Kohlenstoffstahl *m***/reiner [unlegierter]** plain [straight] carbon steel

Kohlenstofftetrachlorid *n* carbon tetrachloride, perchloromethane, *tetrachloromethane

Kohlenstoffverbindung *f* carbon compound

Kohlenstoff-Wasserstoff-Bindung *f* carbon-hydrogen bond

Kohlenstoff-Wasserstoff-Verhältnis *n* C/H ratio

Kohlenstoff-Whisker *m* carbon whisker

Kohlenteer *m* coal tar

Kohlenteerpech *n* coal-tar pitch

Kohlenwassergas *n* coal water gas

Kohlenwasserstoff *m* hydrocarbon

Kohlenwasserstoff *m***/acyclischer** acyclic [aliphatic] hydrocarbon

Kohlenwasserstoff *m***/alicyclischer** alicyclic [cycloaliphatic] hydrocarbon

Kohlenwasserstoff *m***/aliphatischer** aliphatic hydrocarbon

Kohlenwasserstoff *m***/aromatischer** aromatic [benzene] hydrocarbon

Kohlenwasserstoff *m***/chlorierter** chlorinated hydrocarbon

Kohlenwasserstoff *m***/cyclischer** cyclic hydrocarbon

Kohlenwasserstoff *m***/cycloaliphatischer** *s.* Kohlenwasserstoff/alicyclischer

Kohlenwasserstoff *m***/fluorierter** fluorinated hydrocarbon, fluorocarbon

Kohlenwasserstoff *m***/geradkettiger** straight-chain hydrocarbon

Kohlenwasserstoff *m***/gesättigter** saturated [paraffin] hydrocarbon, alkane

Kohlenwasserstoff *m***/höherer** higher hydrocarbon

Kohlenwasserstoff *m***/kettenförmiger** open-chain hydrocarbon, aliphatic hydrocarbon

Kohlenwasserstoff *m***/polycyclischer** polycyclic hydrocarbon

Kohlenwasserstoff *m***/polycyclischer aromatischer** polycyclic aromatic hydrocarbon, PAH

Kohlenwasserstoff *m***/polymerer** hydrocarbon polymer

Kohlenwasserstoff *m***/ringförmiger** cyclic hydrocarbon

Kohlenwasserstoff *m***/verzweigtkettiger** branched-chain hydrocarbon

Kohlenwasserstoffarm *m* spacer *(of a gel in affinity chromatography)*

Kohlenwasserstoff-Fermentation *f (biot)* hydrocarbon fermentation

Kohlenwasserstoffharz *n* hydrocarbon resin

Kohlenwasserstoffkette *f* hydrocarbon chain

Kohlenwasserstoffrest *m* hydrocarbon residue
Kohlenwasserstoffsynthese f/kolbesche Kolbe electrolysis reaction, Kolbe hydrocarbon synthesis
Kohlepyrolyse *f* coal pyrolysis
Kohleraffinerie *f* coal refinery
Kohleschwarz *n* carbon black
kohlestämmig coal-derived
Kohleumwandlung *f* coal conversion
Kohleveredlung *f* coal conversion
Kohleverflüssigung *f* coal liquefaction
Kohlevergasung *f* coal gasification
Kohleverkokung *f* coal carbonization
Kohleverschwelung *f* coal carbonization
Kohlewertstoff *m* coal chemical
Kohlrausch-Brücke *f (phys ch)* Kohlrausch bridge
Kohlsaatöl *n* rape(-seed) oil *(from Brassica rapa L. em. Metzg.)*
Kohlungsbad *n (met)* carburizing bath
Kohlungsmittel *n (met)* carburizing agent, (case-hardening) carburizer
Kohlungsofen *m (met)* carburizing furnace [oven]
Kohlungstiefe *f (met)* case [carburizing] depth
Kohobation *f (distil)* cohobation
kohobieren *v (distil)* cohobate
koinzident coincident
Koinzidenz *f* coincidence
koinzidieren *v* coincide
Kokain *n* cocaine
koken *v* coke
Kokereiabwasser *n* coke plant waste (water)
Kokereibenzol *n* coke-oven benzol
Kokille *f* gravity die
Kokillenguss *m* gravity diecasting, permanent-mould casting
Kokillengussstück *n* gravity die casting
Kokillengussteil *n* gravity die casting
Kokosmilch *f* coconut milk
Koks *m* coke
Koks *m*/**metallurgischer** metallurgical coke
koksbeladen coke-contaminated *(catalyst)*
Koksgenerator *m* coke producer
Kokskohle *f* coking [carbonization] coal
Koksofen *m* coke oven
Koksofenbatterie *f* coke-oven battery
Koksofengas *n* coke-oven gas
Koksofenkammer *f* coking chamber
Koksofenteer *m* coke-oven tar
Koksvergasung *f* coke gasification
Kokungsdestillationsanlage *f* coking still
Kolanuss *f* kola (nut)
Kolatur *f* colature
Kolben *m* 1. *(lab)* flask, bulb; 2. *(tech)* piston, ram
Kolbenkompressor *m* reciprocating compressor
Kolbenpumpe *f* piston pump
Kolbenpumpe *f* **mit hin- und hergehenden Kolben** reciprocating pump
Kolbenpumpe *f* **mit rotierendem Kolben** rotary pump
Kolbenspritzgussmaschine *f (plast)* plunger-type injection machine
Kolbenverdichter *m* reciprocating compressor
Kolbe-Schmitt-Reaktion *f* Kolbe Schmitt reaction [synthesis] *(of aromatic hydroxy acids)*
Kolbe-Schmitt-Synthese *f* Kolbe Schmitt reaction [synthesis] *(of aromatic hydroxy acids)*
Kolbe-Synthese *f* **von Kohlenwasserstoffen** Kolbe hydrocarbon synthesis, Kolbe electrolysis reaction
Kolieren *n* colation, straining
Kollagen *n* collagen *(a scleroprotein)*
Kollektivflotation *f* collective floatation
Kollektor *m (min tech)* collector, promoter, scavenger
Kollektorbürste *f* carbon brush
Kollergang *m* kollergang
Kollermühle *f* kollergang
kollidieren *v* collide
Kollimatorlinse *f (spectr)* collimating lens
Kollision *f* collision
Kollodiummembran *f* collodion membrane
Kollodiumwolle *f* collodion cotton, soluble nitrocellulose, pyroxylin(e), pyrocellulose *(a lower-nitrated cellulose)*
kolloid colloid(al)
Kolloid *n* colloid
Kolloid *n*/**globuläres** globular colloid
Kolloid *n*/**irresolubles [irreversibles]** irreversible colloid

Kolloid *n***/reversibles [wiederauflösbares]** reversible [resoluble] colloid
kolloidal colloidal
Kolloidchemie *f* colloid chemistry
Kolloidchemiker *m* colloid chemist
kolloidchemisch colloid-chemical
kolloiddispers colloid-disperse, colloidally dispersed, colloidal
Kolloidlösung *f* colloidal solution
Kolloidmühle *f* colloid mill
Kolloidpartikel *n* colloidal particle
Kolloidschwefel *m* colloidal sulphur
Kolloidstabilität *f* colloidal stability
Kolloidteilchen *n* colloidal particle
Kolloxylin *n s.* Kollodiumwolle
Kolonne *f* 1. column; 2. *(tech)* tower
Kolonne *f***/leere** wetted-wall column
Kolonne *f***/pulsierte** pulse(d) column
Kolonnendestillation *f* column distillation
Kolonnenfüllung *f (distil)* column packing
Kolonnenfüllung *f***/geordnete** stacked packing
Kolonneninhalt *m***/dynamischer** *(distil)* operating hold-up
Kolonnenkopf *m* still head, top of the column
Kolophonium *n* colophony, pine resin, rosin *(from Pinus specc.)*
Kolorimeter *n* colorimeter
Kolorimeter *n***/visuelles** visual colorimeter
Kolorimetrie *f* colorimetry
Kolorimetrie *f***/visuelle** visual colorimetry
kolorimetrisch colorimetric
kolumnar *(cryst)* columnar
Kolzaöl *n* colza oil, rape(-seed) oil *(from Brassica rapa L. em. Metzg.)*
Kombinationsabbruch *m* coupling termination
Kombinationsdünger *m* compound fertilizer, multinutrient [mixed] fertilizer
Kombinationsprinzip *n***/ritz-rydbergsches** *(phys ch)* Ritz-Rydberg combination principle
Kombinationsreaktion *f* combination reaction
Kombinationsschwingung *f* combination vibration
Kombinationsschwingungsbande *f (spectr)* combination band
kombinieren *v* 1. combine; 2. *(dye)* couple
Kommunalabwasser *n* municipal sewage
kompakt compact, solid
kompaktieren *v* compact
Komparator *m* comparator *(colorimetry)*
Kompartiment *n (bioch)* compartment
Kompartimentierung *f (bioch)* compartment(aliz)ation
kompatibel compatible
Kompatibilität *f* compatibility
Kompensation *f* compensation
Kompensationsmethode *f* compensation method, (null-)balance method, null [zero] method
Kompensationsmethode *f***/poggendorffsche** Poggendorff compensation method
Kompensations-pH-Meter *n* compensation pH-meter
Kompensationsschreiber *m* self-balancing recorder
Kompensationswägung *f* direct weighing
Kompensator *m* compensator
kompensieren *v* 1. compensate, counterbalance; 2. *(dye)* offset
Komplementärfarbe *f* complementary colour
Komplementarität *f* complementarity, complementariness
Komplex *m* complex • **einen Komplex bilden** complex • **im Komplex binden** complex
Komplex *m***/aktivierter** activated complex, transition state *(kinetics)*
Komplex *m***/koordinierter** coordination complex
Komplex *m***/mehrkerniger [polynuklearer]** polynuclear complex
Komplex *m***/schwach gebundener** hypoligated complex
Komplex *m***/stark gebundener** hyperligated complex
1:1-Komplex *m* 1:1 complex
2:1-Komplex *m* 2:1 complex
komplexbildend complex-forming, complexing
Komplexbildner *m* complexing [sequestering] agent
Komplexbildung *f* complex formation, complexation, sequestration

Komplexbildungstitration *f* complexation titration
Komplexchemie *f* coordination chemistry
Komplexdissoziationskonstante *f* instability constant (of complex ions)
komplexgebunden coordinate
komplexieren *v* complex
Komplexierung *f* complexation
Komplexion *n* complex ion
Komplexkatalysator *m* coordination [complexing] catalyst
Komplexkatalyse *f* coordination catalysis
Komplexometrie *f* complexometry
komplexometrisch complexometric
Komplexsalz *n* complex salt
Komplexstabilitätskonstante *f* (complex-)formation constant, stability constant
Komplexstabilitätskonstante *f***/individuelle [konsekutive, sukzessive]** stepwise [successive] formation constant
Komplexverbindung *f* complex [coordination] compound
Komponente *f* component, moiety
Komponente *f***/aktive** *(dye)* diazo [primary] component
Komponente *f***/am schwersten siedende** heavier-than-heavy component
Komponente *f***/dienophile** dienophile
Komponente *f***/flüchtige** volatile constituent
Komponente *f***/höhersiedende** *(distil)* high-boiling component, less volatile component
Komponente *f***/leichtersiedende** *(distil)* low-boiling component
Komponente *f***/leichtest siedende** *(distil)* lighter-than-light component
Komponente *f***/niedrigsiedende** *(distil)* more volatile component, M.V.C., low boiler
Komponente *f***/passive** *(dye)* coupling [secondary] component
Komponente *f***/schwerersiedende** *s.* Komponente/höhersiedende
Kompound *n* *(plast)* compound *(mechanical polymer blend)*
Kompoundmasse *f* *(plast)* compound *(mechanical polymer blend)*
kompressibel compressible
Kompressibilität *f* compressibility
Kompressibilitätsfaktor *m* compressibility factor
Kompressibilitätskoeffizient *m* compressibility coefficient
Kompressibilitätskoeffizient *m***/kubischer** compressibility coefficient
Kompression *f* compression
Kompressionsarbeit *f* work of compression
Kompressionskältemaschine *f* compression refrigerating machine, vapour-compression machine
Kompressionsmanometer *n* **nach McLeod** McLeod gauge
Kompressionsmodul *m* compressive [bulk] modulus
Kompressor *m* compressor
Kompressor *m***/einstufiger** single-stage compressor
komprimierbar compressible
Komprimierbarkeit *f* compressibility
komprimieren *v* compress, *(relating to solids also)* compact
Komproportionierung *f* comproportionation *(kind of redox reaction)*
Kondensat *n* condensate, condensation product
Kondensation *f* 1. condensation *(of gas or vapour)*; 2. *(org ch)* condensation *(as of esters)*; fusion, anellation *(of cyclic compounds)*
Kondensation *f***/claisensche** Claisen condensation *(of esters)*
Kondensation *f***/extramolekulare** self-condensation
Kondensation *f***/kapillare** capillary condensation
Kondensation *f***/partielle** partial condensation
Kondensation *f***/reduktive [reduzierende]** reductive condensation, hydrocondensation
Kondensationsharz *n* condensation resin
Kondensationskalorimeter *n* steam calorimeter
Kondensationskeim *m* condensation nucleus
Kondensationskern *m* condensation nucleus
Kondensationskurve *f* *(distil)* condensation [dew-point] curve

Kondensationslinie *f (distil)* condensation [dew-point] curve
Kondensationspolymer(e) *n* condensation polymer
Kondensationspolymerisation *f* condensation polymerization
Kondensationsprodukt *n* condensation product
Kondensationsreaktion *f* condensation reaction
Kondensationsstelle *f* point of fusion *(of rings)*
Kondensationswärme *f* heat of condensation
Kondensationszentrum *n* condensation [nucleation] centre [site] *(vapour-liquid phase transition)*
Kondensator *m* 1. condenser; 2. *(el ch)* capacitor
Kondensatorkühler *m* condenser
kondensierbar condensable
Kondensierbarkeit *f* condensability
kondensieren *v* 1. condense, precipitate *(gas or vapour)*; 2. *(org ch)* condense *(e.g. esters)*; fuse, anellate *(cyclic compounds)*; 3. *(food)* condense, inspissate; 4. *s.* kondensieren/sich
kondensieren *v*/**sich** condense, precipitate
Kondenswasser *n* condensed [condensate] water
Kondenswasserableiter *m* steam trap
Konditionierapparat *m* conditioner
konditionieren *v* condition
Konduktanz *f (phys ch)* conductance
Konduktometrie *f* conductometry
konfektionieren *v* formulate
Konfektionierung *f* formulation
Konfiguration *f* (spatial) configuration
Konfiguration *f*/**relative** relative configuration
Konfigurationserhalt *m* configuration retention
Konfigurationsformel *f* configurational [space] formula
Konfigurationsisomer(e) *n* configurational isomer
Konfigurationsisomerie *f* configurational isomerism
Konfigurationsumkehr *f* inversion of configuration, Walden inversion
Konfigurationswechsel *m* inversion of configuration, Walden inversion
Konformation *f* conformation *(1. stereochemistry; 2. collectively for the tertiary and quaternary structure of proteins)*
Konformation *f*/**anticlinale** anticlinal conformation, ac *(torsion angle ±120°)*
Konformation *f*/**antiperiplanare** antiperiplanar conformation, ap *(torsion angle ±180°)*
Konformation *f*/**äquatoriale** equatorial conformation
Konformation *f*/**axiale** axial conformation
Konformation *f*/**ekliptische** *s.* Konformation/synperiplanare
Konformation *f*/**gestaffelte** *s.* Konformation/antiperiplanare
Konformation *f*/**gewinkelte** puckered conformation
Konformation *f*/**synclinale** synclinal conformation, sc *(torsion angle ±60°)*
Konformation *f*/**synperiplanare** synperiplanar conformation, sp *(torsion angle 0°)*
***anti*-Konformation** *f* 1. anti conformation, A *(with macromolecules, torsion angle 60°)*; 2. *s.* Konformation/antiperiplanare
***cis*-Konformation** *f* cis conformation, C *(with macromolecules, torsion angle ±180°)*
***trans*-Konformation** *f* trans conformation, T *(with macromolecules, torsion angle 0°)*
Konformationsformel *f* conformational formula
Konformationsgleichgewicht *n* conformational equilibrium
Konformationsisomer(e) *n* conformational isomer, conformer
Konformationsisomerie *f* conformational [rotational] isomerism
Konformationswinkel *m* torsion [rotational] angle
konformer conformational
Konformer(e) *n s.* Konformationsisomer(e)
Konglomerat *n* conglomerate
Kongorot *n* Congo red
Kongorubin *n* Congo rubin(e)
Königswasser *n* aqua regia, chloronitrous acid
konisch conical, cone-shaped

Konjugation *f* conjugation
konjugiert conjugate(d)
konjugiert/nicht non-conjugated *(double bond)*
konkav concave, dished
Konkurrenzhemmung *f* competitive inhibition *(enzyme kinetics)*
Konkurrenzreaktion *f* competitive [concurrent] reaction
Kon(n)ode *f (phys ch)* conode, tie line *(in the equilibrium diagram)*
konrotatorisch *(org ch)* conrotatory
Konserve *f (food)* preserve
Konservendose *f* tin (can), can
konservieren *v* 1. *(tech)* conserve, preserve; 2. *(food)* preserve, *(esp by drying, salting, or smoking:)* cure
Konservierung *f* 1. *(tech)* conservation, preservation; 2. *(food)* preservation, *(esp by drying, salting, or smoking:)* cure, curing
Konservierungsmittel *n (tech, food)* preservative (agent)
Konservierungsmittel *n***/chemisches** chemical preservative
Konservierungsmittel *n* **für Lebensmittel** food preservative
Konservierungsstoff *m s.* Konservierungsmittel
konsistent consistent
Konsistenz *f* 1. consistency, consistence; 2. *(coat)* body
Konsistenz *f***/pastöse [teigige]** doughiness
konstant constant • **konstant halten** maintain [keep, hold] constant • **konstant sein** show a constant reading
Konstante *f* constant
Konstante *f***/boltzmannsche** Boltzmann constant *(thermodynamics)*
Konstante *f***/ebullioskopische** ebullioscopic constant, molal boiling-point(-elevation) constant
Konstante *f***/gyromagnetische** gyromagnetic factor, spectroscopic splitting factor
Konstante *f***/katalytische** catalytic coefficient
Konstante *f***/kryoskopische** cryoscopic constant, molal freezing-point(-depression) constant
Konstante *f***/madelungsche** Madelung constant *(of lattice energy)*
Konstante *f***/physikalische** physical constant
Konstante *f***/plancksche** Planck [action] constant, quantum of action
Konstante *f***/rydbergsche** Rydberg constant [number]
Konstanten *fpl***/kritische** critical constants [data]
konstantsiedend constant-boiling
Konstellation *f s.* Konformation
konstituierend constituent
Konstitution *f* constitution • **die Konstitution aufklären [ermitteln]** elucidate [establish, determine] the structure
Konstitutionsaufklärung *f* structure elucidation [determination], establishment of constitution
Konstitutionsbeweis *m* proof of structure
Konstitutionsermittlung *f s.* Konstitutionsaufklärung
Konstitutionsformel *f* constitutional formula, (valence) structural formula, graphic [line] formula
Konstitutionswasser *n* constitutional water, water of constitution
Konstruktionskunststoff *m* engineering plastics
Konstruktionswerkstoff *m* material of construction
Kontakt *m* 1. contact; 2. heterogeneous [contact] catalyst • **in Kontakt bringen** (bring into) contact • **miteinander in Kontakt kommen** (enter into) contact
Kontaktautoklav *m* contactor
Kontaktbahnentrockner *m* roller dryer
Kontaktbett *n* catalyst bed
Kontaktfläche *f* contact area
Kontaktgift *n* 1. catalyst poison, paralyzer; 2. *(agric)* (direct-)contact poison
Kontaktionenpaar *n* contact [tight] ion pair
Kontaktkatalysator *m s.* Kontakt 2.
Kontaktkatalyse *f* contact catalysis, heterogeneous [surface] catalysis
Kontaktkleber *m* contact adhesive
Kontaktklebstoff *m* contact adhesive
Kontaktkorrosion *f* galvanic [contact] corrosion, two-metal corrosion

Kontaktofen *m* catalytic reactor, converter
Kontaktor *m* contactor *(1. an autoclave; 2. a solvent extractor)*
Kontaktpille *f* catalyst pellet
Kontaktschwefelsäure *f* contact sulphuric acid
Kontaktthermometer *n* contact thermometer
Kontaktträger *m* catalyst support [carrier]
Kontakttrocknung *f* contact drying, conduction [indirect] drying
Kontaktverfahren *n* contact process *(as for manufacturing sulphuric acid)*
Kontaktwinkel *m* contact angle
Kontaktzeit *f* contact time
Kontaminant *m* 1. contaminant; 2. *(biot)* microbial contaminant, contaminant microorganism
Kontaminant *m* **in Lebensmitteln** incidental food additive, food contaminant
Kontamination *f* contamination
Kontamination *f***/radioaktive** radioactive contamination
Kontaminationsgefahr *f* risk of contamination
kontaminationsgefährdet sein *v***/sehr** have a high risk of contamination
kontaminieren *v* contaminate
kontinuierlich continuous • **kontinuierlich arbeitend [betrieben]** continuous
Kontinuitätsgleichung *f* continuity equation
Kontinuum *n* continuum
Kontinuumstrahler *m (anal)* continuum source
kontrahieren *v* contract
Kontraktion *f* contraction
Kontraktionsberührungswinkel *m* receding contact angle *(testing of tension depressors)*
Kontraktionsfaktor *m (plast)* shrinking factor *(branched polymers)*
Kontraktionsstelle *f* vena contracta *(flow measurement)*
Kontrastverbesserung *f s.* Kontrastverstärkung
Kontrastverstärkung *f (spectr)* contrast enhancement
Kontrollanalyse *f* check analysis
Kontrolle *f* control *(of a reaction)*; check, inspection • **außer Kontrolle geraten** get out of control, run wild
Kontrolle *f***/kinetische** kinetic control
Kontrollenzym *n* regulatory enzyme
kontrollieren *v* control *(a reaction)*; check, inspect
Konturlänge *f* contour [extended] length *(of a molecular chain)*
Konus *m* cone
konusartig cone-shaped, conical
Konusmischer *m* conical mixer
Konus-Platte-Viskosimeter *n* cone-and-plate viscometer
Konvektion *f* convection, convective flow of heat
Konvektion *f***/erzwungene** forced convection
Konvektion *f***/freie** natural convection
Konvektionsheizung *f* convection heating
Konvektionsmischen *n* convective mixing
Konvektionstrocknung *f* convection [direct] drying
konvektiv convective
Konversion *f* 1. *(food)* conversion *(of starch)*; 2. *s.* Konvertierung
Konversion *f* **mit Säure** acid conversion
Konverter *m (met)* converter • **im Konverter verblasen** *(met)* convert, bessemerize
Konverter *m***/bodenblasender [normal blasender]** bottom-blown converter
Konverterfrischverfahren *n (met)* converter [Bessemer] process
Konverterstahl *m* converter [Bessemer] steel
Konverterstahl *m***/sauerstoffgefrischter** basic oxygen steel
konvertieren *v* convert; shift *(synthesis gas)*
Konvertierung *f* conversion; shift *(of synthesis gas)*
Konvertierungsgrad *m* degree of conversion
Konvertierungsreaktion *f* conversion reaction
Konvertierungsreaktion *f* **des Kohlenmonoxids** carbon-monoxide-shift reaction, water-gas-shift reaction
konz. *s.* konzentriert
Konzentrat *n* concentrate
Konzentration *f* concentration, *(relating*

to solutions preferably:) strength • **von doppelter Konzentration** double-strength

Konzentration f/bodennahe ground-level concentration *(of chemical species in the air)*

Konzentration *f* **eins [1]** unit concentration

Konzentration f/höchstzulässige *(tox)* maximum allowable [permissible] concentration

Konzentration *f* **null [0]** zero concentration

Konzentration f/(quasi)stationäre steady-state concentration

konzentrationsabhängig concentration-dependent

Konzentrationsabhängigkeit *f* concentration dependence

Konzentrationsänderung *f* change of concentration

Konzentrationsausgleich *m* smoothing-out of concentration

Konzentrationsgefälle *n* concentration gradient

Konzentrationsgradient *m* concentration gradient

Konzentrationsgrenzwerte *mpl* concentration limits

Konzentrationskette *f (el ch)* concentration cell

Konzentrationspolarisation *f* concentration polarization

Konzentrationsschwankung *f* variation [fluctuation] in concentration

Konzentrationsüberspannung *f* concentration overvoltage [overpotential]

Konzentrationsverhältnis *n* ratio of concentrations

Konzentrationszelle *f (el ch)* concentration cell

konzentrieren *v* concentrate

konzentriert concentrated, conc.

konzentriert/mäßig moderately concentrated

Konzentrierung *f* concentration

kooperativ *(bioch)* cooperative

Kooperativität *f (bioch)* cooperativity

Kooperativitätsverhalten *n (bioch)* co-operative behaviour

Koordination *f* coordination *(chemical-bond theory)*

Koordinationsbindung *f* coordinate bond, dipolar [donor-acceptor] bond

Koordinationschemie *f* coordination chemistry, chemistry of coordination compounds

Koordinationsgitter *n* coordination lattice

Koordinationskatalyse *f* coordination catalysis

Koordinationskomplex *m* coordination complex

Koordinationspolymerisation *f* coordination polymerization

Koordinationssphäre *f* coordination sphere

Koordinationsverbindung *f* coordination compound

Koordinationszahl *f* coordination [covalency] number, ligancy

koordinativ coordinate • **koordinativ gebunden werden** coordinate

koordiniert/sechsfach hexacoordinate, six-coordinate

Kopal *m* copal (resin), gum copal *(collectively for high-melting resins esp of fossil origin)*

Kopf *m* 1. *(tech)* head; 2. *(distil)* top • **über Kopf abgehen** *(distil)* leave at the top • **über Kopf abnehmen** *(distil)* take overhead

Kopffraktion *f (petrol)* top fraction

Kopf-Kopf-Polymerisation *f* head-to-head polymerization

Kopfprodukt *n (distil)* overhead (product), top product

Kopfraum *m* headspace *(vapour phase above liquid phase in closed vessels)*

Kopfraumanalyse *f (chromat)* head-space (gas) analysis

Kopf-Schwanz-Polymerisation *f* head-to-tail polymerization

Kopfwaschmittel *n (cosmet)* hair shampoo

Kopie *f (phot)* print; copy *(reprography)*

kopieren *(phot)* print; copy *(reprography)*

Kopierlack *m* resist *(photolithography)*

koplanar coplanar *(bond system)*

Koplanarität *f* coplanarity *(of a bond system)*

Kopplung *f (dye, phys ch)* coupling

Kopplung f/energetische energetic coupling

Kopplung f/galvanische common

electrical connection, direct coupling, galvanic connection
Kopplungsfaktor *m (bioch)* coupling factor
Kopplungskonstante *f (spectr)* coupling constant
Korbflasche *f* carboy *(esp for acids)*
Kork *m* 1. (natural) cork; 2. *s.* Korken • **aus Kork** subereous
korkartig subereous
Korkdichtung *f* cork gasket
Korken *m* cork *(stopper)*
Korkmehl *n* cork powder
Korksäure *f* suberic acid, octanedioic acid
Korkstopfen *m* cork
Korn *n* 1. grain, particle; 2. *(phot, cryst)* grain; bead *(ion exchange)*
Kornbranntwein *m* grain alcohol
Körnchen *n* granule, (small) grain
Korndichte *f* grain [granule] density, apparent density *(of bulk material)*
Korndurchmesser *m* grain [granule] size
körnen *v* grain, granulate
Körnerhaufwerk *n* granular bed, bed of granular solids
Körnerschüttung *f* granular bed, bed of granular solids
Kornform *f* grain [particle] shape
Kornfraktion *f* size fraction *(classifying)*
Kornfraktionen *fpl***/feine [feinste]** fine sizes, fines
Korngrenze *f (cryst)* grain boundary
Korngrenzendiffusion *f* grain-boundary diffusion
Korngrenzenkorrosion *f* intergranular corrosion
Korngröße *f* particle [grain] size, *(in screening also)* grade, screen size
Korngröße *f***/mittlere** average particle size
Korngrößenbereich *m* (particle-)size range
Korngrößenklasse *f* size fraction • **nach Korngrößenklassen einteilen** size • **nach Korngrößenklassen trennen [zerlegen]** size
Korngrößenverteilung *f* particle-size distribution
körnig granular, grainy, gritty • **körnig machen** grain, granulate
Körnigkeit *f* 1. granularity, graininess; 2. *(phot)* graininess
Körnigmachen *n* granulation, graining
Kornkohle *f* granular activated carbon, GAC
Kornkohlefiltration *f* GAC filtration
Kornscheide *f* size of separation, cut size [point], critical diameter *(classifying)*
Körnung *f* grain size, *(screening also)* screen size
Körnungsanalyse *f* size analysis
Koronaentladung *f* corona (discharge)
Körper *m (tech)* body; effect *(of an evaporator)* • **Körper geben** *(coat)* body
Körper *m***/grauer** *(phys ch)* grey body
Körper *m***/schwarzer** black body
Körperfarbe *f* pigment
Körperflüssigkeit *f* body fluid
Körperflüssigkeit *f***/künstliche** simulated body fluid, SBF
Körperflüssigkeit *f***/stimulierte** *s.* Körperflüssigkeit/künstliche
Körperpflegemittel *n* cosmetic
Korpuskel *n (nucl)* corpuscle, particle
Korpuskularstrahlung *f* corpuscular [particle] radiation
Korrekturfaktor *m* 1. correction factor; 2. *(anal)* titrimetric [volumetric] factor
Korrelation *f* correlation
Korrigens *n (pharm)* corrigent, corrective
korrodieren *v* corrode, eat; corrode, undergo corrosion
Korrosion *f* corrosion • **der Korrosion unterliegen** corrode, undergo corrosion
Korrosion *f***/atmosphärische** atmospheric corrosion
Korrosion *f* **durch Konzentrationsketten** concentration-cell corrosion
Korrosion *f***/flächenhafte** general corrosion
Korrosion *f***/galvanische** galvanic [contact] corrosion
Korrosion *f* **in saurem Medium** acid corrosion
Korrosion *f***/interkristalline** intergranular corrosion
Korrosion *f***/lochförmige** pitting corrosion
Korrosion *f***/mikrobiell induzierte** *s.* Biokorrosion
korrosionsanfällig susceptible to corrosion, corrodible

Korrosionsanfälligkeit *f* susceptibility to corrosion, corrodibility
korrosionsbeständig corrosion-resistant, resistant to corrosion, non-corroding
Korrosionsbeständigkeit *f* corrosion resistance, resistance to corrosion
Korrosionselement *n (el ch)* corrosion cell
korrosionsempfindlich *s.* korrosionsanfällig
Korrosionsermüdung *f* corrosion fatigue
korrosionsfest *s.* korrosionsbeständig
Korrosionsgeschwindigkeit *f* corrosion rate [velocity]
korrosionshemmend anticorrosive
Korrosionshemmer *m s.* Korrosionsinhibitor
Korrosionshemmstoff *m s.* Korrosionsinhibitor
korrosionsinaktiv non-corrosive
Korrosionsinhibitor *m* corrosion inhibitor, anticorrosive agent
Korrosionsmedium *n s.* Korrosionsmittel
Korrosionsmilieu *n* corrosive environment
Korrosionsmittel *n* corrosive (agent), corrodent
Korrosionsneigung *f* tendency to corrode
Korrosionsprüfung *f* corrosion testing
Korrosionsrate *f* corrosion rate
Korrosionsschutz *m* corrosion control [protection], protection from corrosion
Korrosionsschutz *m* **mit [durch] Opferanoden/katodischer** sacrificial protection
Korrosionsschutzfett *n* slushing grease
Korrosionsschutzmittel *n* anticorrosive agent, corrosion protective [preventative]
Korrosionsschutzöl *n* slushing oil, corrosion protection oil
Korrosionsstrom *m* corrosion current
Korrosionsversuch *m* corrosion test
Korrosionswiderstand *m* corrosion *(quantitatively)*
korrosiv corrosive
korrosiv/nicht non-corrosive
Korrosivität *f* corrosiveness, corrosivity
Korund *m (min)* corundum *(α-aluminium oxide)*
Kosmetikchemiker *m* cosmetic chemist
Kosmetikum *n* cosmetic
Kosmochemie *f* cosmochemistry
kovalent covalent • **kovalent gebunden** covalently bonded
Kovalenz *f* 1. covalence, covalency; 2. covalent [electron-pair] bond, homopolar bond [linkage], atomic bond *(state)*
Kováts-Index *m (chromat)* Kovats [retention] index
Kp. *s.* Kochpunkt
Krackanlage *f* cracking plant, cracker
Krackbedingungen *fpl (petrol)* cracking conditions
Krackbedingungen *fpl***/scharfe** severe cracking conditions
Krackbehandlung *f (petrol)* cracking treatment
Krackbenzin *n (petrol)* cracked gasoline, cracking fuel
Krackbenzin *n***/katalytisches** cat-cracked gasoline
Krackdestillat *n (petrol)* cracked distillate, cracking distillate
Krackeinsatz *m (petrol)* cracking feedstock
kracken *v* crack
kracken *v***/katalytisch** cat-crack
kracken *v***/mild** give a mild cracking treatment
Kracken *n* cracking
Kracken *n* **auf flüssigen Rückstand** residue cracking
Kracken *n* **auf Koks(rückstand)** non-residue cracking, coking
Kracken *n***/hydrierendes** hydrocracking
Kracken *n* **in der Dampfphase [Gasphase]** vapour-phase cracking
Kracken *n* **in Flüssigphase** liquid-phase cracking
Kracken *n* **in Wirbelschicht/katalytisches** fluid catalytic cracking
Kracken *n***/ionisches [katalytisches]** catalytic cracking
Kracken *n* **mit suspendiertem Katalysator** suspensoid catalytic cracking
Kracken *n***/thermisches** thermal cracking
Krackfraktion *f (petrol)* cracked fraction, cracking fraction

Krackgas *n (petrol)* cracker [cracking] gas
Krackglasur *f (petrol)* cracking glaze
Krackgut *n (petrol)* cracking feed(stock), cracking product
Krackkatalysatoren *mpl (petrol)* cracking catalysts
Krackofen *m (petrol)* cracking furnace
Krackreaktion *f (petrol)* cracking reaction
Krackreaktor *m (petrol)* cracking reactor
Krackröhrenerhitzer *m (petrol)* cracking tube heater
Krackrohstoff *m (petrol)* cracking raw material
Krackschärfe *f (petrol)* cracking sharpness
Krackung *f s.* Kracken
Krackverfahren *n (petrol)* cracking process
Krackvorgang *m (petrol)* cracking process
Krackzone *f (petrol)* cracking zone
Kraft *f (phys ch)* force; power
Kraft *f*/**abstoßende** repulsive force
Kraft *f*/**abweisende** repellency
Kraft *f*/**coulombsche** Coulomb [electrostatic] force
Kraft *f*/**elektromotorische** electromotive force, EMF
Kraft *f*/**elektrostatische** *s.* Kraft/coulombsche
Kraft *f*/**gegenelektromotorische** counter electromotive force, counter [back] e.m.f.
Kraft *f*/**nucleophile** nucleophilicity
Kraft *f*/**rücktreibende** restoring force
Kraft *f*/**treibende** moving force, agency
Kraft *f*/**wasserabweisende** water repellency
Kraft *f*/**zwischenmolekulare** (inter)molecular force
Kraftaufwand *m* power requirements
Kraftbedarf *m* power requirements
Kraftfeld *n* force field
kräftig vigorous *(shaking, agitation)*
Kräftigungsmittel *n (pharm)* tonic, roborant
Kraftkonstante *f* force constant
Kraft-Längenänderungs-Kurve *f* load-elongation [load-extension] curve
Kraft(pack)papier *n* kraft paper
Kraftsprit *m* power [fuel] alcohol
Kraftstoff *m* automotive fuel, (power) fuel
Kraftstoff *m*/**bleifreier** unleaded fuel
Kraftstoff *m*/**fester** solid fuel [propellant]
Kraftstoff *m*/**flüssiger** fluid [liquid] fuel
Kraftstoff *m*/**gasförmiger** gaseous fuel
Kraftstoff *m*/**gebleiter** leaded fuel
Kraftstoffadditiv *n* fuel additive
Kraftwerk *n* power plant
Kraftzellstoff *m (pap)* kraft pulp
Kraftzellstoffverfahren *n (pap)* kraft process
Krählarm *m* raking arm *(of a thickener)*; rabble arm *(of a multiple-hearth furnace)*
krählen *v* rake *(material in a thickener)*; rabble *(material in a multiple-hearth furnace)*
Krählwerk *n* raking mechanism *(of a thickener)*
Krampfgift *n* tetanic poison
krampflindernd anticonvulsant
krampflösend spasmolytic
krankheitserregend pathogenic
Krankheitserreger *m* pathogen
kratzen *v* scrape, scratch
Kratzerförderer *m* flight(ed) conveyor
Kratzfestigkeit *f* scratch resistance
Kratzkühler *m (tech)* scratch cooler
Kratzwärmeaustauscher *m (tech)* scratch heater
Kräuselfaser *f* crimped fibre
Kraut *n* herb
krebsauslösend carcinogenic, carcinogenous, cancerogenic, cancer-producing
krebserregend carcinogenic, carcinogenous, cancerogenic, cancer-producing
krebserzeugend carcinogenic
Krebs-Henseleit-Zyklus *m (bioch)* Krebs-Henseleit cycle, urea [ornithine] cycle
Krebs-Kornberg-Zyklus *m (bioch)* glyoxylate cycle
Krebs-Zyklus *m (bioch)* Krebs (citric acid) cycle, TCA [tricarboxylic-acid] cycle, citric-acid cycle
Kreide *f*/**(fein) gemahlene** whiting
Kreidekalk *m* chalk lime
Kreiselmischer *m* cylindrical batch mixer

Kreiselmolekül *n***/symmetrisches** symmetric-top molecule
Kreiselpumpe *f* centrifugal pump
Kreiselverdichter *m* centrifugal compressor
Kreiselversprüher *m* spinning-top atomizer [sprayer]
Kreiselwäscher *m* centrifugal washer
Kreisfrequenz *f* angular frequency
Kreiskolbenpumpe *f* circular piston pump
Kreislauf *m* 1. cycle, circuit *(scheme)*; 2. circulation *(of liquid or gas)* • **im Kreislauf fahren [führen]** recycle, circulate • **in den Kreislauf zurückführen** return to the circuit, recycle
Kreislauf *m***/biologischer** biological cycle
Kreislauf *m***/geschlossener** closed circuit
Kreislauffahrweise *f* (re)cycling, recirculation
Kreislaufführung *f* (re)cycling, recirculation
Kreislaufgas *n* recycle gas
Kreislaufpumpe *f* circulating pump
Kreislaufreaktor *m* recycle reactor, (tubular) loop reactor
Kreislaufsystem *n* circulation system
Kreislaufwasser *n* circulating [circuit] water, recycling water
Kreislaufwasserführung *f* water recirculation, recycling of water
Kreispolarimeter *n* circular polarimeter
Kreisprozess *m* cycle, cyclic process
Kreisprozess *m***/born-haberscher** Born-Haber (thermochemical) cycle
Kreisprozess *m***/carnotscher** Carnot cycle
Kreisprozess *m***/thermodynamischer** thermodynamic cycle
Kreuzbalgregler *m* cross-belt controller
Kreuzbalgrührer *m* cross-belt mixer
Kreuzbalgscheider *m* cross-belt separator
Kreuzgegenstrom-Wärmeübertrager *m* spiral-tube heat exchanger
Kreuzkopplung *f (org ch)* cross coupling
Kreuzmetathese *f* cross metathesis
Kreuzpolarisation *f (spectr)* cross polarization, CP
Kreuzreaktion *f* cross-reaction
Kreuzschichtstoff *m (plast)* cross-laminate
Kreuzstrom *m* cross flow
Kreuzstromboden *m (distil)* cross-flow tray
Kreuzströmung *f* cross flow
Kreuzstück *n* cross
kreuzteilen *v (anal)* quarter
kriechen *v* creep
Kriechen *n* creep
Kriechfestigkeit *f* creep resistance [strength]
Kriechrheometer *n* creep rheometer
Kriechversuch *m* creep test
Kristall *m* crystal
Kristall *m***/allotriomorpher** *s.* Kristall/xenomorpher
Kristall *m***/flüssiger** liquid crystal, crystalline liquid, mesophase
Kristall *m***/gestörter** imperfect crystal
Kristall *m***/homöopolarer [kovalenter]** covalent crystal
Kristall *m***/lyotroper flüssiger** lyotropic liquid crystal
Kristall *m***/nematischer flüssiger** nematic liquid crystal
Kristall *m***/piezoelektrischer** piezoelectric crystal
Kristall *m***/pseudomorpher** pseudomorph
Kristall *m***/realer** imperfect crystal
Kristall *m***/smektischer flüssiger** smectic liquid crystal
Kristall *m***/thermotroper flüssiger** thermotropic liquid crystal
Kristall *m***/xenomorpher** xenomorphic [allotriomorphic] crystal, anhedron
Kristallachse *f* crystal [crystallographic] axis
Kristallaggregat *n* crystal(line) aggregate
Kristallbau *m* crystal structure
Kristallbaufehler *m* crystal [lattice] defect
Kristallchemie *f* crystal chemistry
kristallchemisch crystallochemical
Kristallebene *f* crystal [crystallographic] plane
Kristallfeldaufspaltung *f (spectr)* crystal-field [zero-field] splitting
Kristallfeldtheorie *f* crystal-field theory
Kristallfläche *f* crystal face

Kristallform *f* crystal shape • **ohne Kristallform** non-crystalline
Kristallgitter *n* crystal lattice
Kristallgittertyp *m* lattice type
Kristallglas *n* crystal glass, *(Am)* rock crystal
Kristallhabitus *m* crystal habit
Kristallhaufwerk *n* crystal(line) aggregate
kristallin crystalline
kristallin/nicht amorphous, non crystalline
Kristallinität *f* crystallinity
Kristallinitätsgrad *m* degree of crystallinity
Kristallisat *n* crystallizate, crop (of crystals), crystalline crop; solid *(in zone melting)*
Kristallisation *f* crystallization • **die Kristallisation anregen [auslösen]** induce crystallization • **zur Kristallisation bringen** crystallize (out)
Kristallisation *f* **durch Animpfen [Impfen]** *(cryst)* seeded crystallization
Kristallisation *f***/fraktionierte** fractional crystallization
Kristallisation *f***/primäre** primary crystallization
Kristallisation *f***/sekundäre** secondary crystallization
Kristallisation *f***/spontane** spontaneous [unseeded] crystallization
kristallisationsfähig crystallizable
Kristallisationsfähigkeit *f* crystallizability
Kristallisationsgeschwindigkeit *f* rate of crystallization
Kristallisationskeim *m* crystal nucleus, nucleus of crystallization • **als Kristallisationskeim wirken** nucleate • **Kristallisationskeime bilden** nucleate
Kristallisationskeimbildung *f* nucleation
Kristallisationskern *m* crystal nucleus, nucleus of crystallization *(consisting of foreign material)*
Kristallisationskinetik *f* crystallization kinetics
Kristallisationskolonne *f* crystallization column
Kristallisationsprodukt *n* crystalline crop, crop (of crystals), crystallizate
Kristallisationszentrum *n* crystallization centre, nucleation centre [site]
Kristallisator *m* crystallizer *(apparatus)*
Kristallisator *m***/offener fest stehender** tank crystallizer
Kristallisierapparat *m* crystallizer
Kristallisierband *n* crystallization conveyor
kristallisierbar crystallizable
Kristallisierbarkeit *f* crystallizability
kristallisieren *v* crystallize (out)
Kristallisierpfanne *f* tank crystallizer
Kristallisierschale *f* crystallizing dish
Kristallisierschnecke *f* crystallization screw conveyor
Kristallisierwalze *f* crystallization roll
Kristallit *m* crystallite
Kristallkeim *m s.* Kristallisationskeim
Kristallkern *m s.* Kristallisationskern
Kristallklasse *f* crystal [symmetry] class, class of crystal symmetry, point group
Kristallographie *f* crystallography
Kristalloid *n* crystalloid
Kristallose *f* sodium [soluble] saccharin, sodium benzosulphimide
Kristall(schütt)schicht *f* bed of crystals
Kristallsoda *f* soda (crystals), salt of soda, washing soda *(sodium carbonate 10-water)*
Kristallstärke *f* crystallizing starch
Kristallstruktur *f* crystal structure
Kristallstrukturanalyse *f* crystal(-structure) analysis
Kristallstrukturanalyse *f***/röntgenographische** X-ray crystal(-structure) analysis
Kristallsymmetrie *f* crystal symmetry
Kristallsystem *n* crystal [crystallographic] system
Kristallversetzung *f* crystal dislocation
Kristallviolett *n* crystal violet
Kristallwachstum *n* crystal growth
Kristallwasser *n* water of crystallization
Kristallzüchtung *f* crystal growth [growing]
Kristallzwilling *m* crystal twin, twin (crystal)
Kroll-Verfahren *n (met)* Kroll process *(a reduction process)*
18-Krone-6 *f* 18-crown 6, *14,7,10,13,16-hexaoxacyclooctadecane

Kronenether *m* crown ether *(a cyclic polyether)*
Kronenverbindung *f (org ch)* crown compound
Kronflintglas *n* crown flint glass, lead crown glass
Kronglas *n* crown (optical) glass
Krümel *mpl(npl)* crumbs
krümelig crumbly, friable
krümeln *v* 1. agglomerate *(using a liquid)*; 2. crumble
Krümeln *n* 1. agglomeration *(using a liquid)*; 2. crumbling
Krümmer *m* 1. *(lab)* bent tube, bent-tube connection; 2. *(tech)* elbow (fitting)
Krümmung *f* curvature, bend
Krümmungsfaktor *m* tortuosity factor *(diffusion theory)*
krumpfen *v (text)* shrink
Krumpfen *n (text)* shrinkage
krumpfend/nicht *(text)* shrink-resistant
krumpffest *(text)* shrink-resistant
Krumpffestigkeit *f (text)* shrink resistance
Krupp-Einschmelzverfahren *n* Krupp melting process
Krupp-Eisenschwammverfahren *n* Krupp iron sponge process
Krupp-Stapelschneide *f* Krupp stacking cutter
Kruste *f* 1. crust, encrustation, incrustation *(as on heated surfaces of dryers)*; 2. *(food)* rind; 3. crust *(of the earth)* • **mit einer Kruste bedecken** encrust • **eine Kruste bilden** encrust, incrust
Kryogenik *f* cryogenics
Kryohydrat *n* cryohydrate
kryohydratisch cryohydric
Kryopumpe *f* cryogenic pump
Kryoskop *n* cryoscope
Kryoskopie *f* cryoscopy
Kryospeicherung *f* cryo storage
Kryotechnik *f* cryogenic [low-temperature] engineering
Kryptand *m (org ch)* cryptand *(spherical crown compound)*
Kryptat *n* cryptate *(complex compound of a kryptand)*
Kryptobase *f* cryptobase
Kryptoion *n* crypto-ion
Kryptoionenreaktion *f* crypto-ionic reaction
kryptoionisch crypto-ionic
kryptokristallin cryptocrystalline
Krypton *n* Kr krypton
Krypton(ionen)-Laser *m* krypton(-ion) laser
K-Schale *f* K shell *(of an atom)*
KU *s.* Kuoxamfaserstoff
Kübel *m* tub, vat; bucket *(esp in conveying)*
kubisch cubic • **kubisch dicht gepackt** *(cryst)* cubic close-packed
kubisch-flächenzentriert *(cryst)* face-centred cubic
kubisch-raumzentriert *(cryst)* body-centred cubic
Kuchen *m (ch, tech)* cake
Kuchenfiltration *f* cake filtration
KUF *s.* Kuoxamfaser
Kufe *f* vat, tub, beck
Kugel *f* 1. sphere; 2. *(lab)* bulb; 3. *(tech)* ball *(as in bearings or mills)*; pellet
Kügelchen *n* spherule; bead; pellet • **zu Kügelchen formen** pelletize, pill; nodulize
Kugeldruckhärte *f* ball-puncture resistance
Kugelfallversuch *m* falling-ball (impact) test
Kugelfallviskosimeter *n* falling-ball viscometer
kugelförmig spherical
Kugelförmigkeit *f* sphericity
Kugelgraphit *m* spheroidal [nodular] graphite
Kugelgraphit(grau)guss *m* nodular cast iron
Kugelkreisel *m* spherical-top molecule
Kugelkühler *m* ball [bulb] condenser
Kugelkühler *m* **nach Allihn** Allihn condenser
kugelmahlen *v* ball-mill
Kugelmühle *f* ball mill, *(specif)* pebble mill *(filled with flint pebbles or porcelain balls)* • **in der Kugelmühle (ver)mahlen** ball-mill
Kugelmühle *f***/schwingende** vibrating (ball) mill
Kugelpackung *f (cryst)* packing (of spheres)
Kugelpackung *f***/dichteste** close(st) packing (of spheres)
Kugelpackung *f***/hexagonal dichteste** hexagonal close(st) packing, hcp

Kugelpackung f/kubisch dichteste cubic close(st) packing, ccp
Kugelprotein *n* globular protein
Kugelschale *f* spherical shell
Kugelschliffverbindung *f* spherical (ground-glass) joint, ball-and-socket joint
kugelsintern *v (min, tech)* pelletize *(ore)*
Kugel-Stäbchen-Modell *n* ball-and-stick model *(of molecules)*
Kugelsymmetrie *f* spherical symmetry
kugelsymmetrisch spherically symmetrical
Kugelventil *n* ball valve
Kühlapparat *m* cooler; chiller *(for temperatures below 0° C)*
Kühlbad *n* cooling bath
Kühlblech *n* heat sink
kühlen *v* 1. cool; *(quickly by immersion:)* chill; 2. *(food)* refrigerate
Kühler *m* 1. *(tech)* cooler, condenser; 2. *(lab)* (steam) condenser
Kühler *m*/**liebigscher** Liebig condenser
Kühler *m* **nach Friedrichs** Friedrichs (reflux) condenser
Kühlfalle *f* cold [cryogenic] trap
Kühlfinger *m (lab)* cold finger, cold-finger condenser [trap], finger-type condenser, acorn condenser
Kühlflüssigkeit *f* cooling liquid
Kühlgeschwindigkeit *f* cooling rate
Kühlkörper *m (tech)* heat sink
Kühlmantel *m* cooling jacket
Kühlmedium *n* coolant, cooling medium
Kühlmittel *n* coolant, cooling medium
Kühlrippe *f* cooling fin
Kühlschlange *f* 1. cooling coil; 2. *(distil)* worm
Kühlschmierstoff *m* cooling lubricant
Kühlschrank *m* refrigerator
Kühlsole *f* (refrigerating) brine
Kühltemperatur f/obere *(glass)* annealing point [temperature]
Kühltemperatur f/untere *(glass)* strain point [temperature]
Kühlturm *m* cooling tower
Kühlung *f* 1. cooling; *(quickly by immersion:)* chilling; 2. *(food)* refrigeration
Kühlwascher *m* washer-cooler
Kühlwasserkreislauf *m* cooling-water circuit, cooling(-water) loop
Küken *n (lab)* plug
kultivieren *v (biot)* cultivate
kultiviert/in Nährlösung *(biot)* solution-grown
Kultivierung *f (biot)* cultivation
Kultur *f (biot)* culture
Kultur f/biologische biological population
Kultur f/frisch eingesäte seed culture
Kultur f/gefriergetrocknete lyophilized culture
Kultur f/kontinuierliche continuous culture
Kultur f/statische steady-state culture
Kultur f/submerse submerged culture
Kulturflüssigkeit *f (biot)* culture solution, nutrient broth
Kulturhefe *f (biot)* culture [industrial] yeast, barm
Kulturkolben *m (biot)* culture [propagating] flask, culture bottle
Kulturkolben *m* **nach Fernbach** Fernbach flask
Kulturmedium *n (biot)* culture medium
kumulativ cumulative
Kumulen *n* cumulene *(any of a class of compounds having three or more cumulated double bonds)*
kumuliert/nicht non-cumulative *(double bond)*
Kunstglas *n* acrylic glass
Kunstharz *n* synthetic resin
Kunstharzionenaustauscher *m* ion-exchange resin, resinous exchanger
Kunstharzlack *m* synthetic-resin varnish
Kunstharzschaum *m* synthetic-resin foam
Kunsthorn *n* casein plastic
Kunstkautschuk *m* synthetic rubber, chemical [artificial, man-made] rubber
Kunstkohlenstoff-Formkörper *m* artificial carbon article [body]
Kunstleder *n* artificial leather
künstlich artificial, non-natural, *(relating to products also)* synthetic, man-made, factitious
Kunstseide *f (text)* artificial silk; viscose, viscose silk, rayon
Kunststoff *m* plastic (material)
Kunststoff *m*/**faserverstärkter** fibre-reinforced plastic, FRP
Kunststoff *m*/**glasfaserverstärkter** glass-fibre reinforced plastic, GRP
Kunststoff *m*/**halbharter** semirigid plastic

Kunststoff *m***/medizinischer** polymer for medical application
Kunststoff *m***/thermoplastischer** thermoplastic (material)
Kunststoff *m***/verstärkter** reinforced plastic
Kunststoff *m***/wärmehärtbarer** thermosetting plastic [resin]
Kunststoff *m***/weicher [weich gestellter]** non-rigid plastic, flexible plastic
Kunststoffabfall *m* plastic waste
Kunststoffalterung *f* plastic ageing
Kunststoffbeton *m* plastic concrete
Kunststofffolie *f* plastic film [foil]
Kunststoffgalvanisierung *f* electroplating of plastics
Kunststoffharz *n* plastic resin
Kunststofflatex *m* polymer latex
Kunststoffrohr *n* plastic pipe
Kunststoffverarbeitung *f* polymer processing
Kunststoff-Verbundmaterial *n* plastic composite material
Kuoxamfaser *f* cuprammonium [cupro] staple
Kuoxamfaserstoff *m* cuprammonium rayon
Küpe *f (dye)* vat *(1. vessel; 2. solution of vat dyes)* • **in der Küpe behandeln** *(dye)* vat
Kupellation *f (met)* cupellation
küpen *v (dye)* vat
Küpenfärberei *f* vat dyeing
Küpenfarbstoff *m* vat dye
Kupfer *n* Cu copper
Kupfer *n***/elementares** elemental copper
Kupfer *n***/gediegenes** native copper
Kupfer(II)-acetat *n* copper(II) acetate, cupric acetate
Kupfer(I)-acetylid *n* copper(I) acetylide, copper(I) carbide, cuprous acetylide
Kupferamalgamelektrode *f* copper amalgam electrode
Kupfer-Anodenschlamm *m* copper anode slime
kupferarm copper-lean, low-copper
Kupferblech *n* copper sheet
Kupferblüte *f* copper blossom
Kupfer(II)-chelat *n* copper(II) chelate, cupric chelate
Kupfer(I)-chlorid *n* copper(I) chloride, copper monochloride, cuprous chloride
Kupfer(II)-chlorid *n* copper(II) chloride, copper dichloride, cupric chloride
Kupfercoulometer *n* copper coulometer
Kupfer(I)-cyanid *n* copper(I) cyanide, cuprous cyanide
Kupfer(II)-cyanid *n* copper(II) cyanide, cupric cyanide
Kupferdampflaser *m* copper-vapour laser
Kupfer(II)-dichromat(VI) *n* copper dichromate
Kupferelektrode *f* copper electrode
Kupferelektrolyt *m* copper-plating bath [solution]
Kupfererz *n* copper ore
Kupferethylendiamin *n (text)* cupriethylenediamine
Kupferglanz *m* copper glance *(a mineral group)*
Kupferglimmer *m* copper mica
Kupfergrün *n* copper green
Kupferhalogenid *n* copper halide
kupferhaltig copper-bearing, cupriferous *(ores)*
Kupfer(I)-hexacyanoferrat(II) *n* copper(I) hexacyanoferrate(II)
Kupfer(II)-hexacyanoferrat(III) *n* copper(II) hexacyanoferrate(III)
Kupfer(II)-hydroxid *n* copper(II) hydroxide, cupric hydroxide
Kupferhydroxidlösung *f***/ammoniakalische** *s.* Kupferoxidammoniak
Kupferkies *m (min)* copper pyrites, chalcopyrite *(Kupfer(I)-Eisen(III)-sulfid)*
Kupferkonverter *m* copper converter
Kupfer-Leaching *n (min tech, biot)* copper leaching
Kupferlegierung *f* copper alloy
Kupfermineral *n* copper mineral
Kupfermonoxid *n s.* Kupfer(II)-oxid
Kupfer-Nickel-Legierung *f* cupronickel alloy
Kupfer-Nickel-Rohstein *m (met)* nickel matte
Kupfer(II)-nitrat *n* copper(II) nitrate, cupric nitrate
Kupfer(I)-oxid *n* copper(I) oxide, cuprous oxide
Kupfer(II)-oxid *n* copper(II) oxide, copper monoxide, cupric oxide
Kupferoxidammoniak *n* cuprammo-

nium (hydroxide) solution, ammoniacal copper oxide solution
Kupferoxidammoniakcellulose *f* cuprammonium cellulose
Kupferoxidammoniak-Spinnverfahren *n* cuprammonia process
Kupfer(II)-oxidchlorid *n* copper(II) chloride oxide, cupric chloride oxide
Kupferphthalocyanin *n* copper phthalocyanine
Kupferrot *n* copper red
Kupfersalzlösungswäsche *f* copper-liquor scrubbing *(for removing carbon monooxide from gases)*
Kupferseide *f* copper rayon, cupro
Kupferseidenelementarfaden *m* cupro filament
Kupfersmaragd *m* copper emerald
Kupfersodabrühe *f* soda bordeaux *(a fungicide)*
Kupferstein *m*/**armer** *(met)* copper matte
Kupfersteinkonverter *m* copper converter
Kupfersteinverblasen *n* copper converting
Kupferstich *m* copper tapping
Kupfer(I)-sulfat *n* copper(I) sulphate, cuprous sulphate
Kupfer(II)-sulfat *n* copper(II) sulphate, cupric sulphate
Kupfer(II)-sulfid *n* copper(II) sulphide, cupric sulphide
Kupfertetramminhydroxidlösung *f* cuprammonium hydroxide solution
Kupfervitriol *m (min)* copper [blue] vitriol *(copper(II) sulphate 5-water)*
Kupfer-Zinn-Legierung *f* copper-tin alloy
kupieren *v (food)* blend
Kupolofen *m* cupola (furnace)
kuppeln *v (dye)* couple
kuppeln *v*/**alkalisch** couple in alkaline solution
kuppeln *v*/**saurer** couple in acid solution
Kupplung *f (dye)* coupling
Kupplung *f*/**oxidative** oxidative coupling
Kupplung *f*/**reduktive** reductive coupling
Kupplungskomponente *f (dye)* coupling [secondary] component
Kuprit *m (min)* cuprite *(copper(I) oxide)*
Kurchialkaloid *n* holarrhena alkaloid
Kurkumapapier *n* turmeric paper *(an indicator paper)*
Kurve *f* curve, graph *(of plotted data)*; *(if recorded automatically also)* trace
Kurve *f*/**binodale** binodal curve, conodal [coexistence] curve
Kurve *f* **gleichen Gehalts an flüchtiger Substanz** isovol
Kurve *f*/**polarographische** polarogram
Kurve *f*/**voltammetrische** voltammogram
Kurvenanpassung *f (anal)* curve fitting
Kurvenschreiber *m* X-Y recorder
Kurzalterung *f* accelerated [artificial] ageing *(materials testing)*
Kurzhalsrundkolben *m (lab)* short-neck round-bottom flask, bolt-head flask
Kurzhalsstehkolben *m* short-neck flat-bottom flask
Kurzkettenverzweigung *f* short-chain branching
kurzkettig short-chain
kurzlebig *(phys ch)* short-lived
Kurzperiode *f* short period *(in the periodic table)*
Kurzperiodensystem *n* short periodic table
Kurzprüfung *f* accelerated testing
Kurzrohrverdampfer *m* short-tube evaparator
kurzschließen *v* short-circuit
Kurzschlussströmung *f* bypass *(in a reactor)*
Kurzwegdestillation *f* short-path (high-vacuum) distillation
Kurzwegkrackung *f (petrol)* short-path cracking
Kurzwegverdampfer *m (tech)* short-path evaporator
Kurzzeitfermentation *f (biot)* short-time fermentation
Kurzzeitpasteurisation *f (food)* high-temperature short-time pasteurization, HTST pasteurization
Kurzzeitversuch *m* accelerated [short-time] test
KUS *s.* Kupferseide
Kuvertkonformation *f* envelope conformation
Küvette *f (anal)* cell, cuvet(te)
Küvette *f*/**geteilte** *s.* Tandemküvette
Küvettenhalter *m (anal)* cell holder
Küvettentest *m (spectr)* cuvette test

Küvettenwechsler *m (anal)* cell changer
Kw. *s.* Königswasser
KW-Stoff *m s.* Kohlenwasserstoff
Kynurenin *n* kynurenine
Kynurensäure *f* kynurenic acid
Kynurin *n* kynurine
Kynursäure *f* kynuric acid
KZ *s.* Koordinationszahl

L

Lab *n* rennet • **mit Lab versetzen** rennet *(milk)*
laben *v* rennet *(milk)*
Labferment *n* rennin, chymosin
labil labile, instable unstable
Labilität *f* lability, instability
Labor *n* laboratory, lab
Laborant *m* laboratory assistant
Laborantin *f* (female) laboratory assistant
Laborapparat *m* laboratory apparatus
Laboratorium *n* laboratory, lab
Laboratorium *n***/chemisches** chemical [chemistry] laboratory
Laborbedingungen *fpl* laboratory conditions
Laborbestimmung *f* laboratory determination
Laborchemikalie *f* laboratory chemical
Laborgerät *n* laboratory apparatus
Laborgeräte *npl* laboratory apparatus, labware
Laborgeräte *f* **aus Glas** laboratory glassware
Labormaßstab/im lab(oratory)-scale, on a [the] laboratory scale
Labormethode *f* laboratory method
Laborprüfung *f* laboratory testing
Laborreaktor *m* laboratory [bench-scale] reactor
labortechnisch 1. laboratory; 2. *s.* Labormaßstab/im
Laboruntersuchung *f* laboratory investigation, bench-scale study
Laborverfahren *n* laboratory procedure
Laborversuch *m* laboratory test
Labung *f* renneting *(of milk)*
Labyrinthfaktor *m* tortuosity factor *(diffusion theory)*
Lachgas *n* laughing gas *(nitrogen(I) oxide)*
Lack *m* (clear) varnish *(chemically drying)*; (clear) lacquer *(physically drying)*
Lack *m* **auf Wasserbasis** *s.* Wasserlack
Lack *m***/ofentrocknender** stoving [baking] varnish, baking finish; stoving lacquer
Lack *m***/wasserlöslicher** water-soluble paint
Lackadditiv *n* varnish additive
Lackbenzin *n* varnish-makers' (and painters') naphtha, painter's naphtha
Lackentferner *m* paint remover; varnish remover
Lackfarbe *f* topcoat [finish] paint; enamel (paint)
Lackfarbe *f***/ofentrocknende** stoving [baking] enamel
Lackglanzmittel *n* paint polish
Lackharz *n* varnish [coating] resin
lackieren *v* lacquer *(using products drying by evaporation)*; *(in a larger sense:)* paint; varnish *(using transparent products)*
Lackmus *n(m)* litmus, lichen blue *(a lichen dye)*
Lackmuspapier *n* litmus (test) paper
Lackrot *n* varnish red
Lackschwarz *n* varnish black
Lack- und Farbenchemie *f* paint chemistry
Lack- und Farbenchemiker *m* paint chemist
Lackwolle *f* varnish wool
Lactalbumin *n* lactalbumin, milk albumin
Lactam *n (org ch)* lactam
Lactat *n* lactate
Lactatdehydrogenase *f* lactic dehydrogenase
Lactatgärung *f (bioch)* lactate fermentation
Lactid *n* lactide
Lactim *n (org ch)* lactim
Lactoglobulin *n* lactoglobulin, milk globulin
Lacton *n* lactone
lactonisieren *v* lactonize
Lactonisierung *f* lactonization
Lactonitril *n* lactonitrile
Lactonregel *f* **(der optischen Drehung)** lactone rule
Lactonsäure *f* lactonic acid *(any of several acids with a lactone ring bearing the carboxyl group)*

Lactose *f* lactose, milk sugar
laden *v* 1. *(phys ch)* charge; 2. *(tech)* load, charge *(bulk material)*
Ladestrom *m* charging [capacity] current *(polarography)*
Ladung *f* 1. *(phys ch)* charge; 2. *(tech)* load, charge, batch *(of bulk material)* • **(mit) gleicher Ladung** like-charged
Ladung f/elektrische electric charge
Ladung f/elektrostatische electrostatic charge
Ladung f/formale formal charge
Ladung f/gleichnamige like charge
Ladung f/partielle partial [fractional] charge
Ladung f/ungleichnamige unlike charge
Ladungsausgleich *m* charge compensation
Ladungsaustausch *m* charge exchange
Ladungsdichte *f* *(phys ch)* charge density
Ladungsdichteverteilung *f* *(phys ch)* charge distribution
Ladungsdoppelschicht *f* electric double layer
Ladungsdurchtrittsfaktor *m* charge-transfer coefficient
Ladungsinjektion *f* charge injection
Ladungskompensation *f s.* Ladungsausgleich
ladungslos uncharged
Ladungsmenge *f* amount of charge
Ladungsneutralität *f* charge neutrality
Ladungsnullpunkt *m* zero point of charge
Ladungsschwerpunkt *m* *(phys ch)* charge centre, centre of charge
Ladungsträger *m* *(phys ch)* charge carrier
Ladungsträgerdichte *f* charge carrier density
Ladungstransport *m* charge transport
Ladungstransportrate *f* charge transport rate
Ladungsübertragung *f* *(phys ch)* charge transfer
Ladungsübertragungskomplex *m* charge-transfer complex, electron-donor-acceptor complex
Ladungsverteilung *f* *(phys ch)* charge distribution
Ladungsvorzeichen *n* charge sign
Ladungswolke *f* *(phys ch)* charge cloud
Ladungszahl *f* *(nomencl)* charge [Ewens-Bassett] number
Lage *f* 1. position; 2. layer; ply *(as of laminated material)*
Lageenergie *f* potential energy
Lagehöhe *f* potential head *(fluid mechanics)*
Lager *n* 1. store(house); storeroom, stock room; 2. *(tech)* bearing *(as of shafts)*; 3. *s.* Lagerstätte
lagerbar storable
Lagerbarkeit *f* storability
Lagerbehälter *m* storage tank
lagerbeständig stable [resistant] in storage
Lagerbeständigkeit *f* storage resistance [stability], stability in storage
Lagerdauer *f* *(food)* storage life, shelf life
Lagerhalle *f* storehouse
Lagerhaus *n* storehouse
Lagermetall *n* bearing metal
lagern *v* store, *(bulk material also)* stockpile
lagern v/kühl store in cool place
Lagerraum *m* storeroom, stock room
Lagerstätte *f* *(geol)* deposit
Lagertank *m* storage tank
Lagertemperatur *f* storage temperature
Lager- und Verarbeitbarkeitsdauer *f* storage [shelf] life
Lagerung *f* 1. storage, storing, *(of bulk material also)* stockpiling; 2. *(ferm)* secondary fermentation; 3. *(tech)* bearing
Lagervorrat *m* stock
Lagerweißmetall *n* white metal
Lag-Phase *f* *(biot)* lag phase [time]
Lake *f* (salt) brine • **mit Lake behandeln** brine
Lakenkonservierung *f* *(tann)* brine curing
lakrimogen lachrymatory
Laktase *f* *(bioch)* lactase
Laktoseintoleranz *f* *(food)* lactose intolerance, lactase deficiency
lamellar lamellar
Lamelle *f* lamella
Lamellenkühler *m* *(tech)* heat sink
Lamellenverdampfer *m* *(tech)* lamella evaporator
Lamellenwärmeaustauscher *m* *(tech)* lamella heater

Laminarströmung *f* laminar [streamline] flow
Laminat *n* laminate
laminieren *v* laminate
Laminieren *n* lamination
LAMS *s.* Laser-Massenspektrometrie
Landé-Faktor *m* **g** Landé [splitting] factor g, Landé g factor
Landwirtschaftschemie *f* agricultural chemistry
längen *v* stretch
Längenänderung *f* change in length, *(increase:)* elongation
Längenzunahme *f* elongation, extension
längerkettig *(org ch)* longer-chain
langfas(e)rig long-fibre
Langhalskolben *m* long-neck flask
Langhalsrundkolben *m* round-bottom long-neck flask
Langhalsstehkolben *m* flat-bottom long-neck flask, Florence [boiling] flask
langkettig *(org ch)* long-chain
Langmuir-Isotherme *f* Langmuir adsorption isotherm
Langperiode *f* long period *(in the periodic table)*
Langperiodensystem *n* long periodic table
langprismatisch *(cryst)* elongate-prismatic
Langrohrverdampfer *m* long-tube evaporator
Langsamfilter *n* slow (sand) filter, low-rate filter
Langsamfiltration *f* slow (sand) filtration
langsamflüchtig slow-evaporating
Langsammischen *n* slow mixing
Langsamrühren *n* slow mixing
langsamziehend *(dye)* slow-striking
Längsdiffusion *f* longitudinal [axial] diffusion
Langsieb(papier)maschine *f* Fourdrinier (paper machine), long sifter
Längsvermischung *f* longitudinal [axial] mixing
Langweggaszelle *f (spectr)* long-path-length cell
Langwegküvette *f (spectr)* long-path-length cell
Langzeitfermentation *f (biot)* long-term fermentation
Langzeitstabilität *f* long-term stability
Langzeitversuch *m* long-term experiment
Lanolin *n* lanolin(e) *(refined wool grease)*
Lanthan *n* La lanthanum
Lanthanoid *n* lanthanoid (element)
Lanthanoidenelement *n* lanthanoid (element)
Lanthanoidenkontraktion *f* lanthanoid contraction
Lanthanoidenreihe *f* lanthanoid series
Lanthan(III)-oxid *n* lanthanum(III) oxide, lanthanum trioxide
Lapislazuli *m (min)* lapis (lazuli) *(a tectosilicate)*
Lardöl *n* lard oil
Lariatether *m* lariat ether *(a crown ether)*
Larmor-Frequenz *f (spectr)* Larmor frequency
Larmor-Präzession *f (spectr)* Larmor precession
Larvengift *n* larvicide
Larvizid *n* larvicide
LAS *s.* Alkylbenzensulfonat/lineares
Laser *m*/**abstimmbarer** *(spectr)* tunable laser
Laser *m*/**chemischer** chemical laser
Laser *m*/**gepulster** pulsed laser
Laser *m*/**kontinuierlicher** CW [continuous-wave] laser
Laser *m*/**phasengekoppelter** mode-locked laser
Laserchemie *f* laser chemistry
Laserdesorptions-/Ionisations-Flugzeit-Massenspektrometrie *f*/**matrixunterstützte** matrix-assisted laser desorption/ionization-time of flight mass spectroscopy, MALDI-TOF
Laser-Desorptions-Massenspektrometrie *f* laser desorption mass spectrometry, LD
Laserlicht-Kleinwinkelstreuung *f* low-angle laser light scattering, LALLS
Laser-Massenspektrometrie *f* laser mass spectrometry, LAMS
Lasermikrosonden-Massenspektrometrie *f* laser microprobe mass analysis, LAMMA
Laser-Oszillator *m* laser
Laser-Raman-Spektroskopie *f* laser Raman spectroscopy
Laser-Raman-Spektroskopie *f*/**oberflächenverstärkte** surface-enhanced Raman spectroscopy, SERS

Laserstrahl *m* laser beam
Laserstrahlverdampfen *n* pulsed laser deposition *(physical vapour deposition technique)*
Lassaigne-Probe *f* Lassaigne test *(for detecting nitrogen)*
Lassoether *m* lariat ether *(a crown ether)*
Last *f* load
Lastschwankungen *fpl (hyd)* load fluctuations
Lasttrum *m(n)* top strand *(of a conveyor)*
Latenzzeit *f (tox, med)* period of latency [induction]
Latex *m* latex
Latex(anstrich)farbe *f* latex paint
latexführend lactiferous, latex-bearing
Latexklebstoff *m* latex adhesive
Latexschaum(gummi) *m* latex foam (rubber)
Laue-Aufnahme *f* Laue pattern
Laue-Diagramm *n* Laue pattern
Laue-Verfahren *n (cryst)* Laue (X-ray) method
Lauf *m* travel *(as of a reactant)*; course *(of a reaction)*
laufen *v* 1. run, flow *(of a liquid)*; 2. *(coat)* sag; 3. run *(of an experiment or a reaction)*; 4. run *(of a machine)*
laufen lassen *v* draw down *(into a vessel)*
laufen lassen *v***/leer** drain (down) *(e.g. a tank)*
laufen *v***/leer** drain
Läufer *m s.* Laufstein
Läuferbildung *f (coat)* sagging
Laufgeschwindigkeit *f* travel rate
Laufmittelfront *f (chromat)* solvent front
Laufstein *m* runner (stone), mill runner, muller
Laufzeit *f* (service) life; running time *(of a chromatogram)*
Lauge *f* 1. lye *(alkaline solution)*; 2. *(tech)* liquor; leach *(solution obtained by leaching)*
Lauge *f***/eingestellte** standard base
Lauge *f***/metallhaltige** *(min tech)* leach liquor [solution]
Lauge-Maßlösung *f (anal)* standard base
Laugemittel *n* leaching agent
laugen *v* 1. *(min tech)* leach, lixiviate; 2. *(text)* mercerize
Laugen *n* 1. *(text)* mercerization; 2. *s.* Laugung
laugenbeständig resistant to alkali(es), alkali-resistant, caustic-proof
Laugenbeständigkeit *f* resistance to alkali(es), alkali resistance
Laugenregeneration *f (pap)* (spent-)liquor recovery
Laugensprödigkeit *f (met)* caustic cracking [embrittlement]
Laugung *f (min tech)* leaching, lixiviation
Laugung *f***/mikrobielle** *(min tech, biot)* microbial leaching
Laugungsflüssigkeit *f (min tech)* leaching solution
Laugungsmittel *n* leaching agent
Lauraldehyd *m* lauraldehyde, *dodecanal
Laurat *n* laurate *(salt or ester of lauric acid)*
Laurinsäure *f* lauric acid, *dodecanoic acid
Laurylalkohol *m* lauryl alcohol, *dodecan-1-ol
Laurylamin *n* laurylamine, *dodecylamine
Läuterboden *m (ferm)* strainer [bottom]
läutern *v* 1. *(filtr)* clarify; 2. *(min tech)* wash
Lavendelöl *n* oil of lavender *(from Lavandula specc.)*
Laves-Phase *f (cryst)* Laves phase *(an intermetallic structure)*
Lävulinsäure *f* laevulinic acid, *4-oxopentanoic acid
Lävulose *f* laevulose, fructose *(a monosaccharide)*
Lawrencium *n* Lr, lawrencium
Laxans *n (pharm)* laxative
Laxativum *n (pharm)* laxative
L-Band *n* conduction band
LCAO-Näherung *f* LCAO approximation, approximation of linear combinations of atomic orbitals
LD *s.* 1. Dosis/letale; 2. Laser-Desorptions-Massenspektrometrie
LD 50 *s.* Dosis/mittlere letale
LDH *s.* Lactatdehydrogenase
leachen *v (biot)* leach
Leaching *n***/mikrobielles** *(biot)* microbial leaching
Leaching-Flüssigkeit *f (min tech)* leaching solution *(containing chemicals or bacteria)*
Lebensdauer *f (phys ch)* lifetime, life

(period), operating [useful, working] life, durability
Lebensdauer f/mittlere *(phys ch)* mean life(time), average life (period)
Lebensgefahr *f* life hazard
Lebensgift *n* biocide *(chemical which kills organisms)*
Lebensmittel *npl* food, edibles
Lebensmittel *npl*/**eiweißreiche** protein food
Lebensmittelanreicherung *f* food fortification *(e.g. with vitamins)*
Lebensmittelbestrahlung *f* food irradiation
Lebensmittelchemie *f* food chemistry
Lebensmittelchemiker *m* food chemist
Lebensmittelfarbstoff *m* food dye
Lebensmittelgesetz *n* food law, food legislation
Lebensmittelindustrie *f* food industry
Lebensmittelkontaminant *m* food contaminant, incidental food additive
Lebensmittelmikrobiologie *f* food microbiology
Lebensmittelrecht *n s.* Lebensmittelgesetz
Lebensmitteltechnologie *f* food technology
Lebensmittelüberwachung *f* food control
Lebensmittelunverträglichkeit *f* food intolerance
Lebensmittelverarbeitung *f* food processing
Lebensmittelverderb *m* food deterioration [spoilage]
Lebensmittelvergiftung *f (food)* food contamination, food intoxication, food poisoning
Lebensmittelverpackung *f* food packaging
Lebensmittelzusatz(stoff) *m* (intentional) food additive
leberschädigend hepatotoxic
Leberstärke *f* animal starch, glycogen
lebhaft 1. vigorous, brisk *(reaction)*; 2. bright, vivid *(colour)*
Leblanc-Soda *f* Leblanc soda
Leblanc-Verfahren *n* Leblanc process *(for obtaining soda)*
Lecithin *n (bioch)* lecithin *(any of a class of phospholipids)*
leck leaking • **leck sein** leak, run
Leck *n* leak
Leckage *f* leakage
lecken *v* leak, run
Leckflüssigkeit *f* leakage
leckfrei leakproof
Leckgas *n* leakage
Leckluft *f* leakage
Leckverlust *m* leakage, slippage loss
Leclanché-Element *n* Leclanché cell
Leder *n* leather
Leder *n*/**pflanzlich gegerbtes** bark leather
lederartig leathery
Lederindustrie-Chemiker *m* leather chemist
leer 1. empty; 2. vacant *(lattice site, orbital)*
leeren *v* empty
Leerlaufen *n* drainage
Leermedikament *n (pharm)* placebo
Leerstelle *f (cryst)* lattice vacancy, hole (position)
Leerstellenpaar *n* vacancy pair
Leerventil *n* outlet [discharge] valve
Leerwert *m (anal)* blank reading, background
legieren *v* alloy
legieren lassen *v*/**sich** alloy
Legierung *f* alloy
Legierung *f*/**binäre** binary alloy
Legierung *f*/**leichtschmelzende** fusible alloy
Legierung *f* **nach Rose** Rose's metal [alloy]
Legierung *f*/**quaternäre** quaternary alloy
Legierung *f*/**ternäre** ternary alloy
Legierungsbestandteil *m* alloying element
Legierungsskelettkatalysator *m* (alloy) skeleton catalyst
Legierungszusatz *m* minor element
Lehm *m* loam, clay
lehmartig loamy, clayey, clayish
lehmhaltig loamy, clayey, clayish
Leichtbenzin *n* light gasoline [naphtha, spirit]
Leichtersiedende(s) *n* more volatile component, low-boiling component
leichtflüchtig highly [readily] volatile
Leichtgut *n (min tech)* light fraction
leichtlöslich readily [freely] soluble
Leichtmetall *n* light metal

Leichtöl *n* light oil
leichtschmelzend low-melting
leichtsiedend low-boiling, light
Leichtwasserreaktor *m (nucl)* light-water reactor
Leim *m* 1. glue; 2. *(pap)* sizing material, size • **mit Leim bestreichen** glue
Leim *m***/tierischer** animal glue
Leimbad *n (pap)* sizing bath
leimen *v* 1. glue; 2. *(pap)* size
Leimung *f (pap)* sizing
Leimungsmittel *n (pap)* sizing material
Leinen(gewebe) *n* linen (fabric)
Leinöl *n* linseed oil
Leinölfirnis *m* boiled linseed oil
Leinöllack *m* linseed-oil varnish
Leistung *f* 1. power *(in strictly physical sense)*; 2. output *(as of a plant)*; 3. performance
Leistungsaufnahme *f* power consumption
leistungsfähig *(tech)* efficient
Leistungsfähigkeit *f (tech)* efficiency, capacity
Leistungsstamm *m (biot)* production strain *(of microorganisms)*
Leistungsverhalten *n (tech)* performance
Leistungsvermögen *n* capacity
Leitblech *n* deflector (plate), baffle (plate) • **mit Leitblechen versehen** baffle
Leitelektrolyt *m (el ch)* supporting electrolyte, base [indifferent] electrolyte
leiten *v* 1. lead *(fluids)*; 2. control *(a reaction)*; 3. conduct *(electricity, heat)*
leiten *v***/durch Rohre** pipe
leitend conducting, conductive
leitend/nicht non-conducting
Leitenzym *n* marker enzyme
Leiter *m* conductor
Leiter *m***/elektrischer** conductor of electricity
Leiter *m***/elektrolytischer** electrolytic [ionic] conductor
Leiter *m* **erster Ordnung** electronic conductor
Leiter *m* **zweiter Ordnung** electrolytic [ionic] conductor
cis-**Leiter-Konformation** *f (bioch)* ladder conformation *(of DNA)*
Leiterpolymer(e) *n* ladder [double-chained] polymer
leitfähig conductive
Leitfähigkeit *f* conductance, conductivity
Leitfähigkeit *f***/elektrische** electrical conductivity
Leitfähigkeit *f***/elektrolytische** electrolytic conductivity
Leitfähigkeit *f***/elektronische** electronic conductivity
Leitfähigkeit *f***/lichtelektrische** photoconductivity
Leitfähigkeit *f***/molare** molar conductance
Leitfähigkeit *f***/spezifische elektrische** specific conductance, conductivity
Leitfähigkeitsmessbrücke *f* conductivity bridge
Leitfähigkeitsmessgerät *n* conductometer
Leitfähigkeitsmesszelle *f* conductivity cell
Leitfähigkeitstitration *f* conductometric titration
Leitisotop *n* (isotopic) tracer
Leitisotopenmethode *f* tracer method, atom tagging method
Leitung *f* 1. line, pipe(line), duct; 2. conduction *(of electricity or heat)*
Leitungsband *n* conduction band
Leitungsband *n***/leeres** empty band
Leitungselektron *n* conduction electron
Leitungsrohr *n* duct, line, pipe(line)
Leitungswasser *n* tap water
Leitwert *m***/elektrischer** electrical conductance
Lektin *n (bioch)* lectin *(any of a class of glycoproteins)*
lenken *v* direct *(a substituent)*; control *(a process)*
Lenkung *f* direction *(of a substituent)*; control *(of a process)*
Lennard-Jones-Potenzial *n* Lennard-Jones potential
Lepton *n* lepton *(any of a family of light elementary particles)*
lesen *v (min tech)* pick
letal *(tox)* lethal, deadly poisonous
Leu leucine
Leuchtdichte *f* luminance, brightness
Leuchtdiode *f* light emitting diode, LED, photodiode
Leuchtdiode *f***/organische** organic light emitting diode, OLED

Leuchtdiode *f***/polymere** polymer light emitting diode, PLED
Leuchten *n***/kaltes** luminescence
leuchtend luminous; bright, brilliant *(colour)*
Leuchtkraft *f* brilliance, brightness *(of colours)*
Leuchtstoff *m* luminophore
Leuchtstoff *m***/aufhellender** *(text)* fluorescent brightener [brightening agent]
Leucin *n* leucine
Leucovorin *n* leucovorin, folinic acid
Leukobase *f* leuco base
Leukochinizarin *n* leucoquinizarine
Leukoindigo *m* leucoindigo
Leukopararosanilin *n* leucopararosaniline
Leukotrien *n (org ch)* leukotriene
Leukoverbindung *f* leuco compound
Lewis-Base *f* Lewis base
Lewis-Säure *f* Lewis acid
LFE-Beziehung *f* linear free energy relation(ship)
LFT *s.* Ligandenfeldtheorie
LH *s.* Luteinisierungshormon
Liberin *n* releasing hormone
Licansäure *f (org ch)* licanic acid
Licht *n* light • **am [im] Licht** on exposure to light, under illumination • **vor Licht geschützt** protected from light • **dem Licht aussetzen** expose to light
Licht *n***/einfallendes** incident light
Licht *n***/gestreutes** scattered light
Licht *n***/kaltes** cold light
Lichtabbau *m* photodegradation, light degradation
lichtabsorbierend light-absorbing
Lichtabsorption *f* light absorption
Lichtatmung *f (bioch)* photorespiration
Lichtausbeute *f* efficiency of luminous source, light [luminous] efficiency, luminosity factor, light yield *(in lm/W)*
lichtbeständig photostable, resistant [fast] to light, light-fast
Lichtbeständigkeit *f* photostability, resistance [fastness] to light, light fastness
Lichtblitz *m* photoflash
Lichtbogen *m* (electric) arc
Lichtbogen *m***/elektrischer** (electric) arc
Lichtbogenentladung *f* arc discharge
Lichtbogenlampe *f* arc lamp
Lichtbogenofen *m* arc(-heated) furnace
Lichtbogenofen *m* **mit indirekter Beheizung** indirect arc(-heated) furnace
Lichtbogenplasmabrenner *m* plasma arc torch
Lichtbogenruß *m* arc black
Lichtbogenschmelzen *n* arc melting
Lichtbogenschweißen *n* arc welding
Lichtbogenverfahren *n* luminous arc process
lichtbrechend refractive
lichtdicht lightproof
lichtdurchlässig transparent, translucent
Lichtdurchlässigkeit *f* transparency, light transmission, translucence
lichtecht fast to light, light-fast
Lichtechtheit *f* resistance [fastness] to light, light fastness
Lichteinwirkung *f***/unter** under illumination, on exposure to light
Lichtemission *f* light emission
lichtempfindlich sensitive to light, light-sensitive, photosensitive • **lichtempfindlich machen** *(phot)* sensitize
Lichtempfindlichkeit *f* sensitivity to light, light sensitivity, photosensitivity
Lichtempfindlichmachen *n (phot)* sensitization
Lichtfilter *n* light filter
lichtgeschützt protected from light • **lichtgeschützt aufbewahren** keep screened from the light
Lichtgeschwindigkeit *f* velocity of light
lichtkatalysiert light-catalyzed
Lichtleitfaser *f* light-conducting fiber
Lichtmikroskop *n* light microscope
Lichtquant *n* photon, light quant(um)
Lichtquelle *f* light source
Lichtreaktion *f* photochemical [light] reaction, photoreaction
Lichtschutz *m* protection from light
Lichtstabilisator *m* photostabilizer *(rubber)*; light stabilizer
Lichtstrahl *m* ray of light, *(if bundled:)* beam of light
Lichtstreuung *f* light scattering
Lichtstreuung *f***/dynamische** dynamic light scattering, DLS
Lichtstreuung *f***/statische** static light scattering, SLS
Lichtstrom *m* luminous flux
lichtundurchlässig opaque, lightproof
Lichtundurchlässigkeit *f* opacity

lichtunempfindlich *s.* lichtbeständig
Lichtunempfindlichkeit *f s.* Lichtbeständigkeit
Lichtunempfindlichmachen *n* photostabilization
Lichtwelle *f* light wave
Lichtwellenleiter *m***/polymerer** plastical optical fibre
Liebig-Kühler *m* Liebig condenser
Lieferdruck *m* discharge pressure
liefern *v* deliver, yield *(as in a reaction)*; donate, furnish, contribute *(electrons)*
Lieferstrom *m* delivery
Liesegang-Ringe *mpl (coll)* Liesegang rings
liften *v* lift *(catalysts)*
Liften *n* lifting *(as of catalysts)*
Ligand *m* ligand
Ligand *m***/anionischer** anionic ligand
Liganden(austausch)chromatographie *f* ligand exchange chromatography
Ligandenfeldtheorie *f* ligand field theory
Ligatoratom *n* ligating atom
ligieren *v (bioch)* ligate *(to join complementary ends of DNA)*
Ligierung *f (bioch)* ligation *(joining complementary ends of DNA)*
Lignan *n* lignan
Lignin *n* lignin
Ligninsulfonsäure *f* ligninsulphonic acid
Lignocellulose *f* lignocellulose *(cellulose associated with lignin)*
Ligroin *n* ligroin(e) *(with boiling range from 90 °C to 120 °C)*
Limonen *n* limonene
Limonit *m (min)* limonite, brown iron ore [stone]
Linde-Verfahren *n* Linde process [cycle] *(for liquefying gases)*
linear unbranched *(chain molecule)* • **linear polarisiert** linearly polarized
Linearbeschleuniger *m* linear accelerator
Linearbeschleuniger *m* **für Elektronen** linear accelerator for electrons
Linearbeschleuniger *m* **für Schwerionen** heavy-ion linear accelerator
Linearbeschleunigung *f* linear acceleration
Linearkolloid *n* fibrous colloid
Linearkombination *f* **von Atomorbitalen** linear combination of atomic orbitals, LCAO
Linearpolyethylen *n* linear [low-pressure] polyethylene, high-density polyethylene
Linearpolymer(e) *n* linear polymer
Lineweaver-Burk-Diagramm *n* Lineweaver-Burk plot, double-reciprocal plot *(of kinetic data)*
Lineweaver-Burk-Methode *f* Lineweaver-Burk method *(for treating kinetic data)*
Linie *f* **gleichen Gehalts an flüchtiger Substanz** isovol
Linienabsorptionsspektrum *n* absorption-line spectrum
Liniendefekt *m (cryst)* line defect
Linienfehler *m (cryst)* line defect
Liniengitter *n (spectr)* line [ruled] grating
Linienspektrum *n* line [discrete] spectrum
Linin *n* linin *(Linum cartharticum principle)*
linksdrehend 1. laevorotatory, laevo; 2. *(cryst)* left-handed
Linksdrehung *f* laevorotation
Linksform *f* laevorotatory form, (-)-form *(of an optically active compound)*
Linksgängigkeit *f* left-handed sense *(of a helix)*
Linksmilchsäure *f* (D)-(-)-lactic acid, laevolactic acid
Linksweinsäure *f (2S,3S)*-(-)-tartaric acid, laevotartaric acid
Linoleat *n* linol(e)ate *(salt or ester of linoleic acid)*
Linolensäure *f* linolenic acid, 9,12,15-octadecatrienoic acid
Linolsäure *f* linoleic acid, 9,12-octadecadienoic acid
Linse *f***/achromatische** achromatic lens
Linse *f***/aplanatische** aplanatic lens
Linse *f***/bifokale** bifocal lens
Linse *f***/bikonkave** biconcave lens
Linse *f***/bikonvexe** biconvex lens
linsenförmig lenticular
Linsenöffnung *f* lens aperture
Lipid *n (bioch)* lipid
Lipid *n***/einfaches** simple lipid
Lipid *n***/komplexes** lipoid, complex [compound] lipid
Lipiddoppelschicht *f (biol)* lipid bilayer
lipidlöslich lipid-soluble

Lipidstoffwechsel *m* lipid metabolism
Lipoid *n (bioch)* lipoid, complex [compound] lipid
Lipolyse *f* lipolysis, fat splitting
lipolytisch lipolytic
lipophil lipophilic, lipophile
Lipopolymer *n* lipopolymer *(liposoluble polymer)*
Lipopolysaccharid *n* lipopolysaccharide
Lipoprotein *n* lipoprotein
Lipoprotein *n* **sehr geringer Dichte** very-low-density lipoprotein, VLDL
Lipoprotein *n* **sehr hoher Dichte** very-high-density lipoprotein, VHDL
Liquiduskurve *f* liquidus curve [line], liquidus
Liquidus-Liquidus-Chromatographie *f* liquid-liquid chromatography, LLC
Liquidus-Solidus-Chromatographie *f* liquid-solid chromatography, LSC
Literaturspektrum *n* library (reference) spectrum
Liter-Molarität *f* molarity
Lithium *n* Li, lithium • **mit Lithium verbinden** lithiate • **mit Lithium versetzen** lithiate
Lithiumaluminiumhydrid *n* lithium aluminium hydride
Lithiumamid *n* lithium amide
Lithiumchlorid *n* lithium chloride
Lithiumhydrid *n* lithium hydride
Lithiumhydroxid *n* lithium hydroxide
Lithiumperchlorat *n* lithium perchlorate
Lithocholsäure *f* lithocholic acid
Lithographie *f* lithography
Lithopon *n* lithophon(e), zinc baryta white *(consisting of ZnS and $BaSO_4$)*
Lithopone *f* lithophon(e), zinc baryta white *(consisting of ZnS and $BaSO_4$)*
Lithosphäre *f (geol)* lithosphere
Littrow-Anordnung *f (spectr)* Littrow mounting
Littrow-Spektrograph *m* Littrow spectrograph
L-Kette *f (bioch)* light chain
L-Konfiguration *f* L configuration
ll *s.* löslich/leicht
LLC *s.* Liquidus-Liquidus-Chromatographie
Lobarsäure *f* lobaric acid
Lobeliaalkaloid *n* lobelia alkaloid
Loch *n* 1. hole; 2. *(plast)* pit *(moulding defect)*; 3. *(phys ch)* (electron) hole, defect electron
Lochblechsieb *n* punched-plate screen
Loch-Elektron-Paar *n* hole-electron pair
Löcherleitung *f (phys ch)* hole [defect] conduction
Lochfraß *m* pitting
Lochfraßkorrosion *f* pitting corrosion
Lochleitungsschicht *f* hole transport layer, HTL *(organic light emitting diodes)*
Lochmaß *n***/lichtes** (size of) aperture *(screening)*
Lochplatte *f (filtr)* strainer
locker fluffy *(bulk material, precipitate)*
lockern *v* loosen (up) *(a chemical bond)*
lockernd antibonding *(chemical-bond theory)*
Lockerung *f* antibonding *(chemical-bond theory)*
Lockmittel *n* attractant
Lockstoff *m* attractant
Löffel *m* bucket *(of an elevator)*
log-Phase *f (biot, hyd)* logarithmic growth phase
Lohe *f (tann)* bark
Lohgerbung *f* bark tannage [tanning]
Lokalanästhetikum *n* local [topical] anaesthetic
Lokalelement *n* local cell [galvanic element]
Lokalkorrosion *f* localized corrosion
London-(van-der-Waals-)Kraft *f* London [dispersion] force
Longitudinaldiffusion *f* longitudinal [axial] diffusion
Lorentz-Profil *n (spectr)* Lorentzian profile
Lorentz-Verbreiterung *f* pressure [collision] broadening *(of spectral lines)*
lösbar 1. *s.* löslich; 2. dissoluble *(joint)*
Löschbrause *f* emergency shower, safety [drench] shower
Lösche *f* breeze
löschen *v* 1. extinguish *(fire)*; quench *(coke)*; 2. slake, hydrate *(lime)*; 3. quench *(fluorescence)*; quench *(an electric arc)*; 4. quench *(a reaction)*
Löscher *m* quencher, quenching agent *(kinetics, photochemistry)*
Löschkalk *m* (water-)slaked lime, lime hydrate *(calcium hydroxide)*
Löschpapier *n* blotting paper

Löschung f/dynamische dynamic quenching *(of fluorescence)*
Löschung f/statische static quenching *(of fluorescence)*
lose (in) bulk
Lösegeschwindigkeit *f* rate of dissolution
lösen *v* 1. dissolve, solve, bring [put] into solution; 2. break, crack, disrupt *(a chemical bond)*
lösen v/sich dissolve, go into solution
lösend dissolving, solvent
Löser *m s.* Lösungsmittel
Lösevermögen *n* solvent power, solvency
löslich soluble • **löslich machen** solubilize
löslich/leicht readily [freely, easily] soluble
löslich/mäßig moderately [reasonably] soluble
löslich/nicht insoluble, non-soluble
löslich/schwer poorly [difficultly] soluble, slightly [sparingly] soluble
löslich/sehr leicht very [highly] soluble
löslich/sehr schwer very [extremely] slightly soluble
Löslichkeit *f* solubility
Löslichkeit f/gegenseitige mutual solubility, intersolubility
Löslichkeit f in festem Zustand solid solubility *(of metals)*
Löslichkeit f in Wasser water solubility
Löslichkeit f/scheinbare apparent solubility
Löslichkeitsdiagramm *n* solubility chart
Löslichkeitserniedrigung *f* decrease in solubility
Löslichkeitsgleichgewicht *n* solubility equilibrium
Löslichkeitskoeffizient *m* solubility coefficient
Löslichkeitskonstante *f* solubility coefficient
Löslichkeitskurve *f* solubility curve, liquidus (curve)
Löslichkeitsprodukt *n* solubility product
Löslichkeitsunterschied *m* difference in solubility
Löslichkeitsvermittler *m* solutizer, solutizing agent
Lossen-Abbau *m* Lossen rearrangement *(of aromatic hydroxamic acids or their derivatives into isocyanates)*
Lösung *f* 1. solution; 2. dissolution *(act or process)* • **in Lösung bringen** bring [put] into solution • **in Lösung gehen** dissolve, go into solution • **in Lösung halten** keep in solution • **in alkalischer Lösung kuppeln** couple in alkaline solution • **in saurer Lösung kuppeln** couple in acid solution
Lösung f/alkalische alkaline solution
Lösung f/alkoholische alcoholic solution
Lösung f/echte true solution
Lösung f/feste solid solution
Lösung f/gepufferte buffered solution
Lösung f/gesättigte saturated solution
Lösung f/interstitielle feste interstitial (solid) solution
Lösung f/isotonische isotonic solution
Lösung f/kolloide colloidal solution
Lösung f/lugolsche Lugol's solution *(aqueous solution of potassium iodide and iodine)*
Lösung f/normale normal solution, N solution
Lösung f/pseudoideale pseudoideal solution
Lösung f/reale real [non-ideal] solution
Lösung f/saure acid solution
Lösung f/standardisierte standard solution
Lösung f/übersättigte supersaturated solution
Lösung f/ungeimpfte *(cryst)* unseeded solution
Lösung f/verdünnte dilute solution, dilution
Lösung f/wässrige aqueous [water] solution
***0,1 m*-Lösung** *f* decimolar [tenth molar] solution
***0,01 m*-Lösung** *f* centimolar solution
***0,001 m*-Lösung** *f* millimolar solution
***m/10*-Lösung** *f* decimolar [tenth molar] solution
***m/100*-Lösung** *f* centimolar solution
***m/1000*-Lösung** *f* millimolar solution
***n*-Lösung** *f* normal solution
***1 n*-Lösung** *f* N solution
***0,1 n*-Lösung** *f* decinormal solution
***0,001 n*-Lösung** *f* millinormal [thousandth normal] solution

***n/10*-Lösung** *f* decinormal solution
***n/1000*-Lösung** *f* millinormal [thousandth normal] solution
Lösungen *fpl* **gleichen Dampfdrucks** isopiestic solutions
Lösungsbenzin *n* mineral [petroleum] spirit(s)
Lösungsdruck *m* solution pressure
Lösungsenthalpie *f* enthalpy of solution
Lösungsgleichgewicht *n* solution equilibrium
Lösungskonzentration *f* solution strength
Lösungsmittel *n* solvent, dissolvent, *(esp for extracting soluble principles from drugs:)* menstruum
Lösungsmittel *n*/**aktives** active [true] solvent
Lösungsmittel *n*/**latentes** latent [indirect] solvent, cosolvent
Lösungsmittel *n*/**leichtflüchtiges** fast solvent
Lösungsmittel *n*/**nicht wässriges** non-aqueous solvent
Lösungsmittel *n*/**organisches** organic solvent
Lösungsmittel *n*/**polares aprot(on)isches** polar-aprotic solvent
Lösungsmittel *n*/**selektives** selective solvent
lösungsmittelabstoßend lyophobic
lösungsmittelanziehend lyophilic
lösungsmittelbeständig solvent-resisting
Lösungsmittelbeständigkeit *f* solvent resistance
Lösungsmitteldampf *m* solvent vapour
Lösungsmittelentparaffinierung *f (petrol)* solvent dewaxing
lösungsmittelfest solvent-resisting
lösungsmittelfrei solventless
Lösungsmittelfront *f (chromat)* solvent front
Lösungsmittelgemisch *n* solvent mixture
Lösungsmittelgerbung *f* solvent tannage
Lösungsmittel-Isotopieeffekt *m* solvent isotope effect
Lösungsmittelkäfig *m* solvent cage
Lösungsmittelklebstoff *m* solvent-based adhesive
Lösungsmittelphase *f* solvent phase
Lösungsmittelraffination *f* solvent refining
Lösungsmittelrückgewinnung *f* solvent recovery
Lösungsmittelverschiebung *f (spectr)* solvent shift
Lösungspolymerisation *f* solvent [solution] polymerization
Lösungspunkt *m*/**kritischer** consolute [critical solution] point
Lösungsreaktion *f* solution reaction
Lösungstemperatur *f*/**kritische** consolute [critical solution] temperature
Lösungstemperatur *f*/**obere kritische** upper consolute [critical solution] temperature
Lösungstemperatur *f*/**untere kritische** lower consolute [critical solution] temperature
Lösungsvermittler *m* solutizer, solutizing agent
Lösungswärme *f* heat of solution
Lösungswärme *f*/**integrale** integral [total] heat of solution
Lösungswärme *f*/**partielle** partial [differential] heat of solution
Lot *n* solder
löten *v* solder
löten lassen *v*/**sich** solder
Lötlampe *f* (blow)torch, blowlamp
Lötmittel *n* flux
Lötrohr *n* blowpipe
Lötrohranalyse *f* blowpipe analysis
Lotuseffekt *m* lotus effect *(self-cleaning of surfaces)*
Lotuspflanze *f (biol)* lotus plant *(lotus effect)*
Lötverbindung *f* soldered joint, soldering
Lötzinn *n* soldering tin, plumber's solder *(with 50 to 75 % Pb)*
Lowry-Methode *f* Lowry method (for protein determination)
LPS *s.* Lipopolysaccharid
l-RNA *f (lösliche RNA)* transfer ribonucleic acid
LSC *s.* Liquidus-Solidus-Chromatographie
L-Schale *f* L shell *(of an atom)*
LSD *s.* Lysergsäurediethylamid
Lsg. *s.* Lösung
Lsgm. *s.* Lösungsmittel
LS-Kopplung *f (spectr)* LS coupling,

spin-orbit coupling [interaction], Russel-Saunders coupling
Lücke *f* gap, interstice • **auf Lücke** staggered, anti-periplanar *(stereochemistry)*
Lückenvolumen *n* void volume
Luft *f* air • **an der Luft** on exposure to air • **der Luft aussetzen** air, expose to air • **mit Luft versetzen** aerate
Luft *f*/**flüssige** liquid air
Luftabschluss *m*/**unter** 1. in the absence of air, out of contact with air, sealed from the air; 2. *(biol)* under anaerobic conditions
Luftanwesenheit *f* presence of air
Luftbad *n* air bath
Luftblase *f* air bubble
Luftdruck *m* air pressure
Luftdurchsatz *m* air rate
Lufteinschluss *m* air pocket [void] *(material fault)*
Lufteintrag *m (hyd)* air entrainment
luftempfindlich air-sensitive
Lüfter *m* fan, blower
Lufterhitzer *m* air heater
Luftfeuchte *f* air humidity, atmospheric moisture
Luftfeuchtemesser *m* hygrometer
Luftfeuchtemessung *f* hygrometry
Luftfilter *n* air filter
Luftförderer *m* air conveyor
luftfrei air-free
Luftführung *f* **im Kreislauf** air recirculation
Luftführung *f* **im Umluftbetrieb** air recirculation
Luftgegenwart *f* presence of air
luftgetrocknet air-dried, air-dry
Luftgüteindex *m* air-pollution index
Luftheber *m* air lift
Lufthülle *f* atmosphere
Luftkalk *m* air-hardening lime
Luftkühler *m (lab)* air condenser
Luftkühlung *f* air cooling
Luftoxidation *f* air [atmospheric] oxidation
Luftpore *f* air void
Luftprobe *f* air sample
Luftqualitätsüberwachung *f* air monitoring
Luftsauerstoff *m* atmospheric oxygen
Luftschadstoff *m* atmospheric contaminant
Luftstickstoff *m* atmospheric nitrogen
Luftstrom *m* air flow, draught [blast] of air
Lufttaupunkt *m* air dew point
lufttrocken air-dry
lufttrocken und mineral(stoff)frei dry and mineral-matter-free
Lufttrockner *m* air dryer
Lufttrocknung *f* air drying
Lüftung *f* aeration
Lüftungsbecken *n (hyd)* aeration tank
Lüftungsöffnung *f* (air) vent, air relief
Luftverflüssigung *f* air liquefaction
Luftverschmutzung *f* air pollution
Luftverteiler *m (hyd, biot)* air diffuser, (air) sparger *(for aerating fermentation tanks)*
Luftverunreinigung *f* air pollution
Luftvorwärmer *m* air preheater
Luftzerlegung *f* air separation
Luftzuführung *f* 1. air supply; 2. *(hyd)* air entrainment; 3. air inlet
Luftzutritt *m* access of air • **unter Luftzutritt** 1. in the presence of air; 2. *(biol)* under aerobic conditions
Luggin-Kapillare *f (phys ch)* capillary salt bridge
Lumineszenz *f* luminescence
Lumineszenzanalyse *f* luminescence analysis
Lumineszenzindikator *m* luminescence indicator
Lumineszenzspektroskopie *f* luminescence spectroscopy
Lumineszenzspektrum *n* luminescence spectrum
Lumineszenzverhalten *n* luminescence behaviour
lumineszieren luminesce
lumineszierend luminescent
Luminophor *m* luminophore
Lumpen *mpl (pap)* rags
Lumpenhalbstoff *m (pap)* rag pulp
Lungengift *n* lung injurant
Lungenreizstoff *m* lung [respiratory] irritant
Lunker *m* pipe *(foundry)*
Lunkerbildung *f* piping *(foundry)*
Lupinenalkaloid *n* lupin(e) alkaloid
Luppe *f (met)* ball • **Luppen bilden** ball
Luppenbildung *f (met)* balling
Lurgi-Druckgasgenerator *m (coal)* Lurgi pressure gasifier [generator]

Lurgi-Druckgasverfahren *n (coal)* Lurgi pressure-gasification process, Lurgi high-pressure process
Lurgi-Spülgas(schwel)verfahren *n (coal)* Lurgi Spülgas low-temperature carbonization process
luteinisieren luteinize
Luteinisierungshormon *n* luteinizing hormone, LH, interstitial-cell-stimulating hormone, ICSH
Lutetium *n* Lu, lutetium
Lutidin *n* lutidine, dimethylpyridine
lutro air-dry, air-dried
Lutropin *n* luteinizing hormone, LH, interstitial-cell-stimulating hormone, ICSH
Luvo *m* air preheater
Lyat-Ion *n* lyate ion
Lycin *n* lycine, betaine, *NNN*-trimethylammonioacetate, oxyneurine
Lycopin *n* lycopene *(a natural dye)*
Lyman-Serie *f (spectr)* Lyman series
Lymphe *f (med)* lymph
Lyoenzym *n* lyo-enzyme
Lyogel *n* lyogel
Lyolysis *f* lyolysis, solvolysis
Lyonium-Ion *n* lyonium ion
lyophil lyophilic • **lyophil trocknen** freeze-dry, lyophilize
Lyophilisat *n (biot)* lyophilized culture
lyophilisieren *v* lyophilize, freeze-dry
Lyophilisierung *f* lyophilization, freeze drying
lyophob lyophobic
Lys *(bioch)* lysine
Lyse *f (biol)* cell lysis, lysis *(death of a cell by bursting)*
Lysergsäure *f* lysergic acid
Lysergsäurealkaloid *n* lysergic acid alkaloid
Lysergsäurediethylamid *n* lysergic acid diethylamide
Lysin *n (bioch)* lysine
Lysozym *n* lysozyme *(a bacteriolytic enzyme)*

M

machen *v*/**brandsicher** fireproof
machen *v*/**keimfrei** sterilize
machen *v*/**körnig** grain, granulate
machen *v*/**lichtempfindlich** *(phot)* sensitize
machen *v*/**löslich** solubilize, render soluble
machen *v*/**luftleer** evacuate, exhaust
machen *v*/**mottenbeständig** mothproof
machen *v*/**weich** 1. soften; 2. *(plast)* plasticize, flexibilize
Madelung-Konstante *f (cryst)* Madelung constant *(of lattice energy)*
Magensäure *f* gastric acid
Magenta *n* magenta, fuchsin(e), rosaniline
Magererz *n* lean [low-grade] ore
Magermilch *f* skim(ed) milk
Magma *n* magma
magmatisch magmatic
Magnesia *f* magnesia, magnesium oxide
Magnesiahärte *f (hyd)* magnesium hardness
Magnesiazement *m* Sorel [magnesia] cement
Magnesiospinell *m (min)* spinel *(magnesium aluminate)*
Magnesitstein *m*/**feuerfester** magnesite brick
Magnesium *n* Mg magnesium
Magnesiumbranntkalk *m* dolomitic lime *(calcium magnesium oxide)*
Magnesiumcarbonat *n* magnesium carbonate
Magnesium-Carbonathärte *f (hyd)* magnesium carbonate hardness
Magnesiumchlorid *n* magnesium chloride
Magnesiumhydrid *n* magnesium hydride
Magnesiumhydroxid *n* magnesium hydroxide
Magnesium-Nichtcarbonathärte *f (hyd)* magnesium non-carbonate hardness
Magnesiumoxid *n* magnesium oxide
Magnesiumsulfat *n* magnesium sulphate
Magnesiumsulfathärte *f (hyd)* magnesium sulphate hardness
Magneteisenerz *n s.* Magnetit

Magnetfeld *n* magnetic field
magnetisch magnetic
magnetisch/nicht non-magnetic
magnetisierbar magnetizable
magnetisieren *v* magnetize
Magnetisierung *f* magnetization
Magnetit *m (min)* magnetite *(iron(II,III) oxide)*
Magnetochemie *f* magnetochemistry
Magneton *n* magneton *(a unit of the magnetic moment)*
Magneton *n*/bohrsches Bohr magneton
Magnetorotation *f* magnetic rotation, Faraday effect
Magnetpulver *n* magnaflux
Magnetquantenzahl *f* magnetic quantum number
Magnetrührer *m* magnetic stirrer
Magnetventil *n* solenoid valve
mahlbar grindable
Mahlbarkeit *f* grindability
mahlen *v* 1. grind, mill; 2. *(pap)* beat *(in a Hollander beater)*
mahlen *v*/auf Staubfeinheit pulverize
mahlen *v*/fein grind finely
mahlen *v*/mit der Kugelmühle ball-mill
Mahlfeinheit *f* fineness of grinding
Mahlhilfsstoff *m* grinding aid
Mahlholländer *m (pap)* Hollander (beater), stuff engine
Mahlkugel *f* grinding ball
Mahlmaschine *f* 1. grinding machine; 2. *(pap)* beater
Mahlscheibenmühle *f* disk mill
Mahlstein *m* grindstone, millstone
Mahltrocknung *f* mill drying
Mahltrocknungsanlage *f* dryer-pulverizer
Mahlung *f* 1. grinding, milling; 2. *(pap)* beating *(in a Hollander beater)*
Mahlung *f*/autogene autogenous grinding
Mahlung *f*/feine fine grinding
Mahlung *f*/grobe coarse grinding
Mahlwalze *f* grinding roll
Maillard Reaktion *f (food)* Maillard reaction, non-enzymatic browning (reaction)
Maischapparat *m* masher
Maische *f (biot)* mash
maischen *v (biot)* mash
Mais(keim)öl *n* maize (germ) oil, *(Am)* corn oil
Maisstärke *f* maize starch, *(Am)* corn starch
Maisvergärung *f (biot)* maize fermentation, *(Am)* corn fermentation
MAk *s.* Antikörper/monoklonaler
MAK *s.* Arbeitsplatzkonzentration/maximale
Makroanalyse *f* macroanalysis
makroanalytisch macroanalytical
Makrobestandteil *m* macroconstituent, macrocomponent
Makrochemie *f* macrochemistry
makrochemisch macrochemical
Makrocyclus *m* macrocycle
makrofibrillär macrofibrillar
Makrofibrille *f* macrofibril
Makrofotografie *f* photomacrograph
Makrogefüge *n* macrostructure
Makroinitiator *m* macroinitiator
Makroion *n* macroion
makrokristallin macrocrystalline
Makrolid-Antibiotikum *n* macrolide antibiotic
Makrolid-Antibiotikum *n*/polyenes polyene macrolide antibiotic
Makromer *n* macromer, macromolecular monomer
Makromethode *f* macromethod
Makromolekül *n* macromolecule, giant molecule
Makromolekül *n*/fadenförmiges [lineares] linear [filamentary] molecule
makromolekular macromolecular
Makronährstoff *m (agric)* macronutrient, major element
Makropore *f* macropore
makroporös macroporous
Makroprobe *f* macro sample *(> 0.1 g)*
Makroradikal *n* macroradical
makroretikulär macroreticular
Makroring *m* macroring, large ring *(consisting of 13 or more members)*
Makrostruktur *f* macrostructure
Malachit *m (min)* malachite *(copper(II) carbonate dihydroxide)*
Malachitgrün *n* benzal green *(a triphenylmethane dye)*
Malaria(bekämpfungs)mittel *n* antimalarial (drug)
Malat *n* malate *(salt or ester of malic acid)*
Malatenzym *n* malic enzyme
Maleat *n* maleate *(salt or ester of maleic acid)*

Maleinat *n* maleate *(salt or ester of maleic acid)*
Maleinsäure *f* maleic acid, **cis*-butenedioic acid
Maleinsäureanhydrid *n* maleic anhydride, 2,5-furandione
Malonat *n* malonate *(salt or ester of malonic acid)*
Malonsäure *f* malonic acid, methanedicarboxylic acid, *propanedioic acid
Malonsäurediethylester *m* diethyl malonate, malonic ester
Malonsäuredinitril *n* malonitrile
***N,N'*-Malonylharnstoff** *m N,N'*-malonyl urea, barbituric acid
Maltose *f* maltose, malt sugar
Malz *n* malt • **mit Malz versetzen** malt
malzartig malty
Malzbereitung *f* malting
mälzen malt
Malzenzym *n* malt enzyme
Mälzerei *f* 1. malting; 2. malt-house
Malzextrakt *m* malt extract
Malztenne *f* (malting) floor
Malzzucker *m* malt sugar, maltose
Mammutpumpe *f* air lift, mammoth pump
Mandelat *n* mandelate *(salt or ester of mandelic acid)*
Mandelsäure *f* mandelic acid, phenylhydroxyacetic acid
Mangan *n* Mn manganese
Manganat *n* manganate
Manganat(IV) *n* manganate(IV)
Manganat(VII) *n* permanganate
Mangancarbid *n* manganese carbide
Mangan(II)-chlorid *n* manganese(II) chloride, manganese dichloride
Mangandioxid *n s.* Mangan(IV)-oxid
Mangangrün *n* manganese green *(barium manganate)*
Manganheptoxid *n s.* Mangan(VII)-oxid
Mangan(II)-hydroxid *n* manganese(II) hydroxide
Manganit *m (min)* manganite *(manganese(III) hydroxide oxide)*
Manganknolle *f (geol)* manganese nodule
Manganmonoxid *n s.* Mangan(II)-oxid
Mangan(II)-oxid *n* manganese(II) oxide, manganese monoxide
Mangan(II,IV)-oxid *n* manganese(II,IV) oxide
Mangan(III)-oxid *n* manganese(III) oxide
Mangan(IV)-oxid *n* manganese(IV) oxide, manganese dioxide
Mangan(VI)-oxid *n* manganese(VI) oxide, manganese trioxide
Mangan(VII)-oxid *n* manganese(VII) oxide, manganese heptoxide
Mangansäure *f* manganic acid
Manganseife *f* manganese soap
Mangansiliciumstahl *m* silicomanganese steel
Manganstahl *m* manganese steel
Mangan(II)-sulfat *n* manganese(II) sulphate
Mangantrioxid *n s.* Mangan(VI)-oxid
Mangel *m* 1. deficiency; 2. *(tech)* fault, defect
Mangelelektron *n (phys ch)* defect electron, electron hole
Mangelkrankheit *f (agric, med)* deficiency disease
Mangelleiter *m (phys ch)* p-type conductor, hole conductor
Mangelleitung *f (phys ch)* p-type conduction, defect [hole] conduction
Manipulation *f*/**genetische [gentechnologische]** *(biot)* genetic [gene] manipulation
Mannan *n* mannan *(a polysaccharide)*
Mannich-Reaktion *f* Mannich reaction *(aminomethylation and variations of it)*
Mannitol *n* mannitol *(a sugar alcohol)*
Mannloch *n* manhole
Mannose *f* mannose *(a monosaccharide)*
Mannuronsäure *f* mannuronic acid
Manometer *n* manometer, pressure gauge
manometrisch manometric
Mantel *m (tech)* casing, sheath, *(esp for heating or cooling:)* mantle, jacket; shell *(of a boiler)*; cover *(of cables)*
Mantelbehälter *m* jacketed vessel
Mantelkessel *m* shell-type boiler
Mantelrohr *n* jacketed pipe [tube]
Margarine *f* margarine
Margarinsäure *f* margaric acid, *heptadecanoic acid
Marihuana *n* marihuana, marijuana, hasheesh, hashish *(from Cannabis indica Lam.)*
Marinade *f (food)* pickle *(e.g. spiced solution of common salt and vinegar)*

Marinemessing *n* naval brass [bronze] *(ca. 60 % Cu, 1 % Sn, remainder Zn)*
Marineöl *n* marine (animal) oil
marinieren *v (food)* pickle
Marke *f* mark *(as of calibration)*
Markenname *m* brand name
Marker-Enzym *n (bioch)* marker enzyme
Marker-Protein *n (bioch)* marker protein
Markersubstanz *f (chromat)* marker
markieren *v* 1. mark; 2. *(anal)* label, tag
markiert/radioaktiv radioactively labelled
Markierung *f* 1. marking; 2. *(anal)* labelling, tagging; 3. mark
Markierungssubstanz *f* 1. tracer; 2. *(chromat)* marker
Markovnikov-Addition *f* Markovnikov addition *(of hydrogen halides to aliphatic double bonds)*
Marmor *m* marble
Martensit *m (met)* martensite
Martensitaushärtung *f (met)* maraging
martensitisch *(met)* martensitic
Masche *f* mesh *(screening)*
Maschenweite *f* mesh size, screen aperture, clear opening
Maschenzahl *f* mesh *(number of openings per linear inch)*
Maschinenausfallzeit *f* downtime
Maschinen(schmier)öl *n* machine oil
maskieren *v* mask, sequester *(ions)*
Maskierung *f* masking, sequestration *(of ions)*
Maskierungsmittel *n* masking [sequestering] agent
Maß *n* measure • **nach Maß aufbauen [herstellen]** tailor(-make), make to measure *(e.g. polymers)*
Maßanalyse *f* volumetric [titrimetric] analysis
maßanalytisch volumetric, titrimetric
Masse *f* 1. mass, *(if loose also)* bulk; 2. *(ceram)* body, paste; 3. *(quantitatively:)* mass
Masse f/atomare atomic mass
Masse f/molare molar mass
Masse f/molekulare molecular mass
Masseäquivalent *n* mass equivalent
Masseeinheit f/atomare atomic mass unit, amu
Maßeinteilung *f* scale *(of a measuring instrument)*
Massekonstanz *f* constant weight • **bis zur Massekonstanz glühen** ignite to constant weight • **bis zur Massekonstanz trocknen** dry to constant weight
Massel *f (met)* pig
Masseleisen *n* pig iron
Massenbilanz *f* mass balance
Massenbilanzgleichung *f* continuity [mass-balance] equation
Massendefekt *m* mass defect
Massendichte *f* mass density
Massenerhaltung *f* conservation of mass [matter]
Massenfluss *m* flux
Massenkonzentration *f* mass per unit volume
Massenkunststoff *m* commodity plastic
Massenquellungsgrad *m* mass swelling degree, mass degree of swelling *(polymer networks)*
Massenskala *f* mass scale
Massenspektrograph m/doppel(t)fokussierender double-focus(s)ing mass spectrograph
Massenspektrometer *n* mass spectrometer
Massenspektrometrie *f* mass spectrometry
massenspektrometrisch mass-spectrometric, by mass spectrometry
Massenspektrum *n* mass spectrum
Massenstrom *m* mass flux
Massenverhältnis *n* mass ratio
Massenwirkungsgesetz *n* law [principle] of mass action, mass-action expression [law], Guldberg and Waage law
Massepolymerisat *n (tech)* bulk polymer
Massepolymerisation *f* mass [bulk] polymerization
Masseprozent *n* percentage by mass, mass percentage
massereich *(nucl)* massive
Masseteil *m* part by weight
Massezahl *f* (nuclear) mass number
mäßig gentle, slight *(heating)*
massiv massive, solid
Maßkolben *m (lab)* measuring [volumetric] flask
Maßlösung *f* standard solution
Maßstab *m* scale • **im Maßstab vergrößern** scale up *(a plant)* • **in den industriellen Maßstab überführen**

scale up to industrial use • **in großem [großtechnischem] Maßstab** on a large scale • **in halbtechnischem Maßstab** on a semicommercial scale • **in präparativem Maßstab** on a preparative scale
Maßstabsvergrößerung *f (tech)* scale-up
Maßzylinder *m* measuring [graduated] cylinder
Masterbatch *m (rubber)* masterbatch • **Masterbatches herstellen** masterbatch • **mit Masterbatches mischen** masterbatch
Mastikation *f (rubber)* mastication
Mastikator *m (rubber)* masticator
Mastix *m* mastic (gum), gum mastic *(from Pistacia lentiscus L.)*
Mastixharz *n s.* Mastix
mastizieren *v (rubber)* masticate
Material *n* material, matter, substance
Material *n***/abgebranntes** burnt-out material
Material *n***/abgeröstetes** *(met)* calcine
Material *n***/basisches feuerfestes** basic refractory (material)
Material *n***/ferroelektrisches** ferroelectric (material, substance)
Material *n***/ferromagnetisches** ferromagnetic (material, substance)
Material *n***/feuerfestes** refractory (material)
Material *n***/filmbildendes** film former
Material *n***/lichtemittierendes** light emitting material
Material *n***/mineralisches** mineral matter
Material *n***/spaltbares** *(nucl)* fissionable material
Material *n***/zu untersuchendes** material to be tested
Material *n***/zu verarbeitendes** (feed-) stock
Materialfehler *m* fault, defect, flaw
Materialfluss *m* flux of material
Materialveränderung *f* material change, change of material
Materialzerstörung *f* **durch Mikroorganismen** biodeterioration of materials
Materie *f* matter
Materiewellenlänge *f* de Broglie wavelength
Matrix *f* 1. matrix; 2. *(geol, ceram)* groundmass
Matrixeffekt *m (phys ch)* matrix (interference) effect
Matrize *f* 1. (mould) cavity; 2. *(bioch)* template
matt flat, dead *(colours)*; mat(t), lustreless, dull *(surfaces)* • **matt werden** dull
Mattblech *n* pickled sheet
Mattheit *f* flatness *(of colours)*; dullness *(of surfaces)*
mattieren *v* flat *(colours)*; dull *(surfaces)*
Mattlack *m* flat varnish
Mauerwerk *n* brickwork
Mauvein *n* mauvein(e), mauve *(a quinone dye)*
Maximaldosis *f* maximum dose
Maximum *n* maximum, peak
Maximumazeotrop *n* maximum-boiling azeotrope, high-boiling azeotrope
Maxwell-Boltzmann-Verteilung *f* Maxwell-Boltzmann distribution *(of particle velocities)*
Mazeral *n (coal)* maceral
Mazeration *f* maceration
mazerieren *v* macerate
McCabe-Thiele-Verfahren *n (distil)* McCabe-Thiele method [construction] *(for estimating the number of plates)*
MCFC *s.* Schmelzkarbonat-Brennstoffzelle
McLafferty-Umlagerung *f* McLafferty rearrangement *(with fragmentation in mass spectrometry)*
McLeod-Manometer *n* McLeod gauge
MCPA *s.* 2-Methyl-4-chlorphenoxyessigsäure
Mechanismus *m***/radikalischer** free-radical mechanism
Mechanismus *m***/sequenzieller** *(bioch)* sequential [single-displacement] mechanism
Mechanochemie *f* mechanochemistry
mechanochemisch mechanochemical
Mediator *m (bioch)* mediator *(substance acting as a messenger)*
Medium *n***/strömendes** fluid medium *(fluid-bed technology)*
Medium *n***/umgebendes** circumambient medium
Medizinalöl *n* medicinal oil
Meersalz *n* sea [marine] salt; solar [bay] salt *(obtained by solar evaporation)*

Meerwasserentsalzung *f* saline water conversion
Meerwasserverunreinigung *f* marine pollution
Meerwein-Ponndorf-Verley-Carbonylreduktion *f* Meerwein-Ponndorf-Verley reduction
M-Effekt *m s.* Mesomerieeffekt
Mehl *n* 1. *(food)* flour, *(if coarsely ground:)* meal; 2. *(tech, ch)* flour, dust, *(if coarser:)* meal, powder
mehlig floury, mealy, farinaceous
Mehlverbesserungsmittel *n* flour improver *(for increasing the baking qualities)*
mehratomig polyatomic
mehrbasig polybasic, multibasic *(acid)*
Mehrbasigkeit *f* polybasicity *(of an acid)*
Mehretagenofen *m* multihearth furnace
Mehretagenröstofen *m* multihearth roaster
Mehrfachapparat *m (tech)* multiple unit
Mehrfachbindung *f* multiple bond
Mehrfachentwicklung *f (chromat)* multiple development
Mehrfachhalogenierung *f* polyhalogenation
Mehrfachion *n* polyion
Mehrfachresistenz *f (biot)* multiple resistance
Mehrfachschicht *f* multilayer, multimolecular layer
Mehrfachstreuung *f (phys ch)* multiple scattering
Mehrfachsubstitution *f* polysubstitution
Mehrfachverdampfer *m* multiple-effect evaporator
Mehrfachzerfall *m (nucl)* multiple disintegration
mehrgängig multipass *(e.g. heat exchanger)*
Mehrhalskolben *m (lab)* multinecked flask
Mehrkammerofen *m* multichamber kiln
Mehrkanalanalysator *m/***optischer** *(spectr)* optical multichannel analyser
Mehrkomponentensystem *n* multicomponent system
Mehrkörperverdampfer *m s.* Mehrstufenverdampfer
mehrlagig multi-ply
Mehrnährstoffdünger *m* multinutrient fertilizer, mixed [compound] fertilizer
Mehrphasensystem *n* multiphase system
Mehrproduktenapparat *m* multiproduct unit
mehrprotonig polybasic *(acid or base)*
mehrsäurig polyacid, multiacidic *(base)*
Mehrscheiben(sicherheits)glas *n* laminated safety (sheet) glass
Mehrschichtenadsorption *f* multilayer adsorption
Mehrschichtenfilter *n* multilayer filter, mixed-media [graded-density] filter
Mehrschichtenfiltration *f* mixed-media filtration
Mehrstoffkatalysator *m* mixed catalyst
Mehrstoffsystem *n* multicomponent system
mehrsträngig multistranded *(e.g. macromolecular chains)*
Mehrstufensynthese *f* many-step synthesis
Mehrstufenverdampfer *m* multieffect evaporator, multistage [multiple-effect, cascade] evaporator
Mehrstufenverdampfung *f* multieffect evaporation, multistage [multiple-effect] evaporation
mehrstufig multistage, multiple-stage, multistep
Mehrweg(e)hahn *m* multiport plug valve
mehrwertig multivalent, polyvalent, polyad; polybasic *(acid or base)*; polyhydric *(alcohol, phenol)*
Mehrwertigkeit *f* multivalence, polyvalence; polybasicity *(of an acid or base)*
Mehrwinkel-Laserlichtstreuung *f* multiangle laser light scattering, MALLS
mehrzählig *s.* mehrzähnig
mehrzähnig multidentate, polydentate *(coordination chemistry)*
Mehrzentrenbindung *f* multicentre bond
Mehrzentrenreaktion *f* multicentre reaction
M-Einheit *f (plast)* M unit, monofunctional unit *(for building up polymers)*
Meisenheimer-Komplex *m (org ch)* Meisenheimer complex [compound, adduct]
Meitnerium *n* Mt meitnerium *(element 109)*
Méker-Brenner *m (lab)* Meker burner
Melamin *n* melamine

Melamin-Formaldehydharz *n* melamine(-formaldehyde) resin
Melaminharz *n* melamine(-formaldehyde) resin
Melanoliberin *n (bioch)* melanotropin releasing hormone, MRH
Melanostatin *n* melanotropin (release) inhibitory hormone, MIH
Melanotropin *n* melanotropin, melanocyte-stimulating hormone, MSH
Melasse *f (sugar)* molasses
Mellithsäure *f* mellitic acid
Membran *f* membrane, diaphragm, partition (wall)
Membran f/biologische *s.* Biomembran
Membran f/selektiv permeable permselective membrane
Membran f/semipermeable semipermeable membrane
Membranfiltration *f* membrane filtration
membrangebunden *(bioch)* membrane-bound
Membrangleichgewicht *n* membrane equilibrium
Membranlipid *n (bioch)* membrane lipid
Membranosmometer *n* membrane osmometer
Membranosmose *f* membrane osmometry *(molar mass determination)*
Membranpotenzial *n* membrane potential
Membranprotein *n* membrane protein
Membranpumpe *f* diaphragm pump
Membranreaktor *m (biot)* ultrafiltration reactor
Membrantransport *m (bioch)* membrane transport
Mendeleev-System *n* Mendeléeff periodic system [table]
Mendelevium *n* Md mendelevium
Menge *f* quantity, quantum, amount
Menge f/adsorbierte amount adsorbed
Menge *f* **an Organohalogenverbindungen/nicht austreibbare** *(hyd)* non-purgeable organic halides
Menge f/äquivalente equivalent
Menge f/aufgenommene intake
mengen *v* mix, mingle, blend
Mengenbestimmung *f* quantitation
Mengenstrom *m* mass flux
Mengenstrommesser *m* flowmeter, rate meter
Mengenstrommessung *f* flow measurement
Meniskus *m* meniscus
Mennige *f* minium, red lead (oxide) *(lead(II) tetraoxoplumbate(IV))*
Mensur *f* measuring [graduated] cylinder, graduate
Menthadien *n (org ch)* mentadiene
***p*-Menthan** *n p*-menthane
Menthen *n (org ch)* menthene
Menthon *n (org ch)* menthone
Mercaptal *n* dithioacetal, *(old term:)* mercaptal *(any of a class of condensation products of thiols with aldehydes)*
Mercaptan *n* thiol, *(org ch deprecated, petrol still often:)* mercaptan
Mercaptid *n* thiolate, *(old term:)* mercaptide *(a metallic derivative of a thiol)*
Mercaptoessigsäure *f* mercaptoacetic acid, thioglycollic acid
Mercaptogruppe *f* –SH thiol group, mercapto [sulphanyl, sulphydryl] group
Mercaptol *n* dithioketal, *(old term:)* mercaptole *(any of a class of condensation products of thiols with ketones)*
mercurieren *v* mercurize *(organic compounds)*
Mercurierung *f* mercurization *(of organic compounds)*
Mercurimetrie *f* mercurimetry *(titration with a mercury(II) nitrate solution)*
mercurimetrisch mercurimetric
Mercurometrie *f* mercurometry *(titration with a mercury(I) nitrate solution)*
mercurometrisch mercurometric
Merrifield-Synthese *f* Merrifield (peptide) synthesis, solid-phase protein synthesis
Merzerisation *f (text)* mercerization
merzerisieren *v (text)* mercerize
Merzerisierhilfsmittel *n (text)* mercerizing assistant
Mescalin *n* mescaline
Mesitylen *n* mesitylene
Mesityloxid *n* mesityl oxide
Mesoform *f (org ch)* meso form
Mesokolloid *n* mesocolloid
mesomer mesomeric
Mesomerie *f* mesomerism, resonance
Mesomerieeffekt *m* mesomeric [electromeric] effect, resonance effect

Mesomerieenergie *f* mesomeric [resonance] energy
mesomeriestabilisiert resonance-stabilized
Mesomeriestabilisierung *f* resonance stabilization
Meson *n (nucl)* meson
Mesophase *f* mesophase, liquid crystal
Mesoprobe *f* *meso sample, semimicro sample *(0.1 to 0.01 g)*
Meso-Spuren-Analyse *f* *meso-trace analysis *(sample weight 0.1 to 0.01 g)*
Mesoverbindung *f (org ch)* meso compound
Mesoxalsäure *f* mesoxalic acid
messbar measurable
Messbehälter *m* measuring vessel
Messbereich *m* measuring [measurable] range, range of measurement
Messelektrode *f* measuring electrode, indicating [indicator] electrode
messen *v* measure
Messenger-Ribonucleinsäure *f (bioch)* messenger ribonucleic acid, messenger RNA
Messenger-RNA *f s.* Messenger-Ribonucleinsäure
Messer *m s.* Messgerät
Messer *n* knife
Messerholländer *m (pap)* Hollander (beater), stuff engine
Messfehler *m* measurement error
Messfühler *m* sensor
Messgefäß *n* measuring vessel, graduate
Messgenauigkeit *f* measuring accuracy
Messgerät *n* measuring device [instrument], meter, *(esp for measuring pressure, volume:)* gauge
Messgerinne *n* flume *(flow measurement)*
Messglas *n* measuring glass
Messgröße *f* quantity being measured; quantity to be measured
Messing *n* brass
Messinstrument *n s.* Messgerät
Messkapillare *f* stem *(of a thermometer)*
Messkolben *m (lab)* measuring [volumetric] flask
Messmethode *f* method of measurement, measurement technique
Messpipette *f* measuring [graduated] pipette
Messung *f* measurement
Messung *f* **nach einer Skale** scaling
Messwagen *m* measuring van
Messwert *m* measured value
Messzelle *f* measuring cell
Messzylinder *m* measuring [graduated] cylinder
Mesylat *n* mesylate
Mesylester *m* mesylate
Mesylierung *f* mesylation
Met *s.* Methionin
metabolisch *(bioch)* metabolic
metabolisierbar *(bioch)* metabolizable
metabolisieren *v (bioch)* metabolize
Metabolismus *m (bioch)* metabolism
Metabolit *m (bioch)* metabolite, product of metabolism
Metabolit *m*/**primärer** primary metabolite
Metabolit *m*/**sekundärer** secondary metabolite
meta-dirigierend meta-directing
Metaldehyd *m* metaldehyde
Metall *n* metal
Metall *n*/**edles** noble metal
Metall *n*/**gediegenes** native (metal)
Metall *n*/**gepulvertes** powder(ed) metal, metal powder
Metall *n*/**Roses [rosesches]** Rose's metal [alloy]
Metall *n*/**schmelzflüssiges** molten metal
Metall *n*/**unedles** base metal
Metallabscheidung *f* metal deposition
Metallamid *n* M^INH_2 (metal) amide, azanide
Metallatom *n* metal atom
Metallbad *n* (molten-)metal bath
Metallbeschichtung *f* electrocoating
Metallbindung *f* metal(lic) bond, metal-metal bond
Metallcarben *n (org ch)* metallocarbene *(metathesis)*
Metallcarbenkomplex *m (org ch)* metallocarbene complex *(metathesis)*
Metallcarbid *n* metal carbide
Metallcarbonyl *n* metal carbonyl, carbonyl (compound)
Metallchelat *n s.* Metallchelatkomplex
Metallchelatkomplex *m* metal chelate (complex), metal chelate compound
Metallchemie *f* metal chemistry

Metall-Dewar-Gefäß *n* Dewar (vessel) *(for holding liquid gases)*
metallen metallic
Metallenzym *n* metalloenzyme
Metallextraktion *f* metal extraction
Metallfaden *m* metallic filament
Metallfaser *f* metallic fibre
Metallfaserstoff *m* metallic fibre
Metallfolie *f* metal foil
Metallgehalt *m*/**biologisch verfügbarer** bioavailable metal level
Metallgewebe *n* metal fabric [gauze], wire cloth
Metallgewinnung *f* metal extraction
Metallgitter *n* metallic lattice
Metallglanz *m* metallic lustre
metallhaltig metal-containing, *(esp ores:)* metalliferous
Metallhydrid *n* metal hydride
Metallierung *f (org ch)* metalation
Metallion *n* metal ion
metallisch metallic
metallisch/nicht non-metallic
metallisieren *v* metallize
Metallisieren *n* metallization
Metall-Isolator-Übergang *m* metal-insulator transition, metal-nonmetal transition
Metallizität *f* metallicity
Metallkatalysator *m* metal(lic) catalyst
Metallketyl *n* ketyl
Metallkomplex *m* metal complex
Metallkomplex-Säurefarbstoff *m* metal mordant dye
Metallkunde *f* metallography
Metalllegierung *f* metallic alloy
Metallmaske *f* metal mask
Metallocen *n* metallocene *(organometallic coordination compound)*
Metallograph *m* metallographer
Metallographie *f* metallography
metallographisch metallographic
Metallorganik *f s.* Organometallchemie
Metallorganyl *n* organometallic [metallo-organic] compound
Metallothionein *n* metallothionein *(protein with high affinity to heavy metals)*
Metallphthalocyanin *n* metal phthalocyanine
Metallpigment *n* metal(lic) pigment
Metallprotein *n* metalloprotein
Metallpulver *n* metal powder, powder(ed) metal
Metallschaum *m* dross
Metallschmelze *f* molten [fused] metal
Metallseife *f* metallic soap
Metallsieb *n* metal [wire] screen
Metallspritzen *n* metal spraying, (spray) metallizing
Metallstearat *n* metal stearate
Metallsulfid *n* metal sulphide
Metallurg(e) *m* metallurgist
Metallurgie *f* metallurgy
Metallurgie *f*/**erzeugende** production [extractive] metallurgy
Metallurgie *f*/**verarbeitende** adaptive metallurgy
metallurgisch metallurgical
Metallzustand *m* metallic state
metamer metameric
Metamer(e) *n* metamer
Metamerie *f* metamerism *(one form of structural isomerism)*
Metametall *n* metametal
Metanilsäure *f* metanilic acid
Metaperiodat *n* $M^{I}IO_4$ metaperiodate
Metaphosphat *n* $(M^{I}PO_3)_n$ metaphosphate
Metaphosphorsäure *f* metaphosphoric acid
metastabil metastable
Metastabilität *f* metastability
metaständig meta, in meta position • **metaständig sein** be (located, situated) meta
meta-Stellung *f* meta position • **in meta-Stellung** meta, in meta position • **sich in meta-Stellung befinden** be (located, situated) meta
metasubstituiert meta-substituted, *m*-substituted
meta-Substitution *f* meta substitution
Metathese *f (org ch)* metathesis, double [mutual] decomposition
Metathesekatalysator *m (org ch)* metathesis catalyst
Metathesereaktion *f* metathesis [metathetical] reaction
metathetisch metathetic(al)
Methacrylat *n* methacrylate
2-Methacrylsäure *f* 2-methacrylic acid, MAA, 2-methylpropenoic acid
Methacrylsäuremethylester *m* methyl methacrylate
Methämoglobin *n* methaemoglobin
Methan *n* methane

Methanal *n* *methanal, formaldehyde
Methanamid *n* methanamide, formamide
Methanbakterien *npl* methane(-forming) bacteria
Methandicarbonsäure *f* *methane-dicarboxylic acid, *propanedioic acid, malonic acid
Methanfermentation *f (biot)* methane fermentation *(production of biomass with methane as carbon source)*
Methangärung *f (biot, hyd)* methane fermentation *(production of methane from waste products)*
Methangärung *f***/alkalische** *(biot, hyd)* methane fermentation *(production of methane from waste products)*
methanisieren *v* methanize *(e.g. synthesis gas)*
Methanisierung *f* methanation *(as of synthesis gas)*
Methanol *n* *methanol, methyl alcohol
Methanolfermentation *f (biot)* methanol fermentation
methanolisch methanolic
Methanolyse *f* methanolysis
Methanoxidierer *mpl (biot)* methane-oxidizing bacteria
Methansäure *f* *methanoic acid, formic acid
Methansulfonat *n* methane sulphonate, mesylate
Methansulfonsäureester *m s.* Methansulfonat
Met-Hb *s.* Methämoglobin
Methenamin *n* methenamine, hexamethylenetetramine
Methid *n (org ch)* methide
Methinfarbstoff *m* methine dye
Methingruppe *f* =CH– methine group
Methionin *n* methionine
Methode *f* method, procedure, technique, approach
Methode *f* **der wandernden Grenzfläche(n)** *(anal)* moving-boundary method, Tiselius method, free electrophoresis
Methode *f* **des quasistationären Zustands** *(phys ch)* steady-state method
Methode *f***/röntgenographische** *(anal)* X-ray method
Methode *f***/standardisierte** standard method
Methode *f***/statische** static method *(for measuring vapour pressures)*
Methode *f***/titrimetrische** *(anal)* titrimetric method
Methodik *f* method; experimental *(in scientific papers)*
Methoxid *n* CH_3OM^I methoxide, methylate
Methoxybenzen *n* *methoxybenzene, anisole
***p*-Methoxybenzoesäure** *f p*-methoxybenzoic acid
Methoxygruppe *f s.* Methoxylgruppe
Methoxylgruppe *f* $-OCH_3$ methoxyl group, methoxy group
Methoxylierung *f* methoxylation
Methoxymethan *n* methoxymethane, dimethyl ether
Methylacetat *n* methyl acetate
Methylal *n* methylal, *dimethoxymethane
Methylalkohol *m* methyl alcohol, *methanol
methylalkoholisch methanolic
Methylamin *n* methylamine
Methylaminoessigsäure *f* methylaminoacetic acid, sarcosine
Methylat *n* CH_3OM^I methylate, methoxide
Methylbenzen *n* *methylbenzene, toluene
Methylbenzoesäure *f* methylbenzoic acid, toluic acid
Methylbernsteinsäure *f* methylsuccinic acid
Methylbromid *n* methyl bromide, bromomethane
2-Methylbutan *n* *2-methylbutane, isopentane
3-Methylbutanal *n* *3-methylbutanal, isovaleraldehyde
3-Methylbutansäure *f* *3-methylbutanoic acid, isovaleric acid
Methylcarboxamidin *n* methyl carboxamidine, acetamidine
Methylcellosolve *n* methylcellosolve, ethylene glycol monomethyl ether
Methylcellulose *f* methyl cellulose
2-Methylchinolin *n* 2-methylquinoline, quinaldine
Methylchlorid *n* methyl chloride, chloromethane
2-Methyl-4-chlorphenoxyessigsäure *f*

2-methyl-4-chlorophenoxyacetic acid, MCPA *(a herbicide)*
Methylchlorsilan *n* methylchlorosilane
Methylcyclohexan *n* methylcyclohexane
Methylen *n* $:CH_2$ methylene, carbene
Methylenbernsteinsäure *f* methylenesuccinic acid, itaconic acid
Methylenblau *n* methylene blue
Methylenbromid *n* *dibromomethane, methylene dibromide
Methylenchlorid *n* *dichloromethane, methylene dichloride
Methylendiphenol *n* methylenediphenol, bisphenol
Methylengruppe *f* $=CH_2$ methylene group, methanediyl group
Methylenhalogenid *n* methylene halide
Methylenierung *f* methylenation
Methyleniodid *n* *diiodomethane, methylene diiodide
Methylester *m* methyl ester
Methylethylketon *n* methyl ethyl ketone, MEK, 2-butanone
***N*-Methylglykokoll** *n N*-methylglycine, sarcosine
Methylglyoxal *n* methylglyoxal, pyruvic aldehyde
Methylgruppe *f* CH_3– methyl group
Methylgruppe *f***/ringständige** ring-methyl group
Methylhalogenid *n* methyl halide
methylieren *v (org ch)* methylate
Methylierung *f (org ch)* methylation
Methylierung *f***/erschöpfende** exhaustive methylation
Methyliodid *n* iodomethane, methyl iodide
Methylmethacrylat *n* methyl methacrylate, MMA
Methylnaphthalen *n* methylnaphthalene
Methylnitren *n* methyl nitrene
Methylorange *n* methyl orange
Methyloxiran *n* 2-methyloxirane, 1,2-epoxypropane, propylene oxide
Methylphenol *n* methylphenol, hydroxytoluene, cresol
Methylphenylether *m* methyl phenyl ether, methoxybenzene, anisole
Methylphenylketon *n* *methyl phenyl ketone, acetophenone
Methylpolysiloxan *n* *polymethylsiloxane, methyl silicone
2-Methylpropan *n* *2-methylpropane, isobutane
2-Methylpropansäure *f* 2-methylpropanoic acid, isobutyric acid
2-Methylpropensäure *f* 2-methylpropenoic acid, methacrylic acid
Methylpropylketon *n* methyl propyl ketone, *pentan-2-one
Methylradikal *n* (free) methyl radical
Methylsalicylat *n* methyl salicylate
Methylsilicon *n* methyl silicone, *polymethylsiloxane
Methylsulfonylgruppe *f* CH_3SO_2– methylsulphonyl group
Methylthiogruppe *f* CH_3S– methylthio group, methylsulphanyl group
Methyltrichlorsilan *n* methyltrichlorosilane
Methylviolett *n* methyl violet
Mevalonsäure *f* mevalonic acid, MVA
MF-Induktionsofen *m* coreless induction furnace
MgH *s.* Magnesiahärte
Mg-KH *s.* Magnesiumcarbonathärte
Mg-NKH *s.* Magnesium-Nichtcarbonathärte
Michael-Addition *f (org ch)* Michael condensation [reaction]
Michaelis-Arbuzov-Reaktion *f* Michaelis-Arbuzov reaction *(for making a phosphorus-carbon bond)*
Michaelis-Komplex *m (bioch)* enzyme-substrate complex
Michaelis-Menten-Komplex *m (bioch)* Michaelis-Menten complex
Michaelis-Menten-Konstante *f (bioch)* Michaelis-Menten constant
Migration *f* migration *(as of ions)*
Migrationsstrom *m* migration current
migrieren *v* migrate *(as of ions)*
Mikroanalyse *f* microanalysis, milligram analysis
mikroanalytisch microanalytic(al)
Mikroarbeitsweise *f s.* Mikromethode
Mikroarray *n* microarray *(molecular biology)*
Mikrobaugruppe *f* micro-assembly
Mikroben *fpl* microbes, microorganisms
Mikrobenaktivität *f* microbial activity
Mikrobestandteil *m* microconstituent, microcomponent
Mikrobestimmung *f* microdetermination
mikrobiell microbial

Mikrobiologie *f* microbiology
Mikrobiologie *f***/industrielle [technische]** industrial microbiology
mikrobiologisch microbiological
Mikrobioreaktor *m* microbioreactor
mikrobizid microbicidal
Mikrobombe *f (lab)* microbomb
Mikrobrenner *m* microburner
Mikrobürette *f* microburet(te)
Mikrochemie *f* microchemistry
mikrochemisch microchemical
Mikrochromatographie *f* microchromatography
Mikrodestillation *f* microdistillation
Mikrodosierspritze *f (anal)* microsyringe
Mikroelektrode *f* microelectrode
Mikroelektrophorese *f* microelectrophoresis
Mikroelement *n s.* Mikronährstoff
Mikroemulsion *f* microemulsion
mikrofibrillär microfibrillar
Mikrofibrille *f* microfibril
Mikrofluidik *f* microfluidic
Mikrofotografie *f* photomicrograph
Mikrogefüge *n* microstructure
Mikrogel *n (coll)* microgel
Mikrogrammbereich *m (anal)* microgram range *(10^{-6} to 10^{-3} g)*
Mikrohärte *f* microhardness
Mikroheterogenität *f* microheterogeneity
Mikrokalorimeter *n* microcalorimeter
mikrokalorimetrisch microcalorimetric
Mikrokanal *m* microchannel
Mikrokapsel *f* microcapsule • **in Mikrokapseln einschließen** microencapsulate
Mikrokatal *n (bioch)* microkatal, µkat *(unit of enzyme activity)*
Mikrokörnungen *fpl* micro grits
mikrokristallin microcrystalline
Mikroküvette *f* microcell
mikrolegiert micro-alloyed
Mikromanipulator *m* micromanipulator
Mikromethode *f* micromethod, microscale procedure
Mikromethode *f* **von Rast** Rast (micro)method, Rast molecular weight method
mikromolekular micromolecular
Mikron *n* micron *(dispersed particle visible in an ordinary microscope)*
Mikronährstoff *m (agric)* micronutrient, trace element
Mikroorganismen *mpl* microorganisms, microbes • **Mikroorganismen abtötend** microbicidal
Mikroorganismen *mpl***/abbauende** decomposers, microconsumers
Mikroorganismen *mpl***/autotrophe** autotrophs
Mikroorganismen *mpl***/cellulolytische [celluloseabbauende]** cellulolytic microorganisms
Mikroorganismen *mpl***/genetisch [gentechnologisch] veränderte** genetically engineered microorganisms
Mikroorganismen *mpl***/heterotrophe** heterotrophs
Mikroorganismen *mpl***/mesophile** mesophiles, mesophilic microorganisms *(growth optimum 20 to 45 °C)*
Mikroorganismen *mpl***/psychrophile** psychrophiles, psychrophilic microorganisms *(growth optimum 5 to 20 °C)*
Mikroorganismeneiweiß *n* microbial protein
Mikroorganismenfermentation *f (biot)* microbial fermentation
Mikroorganismenkonzentration *f* microbial concentration
Mikroorganismenkultur *f* microbial culture, culture of microorganisms
Mikroorganismenmasse *f (hyd)* microbial mass
Mikroorganismenstamm *m* microbial strain
Mikroorganismenzahl *f* microbial count
Mikroparaffin *n* micro(crystalline) wax
mikroperforiert micro-perforated
Mikropipette *f* micropipet(te)
Mikropore *f* micropore
mikroporös microporous
Mikroprobe *f* micro sample *(< 0.01 g)*
Mikroriss *m* micro-crack
Mikrosäule *f (chromat)* microbore [smallbore] column; micropacked [packed capillary] column
Mikroskop *n* microscope
Mikroskopie *f* microscopy
mikrosomal *(bioch)* microsomal
Mikrosonde *f (anal)* microprobe
Mikrospray *n (spectr)* microspray *(electrospray technique)*

Mikrospur *f (anal)* micro-trace *(10^{-7} to 10^{-4} ppm)*
Mikro-Spuren-Analyse *f* micro-trace analysis *(sample weight 10^{-3} to 10^{-2} g)*
Mikrostrom *m* micro-current
Mikrostruktur *f* microstructure
Mikrotitration *f* microtitration
Mikroträger *m (biot)* microcarrier *(for cell cultures)*
Mikrotröpfchen *n* microdroplet
Mikrountersuchung *f* microexamination
Mikroventil *n* microvalve
Mikroverkapselung *f* microencapsulation
Mikrovermischung *f* micromixing, local mixing
Mikroverunreinigung *f* micropollutant
Mikrowachs *n* micro(crystalline) wax
Mikrowelle *f (lab)* microwave
Mikrowellendetektor *m (spectr)* microwave detector
Mikrowellenofen *m (lab)* microwave furnace
Mikrowellenplasma *n*/**kapazitiv gekoppeltes** capacitively coupled microwave plasma, CMP
Mikrowellenreaktor *m (lab)* microwave reactor
Mikrowellenspektroskopie *f* microwave spectroscopy
Mikrowissenschaft *f* microscience
Mikrozustand *m (phys ch)* microstate
Milbengift *n* acaricide, miticide
milbentötend acaricidal, miticidal
Milch *f* milk
Milch *f*/**entrahmte** skim(med) milk
Milch *f*/**ultrahocherhitzte** UHT (processed) milk
Milchalbumin *n* lactalbumin, milk albumin
Milchchemie *f* dairy chemistry
Milcheiweiß *n* milk protein
Milchfett *n* milk fat
Milchglas *n* milk glass
Milchglobulin *n* milk globulin
milchhaltig lactiferous
milchig milky, *(as opposed to translucent also:)* opaque
Milchprotein *n* milk protein
Milchpulver *n* milk powder, dry [dried] milk
Milchsaft *m (bot)* latex, milk, milky sap
milchsaftführend *(bot)* lactiferous, latex-bearing
Milchsäure *f* lactic acid, *2-hydroxy-propionic acid
Milchsäureanhydrid *n* lactic anhydride
Milchsäurebakterien *npl* lactic-acid bacteria
Milchsäuredehydrogenase *f* lactic dehydrogenase
Milchsäuregärung *f* lactic(-acid) fermentation
Milchtrockenmasse *f* milk solids
Milchwirtschaft *f* dairying
Milchzucker *m* lactose, milk sugar
Milchzuckergärung *f* lactose fermentation
mild mild *(reaction conditions)*
Milieu *n* environment
Miller-Indizes *mpl* Miller (crystal) indices
Milli(gramm)äquivalent *n* milliequivalent, mequiv, meq
Milligrammbereich *m (anal)* milligram range *(0.001 to 0.01 g)*
Millimeter *mpl* **Quecksilbersäule** millimetres of mercury *(unit of pressure of fluids; 1 mm Hg = 1 Torr = 133.3 Pa)*
Millival *n s.* Milli(gramm)äquivalent
mindern *v* reduce, decrease, lower
Minderung *f* reduction, decrease, lowering
minderwertig of inferior quality, *(esp of raw materials:)* low-grade
Mindestdosis *f* minimum dose [dosage]
Mindestrücklaufverhältnis *n (distil)* minimum reflux ratio
Mindesttemperatur *f* minimum temperature
Mine *f* mine
Mineral *n* mineral
Mineral *n*/**gesteinsbildendes** rock-forming mineral
Mineral *n*/**primäres** primary mineral
Mineral *n*/**schweres** heavy mineral
Mineral *n*/**sekundäres** secondary mineral
Mineralaufbereitung *f* mineral dressing
Mineralchemie *f* mineral [mineralogical] chemistry
Mineralcorticoid *n* mineralocorticoid *(a hormone)*
Mineraldüngemittel *n* mineral [inorganic] fertilizer

Mineraldünger *m* mineral [inorganic] fertilizer
Mineralfaser *f* mineral fibre
Mineralfaserstoff *m* mineral fibre
Mineralgerbung *f* mineral tanning
Mineralisation *f (geol, soil, hyd)* mineralization
mineralisch mineral
mineralisieren *v (geol, soil, hyd)* mineralize
Mineralisierung *f (geol, soil, hyd)* mineralization
Minerallaugung *f* mineral leaching
Mineralog(e) *m* mineralogist
Mineralogie *f* mineralogy
mineralogisch mineralogical
Mineralöl *n* mineral oil
Mineralsalz *n* mineral salt
Mineralsalzgehalt *m* mineral content
Mineralsäure *f* mineral acid
Mineralstoff *m* mineral matter
Mineralstoffgehalt *m* mineral-matter content
Mineralwasser *n* mineral water
Mineralwolle *f* mineral [rock] wool
Minimaltemperatur *f* minimum temperature
Minimumazeotrop *n* minimum-boiling azeotrope, low-boiling azeotrope
Minoritäts(ladungs)träger *m* minority carrier *(in a semiconductor)*
MIP *s.* Plasma/mikrowelleninduziertes
MIR *s.* Infrarotgebiet/mittleres
Mirbanöl *n (cosmet)* oil of mirbane *(nitrobenzene)*
mischbar miscible
mischbar/beliebig miscible in all proportions
mischbar/nicht immiscible, non-miscible
Mischbarkeit *f* miscibility
Mischbarkeit *f***/begrenzte [teilweise]** incomplete [partial] miscibility
Mischbecken *n***/vollständiges** *(hyd)* well-mixed [flow] reactor, completely mixed tank
Mischbehälter *m* mixing tank [vessel]
Mischbettsäule *f (hyd, chromat)* mixed-bed column
Mischdüse *f* mixing nozzle
Mischelement *n* polyisotopic element
mischen *v* mix, *(esp if so that the constituents cannot be distinguished:)* blend; *(in accordance with a recipe:)* compound
mischen *v***/mit Masterbatches** *(rubber)* masterbatch
mischen *v***/sich** mix, blend
Mischen *n* **von Feststoffkomponenten** dry mixing [blending], solid-solid mixing
Mischer *m* mixer, blender
Mischer *m***/geschlossener** closed mixer
Mischer *m***/satzweise arbeitender** batch mixer
Mischer-Scheider-Extrakteur *m* mixer-settler *(for solvent extraction)*
Mischflügel *m* mixing blade, agitator [stirrer] blade
Mischgas *n* mixed gas
Mischindikator *m* mixed indicator
Mischkatalysator *m* mixed catalyst
Mischkessel *m* mixing vessel
Mischkneter *m* kneader-mixer
Mischkondensation *f (plast)* co-condensation
Mischkristall *m* mixed crystal, solid solution
Mischmaschine *f* mixing machine, mixer, blender
Mischmetall *n* misch metal *(an alloy consisting of rare-earth metals)*
Mischoctanzahl *f* blending octane number
Mischoxid *n* mixed oxide
Mischphase *f* mixed phase
Mischpotenzial *n* mixed potential
Mischreaktor *m* (stirred-)tank reactor, well-mixed (flow) reactor, completely mixed reactor
Mischreaktorkaskade *f* series of stirred-tanks reactors, stirred-tank [well-mixed] reactors in series
Mischsäure *f (tech)* mixed acid, nitrating acid *(consisting of concentrated nitric acid and sulphuric acid)*
Mischschmelzpunkt *m* mixed melting point
Misch-Trenn-Behälter *m* mixer-settler *(for solvent extraction)*
Mischtrommel *f (tech)* rotary barrel mixer
Mischung *f* 1. mixture, mix, blend; 2. *(rubber)* compound, stock, composition; 3. mixing, blending • **eine Mischung herstellen** *(rubber)* compound

Mischung f/azeotrope azeotrope, azeotropic mixture
Mischung f/eutektische eutectic (mixture)
Mischungsherstellung f *(rubber)* compounding
Mischungshöhe f mixing height *(of pollutants in the atmosphere)*
Mischungskalorimeter n water calorimeter
Mischungslücke f miscibility gap
Mischungspunkt m **bei gleich bleibender Temperatur/kritischer** plait [isothermal critical] point *(of a two-component system)*
Mischungspunkt m**/kritischer** 1. consolute [critical solution] point *(of a two-component system)*; 2. *s.* Mischungstemperatur/kritische
Mischungsrezept n *(rubber)* compounding formula [recipe], recipe of mix • **ein Mischungsrezept aufstellen** compound
Mischungstemperatur f/kritische consolute [critical solution] temperature *(of a two-component system)*
Mischungsverhältnis n 1. mixing ratio; 2. abundance ratio *(of isotopes)*
Miszella f miscella *(an extraxt containing an extracted oil or grease)*
mitabscheiden v co-deposit
Mitabscheidung f co-deposition
mitfallen v coprecipitate
mitfällen v coprecipitate
Mitfällung f coprecipitation
Mitfällung f/induzierte coprecipitation
mitführen v entrain, carry over [off] *(as in a gas stream)*
Mitführen n entrainment, carry-over *(as in a gas stream)*
mitlaufen v co-chromatograph
mitlaufen lassen v co-chromatograph
Mitnehmer m *(distil)* entrainer, entraining [azeotroping] agent, azeotrope-former
Mitochondrien-DNA f mitochondrial deoxyribonucleic acid, mtRNA
Mitose f *(bioch)* mitosis • **Mitose auslösend** mitogenic
Mitosegift n *(bioch)* mitotic poison
mitreißen v entrain, carry over [off] *(as in a gas stream)*; carry down *(in a precipitate)*
Mitreißen n entrainment, carry-over *(as in a gas stream)*; carrying down *(in a precipitate)*
Mittel n 1. agent; 2. *(pharm)* preparation; 3. *s.* Mittelwert
Mittel n/absetzverhinderndes antisettling agent
Mittel n/absorptionsbeschleunigendes [absorptionsförderndes] absorbefacient
Mittel n/adstringierendes astringent, styptic
Mittel n/anregendes analeptic, central nervous system stimulant
Mittel n/antikonzeptionelles contraceptive
Mittel n/antiseptisches antiseptic (agent)
Mittel n/antistatisches antistatic (agent)
Mittel n/arithmetisches arithmetic mean *(statistics)*
Mittel n/bakteriostatisches bacteriostat(ic)
Mittel n/blasenziehendes blister agent, vesicant
Mittel n/blutdrucksenkendes antihypertensive [hypotensive] drug, blood-pressure depressant
Mittel n/blutstillendes haemostatic, styptic
Mittel n/chemotherapeutisches chemotherapeutic agent
Mittel n/desinfizierendes disinfectant
Mittel n/desodor(is)ierendes deodorizer
Mittel n/diuretisches diuretic (agent)
Mittel n/entzündungshemmendes antiphlogistic (agent)
Mittel n/feuerhemmendes fire retardant
Mittel n/fiebersenkendes antipyretic
Mittel n/gefäßerweiterndes vasodilator
Mittel n/gefäßkontrahierendes [gefäßverengendes] vasoconstrictor
Mittel n **gegen depressive Verstimmung** antidepressant
Mittel n **gegen Kesselstein** descaling agent
Mittel n **gegen Malariainfektion** antimalarial (drug)
Mittel n **gegen Schaumbildung** foam inhibitor, froth-preventing agent
Mittel n **gegen Schnecken** molluscacide, molluscide

Mittel *n*/**geometrisches** geometric mean *(statistics)*
Mittel *n*/**gerinnungshemmendes [gerinnungsverzögerndes]** anticoagulant
Mittel *n*/**hydrophobierendes** hydrophobing agent, water repellent
Mittel *n*/**insektenabschreckendes [insektenvertreibendes]** insect repellent
Mittel *n*/**keimbildendes** nucleation agent
Mittel *n*/**keimtötendes** germicide, disinfectant
Mittel *n*/**Konvulsionen auslösendes** convulsant
Mittel *n*/**korrodierendes** corrosive (agent)
Mittel *n*/**krampflösendes** spasmolytic, antispasmodic
Mittel *n*/**menstruationsförderndes** emmenagogue
Mittel *n*/**mikrobizides [Mikroorganismen abtötendes]** microbicide
Mittel *n*/**oberflächenaktives** surface-active agent
Mittel *n*/**oxytozisches** oxytocic (agent), uterotonic
Mittel *n*/**photostabilisierendes** *s.* Lichtstabilisator
Mittel *n*/**pilztötendes** fungicide
Mittel *n*/**protektives [protektiv wirkendes]** protective
Mittel *n*/**schmerzlinderndes [schmerzstillendes]** analgesic, pain killer [reliever]
Mittel *n*/**schwangerschaftsverhütendes** contraceptive
Mittel *n*/**schweißhemmendes [schweißlinderndes]** antiperspirant
Mittel *n*/**schweißtreibendes** sudorific
Mittel *n*/**spermienabtötendes [spermizides]** spermatocide
Mittel *n*/**systemisches** systemic (chemical)
Mittel *n*/**virustötendes [viruzides]** virucide
Mittel *n*/**wasserabspaltendes** dehydrating agent
Mittel *n*/**wasserabstoßendes [wasserabweisendes]** water repellent
Mittel *n*/**wasserentziehendes** dehydrating [water-abstracting] agent
Mittel *n*/**wurmabtreibendes** helminthagogue, vermifuge
Mittel *n*/**wurmtötendes** vermicide
Mittel *n*/**wurmwidriges** anthelmint(h)ic
mittelaktiv *(nucl)* medium-level active
Mittelbenzin *n* medium gasoline
Mitteldrucksynthese *f* medium-pressure synthesis
mittelfeinkörnig medium-grained
Mittelfraktion *f (distil)* middle fraction
Mittelgut *n* middlings product *(classifying)*
mittelkörnig medium-grained
Mittelsieder *m* medium boiler *(e.g. solvent)*
mittelständig centrally located
mittelstark moderately strong *(e.g. acid)*
Mitteltemperaturverkokung *f* medium-temperature carbonization
mittelviskos moderately viscous, medium-viscosity
Mittelwert *m* average, mean (value); expected value *(statistics)*
mitten *v* centre
Mixtur *f (pharm)* mixture
mizellar micellar
Mizellbildungskonzentration *f*/**kritische** *s.* Mizellkonzentration/kritische
Mizelle *f* micelle, *(in fibrous material also)* crystallite • **Mizellen bilden** micellize
Mizelle *f*/**inverse** inverted [reverse] micelle
Mizellkolloid *n* micellar [association] colloid
Mizellkonzentration *f*/**kritische** *(coll)* critical micelle concentration, c.m.c.
MO *n s.* Molekülorbital
mobilisieren *v (bioch)* mobilize
Mobilisierung *f (bioch)* mobilization
Möbius-System *n* Möbius system *(a cyclic conjugated hydrocarbon)*
Modacrylfaser *f* modacrylic [modified acrylic] fibre
Modacrylfaserstoff *m* modacrylic [modified acrylic] fibre
Mode *f (spectr)* (transmission) mode, mode of vibration *(propagation of guided waves)*
Modellreaktion *f* model reaction
Modellverbindung *f* model compound
Moderator *m (nucl)* moderator
moderieren *v (nucl)* moderate, slow (down)

Modifikation *f* modification
Modifikation f/allotrope allotropic modification, allotrope *(of an element)*
modifiziert/nicht unmodified
Modulationsfrequenz *f (spectr)* modulation frequency
Modulator *m (bioch)* modulator
Modulator *m***/fördernder** stimulatory [positive] modulator
Modulator *m***/hemmender** inhibitory [negative] modulator
Mohnalkaloid *n* opium alkaloid
Mohs-Skala *f* Mohs' (hardness) scale
Mol *n* mole, mol
molal molal
Molalität *f* molality, molal concentration
molar molar
Molarität *f* molarity, molar concentration
Molekel *f s.* Molekül
Molekül *n* molecule
Molekül *n***/aktives [aktiviertes]** active [activated] molecule
Molekül *n***/angeregtes** excited molecule
Molekül *n***/geknäueltes** coiled molecule
Molekül *n***/gelöstes** solute molecule
Molekül *n***/heteronukleares** heteronuclear molecule
Molekül *n***/homonukleares** homonuclear molecule
Molekül *n***/lineares** linear molecule
Molekül *n* **mit fluktuierenden Bindungen** fluxional molecule
Molekül *n***/polares** polar molecule
Molekül *n***/stäbchenförmiges [stabförmiges]** rod-like molecule
Molekül... *siehe auch: Molekular...*
Molekülaggregat *n* molecular aggregate, aggregate of molecules
Molekülaggregation *f* molecular aggregation, aggregation of molecules
molekular molecular
Molekular... *siehe auch: Molekül...*
Molekularbewegung *f* molecular motion [movement]
Molekulardestillation *f* molecular distillation
Molekulardestillierapparat *m* molecular still
molekulardispers molecularly disperse
Molekulardrehung *f* molecular rotation
Molekulargewicht *n* molecular weight • **von geringem Molekulargewicht** low-molecular-weight • **von hohem Molekulargewicht** high-molecular-weight
Molekularität *f* **(der Reaktion)** molecularity (of reaction), reaction molecularity
Molekularkraft *f* (inter)molecular force
Molekularrotation *f* molecular rotation
Molekularsieb *n* molecular sieve
Molekularstrahl *m* molecular beam
Molekularstrahlen *mpl***/gekreuzte** crossed molecular beams
Molekularstrahlepitaxie *f* molecular beam epitaxy, MBE
Molekülassoziation *f* molecular association
Molekülbande *f (spectr)* molecular band
Moleküleigenfunktion *f* molecular eigenfunction
Moleküleinschlussverbindung *f* molecular inclusion compound
Molekülelektronenspektrum *n* molecular electronic spectrum
Molekülformel *f* molecular formula
Molekülfragment *n* molecular fragment
Molekülgeometrie *f* molecular geometry
Molekülgitter *n* molecular lattice
Molekülion *n* molecular ion
Molekülkolloid *n* molecular colloid
Molekülkomplex *m* molecular complex
Molekülkonformation *f* molecular conformation
Molekülkristall *m* molecular crystal
Molekülmasse *f* molecular mass
Molekülmasse *f***/mittlere relative** average relative molecular mass, average RMM
Molekülmasse *f***/relative** relative molecular mass, RMM
Molekülorbital *n* molecular orbital, MO
Molekülorbital *n***/bindendes** bonding molecular orbital
Molekülorbital *n***/höchstes besetztes** highest(-energy) occupied molecular orbital, HOMO
Molekülorbital *n***/niedrigstes unbesetztes** lowest(-energy) unoccupied molecular orbital, LUMO
Molekülorbitalmethode *f* molecular-orbital method, Hund-Mulliken-Lennard-Jones-Hückel method
Molekülpeak *m (spectr)* molecular peak
Molekülpolarisierbarkeit *f* molecular polarizability

Molekülsäure *f* molecular acid
Molekülschwarm *m* swarm of molecules
Molekülschwingung *f* molecular vibration
Molekülspektroskopie *f* molecular spectroscopy
Molekülspektroskopiker *m* molecular spectroscopist
molekülspektroskopisch by molecular spectroscopy
Molekülspektrum *n* molecular spectrum, band spectrum
Molekülspinorbital *n* molecular spin orbital
Molekülstruktur *f* molecular structure
Molekülverbindung *f* molecular [addition] compound
Molenbruch *m* mole fraction
Molisch-Reaktion *f* Molisch reaction *(for detecting carbohydrates)*
Molkenprotein *n* whey [milk serum] protein
Möller *m (met)* burden
möllern *v (met)* burden
Mollier-Diagramm *n* Mollier [enthalpy-entropy] chart
Molluskizid *n* molluscacide, molluscide
Molmasse *f* molar mass • **von geringer Molmasse** low-molar-mass
Molmasse *f*/**gewichtsmittlere** weight average molecular weight
Molmasse *f*/**mittlere** average molecular weight
Molmasse *f*/**viskositätsmittlere** viscosity average molecular weight
Molmasse *f*/**zahlenmittlere** number average molecular weight
Molmassenverteilung *f* molar mass distribution *(polymers)*
Molozonid *n* molozonide, 1,2,3-trioxolane
Molpolarisation *f* molar polarization
Molsuszeptibilität *f* molar susceptibility
Molvolumen *n* molar volume, gram-molecular volume
Molvolumen *n*/**kritisches** critical molar volume
Molvolumen *n*/**partielles** partial molar volume
Molwärme *f* molar heat capacity
Molybdän *n* Mo molybdenum
Molybdänblau *n* molybdenum blue *(molybdenum(V,VI) oxide)*
Molybdäncarbid *n* molybdenum carbide
Molybdän-Eisen-Protein *n (bioch)* molybdoiron protein
Molybdän(V)-oxid *n* molybdenum(V) oxide, molybdenum pentoxide
Molybdän(VI)-oxid *n* molybdenum(VI) oxide, molybdenum trioxide
Molybdänsäure *f* molybdic acid
Molybdänstahl *m* molybdenum steel
Molybdän(IV)-sulfid *n* molybdenum(IV) sulphide, molybdenum disulphide
Molybdatophosphat *n* molybdophosphate
Molybdatophosphorsäure *f* molybdophosphoric acid
Moment *n*/**atomares magnetisches** atomic magnetic moment
Moment *n*/**bahnmagnetisches** orbital magnetic moment
Moment *n*/**kernmagnetisches** nuclear magnetic moment
Moment *n*/**magnetisches** magnetic (dipole) moment
MO-Methode *f s.* Molekülorbitalmethode
monAk *s.* Antikörper/monoklonaler
Monamid *n* monoamide
Monazit *m (min)* monazite *(cerium phosphate)*
Mondgestein *n* lunar rock
Mond-Niederdruckcarbonylverfahren *n (met)* Mond carbonyl process
Mond-Verfahren *n (tech)* Mond process *(purification of nickel)*
Monoalkylbenzen *n* monoalkylbenzene
monoalkylieren *v* monoalkylate
Monoalkylierung *f* monoalkylation
Monoamin *n* monoamine
Monoarsan *n* monoarsane, arsine
monoatomar mon(o)atomic
Monoazofarbstoff *m* monoazo dye
Monobasizität *f* monobasicity *(of acids)*
Monoboran *n* monoborane
Monoborsäure *f* orthoboric acid
Monobrombenzen *n* (mono)bromobenzene
Monobromid *n* monobromide
Monobrommethan *n* (mono)bromomethane
Monocarbonsäure *f* monocarboxylic acid

Monochloralkan *n* monochloroalkane
Monochlorbenzen *n* (mono)chlorobenzene
Monochlorderivat *n* monochloro derivative
Monochlorethan *n* (mono)chloroethane
Monochlorid *n* monochloride
monochlorieren *v* monochlorinate
Monochlorierung *f* monochlorination
Monochlormethan *n* (mono)chloromethane
Monochlorsilan *n* (mono)chlorosilane
Monochromasie *f* monochromatism
monochromatisch monochromatic
monochromatisieren *v* monochromatize *(light)*
Monochromator *m (spectr)* monochromator
monocyclisch monocyclic
Monoderivat *n* mono derivative
monodispers monodisperse
Monoester *m* monoester
Monoethanolamin *n* (mono)ethanolamine, MEA
Monoethylamin *n* ethylamine
Monofil(garn) *n* monofil, monofilament (yarn)
Monofluorid *n* monofluoride
Monogerman *n* (mono)germane, germanium tetrahydride
Monoglycerid *n* monoglyceride
Monohalogenid *n* monohalide, monohalogenide
monohalogenieren *v* monohalogenate
Monohalogenierung *f* monohalogenation
Monohydrat *n* monohydrate
Monohydrid *n* monohydride
Monohydrogenphosphat *n* monohydrogenphosphate
Monohydrogensalz *n* monohydrogen salt
monoisotop monoisotopic
monokl. *s.* monoklin
monoklin *(cryst)* monoclinic
monoklin-prismatisch monoclinic-prismatic
Monokultur *f (biot)* monoculture
Monolage *f s.* Monoschicht
monomer monomeric
Monomer(e) *n* monomer
monomolekular monomolecular, unimolecular
Monomolekularfilm *m s.* Monoschicht
Mononatriumglutamat *n* monosodium glutamate
mononitrieren *v* mononitrate
Mononitrierung *f* mononitration
Mononitroverbindung *f* mononitro compound
Mononucleotid *n* mononucleotide
Monoolefin *n* monoolefin, alkene, ethylenic hydrocarbon
Monophosphan *n* monophosphane, phosphane, phosphine
Monophosphat *n* monophosphate, orthophosphate
Monophosphorsäure *f* (mono)phosphoric acid, orthophosphoric acid
Monosaccharid *n* monosaccharide
Monosauerstoff *m* monooxygen, atomic oxygen
Monoschicht *f* monolayer, monomolecular [unimolecular] layer [film]
Monoschicht *f*/**selbstangeordnete** self-assembled monolayer
Monosilan *n* monosilane
Monospiro-Verbindung *f* monospiro compound [hydrocarbon]
Monostiban *n* monostibane, stibane, stibine
monosubstituieren *v* monosubstitute
Monosubstitution *f* monosubstitution
Monosulfan *n* monosulphane, hydrogen sulphide
Monosulfid *n* monosulphide
Monosulfonsäure *f* monosulphonic acid
Monoterpen *n (org ch)* monoterpene
Monoterpenalkaloid *n* monoterpenoid alkaloid
Monothioacetal *n* monothioacetal
Monothioketal *n* monothioketal
monotrop monotropic
Monotropie *f* monotropy
monovalent monovalent, univalent
monovariant monovariant, univariant
Monowasserstoff *m* monohydrogen, atomic hydrogen
Monoxid *n* monooxide, monoxide
Montage *f* setting-up, assembling, mounting *(of an apparatus)*
Montanwachs *n* montan wax
Monte-Carlo-Simulation *f* Monte-Carlo simulation *(mathematics)*
montieren *v* set up, assemble, mount *(an apparatus)*

Morphinalkaloid *n* morphine alkaloid
Morpholin *n* morpholine
Morphologie *f (cryst)* morphology
morphologisch *(cryst)* morphological
Mörser *m (lab)* mortar
Mörtel *m (build)* mortar
Mosaikblock *m (cryst)* mosaic bloc
Mosaikstruktur *f (cryst)* mosaic structure [texture]
Mosaiktextur *f (cryst)* mosaic structure [texture]
Moschus *m* musk
moschusartig musk-like
Mößbauer-Effekt *m (phys ch)* Mössbauer effect
Mößbauer-Spektroskopie *f* Mössbauer(-effect) spectroscopy
Mößbauer-Spektrum *n* Mössbauer spectrum
Most *m* must
Motor(en)benzin *n* motor gasoline [spirit]
Motorenkraftstoff *m* motor fuel
Motorenöl *n* motor oil
Motormethode *f* motor method, F2 method *(for octane rating)*
Motor-Octanzahl *f* motor octane number, MON
mottenecht mothproof • **mottenecht machen** mothproof
Motten(schutz)mittel *n* mothproofing agent
moussieren *v (food)* effervesce, sparkle
MPD *s.* Multiphotonendissoziation
MPI *s.* Multiphotonenionisation
MRH *(Melanotropin-Releasinghormon) s.* Melanoliberin
MRIH *(Melanotropin-release-inhibierendes Hormon) s.* Melanostatin
mRNA *s.* Messenger-Ribonucleinsäure
Ms *s.* Messing
M-Schale *f* M shell *(of an atom)*
MT *s.* Metallfaserstoff
mtDNA *s.* Mitochondrien-DNA
mtRNA *s.* RNA/mitochondriale
Mucinsäure *f* mucic acid
Mucopolysaccharid *n* mucopolysaccharide
Mucoproteid *n* mucoprotein
Mucoprotein *n* mucoprotein
Muffe *f* socket
Muffeldrehrohrofen *m* rotary muffle kiln *(as for calcining)*
Muffelofen *m (lab)* muffle furnace
Mühle *f* 1. (grinding) mill, grinder; 2. mill *(works)*
Mühle *f*/**autogen arbeitende** autogenous mill
Mühlstein *m* grindstone, millstone
Mulde *f* trough, pan
Muldenmischer *m* trough mixer
Müll *m* waste (material, product), solid waste, refuse, rubbish, *(Am)* garbage
Müll *m*/**häuslicher** domestic [household] refuse, domestic rubbish
Müll *m*/**kommunaler** municipal solid waste, municipal refuse
Müllaufbereitungsanlage *f* waste-treatment plant
Müllbeseitigung *f* waste [refuse] disposal
Müller-Rochow-Verfahren *n* Müller-Rochow process *(for producing chlorosilanes)*
Mullit *m (min)* mullite *(aluminium silicate)*
Müllkippe *f* refuse [waste] tip, *(Am)* garbage dump
Müllkompostierung *f* garbage composting
Müllverbrennung *f* refuse incineration
Müllverbrennungsanlage *f* incineration plant
Müllverbrennungsofen *m* (refuse, garbage) incinerator
Multienzymkomplex *m* multienzyme complex
Multifil(garn) *n* multifil, multifilament yarn
Multiflash-Verdampfung *f* multiflash [multistage flash] evaporation, MSF evaporation
Multilagen... multilayer
Multiphotonenabsorption *f* multiphoton absorption
Multiphotonendissoziation *f* multiphoton dissociation
Multiphotonenionisation *f* multiphoton ionization
Multiplett *n (spectr)* multiplet
Multiplettaufspaltung *f (spectr)* multiplet splitting
Multiplizität *f* multiplicity *(of an electronic state)*
Mundblaseglas *n* hand-blown glass
Mundstück *n* mouthpiece
Mündung *f* orifice

Muntz-Metall *n* Muntz metal *(60 % Cu, 40 % Zn, up to 0.8 % Pb added)*
Muraminsäure *f* muramic acid
Murein *n* murein *(a polysaccharide-peptide complex in bacterial cell membranes)*
Murexid *n* murexide *(ammonium salt of purpuric acid)*
Murexidprobe *f* murexide test *(for detecting uric acid)*
Muscarin *n* muscarine *(alkaloid from certain mushrooms)*
muschelig *(min)* conchoidal *(surface produced by fracture)*
Muscheltoxin *n* shellfish poison
Muskatnussöl *n*/**ätherisches** nutmeg [myristica] oil *(from Myristica fragrans Houtt.)*
Muskeladenylsäure *f* muscle adenylic acid
Muskovit *m (min)* muscovite, potassium [potash] mica *(a phyllosilicate)*
Muster *n* pattern; sample *(materials testing)*
Musterspektrum *n* library (reference) spectrum
mutagen *(bioch)* mutagenic
Mutagen *n (bioch)* mutagen, mutagenic agent
Mutagenese *f (bioch)* mutagenesis
Mutagenität *f (bioch)* mutagenicity
Mutarotation *f* mutarotation, multirotation *(changed optical rotation of sugars)* • **Mutarotation zeigen** mutarotate
Mutasynthese *f (biot)* mutasynthesis, mutational biosynthesis
Mutation *f*/**spontane** *(biol)* spontaneous mutation
mutationsauslösend *(bioch)* mutagenic
Mutterelement *n (nucl)* parent element
Muttergestein *n (geol)* parent [mother] rock *(as of sediments)*; matrix *(in which larger crystals are embedded)*
Mutterkornalkaloid *n* ergot alkaloid
Mutterlauge *f* mother liquor
Muttersubstanz *f* mother [parent] substance
MVA *s.* Müllverbrennungsanlage
MWG *s.* Massenwirkungsgesetz
Mycolsäure *f* mycolic acid
Myo(hämo)globin *n (bioch)* myoglobin
Myon *n (nucl)* muon, mu meson
Myonchemie *f* muon chemistry
Myoniumchemie *f* muonium chemistry
Myonspinrotation *f (nucl)* muon spin rotation, μSR
Myosin *n* myosin *(a protein of muscle fibres)*
Myristaldehyd *m* myristaldehyde, *tetradecanal
Myristat *n* myristate *(salt or ester of myristic acid)*
Myristinsäure *f* myristic acid, *tetradecanoic acid
Myristylalkohol *m* myristyl alcohol, 1-tetradecanol
Myrrhenharz *n* myrrh gum *(a gum resin from Commiphora specc.)*
myzelgebunden *(biot)* mycelium-bound
Myzelpellets *npl (biot)* mycelium pellets
Myzelsuspension *f (biot)* mycelial suspension
Myzeltrockenmasse *f* mycelial dry weight

N

N_0 *s.* Avogadro-Zahl
N_A *s.* Avogadro-Konstante
NAA *s.* Neutronenaktivierungsanalyse
Nachauflaufherbizid *n* postemergence herbicide
Nachbaratom *n* neighbouring [adjacent] atom
Nachbargruppeneffekt *m* neighbouring group effect
Nachbarschaftshäufigkeit *f (bioch)* frequency of neighbouring, nearest-neighbour (sequence) frequency *(of purine and pyrimidine bases)*
nachbarständig vicinal, adjacent, neighbouring
Nachbarstellung *f* vicinal [adjacent, neighbouring] position
nachbearbeiten *v (tech)* finish
Nachbearbeitung *f (tech)* finish(ing)
nachbehandeln *v* aftertreat, re-treat
Nachbehandlung *f* aftertreatment, re-treatment, secondary treatment, post-treatment
Nachbelichtung *f (phot)* postexposure
nachbilden *v*/**sich** recover *(as of isotopes)*
Nachbildung *f* recovery *(as of isotopes)*
nachchloren *v (hyd)* postchlorinate

nachchlorieren *v (org ch, pap, text)* postchlorinate, afterchlorinate
Nachchlorierung *f (org ch, pap, text)* postchlorination, afterchlorination
Nachchlorung *f (hyd)* postchlorination, final chlorination
Nachchromier(ungs)farbstoff *m* afterchrome dye, chrome-developed dye
nachdosieren *v* make up
Nachdosierung *f* make-up
nachdunkeln *v* darken
Nachfällung *f* postprecipitation, afterprecipitation, delayed precipitation
nachfärben *v (text)* redye, top
Nachfärbung *f (text)* redyeing, topping
Nachfaulbehälter *m (hyd)* secondary digestion tank
nachfüllen *v* replenish, fill up
Nachfüllung *f* replenishment, filling-up
Nachgärung *f* secondary fermentation
Nachglühen *n* afterglow
nachhärten *v (plast)* afterbake
Nachhärtung *f (plast)* afterbake, postcure
Nachklärbecken *n (hyd)* secondary clarifier [settling tank]
Nachklärung *f (hyd)* secondary settling
Nachklassierung *f* secondary screening
Nachkristallisation *f* aftercrystallization, secondary crystallization
nachlassen *v* die down, subside *(of a reaction)*
Nachlauf *m* 1. *(distil)* tailings, tail, foots, back end; 2. *(petrol)* heavy ends
Nachperiode *f* final period, postperiod *(in calorimetric measurements)*
Nachpolymerisation *f* postpolymerization
nachprüfen *v* recheck
Nachprüfung *f* recheck
Nachreinigung *f (hyd)* advanced (tertiary) waste treatment, third-stage treatment
Nachsäulenderivatisierung *f (chromat)* postcolumn derivatization
nachspülen *v* (re)rinse, rewash
nachverarbeiten *v* reprocess
Nachverarbeitung *f* reprocessing
Nachverbrennung *f* afterburning *(of pollutants in waste gas)*
Nachverbrennungsanlage *f* afterburner *(waste incineration)*
Nachweis *m* detection, identification; (confirmatory) test
nachweisbar detectable, identifiable
Nachweisbarkeit *f* detectability
Nachweisempfindlichkeit *f* (detection) sensitivity
nachweisen *v* detect, identify, reaffirm the presence of *(a substance)*
Nachweisgrenze *f* detection [identification] limit, minimum [lower] detection limit, MDL
Nachweisgrenze *f*/absolute absolute detection limit
Nachweismethode *f* detection method, method of detection
Nachweisreagens *n* analytical reagent, detection agent
Nachweisreaktion *f* detection [test] reaction
NAD *s.* Nicotinsäureamid-adenin-dinucleotid
Nadelholz *n* softwood, coniferous wood
nadelig needle-shaped, needle-like, acicular
Nadelkristall *m* needle, (crystal) whisker
Nadelventil *n* needle valve
Nadelwärmeaustauscher *m* bayonet-tube heat exchanger
NADP *s.* Nicotinsäureamid-adenin-dinucleotidphosphat
Nagellack *m (cosmet)* nail laquer
Nagellackentferner *m (cosmet)* nail laquer remover
Näherung *f* approximation
Näherungsformel *f* approximate formula
Näherungsverfahren *n* approximation method, approximate procedure
Näherungsverfahren *n*/empirisches trial-and-error procedure
Näherungswert *m* approximate value
Nahordnung *f (cryst)* short-range order
Nähragar *m(n)* nutrient agar
Nährbouillon *f* nutrient broth
Nährlösung *f (biot)* liquid nutrient medium, nutrient broth; nutrient solution *(for higher plants)* • **in Nährlösung kultiviert** solution-grown
Nährlösung *f*/verbrauchte spent medium [broth]
Nährmedium *n (biot)* nutrient medium, culture [growth] medium, substrate

Nährmedium *n*/**künstliches [synthetisches]** synthetic medium
Nährsalz *n* nutrient salt
Nährstoff *m* nutrient, nutritive, foodstuff
Nährstoffanreicherung *f* nutrient enrichment
Nährstoffarmut *f* nutrient deficiency
Nährstoffaufnahme *f* nutrient uptake
Nährstoffbedarf *m* nutrient demand, nutritional requirements
Nährstoffbilanz *f* nutrient balance
Nährstoffelement *n* nutrient element
Nährstoffentzug *m* nutrient withdrawal
Nährstofflinie *f* nutrient line *(in the periodic table)*
Nährstoffmangel *m* 1. nutrient deficiency; 2. *(agric)* starvation
nährstoffreich nutritious
Nährstoffverhältnis *n* nutrient ratio
Nährstoffzufuhr *f* nutrient supply
Nahrungseiweiß *n* food protein
Nahrungsfett *n* dietary [food] fat
Nahrungskette *f* food chain
Nahrungslipid *n* dietary lipid
Nahrungsmittel *n* food(stuff)
Nahrungsmittelchemie *f* food chemistry
Nahrungsmittelindustrie *f s.* Lebensmittelindustrie
Nahrungsmittelunverträglichkeit *f s.* Lebensmittelunverträglichkeit
Nahrungsmittelvergiftung *f s.* Lebensmittelvergiftung
Nährwert *m* nutritional quality [value], nutrient [food] value
Naht *f* seam; weld
Nalixidinsäure *f (bioch)* nalixidic acid
N-Alkylierung *f* N-alkylation, nitrogen alkylation
Name *m (nomencl)* name, term
Name *m*/**allgemeiner** generic [non-proprietary] name
Name *m*/**falscher** misnomer
Name *m*/**funktioneller** functional name
Name *m*/**geschützter** proprietary name
Name *m*/**halbsystematischer [halbtrivialer]** semisystematic [semitrivial] name
Name *m*/**kommerzieller** trade [commercial] name
Name *m*/**offizieller** official [approved] name
Name *m*/**radikofunktioneller** radicofunctional name
Name *m*/**rationeller [systematischer]** systematic name
Name *m*/**trivialer [unsystematischer]** trivial [unsystematic] name
Namenreaktion *f* name reaction
Nametkin-Umlagerung *f* Nametkin rearrangement *(of a carbenium ion by migration of a methyl group)*
Nanoabdrucklithographie *f* nanoimprint lithography
Nanoarchitektur *f* nanoarchitecture
Nanobereich *m* nano range
Nanobiotechnologie *f* nanobiotechnology
nanobiotechnologisch nanobiotechnological
Nanochemie *f* nanochemistry
nanochemisch nanochemical
Nanocluster *m* nanocluster
Nanodomäne *f* nanodomain
Nanoelektrode *f* nanoelectrode
Nanoelektronik *f* nanoelectronics
Nanofaser *f* nanofibre
Nanofilm *m* nanofilm
Nanofüllstoff *m* nanofiller
Nanogrammbereich *m (anal)* nanogram range *(10^{-9} to 10^{-6} g)*
Nanokanal *m* nanochannel
Nanokapsel *f* nanocapsule
Nanokolloid *n* nanocolloid
Nanokomposit *m s.* Nanoverbundmaterial
Nanokristall *m* nanocrystal
Nanokugel *f* nanosphere
Nanolithographie *f* nanolithography
Nanometertechnik *f s.* Nanotechnik
Nanopartikel *n s.* Nanoteilchen
Nanoperle *f* nanobead
Nanopore *f* nanopore
nanoporös nanoporous
Nanoröhrchen *n* nanotube
Nanoröhre *f s.* Nanoröhrchen
Nanoschicht *f* nanolayer
Nanosensor *m* nanosensor
Nanosonde *f* nanoscale probe
Nanospray *n (spectr)* nanospray *(electrospray technique)*
Nanospur *(anal)* nanotrace *(*10^{-7} to 10^{-10} ppm)*
Nanostab *m s.* Nanostäbchen
Nanostäbchen *n* nanorod
Nanostruktur *f* nanostructure
nanostrukturiert nanostructured

Nanotechnik *f* nanoengineering, nanoscale engineering
Nanotechnologie *f* nanotechnology
Nanoteilchen *n* nanoparticle
Nanoverbundmaterial *n (plast)* nanocomposite
Nanowissenschaften *fpl* nanosciences
Naphtha *n(f)* (heavy) naphtha *(boiling range 150 to 210 °C)*; crude oil
Naphthalen *n* naphthalene
Naphthalencarbonsäure *f* naphthalenecarboxylic acid, naphthoic acid
Naphthalen-1,8-dicarbonsäure *f* 1,8-naphthalene dicarboxylic acid, naphthalic acid
Naphthalen(indol)indigo *m (dye)* naphthindigo
Naphthalensulfonsäure *f* naphthalenesulphonic acid
Naphthalensulfonylchlorid *n* naphthalene sulphonyl chloride
Naphthalin *n s.* Naphthalen
Naphthalol *n* naphthalol
Naphthalsäure *f* naphthalic acid, 1,8-naphthalene dicarboxylic acid
Naphthen *n* naphthene, cycloalkane, cyclic alkane, cycloparaffin
Naphthenat *n* naphthenate
Naphthenbasisöl *n* naphthene-base petroleum
naphthenisch naphthenic
Naphthensäure *f* naphthenic acid
Naphthionsäure *f* naphthionic acid
Naphthochinon *n* naphthoquinone
Naphthoesäure *f* naphthoic acid, naphthalenecarboxylic acid
Naphth-1-ol *n* napht-1-ol, α-naphthol
Naphtholsulfonsäure *f* naphtholsulphonic acid
Naphthylamin *n* naphthylamine
Naphthylaminsulfonsäure *f* naphthylaminesulphonic acid
Naphthylgruppe *f* $C_{10}H_7-$ naphthyl group
Narkotikum *n* narcotic
narkotisch narcotic
Nasen-Rachen-Reizstoff *m (tox)* sternutator, nose irritant, sneeze gas
nass wet, moist
Nassablagerung *f* wet deposition *(of acids from the atmosphere)*
Nassabscheider *m* wet collector, gas washer, scrubber
Nassabscheidung *f* wet collection [scrubbing]
nassaufbereiten *v (min tech)* wet-clean
Nassaufbereitung *f (min tech)* wet cleaning
Nassdampf *m* wet steam
Nässe *f* wetness, moisture
Nasselektroabscheider *m* wet precipitator
Nassfestigkeit *f* wet strength
Nassgas *n* wet gas
Nassgut *n* wet product
Nassherd *m (min tech)* wet table
Nassklassieren *n* wet classification, water sizing
Nassklassierer *m* wet classifier
Nassmahlung *f* wet milling
Nassmetallurgie *f* hydrometallurgy, wet metallurgy
nassmetallurgisch hydrometallurgical
Nassoxidation *f* wet air oxidation *(as a natural process)*
nasspökeln *v (food)* brine, pickle
Nasspökelung *f (food)* brining, brine curing [salting], pickling
Nassprobe *f (met)* wet assay
nassreinigen *v* wet-clean
Nassreiniger *m* wet cleaner
Nassreinigung *f* wet cleaning
Nasssetzmaschine *f (min tech)* jig washer, wet jig
Nasssieben *n* wet screening
Nassspinnen *n* wet spinning
Nassstaubabscheider *m s.* Nassabscheider
Nassverbrennung *f* wet combustion, wet air oxidation *(of waste products)*; burning of wet fuels
Nassverfahren *n* wet process
Nasszyklon *m* liquid cyclone (separator), wet-cyclone classifier, hydraulic cyclone separator, hydroclone
naszierend nascent
nativ native
Natrit *m (min)* natrite, natron, soda *(sodium carbonate 10-water)*
Natrium *n* Na sodium
Natriumacetat *n* sodium acetate
Natriumalaun *m* sodium [soda] alum, aluminium sodium sulphate 12-water
Natriumalginat *n* sodium alginate
Natriumalkoholat *n* sodium alkoxide
Natriumaluminat *n* sodium aluminate

Natriumaluminiumsulfat *n* sodium-aluminium sulphate
Natriumamalgam *n* sodium amalgam
Natriumamid *n* sodium amide
Natrium-*p*-aminobenzoat *n* sodium *para*-aminobenzoate
Natriumammoniumhydrogenphosphat-4-Wasser *n* sodium-ammonium hydrogenphosphate, phosphorus salt
Natriumarsenat(III) *n* sodium arsenite
Natriumarsenit *n* sodium arsenite
Natriumascorbat *n* sodium ascorbate
Natriumaustauscher *m (hyd)* sodium exchanger, Na-form exchanger
Natriumazid *n* sodium azide
Natriumbenzoat *n* sodium benzoate
Natriumbisulfitlösung *f (pap)* sodium bisulphite liquor
Natriumbromat *n* sodium bromate
Natriumbromid *n* sodium bromide
Natriumcarbonat *n* sodium carbonate, soda
Natriumcarbonat ***n*/wasserfreies** 1. anhydrous sodium carbonate; 2. *(tech)* calcined soda, (soda) ash
Natriumcarboxymethylcellulose *f* sodium carboxymethylcellulose
Natriumcaseinat *n* sodium caseinate
Natriumcellulosexanthogenat *n* sodium cellulose xanthate
Natriumchlorat *n* sodium chlorate
Natriumchlorid *n* sodium chloride
Natriumchloridgitter *n* sodium chloride lattice
Natriumchloridlösung *f* **zum Regenerieren** *(hyd)* regenerant brine, brine regenerant *(ion exchange)*
Natriumchlorit *n* sodium chlorite
Natriumchloritbleiche *f (text)* sodium chlorite bleaching
Natriumchromat *n* sodium chromate
Natriumcitrat *n* sodium citrate
Natriumcyanat *n* sodium cyanate
Natriumcyanid *n* sodium cyanide
Natriumdampflampe *f* sodium-vapour (discharge) lamp
Natriumdiacetat *n* sodium diacetate *(commercially for sodium acetate containing additional acetic acid)*
Natriumdichromat *n* sodium dichromate
Natrium-*N,N*-diethyldithiocarbamat *n* sodium diethyldithiocarbamate
Natriumdihydrogenphosphat *n* sodium dihydrogen phosphate
Natriumdiphosphat *n* sodium diphosphate, sodium pyrophosphate
Natriumdithionit *n* sodium dithionite
Natrium-D-Linie *f (anal)* sodium D line
Natriumdodecamolybdatophosphat *n* sodium dodecamolybdophosphate, sodium 12-molybdophosphate
Natriumethylat *n* sodium ethoxide, sodium ethylate
Natriumfluorid *n* sodium fluoride
Natriumfolat *n* sodium folate
Natriumformiat *n* sodium formate
Natriumgluconat *n* sodium gluconate
Natriumglutamat *n* sodium glutamate
Natriumhexachloroplatinat(IV) *n* sodium hexachloroplatinate(IV)
Natriumhexacyanoferrat(III) *n* sodium hexacyanoferrate(III)
Natriumhexafluorosilicat *n* sodium hexafluorosilicate
Natriumhexanitrocobaltat(III) *n* sodium hexanitrocobaltate(III)
Natriumhydrid *n* sodium hydride
Natriumhydrogencarbonat *n* sodium hydrogencarbonate
Natriumhydrogenphosphat *n* sodium hydrogenphosphate
Natriumhydrogensulfat *n* sodium hydrogensulphate
Natriumhydroxid *n* sodium hydroxide
Natriumhypochlorit *n* sodium hypochlorite
Natriumiodat *n* sodium iodate
Natriumiodid *n* sodium iodide
Natriumlactat *n* sodium lactate
Natriumlaurylsulfat *n* sodium lauryl sulphate
Natriummetaborat *n* sodium metaborate
Natriummetaphosphatperle *f* sodium phosphate bead
Natriummetasilicat *n* sodium metasilicate
Natriummethylat *n* sodium methoxide, sodium methylate
Natriummolybdat *n* sodium molybdate
Natriumnitrat *n* sodium nitrate
Natriumnitrit *n* sodium nitrite
Natriumorthophosphat *n* sodium orthophosphate
Natriumoxalat *n* sodium oxalate

Natriumpalmitat *n* sodium palmitate
Natriumperchlorat *n* sodium perchlorate
Natriumperoxid *n* sodium peroxide
Natriumperoxoborat *n* sodium perborate
Natriumperoxodisulfat *n* sodium peroxodisulphate
Natriumphosphat *n* sodium phosphate, *(specif)* sodium orthophosphate
Natriumpolysulfid *n* sodium polysulphide
Natriumpropionat *n* sodium propionate
Natriumpyrophosphat *n* sodium pyrophosphate, sodium diphosphate
Natriumsalicylat *n* sodium salicylate
Natriumseife *f* sodium [soda] soap
Natriumselenat *n* sodium selenate
Natriumselenit *n* sodium selenite
Natriumsilicat *n* sodium silicate
Natriumsilikatglas *n* soda-silica glass
Natriumstearat *n* sodium stearate
Natriumsulfat *n* sodium sulphate
Natriumsulfid *n* sodium sulphide
Natriumsulfit *n* sodium sulphite
Natriumtartrat *n* sodium tartrate
Natriumtetraborat *n* sodium tetraborate
Natriumtetrahydridoborat *n* sodium tetrahydridoborate
Natriumtetrasulfid *n* sodium tetrasulphide
Natriumthiocyanat *n* sodium thiocyanate
Natriumthiosulfat *n* 1. sodium thiosulphate; 2. *(phot)* hyposulphite
Natriumuranylacetat *n* sodium uranyl acetate
Natron *n (min)* natron, soda *(sodium carbonate 10-water)*
Natronaufschluss *m (pap)* soda pulping
Natroncellulose *f* natron cellulose
Natronglas *n* soda glass
Natronkalk *m* soda lime *(a mixture of caustic soda with caustic lime)*
Natronkalkglas *n* soda-lime glass
Natronkochlauge *f (pap)* soda cooking liquor, soda liquor [lye]
Natronlauge *f* sodium hydroxide solution, caustic soda solution
Natronsalpeter *m* soda nitre, Chile saltpetre [nitre] *(sodium nitrate)*
Natronseife *f* soda [sodium] soap
Natronwasserglas *n* (soda) water glass, sodium silicate
Natronzellstoff *m* soda pulp
Naturasphalt *m* native [rock] asphalt
Naturbewitterungsversuch *m* outdoor weathering test
Naturfarbstoff *m* 1. natural colouring matter; 2. *(text)* natural dyestuff; 3. *(bioch)* biochrome
Naturfaser *f* natural fibre
Naturfaserstoff *m* natural fibre
Naturgas *n* natural gas
Naturgas *n***/feuchtes [nasses]** wet natural gas
Naturgas *n***/trockenes** dry natural gas
Naturgasbenzin *n* natural gasoline, casing-head gasoline [spirit]
Naturglas *n* natural glass
Naturgummi *n* natural gum
Naturharz *n* natural resin
Naturkautschuk *m* natural rubber, NR
Naturkohlenwasserstoff *m* natural hydrocarbon
Naturkonstante *f* physical constant
natürlich (vorkommend) naturally occurring, found in nature, from natural sources, native
Naturprodukt *n* natural product
Naturseide *f* natural silk
Naturstoff *m* 1. natural product (material); 2. *(bioch)* natural compound
Naturstoffchemie *f* chemistry of natural products
Naturstoffsynthese *f* natural product synthesis
NBS *s. N*-Brombernsteinsäureimid
n-Dotierung *f* n-doping
Nebel *m* fog, mist
Nebel *m***/saurer** acid fog
Nebelgerät *n (agric)* fog generator, nebulizer
Nebelkammer *f* Wilson (cloud) chamber, cloud chamber
Nebeln *n (agric)* fogging
Nebelspur *f (phys ch)* cloud [fog] track
Nebeltröpfchen *n* fog droplet
Nebenalkaloid *n* companion alkaloid [base]
Nebenbande *f (spectr)* side [subsidiary] band
Nebenbestandteil *m* minor constituent *(*1 to 0.01 %)*

Nebengruppe *f* subgroup, B group *(of the periodic table)*
Nebenkolonne *f* side-stream column
Nebennierenrindenhormon *n* adrenocortical [corticoid] hormone
Nebenprodukt *n* by-product, secondary product *(of a chemical process)*; side product *(of a chemical reaction)*
Nebenproduktenanlage *f* by-product (recovery) plant
Nebenproduktengewinnung *f* by-product recovery
Nebenquantenzahl *f* subsidiary quantum number, azimuthal [orbital] quantum number
Nebenreaktion *f* side [secondary] reaction
Nebenserie *f***/erste** *(spectr)* diffuse series
Nebenserie *f***/zweite** *(spectr)* sharp series
Nebenvalenz *f* secondary valence
Nebenvalenzbindung *f* secondary valence bond
Nebenwirkung *f (pharm)* side effect
Neel-Temperatur *f (cryst)* Neel temperature [point]
Negativ *n (phot)* negative
negativ geladen negatively charged, negative
negativ (geladen)/dreifach trinegative
negativ (geladen)/einfach uninegative
negativ (geladen)/fünffach pentanegative, quinque-negative
negativ (geladen)/vierfach tetranegative
negativ (geladen)/zweifach dinegative
Negativität *f* negativity
Negatron *n* negatron, electron
Neigung *f* 1. slope; 2. propensity, tendency *(as to chemical reaction)*
NE-Legierung *f* non-ferrous alloy
Nelkenöl *n* caryophyllus [clove] oil, oil of cloves *(from Syzygium aromaticum (L.) Merr. et L. M. Perry)*
Nematizid *n* nematocide, nematicide
nematizid nematocidal, nematicidal
NE-Metall *n* non-ferrous metal
N-endständig N-terminal
Neodym *n* Nd neodymium
Neodymlaser *m* neodymium laser
Neodymoxid *n* neodymium oxide
Neon *n* Ne neon
Neonröhre *f* neon discharge tube
Neoprenkautschuk *m* neoprene rubber
Nephelometer *n* nephelometer
Nephelometrie *f* nephelometry
nephelometrisch nephelometric
nephrotoxisch nephrotoxic *(poisonous to the kidneys)*
Neptunium *n* Np neptunium
Neral *n* neral, citral b
Nernst-Gleichung *f* Nernst equation
Nerol *n* nerol
Nervengas *n* nerve gas
Nervengift *n* nerve poison, neurotoxin
nervenschädigend neurotoxic
Nervonsäure *f* nervonic acid
Nesosilicat *n* nesosilicate *(a silicate containing independent SiO_4 tetrahedra)*
Nettogleichung *f* net equation
Nettoladung *f* net charge
Nettoreaktion *f* net reaction
netzbar wettable
Netzbarkeit *f* wettability
Netzebene *f (cryst)* lattice plane
Netzebenenabstand *m (cryst)* lattice distance, spacing of the planes
Netzelektrode *f* gauze electrode
netzen *v* wet, damp(en)
Netzer *m s.* Netzmittel
Netzkette *f (plast)* network chain *(polymer networks)*
Netzkontakt *m* gauze catalyst
Netzmittel *n* wetting agent [aid], spreading agent, humectant
Netzspannung *f* mains voltage
Netzwerk *n***/dreidimensionales** three-dimensional network
Netzwerk *n***/interpenetrierendes** interpenetrating network, IPN
Netzwerkbildner *m (glass)* network former, network-forming ion
Netzwerkwandler *m (glass)* network modifier, network-modifying ion
Neubestimmung *f* redetermination
Neubildungsdauer *f (bioch)* turnover time
Neuraminsäure *f* neuraminic acid *(an amino sugar)*
Neuroleptikum *n (pharm)* neuroleptic (drug), antipsychotic, major tranquilizer, CNS-depressant, central nervous system depressant
neuroleptisch *(pharm)* neuroleptic

Neurotoxin *n* neurotoxin, nerve poison
neurotoxisch neurotoxic
Neurotoxizität *f* neurotoxicity
neutral neutral • **neutral reagierend** neutral-reacting
Neutralaustausch *m (hyd)* sodium cycle exchange, Na-form exchange
Neutralaustauscher *m (hyd)* sodium exchanger, Na-form exchanger
Neutralbereich *m* neutral range
Neutralfett *n* neutral fat
Neutralisation *f* neutralization
Neutralisationsanalyse *f* neutralization titration
Neutralisationsanlage *f* neutralizing plant
Neutralisationsindikator *m* neutralization indicator
Neutralisationskurve *f* neutralization curve
Neutralisationsmittel *n* neutralizer, neutralizing agent
Neutralisationsreaktion *f* neutralization [acid-base] reaction
Neutralisationstitration *f* neutralization [acid-base] titration
Neutralisationszahl *f* neutralization value, acid value [number], AV
neutralisieren *v* neutralize
Neutralität *f* neutrality
Neutralöl *n* neutral oil
Neutralpunkt *m* point of neutrality • **um den Neutralpunkt** around neutrality
Neutralrot *n* neutral [toluylene] red *(an oxidation-reduction indicator)*
Neutralsalz *n* neutral [normal] salt
Neutralsalzfehler *m* (neutral-)salt error *(in pH determinations)*
Neutralsäure *f* molecular acid
Neutralsulfitverfahren *n (pap)* neutral sulphite (semichemical) process, NSSC process, neutral sodium sulphite process
Neutrino *n* neutrino
Neutron *n* neutron
Neutron *n***/kaltes** cold neutron
Neutron *n***/langsames** slow neutron
Neutron *n***/schnelles** fast neutron
Neutron *n***/thermisches** thermal neutron
Neutronenabsorber *m* neutron absorber
Neutronenaktivierungsanalyse *f* neutron-activation analysis, NAA
Neutronenbeschuss *m* neutron bombardment
Neutronenbestrahlung *f* neutron irradiation
Neutronenbeugung *f* neutron diffraction
Neutroneneinfangquerschnitt *m* neutron-capture cross section
Neutronenemission *f* neutron emission
Neutronenquelle *f* neutron source
Neutronenruhemasse *f* neutron (rest) mass
Neutronenstrahl *m* neutron ray, *(if bundled:)* neutron beam
Neutronenstreuung *f***/unelastische** *(anal)* inelastic neutron scattering, INS
n-Halbleiter *m* n-type semiconductor
NHI-Protein *n* non-haem iron protein
Niacin *n s.* Nicotinsäure
Nichtedelmetall *n* base [non-noble] metal
Nichteinheitlichkeit *f* non-uniformity
Nichteisenlegierung *f* non-ferrous alloy
Nichteisenmetall *n* non-ferrous metal
Nichteisenmetallurgie *f* non-ferrous metallurgy
Nichtelektrolyt *m* non-electrolyte
Nichtentflammbarkeit *f* non-flammability, uninflammability, flameproofness
Nichterz *n* gangue (mineral), waste rock, matrix
Nichtflüchtiges *n* non-volatile matter
Nichthämeisen *n (bioch)* non-haem iron
Nichthämeisenprotein *n* non-haem iron protein
Nichtkarbonathärte *f (hyd)* non-carbonate hardness, permanent hardness
Nichtleiter *m* non-conductor, dielectric, electrical insulator
Nichtlöser *m* non-solvent
Nichtlöslichkeit *f* insolubility
Nichtlösungsmittel *n* non-solvent
Nichtmetall *n* non-metal
Nichtproteinanteil *m* non-protein moiety [fraction] *(as of enzymes)*
Nichtproteinstickstoff *m* non-protein nitrogen
Nichtstöchiometrie *f* non-stoichiometry
Nichtzuckeranteil *m* non-sugar portion, aglycon(e)
Nichtzuckerstoff *m* non-sugar (substance)
Nickel *n* Ni nickel
Nickel-Cadmium-Akkumulator *m*

nickel-cadmium battery, cadmium-nickel storage cell
Nickel(II)-chlorid *n* nickel(II) chloride, nickel dichloride
Nickeldimethylglyoxim *n* nickel dimethylglyoxime
Nickel-Eisen-Akkumulator *m* nickel-iron battery, Edison accumulator
Nickelelektrolyt *m* nickel-plating bath
Nickelkatalysator *m* nickel catalyst
Nickellegierung *f* nickel alloy
Nickelnitrat *n* nickel nitrate
Nickel(II)-oxid *n* nickel(II) oxide, nickel monooxide
Nickelphthalocyanin *n* nickel phthalocyanine
Nickelschwammkatalysator *m* spongy-nickel catalyst
Nickelstahl *m* nickel (alloy) steel
Nickelstein *m* *(met)* nickel matte
Nickelsulfat *n* nickel sulphate
Nickeltetracarbonyl *n* nickel tetracarbonyl
Nickeltiegel *m* nickel crucible
Nickschwingung *f* *(spectr)* wagging vibration
Nicol-Prisma *n* Nicol prism
Nicotinat *n* nicotinate *(salt or ester of nicotinic acid)*
Nicotinsäure *f* nicotinic acid, niacin, 3-pyridine carboxylic acid
Nicotinsäureamid *n* nicotinamide
Nicotinsäureamid-adenin-dinucleotid *n* nicotinamide-adenine dinucleotide, NAD
Nicotinsäureamid-adenin-dinucleotidphosphat *n* *(bioch)* nicotinamide-adenine dinucleotide phosphate, NADP
Nicotinvergiftung *f* nicotine poisoning
Niederdruckdampf *m* low-pressure steam
Niederdruckkessel *m* low-pressure boiler
Niederdruckpolyethylen *n* low-pressure [high-density] polyethylene, HDPE
Niederdruckverfahren *n* **nach Ziegler** Ziegler process *(for polymerizing alkanes)*
niederfrequent low-frequency
niedermolekular low-molecular
Niederschachtofen *m* low-shaft furnace
Niederschlag *m* 1. precipitate, sediment, (bottom) settlings, deposit, foots; 2. *(ferm)* lees; condensate *(from vapour)* • **einen Niederschlag bilden** *s.* niederschlagen/sich
Niederschlag *m*/**atmosphärischer** atmospheric precipitation
Niederschlag *m*/**flockiger** flocculate
Niederschlag *m*/**käsiger** curdy precipitate
niederschlagbar condensable *(vapour)*
Niederschlagbarkeit *f* condensability *(of vapour)*
niederschlagen *v* precipitate, sedimentate, deposit; throw [lay] down *(e.g. a coating)*; condense *(vapour)*
niederschlagen *v*/**sich** precipitate, sediment, deposit, settle, subside; condense *(of vapour)*
Niederschlagselektrode *f* precipitating [collecting] electrode
niederschmelzen *v* melt down
Niederspannungselektrophorese *f* low-voltage electrophoresis
Niedertemperatur-Protonenaustauschmembran-Brennstoffzelle *f* proton exchange membrane fuel cell, PEMFC
niederwertig lower-valent, lower-valency *(chemical-bond theory)*
niedrigaktiv *(nucl)* low-level active
niedriggekohlt low-carbon *(e.g. steel)*
niedrigoctan(zahl)ig low-octane
niedrigschmelzend low-melting
niedrigsiedend low-boiling, light
Niedrigsieder *m* low boiler *(solvent)*
niedrigviskos low-viscosity
nierenschädigend nephrotoxic
Ninhydrinreaktion *f* ninhydrin reaction [test] *(for detecting proteins and amino acids)*
Niob *n* Nb niobium
Niob(V)-oxid *n* niobium(V) oxide, niobium pentaoxide
Niobsäure *f* niobic acid
Nioxim *n* nioxime
NIR *s.* Infrarotgebiet/nahes
Nitramid *n* nitramide
Nitramin *n* nitramine *(general formula $RHN{-}NO_2$ or $R_1R_2N{-}NO_2$)*
Nitranilin *n* nitroaniline
Nitranilinrot *n* paranitraniline red
Nitrat *n* nitrate

Nitratbakterien *npl* nitrate bacteria, nitrobacteria *(genus Nitrobacter)*
Nitratentfernung *f (hyd)* nitrate removal
Nitratstickstoff *m* nitrate nitrogen
Nitren *n* :NR nitrene
Nitrid *n* nitride
Nitrierapparat *m* nitrator
nitrieren *v* 1. nitrate; 2. *(met)* nitride
nitrierhärten *v* nitride *(steel)*
Nitrierhärten *n* nitride hardening, nitriding *(of steel)*
Nitriermittel *n* nitrating agent
Nitriersäure *f* nitrating acid, mixed acid *(a mixture of concentrated nitric and sulphuric acid)*
Nitrierung *f* 1. nitration; 2. *s.* Nitrierhärten
Nitrifikation *f (agric)* nitrification
Nitrifikationsbakterien *npl* nitrifying bacteria, nitrobacteria *(collectively for nitrite and nitrate bacteria)*
Nitril *n* RC≡N nitrile
Nitril-Chloroprenkautschuk *m* nitrile-chloroprene rubber, NCR
Nitrilgruppe *f* –C≡N nitrile group
Nitrilium-Ion *n* nitrilium ion
Nitrilkautschuk *m* (acrylo)nitrile-butadiene rubber, NBR
Nitrilotriessigsäure *f* nitrilotriacetic acid, NTA
Nitrit *n* nitrite
Nitritbakterien *npl* nitrite-forming bacteria, ammonia oxidizers
Nitritstickstoff *m* nitrite nitrogen
Nitroalkan *n* nitroalkane, nitroparaffin
Nitrobenzen *n* nitrobenzene
Nitrocobaltat *n* nitrocobaltate
Nitroethan *n* nitroethane
Nitrofarbstoff *m* nitro dye
Nitrogen *n* N nitrogen
Nitrogenbakterien *npl* nitrogen-fixing bacteria
Nitrogruppe *f* $-NO_2$ nitro group
Nitrolack *m* cellulose nitrate lacquer
Nitromethan *n* nitromethane
Nitron *n* nitrone, azomethine oxide
Nitronaphthalen *n* nitronaphthalene
Nitroniumion *n* NO_2^+ nitronium ion
Nitronsäure *f* R=NO(OH) nitronic acid, alkylideneazinic acid, *aci*-nitro compound
Nitroparaffin *n* nitroparaffin, nitroalkane
Nitroprussidnatrium *n* sodium nitroprusside
nitros nitrous *(containing nitrogen oxides)*
Nitrosamid *n* nitrosamide
Nitrosamin *n* nitrosamine
Nitrosat *n (org ch)* nitrosate
Nitrosegase *npl* nitrogen oxides
nitrosieren *v* nitrosate
Nitrosiermittel *n* nitrosating agent
Nitrosierung *f* nitrosation
Nitrosobenzen *n* nitrosobenzene
Nitrosofarbstoff *m* nitroso dye
Nitrosogruppe *f* –N=O nitroso group
Nitrosoverbindung *f* nitroso compound
Nitrosylion *n* NO^+ nitrosyl ion
Nitroverbindung *f* nitro compound
Nitrylion *n* NO_2^+ nitronium ion
Niveau *n (phys ch, tech)* level
Niveaugefäß *n (lab)* levelling bottle
Niveaukugel *f* (gas-)leveling bulb *(gas analysis)*
nivellieren *v* level *(titration)*
Nivellierung *f* 1. levelling *(titration)*; 2. *(phys ch)* levelling effect
NK *s.* Naturkautschuk
NKH *(hyd)* non-carbonate hardness, permanent hardness
nl *s.* löslich/nicht
n-leitend n-type (semi)conducting
NMR *(nuclear magnetic resonance) s.* Resonanz/kernmagnetische
NMR-Spektrometer *n* NMR spectrometer
NMR-Spektroskopie *f* NMR spectroscopy
NMR-Spektroskopie *f***/zweidimensionale** two-dimensional NMR spectroscopy
NMR-Spektrum *n* NMR spectrum
NNR-Hormon *n s.* Nebennierenrindenhormon
Nobelium *n* No nobelium
NOE *(nuclear Overhauser effect) s.* Kern-Overhauser-Effekt
Nomenklatur *f***/chemische** chemical nomenclature
Nomenklatur *f* **der organischen Chemie** organic-chemistry nomenclature
Nomenklaturregel *f* nomenclature rule
Nomenklatursystem *n* nomenclature system
Nonacosan *n (org ch)* nonacosane

Nonandisäure *f* *nonanedioic acid, azelaic acid
Nonansäure *f* *nonanoic acid, pelargonic acid
Non-1-in *n* 1-nonyne
nonvariant non-variant, invariant
Norbornen *n* norbornene
Normal *n* standard
Normalatmosphäre *f* standard atmosphere
Normaldruck *m* normal [standard] pressure
Normaldrucksynthese *f* normal-pressure synthesis
Normalelektrode *f* normal electrode
Normalelement *n* standard cell
normalglühen *v* normalize *(steel)*
Normalisierungsglühen *n* normalizing *(of steel)*
Normalität *f* normality *(of solutions)*
Normalitätsfaktor *m (anal)* normality [volumetric, titrimetric] factor
Normalkalomelelektrode *f* normal calomel electrode
Normallösung *f* standard [normal] solution, *(specif)* N solution, normal solution *(containing one gram equivalent per litre)*
Normalpotenzial *n* normal potential
Normalschwingung *f (phys ch)* fundamental vibration, normal (mode of) vibration
Normalspannung *f* 1. normal voltage; 2. normal stress *(mechanically)*
Normaltemperatur *f* normal temperature
Normalverteilung *f* normal distribution
Normalverteilung *f***/logarithmische** logarithmic normal distribution
Normalvolumen *n* normal [standard] volume
Normalwasserstoffelektrode *f* normal hydrogen electrode, NHE
Normalzustand *m* 1. normal [ground] state, atomic ground state; 2. *s.* Normbedingungen
Normbedingungen *fpl* normal [standard] conditions
Normschliff *m (lab)* standard ground glass joint
Normzustand *m***/physikalischer** normal [standard] temperature and pressure, NTP, STP *(0 °C and 101.325 kPa)*
Normzustand *m***/technischer** atmospheric temperature and pressure *(20 °C and 98.0665 kPa)*
Norrish-Typ-I-Reaktion *f* Norrish type I photoreaction *(alpha-cleavage of a carbonyl compound)*
Norrish-Typ-II-Reaktion *f* Norrish type II photoreaction *(abstraction of a gamma-hydrogen from a carbonyl compound)*
Notation *f (nomencl)* notation
Notationssystem *n (nomencl)* notation system
Notbrause *f* safety [emergency] shower, drench shower
Novolakharz *n* novolak (resin), two-stage resin
NOx *s.* Nitrosegase
N-Oxid *n* amine oxide
N-Schale *f* N-shell *(of an atom)*
NTE *s.* Nitrilotriessigsäure
N-terminal N-terminal
Nuance *f* shade, tint, hue
nuancieren *v* shade, tint
Nuancierung *f* 1. shading, tinting; 2. *s.* Nuance
Nucleinsäure *f (bioch)* nucleic acid
Nucleohiston *n (bioch)* nucleohistone
nucleophil nucleophilic
Nucleophil *n* nucleophile
Nucleophilie *f* nucleophilicity
Nucleoprotein *n* nucleoprotein
Nucleosid-Antibiotikum *n* nucleoside antibiotic
Nucleotid *n* nucleotide
Nucleotid-Exzisionsreparatur *f (biot)* nucleotide-excision repair
Nucleotidsequenz *f (bioch)* nucleotide sequence
nuklear nuclear
Nukleon *n* nucleon, nuclear particle [constituent]
Nukleonik *f* nucleonics
Nuklid *n* nuclide
Nuklid *n***/radioaktives** radioactive nuclide, radionuclide
Nuklid(en)masse *f* nuclidic [nuclide] mass
Nullfeldaufspaltung *f (spectr)* zero-field splitting
Nullinstrument *n* null-balance instrument

Nullladungspunkt *m (phys ch)* point of zero charge, PZC
Nullmethode *f* null method (of measurement), balance method, zero method
Nullpunkt *m* zero (point) *(of a scale)*; null point *(of an instrument)* • **auf den Nullpunkt einstellen** zero
Nullpunkt *m*/**absoluter** absolute zero
Nullpunkteinstellung *f* zero adjustment, zeroing
Nullpunkteinstellung *f*/**automatische** autozero
Nullpunktsenergie *f* zero-point energy
Nullpunktsentropie *f* zero-point entropy, entropy at absolute zero
Nullpunktsschwingung *f* zero-point vibration
Nulltoleranz *f (tox)* zero tolerance
nullwertig zero-valent
Nullwertigkeit *f* zero valence
nummerieren *v (nomencl)* number
Nusselt-Zahl *f* Nusselt number *(of heat transfer)*
Nutsche *f* nutsch(e), nutsch filter
nutschen *v* filter (off) by suction
Nutzarbeit *f (phys ch)* useful [net] work
Nutzarbeit *f* **(der Reaktion)/maximale** *(phys ch)* maximum useful work
nutzbar usable, available *(e.g. resource)*
nutzerfreundlich user-friendly
Nutzleistung *f* efficiency
Nutzungsdauer *f* service life, operating [working, useful] life

O

O-Alkylierung *f* O-alkylation, oxygen alkylation
Oberfläche *f* surface, *(quantitatively:)* surface area
Oberfläche *f*/**äußere** external surface
Oberfläche *f*/**innere** internal [inside, inner] surface
Oberfläche *f*/**spezifische** specific surface
oberflächenaktiv surface-active
Oberflächenaktivität *f* surface activity
Oberflächenanästhetikum *n (med)* surface anaesthetic
Oberflächenantigen *n* surface antigen
Oberflächenbedeckung *f* surface coverage
oberflächenbehandeln *v* surface
Oberflächenbehandlung *f* surface treatment [treating], surfacing
Oberflächenbelastung *f* loading rate per unit area *(sedimentation tanks, trickling filters: $m^3/m^2 \cdot h$)*
Oberflächenbelüftung *f (hyd)* surface aeration
Oberflächenbeschaffenheit *f* surface condition, *(of treated surfaces also)* finish • **die Oberflächenbeschaffenheit verbessern** finish
Oberflächenchemie *f* surface chemistry
Oberflächendenaturierung *f (bioch)* surface denaturation
Oberflächendiffusion *f* surface diffusion
Oberflächendiffusionskoeffizient *m* surface diffusivity
Oberflächeneffekt *m* surface effect
Oberflächeneigenschaften *fpl* surface properties
Oberflächenenergie *f* surface energy
Oberflächenerneuerungstheorie *f* surface-renewal theory *(of mass transfer)*
Oberflächenerscheinung *f* surface phenomenon
Oberflächenfermentation *f (biot)* surface fermentation
Oberflächenfilm *m* surface film
Oberflächenfilm *m*/**monomolekularer** monomolecular [unimolecular] surface film
Oberflächenfilter *n* surface [edge] filter
Oberflächenfiltration *f* surface [edge] filtration
Oberflächengüte *f* (surface) finish • **die Oberflächengüte verbessern** finish
Oberflächenhaftung *f (plast)* surface adhesion
oberflächenhärten *v (met)* surface-harden; cement *(by diffusion)*
Oberflächenhärten *n (met)* surface hardening; cementing *(by diffusion)*
Oberflächenhaut *f* surface film
Oberflächenhelligkeit *f* albedo
Oberflächenheterogenität *f* surface heterogeneity
oberflächeninaktiv surface-inactive
Oberflächenkatalysator *m* heterogeneous catalyst

Oberflächenkatalyse *f* heterogeneous catalysis, surface [contact] catalysis
Oberflächenkraft *f* surface force
Oberflächenkultur *f (biot)* surface culture
Oberflächenladung *f (phys ch)* surface charge
Oberflächenladungsdichte *f* surface charge density
Oberflächenleimung *f (pap)* surface [top] sizing
Oberflächenleitfähigkeit *f* surface conductivity
Oberflächenmodifizierung *f (plast)* surface modification
Oberflächenmusterung *f* surface patterning
Oberflächenoxidation *f* surface oxidation
Oberflächenplasmonresonanzspektroskopie *f (spectr)* surface plasmon resonance spectroscopy, SPR
Oberflächenpotenzial *n* surface potential
Oberflächenrauigkeit *f* surface roughness
Oberflächenreaktion *f* surface reaction
Oberflächenreaktor *m (biot)* surface film reactor
Oberflächenreibung *f (plast)* surface friction
Oberflächenriss *m* surface crack
Oberflächenschicht *f* surface layer
Oberflächenschutz *m* surface protection
Oberflächenschutzschicht *f* surface coating
Oberflächensieden *n* subcooled boiling
Oberflächenspannung *f* surface tension
Oberflächentemperatur *f* surface temperature
Oberflächentextur *f* surface texture
Oberflächenverfahren *n (biot)* surface(-culture) process
Oberflächenwiderstand *m* surface resistance
Oberflächenwiderstand *m***/spezifischer** surface resistivity
oberflächenwirksam surface-active
obergärig top-fermenting, top-fermented
Obergärung *f* top fermentation
Obersäule *f (distil)* enriching section
Oberschwingung *f (spectr)* overtone
Obertrum *m(n)* top strand *(of a conveyor)*
Objekt *n (phys ch)* entity, object
Objekt *n***/chemisches** chemical entity
Objektträger *m* (microscope) slide
Obstessig *m* fruit vinegar
Ocker *m (min)* ochre
Ocker *m***/Gelber** *(min)* yellow ochre *(a mixture of limonite with clay and silica used as a pigment)*
ockerhaltig ochreous
Octadeca-9,12-diensäure *f* *octadeca-9,12-dienoic acid, linoleic acid
Octadecan-1-ol *n* *octadecan-1-ol, stearyl alcohol
Octadecansäure *f* *octadecanoic acid, stearic acid
Octadeca-9,12,15-triensäure *f* *octadeca-9,12,15-trienoic acid, linolenic acid
Octadecensäure *f* octadecenoic acid
Octadec-9-insäure *f* *octadec-9-ynoic acid, stearolic acid
Octaeder *n (cryst)* octahedron
octaedrisch *(cryst)* octahedral
Octafluorid *n* octafluoride
Octahydrat *n* octahydrate
Octahydrid *n* octahydride
Octamer(e) *n* octamer
Octan *n* octane
Octan-1,8-dicarbonsäure *f* *octane-1,8-dicarboxylic acid, *decanedioic acid, sebacic acid
Octandisäure *f* *octanedioic acid, suberic acid
Octan-1-ol *n* *octan-1-ol, octyl alcohol
Octansäure *f* octanoic acid
Octanzahl *f* octane number • **mit hoher Octanzahl** high octane • **mit niedriger Octanzahl** low-octane
Octanzahlbestimmung *f* octane rating
octasubstituiert octasubstituted
octavalent octavalent
Octavalenz *f* octavalency
Oct-1-in *n* oct-1-yne
Octose *f* octose *(any of a class of monosaccharides)*
Octoxid *n* octaoxide
Odorans *n* odorant
odorieren *v* odorize *(toxic gases)*
Odoriermittel *n* odorant
Odorierung *f* odorization *(of toxic gases)*

Ofen *m* 1. *(tech)* furnace; 2. *(ceram)* kiln, *(esp for lower temperatures)* oven
Ofen *m* **für durchlaufenden [kontinuierlichen] Betrieb** continuous furnace
Ofen *m*/**induktionsbeheizter** induction furnace
Ofen *m*/**mehretagiger [mehrherdiger]** multihearth furnace
Ofen *m* **mit Widerstandserhitzung [Widerstandsheizung]** resistance(-heated) furnace
Ofen *m*/**rotierender** rotary furnace; *(for sintering or calcining:)* rotary kiln
Ofenatmosphäre *f* furnace atmosphere
Ofenauskleidung *f* furnace lining
Ofengas *n* furnace gas
ofentrocken kiln-dried, oven-dried
Ofentrocknung *f* kiln (oven) drying
Ofenzug *m* flue
Ofenzustellung *f* *(met)* furnace lining
offizinell official, pharmacopoeial
Öffnung *f* opening, *(esp of a nozzle:)* orifice, *(esp for intake or exhaust of fluids:)* port, *(esp for the passage of radiation:)* aperture
Öffnungsweite *f* **(des Siebs)** sieve size
Öffnungswinkel *m* aperture angle *(of a cone of radiation)*
OH-Austauscher *m* *(hyd)* OH-form [hydroxyl-form] exchanger
OH-Gruppe *f* OH group, hydroxyl group • **mit einer OH-Gruppe** monohydric *(alcohol, phenol)* • **mit zwei OH-Gruppen** dihydric *(alcohol, phenol)* • **mit fünf OH-Gruppen** pentahydric *(alcohol, phenol)*
OH-Radikal *n* OH radical, *(specif)* free OH radical
OHZ *s.* Hydroxylzahl
okkludieren *v* occlude
Okklusion *f* occlusion
Ökosystem *n* ecosystem
Ökotoxikologie *f* ecotoxicology, environmental toxicology
ökotoxikologisch ecotoxicological
Ökotoxizität *f* ecotoxicity, environmental toxicity
Ökotrophologie *f* ecotrophology *(nutrition science)*
Oktett *n* octet
Oktettaufweitung *f* octet expansion
Oktettregel *f* octet rule
Öl *n* oil
Öl *n*/**ätherisches** essential (volatile) oil
Öl *n*/**emulgierbares** emulsifiable [soluble] oil
Öl *n*/**fettes** 1. fatty [fixed] oil *(as opposed to essential oil)*; 2. *(petrol)* rich oil *(in absorption processes)*
Öl *n*/**halbtrocknendes** semidrying oil
Öl *n*/**leichtes** light oil
Öl *n*/**lösliches** soluble oil
Öl *n*/**medizinisches** medicinal oil
Öl *n*/**mineralisches** mineral oil
Öl *n*/**neutrales** neutral oil
Öl *n*/**nicht trocknendes** non-drying oil
Öl *n*/**nicht verharzendes** non-sludging oil
Öl *n*/**pflanzliches** vegetable oil
Öl *n*/**rohes** crude oil
Öl *n*/**schweres** heavy oil
Öl *n*/**trocknendes** drying oil
Ölabscheider *m* oil separator
Ölabscheidung *f* separation of oil, oil removal
Ölabschrecken *n* oil quenching
ölabweisend oil-repellent, oleophobic
ölartig oily, oleaginous
Ölbad *n* oil bath
ölbeständig oil-resistant, resistant to oil
Ölbeständigkeit *f* resistance to oil, oil resistance
Olbrücke *f* ol [olation] bridge *(between metal atoms and OH groups)*
öldicht oiltight *(joint)*; oilproof *(paper)*
Oleat *n* oleate *(salt or ester of oleic acid)*
Olefin *n* olefin, *alkene
Olefinierung *f* olefination
olefinisch olefinic
Olefinmetathese *f* *(org ch)* olefin metathesis
Olefinreihe *f* olefin series, alkene family
Olein *n* 1. olein, glycerol oleate; 2. *s.* Ölsäure/rohe
ölen *v* oil
Oleoresin *n* oleoresin *(a mixture of an essential oil and a resin)*
Oleum *n* oleum, fuming sulphuric acid
Ölfarbe *f* oil(-base) paint
Ölfeld *n* oil field
ölfest *s.* ölbeständig
Ölfirnis *m* boiled oil
Ölfrucht *f* oil plant
ölführend *(geol)* oil-bearing, petroliferous
Ölgas *n* oil gas

Ölgehalt *m* 1. oil content; 2. *(coat)* oil length *(related to resin)*
Ölgrün *n* 1. oil green *(chromium(III) oxide)*; 2. chrome green *(consisting of iron blue and chrome yellow)*
ölhaltig 1. oil-containing, oily, oleaginous; 2. *s.* ölführend
ölhärten *v* oil-harden *(steel)*
Ölhärtung *f* 1. oil hardening *(of steel)*; 2. *(food)* hydrogenation of oil
Ölharz *n s.* Oleoresin
Ölharzlack *m* oleoresinous varnish
Ölhavarie *f* oil spill
ölig oily, oleaginous, *(esp of crystals:)* unctuous
Öligkeit *f* oiliness, *(esp of crystals:)* unctuousness
oligofunktionell oligofunctional
oligomer oligomeric
Oligomer(e) *n* oligomer *(molecule having up to 10 repeating units)*
Oligomerisation *f* oligomerization
Oligonucleotid *n* oligonucleotide
Oligopeptid *n* oligopeptide *(containing up to 10 amino acids)*
Oligosaccharid *n* oligosaccharide
Öl-in-Wasser-Emulsion *f* oil(-in)-water emulsion, O/W emulsion
Olivin *m (min)* olivine *(a nesosilicate)*
Ölkracken *n* oil cracking
Ölkuchen *m (food)* oil cake
Öllack *m* oil varnish
öllöslich oil-soluble
Öl-Naturharz-Lack *m* oleoresinous varnish
Ölpflanze *f* oil plant
Ölquelle *f* oil well
Ölsaat *f* oil-bearing seed
Ölsäure *f* oleic acid, *cis*-octadec-9-enoic acid
Ölsäure *f***/rohe [technische]** olein, commercial oleic acid
Ölspaltung *f (petrol)* oil cracking
Ölvergasung *f* oil gasification
Ölwäsche *f* oil washing [scrubbing] *(of gases)*
OMA *s.* Mehrkanalanalysator/optischer
Ommochrom *n* ommochrome *(any of a class of eye pigments)*
Oniumverbindung *f* onium compound
onkogen oncogenic, oncogenous, tumorigenic
Önologie *f (food)* oenology
önologisch *(food)* oenological
Onsäure *f* aldonic acid *(any of a class of acids derived from aldoses)*
Onyx *m (min)* onyx *(silicon(IV) oxide)*
OPA *s.* o-Phthalaldehyd
opak opaque
Opal *m (min)* opal *(silicon(IV) oxide)*
Opaleszenz *f* opalescence
opaleszieren *v* opalesce
opaleszierend opalescent
Opazität *f* opacity
Operation *f***/grundlegende** basic [fundamental] operation
Operment *n* orpiment (yellow), yellow arsenic *(arsenic(III) sulphide)*
Operon *n (bioch)* operon *(a group of genes which act cooperatively)*
Opferanode *f* sacrificial anode
Opiumalkaloid *n* opium alkaloid
Oppenauer-Reaktion *f* Oppenauer oxidation [reaction] *(for dehydrogenating secondary alcohols)*
Optoakustikspektroskopie *f* photoacoustic spectroscopy, PAS, optoacoustic spectroscopy
optoelektrisch optoelectric
Optoelektronik *f* optoelectronics
Orbital *n* orbital
Orbital *n***/antibindendes** antibonding orbital
Orbital *n***/atomares** atomic orbital
Orbital *n***/nicht bindendes** non-bonding orbital
***p*-Orbital** *n p* orbital
***sp*-Orbital** *n sp* orbital
Orbitaldrehimpuls *m* orbital (angular) momentum
Orbitaldrehimpulsquantenzahl *f* azimuthal [subsidiary] quantum number, orbital (angular momentum) quantum number
Orbitalelektron *n* orbital electron
ORD *s.* Rotationsdispersion/optische
Ordnung *f* order
Ordnungsgrad *m* degree of order [orientation] *(as of polymers)*
Ordnungszahl *f* atomic [ordinal] number, A.N., *(symbol:)* Z
Ordnungszahl *f***/effektive** effective atomic number, E.A.N.
Ordnung-Unordnung-Umwandlung *f* order-disorder transition [transformation]

Organiker *m* organic chemist
organisch organic
organisch-chemisch organic-chemical
Organismen *mpl***/chemoautotrophe** chemoautotrophs
Organismen *mpl***/chemolithotrophe** chemolithotrophs
Organismen *mpl***/chemoorganotrophe** chemoorganotrophs
Organismen *mpl***/photolithotrophe** photolithotrophs
Organismen *mpl***/photoorganotrophe** photoorganotrophs
Organismen *mpl***/phototrophe** phototrophs
Organkultur *f (biot)* organ culture
Organoborverbindung *f* organoboron compound
Organocell-Verfahren *n (pap)* organocell process
Organochemikalie *f* organic chemical
Organochemiker *m* organic chemist
Organochlorsilan *n* organochlorosilane
Organoderivat *n* organoderivative
organofunktionell organofunctional, carbon-functional
Organogel *n* organic gel
Organohalogenverbindungen *fpl* halogenated organic compounds, HOC, organic halides, organohalides, haloorganics
Organoleptik *f* organoleptic [sensory] testing
Organomagnesiumhalogenid *n* organomagnesium halide
Organometallchemie *f* organometallic chemistry
Organometallverbindung *f* organometallic [metallo-organic] compound
Organophosphorverbindung *f* organophosphorus compound
Organopolysiloxan *n* organopolysiloxane, polyorganosiloxane
Organoquecksilberverbindung *f* organomercury compound
Organosiliciumverbindung *f* organosilicon compound
Organosiloxanpolymer(e) *n* organosiloxane polymer
Organozinnverbindung *f* organotin compound
orientieren *v* orient(ate), *(in linear direction also)* align
Orientierung *f* orientation, *(in linear direction also)* alignment
Orientierungsgrad *m* degree of order [orientation]
Orientierungspolarisation *f* orientation polarization
Ornithin *n* ornithine *(intermediate in urea cycle)*
Ornithinzyklus *m (bioch)* ornithine [urea] cycle, Krebs-Henseleit cycle
Orotsäure *f* orotic acid
Orsat-Apparat *m* Orsat (gas analysis) apparatus
Ort *m* site *(of reaction)*; position *(of an elementary particle)* • **an Ort und Stelle** in situ, on-site *(investigation, sampling)* • **vor Ort** in situ, on-site *(investigation, sampling)*
orthoanelliert *(org ch)* ortho-fused
Orthoborsäure *f* orthoboric acid, boric acid
Orthocarbonsäure *n* ortho acid
ortho-dirigierend ortho-directing
Orthokieselsäure *f* orthosilicic acid, silicic acid
Orthoklas *m (min)* orthoclase *(a feldspar)*
orthokondensiert ortho-fused
ortho-Orientierung *f* ortho orientation
ortho-peri-anelliert ortho-peri-fused
ortho-peri-kondensiert ortho-peri-fused
Orthophosphat *n* $M^I_3PO_4$ orthophosphate, phosphate
Orthophosphorsäure *f* orthophosphoric acid, phosphoric acid
Ortho-Positronium *n (nucl)* orthopositronium
orthorhombisch *(cryst)* orthorhombic
Orthosilicat *n* $M^I_4SiO_4$ orthosilicate, silicate
orthoständig ortho, in ortho position • **orthoständig sein** be (located, situated) ortho
ortho-substituiert ortho-substituted, *o*-substituted
ortho-Substitution *f* ortho substitution
Orthotellurat *n* orthotellurate
Orthotellursäure *f* orthotelluric acid, telluric acid
ortho-Umlagerung *f* 1,2-shift
Orthowasserstoff *m* ortho hydrogen
Orton-Umlagerung *f* Orton rearrangement [reaction] *(of N-halogenated ar-*

ylamines into 1,2-halogenated and 1,4-halogenated arylamines)
Osazon *n (org ch)* osazone
Osazonierung *f (org ch)* osazonation
Osmat(VI) *n* $M^{I}_{2}OSO_{4}$ osmate(VI)
Osmium *n* Os osmium
Osmium(VIII)-oxid *n* osmium(VIII) oxide, osmium tetraoxide
Osmiumsäure *f* osmic acid
Osmol *n* osmole *(1 osmol = 1 mol dissolved material per kg of solvent or per l of solution)*
Osmolalität *f* osmolality *(total molar concentration of dissolved material per kg of solvent)*
Osmolarität *f* osmolarity *(total molar concentration of dissolved material per l of solution)*
Osmometer *n* osmometer
Osmometrie *f* osmometry
Osmose *f* osmosis
Osmose f/umgekehrte reverse osmosis, RO
Oson *n (org ch)* osone
Östradiol *n* oestradiol *(a sex hormone)*
Östrogen *n* oestrogen *(any of a class of sex hormones)*
Ostwald-Verfahren *n* Ostwald process *(for obtaining nitric acid)*
Ostwald-Viskosimeter *n* Ostwald viscometer [viscosity pipette]
Oszillation *f* oscillation, vibration
Oszillator *m* oscillator
oszillatorisch oscillatory, vibrational
oszillieren *v* oscillate, vibrate
Oszillometrie *f* oscillometry
Oszillopolarographie *f* oscillographic polarography, oscillopolarography
oszillopolarographisch oscillo-polarographic
Ouabain *n* ouabain, g-strophanthin *(glycoside)*
Ovalbumin *n* egg albumin
Overhauser-Effekt *m (spectr)* Overhauser effect
ovizid ovicidal
Ovizid *n* ovicide *(pest control)*
Ovulationshemmer *m* ovulation inhibitor, antifertility agent
Ö/W *s.* Öl-in-Wasser-Emulsion
Oxalacetatweg *m (bioch)* oxaloacetate pathway
Oxalaldehyd *m* oxalaldehyde, glyoxal
Oxalat *n* oxalate *(salt or ester of oxalic acid)* • **mit Oxalaten behandeln** oxalate *(e.g. blood)*
Oxalbernsteinsäure *f* oxalosuccinic acid
Oxalessigsäure *f* oxalacetic acid
Oxaloacetat *n* oxaloacetate *(intermediate in citric-acid cycle)*
Oxalosuccinat *n* oxalosuccinate *(intermediate in citric-acid cycle)*
Oxalsäure *f* oxalic acid, *ethanedioic acid
Oxalsäurediethylester *m* oxalic acid diethyl ester, oxalic ester
Oxalsäurenitril *n* oxalonitrile
Oxalylharnstoff *m* oxalyl urea, parabanic acid
Oxamid *n* oxamide, oxalic acid diamide
Oxamidsäure *f* oxamic acid
Oxazinfarbstoff *m* oxazine dye
Oxazol *n* oxazole
Oxazolon *n* oxazolone, azlactone
Oxetan *n* oxetane, 1,3-epoxypropane
Oxid *n* oxide
Oxid n/basisches basic oxide
Oxid n/saures acidic oxide
Oxidans *n* oxidant, oxidizing agent
Oxidation *f* oxidation, de-electronation
Oxidation f/aerobe aerobic oxidation
Oxidation f/anodische 1. anodic oxidation; 2. *(met)* anodizing
Oxidation f/biologische biological oxidation, bio-oxidation
Oxidation f durch Luft(sauerstoff) air [atmospheric] oxidation, oxidation by atmospheric oxygen
Oxidation f/elektrochemische [elektrolytische] electrochemical [electrolytic] oxidation
Oxidation f/enzymatische enzymatic oxidation
Oxidation f/oberflächliche surface oxidation
Oxidation f/partielle partial oxidation
Oxidation f/photochemische photochemical oxidation, photooxidation
Oxidation f/schonende mild oxidation
Oxidation f/vorhergehende [vorherige] pre-oxidation
oxidationsanfällig prone to oxidation
Oxidationsanfälligkeit *f* proneness to oxidation

oxidationsbeständig resistant [stable] to oxidation, oxidation-resistant
Oxidationsbeständigkeit *f* resistance [stability] to oxidation, oxidation resistance
oxidationsempfindlich oxidation-sensitive, oxidation-susceptible, susceptible to oxidation
Oxidationsempfindlichkeit *f* sensitivity [susceptibility] to oxidation
Oxidationsenzym *n* oxidative enzyme; *(specif)* flavin [yellow] enzyme
Oxidationsflamme *f* oxidizing flame
oxidationshemmend antioxidant, oxidation-inhibiting
Oxidationsinhibitor *m* 1. oxidation inhibitor, antioxidant, antioxidizing agent; 2. *(petrol)* gum inhibitor
Oxidationsmittel *n* oxidizing agent, oxidant
Oxidationspotenzial *n* oxidation potential
Oxidationsprodukt *n* oxidation product
Oxidationsreaktion *f* oxidation reaction
Oxidationsstufe *f* oxidation state, *(quantitatively:)* oxidation [Stock] number
Oxidationsteich *m (hyd)* oxidation pond
Oxidationszahl *f* oxidation [Stock] number
Oxidationszone *f* zone of oxidation
Oxidchlorid *n* chloride oxide
Oxidfilm *m* oxide film
oxidierbar oxidizable, capable of oxidation
oxidierbar/leicht readily oxidizable
oxidieren *v* oxidize; undergo oxidation
oxidierend/nicht non-oxidizing
Oxidimetrie *f* oxidimetry
oxidimetrisch oxidimetric
oxidisch oxidic
Oxidkeramik *f* oxide ceramics, oxide-ceramic products
Oxidoreductase *f* oxido-reductase, redoxase
Oxidoreduktion *f* oxidoreduction, oxidation-reduction, redox reaction
Oxidsalz *n* oxide salt
Oxidschicht *f* oxide layer, *(if thin:)* oxide film
Oxidschutzschicht *f (met)* oxide coating
Oxidsperrschicht *f* oxide barrier
Oxim *n* oxime *(any of a class of compounds containing the group >C=NOH)*
Oxin *n* oxine, 8-hydroxyquinoline
Oxinat *n* oxinate *(any of the complex compounds of 8-hydroxyquinoline)*
Oxindolalkaloid *n* oxindole alkaloid
Oxiran *n* oxirane, epoxide, ethylene oxide
Oxo-Cyclo-Tautomerie *f* oxo-cyclo tautomerism
Oxoethansäure *f* oxoethanoic acid, glyoxylic acid
Oxoglutarsäure *f* oxoglutaric acid, *(specif)* 2-oxoglutaric acid
Oxoniumion *n* hydronium [oxonium] ion, H_3O^+ ion
Oxoniumsalz *n* oxonium salt
4-Oxopentansäure *f* *4-oxopentanoic acid, laevulinic acid
Oxosalz *n* oxo salt *(salt of an oxo acid)*
Oxosäure *f* oxo acid, oxyacid *(any of the acids having oxygen in the acidic group)*
Oxosynthese *f* oxo synthesis, hydroformylation
Oxosynthesereaktion *f* oxo [hydroformylation] reaction
Oxoverbindung *f* oxo compound
Oxy... *siehe auch: Hydroxy...*
oxybiontisch *(biol)* oxybiotic, aerobiotic *(living in the presence of air oxygen)*
Oxychlorierung *f* oxychlorination
Oxychromatin *n (bioch)* linin *(material in the cell nucleus)*
Oxygen *n* O oxygen
Oxygenase f/mischfunktionelle mixed-function oxygenase
Oxyhämocyanin *n* oxyhaemocyanin *(a copper-containing blood pigment)*
Oxyhämoglobin *n* oxyhaemoglobin
Oxyneurin *n* oxyneurine, betaine, *N,N,N*-trimethylammonioacetate, lycine
Oxytozikum *n (pharm)* oxytocic (agent), uterotonic
OZ *f s.* 1. Ordnungszahl; 2. Octanzahl
Ozokerit *m (min)* ozokerite, earth wax
Ozon *n* ozone, trioxygen • **in Ozon verwandeln** ozonize, convert into ozone • **sich in Ozon verwandeln** ozonize
ozonartig ozonic
Ozonbehandlung *f (hyd)* ozonization

ozonbeständig ozone-resistant, ozone-resisting
Ozonbeständigkeit *f* ozone resistance
Ozongenerator *m* ozone generator, ozonizer
ozonhaltig ozonic, ozoniferous
Ozonid *n* ozonide
Ozonierung *f* ozon(iz)ation
ozonisieren *v* 1. ozonize, ozonate *(to treat, impregnate, or combine with ozone)*; 2. ozonize, convert into ozone
Ozonisierung *f s.* Ozonierung
Ozonloch *n* ozone hole
Ozonolyse *f s.* Ozonspaltung
Ozonosphäre *f s.* Ozonschicht
Ozonschicht *f* ozone layer, ozonosphere
Ozonschutzmittel *n* antiozonant, antiozidant, sunproofing agent
Ozonspaltung *f* ozonolysis, cleavage by ozone
ozonzerstörend ozone-destroying
Ozonzerstörung *f* ozone destruction

P

P *s.* Parachor
p *s.* Proton
¯p *s.* Antiproton
p. a. *(pro analysi) s.* analysenrein
PA *s.* 1. Polyamid; 2. Polyamidfaserstoff
Paarbildung *f* pairing *(of electrons, nucleons)*
paaren *v*/**sich** pair *(as of electrons, nucleons)*
Paarvernichtung *f (phys ch)* annihilation
packen *v* pack(age)
Packung *f* 1. packing *(of a column)*; *(specif)* stacked packing; chromatographic packing; 2. *(cryst)* packing; 3. seal *(as in a valve)*; 4. pack(age) *(a packed quantity)*
Packung *f*/**dichteste** *(cryst)* closest [close] packing
Packung *f*/**hexagonal dichteste** *(cryst)* hexagonal closest [close] packing
Packungsdichte *f* packing density
Packungsmaterial *n* 1. packing *(for a column)*; chromatographic packing; 2. packing material *(as for valves)*
Paddelrührer *m* paddle agitator [mixer], blade mixer
PAK *s.* Kohlenwasserstoff/polycyclischer aromatischer
palisadenartig *(cryst)* columnar
Palladat *n* palladate *(any of a class of complex salts having Pd as a central atom)*
Palladium *n* Pd palladium
Palladiumhydrid *n* palladium hydride
Palladium(II)-iodid *n* palladium(II) iodide, palladium diiodide
Palladiummohr *n* palladium black
Palladium(II)-oxid *n* palladium(II) oxide, palladium monooxide
Palladium(IV)-oxid *n* palladium(IV) oxide, palladium dioxide
Palmenstärke *f* palm starch, sago *(from Metroxylon specc.)*
Palmitat *n* palmitate *(salt or ester of palmitic acid)*
Palmitinsäure *f* palmitic acid, *hexadecanoic acid
Palmkernfett *n* palm kernel (nut) oil *(refined)*
Palmkernöl *n* palm kernel oil *(from Elaeis guineensis Jacq.)*
PAN *s.* Polyacrylnitril
Pantoinsäure *f (bioch)* pantoic acid
Pantothensäure *f* pantothenic acid
Papier *n* paper • **auf [zu] Papier verarbeiten** make into paper
Papier *n*/**chromatographisches** chromatographic paper
Papier *n*/**doppelt logarithmisch geteiltes** loglog paper
Papier *n*/**halblogarithmisches** semilog(arithmic) paper
Papierabfälle *mpl* waste paper
Papierabzug *m (phot)* paper print
Papierausrüstung *f* paper finishing
Papierbahn *f* (paper) web, mat
Papierbild *n (phot)* paper print
Papierbrei *m* paper(making) stock
Papierchromatogramm *n* paper chromatogram
Papierchromatographie *f* paper chromatography
papierchromatographisch paper-chromatographic
Papierelektrophorese *f* paper electrophoresis
Papierfabrikabwasser *n* paper mill waste (water)
Papierfärbung *f* paper dyeing

Papierfaserstoff *m* (paper, raw) stock
Papierfilter *n* paper filter
Papierfüllstoff *m* paper filler
Papierherstellung *f* papermaking
Papierholz *n* pulpwood
Papierkalander *m* calender (machine)
Papierleimung *f* paper sizing
Papiermaschine *f* paper(making) machine
Papiermaschinensieb *n* paper machine wire
Papiermasse *f* paper(making) stock
Papierrohstoff *m* (paper, raw) stock
Papierstoff *m* paper(making) stock, stuff
Papierstoff *m***/fertiger** paper(making) stock, finished stuff
Papierverarbeitung *f (pap)* paper converting
Papierzellstoff *m* paper(making) pulp
Pappe *f* (paper)board
Para *m s.* Parakautschuk
Parabansäure *f* parabanic acid
Paracetaldehyd *m* paraldehyde, paraacetaldehyde
Parachor *m(n) (phys ch)* parachor
Paraffin *n* paraffin (wax)
Paraffin *n***/festes** solid paraffin
Paraffin *n***/flüssiges** *(pharm)* liquid paraffin, paraffin oil
Paraffin *n***/mikrokristallines** microparaffin, micro wax
Paraffinbasisöl *n* paraffin-base petroleum
Paraffingatsch *m* (paraffin) slack wax
paraffinisch paraffinic
Paraffinkohlenwasserstoff *m* paraffin hydrocarbon, saturated hydrocarbon, *alkane
Paraffinöl *n* 1. paraffin oil, liquid paraffin; 2. paraffin-base petroleum
Paraffinreihe *f* alkane family, paraffin series
Paraffinwachs *n* paraffin wax
Paraformaldehyd *m* paraformaldehyde, polyoxymethylene
Parafuchsin *n* parafuchsine
Parakautschuk *m* Para (rubber) *(from Hevea brasiliensis (H. B. K.) Muell Arg.)*
Parakristall *m* paracrystal
parakristallin paracrystalline
Paraldehyd *m* paraldehyde, paraacetaldehyde
Paraleukorosanilin *n* leucopararosaniline
Parallaxenfehler *m (anal)* parallax error
Parallelplattenviskosimeter *n* parallel-disk viscosimeter
Parallelreaktion *f* parallel reaction, competitive [concurrent] reaction
Parallelstrom *m* parallel flow, concurrent [co-current] flow • **im Parallelstrom (geführt)** co-current
Parallelsynthese *f* parallel synthesis
Parallelverschiebung *f (cryst)* translation
Paramagnetikum *n* paramagnetic (material)
paramagnetisch paramagnetic
Paramagnetismus *m* paramagnetism
Parameter *m* 1. parameter; 2. *(cryst)* intercept
Paranitranilinrot *n* paranitraniline red
Para-Positronium *n* parapositronium
Pararosanilin *n* pararosaniline
Pararosolsäure *f* pararosolic acid
paraständig para, in para position • **paraständig sein** be (located, situated) para
para-Stellung *f* para position • **in para-Stellung stehen** be (located, situated) para
parasubstituiert para-substituted, *p*-substituted
para-Substitution *f* para-substitution
Parasympath(ik)omimetikum *n (pharm)* parasympathomimetic agent
parasympath(ik)omimetisch parasympathomimetic
Parathion *n* parathion, diethyl-*p*-nitrophenylphosphorothionate *(insecticide)*
Parathormon *n* parathyroid hormone, parathyrin
Parawasserstoff *m* para hydrogen
Parr-Bombe *f (lab)* Parr bomb
Partialdruck *m* partial pressure
Partialdruckgesetz *n***/daltonsches** Dalton's law of partial pressures
Partialladung *f* partial [fractional] charge
Partialoxidation *f* partial oxidation
Partie *f* batch
Partikel *npl***/suspendierte** suspended particles, particulates
Partikelemission *f* particle emission
Partikelgröße *f* particle size

Partikelgrößenverteilung *f* particle-size distribution
PAS *s.* Photoakustikspektroskopie
Paschen-Runge-Aufstellung *f (spectr)* Paschen-Runge mounting
Paschen-Serie *f (spectr)* Paschen series
Passage *f* pass
passieren *v* pass
passiv passive
Passivator *m* passivator *(for metals)*; inhibitor, negative catalyst, retarder, anticatalyst *(of a chemical reaction)*
passivieren *v (phys ch)* passivate *(metals)*; inhibit, retard *(a chemical reaction)*
Passivierung *f* passivation *(of metals)*; inhibition, retardation *(of a chemical reaction)*
Passivität *f*/**chemische** chemical passivity
Passstück *n* adapter
Paste *f* paste
pastenartig pasty, paste-like
Pasteurisierapparat *m* pasteurizer
pasteurisieren *v* pasteurize
Pasteurisierung *f* pasteurization
Pastille *f (pharm)* pastille
pastös pasty, paste-like
Paterno-Büchi-Reaktion *f* Paterno-Büchi reaction *(photocycloaddition yielding an oxetane)*
pathogen pathogenic
pathogen/nicht non-pathogenic
Patina *f* patina
Patrize *f (plast)* patrix, male mould
Pauli Prinzip *n* Pauli (exclusion) principle
Pauschalanalyse *f* proximate analysis *(determination of main constituents only)*
PC *s.* Polycarbonat
PCB *s.* Biphenyle/polychlorierte
PCDD *s.* Dibenzo-1,4-dioxin/polychloriertes
PCDF *s.* Dibenzofuran/polychloriertes
PCP *s.* Pentachlorphenol
PCTFE *s.* Polychlortrifluorethylen
PCV *s.* Polyvinylcarbazol
p-Dotierung *f* p-doping
PE *s.* 1. Polyethylen; 2. Polyesterfaserstoff
Peakaufweitung *f* peak broadening
Peakbreite *f* **in halber Höhe** peak width at half height
Peakelutionsvolumen *n (chromat)* peak elution volume
Peakfläche *f* peak area
Peakform *f* peak shape
Peakhöhe *f* peak height
Peakintegration *f* peak (area) integration
Peakintensität *f* peak intensity
Peaklage *f* peak position
Peaksignal *n* peak
Peaküberlagerung *f* superposition of peaks
Pech *n* pitch
pechartig pitchy
Pechblende *f (min)* pitchblende *(uranium(IV) oxide)*
Pechkohle *f* pitch coal
Pechkoks *m* pitch coke
Pedologie *f* pedology, soil science
pedologisch pedologic(al)
PEG *s.* Polyethylenglykol
Pegel *m* level
Peierls-Phasenübergang *m* Peierls transition *(a metal-insulator transition)*
Pektat *n* pectate *(salt or ester of pectic acid)*
Pektin *n* pectin
pektinabbauend pect(in)olytic
Pektinat *n* pectinate *(salt or ester of pectinic acid)*
Pektinsäure *f* pectic acid *(a high-molecular-weight polymer of D-galacturonic acid)*
Pektinstoff *m* pectic substance, pectin
Pektisation *f* pectization, flocculation
pektisieren *v* pectize, flocculate
pektolytisch pect(in)olytic
Pelargonaldehyd *m* pelargonic aldehyde, *nonanal
Pelargonsäure *f* pelargonic acid, *nonanoic acid
Pellet *n* pellet
pelletisieren *v* pelletize, nodulize
Pelletisiermaschine *f* pelletizing machine, pelletizer
Pelletisiertrommel *f* pelletizing [balling] drum
Penam *n* penam *(any of a group of antibiotics)*
Pendelbecherwerk *n* pivoted-bucket conveyor
Penem *n* penem *(any of a group of antibiotics)*
Penetration *f* penetration

Penetrationszwillinge *mpl (cryst)* penetration twins
penetrieren *v* penetrate
Penicillamin *n* penicillamine
Penicillansäure *f* penicillanic acid *(a building brick of penicillins)*
Penicillin *n* penicillin *(any of a group of antibiotics)*
Penicillinsäure *f* penicillic acid
Penicilloat *n* penicilloate
Penicillosäure *f* penicilloic acid *(any of a group of amidodicarboxylic acids $RCONH(C_6H_{10}NS)(COOH)_2$)*
Penicillsäure *f* penicillic acid
Penillosäure *f* penilloic acid *(any of a group of acids produced by Penicillum specc.)*
Penillsäure *f* penillic acid *(any of a group of inactivation products of penicillins)*
Penning-Vakuummeter *n* Penning [Philips] gauge, cold-cathode ionization gauge
Pensky-Martens-Gerät *n* Pensky-Martens (flash-point) apparatus
Pentacen *n* pentacene
Pentachlorid *n* pentachloride
Pentachlorphenol *n* pentachlorophenol, PCP
Pentadecan *n* pentadecane
Pentadecanol *n* pentadecanol, *(specif)* pentadecan-1-ol
Pentadecansäure *f* pentadecanoic acid
Pentaerythritol *n* pentaerythritol
Pentahydrat *n* pentahydrate
pentakoordiniert pentacoordinate(d), five-coordinate(d)
Pentamethylendiamin *n* pentamethylenediamine
Pentammin *n* pentammine
Pentan *n* pentane
Pentanal *n* pentanal
Pentancarbonsäure *f* pentanecarboxylic acid
Pentandial *n* *pentanedial, glutaraldehyde
Pentandicarbonsäure *f* pentanedicarboxylic acid
Pentandiol *n* pentanediol
Pentandion *n* pentanedione
Pentandisäure *f* pentanedioic acid, glutaric acid
Pentanol *n* pentanol
Pentanon *n* pentanone
Pentansäure *f* pentanoic acid
Pentapeptid *n* pentapeptide
pentasubstituiert pentasubstituted
Pentasulfid *n* pentasulphide
pentavalent pentavalent, quinquevalent
Pentavalenz *f* pentavalency, quinquevalency
Penten *n* pentene
Pentin *n* pentyne
Pentitol *n* pentite, pentitol
Pentosan *n* pentosan *(any of a class of polysaccharides)*
Pentose *f* pentose, five-carbon sugar
Pentosephosphatweg *m (bioch)* pentose phosphate pathway [cycle]
Pentosephosphatzyklus *m***/oxidativer** s. Pentosephosphatweg
Pentosephosphatzyklus *m***/reduktiver** *(bioch)* C_3 pentose phosphate cycle, ribulose diphosphate cycle, Calvin(-Bassham) cycle *(of CO_2 assimilation)*
Pentoxid *n* pentoxide
Peptid *n* peptide
Peptid *n***/cyclisches** cyclic peptide
Peptidalkaloid *n* peptide alkaloid
Peptidanteil *m* peptide moiety [portion]
Peptid-Antibiotikum *n* peptide antibiotic
Peptidbindung *f* peptide bond
Peptidbindungsort *m (bioch)* peptidyl-tRNA site, P site, donor site
peptisieren *v* 1. peptize; 2. *(rubber)* plasticize; 3. *(coll)* deflocculate
Peptisiermittel *n* 1. peptizer; 2. *(rubber)* (chemical) plasticizer; 3. *(coll)* deflocculant
Peptisierung *f* 1. peptization; 2. *(rubber)* plasticization; 3. *(coll)* deflocculation
Perameisensäure *f* performic acid, peroxyformic acid
Perbenzoesäure *f* perbenzoic acid, peroxybenzoic acid
Perchlorat *n* perchlorate
Perchlormethan *n* perchloromethane, *tetrachloromethane
Perchlorsäure *f* perchloric acid
Peressigsäure *f* peracetic acid, peroxyacetic acid
Perfluorethen *n* perfluoroethylene, tetrafluoroethylene, T.F.E.
perfluorieren *v* perfluorinate
Perfluorierung *f* perfluorination

Perforation *f* 1. *(tech)* perforation; 2. *(lab)* continuos extraction of a liquid
perforieren *v* 1. *(tech)* perforate; 2. *(lab)* extract continuously *(a liquid by another one)*
perhydrieren *v* perhydrogenate
Perhydrierung *f* perhydrogenation
perianelliert peri-fused
perikondensiert peri-fused
Periodat *n* periodate
Periode *f***/kurze** short period *(of the periodic table)*
Periode *f***/lange** long period *(of the periodic table)*
Periodensystem *n* **(der Elemente)** periodic system (of the elements), periodic table
Periodensystem *n***/mendelejewsches** Mendeléeff periodic table
periodisch batch(wise), intermittent *(operation)* • **periodisch arbeitend** batch *(plant)*
Periodizität *f* periodicity
Periodsäure *f* periodic acid
peri-Stellung *f* peri position
Peritektikum *n (phys ch)* peritectic [transition-type] system
Perkin-Kondensation *f* Perkin condensation [reaction]
Perkolat *n* percolate, leachate
Perkolation *f* 1. percolation; 2. *(min tech)* percolation leaching
Perkolator *m* percolator
perkolieren *v* percolate
Perle *f* bead
perlen *v* bubble, *(of beverages also)* sparkle
perlenförmig beaded
Perlenprobe *f (anal)* bead test
Perlit *m (met)* pearlite *(a microconstituent of steel)*
perlitisch *(met)* pearlitic
Perlkontakt *m* bead catalyst
Perlmutter *f* nacre, mother of pearl
Perlpolymerisat *n* bead polymerizate
Perlpolymerisation *f* bead [pearl] polymerization, suspension polymerization
Permanenthärte *f (hyd)* permanent hardness
Permanentmagnet *m* permanent magnet
Permanentweiß *n* permanent white, blanc fixe *(precipitated barium sulphate)*
Permanganat *n* permanganate
Permanganatprobe *f* permanganate test *(for detecting alkenes)*
Permanganatzahl *f* permanganate number
Permangansäure *f* permanganic acid
permeabel permeable
Permeabilität *f* permeability
Permeat *n* permeate
Permeation *f* permeation
permeieren *v* permeate
Permeiervermögen *n* permeativity
Permittivität *f***/relative** *s.* Permittivitätszahl
Permittivitätszahl *f* relative permittivity, specific inductive capacity, SIC
Peroxid *n* peroxide
Peroxidbleiche *f* peroxide bleaching
peroxidieren *v* peroxidize
peroxidisch peroxidic
Peroxidranzigkeit *f* peroxide [oxidative] rancidity, oiliness; tallowiness *(of fats)*
Peroxidzahl *f* peroxide number [value]
Peroxoborat *n* peroxoborate
Peroxoborsäure *f* peroxoboric acid
Peroxocarbonat *n* peroxocarbonate
Peroxochromat *n* peroxochromate
Peroxochromsäure *f* peroxochromic acid
Peroxodischwefelsäure *f* peroxodisulphuric acid
Peroxodisulfat *n* peroxodisulphate
Peroxokohlensäure *f* peroxocarbonic acid
Peroxophosphat *n* peroxophosphate
Peroxophosphorsäure *f* peroxophosphoric acid
Peroxosäure *f* peroxo acid
Peroxoschwefelsäure *f* peroxosulphuric acid
Peroxosulfat *n* peroxosulphate
Peroxycarbonsäure *f* peroxycarboxylic acid
Peroxysäure *f* peroxy acid
Persalz *n* per-salt
Persäure *f* 1. per-acid *(inorganic)*; 2. *(org ch)* peroxy acid
persistent persistent • **persistent sein** persist
Persistenz *f* persistence
Persubstitution *f* per substitution

Perubalsam *m* Peru balsam *(from Myroxylon balsamum (L.) Harms var. pereirae)*
Perylen *n* perylene
PES *s.* Photoelektronenspektroskopie
Pestizid *n* pesticide, biocide
Pestizidrückstand *m* pesticide residue
PETP *s.* Polyethylenterephthalat
Petrischale *f* Petri (culture) dish
Petrochemie *f* 1. petrochemistry, chemistry of rocks; 2. *s.* Petrolchemie
petrochemisch 1. petrochemical; 2. *s.* petrolchemisch
Petrographie *f* petrography
petrographisch petrographic(al)
Petroläther *m* petroleum ether, ligroin(e), light petroleum, benzine *(with a boiling-point range of 40 to 70 °C)*
Petrolatum *n* petrolatum, petroleum jelly
Petrolchemie *f* petrochemistry
Petrolchemikalie *f* petrochemical
petrolchemisch petrochemical
Petroleum *n* kerosine, paraffin oil
Petroleumasphalt *m* petroleum asphalt
Petroleumbenzin *n* petroleum benzin
Petrol(eum)fraktion *f* kerosine fraction
Petrologie *f* petrology
petrologisch petrologic(al)
Pfanne *f* 1. pan; 2. *(met)* ladle; tray; tank
Pfannenkristallisator *m* tank crystallizer
PFIMS *s.* Pyrolyse-Feldionisations-Massenspektrometrie
Pflanze *f***/gerbstoffhaltige** tanniferous plant
Pflanze *f***/kautschukliefernde** rubber-yielding plant
Pflanzenbiochemie *f* plant biochemistry
Pflanzenchemie *f* phytochemistry, plant chemistry
pflanzenchemisch phytochemical
Pflanzeneiweiß *n* phytoprotein
Pflanzenfarbstoff *m* plant pigment
Pflanzenfaser *f* vegetable fibre
Pflanzenfett *n* vegetable [plant] fat
Pflanzengift *n* plant poison
Pflanzengummi *n* (plant) gum
Pflanzenhormon *n* plant hormone, phytohormone
Pflanzenkunde *f* phytology
Pflanzenmasse *f* vegetable matter [material]
Pflanzenöl *n* vegetable oil
Pflanzenprotein *n* vegetable [plant] protein
Pflanzensaft *m* (plant) juice, succus
pflanzenschädigend phytotoxic
Pflanzenschleim *m* mucilage
Pflanzenschutzmittel *n* agricultural pesticide, crop protection chemical, economic poison
Pflanzenschutzmittel *n***/nicht persistentes** non-persistent pesticide
Pflanzenschutzmittel *n***/vorbeugend wirkendes** plant protective, protective toxicant
Pflanzensterol *n* phytosterol, plant sterol
pflanzentötend phytocidal
pflanzenverfügbar *(agric)* available *(nutrients)*
Pflanzenverfügbarkeit *f (agric)* availability *(of nutrients)*
pflanzenverträglich non-phytotoxic
Pflanzenzellkultur *f (biot)* plant cell culture
pflanzlich vegetable
Pfropfcopolymer(e) *n* graft copolymer
Pfropfcopolymerisat *n* graft copolymer
pfropfen *v (plast)* graft
Pfropfenströmung *f* plug [piston] flow
Pfropfenströmungsreaktor *m* plug-flow reactor
Pfropfpolymer(e) *n* graft polymer
Pfropfpolymerisat *n* graft polymer
Pfropfreaktion *f (plast)* grafting reaction
Pfropfung *f (plast)* grafting
Pfund-Serie *f (spectr)* Pfund series
pH-abhängig pH-dependent
phagenresistent *(biot)* phage-resistant
p-Halbleiter *m* p-type semiconductor
Pharmakochemie *f* pharmaceutic [medicinal] chemistry
Pharmakochemiker *m* pharmaceutic [medicinal] chemist
Pharmakodynamik *f* pharmacodynamics
pharmakodynamisch pharmacodynamic
Pharmakognosie *f* pharmacognosy
pharmakognostisch pharmacognostic
Pharmakokinetik *f* pharmacokinetics
Pharmakologie *f* pharmacology
pharmakologisch pharmacologic(al)
Pharmakon *n* pharmacon
Pharmakopöe *f* pharmacop(o)eia

Pharmazeut *m* pharmacist
Pharmazeutik *f* pharmaceutics
Pharmazeutikum *n* pharmaceutic(al)
pharmazeutisch pharmaceutic(al)
Pharmazie *f* pharmacy
Phase *f* 1. phase, stage; 2. *(phys ch)* phase
Phase *f***/cholester(in)ische** cholesteric phase
Phase *f***/dampfförmige** vapour phase
Phase *f* **des aktiven Wachstums** *(biot)* trophophase
Phase *f***/disperse** disperse [internal] phase
Phase *f***/feste** solid phase
Phase *f***/fluide** fluid phase
Phase *f***/flüssige** liquid phase
Phase *f***/gasförmige** gas phase
Phase *f***/gemischte** mixed phase
Phase *f* **geringerer Dichte** V phase *(as with extraction, distillation)*
Phase *f***/geschlossene** continuous [external] phase
Phase *f***/glasige** glassy [vitreous] phase
Phase *f***/intermetallische** intermetallic (compound)
Phase *f***/kristalline** crystalline phase
Phase *f***/letale** *(biot, hyd)* death [endogenous growth] phase
Phase *f***/mobile** *(chromat)* mobile [moving] phase
Phase *f***/nematische** nematic phase
Phase *f***/smektische** smectic phase
Phase *f***/stationäre** 1. *(chromat)* stationary [immobile] phase; 2. *(biot)* stationary [resting] phase *(in the growth of microorganisms)*
Phase *f***/wässrige** aqueous [water] phase
Phasen *fpl***/umgekehrte** reversed phases
Phasenänderung *f* phase change [transition]
Phasendiagramm *n* phase diagram, equilibrium [constitution] diagram
Phasendifferenz *f* phase difference
Phasengleichgewicht *n* phase equilibrium
Phasengrenze *f* phase boundary
Phasengrenzfläche *f* interface, junction
Phasengrenzfläche *f* **flüssig-flüssig** liquid-liquid interface
Phasengrenzreaktion *f* boundary reaction
Phasenkolloid *n* dispersoid, dispersion colloid
Phasenkontrastmikroskop *n* phase-contrast microscope
Phasenregel *f* **(von Gibbs)** (Gibbs) phase rule, phase rule of Gibbs
Phasentransferkatalysator *m* phase-transfer catalyst
Phasentransferkatalyse *f* phase-transfer catalysis, PTC
Phasentrennung *f* phase separation
Phasenübergang *m s.* Phasenumwandlung
Phasenübergangstemperatur *f* phase transition temperature
Phasenumkehr *f* phase inversion, reversal of phases
Phasenumwandlung *f* phase transition (change)
Phasenumwandlung *f***/antiferroelektrische** antiferroelectric transition
Phasenumwandlung *f***/antiferromagnetische** antiferromagnetic transition
Phasenumwandlung *f***/diskontinuierliche** discontinuous [first-order] transition
Phasenumwandlung *f***/displazive** displacive transition
Phasenumwandlung *f***/enantiotrope** enantiotropic transition
Phasenumwandlung *f* **erster Art [Ordnung]** first-order transition, discontinuous transition
Phasenumwandlung *f***/ferrimagnetische** ferrimagnetic transition
Phasenumwandlung *f***/ferroelastische** ferroelastic transition
Phasenumwandlung *f***/ferroelektrische** ferroelectric transition
Phasenumwandlung *f***/ferromagnetische** ferromagnetic transition
Phasenumwandlung *f* **höherer Ordnung** *f* higher-order transition, smooth transition
Phasenumwandlung *f***/irreversible** irreversible transition
Phasenumwandlung *f***/kongruente** congruent transition
Phasenumwandlung *f***/normal supraleitende** superconducting transition

Phasenumwandlung *f*/polymorphe polymorphic transition
Phasenumwandlung *f*/rekonstruktive reconstructive transition
Phasenumwandlung *f*/reversible reversible transition
Phasenumwandlung *f*/strukturelle structural transition
Phasenumwandlung *f* zweiter Art [Ordnung] second-order transition, continuous transition
Phasenumwandlungsentropie *f* entropy of phase transition
Phasenunterschied *m* phase difference
Phasenverschiebung *f* phase shift
pH-Bereich *m* pH range
Phellandren *n* phellandrene *(a monocyclic terpene)*
Phenacetin *n* phenacetin, acetophenetidine
Phenacylchlorid *n* phenacyl chloride, chloroacetophenone
Phenanthren *n* phenanthrene
Phenanthrenchinon *n* phenanthraquinone, *(specif)* 9,10-phenanthraquinone
Phenanthrolin *n* phenanthroline
Phenarsazinchlorid *n* phenarsazine chloride
Phenazin *n* phenazine, azophenylene
Phenazinfarbstoff *m* phenazine-based dye, azine dye
Phenazon *n* phenazone
Phenethylalkohol *m* phenethyl alcohol
Phenethylamin *n* phenethylamine, phenylethylamine
Phenetidin *n* phenetidine, ethoxyaniline
Phenetol *n* phenetole, ethoxybenzene
Phenol *n* phenol, *(specif)* phenol, benzophenol, hydroxybenzene
Phenol *n*/mehrwertiges polyhydric phenol
Phenolaldehyd *m* phenolic aldehyde
Phenolat *n* phenolate, phenoxide
Phenolcarbonsäure *f* phenolic benzoic acid, phenolic acid
Phenolester *m* phenolic ester
Phenol-Formaldehyd-Harz *n* phenol-formaldehyde resin
Phenolfurfuralharz *n* phenol-furfural resin
phenolhaltig phenolic
Phenolharz *n* phenolic resin
Phenolharz *n* im A Zustand resol
Phenolharz *n* im B Zustand resitol
Phenolharz *n* im C Zustand resite
Phenolharzkunststoff *m* phenolic plastic, phenoplast
Phenolharzlack *m* phenolic varnish
Phenolharzpressmasse *f* phenolic moulding compound
Phenolhydroxyl *n* phenolic hydroxyl
phenolisch phenolic
phenolisch/nicht non-phenolic
phenolisieren *v* phenolate
Phenolphthalein *n* phenolphthalein
Phenolrot *n* phenol red *(phenolsulphonephthalein)*
Phenolsulfonphthalein *n* phenolsulphonephthalein
Phenolsulfonsäure *f* phenolsulphonic acid
Phenonium-Ion *n* phenonium ion
Phenoplast *m* phenoplast, phenolic plastic
Phenothiazin *n* phenothiazine
Phenoxyessigsäure *f* phenoxyacetic acid
Phenoxygruppe *f* phenoxy group
***N*-Phenylacetamid** *n* *N*-phenylacetamide
Phenylacetat *n* phenyl acetate
α-Phenylacrylsäure *f* α-phenylacrylic acid, atropic acid
β-Phenylacrylsäure *f* β-phenylacrylic acid, *trans*-cinnamic acid
Phenylalkan *n* phenylalkane
***N*-Phenylamid** *n* *N*-phenyl amide, anilide
Phenylamin *n* *phenylamine, aminobenzene, aniline
Phenylbenzoat *n* phenyl benzoate
Phenylbrenztraubensäure *f* phenylpyruvic acid
Phenylchloroform *n* phenylchloroform, α,α,α-trichlorotoluene
2-Phenylchromenyliumsalz *n* 2-phenylchromenylium salt, flavylium salt
Phenylenblau *n* phenylene blue *(an indamine dye)*
Phenylendiamin *n* phenylenediamine, diaminobenzene
Phenylengruppe *f* $-C_6H_4-$ phenylene group
Phenylessigsäure *f* phenylacetic acid
Phenylester *m* phenyl ester
Phenylethen *n* phenylethene, styrene

Phenylglycin *n* phenylglycine, *(specif)* N-phenylglycine, anilinoacetic acid
Phenylglykolsäure *f* phenylglycollic acid, mandelic acid
Phenylgruppe *f* phenyl group
Phenylhydrazin *n* phenylhydrazine
Phenylhydrazon *n* phenylhydrazone
phenylieren *v* phenylate
Phenylierung *f* phenylation
Phenylkation *n* phenyl cation, phenylium
2-Phenylpropan-2-ol *n* 2-phenyl-2-propanol
Phenylpropansäure *f* phenylpropionic acid
Phenylpropinsäure *f* phenylpropynoic acid
Phenylquecksilberacetat *n* phenylmercuric acetate, PMA *(a pesticide)*
Pheromon *n (bioch)* pheromone
Pheron *n* pheron *(colloid carrier of an enzyme)*
Philips-Vakuummeter *n* Philips [Penning] gauge, cold-cathode ionization gauge
philodien *(org ch)* dienophilic
Philodien *n (org ch)* dienophile
Phiole *f* (glass) vial
Phlegma *n (distil)* less volatile component
Phlobaphen *n* phlobaphene, tanner's red
Phloridzin *n* phlori(d)zin *(β-glucoside)*
Phloroglucinol *n* phloroglucinol, 1,3,5-trihydroxybenzene
pH-Messgerät *n* pH meter [instrument]
pH-Messung *f* pH measurement
pH-Meter *n* pH meter [instrument]
Phonon *n* phonon
Phoron *n* phorone
Phosgen *n* phosgene, carbonyl chloride
Phosgenierung *f* phosgenation
Phosphamsäure *f* phosphamic acid
Phosphan *n* P_nH_{n+2} phosphane, *(specif)* PH_3 phosphane, phosphine
Phosphanoxid *n* phosphine oxide
Phosphat *n* phosphate, *(specif)* orthophosphate
Phosphatase f/saure *(bioch)* acid phosphatase
Phosphatdüngemittel *n* phosphate fertilizer
Phosphatid *n (bioch)* phosphatide, phosphoglyceride
Phosphatidsäure *f (bioch)* phosphatidic acid
Phosphatidylethanolamin *n* 3-phosphatidyl ethanolamine
Phosphatidylserin *n* 3-phosphatidyl serine
phosphatieren *v* phosphate *(for preventing corrosion)*
Phosphatieren *n* phosphating (treatment), phosphation *(for preventing corrosion)*
Phosphatreinigung *f (hyd)* phosphate treatment
Phosphensäure *f* $HOPO_2$ phosphenic acid
Phosphid *n* phosphide
Phosphin *n* phosphine *(PH_3 and ist derivatives)*
Phosphinigsäure *f* $H_2P(OH)$ phosphinous acid
Phosphinoxid *n* phosphine oxide
Phosphinsäure *f* $H_2P(=O)(OH)$ phosphinic acid
Phosphit *n* phosphite
Phosphodiester *m (bioch)* phosphodiester
6-Phosphogluconat *n (bioch)* 6-phosphogluconate
Phosphogluconat-Weg *m (bioch)* phosphogluconate [pentose phosphate] pathway
Phosphoglycerid *n (bioch)* phosphoglyceride, phosphatide
Phospho(hexo)ketolaseweg *m (bioch)* phosphoketolase pathway
Phospholipid *n* phospholipid
Phosphonat *n* phosphonate
Phosphonigsäure *f* $RP(OH)_2$ phosphonous acid
Phosphoniumverbindung *f* phosphonium compound
Phosphoniumylid *n* phosphonium ylide, Wittig reagent
Phosphonsäure *f* $RP(=O)(OH)_2$ phosphonic acid
Phosphonsäureester *m* phosphonate
Phosphonylierung *f* phosphonylation
Phosphoprotein *n* phosphoprotein
Phosphor *m* 1. P phosporus; 2. phosphor *(phosphorescent substance)*
Phosphor *m***/roter** red phosphorus *(commercial product)*

Phosphor *m***/schwarzer** black phosphorus

Phosphor *m***/violetter** violet phosphorus

Phosphor *m***/weißer** white [yellow] phosphorus

Phosphoran *n* phosphorane, λ^5-phosphane

Phosphor(III)-chlorid *n* phosphorus(III) chloride, phosphorus trichloride

Phosphor(V)-chlorid *n* phosphorus(V) chloride, phosphorus pentachloride

Phosphoreszenz *f* phosphorescence

Phosphoreszenzspektroskopie *f* phosphorescence spectroscopy

Phosphoreszenzspektrum *n* phosphorescence spectrum

phosphoreszieren *v* phosphoresce

phosphoreszierend phosphorescent, phosphoric, phosphorous

Phosphorhalogenid *n* phosphorus halide

Phosphor(III)-hydrid *n* phosphorus trihydride, phosphine

Phosphorigsäurealkylesteralkylimid *n* (RO)P=NR alkoxyalkyliminophosphine

Phosphornitrid *n* phosphorus nitride

Phosphorolyse *f (bioch)* phosphorolysis

phosphorolytisch *(bioch)* phosphorolytic

Phosphororganikum *n***/toxisches** *(agric)* O-P toxicant

Phosphor(III)-oxid *n* phosphorus(III) oxide, phosphorus trioxide

Phosphor(V)-oxid *n* phosphorus(V) oxide, phosphorus pentoxide

Phosphorsalz *n* phosphorus salt, ammonium sodium hydrogenphosphate 4-water

Phosphorsalzperle *f* (sodium) phosphate bead

Phosphorsäure *f* phosphoric acid, *(specif)* H_3PO_4 orthophosphoric acid, phosphoric acid

Phosphorsäuretricresylester *m* tricresyl phosphate, TCP

Phosphorsäuretriethylester *m* triethyl phosphate

Phosphorsäuretriphenylester *m* triphenyl phosphate

Phosphorstoffwechsel *m (bioch)* phosphorus metabolism

Phosphortetroxid *n* P_2O_4 phosphorus tetraoxide

Phosphorwasserstoff *m* phosphorus hydride, phosphane

Phosphoryl(V)-chlorid *n* phosphoryl chloride, phosphorus trichloride oxide

Phosphorylid *n* phosphonium ylide, Wittig reagent

phosphorylieren *v (bioch)* phosphorylate

Phosphorylierung *f (bioch)* phosphorylation

Phosphorylierung *f***/oxidative** *(bioch)* oxidative [respiratory-chain] phosphorylation

Phosphorylierungsmittel *n (bioch)* phosphorylating agent

Photoaddition *f* photoaddition

Photoaktivierung *f (bioch)* photoactivation

Photoakustikspektroskopie *f* photoacoustic spectroscopy, PAS

Photoakustikspektrum *n* photoacoustic spectrum

photoakustisch photoacoustic

Photochemie *f* photochemistry

Photochromie *f* photochromism

photocyclisch photocyclic

Photocyclisierung *f* photocyclization

Photocycloaddition *f* photocycloaddition

Photodimerisierung *f* photodimerization

Photodiode *f* photodiode

Photodissoziation *f* photodissociation

Photoeffekt *m* photoelectric effect

Photoelektrochemie *f (el ch)* photoelectrochemistry

photoelektrochemisch *(el ch)* photoelectrochemical

Photoelektron *n* photoelectron

Photoelektronenspektroskopie *f* photoelectron spectroscopy

Photoelektronenspektroskopie *f* **innerer Elektronen** photoelectron spectroscopy of inner-shell electrons

Photoelektronenspektroskopie *f***/röntgenstrahlangeregte** X-ray(-induced) photoelectron spectroscopy, XPS, electron spectroscopy for chemical analysis, ESCA

Photoelektronenvervielfacher *m* photomultiplier (tube), PMT, multiplier phototube, secondary-emission electron multiplier

Photoelement *n* photovoltaic cell, photochemical [barrier-layer] cell
Photoeliminierung *f* photoelimination
Photoemission *f* photoemission
Photo-Fries-Umlagerung *f* photo-Fries rearrangement *(of esters)*
Photografting *n* photografting *(for modification of biomaterial surfaces)*
Photoinaktivierung *f* photoinactivation
Photoinitiierung *f* photoinitiation
Photoionisation *f* **bei Atmosphärendruck** *(spectr)* atmospheric pressure photoionization, APPI
Photoionisationsdetektor *m* photoionization detector, PID
Photoionisierung *f* photoionization
Photokatalysator *m* photocatalyst
Photokatalyse *f* photocatalysis
photokatalysiert light-catalyzed
Photolack *m* photo resist
photoleitend photoconductive
Photoleitfähigkeit *f* photoconductivity
Photoleitwert *m* photoconductance
Photolithographie *f* photolithography
Photolumineszenz *f* photoluminescence
Photolyse *f* photolysis • **durch Photolyse zerlegbar** photolysable
Photolyse *f* **von Ozon** ozone photolysis
Photolyseblitz *m* photolysis flash, photoflash *(kinetics)*
photolytisch photolytic
Photomaske *f* photomask
Photometer *n* photometer
Photometrie *f* photometry
Photomultiplier *m* photomultiplier
Photon *n* photon
Photooxidation *f* photooxidation
photophosphorylieren *v (bioch)* photophosphorylate
Photophosphorylierung *f (bioch)* photophosphorylation
Photopolymer(e) *n* photopolymer
Photopolymerisation *f* photopolymerization
Photoreaktion *f* photochemical reaction, photoreaction
Photoreaktion *f*/**diabatische [nicht adiabatische]** diabatic photoreaction
Photoreaktivierung *f (bioch)* photoreactivation
Photoreaktor *m* photoreactor, photochemical reactor
Photoreduktion *f* photoreduction
Photoresist *n* photoresist *(photolithography)*
Photoresistfarbstoff *m* photoresist dye
Photoresistlack *m* photoresist resin
Photorezeptor *m* photoreceptor
Photoschicht *f* photosensitive surface
photosensibel photosensitive
Photosensibilisator *m* photosensitizer
photosensibilisieren *v* photosensitize
Photosensibilisierung *f* photosensitization
Photosensibilität *f* photosensitivity
Photostrom *m* photocurrent
Photosynthese *f* photosynthesis
Photosynthesezyklus *m (bioch)* ribulose diphosphate cycle, Calvin(-Bassam) cycle
photosynthetisch photosynthetic
Phototropie *f s.* Photochromie
Photo-Volta-Effekt *m* photovoltaic effect
Photovoltaik *f* photovoltaics *(solar power technology)*
photovoltaisch photovoltaic
Photowiderstand *m* photoresistance, photoresistor, photoconductive cell
Photowiderstandszelle *f* photoresistor, photoconductive cell
Photozelle *f* photoelectric cell, photocell, phototube
pH-Papier *n* pH paper
pH-Regelung *f* pH control
pH-Skala *f* pH scale
pH-Standard *m* pH standard
***o*-Phthalaldehyd** *m o*-phthalic aldehyde, OPA
Phthalamid *n* phthalamide
Phthalamidsäure *f* phthalamic acid
Phthalat *n* phthalate *(salt or ester of phthalic acid)*
Phthalein *n (org ch)* phthalein
Phthalid *n* phthalide
Phthalimid *n* phthalimide
Phthalocyaninfarbstoff *m* phthalocyanine dye
Phthalodinitril *n* phthalonitrile
Phthalsäure *f* phthalic acid, *(specif)* o-phthalic acid
Phthalsäureanhydrid *n* phthalic anhydride, PA
pH-unabhängig pH-independent

Phykocyan *n* phycocyan(in) *(pigment of blue-green algae)*
Phykoerythrin *n* phycoerythrin *(pigment of red algae)*
Phyllochinon *n* phylloquinone, anti-haemorrhagic vitamin
Phyllosilicat *n (min)* phyllosilicate, sheet silicate
Physik *f***/chemische** chemical physics
physikalisch-chemisch physicochemical
Physikochemie *f* physical chemistry
Physikochemiker *m* physical chemist
physikochemisch physicochemical
Physisorption *f* physisorption, physical adsorption
Phytin *n* phytin *(Ca-Mg salt of phytic acid)*
Phytinsäure *f* phytic acid
Phytochemie *f* phytochemistry, plant chemistry
phytogen phytogenic, plant-produced
Phytohormon *n* phytohormone, plant hormone
Phytosterol *n* phytosterol, plant sterol
phytotoxisch phytotoxic
Phytotoxizität *f* phytotoxicity
pH-Zahl *f* pH value
PIB *s.* Polyisobuten
Pi-Bindung *f* π bond, pi bond
pichen *v* pitch
Pickeln *n (tann)* pickling process
Pickelverfahren *n s.* Pickeln
Picogrammbereich *m (anal)* picogram range *(10^{-12} to 10^{-9} g)*
Picolinsäure *f* picolinic acid
Picospur *f (anal)* picotrace *(*10^{-13} to 10^{-10} ppm)*
Pictet-Gams-Synthese *f* Pictet-Gams synthesis [reaction] *(of isoquinolines)*
PID *s.* Photoionisationsdetektor
Piezochemie *f* piezochemistry
Piezoelektrizität *f* piezoelectricity
Piezokristall *m* piezoelectric crystal
Pigment *n* pigment
Pigment *n***/metallisches** metal(lic) pigment
Pigment *n***/natürliches** natural pigment
Pigment *n***/natürliches anorganisches** earth [mineral] pigment
Pigment *n***/organisches** organic pigment
Pigment *n***/respiratorisches** respiratory pigment
Pigmentanreibung *f* pigment grinding
Pigmentation *f* pigmentation
Pigment-Bindemittel-Verhältnis *n* pigment-binder ratio
Pigmentfarbstoff *m* pigment (dye, dyestuff)
pigmentieren *v* pigment
pigmentiert/nicht unpigmented
Pigmentkollektiv *n (bioch)* pigment assembly, photosynthetic unit
Pigmentvolumenkonzentration *f* pigment volume concentration, PVC
Pikraminsäure *f* picramic acid
Pikrat *n* picrate *(salt or ester of picric acid)*
Pikrinsäure *f* picric acid, 2,4,6-trinitrophenol
Pille *f* 1. pellet *(as of a catalyst)*; 2. *(pharm)* pill
pillieren *v* pill
Pilotanlage *f* pilot plant, semiworks
Pilzamylase *f* fungal amylase
Pilzbiomasse *f* fungal biomass
Pilzenzym *n* fungal enzyme
Pilzfermentation *f (biot)* fungal fermentation
pilzlich fungal
Pilzlipase *f (biot)* fungal lipase
Pilzmyzel *n (biot)* fungal mycelium
Pilzprotease *f (biot)* fungal protease
Pilzstamm *m (biot)* fungal strain
pilztötend fungicidal
pilzwachstumhemmend fungistatic
pilzwidrig antifungal, fungicidal
Pimelinsäure *f* pimelic acid, heptanedioic acid
Pinakoid *n (cryst)* pinacoid
pinakoidal *(cryst)* pinacoidal
Pinakol *n* pinacol
Pinakolon *n* pinacolone
Pinakol-Pinakolon-Umlagerung *f* pinacol rearrangement
Pinen *n* pinene *(a bicyclic monoterpene)*
Ping-pong-Reaktion *f* ping-pong reaction, double displacement reaction
Pinksalz *n* pink salt, ammonium hexachlorostannate(IV)
Pinsel *m* brush • **mit dem Pinsel aufbringen [auftragen]** brush on
Pinzette *f* tweezers
Pion *n (nucl)* pion, π-meson

Pipeline-Gas *n* pipeline(-quality) gas
Piperazin *n* piperazine
Piperidin *n* piperidine
Piperidinalkaloid *n* piperidine alkaloid
Piperonal *n* piperonal
Pipette *f* pipette
pipettieren *v* pipette
Piranha-Lösung *f* piranha solution *(cleaning of substrates)*
Pirani-Vakuummeter *n* Pirani gauge
Pistill *n (lab)* pestle
Pitot-Rohr *n* Pitot tube
Pitzer-Spannung *f* torsional strain *(in molecules)*
Pivalinsäure *f* pivalic acid
Plagioklas *m (min)* plagioclase *(sodium-calcium feldspar)*
Planck-Konstante *f* Planck [action] constant, (Planck) quantum of action
planieren *v* level
Planschliffverbindung *f* flat flange joint
Plasma *n* 1. *(phys ch)* plasma; 2. *(biol)* plasma
Plasma *n*/**induktiv gekoppeltes** inductively coupled plasma, ICP
Plasma *n*/**mikrowelleninduziertes** microwave-induced plasma, MIP
Plasmabrenner *m* plasma burner, plasma torch, plasma gun
Plasmachemie *f* plasma chemistry
Plasmachemie *f*/**thermische** thermal plasma chemistry
Plasmaexpander *m* blood plasma extender
Plasmalogen *n* plasmalogen *(any of a group of glycerophospholipids)*
Plasmapolymerisation *f* plasma polymerization
Plasmaprotein *n* plasma protein
Plasmaschneiden *n* plasma cutting
Plasmaseparation *f (bioch)* plasma separation
Plasmaspray *n* plasma spray *(coating process)*
Plasmid *n (biot)* plasmid
Plastifikation *f (plast, rubber)* plasticization
plastifizieren *v (plast, rubber)* plasticize
Plastifiziermittel *n (plast, rubber)* plasticizer
Plastigel *n* plastigel
Plastikator *m* 1. *(plast, rubber)* plasticator *(machine)*; 2. *s.* Plastifiziermittel
plastisch plastic • **plastisch machen** soften
Plastisol *n* plastisol
plastizieren *v (plast, rubber)* plasticize, plasticate, soften
Plastiziermaschine *f (plast, rubber)* plasticator
Plastizierung *f (plast, rubber)* plasticization, plastication, softening
Plastizierung *f*/**chemische** chemical plasticization
Plastizierung *f* **mit Peptisiermitteln** peptization
Plastizierungsmittel *n*/**chemisches** 1. chemical plasticizer; 2. *(rubber)* peptizer
Plastizität *f* plasticity
Plastochinon *n (bioch)* plastoquinone
Plastomer(e) *n* plastomer
Platformat *n (petrol)* platformate
Platformieren *n (petrol)* platforming, platinum reforming
Platin *n* Pt platinum
Platinat *n* platinate
Platin(II)-chlorid *n* platinum(II) chloride, platinum dichloride
Platin(IV)-chlorid *n* platinum(IV) chloride, platinum tetrachloride
Platinelektrode *f* platinum electrode
Platinelektrode *f*/**platinierte** platinized platinum electrode
platinhaltig platiniferous
platinieren *v* platinize
Platinieren *n* 1. platinization; 2. *(el ch)* platinum plating
platiniert platinized, platinum-coated
Platinkatalysator *m* platinum catalyst
Platinmohr *n* platinum black
Platin(II)-oxid *n* platinum(II) oxide, platinum monooxide
Platin(IV)-oxid *n* platinum(IV) oxide, platinum dioxide
Platinschwamm *m* platinum sponge, spongy platinum
Platinschwarz *n* platinum black
Platin(II)-sulfid *n* platinum(II) sulphide, platinum monosulphide
Platin(IV)-sulfid *n* platinum(IV) sulphide, platinum disulphide
Platintiegel *m* platinum crucible
Platte *f* 1. plate, *(if thick:)* slab, *(if thin:)* sheet; 2. *(rubber)* sheet, slab; 3. tile *(as*

of fired clay or concrete); board *(as of wood pulp)*; 4. *(filtr)* disk
Plattenbandförderer *m* apron conveyor
Plattenelektrofilter *n* plate precipitator *(gas cleaning)*
Plattengel *n* slab gel *(electrophoresis)*
Plattentrockner *m* shelf dryer
Plattenwärmeaustauscher *m* plate heat exchanger
plattieren *v* clad *(metal by bonding or welding)*
Plattieren *n* **mit Blei** lead cladding
Plattierüberzug *m* cladding
plattstängelig *(cryst)* bladed
platzen *v* burst, break, explode *(by the force of internal pressure)*; crack *(as of a glass plate)*
Platzwechsel *m* shift *(as of atoms)*
Plazebo *n* placebo *(pharmacologically inactive substance)*
PLB-Teilchen *n (chromat)* porous layer bead, PLB
p-leitend p-type (semi)conducting, defect conducting
Pleochroismus *m (cryst)* pleochro(mat)ism
pleochroitisch *(cryst)* pleochro(mat)ic
Plexiglas® *n* acrylic glass, Plexiglas®
PLOT-Säule *f (chromat)* porous layer open tubular column, PLOT column
Plumban *n* plumbane
Plumbat *n* plumbate
Plungerkolben *m* plunger *(of a pump)*
Plungerpumpe *f* plunger pump
plus-Stamm *m (biot)* (+) strain
Plutonium *n* Pu plutonium
PMA *s.* Phenylquecksilberacetat
PMAS *s.* Phenylquecksilberacetat
PMMA *s.* Polymethylmethacrylat
PMP *s.* Poly-4-methylpent-1-en
PMR *s.* Resonanz/paramagnetische
pneumatisch pneumatic
pn-Übergang *m* p-n junction *(in semiconductors)*
PO *s.* Polyolefin
Podbielniak-Zentrifugalextraktor *m* Podbielniak extractor [contactor]
Poiseuille-Gleichung *f* Poiseuille's equation [formula] *(of laminar flow)*
Poiseuille-Strömung *f* Poiseuille flow
Poisson-Verteilung *f* Poisson distribution
Pökeldauer *f (food)* pickling time
Pökellake *f (food)* curing [pickling] solution, pickle
pökeln *v (food)* cure
Pökelsalz *n (food)* curing agent
polar dipolar
polar/nicht non-polar, apolar
Polarimeter *n* polarimeter
Polarimetrie *f* polarimetry
Polarisation *f* polarization
Polarisation *f***/elliptische** elliptical polarization
Polarisation *f***/lineare** linear polarization
Polarisation *f***/zirkulare** circular polarization
Polarisationsebene *f* polarization plane
Polarisationsmikroskop *n* polarizing microscope
Polarisationspotenzial *n* polarization potential
Polarisationsstrom *m* polarization current
Polarisationsstromtitration *f* dead-stop titration
Polarisationswiderstand *m* polarization resistance
Polarisationswinkel *m* angle of polarization, polarizing [Brewster] angle
Polarisator *m* polarizer
polarisierbar polarizable
polarisierbar/nicht non-polarizable
Polarisierbarkeit *f* polarizability
Polarisierbarkeit *f* **des Moleküls** molecular polarizability
polarisieren *v* polarize
polarisiert/linear linearly polarized
polarisiert/linkszirkular left(-hand) circularly polarized
polarisiert/rechtszirkular right(-hand) circularly polarized
polarisiert/zirkular circularly polarized
Polarisierung *f* polarization
Polarität *f* polarity
Polarogramm *n* polarogram
Polarograph *m* polarograph
Polarographie *f* polarography
Polarographie *f* **mit Wechselstrom/oszillographische** *(anal)* multisweep polarography
Polarographie *f***/oszillographische** oscillographic polarography
Polenske-Zahl *f* Polenske number [value] *(of fats and fatty oils)*
polieren *v* polish

Polierfilter *n* polishing [clarifying] filter
Polierfiltration *f* polishing filtration, clarification
polierfiltrieren *v* polish
Poliermittel *n* polishing agent
Polierrot *n* polishing rouge *(iron(III) oxide)*
Polierscheibe *f* polishing wheel
Politur *f* polish
Polizeikampfmittel *n* riot control agent
Polonium *n* Po polonium
Polster *n* pad
Polyacetylen *n* *polyyne, polyacetylene
Polyacrylamid *n* polyacrylamide
Polyacrylat *n* polyacrylate
Polyacrylatelastomer(e) *n* polyacrylate (acrylate) elastomer
Polyacrylnitril *n* polyacrylonitrile
Polyacrylnitrilfaser *f* polyacrylonitrile [acrylic] fibre
Polyacrylnitrilfaserstoff *m* polyacrylonitrile [acrylic] fibre
Polyacrylsäure *f* polyacrylic acid
Polyaddition *f* polyaddition, addition polymerization
Polyaddukt *n* addition polymer
Polyadenylierung *f (bioch)* polyadenylation
Polyalkohol *m* polyalcohol, polyol
Polyallomer(e) *n* polyallomer
Polyamid *n* polyamide, PA
Polyamidfaser *f* polyamide fibre
Polyamidfaserstoff *m* polyamide fibre
Polyampholyt *m* polyampholyte
Polyanion *n* polyanion
Polyarylether *m* polyaryl ether
polyatomar polyatomic
Polyazid *n* polyacid
Polybenzimidazol *n* polybenzimidazole, PBI
Polyblend *n* polyblend, polymer blend *(a mixture of several thermoplastics)*
Polybrombiphenyl *n* polybrominated biphenyl, PBB
Polybutadien *n* polybutadiene
Polybuten *n* polybutene, polybutylene
Polycaprolactam *n* polycaprolactam
Polycarbazol *n* polycarbazole
Polycarbodiimid *n* polycarbodiimide
Polycarbonat *n* polycarbonate, PC
Polycarbonsäure *f* polycarboxylic acid
Polychlordibenzo-1,4-dioxin *n* polychlorinated dibenzo-1,4-dioxin, PCDD
Polychloropren *n* polychloroprene, PCP
Polychlortrifluorethylen *n* polychlorotrifluoroethylene, PCTFE
Polychromasie *f* polychromasia *(of radiation)*
Polycyclus *m* polycyclic compound [ring system]
Poly(dimethylsiloxan) *n* polydimethylsiloxane, PDMS
polydispers polydisperse
Polydispersität *f* polydispersity
Polyeder *n (cryst)* polyhedron
polyedrisch *(cryst)* polyhedral
Polyelektrolyt *m* polyelectrolyte
Polyelektrolytkomplex *m* polyelectrolyte complex
Polyen *n (org ch)* polyene
Polyen-Antibiotikum *n* polyene antibiotic
Polyepoxid *n* polyepoxide
Polyester *m* polyester
Polyester *m***/ungesättigter** unsaturated polyester
Polyesterfaser *f* polyester fibre
Polyesterfaserstoff *m* polyester fibre
Polyesterharz *n* polyester resin
Polyether *m* polyether
Polyethylen *n* polyethylene, PE
Polyethylen *n***/chlorosulfoniertes** chlorosulphonated polyethylene
Polyethylen *n* **hoher Dichte** high-density polyethylene, H.D. polyethylene
Polyethylen *n* **mittlerer Dichte** medium-density polyethylene
Polyethylen *n* **niedriger Dichte** low-density polyethylene, L.D. polyethylene
Polyethylen *n***/verzweigtes** branched [high-pressure] polyethylene
Poly(3,4-ethylendioxythiophen) *n* poly(3,4-ethylenedioxythiophene) *(material for organic light emitting diodes)*
Polyethylenfaser *f* polyethylene fibre
Polyethylenfaserstoff *m* polyethylene fibre
Polyethylenglykol *n* polyethylene glycol, PEG
Polyethylenimin *n* polyethylene imine
Polyethylenoxid *n* polyethylene oxide, PEO
Polyethylenterephthalat *n* polyethylene terephthalate, PETP, PET
Polyformaldehyd *m* polyoxymethylene, POM, polyformaldehyde

Polyformen *n (petrol)* polyforming
polyfunktionell polyfunctional, multifunctional
Polyglykol *n* polyglycol
Polyhalogenid *n* polyhalide, polyhalogenide
Polyharnstoff *m* polyurea
Polyhydrat *n* polyhydrate
Polyhydroxysäure *f* polyhydroxy acid
Polyhydroxyverbindung *f* polyhydroxy compound
Polyimid *n* polyimide, PI
Polyimidazopyrrolon *n* polyimidazopyrrolone, pyrrone
Polyin *n* polyyne, polyacetylene
Polyinsertion *f* insertion polymerization
Polyion *n* polyion
Polyisobuten *n* polyisobutene, polyisobutylene, PIB
Polyisopren *n* polyisoprene
polyisotop polyisotopic
Polykation *n* polycation
Polyketid *n* polyketide, ketide, acetogenin
Polyketon *n* polyketone
Polykieselsäure *f* polysilicic acid
Polykondensat *n* polycondensate, condensation polymer
Polykondensation *f* polycondensation, condensation polymerization
Polykristall *m* polycrystal
polykristallin polycrystalline
polymer polymeric
Polymer *n* 1. polymer; 2. *(tech)* polymerizate
Polymer *n*/**anorganisches** inorganic polymer
Polymer *n*/**ataktisches** atactic polymer
Polymer *n*/**einsträngiges** single-strand polymer
Polymer *n*/**eutaktisches** eutactic polymer
Polymer *n*/**fadenbildendes** fibrous polymer
Polymer *n*/**isotaktisches** isotactic polymer
Polymer *n*/**langkettiges** long-chain polymer
Polymer *n*/**lebendes** living polymer
Polymer *n*/**leitfähiges** conductive polymer
Polymer *n*/**lineares** linear polymer
Polymer *n*/**organisches** organic polymer
Polymer *n*/**syndiotaktisches** syndiotactic polymer
Polymer *n*/**totes** dead polymer
Polymer *n*/**vernetztes** cross-linked polymer
Polymer *n*/**verzweigtes** branched polymer
Polymerase-Kettenreaktion *f* polymerase chain reaction, PCR
Polymerbenzin *n* polymer gasoline
Polymerbeton *m* polymer concrete
Polymerchemie *f* polymer chemistry
Polymere *n s.* Polymer
Polymerisat *n* polymerizate, polymer
Polymerisation *f* polymerization
Polymerisation *f*/**anionische** anionic polymerization
Polymerisation *f*/**cyclisierende** cyclopolymerization
Polymerisation *f* **durch Atomübertragung/radikalische** *s.* Atomübertragungsreaktion
Polymerisation *f* **durch degenerativen Kettentransfer** reversible addition fragmentation chain transfer polymerization, RAFT
Polymerisation *f* **durch Ringöffnung** *s.* Ringöffnungsmetathese
Polymerisation *f*/**elektrochemische** *(plast)* electropolymerisation
Polymerisation *f*/**ionische** ionic polymerization
Polymerisation *f*/**kationische** cationic polymerization
Polymerisation *f*/**koordinative** coordination polymerization
Polymerisation *f*/**lebende** living polymerization
Polymerisation *f*/**photoaktivierte** photoinduced polymerization
Polymerisation *f*/**radikalische** (free-) radical polymerization
Polymerisation *f*/**ringöffnende** ring-opening polymerization
Polymerisation *f*/**stereoselektive** stereoselective polymerization
Polymerisation *f*/**stereospezifische** stereospecific polymerization
Polymerisation *f*/**strahleninduzierte** radiation-induced polymerization
Polymerisationsabstoppmittel *n*

shortstopping agent, shortstop, stopper
Polymerisationsgrad *m* degree of polymerization, DP
Polymerisationsinitiator *m* polymerization initiator
polymerisierbar polymerizable
Polymerisierbarkeit *f* polymerizability
polymerisieren *v* polymerize
Polymerkette *f* polymer chain
Polymermischung *f* polyblend, polymer blend *(mixture of several thermoplastics)*
Polymer-Nanokomposit *m* polymeric nanocomposite
Polymernetzwerk *n (plast)* polymer network
Polymethacrylat *n* polymethacrylate, methacrylate resin
Polymethacrylsäure *f* polymethacrylic acid
Polymethacrylsäureester *m s.* Polymethacrylat
Polymethinfarbstoff *m* polymethine dye
Polymethylen *n* polymethylene
Polymethylmethacrylat *n* polymethyl methacrylate, PMMA
Poly-4-methylpent-1-en *n* poly-4-methyl-1-pentene
Polymilchsäure *f* polylactic acid
polymolekular polymolecular
Polymolekularität *f* polymolecularity
polymorph polymorphic
Polymorphie *f* polymorphism
Polymyxin *n* polymyxin *(protein chemistry)*
Polynosefaser *f* polynosic fibre
Polynosefaserstoff *m* polynosic fibre
Polyol *n* polyol, polyalcohol
Polyolefin *n* polyolefin
Polyorganosiloxan *n* polyorganosiloxane
Polyoxymethylen *n* polyoxymethylene, POM, polyformaldehyde
Polyparabansäure *f* polyparabanic acid
Polypeptidantibiotikum *n* polypeptide antibiotic
Polypeptidkette *f* polypeptide chain
Polyphenylen *n* polyphenylene
Polyphenylenoxid *n* polyphenylene oxide, PPO
Polyphenylensulfid *n* polyphenylene sulphide
Polyphenylensulfon *n* polyphenylene sulphone
Polyphosphat *n* polyphosphate
Polyphosphazen *n* polyphosphazene
Polyphosphorsäure *f* polyphosphoric acid *(any of the phosphoric acids $H_{n+2}[P_nO_{3n+1}]$)*
Polypropylen *n* polypropylene, polypropene, PP
Polypyrrol *n* polypyrrole *(conductive polymer)*
Polyquinan *n* polyquinane *(polycyclic hydrocarbon)*
Polyquinen *n* polyquinene *(unsaturated polycyclic hydrocarbon)*
Polyreaktion *f* polyreaction *(polymerization, polycondensation or polyaddition)*
Polysaccharid *n* polysaccharide
Polysaccharid *n***/mikrobielles** microbial polysaccharide
Polysiloxan *n* polysiloxane
Polysiloxan *n***/organisches** polyorganosiloxane
Polyspiroverbindung *f* polyspiro compound
Polystyren *n* polystyrene, PS
Polystyren *n***/aufschäumbares** expandable polystyrene
Polystyren *n***/geschäumtes** foamed [expanded] polystyrene
Polystyren *n***/hochschlagfestes** high-impact polystyrene
Polystyren *n***/schlagfestes** toughened polystyrene
Polystyrol *n s.* Polystyren
Polystyrolsulfonat *n* poly(styrenesulfonate) *(material for organic light emitting diodes)*
polysubstituiert polysubstituted
Polysubstitution *f* polysubstitution
Polysulfid *n* polysulphide
Polysulfidkautschuk *m* polysulphide rubber
Polysulfon *n* polysulphone, polyphenylene sulphone
Polyterpen *n* polyterpene
Polytetrafluorethylen *n* polytetrafluoroethylene, PTFE
Polythiocarbamid *n* polythiourea
Polythionat *n* $M^I_2[S_xO_6]$ polythionate
Polythionsäure *f* polythionic acid

Polythiophen *n* polythiophene *(conductive polymer)*
Polytriazin *n* polytriazine, triazine polymer
Polyurethan *n* polyurethan(e)
Polyurethanelastomer(e) *n* polyurethane elastomer
Polyurethankautschuk *m* polyurethane [isocyanate] rubber
Polyurethanschaum(stoff) *m* polyurethane foam
Polyuronid *n* polyuronid(e)
Polyuronsäure *f* polyuronic acid
Polyvinylacetal *n* polyvinyl acetal
Polyvinylacetat *n* polyvinyl acetate, PVAC
Polyvinylalkohol *m* polyvinyl alcohol, PVA
Polyvinylbutyral *n* polyvinyl butyral, PVB
Polyvinylcarbazol *n* polyvinyl carbazole
Polyvinylchlorid *n* polyvinyl chloride, PVC
Polyvinylchloridacetat *n* polyvinylchloride acetate
Polyvinylester *m* polyvinyl ester
Polyvinylether *m* polyvinyl ether
Polyvinylfaser *f* polyvinyl fibre
Polyvinylfaserstoff *m* polyvinyl fibre
Polyvinylfluorid *n* polyvinyl fluoride, PVF
Polyvinylformal *n* polyvinyl formal
Polyvinylidenchlorid *n* polyvinylidene chloride
Polyvinylidencyanid *n* polyvinylidene cyanide
Polyvinylidenfluorid *n* polyvinylidene fluoride
Poly-*p*-xylylen *n* poly-*p*-xylylene, poly-*p*-xylene
POM *s.* Polyoxymethylen
Pomeranz-Fritsch-Synthese *f* Pomeranz-Fritsch synthesis [reaction] *(of isoquinolines)*
Popcorn-Polymer *n* popcorn-polymer
Pore *f* 1. pore, void; 2. *(coat)* pinhole • **mit feinen Poren** fine-pore(d) • **mit Poren versehen** poriferous, porous
Porenbeton *m* aerated concrete
Porendiffusion *f* pore diffusion
Porengröße *f* pore size
Porengrößenverteilung *f* pore-size distribution
Porengummi *m* microcellular rubber
Porenvolumen *n* 1. pore [void] volume *(as of a catalyst)*; 2. *(soil)* volume of pore space
Porenvolumen ***n*/relatives** *(filtr)* porosity, voidage
Porenvolumenverteilung *f* pore-volume distribution
Porenweite *f* pore size
porig pored
porös porous, poriferous
Porosimeter *n* porosimeter
Porosität *f* 1. porosity; 2. *s.* Porenvolumen/relatives
Porphin *n* porphin(e) *(cyclic tetrapyrrole)*
Porphobilinogen *n (bioch)* porphobilinogen
Porphyr *m (geol)* porphyry
Porphyrin *n* porphyrin *(substituted porphin)*
portionsweise in portions
Portlandzement *m* portland cement
Porzellan *n* porcelain, *(for non-technical use also)* china • **aus Porzellan** porcelan(e)ous, porcelainous • **zu Porzellan brennen** porcelainize
Porzellan ***n*/chemisch technisches** heavy chemical porcelain
Porzellan ***n*/chemisches** chemical porcelain
porzellanen porcelan(e)ous, porcelainous
Porzellanerde *f (min)* porcelain [china] clay, kaolin, bolus alba, white bole
Porzellanerde ***f*/geschlämmte** (china) clay, kaolin
Porzellangut *n* vitreous china
Porzellantiegel *m (lab)* porcelain crucible
Positionsselektivität *f (org ch)* position selectivity
positiv geladen positively charged
positiv (geladen)/dreifach tripositive
positiv (geladen)/einfach unipositive
positiv (geladen)/fünffach pentapositive, quinque-positive
positiv (geladen)/vierfach tetrapositive
positiv (geladen)/zweifach dipositive
Positiv *n (phot)* positive
Positivform *f (plast)* positive mould, *(in drape and vacuum forming:)* male mould
Positron *n* positron, positive electron
Positronchemie *f* positron chemistry

Positronenzerfall *m* positron [positive beta] decay
Positronium *n* positronium *(a system consisting of a positron and an electron)*
Positroniumchemie *f* positronium chemistry
Posten *m (tech)* batch
Potenz f/nucleophile nucleophilicity, nucleophilic power
Potenzial *n* potential
Potenzial *n*/**äußeres elektrisches** outer [Volta] potential
Potenzial *n*/**chemisches** chemical potential
Potenzial *n*/**elektrokinetisches** electrokinetic potential, zeta [double-layer] potential, ZP
Potenzial *n*/**inneres elektrisches** inner [Galvani] potential
Potenzial *n*/**thermodynamisches** thermodynamic potential
Potenzialabfall *m* potential drop, fall of potential
Potenzialbarriere *f* potential barrier
potenzialbestimmend potential-determining
Potenzialdifferenz *f* potential difference
Potenzialenergiefläche *f* potential-energy (reaction) surface, equipotential surface
Potenzialflächendiagramm *n* potential-energy contour map
Potenzialgradient *m* potential gradient
Potenzialschwelle *f* potential barrier
Potenzialsprung *m* potential jump
Potenzialunterschied *m* potential difference
Potenzialwall *m* potential barrier
Potenziometer *n* potentiometer
Potenziometrie *f* potentiometry
Pottasche *f* potash, potassium carbonate
Po-Z *s.* Polenske-Zahl
PP *s.* Polypropylen
PPO *s.* Polyphenylenoxid
PPS *s.* Polyphenylensulfid
PQ *s.* Plastochinon
Prädioxin *n (tox)* predioxin
Prädissoziation *f (phys ch)* predissociation
Präexponentialfaktor *m* (Arrhenius) pre-exponential factor, frequency factor *(kinetics)*
Prägleichgewicht *n* near-equilibrium
Präionisation *f* preionization, autoionization
praktisch actual *(e.g. test)* • **praktisch erzielbar** actual *(e.g. temperature)* • **praktisch frei** substantially free
Präkursor *m* precursor
Prallfläche *f* baffle (plate), deflector (plate) • **mit Prallflächen ausstatten [versehen]** baffle
Präparat *n* preparation
Präparat *n*/**kosmetisches** cosmetic (preparation)
Präparat *n*/**pharmazeutisches** pharmaceutic(al) preparation
Präparat *n*/**tiermedizinisches [veterinärmedizinisches]** animal health product
Präparat *n*/**virentötendes [viruzides]** virucide
Präparatenglas *n* specimen jar, preservation [museum] jar, show bottle
präparativ preparative • **in präparativem Maßstab** on a preparative scale
präparieren *v* prepare
Präpariersalz *n* preparing salt *(sodium hexahydroxostannate(IV))*
Praseodym *n* Pr praseodymium
Praseodym(III)-chlorid *n* praseodymium(III) chloride, praseodymium trichloride
Praseodym(III)-oxid *n* praseodymium(III) oxide, praseodymium trioxide
Praseodym(IV)-oxid *n* praseodymium(IV) oxide, praseodymium dioxide
Prästationärenzymkinetik *f* pre-steady-state kinetics
Präzession *f (phys ch)* precession
Präzipitat *n*/**gelbes** yellow precipitate, yellow mercuric oxide
Präzipitat *n*/**rotes** red precipitate, red mercuric oxide
Präzipitat *n*/**schmelzbares weißes** $Hg(NH_3)_2Cl_2$ fusible white precipitate, diamminemercury(II) chloride
Präzipitat *n*/**unschmelzbares weißes** $Hg(NH_2)Cl$ infusible white precipitate, aminomercury(II) chloroide
präzipitieren *v* precipitate
Präzision *f (anal)* precision
Präzisionswaage *f* precision balance

Precoatfilter *n* precoat filter
Precoatschicht *f* (filter) precoat
Premix-Masse *f (plast)* premix, premixed moulding compound
Prephensäure *f* prephenic acid *(a cyclohexadiene derivative)*
Prepreg *n (plast)* prepreg *(preimpregnated glass-fibre material)*
Presse *f* 1. press; 2. *(plast)* moulding press
pressen *v* press; press, compact, compress *(e.g. powders)*; mould *(e.g. plastics)*
pressen *v***/kalt** cold-press
pressen *v***/zu Briketts** briquette
Pressform *f* mould; die *(of an extruder)*
Pressharz *n* (compression-)moulding resin
Presskuchen *m* 1. *(tech)* press(ed) cake; 2. *(food)* oil cake
Presslauge *f* press liquor
Pressling *m* 1. *(plast, ceram, glass)* moulding; compact *(powder metallurgy)*; 2. *(coal)* briquette; 3. *(spectr)* pellet, *(for IR spectroscopy:)* pressed disk
Pressluft *f* compressed air
Pressmasse *f (plast)* compression-moulding material
Pressrückstand *m* expressed residue
Presssaft *m* press-juice
Presssintern *n* sintering under pressure, hot pressing *(of metal powder)*
Pressstempel *m (plast)* male mould, patrix, plunger
Presstechnik *f (spectr)* pressed disk technique
Pressteil *n (plast)* moulding
Presstuch *n* press cloth
pressverdichten *v* compact
Pressverdichten *n* compaction
Presswerkzeug *n (plast)* mould
Preußischblau *n* Prussian [Berlin] blue *(a complex iron cyanide)*
PRH *s.* Prolactin-Releasinghormon
prillen *v* prill *(to convert a solution or suspension into pellets)*
Primaquin *n* primaquine *(a quinoline derivative)*
Primärbatterie *f* primary battery
Primärdestillation *f* primary distillation
Primärelektron *n* primary electron
Primärelement *n (el ch)* primary cell *(not capable of being regenerated)*
Primärenergie *f* primary energy
Primärkeimbildung *f (cryst)* primary nucleation
Primärmetabolismus *m (bioch)* primary metabolism
Primärmetabolit *m (bioch)* primary metabolite
Primärreaktion *f* primary [initiation] reaction
Primärreaktion *f***/photochemische** primary photoreaction [photochemical reaction]
Primärschadstoff *m* primary pollutant, precursor
Primärschritt *m* initiation step *(of a reaction)*
Primärstandard *m* primary standard *(pH measurement)*
Primärstoffwechsel *m (bioch)* primary metabolism
Primärstrahl *m* primary ray [beam]
Primärstruktur *f (bioch)* primary structure
Primärteer *m* primary tar
Primärweichmacher *m* primary plasticizer
Prinzip *n* 1. principle, rule; 2. principle, fundamental constituent
Prinzip *n* **der größten Multiplizität** maximum-multiplicity principle [rule]
Prinzip *n* **der harten und weichen Säuren und Basen** HSAB principle
Prinzip *n* **der korrespondierenden Zustände** *(phys ch)* principle of corresponding states
Prinzip *n* **der mikroskopischen Reversibilität** *(phys ch)* principle of microscopic reversibility
Prinzip *n* **des beweglichen Gleichgewichts** *s.* Prinzip des kleinsten Zwangs
Prinzip *n* **des detaillierten Gleichgewichts** *(org ch)* principle of detailed balancing
Prinzip *n* **des kleinsten Zwangs** Le Châtelier principle, Le Châtelier-Braun principle, principle of mobile equilibrium
Prinzip *n* **des quasistationären Zustands** steady-state method
Prinzip *n***/le-chatelier-braunsches** *s.* Prinzip des kleinsten Zwangs
Prinzip *n***/toxisches** toxic principle

Priorität *f (nomencl)* priority, precedence, seniority
Prioritätenfolge *f (nomencl)* order of priority [precedence]
Prisma *n*/**nicolsches** Nicol prism
prismatisch *(cryst)* prismatic
Prismenspektrograph *m* prism spectrograph
PRL *s.* Prolactin
Pro *s.* Prolin
Probe *f* 1. *(anal)* sample; 2. *(chromat)* probe *(a compound having well-defined properties)*; 3. *s.* Probekörper; 4. *s.* Prüfung • **Proben entnehmen** sample, take [collect] samples, (with-)draw samples
Probekörper *m* test piece, specimen
Probelauf *m* experimental run, trial [dry] run
Probelösung *f* test solution; solution to be tested; solution under examination
Probematerial *n* test material
Probenahme *f* sampling, sample collection, withdrawal [collection] of samples
Probenahme *f*/**geschichtete** stratified sampling
Probenahme *f* **mittels Probenstechers** thief sampling
Probenahmefehler *m* sampling error
Probenahmemethode *f* sampling technique
Probenahmeort *m* sampling location
Probenahmestelle *f* sampling site
Probenahmetechnik *f* sampling technique
Probenahmevorrichtung *f* sampling device, sampler
Probenahmevorschrift *f* sampling specification
Probenaufbereitung *f* sample preparation
Probenaufbewahrung *f* sample storage
Probenaufgabe *f* sample application
Probenaufgabe *f* **mit Gasstromteilung** *(chromat)* split sampling
Probenaufgabe *f* **ohne Gasstromteilung** *(chromat)* splitless sampling
Probenbohrer *m* auger sampler
Probendiskriminierung *f* sample discrimination
Probendurchsatz *m* sample throughput
Probenehmer *m* sampling device, sampler, (sample) collector
Probeneinlass *m* sample inlet [port]
Probeneinlasssystem *n* (sample) inlet system
Probengeber *m* 1. (sample) dispenser; 2. *(chromat)* sample injector
Probengeber *m*/**automatischer** autosampler
Probengröße *f* sample size
Probengut *n* test material
Probenkonservierung *f* sample preservation
Probenküvette *f (spectr)* sample cell
Probenmenge *f* sample amount [quantity]
Probenschleife *f (chromat)* sample loop
Probenstecher *m* sampling thief
Probenstecher *m*/**pneumatischer [pneumatisch arbeitender]** pneumatic sampler
Probenteiler *m* sample divider [reducer]
Probenüberführung *f* sample transfer
Probenumfang *m* sample size *(statistically)*
Probenverarbeitung *f* sample reduction *(in a larger sense, relating to volume and grain size)*
Probenverjüngung *f* sample division [reduction]
Probenvorbereitung *f* sample preparation
Probenwechsler *m* sample changer
Probenzerkleinerung *f* sample reduction
Probenzufuhr *f* sample introduction
Probestück *n* sample
Probesubstanz *f* test substance, substance under investigation
Probierglas *n* test tube
prochiral prochiral *(stereochemistry)*
Prochiralität *f* prochirality *(stereochemistry)*
Produkt *n (ch, tech)* product; make *(of a specified factory)*
Produkt *n*/**feuerfestes** refractory
Produkt *n*/**geringveredeltes** low-added-value product
Produkt *n*/**hochveredeltes** high-added-value product
Produkt *n*/**reformiertes** *(petrol)* reformate
Produkt *n*/**verperltes** shot

Produktabtrennung *f* product separation
Produktaufarbeitung *f* product recovery
Produktausbeute *f (biot)* product yield, process [conversion] yield
Produktbildner *mpl (biot)* producers, producing microorganisms
Produktbildung *f* product formation
Produktbildungsphase *f (biot)* product formation phase, idiophase
Produktbildungsrate *f (biot)* rate of product formation; fermentation rate
Produktenhemmung *f (bioch)* product [feedback] inhibition
Produktgas *n* product gas
Produktion *f* 1. *(know-how:)* manufacture; 2. *(in a specified period or scale:)* production; 3. *(of a specified plant:)* make
Produktion *f* **in halbtechnischem Maßstab** pilot-plant(-scale) production
Produktionsabwasser *n* factory effluent, process waste-water
Produktionsanlage *f* production plant, *(as opposed to pilot plant:)* commercial [full-scale] plant
Produktionsbetrieb *m* production plant, *(as opposed to pilot plant:)* commercial [full-scale] plant
Produktionsfermentation *f (biot)* production fermentation
Produktionsfermenter *m (biot)* production fermenter
Produktionskultur *f (biot)* production culture
Produktionsleistung *f* production output
Produktionsprozess *m* manufacturing process
Produktionsreaktor *m* commercial reactor
Produktionsreife *f* production-line status
Produktionsstamm *m (biot)* production strain *(of microorganisms)*
Produktionsverfahren *n* manufacturing [production] process
Produktionsvermögen *n* capacity
Produktionswasser *n* process(ing) water
Produktkühler *m (distil)* product cooler
Produktphase *f* product phase
Produzent *m* 1. manufacturer; 2. *(biot)* producer, producing microorganism
Proenzym *n* zymogen, proenzyme
Profil *n* 1. profile, shape; 2. *(plast)* section
Profil *n*/**Lorentzförmiges** *(spectr)* Lorentzian profile
profilieren *v* shape, profile
Progesteron *n* progesterone *(a hormone)*
Projektionsformel *f* projection formula
Projektionsformel *f* **von Haworth/sterische** Haworth (projection) formula *(for sugars)*
Prokaryot *m (bioch)* prokaryote
Prokaryotenzelle *f (bioch)* prokaryotic cell
Prolactin *n (bioch)* prolactin, lactogenic [luteotrophic] hormone
Prolactin-Releasinghormon *n* prolactin-releasing hormone
Prolactoliberin *n* prolactin-releasing hormone
Prolactostatin *n* prolactin release-inhibiting factor
Prolamin *n* prolamine *(a simple vegetable protein)*
Prolin *n* proline
Promethium *n* Pm promethium
Promotionsenergie *f (phys ch)* promotion energy
Promotor *m* promoter, activator
Prooxidans *n* prooxidant (agent)
Propadien *n* propadiene, allene
Propagierung *f (biot)* multiplication, propagation *(of microorganisms)*
Propan *n* propane
Propanal *n* *propanal, propionaldehyde
Propandisäure *f* propanedioic acid, malonic acid
Propanentparaffinierung *f* propane dewaxing
Propan-1-ol *n* propan-1-ol
Propan-2-on *n* propan-2-one, dimethylketone, acetone
Propan-1,2,3-triol *n* propane-1,2,3-triol, 1,2,3-trihydroxypropane, glycerol
Propargylalkohol *m* propargyl alcohol, prop-2-yn-1-ol
Propellan *n (org ch)* propellane
Propellermischer *m* propeller agitator [mixer]
Propellerrührer *m* propeller agitator [mixer]

Propen *n* propene
Propenal *n* *propenal, acrolein
Propennitril *n* *propenenitrile, acrylonitrile
Prop-2-en-1-ol *n* prop-2-en-1-ol, allyl alcohol
Propensäure *f* propenoic acid, acrylic acid
Propin *n* propyne
Prop-2-in-1-ol *n* prop-2-yn-1-ol, propargyl alcohol
Propiolsäure *f* propiolic acid
Propionaldehyd *m* propionaldehyde, *propanal
Propionat *n* propionate *(salt or ester of propionic acid)*
Propionitril *n* propionitrile
Propionsäure *f* propionic acid
Propionsäureethylester *m* ethyl propionate
Propionylgruppe *f* C_2H_5CO– propionyl group
Propiophenon *n* propiophenone
Propolis *f* propolis, bee glue
Proportionalitätsfaktor *m* proportionality constant
Proportional(itäts)wägung *f* direct weighing
Propylacetat *n* propyl acetate
Propylalkohol *m* *propanol, propyl alcohol
Propylamin *n* propylamine
Propylenglykol *n* 1,2-propylene glycol
Propylenoxid *n* propylene oxide, 1,2-epoxypropane, 2-methyloxirane
Propylierung *f* propylation
prospektieren *v (petrol, mine)* prospect
Prospektierung *f (petrol, mine)* prospecting
Prostaglandin *n* prostaglandin *(any of a group of animal hormones)*
Prostansäure *f (bioch)* prostanoic acid
Protactinium *n* Pa protactinium
Protamin *n* protamine *(any of a class of simple proteins)*
Protease *f* protease, proteolytic enzyme
Proteid *n* proteid(e), conjugated protein
Protein *n* protein
Protein *n*/**acylübertragendes** acyl-carrier protein
Protein *n*/**bakterielles** bacterial protein
Protein *n*/**einfaches** simple protein
Protein *n*/**faserartiges [fibrilläres]** fibrillar protein, fibrous [structural] protein, scleroprotein
Protein *n*/**globuläres** globular protein
Protein *n*/**helixdestabilisierendes** *(bioch)* SSB protein, single-strand (DNA-)binding protein
Protein *n*/**konjugiertes** conjugated protein, proteid(e)
Protein *n*/**mikrobielles** microbial protein
Protein *n*/**natives** native protein
Protein *n*/**pflanzliches** plant [vegetable] protein
Protein *n*/**tierisches** animal protein
Proteinabbau *m* protein degradation [breakdown]
Proteinanteil *m* protein moiety [component]
Proteinaseinhibitor *m* proteinase inhibitor
Proteinbestimmungsmethode *f* **nach Lowry** Lowry method (for protein determination)
Proteinbiosynthese *f (biol)* protein biosynthesis
Proteinchemie *f* protein chemistry
Proteinchemiker *m* protein chemist
Proteindenaturierung *f* protein denaturation
Proteinfaltung *f* protein folding
Proteinfaser *f* protein fibre
Proteinfaserstoff *m* protein fibre
proteingebunden protein-bound
Proteinhydrolysat *n* protein hydrolyzate
proteinisch proteinaceous
Proteinisolat *n* protein isolate
Proteinkomponente *f* protein moiety [component]
Proteinkunststoff *m* protein plastic
Proteinstoffwechsel *m* protein metabolism
Proteinsynthese *f* protein synthesis
Proteohormon *n* proteohormone
Proteolipid *n (bioch)* proteolipid
Proteolyse *f* proteolysis, protein cleavage
Protid *n* $^1H^-$ protide
protisch protic
Protium *n* ^1H protium, light hydrogen
Protoalkaloide *npl* biogenic amines
Protocatechualdehyd *m* protocatechuic aldehyde
Protocatechusäure *f* protocatechuic acid

protogen protogenic
Protohämin *n* protohaemin
Protolyse *f* protolysis
Protolysetitration *f* neutralization [acid-base] titration
Protolyt *m* protolyte
protolytisch protolytic
Protomer(e) *n* protomer, protomeric unit *(protein chemistry)*
Protomerie *f s.* Prototropie
Proton *n* proton
protonenabgebend protogenic
Protonenaffinität *f* proton affinity
Protonenakzeptor *m* proton acceptor, Brønsted-Lowry base
protonenaufnehmend protophilic
Protonenaustausch *m* proton exchange
Protonenaustauschmembran *f* proton exchange membrane
Protonenaustauschreaktion *f* proton exchange reaction *(acid-base reaction)*
Protonendonator *m* proton donor [acid], Brønsted-Lowry acid
Protonen-Kernresonanz *f* proton (magnetic) resonance, PMR
Protonenpumpe *f (bioch)* proton pump
Protonenruhemasse *f* proton (rest) mass
Protonensäure *f s.* Protonendonator
Protonenübertragungsreaktion *f* proton transfer reaction
Protonenverschiebung *f* proton shift
protonieren *v* protonate
Protonierung *f* protonation
protonisch protic
protonophil protophilic
Protoplasma *n* protoplasm
prototrop prototropic
Prototropie *f* prototropy, prototropic rearrangement *(one form of tautomerism)*
Provitamin *n* provitamin, vitamin precursor
Prozentgehalt *m* percentage
Prozessabwasser *n* process wastewater
Prozessanalyse *f* on-line analysis
Prozesschemikalie *f* process chemical
Prozessdampf *m* process steam
Prozessgas *n* process gas
Prozessgestaltung *f* process design
Prozessgröße *f* process quantity
Prozessschritt *m* process(ing) step
Prozesssteuerung *f* process control
Prozessstufe *f* process(ing) stage, treatment stage
Prozessvariable *f* process variable
Prozesswärme *f* process heat
Prozesswasser *n* process(ing) water
Prozessziel *n* process goal
prüfen *v* 1. examine; test *(esp for a specified substance or for purity)*; 2. *(met)* assay *(ores)*; test *(materials)*; check *(esp the correctness or completeness)*
prüfen *v***/nochmals** recheck
Prüfender *m* tester
Prüfer *m s.* 1. Prüfgerät; 2. Prüfender
Prüfgerät *n* testing apparatus, tester
Prüfgut *n* entire [parent] lot
Prüfkörper *m* test piece, specimen
Prüfling *m* test piece, specimen
Prüflösung *f* test(ing) solution
Prüfmethode *f* test(ing) method
Prüfmethode *f***/standardisierte** standard test(ing) method
Prüfpapier *n* test [indicator] paper
Prüfung *f* 1. examination; testing *(of materials)*; 2. test *(esp for a specified substance or for purity)*; 3. *(met)* assay; check *(esp for ensuring the correctness or completeness)*
Prüfung *f***/beschleunigte** accelerated testing
Prüfung *f***/nochmalige** re-examination
Prüfung *f***/organoleptische** organoleptic examination [testing], sensory examination [testing]
Prüfung *f***/orientierende** preliminary test *(as of pesticides)*
Prüfung *f***/zerstörungsfreie** non-destructive testing, NDT
Prüfungsvorschrift *f* specification
Prüfverfahren *n* test(ing) method
Prussiat *n* prussiate, pentacyanoferrate
Prussid *n* prussiate, pentacyanoferrate
PS *s.* Polystyren
PSE *s.* Periodensystem (der Elemente)
p-Serie *f (spectr)* principal series
Pseudoaromaten *pl* pseudo-aromatics
Pseudoasymmetrie *f* pseudo-asymmetry
pseudoasymmetrisch pseudo-asymmetric(al)
Pseudobase *f* pseudo base
Pseudocumen *n* pseudo-cumene

Pseudohalogen *n* pseudo-halogen
Pseudohalogenid *n* pseudo-halide
Pseudokatalyse *f* pseudo-catalysis
pseudokristallin pseudo-crystalline
Pseudolösung *f* pseudo-solution
pseudomorph *(cryst)* pseudomorphic
Pseudomorphie *f (cryst)* pseudomorphism
Pseudopeptid *n* pseudopeptide
Pseudoplastizität *f* pseudo-plasticity
pseudoracemisch pseudo-racemic
Pseudorotation *f* pseudo-rotation
Pseudosalz *n* pseudo salt
Pseudosäure *f* pseudo acid
Pseudosymmetrie *f* pseudo-symmetry
pseudosymmetrisch pseudo-symmetric(al)
Pseudouridylsäure *f (bioch)* pseudouridylic acid
Psychochemikalie *f* psychochemical
Psychodysleptikum *n s.* Psychotomimetikum
Psychogift *n* psychochemical
Psycholeptikum *n (pharm)* neuroleptic (drug), antipsychotic, major tranquilizer
Psychopharmakon *n* psychopharmacologic drug [agent]
Psychosedativum *n (pharm)* (minor) tranquil(l)izer, tranquillizing drug, anxiolytic (sedative)
Psychostimulans *n (pharm)* psychostimulant
Psychotikum *n s.* Psychotomimetikum
Psychotomimetikum *n (pharm)* psychotomimetic (drug), psychodysleptic, hallucinogen
psychotomimetisch *(pharm)* psychotomimetic, hallucinogenic
Psychotonikum *n (pharm)* psychostimulant
Psychotropikon *n* psychopharmacologic drug [agent]
Psychrometer *n* psychrometer, wet-and-dry-bulb hygrometer [thermometer]
PT *s.* Polyethylenfaserstoff
Pteridin *n* pteridine
Pterin *n* pterin *(any of the pteridine derivatives)*
Pteroinsäure *f (bioch)* pteroic acid
Pteroylglutamat *n (bioch)* pteroylglutamate, folate
Pteroylglutaminsäure *f* pteroylglutamic acid, PGA, folic acid
PTFE *s.* Polytetrafluorethylen
PTH *s.* Parathormon
puddeln *v (met)* puddle
Puddelroheisen *n* forge pig iron
Puddelverfahren *n (met)* puddling process
Puder *m (cosmet)* powder
pudern *v* dust, powder
Puffer *m* buffer
Pufferbase *f* buffer base
Pufferbereich *m* buffer region
Puffergemisch *n* buffer(ing) mixture
Pufferkapazität *f* buffering capacity
Pufferlösung *f* buffer solution
Pufferlösung *f*/**standardisierte** standard buffer (solution)
Puffermischung *f* buffer(ing) mixture
puffern *v* buffer
Puffersäure *f* buffer acid
Puffersubstanz *f* buffering agent, buffer reagent
Pulegon *n* pulegone *(a monocyclic ketone)*
Pulfrich-Refraktometer *n* Pulfrich refractometer
Pulpe *f (food)* (fruit) pulp, pomace
Pülpe *f (food)* (fruit) pulp, pomace
Pulper *m (pap)* pulper, pulping engine
Pulsation *f* pulsation
Pulsationskolonne *f* pulsed column
Pulsationsschwingung *f (spectr)* (symmetrical) breathing vibration
pulsieren *v* pulsate
Pulsometer *n* pulsometer (pump), blow case
Pulspolarographie *f (anal)* pulse polarography
Pulsradiolyse *f* pulse(d) radiolysis *(for investigating fast reactions)*
Pultrusion *f (plast)* pultrusion
Pulver *n* powder • **zu Pulver zerfallen** fall to powder
Pulveraufkohlen *n (met)* pack carburizing
Pulverbeschichtung *f* powder coating
Pulverbeschichtung *f*/**elektrostatische** *(coat)* electrostatic powder coating
Pulverdiagramm *n (cryst)* powder diagram, (X-ray) powder pattern, (X-ray) powder photograph
Pulverdiffraktometer *n* X-ray powder

diffractometer, powder (diffraction) camera
Pulverflasche *f* wide-mouth bottle
pulverig powdery, pulverulent
pulverisieren *v* pulverize, reduce to powder, powder, triturate
Pulverisiermühle *f* pulverizing mill
Pulverisolierung *f* powder insulation *(cryogenics)*
Pulverlackierung f/elektrostatische *s.* Pulverbeschichtung/elektrostatische
Pulverlöscher *m* powder extinguisher, dry chemical (fire) extinguisher
Pulvermetallurgie *f* powder metallurgy
pulvermetallurgisch powder-metallurgical
Pulvermethode *f (cryst)* (X-ray) powder method
pulvern *s.* pulverisieren
Pulvertrichter *m* powder funnel, filling [wide-stemmed] funnel
pulvrig powdery, pulverulent
Pummerer-Umlagerung *f* Pummerer rearrangement *(of sulphoxides into α-acyloxysulphides)*
Pumpe *f* pump
Pumpe f/dreistufige three-stage pump
Pumpe f/peristaltische peristaltic pump
Pumpe f/rotierende rotary pump
pumpen *v* pump
pumpen v/voll charge by pumping
pumpfähig pumpable
Pumplaser *m* pump laser
Punkt m/azeotroper azeotropic point
Punkt *m* **bei gleich bleibender Temperatur/kritischer** plait point
Punkt m/eutektischer eutectic point
Punkt m/isoelektrischer isoelectric point, zero point of charge, ZPC
Punkt m/isoionischer isoionic point
Punkt m/kritischer critical point *(as in three-component systems)*
Punkt m/peritektischer peritectic [transition] point
Punktcodon *n (bioch)* terminator codon
Punktdefekt *m (cryst)* point defect
Punktgruppe *f* 1. *(spectr)* point group; 2. Punktsymmetriegruppe
Punktladung *f (phys ch)* point charge
Punktmutation *f (biot)* point mutation
Punktquelle *f* point source (of emission)
Punktsymmetriegruppe *f (cryst)* point [symmetry] group, crystal (symmetry) class
PUR *s.* Polyurethan
Purgans *n (pharm)* purgative
Purgativum *n (pharm)* purgative
Purinalkaloid *n* purine alkaloid
Purinbase *f* purine base
Purinstoffwechsel *m* purine metabolism
Purpurin *n* purpurin
Purpursäure *f* purpuric acid
Putzmittel *n* scouring agent
Putzmörtel *m* plaster (mortar)
PV *s.* Polyvinylfaserstoff
PVA *s.* Polyvinylalkohol
PVAC *s.* Polyvinylacetat
PVB *s.* Polyvinylbutyral
PVC *s.* Polyvinylchlorid
PVCA *s.* Polyvinylchloridacetat
PVC-H unplasticized [rigid] PVC
PVC-hart *n* unplasticized [rigid] PVC
PVC-Schaumstoff *m* cellular PVC, PVC foam
PVC-W plasticized [flexible] PVC
PVC-Weich *n* plasticized [flexible] PVC
PVDC *s.* Polyvinylidenchlorid
PVDF *s.* Polyvinylidenfluorid
PVK *s.* Pigmentvolumenkonzentration
PVY *s.* Polyacrylnitrilfaserstoff
Pyknometer *n* pycnometer, specific-gravity bottle [flask]
pyknometrisch pycnometric
P-Ylid *n s.* Phosphoniumylid
Pyran *n (org ch)* pyran
Pyranose *f* pyranose *(a cyclic monosaccharide)*
Pyranosid *n (bioch)* pyranoside
Pyrazin *n* pyrazine, 1,4-diazine
Pyrazol *n* pyrazole, 1,2-diazole
Pyrazolon *n* pyrazolone
Pyrazolonfarbstoff *m* pyrazolone dye
Pyren *n* pyrene *(a tetracyclic compound)*
Pyrethrin *n* pyrethrine *(insecticide)*
Pyrethroid *n* pyrethroid *(any of a series of insecticidal compounds)*
Pyridazin *n* pyridazine, 1,2-diazine
Pyridin *n* pyridine
Pyridinalkaloid *n* pyridine alkaloid
Pyridinbase *f* pyridine base
Pyridin-3-carbonsäure *f* pyridine-3-carboxylic acid, nicotinic acid
Pyridindicarbonsäure *f* pyridinedicarboxylic acid

Pyridoxalphosphat *n* pyridoxal phosphate *(coenzyme of decarboxylase)*
Pyridoxin *n* pyridoxin, adermin(e) *(vitamin B_6)*
Pyrimidin *n* pyrimidine, 1,3-diazine
Pyrimidinbase *f (bioch)* pyrimidine base
Pyrit *m (min)* (iron) pyrite(s), mundic *(iron(II) sulphide)*
Pyritröstung *f (met)* roasting of pyrite(s)
Pyrogallol *n* pyrogallol, 1,2,3-trihydroxybenzene
Pyrolyse *f* pyrolysis
Pyrolyse-Feldionisations-Massenspektrometrie *f* pyrolysis field ionization mass spectrometry
Pyrolyse-Gaschromatographie *f* pyrolysis gas chromatography
Pyrolyseprodukt *n* pyrolyzate
Pyrolysereaktion *f* pyrolysis [pyrolytic] reaction
pyrolysieren *v* pyrolyze
pyrolytisch pyrolytic
Pyromellit(h)säure *f* pyromellitic acid, benzene-1,2,4,5-tetracarboxylic acid
Pyromellit(h)säuredianhydrid *n* pyromellitic dianhydride, PMDA
Pyrometallurgie *f* pyrometallurgy
pyrometallurgisch pyrometallurgical
Pyrometer *n* pyrometer
Pyrometrie *f* pyrometry
pyrometrisch pyrometric
Pyron *n* pyrone *(any of a class of heterocyclic compounds containing the oxo group)*
pyrophor pyrophoric
Pyrophosphat *n* diphosphate, pyrophosphate
Pyrophosphorsäure *f* diphosphoric acid, pyrophosphoric acid
Pyrophosphorylierung *f (bioch)* pyrophosphorylation
Pyrosäure *f* pyroacid
Pyrotechnik *f* pyrotechnics
pyrotechnisch pyrotechnic(al)
Pyrrol *n* pyrrole
Pyrrolidin *n* pyrrolidine, tetrahydropyrrole
Pyrrolidinalkaloid *n* pyrrolidine alkaloid
Pyrrolidin-2-carbonsäure *f* pyrrolidine-2-carboxylic acid, proline
Pyrrolizidinalkaloid *n* pyrrolizidine [senecio] alkaloid
Pyrrolsynthese *f***/knorrsche** Knorr pyrrole synthesis
Pyrron *n* pyrrone, polyimidazopyrrolone
Pyruvat *n* pyruvate *(salt or ester of pyruvic acid)*
Pyruvinsäure *f* pyruvic acid
PZ *s.* 1. Polenske-Zahl; 2. Portlandzement

Q

Q-Einheit *f* Q unit, tetrafunctional unit *(a structural unit)*
Quadrupeleffektverdampfer *m* quadruple-effect evaporator
Quadrupelpunkt *m (phys ch)* quadruple point
Quadrupol *m (phys ch)* quadrupole
Quadrupolion *n* quadruple ion
Quadrupolmassenspektrometer *n* quadrupole mass spectrometer
Quadrupolmoment *n* quadrupole moment
Quadrupolmoment *n* **des Kerns** nuclear quadrupole moment
Qualität *f* grade, quality • **von minderer Qualität** low-grade
Qualität *f***/handelsübliche** commercial grade
Qualitätsgröße *f* intensive property [quantity]
Qualitätskontrolle *f* quality control
Qualitätsprüfung *f* quality test
Qualitätssicherung *f* quality assurance
Quant *n (phys ch)* quant(um)
quanteln *v* quantize
Quantelung *f* quantization
Quantelung *f* **der Energie** quantization of energy
Quantelung *f***/räumliche** space quantization
Quantenausbeute *f* quantum yield
Quantenchemie *f* quantum chemistry
Quantenhypothese *f***/plancksche** Planck's hypothesis *(of quantized energy)*
Quantenmechanik *f* quantum mechanics
quantenmechanisch quantum-mechanical
Quantenpunkt *m* quantum dot *(semiconductor nanostructure)*

Quantentheorie *f* quantum theory
Quantenzahl *f* quantum number
Quantenzahl *f*/**azimutale** azimuthal [subsidiary, orbital] quantum number
Quantenzahl *f*/**innere** inner quantum number
Quantenzahl *f*/**magnetische** magnetic quantum number
Quantenzahl *f*/**radiale** radial quantum number
Quantitätsgröße *f (phys ch)* extensive property [quantity]
Quantum *n* quantum, quantity, amount; portion
Quartärsalz *n* quaternary salt
Quartärstruktur *f* quaternary structure *(of proteins)*
Quartettaufspaltung *f (phys ch)* quartet splitting
quartieren *v (anal)* quarter
Quarz *m* quartz • **aus Quarz bestehend** quartzose, quartzous, quartzy
quarzartig quartzous, quartzose, quartzy
Quarzglas *n* quartz glass, transparent (vitreous) silica, fused quartz
Quarzgut *n* translucent vitreous silica, fused silica
quarzhaltig quartzose, quartzous, quartzy
Quarzit *m (min)* quartzite
quarzitisch quartzitic
Quarzkristallmikrogravimetrie *f* quartz crystal microgravimetry
Quarzlampe *f* quartz lamp
Quarzmikrowaage *f* quartz crystal microbalance, QMC
Quarzsand *m (geol)* glass sand
Quarz-UV-Bereich *m* near-ultraviolet range
Quasigleichgewicht *n* quasi-equilibrium
quasikristallin quasi-crystalline
quasistationär quasi-stationary
Quasistationaritätsprinzip *n* steady-state method
quaternisieren *v* quaternize *(e.g. amines)*
Quaternisierung *f* quaternization *(as of amines)*
Quayle-Zyklus *m (bioch)* Quayle cycle, ribulose monophosphate cycle
Quecksilber(II)-acetat *n* mercury(II) mercuric acetate
Quecksilberbarometer *n* mercury barometer
Quecksilber(I)-bromid *n* mercury(I) bromide, mercurous bromide
Quecksilber(II)-bromid *n* mercury(II) bromide, mercuric bromide
Quecksilber(I)-chlorid *n* mercury(I) chloride, mercurous chloride, calomel
Quecksilber(II)-chlorid *n* mercury(II) chloride, mercuric chloride, sublimate
Quecksilber(II)-cyanid *n* mercury(II) cyanide, mercuric cyanide
Quecksilberdampflampe *f* mercury-vapour lamp
Quecksilber-Diffusionspumpe *f* mercury diffusion pump, mercury-vapour pump
Quecksilberelektrode *f* mercury electrode
Quecksilberfilmelektrode *f (el ch)* thin mercury film electrode, mercury-film electrode, TMFE
Quecksilber(I)-fluorid *n* mercury(I) fluoride, mercurous fluoride
Quecksilber(II)-fluorid *n* mercury(II) fluoride, mercuric fluoride
Quecksilber(II)-fulminat *n* mercury fulminate, mercuric fulminate
quecksilberhaltig mercurial
Quecksilber(I)-iodid *n* mercury(I) iodide, mercurous iodide
Quecksilber(II)-iodid *n* mercury(II) iodide, mercuric iodide
Quecksilbermanometer *n* mercury manometer
Quecksilberniederdrucklampe *f* low-pressure mercury-vapour lamp
Quecksilber(I)-nitrat *n* mercury(I) nitrate, mercurous nitrate
Quecksilber(II)-nitrat *n* mercury(II) nitrate, mercuric nitrate
Quecksilber(I)-oxid *n* mercury(I) oxide, mercurous oxide
Quecksilber(II)-oxid *n* mercury(II) oxide, mercuric oxide
Quecksilber(II)-oxid *n*/**gefälltes [gelbes]** yellow mercuric oxide, yellow precipitate
Quecksilber(II)-oxid *n*/**rotes** red mercuric oxide, red precipitate
Quecksilberporosimetrie *f* mercury porosimetry
Quecksilbersäule *f* mercury column

Quecksilber(I)-sulfat *n* mercury(I) sulphate, mercurous sulphate
Quecksilber(II)-sulfat *n* mercury(II) sulphate, mercuric sulphate
Quecksilber(II)-sulfid *n* mercury(II) sulphide, mercuric sulphide
Quecksilbersulfid *n***/rotes** red mercuric sulphide
Quecksilberthermometer *n* mercury thermometer
Quecksilber(II)-thiocyanat *n* mercury(II) thiocyanate, mercuric thiocyanate
Quecksilbertropfelektrode *f* dropping mercury electrode, DME
Quecksilbertropfkatode *f* mercury dropping cathode
Quecksilberverbindung *f* mercury compound, mercurial
Quecksilberverfahren *n* mercury-cell process *(electrolysis)*
Quecksilbervergiftung *f* mercury poisoning, mercurialism, hydrargyrism
Quecksilberzange *f* mercury tongs
Quecksilberzelle *f* mercury cell *(electrolysis)*
quellbar capable of swelling, swellable
Quellbarkeit *f* capability of swelling, swellability
quellbeständig resistant to swelling
Quellbeständigkeit *f* resistance to swelling
Quelle *f* 1. source; 2. spring
quellen *v* swell; steep, (cause to) swell
quellfähig capable of swelling
Quellfähigkeit *f* 1. capability of swelling; 2. swelling power
Quellkinetik *f* swelling kinetic
Quellungsgleichgewicht *n* swelling equilibrium
Quellungsgrad *m* swelling degree, degree of swelling *(polymer networks)*
Quellungsmittel *n* swelling agent
quenchen *v* quench *(to slow down rapidly esp a reaction)*
Quencher *m* quenching agent *(photochemistry)*
Querdiffusion *f* lateral diffusion
Querschnitt *m* cross section
Querschnittsfläche *f* cross-sectional area
Querstrom *m* cross flow
Querströmung *f* cross flow
Quervermischung *f* radial mixing
quervernetzen *v* cross-link
Quervernetzung *f* cross-linkage, *(result also)* cross-link
Quetschhahn *m* pinchcock, hose cock, pinch clamp
Quetschhahn *m* **nach Hoffmann** Hoffmann clamp
Quetschwalze *f* press [squeeze] roll, squeegee
Quotient *m***/respiratorischer** *(bioch)* respiratory quotient [ratio]

R

Racemat *n* racemate *(1. optically inactive compound*; *2. salt or ester of DL-tartaric acid)*
Racematspaltung *f (org ch)* resolution
racemisch racemic
racemisieren *v* racemize
Racemisierung *f* racemization
Rad *n* rad *(non-SI unit of absorbed dose of ionizing radiation*; *1 rd = 10^{-2} J/kg)*
Radialgleitlager *n* journal bearing
Radialstromreaktor *m* radial-flow (catalytic) reactor
Radikal *n* (free) radical
Radikal *n***/langlebiges [persistentes]** long-lived radical, persistent radical
Radikalanion *n* anion radical
Radikalbildner *m* radical former
Radikalfalle *f* free-radical trap
Radikalfänger *m* free-radical scavenger
Radikalion *n* radical ion, ion radical
radikalisch radical
Radikalkation *n* cation radical
Radikalkettenpolymerisation *f* (free-) radical polymerization
Radikalkettenreaktion *f* (free-)radical chain reaction
Radikalmechanismus *m* free-radical mechanism
Radikalreaktion *f* (free-)radical reaction
radioaktiv radioactive • **radioaktiv markiert** radio-labelled
radioaktiv/nicht non-radioactive
radioaktiv/schwach feebly radioactive
radioaktiv/stark highly radioactive, hot
Radioaktivität *f* radioactivity
Radioaktivität *f***/induzierte [künstliche]** induced radioactivity

Radioautographie *f* autoradiography, radioautography
Radiochemie *f* radiochemistry
Radiochemiker *m* radiochemist
radiochemisch radiochemical
Radioelement *n* radioelement, radioactive element
Radiographie *f* radiography *(materials testing)*
radiographisch radiographic
Radioimmunoassay *n (bioch)* radioimmunoassay, RIA
Radioindikator *m* radiotracer, radioactive indicator [tracer]
Radiokohlenstoffdatierung *f* radiocarbon dating, C^{14} method
Radiologie *f* radiology
Radiolumineszenz *f* radioluminescence
Radiolyse *f* radiolysis
Radiomarkierung *f* radio labelling
Radiometer *n* radiometer
Radiometrie *f* radiometry
Radiomimetikum *n (biol)* radiomimetic (agent)
Radionuklid *n* radionuclide, radioactive nuclide
Radiotracer *m* radiotracer, radioactive tracer [indicator]
Radium *n* Ra radium
Radiumreihe *f* radium series
Radon *n* Rn radon
Raffinat *n* raffinate *(solvent extraction)*
Raffination *f* refining, purification *(as of metals, oils, sugar)*; refining, (refining) treatment *(of petroleum and its products)*
Raffination *f***/elektrochemische [elektrolytische]** electrolytic refining, electrorefining
Raffination *f***/hydrierende** hydrorefining
Raffination *f* **im Schmelzfluss** fire refining
Raffinerie *f* refinery, refining plant
Raffinerie *f***/carbochemische [kohlechemische]** coal refinery
Raffineriegas *n (petrol)* refinery gas
raffinieren *v* refine, purify *(e.g. metals, oils, sugar)*; refine, treat *(petroleum and its products)*
raffinieren *v***/im Schmelzfluss** fire-refine
Rahmenfilter *n* 1. screen (envelope) filter *(for gases)*; 2. plate-and-frame filter *(for liquids)*
Rahmen(filter)presse *f* plate-and-frame (filter) press
Rainout *n* rain-out, in-cloud scavenging *(of pollutants by raindrops)*
Raketentreibstoff *m* rocket propellant [fuel]
Raketentreibstoff *m***/fester** solid rocket propellant [fuel]
Raketentreibstoff *m***/flüssiger** liquid rocket propellant [fuel]
Raketentreibstoff *m***/hypergoler [selbstzündender]** hypergolic rocket propellant, hypergolic fuel
ramanaktiv Raman-active
Raman-Effekt *m* Raman effect
Raman-Linie *f* Raman line
Raman-Spektroskopie *f* Raman spectroscopy
Raman-Spektroskopie *f***/inverse** inverse Raman spectroscopy, IRS
Raman-Spektrum *n* Raman spectrum
Raman-Streuung *f* Raman scattering
Raman-Verschiebung *f* Raman shift
Rand *m* 1. edge, border, brim; 2. *s.* Randschicht
randaufkohlen *v (met)* case-carburize
Randaufkohlung *f (met)* case carburization, case carburizing
Randbedingung *f* boundary condition
Randhärte *f (met)* case hardness
Randschicht *f (met)* case, surface layer
Randschicht *f***/aufgekohlte** carburized case
Randschicht *f***/gehärtete** hardened case
Randwinkel *m* contact [wetting] angle
Raney-Katalysator *m* Raney catalyst, (alloy) skeleton catalyst
Raney-Nickel *n* Raney nickel (catalyst)
Rang *m (coal)* rank
Rangfolge *f (nomencl)* order of priority [precedence]
Rankine-Skala *f* Rankine (temperature) scale
ranzig *(food)* rancid • **ranzig werden** rancidify
Ranzigkeit *f* rancidity
Ranzigwerden *n (food)* rancidification
Rapsöl *n* rape(-seed) oil *(from Brassica napus L. em. Metzg.)*
Raschig-Ring *m* Raschig ring *(a piece of packing)*
Raschig-Verfahren *n* Raschig process

(catalytic conversion of benzene into phenol)
raschwirkend quick-acting, short-action
Raseneisenerz *n (min)* bog iron (ore)
Rasterelektronenmikroskopie *f* scanning electron microscopy, SEM
Rasterkraftmikroskop *n* atomic force microscope, AFM
Rasterkraftmikroskopie *f* atomic force microscopy, AFM
Rast-Methode *f* Rast method [micromethod], Rast's molecular weight determination
Rationalitätsgesetz *n (cryst)* rational index law, Haüy law
Rattenbekämpfungsmittel *n* rat poison, raticide
Ratteninsulin *n (biot)* rat insulin
rau rough
Rauch *m* smoke *(as of a stack)*; fume *(as of concentrated acids)* • **dem Rauch aussetzen** fumigate
rauchen *v* smoke *(as of a stack)*; fume *(as of concentrated acids)*
Räuchermittel *n* fumigant *(pest control)*
räuchern *v (food)* smoke; fume *(timber)*; fumigate *(e.g. rooms or plants for pest control)*
Rauchfahne *f* (gas) plume
Rauchfahne *f*/**fächerförmige** fanning plume
Rauchfahne *f*/**kegelförmige [konische]** coning plume
Rauchfahne *f*/**schleifenförmige** looping plume
Rauchgas *n* flue gas
Rauchgasanalyse *f* flue-gas analysis
Rauchgasentschwefelung *f* flue-gas desulphurization
Rauchgaskanal *m* breeching
Rauchgasverlust *m* stack loss
Rauchgasvorwärmer *m* flue-gas preheater, economizer
Rauchgaswäscher *m* flue-gas scrubber
Rauchmeldeanlage *f* smoke monitor
Rauchpunkt *m (food)* smoke point
Rauheit *f* roughness
Rauigkeit *f* roughness
Raum *m* space; room, chamber
Raumausdehnungskoeffizient *m* isobaric coefficient of thermal expansion
Raumbelastung *f (hyd)* loading rate per unit volume *(expressed as* $kg/m^3 \cdot d$*)*
Raumelement *n* volume element
Raumformel *f* space [configurational] formula
Raumgitter *n (cryst)* space lattice
Raumgruppe *f (cryst)* space group
Rauminhalt *m* volume
Raumladung *f* space charge
Raumquantelung *f* space quantization
Raumsymmetriegruppe *f (cryst)* space group
Raumteil *m* part by volume
Raumtemperatur *f* room [ordinary] temperature
Raumverhältnis *n* proportion by volume
Raum-Zeit-Ausbeute *f* space-time yield
raumzentriert *(cryst)* space-centred, body-centred
rauscharm *(anal)* low-noise
Rauschen *n (anal)* (analytical) noise
Rauschen *n*/**thermisches** *(anal)* thermal [Johnson] noise
Rauschentkopplung *f (spectr)* broadband decoupling
Rauschgift *n* narcotic, drug
Rauschpegel *m (anal)* noise level
Rauwolfiaalkaloid *n* rauwolfia alkaloid
Rayleigh-Streuung *f (spectr)* Rayleigh scattering
RBW *s.* Wirksamkeit/relative biologische
R.D. *s.* Rotationsdispersion
rDNA *f (bioch)* recombinant DNA
Reagens *n* reagent
Reagens *n*/**chelatbildendes** chelating agent
Reagens *n*/**elektrophiles** electrophilic reagent, electrophile
Reagens *n*/**Millons** Millon's reagent *(for detecting proteins)*
Reagens *n*/**Neßlers** Nessler reagent [solution] *(for detecting ammonia and various amines)*
Reagens *n*/**nucleophiles** nucleophilic reagent, nucleophile
Reagens *n*/**Schweizers** Schweizer reagent, cuprammonium (hydroxide) solution *(for dissolving or detecting cellulose)*
Reagens *n*/**spezifisches** special reagent
Reagenzglas *n* test tube, proof

Reagenzglasversuch *m* test-tube experiment
Reagenzienblindwert *m* reagent blank
Reagenzpapier *n* test [indicator] paper
reagieren *v* react, undergo reaction *(chemically)*; respond *(as to an influence)*
reagieren *v* **gegen/sauer** be acid to
reagieren *v*/**stürmisch** react stormily
reagieren *v*/**träge** react sluggishly
Reaktant *m* reactant
Reaktion *f* 1. reaction; 2. *(bioch, pharm)* response • **die Reaktion in Gang halten** keep the reaction in progress • **eine Reaktion erleiden** undergo reaction • **in Reaktion treten** enter into reaction • **zur Reaktion bringen** cause to react
Reaktion *f*/**abbauende** degradation [breakdown] reaction
Reaktion *f*/**alkalische** alkaline reaction
Reaktion *f*/**anodische** anode reaction
Reaktion *f*/**aufbauende** build-up reaction
Reaktion *f*/**basische** alkaline reaction
Reaktion *f*/**bimolekulare** bimolecular reaction
Reaktion *f*/**blancsche** Blanc chloromethylation reaction
Reaktion *f*/**cannizzarosche** Cannizzaro reaction *(aldehyde dismutation)*
Reaktion *f*/**cheletrope** cheletropic [chelotropic] reaction
Reaktion *f*/**chemische** chemical reaction
Reaktion *f*/**dreistufige** three-step reaction
Reaktion *f* **dritter Ordnung** third-order reaction
Reaktion *f*/**einfache** simple reaction
Reaktion *f*/**einstufige** one-step reaction
Reaktion *f*/**elektrocyclische** electrocyclic reaction
Reaktion *f*/**enzymatische** enzymatic reaction
Reaktion *f* **erster Ordnung** first-order reaction
Reaktion *f*/**étardsche** Étard reaction *(for preparing aromatic aldehydes)*
Reaktion *f*/**gegenläufige** opposing [opposed] reaction
Reaktion *f*/**gekoppelte** coupled reaction
Reaktion *f*/**grundlegende** basic [fundamental] reaction
Reaktion *f*/**heiße** hot-state reaction
Reaktion *f*/**heterogene** heterogeneous reaction
Reaktion *f*/**heterolytische** heterolytic reaction
Reaktion *f* **höherer Ordnung** higher-order reaction
Reaktion *f*/**homogene** homogeneous reaction
Reaktion *f* **im Phasensystem fest flüssig** liquid-solid(-phase heterogeneous) reaction
Reaktion *f* **in (homogener) flüssiger Phase** liquid-phase reaction
Reaktion *f*/**in Lösung** solution reaction
Reaktion *f*/**induzierte** induced [sympathetic] reaction
Reaktion *f*/**ionische** ionic reaction
Reaktion *f*/**irreversible** irreversible [one-way] reaction
Reaktion *f*/**katalysierte [katalytische]** catalytic reaction
Reaktion *f*/**katodische** cathode reaction
Reaktion *f*/**komplexe** composite (complex) reaction
Reaktion *f*/**molischsche** Molisch reaction *(for detecting carbohydrates)*
Reaktion *f*/**monomolekulare** monomolecular [unimolecular] reaction
Reaktion *f*/**monotektische** monotectic reaction
Reaktion *f*/**monotektoide** monotectoid reaction
Reaktion *f*/**nicht katalysierte [katalytische]** non-catalytic reaction
Reaktion *f* **nullter Ordnung** zero-order reaction
Reaktion *f*/**oszillierende** oscillating (periodic) reaction
Reaktion *f*/**pericyclische** pericyclic reaction
Reaktion *f*/**peritektische** peritectic [incongruent] reaction
Reaktion *f*/**peritektoide** peritectoid reaction
Reaktion *f*/**perkinsche** Perkin condensation [reaction] *(for synthesizing unsaturated carboxylic acids)*
Reaktion *f*/**photochemische** photochemical reaction, photoreaction

Reaktion f/protolytische protolytic reaction
Reaktion f pseudo-erster Ordnung pseudo-first-order reaction
Reaktion f/pseudomonomolekulare pseudomonomolecular [pseudounimolecular] reaction
Reaktion f pseudo-nullter Ordnung pseudo-zero-order reaction
Reaktion f/radikalische free-radical reaction
Reaktion f/reversible reversible [balanced] reaction
Reaktion f/schmidtsche Schmidt reaction *(between hydrazoic acid and carbonyl compounds)*
Reaktion f/schnelle fast [rapid] reaction
Reaktion f/schotten-baumannsche Schotten-Baumann reaction *(acylation)*
Reaktion f/stoßinduzierte collision-induced reaction
Reaktion f/synchrone synchronous reaction
Reaktion f/topochemische topochemical reaction
Reaktion f/topotaktische *(cryst)* topotactic [insertion] reaction
Reaktion f/trimolekulare trimolecular [termolecular] reaction
Reaktion f/zweistufige two-step reaction
Reaktion f zweiter Ordnung second-order reaction
Reaktionsablauf *m* course [progress] of reaction
Reaktionsablaufdiagramm *n* progress-of-reaction diagram
Reaktionsaffinität *f (phys ch)* affinity, driving force [potential]
Reaktionsapparat *m*/**chemischer** chemical reactor
Reaktionsbehälter *m* reaction vessel
Reaktionsbestreben *n* tendency to react
Reaktionsdrehofen *m* rotary kiln
Reaktionsenergie *f* reaction energy
Reaktionsenthalpie *f* reaction enthalpy
Reaktionsenthalpie f/freie reaction Gibbs function
Reaktionsentropie *f* reaction entropy
reaktionsfähig reactive
reaktionsfähig/nicht non-reactive, unreactive, inactive
Reaktionsfähigkeit *f* reactivity
Reaktionsfolge *f* reaction sequence, sequence of reactions
reaktionsfreudig reactive
Reaktionsfreudigkeit *f* reactivity
Reaktionsgefäß *n* reaction vessel
Reaktionsgemisch *n* reaction mixture
Reaktionsgeschwindigkeit *f* reaction rate
Reaktionsgeschwindigkeit f/spezifische *s.* Reaktionsgeschwindigkeitskonstante
Reaktionsgeschwindigkeitsgleichung *f* rate equation [law]
Reaktionsgeschwindigkeitskonstante *f* rate coefficient, *(relating to elementary reactions:)* (reaction-)rate constant, specific reaction rate, unitary rate constant
Reaktionsgleichung *f* chemical [reaction] equation, stoichiometric equation
Reaktionsgleichung f/chemische chemical [reaction] equation, stoichiometric equation
Reaktionsisotherme *f* reaction isotherm
Reaktionskessel *m* reaction vessel
Reaktionskette *f* reaction chain
Reaktionskinetik *f* reaction [chemical] kinetics
Reaktionsklebstoff *m (plast)* reactive adhesive, reaction adhesive
Reaktionslaufzahl *f* extent [advancement] of reaction
reaktionslos non-reactive, reactionless
Reaktionslosigkeit *f* lack [absence] of reaction
Reaktionsmechanismus *m* reaction mechanism
Reaktionsmolekularität *f* molecularity (of reaction), reaction molecularity
Reaktionsofen *m* reactor, converter
Reaktionsordnung *f* reaction order
Reaktionsordnung f bezüglich eines Reaktanten partial order (of reaction)
Reaktionspartner *m* reaction partner, (co)reactant
Reaktionsprodukt *n* reaction product
Reaktionsquerschnitt *m* reaction cross section
Reaktionssequenz *f (org ch)* reaction sequence

reaktionsspezifisch *(bioch)* reaction-specific
Reaktionsspezifität *f (bioch)* reaction specificity *(of an enzyme)*
Reaktionstechnik *f*/**chemische** chemical reaction engineering
Reaktionsteilnehmer *m s.* Reaktionspartner
reaktionsträge (chemically) inert, inactive
Reaktionsträgheit *f* inertness
Reaktionsträgheit *f*/**chemische** chemical inertness
reaktionsunfähig non-reactive, unreactive
Reaktionsunfähigkeit *f* non-reactivity
Reaktionsverlauf *m* course [progress] of reaction
Reaktionsvermögen *n* reactivity
Reaktionswärme *f* reaction heat, heat of reaction
Reaktionsweg *m* reaction path(way)
Reaktionszeit *f* contact time
reaktiv reactive
Reaktivfarbstoff *m* reactive dye
reaktivieren *v* reactivate, *(activated carbon also)* revivify
Reaktivierung *f* reactivation, *(of activated carbon also)* revivification
Reaktivität *f* reactivity
Reaktor *m* 1. reactor, converter; 2. nuclear reactor
Reaktor *m*/**adiabat(isch)er** adiabatic reactor
Reaktor *m*/**biochemischer [biologischer]** bioreactor
Reaktor *m*/**chemischer** chemical reactor
Reaktor *m*/**diskontinuierlicher** batch reactor
Reaktor *m*/**festbettkatalytischer** catalytic fixed-bed reactor
Reaktor *m* **für aerobe Verfahren** *(biot)* aerobic reactor
Reaktor *m* **für anaerobe Verfahren** *(biot)* anaerobic reactor
Reaktor *m* **für Fluid-Feststoff-Reaktionen** fluid-solid reactor
Reaktor *m* **für Gas-Feststoff-Reaktionen** gas-solid reactor
Reaktor *m* **für Gas-flüssig-Reaktionen** gas-liquid reactor
Reaktor *m* **für homogene Reaktionssysteme** homogeneous reactor
Reaktor *m* **für katalytische Reaktionen** catalytic reactor
Reaktor *m* **für Oberflächenverfahren** *(biot)* surface film reactor
Reaktor *m* **für Submersverfahren** *(biot)* submerged fermenter
Reaktor *m*/**graphitmoderierter** *(nucl)* graphite-moderated reactor
Reaktor *m*/**halbkontinuierlicher** semi-batch reactor
Reaktor *m*/**homogenphasiger** homogeneous reactor
Reaktor *m*/**idealer** ideal reactor
Reaktor *m*/**isothermer** isothermal reactor
Reaktor *m*/**katalytischer** catalytic reactor
Reaktor *m*/**kontinuierlicher [kontinuierlich durchflossener]** (continuous-) flow reactor, continuous reactor
Reaktor *m* **mit Sauerstoffeintrag** *(biot)* oxygenic reactor
Reaktor *m*/**nicht adiabater** non-adiabatic reactor
Reaktor *m*/**nicht isothermer** non-isothermal reactor
Reaktor *m*/**nicht stationär arbeitender** non-steady-state reactor, non-steady-flow reactor
Reaktor *m*/**photochemischer** photochemical reactor
Reaktor *m*/**stationär arbeitender** steady-state reactor, flowstream reactor
Reaktor *m*/**technischer** commercial reactor
Reaktor *m*/**thermischer** thermal reactor
Reaktorkaskade *f* reactors in series
real real *(e.g. gas)*; actual *(as opposed to theoretical)*
Realgar *m (min)* realgar *(tetraarsenic tetrasulphide)*
Realkristall *m* real [imperfect] crystal
Rechen *m* 1. rake; 2. *(hyd)* (bar) rack
Rechenanlage *f (hyd)* (bar) rack
Rechenarm *m* raking arm
Rechengut *n (hyd)* rakings
Rechenklassierer *m* rake classifier
Rechteckwellenpolarographie *f* square-wave polarography

rechtsdrehend 1. dextrorotatory; 2. *(cryst)* right-handed
Rechtsdrehung *f* 1. dextrorotation; 2. *(cryst)* right-handed rotation
Rechtsform *f* dextro(rotatory) form, (+)-form *(of an optically active compound)*
Rechtsgängigkeit *f* right-handed sense *(of a helix)*
Rechtsmilchsäure *f* L(+)-lactic acid, dextrolactic acid
Rechtsweinsäure *f* *(2R,3R)*-(+)-tartaric acid, dextrotartaric acid
rechtszirkular polarisiert right(-hand) circularly polarized
recken *v* *(text, plast)* stretch, draw
Redestillation *f* redistillation, rerunning
redestillieren *v* redistil, rerun
Redoxase *f* redoxase, oxido-reductase
Redoxgleichgewicht *n* redox equilibrium
Redoxkatalysator *m* redox catalyst
Redoxpaar *n* redox couple
Redoxpotenzial *n* redox [oxidation-reduction] potential, ORP
Redoxreaktion *f* redox [oxidation-reduction] reaction
Redoxtitration *f* redox titration
Reduktion *f* reduction
Reduktion *f***/bouveault-blancsche** Bouveault-Blanc reduction *(of esters to alcohols)*
Reduktion *f***/elektrochemische [elektrolytische]** electrochemical [electrolytic] reduction
Reduktion *f* **nach Wolff Kishner** Wolff-Kishner reduction *(for converting aldehydes and ketones into their corresponding hydrocarbons)*
Reduktionsflamme *f* reducing flame
Reduktionsgas *n* reducing gas
Reduktionsmittel *n* reducing agent, reductant
Reduktions-Oxidations-Potenzial *n s.* Redoxpotential
Reduktionspotenzial *n* reduction potential
reduktiv reductive
Reduzent *m* *(biol)* reducer, decomposer
reduzierbar reducible
Reduzierbarkeit *f* reducibility
reduzieren *v* 1. *(ch)* reduce; 2. reduce, lower *(e.g. pressure or temperature)*
reduzieren lassen *v***/sich** reduce
reduzierend reductive
reduzierend/nicht non-reducing
Reduzierstück *n* 1. reducer, reducing fitting *(for pipes)*; 2. *(lab)* reducing adapter *(small socket to large cone)*
reduziert werden *v* undergo reduction
Reduzierung *f* reduction, lowering *(as of pressure or temperature)*
Reduzierventil *n* reducing valve
Redwood-Viskosimeter *n* Redwood viscometer
Referatezeitschrift *f* abstract(ing) journal
Referenzelektrode *f* reference electrode [element]
Referenzküvette *f* *(spectr)* reference cell
Referenzspannung *f* reference voltage
Referenzstrahl *m* *(spectr)* reference beam
Referenzverbindung *f* reference compound
reflektieren *v* reflect (back)
Reflexion *f* reflection
Reflexionsbeugung *f* **schneller Elektronen** *(anal)* reflection high-energy electron diffraction, RHEED
Reflexionselektronenmikroskopie *f* reflection electron microscopy
Reflexionsfähigkeit *f* reflectivity
Reflexionsgitter *n* *(spectr)* reflection grating
Reflexionsgrad *m* reflectance
Reflexionskraft *f* reflectivity
Reflexionsvermögen *n* reflectivity, *(quantitatively:)* reflectance
Reflexionswinkel *m* reflection angle
Reflexionszahl *f* reflectivity *(radiation of heat)*
Reformat *n* *(petrol)* reformate
Reformatsky-Reaktion *f* Reformatsky reaction *(of 3-hydroxycarboxylic acid esters)*
Reformer *m s.* Reformieranlage
Reformieranlage *f* *(petrol)* reforming plant, reformer
reformieren *v* *(petrol)* reform
Reformieren *n* **an Platinkatalysatoren [Platinkontakten]** *(petrol)* platinum reforming, platforming
Reformieren *n***/katalytisches** catalytic reforming, cat-forming
Reformierung *f* *(petrol)* reforming

Reformingkatalysator *m (petrol)* reforming catalyst
Refraktion *f* refraction
Refraktion *f***/spezifische** specific refraction [refractivity, refractive index]
Refraktionsvermögen *n* refractivity
Refraktometer *n* refractometer
Refraktometrie *f* refractometry
refraktometrisch refractometric
Regel *f* rule, principle
Regel *f* **der größten Multiplizität** maximum-multiplicity rule [principle]
Regel *f* **der IUPAC** IUPAC rule
Regel *f***/stokessche** Stokes rule *(of luminescence)*
Regeleinrichtung *f* controlling device, controller
Regelkreis *m* feedback [closed-loop] control system
regellos random • **regellos verteilt** *(cryst)* randomly distributed
regelmäßig regular • **sich regelmäßig anordnen** align
regeln *v* control
Regelung *f* control; automatic [feedback] control
Regen *m***/saurer** acid rain
Regenbett *n* raining bed *(fluid-bed system consisting of falling particles in upstream fluid)*
Regenerat *n (rubber, plast)* reclaim
Regeneratcellulose *f* regenerated cellulose
Regeneratcellulosefaser *f* regenerated cellulose fibre
Regeneratcellulosefaserstoff *m* regenerated cellulose fibre
Regeneratfaser *f* regenerated [semisynthetic] fibre
Regeneratfaserstoff *m* regenerated [semisynthetic] fibre
Regenerationsbecken *n (hyd)* stabilization tank
Regenerativgummi *m (rubber)* reclaimed rubber, reclaim
Regenerativofen *m* regenerative furnace
Regenerator *m* regenerator, *(in catalytic cracking also)* kiln
regenerierbar regenerable, capable of being regenerated
regenerieren *v* 1. regenerate, *(catalysts also)* revivify, *(of an ion exchanger also)* recharge; 2. *(rubber, plast)* recover, reclaim; remanufacture *(waste paper)*
Regenerierlauge *f* regenerant caustic *(ion exchange)*
Regenerierlösung *f s.* Regeneriermittellösung
Regeneriermittel *n* 1. regenerant *(as for ion exchange)*; 2. *(rubber)* reclaiming agent
Regeneriermittellösung *f* regenerant solution
Regeneriermittellösung *f***/verbrauchte** spent regenerant
Regenerierofen *m* regenerator
Regeneriersäure *f* regenerant acid *(ion exchange)*
Regenerierung *f* 1. regeneration, *(of a catalyst also)* revivification, *(of an ion exchanger also)* recharge; 2. *(rubber, plast)* recovery, reclamation; remanufacture *(as of waste paper)*
Regenerierung *f* **mit Kochsalzsole [Natriumchloridlösung]** brining *(of ion exchange resins)*
Regenwasserrückhaltebecken *n* storm-water retention basin
regioselektiv *(org ch)* regioselective
Regioselektivität *f (org ch)* regioselectivity
regiospezifisch *(org ch)* regiospecific, site-specific
Regiospezifität *f (org ch)* regiospecificity, site specificity
registrieren *v* record *(measuring values)*
Regler *m* 1. (automatic) controller; 2. chain transfer agent *(polymerization)*; 3. modifier, modifying agent *(floatation)*
Regner *m* sprinkler
regulär regular
Regulation *f***/hormonale [humorale]** hormonal control
Regulationsprotein *n (bioch)* regulatory protein
Regulator-Gen *n (bioch)* regulator gene
Regulatorprotein *n* regulatory protein
Regulus *m (lab)* regulus
reiben *v* triturate *(to a fine powder)*; rub
Reibkorrosion *f* fretting corrosion
Reiboxidation *f* fretting corrosion
Reibschale *f* mortar
Reibung *f* friction (force)
Reibung *f***/innere** internal friction, viscous force

Reibung f/kinetische kinetic friction
Reibungskoeffizient *m* frictional constant
Reibungskraft *f* friction force
Reibungslumineszenz *f* triboluminescence
Reibungsverlust *m* friction loss
Reibungsverschleiß *m* abrasive wear, abrasion, attrition
Reibungswärme *f* frictional heat
Reibungswiderstand *m* resistance to friction, friction resistance
Reichert-Meissl-Zahl *f* Reichert-Meissl number [value] *(for evaluating oils and fats)*
Reichgas *n* rich gas
reif mature, ripe • **reif werden** mature, ripen
Reife *f* 1. maturity, ripeness; 2. *s.* Reifung
Reifebeschleuniger *m (food)* ripening agent
reifen *v* mature, ripen; age *(viscose)* • **reifen lassen** mature, ripen
Reifung *f* maturation, ripening; ageing *(of viscose)*; processing *(of ribonucleic acid)*
Reifungsbeschleuniger *m (food)* ripening agent
Reihe f/eluotrope *(chromat)* eluotropic series
Reihe f/homologe *(org ch)* homologous series
Reihe f/isologe *(org ch)* isologous series
Reihe f/lyotrope lyotropic order [series], Hofmeister series
Reimer-Tiemann-Synthese *f* Reimer-Tiemann synthesis [reaction] *(of phenolic aldehydes)*
rein pure, *(of noble metals also)* fine, *(of chemical elements also)* elemental; unalloyed, plain *(steel)*; absolute *(alcohol)* • **rein darstellbar** isolable • **rein darstellen** prepare in pure form, isolate *(natural products)* • **rein erhalten** obtain pure
rein/nicht impure
Reinargon *n* very-high-purity argon
Reinchemikalie *f* pure chemical
Reindarstellung *f* isolation *(of natural products)*
Reineck(e)at *n* reineckate *(a salt of Reinecke acid)*
Reinecke-Salz *n* Reinecke salt, ammonium tetrathiocyanatodiamminechromate(III)
Reinerzeugnis *n* pure product
Reingas *n* clean(ed) gas
Reinhardt-Zimmermann-Lösung *f (anal)* Zimmermann-Reinhardt (preventive) solution
Reinheit *f* purity *(as of chemicals)*; clean(li)ness *(as of surfaces)* • **auf Reinheit prüfen** check for purity • **von höchster Reinheit** superpure
Reinheitsanforderung *f* purity specification
Reinheitsgrad *m* degree [level] of purity, purity level, *(esp of reagents:)* grade
reinigen *v* 1. clean(se) *(surfaces)*; 2. clean-up *(gases esp for removing particulate matter)*; purify *(chemicals, or gases esp for removing gaseous components)*; refine *(metals)*; treat, purify, clean (up) *(water, waste water)*
reinigen v/durch Zonenschmelzen zone-refine
reinigend detergent
Reiniger *m s.* 1. Reinigungsmittel; 2. Reinigungsanlage
Reinigung *f* 1. clean(s)ing *(of surfaces)*; 2. cleaning, clean-up *(of gases esp for removing particulate matter)*; purification *(of chemicals, or of gases esp for removing gaseous components)*; refining *(of metals)*; treatment, purification, cleaning, clean-up *(of water, waste water)*
Reinigung f/elektrische [elektrostatische] electrical [electrostatic] precipitation
Reinigung f/extreme superrefining, ultrapurification *(as of metals)*
Reinigung f/mechanische *(hyd)* primary [first] waste treatment, primary treatment
Reinigung f/vollbiologische *(hyd)* complete biological treatment
Reinigungsanlage *f* 1. purification plant; 2. *(hyd)* treatment plant
Reinigungsapparat *m* purifier
Reinigungsflüssigkeit *f (filtr)* washing liquid [liquor]
Reinigungsgrad *m* 1. degree of purification; 2. *(hyd)* degree of clarification
Reinigungsleistung *f (hyd)* removal efficiency

Reinigungsmittel *n* clean(s)er, cleaning [cleansing] agent, detergent; purifier *(as for chemicals)*
Reinigungsmittel ***n*/synthetisches** synthetic detergent
Reinigungsstufe *f* treatment stage
Reinigungsstufe ***f*/biologische** *s.* Reinigungsstufe/zweite
Reinigungsstufe ***f*/dritte** *(hyd)* tertiary waste treatment stage
Reinigungsstufe ***f*/erste** *(hyd)* primary [first] waste treatment stage
Reinigungsstufe ***f*/zweite** *(hyd)* secondary waste treatment stage
Reinigungszusatz *m* detergent additive
Reinkultur *f (biot)* pure culture, monoculture *(of microorganisms)*
Reinmetall *n* pure metal
Reinprodukt *n* pure (product)
Reinraum *m (anal)* clean room *(provided with filtered air)*
Reinraumtechnik *f* clean room technique
reinst of highest purity, ultrapure, superpure
Reinstaluminium *n* superpure aluminium
Reinstoff *m* pure substance
Reinststoff *m* ultrapure substance
Reinwasser *n (hyd)* clean water; *(as a product of water treatment:)* treated [finished] water; *(obtained by reverse osmosis:)* permeate
Reinwassergüte *f* effluent water quality *(1. water treatment; 2. ion exchange)*
Reinzucht *f (biot)* pure culture [growth]
reißen *v* crack, craze *(of surfaces)*; tear, break *(as of fibres, paper)*
Reissert-Verbindung *f (org ch)* Reissert compound *(obtained by adding cyanide ions to N-benzoylquinoline hydrochloride)*
Reißfestigkeit *f* tear resistance [strength], breaking tenacity [strength]
Reißlänge *f (text)* breaking length
Reißscheibe *f* rupture [bursting] disk
Reisstärke *f* rice starch
Reiter *m* (balance) rider *(of an analytical balance)*
Reiterwägestück *n s.* Reiter
reizen *v* irritate
Reizmittel *n (pharm)* stimulant, *(esp if used externally:)* irritant
Rekombination *f* recombination
Rekombinationsabbruch *m* coupling termination
Rekombinationsreaktion *f* recombination reaction
Rekombinationstechnik *f (bioch)* recombinant DNA technique
rekombinieren *v* recombine
rekombinieren ***v*/sich** recombine
Rekristallisation *f* recrystallization
Rekristallisation ***f*/primäre [verformungsbedingte]** *(met)* (primary) recrystallization
Rekristallisationsglühen *n* recrystallization annealing
rekristallisieren *v* recrystallize
Rektifikation *f* rectification
Rektifikation ***f*/kontinuierliche [stetige]** continuous rectification
Rektifikationsapparat *m* rectification still
Rektifikationskolonne *f* rectifying column
Rektifikationsteil *m* enriching section
Rektifikationszone *f* enriching section
Rektifizieranlage *f* rectification plant
rektifizieren *v* rectify
Rektifiziersäule *f* rectifying column
Rekuperativofen *m* recuperative furnace
Rekuperator *m* recuperator
Relaxation *f* relaxation
Relaxationsmethode *f* relaxation method *(for investigating fast reactions)*
Relaxationsperiode *f* relaxation time
Releasinghormon *n* releasing hormone
REM *s.* Rasterelektronenmikroskopie
Rem-Einheit *f* roentgen equivalent man
renaturieren *v* renature *(e.g. denatured proteins)*
Renaturierung *f* renaturation *(as of denatured proteins)*
Reoxidation *f* reoxidation
Rep *n s.* Rep-Einheit
Reparaturenzym *n (bioch)* repair enzyme
Rep-Einheit *f* roentgen equivalent physical, rep (unit)
Replikation *f (bioch)* replication *(of deoxyribonucleic acid)*
Replikation ***f*/semikonservative** semiconservative replication
replikativ *(bioch)* replicative

replizieren *v* replicate *(deoxyribonucleic acid)*
Reppe-Chemie *f* Reppe chemistry *(industrial chemistry of acetylene)*
Reppe-Synthese *f* Reppe synthesis *(of various compounds from acetylene)*
Repression *f (bioch)* (feedback) repression *(feedback inhibition of enzyme formation)*
Repression *f***/koordinierte** coordinate repression
Repressor-Induktor-Komplex *m (bioch)* repressor-inducer complex
reprimieren *v* repress *(a reaction)*
reproduzierbar reproducible
reproduzierbar/schwer poorly reproducible
Reproduzierbarkeit *f* reproducibility
RER *s.* Reticulum/granuläres [raues] endoplasmatisches
Research-Octanzahl *f* research octane number, RON
Reserve *f* reserve, store
Reservelipid *n (bioch)* storage [depot] lipid
Residualfungizid *n* protective fungicide
Residuum *n* 1. residue; 2. *(petrol)* resid(uum)
Resinat *n* resinate *(resin soap or resin ester)*
Resinoid *n (cosmet)* resinoid *(alcoholic extract from aromatic resins and other odoriferous drugs)*
Resinosäure *f* resin acid
Resist *m* resist *(photolithography)*
resistent resistant
Resistenz *f* resistance
Resistenz *f***/chemische** chemical resistance, resistance to chemical attack
Resistenz *f* **gegen Chemikalien** resistance to chemicals
Resistenzentwicklung *f (biot)* development of (antibiotic) resistance, build-up of resistance
Resit *n* resite, C-stage resin
Resitol *n* resitol, B-stage resin
Resol *n* resol, A-stage resin
Resonanz *f* 1. *(org ch)* resonance, mesomerism; 2. resonance *(physics)*
Resonanz *f***/kernmagnetische** nuclear magnetic resonance, NMR
Resonanz *f***/paramagnetische** paramagnetic resonance, pmr, electron spin resonance, ESR, electron paramagnetic resonance, EPR
Resonanzabsorption *f (nucl)* resonance absorption
resonanzaktiv *(spectr)* resonant
Resonanzenergie *f* resonance energy, delocalization [mesomeric] energy
Resonanzfluoreszenz *f (anal)* resonance fluorescence
Resonanzfrequenz *f* resonance frequency
Resonanzhybrid *n (org ch)* resonance hybrid
Resonanzlinie *f (spectr)* resonance line
Resonanzneutron *n* resonance neutron
Resonanz-Raman-Effekt *m* resonance Raman effect, RRE
Resonanz-Raman-Spektroskopie *f* resonance Raman spectroscopy
Resonanz-Raman-Streuung *f* resonance Raman scattering, RRS
Resonanzsignal *n (spectr)* resonance signal, *(in a graph also)* resonance peak
Resonanzspektroskopie *f***/kernmagnetische** NMR [nuclear magnetic resonance] spectroscopy
Resonanzspektroskopie *f***/magnetische** magnetic resonance spectroscopy
Resonanzspektroskopie *f***/paramagnetische** EPR [electron paramagnetic resonance] spectroscopy, ESR [electron spin resonance] spectroscopy
Resonanzspektrum *n* resonance spectrum
Resonanzspektrum *n***/kernmagnetisches** NMR [nuclear magnetic resonance] spectrum
Resonanzspektrum *n***/paramagnetisches** EPR [electron paramagnetic resonance] spectrum, ESR [electron spin resonance] spectrum
resonanzstabilisiert *(org ch)* resonance-stabilized
Resonanzstabilisierung *f (org ch)* resonance stabilization
Resonanzstrahlung *f* resonance radiation
Resonanzstreuung *f* resonance scattering
Resonanzstruktur *f (org ch)* resonance structure

Resonanzübergang *m (spectr)* resonance transition
Resonanzzustand *m* resonance state
resorbieren *v* resorb
resorbiert werden *v* resorb, undergo resorption
Resorcinharz *n* resorcinol(-formaldehyde) resin
Resorcinol *n* resorcin(ol)
Resorption *f* resorption
Resorption *f* **über die Haut** *(pharm, tox)* skin absorption
Respiration *f* respiration
Respirationsquotient *m (bioch)* respiratory quotient, RQ
respiratorisch respiratory
Rest *m* 1. remnant, remainder *(as of material)*; 2. residue, group *(of a molecule)*; 3. balance *(in analyses)*
Rest *m*/**C-terminaler** C-terminal residue [group] *(in proteins)*
Restaktivität *f (bioch)* remaining activity *(of an enzyme)*
Rest-BSB *m (hyd)* residual [final] BOD
Restchlor *n (hyd)* residual chlorine
Restchlor *n*/**freies wirksames** *s.* Restchlorgehalt
Restchlorgehalt *m (hyd)* chlorine residual, (free) residual chlorine
Restfeuchtebeladung *f* residual moisture content
Restfeuchtegehalt *m* residual moisture content
Restgehalt *m* residual [final] content
Restgehalt *m* **an organischen Inhaltsstoffen** *(hyd)* residual organic content
Resthärte *f (hyd)* residual [final] hardness
Restkohlenwasserstoffe *mpl (hyd)* residual hydrocarbons
Restkörper *m (bioch)* remnant *(remaining from chylomicrons)*
Restmenge *f*/**duldbare [zulässige]** *(tox)* (residue) tolerance
Restriktionsenzym *n (bioch)* restriction enzyme [endonuclease]
Restschwefelgehalt *m* residual sulphur content
Restspannung *f* residual stress
Reststrahlung *f* residual radiation
Reststrom *m (phys ch)* residual current
Restvalenz *f* residual valency
Restwasser *n* residual water
Restwassergehalt *m* residual water content
Retardans *n (bioch)* retarder
Retardiermittel *n* retarder, retarding agent
Retention *f* retention
Retention *f* **der Konfiguration** configuration retention
Retention *f*/**relative** *(chromat)* relative retention
Retentionsindex *m (chromat)* retention index
Retentionsverhältnis *n (chromat)* ratio of retention
Retentionsvolumen *n (chromat)* retention volume
Retentionszeit *f* 1. *(chromat)* retention [hold-up] time; 2. *(tech)* residence [retention, hold-up] time
Reticulum *n*/**agranuläres [glattes] endoplasmatisches** *(bioch)* smooth endoplasmatic reticulum, SER
Reticulum *n*/**granuläres [raues] endoplasmatisches** *(bioch)* rough endoplasmatic reticulum, RER
Retinoid *n (org ch)* retinoid
Retinol *n* retinol, antiinfective [antixerophthalmic] vitamin
Retorte *f* retort • **in der Retorte erhitzen** retort
Retortenofen *m* retort furnace
Retro-En-Reaktion *f (org ch)* retro-ene reaction
Retropinakolinumlagerung *f (org ch)* retropinacolin [Wagner-Meerwein] rearrangement
Retrosynthese *f (org ch)* retrosynthesis
reversibel reversible
Reversibilität *f* reversibility
Reversibilität *f*/**mikroskopische** microscopic reversibility
Reversion *f (rubber, food)* reversion • **Reversion erleiden** revert • **der Reversion unterliegen** revert
Reversosmose *f* reverse osmosis
Revertase *f (bioch)* reverse transcriptase, RNA-directed DNA polymerase
Reynolds-Zahl *f (phys ch)* Reynolds number
rezent recent
Rezept *n* recipe, formula • **nach Rezept aufbauen** formulate
Rezeptaufstellung *f* compounding

Rezeptur *f* recipe, formulation
Rezipientenglocke *f (lab)* bell-jar
Rezirkulationsfermenter *m* fermenter with recycle (loop)
rezirkulieren *v* recirculate, recycle, return to the circuit
Rf Wert *m (chromat)* Rf value
RGK *s.* Reaktionsgeschwindigkeitskonstante
RH *s.* Resthärte
Rhenat(IV) *n* $M^I_2ReO_3$ rhenate(IV)
Rhenat(VI) *n* $M^I_2ReO_4$ rhenate(VI)
Rhenium *n* Re rhenium
Rhenium(VI)-chlorid *n* rhenium(VI) chloride, rhenium hexachloride
Rhenium(VII)-oxid *n* rhenium(VII) oxide, rhenium heptoxide
Rhenium(VII)-säure *f* perrhenic acid
Rheologie *f* rheology
Rheopexie *f (coll)* rheopexy
Rhesus-Faktor *m (bioch)* rhesus factor, rh factor
Rh-Faktor *m (bioch)* rhesus factor, rh factor
Rh-negativ rh-negative
Rhodamin *n (org ch)* rhodamine *(any of a class of fluorescein derivatives)*
Rhodanzahl *f (food)* thiocyanogen number [value] *(a measure of unsaturation of fats)*
Rhodinat *n* rhodate
rhodinieren *v* rhodanize
Rhodinieren *n* rhodanizing, rhodium plating
Rhodium *n* Rh rhodium • **mit Rhodium beschichten** rhodanize
Rhodium(III)-chlorid *n* rhodium(III) chloride, rhodium trichloride
Rhodiummohr *n* rhodium black *(finely divided rhodium)*
Rhodium(III)-oxid *n* rhodium(III) oxide
Rhodiumschwarz *n s.* Rhodiummohr
Rhodium(III)-sulfat *n* rhodium(III) sulphate
Rhodopsin *n* rhodopsin, visual purple
rhombisch *(cryst)* (ortho)rhombic
Rhomboeder *n (cryst)* rhombohedron
rhomboedrisch *(cryst)* rhombohedral
Rh-positiv rh-positive
RhZ *s.* Rhodanzahl
Riboflavin *n* riboflavin(e)
Ribonucleat *n* ribonucleate *(a salt of ribonucleic acid)*
Ribonucleinsäure *f (bioch)* ribonucleic acid, RNA
Ribonucleoproteid *n* ribonucleoprotein
Ribonucleoproteinkerne *mpl (bioch)* core particles *(of ribosomes)*
Ribonucleotid *n (bioch)* ribonucleotide
Ribose *f* ribose *(a pentose)*
Ribosom *n (bioch)* ribosome
Ribosylgruppe *f* ribosyl group
Ribulose-1,5-diphosphatcarboxylase *f (bioch)* ribulose-1,5-bisphosphate carboxylase, Rubisco
Ribulose-Diphosphat-Zyklus *m (bioch)* ribulose diphosphate cycle, C_3 pentose phosphate cycle, Calvin(-Bassham) cycle
Ribulose-Monophosphat-Zyklus *m (bioch)* ribulose monophosphate cycle, Quayle cycle
Richtigkeit *f* (analytical) accuracy
Richtung *f* direction • **in horizontaler Richtung ausschwimmen** *(coat)* flood • **in beiden Richtungen verlaufend** reversible
Richtungsbevorzugung *f* directional preference
Richtungsquantelung *f* space quantization
Richtungsquantenzahl *f* magnetic quantum number
Richtwert *m* guide value
Richtwert *m*/**konventioneller** *(tox)* working level *(for maximum tolerances)*
Richtwerte *mpl* **für Abwasserinhaltsstoffe** *(hyd)* permitted discharge values
Ricinoleat *n* ricinoleate *(a salt of ricinoleic acid)*
Ricinolsäure *f* ricinoleic acid
Ricinusöl *n* castor oil
Ricinusöl *n*/**sulfatiertes [sulfoniertes]** sulphated castor oil
RI-Detektor *m (anal)* RI [refractive index] detector
riechen *v* smell
riechend/stechend pungent(-smelling)
riechend/süßlich sweet-smelling
riechend/übel malodorous, ill-smelling, foul-smelling, noxious-smelling, obnoxious
riechend/unangenehm ill-smelling, foul-smelling, obnoxious
Riechstoff *m* 1. perfume, odoriferous

substance; 2. odour-causing substance
Riegel *m* bar *(as of soap)*
Riemen *m* belt
Rieselabsorber *m* wetted-wall(-column) absorber
Rieselfeld *n (hyd)* drain(age) field, sewage farm
Rieselfilmkolonne *f* wetted-wall column
Rieselfilmreaktor *m (biot)* trickling film reactor
Rieselkühler *m* trickle [spray] cooler
rieseln *v* trickle, run; run *(of bulk material)* • **rieseln lassen** trickle, shower
Rieselreaktor *m* trickle-bed reactor
Rieselvermögen *n* flowability *(of bulk material)*
Riesenmolekül *n* giant molecule, macromolecule
Riffel *f* riffle
Riffelwalze *f* fluted [corrugated] roll
Rille *f* groove, riffle
Rinde *f* 1. rind; 2. *(bot)* bark; 3. crust *(of the earth)*
Rinderserumalbumin *n* bovine serum albumin
Ring *m* 1. *(org ch)* ring, nucleus; 2. *(tech)* ring, *(esp for tightening:)* washer • **sich zum Ring schließen** cyclize
Ring *m*/**chinoider** quinonoid ring
Ring *m*/**großer** large ring, macroring *(13 or more members)*
Ring *m*/**kleiner** small ring *(3 or 4 members)*
Ring *m*/**kondensierter** *(org ch)* fused ring
Ring *m*/**mittelgroßer [mittlerer]** medium ring *(8 to 12 members)*
Ring *m*/**sechsgliedriger** six-membered ring
Ring *m*/**ungeradzahliger** odd-membered ring
Ringaufspaltung *f* ring scission [fission, opening]
Ringbildung *f* ring formation
Ringbrenner *m* ring burner
Ringchromatographie *f* radial [circular] chromatography
Ringdüse *f (plast)* annular die
Ringe *mpl*/**liesegangsche** Liesegang rings
Ringelektrode *f* annular electrode
Ringerweiterung *f* ring enlargement [expansion]
ringförmig annular, cyclic
Ringglied *n* ring member
Ring-Ketten-Tautomerie *f* ring-chain tautomerism
Ringkohlenstoffatom *n* ring-carbon atom
Ringkohlenwasserstoff *m* cyclic hydrocarbon
Ringkohlenwasserstoff *m*/**anellierter [kondensierter]** fused-ring hydrocarbon
Ringkomplex *m (org ch)* ring aggregate
Ringkondensation *f (org ch)* fusion, anellation
Ringöffnung *f* ring opening [scission, fission]
Ringöffnungsmetathese *f* ring-opening metathesis (polymerization), ROMP
Ringöffnungspolymerisation *f* ring-opening polymerization
Ringöffnungsreaktion *f* ring-opening reaction
Ringpresse *f* pot press
Ringprobe *f* brown-ring test *(for detecting nitrate ions)*
Ringrohr *n* annulus
Ringrollenmühle *f* ring-roll mill
Ringschluss *m* ring closure, cyclization • **einem Ringschluss unterwerfen** cyclize
Ringschluss *m*/**doppelter** double ring closure
Ringschlussmetathese *f* ring closing metathesis, RCM
Ringschlussreaktion *f* cyclization reaction
Ringspannung *f* ring [angle] strain
Ringstruktur *f* ring [annular] structure
Ringsynthese *f* ring synthesis
Ringsystem *n*/**anelliertes [kondensiertes]** fused(-ring) system, anellated [condensed-ring] system
Ringverbindung *f* ring [cyclic] compound
Ringverengung *f* ring contraction
Ringversuch *m* interlaboratory test *(for confirming the accuracy of test methods and results)*
Rinne *f* channel, chute, trough
rinnen *v* trickle, run
Rippe *f (tech)* fin

Rippenrohr *n* fin(ned) tube
Rippenrohrwärmeübertrager *m* fin-tube heat exchanger
Rippentrichter *m* fluted funnel
Riss *m* crack, flaw, *(if narrow and deep:)* fissure, crevice
Rissauslösung *f* crack initiation
Rissbildung *f* crack formation, cracking
rissig werden *v* crack
Risskeimbildung *f* crack initiation
Risswachstum *n* crack growth
ritzen *v* scratch
Ritzhärte *f* scratch hardness
RKS *s.* Röntgenkleinwinkelstreuung
RL_{50} *(tox)* residue-life 50 per cent, biological half-life
RMZ *s.* Reichert-Meissl-Zahl
RNA *f* RNA, ribonucleic acid
RNA *f* **des Kerns/heterogene** heterogeneous nuclear RNA
RNA *f*/**lösliche** soluble [transfer] ribonucleic acid, sRNA
RNA *f*/**mitochondriale** mitochondrial ribonucleic acid, mRNA
RNA *f*/**ribosomale** ribosomal ribonucleic acid, rRNA
RNA-Synthetase *f* RNA replicase, RNA-directed RNA polymerase
RNS *f (Ribonucleinsäure) s.* RNA
Roborans *n (pharm)* roborant
Rochellesalz *n* Rochelle salt, potassium sodium tartrate 4-water
Rockwellhärte *f* Rockwell hardness, R.H.
rodentizid rodenticidal
Rodentizid *n* rodenticide
Roelen-Reaktion *f* Roehlen reaction, oxo [hydroformylation] reaction
Roggenstärke *f* rye starch
roh crude *(esp chemicals)*; raw, unprocessed, untreated *(material)*
Rohabwasser *n* raw waste water, untreated sewage
Rohblei *n* crude lead
Rohblock *m (met)* ingot
Rohdestillat *n* crude distillate
Rohdichte *f* bulk density *(of timber)*
Roheisen *n* pig iron
Roherdöl *n* crude oil [petroleum]
Roherdöl *n*/**synthetisches** *(coal)* synthetic crude oil
Roherz *n* crude [raw] ore, run-of-mine ore
Rohextrakt *m* crude extract
Rohfaser *f (bioch, agric)* crude fibre
Rohfördergut *n (mine)* run-of-mine
Rohgas *n* crude [raw] gas
Rohgut *n* crude
Rohholz *n* rough wood
Rohholzgeist *m* natural methanol, wood spirit
Rohkautschuk *m* crude [raw] rubber
Rohkohle *f* raw [run-of-mine] coal
Rohling *m* 1. blank; 2. *(plast)* parison
Rohmaterial *n* raw material
Rohmüll *m* crude refuse
Rohöl *n* 1. crude oil; 2. *(petrol)* crude oil [petroleum], crude
Rohöl *n*/**(ab)getopptes** topped crude (petroleum)
Rohöl *n*/**naphthenbasisches [naphthenisches]** naphthene-base petroleum
Rohöl *n*/**paraffinbasisches [paraffinisches]** paraffin-base petroleum
Rohöl *n*/**synthetisches** *(coal)* synthetic crude oil
Rohopium *n* crude [raw] opium
Rohprodukt *n* crude (product)
Rohprotein *n* raw protein
Rohr *n* tube, pipe
Rohr *n*/**blindes** dummy tube
Rohrbündelreaktor *m* multitube fixed-bed reactor
Rohrbündelwärmeübertrager *m* shell-and-tube heat exchanger
Rohre *npl* tubing • **durch Rohre leiten** tube, pipe • **mit Rohren versehen** tube, pipe
Röhre *f* tube, pipe
Röhre *f*/**pitotsche** Pitot tube
röhrenförmig tubular
Röhrenglas *n* glass piping
Röhrenkühler *m (lab)* tubular condenser
Röhrenofen *m* 1. *(distil)* tube [pipe] still, tube [tubular, pipe] furnace; 2. *(ceram, met)* tube furnace
Röhrenreaktor *m* 1. tubular-flow reactor; 2. multitube fixed-bed reactor
Röhrenströmungsreaktor *m* tubular-flow reactor
Röhrentrockner *m* 1. tube rotary dryer; 2. *(lab)* drying pistol
Röhrenwärmeaustauscher *m* tubular heat exchanger

Rohrkopfbenzin *n* casing-head gasoline [spirit]
Rohrkrümmer *m* elbow (fitting)
Rohrleitung *f* 1. tubing, piping; 2. *(petrol)* pipeline • **durch eine Rohrleitung verbinden** pipe; tube • **in Rohrleitungen führen** pipe; tube • **mit einer Rohrleitung versehen** pipe; tube
Rohrleitungssystem *n* piping system
Rohrmaterial *n* tubing; piping *(heavy-walled and large in diameter)*
Rohrmedium *n* tube-side medium [liquid]
Rohrmühle *f* tube mill
Rohrofen *m (ceram, met)* tube [tubular] furnace
Rohrreaktor *m* tubular(-flow) reactor
Rohrreaktor *m*/**(strömungstechnisch) idealer** ideal tubular-flow reactor, plug-flow reactor
Rohrregister *n* tube bank *(of a heat exchanger)*
Rohrreibung *f* pipe friction
Rohrschelle *f* tube [pipe] clamp
Rohrschlange *f* pipe coil
Rohrspirale *f* pipe coil
Rohrstrang *m* 1. pipeline; 2. *(hyd)* drain
Rohrstutzen *m* socket
Rohrverbindungsstück *n* tube [pipe] joint, union
Rohrverdampfer *m* tubular evaporator
Rohrverteiler *m* manifold
Rohrwand(ung) *f* tube [pipe] wall
Rohrzentrifuge *f* tubular (bowl) centrifuge
Rohrzucker *m* cane sugar
Rohrzuckermelasse *f* cane molasses
Rohsaft *m* raw juice
Rohsalz *n* crude [mine-run] salt
Rohsäure *f* crude [raw] acid
Rohschwefel *m* crude sulphur
Rohseide *f* raw silk
Rohspiritus *m* crude alcohol
Rohstahlblock *m* steel ingot
Rohstärke *f* raw starch
Rohstoff *m* raw material
Rohstoff *m*/**chemischer** chemical raw material
Rohstoff *m*/**pflanzlicher** plant material
Rohstoffquelle *f* source of raw material
Rohstoffquelle *f*/**pflanzliche** plant source
Rohteil *n* blank
Rohton *m* raw clay
Rohvaseline *f* petrolatum, petroleum jelly
Rohvolumen *n (ceram)* bulk volume
Rohwachs *n* crude wax
Rohwasser *n* raw water
Rohweinstein *m* wine stone
Rohwollfett *n* crude wool grease, wool wax
Rohzink *n* spelter
Rohzucker *m* raw sugar
Rohzucker *m*/**brauner** brown sugar
Rolle *f* roll(er)
Rollenlager *n* roller bearing
rollgranulieren *v* pelletize
Rollmühle *f* roller mill
Rommelfass *n (tech)* tumbling barrel
rommeln *v (tech)* tumble
Röntgen *n s.* 1. Röntgen-Einheit; 2. Röntgenuntersuchung
Röntgenabsorption *f* X-ray absorption
röntgenamorph X-amorphous
Röntgenanalyse *f* X-ray analysis
röntgenanalytisch X-ray-analytical
Röntgenaufnahme *f* X-ray photograph (image, picture), radiograph • **eine Röntgenaufnahme anfertigen von** X-ray
Röntgenbestrahlung *f* X-ray irradiation
Röntgenbeugung *f* X-ray diffraction, XRD
Röntgenbeugungsdiagramm *n* X-ray diffractogram
Röntgenbeugungsmethode *f* X-ray diffraction method
Röntgendiagramm *n (cryst)* X-ray diagram [pattern]
Röntgendiagramm *n* **eines Einkristalls** crystallogram
Röntgendiffraktometer *n* X-ray diffractometer
Röntgen-Einheit *f* roentgen (unit)
Röntgenemission *f* X-ray emission
Röntgenemission *f*/**partikelinduzierte** *(anal)* particle-induced X-ray emission
Röntgenfeinstrukturuntersuchung *f s.* Röntgenstrukturanalyse
Röntgenfluoreszenzanalyse *f* X-ray fluorescence analysis
Röntgengebiet *n (anal)* X-ray region [range]
Röntgenkleinwinkelstreuung *f* X-ray small-angle scattering

Röntgenkristallstrukturanalyse *f* X-ray crystal(-structure) analysis
Röntgenmethode *f* X-ray method
Röntgenmikroanalyse *f* X-ray microanalysis
Röntgenmikroskop *n* X-ray microscope
Röntgenmikroskopie *f* X-ray microscopy
Röntgenniveau *n* X-ray level
Röntgenogramm *n s.* Röntgenaufnahme
Röntgenographie *f* X-ray photography, radiography
röntgenographisch X-ray-photographic, radiographic
Röntgenpulverkamera *f* X-ray powder camera
Röntgenspektrometer *n* X-ray spectrometer
Röntgenspektrometrie *f* X-ray spectrometry
Röntgenspektroskopie *f* X-ray spectroscopy
Röntgenspektroskopie ***f*/energiedispersive** energy-dispersive X-ray spectroscopy, EDX
Röntgenspektroskopie ***f*** **nach dem Debye-Scherrer-Verfahren [Pulververfahren]** Debye Scherrer X-ray method, X-ray powder spectroscopy
Röntgenspektrum *n* X-ray spectrum
Röntgenstrahl *m* X-ray
Röntgenstrahlung *f* X-ray radiation, X-radiation
Röntgenstreuung *f* X-ray scattering
Röntgenstrukturanalyse *f* X-ray (structure) analysis
Röntgenterm *m* X-ray level
Röntgenuntersuchung *f* X-ray investigation [examination]
Röntgenverfahren *m* X-ray method
Roots-Gebläse *n* straight-lobe compressor
Rosanilin *n* rosaniline, fuchsin(e)
Rosanilinfarbstoff *m* rosaniline dye
Rosanilinhydrochlorid *n* rosaniline hydrochloride
Rosenmund(-Zajcev)-Reduktion *f* Rosenmund reaction [reduction] *(catalytic hydrogenation of acid chlorides)*
Rosenöl *n* rose oil, oil of roses
Rose-Tiegel *m* Rose crucible
Rosolsäure *f* rosolic acid *(any of several related compounds, esp aurin)*
***p*-Rosolsäure** *f p*-rosolic acid, pararosolic acid
Rost *m* 1. *(ch)* rust; 2. *(tech)* grid, grate, grating
Rost ***m*/weißer** white rust *(consisting of zinc carbonate)*
rostbeständig stainless, rustless, rust-resistant
Rostbeständigkeit *f* stainlessness, rust resistance
Röste *f (text)* ret(ting)
rosten *v* rust, corrode
rösten *v* 1. *(min tech, food)* roast, burn, torrefy, *(relating to ores also)* calcine; 2. *(text)* ret, rot
Rösten *n* 1. *(min tech, food)* roasting, burning, torrefaction, *(of ores also)* calcination; 2. *(text)* ret(ting), rotting
rostend/nicht stainless, rustless, rust-resistant
Rostentferner *m* rust-removing agent
Rostfilm *m* rust film
rostfrei *s.* rostbeständig
Röstherd *m* roasting hearth
Röstofen *m (min tech)* roasting furnace, roaster, calciner
Röstprodukt *n* roasted product
Röstreaktionsverfahren *n* roast-reaction process
Röstreduktionsverfahren *n* roast-reduction process
Rostschutz *m* rust inhibition [protection]
Rostschutzadditiv *n* rust-preventing additive
Rostschutzanstrichstoff *m* rust-protective coating, antirust paint
rostschützend rust-inhibiting, rust-protective, antirust
Rostschutzfarbe *f* rust-protective paint, antirust paint
Rostschutzfett *n* rust-inhibiting grease
Rostschutzmittel *n* rust inhibitor, antirust agent
Rostschutzöl *n* rust-inhibiting oil, slushing oil
Rostsieb *n* bar [rod] screen
Rostsiebmaschine *f* grizzly (screen)
Rostumwandler *m* rust converter *(corrosion)*
Rotation ***f*/behinderte** restricted rotation

Rotation f/freie free rotation
Rotation f/optische optical rotation
Rotationsachse *f* axis of rotation
rotationsangeregt *(spectr)* rotationally excited
Rotationsanregung *f (spectr)* rotational excitation
Rotationsdispersion *f* rotatory dispersion, R.D.
Rotationsdispersion f/magnetooptische *(anal)* magneto-optical rotatory dispersion, MORD
Rotationsdispersion f/optische optical rotatory dispersion, O.R.D.
Rotationsenergie *f* rotational energy, energy of rotation [rotational motion]
Rotationsfilter *n* rotary filter
Rotationsfreiheitsgrad *m (phys ch)* degree of rotational freedom, rotational degree of freedom
Rotationsgebläse *n* rotary blower
Rotationsisomer(e) *n* rotational isomer
Rotationsisomerie *f* rotational isomerism
Rotationskolonne *f* rotary [centrifugal] still
Rotationsnassabscheider *m* rotary washer
Rotationsniveau *n (spectr)* rotational energy level
Rotationsölbrenner *m* rotary-cup (oil) burner
Rotationspolarisation *f* rotary polarization, optical activity
Rotationspumpe *f* rotary pump
Rotationsquantenzahl *f* rotational quantum number
Rotations-Schwingungs-Spektrum *n* rotation-vibration spectrum
Rotationsspektrum *n* rotation(al) spectrum
Rotationsübergang *m (spectr)* rotational transition
Rotationsverdampfer *m* rotary (film) evaporator
Rotationsverdichter *m* rotary compressor
Rotationsverteilungsfunktion *f* rotational partition function
Rotationsviskosimeter *n* rotational viscometer
Rotationswäscher *m* rotary washer
Rotationswinkel *m* angle of rotation, rotational angle
Rotationszerstäuberbrenner *m* rotary-cup (oil) burner
Rotationszustand *m* rotational state
Rotaxan *n (org ch)* rotaxane
Rotenoid *n (org ch)* rotenoid
Rotenon *n* rotenone *(an insecticide)*
rotglühend red-hot
Rotglut *f* red heat, redness • **auf [bis zur] Rotglut erhitzen** heat to redness, make red hot
Rotglut f/dunkle dull red heat
Rotglut f/schwache low redness
Rotgültigerz n/dunkles *(min)* dark red silver ore, silver ruby *(antimony(III) silver sulphide)*
Rotguss *m* red brass *(85 % Cu, 15 % Zn)*
rotieren *v* rotate, spin
Rotkupfererz *n (min)* red copper ore, cuprite *(copper(I) oxide)*
Rotmessing *n* red brass *(85 % Cu, 15 % Zn)*
Rotor m/ausschwingender swinging-bucket rotor *(of a centrifuge)*
Rotverschiebung *f* red-shift, shift to the red
Routineanalyse *f* routine analysis
Roving *m* 1. (glass-fibre) roving; 2. *(text)* roving
Rowland-Anordnung *f (spectr)* Rowland grating mount
RQ *s.* Quotient/respiratorischer
rRNA *f s.* RNA/ribosomale
r-Schwingung *f (spectr)* rocking vibration
RSK *s.* Geruchsschwellenkonzentration
R,S-System *n* RS system, Cahn-Ingold-Prelog system *(for designating absolute configuration)*
Rubeanwasserstoffsäure *f* rubeanic acid, *dithiooxamide
Rübenmelasse *f* beet molasses
Rübensaft *m* beet juice
Rübenschnitzel *npl* beet cossettes
Rübenschnitzel npl/ausgelaugte (exhausted) beet pulp
Rübenzucker *m* beet sugar
Rubidium *n* Rb rubidium
Rubidiumcarbonat *n* rubidium carbonate
Rubidiumchlorid *n* rubidium chloride

Rubidiumhydroxid *n* rubidium hydroxide
Rubidiumperoxid *n* rubidium peroxide
Rubin *m (min)* (oriental) ruby
Rubinlaser *m* ruby laser
Rubinzahl *f* rubin(e) number, Congo rubine number *(a measure for the protective action of a colloid)*
Rüb(sen)öl *n* colza [rape-seed] oil *(from Brassica rapa L. em. Metzg.)*
Rückbildung *f* re-formation
Rückdiffusion *f* back-diffusion
Rückdruck *m* back pressure
Rückfaltung *f (bioch)* refolding *(of a polypeptide chain)*
rückfedern *v (rubber)* recover
Rückfederung *f (rubber)* (elastic) recovery
Rückfluss *m* 1. back [return] flow; 2. *(distil)* reflux • **unter Rückfluss kochen** *(distil)* reflux
Rückflusskühler *m (distil)* reflux condenser
Rückflussrohr *n (distil)* downpipe, downcomer
Rückflussverhältnis *n (distil)* reflux ratio
rückführen *v* return, recycle, feed back
Rückführung *f* return, recycle, feedback
Rückführung *f* **von Abprodukten** recycling
Rückgabebindung *f* back-donation bond
Rückgang *m* reduction, decrease
rückgewinnen *v* recover, reclaim
Rückgewinnung *f* recovery, reclaim
Rückgratkette *f* backbone (chain), skeleton *(of a branched molecule)*
Rückkohlen *n s.* Wiederaufkohlen
Rückkopplung *f***/negative** *s.* Rückkopplungshemmung
Rückkopplung *f***/positive** *(bioch)* positive feedback
Rückkopplungshemmung *f (bioch)* feedback [end-product] inhibition, retroinhibition
Rücklauf *m* 1. back [return] flow; 2. *(distil)* reflux; 3. return flowage
Rücklaufkondensator *m (distil)* reflux condenser
Rücklaufverhältnis *n (distil)* reflux ratio
Rücklaufwasser *n* return water
Rückleitung *f* 1. return (pipe); 2. *s.* Rückführung
Rücklösung *f* redissolution
Rückmutationstest *m* Ames [salmonella] test *(for testing the mutagenicity of chemicals)*
Rückoxidation *f* reoxidation
Rückprall *m* rebound
Rückprallelastizität *f (rubber)* rebound (elasticity), impact resilience
Rückreaktion *f* reverse reaction, back(-ward) reaction
Rückschlag *m* blowback
Rückschlagventil *n* check [non-return] valve
Rücksprunghärteprüfer *m* scleroscope
rückspülen *v* backwash
Rückspülung *f* backwash(ing) • **durch Rückspülung waschen** backwash
Rückstand *m* 1. residue; 2. *(petrol)* resid(uum); 3. *(min tech)* underflow • **Rückstände aufarbeiten** *(petrol)* run resid
Rückstand *m***/atmosphärischer** *(petrol)* long residue [residuum]
Rückstand *m* **einer Vakuumkolonne** *(petrol)* short residue [residuum]
Rückstand *m***/fester** residual solid matter
Rückstand *m***/kutikulärer** cuticular residue *(of pesticides)*
Rückstand *m***/subkutikulärer** subcuticular residue *(of pesticides)*
Rückstandsfilter *n* cake filter
Rückstands-Halbwertszeit *f (tox)* residue-life 50 per cent, RL_{50}, biological half-life
Rückstandsheizöl *n (petrol)* residual fuel oil
Rückstandsmenge *f***/duldbare** *(tox)* residue tolerance
Rückstandstoxizität *f* residual toxicity *(of pesticides)*
Rückstoß *m* rebound, blowback
Rückstoßatom *n* recoil atom
Rückstoßchemie *f* recoil [hot-atom] chemistry
Rückstoßelektron *n* recoil [Compton] electron
Rückstoßkern *m* recoil nucleus
Rückstrahltechnik *f* back reflection method *(for investigating crystal structures)*
Rückstrahlung *f* reflection
Rückstrahlungsvermögen *n* reflectivity

Rückstreuung *f (nucl, anal)* backscatter(ing)
Rückstreuverfahren *n (anal)* backscatter method [radiography]
Rückstrom *m* back flow
Rückstromventil *n* back-flow valve
Rücktitration *f* back titration
rückumwandeln *v* reconvert, convert back
rückumwandeln *v*/**sich** reconvert, convert back
Rückumwandlung *f* reconversion
Rückverdampfung *f* re-evaporation
Rückverformung *f (rubber)* (elastic) recovery, rebound
Rückvermischung *f* back-mixing
Rückvermischung *f*/**axiale** axial [longitudinal] mixing
Rückwäsche *f* backwash(ing)
Rückzugsrandwinkel *m* receding contact angle *(in testing surfactants)*
Ruhemasse *f* rest mass
Ruhemasse *f* **des Elektrons** electron (rest) mass
Ruhemasse *f* **des Neutrons** neutron (rest) mass
Ruhemasse *f* **des Protons** proton (rest) mass
Ruhestoffwechsel *m (bioch)* basal metabolism
Rühranker *m* magnetic [stirring] bar
Rührapparat *m* 1. agitator; 2. *(lab)* stirrer; 3. *(tech)* mixer
Rührarm *m* agitator [stirring] arm; rabble arm *(of a multiple hearth furnace)*
Rührautoklav *m* agitated [stirred] autoclave
rührbar agitable, stirrable
rühren *v* 1. agitate; 2. *(lab)* stir; 3. *(tech)* mix
Rühren *n*/**pneumatisches** air agitation
Rührer *m* 1. agitator; 2. *(lab)* stirrer; 3. *(tech)* mixer
Rührerdrehzahl *f* stirrer speed
Rührfermenter *m (biot)* stirred bioreactor [fermenter], agitated bioreactor
Rührflügel *m* agitator [mixing] blade, stirrer blade
Rührkessel *m* agitated tank, mixing vessel; (stirred-)tank reactor, well-mixed (flow) reactor
Rührkessel *m*/**ideal durchmischter** *(hyd)* completely mixed tank
Rührkesselkaskade *f* series of stirred tanks, stirred-tank [well-mixed] reactors in series
Rührkesselreaktor *m* stirred-tank reactor
Rührkesselreaktor *m*/**diskontinuierlicher** batch-operated tank reactor, agitated batch reactor
Rührkesselreaktor *m*/**kontinuierlicher** continuous-flow stirred-tank reactor, tank-type flow reactor
Rührkolonne *f* agitated extraction tower
Rührlaugung *f (min tech)* agitation leaching
Rührmaschine *f* agitator, stirrer, mixer
Rührorgan *n* mixing element
Rührreaktor *m*/**mehrstufiger** *s.* Rührkesselkaskade
Rührschaufel *f* agitator [mixing] blade
Rührstab *m (lab)* stirring rod
Rührwerk *n* 1. agitator, stirrer, mixer; 2. *(hyd)* raking mechanism
Rührwerk *n*/**magnetisches** *(lab)* magnetic stirrer
Rührwerksautoklav *m* stirred autoclave
Rumpf *m (phys ch)* core, kernel
Rund(absetz)becken *n (hyd)* circular clarifier
Rundfilter *n* filter paper disk, round filter
Rundfilterchromatographie *f* circular (paper) chromatography
rundführen *v* recycle
Rundkolben *m* round-bottom flask
Rundsieb *n* revolving [drum] screen
Ruß *m* soot, *(for technical purposes preferably)* (carbon) black
Russel-Saunders-Kopplung *f (spectr)* Russel-Saunders coupling, LS coupling, spin-orbit coupling
Rußflocke *f* smut
rußig sooty
Rußniederschlag *m* carbon deposit
Rußpunkt *m (petrol)* smoke point
Rußschwarz *n* carbon black
Ruthenat *n* ruthenate
Ruthenium *n* Ru ruthenium
Ruthenium(II)-chlorid *n* ruthenium(II) chloride, ruthenium dichloride
Ruthenium(IV)-chlorid *n* ruthenium(IV) chloride, ruthenium tetrachloride
Ruthenium(III)-hydroxid *n* ruthenium(III) hydroxide

Ruthenium(III)-oxid *n* ruthenium(III) oxide
Ruthenium(IV)-oxid *n* ruthenium(IV) oxide, ruthenium dioxide
Rutherfordium *n* Rf rutherfordium *(element 104)*
Rutil *m (min)* rutile *(titanium(IV) oxide)*
Rutsche *f* chute
rütteln *v* rap, vibrate
Rüttler *m* rapper, vibrator
Rydberg-Atom *n* Rydberg atom
Rydberg-Konstante *f* Rydberg constant [number]
Rydberg-Serie *f (spectr)* Rydberg series
Rydberg-Übergang *m (spectr)* Rydberg transition
Rydberg-Zustand *m (spectr)* Rydberg state
RZ *n s.* Regeneratcellulosefaserstoff

S

s *s.* suprafacial
Saat(gut)beizmittel *n* seed protectant [disinfectant]
Saatschutzmittel *n* seed protectant; bird repellent
Saattechnik *f* seed technique
Säbelkolben *m* **(nach Anschütz)** sausage flask
Saccharase *f* saccharase, sucrase
Saccharat *n* saccharate, sucrate
Saccharid *n* saccharide, carbohydrate
Saccharifikation *f* saccharification
saccharifizieren *v* saccharify
Saccharimeter *n* saccharimeter *(a polarimeter)*
Saccharin *n* saccharin, benzoic sulphimide
Saccharinsäure *f* saccharinic acid
Saccharometer *n* saccharometer *(a hydrometer)*
Saccharose *f* sucrose, saccharose *(a disaccharide)*
Saccharoseester *m* sucrose ester
Saccharoseether *m* sucrose ether
Saccharosephosphat *n* sucrose phosphate
Sachse-Mohr-Theorie *f (org ch)* Sachse-Mohr concept *(of strainless rings)*
Sackfilter *n* bag filter
Safranin *n (dye)* safranin(e)
Safrol *n* safrol(e)
Saft *m* 1. juice, sap; 2. *(pharm)* succus
Sagostärke *f* sago *(from Metroxylon sagu Rottb.)*
Saitengalvanometer *n* string galvanometer
Saizew-Regel *f (org ch)* Saytzeff rule *(of orientation in β-eliminations)*
Salbe *f (pharm)* ointment
salbenartig unctuous
Salbenartigkeit *f* unctuousness
Salicylaldehyd *m* salicylaldehyde
Salicylalkohol *m* salicyl alcohol
Salicylat *n* salicylate *(salt or ester of salicylic acid)*
Salicylsäure *f* salicylic acid, o-hydroxybenzoic acid
Salicylsäureamid *n* salicylamide
Salicylsäureanilid *n* salicylanilide
Salicylsäuremethylester *m* methyl salicylate
Salicylsäure-5-sulfonsäure *f* 5-sulphosalicylic acid
Saline *f* saltern, salt works
Salmiak *m* 1. salmiac, ammonium chloride; 2. *(min)* salmiac *(ammonium chloride)*
Salmiakgeist *m* household [aqua] ammonia, ammonia solution [spirit]
salpeterhaltig containing saltpetre, nitrous
Salpetersäure *f* nitric acid
salpetrig nitrous
Salpetrigsäureamylester *m* amyl nitrite
Salpetrigsäureethylester *m* ethyl nitrite
Salz *n* salt • **Salz [Salze] bilden** salify
Salz *n*/**basisches** oxide [hydroxide] salt
Salz *n*/**buntesches** Bunte salt *(salts of S-alkylthiosulphuric acid, usually sodium salts)*
Salz *n*/**einfaches** simple salt
Salz *n*/**inneres** inner [internal, intramolecular] salt
Salz *n*/**komplexes** complex salt
Salz *n*/**mohrsches** Mohr's salt *(diammonium iron(II) sulphate 6-water)*
Salz *n*/**neutrales** neutral [normal] salt
Salz *n*/**organisches** organic salt
Salz *n*/**quartäres** quaternary salt
Salz *n*/**saures** acid salt
Salz *n*/**schlippesches** Schlippe's salt *(sodium thioantimonate 9-water)*

salzartig saline, salt-like
Salzartigkeit *f* salinity
Salzbad *n* salt bath • **im Salzbad aufkohlen** *(met)* bath-carburize
Salzbad *n*/**aufkohlendes** *(met)* carburizing bath
Salzbadaufkohlen *n (met)* bath carburizing
salzbildend salt-forming
Salzbildung *f* salt formation
Salzbrücke *f (el ch)* salt bridge
salzen *v* salt
Salzfehler *m* (neutral-)salt error *(in pH determinations)*
salzfrei salt-free
salzführend saliferous
Salzgehaltmesser *m* salinimeter *(a hydrometer for salt solutions)*
Salzgeschmack *m* salt [salty] taste
salzhaltig saline, salty, briny *(liquid)*; saliferous *(rock)*
Salzhaltigkeit *f* salinity
salzig salty, saline, briny
Salzigkeit *f* saltiness, salinity
Salzlager *n* salt deposit, saline
Salzlake *f* (salt) brine, pickle • **mit Salzlake behandeln** *(food, tann)* brine
salzlos *(food)* salt-free
Salzlösung *f* salt [saline] solution
Salzpaar *n* salt pair
Salzsäure *f* hydrochloric acid
Salzsäure-Gruppe *f (anal)* insoluble chloride group
Salzschmelze *f* 1. molten [fused] salt; 2. *(met)* salt bath
Salzsole *f* (salt) brine
Salzsprühversuch *m* salt spray test *(a corrosion test)*
Salzwasser *n* salt [saline] water
Salzwerk *n* salt works, saline, saltern
Samarium *n* Sm samarium
Samarium(II)-chlorid *n* samarium(II) chloride, samarium dichloride
Samarium(III)-chlorid *n* samarium(III) chloride, samarium trichloride
Samarium(III)-oxid *n* samarium(III) oxide
Samenöl *n* seed oil
Sämischgerbung *f* chamois [oil] tannage
Sammelbecken *n* **des Stoffwechsels** *(bioch)* metabolic pool
Sammelbehälter *m* 1. collection vessel, receiver, receptacle; sump; 2. *(biot)* collection chamber
Sammelelektrode *f* collecting electrode, collector *(of valves, transistors)*
Sammelgefäß *n* collection vessel, receiver, receptacle
Sammelleitung *f* collecting main, manifold
sammeln *v* collect, accumulate
sammeln *v*/**sich** collect, accumulate
Sammelname *m* generic name [term]
Sammelprobe *f (anal)* cumulative sample
Sammelraum *m (biot)* collection chamber
Sammler *m* 1. *(min tech)* promoter, collector; 2. *(el ch)* (storage) battery, accumulator
sAMP *s.* Adenylosuccinat
Sand *m*/**bitumenhaltiger** bitumnious sand
Sand *m*/**bituminöser** bitumnious sand
Sandbad *n* sand bath
Sandel(holz)öl *n* sandalwood oil *(from Santalum album L.)*
Sandfilter *n* sand filter
Sandmeyer-Reaktion *f* Sandmeyer (diazo) reaction
sandstrahlen *v* sandblast
Sandstrahlreinigung *f* sandblasting
Sandwichbindung *f* sandwich bond
Sandwichkomplex *m* sandwich complex
Sandwichverbindung *f* sandwich compound
Sanitärabwasser *n* sanitary waste (water)
Sanitärkeramik *f* sanitaryware
Sansibarkopal *m* Zanzibar copal *(from Trachylobium specc.)*
Santonin(lacton) *n* santonin
Santoninsäure *f* santoninic acid
Sapphir *m (min)* sapphire *(aluminium oxide)*
saprogen saprogenic, saprogenous
Sapropel *n(m)* sapropel
Sapropel(it)kohle *f* sapropelic coal
Saprophyt *m* saprophyte *(organism feeding on decaying organic matter)*
saprophytisch saprophytic
Sar *s.* Sarkosin
Sarin *n* sarin, isopropyl methylphosphonofluoridate
Sarkosin *n* sarcosine, methylaminoacetic acid

Sassafrasöl *n* sassafras oil *(from Sassafras albidum (Nutt.) Nees)*
Satelliten-DNA *f (bioch)* satellite DNA
Satellitenlinie *f (spectr)* satellite line
satinieren *v* (super)calender
Satinierkalander *m* calender (machine)
Sattdampf *m* saturated steam
Sattel(füll)körper *m (distil)* saddle
sättigen *v* saturate
Sättigung *f* saturation
Sättigungsbeladung *f* saturation capacity
Sättigungsdampfdruck *m* saturation vapour pressure
Sättigungsdefizit *n (hyd)* (dissolved) oxygen deficit
Sättigungsfeuchte *f* saturation humidity
Sättigungskonzentration *f* saturation concentration
Sättigungsmenge *f* amount adsorbed (at saturation)
Saturation *f (sugar)* carbonation *(for removing excess lime)*
saturieren *v (sugar)* carbonate *(for removing excess lime)*
Satz *m* 1. sediment, (bottom) settlings, subsidence, dregs; 2. *(food, ferm)* foots, lees; 3. charge (stock), batch, feed(stock); 4. set *(as of laboratory appliances)*; 5. principle; theorem, law
Satzbetrieb *m* batch operation
satzweise batch(wise) • **satzweise arbeitend** batch *(plant)*
sauber/peinlich scrupulously clean
Sauberkeit *f* cleanliness
säubern *v* clean(se)
Säuberung *f* clean(s)ing
sauer 1. acid; 2. *(soil)* sour; 3. *(petrol)* sour; 4. *(food)* acid, sour, hard *(taste)* • **sauer einstellen** acidify *(solutions)* • **sauer reagieren gegen** be acid to • **sauer werden** sour
sauer/schwach weakly [feebly, faintly] acid, subacid
sauer/sehr stark superacid
sauer/stark strongly acid
Sauergas *n* 1. acid gas *(e.g. carbon dioxide, hydrogen sulphide)*; 2. sour gas *(containing acid components such as hydrogen sulphide, carbon dioxide, and hydrogen cyanide)*
säuerlich *(food)* sourish, subacid
säuern *(food)* sour, acidulate *(milk)*; pickle *(vegetables)*; leaven *(dough)*
Sauerstoff *m* O oxygen • **an Sauerstoff verarmt** oxygen-depleted • **mit Sauerstoff angereichert** oxygenated • **von Sauerstoff befreien** deoxygenate
Sauerstoff *m*/**aktiver** active [available] oxygen *(as in bleaching)*
Sauerstoff *m*/**atmosphärischer** atmospheric oxygen
Sauerstoff *m*/**atomarer** atomic oxygen
Sauerstoff *m*/**flüssiger** liquid oxygen
Sauerstoff *m*/**gelöster** *(hyd)* dissolved oxygen, DO
Sauerstoff *m*/**molekularer** O_2 molecular oxygen
Sauerstoffabscheidung *f* oxygen evolution
Sauerstoffaffinität *f* oxygen affinity, affinity for oxygen
Sauerstoffantransport *m* oxygen supply
sauerstoffarm low-oxygen, oxygen-deficient
Sauerstoffaufblasverfahren *n s.* Sauerstoffblasverfahren
Sauerstoffaufnahme *f* oxygen uptake [absorption]
Sauerstoffausschluss *m* exclusion of oxygen
Sauerstoffbedarf *m* oxygen demand
Sauerstoffbedarf *m*/**biochemischer** *(hyd)* biochemical oxygen demand, BOD
Sauerstoffbedarf *m*/**chemischer** *(hyd)* chemical oxygen demand, COD
Sauerstoffbedarf *m*/**fünftägiger biochemischer** five days BOD [biochemical oxygen demand], BOD_5
sauerstoffbeladen oxygenated *(haemoglobin)*
Sauerstoffblasstahl *m* basic oxygen steel
Sauerstoffblasverfahren *n* oxygen process of steelmaking, top-blown oxygen converter process, basic oxygen converter process
Sauerstoffdefizit *n* 1. oxygen deficiency; 2. *(hyd)* dissolved oxygen deficit
Sauerstoffeintrag *m* 1. *(hyd)* oxygen transport [transfer]; 2. *(hyd)* oxygen transfer *(in g O^2/m^3 h)*

Sauerstoffeintragswert *m (hyd)* oxygen transfer *(in g O^2/m^3 h)*
Sauerstoffentwicklung *f* oxygen evolution
Sauerstofferzeugung *f* oxygen generation
Sauerstofffrischverfahren *n s.* Sauerstoffblasverfahren
Sauerstoffgehalt *m*/**mit geringem** low-oxygen
Sauerstoffgehalt *m*/**mit hohem** high-oxygen
sauerstoffhaltig oxygen-containing, oxygenic
Sauerstoffkonzentrationszelle *f* oxygen concentration cell, differential aeration cell *(corrosion)*
Sauerstoffmangel *m* oxygen deficiency
Sauerstoffpartialdruck *m* oxygen partial pressure
Sauerstoffradikal *n* oxygen-centred radical
sauerstoffreich rich in oxygen
Sauerstoffsäure *f* oxo acid
sauerstofftragend oxygen-carrying
Sauerstoffträger *m* oxygen carrier; oxidizer *(for explosives)*
Sauerstofftransferase *f (bioch)* oxygen transferase, dioxygenase
Sauerstofftransport *m* oxygen transport
Sauerstofftransportpigment *n (bioch)* respiratory pigment
Sauerstoffübergang *m* oxygen transfer
Sauerstoffüberspannung *f (el ch)* oxygen overvoltage
Sauerstoffüberträger *m* oxygen transfer agent
Sauerstoffübertragung *f (hyd)* oxygen transfer [transport]
Sauerstoffverarmung *f* oxygen depletion
Sauerstoffverbindung *f* oxygen compound
Sauerstoffverbrauch *m*/**chemischer** *(hyd)* chemical oxygen demand, COD
sauerstoffverbrauchend oxygen-consuming
sauerstoffzehrend oxygen-depleting, oxygen-consuming
Sauerstoffzehrung *f (hyd)* depletion of dissolved oxygen, dissolved oxygen sag
Sauerstoffzufuhr *f* 1. oxygen supply; 2. *(hyd)* oxygen transfer [transport]
Sauerteig *m* sour(dough), leaven
Säuerung *f (food)* souring, acidulation *(of milk)*; pickling *(of vegetables)*; leavening *(of dough)*
Säuerungsmittel *n (food)* acidulant
Sauerwerden *n* souring *(as of milk)*
Saugdruck *m* suction pressure
saugen *v* suck, draw
Saugen *n* suction, drawing
Saugfähigkeit *f* absorbing capacity, absorbency
Saugfilter *n* suction [vacuum] filter
Saugfiltration *f* suction [vacuum] filtration
Saugflasche *f (lab)* suction [filter] flask, aspirator bottle, Büchner flask
Saugförderer *m* suction conveyor
Saugheber *m* siphon
Saughöhe *f* suction head
Saughub *m* suction stroke
Saugleitung *f* suction line [pipe]
Saugpapier *n* absorbent paper
Saugpumpe *f* suction pump
Saugreibungshöhe *f* suction friction head
Saugrohr *n* suction pipe [line]
Saugseite *f* suction side
Saugstrahlpumpe *f* ejector, eductor, siphon
Saugzellenfilter *n* rotary-drum vacuum filter
Saugzug *m* forced draught, downdraught, suction
Saugzuglüfter *m* exhauster
Säule *f* 1. *(tech)* column, tower; 2. *(chromat)* column; 3. *(cryst)* prism
Säule *f*/**chromatographische** chromatographic column
Säule *f*/**gepackte** *(chromat)* packed column
Säulenchromatographie *f* column chromatography
säulenchromatographisch by column chromatography
säulenförmig *(cryst)* columnar
Säulenfüllung *f* column packing
Säulenpackung *f (chromat)* column packing
Säulenreaktor *m (biot)* column reactor
Säulentemperatur *f (anal)* column temperature

Säure *f* 1. *(ch)* acid; 2. *(food)* sourness
Säure *f*/**anorganische** inorganic [mineral] acid
Säure *f*/**arsenige** arsenious acid
Säure *f*/**bromige** bromous acid
Säure *f*/**carosche** Caro's acid, peroxomonosulphuric acid
Säure *f*/**chlorige** chlorous acid
Säure *f*/**diphosphorige** diphosphorous acid
Säure *f*/**dischweflige** disulphurous acid
Säure *f*/**dithionige** dithionous acid
Säure *f*/**dreiprotonige** triprotic acid, triacid
Säure *f*/**eingestellte** standard acid
Säure *f*/**einprotonige** monoprotic acid, monoacid
Säure *f*/**fuchsinschweflige** fuchsinesulphurous acid
Säure *f*/**harte** hard acid *(according to Pearson's classification)*
Säure *f*/**hochkonzentrierte** high-strength acid
Säure *f*/**hypochlorige** hypochlorous acid
Säure *f*/**hypophosphorige** hypophosphorous acid
Säure *f*/**hyposalpetrige** hyponitrous acid
Säure *f*/**isomere** iso acid
Säure *f*/**konzentrierte** concentrated [strong] acid
Säure *f*/**magische** magic acid *(equimolar mixture of fluorosulphuric acid and antimony pentafluoride)*
Säure *f*/**mehrprotonige** polyprotic acid, polyacid
Säure *f*/**organische** organic acid
Säure *f*/**pektinige** pectinic acid
Säure *f*/**phosphonige** phosphonous acid
Säure *f*/**racemische** racemic acid
Säure *f*/**salpetrige** nitrous acid
Säure *f*/**sauerstofffreie** hydracid
Säure *f*/**schwache** weak acid
Säure *f*/**schweflige** sulphurous acid
Säure *f*/**selenige** selenious acid
Säure *f*/**starke** strong acid
Säure *f*/**tellurige** tellurous acid
Säure *f*/**unterhalogenige** hypohalous acid
Säure *f*/**vierprotonige** tetraprotic acid, tetracid
Säure *f*/**weiche** soft acid *(according to Pearson's classification)*
Säure *f*/**zweiprotonige** diprotic acid, diacid
Säureamid *n* acid amide
Säureanhydrid *n* acid anhydride
Säureanhydrid *n*/**cyclisches** cyclic (acid) anhydride
Säureäquivalent *n* acid equivalent
Säureaustausch *m* acid exchange
Säureazid *n* RCO–N=N=N acid azide
Säureballon *m* acid carboy
Säure-Base-Dissoziationskurve *f* acid-base dissociation curve *(of proteins)*
Säure-Base-Gleichgewicht *n* acid-base equilibrium
Säure-Base-Indikator *m* acid-base indicator
Säure-Base-Katalyse *f* acid-base catalysis
Säure-Base-Katalyse *f*/**allgemeine** general acid-base catalysis
Säure-Base-Paar *n* acid-base pair
Säure-Base-Paar *n*/**konjugiertes** conjugate acid-base pair
Säure-Base-Paar *n*/**korrespondierendes** conjugate acid-base pair
Säure-Base-Reaktion *f* acid-base reaction
Säure-Base-Theorie *f* **von Lewis** Lewis theory of acids and bases
Säure-Base-Titration *f* acid-base titration, neutralization titration
Säurebehandlung *f* acid treatment
säurebeständig acid-resistant, resistant to acid(s), stable [fast] to acid(s)
Säurebeständigkeit *f* resistance [stability] to acid(s), acid resistance
Säurebildner *m (food)* acidogen, acid former
Säurecharakter *m* acidic character
Säurechlorid *n* acid chloride
Säurechloridreduktion *f*/**Rosenmund-Zajcevs** Rosenmund (acid chloride) reduction, Rosenmund reaction
Säuredampf *m* acid vapour [fume]
Säuredenaturierung *f* acid denaturation *(of proteins)*
Säuredissoziationskonstante *f* acid dissociation constant
säureecht *(dye)* acid-fast, fast to acids
Säureechtheit *f (dye)* acid fastness

säureempfindlich sensitive to acids, acid-labile
Säureempfindlichkeit *f* sensitivity to acids
Säurefällung *f* acid precipitation
Säurefarbstoff *m* acid dye
Säureform *f* aci form *(of nitro compounds)*
Säurefunktion *f* acidic [acidity] function
Säuregärung *f*/**gemischte** *(bioch)* mixed [heterolactic] fermentation
Säureglasballon *m* acid carboy
Säuregrad *m* (degree of) acidity
Säurehalogenid *n* acid halide
säurehaltig acidic, acidiferous
Säurehärtung *f (coat, plast)* acid hardening
Säurehydrolyse *f* acid hydrolysis
Säurekatalysator *m* acid(ic) catalyst
Säurekatalyse *f* acid catalysis
Säurekatalyse *f*/**spezifische** specific acid catalysis
säurekatalysiert catalysed by acid(s)
Säurekonstante *f* acidity constant
Säurelösung *f* acid solution
Säure-Maßlösung *f (anal)* standard acid
Säurenebel *m* acid mist
Säurerest *m* acid residue
Säurespaltung *f* acidic cleavage
Säurestärke *f* acid strength
Säureüberschuss *m* excess of acid
säureunbeständig acid-labile
Säure- und Basebegriff *m*/**brönstedscher** Brønsted-Lowry concept of acids and bases
Säurewäsche *f* acid washing
Säurewasserstoff *m* acid hydrogen
Säureweckerkultur *f (food)* starter culture
säurewidrig antacid
Säurezahl *f* acid value, A.V., acid number
Säurezentrum *n* acid site *(in molecules, catalysts)*
Sauter-Durchmesser *m* Sauter [volume-surface] mean diameter
Saybolt-Universalviskosimeter *n (petrol)* Saybolt viscometer
SB *s.* Siedebeginn
SBK *s.* Styren-Butadien-Kautschuk
Sbp. *s.* Sublimationspunkt
SC *s.* Säulenchromatographie
Scandium *n* Sc scandium
Scandiumoxid *n* scandium oxide
scannen *v (anal)* scan
Scanning-Auger-Mikroskopie *f* scanning Auger microscopy
Scanningkalorimeter *n* scanning calorimeter
Schabemesser *n* scraper knife; *(for rolls:)* doctor knife
Schaber *m* scraper; *(for rolls:)* doctor
Schacht *m* shaft *(as of a furnace or kiln)*; stack *(of a blast furnace)*
Schachtelpresse *f* pot press
Schachtofen *m (met)* shaft furnace; shaft [vertical] kiln *(esp for sintering and calcining)*
Schadensanalyse *f* damage analysis
Schadgas *n* pollutant gas
Schädigung *f* **der Atmosphäre** damage to the atmosphere
schädlich damaging, detrimental, *(esp to health:)* harmful, deleterious, noxious
Schädlichkeit *f* detrimentalness, *(esp to health:)* harmfulness, deleteriousness, noxiousness
Schädlingsbekämpfung *f* pest control
Schädlingsbekämpfungsmittel *n* pesticide, (pest) control agent, biocide
Schädlingsbekämpfungsmittelrückstand *m* pesticide residue
Schadraum *m*/**relativer** clearance *(in a reciprocating engine)*
Schadstoff *m* 1. harmful substance [material], agent of damage, *(esp in air and water:)* pollutant; 2. *(hyd)* contaminant
Schadstoff *m*/**biologischer** biological pollutant
Schadstoffbelastung *f* pollutant load(-ing)
Schadstoffemission *f* pollutant emission
Schadstoffgrenzwert *m* pollutant limit value
Schadstoffverminderung *f* pollutant reduction, pollution control
schal stale, flat
Schale *f* 1. shell *(of an atom)*; 2. *(lab)* dish; 3. *(tech)* pan, tray; 4. *(bot)* peel, rind
Schale *f*/**abgeschlossene** closed shell
Schale *f*/**äußere** external shell, outer(-most) shell
Schale *f*/**innere** inner shell
Schale *f*/**offene** open shell

schälen *v* 1. *(pap)* (de)bark; 2. *(tech)* skim *(as in the knife-discharge centrifuge)*; 3. *(food)* shell, hull *(cereals)*
Schalenaufbau *m* shell structure *(of an atom)*
Schalen(hart)guss *m* 1. chill casting; 2. chill-cast iron
Schalenmodell *n* shell model *(of atoms)*
Schälfolie *f* sliced sheet, *(if thin:)* sliced film
Schalheit *f* staleness, flatness
Schallabsorption *f* sound absorption
Schallaerozyklon *m* sonic cyclone *(gas cleaning)*
Schalldämmstoff *m* sound-absorbing material
Schalldruck *m* sound pressure
Schallenergie *f* sound energy
Schallfeld *n* sound field
Schallfrequenz *f* sound frequency
Schallleistung *f* sound power
Schallquant *n* phonon
Schallquelle *f* sound source
Schallwelle *f* sound wave
Schälmesser *n* skimmer *(as of a centrifuge)*
Schälzentrifuge *f* skimmer [knife-discharge] centrifuge
Schamotte *f* chamotte
Schamotteerzeugnis *n*/**SiO_2-reiches** silica refractory *(containing more than 92 % SiO_2)*
Schamottestein *m* refractory brick, firebrick
Schamotteton *m* refractory clay, fireclay
Schardinger-Dextrin *n* Schardinger dextrin, *cyclodextrin
scharf 1. severe, rigorous, drastic *(reaction conditions)*; 2. sharp, acrid, pungent *(taste)*; 3. sharp *(melting point)*
Schärfe *f* 1. severity, rigorousness *(of reaction conditions)*; 2. sharpness, acridity *(of taste)*; 3. sharpness *(of melting point)*
schattieren *v* *(text)* shade
Schattierung *f* 1. *(text)* shade; 2. *(text)* shading
schätzen *v* estimate
Schätzung *f* estimation
Schätzwert *m* *(anal)* estimate
Schaubild *n* chart, graph
Schaufel *f* shovel, blade, paddle, vane
Schaufelgranulator *m* *(tech)* paddle granulator
Schaufelkneter *m* kneading machine, kneader
Schaufelmischer *m* blade mixer
Schaufelradreaktor *m* *(biot)* blade wheel reactor
Schaufelrührer *m* straight-blade(d) turbine
Schauglas *n* 1. inspection glass, viewing window; 2. show bottle, specimen [preservation, museum] jar
Schaukelbecherwerk *n* pivoted-bucket conveyor
schaukeln *v* rock
Schaukelschwingung *f* *(spectr)* rocking vibration
Schauloch *n* inspection hatch [port], spy hole
Schaum *m* 1. foam, froth, *(if undesired:)* scum, skimmings; 2. *(met)* dross; 3. *(cosmet)* lather • **Schaum bilden** foam • **mit Schaum bedeckt** foamy, frothy
Schaum *m*/**fester** 1. solid foam; 2. *(plast)* rigid foam
schaumartig foamy, frothy
Schaumbeständigkeit *f* foam stability
Schaumbildner *m* foaming [frothing] agent
Schaumbildung *f* foam formation, frothing
Schaumdämpfungsmittel *n* foam depressant
schäumen *v* 1. foam, froth, *(by bubbles formed inside a liquid:)* effervesce; 2. foam, expand *(e.g. plastics)*
Schaumentwicklung *f* foam formation, frothing
Schaumflotation *f* froth floatation
Schaumglas *n* foam [sponge] glass
Schaumgummi *m* (latex) foam rubber
schaumig foamy, frothy
Schaumigkeit *f* foaminess, frothiness
Schaumkraft *f* foaminess
Schaumlöscher *m* foam extinguisher
Schäummittel *n* foaming [frothing] agent
Schaumpolystyrol *n* foamed [expanded] polystyrene
Schaumregulator *m* foam-control agent
Schaumschicht *f* *(hyd)* scum layer [mat]
Schaumschwimmaufbereitung *f* froth floatation
Schaumstabilisator *m* foam stabilizer

Schaumstoff *m* foamed plastic, plastic foam, expanded [cellular] plastic
Schaumstoff *m*/**geschlossenzelliger** unicellular [closed-cell] foam
Schaumstoff *m*/**harter** rigid foam
Schaumstoff *m*/**offenzelliger** multicellular [open-cell] foam
Schaumverhütung *f* foam prevention
Schaumverhütungsmittel *n* antifoaming agent, foam inhibitor, froth-preventing agent
Schaumwein *m* sparkling [effervescent] wine
Schaumzerstörung *f* foam breaking *(mechanically)*; foam destruction *(chemically)*
Schaumzerstörungsmittel *n* defoamer, defoaming agent
Schauöffnung *f* inspection hatch [port], spy hole
Scheibe *f* disk; washer *(as for tightening joints)*
Scheibenelektrode *f* disk electrode
Scheibenelektrode *f*/**rotierende** rotating-disk electrode
Scheibenfilter *n* leaf filter; (rotary-)disk filter *(a vacuum filter)*
Scheiben(kolloid)mühle *f* disk (attrition) mill
Scheibenrührer *m* disk agitator
Scheibenversprüher *m* rotary-disk atomizer, spinning-disk atomizer
Scheideerz *n* picked ore
Scheidefiltration *f* cake filtration
Scheideflasche *f* Florentine flask [receiver]
Scheidemittel *n* separating agent
scheiden *v* 1. separate, segregate, *(relating to gold or silver:)* part, *(relating to size:)* make a cut, classify; 2. *(sugar)* defecate, lime
Scheider *m* separator, *(relating to size:)* classifier; settler, separator tank *(fluid extraction)*
Scheidetrichter *m* separatory funnel
Scheidewand *f* partition; *(relating to diffusion:)* diaphragm, membrane
Scheidewand *f*/**semipermeable** semipermeable membrane
Scheidung *f* 1. *(min tech)* separation, *(relating to gold or silver:)* parting, *(relating to size:)* classification; 2. *(sugar)* defecation, liming
Scheinleitwert *m (anal)* admittance
Scheinmedikament *n* placebo
Scheinwiderstand *m (anal)* impedance
Scheitelpunkt *m* peak *(as of a curve)*
Scheitelwert *m* peak
Schellack *m* shellac
Schellbach-Bürette *f* Schellbach burette
Schelle *f* clamp
Schenkel *m* leg *(of an U-tube)*
Scherbeanspruchung *f* shear stress
Scherben *m* (ceramic) body
scheren *v* shear
Scherenbildung *f* chelation *(coordination chemistry)*
Scherenschwingung *f (spectr)* scissor vibration
Scherenverbindung *f* chelate (complex), crab's claw complex
Scherfestigkeit *f* shear strength
Schergeschwindigkeit *f* rate of shear, shear rate
Scherkraft *f* shear force
Schermodul *m* shear modulus (of elasticity)
Scherspannung *f* shear stress
Scherung *f* shear(ing)
scherzerkleinern *v* shear
scheuerbeständig *(text)* abrasion-resistant, resistant to abrasion
Scheuerbeständigkeit *f (text)* abrasion resistance, resistance to abrasion
Scheuermittel *n* scouring agent
scheuern *v* scour; scrub *(using a brush)*
Scheuerprüfung *f (text)* abrasion test
Schicht *f* 1. layer, *(if thin:)* film; coat(ing) *(for protection)*; layer, stratum *(of bulk material)*; bed *(as of a filter)*; ply *(as of laminated material)*; 2. *(geol)* layer, stratum • **in Schichten spalten** delaminate
Schicht *f*/**monomolekulare** monolayer, monomolecular [unimolecular] layer
Schicht *f*/**multimolekulare** multilayer, multimolecular layer
Schichtdicke *f* 1. *(phys ch)* layer [film] thickness; 2. *(spectr)* pathlength, length of path; 3. *(filtr)* media depth
Schichtdicke *f*/**durchstrahlte** *(spectr)* absorption pathlength
Schichtempfindlichkeit *f (phot)* emulsion sensitivity
schichten *v* stratify *(bulk material)*

Schichtenbildung *f (hyd, filtr)* layering
Schichtengitter *n (cryst)* layer lattice
Schichtenplattenfilter *n* sheet filter
Schichtenspaltung *f* delamination
Schichtenströmung *f* laminar [streamline] flow
Schichtentrennung *f* delamination
Schichtfolge *f* layer sequence
schichtförmig layered, lamellar • **schichtförmig einlagern** interleave
Schichtgitter *n (cryst)* layer lattice
Schichthöhe *f (filtr)* media depth
Schichtlänge *f* **eines theoretischen Bodens** *(chromat)* height equivalent to a theoretical plate, HETP
Schichtliniendiagramm *n* **der Potenzialfläche** potential-energy contour map
Schichtpressstoff *m* (plastic) laminate
Schichtsilicat *n* phyllosilicate, sheet silicate
Schichtspaltung *f* delamination
Schichtstruktur *f* layer(ed) structure
Schichtwachstum *m* layer growth
Schieberventil *n* gate valve
Schiedsanalyse *f* referee check *(forensic chemistry)*
Schiefe *f* skewness *(as of a curve)*
Schiefer *m (geol)* schist *(easily splitting crystalline rock)*; slate *(fine-grained clayey metamorphic rock)*; shale *(fine-grained laminated sedimentary rock)*
Schiefer *m*/**bituminöser** bituminous [oil] shale
schieferartig schistose; shaly; slaty
Schieferöl *n* shale oil
Schieferplatte *f* slate
Schieferton *m* shale (clay)
schiefrig *(geol)* schistose
Schießbaumwolle *f* gun-cotton, nitrocotton
Schießmittel *n* low-explosive, propellant (explosive)
Schießpulver *n* gunpowder
Schießrohr *n* **nach Carius** *(lab)* Carius [bomb] tube
Schiffchen *n (lab)* (combustion) boat
Schilddrüsenhormon *n* thyroid hormone
Schillern *n* iridescence
Schimmel *m* mould *(as on food)*; mildew *(as on cloth or leather)*
schimmelbeständig mildew-resistant
Schimmelbeständigkeit *f* mildew resistance
schimmeln *v* mould
Schimmelpilz *m* mould
Schirm *m (tech)* screen
Schlacke *f* 1. *(met)* slag; 2. *(coal)* clinker; 3. *(geol)* scoria • **Schlacke bilden** slag
Schlacke *f*/**basische** basic slag
Schlacke *f*/**saure** acid slag
schlackenartig 1. slaggy; 2. *(geol)* scoriaceous
schlackenbildend slag-forming
Schlackenbildung *f* 1. *(met)* slagging; 2. *(coal)* clinkering
schlackenreich slaggy
Schlackenwolle *f* slag wool
Schlafmittel *n (pharm)* hypnotic (drug), sleeping drug, soporific
Schläger *m* beater *(of a hammer mill)*
Schlägermühle *f* beater mill
schlagfest impact-resistant
Schlagfestigkeit *f* impact resistance
Schlaghärte *f* impact hardness
Schlagkorbmühle *f* cage [bar] mill, (squirrel-)cage disintegrator
Schlagprüfung *f* impact testing
Schlagstiftmühle *f* pinned-disk disintegrator [mill]
Schlagwetter *pl (mine)* firedamp
Schlagzähigkeit *f* impact resistance
Schlamm *m* sludge, mud, slime, *(esp if artificially made:)* slurry
Schlamm *m*/**aktivierter** *(hyd)* activated sludge
Schlamm *m*/**ausgefaulter** *(hyd)* digested sludge
Schlamm *m*/**eingedickter** *(hyd)* concentrated sludge
Schlamm *m*/**getrockneter** *(hyd)* dried sludge
Schlammabsetzung *f (hyd)* sludge settling
Schlammabtrennung *f (hyd)* sludge separation
Schlammanfall *m (hyd)* sludge production
Schlammanreicherung *f* sludge concentration
Schlammaustrag *m* sludge discharge
Schlammbehandlung *f (hyd)* sludge treatment [processing]
Schlammbelastung *f (hyd)* sludge loading

Schlammbelebung *f (hyd)* sludge activation
Schlammbeseitigung *f (hyd)* sludge [solid waste] disposal
Schlammeindicker *m (hyd)* sludge thickener [concentrator]
Schlammeindickung *f (hyd)* sludge thickening [concentration]
schlämmen *v* elutriate
Schlämmen *n* elutriation
Schlammentwässerung *f (hyd)* sludge dewatering
Schlammentwässerungsteich *m (hyd)* sludge lagoon
Schlammfaulbehälter *m (hyd)* sludge digester [digestion tank]
Schlammfaulung *f (hyd)* sludge digestion
Schlammfeststoffe *mpl (hyd)* sludge solids
Schlammfeststoffgehalt *m (hyd)* (dry) sludge solids
Schlammfilter *n* cake filter
Schlammfiltration *f* cake filtration
Schlammflocken *fpl (hyd)* sludge floc
Schlammflockung *f (hyd)* sludge flocculation
schlammig sludgy, slimy
Schlammigkeit *f* sliminess
Schlamminhibitor *m* sludge dispersant, detergent
Schlammkompostierung *f (hyd)* sludge composting
Schlammkonditionierung *f (hyd)* sludge conditioning
Schlammkontaktanlage *f (hyd)* solids contact clarifier [unit]
Schlammkonzentration *f (hyd)* sludge [solids] concentration
Schlammkratzer *m (hyd)* sludge scraper
Schlammkuchen *m (hyd)* sludge cake
Schlammlast *f (hyd)* sludge loading
Schlammproduktion *f (hyd)* sludge production
Schlammpumpe *f* sludge pump, *(esp for useful suspensions:)* slurry pump
Schlammschicht *f (hyd)* sludge layer
Schlammtrockenbeet *n (hyd)* sludge-drying bed
Schlammverbrennung *f (hyd)* incineration [combustion] of sludge, sludge incineration
Schlammverbrennungsofen *m (hyd)* sludge furnace
Schlammzone *f (hyd)* sludge zone
Schlange *f (tech)* coil
Schlangengift *n* snake venom
Schlangenkühler *m (distil)* coil [coiled-tube] condenser
Schlangenrohr-Wärmeübertrager *m* coiled tubular (heat) exchanger
Schlauch *m* 1. hose, *(without textile casing:)* tubing; 2. *(filtr)* bag
Schlauchfilter *n* bag filter, *(gas cleaning also)* bag [cloth-tube] collector
Schlauchfolie *f (plast)* blown tubing
Schlauchkammerfilter *n* bag [cloth-tube] collector *(for gas cleaning)*
Schlauchklemme *f* pinchcock, hose cock
Schlauchklemme *f* **nach Hoffmann** Hoffmann clamp
Schlauchverbindungsstück *n* hose coupling
Schlaufenfermenter *m (biot)* loop-type fermenter, tubular loop reactor
Schlaufenreaktor *m* loop [recycle] reactor, tubular loop reactor
schleichend (wirkend) *(tox)* slow-acting
schleifbar grindable
Schleifbarkeit *f* grindability
schleifen *v* grind, *(finely:)* polish
Schleifholz *n (pap)* pulp wood
Schleifmaschine *f* grinding machine, grinder
Schleifmasse *f (pap)* (mechanical) wood pulp, MWP, groundwood (pulp)
Schleifringbürste *f* carbon brush
Schleifscheibe *f* grinding wheel
Schleifstein *m* grindstone
Schleim *m* 1. *(tech)* slime; 2. *(bot, pharm)* mucilage
schleimhaltig *(bot, pharm)* mucilaginous
schleimig 1. *(tech)* slimy; 2. *(bot, pharm)* mucilaginous
Schleimigkeit *f (tech)* sliminess
Schleimsäure *f* mucic acid, galactosaccharic acid
Schlempe *f* 1. *(ceram)* slip; 2. *(ferm)* stillage, slop
Schlenk-Rohr *n (lab)* Schlenk tube *(for handling air-sensitive materials)*
Schlenk-Technik *f (lab)* Schlenk technique *(for handling air-sensitive materials)*

schleppend sluggish *(reaction)*
Schleppmittel *n (distil)* entrainer, entraining [azeotroping, separating] agent, azeotrope former
Schleuder *f* centrifuge
Schleuderbeschichter *m* spin coater
Schleuderguss *m* 1. *(met)* centrifugal casting; 2. *(plast)* centrifugal moulding
Schleudergussteil *n* 1. *(met)* centrifugal casting; 2. *(plast)* centrifugal moulding
Schleudermühle *f* (squirrel-)cage disintegrator, (squirrel-)cage mill, centrifugal [bar] mill
schleudern *v* centrifuge, centrifugate
Schleudern *n* centrifugation
Schleudertrockner *m* whizzer
Schleuse *f* 1. sluice; (air) lock; 2. *(bioch)* shuttle (system)
Schlichte *f (text)* sizing material, dressing
Schlichtebad *n (text)* sizing bath
schlichten *v (text)* size, dress
Schlick *m* 1. slime, mud; 2. *(geol)* ooze
Schlicker *m (ceram)* slip
Schliere *f* 1. *(glass)* stria; 2. *(geol)* streak; schliere *(in a fluid)*
Schlierenoptik *f* schlieren optics
Schlierenverfahren *n* schlieren method *(for investigating density gradients)*
schließen shut off *(a valve)* • **schließen lassen auf** *(anal)* be suggestive of
schließen v/sich zum Ring cyclize
Schliff *m* 1. grinding; 2. *(glass)* cutting; 3. *(result:)* cut *(of glass or gems)*; 4. section *(of rocks for microscopy)*
Schliffball *m s.* Schliffkugel
Schlifffläche *f* ground surface
Schliffhülse *f (lab)* female tapered joint, socket
Schliffkern *m (lab)* male tapered joint, cone
Schliffkolben *m (lab)* ground-glass flask
Schliffkugel *f (lab)* male spherical joint, ball
Schliffpfanne *f (lab)* female spherical joint, socket
Schliffstopfen *m (lab)* ground-glass stopper
Schliffverbindung *f (lab)* ground-glass joint
Schlitz *m* slit, slot
Schlitzbrenner *m (lab)* bats-wing [flat-flame, fish-tail] burner
Schlitzdüse *f* slot nozzle
Schlitzsiebnutsche *f (lab)* slit sieve funnel
Schlot *m* (smoke)stack, chimney
Schluff *m* silt *(grain size 0.026 to 0.002 mm)*
Schlupf *m* slip(page)
Schlussanstrich *m* 1. *(coat)* finish(ing) *(act)*; 2. *(coat)* finish, topcoat, cover coat(ing) *(result)* • **mit einem Schlussanstrich versehen** finish
Schlüsselenzym *n* key enzyme
Schlüsselfrequenz *f (spectr)* group frequency
Schlüsselintermediäres *n (bioch)* key [crucial] intermediate
Schlüsselreaktion *f* key reaction
Schlüssel-Schloss-Beziehung *f (bioch)* lock-and-key relationship
Schlüsselstoff *m* key chemical
Schlüsselsubstanz *f* 1. key substance; 2. *(bioch)* key intermediate
Schlussreinigung *f (hyd)* final treatment
Schmalz *n* rendered fat, *(esp)* lard
schmecken *v* taste
schmeckend/unangenehm ill-tasting
schmelzbar fusible, meltable
Schmelzbarkeit *f* fusibility, meltability
Schmelzbereich *m* melting range
Schmelzdiagramm *n* melting diagram
Schmelze *f* 1. melt, fusion; 2. *(met)* smelt; 3. *s.* Schmelzen
Schmelze f/alkalische molten alkali (bath)
Schmelzelektrolyse *f s.* Schmelzflusselektrolyse
schmelzen *v* 1. melt, fuse; 2. *(met)* smelt *(ore)*; 3. *(food)* render *(fat)*; thaw *(ice)*; 4. *(bioch)* melt out *(of DNA)*
schmelzen v/auf Stein *(met)* matte-smelt
schmelzen v/nochmals 1. remelt; 2. *(met)* re-fuse
Schmelzen *n* 1. melt(ing), fusion; 2. *(met)* smelting *(of ore)*; 3. *(food)* rendering *(of fat)*; thawing *(of ice)*; 4. *(bioch)* melting-out *(of DNA)*
Schmelzen *n* **auf Stein** *(met)* matte smelting
Schmelzen n/autogenes *(met)* autogenous smelting
Schmelzen n/direktes *(met)* direct smelting
Schmelzen *n* **mit Alkali** alkali fusion

schmelzend/scharf sharp-melting
Schmelzenthalpie *f* enthalpy of fusion [melting]
Schmelzfarbe *f (ceram)* overglaze colour
Schmelzfluss *m* 1. melt fusion; 2. *(met)* smelt, smelting flux
Schmelzflusselektrolyse *f* fused-salt electrolysis
schmelzflüssig molten
Schmelzintervall *n* melting range
Schmelzkarbonat-Brennstoffzelle *f* molten carbonate fuel cell, MCFC
Schmelzkatalysator *m* fused catalyst
Schmelzkegel *m (ceram)* pyrometric [fusion] cone
Schmelzkegel *m* **nach Seger** *(ceram)* Seger cone
Schmelzkleber *m (plast)* hotmelt adhesive
Schmelzkurve *f* melting curve
Schmelzofen *m* 1. melting furnace; 2. *(met)* smelter, smelting furnace
Schmelzpunkt *m* melting point, m.p. • **mit niedrigem Schmelzpunkt** low-melting (point) • **mit scharfem Schmelzpunkt** sharp-melting
Schmelzpunkt(bestimmungs)apparat *m* **nach Thiele** Thiele melting point tube
Schmelzpunkterniedrigung *f* melting-point depression [lowering]
Schmelzpunktsdepression *f s.* Schmelzpunkterniedrigung
Schmelzspinnen *n* melt spinning
Schmelzspinnverbundstoff *m (text)* spun-bonded product
Schmelztauchaluminieren *n* hot-dip aluminizing
Schmelztauchen *n (coat)* hot-melt coating
Schmelztauchmasse *f* hot-melt coating
Schmelztauchverzinken *n* hot-dip galvanizing
Schmelztiegel *m* 1. melting crucible; 2. *(lab)* ignition [fusion] crucible
Schmelztiegel *m* **nach Rose** Rose crucible
Schmelztiegelzange *f (lab)* crucible tongs
Schmelzwanne *f (glass)* melter, melting tank
Schmelzwärme *f* heat of fusion
Schmerzlinderungsmittel *n (pharm)* pain reliever, analgesic
schmerzstillend pain-relieving, analgesic
Schmidt-Reaktion *f* Schmidt reaction *(between hydrazoic acid and carbonyl compounds)*
schmieren *v* lubricate; smear *(of a substance)*
Schmierfett *n* lubricating grease
Schmiergüte *f* oiliness *(of a lubricating oil)*
schmierig greasy, unctuous *(substance)*; slippery *(surface)*
Schmierigkeit *f* greasiness, unctuousness *(of a substance)*; slipperiness *(of a surface)*
Schmiermittel *n* lubricant
Schmiermitteladditiv *n* lubricating additive
Schmieröl *n* lubricating oil
Schmierstoff *m* lubricant
Schmierung *f* lubrication
Schmutz *m* dirt, soil, grime
Schmutzfleck *m* soil, smudge, smear
schmutzig dirty • **schmutzig werden** soil
Schmutzstoff *m* contaminant, *(esp in environment:)* pollutant
Schmutzstoffbelastung *f (hyd)* contaminant load(ing), pollutant load(ing)
Schmutzstoffe *mpl* pollutants • **mit Schmutzstoffen belastet** polluted, contaminated *(water, air)*
Schmutzstoffe *mpl*/**anthropogene** human(-caused) pollutants
Schmutzstoffe *mpl*/**gelöste** dissolved solid pollutants
Schmutzstoffe *mpl* **im Grundwasser** ground-water pollutants
Schmutzstoffe *mpl* **im Wasser** water pollutants
Schmutzstoffe *mpl* **in der Luft** atmospheric contaminants
Schmutzstoffe *mpl*/**organische** organic contaminants [pollutants]
Schmutzstoffe *mpl*/**restliche organische** *(hyd)* residual organic content
Schmutzstoffe *mpl*/**suspendierte** suspended solid pollutants
schmutzstofffrei free from polluting substances *(water, air)*

Schmutzstoffgehalt *m* level of pollutants
Schmutzstoffgehalt *m* **des Abwassers** strength of sewage
Schmutzteilchen *n* dirt particle
Schmutzwasser *n* sewage, dirty water, slops
Schmutzwasserpumpe *f* sewage pump
Schnecke *f (tech)* screw, worm
Schneckenbekämpfungsmittel *n* mollus(ca)cide
Schneckenförderer *m* screw [worm] conveyor, helical [spiral] conveyor
Schneckenkneter *m* screw mixer
Schneckenmischer *m* screw mixer
Schneckenpresse *f* 1. screw press; 2. *(ceram)* (extrusion) auger
Schneckenrührer *m* screw mixer
Schneckenspritzgießmaschine *f (plast)* screw injection (moulding) machine
Schneckenverdichter *m* screw compressor
Schneide *f* knife edge *(as of an analytical balance)*
schneiden *v* cut
Schneider *m* cutter
Schneidflüssigkeit *f* cutting fluid
Schneidkeramik *f* cutting ceramics
Schneidmaschine *f* cutting machine, cutter
Schneidöl *n* cutting oil
Schnellanalyse *f* rapid analysis
schnellbindend rapid-hardening
Schnellfilter *n* rapid [high-rate] filter, HRF
Schnellgerbung *f* rapid tannage
schnellhärtend *(plast)* quick-curing; quick-hardening *(paint)*
Schnellmischen *n* rapid [high-intensity] mixing
schnellrotierend spinning
Schnellrühren *n* rapid [high-intensity] mixing
Schnelltest *m* quick test
schnelltrocknend quick-drying
Schnelltrocknung *f* quick [high-speed] drying
schnellumlaufend spinning
Schnellversuch *m* rapid test
Schnellwaage *f* fast-weighing balance
schnellwirkend quick-acting
Schnitt *m* 1. cut; (micro)section *(microscopy)*; 2. *(distil)* cut
schnitzeln *v* chop
schönen *v (hyd)* polish *(treated waste water)*
schonend mild *(e.g. oxidation)*
Schönung *f* polishing *(of treated waste water)*
Schornstein *m* (smoke)stack, chimney • **durch den Schornstein abgehen** be lost at the stack
Schornsteinhöhe *f*/**effektive** effective chimney height
Schornsteinzug *m* draught
Schotten-Baumann-Reaktion *f* Schotten-Baumann reaction *(of acid chlorides with alcohols or amines)*
Schrägbeziehung *f* diagonal relationship [similarity] *(in the periodic system)*
Schrägblattrührer *m* pitched-blade turbine
Schraubenachse *f (cryst)* screw axis
schraubenförmig spiral, helical
Schraubenkühler *m* Friedrichs (reflux) condenser
Schraubenrührer *m* propeller agitator [mixer]
Schraubenverdichter *m* screw compressor
Schraubenversetzung *f (cryst)* screw dislocation
Schraubklemme *f* **(nach Hoffmann)** *(lab)* screw (compressor) clamp, Hoffmann clamp
Schreiber *m (anal)* (pen) recorder
Schritt *m* step, stage
Schritt *m*/**geschwindigkeitsbestimmender** rate-limiting step *(of a reaction)*
Schrittmacherenzym *n* pacemaker enzyme
Schrittmacherreaktion *f* rate-limiting reaction
Schrock-Katalysatoren *mpl (org ch)* Schrock catalysts *(metathesis)*
Schrödinger-Gleichung *f* Schrödinger (wave) equation
Schrotrauschen *n (anal)* shot noise
Schrott *m* scrap (metal)
schrubben *v* scrub
schrühen *v (ceram)* bake
schrumpfbeständig *(text)* shrink-resistant
Schrumpfbeständigkeit *f (text)* shrink resistance
schrumpfen *v* shrink, contract

Schrumpfriss *m* shrinkage crack
Schrumpfung *f* shrinkage, contraction
Schrumpfungseffekt *m (lab)* shrinking effect
Schubbeanspruchung *f* shear stress
Schubmodul *m* shear modulus (of elasticity)
Schubspannung *f* shear stress
Schuppe *f* scale, flake
Schuppenbildung *f* scaling, flaking
schuppig scaly, foliated
schürfen *v (mine)* prospect, explore
Schürfung *f (mine)* prospecting, exploration
Schüssel *f* bowl, dish
Schüttdichte *f* bulk [apparent] density, B.D.
Schüttelflasche *f* shaking bottle
Schüttelherd *m (min tech)* shaking table
Schüttelmaschine *f* shaking machine
schütteln *v* shake
Schütteln *n* shake, shaking • **unter Schütteln** with shaking • **unter gelegentlichem Schütteln** with occasional shaking • **unter ständigem Schütteln** with constant shaking
Schüttelsieb *n* shaking [shaker] screen
Schüttelsortierer *m* shaking [shaker] screen
Schütteltrichter *m* separatory funnel
schütten *v* pour
Schüttgut *n* bulk material
Schüttgutfilter *n* bed filter
Schüttschicht *f* bed *(as of a filter)*
Schütttrichter *m* (feed) hopper
Schüttung *f* feed (stock), charge *(as of a blast furnace)*; bed *(as of a filter)*; packing *(of a column)*
Schüttung *f***/körnige** granular bed
Schüttung *f***/ruhende [statische]** fixed [static] bed
Schüttvolumen *n* bulk [dry] volume
Schutz *m* protection; safeguard
Schutz *m* **durch Fremdstrom/katodischer** impressed-current cathodic protection
Schutz *m* **durch [mit] Opferanoden/katodischer** sacrificial protection
Schutz *m***/katodischer** cathodic protection
Schutzabdeckung *f* resist *(electroplating)*
Schutzabdeckung *f***/isolierende** resist *(electroplating)*
Schutzanstrich *m* protective coating
Schutzatmosphäre *f* protective atmosphere
Schutzbrille *f* protective [safety] goggles, safety glasses
schützend protective
Schutzfilm *m* protective film
Schutzgas *n* protective gas, *(esp welding:)* shielding gas
Schutzgasatmosphäre *f* protective atmosphere
Schutzgasatmosphäre *f***/kontrollierte** controlled atmosphere
Schutzgasglühen *n* bright anneal(ing)
Schutzglas *n* safety glass
Schutzgruppe *f* protective [protecting] group
Schutzhandschuhe *mpl* protective gloves
Schutzhaut *f* protective skin [film]
Schutzhülle *f* protective shell
Schutzkleidung *f* protective [safety] clothing
Schutzkolloid *n* protective colloid
Schutzmaßnahme *f* protective measure
Schutzmittel *n* protective (agent)
Schutzmittel *n* **mit abstoßender Wirkung** *(text)* repellent
Schutzschicht *f* (protective) coating
Schutzschicht *f***/anorganische nichtmetallische** inorganic coating
Schutzschicht *f***/dekorative** decorative coating
Schutzschicht *f***/dünne** protective film
Schutzschicht *f***/elektrochemisch hergestellte** electroplated coating, electrodeposit
Schutzschicht *f***/metallische** metal(lic) coating
Schutzschicht *f***/oxidische** oxide-coating
Schutzschichtmetall *n* coating metal
schwach 1. weak *(acid or base, bond)*; 2. gentle, slight *(heating)*
schwächen *v* weaken *(e.g. the basicity, a bond)*; attenuate *(e.g. an action)*
schwachfarbig weakly coloured
Schwachgas *n* lean gas
schwachlöslich low-solubility, slightly soluble
Schwächung *f* weakening *(as of basicity*

or of a bond); attenuation *(as of an action)*
schwammartig spongy
Schwammeisen *n* sponge iron
Schwammgummi *m* sponge rubber
schwammig spongy, porous
Schwammkunststoff *m* sponge plastic
schwangerschaftsverhütend contraceptive
Schwankung *f* variation
Schwanz *m* 1. *(chromat)* (solvent) tail; 2. tail *(of a monomer)*
Schwanzbildung *f (chromat)* (peak) tailing
Schwanz-Schwanz-Polymerisation *f* tail-to-tail polymerization
schwärzen *v* blacken
schwärzen *v***/sich** blacken
Schwarzfärbung *f* blackening
Schwarzmetallurgie *f* ferrous [iron] metallurgy
Schwarzpulver *n* gunpowder, black [blasting] powder
Schwärzung *f* 1. blackening; 2. *(phot)* extinction, (optical) density
Schwärzungsdichte *f s.* Schwärzung 2.
Schwärzungsmesser *m (phot)* densitometer
Schwärzungsmessung *f (phot)* densitometry
Schwebedichte *f* buoyant density *(density gradient centrifugation)*
Schwebefilteranlage *f (hyd)* blanket-type clarifier
Schwebekörper-Durchflussmesser *m* rotameter
Schwebemittel *n* antisettling [suspending] agent
Schweberöstung *f (met)* suspension roasting
Schwebeteilchen *npl* particulates, suspended particles
Schwebevergasung *f (coal)* suspension [pulverized-coal] gasification
Schwebstoff *m* suspended material [matter] *(in liquids)*; airborne contaminant
Schwebstoffe *mpl***/feste** *(hyd)* suspended solids, SS *(> 5 · 10^{-4} mm)*
Schwebstofffiltergerät *n* mechanical filter respirator
Schwebstoffgehalt *m (hyd)* suspended-solids content [level]
Schwefel *m* S sulphur • **aus Schwefel bestehend** sulphureous, sulphuric • **mit Schwefel vernetzt [vulkanisiert]** sulphur-cured
Schwefel *m***/elementarer** elemental sulphur
Schwefel *m***/kolloider** colloidal sulphur
Schwefel *m***/monokliner** monoclinic sulphur, β-sulphur
Schwefel *m***/organischer [organisch gebundener]** organic sulphur
Schwefel *m***/perlmuttartiger** mother-of-pearl sulphur *(a modification)*
Schwefel *m***/plastischer** plastic sulphur, γ-sulphur
Schwefel *m***/rhombischer** rhombic sulphur, α-sulphur
Schwefel *m***/sublimierter** sublimed sulphur, sulphur flowers
schwefelarm low-sulphur
Schwefelbakterien *npl* sulphur bacteria
Schwefelblüte *f* sulphur flowers, sublimed sulphur
Schwefelbrücke *f (rubber)* sulphur bridge
Schwefel(II)-chlorid *n* sulphur(II) chloride, sulphur dichloride
Schwefel(IV)-chlorid *n* sulphur(IV) chloride, sulphur tetrachloride
Schwefeldichlorid *n* sulphur dichloride
Schwefeldioxid *n* sulphur dioxide, sulphur(IV) oxide
Schwefelerz *n* sulphide ore
Schwefelfarbstoff *m* sulphur dye, sulphide dye(stuff)
Schwefelfarbstoff *m***/blauer** sulphur blue
Schwefelfarbstoff *m***/gelber** sulphur yellow
Schwefelfarbstoff *m***/schwarzer** sulphur black
Schwefel(IV)-fluorid *n* sulphur(IV) fluoride, sulphur tetrafluoride
Schwefel(VI)-fluorid *n* sulphur(VI) fluoride, sulphur hexafluoride
Schwefelgehalt *m***/mit hohem** high-sulphur
Schwefelgehalt *m***/mit mittlerem** medium-sulphur
Schwefelgehalt *m***/mit niedrigem** low-sulphur
schwefelhaltig sulphur-containing, sulphur(e)ous

Schwefelheptoxid *n* sulphur heptaoxide
Schwefelhexafluorid *n* sulphur hexafluoride, sulphur(VI) fluoride
Schwefelindigoblau *n* sulphur indigo blue
Schwefelkalkbrühe *f* lime sulphur *(a pesticide)*
Schwefelkies *m (min)* iron pyrite(s), pyrite, mundic *(iron disulphide)*
Schwefelkohlenstoff *m* carbon disulphide
Schwefelleber *f* liver of sulphur, sulphurated potash *(technical potassium sulphide)*
Schwefelmilch *f* milk of sulphur
Schwefelmonoxid *n* sulphur monoxide, sulphur(II) oxide
schwefeln *v* sulphur(ize), *(esp with sulphurous acid or sulphites:)* sulphite; thionate *(organic compounds for producing dyestuffs)*
Schwefel(II)-oxid *n* sulphur(II) oxide, sulphur monoxide
Schwefel(IV)-oxid *n* sulphur(IV) oxide, sulphur dioxide
Schwefel(VI)-oxid *n* sulphur(VI) oxide, sulphur trioxide
Schwefelsäure *f* sulphuric acid, vitriol oil • **mit Schwefelsäure beizen** *(met)* vitriol
Schwefelsäure *f*/**hochkonzentrierte** concentrated sulphuric acid, colourless oil of vitriol *(containing 93 to 98 % H_2SO_4)*
Schwefelsäure *f*/**rauchende** fuming sulphuric acid, oleum
Schwefelsäurealkylierung *f (petrol)* sulphuric-acid alkylation
Schwefelsäurediethylester *m* diethyl sulphate
Schwefelsäuredimethylester *m* dimethyl sulphate
Schwefelsol *n* sulphur sol
Schwefeltetrachlorid *n* sulphur tetrachloride, sulphur(IV) chloride
Schwefeltetrafluorid *n* sulphur tetrafluoride, sulphur(IV) fluoride
Schwefeltetroxid *n* sulphur tetr(a)oxide
Schwefeltrioxid *n* sulphur trioxide, sulphur(VI) oxide
Schwefelung *f* 1. sulphurization, *(esp with sulphurous acid or sulphites:)* sulphitation; 2. thionation *(of organic compounds for producing dyestuffs)*
Schwefelverbindung *f* sulphur compound • **mit Schwefelverbindungen versetzen** sulphurize
Schwefelvernetzungsbrücke *f (rubber)* sulphur bridge
Schwefelvulkanisation *f* sulphur vulcanization
Schwefelwasserstoff *m* hydrogen sulphide
Schwefelwasserstoff-Gruppe *f (anal)* insoluble sulphide group
Schweinchen *n (lab)* weighing piggy
Schweineschmalz *n* hog [pig] fat, lard
schweißbar weldable • **schweißbar sein** weld
Schweißbrenner *m* (welding) torch, blowpipe
schweißen *v* 1. weld; 2. *(plast)* seal
schweißen lassen *v*/**sich** weld
Schweißen *n*/**autogenes** autogenous [gas] welding
Schweißflussmittel *n* welding flux
schweißhemmend *(cosmet)* antiperspirant
Schweißhemmungsmittel *n (cosmet)* antiperspirant
Schweißstahl *m* wrought iron
schweißtreibend *(pharm)* sudorific
Schweißung *f* weld(ing)
schwelen *v* carbonize *(e.g. coal at low temperature)*; smoulder
Schwelen *n* low-temperature carbonization
Schweler *m* carbonizer
Schwelgas *n* carbonization gas
Schwelgenerator *m (coal)* carbonizing generator, two-zone gasifier
Schwelkoks *m* low-temperature coke
Schwelle *f (phys ch)* threshold
Schwellenwert *m* threshold (value)
Schwelofen *m* carbonizer
Schwelung *f* low-temperature carbonization
schwenkbar turnable
schwenken *v* 1. swirl *(e.g. a beaker)*; 2. *(tech)* turn
Schwerbenzin *n* heavy gasoline [naphtha]
Schwerchemikalien *fpl* heavy chemicals *(acids, salts, and alkalies produced on a large scale)*

schwererschmelzbar higher-melting
Schwerersiedende *n (distil)* less volatile component, higher-boiling component
Schweretrennung *f* gravity separation
schwerflüchtig slow-evaporating, heavy
Schwerflüssigkeitsaufbereitung *f (min tech)* heavy-medium separation
Schwerflüssigkeitsverfahren *n* sink-float method
Schwerkraftabscheider *m* gravity collector
Schwerkraftabscheidung *f* gravity separation
Schwerkraftaufbereitung *f (min tech)* gravity concentration
Schwerkraft-Eindicker *m (hyd)* gravity thickener
Schwerkraftfilter *n* gravity [hydrostatic head] filter
Schwerkraftfiltration *f* gravity [natural] filtration
Schwerkraftsedimentation *f* gravitational sedimentation
Schwerkraftzuführung *f* gravity feed
schwerlöslich poorly [difficultly] soluble, slightly [sparingly] soluble • **schwerlöslich sein** dissolve with difficulty
Schwermetall *n* heavy metal
Schwermetallsalz *n* heavy-metal salt
Schwermetallvergiftung *f* heavy-metal poisoning
Schwermineral *n* heavy mineral
Schweröl *n* heavy oil
schwerschmelzbar high-melting
schwersiedend high-boiling
Schwerspat *m (min)* heavy spar, baryte *(barium sulphate)*
Schwertkolben *m (lab)* sausage flask
Schwimmanteil *m (min tech)* floating fraction
Schwimmaufbereitung *f* floatation
Schwimmdichte *f* buoyant density
schwimmfähig floatable
Schwimmfähigkeit *f* floatability
Schwimmgut *n (min tech)* floating material, floats, floating fraction
Schwimmmittel *n (min tech)* floatation reagent
Schwimmmittel *n***/regelndes** *(min tech)* modifier
Schwimmschicht *f***/schaumartige** *(hyd)* scum layer [mat]
Schwimmschlamm *m (hyd)* (top) scum, skimmings, float
Schwimmschlammabstreifer *m (hyd)* scum skimmer
Schwimm-Sink-Aufbereitung *f (min tech)* sink-float [heavy-medium] separation
Schwimm-Sink-Verfahren *n (min tech)* sink-float process
Schwimmstoffe *mpl (hyd)* floating solids
schwinden *v* shrink, contract *(of material)*; evanesce *(of a tint)*
schwindend shrinking, contracting *(material)*; evanescent *(tint)*
Schwindung *f* shrinkage, contraction
Schwindungsriss *m (ceram)* shrinkage crack
schwingen *v* vibrate, oscillate
Schwingmühle *f* vibrating mill
Schwingung *f* vibration, oscillation • **in Schwingungen versetzen** vibrate
Schwingung *f***/anharmonische** anharmonic [non-harmonic] vibration
Schwingung *f***/harmonische** harmonic vibration
γ-**Schwingung** *f (spectr)* out-of-plane vibration
δ-**Schwingung** *f (spectr)* in-plane vibration
χ-**Schwingung** *f (spectr)* wagging vibration
ν-**Schwingung** *f (spectr)* stretching [valence] vibration
Schwingungsamplitude *f* vibration amplitude
Schwingungsbande *f (spectr)* vibrational band
Schwingungs-Elektronenenergieverlust-Spektrometrie *f***/hochauflösende** very-high-resolution electron energy loss spectrometry, VHRELS
Schwingungsenergie *f* vibrational energy, energy of vibrational motion
Schwingungsform *f* (vibrational) mode, mode of vibration
Schwingungsfreiheitsgrad *m* degree of vibrational freedom, vibrational degree of freedom
Schwingungsfrequenz *f* vibration(al) frequency
Schwingungskorrosion *f* corrosion fatigue

Schwingungsniveau *n* vibrational (energy) level
Schwingungsquantenzahl *f* vibrational quantum number
Schwingungsrichtung *f* direction of vibration
Schwingungsrisskorrosion *f* corrosion fatigue
Schwingungsrüttler *m* vibrator
Schwingungsspektroskopie *f* vibrational spectroscopy
Schwingungsspektrum *n* vibration(al) spectrum
Schwingungsterm *m* vibrational term
Schwingungsübergang *m* vibrational transition
Schwingungsverteilungsfunktion *f* vibrational partition function
Schwingungszustand *m* vibrational state
schwitzen *v* sweat; bleed *(as of cement)*
Schwitzung *f* sweating *(for refining paraffin)*
Schwitzwasser *n* condensate (water)
Schwund *m* shrinkage, contraction; loss, wastage *(a quantity lost in transportation or storage)*
Schwundriss *m* shrinkage crack
Scopolamin *n* scopolamine *(an alkaloid)*
Scopoletin *n* scopoletin
SCOT-Säule *f* SCOT column, support-coated open tubular column *(gas chromatography)*
SCP *s.* Einzellerprotein
screenen *v (pharm, biot)* screen
SE *s.* 1. Sekundärelektron; 2. Siedeendpunkt
Seaborgium *n* Sg seaborgium *(Element 106)*
Sebacinsäure *f* sebacic acid, *decanedioic acid
sechsatomig hexatomic
sechseckig *(cryst)* hexagonal
sechsfachpositiv hexapositive
Sechsring *m* six-membered ring
Sechsstufenverdampfer *m* sextuple evaporator
sechswertig hexavalent, sexavalent
Sechswertigkeit *f* hexavalency
sechszählig 1. *(cryst)* hexad; 2. sechszähnig
sechszähnig hexadentate, sexadentate *(ligand)*
Sedativum *n (pharm)* sedative
Sediment *n* 1. sediment, subsidence, (bottom) settlings, B.S., dregs; 2. *(food, ferm)* foots
sedimentär sedimentary
Sedimentation *f* sedimentation, settling
sedimentationsfähig *(hyd)* settleable *(flocs)*
Sedimentationsfähigkeit *f (hyd)* settleability *(of flocs)*
Sedimentationsgefäß *n* sedimentation tank, settler
Sedimentationsgeschwindigkeit *f* sedimentation [settling] rate
Sedimentationsgleichgewicht *n* sedimentation equilibrium
Sedimentationskoeffizient *m* sedimentation coefficient
Sedimentationskonstante *f* sedimentation constant [coefficient] *(for determining molecular mass)*
Sedimentationskurve *f* settling curve
Sedimentationsschlamm *m (hyd)* sedimentation [clarification] sludge
sedimentieren *v* sediment(ate), precipitate *(dispersed particles)*; deposit, subside, settle (down) *(of a precipitate)*
Sedimentierzentrifuge *f* sedimentation [solid-bowl] centrifuge, centrifugal decanter
Seesalz *n* marine [sea] salt, *(if obtained by solar evaporation also)* solar [bay] salt
Seetieröl *n* marine (animal) oil
Seger-Kegel *m (ceram)* Seger cone *(a pyrometric cone)*
Segregation *f* segregation
Segregationskonstante *f* segregation [partition] coefficient *(zone melting)*
Sehfarbstoff *m (bioch)* visual pigment, photopigment
Sehpurpur *m (bioch)* visual purple, rhodopsin
Seide *f* silk; continuous-filament yarn *(man-made fibre)*
Seide *f*/**gereckte** drawn yarn
Seide *f*/**monofile** monofil, monofilament (yarn)
Seide *f*/**multifile [polyfile]** multifil, multifilament (yarn)
seidenartig silky
seidig silky

Seifenblasen-Durchflussmesser *m* soap film flowmeter
Seifenchromatographie *f* soap chromatography
Seifenmizelle *f* soap micelle
seigern *v (met)* liquate, segregate *(metal)*; liquate (out), segregate *(said of the metal)*
Seigerung *f (met)* liquation, segregation
Seignettesalz *n* Seignette [Rochelle] salt, potassium-sodium tartrate 4-water
seihen *v* strain
Seiher *m* 1. strainer; 2. cage *(of a cage press)*
Seihfilter *n* bag filter
Seihtuch *n* straining cloth
Seite *f* side *(as of a molecule)*; *(relating to rings also)* face
Seite *f*/**gemeinsame** common face, side of fusion
Seitengruppe *f* side group
Seitenkette *f* side [lateral] chain, branch; spacer *(of a gel in affinity chromatography)*
Seitenkettenabbau *m* breakdown of side chains, side chain degradation
Seitenkettenhalogenierung *f* side chain halogenation
Seitenkettensubstitution *f* side chain substitution
Seitenkolonne *f* 1. stripper column; 2. *(petrol)* side(-stream) stripper, stripper (column)
Seitenprodukt *n (distil)* side product
Seitenschneide *f* terminal knife edge *(as of an analytical balance)*
Seitenstrom *m* side stream
Sekretion *f* secretion
sekretorisch secretory
Sekundärbatterie *f* secondary [storage] battery
Sekundärelektron *n* secondary electron
Sekundärelektronenvervielfacher *m* photomultiplier (tube), PMT, secondary-emission electron multiplier, multiplier phototube
Sekundärelement *n* secondary [storage] cell
Sekundäremission *f* secondary emission
Sekundärionenmassenspektrometrie *f* secondary ion mass spectrometry, SIMS
Sekundärkeimbildung *f (cryst)* secondary [surface] nucleation
Sekundärmetabolismus *m (biot)* secondary metabolism
Sekundärmetabolit *m (biot)* secondary metabolite
Sekundärreaktion *f* secondary reaction
Sekundärreaktion *f*/**photochemische** secondary photochemical reaction
Sekundärrohstoff *m* waste material
Sekundärschadstoff *m* secondary pollutant
Sekundärstandard *m* secondary standard
Sekundärstruktur *f (bioch)* secondary structure
Sekundärweichmacher *m* secondary plasticizer
Selachensäure *f* selacholeic acid, nervonic acid
Selbstabsorption *f* self-absorption
Selbstaggregation *f* self-aggregation
Selbstanordnung *f* self-assembly
selbstauslöschend self-extinguishing
Selbstbeschleunigung *f* autoacceleration
Selbstdiffusion *f* self-diffusion
selbstentzündlich self-igniting, pyrophoric
Selbstentzündung *f* self-ignition, autoignition
selbsthärtend *(plast)* self-curing
Selbstionisation *f* autoionization, preionization
Selbstkondensation *f* self-condensation, autocondensation
selbstlöschend self-quenching *(photochemistry)*; self-extinguishing
Selbstlöschung *f* self-quenching *(photochemistry)*
Selbstorganisation *f* self-organisation
Selbstoxidation *f* autoxidation, autooxidation
Selbstreinigung *f (hyd)* self-purification, self-cleaning *(of surfaces)*
selbstverlöschend self-extinguishing
selbstverstärkend *(plast)* self-reinforcing
Selbstzündtemperatur *f* self-ignition [spontaneous-ignition] temperature, S.I.T., autogenous ignition temperature

Selbstzündung *f* self-ignition, autoignition
Selektivherbizid *n* selective herbicide
Selektivitätskoeffizient *m* selectivity coefficient *(ion exchange)*
Selektivlösungsmittel *n* selective solvent
Selen *n* Se selenium
Selenat *n* $M^I_2SeO_4$ selenate
Selenid *n* M^I_2Se selenide
Selenit *n* $M^I_2SeO_3$ selenite
Selenon *n (org ch)* selenone
Selenonium-Ion *n* selenonium ion
Selenoniumverbindung *f* selenonium compound
Selen(IV)-oxid *n* selenium(IV) oxide, selenium dioxide
Selenoxidchlorid *n* selenium dichloride oxide
Selensäure *f* selenic acid
Selenzelle *f* selenium cell
Seltenerden *pl* rare earths
Seltenerdmetalle *npl* rare-earth elements [metals]
Semicarbazid *n (org ch)* semicarbazide
Semicarbazon *n* semicarbazone
Semichinon *n (org ch)* semiquinone
Semidinumlagerung *f* semidine rearrangement [transformation]
Semiebonit *n* semiebonite, semihard rubber
Semikolloid *n* semicolloid
Semimikroanalyse *f* semimicro analysis
semipermeabel semipermeable
Semipermeabilität *f* semipermeability
semiquantitativ semiquantitative
Semisynthese *f* semisynthesis
semisynthetisch semisynthetical
Semitrivialname semitrivial [semisystematic] name
Senecioalkaloid *n* senecio alkaloid
Senfgas *n* mustard gas, dichlorodiethyl sulphide
Senföl *n* RNCS isothiocyanate
Senke *f* sink *(of pollutants, as opposed to a source of pollutants)*
senken *v* lower, reduce, depress
Senkrechtförderer *m* vertical elevator
Senkrechtofen *m* vertical oven *(as for manufacturing gas)*
Senkspindel *f* areometer, hydrometer
Senkung *f* lowering, reduction, depression
Sensibilisator *m* sensitizer
sensibilisieren *v* sensitize
sensibilisieren *v*/**panchromatisch** panchromatize
Sensibilisierung *f* sensitization
Sensibilisierungsfarbstoff *m* sensitizing dye
Sensor *m* sensor, sensing element
Separator *m* separator
separieren *v* separate (out)
Sequenz *f* sequence
Sequenz *f*/**amphibolische** *(bioch)* amphibolic [central] pathway
Sequenzanalyse *f* 1. sequential analysis *(as opposed to simultaneous analysis)*; 2. sequence analysis [determination] *(protein chemistry)*
Sequenzhomologie *f (bioch)* sequence homology
Sequestiermittel *n* sequestering agent
Ser *s.* Serin
Sericin *n* sericin *(a protein occurring in silk)*
Serie *f* 1. series; 2. battery, bank *(as of apparatus)*
Serienanalyse *f* routine [repetitive] analysis
Serin *n* serine
Serinprotease *f (bioch)* serine protease
Serin-Weg *m (biot)* serine pathway *(methanol fermentation)*
Serotonin *n* serotonin
Serum *n* blood serum
Serumalbumin *n* (blood) serum albumin
Serumglobulin *n* serum globulin
Serumprotein *n* serum protein
Sesquioxid *n* sesquioxide *(deprecated term for an oxide of the type $M^{III}_2O_3$)*
Sesquiterpen *n (org ch)* sesquiterpene
Sesquiterpenoid *n (org ch)* sesquiterpenoid
Sesselform *f* chair form *(stereochemistry)*
Sesselkonformation *f* chair [staircase] conformation
Sesterterpenoid *n (org ch)* sesterterpenoid
Setzapparat *m (min tech)* jig
Setzapparat *m*/**hydraulischer** hydraulic [wet] jig
Setzarbeit *f (min tech)* jigging
setzen *v (min tech)* jig

setzen v/sich settle, deposit, sediment, subside
Setzherd *m (min tech)* table
Setzmaschine *f (min tech)* jig (washer) • **in der Setzmaschine aufbereiten** *(min tech)* jig
Setzmaschine f/hydraulische hydraulic [wet] jig
SEV *s.* Sekundärelektronenvervielfacher
Sexualhormon *n* sex hormone
Sexuallockstoff *m* sex attractant
sezernieren *v (biol)* secrete
sherardisieren *v* sherardize *(to coat with zinc by diffusion coating)*
Shikimisäure *f* shikimic acid
Shikimisäureweg *m (bioch)* shikimic acid pathway
Shine-Delgarno-Sequenz *f (bioch)* Shine-Delgarno sequence *(of the messenger RNA, rich in purines)*
Shore-Härte *f* Shore hardness
Shuttle-Vektor *m (bioch)* shuttle vector
Sial *n* sial *(upper rock layer of the earth's crust)*
Sialinsäure *f* sialic acid *(any of a group of acylated neuraminic acids)*
Sichelzellenhämoglobin *n (bioch)* sickle-cell haemoglobin, haemoglobin S
Sicherheitsbestimmung *f* safety regulation
Sicherheitsglas *n* safety glass
Sicherheitsglas n/geschichtetes laminated safety (sheet) glass
Sicherheitsmaßnahme *f* safety measure, precaution
Sicherheitsmaßregel *f* safety rule [principle]
Sicherheitssprengstoff *m* safety explosive
Sicherheitsventil *n* safety(-relief) valve
Sicherheitsventil n/federbelastetes spring safety valve
Sicherheitsvorkehrungen *fpl* safety precautions
Sicherheitsvorschrift *f* safety instruction
Sicherheitswaschflasche *f* safety wash-bottle
sichern *v* secure; establish *(e.g. the structure of a compound)*
Sichtbares *n (spectr)* visible region [range]
Sichten *n* pneumatic classification, air classifying [sizing], air separation [sweeping]
Sichter *m* pneumatic [air] classifier, air separator
Sichtfeines *n* undersize (material)
Sichtfenster *n* inspection glass, viewing window
Sichtgrobes *n* oversize (material)
Sichtkammer *f* classifying chamber
Sichtprüfung *f* visual examination
Sickerlaugung *f (min tech)* percolation leaching
sickern *v* trickle
Sickerverlust *m* leakage, slippage loss
Sickerwasser *n* 1. *(soil)* percolation water; 2. *(hyd)* seep [drainage] water
Siderit *m* 1. *(min)* siderite, spathic iron (ore) *(iron(II) carbonate)*; 2. (holo)siderite, iron meteorite
Sieb *n* 1. sieve; 2. *(tech)* screen, *(for liquids containing solid particles:)* strainer; 3. *(pap)* wire; 4. *s.* Siebapparat
Sieb *n* **mit bewegter Siebfläche** moving screen
Sieb *n* **mit unbewegter Siebfläche** stationary screen
Siebanalyse *f* sieve [screen] analysis
Siebapparat *m* screening machine, screen (classifier), sifter
Siebboden *m* 1. *(distil)* sieve plate; 2. *(ferm)* strainer bottom
Siebbodenkolonne *f (distil)* sieve-plate column [tower], perforated-plate column [tower]
Siebbodenreaktor *m (biot)* sieve-plate cascade reactor
Siebeinsatz *m* strainer
sieben *v* 1. sieve; 2. *(tech)* screen, sift
Sieben(er)ring *m* seven-membered ring
siebenwertig heptavalent
Siebfeines *n* screen undersize [fines], undersize (product), undersize [minus] material
Siebfilter *n* screen filter
Siebfläche *f* screen area
Siebgewebe *n* gauze
Siebgrobes *n* screen oversize, oversize (product), oversize [plus] material, tailings *(in screening)*
siebklassieren *v* screen
Siebklassieren *n* screen sizing [classification], screening

Siebmaschine *f* screening machine, screen (classifier), sifter
Siebmühle *f* screen-type mill
Siebnummer *f* mesh *(number of openings per linear inch)*
Sieböffnung *f* 1. screen opening [hole]; 2. screen aperture
Sieböffnungsweite *f* screen size [aperture]
Siebtrommel *f* perforated basket *(of a centrifuge)*
Siebweite *f* screen aperture
Siebzentrifuge *f* screen [perforated bowl] centrifuge, filtering centrifuge [centrifugal], centrifugal filter
Siedeanalyse *f* distillation analysis
Siedebeginn *m* initial boiling point, I.B.P.
Siedebereich *m* boiling(-point)-range
Siedediagramm *n* boiling-point diagram
Siedeendpunkt *m* final boiling point, F.B.P., end-point, EP
Siedegrenzen *fpl* distillation range
Siedegrenzenbenzin *n* special boiling-point spirit, SBP spirit
Siedehitze *f* boiling heat
Siedekapillare *f* capillary air bleed, whipping tube
Siedekonstanz *f* regular boiling
Siedekurve *f s.* 1. Siedelinie; 2. Siedepunktskurve
Siedelinie *f* boiling curve *(of a two-component system)*
sieden *v* boil
sieden *v*/**nochmals** reboil
Sieden *n* boil(ing) • **zum Sieden bringen** bring [heat] to boiling • **am Sieden halten** keep at the boil • **nahe am Sieden halten** keep near the boil
Sieden *n*/**konstantes** regular boiling
Sieden *n*/**örtliches** subcooled boiling
siedend/höher higher-boiling, heavier
siedend/konstant constant-boiling
siedend/leichter lower-boiling, light
siedend/niedriger lower-boiling
siedend/schwerer higher-boiling
Siedepunkt *m* boiling point, bp, boiling temperature
Siedepunkt *m*/**mittlerer** mid-boiling point
Siedepunktserhöhung *f* boiling-point elevation, BPE
Siedepunktserhöhung *f*/**molale** molal boiling-point(-elevation) constant, ebullioscopic [boiling-point] constant
Siedepunktserhöhung *f*/**molare [molekulare]** *s.* Siedepunktserhöhung/molale
Siedepunktskurve *f* boiling point curve *(of a multicomponent system)*
Siedepunktsmaximumazeotrop *n* maximum-boiling [high-boiling] azeotrope
Siedepunktsminimumazeotrop *n* minimum-boiling [low-boiling] azeotrope
Siederohr *n* boiling tube
Siederohrdampferzeuger *m* water-tube boiler
Siedeschwanz *m* heavy ends *(boiling analysis)*
Siedestab *m (lab)* bumping stick
Siedestein *m* boiling stone
Siedetemperatur *f* boiling temperature [point]
Siedeverzug *m* delay in boiling
Siedewasserreaktor *m (nucl)* boiling (water) reactor
siegeln *v (plast)* seal
Siemens-Martin-Ofen *m* open-hearth furnace
Siemens-Martin-Stahl *m* open-hearth steel
Siemens-Martin-Verfahren *n* open-hearth process
Sigmabindung *f* sigma-adduct bond, σ bond
Signal *n*/**spitzes** *(anal)* peak
Signalaufspaltung *f (spectr)* signal splitting
Signalerkennungspartikel *n (bioch)* signal recognition particle
Signalfläche *f (anal)* peak area
Signalform *f (anal)* peak shape
Signalhöhe *f (anal)* peak height
Signalintensität *f (anal)* peak intensity
Signallage *f (anal)* peak position
Signal-Rausch-Verhältnis *n (anal)* signal-to-noise ratio, S/N ratio
Signalsequenz *f (bioch)* signal peptide, targeting signal, signal sequence, transit peptide, localization signal *(directs the post-translational transport of a protein)*
Signalüberlagerung *f (anal)* superposition of peaks

Signalverbreiterung *f (anal)* peak broadening
Sikkativ *n (coat)* drier, siccative
Silan *n* Si_nH_{2n+2} silane, *(specif)* SiH_4 monosilane
silanisieren *v (chromat)* silanize
Silanisierung *f (chromat)* silanization
Silanol *n* silanol *(a silicon compound which contains OH groups)*; *(specif)* H_3SiOH silanol
Silazan *n* silazane
Silber *n* Ag silver
Silber *n*/**kolloidales** colloidal silver
Silberazid *n* silver azide
Silberbromid *n* silver bromide
Silberchlorid *n* silver chloride
Silberchloridelektrode *f* silver-chloride electrode
Silbercoulometer *n* silver coulometer
Silbercyanid *n* silver cyanide
Silberelektrode *f* silver electrode
Silberelektrolyt *m* silver-plating bath
Silberfluorid *n* silver fluoride
Silberhalogenid *n* silver halide
silberhaltig argentiferous, argentian *(esp ores)*
Silberiodid *n* silver iodide
Silbernitrat *n* silver nitrate
Silbernitratpapier *n* silver-nitrate paper
Silberoxid *n* silver oxide, *(specif)* silver(I) oxide
Silber-Silberchlorid-Elektrode *f* silver-silver chloride electrode
Silberspiegel *m* silver mirror
Silbersulfat *n* silver sulphate
Silbersulfid *n* silver sulphide
Silberthiocyanat *n* silver thiocyanate
Silicat *n* silicate
silicatisch siliceous
Silicid *n* silicide
Silicium *n* Si silicon
Siliciumbronze *f* silicon bronze
Siliciumcarbid *n* silicon carbide
Silicium(IV)-chlorid *n* silicon(IV) chloride, silicon tetrachloride
Siliciumdioxid *n* silicon dioxide
siliciumdioxidhaltig siliceous
Siliciumeisen *n* silicon iron
Siliciumester *m* silicic-acid ester
Silicium(IV)-fluorid *n* silicon(IV) fluoride, silicon tetrafluoride
siliciumfunktionell silicon-functional
Siliciumgleichrichter *m* silicon rectifier
Siliciumguss *m* silicon cast iron
Siliciumhydrid *n* silicon hydride, silane
Silicium-Kohlenstoff-Bindung *f* silicon-carbon bond
Siliciumoberfläche *f* silica surface
Silicium(IV)-oxid *n* silicon(IV) oxide, silicon dioxide
Siliciumscheibe *f* silicon wafer
Siliciumwasserstoff *m* silicon hydride, silane
Silicon *n* silicone, *polysiloxane
Silicon... *siehe: Silikon...*
silifizieren *v (geoch)* silicify
Silikatglas *n* silicate glass
Silikon *n* silicone • **mit Silikonen behandeln** silicone-treat, siliconize
Silikon *n*/**elastomeres** silicone elastomer
silikonbehandelt silicone-treated, siliconized
Silikonbehandlung *f* silicone treatment, siliconization
silikonbeschichtet silicone-coated
Silikonelastomer(e) *n* silicone elastomer
Silikonfett *n* silicone grease
Silikonfilm *m* silicone film
Silikonflüssigkeit *f* silicone fluid
Silikongummi *m* silicone rubber
Silikonharz *n* silicone resin
Silikonhaut *f* silicone film
silikonisieren *v* siliconize, silicone-treat
Silikonisierung *f* siliconization, silicone treatment
Silikonlack *m* silicone varnish
Silikonöl *n* silicone fluid
Silikonpolymer(e) *n* silicone polymer
Silikonschmiermittel *n* silicone lubricant
Silikose *f (med)* silicosis
Silikospiegel *m* silicospiegel *(an iron-base alloy containing Mn and Si)*
silikothermisch silicothermic
silizieren *v (met)* siliconize *(metals for protecting them)*
Silizieren *n (met)* siliconization *(of metals)*
silizifizieren *v (geoch)* silicify
Silizium... *siehe: Silicium...*
Sillimaniterzeugnis *n (ceram)* sillimanite refractory
Silo *n(m)* silo, bin, (storage-)hopper
Siloxan *n* siloxane
Siloxanpolymer(e) *n* siloxane polymer
Siloxen *n* siloxen(e)

Silt *m (geol)* silt
Silvichemikalie *f* silvichemical
silylieren *v* silylate
Silylierung *f* silylation
Sima *n (geoch)* sima, lower crust, LC
Simazin *n* simazine *(a herbicidal triazine derivative)*
Simmons-Smith-Reaktion *f* Simmons-Smith reaction *(for obtaining cyclopropanes)*
SIMS *s.* Sekundärionenmassenspektrometrie
Simulation *f* simulation *(mathematics)*
Simultanbestimmung *f* simultaneous analysis
Simultanreaktion *f* simultaneous reaction
Sinapinalkohol *m* sinapic alcohol
Sinapinsäure *f* sinapic acid
Singulettbindung *f* single-electron bond
Singulett-Sauerstoff *m* singlet molecular oxygen
Singulett-Singulett-Übergang *m* singlet-singlet transition
Singulett-Triplett-Übergang *m* singlet-triplet transition
Singulett-Übergang *m* singlet-singlet transition
Singulettzustand *m* singlet (electronic) state
sinken *v* 1. sink, fall, subside; *(if bottoms are being formed:)* sediment, settle; 2. decrease, drop *(of physical data)*
Sinken *n* 1. sinking, fall, subsidence; 2. *(if bottoms are being formed:)* sedimentation, settling; decrease, drop *(of physical data)*
Sinkgeschwindigkeit *f* rate of sedimentation [settling], sedimentation rate
Sinkgut *n* 1. settled material; 2. *(min tech)* underflow
Sinkstoffe *mpl* 1. *(hyd)* settleable solids; 2. settled material
Sinter *m* 1. *(ceram, geol)* sinter; 2. *(met)* scale
Sinterapparat *m* sintering machine
Sintergut *n* sinter
Sinterhartmetall *n* cemented [sintered] carbide, hard metal (alloy)
Sintermetall *n* sintered(-powder) metal
sintern *v* sinter
Sinterofen *m* sintering furnace
Sinterrösten *n* sinter roasting
Sinterstahl *m* sintered steel
Sinterwerkstoff *m* sintered material
Siphon *m (lab)* siphon; trap *(as of a sink)*
Sirup *m* 1. syrup; 2. *(sugar)* molasses
sirupartig syrup-like, syrupy
SK *s.* Synthesekautschuk
Skala *f* scale
Skala f/nanoskopische nanoscopic scale
Skale *f* scale *(on a measuring device)*
Skalenteil *m* scale division [interval]
Skalenteilstrich *m* scale division
Skalenwert *m* reading
Skatol *n* skatole
Skelett *n* skeleton
Skelettisomere *n (org ch)* skeletal [chain] isomer
Skelettisomerie *f (org ch)* skeletal [chain] isomerism
Skelettkatalysator *m* skeleton [Raney] catalyst
skimmen *v (petrol)* skim
Skleroprotein *n* scleroprotein, skeletal [structural] protein
Skleroskop *n* scleroscope
Skraup-Synthese *f* Skraup quinoline synthesis
Skrubber *m* scrubber, gas washer, wet collector
slö *s.* schwerlöslich
Smalte *f* smalt *(cobalt(II) potassium silicate)*
Smaragd *m (min)* emerald *(a cyclosilicate)*
Smogalarm *m* smog emergency
SM-Verfahren *n s.* Siemens-Martin-Verfahren
S_N-Reaktion *f* S_N reaction, nucleophilic substitution
S_N1-Reaktion *f* S_N1 reaction, first-order nucleophilic substitution (reaction)
S_N2-Reaktion *f* S_N2 reaction, second-order nucleophilic substitution (reaction)
Soda *f* 1. soda, sodium carbonate; 2. *(min)* soda, natron *(sodium carbonate 10-water)*
Soda f/kalzinierte calcined soda, (soda) ash, anhydrous sodium carbonate
Soda f/kaustische caustic soda, sodium hydroxide *(esp when produced by the lime-soda caustic process)*
Sodaablauge *f* spent soda

Sodaalaun *m* soda alum *(crystalline aluminium sodium sulphate)*
Sodaaufschluss *m (pap)* soda pulping
Sodaauszug *m (anal)* sodium carbonate extract
Sodakalkglas *n* soda-lime glass
Sodastein *m* caustic soda
Sodazellstoff *m* soda pulp
Söderberg-Elektrode *f* Söderberg (continuous) electrode
Sojaöl *n* soybean oil
Sojaprotein *n* soybean protein
Sol *n (coll)* sol • **ein Sol bilden** solate
Solanin *n* solanin(e) *(a steroid alkaloid)*
Solanumalkaloid *n* solanum alkaloid
Sole *f* (salt) brine, saline (water)
Sole *f*/**ablaufende [verbrauchte]** spent brine *(chlor-alkali electrolysis)*
solehaltig briny
Sol-Gel-Umwandlung *f* sol-gel transformation, sol-gel transition [change]
solidensieren *v* desublimate *(to convert a gas directly into the solid state)*
Solidensieren *n* desublimation *(direct conversion of a gas into the solid state)*
Soliduskurve *f* solidus, solidus curve [line]
Soliduspunkt *m* solidus point
Sollwert *m* set point
Solubilisierung *f* solubilization
Solvat *n* solvate *(solvent and solute or dispersed phase and dispersion medium)*; raffinate *(less soluble residue remaining after solvent extraction)*
Solvatation *f* solvation
Solvatationsenergie *f* solvation energy
Solvatationskraft *f* solvating power
Solvatationsreaktion *f* solvation reaction
Solvathülle *f* solvation sheath
solvatisieren *v* solvate
Solvatisierung *f* solvation
Solvay-Verfahren *n* Solvay process, ammonia soda process
Solvens *n* solvent, dissolver
Solvententparaffinierung *f (petrol)* solvent dewaxing
Solventextraktion *f* solvent [liquid-liquid] extraction
Solventnaphtha *n(f)* solvent naphtha
Solventraffination *f* solvent refining
Solvolyse *f* solvolysis, lyolysis • **durch Solvolyse aufspalten** solvolyze
Solvolysereaktion *f* solvolysis reaction
solvolytisch solvolytic
Solvuskurve *f (phys ch)* solvus, solvus curve [line]
Soman *n* soman *(a warfare agent)*
Somatoliberin *n* somatotropin releasing hormone
Somatostatin *n* somatotropin inhibitory hormone
Somatotropin *n* somatotropin, somatotropic hormone
Sommelet-Reaktion *f* Sommelet reaction *(for obtaining aldehydes)*
Sommelet-Umlagerung *f* Sommelet rearrangement *(isomerization of quaternary benzylammonium compounds)*
Sonde *f (anal)* probe
Sonde *f*/**chemische** chemical probe *(molecule which reacts in a specific manner with parts of another one)*
Sonderlegierung *f* specialty alloy
Sondermüll *m* special solid waste
Sonnenblumenöl *n* sunflower (seed) oil
Sonnenbräunungsmittel *n (cosmet)* suntan preparation; suntan make-up
Sonnendestillationsanlage *f (hyd)* solar still
Sonnenofen *m* solar furnace
Sonnenschutzmittel *n* sunscreen (agent)
Sonnenschutzöl *n* sunscreen oil
Sonnenstrahlung *f* solar radiation
Sonolumineszenz *f* sonoluminescence
Sorbend *m* material to be sorbed
Sorbens *n* sorbent, sorption agent
sorbieren *v* sorb
Sorbinsäure *f* sorbic acid, 2,4-hexadienoic acid
Sorbit *m (met)* sorbite
sorbitisch *(met)* sorbitic
Sorel-Zement *m* Sorel cement
Sorosilicat *n (min)* sorosilicate
Sorption *f* sorption
Sorptionsisotherme *f* sorption isotherm
Sorptionsmittel *n* sorbent, sorption agent
sorptiv sorptive
Sorptiv *n* sorbate, sorbed material
Sorte *f* grade
sortieren *v* sorte, grade; pick *(by hand)*
Soxhlet(-Apparat) *m* Soxhlet (extractor)
SOZ *s.* Straßen-Octanzahl
Spachtelmasse *f (coat)* filler, surfacer

Spallation *f (nucl)* spallation
Spalt *m* 1. slit; gap *(between rolls)*; 2. crack, crevice *(fault)*; 3. *(bioch)* trough *(in hydrolase molecules)*
Spaltanlage *f (petrol)* cracking plant
Spaltausbeute *f (nucl)* fission yield
spaltbar 1. divisible, resolvable *(optically active compounds)*; 2. *(nucl)* fissionable, fissile; 3. *(min, cryst)* cleavable, fissile
spaltbar/nicht 1. indivisible *(optically active compounds)*; 2. *(nucl)* non-fissile, non-fissionable
Spaltbarkeit *f* 1. divisibility, resolvability *(of optically active compounds)*; 2. *(nucl)* fissionability; 3. *(min, cryst)* fissility, cleavability, cleavage
Spaltbenzin *n* cracked gasoline
Spaltbreite *f (spectr)* slit width
spalten *v* 1. cleave, crack, decompose *(chemical compounds)*; cleave, break, crack, disrupt *(a chemical bond)*; split, open *(a ring)*; 2. *(nucl)* fission; break, demulsify *(emulsions)*; resolve *(optically active compounds)*
spalten *v***/katalytisch** cat-crack
spalten *v***/sich** 1. decompose, crack, split up *(chemical compounds)*; 2. *(nucl)* split, (undergo) fission
Spalten *n s.* Spaltung
Spaltfaser *f* split fibre
Spaltfilter *n* edge filter
Spaltfläche *f (min, cryst)* cleavage plane [face]
Spaltgut *n (petrol)* cracking feed(stock)
Spaltkette *f (nucl)* fission chain
Spaltkorrosion *f* crevice corrosion
Spaltmaterial *n (nucl)* fissionable material
Spaltneutron *n* fission neutron
Spaltprodukt *n* 1. cleavage product; 2. *(nucl)* fission product
Spaltstück *n* 1. fragment; 2. *(nucl)* fission fragment
Spaltung *f* 1. cleavage, cracking, decomposition *(of chemical compounds)*; cleavage, breaking, disruption, scission *(of a chemical bond)*; splitting, opening *(of a ring)*; 2. *(nucl)* fission; resolution *(of optically active compounds)*
Spaltung *f* **einer C–C Bindung** carbon-carbon scission
Spaltung *f***/enzymatische** enzymatic splitting
Spaltung *f***/hydrierende** *(petrol)* hydrocracking
Spaltung *f***/hydrogenolytische** hydrogenolysis
Spaltung *f***/ionische [katalytische]** catalytic cracking
Spaltung *f***/radikalische [thermische]** thermal cracking
Spaltung *f***/thioklastische [thiolytische]** thiolytic cleavage *(of fatty acids)*
Spaltung *f* **von Emulsionen** breaking of emulsions, emulsion breakdown, demulsification
Spaltungsenergie *f* fission energy
Spaltungsreaktion *f* 1. cleavage [scission] reaction; 2. *(nucl)* fission reaction
Spaltverfahren *n (petrol)* cracking process
Spaltzone *f (nucl)* core
Späne *mpl* shavings
spannen *v* stress, strain, *(esp in length)* stretch
Spannung *f* 1. *(el ch)* voltage; 2. tension *(mechanically)*; *(if unintended:)* stress; 3. strain *(of ring molecules)*
Spannungsabfall *m***/ohmscher** ohmic drop, ohmic overpotential [polarization], IR drop
Spannungsabnahme *f (el ch)* drop in voltage
Spannungsanstieg *m* voltage rise
spannungsarm low-strain *(ring molecule)*
Spannungsarmglühen *n (met, glass, ceram)* (stress relief) annealing
Spannungs-Dehnungs-Diagramm *n* stress-strain diagram
spannungsfrei stress-free, stress-relieved *(materials science)*; strain-free, strainless *(ring molecule)* • **spannungsfrei machen** *(met, glass, ceram)* anneal
Spannungskorrosion *f* stress corrosion
spannungslos *(el ch)* dead
spannungsoptisch stress-optical
Spannungsreihe *f***/elektrochemische** electrochemical series, electromotive (force) series, emf series
Spannungsrelaxation *f* stress relaxation
Spannungsrissbildung *f* stress cracking

Spannungsrisskorrosion *f* stress corrosion cracking, SCC
Spannungsrückgang *m (el ch)* drop in voltage
Spannungstheorie *f (org ch)* strain theory
Spannungstheorie *f***/sachse-mohrsche** Sachse-Mohr concept of strainless rings
Spasmolytikum *n (pharm)* spasmolytic, antispasmodic
spasmolytisch *(pharm)* spasmolytic
spatartig *(min)* spathic
Spatel *m (lab)* spatula
Spatelspitze (voll)/eine a spatula-point, a spatula-tipfull
Spatenwischer *m (lab)* wing shape policeman
Specköl *n* grease [lard] oil
Speckstein *m (min)* steatite *(a variety of talc)*
Speichel *m* saliva
Speicher *m* store(house), reservoir; accumulator *(hydraulics)*
Speicherbecken *n (hyd)* storage basin
Speicherbehälter *m* storage tank [vessel]
Speichergestein *n* reservoir rock
Speichermodul *m* storage modulus
speichern *v* store (up), accumulate
Speicherprotein *n (bioch)* storage protein
Speicherring *m* storage ring *(to keep particle beams or charged particles circulating)*
Speicherung *f* storage, accumulation
Speise *f (met)* speiss, speise
Speiseessig *m* table vinegar
Speisefett *n* dietary [food] fat
speisen *v (tech)* feed, charge; power *(an electrical circuit)*
Speisen *n* feed, charging *(as of reactors, furnaces)*
Speiseöl *n* edible [food] oil
Speiser *m* feeder
Speisesalz *n* table salt
Speisewasser *n* feed water
Speisewasser *n* **für Dampferzeuger** boiler feed water
Speisewasservorwärmer *m* economizer
Speisezone *f (plast)* feed zone
spektral spectral
Spektralanalyse *f* spectral analysis
Spektralbande *f* spectral band
Spektralbereich *m* spectral region [range]
Spektralbereich *m***/kurzwelliger** short-wavelength region
Spektralbereich *m***/langwelliger** long wavelength region
Spektralfarbe *f* spectral colour
Spektrallinie *f* spectral line
Spektralphotometer *n* spectrophotometer
Spektralphotometrie *f* spectrophotometry
spektralphotometrisch spectrophotometric
spektralrein spectro(-quality) grade
Spektralserie *f* spectral series
Spektralterm *m* spectral term
spektrochemisch spectrochemical
Spektrogramm *n* spectrogram
Spektrograph *m* spectrograph
spektrographisch spectrographic
Spektrometer *n* spectrometer
Spektrometrie *f* spectrometry
spektrometrisch spectrometric
Spektrophotometer *n* spectrophotometer
Spektrophotometrie *f* spectrophotometry
spektrophotometrisch spectrophotometric
Spektroskop *n* spectroscope
Spektroskopie *f* spectroscopy
Spektroskopie *f***/dreidimensionale** spin mapping
Spektroskopie *f* **im fernen Infrarot** far-infrared spectroscopy
Spektroskopie *f* **im sichtbaren Bereich** visible spectroscopy
Spektroskopie *f* **im Sichtbaren und Ultraviolett** UV-VIS spectroscopy
Spektroskopie *f* **im sichtbaren und UV Gebiet** UV-VIS spectroscopy
Spektroskopie *f***/optische** optical spectroscopy
Spektroskopie *f***/photoakustische** photoacoustic spectroscopy
Spektroskopie *f***/zeitaufgelöste** time-resolved spectroscopy
Spektroskopiker *m* spectroscopist
spektroskopisch spectroscopic
Spektrum *n* spectrum

Spektrum *n*/**diskretes** discrete spectrum
Spektrum *n*/**elektromagnetisches** electromagnetic spectrum
Spektrum *n*/**kontinuierliches** continuous spectrum
Spektrum *n*/**linienreiches** crowded spectrum
Spektrum *n*/**optisches** optical spectrum
Spekularit *m (min)* specularite, specular iron (ore)
spenden *v* donate *(e.g. electrons)*
Spender *m* dispenser *(for withdrawing materials)*
spermienabtötend *(pharm)* spermatocidal
spermizid *(pharm)* spermatocidal
sperren *v (coat)* seal
Sperrflüssigkeit *f* sealing fluid
Sperrhahn *m* stopcock
sperrig bulky
Sperrschichteffekt *m* photovoltaic effect
Sperrschichtelement *n* photovoltaic [barrier-layer] cell
Sperrschichthöhe *f* inversion height *(in the atmosphere)*
Spezialfermenter *m (biot)* single-purpose fermenter
Spezialzwecke/für special-purpose
Spezies *f*/**chemische** chemical species
Speziesbestimmung *f (anal)* speciation
speziesspezifisch *(bioch)* species-specific
Speziesspezifität *f (bioch)* species specificity
Spezifikation *f* specification
Spezifität *f* specificity
Spezifität *f*/**optische** optical specificity
Spezifität *f*/**stereochemische** stereospecificity
Sphalerit *m (min)* sphalerite, zinc blende
Sphärizität *f* sphericity
sphäroidal spheroidal
Sphärokolloid *n* spherocolloid, globular colloid
Sphäroprotein *n* globular protein
Sphen *m (min)* sphene *(a variety of titanite)*
Sphingolipoid *n (bioch)* sphingolipid(e)
Sphingomyelin *n* sphingomyelin *(any of a group of crystalline phosphatides)*
Sphingosin *n (bioch)* sphingosine
Spiegel *m* level *(of a liquid)*
Spiegelbildisomer(e) *n* enantiomer, optical [mirror-image] isomer, optical antipode
Spiegelbildisomerie *f* enantiomerism, optical [mirror-image] isomerism
Spiegelebene *f (cryst)* plane of symmetry, mirror plane
Spiegeleisen *n* spiegel (iron) *(a pig iron containing manganese)*
spiegeln *v* reflect (back)
Spiegelung *f* reflection
Spin *m* spin, *(of elementary particles also)* intrinsic angular momentum
Spin-Addukt *n* spin adduct
Spin-Bahn-Kopplung *f (spectr)* spin-orbit coupling [interaction], LS coupling, Russel-Saunders coupling
Spindel *f* 1. areometer, hydrometer; 2. stem *(as of a valve)*
Spindrehimpuls *m* spin angular momentum
Spinell *m* spinel *(1. a mineral containing magnesium aluminate; 2. any of a class of aluminates of the type* $M^{II}Al_2O_4$*)*
Spin-Gitter-Relaxation *f (spectr)* spin-lattice relaxation [interaction]
Spin-Kopplungsaufspaltung *f* spin decoupling
Spinmarkierung *f (spectr)* spin labelling
Spinmultiplizität *f* spin multiplicity
Spinnbad *n* spinning bath
Spinndüse *f* spinneret
Spinne *f (distil)* pig, udder-type receiver
spinnen *v* spin
Spinnen *n* **aus der Schmelze** melt spinning [extrusion]
Spinnen *n* **aus Lösungen** solution [solvent] spinning
Spinnfaden *m* strand
Spinnfärbung *f* spin dyeing
Spinnfaser *f* (staple) fibre
spinngefärbt spin-dyed, spun-dyed
Spinnkanne *f* spinning can
Spinnkopf *m* spinning head
Spinnlösung *f* spinning solution
Spinnpumpe *f* spinning [metering] pump
Spinntopf *m* spinning pot
Spinnzwiebel *f* (spinning) bulb
Spinodale *f (phys ch)* spinodal curve
Spinorbital *n* spin orbital, SO
Spinorbital *n*/**molekulares** molecular spin orbital

Spinpopulation *f* spin population
Spinquantenzahl *f* spin quantum number
Spins *mpl***/antiparallele [entgegengesetzte]** antiparallel [opposite] spins
Spin-Spin-Kopplung *f* spin-spin coupling [interaction]
Spin-Spin-Kopplungsaufspaltung *f* spin-spin splitting
Spin-Spin-Relaxation *f* spin-spin relaxation
Spinzustand *m* spin state
Spirale *f* spiral, coil, helix
Spiralkühler *m* **nach Friedrichs** Friedrichs condenser
Spiralrohrbündel-Wärmeübertrager *m* spiral-tube heat exchanger
Spiralrohrschlange *f* spiral coil
Spiralwärmeübertrager *m* spiral-plate heat exchanger
Spiran *n* spiran(e), spiro hydrocarbon
Spirituose *f* spirit
Spiritus *m* spirit *(industrially manufactured ethanol)*
Spiritus *m***/denaturierter** denatured alcohol, methylated spirit
Spiritus *m***/vergällter** methylated spirit, denatured alcohol
Spiritusbrennerei *f* distillery
Spirituslack *m* spirit varnish
Spirokohlenwasserstoff *m* spiro hydrocarbon, spiran(e)
Spiropolymer(e) *n* spiro polymer
Spirosystem *n* spiro ring system
Spitze *f* top; vertex *(as of a tetrahedral molecule)*
Spitzentemperatur *f* peak temperature
Spitzenwert *m* extreme value
Spitzkolben *m* pointed flask
Spitztrichter *m* cone classifier
SPK *s.* Strom-Potenzial-Kurve
spleißen *v (bioch)* splice
Spleißen *n (bioch)* splicing
Spleißen *n* **der RNA** *(bioch)* RNA splicing
splitten *v (chromat)* split *(a gas mixture into two unequal streams)*
Splittereis *n* chipped ice
splittern *v* shatter, spall
splittersicher shatterproof *(glass)*
Spodium *n* spodium, bone char(coal)
Spongiose *f* graphitization *(a type of corrosion)*
Spontanmutation *f (biol)* spontaneous mutation
Spontanzündung *f* autoignition
Sporensuspension *f (biot)* spore suspension
Sporulationsmedium *n (biot)* sporulation medium
spratzen *v (met)* spit, spatter
spreiten *v* spread (out)
Spreitungsdruck *m* spreading pressure
Spreitungskoeffizient *m* spreading coefficient, SC
Spreizschwingung *f* scissor vibration
sprengen *v* 1. break, crack, rupture, disrupt *(a chemical bond)*; 2. blast *(by means of explosives)*
Sprenggelatine *f* blasting gelatin, gelatin dynamite
Sprengkapsel *f* blasting cap, primer
Sprengkraft *f* explosive [shattering] power, brisance
Sprengmittel *n* blasting agent
Sprengpulver *n* blasting [black] powder, gunpowder
Sprengstoff *m* explosive
Sprengstoff *m***/brisanter** high explosive, H.E.
Sprengung *f* 1. breaking, cracking, rupture *(of a chemical bond)*; 2. blast(ing), detonation, shot
springen *v* burst, crack
Sprinkler-Feuerlöschanlage *f* sprinkler system
Sprit *m s.* Spiritus
spritlöslich spirit-soluble
spritzalitieren *v* alumetize
spritzaluminieren *v* alumetize
Spritzbrühe *f (agric)* wash
Spritzdruck *m (plast)* injection pressure
spritzen *v* 1. *(plast)* extrude, inject; 2. *(coat)* spray(-apply) *(e.g. coating material)*; spray(-coat) *(surfaces)*; 3. splash, spatter *(of a liquid)*
Spritzen *n* 1. *(plast)* extrusion (moulding), injection; 2. spraying, spray application *(of coating material)*; spraying, spray coating *(of surfaces)*; 3. splashing, spattering *(of a liquid)*
Spritzer *m* splash
Spritzflasche *f* wash(ing) bottle
Spritzflasche *f* **aus weichem Kunststoff** squeeze bottle
Spritzgießen *n (plast)* injection moulding

Spritzgießmaschine *f (plast)* injection moulding machine
Spritzgussmasse *f* injection-moulding material
Spritzgussteil *n* injection-moulded part
Spritzgussverfahren *(plast)* injection moulding
Spritzkammer *f* spray chamber
Spritzmaschine *f (plast)* extruder
Spritzmittel *n* spray
Spritzpistole *f (coat)* spray(ing) gun, paint sprayer
Spritzprägen *n (plast)* injection embossing
Spritzrückstand *m (agric)* spray residue
Spritzschaden *m (agric)* spray injury
Spritzschicht *f* spray coat(ing)
Spritzschleier *m* spray swathe
Sprödbruch *m* brittle fracture [failure]
Sprödbruchtemperatur *f* brittle(-point) temperature
spröde brittle • **spröde machen** embrittle, make brittle • **spröde werden** embrittle, become brittle
Sprödigkeit *f* brittleness
Sprödwerden *n* embrittlement
S-Protein *n* S protein *(breakdown product of ribonuclease)*
Sprudel *m* sparkling water
sprudeln *v* bubble
Sprudelschicht *f* spouted bed *(fluid-bed technology)*
Sprühabsorber *m* spray absorber
Sprühdüse *f* spray [atomizing] nozzle
Sprühelektrode *f* discharge electrode
sprühen *v* spray
Sprüher *m* sprayer, atomizer
Sprühflüssigkeit *f* spray
Sprühkegel *m* spray cone
Sprühkolonne *f* spray column *(gas cleaning)*
sprühkristallisieren *v* prill
Sprühnebel *m (agric)* spray mist
Sprühreagens *n* spray reagent
Sprühtrockner *m* spray dryer
Sprühtrocknung *f* spray drying
Sprühturm *m* spray tower
Sprühwascher *m* spray scrubber [washer] *(gas cleaning)*
Sprung *m* 1. crack; 2. *(ceram)* flaw
Sprungtemperatur *f* superconductive [transition] temperature *(of a superconductor)*
spucken *v* 1. prime *(of kettles or distillation columns)*; 2. *(distil)* puke
Spuckgrenze *f (distil)* flooding point
Spülbad *n* rinsing [scouring] bath
Spülbecken *n* (bench) sink
spülen *v* rinse, scour, wash, swill, flush; purge *(with inert gas)*
Spülflüssigkeit *f* rinsing liquid, rinsing(s)
Spülgas *n (chromat)* purge gas
Spülmaschine *f (lab)* dishwasher
Spülmittel *n* rinsing [scouring] agent, rinse, scavenger
Spülwasser *n* rinse [swill] water, wash(-ing) water
Spur *f* 1. *(anal)* trace (amount) *($^*10^2$ to 10^{-4} ppm)*; 2. trace *(of a recording instrument)*
Spurenanalyse *f* trace analysis
Spurenbestandteil *m* trace component [constituent]
Spurenelement *n (biol, agric)* trace element, microelement
Spurenmetall *n* trace metal
Spurennährstoff *m* micronutrient
Spurstein *m* white metal *(product obtained by removing iron from copper matte)*
sputtern *v* sputter
S-PVC *s.* Suspensionspolyvinylchlorid
Squalen *n* squalene *(an aliphatic triterpene)*
SR-Benzin *n* straight-run gasoline [naphtha, spirit]
Sre. *s.* Säure
–S–S-Bindung *f* disulphide bond
ssb-Protein *n (bioch)* SSB protein, single-stranded (DNA-)binding protein
–S–S-Brücke *f* disulphide bridge
ssl. *(sehr schwer löslich)* extremely slightly soluble
–S–S-Spange *f (bioch)* intrachain disulphide bond
stabil stable, resistant, *(esp relating to pesticides:)* persistent; stable, rigid *(physically)* • **stabil sein** persist *(pesticides)* • **stabil werden** come into stability
stabilglühen *v (met)* stabilize
Stabilglühen *n (met)* stabilizing anneal
Stabilisator *m* 1. stabilizer, stabilizing agent, *(relating to suspensions also)* suspending agent; 2. *s.* Stabilisierkolonne

stabilisieren *v* stabilize
stabilisieren *v***/sich** stabilize
Stabilisierkolonne *f (petrol)* stabilizer
Stabilisierung *f* stabilizing
Stabilisierungsbecken *n (hyd)* stabilization tank
Stabilisierungsenergie *f* stabilizing energy
Stabilisierungsmittel *n* stabilizing agent, stabilizer
Stabilisierungsteich *m (hyd)* stabilization pond [lagoon]
Stabilität *f* stability, resistance, *(esp relating to pesticides:)* persistence; stability, rigidity *(physically)* • **Stabilität erlangen** come into stability
Stabilität *f***/thermische** thermal stability *(as of plastics)*
Stabilitätskonstante *f* (complex) formation constant, stability constant
Stabilitätskonstante *f***/individuelle [konsekutive]** successive formation constant
Stabilitätskonstante *f***/kumulative** overall formation constant
Stabilitätskonstante *f***/sukzessive** successive formation constant
Stabmühle *f* rod mill
Stabrostsieb *n* bar [rod] screen
Stabrostsiebmaschine *f* grizzly screen
Stabsiebrost *m* (bar, rod) grizzly
Stadium *n* stage
Stadium *n* **der großtechnischen Erprobung** preproduction stage
Stadtentwässerung *f (hyd)* sewerage
Stadtgas *n* town [city] gas
Stadtmüll *m* town refuse
Stadtnebel *m* smog *(a type of air pollution)*
Stahl *m* steel
Stahl *m***/basischer** basic steel
Stahl *m***/beruhigter [beruhigt vergossener]** killed steel
Stahl *m***/einsatzgehärteter** case-hardened steel
Stahl *m* **für Einsatzhärtung** case-hardening steel, carburizing steel
Stahl *m***/halbberuhigter** semikilled steel
Stahl *m***/herdgefrischter** open-hearth steel
Stahl *m***/hochlegierter** high-alloy steel
Stahl *m***/inchromierter** chromized steel
Stahl *m***/legierter** alloy steel
Stahl *m***/lufthärtender** air-hardening steel
Stahl *m***/niedriglegierter** low-alloy steel
Stahl *m***/sauerstoffgefrischter** basic oxygen steel
Stahl *m***/saurer** acid steel
Stahl *m***/unberuhigter [unberuhigt vergossener]** rimming [wild] steel
Stahl *m***/unlegierter** (plain) carbon steel
Stahl *m***/weicher** mild [soft] steel
Stahl *m***/windgefrischter** Bessemer [converter] steel
Stahlakkumulator *m* Edison accumulator [cell]
Stahlbeton *m* reinforced concrete
Stahlblock *m* steel ingot
Stahlflasche *f* steel cylinder
Stahlformguss *m* cast steel
Stahlgewinnung *f* steelmaking
Stahlguss *m* 1. steel casting; 2. cast steel *(products)*
Stahlgussstück *n* steel casting
Stahlmantel *m* steel jacket
Stahlpanzer *m* steel shell
Stahlschrott *m* steel scrap
Stalagmometer *n* stalagmometer *(for measuring surface tension)*
Stalagmometrie *f* stalagmometry *(a method for measuring surface tension)*
Stamm *m* 1. *(nomencl)* parent; 2. strain *(of microorganisms)*
Stammemulsion *f* stock emulsion
Stammentwicklung *f (biot)* strain development
Stammflotte *f (text)* stock liquor
Stammkohlenwasserstoff *m* parent hydrocarbon
Stammkörper *m* parent [mother] substance
Stammkultur *f (biot)* stock culture
Stammküpe *f (dye)* stock vat
Stammlauge *f* mother liquor
Stammlösung *f* stock solution
Stammname *m (nomencl)* parent name
Stammverbindung *f* parent compound
stampfen *v* stamp, ram
Stampfer *m* beater *(of a hammer mill)*
Stand *m* level; height *(as of a liquid in a vessel)*
Standard *m***/innerer [interner]** *(anal)* internal standard
Standard *m***/sekundärer** secondary standard

Standardabweichung *f* standard deviation, root-mean-square error
Standardaktivierungsenthalpie *f*/freie standard free energy of activation
Standardbedingungen *fpl* standard [asterisk] conditions
Standardbildungsenthalpie *f* standard enthalpy [heat] of formation
Standardbildungsenthalpie *f*/freie standard free energy of formation, standard Gibbs function of formation
Standarddruck *m* standard [normal] pressure
Standardeinheit *f* standard unit
Standardelektrodenpotenzial *n* standard electrode potential, S.E.P.
Standardelement *n (el ch)* standard cell
Standard-EMK *f* standard emf [electromotive force]
Standardenthalpie *f* standard enthalpy
Standardenthalpie *f*/freie standard Gibbs [free] energy
Standardenthalpie *f*/molare freie standard molar Gibbs function
Standardentropie *f* standard entropy
Standardentropie *f*/molare standard third-law entropy
Standardkunststoff *m* commodity plastic
Standardlösung *f* standard solution
Standardmethode *f* standard method [procedure, technique]
Standardnormalpotenzial *n* standard potential
Standardoxidationspotenzial *n* standard oxidation potential
Standardpotenzial *n* standard [normal] potential
Standardpotenzial *n*/chemisches standard chemical potential
Standardpotenzial *n* des Halbelements standard half-cell potential
Standardpotenzial *n* des Redoxsystems standard redox potential
Standardpufferlösung *f* standard buffer (solution)
Standardreagens *n* standard reagent
Standardreaktionsenthalpie *f* standard reaction enthalpy
Standardreaktionsentropie *f* standard reaction entropy
Standardredoxpotenzial *n* standard redox potential
Standardreduktionspotenzial *n* standard reduction potential
Standardschliffverbindung *f* standard ground glass joint
Standardsiebreihe *f* standard sieve scale [series]
Standardspannung *f* standard voltage
Standardsubstanz *f* standard (substance)
Standardverfahren *n* standard method [procedure, technique]
Standardwasserstoffelektrode *f* standard hydrogen electrode, SHE
Standardwert *m* standard value
Standardzelle *f (el ch)* standard cell
Standardzustand *m* standard (reference) state
Standflasche *f* storage [laboratory] bottle
***m*-ständig** in meta position, meta • ***m*-ständig (angeordnet) sein** be (located, situated) meta
***o*-ständig** in ortho position, ortho • ***o*-ständig sein** be (located, situated) ortho
***p*-ständig** in para position, para • ***p*-ständig sein** be (located, situated) para
Standöl *n* stand oil
Standrohr *n* standpipe
Standzeit *f* 1. operating life, useful [working] life *(of a machine)*; 2. residence time, retention [hold-up] time
Standzylinder *m* 1. *(lab)* plain cylinder, gas jar; 2. *(hyd)* settling cylinder
Stange *f (tech)* rod, bar
Stängel *m (cryst)* column
Stängel *m*/flacher blade
stängelförmig *(cryst)* columnar
Stängelkristall *m* columnar crystal
stängelkristallinisch *(cryst)* columnar
Stangenrostsieb *n* bar [rod] screen
Stangensiebrost *m* (bar, rod) grizzly
Stannan *n* stannane, tin [stannic] hydride
Stannat *n* stannate
Stanton-Zahl *f* Stanton number *(fluid mechanics)*
stanzen *v* punch
Stanzmaschine *f (met)* punching machine
Stanzwerkzeug *n* die
Stapel *m* 1. stack; 2. *(text)* staple (length), fibre staple

Stapelfaser *f (text)* staple (fibre)
Stapelfehler *m (cryst)* stacking fault
Stapellänge *f (text)* staple (length), fibre staple
stapeln *v* stack
Stapelung *f (cryst)* stacking
stark 1. strong *(acid, base)*; 2. vigorous, intense *(reaction)*; 3. intense *(irradiation)*
Stark-Aufspaltung *f (spectr)* Stark-splitting
Stärke *f* 1. starch *(a polysaccharide)*; 2. strength *(of an acid or base)*; 3. intensity, power *(of a reaction)*; 4. intensity *(of irradiation)*
Stärke *f* **des Abwassers** *(hyd)* strength of sewage
Stärke *f*/**dünnkochende [lösliche]** soluble starch
Stärke *f*/**modifizierte** modified starch
Stärke *f*/**tierische** animal starch
Stärkeabbau *m* starch degradation
stärkeartig amylaceous
Stärkeblockelektrophorese *f* starch block electrophoresis
Stark-Effekt *m* Stark effect, electrochromic [electric field] effect *(splitting of spectral lines)*
Stärkegelelektrophorese *f* starch gel electrophoresis
Stärkegummi *n* dextrin, starch gum *(a polysaccharide)*
stärkehaltig starchy, amyliferous, farinaceous
Stärkehydrolysat *n* starch hydrolyzate
Stärkehydrolyse *f* starch hydrolysis
Stärkekette *f* starch chain
Stärkekorn *n* starch grain [granule]
Stärkeleim *m* starch glue [adhesive]
Stärkemehl *n* starch flour
Stärkenitrat *n* starch nitrate, nitrostarch
Stärkepflanze *f* starch plant *(a plant which converts excess assimilation products into starch)*
stärkereich rich in starch, farinaceous
Stärkeretrogradation *f* starch retrogradation
Stärkeumwandlung *f* starch conversion
Stärkeverflüssigung *f* starch liquefaction
Stärkeverzuckerung *f* starch saccharification [conversion] *(by acids and enzymes)*
Starkgas *n* rich gas
starr rigid *(e.g. molecule, gel)*
Starrheit *f* rigidity *(as of molecules or gels)*
Start *m* initiation *(as of a reaction)*
Startcodon *n (bioch)* (chain) initiation codon
Startdünger *m* starter (fertilizer)
starten *v* initiate *(e.g. a reaction)*
Starter *m* 1. *(bioch)* primer; 2. *(plast)* polymerization initiator
Starterkultur *f (biot)* starter culture
Startermolekül *n (bioch)* primer
Startreaktion *f* initiation reaction
Startschritt *m (plast)* initiation step
Stationärenzymkinetik *f (bioch)* steady-state kinetics
Stationaritätsprinzip *n* Bodenstein (steady-state) approximation, steady-state method
Stationaritätsprinzip *n*/**bodensteinsches** Bodenstein (steady-state) approximation, steady-state method
Stationärzustand *m* steady [stationary] state *(kinetics)*
Statistik *f*/**boltzmannsche** Boltzmann statistics
Stativ *n (lab)* (retort) stand, ring stand
Stativklemme *f* stand clamp
Staub *m* dust • **mit Staub beladen** dust-laden
Staubablagerung *f* 1. dust deposition [fall]; 2. dust deposit
Staubabscheider *m* dust collector [catcher]
Staubabscheidung *f* dust collection
Staubbekämpfung *f* dust control
staubbeladen dust-laden
Staubbeladung *f* dust loading
Staubbildung *f* dust formation
Staubbrenner *m* pulverized-coal burner
staubdicht dustproof
Stäubemittel *n* dusting agent, dust
stäuben *v (agric)* dust *(with pesticides)*
Staubexplosion *f* dust explosion
Staubfeuerung *f* suspension firing
Staubfilter *n* dust(-control) filter, filter-type dust collector
Staubfließkatalysator *m* fluid(ized) catalyst
Staubfließtechnik *f* fluid-bed technique
Staubfließverfahren *n* fluid(ized) process

staubfrei dust-free, dustless
Staubgehalt *m* dust loading
staubhaltig dust-laden
staubig dusty, pulverulent
Staubkammer *f* settling chamber *(gas cleaning)*
Staubkonzentration *f* dust concentration
Staubkorn *n* dust particle
Staubmühle *f* pulverizing mill
Staubschutzmaske *f* aspirator, mechanical filter respirator
Staubteilchen *n* dust particle
Staubvergasung *f (coal)* pulverized-coal gasification, suspension gasification
Staudinger-Index *m* intrinsic viscosity, limiting viscosity number
Staudruck *m* stagnation pressure
Stauhöhe *f (hyd)* head of water, liquid head *(above a filter medium)*
Staupunkt *m* 1. *(distil)* loading point; 2. stagnation point *(fluid mechanics)*
Staurohr *n* Pitot tube *(flow measurement)*
Stearamid *n* stearamide
Stearat *n* stearate
Stearin *n* stearin, *(specif)* tristearin
Stearinöl *n* olein
Stearinsäure *f* stearic acid, *octadecanoic acid
Stearolsäure *f* stearolic acid, octadec-9-ynoic acid
Stearylalkohol *m* stearyl alcohol, octadecan-1-ol
Steatit *m (min)* steatite *(a variety of talc)*
stechend pungent *(smell)*
Stechpipette *f* dropping [teat] pipette, medicine dropper
Stefan-Boltzmann-Gesetz *n* Stefan-Boltzmann law *(emissive power of black bodies)*
Steffen-Verfahren *n* Steffen process *(for desugarizing molasses)*
Stegkettenförderer *m* scraper(-chain) conveyor
stehen lassen *v* allow to stand
Stehkolben *m* flat-bottom(ed) flask
Stehzentrifuge *f* underdriven [bottom-driven] centrifuge
steif rigid, stiff • **steif werden** *(coll)* gel
Steifheit *f* rigidity, stiffness
Steifwerden *n (coll)* gelation
steigen *v* 1. rise, increase *(as of temperature)*; 2. rise *(as of fluid)*
steigern *v* increase *(e.g. number, quantity)*; elevate, raise *(e.g. temperature, boiling point)*; potentiate *(the effect esp of drugs or pesticides by adding another agent)*
Steigerung *f* increase *(as of number, quantity)*; elevation, raise *(as of temperature, boiling point)*; potentiation *(the effect esp of drugs or pesticides by adding another agent)*
Steigrohr *n* riser (pipe), riser tube
Steigung *f* pitch *(as of the blades of a propeller turbine)*
Steilheit *f* steepness *(of a curve or an electrode)*
Stein *m* 1. stone; 2. *(ceram)* brick; 3. matte *(smelted mixture of metal sulphides)* • **auf Stein (ver)schmelzen** *(met)* matte-smelt
Stein *m*/**feuerfester** refractory brick, firebrick
Steinbildung *f (hyd)* scaling, scale formation *(in a vessel)*
Steingitterwerk *n* chequerwork
Steingut *n* earthenware
Steinkohle *f* (hard) coal
Steinkohlenaufbereitung *f* coal preparation [cleaning]
Steinkohlengas *n* coal gas
Steinkohlenhydrierung *f* coal hydrogenation
Steinkohlenkoks *m* coal coke
Steinkohlenteer *m* coal tar
Steinkohlenteerpech *n* coal-tar pitch
Steinkohlenverkokung *f* coal carbonization
Steinkohlenverschwelung *f* coal carbonization
Steinsalz *n (min)* rock salt, halite *(sodium chloride)*
Steinsalzgitter *n (cryst)* rock-salt lattice, sodium chloride lattice
Steinschmelzen *n (met)* matte smelting
Steinwolle *f* rock wool
Steinzeug *n (ceram)* stoneware
Steinzeug *n*/**braunes** brown ware
Steinzeug *n*/**chemisches** chemical stoneware
Stelle *f*/**aktive** activator site [centre] *(catalysis)*

Stellglied *n* controller *(control engineering)*
Stellmotor *m* actuator *(control engineering)*
Stellschraube *f* adjusting screw *(of a balance)*
Stellung *f* position
***anti*-Stellung** *f* anti position • **in *anti*-Stellung stehen** be anti
***cis*-Stellung** *f* cis position • **in *cis*-Stellung** cis
***m*-Stellung** *f* meta position • **in *m*-Stellung** in meta position, meta • **sich in *m*-Stellung befinden** be (located, situated) meta
***o*-Stellung** *f* ortho position • **in *o*-Stellung** in ortho position, ortho • **sich in *o*-Stellung befinden** be (located, situated) ortho
***p*-Stellung** *f* para position • **in *p*-Stellung** in para position, para • **sich in *p*-Stellung befinden** be (located, situated) para
***syn*-Stellung** *f* syn position • **in *syn*-Stellung stehen** be syn
***trans*-Stellung** *f* trans position • **in *trans*-Stellung** trans
1,2-Stellung *s. o*-Stellung
1,2,3-Stellung vicinal [neighbouring] position
1,3-Stellung *s. m*-Stellung
1,4-Stellung *s. p*-Stellung
Stellungsisomer(e) *n* position(al) isomer
Stellungsisomerie *f* position(al) isomerism, place [substitution] isomerism
Stellungsziffer *f (nomencl)* position [locator] number
Stempel *m* 1. *(tech)* piston, plunger; 2. *(plast)* male mould, stamp; 3. *(ceram)* punch
Stengel-Verfahren *n* Stengel process *(for making ammonium nitrate fertilizer)*
Stephen-Reduktion *f* Stephen reduction [reaction] *(for obtaining aldehydes from nitriles)*
Stereoblockpolymer(e) *n* stereoblock polymer
Stereochemie *f* stereochemistry
stereochemisch stereochemical
stereoelektronisch stereoelectronic
Stereoformel *f* stereochemical formula, space [configurational] formula
stereoisomer stereoisomeric
Stereoisomer(e) *n* stereo(iso)mer, space isomer
Stereoisomerie *f* stereoisomerism, space isomerism
Stereomer(e) *n* stereo(iso)mer
stereoregulär stereoregular
Stereoregularität *f* stereoregularity
stereoselektiv stereoselective
Stereoselektivität *f* stereoselectivity
stereospezifisch stereospecific
Stereospezifität *f* stereospecificity
steril 1. sterile, *(if achieved by physical methods also)* aseptic; 2. *(soil)* barren, infertile
Sterilbank *f (biol)* clean bench, laminar flow
Sterilbox *f s.* Sterilbank
Sterilfiltration *f* sterile filtration, filter sterilization
sterilfiltrieren *v* sterilize by filtration, filter aseptically
Sterilisation *f* sterilization
Sterilisation *f*/thermische heat sterilization
Sterilisator *m* sterilizer
sterilisieren *v* sterilize
Sterilisierung *f* sterilization
Sterilität *f* 1. sterility, *(if achieved by physical methods also)* asepsis; 2. *(soil)* barrenness, infertility
sterisch steric • **sterisch gehindert** sterically hindered • **sterisch geordnet** stereoregular • **sterisch möglich** sterically feasible
Stern-Gerlach-Versuch *m (phys ch)* Stern-Gerlach experiment
Sternpolymer *n (plast)* star polymer *(polymeric architecture)*
Sternutatorium *n (pharm)* sternutator, sternutative
Stern-Volmer-Beziehung *f* Stern-Volmer kinetic relationship *(of fluorescence quenching)*
Stern-Volmer-Gleichung *f* Stern-Gerlach-Volmer kinetic relationship *(of fluorescence quenching)*
Steroid *n* steroid
Steroidalkaloid *n* steroid(al) alkaloid
Steroidchemie *f* steroid chemistry
Steroidhormon *n* steroid hormone
Sterol *n (bioch)* sterol
Sterol *n*/pflanzliches plant sterol, phytosterol

stetig continuous *(operation)*; constant *(e.g. flow)*
Stetigförderer *m* conveyor
steuerbar controllable *(e.g. a chemical reactor)*
Steuerbarkeit *f* controllability *(as of a chemical reactor)*
Steuerdruck *m* control pressure
Steuergerät *n* controller
Steuermetabolit *m (bioch)* effector
steuern *v* control *(a process)*
Steuerung *f* 1. control; 2. control equipment
Steuerung *f***/automatische** automatic control
Steuerung *f***/hormonale** hormonal control
Steuerung *f***/kinetische** kinetic control
Stevens-Umlagerung *f* Stevens rearrangement *(of a benzyl group in a quaternary ammonium salt)*
STH *s.* Hormon/somatotropes
Stiban *n* 1. Sb_nH_{n+2} stibane; 2. *s.* Stibin 1.
Stibin *n* 1. SbH_3 stibine, antimony hydride; 2. *(org ch)* SbR_3 stibine
Stibinsäure *f* RR′SbOOH stibinic acid
Stibnit *m (min)* stibnite, antimonite *(antimony(III) sulphide)*
Stich *m* 1. *(dye)* hue, shade, tinge, tint; 2. *(met)* tapping; 3. *(met)* taphole
Stichkultur *f* stab culture *(microbiology)*
Stichloch *n (met)* taphole
Stichprobe *f* random sample
Stickoxid *n* nitrogen oxide, *(specif)* NO nitrogen(II) oxide, nitrogen monooxide
Stickstoff *m* N nitrogen
Stickstoff *m***/atmosphärischer** atmospheric nitrogen
Stickstoff *m***/flüssiger** liquid nitrogen
Stickstoff *m***/gebundener** fixed nitrogen
Stickstoffbestimmung *f* **nach Dumas** Dumas nitrogen analysis
Stickstoffbestimmung *f* **nach Kjeldahl** Kjeldahl nitrogen analysis, Kjeldahl determination
Stickstoffbestimmungsapparat *m* **nach Kjeldahl** Kjeldahl digestion apparatus
Stickstoffbilanz *f (bioch)* nitrogen balance
Stickstoffbindung *f* nitrogen fixation *(chemically or biologically)*
Stickstoffbindung *f***/nicht symbio(n)tische** non-symbiotic nitrogen fixation, free nitrogen fixation, azofication
Stickstoffbindung *f***/symbio(n)tische** *(agric)* symbiotic nitrogen fixation *(esp by nodule-forming bacteria)*
Stickstoffdioxid *n* nitrogen dioxide
Stickstoffdüngemittel *n* nitrogen fertilizer
Stickstoffgruppe *f* nitrogen family *(in the periodic table)*
stickstoffhaltig nitrogenous
Stickstoff-Laser *m* nitrogen laser
Stickstoffmonoxid *n* nitrogen monooxide
Stickstoffoxid *n* nitrogen oxide
Stickstoff(I)-oxid *n* nitrogen(I) oxide, dinitrogen monooxide
Stickstoff(II)-oxid *n* nitrogen(II) oxide, nitrogen monooxide
Stickstoff(III)-oxid *n* nitrogen(III) oxide, dinitrogen trioxide
Stickstoff(IV)-oxid *n* nitrogen(IV) oxide, nitrogen dioxide
Stickstoff(V)-oxid *n* nitrogen(V) oxide, dinitrogen pentaoxide
Stickstoffprobe *f* **nach Lassaigne** Lassaigne test
Stickstoffradikal *n* nitrogen-centred radical
Stickstoffstoffwechsel *m (bioch)* nitrogen metabolism
Stickstofftetroxid *n* dinitrogen tetraoxide
Stickstoffwasserstoffsäure *f* hydrazoic acid, azoimide
Stickstoffyperit *n* nitrogen mustard
Stiel *m* stem *(of a funnel)*
Stift *m (tech)* pin
Stiftloch *n* pinhole *(a coating defect)*
Stiftscheibenmühle *f* pinned-disk disintegrator [mill]
Stilben *n* stilbene
Stilbenfarbstoff *m* stilbene dye
stilllegen *v* shut down
Stilllegung *f* shutdown
Stillstandszeit *f* downtime, outage time
Stimulans *n* stimulant
Stobbe-Kondensation *f* Stobbe condensation [reaction] *(of aldehydes or ketones with esters of succinic acid)*
Stöchiometrie *f* stoichiometry

Stöchiometriezahl *f* stoichiometric number
stöchiometrisch stoichiometric
stöchiometrisch/nicht non-stoichiometric
Stockpunkt *m* setting point, s.p., solid [solidification] point *(of oils)*
Stock-Zahl *f* Stock number *(of valence)*
Stoff *m* 1. substance, material, matter; 2. *(pap)* (fibrous) pulp; 3. *(text)* fabric, (fabric) cloth, tissue, textile
Stoff *m*/**absorbierender** absorbent
Stoff *m*/**absorbierter** absorbate
Stoff *m*/**adsorbierender** adsorbent
Stoff *m*/**adsorbierter** adsorbate
Stoff *m*/**amphoterer** ampholyte
Stoff *m*/**anion(en)aktiver** anionic surfactant
Stoff *m*/**antibiotischer** *(pharm)* antibiotic (agent)
Stoff *m*/**arsenhaltiger** arsenical
Stoff *m*/**aufgenommener** sorbate
Stoff *m*/**aufnehmender** sorbent
Stoff *m*/**basischer** basic substance
Stoff *m*/**diamagnetischer** diamagnetic (substance)
Stoff *m*/**einheitlicher** uniform substance
Stoff *m*/**färbender [farbgebender]** colouring matter [substance]
Stoff *m*/**ferroelektrischer** ferroelectric (substance)
Stoff *m*/**ferromagnetischer** ferromagnetic (substance)
Stoff *m*/**fester** solid (substance)
Stoff *m*/**filmbildender** film former
Stoff *m*/**flüchtiger** volatile substance
Stoff *m*/**fremder** foreign [extraneous] substance
Stoff *m*/**gefährlicher** hazardous substance
Stoff *m*/**gelöster** dissolved substance, solute
Stoff *m*/**geruchsaktiver [geruchsbildender]** odour-causing substance
Stoff *m*/**geschmacksverstärkender** flavour enhancer
Stoff *m*/**geschmacksverursachender** taste-causing substance
Stoff *m*/**glasiger** glassy material
Stoff *m*/**glaskeramischer** glass ceramic, vitroceramic, devitrified glass
Stoff *m*/**grenzflächenaktiver** surface-active agent, surfactant
Stoff *m*/**gummierter** rubber-coated fabric, proofing
Stoff *m*/**hautreizender** skin irritant
Stoff *m*/**indifferenter** inert substance
Stoff *m*/**ionogener grenzflächenaktiver** ionic surfactant
Stoff *m*/**karzinogener** carcinogen
Stoff *m*/**komplexbildender** complexing agent
Stoff *m*/**maskierender** masking [sequestering] agent
Stoff *m*/**mineralischer** mineral matter
Stoff *m*/**nicht flüchtiger** non-volatile substance
Stoff *m*/**nicht ionogener grenzflächenaktiver** non-ionic (surfactant)
Stoff *m*/**oberflächenaktiver** surface-active agent [substance], surfactant
Stoff *m*/**paramagnetischer** paramagnetic (substance)
Stoff *m*/**partikulärer** particulate matter *(solid or liquid)*
Stoff *m*/**pflanzenschädigender** phytotoxicant
Stoff *m*/**polymorpher** polymorph
Stoff *m*/**reagierender** reactant
Stoff *m*/**reiner** pure substance
Stoff *m*/**sorbierter** sorbate
Stoff *m*/**therapeutisch wirksamer** therapeutic (agent)
Stoff *m*/**thixotroper** thixotropic substance
Stoff *m*/**wachstumsfördernder** growth substance
Stoff *m*/**waschaktiver** detergent surfactant
Stoff *m*/**zu absorbierender** material to be absorbed
Stoff *m*/**zu adsorbierender** material to be adsorbed
Stoff *m*/**zu sublimierender** sublimand
Stoff *m*/**zusammengesetzter** compound substance
Stoffänderung *f* material change
Stoffauflöser *m (pap)* pulper
Stoffaustausch *m* exchange of materials
Stoffbahn *f* (paper) web, mat
Stoffbilanz *f* mass [material] balance
Stoffbrei *m (pap)* (fibrous) pulp
Stoffdichte *f (pap)* stock density
Stoffdurchgang *m* mass transfer between phases

Stoffdurchgangskoeffizient *m* overall mass-transfer coefficient
Stoffe *mpl*/**absetzbare** settleable solids
Stoffe *mpl*/**gelöste** dissolved solids
Stoffe *mpl*/**halbsynthetisch hergestellte** semisynthetics
Stoffe *mpl*/**schwimmfähige** floatables
Stoffe *mpl*/**trübende** turbidity
Stoffe *mpl*/**vergärbare** fermentables
Stofffilter *n* cloth [fabric, woven] filter
Stoffgemisch *n* mixture of substances
Stoffgleichung *f* stoichiometric equation
Stoffgruppe *f* group of substances
Stoffklasse *f* family of compounds, class of substances
Stoffmahlung *f (pap)* stock disintegration, beating
Stoffmenge *f* amount [quantity] of substance
Stoffmengeneinheit *f* unit of amount of substance
Stoffmühle *f (pap)* beater
Stoffproduktion *f* product formation
Stoffproduktion *f*/**biotechnologische** bioprocess production
Stoffproduzent *m (biot)* producer, producing microorganism
Stoffstrom *m* mass flux
Stoffsystem *n*/**einschleusendes** *(bioch)* shuttle (system)
Stofftransport *m* mass transport
Stofftrübe *f (min tech)* pulp of ore
Stoffübergang *m* mass transfer
Stoffübergangskoeffizient *m* mass-transfer coefficient
Stoffübergangswiderstand *m* mass-transfer resistance
Stoffübertragung *f* mass transfer
Stoffumsatz *m (bioch)* turnover
Stoffumwandlung *f* conversion [transformation] of materials, materials conversion
Stoffveränderung *f* material change
Stoffwandlungsleistung *f (biot)* (substrate) conversion efficiency
Stoffwechsel *m* metabolism
Stoffwechselkette *f* metabolic chain
stoffwechseln *v* metabolize
Stoffwechselpool *m* metabolic pool, central area of metabolism
Stoffwechselprodukt *n* metabolite, product of metabolism
Stoffwechselprodukt *n*/**primäres** primary metabolite
Stoffwechselprodukt *n*/**sekundäres** secondary metabolite
Stoffwechselraum *m (bioch)* compartment *(of a cell)*
Stoffwechselweg *m* metabolic pathway [route], pathway of metabolism
Stoffwechselweg *m*/**kataboler** catabolic pathway [route]
Stokes-Linie *f (spectr)* Stokes line
Stokes-Verschiebung *f (spectr)* Stokes shift
Stopfbuchse *f* stuffing box
Stopfen *m* stopper, plug, cork • **mit einem Stopfen verschließen** stopper
Stoppcodon *n (bioch)* terminator codon
stoppen *v* terminate, shortstop *(polymerization)*; arrest *(a reaction)*
Stopper *m (plast)* chain terminator
störanfällig susceptible to failure
Störanfälligkeit *f* susceptibility to failure
Störatom *n* foreign [impurity] atom
Störelement *n (anal)* interfering element
stören *v (anal)* interfere with
Störion *n (anal)* interfering ion
Störpeak *m* ghost [spurious] peak
Störstelle *f* lattice [crystal] defect, lattice imperfection
Störstellen(halb)leiter *m* impurity semiconductor
Störsubstanz *f (anal)* interferent
Störung *f (cryst)* imperfection
Stoß *m* 1. impact, *(esp relating to particles:)* impingement; 2. *(phys ch)* collision
Stoß *m*/**aktivierender [anregender]** activating collision, collision of the first kind
Stoß *m*/**bimolekularer** bimolecular collision
Stoß *m*/**desaktivierender** deactivating collision, collision of the second kind
Stoß *m*/**elastischer** elastic collision
Stoß *m* **mit der Gefäßwand** wall collision *(with chain reactions)*
Stoß *m*/**unelastischer** inelastic collision
Stoßaktivierung *f (spectr)* collision activation, CA
Stoßanregung *f* collision activation, CA
stoßdämpfend shock-absorbing
Stoßelastizität *f* (impact) resilience
Stoßelektron *n* impact electron

stoßen *v* 1. chop, crush *(ice)*; 2. bump *(of boiling liquids)*
stoßfest impact-resistant, shockproof
Stoßfestigkeit *f* impact resistance, resistance to shock
Stoßfrequenz *f* collision number, collision(al) frequency
Stoßfront *f* shock front *(kinetics)*
Stoßgalvanometer *n* ballistic galvanometer
Stoßgleichung *f* momentum (balance) equation
Stoßhäufigkeit *f* collision number, collision(al) frequency
Stoßherd *m (min tech)* shaking table
Stoßionisation *f* collision [impact] ionization
Stoßkomplex *m (phys ch)* collision complex
Stoßquerschnitt *m (phys ch)* collision cross section
stoßsicher shockproof
Stoßverbreiterung *f (spectr)* collision broadening
Stoßverdampfung *f* flash evaporation
Stoßwahrscheinlichkeit *f* probability [likelihood] of collision
Stoßwelle *f* shock wave
Stoßwellenverfahren *n* shock tube technique *(for investigating fast reactions)*
Stoßzahl *f* **(je Zeiteinheit)** *(phys ch)* collision number, collision(al) frequency
Strahl *m* 1. ray *(of light or other radiation)*; *(if bundled:)* beam; 2. jet *(of liquid or gas)*
Strahl *m*/**positiver** positive [canal] ray
strahlen *v* radiate
Strahlen... *siehe auch: Strahlungs...*
strahlenbeständig radiation-resistant
Strahlenbeständigkeit *f* radiation resistance
Strahlenbündel *n* beam of rays
Strahlenchemie *f* radiation chemistry
strahlenchemisch radiation-chemical
Strahlendetektor *m* radiation detector
Strahlendosimetrie *f* radiation dosimetry
Strahlendosis *f* radiation dosage [dose]
strahlendurchlässig radiolucent
Strahlendurchlässigkeit *f* radiolucency
strahlenempfindlich radiosensitive
Strahlenempfindlichkeit *f* radiosensitivity
Strahlengefährdung *f* radiation hazard
strahleninitiiert initiated by irradiation *(e.g. polymerization)*
Strahlenkrankheit *f* radiation sickness
Strahlenkunde *f* radiology
Strahlenpasteurisierung *f* radiopasteurization
Strahlenquelle *f* radiation source
strahlenresistent radiation-resistant
Strahlenresistenz *f* radiation resistance
Strahlenschaden *m* radiation damage
Strahlenschutz *m* radiation protection
Strahlenvernetzung *f* irradiation crosslinking
Strahlenzählrohr *n* radiation counter
Strahler *m* 1. *(phys ch)* radiator; 2. *(nucl)* emitter, radiator
Strahler *m*/**grauer** grey body
Strahler *m*/**schwarzer** black body, black-body radiator
Strahlmühle *f* jet mill
Strahlpumpe *f* jet pump
Strahlstärke *f* radiation intensity
Strahlung *f* radiation
Strahlung *f*/**durchdringende** penetrating radiation
Strahlung *f*/**durchgelassene** *(spectr)* transmitted radiation
Strahlung *f*/**energiearme** soft [low-energy] radiation
Strahlung *f*/**energiereiche [harte]** hard [high-energy] radiation, penetrating radiation
Strahlung *f*/**infrarote** infrared radiation
Strahlung *f*/**inkohärente** incoherent radiation
Strahlung *f*/**ionisierende** ionizing radiation
Strahlung *f*/**kohärente** coherent radiation
Strahlung *f*/**photosynthetisch ausnutzbare** photosynthetically active radiation, PhAR
Strahlung *f*/**schwarze** black-body radiation
Strahlung *f*/**solare** solar radiation
Strahlung *f*/**thermische** heat [thermal] radiation
Strahlung *f*/**weiche** *s.* Strahlung/energiearme
Strahlungs... *siehe auch: Strahlen...*

Strahlungsdesaktivierung *f* radiative deactivation
Strahlungsdetektor *m* radiation detector
Strahlungsdichte *f* (spectral) radiance
Strahlungseinfang *m* radiative capture
Strahlungsempfänger *m (anal)* detector
Strahlungsenergie *f* radiant energy, energy of radiation
Strahlungsfeld *n* radiation field
Strahlungsfilter *n (spectr)* monochromator
Strahlungsfluss *m* radiant flux
Strahlungsflussdichte *f* irradiance, radiant flux density
Strahlungsflussdichte *f***/spektrale** spectral irradiance
strahlungsfrei radiationless, non-radiative
Strahlungsgesetz *n***/rayleigh-jeanssches** Rayleigh-Jeans radiation law
Strahlungsintensität *f***/spektrale** spectral radiant intensity
Strahlungslebensdauer *f* radiative lifetime
Strahlungslebensdauer *f***/mittlere** radiative lifetime
Strahlungslebensdauer *f***/natürliche** radiative lifetime
Strahlungsleistung *f* radiant power
strahlungslos radiationless, non-radiative
Strahlungsschutz *m* radiation shield
Strahlungsspektrum *n* radiation spectrum
Strahlungsteiler *m (spectr)* beam splitter
Strahlungsthermometer *n* pyrometer
Strahlungsübergang *m* radiative transition
Strahlungsverlust *m* radiation loss
Strahlungswärme *f* radiant heat
Strähn *m (text)* skein, hank
Straight-run-Benzin *n* straight-run gasoline
Straight-run-Destillation *f* straight(-run) distillation
Strang *m* 1. strand *(as of molecules)*; 2. *(text)* hank
Strangaufweitung *f (plast)* swelling from the die *(extrusion)*
Strangaufweitungsverhältnis *n (plast)* swell ratio *(extrusion)*
Strangfärben *n (text)* rope dyeing *(of cloth)*; hank dyeing *(of yarn)*
Stranggießen *n (tech)* continuous casting
Strangguss *s.* Stranggießen
Strangpresse *f* extruder
strangpressen *v* extrude
Strangpressen *n* extrusion (moulding)
Strangpresserzeugnis *n* extrudate
Strangpressmischung *f* extrusion compound
Strass *m* strass *(a glass of high lead content)*
Straßen-Octanzahl *f* road octane number
Straßentankwagen *m* road tanker
streckbar extensible, *(esp relating to metal:)* ductile
Streckbarkeit *f* extensibility, *(esp relating to metal:)* ductility
strecken *v* 1. stretch, extend *(mechanically)*; 2. extend *(by adding additives)*; dilute *(liquids)*; 3. *(text, plast)* stretch, draw
Strecker-Abbau *m* Strecker degradation *(of amino acids to aldehydes)*
Strecker-Synthese *f* Strecker synthesis *(of α-amino acids)*
Streckformen *n (plast)* stretch forming
Streckgrenze *f (plast, met)* yield point
Streckgrenze *f***/untere** yield value
Streckmittel *n* extender, (extending) filler, cheapener, *(relating to liquids also)* diluent
Streckschwingung *f* stretching [valence] vibration
Streckspannung *f* yield stress
Streckspinnen *n* stretch spinning
Streckung *f* 1. stretching, extension *(mechanically)*; 2. extension *(by adding additives)*; dilution *(of liquids)*; 3. *(text, plast)* stretching, drawing
Streichauftrag *m (coat)* brush application
streichen *v* 1. *(plast, pap)* coat; 2. *(coat)* paint; brush *(the paint onto a surface)*
Streichfarbe *f (coat)* brushing paint
Streifen *m* band *(in a spectrum)*
streifig banded
Streudiffusion *f* eddy [turbulent] diffusion
Streuelektron *n* scattered electron

streuen *v* 1. scatter *(e.g. rays, electrons)*; 2. sprinkle *(e.g. powder)*
Streulicht *n* scattered [stray] light
Streulichtphotometer *n* light-scattering photometer
Streuquerschnitt *m* scattering cross section
Streustrahlung *f* scattered radiation
Streuung *f (phys ch)* scatter(ing)
Streuung *f*/**anomale elastische** resonance scattering
Streuung *f* **der Röntgenstrahlen** X-ray scattering
Streu(ungs)winkel *m* scattering angle
Strich *m (min)* streak colour
Strichfarbe *f (min)* streak colour
Strichgitter *n (spectr)* line [ruled] grating
Strichmarke *f* mark *(as of a volumetric flask)*
Strichspektrum *n* stick spectrum
Strichteilung *f* division *(of a gauge)*
strippen *v (distil)* strip (off, out)
Stripper *m (distil)* stripper
Strippgas *n* stripping gas
Strippkolonne *f (distil)* stripping [stripper] column
Strohstoff *m* 1. straw pulp; 2. *s.* Strohzellstoff
Strohzellstoff *m* straw cellulose, fine straw pulp
Strohzellstoff *m*/**vollaufgeschlossener** straw cellulose, fine straw pulp
Strom *m* 1. *(tech)* current, stream *(of a fluid)*; 2. (electric) current; 3. *s.* Strömung
Strom *m*/**abwärtsgerichteter** downward [descending] current
Strom *m*/**aufsteigender** upward [ascending] current, upflow
Strom *m*/**kapazitiver** capacity current
Stromanschluss *m* power connection
Stromausbeute *f* current efficiency
Stromdichte *f* current density, C.D.
strömen *v* stream, flow, run, *(if swiftly:)* flush
Stromfaden *m* stream filament *(rheology)*
Stromfluss *m* current flow
stromführend current-carrying
Stromkreis *m* electric circuit
Stromleiter *m* conductor (of electricity), electric conductor
stromlos electroless, dead
Strom-Potenzial-Kurve *f* current-potential curve
Stromquelle *f* power [current] source
Stromrichtung *f* current direction; direction of flow
Stromschlüssel *m* salt bridge
Stromschlüssel *m*/**elektrolytischer** salt bridge
Strom-Spannungs-Kurve *f* current-potential curve
Stromstärke *f* current strength
Strömung *f* flow
Strömung *f*/**abwärtsgerichtete** downward flow
Strömung *f*/**aufwärtsgerichtete** upward flow
Strömung *f*/**axiale** axial flow
Strömung *f*/**laminare** laminar [streamline] flow
Strömung *f*/**nichtideale** non-ideal flow
Strömung *f*/**radiale** radial flow
Strömung *f*/**reale** non-ideal flow
Strömung *f*/**reibungsfreie** frictionless flow, inviscid [non-viscous] flow
Strömung *f*/**stationäre** steady(-state) flow
Strömung *f*/**turbulente** turbulent flow
Strömungsdoppelbrechung *f (anal)* flow birefringence, double refraction of flow
Strömungsdruck *m (hyd)* head of water
Strömungsform *f* flow pattern *(in a reactor)*
Strömungsgeschwindigkeit *f* flow rate [velocity]
Strömungsgrenzschicht *f*/**prandtlsche** Prandtl boundary layer
Strömungskanal *m* flow channel
Strömungskorrosion *f* flow [erosion] corrosion
Strömungsmesser *m* flowmeter, fluid meter
Strömungsmesser *m* **für Gase** gas flowmeter
Strömungsmethode *f* (continuous-)flow method *(kinetics, calorimetry)*
Strömungspotenzial *n* streaming potential
Strömungsrichtung *f* direction of flow, flow direction
Strömungsrohr *n* (continuous-)tubular-flow reactor

Strömungsrohr *n*/**ideales** longitudinal reactor
Strömungsteiler *m (chromat)* (stream, sample) splitter
Strömungsteilung *f (chromat)* (stream) splitting
Strömungswiderstand *m* resistance to fluid flow
Stromversorgung *f* power supply
Stromzufuhr *f* power supply
Strontium *n* Sr strontium
Strontiumcarbonat *n* strontium carbonate
Strontiumchlorid *n* strontium chloride
Strontiumhydroxid *n* strontium hydroxide
Strontiumnitrat *n* strontium nitrate
Strontiumoxid *n* strontium oxide
Strontiumsulfat *n* strontium sulphate
Strophanthin *n* strophanthin *(any of a group of glycosides)*
Strophanthin *n* **G** G-strophanthin, ouabain
Strudel *m* vortex, eddy
Struktur *f* structure, *(relating to chemical compounds also)* constitution
Struktur *f*/**hochgeordnete** highly-ordered structure
Struktur *f*/**körnige** granularity
Struktur *f*/**nanodimensionierte** *s.* Nanostruktur
Struktur *f*/**ringförmige** annular structure
Struktur *f*/**übermolekulare** supermolecular structure
Struktur *f*/**wabenartige [wabenförmige]** honeycomb structure
Struktur *f*/**wahrscheinliche** likely structure
Struktur *f*/**zybotaktische** cybotaxis *(of liquids)*
Strukturanalyse *f* structure [structural] analysis
Strukturänderung *f* change in structure
Strukturaufklärung *f* structure elucidation [determination]
Strukturbeweis *m* proof of structure
Strukturchemie *f* structural chemistry
Struktureinheit *f* structural unit
Struktureinheit *f*/**sich wiederholende** constitutional repeating unit, CRU *(of polymers)*
Strukturelement *n* structural element
strukturell structural
Strukturformel *f* structural formula
Strukturformel *f* (valence) structural formula, constitutional [line, graphic] formula
strukturidentisch structurally identical
Strukturisomer(e) *n* structural isomer
Strukturisomerie *f* structural isomerism
strukturlos structureless, devoid of structure
Strukturparachor *m* structural parachor
Strukturprotein *n* skeletal [fibrous] protein, scleroprotein
Strukturresonanz *f* resonance, mesomerism
Strukturtyp *m (cryst)* lattice [structure] type
Strukturuntersuchung *f* investigation of structure, structural study
strukturviskos pseudoplastic
Strukturviskosität *f* structural viscosity, shear thinning, pseudoplasticity
Strychnosalkaloid *n* strychnos alkaloid
Stuart-Modell *n* Stuart model *(for illustrating molecular structures)*
Stückerz *n* lump ore
Stuckgips *m* stucco
Stückgröße *f* particle size
stückig lumpy • **stückig machen** agglomerate
Stückigmachen *n* agglomeration
Stufe *f* 1. stage, step *(as of a reaction)*; 2. effect *(of an evaporator)*
Stufe *f*/**erste** initiating step *(of a reaction)*
Stufe *f*/**letzte** completing step *(of a reaction)*
Stufe *f*/**polarographische** polarographic wave
Stufeneluierung *f (chromat)* stepwise elution
Stufengitter *n (spectr)* echelon grating
Stufenhöhe *f (chromat)* step height; wave height *(polarography)*
Stufenpolymerisation *f* step-growth polymerization
Stufenreaktion *f* step(wise) reaction
Stufenregel *f*/**ostwaldsche** Ostwald rule, successive reactions law
Stufenversetzung *f (cryst)* edge dislocation
Stufenwachstumspolymerisation *f* step-growth polymerization
Stuhlzäpfchen *n (pharm)* suppository

stumpf matt, dull *(surface)*; dead, dull, flat *(colour)* • **stumpf werden** dull
Stumpfheit *f* mattness, dullness *(of a surface)*; dullness, flatness *(of a colour)*
Stumpfwerden *n* dulling *(of a surface or colour)*
15-Stunden-Entspannungstemperatur *f (glass)* strain point [temperature]
stürmisch vigorous *(reaction)*
Sturzmühle *f* tumbling mill
Stutzen *m (tech)* port
Stutzenflasche *f (lab)* aspirator
Stützgewebe *n (hyd)* septum *(of a precoat filter)*
Styphnat *n* styphnate *(salt or charge-transfer complex of styphnic acid)*
Styphninsäure *f* styphnic acid
Styptikum *n (pharm)* styptic
styptisch styptic
Styren *n* styrene, vinylbenzene
Styrenalkydharz *n* styrenated alkyd
Styren-Butadien-Kautschuk *m* styrene-butadiene rubber, SBR
Styren-Chloropren-Kautschuk *m* styrene-chloroprene rubber, SCR
Styren-Isopren-Kautschuk *m* styrene-isoprene rubber, SIR
Styrol *n s.* Styren
styrolisieren *v* styrenate
subatomar subatomic
subazid subacid
Subazidität *f* 1. subacidity; 2. *(med)* hypoacidity *(of gastric juice)*
Suberan *n* suberane, *cycloheptane
Suberin *n (bioch)* suberin
Suberon *n* suberone, *cycloheptanone
Suberylalkohol *m* suberyl alcohol, *cycloheptanol
Subkeim *m (cryst)* cluster
Subklonieren *n (bioch)* subcloning
Sublimand *m* sublimand
Sublimat *n* 1. sublimate *(a product obtained by sublimation)*; 2. sublimate, mercury(II) chloride, mercury dichloride
Sublimation *f* sublimation
Sublimations(druck)kurve *f* sublimation curve
Sublimationspunkt *m* sublimation point
sublimierbar sublimable
sublimieren *v* sublimate
Sublimiergut *n* sublimand, material being [to be] sublimated
submers submerged
Submersfermentation *f (biot)* submerged(-culture) fermentation
Submersfermenter *m (biot)* submerged fermenter
Submerskultur *f (biot)* submerged culture
Submikrogefüge *n* submicrostructure
Submikron *n* submicron
Submikroprobe *f* submicro sample *(*10^{-3} to 10^{-4} g)*
submikroskopisch submicroscopic
Submikrostruktur *f* submicrostructure
Submolekül *n* submolecule
Suboxid *n* suboxide
Sub-Spuren-Analyse *f* sub-trace analysis *(*sample weight 10^{-3} to 10^{-4} g)*
Substantivfarbstoff *m* substantive [direct] dye [dyestuff]
Substanz *f* substance, matter, material
Substanz f/abgewogene weighed substance
Substanz f/abzuwägende substance to be weighed
Substanz f/adstringierende astringent, styptic
Substanz f/bakterientötende [bakterizide] bactericide
Substanz f/giftige toxicant
Substanz f/inkrustierende incrustant
Substanz f/ionophore *(bioch)* ionophore
Substanz f/mikrobielle *(biot)* microbial product
Substanz f/pflanzliche vegetable matter
Substanz f/standardisierte standard substance
Substanz f/synthetisch gewonnene synthetic
Substanz f/unbekannte unknown (substance)
Substanz f/unlösliche insoluble
Substanz f/wirksame active substance, (active) agent
Substanzformel *f* stoichiometric formula, simplest formula
Substanzklasse *f* class of substances
Substanzmenge *f* amount of substance
Substanzpolymerisation *f* mass [bulk] polymerization
Substanzprobe *f* sample
Substanzschiffchen *n (lab)* boat

Substanzschiffchen *n* **aus Tonerde** alumina boat
Substanzverlust *m* loss of material
Substituent *m* substituent
Substituent *m* **erster Ordnung** ortho-para-directing group
Substituent *m* **in meta-Stellung [*m*-Stellung]** meta substituent
Substituent *m***/meta-ständiger** meta substituent
Substituent *m* **mit aktivierender Wirkung** activating group
Substituent *m***/ortho-para-dirigierender** ortho-para-directing group
***m*-Substituent** *m* meta substituent
***o*-Substituent** *m* ortho substituent
***p*-Substituent** *m* para substituent
Substituenteneffekt *m* substituent effect
Substituentenkonstante *f (org ch)* substituent constant, sigma-constant, σ-constant
substituierbar substitutable, replaceable
Substituierbarkeit *f* substitutability
substituieren *v* substitute, replace
substituiert/mehrfach polysubstituted
substituiert/zweifach disubstituted
1,2 substituiert *o*-substituted, ortho-substituted
1,3 substituiert *m*-substituted, meta-substituted
1,4 substituiert *p*-substituted, para-substituted
***m*-substituiert** meta-substituted, *m*-substituted
***o*-substituiert** ortho-substituted, *o*-substituted
***p*-substituiert** para-substituted, *p*-substituted
Substitution *f* substitution, (simple) replacement
Substitution *f***/bimolekulare nucleophile** second-order nucleophilic substitution (reaction), S_N2 reaction
Substitution *f***/elektrophile** electrophilic substitution
Substitution *f***/monomolekulare nucleophile** first-order nucleophilic substitution (reaction), S_N1 reaction
Substitution *f***/nucleophile** nucleophilic substitution, S_N reaction
Substitutionsbereitschaft *f* susceptibility to substitution
Substitutionsgrad *m* degree of substitution, D.S.
Substitutionsisomerie *f* substitution [place] isomerism, position(al) isomerism
Substitutionslegierung *f* substitutional alloy
Substitutionsmischkristall *m* substitutional solid solution
Substitutionsname *m* substitutive name
Substitutionsreaktion *f* substitution reaction
Substitutionstautomerie *f* substitution tautomerism
Substitutionswägung *f* substitution weighing
Substrat *n* 1. substrate; 2. *(biot)* medium, *(enzymology also)* reactant
Substrat *n***/führendes [obligatorisches]** leading [obligatory] substrate [reactant]
Substratbindung *f (bioch)* substrate binding
Substratbindungsmechanismus *m***/kooperativer** *(bioch)* cooperativity
Substrathemmung *f (bioch)* substrate inhibition
Substrat(ketten)phosphorylierung *f (bioch)* substrate-level phosphorylation
Substratsättigung *f* substrate saturation *(enzyme kinetics)*
substratspezifisch *(bioch)* substrate-specific
Substratspezifität *f (bioch)* substrate specificity
substratvermittelt *(bioch)* substrate-induced
Subtraktionsmischkristall *m* subtraction solid solution
Subtraktionsname *m* subtractive name
Succinat *n* succinate *(salt or ester of succinic acid)*
Succinatdehydrogenase *f* succinate dehydrogenase
Succinatoxidase *f* succinoxidase
Succindialdehyd *m* succindialdehyde
Succinimid *n* succinimide
Succinit *m* succinite, amber
suchterzeugend *(pharm)* addictive, habit-forming

Sucrochemie *f* sucrochemistry
Sud *m (ferm)* brew(ing)
Sudhaus *n* brewhouse
sukzinylieren *v* succinylate
Sukzinylierung *f* succinylation
Sulfacetamid *n* sulphacetamide *(a sulphonamide)*
Sulfadiazin *n* sulphadiazine
Sulfamat *n s.* Sulfamidat
Sulfamid *n s.* Sulfonamid
Sulfamidat *n* amidosulphate, sulphamate
Sulfamidsäure *f* sulphamic acid
Sulfaminierung *f* sulphamation, *N*-sulphonation
Sulfan *n* sulphane
Sulfanilamid *n* sulphanilamide
Sulfanilsäure *f* sulphanilic acid, aniline-*p*-sulphonic acid
Sulfapyridin *n* sulphapyridine
Sulfat *n* sulphate
Sulfataufschluss *m (pap)* sulphate pulping
Sulfathärte *f (hyd)* sulphate hardness
Sulfathiazol *n* sulphathiazole
sulfatieren *v* sulphate *(to esterify alcohols with sulphuric acid)*
Sulfatierung *f* sulphation *(esterification of alcohols with sulphuric acid)*
Sulfat-Ion *n* SO_4^{-2} sulphate ion
sulfatisieren *v* 1. sulphatize *(e.g. sulphide ores by roasting)*; 2. sulphate *(the plates of a lead-acid accumulator)*
Sulfatisierung *f* 1. sulphatizing *(as of sulphide ores by roasting)*; 2. sulphation *(of the plates of a lead-acid accumulator)*
Sulfatstein *m (hyd)* sulphate scale
Sulfatzellstoff *m* sulphate pulp
Sulfen *n (org ch)* sulphene
Sulfenamid *n* sulphenamide
Sulfensäure *f* sulphenic acid
Sulfenylgruppe *n* sulphenyl [hydrocarbylsulphanyl] group
Sulfenylium-Ion *n* sulphenylium [sulphenium] ion
Sulfid *n* sulphide
Sulfid *n*/**primäres [saures]** hydrogensulphide
Sulfiderz *n* sulphide ore
sulfidieren *v* 1. sulphidize; 2. *(text)* xanthate, churn
Sulfidieren *n* 1. sulphidizing; 2. *(text)* xanthation, churning
sulfidisch sulphidic
Sulfidschwefel *n* sulphide sulphur
Sulfid-Spannungsrisskorrosion *f* sulphide stress corrosion
sulfieren *v* sulphonate *(to treat an organic compound with sulphuric acid or a related agent)*
Sulfin *n (org ch)* sulphine
Sulfinsäure *f (org ch)* sulphinic acid
Sulfit *n* sulphite
Sulfitablauge *f (pap)* spent sulphite liquor
Sulfitaufschluss *m (pap)* sulphite pulping
sulfitieren *v* sulphite
Sulfitieren *n* sulphitation
Sulfitkochsäure *f (pap)* sulphite [bisulphite] cooking liquor
Sulfitkraftpapier *n* sulphite kraft paper
Sulfit(zell)stoff *m* sulphite pulp
Sulfochlorid *n (org ch)* sulphonyl chloride
sulfochlorieren *v* sulphochlorinate
Sulfochlorierung *f* sulphochlorination
Sulfogruppe *f* $-SO_3H$ sulpho group
Sulfon *n* sulphone
Sulfonamid *n* sulphonamide
Sulfonamidpräparat *n* sulpha drug
Sulfonat *n* sulphonate
sulfonieren *v* sulphonate
sulfonieren lassen *v*/**sich** sulphonate
Sulfonierung *f* sulphonation
***N*-Sulfonierung** *f* *N*-sulphonation, sulphamation
Sulfonierungsmittel *n* sulphonating agent
Sulfonium-Ion *n* sulphonium ion
Sulfonphthalein *n* sulphonephthalein
Sulfonsäure *f* sulphonic acid
Sulfonsäureamid *n* sulphonamide
Sulfonsäurechlorid *n* sulphonyl chloride
Sulfonsäuregruppe *f s.* Sulfogruppe
Sulfonylgruppe *f* $=SO_2$ sulphonyl group
Sulfonylierung *f* sulphonylation
Sulfonyloxygruppe *f* $-OSO_2R$ sulphonyloxy group
Sulfosalicylsäure *f* sulphosalicylic acid
Sulfoxid *n* R_2SO sulphoxide
Sulfoxidation *f* sulphoxidation
Sulfoxylat *n* $M^I_2SO_2$ sulphoxylate
Sulfoxylsäure *f* sulphoxylic acid

Sulfur *m s.* Schwefel
sulfurieren *v* sulphonate
sulfurieren lassen *v*/**sich** sulphonate
Sulfurierung *f* sulphonation
Sulfurikation *f (soil)* sulphofication *(microbial oxidation of organically bound sulphur to sulphate)*
Sulfurylchlorid *n* sulphuryl chloride
Summenbande *f (spectr)* sum band
Summenformel *f (ch)* empirical formula
Summenformel *f*/**einfachste** simplest (possible) formula, stoichiometric formula
Summenformel *f*/**wahre** true formula
Sumpf *m* 1. *(tech)* pond, sump; 2. *(distil)* bottom *(of a column)*
sumpfen *v (ceram)* soak
Sumpfgas *n* marsh gas
Sumpfphase *f* liquid phase *(in coal hydrogenation)*
Sumpfprodukt *n (distil)* bottoms, bottom product
Superabsorber *m* superabsorber, superabsorbent polymer, SAP
superabsorbierend superabsorbent
superazid superacid
Superazidität *f* 1. superacidity; 2. *(med)* hyperacidity *(of gastric juice)*
superfluid superfluid, superliquid
Superfluidität *f* superfluidity, superfluid state
superflüssig superfluid, superliquid
superhelikal *(bioch)* superhelical
Superhelix *f (bioch)* superhelix, triple-stranded helix, coiled-coil
Superisolierung *f* superinsulation, multiple-layer insulation *(cryogenic engineering)*
Superknäuelung *f (bioch)* supercoiling *(of DNA)*
Superlegierung *f* superalloy
Superphosphat *n* superphosphate
supersauer superacid
Supersäure *f* superacid
Supersekundärstruktur *f (bioch)* supersecondary structure
Suppositorium *n (pharm)* suppository
Suppressormutation *f (biot)* suppressor mutation
suprafacial *(org ch)* suprafacial, s
suprafluid superfluid, superliquid
Suprafluidität *f* superfluidity, superfluid state
supraflüssig superfluid, superliquid
supraleitend superconducting
Supraleiter *m* superconductor
supraleitfähig superconducting
Supraleitfähigkeit *f* superconductivity • **Supraleitfähigkeit zeigen** superconduct
Supraleitung *f s.* Supraleitfähigkeit
Surrogat *n* substitute
suspendierbar suspensible
Suspendierbarkeit *f* suspensibility
suspendieren *v* suspend
Suspendiermittel *n* suspending agent
Suspendiervermögen *n* suspending capacity
Suspension *f* suspension • **in Suspension halten** keep [maintain] in suspension
Suspension *f*/**erneute** resuspension
Suspensionskolloid *n* suspensoid (colloid)
Suspensionskultur *f (biot)* submerged culture
Suspensionsmittel *n* suspending agent [medium]
Suspensionspartikel *npl* suspended particles, particulates
Suspensionspolymerisation *f* suspension [bead] polymerization
Suspensionspolyvinylchlorid *n* suspension polyvinyl chloride
Suspensionsreaktor *m* (catalyst-)slurry reactor
Suspensionsröstung *f (met)* suspension roasting
Suspensionstechnik *f* mull technique *(IR spectroscopy)*
Suspensoid *n* suspensoid (colloid)
Suspensoid-Kracken *n (petrol)* suspensoid (catalytic) cracking
süß 1. sweet *(taste)*; 2. *(petrol)* sweet(-ened)
Süße *f* sweetness
süßen *v* 1. *(food)* sweeten; 2. *(petrol)* sweeten *(to remove malodorous sulphur compounds)*
süßlich sweetish
Süßmittel *n (food)* sweetener
Süßstoff *m* non-nutritional [non-nutritive] sweetener
Süßungsmittel *n* 1. *(petrol)* sweetening agent; 2. *(food)* sweetener

Süß(ungs)verfahren *n (petrol)* sweetening process
Süßwasser *n* sweet [fresh] water
Süßwurzel *f* sugar wort
Suszeptibilität *f* susceptibility
Suszeptibilität f/diamagnetische diamagnetic susceptibility
Suszeptibilität f/ferromagnetische ferromagnetic susceptibility
Suszeptibilität f/magnetische magnetic susceptibility
Suszeptibilität f/molare molar susceptibility
Suszeptibilität f/paramagnetische paramagnetic susceptibility
Suszeptibilität f/spezifische specific susceptibility
Svedberg-Einheit *f* Svedberg unit *(determination of molecular weight by sedimentation)*
Swarts-Reaktion *f* Swarts reaction *(for obtaining fluoroalkanes)*
Sweep-Generator *m (spectr)* sweep generator
Sweep-Spule *f (spectr)* sweep coil
Sweetland-Filter *n* Sweetland filter, Sweetland (filter) press
swl. *(sehr wenig löslich)* extremely slightly soluble
Sydnon *n (org ch)* sydnone
Sydnonimin *n* sydnone imine
Sylvin *n (min)* sylvite, sylvin(e) *(potassium chloride)*
Symbol *n* symbol, sign, *(for representing chemical elements also)* chemical symbol [sign]
Symmetrieachse *f* axis of symmetry
Symmetrieachse f/sechszählige sixfold axis of symmetry
Symmetrieachse f/zweizählige twofold axis of symmetry
Symmetrieebene *f* plane of symmetry, symmetry plane
Symmetrieelement *n* symmetry element
Symmetrieklasse *f* class of crystal symmetry, symmetry [crystal] class, point group
Symmetrieoperation *f* symmetry operation
Symmetriezentrum *n* centre of symmetry, inversion centre
symmetrisch symmetric(al)
Sympath(ik)olytikum *n (pharm)* sympatholytic (agent)
sympath(ik)olytisch sympatholytic
Sympath(ik)omimetikum *n (pharm)* sympathomimetic (agent)
sympath(ik)omimetisch sympathomimetic
Symport *m (bioch)* symport *(one form of carrier transport)*
Symproportionierung *f* comproportionation
synantetisch *(geoch)* synante(c)tic *(formed by the reaction of two minerals)*
Synärese *f (coll)* syneresis • **Synärese zeigen** synerize *(of gels)*
Synchrotron *n (nucl)* synchrotron
Synchrotronstrahlung *f* synchrotron radiation
Synchrozyklotron *n* synchrocyclotron
syndiotaktisch syn(dio)tactic *(polymers)*
Syndiotaktizität *f* syndiotacticity *(of polymers)*
Synergismus *m* synergism
Synergist *m* synergist
synergistisch synergistic
±synperiplanar synperiplanar, eclipsed, opposed *(stereochemistry)*
Syntan *n* syntan, synthetic tannin
Synthese *f* synthesis
Synthese *f* **an polymeren Trägern** solid-phase protein synthesis *(according to Merrifield)*
Synthese f/asymmetrische asymmetric synthesis
Synthese f/biologische biosynthesis, biogenesis
Synthese f/conrad-limpachsche Conrad-Limpach synthesis *(of 4-hydroxyquinoline)*
Synthese f/enzymatische enzymatic synthesis
Synthese f/knoevenagelsche Knoevenagel condensation [synthesis] *(of α,β-unsaturated acids or esters)*
Synthese f/perkinsche Perkin condensation [synthesis] *(of unsaturated carboxylic acids)*
Synthese f/reimer-tiemannsche Reimer-Tiemann synthesis *(of hydroxyaldehydes or hydroxy acids)*
Synthese f/skraupsche Skraup (quinoline) synthesis

Synthese *f*/streckersche Strecker synthesis *(of α-amino acids)*
Synthese *f*/technische commercial synthesis
Syntheseammoniak *n* synthetic ammonia
Synthesechemie *f* synthetic chemistry
Synthesechemie *f*/organische synthetic organic chemistry
Synthesechemiker *m* synthetic chemist
Synthesefaser *f* synthetic fibre
Synthesefaserstoff *m* synthetic fibre
Synthesegas *n* synthesis gas
Synthesekautschuk *m* synthetic [artificial] rubber, chemical [man-made] rubber
Syntheseort *m (bioch)* site of synthesis
Syntheseprodukt *n* synthetic
Syntheseschritt *m* synthesis step
Syntheseweg *m (bioch)* synthesis path(way)
synthetisch 1. synthetic(al), non-natural; 2. *(bioch)* anabolic • **synthetisch herstellen** synthesize
synthetisieren *v* synthesize
System *n*/abgeschlossenes *(phys ch)* isolated system
System *n*/binäres binary [two-component] system
System *n*/disperses disperse system
System *n*/einphasiges *s.* System/homogenes
System *n*/endocyclisches endocyclic [caged ring] system
System *n*/heteroaromatisches [heterocyclisches konjugiertes] heteroconjugated system
System *n*/heterogenes heterogeneous system
System *n*/homogenes homogeneous [one-phase] system
System *n*/isodisperses *(coll)* isodisperse system, isodispersion
System *n*/isoliertes *(phys ch)* isolated system
System *n*/kristallographisches crystal [crystallographic] system
System *n*/mehrphasiges multiphase system
System *n*/offenes *(phys ch)* open system
System *n*/polydisperses *(coll)* polydisperse system, polydispersion
System *n*/polynäres multicomponent system
System *n*/quaternäres quaternary [four-component] system
System *n*/sich selbst organisierendes *(bioch)* self-assembling system
System *n*/spirocyclisches spiro ring system
System *n*/ternäres ternary [three-component] system
System *n*/ungeordnetes *(phys ch)* random system
System *n*/unitäres unary [one-component] system
System *n*/zweiphasiges *s.* System/binäres
systemisch/nicht *(agric)* non-systemic
SZ *s.* Säurezahl
Szent-Györgyi-Krebs-Zyklus *m (bioch)* Krebs cycle, citric-acid [tricarboxylic-acid] cycle, TCA cycle
Szilard-Chalmers-Effekt *m (nucl)* Szilard-Chalmers effect
Szilard-Chalmers-Methode *f (nucl)* Szilard-Chalmers method
Szintillation *f* scintillation
Szintillationsspektrometer *n* scintillation spectrometer
Szintillationszähler *m* scintillation counter
Szintillator *m* scintillator
szintillieren *v* scintillate
S-Zustand *m* singlet (electronic) state

T

2,4,5-T *s.* 2,4,5-Trichlorphenoxyessigsäure
TA *s.* Triacetatfaserstoff
Tabakalkaloid *n* tobacco alkaloid
Tablette *f* 1. tablet; 2. *(spectr)* pellet, pressed disk
tablettieren *v* tablet
Tablettiermaschine *f* tabletting [tablet-compressing] machine, tablet press
Tafel *f* 1. plate, slab, *(if thin:)* sheet; 2. *(cryst)* plate
Tafel-Diagramm *n (el ch)* Tafel diagram [plot]
tafelförmig 1. tabular; 2. *(cryst)* platy
Tafelglas *n* sheet glass

Tafelglasziehverfahren *n (glass)* (flat) sheet drawing process
tafelig *s.* tafelförmig
Tafelsalz *n* table salt
Tagebaubetrieb *m* surface [open-cast, open-cut] mining, open-pit method, *(esp relating to ores)* (surface) quarrying
Tagebauförderung *f* surface [open-cast, open-cut] mining, open-pit method, *(esp relating to ores)* (surface) quarrying
Tagesaufnahme *f***/empfohlene** *(food)* recommended daily allowance, RDA *(of nutrients)*
Tagesdosis *f (tox)* daily intake (dose)
Tagesdosis *f***/bedingt duldbare [zulässige]** conditional acceptable daily intake
Tagesdosis *f***/duldbare [zulässige]** acceptable daily intake, ADI, permissible daily body intake
Taktizität *f* tacticity *(of polymers)*
Talg *m* tallow
talgartig tallowy, sebaceous
Talggeschmack *m* tallowy flavour, tallowiness *(of spoiled fats)*
talgig tallowy, sebaceous
Talgigkeit *f* tallowiness
Talk *m* 1. talc(um) *(magnesium dihydrogentetrasilicate as a mineral or synthetic product)*; 2. *s.* Talkum
talkig talcose, talcous, talcky
Talkum *n* talcum (powder), talc
Tallöl *n* tall oil, liquid resin [rosin]
Tandemküvette *f (spectr)* tandem cuvette
Tank *m* tank, reservoir
Tanklager *n* tank farm
Tankschiff *n* tank ship, tanker
Tankwagen *m* tank truck *(road)*; tank car [wagon] *(railway)*
Tannat *n* tannate *(salt or ester of tannic acid)*
Tannin *n* tannic acid, gallotannic acid, gallotannin, tannin; *(broadly)* vegetable tannin *(any of a large number of substances used in leather tanning)*
Tantal *n* Ta tantalum
Tantalcarbid *n* tantalum carbide
Tantal(V)-chlorid *n* tantalum(V) chloride, tantalum pentachloride
Tantal(V)-oxid *n* tantalum(V) oxide, ditantalum pentaoxide, tantalum pentaoxide
Tantalsäure *f* tantalic acid
Tapioka(stärke) *f* tapioca (starch) *(from Manihot utilissima Pohl)*
tarieren *v* tare
Tartrat *n* tartrate *(salt or ester of tartaric acid)*
Tartrazin *n* tartrazine, hydrazine yellow *(a pyrazole derivative)*
Tartronsäure *f* tartronic acid
Tastpolarographie *f* tast polarography
tato *(Tagestonnen) s.* Tonnen je Tag
Tato *(Tagestonnen) s.* Tonnen je Tag
taub *(mine)* barren, dead
tauchaluminieren *v* dip-aluminize
Tauchbad *n* 1. dipping bath; 2. *(agric)* dip
Tauchbehälter *m* dip(ping) tank
Tauchbeschichtung *f* dip coating
Tauchbrenner *m* submerged burner, submerged combustion burner [heater]
Tauchelektrode *f* immersion electrode, dipping [dipped] electrode
tauchen *v* immerse, immerge, plunge, *(for a short time:)* dip, *(totally:)* submerge
Tauchen *n* immersion, plunge, *(for a short time:)* dip(ping), *(totally:)* submergence
Tauchkolbenpumpe *f* plunger [ram] pump
Tauchlack *m* dipping varnish
Tauchlackieren *n* dip painting [coating], dipping
Tauchlackieren *n***/elektrophoretisches** electrophoretic coating [painting, dipping], electrocoating, electropainting
Tauchlackierung *f s.* Tauchlackieren
Tauchlackierung *f***/katodische** cathodic dip painting *(automotive coatings)*
Tauchschicht *f* dip coat
Tauchsieder *m* immersion heater
Tauchtiefe *f* submergence
Tauchtränkung *f* steeping
Tauchwanne *f* dip(ping) tank
tauen *v* thaw
Taukurve *f (distil)* dew-point curve, condensation curve
Taulinie *f (distil)* dew-point curve, condensation curve
Taupunkt *m* dew point, saturation temperature

Taurocholsäure *f (bioch)* taurocholic acid
tautomer tautomeric
Tautomer(e) *n* tautomer(ide), dynamic isomer
Tautomerie *f* tautomerism, tautomery, dynamic isomerism
tautomerisieren *v* tautomerize
Tautomerisierung *f* tautomerization
TBA *s.* 1. Trichlorbenzoesäure; 2. Thiobarbitursäure
TCC *s.* Tricarbonsäurezyklus
TCDD *n s.* 2,3,7,8-Tetrachlordibenzo-1,4-dioxin
2,3,7,8-TCDD *n s.* 2,3,7,8-Tetrachlordibenzo-1,4-dioxin
Technetium *n* Tc technetium
Technik *f* 1. engineering *(branch of science)*; 2. technology *(totality of means employed in the production of material goods)*; 3. technique, procedure, method *(manner of performing technical details, esp in the laboratory)*
Technik f/chemische chemical engineering
Techniker *m* technologist, technician
technisch durchführbar feasible, practicable
technisch rein technical
Technologie *f* technology *(1. of the manufacture of specified products; 2. branch of science)*
Technologie f/organisch-chemische organic-chemical technology
Teelöffel voll/einen a teaspoonful, tsp, tspn
Teelöffel voll/einen gestrichenen a level teaspoon
Teer *m* tar
Teerabscheider *m* tar separator [extractor], detarrer
teerartig tarry
Teerbase *f* tar base
Teercresol *n* coal-tar-derived cresylic acid (*a mixture of o-, m-, and p-cresol and other phenolic compounds*)
Teerdestillation *f* tar distillation
teeren *v* tar
Teerfarbstoff *m* coal-tar dye(stuff), *(Am)* coaltar color
Teerhydrierung *f* tar hydrogenation
teerig tarry
Teeröl *n* tar oil
Teerpech *n* tar pitch
Teersand *m* tar sand
Teersäure *f* tar acid
Teichonsäure *f (bioch)* teichoic acid
Teichreinigung *f (hyd)* pond treatment, ponding, lagooning *(of waste water)*
Teig *m* 1. *(tech)* dough, paste; 2. *(food)* dough
teigartig dough-like, doughy
teigig dough-like, doughy
Teigkneter *m* dough kneader [mixer], dough kneading [mixing] machine
Teigknetmaschine *f* dough kneader [mixer], dough kneading [mixing] machine
Teigtriebmittel *n (food)* leaven, leavening (agent)
Teil *m* 1. part, portion *(as of bulk material)*; *(esp relating to approximately equal quantities:)* moiety; 2. section *(as of an industrial plant)*
Teil m/aliquoter aliquot (part, portion)
Teil *n* part
Teil n/grünes green compact *(in powder metallurgy)*
Teilanalyse *f* partial analysis
Teilchen *n* particle
Teilchen n/energiereiches high-energy particle
Teilchen n/seltsames *(nucl)* strange particle
α-Teilchen *n* alpha particle, α-particle
β-Teilchen *n* beta particle, β-particle
Teilchen npl/suspendierte suspended particles, particulates, particulate matter
Teilchenaggregat *n* particle [particulate] aggregate
Teilchenbeschleuniger *m* particle accelerator
Teilchenfluss *m* flux
Teilchenform *f* particle shape
Teilchengröße *f* particle size, *(screen classification also)* screen size
Teilchengrößenbereich *m* range of particle sizes, particle-size range, *(screen classification also)* range of screen sizes
Teilchengrößenverteilung *f* particle-size distribution
Teilchenkollision *f* particle collision
Teilchenstrahlung *f* corpuscular [particle] radiation

Teilchenwachstum *n (hyd)* particle growth
Teildruck *m* partial pressure
Teile *mpl* **je hundert Millionen Teile** parts per hundred million, pphm
Teile *mpl* **je Million Teile** parts per million, ppm
Teile *mpl* **je tausend Teile** parts per thousand, ppt
teilen *v***/in Grade** graduate
Teilentsalzungsanlage *f (hyd)* dealkalizer
Teilfäserchen *n (text)* fibril(la)
Teilkondensation *f* partial condensation, dephlegmation
Teilkondensator *m (distil)* (countercurrent) partial condenser, partial-condensation head
Teilladung *f* partial [fractional] charge
teilnehmen *v* participate *(as in a reaction)*
Teilordnung *f* partial order (of reaction), individual order
Teilpipette *f* graduated [measuring] pipette
Teilprobe *f (anal)* sub-sample
Teilreaktion *f* partial reaction, half-reaction
Teilreinigung *f* partial purification [treatment] *(of waste water)*
Teilschritt *m* individual step *(of a reaction)*
Teilstrich *m* graduation [division] mark [line], graduation, division
Teilverbrennung *f* partial combustion
Tein *n* caffeine, theine, 1,3,7-trimethylxanthine
T-Einheit *f* trifunctional unit, T unit *(for building up polymers)*
Tektosilicat *n (min)* tectosilicate *(any of a class of polymeric silicates)*
tele-Substitution *f (org ch)* tele-substitution
Tellermühle *f* disk (attrition) mill, disk grinder
Tellerwäscher *m* plate scrubber *(gas cleaning)*
Tellerwäscher *m* **nach Theisen** Theisen disintegrator
Tellur *n* Te tellurium
Tellurat *n* tellurate
Tellur(II)-chlorid *n* tellurium(II) chloride, tellurium dichloride
Tellur(VI)-fluorid *n* tellurium(VI) fluoride, tellurium hexafluoride
Tellurid *n* M^I_2Te telluride
Tellurit *n* $M^I_2TeO_3$ tellurite
Telluroniumverbindung *f* telluronium compound
Tellur(II)-oxid *n* tellurium(II) oxide, tellurium monooxide
Tellur(IV)-oxid *n* tellurium(IV) oxide, tellurium dioxide
Tellur(VI)-oxid *n* tellurium(VI) oxide, tellurium trioxide
Tellursäure *f* orthotelluric acid, telluric acid, hexaoxotelluric acid
Tellurwasserstoff *m* hydrogen telluride, tellurium hydride
Telogen *n* telogen *(agent for controlling the degree of polymerization)*
Telomer(e) *n* telomer
Telomerisat *n* telomer
Telomerisation *f (org ch)* telomerization
Temperatur *f***/absolute** *s.* Temperatur/thermodynamische
Temperatur *f***/charakteristische** characteristic [Debye] temperature
Temperatur *f***/gewöhnliche** normal [ordinary] temperature
Temperatur *f***/isokinetische** isokinetic temperature
Temperatur *f***/kritische** critical temperature
Temperatur *f***/thermodynamische** thermodynamic [absolute] temperature
Θ-Temperatur *f* Flory theta temperature
Temperaturabfall *m* temperature drop
temperaturabhängig temperature-dependent
Temperaturabhängigkeit *f* temperature dependence
Temperaturanstieg *m* rise [increase] in temperature
Temperaturbeiwert *m* temperature coefficient
Temperaturbereich *m* temperature range, range of temperatures
temperaturbeständig temperature-resistant, temperature-resisting, temperature-stable
Temperaturbeständigkeit *f* temperature resistance [stability]
temperaturempfindlich temperature-sensitive

Temperaturempfindlichkeit *f* temperature sensitivity
Temperatur-Entropie-Diagramm *n* (temperature-)entropy chart, entropy diagram
Temperaturerhöhung *f* temperature raising; rise [increase] in temperature
Temperaturgleichgewicht *n* temperature equilibrium
Temperaturgradient *m* temperature gradient; (temperature) lapse rate *(in the atmosphere)*
Temperaturkoeffizient *m* **der Viskosität [Zähigkeit]** viscosity-temperature coefficient
Temperaturleitfähigkeit *f s.* Temperaturleitkoeffizient
Temperaturleitkoeffizient *m* (thermal) diffusivity, thermometric conductivity
Temperaturleitzahl *f* (thermal) diffusivity, thermometric conductivity
Temperaturregelung *f* temperature control
Temperaturschwankung *f* temperature fluctuation
Temperaturskala *f* temperature scale
Temperaturskala *f*/**absolute** *s.* Temperaturskala/thermodynamische
Temperaturskala *f*/**ideale gasthermometrische** perfect gas temperature scale
Temperaturskala *f*/**thermodynamische** absolute [thermodynamic] temperature scale
Temperatursprung *m* temperature jump, T-jump
temperaturunabhängig temperature-independent
Temperaturunabhängigkeit *f* temperature independence
temperaturwechselbeständig thermal-shock resistant
Temperaturwechselbeständigkeit *f* thermal-shock resistance
Temperguss *m* malleable (cast) iron
Temperierbad *n* (constant-)temperature bath
temperieren *v* bring to a specified temperature, keep at a specified temperature, *(esp)* thermostat
tempern *v* 1. malleabl(e)ize, anneal *(cast iron)*; 2. *(plast)* anneal, temper; 3. *(glass)* anneal

Tensammetrie *f (el ch)* tensammetry
tensammetrisch *(el ch)* tensammetric
Tensid *n* surfactant, surface-active agent
Tensid *n*/**anionisches** anionic surfactant
Tensid *n*/**kationisches** cationic surfactant
Tensid *n*/**nicht ionogenes** non-ionic (surfactant)
Terbium *n* Tb terbium
Terbiumchlorid *n* terbium chloride
Terbiumoxid *n* terbium oxide
Terephthalsäure *f* terephthalic acid, TPA, *p*-phthalic acid, benzene-1,4-dicarboxylic acid
Term *m (phys ch)* term (value), energy level
Termdiagramm *n* energy(-level) diagram, level scheme [diagram]
terminal terminal
Terminationscodon *n (bioch)* termination codon, terminator [chain-terminating] codon
termolekular termolecular, trimolecular
Termschema *n s.* Termdiagramm
ternär ternary
Terpen *n* terpene, *(broadly)* terpenoid
Terpenalkaloid *n* terpenoid alkaloid
Terpenalkohol *m* terpene alcohol
Terpenkohlenwasserstoff *m* terpenoid hydrocarbon, terpene
Terpenoid *n* terpenoid
Terpentin *n* turpentine (oleoresin) *(balsam obtained from coniferous trees)*
Terpentinharz *n* rosin, colophony, pine resin [rosin]
Terpentinharzöl *n* rosin oil, rosinol *(by fractional distillation of rosin)*
Terpentinöl *n* oil [spirit] of turpentine
Terpenverbindung *f* terpenoid
Terpin *n* terpin *(a cyclohexanol derivative)*
Terpinen *n (org ch)* terpinene
Terpineol *n* terpineol *(any of three isomeric terpene alcohols)*
Terpinolen *n* terpinolene, 4-isopropylidene-1-methylcyclohexane
Terpolymer(e) *n* terpolymer
Terrakotta *f (ceram)* terra cotta
tertiär tertiary
Tertiärstruktur *f (bioch)* tertiary structure
Test... *siehe: Prüf...*
testen *v s.* prüfen
Testosteron *n* testosterone, 17b-

hydroxyandrost-4-en-3-one *(a sex hormone)*
Tetra *m s.* Tetrachlorkohlenstoff
Tetraacetat *n* tetraacetate
Tetrabasizität *f* tetrabasicity *(of acids)*
Tetraborat *n* $M^{I}_{2}B_4O_7$ tetraborate, heptaoxotetraborate
Tetraborsäure *f* tetraboric acid, heptaoxotetraboric acid
Tetrabromfluorescein *n* tetrabromofluorescein, eosin
Tetrachlor-*p*-benzochinon *n* tetrachloro-*p*-benzoquinone, tetrachloroquinone, chloranil
2,3,7,8-Tetrachlordibenzo-1,4-dioxin *n* 2,3,7,8-tetrachlorodibenzo-1,4-dioxin, 2,3,7,8-tetrachlorodibenzo-*p*-dioxin, 2,3,7,8-TCDD, TCDD, dioxin
Tetrachlorethen *n* tetrachloroethylene, perchloroethylene
Tetrachlorethylen *n* tetrachloroethylene, perchloroethylene
Tetrachlorid *n* tetrachloride
Tetrachlorkohlenstoff *m* carbon tetrachloride, *tetrachloromethane, perchloromethane
Tetrachlormethan *n* carbon tetrachloride, *tetrachloromethane, perchloromethane
Tetrachlorodioxin *n s.* 2,3,7,8-Tetrachlorodibenzo-1,4-dioxin
Tetrachlorsilan *n* tetrachlorosilane
Tetracyanoaurat(III) *n* $M^{I}[Au(CN)_4]$ tetracyanoaurate(III), cyanoaurate(III)
Tetracyanoplatinat(II) *n* $M^{I}_{2}[Pt(CN)_4]$ tetracyanoplatinate(II), cyanoplatinate(II)
Tetradecanal *n* *tetradecanal, tetradecylaldehyde, myristaldehyde
Tetradecan-1-ol *n* *tetradecan-1-ol, myristyl alcohol
1-Tetradecanol *n* *tetradecan-1-ol, myristyl alcohol
Tetradecansäure *f* tetradecanoic acid
Tetraederanordnung *f* tetrahedral arrangement [orientation]
Tetraen *n (org ch)* tetraene
Tetraethylblei *n* tetraethyl lead, TEL
• **mit Tetraethylblei versetzen** lead *(motor fuel)*
Tetrafluorethen *n* tetrafluoroethylene, TFE, perfluoroethylene
Tetrafluorethylen *n* tetrafluoroethylene, TFE, perfluoroethylene
Tetrafluorsilan *n* tetrafluorosilane
tetrafunktionell tetrafunctional
tetragonal *(cryst)* tetragonal
Tetrahalogenid *n* *tetrahalide, *tetrahalogenide
Tetrahydrat *n* tetrahydrate
Tetrahydridoborat *n* $M^{I}[BH_4]$ tetrahydridoborate
Tetrahydrobenzen *n* tetrahydrobenzene, cyclohexene
Tetrahydrobenzol *n* tetrahydrobenzene, cyclohexene
Tetrahydrofuran *n* tetrahydrofuran, THF
Tetrahydrofurfuran *n* tetrahydrofuran, THF
Tetrahydronaphthalen *n* tetrahydronaphthalene
Tetrakisazofarbstoff *m* tetrakisazo dye
Tetramer(e) *n* tetramer
Tetramethylblei *n* tetramethyl lead, TML
Tetramethyldiarsin *n* tetramethyldiarsine, cacodyl
Tetramethylenimin *n* pyrrolidine, tetramethyleneimine
Tetramethylethylenglykol *n* tetramethylethylene glycol, *2,3-dimethylbutane-2,3-diol, pinacol
Tetramethylsilan *n* tetramethylsilane, TMS
Tetrammin *n* tetraammine
tetramolekular tetramolecular, quadrimolecular
Tetranatriumsalz *n* tetrasodium salt
Tetrapeptid *n* tetrapeptide
Tetrapyrrol *n (org ch)* tetrapyrrole
Tetrasilan *n* tetrasilane
tetrasubstituiert tetrasubstituted
Tetrasubstitution *f* tetrasubstitution
Tetrasulfid *n* tetrasulphide
Tetraterpen *n (bioch)* tetraterpene
Tetraterpenoid *n (bioch)* tetraterpenoid
tetravalent tetravalent, quadrivalent, *(relating to molecules also)* tetraatomic
Tetravalenz *f* tetravalence, quadrivalence
Tetrazin *n* tetrazine
tetrazotieren *v (dye)* tetraazotize
Tetrazotierung *f (dye)* tetraazotization
Tetrazoverbindung *f* tetraazo compound

Tetritol *n* tetritol *(a tetrahydroxy alcohol derived from tetrose)*
Tetrose *f* tetrose *(monosaccharide containing four carbon atoms per molecule)*
Tetroxid *n* tetraoxide
Tetroxokieselsäure *f* tetraoxosilicic acid, orthosilicic acid
Tetroxosilicat *n* $M^I_4SiO_4$ tetraoxosilicate, orthosilicate, silicate
Textilausrüstung *f* textile finish
Textilchemie *f* textile chemistry
Textilchemiker *m* textile chemist
Textilfaser *f* textile fibre
Textilfaserstoff *m* textile fibre
Textilgewebe *n* fabric (cloth), tissue
Textilhilfsmittel *n* textile auxiliary
Textilien *pl* textiles
Textilverbundstoffe *mpl* non-woven textiles, non-wovens
Textilvered(e)lung *f* textile finishing
Textilveredlungsmittel *n* textile auxiliary
Textur *f* texture
texturieren *v (text)* texture, bulk
Thallium *m* Tl thallium
Thallium(I)-chlorid *n* *thallium(I) chloride, thallium monochloride, thallous chloride
Thallium(III)-chlorid *n* *thallium(III) chloride, thallium trichloride, thallic chloride
Thallium(I)-hydroxid *n* *thallium(I) hydroxide, thallous hydroxide
Thallium(I)-oxid *n* *thallium(I) oxide, thallium monooxide, thallous oxide
Thallium(III)-oxid *n* *thallium(III) oxide, thallium trioxide, thallic oxide
Thallium(I)-sulfat *n* *thallium(I) sulphate, thallous sulphate
Thallium(III)-sulfat *n* *thallium(III) sulphate, thallic sulphate
Thein *n s.* Tein
Theisen-Desintegrator(wäscher) *m* Theisen disintegrator *(gas cleaning)*
Theisen-Wäscher *m* Theisen disintegrator *(gas cleaning)*
Theobromin *n* theobromine, 2,6-dihydroxy-3,7-dimethylpurine *(alkaloid)*
Theophyllin *n* theophylline, 2,6-dihydroxy-1,3-dimethylpurine *(alkaloid)*
Theorie *f* theory • **etwas mehr als der Theorie entspricht** in slight excess of theory
Theorie *f* **der Valenzstrukturen** valence-bond [electron-pair] theory, VB theory, Heitler-London-Slater-Pauling theory, HLSP theory
Theorie *f* **des Übergangszustandes** transition-state theory, activated-complex theory, absolute-(reaction-) rate theory
Theorie *f* **spannungsfreier Ringe** Sachse-Mohr concept (of strainless rings)
Theorie *f* **spannungsfreier Ringe/sachse-mohrsche** Sachse-Mohr concept (of strainless rings)
Therapeutikum *n* therapeutic agent
therapeutisch therapeutic(al)
thermionisch thermionic
thermisch thermal
Thermitverfahren *n* aluminothermic [thermite, Goldschmidt's] process, aluminothermics, aluminothermy
Thermoanalyse *f* thermal analysis, thermoanalysis, TA
Thermochemie *f* thermochemistry
Thermochemiker *m* thermochemist
thermochemisch thermochemical
thermochrom thermochromic
Thermochromie *f* thermochromism
Thermodiffusion *f* thermal diffusion, thermodiffusion
Thermodiffusionsverfahren *n* thermal-diffusion process
Thermodynamik *f* thermodynamics
Thermodynamik *f***/chemische** chemical thermodynamics
Thermodynamik *f* **irreversibler Prozesse** thermodynamics of irreversible processes, non-equilibrium thermodynamics
Thermodynamik *f***/statistische** statistical thermodynamics
Thermodynamik *f***/technische** engineering thermodynamics
thermodynamisch thermodynamic(al)
Thermoelement *n* thermocouple
Thermogramm *n* thermogram
Thermogravimetrie *f* thermogravimetry
Thermogravimetrie *f***/derivative** derivative [differential] thermogravimetry, DTG

thermogravimetrisch thermogravimetric
Thermoionisationsdetektor *m (anal)* thermionic detector
thermolabil thermolabile, heat-labile
Thermolabilität *f* thermolability, thermal lability
Thermolumineszenz *f* thermoluminescence, thermal luminescence
Thermolyse *f* thermolysis
thermolytisch thermolytic
thermomechanisch thermomechanical
Thermometer *n*/**beckmannsches** Beckmann thermometer
Thermometer *n*/**feuchtes** wet-bulb thermometer
Thermometer *n*/**trockenes** dry-bulb thermometer
Thermometerfehler *m* thermometer error
Thermometrie *f* thermometry
thermometrisch thermometric(al)
thermooxidativ thermal-oxidative
Thermopaar *n* thermocouple
thermophil thermophilic, thermophilous
Thermoplast *m* thermoplastic
thermoplastisch thermoplastic
thermoresistent thermoduric, heat-resistant *(microorganisms)*
Thermoresistenz *f* heat resistance *(of microorganisms)*
Thermospray *n* thermal spray *(coating process)*
thermostabil thermostable, heat-stable *(enzymes, vitamins)*
Thermostabilität *f* thermostability, heat stability *(of enzymes, vitamins)*
Thermostat *m (lab)* constant-temperature chamber, thermostat; (constant-) temperature bath
thermostat(is)ieren *v* thermostat
Thermostat(is)ierung *f* thermostatting
Theta-Lösung *f* pseudoideal solution
Theta-Lösungsmittel *n* theta solvent
Theta-Temperatur *f* theta temperature
Thiamin *n* thiamin(e), aneurin(e) *(vitamin* B_1*)*
Thiazinfarbstoff *m* thiazine dye
Thiazolfarbstoff *m* thiazole dye
Thiazolidin *n* thiazolidine, tetrahydrothiazole
Thioacetal *n (org ch)* thioacetal
Thioaldehyd *n* thioaldehyde
Thioalkohol *m* *thiol, thioalcohol, mercaptan
Thioamid *n (org ch)* thio-amide
Thiobakterien *npl* sulphur bacteria
Thiobarbitursäure *f* thiobarbituric acid, TBA
Thiocarbamid *n s.* Thioharnstoff
Thiocarbanilid *n* thiocarbanilide
Thiocarbonat *n* $M^I_2CS_3$ thiocarbonate, trithiocarbonate
Thiocarbonsäure *f* thiocarboxylic acid
Thiocarbonyldichlorid *n* thiocarbonyl chloride, thiophosgene
Thiocyanat *n* M^ISCN thiocyanate, rhodanide
Thiocyansäure *f* thiocyanic acid
Thioderivat *n* thio derivative
Thioessigsäure *f* thioacetic acid
Thioglykolsäure *f* thioglycollic acid
Thioharnstoff *m* thiourea, thiocarbamide
Thioharnstoffharz *n* thiourea resin, polythiourea
Thioketal *n (org ch)* thioketal
Thiokohlensäure *f* thiocarbonic acid, trithiocarbonic acid
Thiol *n (org ch)* *thiol, mercaptan
Thiolgruppe *f* –SH thiol group, sulphanyl [sulphydryl, mercapto] group
Thiolthionkohlensäure *f* thionothiolcarbonic acid, dithiocarbonic acid, xanthic acid
Thiolyse *f (bioch)* thiolysis
thiolytisch *(bioch)* thiolytic
Thion-Thiol-Umlagerung *f* thion-thiol rearrangement
Thionylchlorid *n* thionyl chloride
Thiophen *n* thiophene
Thiophenol *n* thiophenol, phenyl mercaptan
Thiophosphat *n* thiophosphate
Thioplast *m* thioplast, polysulphide rubber
Thiosäure *f* thioacid, monothiocarboxylic acid
Thiosemicarbazid *n* thiosemicarbazide, aminothiourea
Thiosulfat *n* $M^I_2S_2O_3$ thiosulphate
Thiosulfit *n* $M^I_2S_2O_2$ thiosulphite
thixotrop *(coll)* thixotropic
Thixotropie *f (coll)* thixotropy
Thixotropier(ungs)mittel *n (coll)* thixotroping [thixotropic] agent

Thomas-Stahl *m* Thomas steel, basic (Bessemer, converter) steel
Thomas-Verfahren *n* Thomas(-Gilchrist) process, basic (Bessemer, converter) process
thorieren *v* thoriate *(e.g. tungsten filaments)*
Thorium *n* Th thorium • **mit Thorium beschichten** thoriate *(e.g. tungsten filaments)*
Thoriumchlorid *n* thorium chloride
Thoriumdioxid *n s.* Thoriumoxid
Thoriumoxid *n* thorium oxide, thorium dioxide
Thoriumsulfat *n* thorium sulphate
Thoriumzerfallsreihe *f (nucl)* thorium (decay) series
Thorpe-Reaktion *f* Thorpe reaction *(for obtaining cyclic alkanones)*
Thr *s.* Threonin
Threonin *n* threonine, 2-amino-3-hydroxybutyric acid
Thulium *n* Tm thulium
Thymiancampher *m* thyme camphor, thymol, thymic acid, 1-methyl-3-hydroxy-4-isopropyl benzene
Thymianöl *n* thyme oil *(from Thymus vulgaris L. and Th. zygis L.)*
Thymiansäure *f s.* Thymiancampher
Thymidylsäure *f (bioch)* thymidylic acid, ribothymidylic acid
Thymolblau *n* thymol blue, thymolsulphonephthalein *(a pH indicator)*
Thymolphthalein *n* thymolphthalein *(a pH indicator)*
Thymolsulfophthalein *n s.* Thymolblau
Thyreoglobulin *n (bioch)* thyroglobuline
thyreotrop thyrotropic
Thyreotropin *n* thyrotropin, thyrotropic hormone, thyroid-stimulating hormone
Thyreotropin-Releasinghormon *n* thyrotropin-releasing hormone [factor], TRH, TRF
Tiefbau(betrieb) *m* deep [underground] mining, underground work(ing)
Tiefbauförderung *f* deep [underground] mining, underground work(ing)
Tiefenfilter *n (hyd)* depth filter
Tiefenfiltration *f* 1. filter-medium filtration [operation]; 2. *(hyd)* deep-bed filtration
Tiefkühltrocknung *f* 1. freeze drying, lyophilization; 2. *(food)* dehydrofreezing
tiefschmelzend low-melting(-point), low-fusion
tiefsiedend low-boiling, light
tieftemperaturbeständig low-temperature-resistant, resistant to cold, cold-resistant
Tieftemperaturbeständigkeit *f* low-temperature resistance, resistance to cold
Tieftemperatureigenschaften *fpl* low-temperature properties [characteristics]
Tieftemperaturhydrierung *f* low-temperature hydrogenation
Tieftemperaturkautschuk *m* cold(-polymerized) rubber, low-temperature rubber
Tieftemperaturkoks *m* low-temperature coke
Tieftemperaturpolymer(e) *n* low-temperature polymer, cold polymer
Tieftemperaturpolymerisat *n* low-temperature polymer, cold polymer
Tieftemperaturpolymerisation *f* low-temperature polymerization, cold polymerisation
Tieftemperaturreaktion *f* low-temperature reaction
Tieftemperaturtechnik *f* cryogenic [low-temperature] engineering, cryogenics
Tieftemperaturverhalten *n* low-temperature behaviour
Tieftemperaturverkokung *f* low-temperature carbonization
Tieftemperaturwerkstoff *m* low-temperature material
Tiegel *m* crucible
Tiegelkoks *m* crucible coke
Tiegelofen *m (met, lab)* crucible furnace
Tiegelstahl *m* drill steel, *(obsolete:)* crucible (cast) steel
Tiegelzange *f* (pair of) crucible tongs
Tierarzneimittel *n* animal health product
Tiereiweiß *n* animal protein
Tierfaser *f* animal fibre
Tierfett *n* animal fat
tierisch animal
Tierkohle *f* animal char(coal)
Tierleim *m* animal (protein) glue, animal adhesive
Tieröl *n* animal oil

Tieröl *n*/**ätherisches [dippelsches]** hartshorn oil, Dippel's oil
Tierstärke *f* animal starch
Tierversuch *m (pharm, tox)* animal assay [test], test on animals
Tiffeneau-Reaktion *f* Tiffeneau rearrangement [reaction] *(for converting amino alcohols into carbonyl compounds)*
Tiffeneau-Umlagerung *f* Tiffeneau rearrangement [reaction] *(for converting amino alcohols into carbonyl compounds)*
Tiglinsäure *f* tiglic acid, 2-methylcrotonic acid, trans-2-methylbuten-2-oic acid
Tinkal *m (min)* tincal, borax *(sodium tetraborate 10-water)*
Tinktur *f* 1. *(pharm)* tincture; 2. *(food)* miscella *(an extractant containing an extracted oil or grease)*
Tischabzug(sschrank) *m (lab)* local exhaust hood, fume closet
Tischtschenko-Reaktion *f* Tishchenko reaction *(for obtaining esters from aldehydes)*
Tiselius-Elektrophorese *f (anal)* free [moving-boundary] electrophoresis, Tiselius method
Titan *n* Ti titanium
Titanat *n* titanate
Titan(II)-chlorid *n* titanium(II) chloride, titanium dichloride
Titan(III)-chlorid *n* titanium(III) chloride, titanium trichloride
Titan(IV)-chlorid *n* titanium(IV) chloride, titanium tetrachloride
titanhaltig titaniferous
Titanhydrid *n* titanium hydride
Titannitrid *n* titanium nitride
Titanometrie *f* titanometry
Titan(II)-oxid *n* titanium(II) oxide, titanium monooxide
Titan(III)-oxid *n* titanium(III) oxide, dititanium trioxide
Titan(IV)-oxid *n* titanium(IV) oxide, titanium dioxide
Titansäure *f* titanic acid *(any of various hydrates of titanium oxide)*
Titan(IV)-sulfat *n* titanium(IV) sulphate
Titan(IV)-sulfid *n* titanium(IV) sulphide, titanium disulphide
Titanweiß *n* titanium white *(a pigment consisting mainly of TiO_2)*
Titer *m* 1. titre, titer *(strength of a solution)*; 2. titre, titer *(for defining the fineness of yarn)*; 3. titre (value) *(the solidifying point of oils)*
Titersubstanz *f* standard reagent [titrant, titrimetric substance]
Titrand *m* solution to be titrated, solution being titrated
Titrand-Titrator-System *n* titration system
Titrans *n s.* Titrator
Titrant *m s.* Titrator
Titration *f* titration
Titration *f*/**amperometrische** amperometric titration
Titration *f*/**chelatometrische** chelatometric titration
Titration *f*/**coulometrische** coulometric titration
Titration *f*/**direkte** direct titration
Titration *f*/**elektrometrische** *s.* Titration/potenziometrische
Titration *f* **in nicht wässriger Lösung** non-aqueous titration
Titration *f* **in wasserfreiem Medium** non-aqueous titration
Titration *f*/**komplexometrische** complexometric titration
Titration *f*/**konduktometrische** conductometric [conductance] titration
Titration *f*/**manganometrische** permanganate titration
Titration *f* **nach Mohr** Mohr titration [method] *(argentometry)*
Titration *f* **nach Volhard** Volhard titration [method] *(for determining chlorine, bromine, or iodine)*
Titration *f*/**nephelometrische** nephelometric titration
Titration *f*/**oszillometrische** high-frequency titration
Titration *f*/**photometrische** photometric titration
Titration *f*/**potenziometrische** potentiometric [electrometric] titration
Titration *f*/**turbidimetrische** turbidimetric [turbidity] titration
Titrationsfehler *m* titration error
Titrationskurve *f* titration curve
Titrationsmethode *f* titrimetric method
Titrationsmittel *n s.* Titrator

Titrator *m* titrant, titrator
Titrieranalyse *f* titrimetric [volumetric, mensuration] analysis
Titrierautomat *m* automatic titrator, autotitrator
titrierbar titr(at)able
titrieren *v* titrate
Titrierfehler *m* titration error
Titriergerät *n* titrator
Titrierkolben *m* titration flask
Titrimeter *n* titrimeter, titrator
Titrimetrie *f* titrimetry, volumetry
titrimetrisch titrimetric, volumetric
TMS *s.* Tetramethylsilan
TNT *s.* Trinitrotoluen
TOC-Gehalt *m (hyd)* TOC content
Tochterstrang *m (bioch)* daughter strand *(of DNA)*
tödlich deadly poisonous; lethal
Tödlichkeitsdosis *f (tox)* lethal dose, LD, fatal dose
Tödlichkeitsindex *m (tox)* ct product *(product of concentration and survival time)*
Tolan *n* tolan, diphenyl acetylene, *diphenylethyne
Toleranz *f* 1. tolerance, *(relating to dimensions also)* allowance; 2. *s.* Toleranzwert
Toleranzdosis *f s.* Toleranzwert
Toleranzgrenze *f* maximum allowable content, maximum permissible concentration, maximum permitted level *(of one component in a material)*
Toleranzwert *m (tox)* (maximum) tolerance, maximum permissible concentration, maximum permitted level
Tollens-Reagens *n* Tollens' reagent *(for detecting aldehydes)*
Toluen *n* toluene, methylbenzene
Toluendiisocyanat *n* toluene diisocyanate, TDI
***p*-Toluensulfochlorid** *n p*-toluenesulphonyl chloride, tosyl chloride
***p*-Toluensulfonat** *n p*-toluenesulphonate, tosylate
***p*-Toluensulfonsäure** *f* toluene-*p*-sulphonic acid
***p*-Toluensulfonsäureester** *m s. p*-Toluensulfonat
***p*-Toluensulfonylgruppe** *f* $CH_3C_6H_4SO_2$– *p*-toluenesulphonyl group, tosyl group
Toluidin *n* toluidine, aminotoluene
Toluol *n s.* Toluen
Toluylenrot *n* toluylene [neutral] red *(an oxidation-reduction indicator)*
Toluylsäure *f* toluic acid, methylbenzoic acid
α-Tolylsäure *f* α-toluic acid, toluylic acid, phenylacetic acid
1′-Tolylsäure *f* α-toluic acid, toluylic acid, phenylacetic acid
Tombak *m* tombac, tombak *(an alloy chiefly consisting of copper and zinc)*
Ton *m* 1. clay, argil(la); 2. shade, hue *(of colour)* • **einen Ton treffen** *(dye)* match a shade
Ton *m*/**aktivierter** activated clay
Ton *m*/**feuerfester** refractory clay, fireclay
Ton *m*/**weißer** white [china, porcelain] clay, kaolin(e), white bole, bolus alba
tonartig clayey, clayish, argillaceous
tönen *v* tint, tinge, tone, shade
Tonerde *f* alumina *(aluminium oxide)*
Tonerdegel *n* alumina gel, gelatinous aluminium hydroxide
tonerdehaltig aluminiferous
Tonerdeschiffchen *n* alumina boat
Tongut *n* earthenware
tonhaltig 1. containing clay, clayey; 2. *(geol)* argilliferous, argillaceous
tonig 1. clayey, clayish; 2. *(geol)* argillaceous
Tonkabohnencampher *m* tonka bean camphor, coumarin
Tonne *f* cask, barrel, *(if large)* tun, *(if small)* keg
Tonnen *fpl* **je Jahr** tons per year [annum], tpy, tpa, metric tons annually, mta
Tonnen *fpl* **je Tag** tons per day, tpd
Tonschiefer *m (geol)* (clay) slate
Tontiegel *m* clay crucible
Tönung *f* 1. tinting, tinging, toning, shading; 2. tint, tinge, tone, shade, cast
Tonware *f* clay ware
Topf *m* 1. pot; 2. *(lab)* jar; 3. *(text)* (spinning) can; pot *(for pot spinning)*
Topf *m*/**wittscher** Witt jar
Topfcurare *n* pot curare *(from Chondrodendron specc.)*
Töpferton *m* potter's clay
Töpferware *f* pottery, earthenware
Topfzeit *f* (liquid) pot life *(as of adhesives and organic coatings)*

Topochemie *f* topochemistry
topochemisch topochemical
topotaktisch *(cryst)* topotactic
Topotaxie *f (cryst)* topotaxy
Toppdestillation *f* topping
toppen *v (petrol)* top, skim
Toppprodukt *n (petrol)* tops, top [overhead] product, overhead(s)
Topprückstand *m (petrol)* long residue [residuum] • **Topprückstände aufarbeiten** run resid
Torf *m* peat
Torfgas *n* peat gas
torfig peaty
Torfkoks *m* peat coke
Torfteer *m* peat tar
Torsion *f* torsion, twist
Torsionsschwingung *f* 1. torsional vibration; 2. *(spectr)* torsional oscillation [mode], twisting vibration
Torsionsspannung *f* torsional strain *(in molecules)*
Torsionssteifigkeit *f* stiffness in torsion
Torsionswaage *f* torsion balance
Torsionswinkel *m* torsion [rotational] angle, angle of rotation
Tosylat *n* tosylate, *p*-toluenesulphonate *(salt or ester of p-toluenesulphonic acid)*
Tosylchlorid *n* tosyl chloride, *p*-toluenesulphonyl chloride
Tosylester *m* tosylate, *p*-toluenesulphonate
Tosylgruppe *f* $CH_3C_6H_4SO_2$– tosyl group, *p*-toluenesulphonyl group
Tosylierung *f* tosylation *(introduction of the p-toluenesulphonyl group)*
Totalherbizid *n* general herbicide, non-selective herbicide [weed-killer], soil sterilant
Totalkondensation *f (distil)* total condensation
Totalreflexion *f* total reflection
Totalreflexion *f***/abgeschwächte** *(anal)* attenuated total reflectance, ATR
Totalsynthese *f* total synthesis
Totraum *m* dead space
totrösten *v (met)* dead-roast
Totvolumen *n* dead volume
totvolumenarm low-dead-volume
Totzeit *f* 1. dead time *(of a counting tube)*; 2. *(chromat)* (gas) hold-up time, time of passage *(period between injection and detection)*
Totzone *f* dead zone [spot]
Toxikologe *m* toxicologist
Toxikologie *f* toxicology
toxikologisch toxicologic(al)
Toxikum *n* toxic (substance), toxicant
Toxin *n (bioch)* toxin • **Toxin erzeugend** toxigenic
Toxin *n***/pflanzliches** phytotoxin
toxisch toxic(al), poisonous
toxisch/nicht non-toxic, non-poisonous
toxisch/schwach mildly toxic
toxisch/stark highly toxic
Toxizität *f* toxicity
Toxizität *f***/akute** acute toxicity
Toxizität *f***/chronische** chronic toxicity
Toxizität *f* **für Säugetiere** mammalian toxicity
Toxizität *f***/orale** oral toxicity
Tracer *m***/radioaktiver** radioactive tracer [indicator], radiotracer
Tracermethode *f* tracer [indicator] method
Tracersubstanz *f* tracer
Tragant *m* tragacanth gum, gum tragacanth *(any of various gums esp from Astragalus specc.)*
Tragantgummi *n* tragacanth gum, gum tragacanth *(any of various gums esp from Astragalus specc.)*
träge inert, inactive, passive, indifferent *(chemical)*; sluggish, slow *(reaction)*
Träger *m* 1. carrier, *(esp relating to heat transfer also)* medium; 2. support, supporting material *(for a layer containing active substances, as catalysts or photosensitive compounds)*; 3. substrate *(for pigments)*
Trägerbindung *f (biot)* enzyme bonding [binding attachment] to a carrier
Trägerdampfdestillation *f* carrier distillation, codistillation
Trägerflüssigkeit *f* carrier liquid
Trägergassublimation *f* entrainer [carrier] sublimation
Trägerkatalysator *m* supported [high-area] catalyst
Trägerstoffdestillation *f* carrier distillation
Trägerstrom *m* fluid medium *(fluid-bed technology)*
Trägheit *f* 1. inertness, inactivity, pas-

sivity, indifference *(of a chemical)*; slowness *(of a reaction)*; 2. inertia *(physics)*
Trägheitsgesetz *n* law of inertia
Trägheitsmoment *n* moment of inertia
Trägheitsradius *m* radius of gyration
Tränengas *n* tear gas, lachrymator
tränenreizend lachrymatory, lacrimatory, causing [prompting] tears
Tränenreizstoff *m* lachrymator, lacrimator
tränken *v* impregnate, saturate, imbibe, *(using an aqueous solution)* steep, soak, water
tränken *v*/**mit Harz** *(plast)* resin
Tränkflüssigkeit *f* 1. impregnation solution; 2. *(dye)* steeping liquor
Tränkmittel *n* impregnating [impregnation] material
Tränkstoff *m* impregnating [impregnation] material
Tränkung *f* impregnation, saturation, imbibition, *(using an aqueous solution)* steep(age), soak(age), watering
Tranquilizer *m (pharm)* (minor) tranquil(l)izer, tranquillizing drug, anxiolytic (sedative)
Transaminierung *f* transamination
Transduktion *f (bioch)* transduction *(transfer of genetic information by bacteriophages)*
Transfektion *f* transfection *(introduction of DNA in eukaryotic cells)*
Transferfaktor *m (bioch)* transfer [elongation, propagation] factor
Transfer-Ribonucleinsäure *f* transfer [soluble] ribonucleic acid, tRNA, s-RNA
Transfer-RNA *f* transfer [soluble] ribonucleic acid, tRNA, s-RNA
Transformation *f (phys ch)* transformation
Transformation *f*/**mikrobielle** *s.* Biokonversion
Transformationsintervall *n (phys ch)* transition interval, *(relating to glass:)* transformation range
Transformationstemperatur *f (glass)* glass-transition temperature *(where the viscosity of the glass is 10^{13} dPa · s)*
Transfusion *f* diffusion *(of gases through a porous diaphragm)*
Transglykosidierung *f (bioch)* transglycosylation
Transglykosylierung *f (bioch)* transglycosylation
Transient *m* (reaction) intermediate
Transition *f (bioch)* transition
trans-Konformation *f* trans conformation, T *(torsion angle 0°)*
Transkription *f (bioch)* transcription *(enzymatic transfer of genetic information from DNA onto RNA)*
Translation *f* 1. *(phys ch)* translation; translation (gliding) *(crystal gliding along a crystal plane)*; 2. *(bioch)* translation *(esp of genetic information into an amino-acid sequence)*
Translationsenergie *f* translational (kinetic) energy
Translationsfreiheitsgrad *m* translational degree of freedom, degree of translational freedom
Translationsgitter *n (cryst)* space lattice
Translationsverteilungsfunktion *f* translational partition function
translokal systemic *(pesticide)*
Transmembranprotein *n (bioch)* transmembrane protein
Transmetallierung *f (org ch)* transmetalation
Transmissions-Elektronenenergieverlust-Spektroskopie *f* transmission electron energy loss spectrometry, TEELS
Transmissionselektronmikroskop *n* transmission electron microscope, TEM
Transmissionselektronmikroskopie *f* transmission electron microscopy, TEM
Transmissionsgrad *m* transmission ratio, transmittance, transmittancy *(optics)*
Transmutation *f* transmutation
transparent transparent
transparent/unvollkommen translucent, translucid
Transparenz *f* transparency, light transmittance
Transparenz *f*/**unvollständige** translucence, translucency
transpirationshemmend *(cosmet)* antiperspirant
Transport *m* 1. transport, transfer *(as of electrons or heat)*; 2. *(bioch)* transport

Transport *m*/**chemischer** chemical transport
Transport *m*/**katalysierter** *(bioch)* mediated transport
Transport *m*/**passiv katalysierter** *(bioch)* facilitated diffusion
transportabel (trans)portable
Transportband *n* 1. conveyor [conveying] belt; 2. *(pap)* delivery tape *(of a cross-cutter)*
Transportbehälter *m* container
transportieren *v* 1. transport, transfer *(e.g. electrons or heat)*; 2. *(bioch)* transport
Transportkübel *m* skip (car)
Transportreaktion *f* transport reaction
Transportreaktor *m* transport [entrained-bed] reactor
Transportschnecke *f* conveyor [conveying] screw [worm]
Transportsystem *n* 1. *(phys ch, bioch)* transport system; 2. *(bioch)* carrier, porter, translocase
trans-ständig trans • **trans-ständig (angeordnet) sein** be trans
trans-taktisch transtactic *(polymer)*
trans-Taktizität *f* transtacticity *(of polymers)*
Transuran *n* transuranium element, transuranic [transuranian] element, uranoid
traubenförmig botryoidal
Traubensäure *f* racemic acid, racemic [inactive] tartaric acid, (±)-tartaric acid
Traubenzucker *m* grape sugar, *D*-glucose, dextrose *(a monosaccharide)*
Traube-Reaktion *f* Traube reaction *(for obtaining purines)*
traubig botryoidal
träufeln *v* trickle
Treibarbeit *f (met)* cupellation
Treibdampf *m* operating [motive] steam *(as in injector-type jet pumps)*
treiben *v (met)* cupel *(to extract gold and silver from lead)*
treiben *v*/**an die Oberfläche** buoy up
Treiben *n (met)* cupellation *(extraction of gold and silver from lead)*
Treibgas *n* 1. fuel gas; 2. propellant (gas), propellent *(for liquids)*
Treibgas *n* **für Aerosole** aerosol propellant
Treibhauseffekt *m* greenhouse effect
Treibhausgas *n* greenhouse gas
Treibmittel *n* 1. expanding agent; 2. *(plast, rubber)* blowing agent; gasifying agent *(for producing foam glass)*; 3. *(food)* leavening (agent), leaven; 4. pumping [motivating] fluid *(of a jet pump)*; 5. aerosol propellant; 6. *s.* Treibstoff 2.
Treibmittelpumpe *f* jet pump
Treibspiritus *m* power [fuel] alcohol
Treibstoff *m* 1. fuel, *(esp for rockets:)* propellant; 2. propellant (explosive), propellent, low explosive, deflagrating powder
Treibstoff *m*/**fester** solid fuel, *(esp for rockets:)* solid propellant
Treibstoff *m*/**hypergoler** hypergolic fuel [rocket propellant], hypergol
Trennaufgabe *f* separation task
trennbar separable
Trennbarkeit *f* separability
trennen *v* 1. *(ch)* separate; break, crack, demulsify *(emulsions)*; resolve *(racemates)*; 2. *(tech)* disconnect *(e.g. the joints of an apparatus)*; release *(mouldings from the mould)*
trennen *v*/**in Schichten** delaminate
trennen *v*/**nach Korn(größen)klassen** size(-separate), grade
trennen *v*/**sich** separate
trennen *v*/**sich nach Korn(größen)klassen** size-fractionate
Trennerfolg *m* separation efficiency
Trennfaktor *m (anal)* separation factor; relative centrifugal force *(of a centrifuge)*
Trennfiltration *f* solids recovery filtration
Trenngrad *m* degree of separation, separation efficiency
Trennkapillare *f* open tubular (capillary) column *(gas chromatography)*
Trennkorngröße *f* size [mesh] of separation, critical diameter, cut size [point]
Trennleistung *f* separation efficiency
Trennmethode *f* separation method
Trennmittel *n* 1. release agent, parting [separating, antitack] agent, abherent *(for mouldings)*; 2. *(distil)* separating agent; 3. *(min tech)* separating fluid; 4. *(chromat)* mobile solvent
Trennproblem *n* separation task
Trennrohrverfahren *n* thermal-diffusion method *(for separating isotopes)*

Trennrohrverfahren *n*/**clusiussches** thermal-diffusion method *(for separating isotopes)*
Trennsäule *f* 1. *(chromat)* analytical [separation] column; 2. *(distil)* rectifying [rectification] column
Trennschärfe *f* sharpness [degree] of separation, separation efficiency, selectivity
Trennstufe *f* 1. separation stage; 2. *(distil)* distillation stage
Trennstufe *f*/**praktische** *(distil)* actual plate [tray]
Trennstufe *f*/**theoretische** *(distil)* theoretical plate [tray], ideal [perfect] plate
Trennstufenhöhe *f* 1. *(anal, distil)* plate height; 2. *(distil)* height equivalent to a theoretical plate, HETP
Trennstufenzahl *f*/**theoretische** number of theoretical plates [trays]
Trenntechnik *f* separation technique
Trenntrichter *m* separatory [separating] funnel
Trennung *f* 1. *(ch)* separation; breaking, cracking, demulsification *(of emulsions)*; resolution *(of racemates)*; 2. *(tech)* disconnection *(as of the joints of an apparatus)*; release *(of mouldings from the mould)*
Trennung *f*/**elektrolytische** electrolytic separation
Trennung *f*/**flotative** *(min tech)* floatation separation
Trennung *f* **in Gruppen** *(anal)* group separation, separation into groups
Trennung *f* **in Schweretrüben** *(min tech)* heavy-medium [dense-medium] separation [cleaning]
Trennung *f* **mittels semipermeabler Membranen** *(hyd)* membrane separation
Trennung *f* **nach der Dichte** density separation [cut]
Trennung *f* **nach Korn(größen)klassen** grading (into size), size classification [grading, separation], sizing
6σ-Trennung *f* *(chromat)* 6σ separation
Trennungsgang *m* qualitative analysis scheme, separation scheme
Trennverfahren *n* 1. separation process; 2. *(lab)* separation procedure
Trennwand *f* separating [dividing] wall, partition; *(relating to diffusion:)* diaphragm, membrane, barrier, *(esp osmosis, dialysis:)* membrane, *(esp electrolysis:)* diaphragm
Trennwand *f*/**halbdurchlässige [semipermeable]** semipermeable membrane
Trennwirksamkeit *f* separation efficiency
Trester *pl (ferm)* marc, rape, (grape) pomace, pummace
treten *v*/**in Reaktion** enter into reaction
treten *v*/**in Wechselwirkung** interact
TRH *s.* Thyreotropin-Releasinghormon
Triacetat *n* 1. triacetate; 2. *(text)* cellulose triacetate, primary (cellulose) acetate
Triacetatfaser *f* cellulose triacetate fibre
Triacetatfaserstoff *m* cellulose triacetate fibre
Triade *f* triad *(in the periodic table)*
Triadenregel *f*/**döbereinersche** Döbereiner's law of triads
Trialkylaluminium *n* trialkylaluminium
Triarylmethanfarbstoff *m* triarylmethane dye
Triazin *n (org ch)* triazine
Triazinpolymer(e) *n* triazine polymer, polytriazine
Triazol *n (org ch)* triazole
Tribasizität *f* tribasicity *(of acids)*
Triblockcopolymer *n (plast)* triblock copolymer
Tribochemie *f* tribochemistry
Tribolumineszenz *f* triboluminescence
tribolumineszierend triboluminescent
Tribrommethan *n* tribromomethane, bromoform
Tributylzinnverbindung *f* tributyl tin (compound), TBT (compound)
Tricarbonsäure *f* tricarboxylic acid
Tricarbonsäurezyklus *m (bioch)* tricarboxylic-acid cycle, TCA cycle, citric-acid cycle, Krebs cycle
Trichlorbenzen *n* trichlorobenzene, TCB
Trichlorbenzoesäure *f* trichlorobenzoic acid
Trichlorderivat *n* trichloro derivative
Trichloressigsäure *f* trichloroacetic acid, TCA
Trichlorethan *n* trichloroethane, TCE
Trichlorethanal *n* *trichloroethanal, trichloroacetaldehyde, chloral
Trichlorid *n* trichloride

Trichlormethan *n* *trichloromethane, chloroform
2,4,5-Trichlorphenoxyessigsäure *f* 2,4,5-trichlorophenoxyacetic acid, 2,4,5-T *(a herbicide)*
2-(2,4,5-Trichlorphenoxy)propionsäure *f* 2-(2,4,5-trichlorophenoxy)propionic acid, fenoprop, 2,4,5-TP *(a herbicide)*
2,3,6-Trichlorphenylessigsäure *f* 2,3,6-trichlorophenylacetic acid, fenac *(a herbicide)*
Trichlorsilan *n* trichlorosilane
α-**Trichlortoluen** *n* α,α,α-trichlorotoluene, benzotrichloride
α-**Trichlortoluol** α,α,α-trichlorotoluene, benzotrichloride
Trichlor-*sym*-triazin *n* 2,4,6-trichloro-1,3,5-triazine, cyanuric chloride
Trichroismus *m (cryst)* trichroism
Trichter *m* 1. funnel; 2. *(tech)* hopper; cup *(of a blast furnace)*
Trichter *m* **mit langem Rohr [Stiel]** long-stem(med) funnel, Bunsen funnel
Trichter *m* **nach Hirsch** Hirsch funnel
Tricresylphosphat *n* tritolyl phosphate, tricresyl phosphate, TCP
Triebkraft *f* driving force [potential], (chemical) affinity *(of a reaction)*
Trien *n* triene *(any of a class of hydrocarbons containing three carbon double bonds)*
Triester *m (org ch)* triester
Triethanolamin *n* triethanolamine, TEA, tri(2-hydroxyethyl)amine
Triethylaluminium *n* triethylaluminium
Triethylamin *n* triethylamine
Triethylcellulose *f* ethylcellulose
Triethylenglykol *n* triethylene glycol, TEG
Triethylphosphat *n* triethyl phosphate
Trifluoressigsäure *f* trifluoroacetic acid
Trifluormethansulfonsäure *f* trifluoromethane sulphonic acid, triflic acid
trifunktionell trifunctional
trig. *s.* trigonal
Triglycerid *n* triglyceride
Triglycerid *n*/**einsäuriges** simple triglyceride
Triglycerid *n*/**gemischtes [gemischtsäuriges]** mixed triglyceride
Triglykol *n s.* Triethylenglykol
trigonal *(cryst)* trigonal, trig.
Trihalogenid *n* trihalide, trihalogenide
Trihalo(gen)methan *n* 1. trihalomethane; 2. *(hyd)* THM
Trihydrat *n* trihydrate
Trihydroxid *n* trihydroxide
Triiodmethan *n* triiodomethane, iodoform
Trikaliumorthophosphat *n* tripotassium orthophosphate, potassium phosphate
Trikaliumphosphat *n* tripotassium orthophosphate, potassium phosphate
trikl. *s.* triklin
triklin *(cryst)* triclinic, tric., anorthic
Trimellithsäure *f* trimellitic acid, *benzene-1,2,4-tricarboxylic acid
trimer trimeric
Trimer(e) *n* trimer
Trimerisation *f* trimerization
trimerisieren *v* trimerize; trimerize, undergo trimerization
Trimethylaluminium *n* trimethylaluminium
Trimethylamin *n* trimethylamine
Trimethylammonioacetat *n NNN*-trimethylammonioacetate, betaine, oxyneurine, lycine
Trimethylsilylgruppe *f* $-Si(CH_3)_3$ trimethylsilyl group
Trimethylsilylierung *f* trimethylsilylation
trimolekular trimolecular
Trinatriumorthophosphat *n* trisodium orthophosphate, sodium phosphate
Trinatriumphosphat *n* trisodium orthophosphate, sodium phosphate
Trinatriumphosphatverfahren *n (hyd)* phosphate treatment
Trinatriumsalz *n* trisodium salt
Trinitrat *n* trinitrate
Trinitrid *n* trinitride
Trinitrierung *f* trinitration
2,4,6-Trinitrophenol *n* 2,4,6-trinitrophenol, picric acid
2,4,6-Trinitroresorcinol *n* 2,4,6-trinitroresorcinol, styphnic acid
Trinitrotoluen *n* trinitrotoluene, TNT
Trinitrotoluol *n* trinitrotoluene, TNT
trinkbar potable
Trinkbranntwein *m* potable spirit, (distilled) beverage spirit, beverage alcohol
Trinkwasser *n* drinking [potable] water, *(specif)* municipal (drinking) water, city water

Trinkwasseraufbereitung *f* drinking-water treatment
Trinkwasserentkeimung *f* drinking-water disinfection
Trinkwasser-Pasteurisieranlage *f* water pasteurizer
Trinkwasserschutzzone *f* aquifer protection zone
Trinkwasser-Standard *m* drinking-water standards [regulations]
Triol *n* triol, trihydric alcohol
Triolein *n* triolein, glycerol trioleate, olein
Triose *f* triose *(monosaccharide containing three carbon atoms per molecule)*
1,3,5-Trioxan *n* 1,3,5-trioxan, trioxymethylene, metaformaldehyde
Trioxid *n* trioxide
Trioxoborat *n* $M^I_3BO_3$ trioxoborate, orthoborate, borate
Trioxoborsäure *f* trioxoboric acid, orthoboric acid, boric acid
Trioxokieselsäure *f* trioxosilicic acid, metasilicic acid
Trioxosilicat *n* M^I_2SiO trioxosilicate, metasilicate
Tripalmitin *n* tripalmitin, glycerol tripalmitate
Tripeleffektverdampfer *m* triple-effect evaporator [evaporating unit]
Tripelhelix *f (bioch)* triple(-stranded) helix, superhelix
Tripelpunkt *m* triple point
Tripelsalz *n* triple salt
Tripeptid *n* tripeptide
Triphenylaminfarbstoff *m* triphenylamine dye
Triphenylmethanfarbstoff *m* triphenylmethane dye
Triphenylphosphat *n* triphenyl phosphate
Triphosphat *n* $M^I_5P_3O_{10}$ triphosphate
Triplett *n* 1. *(phys ch, spectr)* triplet; 2. *(bioch)* triplet (codon)
Triplett-Singulett-Übergang *m (phys ch)* triplet-singlet transition
Triplett-Triplett-Übergang *m (phys ch)* triplet-triplet transition
Triplettzustand *m* triplet (electronic) state
Trisaccharid *n* trisaccharide
Trisauerstoff *m s.* Ozon
Trisilicat *n* trisilicate
Trispiro-Verbindung *f* trispiro compound [hydrocarbon]
Trispuffer *m (bioch)* tris buffer
Tristearin *n* tristearin, glyceryl tristearate
trisubstituiert trisubstituted
Trisubstitutionsprodukt *n* trisubstitution product
Trisulfid *n* trisulphide
Trisulfonsäure *f* trisulphonic acid
Triterpen *n (org ch)* triterpene
Triterpenoid *n (org ch)* triterpenoid
Trithionat *n* $M^I_2S_3O_6$ trithionate
tritiieren *v* tritiate
Tritiierung *f* tritiation
Tritium *n* T, ^3_1H, tritium • **mit Tritium markiert** tritium-labelled
Triton *n (nucl)* triton
Tritylfarbstoff *m* triarylmethane dye
Tritylierung *f* tritylation, triphenylmethylation
trivalent trivalent, tervalent
Trivalenz *f* trivalency, tervalency
trivariant trivariant
Trivialname *m* trivial [common, unsystematic] name
t-RNA *s.* Transfer-Ribonucleinsäure
trocken dry
trocken/absolut bone-dry, oven-dry, oven-dried, OD
trocken und mineral(stoff)frei dry and mineral-matter-free, d.m.m.f., D.M.F.
Trockenablagerung *f* dry deposition *(of pollutants from the atmosphere)*
Trockenanlage *f* drying plant
Trockenapparat *m* 1. dryer; 2. *(lab)* drying apparatus
Trockenaufbereitung *f* 1. *(min tech)* dry [pneumatic] cleaning; 2. *(ceram)* dry preparation [mixing, mix]
Trockenbatterie *f* dry battery
Trockenbestandteil *m* solid constituent
Trockenbiomasse *f (biot)* dry weight of biomass
Trockenblech *n* (drying) tray
trockencyanieren *v* dry-cyanide, gas-cyanide, carbonitride *(steel)*
Trockencyanieren *n* dry [gas] cyaniding [cyanization], carbonitriding, nicarbing, nitrocementation *(of steel)*
Trockendampf *m* dry (saturated) steam
Trockendestillation *f* dry [destructive] distillation

Trockeneis *n* dry ice, carbon dioxide ice [snow], solid carbon dioxide
Trockenelement *n* dry cell
Trockenentschwefelung *f* dry desulphurization *(of gas)*
Trockenfilter *n* dry filter
Trockengas *n* dry (natural) gas *(main constituent methane)*
Trockengeschwindigkeit *f* drying rate
Trockengewicht *n* dry weight
Trockengut *n* material being [to be] dried; dry product
Trockenhefe *f* dried [dry] yeast, yeast powder
Trockenheit *f* dryness
Trockenhorde *f* (drying) tray
Trockenkammer *f* drying room [chamber], *(if small:)* drying cabinet [box]
Trockenkugeltemperatur *f* dry-bulb temperature
Trockenlöscher *m* dry chemical (fire) extinguisher, powder extinguisher
Trockenmahlung *f* dry milling [grinding]
Trockenmasse *f* 1. solids; 2. (bone-)dry weight, moisture-free weight
Trockenmasse *f***/fettfreie** *(food)* non-fat(ty) solids, solids-non-fat, S.N.F.
Trockenmilch *f* dried [dry] milk, milk powder
Trockenmittel *n* desiccant, desiccating agent, drying [dehydrating] agent [medium], dehydrator, dehumidifier
Trockenofen *m* 1. drying oven; 2. *(ceram)* drying kiln; 3. *(coat)* stoving [baking] oven
Trockenpistole *f* (vacuum) drying pistol
Trockenreinigung *f* dry cleaning *(1. of gases; 2. of clothes)*
Trockenrohr *n s.* Trockenpistole
Trockenschale *f* drying tray
Trockenschlamm *m (hyd)* dry [dried] sludge
Trockenschrank *m (lab)* drying oven, (air) oven • **im Trockenschrank getrocknet** oven-dry, oven-dried, OD • **im Trockenschrank trocknen** oven-dry
Trockensieben *n* dry screening [sieving]
Trockensubstanz *f* dry [solid] matter, solids, dry residue [substance]
Trockentemperatur *f* drying temperature
Trockenturm *m* 1. drying tower; 2. *(lab)* (gas) drying jar
Trockenvermahlung *f* dry milling [grinding]
Trockenzentrifuge *f* centrifugal dryer, whizzer, hydroextractor
Trockne *f* dryness • **zur Trockne eindampfen** evaporate to dryness
Trockne *f***/beginnende** incipient [moist] dryness
trocknen *v* dry, dehydrate, desiccate, *(esp relating to gases:)* dehumidify; season *(wood)*
trocknen *v***/an der Luft** air-dry
trocknen *v* **bis zur Gewichtskonstanz [Massekonstanz]** dry to constant weight
trocknen *v***/im Ofen** *(coat)* stove
trocknen *v***/im Sprühverfahren** spray-dry
trocknen *v***/im Trockenschrank** oven-dry
trocknen *v***/im Vakuum** vacuum-dry
trocknen *v***/thermisch** stove, bake
trocknen *v***/unvollständig [unzureichend]** underdry
trocknend/chemisch [durch chemische Reaktion] convertible *(organic coating)*
trocknend/physikalisch non-convertible *(organic coating)*
Trockner *m* 1. dryer; 2. *(lab)* drying apparatus; 3. *s.* Trockenmittel
Trocknung *f* drying, dehydration, desiccation, *(esp relating to gases:)* dehumidification; seasoning *(of wood)*
Trocknung *f* **an der Luft** air drying
Trocknung *f* **bis zur Gewichtskonstanz [Massekonstanz]** drying to constant weight
Trocknung *f***/thermische** thermal drying, stoving
Trocknung *f***/unvollständige [unzureichende]** underdrying
Trocknungskurve *f* drying curve
Trog *m* trough, vat, tank, tub
Trogkneter *m* trough [open-pan] mixer
Trogmischer *m* trough [open-pan] mixer
Trogpresse *f* pot press
Trogtränkung *f* steeping *(of timber)*
Trombenbildung *f* vortex formation
Trommel *f* drum, cylinder; bowl, basket *(of a centrifuge)*

Trommelfilter *n* (rotary-)drum filter
Trommelmischer *m* drum mixer, barrel blender [mixer]
Trommelmühle *f* drum mill
trommeln *v (plast, coal)* tumble *(mass-production parts)*
Trommelsaugfilter *n* rotary-drum vacuum filter, (rotary) vacuum drum filter
Trommelsieb *n* drum [trommel, revolving] screen
Trommeltrockner *m* rotary (drum) dryer, rotatory dryer • **im Trommeltrockner trocknen** drum-dry
Trommelzentrifuge *f* bowl [basket] centrifuge
Trommsdorff-(Norrish-)Effekt *m (plast)* Trommsdorff effect, gel effect
Tropa(n)alkaloid *n* tropane alkaloid
Tröpfchen *n* droplet
Tröpfchengröße *f* droplet size
Tropfelektrode *f* drop(ping) electrode
tröpfeln *v* trickle
tropfen *v* drop, drip
Tropfen *m* 1. drop; 2. *(coat)* tear, drip
Tropfen *m***/hängender** pendant drop
Tropfen *m***/liegender** captive drop
Tropfenformanalyse *f* drop shape analysis
Tropfenformanalyse *f***/asymmetrische** asymmetric drop shape analysis, ADSA
tropfenweise dropwise, drop by drop
Tropfenzähler *m* 1. *(lab)* drop counter; 2. *s.* Tropfflasche; 3. *s.* Tropfpipette
Tropfer *m s.* Tropfpipette
Tropfflasche *f (lab)* dropping bottle
Tropfglas *n (lab)* dropping bottle
Tropfkörper *m (hyd)* trickling [percolating] filter, biological filter, biofilter, fixed-film reactor, packed-bed biological film reactor
Tropfkörperreaktor *m (biot)* trickling film reactor, trickle-flow fermenter [reactor]
Tropfpipette *f (lab)* dropping [teat] pipette, (medicine) dropper
Tropfpunkt *m* drop(ping) point
Tropftrichter *m (lab)* dropping [tap] funnel
tropfwassergeschützt drip-proof
Trophophase *f (biot)* trophophase *(growth of microorganisms)*
Tropiliden *n* tropilidene, cyclohepta-1,3,5-triene
Tropinalkaloid *n* tropane alkaloid
Tropolon *n (org ch)* tropolone
Tropon *n* tropone, cyclohepta-2,4,6-trienone
Tropylium-Ion *n* tropylium ion, cycloheptatrienylium ion
Trp *s.* Tryptophan
Trub *m (ferm)* sediment, settling(s), *(relating to wine also)* lees, *(brewing also)* trub
trüb(e) opaque, *(esp relating to liquids:)* turbid, hazy, cloudy, feculent; dull, dusky *(shade)*
Trübe *f* 1. slurry, pulp; 2. *(filtr)* prefilt, prefilt slurry [feed], material being [to be] filtered; fluid pulp *(classifying)*; 3. *(min tech)* pulp (of ore); 4. *s.* Trub; 5. *s.* Trübheit
Trübglas *n* opaque [opal] glass
Trübheit *f* opacity, *(esp of liquids:)* turbidity, haziness, cloudiness, feculence; dullness, duskiness *(of a colour)*
Trüblauf *m (filtr)* bleeding
Trübstoffe *mpl (hyd)* turbidity-causing solids, turbidity, turbidities
Trübstoffelimination *f (hyd)* removal of turbidity, turbidity removal
Trübstoffentfernung *f (hyd)* removal of turbidity, turbidity removal
Trübung *f* 1. *s.* Trübheit; 2. haze *(in beer)*; casse *(in wine)*; 3. opacifying, *(esp of liquids:)* clouding
Trübung *f***/milchige** milkiness
Trübungsanalyse *f* 1. turbidimetric analysis; 2. *s.* Turbidimetrie
Trübungsmesser *m* turbidimeter, turbidometer, nephelometer
Trübungspunkt *m* 1. turbidity [cloud] point *(of solutions or oils)*; 2. *(petrol)* cloud point; titre *(of a soap)*
Trübungstitration *f* turbidimetric [turbidity] titration
Trum *m(n)* strand *(of a conveyor)*
trypanozid (wirkend) *(pharm)* trypanocidal *(tending or used to destroy trypanosomes)*
Trypanrot *n* trypan red *(a biological stain)*
Trypsin *n* trypsin *(a proteolytic enzyme or a preparation containing proteolytic enzymes)*
Tryptophan *n* tryptophane, 2-amino-3,3′-indolylpropionic acid
TS *s.* Trockensubstanz

Tschitschibabin-Kohlenwasserstoff *m* Chichibabin hydrocarbon
Tschugajew-Reaktion *f* Chugaev [Tschugaeff] reaction *(for obtaining alkenes)*
T-s-Diagramm *n* temperature-entropy chart, entropy chart [diagram]
TSH *s.* Hormon/thyreotropes
T-Stück *n* T-shape connecting tube, T-piece, T-type connector, tee (connector) • **mit einem T-Stück verbinden** tee
tt *s.* trans-taktisch
Tuberkulosemittel *n* antitubercular [antituberculosis] drug
Tuberkulostatikum *n* antitubercular [antituberculosis] drug
Tubocurare *n* tubocurare, tube [pot] curare
Tubus *m* beak *(of a retort)*
Tuch *n* (woven) fabric, *(if of definite size:)* cloth
Tuchfilter *n* cloth [fabric] filter, woven(-fabric) filter
Tulpe *f (lab)* crucible adapter
tumorerzeugend tumour-causing, tumorigenic, oncogenic, oncogenous
tumorhemmend antitumour
Tunneleffekt *m (phys ch)* tunnel effect
Tunnelofen *m* 1. tunnel furnace; 2. *(ceram)* tunnel kiln
Tunnelung *f (phys ch)* tunnel(l)ing *(penetration of the energy barrier)*
Tüpfelanalyse *f* spot-test analysis, drop analysis
tüpfeln *v* spot, test by spotting
Tüpfelplatte *f* spot [spotting, cavity] plate
Tüpfelprobe *f* spot (plate) test, drop test
Tüpfelreaktion *f* spot [drop] reaction
Turbidimeter *n* turbidimeter, turbidometer, nephelometer
Turbidimetrie *f* turbidimetry, turbidimetric analysis
turbidimetrisch turbidimetric
Turbinenrührer *m* turbine mixer [agitator, impeller]
Turbinentreibstoff *m* turbine fuel
Turbinentreibstoff *m* **für Flugturbinen** aviation [aircraft] turbine fuel
Turbomischer *m* turbine mixer [agitator, stirrer]
Turborührer *m* turbine mixer [agitator, stirrer]
turbulent turbulent
Turbulenz *f* turbulence
Türkischrotöl *n* Turkey-red oil, sulphated castor oil
Turm *m* tower
Turmfermentation *f (biot)* tower fermentation
Turmfermenter *m (biot)* tower reactor [fermenter]
Turmkühler *m* cooling tower
Turmreaktor *m s.* Turmfermenter
Turmwäscher *m* tower scrubber, scrubbing [gas-washing] tower, (gas) scrubber
Turnover *n (bioch)* molar [molecular] activity, (metabolic) turnover
Twistbootkonformation *f* twist conformation
Twisting-Schwingung *f (spectr)* twisting vibration, torsional oscillation [mode]
Tyndall-Effekt *m (coll)* Tyndall effect [phenomenon]
Tyndall-Kegel *m (coll)* (Faraday-)Tyndall cone
Tyndall(o)meter *n (coll)* tyndallimeter, tyndallometer, Tyndall meter
Tyndallometrie *f (coll)* tyndallimetry, tyndallometry
Tyr *s.* Tyrosin
Tyramin *n* tyramine, 2-(*p*-hydroxyphenyl)ethylamine
Tyrosin *n* tyrosine, 2-amino-3-(*p*-hydroxyphenyl)propionic acid
T-Zustand *m s.* Triplettzustand

U

Ubbelohde-Viskosimeter *n* Ubbelohde viscometer
überbelichten *v (phot)* overexpose
Überbelichtung *f (phot)* overexposure
überbleichen *v* overbleach
überbrennen *v* 1. overburn *(e.g. cement clinker)*; 2. *(ceram)* overfire; 3. *(coat)* overstove
überbrücken *v* bridge
Überbrückung *f* bridging
überdestillieren *v* distil over *(of human agent)*; distil (over), pass [come] over *(of a distillate)*

überdosieren *v* overdose, overfeed
Überdosierung *f* overdosage, overfeed(ing)
Überdosis *f* overdose
Überdruck *m* excess(ive) pressure, overpressure
Überdruckventil *n* pressure relief valve
überdüngen *v* fertilize excessively, overfertilize
übereinanderlagern *v* superimpose *(stereochemistry)*
überempfindlich *(med, phot)* hypersensitive, supersensitive
Überempfindlichkeit *f (med, phot)* hypersensitivity, hypersensitiveness, supersensitivity
überentwickeln *v (phot)* overdevelop
übereutektisch hypereutectic
übereutektoid(isch) hypereutectoid
überfärben *v* 1. *(text)* overdye *(unintentionally)*; 2. *(text)* dye over, overdye; cross-dye *(a second component in fibre mixtures)*; 3. *(text)* top *(by applying further dye to achieve a desired final shade)*
überfluten *v* flood *(a column unintentionally)*
überführen *v* 1. *(chemically:)* convert; 2. *(mechanically:)* transfer
überführen *v*/**in den industriellen Maßstab** scale up to industrial use
überführen *v*/**in ein Salz** salify *(an acid or a base)*
überführen *v*/**in einen Komplex** complex
Überführung *f* 1. *(chemically:)* conversion; 2. *(el ch)* transference; 3. *(mechanically:)* transfer
Überführungszahl *f (el ch)* transference [transport] number
Überführungszahl *f*/**anomale** abnormal transference number
Überführungszahl *f*/**hittorfsche** Hittorf number
Überführungszahl *f*/**wahre** true transference number
Übergang *m* 1. transition, conversion, change *(into another state or modification)*; 2. *(el ch)* transference; change *(of colour)*; 3. junction *(in semiconductors)*
Übergang *m*/**antiferroelektrischer** antiferroelectric transition
Übergang *m*/**erlaubter** allowed transition *(of electrons)*
Übergang *m*/**ferroelektrischer** ferroelectric transition
Übergang *m* **in den Spinglaszustand** spin-glass transition
Übergang *m*/**kongruenter** congruent transition
Übergang *m*/**konzertierter** *(bioch)* concerted transition [transfer]
Übergang *m*/**magnetischer** magnetic transition
Übergang *m*/**spinverbotener** spin-forbidden transition
Übergang *m*/**strahlender** radiative transition
Übergang *m*/**strahlungsfreier [strahlungsloser]** radiationless transition
Übergang *m*/**verbotener** forbidden transition *(of electrons)*
Übergang *m* **vom flüssigen in den festen Zustand** fluid-solid transition
***pn*-Übergang** *m* p-n junction *(in semiconductors)*
Übergangsbereich *m* transition region; transition interval *(of a colour indicator)*
Übergangsmetall *n* transition(al) metal, transition element
Übergangsphase *f (bioch)* transient state
Übergangsreaktion *f*/**eutektische** eutectic reaction
Übergangsreaktion *f*/**eutektoide** eutectoid reaction
Übergangsreaktion *f*/**monotektische** monotectic reaction
Übergangsreaktion *f*/**monotektoide** monotectoid reaction
Übergangsreaktion *f*/**peritektische** peritectic [incongruent] reaction
Übergangsreaktion *f*/**peritektoide** peritectoid reaction
Übergangsreihe *f* transition series *(in the periodic table)*
Übergangsstück *n (distil)* adapter
Übergangsstück *n* **von Kegelschliff auf Kugelschliff** conical-spherical adapter
Übergangsstück *n* **von Kugelschliff auf Kegelschliff [Normschliff]** spherical-conical adapter
Übergangswiderstand *m* transfer resistance, resistance to transfer
Übergangszustand *m* transition state,

(reaction theory also) activated complex
übergehen *v* 1. pass, be converted; change *(into another state or modification)*; 2. *(distil)* come [pass] over, distil (over); 3. *(el ch)* transfer; change *(in colour)*
übergehen v/gemeinsam distil together
übergehen v/in den Fließbettzustand fluidize
übergehen v/in den Gelzustand *(coll)* gel
übergehen v/in den Solzustand *(coll)* solate
Übergerbung *f* overtannage
überhärten *v (plast)* overcure
Überhärtung *f (plast)* overcure, overcuring
überhitzen *v* superheat, overheat
Überhitzung *f* superheating, overheating
überkochen *v* boil over
Überkochen *n* boilover
überkohlen *v* supercarburize, overcarburize *(steel)*
Überkopfprodukt *n (distil)* overhead [top] product, overhead(s)
Überkorn *n* oversize (material, product), tailings, tails, plus material *(classifying)*; *(screening also)* screen oversize, plus mesh
überkritisch *(nucl)* supercritical
Überlack *m (coat)* clearcoat
überlagern *v* superpose, superimpose
Überlagerung *f* superposition, *(act)* superimposition; spectral overlap
überlappen *v* overlap *(1. orbital; 2. genetic code)*
überlappen v/sich overlap
Überlappung *f* overlap(ping) *(1. of orbitals; 2. of genetic code)*
Überlappungskonzentration *f* overlapping concentration *(in polymer solutions)*
überlasten *v* overload
Überlauf *m* 1. overflow (weir, lip, port), weir; 2. overflow (fraction, product), overs, tailings, tails *(classifying)*; 3. overflow rate
überlaufen *v* run over, overflow; *(with heat:)* boil over
Überlaufgut *n s.* Überlauf 2.
Überlaufprodukt *n s.* Überlauf 2.
Überlebensdauer *f (tox)* survival time
Überlebensfaktor *m (biot)* survival factor
überleiten *v* pass [carry] over *(e.g. vapour)*
übermahlen *v* overgrind
Überproduktion *f* over-production, excess production
übersättigen *v* supersaturate
Übersättigung *f* supersaturation, supersolubility
Übersäure *f* peracid
überschäumen *v* froth [foam] over
überschichten *v* cover with a layer *(of another liquid)*
überschreiten *v* exceed, overrun, overshoot *(a specified value)*
Überschuss *m* excess, surplus • **mit einem geringen Überschuss an** in slight excess of
Überschusschlor(ier)ung *f* excess chlorination, superchlorination *(of water)*
Überschusselektron *n* excess electron
Überschussenergie *f* excess energy
Überschusshalbleiter *m* n-type semiconductor
überschüssig excess(ive)
Überschussladung *f* excess charge
überschussleitend n-type (semi)conducting
Überspannung *f* overvoltage, overpotential, excess potential
Überspannung f/elektrochemische [elektrolytische] overvoltage, overpotential, excess potential
Überstand *m* 1. supernatant (liquid, liquor); 2. *(biot)* culture supernatant
überstöchiometrisch hyperstoichiometric
Überstruktur *f (cryst)* superstructure, hyperstructure, superlattice
übertitrieren *v* overtitrate, overrun [overshoot] the end point
übertragen *v* transfer
Überträger *m* carrier; chain-transfer agent *(polymerization)*
Übertragung *f* transfer(ence)
Übertragungsreaktion *f* transfer reaction *(polymerization)*
übertreiben *v (distil)* carry [distil, pass] over
übertrocknen *v* overdry

Übertrocknung *f* overdrying, excessive drying
Übervulkanisation *f* overcure, overcuring, over-vulcanization
übervulkanisieren *v* overcure, over-vulcanize
überwachen *v* monitor *(e.g. an experiment)*; supervise *(by an authority)*
Überwachsung *f (cryst)* overgrowth
Überwachung *f* monitoring *(as of an experiment)*; supervision *(by an authority)*
überziehen *v* 1. cover *(with prefabricated sheet)*; *(esp with sheet metal:)* clad; 2. *s.* beschichten
überziehen *v*/**sich** become covered
Überzug *m* 1. cover(ing) *(consisting of prefabricated sheet)*; *(esp consisting of sheet metal:)* cladding; 2. *s.* Schutzschicht
UDP *s.* Uridindiphosphat
Ugi-Reaktion *f* Ugi reaction *(for obtaining α-amino acids or peptides)*
Uhrglas *n (lab)* watch [clock] glass
Uhrglasschale *f (lab)* watch [clock] glass
Uhrzeigersinn/entgegen dem counterclockwise
Uhrzeigersinn/im clockwise
UHT-Verfahren *n s.* Ultrahocherhitzungsverfahren
UKS = Ultrakurzzeitspektroskopie
Ullmann-Reaktion *f* Ullmann reaction *(for synthesizing diaryls)*
Ultrafilter *n* ultrafilter
Ultrafiltration *f* ultrafiltration, UF
ultrafiltrieren *v* ultrafilter
Ultrahocherhitzungsverfahren *n (food)* ultra-high temperature process
Ultrahochvakuum *n* ultra-high vacuum
Ultramikroanalyse *f* ultramicroanalysis
Ultramikrochemie *f* ultramicrochemistry
Ultramikroprobe *f* ultramicro sample *(*< 10^{-4} g)*
Ultramikroskop *n* ultramicroscope
ultramikroskopisch ultramicroscopic
ultrarein ultrapure, extremely pure
Ultraschallaufschluss *m (bioch)* cell breakage by sonication
Ultraschallbad *n* ultrasonic bath
Ultraschallprüfung *f* ultrasonic [supersonic] testing
Ultraschallreinigung *f* ultrasonic cleaning
Ultra-Spuren-Analyse *f* ultra-trace analysis *(*sample weight ≤ 10^{-4} g)*
Ultraviolett *n* ultraviolet (radiation), uv
Ultraviolett *n*/**nahes** near ultraviolet
Ultraviolett... *siehe: UV-...*
Ultrazentrifuge *f* ultracentrifuge, high-speed centrifuge
Ultrazentrifugieren *n* ultracentrifugation, high-speed centrifugation
umestern *v* interesterify, transesterify, re-esterify
Umesterung *f* interesterification, transesterification, re-esterification, interchange esterification, ester interchange
umfällen *v* reprecipitate
Umfällung *f* reprecipitation
Umfangsgeschwindigkeit *f* peripheral speed
umformen *v* form *(materials)*; change *(a structure)*; convert, transform *(substances)*
umformen *v*/**sich** change *(of a structure)*
Umformung *f* forming *(of materials)*; change *(of a structure)*; conversion, transformation *(of substances)*
Umformverfahren *n (tech)* sheet metal forming *(manufacturing process)*
umfüllen *v* transfer *(substances into another vessel)*
Umfüllen *n* transfer *(of substances)*
Umgang *m* handling *(as of dangerous materials)*
umgebend ambient
Umgebung *f* environment; vicinity
Umgebungstemperatur *f* ambient temperature
umgesetzt/nicht unreacted, unconverted
umgesetzt werden *v* undergo conversion
umhausen *v* house *(a treatment unit)*
umkehrbar reversible, invertible
umkehrbar/nicht non-reversible, irreversible
Umkehrbarkeit *f* reversibility, invertibility
umkehren *v* reverse, invert *(a process)*
Umkehrosmose *f* reverse osmosis, RO
Umkehrphasenchromatographie *f* reversed-phase chromatography, RPC
Umkehrung *f* reversion, reversal, inversion

Umkehrung f/waldensche *(org ch)* Walden inversion
Umklonieren *n (bioch)* subcloning
Umkristallisation *f* recrystallization
umkristallisieren *v* recrystallize
umlagern *v* rearrange
umlagern v/sich rearrange
Umlagerung *f* (molecular) rearrangement
Umlagerung *f* **am Aromaten** aromatic rearrangement
Umlagerung f/beckmannsche Beckmann rearrangement [molecular transformation] *(of a ketoxime into an amide derivative)*
Umlagerung f/friessche Fries rearrangement [migration] *(as for synthesizing phenolic ketones)*
Umlagerung f/innermolekulare [intramolekulare] intramolecular rearrangement
Umlagerung f/lossensche Lossen rearrangement *(of aromatic hydroxamic acids into isocyanates)*
Umlagerung f/photochemische photorearrangement
Umlagerung f/sigmatrope *(org ch)* sigmatropic rearrangement
1,2-Umlagerung *f* 1,2-shift
Umlauf *m* (re)circulation
Umlauf *m*/**natürlicher** natural circulation
Umlaufbahn *f* orbit
umlaufen *v* circulate *(in a closed cycle)*; rotate, revolve, *(if rapidly:)* spin
Umlaufgas *n* recycle gas
Umlaufkolbenpumpe *f* rotary pump
Umlaufsystem *n* circulation system
Umlaufverdampfer *m* 1. circulation evaporator; 2. *(distil)* forced-circulation reboiler
Umlenkblech *n* baffle (plate) • **mit Umlenkblechen (versehen)** baffled • **ohne Umlenkbleche** unbaffled
ummanteln *v* jacket, sheathe
Ummantelung *f* jacket, sheath(ing)
UMP = Uridin-5-monophosphat
umpumpen *v* recirculate, recycle
Umrechnungsfaktor *m* conversion factor
Umrechnungstabelle *f* conversion [converting] table
umrühren *v* agitate, *(esp relating to molten metal:)* puddle
Umrühren *n* agitation, *(esp relating to molten metal:)* puddling
Umsatz *m* conversion
Umsatzgeschwindigkeit *f* conversion rate
Umsatzgleichung *f* chemical equation, reaction equation
Umsatzgleichung f/chemische chemical equation, chemical-reaction equation
Umsatzkurve *f* conversion curve
Umsatzrate *f (biot)* conversion [fermentation] rate, rate of product formation
Umsatzreaktion f/doppelte double-decomposition reaction, metathesis [metathetical] reaction
Umschlag *m* break *(as in titration)*; change(over), transition *(of an indicator)*
umschlagen *v* 1. change *(of colours or reactions)*; 2. reverse *(of emulsions)*
Umschlagen *n* 1. change *(of colours or reactions)*; 2. reversion *(of emulsions)*
Umschlagsintervall *n* transition [colour-change] interval *(of a pH indicator)*
Umschlagspunkt *m* point of change *(titration)*
umschmelzen *v* 1. remelt; 2. *(met)* refuse
umschwenken *v* swirl
umsetzen *v* 1. convert; 2. *(bioch)* metabolize
umsetzen v/sich 1. react, convert; 2. *(bioch)* metabolize
Umsetzung *f* conversion • **(eine) Umsetzung erleiden** undergo conversion
Umsetzung f/doppelte double [mutual] decomposition, metathesis
Umsetzung f/enzymatische enzym(at)ic conversion
Umsetzung f/gegenseitige interconversion (reaction)
Umsetzung f/stoffliche conversion of materials, materials conversion
Umsetzungsgeschwindigkeit *f* conversion rate [velocity]
Umsetzungsgrad *m* degree of conversion
Umsetzungsprodukt *n* 1. conversion product; 2. *(biot)* fermentation product
Umsetzungsreaktion *f* conversion reaction
umstellen *v* 1. rearrange; 2. reverse *(e.g. the direction of flow)*

Umstellung *f* 1. rearrangement; 2. reversal *(as of the direction of flow)*
umwälzen *v* circulate
Umwälzpumpe *f* circulating pump, recirculation [recycle] pump, circulator
Umwälzung *f* circulation
umwandelbar convertible, transformable
umwandelbar/ineinander [wechselseitig] (mutually) interconvertible, reversibly convertible; enantiotropic *(modifications)*
Umwandelbarkeit *f* convertibility, transformability
Umwandelbarkeit *f***/wechselseitige** interconvertibility; enantiotropy *(of modifications)*
umwandeln *v* 1. transform, convert, change; reform *(hydrocarbons)*; 2. *(nucl)* transform, transmute; 3. *(cryst)* invert
umwandeln *v***/gegenseitig** interconvert
umwandeln *v***/in Dextrin** dextrinate, dextrinize
umwandeln *v***/in ein Carbonat** carbonatize, carbonate
umwandeln *v***/in Kohlenstoff** carbonize
umwandeln *v***/sich** 1. convert, transform, revert, undergo change, be [become] converted; 2. *(nucl)* transform, transmute, undergo transformation [transmutation]; 3. *(cryst)* invert
umwandeln *v***/sich gegenseitig** interconvert
umwandeln *v***/sich in die metallische Form [Modifikation]** metallize
umwandeln *v***/sich ineinander [wechselseitig]** interconvert
Umwandlung *f* 1. transformation, conversion, change; reforming *(of hydrocarbons)*; 2. *(nucl)* transformation, transmutation, *(by decay also)* devolution; transition *(into another phase)*; 3. *(cryst)* transition, inversion • **(eine) Umwandlung erleiden** undergo conversion
Umwandlung *f***/allotrope** allotropic [allotriomorphic] transition *(of a pure element)*
Umwandlung *f***/biochemische** biochemical transformation
Umwandlung *f***/diffusionslose** diffusionless transition
Umwandlung *f***/diskontinuierliche** *s.* Umwandlung 1. Ordnung
Umwandlung *f***/displazive** displacive transition
Umwandlung *f***/druckinduzierte** pressure-induced transition
Umwandlung *f***/enantiotrope** enantiotropic transition
Umwandlung *f***/enzymatische** enzym(at)ic conversion
Umwandlung *f***/ferrimagnetische** ferrimagnetic transition
Umwandlung *f***/ferroelastische** ferroelastic transition
Umwandlung *f***/gegenseitige** interconversion (reaction)
Umwandlung *f* **höherer Ordnung** higher-order transition, smooth transition
Umwandlung *f***/irreversible** irreversible transition
Umwandlung *f***/isotherme** isothermal [constant-temperature] transformation
Umwandlung *f***/kontinuierliche** *s.* Umwandlung 2. Ordnung
Umwandlung *f***/martensitische** martensitic transition
Umwandlung *f***/metamagnetische** metamagnetic transition
Umwandlung *f***/monotrope** monotropic transition
Umwandlung *f* **normal-supraleitend** superconducting transition
Umwandlung *f* **1. Ordnung** first-order transition, discontinuous transition
Umwandlung *f* **2. Ordnung** second-order transition, continuous transition
Umwandlung *f***/polymorphe** polymorphic transition
Umwandlung *f***/radioaktive** radioactive transformation [transmutation]
Umwandlung *f***/rekonstruktive** reconstructive transition
Umwandlung *f***/reversible** reversible transition
Umwandlung *f***/stoffliche** conversion [transformation] of materials, materials conversion
Umwandlung *f***/strukturelle** structural transition
Umwandlungsgeschwindigkeit *f* rate of transformation, conversion rate, rate of change

Umwandlungsgrad *m* degree of conversion
Umwandlungsintervall *n* transition interval *(of a double salt)*
Umwandlungsprodukt *n* transformation [conversion] product
Umwandlungspunkt *m* 1. transition point [temperature]; 2. *(cryst)* inversion point [temperature]
Umwandlungspunkt *m* **erster Ordnung** first-order transition point [temperature]
Umwandlungspunkt *m* **zweiter Ordnung** 1. second-order transition point [temperature]; 2. *(plast, rubber)* glass transition point, Tg (point)
Umwandlungsrate *f* disintegration rate, activity *(of a radioactive substance)*
Umwandlungsreaktion *f* conversion reaction
Umwandlungswärme *f* heat of transition, latent heat *(of phases or crystals)*
Umweltanalytik *f* environmental analytical chemistry, environmental analysis
Umweltbelastung *f* environmental load(ing), environmental burden
Umweltbiochemie *f* environmental biochemistry
Umweltblei *n* environmental lead
Umweltchemie *f* environmental chemistry
Umweltchemikalie *f* environmental chemical
Umweltchemiker *m* environmental chemist
umweltfreundlich environmentally acceptable [beneficial, friendly, compatible]
Umweltgefahr *f* environmental hazard [danger]
umweltgefährdend environmentally hazardous [dangerous]
umweltgefährlich environmentally hazardous [dangerous]
Umweltgift *n* environmental toxicant
Umweltprobe *f* environmental sample
Umweltradioaktivität *f* environmental radioactivity
umweltschädigend enviromentally damaging [harmful]
Umweltschadstoff *m* environmental pollutant
Umweltschmutzstoff *m* environmental contaminant
Umweltschutz *m* environmental protection
Umweltschutztechnik *f* environmental technology
Umweltstoff *m*/**toxischer** environmental toxicant
Umwelttoxikologie *f* environmental toxicology
Umwelttoxizität *f* environmental toxicity
Umweltverschmutzung *f* environmental contamination
Umweltverschmutzung *f* **durch Schadstoffe** environmental pollution
umweltverträglich environmentally compatible [acceptable]
Unabhängigkeitsverbundwirkung *f* *(tox)* independent joint action *(without reciprocal influence of the components)*
unabtrennbar inseparable
unangegriffen unchanged, unattacked, unaffected
unangenehm disagreeable, disgusting, unpleasant, objectionable *(e.g. smell)*
unangreifbar unattackable, stable, resistant
unangreifbar/chemisch stable [resistant] to chemical attack, chemically stable [resistant]
Unangreifbarkeit *f* unattackability, stability, resistance
Unangreifbarkeit *f*/**chemische** stability [resistance] to chemical attack, chemical stability [resistance]
unbehandelt untreated, unprocessed, unfinished, raw, virgin
unbenetzbar non-wettable
unbesetzt vacant *(lattice position)*; unoccupied *(subshell)*
unbeständig unstable, instable, labile, transient, *(relating to isotopes also)* evanescent
unbeständig/thermisch thermolabile, heat-labile
Unbeständigkeit *f* instability, lability, transience
Unbeständigkeit *f*/**thermische** thermolability, thermal lability
Unbestimmtheitsrelation *f* uncertainty relation
Unbestimmtheitsrelation *f*/**heisen-**

bergsche Heisenberg uncertainty relation
unbrennbar incombustible, non-combustible, fireproof • **unbrennbar machen** fireproof
Unbrennbarkeit *f* incombustibility
Undecan *n* undecane
Undecanal *n* undecanal, undecyl aldehyde
Undecanon *n* undecanone *(any of several isomeric ketones $C_{11}H_{22}O$)*
Undecansäure *f* undecanoic acid
Undecen *n* undecene *(any of several isomeric alkenes $C_{11}H_{22}$)*
Undecensäure *f* undecenoic acid *(any of several isomeric alkenoic acids $C_{11}H_{20}O_2$)*
Undecin *n* undecyne *(any of several isomeric alkynes $C_{11}H_{20}$)*
Undecinsäure *f* undecynoic acid *(any of several isomeric alkynoic acids $C_{11}H_{18}O_2$)*
Undecylalkohol *m* undecyl alcohol, *undecan-1-ol
undicht leaky, pervious, porous, porose • **undicht sein** leak, run • **undicht werden** spring a leak
Undichtheit *f* leakiness
Undichtigkeit *f* leakiness
undissoziiert undissociated
undurchlässig impermeable, impenetrable, impervious, (leak)tight, (leak-)proof; *(optically:)* opaque • **undurchlässig machen** make impervious
Undurchlässigkeit *f* impermeability, impenetrability, imperviousness, tightness, proofness; *(optically:)* opacity
undurchsichtig opaque, non-transparent
Undurchsichtigkeit *f* opacity, non-transparency
unedel base, non-noble, (re)active *(metal)*
Unedelmetall *n* base [non-noble] metal
uneinheitlich non-uniform
uneinheitlich/innerlich inhomogeneous
Uneinheitlichkeit *f* non-uniformity
Uneinheitlichkeit *f***/innerliche** inhomogeneity
unempfindlich insensitive
Unempfindlichkeit *f* insensitivity, insensitiveness
unentbehrlich *(bioch, agric)* essential
Unentbehrlichkeit *f (bioch, agric)* essentiality
unentflammbar non-(in)flammable, uninflammable, flameproof
Unentflammbarkeit *f* non-flammability, uninflammability, flameproofness
ungebleicht unbleached
ungebrannt *(ceram)* unfired, unburned, green, raw
ungebunden unbonded, unbound, uncombined *(atoms, ions)*
ungefährlich harmless
Ungefährlichkeit *f* harmlessness
ungefüllt filler-free, unfilled, unloaded
ungeladen uncharged
ungelöst undissolved
ungeordnet unordered, disordered, random
ungepuffert unbuffered
ungeradzahlig odd-number(ed)
ungesättigt unsaturated *(solution or compound)*
ungesättigt/dreifach triply unsaturated
ungesättigt/einfach monounsaturated, monoenoic
ungesättigt/mehrfach polyunsaturated, polyenoic
ungesättigt/zweifach diunsaturated, dienoic, doubly unsaturated
Ungesättigtheit *f* unsaturation
ungiftig non-toxic, non-poisonous
Ungiftigkeit *f* non-toxicity, freedom from toxicity
Ungleichgewicht *n* imbalance, unbalance, non-equilibrium
unimolekular unimolecular, monomolecular
Union *f* **für reine und angewandte Chemie/Internationale** International Union of Pure and Applied Chemistry, IUPAC
unionisiert unionized
Unipolarzelle *f* monopolar cell
Unipolymer(e) *n* homopolymer
unitär unary *(system)*
univariant univariant, monovariant
Universaldoppelmuffe *f (lab)* swivel clamp holder
Universalfermenter *m (biot)* multi-purpose fermenter
Universalindikator *m* universal indicator
unkatalysiert uncatalyzed

Unkrautbekämpfungsmittel *n* herbicide, weed-killer, weed-control agent
unl. *s.* unlöslich
unlegiert unalloyed
unlöslich insoluble, i.s., non-soluble, indissoluble • **unlöslich machen** insolubilize
Unlösliches *n* insoluble, i.
Unlöslichkeit *f* insolubility, insolubleness
unmagnetisch non-magnetic
unmischbar immiscible
Unmischbarkeit *f* immiscibility
unmodifiziert unmodified
Unordnung *f* disorder, randomness
unpigmentiert unpigmented
unpolar non-polar, apolar; non-polar, homopolar, covalent *(chemical bond)*
unpolarisierbar non-polarizable
unrein impure
Unreinheit *f* impurity, *(if localized also)* speck, spot
uns. *s.* unsymmetrisch
unschädlich harmless, innocuous, benign
Unschädlichkeit *f* harmlessness, innocuousness, innocuity, benignity
unschmackhaft unpalatable, distasteful
unschmelzbar infusible
Unschmelzbarkeit *f* infusibility
unspezifisch non-specific
Unstöchiometrie *f* non-stoichiometry
unstöchiometrisch non-stoichiometric
Unsymmetrie *f* asymmetry, dissymmetry
unsymmetrisch asymmetric(al), dissymmetric(al), unsymmetric(al)
Untenaustrag *m* bottom discharge
Untenentleerung *f* bottom discharge
unterbelasten *v* underload
Unterbleiche *f* *(pap)* underbleaching
unterbleichen *v* *(pap)* underbleach
unterbringen *v* locate *(e.g. atoms in interstitial lattice sites)*
Unterbringung *f* location *(as of atoms in interstitial lattice sites)*
unterdosieren *v* underdose, underfeed
Unterdosierung *f* underdosage, underfeed(ing)
Unterdruckdestillation *f* distillation under vacuum [reduced pressure]
unterdrücken *v* suppress, repress *(e.g. a reaction)*
Unterdruckfilter *n* vacuum [suction] filter
Unterdruckfiltration *f* vacuum filtration
Untereinheit *f* subunit *(protein chemistry)*
untereutektisch hypoeutectic
untereutektoid(isch) hypoeutectoid
untergärig bottom-fermenting, bottom-fermented
Untergärung *f* bottom fermentation
Unterglasurfarbe *f* *(ceram)* underglaze colour
Untergrund *m* 1. *(coat)* ground, undersurface; 2. *(anal, nucl)* background
Untergrundrauschen *n* *(anal)* background [base-line] noise
Untergrundspektrum *n* background [residual] spectrum
Untergrundstörung *f* *(spectr)* background interference
Untergruppe *f* subgroup; sub-subclass
unterhalten *v* 1. *(tech)* maintain; 2. support, sustain *(e.g. combustion)*
unterhaltend/sich selbst self-supporting, self-sustaining, self-sustained *(reaction)*
Unterhaltung *f* 1. *(tech)* maintenance, upkeep; 2. support, sustainment *(as of combustion)*
Unterhaltungskosten *pl* maintenance cost(s), cost of upkeep
Unterhärtung *f* *(plast)* undercure
Unterkeim *m* *(cryst)* cluster
Unterklasse *f* subclass
Unterkorn *n* undersize (material, product), fines, minus material *(classifying)*; *(screening also)* screen undersize [fines]
unterkritisch *(nucl, cryst)* subcritical
unterkühlen *v* 1. supercool, subcool; 2. *(glass)* overcool
Unterlaufprodukt *n* underflow *(classifying)*
Unterniveau *n* sublevel *(of electrons)*
untersättigt subsaturated, undersaturated
Untersättigung *f* subsaturation, undersaturation
Unterschale *f* (electronic) subshell
unterschichten *v* add to form a lower layer
unterstöchiometrisch hypostoichiometric
Unterstruktur *f* *(cryst)* substructure
Unterstützung *f*/**anchimere** anchimeric

assistance *(of neighbouring groups in accelerating the reaction rate)*
untersuchen *v* investigate, examine, assay, analyse
untersucht werden *v* undergo examination
Untersuchung *f* investigation, research, examination, assay, analysis
Untersuchung *f*/organoleptische organoleptic [sensory] examination
Untersuchung *f*/röntgenographische X-ray investigation
Untersuchungslösung *f* solution for [under] investigation [examination], solution being [to be] tested, test solution
Untersuchungsmaterial *n* material for [under] investigation [examination], experimental [test] material
Untersuchungsmethode *f* investigational method, test(ing) method
Untersuchungsmethode *f*/standardisierte standard test(ing) method
Untersuchungsprobe *f* test sample *(product of sample reduction)*
Untertagebau *m* underground mining [working], deep mining
Untervulkanisation *f* undercure, undercuring, undervulcanization
untervulkanisieren *v* undercure, undervulcanize
Unterwerkzeug *n (plast)* die
Unterwind *m* downdraught
Unterwindfrischkonverter *m (met)* bottom-blown converter
untrennbar inseparable
unumgesetzt unreacted, unconverted
unveränderlich invariable
unverändert unchanged, unaltered
unverbrannt unburned, unburnt
unverbraucht unspent *(chemicals)*; unreacted *(reactant)*
unverdünnt undiluted
unvergärbar unfermentable
unvernetzt 1. uncross-linked, uncured; 2. *(rubber)* unvulcanized
unverschmutzt uncontaminated, noncontaminated, unpolluted, free from polluting substances *(water, air)*
unverseifbar unsaponifiable, non-saponifiable
Unverseifbares *n* unsaponifiable matter [residue]
unverstärkt unreinforced *(e.g. plastic material)*
unverträglich incompatible
Unverträglichkeit *f* incompatibility
unverzweigt unbranched(-chain)
unvollständig non-quantitative *(e.g. precipitation)*
Unvollständigkeit *f* non-quantitativeness *(as of a precipitation)*
unvulkanisiert *(rubber)* uncured, unvulcanized
unwirksam ineffective • **unwirksam machen** deactivate, inactivate, block
UP *s.* Polyester/ungesättigter
Uracil *n* uracil, pyrimidine-2,4-dione
Uracil-4-carbonsäure *f* uracil-4-carboxylic acid, orotic acid
Uramil *n* uramil, 5-aminobarbituric acid
Uran *n* U uranium • **an Uran verarmt** uranium-barren
Uran-Actinium-Reihe *f* actinium (decay) series, uranium-actinium (disintegration) series
Uran-Actinium-Zerfallsreihe *f* actinium (decay) series, uranium-actinium (disintegration) series
Uranat *n* uranate *(salt or ester of uranic acid)*
Uran(VI)-fluorid *n* uranium(VI) fluoride, uranium hexafluoride
Uranhydrid *n* uranium hydride
Uranin *n* uranin(e), sodium fluorescein
Uran(IV)-oxid *n* uranium(IV) oxide, uranium dioxide
Uranpechblende *f (min)* pitchblende, uraninite *(uranium(IV) oxide)*
Uranpecherz *n (min)* pitchblende, uraninite *(uranium(IV) oxide)*
Uran-Radium-Reihe *f* uranium (decay) series, uranium-radium (disintegration) series
Uran-Radium-Zerfallsreihe *f* uranium (decay) series, uranium-radium (disintegration) series
Uransäure *f* uranic acid
Uranspaltung *f* uranium fission
Uranylacetat *n* uranyl acetate
Uranylchlorid *n* uranyl chloride
Uranylnitrat *n* uranyl nitrate
Uranylsalz *n* uranyl salt
Uranylverbindung *f* uranyl compound
Urat *n* urate *(a salt of uric acid)*

p-Urazin *n* p-urazine, hexahydro-s-tetrazine-3,6-dione
Ureid *n* ureide *(an acyl derivative of urea)*
Ureidosäure *f* ureido acid
ureotel ureotelic *(excreting nitrogen as urea)*
Urethan *n* urethane
Urethanelastomer(e) *n* (poly)urethane elastomer
Urethankautschuk *m* (poly)urethane rubber, isocyanate rubber
Urethanschaumstoff *m* urethane foam
Uricolyse *f* uricolysis *(the conversion of uric acid to urea)*
uricolytisch uricolytic
Uridin *n (org ch)* uridine
Uridindiphosphat *n* uridine diphosphate, UDP
Uridinphosphat *n* uridine phosphate, *(esp)* uridine monophosphate, uridine phosphoric acid, UMP
Uridinphosphorsäure *f* uridine phosphoric acid
Uridintriphosphat *n* uridine triphosphate, UTP
Uridintriphosphorsäure *f* uridine triphosphate, UTP
Uridylsäure *f (bioch)* uridylic acid, uridine 5′-monophosphate, UMP
urikotel uricotelic *(excreting nitrogen as uric acid)*
U-Rohr-Manometer *n* U-tube manometer
U-Rohr-Wärmeaustauscher *m* U-tube heat exchanger
U-Rohr-Wärmeübertrager *m* U-tube heat exchanger
Uronsäure *f* $OHC[CHOH]_nCOOH$ uronic acid *(any of a series of aldehyde acids)*
Urspannung *f* electromotive force, emf
Urspannung *f*/**elektrische** electromotive force, emf
Urspannungsnormal *n* standard of emf
Ursprung *m* origin, source • **pflanzlichen Ursprungs** of vegetable [plant] origin, plant-derived • **tierischen Ursprungs** of animal origin • **vulkanischen Ursprungs** of volcanic origin, *(esp of gases)* plume-borne
Ursprungsgestein *n* parent [mother, source] rock
Ursprungssubstanz *f* parent substance
Urteer *m* primary [low-temperature] tar
Urtitersubstanz *f* primary-standard chemical, primary standard *(in titrimetric analysis)*
Uterotonikum *n (pharm)* uterotonic, oxytocic (agent)
Uterustonikum *n (pharm)* uterotonic, oxytocic (agent)
UV UV, ultraviolet
UV-Bereich *m* UV region, ultraviolet (spectral) region
UV-Gebiet *n* UV region, ultraviolet (spectral) region
UVS-Bereich *m* UV-VIS range [region], ultraviolet-visible [UV-visible] region
UVS-Gebiet *n* UV-VIS range [region], ultraviolet-visible [UV-visible] region
UV-Spektroskopie *f* UV spectroscopy
UV-Spektrum *n* UV spectrum
UV-Strahlung *f* UV radiation
UV-VIS-Spektroskopie *f* UV-VIS spectroscopy

V

Vaccensäure *f* vaccenic acid, *octadec-11-enoic acid *(specif trans form)*
Vakuum *n* vacuum • **Vakuum herstellen** create a vacuum • **unter Vakuum arbeiten** operate under a vacuum
Vakuumaufdampfung *f* vacuum deposition
Vakuumbandfilter *n* rotary-drum belt-type vacuum filter
Vakuumdestillat *n* vacuum distillate, VD
Vakuumdestillation *f* vacuum distillation
Vakuumdestillierapparat *m* vacuum still
vakuumdestilliert vacuum-distilled, distilled in vacuo
Vakuumdrehfilter *n* rotary vacuum filter
Vakuumeindampfung *f* vacuum concentration
Vakuumexsikkator *m* vacuum desiccator
Vakuumfilter *n* vacuum [suction] filter
Vakuumfiltration *f* vacuum [suction] filtration
Vakuumkolonne *f (distil)* vacuum column [still]
Vakuummessgerät *n* vacuum gauge, vacuometer
Vakuumpumpe *f* vacuum pump

Vakuumrückstand *m (petrol)* vacuum residue, VR, short residue [residuum]
Vakuumtrockner *m* vacuum dryer
Vakuumtrocknung *f* vacuum drying [dehydration]
Vakuumtrommelfilter *n* (rotary) vacuum drum filter, rotary-drum vacuum filter
Vakuumtrommelzellenfilter *n* multicompartment drum filter
Vakuumverdampfapparat *m* vacuum evaporator
Vakuumverdampfer *m* vacuum evaporator
Vakuumverdampfung *f* vacuum evaporation
Vakuum-Vielschichtisolierung *f* multiple-layer insulation, superinsulation *(cryogenic engineering)*
Vakuumzerstäuben *n* sputtering
Val *n* gram equivalent, g. equiv.
Valenz *f* 1. valence, valency *(unit of combining power)*; 2. *s.* Wertigkeit; 3. *s.* Bindungskraft
Valenzabsättigung *f* valence saturation
Valenzband *n* valence band
Valenzbindung *f* valence bond
Valenzbindungstheorie *f* valence-bond theory, VB theory, electron-pair theory, HLSP [Heitler-London-Slater-Pauling] theory
Valenzelektron *n* valence electron, outermost [valence shell, optical] electron
Valenzelektronenpaar *n* valence [outer-shell] electron pair, valence-shell (electron) pair
Valenzorbital *n* valence [bond] orbital, outer-shell orbital
Valenzschale *f* valence shell
Valenzschwingung *f* valence [stretching] vibration, stretching mode
Valenzstrichformel *f* (valence) structural formula, line [constitutional, graphic] formula
Valenztheorie *f* theory of valence, valence theory
Valenzwinkel *m* valence angle, bond [interbonding] angle
Valeraldehyd *m* valeraldehyde, *pentanal
Valerat *n* valerate *(salt or ester of a valeric acid)*
Valeriansäure *f* valeric acid, *(specif)* $CH_3[CH_2]_3COOH$ *pentanoic acid
Valin *n* valine, *2-amino-3-methylbutanoic acid
Vanadat *n* vanadate
Vanadium *n* V vanadium
Vanadium(III)-chlorid *n* vanadium(III) chloride, vanadium trichloride
Vanadium(III)-fluorid *n* vanadium(III) fluoride, vanadium trifluoride
Vanadium(V)-fluorid *n* vanadium(V) fluoride, vanadium pentafluoride
Vanadium(III)-oxid *n* vanadium(III) oxide, vanadium trioxide, divanadium trioxide
Vanadium(IV)-oxid *n* vanadium(IV) oxide, vanadium dioxide
Vanadium(V)-oxid *n* vanadium(V) oxide, vanadium pentaoxide, divanadium pentaoxide
Vanadium(III)-oxidchlorid *n* vanadium(III) monochloride oxide
Vanadium(V)-oxidchlorid *n* vanadium(V) trichloride oxide
Vanadiumsäure *f* vanadic acid
Vanadiumtri... *siehe: Vanadium(III)-...*
Van-der-Waals-Bindung *f* van der Waals bond
Van-der-Waals-Kräfte *fpl* van der Waals forces (of attraction)
Vanillal *n* vanillal, bourbonal, 3-ethoxy-4-hydroxybenzaldehyde
Vanillaldehyd *m* vanillin, 4-hydroxy-3-methoxybenzaldehyde
Vanillin *n* vanillin, 4-hydroxy-3-methoxybenzaldehyde
Vanillinsäure *f* vanillic acid, 4-hydroxy-3-methoxybenzoic acid
Van't-Hoff-Gleichung *f* van't Hoff equation *(osmosis)*
Varianz *f* variance *(statistics)*
Varietät *f (min)* variety
Vasodilatans *n (pharm)* vasodilator
Vasodilat(at)or *m (pharm)* vasodilator
vasodilatatorisch vasodilating
Vasokonstriktor *m (pharm)* vasoconstrictor
vasokonstriktorisch vasoconstrictive
VB-Theorie *f s.* Valenzbindungstheorie
vegetabilisch vegetable
vektoriell *(bioch)* vectorial *(transport mechanism)*
Ventil *n* valve
Ventilator *m* blower, fan
Ventilboden *m* valve tray

Venturi-Düse *f* Venturi tube
Venturi-Rohr *n* Venturi tube
Venturi-Wascher *m* Venturi scrubber [washer], Venturi-type water jet scrubber
verabreichen *v* 1. apply; 2. *(pharm)* administer
Verabreichung *f* 1. application; 2. *(pharm)* administration
verändern *v***/sich** undergo change
verändern *v***/sich chemisch** undergo chemical change
Veränderung *f* change • **eine chemische Veränderung erfahren [erleiden]** undergo chemical change
Veränderung *f***/stoffliche** change of material, material change
verankern *v* anchor
verarbeitbar processable, processible, workable
Verarbeitbarkeit *f* processability, processibility, workability
verarbeiten *v* process, work
Verarbeitung *f* 1. processing, working; 2. *(petrol)* refining
Verarbeitungsbereich *m* working range
Verarbeitungseigenschaften *fpl* processing characteristics [properties]
Verarbeitungshilfsmittel *n* processing agent [aid]
Verarbeitungstechnik *f* processing technology *(branch of science or its application)*
Verarbeitungstechnologie *f* processing technology *(sequence of processing steps)*
Verarbeitungsverfahren *n* manufacturing process
Verarbeitungsvorschrift *f* (processing) specification
verarmen *v* become poor [depleted, impoverished]
verarmen lassen *v* deplete, impoverish
verarmt/an Sauerstoff oxygen-depleted
Verarmung *f* depletion, impoverishment
Verarmung *f* **an Sauerstoff** oxygen depletion
veraschen *v* 1. ash, reduce to ashes, incinerate; 2. *(filtr)* burn off
Veraschen *n* ashing, incineration
Veraschen *n***/nasses** wet ashing
Veraschen *n***/trockenes** dry ashing
Veraschungsschälchen *n* incinerating dish
verästelt *(cryst)* dendritic(al)
Veratrumaldehyd *m* veratric aldehyde, 3,4-dimethoxybenzaldehyde
Veratrumsäure *f* veratric acid, dimethoxybenzoic acid, *(specif)* 3,4-dimethoxybenzoic acid
verätzen *v* 1. burn; 2. *(agric)* scorch; 3. *(med)* cauterize, burn *(e.g. a wound)*
Verätzung *f* 1. *(act:)* burning; 2. *(agric)* scorching; 3. *(med)* cauterization, cautery *(as of wounds)*; 4. *(result:)* burn; 5. *(agric)* scorch; 6. *(med)* chemical burn
Verband *m* union *(of atoms, molecules)*
verbessern *v* improve; upgrade *(the quality of a material)*
verbessern *v***/die Oberflächenbeschaffenheit [Oberflächengüte]** finish
Verbesserung *f* improvement; upgrading *(of the quality of a material)*
verbinden *v* 1. combine *(elements, compounds)*; bond (together), link (together) *(atoms)*; 2. connect, join, link (together), *(esp by means of an adhesive:)* bond
verbinden *v***/durch eine Rohrleitung** pipe
verbinden *v***/mit einem T-Stück** tee
verbinden *v***/miteinander** *s.* verbinden
verbinden *v***/sich (miteinander)** combine *(of elements, compounds)*; bond, link (together) *(of atoms)*
Verbindung *f* 1. *(act or process:)* combination *(of elements or compounds)*; bonding, linking, linkage *(of atoms)*; *(mechanically:)* connection, joining, linkage; 2. *(substance:)* compound; 3. *(piece:)* connection, joint, link, union, *(esp for pipes and hoses:)* fitting; 4. *(state:)* connection, *(esp by an adhesive:)* bond • **eine (chemische) Verbindung eingehen** enter into (chemical) combination [union], combine
Verbindung *f***/abbauresistente** persistent [refractory] compound
Verbindung *f***/alicyclische** alicyclic compound
Verbindung *f***/aliphatische** aliphatic compound
Verbindung *f***/analoge** analogue

Verbindung *f*/**anorganische** inorganic (compound)
Verbindung *f*/**antiaromatische** antiaromatic compound
Verbindung *f*/**aromatische** aromatic compound
Verbindung *f*/**berthollide** berthollide compound, non-stoichiometric [non-Daltonian, non-daltonide] compound
Verbindung *f*/**binäre** binary compound
Verbindung *f*/**biologisch nicht abbaubare** non-biodegradable compound, biorefractory compound
Verbindung *f*/**bororganische** organoboron compound
Verbindung *f*/**chemische** chemical compound
Verbindung *f*/**cyclische** cyclic [ring] compound
Verbindung *f*/**daltonide** Daltonian [daltonide] compound
Verbindung *f*/**diastereomere** diastereo(iso)mer
Verbindung *f*/**herzsche** *(dye)* Herz compound
Verbindung *f*/**heterocyclische** heterocycle, heterocyclic (compound)
Verbindung *f*/**heteropolare** heteropolar [polar, ionic] compound
Verbindung *f*/**homologe** homolog(ue)
Verbindung *f*/**homöopolare** homopolar [non-polar, non-ionic] compound
Verbindung *f*/**hydroaromatische** hydroaromatic compound
Verbindung *f*/**intermetallische** intermetallic [metal] compound, intermetallic, electron compound, intermediate constituent [phase]
Verbindung *f*/**isocyclische** isocyclic compound, homocyclic [carbocyclic] compound, homocycle
Verbindung *f*/**isomere** isomeric compound
Verbindung *f*/**isostere** isosteric compound, isostere
Verbindung *f*/**lamellare** lamellar compound, intercalation [intercalate] compound
Verbindung *f*/**makrocyclische** macrocyclic compound, macroring [large-ring] compound
Verbindung *f*/**markierte** (isotopically-) labelled compound, tagged compound
Verbindung *f*/**mesoionische** mesoionic compound
Verbindung *f*/**mesomere** resonance hybrid
Verbindung *f*/**metallorganische** organometallic [metallo-organic] compound, organometal
Verbindung *f* **mit kleinem Ring** small-ring compound
Verbindung *f*/**oberflächenaktive** surface-active compound, surfactant
Verbindung *f*/**organische** organic (compound)
Verbindung *f*/**persistente** persistent [refractory] compound
Verbindung *f*/**phosphororganische** organophosphorus compound
Verbindung *f*/**polycyclische** polycyclic compound
Verbindung *f*/**quecksilberorganische** organomercury compound, organomercurial
Verbindung *f*/**siliciumorganische** organosilicon compound
Verbindung *f*/**spirocyclische** spiro hydrocarbon [compound], spiran(e), spirocyclane
Verbindung *f*/**ternäre** ternary compound
Verbindung *f*/**trimere** trimer
Verbindung *f*/**unpolare** non-polar compound
Verbindung *f*/**unsymmetrische** unsymmetrical compound
Verbindung *f*/**zinnorganische** organotin compound
***m*-Verbindung** *f* meta compound
***o*-Verbindung** *f* ortho compound
***p*-Verbindung** *f* para compound
Verbindungsklasse *f* family, class of compounds
Verbindungspeptid *n (bioch)* connecting peptide
Verbindungsstück *n* connection, connector; *(for pipes and hoses:)* fitting
verblasen *v (glass, met)* blow
verblasen *v*/**im Konverter** *(met)* convert, bessemerize
verblassen *v* fade, discolour
Verblassen *n* fading, discoloration
verbleichen *v s.* verblassen
verbleien *v* 1. lead *(motor fuel)*; 2. lead(-

coat) *(surfaces for protection)*; *(relating to the inside also)* lead-line; *(using prefabricated material:)* lead-clad
Verbrauch *m* consumption
verbrauchen *v* consume; spend, exhaust *(e.g. dye liquor)* • **zu verbrauchen bis ...** use by ...
Verbrauchsgeschwindigkeit *f* rate of consumption [disappearance] *(of a reactant)*
verbraucht 1. spent, exhausted *(e.g. solution)*; 2. *(tech)* worn out
verbrennbar combustible
Verbrennbarkeit *f* combustibility
verbrennen *v* 1. burn; incinerate *(garbage, sludge)*; 2. *(agric)* burn, scorch *(e.g. leaves by excess fertilization)*; burn, undergo combustion; 3. *(agric)* be burnt [scorched]
Verbrennung *f* 1. combustion, burning; incineration *(of garbage, sludge)*; 2. *(agric)* burn, scorch *(as of leaves by excess fertilization)* • **die Verbrennung unterhalten** sustain combustion
Verbrennung *f***/partielle** partial combustion
Verbrennung *f***/unvollkommene [unvollständige]** incomplete combustion
Verbrennungsanalyse *f* combustion analysis
Verbrennungsapparat *m* combustion train *(for elementary organic analysis)*
Verbrennungsgas *n* combustion gas
Verbrennungsgeschwindigkeit *f* rate of combustion, burning rate
Verbrennungskammer *f* combustion [furnace] chamber
Verbrennungsofen *m* combustion furnace, burner
Verbrennungsraum *m* combustion space; incineration chamber *(for refuse)*
Verbrennungsreaktion *f* combustion reaction
Verbrennungsrohr *n* combustion tube
Verbrennungsröhre *f* combustion tube
Verbrennungsrückstand *m* residue of combustion
Verbrennungsschiffchen *n* combustion boat
Verbrennungswärme *f* 1. heat of combustion; 2. *s.* Brennwert/spezifischer
Verbrennungswärme *f***/molare** *s.* Brennwert/stoffmengenbezogener
Verbrennungszone *f* combustion [oxidation] zone
verbunden/über Wasserstoff hydrogen-bonded
Verbundsicherheitsglas *n* laminated safety (sheet) glass
Verbundwerkstoff *m* composite
Verbundwirkung *f* joint action
verchromen *v* *(el ch)* plate with chromium, chrome; chromize *(by thermal diffusion)*
verdampfbar (e)vaporable, vaporizable
verdampfbar/leicht volatile, volatilizable
Verdampfbarkeit *f* (e)vaporability
verdampfen *(of human agent:)* vaporize, boil away [down], evaporate, volat(il)ize, *(suddenly as by expansion:)* flash; *(of a liquid:)* evaporate, vaporize, volat(il)ize, *(suddenly as by expansion:)* flash
Verdampfer *m* evaporator, vaporizer; expander *(cryogenics)*
Verdampfer *m***/mehrstufiger** multistage evaporator
Verdampferkörper *m* body, *(in a series of evaporators preferably)* effect
Verdampferstufe *f* effect
Verdampfung *f* *(esp process:)* evaporation, *(esp act:)* vaporization, *(esp if readily:)* volatilization; *(rapidly as by expansion:)* flash
Verdampfung *f* **durch Sonnenbestrahlung** *(hyd)* solar evaporation [distillation]
Verdampfungsgeschwindigkeit *f* evaporation rate, rate of vaporization
Verdampfungskristallisator *m* evaporative crystallizer, crystallizing [salting-out] evaporator
Verdampfungsofen *m* *(distil)* still pot [body], reboiler
Verdampfungstrocknung *f* drying by evaporation, evaporation drying
Verdampfungswärme *f* heat of evaporation [vaporization]
verdauen *v* digest
verdaulich digestible
Verdaulichkeit *f* digestibility
Verdauung *f* digestion
Verdauungsenzym *n* *(biot)* digestive enzyme

Verdauungsferment *n (biot)* digestive enzyme
verdeckt opposed, eclipsed *(stereochemistry)*
Verderb *m (food)* decay, deterioration, spoilage
Verderb *m*/**oxidativer** oxidative deterioration
verderben *v* decay, deteriorate, spoil, perish
Verderbniserreger *mpl*/**mikrobielle** spoilage microorganisms
verdichtbar compressible; condensable *(vapour)*
Verdichtbarkeit *f* compressibility; condensability *(of vapour)*
verdichten *v* densify, compress, *(esp relating to bulk material:)* compact, consolidate; condense *(vapours)*
verdichten *v*/**sich** condense *(of vapours)*
Verdichter *m* compressor, *(relating to refrigerating machines also)* vapour compressor
Verdichter *m*/**einstufiger** single-stage compressor
Verdichtung *f* densification, compression, *(esp relating to bulk material:)* compacting, consolidation; condensation *(of vapours)*
Verdichtungsarbeit *f* work of compression *(as of a pump)*
verdicken *v s.* eindicken
Verdicker *m* thickener
Verdickungsmittel *n* thickening agent
Verdopplungszeit *f (biot)* (cell) doubling time *(growth of microorganisms)*
verdrängbar *(ch)* displaceable
verdrängen *v (ch, tech)* displace; expel, push [sweep] out, eject *(e.g. gases)*
Verdrängerpumpe *f* positive-displacement pump
Verdrängung *f (ch, tech)* displacement; expulsion, ejection *(as of gases)*
Verdrängungsanalyse *f* displacement analysis [chromatography]
Verdrängungschromatographie *f* displacement analysis [chromatography]
Verdrängungsreaktion *f* displacement reaction
Verdrängungsreaktion *f*/**doppelte** double displacement reaction, ping-pong reaction *(enzyme kinetics)*
Verdrängungstitration *f* displacement [replacement] titration
verdrillt twisted, interwound, intertwined *(macromolecules)*
verdünnbar dilutable *(liquid)*
verdünnbar/mit Wasser water-dilutable
Verdünnbarkeit *f* dilutability
verdünnen *v* dilute, thin, attenuate, *(esp relating to gases:)* rarefy; potentiate *(homoeopathy)*
Verdünnen *n* dilution, thinning, attenuation, *(esp relating to gases:)* rarefaction; potentiation *(homoeopathy)*
Verdünner *m s.* Verdünnungsmittel
verdünnt/unendlich infinitely dilute
Verdünnung *f* 1. dilution *(diluted liquid)*; potency *(homoeopathy)*; 2. *s.* Verdünnungsmittel; 3. *s.* Verdünnen
Verdünnung *f*/**unendliche** infinite dilution
Verdünnungsgesetz *n*/**ostwaldsches** Ostwald dilution law
Verdünnungsgrad *m* degree of dilution
Verdünnungsmittel *n* 1. diluting agent, diluent; 2. *(coat)* thinner, thinning agent
Verdünnungsprinzip *n*/**ruggli-zieglersches** Ruggli high-dilution principle
Verdünnungsreihe *f* dilution series
Verdünnungswasser *n* dilution water, water of dilution
verdunsten *v* evaporate, vaporize, volat(il)ize *(below normal boiling point)*
Verdunstung *f* evaporation, vaporization, volatilization *(below normal boiling point)*
Verdunstungsgeschwindigkeit *f* rate of evaporation
Verdunstungstrocknung *f* drying by evaporation
Verdunstungswärme *f* heat of evaporation
verdüsen *v* atomize, spray
Verdüsung *f* (nozzle) atomization, spraying
veredeln *v* 1. refine, improve; finish *(a surface)*; 2. *(text)* finish, process; 3. *(pap)* refine *(pulp)*
Vered(e)lung *f* 1. refining (treatment), refinement, improvement; finish(ing) *(of a surface)*; 2. *(text)* finish(ing), processing; 3. *(pap)* (alkali) refining *(of pulp)*

vereinigen *v* 1. combine *(elements)*; 2. assemble *(parts of an apparatus)*
vereinigen *v***/sich** combine *(of elements)*; coalesce *(as of gas bubbles)*
Vereinigung *f* 1. combination, union *(of elements or chemical compounds)*; 2. mixing *(of filtrates)*; coalescence *(as of gas bubbles)*; 3. assembly *(of parts of an apparatus)*
Verengung *f* throat, restriction *(as of a nozzle)*
verestern *v* esterify
Veresterung *f* esterification
Veresterung *f***/extraktive** extractive esterification
verethern *v* etherify
Veretherung *f* etherification
Verfahren *n* process, method, approach; *(manner of performing technical details)* method, technique, procedure
Verfahren *n***/aluminothermisches** aluminothermic process
Verfahren *n***/diskontinuierliches** batch process
Verfahren *n***/großtechnisches** large-scale process
Verfahren *n***/kontinuierliches** continuous process
Verfahren *n***/nasses** wet process
Verfahren *n***/röntgenographisches** X-ray method *(as in crystallography)*
Verfahren *n***/standardisiertes** standard method
Verfahren *n***/technisches** commercial process *(as opposed to lab-scale techniques or pilot-plant-scale processes)*
Verfahren *n***/trockenes** dry process
Verfahren *n***/zweistufiges** two-stage [two-step] process
Verfahrenschemie *f* process chemistry
Verfahrenschemiker *m* process chemist
Verfahrensentwicklung *f* process development
Verfahrensindustrie *f* process industry
Verfahrensingenieur *m* process engineer
Verfahrensschritt *m* process(ing) step, treatment step
Verfahrenssteuerung *f* process control
Verfahrensstufe *f* process(ing) stage, treatment stage
Verfahrenstechnik *f* process engineering
Verfahrenstechnik *f***/biochemische** biochemical engineering
Verfahrenstechnik *f***/biotechnologische** bioprocess engineering
Verfahrenstechnik *f***/chemische** chemical engineering
Verfahrenstechniker *m* process engineer
Verfahrensziel *n* process goal
verfärben *v* discolour, *(esp if locally:)* stain
verfärben *v***/sich** discolour, *(esp if locally:)* stain; fade *(to lose intensity of colour)*
Verfärbung *f* 1. *(process:)* discoloration, *(esp if locally:)* staining; fading *(loss of colour intensity)*; 2. *(state:)* discoloration, stain
verfaulen *v* putrefy
verfestigen *v* 1. compact, consolidate *(bulk material)*; solidify *(a fluid)*; 2. *(met)* work-harden *(by cold forming)*; 3. *(glass)* strengthen
verfestigen *v***/sich** 1. compact, consolidate *(as of bulk material)*; *(relating to a fluid)* solidify, set; 2. *(met)* harden *(as by cold forming)*
Verfestigen *n* 1. *(met)* (work-)hardening *(as by cold forming)*; 2. *(glass)* strengthening
Verfestigung *f* 1. compaction, consolidation *(of bulk material)*; solidification, set(ting) *(of a fluid)*; 2. *s.* Verfestigen
Verfestigungsmittel *n* solidifying agent
verfilzen *v***/sich** felt (together), mat (together) *(as of fibres)*
verflüchtigen *v* volat(il)ize
verflüchtigen *v***/sich** volat(il)ize
Verflüchtigung *f* volatilization
verflüssigbar liquefiable
verflüssigen *v* liquefy *(a solid or gas)*
verflüssigen *v***/sich** liquefy *(of a solid or gas)*; run *(of metal)*
Verflüssiger *m* 1. liquefier *(in gas liquefaction)*; 2. condenser *(of a refrigerating machine)*
Verflüssigung *f* liquefaction *(of a solid or gas)*
Verfolgung *f* monitoring *(of chemical reactions)*
Verformbarkeit *f* deformability

verformen *v* 1. deform, *(relating to elastic material also)* strain; 2. *s.* umformen
verformen v/sich deform
Verformung *f* 1. deformation, *(relating to elastic material also)* strain; 2. *s.* Umformung
Verformung f/bleibende permanent [residual, plastic] deformation, permanent set
Verformung f/elastische elastic deformation
Verformung f/nicht elastische [plastische] *s.* Verformung/bleibende
Verformungsarbeit *f* resilience, resiliency
Verformungsverhalten *n (tech)* deformability behaviour
verfügbar available
verfügbar/nicht unavailable
Verfügbarkeit *f* availability
verfüttern *v (biot)* feed *(nutrients to microorganisms)*
vergällen *v* 1. denature, denaturize; 2. *(ferm)* methylate
Vergällung *f* 1. denaturation, denaturing; 2. *(ferm)* methylation
Vergällungsmittel *n* denaturant, denaturing agent
vergärbar fermentable
Vergärbarkeit *f* fermentability
vergären *v* ferment
vergasen *v* 1. gasify *(e.g. coal)*; 2. carburet *(motor fuel)*; volat(il)ize *(pesticides)*
vergasen v/hydrierend *(coal)* hydrogasify
Vergaser *m* 1. *(coal)* gasifier; 2. carburet(t)or, carburetter *(of an internal-combustion engine)*
Vergaserkraftstoff *m* carburetting fuel, carburant
Vergasung *f* 1. gasification *(as of coal)*; 2. carburetting *(of motor fuel)*; volatilization *(of pesticides)*
Vergasung f/hydrierende *(coal)* hydrogasification
Vergasung *f* **in der Flugstaubwolke [Staubwolke]** *(coal)* suspension gasification, dilute-phase [entrained-bed, entrainment] gasification, pulverized-coal gasification
Vergasungsmittel *n* fumigant, fumigator *(crop protection)*
vergesellschaften v/sich *(min)* associate
Vergesellschaftung *f (min)* association
vergießbar castable, pourable
Vergießbarkeit *f* castability, pourability
vergießen *v* cast, pour *(molten material into a mould)*
Vergießen *n (plast)* encapsulation
vergiften *v* poison, toxify
Vergiftung *f* poisoning, toxification
Vergiftung f/absichtliche deliberate poisoning *(of a catalyst)*
Vergiftung f/tödliche fatal poisoning
Vergiftungsgefahr *f* poisoning hazard
vergilben *v* 1. (go) yellow; 2. *(pap)* discolour, age
vergilbend/nicht non-yellowing
Vergilbung *f* 1. yellowing; 2. *(pap)* discoloration, ag(e)ing
vergilbungsbeständig non-yellowing
verglasen *v (ceram, geoch)* vitrify
Verglasung *f (ceram, geoch)* vitrification
Vergleichselektrode *f* reference electrode [element]
Vergleichsküvette *f* reference cell
Vergleichslösung *f* standard [reference] solution
Vergleichsmassestück *n* balance weight
Vergleichsprobe *f* 1. standard [blank] sample, control, reference (sample, material); 2. *s.* Vergleichstest
Vergleichsspektrum *n* reference [comparison] spectrum
Vergleichsstrahl *m (spectr)* reference beam
Vergleichstest *m* blank test
Vergleichsverbindung *f* reference [comparison] compound
vergolden *v* gild
Vergraben *n* burial *(as of industrial wastes)*
vergrauen *v (text)* grey
vergrößern *v* scale up *(a plant)*
vergrößern v/im Maßstab scale up *(a plant)*
Vergrößerung *f* **(im Maßstab)** scale-up *(of a plant)*
Verhakung *f* entanglement *(of polymer chains)*
Verhakungsnetzwerk *n* entanglement [physical] network
verhalten v/sich behave *(of material)*

verhalten *v*/**sich ideal** behave perfectly
Verhalten *n* behaviour, performance *(of material)*
Verhalten *n* **bei hohen Temperaturen** high-temperature behaviour
Verhalten *n* **bei Kälte** low-temperature behaviour
Verhalten *n*/**chemisches** chemical behaviour
Verhalten *n*/**gummielastisches** rubber-elasticity behaviour
Verhalten *n*/**hookesches** Hookean behaviour *(of elastic material)*
Verhalten *n*/**newtonsches** Newtonian behaviour *(of viscous fluids)*
Verhalten *n*/**nicht-newtonsches** non-Newtonian behaviour *(of viscous fluids)*
Verhalten *n*/**rheologisches** rheological behaviour
Verhalten *n*/**rheopexes** rheopectic behaviour, antithixotrophy
Verhalten *n*/**strukturviskoses** shear thinning, pseudoplasticity
Verhalten *n*/**wasserabweisendes** water repellency
Verhältnis *n* relation; *(quantitatively:)* ratio, proportion • **in jedem Verhältnis mischbar** miscible in all proportions • **in molekularem Verhältnis** in molecular proportions
Verhältnis *n*/**atomares** *(anal)* atomic (combining) ratio, atomic proportion, proportion by atoms
Verhältnis *n*/**gyromagnetisches** (nuclear) gyromagnetic ratio
Verhältnisformel *f* stoichiometric formula, simplest (possible) formula
verhärten *v* harden, solidify
Verhärtung *f* hardening, solidification
verharzen *v* resinify, *(specif petrol)* gum
Verharzung *f* 1. resinification; 2. *(petrol)* gum formation, gumming
verhütten *v* smelt *(ores)*
verkapseln *v* encapsulate
Verkapselung *f* encapsulation
verketten *v* link *(molecules)*
Verkettung *f* linking, linkage *(of molecules)*; concatenation *(of reactions)*
verkieseln *v* *(geoch)* silicify
Verkieselung *f* *(geoch)* silicification
verkippen *v* dump, dispose of to tip, dispose of by tipping, deposit [place] on a tip *(waste products)*
verkippen *v*/**ins Meer** dump at sea, dispose of to sea, dispose of in the ocean
Verkippen *n* dumping, disposal to tip, tipping *(of waste products)*
Verkippen *n* **ins Meer** dumping at sea, sea [ocean] dumping, sea burial
verkitten *v* putty, lute, cement, seal *(e.g. joints of tubes)*
verkleben *v* cement, glue, bond, agglutinate; stick together, agglutinate
Verklebung *f* cementation, gluing, bond, bonding, agglutination *(act)*; agglutination *(process)*
verkleiden *v* cover, case, jacket, sheathe, *(esp relating to metal:)* clad
Verkleidung *f* cover(ing), casing, jacket, sheathing, *(esp relating to metal:)* cladding
verklumpen *v* agglomerate, clog, clot, lump, agglutinate, curd(le)
Verklumpung *f* agglomeration, clogging, clotting, lumping, agglutination, curd(l)ing
verknappen *v* squeeze *(of resources)*
Verknappung *f* squeeze *(of resources)*
verknüpfen *v* link (together), bond (together), couple *(atoms)*
verknüpfen *v*/**miteinander** link (together), bond (together), couple *(atoms)*
Verknüpfung *f* 1. linking, linkage, bonding, coupling *(of atoms)*; 2. concatenation *(of reactions)*
verkochen *v* boil down, concentrate; boil away
Verkochen *n* boildown, boiling-down, concentration; boiling-away
verkohlen *v* char, carbonize
Verkohlung *f* charring, carbonization
verkoken *v* coke, carbonize
Verkokung *f* coking, carbonization, *(specif)* high-temperature coking [carbonization]
Verkokungsgas *n* coke-oven gas, carbonization gas
Verkokungskammer *f* *(petrol, coal)* coking [coke] chamber
Verkokungsofen *m* coke [coking] oven
Verkokungsrückstand *m* carbon [coke] residue
Verkokungswert *m* carbon-residue value, coke value [number]

Verkokungszahl *f* carbon-residue value, coke value [number]
verkrusten *v* encrust, incrust, scale
Verkrustung *f* encrustation, incrustation, scaling
verküpen *v (dye)* vat
verkupfern *v* copper
verkupfern *v*/**galvanisch [elektrochemisch]** plate with copper, copper-plate
Verkupferungselektrolyt *m* copper plating bath [solution]
Verküpung *f (dye)* vatting
verlaben *v* rennet *(milk)*
verlagern *v* displace; shift *(e.g. an equilibrium)*; translocate *(e.g. nutrients in plants)*
verlagern *v*/**sich** shift *(as of an equilibrium)*
Verlagerung *f* displacement; shift *(as of an equilibrium)*; translocation *(as of nutrients in plants)*
verlängern *v* lengthen *(a chain)*
verlangsamen *v* 1. slow down, decelerate, retard; 2. *(nucl)* moderate
verlangsamen *v*/**sich** slow down, decelerate
Verlangsamung *f* 1. slowing-down, deceleration, retardation; 2. *(nucl)* moderation
Verlauf *m* 1. course, progress *(as of a reaction)*; 2. *(coat)* flow
Verlauf *m*/**zeitlicher** time course
verlaufen *v* 1. proceed, go, run *(as of reactions)*; 2. *(coat)* flow
verlaufen *v*/**glatt** proceed smoothly
verlaufen *v*/**schleppend** proceed poorly [sluggishly]
verlaufen *v*/**stürmisch** proceed vigorously
verlaufen *v*/**träge** *s.* verlaufen/schleppend
verleimen *v* cement, glue
verlieren *v* lose *(e.g. hydrate water)*
Verlust *m* loss
Verlust *m* **durch Mitreißen** entrainment loss
Verlustfaktor *m* loss tangent, dissipation factor
Verlustfaktor *m*/**dielektrischer** loss tangent, dissipation factor
Verlustwinkel *m* loss angle
vermahlen *v* grind, mill, *(relating to soft material also)* disintegrate, *(relating to hard material also)* pulverize
vermahlen *v*/**fein** grind finely
Vermahlen *n* grinding, milling, *(relating to soft material also)* disintegration, *(relating to hard material also)* pulverization
vermälzen *v* malt
vermehren *v* 1. *(nucl)* breed *(fissionable material)*; multiply *(neutrons)*; 2. *(biot)* propagate, multiply
vermehren *v*/**sich** 1. *(nucl)* multiply *(of neutrons)*; 2. *(biot)* propagate, multiply
Vermehrung *f* 1. *(nucl)* breeding *(of fissionable material)*; multiplication *(of neutrons)*; 2. *(biot)* propagation, multiplication
vermengen *v* mingle, mix *(solids)*
vermischen *v* mix, *(esp relating to solids:)* mingle, *(specif if so that the components cannot be distinguished:)* blend
vermischen *v*/**innig** mix intimately, blend
vermischen *v*/**sich** (inter)mix, blend
Vermischen *n* mix(ing), mixture, *(esp relating to solids:)* mingling, *(specif if so that the components cannot be distinguished:)* blending
Vermischen *n*/**vollständiges** complete mixing
Vermischen *n* **von Feststoffkomponenten** dry mixing [blending]
Vermizid *n* vermicide
vernachlässigbar negligible
vernebeln *v* nebulize, aerosolize
Verneb(e)lung *f* nebulization, aerosolization
vernetzbar 1. cross-linkable, curable; 2. *(rubber)* vulcanizable
vernetzen *v* 1. cross-link, cure; 2. *(rubber)* vulcanize
vernetzen *v*/**räumlich** enmesh *(e.g. starch molecules)*
Vernetzer *m s.* Vernetzungsmittel
vernetzt/nicht 1. uncrosslinked, uncured; 2. *(rubber)* unvulcanized
Vernetzung *f* 1. cross-linking, crosslinkage, cure; 2. *(rubber)* vulcanization
Vernetzung *f* **mit Schwefel** *(rubber)* (elemental-)sulphur vulcanization
Vernetzung *f*/**photochemische** photocrosslinking

Vernetzung f/räumliche enmeshing *(as of starch molecules)*
Vernetzung f/schwefelfreie *(rubber)* sulphurless [non-sulphur] vulcanization
Vernetzungsdichte *f* cross-link density, degree of cross-linking
Vernetzungsgrad *m* cross-link density, degree of cross-linking
Vernetzungsmittel *n* 1. cross-linking agent, curing agent; 2. *(rubber)* vulcanizing agent
Vernetzungspunkt *m (plast)* junction *(polymer networks)*
Vernetzungsstelle *f* cross-link(age)
vernichten *v (phys ch)* annihilate
Vernichtung *f (phys ch)* annihilation
Vernichtungsstrahlung *f (phys ch)* annihilation radiation
vernickeln *v* nickel(ize)
vernickeln v/elektrochemisch [galvanisch] plate with nickel, nickel-plate
Vernickeln *n* nickelization
Vernickeln n/elektrochemisches [galvanisches] nickel plating
Vernickelungsbad *n* nickel plating bath [solution]
Vernickelungselektrolyt *m* nickel plating bath [solution]
verolen *v* olate *(of hydroxo compounds)*
Verolung *f* olation *(of hydroxo compounds)*
verpacken *v* 1. pack(age); 2. *(met)* pack *(with a carburizing powder)*
Verpacken *n* 1. pack(ag)ing, package; 2. *(met)* packing *(in a carburizing powder)*
Verpackungsmaterial *n* pack(ag)ing material, package, packing, *(relating to paper, plastic film or cloth also)* wrapping
verperlen *v* prill
verpressen v/heiß hot-press, *(relating to powders also)* sinter under pressure
verpressen v/zu Briketts briquette
verprillen *v* 1. *(biot)* prill, microencapsulate *(an enzyme)*; 2. *s.* verperlen
Verprillung *f (biot)* prilling, (micro)encapsulation *(of an enzyme)*
verpuffen *v* deflagrate
Verpuffung *f* deflagration
Verrauchungsfahne *f* fumigating plume
verreiben *v* 1. grind, powder, pulverize; 2. *(pharm)* triturate
verreiben v/gemeinsam powder together
Verreiben *n* 1. grinding, powdering, pulverization; 2. *(pharm)* trituration
Verreibung *f* trituration *(medicament)*
verrohren *v* pipe, furnish [equip] with pipes
Verrohrung *f* piping, tubing *(act or material)*
verrosten *v* rust, corrode
verrotten *v* rot
verrottungsfest rot-resistant, rotproof
Verrottungsfestigkeit *f* rot resistance, rotproofness
Verrottungsschutzmittel *n* rotproofing agent
verrühren *v* 1. *(lab)* stir up; 2. *(tech)* mix up
Versand *m* shipment
Versandvorschrift *f* shipping regulation
Versatz *m* 1. *(ceram)* batch (composition); 2. blinding *(as of screens, filter cloth)*
versauern *v (agric)* acidify *(a soil, as of fertilizers)*; sour, become acid *(of soils)*
Versauerung *f (agric)* souring *(of soils)*
verschäumbar *(plast)* foamable, expandable, expandible
verschäumen *v (plast)* foam, expand, froth
verschiebbar shiftable, *(relating to atoms or radicals also)* displaceable
verschieben *v* shift, *(relating to atoms or radicals also)* displace
verschieben v/sich shift *(frequency, electron, equilibrium)*
Verschiebung *f* shift, *(relating to atoms or radicals also)* displacement
Verschiebung f/bathochrome bathochromic shift
Verschiebung f/chemische *(spectr)* chemical shift, nmr shift
Verschiebung f/hypsochrome hypsochromic shift, blue-shift
Verschiebungsgesetz n/wiensches Wien displacement law *(a radiation law)*
Verschiebungspolarisation *f* induced [distortion] polarization
Verschiebungssatz *m (nucl)* (radioactive-)displacement law, group displacement law
Verschiebungssatz m/sommerfeld-

kosselscher *s.* Verschiebungssatz/spektroskopischer
Verschiebungssatz *m*/**spektroskopischer** spectroscopic displacement law (of Kossel and Sommerfeld)
verschießen *v (text)* fade
verschimmeln *v* mould
verschlacken *v* slag, scorify *(impurities)*; slag *(of impurities)*; clinker *(of ashes)*
Verschlackung *f* slagging, scorification; clinkering *(of ashes)*
verschlechtern *v* deteriorate
verschlechtern *v*/**sich** deteriorate
Verschlechterung *f* deterioration
Verschleiß *m* wear
Verschleiß *m*/**reibender** abrasive wear, abrasion
verschleißen *v* wear (out)
verschleißfest 1. wear-resistant, resistant to wear; 2. *(tech)* abrasion-resistant, resistant to abrasion
Verschleißfestigkeit *f* 1. wear resistance, resistance to wear; 2. *(tech)* abrasion resistance, resistance to abrasion
verschleppen *v* carry over [through]
Verschleppen *n* carry-over
verschließen *v* close; stopper, plug *(e.g. a flask)*; *(if gas-tight or water-tight:)* seal
verschlucken *v* absorb *(rays)*
Verschluss *m* 1. *(appurtenance:)* closure, *(if gas-tight or water-tight:)* seal; 2. *(act:)* closure; sealing
verschmelzen *v* 1. fuse, melt (together); seal *(as in working glass)*; coalesce *(as of gas bubbles)*; 2. *(nucl)* fuse, merge
verschmelzen *v*/**auf Stein** *(met)* matte-smelt
Verschmelzung *f* coalescence *(as of gas bubbles)*
verschmutzen *v* soil, *(esp relating to the environment:)* pollute, *(esp with poisonous matter:)* contaminate; be soiled
verschmutzt/durch Abwässer sewage-contaminated
verschmutzt/nicht unpolluted, uncontaminated, non-contaminated, free from polluting substances *(water, air)*
Verschmutzung *f* 1. *(act or process:)* soiling, contamination, *(esp relating to the environment:)* pollution; 2. *(state:)* pollution, contamination, dirtiness; 3. contaminant, soil, *(esp in the environment:)* pollutant, polluting matter
Verschmutzung *f*/**anthropogene** human(-induced) pollution
Verschmutzung *f* **durch Abwässer** *(hyd)* sewage pollution [contamination]
Verschmutzung *f* **durch Schadstoffe** pollution
Verschmutzungsgrad *m* 1. degree of soiling; degree of pollution *(of air or water)*; 2. *(hyd)* contaminant [pollutant] concentration
verschneiden *v* 1. *(food, rubber)* blend; 2. *(coat)* dilute *(solvents)*; 3. *(coat)* extend *(pigments)*; cut back, flux *(high-boiling petroleum fractions)*
Verschneidmittel *n (coat)* diluent, diluting agent, indirect [latent] solvent; (paint) extender, (extending) filler *(for pigments)*
Verschnitt *m* 1. *(food, rubber)* blending *(process)*; 2. *(food, rubber)* blend *(product)*
Verschnittbitumen *n* cutback (bitumen), bitumen cutback
Verschnittprodukt *n (petrol)* cutback product
verschreiben *v* prescribe *(a medicine)*
verschütten *v* spill
verschwelen *v* carbonize *(e.g. coal at low temperatures)*; char *(organic matter by heat)*; char *(of organic matter under the influence of heat)*
Verschwelung *f (coal)* (low-temperature) carbonization, low-temperature coking
versehen *v*/**mit einem Deckanstrich [Schlussanstrich]** *(coat)* finish
verseifbar saponifiable
verseifbar/nicht unsaponifiable, non-saponifiable
Verseifbares *n* saponifiable matter
verseifen *v* saponify
verseift/partiell partly saponified
Verseifung *f* saponification
Verseifung *f*/**alkalische** alkaline saponification
Verseifungszahl *f* saponification value [number], sap. value, S.V., Koettstorfer number [value]
versengen *v* scorch
versenken *v* sink, submerge, *(relating to industrial wastes:)* bury
Versenken *n* sinking, submergence,

submersion, *(relating to industrial wastes:)* burial
Versenken *n* **ins Meer** sea disposal [burial]
versetzen *v* 1. *(filtr)* clog, blind; 2. lace *(with a minor quantity of something)*; add
versetzen *v*/**in Schwingungen** vibrate
versetzen *v*/**mit Citrat** citrate
versetzen *v*/**mit Hefe** yeast
versetzen *v*/**mit Luft** aerate
versetzen *v*/**mit Malz** malt
versetzt/gegeneinander *(cryst)* dislocated; staggered *(stereochemistry)*
Versetzung *f (cryst)* dislocation
verseuchen *v* contaminate
verseuchen *v*/**radioaktiv** contaminate
Verseuchung *f* (radioactive) contamination
Verseuchung *f*/**radioaktive** (radioactive) contamination
versickern *v* seep (away), soak (away), percolate, *(esp if slowly:)* ooze (away)
Versickerung *f* seepage, soaking, percolation, *(esp if slowly:)* oozing; floor drain *(waste-water treatment)*
versiegeln *v* seal
versilbern *v* silver
versilbern *v*/**elektrochemisch [galvanisch]** silver-plate
Versilbern *n* silvering
Versilbern *n*/**elektrochemisches [galvanisches]** silver plating
Versilberungsbad *n* silver-plating bath [solution]
Versilberungselektrolyt *m* silver-plating bath [solution]
verspinnbar spinnable
Verspinnbarkeit *f* spinnability
verspinnen *v* spin *(a dope)*
verspritzen *v* splash, spatter, spill
verspröden *v* embrittle
Versprödung *f* embrittlement
Versprödungstemperatur *f* brittle(ness) temperature, brittle-point temperature
versprühen *v* atomize *(liquids)*; spray
Versprüher *m* (liquid) atomizer, sprayer
Versprüher *m*/**rotierender** rotating [rotary] atomizer
Versprühung *f* atomization, spraying
verstärken *v* 1. fortify *(e.g. a construction)*; reinforce *(e.g. plastics)*; 2. intensify, potentiate *(an action)*; 3. fortify, strengthen *(solutions)*; *(esp by evaporation:)* graduate
Verstärker *m*/**synergetischer** activator, (catalyst) promoter, activating [promoting] agent
Verstärkerfüllstoff *m (rubber)* reinforcing filler [ingredient, pigment], active filler
Verstärkung *f* 1. fortification *(as of a construction)*; reinforcement *(as of plastics)*; 2. intensification, potentiation *(as of an action)*; 3. fortification, strengthening *(of solutions)*; *(esp by evaporation:)* graduation
Verstärkungsgerade *f (distil)* enrichment [rectifying operating] line
Verstärkungslinie *f (distil)* enrichment [rectifying operating] line
Verstärkungsmaterial *n* reinforcing agent [material] *(as for plastics)*
Verstärkungsteil *m (distil)* enriching section, rectifying section
Verstärkungsverhältnis *n*/**mittleres** *(distil)* overall column [plate] efficiency *(ratio of number of theoretical plates to number of actual plates)*
verstäuben *v* dust *(e.g. pesticides)*
versteifen *v* stiffen
verstopfen *v* clog (up), choke (up), blind, plug *(e.g. screen openings)*
verstopfen *v*/**sich** clog, choke, plug *(as screen openings)*
Versuch *m* experiment, trial, assay, *(if conducted under specified conditions:)* test • **einen Versuch durchführen** run [carry out] an experiment, perform [conduct] an experiment • **Versuche anstellen** experiment(alize), run [carry out] experiments
Versuchsanlage *f* experimental plant
Versuchsanlage *f*/**halbtechnische** pilot plant, semiworks
Versuchsanordnung *f* experimental [test] arrangement
Versuchsdurchführung *f* performance of the experiment; experimental procedure, test practice *(know-how)*
Versuchsfehler *m* experimental error
Versuchssubstanz *f* experimental [test] substance [material], substance for [under] investigation
Vertauschungswägung *f* **(nach Gauß)** (Gauss's method of) double weighing

verteilen *v* distribute, partition; spread (out) *(over a surface)*; disperse *(e.g. solid particles in a liquid)*; homogenize *(e.g. for forming an emulsion)*
Verteiler *m* distributor, distributing device
Verteilerwirksamkeit *f* spreading efficiency *(of surfactants)*
verteilt/fein finely dispersed [divided]
verteilt/regellos [statistisch, zufällig] randomly distributed
Verteilung *f* distribution, partition; spreading *(over a surface)*; dispersion *(as of solid particles in a liquid)*; homogenization *(as for forming an emulsion)* • **in feiner Verteilung** finely divided, disperse
Verteilung *f*/bimodale bimodal [double-peaked] distribution *(the occurrence of two maxima in a frequency distribution)*
Verteilung *f*/fraktionierte countercurrent extraction [separation]
Verteilung *f*/klassische *(phys ch)* classical distribution
Verteilung *f*/regellose [statistische, zufällige] random distribution
Verteilung *f*/zweigipflige *s.* Verteilung/bimodale
Verteilungschromatographie *f* partition [extraction] chromatography
verteilungschromatographisch by partition chromatography, partition-chromatographic
Verteilungsfunktion *f* 1. distribution function; 2. *(phys ch)* partition function
Verteilungsfunktion *f*/gaußsche [normale] Gaussian [normal] distribution function
Verteilungsfunktion *f*/radiale radial distribution [probability] function *(for electrons)*
Verteilungsgesetz *n* distribution [partition] law
Verteilungsgleichgewicht *n* distribution equilibrium
Verteilungskoeffizient *m* distribution [partition] coefficient [ratio], *(in zone refining also)* segregation coefficient
Verteilungskonstante *f* distribution [partition] coefficient [ratio], *(in zone refining also)* segregation coefficient
Verteilungskurve *f* distribution curve
Verteilungskurve *f*/gaußsche Gaussian curve *(statistics)*
Verteilungssatz *m*/nernstscher Nernst distribution [partition] law
Verteilungsverhältnis *n* (mass) distribution ratio, partition ratio, capacity factor
vertiefen *v* intensify *(a colour)*
Vertiefung *f* intensification *(of a colour)*
Vertikalbeziehung *f* family relationship *(in the periodic system)*
Vertikalofen *m* vertical oven *(as for gas production)*
Vertikalrohrverdampfer *m* vertical (tube) evaporator
Vertikalspalte *f* vertical column *(in the periodic system)*
verträglich compatible
verträglich/biologisch biocompatible
Verträglichkeit *f* compatibility
Verträglichkeit *f*/biologische *s.* Biokompatibilität
Verträglichkeitsmacher *m s.* Verträglichkeitsvermittler
Verträglichkeitsvermittler *m* compatibilizer
Vertrauensbereich *m* confidence interval *(statistics)*
Vertrauensgrenze *f* confidence limit *(statistics)*
vertreiben *v* drive off [out], expel, dispel *(volatile matter)*; repel *(e.g. insects by repellents)*
Vertreibung *f* expulsion *(of volatile matter)*; repulsion *(as of insects by repellents)*
verunreinigen *v* soil, *(esp relating to the environment:)* pollute, *(esp with poisonous matter:)* contaminate
Verunreinigung *f* 1. soiling, contamination, *(esp relating to the environment:)* pollution; 2. impurity, foreign matter *(in materials)*; contaminant, soil, *(esp in environment:)* pollutant, polluting matter, *(esp in zone melting:)* solute
Verunreinigung *f*/biologische biological pollution
Verunreinigung *f*/zufällige chance contaminant
Verunreinigungen *fpl*/gelöste dissolved solid pollutants
Verunreinigungen *fpl*/suspendierte suspended (solid) pollutants

verwandt related, allied *(chemical substances)*
Verweilzeit *f* 1. residence time, retention [hold-up, holding, detention] time; 2. *(bioch)* turnover time
Verweilzeit *f***/mittlere** average [mean] residence time
Verweilzeitverteilung *f* residence-time distribution, RTD
verwendbar applicable
verwendbar/allgemein [universell] generally applicable, general-purpose
Verwendbarkeit *f* applicability
verwenden *v* apply, use
verwerfen *v (lab)* discard, reject
verwesen *v* decay, decompose, undergo decomposition
Verwesung *f* decay, decomposition
verwirbeln *v* vortex *(material in a cyclone)*
Verwirbelung *f* vortexing *(of material in a cyclone)*
verwittern *v* weather
verwittern *v***/unter Kristallwasserverlust** *(cryst)* effloresce
Verwitterung *f* weathering, decay, disintegration
Verwitterung *f* **unter Kristallwasserverlust** *(cryst)* efflorescence
Verwitterungsprodukt *n* weathered [weathering] product
verzerren *v* distort *(e.g. an electron cloud)*
Verzerrung *f* distortion *(as of an electron cloud)*
verzinken *v* zinc, *(esp by hot-dipping:)* galvanize
verzinken *v***/elektrochemisch [galvanisch]** electrogalvanize, zinc-plate
verzinken *v***/im Pulver** sherardize
Verzinken *n* zinc coating, zinc(k)ing, *(esp by hot-dipping:)* galvanizing, galvanization
Verzinken *n***/elektrochemisches [galvanisches]** electrogalvanizing, zinc plating, wet [cold] galvanizing
Verzinken *n* **im Pulver** sherardizing
Verzinkung *f s.* Verzinken
Verzinkungsbad *n* zinc plating bath [solution]
Verzinkungselektrolyt *m* zinc plating bath [solution]
verzinnen *v* tin
verzinnen *v***/elektrochemisch [galvanisch]** tin-plate, electrotin
Verzinnen *n* tinning
Verzinnen *n***/elektrochemisches [galvanisches]** tin plating, electrotinning
Verzinnungsbad *n* tin plating bath [solution]
Verzinnungselektrolyt *m* tin plating bath [solution]
Verzögerer *m* retarder, retarding agent, inhibitor, anticatalyst, negative catalyst
verzögern *v* retard, inhibit, delay
verzögern *v***/sich** retard
verzögernd retardant, inhibitory, inhibitive
Verzögerung *f* retardation, inhibition, delay
Verzögerung *f***/sterische** *(org ch)* steric retardation, F strain effect
Verzögerungsphase *f* lag phase [time]
verzuckern *v* saccharify
Verzuckerung *f* saccharification
Verzuckerung *f***/enzymatische** *(food)* enzymatic [enzyme] conversion
Verzuckerung *f* **mit Säure** *(food)* acid conversion
verzundern *v (met)* scale
Verzunderung *f (met)* scaling
verzweigen *v***/sich** branch *(molecules and reaction chains)*
verzweigt 1. branched; 2. *(cryst)* dendritic(al)
verzweigt/schwach lightly branched
verzweigt/stark highly branched
verzweigtkettig branched-chain
Verzweigung *f* branching *(process)*; branch *(state)*
Verzweigungsenzym *n (bioch)* branching enzyme
Verzweigungsgrad *m* degree of branching
Verzweigungsstelle *f* branch point *(of a molecule or of a metabolic pathway)*
Vetiveröl *n* vetiver [cuscus] oil, vetivert *(from Vetiveria zizanioides (L.) Nash)*
VI *s.* 1. Viskositätsindex; 2. Viskosefaserstoff
Vibration *f* vibration, oscillation
Vibrationssieb *n* vibrating [reciprocating] screen
Vibrationssortierer *m* vibrating [reciprocating] screen
vibrieren *v* vibrate, oscillate

vibrieren lassen *v* vibrate, oscillate
vic. *s.* vicinal
Vicat-Zahl *f (plast)* Vicat softening point [temperature], V.S.P., Vicat needle point
vicinal vicinal, in 1,2,3 position, in 1,2,3,4 position
Vickers-Härte *f* Vickers hardness, diamond pyramid hardness, DPH
vielatomig polyatomic
Vielfachapparat *m* multiple (unit)
vielflächig *(cryst)* polyhedral
vielgestaltig polymorphic, polymorphous
vielgliedrig multimembered, many-membered *(ring)*
Vielkristall *m* polycrystal
vielstufig multistep *(process)*
vieratomig tetra-atomic
vierbasig tetrabasic, tetraprotic *(acid)*; tetrabasic, tetra-acid *(base)*
vierfach positiv (geladen) tetrapositive
viergliedrig four-membered
Vierhalskolben *m* four-neck(ed) flask
Vierkohlenstoffverbindung *f (bioch)* four-carbon compound
Vierkomponentensystem *n* four-component system, quaternary system
Vierkörperverdampfer *m* quadruple-effect (evaporator)
vierprotonig tetraprotic, tetrabasic *(acid)*
Vierring *m* four-membered ring
viersäurig tetra-acid, tetrabasic *(base)*
Vierstufenreaktion *f* four-step reaction
vierteln *v (anal)* (cone and) quarter
Viertelung *f (anal)* (coning and) quartering
Vierwegehahn *m* four-way valve
Vierwegeventil *n* four-way valve
vierwertig tetravalent, quadrivalent *(element)*; tetrabasic, tetraprotic *(acid)*; tetrabasic, tetra-acid *(base)*; tetrahydric *(alcohol)*
vierwertig/koordinativ tetracovalent
Vierwertigkeit *f* tetravalence, tetravalency, quadrivalence, quadrivalency *(of an element)*; tetrabasicity *(of an acid or base)*
vierzählig 1. *(cryst)* tetrad; 2. *s.* vierzähnig
vierzähnig tetradentate, quadridentate *(ligand)*
Vigreux-Kolonne *f (distil)* Vigreux column
Vilsmeier-Haack-Reaktion *f* Vilsmeier-Haack synthesis [reaction] *(for obtaining aldehydes by formylation)*
Vilsmeier-Haack-Synthese *f* Vilsmeier-Haack synthesis [reaction] *(for obtaining aldehydes by formylation)*
Vinylacetat *n* vinyl acetate
Vinylalkohol *m* vinyl alcohol, *ethenol
Vinylbenzen *n* vinylbenzene, phenylethylene, styrene
Vinylbenzol *n* vinylbenzene, phenylethylene, styrene
Vinylbromid *n* vinyl bromide, *bromoethene
Vinylchlorid *n* vinyl chloride, *chloroethene
Vinylcyanid *n* vinyl cyanide, *cyanoethene, acrylonitrile
Vinylengruppe *f* –HC=CH– vinylene group
Vinylester *m* vinyl ester
Vinylgruppe *f* CH_2=CH– vinyl [vinylic] group [residue]
Vinylhalogenid *n* vinyl halide, vinyl halogenide
Vinylharz *n* vinyl [ethenoic] resin
Vinylharzkunststoff *m* vinyl plastic
vinylhomolog *s.* vinylog
Vinylidenchlorid *n* 1,1-dichloroethylene, vinylidene chloride
Vinylkation *n* vinyl(ic) cation
vinylog vinylogous
Vinyloge *n* vinylog(ue), vinyl homologue
Vinylogie *f* vinylogy
Vinylpolymerisation *f* vinyl polymerization
Vinylradikal *n* vinyl radical, *(specif)* CH=CH_2 free vinyl radical
Vinylsulfonfarbstoff *m* vinyl sulphone dye
Vinylverbindung *f* vinyl compound
Violursäure *f* violuric acid, 5-isonitrosobarbituric acid
Virialgleichung *f (phys ch)* virial equation (of state)
Virialkoeffizient *m (phys ch)* virial coefficient
Viruseiweiß *n* viral protein
Virushüllprotein *n* (viral) coat protein, capsid protein
Virusprotein *n* viral protein

virustötend virucidal, viricidal, antiviral
viruzid virucidal, viricidal, antiviral
viskoelastisch viscoelastic
Viskoelastizität *f (plast)* viscoelasticity, viscous elasticity
viskos viscous
Viskose *f* viscose
Viskoseerspinnlösung *f* viscose (spinning) solution, viscose dope
Viskosefaser *f* viscose (staple) fibre
Viskosefaserstoff *m* viscose fibre
Viskoseseide *f (text)* viscose rayon, viscose silk
Viskosimeter *n* viscometer, viscosimeter
Viskosimetrie *f* viscometry, viscosimetry
viskosimetrisch viscometric, viscosimetric
Viskosität *f* viscosity
Viskosität *f*/**dynamische** dynamic [absolute] viscosity; *(quantitatively:)* viscosity coefficient
Viskosität *f*/**kinematische** kinematic viscosity, viscosity/density ratio
Viskositätsbrechen *n (petrol)* viscosity breaking, visbreaking
Viskositäts-Dichte-Verhältnis *n* viscosity/density ratio, kinematic viscosity
Viskositätsindex *m* viscosity index, V.I. *(of lubricating oils)*
Viskositätskoeffizient *m* viscosity coefficient
Viskositätskonstante *f* viscosity coefficient
Viskositätsmittel *n* viscosity average *(of relative molecular mass)*
Viskositäts-Temperatur-Koeffizient *m* viscosity-temperature coefficient, temperature coefficient of viscosity
Vitamere *n* vitamer
Vitamin *n* vitamin • **mit Vitaminen anreichern** *s.* vitaminisieren
Vitamin *n*/**antihämorrhagisches** antihaemorrhagic vitamin, phylloquinone, vitamin K
Vitamin *n*/**antirachitisches** antirachitic vitamin, vitamin D
Vitamin *n*/**antiskorbutisches** antiscorbutic vitamin, vitamin C
Vitamin *n*/**antixerophthalmisches** antixerophthalmic vitamin, vitamin A
Vitaminaufnahme *f* vitamin intake
vitaminisieren *v* vitaminize, fortify [enrich] by vitamins
Vitaminisierung *f* vitaminization
Vitaminkonzentrat *n* vitamin concentrate
Vitaminmangel *m* vitamin deficiency
vitaminreich rich in vitamins, vitamined
Vitriol *n* vitriol *(a hydrated sulphate of a divalent metal)*
Vitrokeram *n* vitroceramic, glass ceramic, devitrified [neo-ceramic] glass
Vitrokeramik *f* vitroceramic, glass ceramic, devitrified [neo-ceramic] glass
VK *s.* Vergaserkraftstoff
Vlies *n* 1. *(text)* fleece; 2. *(glass)* mat
V-Mischer *m* vee-type [twin-shell] blender [mixer]
Vogelabschreckmittel *n* bird repellent
Vogelbekämpfungsmittel *n* avicide
Vol.% *s.* Volumenprozent
Vollanalyse *f* complete analysis
vollenthärten *v (hyd)* soften completely
Vollenthärtung *f (hyd)* complete softening, softening to zero hardness
vollentsalzen *v (hyd)* demineralize, *(by ion exchange)* deionize (completely)
Vollentsalzung *f (hyd)* demineralization *(by ion exchange)*; (complete) deionization, DI
Vollentsalzungsanlage *f (hyd)* demineralizing system, demineralizer, deionization unit
Vollpipette *f* transfer pipette, volumetric [bulb] pipette
Vollreinigung *f (hyd)* complete purification [treatment], complete [overall] removal
vollsaugen *v*/**sich** soak
Voltameter *n* voltameter, coulo(mb)meter
Voltammetrie *f (anal)* voltammetry
voltammetrisch voltammetric
Volta-Potenzial *n* Volta [outer] potential
Volumen *n* volume • **mit konstantem Volumen** constant-volume
Volumen *n*/**ausgeschlossenes** excluded volume *(of a coiled macromolecule)*
Volumen *n*/**kritisches** critical volume
Volumen *n*/**partielles molares** partial molar [molal] volume
Volumen *n*/**spezifisches** specific volume
Volumen *n*/**stoffmengenbezogenes** molar volume

Volumenabnahme *f* decrease [diminution] in volume, volume decrease
Volumenänderung *f* change in volume, volume change
Volumenarbeit *f* work of expansion, expansion work, p,V-work
Volumenarbeit *f*/**reversible isotherme [technische]** reversible isothermal work of expansion
Volumenausdehnung *f* cubic(al) expansion, volume expansion
Volumenausdehnungskoeffizient *m* coefficient of cubic(al) expansion, coefficient of volume expansion
Volumenausdehnungskoeffizient *m*/**thermischer** isobaric coefficient of thermal expansion, isobaric thermal expansivity
Volumendurchsatz *m* volume flux
Volumeneinheit *f* unit of volume • **je Volumeneinheit** per unit volume
Volumenmolarität *f* molarity, molar concentration
Volumenprozent *n* percentage [per cent] by volume
Volumenquellungsgrad *m* volume swelling degree, volume degree of swelling *(polymer networks)*
Volumenstrom *m* volumetric flow rate
Volumenteil *m* part by volume
Volumenverhältnis *n* proportion by volume, volume ratio
Volumenzunahme *f* increase in volume, volume increase
Volumetrie *f* volumetric analysis, *(using solutions also)* titrimetry, titrimetric analysis
volumetrisch volumetric, *(using solutions also)* titrimetric
Vorabscheider *m* precleaner *(air cleaning)*
Voranstrichstoff *m* undercoat paint
Voraufbereitung *f* pretreatment, preconditioning *(of water)*
Vorauflaufherbizid *n* pre-emergence herbicide [weed-killer]
vorbehandeln *v* pretreat, prepare, (pre)condition
Vorbehandlung *f* pretreatment, preparation, (pre)conditioning, preparatory [preliminary] treatment
vorbeiführen *v* by-pass
vorbereiten *v* prepare, prime
Vorbereitung *f* preparation, priming
vorbestrahlen *v (plast)* pre-irradiate
Vorbestrahlung *f (plast)* pre-irradiation
Vorbleiche *f (pap)* prechlorination
vorbleichen *v (pap)* prechlorinate
Vorblütenspritzung *f (agric)* pre-bloom spray
Vorbrenne *f s.* Ätzbad
vorchloren *v (hyd)* prechlorinate
vorchlorieren *v (pap)* prechlorinate
Vorchlorierung *f (pap)* prechlorination
Vorchlorung *f (hyd)* prechlorination
Vorderflanke *f (anal)* front *(of a peak)*
Vorderfront *f (anal)* front *(of a peak)*
Vordestillationskolonne *f* primary column
Vorelektrolyse *f* pre-electrolysis
vorerhitzen *v* preheat
Vorfermentation *f (biot)* prefermentation
Vorfermenter *m (biot)* prefermenter
Vorfilter *n* prefilter, roughing filter
Vorfluter *m (hyd)* receiving (body of) water, *(specif)* receiving stream
vorflutergerecht *(hyd)* acceptable for discharge to a receiving stream
vorfluterwürdig *(hyd)* acceptable for discharge to a receiving stream
Vorformling *m (plast)* preform, *(if hollow:)* parison, *(if massive:)* pill
vorfraktionieren *v (petrol, chromat)* prefractionate
Vorfraktionierung *f (petrol, chromat)* prefractionation
Vorgang *m* process
Vorgang *m*/**irreversibler [nicht umkehrbarer]** irreversible process
Vorgang *m*/**reversibler [umkehrbarer]** reversible process
Vorgarn *n* 1. roving; 2. *(glass)* sliver
Vorgärung *f* prefermentation
vorgerben *v* pretan
Vorgerbung *f* pretannage
vorhärten *v (plast)* precure
Vorhärtung *f (plast)* precure, precuring
vorheizen *v* 1. preheat; 2. *(rubber)* prevulcanize, precure
Vorheizer *m* preheater
Vorheizofen *m* preheater
Vorheizung *f* 1. preheating; 2. *(rubber)* prevulcanization, precure, set cure, semicure
Vorheizzone *f* preheating zone [compartment]

Vorhydrolyse *f* 1. prehydrolysis; 2. *(pap)* pre-impregnation, preliminary impregnation [penetration], presoaking, steeping, steepage *(of chips)*
vorimprägnieren *v* 1. pre-impregnate; 2. *(plast)* precompound; 3. *(pap)* presoak, steep
Vorimprägnierung *f* 1. pre-impregnation, preliminary impregnation [penetration]; 2. *(plast)* precompounding; 3. *s.* Vorhydrolyse
Vorklärbecken *n (hyd)* primary clarifier [tank], primary settling [sedimentation] tank, presedimentation [primary sedimentation] basin
vorklären *v (hyd, filtr)* preclarify
Vorklärung *f* 1. *(hyd, filtr)* preclarification; 2. *(hyd)* primary sedimentation, presedimentation
Vorkommen *n* 1. occurrence; 2. *(geol)* deposit
Vorkommen *n***/natürliches** occurrence in nature, *(esp of a chemical element:)* natural abundance
vorkommend/natürlich naturally occurring, found in nature, from natural sources, native
vorkonzentrieren *v* preconcentrate
Vorkonzentrierung *f* preliminary concentration, preconcentration
vorkühlen *v* precool, prechill
Vorkühlung *f* precooling, preliminary cooling, prechill
vorkultivieren *v (biot)* precultivate
Vorkultur *f (biot)* preculture, preliminary culture
Vorlage *f* 1. (distillate, distillation, still) receiver; 2. *(lab)* receiving flask
Vorlauf *m* 1. *(distil)* first runnings, forerun(s), forerunning(s), *(esp in distilling alcohol:)* heads, foreshot(s); 2. *(sugar)* green syrup
Vorläufer *m* precursor, progenitor
Vorläufersubstanz *f* precursor, progenitor
vormetallisieren *v* premetallize
vormischen *v* premix, preblend
Vormischung *f* 1. premix, preblend, initial mixture; 2. *(rubber)* masterbatch, mother stock • **Vormischungen herstellen** *(rubber)* masterbatch, mix into a masterbatch
Vor-Ort-Analyse *f* on-site analysis
Vorozon(is)ierung *f (hyd)* pre-ozonation
Vorperiode *f* pre-period, initial period *(as of calorimetric measurement)*
vorplastizieren *v (plast)* preplasticate, preplasticize
Vorplastizierung *f (plast)* preplastication, preplasticating, preplasticizing
Vorpolymer(e) *n* prepolymer
Vorpolymerisat *n* prepolymer
Vorprobe *f* preliminary test
Vorprodukt *n* intermediate (product)
Vorprüfung *f* preliminary examination
Vorprüfung *f***/trockene** dry analysis
Vorpumpe *f* forepump, backing pump *(vacuum technology)*
Vorrang *m* (nomenclature) priority, seniority, precedence
Vorrat *m* stock, store
Vorratsbehälter *m* 1. storage vessel; 2. *(tech)* storage tank [bin], hold(ing) tank, stock bin
Vorratsbunker *m* storage [stock] bin
Vorratsbürette *f* dispensing burette
Vorratsflasche *f* storage bottle, stock [dispensing] bottle
Vorratshaltung *f* stockpiling, storing
Vorratslösung *f* stock solution
Vorratsraum *m* store [stock] room, storage, store
vorreinigen *v* 1. preclean; 2. *(hyd)* pretreat, precondition *(water)*; preclarify *(waste water)*
Vorreiniger *m* precleaner
Vorreinigung *f* 1. precleaning, preliminary cleaning [purification]; 2. *(hyd)* pretreatment, preconditioning *(of water)*; pretreatment, primary treatment, primary [first] waste treatment stage, presedimentation, preclarification *(of waste water)*
Vorrichtung *f* contrivance, device
vorrösten *v (met)* preroast
Vorröstung *f (met)* preroast
Vorsaatanwendung *f* pre-sowing application *(of pesticides)*
Vorsaatherbizid *n* pre-sowing herbicide *(weed-killer)*
Vorsäule *f (chromat)* precolumn, guard column
Vorsäulenderivatisierung *f (chromat)* precolumn derivatization
vorschäumen *v (plast)* pre-expand, prefoam

vorschreiben *v* order, prescribe; specify *(the qualities of a product)*; formulate *(e.g. the mixture of a batch)*
Vorschrift *f* 1. instruction, direction; specification *(relating to the qualities of a product)*; formula *(as for mixing a batch)*; 2. *(pharm)* prescription • **den Vorschriften entsprechen** meet the specifications
vorsichtig gentle, slight *(heating)*
Vorsichtsmaßnahme *f* (safety) precaution • **Vorsichtsmaßnahmen treffen** take precautions
vorsintern *v* presinter
Vorsinterung *f* presintering
vorspannen *v* 1. prestress; 2. *(glass)* temper
Vorstoß *m* 1. adapter; 2. *(distil)* (condenser) adapter, receiver adapter [tube], delivery tube
Vorstoß *m* **für Frittentiegel** *(lab)* crucible adapter
Vorstreichfarbe *f* undercoat paint
Vorstufe *f* precursor, progenitor
Vortrennung *f (chromat)* preliminary separation
vortrocknen *v* predry
Vortrockner *m* predryer
Vortrocknung *f* predrying, preliminary drying
Voruntersuchung *f* preliminary examination [investigation]
Vorvakuum *n* forevacuum
Vorvakuumpumpe *f* forepump, backing pump
Vorverdampfer *m* pre-evaporator
Vorverstärker *m (anal)* pre-amplifier
Vorversuch *m* preliminary test
Vorversuch *m* **mit kleinen Substanzmengen** small-scale (preliminary) test
vorwärmen *v* preheat, warm (up)
Vorwärmer *m* preheater, feed heater
Vorwärmung *f* preheating, forewarming, warming(-up)
Vorwärmzone *f* preheating zone [compartment]; non-boiling zone *(of an evaporator)*
Vorwärtskontrolle *f (bioch)* feed-forward regulation
Vorwäsche *f* prewashing
vorwaschen *v* prewash
Vorzeichen *n*/**mit umgekehrtem** opposite in sign
Vorzugsbenennung *f (nomencl)* preferred name
VTC *s.* Viskositäts-Temperatur-Koeffizient
Vulkanisat *n (rubber)* vulcanizate, vulcanized product
Vulkanisation *f (rubber)* vulcanization, *(in compounds usually)* cure, curing
Vulkanisation *f* **bei Raumtemperatur** room-temperature cure
Vulkanisation *f* **in Dampf** steam cure
Vulkanisation *f* **in Formen** mould cure, moulding
Vulkanisationsbeschleuniger *m* vulcanization [cure] accelerator
Vulkanisationshilfsmittel *n* vulcanization assistant
Vulkanisationskurve *f* vulcanization [curing] curve
Vulkanisationsmittel *n* vulcanizing [curing] agent
Vulkanisationsreaktion *f* vulcanization [curing] reaction
Vulkanisierapparat *m* vulcanizer, vulcanizing apparatus, heater
vulkanisierbar vulcanizable, curable
vulkanisieren *v* vulcanize, cure
vulkanisieren *v*/**bei Raumtemperatur** vulcanize [cure] at room temperature
vulkanisieren *v*/**in Formen** mould
Vulkanisierform *f* vulcanizing [curing] mould
Vulkanisierkessel *m* vulcanizing autoclave [boiler], open-steam vulcanizer
Vulkanisierpresse *f* vulcanizing press, curing [moulding] press
vulkanisiert/mit Schwefel sulphur-cured
VZ *s.* Verseifungszahl

W

Waage *f* balance, *(of simple construction also)* scales
Waage *f*/**chemische** chemical [assay] balance
Waage *f* **mit Luftdämpfung** air-damped balance
Waage *f*/**Westphalsche** Westphal balance
Waagebalken *m* balance beam
Waagschale *f* balance pan

Wabenstruktur *f* honeycomb structure • **mit Wabenstruktur** honeycombed
Wacholder(beer)öl *n* juniper [Jupiter] oil, oil of juniper *(from Juniperus communis L.)*
Wachs *n* wax
Wachs *n***/mikrokristallines** micro(crystalline) wax, microparaffin
Wachs *n***/mineralisches** mineral [fossil] wax
wachsartig wax-like, waxy
wachsen *v* 1. wax; 2. grow *(as of crystals, microorganisms)*; grow, propagate *(of polymer chains)*; increase *(of physical parameters)*
wächsern waxy
Wachstuch *n* oilcloth
Wachstum *n* growth *(as of crystals or microorganisms)*; growth, propagation *(of polymer chains)*; increase *(of physical parameters)*
Wachstum *n***/diauxisches** diauxic growth *(of microorganisms)*
Wachstumsgeschwindigkeit *f* growth rate, rate of growth, *(relating to polymer chains also)* rate of propagation
Wachstumshormon *n* growth hormone, GH, somatotropin
Wachstumsinhibitor *m* growth inhibitor
Wachstumsmechanismus *m* growth mechanism
Wachstumsperiode *f* period of growth; period of chain growth [propagation] *(polymerization)*
Wachstumsphase *f (biot, hyd)* growth phase
Wachstumsphase *f***/abnehmende** declining [declined] growth phase
Wachstumsphase *f***/endogene** death [endogenous growth] phase
Wachstumsphase *f***/logarithmische** logarithmic growth phase, log phase
Wachstumsrate *f* growth rate, rate of growth
Wachstumsreaktion *f* growth [propagation] reaction *(polymerization)*
Wachstumsregulator *m* growth-regulating substance, growth regulator [modifier]
Wachstumsschritt *m* growth [propagation] step *(polymerization)*
Wächter *m (tech)* safeguard
Wadsworth-Anordnung *f (spectr)* Wadsworth mounting
Wadsworth-Aufstellung *f (spectr)* Wadsworth mounting
waf *s.* wasser- und aschefrei
Wägefehler *m* weighing error, error in weighing
Wägeform *f* weighing form *(gravimetric analysis)*
Wägeglas *n* weighing bottle
Wägegläschen *n* weighing bottle
Wägegut *n* material being [to be] weighed
wägen *v* weigh
wägen *v***/auf ein Milligramm genau** weigh to the nearest milligram, weigh to 1 mg accuracy
wägen *v***/nochmals** reweigh
wägen *v***/vorher** preweigh
Wägeröhrchen *n* weighing piggy
Wägesatz *m* set of weights
Wägeschiffchen *n* weighing scoop, balance dish
Wägeschweinchen *n s.* Wägeröhrchen
Wägestück *n* balance weight
Wagner-Meerwein-Umlagerung *f (org ch)* Wagner-Meerwein rearrangement
Wägung *f* weighing
Wägung *f***/direkte [einfache]** direct weighing
Wägung *f***/gaußsche** Gauss's method of double weighing
wahrnehmbar/geruchlich detectable by odour
wahrnehmbar/geschmacklich detectable by taste
Wahrscheinlichkeit *f***/thermodynamische** thermodynamic probability
Wahrscheinlichkeitsdichte *f* probability density
Wahrscheinlichkeitsfaktor *m* probability [steric] factor *(reaction kinetics)*
Wahrscheinlichkeitsverteilung *f* probability distribution, distribution of a random variable
Walden-Umkehr *f* Walden inversion, inversion of configuration
walken *v* 1. *(text, tann)* mill, full; 2. *(ceram)* wedge
Walkmaschine *f* milling [fulling] machine, mill, beater
Wallach-Umlagerung *f* Wallach rear-

rangement *(of azoxybenzene into 4-hydroxyazobenzene)*
Walze *f* roll(er), *(if hollow:)* cylinder, drum • **mit mehreren Walzen (versehen)** multiroll(er), multiple-roll
Walze f/geriffelte corrugated [fluted] roll
walzen *v* roll, mill
Walzenbrecher *m* roll(er) crusher
Walzenmühle *f* roll(er) mill, roller-milling system
Walzenpresse *f* roll(er) press
Walzenspalt *m* roll nip, nip of the rolls, bite *(clearance between the rolls)*
Walzentrockner *m* drum dryer
Walzentrocknung *f* drum drying
Wälzkolbenverdichter *m* straight-lobe compressor
Wälzmühle *f* roll(er) mill
Wälzsieb *n* revolving [drum] screen, trommel (screen)
Walzwerk *n* rolling mill
Walzzunder *m (met)* roll(ing) scale, mill scale
Wand f/blochsche *(cryst)* Bloch [domain] wall
Wand f/halbdurchlässige [semipermeable] semipermeable membrane
Wandeinfluss *m* wall effect *(on the course of a reaction)*
Wanderbettreaktor *m* moving-bed reactor
wandern *v* travel *(as of bulk material being processed)*; migrate *(as of ions, groups, or pigments)*; move *(as in a chromatogram)*; drift *(as of zero point)*
Wanderung *f* travel *(as of bulk material being processed)*; migration *(as of ions, groups, or pigments)*; movement *(as in a chromatogram)*; drift *(as of zero point)*
Wanderungsgeschwindigkeit *f* travel rate *(as of bulk material being processed)*; migration rate [velocity] *(as of ions)*
Wanderungsstrom *m (el ch)* migration current
Wandreaktion *f* wall reaction
Wandung *f* wall
Wanne *f* 1. *(tech)* trough, vat; back *(brewing, dyeing, soap manufacture)*; 2. *(glass)* tank; crucible *(of an arc furnace)*; 3. *s.* Wannenform
Wannenform *f* boat (form) *(stereochemistry)*
Wannenglas *n* tank glass
Wannenkonformation *f* boat [tub] conformation *(stereochemistry)*
Wannenofen *m (glass)* tank furnace
Warburg-Dickens-Horecker-Schema *n s.* Pentosephosphatweg
Warendichte *f (text)* gauge, gg.
Warenname m/freier non-proprietary name
Warenzeichen n/eingetragenes [registriertes] registered trademark
Wärme *f (phys ch, tech)* heat; *(sensation:)* warmness, warmth
Wärme f/differenzielle differential heat
Wärme f/integrale integral heat
Wärme f/latente latent heat
Wärme f/mäßige moderate heat
wärmeabgebend exothermic, heat-giving
Wärmeableitung *f* dissipation of heat, thermal dissipation • **mit geringer Wärmeableitung** low-thermal-dissipation
Wärmeänderung *f* heat change
Wärmeäquivalent n/elektrisches electrical equivalent of heat
Wärmeäquivalent n/mechanisches mechanical equivalent of heat
wärmeaufnehmend endothermic
Wärmeausdehnung *f* thermal expansion
Wärmeausdehnungskoeffizient *m* coefficient of thermal expansion
Wärmeausdehnungszahl *f* coefficient of thermal expansion
Wärmeaustausch *m* heat exchange [interchange], thermal transfer
Wärmeaustauscher *m s.* Wärmeübertrager
Wärmeaustauschkoeffizient *m* heat exchange coefficient
wärmebehandeln *v* heat-treat
Wärmebehandlung *f* heat treatment [processing]
wärmebeständig heat-resistant, thermally stable *(e.g. plastics)*
Wärmebeständigkeit *f* heat resistance, thermal stability *(as of plastics)*
Wärmebewegung *f* thermal motion [agitation]
Wärmebilanz *f* heat balance
Wärmedämmstoff *m* heat-insulating material, thermal [heat] insulator

Wärmedämmung *f* thermal [heat] insulation • **mit Wärmedämmung** heat-insulated
Wärmedurchgangskoeffizient *m* overall heat-transfer coefficient
Wärmeeintrag *m* heat input [addition]
wärmeempfindlich heat-sensitive
Wärmeempfindlichkeit *f* heat sensitivity
Wärmeenergie *f* thermal [heat] energy
Wärmeentwicklung *f* evolution of heat
wärmeerzeugend heat-generating, calorific, calorigenic
Wärmefluss *m* heat flow
Wärmefreisetzung *f* heat release
Wärmefunktion *f*/**gibbssche** Gibbs function [free energy], free energy G, free enthalpy
wärmehärtbar *(plast)* thermosetting, thermoreactive, heat-curable
wärmehärtend *(plast)* thermosetting, thermoreactive, heat-curable
Wärmeinhalt *m* heat content, *(per unit mass also)* enthalpy, heat function at constant pressure
Wärmeinhalts-Temperatur-Diagramm *n* enthalpy/temperature diagram, heat-content/temperature diagram [chart], It diagram
Wärmekapazität *f* heat [thermal] capacity
Wärmekapazität *f* **bei konstantem Druck** isobaric heat capacity
Wärmekapazität *f* **bei konstantem Volumen** isochoric [constant volume] heat capacity
Wärmekapazität *f*/**molare** molar heat capacity
Wärmekapazität *f*/**spezifische** specific heat capacity
Wärmekonvektion *f* thermal convection [siphoning]
Wärmekonvektion *f*/**erzwungene** forced convection
Wärmekonvektion *f*/**freie** natural convection
Wärmeleistung *f* heat flow
Wärmeleiter *m* heat conductor
Wärmeleitfähigkeit *f* 1. thermal [heat] conductivity; 2. *s.* Wärmeleitzahl
Wärmeleitfähigkeitsdetektor *m* thermal-conductivity detector [cell], katharometer
Wärmeleitfähigkeits(mess)zelle *f* thermal-conductivity detector [cell], katharometer
Wärmeleitung *f* heat [thermal] conduction
Wärmeleitzahl *f* coefficient of thermal conduction, specific thermal conductivity
wärmeliefernd exothermic, heat-giving
Wärmemenge *f* quantity [amount] of heat
Wärmepumpe *f* heat pump
Wärmequelle *f* heat source
Wärmerückgewinnung *f* heat recovery
Wärmesatz *m*/**nernstscher** *s.* Wärmetheorem/nernstsches
Wärmestein *m* pebble *(of a pebble heater)*
Wärmesteinerhitzer *m* pebble heater [stove]
Wärmestrahlung *f* heat radiation, thermal [calorific, caloric] radiation
Wärmestrom *m* rate of heat flow *(heat conduction)*; rate of heat transfer
Wärmestromdichte *f* heat [thermal] flux
Wärmestromintensität *f* heat [thermal] flux
Wärmetauscher *m s.* Wärmeübertrager
Wärmetheorem *n* heat theorem, third law of thermodynamics
Wärmetheorem *n*/**nernstsches** Nernst heat theorem, third law of thermodynamics
Wärmetönung *f* heat tonality
Wärmeträger *m* heat carrier, heat-exchanging [heat-transfer] medium, heating medium
Wärmeübergang *m* heat [thermal] transfer
Wärmeübergangskoeffizient *m* heat-transfer coefficient
Wärmeübergangszahl *f* heat-transfer coefficient
Wärmeübertrager *m* heat exchanger [interchanger]
Wärmeüberträger *m s.* 1. Wärmeübertrager; 2. Wärmeträger
Wärmeübertragung *f* heat [thermal] transfer, heat exchange [interchange]
Wärmeübertragungsmittel *n s.* Wärmeträger
wärmeunbeständig thermolabile, heat-labile
wärmeverbrauchend endothermic

Wärmeverlust *m* heat loss
Wärmezufuhr *f* heat addition [input]
warmfest heat-resistant *(steel up to 600 °C)*
Warmfestigkeit *f* heat resistance *(of steel up to 600 °C)*
Warmformen *n* 1. hot forming [working]; 2. *(plast)* thermoforming
Warmhärte *f* hot hardness
warmpressen *v* hot-press
Warmpressen *n* hot pressing
Warmumformen *n* hot forming [working]
warmverarbeiten *v* hot-work
Warmverarbeitung *f* hot working
Wartezeit *f* 1. downtime, down period; 2. *(agric)* preharvest interval *(after application of pesticides)*
Wartungsaufwand *m* maintenance requirements
WAS *s.* Stoff/waschaktiver
Wäsche *f (tech)* wash(ing), wet cleaning, wash-up
waschecht fast to washing, wash-fast, washable
Waschechtheit *f* fastness to washing, wash(ing) fastness, washability
waschen *v* wash, wet-clean; scrub *(gases)*; launder *(textiles)*
waschen *v*/**im Gegenstrom** wash countercurrently
Waschen *n* wash(ing), wet cleaning, (gas) scrubbing, gas washing; laundering *(of textiles)*
Wäscher *m* (gas) washer, scrubber, wet collector
Waschflasche *f (lab)* wash(ing) bottle, absorption bottle, bubbler
Waschflotte *f (text)* washing [scouring] liquor, washing bath [liquid], detergent solution
Waschflüssigkeit *f* 1. wash(ing) liquid, washing liquor [solvent], *(if spent also)* washings; scrubbing liquid *(for gases)*; 2. *s.* Waschflotte
Waschkolonne *f* scrubber [wash] column
Waschkühler *m* washer-cooler
Waschlauge *f s.* Waschflotte
Waschmittel *n* detergent, washing agent, *(specif)* laundry [fabric] detergent, laundering formulation, (fabric) laundering composition, washing composition
Waschmittel *n*/**synthetisches** (synthetic) detergent, syndet
Waschöl *n* wash oil, absorption [scrubbing, stripping] oil
Waschpulver *n* washing powder
Waschsoda *f* washing soda, salt of soda, (sal) soda, natron *(sodium carbonate 10-water)*
Waschturm *m* scrubbing [gas-washing] tower, tower scrubber, (gas) scrubber
Waschung *f* wash(ing)
Waschwasser *n* wash(ing) water, wash (liquid), rinsing [rinse] water, rinse, *(if spent also)* washings
Waschwirkung *f* cleaning [detergent] effect
Washout *m* wash out *(the air pollutants collected by falling raindrops)*
Washout *n* wash out, below-cloud scavenging *(of air pollutants by falling raindrops)*
Wasser *n* 1. H_2O water; 2. *(hyd, tech)* water; 3. *(tech)* liquor, liquid; 4. *(cosmet)* wash, waters; 5. *(pharm)* waters; 6. *(min)* water *(limpidity and lustre of gems)* • **in Wasser quellbar** water-swellable • **mit Wasser verdünnbar** water-dilutable
Wasser *n*/**ablaufendes** *(hyd)* effluent water *(as from a treatment plant)*; underflow *(oil removal)*
Wasser *n*/**aufbereitetes** *(hyd)* finished [treated] water, treated [plant] effluent
Wasser *n*/**chemisch gebundenes** combined [chemically bound] water
Wasser *n*/**destilliertes** distilled water
Wasser *n*/**doppelt destilliertes** double-distilled water
Wasser *n*/**fließendes** running water
Wasser *n*/**gebundenes** bound water [moisture]
Wasser *n*/**hartes** hard water
Wasser *n*/**hochreines** ultrapure [ultra-high-purity] water
Wasser *n*/**hygroskopisches** hygroscopic water, hygroscopic(al) moisture
Wasser *n*/**konstitutiv gebundenes** constitutional water, water of constitution
Wasser *n*/**salzhaltiges** saline [salt] water
Wasser *n*/**schweres** D_2O heavy water, deuterium oxide

Wasser *n*/**stark belastetes [verschmutztes]** heavily polluted water
Wasser *n*/**vollentsalztes** *(hyd)* deionized [demineralized] water
Wasser *n*/**weiches** soft water
Wasser *n*/**zeolithisches [zeolithisch gebundenes]** zeolitic water
Wasserabscheider *m* water separator
Wasserabschreckung *f (met)* water quenching
wasserabsorbierend water-absorbing
Wasserabsorption *f* water absorption
wasserabspaltend dehydrating
Wasserabspaltung *f* elimination of water, dehydration
wasserabstoßend water-repellent, hydrophobic, hydrophobe
wasserabweisend water-repellent, hydrophobic, hydrophobe
Wasserabweisungsvermögen *n* water repellency
Wasseranalyse *f* water(-quality) analysis
Wasseranlagerung *f* water addition, hydrate formation
wasseranziehend water-attracting, hygroscopic(al)
wasserartig aqueous
Wasseraufbereitung *f (hyd)* (fresh) water treatment, water conditioning [purification]
Wasseraufbereitungsanlage *f (hyd)* water treatment plant [works], water (purification) plant, waterworks
Wasseraufbereitungsverfahren *n* water treatment process
Wasseraufnahme *f* water absorption, uptake of water
wasseraufnehmend water-absorbing, hydrophilic, hydrophile
Wasserbad *n* (hot-)water bath
wasserbeständig water-resistant
Wasserbeständigkeit *f* water resistance
Wasserchemie *f* water [aquatic] chemistry, hydrochemistry
Wasserchemiker *m* water chemist
Wasserdampf *m* water vapour, *(if generated at boiling point of water:)* steam • **mit Wasserdampf behandeln** steam
Wasserdampf *m*/**gesättigter** saturated steam
Wasserdampf *m*/**überhitzter** superheated steam
Wasserdampfaufnahme *f* water vapour absorption
Wasserdampfdestillation *f* steam distillation [stripping]
wasserdampfdurchlässig permeable to water vapour
Wasserdampfdurchlässigkeit *f* water-vapour transmission [permeability], WVT, moisture-vapour transmission [permeability], MVT
wasserdampfflüchtig steam-volatile
wasserdampfundurchlässig impervious to water vapour, moisture-resistant, moistureproof
Wasserdampfundurchlässigkeit *f* imperviousness to water vapour, moisture resistance [proofness]
wasserdicht water-tight, waterproof, impermeable to water • **wasserdicht abschließen** waterproof • **wasserdicht machen** *(text)* waterproof
Wasserelektrolyse *f* water electrolysis
Wasserenthärtung *f* water softening
Wasserenthärtungsanlage *f* water-softening plant
Wasserenthärtungsmittel *n* water-softening agent, water softener
Wasserentsalzung *f* water desalination, demineralization of water
wasserentziehend dehydrating
Wasserentzug *m* water removal, abstraction [removal] of water, dehydration
Wasserfarbe *f* water(-base) paint, *(Am)* watercolor; water colour *(art)*
wasserfest water-resistant
Wasserfestigkeit *f* water resistance
Wasserfilter *n* water filter
Wasserfiltration *f* water filtration
wasserfrei free from water, *(esp relating to chemical substances:)* anhydrous, anh., *(esp relating to coal:)* moisture-free, mf
Wassergas *n* water gas
Wassergas *n*/**karburiertes** carburetted [enriched] water gas
Wassergasgenerator *m* water-gas generator
Wassergasgleichgewicht *n* water-gas(-shift) equilibrium
Wassergasreaktion *f* water-gas [steam-carbon] reaction *(reduction of water by means of carbon)*

Wassergasreaktion *f***/heterogene** water-gas [steam-carbon] reaction *(reduction of water by means of carbon)*
Wassergasreaktion *f***/homogene** water-gas-shift reaction, carbonmonoxide-shift reaction
Wassergehalt *m* water content, *(esp of gases and hygroscopic substances:)* moisture content
wassergekühlt water-cooled
wassergeschützt waterproof
Wasserglas *n* water glass, liquid [soluble] glass *(solution of alkali silicates in water)*
Wassergüte *f* water quality
Wassergütebedingungen *fpl* water-quality standards
Wassergütemerkmale *npl* water(-quality) characteristics, water-quality criteria [parameters]
wasserhaltig hydrous
Wasserhärte *f* water hardness
wasserhell water-white, water-clear
Wasserhülle *f* water sheath *(as of ions)*
Wasser-in-Fett-Emulsion *f* water-in-fat emulsion
Wasserinhaltsstoffe *mpl* water(-quality) constituents, impurities in water
Wasserinhaltsstoffe *mpl***/flüchtige organische** volatile organic compounds
Wasserinhaltsstoffe *mpl***/grobdisperse** suspended solids, SS, suspended solid matter *(> 5 · 10^{-4} mm)*
Wasserinhaltsstoffe *mpl***/schädliche** water pollutants [contaminants]
Wasser-in-Öl-Emulsion *f* water-in-oil emulsion, W/O emulsion
Wasserkalorimeter *n* water calorimeter
wasserklar water-clear, water-white
Wasserkreislauf *m* water recycle [cycle, circuit], circulation of water
Wasserkreislaufführung *f* recycling [recirculation] of water, water recirculation
Wasserkühlung *f* water cooling
Wasserlack *m (coat)* waterborne coating
wasserlöslich water-soluble
wasserlöslich/nicht water-insoluble
Wasserlöslichkeit *f* water [aqueous] solubility
Wassermantel *m* water jacket • **mit Wassermantel** water-jacketed
Wassermehrfachnutzung *f* water reuse
Wassermolekül *n* water molecule
wässern *v (of human agent:)* steep, soak; *(in running water:)* rinse; *(of material:)* soak
Wasseroxidation *f***/überkritische** supercritical water oxidation, SCWO
Wasserprobe *f* water sample, sample of water
Wasserqualität *f s.* Wassergüte
Wasserreinhaltung *f (hyd)* water pollution control, water protection
Wasserreinigung *f* 1. water purification [clean-up]; 2. *s.* Wasseraufbereitung
Wasserrohrdampferzeuger *m* water-tube boiler
Wasserrohrkessel *m* water-tube boiler
Wasserrückgewinnung *f* water recovery
Wasserschadstoff *m* water pollutant
Wasserschlepper *m (distil)* water entrainer
Wasserschöpfer *m* water sampler
Wasserstein *m* (hardness) scale, mineral scale, hardwater depositions [residue]
Wassersteinbildung *f* formation of hardness scale, scale formation [build-up], scaling
Wasserstoff *m* H hydrogen • **über Wasserstoff verbunden** hydrogen-bonded
Wasserstoff *m***/aktiver** active hydrogen
Wasserstoff *m***/atomarer [einatomiger]** atomic hydrogen, monohydrogen
Wasserstoff *m***/leichter** ^{1}H protium, light hydrogen
Wasserstoff *m***/schwerer** $^{2}_{1}$H, D deuterium, heavy hydrogen
Wasserstoff *m***/überschwerer** $^{3}_{1}$H, T tritium
Wasserstoffabscheidung *f* hydrogen evolution [liberation, generation]
wasserstoffabspaltend hydrogen-abstracting
Wasserstoffabspaltung *f* abstraction [elimination] of hydrogen, dehydrogenation
Wasserstoffakzeptor *m* hydrogen acceptor
wasserstoffarm poor in hydrogen
Wasserstoffatom *n* hydrogen atom
Wasserstoffatom *n***/allylständiges** allylic hydrogen atom

Wasserstoffatom *n*/**bewegliches** mobile hydrogen atom
Wasserstoffatom *n* **in** γ**-Stellung** gamma hydrogen, γ-hydrogen
Wasserstoffaustauscher *m s.* Wasserstoffionenaustauscher
Wasserstoffbindung *f* 1. hydrogen bond *(state)*; 2. hydrogen bonding *(process)*
Wasserstoffbrüchigkeit *f* acid brittleness
Wasserstoffbrücke *f* hydrogen bridge
Wasserstoffbrückenbindung *f* hydrogen bridge bond
Wasserstoffdon(at)or *m* hydrogen donor
Wasserstoffelektrode *f* hydrogen (gas) electrode
Wasserstoffelektrode *f*/**reversible** reversible hydrogen electrode, RHE
Wasserstoffentwicklung *f* hydrogen evolution [liberation, generation]
Wasserstoffentzug *m* abstraction [elimination] of hydrogen, dehydrogenation
Wasserstofffluoridlaser *m* hydrogen-fluoride laser, HF laser
Wasserstoffgas *n* hydrogen gas
Wasserstoffhalbelement *n* hydrogen half-cell
wasserstoffhaltig containing hydrogen, hydrogenous
Wasserstoffion *n* hydrogen ion, H^+ ion
Wasserstoffionenaktivität *f* hydrogen-ion activity
Wasserstoffionenaustauscher *m (hyd)* H-form exchanger, hydrogen(-form) exchanger, hydrogen-cation exchanger
Wasserstoffionenexponent *m* hydrogen-ion exponent, pH value [number, level]
Wasserstoffionenkonzentration *f* hydrogen-ion concentration
Wasserstoffkern *m* hydrogen nucleus
wasserstoffliefernd hydrogen-donating
Wasserstoffnormalelektrode *f* normal hydrogen electrode, N.H.E.
Wasserstoffperoxid *n* hydrogen peroxide, dioxidane
Wasserstoff-Sauerstoff-Brennstoffzelle *f* hydrogen-oxygen fuel cell
Wasserstoffsäure *f* hydrogen acid, hydracid
Wasserstoffspektrum *n* hydrogen spectrum
Wasserstoffüberspannung *f (el ch)* hydrogen overvoltage
wasserstoffübertragend hydrogen-transferring
Wasserstoffübertragung *f* hydrogen transfer
Wasserstoffverbindung *f* hydrogen compound
Wasserstoffvergasung *f (coal)* hydrogasification
1,5-Wasserstoff-Verschiebung *f* 1,5 sigmatropic shift of hydrogen
Wasserstoffversprödung *f* hydrogen embrittlement
Wasserstoffwechsel *m (biol)* water metabolism
Wasserstrahl *m* water jet
Wasserstrahlpumpe *f* 1. *(lab)* water (suction) pump, water aspirator; 2. *(tech)* water(-operated) ejector
Wassertechnologe *m* water technologist
Wassertechnologie *f* water (treatment) technology
Wassertröpfchen *n* water droplet
Wasserüberstau *m (hyd, filtr)* head of water, liquid head
Wasserüberstauhöhe *f (hyd, filtr)* head of water, liquid head
wasser- und aschefrei moisture-and-ash-free, maf
Wässerung *f* steep(ing), steepage, soaking; *(in running water:)* rinse, rinsing
wasserunlöslich insoluble in water, water-insoluble
Wasserunlöslichkeit *f* water-insolubility
wasserverdünnbar water-dilutable
Wasserverschmutzung *f* water pollution, *(esp relating to poisons or bacteria:)* water contamination
Wasserverunreinigung *f* 1. water pollutant [contaminant, impurity]; 2. *s.* Wasserverschmutzung
Wasserwäsche *f* water wash(ing), *(relating to gases also)* water scrubbing
Wasserwerk *n* waterworks, water-treatment plant [works], water(-purification) plant
Wasserwerksschlamm *m* waterworks sludge, water treatment (plant) sludge

Wasserwert *m* water equivalent *(of a calorimeter)*
Wasserwiederverwendung *f* water reuse
Wasserzersetzung *f* water electrolysis
Wasserzersetzungszelle *f* water electrolyzer
Wasserzusatz *m* water addition
wässrig aqueous, watery
wässrig/nicht non-aqueous
Watson-Crick-Modell *n (bioch)* Watson-Crick model *(of DNA)*
Watte *f* 1. wadding; 2. *(med)* cotton wool
Watte *f* **in Lagen** batting
wdfl. *s.* wasserdampfflüchtig
Wechsel *m* change(over), alteration, *(if periodically:)* alternation
Wechselbeziehung *f* correlation
wechseln *v* change, *(if periodically:)* alternate
Wechselspannung *f*/**periodische** periodic voltage
Wechselspannung *f*/**sinusförmige** alternating voltage
Wechselstrom *m* alternating current, a.c.
Wechselstrom *m*/**periodischer** periodic current
Wechselstrompolarographie *f* a.c. polarography
Wechselstromwiderstand *m* impedance
Wechselumsetzung *f* metathesis, double [mutual] decomposition
wechselwirken *v* interact
Wechselwirkung *f* interaction • **in Wechselwirkung stehen** interact • **in Wechselwirkung treten** interact
Wechselwirkung *f* **durch Stapelkräfte** *(bioch)* stacking interaction
Wechselwirkung *f*/**hydrophobe** hydrophobic interaction
Wechselwirkung *f*/**interionische** ionic [ion-ion] interaction, interionic action
Wechselwirkung *f*/**intermolekulare** (inter)molecular interaction
Wechselwirkungskraft *f* interaction force
Wechselwirkungskraft *f*/**interionische** interionic force
Wechselzahl *f (bioch)* turnover number
Wechselzersetzung *f s.* Wechselumsetzung
Wedge-Ofen *m* Wedge (roasting) furnace, Wedge roaster [burner]
Weg *m* 1. path(way) *(of a reaction)*; 2. trajectory, path *(of a particle)*
wegabhängig dependent on the path
Wegabhängigkeit *f* path dependence
wegbrennen *v* burn off
wegdiffundieren *v* diffuse away
Wegdiffusion *f* diffusing-away
Weglänge *f (phys ch)* length of path
Weglänge *f*/**freie** free path
Weglänge *f*/**mittlere freie** mean free path
Weglänge *f*/**optische** *(spectr)* optical pathlength, light path
wegunabhängig independent of the path
Wegunabhängigkeit *f* path independence
wehenerregend oxytocic
Wehenmittel *n (pharm)* uterotonic (agent), oxytocic (agent)
weich soft • **weich machen** soften *(e.g. water, plastics)*
weichen *v* steep, soak; soak *(of material)*
Weichen *n* steep(ing), steepage, soaking *(act)*; soaking *(process)*
Weichglas *n* soft (sealing) glass
weichglühen *v* soft-anneal, spheroidize
Weichglühen *n* soft annealing, softening anneal, spheroidizing (anneal), spheroidization
Weichgummi *m* soft rubber
Weichharz *n* soft resin
Weichlot *n* soft solder (alloy)
weichlöten *v* soft-solder
Weichlöten *n* soft soldering
Weichmachen *n* 1. softening; 2. *(plast)* plasticization, flexibilization
Weichmacher *m* 1. softener, softening agent; 2. *(plast)* plasticizer, plasticizing agent, flexibilizer
Weichmacher *m*/**äußerer** external plasticizer
Weichmacher *m*/**innerer** internal plasticizer
Weichmacher *m*/**lösender** primary plasticizer
Weichmacher *m*/**nicht lösender** secondary plasticizer
weichmacherfrei unplasticized
Weichporzellan *n* soft(-paste) porcelain
Weich-PVC *n* flexible [plasticized] PVC

Weichstahl *m* soft [mild] steel
Weichwachs *n* soft wax
Weichwasser *n (hyd)* soft(ened) water
Weichwassergüte *f* effluent (water) quality *(ion exchange)*
Weingeist *m* spirit(s) of wine *(aqueous ethyl alcohol)*
Weinhold-Gefäß *n* Dewar (flask, vessel) *(for holding liquid gases)*
Weinkunde *f* oenology
weinkundlich oenological
Weinsäure *f* tartaric acid
Weinsäure *f***/linksdrehende** *s. (2S,3S)*-(-)-Weinsäure
Weinsäure *f***/racemische** *s. DL*-Weinsäure
Weinsäure *f***/rechtsdrehende** *s. (2S,3S)*-(+)-Weinsäure
***DL*-Weinsäure** *f* (±)-tartaric acid, racemic [inactive] tartaric acid
***(2R,3R)*-(+)-Weinsäure** *f (2R,3R)*-(+)-tartaric acid, L-tartaric acid, dextrotartaric acid, dextrorotatory tartaric acid
***(2S,3S)*-(-)-Weinsäure** *f (2S,3S)*-(-)-tartaric acid, D-tartaric acid, laevotartaric acid, laevorotatory tartaric acid
Weinstein *m* tartar
Weinsteinsalz *n* salt of tartar, potassium carbonate
Weisel(zellen)futtersaft *m (pharm)* royal jelly, queen-bee's nutrient jelly
Weiß *n* white *(sensation or substance)*
Weiß-Bezirk *m (cryst)* Weiss (molecular magnetic) field, ferromagnetic domain
Weißei *n* egg white, (egg) albumen
Weissenberg-Effekt *m* Weissenberg effect *(with viscoelastic liquids)*
Weißgehalt *m* 1. degree of white(ness); 2. *(pap)* brightness (level)
weißglühen *v* incandesce
weißglühend incandescent, white-hot
Weißglut *f* incandescence, incandescency, white heat, W.H. • **auf [bis zur] Weißglut erhitzen** incandesce
Weißguss *m* white cast iron
Weißmacher *m* (detergent) brightener, whitening agent
Weißmetall *n* white metal *(1. any of several bearing metals; 2. copper matte having its iron removed)*
Weißöl *n* (petroleum) white oil, white mineral [petroleum] oil
Weißtöner *m* optical brightener [bleach], fluorescent brightener, optical brightening [whitening, bleaching] agent
Weißwerden *n* whitening, *(esp of nitrocellulose lacquer:)* blushing
Weitenverteilung *f* size distribution *(as of pores)*
weiterreagieren *v* react further, undergo further reaction
Weiterreaktion *f* further reaction
weitertragen *v* propagate *(e.g. a reaction)*
Weitertragen *n* propagation *(as of a reaction)*
Weithalsflasche *f* wide-mouth [wide-neck] bottle
Weithalskolben *m* wide-mouth [wide-neck] flask
Weithalsrundkolben *m* round-bottom wide-mouth flask
weitmaschig coarse-mesh(ed)
Weitwegkopplung *f (phys ch)* long-range coupling [interaction]
Weitwinkelaufnahme *f (cryst)* wide-angle X-ray pattern
Weizenmehl *n* wheat flour, *(if coarse:)* wheat meal
Weizenstärke *f* wheat starch
Wellasbest *m* corrugated asbestos
Welle *f* 1. wave; 2. *(tech)* shaft
Welle *f***/polarographische** polarographic wave
Wellenfunktion *f* wave function
Wellengleichung *f* wave equation
Wellengleichung *f***/schrödingersche** Schrödinger (wave) equation
Wellenlänge *f* wavelength
Wellenmechanik *f* wave mechanics
Wellenzahl *f* wave number
Wellplatte *f (plast)* corrugated sheet
Wellrohr(dehnungs)ausgleicher *m* corrugated [bellows-type] expansion joint
Wellrohrkompensator *m* corrugated [bellows-type] expansion joint
Wendel *f (distil)* helix
Wendepunkt *m* inflection point *(of a curve)*
Werkblei *n* crude [pig] lead
Werksabwasser *n* plant waste water, plant [works, factory] effluent
Werkschemiker *m* industrial chemist
Werkstoff *m* material, matter

Werkstoff *m* **aus Kohlenstoff** carbon material
Werkstoff *m*/**glasiger** glassy material
Werkstoff *m*/**halbleitender** semiconducting material
Werkstoff *m*/**hochtemperaturbeständiger** high-temperature material
Werkstoff *m*/**keramischer** ceramic material
Werkstoff *m*/**metallkeramischer [mischkeramischer]** cermet, ceramal, ceramel, ceramet
Werkstoffkunde *f* materials science
Werkstoffprüfung *f* materials testing
Werkstoffwissenschaft *f* materials science
Werkzeug *n* 1. *(plast)* mould, form; 2. tool
Werkzeugstahl *m* tool steel
Wert *m* value • **einen konstanten Wert aufweisen** show a constant reading • **etwas über dem theoretischen Wert** in slight excess of theory, slightly over the theoretical • **etwas unter dem theoretischen Wert** slightly short of theory
Wert *m*/**abweichender** *(anal)* outlier, maverick
Wert *m*/**kalorischer** *(food)* calorific value, c.v.
Wert *m*/**quasistationärer** steady-state value
Wert *m*/**vorgegebener** preset value
ν-**Wert** *m* Abbe number [value], ν-value, constringence *(reciprocal relative dispersion)*
Wertbestimmung *f*/**biologische** biological assay, bioassay
Wertigkeit *f* valence, valency
Wertigkeit *f*/**elektrochemische** electrochemical valence, electrovalence
Wertigkeit *f*/**koordinative** coordination [covalence] number, coordination, ligancy
Wertminderung *f* deterioration
Wertsteigerung *f* upgrading
Wertstoff *m* valuable product, product of value, useful material
Wertstoffe *mpl* values
Wertstoffrückgewinnung *f* recovery [reclaim] of values [useful material]
Weston-Element *n* Weston cell, Weston normal [standard] cell, Weston saturated cadmium cell
Weston-Normalelement *n* Weston cell, Weston normal [standard] cell, Weston saturated cadmium cell
Wetter *pl*/**schlagende** *(mine)* firedamp
wetterbeständig 1. weather-resistant; 2. *(text)* weatherproof
Wetterbeständigkeit *f* 1. weather(ing) resistance, resistance to weathering, weatherability; 2. *(text)* weatherproofness
Wettersprengstoff *m (mine)* safety explosive
wf *s.* wasserfrei
Wheatstone-Brücke *f* Wheatstone bridge
Wheland-Intermediat *n (org ch)* Wheland intermediate, arenium σ-adduct
Whisker *m* (crystal) whisker
Wichte *f* specific weight *(the weight of a substance per unit volume)*
widerlich disagreeable, obnoxious, repulsive, vile *(odour)*
Widerstand *m* 1. resistance; 2. resistor
Widerstand *m*/**aerodynamischer** drag (force)
Widerstand *m*/**elektrischer** 1. electrical resistance; 2. resistor
Widerstand *m*/**hydrodynamischer** fluid friction, drag (force)
Widerstand *m*/**spezifischer** specific resistance, resistivity
Widerstand *m*/**lichtelektrischer** *s.* Photowiderstand
Widerstand *m*/**piezoelektrischer** piezoelectric resistor
Widerstand *m*/**thermischer** thermal resistance
Widerstandsbrücke *f* resistance bridge
Widerstandsbrücke *f*/**elektrische** resistance bridge
widerstandsfähig resistant, stable, fast
Widerstandsfähigkeit *f* resistance, stability, fastness
Widerstandsofen *m* resistance [resistance-heated, resistor] furnace
Widerstandsofen *m*/**elektrischer** resistance [resistance-heated, resistor] furnace
Widerstandspolarisation *f* ohmic polarization [overpotential, drop], iR drop
Widerstandsthermometer *n* resistance thermometer

Widerstandsüberspannung *f s.* Widerstandspolarisation
wiederaufarbeiten *v* reprocess, rework, reclaim
wiederaufbereiten *v* reprocess *(e.g. nuclear fuel)*; remanufacture *(e.g. waste paper)*
Wiederaufbereitung *f* reprocessing *(as of nuclear fuel)*; remanufacture *(as of waste paper)*
Wiederaufbereitung *f*/**chemische** *(nucl)* chemical reprocessing
Wiederaufbereitungsanlage *f* 1. reprocessing plant; 2. *(nucl)* fuel-reprocessing plant
wiederaufkohlen *v (met)* recarburize
Wiederaufkohlen *n (met)* carbon restoration, recarburization
Wiederaufkohlung *f s.* Wiederaufkohlen
wiederauflösen *v* redissolve
Wiederauflösung *f* redissolution
wiederaufschwemmen *v* repulp, reslurry
wiederausstrahlen *v* reradiate, re-emit
Wiederausstrahlung *f* reradiation, reemission
wiederbeleben *v* revivify, regenerate *(a sorbent or a catalyst)*
Wiederbelebung *f* revivification, regeneration *(of a sorbent or a catalyst)*
Wiederbelebungsmittel *n* regenerant *(for sorbents)*
Wiedereinfangen *n* recapture *(as of ions)*
Wiederfindungsrate *f (anal)* percentage recovery
wiedergewinnbar recoverable, reclaimable
wiedergewinnen *v* recover, reclaim
Wiedergewinnung *f* recovery, reclaim(-ing)
wiederherstellen *v* re-establish, restore
wiederherstellen *v*/**das Gleichgewicht** re-establish equilibrium
Wiederherstellung *f* re-establishment, restoration
Wiederherstellung *f* **des Gleichgewichts** re-equilibration
wiederholbar repeatable *(experiment)*
Wiederholbarkeit *f* repeatability *(of an experiment)*
Wiederholstreubereich *m* repeatability *(statistics)*
Wiederholung *f* replicate, replication *(statistics)*
Wiederholungseinheit *f* repeat(ing) unit *(as in a macromolecule)*
wiederverarbeiten *v* reprocess
Wiederverarbeitung *f* reprocessing
wiedervereinigen *v (ch)* recombine
wiedervereinigen *v*/**sich** recombine
Wiedervereinigung *f (ch)* recombination
Wiederverfestigung *f* resolidification
wiederverwendbar reusable
wiederverwenden *v* reuse
Wiederverwendung *f* reuse
wiegen *v* weigh
Wien-Effekt *m (el ch)* Wien effect
Wien-Verschiebungsgesetz *n* Wien displacement law
Wijs-Lösung *f* Wijs (iodine monochloride) solution *(for determining the iodine number of fats)*
Williamson-Synthese *f* Williamson (ether) synthesis
Wilson-Kammer *f (nucl)* cloud [Wilson] chamber, (Wilson) cloud-track apparatus
Wind *m* 1. wind; 2. *(met)* (air) blast
winden *v (text)* reel (up)
Winderhitzer *m (met)* hot-blast stove, air-blast [blast-furnace] stove, air heater
Windform *f (met)* air-blast tuyère
windfrischen *v (met)* bessemerize, air-refine, convert
Windfrischen *n (met)* bessemerizing, air refining
Windfrischkonverter *m (met)* Bessemer converter
Windfrischstahl *m (met)* Bessemer [converter] steel
Windfrischverfahren *n (met)* Bessemer [converter] process
Windfrischverfahren *n*/**basisches** basic Bessemer [converter] process, Thomas(-Gilchrist) process
Windfrischverfahren *n*/**saures** acid Bessemer [converter] process
Windleitung *f (met)* (air-)blast main
Windsichten *n s.* Sichten
Windung *f (tech, bioch)* turn *(of a screw or an alpha helix)*
Winkel *m*/**braggscher** *(cryst)* Bragg angle

Winkel *m*/**brewsterscher** Brewster [polarizing] angle
Winkelfrequenz *f* angular frequency
Winkelgeschwindigkeit *f* angular velocity
Winkelkopf *m* (fixed-)angle head, (fixed-) angle rotor, angular rotor *(of a centrifuge)*
Winkelrotor *m* (fixed-)angle head, (fixed-)angle rotor, angular rotor *(of a centrifuge)*
Winkelspannung *f* angle strain
Winkelstück *n* 1. *(tech)* elbow (fitting); 2. *(lab)* angle connector, bent-tube connection, bent tube, elbow
Winkelthermometer *n* angle thermometer
Winkler-Generator *m* Winkler generator *(a fluidized-bed gasifier)*
Winkler-Vergasungsverfahren *n* Winkler process *(for gasifying lignite in a fluidized bed)*
Wintergrünöl *n* oil of wintergreen *(from Gaultheria procumbens L.)*
Wirbel *m* eddy, whirl, swirl, *(if large:)* vortex
Wirbelbett *n* fluid(ized) bed
Wirbelbett... *siehe auch: Wirbelschicht...*
Wirbelbettkatalysator *m* fluid(ized) catalyst, fluid-bed catalyst
Wirbeldiffusion *f* eddy [turbulent] diffusion
Wirbeldispersion *f* eddy dispersion
Wirbelgas *n* fluidizing [fluidization] gas
Wirbelschicht *f* fluid(ized) bed
Wirbelschicht *f*/**brodelnde** bubbling bed
Wirbelschicht *f*/**kochende** boiling bed
Wirbelschicht *f*/**stoßende** slugging bed
Wirbelschichtkracken *n*/**katalytisches** fluid catalytic cracking, FCC
Wirbelschichtofen *m (hyd)* fluid(ized-bed) incinerator
Wirbelschichtreaktor *m* 1. fluid(ized)-bed reactor; 2. *(biot)* fluid(ized)-bed fermenter [bioreactor]
Wirbelschichttechnik *f* fluid(ized)-bed technology, fluid(ized)-bed technique, fluidized(-solid) technique
Wirbelschichtverfahren *n* fluid(-bed) process, fluidized process
Wirbelschichtvergasung *f (coal)* fluid(ized)-bed gasification
Wirbelströmung *f* eddy current, turbulent [sinuous] flow
Wirbeltrockner *m* vortex dryer
Wirkdosis *f s.* Wirkungsdosis
Wirkdruck *m* differential pressure [head]
wirken *v* act
wirken *v*/**als Kristallisationskeim** form a nucleus, nucleate
wirken *v*/**antifungal [fungizid]** have antifungal activity
wirkend/langsam slow-acting
wirkend/schleichend *(tox)* slow-acting
wirkend/schnell quick-acting
wirkend/trypanozid *(pharm)* trypanocidal *(tending or used to destroy trypanosomes)*
Wirkgruppe *f* active group, coenzyme, prosthetic group, agon
Wirkkraft *f (pharm)* potency
wirksam active *(substance, component)*; efficient, effective *(method)*; *(with emphasis:)* potent *(agent, method)*
wirksam/anodisch galvanic, sacrificial *(coating)*
Wirksamkeit *f* activity *(of a substance, component)*; efficiency *(of a method)*
Wirksamkeit *f*/**biologische** biological activity
Wirksamkeit *f*/**fungizide** fungicidal activity, fungitoxicity
Wirksamkeit *f*/**katalytische** catalytic activity
Wirksamkeit *f*/**relative biologische** relative biological effectiveness, RBE *(radiation biology)*
Wirkstoff *m* 1. (active) agent, active substance [material]; 2. *(biol)* ergone, *(if part of a mixture:)* active ingredient [component, constituent]
Wirkstoff *m*/**antibakterieller** bactericide
Wirkstoff *m*/**antibiotischer** antibiotic
Wirkstoff *m*/**antimikrobieller** antimicrobial agent
Wirkstoff *m*/**chemischer** chemical agent
Wirkstoff *m*/**giftiger** toxicant, toxic (substance)
Wirkstoff *m*/**therapeutischer** therapeutic agent
Wirkstoffabgabe *f* drug delivery

Wirkstoffgehalt *m* content of active ingredient [component]
Wirkstoffkonzentration *f* concentration of active ingredient [component]
Wirkung *f* 1. action; 2. effect *(result)*
Wirkung *f*/**adstringierende** *(pharm)* stypticity
Wirkung *f*/**anästhe(ti)sierende** anaesthetic action
Wirkung *f*/**antagonistische** antagonistic action
Wirkung *f*/**dirigierende** directive influence *(as on substituents)*
Wirkung *f* **fremdioniger Zusätze** diverse ion effect
Wirkung *f*/**geschwulsterregende** tumorigenicity
Wirkung *f* **gleichioniger Zusätze** common-ion effect
Wirkung *f*/**heilende** curative action
Wirkung *f*/**korrodierende** corrosiveness, corrosivity
Wirkung *f*/**krebserregende** carcinogenicity
Wirkung *f*/**kumulative** cumulative action
Wirkung *f*/**lösungsvermittelnde** solubilizing action
Wirkung *f*/**selektive** selectivity
Wirkung *f*/**synergistische** synergistic action
Wirkung *f*/**toxische** toxic action
Wirkungsbreite *f* spectrum of activity *(as of pesticides)* • **mit großer Wirkungsbreite** broad-spectrum, broad-scale
Wirkungsdosis *f (tox, pharm)* effective dose
Wirkungsgrad *m* 1. *(ch)* strength *(of a reagent)*; 2. *(tech)* efficiency (factor), efficiency of action • **mit hohem Wirkungsgrad** *(tech)* efficient
Wirkungsgrad *m*/**carnotscher** *(phys ch)* Carnot efficiency
Wirkungsgrad *m*/**elastischer** *(rubber)* resilience, resiliency
Wirkungsgrad *m*/**thermischer** thermal efficiency
Wirkungsgrad *m*/**thermodynamischer** thermodynamic efficiency
wirkungslos ineffective
Wirkungsmechanismus *m* mechanism of action
Wirkungsquantum *n*/**plancksches** Planck (action) constant, (Planck) quantum of action
Wirkungsquerschnitt *m* 1. *(anal)* activation cross section; 2. *(nucl)* cross-section target area, (nuclear) cross section
Wirkungsspektrum *n* spectrum of activity *(as of pesticides)* • **mit breitem Wirkungsspektrum** broad-spectrum, broad-scale
Wirkungsspektrum *n*/**antimikrobielles** antimicrobial spectrum *(of an antibiotic)*
wirkungsspezifisch reaction-specific *(enzyme)*
Wirkungsspezifität *f* reaction specificity *(of an enzyme)*
Wirkungsstärke *f (pharm)* strength, potency
Wirkungssteigerung *f* fortification
Wirkungsweise *f* mode [mechanism] of action
Wirt *m* host (component, compound) *(of an inclusion compound)*
Wirtschaftsabwasser *n (hyd)* sanitary waste water *(from industrial plants)*
Wirtsgitter *n* host (crystal) lattice
Wirtskomponente *f* host (component) *(of an inclusion compound)*
Wirtsmolekül *n* host molecule
Wirtszelle *f (bioch)* host cell
Wismut *n s.* Bismut
Wismutglanz *m (min)* bismuth glance, bismuthin(it)e *(bismuth(III) sulphide)*
Wismutocker *m (min)* bismuth ochre
Witherit *m (min)* witherite *(barium carbonate)*
witterungsbeständig *s.* wetterbeständig
Wittig-Reaktion *f* Wittig reaction *(between alkylidene phosphoranes and carbonyl compounds)*
Wittig-Umlagerung *f* Wittig (ether) rearrangement
WLD *s.* Wärmeleitfähigkeitsdetektor
Wobblebase *f (bioch)* wobble base
Wohl-Abbau *m* Wohl degradation *(of sugars)*
Wohlgeruch *m* fragrance, aroma, scent, odour
Wohlgeschmack *m* tastiness, palatability

wohlriechend fragrant, aromatic, flavourful
wohlschmeckend tasty, palatable, aromatic, flavourful
Wolff-Kishner-Reaktion *f (org ch)* Wolff-Kishner reaction [reduction]
Wolfram *n* W tungsten, wolfram
Wolframat *n* *wolframate, tungstate
Wolframatophosphat *n* *wolframophosphate, tungstophosphate
Wolframatophosphorsäure *f* *wolframophosphoric acid, tungstophosphoric acid
Wolframbandlampe *f (spectr)* tungsten source
Wolframbronze *f* tungsten bronze
Wolframcarbid *n* tungsten carbide
Wolframcarbidhartmetall *n* cemented tungsten carbide
Wolfram(II)-chlorid *n* tungsten(II) chloride, tungsten dichloride, wolfram(II) chloride
Wolfram(IV)-chlorid *n* tungsten(IV) chloride, tungsten tetrachloride, wolfram(IV) chloride
Wolfram(V)-chlorid *n* tungsten(V) chloride, tungsten pentachloride, wolfram(V) chloride
Wolfram(VI)-chlorid *n* tungsten(VI) chloride, tungsten hexachloride, wolfram(VI) chloride
Wolframit *m (min)* wolframite *(iron manganese wolframate)*
Wolfram(IV)-oxid *n* tungsten(IV) oxide, tungsten dioxide, wolfram(IV) oxide
Wolfram(VI)-oxid *n* tungsten(VI) oxide, tungsten trioxide, wolfram(VI) oxide
Wolframsäure *f* *wolframic acid, tungstic acid
Wolfram(IV)-sulfid *n* tungsten(IV) sulphide, tungsten disulphide, wolfram(IV) sulphide
Wolke *f* cloud
wolkig cloudy
Wollastonit *m (min)* wollastonite *(calcium metasilicate)*
Wollfarbstoff *m* wool dye
Wollfett *n* 1. wool grease; 2. *(text)* yolk, suint
Wollfett *n***/gereinigtes** lanolin, wool fat
Wollfettalkohol *m* wool alcohol
Wollwachs *n* wool wax, crude wool grease
Woodward-Regel *f (spectr)* Woodward's rule
WSE *s.* Struktureinheit/sich wiederholende
wss. *s.* wässrig
wssr. *s.* wässrig
Wuchshormon *n (bot)* growth hormone
Wuchsstoff *m* growth substance, *(specif)* growth-promoting [growth-stimulating] substance, growth promotant
Wuchsstoffherbizid *n* hormone weedkiller, (plant-)growth regulator
Wulst *m(f)* bead, boss
würfelförmig cubic(al)
Wurmmittel *n (pharm)* vermifuge, anthelmint(h)ic, helminthagogue
wurmvertreibend anthelmint(h)ic
wurmwidrig anthelmint(h)ic
Wurtz-Fittig-Synthese *f* Wurtz-Fittig synthesis *(of hydrocarbons)*
Wurtz-Synthese *f* Wurtz reaction, Wurtz synthesis *(of hydrocarbons)*
Würze *f* 1. *(ferm)* (beer) wort; 2. aroma, flavour *(of wine)*; 3. *s.* Gewürz
Würze *f***/gehopfte** *(ferm)* hopped wort
Würze(koch)kessel *m (ferm)* wort [brew] kettle, (wort) copper
Wurzelharz *n* wood rosin
Wurzelkolophonium *n* wood rosin
würzig aromatic, spicy
Würzigkeit *f* aromaticity, spiciness
Würzmittel *n* condiment, seasoning
WW *s.* Wasserwerk

X

Xanthan *n* 1. *(biot)* xanthan gum *(a microbial polysaccharide)*; 2. *s.* Xanthen
Xanthat *n s.* Xanthogenat
Xanthen *n* xanthene, *(specif)* dibenzo[a,e]pyran
Xanthenfarbstoff *m* xanthene dye(stuff)
Xanthin *n* xanthine, *(specif)* 2,6-dihydroxypurine
Xanthogenat *n* xanthate, xanthogenate *(salt or ester of a xanthic acid)*
xanthogenieren *v (text)* xanthate, sulphidize, sulphide
Xanthogenierung *f (text)* xanthation, sulphidizing
Xanthogensäure *f* 1. RO–CS–SH

xanthic acid, xanthogenic acid, *O*-ester of dithiocarbonic acid, *(specif)* C_2H_5O–CS–SH dithiocarbonic *O*-ethyl ester, ethylxanthogenic acid; 2. HO–CS–SH dithiocarbonic acid
Xanthon *n* xanthone, xanthen-9-one, dibenzopyrone
Xanthophyll *n* xanthophyll *(any of a group of carotenoid pigments)*
Xanthoprotein *n* xanthoprotein
Xanthoproteinreaktion *f* xanthoproteic [xanthoprotein] reaction
Xanthoproteinsäure *f* xanthoproteic acid
Xanthosin *n* xanthosine, xanthine 9-ribofuranoside *(a nucleoside)*
Xanthurensäure *f* xanthurenic acid, 4,8-dihydroxyquinoline-2-carboxylic acid
Xanthyliumsalz *n* xanthylium salt
Xanthylsäure *f (bioch)* xanthylic acid, xanthosine 5′-monophosphate, XMP
Xenat *n* xenate *(salt or ester of xenic acid)*
xenomorph *(cryst)* xenomorphic, allotriomorphic, anhedral
Xenon *n* Xe xenon
Xenonlampe *f* xenon (arc) lamp
Xenonsäure *f* xenic acid
Xenylamin *n* xenylamine, *p*-aminobiphenyl
Xerogel *n (coll)* xerogel
Ximeninsäure *f* ximenynic acid, santalbic acid, *octadec-11-en-9-ynoic acid
Xylan *n* xylan *(a polysaccharide)*
Xylen *n* xylene, dimethylbenzene, *(esp commercially for impure products:)* xylol
Xylenol *n* xylenol, hydroxyxylene, dimethylphenol
Xylidin *n* xylidine, aminodimethylbenzene
Xylit *m* 1. xylite, woody lignite *(brown coal)*; 2. *s.* Xylitol
Xylitol *n* xylitol *(a pentanepentol)*
Xyloketose *f s.* Xylulose
Xylol *n s.* Xylen
Xylonsäure *f* xylonic acid *(a tetrahydroxypentanoic acid)*
Xylose *f* xylose, wood sugar *(a monosaccharide)*
Xylulose *f* xylulose, xyloketose
Xylylbromid *n* α-bromoxylene, xylyl bromide
Xylylenbromid *n* α,α′-dibromoxylene, xylylene dibromide
Xylylsäure *f* xylylic acid, dimethylbenzoic acid
x-y-Schreiber *m* X-Y recorder, function plotter

Y

Ylid *n* ylide *(any of a class of internal salts)*
***P*-Ylid** *n* phosphonium [phosphorus] ylide, Wittig reagent
Yohimbealkaloid *n* yohimbé alkaloid
Yperit *n* yperite, *di-2-chloroethylsulphide
Ypsilonstück *n* Y-shape connecting tube, Y-piece, Y-type connector
Y-Stück *n* Y-shape connecting tube, Y-piece, Y-type connector
Ytterbium *n* Yb ytterbium
Ytterbium(II)-chlorid *n* ytterbium(II) chloride, ytterbium dichloride
Ytterbium(III)-chlorid *n* ytterbium(III) chloride, ytterbium trichloride
Ytterbiumoxid *n* ytterbium oxide
Ytterbium(III)-sulfat *n* ytterbium(III) sulphate
Yttererde *f* 1. yttrium earth *(class name)*; 2. *s.* Yttriumoxid
Yttrium *n* Y yttrium
Yttrium-Aluminium-Granat *m* yttrium-aluminium garnet, YAG
Yttriumcarbonat *n* yttrium carbonate
Yttriumchlorid *n* yttrium chloride
Yttriumhydroxid *n* yttrium hydroxide
Yttriumoxid *n* yttrium oxide

Z

Z *s.* Ordnungszahl
Zähbruch *m* ductile fracture, tough fracture
zäh(e) 1. tough, tenacious *(material)*; 2. *s.* zähflüssig
zähflüssig high-viscosity, viscid, viscous, semiliquid, ropy • **zähflüssig machen** thicken • **zähflüssig werden** thicken, inspissate

Zähflüssigkeit *f* viscosity, ropiness, thickness
Zähigkeit *f* 1. toughness, tenacity; 2. *s.* Viskosität
Zähler *m* 1. meter *(for fluids)*; 2. *(nucl)* counter, counting tube
Zähligkeit *f* ligancy, coordination, coordination [covalence, covalency] number
Zählrohr *n (nucl)* counting tube, counter
Zählrohr *n*/**selbstlöschendes** self-quenched [self-quenching] counter
Zahngold *n* dental gold (alloy)
Zahnradpumpe *f* gear pump
Zahnzement *m* dental cement, cementum
Zajcev-Regel *f* Zaitsev rule *(of orientation in β-eliminations)*
Zapfen *m* 1. trunnion, pivot, journal, neck *(as of a shaft or cylinder)*; 2. bung, spigot, plug, plug cock [bib] *(on a cask)*
Zapfenlager *n* journal bearing
Zartmachersalz *n (food)* meat tenderizer
Zeeman-Atomabsorptionsspektrophotometrie *f* Zeeman atomic absorption spectrophotometry, ZAAS
Zeeman-Aufspaltung *f* **(der Energieniveaus)** *s.* Zeeman-Effekt
Zeeman-Effekt *m* Zeeman effect
Zeeman-Niveau, Zeeman-Term *m* Zeeman term [energy level]
Zeeman-Verbreiterung *f* Zeeman broadening *(of spectral lines under the influence of magnetic fields)*
Zehntelnormallösung *f* decinormal solution
zehnzählig decadentate *(coordination chemistry)*
zehnzähnig decadentate *(coordination chemistry)*
Zeichen *n*/**chemisches** chemical symbol [sign]
Zeiger *m* pointer, indicator, index *(of an apparatus)*
Zein *n* zein *(a prolamin obtained from maize)*
Zeinfaser *f* zein staple (fibre), maize protein staple
Zeinkunststoff *m* zein plastic
Zeiß-Abbe-Refraktometer *n* Abbe refractometer
Zeitdomäne *f (spectr)* time domain
Zeiteinheit *f* unit of time • **in der Zeiteinheit** per unit time
Zeitgesetz *n s.* Geschwindigkeitsgleichung
Zeitkonstante *f* time constant
Zeitreaktion *f* clock [time] reaction
Zeitstandfestigkeit *f* creep strength [resistance]
Zeitstandversuch *m* creep test [experiment]
Zeit-Temperatur-Umwandlungsdiagramm *n* time-temperature-transformation diagram, T.T.T. diagram
Zellabtrennung *f (bioch)* cell detachment
Zellaufschluss *m (bioch)* cell disruption
Zellausbeute *f (biot)* cell yield
Zellbeweglichkeit *f* cell motility
Zelle *f* 1. cell; 2. *(tech)* compartment, chamber
Zelle *f*/**bipolare** *(el ch)* bipolar cell
Zelle *f*/**elektrochemische** *s.* Zelle/galvanische
Zelle *f*/**elektrolytische** electrolytic [electrolysis] cell
Zelle *f*/**galvanische** voltaic cell, galvanic [electrical, chemical] cell
Zelle *f*/**künstliche** artificial cell
Zelle *f*/**lichtelektrische [photoelektrische]** photoelectric cell, photocell, phototube, photovalve
Zelle *f*/**photogalvanische** photogalvanic cell
Zelle *f*/**unipolare** *(el ch)* monopolar cell
zellenartig cellular
Zellenfilter *n* cellular filter
zellgebunden *(biol)* cell-bound
Zellgift *n* cytotoxin
Zellglas *n* cellophane, cellulose film
Zellhaftung *f (bioch)* cell attachment
Zellhorn *n* celluloid
zellig cellular
Zellimmobilisierung *f (biot)* cell immobilization
Zellkern *m (biol)* nucleus
Zellkultur *f (biot)* cell culture
Zellprotein *n* cell(ular) protein
Zellreaktion *f (phys ch)* cell reaction
zellschädigend cytotoxic
Zellspannung *f (el ch)* cell voltage
Zellstoff *m* 1. *(tech)* (chemical) pulp, *(specif)* (chemical) wood pulp, CWP; 2. *s.* Cellulose

Zellstoff *m*/**alkalisch aufgeschlossener [erkochter, gekochter]** alkaline-cooked pulp, alkali pulp
Zellstoff *m* **für die Papierindustrie** paper(making) pulp
Zellstoff *m*/**klassischer** chemical pulp
Zellstoffaufschluss *m* chemical pulping
Zellstoffbahn *f* web of pulp
Zellstoffbleiche *f* pulp bleaching
Zellstoffindustrie *f* pulp industry
Zellstoffwechsel *m (biol)* cell metabolism
Zellsuspension *f (biot)* cell suspension [slurry]
Zellteilung *f (biol)* cell division
Zelltod *m (biol)* cell death
Zelltod *m*/**programmierter** *(biol)* programmed cell death
Zelltrockenmasse *f (biot)* cell dry weight, dry weight of cells
Zelltrockensubstanz *f (biot)* cell dry weight, dry weight of cells
Zell-TS *f s.* Zelltrockenmasse
zellulär cellular
Zelluloid *n* celluloid
Zellulose *f s.* Cellulose
Zellwolle *f* (viscose) staple fibre
Zellvermehrung *f (biol)* cell proliferation
Zellwachstum *m (biol)* cell growth
Zement *m* 1. *(build, rubber)* cement; 2. *(geoch)* cement, cementing agent, binder, agglutinant
Zement *m*/**hydraulischer [wasserbindender]** hydraulic cement
zementartig cementitious
Zementation *f* 1. *(met, geol)* cementation *(precipitation of a metal from salt solutions by the action of a less noble one)*; 2. *(met)* carburization
Zementationsbad *n (met)* carburizing bath
Zementationshärten *n (met)* case-hardening *(carburizing and subsequent hardening)*
Zementationsmittel *n (met)* carburizing medium [agent], (case-hardening) carburizer
Zementationsschicht *f (met)* carburized case
Zementationstiefe *f (met)* carburizing [carburization] depth, depth of case
Zementier... *siehe auch: Zementations...*
zementieren *v* 1. *(build)* cement; 2. *(met, geoch)* cement *(precipitate a metal from salt solutions by the action of a less noble one)*; 3. *(met)* carburize
zementieren *v* **in festem Einsatz** *(met)* pack-carburize
zementieren *v* **in festen Mitteln** *(met)* pack-carburize
zementieren *v*/**in flüssigen Mitteln** *(met)* liquid-carburize, bath-carburize
Zementieren *n* 1. *(build)* cementing; 2. *(met)* cementing *(precipitating a metal from salt solutions by the action of a less noble one)*; 3. *(met)* carburizing
Zementieren *n* **in festem Einsatz** *(met)* pack carburizing, solid(-pack) carburizing
Zementieren *n* **in festen Mitteln** *(met)* pack carburizing, solid(-pack) carburizing
Zementieren *n* **in flüssigen Mitteln** *(met)* liquid(-salt) carburizing, bath carburizing
Zementierofen *m (met)* carburizing oven
Zementit *m (met)* cementite, iron carbide
Zementkupfer *n* cement [precipitated] copper
Zentigrammbereich *m (anal)* centigram range
Zentralatom *n* central atom
Zentralbahn *f (bioch)* central (metabolic) pathway, amphibolic pathway
zentralsymmetrisch centrosymmetric
Zentrifugalabscheider *m* centrifugal collector [separator]
Zentrifugalextraktor *m* centrifugal extractor
Zentrifugalgießen *n* centrifugal casting
Zentrifugalguss *m* centrifugal casting
Zentrifugalklassierer *m* centrifugal classifier
Zentrifugalkraft *f* centrifugal force
Zentrifugalpumpe *f* centrifugal pump
Zentrifugalscheider *m* centrifugal collector [separator]
Zentrifugalversprüher *m* centrifugal atomizer
Zentrifugat *n* centrifugate
Zentrifuge *f* centrifuge, *(tech esp for filtering:)* centrifugal
Zentrifuge *f*/**stehende** underdriven [bottom-driven] centrifuge, base-bearing centrifuge

Zentrifugenkuchen *m* (solids) cake, filter cake *(of a centrifuge)*
Zentrifugenmittel *n* z average *(of relative molecular mass)*
Zentrifugenrotor *m* centrifuge rotor [head]
Zentrifugentrommel *f* bowl, centrifuge basket
zentrifugieren *v* centrifugate, centrifuge
Zentrifugierung *f* centrifugation
Zentripetalkraft *f* centripetal force
zentrosymmetrisch centrosymmetric
Zentrum *n*/**aktives** active site *(in molecules, catalysts)*
Zentrum *n*/**reaktives** reactive site *(in molecules)*
Zentrum *n*/**saures** acid site *(in molecules, catalysts)*
***sp*^{3}-Zentrum** *n* sp^3 centre *(chemical bond theory)*
Zentrumsaktivität *f*/**katalytische** *(bioch)* turnover number
Zeolith *m (min)* zeolite
Zeolith *m*/**künstlicher** artificial [synthetic] zeolite, zeolite exchange resin *(ion exchange)*
zeolithisch zeolitic
Zeolithwasser *n* zeolitic water
zerbrechen *v* break
zerbrechlich breakable, fragile, brittle
Zerbrechlichkeit *f* fragility, brittleness
zerbröckeln *v* crumble; crumble (away), slake *(of material)*
Zerfall *m* 1. *(ch)* decomposition, *(into ions:)* dissociation; 2. *(nucl)* disintegration, fission, decay; 3. *(biol)* decomposition, decay; 4. *(tech)* disintegration, breakdown of size • **durch radioaktiven Zerfall entstanden** radiogenic • **zum Zerfall bringen** disintegrate
Zerfall *m*/**dualer** *(nucl)* branched [multiple] disintegration [decay], branching
Zerfall *m*/**radioaktiver** radioactive disintegration [decay]
Zerfall *m*/**verzweigter** *s.* Zerfall/dualer
β-**Zerfall** *m (nucl)* beta [β-ray] disintegration [decay]
β^+-**Zerfall** *m (nucl)* positive beta decay, positron decay
β^--**Zerfall** *m (nucl)* negative beta decay
zerfallen *v* 1. *(ch)* decompose, undergo decomposition, *(of molecules:)* fragment, *(into ions:)* dissociate; 2. *(nucl)* disintegrate, decay; 3. *(biol)* decompose, decay; 4. *(tech)* disintegrate, break down, crumble (away), *(esp of coal)* slake
Zerfallsenergie *f (nucl)* disintegration [decay] energy
Zerfallsgeschwindigkeit *f* 1. *(ch)* decomposition rate; 2. *(nucl)* rate of disintegration [decay]
Zerfallskonstante *f (nucl)* disintegration constant, decay [transformation] constant, decay coefficient [factor]
Zerfallsprodukt *n* 1. *(ch)* decomposition product; 2. *(nucl)* disintegration [decay] product
Zerfallsrate *f (nucl)* rate of disintegration [decay]
Zerfallsreaktion *f (ch)* decomposition reaction
Zerfallsreihe *f* radioactive (decay) series
Zerfallsreihe *f*/**radioaktive** radioactive (decay) series
Zerfallswärme *f (nucl)* decay heat
Zerfallsweg *m* fragmentation path *(of molecules in mass spectrometry)*
zerfasern *v (pap)* defibre, defibrate, reduce to fibres, shred, *(Am)* (de)fiberize
zerfließen *v* deliquesce, liquefy
zerfließlich deliquescent
Zerfließlichkeit *f* deliquescence
zerfressen *v* corrode, eat
zerhacken *v* 1. chop, chip *(mechanically)*; 2. *(spectr)* chop
Zerhacker *m (spectr)* chopper
zerkleinern *v* comminute, reduce (in size); mill, grind; chip, chop; crush *(brittle material)*; shred *(garbage)*
zerkleinern *v*/**fein** comminute
Zerkleinerung *f* comminution, (size) reduction; milling, grinding; chipping, chopping; crushing *(of brittle material)*; shredding *(of garbage)*
zerknallen *v* explode *(of vessels)*
zerkrümeln *v* crumble; crumble (away), slake *(of material)*
zerlegbar decomposable
zerlegen *v* 1. *(ch)* decompose, split (up), break down; 2. *(bioch)* catabolize; *(physically:)* fractionate, split (up), break down, separate
Zerlegung *f (ch)* decomposition, breakdown, splitting(-up); *(physically:)* frac-

tionation, splitting(-up), breakdown, separation
Zerlegungsprodukt *n* decomposition [breakdown] product
zermahlen *v* grind, mill, triturate
zerplatzen *v* crack, *(violently:)* explode, blow up
zerpulvern *v* pulverize, powder, triturate
zerquetschen *v* crush, mash
zerreiben *v* triturate, grind, mill
zerreiblich friable, triturable
Zerreiblichkeit *f* friability
zerreißen *v* tear, rupture, break; macerate, tear *(fibrous material)*; break, crack *(of material)*
Zerreißfestigkeit *f* resistance to tear(-ing), tearing [tensile, bursting] strength
zerschmelzen *v* melt
zersetzen *v (ch, biol)* decompose
zersetzen *v***/sich** 1. *(ch)* undergo decomposition, decompose; 2. *(biol)* decay
zersetzend destructive
Zersetzer *m* decomposer *(electrolysis)*
Zersetzerzelle *f* decomposer *(electrolysis)*
zersetzlich decomposable
zersetzlich/leicht readily [easily] decomposing
Zersetzlichkeit *f* decomposability
Zersetzung *f* 1. *(ch)* decomposition; 2. *(biol)* decay, decomposition • **Zersetzung erleiden** undergo decomposition
Zersetzung *f* **durch Licht** photolysis, photodecomposition
Zersetzung *f***/elektrolytische** electrolysis, electrolytic dissociation
Zersetzung *f***/strahlenchemische** radiolysis
Zersetzung *f***/thermische** thermal decomposition [degradation]
Zersetzungsdestillation *f* destructive distillation
Zersetzungspotenzial *n* decomposition potential
Zersetzungsreaktion *f* decomposition reaction
zersplittern *v* 1. spall, shatter; 2. *(min)* delaminate
zerspringen *v* crack, shatter
zerstäuben *v* atomize
Zerstäuber *m* 1. atomizer; 2. *(spectr)* nebulizer
Zerstäuberdüse *f* atomizing nozzle
Zerstäubung *f* atomization
zerstören *v* 1. destroy; 2. *(bioch)* decompose; break, crack, demulsify *(emulsions)*
zerstört werden *v* break down *(of emulsions)*
Zerstörung *f* 1. destruction; 2. *(bioch)* decomposition; 3. *(el ch)* breakdown *(of passivity or a passive film)*; breaking, cracking *(of emulsions)*
zerstörungsfrei non-destructive
zerstoßen *v* crush, triturate
zerstrahlen *v* undergo annihilation
Zerstrahlung *f* annihilation, energy pooling
zerteilen *v* 1. disperse; 2. *(coll)* deflocculate; *(relating to solids:)* comminute, reduce, *(relating to droplets:)* reduce in size, subdivide, atomize
Zerteilung *f* 1. dispersion, dispersal; 2. *(coll)* deflocculation; *(relating to solids:)* comminution, (size) reduction, *(relating to droplets:)* size reduction, subdivision, atomization
zertrümmern *v* shatter, destroy; *(esp for further processing:)* break up *(e.g. ore)*
Zertrümmerung *f* shatter(ing), destruction; *(esp for further processing as of ore:)* breaking-up
Zeta-Potenzial *n* zeta potential, ZP, electrokinetic [double-layer] potential
Zeugmatographie *f (spectr)* spin mapping
Zickzackkette *f* zigzag chain
Ziegel *m* brick
Ziegeleierzeugnis *n* brickware
Ziegelofen *m* brick kiln
Ziegelton *m* brick(-making) clay
Ziegler-Natta-Katalysator *m* Ziegler-Natta catalyst
Ziegler-Verfahren *n* Ziegler process *(for polymerizing alkenes)*
ziehen *v* 1. draw, pull *(e.g. glass or plastics)*; 2. pull *(a reaction)*; 3. *(of viscous material:)* slide *(e.g. to the discharge point)*
Ziehen *n* **aus der Schmelze** crystal pulling *(of single crystals)*
Ziehzwiebel *f* bulb *(for producing glass silk)*
Zimm-Diagramm *n* Zimm plot *(for de-*

termining the molecular weights of polymers)
Zimmertemperatur *f* room temperature, normal [ordinary] temperature
Zimtaldehyd *m* cinnamaldehyde, cinnamic aldehyde, *3-phenylpropenal
Zimtalkohol *m* cinnamyl alcohol, *3-phenyl-2-propen-1-ol
Zimtsäure *f* cinnamic acid, 3-phenylacrylic acid
***trans*-Zimtsäure** *f trans*-cinnamic acid, ordinary cinnamic acid
Zineb *n* zineb, zinc ethylenebisdithiocarbamate
Zingeron *n* zingerone, 4-hydroxy-3-methoxybenzylacetone
Zingiberen *n* zingiberene *(a monocyclic sesquiterpene)*
Zink *n* Zn zinc
Zinkamalgam *n* zinc amalgam
Zinkat *n* zincate
Zinkblende *f (min)* zinc blende, sphalerite *(zinc sulphide)*
Zinkblendegitter *n (cryst)* zinc-blende lattice
Zinkchlorid *n* zinc chloride
Zinkgelb *n* zinc yellow, zinc(-potassium) chromate
Zinkgranalien *fpl* mossy zinc
Zinkhexafluorosilicat *n* zinc hexafluorosilicate
Zinkoxid *n* zinc oxide
Zinkoxidkatalysator *m* zinc-oxide catalyst
Zinkrost *m* white rust
Zinkspat *m (min)* zinc spar, smithsonite, dry-bone ore *(zinc carbonate)*
Zinkstaubdestillation *f* zinc-dust distillation
Zinkvitriol *m (min)* goslarite, zinc vitriol, white vitriol [copperas] *(zinc sulphate 7-water)*
Zinkvitriol *n* zinc vitriol, zinc sulphate 7-water
Zinkweiß *n* zinc white *(crude zinc oxide)*
Zinn *n* Sn tin
α-Zinn *n* alpha [grey] tin, α tin
β-Zinn *n* beta [white] tin, β tin
γ-Zinn *n* gamma [brittle] tin, γ tin
Zinnbronze *f* tin bronze
Zinn(II)-chlorid *n* tin(II) chloride, tin dichloride, stannous chloride
Zinn(IV)-chlorid *n* tin(IV) chloride, tin tetrachloride, stannic chloride
Zinndi... *siehe auch: Zinn(II)-...*
Zinndioxid *n s.* Zinn(IV)-oxid
Zinndisulfid *n s.* Zinn(IV)-sulfid
Zinnelektrolyt *m* tin plating bath [solution]
Zinngeschrei *n* tin cry
Zinnhydrid *n* tin hydride, stannic hydride, *stannane
Zinnlegierung *f* tin alloy
Zinnmonosulfid *n s.* Zinn(II)-sulfid
Zinnober *m* 1. cinnabar, cinebarite, red mercuric sulphide; 2. *(min)* liver ore *(mercury(II) sulphide)*
Zinn(IV)-oxid *n* tin (IV) oxide, tin dioxide, stannic oxide
Zinnpest *f* tin pest [plague, disease]
Zinnsäure *f* stannic acid
Zinnstein *m (min)* tin stone, cassiterite *(tin(IV) oxide)*
Zinn(II)-sulfid *n* tin(II) sulphide, tin monosulphide, stannous sulphide
Zinn(IV)-sulfid *n* tin(IV) sulphide, tin disulphide, stannic sulphide
Zinntetra ... *siehe: Zinn(IV)-...*
Ziram *n* ziram, zinc dimethyldithiocarbamate
Zirconat *n* zirconate *(salt of zirconic acid)*
Zirconium *n* Zr zirconium
Zirconiumcarbid *n* zirconium carbide
Zirconium(II)-chlorid *n* zirconium(II) chloride, zirconium dichloride
Zirconium(III)-chlorid *n* zirconium(III) chloride, zirconium trichloride
Zirconium(IV)-chlorid *n* zirconium(IV) chloride, zirconium tetrachloride
Zirconiumhydrid *n* zirconium hydride
Zirconium(IV)-oxid *n* zirconium(IV) oxide, zirconium dioxide
Zirconiumoxidchlorid *n* zirconium dichloride oxide
Zirconiumsilicat *n* zirconium silicate, *(specif)* zirconium orthosilicate
Zirconium(IV)-sulfat *n* zirconium(IV) sulphate
Zirconsäure *f* zirconic acid
Zirkon *m (min)* zircon *(zirconium orthosilicate)*
Zirkonporzellan *n* zircon porcelain
Zirkularchromatographie *f* radial [circular] chromatography

Zirkulardichroismus *m* circular dichroism, CD
Zirkularpolarisation *f* circular polarization
zirkularpolarisiert circularly polarized
Zirkulation *f* (re)circulation
zirkulieren *v* circulate
zirkulieren lassen *v* circulate
Zitronensäure *f s.* Citronensäure
Zitronensäurezyklus *m s.* Citratzyklus
Zittersieb *n* vibrating [reciprocating] screen
z-Mittel *n* z average *(of relative molecular mass)*
Zone *f* zone, region
Zone f/stagnierende stagnant zone, dead space *(in a reactor)*
Zonenelektrophorese *f* zone electrophoresis
zonenreinigen *v* zone-refine, zone-purify
Zonenreinigung *f* zone refining [purification]
zonenschmelzen *v* zone-melt
Zonenschmelzen *n* zone melting
Zonenschmelzverfahren *n* float zone process *(semiconductor materials processing)*
Zonensedimentation *f (hyd)* zone settling
Zonenzentrifugation *f (anal)* zonal centrifugation
Zoosterin *n (bioch)* zoosterol, animal sterol
Zoosterol *n (bioch)* zoosterol, animal sterol
ZT *s.* Zimmertemperatur
ZTU-Diagramm *n* time-temperature-transformation diagram, T.T.T. diagram
zubereiten *v* prepare, make up
zubereitet/frisch freshly prepared
Zubereitung *f* preparation, formulation *(act or substance)*
züchten *v* 1. *(cryst)* grow; 2. *(biot)* culture, grow
Zuchthefe *f (biot)* culture(d) yeast, industrial yeast, barm
Züchtung *f* 1. *(cryst)* growth; 2. *(biot)* cultivation, growth
Zucker *m* sugar • **in Zucker umwandeln** saccharify, saccharize
Zucker *m***/brauner** brown [soft] sugar
Zucker *m***/einfacher** simple sugar, monosaccharide
Zucker *m***/flüssiger** liquid sugar
Zucker *m***/gebrannter [karamellisierter]** caramelized sugar, caramel
Zucker *m***/linksdrehender** laevorotatory sugar
Zucker *m***/reduzierender** reducing sugar
Zuckeralkohol *m* sugar alcohol
Zuckeranhydrid *n* sugar anhydride
Zuckeranteil *m* sugar moiety *(as of a glycoside)*
zuckerartig sugar-like, saccharine, saccharoid
Zuckercouleur *f* sugar colouring [dye]
Zuckerderivat *n* sugar derivative
Zuckerdicarbonsäure *f* aldaric acid
Zuckerersatzstoff *m* sugar substitute
Zuckerester *m* sugar ester
zuckerhaltig sacchariferous, sugary, sugar-bearing, sugar-containing
Zuckerkomponente *f* sugar moiety *(as of a glycoside)*
zuckerliefernd 1. *(bioch)* glycogenic *(e.g. amino acid)*; 2. *s.* zuckerhaltig
Zuckerpflanze *f* sugar plant *(a plant not converting its assimilation products into starch)*
Zuckerraffination *f* sugar refining
Zuckerreihe *f* sugar series
Zuckerrohrbagasse *f* (sugar-cane) bagasse, begass(e), megass(e)
Zuckerrohrmelasse *f* (sugar-)cane molasses, cane blackstrap (molasses)
Zuckerrübenmelasse *f* (sugar-)beet molasses, beet discard molasses
Zuckerrübenschnitzel *npl* (beet) cossettes, beet slices
Zuckersäure *f* sugar acid, *(specif)* $HOOC[CHOH]_4COOH$ saccharic acid, glucosaccharic acid *(one form of 2,3,4,5-tetrahydroxyhexanedioic acid)*
Zuckerumwandlung *f* sugar conversion
zudosieren *v* charge, apportion
Zufallsfehler *m* random [accidental] error
Zufallsknäuel *n* random coil *(chemistry of macromolecules)*
Zufallsprobe *f* random sample
Zufallsprobe(nent)nahme *f* random sampling
Zufallsvariable *f* variate *(statistics)*

zufließen *v* flow in, run in
zufließen lassen *v* run in
Zufluss *m* 1. *s.* Zulauf; 2. *(bioch)* influx
zufügen *v s.* zugeben
Zufuhr *f* feed(ing)
zuführen *v* feed *(a material)*
zuführen *v*/**wieder** return
Zuführung *f* 1. feed(ing); 2. (feed) inlet
Zuführungsrohr *n* feed pipe
zufüttern *v (biot)* feed *(nutrients to microorganisms)*
Zug *m* 1. flue *(passageway as of an oven)*; 2. draught, pull, suction *(as caused by a stack or fan)*
Zug *m*/**künstlicher** forced draught
Zugabe *f* addition, supplementation
zugänglich *(agric)* available *(nutrients)*
Zugänglichkeit *f (agric)* availability *(of nutrients)*
Zugdehnungseigenschaften *fpl* tensile (stress-strain) characteristics
Zugdehnungsverhalten *n* tensile (stress-strain) characteristics
zugeben *v* add
zugeben *v*/**chargenweise [portionsweise]** charge
zugeben *v*/**tropfenweise** add dropwise
Zugeben *n* addition
zügeln *v* control *(e.g. a reaction)*
zugestellt/basisch *(met)* basic-lined
zugestellt/sauer acid-lined
zugfest tenacious
Zugfestigkeit *f* 1. tensile strength, tenacity; 2. *(pap)* breaking strength
Zugkraft *f* tensile force
Zugspannung *f* tensile stress
Zugversuch *m* tensile [tension] test, stress-strain experiment
Zulauf *m* 1. *(liquid:)* influent, affluent, afflux, inflow; 2. *(distil)* feed (stream); 3. *(site:)* feed (liquor) inlet; 4. *(process:)* inflow, afflux
zulaufen *v* run in, flow in
zulaufen lassen *v* run in
Zulaufrohr *n* feed pipe [tube], charging [influent] pipe
Zulaufventil *n* feed valve
Zuleitung *f* 1. feed(ing); 2. feed pipe [line]
Zuluft *f* incoming [inlet] air
Zuluftfilter *n (biot)* incoming air filter
zumessen *v* meter, dose, proportion
Zumesspumpe *f* metering pump, dosing [proportioning, controlled-volume] pump
Zumessung *f* metering, dosing, dosage, proportioning
zumischen *v* admix
Zumischung *f* admixture
Zunahme *f* increase, growth, rise
zünden *v* ignite, fire
Zunder *m* 1. (oxide, oxidation) scale; 2. *s.* Walzzunder
Zünder *m* igniter, primer
zunderbeständig non-scaling
Zunderbildung *f* scale formation [build-up]
zunderfrei non-scaling
zundern *v* scale
Zündhütchen *n* ignition cap, primer
Zündpunkt *m* spontaneous-ignition temperature, S.I.T.
Zündquelle *f* ignition source
Zündsprengstoff *m* initiating [primary] explosive, initiator, primer, detonator
Zündtemperatur *f* ignition temperature
Zündung *f* 1. ignition, inflammation, *(relating to explosives also)* firing; 2. *(bioch)* sparking *(of fatty-acid oxidation)*
zunehmen *v* increase, grow, rise
zurückbilden *v* re-form
zurückbilden *v*/**sich** re-form
zurückfließen *v* flow back, return
zurückführen *v* return
zurückführen *v*/**in den Kreislauf** return to the circuit, recirculate, recycle
zurückgewinnen *v* recover, reclaim
zurückhalten *v* retain, trap
Zurückhaltung *f* retention
zurücklaufen *v* run back
zurückoxidieren *v* reoxidize, oxidize back
zurückschlagen *v* strike [flash] back *(of flame in a burner)*; suck back *(of liquid in a pipe)*
zurückspringen *v* 1. rebound, spring back; 2. *(rubber)* recover; drop, return, fall *(of electrons to a lower energy level)*
zurückstrahlen *v* reflect (back)
zurücktitrieren *v* back-titrate
Zurücktitrieren *n* back-titration
zurückverwandeln *v* reconvert, convert back
zurückverwandeln *v*/**sich** reconvert, convert back
zurückwerfen *v* reflect (back)

zusammenbacken *v* agglomerate, set up, stick together, *(esp under the influence of heat:)* cake
zusammenballen *v* agglomerate, agglutinate, ball (up)
zusammenballen **v/sich** agglomerate, aggregate, agglutinate, ball (up), clog, clot, lump
zusammenbauen *v* 1. assemble; 2. *(rubber)* assemble, build
zusammenbrechen *v* 1. collapse, break down *(as of foam)*; 2. *(el ch)* break down
Zusammenbruch *m* 1. collapse, breakage *(as of foam)*; 2. *(el ch)* breakdown *(of passivity, a passive film, or potential)*
zusammendrücken *v* compress, squeeze
zusammenfließen *v* flow together, coalesce
zusammenfritten *v* agglomerate, cake *(under the influence of heat)*
zusammengesetzt/gleichartig homogeneous
zusammengesetzt/ungleichartig heterogeneous
zusammenhalten *v* cohere, hold together
zusammenkleben *v* stick together
zusammenlagern **v/sich** congregate
zusammenmischen *v* mix, *(esp relating to bulk material)* blend
Zusammenprall *m* collision, impact, impingement
zusammenprallen *v* collide, impact, impinge
zusammenpressen *v* compact, compress, squeeze
zusammenrühren *v* stir together
zusammenschmelzen *v* melt [fuse] together; melt down
zusammensetzen *v* arrange, set up *(mechanically)*
zusammensetzen *v* **aus/sich** be made up of, consist of
Zusammensetzung *f* composition, make-up, analysis *(nature of a chemical compound or mixture)* • **von unklarer Zusammensetzung** ill-defined, of ill-defined composition
Zusammensetzung **f/chemische** chemical composition
Zusammensetzung **f/mittlere** average analysis
Zusammensetzung **f/prozentuale** percent(age) composition
Zusammenstoß *m* collision, impact, impingement
zusammenstoßen *v* collide, impact, impinge
zusammentreffen *v* encounter *(of molecules in solutions)*
Zusammentreffen *n* encounter *(of molecules in solutions)*
zusammenziehen **v/sich** contract
Zusatz *m* 1. addition, admixture *(act)*; 2. additive (substance), admixture
Zusatzchemikalie *f* make-up (chemical) *(for compensating losses)*
Zusatzelektrolyt *m* support(ing) electrolyte, base [indifferent] electrolyte *(polarography)*
Zusatzkesselspeisewasser *n* make-up boiler (feed) water, boiler make-up (water), feed-water make-up
Zusatzkomponente *f* *(distil)* codistillant
Zusatzstoff *m* 1. additive (substance), accessory agent, *(esp for cheapening or weighting:)* loading material, load; conditioner *(for improving physical properties)*; 2. *(distil)* codistillant, separating agent; 3. *(food)* intentional food additive; blending agent *(for improving the octane number of motor fuel)*
Zusatzwasser *n* make-up water, *(specif)* make-up boiler water, boiler make-up
Zuschlag *m* 1. loading *(as of filling materials)*; 2. *s.* Zuschlagstoff
Zuschlagstoff *m* 1. *(met)* flux; 2. *(build)* aggregate; loading material *(as for filling rubber)*
zuschmelzen *v* seal (up)
zusetzen *v* 1. add, admix; 2. clog (up), blind *(e.g. screen openings)*
zusetzen **v/sich** clog, plug *(as of screen openings)*
zuspeisen *v* feed
Zustand *m* state, condition • **in fein verteiltem Zustand** in a fine state • **in freiem Zustand** in the free state • **in gebundenem Zustand** in the combined state • **in reinem Zustand darstellen** prepare in pure form, prepare in a pure condition [state], isolate *(natural products)* • **in reinem Zustand gewinnen** obtain pure [in pure form], obtain in a pure condition [state]

Zustand *m*/**angeregter** excited state
Zustand *m*/**atomarer** atomic state
Zustand *m*/**fester** solid state
Zustand *m*/**flüssig-kristalliner** *s.* Zustand/mesomorpher
Zustand *m*/**glasartiger [glasiger]** vitreous [glass-like, glassy] state
Zustand *m*/**mesomerer** *(org ch)* resonance state
Zustand *m*/**mesomorpher** mesomorphic [liquid crystalline] state, mesomorphism
Zustand *m*/**metallischer** metallic state
Zustand *m*/**metastabiler** metastable state
Zustand *m*/**plastischer** 1. plastic state; 2. *(coal)* stage of plasticity
Zustand *m*/**quasistationärer** steady [stationary] state *(kinetics)*
Zustand *m*/**stabiler [stationärer]** steady [stationary] state
Zustand *m*/**superangeregter** superexcited state
Zustand *m*/**superfluider [superflüssiger]** superfluid state, superfluidity
Zustände *mpl*/**übereinstimmende** *(phys ch)* corresponding states
Zustandsänderung *f* change of state
Zustandsdiagramm *n* constitution(al) diagram, equilibrium [phase] diagram
Zustandsenergie *f* potential energy
Zustandsfunktion *f* state function
Zustandsgleichung *f* equation of state
Zustandsgleichung *f* **idealer Gase** ideal gas equation, ideal [perfect] gas law
Zustandsgleichung *f* **idealer Gase/thermische** ideal gas equation, ideal [perfect] gas law
Zustandsgleichung *f*/**reduzierte** reduced equation of state
Zustandsgleichung *f*/**thermodynamische** thermodynamic equation of state
Zustandsgröße *f* property of state
Zustandsgröße *f*/**reduzierte** reduced property
Zustandsvariable *f* state variable, variable of state
Zustandsvariable *f*/**reduzierte** reduced variable
zustellen *v* *(met)* line
Zustellung *f* *(met)* lining *(act or material)*
Zustellung *f*/**basische** basic lining
Zustellung *f*/**saure** acid lining
zustreben *v*/**dem Gleichgewicht** go toward equilibrium
Zuteileinrichtung *f* dosing [proportioning] apparatus, feeder
zuteilen *v* dose, proportion, charge, batch
zutropfen *v* add dropwise, add drops of
Zutropfen *n* dropwise addition
Zwangumlauf *m* forced circulation
Zwangumlaufverdampfer *m* forced-circulation evaporator
zweiachsig *(cryst)* biaxial
zweiatomig diatomic
zweibasig dibasic, diprotic *(acid)*; dibasic, diacid *(base)*
zweibindig divalent, bivalent *(relating to homopolar bonds)*
Zweielektronenbindung *f* two-electron bond
Zweierstoß *m* *(phys ch)* bimolecular collision, dual [binary, two-body] collision
zweifach negativ (geladen) dinegative
zweifach positiv (geladen) dipositive
Zweifachbindung *f* double bond [link, linkage]
Zweifachfilter *n* dual media filter
zweifachfrei divariant, bivariant
Zweifachzucker *m* disaccharide
Zweiflächner *m* *(cryst)* pinacoid
Zweig *m* 1. *(spectr)* branch; 2. bridge *(of a bridge-ring compound)*
zweigängig two-pass *(e.g. heat exchanger)*
Zweigutapparat *m* two-product unit
Zweihalskolben *m* two-neck(ed) flask
zweikernig binuclear, binucleate
Zweikomponentensystem *n* 1. *(phys ch)* two-component system, binary system; 2. *(coat)* two-can [two-pack] system
Zweikörperverdampfer *m* double-effect evaporator
Zweiphasensystem *n* two-phase system
Zweiphotonenabsorption *f* two-photon absorption, TPA
Zweiphotonenanregung *f* two-photon excitation, biphotonic excitation
Zweiproduktenapparat *m* two-product unit
zweiprotonig diprotic, dibasic *(acid)*

Zweipunktregelung *f* two-position [on-off] control
zweisäurig diacid, dibasic *(base)*
Zweischichtenfilter *n* dual-media filter
Zweischneckenextruder *m* double-screw [twin-screw] extruder
Zweistoffgemisch *n* binary mixture
Zweistofflegierung *f* binary alloy
Zweistoffsystem *n* two-component system, binary system
Zweistrahlspektralphotometer *n* double-beam spectro(photo)meter
Zweistrahlspektrometer *n* double-beam spectro(photo)meter
zweisträngig double-strand *(e.g. conveyor)*
Zweistufenreaktion *f* two-step reaction
Zweistufenverdampfer *m* double-effect evaporator
Zweistufenverfahren *n* 1. two-stage [two-step] process; 2. *(met)* duplex process
zweistufig two-stage, two-step, double-stage
Zwei-Substrat-Reaktion *f* bisubstrate [two-substrate] reaction *(enzyme kinetics)*
Zweitdestillation *f* redistillation, rerun(-ning)
Zweitkomponente *f (dye)* secondary component
Zweitsubstituent *m* second substituent
Zweiwalzen-Walzwerk *n* two-roll mill
Zweiweghahn *m* two-way stopcock
zweiwertig divalent *(element)*; dibasic, diprotic *(acid)*; dibasic, diacid *(base)*; dihydric *(alcohol)*
Zweiwertigkeit *f* divalency, bivalency *(of an element)*; dibasicity *(of an acid or base)*
zweizählig 1. *(cryst)* diadic; 2. *s.* zweizähnig
zweizähnig bidentate *(ligand)*
Zweizentrenbindung *f* two-centre bond *(chemical-bond theory)*
Zweizentren(molekül)orbital *n* two-centre (molecular) orbital
Z-Wert *m* (Kossower) Z-value *(for characterizing the polarity of solvents)*
Zwiebel *f (tech)* bulb
Zwillingsbildung *f (cryst)* twinning, twin formation
Zwillingskristall *m* twin (crystal), crystal twin
Zwischenanstrich *m* undercoat
Zwischenanstrichstoff *m* undercoat (paint), undercoater
Zwischenerzeugnis *n* intermediate (product)
Zwischenfraktion *f* 1. *(distil)* intermediate fraction [cut]; 2. intermediate material, middlings product *(classifying)*
Zwischengitteratom *n* interstitial atom, (lattice) interstitial
Zwischengitterion *n* interstitial ion
Zwischengitterplatz *m* interstitial (lattice) site, interstitial position
Zwischenglühen *n* process [commercial] annealing *(of steel)*
Zwischenkornvolumen *n* intergranular [void] volume
Zwischenlage *f* intermediate layer, interlayer, interlining, ply
Zwischenlauf *m (distil)* intermediate cut [fraction]
zwischenmolekular intermolecular
Zwischenprodukt *n* 1. intermediate (product); 2. *s.* Zwischenstoff
Zwischenraum *m* interstice, spacing
Zwischenschicht *f* intermediate layer, interlayer, interlining, ply
Zwischenstoff *m* (reaction) intermediate, intermediate substance [product]
Zwischenstoffwechsel *m (bioch)* intermediary metabolism
Zwischenstück *n* spacer; adapter, connector, *(if tubular:)* connecting tube
Zwischenstufe *f* intermediate stage
Zwischenstufengefüge *n (met)* bainite (structure), bainitic structure
Zwischenverbindung *f* intermediate (compound)
Zwischenwand *f* 1. partition (wall), *(if thin:)* diaphragm; 2. *(pap)* midfeather, midwall, midriff, centre division *(of a Hollander beater)*
Zwischenzustand *m* intermediate [transitory] state [condition]
Zwitterion *n* zwitterion, zwitterionic compound, inner [internal, intramolecular] salt, amphion, dipolar ion
zwitterionisch zwitterionic
Zwölfflächner *m (cryst)* dodecahedron
zybotaktisch *(phys ch)* cybotactic
zyklisch cyclic

Zyklo... *siehe auch: Cyclo...*
Zyklon(abscheider) *m* (centrifugal) cyclone separator, centrifugal [cyclonic] separator [collector], cyclone
Zyklonieren *n* cyclonic [cyclone] separation
Zyklotron *n* cyclotron
Zyklus *m* cycle
zylinderförmig cylindrical
Zylindertrockner *m (pap, text)* cylinder [can] dryer
zylindrisch cylindrical
zymogen zymogenic, zymogenous
Zymogen *n (bioch)* zymogen, pre-enzyme
zymotisch *s.* zymogen
Zytostatikum *n (pharm)* cytostatic agent
zytostatisch cytostatic

Appendix
Anhang

Appendix / Anhang

Table of elements, sorted alphabetically by name / Tabelle der Elemente, alphabetisch nach den Namen geordnet

English term / Bezeichnung in englischer Sprache	*German term / Bezeichnung in deutscher Sprache*	*Symbol / Symbol*
actinium	Actinium	Ac
aluminium	Aluminium	Al
americium	Americium	Am
antimony	Antimon	Sb
argon	Argon	Ar
arsenic	Arsen	As
astatine	Astat	At
barium	Barium	Ba
berkelium	Berkelium	Bk
beryllium	Beryllium	Be
bismuth	Bismut	Bi
bohrium	Bohrium	Bh
boron	Bor	B
bromine	Brom	Br
cadmium	Cadmium	Cd
caesium	Caesium	Cs
calcium	Calcium	Ca
californium	Californium	Cf
carbon	Kohlenstoff	C
cerium	Cer	Ce
chlorine	Chlor	Cl
chromium	Chrom	Cr
cobalt	Cobalt	Co
copper	Kupfer	Cu
curium	Curium	Cm
dubnium	Dubnium	Db
dysprosium	Dysprosium	Dy
einsteinium	Einsteinium	Es
erbium	Erbium	Er
europium	Europium	Eu
fermium	Fermium	Fm
fluorine	Fluor	F
francium	Francium	Fr
gadolinium	Gadolinium	Gd
gallium	Gallium	Ga
germanium	Germanium	Ge
gold	Gold	Au
hafnium	Hafnium	Hf
hassium	Hassium	Hs
helium	Helium	He

English term / Bezeichnung in englischer Sprache	*German term / Bezeichnung in deutscher Sprache*	*Symbol / Symbol*
holmium	Holmium	Ho
hydrogen	Wasserstoff	H
indium	Indium	In
iodine	Iod	I
iridium	Iridium	Ir
iron	Eisen	Fe
krypton	Krypton	Kr
lanthanum	Lanthan	La
lawrencium	Lawrentium	Lr
lead	Blei	Pb
lithium	Lithium	Li
lutetium	Lutetium	Lu
magnesium	Magnesium	Mg
manganese	Mangan	Mn
meitnerium	Meitnerium	Mt
mendelevium	Mendelevium	Md
mercury	Quecksilber	Hg
molybdenum	Molybdän	Mo
neodymium	Neodym	Nd
neon	Neon	Ne
neptunium	Neptunium	Np
nickel	Nickel	Ni
niobium	Niob	Nb
nitrogen	Stickstoff	N
nobelium	Nobelium	No
osmium	Osmium	Os
oxygen	Sauerstoff	O
palladium	Palladium	Pd
phosphorus	Phosphor	P
platinum	Platin	Pt
plutonium	Plutonium	Pu
polonium	Polonium	Po
potassium	Kalium	K
praseodymium	Praseodym	Pr
promethium	Promethium	Pm
protactinium	Protaktinium	Pa
radium	Radium	Ra
radon	Radon	Rn
rhenium	Rhenium	Re
rhodium	Rhodium	Rh
rubidium	Rubidium	Rb
ruthenium	Ruthenium	Ru
rutherfordium	Rutherfordium	Rm
samarium	Samarium	Sm
scandium	Scandium	Sc
seaborgium	Seaborgium	Sg

English term / Bezeichnung in englischer Sprache	*German term / Bezeichnung in deutscher Sprache*	*Symbol / Symbol*
selenium	Selen	Se
silicon	Silicium	Si
silver	Silber	Ag
sodium	Natrium	Na
strontium	Strontium	Sr
sulphur	Schwefel	S
tantalum	Tantal	Ta
technetium	Technetium	Tc
tellurium	Tellur	Te
terbium	Terbium	Tb
thallium	Thallium	Tl
thorium	Thorium	Th
thulium	Thulium	Tm
tin	Zinn	Sn
titanium	Titan	Ti
tungsten / wolfram	Wolfram	W
uranium	Uran	U
vanadium	Vanadin	V
xenon	Xenon	Xe
ytterbium	Ytterbium	Yb
yttrium	Yttrium	Y
zinc	Zink	Zn
zirconium	Zirconium	Zr

Table of elements, sorted alphabetically by symbol / Tabelle der Elemente, alphabetisch nach den Symbolen geordnet

Symbol / Symbol	*German term / Bezeichnung in deutscher Sprache*	*English term / Bezeichnung in englischer Sprache*
Ac	Actinium	actinium
Ag	Silber	silver
Al	Aluminium	aluminium
Am	Americium	americium
Ar	Argon	argon
As	Arsen	arsenic
At	Astat	astatine
Au	Gold	gold
B	Bor	boron
Ba	Barium	barium
Be	Beryllium	beryllium
Bh	Bohrium	bohrium
Bi	Bismut	bismuth
Bk	Berkelium	berkelium
Br	Brom	bromine
C	Kohlenstoff	carbon
Ca	Calcium	calcium
Cd	Cadmium	cadmium
Ce	Cer	cerium
Cf	Californium	californium
Cl	Chlor	chlorine
Cm	Curium	curium
Co	Cobalt	cobalt
Cr	Chrom	chromium
Cs	Caesium	caesium
Cu	Kupfer	copper
Db	Dubnium	dubnium
Dy	Dysprosium	dysprosium
Er	Erbium	erbium
Es	Einsteinium	einsteinium
Eu	Europium	europium
F	Fluor	fluorine
Fe	Eisen	iron
Fm	Fermium	fermium
Fr	Francium	francium
Ga	Gallium	gallium
Gd	Gadolinium	gadolinium
Ge	Germanium	germanium
H	Wasserstoff	hydrogen
He	Helium	helium

Symbol / Symbol	*German term / Bezeichnung in deutscher Sprache*	*English term / Bezeichnung in englischer Sprache*
Hf	Hafnium	hafnium
Hg	Quecksilber	mercury
Ho	Holmium	holmium
Hs	Hassium	hassium
I	Iod	iodine
In	Indium	indium
Ir	Iridium	iridium
K	Kalium	potassium
Kr	Krypton	krypton
La	Lanthan	lanthanum
Li	Lithium	lithium
Lr	Lawrentium	lawrencium
Lu	Lutetium	lutetium
Md	Mendelevium	mendelevium
Mg	Magnesium	magnesium
Mn	Mangan	manganese
Mo	Molybdän	molybdenum
Mt	Meitnerium	meitnerium
N	Stickstoff	nitrogen
Na	Natrium	sodium
Nb	Niob	niobium
Nd	Neodym	neodymium
Ne	Neon	neon
Ni	Nickel	nickel
No	Nobelium	nobelium
Np	Neptunium	neptunium
O	Sauerstoff	oxygen
Os	Osmium	osmium
P	Phosphor	phosphorus
Pa	Protaktinium	protactinium
Pb	Blei	lead
Pd	Palladium	palladium
Pm	Promethium	promethium
Po	Polonium	polonium
Pr	Praseodym	praseodymium
Pt	Platin	platinum
Pu	Plutonium	plutonium
Ra	Radium	radium
Rb	Rubidium	rubidium
Re	Rhenium	rhenium
Rh	Rhodium	rhodium
Rm	Rutherfordium	rutherfordium
Rn	Radon	radon
Ru	Ruthenium	ruthenium
S	Schwefel	sulphur
Sb	Antimon	antimony

Symbol / Symbol	*German term / Bezeichnung in deutscher Sprache*	*English term / Bezeichnung in englischer Sprache*
Sc	Scandium	scandium
Se	Selen	selenium
Sg	Seaborgium	seaborgium
Si	Silicium	silicon
Sm	Samarium	samarium
Sn	Zinn	tin
Sr	Strontium	strontium
Ta	Tantal	tantalum
Tb	Terbium	terbium
Tc	Technetium	technetium
Te	Tellur	tellurium
Th	Thorium	thorium
Ti	Titan	titanium
Tl	Thallium	thallium
Tm	Thulium	thulium
U	Uran	uranium
V	Vanadin	vanadium
W	Wolfram	tungsten / wolfram
Xe	Xenon	xenon
Y	Yttrium	yttrium
Yb	Ytterbium	ytterbium
Zn	Zink	zinc
Zr	Zirconium	zirconium

Reading of some chemical formulae and equations / Lesart einiger chemischer Formeln und Gleichungen

Description of the valency and quantity ratios of constituents / Kennzeichnung der Wertigkeit und der Mengenverhältnisse der Bestandteile

R^{I}	any univalent radical	ein einwertiges Radikal
Me^{II}	any bivalent metal	ein zweiwertiges Metall
Fe^{II}	f e bivalent, iron bivalent, ferrous [bivalent] iron	zweiwertiges f e, bivalentes Eisen, zweiwertiges [bivalentes] Eisen
Fe^{III}	f e tervalent, iron tervalent, ferric [tervalent] iron	dreiwertiges f e, trivalentes Eisen, dreiwertiges [trivalentes] Eisen
$Me^{I}_{2}SO_4$	a metal sulphate, where the metal has a valency of one	ein Metallsulfat, bei dem das Metall eine Valenz von eins besitzt
$Fe^{III}_{4}Fe^{II}_{3}[Fe^{III}(CN)_6]_6$	f e tervalent four / f e bivalent three / open square bracket / f e tervalent / c n / six times / close square bracket / all six times*)	dreiwertiges f e vier / zweiwertiges f e drei / eckige Klammer auf / dreiwertiges f e / in Klammern c n / eckige Klammer zu / sechs*)
H_2O	h two o	h zwei o
P_2O_5	p two o five	p zwei o fünf
H_2SO_4	h two s o four	h zwei s o vier
NaOH	n a o h	n a o h
$Ca(OH)_2$	c a / o h / twice	c a / o h / zwei
$(C_6H_{10}O_5)_n$	c six h ten o five / n times	n mal c 6 h zehn o fünf

General formulae / Allgemeine Formeln

$C_x(H_2O)_y$	c x / h two o / y times	c x / h zwei o / y
C_nH_{2n+2}	c n / h two n plus two	c n / h zwei n plus zwei
C_nH_{2n-2}	c n / h two n minus two	c n / h zwei n minus zwei
$R_1–O–R_2$	r one o r two	r eins o r zwei
$H_{n+2}P_nO_{3n+1}$	h n plus two / p n / o three n plus one	h n plus zwei / p n / o drei n plus eins

---

*) The strokes indicate pauses in the reading.
Die Schrägstriche bedeuten Sprechpausen.

Description of the bonding / Kennzeichnung der Bindung

$CH_2{:}CH{\cdot}C{\vdots}CH$	c h two / double bond / c h / single bond / c / triple bond / c h	c h zwei / Doppelbindung / c h / Einfachbindung / c / Dreifachbindung / c h

Description of the number of molecules of water of crystallization / Kennzeichnung der Anzahl der Moleküle Kristallwasser

$CuSO_4 \cdot H_2O$	c u s o four / h two o	c u s o vier / h 2 o
$KBO_3 \cdot 1/2\ H_2O$	k b o three / (one) half h two o	k b o drei / ein halb h zwei o
$Na_2Mg(SO_4)_2 \cdot 2.5\ H_2O$	n a two m g / s o four / twice / two point five h two o	n a zwei m g / s o vier / zwei / zwei Komma 5 h zwei o

Description of the ionic charge / Kennzeichnung der Ionenladung

Cu^{+}	c u plus, cuprous ion	c u plus, Kupfer-eins-Ion
Cu^{2+}	c u two plus, cupric ion	c u zwei plus, Kupfer-zwei-Ion
Cl^{-}	c l minus, chloride ion	c l minus, Chlor-Ion
O^{2-}	o two minus, oxide ion	o zwei minus, Sauerstoff-Ion
Si^{4-}	s i four minus, silicide ion	s i vier minus, Silicium-Ion

Description of mass, atomic number, number of atoms and ionic charge on the element symbol / Kennzeichnung der Masse, der Ordnungszahl, der Anzahl der Atome und der Ionenladung am Elementsymbol

${}^{32}_{16}S^{2+}_{2}$	a doubly ionized molecule containing two atoms of sulphur, each of which has the atomic number sixteen and mass number thirty-two	ein doppelt ionisiertes Molekül, das zwei Atome Schwefel enthält, von denen jedes die Ordnungszahl sechzehn und die Massenzahl zweiunddreißig hat

Description of the isotopes of an element / Kennzeichnung der Isotope eines Elements

${}^{18}O$	oxygen-eighteen	Sauerstoff achtzehn

Description of isotope-labelled compounds / Kennzeichnung von durch Isotope markierten Verbindungen

${}^{32}PCl_3$	phosphorus-thirty-two trichloride	Phosphor-zweiunddreißig-trichlorid
$H^{36}Cl$	hydrogen chloride-thirty-six	Hydrogenchlorid sechsunddreißig
${}^{15}NH_3$	ammonia nitrogen-fifteen	Ammoniakstickstoff-fünfzehn

$^{15}N^{2}H_3$	ammonia nitrogen-fifteen hydrogen-two	Ammoniakstickstoff-fünfzehn Wasserstoff-zwei
$^{2}H_2{}^{35}SO_4$	sulphuric acid sulphur-thirty-five hydrogen-two	Schwefelsäure Schwefel-fünfunddreißig Wasserstoff-zwei

Equations / Gleichungen

$NaOH + HCl = NaCl + H_2O$

n a o h plus h c l equals n a c l plus h two o	n a o h plus h c l gleich n a c l plus h zwei o

$CaO + H_2O = Ca(OH)_2.\ \Delta H = -16$ kcal

c a o plus h two o equals c a / o h / twice / full stop / delta h equals minus sixteen kilocalories	c a o plus h plus h zwei o gleich c a / o h / zwei Punkt delta h gleich minus sechzehn Kilokalorien

$Zn + H_2SO_4 = ZnSO_4 + H_2\uparrow$

z n plus h two s o four equals z n s o four plus h two (**or**: plus hydrogen gas, plus hydrogen in the gaseous state)	z n plus h zwei s o vier gleich z n s o vier plus h zwei (**oder**: plus Wasserstoffgas, plus gasförmigen Wasserstoff)

$Ba(OH)_2 + H_2SO_4 = BaSO_4\downarrow + 2\ H_2O$

b a / o h / twice / plus h two s o four equals b a s o four / precipitated / plus two h two o	b a / o h / zwei / plus h zwei s o vier gleich b a s o vier / gefällt / plus zwei h zwei o

Reading of some mathematical expressions / Lesart einiger mathematischer Ausdrücke

Mathematical symbols / Mathematische Zeichen

$+$	plus	plus
$-$	minus	minus
$\pm$	plus or minus	plus minus
$\mp$	minus or plus	minus plus
$\times$, $\cdot$	multiplied by, times*)	multipliziert mit, mal*)
$\div$, /	divided by	dividiert durch, geteilt durch
$=$	is equal to, equals	gleich, ist gleich
$\neq$	is not equal to	ungleich, nicht gleich
$\equiv$	is identical with	identisch
$\not\equiv$	is not identical with	nicht identisch
$\approx$	is approximately equal to	annähernd gleich
$\sim$	is asymptotically equal to	asymptotisch gleich
$>$	is greater than	größer als
$\ngtr$	is not greater than	nicht größer als
$<$	is less than	kleiner als
$\nless$	is not less than	nicht kleiner als
$\geqq$, $\geq$	is greater than or equal to	größer als oder gleich
$\leqq$, $\leq$	is less than or equal to	kleiner als oder gleich
$\gg$	is much greater than	viel größer als
$\ll$	is much less than	viel kleiner als
$\therefore$	therefore	deswegen
$\because$	because, since	weil, da
$\ldots$	dot dot dot	Punkt, Punkt, Punkt
$\propto$	is proportional to	proportional zu
∞	infinity	unendlich
0 °C	nought degrees c (Celsius)	null Grad (Celsius)
%	per cent, per hundred	Prozent
‰	per thousand	Promille
$\sqrt{}$	(square) root of	Wurzel aus, Quadratwurzel aus
$\sum$	capital sigma; sum of	Summenzeichen
$\prod$	capital pi; product of	Produktzeichen
lim	limit	Grenzwert, Limes
$\rightarrow$	approaches, tends to	strebt nach, geht nach
$\int$	intregral of	Intregral von

*) In UK usage, the multiplying dot is often on the line, while the decimal point is halfway up.
Im Zeichengebrauch des Vereinigten Königreiches steht der Multiplikationspunkt häufig auf der Grundlinie, während der Dezimalpunkt hochgestellt ist.

( )	(ordinary) brackets (in printers' terminology: parentheses)	(runde) Klammern
(	open [initial] bracket, open [initial] parenthesis	runde Klammer auf
)	close [final] bracket, close [final] parenthesis	runde Klammer zu
[ ]	square brackets (in printers' terminology: brackets)	eckige Klammern
{ }	curly brackets (in printers' terminology: braces)	geschweifte Klammern
x'	x dash, x prime	x Strich
x''	x double dash [prime]	x zweigestrichen
x'''	x triple dash [prime]	x dreigestrichen

Decimals / Dezimalzahlen

0.5	zero point five, o point five, nought point five	null Komma fünf
2.87	two point eight seven	zwei Komma acht sieben
15.3 °C	fifteen point three degrees c [Celsius]	fünfzehn Komma drei Grad (Celsius)
–273.1 °C	minus two seven three point one degrees c [Celsius]	minus zweihundertdreiundsiebzig Komma ein Grad (Celsius)

Fractions / Brüche

1/2	half, one half, a half	ein Halb
1/3	one third, a third	ein Drittel
2/3	two thirds	zwei Drittel
1/4	one [a] fourth, one [a] quarter	ein Viertel
3/4	three fourths, three quarters	drei Viertel
1 1/2	one and a half	eineinhalb, anderthalb
$\frac{a}{b}$, a/b	a over b, a divided by b	a durch b, a geteilt durch b
$\frac{ab^2}{4b}$	a b squared over [divided by] four b	a b quadrat durch [geteilt durch] vier b
$\frac{1}{n}$	one over n, one nth	eins geteilt durch n, ein n-tel

Letters with subscripts and superscripts / Buchstaben mit Indizes

a_b	a subscript b, a sub b	a unten b
S_n	capital s sub n	groß s unten n

a^b	a superscript b, a super b	a hoch b
x_0	x subscript zero, x sub zero, x zero, x nought	x null

Powers / Potenzausdrücke

a^n	a to the power (of) n, a to the nth (power), a to the n	a hoch n, a zur n-ten Potenz, a zur n-ten
a^{-n}	a to the power (of) minus n, a to the minus nth (power), a to the minus n	a hoch minus n
a^{2x}	a to the power of two x	a hoch x
x^2	x squared	x quadrat
x^3	x cubed	x hoch drei, x kubik
$(a + b)^2$	a plus b / all squared*)	a plus b / in Klammern / zum Quadrat*)
$(x + 4)^{-\frac{x}{6}}$	x plus four / in brackets / to the power of minus one sixth x	x plus vier / in Klammern / hoch minus x Sechstel
10^2	ten to the power (of) two, ten to the second (power), ten squared	zehn hoch zwei, zehn quadrat
10^3	ten to the third (power), ten cubed	zehn hoch drei, zehn kubik
10^{-2}	ten to the power (of) minus two, ten to the minus two, ten to the minus second (power)	zehn hoch minus zwei

Root expressions / Wurzelausdrücke

$\sqrt{a}$	root a, square root of a	Wurzel a, Quadratwurzel aus a
$\sqrt[3]{a}$	cube root of a, third root of a	Kubikwurzel aus a, dritte Wurzel aus a
$\sqrt[n]{a}$	nth root of a	n-te Wurzel aus a
$2\sqrt{x}$	twice the square root of x	zwei Mal Quadratwurzel (aus) x
$3\sqrt[4]{x}$	three times the fourth root of x	drei Mal vierte Wurzel x
$\sqrt{(a+b)}$	square root of the sum (of) a plus b, square root of the sum of a and b	Quadratwurzel aus der Summe (von) a plus b

*) The strokes indicate pauses in the reading.
Die Schrägstriche bedeuten Sprechpausen.

Logarithms / Logarithmen

log	log, logarithm	log, Logarithmus
$\log_{10}$, lg	common [Briggsian] logarithm, log to the base ten	gewöhnlicher [briggscher] Logarithmus, Logarithmus zur Basis zehn
$\log_e$, ln	natural [Neperian] logarithm, l n	natürlicher [napierscher] Logarithmus, l n
$\log_{10} 2$	log two to the base ten	Logarithmus von zwei zur Basis zehn
$\log_a m$	log of m to the base a	Logarithmus von m zur Basis a
ln x	natural logarithm [log] of x, l n x	natürlicher Logarithmus von x, l n x

Differential calculus / Differenzialrechnung

f(x)	function of x, f of x	Funktion von x, f von x
f′(x)	f dash [prime] of x	f Strich von x
f″(x)	f double dash [prime] of x	f zweigestrichen von x
lim f(x)	the limit of the function of x, the limit of f of x, the limit f x	Limes [Grenzwert] der Funktion f von x, Limes von f x, Limes f x
∂ f(x)	a variant or increment in f x	Inkrement in f x
Δx	delta x	delta x
dx	d x, differential of x	d x, Differenzial von x
$\frac{dy}{dx}$, dy/dx	d y by d x, derivative of y with respect to x	d y nach d x, Ableitung von y nach x
$\frac{\partial z}{\partial x}$	dif z to dif x, partial derivative of z with respect to x	del z nach del x, partielle Ableitung von z nach x
$x \to 0$	x tends to zero, x approaches zero	x strebt nach null, x gegen null
$\Delta x \to \infty$	delta x tends to infinity	delta x strebt nach unendlich, delta x gegen unendlich

Integral calculus / Integralrechnung

$\int$	integral	Integral
$\iint$	double integral	Doppelintegral
$\iiint$	triple integral	dreifaches Integral
$\int_a^b x\,dx$	integral of x dx between the limits a and b	Integral x d x von a bis b

$\int_{a_0=m}^{\infty} xy^2 \, da_0$	integral / between the limits a subscript zero equal to m and to infinity / of x y squared d a subscript zero	Integral x y quadrat d a null / für a null gleich m bis unendlich

Some examples / Einige Beispiele

$a + b - c = d$	a plus b minus c equals [is equal to] d	a plus b minus c gleich [ist gleich] d
$5 \times 3 = 15$	five times [multiplied by] three is equal to fifteen	fünf mal [multipliziert mit] drei (ist) gleich fünfzehn
$15 : 3 = 5$	fifteen divided by three is equal to five	fünfzehn geteilt durch drei ist (gleich) fünf
$a : b :: c : d$	a is to b as c is to d	a verhält sich zu b wie c zu d
$t_1 : t_2 = \sqrt{m_1} : \sqrt{m_2}$	the ratio of t sub one to t sub two is equal to the ratio of root of m sub one to root of m sub to, the ratio of t one to t two is equal to the ratio of root m one to root m two	das Verhältnis von t eins zu t zwei ist gleich dem Verhältnis von Wurzel m eins zu Wurzel m zwei
$\frac{a}{b} = c$	a over b is equal to c	a durch b gleich c, a geteilt durch b ist gleich c
$\frac{8 \times 79.9}{20.1} = 31.8$	eight times seventy-nine point nine over twenty point one equals thirty-one point eight	acht mal neunundsiebzig Komma neun geteilt durch zwanzig Komma eins ist gleich einunddreißig Komma acht
10–20%	ten to twenty per cent	zehn bis zwanzig Prozent
$1 + 3 + 5 \ldots$	one plus three plus five / dot dot dot	eins plus drei plus fünf / Punkt Punkt Punkt
$a(b + c)$	a times / b plus c / in brackets, a times / b plus c	a mal / b plus c / in Klammern, a mal / Klammer auf / b plus c / Klammer zu
$a[b + c]$	a times / square bracket / b plus c / close the bracket	a mal / b plus c / in eckigen Klammern, a mal / eckige Klammer auf / b plus c / eckige Klammer zu

$v_t = v_0\left(1+\frac{t}{273}\right) = v_0(1+\alpha t)$	v subscript t equals v subscript zero / bracket / one plus t over two seven three / close the bracket / equals v subscript zero / bracket / one plus alpha t / close bracket	v t gleich v null mal / Klammer auf / eins plus t durch zweihundertdreiundsiebzig / Klammer zu / gleich v null mal / Klammer auf / 1 plus Alpha t / Klammer zu
$\left[(a+b)^x - \sqrt[n]{c}\,\right]^{-y} = 0.503$	square bracket / ordinary bracket into a plus b / close ordinary bracket / to the power x / minus the nth root of c / close the square bracket / all to the power minus y / equals 0 point five 0 three	eckige Klammer auf / runde Klammer auf / a plus b / runde Klammer zu / hoch x minus n-te Wurzel aus c / eckige Klammer zu / hoch minus y gleich null Komma fünf null drei
$2\sqrt[3]{(x-a)} = 0$	twice the cube root of / x minus a / equals zero	zwei Mal Kubikwurzel aus / x minus a / gleich null
$\lim\limits_{\Delta x \to \infty} \sum\limits_{a_b=0}^{n} f(a_k x^m)$	the limit / as delta x tends to infinity / of the sum of f / bracket / a subscript k times x to the m / from a sub b equals zero to a sub b equals n	Limes / für delta x gegen Unendlich / von Summe über f von a k mal x hoch m / von a unten b gleich null bis n
$y = \prod\limits_{a=1}^{b} x_n$	y equals the product / from a equals one to a equals b / of x subscript n	y gleich Produkt aller x n / von x gleich eins bis b /

Comparison of SI-units with previously used units / Gegenüberstellung von SI-Einheiten und früher gebräuchlichen Einheiten

Force / Kraft

N	kp
1	0,102
9,81	1

1 mN = 0,102 p   1 kN = 0,102 Mp
9,81 mN = 1 p   9,81 kN = 1 Mp

Pressure, stress / Druck, mechanische Spannung

N/m^2 = Pa		mmWS
1		0,102
9,81		1
kN/m^2 = kPa		mWS
1		0,102
9,81		1
		Torr
1		7,50
0,133		1
N/mm^2 = MPa		at = kp/cm^2
1		10,2
0,0981		1
		atm
1		9,87
0,1013		1
	bar	kp/mm^2
1	10	0,102
0,1	1	0,0102
9,81	98,1	1

Stress / mechanische Spannungen

1 N/mm^2 = 10^3 kN/m^2 = 10^6 N/m^2

Pressure in fluids / Drücke in Fluiden

1 MPa = 10^3 kPa = 10^6 Pa
1 bar = 10^3 mbar = 10^5 Pa

1 N/m^2 = 1 Pa = 0,01 mbar
1 kN/m^2 = 1 kPa = 10 mbar

1 kp/cm^2 = 1 at
= 0,0981 N/mm^2 = 0,981 bar
1 kp/mm^2 = 9,81 N/mm^2 = 98,1 bar
1 atm = 0,1013 N/mm^2 = 1,013 bar
1 mWS = 9,81 kN/m^2 = 0,0981 bar
1 mmWS = 9,81 N/m^2 = 0,0981 mbar
1 Torr = 1 mmHg
= 0,133 kN/m^2 = 1,33 mbar

Energy, work, quantity of heat / Energie, Arbeit, Wärmemenge

J = W · s = N · m	kp · m	cal
1	0,102	0,239
9,81	1	2,342
4,19	0,427	1

1 kcal = 4,19 kJ = 4,19 kW · s
1 kW · h = 3,6 MJ = 3,6 MW · s

Power / Leistung

W = J/s = N · m/s	kp · m/s	kcal/h
1	0,102	0,860
9,81	1	8,43
1,16	0,119	1

1 kW = 1,360 PS
1 PS = 0,7355 kW

Quantity / Basisgrößenarten			Si unit / Basiseinheiten		
length	Länge	*l*, *s*	meter	Meter	m
times	Zeit	*t*	second	Sekunde	s
mass	Masse	*m*	kilogram	Kilogramm	kg
temperature	Temperatur	*T*	kelvin	Kelvin	K
amperage	Elektrische Stromstärke	*I*	ampere	Ampere	A
amount of substance	Stoffmenge	*n*	mol	Mol	mol
luminous intensity	Lichtstärke	I_v	candela	Candela	cd

Prefixes / Vorsätze							
exa	Exa	E	10^{18}	atto	Atto	a	10^{-18}
peta	Peta	P	10^{15}	femto	Femto	f	10^{-15}
tera	Tera	T	10^{12}	pico	Piko	p	10^{-12}
giga	Giga	G	10^{9}	nano	Nano	n	10^{-9}
mega	Mega	M	10^{6}	micro	Mikro	µ	10^{-6}
kilo	Kilo	k	10^{3}	milli	Milli	m	10^{-3}
hecto	Hekto	h	10^{2}	centi	Zenti	c	10^{-2}
deca	Deka	da	10	deci	Dezi	d	10^{-1}

Hecto, deca, deci and centi should only be used as hitherto customary. / Hekto, Deka, Dezi und Zenti sollen nur noch verwendet werden, soweit es bisher üblich war.